2020

AutoCAD®
and Its Applications
COMPREHENSIVE

by

Terence M. Shumaker
Faculty Emeritus
Former Chairperson
Drafting Technology
Autodesk Premier Training Center
Clackamas Community College
Oregon City, OR

David A. Madsen
President, Madsen Designs Inc.
Faculty Emeritus
Former Department Chairperson
Drafting Technology
Autodesk Premier Training Center
Clackamas Community College
Oregon City, OR
Director Emeritus, American Design Drafting Association

David P. Madsen
President, Engineering Drafting & Design, Inc.
Vice President, Madsen Designs Inc.
Computer-Aided Design and Drafting Consultant and Educator
Autodesk Developer Network Member
American Design Drafting Association Member

Jeffrey A. Laurich
Instructor, Mechanical D
Technology
Fox Valley Technical Coll
Appleton, WI

J.C. Malitzke
President, Digital JC CAD Services Inc.
Former Department Chair
Computer Integrated Technologies
Former Manager and Instructor
Autodesk Authorized Training Center
Moraine Valley Community College
Palos Hills, IL

Craig P. Black
Instructor, Mechanical Design Technology
Former Manager and Instructor
Autodesk Premier Training Center
Fox Valley Technical College
Appleton, WI

Publisher
The Goodheart-Willcox Company, Inc.
Tinley Park, IL
www.g-w.com

Manufactured in the United States of America.

ISBN 978-1-63563-866-0

1 2 3 4 5 6 7 8 9 – 20 – 24 23 22 21 20 19

Cataloging Data

Names: Shumaker, Terence M., author. | Madsen, David A., author. | Madsen, David P., author. | Laurich, Jeffrey A., author. | Malitzke, J.C., author. | Black, Craig P., author.
Title: AutoCAD and Its Applications: Comprehensive 2020 / by Terence M. Shumaker, David A. Madsen, David P. Madsen, Jeffrey A. Laurich, J.C. Malitzke, Craig P. Black.
Description: 27th edition. | Tinley Park, IL : The Goodheart-Willcox Company, Inc., [2020] | Includes index.
Identifiers: ISBN 9781635638660
Subjects: LCSH: Computer graphics. | AutoCAD.
Classification: LCC T385.S4616445 2020 | DDC 620'.0042028553--dc23

2020

AutoCAD®

and Its Applications

BASICS

by

Terence M. Shumaker
Faculty Emeritus
Former Chairperson
Drafting Technology
Autodesk Premier Training Center
Clackamas Community College
Oregon City, OR

David A. Madsen
President
Madsen Designs Inc.
Faculty Emeritus
Former Department Chairperson
Drafting Technology
Autodesk Premier Training Center
Clackamas Community College
Oregon City, OR
Director Emeritus
American Design Drafting Association

David P. Madsen
President
Engineering Drafting & Design, Inc.
Vice President
Madsen Designs Inc.
Computer-Aided Design and Drafting Consultant and Educator
Autodesk Developer Network Member
American Design Drafting Association Member

Publisher
The Goodheart-Willcox Company, Inc.
Tinley Park, IL
www.g-w.com

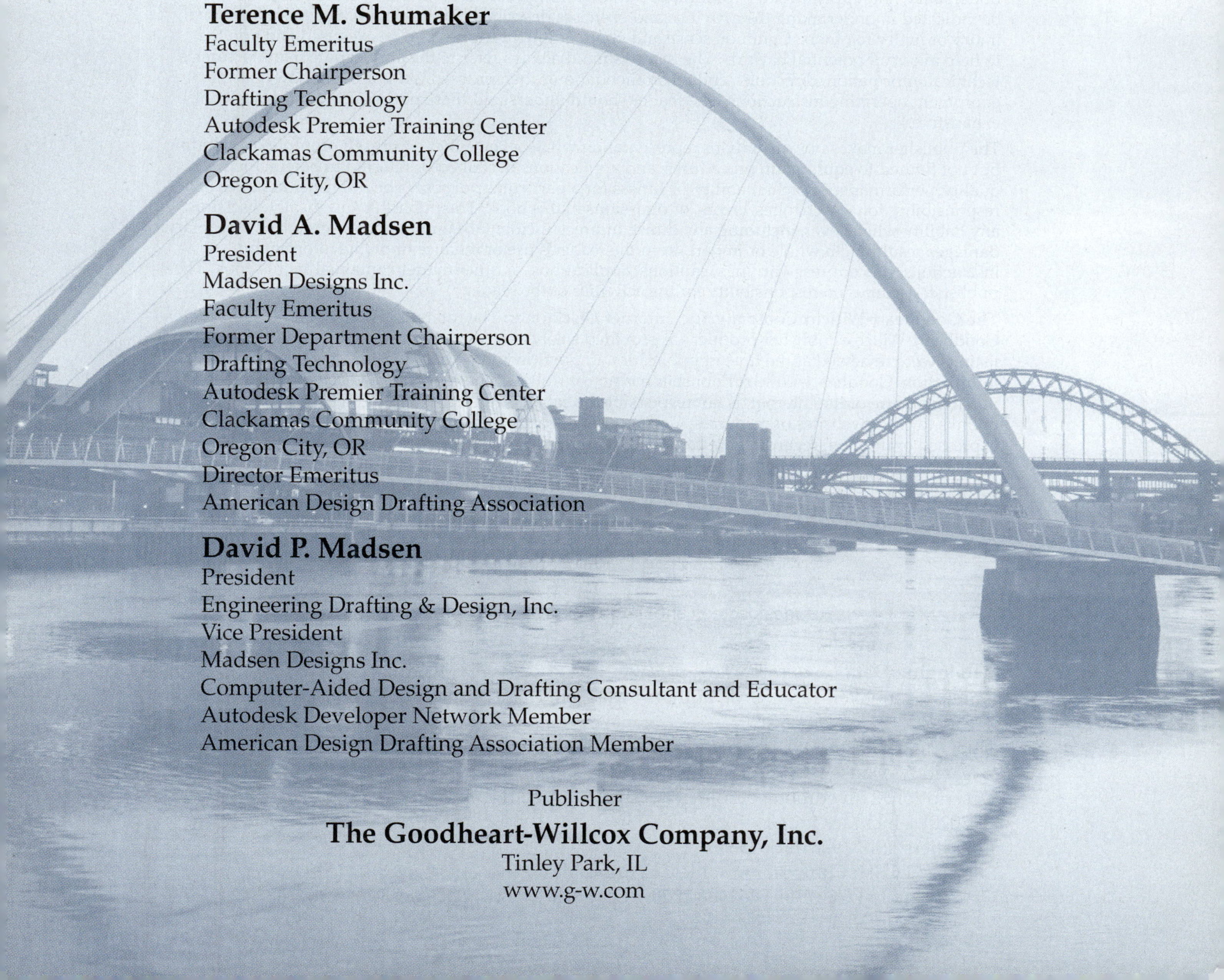

Manufactured in the United States of America.

ISBN 978-1-63563-864-6

1 2 3 4 5 6 7 8 9 – 20 – 24 23 22 21 20 19

Cataloging Data
Names: Shumaker, Terence M., author. | Madsen, David A., author. | Madsen, David P., author.
Title: AutoCAD and Its Applications: Basics 2020 / by Terence M. Shumaker, David A. Madsen, David P. Madsen.
Description: 27th edition. | Tinley Park, IL : The Goodheart-Willcox Company, Inc., [2020] | Includes index.
Identifiers: ISBN 9781635638646
Subjects: LCSH: Computer graphics. | AutoCAD.
Classification: LCC T385.S461466 2020 | DDC 620'.0042028553--dc23

INTRODUCTION

AutoCAD and Its Applications—Basics is a textbook providing complete instruction in mastering fundamental AutoCAD® 2020 commands and drawing techniques. Typical applications of AutoCAD are presented with basic drafting and design concepts. The topics are covered in an easy-to-understand sequence and progress in a way that allows you to become comfortable with the commands as your knowledge builds from one chapter to the next. ***AutoCAD and Its Applications—Basics*** offers the following features:

- Step-by-step use of AutoCAD commands
- In-depth explanations of how and why commands function as they do
- Extensive use of font changes to specify certain meanings
- Examples and descriptions of industry practices and standards
- Detailed illustrations of AutoCAD features and functions
- Professional tips explaining how to use AutoCAD effectively and efficiently
- More than 280 exercises to reinforce the chapter topics and build on previously learned material
- Chapter reviews for review of commands and key AutoCAD concepts
- Practice questions and problems for the AutoCAD Certified Professional Exam
- A large selection of drafting problems supplementing each chapter

With ***AutoCAD and Its Applications—Basics***, you learn AutoCAD commands and become acquainted with information in other areas:

- Preliminary planning and sketches
- Drawing geometric shapes and constructions
- Editing operations that increase productivity
- Making multiview drawings (orthographic projection)
- Placing text according to accepted industry practices
- Dimensioning techniques and practices based on accepted standards
- Parametric drawing techniques
- Drawing section views and designing graphic patterns
- Creating shapes and symbols
- Creating and managing symbol libraries
- Plotting and printing drawings

AutoCAD and Its Applications—Basics covers basic AutoCAD applications. For detailed coverage on advanced AutoCAD applications, please refer to the ***AutoCAD and Its Applications—Advanced*** portion of ***AutoCAD and Its Applications—Comprehensive***.

Features of the textbook

Learning Objectives identify key items you will learn in the chapter.

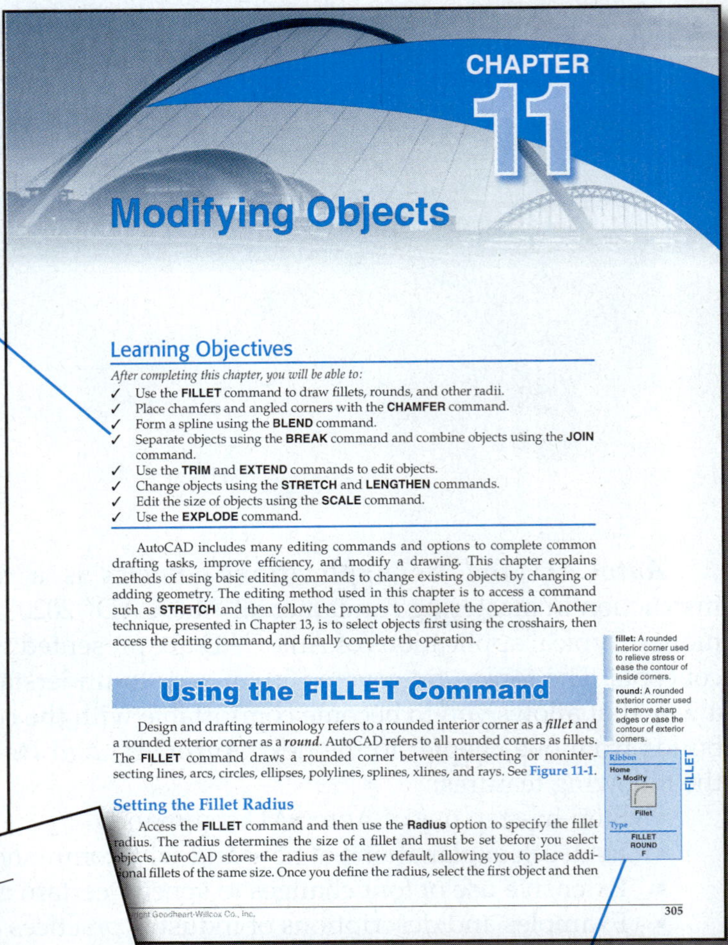
CHAPTER 11

Modifying Objects

Learning Objectives

After completing this chapter, you will be able to:

- ✓ Use the **FILLET** command to draw fillets, rounds, and other radii.
- ✓ Place chamfers and angled corners with the **CHAMFER** command.
- ✓ Form a spline using the **BLEND** command.
- ✓ Separate objects using the **BREAK** command and combine objects using the **JOIN** command.
- ✓ Use the **TRIM** and **EXTEND** commands to edit objects.
- ✓ Change objects using the **STRETCH** and **LENGTHEN** commands.
- ✓ Edit the size of objects using the **SCALE** command.
- ✓ Use the **EXPLODE** command.

AutoCAD includes many editing commands and options to complete common drafting tasks, improve efficiency, and modify a drawing. This chapter explains methods of using basic editing commands to change existing objects by changing or adding geometry. The editing method used in this chapter is to access a command such as **STRETCH** and then follow the prompts to complete the operation. Another technique, presented in Chapter 13, is to select objects first using the crosshairs, then access the editing command, and finally complete the operation.

fillet: A rounded interior corner used to relieve stress or ease the contour of inside corners.

round: A rounded exterior corner used to remove sharp edges or ease the contour of exterior corners.

Using the FILLET Command

Design and drafting terminology refers to a rounded interior corner as a *fillet* and a rounded exterior corner as a *round*. AutoCAD refers to all rounded corners as fillets. The **FILLET** command draws a rounded corner between intersecting or nonintersecting lines, arcs, circles, ellipses, polylines, splines, xlines, and rays. See Figure 11-1.

Setting the Fillet Radius

Access the **FILLET** command and then use the **Radius** option to specify the fillet radius. The radius determines the size of a fillet and must be set before you select objects. AutoCAD stores the radius as the new default, allowing you to place additional fillets of the same size. Once you define the radius, select the first object and then

305

New Feature Graphics identify new and updated features for AutoCAD 2020.

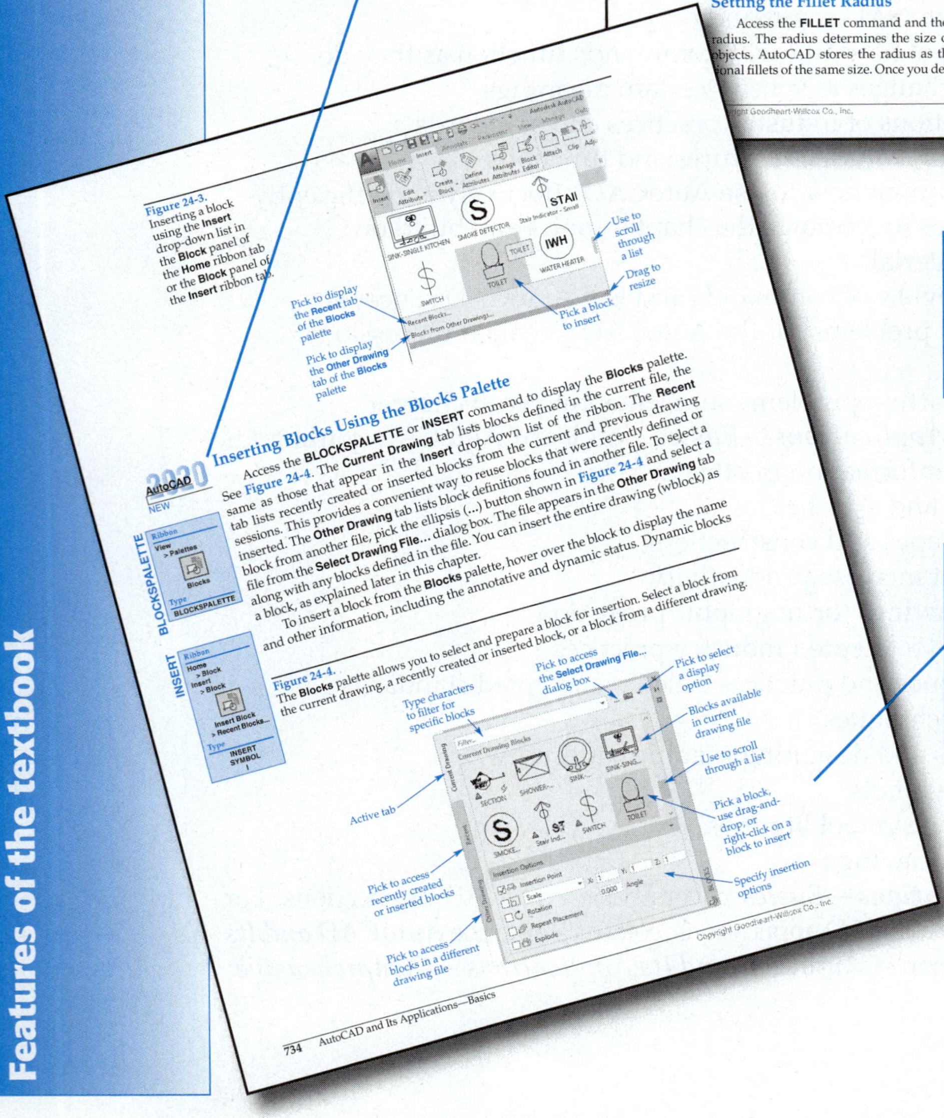
Figure 24-3. Inserting a block using the **Insert** drop-down list in the **Block** panel of the **Home** ribbon tab or the **Block** panel of the **Insert** ribbon tab.

AutoCAD 2020 NEW

Inserting Blocks Using the Blocks Palette

Access the **BLOCKSPALETTE** or **INSERT** command to display the **Blocks** palette. See Figure 24-4. The **Current Drawing** tab lists blocks defined in the current file, the same as those that appear in the **Insert** drop-down list of the ribbon. The **Recent** tab lists recently created or inserted blocks from the current and previous drawing sessions. This provides a convenient way to reuse blocks that were recently defined or inserted. The **Other Drawing** tab lists block definitions found in another file. To select a block from another file, pick the ellipsis (...) button shown in Figure 24-4 and select a file from the **Select Drawing File...** dialog box. The file appears in the **Other Drawing** tab along with any blocks defined in the file. You can insert the entire drawing (wblock) as a block, as explained later in this chapter.

To insert a block from the **Blocks** palette, hover over the block to display the name and other information, including the annotative and dynamic status. Dynamic blocks

Figure 24-4. The **Blocks** palette allows you to select and prepare a block for insertion. Select a block from the current drawing, a recently created or inserted block, or a block from a different drawing.

734 AutoCAD and Its Applications—Basics

Command Entry Graphics show ribbon, **Application Menu**, and keyboard entry options.

Illustrations, including line art illustrations, AutoCAD "screen shots," and illustrated tables, make learning easy.

Features of the textbook

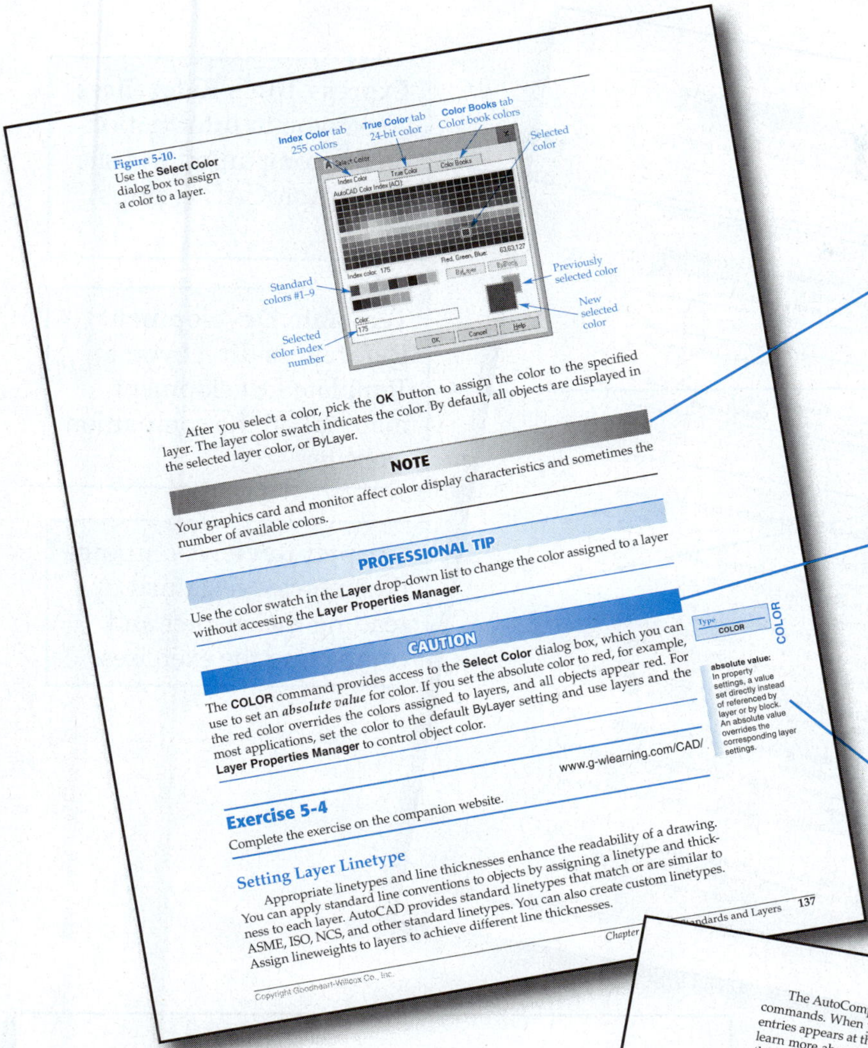

Figure 5-10. Use the **Select Color** dialog box to assign a color to a layer.

After you select a color, pick the **OK** button to assign the color to the specified layer. The layer color swatch indicates the color. By default, all objects are displayed in the selected layer color, or ByLayer.

NOTE

Your graphics card and monitor affect color display characteristics and sometimes the number of available colors.

PROFESSIONAL TIP

Use the color swatch in the **Layer** drop-down list to change the color assigned to a layer without accessing the **Layer Properties Manager**.

CAUTION

The **COLOR** command provides access to the **Select Color** dialog box, which you can use to set an ***absolute value*** for color. If you set the absolute color to red, for example, the red color overrides the colors assigned to layers, and all objects appear red. For most applications, set the color to the default ByLayer setting and use layers and the **Layer Properties Manager** to control object color.

Type COLOR

absolute value: In property settings, a value set directly instead of referenced by layer or by block. An absolute value overrides the corresponding layer settings.

www.g-wlearning.com/CAD/

Exercise 5-4

Complete the exercise on the companion website.

Setting Layer Linetype

Appropriate linetypes and line thicknesses enhance the readability of a drawing. You can apply standard line conventions to objects by assigning a linetype and thickness to each layer. AutoCAD provides standard linetypes that match or are similar to ASME, ISO, NCS, and other standard linetypes. You can also create custom linetypes. Assign lineweights to layers to achieve different line thicknesses.

Copyright Goodheart-Willcox Co., Inc. Chapter 5 Drawing Standards and Layers 137

Notes explain important aspects of a topic.

Cautions alert you to potential problems.

Running Glossary Entries define key terms.

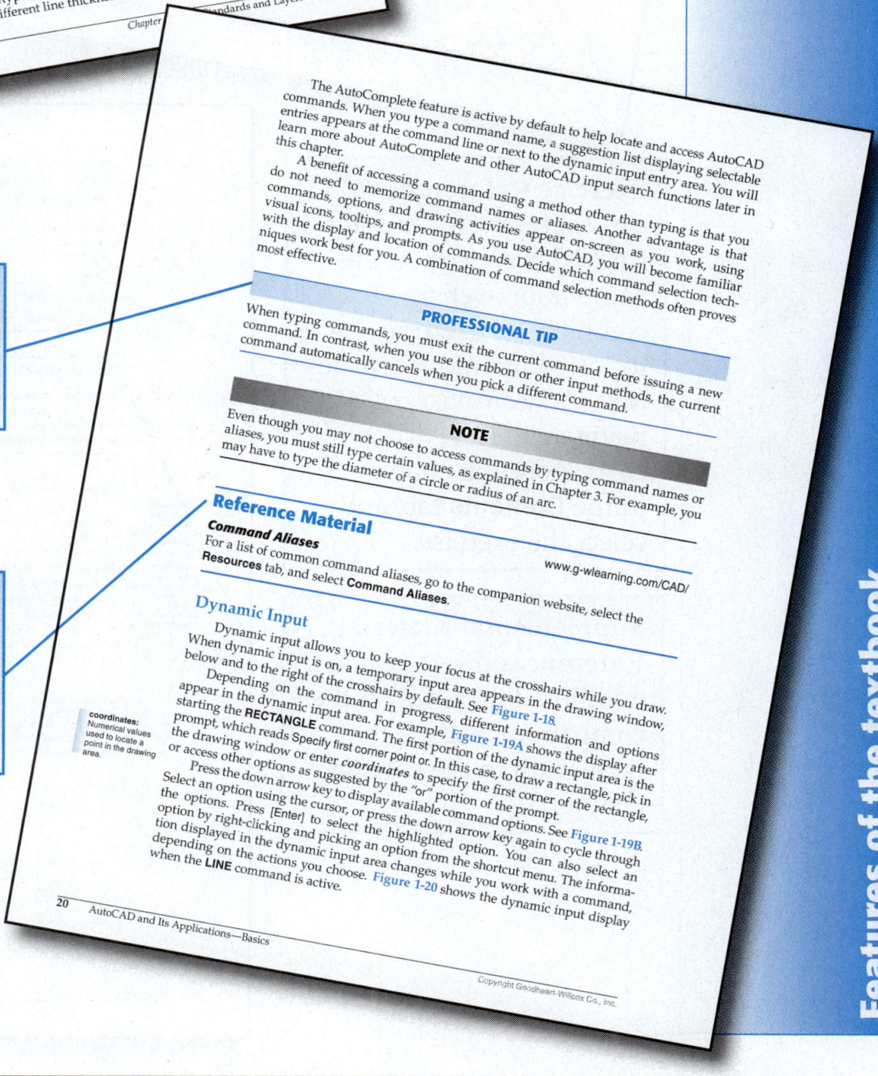

The AutoComplete feature is active by default to help locate and access AutoCAD commands. When you type a command name, a suggestion list displaying selectable entries appears at the command line or next to the dynamic input entry area. You will learn more about AutoComplete and other AutoCAD input search functions later in this chapter.

A benefit of accessing a command using a method other than typing is that you do not need to memorize command names or aliases. Another advantage is that commands, options, and drawing activities appear on-screen as you work, using visual icons, tooltips, and prompts. As you use AutoCAD, you will become familiar with the display and location of commands. Decide which command selection techniques work best for you. A combination of command selection methods often proves most effective.

PROFESSIONAL TIP

When typing commands, you must exit the current command before issuing a new command. In contrast, when you use the ribbon or other input methods, the current command automatically cancels when you pick a different command.

NOTE

Even though you may not choose to access commands by typing command names or aliases, you must still type certain values, as explained in Chapter 3. For example, you may have to type the diameter of a circle or radius of an arc.

Reference Material

Command Aliases

For a list of common command aliases, go to the companion website, select the **Resources** tab, and select **Command Aliases**.

www.g-wlearning.com/CAD/

Dynamic Input

Dynamic input allows you to keep your focus at the crosshairs while you draw. When dynamic input is on, a temporary input area appears in the drawing window, below and to the right of the crosshairs by default. See Figure 1-18.

Depending on the command in progress, different information and options appear in the dynamic input area. For example, Figure 1-19A shows the display after starting the **RECTANGLE** command. The first portion of the dynamic input area is the prompt, which reads Specify first corner point or. In this case, to draw a rectangle, pick in the drawing window or enter ***coordinates*** to specify the first corner of the rectangle, or access other options as suggested by the "or" portion of the prompt.

Press the down arrow key to display available command options. See Figure 1-19B. Select an option using the cursor, or press the down arrow key again to cycle through the options. Press [Enter] to select the highlighted option. You can also select an option by right-clicking and picking an option from the shortcut menu. The information displayed in the dynamic input area changes while you work with a command, depending on the actions you choose. Figure 1-20 shows the dynamic input display when the **LINE** command is active.

coordinates: Numerical values used to locate a point in the drawing area.

20 AutoCAD and Its Applications—Basics

Copyright Goodheart-Willcox Co., Inc.

Professional Tips increase your productivity in using AutoCAD commands and techniques.

Reference Material References direct you to charts, tables, and other useful references available on the companion website.

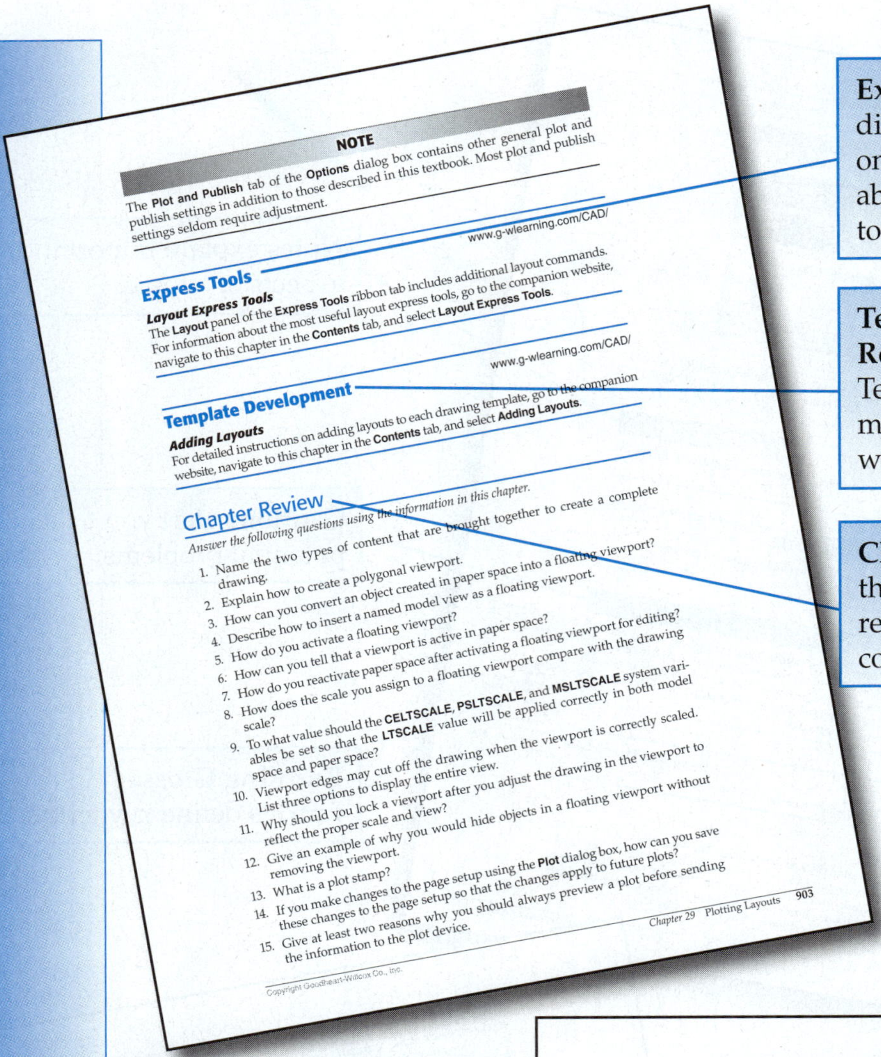

NOTE

The **Plot and Publish** tab of the **Options** dialog box contains other general plot and publish settings in addition to those described in this textbook. Most plot and publish settings seldom require adjustment.

Express Tools www.g-wlearning.com/CAD/

Layout Express Tools

The **Layout** panel of the **Express Tools** ribbon tab includes additional layout commands. For information about the most useful layout express tools, go to the companion website, navigate to this chapter in the **Contents** tab, and select **Layout Express Tools**.

Template Development www.g-wlearning.com/CAD/

Adding Layouts

For detailed instructions on adding layouts to each drawing template, go to the companion website, navigate to this chapter in the **Contents** tab, and select **Adding Layouts**.

Chapter Review

Answer the following questions using the information in this chapter.

1. Name the two types of content that are brought together to create a complete drawing.
2. Explain how to create a polygonal viewport.
3. How can you convert an object created in paper space into a floating viewport?
4. Describe how to insert a named model view as a floating viewport.
5. How do you activate a floating viewport?
6. How can you tell that a viewport is active in paper space?
7. How do you reactivate paper space after activating a floating viewport for editing?
8. How does the scale you assign to a floating viewport compare with the drawing scale?
9. To what value should the **CELTSCALE**, **PSLTSCALE**, and **MSLTSCALE** system variables be set so that the **LTSCALE** value will be applied correctly in both model space and paper space?
10. Viewport edges may cut off the drawing when the viewport is correctly scaled. List three options to display the entire view.
11. Why should you lock a viewport after you adjust the drawing in the viewport to reflect the proper scale and view?
12. Give an example of why you would hide objects in a floating viewport without removing the viewport.
13. What is a plot stamp?
14. If you make changes to the page setup using the **Plot** dialog box, how can you save these changes to the page setup so that the changes apply to future plots?
15. Give at least two reasons why you should always preview a plot before sending the information to the plot device.

Chapter 29 Plotting Layouts 903

Copyright Goodheart-Willcox Co., Inc.

Express Tools References direct you to information on the companion website about AutoCAD express tools.

Template Development References direct you to Template Development material on the companion website.

Chapter Reviews reinforce the knowledge gained by reading the chapter and completing the exercises.

Exercise References direct you to step-by-step tutorial exercises on the companion website. To complete an exercise, go to the companion website (www.g-wlearning.com/CAD/), navigate to the corresponding chapter in the **Contents** tab, and select the exercise.

Supplemental Material References direct you to additional material on the companion website that is relevant to the current chapter.

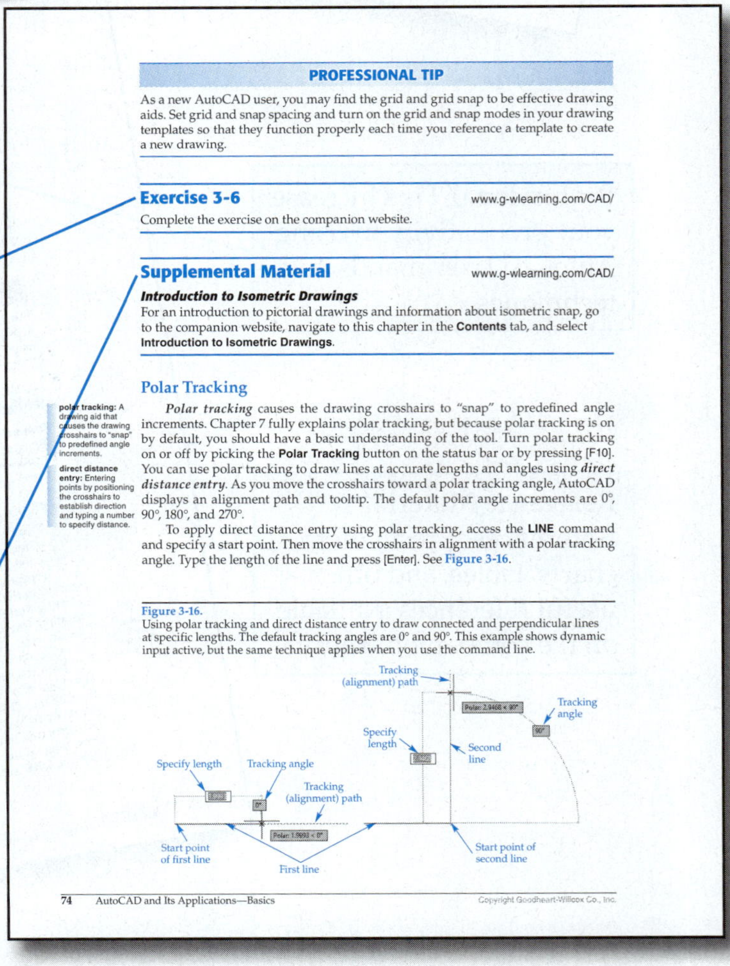

PROFESSIONAL TIP

As a new AutoCAD user, you may find the grid and grid snap to be effective drawing aids. Set grid and snap spacing and turn on the grid and snap modes in your drawing templates so that they function properly each time you reference a template to create a new drawing.

Exercise 3-6 www.g-wlearning.com/CAD/

Complete the exercise on the companion website.

Supplemental Material www.g-wlearning.com/CAD/

Introduction to Isometric Drawings

For an introduction to pictorial drawings and information about isometric snap, go to the companion website, navigate to this chapter in the **Contents** tab, and select **Introduction to Isometric Drawings**.

Polar Tracking

polar tracking: A drawing aid that causes the drawing crosshairs to "snap" to predefined angle increments.

direct distance entry: Entering points by positioning the crosshairs to establish direction and typing a number to specify distance.

Polar tracking causes the drawing crosshairs to "snap" to predefined angle increments. Chapter 7 fully explains polar tracking, but because polar tracking is on by default, you should have a basic understanding of the tool. Turn polar tracking on or off by picking the **Polar Tracking** button on the status bar or by pressing [F10]. You can use polar tracking to draw lines at accurate lengths and angles using *direct distance entry*. As you move the crosshairs toward a polar tracking angle, AutoCAD displays an alignment path and tooltip. The default polar angle increments are 0°, 90°, 180°, and 270°.

To apply direct distance entry using polar tracking, access the **LINE** command and specify a start point. Then move the crosshairs in alignment with a polar tracking angle. Type the length of the line and press [Enter]. See Figure 3-16.

Figure 3-16.
Using polar tracking and direct distance entry to draw connected and perpendicular lines at specific lengths. The default tracking angles are 0° and 90°. This example shows dynamic input active, but the same technique applies when you use the command line.

74 AutoCAD and Its Applications—Basics

Copyright Goodheart-Willcox Co., Inc.

Features of the textbook

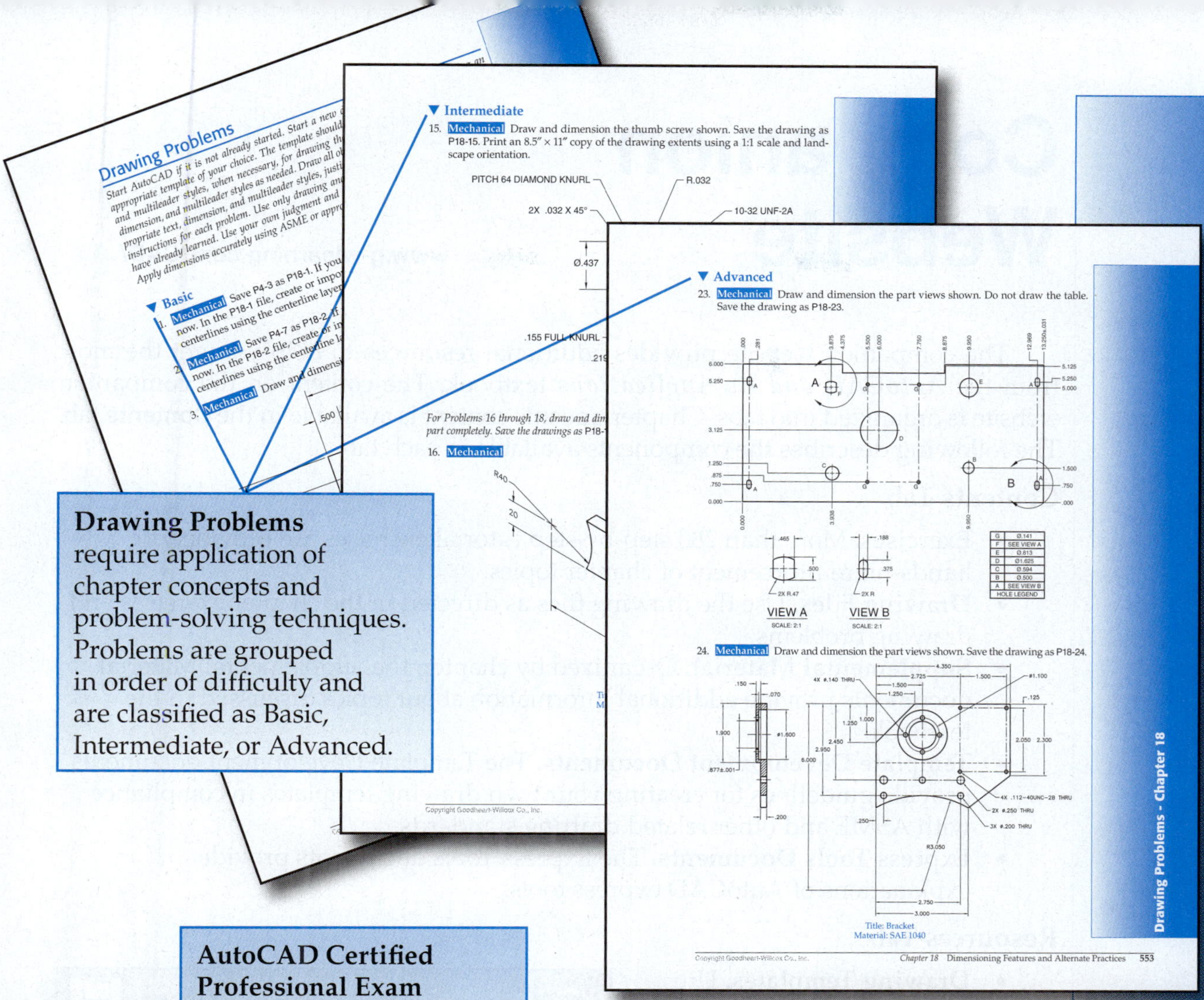

Drawing Problems require application of chapter concepts and problem-solving techniques. Problems are grouped in order of difficulty and are classified as Basic, Intermediate, or Advanced.

AutoCAD Certified Professional Exam Practice questions help prepare you for the AutoCAD professional level certification exam.

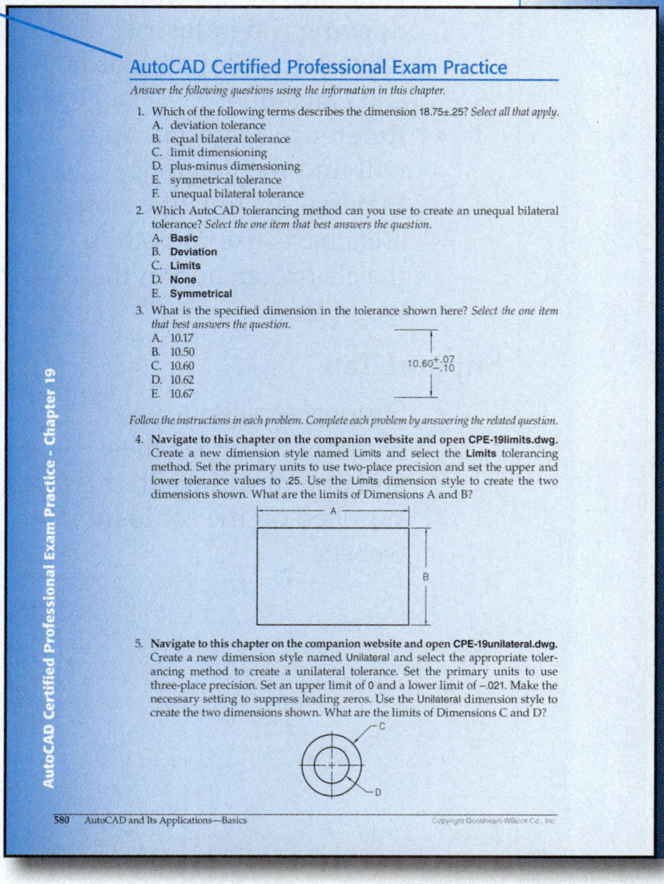

Discipline labels identify problems from different drafting disciplines.

Discipline Labels

Mechanical	Civil
Architectural	Graphic Design
Electronics	Piping
General	

Features of the textbook

Companion Website

www.g-wlearning.com/CAD/

The companion website provides additional resources to help you get the most from the ***AutoCAD and Its Applications*** textbook. The content on the companion website is organized into tabs. Chapter-specific content is available in the **Contents** tab. The following describes the components available in each tab.

Contents Tab

- **Exercises.** More than 280 step-by-step tutorial exercises are provided for hands-on reinforcement of chapter topics.
- **Drawing Files.** Use the drawing files as directed in the chapter exercises and drawing problems.
- **Supplemental Material.** Organized by chapter, the Supplemental Material documents provide additional information about topics discussed in the textbook.
- **Template Development Documents.** The Template Development documents provide guidelines for creating your own drawing templates in compliance with ASME and other related drafting standards.
- **Express Tools Documents.** The Express Tools documents provide explanations of AutoCAD express tools.

Resources Tab

- **Drawing Templates.** Use the predefined drawing templates to base your drawings on industry-related drawing standards and conventions.
- **Reference Material.** You will find the Reference Material documents, tables, and charts useful both in the classroom and in the workplace.

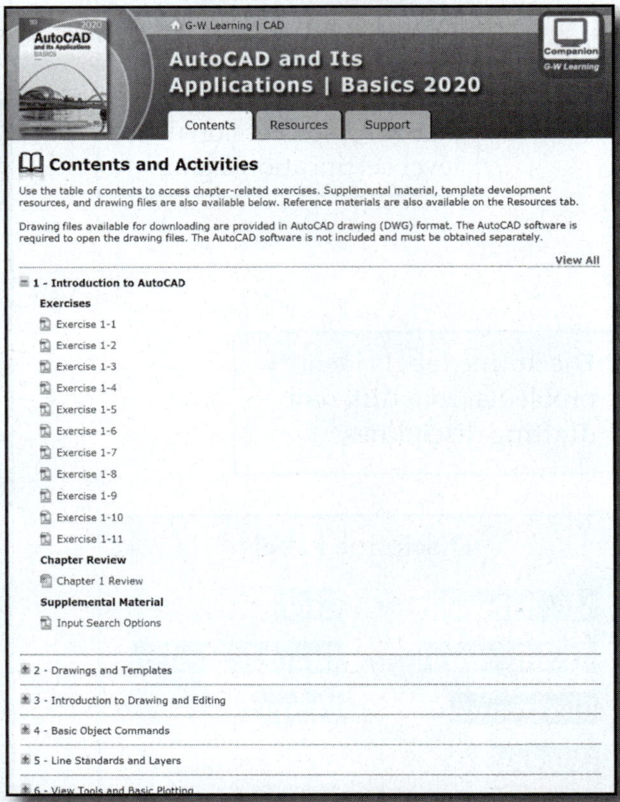

Support Tab

- Support topics address working with PDF files and downloading ZIP files supplied on the companion website.

Online Instructor Resources

The Online Instructor Resources provide support materials designed to help instructors effectively plan and teach AutoCAD courses. Support materials are available in one accessible location and include tips for the instructor, answer keys, drawing problem solution files, course syllabi, final exams, and practice certification exam questions. Also included is assessment software with question banks featuring hundreds of questions corresponding to each chapter in the text. The Online Instructor Resources are available as a subscription and can be accessed at school or at home.

Fonts Used in This Textbook

Different typefaces are used throughout this textbook to define terms and identify AutoCAD commands. The following typeface conventions are used in this textbook:

Text Element	Example
AutoCAD commands	**LINE** command
AutoCAD menu and ribbon selections	**Draw** > **Arc** > **3 Points**
AutoCAD system variables	**LTSCALE** system variable
AutoCAD toolbars and buttons	**Quick Access** toolbar, **Undo** button
AutoCAD dialog boxes	**Insert Table** dialog box
Keyboard entry (in text)	Type LINE
Keyboard keys	[Ctrl]+[1] key combination
File names, folders, and paths	C:\Projects\Blocks\Desk.dwg
Microsoft Windows features	Start menu, Start screen, Programs folder
Prompt sequence	Command:
Keyboard input at prompt sequence	Command: **L** *or* **LINE**↵
Comment at a prompt sequence	Specify first point: (*pick a point or press* [Enter])

Other Text References

For additional information, standards from organizations such as ANSI (American National Standards Institute) and ASME (American Society of Mechanical Engineers) are referenced throughout the textbook. Use these standards to create drawings that follow industry, national, and international practices.

Trademarks

Autodesk, the Autodesk logo, DWG, the DWG logo, and AutoCAD are registered trademarks or trademarks of Autodesk, Inc., and/or its subsidiaries and/or affiliates in the USA and other countries.

Microsoft, Windows, Windows 10, Windows 8, and Windows 7 are registered trademarks of Microsoft Corporation in the United States and/or other countries.

About the Authors

Terence M. Shumaker is Faculty Emeritus, the former Chairperson of the Drafting Technology Department, and former Director of the Autodesk Premier Training Center at Clackamas Community College in Oregon City, Oregon. Terence taught at the community college level for over 28 years. He worked as a training consultant for Autodesk, Inc., and conducted CAD program development workshops around the country. He has professional experience in surveying, civil drafting, industrial piping, and technical illustration. He is the author of Goodheart-Willcox's ***Process Pipe Drafting*** and coauthor of the ***AutoCAD and Its Applications*** series.

David A. Madsen is the president of Madsen Designs Inc. (www.madsendesigns.com). David is Faculty Emeritus and the former Chairperson of Drafting Technology and the Autodesk Premier Training Center at Clackamas Community College in Oregon City, Oregon. David was an instructor and department Chairperson at Clackamas Community College for nearly 30 years. In addition to community college experience, David was a Drafting Technology instructor at Centennial High School in Gresham, Oregon. David is a former member of the American Design Drafting Association (ADDA) Board of Directors, and was honored by the ADDA with Director Emeritus status at the annual conference in 2005. David has extensive experience in mechanical drafting, architectural design and drafting, and building construction. David holds a Master of Education degree in Vocational Administration and a Bachelor of Science degree in Industrial Education. David is the author of ***Geometric Dimensioning and Tolerancing*** and coauthor of ***Architectural Drafting Using AutoCAD***, the ***AutoCAD and Its Applications*** series, and other textbooks in the areas of architectural drafting, mechanical drafting, engineering drafting, civil drafting, architectural print reading, and mechanical print reading.

David P. Madsen is the president of Engineering Drafting & Design, Inc., and the vice president of Madsen Designs Inc. (www.madsendesigns.com). Dave provides drafting and design consultation and training for all disciplines. Dave has been a professional design drafter since 1996, and has extensive experience in a variety of drafting, design, and engineering disciplines. Dave has provided drafting and computer-aided design and drafting instruction to secondary and postsecondary learners since 1999, and has considerable curriculum, program coordination, and development experience. Dave holds a Master of Science degree in Educational Policy, Foundations, and Administrative Studies with a specialization in Postsecondary, Adult, and Continuing Education; a Bachelor of Science degree in Technology Education; and an Associate of Science degree in General Studies and Drafting Technology. Dave is the author of ***Inventor and Its Applications*** and coauthor of ***Architectural Drafting Using AutoCAD, AutoCAD and Its Applications—Basics, AutoCAD and Its Applications—Comprehensive, Geometric Dimensioning and Tolerancing***, and other textbooks in the areas of architectural drafting, mechanical drafting, engineering drafting, civil drafting, architectural print reading, and mechanical print reading.

ADDA Technical Publication

The content of this text is considered a fundamental component to the design drafting profession by the American Design Drafting Association. This publication covers topics and related material relevant to the delivery of the design drafting process. Although this publication is not conclusive, it should be considered a key reference tool in furthering the knowledge, abilities, and skills of a properly trained designer or drafter in the pursuit of a professional career.

BRIEF CONTENTS

1 Introduction to AutoCAD . . . 1
2 Drawings and Templates . . . 33
3 Introduction to Drawing and Editing . . . 61
4 Basic Object Commands . . . 95
5 Line Standards and Layers . . . 127
6 View Tools and Basic Plotting . . . 163
7 Object Snap and AutoTrack . . . 193
8 Construction Tools and Multiview Drawings . . . 221
9 Text Styles and Multiline Text . . . 249
10 Single-Line Text and Additional Text Tools . . . 285
11 Modifying Objects . . . 305
12 Arranging and Patterning Objects . . . 333
13 Grips, Properties, and Additional Selection Techniques . . . 365
14 Polyline and Spline Editing Tools . . . 401
15 Obtaining Drawing Information . . . 427
16 Dimension Standards and Styles . . . 455
17 Linear and Angular Dimensioning . . . 487
18 Dimensioning Features and Alternate Practices . . . 511
19 Dimensioning with Tolerances . . . 559
20 Editing Dimensions . . . 581
21 Tables . . . 615
22 Parametric Drafting . . . 649
23 Section Views and Graphic Patterns . . . 691
24 Standard Blocks . . . 729
25 Block Attributes . . . 761
26 Introduction to Dynamic Blocks . . . 783
27 Additional Dynamic Block Tools . . . 819
28 Layout Setup . . . 845
29 Plotting Layouts . . . 875
30 Annotative Objects . . . 909
31 External References . . . 941
32 Introduction to Sheet Sets . . . 969
33 Additional Sheet Set Tools . . . 989

EXPANDED CONTENTS

Chapter 1

Introduction to AutoCAD 1

AutoCAD Applications 1
Before You Begin 3
Starting AutoCAD 5
Exiting AutoCAD 6
The AutoCAD Interface 6
Accessing Commands 19
Getting Help 27

Chapter 2

Drawings and Templates 33

Starting a New Drawing 33
Saving Your Work 36
Closing Files 40
Opening a Saved File 40
Managing Multiple Documents 45
Drawing Properties 49
Basic Drawing Settings 51
Introduction to Templates 56

Chapter 3

Introduction to Drawing and Editing 61

Responding to Prompts 61
Introduction to Drawing 63
Introduction to Drawing Aids 70
Introduction to Editing 77
Object Selection 79

Chapter 4

Basic Object Commands 95

Drawing Circles 95
Drawing Arcs 99
Drawing Ellipses 102
Drawing Polylines 105
Drawing Regular Polygons 108
Drawing Rectangles 109
Drawing Donuts and Filled Circles 112
Drawing True Splines 112

Chapter 5
Line Standards and Layers 127
Line Standards 127
Introduction to Layers 131
Layers Used in Drafting Fields 131
Creating and Using Layers 131
Reusing Drawing Content 149

Chapter 6
View Tools and Basic Plotting 163
View Tools 163
Using Commands Transparently 175
Controlling Draw Order 175
Named Views 176
Model Viewports 180
Introduction to Printing and Plotting 184

Chapter 7
Object Snap and AutoTrack 193
Object Snap 193
AutoTrack 206

Chapter 8
Construction Tools and Multiview Drawings 221
Parallel Offsets 221
Drawing Points 223
Construction Lines 226
Multiview Drawings 228
Multiview Drawing Construction 232

Chapter 9
Text Styles and Multiline Text 249
Text Standards and Composition 249
Drawing Scale and Text Height 252
Text Styles 255
Multiline Text 259

Chapter 10
Single-Line Text and Additional Text Tools 285
Single-Line Text 285
Fields 289
Checking Spelling 291
Revising Text 294

Chapter 11
Modifying Objects . . . 305
Using the FILLET Command . . . 305
Using the CHAMFER Command . . . 308
Using the BLEND Command . . . 310
Using the BREAK Command . . . 310
Using the JOIN Command . . . 312
Trimming . . . 313
Extending . . . 315
Stretching . . . 317
Using the LENGTHEN Command . . . 319
Using the SCALE Command . . . 321
Exploding Objects . . . 322

Chapter 12
Arranging and Patterning Objects . . . 333
Moving Objects . . . 333
Rotating Objects . . . 335
Aligning Objects . . . 337
Copying Objects . . . 338
Mirroring Objects . . . 339
Reversing an Object's Point Calculation . . . 340
Arraying Objects . . . 341

Chapter 13
Grips, Properties, and Additional Selection Techniques . . . 365
Grips . . . 365
Editing Associative Arrays . . . 373
Object Properties . . . 380
Matching Properties . . . 387
Editing between Drawings . . . 387
Add Selected . . . 388
Select Similar . . . 389
Quick Select . . . 390

Chapter 14
Polyline and Spline Editing Tools . . . 401
Using the PEDIT Command . . . 401
Polyline Grip Commands . . . 407
Creating a Polyline Boundary . . . 409
Using the SPLINEDIT Command . . . 410
Spline Grip Commands . . . 414

Chapter 15
Obtaining Drawing Information 427
Taking Measurements..427
Listing Drawing Data..434
Reviewing the Drawing Status..................................435
Checking the Time..435
Using QuickCalc...437
Comparing Drawings..443

Chapter 16
Dimension Standards and Styles 455
Dimension Standards and Practices..............................455
Drawing Scale and Dimensions..................................460
Dimension Styles..463

Chapter 17
Linear and Angular Dimensioning 487
Introduction to Placing Dimensions.............................487
Linear Dimensions..488
Dimensioning Angles..494
Baseline, Chain, and Direct Dimensioning........................497
Adjusting Overlapping Dimensions...............................501
Dimension Layer Override.......................................502
Using QDIM to Dimension..503

Chapter 18
Dimensioning Features and Alternate Practices 511
Dimensioning Circles, Arcs, and Other Curves....................511
Adding Center Dashes and Centerlines...........................517
Drawing Leader Lines...519
Multileader Styles..519
Inserting Multileaders..530
Alternate Dimensioning Practices...............................534
Creating Ordinate Dimension Objects............................536
Marking Up Drawings..539

Chapter 19
Dimensioning with Tolerances 559
Tolerancing Fundamentals.......................................559
Dimensioning Units...561
Setting Primary Units...562
Setting Tolerance Methods.......................................564
Specifying Tolerance Settings...................................568
Introduction to GD&T..570

Chapter 20
Editing Dimensions . . . 581
Associative Dimensioning . . . 581
Dimension Editing Tools . . . 583
Overriding Dimension Style . . . 591
Spacing and Aligning Dimensions . . . 593
Using the DIMBREAK Command . . . 596
Creating Inspection Dimensions . . . 597
Center Mark and Centerline Editing Tools . . . 599
Multileader Editing Tools . . . 601

Chapter 21
Tables . . . 615
Table Styles . . . 615
Inserting Tables . . . 623
Editing Tables . . . 628
Calculating Values in Tables . . . 638

Chapter 22
Parametric Drafting . . . 649
Parametric Fundamentals . . . 649
Geometric Constraints . . . 657
Dimensional Constraints . . . 668
Parametric Editing . . . 678

Chapter 23
Section Views and Graphic Patterns . . . 691
Section Views . . . 691
Using the HATCH Command . . . 696
Hatching Using DesignCenter . . . 709
Hatching Using Tool Palettes . . . 710
Editing Hatch Patterns . . . 711

Chapter 24
Standard Blocks . . . 729
Constructing Blocks . . . 729
Inserting Blocks . . . 733
Editing Blocks . . . 740
Creating Blocks as Drawing Files . . . 746
Symbol Libraries . . . 747
Purging Named Objects . . . 749

Chapter 25

Block Attributes 761

Defining Attributes 761
Creating Blocks with Attributes 764
Inserting Blocks with Attributes 766
Editing Attribute References 767
Editing Attribute Definitions 770
Automating Drafting Documentation 772
Using Fields to Reference Attributes 777

Chapter 26

Introduction to Dynamic Blocks 783

Dynamic Block Fundamentals 783
Point Parameters 786
Testing and Adjusting Dynamic Properties 789
Linear Parameters 791
Polar Parameters 796
Rotation Parameters 799
Alignment Parameters 801
Flip Parameters 803
XY Parameters 804
Base Point Parameters 806
Parameter Value Sets 806
Chain Actions 808

Chapter 27

Additional Dynamic Block Tools 819

Visibility Parameters 819
Lookup Parameters 823
Parameter Sets 828
Constraining Block Geometry 828
Block Properties Tables 834

Chapter 28

Layout Setup 845

Introduction to Layouts 845
Working with Layouts 846
Initial Layout Setup 852
Page Setups 852

Chapter 29

Plotting Layouts 875

Layout Content 875
Floating Viewports 878
Plotting 899

Chapter 30
Annotative Objects 909
Introduction to Annotative Objects 909
Defining Annotative Objects 910
Drawing Annotative Objects 914
Displaying Annotative Objects in Layouts 917
Changing Drawing Scale 919
Preparing Multiview Drawings. 926

Chapter 31
External References 941
Introduction to Xrefs. 941
Preparing Xref and Host Drawings 945
Placing Xref Drawings 945
Working with Xref Objects. 949
Managing Xrefs 952
Binding an Xref 960
Editing Xref Drawings 961

Chapter 32
Introduction to Sheet Sets 969
Sheet Set Fundamentals 969
Creating Sheet Sets 973
Managing Sheet Sets. 980
Subsets 981
Sheets 982

Chapter 33
Additional Sheet Set Tools. 989
Sheet Views 989
Model Views. 997
Sheet List Tables. 1000
Sheet Set Fields 1003
Publishing a Sheet Set. 1008
Archiving a Sheet Set 1010

Index 1017

Companion Website Contents

www.g-wlearning.com/CAD/

Contents Tab

- Exercises
- Drawing Files
- Supplemental Material
- Template Development Documents
- Express Tools Documents

Resources Tab

- Drawing Templates
- Reference Material

Support Tab

- Support and Information

Students can download the AutoCAD software and other Autodesk software products for free from the Autodesk Education Community website. A user account is required. The Autodesk Education Community website provides many other resources, including discussion groups, help videos, and career information.

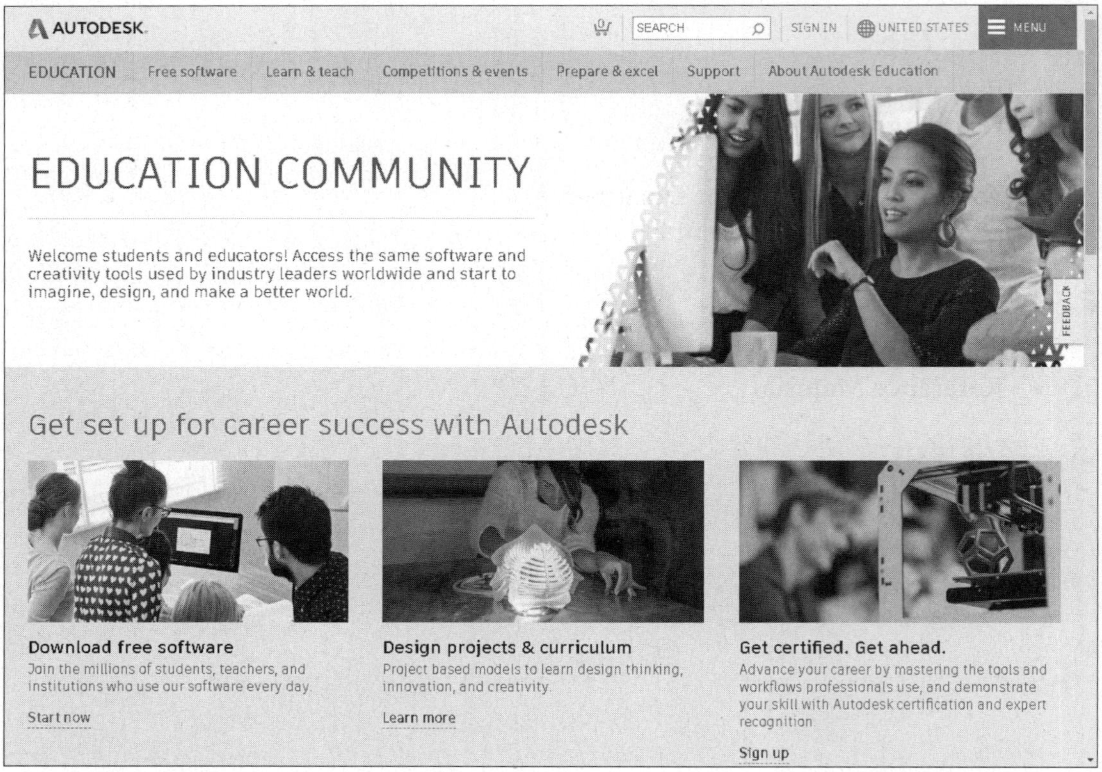

CHAPTER 1

Introduction to AutoCAD

Learning Objectives

After completing this chapter, you will be able to:

- ✓ Define computer-aided design and drafting.
- ✓ Describe typical AutoCAD applications.
- ✓ Explain the value of planning your work and system management.
- ✓ Describe the purpose and importance of drawing standards.
- ✓ Demonstrate how to start and exit AutoCAD.
- ✓ Recognize the AutoCAD interface and access AutoCAD commands.
- ✓ Use help resources.

Computer-aided design and drafting (CADD) is the process of using a computer with CADD software to design and produce drawings and models according to specific industry and company standards. The terms *computer-aided design (CAD)* and *computer-aided drafting (CAD)* refer to specific aspects of the CADD process. This chapter introduces the AutoCAD CADD system. You will begin working with AutoCAD and learn to control the AutoCAD environment.

computer-aided design and drafting (CADD): The process of using a computer with CADD software to design and produce drawings and models.

AutoCAD Applications

AutoCAD ***commands*** and ***options*** allow you to draw objects of any size and shape. Use AutoCAD to prepare two-dimensional (2D) drawings, three-dimensional (3D) models, and animations. AutoCAD is a universal CADD software program that applies to any drafting, design, or engineering discipline. For example, use AutoCAD to design and document mechanical parts and assemblies, architectural buildings, civil and structural engineering projects, and electronics.

command: An instruction issued to the computer to complete a specific task. For example, use the **LINE** command to draw line objects.

option: A choice associated with a command or an alternative function of a command.

2D Drawings

2D drawings display object length and width, width and height, or height and length in a flat (2D) form. 2D drawings are the established design and drafting format and are common in all engineering and architectural industries and related disciplines.

A complete 2D drawing typically includes dimensions, notes, and symbols that describe view features and information. This practice results in a document used to manufacture or construct a product. 2D drawings are the conventional and often required method of communicating a project. Figure 1-1 shows an example of a 2D architectural floor plan created using AutoCAD. Use this textbook to learn how to construct, design, dimension, and annotate 2D AutoCAD drawings.

3D Models

3D models allow for advanced visualization, simulation, and analysis typically not possible with 2D drawings. AutoCAD provides commands and options for developing ***wireframe***, ***surface***, and ***solid models***. An accurate solid model is an exact digital representation of a product. Add color, lighting, and texture to display a realistic view of the model. See Figure 1-2A. Use view tools to view the model from any angle. See Figure 1-2B. Apply animation to a model to show product design or function. For example, you can perform a ***walkthrough*** of a model home or a ***flythrough*** of a model civil engineering project. *AutoCAD and Its Applications—Advanced* provides detailed instruction on 3D modeling and rendering.

wireframe model: The most basic 3D model—contains only information about object edges and the points where edges intersect, known as *vertices*; describes the appearance of the model as if it were constructed from wires.

surface model: A 3D model that contains information about object edges, vertices, and the outer boundaries of the object, known as *surfaces*; surface models have zero thickness, lack mass, and may not enclose a volume.

solid model: The most complex 3D model—contains information about object edges, vertices, surfaces, and mass; solid models enclose a volume.

walkthrough: A computer simulation that replicates walking through or around a 3D model.

flythrough: A computer simulation that replicates flying through or around a 3D model.

Reference Material

www.g-wlearning.com/CAD/

Glossary of CADD Terms

For a detailed glossary of CADD and AutoCAD terms, go to the companion website, select the **Resources** tab, and select **Glossary of CADD Terms**.

Figure 1-1.
AutoCAD provides commands and options to create accurate 2D drawings for building design and construction, such as this architectural floor plan of a home.

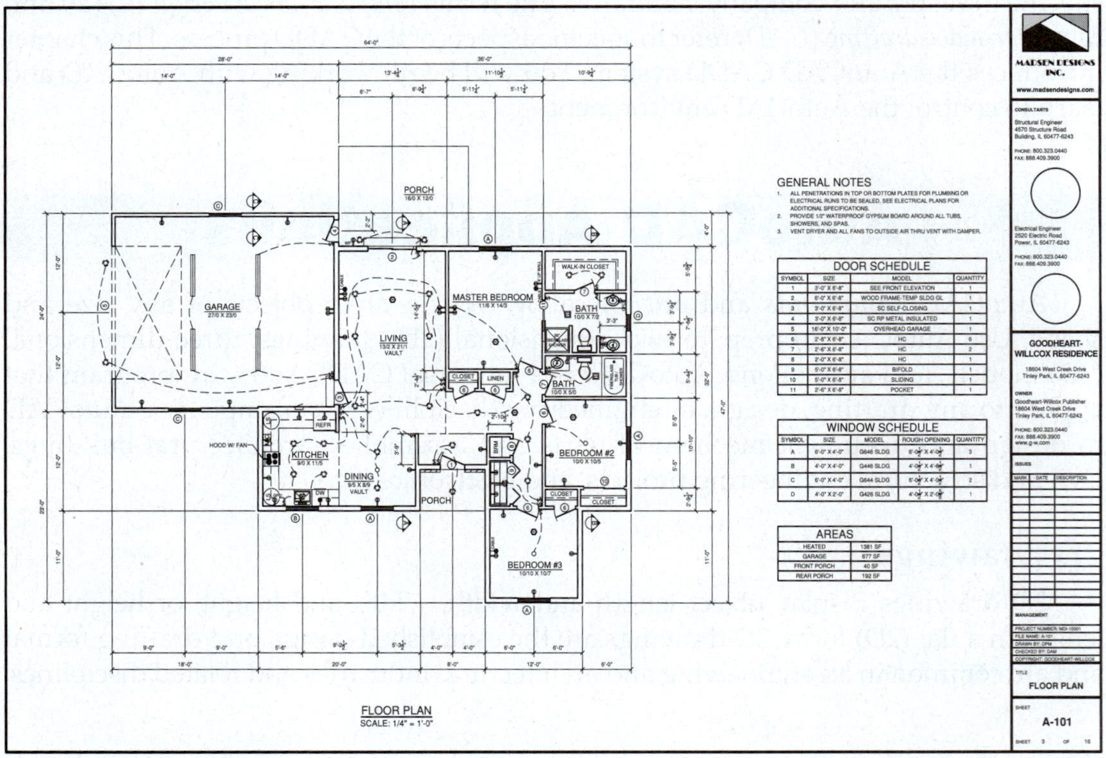

Figure 1-2.
A 3D AutoCAD model of a mechanical assembly. A—A wireframe visual style (left) and a conceptual visual style with color, lighting, and texture (right). B—A hidden visual style (left) and the same display viewed at an alternate angle and zoomed in (right).

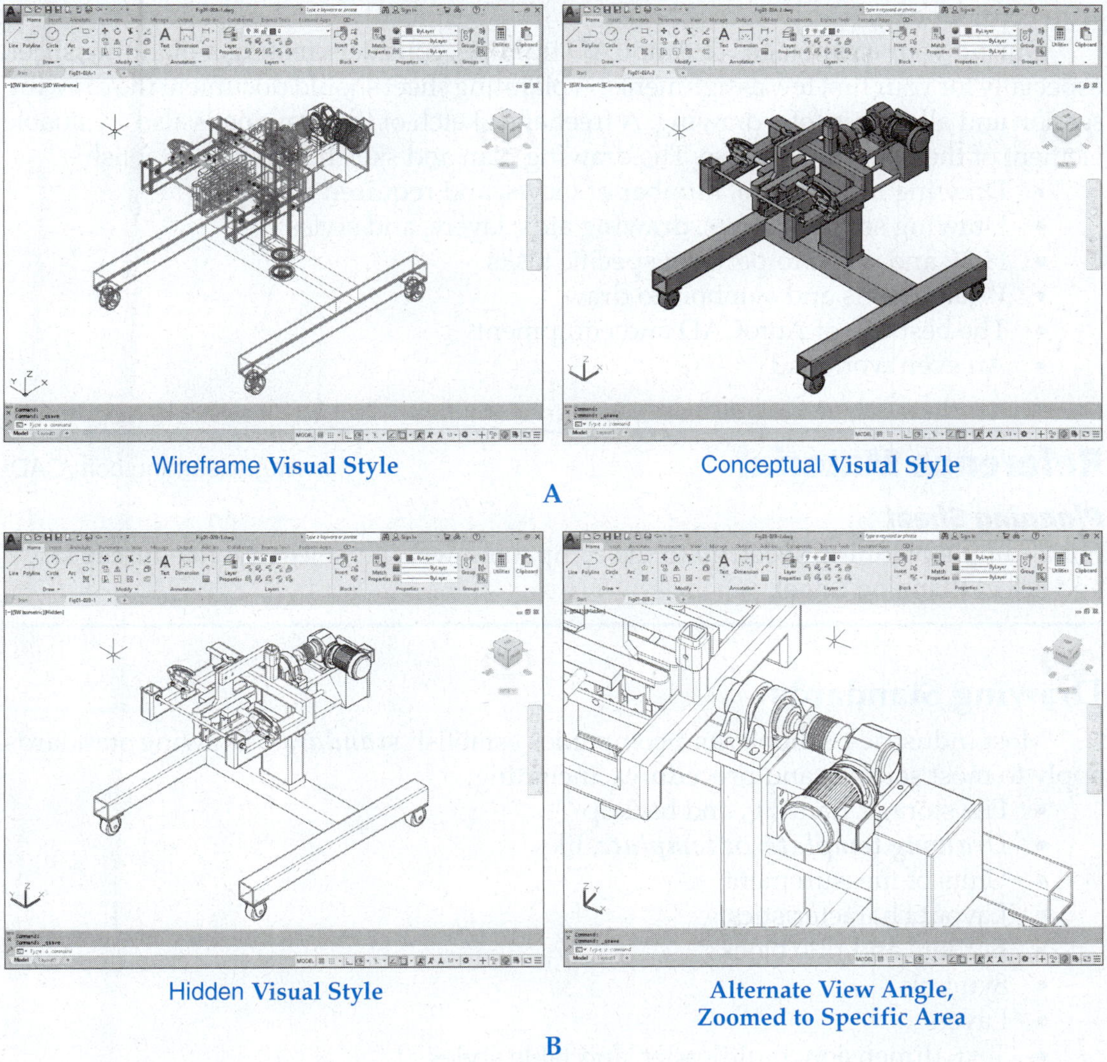

Before You Begin

Designing and drafting effectively with a computer requires a skilled CADD operator. To be a proficient AutoCAD user, you must have detailed knowledge of AutoCAD commands and processes, and know which command and process is best suited for a specific task. You must also understand and be able to apply design and drafting systems and conventions when using AutoCAD.

As you begin your CADD training, develop effective methods for managing your work. First, plan your ***drawing sessions*** thoroughly to organize your thoughts. Second, learn and use industry, classroom, or office standards. Third, save your work often. If you follow these procedures, you will find it easier to use AutoCAD commands and methods, and your drawing experience will be more productive and enjoyable.

drawing sessions: Time spent working on a drawing project, including analyzing design parameters and using AutoCAD.

Planning Your Work

A drawing plan involves thinking about the entire process or project in which you are involved. Your drawing plan focuses on the content you want to present, the objects and symbols you intend to create, and the appropriate use of standards. You may want processes to be automatic or to happen immediately, but if you hurry and do little or no planning, you may become frustrated and waste time while drawing. Take as much time as needed to develop drawing and project goals so that you can proceed with confidence.

During your early stages of AutoCAD training, consider creating a planning sheet, especially for your first few assignments. A planning sheet should document the drawing session and all aspects of a drawing. A freehand sketch of the drawing is also a valuable element of the planning process. The drawing plan and sketch help you establish:

- Drawing layout: area, number of views, and required free space
- Drawing settings: units, drawing aids, layers, and styles
- How and when to perform specific tasks
- What objects and symbols to draw
- The best use of AutoCAD and equipment
- An even workload

Reference Material

www.g-wlearning.com/CAD/

Planning Sheet

For a sample planning sheet, go to the companion website, select the **Resources** tab, and select **Planning Sheet**.

Drawing Standards

standards: Guidelines that specify drawing requirements, appearance, techniques, operating procedures, and record-keeping methods.

drawing template (template): A file that contains standard drawing settings and objects for use in new drawings.

Most industries, schools, and companies establish ***standards***. Drawing standards apply to most settings and procedures, including:

- File storage, naming, and backup
- ***Drawing template***, or ***template***, files
- Units of measurement
- Layout characteristics
- Borders and title blocks
- Symbols
- Layers
- Text, dimension, multileader, and table styles
- Plot styles and plotting

Company or school drawing standards should follow appropriate national industry standards whenever possible. Although standards vary in content, the most important aspect is that standards exist and are understood and used by all CADD personnel. When you follow drawing standards, your drawings are consistent, you become more productive, and the classroom or office functions more efficiently.

This textbook presents mechanical drafting standards developed by the American Society of Mechanical Engineers (ASME) and accredited by the American National Standards Institute (ANSI). This textbook also references International Standards Organization (ISO) mechanical drafting standards and discipline-specific standards when appropriate, including the United States National CAD Standard® (NCS) and American Welding Society (AWS) standards.

Reference Material

www.g-wlearning.com/CAD/

Drawing Standards

For more information about drawing standards, go to the companion website, select the **Resources** tab, and select **Drawing Standards**.

NOTE

You may consider other drafting standards when preparing drawings, such as the ***BSI***, ***DIN***, ***GB***, ***GOST***, and ***JIS*** standards.

BSI: British Standards Institution.

DIN: Deutsches Institut Für Normung, established by the German Institute for Standardization.

GB: Chinese Guóbiāo.

GOST: Gosudarstvennyy Standart, maintained by the Euro-Asian Council for Standardization, Metrology, and Certification.

JIS: Japanese Industrial Standards.

Saving Your Work

Drawings are lost due to software error, hardware malfunction, power failure, or accident. Prepare for such an event by saving your work frequently. Develop a habit of saving your work at least every 10 to 15 minutes. You can set the automatic save option, described in Chapter 2, to save drawings automatically at set intervals. However, you should also frequently save your work manually.

Working Procedures Checklist

Proficient use of AutoCAD requires several skills. Use the following checklist to become comfortable with AutoCAD, and to help you work quickly and efficiently:

- ✓ Carefully plan your work.
- ✓ Frequently check object and drawing settings, such as layers, styles, and properties, to see which object characteristics and drawing options are in effect.
- ✓ Follow the prompts, tooltips, notifications, and ***alerts*** that appear as you work.
- ✓ Constantly check for the correct options, instructions, or keyboard entry.
- ✓ ***Right-click*** to access shortcut menus and review available options.
- ✓ Think ahead to prepare for each stage of the drawing session.
- ✓ Learn commands, tools, and options that increase your speed and efficiency.
- ✓ Save your work at least every 10 to 15 minutes.
- ✓ Learn to use available resources, such as this textbook, to help solve problems and answer questions.

alert: A pop-up that indicates a required action or potential problem.

right-click: Press the right mouse button.

Exercise 1-1

www.g-wlearning.com/CAD/

Complete the exercise on the companion website.

Starting AutoCAD

Start AutoCAD using the same methods you use to launch other programs within the Windows operating system. To launch AutoCAD from the Windows desktop, ***double-click*** on the AutoCAD 2020 ***icon***. To start AutoCAD from the Start menu, ***pick*** the Start ***button***, then pick All apps or All Programs. Then select AutoCAD 2020, followed by AutoCAD 2020. To start AutoCAD from the Start screen, pick the AutoCAD 2020 tile or use the search function to search for AutoCAD 2020.

double-click: Quickly press the left mouse button twice.

icon: Small graphic representing an application, file, or command.

pick (click): Press the left mouse button.

button: A "hot spot" button on the screen that you pick to access an application, command, or option.

NOTE

AutoCAD 2020 operates with Windows 10, Windows 8.1, Windows 8, and Windows 7. Do not be concerned if you see illustrations in this textbook that appear slightly different from those on your screen.

Exiting AutoCAD

Use the **EXIT** command to end an AutoCAD session. A common technique to access the **EXIT** command is to pick the program **Close** button, located in the upper-right corner of the AutoCAD window. Other ways to close AutoCAD are to double-click the **Application Menu** button in the upper-left corner of the AutoCAD window, select the **Exit Autodesk AutoCAD 2020** button in the **Application Menu**, or, with a file open, type EXIT or QUIT and press [Enter]. See Figure 1-3.

NOTE

If you attempt to exit before saving your work, AutoCAD prompts you to save or discard changes.

Exercise 1-2

www.g-wlearning.com/CAD/

Complete the exercise on the companion website.

The AutoCAD Interface

interface: Items that allow you to input data to and receive outputs from a computer system.

graphical user interface (GUI): On-screen features that allow you to interact with software.

Interface items include devices to input data, such as the keyboard and mouse, and devices to receive computer outputs, such as the monitor. AutoCAD uses a Windows-style ***graphical user interface (GUI)*** with an **Application Menu**, ribbon, dialog boxes, and AutoCAD-specific items. You will explore specific elements of the

Figure 1-3.
Use any of several techniques to exit AutoCAD when you finish a drawing session.

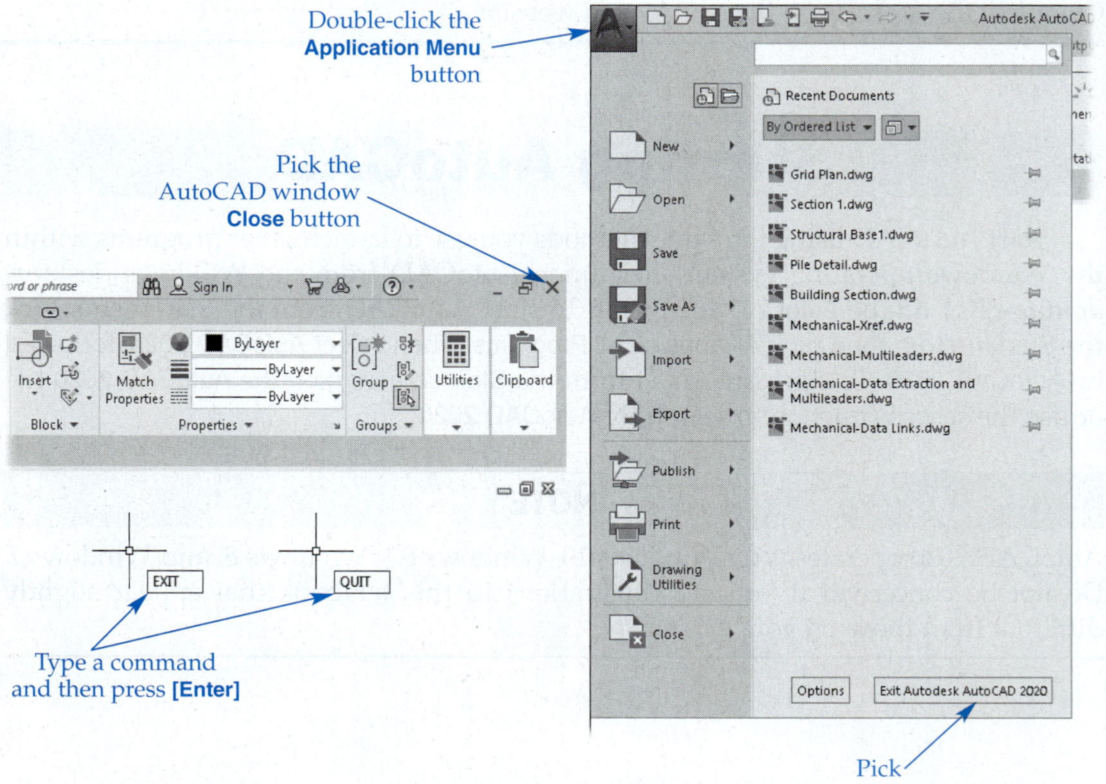

unique AutoCAD interface in this chapter and throughout this textbook. Learn the format, appearance, and proper use of interface items to help quickly master AutoCAD.

NOTE

As you learn AutoCAD, you may want to customize the graphical user interface according to common tasks and specific applications. Use the **Options** dialog box to change display colors and window elements. To display the **Options** dialog box, pick the **Options** button at the bottom of the **Application Menu**, or right-click in the drawing area and select **Options...**. Use the **Customize User Interface** dialog box, accessed with the **CUI** command, to customize the ribbon and other AutoCAD-specific items.

Start Tab

The **Start Tab** is an element of the AutoCAD file tab system and is displayed by ***default***. Chapter 2 explains file tabs. The **Create** page appears by default when you first launch AutoCAD. See Figure 1-4. The **Get Started** section includes options for starting a new drawing or opening a saved file. Pick the **Start Drawing** button to start a new drawing. The **Templates** ***drop-down list*** below the **Start Drawing** button lists drawing templates currently found in the default template folder. The highlighted template is used to start a new drawing file when you pick the **Start Drawing** button. Pick a different template from the list to start a new drawing using the selected template. The template you select becomes the new default assigned to the **Start Drawing** button.

default: A value maintained by the computer until changed.

drop-down list: A list of options that appears when you pick a button that contains a down arrow.

Figure 1-4.
The **Create** page of the **Start Tab** appears by default when you first launch AutoCAD.

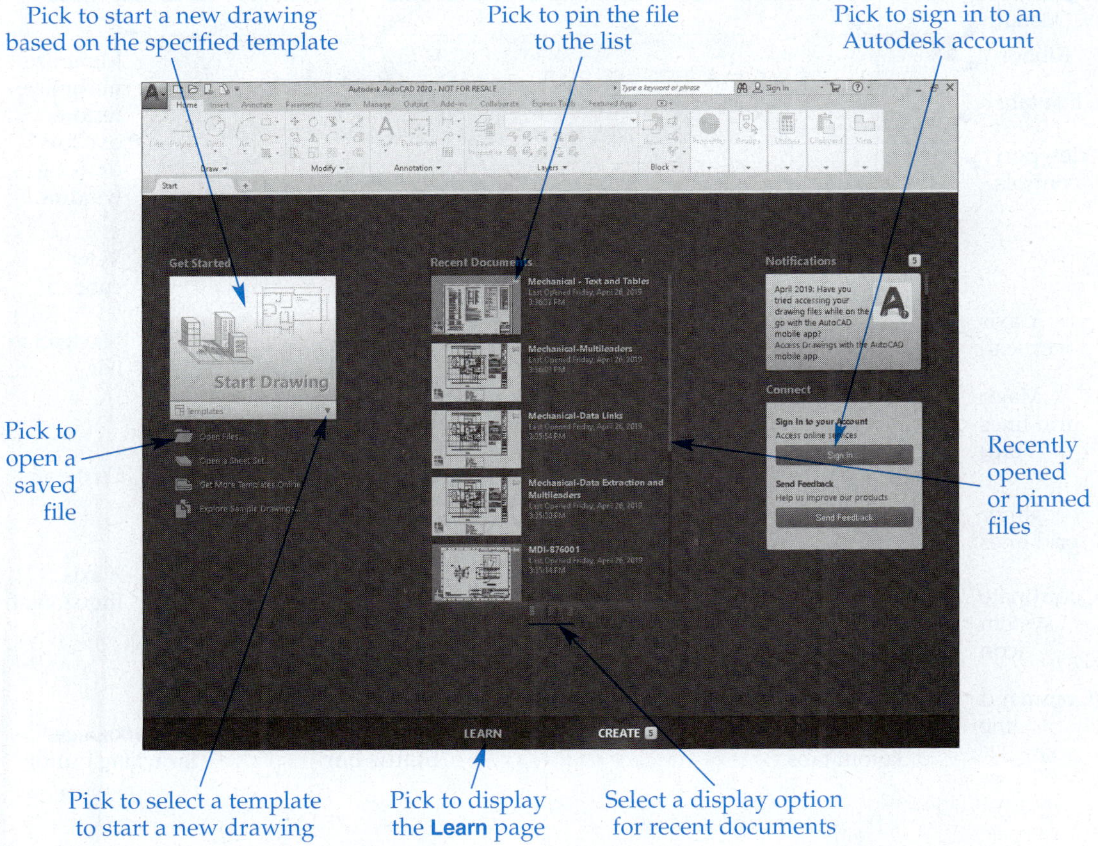

Use the **Recent Documents** section to open a recently saved file or a file that has been pinned to the **Recent Documents** list. Select from the display options below the list to specify how recent files appear. You will learn more about starting new drawings and opening existing files in Chapter 2. The **Notifications** section contains announcements and links to a variety of resources online. The **Connect** section offers access to additional AutoCAD and Autodesk tools and services online. Pick the **Sign In...** button to log in to or create an Autodesk account. Access the **Learn** page to explore AutoCAD resources online, including the AutoCAD help system.

NOTE

The **Try** page appears if AutoCAD has not been activated. The **Start Tab** disappears when you turn off file tabs, but the **Create** and **Learn** pages remain.

Workspaces

workspace: A preset work environment containing specific interface items.

drawing window (graphics window): The largest area in the AutoCAD window, where drawing and modeling occurs.

The Drafting & Annotation *workspace* is active by default when you launch AutoCAD and begin a new drawing. See Figure 1-5. The Drafting & Annotation workspace displays interface features above and below a large *drawing window,* also called the *graphics window,* and contains the commands and options most often used for 2D drawing.

To activate a different workspace, pick the **Workspace Switching** button on the status bar and select a different workspace. See Figure 1-6. The 3D Basics and 3D Modeling workspaces provide commands and options appropriate for 3D modeling. The list also includes saved custom workspaces.

Figure 1-5.
A new drawing started from the acad.dwt template installed with AutoCAD. The default AutoCAD window appears with the Drafting & Annotation workspace active.

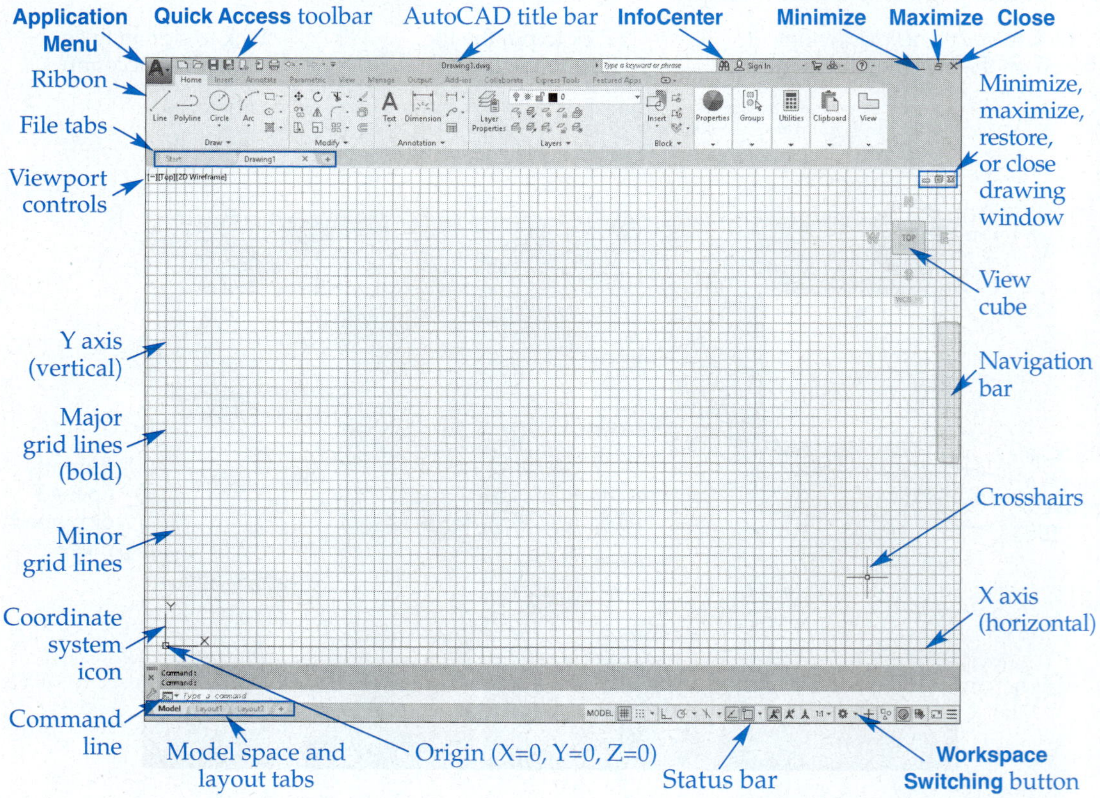

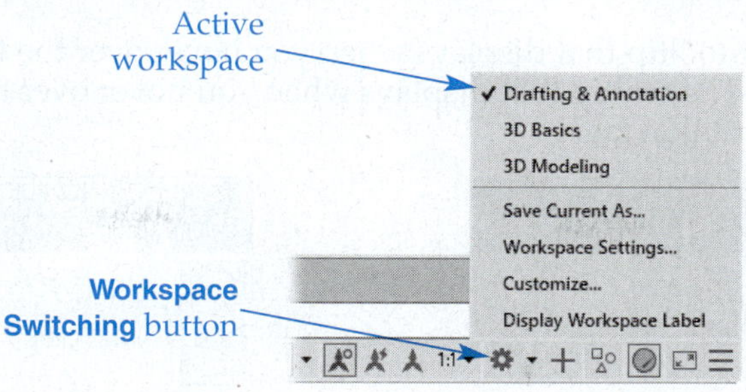

Figure 1-6. Use the **Workspace Switching** button on the status bar to change to a different workspace, create a new workspace, or customize the user interface.

NOTE

This textbook focuses on the default Drafting & Annotation workspace, except in specific situations that require additional interface items. The default model space drawing window background color is dark blue, but this textbook shows a white background for clarity. Add items and AutoCAD tools to the interface as needed. ***AutoCAD and Its Applications—Advanced*** details the 3D Modeling workspace.

PROFESSIONAL TIP

Reload the Drafting & Annotation workspace to return interface items to their default locations. To reload the current workspace, use the **setCurrent** option of the **WORKSPACE** command.

Exercise 1-3

www.g-wlearning.com/CAD/

Complete the exercise on the companion website.

Crosshairs and Cursor

The AutoCAD crosshairs is the primary means of pointing to and selecting objects or locations within the drawing window. The crosshairs changes to the familiar Windows cursor when you move it outside of the drawing area or over an interface item, such as the status bar.

PROFESSIONAL TIP

Control crosshairs length using the ***text box*** or ***slider*** found in the **Crosshair size** area on the **Display** tab of the **Options** dialog box. Longer crosshairs can help to reference alignment between objects.

text box: A box in which you type a name, number, or single line of information.

slider: A movable bar that increases or decreases a value when you slide the bar.

Tooltips

A ***tooltip*** displays when you ***hover*** over most interface items. See Figure 1-7. Tooltip content varies depending on the item. Many tooltips expand as you continue to hover. The initial tooltip might display the command name and a brief description of the command. As you continue to hover, an explanation, illustration, or short video on how to use the command may appear. Use the tooltip to help confirm and learn more about your desired selection.

tooltip: A pop-up that provides information about the item over which you hover.

hover: Pause the cursor over an item to display information or options.

Figure 1-7.
A—The basic tooltip that displays when you hover over the **Close** button of the AutoCAD window. B—The tooltip that displays when you hover over the **Line** button on the **Draw** panel in the **Home** ribbon tab.

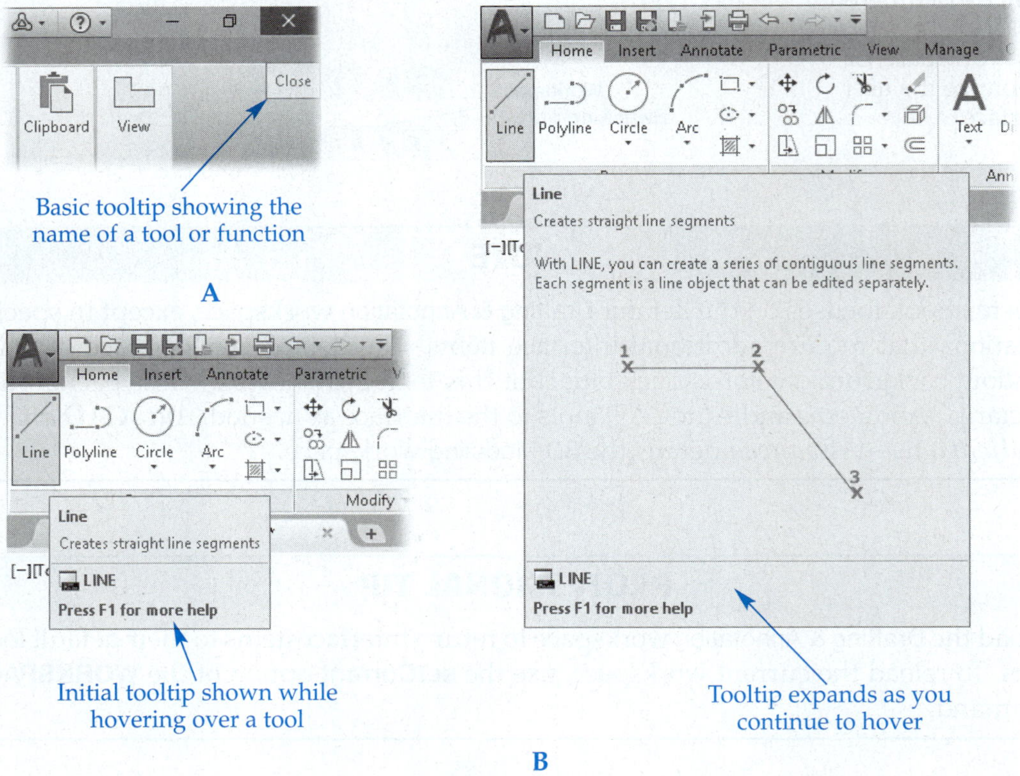

Shortcut Menus

shortcut menu (cursor menu, right-click menu, pop-up menu): A general or context-sensitive menu available by right-clicking on interface items or objects.

context-sensitive: Specific to the active command or option.

cascading menu (cascading submenu): A menu of options related to the chosen menu item.

AutoCAD uses ***shortcut menus***, also known as ***cursor menus***, ***right-click menus***, or ***pop-up menus***, to simplify and accelerate command and option access. When you right-click in the drawing area while a command is not active, the first item in the shortcut menu is typically an option to repeat the previous command or operation. If you right-click while a command is active, the shortcut menu contains ***context-sensitive*** menu options. See Figure 1-8. Some menu options have a small arrow to the right of the option name. Hover over the option to display a ***cascading menu***, also known as a ***cascading submenu***. The **Recent Input** cascading menu shows a list of recently used commands, options, or values, depending on the shortcut menu. Pick from the list to reuse a function or value.

Controlling Windows

Control the AutoCAD and drawing windows using the same methods you use to control other windows within the Windows operating system. To minimize, maximize, restore, or close the AutoCAD window or individual drawing windows, pick the appropriate button in the upper-right corner of the window. You can also adjust the AutoCAD window by right-clicking on the title bar and choosing from the standard window control menu. Window sizing operations are also the same as those for other windows within the Windows operating system.

Floating and Docking

float: Describes interface items that appear within a frame and that you can resize or move.

dock: Describes interface items set into position on an edge of the AutoCAD window (top, bottom, left, or right).

Several interface items, including the AutoCAD and drawing windows, can ***float*** or ***dock***. Some items, such as the drawing window, have a title bar at the top or side. You can move and resize floating windows in the same manner as other windows.

Figure 1-8.
Shortcut menus provide direct access to general or context-sensitive commands and options.

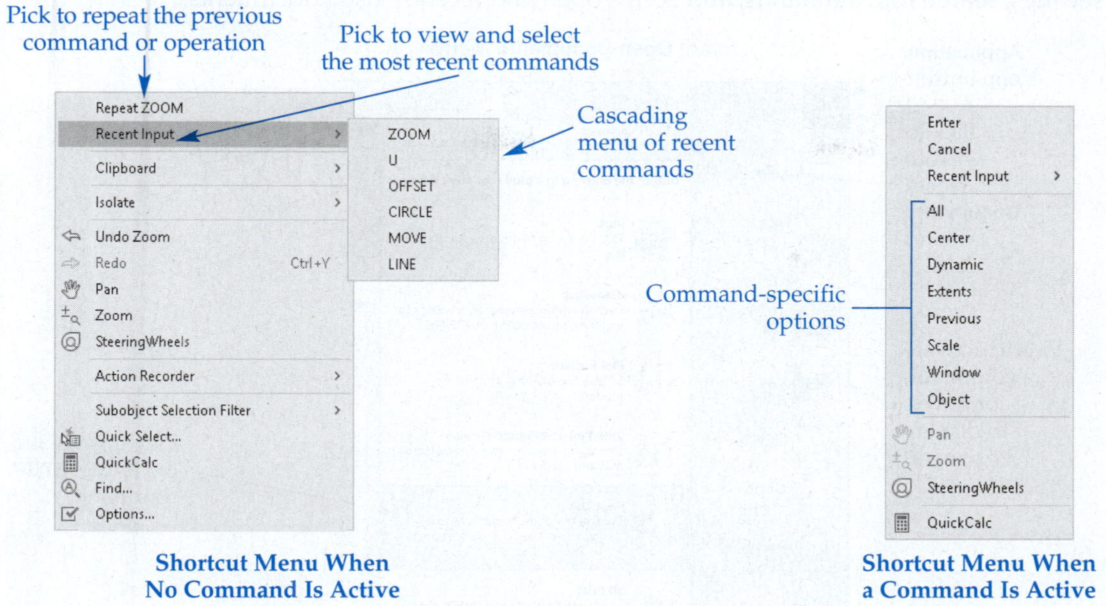

However, you can only move and resize drawing windows within the AutoCAD window. Different options are available depending on the particular interface item and the float or dock status of the item. Typically, the close and minimize or maximize options are available. Some floating items, such as ***sticky panels***, include ***grab bars***.

sticky panel: A ribbon panel moved out of a tab and made to float in the drawing window.

grab bar: Thin bar at the edge of a docked or floating interface item that you can use to move the item.

Locking

Use the **LOCKUI** ***system variable*** to prevent floating or docked ***toolbars***, panels, and windows from moving accidentally. The setting applies the sum of the following values. For example, a setting of 15 locks all items. The default setting is 0, which unlocks all items.

1 Docked toolbars and panels are locked.
2 Docked windows are locked.
4 Floating toolbars and panels are locked.
8 Floating windows are locked.

To move a locked item without unlocking it, hold down [Ctrl] while moving the item.

system variable: A named definition that stores a value and configures AutoCAD to accomplish a specific task or exhibit a certain behavior.

toolbars: Interface items that contain tool buttons or drop-down lists.

Use the **Lock User Interface** flyout on the status bar to manage the **LOCKUI** system variable setting. The status bar is described later in this chapter. The **Lock User Interface** flyout is not displayed by default and can be added to the status bar by selecting the **Lock UI** option in the status bar **Customization** flyout.

Exercise 1-4

www.g-wlearning.com/CAD/

Complete the exercise on the companion website.

Application Menu

The **Application Menu** is a menu system that provides access to application- and file-related commands and settings. The **Application Menu** displays when you pick the **Application Menu** button, located in the upper-left corner of the AutoCAD window. See **Figure 1-9**.

Figure 1-9.
Use the **Application Menu** to access common application and file management commands and settings, search for commands, and access open and recently used documents.

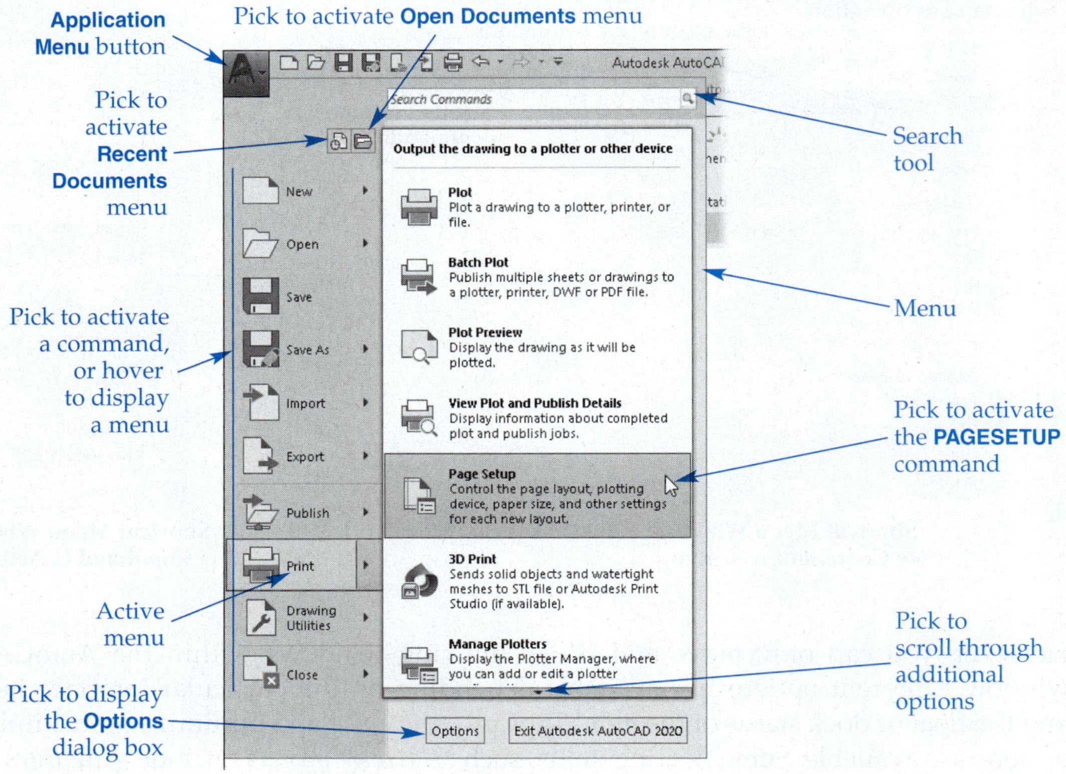

Using the Buttons and Menus

Items on the left side of the **Application Menu** function as buttons to activate common application commands, display menus, or do both. For example, pick the **New** button to begin a new file using the **NEW** command. To display a menu, hover over the menu name, or pick the arrow on the right side of the button. Long menus include small arrows at the top and bottom for scrolling through selections. Some options have a small arrow to the right of the item name that, when selected or hovered over, expands to provide a submenu. Pick an option from the list to activate the command.

PAGESETUP
Application Menu
Print
> Page Setup

A command or option accessible from the **Application Menu** appears as a graphic in the margin of this textbook. The graphic represents the process of picking the **Application Menu** button, then selecting a menu button, or hovering over a menu, and picking a menu or submenu option. The example shown in this margin illustrates picking the **Page Setup** option of the **Print** menu, as shown in **Figure 1-9**, to access the **PAGESETUP** command.

Searching for Commands

Use the **Application Menu** search tool to locate and access any AutoCAD command listed in the Customize User Interface (CUIx) file. Type a command name in the **Search** text box. Commands that match the letters you enter appear as you type. Type additional letters to narrow the search, with the best-matched command listed first. **Figure 1-10** shows using the **Search** text box to locate the **SAVE** command for saving a file. Pick a command from the list to activate the command.

NOTE

The **Recent Documents** and **Open Documents** menus of the **Application Menu** provide access to recent and active files, as described in Chapter 2.

Figure 1-10.
Use the **Application Menu** to search for a command. Pick the command from the list to activate the command.

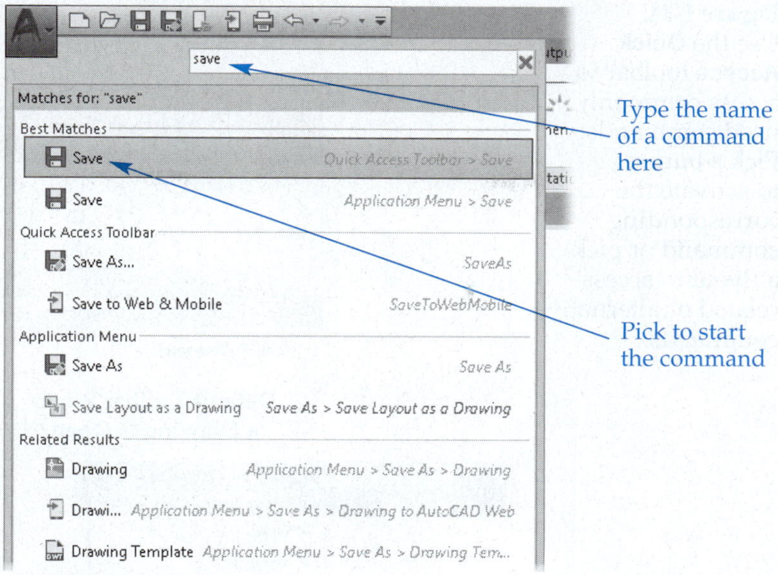

Exercise 1-5

www.g-wlearning.com/CAD/

Complete the exercise on the companion website.

Quick Access Toolbar

Toolbars contain ***tool buttons***. Each tool button includes an icon that represents an AutoCAD command or option. As you move the cursor over a tool button, the button highlights and may display a border and tooltip. Use the tooltip to become familiar with the command. Select a tool button to activate the associated command. Some tool buttons include ***flyouts***. Select a flyout and then pick from the list to activate the command.

tool buttons: Interface items used to start commands.

flyout: A set of related buttons that appears when you pick the arrow next to certain command buttons.

The default **Quick Access** toolbar appears on the title bar in the upper-left corner of the AutoCAD window, to the right of the **Application Menu** button. See Figure 1-11. The **Quick Access** toolbar provides fast, convenient access to several common commands using one or two picks. Most other interface items require two or more picks to activate a command.

When a drawing is open and the default Drafting & Annotation workspace is active, the **Quick Access** toolbar contains the **New**, **Open**, **Save**, **Save As...**, **Open from Web & Mobile**, **Save to Web & Mobile**, **Plot**, **Undo**, and **Redo** buttons. When no drawings are open, the **New**, **Open**, **Open from Web & Mobile**, and **Sheet Set Manager** buttons display.

A command or option accessible from the **Quick Access** toolbar appears as a graphic in the margin of this textbook. The graphic represents the process of picking a **Quick Access** toolbar button from the toolbar or flyout. The example shown in this margin illustrates picking the **Redo** button to access the **REDO** command.

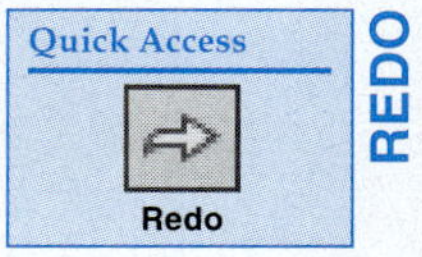

NOTE

Pick the **Customize Quick Access Toolbar** flyout on the right side of the **Quick Access** toolbar to add, remove, and relocate tool buttons.

Exercise 1-6

www.g-wlearning.com/CAD/

Complete the exercise on the companion website.

Figure 1-11.
Use the **Quick Access** toolbar to access commonly used commands. Pick a button to activate the corresponding command or pick a flyout to access related or alternative commands.

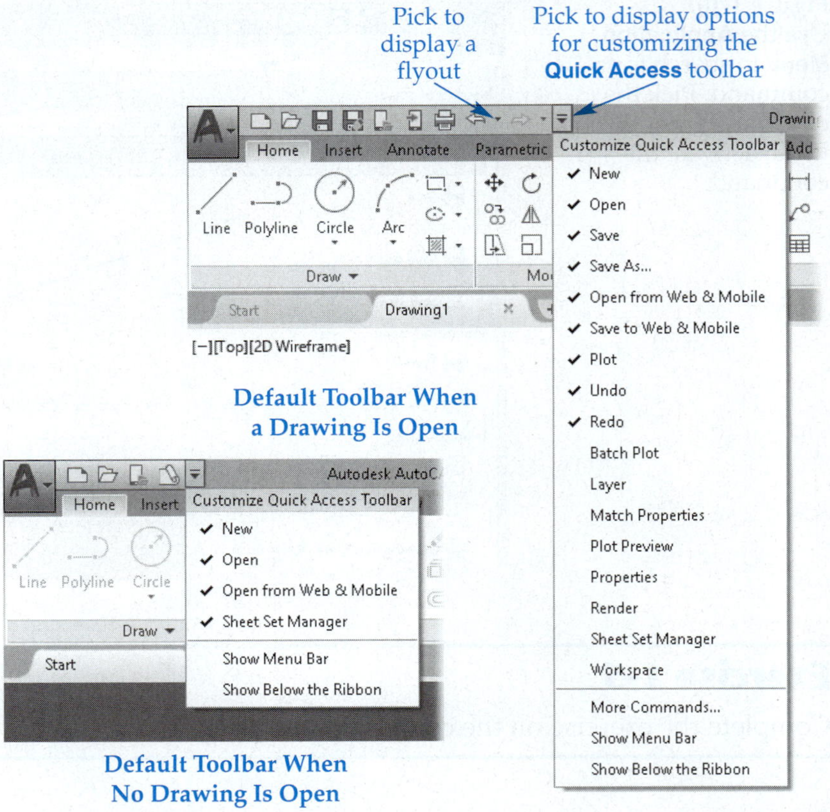

Ribbon

The ribbon docks horizontally below the AutoCAD window title bar by default. See **Figure 1-12**. The ribbon provides a convenient location from which to select commands and options that traditionally would require access by extensive typing, multiple toolbars, or several menus.

tab: A small stub at the top or side of a page, window, dialog box, or palette that allows you to access other portions of the item.

ribbon panels (panels): Divided areas in the ribbon that group commands.

Use the ***tabs*** along the top of the ribbon to access collections of related ***ribbon panels***, or ***panels***. Each panel houses groups of similar commands. For example, the **Annotate** tab includes several panels, each with specific commands for creating, modifying, and formatting annotations, such as text. The tabs and panels shown when the Drafting & Annotation workspace is active provide access to 2D drawing commands. Highlighted, context-sensitive tabs appear when some commands, such as the **HATCH** command, are active or when you work in a unique environment, such as the **Block Editor**.

Figure 1-12.
The ribbon is docked at the top of the drawing window by default and provides access to commands and command options.

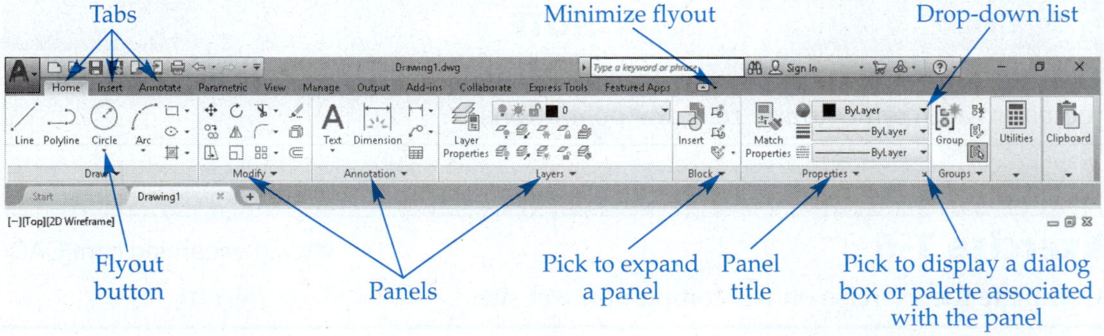

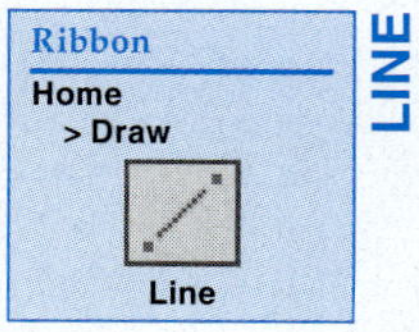

A command or option accessible from the ribbon appears in a graphic located in the margin of this textbook. The graphic identifies the tab and panel where the command is located. You may need to expand the panel or pick a flyout to locate the command. The example shown in this margin illustrates picking the **Line** button on the **Draw** panel in the **Home** ribbon tab to access the **LINE** command.

Ribbon Panels

The large tool button of a panel typically signifies the most often used panel command. In addition to tool buttons, panels can contain flyouts, drop-down lists, and other items. Some panels have a solid-filled triangle, or down arrow, next to the panel name. If you see this down arrow, pick the title at the bottom of the panel to display additional related commands and functions. See Figure 1-13. To show the expanded list on-screen at all times, select the pushpin button in the lower-left corner of the expanded panel.

NOTE

When you pick an option from a ribbon flyout, the option becomes the new default and appears in the ribbon. This makes it easier to select the same option the next time you use the command.

Some panels include a small arrow in the lower-right corner of the panel. Pick this arrow to access a dialog box or palette associated with the panel. For example, pick the arrow in the lower-right corner of the **Dimensions** panel in the **Annotate** tab, as shown in Figure 1-13, to display the **Dimension Style Manager** dialog box used to format dimensions.

Basic Adjustment

The ribbon appears maximized by default. You can minimize the display to show only tabs, panel titles, or panel buttons by repeatedly picking the **Minimize** button to the right of the tabs, or by selecting the appropriate option from the **Minimize** flyout. Picking the **Minimize** button corresponds to the **Cycle through All** flyout selection. When **Minimize to Tabs** is active, pick a tab to show all panels in the tab. When **Minimize to Panel Titles** is active, pick a panel title to display the panel. When **Minimize to Panel Buttons** is active, pick a panel button to display the panel. Right-click on a portion of the ribbon unoccupied by a panel to access the options described in Figure 1-14.

NOTE

The **Application Menu**, **Quick Access** toolbar, and ribbon replace the traditional menu bar. To display the menu bar, pick the **Customize Quick Access Toolbar** flyout on the right side of the **Quick Access** toolbar and select **Show Menu Bar**.

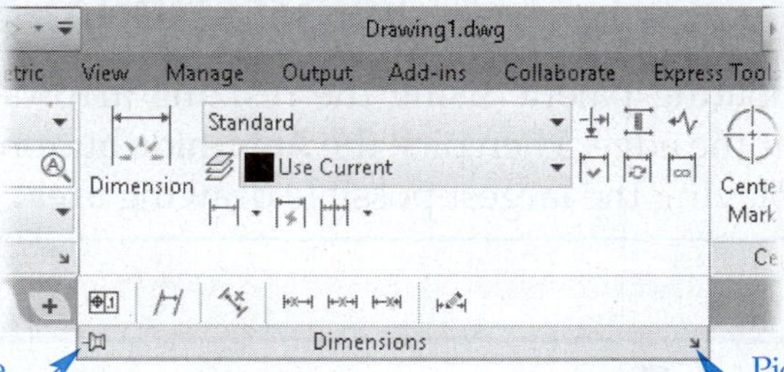

Figure 1-13. An expanded panel provides additional, related commands and functions. This example shows the expanded list of dimensioning commands found in the **Dimensions** panel in the **Annotate** ribbon tab.

Figure 1-14.
Right-click options for displaying and organizing ribbon elements.

Selection	Result
Show Related Tool Palette Group	Displays tool palette groups customized to associate with a ribbon tab.
Tool Palette Group	Allows you to select which related tool palette groups to show.
Show Tabs	Allows you to choose which tabs to display; also available by right-clicking on a panel.
Show Panels	Allows you to select which panels to display; also available by right-clicking on a panel.
Show Panel Titles	Uncheck to hide panel titles.
Undock	Changes the ribbon to a floating state. Double-click the ribbon title bar or drag and drop to dock the floating ribbon.
Close	Closes the ribbon. Use the **RIBBON** command to redisplay the ribbon.

Palettes

palette (modeless dialog box): Special type of window containing tool buttons and features common to dialog boxes. Palettes can remain open while other commands are active.

list box: A framed area that contains a list of items or options from which to select.

scroll bar: A bar tipped with arrow buttons used to scroll through a list of options or information.

Palettes, also known as ***modeless dialog boxes***, control many AutoCAD functions. Palettes contain tool buttons, flyouts, drop-down lists, and many other features, such as ***list boxes*** and ***scroll bars***. As with the ribbon, panels divide some palettes into groups of commands. Large palettes are divided into separate pages or windows, which you commonly access using tabs. Unlike dialog boxes, you do not need to close palettes in order to use other commands and work on the drawing.

To display a palette, pick a palette button from the **Palettes** panel in the **View** ribbon tab. You can also display most palettes using palette-specific access techniques. For example, to access the **Properties** palette, pick the arrow in the lower-right corner of the **Properties** panel in the **Home** ribbon tab; select an object, right-click, and select **Properties**; or type PROPERTIES and press [Enter].

When you display a palette for the first time, it is often in a floating state, although you can dock some palettes. Right-click on the palette title bar or pick the **Properties** button to select from a list of undocked palette control options. The **Auto-hide** option allows the palette to minimize when the cursor is away from the palette, conserving drawing space. Deselect the **Allow Docking** palette property or menu option to disable the ability to dock palettes. The **Properties** button or shortcut menu on some palettes includes other functions, such as the **Transparency...** option. Use the **Transparency...** option to make the palette transparent in order to view drawing geometry behind the palette. See Figure 1-15.

PROFESSIONAL TIP

Resize a floating palette using the resizing arrows that appear when you move the cursor over the edge. Then pick the **Auto-hide** button to have quick access to the palette while displaying the largest possible drawing area.

Exercise 1-7

www.g-wlearning.com/CAD/

Complete the exercise on the companion website.

Figure 1-15.
A—Pick the **Properties** button or right-click on the title bar and select **Transparency...** to access the **Transparency** dialog box. B—The transparent **Layer Properties Manager** palette positioned over a commercial building floor plan.

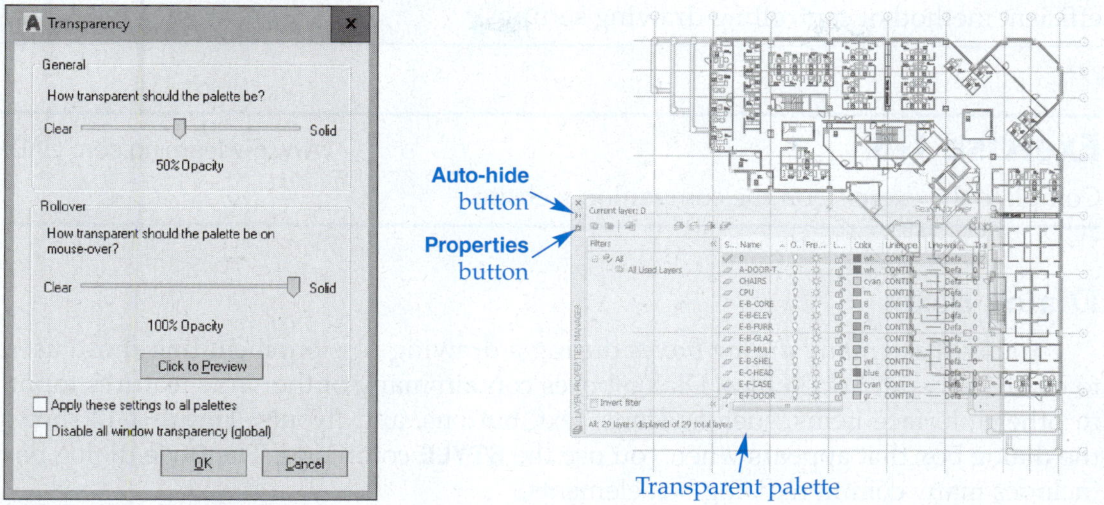

A B

Status Bar

The status bar appears along the lower right edge of the AutoCAD window, below the ***command line***. See **Figure 1-16**. The status bar includes controls for a variety of drawing aids and commands. Most of the controls are ***status toggle buttons***. The status bar is often the quickest and most effective way to manage certain drawing settings. By default, the status bar provides notifications for certain AutoCAD processes, such as plotting.

command line: Area where you can type commands (command names) and type or select command options.

status toggle buttons: Buttons that toggle drawing aids and commands on and off.

NOTE

Use the **Customization** flyout to add tools to or hide tools from the status bar. AutoCAD forms the status bar into two rows if not all of the status bar controls fit into a single row.

Figure 1-16.
The default status bar. Pick the **Customization** flyout to add tools to or remove tools from the status bar.

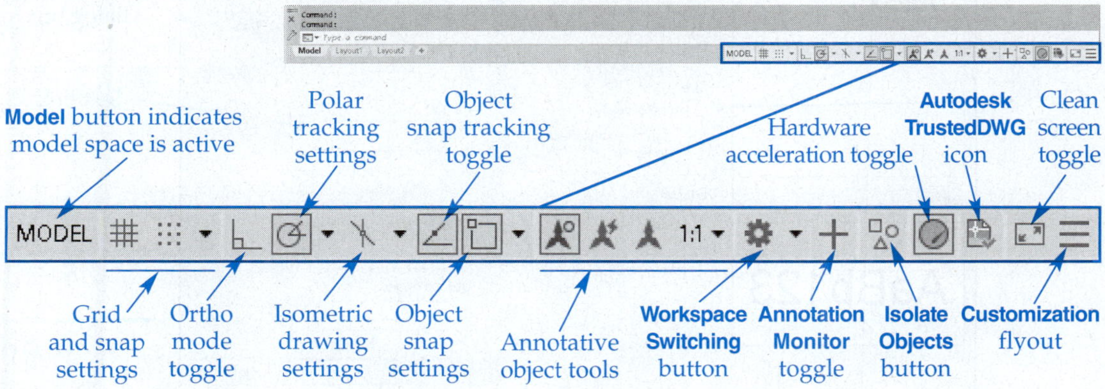

PROFESSIONAL TIP

Right-click on a button or pick a flyout on the status bar to view a shortcut menu specific to the item. Picking options from a status bar shortcut menu is often the most efficient method of controlling drawing settings.

Exercise 1-8

www.g-wlearning.com/CAD/

Complete the exercise on the companion website.

Dialog Boxes

dialog box: A window-like item that contains various settings and information.

You will see many ***dialog boxes*** during a drawing session, including those used to create, save, and open files. Dialog boxes contain many of the same features found in other interface items, such as icons, text, buttons, and flyouts. Figure 1-17 shows the dialog box that appears when you use the **STYLE** command. The **Style** dialog box includes many common dialog box elements.

NOTE

A dialog box appears when you pick any menu selection or button displaying an ellipsis (…).

preview box: An area in a dialog box that shows the results of the options and settings you select.

Use the cursor to set variables and select items in a dialog box. Many dialog boxes include icons, images, ***preview boxes***, or other cues to help you to select appropriate options. When you pick a button in a dialog box that includes an ellipsis (…), another dialog box appears. You must make a selection from the second dialog box before returning to the original dialog box. A button with a crosshairs icon requires you to select in the drawing area.

Figure 1-17.
This dialog box displays when you issue the **STYLE** command.

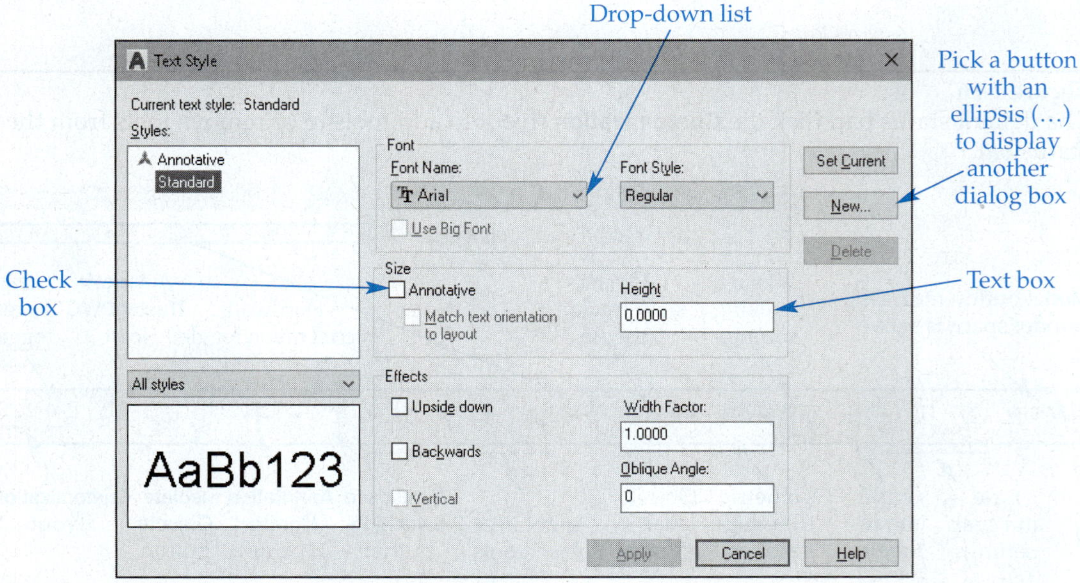

PROFESSIONAL TIP

Certain AutoCAD dialog boxes can be resized. Resize dialog boxes as needed to view hidden content and reduce the need to scroll through long lists. The size of the dialog box is maintained when you exit and then restart AutoCAD.

CAUTION

The AutoCAD interface includes several other unique items, such as file tabs, the viewport controls, the navigation bar, the view cube, and **Model** and **Layout** tabs. Refer to Figure 1-5 to recognize these features. You will explore these features and their specific controls throughout this textbook. Do not use or adjust these tools until you learn about their functions, because doing so can unexpectedly change the interface display and operation. Ensure that the **Model** tab is active.

Exercise 1-9

www.g-wlearning.com/CAD/

Complete the exercise on the companion website.

AutoCAD Options

The **Options** dialog box contains AutoCAD system and drawing file options. System options apply to the entire program. Drawing file options, identified by the AutoCAD drawing icon, are file-specific. Many system options are available to help configure the work environment, such as the background color of the drawing window. This textbook focuses on the default system options and references the **Options** dialog box when applicable.

OPTIONS
Application Menu
Options
Type
OPTIONS
SETTINGS
OP

NOTE

You can also access the **Options** dialog box by right-clicking when no command is active and selecting **Options...**.

Accessing Commands

Commands are available by direct access from the ribbon, shortcut menus, **Application Menu**, **Quick Access** toolbar, palettes, status bar, file tabs, viewport controls, navigation bar, and view cube. An alternative is to enter the command using ***dynamic input*** or the command line. The basic method to activate a command by typing is to type the single-word command name or the ***command alias*** and press [Enter] or the space bar, or right-click. You can use uppercase, lowercase, or a combination of uppercase and lowercase letters. You can only issue one command at a time.

You can activate any command or option by typing. Each command name and alias, along with other access techniques available in the Drafting & Annotation workspace, appears in a graphic in the margin of this textbook. The example displayed in this margin shows the command name (**LINE**) and alias (**L**) you can enter to access the **LINE** command.

dynamic input: Area near the crosshairs where you can type commands (command names), type or select command options, and view context-oriented information.

command alias: Abbreviated command name entered at the keyboard.

LINE
Type
LINE
L

The AutoComplete feature is active by default to help locate and access AutoCAD commands. When you type a command name, a suggestion list displaying selectable entries appears at the command line or next to the dynamic input entry area. You will learn more about AutoComplete and other AutoCAD input search functions later in this chapter.

A benefit of accessing a command using a method other than typing is that you do not need to memorize command names or aliases. Another advantage is that commands, options, and drawing activities appear on-screen as you work, using visual icons, tooltips, and prompts. As you use AutoCAD, you will become familiar with the display and location of commands. Decide which command selection techniques work best for you. A combination of command selection methods often proves most effective.

PROFESSIONAL TIP

When typing commands, you must exit the current command before issuing a new command. In contrast, when you use the ribbon or other input methods, the current command automatically cancels when you pick a different command.

NOTE

Even though you may not choose to access commands by typing command names or aliases, you must still type certain values, as explained in Chapter 3. For example, you may have to type the diameter of a circle or radius of an arc.

Reference Material

www.g-wlearning.com/CAD/

Command Aliases

For a list of common command aliases, go to the companion website, select the **Resources** tab, and select **Command Aliases**.

Dynamic Input

Dynamic input allows you to keep your focus at the crosshairs while you draw. When dynamic input is on, a temporary input area appears in the drawing window, below and to the right of the crosshairs by default. See Figure 1-18.

Depending on the command in progress, different information and options appear in the dynamic input area. For example, Figure 1-19A shows the display after starting the **RECTANGLE** command. The first portion of the dynamic input area is the prompt, which reads Specify first corner point or. In this case, to draw a rectangle, pick in the drawing window or enter ***coordinates*** to specify the first corner of the rectangle, or access other options as suggested by the "or" portion of the prompt.

coordinates: Numerical values used to locate a point in the drawing area.

Press the down arrow key to display available command options. See Figure 1-19B. Select an option using the cursor, or press the down arrow key again to cycle through the options. Press [Enter] to select the highlighted option. You can also select an option by right-clicking and picking an option from the shortcut menu. The information displayed in the dynamic input area changes while you work with a command, depending on the actions you choose. Figure 1-20 shows the dynamic input display when the **LINE** command is active.

Figure 1-18.
Use dynamic input to type or select commands and values from a temporary input area near the crosshairs. You can use the command line as an alternative to, or with, dynamic input. The command line also offers utility not available with dynamic input.

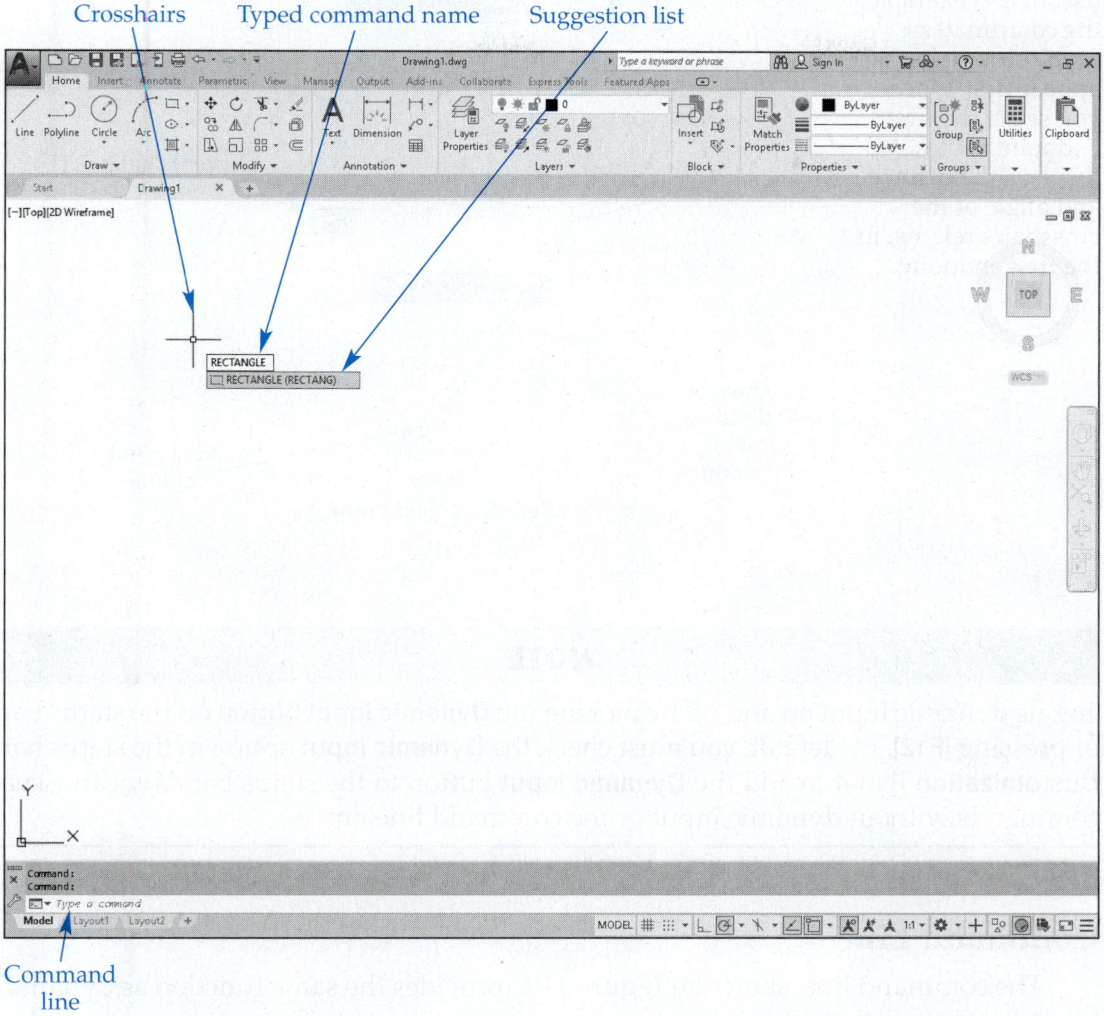

Figure 1-19.
A—The dynamic input fields that appear after you first enter the **RECTANGLE** command. B—Press the down arrow key to display command options. Pick an option with the cursor, or use the up and down arrow keys to highlight the desired option and press [Enter] to select.

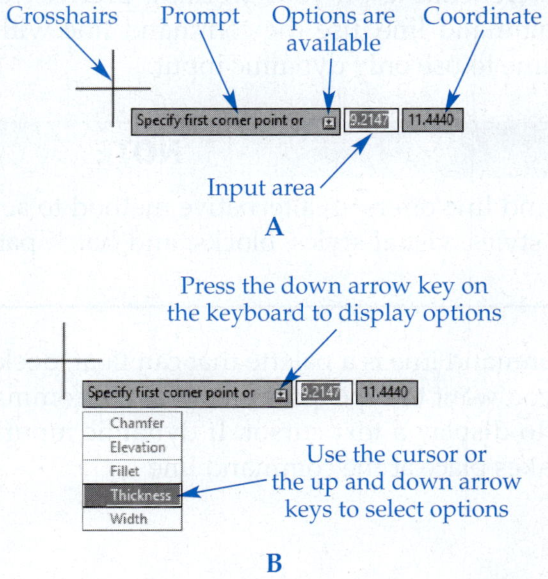

Figure 1-20. Dynamic input fields change while a command is in use. In this example, the coordinates of the crosshairs appear first. Once you select the first endpoint, AutoCAD displays the distance and angle of the crosshairs relative to the first endpoint.

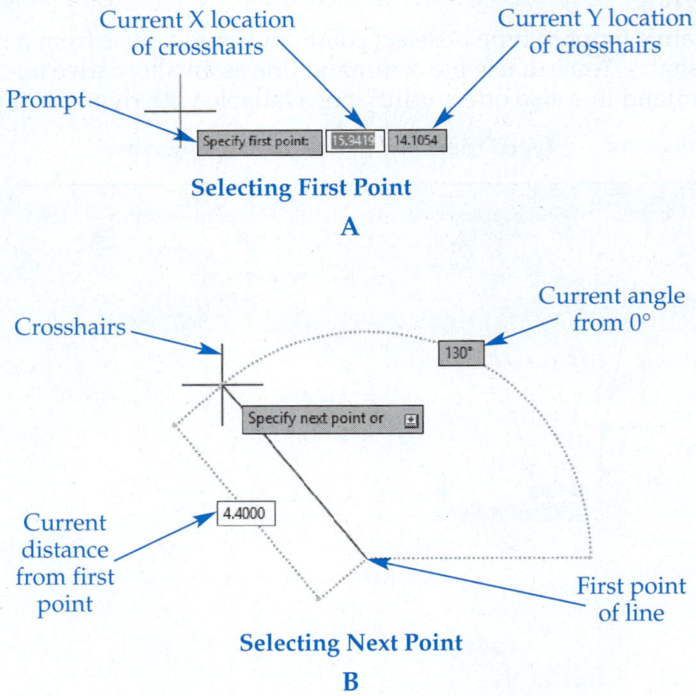

NOTE

Toggle dynamic input on and off by picking the **Dynamic Input** button on the status bar or pressing [F12]. By default, you must check the **Dynamic Input** option in the status bar **Customization** flyout to add the **Dynamic Input** button to the status bar. You can issue commands without dynamic input or the command line on.

Command Line

COMMANDLINE
Ribbon
View
> Palettes
Command Line
Type
COMMANDLINE
[Ctrl]+[9]

The command line, shown in **Figure 1-18**, provides the same function as dynamic input but uses a more traditional command input format. The command line also allows you to display extensive command history and search for information about commands. Depending on your working preference, disable dynamic input to use only the command line, use the command line with dynamic input, or disable the command line to use only dynamic input.

NOTE

The command line offers an alternative method to activate and use layers, text styles, dimension styles, visual styles, blocks, and hatch patterns, as explained later in this textbook.

The command line is a palette that can float, dock, or be resized. If dynamic input is on, and you want to type a command at the command line, pick the command line input area to display a text cursor. If dynamic input is off, typing a command automatically takes place at the command line.

NOTE

Illustrations in this textbook that display the command line show the command line docked at the bottom of the AutoCAD window, above the status bar. This may be different from the appearance you see on-screen. Where appropriate in this textbook, the command line is shown floating to explain features associated with the floating command line or to show additional command line history.

Depending on the command in progress, different information and options appear at the command line. For example, Figure 1-21A shows the display after starting the **CIRCLE** command. The prompt line shows an icon and the name of the active command followed by a prompt, which reads Specify center point for circle or. In this case, to draw a circle, pick in the drawing window or enter coordinates to specify the center point of the circle, or select a different option as suggested by the "or" portion of the prompt. The square brackets ([]) contain available options. Select an option using the cursor or type the name of the option. Each option has an alias, or unique highlighted uppercase character(s), that you can enter at the prompt rather than typing the entire option name. You can also select an option by right-clicking in the drawing window and picking an option from the shortcut menu.

As you work with a command, the information displayed at the command line changes. The active prompt background is white by default to distinguish it from prompt and command line history, which is gray by default. When the command line is docked, one line of command history appears by default. Figure 1-21B shows the display after specifying the center point of the circle and entering the **Diameter** option. At this prompt, you must specify the diameter of the circle. Some command prompts include angle brackets (<>) surrounding the default option or value; press [Enter] to accept the default instead of typing the value again.

To move the command line to a different location in the drawing window, pick and hold the grab bar at the left edge and drag it. When the command line is floating, three lines of command history appear by default. See Figure 1-22. Some commands, such as the **FILLET** command active in Figure 1-22, display current command settings as the first line of prompt history after the command name for reference.

Figure 1-21.
A—The command line after you first enter the **CIRCLE** command. Follow the prompt or select an available option shown in square brackets ([]). B—Prompt history appears as you progress through a command sequence. Adjust the height of the docked command line to change the number of lines of command history. Press [Enter] to accept a default value shown in angle brackets (<>).

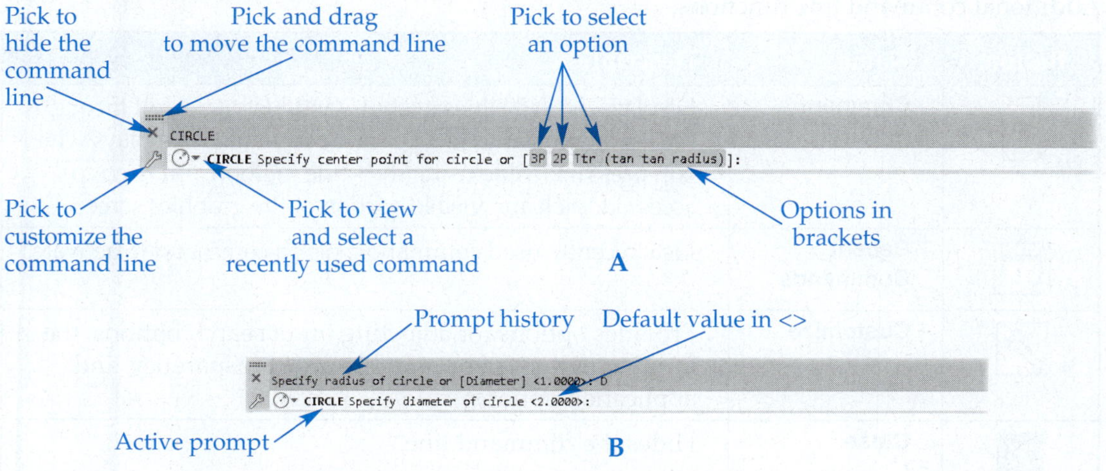

Figure 1-22.
The floating command line displays additional lines of command history for reference.

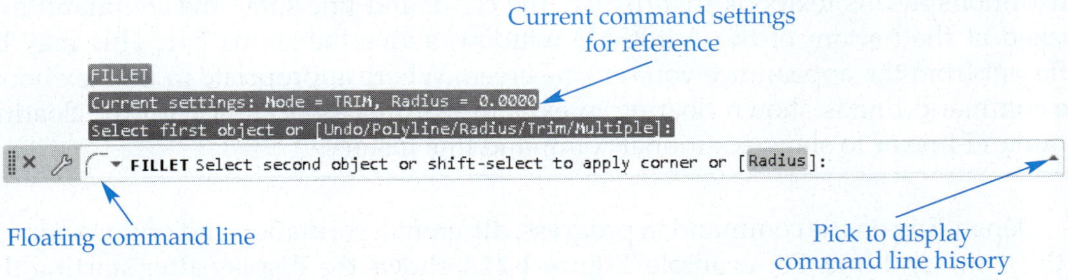

Figure 1-23 describes other elements of the command line. Note that scroll arrows replace the **Command History** button when the command line is docked. You can also access many of these functions by right-clicking on the command line.

NOTE

The floating command line attaches to framed interface items, such as the drawing window or ribbon, when you drag it near the item. Hold [Ctrl] to float the command line near or over items without attaching or docking.

PROFESSIONAL TIP

While learning AutoCAD, pay close attention to prompts in the dynamic input area and at the command line. Prompts guide you through the operation.

Input Search Tools

Input search tools are active by default to help you access and research AutoCAD commands and system variables. The AutoCAD input search tools include AutoComplete, AutoCorrect, and additional functions available when accessing commands by typing. Begin typing a command name using dynamic input or the command line. Commands that match the letters you enter appear in a suggestion list as you type. Type additional letters to narrow the search, with the best-matched

Figure 1-23.
Additional command line functions.

Button	Name	Description
	Command History	Displays up to 50 lines of command history; scroll through the list as needed to view earlier commands. Displays when you press [F2]; hide using the same method you used to access, or pick any visible portion of the graphics screen.
	Recent Commands	Lists recently used commands; pick a command to activate.
	Customize	Provides options for adjusting input search options, the number of lines of prompt history, transparency, and application options.
	Close	Hides the command line.

command listed first. Figure 1-24A shows the results of typing a lowercase letter l. The lowercase l changes to the preferred format of an uppercase L by default.

The suggestion list identifies commands by alias followed by the command name in parentheses, such as L (LINE) shown in Figure 1-24A. If you type a word that does not exist as a command, AutoCAD references a list of synonyms in an effort to match the word you type. For example, the word *curve* is not a command, but if you type CURVE, you will see CURVE (SPLINE) in the list. Select this suggestion to access the **SPLINE** command.

An easy way to toggle input search settings on and off is to right-click on the dynamic input suggestion list and select from the menu. You can also right-click on the command line, or pick the **Customize** button on the command line, and select from the **Input Settings** cascading menu. Figure 1-24B describes the input search settings.

The command line offers input search tools not available using dynamic input. Content search, explained later in this textbook, is only available from the command line. In addition, by default, the command line suggestion list uses categories to group commands, system variables, and content. Pick the expand icon to expand a category, as shown in Figure 1-24A, or press [Tab] to cycle through the categories. Pick the **Search in Help** button to search the AutoCAD help system for information about the command. Pick the **Search on Internet** button to search the Internet for information about the command. The word AutoCAD followed by the command name appears in the search bar of your Internet browser.

NOTE

By default, AutoCAD sorts the suggestion list according to commonly used commands. As you work, the suggestion list adapts to sort suggestions based on previously used commands. You can customize command aliases, suggestion list synonyms, and AutoCorrect spellings. Tools for customizing these functions are accessed in the **Edit Aliases** flyout in the **Customization** panel of the **Manage** ribbon tab.

Supplemental Material

www.g-wlearning.com/CAD/

Input Search Options

Use the **INPUTSEARCHOPTIONS** command to adjust input search preferences using the **Input Search Options** dialog box. For information about the **Input Search Options** dialog box, go to the companion website, navigate to this chapter in the **Contents** tab, and select **Input Search Options**.

Keyboard Shortcuts

Many keys on the keyboard, known as ***shortcut keys*** or ***keyboard shortcuts***, allow you to perform AutoCAD functions quickly. Become familiar with these keys to improve your AutoCAD performance. To cancel a command or exit a dialog box, press the ***escape key*** [Esc]. Some command sequences require that you press [Esc] twice to cancel the operation.

shortcut key (keyboard shortcut): Single key or key combination used to issue a command or select an option.

escape key: Keyboard key used to cancel a command or exit a dialog box.

When no command is active, press the up arrow key as many times as necessary to cycle through the sequence of previously used command names. Use the down arrow key to return to a later command in the list. If dynamic input is active, previously used commands appear in the dynamic input box near the crosshairs by default. To display previously used commands at the command line, pick the command line before pressing the up arrow or turn off dynamic input. Press [Enter] to activate the displayed command. You can also use the up and down arrow keys to reference previously used values while a command is active.

Figure 1-24.
A—Use AutoCAD input search tools to help access and research commands when typing. Notice that the suggestion list displays the command alias and name, or a synonym and name. B—Basic input search tool settings. The **Search Content** setting is only available from the command line.

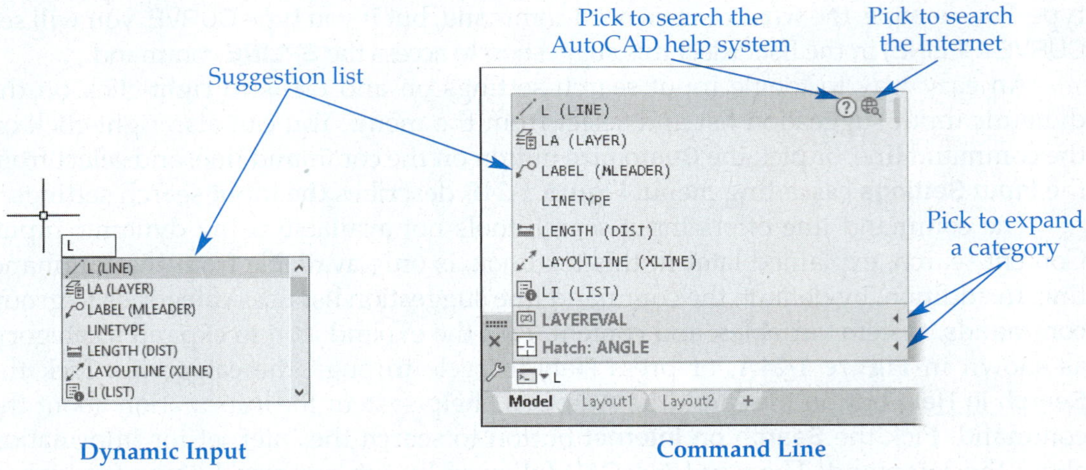

A

Selection	Result	Example
AutoComplete	Displays a list of suggested commands and system variables.	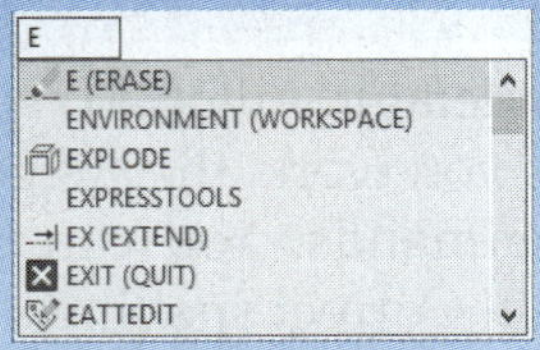
AutoCorrect	Suggests commands and system variables even if you type the name incorrectly; lowercase letters identify the error.	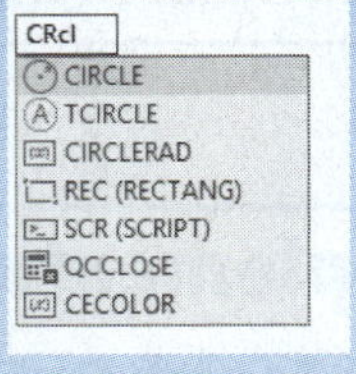
Search System Variables	Includes system variables in the suggestion list; identified by the system variable icon.	
Search Content	An alternative method to activate and use layers, text styles, dimension styles, visual styles, blocks, and hatch patterns, as explained later in this textbook.	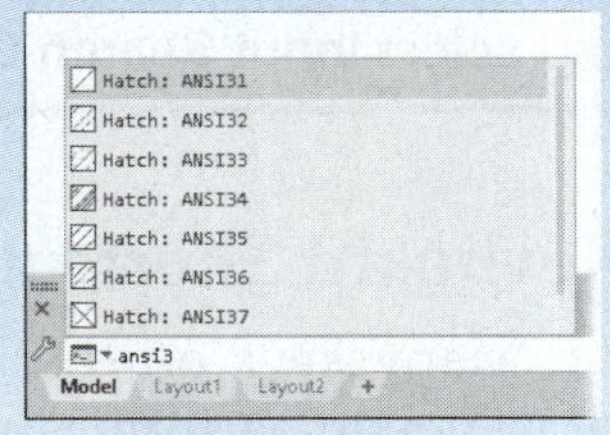
Mid-String Search	Suggests all commands and system variables that include the letters you type; disable to show only suggestions that begin with the letters you type.	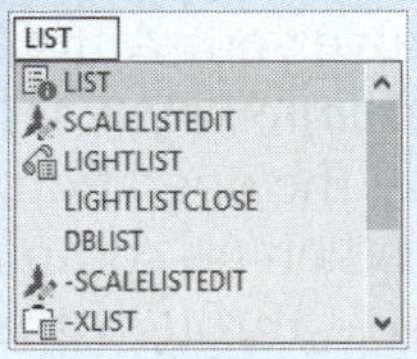
Delay Time	Specifies the number of milliseconds before the suggestion list appears.	300 millisecond delay is the default.

B

Function keys provide instant access to commands and are programmable to perform a series of commands. Control and shift key combinations require that you press and hold [Ctrl] or [Shift] and then press a second key. You can activate several commands using [Ctrl] combinations. A tooltip or a display in a shortcut menu typically indicates if a key combination is available for a command.

function keys: The keys labeled [F1] through [F12] along the top of the keyboard.

Reference Material

www.g-wlearning.com/CAD/

Shortcut Keys

For a complete list of keyboard shortcuts, go to the companion website, select the **Resources** tab, and select **Shortcut Keys**.

Exercise 1-10

www.g-wlearning.com/CAD/

Complete the exercise on the companion website.

Getting Help

InfoCenter
Help
Type
HELP
?
[F1]
HELP

If you need help with a specific command, option, or AutoCAD feature, use this textbook as a guide. AutoCAD also includes learning tools, such as the **Learn** page of the **Start Tab**, and a help system that you access online, by default, through the **Help** window. See Figure 1-25. The graphic shown in this margin identifies ways to access help and the **Help** window. You can also open the **Help** window from the **InfoCenter**, described in the next section. You can also select **Help** from a shortcut menu or use the command line **Search in Help** tool.

You can download and install help system files from Autodesk in order to view the **Help** window offline. If you cannot view the online version, force AutoCAD to display the installed help system by deselecting the **Access online content when available** ***check box*** in the **Help** area on the **System** tab of the **Options** dialog box.

check box: A selectable box that turns an item on (when checked) or off (when unchecked).

The **Help** window uses a format similar to a typical website with menus of links, navigation options, and a search function. To search the help system index for a specific topic, such as a command, type in the **Search** text box, located near the upper-right corner of the **Help** window. Select the desired command topic from the search results. Once you see the description of the command in the help documentation, pick the **Find** button or click the **Find** link to display an arrow pointing to the item in the ribbon, **Application Menu**, **Quick Access** toolbar, or status bar. If the item is hidden, a tooltip appears to identify where the item is located. Menus available in the **Help** window provide direct access to content such as tutorials, sample files, and Autodesk support.

PROFESSIONAL TIP

Press [F1] while using a command to display ***context-oriented help***. This saves time when you are looking for help with the current command and drawing task.

context-oriented help: Help information for the active command.

Figure 1-25.
Learning and help resources available in AutoCAD. A—The **Learn** page of the **Start Tab**. B—The **Help** window and the **InfoCenter** displayed in the AutoCAD window.

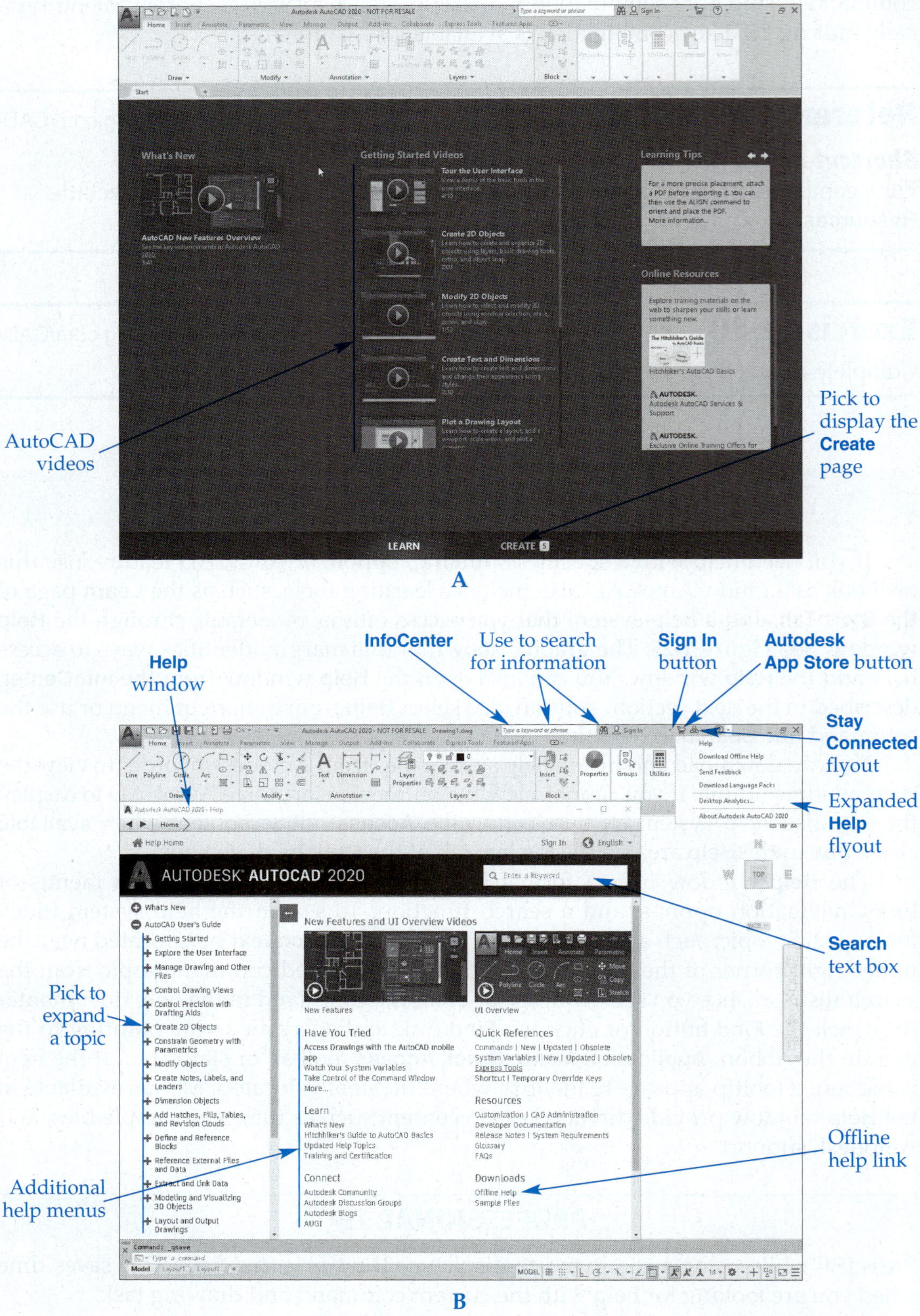

InfoCenter

The **InfoCenter**, located on the right side of the title bar in **Figure 1-25**, allows you to search for help topics without first displaying the **Help** window. Enter a topic in the text box to search for related information. The **Help** window appears with the search results. Pick the **Sign In** button to log in to or create an Autodesk account. Pick the **Autodesk App Store** button to display the Autodesk App Store, where you can download applications to complement Autodesk software. Use the **Stay Connected** flyout to connect with Autodesk and AutoCAD services. Pick the **Help** button to access the **Help** window, or select a help system option from the **Help** flyout.

Exercise 1-11

www.g-wlearning.com/CAD/

Complete the exercise on the companion website.

Chapter Review

Answer the following questions using the information in this chapter.

1. Describe at least one application for AutoCAD software.
2. Explain what is involved in planning a drawing.
3. What are drawing standards?
4. Why should you save your work every 10 to 15 minutes?
5. Name one method of starting AutoCAD.
6. Name one method of exiting AutoCAD.
7. What is the name for an interface that includes on-screen features?
8. Define or explain the following terms:
   A. Default
   B. Pick (or click)
   C. Hover
   D. Button
   E. Function key
   F. Option
   G. Command
9. What is a workspace?
10. What is a flyout?
11. How do you change from one workspace to another?
12. How do you access a shortcut menu?
13. What does it mean when a shortcut menu is described as context-sensitive?
14. What is the difference between a docked interface item and a floating interface item?
15. Explain the basic function of the **Application Menu**.
16. Describe the **Application Menu** search tool and explain how to use it.
17. Describe an advantage of using the ribbon.
18. What is the function of tabs in the ribbon?
19. What is another name for a palette?
20. Describe the function of the status bar.
21. What is the meaning of the … (ellipsis) in a menu option or button?
22. List two methods for accessing AutoCAD commands.

23. Identify two ways to access AutoCAD input search settings from the command line.
24. Describe the function of dynamic input.
25. Explain the function of the [Esc] key.
26. How do you access previously used commands when dynamic input is on?
27. Name the function keys that execute the following tasks. (Refer to the **Shortcut Keys** document in the **Reference Material** section on the companion website.)
    A. Snap mode (toggle)
    B. Grid mode (toggle)
    C. Ortho mode (toggle)
28. Describe two ways to access the **Help** window.
29. What is context-oriented help, and how is it accessed?
30. Describe the purpose of the **InfoCenter**, and explain how to use the **InfoCenter** text box.

Problems

Start AutoCAD if it is not already started. Follow the specific instructions for each problem.

▼ Basic

1. General Perform the following tasks:
    A. Open the **Help** window.
    B. Type InfoCenter into the **Search** text box.
    C. Pick the **About InfoCenter** link from the search results.
    D. Read the information provided.
    E. Close the **Help** window, and then close AutoCAD.

2. General Perform the following tasks:
    A. Start a new drawing.
    B. Move the cursor over the buttons in the status bar and read the tooltip for each.
    C. Slowly move the cursor over each of the ribbon panels and read the tooltips.
    D. Pick the **Application Menu** button to display the **Application Menu**. Hover over the **Open** menu, and then use the up and down arrow keys to cycle through the options.
    E. Press [Esc] once to hide the **Open** menu options and a second time to hide the **Application Menu**.
    F. Close AutoCAD.

▼ Intermediate

3. General Interview your drafting instructor or supervisor and try to determine what types of drawing standards exist at your school or company. Write them down and keep them with you as you learn AutoCAD. Make notes as you progress through this textbook on how you use these standards. Also, note how you could change the standards to match the capabilities of AutoCAD.

4. General Research your drawing department standards. If you do not have a copy of the standards, acquire one. If AutoCAD standards exist, make notes as to how you can use these in your projects. If no standards exist in your department or company, make notes about how you can help develop standards. Write a report on why your school or company should create CAD standards and how to use the standards. Describe who should be responsible for specific tasks. Recommend procedures, techniques, and forms, if necessary. Develop this report as you progress through your AutoCAD instruction and as you read this textbook.

5. General Develop a drawing planning sheet for use in your school or company. List items you think are important for planning a CAD drawing. Make changes to this sheet as you learn more about AutoCAD.

6. General Create a freehand sketch of the default AutoCAD window with the Drafting & Annotation workspace active and a new drawing started using the acad.dwt template. Label each of the screen areas. To the side of the sketch, write a short description of the function of each screen area.

7. General Create a freehand sketch showing three examples of tooltips displayed as you hover over an item. To the side of the sketch, write a short description of each example's function.

8. General Using the **Application Menu** search tool, type the letter C and review the information provided in the **Application Menu**. Then add the letter L. How does the information change? Continue typing O, S, and E to complete the **CLOSE** command. Write a short paragraph explaining how you might use this search tool to find a command if you are unsure how the command is spelled or where it is located.

▼ Advanced

9. General Research and write a report of approximately 250 words covering the American Society of Mechanical Engineers (ASME) standards accredited by the American National Standards Institute (ANSI).

10. General Research and write a report of approximately 250 words covering the International Standards Organization (ISO) drafting standards.

11. General Research and write a report of approximately 250 words covering the United States National CAD Standard (NCS).

12. General Research and write a report of approximately 250 words covering workplace ethics, especially as it applies to CADD applications and CADD-related software. Prepare a slide show presentation of your research and present it to your class or office.

13. General Research and write a report of approximately 150 words covering an ergonomically designed CADD workstation. Include a freehand sketch of what you consider a high-quality design for a workstation and label its characteristics. Prepare a slide show presentation of your research and present it to your class or office.

14. General Go to the Autodesk Education Community website and register to join the Autodesk Education Community. After you register, download a student version of AutoCAD to your home or laptop computer. Use your copy of AutoCAD to complete assignments and study AutoCAD when you are unable to access a CADD lab. The Autodesk Education Community website provides complete information on the registration and download process.

AutoCAD Certified Professional Exam Practice

Answer the following questions using the information in this chapter.

1. Which workspaces are available by default in AutoCAD? *Select all that apply.*
   A. 3D Basics
   B. 3D Animation
   C. 3D Modeling
   D. 2D Modeling
   E. Drafting & Annotation
2. Which of the following is a method of starting the AutoCAD software? *Select the one item that best answers the question.*
   A. Access the Windows Start menu, select Run..., and enter autocad.exe
   B. Right-click on the Windows desktop and select AutoCAD 2020 from the shortcut menu
   C. Double-click the AutoCAD 2020 icon on the Windows desktop
   D. Navigate to the Autodesk website and double-click AutoCAD 2020
3. If you cannot view the online version of the **Help** window, you can force AutoCAD to display the installed help system by deselecting the **Access online content when available** check box in which of the following locations? *Select the one item that best answers the question.*
   A. **InfoCenter**
   B. **Options** dialog box, **System** tab
   C. **Application Menu**, **Drawing Utilities** menu
   D. Status bar

Follow the instructions in each problem. Complete each problem by answering the related question.

4. **Start AutoCAD and a new drawing using one of the methods described in the chapter.**
   Turn off dynamic input. At the Command: prompt, type the letters LA and press [Enter]. What command does AutoCAD execute?
5. **Start AutoCAD and a new drawing using one of the methods described in the chapter.**
   Access the **Application Menu** search tool and type the letters C, L, and O. Review the results of the search. What entry appears in the Ribbon Tab: Home category?

CHAPTER 2

Drawings and Templates

Learning Objectives

After completing this chapter, you will be able to:

- ✓ Start a new drawing.
- ✓ Save files.
- ✓ Close files.
- ✓ Open saved files.
- ✓ Work with multiple open documents.
- ✓ View and adjust drawing properties.
- ✓ Determine and specify drawing units and limits.
- ✓ Create drawing template files.

In this chapter, you will learn how to start new drawings, save drawings, open existing drawings, and begin the process of preparing drawing template files. This chapter also explains commands and options that assist in organizing and setting up a drawing session, including basic drawing settings. You will find the drawing settings described in this chapter very useful as you begin working with drawing and drawing template files.

Starting a New Drawing

AutoCAD uses several different types of files for specific functions. The primary file types are ***drawing files***, which have a .dwg extension, and ***drawing template files***, also known as ***templates***, which have a .dwt extension. You typically begin a new drawing by referencing a template. A new drawing based on a template includes all of the template settings and content. To help avoid confusion as you learn AutoCAD, remember that a new drawing file references a drawing template file, but the drawing file is where you prepare the new drawing. You can also start a new drawing from scratch.

drawing files: Files you use to create and store drawings.

drawing template files (templates): Files you reference to develop new drawings; contain standard drawing settings and objects.

Starting from a Template

Use the **QNEW** command to begin a new drawing. By default, the **Select template** dialog box appears when you access the **QNEW** command. See Figure 2-1. You can use the **QNEW** command to start a drawing without displaying the **Select template** dialog box by setting up the quick start template, as described in this chapter.

Quick Access
New
QNEW

NEW

Application Menu
New
> Drawing
New
Type
NEW
[Ctrl]+[N]

The **NEW** command can also be used to start a new drawing and displays the **Select template** dialog box by default.

The **Select template** dialog box lists the templates found in the specified drawing template folder. The default template folder shown in Figure 2-1 includes a variety of templates supplied with AutoCAD.

The tutorial templates for manufacturing and architecture include a border and title block. All of the other supplied templates are blank, but include drawing settings specific to the requirements of a certain industry or drawing. For general 2D drawing applications, select the default acad.dwt template, which uses decimal unit settings with a US Customary (inch) unit preference. Select the acadiso.dwt template to use basic decimal unit settings according to metric ISO standards.

drawing standards file (DWS): A file used to check the standards of another file using AutoCAD standards-checking tools.

You can start a new drawing file by referencing any available template, drawing, or ***drawing standards file (DWS)***. To select a file not listed in the default templates folder, first select the type of files to locate from the **Files of type:** drop-down list. The default Drawing Template (*.dwt) filter shows only template files. Select Drawing (*.dwg) to list only drawing files or select Standards (*.dws) to list only drawing standards files. Next, select the folder that contains the desired files from the **Look in:** drop-down list. To use a file from the list to begin a new drawing, double-click on the file name, right-click on the file and pick **Select**, or select the file and pick the **Open** button.

NOTE

You can also start a new drawing from the **Create** page of the **Start Tab**, which appears by default when you first launch AutoCAD. The **Create** page of the **Start Tab** can also be accessed with the **GOTOSTART** command. File tabs, described later in this chapter, provide quick access to the **QNEW** command.

GOTOSTART

Type
GOTOSTART

Figure 2-1.
Use the **Select template** dialog box to start a new drawing. Select a template to reference, or pick the arrow next to the **Open** button and select an **Open with no Template** option to start a drawing from scratch.

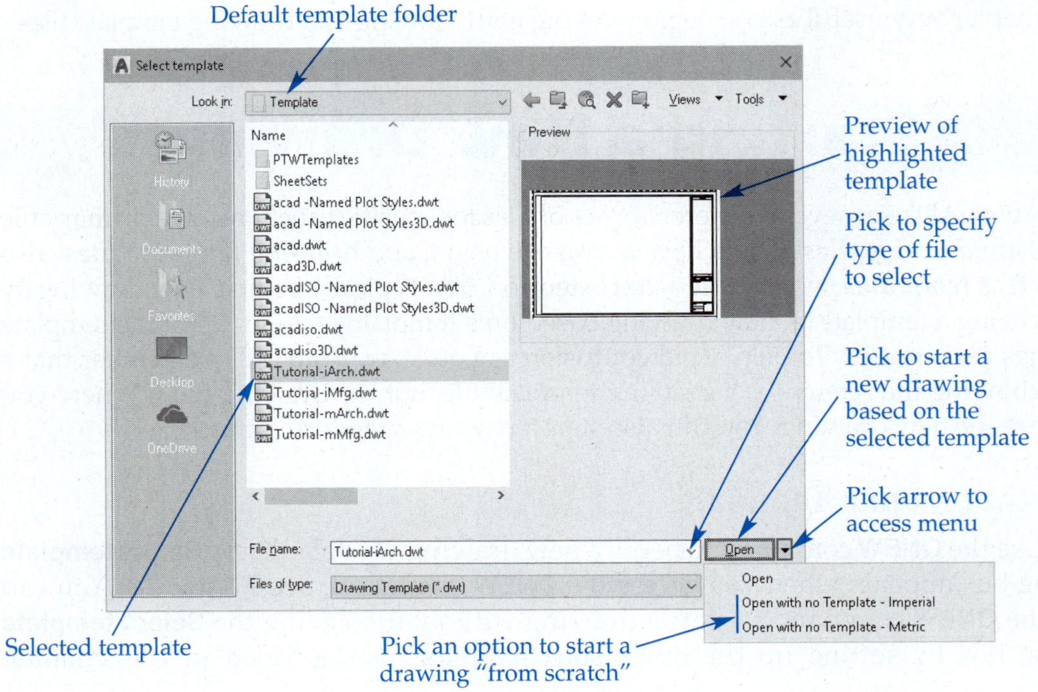

PROFESSIONAL TIP

Use the **Options** dialog box to change the default drawing template folder displayed in the **Select template** dialog box. In the **Files** tab, expand **Template Settings**, and then expand **Drawing Template File Location**. Pick the **Browse...** button to select a folder.

Starting from Scratch

For a new AutoCAD user, an effective way to begin a new drawing is to start "from scratch" using a blank drawing file without a border, title block, modified layouts, or customized drawing settings. Start from scratch when you just begin to learn AutoCAD, when you do not yet know specific units and drawing settings to use, and when you create your own templates. To start from scratch, pick the flyout next to the **Open** button in the **Select template** dialog box. See Figure 2-1. Then, to begin a drawing using basic inch unit settings, pick **Open with no Template – Imperial**, or to begin a drawing using basic metric unit settings, pick **Open with no Template – Metric**.

NOTE

The default **Templates** drop-down list in the **Get Started** section of the **Create** page of the **Start Tab** includes **No Template – Imperial** and **No Template – Metric** options to start from scratch.

Setting the Quick Start Template

The **Options** dialog box provides a quick start feature that allows you to begin a drawing using the **QNEW** command and a specific template, skipping the **Select template** dialog box. Pick the **Files** tab, expand the **Template Settings** option, and then expand the **Default Template File Name for QNEW** function. See Figure 2-2.

Figure 2-2.
Specifying a template for the **QNEW** command.

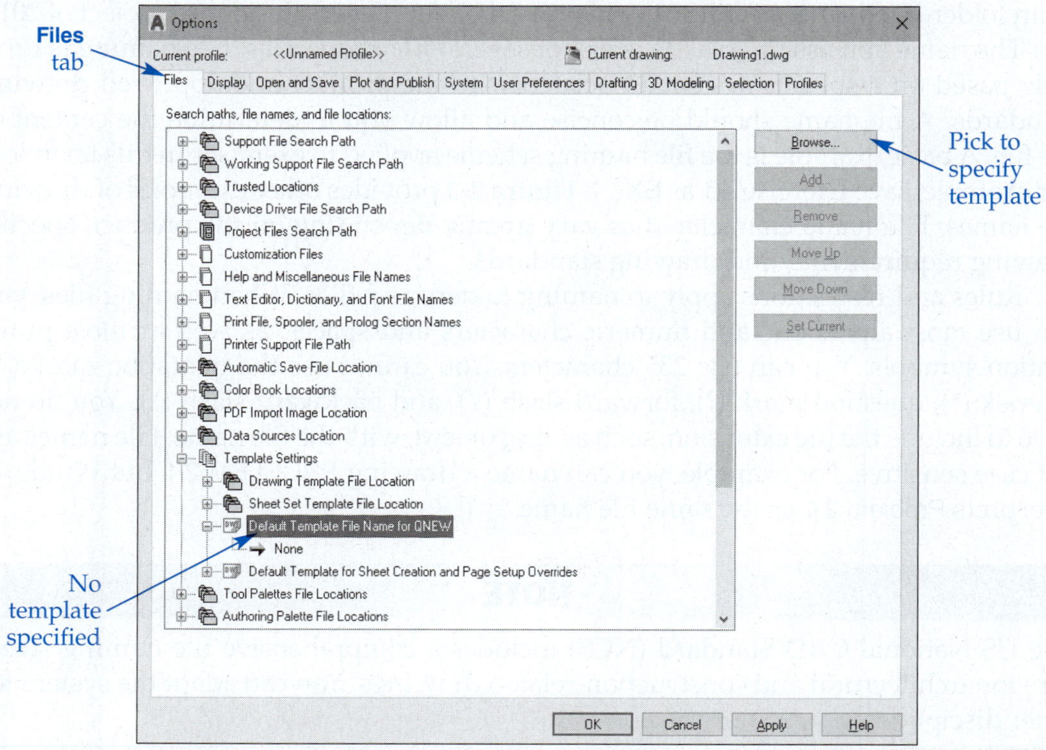

None displays by default and causes the **Select template** dialog box to appear. Pick the **Browse...** button to select a specific template to launch each time you use the **QNEW** command.

PROFESSIONAL TIP

Use the **NEW** command to override the quick start template and display the **Select template** dialog box. This allows you to use the **QNEW** command to begin a drawing by referencing the template you use most frequently, but access other templates as needed.

Saving Your Work

You should save your work immediately after you start a new file. Then save at least every 10 to 15 minutes while working to avoid the possibility of losing your work. Several AutoCAD commands allow you to save your work. In addition, when you close a file or exit AutoCAD, an alert appears asking if you want to save changes. This gives you a final opportunity to save or discard changes.

File Storage and Naming

Store drawings on your computer or a network drive according to school or company practice. Back up files to a removable storage device, online service, or other system to help reduce the possibility of permanently losing information during a system failure.

Develop an organized structure of file folders, and use subfolders as needed to help organize each project. Using a standard naming system allows you to find drawings quickly and easily by content, category, or other criteria. For example, you might save all of your textbook exercises to an Exercises folder, or create chapter-specific folders to store exercises, tests, and problems. An example in a mechanical drafting project is a main folder titled ACME.4001 based on the company name ACME, Inc. and project and assembly number 4001. An example in an architectural drafting project is a main folder titled 2019MEYERS01 to identify the first Meyers Residence project of 2019.

file properties: Values used to define a variety of file and design characteristics.

The name you assign to a file is one of several ***file properties***. File naming is typically based on a specific system associated with the product and approved drawing standards. A file name should be concise and allow you to determine the content of the file. A basic example is the file naming scheme applied to exercises in this textbook. For example, save Exercise 2-1 as EX2-1. Figure 2-3 provides other examples of drawing file names. File name characteristics vary greatly depending on the product, specific drawing requirements, and drawing standards.

Rules and restrictions apply to naming folders and files. When naming files, you can use most alphabetic and numeric characters and spaces, as well as most punctuation symbols. You can use 255 characters. You cannot use the quotation mark ("), asterisk (*), question mark (?), forward slash (/), and backward slash (\). You do not have to include the file extension, such as .dwg or .dwt, with the file name. File names are not case sensitive. For example, you can name a drawing PROBLEM 2-1, but Windows interprets Problem 2-1 as the same file name.

NOTE

The US National CAD Standard (NCS) includes a comprehensive file naming structure for architectural and construction-related drawings. You can adapt the system for other disciplines.

Figure 2-3. Examples of mechanical, architectural, and structural drawing file naming schemes. The architectural and structural examples follow the US National CAD Standard (NCS) file naming format.

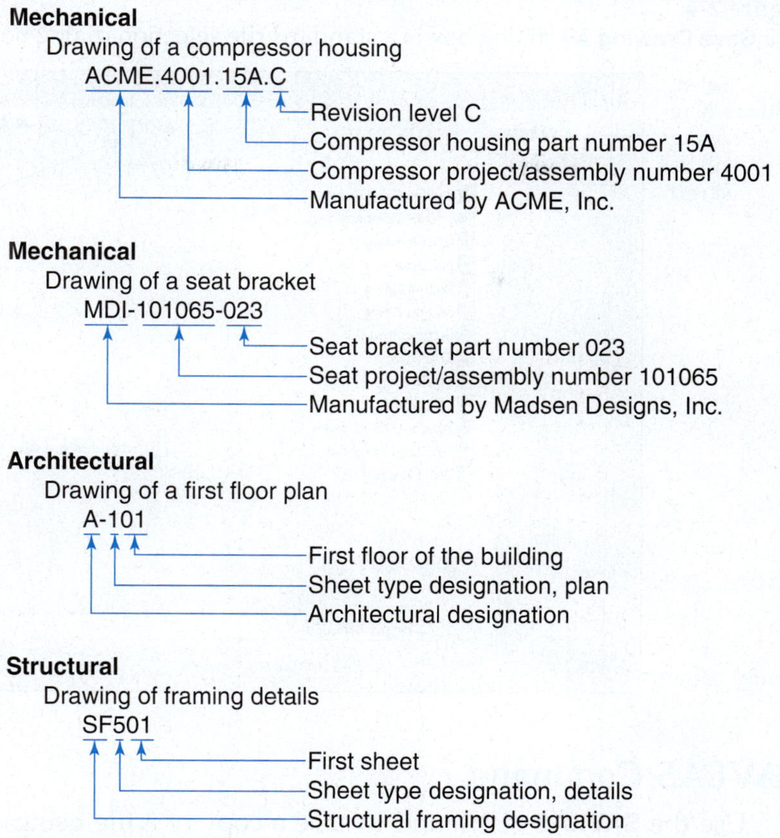

QSAVE Command

If you have not saved a file, the **QSAVE** command displays the **Save Drawing As** dialog box. See Figure 2-4. To save a file, first select the type of file to save from the **Files of type:** drop-down list. Select AutoCAD 2018 Drawing (*.dwg) for most applications. Use AutoCAD Drawing Template (*.dwt) to save a template file. To share a file with someone using an older version of AutoCAD, select the appropriate older AutoCAD version from the list. You also have the option to save a ***drawing exchange file (DXF)*** or drawing standards file (DWS).

QSAVE
Quick Access
Save
Application Menu
Save
Type
QSAVE
[Ctrl]+[S]

drawing exchange file (DXF): A common file format recognized by other CADD systems.

Next, select the folder in which to store the file from the **Save in:** drop-down list. To move upward from the current folder, pick the **Up one level** button. To create a new folder in the current location, pick the **Create New Folder** button and type a new folder name.

The name Drawing1 appears in the **File name:** text box if the file is the first file since you launched AutoCAD. Change the name to the desired file name. The **Options** area contains the **Update sheet and view thumbnails now** check box. Selecting this check box forces AutoCAD to update thumbnail previews in the **Sheet Set Manager**, described later in this textbook, for the file you are saving. After you specify the correct folder location and file name, pick the **Save** button to save the file. You can also press [Enter] to activate the **Save** button.

If you saved the file previously, the **QSAVE** command updates, or resaves, the file based on the current file state. In this situation, **QSAVE** issues no prompts and displays no dialog box.

NOTE

You can also access the **QSAVE** command from file tabs, as described later in this chapter.

Figure 2-4.
The **Save Drawing As** dialog box is a standard file selection dialog box.

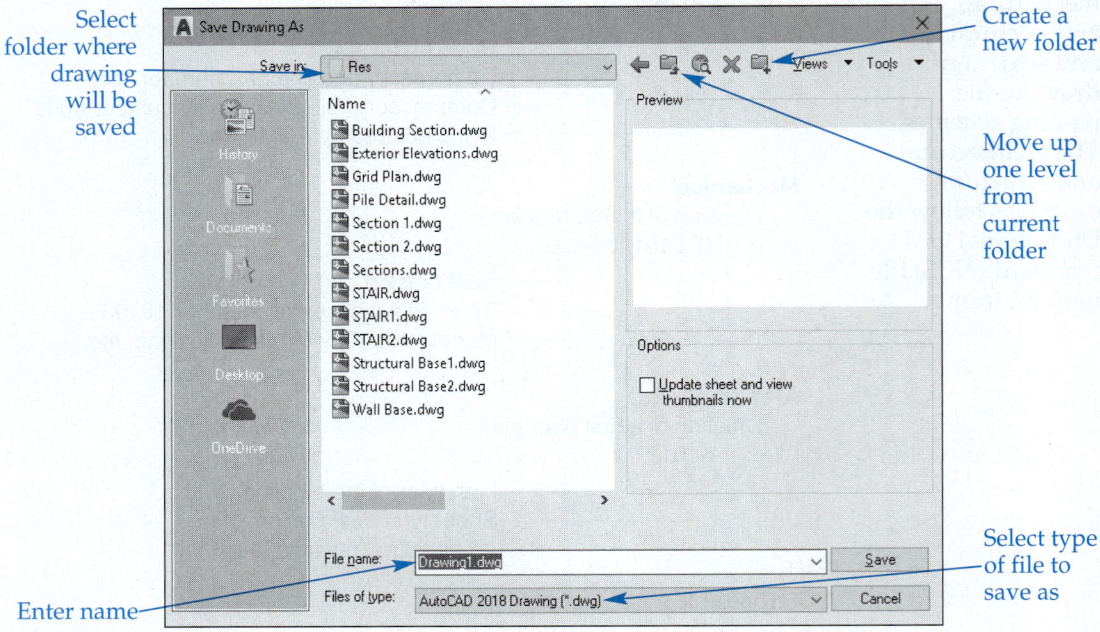

SAVEAS Command

SAVEAS

Quick Access
Save As...
Application Menu
Save As
Type
SAVEAS
[Ctrl]+[Shift]+[S]

Use the **SAVEAS** command to save a copy of a file using a different name or file type. You can also use the **SAVEAS** command when you open a drawing template file to use as a basis for another drawing. This leaves the template unchanged and ready to use for starting other drawings.

The **SAVEAS** command always displays the **Save Drawing As** dialog box. The location and name of an existing file appear. Confirm that the **Files of type:** drop-down list displays the desired file type and that the **Save in:** drop-down list displays the correct drive and folder. Type the new file name in the **File name:** text box and pick the **Save** button.

NOTE

You can also access the **SAVEAS** command from file tabs, as described later in this chapter. Pick an option from the **Save As** menu in the **Application Menu** to preset the file type in the **Save Drawing As** dialog box.

PROFESSIONAL TIP

When you save a version of a drawing in an earlier format, give the file a name that is different from the current version to prevent accidentally overwriting your drawing with the older format.

SAVETOWEBMOBILE

Quick Access
Save to Web & Mobile
Type
SAVETOWEBMOBILE

NOTE

Use the **SAVETOWEBMOBILE** command to save a file to online storage for use with the AutoCAD web app or the AutoCAD mobile app. The AutoCAD web app is an online version of the AutoCAD software. The AutoCAD mobile app is an app version of the software designed for use on mobile devices.

Automatic Saves

AutoCAD provides an ***automatic save*** function that automatically creates a temporary backup file while you work. Settings in the **File Safety Precautions** area of the **Open and Save** tab in the **Options** dialog box control automatic saves. See Figure 2-5. Automatic save is on by default and saves every 10 minutes. Type the number of minutes between saves in the **Minutes between saves** text box. By default, AutoCAD names automatically saved files *FileName_n_n_nnnn.sv$* in the C:\Users*user*\AppData\Local\Temp folder.

automatic save: A save procedure that occurs at specified intervals without your input.

The automatic save timer starts when you make a change to the file and resets when you save the file. The file saves automatically when you start the first command after reaching the automatic save time. Keep this in mind if you let the computer remain idle; an automatic save does not execute until you return and issue a command. Be sure to save your file manually if you plan to be away from your computer for an extended period.

Drawing Recovery Manager

The automatic save file is available for use if AutoCAD shuts down unexpectedly. Therefore, when you close a file normally, AutoCAD deletes the automatic save file. However, after a system failure, the **Drawing Recovery Manager** displays the next time you open AutoCAD. The **Drawing Recovery Manager** contains a node for every file that was open at the time of the system failure. It displays all of the available versions of each file: the original file, the recovered file saved at the time of the system failure, the automatic save file, and the backup file. Pick a version in the **Drawing Recovery Manager** to view, determine which version you want to save, and then save that file. You can save the recovered file over the original file name.

Application Menu
Drawing Utilities > Open the Drawing Recovery Manager

Drawing Recovery Manager...

DRAWINGRECOVERY

NOTE

The **Automatic Save File Location** path in the **Files** tab of the **Options** dialog box determines the folder in which automatic save files are stored.

Backup Files

By default, AutoCAD saves a backup file in the same folder as the drawing or template file. Backup files have a .bak extension. When you save a drawing or template,

Figure 2-5.
Use the **Open and Save** tab in the **Options** dialog box to set up the automatic save feature and backup files.

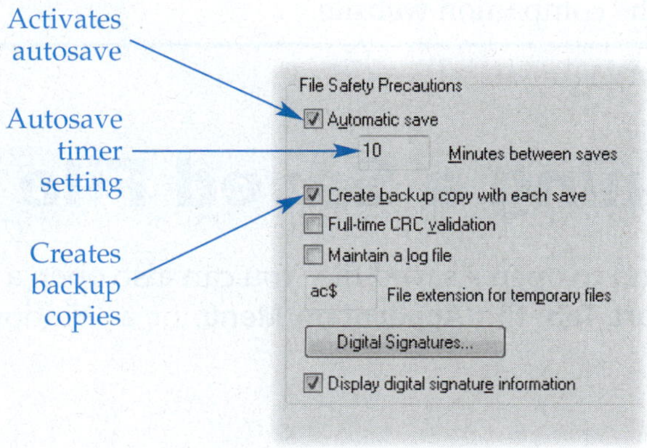

the file updates, and the old file overwrites the backup file. Therefore, the backup file is always one save behind the drawing or template file.

The backup feature is on by default and is controlled using the **Create backup copy with each save** check box in the **Open and Save** tab of the **Options** dialog box. See Figure 2-5. If AutoCAD shuts down unexpectedly, you may be able to recover a file from the backup version using the **Drawing Recovery Manager**. You can also use a backup file by changing the .bak extension to .dwg to restore a drawing, or to .dwt to restore a template. This method allows you to return to an earlier version of the file.

Supplemental Material

www.g-wlearning.com/CAD/

Recovering a Damaged File

For information about recovering a damaged file, go to the companion website, navigate to this chapter in the **Contents** tab, and select **Recovering a Damaged File**.

Closing Files

CLOSE
Application Menu
Close
> Current Drawing
Close
Type
CLOSE

CLOSEALL
Application Menu
Close
> All Drawings
Close All
Type
CLOSEALL

CLOSEALLOTHER
Type
CLOSEALLOTHER

Use the **CLOSE** command to close the current file without exiting AutoCAD. One of the quickest methods of closing a file is to pick the **Close** button from the title bar of a drawing window. If you close a file before saving, AutoCAD prompts you to save or discard changes. Pick the **Yes** button to save the file, or pick the **No** button to discard any changes made to the file since the previous save. Pick the **Cancel** button if you decide not to close the drawing.

AutoCAD allows you to have multiple files open at the same time. Use the **CLOSEALL** command to close all open files. Use the **CLOSEALLOTHER** command to close all open files except the current file. AutoCAD prompts you to save each file to which you made changes.

NOTE

You can also close files using file tabs, as described later in this chapter.

Exercise 2-1

www.g-wlearning.com/CAD/

Complete the exercise on the companion website.

Opening a Saved File

OPEN
Quick Access
Open
Application Menu
Open
> Drawing
Type
OPEN
[Ctrl]+[O]

Use the **OPEN** command to open a saved file. You can also open a saved file using the **Create** page of the **Start Tab**, the **Application Menu**, or a Windows tool such as Windows Explorer.

OPEN Command

Access the **OPEN** command to display the **Select File** dialog box shown in Figure 2-6. The buttons in the Places list on the left side of the dialog box provide

Figure 2-6.
The **Select File** dialog box is a standard file selection dialog box that allows you to open an AutoCAD file. The AutoCAD 2020\Sample\Mechanical Sample folder is open in this illustration. The Mechanical – Multileaders drawing is selected and appears in the **File name:** text box and in the **Preview** area.

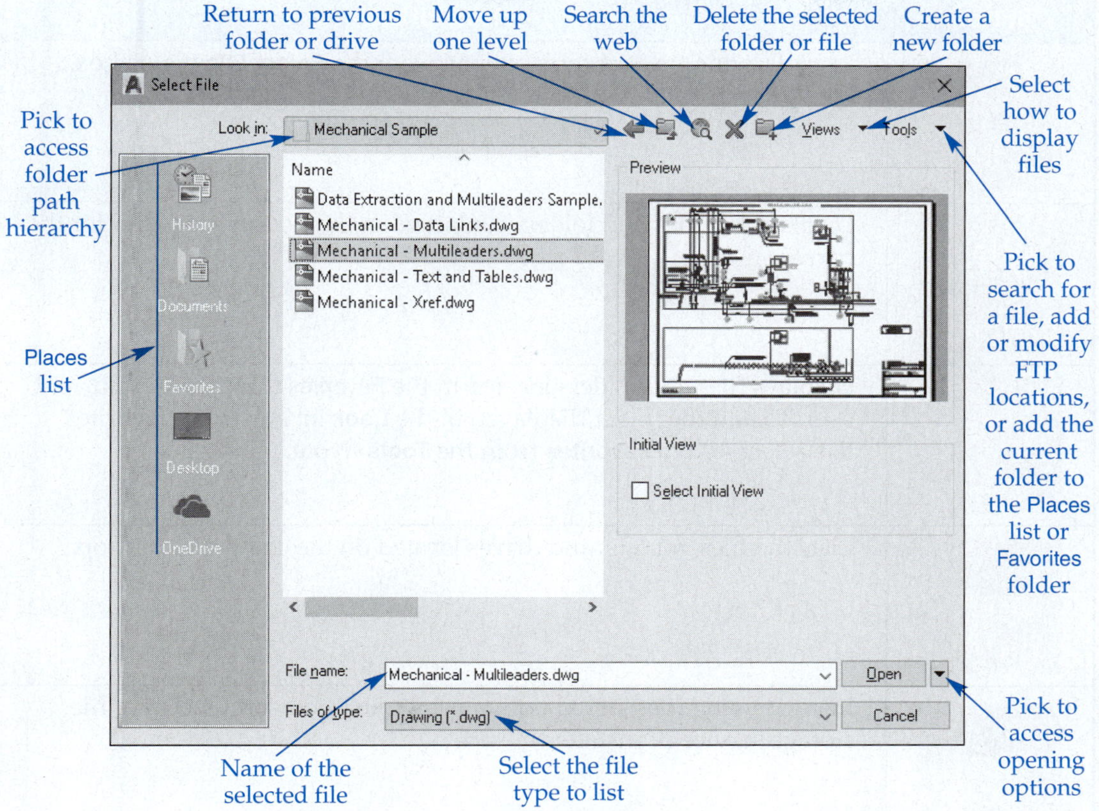

instant access to common items. See **Figure 2-7**. Double-click on a folder to display the contents. Pick a file to view an image of the file in the **Preview** area. The preview provides an easy way for you to identify the content of a file without loading the file into AutoCAD. You can quickly preview other files using the up and down arrow keys to move vertically, and the left and right arrow keys to move horizontally. To open the highlighted file, double-click on the file, pick the **Open** button, or press [Enter].

NOTE

You can also access the **OPEN** command from file tabs, as described later in this chapter.

Exercise 2-2

www.g-wlearning.com/CAD/

Complete the exercise on the companion website.

Finding Files

Pick **Find...** from the **Tools** flyout at the top of the **Select File** dialog box to search for files using the **Find** dialog box. See **Figure 2-8**. If you know the file name, type it in the **Named:** text box. If you do not know the name, use wildcard characters, such as *, to narrow the search. Use the **Type:** drop-down list to search specifically for DWG, DWS, DXF, or DWT files. Use Windows Explorer or Windows Search to search for other file types.

Figure 2-7.
Buttons found in the Places list of the **Select File** dialog box. Additional items may be available in the Places list depending on the programs installed on your computer. Right-click on the list to access options for adding, removing, and adjusting items, and for restoring the original Places list.

Button	Description
	Lists drawing files recently opened from the **Select File** dialog box.
	Displays the files and folders contained in the Documents folder for the current user.
	Displays files and folders located in the Favorites folder for the current user. To add the folder displayed in the **Look in:** box to the favorites list, select **Add to Favorites** from the **Tools** flyout.
	Lists the files, folders, and drives located on the computer desktop.
	Lists drawing files stored online using the Microsoft OneDrive file storage service (if installed).

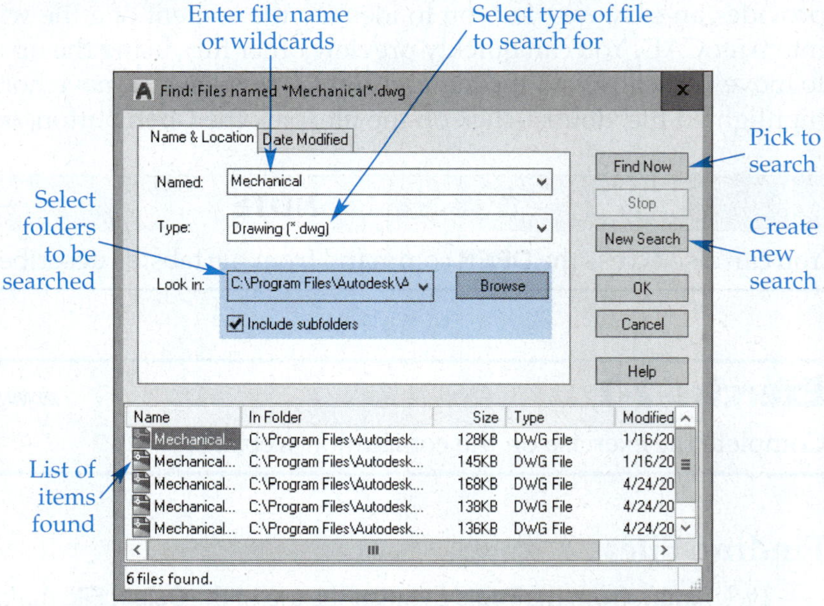

Figure 2-8.
Use the **Find** dialog box to locate AutoCAD files.

If you know the folder in which the file is located, specify the folder in the **Look in:** text box, or pick the **Browse** button to select a folder from the **Browse for Folder** dialog box. Check the **Include subfolders** check box to search the subfolders within the selected folder. The **Date Modified** tab provides options to search for files modified within a certain time. This function is useful if you want to list all drawings modified within a specific week or month.

NOTE

Right-click on a folder or file in the **Select File** dialog box to display a shortcut menu of options. Pick somewhere off the menu to close the menu. The **Initial View** area of the **Select File** dialog box houses a **Select Initial View** check box that, when selected, allows you to select a specific named view display when the file opens. You will learn more about named views later in this textbook.

PROFESSIONAL TIP

Certain file management capabilities are available in file dialog boxes, similar to those in Windows Explorer. For example, to rename an existing file or folder, slowly double-click on the name. This places the name in a text box for editing. Type the new name and press [Enter].

CAUTION

Use caution when deleting or renaming files and folders. Never delete or rename a file if you are not certain you should. If you are unsure, ask your instructor or system administrator for assistance.

Exercise 2-3

www.g-wlearning.com/CAD/

Complete the exercise on the companion website.

Recent Documents

The **Recent Documents** section on the **Create** page of the **Start Tab** and the **Recent Documents** menu in the **Application Menu** provide quick access to recently used files. Pick the **Recent Documents** button in the **Application Menu** to display the list of recently opened documents. See Figure 2-9. Nine of the most recent AutoCAD files appear by default.

Use the **Ordered List** flyout to organize the files. Select **By Ordered List**, **By Access Date**, **By Size**, or **By Type** according to how you want files arranged. Select from the display options flyout to specify how recent files appear. Select the appropriate option to display files as small icons, large icons, small images, or large images.

Hover over a file in the list to display a tooltip with file information and a preview of the file. Pick a file from the list to open. Pick the pushpin icon to the right of the file to keep the file in the recent documents list. AutoCAD eventually removes unpinned files from the recent documents list as you open other files.

PROFESSIONAL TIP

Specify the number of files displayed in the **Recent Documents** list by accessing the **Open and Save** tab of the **Options** dialog box. The **Number of recently-used files** text box in the **Application Menu** area controls this function.

Figure 2-9.
Use the **Recent Documents** menu in the **Application Menu** to open a recent file.

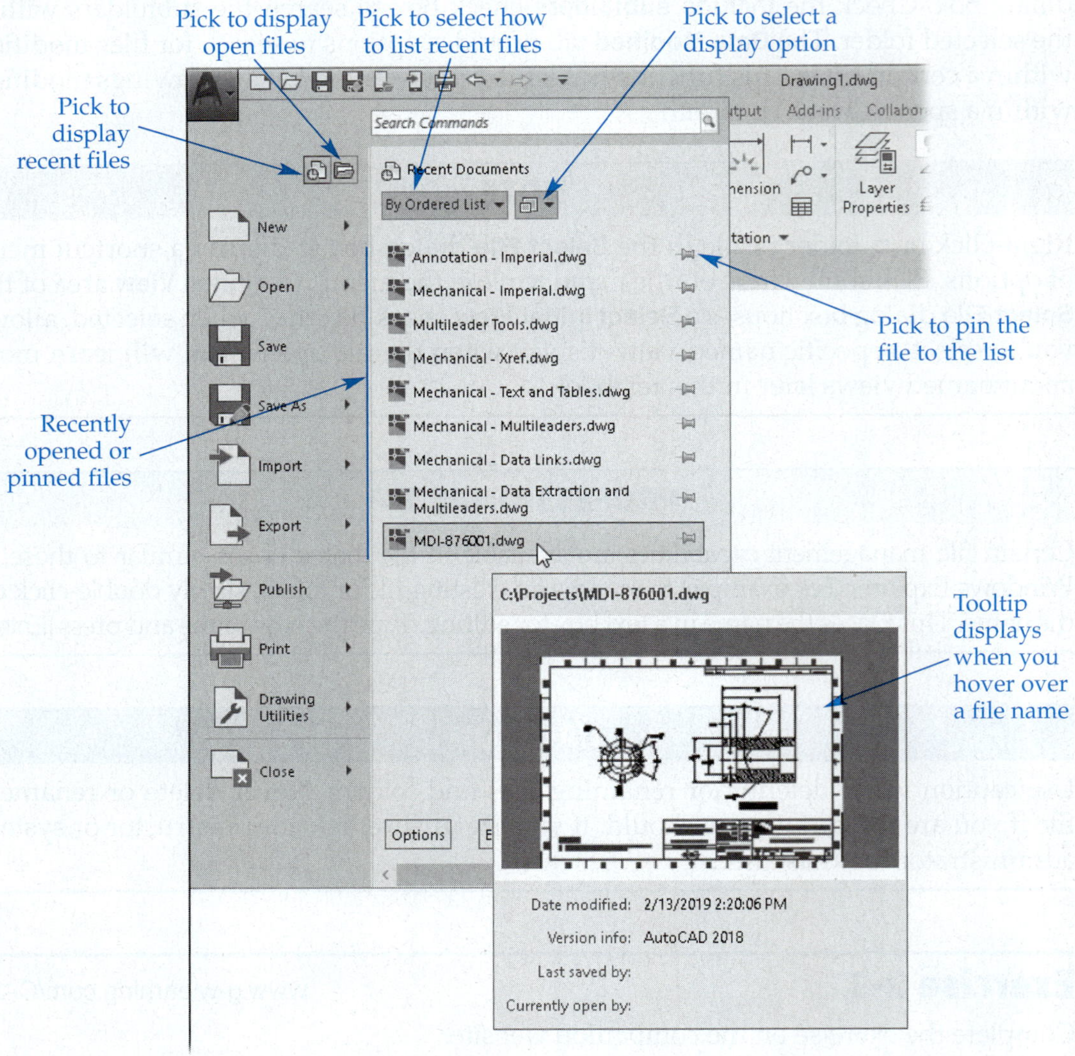

NOTE

If you try to open a deleted or moved file, AutoCAD displays the message Cannot find the specified drawing file. Please verify that the file exists. AutoCAD then opens the **Select File** dialog box.

Windows Explorer

Double-click on an AutoCAD file from Windows Explorer to open the file. If AutoCAD is not already running, it starts and the file opens. You can also launch AutoCAD, if it is not running, and open a file by dragging and dropping the file from Windows Explorer onto the AutoCAD 2020 desktop icon. If AutoCAD is running, you can drag and drop a file onto the command line to open the file.

CAUTION

Dragging and dropping an AutoCAD file into the drawing area inserts the file as a block into the existing file. You must drop the file onto the command line to open the file.

Opening Old Files

When using AutoCAD 2020, you can open AutoCAD files created in any previous release. When you save a file from a previous release in AutoCAD 2020, AutoCAD automatically updates the file to the current file format. A file saved in AutoCAD 2020 displays in the **Preview** area in the **Select File** dialog box.

NOTE

If you want to keep the original format of an older release file opened in AutoCAD 2020, close the drawing without saving or use the **SAVEAS** command to save it back to the appropriate file format.

Opening as Read-Only or Partial Open

You can open files in various modes by selecting the appropriate option from the **Open** flyout in the **Select File** dialog box. When you open a file as ***read-only***, you cannot save changes to the original file. However, you can make changes to the file and then use the **SAVEAS** command to save changes using a different file name. This technique ensures that the original file remains unchanged.

read-only: Describes a drawing file intended for viewing only.

When opening a large drawing, you may choose to issue a ***partial open***. This allows you to open a portion of a drawing by selecting specific views and layers to open. Views and layers are described later in this textbook. You can also partially open a drawing in the read-only mode.

partial open: Describes opening a portion of a file by specifying only the views and layers you want to see.

NOTE

Use the **OPENFROMWEBMOBILE** command to open drawing files from online storage.

Quick Access

Open from Web & Mobile

Type

OPENFROMWEB-MOBILE

OPENFROMWEBMOBILE

Managing Multiple Documents

Most projects include several closely related files. Each file presents or organizes a different aspect of the project. Examples of drawings for a mechanical drafting project include an assembly drawing, subassembly drawings, and detail drawings of each part. Examples of drawings for a basic architectural drafting project include a site plan, floor plans, elevations, sections, and details. You can open multiple AutoCAD files at the same time to work with different portions of a project. Drag and drop and similar operations allow you to share content between documents.

Controlling Windows

Each file you start or open in AutoCAD appears in its own drawing window. By default, AutoCAD maximizes each new drawing window and you only see the active file. See Figure 2-10A. Control drawing windows using the minimize, maximize, restore, and close buttons located in the upper-right corner of each window. Figure 2-10B shows minimized windows. When windows are minimized, the name of the active file appears on the AutoCAD window title bar, and each window includes a title bar that displays the name of the file.

The **Interface** panel of the **View** ribbon tab provides one method to access additional window management commands. Pick the **Switch Windows** flyout to display a list of all open files. Select a file from the flyout to activate. The **Tile Horizontally**, **Tile Vertically**, and **Cascade** buttons allow you to control the arrangement of floating drawing windows. Select the **Tile Vertically** button to tile drawings in a vertical arrangement, with the active window placed on the left. Select the **Tile Horizontally** button

Figure 2-10.
By default, AutoCAD maximizes each new drawing window and you only see the active file, even if multiple files are open. In this figure, three files are open. A—Use the window control buttons to manage the display of multiple windows. B—Minimize drawing windows to display title bars only. Pick the title bar to display a window control menu.

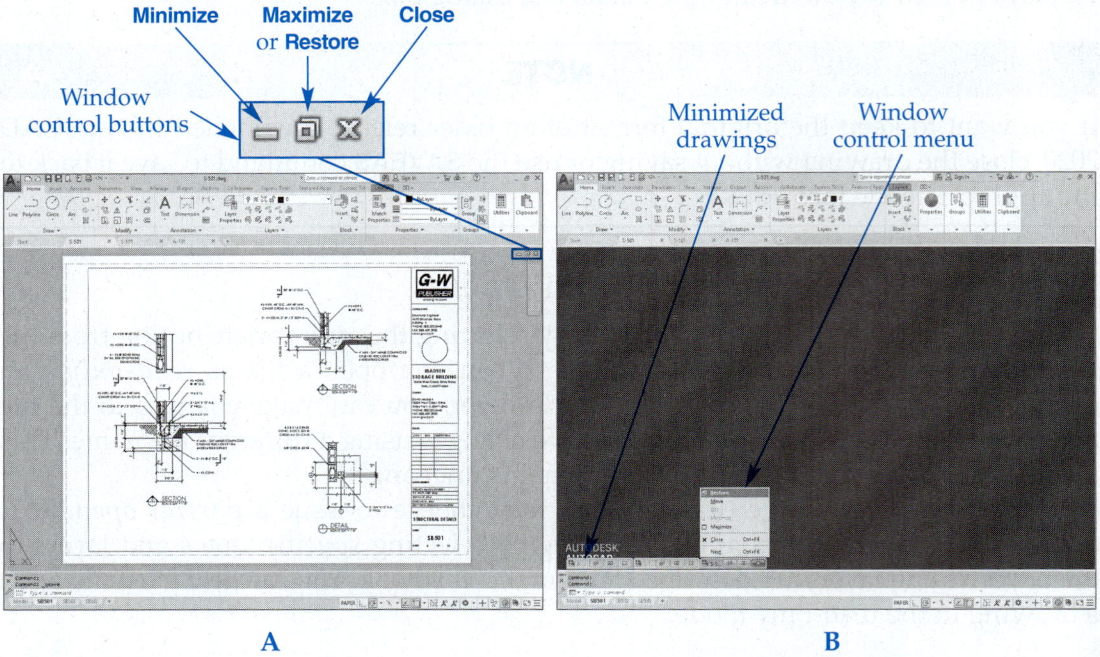

to tile drawings in a horizontal arrangement, with the active window placed on top. See **Figure 2-11A**. Select the **Cascade** button to arrange drawing windows in a cascading style. See **Figure 2-11B**.

You can also switch drawing windows from the **Application Menu** by picking the **Open Documents** button. Refer to **Figure 2-9**. The **Open Documents** menu lists files in the order you opened them. You can display open files as small icons, large icons, small images, or large images by picking the appropriate option from the display options flyout. Pick a file from the list to activate the corresponding drawing window.

PROFESSIONAL TIP

Another technique for switching between open drawings is to press the [Ctrl]+[F6] key combination. This is an effective way to cycle through open drawings quickly.

NOTE

You can change the active drawing window as desired. However, you cannot activate a different window during certain operations, such as while a dialog box is open. You must complete or cancel the operation before switching.

File Tabs

File tabs are displayed by default and are arranged in a horizontal bar above the drawing area. See **Figure 2-12**. The **Start Tab**, introduced in Chapter 1, is an element of the file tabs system. File tabs provide an effective way to see and switch between open AutoCAD files. A quick way to toggle the display of file tabs on

Figure 2-11.
Examples of tiled and cascading floating drawing windows. A—The effect of tiling drawing windows varies depending on the number of windows and the tile option you select. B—A cascading arrangement.

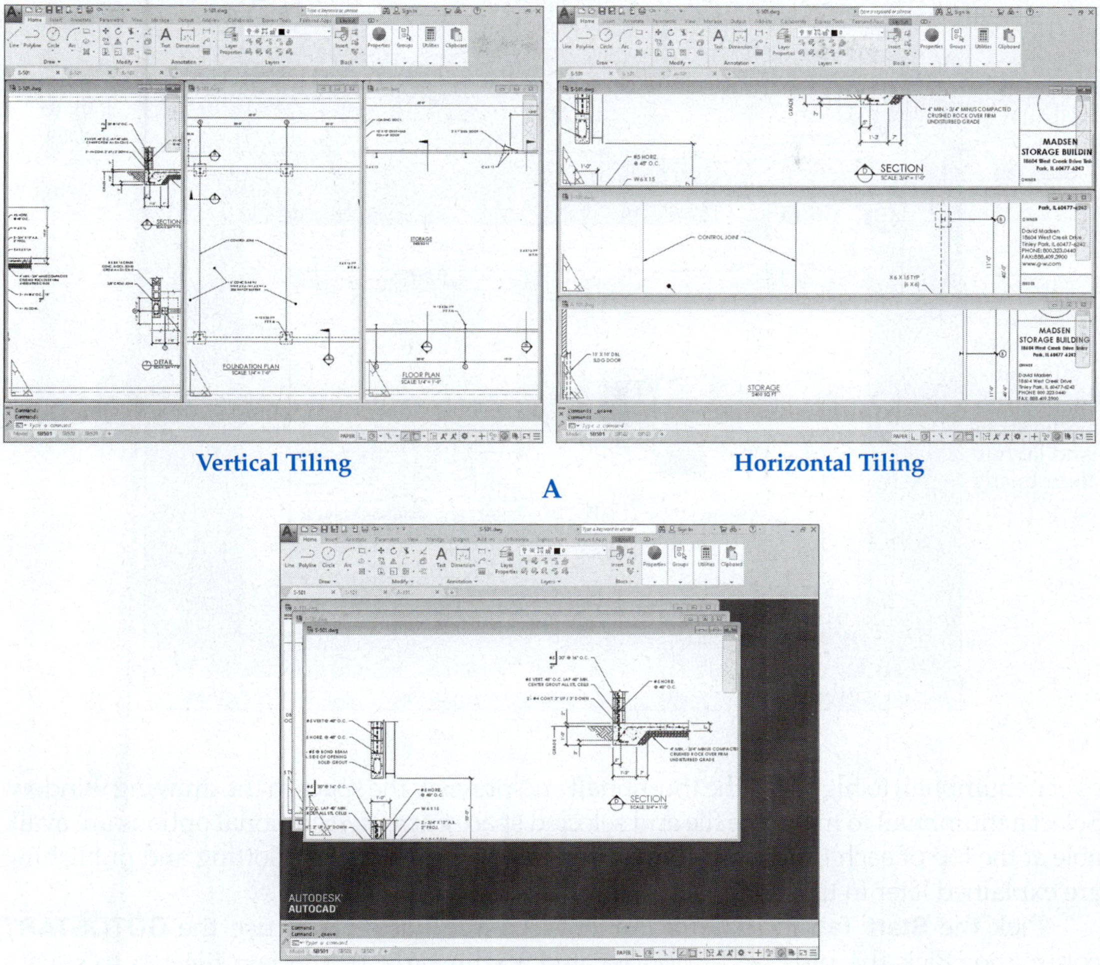

Vertical Tiling

Horizontal Tiling

A

Cascading

B

and off is to pick the **File Tabs** button in the **Interface** panel of the **View** ribbon tab. You can also use the **FILETAB** command to display file tabs or the **FILETABCLOSE** command to hide file tabs.

Type
FILETAB

Type
FILETABCLOSE

A tab and file name identify each open file. See Figure 2-13A. Files are arranged in the order opened, with the file opened first on the left side of the file tab bar, next to the **Start Tab**. The maximized, minimized, or floating state of multiple windows does not change the position of the tabs. If there are so many open files that the tabs spread past the screen, pick the button on the right edge of the file tabs bar to access a menu of open files. See Figure 2-13B. Drag and drop tabs left or right to rearrange if necessary.

The tab associated with the current file is highlighted and is set in front of the other tabs. An asterisk (*) next to a file name indicates that the file has been changed since it was last saved, and a lock icon indicates that the file is open as read-only. Pick a file tab to make the file current. Pick the **Close** button on the right edge of a file tab to close the file. Move the cursor over a tab to display the file path above the tab and model space and layout thumbnails below the tab. Model space and layouts are introduced later in this chapter. The bold frame around a thumbnail indicates the current file and space. Move the cursor

FILETAB FILETABCLOSE

Figure 2-12.
File tabs offer an effective visual method for switching between and controlling open files. File tabs also provide access to basic functions such as beginning a new drawing.

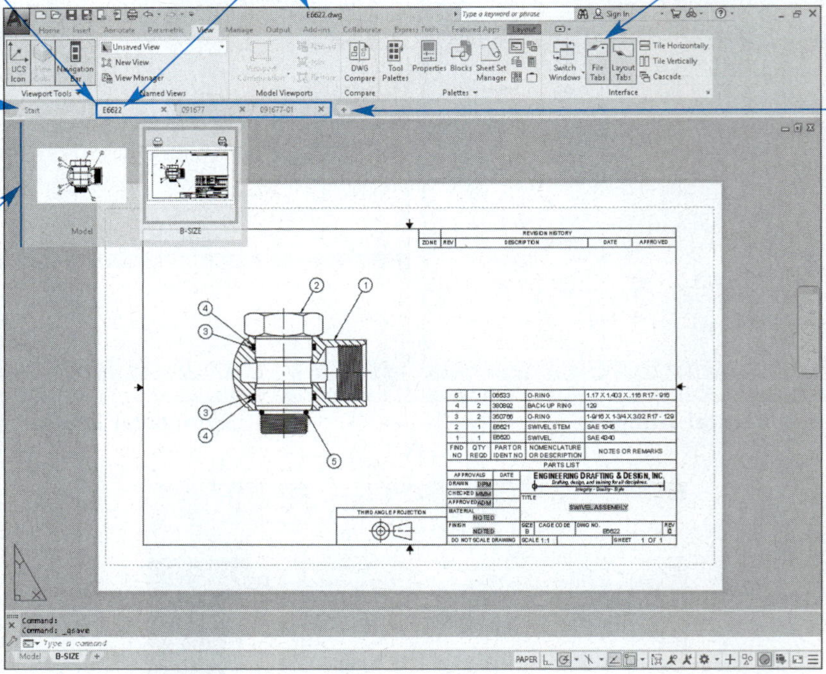

over a thumbnail to highlight the thumbnail and preview the space in the drawing window. Select a thumbnail to make the file and selected space current. Additional options are available at the top of each thumbnail for plotting and publishing files. Plotting and publishing are explained later in this textbook.

Type
GOTOSTART
[Ctrl]+[Home]

GOTOSTART

Pick the **Start Tab** to make it current. An alternative is to use the **GOTOSTART** command. Pick the plus (+) button located to the right of the last file tab to start a new drawing that references the most recently used template. This is especially useful for starting a new drawing that relates to the current project. Right-click on a file tab or the file tab bar to display a shortcut menu with options to access the **QNEW** command, open a drawing file, save all open documents, or close all open documents. The shortcut menu that appears when you right-click on a file tab also includes save and close options for the file associated with the tab. Select **Close All Other Drawings** to close all files except the selected file tab. Select **Copy Full File Path** to copy the file path to the Windows Clipboard. Select **Open File Location** to display the file in Windows Explorer.

NOTE

Type
QVDRAWING

QVDRAWING

The **Quick View Drawings** tool provides access to the same controls available with file tabs. Use the **QVDRAWING** command to access the **Quick View Drawings** tool.

Exercise 2-4

www.g-wlearning.com/CAD/

Complete the exercise on the companion website.

Figure 2-13.
A—Hover over a file tab to display the file path and model space and layout thumbnails, and then hover over a thumbnail to preview the space in the drawing window. B—Use the file tabs menu to access additional open files.

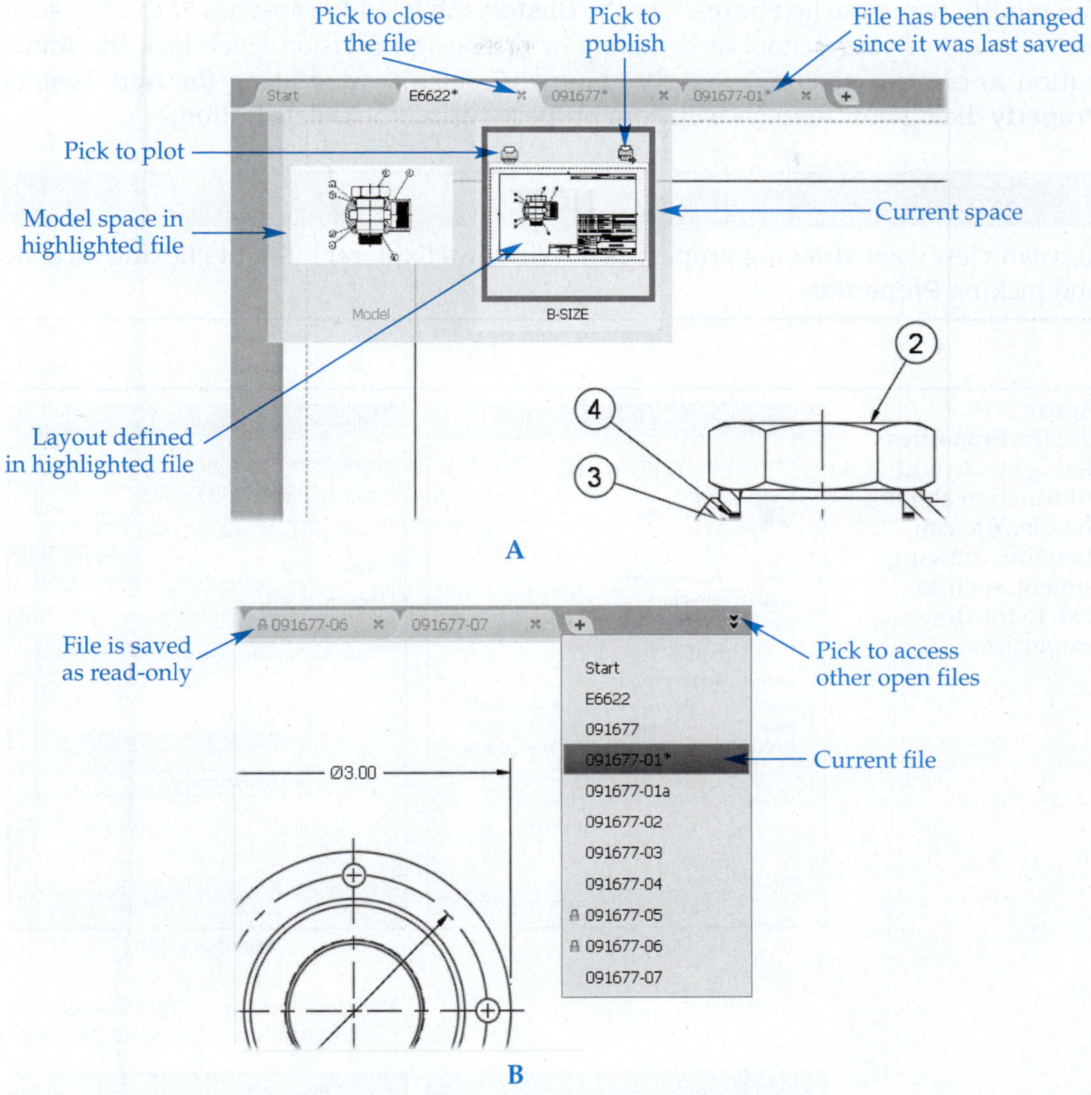

Drawing Properties

An AutoCAD file stores drawing properties that identify the file. The file type and name are examples of file properties. Add values to drawing properties to record common drawing information, such as the title of the product represented by the drawing. You can then reference drawing properties for a variety of purposes, such as adding text to a title block, or creating a bill of materials or report. Additionally, content in the drawing that is linked to drawing properties updates when you make changes to values in the **Properties** dialog box.

NOTE

The **Properties** dialog box is different from the **Properties** palette, which is explained later in this textbook. The **Properties** dialog box contains properties of the drawing file. The **Properties** palette provides access to the properties of objects created in the drawing file.

DWGPROPS

Application Menu
Drawing Utilities > Drawing Properties
Drawing Properties...
Type
DWGPROPS

Access the **Properties** dialog box to view and make changes to drawing properties. See Figure 2-14. Values form in the **General** and **Statistics** tabs when you save the file. You cannot edit general and statistic properties using the **Properties** dialog box. The **Summary** tab includes basic file properties, such as the title and author, that you can modify using the text boxes. Use the **Custom** tab to add properties to the file, such as the name of your school or company or the design revision level. Pick the **Add...** button to create a custom property name and default value using the **Add Custom Property** dialog box. Remove a custom property using the **Delete** button.

NOTE

You can view some drawing properties in Windows Explorer by right-clicking on a file and picking **Properties**.

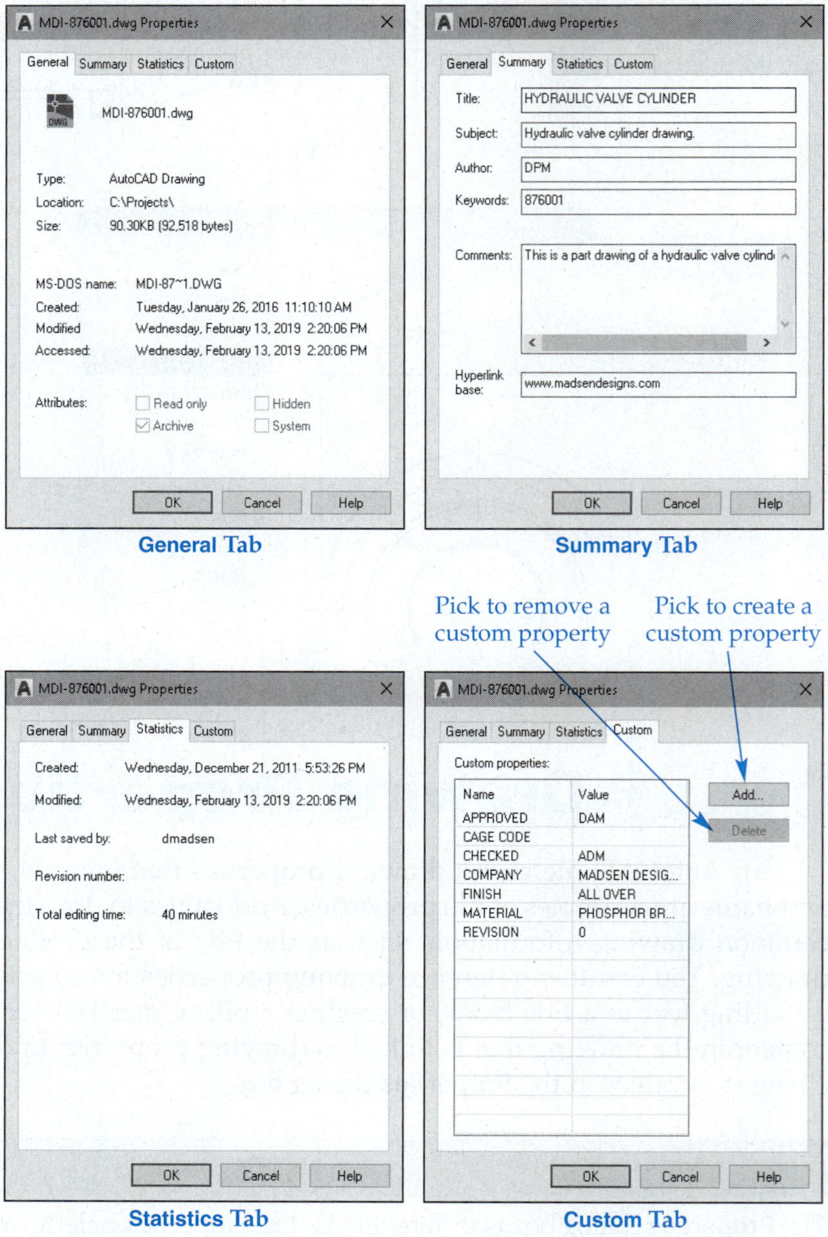

Figure 2-14.
Use the **Properties** dialog box to add information about the file. You can then link drawing content, such as text, to the drawing properties.

PROFESSIONAL TIP

Specify properties that will be used throughout a project and those that are common to multiple projects in template files. These properties will then be predefined every time you begin a new file. You will create template files later in this chapter.

Exercise 2-5

www.g-wlearning.com/CAD/

Complete the exercise on the companion website.

Basic Drawing Settings

Drawing settings determine the general characteristics of a drawing, such as the units of measurement and limits of the drawing area. You can change drawing settings as needed throughout the drawing process. However, you can also specify appropriate drawing settings in template files to preset common drawing characteristics every time you begin a new file. You will create templates later in this chapter.

Introduction to Model Space

Before you adjust drawing settings and begin drawing, you should understand two environments in which you can work. ***Model space*** is where you design and draft the ***model*** of a product. In mechanical drafting, for example, use model space to draw part and assembly views. In architectural drafting, use model space to draw building plans, elevations, sections, and details.

Once you complete the drawing or model in model space, you switch to ***paper (layout) space***, where you prepare a ***layout***. A layout represents the ***sheet*** used to organize and scale, or lay out, a drawing or model to be plotted or exported. Layouts typically include items such as a border, title block, schedules, and general notes. A single drawing can have multiple layouts.

You know you are in model space when you see the model space coordinate system icon, the active **Model** tab, the **MODEL** button on the status bar, and no representation of a sheet in the drawing area. See Figure 2-15. You know you are working with a layout when you see the paper space coordinate system icon, an active layout tab, the **PAPER** button on the status bar, and a representation of a sheet in the drawing area. See Figure 2-16. If you find that you are not in model space, pick the **Model** tab, pick the **PAPER** button on the status bar, or use file tabs.

model space: The environment in AutoCAD where the majority of drawing usually occurs, including the design and drafting of drawing views.

model: A term that usually describes a 3D model, but in AutoCAD also refers to 2D drawing geometry, typically created at full size.

paper (layout) space: The environment in AutoCAD where you create layouts for plotting and display purposes.

layout: An arrangement in paper space of sheet elements, typically including a border, title block, general notes, and a display of items drawn in model space.

sheet: The paper used to lay out and plot drawings.

CAUTION

This textbook explains paper space when appropriate. Until you are ready to lay out a drawing, all of your drawing and drawing setup should occur in model space. Model space is active by default when you start a drawing from scratch and when you begin a file using one of several available templates. Activate model space in all current drawings and custom templates.

Figure 2-15.
Model space is the environment in which you design and draft product geometry, such as the views of this hydraulic valve cylinder.

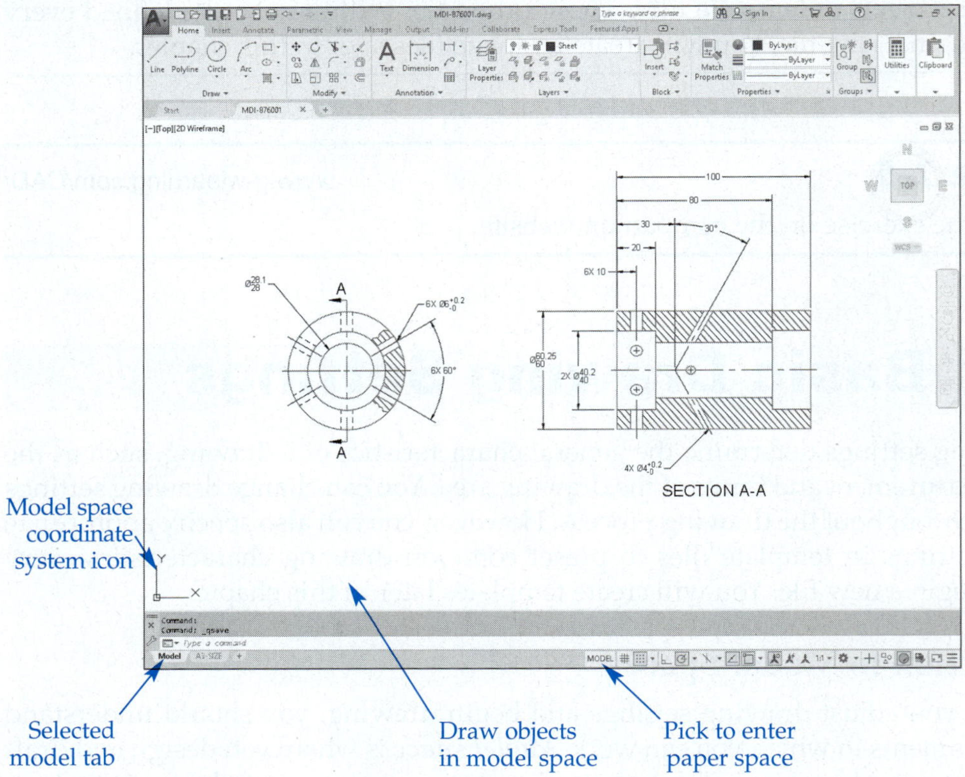

Figure 2-16.
Paper space is the environment in which you lay out model space geometry to complete the drawing and prepare for plotting or export. This example shows common items added to a layout sheet, including an ASME border, title block, revision block, angle of projection block, and general notes.

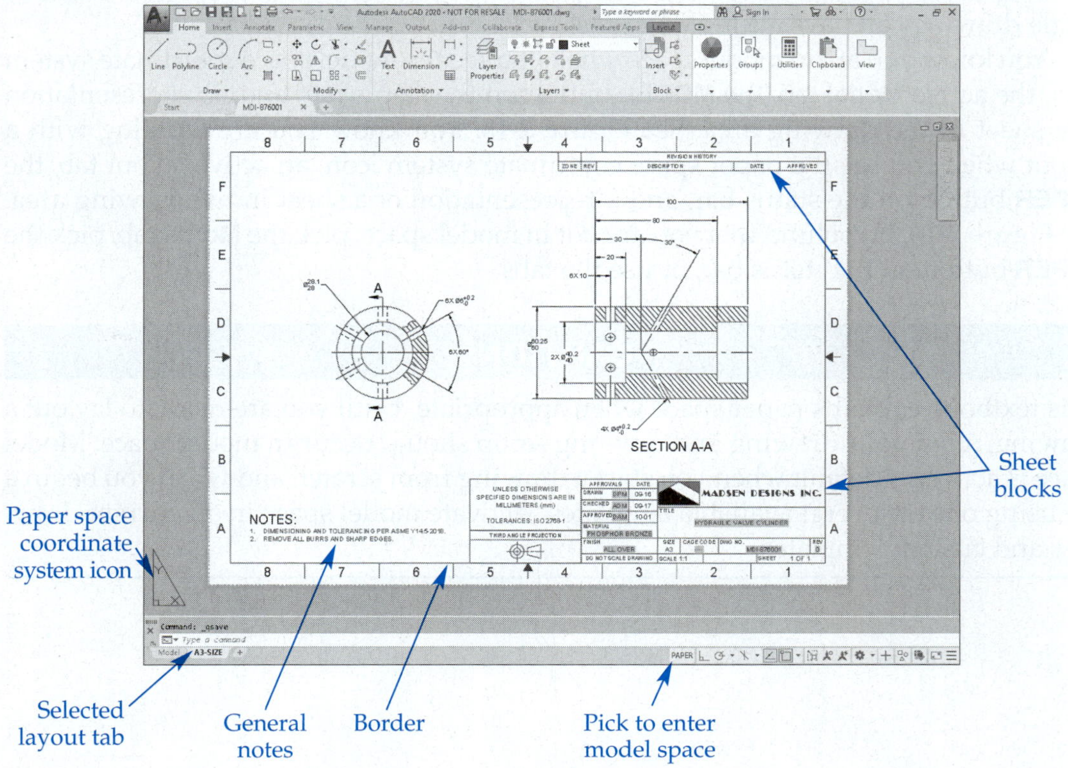

Drawing Units

Drawing units define the linear and angular units of measurement used while drawing and the precision to which these measurements display. *You* determine what "1 unit" means in AutoCAD. For example, 1 unit can mean 1 inch, 1 millimeter, 1 meter, or 1 mile. Most AutoCAD users generally think of 1 unit to be 1 inch or 1 millimeter. Access the **Drawing Units** dialog box to set linear and angular units. See Figure 2-17. Specify linear unit characteristics in the **Length** area. Use the **Type:** drop-down list to set the linear units format, and use the **Precision:** drop-down list to specify the precision of linear units. Figure 2-18 describes linear unit formats.

drawing units: The standard linear and angular units and measurement precision.

Application Menu
Drawing Utilities > Units
Units...
Type
UNITS
UN

UNITS

Set the angular unit format and precision in the **Type:** and **Precision:** drop-down lists in the **Angle** area of the **Drawing Units** dialog box. To change the direction for angular measurements to clockwise from the default setting of counterclockwise, select the **Clockwise** check box.

Figure 2-17.
Use the **Drawing Units** dialog box to set linear and angular unit values.

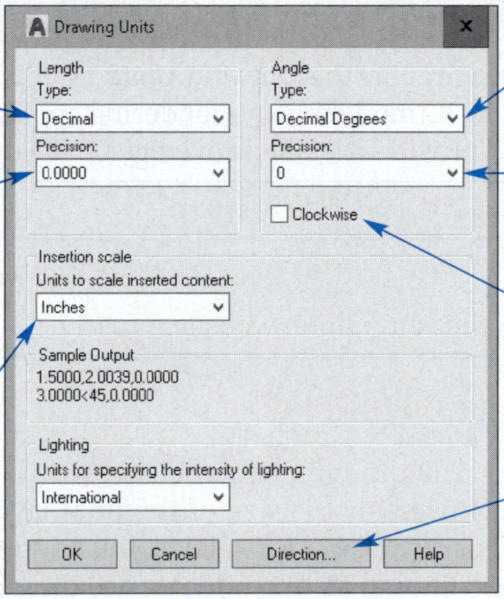

Figure 2-18.
Linear unit formats available in the **Drawing Units** dialog box. Select the appropriate type and precision for the specific drawing application.

Type	Typical Applications	Characteristics	Example
Decimal	Mechanical, architectural, structural, civil (inch or metric)	• Decimal inches or millimeters. • Conforms to the ASME Y14.5 dimensioning and tolerancing standard. • Four decimal place default precision.	14.1655
Engineering	Civil—feet and inch	• Feet and decimal inches. • Four decimal place default precision.	1'-2.1655"
Architectural	Architectural, structural—feet and inch	• Feet, inches, and fractional inches. • 1/16" default precision.	1'-2 3/16"
Fractional	Mechanical—fractional	• Fractional parts of any common unit of measure. • 1/16" default precision.	14 3/16
Scientific	Chemical engineering, astronomy	• E+01 means the base number is multiplied by 10 to the first power. • Used when very large or small values are required. • Four decimal place default precision.	1.4166E+01

radio button: A selection that activates a single item in a group of options.

Pick the **Direction...** button to access the **Direction Control** dialog box. See Figure 2-19. Pick the **East**, **North**, **West**, or **South** *radio button* to set the compass orientation. Picking the **Other** radio button activates the **Angle:** text box and the **Pick an angle** button. Enter an angle for zero direction in the **Angle:** text box. The **Pick an angle** button allows you to pick two points on-screen to establish the angle zero direction.

CAUTION

Use the default direction of 0° **East** at all times, unless you have a specific need to change the compass direction angle, such as when measuring direction using azimuths (0° **North**). This textbook uses the 0° **East** direction, and you should use this default in order to complete most exercises and problems correctly.

Figure 2-20 describes angular unit formats. After selecting the linear and angular units and precision, pick the **OK** button to exit the **Drawing Units** dialog box.

NOTE

The setting in the **Lighting** area in the **Drawing Units** dialog box specifies the default lighting units and is used in 3D modeling and rendering applications. *AutoCAD and Its Applications—Advanced* provides detailed coverage on 3D modeling and rendering.

Exercise 2-6

www.g-wlearning.com/CAD/

Complete the exercise on the companion website.

Drawing Limits

limits: The size of the virtual drawing area in model space.

You prepare an AutoCAD drawing at actual size, or full scale, regardless of the type of drawing, the units used, or the size of the final layout on paper. Use model space to draw full-scale objects. AutoCAD allows you to specify the size of a virtual model space drawing area, known as the model space drawing limits, or *limits*. You typically set limits in a template, but you can change limits as necessary.

The concept of limits is somewhat misleading, because the AutoCAD drawing area is infinite in size. For example, if you set limits to 17″ × 11″, you can still create objects that extend past the 17″ × 11″ area, such as a line that is 1200′ long. Therefore, you can choose not to consider limits while developing a template or creating a drawing. Conversely, as you learn AutoCAD, you may decide that setting appropriate drawing limits is helpful, especially when you are drawing large objects. Regardless of whether you choose to acknowledge limits, you should be familiar with the concept of limits and recognize that some AutoCAD commands, such as **ZOOM** and **PLOT**, provide options associated with limits. The **ZOOM** and **PLOT** commands are explained later in this textbook.

Figure 2-19.
The **Direction Control** dialog box.

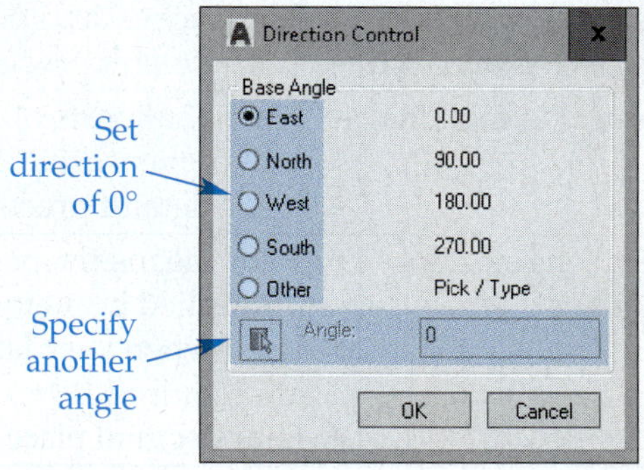

Figure 2-20.
Angular unit formats available in the **Drawing Units** dialog box. Select the appropriate type and precision for the specific drawing application.

Type	Applications and Characteristics	Example	AutoCAD Entry
Decimal Degrees	• Mechanical, architectural, and structural applications. • Degrees (d) and decimal parts of a degree. • Default setting.	45.5°	45.5
Deg/Min/Sec	• Civil and some mechanical, architectural, and structural applications. • Degrees (d), minutes (′), and seconds (″). • 1 degree = 60 minutes; 1 minute = 60 seconds.	45°20′15″	45d20'15"
Grads	• Grad (g) is the abbreviation for *gradient*. • One-quarter of a circle has 100 grads; a full circle has 400 grads.	50.05g	50.05
Radians	• A radian (r) is an angular unit of measurement in which 2π radians = 360° and π radians = 180°. Pi (π) is approximately equal to 3.1416. • A 90° angle has π/2 radians and an arc length of π/2. • Changing the precision displays the radian value rounded to the specified decimal place.	0.785r	.785
Surveyor's Units	• Mostly civil applications. • Uses bearings. A bearing is the direction of a line with respect to one of the quadrants of a compass. Bearings are measured clockwise or counterclockwise, depending on the east (e) or west (w) quadrant, beginning from either north (n) or south (s). • Degrees (d), minutes (′), and seconds (″). • Set precision to degrees, degrees/minutes, degrees/minutes/seconds, or decimal display accuracy of the seconds part of the measurement.	N35°15′E (35°15′ from north toward east)	n35d15'e

Type
LIMITS

LIMITS

Access the **LIMITS** command to set model space drawing limits. The first prompt asks you to specify the coordinates for the lower-left corner of the drawing limits. Press [Enter] to accept the default 0,0 value or enter a new value and press [Enter]. The next prompt asks you to specify the coordinates for the upper-right corner of the virtual drawing area. For example, type 17,11 and then press [Enter]. The first value is the horizontal measurement of the limits, and the second value is the vertical measurement. A comma separates the values. This example creates a limits of 17 units × 11 units if you specify 0,0 as the lower-left corner.

In general, you should set limits larger than the objects you plan to draw. You can determine limits accurately by identifying the drawing scale, converting the scale to a scale factor, and then multiplying the scale factor by the size of sheet on which you plan to plot the drawing. For now, calculate the approximate total length and width of all objects you plan to draw, and add extra space for dimensions and notes. For example, when drawing a 48′ × 24′ building floor plan, allow 10′ on each side for dimensions and notes to make a total virtual drawing area of 68′ × 44′.

NOTE

The **LIMITS** command provides a limits-checking feature that, when turned on, restricts your ability to draw outside of the drawing limits. Use the **ON** option of the **LIMITS** command to turn on limits checking and the **OFF** option to turn off limits checking.

Introduction to Templates

A drawing template file automates the process of using an existing drawing as a starting point for a new drawing. A template includes drawing settings for specific applications that are preset each time you begin a new drawing. Templates are incredible productivity boosters, and can help ensure that everyone at your school or company uses the same drawing standards.

Templates usually include the following, set according to the drawing requirements:

- Units and limits
- Drawing aids, such as dynamic input, grid display, and snaps
- Layers with specific line standards
- Text, dimension, multileader, table, and plot styles
- Common symbols and blocks
- Layouts with a border, sheet blocks, and general notes

Reference Material

www.g-wlearning.com/CAD/

Drawing Sheets

sheet size: Size of the paper used to lay out and plot drawings.

For tables describing sheet characteristics, including ***sheet size***, drawing scale, and drawing limits, go to the companion website, select the **Resources** tab, and select **Drawing Sheets**. Additional information regarding sheet parameters and selection is provided later in this textbook.

Template Development

Template Development
Chapter 2

Starting from scratch or using a template supplied with AutoCAD is an appropriate way to begin drawing as you first learn AutoCAD. However, you will soon find it necessary to create your own templates. Creating your own templates helps you reduce setup time significantly, increase productivity, and consistently adhere to drafting standards. Begin the basic process of template development now, and continue to add content to your templates while you learn AutoCAD.

You can save any drawing or template file as a template using the **SAVEAS** command and **Save Drawing As** dialog box. Select AutoCAD Drawing Template (*.dwt) from the **Files of type:** drop-down list, and then specify a template name and storage location. Template names typically relate to the drawing application, such as Mechanical Template for mechanical part and assembly drawings, or Architectural Floor Plans for architectural floor plan drawings.

By default, the file list box in the **Save Drawing As** dialog box shows the drawing templates currently found in the default Template folder. See Figure 2-21A. You can store templates in any appropriate location, but if you place them in the Template folder, they automatically appear in the **Select template** dialog box and in the **Templates** drop-down list on the **Create** page of the **Start Tab**. Pick the **Save** button to save the template and display the **Template Options** dialog box. See Figure 2-21B. Type a brief description of the template file in the **Description** area. Select **English** or **Metric** units in the **Measurement** drop-down list and pick the **OK** button.

NOTE

Select **Drawing Template** from the **Save As** menu of the **Application Menu** to preset the template file type in the **Save Drawing As** dialog box. Select the **Save all layers as unreconciled** radio button in the **Template Options** dialog box until you are familiar with layers, described in Chapter 5.

Figure 2-21.
A—Save a template using the .dwt file extension. If desired, save it in the AutoCAD Template folder. B—Type a description of the new template in the **Template Options** dialog box.

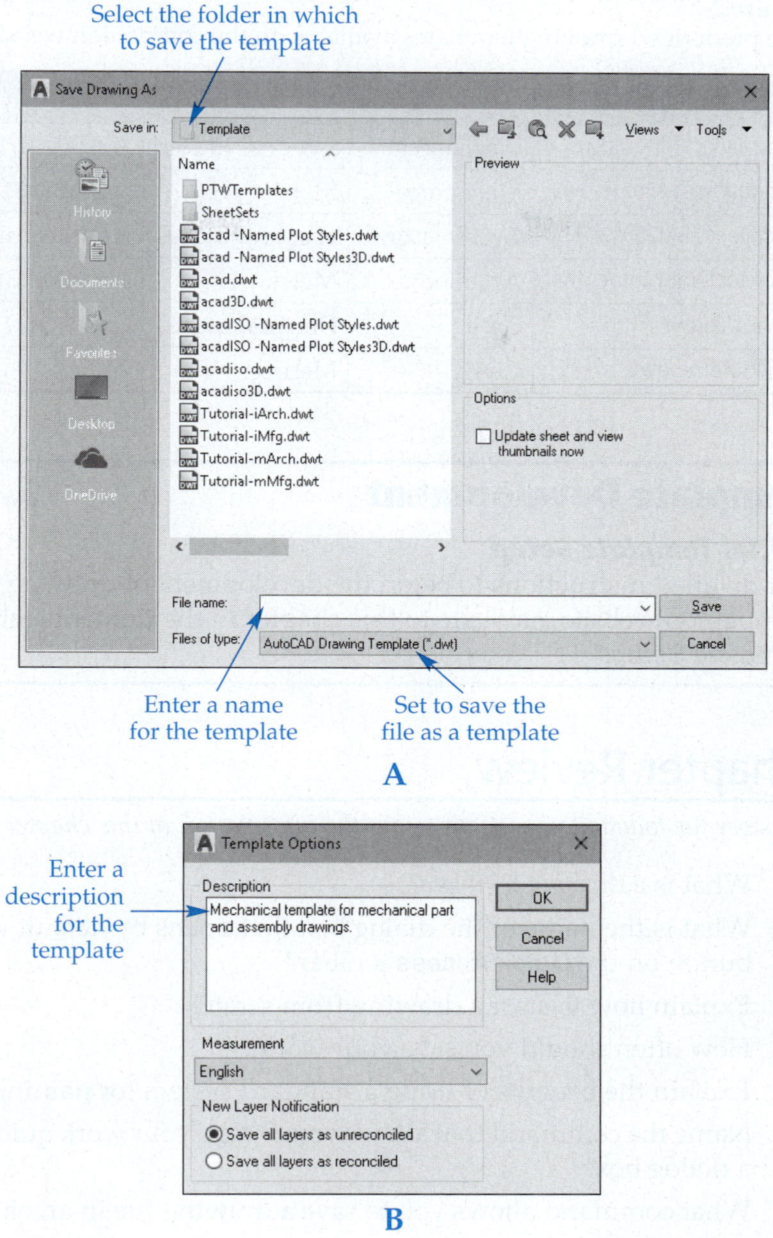

PROFESSIONAL TIP

As you refine your setup procedure, open, revise, and then resave your template files. Use the **SAVEAS** command to save a new template from an existing template.

The companion website for this textbook contains several predefined templates that you can use to create drawings in accordance with correct mechanical, architectural, and civil drafting standards. Figure 2-22 describes each available template. The mechanical drafting templates follow ASME and ISO drafting standards. The architectural and civil drafting templates follow appropriate architectural and civil drafting standards, including standards specified in the US National CAD Standard.

In addition to the complete, ready-to-use templates on the companion website, the Template Development sections at the end of several chapters refer you to the companion website for important template creation topics and procedures. Use the Template Development feature of this textbook to learn gradually how to prepare templates in accordance with correct mechanical, architectural, and civil drafting standards.

Figure 2-22.
The predefined drawing templates available on the companion website.

Template File	Discipline	Units	Layout Sheet Sizes
Mechanical-Inch.dwt	Mechanical	Decimal inches	A, B, C, D, E, F
Mechanical-Metric.dwt	Mechanical	Metric	A4, A3, A2, A1, A0
Architectural-US.dwt	Architectural	Feet and inches	Architectural C, Architectural D
Architectural-Metric.dwt	Architectural	Metric	Architectural A2, Architectural A1
Civil-US.dwt	Civil	Decimal inches	C, D
Civil-Metric.dwt	Civil	Metric	A2, A1

Template Development

www.g-wlearning.com/CAD/

Initial Template Setup

For detailed instructions to begin the development of drawing templates, go to the companion website, navigate to this chapter in the **Contents** tab, and select **Initial Template Setup**.

Chapter Review

Answer the following questions using the information in this chapter.

1. What is a *drawing template*?
2. What is the name of the dialog box that opens by default when you pick the **New** button on the **Quick Access** toolbar?
3. Explain how to start a drawing from scratch.
4. How often should you save your work?
5. Explain the benefits of using a standard system for naming drawing files.
6. Name the command that allows you to save your work quickly without displaying a dialog box.
7. What command allows you to save a drawing file in an older AutoCAD format?
8. How do you set AutoCAD to save your work automatically at designated intervals?
9. Identify the command you would use to exit a drawing file, but remain in the AutoCAD session.
10. How can you close all open drawing windows at the same time?
11. How can you set the number of recently opened files listed in the **Application Menu**?
12. What does the term *read-only* mean?
13. Describe the purpose of file tabs.
14. What occurs when you pick the plus (+) button to the right of the last file tab?
15. What is the function of drawing properties?
16. What is the advantage of adding values to drawing properties?
17. Name three settings you can specify in the **Drawing Units** dialog box.
18. What is *sheet size*?
19. Explain the benefit of using a drawing template file to create a new drawing.
20. How can you convert a drawing file into a drawing template?

Drawing Problems

Start AutoCAD if it is not already started. Follow the specific instructions for each problem.

▼ Basic

1. **General** Start a new drawing using the acad-Named Plot Styles template supplied by AutoCAD. Save the file as a drawing named P2-1.dwg.

2. **General** Start a new drawing using the Tutorial-iArch template supplied by AutoCAD. Save the file as a drawing named P2-2.dwg.

▼ Intermediate

3. **General** Open the Tutorial-iMfg.dwt template file in the AutoCAD Template folder. Type Z and press [Enter], then type A and press [Enter] to view the entire drawing. Describe the drawing. Close the file without saving.

4. **General** Open the chroma.dwg file in the AutoCAD Support folder. Type Z and press [Enter], then type A and press [Enter] to view the entire drawing. Describe the drawing. Close the file without saving.

▼ Advanced

For Problems 5–7, create the specified template for possible future use.

5. **Mechanical** Create a template with the following settings: decimal units with 0.0 precision, decimal degrees with 0.0 precision, default angle measure and orientation, and limits of 0,0 × 17,11. Save the file as a template named P2-5.dwt. Enter an appropriate description for the template. Close the file.

6. **Mechanical** Create a template with the following settings: metric units with 0.0 precision, decimal degrees with 0.0 precision, default angle measure and orientation, and limits of 0,0 × 22,17. Save the file as a template named P2-6.dwt. Enter an appropriate description for the template. Close the file.

7. **Architectural** Create a template with architectural units with 0'-0" precision, decimal degrees with 0 precision, and limits of 0,0 × 1632,1056. Save the file as a template named P2-7.dwt. Enter an appropriate description for the template. Close the file.

8. **Mechanical** Research the requirements for a set of working drawings of a mechanical assembly. Identify a mechanical assembly that consists of several parts and possibly subassemblies. Describe the product and provide a detailed list of each drawing required in the set of working drawings. Create, in writing only, a complete directory of folders for storing and organizing the drawing files. Use an appropriate system to name each folder and file. Prepare a slide show presentation of your research and present it to your class or office.

9. **Architectural** Research the requirements for a set of working drawings for a single-family home according to national and local codes. Provide a detailed list of each drawing required in the set of working drawings. Create, in writing only, a complete directory of folders for storing and organizing the drawing files. Use an appropriate system to name each folder and file. Refer to the US National CAD Standard if available. Prepare a slide show presentation of your research and present it to your class or office.

AutoCAD Certified Professional Exam Practice

Answer the following questions using the information in this chapter.

1. Which of the following methods can you use to close an AutoCAD drawing file? *Select all that apply.*
   A. Enter **CLOSEFILE** at the keyboard
   B. Pick the **Close** button on the drawing window title bar
   C. Pick **Close** in the **Application Menu**
   D. Right-click in the drawing area and pick **Close...**
   E. Select **Close** and then select **Current Drawing** in the **Application Menu**
2. Which of the following tasks can you accomplish using the **Application Menu**? *Select all that apply.*
   A. Access the **Options** dialog box
   B. Open a new drawing
   C. Specify the limits for a drawing
   D. Specify the units and precision for a drawing
   E. Toggle the display of file tabs
3. Open the Floor Plan Sample.dwg file in the Database Connectivity subfolder of the AutoCAD Sample folder.
   Access the **LIMITS** command and press [Enter] repeatedly to cycle through the prompts. Do not change any of the values. What are the coordinates for the upper-right corner of the drawing limits for this drawing?
4. Open the Mechanical - Multileaders.dwg file in the Mechanical Sample subfolder of the AutoCAD Sample folder.
   Access the **Drawing Units** dialog box. Do not change any of the values. What type of linear drawing units are specified for this drawing, and at what precision?

CHAPTER 3

Introduction to Drawing and Editing

Learning Objectives

After completing this chapter, you will be able to:

- ✓ Use appropriate values when responding to prompts.
- ✓ Apply basic viewing methods.
- ✓ Draw given objects using the **LINE** command.
- ✓ Describe and use several point entry methods.
- ✓ Describe and use basic drawing aids.
- ✓ Use the **ERASE**, **UNDO**, **U**, **REDO**, and **OOPS** commands.
- ✓ Create selection sets using various selection options.

This chapter introduces a variety of fundamental drawing and editing concepts and processes. You will learn to specify the location of points, draw lines, select objects, and erase objects. You will also use basic drawing aids and explore several other primary AutoCAD commands and operations.

Responding to Prompts

AutoCAD commands prompt, or ask, you to perform a specific task. For example, when you draw a line, prompts ask you to specify line endpoints. When you erase an object, a prompt asks you to select objects to erase. You must understand how to respond to prompts before you begin drawing. Many prompts provide options that you can select instead of responding to the immediate request. For example, after you pick the first point of a line, a prompt asks you to select the next point, or you can select the **Undo** option to remove the previous selection.

Responding with Numbers

Many commands require you to enter specific numerical data, such as the location of a point or the radius of a circle. Acceptable values vary depending on the command and prompt. AutoCAD understands that a number is positive without the plus sign (+) in front of the value. However, you must add a hyphen (-) in front of a negative number, such as -6.375. Do not include a space when responding to a prompt, because the space bar functions like [Enter] to enter the input.

The drawing units have some effect on values that you can enter when responding to prompts. All drawing length units accept decimal or fractional values. For example, you can type 2.5 or 2-1/2 to specify two and one half units. Remember, you are responsible for applying the appropriate units. Architectural units recognize two and one half units as 2 1/2", decimal units as 2.5 units, engineering units as 2.5", fractional units as 2 1/2 units, and scientific units as 2.5E+0 units, depending on the precision.

When a drawing is set up with architectural or engineering units, AutoCAD accepts only the inch (″) and foot (′) symbols and does not recognize other suffixes, such as mm. The numerator and denominator of fractions must be whole numbers greater than zero, such as 1/2 or 2/3. You must include a hyphen between a whole number and a fraction for values greater than one, such as 2-3/4, because of the function of the space bar. The numerator can be larger than the denominator, as in 3/2, but only if you do not include a whole number with the fraction. For example, 1-3/2 is not a valid input.

AutoCAD assumes inches when a drawing is set up with architectural and engineering length units. Do not add inch marks (″) when specifying an inch value; they are unnecessary and time-consuming. Values can be whole numbers, decimals, or fractions. For measurements in feet, place the foot symbol (′) after the number, as in 24'. Do not include a space or hyphen when specifying a value in feet and whole inches. For example, 24'6 is the proper input for the value 24′-6″. Separate fractional inches with a hyphen, such as 24'6-1/2. Never mix feet with inch values greater than one foot. For example, 24'18 is invalid; type 25'6 instead.

Angular units also have an effect on values that you can enter when responding to prompts. AutoCAD recognizes *d* for degrees, *g* for grads, and *r* for radians. However, it is not necessary to type a suffix when using decimal degree, grad, or radian units, or if the angle is a whole number when using deg/min/sec units. When a drawing is set up with deg/min/sec units, follow degrees with *d*, minutes with the foot symbol (′), and seconds with the inch symbol (″). For example, 32d24'5" is the proper input for the angle 32°24′5″.

When a drawing is set up with surveyor's units, enter a bearing to specify an angle. Start with the meridian (*n* or *s*); then the angle with degrees followed by *d*, minutes followed by the foot symbol (′), and seconds followed by the inch symbol (″); and finally the direction from the quadrant (*e* or *w*). For example, s25d30'10"e is the proper input for the bearing S25°30′10″E.

Ending and Canceling Commands

Some AutoCAD commands remain active until stopped. For example, you can continue to pick points to create new line segments until you end the **LINE** command. This textbook focuses on ending a command by pressing [Enter]. However, you can also press the space bar or right-click and select **Enter** to end a command. Choose the method that you prefer to end a command. Press [Esc] to cancel an active command or abort data entry. It may be necessary to press [Esc] twice to cancel certain commands completely.

If you press the wrong key or misspell a word when answering a prompt, use [Backspace] to correct the error. This works only if you notice your mistake *before* you accept the value. If you enter an invalid value or option, AutoCAD usually responds with an error message. Pick the **Command History** button on the command line to view lengthy error messages or to review entries.

TEXTSCR

Ribbon
View
> Palettes
Text Window
Type
TEXTSCR
[Ctrl]+[F2]
[F2]

NOTE

You can also access the **AutoCAD Text Window** to view lengthy error messages or to review entries. Return to the graphics screen using the same method you used to access the text window, or pick any visible portion of the graphics window.

PROFESSIONAL TIP

You can cancel the active command and access a new command at the same time by picking a ribbon button or an **Application Menu** option.

Introduction to Drawing

Most object drawing commands use ***point entry*** to locate and size geometry. Examples of point entry include specifying the two endpoints of a line, locating the center and a point on the edge of a circle, and specifying the points at opposite corners of a rectangle. The most basic point entry technique is to pick a random location in space using the left mouse button. Specific coordinate entry methods and drawing aids allow for accurate point entry, as explained in this chapter and later chapters.

point entry: Locating a point, such as the endpoint of a line, on the AutoCAD coordinate system.

AutoCAD uses the ***Cartesian (rectangular) coordinate system*** of X, Y, and Z coordinate values. These values, called ***rectangular coordinates***, locate any point in 3D space. In 2D drafting, the ***origin*** divides the coordinate system into four quadrants on the XY plane. See Figure 3-1. The origin is usually positioned at the lower-left corner of the drawing. This setup places all points in the upper-right quadrant of the XY plane, where both X and Y coordinate values are positive. See Figure 3-2. To locate a point in 3D space, a third dimension rises up from the surface of the XY plane along the Z axis.

Cartesian (rectangular) coordinate system: A system that locates points in space according to distances from three intersecting axes.

rectangular coordinates: A set of numerical values that identify the location of a point on the X, Y, and Z axes of the Cartesian coordinate system.

Describe a coordinate location using the X value first, followed by the Y value, and finally the Z value. A comma separates each value. For example, the coordinate location of 3,1,6 represents a point that is three units from the origin in the X direction, one unit from the origin in the Y direction, and six units from the origin in the Z direction. ***AutoCAD and Its Applications—Advanced*** describes how to use the Z axis in 3D modeling.

origin: The intersection point of the X, Y, and Z axes. The position of the default 2D origin is 0,0, where X = 0 and Y = 0.

Exercise 3-1

www.g-wlearning.com/CAD/

Complete the exercise on the companion website.

Basic Viewing Methods

View tools allow you to navigate the infinite AutoCAD drawing area, and observe and work more efficiently with a specific portion of a drawing. Chapter 6 explains the many view tools available to adjust the display of a 2D drawing. However, as you begin drawing, you should be able to apply basic ***zoom*** and ***pan*** operations.

zoom: Make objects appear bigger (zoom in) or smaller (zoom out) on the screen without affecting their actual size.

pan: Change the drawing display so that different portions of the drawing are visible on-screen.

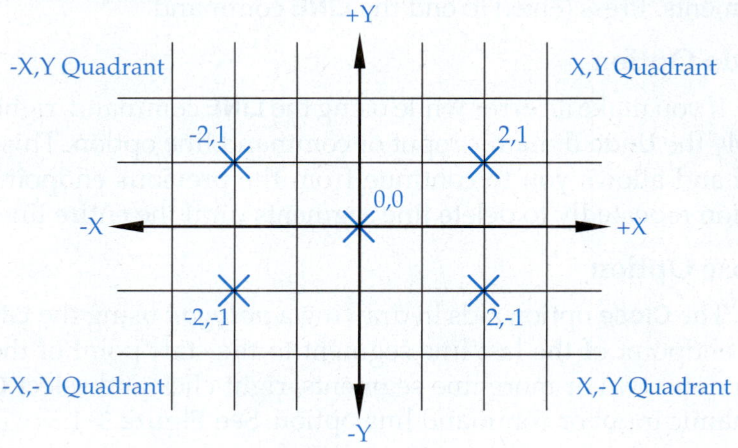

Figure 3-1. The 2D Cartesian coordinate system consists of X and Y axes. The origin is located at the intersection of the axes.

Figure 3-2.
By default, the upper-right quadrant of the Cartesian coordinate system fills the screen. The model space coordinate system icon identifies the coordinate system orientation.

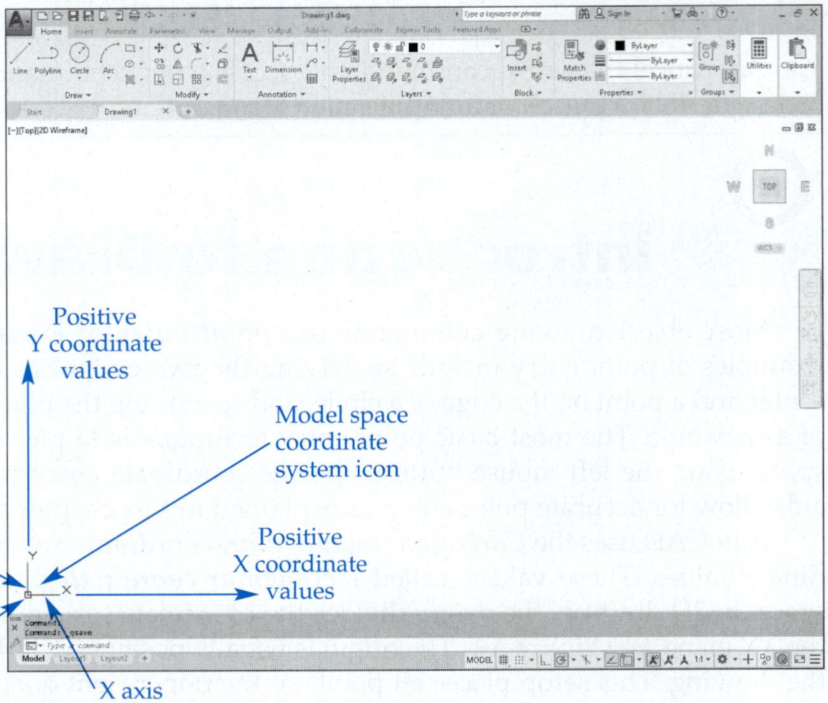

zoom in: Change the display area to show a smaller part of the drawing at a higher magnification.

zoom out: Change the display area to show a larger part of the drawing at a lower magnification.

To ***zoom in***, roll the mouse wheel forward. To ***zoom out***, roll the mouse wheel back. Double-click the wheel to zoom to the furthest extents of objects in the drawing. To pan, press and hold the mouse wheel while you move the mouse. For now, these basic zoom and pan functions will allow you to view drawings effectively.

NOTE

Refer to the X and Y axis lines in the drawing window to help identify the location you are viewing and the location of objects in space. If the origin is visible in the drawing window, the model space coordinate system icon appears at the origin, collinear with the axes. If the current view does not show the origin, the coordinate system icon floats in space near the lower-left corner of the drawing window.

Drawing Lines

LINE
Ribbon
Home
> Draw
Line
Type
LINE
L

To draw a line, access the **LINE** command and specify a start point, which is the first line endpoint. As you move the crosshairs, a ***rubberband line*** appears connecting the first point and the crosshairs. Continue locating points to connect a series of line segments. Press [Enter] to end the **LINE** command.

rubberband line: A reference line that extends from the crosshairs in certain drawing commands after you make the first selection.

Undo Option

If you make an error while using the **LINE** command, right-click and select **Undo**, or apply the **Undo** dynamic input or command line option. This removes the most recent line and allows you to continue from the previous endpoint. You can use the **Undo** option repeatedly to delete line segments until the entire line is gone. See Figure 3-3.

Close Option

polygon: A closed plane figure with at least three sides, such as a triangle or rectangle.

The **Close** option aids in drawing a ***polygon*** using the **LINE** command. To connect the endpoint of the last line segment to the start point of the first line segment after you draw two or more line segments, right-click and select **Close**, or apply the **Close** dynamic input or command line option. See Figure 3-4.

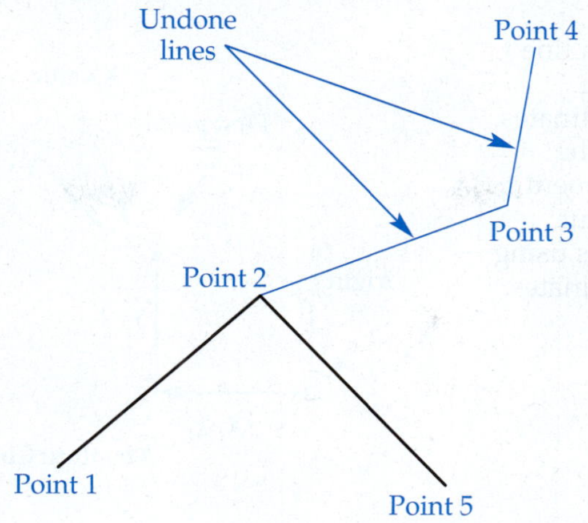

Figure 3-3. Using the **Undo** option of the active **LINE** command. The lines in color represent the undone lines and do not appear when you use the **Undo** option.

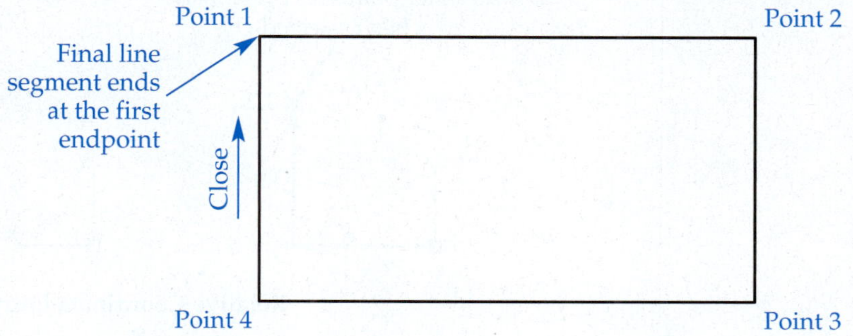

Figure 3-4. Constructing the final segment of a rectangle using the **Close** option of the **LINE** command.

Exercise 3-2

www.g-wlearning.com/CAD/

Complete the exercise on the companion website.

Coordinate Entry Methods

Creating an object by picking random points in space with the left mouse button is generally not accurate. Specific coordinate input is one way to locate points precisely to create accurate geometry. Another option is to use drawing aids, as explained later in this chapter.

Absolute Coordinates

absolute coordinates: Coordinate distances measured from the origin.

Points located using ***absolute coordinates*** are measured from the origin (0,0). For example, a point located two units horizontally (X = 2) and two units vertically (Y = 2) from the origin is at the absolute coordinate 2,2. Figure 3-5A shows using absolute coordinates to draw a line starting at 2,2 and ending at 4,4. The first point of a line is often positioned using absolute coordinates. Remember, when using the absolute coordinate system, you locate each point from 0,0. If you enter negative X and Y values, the selection occurs outside of the upper-right XY plane quadrant.

Relative Coordinates

relative coordinates: Coordinates specified from, or relative to, the previous coordinate, rather than from the origin.

When using ***relative coordinates***, think of the previous point as the "temporary origin." The @ symbol specifies coordinates as relative. For example, input @2,2 to locate the second point shown in Figure 3-5B at the absolute coordinate 4,4.

Figure 3-5.
A—Drawing a line segment using absolute coordinates. B—Drawing the same line segment shown in A, but locating points using relative coordinates.

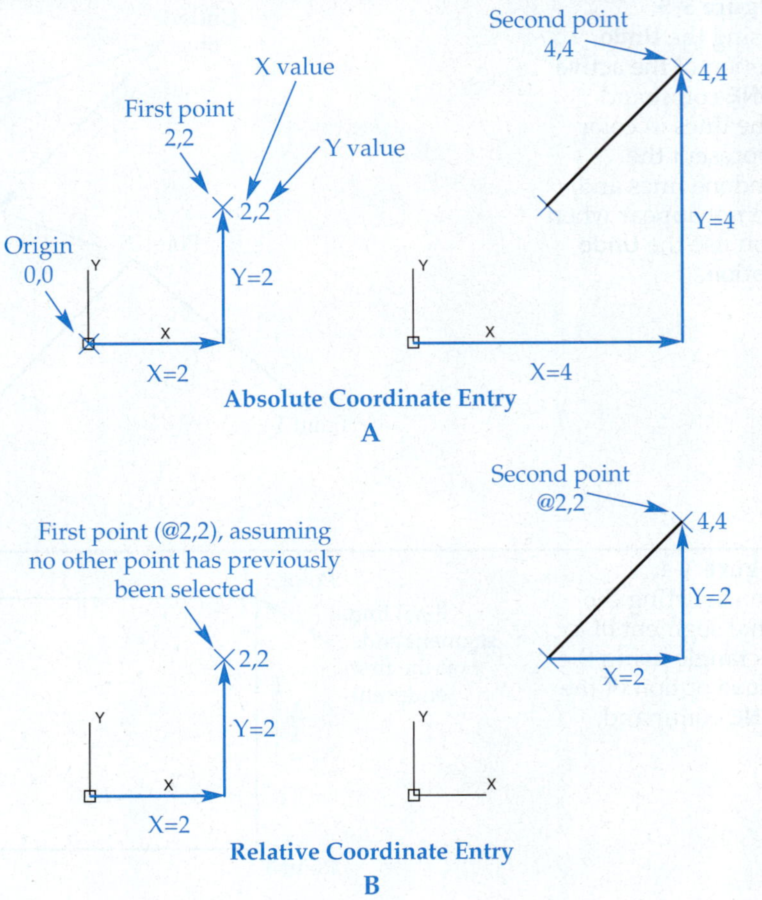

Polar Coordinates

polar coordinates: Coordinates based on the distance from a fixed point at a given angle.

To apply ***polar coordinates***, specify the length of the line, followed by the less than (<) symbol, and then the angle at which the line is drawn. **Figure 3-6** shows the default angular values used for polar coordinate entry. Precede a polar coordinate with the @ symbol to locate the point relative to the previous point. Otherwise, the coordinate is located relative to the origin. For example, to draw a line 2 units long at a 45° angle, starting 2 units from 0,0 at a 45° angle, type 2<45 for the first point and @2<45 for the second point. See **Figure 3-7**.

Figure 3-6.
The default angles used when entering polar coordinates. 0° is to the right, or east, and angles are measured counterclockwise. Use the **Drawing Units** dialog box to adjust the direction and rotation for specific applications, such as when measuring azimuths.

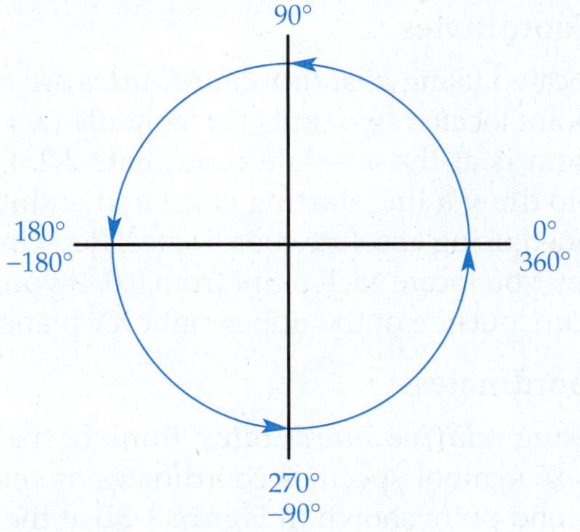

Figure 3-7.
Locating points using polar coordinates.

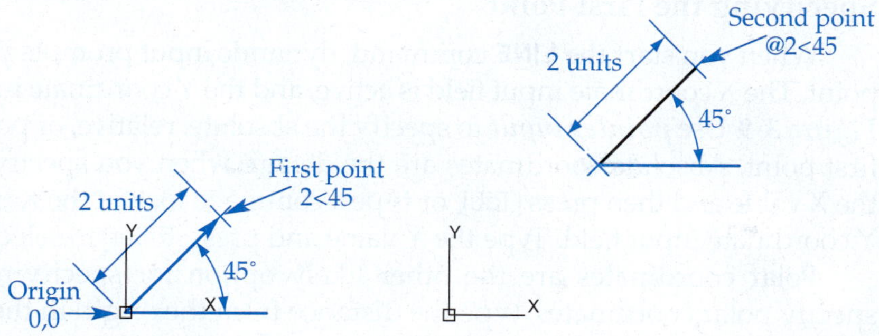

PROFESSIONAL TIP

Polar coordinates with angles measured as bearings are used for civil engineering applications such as laying out property lines or the centerline of a road. For example, to draw a property line 46.20′ long at a S25°30′10″E bearing (25°30′10″ from south toward east), from a previous point, type @46.20'<s25d30'10"e. This method requires architectural or engineering linear units and surveyor's angular units, as described in Chapter 2.

Coordinate Display

The status bar offers a coordinate display tool that can help you identify the location of the crosshairs and selections in space. See Figure 3-8. The coordinate display does not appear on the status bar by default. Select the **Coordinates** option in the status bar **Customization** flyout to add the coordinate display to the status bar. The drawing units determine the format and precision shown. By default, the coordinates constantly change as the crosshairs moves. Pick the coordinates to switch this function on or off. When off, the coordinates are "grayed out" but update to identify the location of the last point you select.

You can switch the coordinate display mode when a command such as **LINE** is active and you have selected a point. Right-click on the coordinate display and select **Relative** to view the coordinates of the crosshairs as polar coordinates relative to the previously picked point. The coordinates update each time you pick a new point. Select **Absolute** to view the coordinates of the crosshairs relative to the origin. Select **Specific** to view the coordinates of the last point you select. The **Geographic** option is available if you specify the geographic drawing location, as explained in *AutoCAD and Its Applications—Advanced*.

Dynamic Input

Dynamic input is on by default and is a very effective way to enter coordinates. A quick way to toggle dynamic input on and off is to press [F12]. You can also pick the **Dynamic Input** button on the status bar. The **Dynamic Input** button does not appear by default. Select the **Dynamic Input** option in the status bar **Customization** flyout to add the **Dynamic Input** button to the status bar. Dynamic input provides the same coordinate entry functions as the command line, but allows you to keep your focus at the crosshairs while you draw. Dynamic input also offers additional coordinate entry techniques. When dynamic input is active, point entry functions differently, according to dynamic input settings.

Figure 3-8.
Turn on the status bar **Coordinates** option to view the coordinate display.

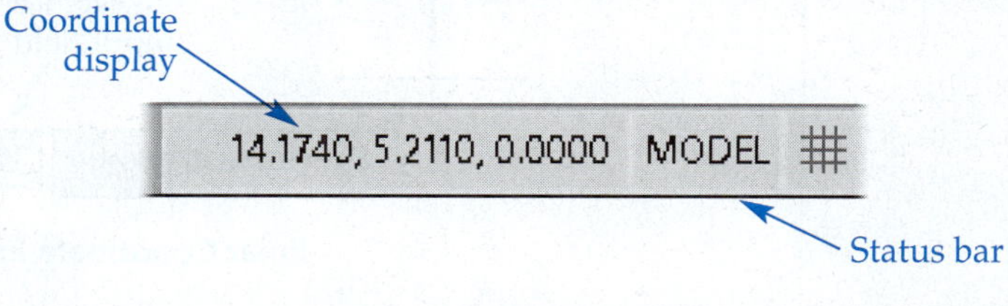

Specifying the First Point

pointer input: The process of entering points using dynamic input.

When you start the **LINE** command, dynamic input prompts you to specify the first point. The X coordinate input field is active, and the Y coordinate input field appears. See **Figure 3-9**. Use ***pointer input*** to specify the absolute, relative, or polar coordinates of the first point. Absolute coordinates are the default when you specify the first point. Type the X value and then press [Tab], or type a comma to lock in the X value, and move to the Y coordinate input field. Type the Y value and press [Enter] to select the point.

Polar coordinates are the other likely option for specifying the first point. To specify polar coordinates, type the distance from the origin in the distance input field, followed by the less than symbol (<) to move to the angle input field. Type the angle from the origin and press [Enter] to select the point. Dynamic input fields automatically change to anticipate the next entry.

Specifying the Next Point

dimensional input: An instinctive dynamic input point entry technique, similar to polar coordinate entry.

Dynamic input also provides a ***dimensional input*** feature that allows you to enter the length of a line and the angle at which the line is drawn. To use dimensional input, access the **LINE** command and specify a start point. Distance and angle input fields appear by default. See **Figure 3-10A**. Move the crosshairs toward where you want to locate the endpoint, type the length of the line in the distance input field, and press [Tab]. This locks in the distance and moves the cursor to the angle input field. Type the angle of the line and press [Enter] to select the point. Dimensional input does not follow the same rules as polar coordinates, and drawing units do not control direction and rotation. All angles relate to the location of the crosshairs from 0° east. **Figure 3-10B** shows additional examples of specifying dimensional input angles.

You can also use pointer input to pick additional points once you select the start point of the line. Relative coordinates are active by default, which means you do not need to type @ before entering the X and Y values. To select the second point using a relative coordinate entry, type the X value in the distance input field, then type a comma to lock in the X value and move to the Y coordinate input field. Type the Y value and press [Enter] to select the point.

Dimensional input is on by default, but you can turn it off temporarily by typing the pound symbol (#) before entering values. When dimensional input is off, dynamic input defaults to polar format. Type the length of the line in the distance input field,

Figure 3-9.
Dynamic input displays these fields after you start the **LINE** command. When you type @ to use relative coordinates, the symbol appears in a field to the right of the prompt. When you use polar coordinates, the less than symbol (<) appears before the angle input field.

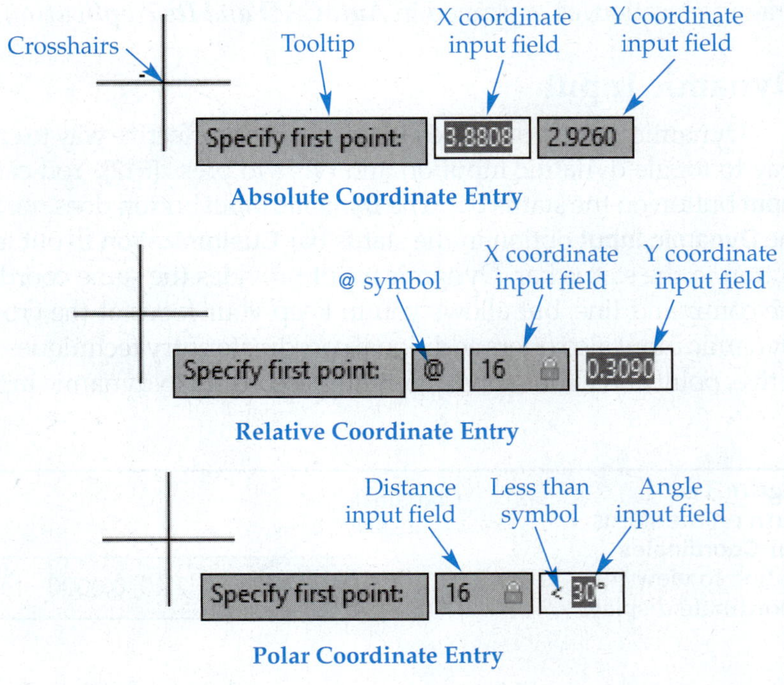

Figure 3-10.
A—Steps for using dimensional input to define the length and angle of a line. B—Pay close attention to the location of the crosshairs and the reading in the angle input field. All angles relate to the location of the crosshairs from 0° east.

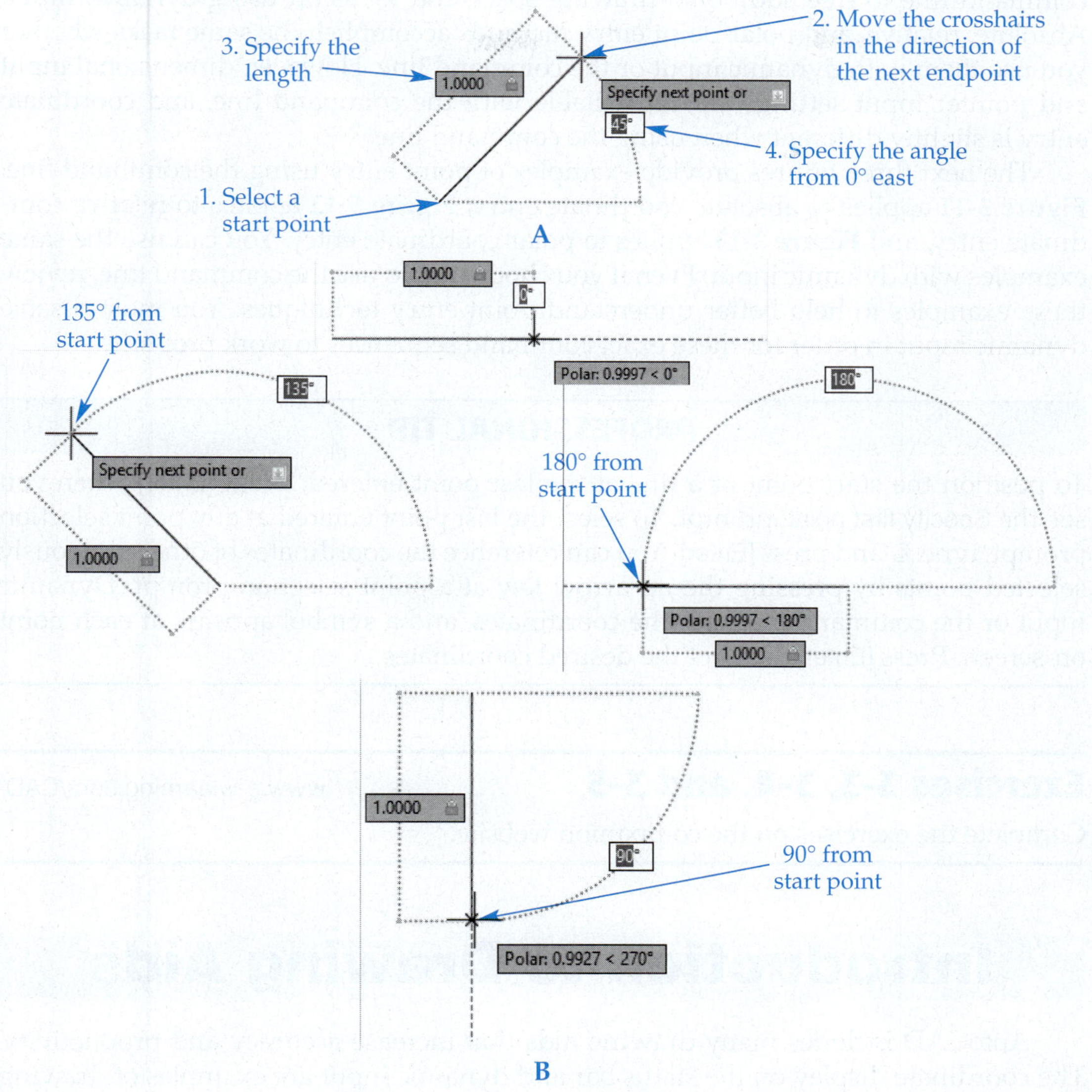

and press [Tab] to lock in the length and move to the angle input field. Enter the angle of the line and press [Enter] to select the point. In order to use an absolute coordinate entry with the default settings, type #, enter the X coordinate in the active field, type a comma, type the Y value, and press [Enter] to select the point.

PROFESSIONAL TIP

Use [Tab] to cycle through dynamic input fields. You can make changes to values before accepting the coordinates.

NOTE

Use the **Dynamic Input** tab of the **Drafting Settings** dialog box to configure dynamic input to meet your personal, school, or company standards. A quick way to access the **Dynamic Input** tab is to right-click on the **Dynamic Input** button on the status bar and select **Dynamic Input Settings...**.

Type
DSETTINGS
DS
SE

DSETTINGS

Command Line

You can use the command line at the same time as dynamic input, or you can disable dynamic input to use only the command line. Another option is to hide the command line to free additional drawing space and focus on using dynamic input. Absolute, relative, and polar point entry methods accomplish the same tasks whether you use them with dynamic input or the command line. However, dimensional input and pointer input settings are unavailable with the command line, and coordinate entry is slightly different when using the command line.

The next three figures provide examples of point entry using the command line. Figure 3-11 applies to absolute coordinate entry, Figure 3-12 applies to relative coordinate entry, and Figure 3-13 applies to polar coordinate entry. You can use the same examples with dynamic input. Even if you choose not to use the command line, review these examples to help better understand point entry techniques. You must disable dynamic input in order for these exact command sequences to work properly.

PROFESSIONAL TIP

To position the start point of a line at the last point entered, press [Enter] when you see the Specify first point: prompt. To select the last point entered at any point selection prompt, type @ and press [Enter]. You can reference the coordinates of other previously selected points by pressing the up arrow key at a point selection prompt. Dynamic input or the command line lists the coordinates, and a symbol appears at each point on-screen. Press [Enter] to select the desired coordinates.

Exercises 3-3, 3-4, and 3-5

www.g-wlearning.com/CAD/

Complete the exercises on the companion website.

Introduction to Drawing Aids

AutoCAD includes many drawing aids that increase accuracy and productivity. The coordinate display on the status bar and dynamic input are examples of drawing aids. This chapter describes other basic drawing aids and provides an overview of several important drawing aids that are explained in detail later in this textbook.

Figure 3-11.
Drawing a shape using the **LINE** command and absolute coordinates.

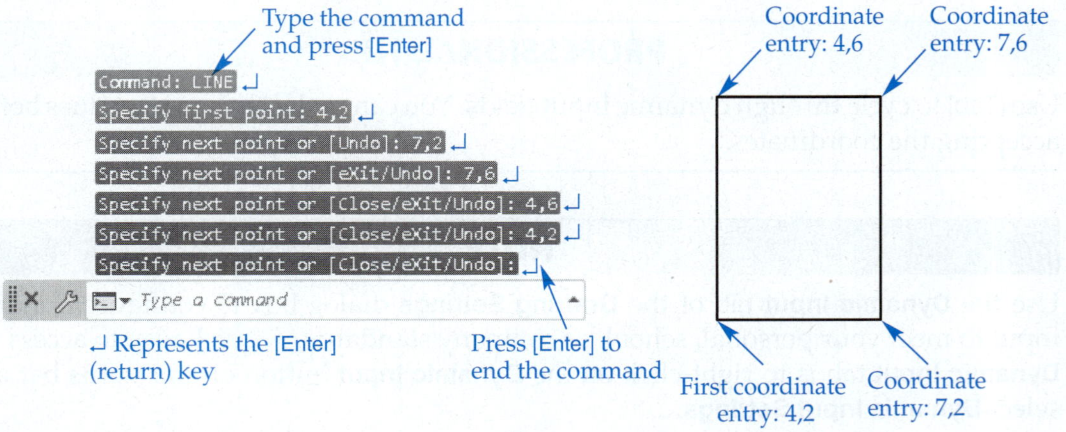

Figure 3-12.
Drawing a shape using the **LINE** command and a combination of absolute and relative coordinates. Notice that negative (–) values are used and the coordinates are entered counterclockwise from the first point, in this case 2,2.

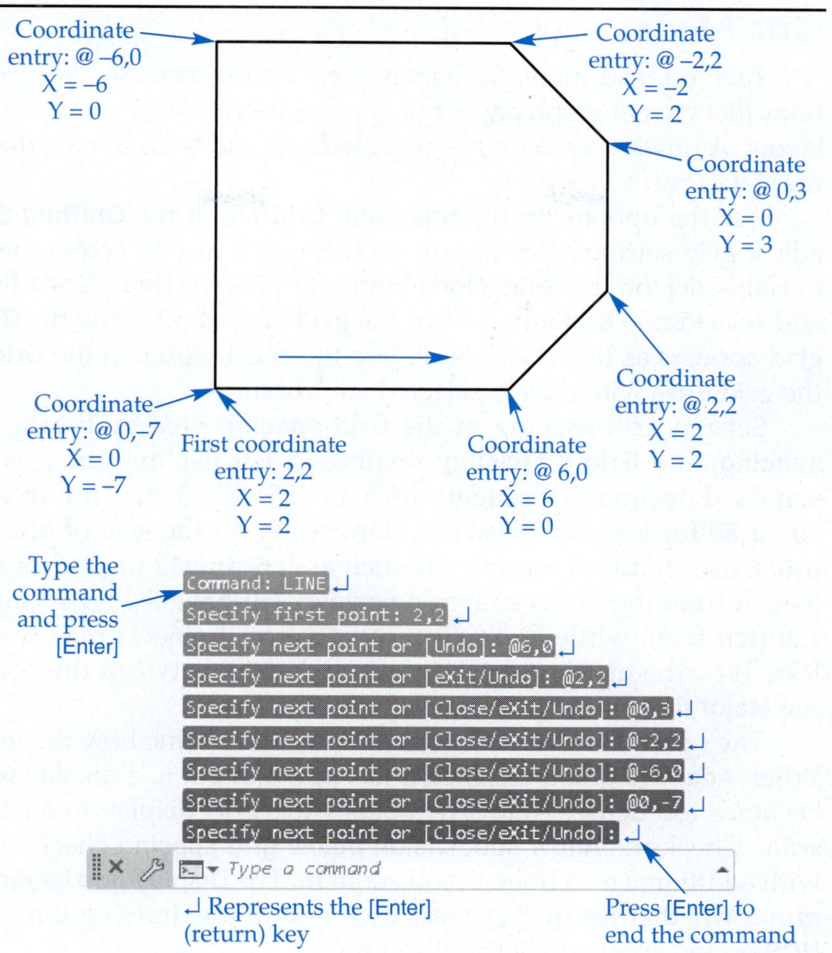

Figure 3-13.
Drawing shapes using the **LINE** command and a combination of absolute and polar coordinates.

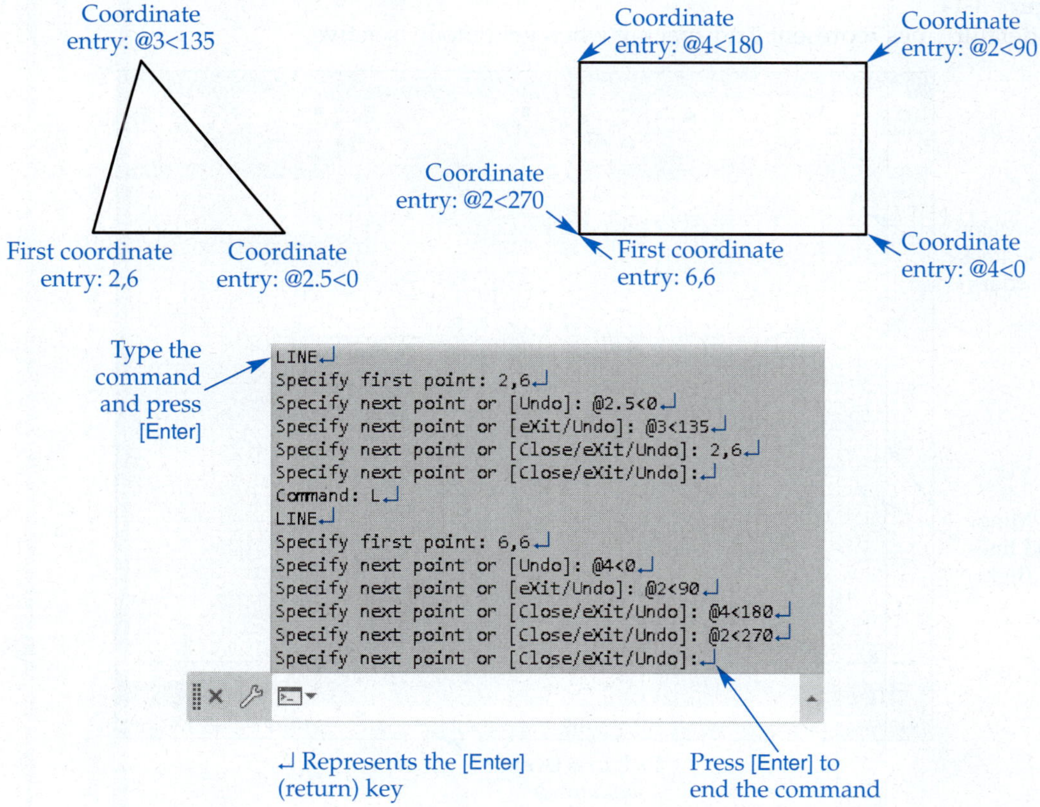

grid: A pattern of lines that appears on-screen for reference, analogous to graph paper.

GRID

Type
GRID
[Ctrl]+[G]
F7

DSETTINGS

Type
DSETTINGS
DS
SE

Grid Mode

Turn on grid mode to display a ***grid*** on-screen. See Figure 3-14. The grid functions like virtual graph paper but appears for reference only, as a visual aid to drawing layout. A quick way to toggle grid mode on and off is to pick the **Grid Mode** button on the status bar.

Use the options on the **Snap and Grid** tab of the **Drafting Settings** dialog box to adjust grid settings. See Figure 3-15. A quick way to access the **Snap and Grid** tab is to right-click on the **Snap Mode** button or pick the **Snap Mode** flyout on the status bar and select **Snap Settings...**. Turn the grid on and off using the **Grid On** check box. The grid appears as lines by default. Use the check boxes in the **Grid style** area to replace the grid lines with dots in selected environments.

Set the grid spacing in the **Grid spacing** area by typing values in the **Grid X spacing:** and **Grid Y spacing:** text boxes. For decimal units, set the grid spacing to standard decimal increments such as .125, .25, .5, or 1 for an inch drawing or 1, 10, 20, or 50 for a metric drawing, depending on the size of objects. For architectural units, use standard increments such as 1, 6, and 12 for inches or 1, 2, 4, 5, and 10 for feet. A drawing of very large objects might have a grid spacing of 12 (one foot) or 120 (ten feet), while a drawing of very small objects may use a spacing of .125 or less. Type the number of grid rows to display between the bold, major grid lines in the **Major line every:** text box.

The options in the **Grid behavior** area determine how the grid appears on-screen. When **Adaptive grid** is checked and you zoom out to a display where the grid spacing becomes too dense, AutoCAD adjusts the grid display to show the grid at a larger scale. Check the **Allow subdivision below grid spacing** check box to divide each grid with additional grid lines as you zoom in. The **Display grid beyond Limits** option determines whether the grid appears only within the drawing limits. The **Follow Dynamic UCS** option applies to 3D applications.

Figure 3-14.
By default, lines represent grid spacing when grid mode is active.

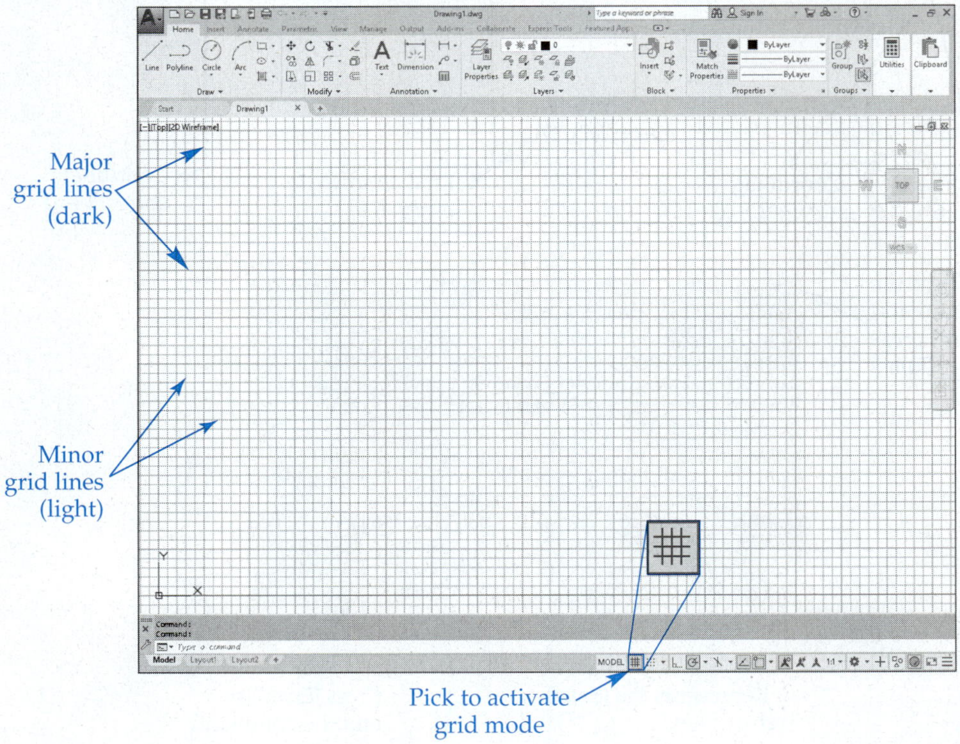

Figure 3-15. Use the **Snap and Grid** tab of the **Drafting Settings** dialog box to specify grid and snap grid settings.

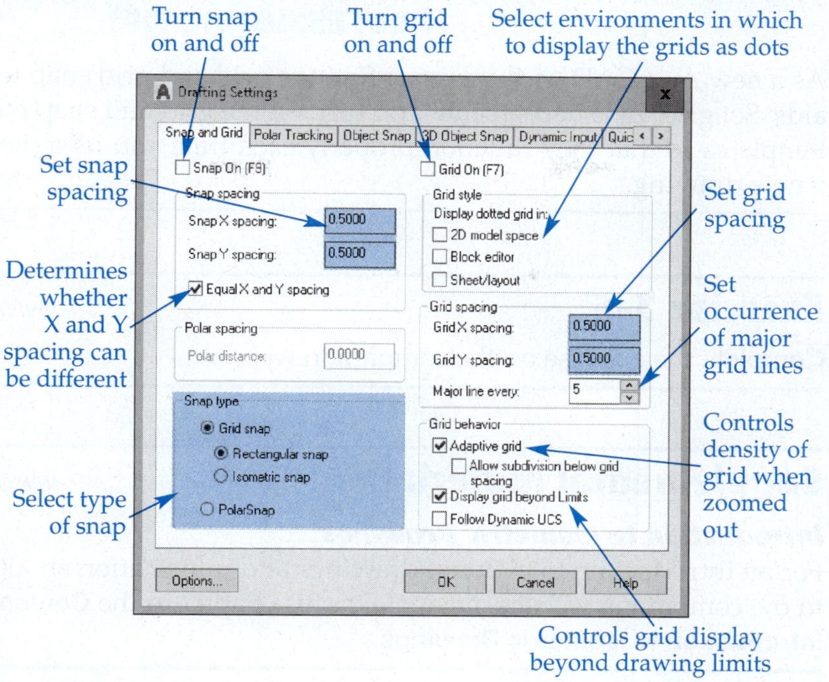

Snap Mode

Turn on snap mode to activate the ***snap grid***, also known as ***snap resolution*** or ***snap***. A quick way to toggle snap on and off is to pick the **Snap Mode** button on the status bar. By default, snap is only noticeable when you are drawing or editing objects, such as constructing a line or moving a circle. Snap is off by default, allowing the crosshairs to move freely on-screen. Turn snap on to move the crosshairs in specific increments. Snap is different from the grid, because snap controls the movement of the crosshairs, while the grid is only a visual guide. However, grid and snap settings typically complement each other.

SNAP

Type
SNAP
[Ctrl]+[B]
[F9]

snap grid (snap resolution, snap): An invisible grid that allows the crosshairs to move in, or snap to, specified increments during the drawing or editing process.

The **Snap and Grid** tab of the **Drafting Settings** dialog box includes options for setting snaps. Refer to **Figure 3-15**. Turn snap on or off using the **Snap On** check box. Set the snap increment in the **Snap spacing** area by typing values in the **Snap X spacing:** and **Snap Y spacing:** text boxes. For example, if you set the X and Y grid spacing to .5, an appropriate X and Y snap spacing is .125 or .25. With these settings, each mode plays a separate role in assisting drafting. Often the most effective use of grid and snap is to set equal X and Y spacing. However, if many horizontal features conform to one increment and most vertical features correspond to another, you may choose to set different X and Y values.

Use the **Snap type** area to control how snaps function. The default, previously described snap type is **Grid snap** with the **Rectangular snap** style. Select the **Grid snap** type and **Isometric snap** style to aid in creating isometric drawings. The **PolarSnap** type allows you to snap to precise distances along alignment paths when you use polar tracking, as explained in Chapter 7. You can also change the snap type by selecting **Polar Snap** or **Grid Snap** from the **Snap Mode** flyout on the status bar.

NOTE

Adjust the grid and snap settings as needed, such as when larger or smaller values would assist you with a certain drawing task. Changing grid and snap settings does not affect the location of existing points or objects. Change the **Legacy** option of the **SNAP** command to **Yes** to snap the crosshairs, even if a drawing or editing command is inactive.

PROFESSIONAL TIP

As a new AutoCAD user, you may find the grid and grid snap to be effective drawing aids. Set grid and snap spacing and turn on the grid and snap modes in your drawing templates so that they function properly each time you reference a template to create a new drawing.

Exercise 3-6

www.g-wlearning.com/CAD/

Complete the exercise on the companion website.

Supplemental Material

www.g-wlearning.com/CAD/

Introduction to Isometric Drawings

For an introduction to pictorial drawings and information about isometric snap, go to the companion website, navigate to this chapter in the **Contents** tab, and select **Introduction to Isometric Drawings**.

Polar Tracking

polar tracking: A drawing aid that causes the drawing crosshairs to "snap" to predefined angle increments.

direct distance entry: Entering points by positioning the crosshairs to establish direction and typing a number to specify distance.

Polar tracking causes the drawing crosshairs to "snap" to predefined angle increments. Chapter 7 fully explains polar tracking, but because polar tracking is on by default, you should have a basic understanding of the tool. Turn polar tracking on or off by picking the **Polar Tracking** button on the status bar or by pressing [F10]. You can use polar tracking to draw lines at accurate lengths and angles using ***direct distance entry***. As you move the crosshairs toward a polar tracking angle, AutoCAD displays an alignment path and tooltip. The default polar angle increments are 0°, 90°, 180°, and 270°.

To apply direct distance entry using polar tracking, access the **LINE** command and specify a start point. Then move the crosshairs in alignment with a polar tracking angle. Type the length of the line and press [Enter]. See Figure 3-16.

Figure 3-16.
Using polar tracking and direct distance entry to draw connected and perpendicular lines at specific lengths. The default tracking angles are 0° and 90°. This example shows dynamic input active, but the same technique applies when you use the command line.

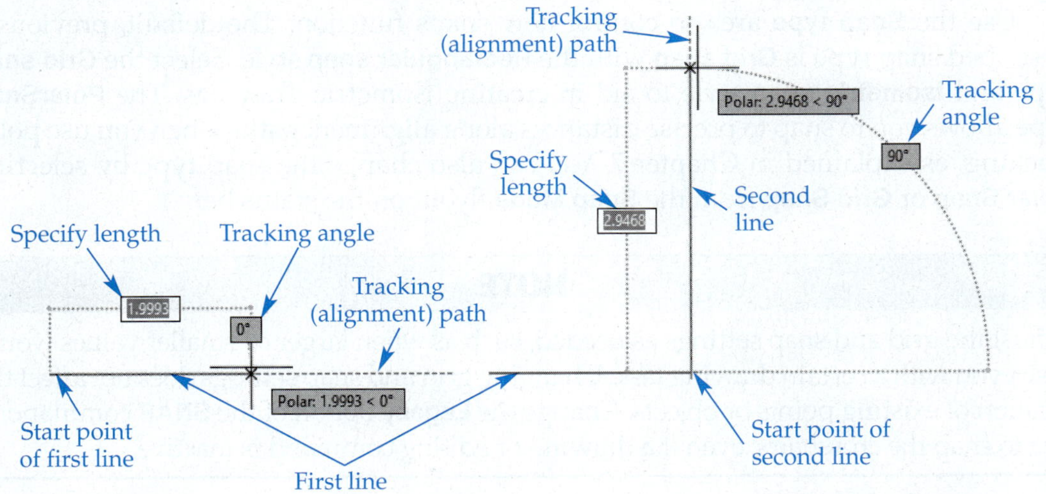

Exercise 3-7

www.g-wlearning.com/CAD/

Complete the exercise on the companion website.

Ortho Mode

ortho: From *orthogonal*, which means "at right angles."

Type
ORTHO
[Ctrl]+[L]
[F8]

ORTHO

Ortho forces a horizontal or vertical line. Ortho mode is off by default. Pick the **Ortho Mode** button on the status bar to toggle ortho mode on and off. If ortho mode is off, you can temporarily turn it on while drawing by holding down [Shift]. You can use ortho mode to draw accurate lengths of horizontal and vertical lines using direct distance entry.

To apply direct distance entry using ortho mode, access the **LINE** command and specify a start point. Move the crosshairs to display a horizontal or vertical rubberband line in the direction you want to draw. Then type the length of the line and press [Enter]. See Figure 3-17.

NOTE

You can use direct distance entry to specify the length of a line at any angle. However, direct distance entry itself is not very useful unless you incorporate drawing aids such as polar tracking or ortho mode.

Figure 3-17.
Using ortho mode and direct distance entry to construct a rectangle. Notice that the crosshairs specifies the general direction of the line endpoint and does not attach to the rubberband line. This example shows dynamic input active. The same technique applies when you use the command line.

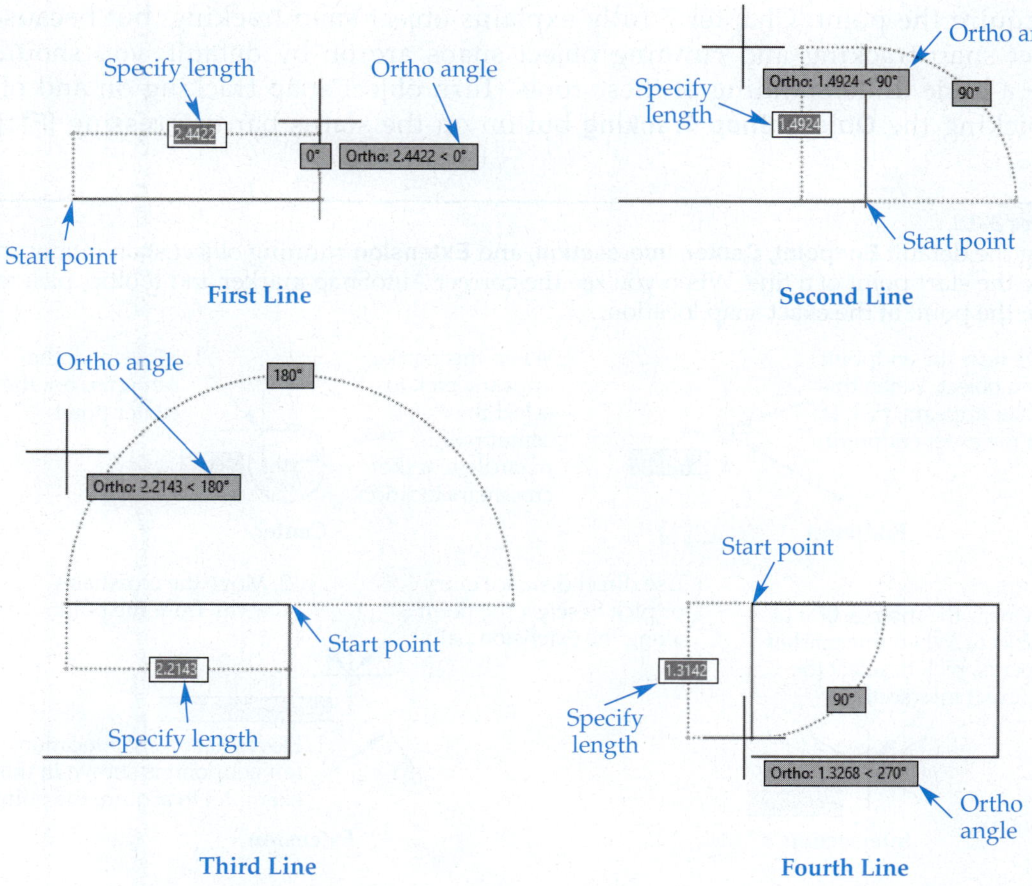

Exercise 3-8

www.g-wlearning.com/CAD/

Complete the exercise on the companion website.

Object Snap

object snap: A tool that snaps to exact points on or in relation to existing objects, such as endpoints or midpoints.

snapping: Picking a point near the intended position to have the crosshairs "snap" exactly to the specific point.

running object snaps: Object snap modes that run in the background during all drawing and editing procedures.

markers: Visual cues to confirm object snap points.

Object snap increases drafting performance and accuracy through the concept of ***snapping***. Chapter 7 fully explains object snap, but because ***running object snaps*** are on by default, you should have a basic understanding of how to use them. Turn running object snaps on or off by picking the **Object Snap** button on the status bar or pressing [F3]. Object snap modes identify the points on objects to which the crosshairs snap. The **Endpoint**, **Center**, **Intersection**, and **Extension** running object snap modes are active by default. The AutoSnap feature is also on by default and displays ***markers*** at each active snap point. After a brief pause, a tooltip appears to indicate the object snap mode. See Figure 3-18.

You can snap to a point at any point selection prompt, for example, when picking the start point or endpoint of a line segment. Figure 3-18 shows a basic example of locating the start point of a line using each default running object snap mode. Follow the instructions shown to snap to the corresponding point.

Exercise 3-9

www.g-wlearning.com/CAD/

Complete the exercise on the companion website.

Object Snap Tracking

object snap tracking: A drawing aid that provides horizontal and vertical alignment paths for locating points after a point is acquired with object snap.

Object snap tracking has two requirements: running object snaps must be active, and the crosshairs must hover over the intended selection long enough to acquire the point. Chapter 7 fully explains object snap tracking, but because object snap tracking and running object snaps are on by default, you should have a basic understanding of these tools. Turn object snap tracking on and off by picking the **Object Snap Tracking** button on the status bar or pressing [F11].

Figure 3-18.
Using the default **Endpoint**, **Center**, **Intersection**, and **Extension** running object snap modes to locate the start point of a line. When you see the correct AutoSnap marker and tooltip, pick to locate the point at the exact snap location.

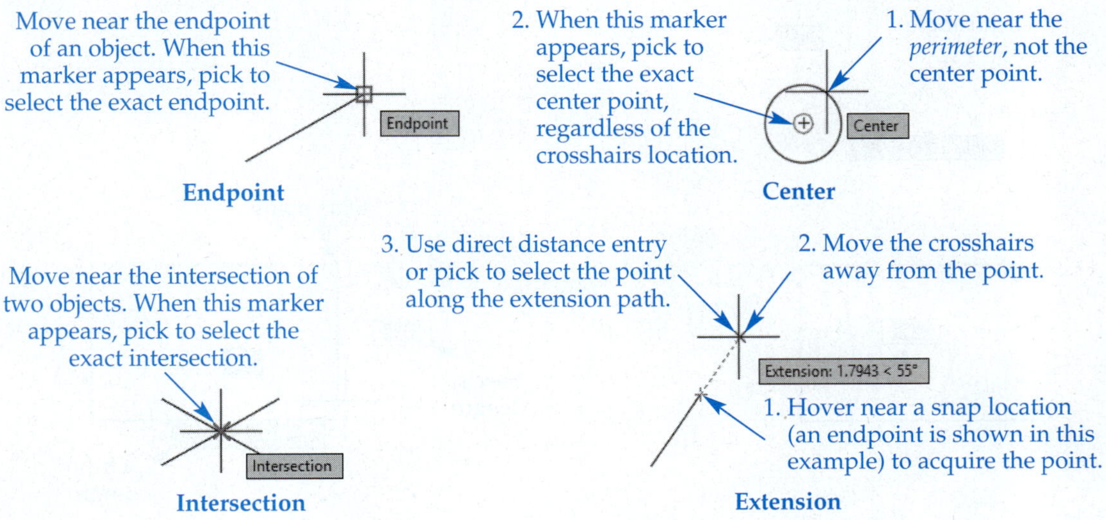

Figure 3-19 shows an example of using the **Endpoint** running object snap with object snap tracking to locate the endpoint of a line exactly vertical to another endpoint.

PROFESSIONAL TIP

Practice using different point entry techniques and drawing aids, and decide which method works best for specific situations.

Exercise 3-10

www.g-wlearning.com/CAD/

Complete the exercise on the companion website.

Introduction to Editing

AutoCAD includes many ***editing*** commands for making changes to a drawing and increasing productivity. One of the most basic ways to edit a drawing is to remove objects using the **ERASE** command. In this chapter, you will also learn to use the **OOPS**, **UNDO**, **U**, and **REDO** commands, which are often needed in the drawing and editing processes.

editing: A procedure used to modify an existing object.

A common approach to editing is to access a command, such as **ERASE**, select the objects to modify, and then close the command by pressing [Enter]. Another approach is to select objects first using the crosshairs, access the editing command, modify objects, and finally close the command. The process of selecting objects is generally the same for both methods. Choose the technique you prefer, but selecting objects first is most appropriate when using ***grips*** to edit objects. Chapter 13 explains grip editing.

grips: Small boxes that appear at strategic points on a selected object, allowing you to edit the object directly.

ERASE Command

Access the **ERASE** command to remove objects from the drawing. The Select objects: prompt appears and an object selection target, or ***pick box***, replaces the screen crosshairs. Move the pick box over the object to erase. When the object becomes faded and you see the erase icon next to the pick box, pick the object. The object remains faded and the Select objects: prompt remains active, allowing you to select additional objects to erase.

ERASE

Ribbon

Home
> Modify

Erase

Type

ERASE
E

pick box: A small box that replaces the crosshairs when the Select objects: prompt is active.

Figure 3-19.
Using the **Endpoint** running object snap and object snap tracking to construct the top side of a rectangle using the **LINE** command. Polar tracking is active to create a horizontal line.

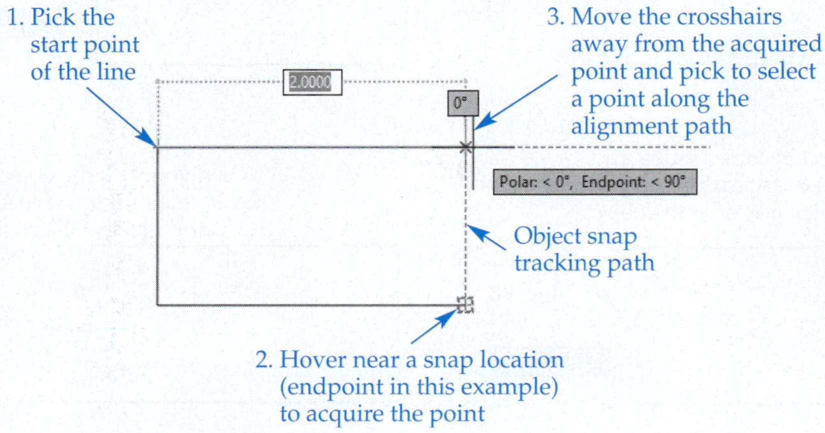

When you finish selecting objects, press [Enter] to erase the selection set. See **Figure 3-20**. If you choose to select objects before accessing the **ERASE** command, you can erase the selected objects by pressing [Delete].

NOTE

By default, when you hover over an object, the object becomes highlighted. Basic object properties also appear. When you move the crosshairs or pick box off the object, the object display returns to normal. This allows you to preview the object before you select. When a small area contains many objects, this feature helps you select the correct object and often eliminates the need to cycle through stacked objects.

Exercise 3-11

www.g-wlearning.com/CAD/

Complete the exercise on the companion website.

U Command

U
Type
U

The **U** command undoes the effect of the last command you entered. You can reissue the **U** command to continue undoing command actions, but you can only undo one command at a time. Commands are undone in the order you used them, starting with the most recent.

NOTE

You can also activate the **U** command by right-clicking in the drawing window and selecting **Undo** *current*.

Figure 3-20.
Using the **ERASE** command to erase a single object. A—The initial display before you access the **ERASE** command. B—The pick box that appears when you access the **ERASE** command. C—Move over an object to select it. D—The completed erase operation.

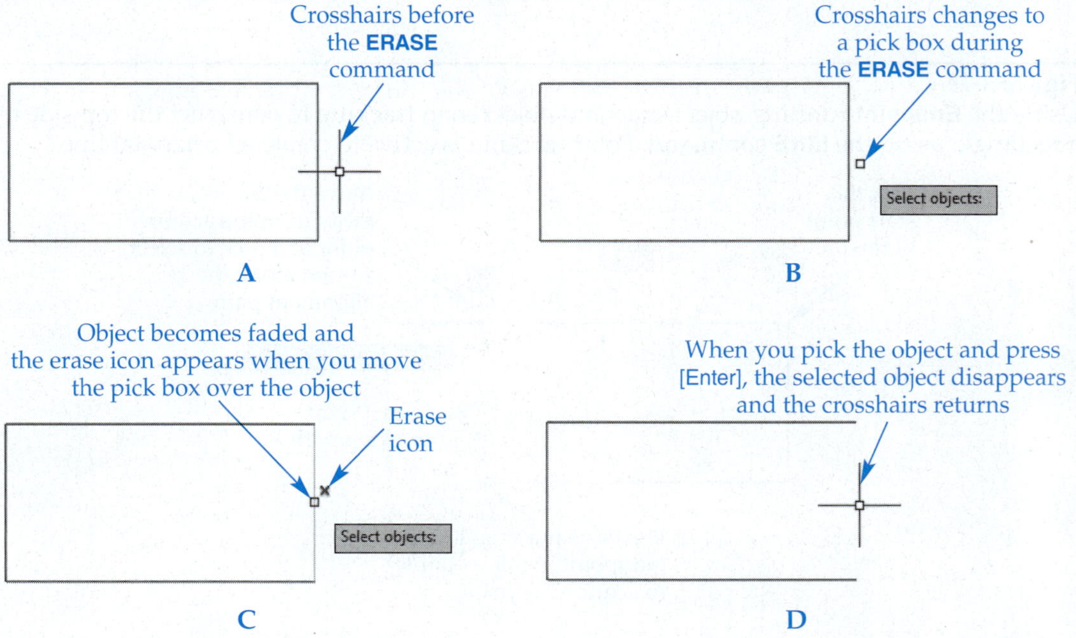

UNDO Command

The **UNDO** command allows you to undo a single operation or a number of operations at once. The **UNDO** command is different from the **Undo** option of certain commands, such as the **LINE** command. A quick way to use the **UNDO** command is to pick the **Undo** button on the **Quick Access** toolbar. Select the button as many times as needed to undo multiple operations. An alternative is to pick the flyout and select all of the commands to undo from the list.

> Quick Access
> **Undo**
> Type
> **UNDO**
>
> **UNDO**

Supplemental Material

www.g-wlearning.com/CAD/

UNDO Options

For information about the options available when you access the **UNDO** command from a source other than the **Quick Access** toolbar, go to the companion website, navigate to this chapter in the **Contents** tab, and select **UNDO Options**.

REDO Command

Use the **REDO** command to reverse the action of the **UNDO** and **U** commands. The **REDO** command works only *immediately* after you have undone something. A quick way to use the **REDO** command is to pick the **Redo** button on the **Quick Access** toolbar. Select the button as many times as needed to redo multiple undone operations. An alternative is to pick the flyout and select one or more undone operations from the list to redo. The **REDO** command does not bring back line segments removed using the **Undo** option of the **LINE** command.

> Quick Access
> **Redo**
> Type
> **REDO**
>
> **REDO**

Exercise 3-12

www.g-wlearning.com/CAD/

Complete the exercise on the companion website.

OOPS Command

The **OOPS** command brings back the last object you *erased*. Unlike the **UNDO** and **U** commands, **OOPS** only returns objects erased in the most recent procedure. It has no effect on other modifications. If you erase several objects in the same command sequence, all of the objects return to the screen.

> Type
> **OOPS**
>
> **OOPS**

Object Selection

Editing commands and similar operations prompt you to select the objects to modify. So far, you have selected objects individually to create a ***selection set***. However, AutoCAD includes more efficient options for selecting multiple objects. This chapter describes several basic options for creating selection sets. You will explore additional selection techniques later in this textbook.

> **selection set:** A group of one or more selected objects, typically created to perform an editing operation on the selected objects.

Window and Crossing Selection

Window selection allows you to select objects by creating a selection rectangle around the objects. Only objects entirely within the window are selected. Crossing selection also requires you to create a selection rectangle, but all objects within *and crossing* the rectangle are selected. A quick way to use window or crossing selection

automatic windowing (implied windowing): A selection method that allows you to select multiple objects at one time without entering a selection option.

is through a feature known as ***automatic windowing***, or ***implied windowing***, which is on by default. You can apply automatic windowing at the Select objects: prompt or when no command is active, such as when grip editing.

To apply automatic window selection, pick a point clearly above or below and to the *left* of the objects to be selected. Then move the crosshairs to the right and up or down to enclose the objects to be selected. When you move the crosshairs in this direction, the window selection icon appears and the window selection rectangle has a solid outline and light blue background. Objects to be included in the selection set are highlighted by default. Pick to locate the second corner and create the selection set. Objects in the selection set remain highlighted. Press [Enter] to complete the operation. See **Figure 3-21**.

To apply automatic crossing selection, pick a point clearly above or below and to the *right* of the objects to be selected. Then move the crosshairs to the left and up or down, across the objects to be selected. When you move the crosshairs in this direction, the crossing selection icon appears and the crossing selection rectangle has a dotted outline and light green background. Objects to be included in the selection set are highlighted by default. Pick to locate the second corner and create the selection set. Objects in the selection set remain highlighted. Press [Enter] to complete the operation. See **Figure 3-22**.

NOTE

You can type W or WINDOW at the Select objects: prompt to use manual window selection, or type C or CROSSING to use manual crossing selection. When you use manual window or crossing selection, the selection box uses the window or crossing format regardless of where you pick to create the box.

Figure 3-21.
Using window selection on a mechanical part drawing view. A—The window selection icon appears next to the crosshairs when drawing the selection rectangle. B—A window selection is used to select and erase all objects that lie completely inside the window selection box. Notice that the selection set does not include the circles that are only partially inside the window.

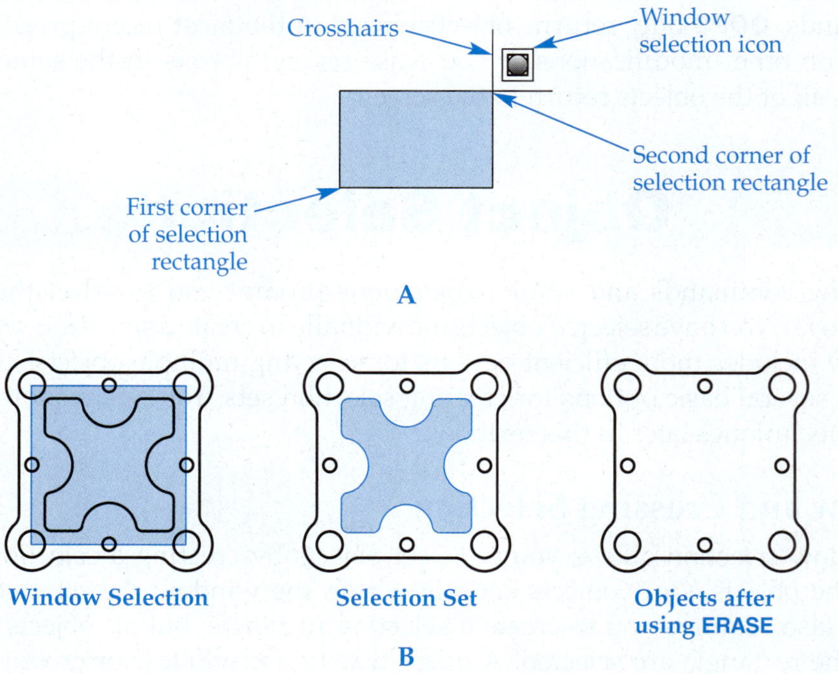

Figure 3-22.
Using crossing selection on a portion of an architectural floor plan. A—The crossing selection icon appears next to the crosshairs when drawing the selection rectangle. B—A crossing selection is used to select all objects inside the selection box and all objects that touch the sides of the box.

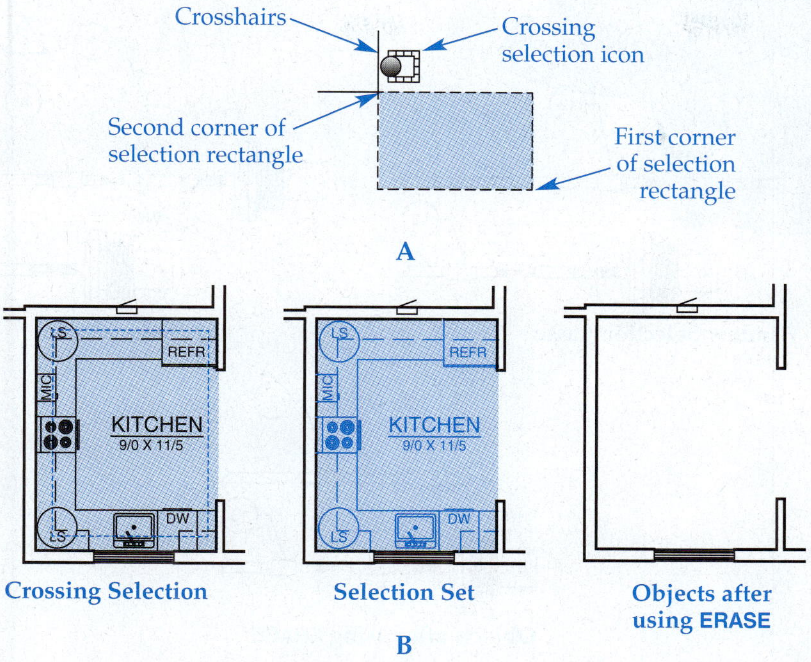

Exercise 3-13

www.g-wlearning.com/CAD/

Complete the exercise on the companion website.

Lasso Selection

Lasso selection is another method of implied windowing. Use a lasso to create an irregular selection shape that would otherwise require multiple individual or rectangular selections. A lasso selection can be created as a window or crossing selection. To apply a window selection lasso, pick and hold down the left mouse button at a point clearly above or below and to the *left* of the objects to be selected. Then drag the crosshairs to the right and up or down as necessary to enclose the objects to be selected. Release the left mouse button to make the selection. Press [Enter] to complete the operation. See Figure 3-23.

To apply a crossing selection lasso, pick and hold down the left mouse button at a point clearly above or below and to the *right* of the objects to be selected. Then drag the crosshairs to the left and up or down as necessary, across the objects to be selected. Release the left mouse button to make the selection. Press [Enter] to complete the operation. See Figure 3-24.

Exercise 3-14

www.g-wlearning.com/CAD/

Complete the exercise on the companion website.

Figure 3-23.
Using a window selection lasso on a portion of an electronic wire harness assembly drawing to select and erase all objects that lie completely inside the lasso.

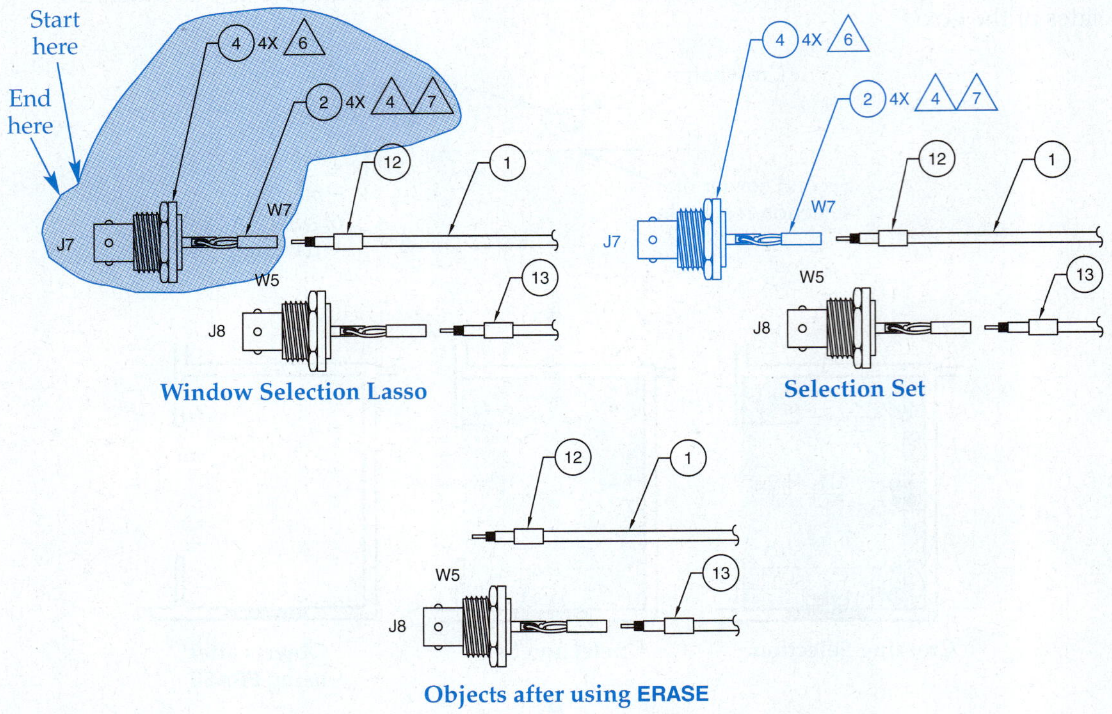

Figure 3-24.
Using a crossing selection lasso on an architectural isometric drawing of a fireplace to select and erase all objects that make up the hearth.

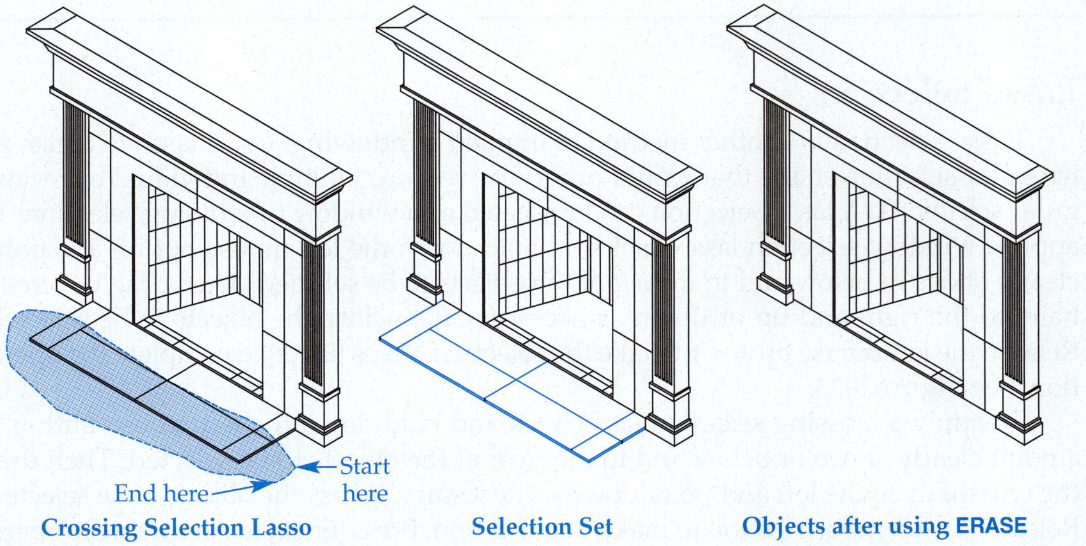

Window and Crossing Polygon Selection

The window and crossing polygon selection methods are also useful for creating a selection set that would otherwise require multiple individual or rectangular selections. To use window polygon selection, type WP or WPOLYGON at the Select objects: prompt. Then pick points to draw a polygon enclosing the objects to select. See **Figure 3-25A**. To use crossing polygon selection, type CP or CPOLYGON at the Select objects: prompt. Then pick points to draw a polygon around and through the objects to select. See **Figure 3-25B**.

Figure 3-25.
A—Using window polygon selection to erase only the wire reinforcement symbol from a structural slab detail. B—Using crossing polygon selection to erase multiple objects from the structural slab detail. Everything within and contacting the crossing polygon is selected.

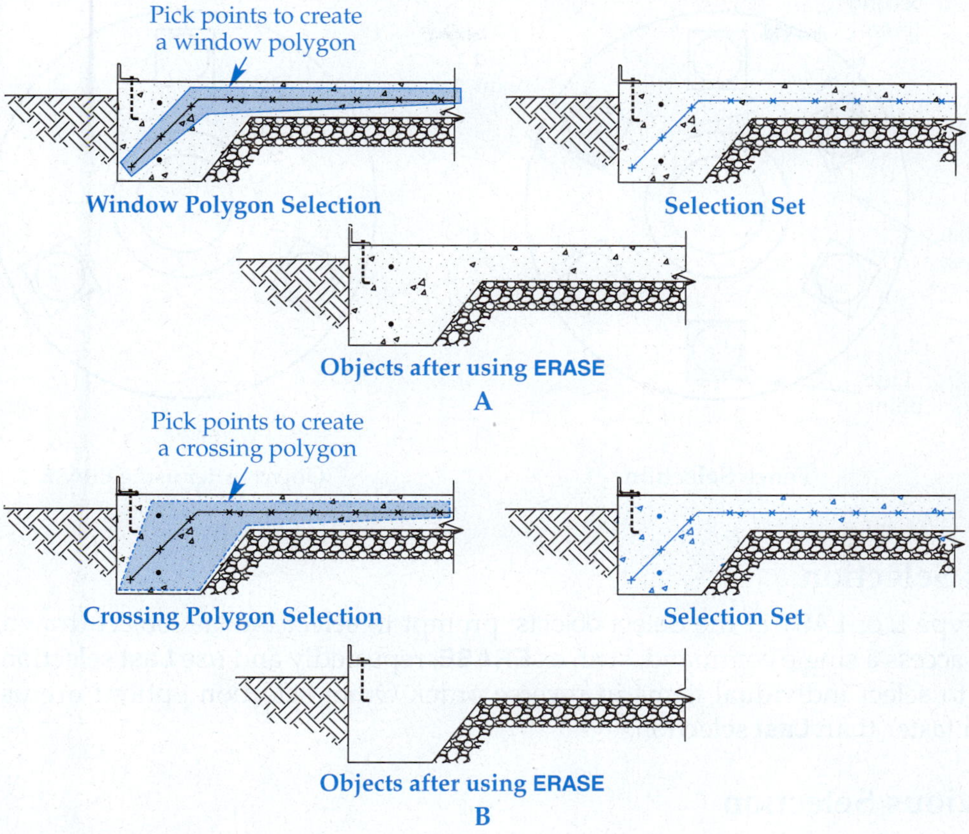

PROFESSIONAL TIP

In window or crossing polygon selection, AutoCAD does not allow you to select a point that causes the lines of the selection polygon to intersect each other. Pick locations that do not result in an intersection. Use the **Undo** option if you need to go back and relocate a previous pick point.

Fence Selection

Fence selection allows you to select all objects that contact a "fence" of connected segments you draw while using an editing command. To use fence selection, type F or FENCE at the Select objects: prompt. Then pick points to draw a fence through the objects to include in the selection set. See Figure 3-26. Often you only need to draw a single segment to create a useful selection set.

PROFESSIONAL TIP

Fence, **WPolygon**, and **CPolygon** options are also available after you specify the first corner when you use implied windowing before accessing a command, such as when grip editing.

Exercise 3-15

www.g-wlearning.com/CAD/

Complete the exercise on the companion website.

Figure 3-26.
Using fence selection to erase specific objects from a mechanical part drawing view. The fence can be staggered, as shown, or a single, straight segment.

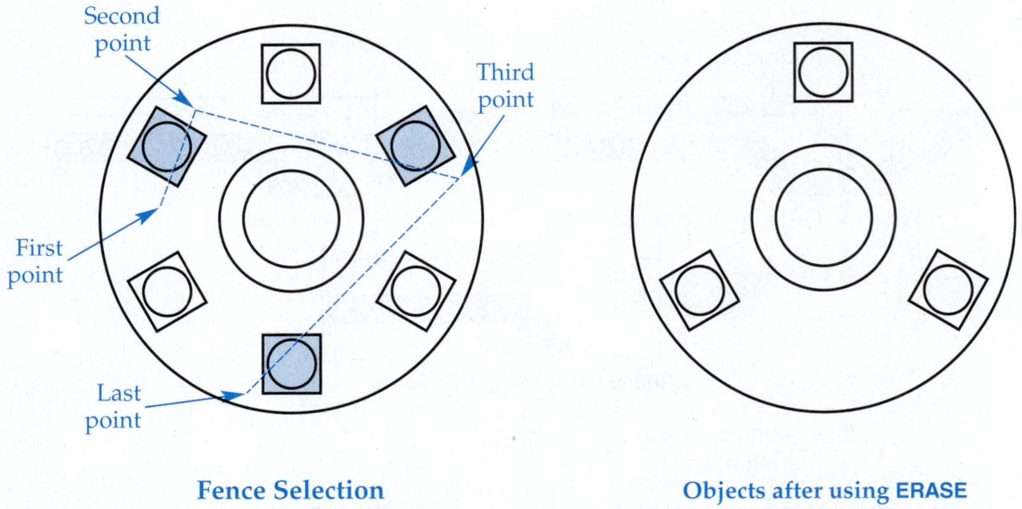

Last Selection

Type L or LAST at the Select objects: prompt to select the last object drawn. You must access a single command, such as **ERASE**, repeatedly and use **Last** selection each time to select individual items in reverse order. Other selection options are usually much faster than **Last** selection.

Previous Selection

Type P or PREVIOUS at the Select objects: prompt to reselect all the objects selected in the previous selection set. **Previous** selection is especially useful when you need to carry out more than one editing operation on a specific group of objects. Use **Previous** selection to reselect the objects you just edited.

NOTE

Previous selection does not reselect erased objects.

Selecting All Objects

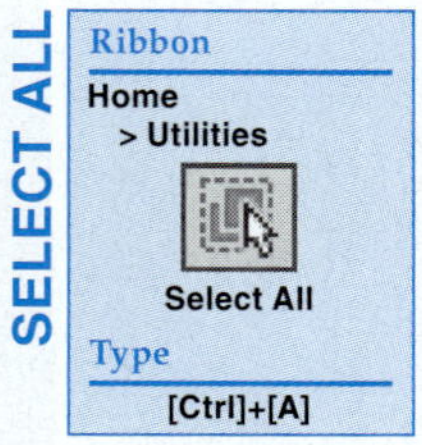

Use the **Select All** option to select every object in the drawing that is not on a frozen layer, including objects that are outside of the current drawing window display. Chapter 5 describes layers.

Changing the Selection Set

A quick way to remove one or more objects from a selection set is to hold down [Shift] and reselect the objects. This is possible only for individual picks and automatic windowing. To change the selection set using automatic windowing, hold down [Shift] and pick the first corner, then release [Shift] and pick the second corner. You can also drag a lasso as previously described. Select objects as usual to add them back to the selection set.

Another option for removing objects from a selection set is to type R or REMOVE at the Select objects: prompt. This enters the **Remove** option and changes the Select objects: prompt to Remove objects:, allowing you to pick objects to remove from the selection set. To switch back to selection mode, type A or ADD at the Remove objects: prompt. This enters the **Add** option and restores the Select objects: prompt, allowing you to select additional objects.

PROFESSIONAL TIP

Removing items from a selection set is especially effective if you first use the **Select All** selection option. This allows you to keep a few specific objects while erasing everything else.

Exercise 3-16

www.g-wlearning.com/CAD/

Complete the exercise on the companion website.

Cycling through Stacked Objects

stacked objects: Objects that overlap in a drawing. When you pick with the mouse, the topmost object is selected by default.

cycle: Repeatedly select a series of stacked objects until the desired object is highlighted.

While drawing, you will sometimes create ***stacked objects***, intersecting objects, or objects that are very close together. To ***cycle*** through overlapping objects to find the object to select, move the crosshairs over the intersecting objects. Then, hold down [Shift] and press the space bar repeatedly to cycle through the stacked objects. When the object you want to select is highlighted, release [Shift] and pick to select. This method can be used at the Select objects: prompt or before accessing an editing command. See Figure 3-27.

AutoCAD also includes a **Selection Cycling** tool for cycling through stacked objects. This tool can be used at the Select objects: prompt or before accessing an editing command. To access the **Selection Cycling** tool, first check the **Selection Cycling** option in the status bar **Customization** flyout to add the **Selection Cycling** button to the status bar. Then pick the **Selection Cycling** button to activate the tool. Move the crosshairs over stacked objects. When you see the **Selection Cycling** icon, pick to display a list of stacked objects. Move the cursor over an object in the list to highlight the corresponding object in the drawing. Select an object from the list box or select **None** to end the tool. See Figure 3-28A. Use the options on the **Selection Cycling** tab of the **Drafting Settings** dialog box to adjust selection cycling settings. See Figure 3-28B. A quick way to access the **Selection Cycling** tab is to right-click on the **Selection Cycling** button on the status bar and select **Selection Cycling Settings...**.

Figure 3-27.
Cycling through a series of stacked bushes on a landscaping plan to select a specific bush.

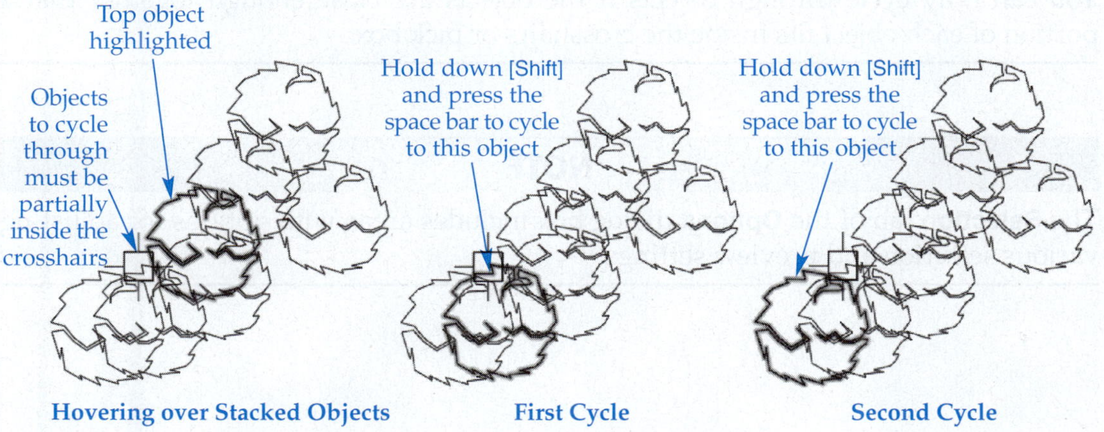

Figure 3-28.
A—Turn on selection cycling to cycle through stacked objects. B—Use the **Selection Cycling** tab of the **Drafting Settings** dialog box to specify selection cycling preferences.

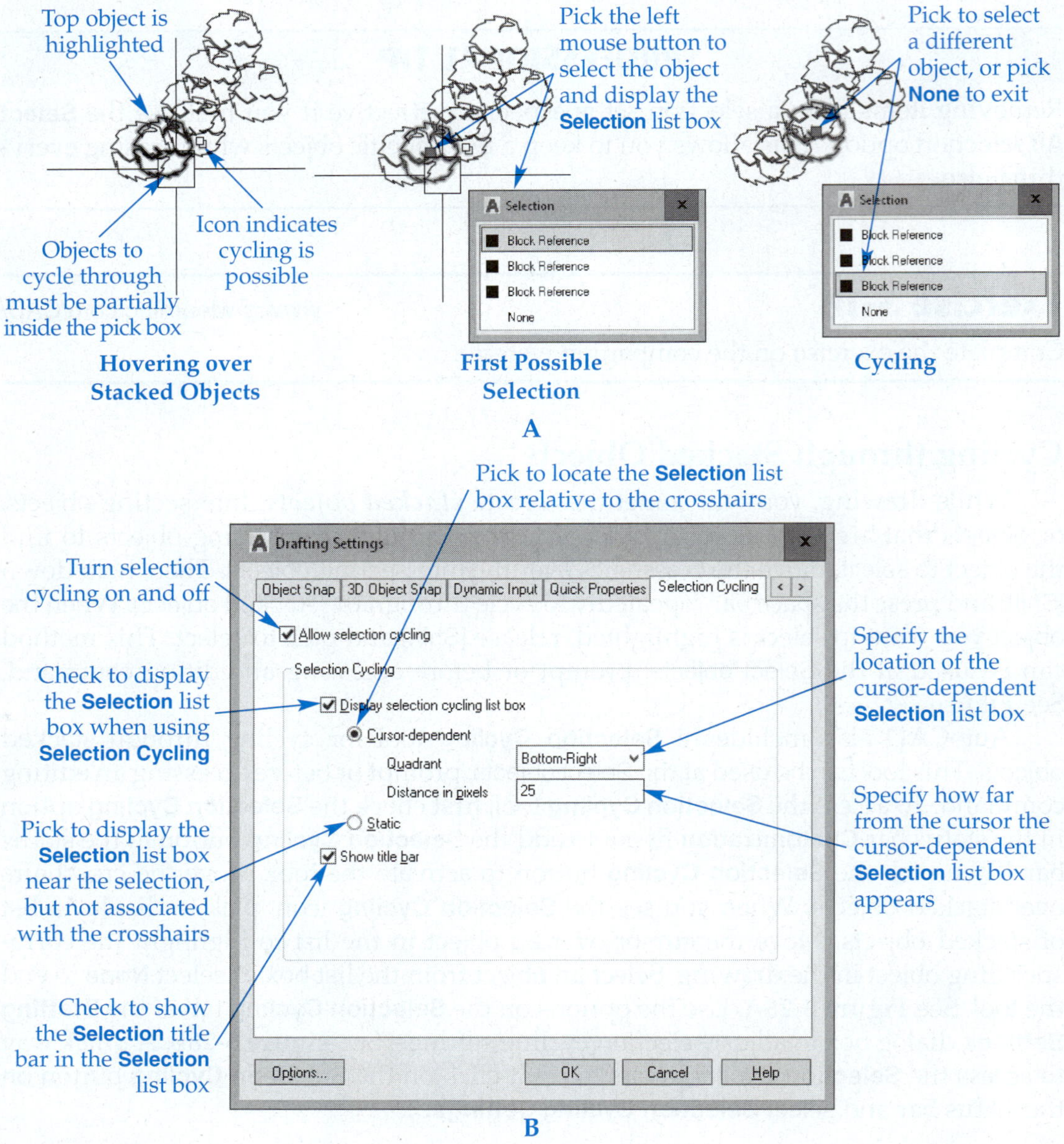

NOTE

You can only cycle through objects if the objects are close enough together that a portion of each object fits inside the crosshairs or pick box.

NOTE

The **Selection** tab of the **Options** dialog box includes areas with options for adjusting various selection and preview settings.

Chapter Review

Answer the following questions using the information in this chapter.

1. When you enter a fractional number in AutoCAD, why is a hyphen required between a whole number and its associated fraction?
2. Describe the Cartesian coordinate system.
3. Explain how to use the wheel on a mouse to zoom in, zoom out, and pan.
4. List two ways to discontinue drawing a line.
5. Explain what is meant by the absolute coordinate entry 5.250,7.875 in relationship to the origin.
6. Explain what is meant by the relative coordinate entry @2,2.
7. Explain what is meant by the polar coordinate entry @2.750<90.
8. What two general methods of point entry are available when dynamic input is active?
9. Explain how you can continue drawing another line segment from a previously drawn line.
10. Name two ways to quickly access the **Snap and Grid** tab of the **Drafting Settings** dialog box.
11. How do you activate snap mode?
12. How do you set a grid spacing of .25?
13. Explain, in general terms, how direct distance entry works.
14. What are the default angle increments for polar tracking?
15. How can you turn on ortho mode?
16. Which running object snap modes are active by default?
17. Name the drawing aids that must be active for object snap tracking to function.
18. When you access the **ERASE** command, what replaces the screen crosshairs?
19. How many command sequences can you undo at one time with the **U** command?
20. Name the command used to bring back an object that was previously removed using the **UNDO** command.
21. Name the command used to bring back objects erased in the most recent procedure.
22. How does the appearance of window and crossing selection boxes differ?
23. List five ways to select an object to erase.
24. Define *stacked objects*.

Drawing Problems

Start AutoCAD if it is not already started. Follow the specific instructions for each problem. Use only drawing commands and techniques you have already learned. Do not draw dimensions or text. Use your own judgment and approximate dimensions when necessary. Several problems in this chapter and throughout this textbook show dimension values that include the X symbol. This practice is used to identify multiple features of the same size. For example, a 4X ∅.312 dimension value indicates that there are four circles with a .312 diameter. A 2X R.625 dimension value indicates that there are two arcs with a .625 radius. A 2X .750 dimension value indicates that there are two of the same .750 features in the drawing. These practices are explained later in this textbook.

▼ Basic

1. **General** Start a new drawing from scratch or use a template of your choice. Use the status bar to turn off all drawing aids, including grid, snap, polar tracking, object snap tracking, and ortho. Use the **LINE** command to draw the following objects as accurately as possible.
   - Right triangle
   - Isosceles triangle
   - Rectangle
   - Square

   Save the drawing as P3-1.

2. **General** Start a new drawing from scratch or use a template of your choice. Draw the same objects specified in Problem 1, but this time, turn the snap grid on. Observe the difference between having snap mode on for this problem and off for the previous problem. Save the drawing as P3-2.

3. **General** Start a new drawing from scratch or use a decimal-unit template of your choice. Draw an object by connecting the following point coordinates. Use dynamic input to enter the coordinates. Save the drawing as P3-3.

Point	Coordinates	Point	Coordinates
1	2,2	8	@-1.5,0
2	@1.5,0	9	@0,1.25
3	@.75<90	10	@-1.25,1.25
4	@1.5<0	11	@2<180
5	@0,-.75	12	@-1.25,-1.25
6	@3,0	13	@2.25<270
7	@1<90		

4. **Architectural** Start a new drawing from scratch or use an architectural-unit template of your choice. Draw the front and side views of a wide flange, similar to the wide flange shown. Use grid and snap modes, default running object snaps, and object snap tracking when possible. Save the drawing as P3-4.

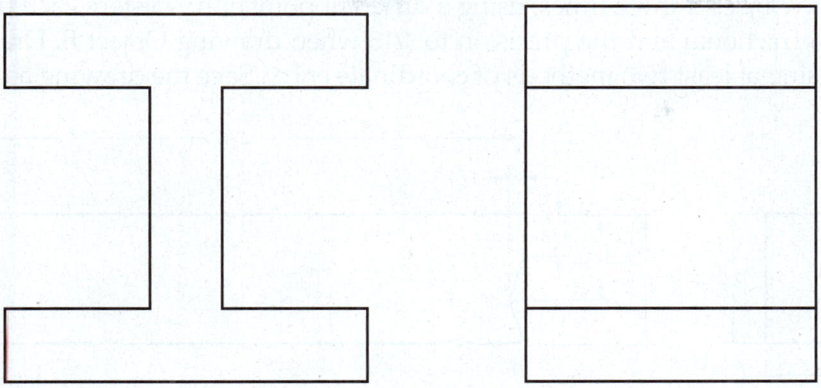

5. **General** Start a new drawing from scratch or use a template of your choice. Draw the bar graph shown using direct distance entry and polar tracking. Each grid square represents one unit. Do not draw the grid lines. Save the drawing as P3-5.

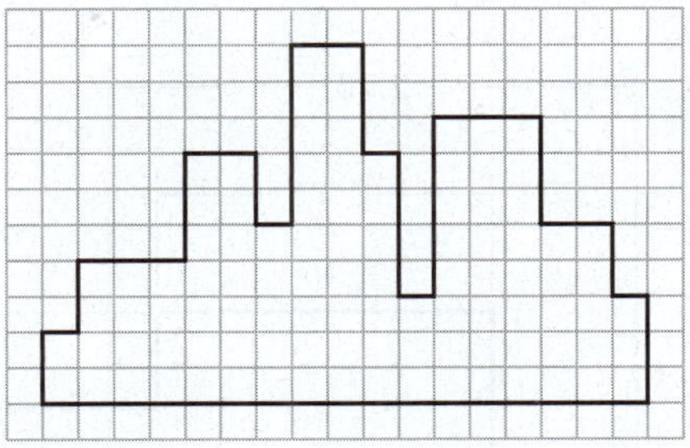

6. **General** Start a new drawing from scratch or use a template of your choice. Draw the hexagon shown using the dimensional input feature of dynamic input. Each side of the hexagon is 2 units. Begin at the start point and draw the lines in the direction indicated by the arrows. Save the drawing as P3-6.

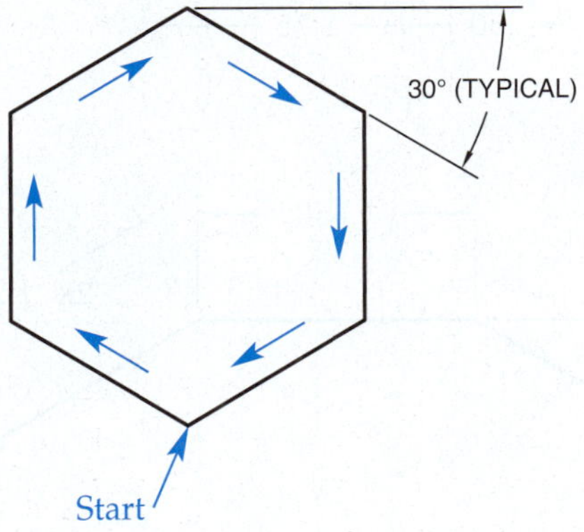

Drawing Problems - Chapter 3

▼ Intermediate

7. **Mechanical** Start a new drawing from scratch or use a fractional-unit template of your choice. Draw the part views shown using absolute, relative, and polar coordinate entry methods. Set the units to decimal and the precision to 0.0 when drawing Object A. Draw Object A three times, using a different point entry system each time. Set the units to fractional and the precision to 1/16 when drawing Object B. Draw Object B once, using at least two methods of coordinate entry. Save the drawing as P3-7.

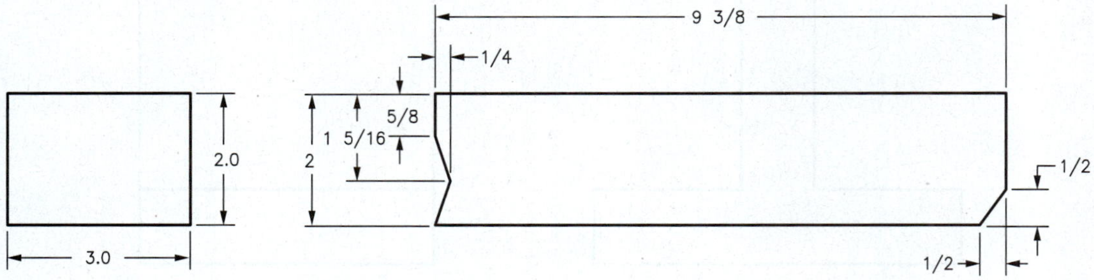

A B

For Problems 8–9, start a new drawing from scratch or use a decimal-unit template of your choice. Draw the part view shown. Save the drawings as P3-8 *and* P3-9.

8. **Mechanical**

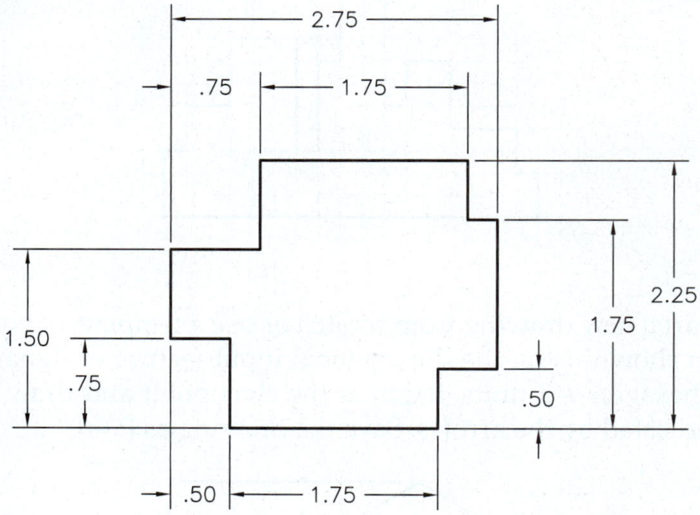

9. **Mechanical**

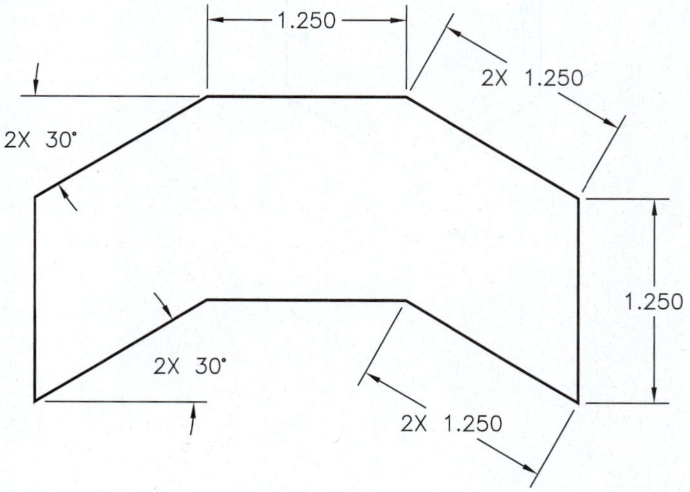

10. **General** Start a new drawing from scratch or use a decimal-unit template of your choice. Draw the part views shown in A and B. Begin at the start point and then discontinue the **LINE** command at the point shown. Complete each view by continuing from the previous endpoint. Save the drawing as P3-10.

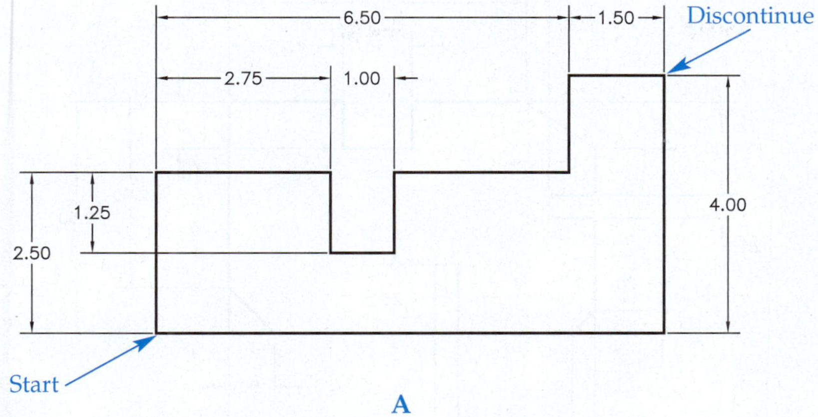

A

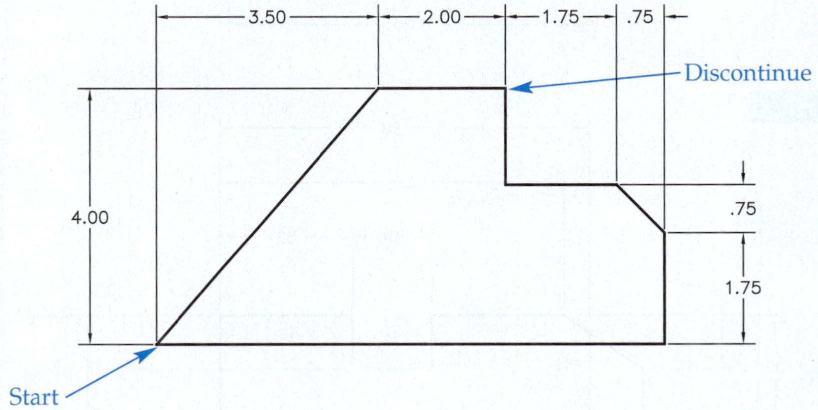

B

11. **General** Create a freehand sketch of the X and Y axes on a sheet of paper. Label the origin, the positive values for X = 1 through X = 10, and the positive values for Y = 1 through Y = 10. Then sketch the object described by the following coordinate points:

Point 1:	2,2	Point 8:	4,6
Point 2:	8,2	Point 9:	4,3
Point 3:	8,7	Point 10:	3,3
Point 4:	7,7	Point 11:	3,7
Point 5:	7,3	Point 12:	2,7
Point 6:	6,3	Point 13:	2,2
Point 7:	6,6		

▼ Advanced

12. **General** Create a freehand sketch of the X and Y axes as you did for the previous problem. Then sketch an object outline of your choice within the axes. List, in order, the rectangular coordinates of the points a drafter would need to specify in AutoCAD to recreate the object in your sketch.

For Problems 13–15, start a new drawing from scratch or use a decimal-unit template of your choice. Draw the part view shown. Save the drawings as P3-13, P3-14, *and* P3-15.

13. **Mechanical**

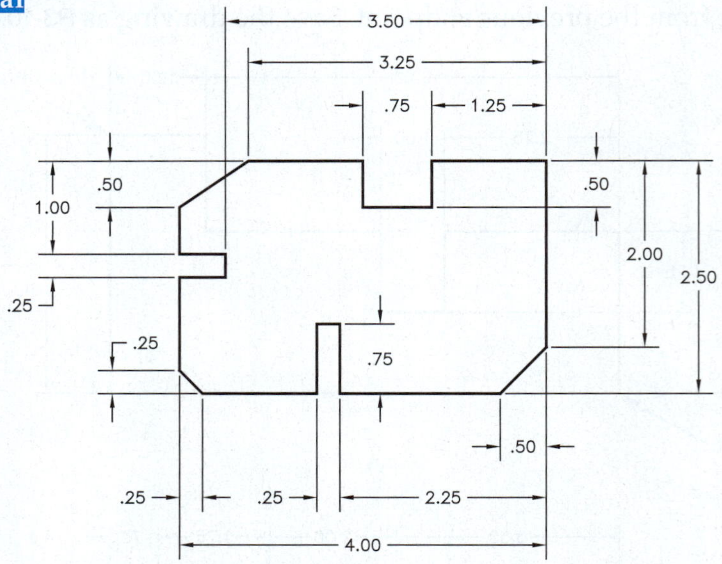

14. **Mechanical**

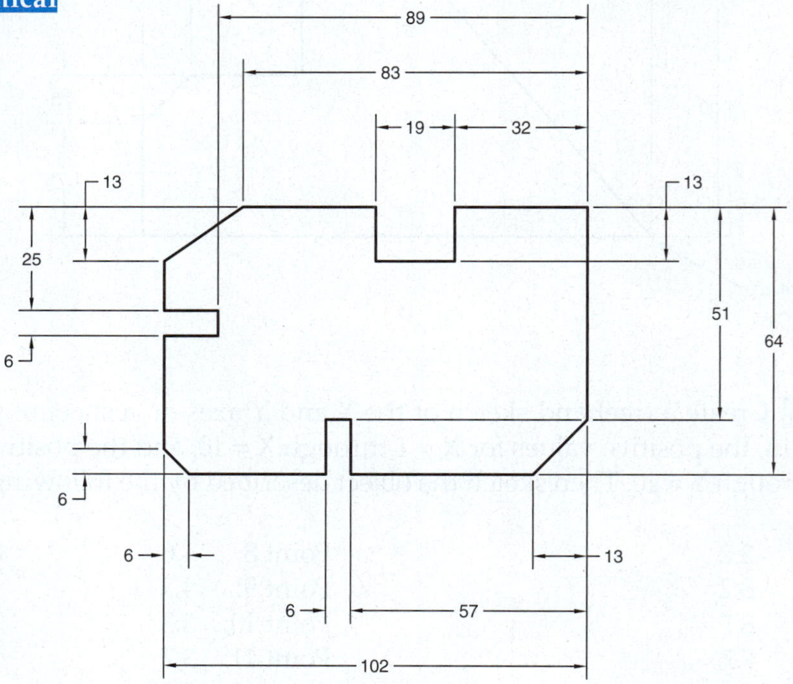

15. **Mechanical**

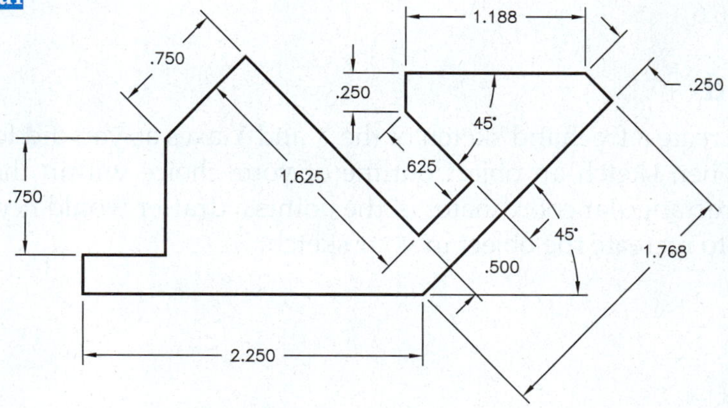

16. **Architectural** Start a new drawing from scratch or use an architectural-unit template of your choice. Draw the window elevation symbol shown. Save the drawing as P3-16.

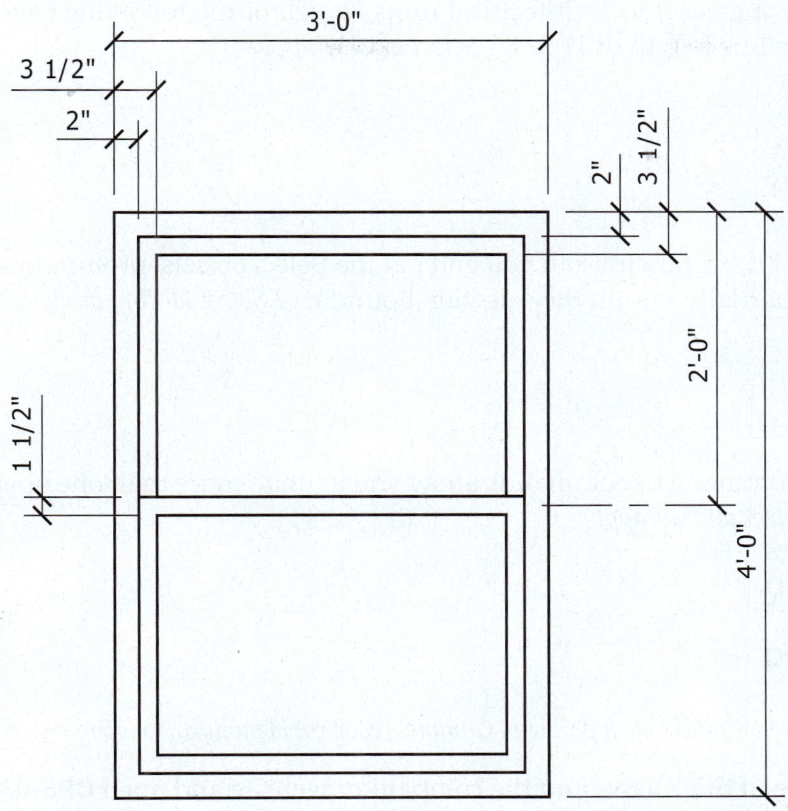

17. **Mechanical** Create a freehand 2D sketch of the front of your computer monitor. Use available measuring devices, such as a tape measure and caliper, to dimension the size and location of each feature accurately. Convert any round objects to rectangular shapes that you can draw using the **LINE** command. Start a new drawing from scratch or use a decimal-unit template of your choice. Draw the monitor from your sketch. Save the drawing as P3-17.

18. **Architectural** Create a freehand, dimensioned 2D sketch of the floor plan of a room in your school or company, complete with furniture. Use a tape measure to dimension the size and location of walls, doors, windows, and furniture accurately. Convert any round objects to rectangular shapes that you can draw using the **LINE** command. Start a new drawing from scratch or use an architectural-unit template of your choice. Draw the room from your sketch. Save the drawing as P3-18.

AutoCAD Certified Professional Exam Practice

Answer the following questions using the information in this chapter.

1. If a drawing is set to architectural units, which of the following can you enter to specify a line length of 11 3/4"? *Select all that apply.*
   A. 11.75"
   B. 11.75
   C. 11 3/4
   D. 11-3/4
   E. 11 3/4"
2. Which of the following can you enter at the Select objects: prompt to select objects that lie partially within the selection boundary? *Select all that apply.*
   A. C
   B. W
   C. CP
   D. WP
3. Which command or commands allow you to undo more than one operation at one time? *Select all that apply.*
   A. **OOPS**
   B. **REDO**
   C. **U**
   D. **UNDO**

Follow the instructions in each problem. Complete each problem by answering the related question.

4. **Navigate to this chapter on the companion website and open CPE-03line.dwg.** With dynamic input off, use the coordinates below to create the object shown. Do not change any settings in the drawing file. When you finish, use the **Endpoint** object snap mode to select Point A. According to the coordinate display in the status bar, what are the coordinates of this point? If necessary, select the **Coordinates** option in the status bar **Customization** flyout to add the coordinate display to the status bar.

   Coordinate Values:
   2,3
   @6.25<0
   @4<50
   @3.5<90
   @4<130
   @6.25<180
   CLOSE

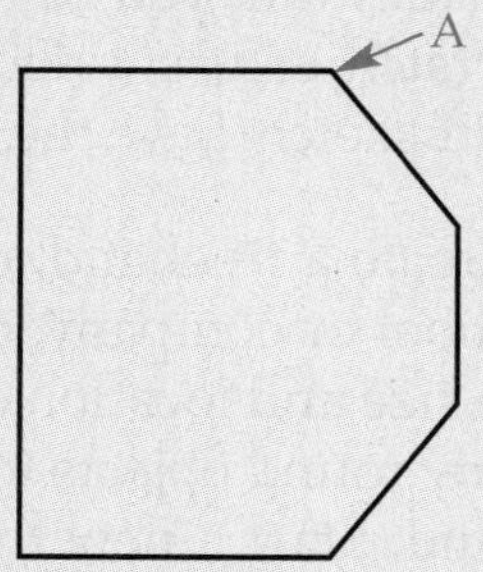

5. **Navigate to this chapter on the companion website and open CPE-03intersect.dwg.** With dynamic input off, use the coordinates below to create the object shown. Do not change any settings in the drawing file. Use the **Intersection** object snap mode to select Point B. According to the coordinate display in the status bar, what are the coordinates of this point? If necessary, select the **Coordinates** option in the status bar **Customization** flyout to add the coordinate display to the status bar.

   Coordinate Values:
   5,2
   @3.45<60
   @4.75<180
   @8<–15

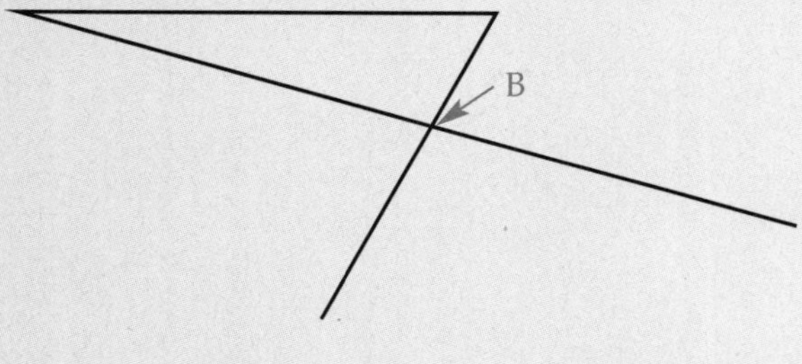

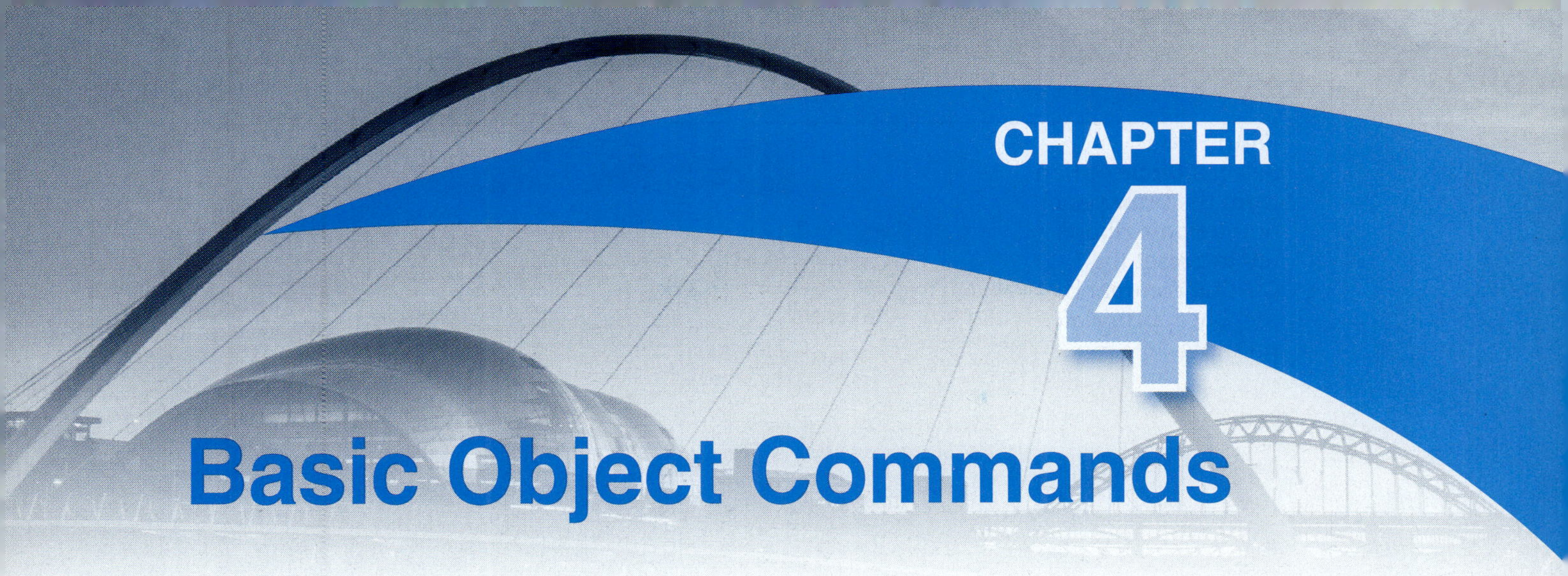

CHAPTER

4

Basic Object Commands

Learning Objectives

After completing this chapter, you will be able to:

✓ Draw circles using the **CIRCLE** command.
✓ Draw arcs using the **ARC** command.
✓ Use the **ELLIPSE** command to draw ellipses and elliptical arcs.
✓ Use the **PLINE** command to draw polylines.
✓ Draw regular polygons using the **POLYGON** command.
✓ Draw rectangles using the **RECTANGLE** command.
✓ Draw donuts and filled circles using the **DONUT** command.
✓ Draw true spline curves using the **SPLINE** command.

This chapter describes several object commands and their options. It presents the ribbon as the primary way to access object commands, because the ribbon provides a direct link to many command options. Prompts associated with the options appear when you draw to automate the process. In contrast, when you issue a command using dynamic input or the command line, you must select specific options while you draw to receive appropriate prompts for constructing the object. You can draw objects using point entry or drawing aids, similar to locating endpoints while using the **LINE** command. Several object commands also offer the option to input a direct value, such as for the radius of a circle.

circle: A closed curve with a constant radius (R) around a center point; usually dimensioned according to the diameter (∅).

radius: The distance from the center of a circle to its circumference; always one-half the diameter; usually represented on a drawing with the R symbol.

Drawing Circles

The **CIRCLE** command provides several options for drawing *circles*. Select the appropriate option based on the information you know about locating and constructing the circle. The ribbon is an effective way to access **CIRCLE** command options. See Figure 4-1.

CIRCLE

Ribbon
Home
> Draw
Circle
Type
CIRCLE
C

Center, Radius Option

Access the **Center, Radius** option to specify the center of the circle, followed by the *radius*. Use point entry or drawing aids to locate the center point. If you know the radius, type a value and press [Enter]. You can also define the radius using point entry or drawing aids. See Figure 4-2.

Figure 4-1.
Select a **Circle** option from the **Draw** panel of the **Home** ribbon tab to preset the **CIRCLE** command to display appropriate prompts.

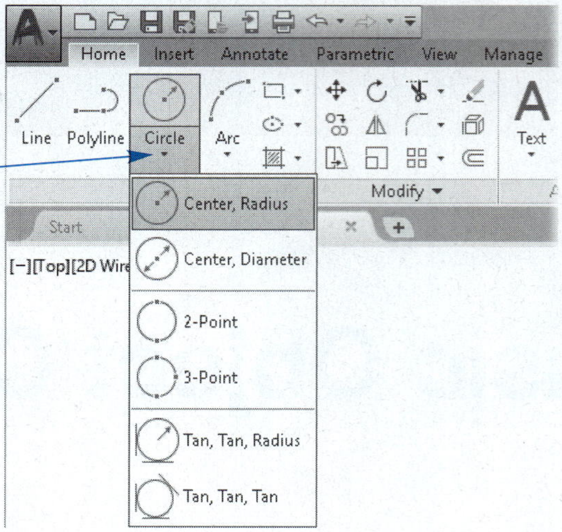

Figure 4-2.
Drawing a circle by specifying the center point and radius. Notice the rubberband line that appears when you move the crosshairs away from the center point.

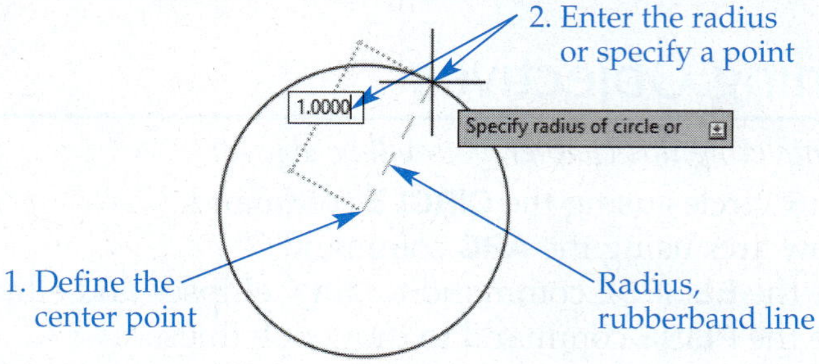

NOTE

AutoCAD stores the radius of the circle you draw as the new default radius setting, allowing you to draw another circle quickly with the same radius.

Center, Diameter Option

diameter: The distance across a circle measured through the center; usually represented on a drawing with the ∅ symbol.

Access the **Center, Diameter** option to specify the center of the circle, followed by the ***diameter***. The **Center, Diameter** option is convenient for designs such as circular holes, shafts, and features sized according to diameter. Use point entry or drawing aids to locate the center point. If you know the diameter, type a value and press [Enter]. You can also define the diameter using point entry or drawing aids. See **Figure 4-3**.

NOTE

If you use the **Center, Radius** option to draw a circle after using the **Diameter** option, AutoCAD changes the default to a radius measurement based on the previous diameter.

2-Point Option

quadrant: A point on the circumference at the horizontal or vertical quarter of a circle, arc, donut, or ellipse.

Access the **2-Point** option to specify diameter using two points at opposite ***quadrants*** of the circle. The **2-Point** option is useful for defining the diameter of a circle without specifying the center point, such as when drawing a circle between two existing objects. Specify the points in any order using point entry or drawing aids. See **Figure 4-4**.

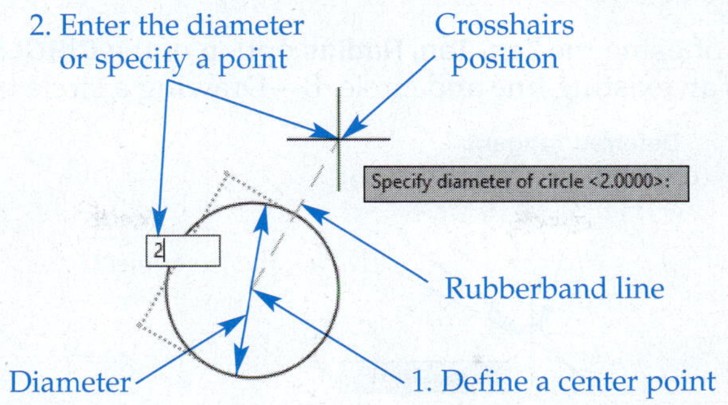

Figure 4-3. Drawing a circle by specifying the center point and diameter. Notice that the crosshairs measures the diameter, but the rubberband line passes midway between the center and the crosshairs.

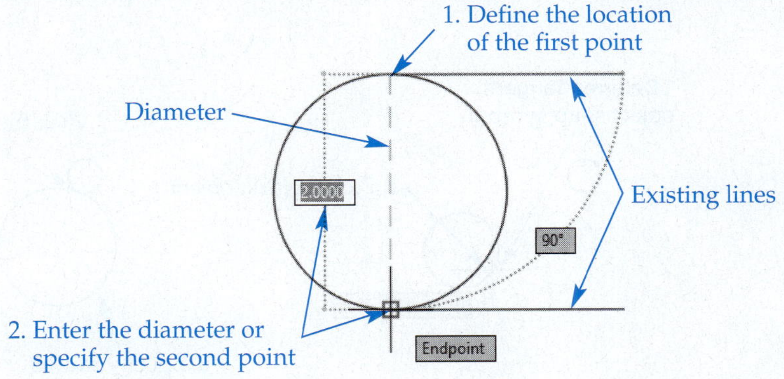

Figure 4-4. Using the **2-Point** option of the **CIRCLE** command. A common application is drawing a circle between two existing objects, such as these lines.

3-Point Option

Access the **3-Point** option to draw a circle according to three known points on the circumference of the circle. The **3-Point** option is useful for constructing a circle without specifying the center point, the radius, or the diameter, such as when drawing a circle between three existing objects. Specify the points in any order using point entry or drawing aids. See **Figure 4-5**.

Tan, Tan, Radius Option

Access the **Tan, Tan, Radius** option to pick two objects ***tangent*** to a circle and specify the circle radius. Hover the crosshairs over the first line, arc, or circle to which the new circle will be tangent. When you see the **Deferred Tangent** object snap marker, pick to select the first ***point of tangency***. Repeat the process to select the second object to which the new circle will be tangent. The order in which you pick is not critical. If you know the radius, type a value and press [Enter]. You can also define the radius using point entry or drawing aids. See **Figure 4-6**.

tangent: A line, circle, or arc that meets another circle or arc at only one point.

point of tangency: The point shared by tangent objects.

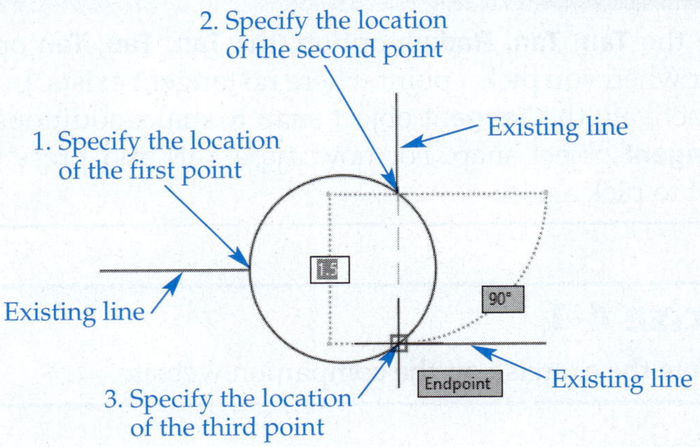

Figure 4-5. Using the **3-Point** option of the **CIRCLE** command. A common application is drawing a circle by referencing three known points, such as the endpoints of these lines.

Figure 4-6.
Examples of using the **Tan, Tan, Radius** option of the **CIRCLE** command. A—Drawing a circle tangent to an existing line and circle. B—Drawing a circle tangent to two existing circles.

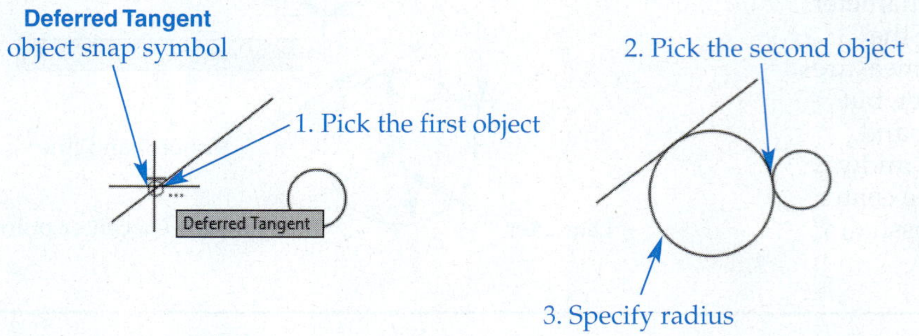

A

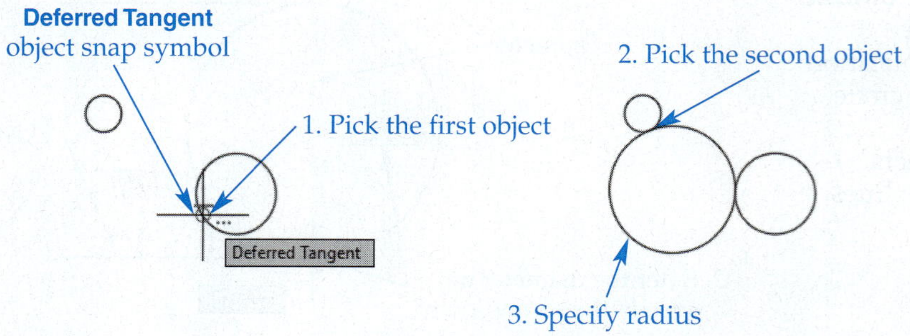

B

NOTE

If the radius you enter while using the **Tan, Tan, Radius** option is too small, AutoCAD displays the message Circle does not exist.

Tan, Tan, Tan Option

Select the **Tan, Tan, Tan** option to draw a circle tangent to three existing objects. Hover the crosshairs over the first line, arc, or circle to which the new circle will be tangent. When you see the **Deferred Tangent** object snap marker, pick to select the first point of tangency. Repeat the process to select the second and third objects to which the new circle will be tangent. You must make selections when you see the **Deferred Tangent** object snap marker, but the order in which you pick is not critical. See Figure 4-7.

NOTE

Unlike the **Tan, Tan, Radius** option, the **Tan, Tan, Tan** option does not automatically recover when you pick a point where no tangent exists. In such a case, you must manually reactivate the **Tangent** object snap to make additional picks. Chapter 7 describes the **Tangent** object snap. For now, type TAN and press [Enter] at the point selection prompt to pick again.

Exercise 4-1

www.g-wlearning.com/CAD/

Complete the exercise on the companion website.

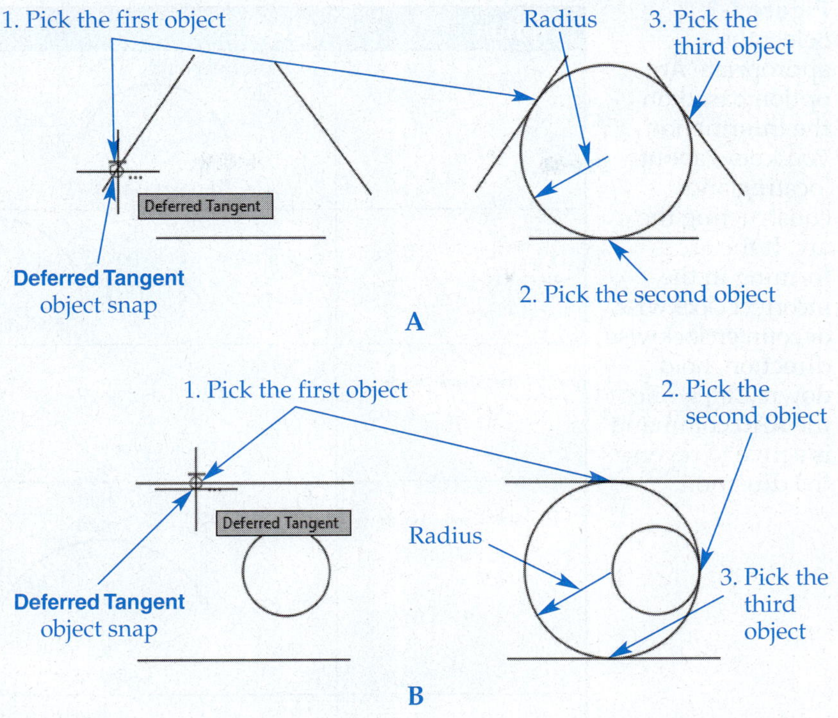

Figure 4-7. Examples of using the **Tan, Tan, Tan** option of the **CIRCLE** command. A—Drawing a circle tangent to three existing lines. B—Drawing a circle tangent to two existing lines and a circle.

Drawing Arcs

The **ARC** command offers multiple options for drawing ***arcs***. Select the appropriate option based on the information you know about locating and constructing the arc. The ribbon is an effective way to access arc command options. See **Figure 4-8**.

Figure 4-9 provides a step-by-step example of using each **ARC** command option. The selections and values you enter determine arc placement. Some options prompt for the ***included angle***, and others prompt for the ***chord length***. Locating points in a clockwise or counterclockwise pattern affects the construction of most arcs. If the arc

ARC

Ribbon
Home
> Draw
Arc
Type
ARC
A

arc: Any portion of a circle, usually dimensioned according to the radius (R).

included angle: The angle formed between the center, start point, and endpoint of an arc.

chord length: The linear distance between two points on a circle or arc.

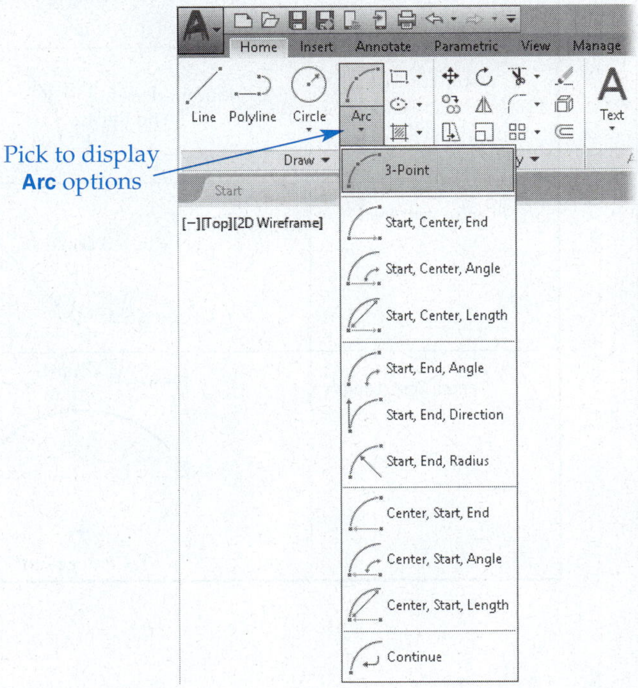

Figure 4-8. Select an **Arc** option from the **Draw** panel of the **Home** ribbon tab to preset the **ARC** command to display appropriate prompts.

Figure 4-9.
Select the appropriate **Arc** option based on the information you know about locating and constructing the arc. If the arc is forming in the incorrect clockwise or counterclockwise direction, hold down [Ctrl] while the **ARC** command is active to reverse the direction.

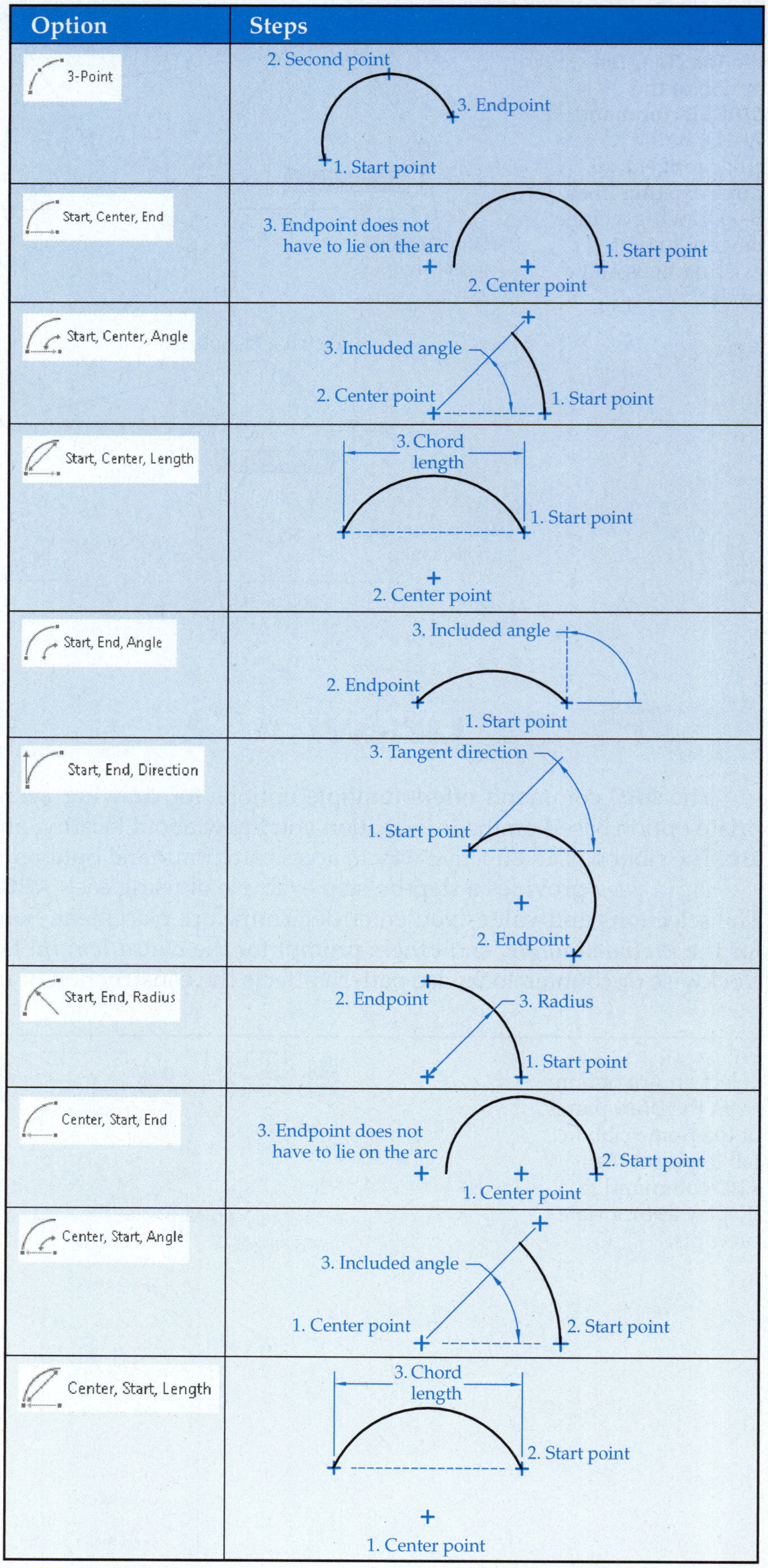

is forming in the incorrect clockwise or counterclockwise direction, hold down [Ctrl] while the **ARC** command is active to reverse the direction. The values you specify, including the use of positive or negative numbers, also affects the result.

NOTE

The **3-Point** option is default when you enter the **ARC** command at the keyboard.

Reference Material

www.g-wlearning.com/CAD/

Chord Length Table

For a chord length table and other reference tables, go to the companion website, select the **Resources** tab, and select **Standard Tables**.

Exercises 4-2 and 4-3

www.g-wlearning.com/CAD/

Complete the exercises on the companion website.

Continue Option

Use the **Continue** option to continue an arc from the endpoint of a previously drawn line or arc. The arc automatically attaches to the endpoint of the previously drawn line or arc, and the Specify end point of arc (hold Ctrl to switch direction): prompt appears. Pick the second endpoint of the new arc to create the arc.

The **Continue** option is a quick way to draw an arc beginning at the endpoint of a previously drawn line, tangent to the line. See **Figure 4-10A**. Use this technique for applications such as drawing slots. When you draw a series of arcs using the **Continue** option, each arc is tangent to the previous arc. The start point and direction are based on the endpoint and direction of the previous arc. See **Figure 4-10B**.

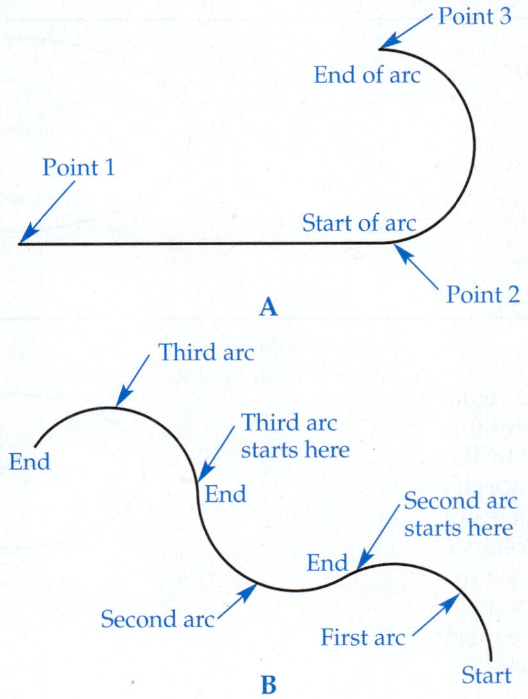

Figure 4-10. A—Continuing an arc from the previous line. Point 2 is the start of the arc, and Point 3 is the end of the arc. The arc and line are tangent at Point 2. If the arc is forming in the incorrect clockwise or counterclockwise direction, hold down [Ctrl] while the **ARC** command is active to reverse the direction. B—Using the **Continue** option to draw three tangent arcs.

NOTE

You can also access the **Continue** option by beginning the **ARC** command and pressing [Enter] when prompted to specify the start point of the arc.

Exercise 4-4

www.g-wlearning.com/CAD/

Complete the exercise on the companion website.

Drawing Ellipses

ellipse: An oval shape that contains two centers of equal radius.

ELLIPSE

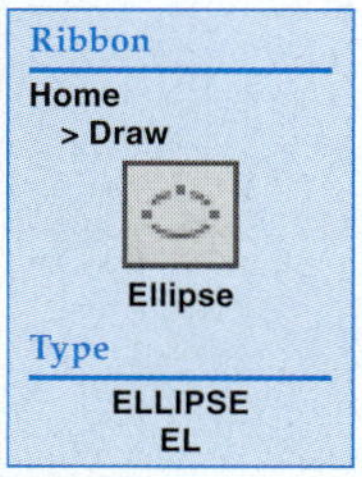

major axis: The longer of the two axes in an ellipse.

minor axis: The shorter of the two axes in an ellipse.

An ***ellipse*** has a ***major axis*** and a ***minor axis***. See Figure 4-11. A circle appears as an ellipse when you view the circle at an angle. For example, a 30° ellipse is a circle rotated 30° from the line of sight.

The **ELLIPSE** command offers several options for drawing elliptical shapes. Select the appropriate option based on the information you know about locating and constructing the ellipse and on whether the ellipse is whole or an elliptical arc.

Center Option

Use the **Center** option to specify the center of the ellipse, then an endpoint of the first axis, and finally an endpoint of the second axis. Axis endpoints originate from the center of the ellipse, forming half of the major and minor axes. See Figure 4-12.

Axis, End Option

Use the **Axis, End** option to specify the first endpoint of an axis, then the second endpoint of the same axis, and finally one endpoint of the second axis. The first axis can be the major or minor axis. See Figure 4-13.

Figure 4-11. The parts of an ellipse.

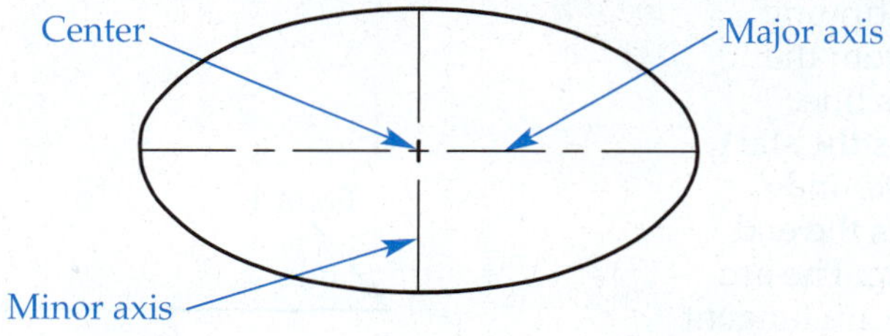

Figure 4-12. Drawing an ellipse by picking the center and an endpoint for each axis. The order in which you specify axis endpoints is not critical. The distance from each endpoint to the center point determines the major and minor axes.

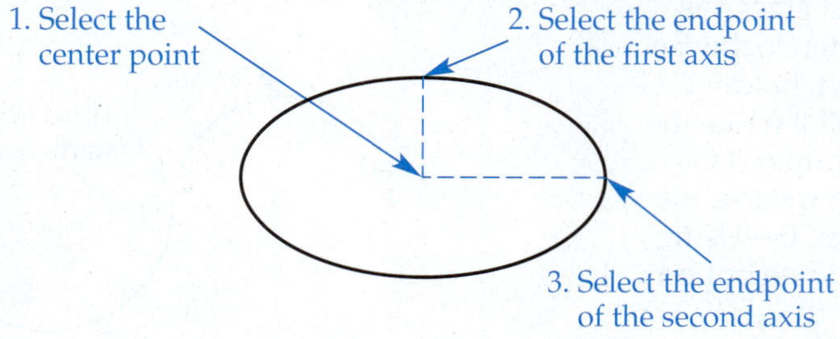

Figure 4-13.
Constructing the same ellipse by selecting different axis endpoints. Select points based on known information or the location of existing objects.

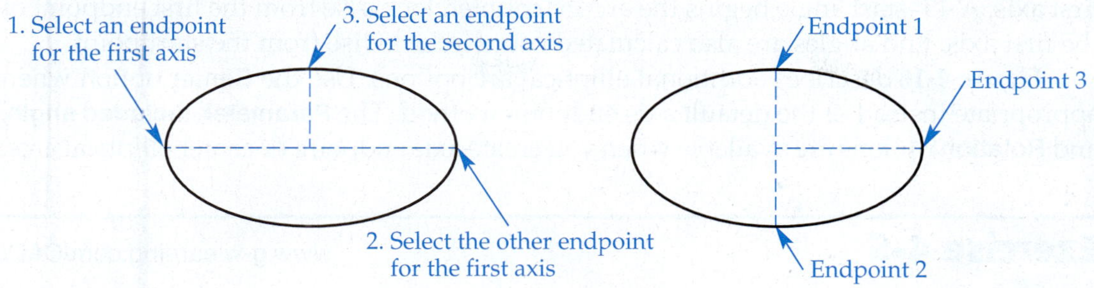

Rotation Option

Use the **Rotation** option to create an ellipse by specifying the angle at which a circle rotates from the line of sight. Begin by constructing an ellipse as usual, but be sure to create the major axis when you specify the first axis endpoint. Then, when the Specify distance to other axis or [Rotation]: prompt appears, select the **Rotation** option instead of picking the second axis endpoint. Finally, enter the angle at which the circle rotates from the line of sight, such as 30 for a 30° rotation. Figure 4-14 shows examples of rotation angles.

NOTE

The **Rotation** option works with the **Center** and **Axis, End** options. A 0 response draws an ellipse with the minor axis equal to the major axis, which is a circle. AutoCAD rejects values of 90° and 270°.

Exercise 4-5

www.g-wlearning.com/CAD/

Complete the exercise on the companion website.

Drawing Elliptical Arcs

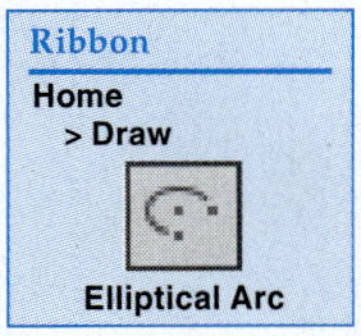

Use the **Arc** option of the **ELLIPSE** command to draw elliptical arcs. Drawing an elliptical arc is just like drawing an ellipse, but with two additional steps that define the beginning and end of the elliptical arc. Several options are available for defining the size and shape of an elliptical arc.

The default elliptical arc option is similar to the **Axis, End** ellipse option. Specify the first endpoint of an axis, then the second endpoint of the same axis, and then one endpoint of the second axis. Finally, select the start and end angles for the elliptical arc.

Figure 4-14.
The relationship among several ellipses having the same major axis length but different rotation angles.

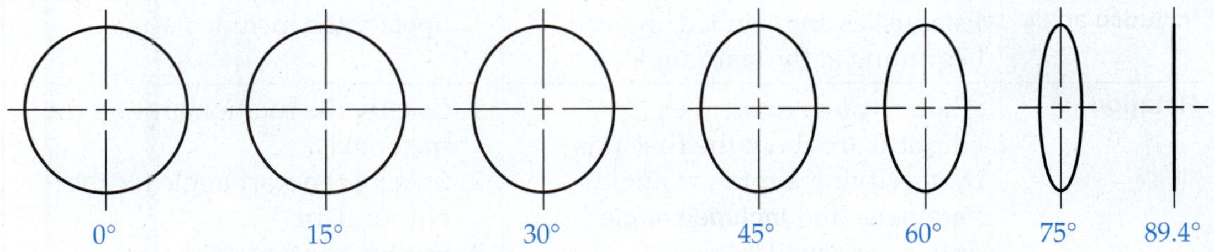

See Figure 4-15. The start and end angles are the angular relationships between the center of the ellipse and the arc endpoints. The angle of the first axis establishes the angle of the elliptical arc. For example, a 0° start angle begins the arc at the first endpoint of the first axis. A 45° start angle begins the arc 45° counterclockwise from the first endpoint of the first axis. End angles are also calculated counterclockwise from the start point.

Figure 4-16 describes additional elliptical arc options. Use the **Center** option when appropriate instead of the default axis endpoint method. The **Parameter, Included angle,** and **Rotation** options are available when you create axis endpoint or center elliptical arcs.

Exercise 4-6

www.g-wlearning.com/CAD/

Complete the exercise on the companion website.

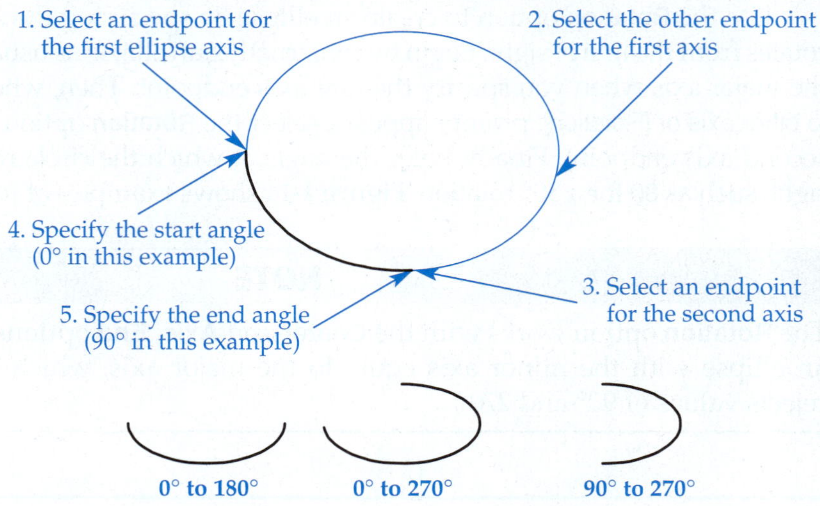

Figure 4-15. The steps required to draw an elliptical arc using a 0° start angle and a 90° end angle. The three examples at the bottom were created using the same steps but with different start and end angles.

Figure 4-16. Additional options for drawing elliptical arcs.

Option	Application	Process
Center	Lets you establish the center of the elliptical arc. **Rotation**, **Parameter**, and **Included angle** options are available.	1. Select the ellipse center point. 2. Select the endpoint of one of the ellipse axes. 3. Pick the endpoint of the other axis to form the ellipse. 4. Enter the start angle for the elliptical arc. 5. Select the end angle.
Parameter	Use instead of picking the start angle of the elliptical arc. AutoCAD uses a different means of vector calculation to create the elliptical arc.	1. Specify the start parameter point. 2. Specify the end parameter point.
Included angle	Establishes an included angle beginning at the start angle.	1. Specify the included angle.
Rotation	Allows you to rotate the elliptical arc about the first axis by specifying a rotation angle. **Parameter** and **Included angle** options are available.	1. Specify the rotation around the major axis. 2. Specify the start angle for the elliptical arc. 3. Specify the end angle.

Supplemental Material

www.g-wlearning.com/CAD/

Isometric Circles and Arcs

For information about using the **Isocircle** option of the **ELLIPSE** command to draw isometric circles and arcs, go to the companion website, navigate to this chapter in the **Contents** tab, and select **Isometric Circles and Arcs**.

Drawing Polylines

Ribbon	
Home > Draw	Polyline
Type	PLINE, PL

PLINE

Use the **PLINE** command to draw ***polylines***. When you use the default polyline settings, drawing polyline segments is identical to drawing line segments using the **LINE** command. Access the **PLINE** command and use point entry or drawing aids to locate polyline endpoints. Press [Enter] to exit. The difference between a polyline and a line is that all of the segments of a polyline act as a single object. The **PLINE** command also provides more flexibility than the **LINE** command, allowing you to draw a single object composed of straight lines and arcs of varying thickness.

polyline: A series of lines and arcs that constitute a single object.

The **PLINE** command includes the same **Undo** and **Close** options available with the **LINE** command. Use the **Undo** option to remove the last segment of a polyline and continue from the previous endpoint without leaving the **PLINE** command. You can use the **Undo** option repeatedly to delete polyline segments until the entire object is gone. Use the **Close** option to connect the endpoint of the last polyline segment to the start point of the first polyline segment.

Setting Polyline Width

The default polyline settings create a polyline with a constant width of 0. A polyline with a constant width of 0 is similar to a standard line and accepts the lineweight applied to the layer on which the polyline is drawn. Chapter 5 explains layers. Adjust the polyline width to create thick or tapered polyline objects.

To change the width of a polyline segment, access the **PLINE** command, pick the first point, and select the **Width** option. AutoCAD prompts you to specify the starting width of the line, followed by the ending width of the line. Enter the same starting and ending width value to draw a polyline with constant width. See Figure 4-17A. The rubberband line from the first point reflects the width settings. The location of the start point and endpoint is at the center of the segment width.

To create a tapered line segment for applications such as an arrowhead, enter different values for the starting and ending widths. See Figure 4-17B. To draw an

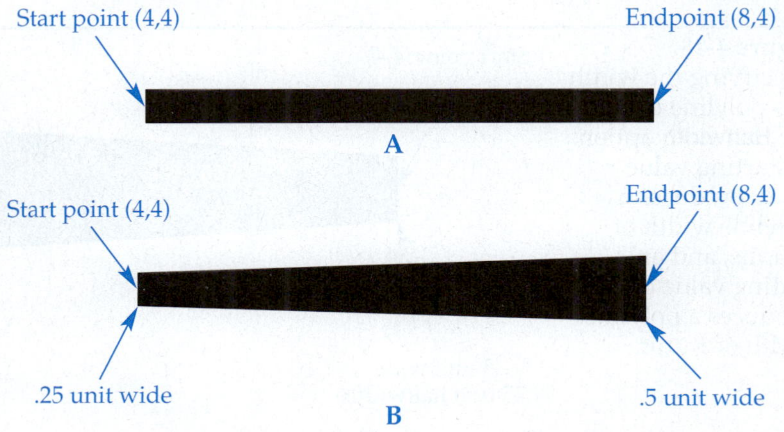

Figure 4-17. A—A thick polyline drawn using the **Width** option of the **PLINE** command. B—Using the **Width** option of the **PLINE** command to draw a tapered polyline.

arrowhead with a sharp point, use the **Width** option and specify 0 as the starting or ending width and use an appropriate value greater than 0 for the opposite width.

NOTE

A starting or ending width value other than 0 overrides the lineweight, typically assigned to the layer, you apply to the polyline.

Setting Polyline Halfwidth

Select the **Halfwidth** option to specify the width of the polyline from the center to one side, as opposed to the total width of the polyline defined using the **Width** option. Access the **PLINE** command, pick the first polyline endpoint, and then select the **Halfwidth** option. Specify starting and ending values at the appropriate prompts. Figure 4-18 shows a polyline drawn using the **Halfwidth** option and the same width values applied in Figure 4-17B, resulting in a polyline that is twice as wide.

NOTE

REGEN FILL

Type **FILL**

Type **REGEN**

All polyline objects with width—including polylines, polygons drawn using the **POLYGON** command, rectangles drawn using the **RECTANGLE** command, and donuts—can appear filled or empty. The **Apply solid fill** setting in the **Display performance** area of the **Display** tab in the **Options** dialog box controls the appearance. You can also enter FILL and use the **On** or **Off** option. Polyline objects are filled by default. The fill display for previously drawn polyline objects updates when the drawing regenerates. Use the **REGEN** command to regenerate the drawing manually.

Length Option

The **Length** option allows you to draw a polyline parallel to a previously drawn line or polyline. After you draw a line or polyline, access the **PLINE** command and pick a start point. Select the **Length** option and specify the length. The resulting polyline is automatically drawn parallel to the previous line or polyline using the specified length.

Exercises 4-7 and 4-8

www.g-wlearning.com/CAD/

Complete the exercises on the companion website.

Figure 4-18.
Specifying the width of a polyline using the **Halfwidth** option. A starting value of .25 produces a polyline width of .5 units, and an ending value of .5 produces a polyline width of 1 unit.

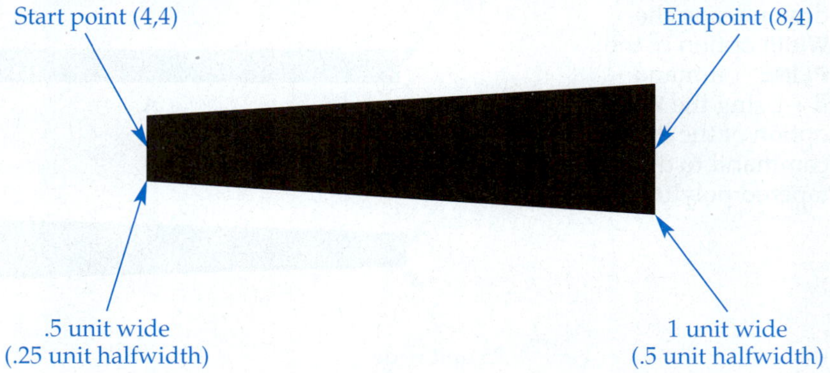

Drawing Polyline Arcs

Use the **Arc** option to draw polyline arcs. Polyline arcs can continue from or to polyline segments drawn during the same operation to form a single object. You can use the **Width** or **Halfwidth** option to add width to a polyline arc, ranging from 0 to the radius of the arc. You can also set different starting and ending arc widths. See Figure 4-19. Enter the **Width** or **Halfwidth** and **Arc** options in either order. Use the **Line** option to return to the straight-line segment mode of the **PLINE** command.

In addition to the **Close**, **Undo**, **Width**, and **Halfwidth** options, the polyline **Arc** option includes functions for controlling the size and location of polyline arcs. Many of the polyline **Arc** options allow you to create polyline arcs using the same methods available for drawing arcs using the **ARC** command. Select the appropriate option and follow the prompts to create the polyline arc. Review the **ARC** command options described in Figure 4-20 to help recognize the function of similar polyline **Arc** options.

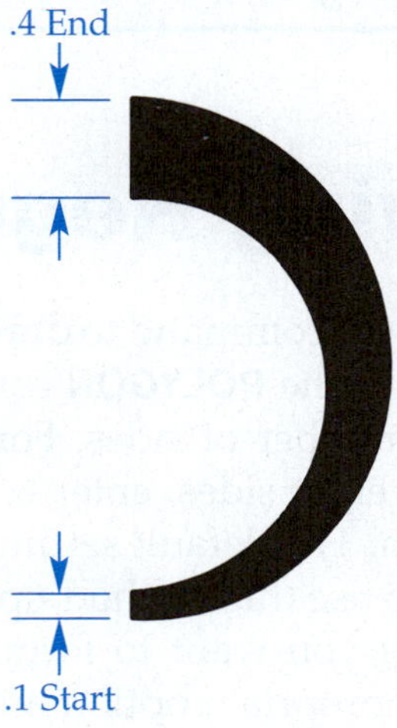

Figure 4-19.
An example of a polyline arc with different starting and ending widths.

Figure 4-20.
Additional options available for drawing polyline arcs.

Option	Application	Options for Completion
Angle	Specify the polyline arc size according to an included angle.	1. Specify an endpoint. 2. Use the **Center** option to select the center point. 3. Use the **Radius** option to enter the radius.
Center	Specify the location of the polyline arc center point, instead of allowing AutoCAD to calculate the location automatically.	1. Specify an endpoint. 2. Use the **Angle** option to specify the included angle. 3. Use the **Length** option to specify the chord length.
Direction	Alter the polyline arc bearing, or tangent direction, instead of allowing the polyline arc to form tangent to the last object drawn.	1. Specify an endpoint.
Radius	Specify the polyline arc radius.	1. Specify an endpoint. 2. Use the **Angle** option to specify the included angle.
Second point	Draw a three-point polyline arc.	1. Pick the second point, followed by the endpoint.

If the polyline arc is forming in the incorrect clockwise or counterclockwise direction, hold down [Ctrl] while the **Arc** option is active to reverse the direction. The values you specify, including the use of positive or negative numbers, also affects the result.

Exercise 4-9

www.g-wlearning.com/CAD/

Complete the exercise on the companion website.

Supplemental Material

www.g-wlearning.com/CAD/

Multilines

AutoCAD also includes commands for working with objects made up of multiple lines. For information about drawing and editing multiline objects, go to the companion website, navigate to this chapter in the **Contents** tab, and select **Multilines**.

Drawing Regular Polygons

POLYGON

Ribbon
Home
> Draw
Polygon
Type
POLYGON
POL

regular polygon: A closed geometric figure with three or more equal sides and equal angles.

inscribed polygon: A polygon drawn inside an imaginary circle so that the corners of the polygon touch the circle.

circumscribed polygon: A polygon drawn outside an imaginary circle so that the sides of the polygon are tangent to the circle.

Access the **POLYGON** command to draw any ***regular polygon*** with up to 1024 sides. A polygon drawn using the **POLYGON** command is a single polyline object. The first prompt asks for the number of sides. For example, to draw an octagon, which is a regular polygon with eight sides, enter 8. Next, decide how to describe the size and location of the polygon. The default setting involves selecting the center and radius of an imaginary circle. To use this method, specify a location for the polygon center point. A prompt then asks if you want to form an ***inscribed polygon*** or a ***circumscribed polygon***. Select the appropriate option and specify the radius to create the polygon. See Figure 4-21.

NOTE

The number of polygon sides you enter, the **Inscribed in circle** or **Circumscribed about circle** option you select, and the radius you specify are stored as the new default settings, allowing you to draw another polygon quickly with the same characteristics.

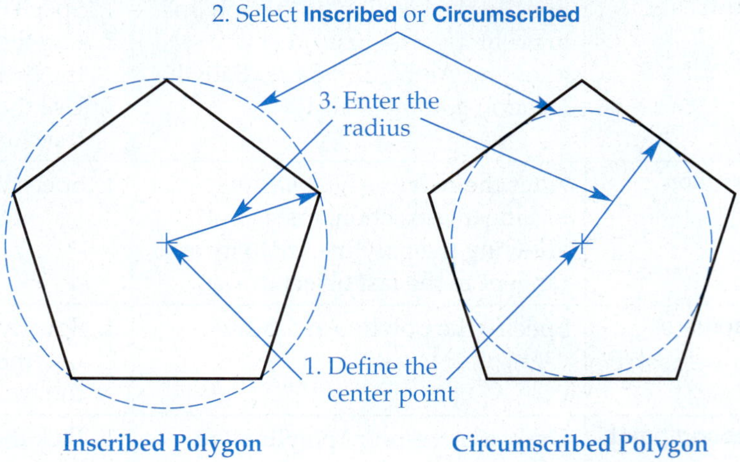

Figure 4-21. Regular polygons, such as these pentagons, can be inscribed in a circle or circumscribed around a circle.

PROFESSIONAL TIP

Regular polygons, such as the ***hexagons*** commonly drawn to represent bolt heads and nuts on mechanical drawings, are normally dimensioned across the flats. Use the **Circumscribed about circle** option to draw a polygon dimensioned across the flats. The radius you enter is equal to one-half the distance across the flats. Use the **Inscribed in circle** option to draw a polygon dimensioned across the corners or to confine the polygon within a circular area.

hexagon: A six-sided regular polygon.

Edge Option

Use the **Edge** option to construct a polygon if you do not know the location of the center point or the radius of the imaginary circle, but you do know the size and location of a polygon edge. After you access the **POLYGON** command and enter the number of sides, select the **Edge** option at the Specify center of polygon or [Edge]: prompt. Specify a point for the first endpoint of one side, followed by the second endpoint of the side. See Figure 4-22.

Exercise 4-10

www.g-wlearning.com/CAD/

Complete the exercise on the companion website.

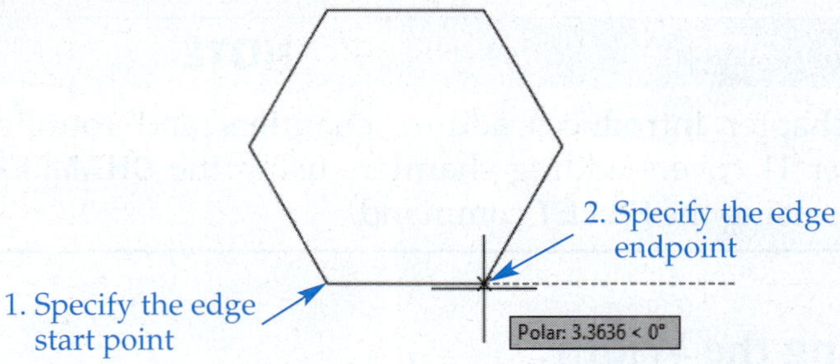

Figure 4-22. Use the **Edge** option of the **POLYGON** command to construct a regular polygon according to the location and size of an edge.

Drawing Rectangles

Use the **RECTANGLE** command to draw rectangles easily. A rectangle drawn using the **RECTANGLE** command is a single polyline object. To draw a rectangle using default settings, specify the point of one corner, followed by the point of the diagonally opposite corner. See Figure 4-23. By default, the **RECTANGLE** command draws a rectangle at a 0° angle with sharp corners.

RECTANGLE
Ribbon
Home
> Draw
Rectangle
Type
RECTANGLE
RECTANG
REC

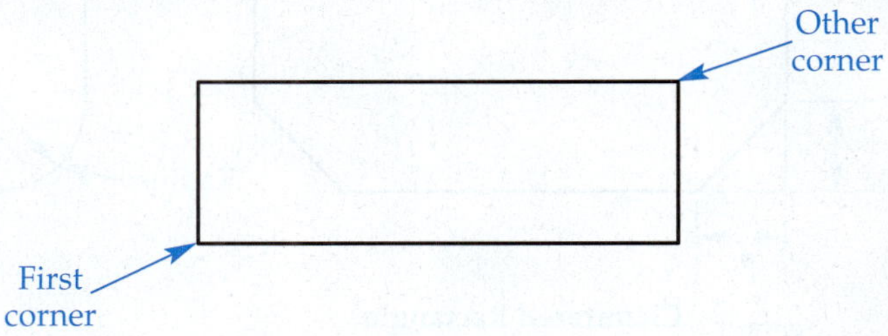

Figure 4-23. Using the **RECTANGLE** command and point entry or drawing aids to construct a rectangle.

Adding Chamfered Corners

chamfer: In mechanical drafting, a small, angled surface used to relieve a sharp corner.

Use the **Chamfer** option to include angled corners during rectangle construction. See Figure 4-24A. When prompted, enter the first *chamfer* distance, followed by the second chamfer distance. Entering 0 at the first or second chamfer distance prompt creates a rectangle with sharp corners. After setting the distances, you can either draw the rectangle or set additional options. However, using the **Fillet** option overrides the **Chamfer** option.

The rectangle you draw must be large enough to accommodate the specified chamfer distances. Otherwise, the rectangle will have sharp corners. New rectangles are drawn with the specified chamfer until you reset the chamfer distances to 0 or use the **Fillet** option to create rounded corners.

Adding Rounded Corners

fillet: A rounded interior corner.

round: A rounded exterior corner.

Use the **Fillet** option to include rounded corners during rectangle construction. See Figure 4-24B. AutoCAD uses the term *fillet* to describe both ***fillets*** and ***rounds***. When prompted, enter the radius for all fillets and rounds. Entering a radius of 0 creates a rectangle with sharp corners. After setting the radius, you can either draw the rectangle or set additional options. However, using the **Chamfer** option overrides the **Fillet** option.

The rectangle you draw must be large enough to accommodate the specified fillet radii. Otherwise, the rectangle will have sharp corners. New rectangles are drawn with the specified fillets until you reset the fillet radius to 0 or use the **Chamfer** option to create chamfered corners.

NOTE

This chapter introduces adding chamfers and rounds while creating rectangles. Chapter 11 covers adding chamfers using the **CHAMFER** command, and fillets and rounds using the **FILLET** command.

Setting the Width

The default rectangle settings create a polyline object with a constant width of 0. Select the **Width** option to adjust rectangle line width. As previously explained, do not confuse width with lineweight, described in Chapter 5. AutoCAD prompts you to

Figure 4-24.
A—Use the **Chamfer** option of the **RECTANGLE** command to add chamfers to rectangle corners during construction. B—Use the **Fillet** option of the **RECTANGLE** command to add rounds to rectangle corners during construction.

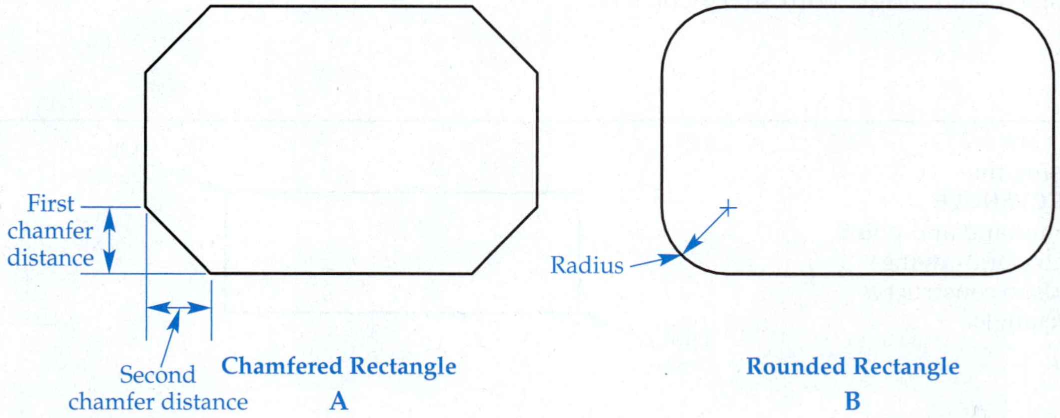

specify the line width. For example, to draw a rectangle with sides one-half unit wide, enter .5. After setting the rectangle width, you can either draw the rectangle or set additional options. All new rectangles are drawn using the specified width. Reset the **Width** option to 0 to create new rectangles using a standard "0-width" line.

Specifying the Area

The **Area** option is available after you pick the first corner point and is useful for drawing a rectangle when you know the area of the rectangle and the length of one side. Select the **Area** option, and then specify the total area for the rectangle using a value that corresponds to the current units. For example, enter 45 to draw a rectangle with an area of 45 units. Next, select the **Length** option if you know the length of a side (the X value), or select the **Width** option if you know the width of a side (the Y value). When prompted, enter the length or width to complete the rectangle. AutoCAD calculates the unspecified dimension and draws the rectangle.

Specifying Rectangle Dimensions

The **Dimensions** option is available after you pick the first corner of the rectangle and allows you to specify the length and width of the rectangle. Select the **Dimensions** option and specify the length of a side to indicate the X value. Next, enter the width of a side to indicate the Y value. AutoCAD then prompts for the other corner point. To change the dimensions, select the **Dimensions** option again. If the dimensions are correct, specify another point to complete the rectangle. The second point determines which of four possible rectangles you draw. See **Figure 4-25**.

Drawing a Rotated Rectangle

The **Rotation** option is available after you pick the first corner of the rectangle and allows you to draw a rectangle at an angle other than 0°. When prompted, specify the angle to rotate the rectangle from the default of 0°. Then locate the opposite corner of the rectangle. An alternative is to use the **Pick points** option at the Specify rotation angle or [Pick points]: prompt. If you select the **Pick points** option, the prompt asks you to select two points to define the angle. The rotation value becomes the new default angle for using the **Rotation** option.

Figure 4-25.
The second corner point, or quadrant, determines the orientation of the rectangle relative to the first corner point.

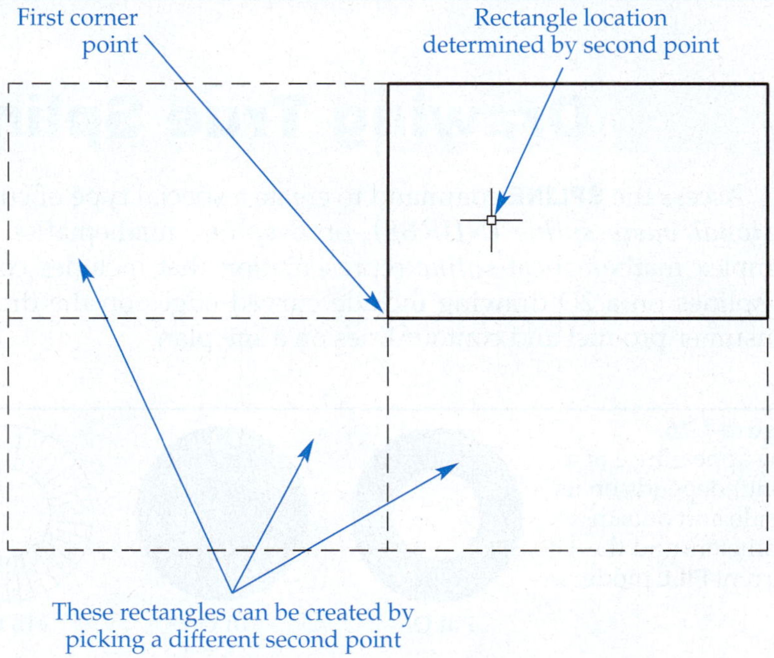

NOTE

The **Elevation** and **Thickness** options of the **RECTANGLE** command are appropriate for 3D applications. ***AutoCAD and Its Applications—Advanced*** explains 3D drawing applications.

PROFESSIONAL TIP

You can use a combination of rectangle settings to draw a single rectangle. For example, you can enter a width value, chamfer distances, and length and width dimensions to create a rectangle.

Exercise 4-11

www.g-wlearning.com/CAD/

Complete the exercise on the companion website.

Drawing Donuts and Filled Circles

DONUT

Ribbon
Home
> Draw
Donut
Type
DONUT
DO

The **DONUT** command allows you to draw a thick or filled circle. See Figure 4-26. A donut is a single polyline object. After activating the **DONUT** command, enter the inside diameter and then the outside diameter of the donut. Enter a value of 0 for the inside diameter to create a closed donut.

The center point of the donut attaches to the crosshairs, and the Specify center of donut or <exit>: prompt appears. Pick a location to place the donut. The **DONUT** command remains active until you press [Enter] or [Esc]. This allows you to place multiple donuts of the same size using a single instance of the **DONUT** command.

non-uniform rational basis spline (NURBS, B-spline): The mathematics used by most surface modeling CADD systems to produce accurate curves and surfaces.

spline: A curve that uses a series of control points and other mathematical principles to define the location and form of the curve.

Exercise 4-12

www.g-wlearning.com/CAD/

Complete the exercise on the companion website.

Drawing True Splines

SPLINE

Ribbon
Home
> Draw
Spline Fit
Type
SPLINE
CURVE
B-SPLINE
SPL

Access the **SPLINE** command to create a special type of curve using ***non-uniform rational basis spline (NURBS)***, or ***B-spline***, mathematics. A NURBS curve is a complex mathematical ***spline*** representation that includes control points. Examples of splines on a 2D drawing include curved edges on the drawing of an ergonomic consumer product and contour lines on a site plan.

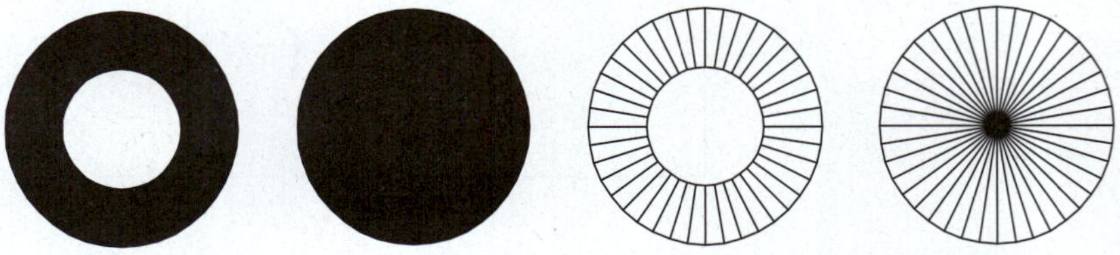

Figure 4-26. The appearance of a donut depends on its inside and outside diameters and the current **FILL** mode.

To draw a default spline, specify fit points using point entry or drawing aids. By default, the spline forms, or fits, through the points. When you finish locating points, press [Enter] to create the spline and end the command. Figure 4-27 shows a default spline drawn using absolute coordinates. Use the **Undo** option to remove the last segment of a spline without leaving the **SPLINE** command.

NOTE

If you specify only two points for a spline curve and use default settings, an object that looks like a line forms, but the object is a spline.

Drawing Closed Splines

Use the **Close** option after locating at least three control points to connect the last point to the first point. See Figure 4-28. The **Close** option forms a smooth curve by creating what is known as a "periodic spline with C2 geometric continuity."

Exercise 4-13

www.g-wlearning.com/CAD/

Complete the exercise on the companion website.

Supplemental Material

www.g-wlearning.com/CAD/

Spline Options

For information about options available for drawing splines, go to the companion website, navigate to this chapter in the **Contents** tab, and select **Spline Options**.

Figure 4-27.
A spline drawn using the default settings of the **SPLINE** command and three fit points.

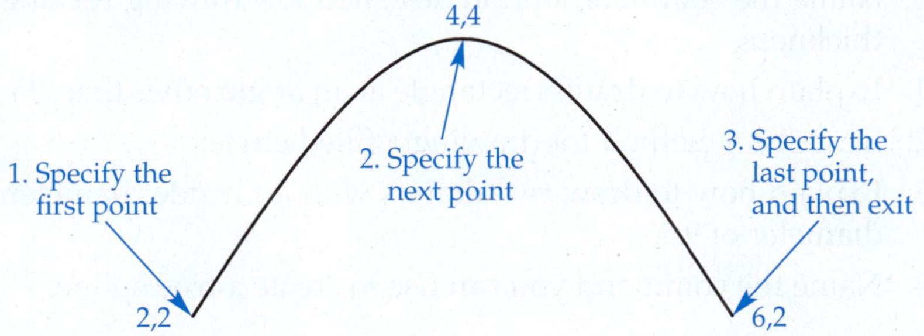

Figure 4-28.
Using the **Close** option of the **SPLINE** command with AutoCAD default tangents to draw a closed spline. Compare this spline to the spline shown in Figure 4-27.

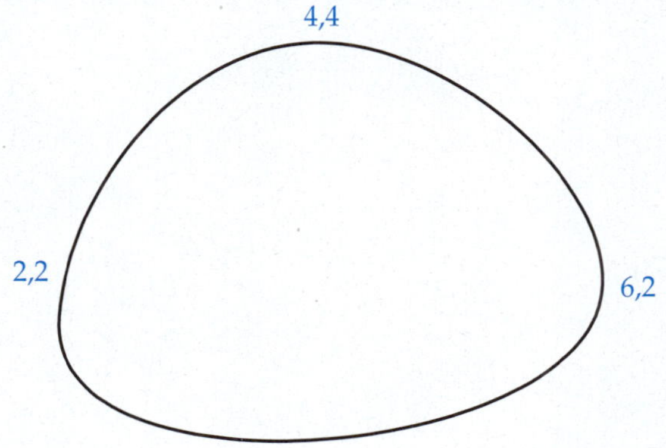

Chapter Review

Answer the following questions using the information in this chapter.

1. When you use the **Center, Radius** option of the **CIRCLE** command, what are the options for specifying the radius?
2. Explain how to create a circle with a diameter of 2.5 units.
3. What option of the **CIRCLE** command creates a circle of a specific radius that is tangent to two existing objects?
4. Define the term *point of tangency*.
5. What option of the **CIRCLE** command draws a circle tangent to three objects?
6. Explain the procedure to draw an arc beginning with the center point and having a 60° included angle.
7. Define the term *included angle* as it applies to an arc.
8. What is the default option if you enter the **ARC** command at the keyboard?
9. List two input options that you can use to draw an arc tangent to the endpoint of a previously drawn arc.
10. Name the two axes found on an ellipse.
11. Describe the procedure to draw an ellipse using the **Axis, End** option.
12. What **ELLIPSE** rotation angle results in a circle?
13. How do you draw an arrowhead with a sharp point using the **PLINE** command?
14. Which **PLINE** option allows you to specify the width from the center to one side?
15. Explain how to turn off the fill display for polyline objects.
16. Describe how to create a polyline parallel to a previously drawn line or polyline.
17. Explain how to draw a hexagon measuring 4″ (102 mm) across the flats.
18. Name at least three commands you could use to create a rectangle.
19. Name the command option used to draw rectangles with rounded corners.
20. Name the command option designed for drawing rectangles with a specific line thickness.
21. Explain how to draw a rectangle at an angle other than 0°.
22. Describe a method for drawing a filled circle.
23. Explain how to draw two donuts with an inside diameter of 6.25 and an outside diameter of 9.50.
24. Name the command you can use to create a true spline.

Drawing Problems

Start AutoCAD if it is not already started. Start a new drawing from scratch or use an appropriate template of your choice. Follow the specific instructions for each problem. Use only drawing commands and techniques you have already learned. Do not draw dimensions or text. Use your own judgment and approximate dimensions when necessary.

▼ Basic

1. **General** Use the **LINE**, **CIRCLE**, and **RECTANGLE** commands to draw the objects shown. Save the drawing as P4-1.

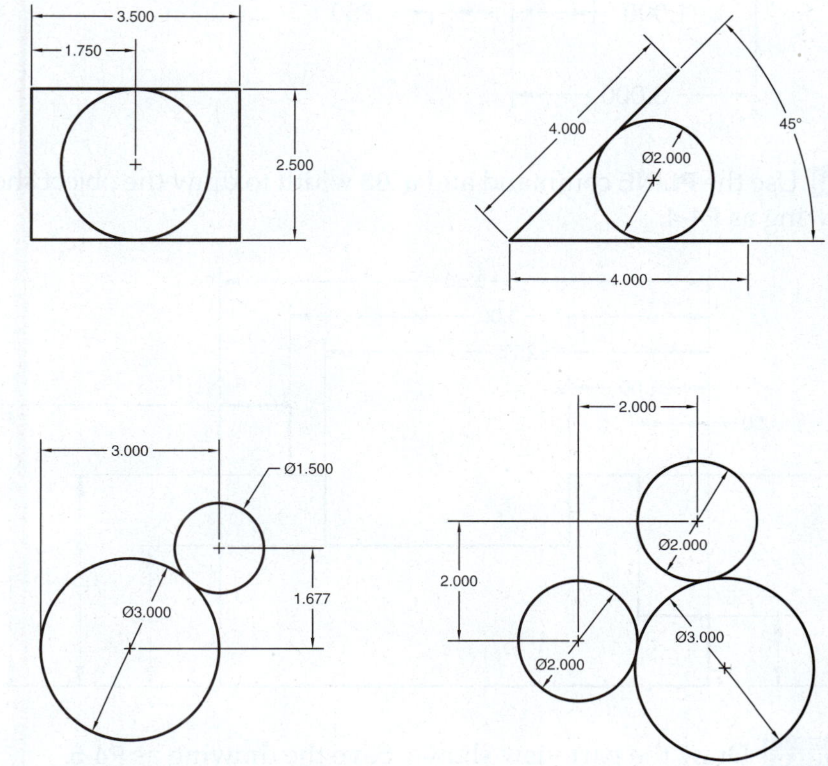

2. **Mechanical** Use the **CIRCLE** and **ARC** commands to draw the object shown. Save the drawing as P4-2.

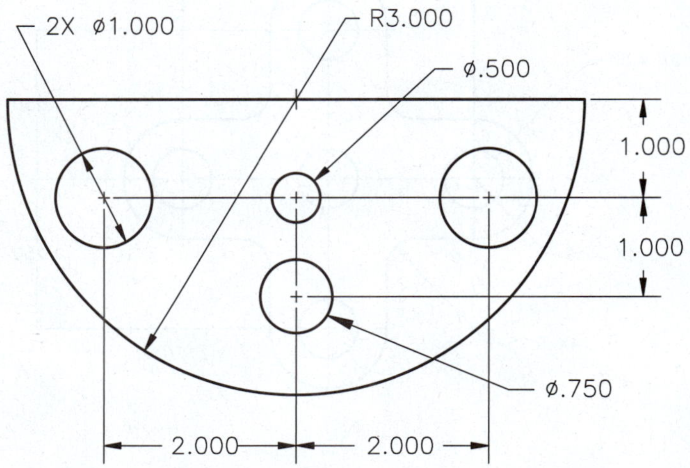

3. **Mechanical** Draw the spacer shown. Save the drawing as P4-3.

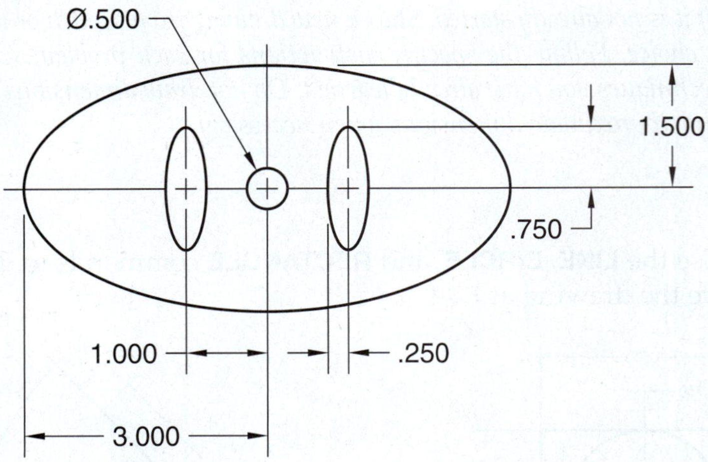

4. **General** Use the **PLINE** command and a .03 width to draw the object shown. Save the drawing as P4-4.

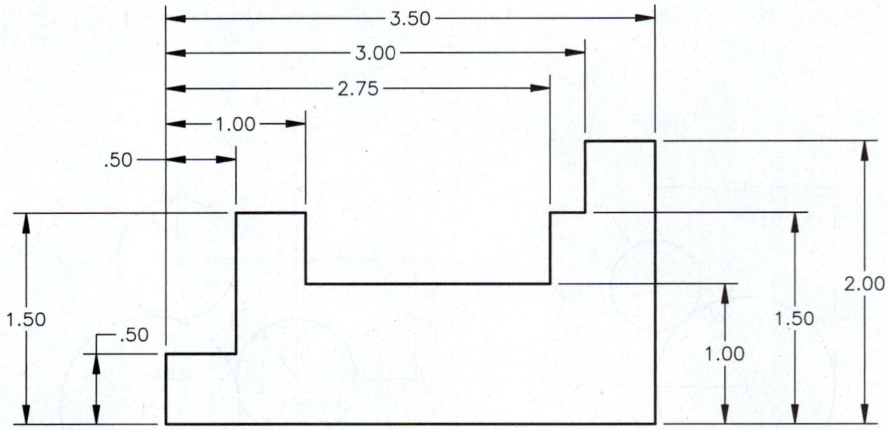

5. **Mechanical** Draw the part view shown. Save the drawing as P4-5.

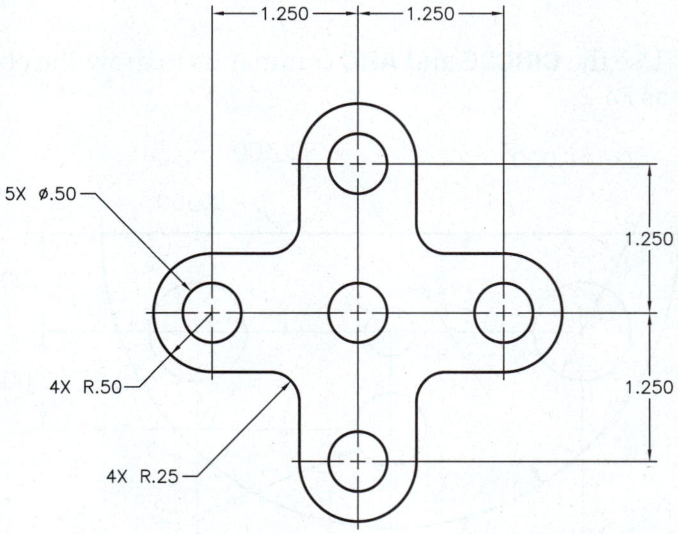

Drawing Problems - Chapter 4

6. **Mechanical** Draw the part view shown. Save the drawing as P4-6.

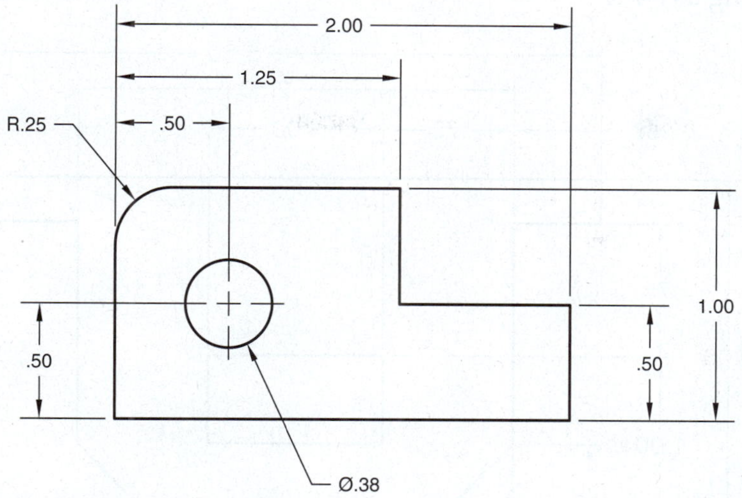

7. **Mechanical** Draw the pipe spacer shown. Save the drawing as P4-7.

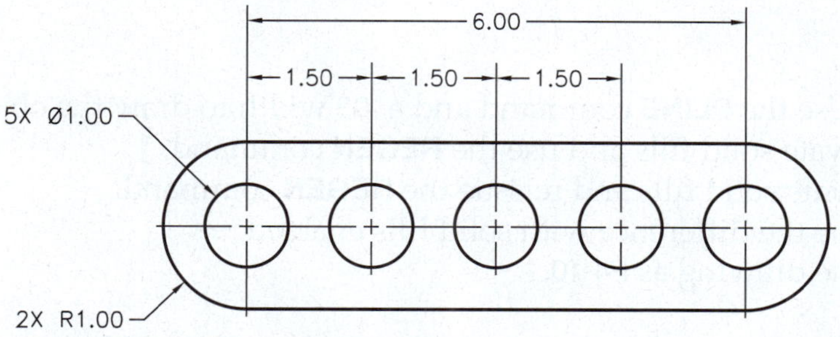

8. **Mechanical** Draw the part view shown. Save the drawing as P4-8.

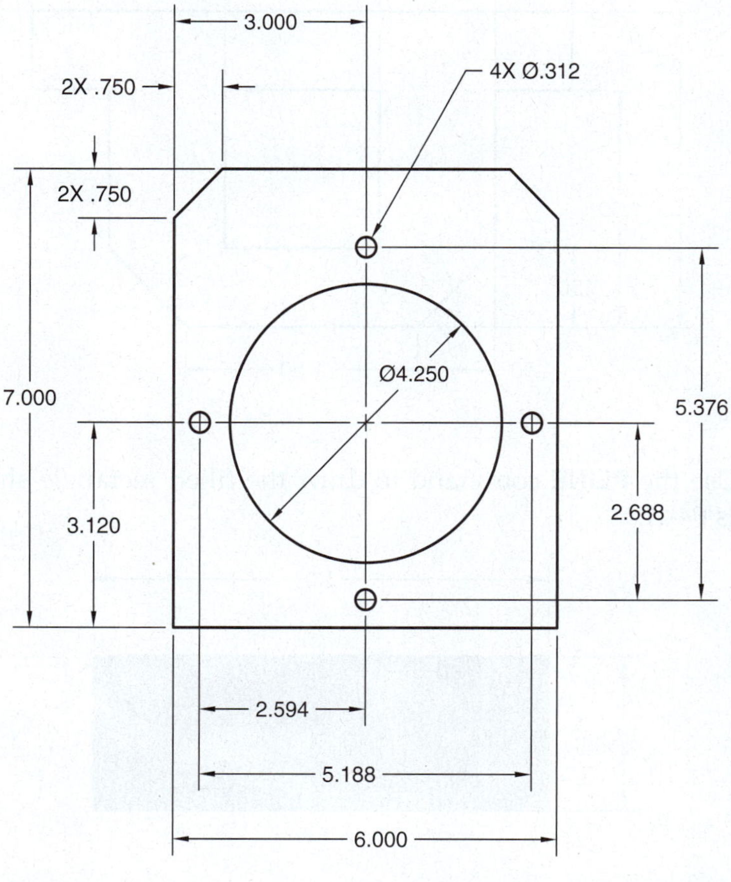

Drawing Problems - Chapter 4

9. **General** Use the **PLINE** command and a .03 width to draw the object shown. Save the drawing as P4-9.

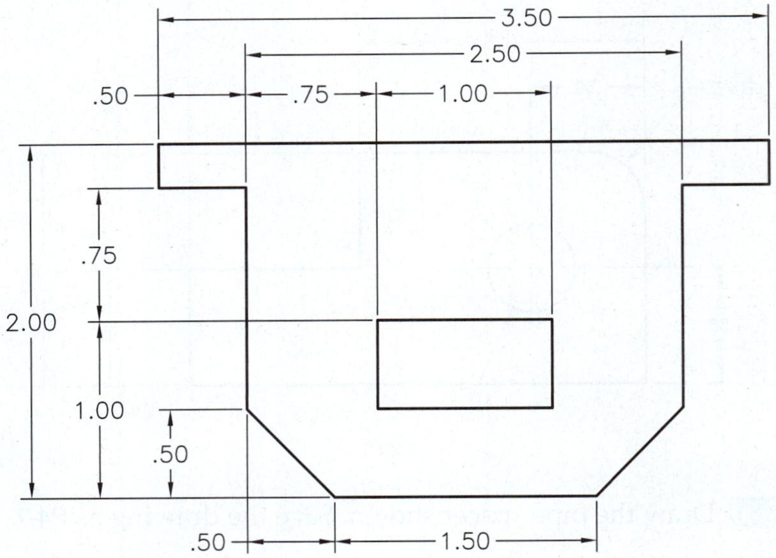

10. **General** Use the **PLINE** command and a .03 width to draw the object shown.
    A. Deactivate solid fills and use the **REGEN** command.
    B. Reactivate solid fills and reissue the **REGEN** command.
    C. Observe the difference with solid fills enabled.
    D. Save the drawing as P4-10.

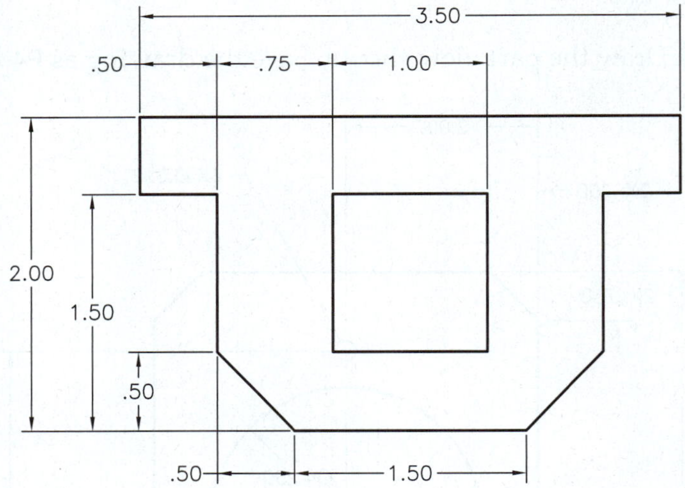

11. **General** Use the **PLINE** command to draw the filled rectangle shown. Save the drawing as P4-11.

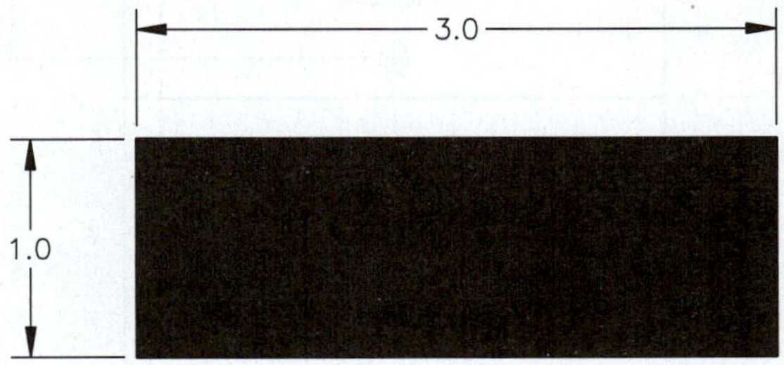

12. **General** Use the **PLINE** command to draw the arrowheads shown. Save the drawing as P4-12.

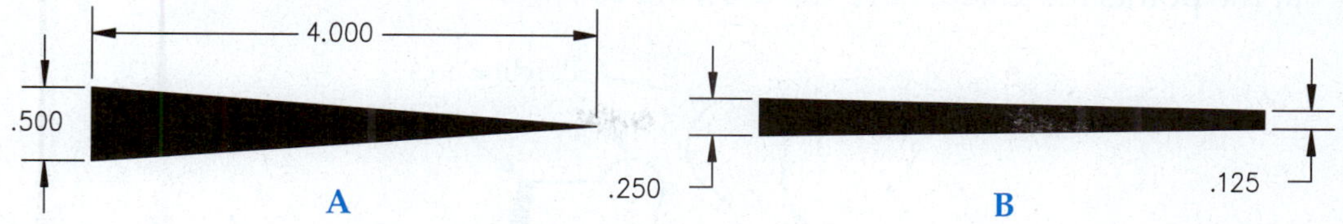

13. **General** Use the **PLINE** command to draw the arrow shown. Set decimal units, .25 grid spacing, .0625 snap spacing, and limits of 11,8.5. Save the drawing as P4-13.

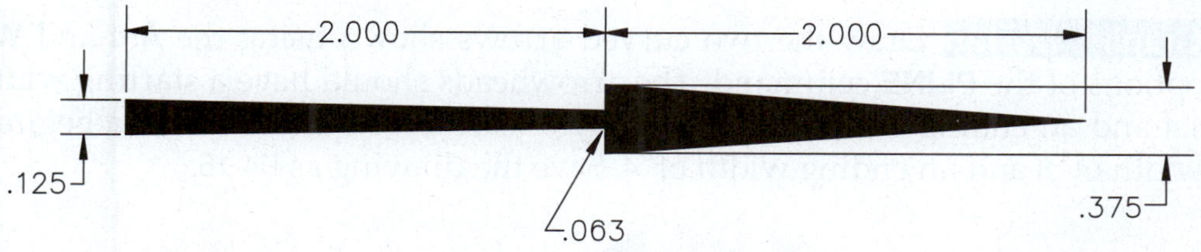

▼ Intermediate

14. **Architectural** Use the **RECTANGLE** and **CIRCLE** commands to draw the single kitchen sink shown. Save the drawing as P4-14.

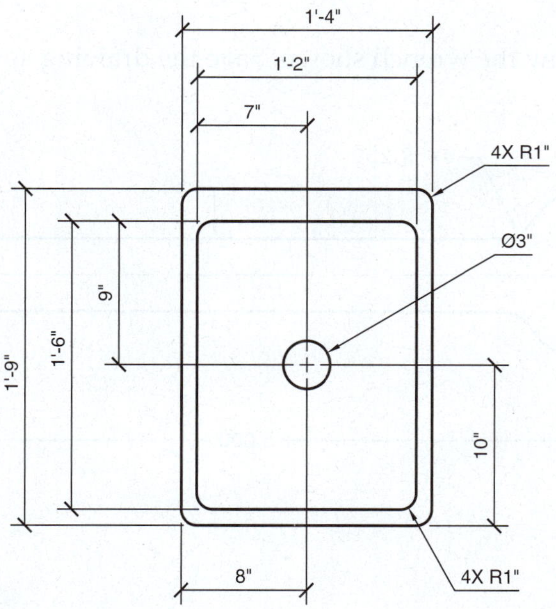

Drawing Problems - Chapter 4

15. **General** Draw the single polyline shown. Use the **Arc**, **Width**, and **Close** options of the **PLINE** command to complete the shape. Set the polyline width to 0, except at the points indicated. Save the drawing as P4-15.

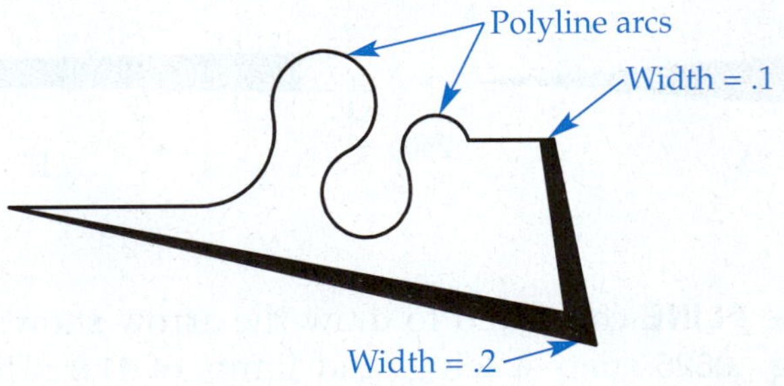

16. **Graphic Design** Draw the two curved arrows shown using the **Arc** and **Width** options of the **PLINE** command. The arrowheads should have a starting width of 1.4 and an ending width of 0. The body of each arrow should have a beginning width of .8 and an ending width of .4. Save the drawing as P4-16.

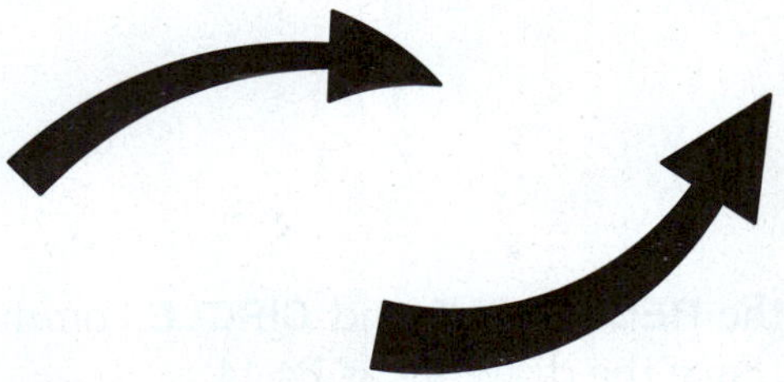

17. **Mechanical** Draw the wrench shown. Save the drawing as P4-17.

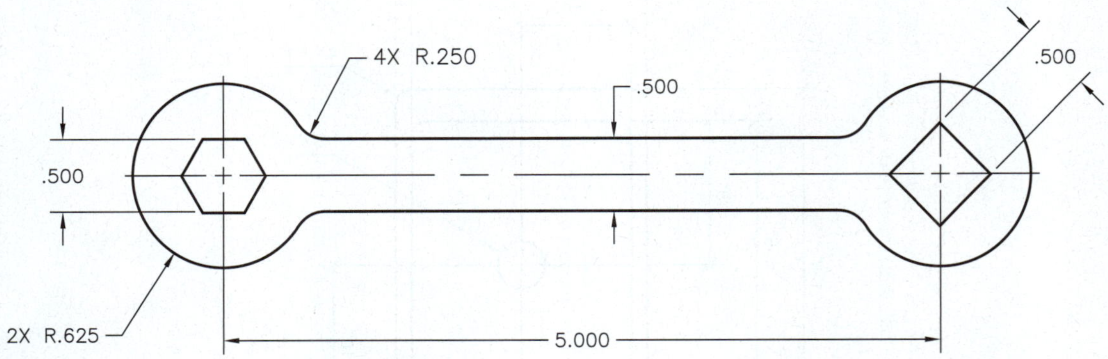

18. **Piping** Draw the pipe fitting shown. Save the drawing as P4-18.

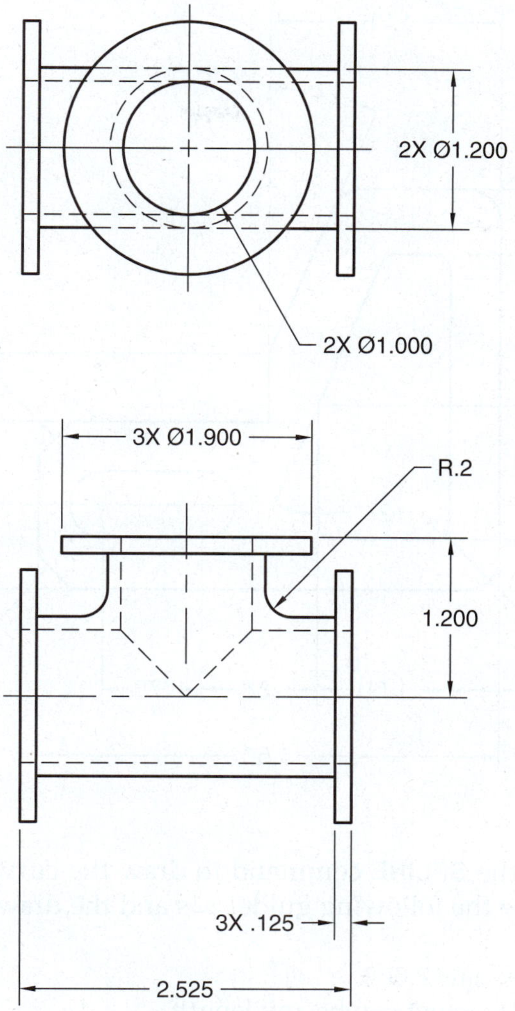

19. **General** Draw the ellipse template shown. Save the drawing as P4-19.

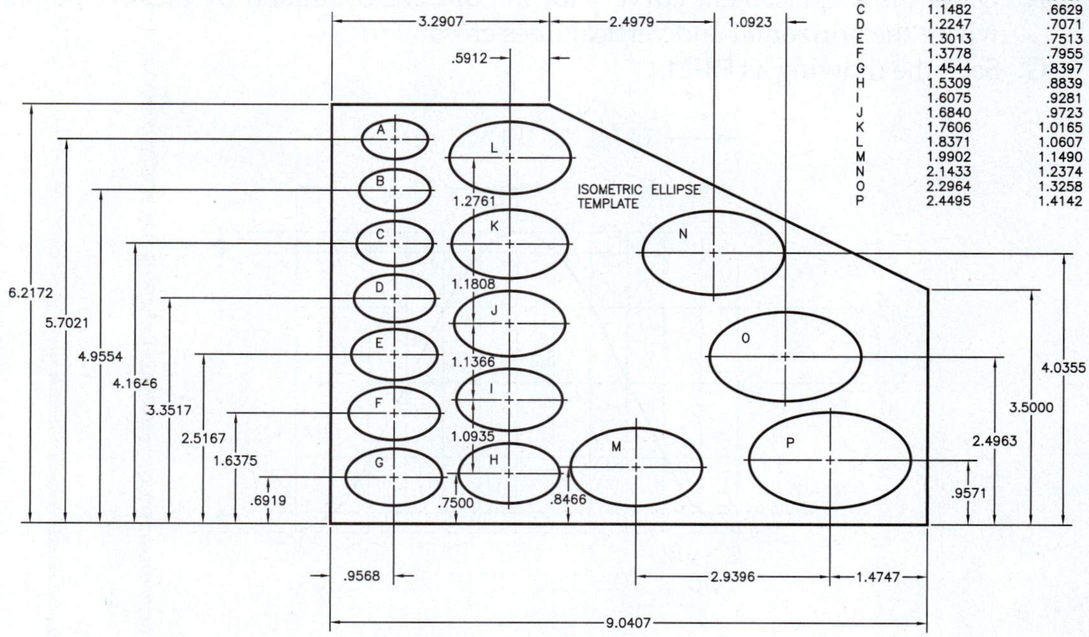

ELLIPSE TABLE

KEY	MAJOR DIA	MINOR DIA
A	.9951	.5745
B	1.0717	.6187
C	1.1482	.6629
D	1.2247	.7071
E	1.3013	.7513
F	1.3778	.7955
G	1.4544	.8397
H	1.5309	.8839
I	1.6075	.9281
J	1.6840	.9723
K	1.7606	1.0165
L	1.8371	1.0607
M	1.9902	1.1490
N	2.1433	1.2374
O	2.2964	1.3258
P	2.4495	1.4142

20. **Mechanical** Draw the gasket shown. Save the drawing as P4-20.

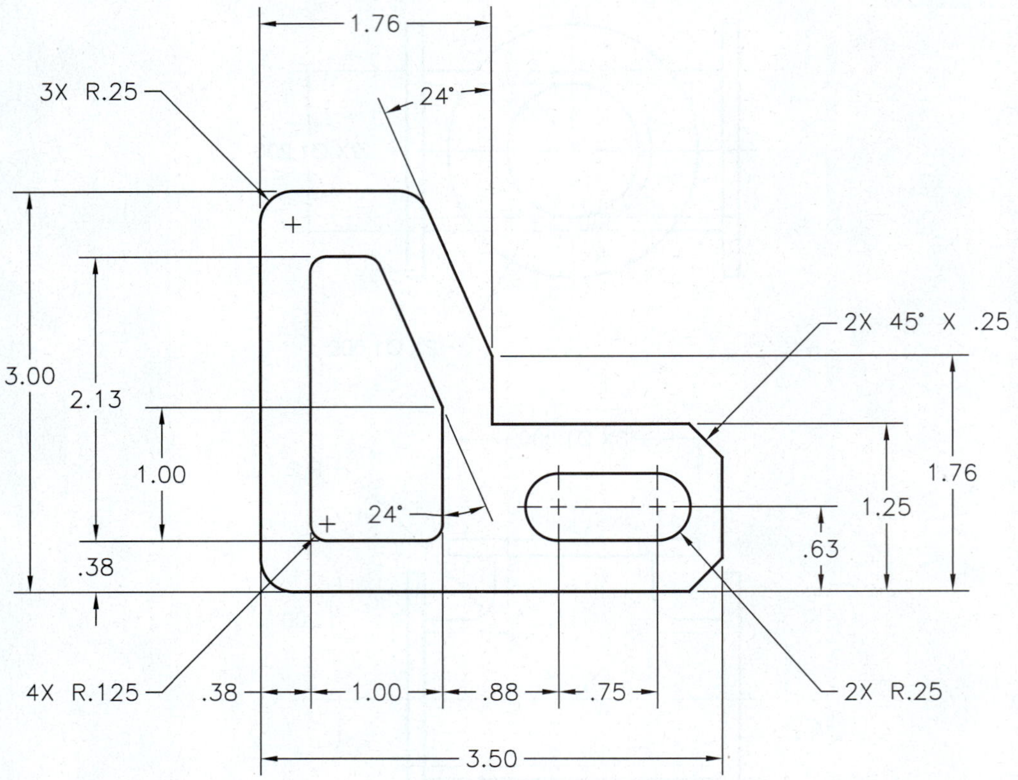

21. **Mechanical** Use the **SPLINE** command to draw the curve for the cam displacement diagram. Use the following guidelines and the drawing shown to complete this problem.

A. The total rise equals 2.000.
B. The total displacement can be any length.
C. Divide the total displacement into 30° increments.
D. Draw a half circle divided into 6 equal parts on one end.
E. Draw a horizontal line from each division of the half circle to the other end of the diagram.
F. Draw the displacement curve with the **SPLINE** command by picking points where the horizontal and vertical lines cross.
G. Save the drawing as P4-21.

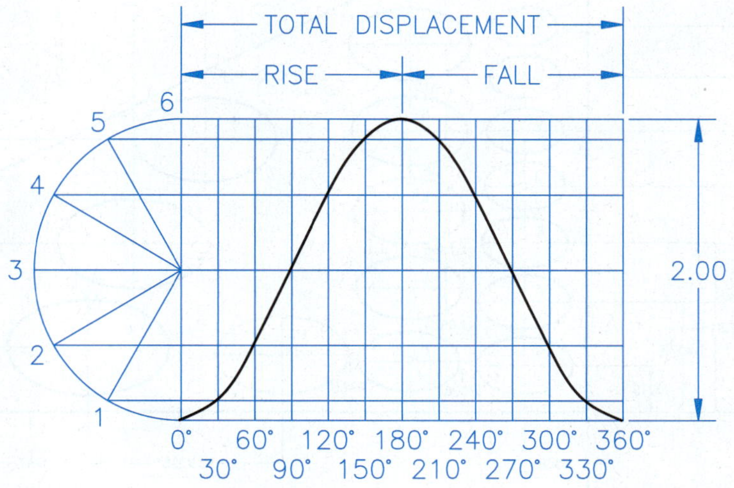

▼ Advanced

22. **Mechanical** Draw the part view shown. Save the drawing as P4-22.

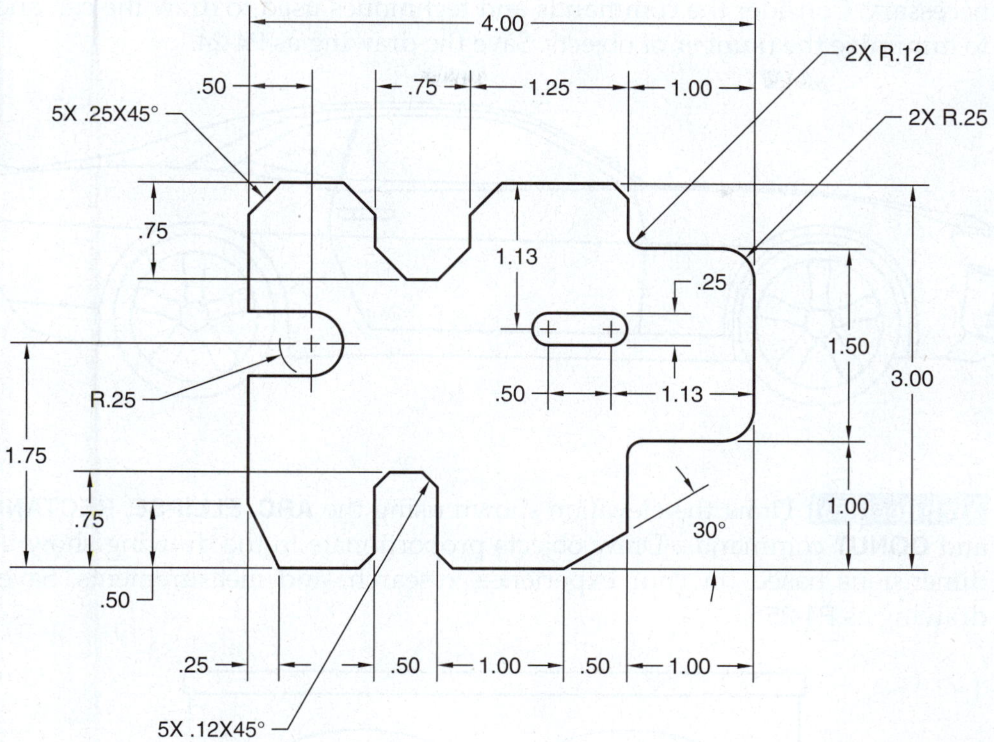

23. **Mechanical** Draw the part view shown. Save the drawing as P4-23.

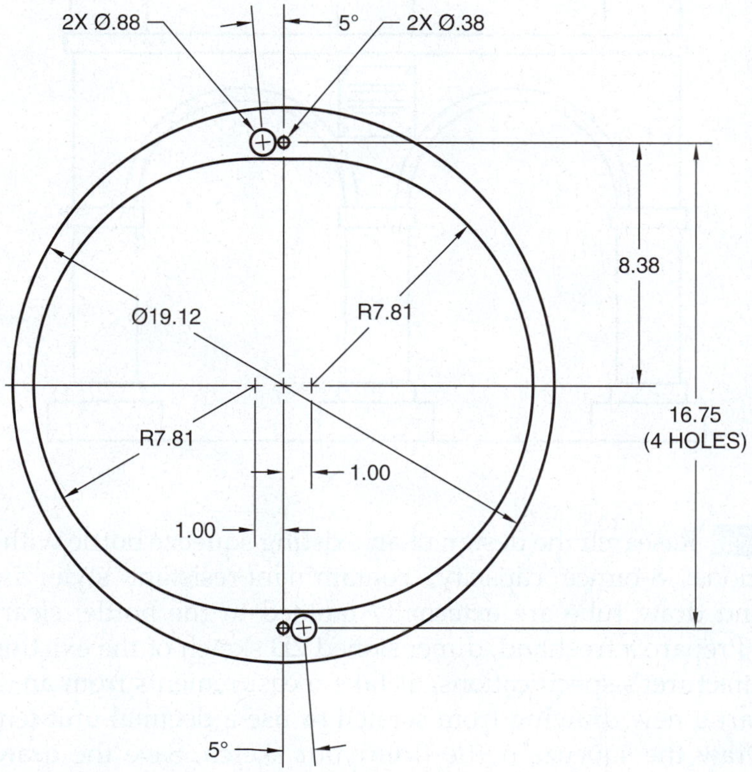

24. General Create a drawing from the sketch of a car design shown. Use the **LINE** command and selected shape commands to draw the car using appropriate size and scale features. Use a tape measure to measure an actual car for reference if necessary. Consider the commands and techniques used to draw the car, and try to minimize the number of objects. Save the drawing as P4-24.

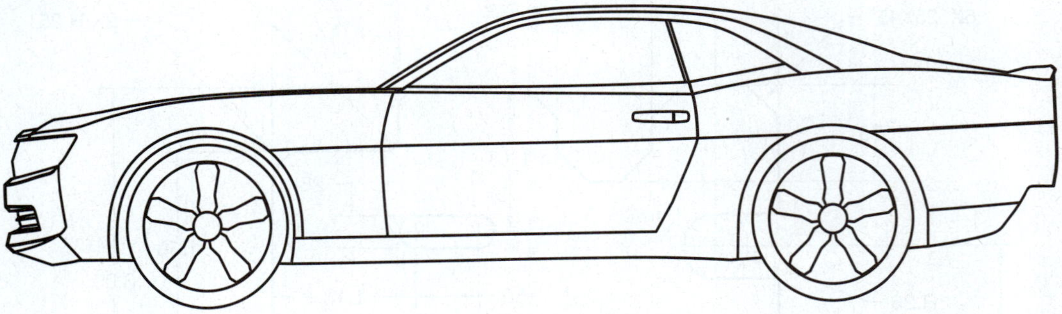

25. Architectural Draw the elevation shown using the **ARC**, **ELLIPSE**, **RECTANGLE**, and **DONUT** commands. Draw objects proportionate to the drawing shown. Use dimensions based on your experience, research, and measurements. Save the drawing as P4-25.

26. Mechanical Research the design of an existing squeeze bottle with the following specifications: 8-ounce capacity; contaminant-resistant style; integral spout, nozzle, and draw tube are externally molded to the bottle; clear polyethylene material. Prepare a freehand, dimensioned 2D sketch of the existing design from the manufacturer's specifications, or take measurements from an actual squeeze bottle. Start a new drawing from scratch or use a decimal-unit template of your choice. Draw the squeeze bottle from your sketch. Save the drawing as P4-26. Prepare a slide show presentation of your research, design process, and drawing and present it to your class or office.

27. **Architectural** Find a door at your school, company, or home that includes features with several different shapes. Prepare a freehand 2D sketch of the door elevation complete with casework and hardware. Use measuring devices such as a tape measure and caliper to dimension the size and location of door features accurately. Start a new drawing from scratch or use an architectural template of your choice. Draw the door from your sketch. Save the drawing as P4-27. Prepare a slide show presentation of your research, design process, and drawing and present it to your class or office.

AutoCAD Certified Professional Exam Practice

Answer the following questions using the information in this chapter.

1. Which command allows you to draw a rectangle? *Select all that apply.*
   A. **LINE**
   B. **PLINE**
   C. **RECTANGLE**
   D. **DONUT**
   E. **POLYGON**
2. If you use the **Rotation** option of the **ELLIPSE** command and specify a rotation of 45, AutoCAD creates a circle rotated 45° from which of the following? *Select the one item that best answers the question.*
   A. line of sight
   B. major axis of the ellipse
   C. minor axis of the ellipse
   D. X axis
   E. Z axis
3. When you use the **2-Point** option of the **CIRCLE** command to create a circle, which of the following is defined by the points you specify? *Select the one item that best answers the question.*
   A. radius of the circle
   B. circumference of the circle
   C. diameter of the circle
   D. area of the circle

Follow the instructions in each problem. Complete each problem by answering the related question.

4. **Navigate to this chapter on the companion website and open CPE-04circle.dwg.** Use the **CIRCLE** command to create a circle that is tangent to both of the lines in the drawing and has a radius of 7.3. Do not change any settings in the drawing file. When you finish, type **LIST**, press [Enter], select the circle, and press [Enter] to display a list of information about the circle in a text window. (Chapter 15 explains the **LIST** command in more detail.) According to the listed information, what are the coordinates of the center point of the circle?
5. **Navigate to this chapter on the companion website and open CPE-04arc.dwg.** Use the **ARC** command to create the arc shown below. Specify the center point and start point of the arc using the **Endpoint** object snap, and specify an angle of 37°. Do not change any settings in the drawing file. Then use the **Endpoint** object snap mode to select the upper endpoint of the arc (Point A). According to the coordinate display, what are the coordinates of this point? If necessary, select the **Coordinates** option in the status bar **Customization** flyout to add the coordinate display to the status bar.

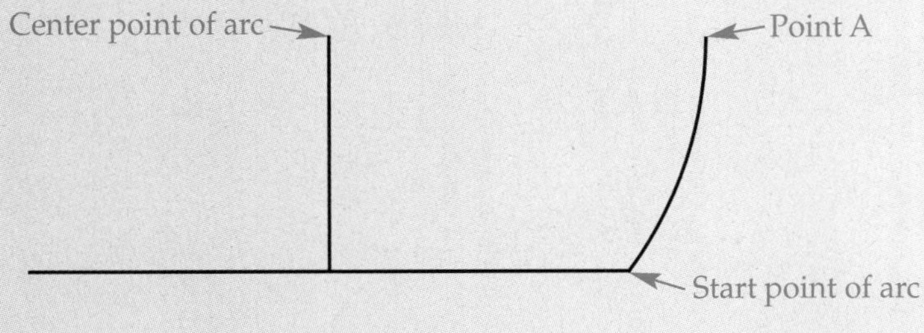

CHAPTER

5

Line Standards and Layers

Learning Objectives

After completing this chapter, you will be able to:

- ✓ Describe basic line conventions.
- ✓ Explain the concept of layers in a CADD drawing.
- ✓ Describe how layers are used in various drafting fields.
- ✓ Create and manage layers.
- ✓ Use **DesignCenter** to copy layers and linetypes between drawings.

AutoCAD has a layer system that allows you to organize several properties and assign them to objects. ***Layers*** help you conform to drawing standards and conventions and create various displays, views, and sheets. This chapter introduces line conventions and the AutoCAD layer system. It also introduces **DesignCenter** as a tool for reusing drawing content.

layers: Components of the AutoCAD overlay system that allow you to separate objects into logical groups for formatting and display purposes.

Line Standards

Drafting is a graphic language that uses lines, symbols, and text to describe how to manufacture or construct a product. ***Line conventions*** provide a way to classify the content of a drawing to enhance readability. Layers allow you to apply line conventions while drawing with AutoCAD. Use line conventions as a guide to develop layers.

line conventions: Standards related to line thickness, type, and purpose.

The ASME Y14.2 standard, *Line Conventions and Lettering*, recommends two line thicknesses to establish contrasting lines: lines are thick or thin. Thick lines are twice as thick as thin lines. The recommended thicknesses are 0.6 mm for thick lines and 0.3 mm for thin lines. **Figure 5-1** describes common linetypes. **Figure 5-2** shows an example of a drawing with several common linetypes.

The US National CAD Standard (NCS) recommends specific line thicknesses and characteristics for architectural and similar drawings. Thicknesses range from *extra fine* at 0.13 mm to *4X* at 2 mm. Use the range of NCS-recommended line thicknesses to provide accents to drawings as needed. A common practice is to select an assortment of line thicknesses that correlate best to specific applications, such as those listed in **Figure 5-3**.

Figure 5-1.
Line conventions adapted from ASME Y14.2, *Line Conventions and Lettering*. Line characteristics and spacing are measured at full scale. Specifications vary according to drawing size.

Line Conventions			
Appearance	**Purpose**	**Characteristics**	**AutoCAD Linetype**
Object line			
	Show the visible contour or outline of objects.	Thick (0.6 mm), solid.	Continuous
Hidden line			
	Represent features hidden in a view.	Thin (0.3 mm), dashed. Dashes are .125″ (3 mm) long and are spaced .06″ (1.5 mm) apart.	HIDDEN or DASHED
Centerline			
	Identify centers and axes of circular features and lines of symmetry for symmetrical objects.	Thin (0.3 mm). Consists of one .125″ (3 mm) dash alternating with one .75″ to 1.5″ (19 mm to 38 mm) dash. A .06″ (1.5 mm) space separates dashes. Extends .125″ to .25″ (3 mm to 6 mm) past objects. See Chapter 16.	CENTER
Symmetry line			
Symmetry symbol	Identify lines of symmetry for symmetrical objects.	Thin (0.3 mm) centerline with thick (0.6 mm) symmetry symbol. Symbol lines are 1.5 times longer than and spaced half the height of dimension text.	CENTER for line of symmetry and Continuous for symmetry symbol
Section line			
	In a section view, show where material has been cut away.	Thin (0.3 mm), usually drawn in a pattern. Different types of lines and patterns can be used to indicate specific materials or components. See Chapter 23.	Continuous
Phantom line			
	Identify repetitive details, show alternate positions of moving parts, and locate adjacent positions of related parts.	Thin (0.3 mm). Two .125″ (3 mm) dashes alternating with one .75″ to 1.5″ (19 mm to 38 mm) dash. Spaces between dashes are .06″ (1.5 mm).	PHANTOM

Line Conventions (Cont.)			
Appearance	**Purpose**	**Characteristics**	**AutoCAD Linetype**
Chain line			
	Indicate special features or unique treatment for a surface.	Thick (0.6 mm). Consists of one .125″ (3 mm) dash alternating with one .75″ to 1.5″ (19 mm to 38 mm) dash. A .06″ (1.5 mm) space separates the dashes.	CENTER
Extension line			
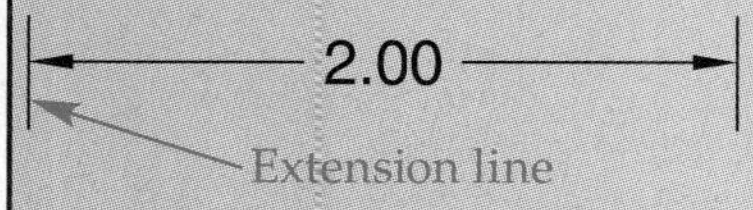	Show the extent of a dimension.	Thin (0.3 mm), usually solid. Begins .06″ (1.5 mm) from an object and extends .125″ (3 mm) beyond the last dimension line. See Chapter 16.	Continuous
Dimension line			
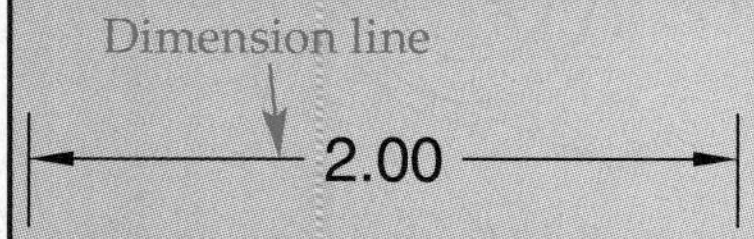	Show the distance being measured.	Thin (0.3 mm), usually solid. See Chapter 16.	Continuous
Leader line			
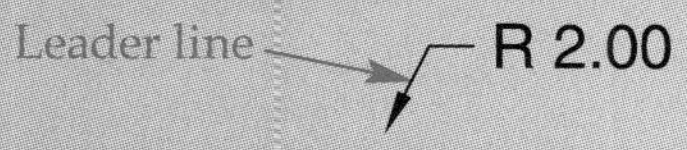	Connect a specific note to a feature on a drawing.	Thin (0.3 mm), solid. See Chapter 16.	Continuous
Cutting-plane line			
	Identify the location and viewing direction of a section view.	Thick (0.6 mm). See Chapter 23.	PHANTOM, Continuous, or DASHED
Viewing-plane line			
	Identify the location and viewing direction of a removed view.	Thick (0.6 mm). See Chapter 8.	PHANTOM, Continuous, or DASHED
Break line			
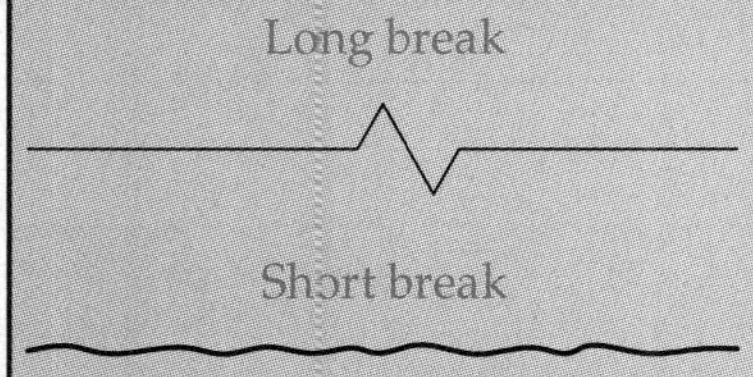	Show where a portion of an object has been removed for clarity or convenience.	Thin (0.3 mm) for long break lines and thick (0.6 mm) for short break lines, solid. Break representation is based on object or material being broken.	Continuous
Stitch line			
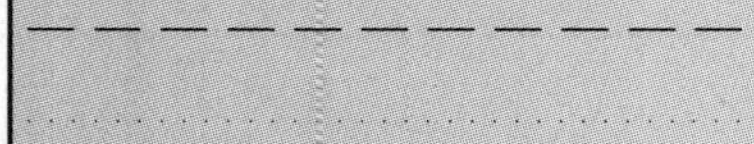	Identify the location of a stitching or sewing process.	Thin (0.3 mm), dashed or .01″ (0.3 mm) diameter dots.	HIDDEN, DASHED, or DOT

Figure 5-2.
An example of a mechanical assembly drawing with several common types of lines.

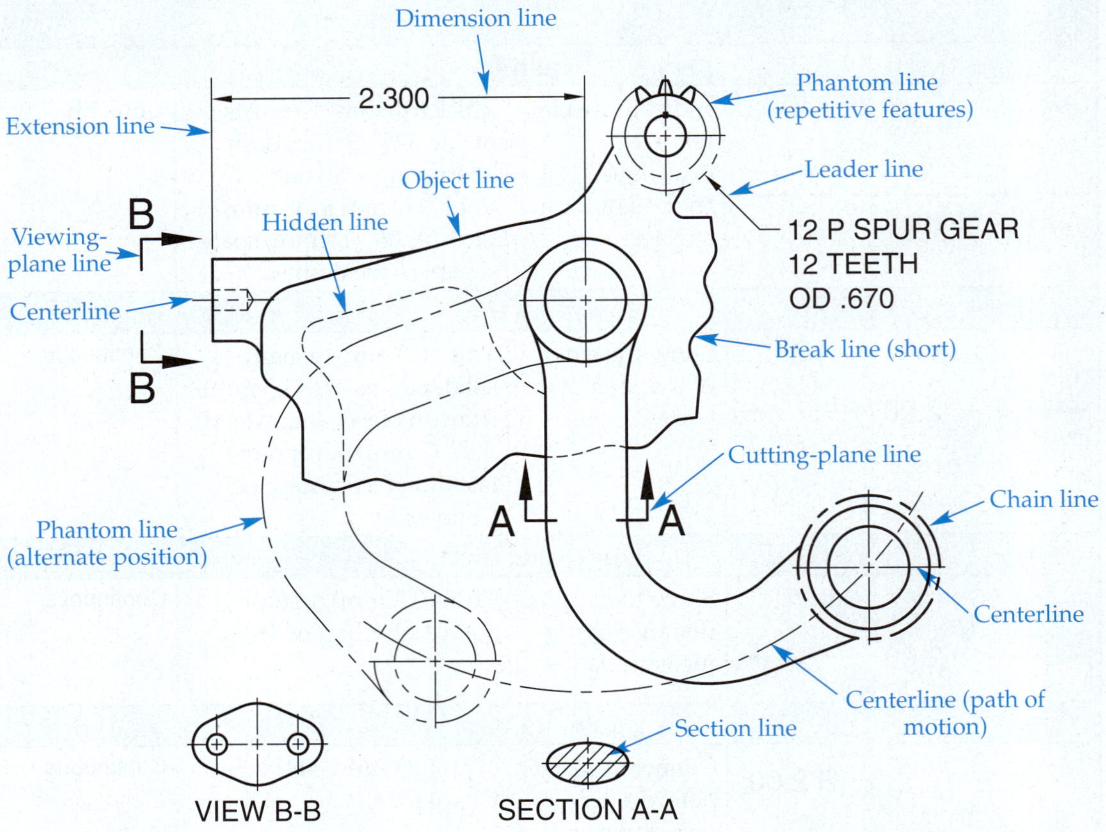

Figure 5-3.
An example of applications for common US National CAD Standard line thicknesses.

Thickness	Application
Thin, 0.25 mm	Dimension elements, phantom lines, hidden lines, centerlines, long break lines, schedule grid lines, and background objects.
Medium, 0.35 mm	Object lines, text for dimension values, notes, and schedules, terminator marks, door and window elevations, and schedule grid accent lines.
Wide, 0.5 mm	Major object lines at elevation edges, cutting-plane lines, short break lines, title text, minor title underlines, and border lines.
Extra wide, 0.7 mm	Major title underlines, schedule outlines, large titles, special emphasis object lines, elevation and section grade lines, property lines, sheet borders, and schedule borders.

NOTE

Many AutoCAD tools simplify and automate the process of applying correct line standards. You will learn applications and techniques for drawing specific types of lines in this textbook.

Introduction to Layers

In AutoCAD, you can use an ***overlay system*** of layers to separate different objects and elements of a drawing. For example, you might choose to draw all visible object lines on an Object layer and all dimension objects on a Dimension layer. You can display both layers to show the complete drawing with dimensions or hide the Dimension layer to show only the visible objects. The following is a list of ways you can use layers to increase productivity and add value to a drawing:

overlay system: A system of separating drawing components by layer.

- Assign each layer a different color, linetype, and lineweight to correspond to line conventions and to help improve clarity.
- Make changes to layer properties to update all objects drawn on the layer.
- Turn off or freeze selected layers to decrease the amount of information displayed on-screen or to speed up screen regeneration.
- Plot each layer in a different color, linetype, or lineweight, or set a layer not to plot.
- Use separate layers to group specific information. For example, draw a floor plan using floor plan layers, an electrical plan using electrical layers, and a plumbing plan using plumbing layers.
- Create several sheets from the same drawing file by controlling layer visibility to separate or combine drawing information. For example, use layers to display a floor plan and electrical plan together to send to an electrical contractor or to display a floor plan and plumbing plan together to send to a plumbing contractor.

Layers Used in Drafting Fields

The type of drawing typically determines the function of each layer. In mechanical drafting, you usually assign a specific layer to each different type of line or object. For example, draw object lines on an Object layer that is black in color and uses a 0.6 mm solid (Continuous) linetype. Draw hidden lines on a green Hidden layer that uses a 0.3 mm hidden (HIDDEN or DASHED) linetype.

Architectural and civil drawings may require hundreds of layers, each used to produce a specific item. For example, draw full-height walls on a floor plan using a black A-WALL-FULL layer that has a 0.5 mm solid (Continuous) linetype. Add plumbing fixtures to a floor plan using a blue (150) P-FLOR-FIXT layer that has a 0.35 mm solid (Continuous) linetype. Figure 5-4 lists examples of typical mechanical, architectural, civil, and electronic drafting layers.

Creating and Using Layers

Use the **LAYER** command to open the **Layer Properties Manager**, where you can create and control layers. See Figure 5-5. The columns in the list view pane on the right side of the **Layer Properties Manager** list layers and provide layer property controls. Properties in each column appear as an icon or as an icon and a name. Pick a property to change the corresponding layer setting. The tree view pane on the left side of the **Layer Properties Manager** displays filters for limiting the number of layers displayed in the list view pane.

LAYER

Ribbon
Home
> Layers
View
> Palettes
Layer Properties
Type
LAYER
LA

The default 0 layer is the only required layer in an AutoCAD drawing. However, the 0 layer is primarily reserved for drawing blocks, as described later in this textbook.

Figure 5-4.
Examples of typical layers in common drafting fields.

Mechanical				
Name	**Color**	**Linetype**	**Lineweight**	**Description**
Center	red	CENTER	0.30 mm	Centerlines, symmetry lines
Construction	cyan	Continuous	0.30 mm	Construction lines
Dimension	red	Continuous	0.30 mm	Dimensions
Hidden	green	HIDDEN	0.30 mm	Hidden lines
Object	white	Continuous	0.60 mm	Object lines
Sheet	blue	Continuous	0.60 mm	Borders, sheet blocks
Architectural				
Name	**Color**	**Linetype**	**Lineweight**	**Description**
A-DOOR	blue	Continuous	0.35 mm	Doors
A-WALL-FULL	white	Continuous	0.50 mm	Full height walls
E-LITE-CLNG	160	Continuous	0.35 mm	Ceiling mounted lights
M-HVAC	white	Continuous	0.35 mm	HVAC system
P-FLOR-FIXT	150	Continuous	0.35 mm	Plumbing fixtures
S-FNDN	magenta	Continuous	0.35 mm	Foundation
Civil				
Name	**Color**	**Linetype**	**Lineweight**	**Description**
C-BLDG	white	Continuous	0.60 mm	Buildings, primary structures
C-NGAS-UNDR	yellow	GAS_LINE	0.50 mm	Underground natural gas
C-PROP	white	PHANTOM	0.70 mm	Property boundary
C-SSWR-UNDR	green	DASHED	0.50 mm	Underground sanitary sewer
C-TOPO-MAJR	44	CENTER	0.60 mm	Index contour lines
C-WATR-UNDR	blue	DASHED	0.50 mm	Underground water supply
Electronic				
Name	**Color**	**Linetype**	**Lineweight**	**Description**
Capacitor	green	Continuous	0.30 mm	Capacitors
Diode	blue	Continuous	0.30 mm	Diodes
Inductor	yellow	Continuous	0.30 mm	Inductors
Resistor	red	Continuous	0.30 mm	Resistors
Switch	white	Continuous	0.30 mm	Switches
Transistor	magenta	Continuous	0.30 mm	Transistors

purge: To remove unused items, such as block definitions and layers, from the drawing.

You cannot delete, rename, or ***purge*** the 0 layer. Draw each object on a layer specific to the object. For example, draw object lines on an Object layer, draw walls on a floor plan using an A-WALL layer, and draw construction lines on a Construction or A-ANNO-NPLT layer. Make a conscious effort to assign an appropriate layer to each object.

Figure 5-5.
The **Layer Properties Manager**. Layer 0 is the default layer. Pick an icon to change the corresponding layer setting.

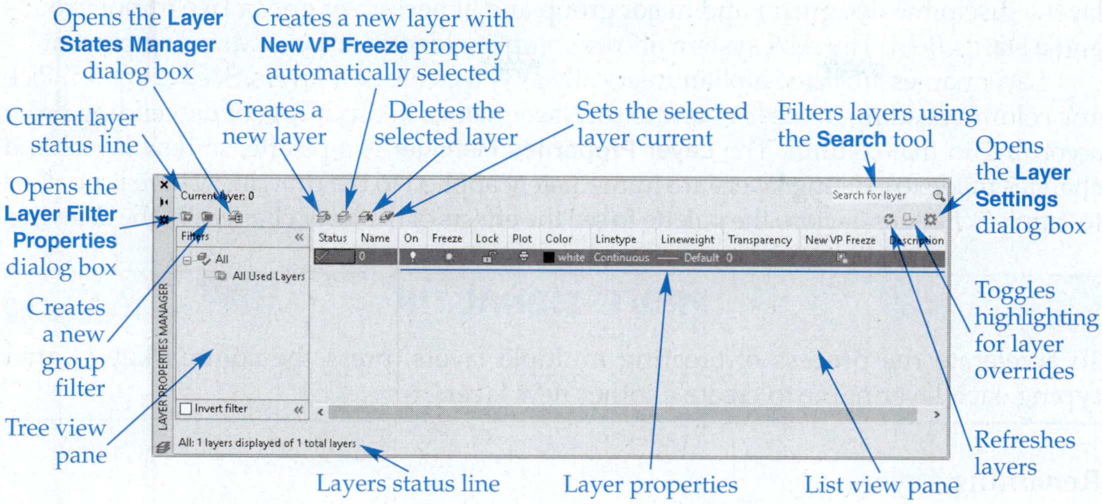

Adding Layers

Add layers to a drawing to meet the needs of the drawing project. Use the **Layer Properties Manager** to add a new layer. Select an existing layer with properties similar to those you want to assign to the new layer. Reference the 0 layer to create the first new layer using a default template. Then pick the **New Layer** button, right-click in the list view and select **New Layer**, or press the [Alt]+[N] key combination. A new layer appears, using a default name. The layer name is highlighted, allowing you to type a new name. See **Figure 5-6**. Pick away from the layer in the list or press [Enter] to accept the layer.

Layer Names

Name layers to reflect drawing content. Layer names can include letters, numbers, and certain other characters, including spaces. Layer names are usually set according to specific industry or company standards. Refer to **Figure 5-4**.

However, simple or generic drawings may use a more basic naming system. For example, the name Continuous-White indicates a layer assigned a continuous linetype and white color. The name Object-7 identifies a layer for drawing object lines, assigned color 7. Another option is to assign the linetype a numerical value. For example, name object lines 1, hidden lines 2, and centerlines 3. If you use this method, keep a written record of the numbering system for reference.

More complex layer names are appropriate for some applications, and may include items such as drawing number, color code, and layer content. For example, the name Dwg100-2-Dimen refers to drawing DWG100, color 2, for use when adding dimensions.

Figure 5-6.
AutoCAD names a new layer Layer*n* by default and provides the opportunity for you to change the name immediately.

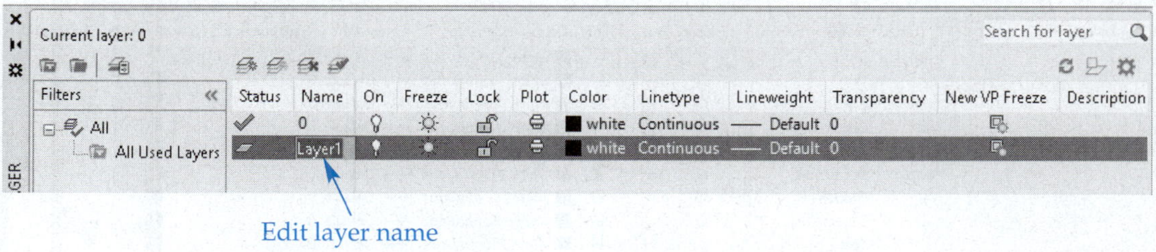

The American Institute of Architects (AIA) *CAD Layer Guidelines*, associated with the US National CAD Standard, specifies a layer naming system for architectural and related drawings. The system uses a highly detailed layer naming process that assigns each layer a discipline designator and major group and, if necessary, one or two minor groups and a status field. The AIA system allows complete identification of drawing content.

Layer names are listed alphanumerically as you create new layers. See Figure 5-7. Pick any column heading in the list view to sort layer names in ascending or descending order according to that column. The **Layer Properties Manager** is a palette, so new layers and changes made to existing layers are immediately applied to the drawing. There is no need to "apply" changes or close the palette to see the effects of the layer changes in the drawing.

PROFESSIONAL TIP

To accelerate the process of creating multiple layers, press the comma key [,] after typing each layer name to create another new layer.

Renaming Layers

To change a layer name using the **Layer Properties Manager**, slowly double-click on the existing name in the **Name** column to highlight it. Type the new name and press [Enter] or pick outside of the text box. You can also rename a layer by picking the name once to highlight it and then pressing [F2], or by right-clicking and selecting **Rename Layer**. You cannot rename layer 0 or layers associated with an external reference.

Exercise 5-1

www.g-wlearning.com/CAD/

Complete the exercise on the companion website.

Selecting Multiple Layers

Select multiple layers to speed the process of deleting or applying the same properties to several layers. Use standard selection practices or the shortcut menu to select multiple layers. Hold [Shift] to select several consecutive layers, or hold [Ctrl] to select several nonconsecutive layers. You can also use a window to select all the layers that contact the window. The following selection options are available when you right-click in the list view:

- **Select All.** Selects all layers.
- **Clear All.** Deselects all layers.
- **Select All but Current.** Selects all layers except the current layer.
- **Invert Selection.** Deselects all selected layers and selects all deselected layers.

Figure 5-7.
Layer names are automatically listed in alphanumeric order when you create new layers or change layer names.

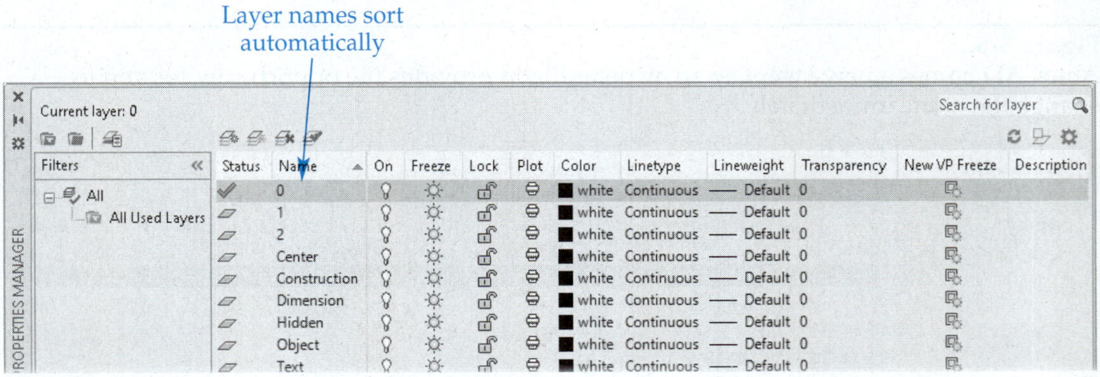

Exercise 5-2

www.g-wlearning.com/CAD/

Complete the exercise on the companion website.

Layer Status

current layer: The active layer. Whatever you draw is placed on the current layer.

The icon in the **Status** column describes the status, or use, of a layer. A green check mark indicates the ***current layer***. The status line at the top of the **Layer Properties Manager** also identifies the current layer.

Current

A gray sheet of paper, or **Not In Use** icon, in the **Status** column indicates a non-current layer that is not used by the drawing. A blue sheet of paper, or **In Use** icon, in the **Status** column means the layer is assigned to objects, but the layer is not current. The **In Use** icon also means that you cannot delete or purge the layer, even if no objects are assigned to the layer.

Not in Use

In Use

NOTE

If the **Layer Properties Manager** does not indicate layers in use, pick the **Settings** button in the upper-right corner to display the **Layer Settings** dialog box and select the **Indicate layers in use** check box.

Setting the Current Layer

To set a layer current using the **Layer Properties Manager**, double-click the layer name, pick the layer name and select the **Set Current** button, or right-click on the layer and select **Set current**. You can also make a layer current without using the **Layer Properties Manager** by picking the layer name from the **Layer** drop-down list in the **Layers** panel of the **Home** ribbon tab. See Figure 5-8. Use the vertical scroll bar to move up and down through a long list. The **Layer** drop-down list is a very effective way to activate and manage layers while drawing. The **Layer** drop-down list is also available from the **Quick Access** toolbar, but does not appear by default. Pick the **Customize Quick Access Toolbar** flyout on the right side of the **Quick Access** toolbar and check **Layer** to add the **Layer** drop-down list to the **Quick Access** toolbar.

Command line content search offers another way to find a layer and make a layer current without using the **Layer Properties Manager**. Begin typing the letters in a layer name on the command line. Layers that match the letters you enter appear in the suggestion list as you type. Figure 5-9 shows the results of typing LITE at the command line to locate layers used to draw lighting symbols in an architectural drawing.

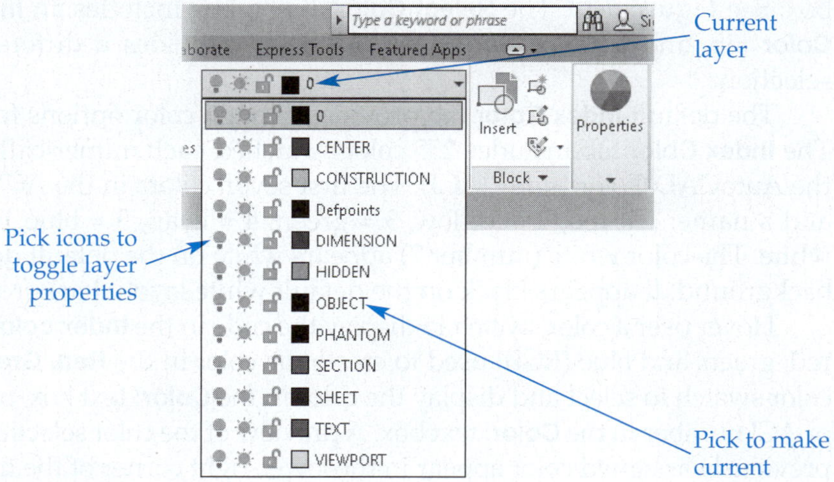

Figure 5-8. The **Layer** drop-down list allows you to change the current layer and adjust specific layer properties.

Figure 5-9.
Using command line content search to locate layers and set a specific layer current.

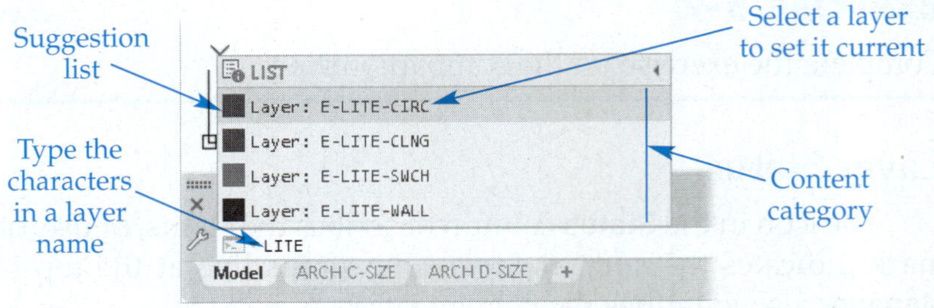

You may need to search through the content category in the suggestion list to find the desired layer. Select a layer from the suggestion list to set current.

NOTE

You can use the **Layer Properties Manager** or the **Layer** drop-down list to change the current layer or layer properties while a command is active. For example, draw a line segment using the current layer, and then, without exiting the **LINE** command, make a different layer current to draw the next line segment on a different layer.

PROFESSIONAL TIP

To assign a different layer to existing objects, select the objects and then select the layer using the **Layer** drop-down list.

Exercise 5-3

www.g-wlearning.com/CAD/

Complete the exercise on the companion website.

Setting Layer Color

Assign a unique color to each layer to help distinguish objects on-screen. You can also use layer colors when it is necessary to plot a drawing in color. Assigning colors to layers is very important for on-screen drawing clarity, organization, workability, and format. Layer colors should emphasize important features and symbols and not cause eyestrain.

Use the **Color** column of the **Layer Properties Manager** to assign a color to each layer. Pick the existing color swatch to change the color using the **Select Color** dialog box. See Figure 5-10. The **Select Color** dialog box includes an **Index Color** tab, a **True Color** tab, and a **Color Books** tab. Each tab provides a different method for color selection.

The default **Index Color** tab provides enough color options for most 2D drawings. The **Index Color** tab includes 255 color swatches, each numerically coded according to the AutoCAD Color Index (ACI). The first seven colors in the ACI have both a number and a name: 1 = red, 2 = yellow, 3 = green, 4 = cyan, 5 = blue, 6 = magenta, and 7 = white. The color white (number 7) appears white on the default dark-gray model space background. It appears black on the default white layout background.

Hover over a color swatch to display the code in the **Index color:** field and the mix of red, green, and blue (RGB) used to create the color in the **Red, Green, Blue:** field. Pick a color swatch to select and display the color in the **Color:** text box, or type the color name or ACI number in the **Color:** text box. A preview of the color selection and a sample of the previously assigned color appear in the lower-right corner of the dialog box.

Figure 5-10.
Use the **Select Color** dialog box to assign a color to a layer.

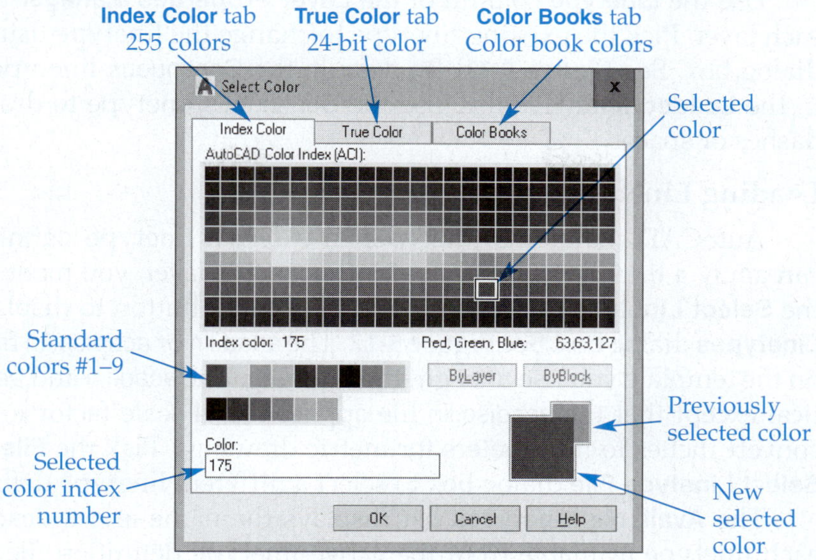

After you select a color, pick the **OK** button to assign the color to the specified layer. The layer color swatch indicates the color. By default, all objects are displayed in the selected layer color, or ByLayer.

NOTE

Your graphics card and monitor affect color display characteristics and sometimes the number of available colors.

PROFESSIONAL TIP

Use the color swatch in the **Layer** drop-down list to change the color assigned to a layer without accessing the **Layer Properties Manager**.

CAUTION

The **COLOR** command provides access to the **Select Color** dialog box, which you can use to set an ***absolute value*** for color. If you set the absolute color to red, for example, the red color overrides the colors assigned to layers, and all objects appear red. For most applications, set the color to the default ByLayer setting and use layers and the **Layer Properties Manager** to control object color.

Type
COLOR

COLOR

absolute value: In property settings, a value set directly instead of referenced by layer or by block. An absolute value overrides the corresponding layer settings.

Exercise 5-4

www.g-wlearning.com/CAD/

Complete the exercise on the companion website.

Setting Layer Linetype

Appropriate linetypes and line thicknesses enhance the readability of a drawing. You can apply standard line conventions to objects by assigning a linetype and thickness to each layer. AutoCAD provides standard linetypes that match or are similar to ASME, ISO, NCS, and other standard linetypes. You can also create custom linetypes. Assign lineweights to layers to achieve different line thicknesses.

Use the **Linetype** column of the **Layer Properties Manager** to assign a linetype to each layer. Pick the existing linetype to change the linetype using the **Select Linetype** dialog box. See Figure 5-11. By default, the Continuous linetype is the only linetype in the **Loaded linetypes** list. Use the Continuous linetype to draw solid lines with no dashes or spaces.

Loading Linetypes

AutoCAD maintains linetypes in external linetype definition files. Before you can apply a linetype other than Continuous to a layer, you must load the linetype into the **Select Linetype** dialog box. Pick the **Load...** button to display the **Load or Reload Linetypes** dialog box. See Figure 5-12. The acad.lin or acadiso.lin file is active, depending on the template you use to begin the drawing. The acad.lin and acadiso.lin files are identical except that the acadiso.lin file applies a 25.4 scale factor to non-ISO linetypes to convert inches to millimeters for metric drawings. Pick the **File...** button and use the **Select Linetype File** dialog box to select a different linetype definition file.

The **Available Linetypes** list displays the name and a description and image of each linetype available from the active linetype definition file. Use the scroll bar to view all available linetypes, and use the image in the **Description** column to aid in selecting the appropriate linetypes to load. Select a single linetype, or select multiple linetypes using standard selection practices or the shortcut menu. Pick the **OK** button to return to the **Select Linetype** dialog box, where the linetypes you selected now appear. See Figure 5-13. In the **Select Linetype** dialog box, pick the linetype to assign to the layer, and then pick the **OK** button. Figure 5-14 shows the HIDDEN linetype selected in Figure 5-13 assigned to the layer named Hidden.

Figure 5-11.
The **Select Linetype** dialog box allows you to load linetypes for use in the current drawing.

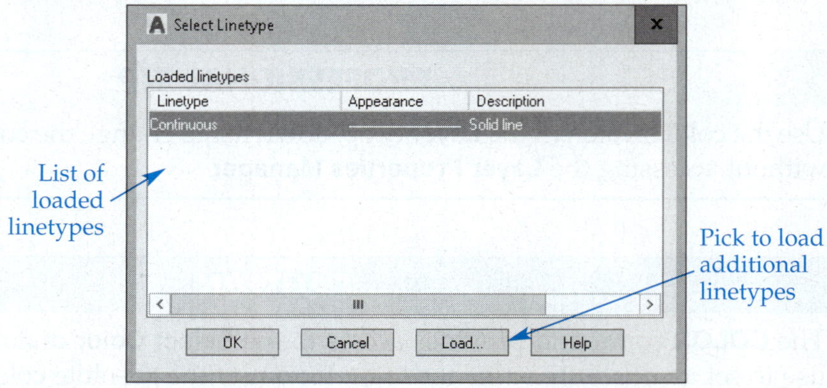

Figure 5-12.
The **Load or Reload Linetypes** dialog box displays linetypes available for loading.

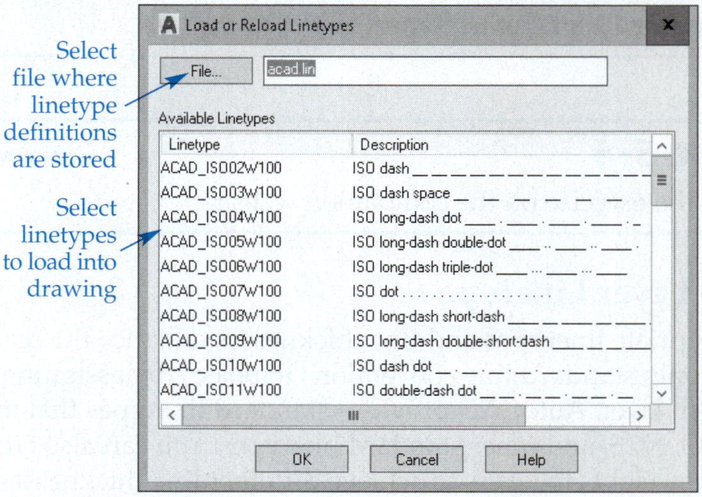

Figure 5-13.
Linetypes loaded into the drawing using the **Load or Reload Linetypes** dialog box appear in the **Loaded linetypes** list.

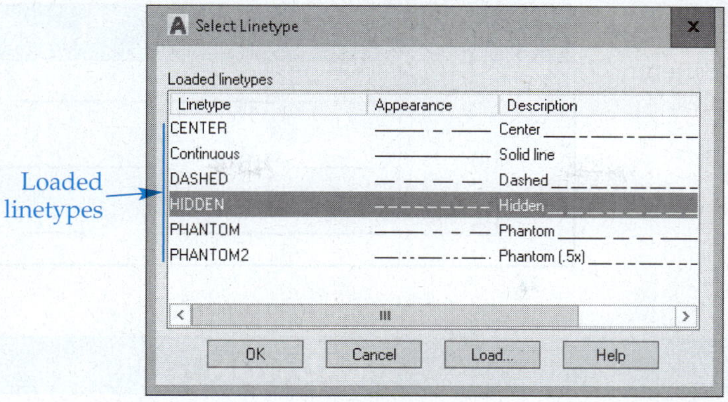

Figure 5-14.
Objects drawn on the Hidden layer will have the HIDDEN linetype.

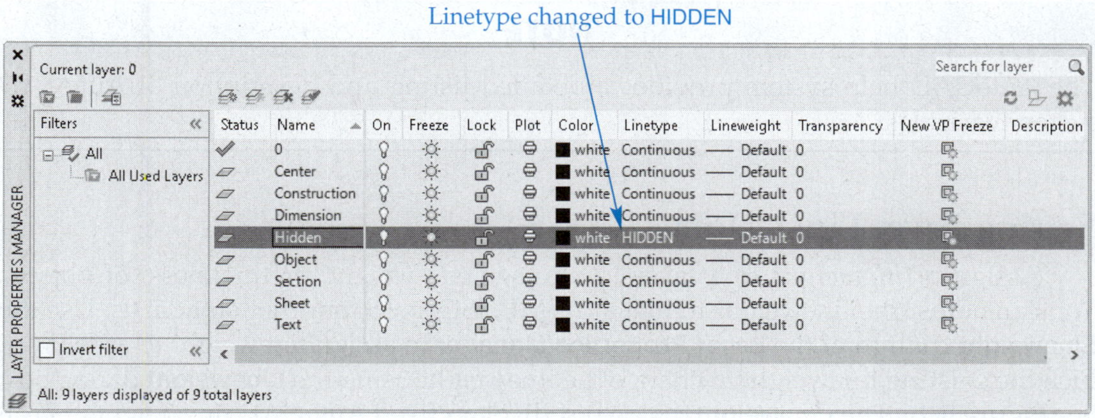

CAUTION

The **LINETYPE** command provides access to the **Linetype Manager**, which you can use to set an absolute value for linetype. If you set the absolute linetype to HIDDEN, for example, the HIDDEN linetype overrides the linetypes assigned to layers, and all objects appear in the HIDDEN linetype. For most applications, set linetype to ByLayer and use layers and the **Layer Properties Manager** to control object linetype.

Type
LINETYPE   **LT**

LINETYPE

Exercise 5-5

www.g-wlearning.com/CAD/

Complete the exercise on the companion website.

Setting Linetype Scale

You can change the ***linetype scale*** to adjust the lengths of dashes and spaces in linetypes to make a drawing more closely match standard drafting practices. Changing the ***global linetype scale*** is the preferred method for adjusting linetype scale, but it is possible to change the linetype scale of specific objects.

linetype scale: The lengths of dashes and spaces in linetypes.

global linetype scale: A linetype scale applied to every linetype in the current drawing.

Use the **LTSCALE** system variable to make a global change to the linetype scale. Enter LTS or LTSCALE at the keyboard and then enter a new value. The default global linetype scale factor is 1. A value less than 1 makes dashes and spaces smaller, and a value greater than 1 makes them larger. See Figure 5-15. After you enter a value and press [Enter], the drawing regenerates and the global linetype scale changes for all lines on the drawing. Experiment with different linetype scales until you achieve the desired result.

Figure 5-15.
The CENTER linetype at different linetype scales.

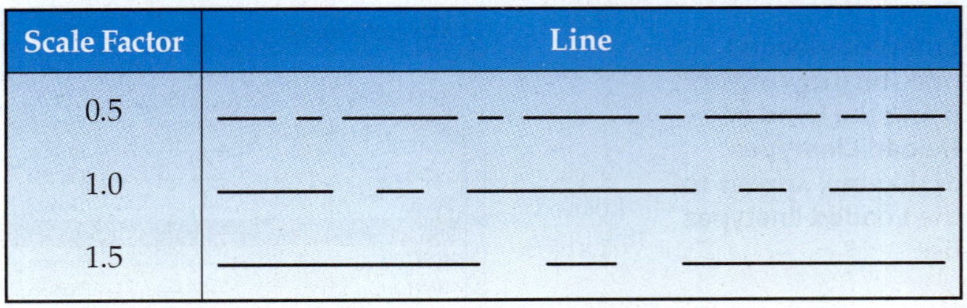

CAUTION

Be careful when changing linetype scales to avoid making your drawing look odd and not in accordance with drafting standards.

NOTE

The AutoCAD help system provides a detailed listing and description of AutoCAD system variables.

Setting Layer Lineweight

lineweight: The assigned width of lines for display and plotting.

Assign a ***lineweight*** to a layer to manage the weight, or thickness, of objects. You can adjust the lineweight to match ASME, ISO, NCS, or other standards. Use the **Lineweight** column of the **Layer Properties Manager** to assign lineweight to each layer. Pick the existing lineweight to change the lineweight using the **Lineweight** dialog box. See Figure 5-16. The **Lineweight** dialog box displays fixed AutoCAD lineweights. Scroll through the **Lineweights:** list and select the lineweight to assign to the layer. Pick the **OK** button to apply the lineweight and return to the **Layer Properties Manager**.

Lineweight Settings

LINEWEIGHT

Type
LINEWEIGHT
LWEIGHT
LW

The **LINEWEIGHT** command provides access to the **Lineweight Settings** dialog box, shown in Figure 5-17. Use the **Units for Listing** area to set the lineweight thickness to **Millimeters (mm)** or **Inches (in)**. The units apply only to values in the **Lineweight** and **Lineweight Settings** dialog boxes, helping you to select lineweights based on a known unit of measurement.

Figure 5-16.
Use the **Lineweight** dialog box to assign a lineweight to a layer.

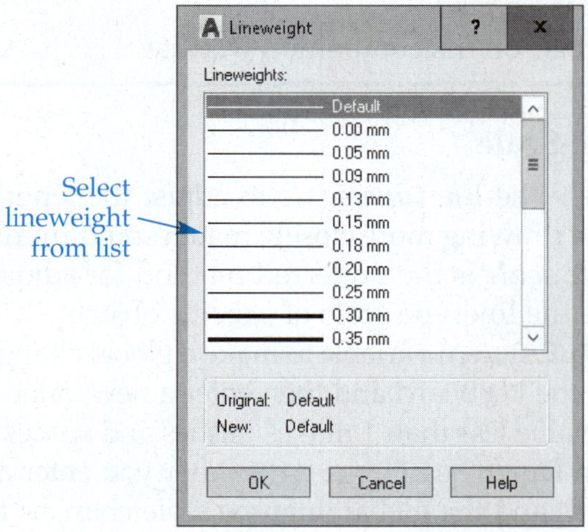

Figure 5-17.
The **Lineweight Settings** dialog box.

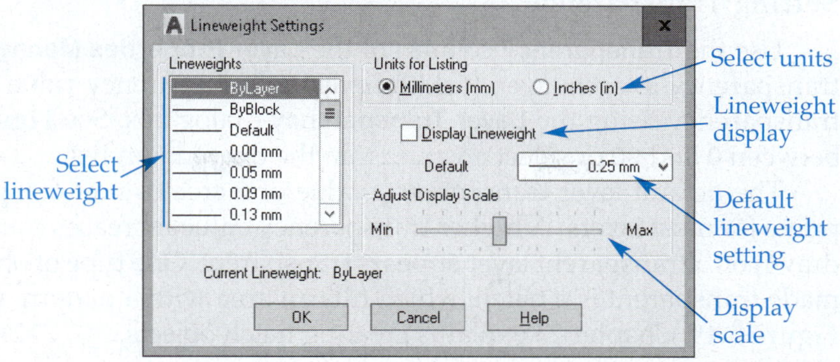

NOTE

You can also access the **Lineweight Settings** dialog box by right-clicking on the **Show/Hide Lineweight** button on the status bar and selecting **Lineweight Settings...**. The **Show/Hide Lineweight** button does not appear on the status bar by default. Select the **LineWeight** option in the status bar **Customization** flyout to add the **Show/Hide Lineweight** button to the status bar.

Check the **Display Lineweight** check box to display object lineweight on-screen. Use the **Adjust Display Scale** slider to adjust the lineweight display scale to improve the appearance of lineweights when lineweight display is on. When lineweight display is off, all objects display a 0, or one-pixel, thickness regardless of the lineweight assigned to the layer. You can also toggle screen lineweight on and off using the **Show/Hide Lineweight** button on the status bar.

The **Default** drop-down list sets the value used when you assign the Default lineweight to a layer. The Default lineweight is an application setting and applies to any drawing you open. It is not template-specific and remains set until you change the value. Do not assign the Default lineweight to layers if you anticipate using a different default lineweight for different drawing applications. Assign a specific lineweight, other than Default, to each layer to maintain flexibility and consistency between drawings.

CAUTION

You can use the **Lineweights** list in the **Lineweight Settings** dialog box to set an absolute value for lineweight. If you set the absolute lineweight to 0.30 mm, for example, the 0.30 mm lineweight overrides the lineweights assigned to layers and all objects appear 0.30 mm thick. For most applications, set lineweight to ByLayer and use layers and the **Layer Properties Manager** to control object lineweight.

Exercise 5-6

www.g-wlearning.com/CAD/

Complete the exercise on the companion website.

Layer Transparency

ASME standards recommend that all objects be opaque and dark for most applications. However, you can choose to draw transparent objects for specific drawing requirements, usually for architectural, civil, or technical illustration applications. For example, draw an existing building using transparent objects to highlight a proposed structure drawn using nontransparent objects.

Setting Transparency

Use the **Transparency** column of the **Layer Properties Manager** to assign a level of transparency to each layer. Pick the existing transparency value to change the level of transparency using the **Layer Transparency** dialog box. See Figure 5-18. Type a value between 0 and 90 or select a value from the drop-down list.

The default layer transparency value of 0 creates nontransparent objects, appropriate for most layers. A higher transparency value increases transparency. Any object drawn on a transparent layer appears transparent. One type of object that is commonly made transparent is a hatch, which fills an area with a pattern, solid, or gradient. See Figure 5-19. Chapter 23 explains creating hatch objects.

Showing and Hiding Transparency

Show or hide transparency using the **Transparency** button on the status bar. The **Transparency** button does not appear on the status bar by default. Select the **Transparency** option in the status bar **Customization** flyout to add the **Transparency** button to the status bar. Transparency is on by default, and all objects drawn using a transparent layer appear at their transparent level. Disabling transparency using the **Transparency** button makes all transparent objects appear nontransparent, but does not change the layer transparency property.

Figure 5-18.
Use the **Layer Transparency** dialog box to assign a level of transparency to a layer.

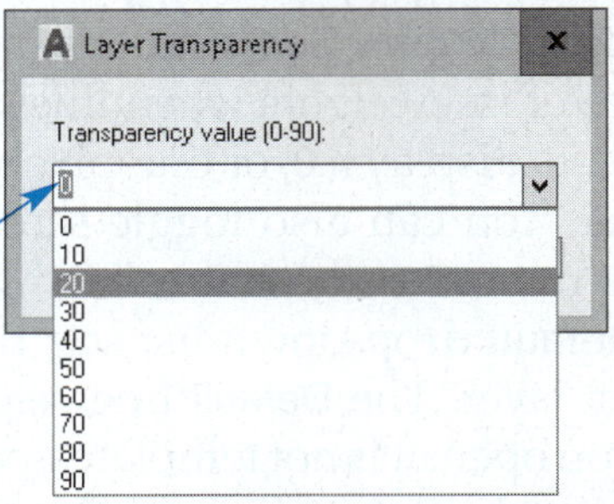

Figure 5-19.
An example of a portion of a storm water pollution control plan with transparent solid hatch objects that represent different impervious and non-impervious surfaces.

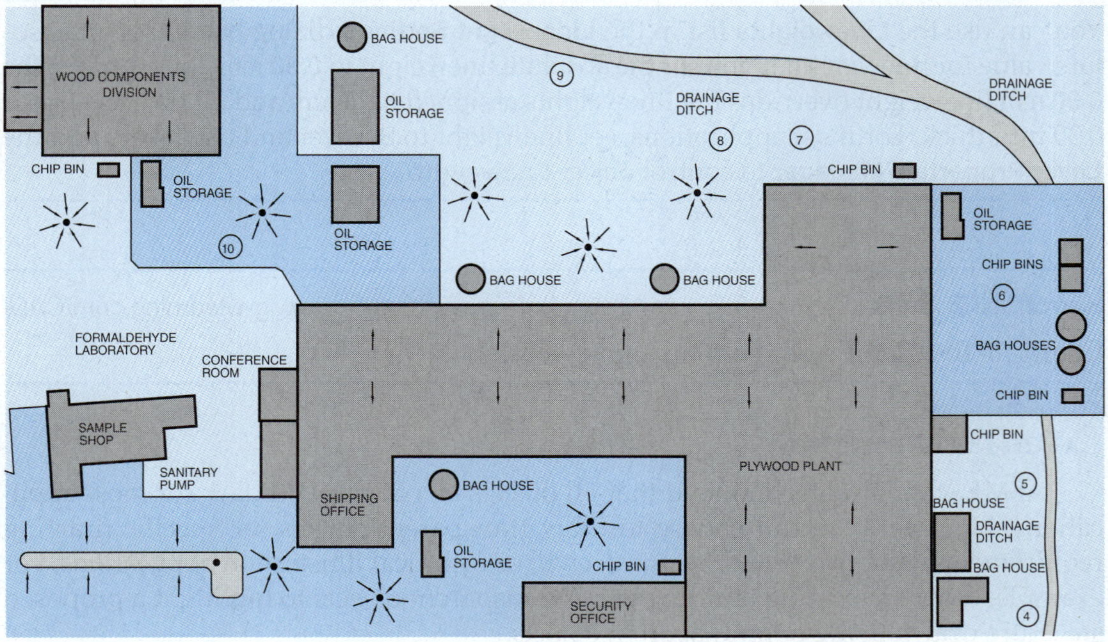

CAUTION

You can override layer transparency for specific objects, but for most applications, set transparency to ByLayer and use layers and the **Layer Properties Manager** to control object transparency.

PROFESSIONAL TIP

Layers should simplify and support drafting. Set color, linetype, lineweight, and transparency using layers and do not override these properties for individual objects. Also, once you establish layers, avoid resetting and mixing color, linetype, lineweight, and transparency, which can lead to confusion and disorder.

Plotting Layers

Use the **Plot** column of the **Layer Properties Manager** to disable a layer from printing or plotting. Pick the printer, or **Plot**, icon to change it to the **No Plot** icon if the layer should not plot. The layer displays on-screen and is selectable, but does not plot.

Plot No Plot

Adding a Layer Description

Use the **Description** column of the **Layer Properties Manager** to describe each layer. To add or change a description, slowly double-click on the blank area or existing description, type a description, and press [Enter] or pick outside of the **Description** text box. Another method is to right-click and select **Change Description**.

Turning Layers On and Off

Use the **On** column of the **Layer Properties Manager** to turn a layer on or off. The lit light bulb, or **On** icon, indicates that the layer is turned on. Objects assigned to an "on" layer display on-screen and can be selected, regenerated, and plotted. Pick the lit light bulb to turn the layer off, indicated by the unlit light bulb, or **Off** icon. Objects assigned to an "off" layer do not display on-screen and do not plot. You can select and edit objects that are off using advanced selection techniques. Objects that are off can be regenerated.

On Off

NOTE

Turn layers on or off using the **Layer** drop-down list in the **Layers** panel of the **Home** ribbon tab.

Freezing and Thawing Layers

Use the **Freeze** column of the **Layer Properties Manager** to freeze or thaw a layer. The sun, or **Thaw** icon, indicates a thawed layer. Objects assigned to a thawed layer display on-screen and can be selected, regenerated, and plotted. Pick the sun to freeze the layer, indicated by the ice crystal, or **Freeze** icon. Objects assigned to a frozen layer do not display on-screen, and cannot be plotted or regenerated. You cannot select or edit objects that are frozen. Freeze layers to hide objects and ensure that you do not accidentally modify the objects.

Freeze Thaw

VP Freeze

Use the **New VP Freeze** column to control thawing or freezing of layers when you create a new viewport. Additional layer functions also apply to layouts and viewports. This textbook explains layouts and viewports when appropriate.

VP Thaw

NOTE

You can also freeze or thaw layers using the **Layer** drop-down list in the **Layers** panel of the **Home** ribbon tab. You cannot freeze the current layer or make a frozen layer current.

CAUTION

You cannot modify frozen objects, but you can modify objects that are off. For example, if you turn off layers and use the **All** selection option with the **ERASE** command, even the objects assigned to the off layers are erased. However, if you freeze the layers, the objects are not selected or erased.

Locking and Unlocking Layers

Lock Unlock

Use the **Lock** column of the **Layer Properties Manager** to lock or unlock a layer. The unlocked padlock, or **Unlock** icon, indicates an unlocked layer. Pick the unlocked padlock to lock the layer, indicated by the locked padlock, or **Lock** icon. A **Lock** icon also appears next to the cursor when you hover over an object on a locked layer. Objects assigned to a locked layer display on-screen, and you can use a locked layer to draw new objects. However, you cannot select or edit locked objects. Lock layers to display objects but eliminate the possibility of selecting the objects.

NOTE

You can also lock or unlock layers using the **Layer** drop-down list in the **Layers** panel of the **Home** ribbon tab.

Locked Layer Fading

By default, all locked layers fade, allowing unlocked layers to stand out on-screen. A quick way to control locked layer fading is to use the options available in the expanded **Layers** panel of the **Home** ribbon tab. See Figure 5-20. Pick the **Locked Layer Fading** button to allow or disable locked layer fading. Use the **Locked Layer Fading** slider to increase or decrease fading, or type a fading percentage between 0 and 90. The default fade value is 50%. A higher fade value increases fading. Figure 5-21 shows an example of using locked layer fading on an architectural drawing.

NOTE

Transparency is a layer property that makes objects transparent. Locked layer fading is a function of the lock state, intended for on-screen drawing purposes only. For example, you can plot transparent objects, but you cannot plot the display created by locked faded layers.

Figure 5-20.
Locked layers fade by default. Increase or decrease fading and enable or disable locked layer fading as needed.

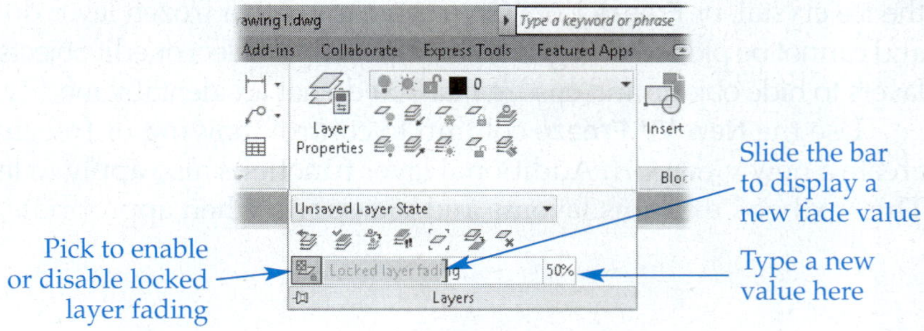

Figure 5-21.
An example of a floor plan with all layers locked except the A-WALL-FULL layer, which contains the walls. A—Locked layer fading disabled. B—Locked layer fading enabled and set to a fade value of 75.

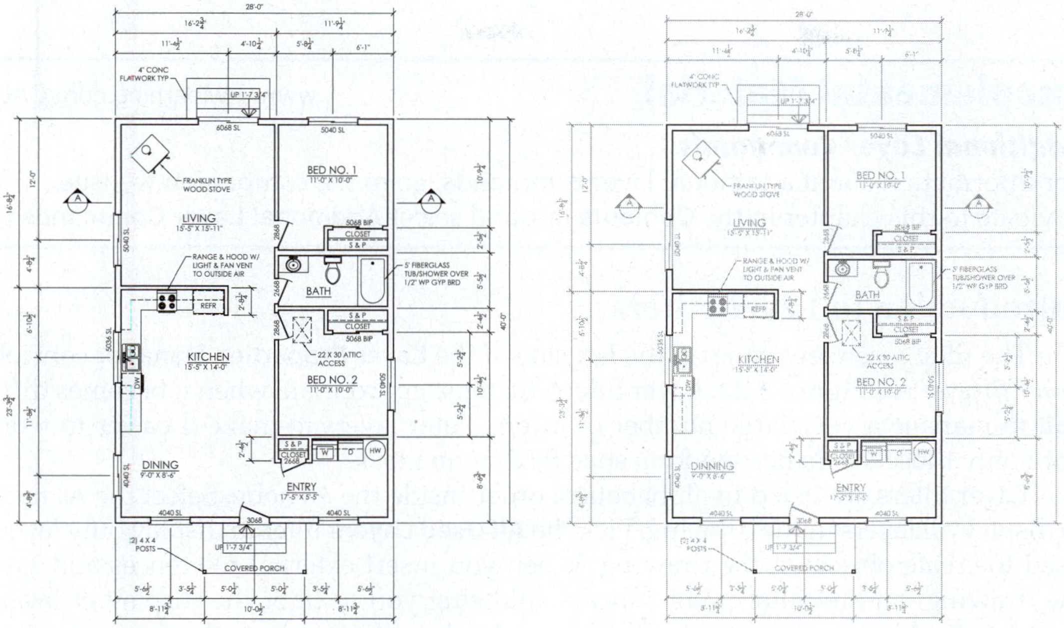

Locked Layer Fading Disabled

A

Locked Layer Fading Enabled
Fade Value: 75

B

Deleting Layers

To delete a layer using the **Layer Properties Manager**, select the layer and pick the **Delete Layer** button, or right-click on the layer and select **Delete Layer**. You cannot delete or purge the 0 layer, the current layer, layers containing objects, or layers associated with external references.

Adjusting Property Columns

To resize a column in the **Layer Properties Manager**, move the cursor over the column edge to display the resize icon and drag the column width. Maximize column width to show the full heading or longest value in the column. To maximize the width of a specific column, double-click on the column edge, or right-click on the heading and select **Maximize column**. To maximize the width of all columns, right-click on any heading and select **Maximize all columns**.

Optimize column width to show the longest value in columns that list properties as text and reduce the width of columns that list properties as icons. To optimize the width of a specific column, right-click on the heading and select **Optimize column**. To optimize the width of all columns, right-click on any heading and select **Optimize all columns**.

By default, a vertical bar appears on the right side of the **Name** column. Any column to the left of the bar is "frozen" to remain in position when you move the scroll bar near the bottom of the **Layer Properties Manager**. Scroll columns to the right of the vertical bar using the horizontal scroll bar. To disable the column freeze function, right-click on a heading and select **Unfreeze column**. Right-click on a heading and select **Freeze column** to turn on the freeze function for the selected column and every column to its left.

To hide a column in the **Layer Properties Manager**, right-click on a heading and deselect the column name. Another option is to right-click on a heading and select **Customize...** to display the **Customize Layer Columns** dialog box. Deselect the check boxes corresponding to the columns to hide. To move a column left or right in the **Layer Properties Manager**, pick a column name and select the **Move Up** or **Move Down** button. Reset the display of all property columns to their default settings by right-clicking on a heading and selecting **Restore all columns to defaults**.

Supplemental Material

www.g-wlearning.com/CAD/

Additional Layer Commands

For information about additional layer commands, go to the companion website, navigate to this chapter in the **Contents** tab, and select **Additional Layer Commands**.

Introduction to Layer Filters

layer filters: Settings that screen out, or filter, layers you do not want to display in the list view pane of the **Layer Properties Manager**.

The filter tree view pane on the left side of the **Layer Properties Manager** controls *layer filters*. See Figure 5-22. Layer filters are most appropriate when it becomes difficult to manage a very large number of layers. Filter layers to make it easier to work with only those layers needed for a specific drawing task.

Layer filters are listed in alphabetical order inside the All node. Select the All node to display all layers in the drawing. Pick the **All Used Layers** filter to display only layers used to create objects in the drawing. When you insert external references and save the drawing, an Xref filter node appears, allowing you to filter the display of layers associated with external references. This textbook explains external references when appropriate. You can create custom filters as needed.

NOTE

Collapse the filter tree view of the **Layer Properties Manager** by picking the **Collapse Layer filter tree** button. To display all filters and layers in the list view, right-click in the layer list area and select **Show Filters in Layer List**.

Supplemental Material

www.g-wlearning.com/CAD/

Creating Layer Filters

For information about creating and managing layer filters, go to the companion website, navigate to this chapter in the **Contents** tab, and select **Creating Layer Filters**.

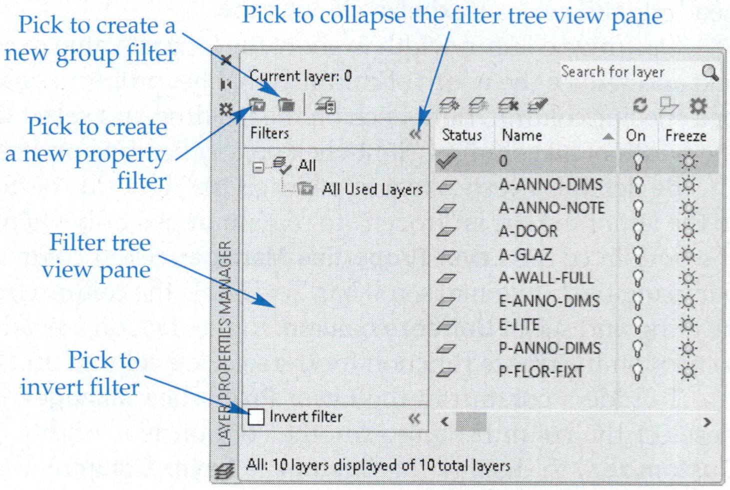

Figure 5-22. Create and restore layer filters using the filter tree view of the **Layer Properties Manager**.

Layer States

Once you save a ***layer state***, you can readjust layer settings to meet drawing tasks, with the option to restore a saved layer state when needed. For example, a basic architectural drawing might use the layers shown in Figure 5-22. You can use the drawing file to prepare a floor plan, a plumbing plan, and an electrical plan. Figure 5-23 shows the layer settings for each of the three plans.

layer state: A saved setting, or state, of layer properties for all layers in the drawing.

Save each of the three groups of settings as an individual layer state. Then restore a layer state to return the layer settings for a specific drawing. This method is easier than changing the settings for each layer individually.

Use the **Layer States Manager**, shown in Figure 5-24, to create a new layer state. Pick the **New...** button to display the **New Layer State to Save** dialog box. See Figure 5-25. Type a name in the **New layer state name:** text box and a description in the **Description** text box. Pick the **OK** button to save the new layer state. Once you create a layer state, adjust layer properties as needed. Figure 5-26 describes the areas, options, and buttons available in the **Layer States Manager**.

LAYERSTATE
Ribbon
Home
> Layers
> Layer State
Manage Layer States...
Type
LAYERSTATE

Figure 5-23.
A basic example of using layer states to manage the layers needed for several different architectural plans.

Layer	Description	Floor Plan	Plumbing Plan	Electrical Plan
0		Off	Off	Off
A-ANNO-DIMS	Floor Plan Dimensions	On	Frozen	Frozen
A-ANNO-NOTE	Floor Plan Notes	On	Frozen	Frozen
A-DOOR	Doors	On	Frozen	Locked
A-GLAZ	Windows	On	Frozen	Locked
A-WALL-FULL	Full Height Walls	On	Locked	Locked
E-ANNO-DIMS	Electrical Panel Dimensions	Frozen	Frozen	On
E-LITE	Electrical Plan Lights	Frozen	Frozen	On
P-ANNO-DIMS	Plumbing Plan Dimensions	Frozen	On	Frozen
P-FLOR-FIXT	Plumbing Plan Fixtures	Locked	On	Locked

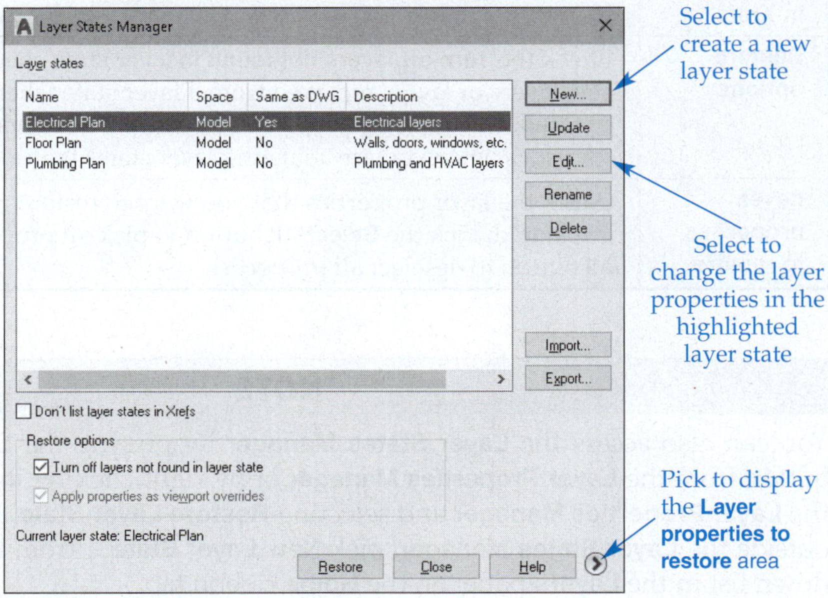

Figure 5-24.
The **Layer States Manager** allows you to save, restore, and manage layer settings.

Figure 5-25.
Creating a new layer state.

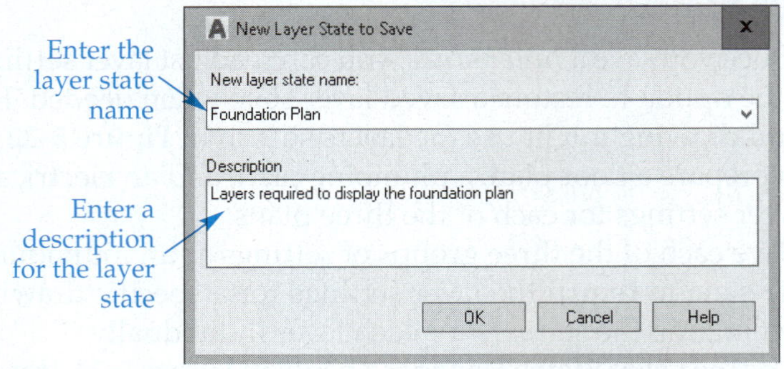

Figure 5-26.
Layer state options available in the **Layer States Manager**.

Item	Description
Layer states	Displays saved layer states. The **Name** column provides the name of the layer state. The **Space** column indicates whether the layer state was saved in model space or paper space. The **Same as DWG** column indicates whether the layer state is the same as the current layer properties. The **Description** column lists the layer state description added when the layer state was saved.
Update	Pick to resave and override the selected layer state with the current layer properties.
Edit...	Opens the **Edit Layer State** dialog box, where you can adjust the properties of each layer state without exiting the **Layer States Manager**.
Rename	Activates a text box that allows you to rename the current layer state.
Delete	Deletes the selected layer state.
Import...	Opens the **Import layer state** dialog box, used to import an LAS file containing an existing layer state into the **Layer States Manager**.
Export...	Opens the **Export layer state** dialog box, used to save a layer state as an LAS file. The file can be imported into other drawings, allowing you to share layer states between drawings containing identical layers.
Don't list layer states in Xrefs	Hides layer states associated with external reference drawings. External references are described in Chapter 31.
Restore options	Check the **Turn off layers not found in layer state** check box to turn off new layers or layers removed from a layer state when the layer state is restored. Check **Apply properties as viewport overrides** to apply property overrides when you are adjusting layer states within a layout.
Layer properties to restore	Check the layer properties that you want to restore when the layer state is restored. Pick the **Select All** button to pick all properties. Pick the **Clear All** button to deselect all properties.

NOTE

You can also access the **Layer States Manager** by picking the **Layer States Manager** button from the **Layer Properties Manager** or by right-clicking in the layer list view of the **Layer Properties Manager** and selecting **Restore Layer State...**. To save a layer state outside the **Layer States Manager**, pick **New Layer State...** from the **Layer State** drop-down list in the **Layers** panel on the **Home** ribbon tab.

PROFESSIONAL TIP

Importing a layer state file (.las file) to a drawing that does not contain any layers other than 0 adds the layers from the layer state file to the drawing.

After you create a layer state, you can restore layer properties to the settings saved in the layer state at any time. To activate a layer state using the ribbon, select it from the **Layer State** drop-down list in the **Layers** panel on the **Home** ribbon tab. You can also restore a layer state using the **Layer States Manager** by selecting the layer state from the list and picking the **Restore** button.

Supplemental Material

www.g-wlearning.com/CAD/

Layer Settings

For information about options available in the **Layer Settings** dialog box, go to the companion website, navigate to this chapter in the **Contents** tab, and select **Layer Settings**.

Reusing Drawing Content

The drawings created for a given project generally have the same set of standards, and may share a number of common elements. You often duplicate ***drawing content***, such as layers, text and dimension characteristics, symbols, layouts, and details, in many different drawings. One of the fundamental advantages of CADD is the ease with which you can share content between files. Once you create a common drawing element, you can reuse the item as needed in any number of drawings.

drawing content: All of the objects, settings, and other components that make up a drawing.

Drawing templates provide one way to reuse drawing content. Customized templates provide an effective way to start each new drawing using standard settings. Another way to reuse drawing content is to seek out data from existing files. This is a common requirement when you develop related drawings for a specific project or work on similar projects. Sharing drawing content is also common when revising drawings and when duplicating standards used by a consultant, vendor, or client. AutoCAD provides several ways to share drawing content, as described throughout this textbook. **DesignCenter** is one of the most useful tools.

Introduction to DesignCenter

DesignCenter is a palette for managing drawing content between files. See Figure 5-27. You can use **DesignCenter** to locate and reuse a variety of drawing content you will use with this textbook, including layers, linetypes, text styles, dimension styles, multileader styles, table styles, blocks, and external references. **DesignCenter** allows you to load content from a file without actually opening the file.

ADCENTER
Ribbon
View
> Palettes
DesignCenter
Type
ADCENTER
ADC

NOTE

DesignCenter floats by default. You can resize, move, autohide, or dock the floating **DesignCenter**.

Figure 5-27.
Use **DesignCenter** to copy content from one drawing to another.

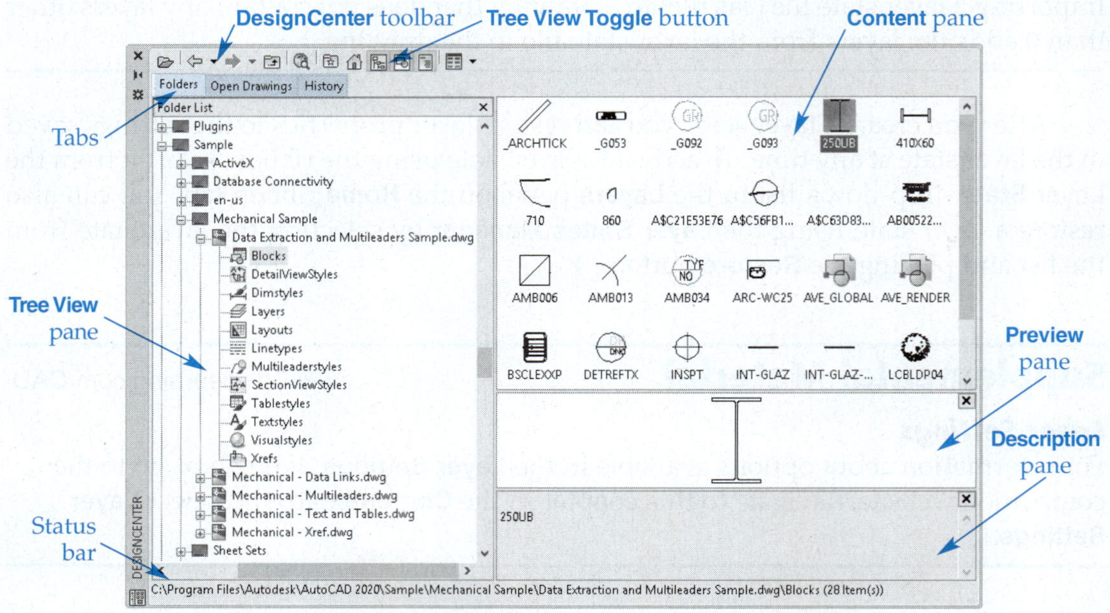

Copying Layers and Linetypes

To copy content using **DesignCenter**, first pick a tab below the **DesignCenter** toolbar to locate a file with drawing content. Pick the **Folders** tab to explore the folders and files found on the hard drive and network, similar to using Windows Explorer. Pick the **Open Drawings** tab to list only drawings that are currently open. The **History** tab lists recently opened drawings and templates. Double-click on a recent file, or right-click on the file and select **Explore**, to navigate to the file in the **Folders** tab.

The **Tree View** pane is displayed directly below the tabs by default. If the **Tree View** pane is not visible, toggle it on by picking the **Tree View Toggle** button from the **DesignCenter** toolbar. Use the **Tree View** pane with the **Folders** or **Open Drawings** tab to select a file with the content to reuse. Double-click the file or pick the plus sign (+) to view content categories. Pick **Layers** to load the **Content** pane with layers found in the selected drawing. See Figure 5-28.

Figure 5-28.
Using **DesignCenter** to display the layers found in a drawing.

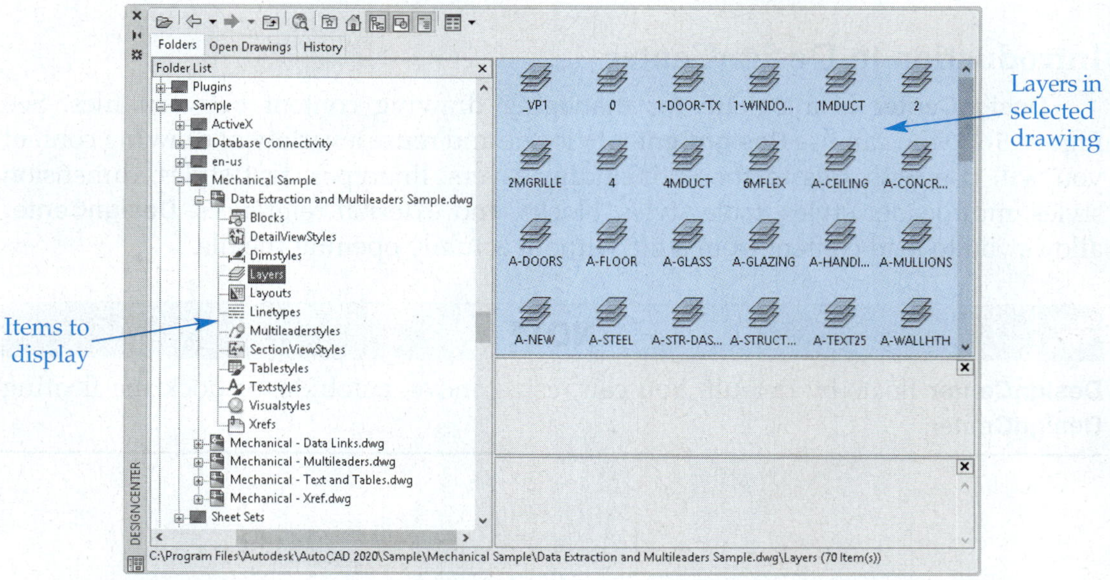

Use a drag-and-drop operation to import content into the current drawing. Use standard selection practices to select multiple layers. Press and hold down the pick button on the layers to import, and then drag the cursor to the drawing window. See Figure 5-29. Release the pick button to add the layers to the current file. An alternative is to right-click on the layers in the **Content** pane and pick **Add Layer(s)**.

Use **DesignCenter** to copy linetypes from one file to another using the same procedure as copying layers. In the **Tree View** pane, select and expand the file containing the linetypes to copy. Pick the **Linetypes** category to display the linetypes in the **Content** pane. Use drag and drop or the shortcut menu to add the linetypes to the current drawing. Figure 5-30 describes several additional **DesignCenter** features.

NOTE

You cannot import drawing content if the content uses the same name as existing content. For example, AutoCAD ignores a layer you try to import if the layer name already exists in the destination drawing. The existing settings for the layer are preserved, and a message at the command line indicates that duplicate settings were ignored. Purge the existing content, as described in Chapter 24, in order to import content with the same name.

Exercise 5-7

www.g-wlearning.com/CAD/

Complete the exercise on the companion website.

Template Development

www.g-wlearning.com/CAD/

Adding Layers

For detailed instructions on adding layers to each of your drawing templates, go to the companion website, navigate to this chapter in the **Contents** tab, and select **Adding Layers**.

Figure 5-29.
To copy layers shown in **DesignCenter** into the current drawing, drag and drop the layers into the drawing area.

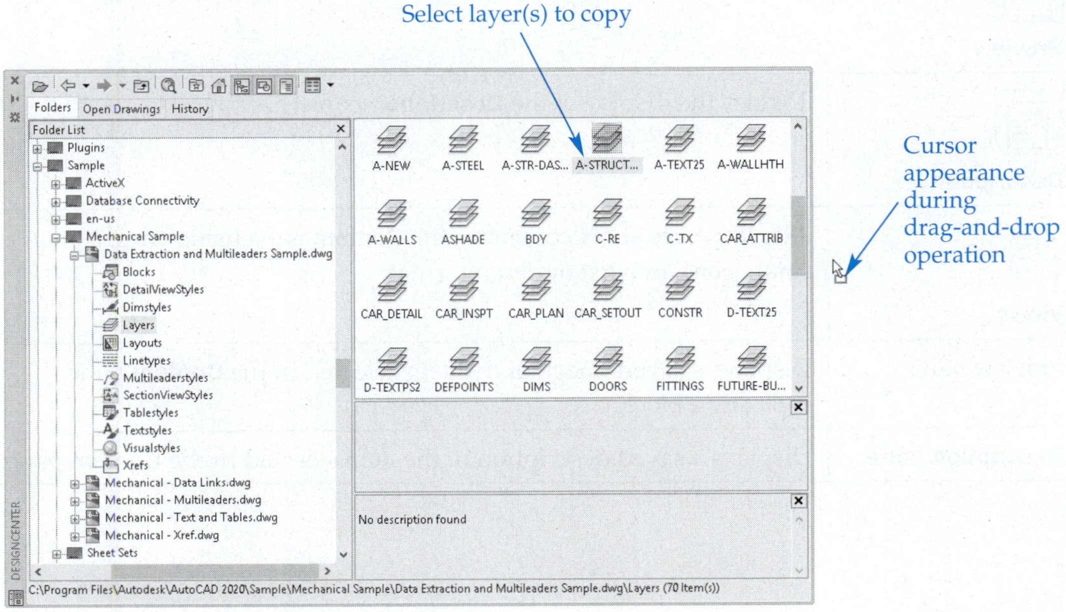

Figure 5-30.
Additional items available in **DesignCenter**.

Item	Function
Load	Displays the **Load** dialog box, which you can use to select a file with drawing content or download online content to **DesignCenter**.
Back	Select to show the content of the last selected file, or select the flyout to access a list of previously viewed content.
Forward	Pick after using the **Back** button to show the content of the last selected file, or select the flyout to access a list of previously viewed content.
Up	Moves up one level in the **Folder List** tree to a file, folder, drive, This PC, or the desktop.
Search	Displays the **Search** dialog box, which you can use to locate drawing content on any drive that meets search criteria and load the content to **DesignCenter**.
Favorites	Navigates to the Autodesk folder in your Favorites folder. To add commonly used content to the Favorites folder, right-click on an item in the **Folder List** area and select **Add to Favorites**.
Home	Navigates to the AutoCAD Sample folder. To change the "home" location, right-click on an item in the **Folder List** area and select **Set as Home**.
Tree View Toggle	Toggles the display of the **Tree View** pane.
Preview	Toggles the display of the **Preview** pane.
Description	Toggles the display of the **Description** pane.
Views	Allows you to show content in the **Content** pane using large icons, small icons, or a list or detail format.
Preview pane	Displays a saved image of the item selected in the **Content** pane, typically a block.
Description pane	Displays a saved description of the item selected in the **Content** pane.

Chapter Review

Answer the following questions using the information in this chapter.

1. Identify the following linetypes.

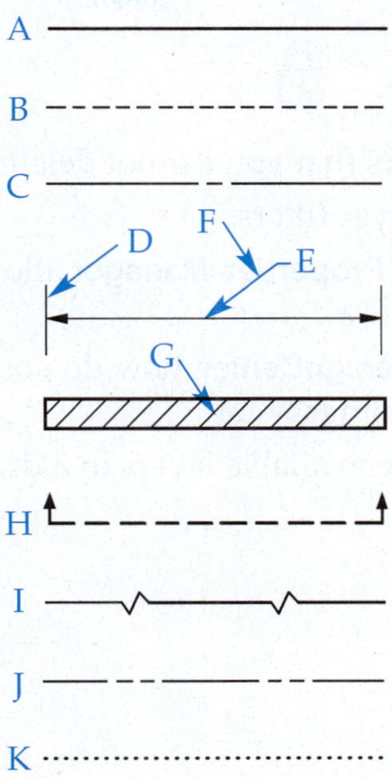

2. How can you tell if a layer is off, thawed, or unlocked by looking at the **Layer Properties Manager**?
3. Should you draw on the 0 layer? Explain.
4. Identify two ways to access the **Layer Properties Manager**.
5. How can you enter several new layer names consecutively in the **Layer Properties Manager** without using the **New Layer** button?
6. How do you make another layer current using the **Layer Properties Manager**?
7. How do you make another layer current using the ribbon?
8. How do you make another layer current using the command line?
9. How can you display the **Select Color** dialog box from the **Layer Properties Manager**?
10. List the names and numbers of the first seven colors in the AutoCAD Color Index (ACI).
11. How do you use the **Layer Properties Manager** to change the linetype assigned to a layer?
12. What is the default loaded linetype in AutoCAD?
13. What condition must exist before you can assign a linetype to a layer?
14. Describe the basic procedure to change a layer's linetype to HIDDEN.
15. What is the function of the linetype scale?
16. Explain the effects of using a global linetype scale.
17. Why do you have to be careful when changing linetype scales?
18. What is the state of a layer not displayed on-screen and not calculated by the computer when you regenerate the drawing?
19. Explain the purpose of locking a layer.
20. Explain the difference between a locked layer state and layer transparency.

21. Identify the following layer status icons.

A. 

B. 

C. 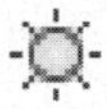

D. 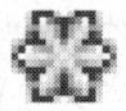

E. 

F. 

22. Identify at least three layers that you cannot delete from a drawing.
23. Describe the purpose of layer filters.
24. Which button in the **Layer Properties Manager** allows you to save layer settings so they can be restored later?
25. In the **Tree View** pane of **DesignCenter**, how do you view the content categories of one of the listed open drawings?
26. How do you display all the available layers in a drawing in **DesignCenter**?

Drawing Problems

Start AutoCAD if it is not already started. For each problem, start a new drawing from scratch or use an appropriate template of your choice. The template should include layers for drawing the given objects. Add layers as needed. Draw all objects using appropriate layers. Follow the specific instructions for each problem. Use only drawing commands and techniques you have already learned. Do not draw dimensions or text. Use your own judgment and approximate dimensions when necessary.

▼ Basic

1. **Mechanical** Draw the hex head bolt pattern shown. Save the drawing as P5-1.

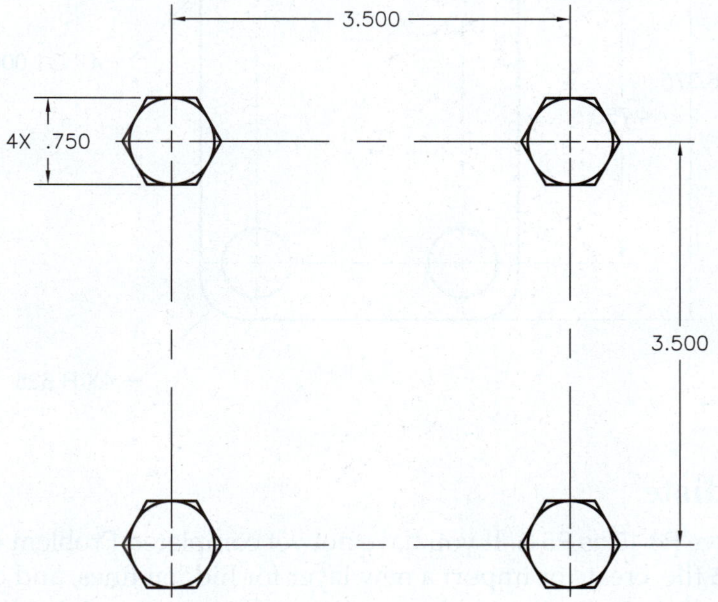

2. **Mechanical** Draw the 1/2″ hex nut with 3/4″ across the flats and a .422″ minor diameter. Save the drawing as P5-2.

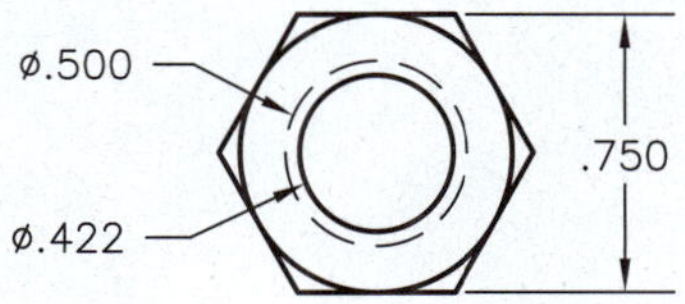

3. **Mechanical** Draw the part view shown. Save the drawing as P5-3.

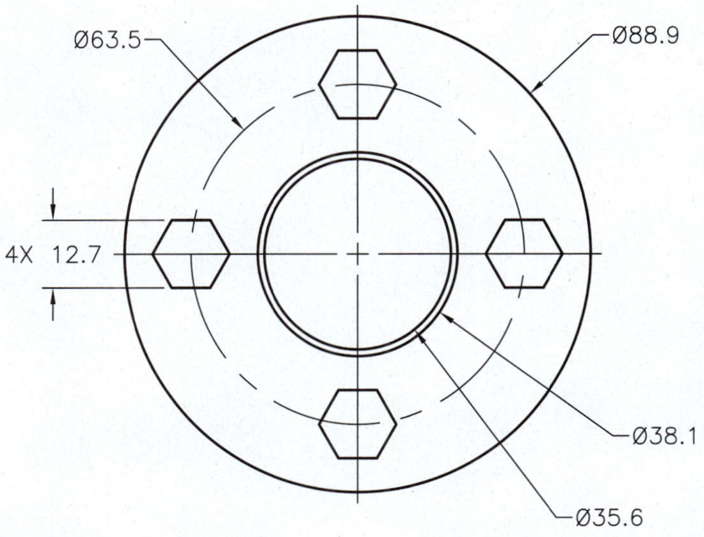

Drawing Problems - Chapter 5

4. **Mechanical** Draw the part view shown. Save the drawing as P5-4.

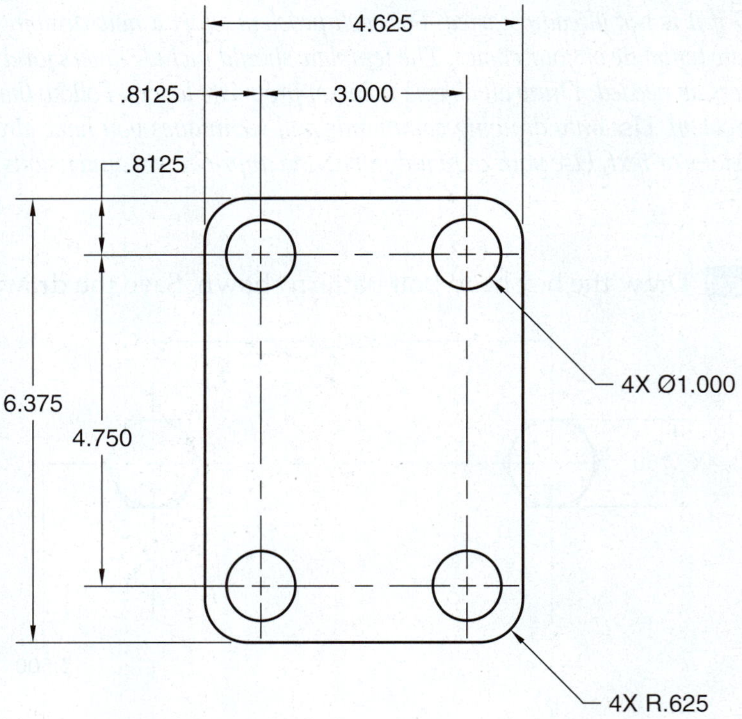

▼ Intermediate

5. **Piping** Save P4-18 as P5-5. If you have not yet completed Problem 4-18, do so now. In the P5-5 file, create or import a new layer for hidden lines, and draw or redraw the hidden lines using the hidden line layer. Change the global linetype scale to achieve an effect similar to the hidden lines shown in Chapter 4.

▼ Advanced

6. **Civil** Draw the plot plan shown using the layers shown in the table. Draw objects proportionate to the drawing shown. Use dimensions based on your experience, research, or measurements. Save the drawing as P5-6.

Name	Color	Linetype	Lineweight	Description
C-BLDG	white	Continuous	0.60 mm	Buildings, primary structures
C-DRIV	8	Continuous	0.35 mm	Driveway
C-FENC	134	FENCELINE2	0.35 mm	Fences
C-NGAS-UNDR	30	GAS_LINE	0.50 mm	Underground natural gas
C-PROP	white	PHANTOM	0.70 mm	Property boundary
C-ROAD-CNTR	white	CENTER	0.35 mm	Road center
C-ROAD-CURB	9	Continuous	0.35 mm	Road curb
C-SSWR-UNDR	54	DASHED	0.50 mm	Underground sanitary sewer
C-SWLK	252	Continuous	0.35 mm	Sidewalk

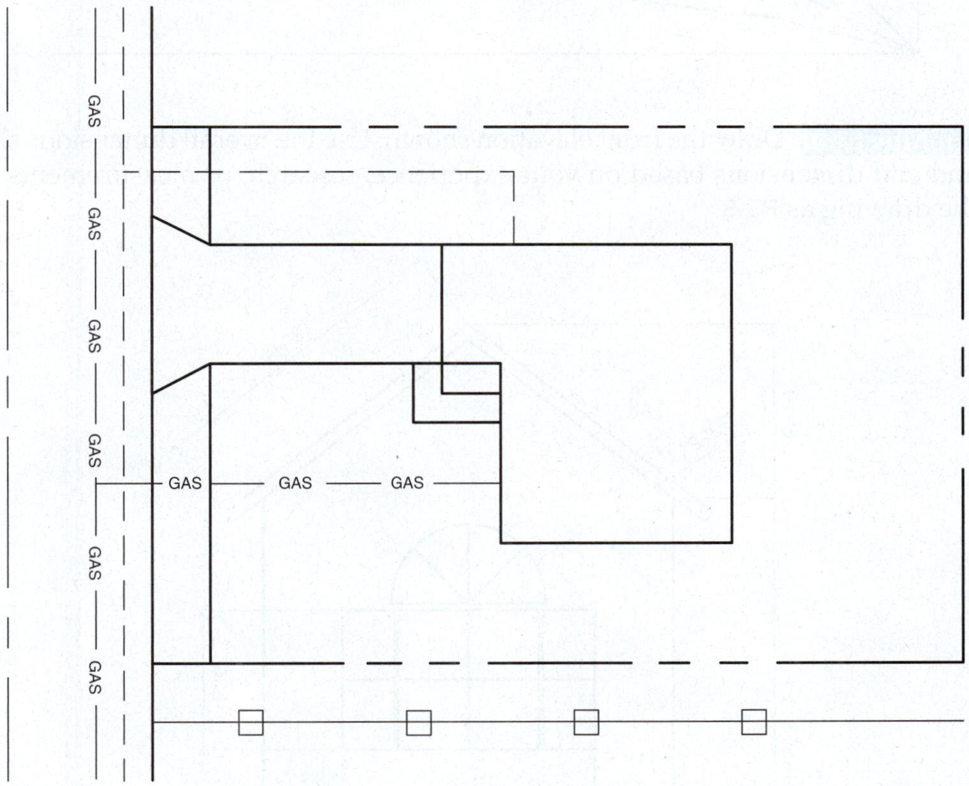

7. **Graphic Design** Draw the line chart shown. Use the linetypes shown, which include Continuous, HIDDEN, PHANTOM, CENTER, FENCELINE1, and FENCELINE2. Draw objects proportionate to the drawing. Save the drawing as P5-7.

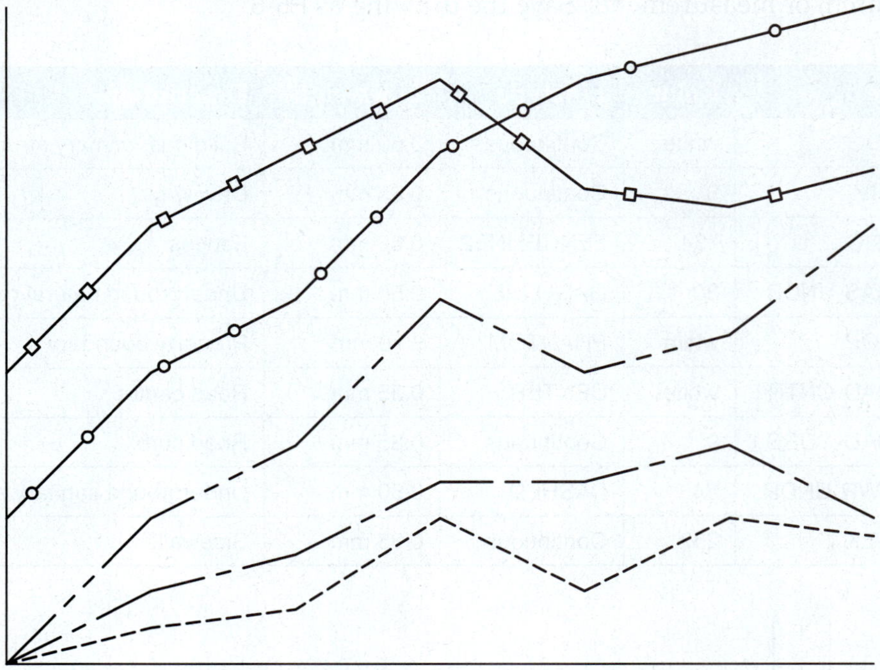

8. **Architectural** Draw the front elevation shown. Use the overall dimensions shown and add dimensions based on your experience, research, or measurements. Save the drawing as P5-8.

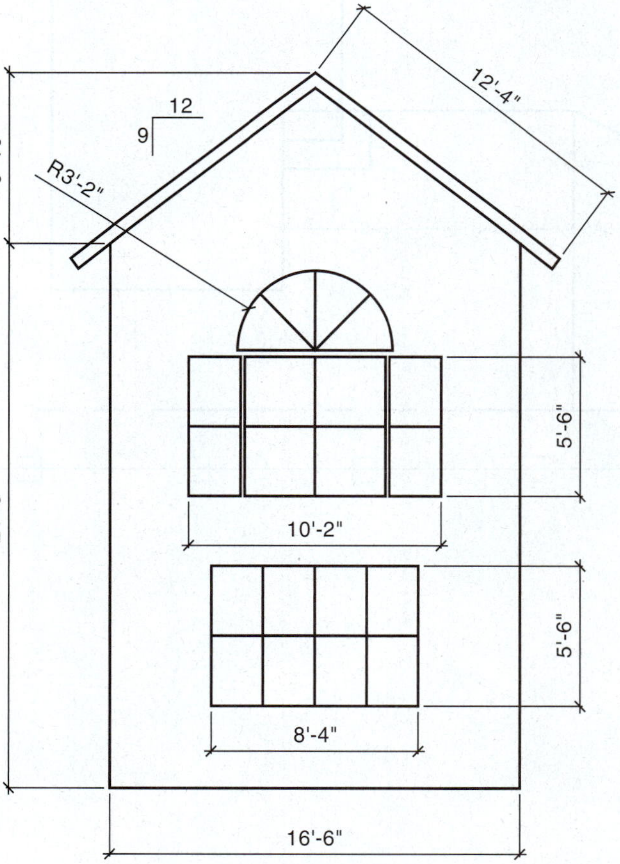

9. **Electronics** Create the controller integrated circuit diagram shown. Use a ruler or scale to keep the proportion as close as possible. Save the drawing as P5-9.

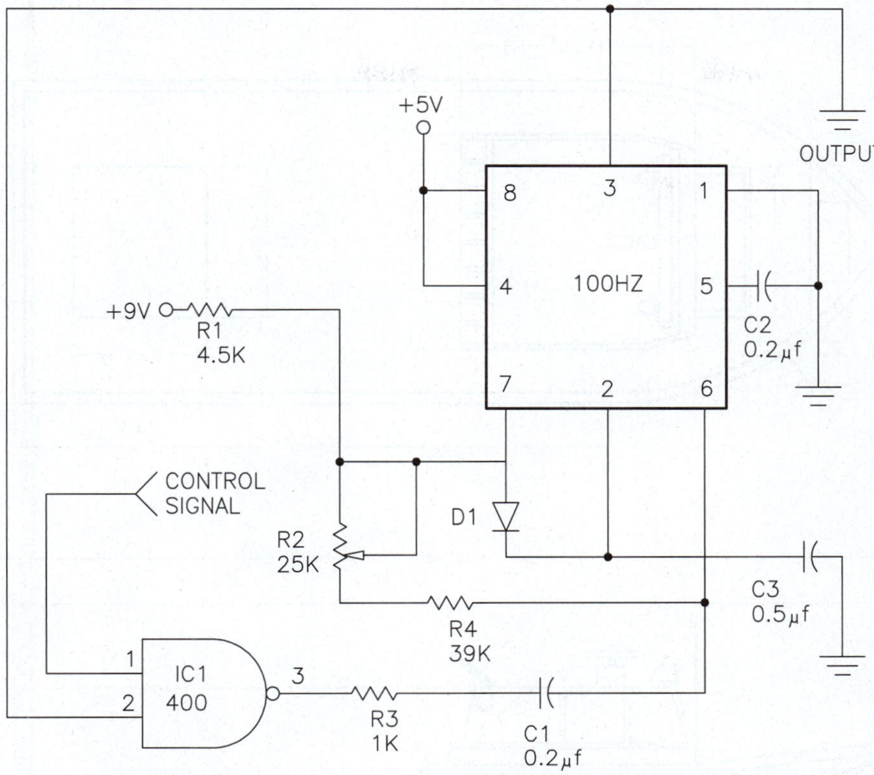

10. **General** Draw a drift boat similar to the drift boat shown. Use dimensions based on your experience, research, or measurements. Use layers to group specific objects. Save the drawing as P5-10.

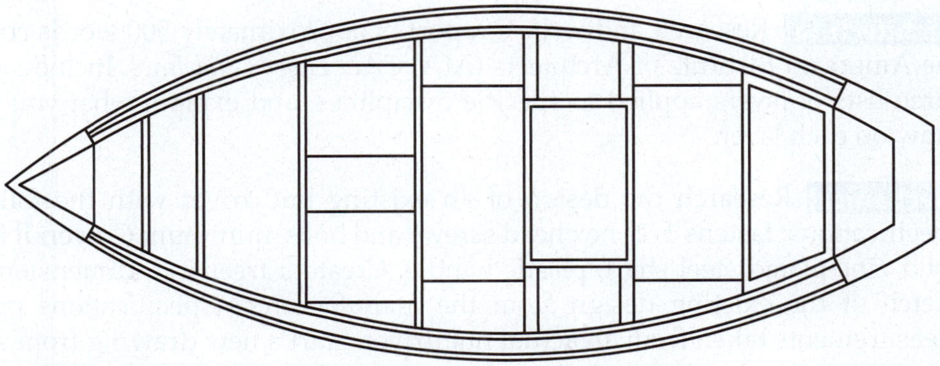

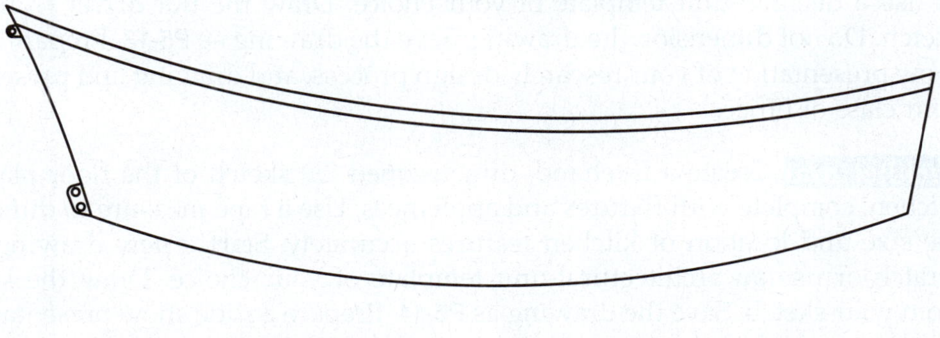

Drawing Problems - Chapter 5

11. **General** Draw a fishing boat similar to the fishing boat shown. Use the overall dimensions shown and add dimensions based on your experience, research, or measurements. Use layers to group specific objects. Save the drawing as P5-11.

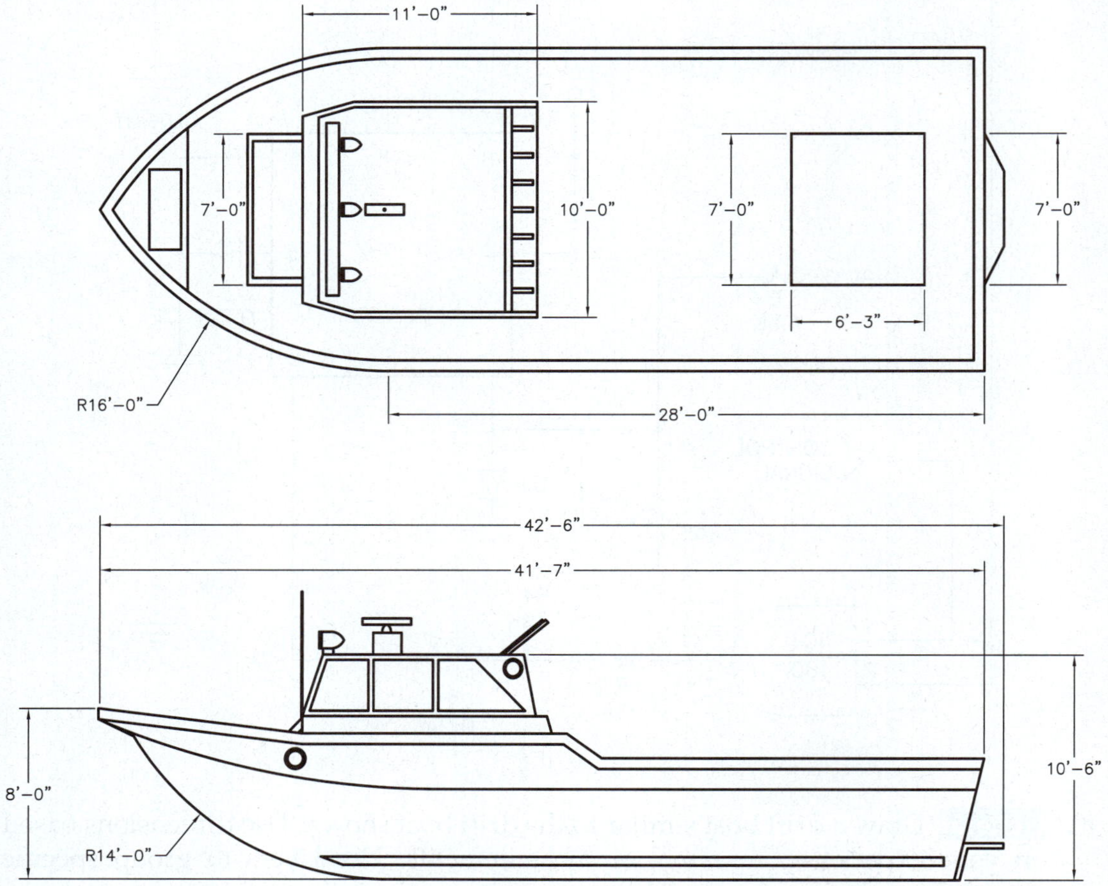

12. **Architectural** Research and write a report of approximately 500 words covering the American Institute of Architects (AIA) *CAD Layer Guidelines*. Include at least three lists of layers applied to specific disciplines, and explain what you would draw on each layer.

13. **Mechanical** Research the design of an existing nut driver with the following specifications: fastens 1/2″ hex head screws and bolts, minimum 6″ overall length, solid zinc-plated steel shaft, plastic handle. Create a freehand, dimensioned 2D sketch of the existing design from the manufacturer's specifications or from measurements taken from an actual nut driver. Start a new drawing from scratch or use a decimal-unit template of your choice. Draw the nut driver from your sketch. Do not dimension the drawing. Save the drawing as P5-13. Prepare a slide show presentation of your research, design process, and drawing and present it to your class or office.

14. **Architectural** Create a freehand, dimensioned 2D sketch of the floor plan of a kitchen, complete with fixtures and appliances. Use a tape measure to dimension the size and location of kitchen features accurately. Start a new drawing from scratch or use an architectural-unit template of your choice. Draw the kitchen from your sketch. Save the drawing as P5-14. Prepare a slide show presentation of your research, design process, and drawing and present it to your class or office.

AutoCAD Certified Professional Exam Practice

Answer the following questions using the information in this chapter.

1. Which line widths are recommended by ASME Y14.2? *Select all that apply.*
   A. extra fine
   B. thin
   C. thick
   D. 3X
   E. 4X
2. Which of the following line characteristics would you typically assign to a layer for drawing object lines on a mechanical part view drawing? *Select the one item that best answers the question.*
   A. thick, HIDDEN linetype
   B. 4X, Continuous linetype
   C. extra fine, Continuous linetype
   D. 0.3 mm, Continuous linetype
   E. 0.6 mm, Continuous linetype
3. Which of the following operations are possible on a locked layer? *Select all that apply.*
   A. drawing new objects
   B. editing objects
   C. erasing objects
   D. selecting objects
   E. displaying objects on-screen

Follow the instructions in each problem. Complete each problem by answering the related question.

4. **Navigate to this chapter on the companion website and open CPE-05linetype.dwg.** Create a new layer named Centerline and assign it the color blue. Load the CENTER linetype and apply it to the Centerline layer. Then select the existing line and change it to the Centerline layer. Enter **LTSCALE** and specify a linetype scale of 2.0. The centerline linetype consists of long dashes alternating with short dashes. How many short dashes appear on the line at the current linetype scale?

5. **Navigate to this chapter on the companion website and open CPE-05layers.dwg.** Use the **Layer** drop-down list and the **Layer Properties Manager** to find the following information: On which layer was Line A drawn, and what lineweight is assigned to that layer?

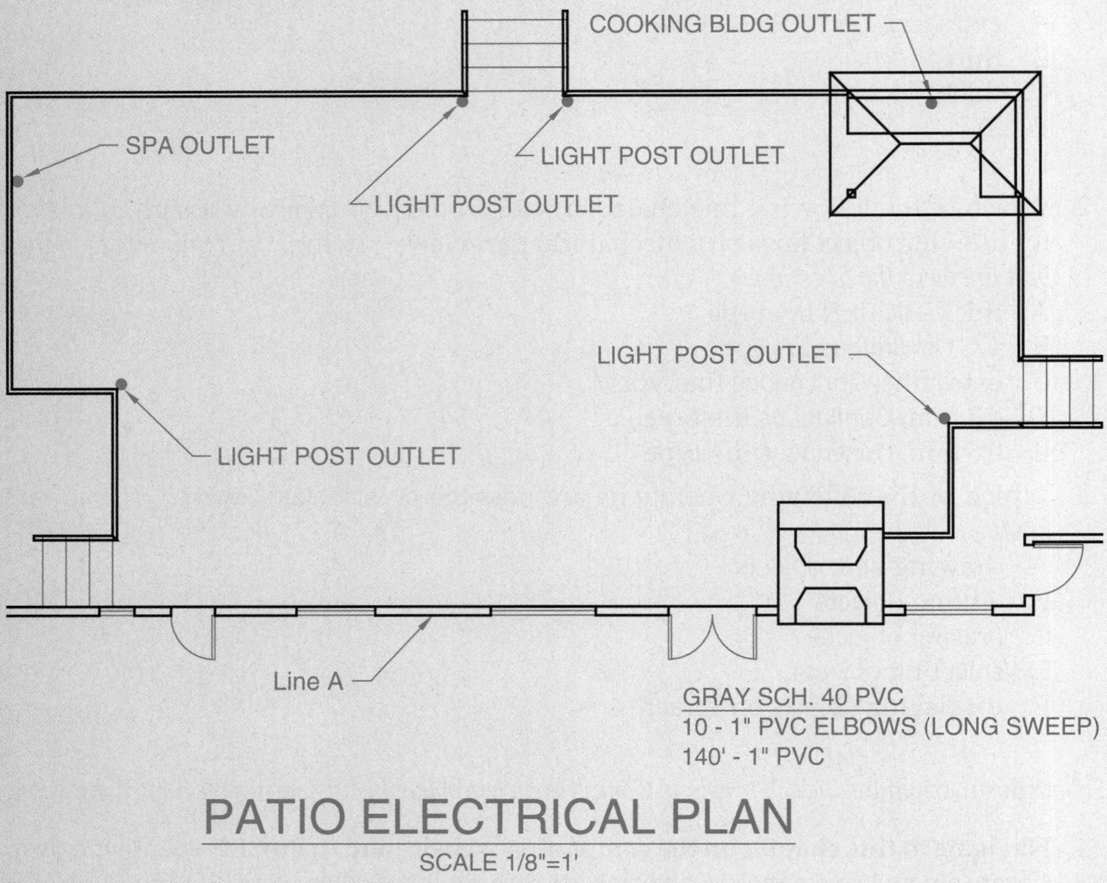

AutoCAD Certified Professional Exam Practice - Chapter 5

CHAPTER 6

View Tools and Basic Plotting

Learning Objectives

After completing this chapter, you will be able to:

- ✓ Adjust the graphics window to view specific portions of a drawing.
- ✓ Use display commands transparently.
- ✓ Control object display order.
- ✓ Create named views for direct recall.
- ✓ Create multiple model viewports in the drawing window.
- ✓ Print and plot drawings.

As you create drawings that are more complex and begin drawing large and small objects, you will realize the importance of using view tools to adjust the drawing display. This chapter describes several view commands that you will use frequently during the drawing process. This chapter also introduces printing and plotting so that you can begin printing your drawings. Later chapters describe printing and plotting in more detail.

View Tools

View tools help you to navigate in drawings. They adjust the display without changing the objects in a drawing. View tools are available from the ribbon, keyboard entry, shortcut menus, the viewport controls, the navigation bar, steering wheels, and the view cube.

view tools: AutoCAD display commands, options, and settings.

Navigation Bar

The navigation bar is a view toolbar positioned near the upper-right corner of the drawing window by default. See **Figure 6-1**. The navigation bar includes tool buttons for accessing common view commands. Right-click on a tool button or pick the **Settings** flyout to access options for changing the navigation bar.

A command or option accessible from the navigation bar appears as a graphic in the margin of this textbook. The graphic indicates picking a navigation bar button. The example shown in this margin illustrates using the navigation bar to access the **PAN** command.

NAVBAR
Ribbon: View > Viewport Tools > Navigation Bar
Type: NAVBAR

PAN
Navigation Bar: Pan

Figure 6-1.
The default navigation bar positioned below the view cube in model space.

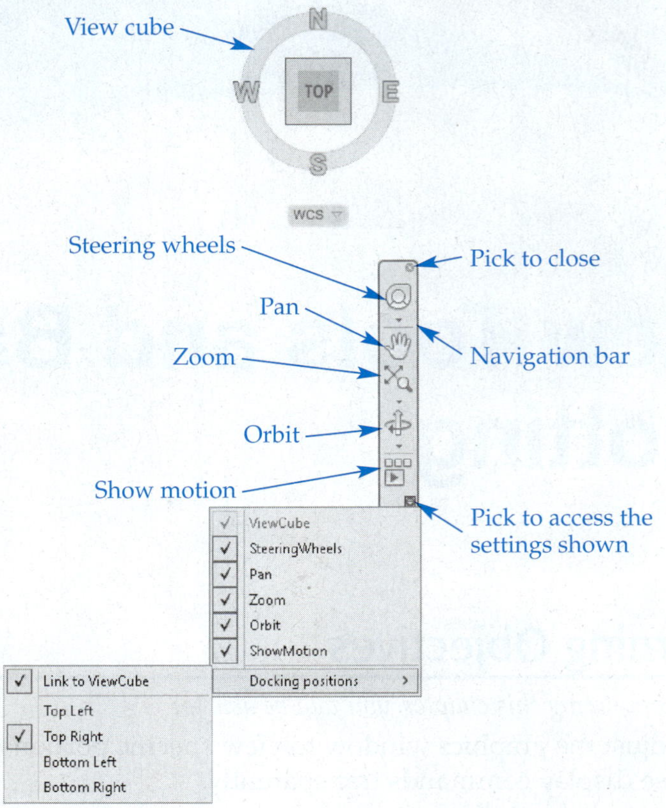

zoom: Makes objects appear bigger (zoom in) or smaller (zoom out) on the screen without affecting their actual size.

zoom in: Changes the display area to show a smaller portion of the drawing at a higher magnification.

zoom out: Changes the display area to show a larger portion of the drawing at a lower magnification.

ZOOM

Ribbon
View
> Navigate

Zoom flyout

Type
ZOOM
Z

NOTE

Some of the view commands available in the default Drafting & Annotation workspace are appropriate for 3D applications only. ***AutoCAD and Its Applications—Advanced*** explains 3D and animation view commands.

Zooming

AutoCAD provides several methods for ***zooming***. As described in Chapter 3, an easy way to zoom is to use a mouse with a scroll wheel. To ***zoom in***, roll the mouse wheel forward. To ***zoom out***, roll the mouse wheel back. This technique also pans to the location of the crosshairs while zooming. Double-click the wheel to zoom to the furthest extents of objects in the drawing.

Additional zooming options are available from the **ZOOM** command. The ribbon and navigation bar provide a direct link to **ZOOM** command options. See Figure 6-2. In contrast, when you issue the **ZOOM** command using dynamic input or the command line, you must select a specific option to receive appropriate prompts.

NOTE

The **Navigate** panel of the **View** ribbon tab is not displayed by default. To display the **Navigate** panel, right-click on a portion of the ribbon unoccupied by a panel and select **Navigate** from the **Show Panels** cascading menu.

Realtime Zooming

realtime zoom: A zoom that you view as it occurs.

Access the **Realtime** option to apply ***realtime zooming***. The zoom cursor appears as a magnifying glass with a plus sign icon when zooming in and a minus sign icon when zooming out. Press and hold the left mouse button and move the cursor up to

Figure 6-2.
ZOOM options in the **Zoom** flyout on the **Navigate** panel of the **View** ribbon tab and on the navigation bar.

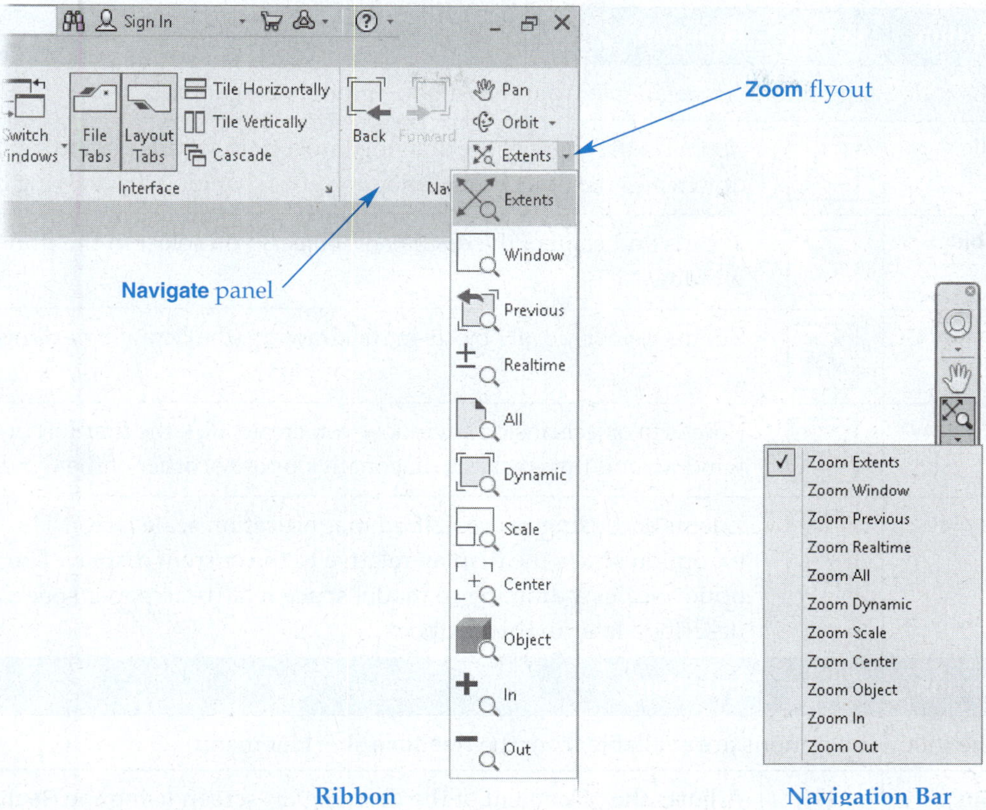

zoom in or down to zoom out. Release the mouse button when you achieve the appropriate display. The **Realtime** option includes a shortcut menu that you can access while the zoom cursor is active. Use the menu to select a different zoom option or view tool. Press [Enter] or [Esc] to exit realtime zooming.

Additional Zoom Options

The mouse wheel and the **Realtime** option of the **ZOOM** command are effective for most zooming requirements. Apply one of the other zoom options to achieve a specific display. Figure 6-3 provides a description of the most useful additional zoom options.

NOTE

The **Previous**, **In**, **Out**, **Center**, and **Dynamic** options of the **ZOOM** command provide functions that you can achieve more easily using other viewing methods.

PROFESSIONAL TIP

Apply the **All** option of the **ZOOM** command in each of your drawing templates to zoom to the edge of the drawing limits. When you start a new drawing using one of your templates, the drawing will automatically display the entire drawing area.

Exercise 6-1

www.g-wlearning.com/CAD/

Complete the exercise on the companion website.

Figure 6-3.
Zoom options available from the ribbon, the navigation bar, and the shortcut menu that appears when you right-click while using the **Realtime** option of the **ZOOM** command.

Option	Button	Function
The following options are available from the ribbon and the navigation bar.		
All		Zooms to the edges of the drawing limits, or to the edge of the geometry drawn past the limits. Use this option after changing the drawing limits.
Object		Zooms and centers the display on objects you select in the drawing window.
Extents*		Zooms to include all objects in the drawing (the "drawing extents").
Window*		Zooms to objects inside a window you create. Pick the first corner of the window, and then pick the diagonally opposite corner of the window.
Scale		Zooms according to a specified magnification scale factor. The **nX** option scales the display relative to the current display. The **nXP** option scales a drawing in model space relative to paper space, as described later in this textbook.
Option	**Cursor**	**Function**
The following options are available from the **Realtime** shortcut menu.		
Pan		Adjusts the placement of the drawing on-screen using the **Realtime** option of the **PAN** command.
Zoom		Returns from the **Realtime** option of the **PAN** command to the **Realtime** option of the **ZOOM** command.
3D Orbit		Rotates the view of a 3D object as described in *AutoCAD and Its Applications—Advanced*.
Zoom Original		Restores the previous display before any realtime zooming or panning occurred; useful if the modified display is not appropriate.

*Also available from the **Realtime** shortcut menu.

pan: Move the drawing display to view different portions of the drawing without changing the magnification.

PAN
Ribbon
View
> Navigate

Pan
Type
PAN
P

realtime panning: A panning operation in which you can see the drawing move on-screen as you pan.

Panning

An easy way to ***pan*** is to use a mouse with a scroll wheel. Press and hold the mouse wheel and then move the mouse. An alternative is to access the **PAN** command to apply ***realtime panning***. The pan cursor appears as a hand. Press and hold the left mouse button and move the cursor in the direction to pan. Right-click to display the same shortcut menu available for realtime zooming. Press [Enter] or [Esc] to exit realtime panning.

Another, but less effective, way to pan involves using drawing window scroll bars. To display the scroll bars, check the **Display scroll bars in drawing window** check box in the **Window Elements** area of the **Display** tab in the **Options** dialog box.

NOTE

AutoCAD supports some high-performance mice and touch screen monitors that include additional buttons and functions.

NOTE

By default, when you use the **UNDO** command after multiple zooming and panning operations, zooms and pans are grouped together, allowing you to return to the original view. To make each zoom and pan operation count individually, deselect the **Combine zoom and pan commands** check box in the **Undo/Redo** area of the **User Preferences** tab in the **Options** dialog box.

View Back and Forward

Ribbon
View
> Navigate
View Back
Type
VIEWBACK
VIEWBACK

Ribbon
View
> Navigate
View Forward
Type
VIEWFORWARD
VIEWFORWARD

Issue the **VIEWBACK** command as needed to cycle back to prior views created by an operation such as zooming or panning. The **VIEWFORWARD** command is available after you issue the **VIEWBACK** command, and can be used to return to the previous view. The **VIEWBACK** and **VIEWFORWARD** commands are most effective in basic 2D drafting for working between specific displays quickly, such as viewing fine detail while zoomed in and then returning to the previous view of the drawing extents to view the overall design.

Exercise 6-2

www.g-wlearning.com/CAD/

Complete the exercise on the companion website.

Introduction to Steering Wheels

AutoCAD steering wheels provide an alternate way to access and use certain view commands. A ***navigation wheel*** is a specific steering wheel. Some navigation wheels and many of the commands available from navigation wheels are most appropriate for 3D modeling. This textbook focuses on the default navigation wheels and the **ZOOM**, **CENTER**, **PAN**, and **REWIND** commands for 2D applications. *AutoCAD and Its Applications—Advanced* explains additional navigation wheel commands and settings.

navigation wheel: A steering wheel designed for use in a specific drawing setting or with a particular type of drawing.

Type
NAVSWHEEL
NAVSWHEEL

The Full Navigation wheel is the default model space steering wheel. When the Full Navigation wheel is displayed, the UCS icon changes to a 3D display. The 2D wheel is the only steering wheel available in layout space. See Figure 6-4. To access a different navigation wheel, select from the **Steering Wheels** flyout on the navigation bar, the shortcut menu that appears when you right-click while using a navigation wheel, or the flyout in the lower-right corner of a navigation wheel. The default format is big, but mini navigation wheels are available.

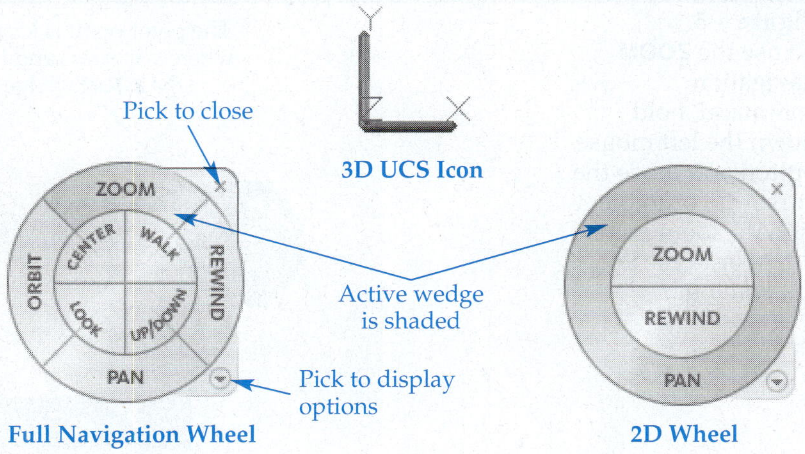

Figure 6-4.
The Full Navigation wheel appears by default in model space. The 2D wheel is the only navigation wheel available in a layout. You can also display the 2D wheel in model space.

wedges: The parts of a navigation wheel that contain navigation commands.

Navigation wheels display next to the cursor. ***Wedges*** contain the individual navigation commands and function similar to tool buttons. Hover over a wedge to highlight it. Some commands activate when you pick a wedge, and other commands require that you hold down the left mouse button on the wedge.

A navigation wheel remains on-screen until you close it. This allows you to use multiple navigation commands. To close a navigation wheel, pick the **Close** button in the upper-right corner of the wheel, press [Esc] or [Enter], or right-click and pick **Close Wheel**.

NOTE

The flyout in the lower-right corner of a navigation wheel provides access to other navigation wheels and steering wheel settings. The flyout includes a **Close Wheel** option and may include additional options, such as **Fit to Window**, which zooms and pans to show all objects centered in the drawing window.

Zooming with the Navigation Wheel

The **ZOOM** navigation command offers realtime zooming. Press and hold the left mouse button on the **ZOOM** wedge on the Full Navigation wheel to display the pivot point icon and zoom navigation cursor. See Figure 6-5. The pivot point is the location at which you press the **ZOOM** wedge. Move the zoom navigation cursor up or to the right to zoom in. Move the cursor down or to the left to zoom out. The pivot point icon also zooms in or out as a visual aid to zooming. Release the left mouse button when you achieve the appropriate display.

Using the Center Navigation Command

The **CENTER** navigation command on the Full Navigation wheel centers the display screen at a picked point, without zooming. Press and hold the left mouse button on the **CENTER** wedge. The pivot point icon appears when you move the cursor over an object. Release the mouse button to pan so the object at the pivot point is displayed in the center of the drawing window when you release the mouse button. See Figure 6-6.

Panning with the Navigation Wheel

The **PAN** navigation command offers realtime panning. Press and hold the left mouse button on the **PAN** wedge to display the pan navigation cursor. Move the pan navigation cursor in the direction to pan. Release the mouse button when you achieve the desired display.

Rewinding

The **REWIND** navigation command allows you to observe display changes and return to a previous display. For example, if you use the navigation wheel to zoom in,

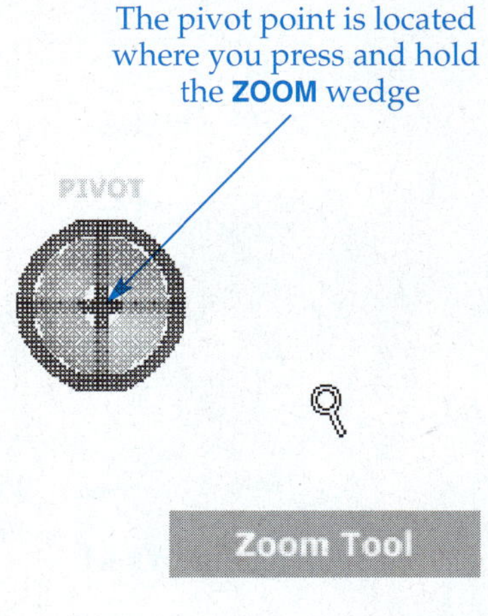

Figure 6-5. To use the **ZOOM** navigation command, hold down the left mouse button and move the cursor up or to the right to zoom in, or down or to the left to zoom out.

Figure 6-6.
To use the **CENTER** navigation command, hold down the left mouse button, move the cursor over an object at the point you want to center in the drawing window, and release the mouse button.

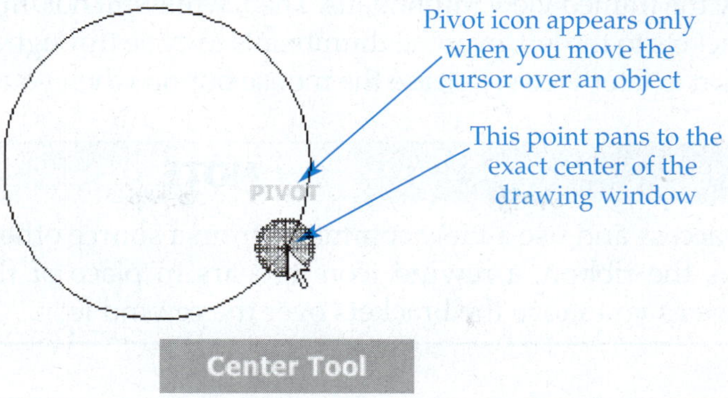

then pan, then zoom out, you can rewind through each action and return to the original zoomed-in display, the panned display, and then back to the current zoomed-out display. You can rewind through view actions created using most view commands, not just navigation wheel commands.

Pick the **REWIND** command once to return to the previous display. Thumbnail images appear in frames as the previous view is restored. The thumbnail with an orange frame surrounded by brackets indicates the restored display and its location in the sequence of events. See **Figure 6-7**. Pick the **REWIND** button repeatedly to cycle back through prior views. Another option is to press and hold the left mouse button on the **REWIND** wedge to

Figure 6-7.
Use the **REWIND** navigation command to step back through and restore previous display configurations. This example shows rewinding through the views of a detail drawing of a mechanical part.

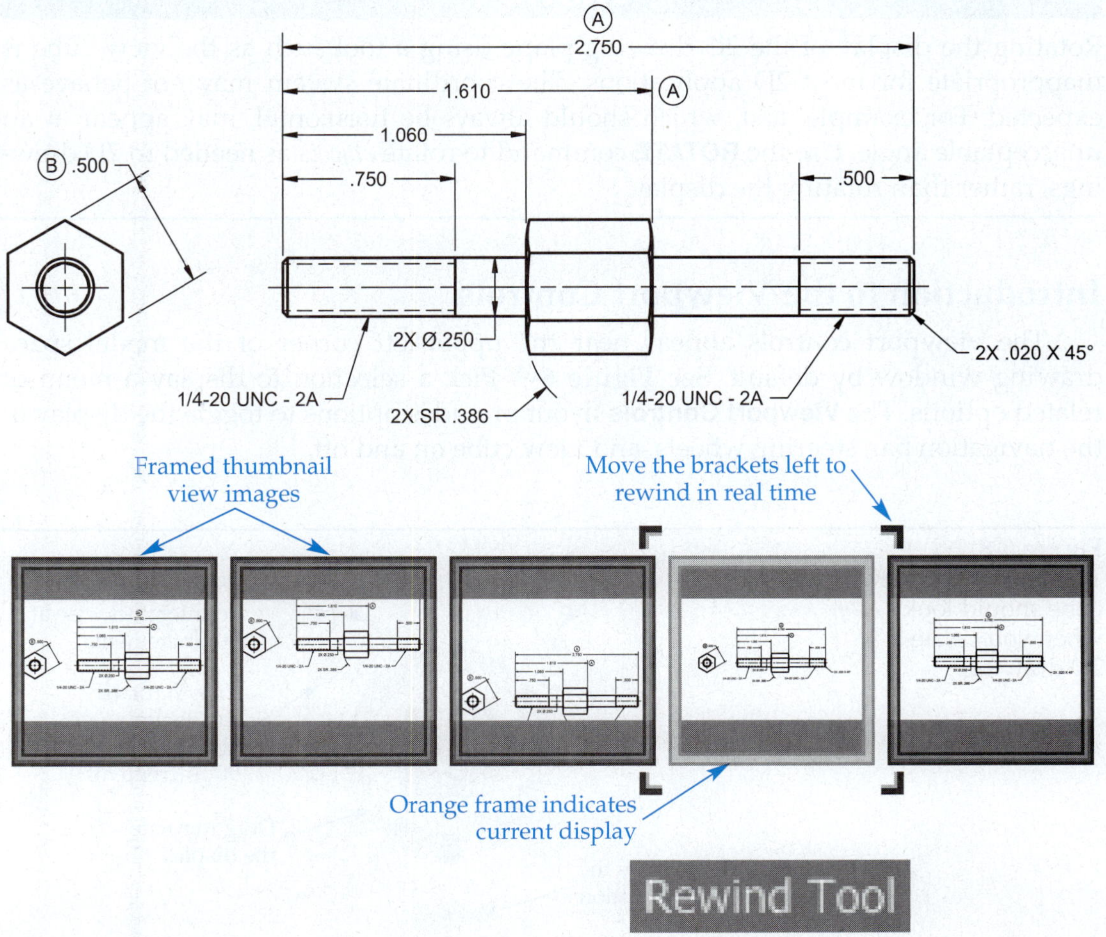

display the framed view thumbnails. Then, while still holding the left mouse button, move the brackets to the left, over the thumbnails to cycle through earlier views, and to the right to return to later views. Release the mouse button when you achieve the desired display.

NOTE

If you access and use a view command from a source other than the navigation wheel, such as the ribbon, a rewind icon appears in place of the thumbnail. A thumbnail displays as you move the brackets over the rewind icon.

Exercise 6-3

www.g-wlearning.com/CAD/

Complete the exercise on the companion website.

Introduction to the View Cube

The view cube appears near the upper-right corner of the drawing window. This is the default display when you start a new drawing using the default acad.dwt template. See Figure 6-8. The view cube includes a WCS menu that is useful for some 2D drawing applications. However, the primary purpose of the view cube is to view a 3D model precisely using a labeled cube. Until you are ready to create 3D models, use the **NAVVCUBE** command to turn off the view cube. If you change the view orientation, pick the **Home** button and then the **TOP** face to return to the default 2D view. ***AutoCAD and Its Applications—Advanced*** provides complete information on the view cube.

NAVVCUBE
Ribbon
View
> Viewport Tools
View Cube
Type
NAVVCUBE

CAUTION

Rotating the display of the 2D drawing plane using a tool such as the view cube is inappropriate for most 2D applications. The coordinate system may not behave as expected. For example, text, which should always be horizontal, may appear at an unacceptable angle. Use the **ROTATE** command to rotate *objects* as needed in 2D drawings, rather than rotating the display.

Introduction to the Viewport Controls

The viewport controls appear near the upper-left corner of the model space drawing window by default. See Figure 6-9. Pick a selection to display a menu of related options. The **Viewport Controls** flyout provides options to toggle the display of the navigation bar, steering wheels, and view cube on and off.

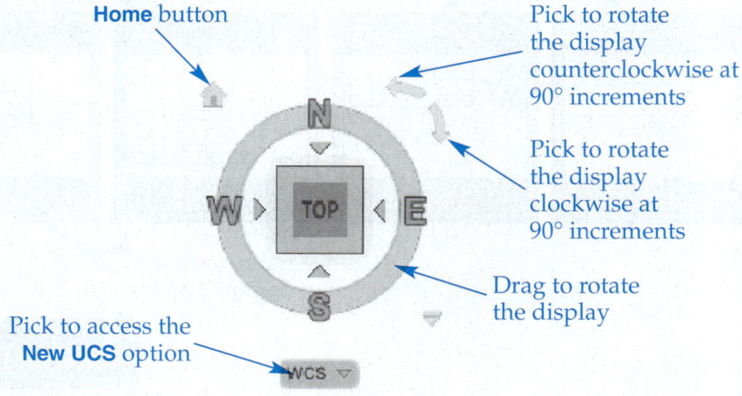

Figure 6-8. This is how the view cube should look when you create a 2D drawing.

Figure 6-9.
The viewport controls in the model space drawing window, with the **Viewport Controls** flyout selected.

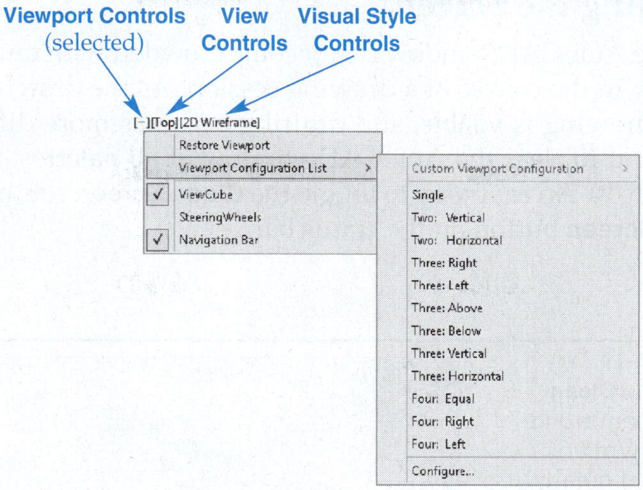

The **Viewport Controls** flyout also offers convenient access to model viewport controls, as explained later in this chapter. The **View Controls** flyout contains options that are appropriate for viewing a 3D model, as explained in ***AutoCAD and Its Applications—Advanced***. The **View Controls** flyout also includes an option for accessing the **View Manager** described later in this chapter. The **Visual Style Controls** flyout lists options typically associated with 3D model display, as explained in ***AutoCAD and Its Applications—Advanced***.

NOTE

Use the **Display the Viewport Controls** check box in the **Display Tools in Viewport** area of the **3D Modeling** tab in the **Options** dialog box to control the display of the viewport controls.

Regenerating the Screen

AutoCAD stores all drawing information in the file database, but only displays data on-screen that is necessary for you to work. ***Regenerating*** evaluates the drawing to correct display issues that result when the file database does not match exactly what you see on-screen. For example, if curved objects appear as straight segments when you zoom in, regenerate the display to smooth the curves.

regenerating: Recalculating and redisplaying all objects on-screen to correspond to the information in the file database.

The **REGENMODE** system variable is set to 1 (on) by default. AutoCAD performs an automatic regeneration when you use specific commands that change the display. Examples of actions that cause automatic regeneration include thawing layers and switching layouts. If automatic regeneration takes too much time, especially on large, complex drawings, set the **REGENMODE** system variable to 0 (off). AutoCAD then issues prompts when regeneration is appropriate.

You also have the option of regenerating the display manually. Use the **REGEN** command to regenerate the display in the current viewport only. Use the **REGENALL** command to regenerate the display in all viewports.

Type
REGEN
RE

Type
REGENALL
REA

NOTE

Redrawing is different from regenerating. Redrawing is an old function that was important when computers and graphics were slower and less advanced. The **REDRAW** command redraws the display of the current viewport only. The **REDRAWALL** command redraws the display in all viewports. Neither of these commands recalculates the display based on the file database.

redrawing: A process that was once useful for refreshing the screen display without regenerating the drawing.

Cleaning the Screen

Type
[Ctrl]+[0] (zero)

The AutoCAD window can become crowded with multiple interface items, such as palettes, in the course of a drawing session. As the drawing window gets smaller, less of the drawing is visible, and drafting becomes more difficult. Use the **Clean Screen** command to clear the AutoCAD window of all palettes, toolbars, and the ribbon. See Figure 6-10. An easy way to toggle the **Clean Screen** command on and off is to pick the **Clean Screen** button on the status bar.

Figure 6-10. Using the **Clean Screen** command on a drawing of structural details. A—Initial display with the **Properties** palette and **Layer Properties Manager** displayed. B—Display after using the **Clean Screen** command.

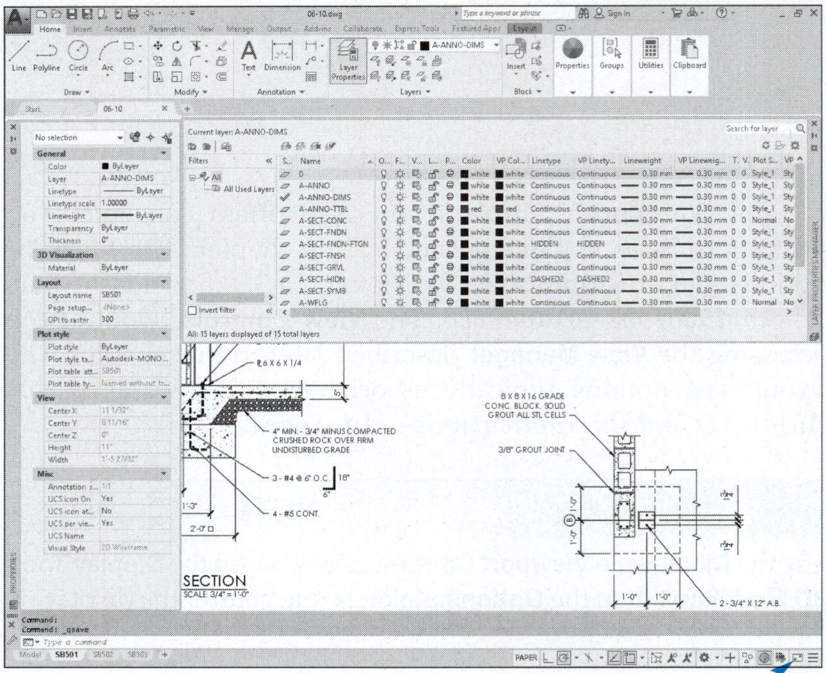

A

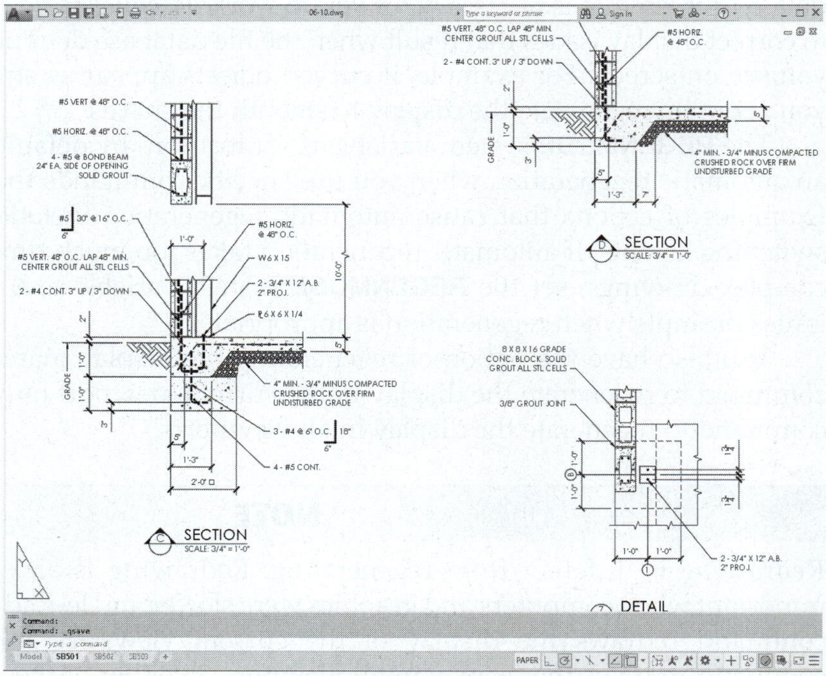

B

PROFESSIONAL TIP

The **Clean Screen** command can be helpful for displaying multiple drawings. The active drawing appears when you use the **Clean Screen** command. This allows you to work more efficiently within one of the drawings.

Supplemental Material

www.g-wlearning.com/CAD/

View Transitions and Resolution

For information about view transitions and view resolution, go to the companion website, navigate to this chapter in the **Contents** tab, and select **View Transitions and Resolution**.

Isolating or Hiding Objects

Layer states allow you to use layers to control the display of objects. AutoCAD includes other commands that allow you to display (isolate) or hide specific objects. These functions are view commands for temporarily isolating or hiding objects to clarify the drawing and focus on particular items.

Type: ISOLATEOBJECTS, ISOLATE

Use the **ISOLATEOBJECTS** command to hide all objects except the objects you select. An easy way to access the command is to pick the **Isolate Objects** button on the status bar and pick the **Isolate Objects** option, or right-click away from the status bar and select **Isolate Objects** from the **Isolate** cascading menu. See Figure 6-11A. Pick the objects to isolate as shown in Figure 6-11B.

Type: HIDEOBJECTS

Use the **HIDEOBJECTS** command to hide selected objects. An easy way to access the command is to pick the **Isolate Objects** status bar button followed by the **Hide Objects** option, or right-click away from the status bar and select **Hide Objects** from the **Isolate** cascading menu. Pick the objects to hide as shown in Figure 6-11C.

You can use the **ISOLATEOBJECTS** and **HIDEOBJECTS** commands at the same time to isolate and hide objects. You can also add objects to the set of isolated or hidden objects as needed.

Type: UNISOLATEOBJECTS, UNISOLATE

Use the **UNISOLATEOBJECTS** command to redisplay all objects hidden by the **ISOLATEOBJECTS** and **HIDEOBJECTS** commands. An easy way to access the command is to pick the **Isolate Objects** status bar button followed by the **End Object Isolation** option, or right-click away from the status bar and select **End Object Isolation** from the **Isolate** cascading menu.

NOTE

The **Isolate Objects** status bar button indicates whether the drawing includes isolated or hidden objects. Use the **OBJECTISOLATIONMODE** system variable to control whether isolation and hiding remains active when you close and reopen a drawing.

Exercise 6-4

www.g-wlearning.com/CAD/

Complete the exercise on the companion website.

Figure 6-11.
A—Use the status bar or shortcut menu to access object isolation and hiding commands. B—An example of using the **ISOLATEOBJECTS** command and window selection to show only objects within a portion of a sheet metal detail for a flat pattern drawing. C—An example of using the **HIDEOBJECTS** command to hide several dimension and multileader objects on a structural detail.

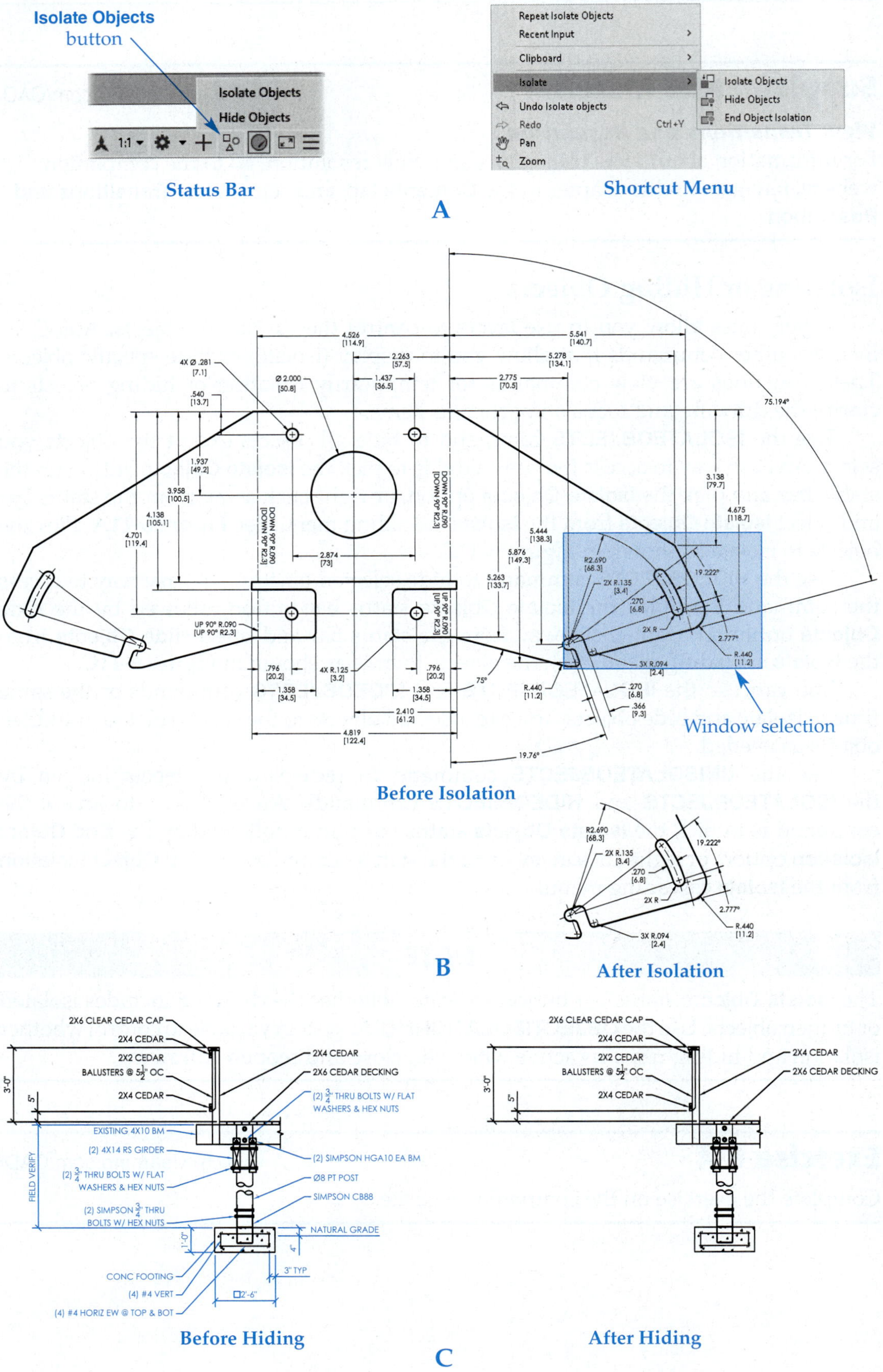

Using Commands Transparently

Activating a command usually cancels the command in progress and starts the new command. However, you can use some commands ***transparently***. After completing the transparent operation, the interrupted command resumes. Therefore, it is not necessary to cancel the initial command. You can use many display commands transparently, including **ZOOM** and **PAN**.

transparently: When referring to command access, describes temporarily interrupting the active command to use a different command.

An example of when transparent commands are useful is drawing a line when one end of the line is somewhere off the screen. One option is to cancel the **LINE** command, zoom out, and then reactivate the **LINE** command. A more efficient method is to use **PAN** or **ZOOM** transparently with the **LINE** command. To do so, begin the **LINE** command and pick the first point. At the Specify next point or [Undo]: prompt, pan or zoom to display the next line endpoint. You can use any access method, but using the mouse wheel is often quickest. When the correct view appears, pick the second line endpoint. You can also activate commands transparently by entering an apostrophe (') before the command name. For example, to enter the **ZOOM** command transparently, type 'Z or 'ZOOM.

PROFESSIONAL TIP

You can activate and adjust drawing aids such as grid, snap, ortho, and object snap transparently. The quickest activation and adjustment method is to use the appropriate button on the status bar or press a function key.

Controlling Draw Order

Drawings often include overlapping objects. The overlap is difficult to see when all objects have a thin lineweight and when lineweight display is off. To help understand display order, view an object that has width, such as the donuts shown in Figure 6-12. The donuts in this example were drawn after the other objects. You can change the drawing order of objects to place them above or below selected objects or to the front or back of all objects.

Use the **DRAWORDER** command to change the order of objects in a drawing. You can also set certain draw order options by selecting an object, right-clicking, and selecting an option from the **Draw Order** cascading menu. Figure 6-13 describes draw order options.

DRAWORDER

Ribbon
Home
> Modify

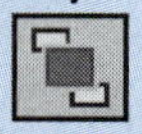

Bring to Front

Type
DRAWORDER
DR

Figure 6-12.
Change the order of objects to place selected objects under or above other objects.

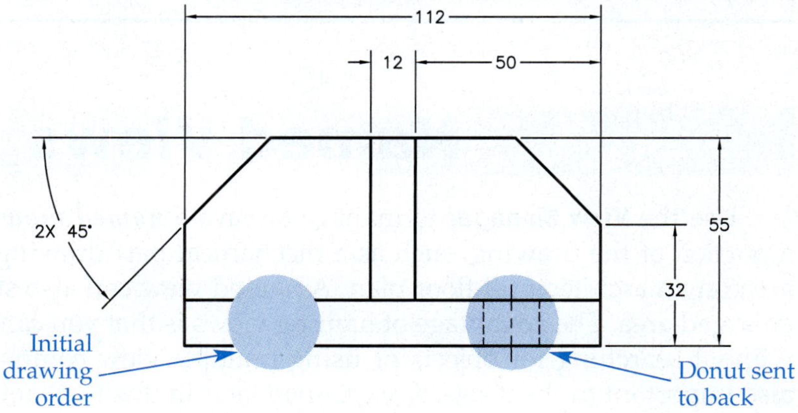

Figure 6-13.
The options available for changing the order of objects in a drawing. The text, dimension, leader, annotation, and hatch options are available from the ribbon and apply to all corresponding objects in the drawing.

Option	Function
Bring to Front	Places the selected objects at the front of the drawing.
Send to Back	Places the selected objects at the back of the drawing.
Bring Above Objects	Moves the selected objects above the reference object.
Send Under Objects	Moves the selected objects below the reference object.
Bring Text to Front	Places all text objects above other objects.
Bring Dimensions to Front	Places all dimension objects above other objects.
Bring Leaders to Front	Places all leader and multileader objects above other objects.
Bring All Annotations to Front	Places all annotation objects, such as text, dimension, and multileader objects, above other objects.
Send Hatches to Back	Moves all hatch objects below other objects.

NOTE

You may need to use the **DRAWORDER** command on several objects until the objects display correctly. Objects move to the front of the drawing when you modify them.

PROFESSIONAL TIP

Use the **DRAWORDER** command to help display and select objects that are hidden by other objects.

Exercise 6-5

www.g-wlearning.com/CAD/

Complete the exercise on the companion website.

named view: A specific drawing display saved for easy recall and future use, analogous to taking a picture.

VIEW

Ribbon
View
> Named Views
View Manager
Type
VIEW
V

Named Views

Use the **View Manager** to manage or save a ***named view***. A named view can be a portion of the drawing, such as a mechanical part drawing or detail, or a room or area on an architectural floor plan. A named view can also show an entire design or enlarged area. The advantage of naming views is that you can recall a specific display without searching for objects or using multiple view commands. Named views are also important to sheet sets, as explained later in this textbook.

Figure 6-14 shows the **View Manager** and an architectural plan that includes custom named views. The tree on the left side of the **View Manager** lists each type of view. Pick a view type to display its information. The **Current** node lists the properties of the current display in the drawing window. Pick the plus sign (+) to list related views. The **Model Views** node lists named model views, and the **Layout Views** node lists named layout views. The **Preset Views** node lists all preset orthographic and isometric views. This chapter focuses on the use of named model views. You will learn about layout views later in this textbook. ***AutoCAD and Its Applications—Advanced*** explains preset views.

Pick a model view to list its properties in the middle portion of the **View Manager**. You can modify some properties, such as **Name**. Other properties are read-only and are set during view configuration. The right side of the **View Manager** contains buttons to create and control views. The options are also available from the shortcut menu that appears when you right-click on a view. The lower-right corner of the **View Manager** shows a preview of the selected model or layout view.

Figure 6-14.
Use the **View Manager** to assign new named views and to control existing named views. This example shows a named view of an overall floor plan titled FLOOR PLAN. The other named views display specific remodel areas for easy recall and future use.

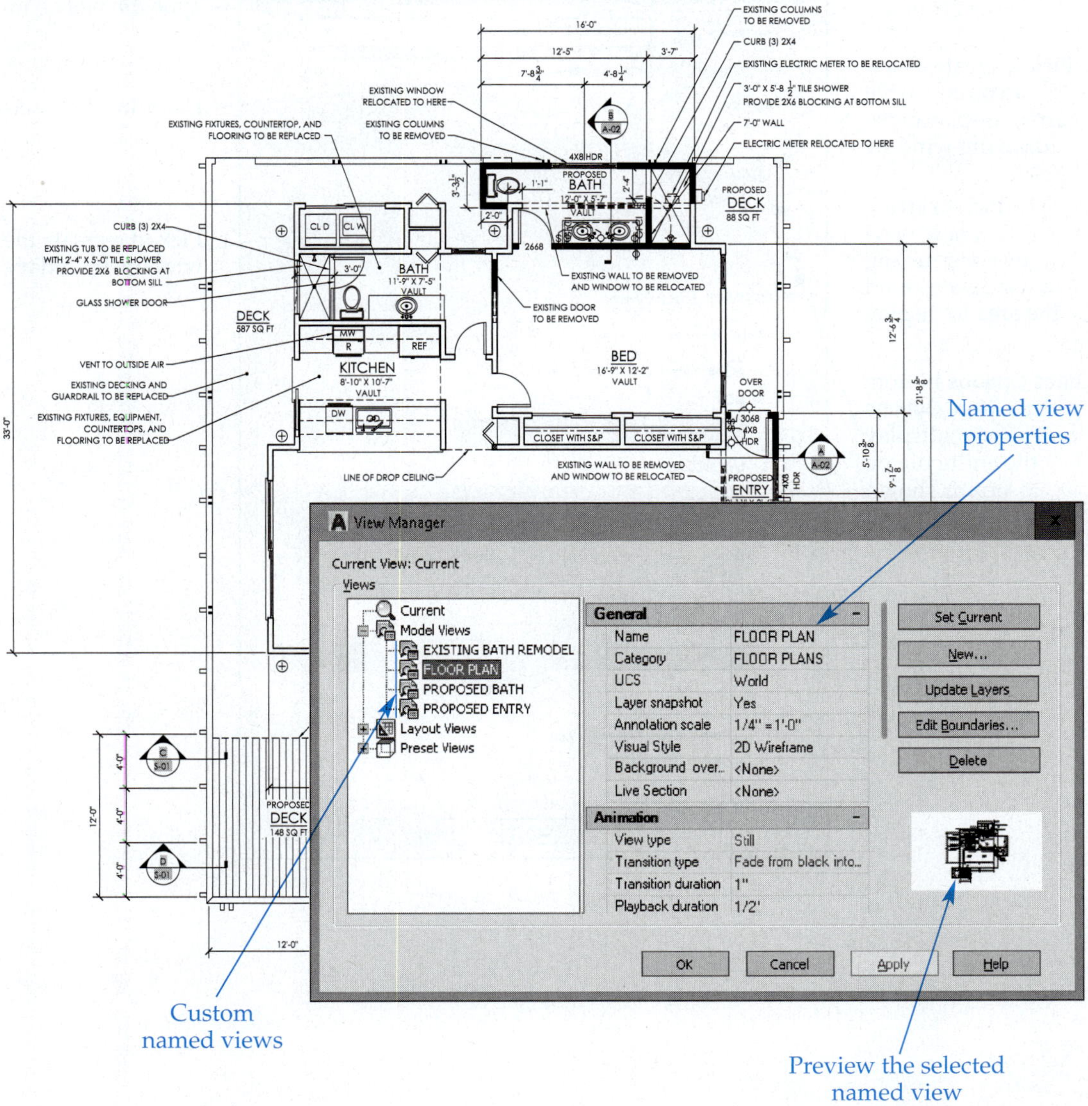

New Views

NEWVIEW

Ribbon
View
> Named Views

New View

Type
NEWVIEW
NVIEW

To save a display as a named view, access the **NEWVIEW** command or pick the **New...** button in the **View Manager** to open the **New View/Shot Properties** dialog box. See **Figure 6-15**. Pick the down arrow to expand the dialog box. Type the view name in the **View name:** text box.

To create a basic 2D view, select **Still** from the **View type:** drop-down list. The **Current display** radio button option is set by default and prepares the view boundary according to the current display in the drawing window. To construct a view boundary, pick the **Define window** radio button. The **New View/Shot Properties** dialog box closes temporarily so you can pick two opposite corners to define a window around the area to display. See **Figure 6-16**. Press [Enter] to return to the **New View/Shot Properties** dialog box.

Select the **Save layer snapshot with view** check box to save the current layer settings when you save the new view. AutoCAD recalls saved layer settings each time the view is set current. Pick the **OK** button to finish the new view.

Figure 6-15.
Use the **New View/Shot Properties** dialog box to save the current display as a view or define a window to create the view. This example shows creating a named view for viewing a proposed bathroom on an architectural floor plan.

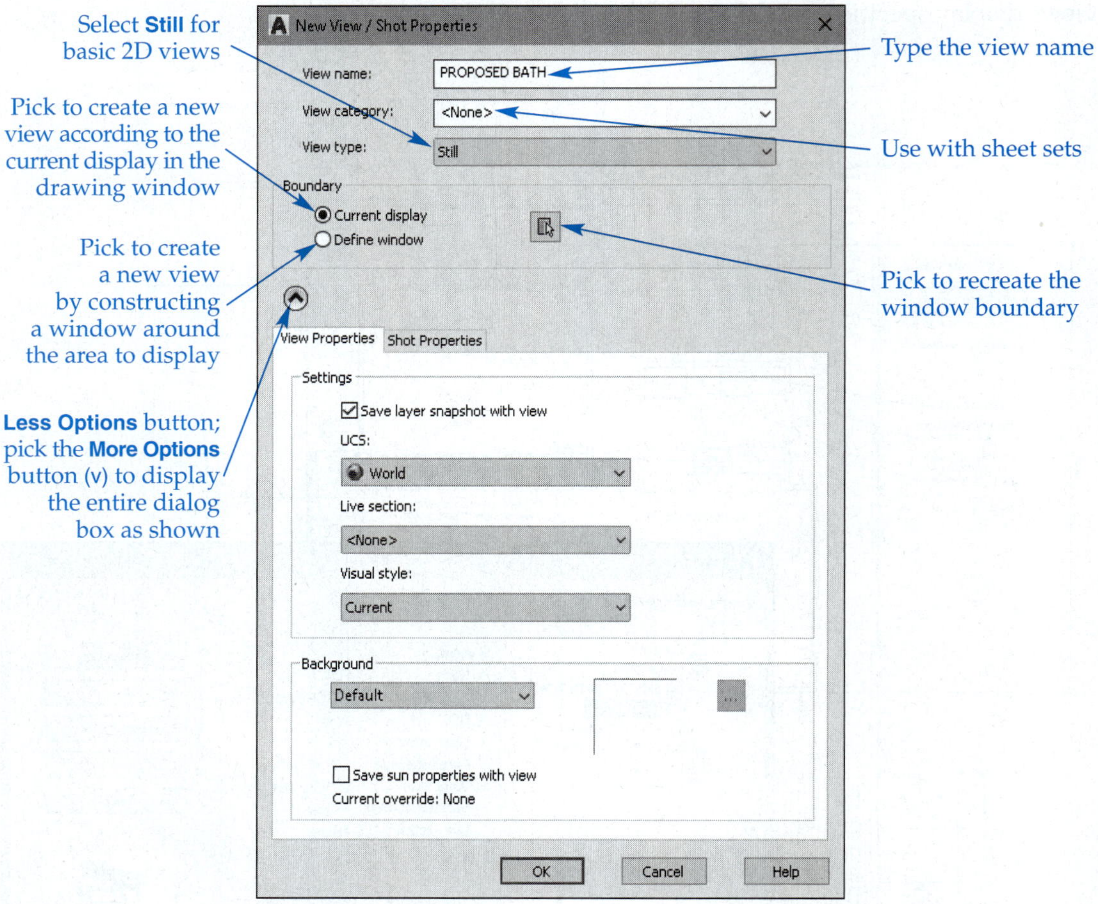

Figure 6-16.
Specifying a view boundary using the **Define window** option. This example shows creating a view boundary around the proposed bathroom that corresponds to the PROPOSED BATH named view.

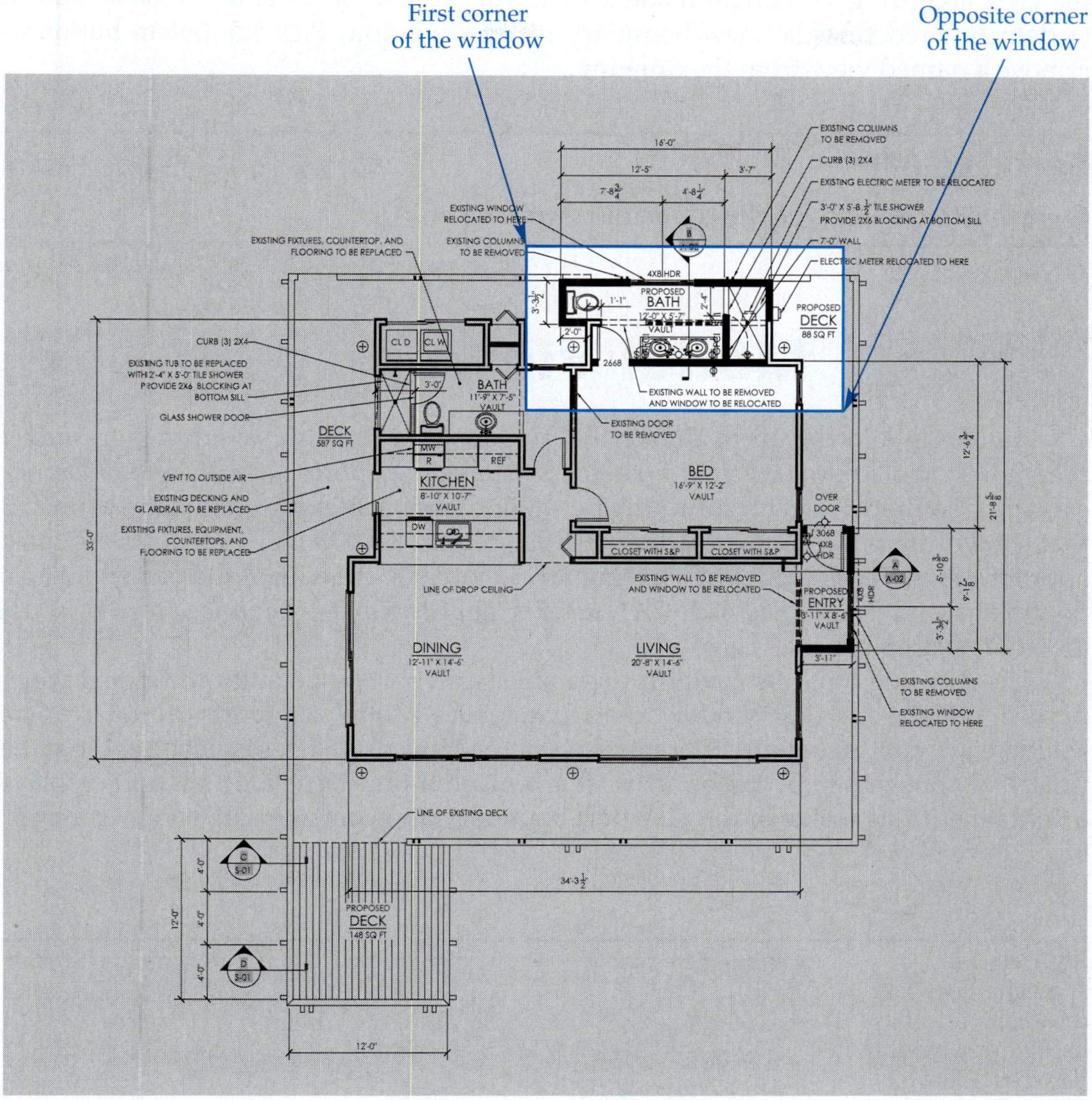

NOTE

AutoCAD and Its Applications—Advanced describes the other options in the **New View/Shot Properties** dialog box, which are used for animating 3D model views.

Activating a View

To display a named view in the drawing window without accessing the **View Manager**, select the view from the list in the **Named Views** panel of the **View** ribbon tab, or the **Custom Model Views** cascading menu of the **View Controls** flyout in the viewport controls. To make a named view current from inside the **View Manager**, select the view from the list and pick the **Set Current** button or right-click and select **Set Current**. Pick the **Apply** button to display the view, or pick the **OK** button to display the view and exit the dialog box.

Managing Views

Return to the **View Manager** to adjust named views using the appropriate button or shortcut menu option. Pick the **Update Layers** button to update layers displayed in the view according to changes made to layer states. Pick the **Edit Boundaries...** button to define or redefine the view boundary using a window. Pick the **Delete** button to remove a named view from the drawing.

Exercise 6-6

www.g-wlearning.com/CAD/

Complete the exercise on the companion website.

Model Viewports

model (tiled) viewport: A window or frame within which a drawing is visible in model space.

AutoCAD allows you to divide the model space drawing window into ***model viewports***, also known as ***tiled viewports***. Model viewports have 2D and 3D applications. Use model viewports to divide a 2D drawing window into compartments that show different aspects of a drawing. See Figure 6-17. Model viewports are most appropriate in 2D drafting for viewing large, complex drawings, such as site plans, floor plans, and details. See ***AutoCAD and Its Applications—Advanced*** for 3D model viewport examples.

The drawing window contains one model viewport by default. Additional viewports divide the drawing window into separate tiles. Model viewports cannot overlap. Multiple viewports contain different views of the same drawing, displayed at the same time. Only one viewport can be active at any given time. The active viewport displays a bold blue frame inside of the viewport boundary and contains active view controls. See Figure 6-17.

Figure 6-17.
A mechanical part drawing viewed using three model viewports. Each viewport contains the same drawing, but can present a different display. Model viewports are most useful in 2D drafting for viewing large, complex drawings.

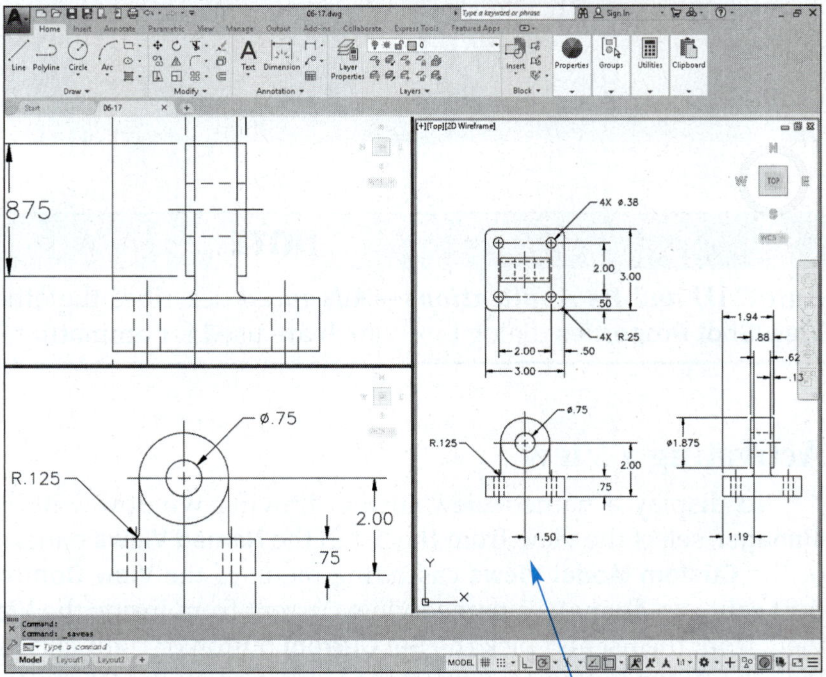

Active viewport

Configuring Model Viewports

The **Viewports** dialog box provides one method for configuring model viewports. Figure 6-18 shows the **New Viewports** tab of the **Viewports** dialog box. The **Standard viewports:** list contains preset viewport configurations. The configuration name identifies the number of viewports and the arrangement or location of the largest viewport. Select a configuration to see a preview of the model viewports in the **Preview** area. Select *Active Model Configuration* to preview the current configuration. Pick the **OK** button to divide the drawing window into the selected model viewports.

VIEWPORTS

Ribbon
View
> Model Viewports
Named
Type
VIEWPORTS
VPORTS

NOTE

You can activate the preset viewport configurations using the **Viewport Configuration** flyout on the **Model Viewports** panel of the **View** ribbon tab or the **Viewport Configuration List** cascading menu of the **Viewport Controls** flyout in the viewport controls.

The default setting in the **Setup:** drop-down list is **2D**, and all viewports show the 2D drawing plane, or top view. If you select the **3D** option, the different viewports display various 3D views of the drawing. At least one viewport shows a 3D isometric view. The other viewports show different views, such as a top view or side view. The viewport configuration is displayed in the **Preview** image. To change a view in a viewport, pick the viewport in the **Preview** image and then select the new viewpoint from the **Change view to:** drop-down list.

The options in the **Apply to:** drop-down list specify how the viewport configuration is applied. Select **Display** to apply the viewport configuration to the entire drawing window. Select **Current Viewport** to apply the new configuration in the active viewport only. See Figure 6-19.

Create a new viewport configuration if none of the preset configurations is acceptable. Enter a descriptive name in the **New name:** text box. When you pick the **OK** button, the new viewport configuration is displayed in the **Named Viewports** tab the next time you access the **Viewports** dialog box. See Figure 6-20. Select a different named viewport configuration and pick **OK** to apply changes to the drawing area.

Figure 6-18.
Specify the number and arrangement of model viewports in the **New Viewports** tab of the **Viewports** dialog box.

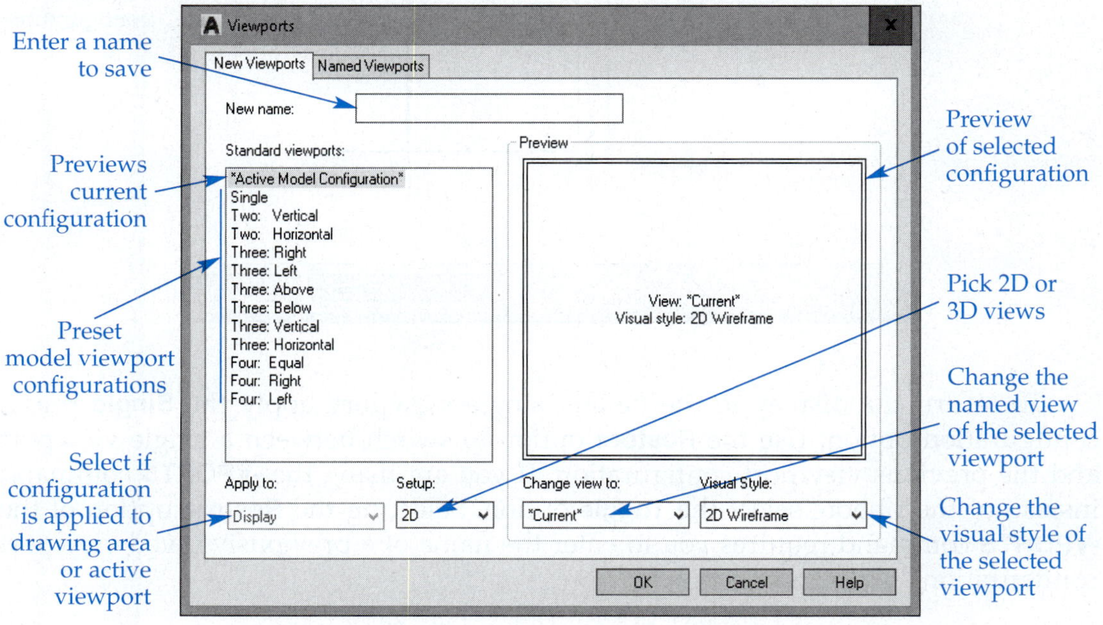

Figure 6-19.
You can subdivide a viewport by selecting **Current Viewport** in the **Apply to:** drop-down list. This example shows dividing the top-left viewport using the **Two: Vertical** configuration.

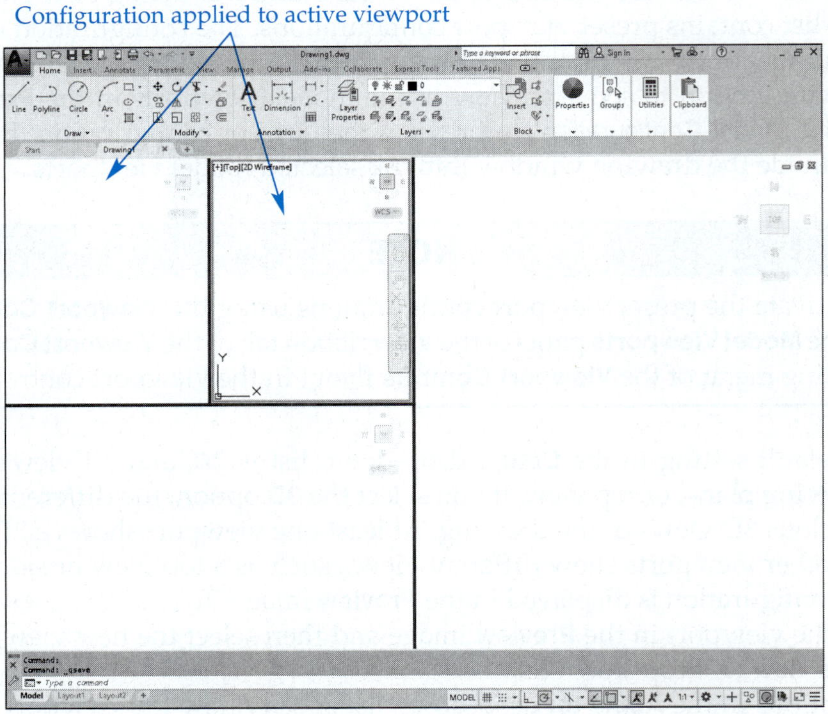

Figure 6-20.
The **Named Viewports** tab displays custom viewports.

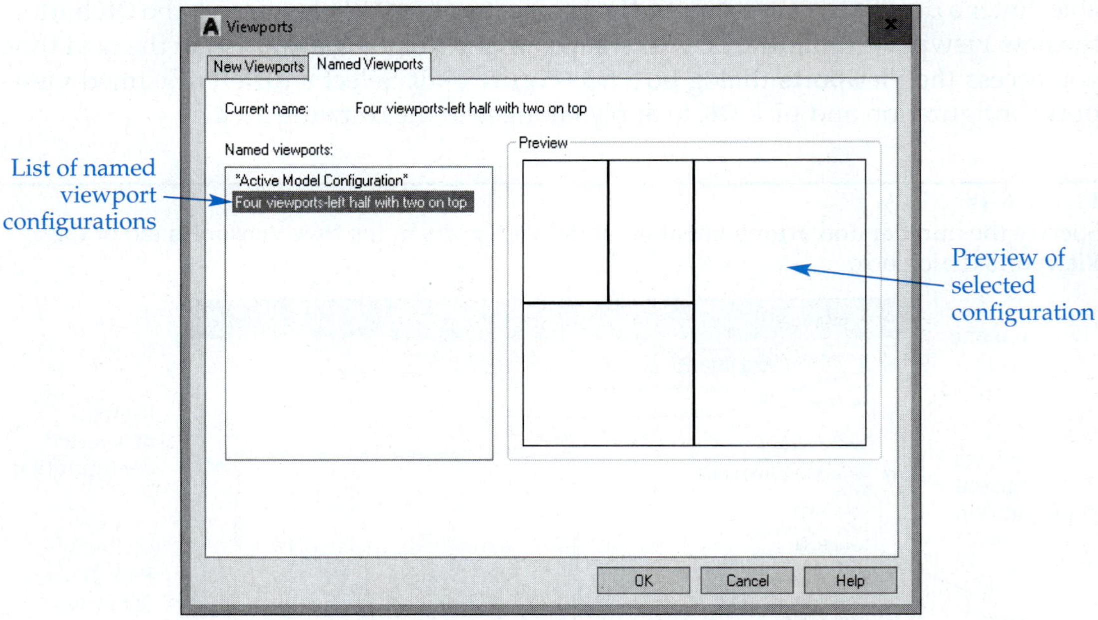

-VPORTS

Ribbon
View
> Model Viewports
Restore Viewports
Type
-VPORTS

To return the display to the default single viewport, apply the **Single** preset configuration option. Use the **Restore** option to switch between a single viewport and the previous viewport configuration. If you are using the **-VPORTS** command instead of the ribbon, select the **Toggle** option. Selecting the **Restore** option of the **-VPORTS** command requires you to enter the name of a previously saved viewport configuration.

Working in Model Viewports

After you select the viewport configuration and return to the drawing window, move the mouse and notice that only the active viewport contains crosshairs. The cursor is an arrow in the other viewports. To make an inactive viewport active, move the cursor into the inactive viewport and pick.

Depending on the zoom level, as you draw in one viewport, objects appear in other viewports. Try drawing objects and notice how drawing affects the viewports. Use a display command, such as **ZOOM**, in the active viewport and notice the results. Only the active viewport reflects the use of the **ZOOM** command.

Joining Model Viewports

Ribbon
View
> Model Viewports
Join Viewports
Type
-VPORTS

-VPORTS

Use the **Join** option to join two viewports. Select the dominant viewport, which is the viewport that shows the view you want to display in the joined viewport, or press [Enter] to select the active viewport. Then select the viewport to join with the dominant viewport. AutoCAD joins the two viewports together and retains the dominant view. The two viewports you join cannot create an L-shape viewport. The adjoining edges of the viewports must be the same size in order to be joined.

Adjusting Model Viewports

Viewport boundaries provide options for resizing, creating, joining, and removing viewports quickly. Boundary adjustments are made by picking a boundary with the left mouse button and dragging to a specific location. Release the mouse button to complete the operation. Drag a horizontal or vertical viewport boundary, or the intersection of three or four viewports, to resize viewports. See **Figure 6-21A**. To add a viewport, pick the small plus sign (+) near the edge of a viewport boundary and drag to display a green splitter bar. See **Figure 6-21B**. Drag the splitter bar to locate the new viewport boundary. You can also add a viewport by pressing [Ctrl] while dragging a viewport boundary. To join or remove viewports, drag a viewport boundary near another viewport boundary.

Figure 6-21.
Adjusting viewport configurations. A—Drag a horizontal or vertical viewport boundary or drag a boundary intersection to resize viewports. B—Pick the small plus sign (+) and drag or press [Ctrl] while dragging a boundary to add a viewport.

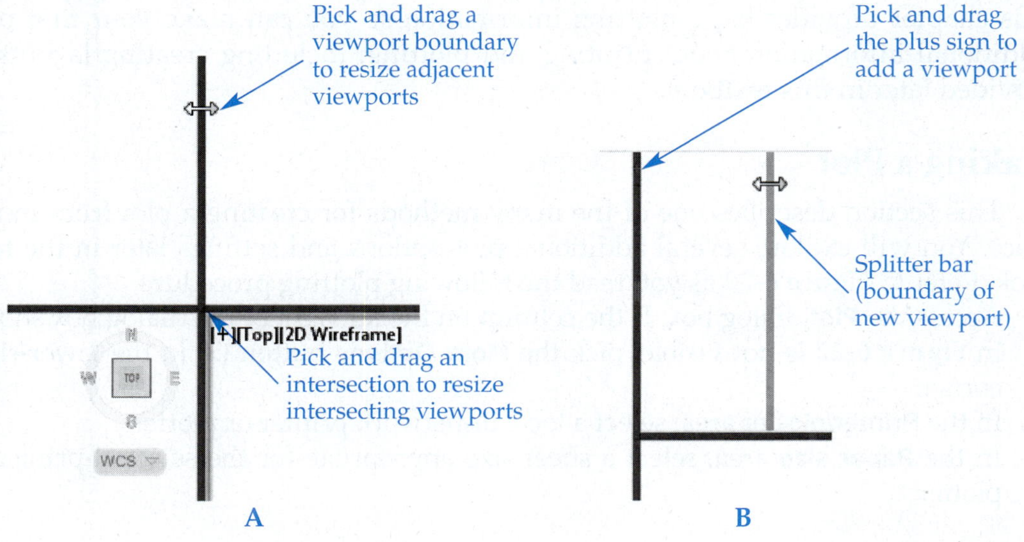

PROFESSIONAL TIP

Add named views to templates to recall common displays for repeated use. Apply a model viewport arrangement to templates to organize the drawing window for repeated use.

Exercise 6-7

www.g-wlearning.com/CAD/

Complete the exercise on the companion website.

Introduction to Printing and Plotting

soft copy: The electronic data file of a drawing.

hard copy: A physical drawing produced by a printer or plotter.

A ***soft copy*** appears on-screen, often making a drawing inconvenient to use for manufacturing and construction purposes. A ***hard copy*** is useful on the shop floor or at a construction site. A design team can check and redline a hard copy without a computer or CADD software. CADD is the standard throughout the world for generating drawings, and electronic data exchange is popular. However, hard-copy drawings are still a vital tool for communicating a design.

A printer or plotter transfers soft copy images onto paper. The terms *printer* and *plotter* are interchangeable, although *plotter* typically refers to a large-format printer. Desktop printers are commonly used at computer workstations. They generally print 8 1/2″ × 11″ and sometimes 11″ × 17″ sheets. Desktop printers are most appropriate for printing small drawings and reduced-size test prints. Large-format printers print larger drawings, such as C-size and D-size drawings. The most common types of both desktop and large-format printers are inkjet and laser printers. Traditional pen plotters, which "draw" with actual ink pens, are much less common.

Plotting in Model Space

You typically plot final drawings using a layout in paper space. A layout represents the sheet of paper used to organize and scale, or lay out, a drawing or model. The layout can be printed or plotted on paper or exported to another format. However, you can also plot from model space.

Plotting from model space is common when a layout is unnecessary. It allows you to make quick hard copies for viewing how model space objects will appear on paper, and for submitting basic assignments to your instructor or supervisor. This chapter provides basic plotting information so you can make your first plot. Additional information about printing and plotting, including creating layouts, is provided later in this textbook.

Making a Plot

PLOT

Quick Access
Plot

Ribbon
Output > Plot
Plot

Application Menu
Print > Plot

Type
PLOT
[Ctrl]+[P]

This section describes one of the many methods for creating a plot from model space. You will explore several additional plot options and settings later in the textbook. Refer to Figure 6-22 as you read the following plotting procedure.

1. Access the **Plot** dialog box. If the column on the far right of the dialog box shown in Figure 6-22 is not visible, pick the **More Options** button (>) in the lower-right corner.
2. In the **Printer/plotter** area, select a local or network printer or plotter.
3. In the **Paper size** area, select a sheet size appropriate for the selected printer or plotter.

Figure 6-22.
Use the **Plot** dialog box to plot or print a drawing.

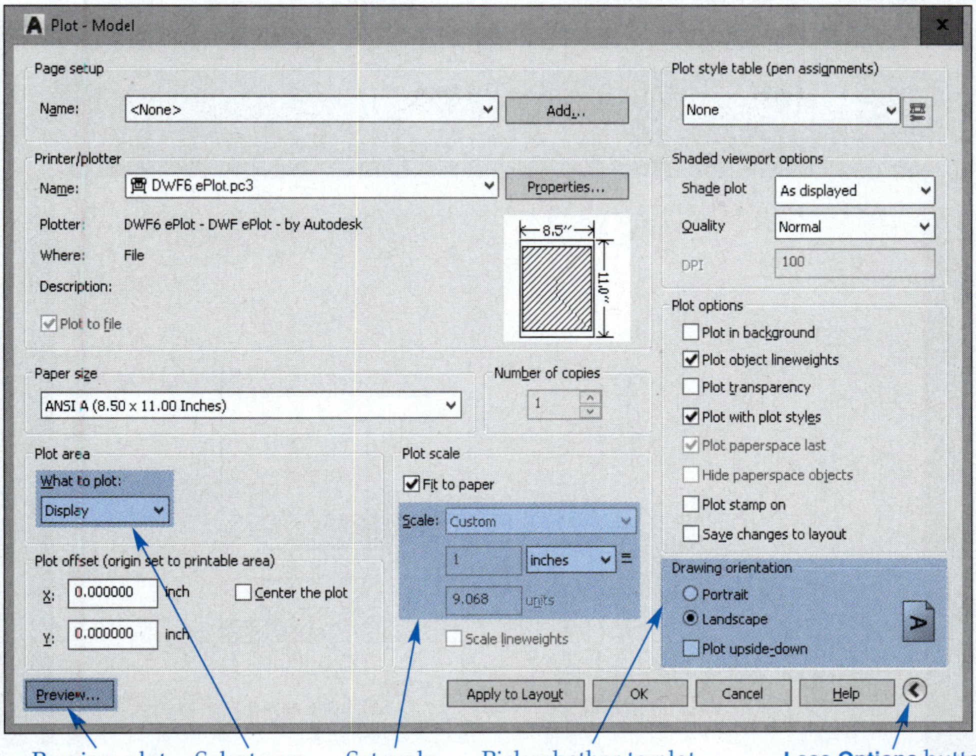

4. Select the area to plot from the **Plot area** section. Select the **Display** option to plot the current screen display, exactly as shown. Pick the **Extents** option to plot the furthest extents of objects in the drawing. The **Limits** option allows you to plot everything inside the specified drawing limits. The **View** option is available if the file includes named views. Select a named view from the list that appears when you select the **View** option. When you select the **Window** option, the **Plot** dialog box closes temporarily so you can pick two opposite corners of a window around the area to plot. Once you create the window, a **Window<** button appears in the **Plot area** section. Pick the button to redefine the window around the area to plot.
5. Select an option in the **Drawing orientation** area. Select **Portrait** to orient the drawing vertically (***portrait***) or select **Landscape** to orient the drawing horizontally (***landscape***). The **Plot upside-down** option rotates the drawing 180° on the paper.
6. Set the scale in the **Plot scale** area. The scale is a ratio of inches or millimeters to drawing units. Select a scale from the **Scale:** drop-down list or type values in the custom fields. Select the **Fit to paper** check box to let AutoCAD increase or decrease the plot scale to fill the paper.
7. If necessary, use the **Plot offset (origin set to printable area)** section to set additional left and bottom margins around the plot, or to center the plot.
8. Pick the **Preview...** button to display the sheet as it will look when it plots. See Figure 6-23. The realtime zoom cursor appears. Hold down the left mouse button and move the cursor to increase or decrease the displayed image to view more or less detail. Press [Esc] to exit the preview.
9. Pick the **OK** button in the **Plot** dialog box to send the data to the plotting device.

portrait: A vertical paper orientation.

landscape: A horizontal paper orientation.

Figure 6-23.
A plot preview of mechanical part views drawn in model space. The preview shows exactly how the drawing will appear on paper.

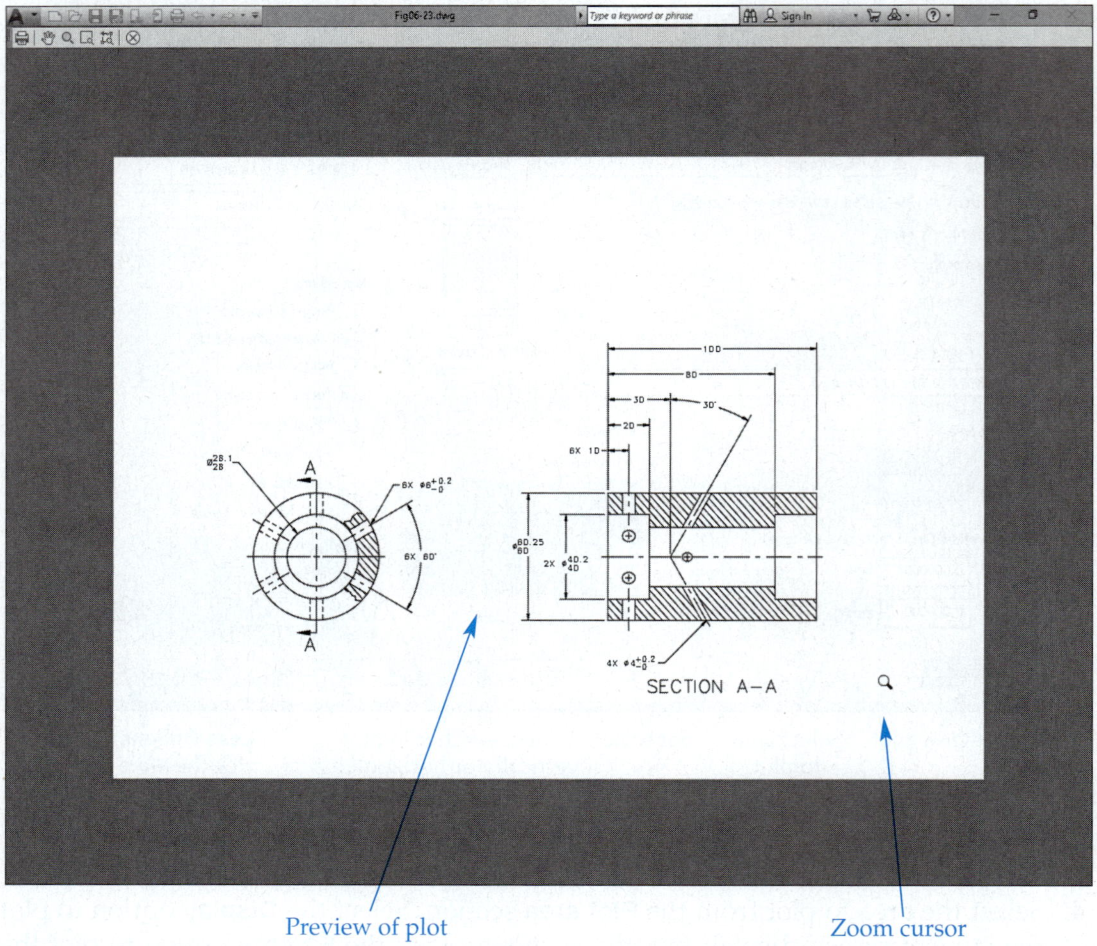

NOTE

AutoCAD may display a **Batch Plot** alert when you access the **PLOT** command and have more than one drawing file open. Pick **Continue to plot a single sheet** to display the **Plot** dialog box. Batch plotting, also referred to as ***publishing*** in AutoCAD, is used to plot multiple layouts and/or sheets. Publishing is explained later in this textbook.

publishing: Preparing a sequential set of multiple drawings for hard copy or electronic plotting of the set.

Exercise 6-8

www.g-wlearning.com/CAD/

Complete the exercise on the companion website.

Chapter Review

Answer the following questions using the information in this chapter.

1. Explain how to use the **Realtime** zoom option.
2. What is the difference between zooming and panning?
3. What two commands allow you to cycle back and forth quickly between prior views that were created by zooming or panning?
4. Which steering wheels navigation commands are most appropriate for 2D drafting applications?
5. Explain how to use the **CENTER** command on the Full Navigation wheel.
6. What feature of the Full Navigation wheel allows you to return to previous display settings?
7. Why should you avoid using the view cube to rotate the display of a 2D drawing?
8. Provide an example of when regenerating the display is necessary.
9. Which command regenerates all of the viewports?
10. Explain the difference between using layer states and object isolation or object hiding to hide objects.
11. How do you enter a display command transparently at the keyboard?
12. Which command changes the order in which objects are displayed in a drawing?
13. How do you create a 2D named view of the current screen display?
14. How do you display an existing named view?
15. By default, how many model viewports appear in the drawing window?
16. How can you specify whether a new model viewport configuration applies to the entire drawing window or the active viewport?
17. Explain the procedure for joining model viewports.
18. Define *hard copy* and *soft copy*.
19. Identify four ways to access the **Plot** dialog box.
20. Describe the difference between the **Display** and **Window** options in the **Plot area** section of the **Plot** dialog box.

Drawing Problems

Start AutoCAD if it is not already started. Start a new drawing from scratch or use an appropriate template of your choice. The template should include layers for drawing the given objects. Add layers as needed. Draw all objects using appropriate layers. Follow the specific instructions for each problem. Use only drawing commands and techniques you have already learned. Do not draw dimensions or text. Use your own judgment and approximate dimensions when necessary.

▼ Basic

1. Architectural Draw the surface-mounted fluorescent light fixture shown below. Zoom in and out on the drawing. Pan the screen display. Save the drawing as P6-1. Print an 8.5″ × 11″ copy of the drawing extents, fit to paper, using a portrait orientation.

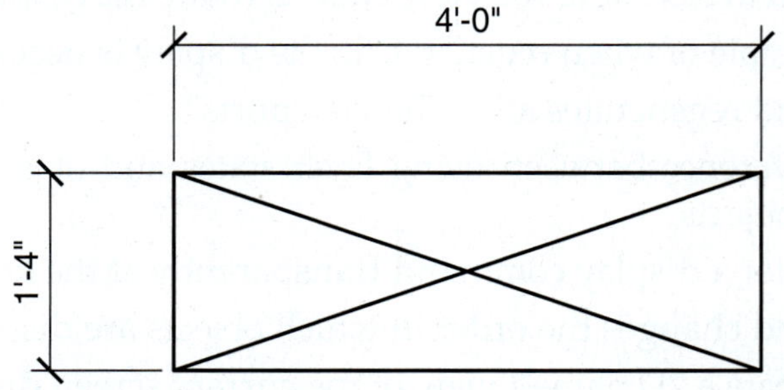

Surface-Mounted Fluorescent Light Fixture

2. Architectural Draw the surface-mounted light fixture shown below. Zoom in and out on the drawing. Pan the screen display. Save the drawing as P6-2. Print an 8.5″ × 11″ copy of the drawing extents, fit to paper, using a portrait orientation.

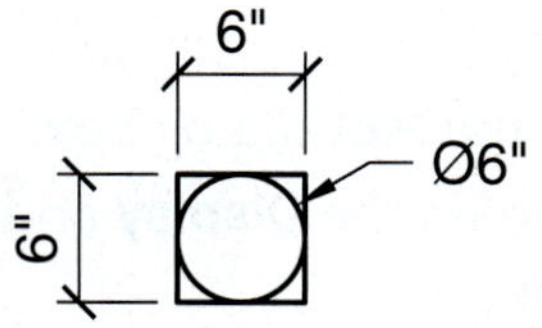

Surface-Mounted Light Fixture

3. Mechanical Open the drawing named VW252-03-1200.dwg in the AutoCAD 2020\Sample\Sheet Sets\Manufacturing folder. Pick the **Model** tab below the drawing window to activate model space. Zoom to the extents of the drawing. Zoom in on each drawing view. Use a variety of view commands to view the gears in the full side view, the section view, and the isometric view. Close the drawing without saving.

4. Architectural Open the drawing named A-02.dwg in the AutoCAD 2020\Sample\Sheet Sets\Architectural folder. Pick the **Model** tab below the drawing window to activate model space. Zoom to the extents of the drawing and print using the **Fit to paper** option. Use a variety of view commands to locate and view the following items.
   A. Front elevation
   B. Doors on the right elevation
   C. Downspouts on the rear elevation
   D. Column on the left elevation
   E. Main entrance

   Mark the location of each item on your hard copy. Close the drawing without saving.

Drawing Problems - Chapter 6

5. **Civil** Open the drawing named Erosion Control Plan.dwg in the AutoCAD 2020\Sample\Sheet Sets\Civil folder. Pick the **Model** tab below the drawing window to activate model space. Zoom to the extents of the drawing and print using the **Fit to paper** option. Use a variety of view commands to locate and view the following items on the plan.
   A. Coburn Avenue
   B. Proposed track and field
   C. Proposed high school
   D. Proposed tennis courts
   E. Two proposed baseball fields
   F. Two proposed softball fields
   G. Accessible parking near the northeast corner of the high school

   Mark the location of each item on your hard copy. Close the drawing without saving.

▼ Intermediate

6. **General** Create a freehand sketch of the default navigation bar that appears in model space. Label each command button and describe the commands that are most appropriate for 2D drafting applications.

7. **General** Create a freehand sketch of the Full Navigation wheel. Label each wedge and describe the commands that are most appropriate for 2D drafting applications.

8. **Mechanical** Draw the gasket shown. Save the drawing as P6-8. Print an 8.5″ × 11″ copy of the drawing extents, using a 1:1 scale and a landscape orientation.

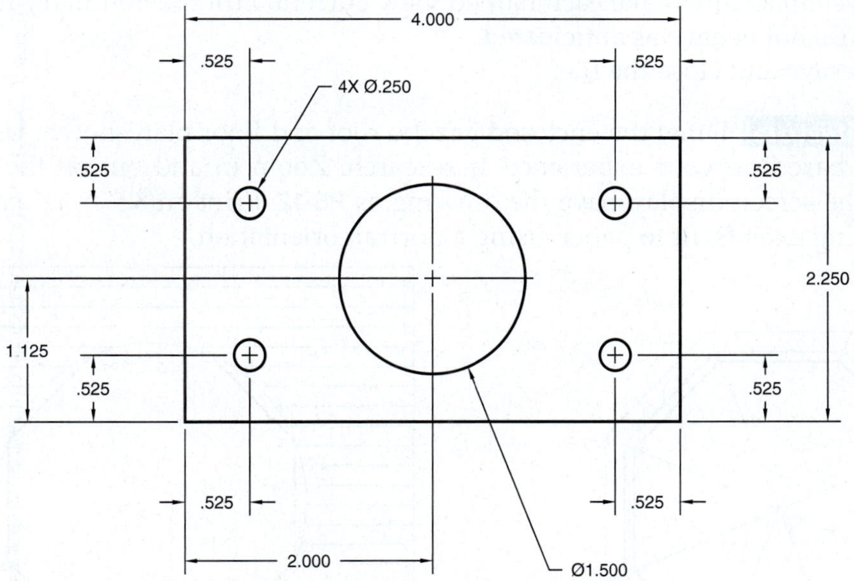

9. **Architectural** Open the drawing named Floor Plan.dwg available on the companion website. Save a copy of the drawing as P6-9. The P6-9 file should be active. Perform the following display functions on the drawing.
   A. Zoom to the drawing extents.
   B. Zoom to create a display of the dining room and create a view of this display named Dining Room.
   C. Zoom to create a display of the kitchen and create a view of this display named Kitchen.
   D. Zoom to create a display of the living room and create a view of this display named Living Room.
   E. Display the view named Kitchen.
   F. Resave and close the file.

▼ Advanced

10. **Architectural** Open the drawing named Kitchens.dwg in the AutoCAD 2020\Sample\en-us\DesignCenter folder. Save a copy of the drawing as P6-10. The P6-10 file should be active. Perform the following display functions on the drawing.
    A. Zoom to the drawing extents.
    B. Zoom in on each symbol and research if necessary to identify what the symbol represents.
    C. Create a named view of the extents of each symbol using the **Define window** option in the **New View/Shot Properties** dialog box. Use an appropriate name for each view.
    D. Systematically make each named view current. Edit the boundary if the view does not behave as anticipated.
    E. Resave and close the file.

11. **Piping** Open the drawing named Pipe Fittings.dwg in the AutoCAD 2020\Sample\en-us\DesignCenter folder. Save a copy of Pipe Fittings as P6-11. The P6-11 file should be active. Perform the following display functions on the drawing.
    A. Zoom to the drawing extents.
    B. Zoom in on each symbol and research if necessary to identify what the symbol represents.
    C. Create a named view of the extents of each symbol using the **Define window** option in the **New View/Shot Properties** dialog box. Use an appropriate name for each view.
    D. Systematically make each named view current. Edit the boundary if the view does not behave as anticipated.
    E. Resave and close the file.

12. **Architectural** Draw the enclosed gazebo roof and floor plan shown. Use dimensions based on your experience or research. Zoom in and out on the drawing. Pan the screen display. Save the drawing as P6-12. Print an 8.5″ × 11″ copy of the drawing extents, fit to paper, using a portrait orientation.

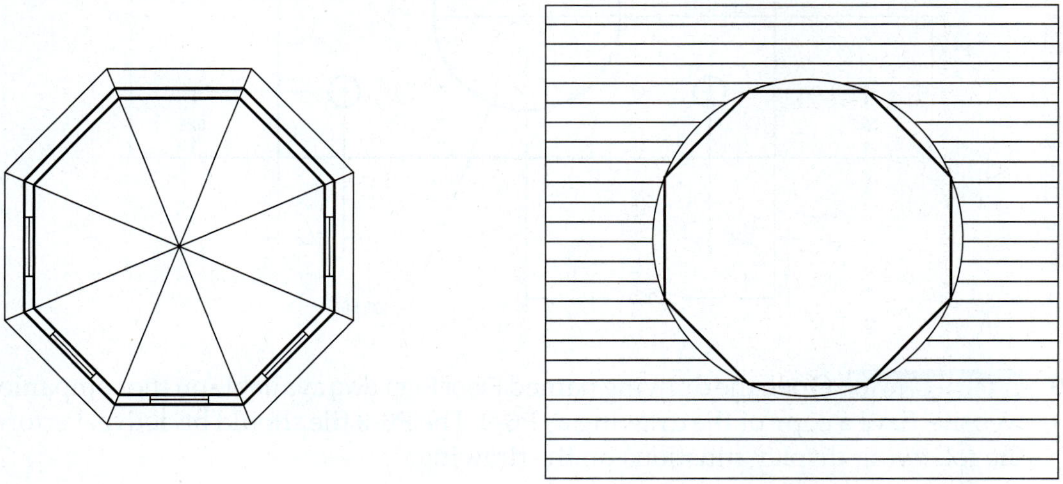

13. **Mechanical** Research the design of an existing exercise dumbbell with the following specifications: steel, one-piece hexagon shape, five pounds. Create a freehand, dimensioned 2D sketch of the existing design from manufacturer's specifications or from measurements taken from an actual dumbbell. Research the mass properties of steel to design the dumbbell to weigh exactly five pounds. Start a new drawing from scratch or use a decimal template of your choice. Draw the dumbbell from your sketch. Save the drawing as P6-13. Print an 8.5″ × 11″ copy of the drawing extents, using a 1:1 scale and a landscape orientation. Do not dimension the drawing.

AutoCAD Certified Professional Exam Practice

Answer the following questions using the information in this chapter.

1. Which of the following actions enters a view command or drawing aid transparently? *Select all that apply.*
   A. picking a button on the status bar
   B. picking a button on the ribbon
   C. picking a button on the navigation bar
   D. typing a colon (:) before the command name
   E. using the mouse wheel
2. How can you display a named view in a drawing? *Select all that apply.*
   A. select the view from the list in the **Named Views** panel of the **View** ribbon tab
   B. select the view from the flyout in the navigation bar
   C. double-click the view name in the view cube
   D. right-click and select the view name
3. When a drawing contains two or more model viewports, how can you activate a different viewport? *Select all that apply.*
   A. press the [F3] key
   B. press the space bar
   C. pick anywhere in the viewport to activate
   D. right-click and select the viewport to activate

Follow the instructions in each problem. Complete each problem by answering the related question.

4. **Navigate to this chapter on the companion website and open CPE-06zoom.dwg.** Use the view command of your choice to zoom in on the notes in the lower-left corner of the drawing. Which ANSI standard should be used to interpret the graphic symbols for the electrical and electronic elements in this drawing?

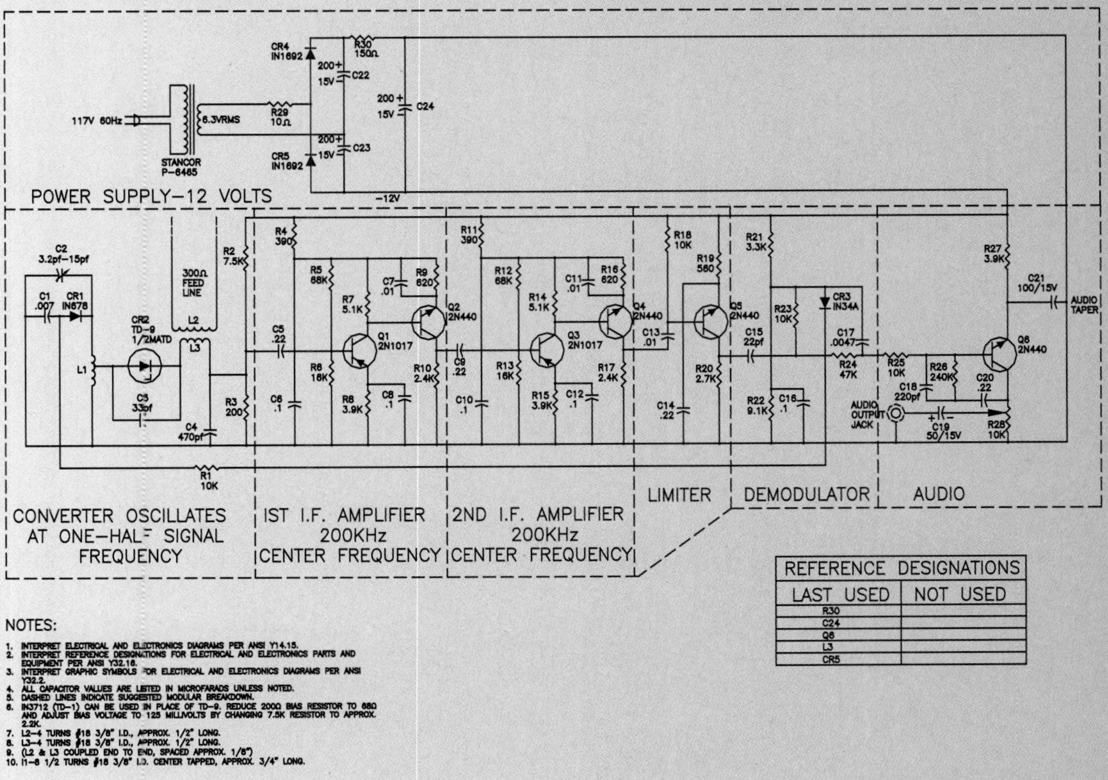

5. **Navigate to this chapter on the companion website and open CPE-06viewport.dwg.** Create model viewports using the **Three: Left** viewport configuration. In the left viewport, zoom to the drawing extents. In the top-right viewport, zoom to show the general drawing notes. Answer the following questions.
   A. What is the scale of VIEW A?
   B. What is the current revision level for this drawing?

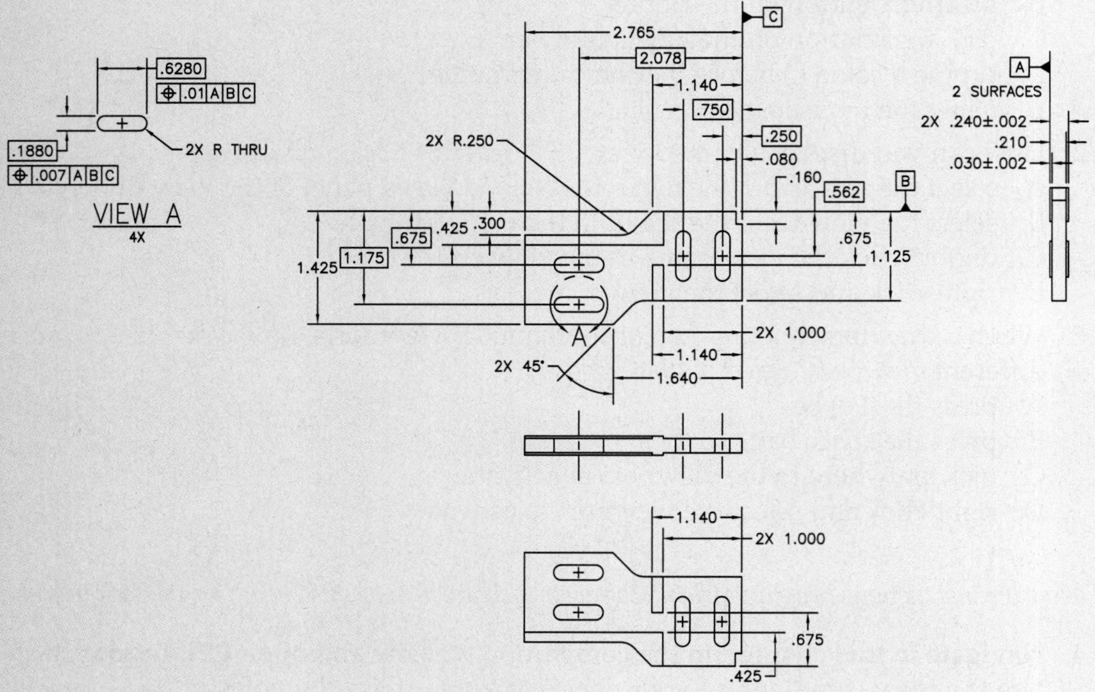

CHAPTER 7

Object Snap and AutoTrack

Learning Objectives

After completing this chapter, you will be able to:

- ✓ Set running object snap modes for continuous use.
- ✓ Use object snap overrides for single point selections.
- ✓ Select appropriate object snaps for various drawing tasks.
- ✓ Use AutoSnap features to speed up point specifications.
- ✓ Use AutoTrack to locate points relative to other points in a drawing.

This chapter explains how to use object snap and AutoTrack tools and options to draw accurately. Object snaps and AutoTrack are very useful and efficient drawing aids. Object snaps are also an important aid to parametric drafting. Chapter 22 explains how to use parametric drafting tools to constrain objects.

Object Snap

Object snap increases drafting performance and accuracy with ***snapping***. See Figure 7-1. You can use object snap with any command that requires point selection. Object snap modes identify the object snap point. The AutoSnap feature, which controls object snap, is on by default and displays ***markers*** while you draw. After a brief pause, a tooltip appears, indicating the object snap mode. See Figure 7-2. Refer to the list of standard object snap modes in Figure 7-3 to identify the appropriate object snap for each drafting task. Object snap use becomes second nature with practice, and greatly increases productivity and accuracy.

object snap: A tool that locates an exact point, such as an endpoint, midpoint, or center point, on or in relation to an existing object.

snapping: Picking a point near the intended position to have the crosshairs "snap" exactly to the specific point.

markers: Visual cues that appear at the snap point to confirm object snap mode and location.

NOTE

If you cannot see an AutoSnap marker because of the size of the current screen display, you can still confirm the point before picking by reading the tooltip, which indicates if a point is acquired beyond the visible area.

Figure 7-1.
An example of object snaps used to aid construction of specific objects in a part drawing view. Object snaps aid geometric construction for any drawing application.

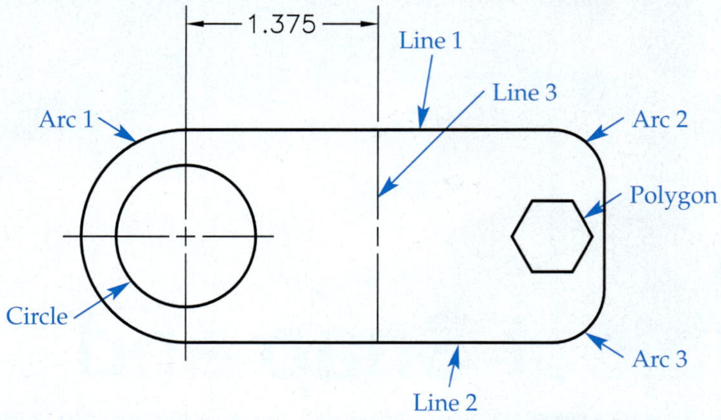

Object Snap	Construction Example
Parallel	Draw Line 2 parallel to Line 1
Endpoint	Begin and end Arc 1 at the endpoints of Line 1 and Line 2
Center	Locate the center of the Circle at the center point of Arc 1
Midpoint	Begin Line 3 at the midpoint of Line 1
Perpendicular	End Line 3 perpendicular to Line 2
Mid Between 2 Points	Locate the center of the Polygon at the midpoint between the center points of Arc 2 and Arc 3

Figure 7-2.
The AutoSnap marker and the related tooltip that appears when you snap to an endpoint and a point of tangency.

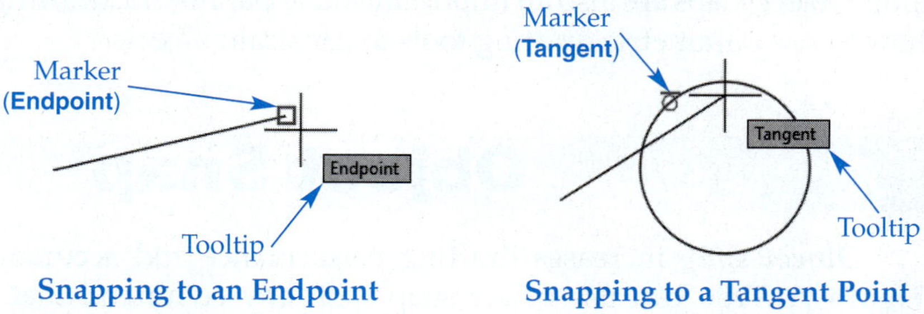

Running Object Snaps

Type
DSETTINGS
DS
SE

running object snaps: Automatic object snap modes that run in the background while you work.

Running object snaps are on by default and are often the quickest and most effective way to use object snap. The **Endpoint**, **Center**, **Intersection**, and **Extension** running object snap modes are active by default. Right-click on the **Object Snap** button or pick the **Object Snap** flyout on the status bar to select modes to activate or deactivate. See Figure 7-4A. You can also set running object snaps using the **Object Snap** tab of the **Drafting Settings** dialog box. See Figure 7-4B. A quick way to access the **Object Snap** tab is to right-click on the **Object Snap** button or pick the **Object Snap** flyout on the status bar and select **Object Snap Settings...**.

To use most running object snaps, move the crosshairs near the location on an existing object where the object snap should occur. When you see the appropriate marker and tooltip, pick to locate the point at the exact position on the object. See Figure 7-2.

Figure 7-3.
Standard object snap modes. You can snap to appropriate geometry of many different objects in addition to the objects listed in the description, such as dimension, mutileader, and table objects.

Mode	Marker	Alias	Description
Endpoint	□	END	Locates the nearest endpoint of an object, such as the endpoint of a line, arc, elliptical arc, polyline, spline, ray, solid, or multiline.
Midpoint	△	MID	Finds the point halfway between the endpoints of an object, such as the midpoint of a line, arc, elliptical arc, polyline, spline, or multiline, or finds the root point of an xline.
Center	○	CEN	Finds the center point of a circular object, such as the center of a circle, arc, ellipse, elliptical arc, polyline, or radial solid.
Geometric Center	○	GCEN	Finds the geometric center of a closed polyline, polygon, or spline.
Node	⊗	NOD	Locates a point object, such as an object drawn with the **POINT**, **DIVIDE**, or **MEASURE** command, or a dimension definition point.
Quadrant	◇	QUA	Locates the closest of the four quadrant points of an object, such as the quadrant of a circle, arc, ellipse, elliptical arc, polyline, or radial solid. (Some objects may not have all four quadrants.)
Intersection	×	INT	Locates the closest intersection of two objects.
Extension	+	EXT	Finds a point along the imaginary extension of an existing object, such as a line, arc, polyline, elliptical arc, ray, solid, or multiline.
Insert		INS	Locates the insertion point of objects, such as text objects and blocks.
Perpendicular	∟	PER	Finds a point that is perpendicular to an object from the previously picked point.
Tangent		TAN	Finds points of tangency between circular and linear objects.
Nearest	⧖	NEA	Locates the point on an object closest to the crosshairs.
Apparent Intersect	⊠	APP	Locates the intersection between two objects that appear to intersect on-screen in the current view, but may not actually intersect in 3D space. *AutoCAD and Its Applications—Advanced* describes creating and editing 3D objects.
Parallel	//	PAR	Finds any point along an imaginary line parallel to an existing linear object, such as a line or polyline.
None		NON	Temporarily turns running object snap off during the current selection.

Figure 7-4.
A—Right-click on the **Object Snap** button or pick the **Object Snap** flyout on the status bar to activate or deactivate running object snap modes. B—You can also set running object snap modes using the **Drafting Settings** dialog box.

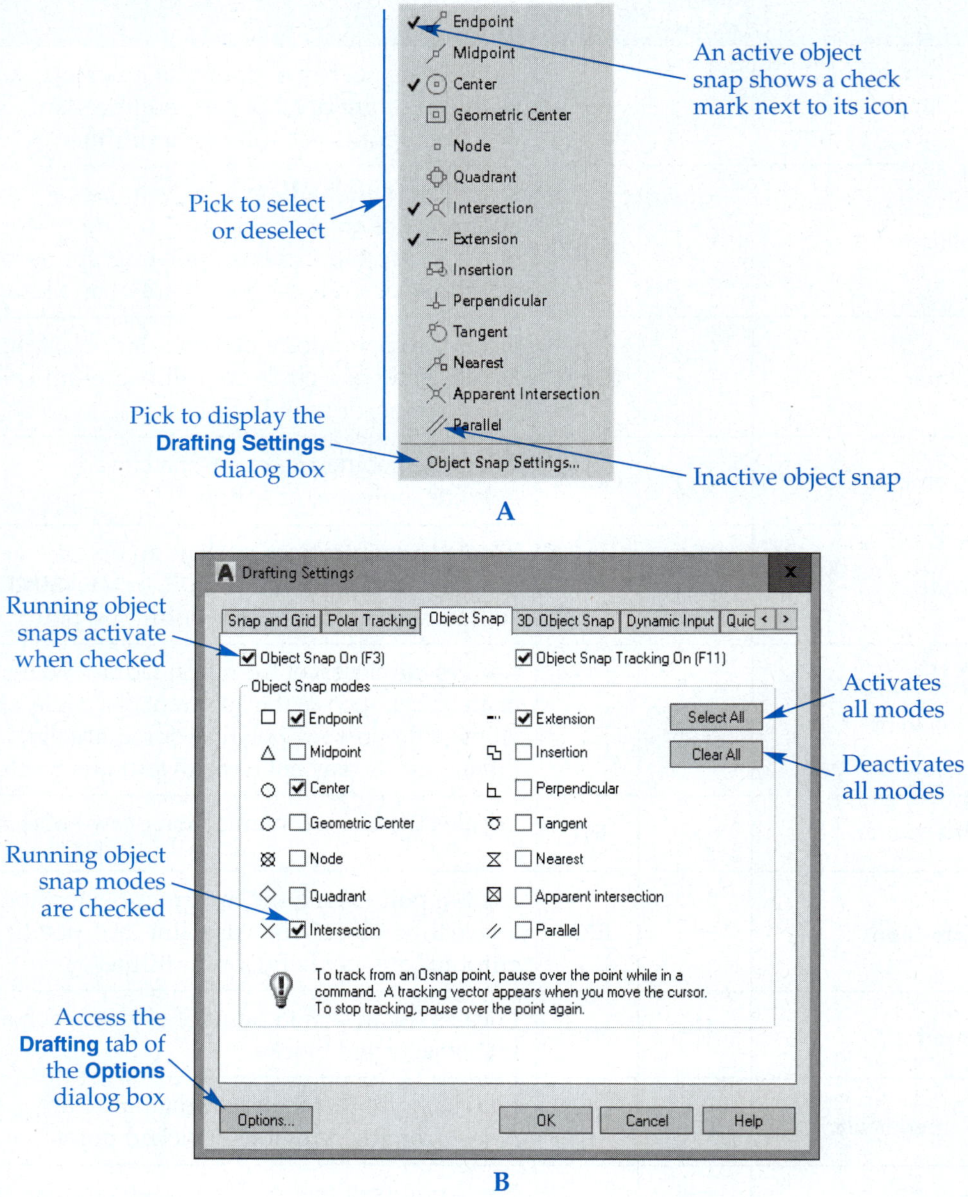

Toggle running object snaps off and on by picking the **Object Snap** button on the status bar, pressing [F3], or using the **Object Snap On (F3)** check box on the **Object Snap** tab of the **Drafting Settings** dialog box. Turn off running object snaps to locate points without the aid, or to avoid possible confusion of object snap modes. AutoCAD restores the selected running object snap modes when you reactivate running object snaps.

PROFESSIONAL TIP

Activate only the running object snap modes that you use most often. Too many running object snaps can make it difficult to snap to the appropriate location, especially on detailed drawings with several objects near each other. Use object snap overrides to access object snap modes that you use less often.

NOTE

By default, a keyboard point entry overrides running object snaps. Use the **Priority for Coordinate Data Entry** area on the **User Preferences** tab of the **Options** dialog box to adjust the default setting.

Object Snap Overrides

Use an ***object snap override*** to use an object snap that is not active, or to select a specific point if you experience conflicting running object snaps. Running object snaps return after you make the object snap override selection. All running object snap modes are available as object snap overrides, but some object snap options are only available as object snap overrides.

object snap override: A method of isolating a specific object snap mode while using a drawing or editing command. The selected object snap temporarily overrides the running object snap modes.

After you access a command and are ready to apply an object snap, hold [Shift] or [Ctrl] and then right-click and select an object snap override from the shortcut menu shown in Figure 7-5. An alternative, when you right-click and AutoCAD does not select the previous point, is to right-click without holding [Shift] or [Ctrl] and select from the **Osnap Overrides** cascading menu. Once you activate an object snap override, move the crosshairs near the location on an existing object where the object snap should occur. When you see the corresponding marker and tooltip, pick to locate the point at the exact position on the object.

NOTE

You can activate an object snap override by entering the first three letters of the object snap name. For example, enter END to activate the **Endpoint** object snap or CEN to activate the **Center** object snap.

PROFESSIONAL TIP

Remember that object snap modes function with the active command. An error message appears if you try to apply an object snap when no command is active.

Figure 7-5.
The **Object Snap** shortcut menu provides quick access to object snap overrides.

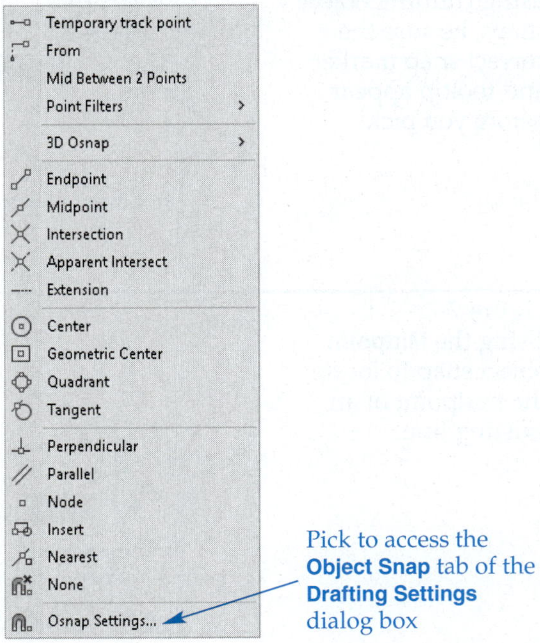

Endpoint Object Snap

To snap to an endpoint using the **Endpoint** object snap mode, move the crosshairs near the endpoint of an object. When the endpoint marker and tooltip appear, pick to locate the point at the exact endpoint. See Figure 7-6.

Midpoint Object Snap

To snap to a midpoint using the **Midpoint** object snap mode, move the crosshairs near the point halfway between the endpoints of an object. When the midpoint marker and tooltip appear, pick to locate the point at the exact midpoint. See Figure 7-7.

Exercise 7-1

www.g-wlearning.com/CAD/

Complete the exercise on the companion website.

Center Object Snap

To snap to a center point using the **Center** object snap mode, move the crosshairs near the *perimeter*, not the center point, of a circular object. You must move the crosshairs near the perimeter of a circular object, especially if the object is large, to acquire the center point. When you see the center marker and tooltip, pick to locate the point at the exact center. See Figure 7-8.

Geometric Center Object Snap

To snap to a ***centroid*** using the **Geometric Center** object snap mode, move the crosshairs near the perimeter of a closed polyline or spline. When you see the geometric center marker and tooltip, pick to locate the point at the exact centroid. See Figure 7-9.

centroid: The mathematical center of a geometric figure.

Figure 7-6.
Using the **Endpoint** object snap to locate the endpoint of an existing line. When using running object snaps, be sure the correct snap marker and tooltip appear before you pick.

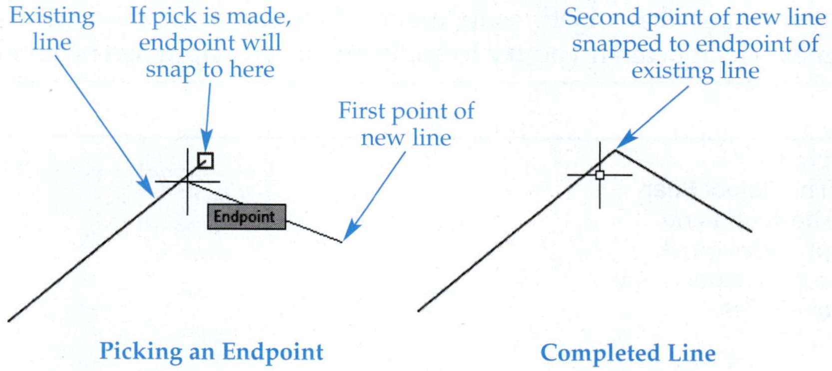

Figure 7-7.
Using the **Midpoint** object snap to locate the midpoint of an existing line.

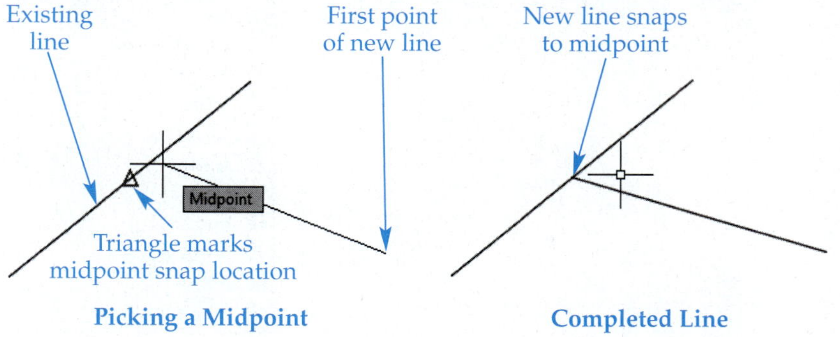

Figure 7-8.
Using the **Center** object snap to locate the center point of existing circles.

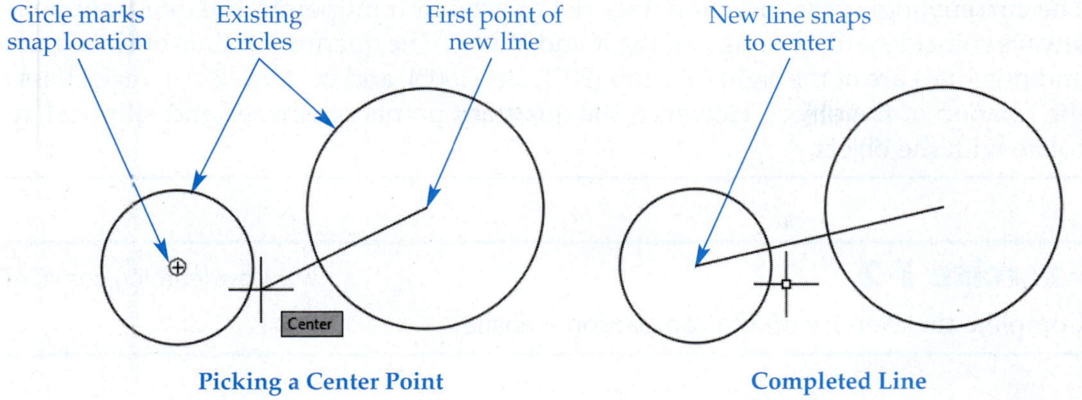

Figure 7-9.
Using the **Geometric Center** object snap to locate the geometric center point of a polygon.

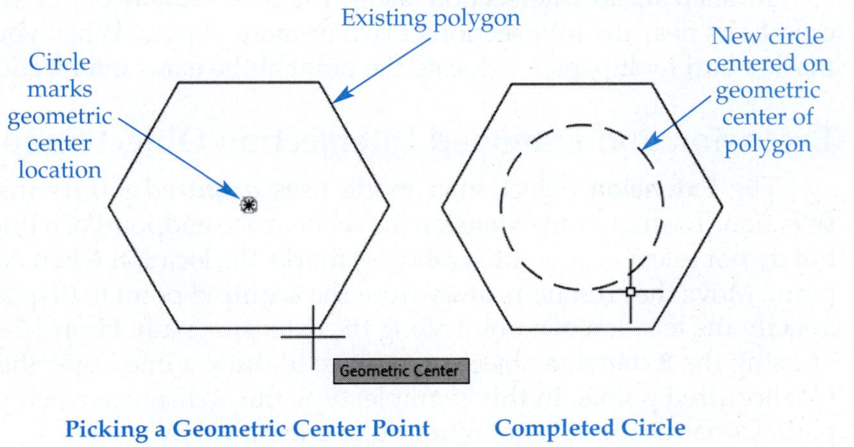

Quadrant Object Snap

To snap to a ***quadrant*** using the **Quadrant** object snap mode, move the crosshairs near the appropriate 0°, 90°, 180°, or 270° point on the circumference of a circular object. When you see the quadrant marker and tooltip, pick to locate the point at the exact quadrant position. See **Figure 7-10**.

quadrant: A point on the circumference at the horizontal or vertical quarter of a circular object, such as a circle, arc, ellipse, or polyline.

Figure 7-10.
Using the **Quadrant** object snap to locate a quadrant point on the circumference of a circle.

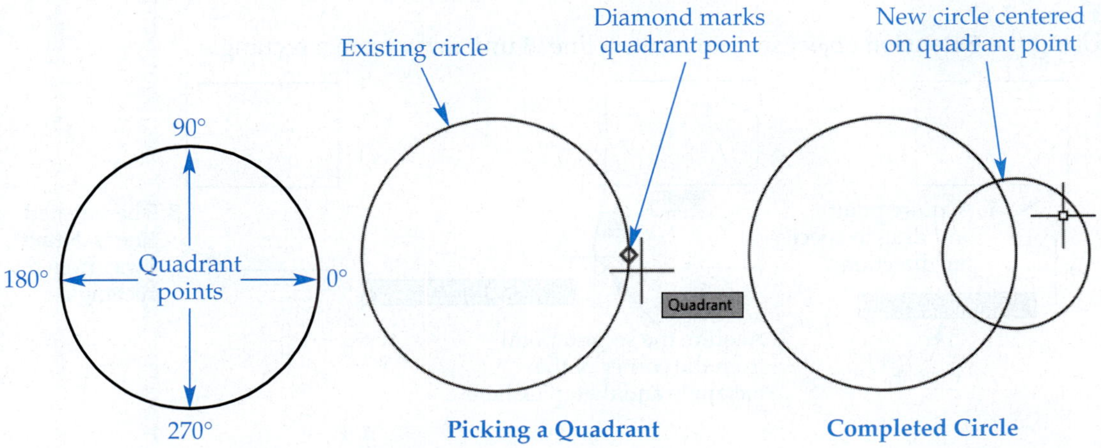

NOTE

The current angle zero direction does not affect quadrant points, but quadrant points always coincide with the angle of the X and Y axes. The quadrant points of circles, arcs, and polylines are at the right (0°), top (90°), left (180°), and bottom (270°), regardless of the rotation of the object. However, the quadrant points of ellipses and elliptical arcs rotate with the object.

Exercise 7-2

www.g-wlearning.com/CAD/

Complete the exercise on the companion website.

Intersection Object Snap

To snap to an intersection using the **Intersection** object snap mode, move the crosshairs near the intersection of two or more objects. When you see the intersection marker and tooltip, pick to locate the point at the exact intersection. See Figure 7-11.

Extension and Extended Intersection Object Snaps

acquired point: A point found by moving the crosshairs over a point on an existing object to reference the point when picking a new point.

extension path: A dashed line or arc that extends from an acquired point to the current location of the crosshairs.

The **Extension** object snap mode uses ***acquired points*** instead of direct point selection. To snap to an extension, hover near the endpoint of a linear or circular object, but do not select it. A point symbol (+) marks the location when AutoCAD acquires the point. Move the crosshairs away from the acquired point to display an ***extension path***. Specify the location of a point along the extension path. Figure 7-12 shows an example of using the **Extension** object snap twice to draw a line a specific distance away from two acquired points. In this example, type the .8 distance when you see the extension path. Dynamic input is not required to enter a value.

Figure 7-11.
Using the **Intersection** object snap to locate the intersection of a line and an arc.

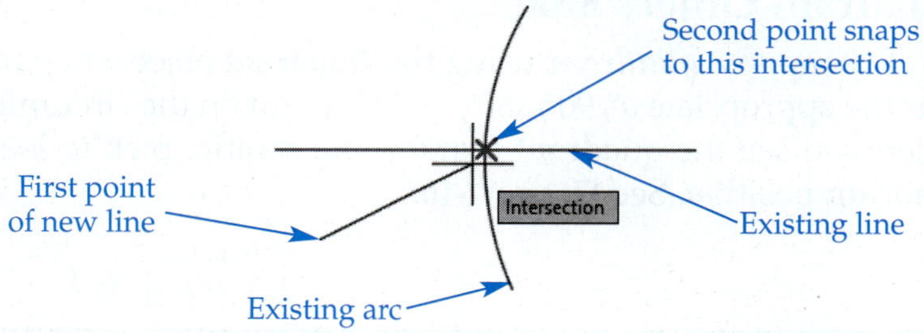

Figure 7-12.
Using the **Extension** object snap to create a line .8 unit away from a rectangle.

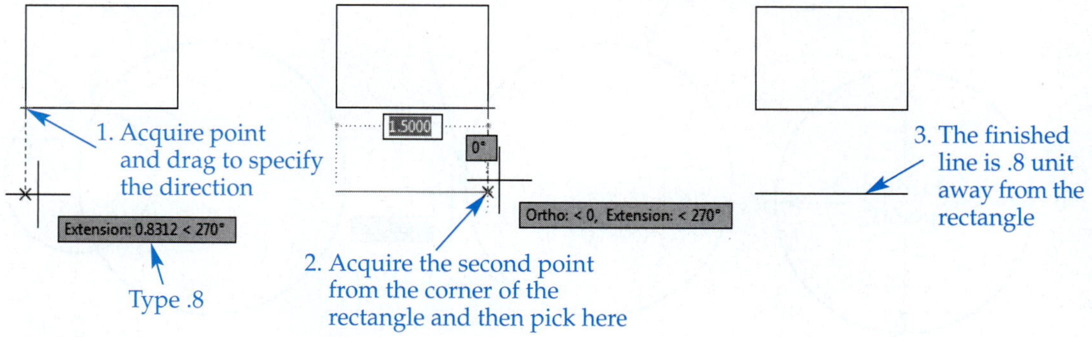

Exercise 7-3

www.g-wlearning.com/CAD/

Complete the exercise on the companion website.

Use the **Extension** or **Extended Intersection** object snap override to snap to the location where objects would intersect if they were long enough. To use the **Extension** object snap mode, hover near the endpoint of one object to acquire the first point, and then hover near the endpoint of another object to acquire the second point. Then move the crosshairs away from the acquired point, near the location of where the objects would intersect. When you see two extension paths and an intersection icon, pick to locate the point. See Figure 7-13A.

To use the **Extended Intersection** object snap override, select objects one at a time using the **Intersection** object snap override. Move the cursor over one of the objects to display the intersection marker with an ellipsis (...), and pick the object. Then move the cursor over the other object to display the intersection marker at the extended intersection, and pick. See Figure 7-13B.

Exercise 7-4

www.g-wlearning.com/CAD/

Complete the exercise on the companion website.

Figure 7-13.
Locating the center of a circle at the extended intersection of a line and an arc. A—Using the **Extension** object snap. B—Using the **Extended Intersection** object snap.

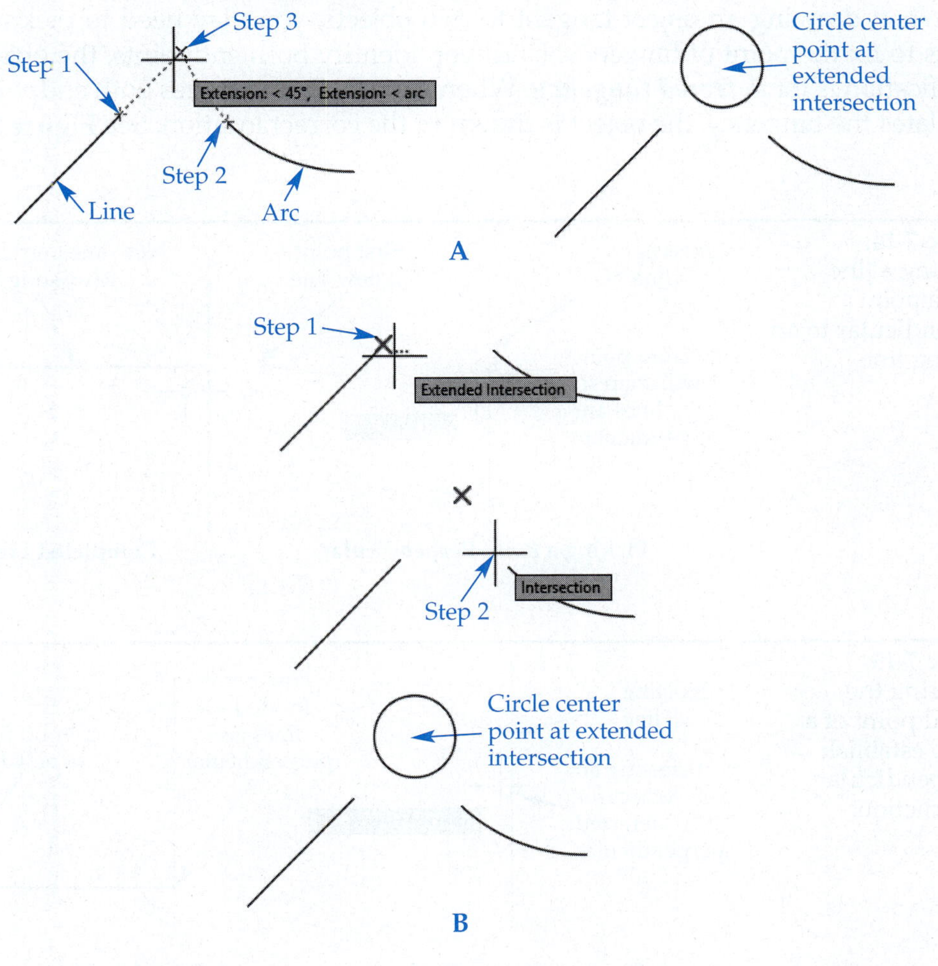

Perpendicular Object Snap

To snap to a perpendicular point using the **Perpendicular** object snap mode, move the crosshairs near the point of perpendicularity of an object. When you see the perpendicular marker and tooltip, pick to locate the point exactly perpendicular to the existing object. See Figure 7-14.

Figure 7-15 shows using the **Perpendicular** object snap mode to begin a line perpendicular to an existing object. The tooltip reads Deferred Perpendicular, and the perpendicular marker includes an ellipsis (...). The second endpoint determines the location of the line in a ***deferred perpendicular*** condition.

deferred perpendicular: A calculation of the perpendicular point that is delayed until you pick another point.

NOTE

Perpendicularity is determined from the point of intersection. Therefore, it is possible to draw a line perpendicular to a circular object, such as a circle or arc.

Exercise 7-5

www.g-wlearning.com/CAD/

Complete the exercise on the companion website.

Tangent Object Snap

To snap to the point of tangency using the **Tangent** object snap mode, move the crosshairs near a point of tangency of an object. When you see the tangent marker, pick to locate the point at the exact point of tangency. See Figure 7-16.

When drawing an object tangent to two objects, you may need to pick multiple points to fix the point of tangency. Until you identify both endpoints, the object snap specification is for ***deferred tangency***. When AutoCAD recognizes both endpoints and calculates the tangency, the object is drawn in the correct location. See Figure 7-17.

deferred tangency: A calculation of the point of tangency that is delayed until you pick both points.

Figure 7-14. Drawing a line from a point perpendicular to an existing line.

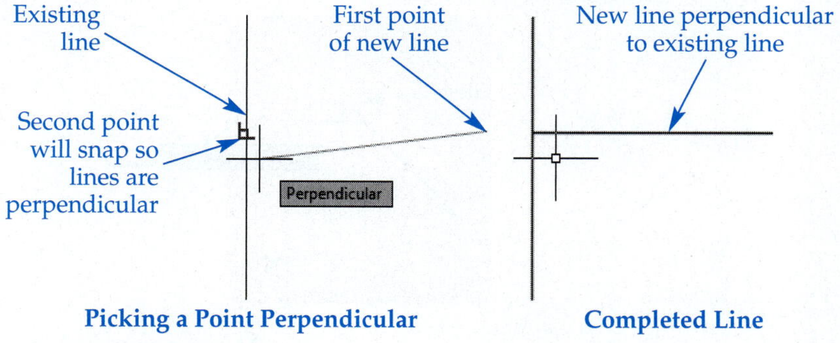

Figure 7-15. Deferring the second point of a line to establish a perpendicular construction.

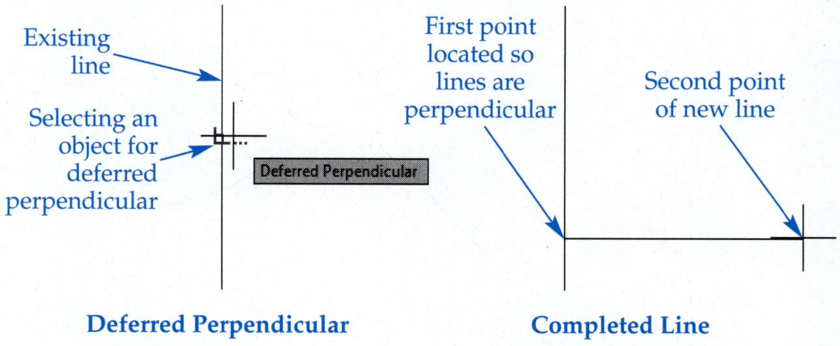

Figure 7-16.
Using the **Tangent** object snap to end a line tangent to a circle.

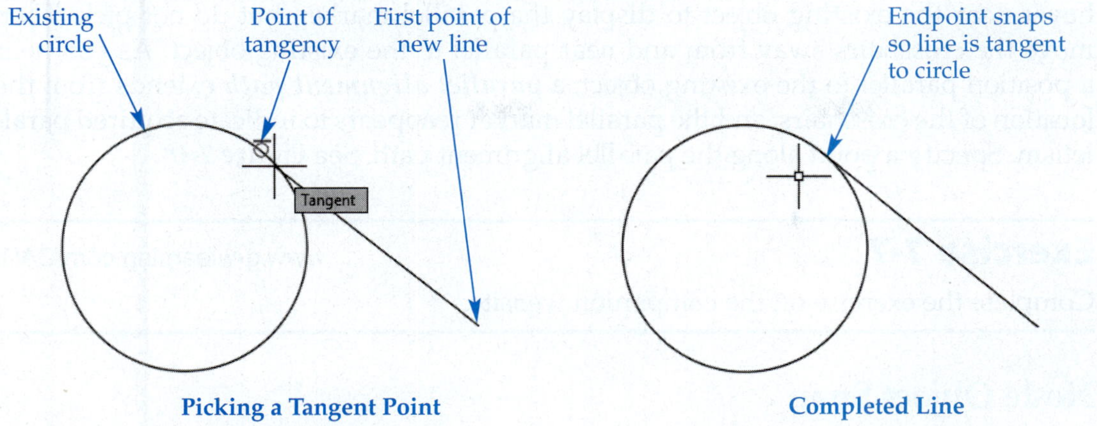

Figure 7-17.
Drawing a line tangent to two circles.

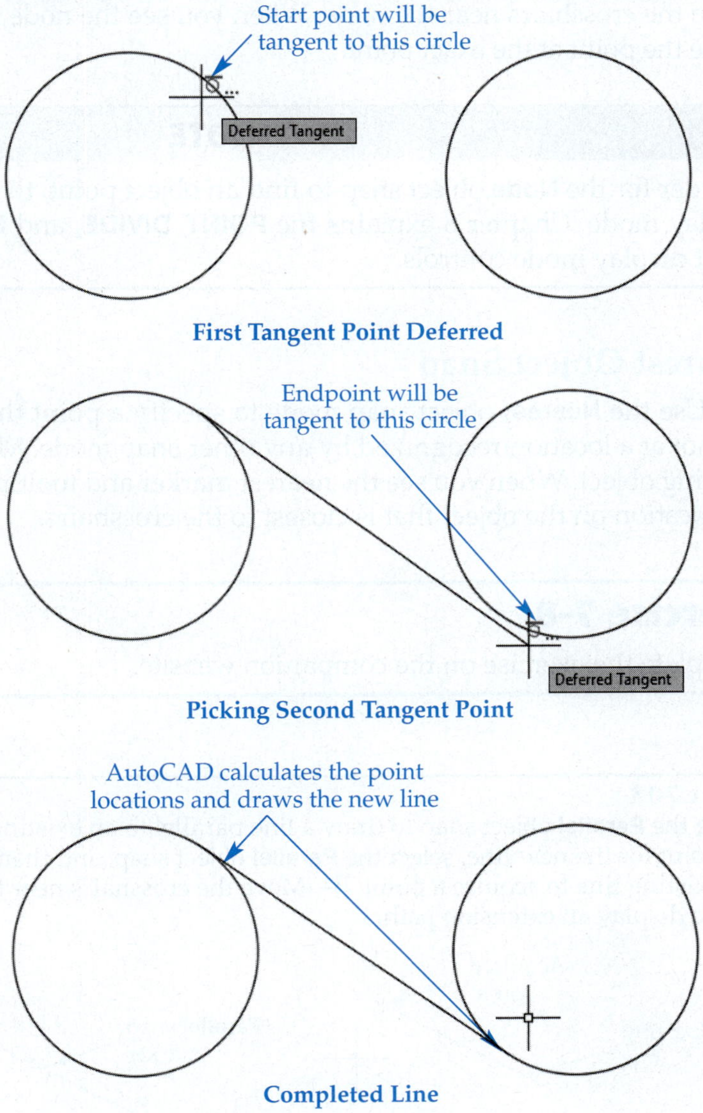

Exercise 7-6

www.g-wlearning.com/CAD/

Complete the exercise on the companion website.

Parallel Object Snap

To snap to a point parallel to a line or polyline using the **Parallel** object snap mode, hover near the existing object to display the parallel marker, but do not pick. Then move the crosshairs away from and near parallel to the existing object. As you near a position parallel to the existing object, a ***parallel alignment path*** extends from the location of the crosshairs, and the parallel marker reappears to indicate acquired parallelism. Specify a point along the parallel alignment path. See Figure 7-18.

parallel alignment path: A dashed line parallel to an existing line that extends from the location of the crosshairs.

Exercise 7-7

www.g-wlearning.com/CAD/

Complete the exercise on the companion website.

Node Object Snap

Use the **Node** object snap mode to snap to a point, such as a point drawn using the **POINT**, **DIVIDE**, or **MEASURE** command, or to the origin of a dimension ***definition point***. Move the crosshairs near the node. When you see the node marker and tooltip, pick to locate the point at the exact point.

definition point: A point used to specify the location of a dimension.

NOTE

In order for the **Node** object snap to find an object point, the point must be in a visible display mode. Chapter 8 explains the **POINT**, **DIVIDE**, and **MEASURE** commands and point display mode controls.

Nearest Object Snap

Use the **Nearest** object snap mode to specify a point that is directly on an object, but not at a location recognized by any other snap mode. Move the crosshairs near an existing object. When you see the nearest marker and tooltip, pick to locate the point at the location on the object that is closest to the crosshairs.

Exercise 7-8

www.g-wlearning.com/CAD/

Complete the exercise on the companion website.

Figure 7-18.
Using the **Parallel** object snap to draw a line parallel to an existing line. A—Select the first endpoint for the new line, select the **Parallel** object snap, and then move the crosshairs near the existing line to acquire a point. B—Move the crosshairs near the location of the parallel line to display an extension path.

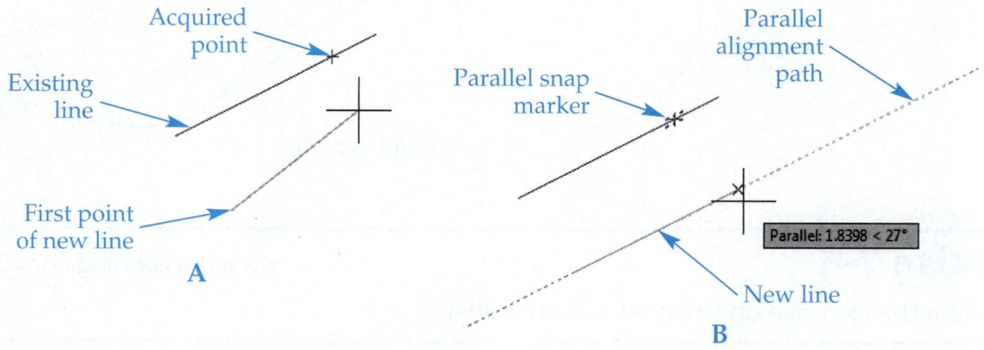

Temporary Track Point Snap

The **Temporary track point** snap mode is available only as an object snap override and allows you to locate a point aligned orthogonally with or relative to another point. For example, use the **Temporary track point** override to place the center of a circle a specified distance from, and horizontally aligned with, the midpoint of a line. At the Specify center point for circle or [3P/2P/Ttr (tan tan radius)]: prompt, select the **Temporary track point** override. Then use the **Midpoint** object snap to pick the midpoint of the line. The direction in which you move the crosshairs from the temporary tracking point determines the alignment of ***tracking vectors***. See Figure 7-19A. Next, specify the distance from the midpoint using the Specify center point for circle or [3P/2P/Ttr (tan tan radius)]: prompt. See Figure 7-19B. Finally, specify the circle radius. See Figure 7-19C.

tracking vectors: Temporary lines that display at specific angles, 0°, 90°, 180°, and 270° by default.

NOTE

You can typically use object snap tracking or polar tracking, explained later in this chapter, to accomplish the same tasks performed with the **Temporary track point** override.

Snap From

The **From** snap mode is available only as an object snap override and allows you to locate a point using coordinate entry from a specified reference base point. For example, use the **From** snap to place the center of a circle using a polar coordinate entry from the midpoint of an existing line. At the Specify center point for circle or [3P/2P/Ttr (tan tan radius)]: prompt, select the **From** snap mode, and then use the **Midpoint** object snap to pick the midpoint of the line. At the <Offset>: prompt, enter the polar coordinate @2<45 to establish the center of the circle 2 units and at a 45° angle from the midpoint of the line. Specify the radius of the circle to complete the operation. See Figure 7-20.

Mid Between 2 Points Snap

The **Mid Between 2 Points** snap mode is available only as an object snap override and is effective for locating a point halfway between two specified points. Use object snaps or coordinate point entry to pick reference points accurately. Figure 7-21 shows locating the center of a circle between two line endpoints.

Figure 7-19.
Using temporary tracking to position the center of a circle horizontal to the midpoint of a line. A—Acquire the midpoint of the line and move the crosshairs to form the desired alignment. B—Specify the distance from the midpoint along the tracking vector. C—Specify the size of the circle.

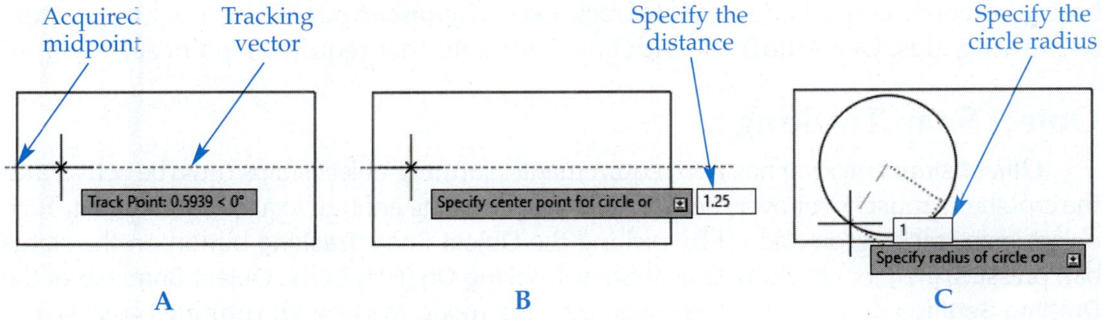

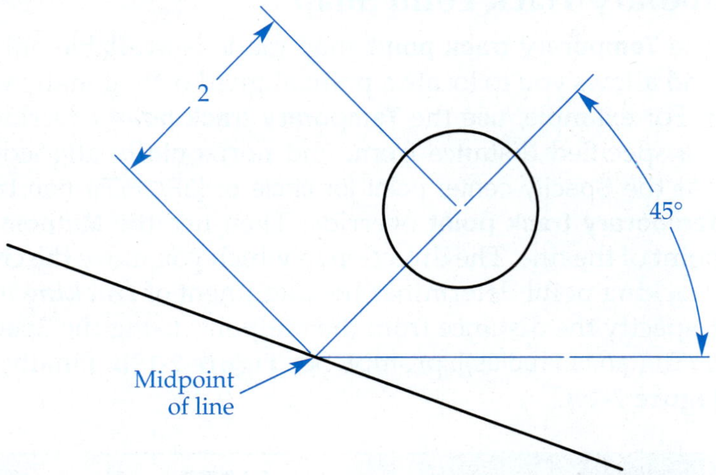

Figure 7-20. An example of using the **From** snap mode to locate the center of a circle using the **Midpoint** object snap and polar coordinate entry.

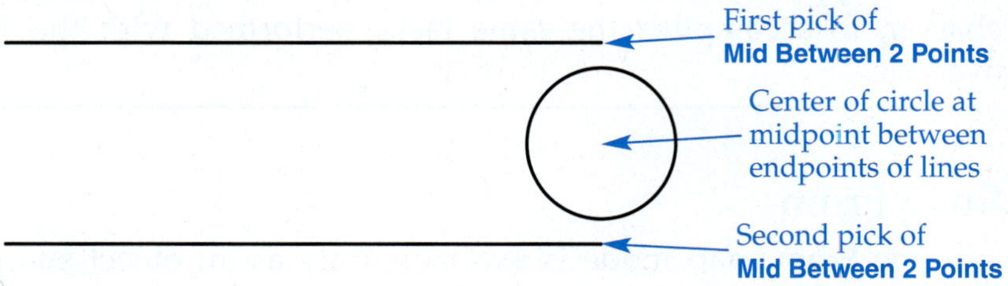

Figure 7-21. Using the **Mid Between 2 Points** snap mode to create a circle with the center point located at an exactly equal distance between two line endpoints.

NOTE

Object snaps are also important when constructing a parametric drawing, especially for inferring geometric constraints. Chapter 22 explains parametric drafting. 3D object snaps are available for 3D applications, as described in ***AutoCAD and Its Applications—Advanced***.

Exercise 7-9

www.g-wlearning.com/CAD/

Complete the exercise on the companion website.

AutoTrack

AutoTrack offers an object snap tracking mode and a polar tracking mode. Object snap tracking and polar tracking are helpful for common drafting tasks, including basic geometric constructions. AutoTrack uses ***alignment paths*** and tracking vectors as drawing aids. Use AutoTrack with any command that requires a point selection.

alignment paths: Temporary lines and arcs that coincide with the position of existing objects.

Object Snap Tracking

Object snap tracking has two requirements: running object snaps must be active, and the crosshairs must hover over the intended selection long enough to acquire the point. Turn object snap tracking on and off by picking the **Object Snap Tracking** button on the status bar, pressing [F11], or checking **Object Snap Tracking On (F11)** in the **Object Snap** tab of the **Drafting Settings** dialog box. Object snap tracking mode works with running object snaps.

object snap tracking: Mode that provides horizontal and vertical alignment paths for locating points after a point is acquired with object snap.

You must activate object snap tracking, running object snaps, and the appropriate running object snap modes in order for object snap tracking to function properly.

Figure 7-22 shows an example of using object snap tracking with the **Perpendicular** and **Midpoint** running object snaps to draw a line 1 unit long, perpendicular to the existing diagonal line. Running object snaps, the **Perpendicular** and **Midpoint** running object snap modes, and object snap tracking must be active before you use the **LINE** command to draw the line.

Figure 7-23 shows an example of using object snap tracking with the **Midpoint** running object snap to locate the center point of a circle directly above the midpoint

Figure 7-22.
Using object snap tracking to draw a line perpendicular to and at the midpoint of an existing line.

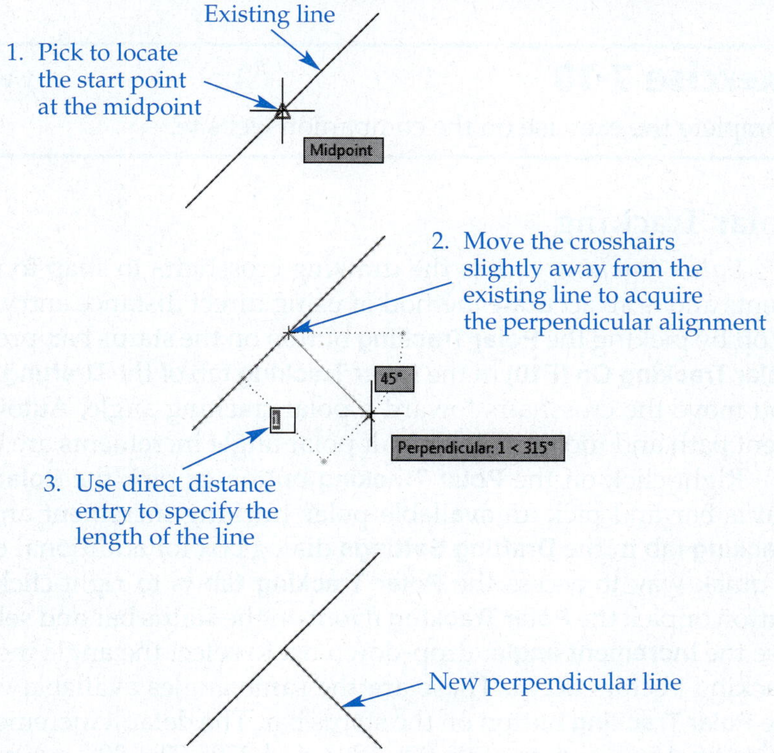

Figure 7-23.
Using object snap tracking to locate the center point of a circle in line with the midpoints of two lines.

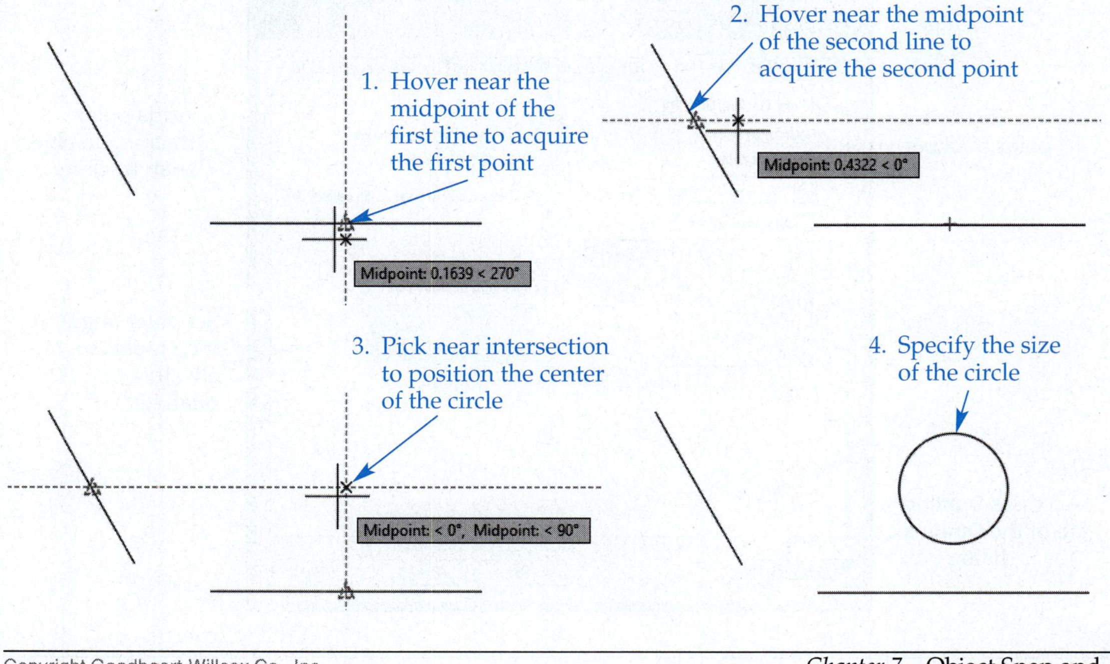

of a horizontal line and to the right of the midpoint of an angled line. Running object snaps, the **Midpoint** running object snap mode, and object snap tracking must be active before you use the **CIRCLE** command to draw the circle.

PROFESSIONAL TIP

Use object snap tracking whenever possible to complete tasks that require you to reference locations on existing objects. Often the combination of running object snaps and object snap tracking is the quickest way to construct geometry.

Exercise 7-10

www.g-wlearning.com/CAD/

Complete the exercise on the companion website.

Polar Tracking

polar tracking: Mode that allows the crosshairs to snap to preset incremental angles when you locate a point relative to another point.

Polar tracking causes the drawing crosshairs to snap to predefined angle increments and is an accurate method of using direct distance entry. Turn polar tracking on or off by picking the **Polar Tracking** button on the status bar, pressing [F10], or checking **Polar Tracking On (F10)** in the **Polar Tracking** tab of the **Drafting Settings** dialog box. As you move the crosshairs toward a polar tracking angle, AutoCAD displays an alignment path and tooltip. The default polar angle increments are 0°, 90°, 180°, and 270°.

Right-click on the **Polar Tracking** button or pick the **Polar Tracking** flyout on the status bar and pick an available polar tracking increment angle, or access the **Polar Tracking** tab in the **Drafting Settings** dialog box for additional control. See Figure 7-24. A quick way to access the **Polar Tracking** tab is to right-click on the **Polar Tracking** button or pick the **Polar Tracking** flyout on the status bar and select **Tracking Settings...**. Use the **Increment angle:** drop-down list to select the angle increments at which polar tracking vectors occur. These are the same angles available when you right-click on the **Polar Tracking** button on the status bar. The default increment is 90, which displays polar tracking vectors at 0°, 90°, 180°, and 270°. The 30° setting shown in Figure 7-24

Figure 7-24.
The **Polar Tracking** tab of the **Drafting Settings** dialog box.

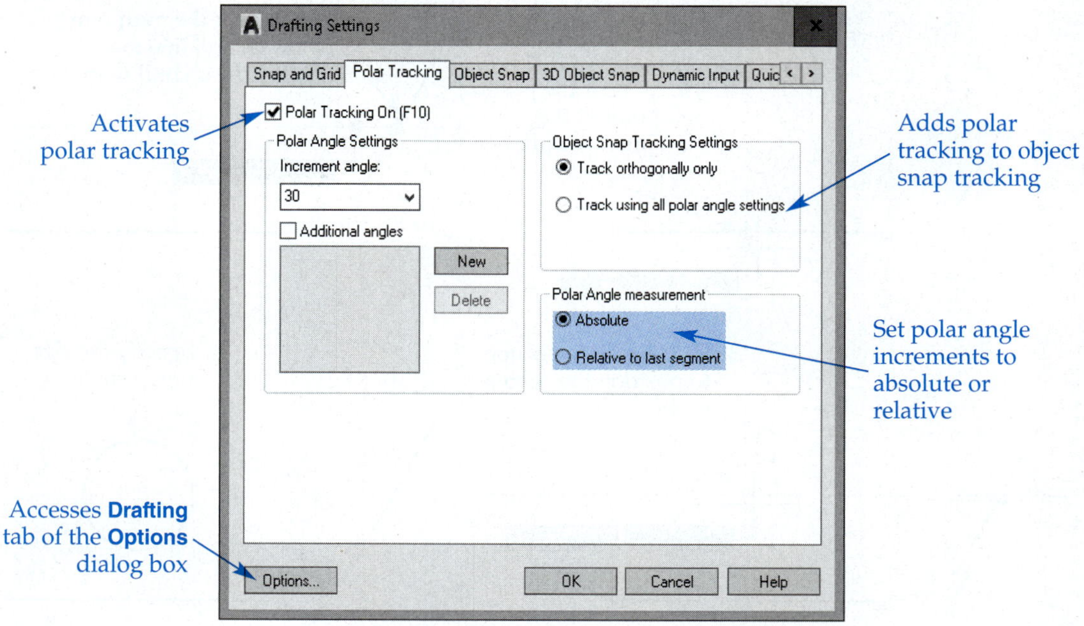

provides polar tracking at 30° increments. Figure 7-25 shows an example of drawing a parallelogram using the **LINE** command, polar tracking set to 30° angle increments, and direct distance entry.

To add specific polar tracking angles that are not associated with the increment angle, pick the **New** button in the **Polar Angle Settings** area and type an angle in the text box that appears in the **Additional angles** window. The added angles work with the increment angle setting when polar tracking is used. AutoCAD recognizes only the specific additional angles you enter, not each increment of the angle. Use the **Delete** button to remove angles from the list. Uncheck **Additional angles** to deactivate the additional angles.

The **Object Snap Tracking Settings** area sets the angles available with object snap tracking. If you select **Track orthogonally only**, only horizontal and vertical alignment paths are active. If you select **Track using all polar angle settings**, alignment paths are active for all polar tracking angles.

The **Polar Angle measurement** setting determines whether the polar angle increments are constant or relative to the previous segment. If you select **Absolute**, polar angles are measured from the base angle of 0° set for the drawing. If you select **Relative to last segment**, each increment angle is measured from a base angle established by the previously drawn segment.

NOTE

AutoCAD automatically turns ortho off when polar tracking is on, and it turns polar tracking off when ortho is on.

Exercise 7-11

www.g-wlearning.com/CAD/

Complete the exercise on the companion website.

Polar Tracking with Polar Snaps

You can also use polar tracking with polar snaps. For example, if you use polar tracking and polar snaps to draw the parallelogram in Figure 7-25, there is no need to type the length of the line, because you set the angle increment with polar tracking and a length increment with polar snaps. Establish the length increment using the **Snap and Grid** tab of the **Drafting Settings** dialog box. See Figure 7-26.

Figure 7-25.
Using polar tracking with 30° angle increments to draw a parallelogram.

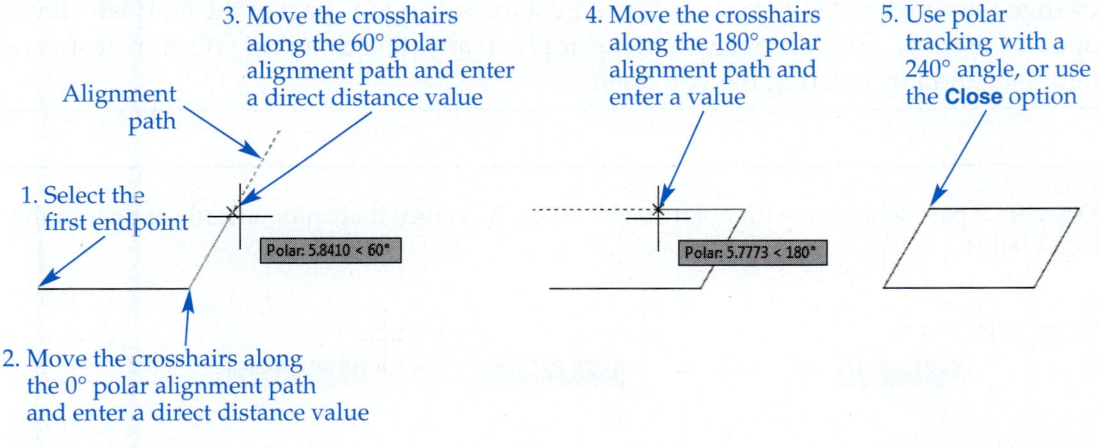

Figure 7-26.
Use the **Snap and Grid** tab of the **Drafting Settings** dialog box to set polar snap distance.

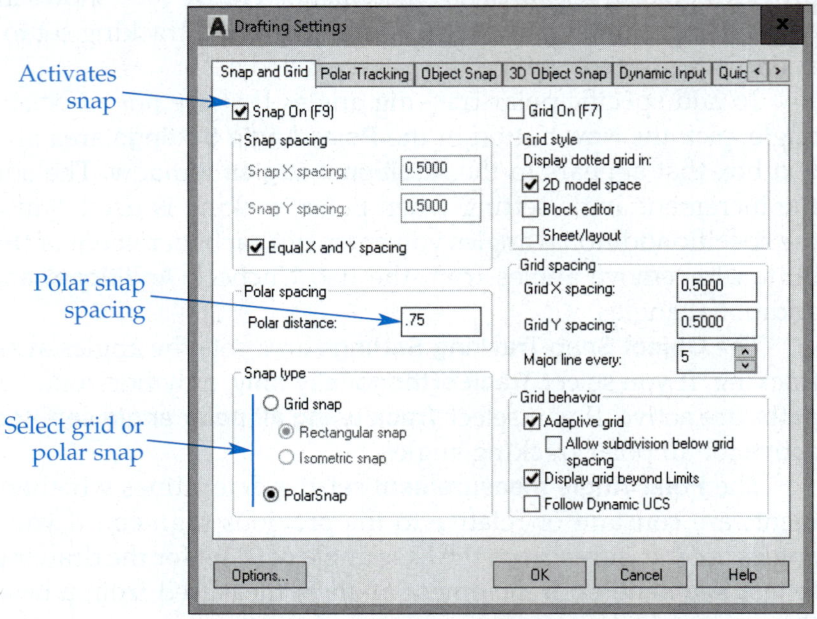

To activate polar snap, pick the **PolarSnap** radio button in the **Snap type** area of the dialog box. This activates the **Polar spacing** area and deactivates the **Snap spacing** area. Set the length of the polar snap increment in the **Polar distance:** text box. If the **Polar distance:** setting is 0, the polar snap distance is the orthogonal snap distance. Figure 7-27 shows a parallelogram drawn with 30° angle increments and length increments of .75. The lengths of the parallelogram sides are 1.5 and .75.

Polar Tracking Overrides

It takes time to set up polar tracking and polar snap options, but it is worth the effort if you intend to draw several objects that can take advantage of these aids. Use polar tracking overrides to define unique polar tracking angles. Polar tracking overrides work for the specified angle whether polar tracking is on or off. To activate a polar tracking override when AutoCAD asks you to specify a point, type a less than symbol (<) followed by the angle. For example, after you access the **LINE** command and pick a first point, enter <30 to set a 30° override. Then move the crosshairs in one of the possible directions and enter a length.

PROFESSIONAL TIP

When appropriate, set running object snaps and activate polar tracking for use in your drawing templates to increase drawing efficiency. Running object snaps, polar tracking, and polar tracking values are not template-specific and remain set until you change the settings. However, you can adjust the settings as part of the template development process. The settings you make apply to any new drawing you start from any template and any existing file you open.

Figure 7-27.
Drawing a parallelogram with polar snap. Notice the values that automatically appear in the input fields.

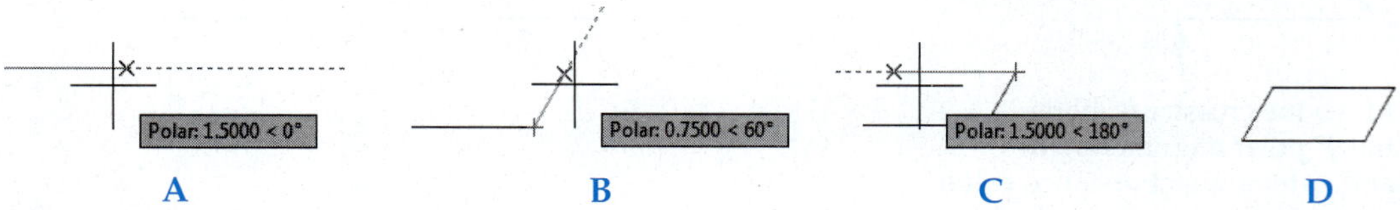

Exercise 7-12

www.g-wlearning.com/CAD/

Complete the exercise on the companion website.

Supplemental Material

www.g-wlearning.com/CAD/

AutoSnap and AutoTrack Options

For information about options for controlling the appearance and function of AutoSnap and AutoTrack, go to the companion website, select this chapter, and select **AutoSnap and AutoTrack Options**.

Chapter Review

Answer the following questions using the information in this chapter.

1. Define the term *object snap*.
2. Name the following AutoSnap markers.

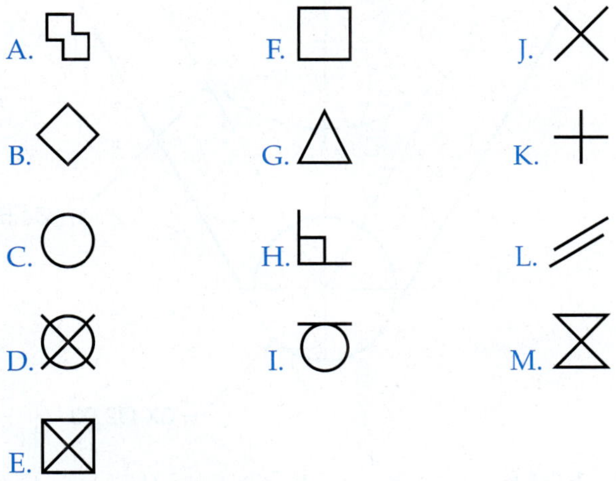

3. Define the term *running object snap*.
4. How do you access the **Drafting Settings** dialog box to change object snap settings?
5. How do you set running object snaps?
6. Why should you activate only the running object snaps you use most often?
7. What is an *object snap override*?
8. How do you access the **Object Snap** shortcut menu?
9. Where are the four quadrant points on a circle?
10. What does it mean when the tooltip reads Deferred Perpendicular?
11. What is a *deferred tangency*?
12. What is an *acquired point*?
13. What two display features does AutoTrack use as drawing aids to help you align new objects with existing geometry?
14. What are the two requirements to use object snap tracking?
15. What would you enter to specify a 40° polar tracking override?

Drawing Problems

Start AutoCAD if it is not already started. Start a new drawing from scratch or use an appropriate template of your choice. The template should include layers for drawing the given objects. Add layers as needed. Draw all objects using appropriate layers. Follow the specific instructions for each problem. Use only drawing commands and techniques you have already learned. Use object snap and AutoTrack when possible. Do not draw dimensions or text. Use your own judgment and approximate dimensions when necessary.

▼ Basic

1. **Mechanical** Draw the part view shown using object snap modes. Save the drawing as P7-1.

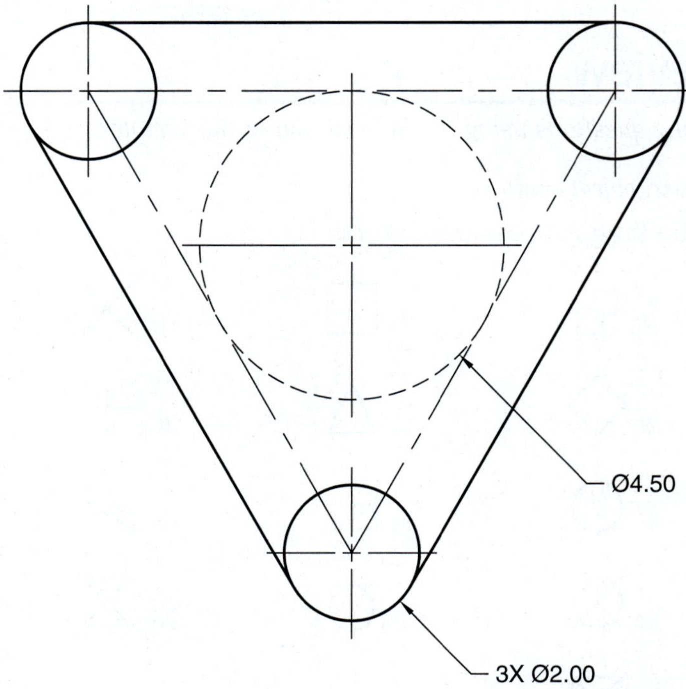

2. **General** Draw the highlighted objects shown, and then use the object snap modes indicated to draw the remaining objects. Save the drawing as P7-2.

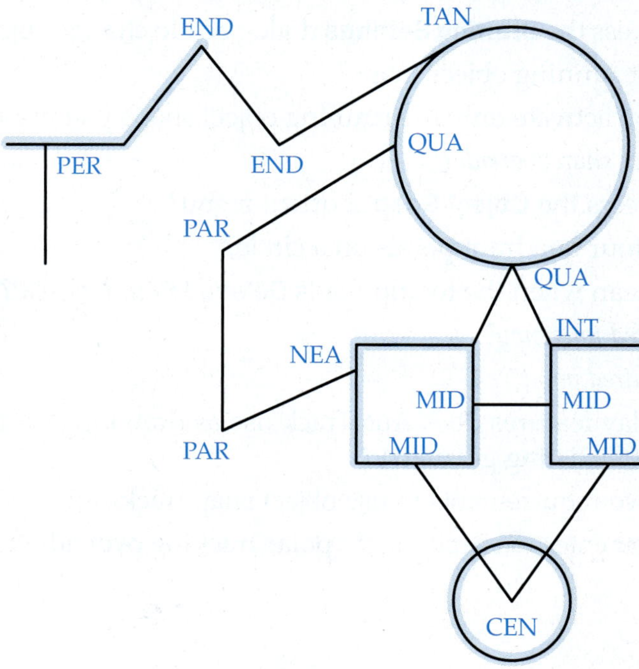

3. General Draw the schematic shown using the **Endpoint**, **Tangent**, **Perpendicular**, and **Quadrant** object snap modes. Save the drawing as P7-3.

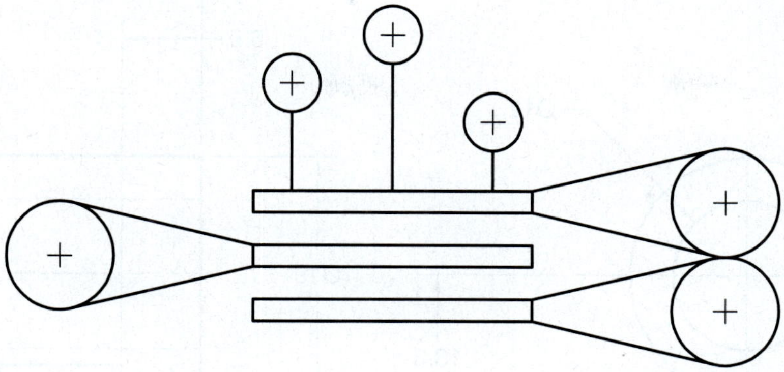

4. Piping Draw the pipe separator shown. Save the drawing as P7-4.

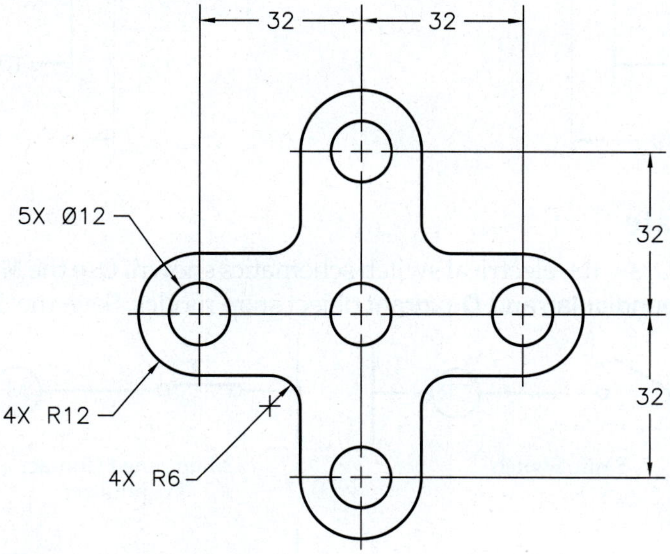

5. Architectural Draw the column base detail shown. The thickness of the column is 1/2″. Save the drawing as P7-5.

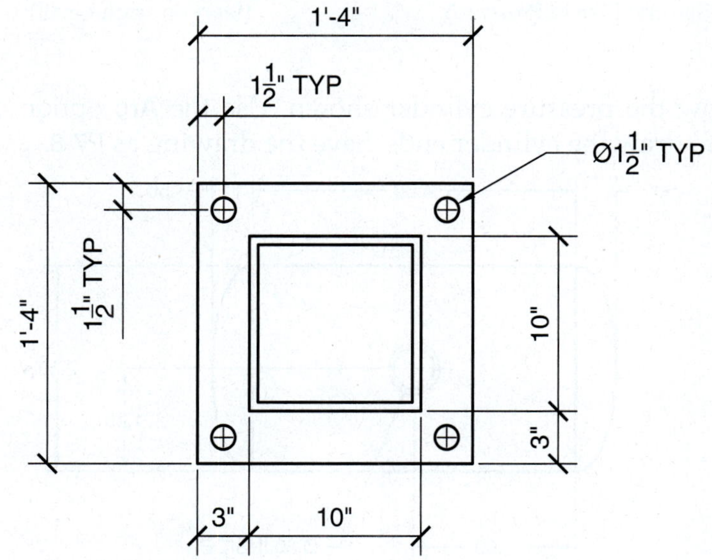

Drawing Problems - Chapter 7

6. **Piping** Draw the views of the tube hanger shown. Save the drawing as P7-6.

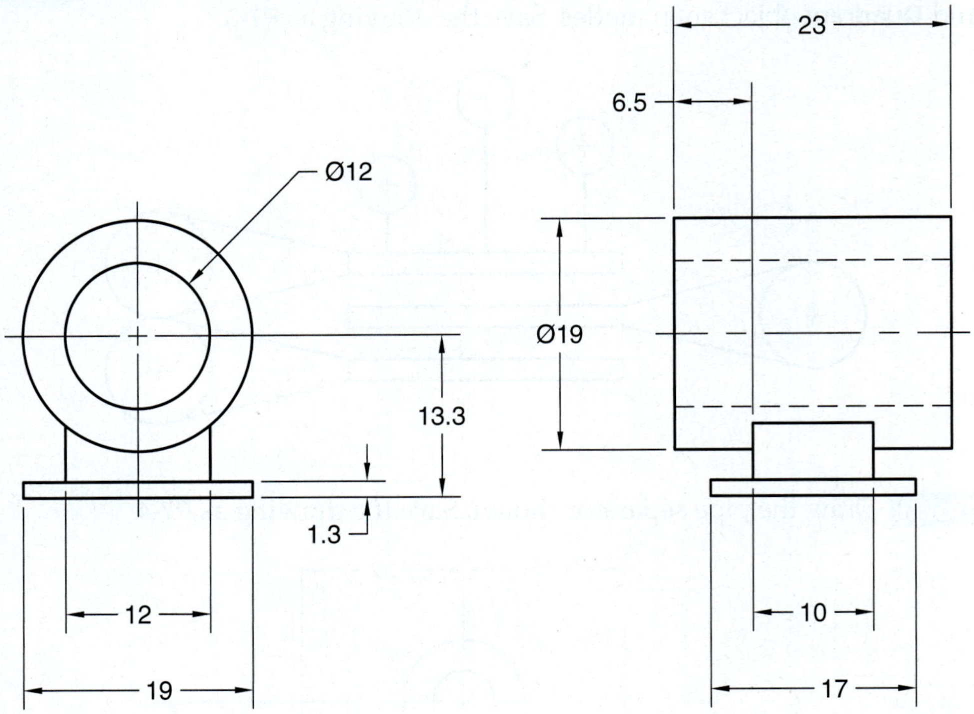

▼ Intermediate

7. **Electronics** Draw the electrical switch schematics shown. Use the **Midpoint**, **Endpoint**, **Tangent**, **Perpendicular**, and **Quadrant** object snap modes. Save the drawing as P7-7.

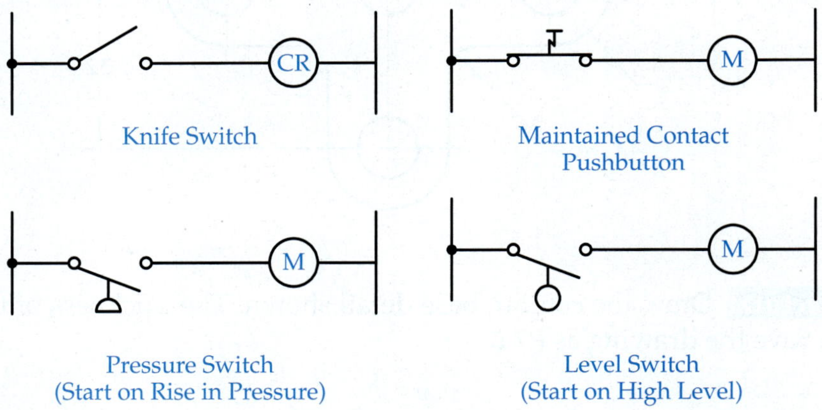

8. **Piping** Draw the pressure cylinder shown. Use the **Arc** option of the **ELLIPSE** command to draw the cylinder ends. Save the drawing as P7-8.

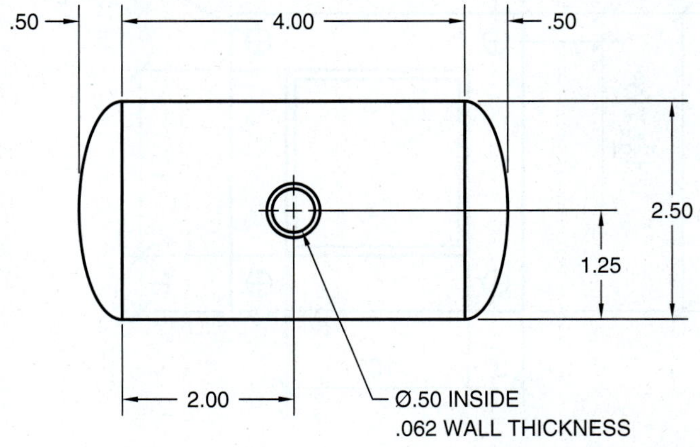

9. Mechanical Draw the stud shown. Save the drawing as P7-9.

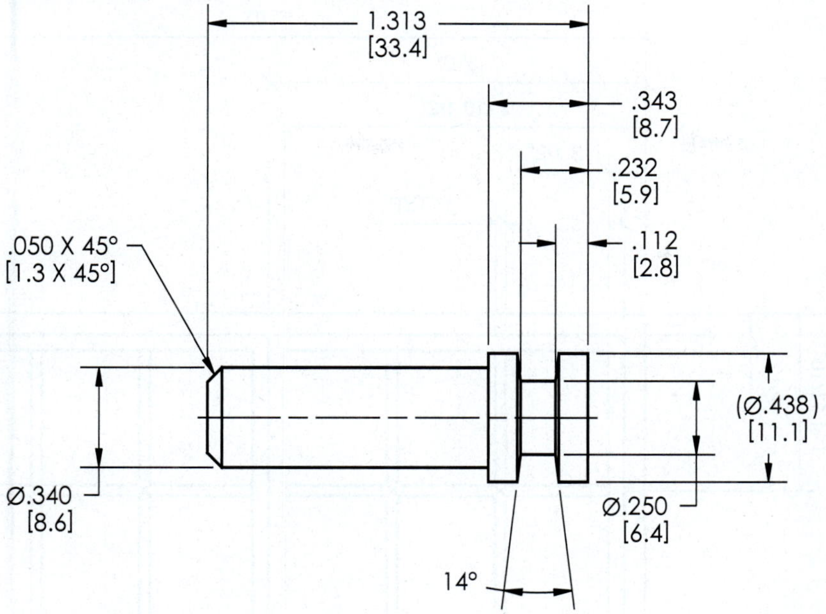

10. Mechanical Draw the part view shown. Save the drawing as P7-10.

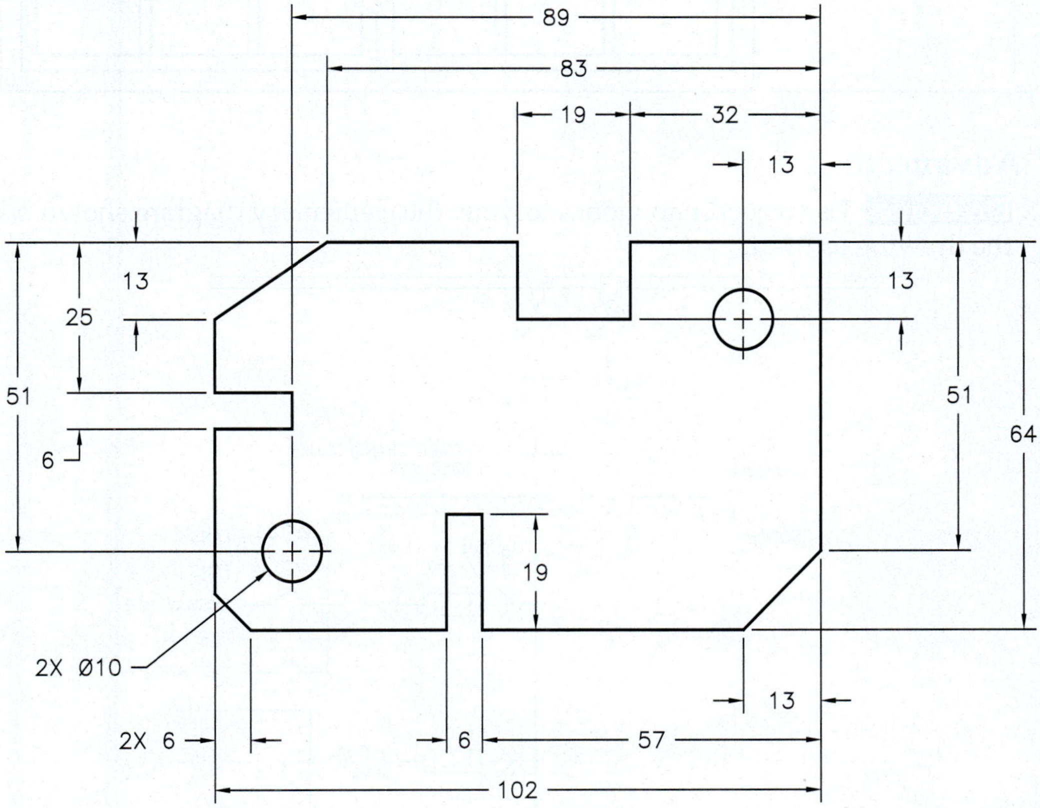

Drawing Problems - Chapter 7

11. **Architectural** Draw the window elevation symbol shown. Save the drawing as P7-11.

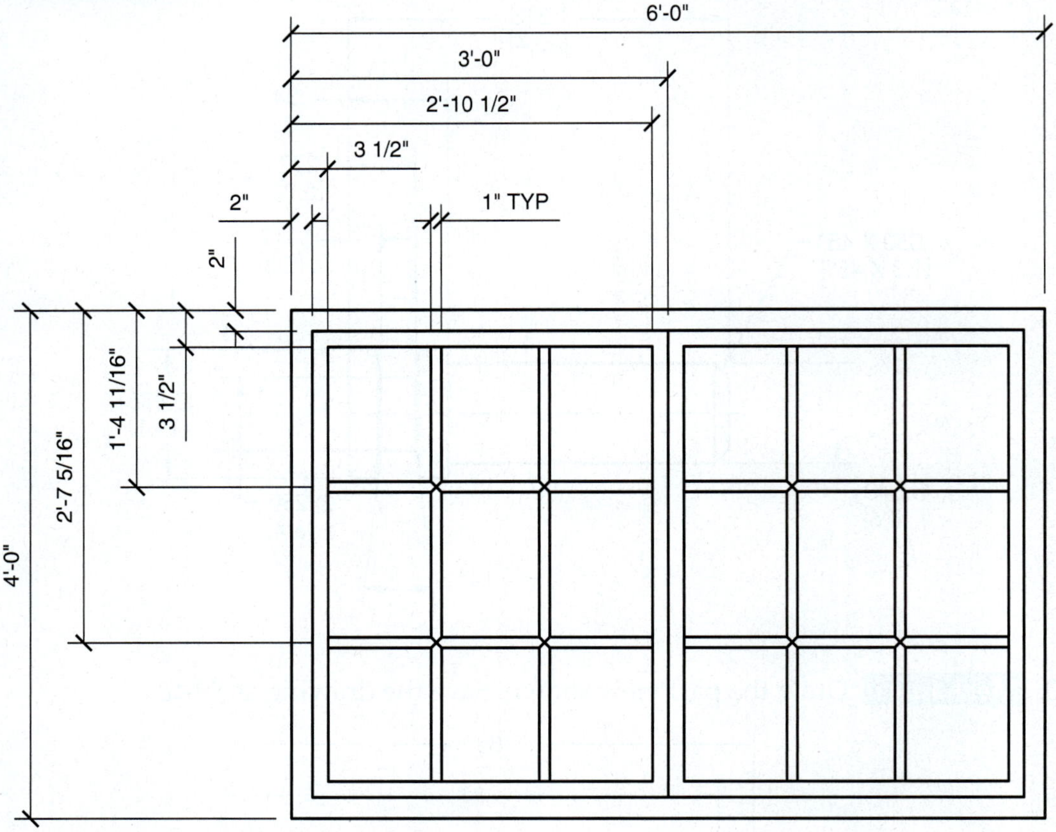

▼ Advanced

12. **Electronics** Use object snap modes to draw the elementary diagram shown. Save the drawing as P7-12.

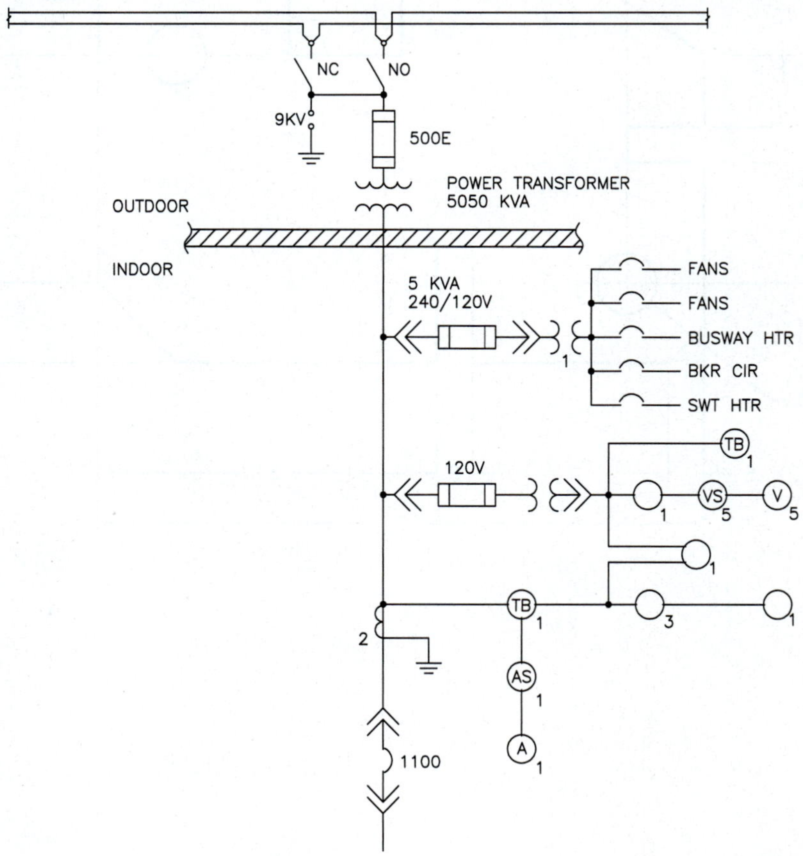

13. Mechanical Design and draft a hammer similar to the hammer shown. Use the overall dimensions shown, and add dimensions based on your experience, research, and measurements. Save the drawing as P7-13.

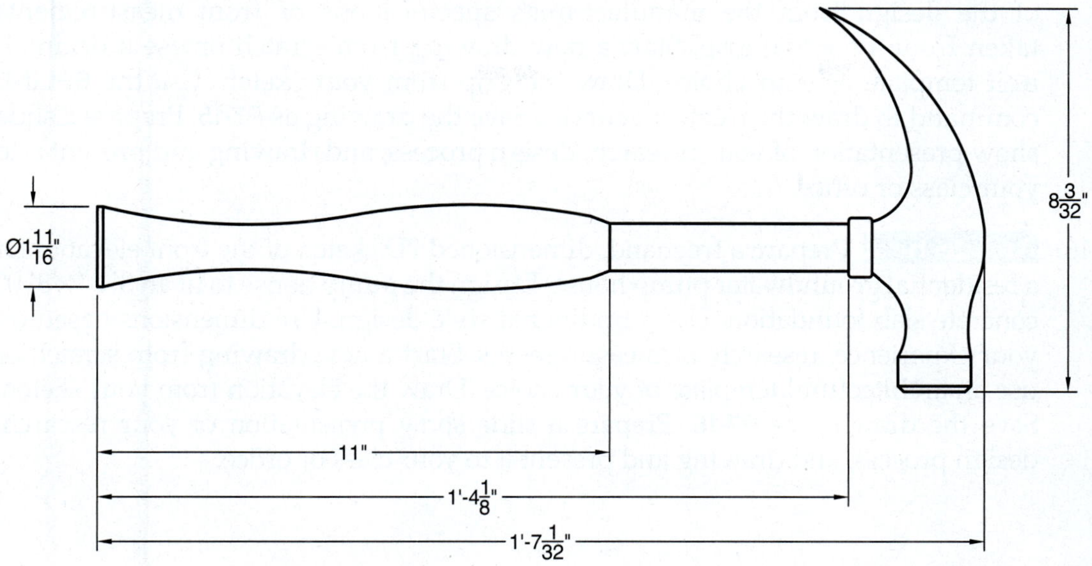

14. General Create a drawing from the sketch of a truck design shown. Draw the truck using appropriate size and scale features. Use a tape measure to measure an actual truck for reference if instructed. Consider the commands and techniques used to draw the truck, and try to minimize the number of objects. Save the drawing as P7-14.

Drawing Problems - Chapter 7

15. Mechanical Research the design of an existing handlebar grip with the following specifications: slips onto a Ø1″ pipe, freeform ergonomic design with finger grooves, silicone rubber material. Prepare a freehand, dimensioned 2D sketch of the design from the manufacturer's specifications or from measurements taken from an actual grip. Start a new drawing from scratch or use a decimal-unit template of your choice. Draw the grip from your sketch. Use the **SPLINE** command to draw the freeform curves. Save the drawing as P7-15. Prepare a slide show presentation of your research, design process, and drawing and present it to your class or office.

16. Architectural Prepare a freehand, dimensioned 2D sketch of the front elevation of a residential groundwater pump house. Design the pump house to fit an 8′-0″ × 8′-0″ concrete slab foundation. Use a traditional style design. Use dimensions based on your experience, research, or measurements. Start a new drawing from scratch or use an architectural template of your choice. Draw the elevation from your sketch. Save the drawing as P7-16. Prepare a slide show presentation of your research, design process, and drawing and present it to your class or office.

AutoCAD Certified Professional Exam Practice

Answer the following questions using the information in this chapter.

1. Which of the labeled points shown can you select using the AutoSnap feature, without entering an object snap override, if the **Midpoint**, **Endpoint**, and **Center** object snaps are set as running object snaps? *Select the one item that best answers the question.*

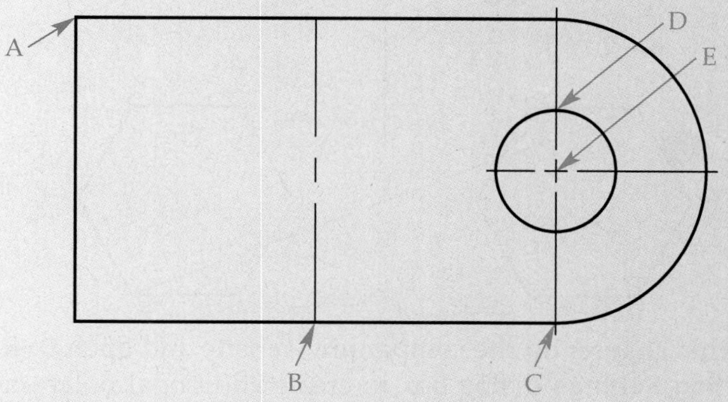

   A. points A, B, and C
   B. points A, C, and D
   C. points A, B, D, and E
   D. points A, B, C, and E
   E. points A and E only

2. Which of the following positions describe quadrant points on a circle? *Select all that apply.*
   A. 0°
   B. 30°
   C. 60°
   D. 90°
   E. 180°

3. Which of the following points can be selected using the **Endpoint** object snap? *Select all that apply.*
   A. a corner of a rectangle
   B. the center of a circle
   C. the end of an arc
   D. a vertex of a polygon
   E. the midpoint of a line

Follow the instructions in each problem. Complete each problem by answering the related question.

4. **Navigate to this chapter on the companion website and open CPE-07snaps.dwg.** Use running object snaps or object snap overrides to create Line A as shown. Restart the **LINE** command and create Line B from the midpoint of Line A, extending 1.5 units upward vertically. What are the coordinates of the upper endpoint of Line B?

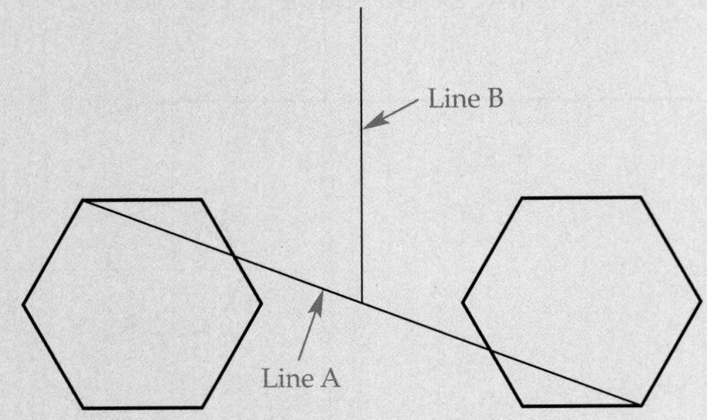

5. **Navigate to this chapter on the companion website and open CPE-07polar.dwg.** Use the **Drafting Settings** dialog box to create additional polar angles of 20, 40, and 80. Make sure **Polar Angle measurement** is set to **Absolute.** Then use polar tracking to create the following line segments.
   A. A line starting at the right endpoint of the existing line and extending 3 units at 80°
   B. A line starting at the upper endpoint of the previous line and extending 4.5 units at 20°
   C. A line starting at the right endpoint of the previous line and extending 3 units upward vertically
   D. A line starting at the upper endpoint of the previous line and extending 6.3 units at 40°

   What are the coordinates of the upper endpoint of the last line you drew?

AutoCAD Certified Professional Exam Practice - Chapter 7

CHAPTER 8

Construction Tools and Multiview Drawings

Learning Objectives

After completing this chapter, you will be able to:

- ✓ Use the **OFFSET** command to draw parallel and concentric objects.
- ✓ Place construction points.
- ✓ Mark points on objects at equal lengths using the **DIVIDE** command.
- ✓ Mark points on objects at designated increments using the **MEASURE** command.
- ✓ Create construction lines using the **XLINE** and **RAY** commands.
- ✓ Create multiview drawings.

This chapter explains how to create parallel offsets, divide objects, place point objects, and use construction lines. You can use these skills and your existing geometric construction ability to create multiview drawings. This chapter describes tools and methods for producing accurate geometric constructions. Construction tools do not constrain, or apply relationships between, objects. Chapter 22 explains how to use parametric drafting tools to constrain objects.

Parallel Offsets

The **OFFSET** command is a common geometric construction tool that is useful for many different drafting tasks. For example, you can offset lines or polylines to construct multiview drawings or form the thickness of architectural floor plan walls. Offset circles, arcs, or other curves to form concentric objects. For example, you can offset a circle to create the wall thickness of a pipe.

OFFSET

Ribbon
Home
> Modify
Offset

Type
OFFSET
PARALLEL
STEPPED
WALL
O

Specifying the Offset Distance

Often the best way to use the **OFFSET** command is to enter an offset value at the Specify offset distance or [Through/Erase/Layer] <*current*>: prompt. For example, to draw a circle concentric to and 1 unit from an existing circle, access the **OFFSET** command and specify an offset distance of 1. Pick the circle to offset, and then pick the side of the circle where the offset will occur. Use the offset preview and dynamic input to confirm the correct offset. See Figure 8-1. The **OFFSET** command remains active, allowing you to pick another object to offset using the same offset distance. Press [Enter] or select the **Exit** option to end the command.

Figure 8-1.
Drawing an offset circle using a designated distance and offset preview.

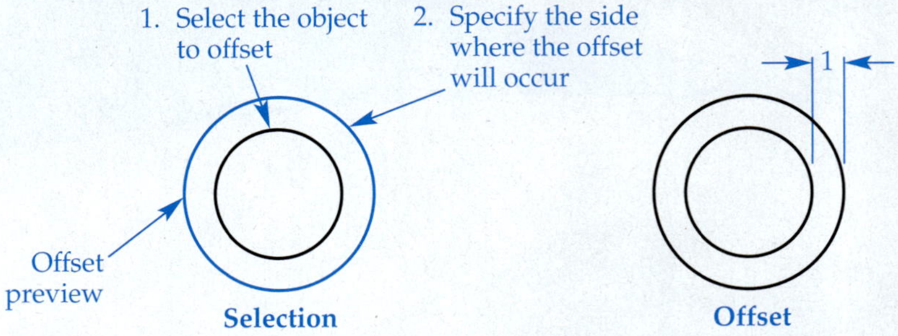

NOTE

Select objects to offset individually. No other selection option, such as window or crossing selection, works for selecting objects to offset. The offset preview appears by default.

PROFESSIONAL TIP

When using most commands that prompt you to specify a value, such as distance or height, and you do not know the numeric value, pick two points as an alternative. The distance between the points sets the value. You typically pick two points on existing objects to specify the appropriate value. Use object snaps, AutoTrack, or coordinate entry to make accurate selections.

Through Option

Another option to specify the offset distance is to pick a point through which the offset occurs. Access the **OFFSET** command and, instead of specifying the offset distance, select the **Through** option at the Specify offset distance or [Through/Erase/Layer] <*current*>: prompt. Then pick the object to offset, and pick the point through which the offset occurs. Use the offset preview and dynamic input to confirm the correct offset. See Figure 8-2. The **OFFSET** command remains active, allowing you to pick another object to offset using the **Through** option. Exit when you are finished offsetting.

Erasing the Source Object

Use the **Erase** option of the **OFFSET** command to erase the source, or original, object during the offset. Start the **OFFSET** command, activate the **Erase** option, and select **Yes** at the Erase source object after offsetting? prompt. The **Yes** option remains set as the default until you change the setting to **No**. Be sure to change the **Erase** setting back to **No** if the source object should remain the next time you use the **OFFSET** command. Exit when you are finished offsetting.

Figure 8-2.
Drawing an offset of a polyline object, constructed using the **POLYGON** command, through a given point.

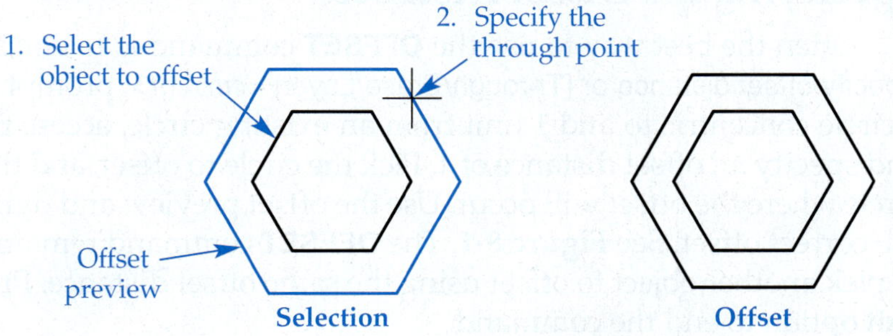

Layer Option

By default, offsets use the same properties as the source object, including layer. Use the **Layer** option of the **OFFSET** command to place the offset object on the current layer, regardless of the layer assigned to the source object. First, make the layer to apply to the offset current. Then start the **OFFSET** command, activate the **Layer** option, and select **Current** at the Enter layer option for offset objects prompt. The **Current** option remains set as the default until you change the setting to **Source**. Be sure to change the **Layer** setting back to **Source** if the layer assigned to the source object should apply to the offset the next time you use the **OFFSET** command. Exit when you are finished offsetting.

Multiple Option

The **Multiple** option is useful for offsetting more than once using the same distance between objects, without having to reselect the object to offset. Access the **OFFSET** command, specify the offset distance, and pick the source object. Then select the **Multiple** option and begin picking to specify the offset direction. See **Figure 8-3**. Exit when you are finished offsetting.

NOTE

You can use the **Undo** option, when available, to undo the last offset without exiting the **OFFSET** command.

Exercise 8-1

www.g-wlearning.com/CAD/

Complete the exercise on the companion website.

Drawing Points

Use the **POINT** command to draw point objects. Points are useful for marking positions anywhere in space or on objects. To place a single point object and then exit the **POINT** command, enter PO or POINT at the keyboard. Use any appropriate method to specify the location of a point. To draw multiple points without exiting the **POINT** command, select the **Multiple Points** button from the ribbon. Press [Esc] to exit the command.

POINT

Ribbon
Home
> Draw

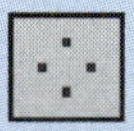

Multiple Points

Type
POINT
PO

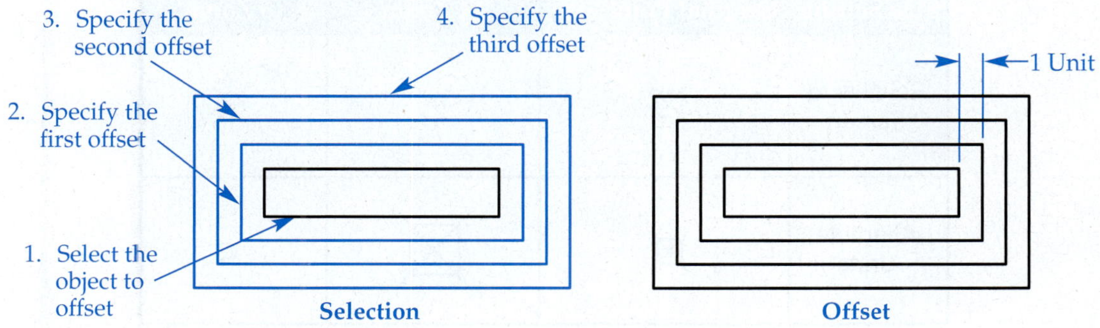

Figure 8-3.
Using the **Multiple** option to create three offsets of a polyline object, constructed using the **RECTANGLE** command, the same distance, without having to reselect the source object. The offset distance in this example is 1 unit.

Setting Point Style

Points appear as one-pixel dots by default. The default appearance is functional and does not interfere with objects. However, the one-pixel style is difficult to see and can get lost on-screen. Change the point style and size using the **Point Style** dialog box, shown in Figure 8-4. Pick the image of the point style to use.

PTYPE
Ribbon
Home
> Utilities

Point Style...
Type
PTYPE

Enter a value in the **Point Size:** text box to set the point size. Pick the **Set Size Relative to Screen** radio button to specify the point size as a percentage of the screen size. You may need to regenerate the display to view the relative sizes. Pick the **Set Size in Absolute Units** radio button to specify an absolute value for the point size. At this setting, points appear larger or smaller at different screen magnifications. See Figure 8-5. Pick the **OK** button to exit the **Point Style** dialog box. All existing and new points change to the current style and size.

Exercise 8-2

www.g-wlearning.com/CAD/

Complete the exercise on the companion website.

block: A symbol or shape saved for repeated use.

Marking an Object at Specified Increments

DIVIDE
Ribbon
Home
> Draw
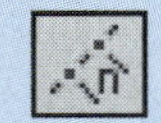
Divide
Type
DIVIDE
DIV

Use the **DIVIDE** command to place point objects or ***blocks*** at equally spaced locations on a line, circle, arc, ellipse, polyline, or spline. AutoCAD calculates the distance between marks based on the number of segments you specify. The **DIVIDE** command does not remove geometry or break an object into segments. Use the **BREAK** command, described in Chapter 11, to remove geometry or break an object into segments.

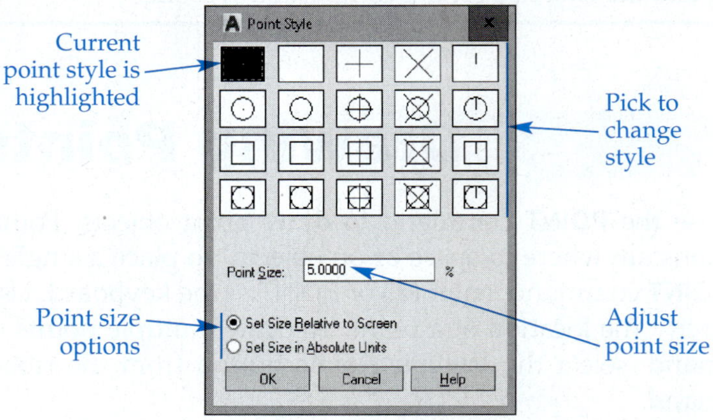

Figure 8-4.
The **Point Style** dialog box provides a quick way to select the point style and change the point size.

Figure 8-5.
Points sized with the **Set Size Relative to Screen** setting remain at a constant size regardless of the zoom magnification. Points sized with the **Set Size in Absolute Units** setting have an absolute size and change size as you zoom in and out.

Size Setting	Original Point Size	2X Zoom	.5 Zoom
Relative to Screen			
Absolute Units		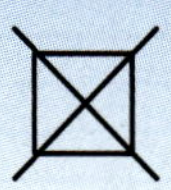	

Access the **DIVIDE** command and select the object to mark. Enter the number of divisions and then exit the command. The point style determines the appearance of the point objects. Figure 8-6 shows marking objects with points at seven segments.

The **Block** option of the **DIVIDE** command allows you to place a block at each increment, instead of a point object. Select the **Block** option at the Enter the number of segments or [Block]: prompt to insert a block. AutoCAD asks if the block should align with the object, such as rotate around a circle. You will learn about blocks later in this textbook.

Marking an Object at Specified Distances

MEASURE

Ribbon
Home
> Draw

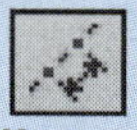

Measure

Type
MEASURE
ME

Use the **MEASURE** command to place point objects or blocks a specified distance apart on a line, circle, arc, ellipse, polyline, or spline. In contrast to the **DIVIDE** command, the measurement length you specify and the total length of the object determine the number of marks, or divisions. The **MEASURE** command does not remove geometry or break an object into segments. Use the **BREAK** command, described in Chapter 11, to remove geometry or break an object into segments.

Access the **MEASURE** command and select the object to mark. Measurement begins at the end closest to where you pick the object. Enter the distance between points and then exit the command. All increments are equal to the specified segment length except the last segment, which may be shorter. The point style determines the appearance of point objects. Figure 8-7 shows marking points at .75 unit intervals. Use the **Block** option of the **MEASURE** command to place a block at each interval.

PROFESSIONAL TIP

Specify an appropriate point style in your drawing templates for use with the **POINT**, **DIVIDE**, and **MEASURE** commands. The × style is a good general-purpose point style that is easy to see and select in most drawings.

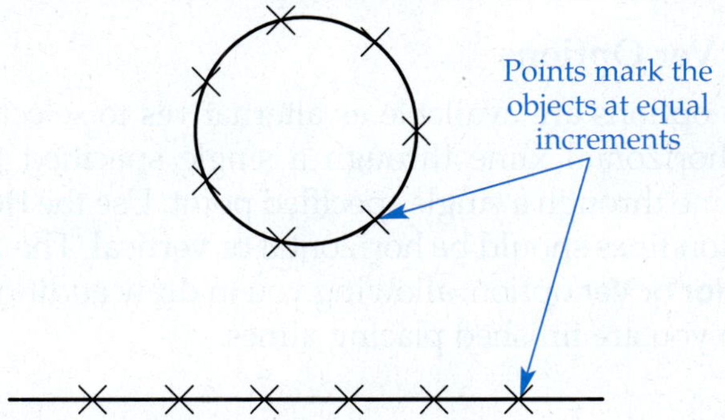

Figure 8-6. Marking seven equal divisions with points on a circle and a line using the **DIVIDE** command. An × point style replaces the default point appearance in these examples. Notice that points do not appear at the endpoints of open objects, such as the line.

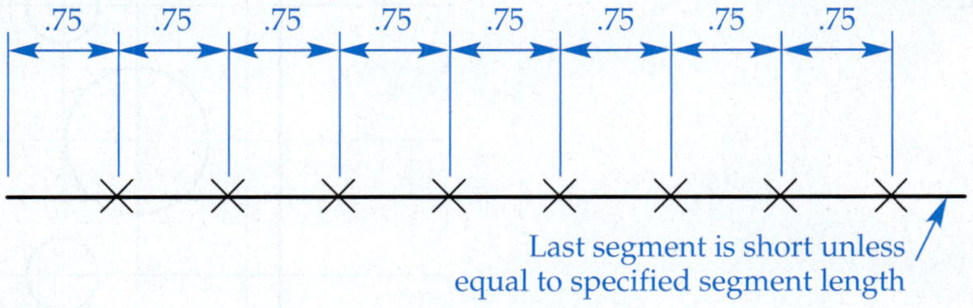

Figure 8-7. Using the **MEASURE** command to place point objects on a line at .75 unit intervals. Notice that the last segment may be short, depending on the specified interval and total length of the object.

Exercise 8-3

www.g-wlearning.com/CAD/

Complete the exercise on the companion website.

Construction Lines

construction lines: Lines commonly used to lay out a drawing.

The tracking vectors and alignment paths available with the object snap and AutoTrack tools are examples of ***construction lines*** generated by AutoCAD. Object snap and AutoTrack are efficient for constructing geometry because vector and alignment lines appear as you draw. Often, however, drawings require construction lines that remain on-screen for reference and future use. You can draw construction geometry using any drawing command, such as **LINE**, **CIRCLE**, or **ARC**. The **XLINE** and **RAY** commands are specifically designed for adding construction lines to help lay out a drawing. See Figure 8-8.

PROFESSIONAL TIP

Create construction geometry on a separate construction layer. Use an appropriate layer name, such as CONST, CONSTRUCTION, or A-ANNO-NPLT. Turn off or freeze the construction layer when you do not need it, or erase objects drawn on the construction layer if necessary.

Using the XLINE Command

XLINE

Ribbon
Home
> Draw
Construction Line
Type
XLINE
LAYOUTLINE
XL

Use the **XLINE** command to draw an infinitely long AutoCAD construction line object, or ***xline***. To draw an xline, specify the location of the first point through which the xline passes, or ***root point***. Then select a second point through which the xline passes. The **XLINE** command remains active, and the initial root point acts as an axis point. Continue locating additional points from the same root point to create additional xlines. See Figure 8-9. To exit, press [Enter] or [Esc].

xline: A construction line in AutoCAD that is infinite in both directions; helpful for creating accurate geometry and multiview drawings.

root point: The first point specified to create a construction line or ray.

Hor and Ver Options

Xline options are available as alternatives to selecting two points. The **Hor** option draws a horizontal xline through a single specified point. The **Ver** option draws a vertical xline through a single specified point. Use the **Hor** or **Ver** option when you know construction lines should be horizontal or vertical. The **XLINE** command remains active with the **Hor** or **Ver** option, allowing you to draw additional horizontal or vertical xlines. Exit when you are finished placing xlines.

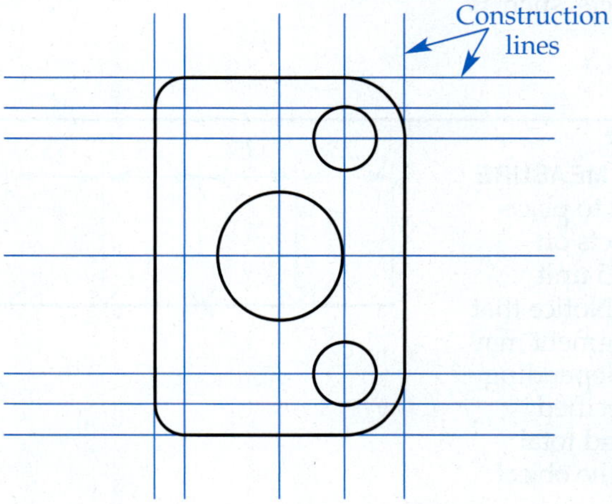

Figure 8-8.
An example of a drawing laid out using construction lines.

Figure 8-9.
A—Using the **XLINE** command to draw an infinitely long construction line by specifying two points through which the line passes. B—Constructing multiple xlines that intersect the root point before exiting the command.

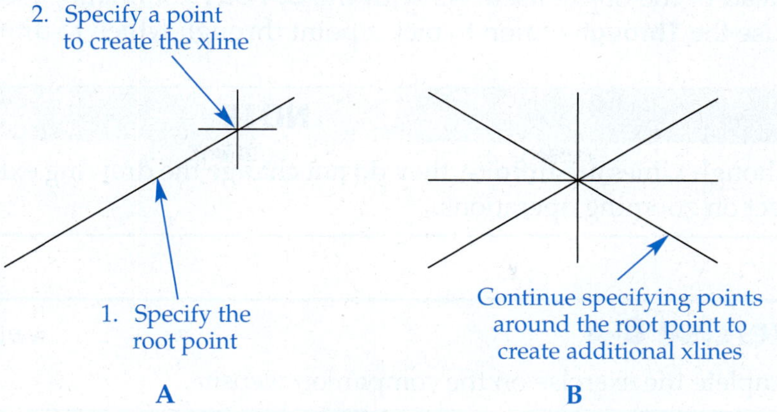

Ang Option

Use the **Ang** option to draw an xline at a specified angle through a selected point. Define an angle by entering a value, or picking two points, and then pick a point through which the xline passes. The **Reference** option of the **Ang** option allows you to reference the angle of an existing line object to use as the xline angle. The **Reference** option is useful when you do not know the angle of the xline, but you do know the angle between an existing object and the xline. See Figure 8-10.

Bisect Option

The **Bisect** option draws an xline that bisects a specified angle, using the root point as the vertex. See Figure 8-11. The **Bisect** option is useful for bisecting an angle and similar applications.

Offset Option

The **Offset** option draws an xline parallel to a selected linear object. The **Offset** option of the **XLINE** command functions much like the **OFFSET** command, except that the **Offset** option of the **XLINE** command offsets an xline from the selected object,

Figure 8-10.
Using the **Reference** option of the **Ang** option of the **XLINE** command. This example shows selecting an existing line and entering a value of 90° to make an xline perpendicular to the line.

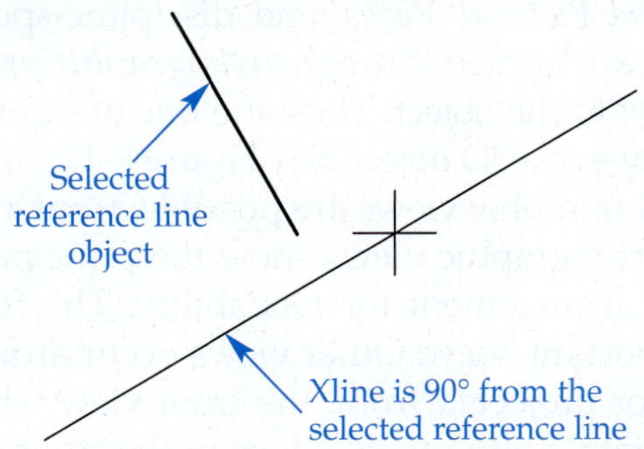

Figure 8-11.
Using the **Bisect** option of the **XLINE** command.

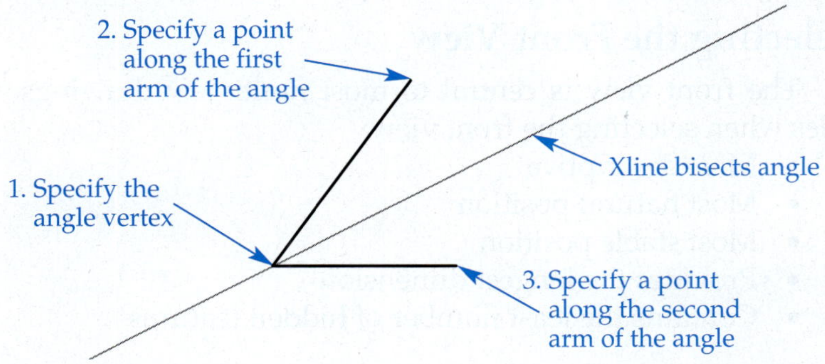

instead of the object itself. As with the **OFFSET** command, specify an offset distance or use the **Through** option to pick a point through which to draw the xline.

NOTE

Although xlines are infinite, they do not change the drawing extents, and they have no effect on zooming operations.

Exercise 8-4

www.g-wlearning.com/CAD/

Complete the exercise on the companion website.

Using the RAY Command

RAY

Ribbon
Home
> Draw
Ray

Type
RAY
CONSTRUCTIONLINE

Use the **RAY** command to draw an AutoCAD ***ray*** object. A ray begins at an origin point and extends through another point, similar to an xline drawn with the default option of the **XLINE** command. However, a ray is infinite only in the direction of the second point. To draw a ray, specify the location of the root point, followed by a second point through which the ray passes. Continue locating points to create additional rays from the same root point. To exit, press [Enter] or [Esc].

ray: A linear AutoCAD object that is infinite in one direction only; considered semi-infinite.

Multiview Drawings

multiview drawing: A presentation of drawing views created through orthographic projection.

A ***multiview drawing*** presents views of an object. The general multiview concept applies to most drafting fields. Mechanical drafting uses multiview drawings to describe parts and assemblies. Architectural drafting uses plans, elevations, and sections to show building construction and layout. Electronics drafting uses schematic diagrams with electronic symbols to show circuit layout. Civil drafting uses plans and profiles to show the topography of land.

This textbook explains multiview drawings based on the ASME Y14.3 standard, *Orthographic and Pictorial Views*, and discipline-specific drafting practices. Multiview drawing views are formed through ***orthographic projection***. The imaginary ***projection plane*** is parallel to the object. Thus, the line of sight is perpendicular to the object. The result is 2D views of a 3D object. See Figure 8-12.

orthographic projection: Projecting object features onto an imaginary plane.

projection plane: An imaginary projection plane parallel to the surface of an object.

Six 2D orthographic views are possible: front, right side, left side, top, bottom, and rear. The six orthographic views show the principal sides of an object and are drawn in a standard arrangement for readability. The front view is the central, and typically most important, view. Other views occur around the front view and are directly aligned with, or projected from, the front view when possible. Notice in Figure 8-13 that the horizontal and vertical edges in the front view align with the corresponding edges in the other views.

Selecting the Front View

The front view is central to most multiview drawings. Consider the following rules when selecting the front view:

- Most descriptive
- Most natural position
- Most stable position
- Provides the longest dimension
- Contains the least number of hidden features

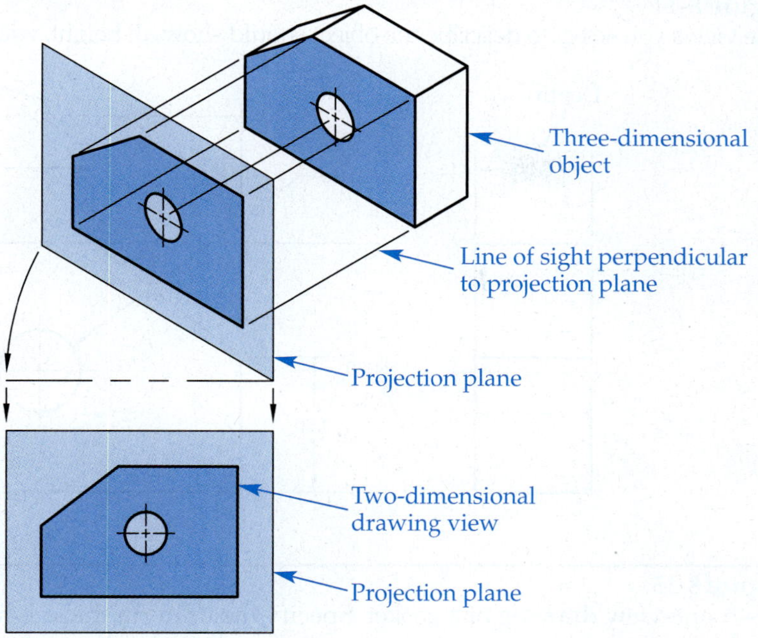

Figure 8-12. Obtaining a front view with orthographic projection.

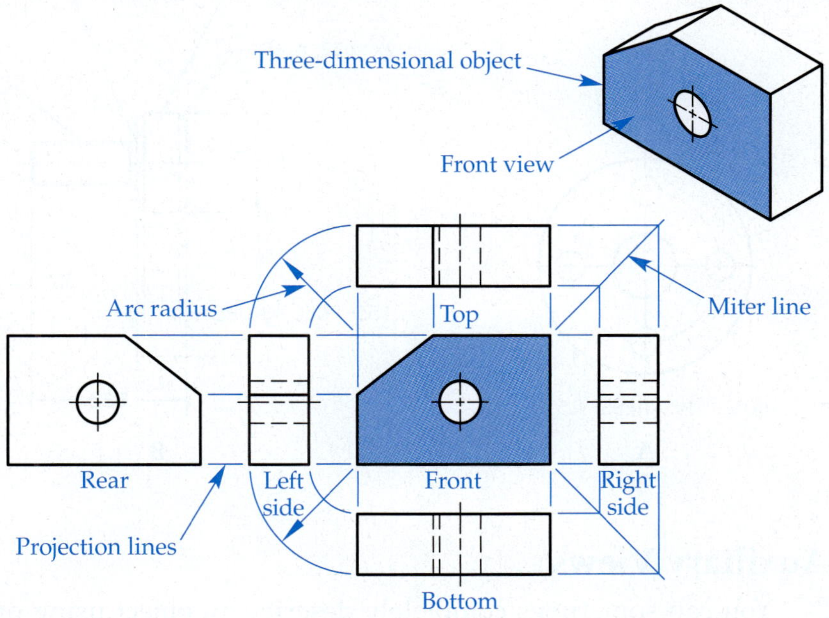

Figure 8-13. Arrangement of the six orthographic views. The front view is typically central.

Selecting Additional Views

Select additional views relative to the front view. Few products require all six views. The required number of views depends on the complexity of the object. Use only enough views to describe the object completely. Drawing too many views is time consuming and can clutter the drawing. **Figure 8-14** shows an example of an object represented by two views. The two views completely describe the width, height, depth, and features of the object. In some cases, a single view is enough to describe the object. The example shown in **Figure 8-14** could be a one-view drawing with a note specifying the uniform depth. You can often draw a thin part that has a uniform thickness, such as a gasket, with one view. If necessary, provide the thickness or material specification as a general note or in the title block. See **Figure 8-15**.

Figure 8-14.
The views you select to describe the object should show all height, width, and depth dimensions.

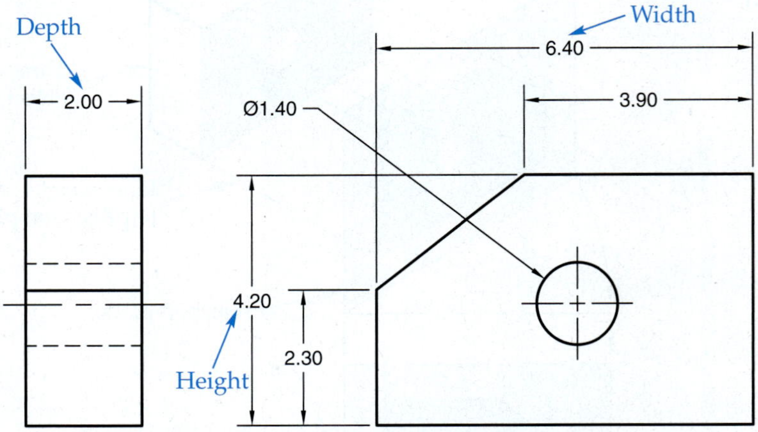

Figure 8-15.
A—A one-view drawing of a gasket. Specify the uniform thickness in a general note.
B—A one-view drawing of a thumbscrew. The diameter dimensions indicate cylindrical features, eliminating the need for a side view.

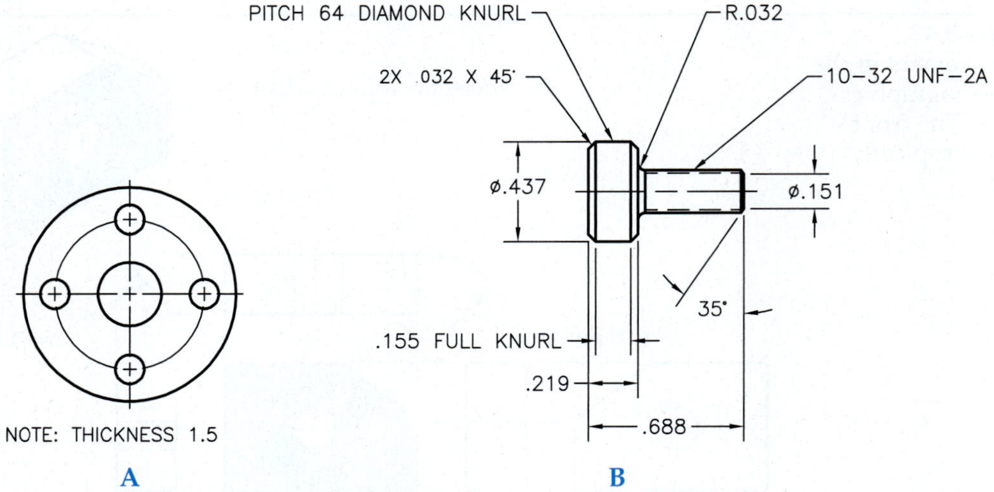

auxiliary view: View used to show the true size and shape of a foreshortened surface.

foreshortened: Describes a surface at an angle to the line of sight. Foreshortened surfaces appear shorter than their true size and shape.

partial auxiliary view: An auxiliary view that shows a specific inclined surface of an object, rather than the entire object.

removed view: A view removed from alignment with other views when drawing space is unavailable.

Auxiliary Views

You can sometimes completely describe an object using one or more of the six standard views. However, you must use an ***auxiliary view*** to describe a surface that appears ***foreshortened*** in standard orthographic views. Draw an auxiliary view using projection lines perpendicular to a slanted surface. One projection line is sometimes included on the drawing to connect the auxiliary view to the view where the slanted surface appears as a line. The resulting auxiliary view shows the surface at its true size and shape. A ***partial auxiliary view*** is enough for most applications. See Figure 8-16.

Removed Views

Sometimes there is not enough room on a drawing to project directly from one view to another. This requires that you create a ***removed view*** to locate a view elsewhere on the drawing. An auxiliary view is a common example of a removed view in mechanical drafting, but you can relocate any view if necessary. See Figure 8-17.

Figure 8-16.
Auxiliary views show the true size and shape of an inclined surface. Use a partial auxiliary view to show only the inclined surface, because the other features appear foreshortened and reduce clarity.

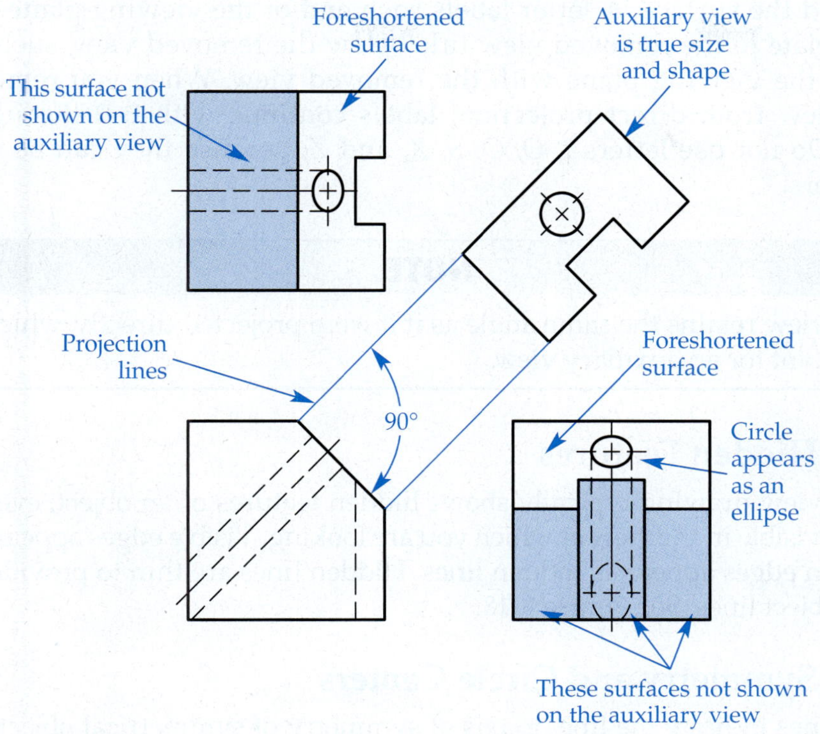

Figure 8-17.
A viewing-plane line creates the flexibility to move a view to a location where there is enough space for the view. This process creates a removed view.

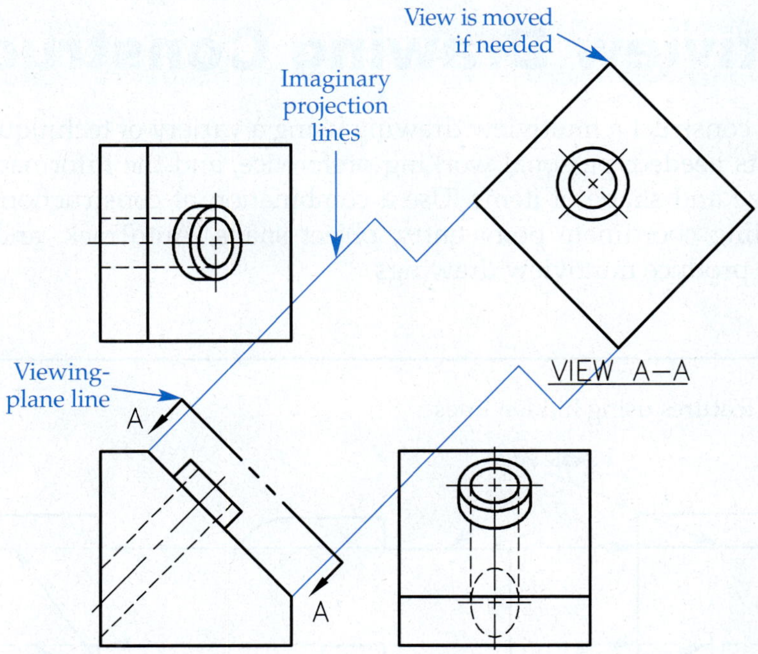

In other disciplines, such as architectural drafting, views may not align and often occur on different sheets.

viewing-plane line: A thick dashed or phantom line identifying the viewing direction of a related view.

Draw a ***viewing-plane line*** parallel to the view in which the surface appears as a line. The standard viewing-plane line terminates with bold arrowheads that point toward the surface. A letter labels each end of the viewing-plane line. The letters correlate to the removed view title below the removed view, such as VIEW A-A, to key the viewing plane with the removed view. When you remove more than one view from direct projection, labels continue with B-B through Z-Z, if necessary. Do not use letters *I*, *O*, *Q*, *S*, *X*, and *Z*, because they can be confused with numbers.

NOTE

A removed view retains the same angle as if it were projected directly, which is especially important for an auxiliary view.

Showing Hidden Features

A multiview drawing typically shows hidden features of an object, even though they are not visible in the view at which you are looking. Visible edges appear as object lines. Hidden edges appear as hidden lines. Hidden lines are thin to provide contrast with thick object lines. See **Figure 8-18**.

Showing Symmetry and Circle Centers

Centerlines indicate the line or axis of symmetry of symmetrical objects and the centers of circles. For example, in the circular view of a cylinder, centerlines cross to show the center of the cylinder. In the other view, a centerline identifies the axis. See **Figure 8-19**. The only place the small centerline dashes should cross is at the center of a circle, arc, ellipse, or other circular feature.

Multiview Drawing Construction

You can construct a multiview drawing using a variety of techniques, depending on the objects needed, personal working preference, and the information you know about the size and shape of items. Use a combination of construction methods and tools, including coordinate point entry, object snaps, AutoTrack, and construction geometry, to produce multiview drawings.

Figure 8-18.
Draw hidden features using hidden lines.

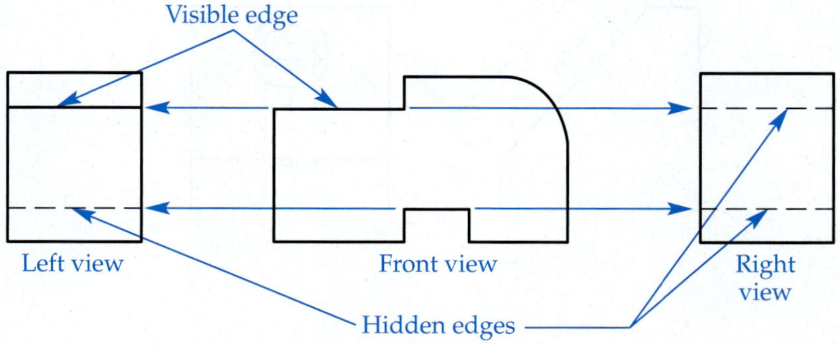

Figure 8-19.
Using centerlines in multiview drawings.

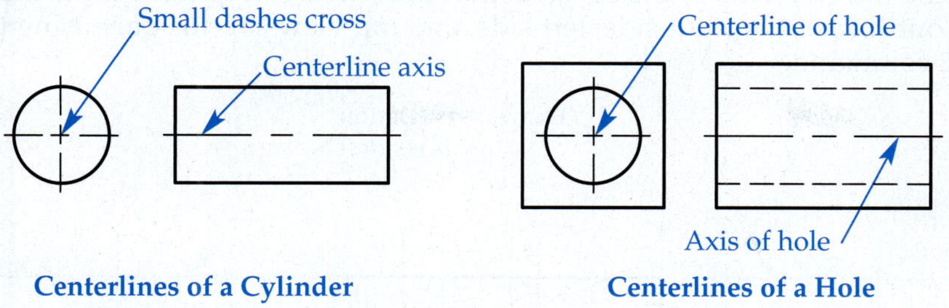

Orthographic Views

Figure 8-20 shows an example of using object snap tracking and a running **Endpoint** object snap mode to locate points for a left-side view by referencing points on the existing front view. Notice that the AutoTrack alignment path in Figure 8-20A provides a temporary construction line. Polar tracking vectors offer a similar temporary construction line. Figure 8-20B shows the complete front and left-side views.

Figure 8-21 shows an example of using construction lines to form three views. This example shows offsetting vertical and horizontal xlines to form an xline grid. Use the **Intersection** object snap mode to select the intersecting xlines when locating line and arc endpoints and the center point of the arc. Notice that a single infinitely long xline can provide construction geometry for multiple views.

Exercise 8-5

www.g-wlearning.com/CAD/

Complete the exercise on the companion website.

Figure 8-20.
An example of using object snap tracking to create an additional view. A—Referencing a point from the front view to establish the first line of the left-side view. B—The completed front and left-side views.

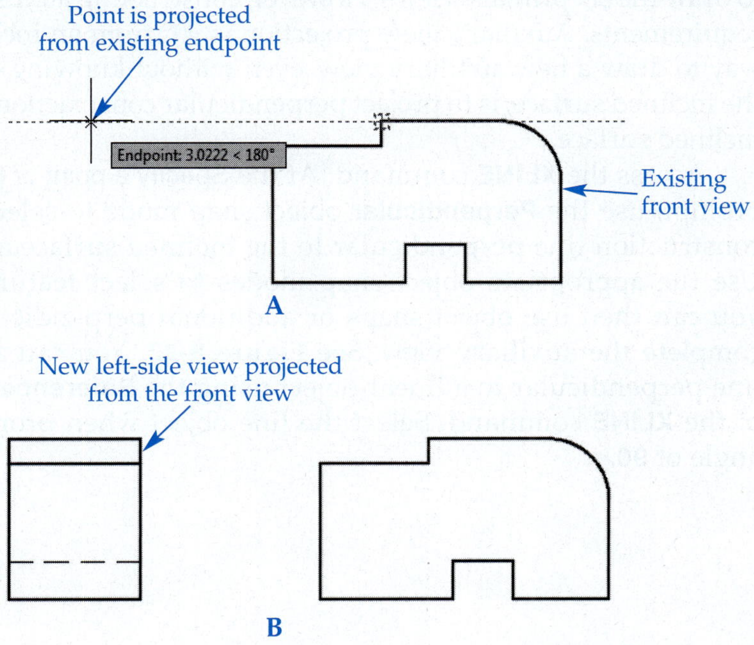

Figure 8-21.
Using a complete grid of construction lines to form a multiview drawing by "connecting the dots" at the intersections of the construction lines. You can quickly draw the rectangular outlines of the right-side, left-side, and top views of this part using the **RECTANGLE** command.

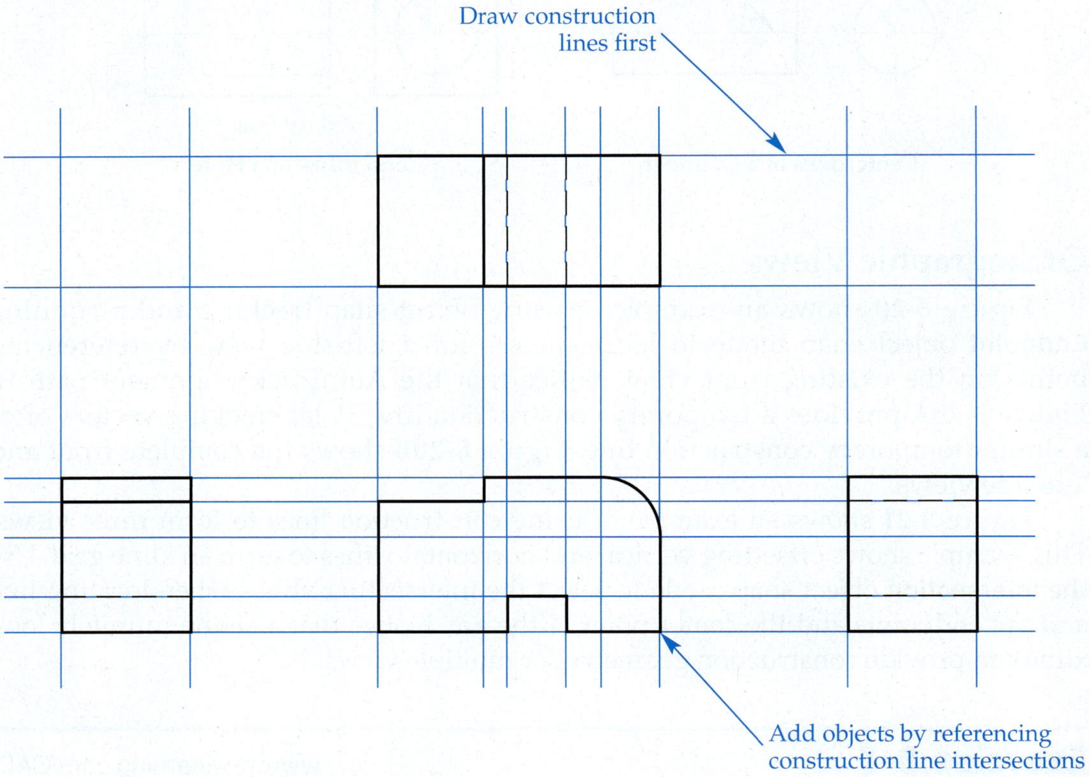

Auxiliary Views

You can construct auxiliary views using the same commands and options you use to draw the six primary views. However, constructing auxiliary views presents unique requirements. Auxiliary view projection is 90° from an inclined surface. An effective way to draw a new auxiliary view, even without knowing or calculating the angle of the inclined surface, is to project perpendicular construction lines from features on the inclined surface.

Access the **XLINE** command. At the Specify a point or [Hor/Ver/Ang/Bisect/Offset]: prompt, use the **Perpendicular** object snap mode to select the inclined surface. A construction line perpendicular to the inclined surface attaches to the crosshairs. Use the appropriate object snap modes to select features on the existing view. You can then use object snaps or additional perpendicular construction lines to complete the auxiliary view. See **Figure 8-22**. You can also make a construction line perpendicular to a linear object using the **Reference** option of the **Ang** option of the **XLINE** command. Select the line object when prompted and enter an xline angle of 90.

Figure 8-22.
Using construction lines drawn perpendicular to an inclined surface on an existing view to construct an auxiliary view. The completed top and auxiliary views are shown for reference.

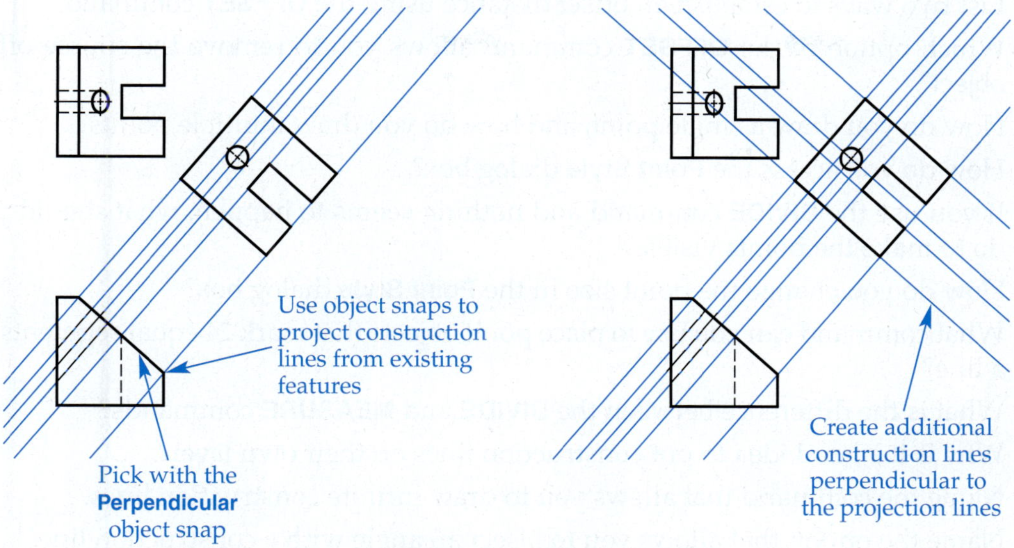

NOTE

You can also use parametric tools, explained in Chapter 22, in addition or as an alternative to other techniques for constructing multiview drawings.

Exercise 8-6

www.g-wlearning.com/CAD/

Complete the exercise on the companion website.

Chapter Review

Answer the following questions using the information in this chapter.

1. List two ways to establish an offset distance using the **OFFSET** command.
2. Which option of the **OFFSET** command allows you to remove the source offset object?
3. How do you draw a single point, and how do you draw multiple points?
4. How do you access the **Point Style** dialog box?
5. If you use the **DIVIDE** command and nothing seems to happen, what should you do to make the points visible?
6. How do you change the point size in the **Point Style** dialog box?
7. What command can you use to place point objects that mark 24 equal segments on a line?
8. What is the difference between the **DIVIDE** and **MEASURE** commands?
9. Why is it a good idea to put construction lines on their own layer?
10. Name the command that allows you to draw infinite construction lines.
11. Name the option that allows you to bisect an angle with a construction line.
12. What is the difference between the construction lines drawn with the command identified in Question 10 and rays drawn with the **RAY** command?
13. Name the six 2D views possible in orthographic projection.
14. Provide at least four guidelines for selecting the front view of an orthographic multiview drawing.
15. How do you determine how many views of an object are necessary in a multiview drawing?
16. When can you describe a part with only one view?
17. When does a drawing require an auxiliary view?
18. What is a *removed view*?
19. What is the angle of projection from the inclined surface to the auxiliary view?
20. Describe an effective method of constructing an auxiliary view even if you do not know the angle of the inclined surface.

Drawing Problems

Start AutoCAD if it is not already started. Start a new drawing from scratch or use an appropriate template of your choice. The template should include layers for drawing the given objects. Add layers as needed. Draw all objects using appropriate layers. Follow the specific instructions for each problem. Use only drawing commands and techniques you have already learned. Use object snaps and AutoTrack when possible. Do not draw dimensions or text. Use your own judgment and approximate dimensions when necessary.

▼ Basic

1. **Mechanical** Draw the front and side views of the offset support shown. Save the drawing as P8-1.

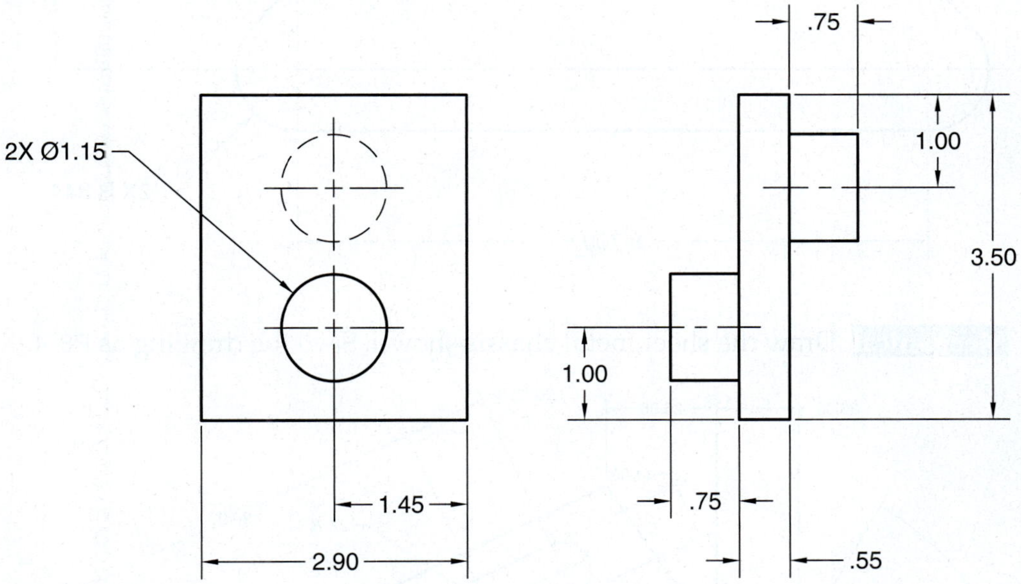

2. **Mechanical** Draw the front and top views of the hitch bracket shown. Save the drawing as P8-2.

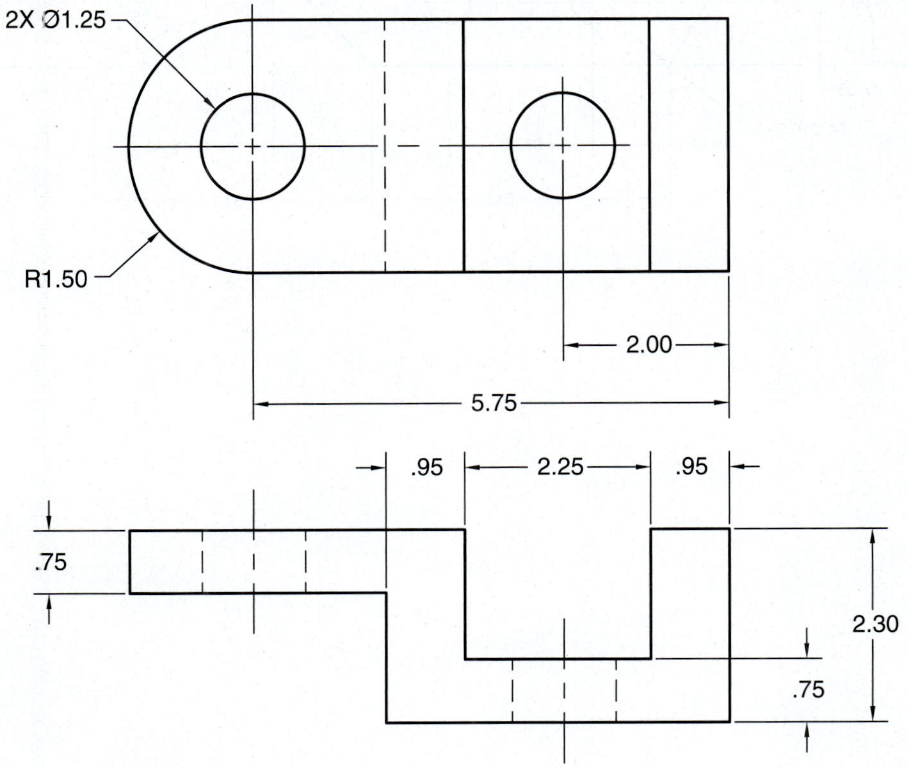

3. **Mechanical** Draw the spring shown using the **PLINE** command with a width of .024. Save the drawing as P8-3.

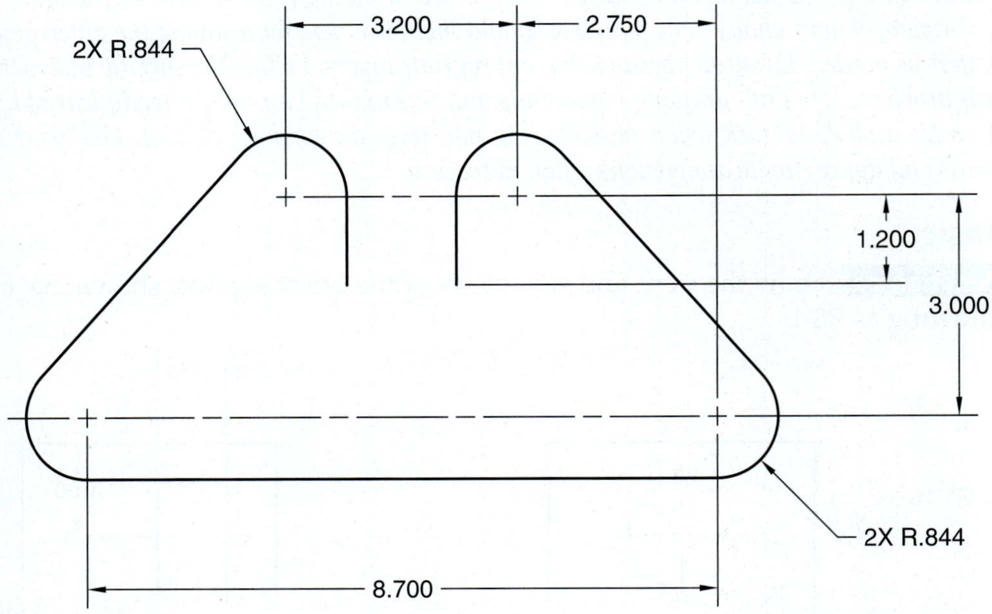

4. **Mechanical** Draw the sheet metal chassis shown. Save the drawing as P8-4.

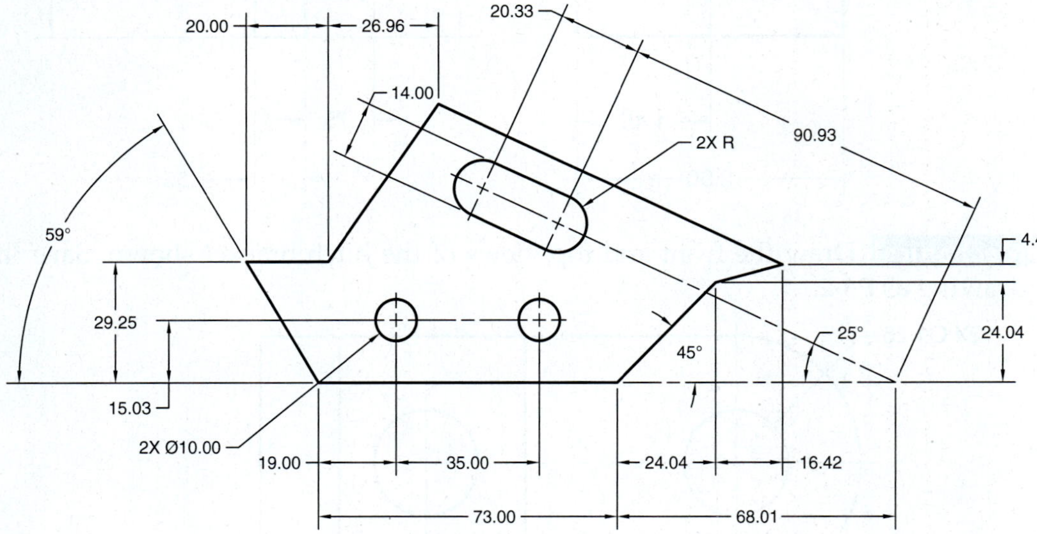

▼ Intermediate

5. **Piping** Draw the views of the elbow shown. Save the drawing as P8-5.

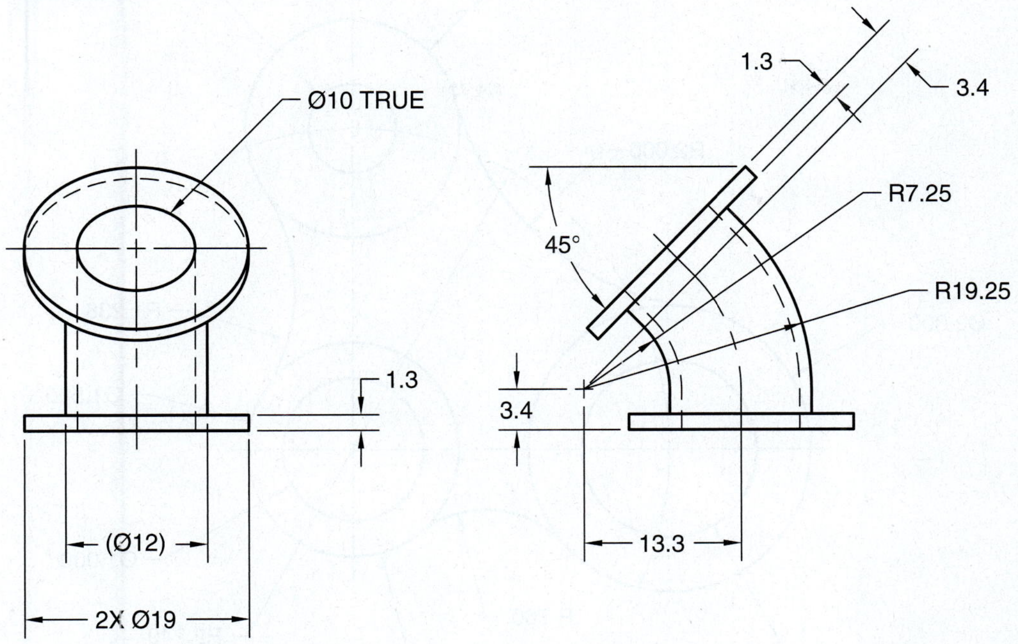

6. **Mechanical** Draw the part view shown. Save the drawing as P8-6.

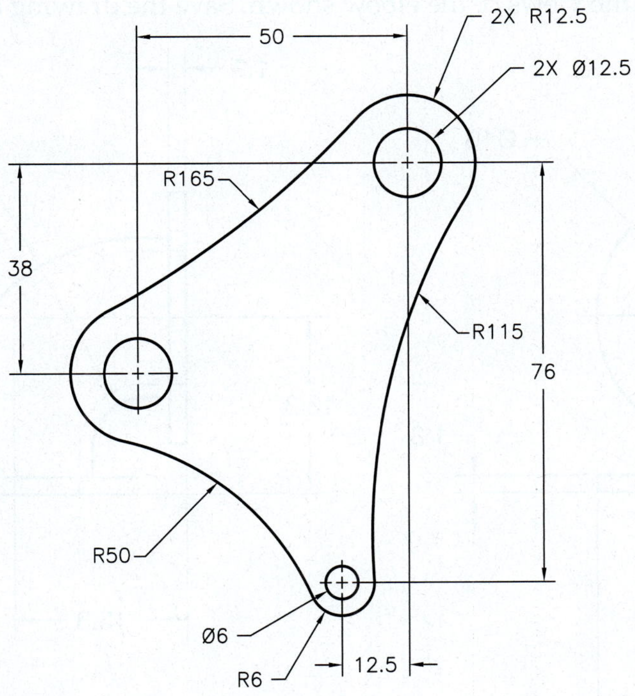

7. **Mechanical** Draw the part view shown. Save the drawing as P8-7.

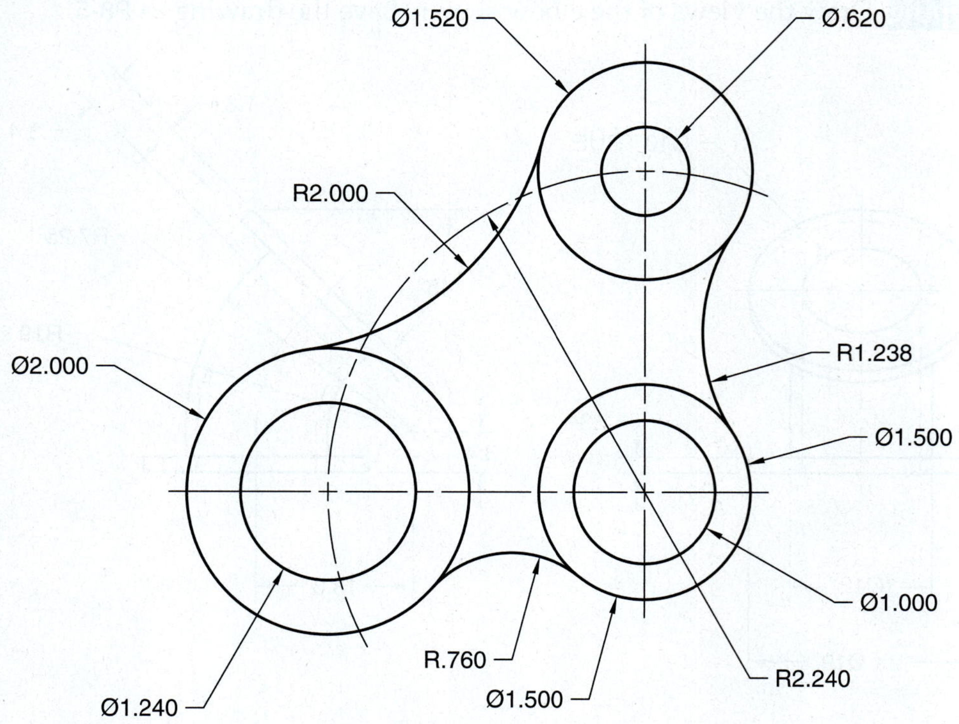

8. **Piping** Draw the views of the elbow shown. Save the drawing as P8-8.

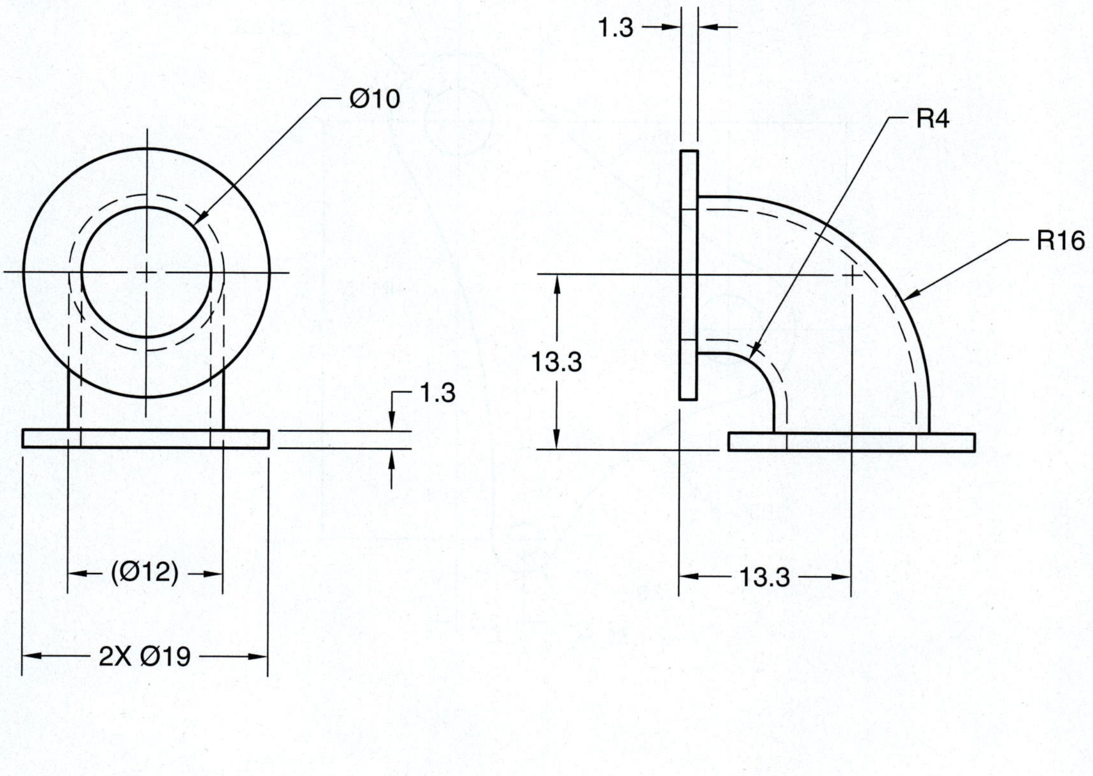

9. **Architectural** Use the **OFFSET** command to draw the elevation of the desk shown. Center 1″ × 4″ rectangular drawer handles 2″ below the top of each drawer. The tops of the legs begin 1″ from the side edges at the bottom of the desk. Save the drawing as P8-9.

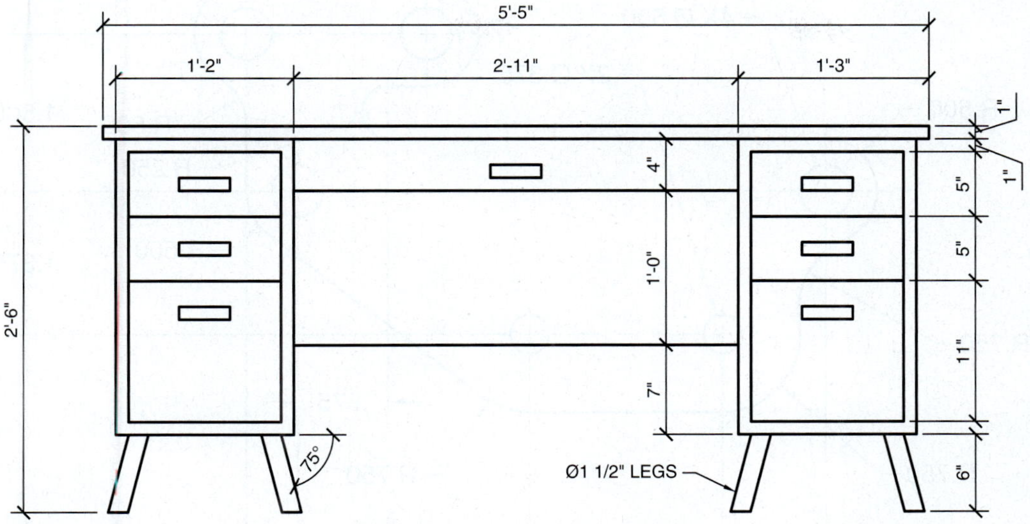

10. **Mechanical** Draw the aluminum spacer shown. Save the drawing as P8-10.

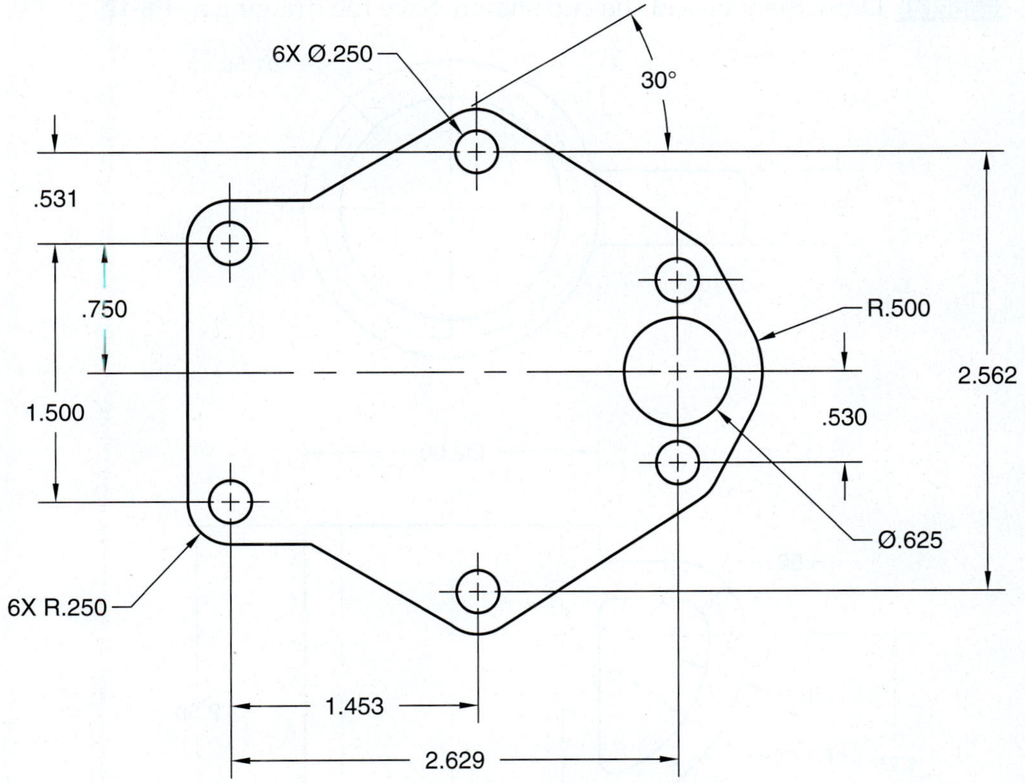

11. **Mechanical** Draw the gasket shown. Save the drawing as P8-11.

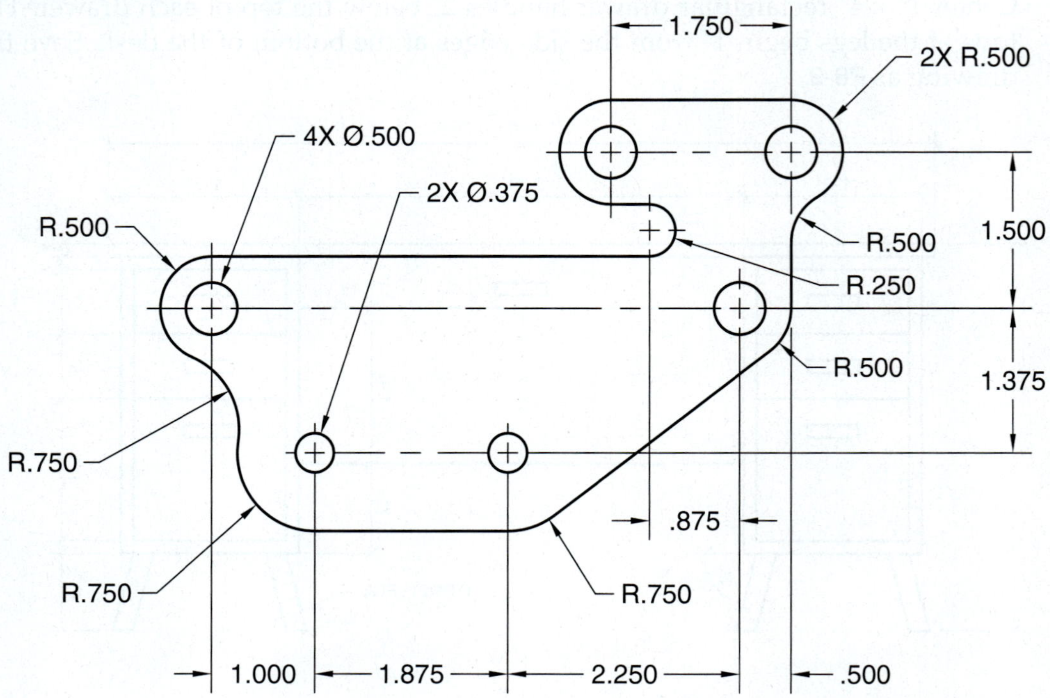

12. **General** Draw the views of the cup shown. Save the drawing as P8-12.

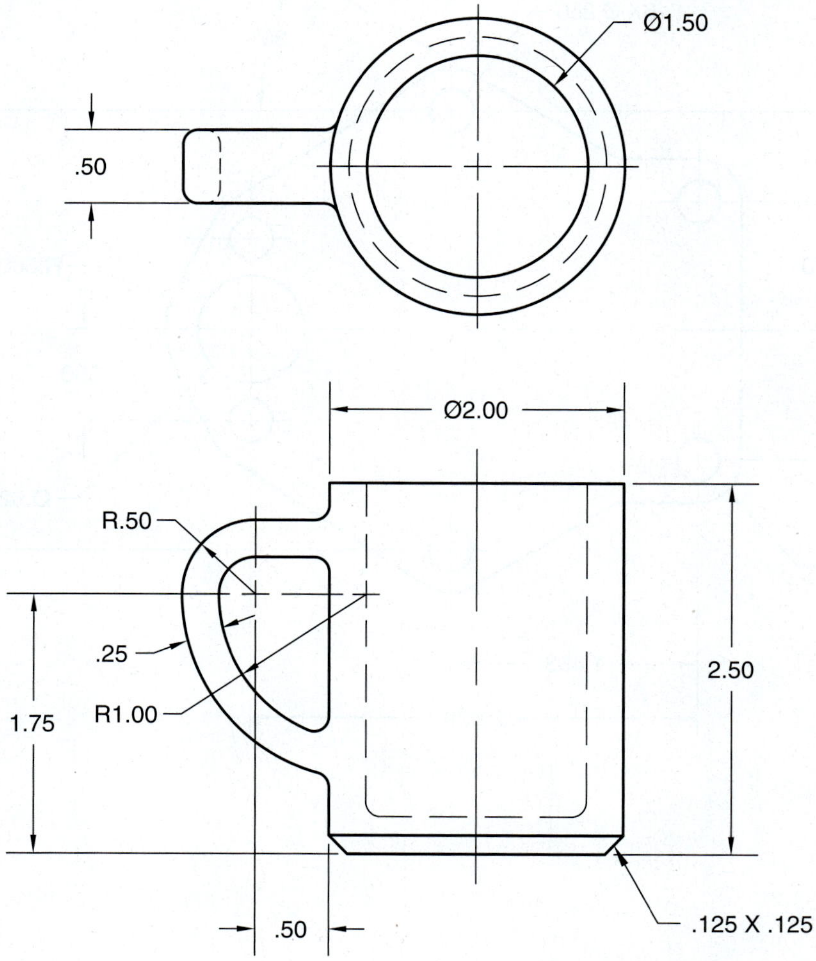

13. **Mechanical** Draw the views of the bushing shown. Save the drawing as P8-13.

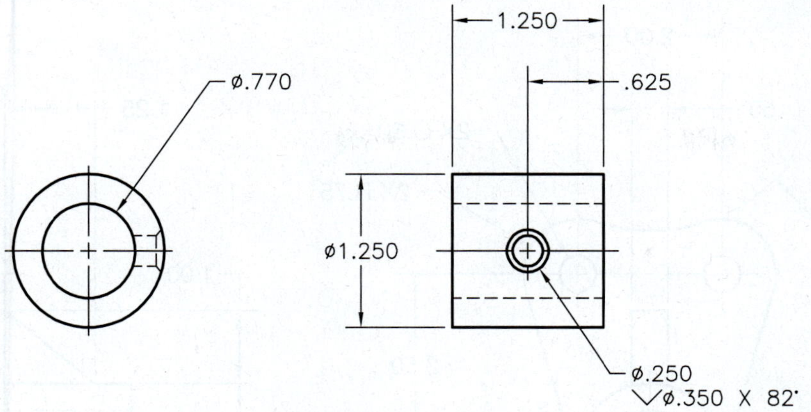

14. **Mechanical** Draw the views of the wrench shown. Save the drawing as P8-14.

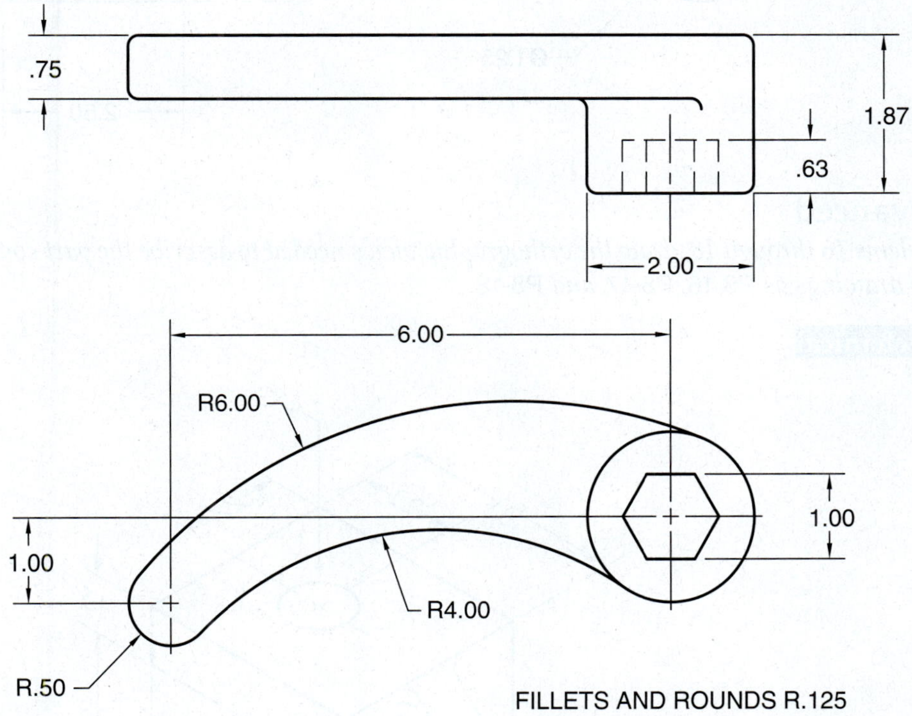

15. **Mechanical** Draw the views of the support shown. Save the drawing as P8-15.

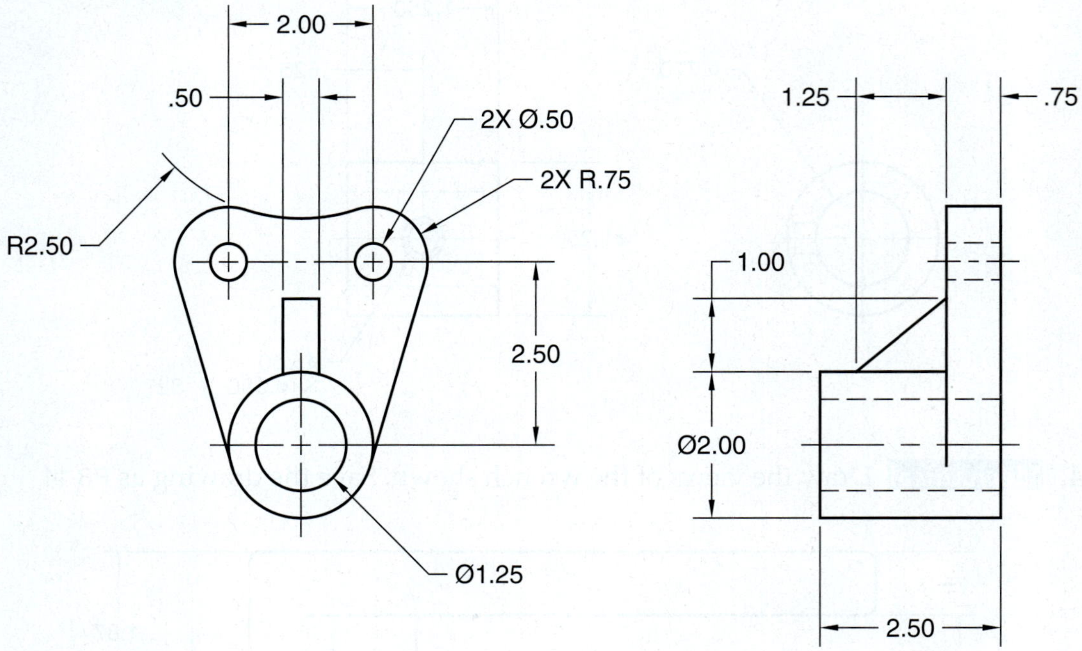

▼ Advanced

For Problems 16 through 18, draw the orthographic views needed to describe the part completely. Save the drawings as P8-16, P8-17, *and* P8-18.

16. **Mechanical**

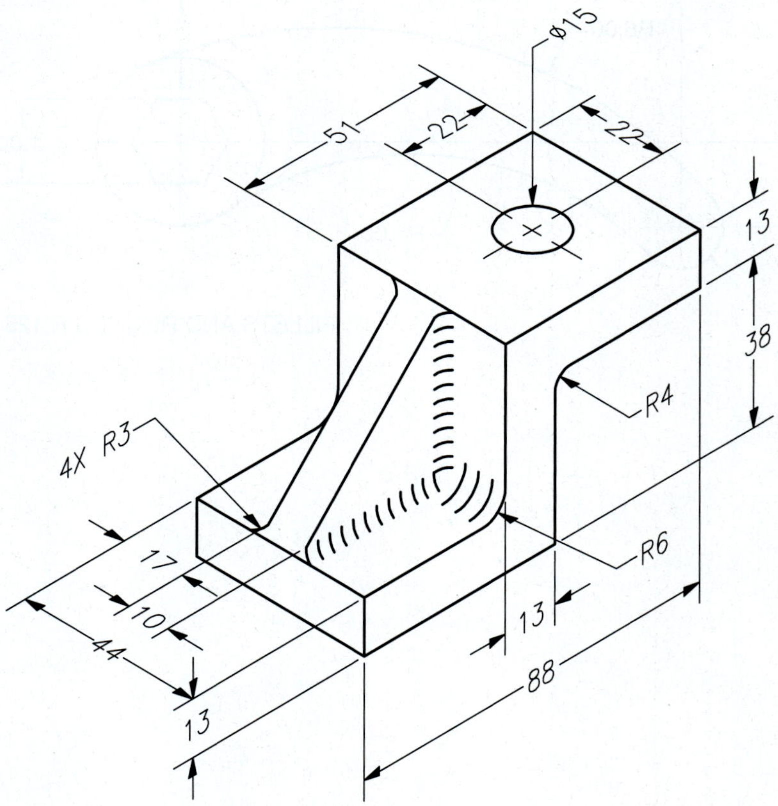

Brace

17. **Mechanical**

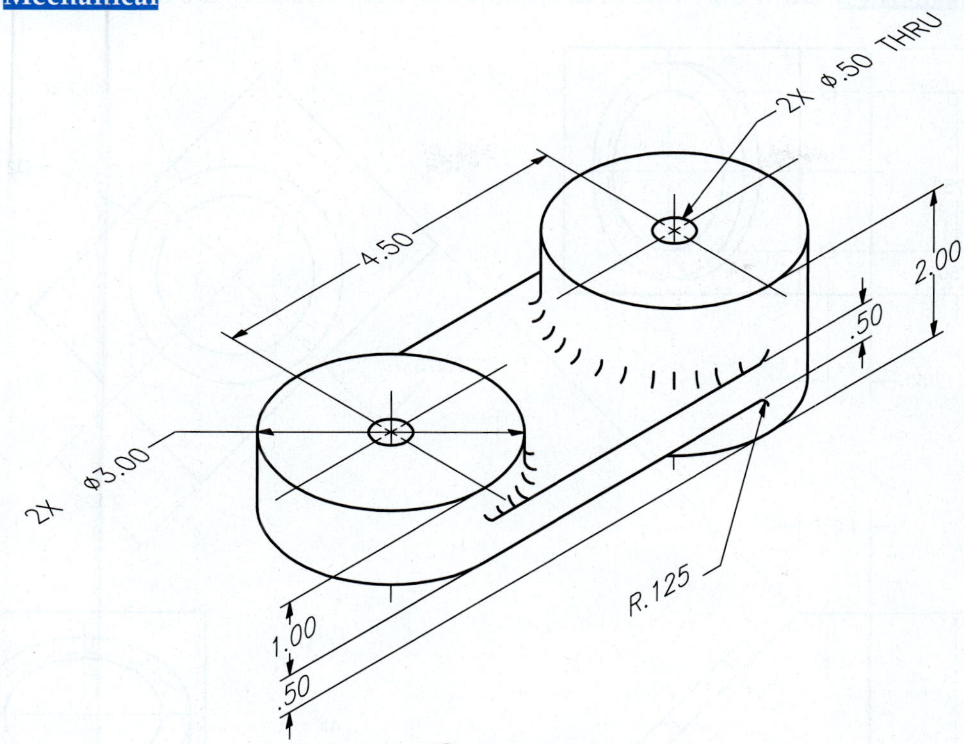

Connector

18. **Mechanical**

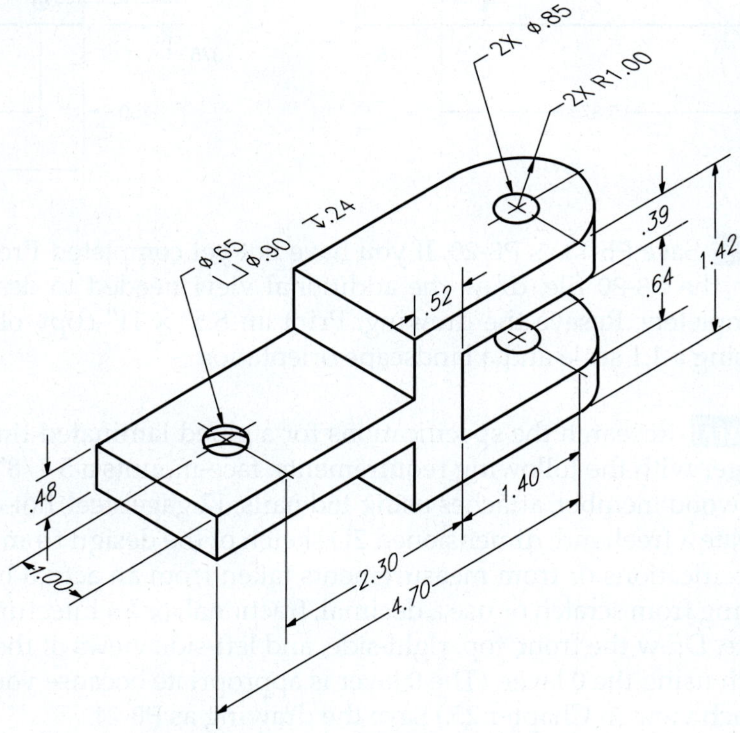

Hitch Bracket

19. **Mechanical** Draw all views of the pillow block shown. Save the drawing as P8-19.

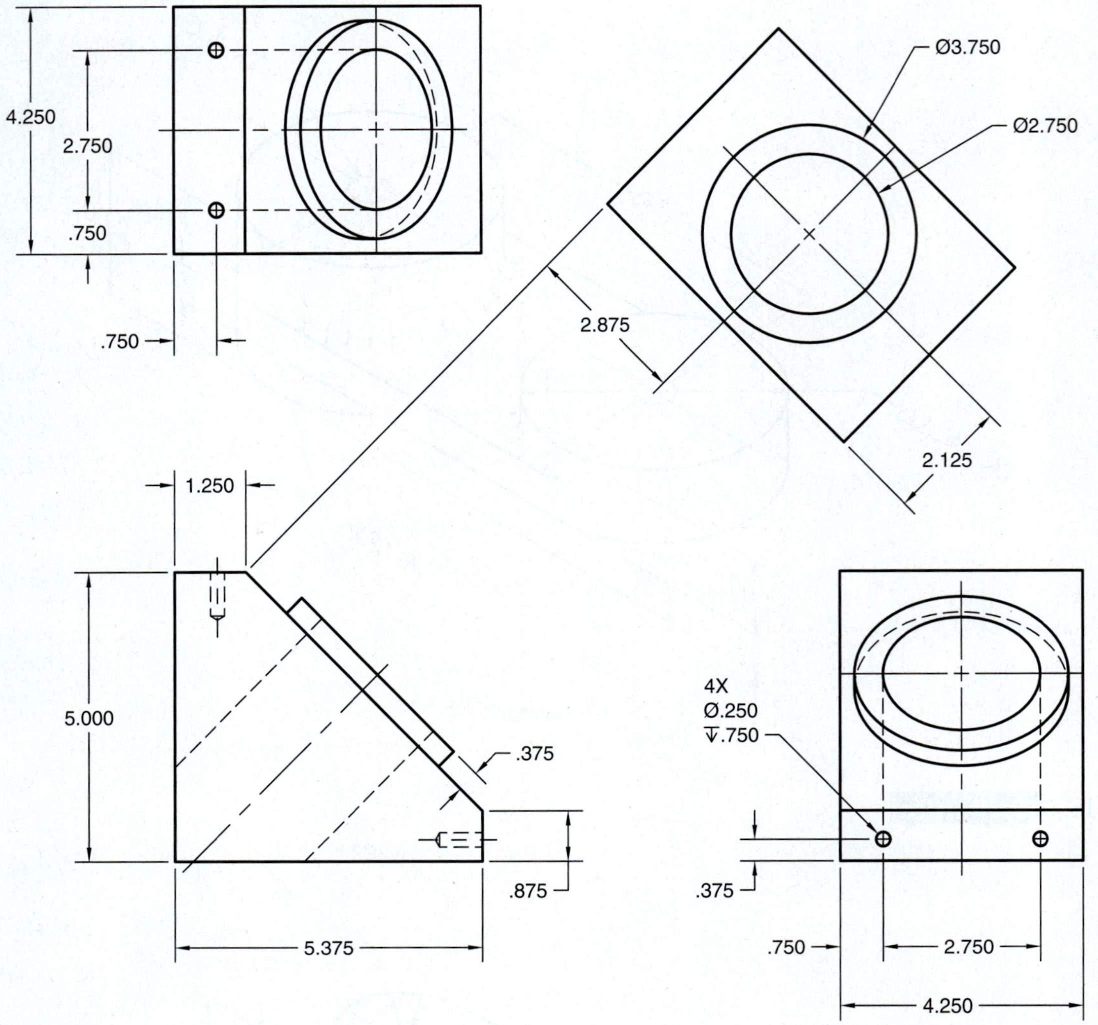

20. **Mechanical** Save P5-13 as P8-20. If you have not yet completed Problem 5-13, do so now. In the P8-20 file, draw the additional view needed to describe the nut driver completely. Resave the drawing. Print an 8.5″ × 11″ copy of the drawing extents, using a 1:1 scale and a landscape orientation.

21. **Architectural** Research the specifications for a glued laminated timber (glulam) beam hanger with the following requirements: face-mounts a 5-1/8″ × 10″ glulam beam to a wood member, attaches using 16d nails, 12-gage steel, hot-dipped galvanized. Create a freehand, dimensioned 2D sketch of the design from the manufacturer's specifications or from measurements taken from an actual hanger. Start a new drawing from scratch or use a decimal, fractional, or architectural template of your choice. Draw the front, top, right-side, and left-side views of the hanger from your sketch using the 0 layer. (The 0 layer is appropriate because you will create a block of each view in Chapter 25.) Save the drawing as P8-21.

AutoCAD Certified Professional Exam Practice

Answer the following questions using the information in this chapter.

1. Which of the following operations can you complete using the **OFFSET** command? *Select all that apply.*
   A. create an exact copy of a line at a specified distance from the original
   B. create an exact copy of an arc at a specified distance from the original
   C. create a circle concentric with an existing circle
   D. create an arc concentric with an existing arc
   E. create an exact copy of a line through a specified point
2. Which of the following operations can you complete using the **MEASURE** command? *Select all that apply.*
   A. place point objects anywhere in the drawing area
   B. convert point objects into blocks
   C. break a circle into a specified number of pieces
   D. place point objects at specified intervals along a line
   E. place point objects at a specified number of equally spaced locations on an arc
3. Which of the following are primary 2D orthographic views for a mechanical part drawing? *Select all that apply.*
   A. front
   B. rear
   C. auxiliary
   D. central
   E. top

Follow the instructions in each problem. Complete each problem by answering the related question.

4. **Navigate to this chapter on the companion website and open CPE-08offset.dwg.** Use the **Multiple** option of the **OFFSET** command to offset the existing circle to the outside, using an offset distance of 2, to create Circle 2, Circle 3, and Circle 4. Then offset the outermost circle to the inside using an offset distance of .5 to create Circle 5. What are the coordinates of the top quadrant point of Circle 5?

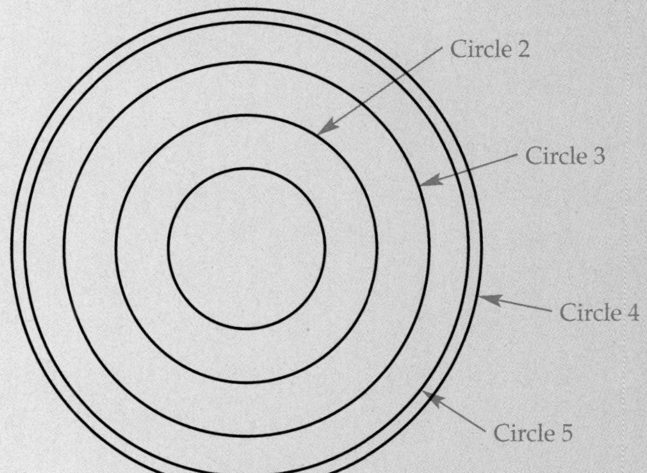

AutoCAD Certified Professional Exam Practice - Chapter 8

5. **Navigate to this chapter on the companion website and open CPE-08multiview.dwg.** This file contains the front view of a cylindrical spacer, as well as the starting point for the lower-left corner of the right-side view. Change the point display to an × to make it more visible. Then create the right-side view, assuming that the spacer has a total length of 4 units. Use any method of construction described in this chapter to create the view accurately. Use hidden lines to show the inside diameter of the spacer and include a centerline to show the axis. Place the hidden lines and centerlines on the correct layers. What are the coordinates of Point A?

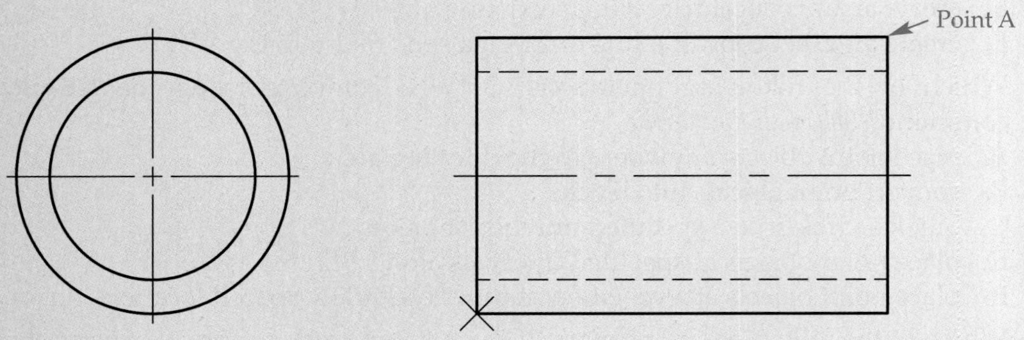

CHAPTER 9

Text Styles and Multiline Text

Learning Objectives

After completing this chapter, you will be able to:

- ✓ Describe and use proper text standards.
- ✓ Calculate drawing scale and text height.
- ✓ Develop and use text styles.
- ✓ Use the **MTEXT** command to create multiline text objects.

Annotation, dimensions, and symbols with ***text*** provide necessary information about features on a drawing. AutoCAD provides commands and settings to create uniform, easy-to-read text according to drafting standards. This chapter introduces text standards and composition and explains how to use the **MTEXT** command to create a text object that can include multiple lines of text, such as paragraphs or a list of general notes. Chapter 10 explains how to create single-line text objects using the **TEXT** command, as well as additional text tools.

annotation: Textual information presented in notes, specifications, comments, and symbols.

text: Lettering on a CADD drawing.

Text Standards and Composition

Industry and company standards dictate how text should appear on a drawing. Consistent text type, format, height, and spacing are critical to legibility and drawing clarity. A drawing should use the same text ***font*** and format throughout, except for specific cases such as text in traditional architectural title blocks or on maps. Refer to Figure 9-1 as you read the following information about text type and format.

font: The face design of a letter or number.

The ASME Y14.2 standard, *Line Conventions and Lettering*, applies to the process of hand lettering each character using one or more single straight or curved elements, in a Gothic design. For example, the letter A has three single-stroke lines. You can achieve the ASME standard with AutoCAD using a font such as Arial, Romans (roman simplex), or Century Gothic. The US National CAD Standard (NCS) recommends a sans serif font, which can be achieved with a font such as the Arial, Romans, or SansSerif font. Some architectural companies prefer the Stylus BT or CountryBlueprint font, because these fonts provide a more traditional architectural hand-lettering appearance.

Figure 9-1.
Examples of common text typefaces and formats.

DIMENSIONING AND TOLERANCING PER ASME Y14.5-2018.
ASME Y14.2 standard: vertical UPPERCASE, Arial font

BEND DOWN 90° R.50
ASME Y14.2 standard: vertical UPPERCASE, Romans font

REMOVE ALL BURRS AND SHARP EDGES.
ASME Y14.2 standard: vertical UPPERCASE, Century Gothic font

SIMPSON LCC5.25-3.5 TYP
US National CAD standard: vertical UPPERCASE, SansSerif font

HOOD W/FAN, VENT TO OUTSIDE AIR
Traditional architecture format: vertical UPPERCASE, Stylus BT font

ALL FRAMING LUMBER TO BE DFL #2 OR BETTER
Traditional architecture format: vertical UPPERCASE, CountryBlueprint font

TYP EACH END TWO FLANGES
ASME Y14.2 standard variation: inclined UPPERCASE, Arial font

Mississippi River
River identification on a map: inclined lowercase, SansSerif font

Vertical text is most common. However, inclined text is an approved ASME standard and is used by some companies, most often in structural drafting and for specific drawing requirements, such as water-feature labels on maps. The recommended slant for inclined text is 68° from horizontal, although some drafters find 75° more appropriate. Uppercase text is standard. However, some companies use lowercase letters for applications such as civil engineering drawings or maps.

A drawing displays specific text heights for different purposes. **Figure 9-2** lists minimum letter heights based on the ASME Y14.2 standard, *Line Conventions and Lettering.*

Figure 9-2.
Minimum letter heights based on the ASME Y14.2 standard, *Line Conventions and Lettering.*

Application	Height INCH	Height METRIC (mm)
Most text (dimension values, notes)	.12	3
Drawing title, drawing size, CAGE code, drawing number, revision letter	.24* .12**	6* 3**
Section and view letters	.24	6
Zone letters and numerals in borders	.24	6
Drawing block headings	.10	2.5

*D, E, F, H, J, K, A0, and A1 size sheets
**A, B, C, G, A2, A3, and A4 size sheets

The NCS and many companies, especially those that produce architectural and civil drawings, depart slightly from the ASME standard. The NCS specifies a minimum text height of 3/32″ (2.4 mm). Most text is 1/8″ (3 mm) high, with titles and similar text 1/4″ (6 mm) high.

Numbers in dimensions and notes are the same height as standard text. AutoCAD provides several methods for stacking text in fractions. When dimensions contain fractions, the fraction bar should usually appear horizontally between the numerator and denominator. However, many notes have fractions displayed with a diagonal fraction bar (/). In this case, use a dash or space between the whole number and the fraction. **Figure 9-3** shows examples of text for numbers and fractions in different unit formats.

AutoCAD text commands provide great control over text ***composition***. You can lay out text horizontally, as is typical when adding notes, or draw text at any angle according to specific requirements. AutoCAD automatically spaces letters and lines of text to help maintain the identity of individual notes. See **Figure 9-4**.

composition: The spacing, layout, and appearance of text.

PROFESSIONAL TIP

Text presentation is important. Refer to appropriate industry, company, or school standards and consider the following tips when adding text.

- Plan your drawing using rough sketches to allow room for text and notes.
- Arrange text to avoid crowding.
- Place related notes in groups to make the drawing easier to read.
- Place all general notes in a common location. Locate notes in the lower-left corner or above the title block when using ASME standards, in the upper-left corner when using military (MIL) standards, or in the note block when using the NCS.
- Review all text and use the spell checker.

Figure 9-3.
Examples of fractional text for different unit formats.

Decimal Inch	Fractional Inch	Millimeter
2.750 .25	$2\frac{3}{4}$ 2–3/4 2 3/4	2.5 3 0.7

Figure 9-4.
An example of general notes on a mechanical part drawing, typed according to the ASME Y14.2 standard, *Line Conventions and Lettering*. The standard spacing is also appropriate for use on drawings for other disciplines.

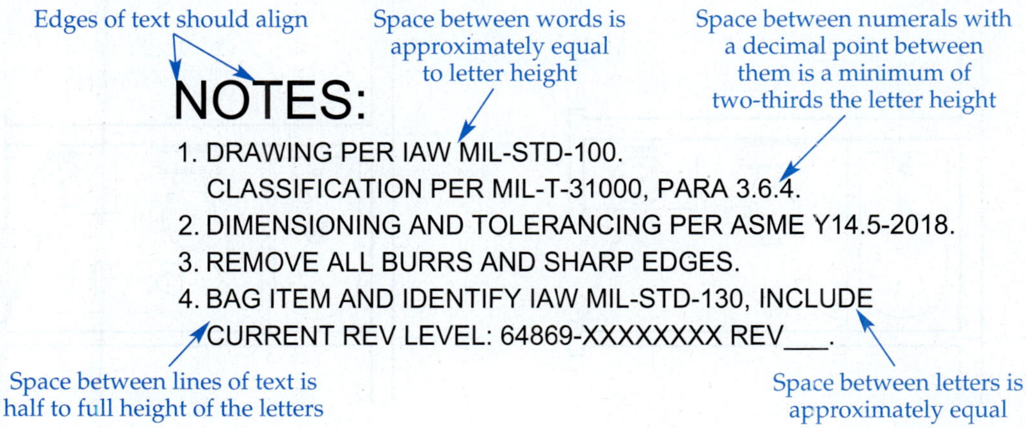

Drawing Scale and Text Height

Ideally, you should determine drawing scale, scale factors, and text heights before you begin drawing. Incorporate these settings into drawing template files and make changes when necessary. The drawing scale factor determines how text height appears on-screen and plots.

To understand the concept of drawing scale, look at the portion of a floor plan shown in Figure 9-5. You should draw everything in model space at full scale. This means that the bathtub, for example, is actually drawn 5′ long. However, at this scale, text size becomes an issue, because full-scale text that is 1/8″ high is too small compared to the other full-scale objects. See Figure 9-5A. As a result, you must adjust the text height according to the drawing scale, as shown in Figure 9-5B. You can calculate the scale factor manually and apply it to text height, or you can allow AutoCAD to calculate the scale factor using annotative text.

Scaling Text Manually

text height: The specified height of text, which may be different from the plotting size for text scaled manually.

scale factor: The reciprocal of the drawing scale.

paper text height: The plotted text height.

To adjust ***text height*** manually according to a specific drawing scale, you must calculate the drawing ***scale factor***. Figure 9-6 provides examples of calculating scale factor. You then multiply the scale factor by the ***paper text height*** to get the model space text height.

For example, a site plan plotted at a 1″ = 60′ scale has a scale factor of 720. Text drawn 1/8″ high is almost invisible, because the drawing is 720 times larger than the text is when plotted at the proper scale. Therefore, multiply the scale factor of 720 by the text height of 1/8″ (.125″) to find the 90″ scaled text height for model space. The proper height of 1/8″ text in model space at a 1″ = 60′ scale is 90″.

Annotative Text

annotative text: Text scaled by AutoCAD according to the specified annotation scale.

annotation scale: The drawing scale AutoCAD uses to calculate the height of annotative text.

AutoCAD scales ***annotative text*** according to the ***annotation scale*** you select, which reduces the need for you to calculate the scale factor. Once you select an annotation scale, AutoCAD applies the corresponding scale factor to annotative text and all other annotative objects. For example, if you manually scale 1/8″ text for a drawing with a 1/4″ = 1′-0″ scale, or a scale factor of 48, you must draw the text using a text height of 6″ (1/8″ × 48 = 6″) in

Figure 9-5.
An example of a portion of a floor plan drawn at full scale in model space. A—Text drawn at full scale (1/8″ high) is too small compared to the large full-scale objects. B—Text scaled (6″ high) to display and plot correctly relative to the size of features on the drawing.

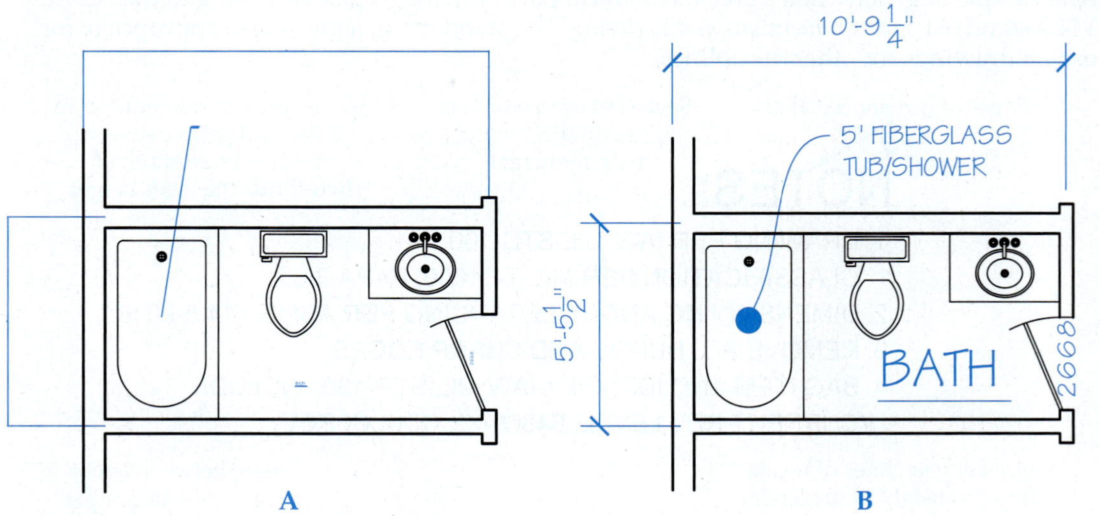

Figure 9-6.
Examples of calculating scale factor.

Example	Scale	Conversion	Calculation	Scale Factor
Mechanical	1:2	None	2 ÷ 1 = 2	2
Civil	1″ = 60′	1″ = 720″ (60′ = 720″)	720 ÷ 1 = 720	720
Architectural	1/4″ = 1′-0″	1/4″ (.25″) = 12″ (1′ = 12″)	12 ÷ .25 = 48	48
Metric to Inch	1:1	1″ = 25.4 mm	25.4 ÷ 1 = 25.4	25.4
Metric to Inch	1:2	1″ = 25.4 mm × 2 (50.8)	50.8 ÷ 1 = 50.8	50.8

model space. When placing annotative text, using this example, you set an annotation scale of 1/4″ = 1′-0″. Then you draw the text using a paper text height of 1/8″ in model space. The 1/8″ text is scaled to 6″ automatically, according to the preset 1/4″ = 1′-0″ annotation scale.

Annotative text offers several advantages over manually scaled text, including the ability to control text appearance based on the drawing scale and paper text height, while reducing the need to focus on the scale factor. Annotative text is especially effective when the drawing scale changes or when a single sheet includes views at different scales.

PROFESSIONAL TIP

If you anticipate preparing scaled drawings, you should use annotative text and other annotative objects instead of manual scaling. However, scale factor does influence non-annotative items and is still an important value to identify and use throughout the drawing process.

Setting the Annotation Scale

The annotation scale is typically the same as the drawing scale. You should usually set the annotation scale before you begin typing text so that the text height is scaled automatically. However, this is not always possible. It may be necessary to adjust the annotation scale throughout the drawing process, especially if you prepare views at different scales on one sheet. This textbook approaches annotation scaling in model space, using the process of selecting the appropriate annotation scale before typing text. To create text using a different scale, pick the new annotation scale and then type the text.

The **Annotation Scale** flyout on the status bar provides a quick way to set the annotation scale. See Figure 9-7A. Scroll through a long list as necessary to find the appropriate annotation scale. The **Annotation Scale** flyout shows the current annotation scale as a ratio by default. Check the **Percentages** option to also display corresponding percentages.

NOTE

If you access the **MTEXT** command and an annotative text style is current, the **Select Annotation Scale** dialog box appears. This dialog box provides a convenient way to set annotation scale before creating multiline text.

NOTE

This textbook describes many additional annotative object tools. Some of these tools are more appropriate for working with layouts, as explained later in this textbook.

Figure 9-7.
A—Use the **Annotation Scale** flyout on the status bar to set the annotation scale. B—Use the **Edit Drawing Scales** dialog box to make changes within the list of existing annotation scales and to create new annotation scales.

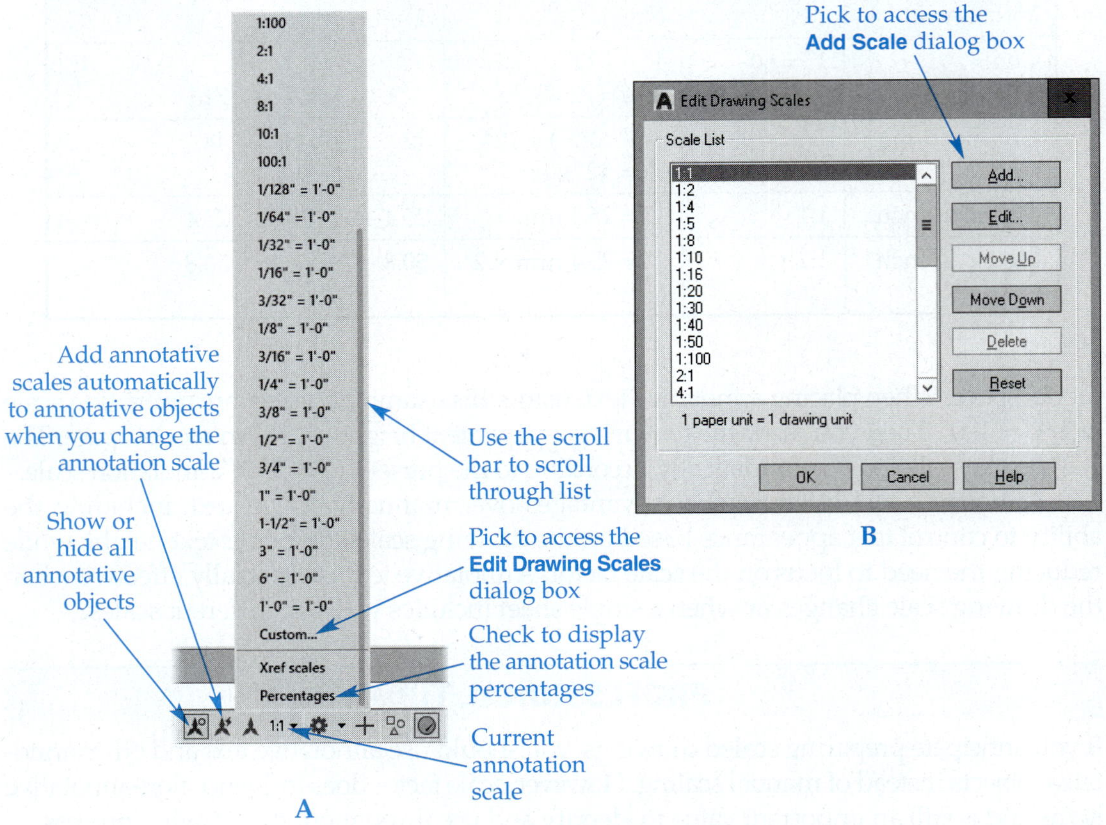

Editing Annotation Scales

SCALELISTEDIT

Ribbon
Annotate
> Annotation Scaling
Scale List
Type
SCALELISTEDIT

Access the **Edit Drawing Scales** dialog box, shown in **Figure 9-7B**, to make changes within the list of existing annotation scales and to create new annotation scales. A quick way to access the **Edit Drawing Scales** dialog box is to select the **Custom...** option from the **Annotation Scale** flyout on the status bar. Move the highlighted scale up or down in the list using the **Move Up** or **Move Down** button. To remove the highlighted scale from the list, pick the **Delete** button.

Select the **Edit...** button to open the **Edit Scale** dialog box, where you can change the name of the scale and adjust the scale by entering the paper and drawing units. For example, a scale of 1/4″ = 1′-0″ has a paper units value of .25 or 1 and a drawing units value of 12 or 48.

To create a new annotation scale, pick the **Add...** button to display the **Add Scale** dialog box, which provides the same options as the **Edit Scale** dialog box. Pick the **Reset** button to restore the list to display the default annotation scales. Once you select an annotation scale, you are ready to type annotative text.

NOTE

Changes you make in the **Edit Drawing Scales** dialog box are stored with the drawing and are specific to the drawing. To make changes to the default scale list saved to the system registry, pick the **Default Scale List...** button in the **User Preferences** tab of the **Options** dialog box to access the **Default Scale List** dialog box. The options are the same as those in the **Edit Drawing Scales** dialog box, but changes are saved as the default for new drawings.

Text Styles

A ***text style*** presets many text characteristics. Create a text style for each different text appearance or function. For example, use an annotative text style to draw annotative text and a non-annotative text style to draw non-annotative text. Another example is creating text styles that correspond to a specific text height or other characteristics. Add text styles to your drawing templates for repeated use. Avoid adjusting text format independently of the text style assigned to the text.

text style: A saved collection of settings for text height, width, oblique angle (slant), and other text effects.

Text Style Dialog Box

Create, modify, and delete text styles using the **Text Style** dialog box. See Figure 9-8. The **Styles:** list box displays existing text styles. The Annotative text style allows you to create annotative text, as indicated by the icon to the left of the style name. The Standard text style does not use the annotative function.

To make a text style current, double-click the style name, right-click the name and select **Set current**, or pick the name and select the **Set Current** button. Below the **Styles:** list box is a drop-down list that you can use to filter the number of text styles displayed in the **Text Style** dialog box. Pick the **All styles** option to show all text styles in the file, or pick the **Styles in use** option to show only the current style and styles used in the drawing.

STYLE

Ribbon
Home
> Annotation
Text Style
Annotate
> Text
Text Style
Annotate
> Text
> Text Style
Manage Text
Styles...
Type
STYLE
ST

Creating New Text Styles

To create a new text style, select an existing text style from the **Styles:** list box to use as a base for formatting the new text style. Then pick the **New...** button to open the **New Text Style** dialog box. See Figure 9-9. Notice that style1 appears in the **Style Name:** text box. Replace the default name with a more descriptive name.

Figure 9-8.
Use the **Text Style** dialog box to create, rename, delete, and set the characteristics of a text style. The **Big Font:** drop-down list replaces the **Font Style:** drop-down list when you select the **Use Big Font** check box. The **Height** text box replaces the **Paper Text Height** text box when you deselect the **Annotative** check box.

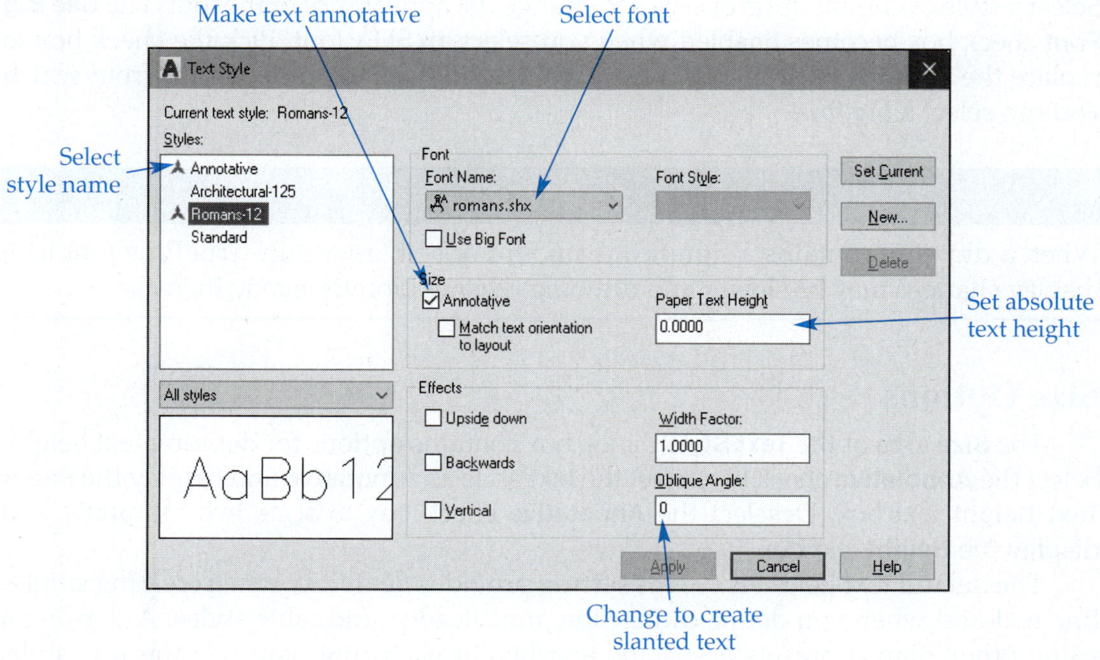

Figure 9-9. Enter a descriptive name for the new text style in the **New Text Style** dialog box.

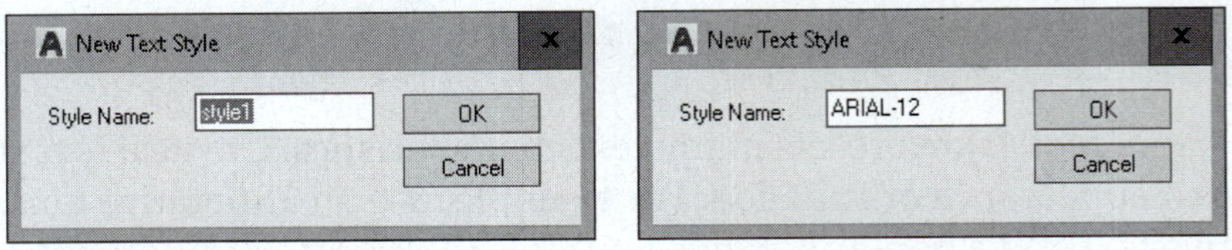

Default Style   New Style

For example, the name ARIAL-12 describes a text style that uses the Arial font and characters .12″ high. The name ARCHITECTURAL-125 describes a text style that uses the Stylus BT font and characters 1/8″ high.

Text style names can have up to 255 characters, including uppercase and lowercase letters, numbers, dashes (–), underlines (_), and dollar signs ($). After typing the text style name, pick the **OK** button. The new text style appears in the **Styles:** list box of the **Text Style** dialog box, and you are ready to adjust text style characteristics. Pick the **Apply** button to apply changes, and pick the **Close** button to exit the **Text Style** dialog box.

PROFESSIONAL TIP

Record the names and details of the text styles you create and keep the information in a log for future reference.

Font Options

scalable fonts: Fonts that can be displayed or printed at any size while retaining proportional letter thickness.

Use the **Font Name:** drop-down list in the **Font** area of the **Text Style** dialog box to select a font. The list includes TrueType fonts installed on your computer and fonts linked to AutoCAD shape files. TrueType fonts are ***scalable fonts*** and have an outline. Most TrueType fonts appear and plot filled by default. You can recognize fonts from AutoCAD shape files (SHX fonts) by the .shx file extension and the AutoCAD compass icon.

The **Font Style:** drop-down list is active if the selected font includes options, such as bold or italic. SHX fonts do not provide style options, but some TrueType fonts do. For example, the SansSerif font has Regular, Bold, BoldOblique, and Oblique options. Select a style or combination of styles to change the appearance of the font. The **Use Big Font** check box becomes enabled when you select an SHX font. Pick the check box to replace the **Font Style:** drop-down list with the **Big Font:** drop-down list, from which you can select a ***big font***.

big font: A supplement that provides Asian and other large-format fonts that have characters and symbols not present in other font files.

CAUTION

When a drawing contains a significant amount of text, especially TrueType font text, display changes may be slower and drawing regeneration time may increase.

Size Options

The **Size** area of the **Text Style** dialog box contains options for defining text height. Select the **Annotative** check box to set the text style as annotative and display the **Paper Text Height** text box. Deselect the **Annotative** check box to scale text manually and display the **Height** text box.

The default text height is 0. This setting provides flexibility when creating single-line text and when you define dimension, multileader, and table styles. Assigning a value other than 0 presets a specific text height each time you use the text style.

When you assign a value other than 0, you can only initially create single-line text at the specified height, and you do not have the option of assigning a different text height to dimension, multileader, and table styles. However, once created, you can use an editing command to change the initial text height. Single-line text and editing text are explained in detail in Chapter 10. Dimension, multileader, and table styles are explained later in this textbook.

PROFESSIONAL TIP

A text style height of 0 provides the greatest flexibility when drawing. A prompt asks you to specify a text height when you create single-line text, and you can assign a text height to dimension, multileader, and table styles. As an alternative, assigning a specific text height to a text style can increase productivity and accuracy, but requires that you create a text style for each unique height or edit text objects to change text height.

The **Match text orientation to layout** check box becomes enabled when you pick the **Annotative** check box. Check **Match text orientation to layout** to match the orientation of text in layout viewports with the layout orientation. Layouts are described later in this textbook.

Effects

The **Effects** area of the **Text Style** dialog box offers text format options, including settings for drawing text upside-down, backwards, and vertically. See Figure 9-10. The **Vertical** check box is available for SHX fonts. Text on drawings is normally horizontal. Vertical text is appropriate for special effects and graphic designs and usually works best with a 270° rotation angle.

Use the **Width Factor:** text box to specify the text character width relative to its height. You can set the width factor between 0.01 and 100. A width factor of 1 is default and is recommended for most drawing applications. A width factor greater than 1 expands characters, and a factor less than 1 compresses characters. If necessary, reduce the width slightly, and only if you cannot reduce the height. See Figure 9-11.

The **Oblique Angle:** text box allows you to set the angle at which text inclines. The 0 default draws vertical characters. A value greater than 0 slants characters to the right, and a negative value slants characters to the left. See Figure 9-12. Some fonts, such as the italic SHX font, are already angled.

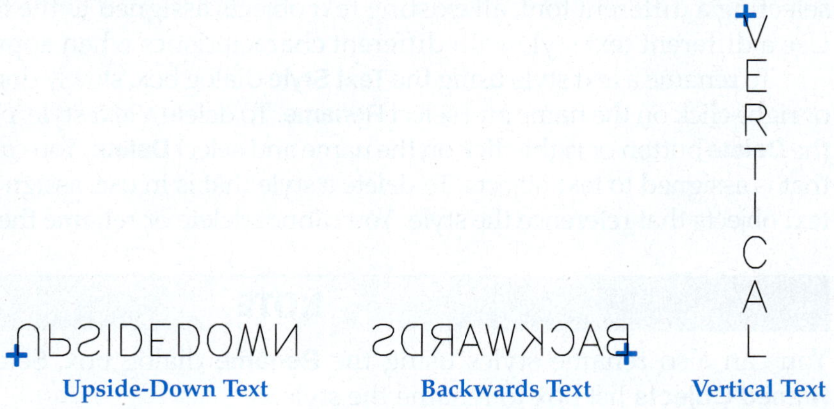

Figure 9-10. Special effects for text styles can be set in the **Effects** area of the **Text Style** dialog box. The + symbol indicates the text justification start point.

Figure 9-11.
Examples of width factor settings for text. A width factor of 1 is the default and is recommended for most drawing applications.

Width Factor	Text
1	ABCDEFGHIJKLM
.5	ABCDEFGHIJKLMNOPQRSTUVWXY
1.5	ABCDEFGHI
2	ABCDEFG

Figure 9-12.
Examples of oblique angle settings for text.

Obliquing Angle	Text
0	ABCDEFGHIJKLM
15	*ABCDEFGHIJKLM*
–15	ABCDEFGHIJKLM

PROFESSIONAL TIP

AutoCAD text slant is measured from vertical. Use a 22° oblique angle to slant text according to the 68° horizontal incline standard.

NOTE

The preview area of the **Text Style** dialog box displays an example of the selected font and font effects. This is a convenient way to see how the text appears before using it in a new text style.

Exercise 9-1

www.g-wlearning.com/CAD/

Complete the exercise on the companion website.

Changing, Renaming, and Deleting Text Styles

Select a text style from the **Styles:** list box to edit. Make the necessary changes and pick the **Apply** button to apply the changes. If you make changes to a text style, such as selecting a different font, all existing text objects assigned to the text style are updated. Use a different text style with different characteristics when appropriate.

To rename a text style using the **Text Style** dialog box, slowly double-click on the name or right-click on the name and select **Rename**. To delete a text style, pick the style and select the **Delete** button or right-click on the name and select **Delete**. You cannot delete a text style that is assigned to text objects. To delete a style that is in use, assign a different style to the text objects that reference the style. You cannot delete or rename the Standard style.

NOTE

RENAME

Type
RENAME
REN

You can also rename styles using the **Rename** dialog box. Select **Text styles** in the **Named Objects** list box to rename the style.

Setting a Text Style Current

Set a text style current using the **Text Style** dialog box, as described earlier in this chapter. To set a text style current without opening the **Text Style** dialog box, use the **Text Style** drop-down list on the expanded **Annotation** panel of the **Home** ribbon tab or on the **Text** panel of the **Annotate** ribbon tab. See Figure 9-13.

Command line content search offers another way to find a text style and make a text style current. Begin typing a text style name using the command line. Text styles that match the letters you enter appear in the suggestion list as you type. Select a text style from the suggestion list to set current.

PROFESSIONAL TIP

You can import text styles from existing drawings using **DesignCenter**. See Chapter 5 for more information about using **DesignCenter** to reuse drawing content.

Multiline Text

The **MTEXT** command draws a single multiline text, or mtext, object that can include extensive paragraph formatting, lists, symbols, and columns. It is good practice to set the appropriate text style current before you access the **MTEXT** command. Activate the **MTEXT** command to display letters near the crosshairs that indicate the current text style and height. Pick the first corner of the ***text boundary***. A prompt then asks you to specify the opposite corner or select an option. You can use the options to preset mtext layout. However, it is typically easier to set mtext options while you are typing. Rotation is the only option you cannot adjust while typing. Use the **Rotation** option before selecting the opposite corner to rotate the text boundary, or use an editing command to rotate the mtext object after you complete the **MTEXT** command.

MTEXT

Ribbon
Annotate
> Text
Home
> Annotation
Multiline Text
Type
MTEXT
MT
T

text boundary: An imaginary box that sets the location and width for multiline text.

Mtext uses dynamic columns by default to help you organize multiple text columns in an mtext object. Disable columns for typical text requirements without columns. Before selecting the second corner of the text boundary, select the **Columns** option and then the **No columns** option. An arrow in the boundary shows the direction of text flow and where the boundary will expand as you type, if necessary. Pick the opposite corner of the text boundary to continue. See Figure 9-14. You will learn to format columns later in this chapter.

Figure 9-13.
The fastest way to set a text style current is to use one of the drop-down lists on the ribbon.

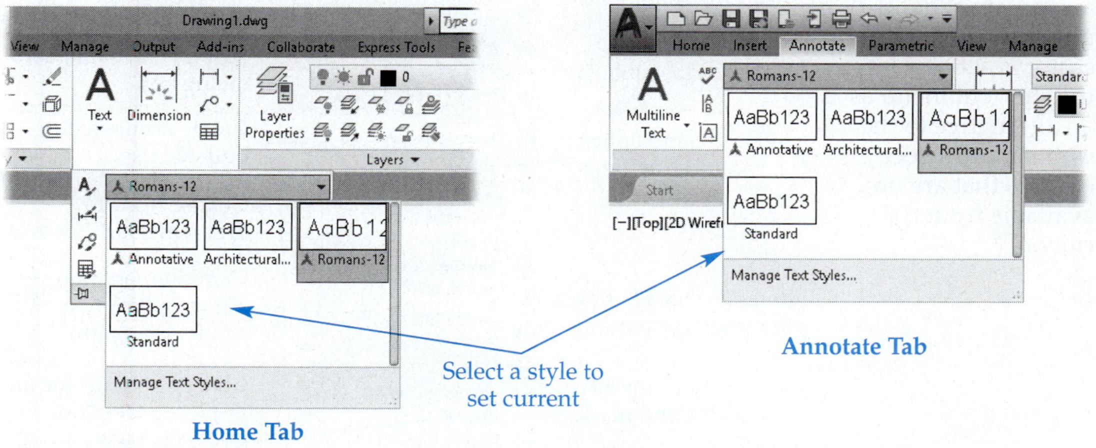

Figure 9-14.
The text boundary is the box in which you type. Consider the text boundary the extents of the text object. Long words extend past the boundary in the direction the arrow indicates.

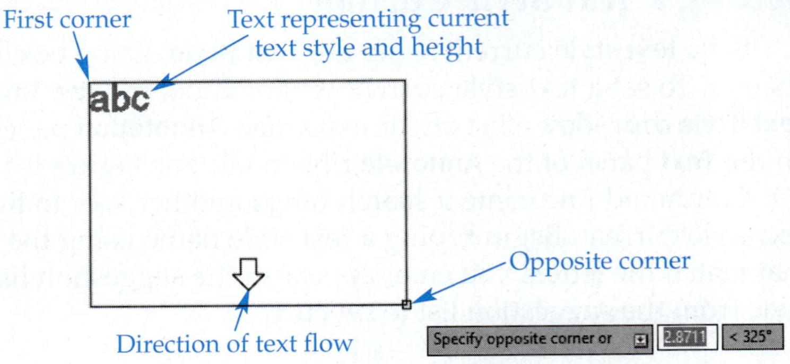

Using the Text Editor

text editor: The area of the multiline or single-line text system where you type text.

When you select the opposite corner of the text boundary, the **Text Editor** contextual ribbon tab and the ***text editor*** appear. Figure 9-15 shows the display with columns disabled. Typing mtext is similar to typing using word processing software such as Microsoft® Word. The **Text Editor** ribbon tab provides options for adjusting and formatting text in the text editor. You can access many of the same options found in the **Text Editor** ribbon tab, as well as Windows Clipboard functions, from the shortcut menu that appears when you right-click away from the ribbon. See Figure 9-16.

Figure 9-15.
The **Text Editor** contextual ribbon tab provides options for working with mtext.

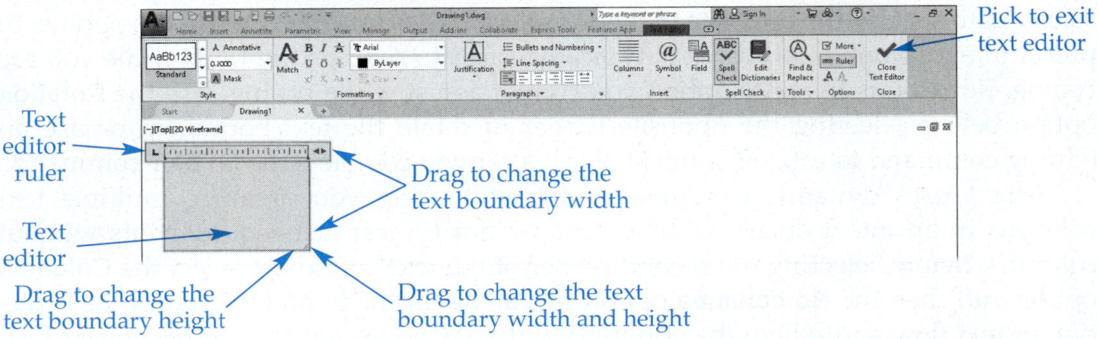

Figure 9-16.
Display the text editor shortcut menu by right-clicking away from the ribbon while the text editor is active. Use the shortcut menu as an alternative to the ribbon or to select options that are not available from the ribbon.

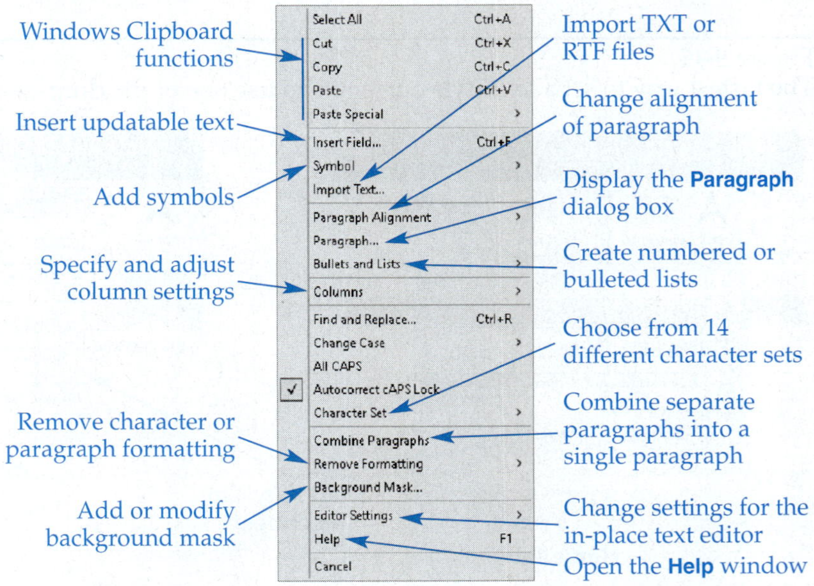

NOTE

If you close the ribbon, the **Text Formatting** toolbar appears instead of the **Text Editor** ribbon tab. This textbook focuses on using the **Text Editor** ribbon tab to add mtext. The **Text Formatting** toolbar provides the same functions.

The text editor indicates the initial size of the area in which you type. When columns are not active, long words and paragraphs extend past the text editor limits. The text editor displays a semitransparent background by default. Deselect the **Show Background** option to make the text editor transparent. The text editor includes a ruler that displays indent and tab stops and indent and tab markers. Use the **Ruler** option to turn the ruler on or off.

Ribbon
Text Editor
> Options
> More
> Editor Settings
> Show Background

Ribbon
Text Editor
> Options

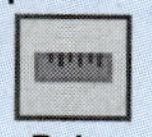

Ruler

NOTE

Use the **Undo** and **Redo** options in the **Options** panel of the **Text Editor** tab to undo or redo text editor operations.

To change the width of the text editor, drag the far end of the ruler or the far vertical edge of the text boundary. An alternative, available when not initially using columns, is to right-click on the ruler and select **Set Mtext Width...** to use the **Set Mtext Width** dialog box. To change the height of the text editor, drag the far horizontal edge of the text editor, or, when not initially using columns, right-click on the ruler and select **Set Mtext Height...** to use the **Set Mtext Height** dialog box. Drag the far corner of the text boundary to change the width and height.

CAUTION

Change the width and height of the text editor to increase or decrease the number of lines of text. Do not press [Enter] to form lines of text, unless you are specifically creating a new paragraph or a new item in a list.

The familiar Windows text editor cursor is displayed at the current text height within the text editor. Begin typing or editing text. The procedure for selecting and editing existing text is the same as in standard Windows text editors. To select all text in the text editor, right-click inside the text editor and select **Select All**. To change the selected text highlight color, use the **Text Highlight Color...** option to access the **Select Color** dialog box.

Ribbon
Text Editor
> Options
> More
> Editor Settings
> Text Highlight Color...

When you finish typing, exit the mtext system using the **Close Text Editor** option, or pick outside of the text editor. You can also press [Esc] or right-click and select **Cancel**, but AutoCAD prompts you to save changes to the mtext. The easiest way to reopen the text editor to make changes to text content is to double-click on the mtext object.

Ribbon
Text Editor
> Close

Close Text Editor

Reference Material

www.g-wlearning.com/CAD/

Shortcut Keys

For a complete list of keyboard shortcuts for text editing, go to the companion website, select the **Resources** tab, and select **Shortcut Keys**.

Exercise 9-2

www.g-wlearning.com/CAD/

Complete the exercise on the companion website.

Stacking Text

The AutoStacking feature is active by default and automatically stacks numerical characters when you type a specific character sequence. For example, to create a stacked fraction with a horizontal fraction bar, type a forward slash (/) between characters. To create a stacked fraction with a diagonal fraction bar, type a number sign (#) between characters. To create a ***tolerance stack*** or a decimal stack, type the caret (^) character between characters. Type a space or a nonnumerical character, such as an inch mark ("), after the characters to form the stack. See **Figure 9-17**. Press [Backspace] immediately after characters stack to remove stacking.

tolerance stack: Text stacked vertically without a fraction bar.

The icon shown in **Figure 9-17A** appears as soon as characters stack and when you pick stacked characters. Select the icon to access a shortcut menu of stack options. Select **Diagonal** to convert the stack to use a diagonal fraction bar or select **Horizontal** to convert the stack to use a horizontal fraction bar. Select **Unstack** to remove stacking.

Select **Stack Properties** to display the **Stack Properties** dialog box, described in **Figure 9-18**. Select **100%** from the **Text size** drop-down list to conform to ASME standards. This is an appropriate standard to follow for all disciplines. Pick the **AutoStack...** button to access the **AutoStack Properties** dialog box shown in **Figure 9-19**. Use the **AutoStack Properties** dialog box to enable or disable AutoStacking and control the stack format. You can also remove the leading space between a whole number and the fraction.

Figure 9-17.
Using AutoStacking to create stacked text. A—Typing a specific sequence of characters causes the characters to stack by default. Type a space or a nonnumerical character following the characters. B—Sequences used to stack characters. You must adjust stack properties after typing to create the decimal stack shown.

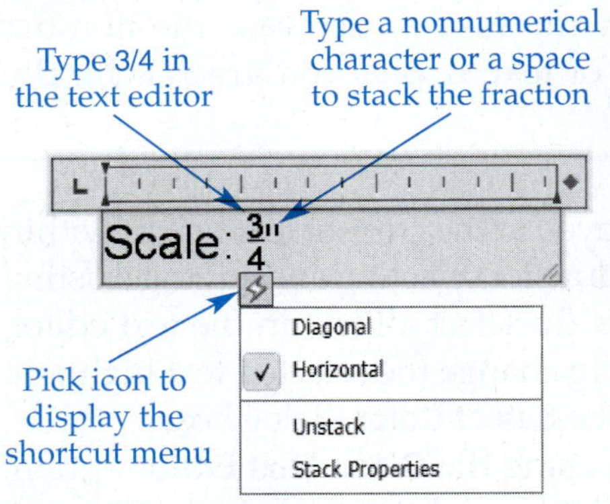

A

Stack Type	Character Sequence	Stacked Text
Vertical Fraction	1/2	$\frac{1}{2}$
Diagonal Fraction	1#2	1/2
Tolerance Stack	1^2	1 2
Decimal Stack	1.10^.90	1.10 .90

B

Figure 9-18.
The **Stack Properties** dialog box. ASME standards recommend that the text height of stacked fraction numerals be the same as the height of other dimension numerals. Select **100%** from the **Text size** drop-down list to conform to ASME standards.

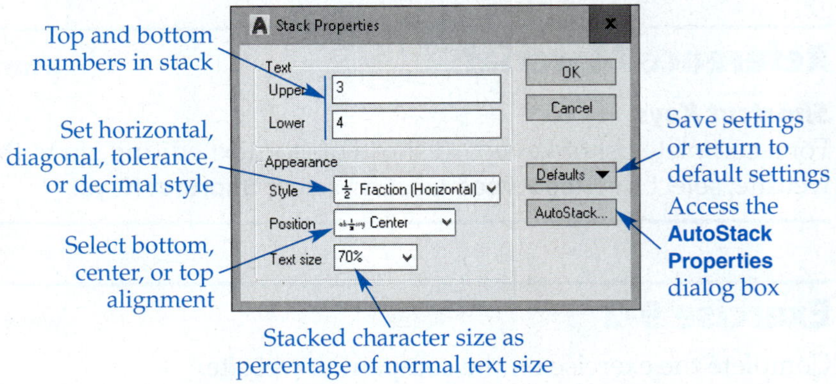

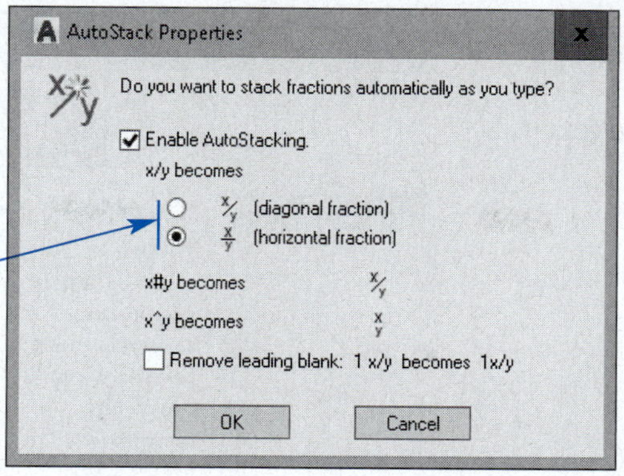

Figure 9-19. The **AutoStack Properties** dialog box.

Manual Stacking

Use the **Stack** option to manually stack selected characters using the appropriate format shown in **Figure 9-17B**. To return stacked text to the original unstacked format, select the stacked text and select the **Stack** option, or use the **Unstack** shortcut menu option.

Ribbon
Text Editor
> Formatting
Stack

Superscript and Subscript Text

The **Superscript** and **Subscript** options make it easy to create superscript or subscript text. Select the characters to convert to superscript or subscript text and select the **Superscript** or **Subscript** button in the **Formatting** panel of the **Text Editor** ribbon tab. See **Figure 9-20**. Use the **Stack Properties** dialog box to make changes to superscript and subscript format.

Ribbon
Text Editor
> Formatting
Superscript

Ribbon
Text Editor
> Formatting
Subscript

Adding Symbols

Use the **Symbol** option to insert a common drafting symbol or other unique character not found on a typical keyboard. See **Figure 9-21**. The first two sections in the **Symbol** flyout menu contain common symbols. The third section contains the option to add a ***non-breaking space***. Pick a symbol to insert at the current location of the text cursor. Selecting the **Other...** option opens the **Character Map** dialog box, shown in **Figure 9-22**. Use the following steps to insert a symbol from the **Character Map** dialog box.

1. Pick a font from the **Font:** drop-down list to display symbols associated with the font.
2. Locate and pick a symbol and then pick the **Select** button. The symbol appears in the **Characters to copy:** box. You can copy multiple symbols to the box.
3. Pick the **Copy** button to copy the symbols to the Clipboard.
4. Close the **Character Map** dialog box.
5. In the text editor, place the text cursor at the location where the symbols are to be inserted.
6. Right-click and select **Paste** to paste the symbols at the cursor location.

Ribbon
Text Editor
> Insert
Symbol

non-breaking space: A symbol that you insert in place of a space to keep separate words together on one line.

Exercise 9-3

www.g-wlearning.com/CAD/

Complete the exercise on the companion website.

Figure 9-20. Examples of superscript and subscript text.

Figure 9-21.
Use the **Symbol** option to add symbols to the text editor.

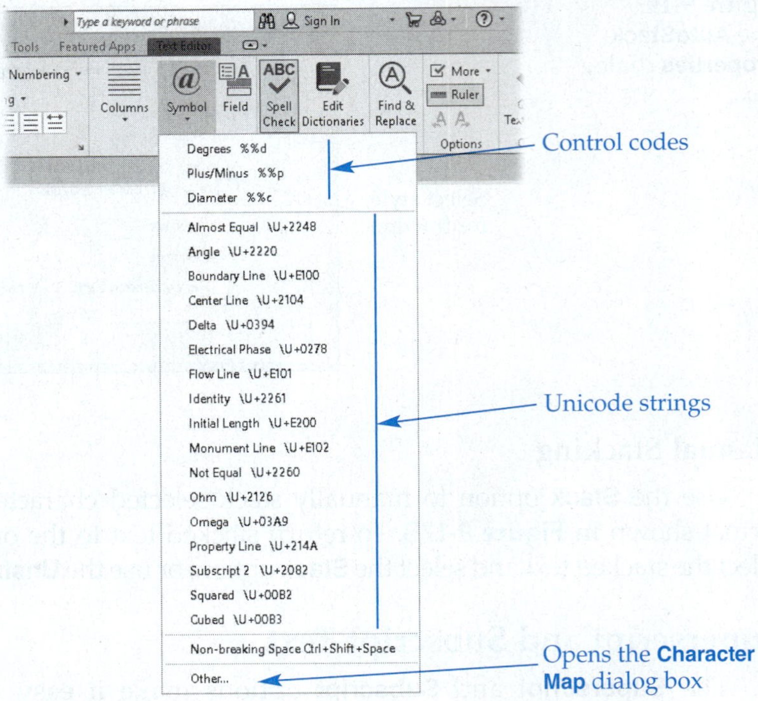

Figure 9-22.
The **Character Map** dialog box.

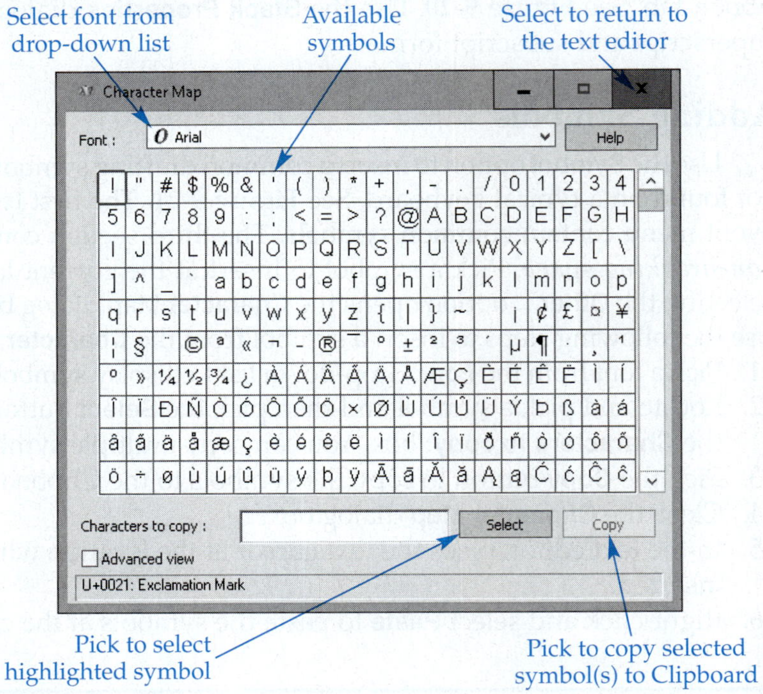

Style Settings

The **Style** panel of the **Text Editor** ribbon tab contains options for changing the text style and for overriding the annotative setting and text height. A single mtext object can use a combination of character formatting and text heights, but it can only be assigned one text style and must be entirely annotative or non-annotative. Use the scroll buttons to the right of the text styles to locate styles, or pick the expansion arrow to display styles in a temporary window. Pick a style different from the current style to apply to all text in the text editor.

Use the text height drop-down list to set the text height. The text height for annotative text is the paper text height. The text height for non-annotative text is the text height multiplied by the scale factor. You should usually only change the text height if the current text style uses a 0 height. Otherwise, you override the specified text style height. You can use the **Annotative** button to override the annotative setting of the current text style, but this is typically not appropriate.

Using a Background Mask

Sometimes drawings display text over existing objects, such as graphic patterns, making the text difficult to read. A ***background mask*** can solve this problem, and can be used to display a background color behind the mtext. Access the **Background Mask** option to display the **Background Mask** dialog box. See Figure 9-23. To mask the current mtext object, check **Use background mask**.

Ribbon
Text Editor
> Style
Background

The **Border offset factor:** text box sets the amount of mask, from 1 to 5. The border offset factor works with the text height value according to the following formula: border offset factor × text height = total masking distance from the bottom of the text. If you set the border offset factor to 1, the mask occurs directly within the boundary of the text. Use a value greater than 1 to offset the mask beyond the text boundary. See Figure 9-24. Use the options in the **Fill Color** area to set the mask color. Select a color from the **Fill Color** drop-down list or check the **Use drawing background color** check box to use the drawing background color.

background mask: A mask that hides a portion of objects behind and around text so that the text is unobstructed.

NOTE

You may need to adjust the draw order to display masked mtext in front of or above other objects correctly.

Figure 9-23.
The **Background Mask** dialog box controls text mask settings.

Figure 9-24.
The border offset factor determines the size of the background mask. The text in the figure is 1/8″ with different border offset factors.

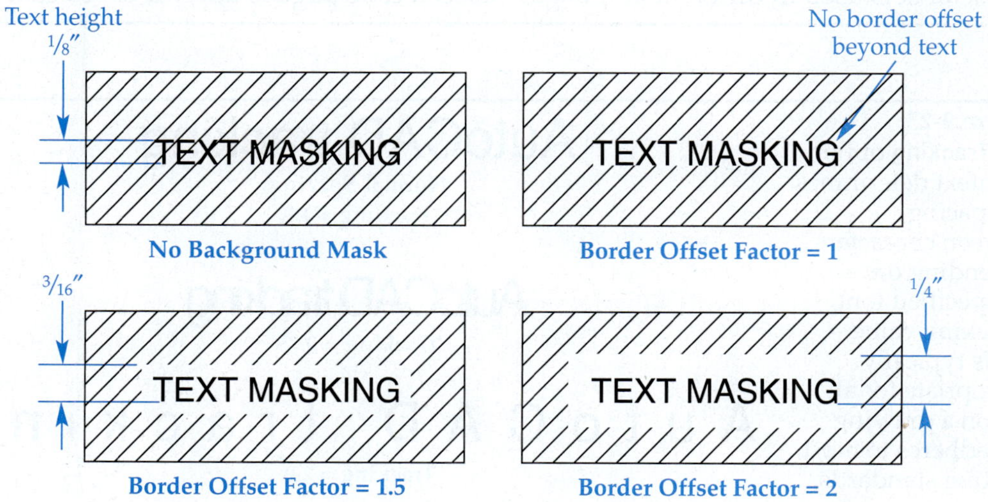

Character Formatting

Use the **Formatting** panel of the **Text Editor** ribbon tab to adjust character format. Some of the same settings are also available from the shortcut menu. A single mtext object can use a combination of character formats. Remember, however, that making changes to character formatting overrides specified text style format and preset object properties, such as color. You should usually avoid this practice.

The **Bold** and **Italic** buttons are enabled for some TrueType fonts. Select the appropriate button(s) to make text bold, italic, or both. Select the **Underline** button to underline text. Select the **Overline** button to place a line over text. Use the **Uppercase** button to make all selected text uppercase, or use the **Lowercase** button to make all selected text lowercase. Use the **Strikethrough** button to draw a line through, or strike-through, characters. The **Font** drop-down list allows you to override the text font. The text color is set to ByLayer by default, but you can change the color by picking a color or option from the **Color** drop-down list. Although you should usually define color as ByLayer, a single mtext object can have a combination of text colors.

NOTE

Ribbon
Text Editor
> Tools
> All CAPS

The **Autocorrect cAPS Lock** and **All CAPS** options provide additional controls for use with uppercase text. The **Autocorrect cAPS Lock** option, available from the shortcut menu, is used to correct typing errors. For example, if [Caps Lock] is on, and you incorrectly type nOTE using [Shift] to type the letter N, AutoCAD changes the word to Note after you type a space, and turns [Caps Lock] off. The **Autocorrect cAPS Lock** option is active by default. Activate the **All CAPS** option to type all text uppercase regardless of the [Caps Lock] status.

Additional character formatting options are available from the expanded **Formatting** panel. The **Oblique Angle** text box overrides the angle at which text is inclined. The value in the **Tracking** text box determines the amount of space between text characters. The default tracking value is 1, which results in normal spacing. Increase the value to add space between characters, or decrease the value to tighten the spacing between characters. You can enter any value between 0.75 and 4.0. See Figure 9-25. The value in the **Width Factor** text box overrides the text character width.

NOTE

The **Character Set** cascading menu, available from the **More** flyout in the **Options** panel, or the shortcut menu, displays a menu of code pages. A code page provides support for character sets used in different languages. Select a code page to apply it to selected text.

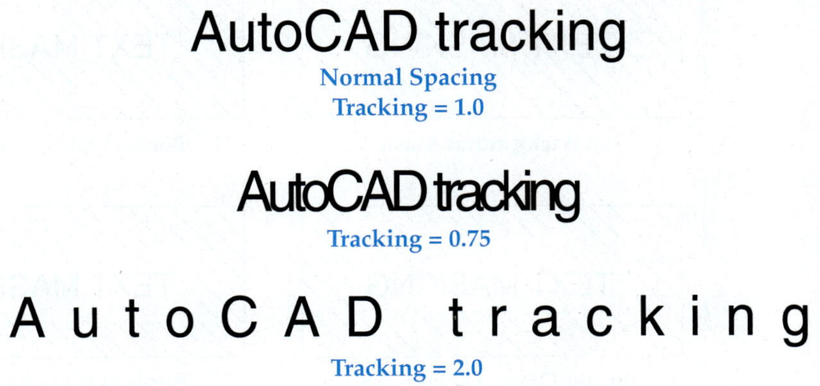

Figure 9-25.
The **Tracking** option for mtext determines the spacing between characters. Depending on the specified font, a tracking value of 1 is typically appropriate for all text on a drawing and adheres to most drafting standards.

Exercise 9-4

www.g-wlearning.com/CAD/

Complete the exercise on the companion website.

Paragraph Formatting

Use the **Paragraph** panel of the **Text Editor** ribbon tab to adjust paragraph formatting. Some of the same settings are also available from the shortcut menu. ***Justify*** the text boundary to control the arrangement and location of text within the text editor. You can also justify the text within the boundary independently of the text boundary justification. This provides flexibility for determining the location and arrangement of text. Justification also determines the direction of text flow. To justify the text boundary, as shown in Figure 9-26, select an option from the **Justification** flyout.

justify: Align the margins or edges of text. For example, left-justified text aligns along an imaginary left border.

Paragraph alignment occurs inside the text boundary. For example, when you apply the **Middle Center** text boundary justification, then set the paragraph alignment to **Left**, the text inside the boundary aligns to the left edge of the text boundary, while the text boundary remains positioned according to the **Middle Center** justification. See Figure 9-27.

paragraph alignment: The alignment of multiline text inside the text boundary.

To adjust paragraph alignment, select a paragraph alignment button on the **Paragraph** panel, or pick from the **Paragraph Alignment** cascading menu of the shortcut menu. You can also control paragraph alignment using the **Paragraph** dialog box, shown in Figure 9-28. To set paragraph alignment in the **Paragraph** dialog box, pick the **Paragraph Alignment** check box, and then select the appropriate radio button. Figure 9-29 shows paragraph alignment options.

Ribbon
Text Editor
> Paragraph
Paragraph

Figure 9-26.
Options for justifying the mtext boundary.

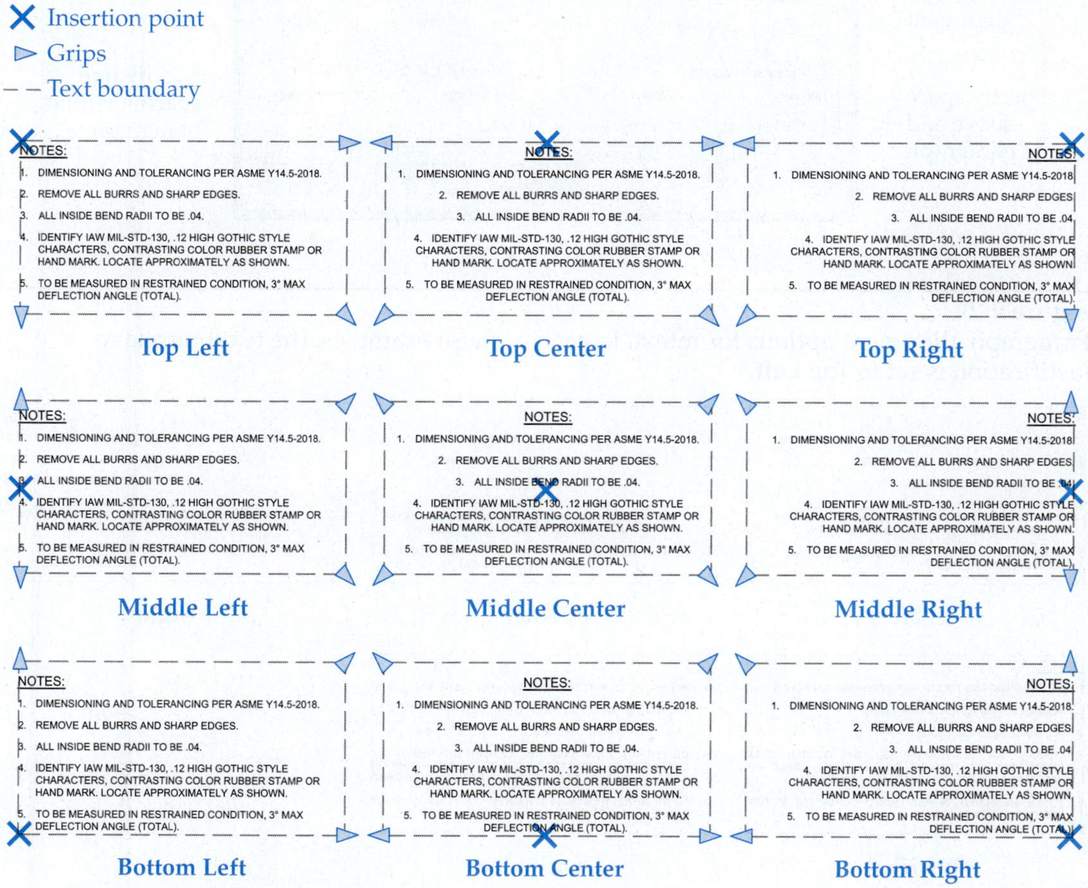

Figure 9-27.
You can adjust paragraph alignment independently of text boundary justification.

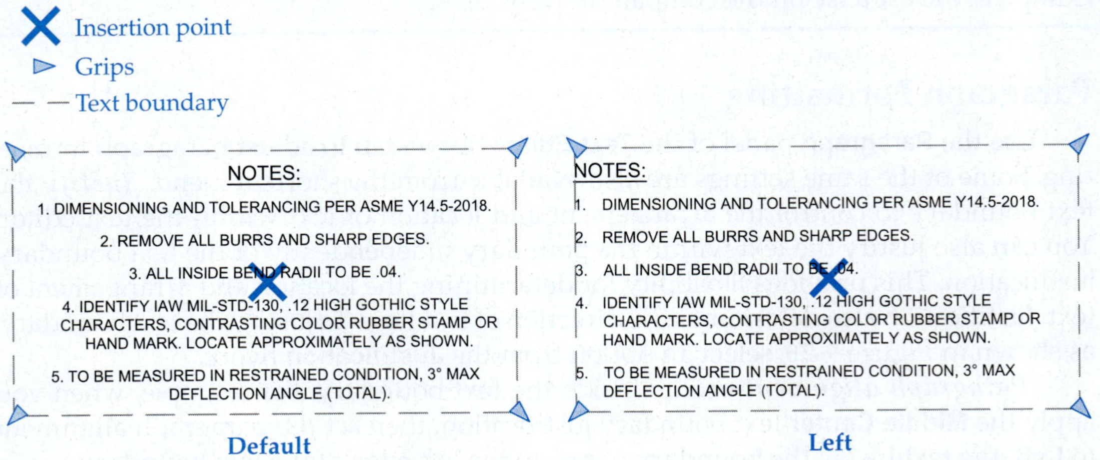

Figure 9-28.
The **Paragraph** dialog box.

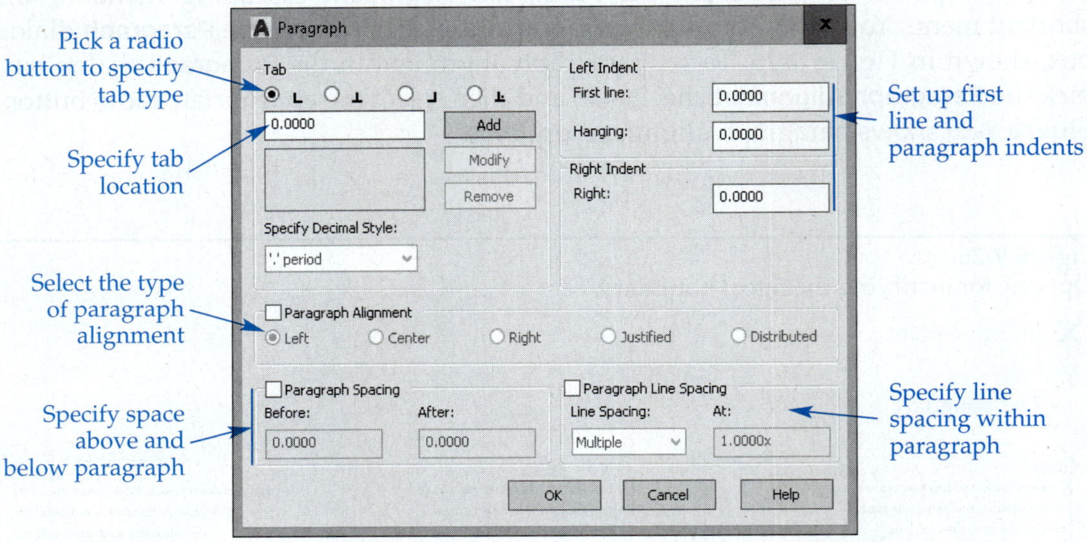

Figure 9-29.
Paragraph alignment options for mtext. In each of these examples, the text boundary justification is set to **Top Left**.

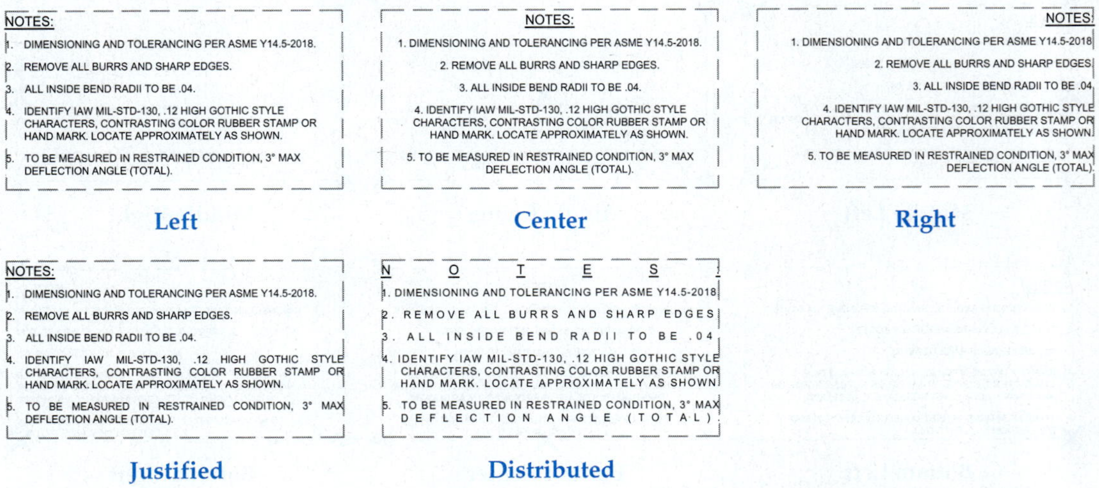

The **Paragraph** dialog box also includes tab, indent, paragraph spacing, and paragraph line spacing settings. Use the **Tab** area to set custom tab stops. Pick a tab type radio button, enter a value for the tab in the text box, and then pick the **Add** button to add the tab to the list and ruler. Figure 9-30 describes each tab option. Add as many custom tabs as necessary. Use the **Modify** and **Remove** buttons as necessary to edit or remove tabs. You can also add custom tabs to the ruler by picking the tab button on the far left side of the ruler until the desired tab symbol appears. Then pick a location on the ruler to insert the tab.

The options in the **Left Indent** area set the indentation for the first line of a paragraph of text and the remaining portion of a paragraph. The **First line:** indent applies each time you start a new paragraph. As text wraps to the next line, the **Hanging:** indent is used. The **Right:** option in the **Right Indent** area sets the indentation for the right side of a paragraph. As you type, the right indent value, not the right edge of the text boundary, determines when the text wraps to the next line.

The options in the **Paragraph Spacing** area define the amount of space before and after paragraphs. To set paragraph spacing, pick the **Paragraph Spacing** check box. Then enter the spacing above a paragraph in the **Before:** text box, and the spacing below a paragraph in the **After:** text box. Figure 9-31 shows examples of paragraph spacing settings.

Figure 9-30.
Using custom tabs to position text in the text editor. When you press [Tab], the cursor moves to the tab position. The type of tab determines text behavior.

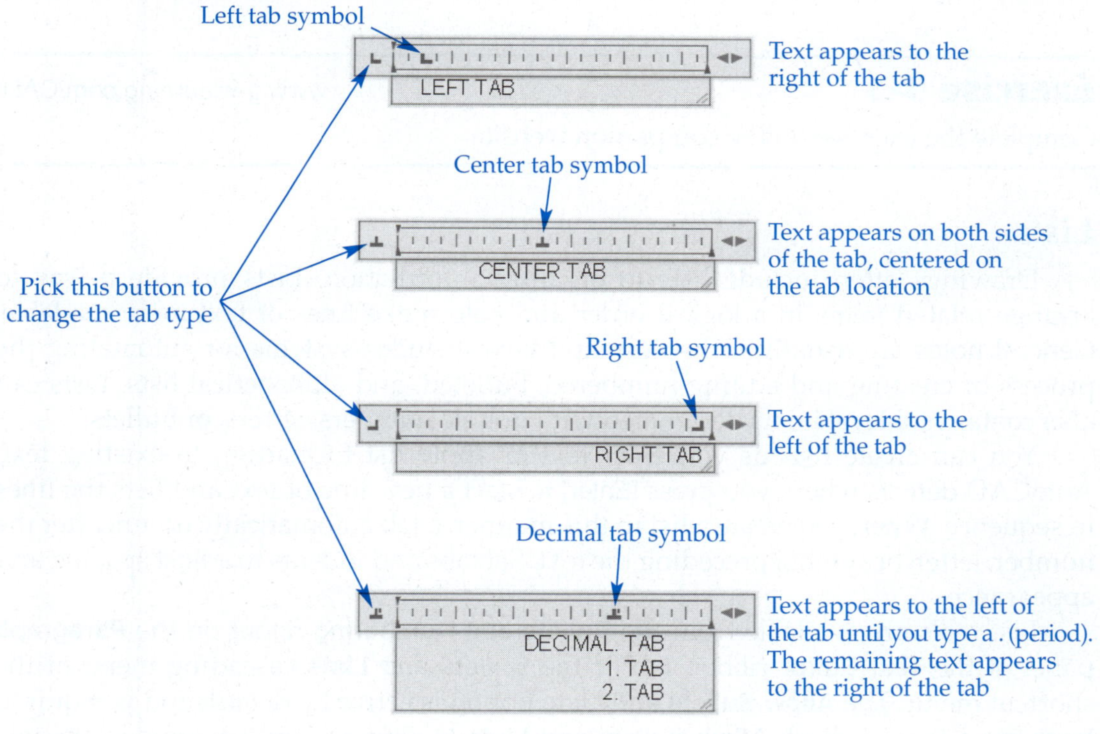

Figure 9-31.
Examples of paragraph spacing. Each example uses a text height of .1875″ and a first-line left indent of .5″.

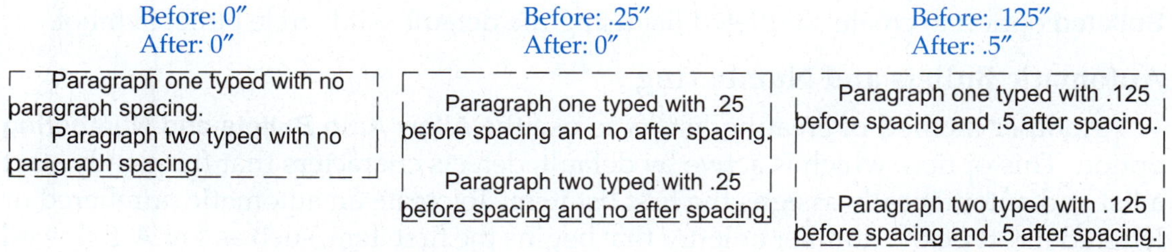

line spacing: The vertical distance from the bottom of one line of text to the bottom of the next line.

The options in the **Paragraph Line Spacing** area adjust *line spacing*. Default line spacing for single lines of text is equal to 1.66 times the text height. To adjust the line spacing, pick the **Paragraph Line Spacing** check box. Select the **Multiple** option from the **Line Spacing:** drop-down list to enter a multiple of the default line spacing in the **At:** text box. For example, to double-space lines, enter a value of 2x.

To force the line spacing to be the same for all lines of text, select the **Exactly** option from the **Line Spacing:** drop-down list and enter a value in the **At:** text box. If you enter an exact line spacing that is less than the text height, lines of text stack on top of each other. To add spaces between lines automatically based on the height of the characters in the line, select the **At least** option from the **Line Spacing:** drop-down list and enter a value in the **At:** text box. The result is an equal spacing between lines, even if the text has different heights.

Ribbon
Text Editor
> Paragraph
Line Spacing

You can also set line spacing without opening the **Paragraph** dialog box using the **Line Spacing** flyout of the ribbon. Select an available spacing, select the **More...** option to display the **Paragraph** dialog box, or select the **Clear Line Space** option, which restores the default line spacing.

NOTE

Ribbon
Text Editor
> Paragraph
> Combine Paragraphs

Combine multiple selected paragraphs to form a single paragraph using the **Combine Paragraphs** option.

Exercise 9-5

www.g-wlearning.com/CAD/

Complete the exercise on the companion website.

Lists

Drawings often include lists to organize information. Lists provide a way to arrange related items in a logical order and help make lines of text more readable. General notes are usually in list format. Mtext includes systems for automating the process of creating and editing numbered, bulleted, and alphabetical lists. Lists can also contain sublevel items designated with double numbers, letters, or bullets.

You can create lists as you enter text or apply list formatting to existing text. AutoCAD detects where you press [Enter] to start a new line of text and lists the lines in sequence. When you create a list in this manner, a tab automatically occurs after the number, letter, or symbol preceding the text. Set tabs and indents to adjust spacing and appearance.

Ribbon
Text Editor
> Paragraph
Bullets and Numbering

List tools are available from the **Bullets and Numbering** flyout on the **Paragraph** panel of the **Text Editor** ribbon tab or the **Bullets and Lists** cascading menu of the shortcut menu. The **Allow Bullets and Lists** option is active by default and is required to create a list. Uncheck **Allow Bullets and Lists** to convert any list items in the text object to plain text characters and disable bullet and list options.

Select an option from the **Lettered** cascading menu to create an alphabetical list. Select the **Uppercase** option to use uppercase lettering, or select **Lowercase** to use lowercase lettering. Select the **Numbered** option to create a numbered list. Select the **Bulleted** option to create a bulleted list using the default solid circle bullet symbol.

Automatic Bullets and Numbering

Another method of creating lists is to use the **Allow Auto Bullets and Numbering** option. This option, which is active by default, detects characters that frequently start a list and automatically assigns the first list item. To create an automatic numbered or lettered list, type the number or letter that begins the first item, such as 1 or A, followed by a period [.], colon [:], parenthesis [)], bracket []], angle bracket [<], or curly bracket [}]. Then press the space bar or [Tab] and type the line of text. When you press [Enter] to start a new line of text, the new line has the same formatting as the previous line, and the next consecutive number or letter appears. To end the list, press [Enter] twice. Figure 9-32 shows an example of a numbered list.

When creating an automatic bulleted list, use typical keyboard characters, such as a hyphen [-], tilde [~], angle bracket [>], or asterisk [*], at the beginning of a line. Another option is to insert a symbol at the beginning of a line. Then, press the space bar or [Tab] and type the line of text. When you press [Enter] to start a new line of text, the new line uses the same formatting bullet symbol as the previous line. To end the list, press [Enter] twice. See Figure 9-33.

Additional List Options

Additional options are available for controlling lists. Select the **Off** option to remove list characters or bulleting from selected text. Select the **Start** option to renumber or re-letter selected items to create a new list. The numbering or lettering restarts from the beginning, using 1 or A, for example. Select the **Continue** option to add selected items to a list that exists above the selected lines of text. The number of the first selected line of text continues from the previous list. Items below the first selected line are also renumbered.

NOTE

The automatic list icon appears as soon as AutoCAD initiates the list. Select the icon to access a shortcut menu of options, including **Start** and **Continue**. Select **Remove Bullets or Numbering** to remove the list. You can also press [Backspace] immediately after the first list item forms to remove the list. Select **Stop Auto Bullets and Numbering** to disable the **Allow Auto Bullets and Numbering** option.

Matching and Removing Formatting

Ribbon
Text Editor
> Formatting
Match Text Formatting

The **Match Text Formatting** option allows you to match, or "paint," character and paragraph formatting from one selection in the text editor to another. Select the text with the format you want to match, select the **Match Text Formatting** option, and then use the text cursor with the paintbrush icon to select the text to receive the formatting.

FRAMING NOTES:
1. ALL FRAMING LUMBER TO BE DFL #2 OR BETTER.
2. ALL HEATED WALLS @ HEATED LIVING AREA TO BE 2 X 6 @ 16" OC.
FRAME ALL EXTERIOR NON-BEARING WALLS W/2 X 6 STUDS @ 24" OC.
3. USE 2 X 6 NAILER AT THE BOTTOM OF ALL 2-2 X 12 OR 4 X HEADERS @ EXTERIOR WALLS, BACK HEADER W/2" RIGID INSULATION.
4. BLOCK ALL WALLS OVER 10'-0" HIGH AT MID HEIGHT.

Figure 9-32. Framing notes arranged in a numbered list, using the traditional architectural Stylus BT font.

- An elevation of the beam with end views or sections
- Complete locational dimensions for holes, plates, and angles
- Length dimensions

Bulleted List with Bullet Symbols

~ Connection specifications
~ Cutouts
~ Miscellaneous notes for the fabricator

Bulleted List with Tilde Characters

Figure 9-33. In addition to the standard solid circle bullet symbol, you can use other keyboard characters or symbols to create a bulleted list.

Ribbon
Text Editor
> Formatting

Clear

Use the **Clear** flyout of the ribbon or the **Remove Formatting** cascading submenu of the shortcut menu to remove formatting from selected text. Select the **Remove Character Formatting** option to remove character formatting such as bold, italic, or underline. Select the **Remove Paragraph Formatting** option to remove paragraph formatting, including lists. Pick the **Remove All Formatting** option to remove all character and paragraph formatting.

Exercise 9-6

www.g-wlearning.com/CAD/

Complete the exercise on the companion website.

Columns

Sometimes it is necessary to group text into multiple columns. A common example is dividing lengthy general notes into columns. See Figure 9-34. Mtext with columns is still a single mtext object. This eliminates the need to create multiple text objects to form separate columns of text. You can create columns as you enter text or apply column formatting to existing text.

dynamic columns: Columns calculated automatically by AutoCAD according to the amount of text and the specified height and width of the columns.

Earlier in this chapter, you were told to turn columns off before accessing the text editor. This approach is appropriate for drawing text without columns, especially as you learn to create mtext. However, mtext is set to form ***dynamic columns*** using the **Manual height** option by default. Column options are also available while the text editor is active, from the **Columns** flyout on the **Insert** panel of the **Text Editor** ribbon tab or from the **Columns** cascading menu in the shortcut menu.

Ribbon
Text Editor
> Insert

Columns

Figure 9-34.
An example of general construction notes created as a single mtext object and divided into three columns.

1.01 RELATED WORK

A. REQUIREMENTS: PROVIDE METAL FABRICATION IN ACCORDANCE WITH CONTRACT DOCUMENTS.

1.02 SUBMITTALS

A. SHOP DRAWINGS: INCLUDE PLANS AND ELEVATIONS AT NOT LESS THAN 1" = 1'-0" SCALE, AND INCLUDE DETAILS OF SECTIONS AND CONNECTIONS AT NOT LESS THAN 3" = 1'-0" SCALE. SHOW ANCHORAGE AND ACCESSORY ITEMS. SHOP DRAWINGS FOR ITEMS SPECIFIED BY DESIGN LOAD SHALL INCLUDE ENGINEERING CALCULATIONS AND SHALL BEAR SEAL AND SIGNATURE OF PROFESSIONAL ENGINEER REGISTERED IN STATE IN WHICH PROJECT IS LOCATED.

1.03 DELIVERY, STORAGE AND HANDLING

A. DELIVERY: DELIVER ITEMS, WHICH ARE TO BE BUILT INTO WORK OF OTHER SECTIONS IN TIME TO NOT DELAY WORK.

B. STORAGE: STORE IN UNOPENED CONTAINERS. STORE OFF GROUND AND UNDER COVER, PROTECTED FROM DAMAGE.

C. HANDLING: HANDLE IN MANNER TO PROTECT SURFACES. PREVENT DISTORTION OF, AND OTHER DAMAGE TO, FABRICATED PIECES.

2.01 MATERIALS

A. STRUCTURAL STEEL SHAPES: ASTM A36.

B. STEEL PLATES: ASTM A283 GRADE C FOR BENDING OR FORMING COLD.

C. STEEL TUBING: ASTM A500, GALVANIZED WHERE INDICATED.

D. STEEL BARS AND BAR SHAPES: ASTM A675, GRADE 65; OR ASTM A36.

E. COLD FINISHED STEEL BARS: ASTM A108, GRADE AS SELECTED BY FABRICATOR.

F. STAINLESS STEEL: ASTM A167, TYPE 302 OR 304; NUMBER 4 FINISH.

G. BOLTS AND NUTS: ASTM A307, GRADE A BOLTS.

H. MACHINE SCREWS: F FF-S-92, TYPE COMPATIBLE WITH METALS BEING FASTENED AND FINISHED TO MATCH METAL FINISH.

2.02 FABRICATION, GENERAL

A. INCLUDE SUPPLEMENTARY BARS NECESSARY TO COMPLETE METAL FABRICATION WORK THOUGH NOT DEFINITELY INDICATED.

B. USE MATERIALS OF SIZE AND THICKNESS INDICATED, OR IF NOT INDICATED, OF REQUIRED SIZE AND THICKNESS TO PRODUCE ADEQUATE STRENGTH AND DURABILITY IN FINISHED PRODUCT FOR INTENDED USE.

C. FORM EXPOSED WORK TRUE TO LINE AND LEVEL WITH ACCURATE ANGLES AND SURFACES AND STRAIGHT SHARP EDGES. EASE EXPOSED EDGES TO RADIUS OF APPROXIMATELY 1/32 INCH UNLESS OTHERWISE INDICATED. FORM BENT-METAL CORNERS TO SMALLEST RADIUS WITHOUT CAUSING GRAIN SEPARATION OF OTHERWISE IMPAIRING WORK.

3.01 INSPECTION

A. EXAMINATION: EXAMINE SUBSTRATES, ADJOINING CONSTRUCTION AND CONDITIONS UNDER WHICH WORK IS TO BE INSTALLED. DO NOT PROCEED WITH WORK UNTIL UNSATISFACTORY CONDITIONS HAVE BEEN CORRECTED.

3.02 PREPARATION

A. FIELD MEASUREMENTS: VERIFY DIMENSIONS BEFORE PROCEEDING WITH WORK. OBTAIN FIELD MEASUREMENTS FOR WORK REQUIRED TO BE ACCURATELY FITTED TO OTHER CONSTRUCTION. BE RESPONSIBLE FOR ACCURACY OF SUCH MEASUREMENTS AND PRECISE FITTING AND ASSEMBLY OF FINISHED WORK.

3.03 INSTALLATION

A. INSTALL WORK IN LOCATIONS INDICATED, PLUMB, LEVEL, AND IN LINE WITH ADJACENT MATERIALS WHERE REQUIRED. PROVIDE FASTENINGS INDICATED.

B. FILL SPACE BETWEEN SLEEVES AND POSTS OF RAILINGS WITH SETTING COMPOUND. SLOPE TOP SURFACE TO DRAIN AWAY FROM POSTS.

3.04 ADJUSTING AND CLEANING

A. TOUCH-UP MARRED AND ABRADED SURFACES WITH SPECIFIED PAINT AFTER FIELD ERECTION.

3.05 PROTECTION

A. PROTECT FINISHED SURFACES AGAINST DAMAGE DURING SUBSEQUENT CONSTRUCTION OPERATIONS. REMOVE PROTECTION AT TIME OF SUBSTANTIAL COMPLETION.

Dynamic Columns

To form dynamic columns, select an option from the **Dynamic Columns** cascading menu. The default **Manual height** option creates columns you can adjust individually for height to produce distinct groups of information. Figure 9-35 shows methods for adjusting dynamic columns using the **Manual height** option. Select the **Auto height** option to produce columns of equal height. Figure 9-36 shows methods for adjusting dynamic columns using the **Auto height** option. Increase column width or height to reduce the number of columns, or decrease column width or height to produce more columns.

Static Columns

To form ***static columns***, select the number of columns from the **Static Columns** cascading menu. The display of text in static columns depends on how much text is in the text editor and the height and width of the columns. However, the selected number of columns does not change even if text is not completely filled or extends past a column. Figure 9-37 shows methods for adjusting static columns. Increasing column width or height rearranges the text in the specified number of columns, but the number of static columns does not change based on column width or height.

static columns: Columns in which you divide the text into a specified number of columns.

NOTE

To create more than six static columns, pick the **More...** option to access the **Column Settings** dialog box and enter the number of columns in the **Column Number** text box.

Using the Column Settings Dialog Box

You can use the **Column Settings** dialog box as an alternative method to create columns. To create dynamic columns, select the **Dynamic Columns** radio button, and then select the **Auto Height** or **Manual Height** radio button. To create static columns, select the **Static Columns** radio button and enter the number of static columns in the **Column Number** text box.

Figure 9-35.
Controlling the length of dynamic columns individually, or manually, using the **Manual height** option. When you drag a text boundary edge, the column text reflows automatically from one column to the next.

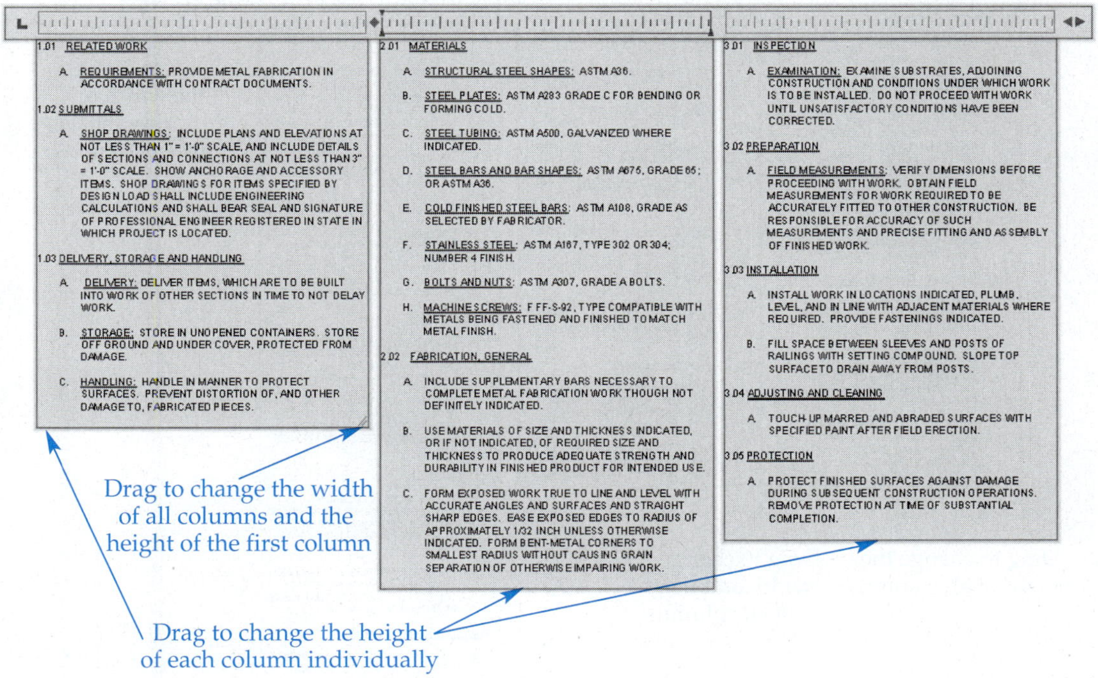

Figure 9-36.
Controlling columns using the dynamic column **Auto height** option.

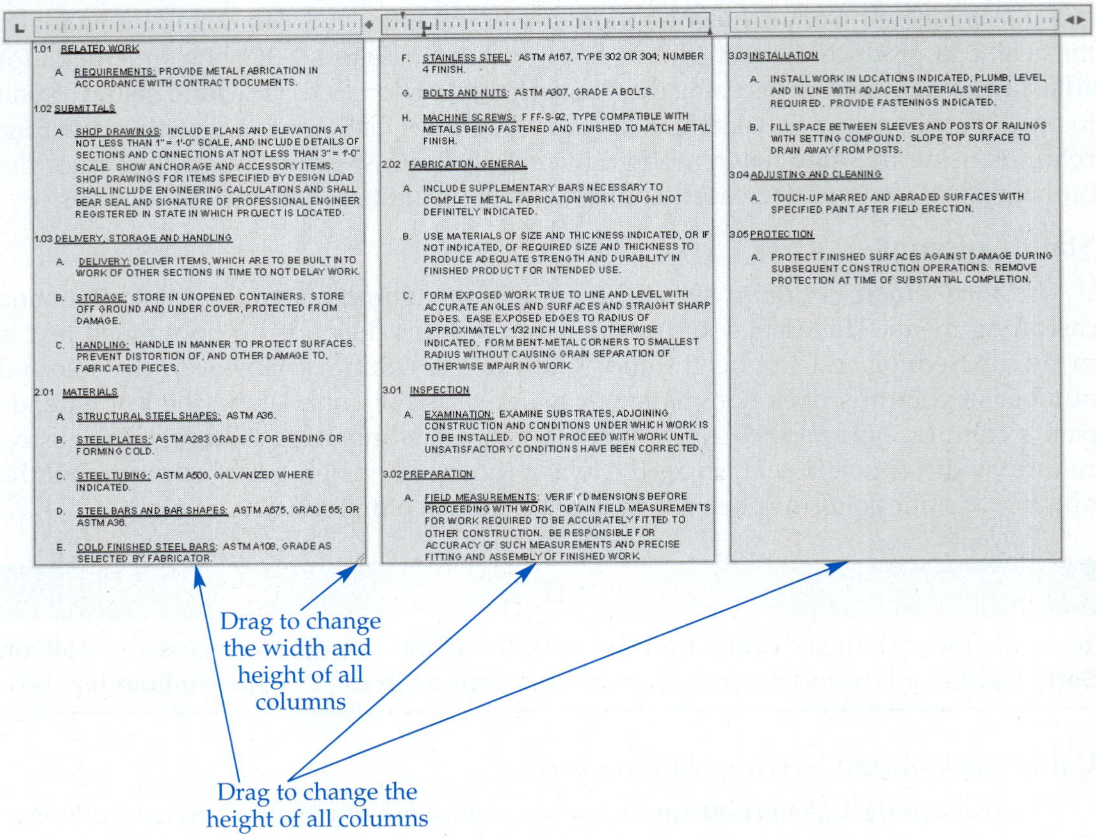

Figure 9-37.
Controlling static columns. You can also right-click on the ruler and select **Set Mtext Width...** to use the **Set Mtext Width** dialog box, or **Set Mtext Height...** to use the **Set Mtext Height** dialog box.

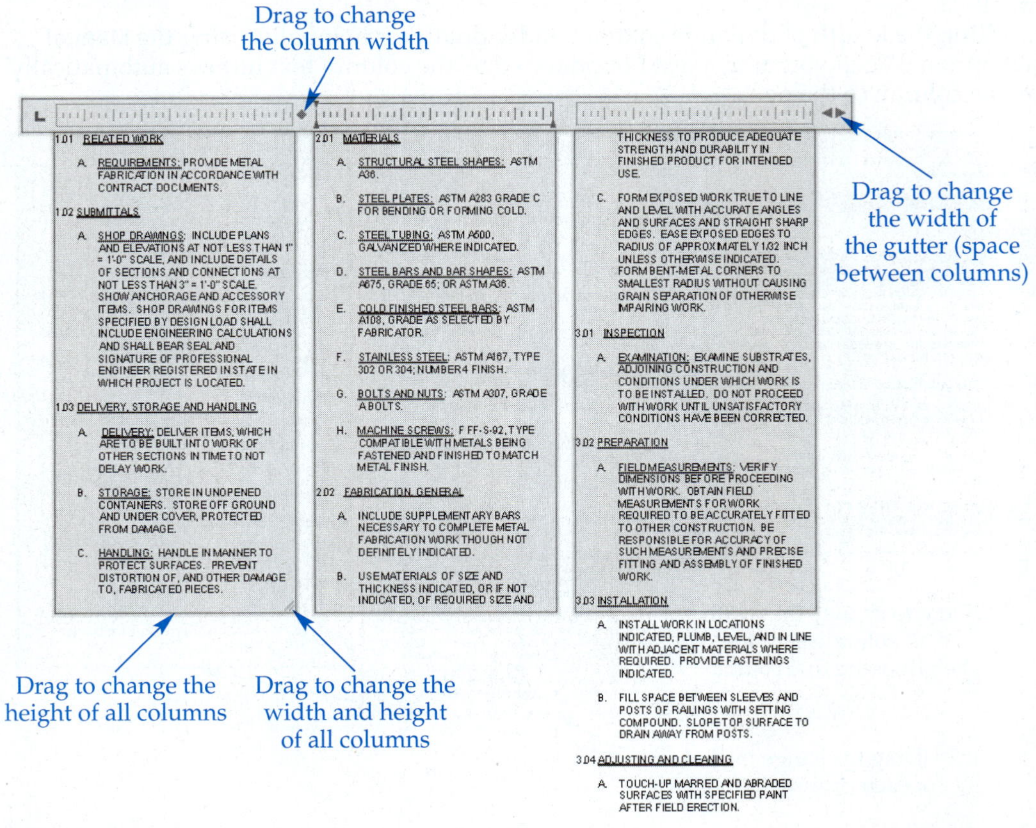

Additional controls become available depending on the selected **Column Type** radio buttons. The **Height** text box allows you to enter the height for all static or dynamic columns. The **Width** area allows you to set column width and the ***gutter***. Enter the column width in the **Column:** text box and the gutter width in the **Gutter:** text box. The **Total:** text box, available only with static columns, allows you to enter the total width of the text editor, which is the sum of the width of all columns and the gutter spacing between columns. To eliminate columns, pick the **No Columns** radio button.

gutter: The space between columns of text.

NOTE

If you apply a background mask to an mtext object with multiple columns, AutoCAD does not mask the gutter.

Controlling Column Breaks

Use the **Insert Column Break** option to specify the line of text at which a new column begins. To assign a column break, first form a dynamic or static column. Then place the cursor at a location in the text editor where a new column is to start, such as the start of a paragraph. Pick the **Insert Column Break** option to form the break. The text shifts to the next column at the location of the break. Continue applying column breaks as needed to separate sections of information.

NOTE

If you remove columns using the **No Columns** option, any column breaks added using the **Insert Column Break** option remain set. Backspace to remove column breaks.

Exercise 9-7

www.g-wlearning.com/CAD/

Complete the exercise on the companion website.

Importing Text

The **Import Text** option, also available from the shortcut menu, allows you to import text from an existing text file directly into the text editor. The text file can be either a standard ASCII text file (TXT) or a rich text format (RTF) file. The **Select File** dialog box appears when you access the **Import Text** option. Select the text file to import and pick the **Open** button. The text is inserted at the location of the text cursor. Imported text becomes part of the mtext object.

Ribbon
Text Editor
> Tools
> Import Text

PROFESSIONAL TIP

Importing text is useful if someone has already created specifications or notes in a program other than AutoCAD and you want to place the same notes in your drawings.

Template Development

www.g-wlearning.com/CAD/

Adding Text Styles

For detailed instructions on adding text styles to each drawing template, go to the companion website, navigate to this chapter in the **Contents** tab, and select **Adding Text Styles**.

Chapter Review

Answer the following questions using the information in this chapter.

1. Define *font*.
2. Which ASME standard contains guidelines for lettering?
3. What is text *composition*?
4. Determine the AutoCAD text height for text to be plotted .188″ high using a half (1:2) scale. Show your calculations.
5. Determine the AutoCAD text height for text to be plotted .188″ high using a scale of 1/4″ = 1′-0″. Show your calculations.
6. Explain the function of annotative text and give an example.
7. What is the relationship between the drawing scale and the annotation scale for annotative text?
8. Define *text style*.
9. Describe how to create a text style that has the name ROMANS-12_15, uses the romans.shx font, and has a fixed height of .12, a text width of 1.25, and an oblique angle of 15.
10. What are *big fonts*?
11. When setting text height in the **Text Style** dialog box, what value do you enter so that you can alter the text height when you create single-line text?
12. How would you specify text to display vertically on-screen?
13. What does a width factor of .5 do to text compared to the default width factor of 1?
14. How can you set a text style current without opening the **Text Style** dialog box?
15. Name the command that lets you create multiline text objects.
16. How can you change the width of the text editor when using the **MTEXT** command?
17. What character sequence do you type in the text editor to automatically create the fraction $\frac{1}{2}$ with a horizontal fraction bar?
18. How can you draw stacked fractions manually when using the **MTEXT** command?
19. What is the purpose of tracking?
20. What is the purpose of a background mask?
21. What is the difference between text boundary justification and paragraph alignment?
22. Explain how to create an automatic numbered list.
23. Describe the difference between dynamic columns and static columns.
24. From what two file formats can you import text?

Drawing Problems

Start AutoCAD if it is not already started. Start a new drawing for each problem using an appropriate template of your choice. The template should include layers and text styles for drawing the given objects. Add layers and text styles as needed. Draw all objects using appropriate layers. Use appropriate text styles, justification, and format. Follow the specific instructions for each problem. Use only drawing commands and techniques you have already learned. Use your own judgment and approximate dimensions when necessary.

▼ Basic

1. **General** Use the **MTEXT** command to type your name using a text style of your choosing and a text height of 1″. Save the drawing as P9-1. Print an 8.5″ × 11″ copy of the drawing extents using a 1:1 scale and a landscape orientation. Construct the sheet into a name tag.

2. **General** Use the **MTEXT** command to type the definitions of the following terms. Use a text style with the Arial font and a .12 text height. Save the drawing as P9-2.
   - annotation
   - text
   - composition
   - font
   - justify

3. **General** Use the **MTEXT** command to type the definitions of the following terms. Use a text style with the Romand font and a .12 text height. Save the drawing as P9-3.
   - scale factor
   - annotative text
   - annotation scale
   - text height
   - paper text height

4. **Mechanical** Use the **MTEXT** command to type the key notes shown. Use a text style with the Stylus BT font. The heading text height is .25 and the note text height is .125. Save the drawing as P9-4.

KEY NOTES

1. SLOPING SURFACE
2. DIAGONAL SUPPORT STRUT
3. VENT- PROVIDE NEW CANT FLASHING
4. BRICK CHIMNEY- REMOVE TO BELOW DECK SURFACE

5. **Architectural** Use the **MTEXT** command to type the general floor plan notes shown. Use a text style with the Century Gothic font. The heading text height is .25 and the note text height is .125. Save the drawing as P9-5.

GENERAL NOTES

1. ALL PENETRATIONS IN TOP OR BOTTOM PLATES FOR PLUMBING OR ELECTRICAL RUNS TO BE SEALED. SEE ELECTRICAL PLANS FOR ADDITIONAL SPECIFICATIONS.
2. PROVIDE 1/2" WATERPROOF GYPSUM BOARD AROUND ALL TUBS, SHOWERS, AND SPAS.
3. VENT DRYER AND ALL FANS TO OUTSIDE AIR THRU VENT WITH DAMPER.

▼ Intermediate

6. **Graphic Design** Draw the basic organizational chart shown below. Save the drawing as P9-6.

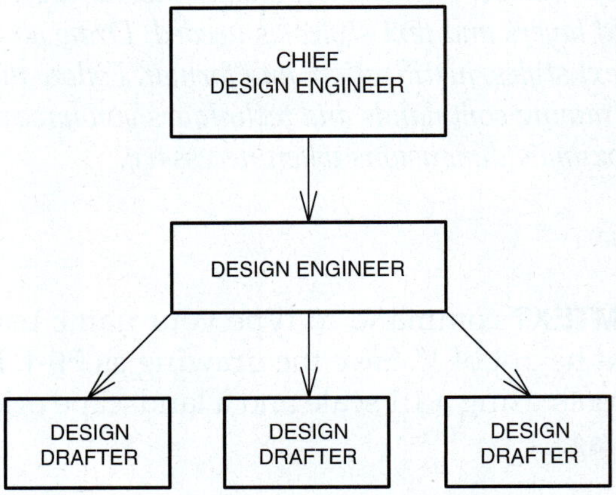

7. **Mechanical** Use the **MTEXT** command to type the general notes shown for a part drawing. Use a text style with the Romans font. The heading text height is 6 mm and the note text height is 3 mm. Save the drawing as P9-7.

NOTES:

1. DIMENSIONING AND TOLERANCING PER ASME Y14.5–2018.
2. REMOVE ALL BURRS AND SHARP EDGES.

CASTING NOTES UNLESS OTHERWISE SPECIFIED:
1. .31 WALL THICKNESS
2. R.12 FILLETS
3. R.06 CORNERS
4. 1.5°–3.0° DRAFT
5. TOLERANCES
   ± 1° ANGULAR
   ± .03 TWO–PLACE DIMENSIONS
6. PROVIDE .12 THK MACHINING STOCK ON ALL MACHINED SURFACES.

8. **Architectural** Use the **MTEXT** command to type the common framing notes shown. Use a text style with the Stylus BT font. The heading text height is .188 and the note text height is .125. After typing the text exactly as shown, edit the text with the following changes.

A. Change the \ in item 7 to 1/2.
B. Change the [in item 8 to 1.
C. Change the 1/2 in item 8 to 3/4.
D. Change the ^ in item 10 to a degree symbol.

Save the drawing as P9-8.

COMMON FRAMING NOTES:

1. ALL FRAMING LUMBER TO BE DFL #2 OR BETTER.
2. ALL HEATED WALLS @ HEATED LIVING AREAS TO BE 2 X 6 @ 24" OC.
3. ALL EXTERIOR HEADERS TO BE 2-2 X 12 UNLESS NOTED, W/ 2" RIGID INSULATION BACKING UNLESS NOTED.
4. ALL SHEAR PANELS TO BE 1/2" CDX PLY W/8d @ 4" OC @ EDGE, HDRS, ¢ BLOCKING AND 8d @ 8" OC @ FIELD UNLESS NOTED.
5. ALL METAL CONNECTORS TO BE SIMPSON CO. OR EQUAL.
6. ALL TRUSSES TO BE 24" OC. SUBMIT TRUSS CALCS TO BUILDING DEPT. PRIOR TO ERECTION.
7. PLYWOOD ROOF SHEATHING TO BE \ STD GRADE 32/16 PLY LAID PERP TO RAFTERS. NAIL W/8d @ 6" OC @ EDGES AND 12" OC @ FIELD.
8. PROVIDE [1/2" STD GRADE T¢G PLY FLOOR SHEATHING LAID PERP TO FLOOR JOISTS. NAIL W/10d @ 6" OC @ EDGES AND BLOCKING AND 12" OC @ FIELD.
9. BLOCK ALL WALLS OVER 10'-0" HIGH AT MID.
10. LET-IN BRACES TO BE 1 X 4 DIAG BRACES @ 45^ FOR ALL INTERIOR LOAD-BEARING WALLS.

9. **Architectural** Use the **MTEXT** command to type the caulking notes shown. Use a text style with the SansSerif font. The heading text height is .188 and the note text height is .125. Save the drawing as P9-9.

CAULKING NOTES:

CAULKING REQUIREMENTS BASED ON
OREGON RESIDENTIAL ENERGY CODE

1. SEAL THE EXTERIOR SHEATHING AT CORNERS, JOINTS, DOORS, WINDOWS, AND FOUNDATION SILL WITH SILICONE CAULK.
2. CAULK THE FOLLOWING OPENINGS W/ EXPANDED FOAM, BACKER RODS, OR SIMILAR:
   - ANY SPACE BETWEEN WINDOW AND DOOR FRAMES
   - BETWEEN ALL EXTERIOR WALL SOLE PLATES AND PLY SHEATHING
   - ON TOP OF RIM JOIST PRIOR TO PLYWOOD FLOOR APPLICATION
   - WALL SHEATHING TO TOP PLATE
   - JOINTS BETWEEN WALL AND FOUNDATION
   - JOINTS BETWEEN WALL AND ROOF
   - JOINTS BETWEEN WALL PANELS
   - AROUND OPENINGS

▼ Advanced

10. **Architectural** Using the **MTEXT** command, create the electrical notes shown below. Format the notes properly to make them easier to read.
    A. Use a text style with the Stylus BT font.
    B. Add the heading ELECTRICAL NOTES: on a separate line.
    C. Number each item separately.
    D. Use all capital letters and set the notes in two columns.
    Save the drawing as P9-10.

    All garage and exterior plugs and light fixtures to be on GFCI circuit. All kitchen plugs and light fixtures to be on GFCI circuit. Provide a separate circuit for microwave oven. Provide a separate circuit for personal computer. Verify all electrical locations with owner. Exterior spotlights to be on photoelectric cell with timer. All recessed lights in exterior ceilings to be insulation cover rated. Electrical outlet plate gaskets to be insulated on receptacle, switch, and any other boxes in exterior wall. Provide thermostatically controlled fan in attic with manual override; verify location with owner. All fans vent to outside air; all fan ducts to have automatic dampers. Hot water tanks to be insulated to R-11 minimum. Insulate all hot water lines to R-4 minimum; provide alternate bid to insulate all pipes for noise control. Provide 6 sq. ft. of vent for combustion air to outside air for fireplace connected directly to firebox; provide fully closable air inlet. Heating to be electric heat pump; provide bid for single unit near garage or for a unit each floor (in attic). Insulate all heating ducts in unheated areas to R-11; all HVAC ducts to be sealed at joints and corners.

11. **Architectural** Draw the general construction notes exactly as shown in **Figure 9-34** using columns. Save the drawing as P9-11.

Drawing Problems - Chapter 9

12. **Graphic Design** Draw the flowchart shown. Save the drawing as P9-12.

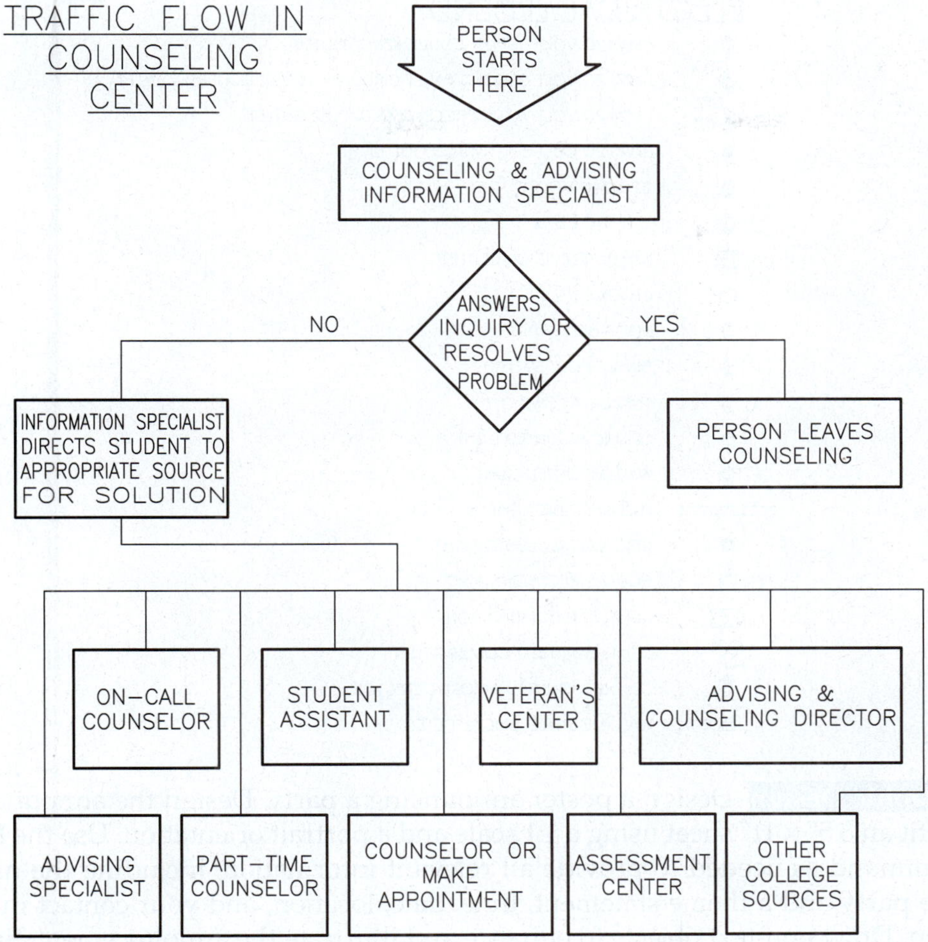

13. **Electronics** Draw the controller schematic shown. Save the drawing as P9-13.

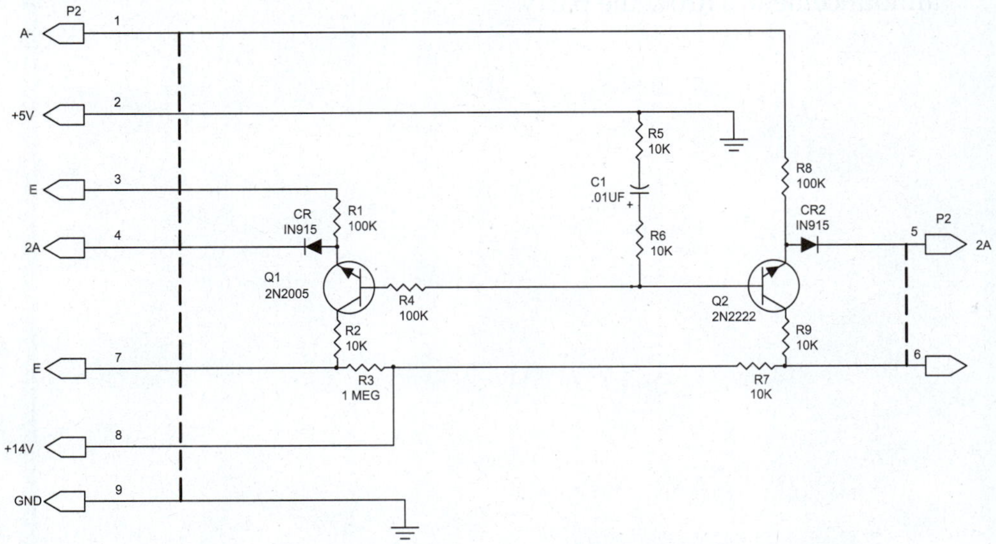

NOTES:

1. INTERPRET ELECTRICAL AND ELECTRONICS DIAGRAMS PER ANSI Y14.15.
2. UNLESS OTHERWISE SPECIFIED:

   RESISTANCE VALUES ARE IN OHMS.
   RESISTANCE TOLERANCE IS 5%.
   RESISTORS ARE 1/4 WATT.
   CAPACITANCE VALUES ARE IN MICROFARADS.
   CAPACITANCE TOLERANCE IS 10%.
   CAPACITOR VOLTAGE RATING IS 20V.
   INDUCTANCE VALUES ARE IN MICROHENRIES.

REFERENCE DESIGNATIONS
LAST USED
R9
C1
CR2
Q2

14. **Architectural** Draw the electrical legend shown. Save the drawing as P9-14.

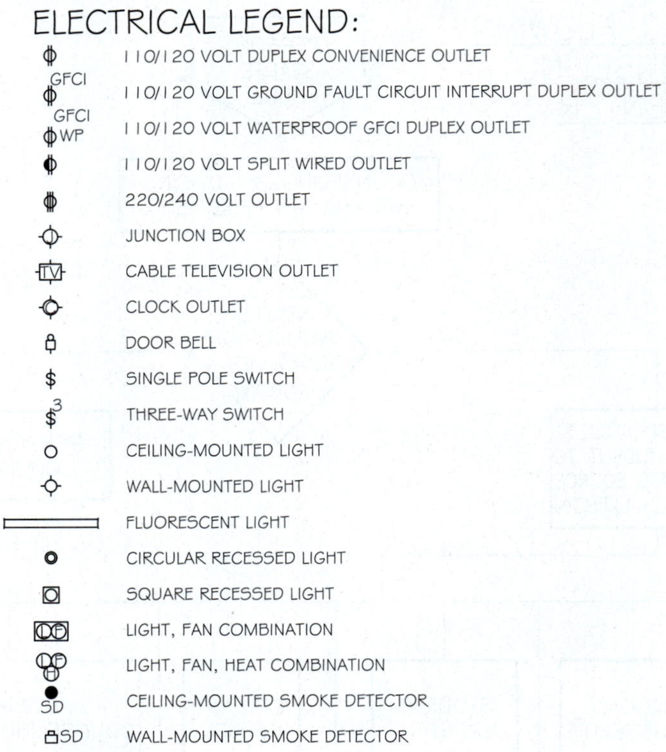

15. **Graphic Design** Design a poster announcing a party. Design the announcement to fit an 8.5″ × 11″ sheet using a 1:1 scale and a portrait orientation. Use the **MTEXT** command as needed to provide all relevant information, including the name of the party and a theme statement, time, date, location, and your contact information. Draw non-text objects to enhance and illustrate the announcement. Save the drawing as P9-15. Print 8.5″ × 11″ copies of the drawing extents, centered on the sheet, using a 1:1 scale and a portrait orientation. Distribute and post copies of the announcement. Throw the party.

16. Graphic Design Design a greeting card. Design the card to fit an 8.5″ × 11″ sheet using a 1:1 scale, a portrait orientation, and the format shown. Do not draw the dimensions, sheet edges, or fold lines. Use the **MTEXT** command as needed to provide all relevant information, including the occasion, name of the card recipient, and a greeting. Draw non-text objects to enhance and illustrate the card. Save the drawing as P9-16. Print an 8.5″ × 11″ copy of the drawing extents, centered on the sheet, using a 1:1 scale and a portrait orientation. Fold the card and give it to the recipient.

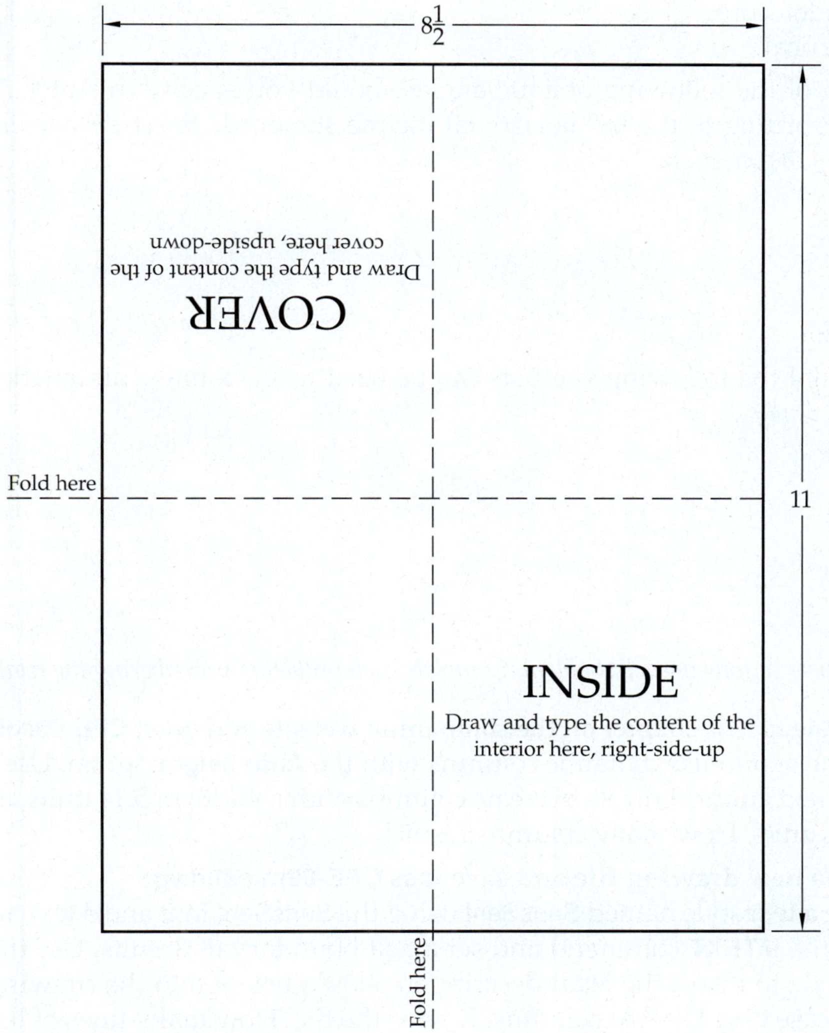

Drawing Problems - Chapter 9

AutoCAD Certified Professional Exam Practice

Answer the following questions using the information in this chapter.

1. Which of the following options are available for AutoCAD SHX (shape) fonts? *Select all that apply.*
   A. backwards
   B. bold
   C. italic
   D. underscore
   E. vertical
2. Which of the following oblique angles should you specify in AutoCAD to slant text according to the 68° horizontal incline standard? *Select the one item that best answers the question.*
   A. 0°
   B. –22°
   C. 22°
   D. 68°
   E. –68°
3. Which of the following symbols can be used to stack mtext automatically? *Select all that apply.*
   A. *
   B. ^
   C. /
   D. #
   E. ~

Follow the instructions in each problem. Complete each problem by answering the related question.

4. **Navigate to this chapter on the companion website and open CPE-09columns.dwg.** Edit the text to use dynamic columns with the **Auto height** option. Use the appropriate text editor controls to create columns with a width of 5.44 units and a height of 2.41 units. How many columns result?
5. **Open a new drawing file and save it as CPE-09mtext.dwg.**
   Create a text style named SansSerif using the SansSerif font and a text height of .12. Enter the **MTEXT** command and set a text boundary of 3 units. Use the SansSerif text style to insert the legal description shown below into the drawing using all uppercase text. Disable columns. Resave the file. How many lines of text does the legal description occupy with these settings?

   > Beginning at a point 20 feet north of center line of Unger Road, north for a length of 655.18′, thence S89°33′10″E for a length of 447.75′, thence south for a length of 475.53′, thence N84°53′42″W for a length of 224.77′, thence north for a length of 92.02′, thence S81°12′15″W for a length of 196.19′, thence south for a length of 256.56′, thence S86°55′3″W for a length of 30.23′ to the point of beginning.

CHAPTER 10

Single-Line Text and Additional Text Tools

Learning Objectives

After completing this chapter, you will be able to:

✓ Use the **TEXT** command to create single-line text.
✓ Insert and use fields.
✓ Check spelling.
✓ Edit existing text.
✓ Search for and replace text automatically.

This chapter describes how to use the **TEXT** command to place a single-line text object. The **TEXT** command is most useful for adding a single character, word, or line of text. Use the **MTEXT** command for all other text requirements. This chapter also explains methods for editing text and other valuable text tools, such as fields, the spell checker, and ways to find and replace text. You can apply most of the additional text tools described in this chapter to mtext or single-line text objects.

Single-Line Text

It is good practice to set the appropriate text style current before you access the **TEXT** command. Then activate the **TEXT** command to create a single-line text object. **Left** is the default justification. To use a different justification, select the **Justify** option at the Specify start point of text or [Justify/Style]: prompt before you pick the start point of the text. Figure 10-1 shows all of the justification options except the **Align** and **Fit** options, which are described in the next section. Select a justification option according to where and how you want text to form.

TEXT

Ribbon
Home
> Annotation
Annotate
> Text
Single Line Text
Type
TEXT
DT

Pick a start point to locate the text according to the specified justification, known as the ***justification point***. Next, if the current text style uses a height of 0, enter the text height. If the current text style is annotative, enter the paper text height. If the current text style is not annotative, enter the text height multiplied by the scale factor. The next prompt asks for the text rotation angle. The default value is 0, which draws horizontal text. Specify a rotation angle to pivot the text around the start point in a counterclockwise direction. See Figure 10-2.

justification point: The point from which text is justified according to the current justification option.

Figure 10-1.
Standard justification options available for drawing single-line text. The difference between some options is evident only when you use characters that extend below the baseline, such as the "y" in the word "Justify."

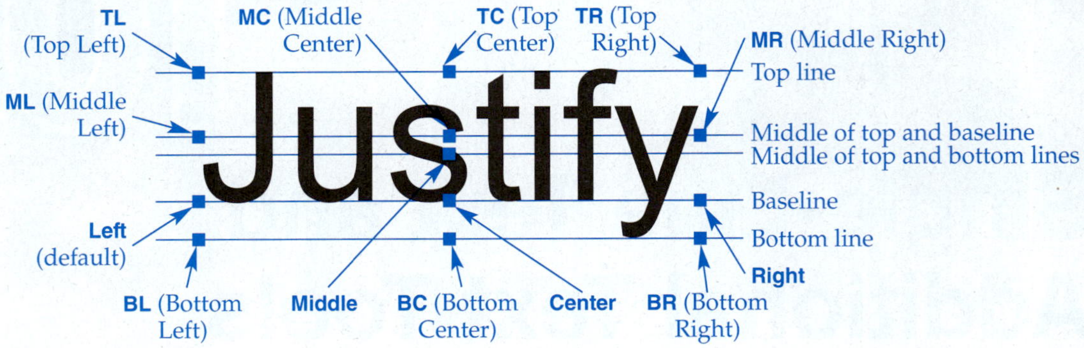

Figure 10-2.
Examples of rotating single-line text. The plus sign indicates the start point.

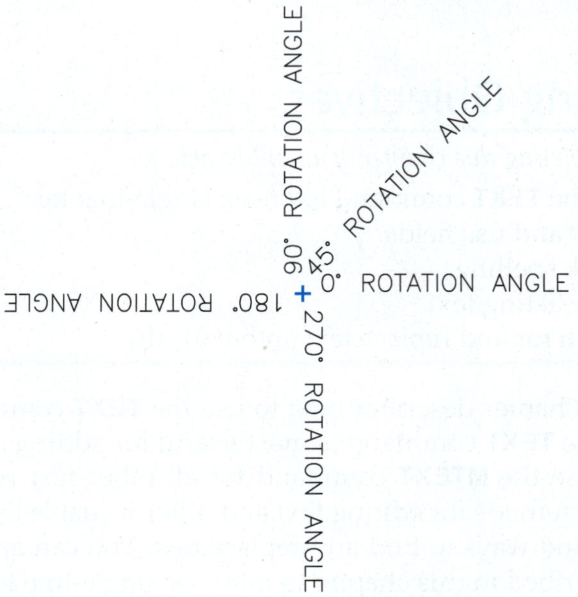

NOTE

AutoCAD applies the specified justification to single-line text until you change the justification setting. The text rotation is affected if you change the default angular measurement direction or zero direction.

Once you set the justification, height, and rotation angle, a text editor appears on-screen with a cursor equal in height to the text height at the start point. As you type, the text editor expands in size to display the characters. See **Figure 10-3**. Right-click to display a shortcut menu of text options, similar to those available for the **MTEXT** command.

Press [Enter] at the end of each line of text to move the cursor to a start point one line below the preceding line. Each new line of text is a new single-line text object, not

Figure 10-3.
Typing text using the **TEXT** command and default left justification.

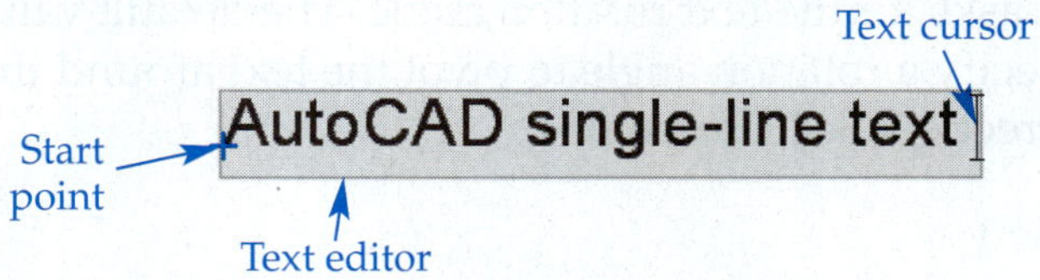

text in a grouped paragraph. When you are finished typing, press [Enter] twice to exit. Press [Esc] to cancel the **TEXT** command and remove incomplete lines of text.

NOTE

To change selected text to uppercase, right-click and select **UPPERCASE** from the **Change Case** cascading menu. Use the **lowercase** option to change all text to lowercase.

PROFESSIONAL TIP

Set the text style current before accessing the **TEXT** command. A **Style** option is available before you pick the text start point to set the text style, but it is difficult to use.

Align and Fit Justification

Align and **Fit** justification options are available from the **Justify** option of the **TEXT** command as an alternative to the previously described method of creating single-line text. The **Align** option requires you to select the start point and endpoint of the line of text. AutoCAD adjusts the text height to form between the start point and endpoint. The height varies according to the distance between the points and the number of characters. The **Fit** option requires you to select the start point and endpoint of the line of text and the text height. AutoCAD adjusts character width to fit between the specified points, while keeping text height constant. Figure 10-4 shows the effects of the **Align** and **Fit** options.

CAUTION

Avoid using the **Align** and **Fit** options of the **TEXT** command for standard drafting practices, because the text height or width is inconsistent from one line of text to another.

Exercise 10-1

www.g-wlearning.com/CAD/

Complete the exercise on the companion website.

Adding Symbols

Type a ***control code sequence*** to add a symbol with the **TEXT** command. For example, type %%C2.75 to create the note ∅2.75. In this example, %%C is the code used to add the diameter symbol. Figure 10-5 shows several symbol codes and the symbol the code creates.

control code sequence: A key sequence beginning with %% that defines symbols in text.

Figure 10-4.
Examples of using the **Align** and **Fit** justification options to create aligned and fit text. The plus signs indicate the start point and endpoint.

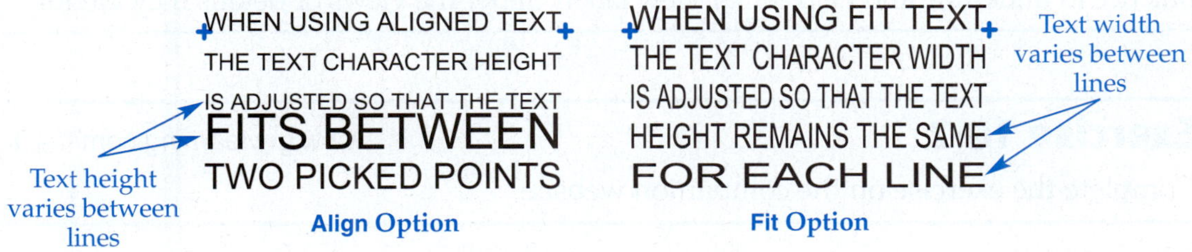

Figure 10-5.
Common control code sequences used to add symbols to single-line text.

Control Code or Unicode	Type of Symbol	Appearance	Control Code or Unicode	Type of Symbol	Appearance
%%d	Degrees	°	\U+2261	Identity	≡
%%p	Plus/Minus	±	\U+E200	Initial length	○→
%%c	Diameter	Ø	\U+E102	Monument line	ML
%%%	Percent	%	\U+2260	Not equal	≠
\U+2248	Almost equal	≈	\U+2126	Ohm	Ω
\U+2220	Angle	∠	\U+03A9	Omega	Ω
\U+E100	Boundary line	BL	\U+214A	Property line	⅊
\U+2104	Centerline	℄	\U+2082	Subscript 2	$_2$
\U+0394	Delta	Δ	\U+00B2	Squared	^{2}
\U+0278	Electrical phase	ɸ	\U+00B3	Cubed	^{3}
\U+E101	Flow line	FL			

Underscoring or Overscoring Text

Type %%U in front of text to underscore and %%O in front of text to overscore. For example, type %%UUNDERSCORING TEXT to create the note UNDERSCORING TEXT. Use both control code sequences to underscore and overscore text. For example, the control code sequence %%O%%ULINE OF TEXT produces LINE OF TEXT.

The %%U and %%O control codes are toggles that turn underscoring and overscoring on and off. Type %%U before a word or phrase to be underscored, and then type %%U after the word or phrase to turn underscoring off. Text following the second %%U appears without underscoring. For example, type %%UDETAIL A%%U HUB ASSEMBLY to create DETAIL A HUB ASSEMBLY.

PROFESSIONAL TIP

Underline labels such as SECTION A-A or DETAIL B instead of drawing a line or polyline object under the text. In addition, use the **Middle** or **Center** justification option. Follow this practice to underline and help center view labels under the views or details they identify.

Exercise 10-2

www.g-wlearning.com/CAD/

Complete the exercise on the companion website.

Fields

A ***field*** references drawing properties, AutoCAD functions, or object properties, and then displays the content in a text object. For example, insert the **Title** field below a drawing view to display the title of the drawing. You can update fields when changes occur to the reference data. For example, insert the **Date** field into a title block to update the field with the current date as necessary. Fields allow you to create "intelligent" text that uses existing drawing data and displays current information that changes throughout the course of a project.

field: A text object that can display a specific property value, setting, or characteristic.

Inserting Fields

To insert a field in an active mtext editor, access the **Field** option from the ribbon, also available by right-clicking and selecting **Insert Field...** or by pressing [Ctrl]+[F]. To insert a field in an active single-line text editor, right-click and select **Insert Field...** or press [Ctrl]+[F].

Ribbon
Insert
> Data
Text Editor
> Insert
Field
Type
FIELD

FIELD

The **Field** dialog box appears, allowing you to select a field. See Figure 10-6. The **Field** dialog box includes many preset fields. Filter the list of fields by selecting a category from the **Field category:** drop-down list. Fields related to the category appear in the **Field names:** list box. Pick a field from the list and then select a format from the list box to the right of the **Field names:** list box.

Pick the **OK** button to insert the field. The field assumes the current text style. By default, field text displays a light gray background. See Figure 10-7. This keeps you aware that the text is a field and that the value may change. You can deactivate the background in the **Fields** area of the **User Preferences** tab of the **Options** dialog box. See Figure 10-8.

Figure 10-6.
Select fields using the **Field** dialog box.

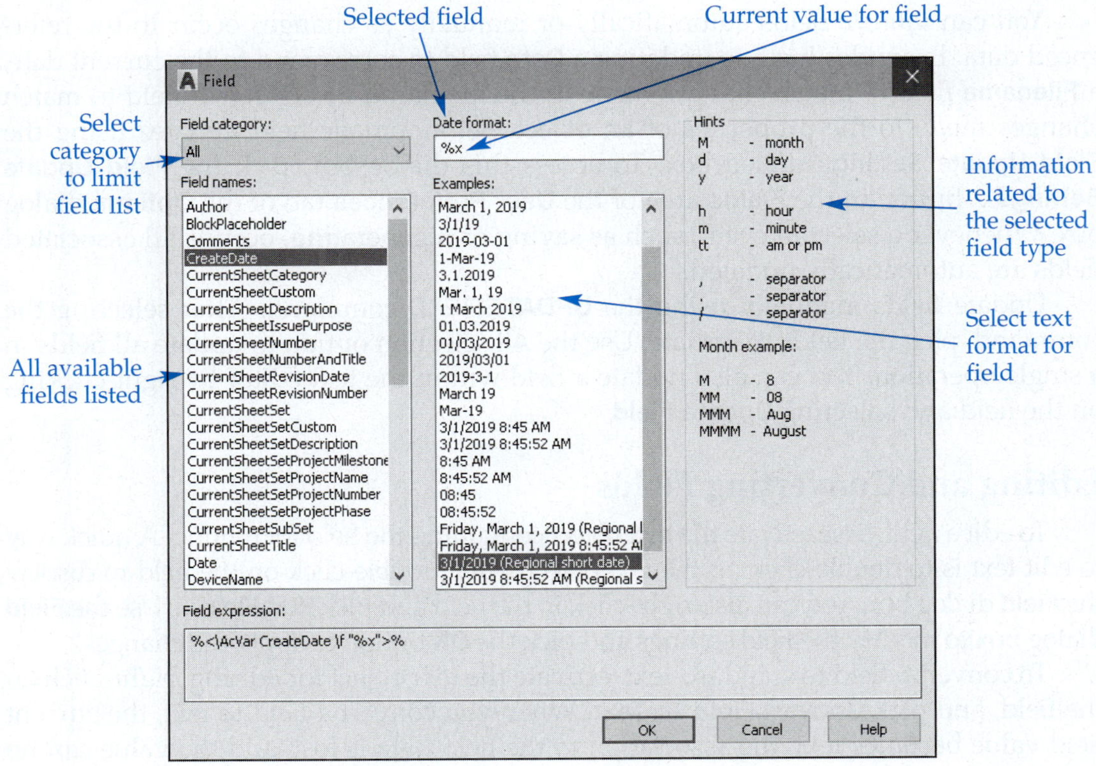

Figure 10-7.
A date field added to an mtext object. The gray background identifies the text as a field. The date field references the current date and time.

Figure 10-8.
Control the background display for fields in the **User Preferences** tab of the **Options** dialog box.

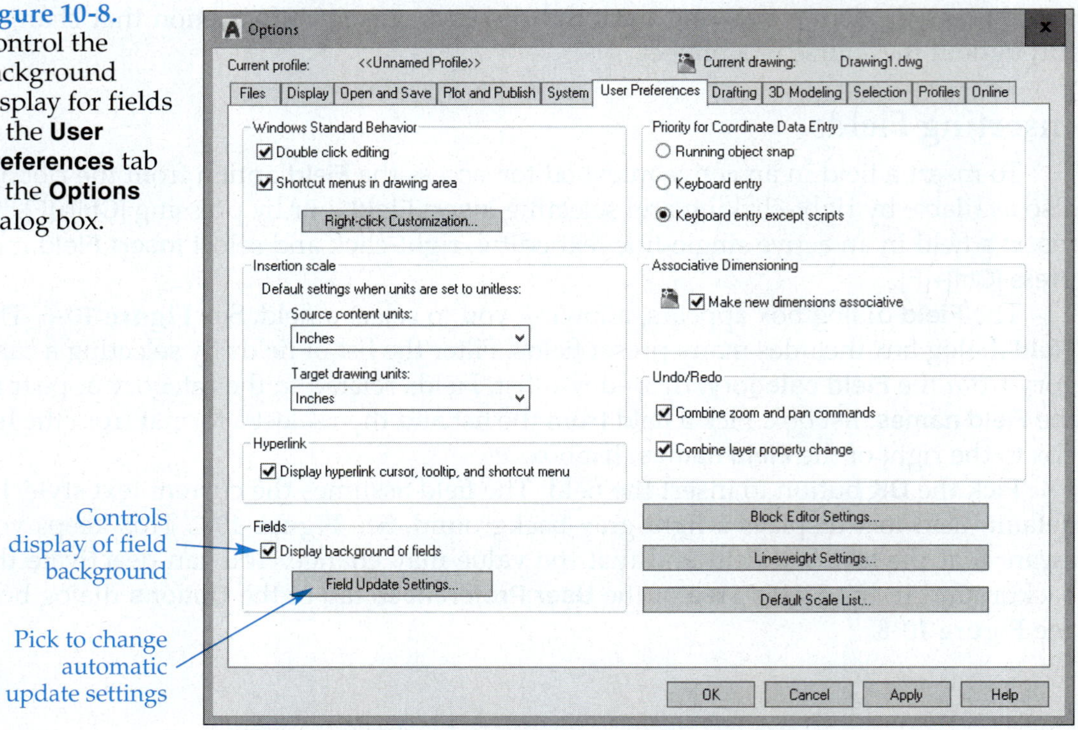

Updating Fields

update: The AutoCAD procedure for changing text in a field to reflect the current value.

You can *update* fields automatically or manually as changes occur to the referenced data. Examples include updating a **Date** field to correspond to the current date, a **Filename** field to match changes made to the file name, or an **Object** field to match changes made to the properties of an object. Set automatic field updates using the **Field Update Settings** dialog box. To access this dialog box, pick the **Field Update Settings...** button in the **Fields** area of the **User Preferences** tab of the **Options** dialog box. Whenever a selected event, such as saving or regenerating, occurs, all associated fields are automatically updated.

UPDATEFIELD
Ribbon
Insert
> Data

Update Fields
Type
UPDATEFIELD

Update fields manually using the **UPDATEFIELD** command. After selecting the command, pick the fields to update. Use the **All** selection option to update all fields in a single operation. You can also update a field within the text editor by right-clicking on the field and selecting **Update Field**.

Editing and Converting Fields

To edit a field, first activate the text object containing the field for editing. A quick way to edit text is to double-click on the text object. Then double-click on the field to display the **Field** dialog box. You can also right-click in the field and pick **Edit Field...**. Use the **Field** dialog box to modify the field settings and pick the **OK** button to apply the changes.

To convert a field to standard text, activate the text object for editing, right-click on the field, and pick **Convert Field To Text**. When you convert a field to text, the current field value becomes text, the association to the field data is lost, and the value can no longer be updated automatically.

NOTE

You can use fields with many AutoCAD functions and tools, including object properties, drawing properties, attributes, and sheet sets. You will learn specific field applications throughout this textbook.

Exercise 10-3

www.g-wlearning.com/CAD/

Complete the exercise on the companion website.

Checking Spelling

> **Ribbon**
> Text Editor
> > Spell Check
>
> 
>
> Spell Check

A quick way to check for correct spelling in text objects is to use the **Spell Check** tool available in text editors. The **Spell Check** tool is active by default. To toggle **Spell Check** on or off, select the **Check Spelling** option from the **Editor Settings** cascading menu available from the text editor shortcut menu. When using the mtext editor, a quick way to toggle **Spell Check** on or off is to select **Spell Check** from the **Spell Check** panel on the **Text Editor** ribbon tab.

A red dashed line appears under a word that the AutoCAD dictionary does not recognize. Right-click on the underlined word to display options for adjusting the spelling. See Figure 10-9. The first section at the top of the shortcut menu provides suggested replacements for the word. Pick a word to change the spelling in the text editor. If you do not see the correct suggestion, you may be able to find the correct spelling from the **More Suggestions** cascading menu.

Figure 10-9.
Checking spelling using the **Spell Check** tool in the mtext editor. The tool functions the same in the single-line text editor.

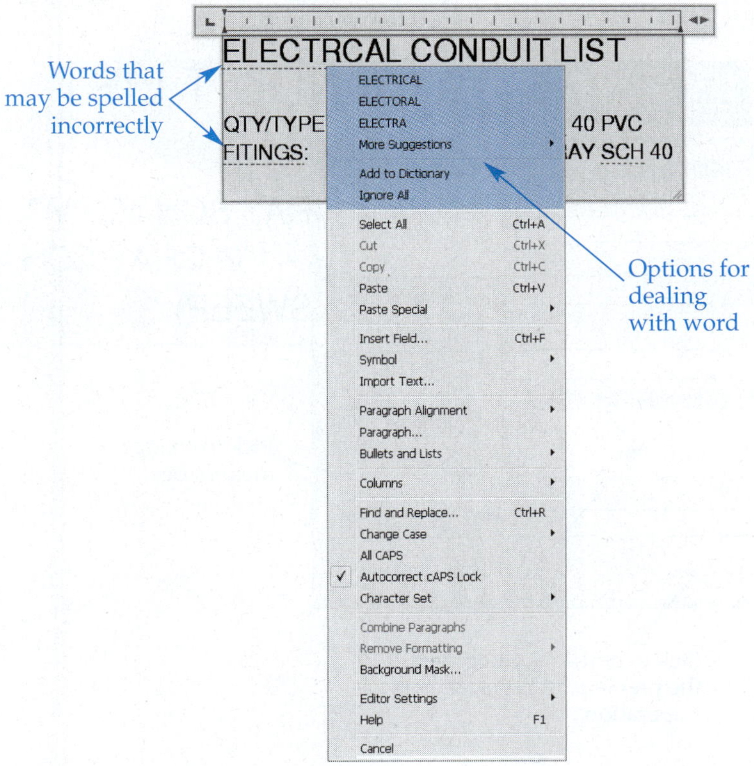

If you still cannot find a correct spelling from the suggestions, either the word is spelled correctly but not found in the dictionary, or it is spelled so incorrectly that AutoCAD cannot suggest a replacement. If the word is spelled correctly, pick the **Add to Dictionary** option to add the current word to the custom dictionary. You can add words with up to 63 characters. To use the current spelling without adding the word to the dictionary, pick the **Ignore All** option. Spell checking ignores all words that match the spelling in the active text editor and hides the underline. Add common drafting words and abbreviations to the dictionary, such as the abbreviation for the word SCHEDULE (SCH) in Figure 10-9, or ignore the words.

Using the SPELL Command

SPELL

Ribbon
Annotate > Text
Spell Check
Type
SPELL
SP

The **SPELL** command uses the **Check Spelling** dialog box, shown in Figure 10-10, to check the spelling of text objects without activating a text editor. To check spelling, first identify the portion of the drawing to check by selecting an option from the **Where to check:** drop-down list. Select the **Entire drawing** option to check the spelling of all text objects in the drawing file, including model space and layouts. Select **Current space/layout** to check spelling only of text objects in the active layout or in model space, if model space is active. You can also pick the **Select objects** button next to the **Where to check:** drop-down list to enter the drawing window and select specific text objects to check. You do not need to select the **Selected objects** option from the **Where to check:** drop-down list to check selected objects.

After you define what and where to check, pick the **Start** button to begin spell checking. The first word that may be misspelled becomes highlighted in the drawing window and is active in the **Check Spelling** dialog box. Pick the appropriate button to add, ignore, or change the spelling. Use the **Suggestions:** text box to type a different spelling that is not available from the list and then pick the **Change** or **Change All** button to insert the new word.

Figure 10-10.
The **Check Spelling** dialog box.

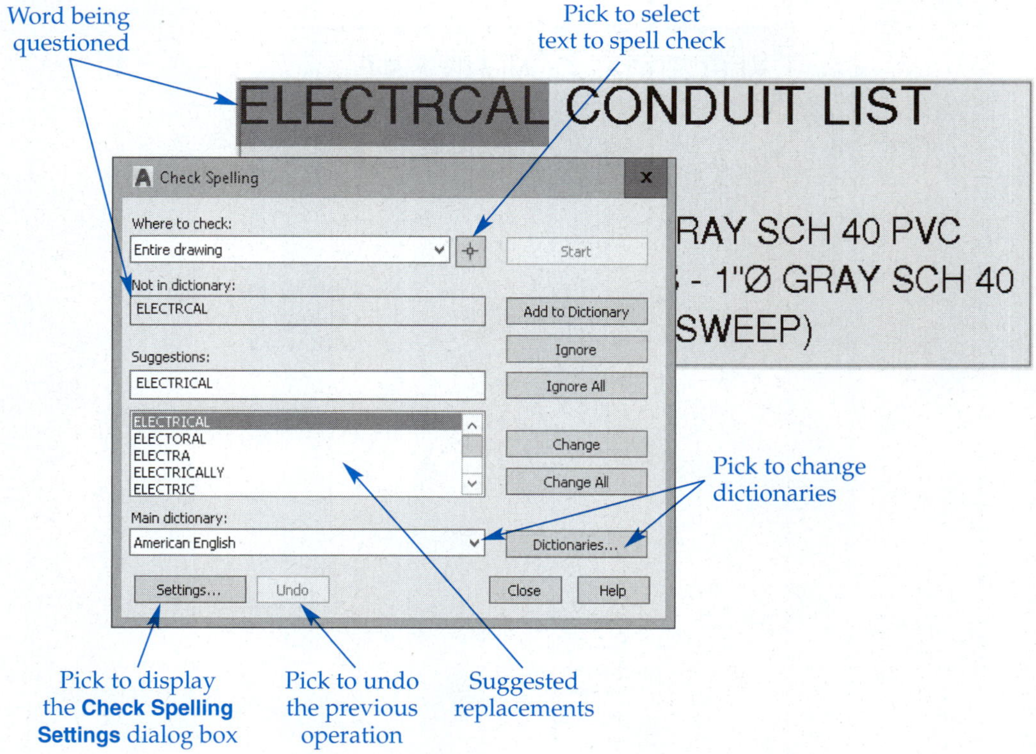

PROFESSIONAL TIP

Before you check spelling, you may want to adjust some of the spell-checking preferences provided in the **Check Spelling Settings** dialog box. Access this dialog box by picking the **Settings...** button in the **Check Spelling** dialog box, or select the **Check Spelling Settings...** option from the **Editor Settings** cascading menu available from the text editor shortcut menu. If the mtext editor is already open, pick the small arrow in the lower-right corner of the **Spell Check** panel on the **Text Editor** ribbon tab. The settings apply to spelling checked using the **Spell Check** tool in the active text editor and using the **Check Spelling** dialog box.

Changing Dictionaries

Ribbon
Text Editor
> Spell Check

Edit Dictionaries

To change the dictionary used to check spelling, pick the **Dictionaries...** button in the **Check Spelling** dialog box. You can also select the **Dictionaries...** option from the **Editor Settings** cascading menu available from the text editor shortcut menu to access the **Dictionaries** dialog box. See **Figure 10-11**.

Use the drop-down list in the **Main dictionary** area to select a language dictionary to use as the current main dictionary. You cannot add definitions to the main dictionary. Use the drop-down list in the **Custom dictionaries** area to select the active custom dictionary. The default custom dictionary is sample.cus. Type a word in the **Content:** text box to add or delete from the custom dictionary. For example, ASME is custom text used in engineering drafting. Pick the **Add** button to accept the custom words in the text box, or pick the **Delete** button to remove the words from the custom dictionary. Custom dictionary entries can use up to 63 characters.

To create and manage a custom dictionary, pick the **Manage custom dictionaries...** option from the drop-down list to access the **Manage Custom Dictionaries** dialog box. Pick the **New...** button to create a new custom dictionary by entering a new file name with a .cus extension. Use any standard text editor to edit the file. If you use a word processor such as Microsoft® Word, be sure to save the file as *text only*, with no special text formatting or printer codes. Add a custom dictionary by picking the **Add...** button, and pick the **Remove** button to delete a custom dictionary from the list. To add words from a different dictionary to the current custom dictionary, pick the **Import...** button from the **Custom dictionaries** area in the **Dictionaries** dialog box.

Figure 10-11.
The **Dictionaries** dialog box.

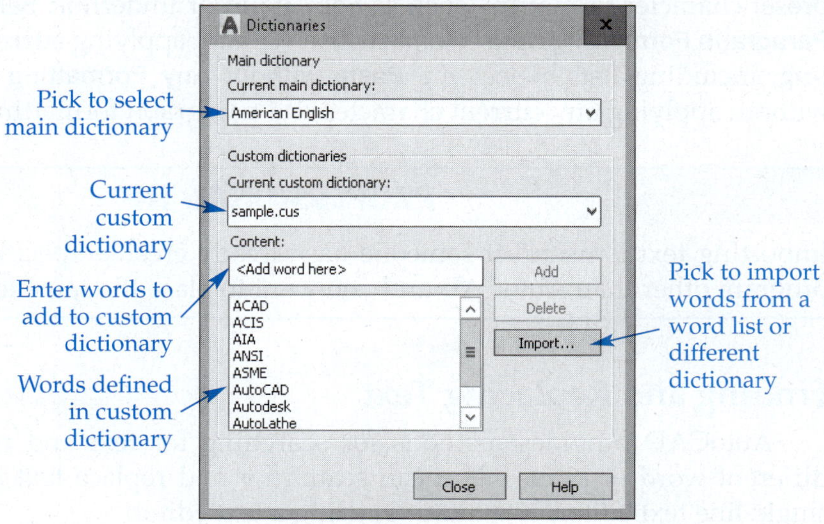

PROFESSIONAL TIP

Create custom dictionaries for various disciplines. For example, add common abbreviations and brand names for mechanical drawings to a mech.cus file. A separate file named arch.cus might contain common architectural abbreviations and brand names.

Revising Text

MTEDIT TEXTEDIT

Type
TEXTEDIT
ED

Type
MTEDIT

Use the **TEXTEDIT** command to revise selected mtext or single-line text, or the **MTEDIT** command to revise selected mtext. An easy way to issue the **MTEDIT** command is to double-click on an mtext object. Double-click on a text object to issue the **TEXTEDIT** command. Another technique to re-enter the text editor is to pick the text to modify and then right-click and select **Mtext Edit...** to revise mtext, or **Edit...** to modify single-line text.

The **TEXTEDIT** command includes a **Mode** option that is set to **Multiple** by default. As a result, the **TEXTEDIT** command remains active after you close the text editor, allowing you to select another annotation object to revise. When in **Multiple** mode, an **Undo** option is available to undo changes made during the previous revision. Change the **Mode** option to the **Single** setting to exit the **TEXTEDIT** command when you close the text editor.

Exercise 10-4

www.g-wlearning.com/CAD/

Complete the exercise on the companion website.

Cutting, Copying, and Pasting Text

Windows Clipboard functions allow you to cut, copy, and paste text from any text-based application, such as Microsoft® Word, into a text editor. You can also cut or copy and paste text from the text editor into other text-based applications. Access Clipboard functions from the shortcut menu when an mtext or single-line text editor is active. Text that you paste retains the original text properties.

AutoCAD provides additional paste options for pasting text into the active mtext text editor. Right-click and select an option from the **Paste Special** cascading menu. Select the **Paste without Character Formatting** option to paste text without applying preset character formatting such as bold, italic, or underline. Select the **Paste without Paragraph Formatting** option to paste text without applying current paragraph formatting, including lists. Select the **Paste without Any Formatting** option to paste text without applying any current character and paragraph formatting.

PROFESSIONAL TIP

Importing text is useful if someone has already created specifications or notes in a program other than AutoCAD and you want to place the same text in your drawings.

Finding and Replacing Text

AutoCAD provides methods for searching for text and replacing text with a different word or phrase. You can search for and replace text in an active mtext or single-line text editor, or without opening a text editor.

Using the Find and Replace Tool

Ribbon
Text Editor
> Tools
Find & Replace

Use the **Find and Replace** tool to find and replace text in the active mtext or single-line text editor. Right-click and select **Find and Replace...** to display the **Find and Replace** dialog box. The dialog box is also available in the mtext editor using the **Find & Replace** button on the **Tools** panel of the **Text Editor** ribbon tab. See Figure 10-12.

Type the text to search for in the **Find what:** text box. Type the text to substitute in the **Replace with:** text box. Then pick the **Find Next** button to highlight the next instance of the search text. You can then pick the **Replace** or **Replace All** button to replace just the highlighted text or all words that match the search criteria. Check boxes control the characters and words that are recognized.

Using the FIND Command

Ribbon
Annotate
> Text
Find...
Type
FIND

FIND

The **FIND** command uses the **Find and Replace** dialog box, shown in Figure 10-13, to find and replace text objects without activating a text editor. Another method for accessing the **Find and Replace** dialog box is to enter the text string to find in the **Find text** text box in the **Text** panel of the **Annotate** ribbon tab and press [Enter]. You can also activate the **FIND** command by right-clicking when no command is active and selecting **Find...**.

To find and replace text, first identify the portion of the drawing to search by selecting an option from the **Find where:** drop-down list. The options are similar to those in the **Where to check:** drop-down list in the **Check Spelling** dialog box. The **Find and Replace** dialog box is much like the dialog box of the same name that appears when you find and replace text within a text editor. However, the **FIND** command version allows you to display the search results in a table within the dialog box and provides more search options. Pick the **More Options** button to display check boxes used to control the characters and words recognized when you are finding and replacing text.

NOTE

The find and replace strings are saved with the drawing file for future use.

Figure 10-12.
Using the **Find and Replace** dialog box in the mtext editor. The tool functions the same in the single-line text editor.

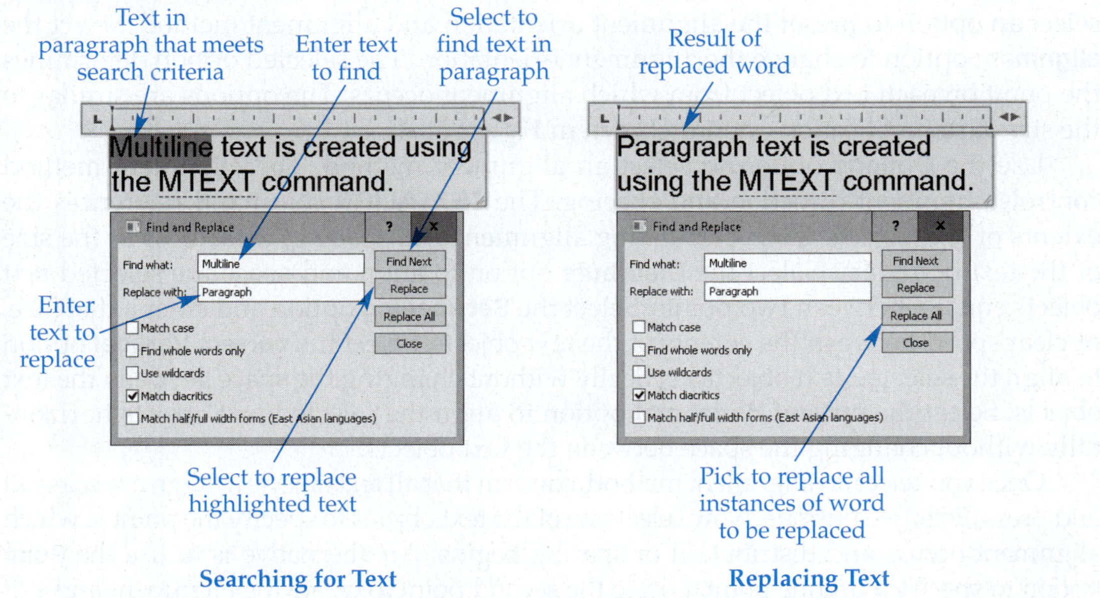

Figure 10-13.
Using the version of the **Find and Replace** dialog box that appears when you use the **FIND** command.

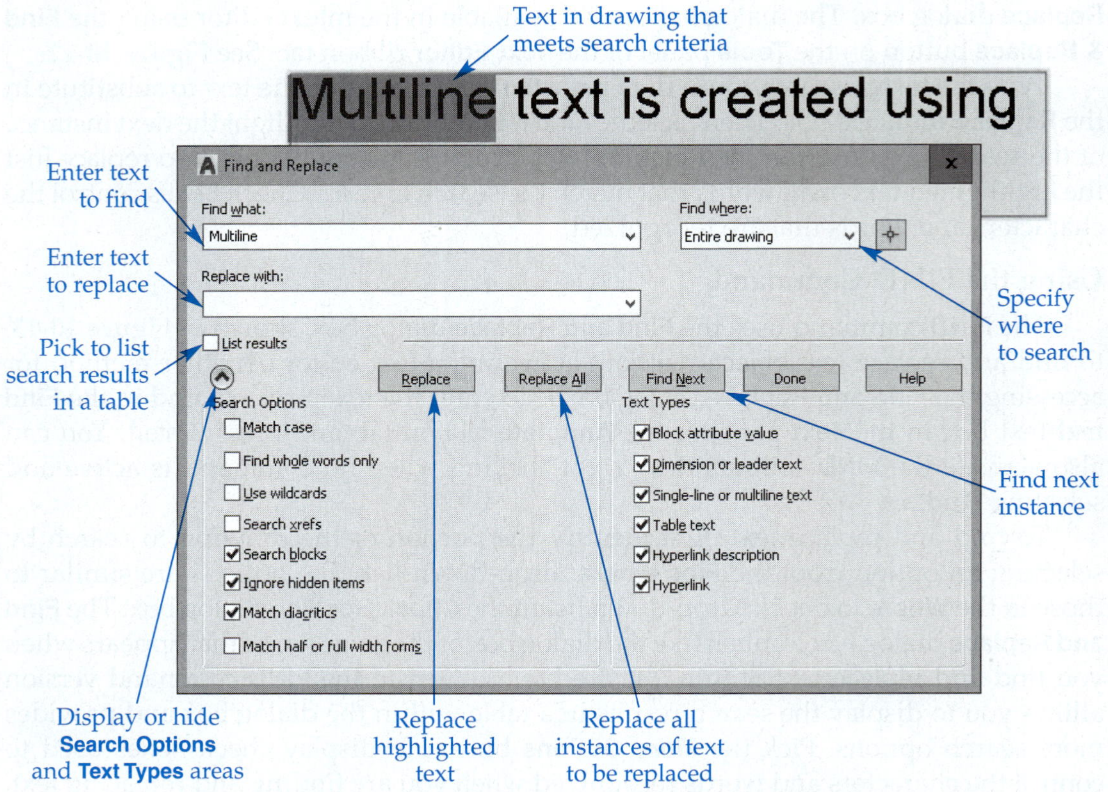

Aligning Text

One option to change the location and organization of text objects is to use editing commands, such as **MOVE**, as described in Chapter 12. An alternative is to use the **TEXTALIGN** command to align and space any combination of mtext, single-line text, and attribute objects. Attributes are described in Chapter 25. Aligning text objects is a common requirement when the drawing scale changes, or when text is misaligned or unequally spaced.

TEXTALIGN
Ribbon
Annotate > Text
Text Align
Type
TEXTALIGN
TA

Access the **TEXTALIGN** command and select the text objects to align and space, or select an option to preset the alignment orientation and alignment method. Select the **alignment** option to change the alignment orientation. The selected option determines the point on each text object from which alignment occurs. The options are similar to the standard justification options shown in Figure 10-1.

Use the **Options** option to select an alignment method. The alignment method controls alignment direction and spacing. The **TEXTALIGN** command references the extents of text objects when calculating alignment and spacing, regardless of the size of the text boundary. Select the **Distribute** option to align and space the selected text objects equally between two points. Select the **Set spacing** option and enter a distance, or clear space, between the extents of the text objects. Select the **current Vertical** option to align the selected text objects vertically without changing the space between the text objects. Select the **current Horizontal** option to align the selected text objects horizontally without changing the space between the text objects.

Once you select an alignment method, confirm that all text objects to align are selected and press [Enter] to continue. Now select one of the text objects to specify the point at which alignment occurs and distribution or spacing begins. An alternative is to use the **Point** option to specify a unique point. Locate the second point to create the alignment and exit the **TEXTALIGN** command. Figure 10-14 shows examples of using each alignment method.

Figure 10-14.
Examples of using the **TEXTALIGN** command to align text. The text object in color is selected as the object to align to. The plus signs indicate the first and second points used to define the alignment. Locate the first point by selecting the object to align to or use the **Point** option. Pick the second point at an appropriate location to create the alignment. Use drawing aids as needed for accurate placement.

Alignment Method	Orientation	Original Arrangement	Result
Distribute	**BL** (Bottom Left)	1 CONSULTANTS Structural Engineer 4570 Structure Road Building, IL 60477-6243 PHONE: 800.323.0440 FAX: 888.409.3900 2	CONSULTANTS Structural Engineer 4570 Structure Road Building, IL 60477-6243 PHONE: 800.323.0440 FAX: 888.409.3900
Set spacing	**BC** (Bottom Center)	ENGINEERING DRAFTING & DESIGN, INC. 1 2 *Drafting, design, and training for all disciplines.* *Integrity - Quality - Style*	ENGINEERING DRAFTING & DESIGN, INC. *Drafting, design, and training for all disciplines.* *Integrity - Quality - Style*
current Vertical	**ML** (Middle Left)	1 EXISTING PARKING AREA EXISTING SIDEWALK 2 EXISTING GARDEN	EXISTING PARKING AREA EXISTING SIDEWALK EXISTING GARDEN
current Horizontal	**TC** (Top Center)	W1 = 15.5"±1" 1 2 W2 = 14"±1"	W1 = 15.5"±1" W2 = 14"±1"

NOTE

The second point you locate with the **TEXTALIGN** command only determines alignment direction, except when using the **Distribute** option, in which case the second point controls alignment direction and spacing.

Scaling Text

The **SCALETEXT** command is one option for changing the height of text objects. Access the **SCALETEXT** command and select the text objects to be scaled. Next, specify the base point from which the increase or decrease in size occurs. The base point options are similar to the standard justification options shown in Figure 10-1. The **Existing** option scales text objects using their existing justification setting as the base point.

Ribbon
Annotate
> Text
Scale
Type
SCALETEXT

SCALETEXT

After you specify the base point, AutoCAD prompts for the scaling type. The default **Specify new model height** option allows you to type a new value for the text height of non-annotative objects. If the selected text is annotative, AutoCAD ignores the value you enter. Use the **Paper height** option to type a new paper text height for the annotative objects. If the selected text is non-annotative, AutoCAD ignores the paper height value you enter.

Use the **Match object** option to match the height of the text to the height of a different selected text object. Use the **Scale factor** option to scale text objects relative to their current heights. For example, a scale factor of 2 scales all selected text objects to twice their current size.

CAUTION

Only use the **SCALETEXT** command to scale non-annotative text.

JUSTIFYTEXT

Ribbon
Annotate
> Text
Justify
Type
JUSTIFYTEXT

Changing Justification

Use the **JUSTIFYTEXT** command to change the justification point without moving the text. Pick the text for which you want to change the justification and then select a new justification option.

Exercise 10-5

www.g-wlearning.com/CAD/

Complete the exercise on the companion website.

Supplemental Material

www.g-wlearning.com/CAD/

Isometric Text

For information about constructing text for isometric views, go to the companion website, navigate to this chapter in the **Contents** tab, and select **Isometric Text**.

Express Tools

www.g-wlearning.com/CAD/

Text Express Tools

The **Text** panel of the **Express Tools** ribbon tab includes additional text commands. For information about the most useful text express tools, go to the companion website, navigate to this chapter in the **Contents** tab, and select **Text Express Tools**.

Chapter Review

Answer the following questions using the information in this chapter.

1. List two ways to access the **TEXT** command.
2. Write the control code sequence required to draw the following symbols using the **TEXT** command.
   A. 30°
   B. 1.375±.005
   C. ∅24
   D. <u>NOT FOR CONSTRUCTION</u>
3. Explain the function and purpose of fields.
4. What is different about the default on-screen display of fields compared to that of text?
5. How can you access the **Field Update Settings** dialog box?
6. Explain how to convert a field to text.
7. What is a quick way to check your spelling within a current text editor?
8. How do you change the current word in the **Check Spelling** dialog box if the correct spelling is not displayed in the **Suggestions:** text box or the list of suggested words?
9. Identify two ways to access the **Check Spelling** dialog box.
10. How do you change the main dictionary for use in the **Check Spelling** dialog box?
11. Why might you want to create more than one custom dictionary?
12. What happens if you double-click on multiline text?
13. Describe how to find and replace text when an mtext or single-line text editor is open.
14. Name the command that allows you to find text and replace it with different text without activating a text editor.
15. When using the **SCALETEXT** command, which base point option would you select to scale a text object using the current justification setting of the text as the base point?

Drawing Problems

Start AutoCAD if it is not already started. Start a new drawing using an appropriate template of your choice. The template should include layers and text styles for drawing the given objects. Add layers and text styles as needed. Draw all objects using appropriate layers. Use appropriate text styles, justification, and format. Follow the specific instructions for each problem. Use only drawing commands and techniques you have already learned. Use your own judgment and approximate dimensions when necessary.

▼ Basic

1. General Create text styles according to the list shown. Use the **TEXT** command and the appropriate text style to type the text shown. Use a .25-unit text height and 0° rotation angle. Save the drawing as P10-1.

ARIAL - A VERY BASIC FONT USED FOR GENERAL-PURPOSE TEXT.

ROMANS – A FONT THAT CLOSELY DUPLICATES THE SINGLE–STROKE LETTERING THAT HAS BEEN THE STANDARD FOR DRAFTING.

ROMANC – A MULTISTROKE DECORATIVE FONT THAT IS GOOD FOR USE IN DRAWING TITLES.

ITALICC – AN ORNAMENTAL FONT THAT IS SLANTED TO THE RIGHT AND HAS THE SAME LETTER DESIGN AS THE COMPLEX FONT.

2. General Create text styles according to the list shown. Use the **TEXT** command and the appropriate text style to type the text shown. Use a .25-unit text height. Save the drawing as P10-2.

ARIAL–EXPAND THE WIDTH BY THREE.

MONOTXT–SLANT TO THE LEFT –30°.

ROMANS–SLANT TO THE RIGHT 30°.

ROMAND–BACKWARDS.

ITALICC–UNDERSCORED AND OVERSCORED.

ROMANS–USE 16d NAILS @ 10"OC.

ROMANT–ø32 (812.8).

ROMANC–VERTICAL

3. Electronics Save P7-7 as P10-3. If you have not yet completed Problem 7-7, do so now. In the P10-3 file, use a text style with the Arial font to add text and titles to the electrical switch schematics as shown in Problem 7-7.

4. Electronics Save P5-9 as P10-4. If you have not yet completed Problem 5-9, do so now. In the P10-4 file, use a text style with the Romans font to add text to the circuit diagram as shown in Problem 5-9.

5. Electronics Save P7-12 as P10-5. If you have not yet completed Problem 7-12, do so now. In the P10-5 file, use a text style with the Romans font to add text to the elementary diagram as shown in Problem 7-12.

▼ Intermediate

6. **General** Create text styles with a .375 height and the following fonts: Arial, BankGothic Lt BT, CityBlueprint, Stylus BT, Swis721 BdOul BT, Vineta BT, and Wingdings. Use each text style with the **TEXT** command to type the complete alphabet and numbers 0–10. In addition, type all symbols available on the keyboard and the diameter, degree, and plus/minus symbols. Save the drawing as P10-6.

7. **Architectural** Draw the interior finish schedule shown. Use a text style with the Stylus BT font to add the text. Save the drawing as P10-7. You will learn to create tables using the **TABLE** command in Chapter 21. The purpose of this problem is to practice using the **TEXT** command to place text objects in specific areas, using appropriate justification and format. In general, the **TABLE** command is more appropriate for drawing schedules.

INTERIOR FINISH SCHEDULE												
ROOM	FLOOR					WALLS				CEILING		
	VINYL	CARPET	TILE	HARDWOOD	CONCRETE	PAINT	PAPER	TEXTURE	SPRAY	SMOOTH	BROCADE	PAINT
ENTRY					•							
FOYER			•			•			•			•
KITCHEN			•				•			•		•
DINING				•		•			•		•	•
FAMILY		•				•			•		•	•
LIVING		•				•		•			•	•
MSTR. BATH			•			•				•		•
BATH #2			•			•			•	•		•
MSTR. BED		•				•		•			•	•
BED #2		•				•			•		•	•
BED #3		•				•			•		•	•
UTILITY	•					•			•	•		•

8. **Electronics** Draw the block diagram shown. Use a text style with the Romans font to add the text. Save the drawing as P10-8.

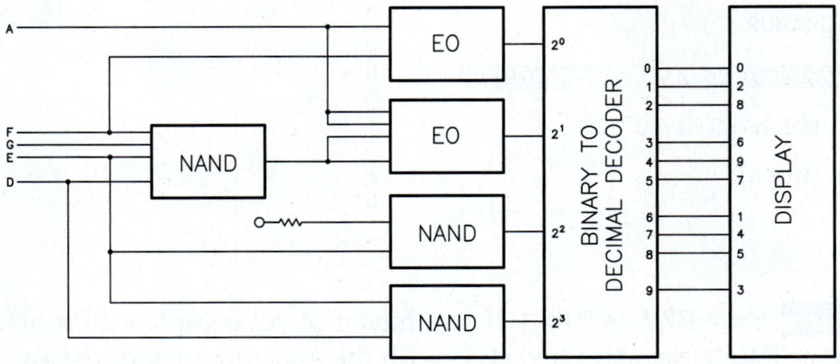

9. **Electronics** Draw the block diagram shown. Use a text style with the Romans font to add the text. Use polylines to create the arrowheads. Save the drawing as P10-9.

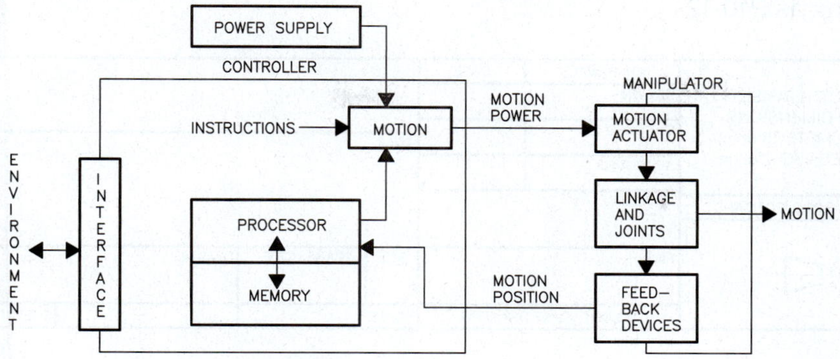

10. **Electronics** Draw the AND/OR schematic shown. Save the drawing as P10-10.

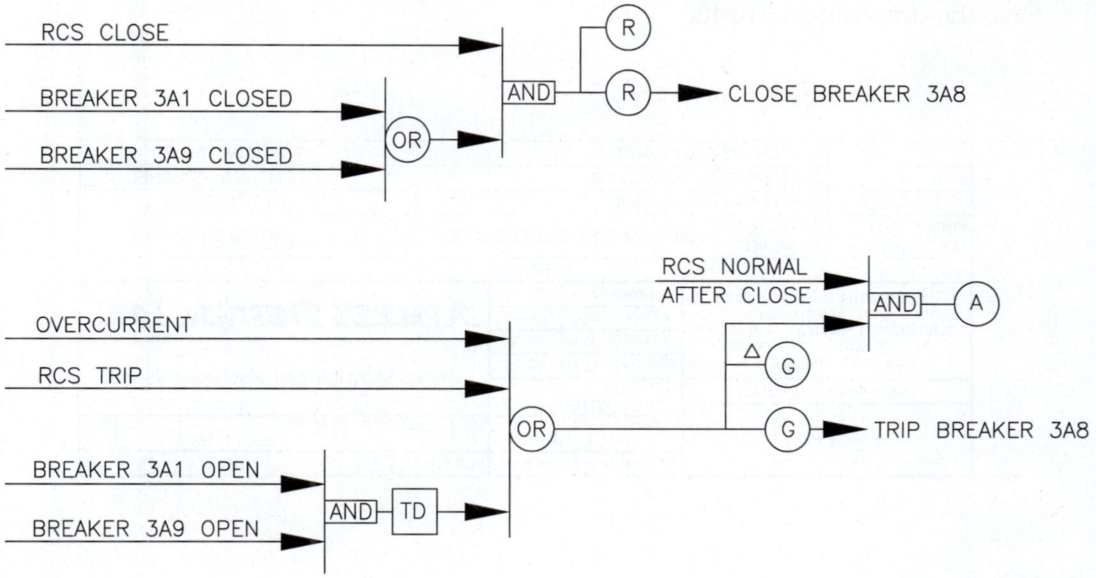

▼ Advanced

11. **General** Save EX2-5 as P10-11. If you have not yet completed Exercise 2-5, do so now. In the P10-11 file, use a text style with the Century Gothic font and the **MTEXT** command to type the text shown. Use a .25-unit text height. Type text for the titles, and insert fields to add the values. Use appropriate field options to create the list exactly as shown, except that the Author, Approved by, and Checked by values will be specific to your drawing properties. Next, change the Title, Subject, Keywords, Comments, and Revision drawing file properties to values according to this problem. Use a revision value of A. Using the **UPDATEFIELD** command, apply the **All** selection option to update the fields. Resave the drawing.

Filename: P10-11.dwg
Date: 03-01-2019
Title: EXERCISE 2-5
Subject: Exercise 2-5
Author: ABC
Keywords: Exercise 2-5
Comments: This is Chapter 2, exercise number five.
Approved by: ABC
Checked by: ABC
Revision level: 0
Web site: www.g-w.com

12. **Mechanical** Draw the metric-unit mechanical drafting sheet blocks shown. Enter your school or company name in the company block (upper-right cell). Save the drawing as P10-12.

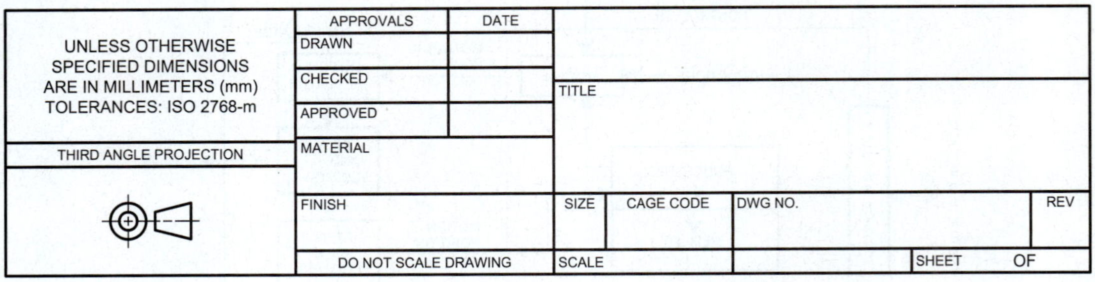

13. **Mechanical** Draw the mechanical drafting sheet blocks and parts list shown. Save the drawing as P10-13.

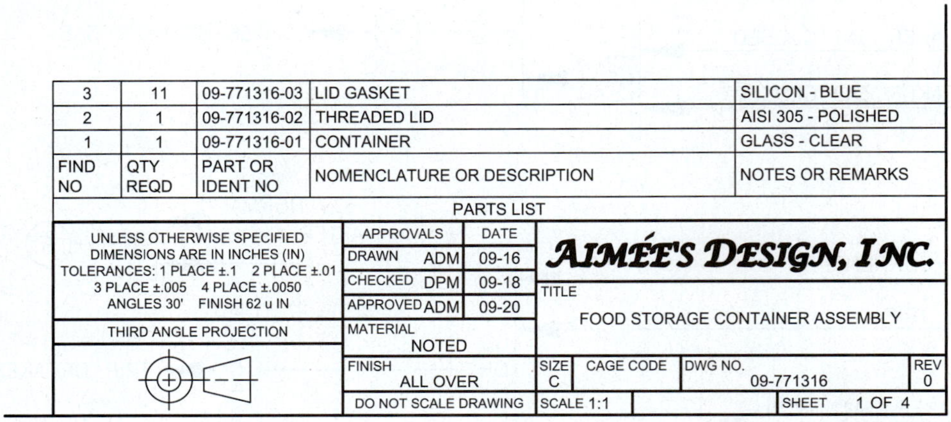

14. **Civil** Create a freehand 2D sketch of a map providing driving or walking directions from your home to your school or office. Label all features, including roads and distances. Use available measuring devices such as the odometer in your car, a tape measure, or surveying equipment to measure features. Draw the map from your sketch using real-world units. Add text to label features and distances, but do not dimension the drawing. Save the drawing as P10-14.

15. **Architectural** Draw the architectural title block shown. Save the drawing as P10-15.

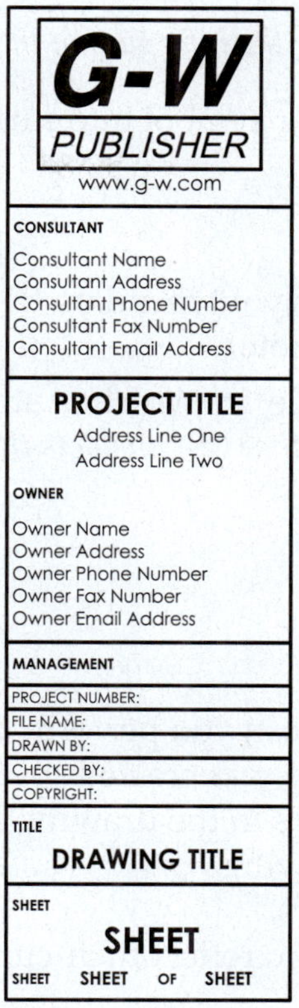

16. **General** Draw title blocks with borders for electrical, piping, and general drawings. Research sample title blocks to come up with your designs. Save the drawings as P10-16A, P10-16B, and P10-16C.

17. **Mechanical** Draw the engineering change notice shown. Use fields as appropriate for areas highlighted in gray. Save the drawing as P10-17.

Engineering Change Notice

ECN# enter the ECN number

Disposition of production stock:
A = Alter or rework, U = Use in production, T = Transfer to service stock, S = Scrap

	Qty.	Drawing Size Part Number	R/N	Description	Change	Other Usage in Production	D/S
01							
02							
03							
04							
05							
06							
07							
08							
09							
10							
11							
12							
13							
14							
15							
16							
17							
18							

Reason: enter the reason for the change

Castings & forgings affected? enter YES or NO	Design engineer: enter design engineer's initials	Supervisor approval: enter supervisor's initials	Release date: enter the date	Page: enter # of pages

Drawing Problems - Chapter 10

AutoCAD Certified Professional Exam Practice

Answer the following questions using the information in this chapter.

1. How can you insert a diameter symbol into single-line text? *Select all that apply.*
   A. type %%D
   B. type %%C
   C. type (D)
   D. select it from the **Symbol** flyout menu on the **Text Editor** ribbon tab
   E. right-click and select **Diameter**
2. What is the name of the file in which AutoCAD stores the default custom dictionary? *Select the one item that best answers the question.*
   A. dictionary.smp
   B. sample.cus
   C. sample.dct
   D. sample.dic
   E. sample.dwt
3. Which of the following tasks can you perform using a field? *Select all that apply.*
   A. display the date a drawing was created
   B. display drawing properties in the drawing
   C. change the name of a drawing
   D. change the names of layers
   E. update property displays to reflect their current status

Follow the instructions in each problem. Complete each problem by answering the related question.

4. **Navigate to this chapter on the companion website and open CPE-10spell.dwg.** Use the default spell checker or the **SPELL** command to check the text in the drawing. What are the first three alternate spellings AutoCAD suggests for "ASME"?
5. **Navigate to this chapter on the companion website and open CPE-10scale.dwg.** Use the **SCALETEXT** command to scale the text using the existing justification point and a scale factor of 2.67. What is the height of the text after the **SCALETEXT** operation?

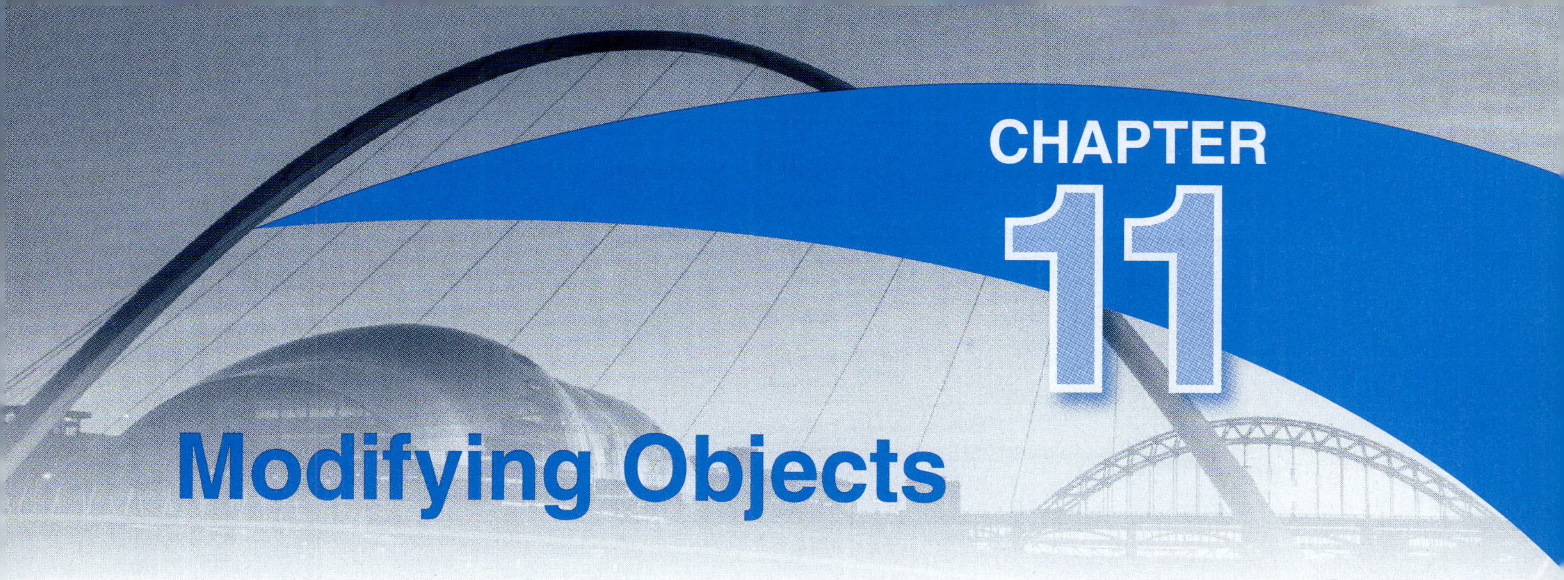

CHAPTER 11

Modifying Objects

Learning Objectives

After completing this chapter, you will be able to:

- ✓ Use the **FILLET** command to draw fillets, rounds, and other radii.
- ✓ Place chamfers and angled corners with the **CHAMFER** command.
- ✓ Form a spline using the **BLEND** command.
- ✓ Separate objects using the **BREAK** command and combine objects using the **JOIN** command.
- ✓ Use the **TRIM** and **EXTEND** commands to edit objects.
- ✓ Change objects using the **STRETCH** and **LENGTHEN** commands.
- ✓ Edit the size of objects using the **SCALE** command.
- ✓ Use the **EXPLODE** command.

AutoCAD includes many editing commands and options to complete common drafting tasks, improve efficiency, and modify a drawing. This chapter explains methods of using basic editing commands to change existing objects by changing or adding geometry. The editing method used in this chapter is to access a command such as **STRETCH** and then follow the prompts to complete the operation. Another technique, presented in Chapter 13, is to select objects first using the crosshairs, then access the editing command, and finally complete the operation.

fillet: A rounded interior corner used to relieve stress or ease the contour of inside corners.

round: A rounded exterior corner used to remove sharp edges or ease the contour of exterior corners.

Using the FILLET Command

Design and drafting terminology refers to a rounded interior corner as a ***fillet*** and a rounded exterior corner as a ***round***. AutoCAD refers to all rounded corners as fillets. The **FILLET** command draws a rounded corner between intersecting or nonintersecting lines, arcs, circles, ellipses, polylines, splines, xlines, and rays. See Figure 11-1.

FILLET

Ribbon
Home
> Modify
Fillet

Type
FILLET
ROUND
F

Setting the Fillet Radius

Access the **FILLET** command and then use the **Radius** option to specify the fillet radius. The radius determines the size of a fillet and must be set before you select objects. AutoCAD stores the radius as the new default, allowing you to place additional fillets of the same size. Once you define the radius, select the first object and then

Figure 11-1.
Using the **FILLET** command. A preview of the operation appears before selecting the second object, allowing you to adjust the fillet size.

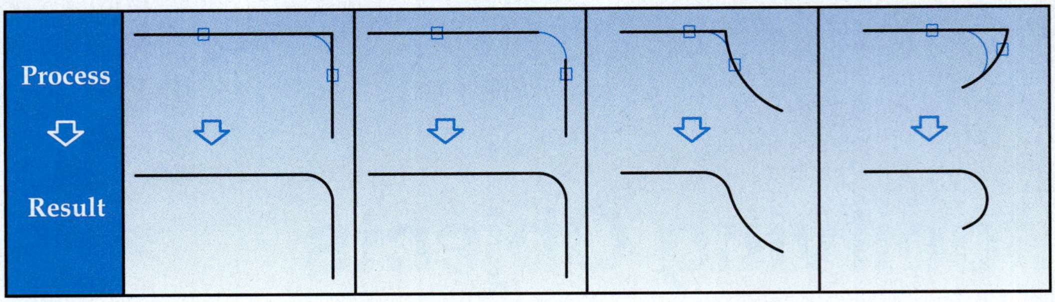

hover the pick box over the second object to preview the operation. The **Radius** option, available after you pick the first object, provides another chance to modify the radius before drawing the fillet or round. If the preview looks acceptable, pick the second object to apply the fillet or round.

PROFESSIONAL TIP

Use the **FILLET** command to form sharp corners by specifying a radius of 0 or by holding down [Shift] when you pick the second object.

NOTE

Only corners large enough to accept the specified fillet radius are eligible for filleting.

Exercise 11-1

www.g-wlearning.com/CAD/

Complete the exercise on the companion website.

Creating Full Rounds

You can use the **FILLET** command to draw a full round between parallel lines. The radius of a fillet between parallel lines is always half the distance between the two lines, regardless of the radius setting. Use this method to create a full round, such as the end radii of a slot.

Polyline Option

Use the **Polyline** option to fillet all corners of a closed polyline. See Figure 11-2. Set the appropriate radius before selecting the **Polyline** option or use the **Radius** option after you select the **Polyline** option to modify the radius.

Closing Polylines

You can use the **FILLET** command to extend, or close, nonparallel and nonintersecting polylines to the ***implied intersection***. Use this technique to close separate or linked polyline or polyline arc segments. The specified radius applies and the result is a single polyline object.

implied intersection: The point at which objects would meet if they were extended.

Figure 11-2.
Using the **Polyline** option of the **FILLET** command. Use the preview to verify the result before completing the operation.

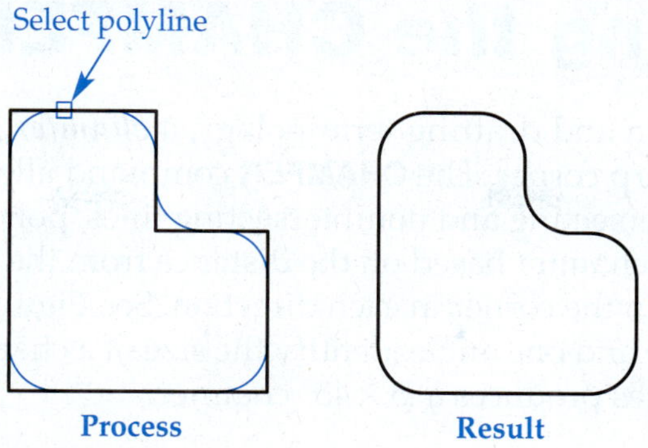

Trim Settings

The **Trim** option controls whether the **FILLET** command trims object segments that extend beyond the fillet radius point of tangency. See Figure 11-3. Use the default **Trim** setting to trim objects. When you set the **Trim** option to **No trim**, the fillet occurs, but the filleted objects do not change.

PROFESSIONAL TIP

You can fillet objects even when the corners do not meet. If the **Trim** option is set to **Trim**, objects extend as required to generate the fillet and complete the corner. If the **Trim** option is set to **No trim**, objects do not extend to complete the corner.

Multiple Option

Use the **Multiple** option to make several fillets without exiting the **FILLET** command. The prompt for a first object repeats after each fillet is drawn. To exit, press [Enter]. While using the **Multiple** option, select the **Undo** option to discard the previous fillet if necessary.

Exercise 11-2

www.g-wlearning.com/CAD/

Complete the exercise on the companion website.

Figure 11-3.
Comparison of the **Trim** and **No trim** options of the **FILLET** command.

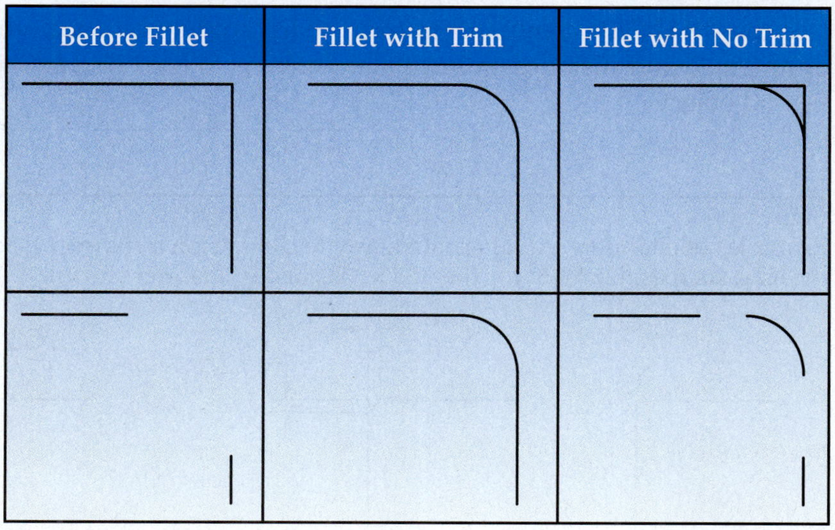

chamfer: A small, angled surface used to relieve a sharp corner.

CHAMFER

Ribbon
Home
> Modify
Chamfer
Type
CHAMFER
CHA

Using the CHAMFER Command

In design and drafting terminology, a ***chamfer*** is a small, angled surface used to relieve a sharp corner. The **CHAMFER** command allows you to draw an angled corner between intersecting and nonintersecting lines, polylines, xlines, and rays. Determine the size of a chamfer based on the distance from the corner. A 45° chamfer is the same distance from the corner in each direction. See Figure 11-4. Typically, two distances or one distance and one angle identify the size of a chamfer. For example, a value of .5 for both distances produces a .5 × 45° chamfer.

Setting Chamfer Distances

Access the **CHAMFER** command and then use the **Distance** option to specify the chamfer distances from a corner. You must set chamfer distances before you select objects. AutoCAD stores the distances as the new defaults, allowing you to place additional chamfers of the same size. Once you define the distances, select the first object and then hover the pick box over the second object to preview the operation. The **Distance** option, available after you pick the first object, provides another chance to modify the distances before drawing the chamfer. If the preview looks acceptable, pick the second object to apply the chamfer. See Figure 11-5.

Setting the Chamfer Angle

Use the **Angle** option as an alternative to setting two chamfer distances. Specify the chamfer distance along the first selected object, followed by the chamfer angle. Then select the objects to chamfer. See Figure 11-6. AutoCAD stores the distance and angle settings as the new defaults, allowing you to place additional chamfers of the same size. The **Angle** option, available after you pick the first object, provides another chance to modify the distance and angle before drawing the chamfer.

Method Option

AutoCAD maintains the specified chamfer distances and the distance and angle until you change the values. You can set the values for each method without affecting the other. Use the **Method** option to toggle between drawing chamfers using the **Distance** and **Angle** options. The **Method** option, available after you pick the first object, provides another chance to modify the method before drawing the chamfer.

PROFESSIONAL TIP

Use the **CHAMFER** command to form sharp corners by specifying chamfer distances or an angle and distance of 0. An alternative is to hold down [Shift] when you pick the second object.

Figure 11-4.
Examples of chamfers. An alternative way to dimension a chamfer is to specify the angle by the size, such as .125 × 45° for the .125 × .125, or 45°, chamfer shown.

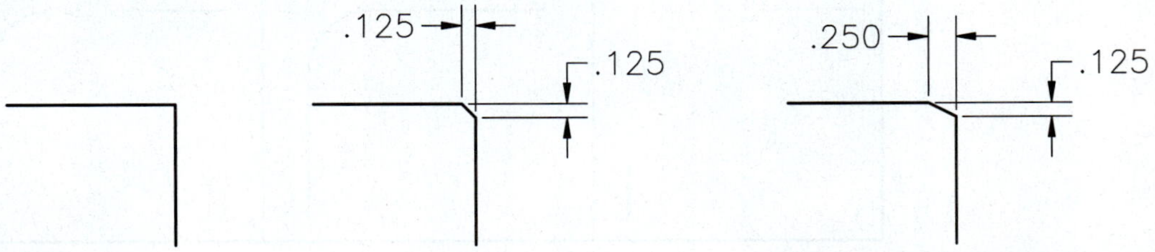

"0" Chamfer 45° Chamfer Unequal Chamfer

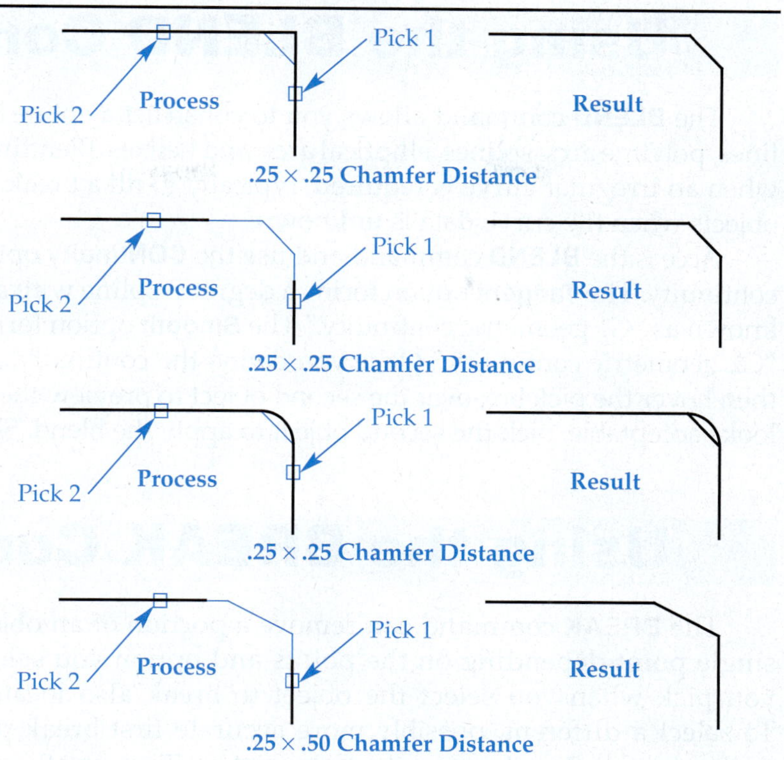

Figure 11-5. Examples of using the **CHAMFER** command to create chamfered corners. Use the preview to verify the result before making the second pick.

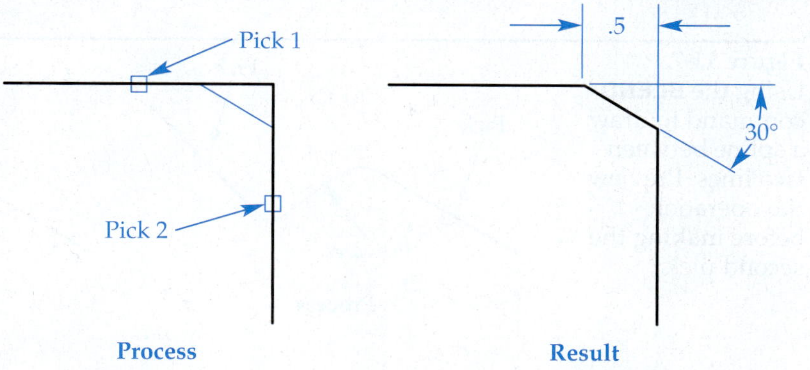

Figure 11-6. Using the **Angle** option of the **CHAMFER** command with the chamfer length set at .5 and the angle set at 30°. Use the preview to verify the result before making the second pick.

NOTE

Only corners large enough to accept the specified chamfer size are eligible for chamfering.

Exercise 11-3

www.g-wlearning.com/CAD/

Complete the exercise on the companion website.

Additional Chamfer Options

The **CHAMFER** command includes the same **Polyline**, **Trim**, and **Multiple** options, and the same ability to close polylines, as the **FILLET** command. Similar rules apply for using these options with the **CHAMFER** command.

Exercise 11-4

www.g-wlearning.com/CAD/

Complete the exercise on the companion website.

Using the BLEND Command

BLEND

Ribbon
Home
> Modify

Blend Curves

Type
BLEND

The **BLEND** command allows you to construct a spline between lines, arcs, polylines, polyline arcs, splines, elliptical arcs, and helixes. Blending a spline can be effective when an irregular curve is required, typically to fill a break between nonintersecting objects when the curve data is unknown.

Access the **BLEND** command and use the **CONtinuity** option to specify the type of continuity. The **Tangent** option forms a degree 3 spline with a mathematical algorithm known as "G1 geometric continuity." The **Smooth** option forms a degree 5 spline with "G2 geometric continuity." Once you define the continuity, select the first object and then hover the pick box over the second object to preview the operation. If the preview looks acceptable, pick the second object to apply the blend. See Figure 11-7.

Using the BREAK Command

BREAK

Ribbon
Home
> Modify

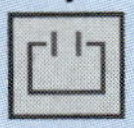

Break

Type
BREAK
BR

The **BREAK** command can remove a portion of an object or split an object at a single point depending on the points and option you select. By default, the point you pick when you select the object to break also locates the first break point. To select a different, possibly more accurate first break point, use the **First point** option at the Specify second break point or [First point]: prompt. See Figure 11-8.

Figure 11-7.
Using the **BLEND** command to draw a spline between two lines. Preview the operation before making the second pick.

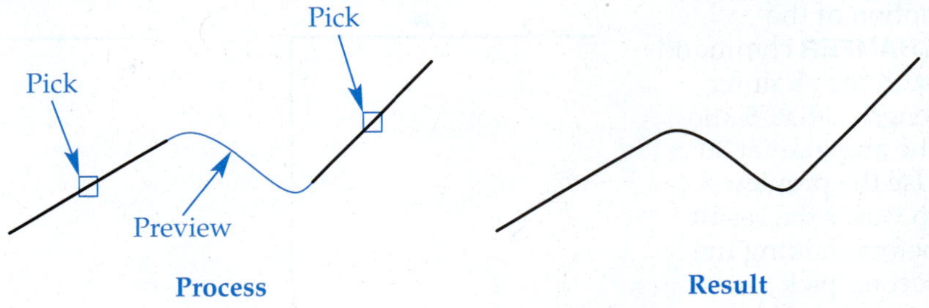

Figure 11-8.
Using the **BREAK** command to break an object. When using the default method, the first pick selects the object and the first break point. Use the **First point** option to select the object to break and then specify accurate break points using object snaps, such as the **Endpoint** object snap as shown here.

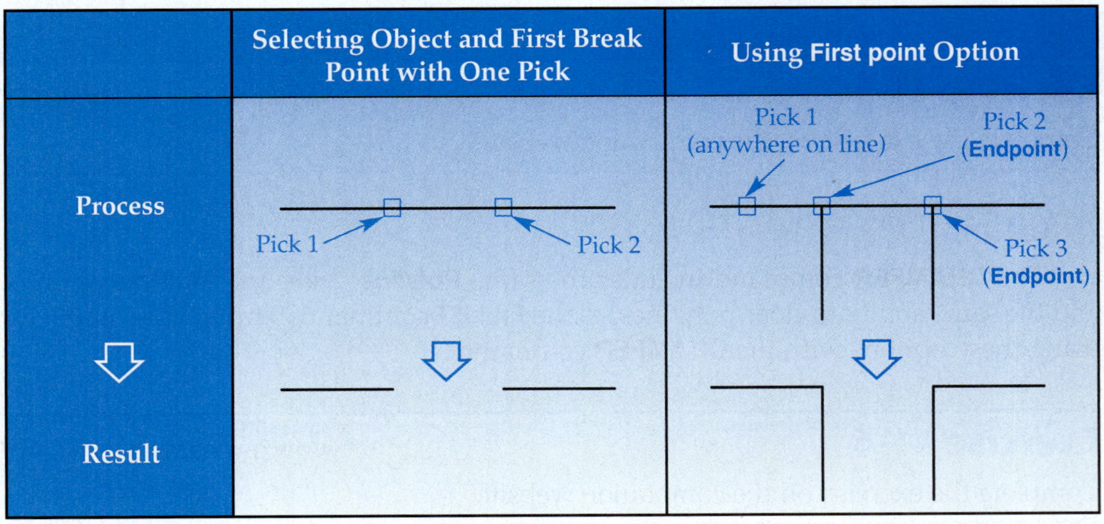

Pick a second break point away from the first break point to break the part of the object between the two points. Preview the operation before making the second pick.

If you select the same point for the first and second break points, the **BREAK** command splits the object into separate segments without removing a portion. At the Specify second break point or [First point]: prompt, use object snaps or coordinate entry, or enter @, to select the same coordinates as the first break point. See **Figure 11-9**. To automate the process, use the **Break at Point** command instead of the standard **BREAK** command.

Ribbon
Home
> Modify
Break at Point

Select points in a counterclockwise direction when breaking circular objects to ensure that you remove the correct portion of the object. See **Figure 11-10**. Notice in **Figure 11-10** that you can break off the end of an open object by picking the first point on the object and the second point slightly beyond the end to be cut off. AutoCAD selects the endpoint nearest the point you pick.

PROFESSIONAL TIP

Use object snaps to pick a point accurately when using the **First point** option of the **BREAK** command. However, it may be necessary to turn running object snaps off if they conflict with points you try to pick.

Exercise 11-5

www.g-wlearning.com/CAD/

Complete the exercise on the companion website.

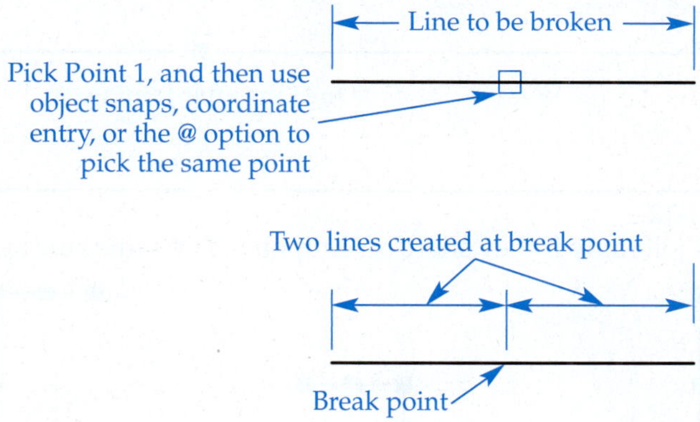

Figure 11-9. Using the **BREAK** command to break an object at a single point without removing any part of the object. Select the same point for the first and second break points. You can use the **Break at Point** command to automate the process.

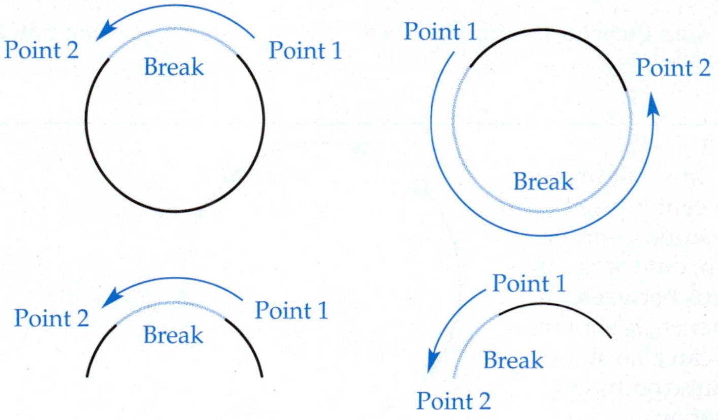

Figure 11-10. Select points in a counterclockwise direction when using the **BREAK** command on circular objects, such as circles and arcs.

Using the JOIN Command

While drawing and editing, you sometimes create multiple objects that should be one object. These multiple objects are cumbersome to work with and increase the size of the drawing file. The **JOIN** command is one option for combining lines, arcs, polylines, polyline arcs, splines, and elliptical arcs. You can join objects of the same type or specific combinations of objects.

JOIN

Ribbon
Home
> Modify

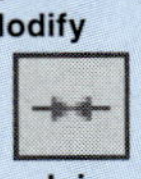

Join

Type
JOIN
UNITE
RECOMBINE
J

Joining the Same Object Type

Access the **JOIN** command and select objects of the same type to join. Lines must be collinear in order to be joined. Lines can share the same endpoint, overlap, or have gaps between segments. See Figure 11-11. Polylines must share a common endpoint and cannot overlap or have gaps between segments. See Figure 11-12. The same rules apply for joining splines.

Arcs must share the same center point and radius. They can overlap, share the same endpoint, or have gaps between segments. See Figure 11-13. Similar rules apply for joining elliptical arcs, except elliptical arcs must share the same axes. Pick arcs or elliptical arcs in a counterclockwise direction to close the nearest gap.

Figure 11-11.
Lines must be collinear to join, but there can be gaps between segments, as shown. Lines can also share the same endpoint or overlap.

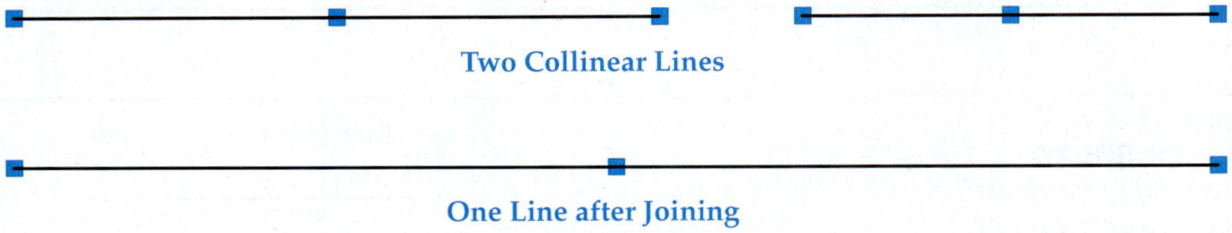

Figure 11-12.
Polylines join only if they share an endpoint. The same rule applies for joining splines.

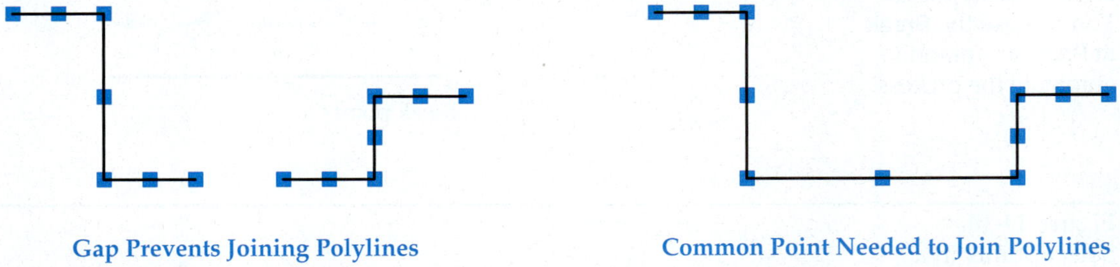

Figure 11-13.
Arcs must share the same center point and radius in order to join, but there can be gaps between segments, as shown. Arcs can also share the same endpoint or overlap.

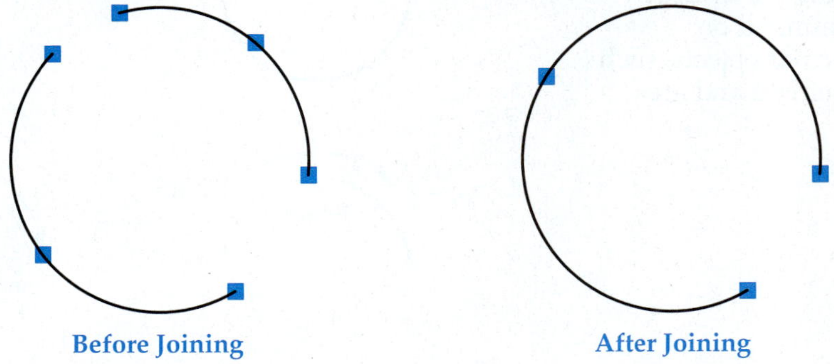

Depending on your selections, you may be prompted to convert arcs to a circle. To maintain the original object type, select the **No** option.

NOTE

After selecting a single arc or a single elliptical arc segment to join and pressing [Enter], you can use the **cLose** option to form a circle from the arc, or to form an ellipse from an elliptical arc. This option closes the two ends of the selected segment and does not join the segment with any other objects.

Joining Different Object Types

You can join lines and arcs to polylines, or join lines, arcs, and polylines to splines. Access the **JOIN** command and select the objects to join. Apply the rules of joining objects of the same type to joining a combination of objects. However, segments cannot overlap or have gaps. The final object becomes the most complex of the original objects.

Trimming

Use the **TRIM** command to cut lines, polylines, circles, arcs, ellipses, splines, xlines, and rays that extend beyond an intersection. Access the **TRIM** command, pick as many ***cutting edges*** as necessary, and then press [Enter]. Then pick the objects to trim to the cutting edges. Preview the operation before trimming. To exit, press [Enter]. See **Figure 11-14**.

Automatic windowing is often a quick and effective method for trimming multiple objects. The **TRIM** command also offers **Crossing** and **Fence** options that provide standard crossing and fence selection. **Figure 11-15** shows using the **Crossing** option to trim multiple objects. **Figure 11-16** shows using the **Fence** option to trim multiple objects.

TRIM

Ribbon
Home > Modify

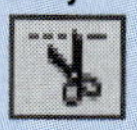

Trim

Type
TRIM
TR

cutting edge: An object such as a line, arc, or text that defines the point (edge) at which the object you trim will be cut.

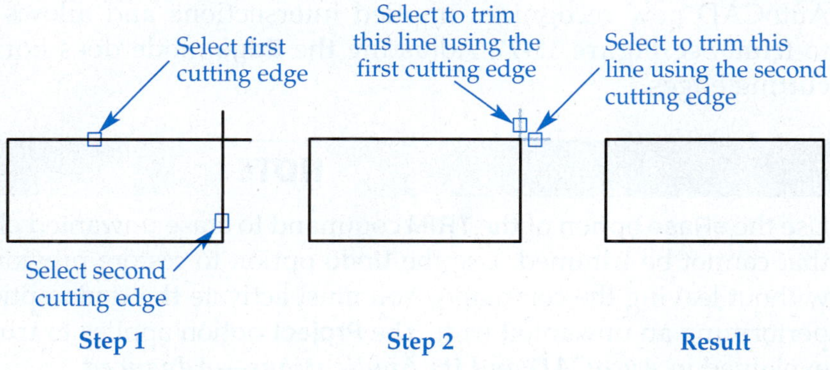

Figure 11-14.
Using the **TRIM** command. Note the cutting edges.

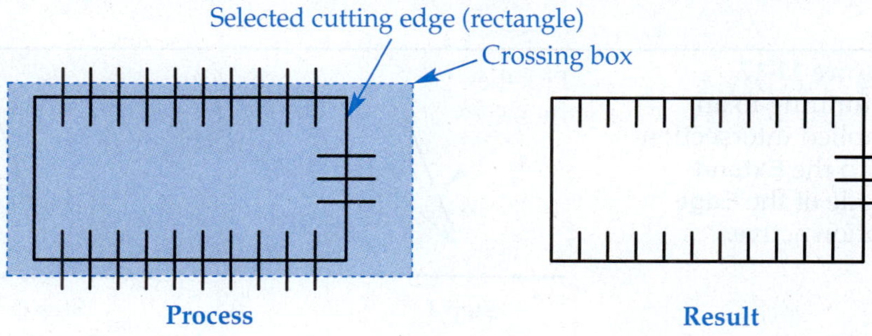

Figure 11-15.
The only objects trimmed with the **Crossing** option are those that cross the edges of the crossing window. Automatic windowing accomplishes the same task.

Figure 11-16.
The **Fence** option allows you to select around objects. In this example, the cutting edge consists of a rectangle drawn using the **RECTANGLE** command.

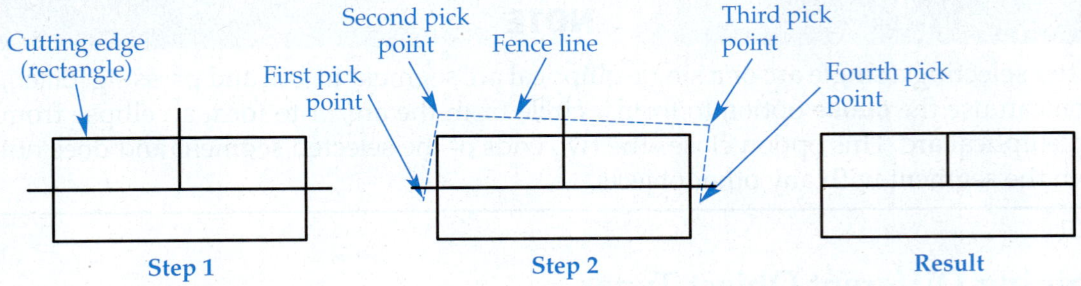

NOTE

To access the **EXTEND** command while using the **TRIM** command, after selecting the cutting edge(s), hold down [Shift] and pick objects to extend to the cutting edge. The **EXTEND** command is described later in this chapter.

Trimming without Selecting a Cutting Edge

To trim objects to the nearest intersection without selecting a cutting edge, access the **TRIM** command. At the Select objects or <select all>: prompt, press [Enter] instead of picking a cutting edge. Then pick the objects to trim. You can continue selecting objects to trim without restarting the **TRIM** command. To exit, press [Enter].

Trimming to an Implied Intersection

Use the **Edge** option of the **TRIM** command to trim to an implied intersection. The **No extend** mode is active by default, and does not allow you to trim objects that do not intersect. To trim nonintersecting objects, access the **TRIM** command, pick the cutting edges, and select the **Edge** option followed by the **Extend** option. AutoCAD now recognizes implied intersections and allows you to pick objects to trim. See Figure 11-17. Adjusting the **Edge** mode does not change the selected cutting edges.

NOTE

Use the **eRase** option of the **TRIM** command to erase unwanted objects, usually objects that cannot be trimmed. Use the **Undo** option to restore previously trimmed objects without leaving the command. You must activate the **Undo** option immediately after performing an unwanted trim. The **Project** option applies to trimming 3D objects, as explained in *AutoCAD and Its Applications—Advanced*.

Figure 11-17.
Trimming to an implied intersection with the **Extend** mode of the **Edge** option active.

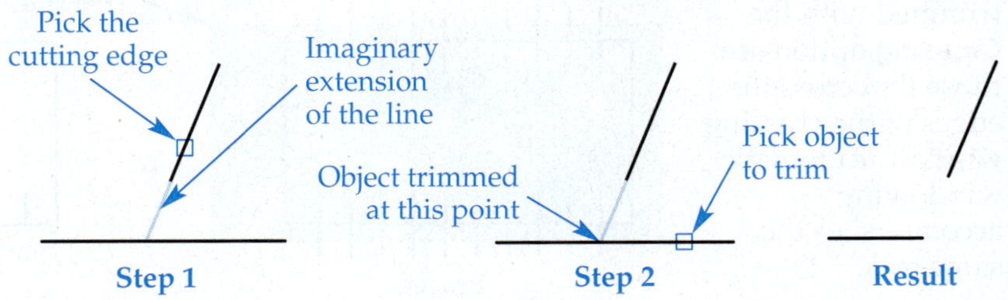

Extending

Use the **EXTEND** command to extend lines, elliptical arcs, rays, open polylines, arcs, and splines to meet other objects. You cannot extend a closed object because an unconnected endpoint does not exist. Access the **EXTEND** command, pick as many ***boundary edges*** as necessary, and then press [Enter]. Now pick the objects to extend to the boundary edges. Preview the operation before extending. To exit, press [Enter]. See Figure 11-18.

Automatic windowing is a quick and effective method for extending multiple objects. The **EXTEND** command also offers **Crossing** and **Fence** options that provide standard crossing and fence selection. Figure 11-19 shows using the **Crossing** option to extend multiple objects. Figure 11-20 shows an example of using the **Fence** option to extend multiple objects.

Ribbon
Home
> Modify
Extend
Type
EXTEND
EX

EXTEND

boundary edge: The edge to which objects such as lines, arcs, and polylines extend.

Extending without Selecting a Boundary Edge

To extend objects to the nearest intersection without selecting a boundary edge, access the **EXTEND** command and, at the Select objects or <select all>: prompt, press [Enter] instead of picking a boundary edge. Then pick the objects to extend. You can continue selecting objects to extend without restarting the **EXTEND** command. To exit, press [Enter].

Figure 11-18.
Using the **EXTEND** command. Note the boundary edge.

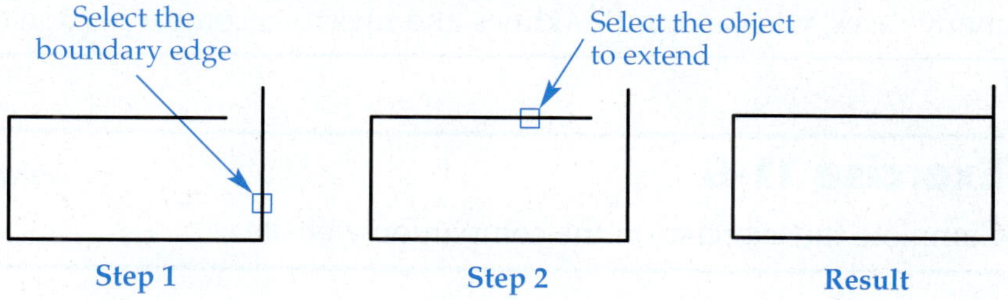

Figure 11-19.
Selecting objects to extend using the **Crossing** option. Automatic windowing accomplishes the same task.

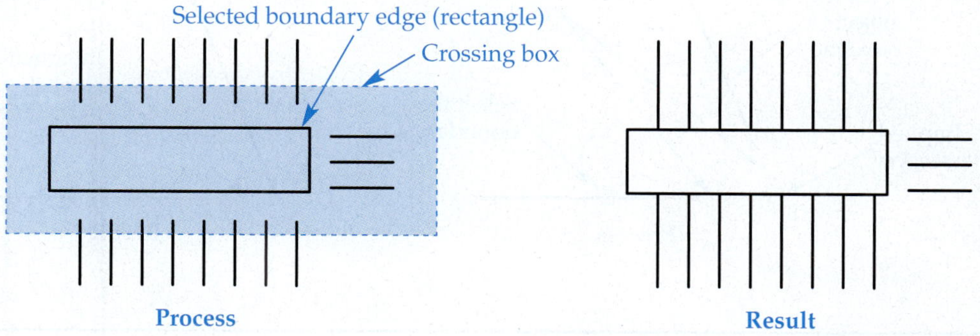

Figure 11-20.
Extending multiple lines to a boundary edge using the **Fence** option.

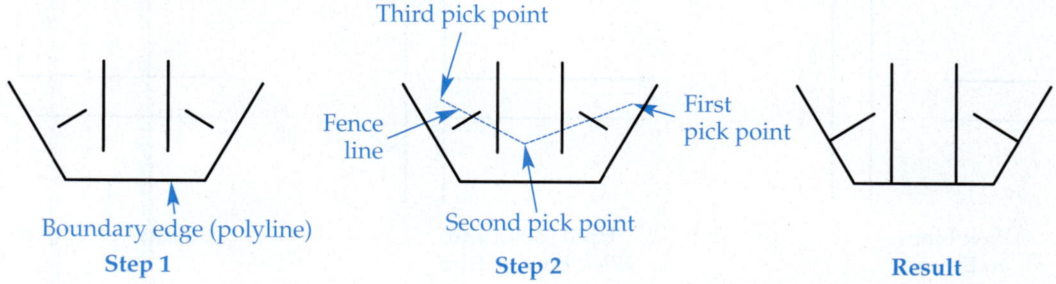

Additional Extend Options

The **EXTEND** command includes the same **Edge**, **Undo**, and **Project** options as the **TRIM** command, and similar rules apply when using these options with the **EXTEND** command. Use the **Extend** mode of the **Edge** option to extend to an implied intersection, as shown in Figure 11-21. Select the **Undo** option immediately after performing an unwanted extend to restore previous objects without leaving the command. The **Project** option applies to extending 3D objects, as explained in *AutoCAD and Its Applications—Advanced*.

PROFESSIONAL TIP

Figure 11-22 illustrates how to combine the **EXTEND** and **TRIM** commands to draw a wall on a floor plan without selecting a boundary edge. You can apply this process to a variety of applications.

NOTE

When you trim one infinite end of an xline, the object becomes a ray. When you trim both infinite ends of an xline or the infinite end of a ray, the object becomes a line. Therefore, in many cases, you can modify xlines and rays to become a portion of the actual drawing.

Exercise 11-6

www.g-wlearning.com/CAD/

Complete the exercise on the companion website.

Figure 11-21.
Extending to an implied intersection with the **Extend** mode.

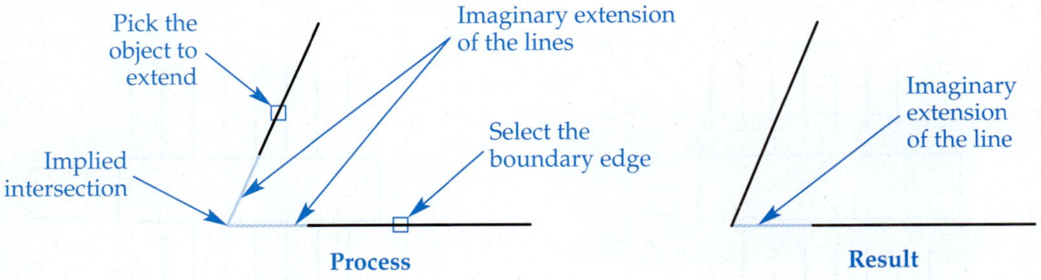

Figure 11-22.
To extend objects to the nearest intersection without selecting a boundary edge, press [Enter] instead of picking a boundary edge. Then pick objects to extend. Hold down [Shift] to toggle between extending and trimming.

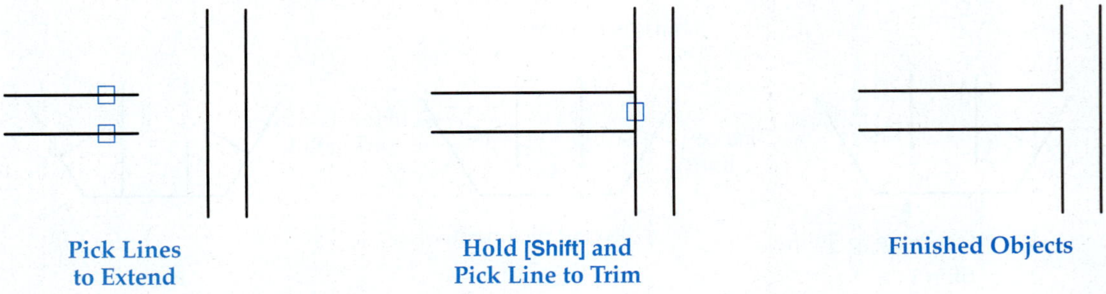

Stretching

The **STRETCH** command allows you to modify certain dimensions of an object without changing other dimensions. In mechanical drafting, for example, you can stretch a screw body to create a longer or shorter screw. In architectural design, you can stretch room sizes to increase or decrease square footage.

Once you access the **STRETCH** command, you must use a crossing box, crossing lasso, or crossing polygon to select only the objects to stretch. This is a very important requirement and is different from object selection for other editing commands. See Figure 11-23. If you select using the pick box or a window, the **STRETCH** command works like the **MOVE** command, as described in Chapter 12.

STRETCH

Ribbon
Home
> Modify

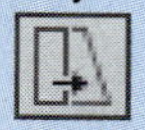

Stretch

Type
STRETCH
S

After selecting the objects to stretch, specify the ***base point*** from which the objects will stretch. Although the position of the base point is often not critical, you may want to select a point on an object. For example, select the corner of a view or the center of a circle. The selection stretches or compresses as you move the cross-hairs. Specify a second point to complete the stretch. Preview the operation before stretching.

base point: The initial reference point AutoCAD uses when stretching, moving, copying, and scaling objects.

PROFESSIONAL TIP

Use object snap modes while editing. For example, to stretch a rectangle to make it twice as long, use the **Endpoint** object snap to select the endpoint of the rectangle as the base point, and another **Endpoint** object snap to select the opposite endpoint of the rectangle as the second point.

Displacement Option

The **Displacement** option allows you to stretch objects relative to their current location. To stretch using a ***displacement***, access the **STRETCH** command and use crossing selection to select only the objects to stretch. Then select the **Displacement** option instead of defining the base point. At the Specify displacement <0,0,0>: prompt, enter a coordinate or pick a point to specify the displacement. See Figure 11-24.

displacement: The direction and distance in which an object moves.

Figure 11-23. Using the **STRETCH** command to edit specific dimensions. Preview the operation before making the second pick.

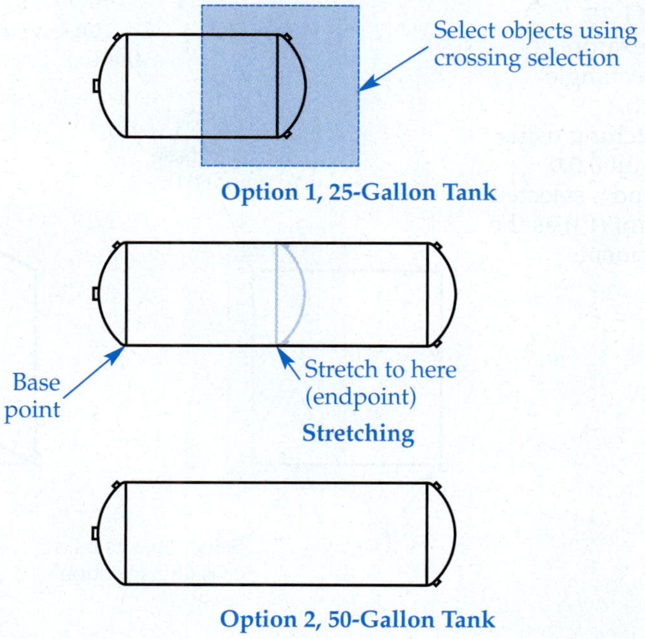

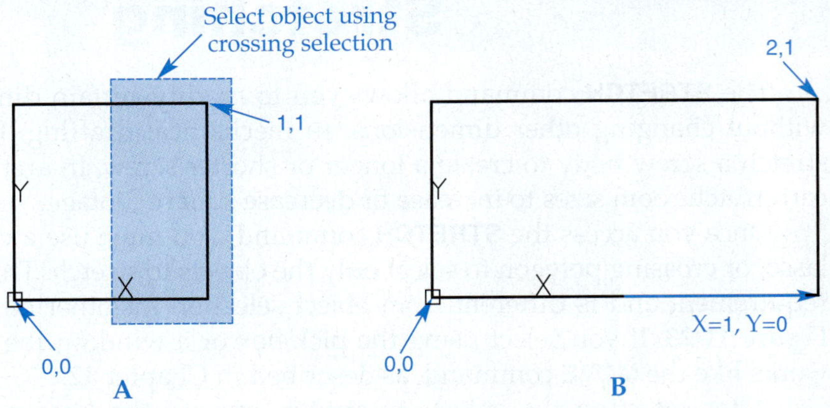

Figure 11-24. Using the **Displacement** option of the **STRETCH** command. A—An example of a 1 × 1 rectangle to stretch. B—Stretching the rectangle using a 1,0 displacement.

Using the First Point as Displacement

Another method for stretching an object is to use the first point as the displacement. The coordinates you use to select the base point automatically define the coordinates for the direction and distance for stretching the object. Access the **STRETCH** command and use crossing selection to select only the objects to stretch. Then specify the base point, and instead of locating the second point, press [Enter] to accept the <use first point as displacement> default. See Figure 11-25.

PROFESSIONAL TIP

Objects often do not align conveniently for using crossing selection to pick objects to stretch. Consider using a crossing lasso or crossing polygon to make selection easier. If the stretch is not as expected, press [Esc] to cancel. The **STRETCH** command and other editing commands work well with polar tracking or **Ortho** mode.

Exercise 11-7

www.g-wlearning.com/CAD/

Complete the exercise on the companion website.

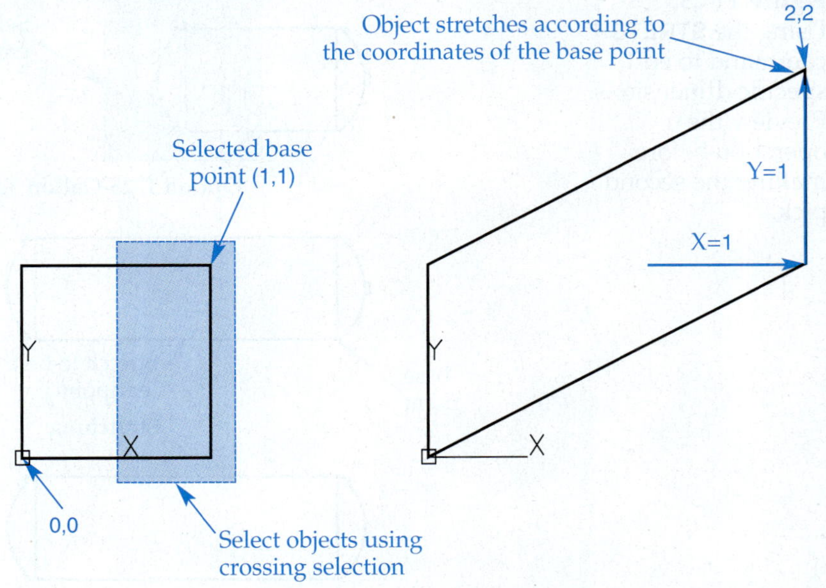

Figure 11-25. A—An example of a 1 × 1 rectangle to stretch. B—Stretching using the default 0,0,0 origin and a selected base point (1,1) as the displacement.

Using the LENGTHEN Command

Ribbon
Home
> Modify

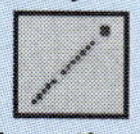

Lengthen

Type
LENGTHEN
LEN

LENGTHEN

Use the **LENGTHEN** command to change the length of a line, arc, polyline, spline, or elliptical arc, or the included angle of an arc. You can only lengthen one object at a time. You cannot lengthen a closed object because an unconnected endpoint does not exist. Access the **LENGTHEN** command and select the object to change. AutoCAD identifies the current length. If the selected object is an arc, the length and included angle are identified. The initial selection is for reference only. Select an option and follow the prompts to complete the operation. Preview the operation before completing the command.

Use the **DElta** option to specify a positive or negative change in length, measured from the endpoint of the selected object. The change in length occurs at the end closest to the selection point. See Figure 11-26. Use the **Angle** option of the **DElta** option to change the included angle of an arc according to a specified angle. See Figure 11-27.

Select the **Percent** option to change the length of an object or the angle of an arc by a specified percentage. The original length is 100 percent. Specify a percentage more than 100 to increase the length or less than 100 to decrease the length. See Figure 11-28.

The **Total** option allows you to set the total length or angle of the object after the **LENGTHEN** operation. See Figure 11-29. The **DYnamic** option lets you drag the endpoint of the object to the desired length or angle using the crosshairs. See Figure 11-30. It is helpful to use dynamic input with polar tracking or **Ortho** mode or to have the grid and snap set to usable increments when using the **DYnamic** option.

Figure 11-26.
Using the **DElta** option of the **LENGTHEN** command with values of .75 and –.75.

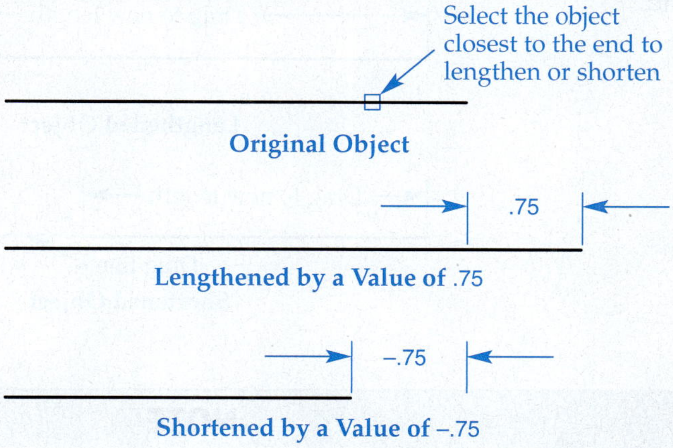

Figure 11-27.
Using the **Angle** option of the **DElta** option to increase or decrease the included angle of an arc by 45°.

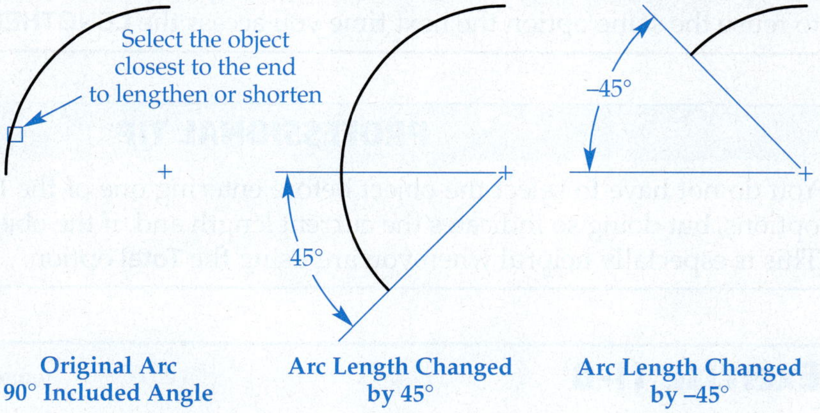

Figure 11-28.
Examples of changing the length of a line using the **Percent** option of the **LENGTHEN** command.

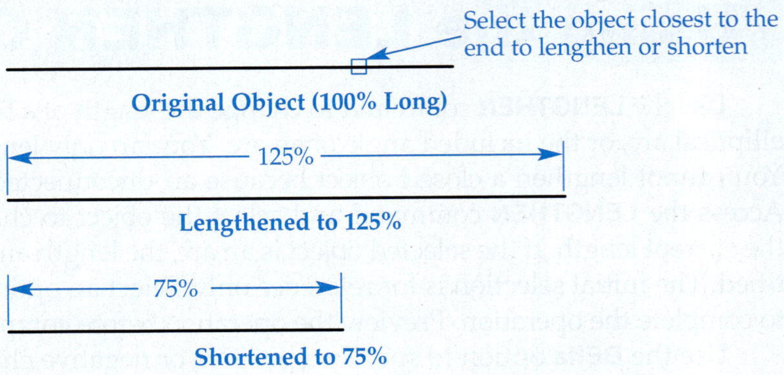

Figure 11-29.
Using the **Total** option of the **LENGTHEN** command.

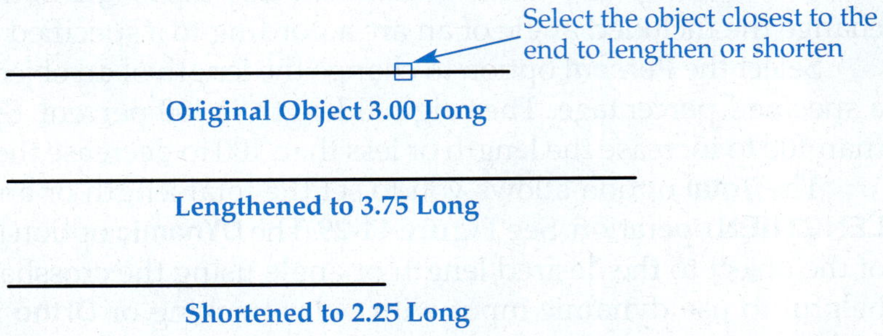

Figure 11-30.
Using the **DYnamic** option of the **LENGTHEN** command.

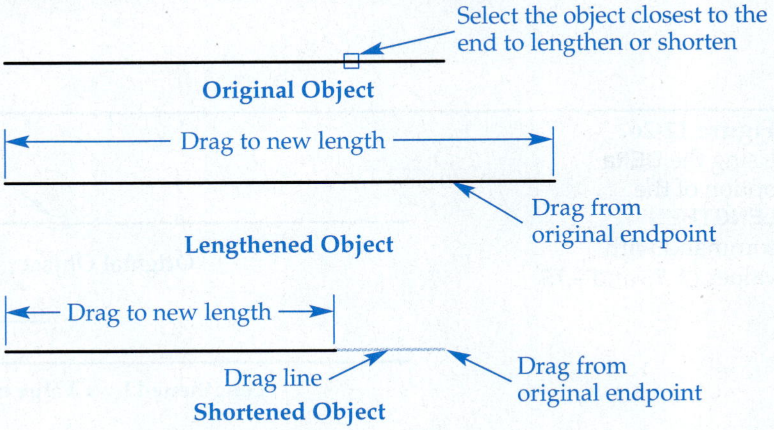

NOTE

You can only lengthen lines and arcs dynamically, and you can only decrease the length of a spline. AutoCAD stores the last lengthen option as the new default, allowing you to reuse the same option the next time you access the **LENGTHEN** command.

PROFESSIONAL TIP

You do not have to select the object before entering one of the **LENGTHEN** command options, but doing so indicates the current length and, if the object is an arc, the angle. This is especially helpful when you are using the **Total** option.

Exercise 11-8

www.g-wlearning.com/CAD/

Complete the exercise on the companion website.

Using the SCALE Command

Access the **SCALE** command to proportionately enlarge or reduce the size of objects. Select objects to scale and then proceed to the next prompt to specify the base point from which the increase or decrease in size should occur. Scaling occurs away from the base point for enlargement or toward the base point for reduction. After you specify the base point, the scale icon appears next to the crosshairs and the selected objects scale in size as you move the crosshairs. The next step is to specify the scale factor. Enter a number to indicate the amount of enlargement or reduction. For example, to make the objects twice the current size, type 2 at the Specify scale factor or [Copy/Reference]: prompt. See **Figure 11-31**. Use fractions or decimal values to reduce scale. For example, enter 3/4 or .75 to scale objects three-quarters their original size or 1/2 or .5 to scale objects half their original size. You can also use coordinate entry or drawing aids to define the scale factor. Preview the operation before completing the command.

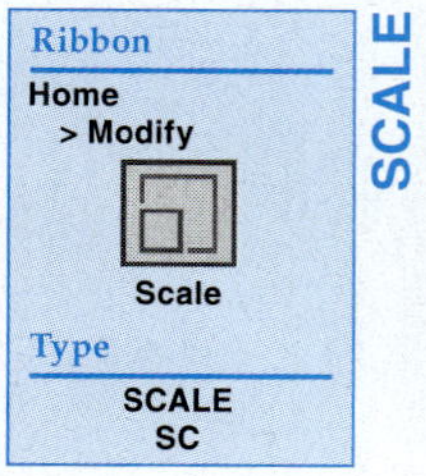

Reference Option

The **Reference** option is an alternative to entering a scale factor that allows you to specify a new size in relation to an existing dimension. For example, to change the size of a part with an overall dimension of 2.50″ to an overall dimension of 3.00″ proportionately, select the **Reference** option. Specify the current length, 2.5 in the example, at the Specify reference length <current>: prompt. Next, specify the new length for the dimension, 3 in the example. See **Figure 11-32**.

PROFESSIONAL TIP

Specify the reference length and new length using specific values or select points on existing objects. Picking points is especially effective when you do not know the exact reference and new lengths.

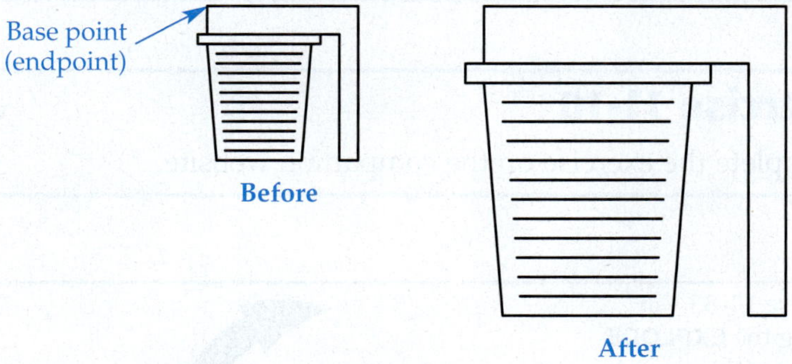

Figure 11-31. Using the **SCALE** command to make objects twice the previous size. The points associated with the selected objects adjust from the location of the base point.

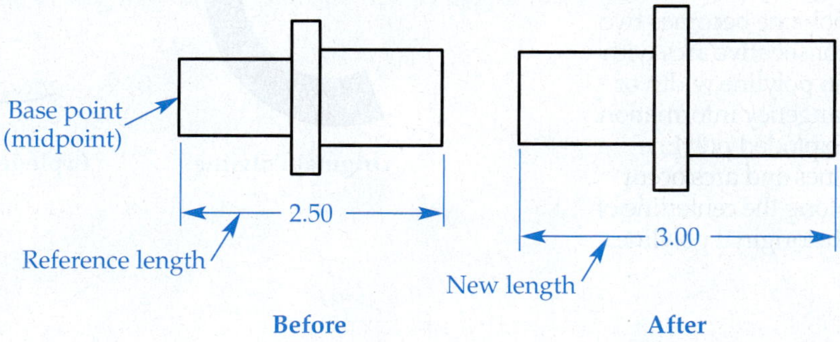

Figure 11-32. Using the **Reference** option of the **SCALE** command.

Copying While Scaling

The **Copy** option of the **SCALE** command copies and scales the selected objects, leaving the original objects unchanged. The copy is created at the specified location.

NOTE

The **SCALE** command changes all dimensions of an object proportionately. Use the **STRETCH** or **LENGTHEN** command to change only the length, width, or height.

Exercise 11-9

www.g-wlearning.com/CAD/

Complete the exercise on the companion website.

Exploding Objects

EXPLODE

Ribbon
Home
> Modify
Explode

Type
EXPLODE
BREAKUP
DISASSEMBLE
X

The **EXPLODE** command allows you to change a single object that consists of multiple items into a series of individual objects. For example, you can explode a polyline object into individual line and arc objects. See Figure 11-33. Another example is exploding a multiline text object to create several single-line text objects. You can explode a variety of other objects, including multilines, regions, dimensions, leaders, and blocks. This textbook explains these objects when appropriate. Access the **EXPLODE** command, pick the object to explode, and press [Enter] to cause the explosion.

CAUTION

Exploding eliminates the original object properties, characteristics, and associations. The need to explode objects is usually rare. Exploding is only appropriate if no other command or option can produce the desired effect.

Exercise 11-10

www.g-wlearning.com/CAD/

Complete the exercise on the companion website.

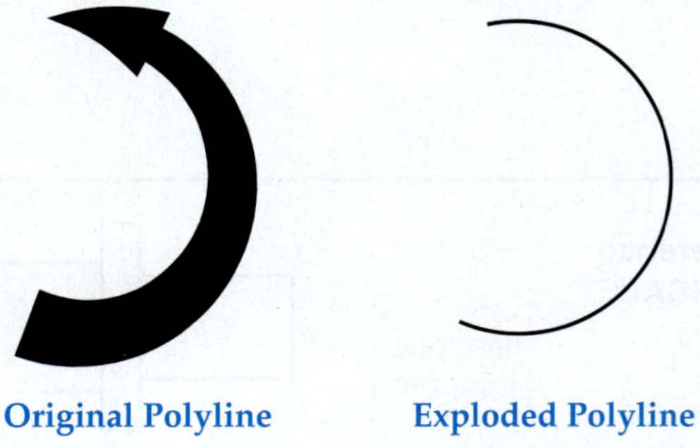

Figure 11-33. Using the **EXPLODE** command to explode a polyline. In this example, the polyline becomes two consecutive arcs with no polyline width or tangency information. Exploded polyline lines and arcs occur along the centerline of the original polyline.

Supplemental Material

www.g-wlearning.com/CAD/

Deleting Duplicate Objects

For information about using the **OVERKILL** command to remove duplicate and/or unnecessary objects, go to the companion website, navigate to this chapter in the **Contents** tab, and select **Deleting Duplicate Objects**.

Chapter Review

Answer the following questions using the information in this chapter.

1. What determines the size of a fillet?
2. Explain how to set the radius of a fillet to .50.
3. Describe the difference between the **Trim** and **No trim** options of the **FILLET** command.
4. Which option of the **CHAMFER** command would you use to specify a chamfer using two distances?
5. What is the purpose of the **Method** option of the **CHAMFER** command?
6. Which command allows you to create an irregular curve to fill a break between nonintersecting objects when the curve data is unknown?
7. How can you split an object in two without removing a portion?
8. In what direction should you pick points to break a portion out of a circle or arc?
9. What command can you use to combine two collinear lines into a single line object?
10. What two requirements must be met before you can join two arcs?
11. Name the command that trims an object to a cutting edge.
12. Which command performs the opposite function of the **EXTEND** command?
13. Name the command associated with boundary edges.
14. Name the **TRIM** and **EXTEND** command option that allows you to trim or extend to an implied intersection.
15. When using the **STRETCH** command, what must you use to select the objects to stretch?
16. Define the term *displacement* as it relates to the **STRETCH** command.
17. Identify the **LENGTHEN** command option that corresponds to each of the following descriptions:
    A. Allows a positive or negative change in length from the endpoint
    B. Changes a length or an arc angle by a percentage of the total
    C. Sets the total length or angle to the value specified
    D. Drags the endpoint of the object to the desired length or angle
18. Write the command aliases for the following commands:
    A. **CHAMFER**
    B. **FILLET**
    C. **BREAK**
    D. **TRIM**
    E. **EXTEND**
    F. **SCALE**
    G. **LENGTHEN**
19. What command would you use to reduce the size of an entire drawing by one-half?
20. Which command removes all width characteristics and tangency information from a polyline?

Drawing Problems

Start AutoCAD if it is not already started. Start a new drawing for each problem using an appropriate template of your choice. The template should include layers and text styles, when necessary, for drawing the given objects. Add layers and text styles as needed. Draw all objects using appropriate layers. Use appropriate text styles, justification, and format. Follow the specific instructions for each problem. Use only drawing and editing commands and techniques you have already learned. Do not draw dimensions. Use your own judgment and approximate dimensions when necessary.

▼ Basic

1. General Draw Object A using the **LINE** and **ARC** commands. Make sure the corners overrun and center the arc on the lines, but the arc should not touch the lines. Use the **TRIM**, **EXTEND**, and **STRETCH** commands to make Object B. Save the drawing as P11-1.

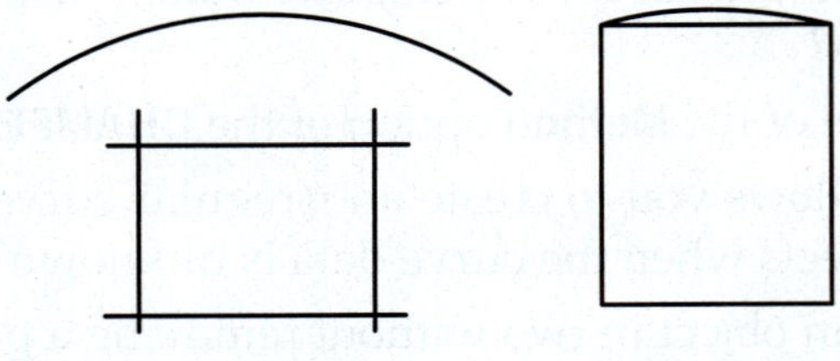

2. General Save P11-1 as P11-2. In the P11-2 file, use the **STRETCH** command to convert Object A to Object B. Resave the drawing.

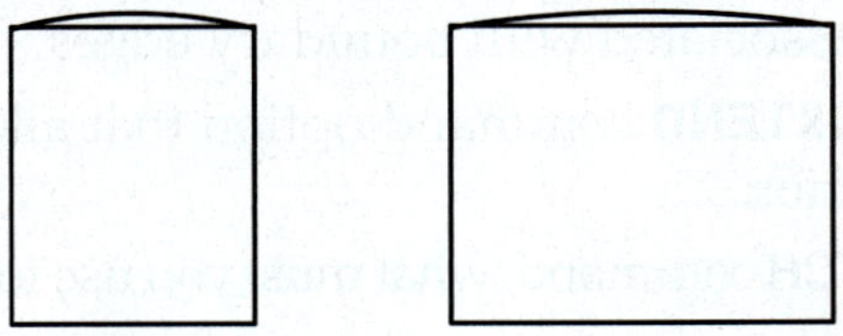

3. Piping Refer to **Figure 11-23** in this chapter. Draw the object shown in Option 1. Stretch the object to twice its length, as shown in Option 2. Stretch the object again to one and a half times its length. *Hint:* Use endpoint and midpoint object snap modes to stretch accurately. Save the drawing as P11-3.

Drawing Problems - Chapter 11

4. **Mechanical** Draw the part view shown. Use the **CHAMFER** command to create the inclined surface. Save the drawing as P11-4.

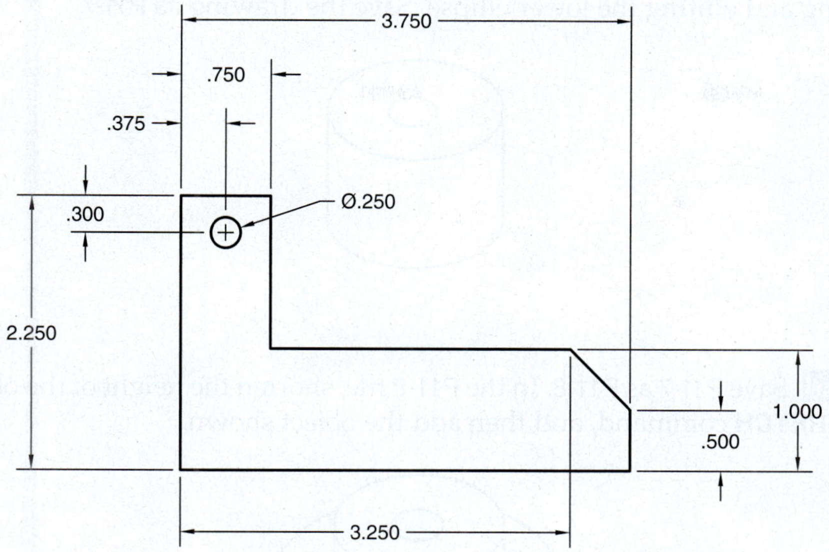

▼ Intermediate

5. **Mechanical** Draw the wrench shown. Save the drawing as P11-5.

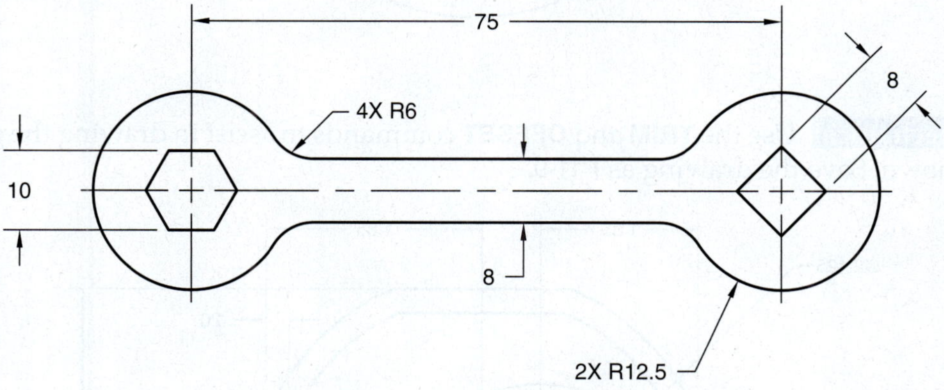

6. **Mechanical** Draw the bracket shown. Save the drawing as P11-6.

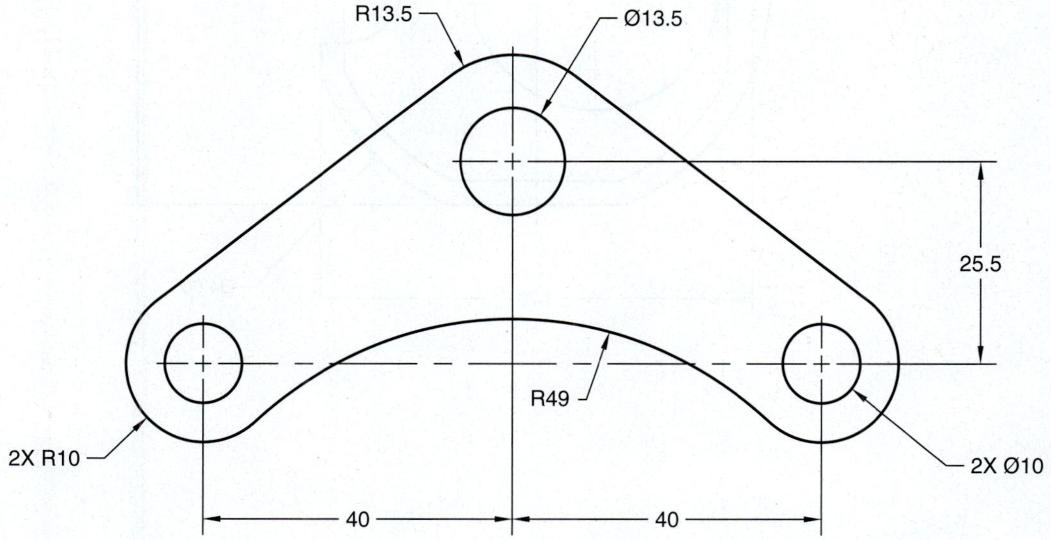

7. General Draw the bushing shown using the **ELLIPSE** and **LINE** commands. Draw the ellipses using a 30° rotation angle. Use the **BREAK** or **TRIM** command when drawing and editing the lower ellipse. Save the drawing as P11-7.

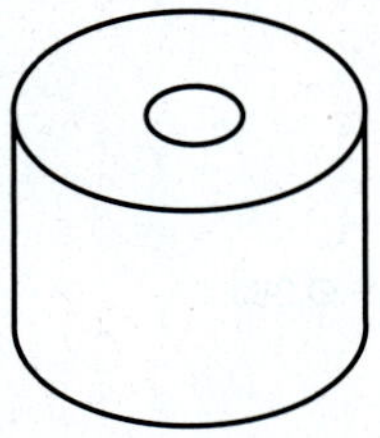

8. General Save P11-7 as P11-8. In the P11-8 file, shorten the height of the object using the **STRETCH** command, and then add the object shown.

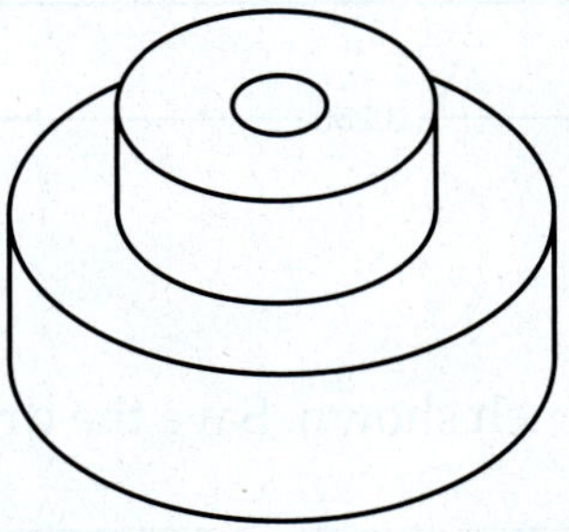

9. Mechanical Use the **TRIM** and **OFFSET** commands to assist in drawing the gasket shown. Save the drawing as P11-9.

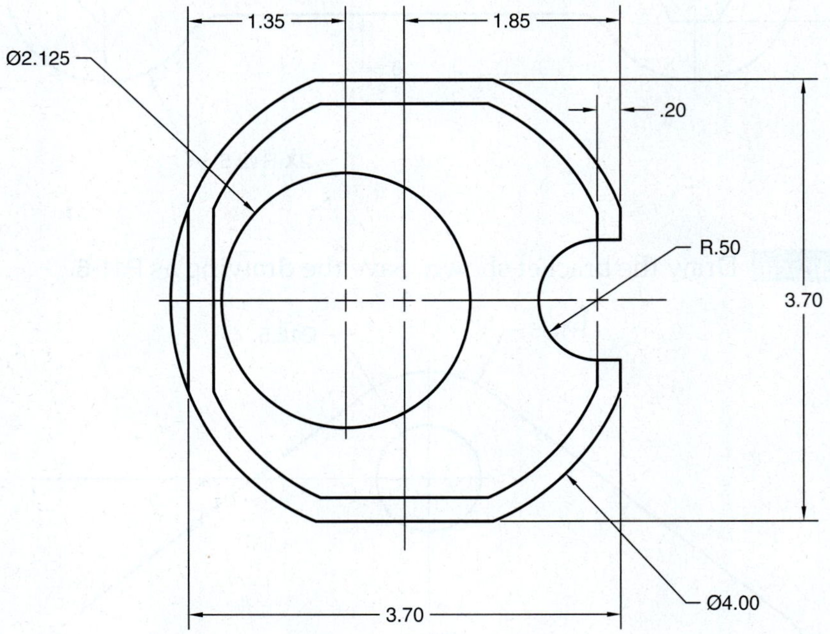

10. **Mechanical** Draw the plate shown. Use the **FILLET** command where appropriate. Save the drawing as P11-10.

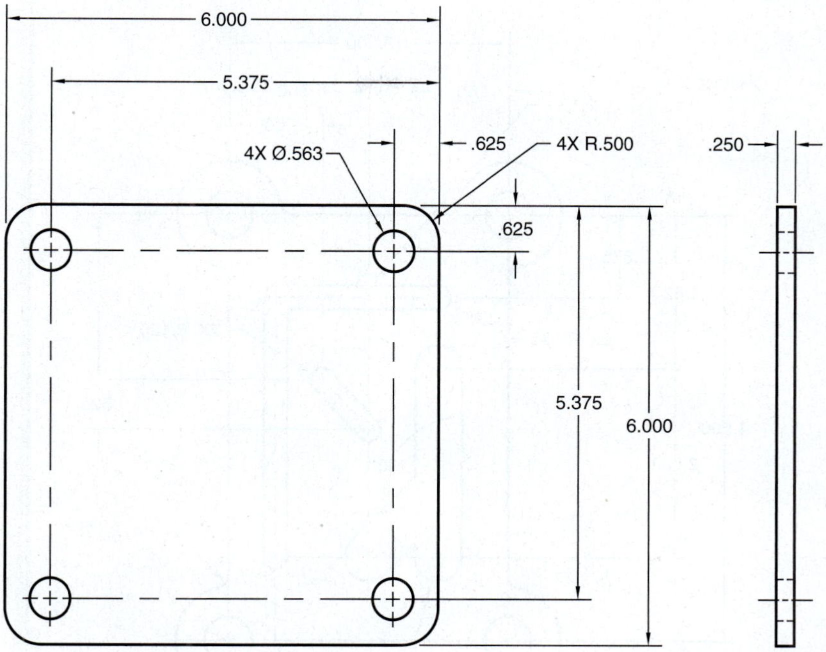

11. **Architectural** Draw the toilet shown. Use dimensions of your choice for objects not dimensioned. Save the drawing as P11-11.

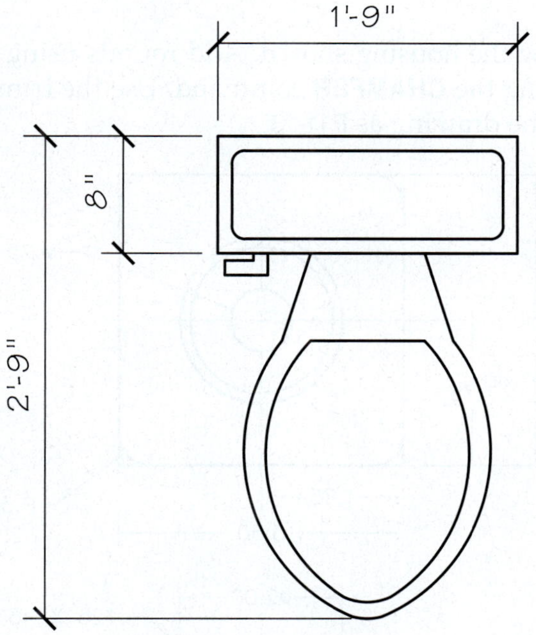

▼ Advanced

12. **Mechanical** Draw the part view shown. Save the drawing as P11-12.

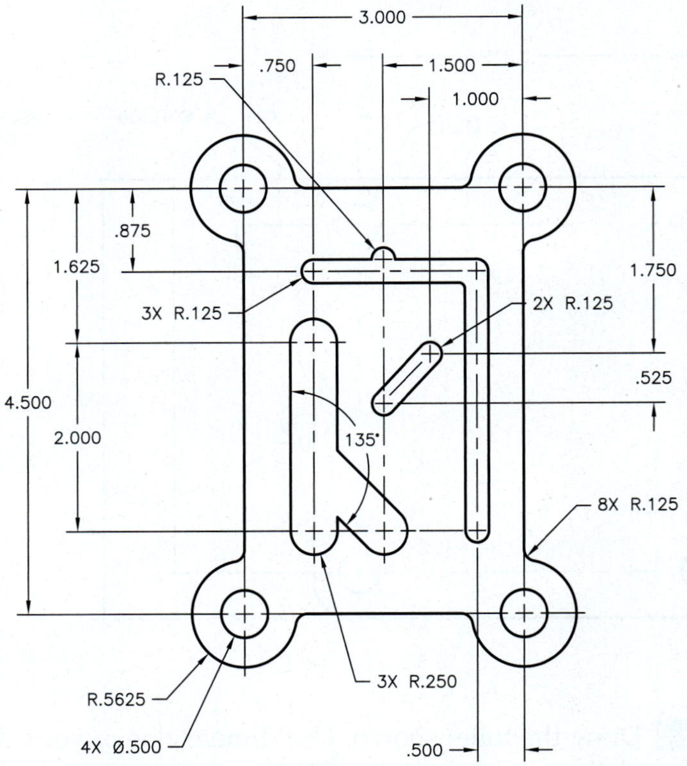

13. **Mechanical** Draw the housing shown. Add rounds using the **FILLET** command and chamfers using the **CHAMFER** command. Use the trim mode setting to your advantage. Save the drawing as P11-13.

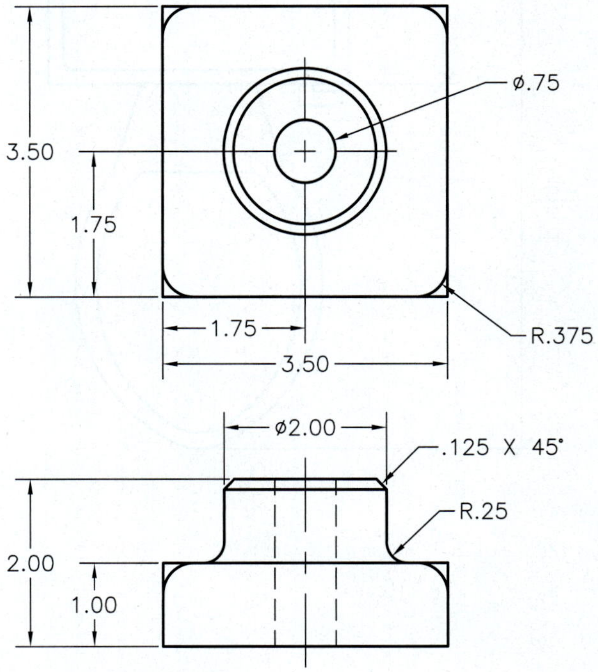

Drawing Problems - Chapter 11

14. **Architectural** Draw the beam wrap detail shown. Save the drawing as P11-14.

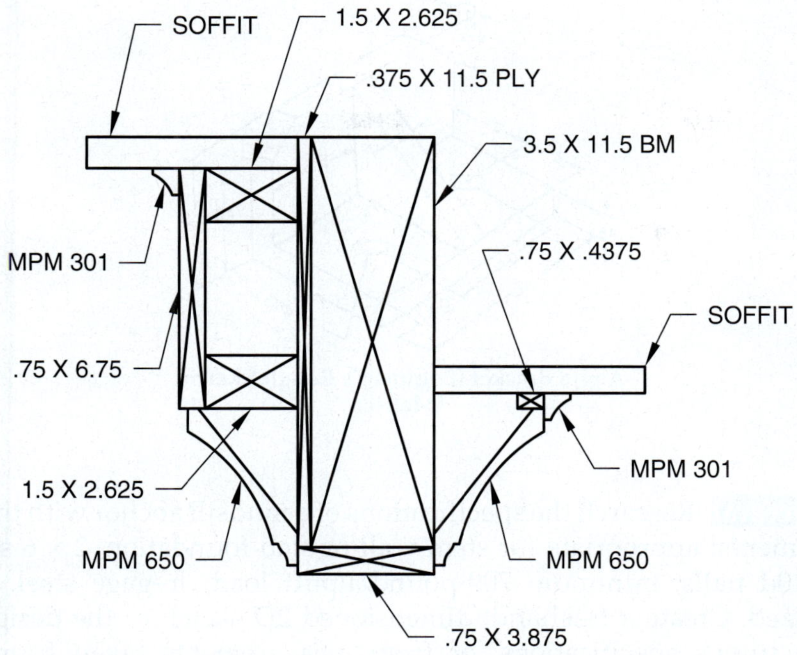

For Problems 15 and 16, draw the orthographic views needed to describe each part completely. Save the drawings as P11-15 *and* P11-16.

15. **Mechanical**

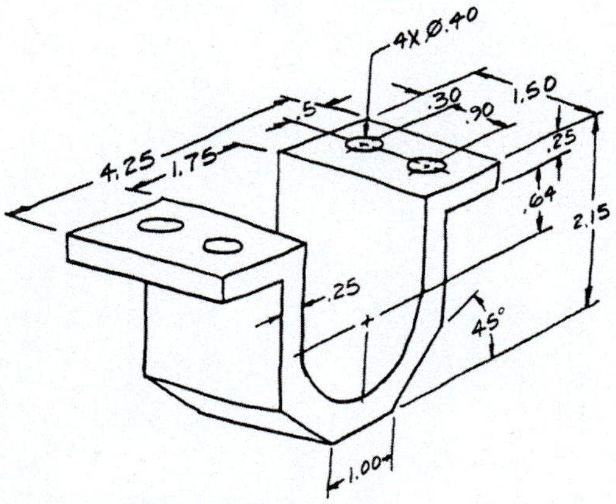

Journal Bracket (Engineer's Rough Sketch)

16. **Mechanical**

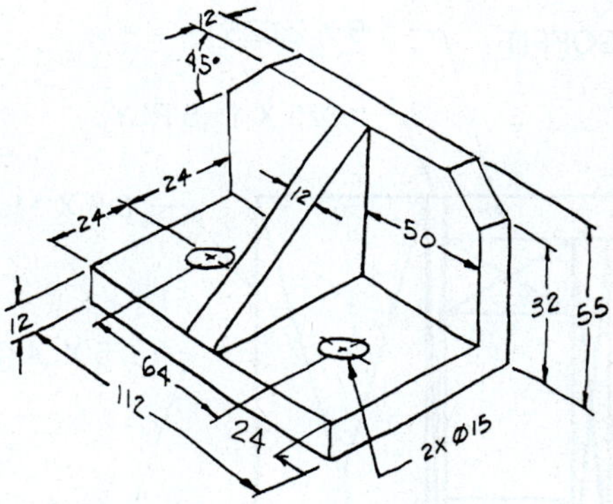

Angle Bracket (Engineer's Rough Sketch)
(Metric)

17. **Architectural** Research the specifications of a mudsill anchor with the following requirements: appropriate for stem wall or slab foundation, 2 × 6 sill, attaches using 10d nails, minimum 700-pound uplift load, 16-gage steel, hot-dipped galvanized. Create a freehand, dimensioned 2D sketch of the design from the manufacturer's specifications, or from measurements taken from an actual anchor. Start a new drawing from scratch or use a decimal, fractional, or architectural template of your choice. Draw the front, top, right-side, and left-side views of the anchor from your sketch using the 0 layer. The 0 layer is appropriate because you will create a block of each view in Chapter 25. Do not draw dimensions. Save the drawing as P11-17.

AutoCAD Certified Professional Exam Practice

Answer the following questions using the information in this chapter.

1. Which of the following can you use to extend two line segments to meet exactly at a sharp point? *Select all that apply.*
   A. **CHAMFER** command with a distance of 0
   B. **Extend** mode of the **EXTEND** command
   C. **Extend** mode of the **TRIM** command
   D. **FILLET** command with a radius of 0
   E. **LENGTHEN** command with a delta of 0
2. Which of the following commands can resize a rectangle from 2″ × 4″ to 2″ × 8″ in a single operation? *Select all that apply.*
   A. **EXTEND**
   B. **JOIN**
   C. **LENGTHEN**
   D. **SCALE**
   E. **STRETCH**
3. Which keyboard key or keys can you press to trim an object while the **EXTEND** command is active? *Select the one item that best answers the question.*
   A. [Alt]
   B. [Ctrl]
   C. [F2]
   D. [Shift]
   E. [Shift]+[Ctrl]

Follow the instructions in each problem. Complete each problem by answering the related question.

4. **Navigate to this chapter on the companion website and open CPE-11fillet.dwg.** Use an appropriate option of the **FILLET** command to round all of the corners of the polyline using a fillet radius of 1.5, as shown in the figure. What are the coordinates of Point 1 (the center of the bottom-right radius)?

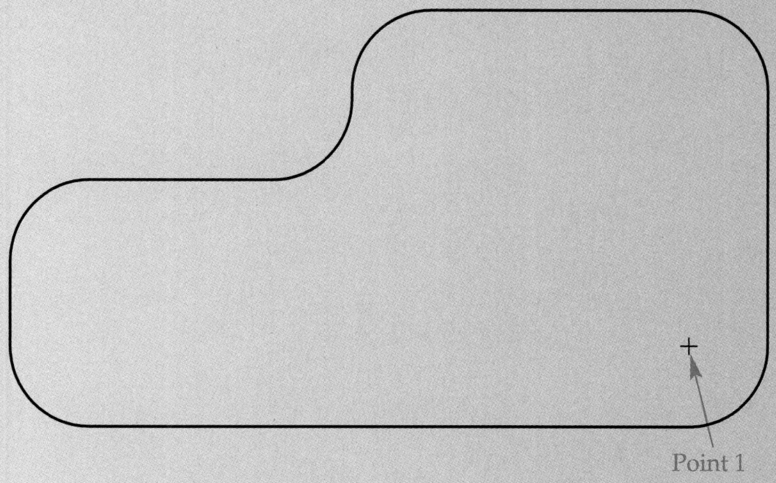

5. **Navigate to this chapter on the companion website and open CPE-11scale.dwg.** Enlarge the entire object using the **Reference** option of the **SCALE** command. Use Point 1 as the base point, and use Line A to set the reference length. Specify a new length of 1.35. Zoom out to see the result. What are the coordinates of Point 2?

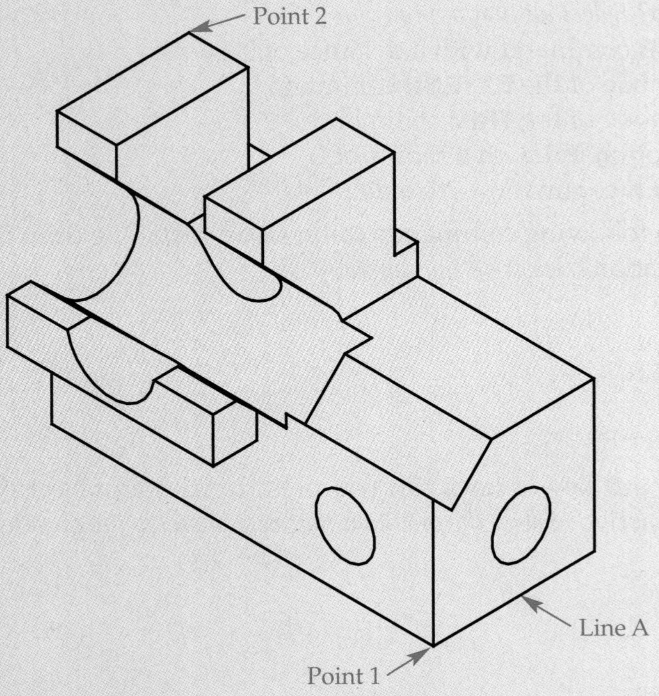

CHAPTER 12

Arranging and Patterning Objects

Learning Objectives

After completing this chapter, you will be able to:

- ✓ Relocate objects using the **MOVE** command.
- ✓ Change the angular position of objects using the **ROTATE** command.
- ✓ Use the **ALIGN** command to move and rotate objects at the same time.
- ✓ Make copies of objects using the **COPY** command.
- ✓ Draw mirror images of objects using the **MIRROR** command.
- ✓ Use the **REVERSE** command.
- ✓ Create patterns of objects using array commands.

This chapter explains methods for arranging and patterning existing objects using basic editing commands. The approach to editing presented in this chapter is to access a command, such as **MOVE**, and then follow the prompts to complete the operation. Another technique, presented in Chapter 13, is to select objects first using the crosshairs, then access the editing command, and finally complete the operation.

Moving Objects

Access the **MOVE** command to move objects to a different location. Select objects to move and then proceed to the next prompt to specify the base point from which the objects will move. Although the position of the base point is often not critical, you may want to select a point on an object, the corner of a view, or the center of a circle, for example. After you specify the base point, the move icon appears next to the crosshairs and the selected objects move as you move the crosshairs. Specify a second point to complete the move. Preview the operation before moving. See **Figure 12-1**.

MOVE

Ribbon
Home
> Modify
Move
Type
MOVE
M

Displacement Option

The **Displacement** option allows you to move objects relative to their current location. To move using a displacement, access the **MOVE** command and select objects to move. Then select the **Displacement** option instead of defining the base point. At the Specify displacement <0.0000,0.0000,0.0000>: prompt, enter a coordinate or pick a point to specify the displacement. See **Figure 12-2**.

Figure 12-1.
Using the **MOVE** command to relocate objects.

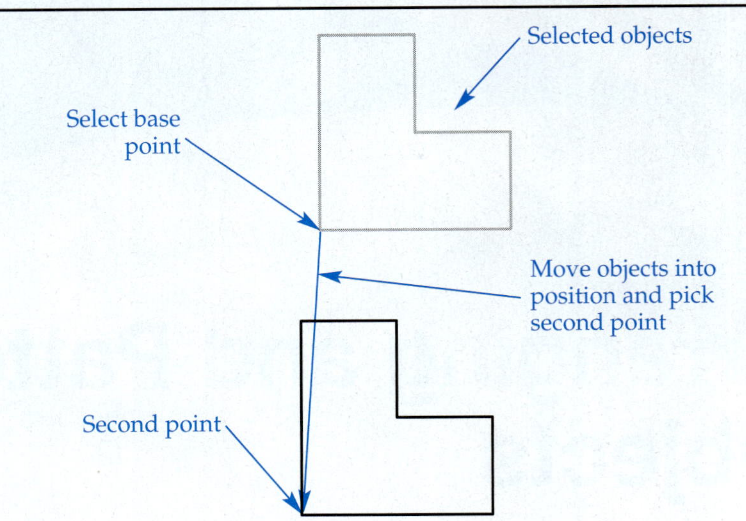

Figure 12-2.
Using the **Displacement** option of the **MOVE** command. In this example, the bottom-left corner of the original polygon is located at the 0,0 origin. The coordinate point 2,2 is the displacement.

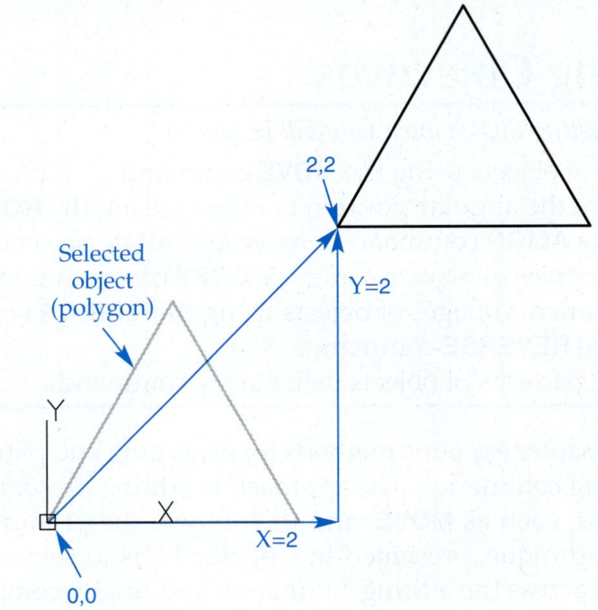

Using the First Point As Displacement

Another method for moving an object is to use the first point as the displacement. The coordinates you use for the base point automatically define the coordinates for the direction and the distance to move the object. Access the **MOVE** command and select objects to move. Then specify the base point, and instead of locating the second point, press [Enter] to accept the <use first point as displacement> default. See Figure 12-3.

PROFESSIONAL TIP

Use object snap modes while editing. For example, to move an object to the center of a circle, use the **Center** object snap mode to select the center of the circle.

Figure 12-3.
Moving a circle using the selected base point, 1,1 in this example, as the displacement.

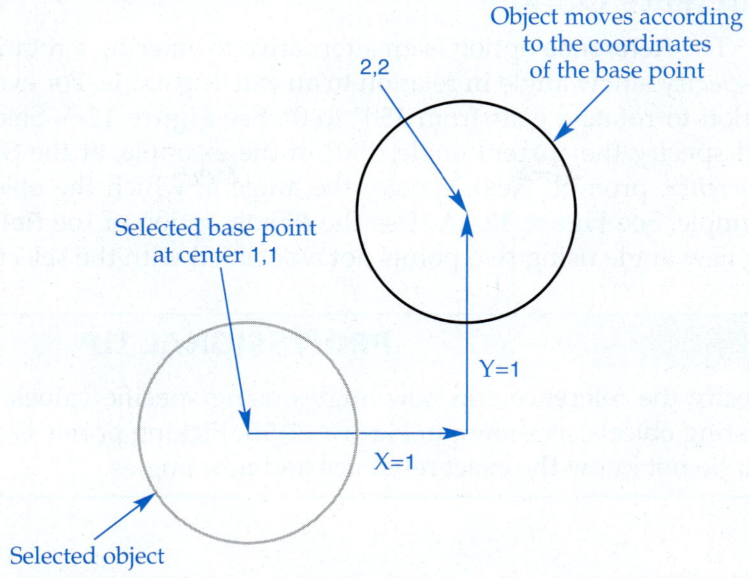

Exercise 12-1

www.g-wlearning.com/CAD/

Complete the exercise on the companion website.

Rotating Objects

Ribbon
Home
> Modify
Rotate
Type
ROTATE
RO

ROTATE

Use the **ROTATE** command to rotate objects. For example, rotate furniture to adjust an interior design plan, or rotate the north arrow on a site plan. Access the **ROTATE** command and select objects to rotate. Proceed to the next prompt and specify the base point, or axis of rotation, around which the objects rotate. After you specify the base point, the rotate icon appears next to the crosshairs and the selected objects rotate as you move the crosshairs. Next, enter a value or specify a point to define a rotation angle at the Specify rotation angle or [Copy/Reference] <*current*>: prompt. Objects rotate counterclockwise by default. To rotate an object clockwise, use a negative value. Preview the operation before rotating. See **Figure 12-4**.

Figure 12-4.
Rotating a north arrow on a site plan –30° (330°) and 30°.

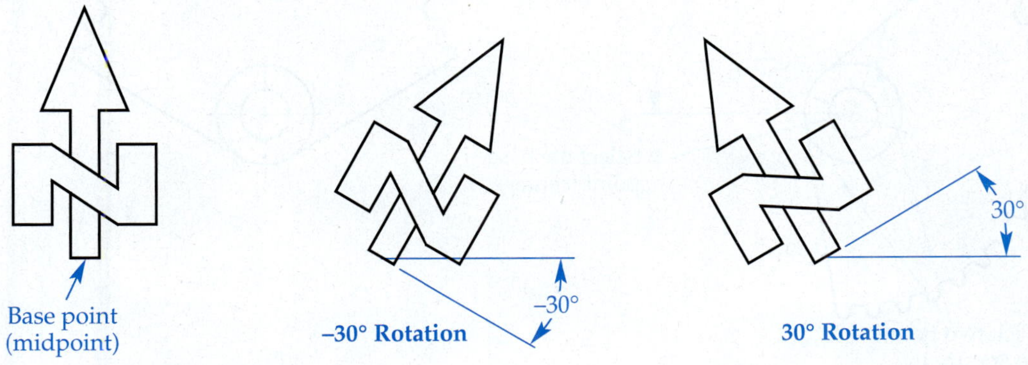

Reference Option

The **Reference** option is an alternative to entering a rotation angle and allows you to specify a new angle in relation to an existing angle. For example, use the **Reference** option to rotate a gear from 150° to 0°. See Figure 12-5. Select the **Reference** option and specify the current angle, 150° in the example, at the Specify the reference angle <*current*>: prompt. Next, specify the angle at which the objects should be, 0° in the example. See Figure 12-5A. Use the **Points** option of the **Reference** option to specify the new angle using two points not associated with the selected base point.

PROFESSIONAL TIP

Specify the reference and new angles using specific values; or pick points, often on existing objects, as shown in Figure 12-5B. Picking points is especially effective when you do not know the exact reference and new angles.

Figure 12-5.
Using the **Reference** option of the **ROTATE** command to rotate a gear on an assembly drawing according to the current angle of the objects. A—Entering reference angles. B—Selecting points on a reference line.

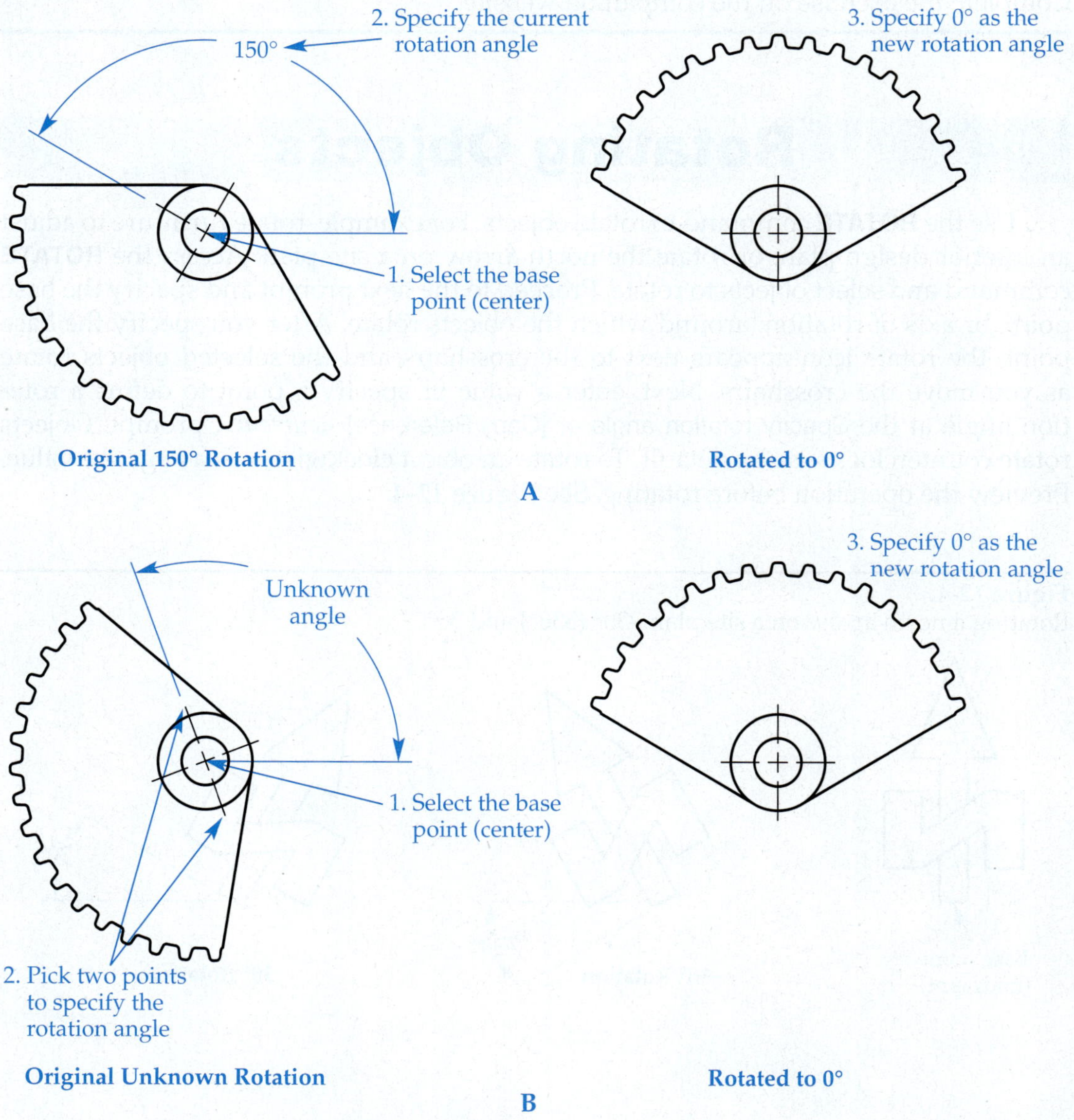

Copying While Rotating

The **Copy** option of the **ROTATE** command copies and rotates the selected objects, leaving the original objects unchanged. The copy rotates to the specified angle.

Exercise 12-2

www.g-wlearning.com/CAD/

Complete the exercise on the companion website.

Aligning Objects

Ribbon
Home
> Modify

Align

Type
ALIGN
AL

The **ALIGN** command moves and rotates objects in one operation. The **ALIGN** command can be used to align 2D or 3D objects. Aligning 3D objects is described in ***AutoCAD and Its Applications—Advanced***. Access the **ALIGN** command and select objects to align. Proceed to the next prompt to specify ***source points*** and ***destination points***. Pick the first source point, followed by the first destination point. Then pick the second source point and the second destination point. Press [Enter] when the prompt requests the third source and destination points, as these points are not necessary for aligning 2D objects. See Figure 12-6. The last prompt allows you to change the size of the source objects. Select **Yes** to scale the source objects if the distance between the source points is different from the distance between the destination points. See Figure 12-7.

source points: Points to define the original position of objects during an **ALIGN** operation.

destination points: Points to define the new location of objects during an **ALIGN** operation.

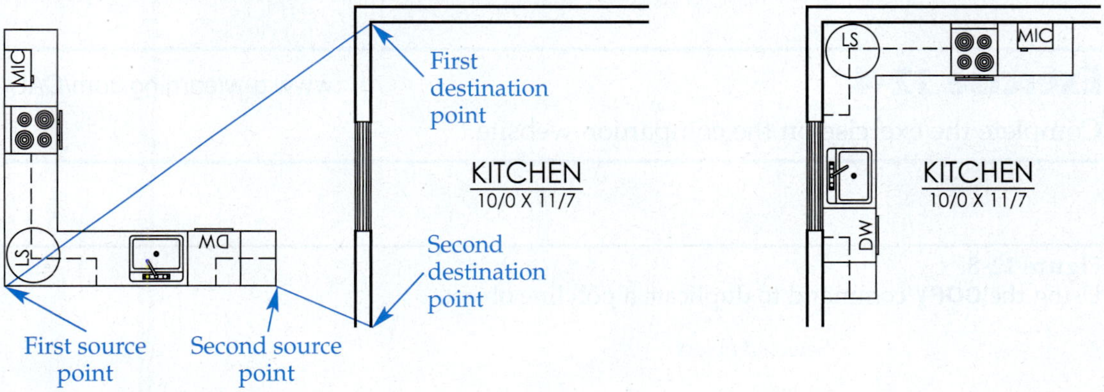

Figure 12-6.
Using the **ALIGN** command to move and rotate a kitchen cabinet layout against a wall. Select the **No** option when prompted to scale objects to apply this example.

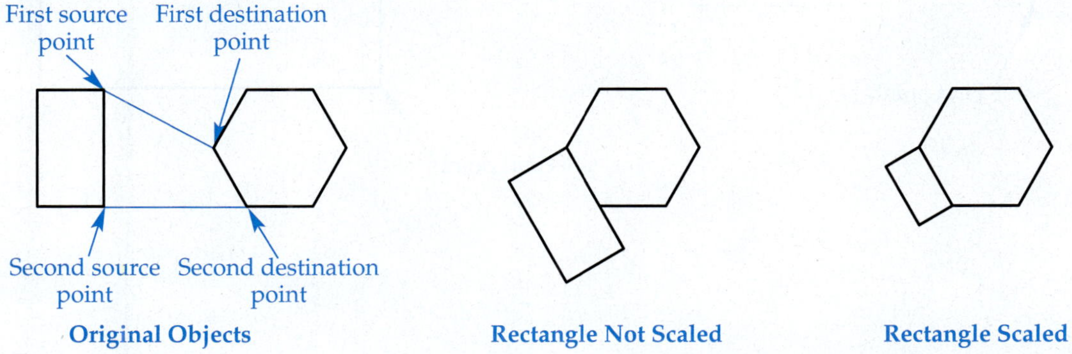

Figure 12-7.
Select the **Yes** option when prompted to scale objects to change the size of an object during the alignment. The rectangle in this example is a polyline object created using the **RECTANGLE** command.

Exercise 12-3

www.g-wlearning.com/CAD/

Complete the exercise on the companion website.

Copying Objects

COPY

Ribbon
Home
> Modify
Copy

Type
COPY
CO
CP

Access the **COPY** command to duplicate objects. Select objects to copy and then proceed to the next prompt to specify the base point from which the objects will copy. After you specify the base point, the copy icon appears next to the crosshairs and the selected objects move as you move the crosshairs. Specify a second point to complete the copy. Preview the operation before copying. See Figure 12-8. You can continue creating copies of the selected objects by specifying additional points. Use the **Undo** option to remove copies without exiting the **COPY** command. Press [Enter] to exit.

The **COPY** command provides the same options as the **MOVE** command, allowing you to specify a base point and a second point, select a displacement using the **Displacement** option, or define the first point as the displacement.

NOTE

The **Multiple** copy mode is active by default and allows you to create several copies of the same object using a single **COPY** operation. To make a single copy and exit the command after placing the copy, select the **mOde** option and activate the **Single** option.

Exercise 12-4

www.g-wlearning.com/CAD/

Complete the exercise on the companion website.

Figure 12-8.
Using the **COPY** command to duplicate a polyline object.

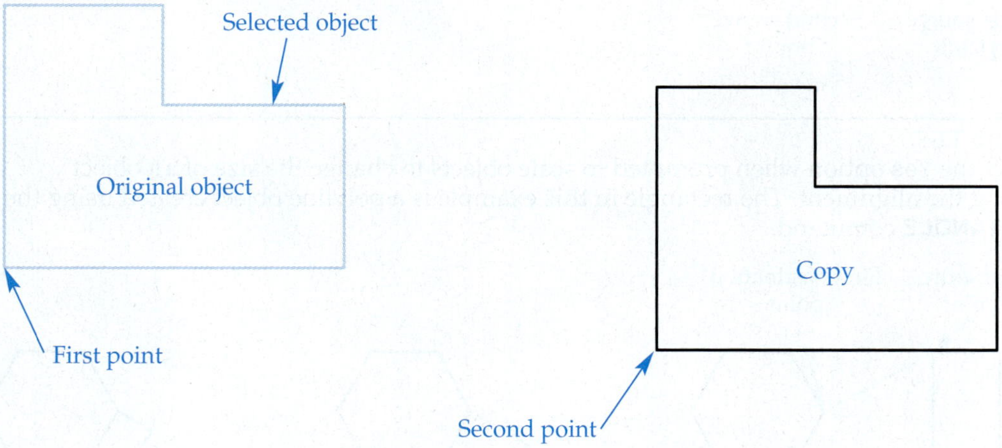

Array Option

Use the **Array** option of the **COPY** command, available after you specify the base point, to create a linear pattern of the selected objects. AutoCAD includes other **ARRAY** commands, described later in this chapter, which are usually more appropriate for patterning objects. However, the **Array** option of the **COPY** command is effective for copying multiple, equally spaced objects quickly.

Select the **Array** option and then enter the total number of copies, including the selected objects, to create. Specify the location of the first copy, which also defines the spacing between copies. An alternative is to use the **Fit** option and then specify the location of the last copy, which divides the number of items equally between the base point and second point. See Figure 12-9.

Exercise 12-5

www.g-wlearning.com/CAD/

Complete the exercise on the companion website.

Mirroring Objects

The **MIRROR** command allows you to reflect, or mirror, objects. For example, in mechanical drafting, mirror a part to form the opposite component of a symmetrical assembly. In architectural drafting, mirror a floor plan to create a duplex residence or to accommodate a different site orientation. Access the **MIRROR** command and select the objects to mirror. Proceed to the next prompt and specify two points to define an imaginary ***mirror line***. After you locate the second mirror line point, you have the option to delete the original objects. Preview the operation before mirroring. See Figure 12-10.

MIRROR

Ribbon
Home
> Modify
Mirror

Type
MIRROR
MI

mirror line: The line of symmetry across which objects are mirrored.

NOTE

The **MIRRTEXT** system variable, which is set to 0 by default, prevents text from reversing during a mirror operation. Change the **MIRRTEXT** value to 1 to mirror text in relation to the original object. See Figure 12-11. Backward text is generally not acceptable, except for in applications such as reverse imaging.

Figure 12-9.
Using the **Array** option of the **COPY** command to draw a linear pattern of rollers along the frame of a conveyor.

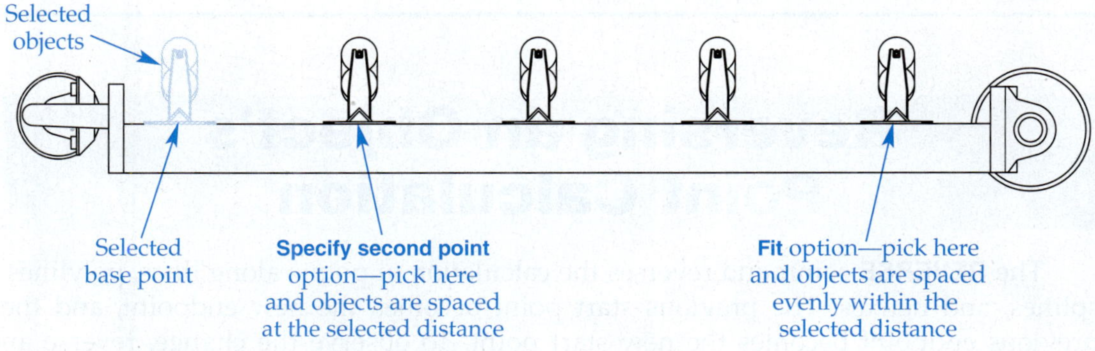

Figure 12-10.
Using the **MIRROR** command to reflect objects over an imaginary mirror line. You have the option of erasing the original objects. A—To specify the mirror line, pick two points. B—To erase the source objects, enter Y at the Erase source objects? prompt. C—To keep the source objects, enter N at the Erase source objects? prompt. D—The final object after using the **TRIM** and **FILLET** commands.

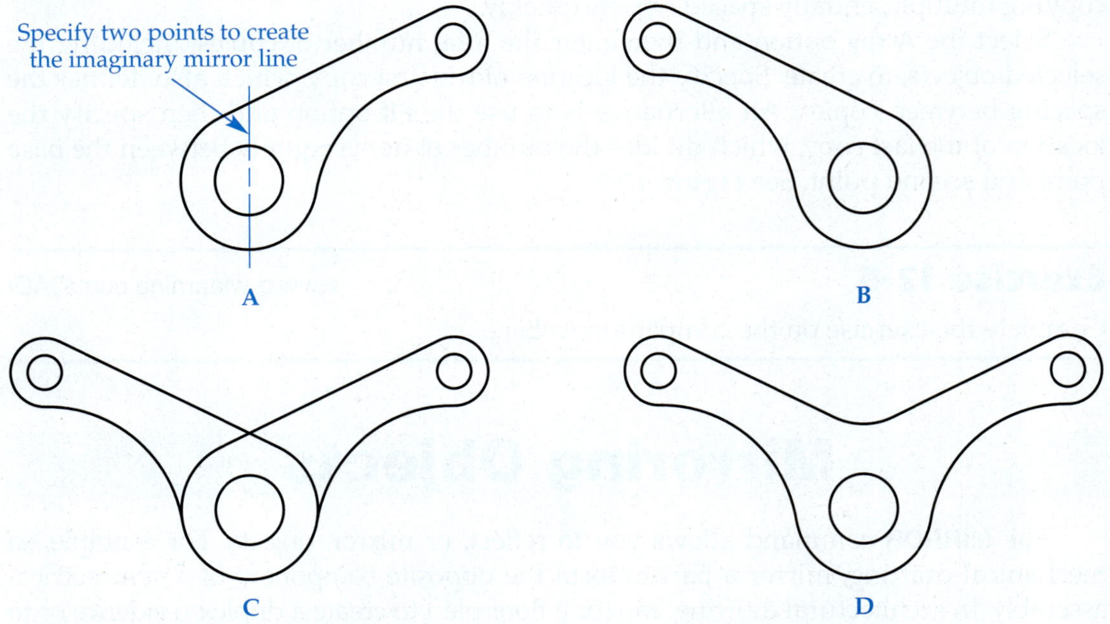

Figure 12-11.
The **MIRRTEXT** system variable options.

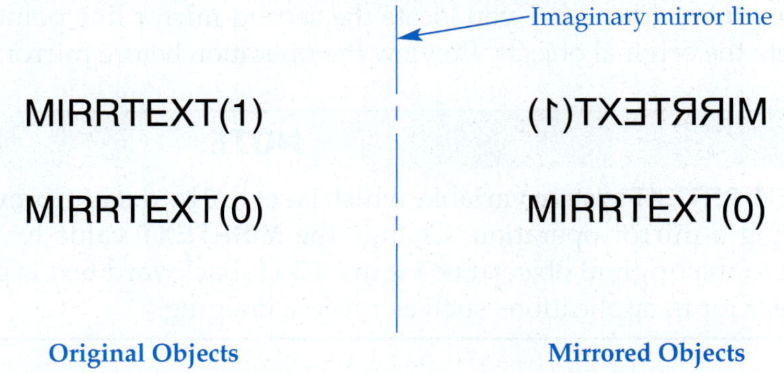

Exercise 12-6

www.g-wlearning.com/CAD/

Complete the exercise on the companion website.

Reversing an Object's Point Calculation

REVERSE

Ribbon
Home
> Modify
Reverse
Type
REVERSE

The **REVERSE** command reverses the calculation of points along lines, polylines, splines, and helixes. The previous start point becomes the new endpoint, and the previous endpoint becomes the new start point. To observe the change, reverse an object assigned a linetype that includes text. See Figure 12-12A. However, AutoCAD attempts to orient text included with linetypes correctly by default for all objects. You should typically avoid reversing text included with linetypes.

Figure 12-12.
A—Reversing a polyline assigned a linetype that includes text. B—Using the **REVERSE** command to reverse a polyline with varying width. You must change the **PLINEREVERSEWIDTHS** system variable to 1 in order to make this edit.

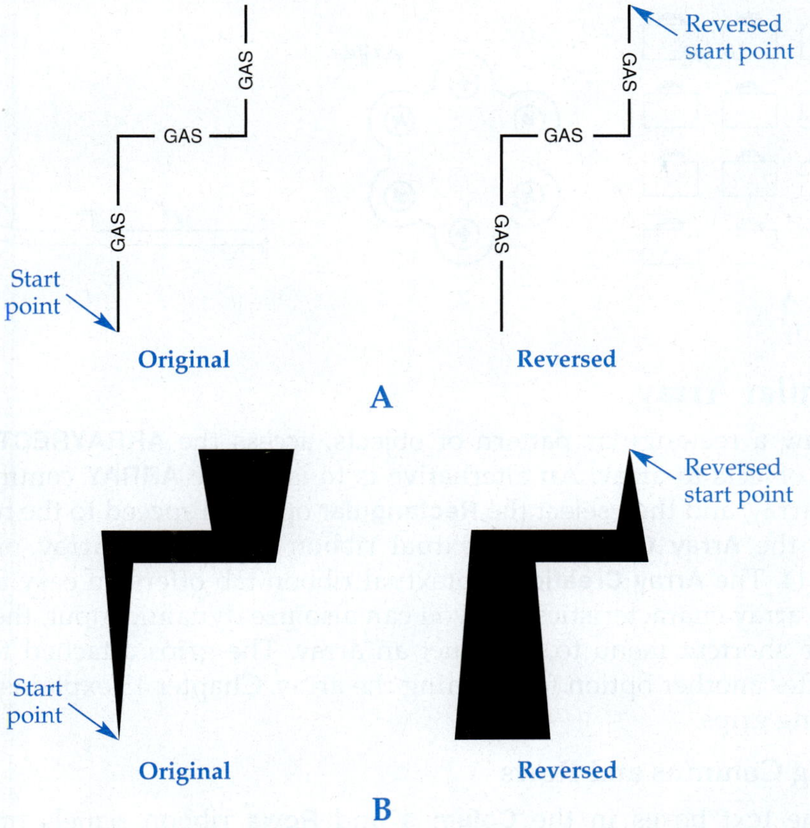

The **PLINEREVERSEWIDTHS** system variable controls how the **REVERSE** command affects polyline objects. The default setting of 0 only allows you to reverse start and end points. Polylines with varying width appear unchanged. Set the **PLINEREVERSEWIDTHS** system variable to 1 to also reverse varying polyline width. See Figure 12-12B.

NOTE

You can also use the **Reverse** option of the **PEDIT** command to reverse polylines and the **Reverse** option of the **SPLINEDIT** command to reverse splines. The **PEDIT** and **SPLINEDIT** commands are explained later in this textbook. Reversing affects the vertex options of the **PEDIT** command and control point options of the **SPLINEDIT** command. The **PLINEREVERSEWIDTHS** system variable also applies to the **PEDIT** command.

Arraying Objects

AutoCAD includes methods for creating and modifying a pattern, or ***array***, of existing objects. For example, create a ***rectangular array*** of computer workstations on a classroom design plan, or create a ***polar array*** (also called a ***circular array***) of screws on a mechanical assembly drawing. You can also pattern objects in reference to an existing object using a ***path array***. Figure 12-13 shows examples of arrays.

array: Multiple copies of an object arranged in a pattern.

rectangular array: A pattern made up of columns and rows of objects.

polar (circular) array: A circular pattern of objects.

path array: A pattern of objects drawn in reference to another object, or path.

Figure 12-13.
A—A rectangular array of computer workstations on a classroom design plan. B—A polar array of arcs and screws on a mechanical assembly drawing. C—A path array of angle brackets along a polyline on the drawing of a steel structure.

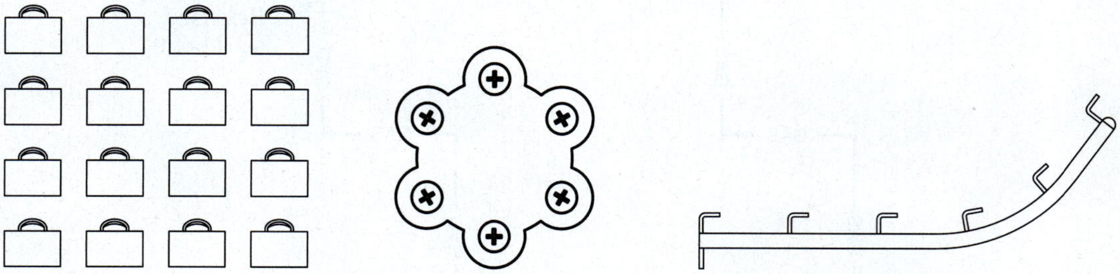

A B C

Rectangular Array

ARRAYRECT
Ribbon
Home
> Modify
Rectangular Array
Type
ARRAYRECT

ARRAY
Type
ARRAY
AR

To draw a rectangular pattern of objects, access the **ARRAYRECT** command and select objects to array. An alternative is to issue the **ARRAY** command, select objects to array, and then select the **Rectangular** option. Proceed to the next prompt to display the **Array Creation** contextual ribbon tab and an array preview. See Figure 12-14. The **Array Creation** contextual ribbon tab offers an easy approach to specifying array characteristics, but you can also use dynamic input, the command line, or the shortcut menu to construct an array. The grips attached to the array preview offer another option for defining the array. Chapter 13 explains modifying objects using grips.

Specifying Columns and Rows

Use the text boxes in the **Columns** and **Rows** ribbon panels or apply the **COLumns** and **Rows** options to adjust the number of columns and rows, column and row spacing, and total column and row distance. Column and row count determines the total number of columns and rows in the array. Figure 12-14B shows the default array of four columns and three rows. If you do not know exact counts, enter equations or formulas in the **Column Count** and **Row Count** ribbon text boxes, or apply the **Expression** option at the appropriate prompt of the **COLumns** and **Rows** options.

You can specify column spacing or total column distance and row spacing or total row distance. Column or row spacing is the distance from a point on an item in a column or row to the corresponding object point in the next column or row. See Figure 12-14B. If you do not know exact spacing values, enter equations or formulas in the **Column Spacing** and **Row Spacing** ribbon text boxes, or apply the **Expression** option at the appropriate prompt of the **COLumns** and **Rows** options.

The total column or row distance is the distance from a point on an item in the first column or row to the corresponding object point in the last column or row. See Figure 12-14B. Specify total column and row distances using the **Total Column Distance** and **Total Row Distance** ribbon text boxes, or apply the **Total** option at the appropriate prompt of the **COLumns** and **Rows** options. Adjust the direction of the array using positive or negative values, as shown in Figure 12-15.

NOTE

The **COLumns** and **Rows** options allow you to enter a numeric value or specify points for column and row spacing or total distance.

Figure 12-14.
Using the **ARRAYRECT** command to create a rectangular array of a circle. A—The **Array Creation** contextual ribbon tab. B—The initial array preview and grips. Use the preview for reference as you construct the array.

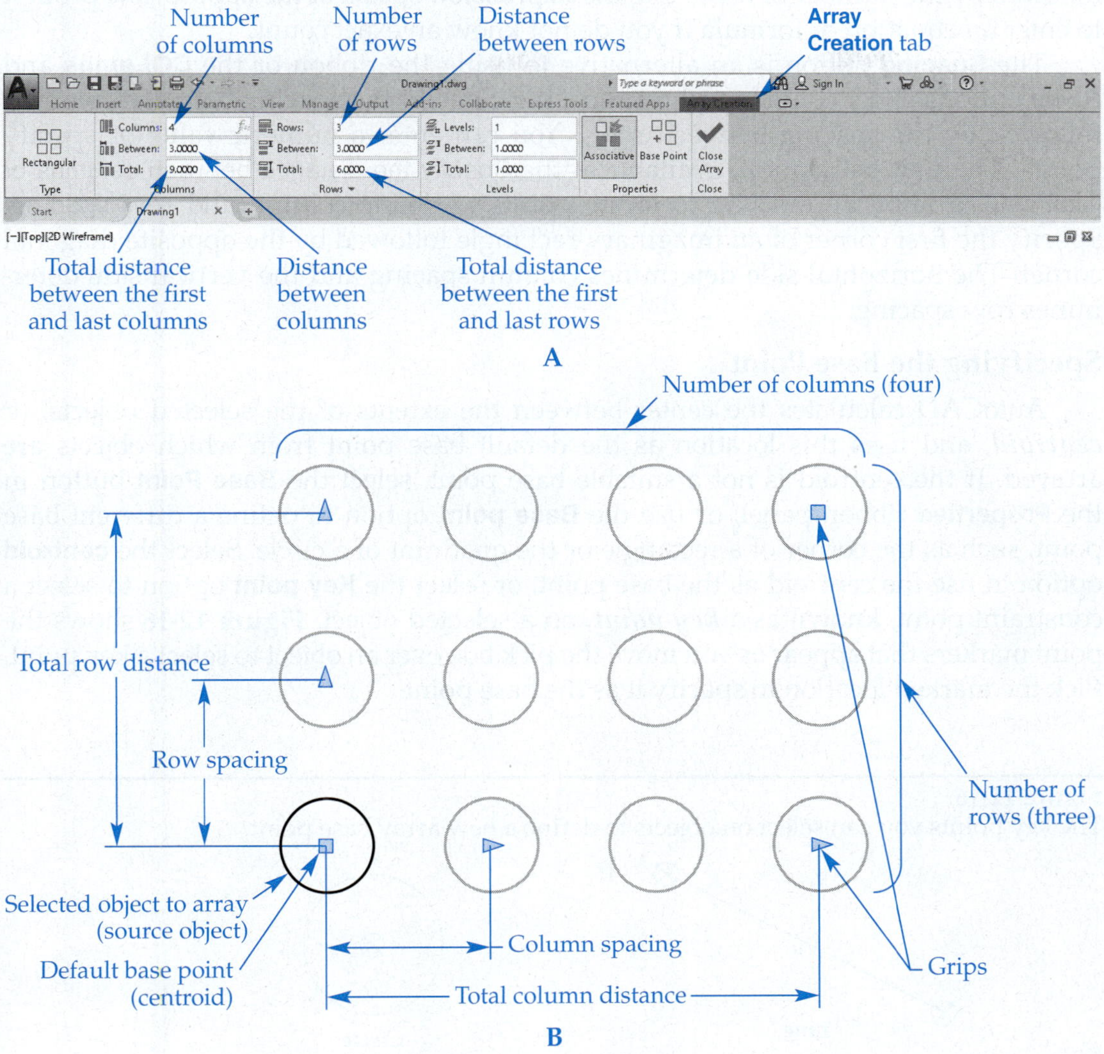

Figure 12-15.
Positive or negative column and row spacing and total column and row distance determine the direction in which an array will grow from the source objects.

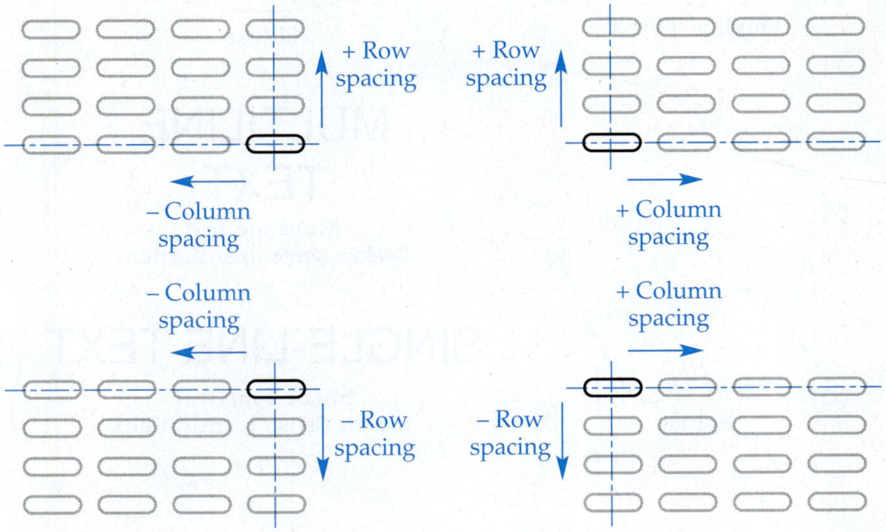

Count and Spacing Options

The **COUnt** option is an alternative to using the ribbon or the **COLumns** and **Rows** options to specify the number of columns and rows. Enter the number of columns followed by the number of rows. Use the **Expression** option at the appropriate prompt to enter an equation or formula if you do not know an exact count.

The **Spacing** option is an alternative to using the ribbon or the **COLumns** and **Rows** options to set column and row spacing. Specify the spacing between columns followed by the spacing between rows. You can enter a numeric value or specify points. The **Unit cell** option, available at the Specify the distance between columns or [Unit cell] prompt, allows you to locate points to define column and row spacing. Specify the first corner of an imaginary rectangle followed by the opposite diagonal corner. The horizontal side determines column spacing and the vertical side determines row spacing.

Specifying the Base Point

centroid: The mathematical center of a geometric figure.

Ribbon
Array Creation > Properties

Base Point

key point: The point on a selected object that you use to manipulate the object.

AutoCAD calculates the center between the extents of the selected objects, or ***centroid***, and uses this location as the default base point from which objects are arrayed. If the centroid is not a suitable base point, select the **Base Point** button in the **Properties** ribbon panel, or use the **Base point** option to define a different base point, such as the corner of a rectangle or the quadrant of a circle. Select the **centroid** option to use the centroid as the base point, or select the **Key point** option to select a constraint point, known as a ***key point***, on a selected object. Figure 12-16 shows the point markers that appear as you move the pick box over an object to select a key point. Pick the marked location to specify it as the base point.

Figure 12-16.
The key points you can select on objects to define a new array base point.

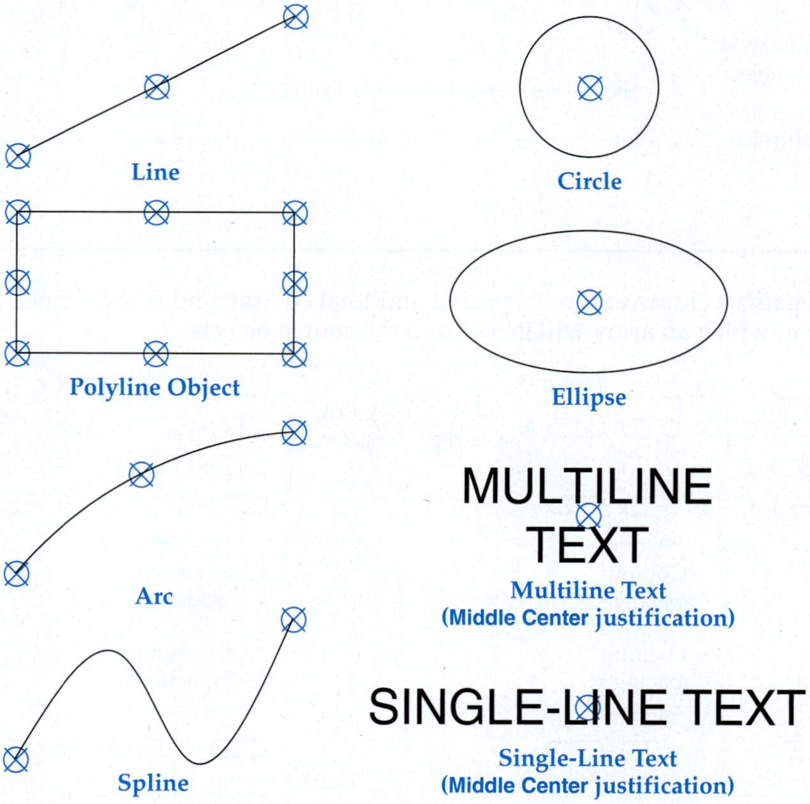

Finalizing the Array

AutoCAD creates an ***associative array*** by default. An associative array is a single array object. The major advantage of using associative arrays is the ability to modify the parameters of the array and to edit or replace the source objects without recreating the array. Chapter 13 describes working with associative arrays. To create a ***non-associative array***, deselect the **Associative** button in the **Properties** ribbon panel or select **No** at the **ASsociative** option. To create the array and exit the command, pick the **Close Array** button in the **Close** ribbon panel; press [Enter] or [Esc]; or select the **eXit** option.

Ribbon
Array Creation
> Properties
Associative

Ribbon
Array Creation
> Close
Close Array

associative array: An adjustable array object; all items are grouped to form a single object, allowing you to make modifications such as changing the number of items and spacing between items.

non-associative array: An array of copied, or static, source objects that do not form a single adjustable array object.

PROFESSIONAL TIP

Form associative arrays whenever possible to increase efficiency when making changes to arrayed objects. AutoCAD stores the current associative setting as the default for all arrays. Remember to change the associative setting as necessary before finalizing an array. If necessary, use the **EXPLODE** command to remove the associative property from existing arrays.

CAUTION

The **Incremental Elevation** text box in the expanded **Rows** panel, the Specify the incrementing elevation between rows or [Expression] <0.0000>: prompt associated with the **Rows** option, the **Levels** panel, and the **Levels** option are appropriate only for 3D applications. *AutoCAD and Its Applications—Advanced* provides complete information on 3D modeling. Use an incremental elevation of zero and one level for all 2D applications.

Exercise 12-7

www.g-wlearning.com/CAD/

Complete the exercise on the companion website.

Polar Array

Ribbon
Home
> Modify
Polar Array
Type
ARRAYPOLAR

ARRAYPOLAR

To draw a circular pattern of objects, access the **ARRAYPOLAR** command and select objects to array. An alternative is to issue the **ARRAY** command, select objects to array, and then select the **POlar** option. Proceed to the next prompt to specify the center point around which the objects will be arrayed. The array forms around the Z axis of the UCS. You can apply the **Base point** option before locating the center point if necessary. The **Base point** option operates the same as when creating a rectangular array. However, the purpose of relocating the base point for a polar array is most apparent when you edit an associative polar array, as explained in Chapter 13.

Once you specify a center point, the **Array Creation** ribbon tab and an array preview appear. See Figure 12-17. Use the **Array Creation** ribbon tab, dynamic input or the command line, or the shortcut menu to construct the array. Grips offer another option for defining the array, as described in Chapter 13.

Specifying Items

You can specify the total number of items and angle between items or the total number of items and angle to fill. Use the text boxes in the **Items** ribbon panel, or apply the **Items**, **Angle between**, and **Fill angle** options. Item count determines the total number of items in the array. Figure 12-17B shows the default array of six items. If you do not know an exact count, enter an equation or formula in the **Item Count** ribbon text box, or apply the **Expression** option at the appropriate prompt of the **Items** option.

Figure 12-17.
Creating a polar array of a polyline object. A—The **Array Creation** contextual ribbon tab. B—The initial array preview and grips. Use the preview for reference as you construct the array.

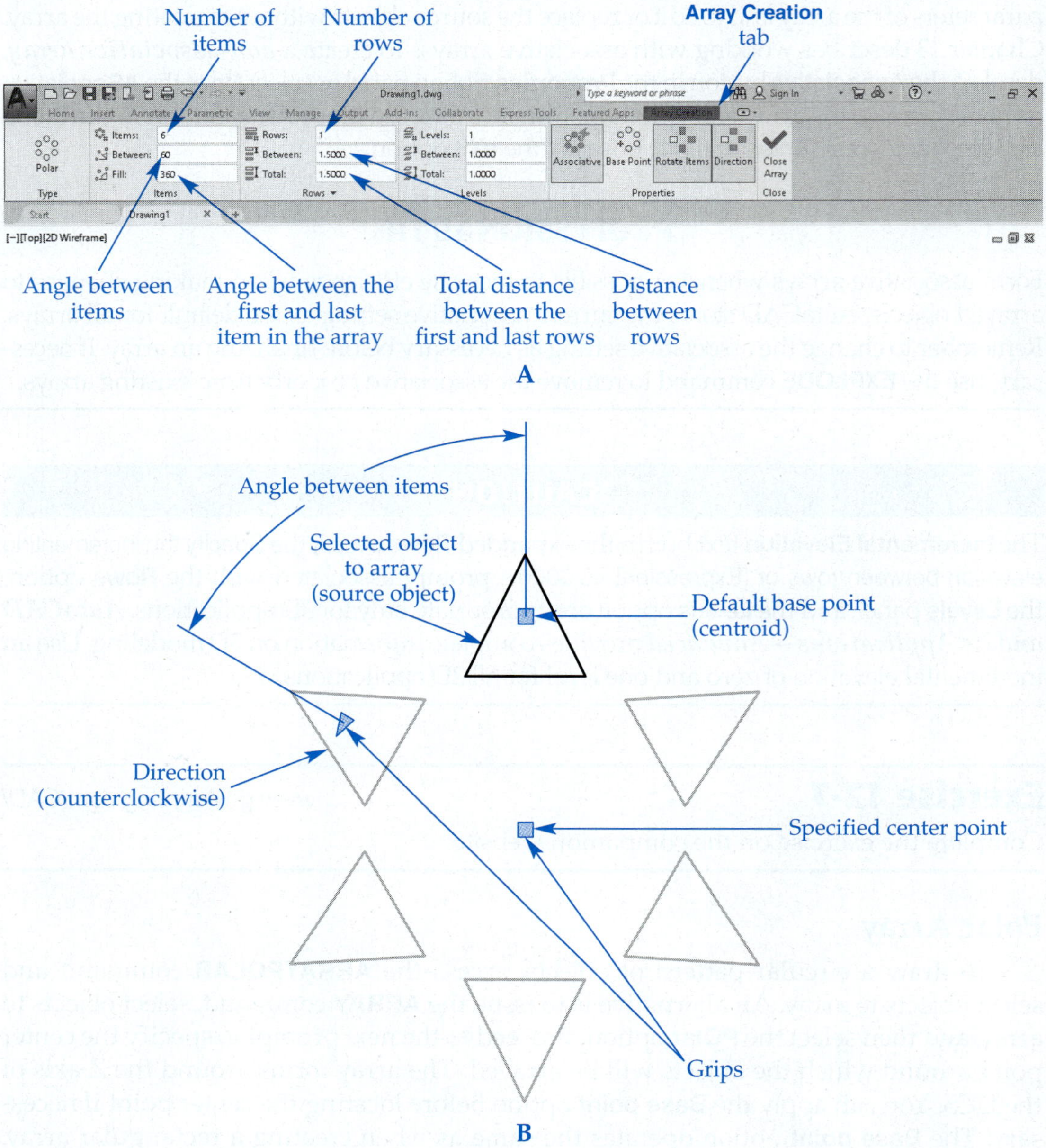

Total Number of Items and Angle between Items

Specify the included angle between adjacent items, as shown in Figure 12-17B, in the **Angle between Items** ribbon text box, or use the **Angle between** option. Enter a positive, nonzero value. If you do not know an exact angle, enter an equation or formula in the **Angle between Items** ribbon text box, or apply the **Expression** option at the appropriate prompt of the **Angle between** option.

Total Number of Items and Angle to Fill

Specify the included angle to fill with all items in the **Fill Angle** ribbon text box or use the **Fill angle** option. For example, enter 360 to create a complete circular array, as shown in Figure 12-17B, or enter 180 to create a half circle array. AutoCAD converts a negative value entered in the **Fill Angle** ribbon text box to positive. However, a negative value entered while using the **Fill angle** option changes the array to occur in a clockwise direction. You will learn more about polar array direction later in this chapter.

If you do not know an exact angle, enter an equation or formula in the **Fill Angle** ribbon text box, or apply the **Expression** option at the appropriate prompt of the **Fill angle** option.

NOTE

The **Angle between** and **Fill angle** options allow you to enter a numeric value or specify points for the angle between items or fill angle.

Adding Rows

Use the text boxes in the **Rows** ribbon panel, or apply the **ROWs** option, to add multiple rows during the array. See Figure 12-18. Row count determines the total number of rows in the array. Figure 12-17B shows the default array with one row. Figure 12-18 shows an array with three rows. If you do not know an exact count, enter an equation or formula in the **Row Count** ribbon text box, or apply the **Expression** option at the appropriate prompt of the **ROWs** option.

You can specify row spacing or total row distance. Row spacing is the distance from a point on an item in a row to the corresponding object point in the adjacent row. Enter a positive, nonzero value. If you do not know an exact spacing, enter an equation or formula in the **Row Spacing** ribbon text box, or apply the **Expression** option at the appropriate prompt of the **ROWs** option.

The total row distance is the total spacing from a point on an item in the first row to the corresponding object point in the last row. To specify the total row distance, use the **Total Row Distance** ribbon text box or apply the **Total** option at the appropriate prompt of the **ROWs** option. Enter a positive, nonzero value.

NOTE

The **ROWs** option allows you to enter a numeric value or specify points for row spacing or total distance.

Direction and Rotation Controls

To array items in a counterclockwise direction, activate the **Direction** button in the **Properties** ribbon panel, or enter a positive value while using the **Fill angle** option. To array items in a clockwise direction, deselect the **Direction** button, or enter a negative value while using the **Fill angle** option.

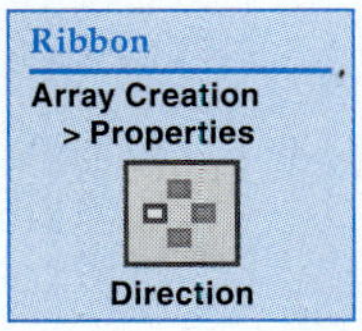

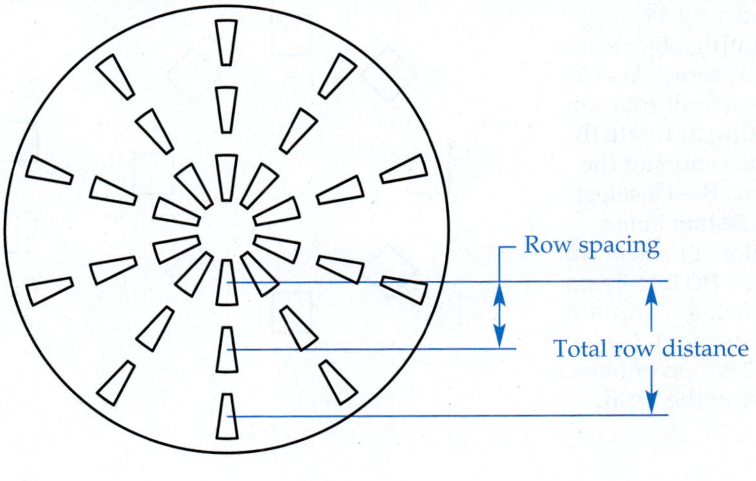

Figure 12-18. A nozzle design drawn using a polar array with 10 items, 360° angle to fill, and three rows.

AutoCAD rotates items perpendicular to the center point by default. Deselect the **Rotate Items** button in the **Properties** ribbon panel, or select **No** at the **ROTate items** option, to position items in the same orientation as the source objects. See Figure 12-19.

Ribbon
Array Creation
> Properties
Rotate Items

Finalizing the Array

The **Base Point** button and **Base point** option allow you to define a different base point and operate the same as when creating a rectangular array. Use the **Associative** button or **ASsociative** option to specify an array as associative or non-associative. To generate the array and exit the command, pick the **Close Array** button in the **Close** ribbon panel; press [Enter] or [Esc]; or select the **eXit** option.

CAUTION

The **Incremental Elevation** text box in the expanded **Rows** panel, the Specify the incrementing elevation between rows or [Expression] <0.0000>: prompt associated with the **ROWs** option, the **Levels** panel, and the **Levels** option are appropriate only for 3D applications. *AutoCAD and Its Applications—Advanced* provides complete information on 3D modeling. Use an incremental elevation of zero and one level for all 2D applications.

Exercise 12-8

www.g-wlearning.com/CAD/

Complete the exercise on the companion website.

Path Array

ARRAYPATH

Ribbon
Home
> Modify
Path Array
Type
ARRAYPATH

To array objects along a path, access the **ARRAYPATH** command and select objects to array. An alternative is to issue the **ARRAY** command, select objects to array, and then select the **PAth** option. Proceed to the next prompt to pick a line, circle, arc, ellipse, polyline, spline, or helix to use as the path along which the source objects will be arrayed. Pick the object near where you want the array to begin, because the default base point is the endpoint of the path closest to where you select the path.

Once you select the path, the **Array Creation** ribbon tab and an array preview appear. See Figure 12-20. Use the **Array Creation** ribbon tab, dynamic input or the command line, or the shortcut menu to construct the array. Grips offer another option for defining the array, as described in Chapter 13.

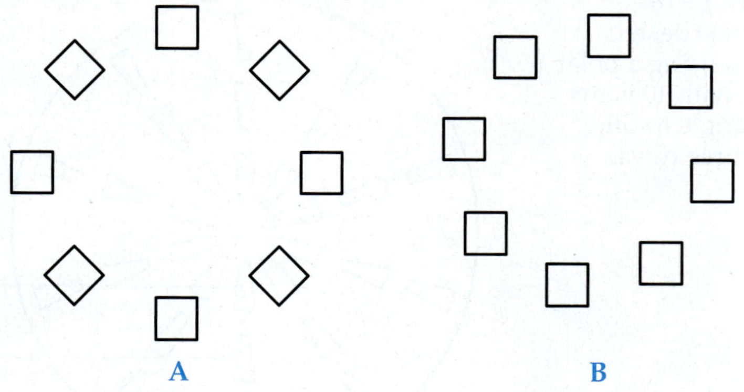

Figure 12-19. Rotating objects in a polar array. A—Use the default rotation setting to rotate the square during the array. B—Deselect the **Rotate Items** button, or select **No** at the **ROTate items** option, to maintain the original orientation of objects during the array.

Measure Path Method

The measure path method is active by default and allows you to fill the entire path with items a specific distance apart, specify the total number of items and item spacing, or specify the total number of items and total item distance. Use the text boxes in the **Items** ribbon panel or apply the **Items** option.

Ribbon
Array Creation > Properties
Measure Method

By default, AutoCAD fills the entire path with items and allows you to set item spacing, as shown in **Figure 12-20B**. Item spacing is the distance between a point on an item and the corresponding object point on the adjacent item. If you do not know an exact spacing, enter an equation or formula in the **Item Spacing** ribbon text box, or apply the **Expression** option at the appropriate prompt of the **Items** option.

In order to specify the total number of items and item spacing, or the total number of items and total item distance, deselect the **Item Count** ribbon button, or enter the item count at the appropriate prompt of the **Items** option. If you do not know an exact count, enter an equation or formula in the **Item Count** ribbon text box, or apply the **Expression** option at the appropriate prompt of the **Items** option. Set item spacing as previously described, or specify the total item distance. The total item distance is the distance between a point on the first item and the corresponding object point on the last item. See **Figure 12-20B**.

Divide Path Method

Select the **Divide Method** button from the flyout in the **Properties** ribbon panel or select **Divide** at the **Method** option to divide the item count equally along the path. Use the **Item Count** ribbon text box or the **Items** option to specify the total number of items.

Ribbon
Array Creation > Properties
Divide Method

Figure 12-20.
Creating a path array of a polyline object. A—The **Array Creation** contextual ribbon tab. B—The initial array preview and grips. Use the preview for reference as you construct the array.

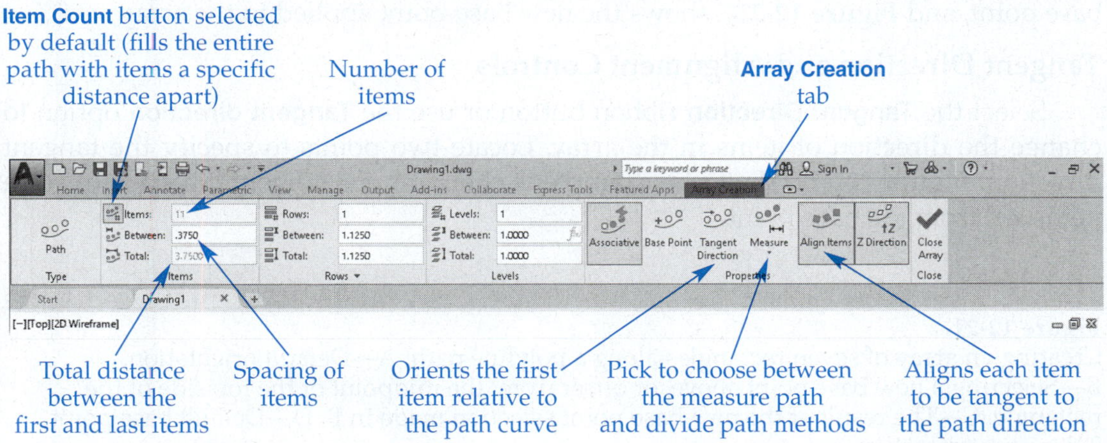

A

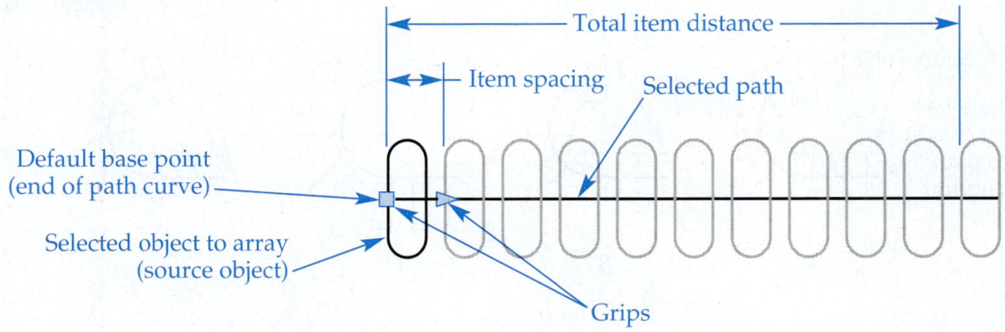

B

If you do not know an exact count, enter an equation or formula in the **Item Count** ribbon text box, or apply the **Expression** option at the appropriate prompt of the **Items** option.

NOTE

To return to the default measure path method, select the **Measure Method** button from the flyout in the **Properties** ribbon panel, or select **Measure** at the **Method** option. To return to the default technique of filling the entire path with items a specific distance apart, select the **Item Count** ribbon button, or select **Fill entire path** at the appropriate prompt of the **Items** option.

PROFESSIONAL TIP

Select the appropriate path method and specify path properties based on the information you know about the array. Use the divide path method to equally distribute a specific number of items along the length of the path.

Specifying the Base Point

The default base point from which objects are arrayed is the endpoint of the path closest to where you select the path, not a point associated with the source objects. See Figure 12-21A. If the end of the path curve is not a suitable base point, select the **Base Point** button in the **Properties** ribbon panel, or use the **Base point** option to define a different base point. Select the **end of path curve** option to reselect the default base point, or apply the **Key point** option to select a key point for the base point, as explained for creating a rectangular array. Figure 12-21B shows an example of a likely base point, and Figure 12-21C shows the new base point applied to the array.

Tangent Direction and Alignment Controls

Ribbon
Array Creation > Properties
Tangent Direction

Select the **Tangent Direction** ribbon button or use the **Tangent direction** option to change the direction of items in the array. Locate two points to specify the tangent direction. Figure 12-21D shows an example of changing the tangent direction to align items 20° from the base point.

Figure 12-21.
Creating an array of seven rectangles along a polyline path. A—Default orientation. B—Selecting a new base point above, or offset from, the midpoint of the top side of the rectangle. C—The result of the new base point selection made in B. D—Default base point, 20° tangent direction.

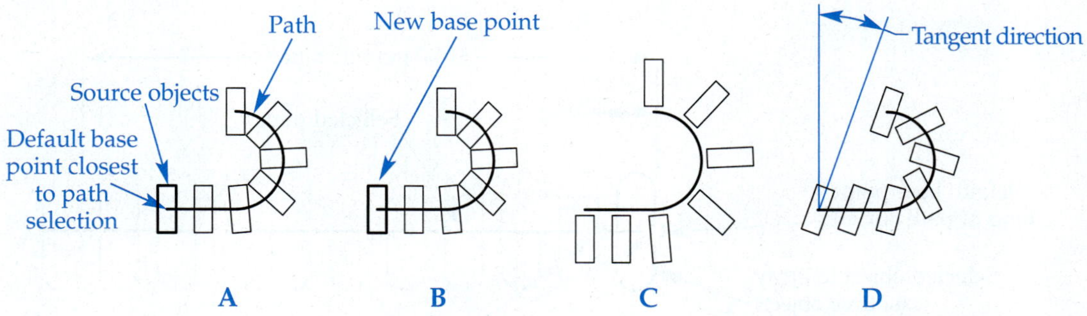

Figure 12-22.
Creating an array of four coconut palm trees along a spline path for a landscape elevation. Use the **Align Items** control to set alignment of items along the path.

Ribbon
Array Creation > Properties

Align Items

Use the **Align Items** ribbon button or the **Align items** option to control the alignment of items along the path. Items align with the path by default. See Figure 12-22A. Deselect the **Align Items** button, or select **No** at the **Align items** option, to apply the alignment of the source objects to all items along the path. See Figure 12-22B.

Finalizing the Array

Use the text boxes in the **Rows** ribbon panel or apply the **Rows** option to add multiple rows during the array, as explained for a polar array. Use the **Associative** button or **ASsociative** option to specify an array as associative or non-associative. To generate the array and exit the command, pick the **Close Array** button in the **Close** ribbon panel; press [Enter] or [Esc]; or select the **eXit** option.

CAUTION

The **Incremental Elevation** text box in the expanded **Rows** panel, the Specify the incrementing elevation between rows or [Expression] <0.0000>: prompt associated with the **Rows** option, the **Levels** panel, the **Levels** option, the **Normal** setting of the **Tangent Direction** option, and the **No** setting of the **Z Direction** option are appropriate only for 3D applications. *AutoCAD and Its Applications—Advanced* provides complete information on 3D modeling. Use an incremental elevation of zero and one level, do not use normal orientation, and maintain Z direction for all items for all 2D applications.

Exercise 12-9

www.g-wlearning.com/CAD/

Complete the exercise on the companion website.

Chapter Review

Answer the following questions using the information in this chapter.

1. How would you rotate an object 45° clockwise?
2. Describe the two methods of using the **Reference** option of the **ROTATE** command.
3. Name the command that you can use to move and rotate an object at the same time.
4. How many points must you select to align an object in a 2D drawing?
5. Explain the difference between the **MOVE** and **COPY** commands.
6. Which command allows you to draw a reflected image of an existing object?
7. What is the purpose of the **REVERSE** command?
8. What is the difference between polar and rectangular arrays?
9. What does AutoCAD use as the default base point for a rectangular array?
10. Suppose an object is 1.5″ (38 mm) wide and you want to create a rectangular array with .75″ (19 mm) spacing between objects. What should you specify for the distance between columns?
11. Define *associative array*.
12. How do you specify a clockwise circular array rotation?
13. What value should you specify for the angle to fill to create a complete circular array?
14. List the objects you can use as a path for a path array.
15. Describe the spacing between items that occurs when you use the divide path method available with the **ARRAYPATH** command.

Drawing Problems

Start AutoCAD if it is not already started. Start a new drawing for each problem using an appropriate template of your choice. The template should include layers and text styles, when necessary, for drawing the given objects. Add layers and text styles as needed. Draw all objects using appropriate layers. Use appropriate text styles, justification, and format. Follow the specific instructions for each problem. Use only drawing and editing commands and techniques you have already learned. Do not draw dimensions. Use your own judgment and approximate dimensions when necessary.

▼ Basic

1. **General** Save P11-1 as P12-1. If you have not yet completed Problem 11-1, do so now. In the P12-1 file, rotate the object 90° to the right and mirror the object to the left. Use the vertical base of the object as the mirror line. The final drawing should look like the example below.

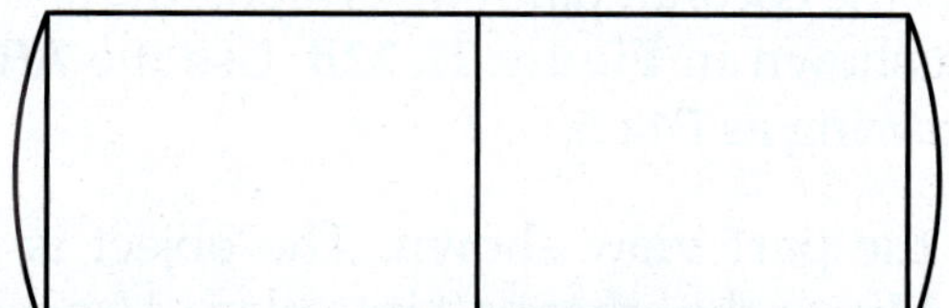

2. **General** Save P11-2 as P12-2. If you have not yet completed Problem 11-2, do so now. In the P12-2 file, make two copies of the object to the right of the original object. Scale the first copy 1.5 times the size of the original object. Scale the second copy 2 times the size of the original object. Move the objects so they are approximately centered in the drawing area. Move the objects as needed to align the bases of all objects and provide an equal amount of space between the objects. The final drawing should look like the example below.

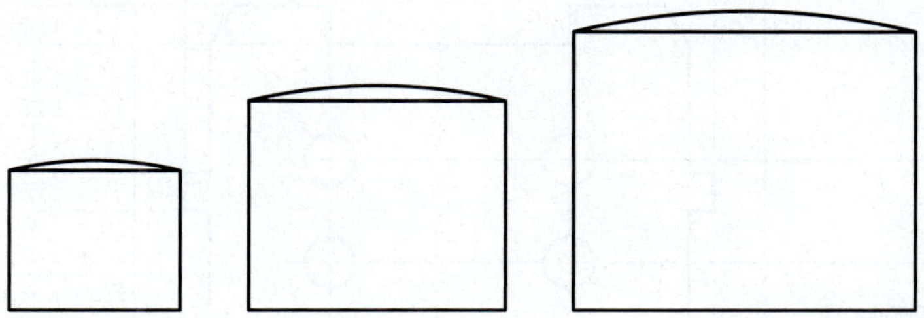

3. **General** Draw Objects A, B, and C. Make a copy of Object A two units up. Make four copies of Object B three units up, center to center. Make three copies of Object C three units up, center to center. Save the drawing as P12-3.

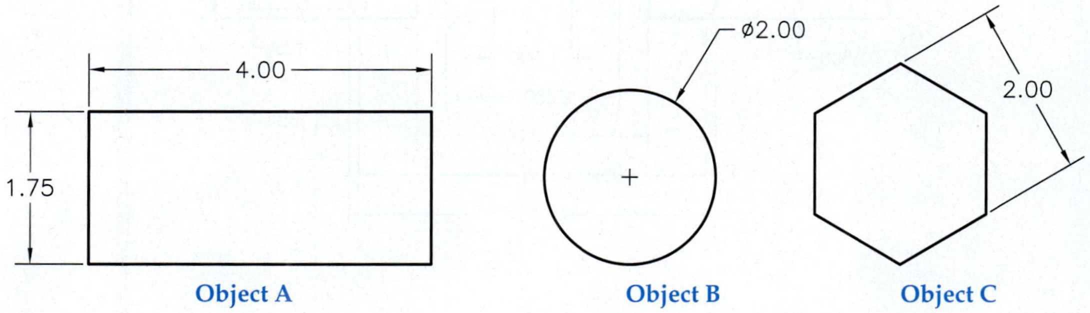

4. **Mechanical** Save P11-4 as P12-4. If you have not yet completed Problem 11-4, do so now. In the P12-4 file, draw a mirror image as Object B. Then remove the original view and move the new view so that Point 2 is at the original Point 1 location.

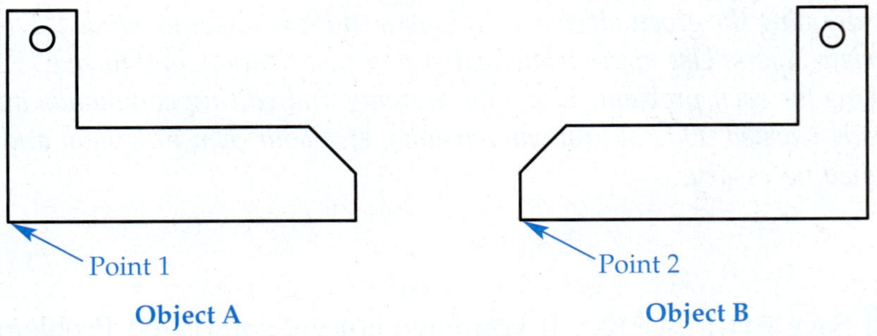

▼ Intermediate

5. **Architectural** Draw the coconut palm trees along the spline for the portion of the landscape elevation shown in **Figure 12-22B**. Use the **ARRAYPATH** command as needed. Save the drawing as P12-5.

6. **Mechanical** Draw the part view shown. The object is symmetrical; therefore, draw only one half. Mirror the other half into place. Use the **CHAMFER** and **FILLET** commands to your best advantage. All fillets and rounds are .125. Use the **JOIN** command where necessary. Use the **Array** option of the **COPY** command to array the row of ∅.500 holes. Save the drawing as P12-6.

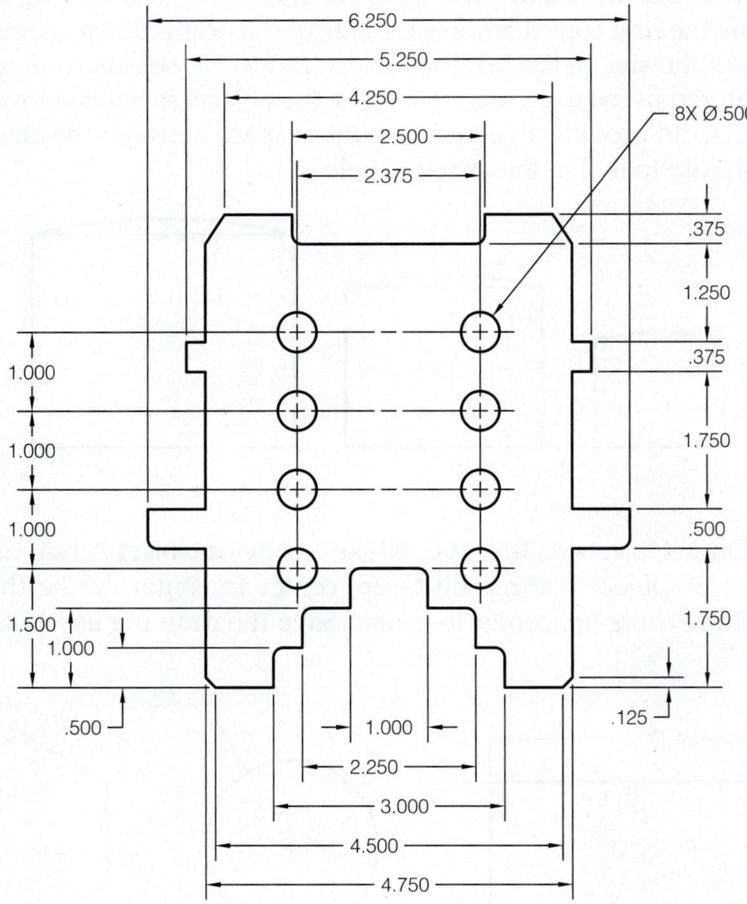

7. Mechanical Draw the portion of the part view shown. Mirror the right half into place. Use the **CHAMFER** and **FILLET** commands to your best advantage. Save the drawing as P12-7.

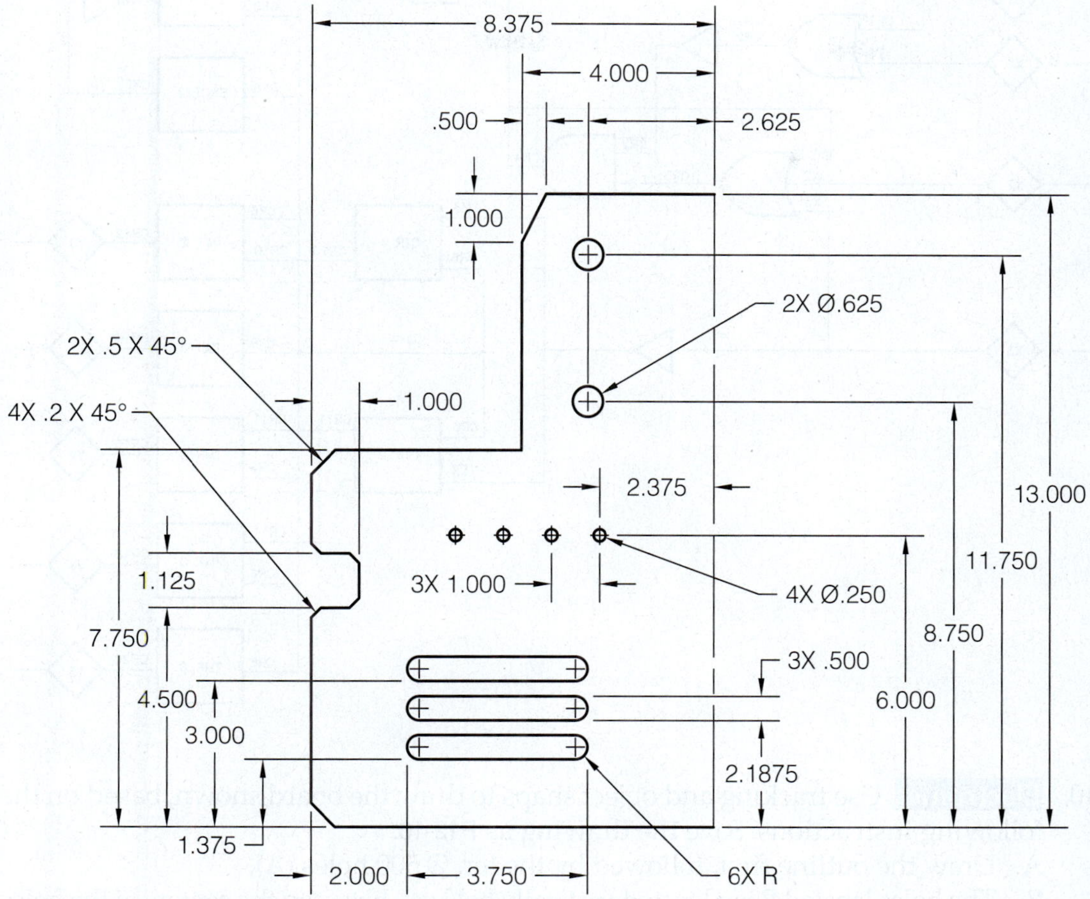

8. Electronics Draw the electronic schematic symbols shown. Mirror the drawing, but make sure the text remains readable. Delete the original image during the mirroring process. Save the drawing as P12-8.

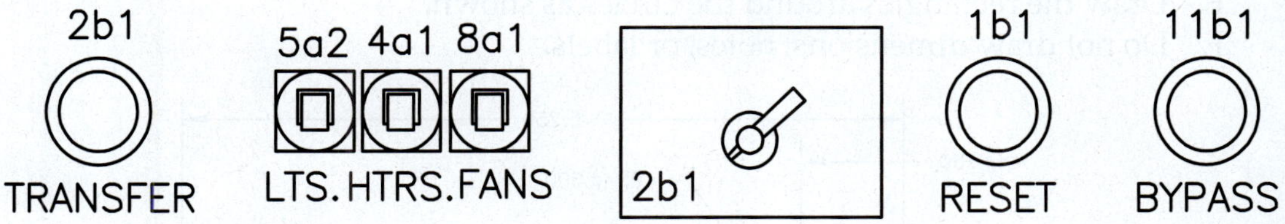

Drawing Problems - Chapter 12

9. **Electronics** Draw the timer schematic shown. Save the drawing as P12-9.

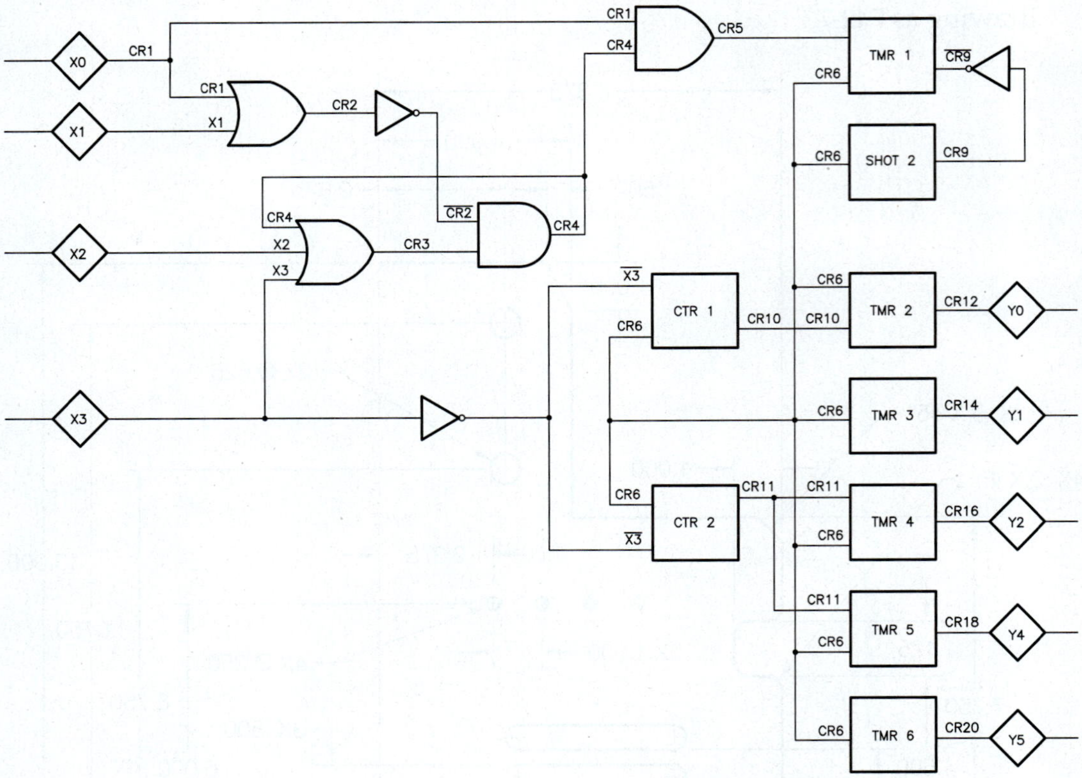

10. **Electronics** Use tracking and object snaps to draw the board shown, based on the following instructions. Save the drawing as P12-10.
   A. Draw the outline first, followed by the ten ∅.500 holes (A).
   B. The holes labeled B are located vertically halfway between the centers of the holes labeled A. They have a diameter one-quarter the size of the holes labeled A.
   C. The holes labeled C are located vertically halfway between the holes labeled A and B. Their diameter is three-quarters of the diameter of the holes labeled B.
   D. The holes labeled D are located horizontally halfway between the centers of the holes labeled A. These holes have the same diameter as the holes labeled B.
   E. Draw the rectangles around the circles as shown.
   F. Do not draw dimensions, notes, or labels.

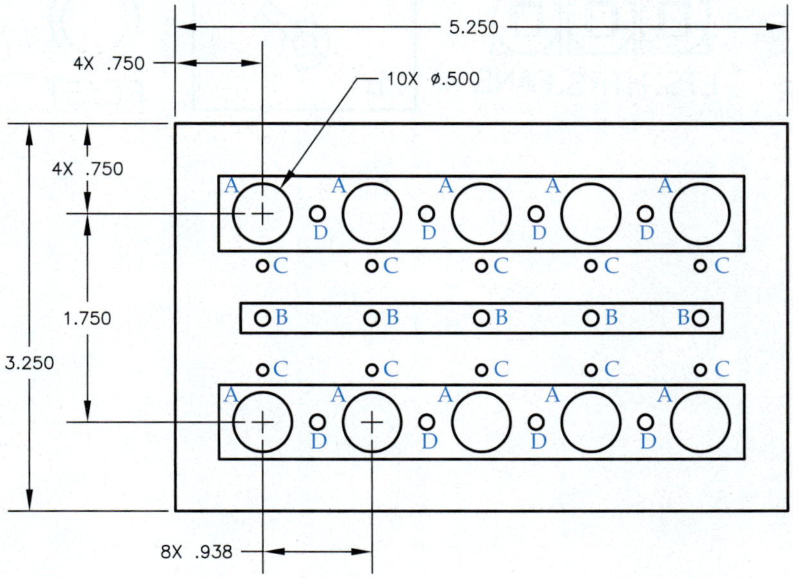

11. Mechanical Draw the portion of the gasket shown on the left. Use the **MIRROR** command to complete the gasket as shown on the right. Save the drawing as P12-11.

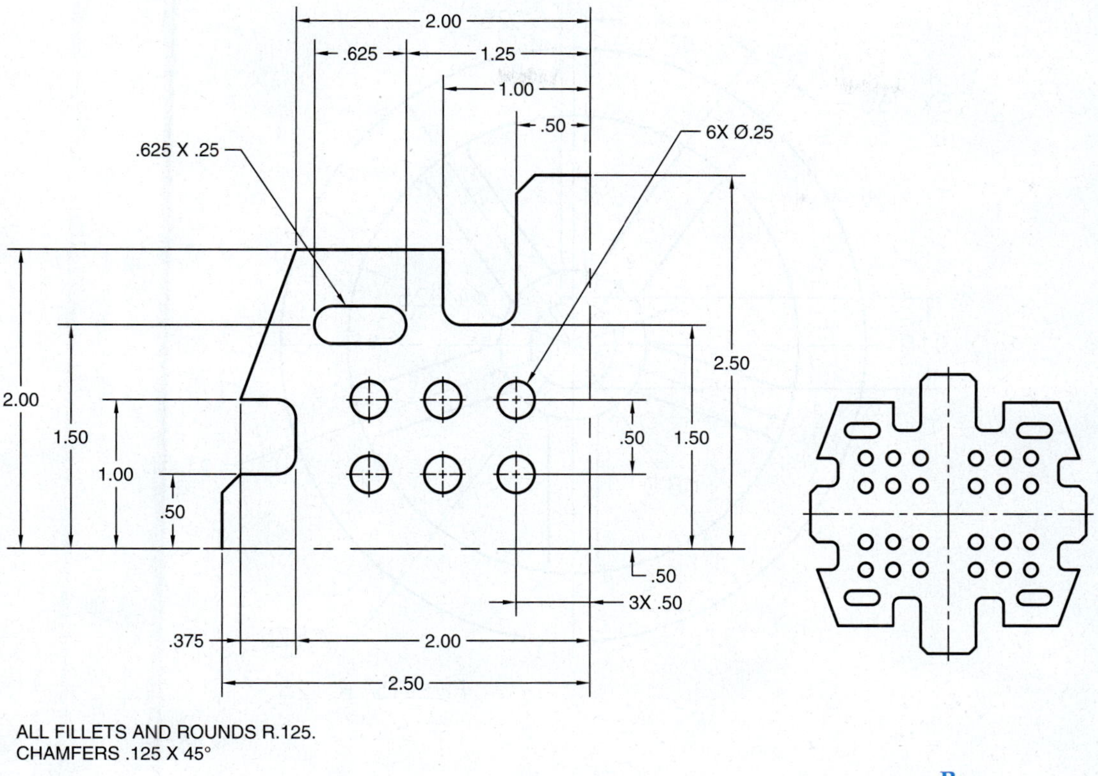

A

B

12. Architectural Draw the padded bench shown. Use the **COPY** and **ARRAY** commands as needed. Save the drawing as P12-12.

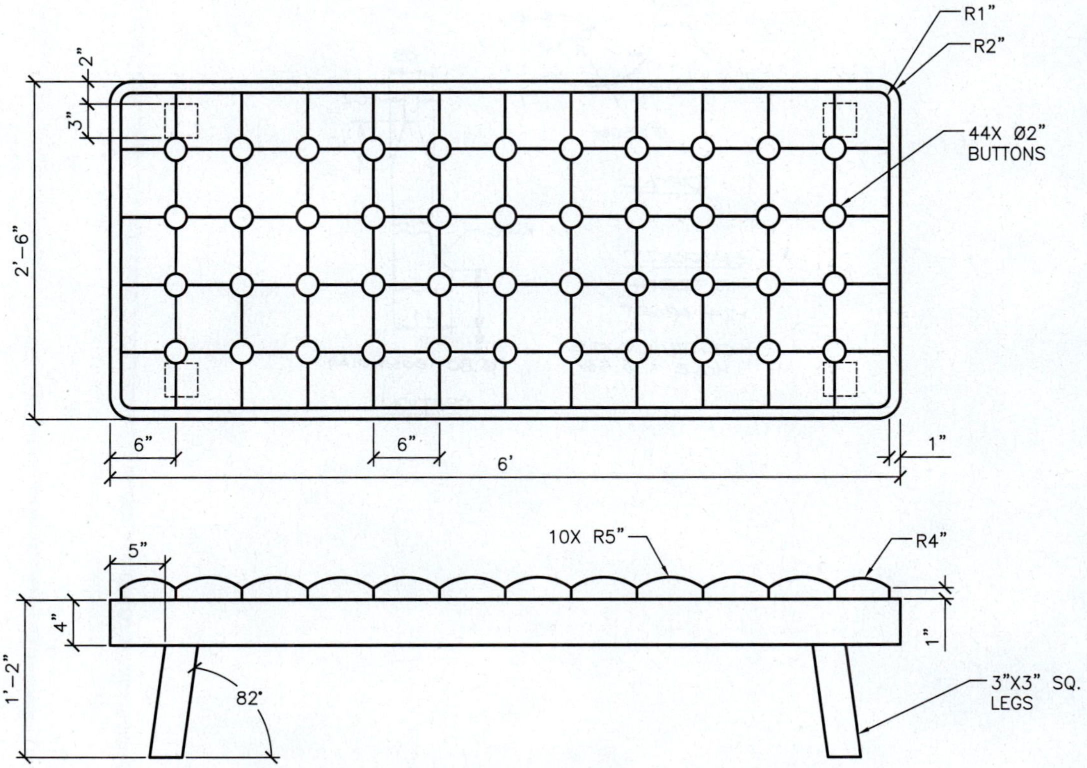

13. **Mechanical** Draw the hand wheel shown. Use the **ARRAYPOLAR** command to draw the spokes. Save the drawing as P12-13.

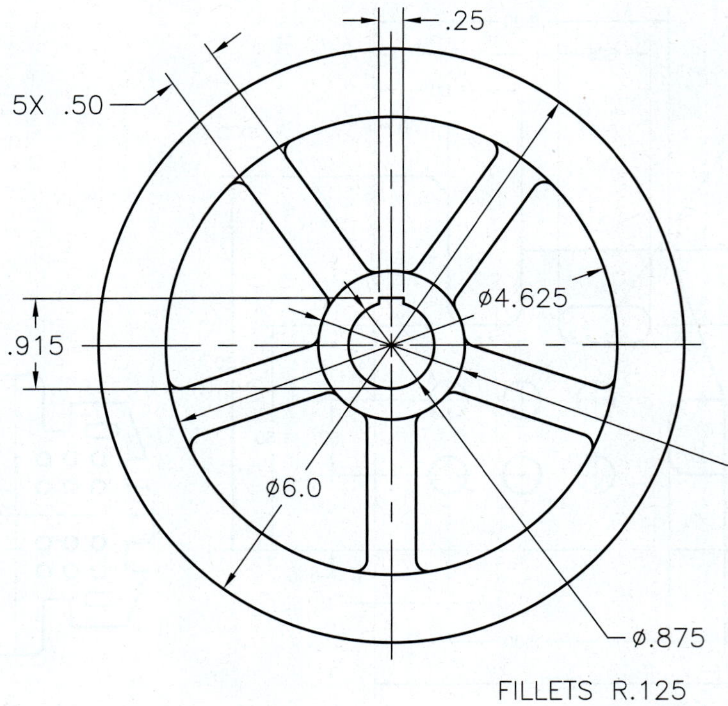

▼ Advanced

14. **Mechanical** Use the engineer's sketch and notes shown to draw the sprocket. Create a front and side view of the sprocket. Use the **ARRAYPOLAR** command as needed. Save the drawing as P12-14.

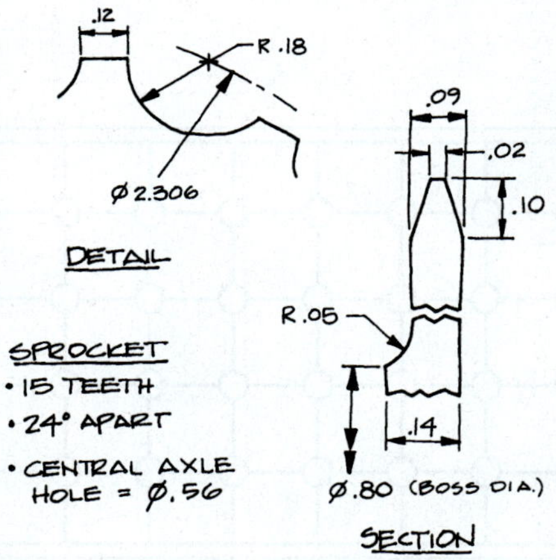

15. **Mechanical** Draw the views of the sprocket shown. Use **ARRAYPOLAR** to construct the hole and tooth arrangements. Save the drawing as P12-15.

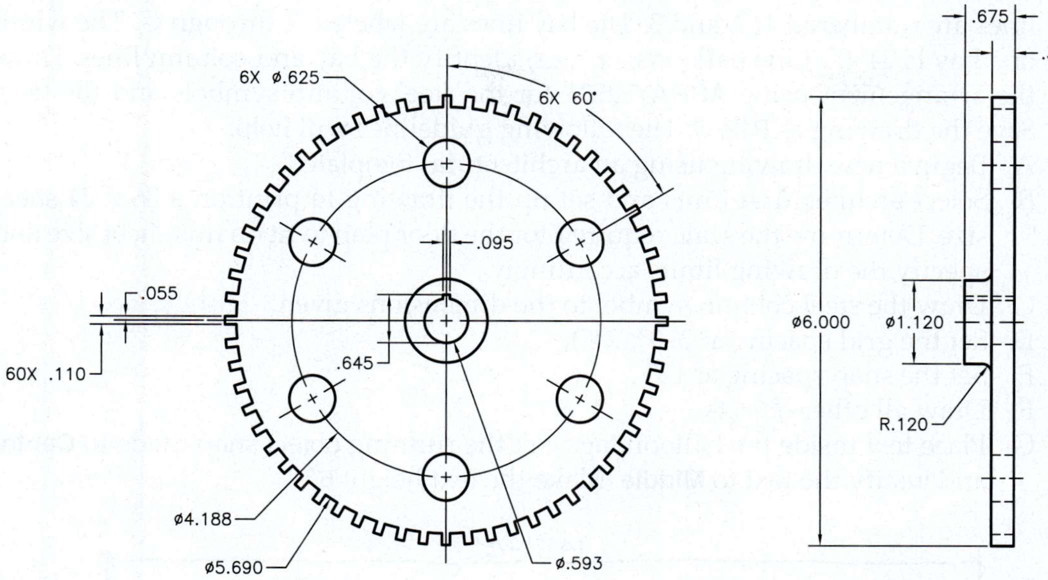

16. **Electronics** Draw the refrigeration system schematic shown. Save the drawing as P12-16.

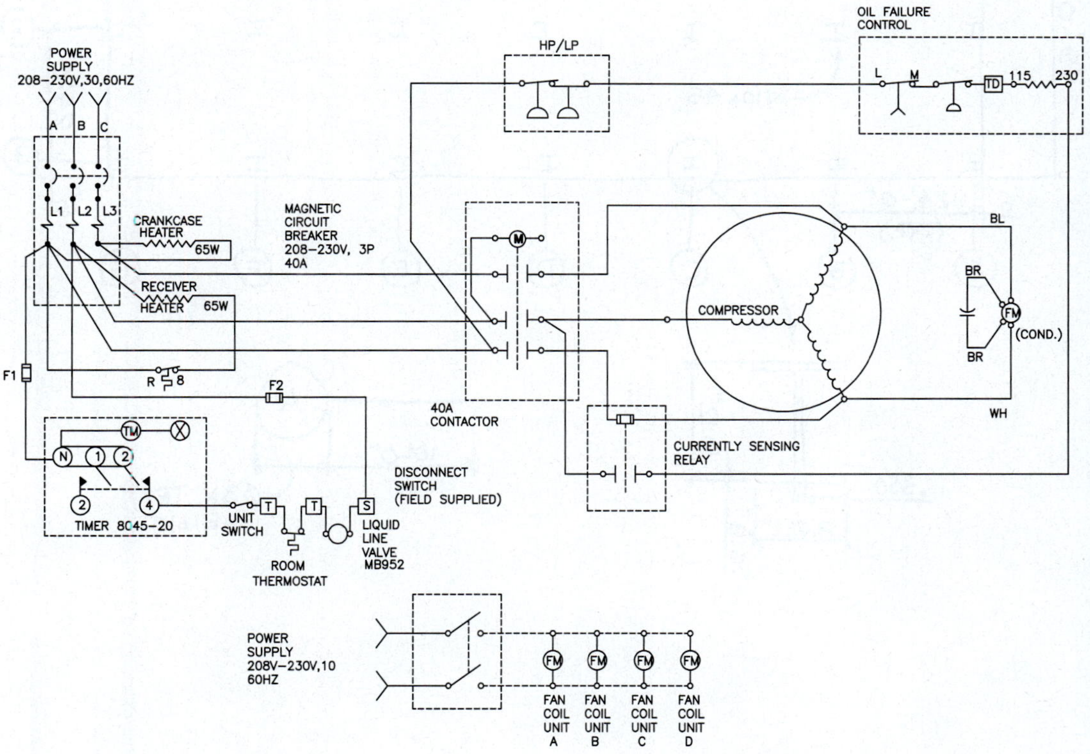

17. **Architectural** The structural sketch shown is a steel column arrangement on a concrete floor slab for a new building. The I-shaped symbols represent the steel columns. The columns are arranged in "bay lines" and "column lines." The column lines are numbered 1, 2, and 3. The bay lines are labeled A through G. The width of a bay is 24′-0″. Line balloons, or tags, identify the bay and column lines. Draw the arrangement using **ARRAYRECT** for the steel column symbols and the tags. Save the drawing as P12-17. The following guidelines will help.
    A. Begin a new drawing using an architectural template.
    B. Select architectural units and set up the drawing to print on a 36 × 24 sheet size. Determine the scale required for the floor plan to fit on this sheet size and specify the drawing limits accordingly.
    C. Draw the steel column symbol to the dimensions given.
    D. Set the grid spacing at 2′-0″ (24″).
    E. Set the snap spacing at 12″.
    F. Draw all other objects.
    G. Place text inside the balloon tags. Set the running object snap mode to **Center** and justify the text to **Middle**. Make the text height 6″.

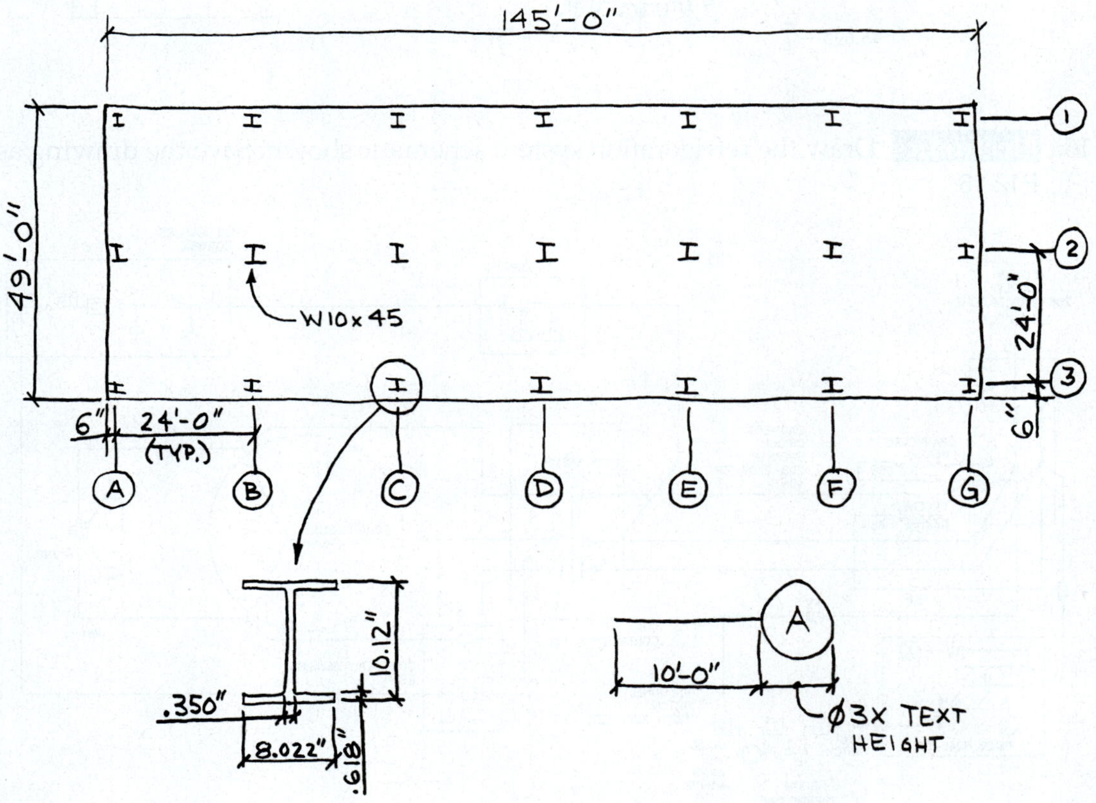

18. **Architectural** The sketch shown is a proposed classroom layout of desks and chairs. One desk is shown with the layout of a chair, keyboard, monitor, and tower-mounted computer (drawn with dotted lines). All of the desk workstations should have the same configuration. The exact sizes and locations of the doors and windows are not important for this problem. Save the drawing as P12-18. Use the following guidelines to complete this problem.

   A. Begin a new drawing.
   B. Specify architectural units.
   C. Set up the drawing to print on a C-size sheet, and be sure to create the drawing in model space.
   D. Use the appropriate drawing and editing commands to complete this problem quickly and efficiently.
   E. Draw the desk and computer hardware to the dimensions given.
   F. Do not dimension the drawing.

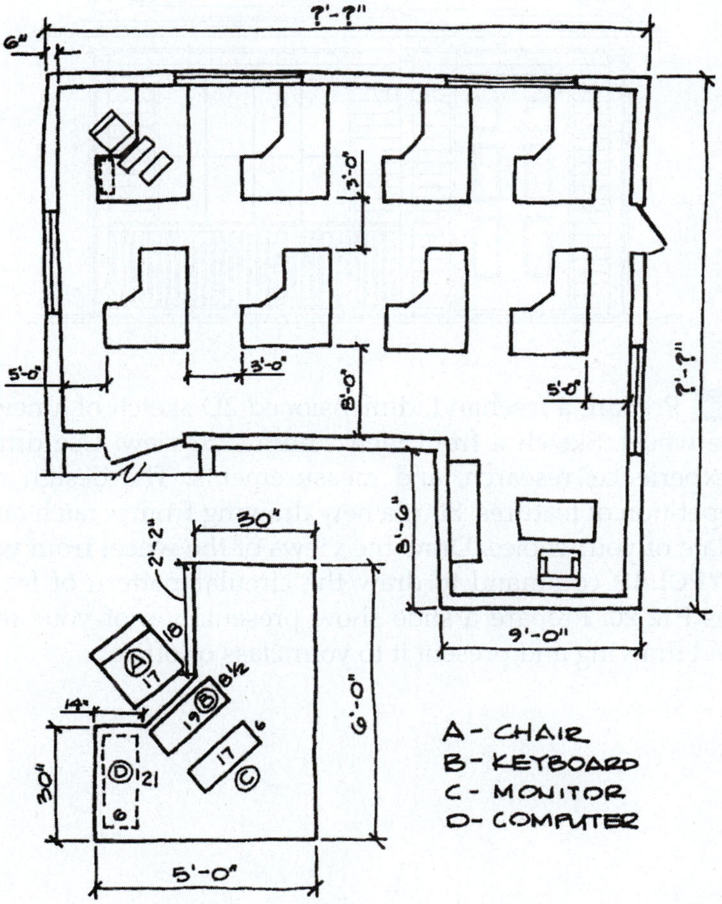

19. Architectural Draw the front elevation of this house. Create the features proportional to the given drawing. Use the **ARRAYRECT** and **TRIM** commands to place the siding and porch rails evenly. Save the drawing as P12-19.

20. Mechanical Prepare a freehand, dimensioned 2D sketch of a new design for an automobile wheel. Sketch a front view and a side view. Use dimensions based on your experience, research, and measurements. The design must include a circular repetition of features. Start a new drawing from scratch or use a decimal-unit template of your choice. Draw the views of the wheel from your sketch. Use the **ARRAYPOLAR** command to draw the circular pattern of features. Save the drawing as P12-20. Prepare a slide show presentation of your research, design process, and drawing and present it to your class or office.

AutoCAD Certified Professional Exam Practice

Answer the following questions using the information in this chapter.

1. Which of the following can you do using the **MOVE** command? *Select all that apply.*
   A. move objects relative to their current location
   B. move objects from a base point to a second specified point
   C. rotate objects during the move operation
   D. use the first point you pick as the point of displacement
   E. use a scale factor
2. In which order do you pick the source and destination points when using the **ALIGN** command? *Select the one item that best answers the question.*
   A. destination point 1, destination point 2, source point 1, source point 2
   B. destination point 1, source point 1, destination point 2, source point 2
   C. destination point 2, source point 2, destination point 1, source point 1
   D. source point 1, destination point 1, source point 2, destination point 2
   E. source point 1, source point 2, destination point 1, destination point 2
3. In the array shown below, what would you enter for the row and column spacing? *Select the one item that best answers the question.*
   A. row spacing 1.00, column spacing 2.00
   B. row spacing 2.00, column spacing 1.00
   C. row spacing 2.00, column spacing 4.00
   D. row spacing 4.00, column spacing 2.00
   E. row spacing 5.00, column spacing 6.00
   F. row spacing 6.00, column spacing 5.00

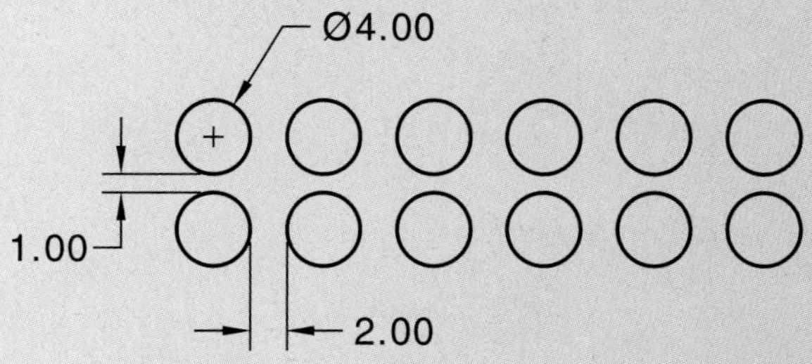

Follow the instructions in each problem. Complete each problem by answering the related question.

4. **Navigate to this chapter on the companion website and open CPE-12array.dwg.** Use the **ARRAYPOLAR** command to finish the view of a fan plate as shown. Analyze the drawing and use the most appropriate options for the polar array. What are the coordinates of Point 1?

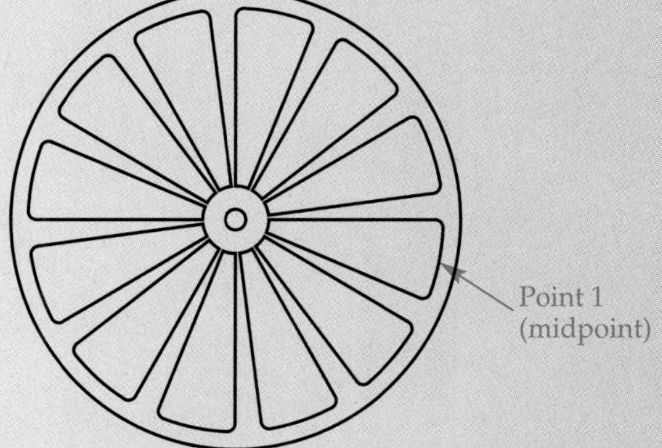

5. **Navigate to this chapter on the companion website and open CPE-12mirror.dwg.** Create a mirror line starting at absolute coordinates 13′,8′ and extending 5′ at 120°. Mirror the couch across this line. What are the coordinates of Point 1?

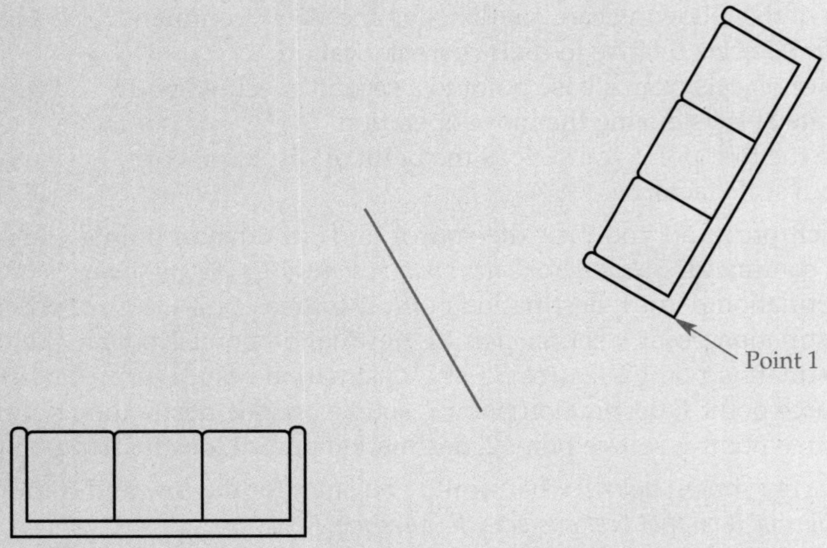

CHAPTER 13

Grips, Properties, and Additional Selection Techniques

Learning Objectives

After completing this chapter, you will be able to:

- ✓ Use grips to stretch, move, rotate, scale, mirror, and copy objects.
- ✓ Edit associative arrays.
- ✓ Adjust object properties using the **Quick Properties** palette and the **Properties** palette.
- ✓ Use the **MATCHPROP** command to match object properties.
- ✓ Edit between drawings.
- ✓ Use the **ADDSELECTED** command to draw an object based on an existing object.
- ✓ Create selection sets using the **SELECTSIMILAR** and **QSELECT** commands.

One approach to editing is to access a command, such as **ERASE**, **FILLET**, **MOVE**, or **COPY**, select the objects to modify, and follow prompts to complete the operation. This chapter explains the alternative approach of selecting objects first and then using editing commands or object properties to make changes. This chapter also describes additional selection options, selection set filters, and related tools.

Grips

Use the crosshairs to select objects and display ***grips***. See Figure 13-1. Selected objects become highlighted, and grips initially appear as ***unselected grips***. Unselected grips are blue (Color 150) by default. Grips are specific to object type. Most objects include the standard filled-square grips at critical and editable points on the object. Several objects, including elliptical arcs, mtext, polylines, splines, associative arrays, tables, hatches, and blocks, also have specialized grips. For example, elliptical arcs include filled-arrow grips for adjusting the length of the elliptical arc. This textbook explains the grips specific to various object types when applicable.

grips: Small boxes that appear at strategic points on a selected object, allowing you to edit the object directly.

unselected grips: Grips that you have not yet picked to perform an operation.

Move the crosshairs over an unselected grip to snap to the grip. Then pause to change the color of the grip to pink (Color 11). Hovering over an unselected grip and allowing it to change color helps you select the correct grip, especially when multiple grips are close together. A tooltip or options may appear, depending on the object and grip.

Figure 13-1.
Grips appear at specific locations on objects when you select the objects while no drawing or editing command is active.

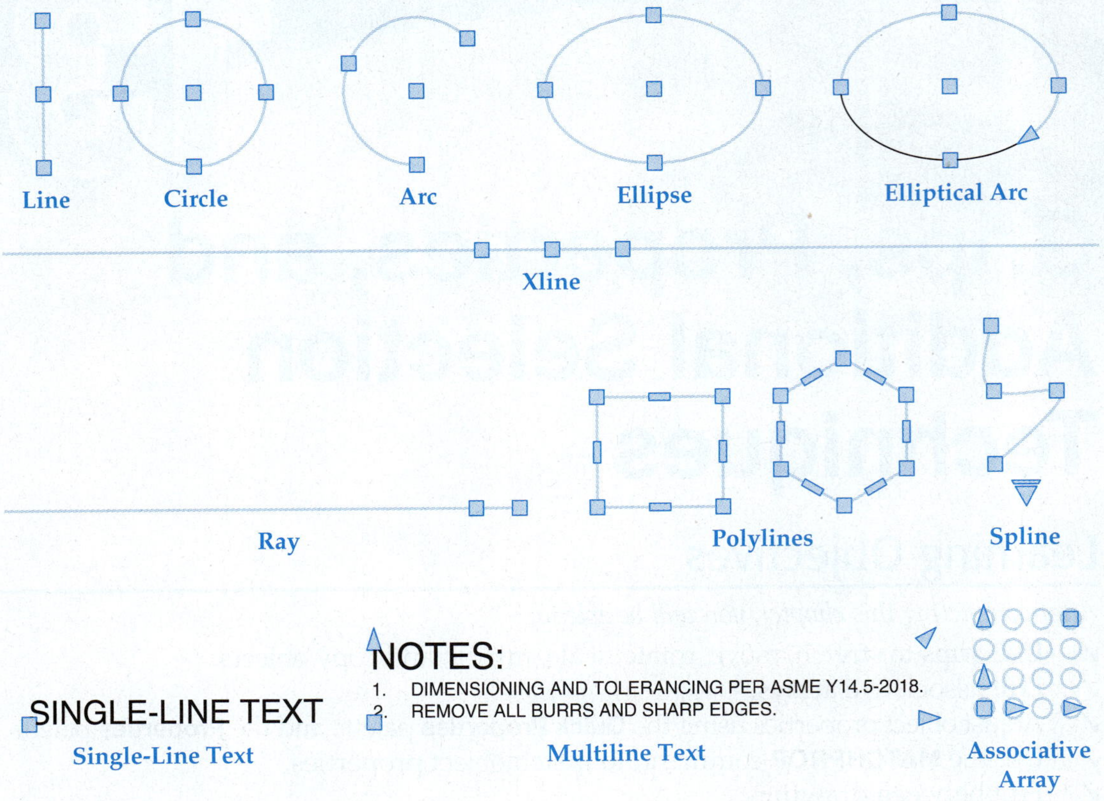

selected grip: A grip that you have picked to perform an operation.

Pick a grip to perform an editing operation at the location of the grip. A ***selected grip*** appears red (Color 12) by default. If you select grips on more than one object, what you do with the selected grips affects all of the selected objects. Objects that have unselected and selected grips become part of the current selection set.

To remove objects from a selection set, hold down [Shift] and pick the objects to deselect. Select additional objects without pressing [Shift] to add to the selection set. [Shift] also allows you to select multiple grips. Hold down [Shift] and then select each grip. Remember not to release [Shift] until you pick all the grips you want to activate. While still holding down [Shift], you can pick a selected (red) grip to return it to the unselected (blue) state. **Figure 13-2** shows an example of modifying two circles at the same time using selected grips.

Figure 13-2.
You can modify multiple objects at the same time by pressing [Shift] to select additional grips. In this example, if you only select one grip, you can only edit one circle.

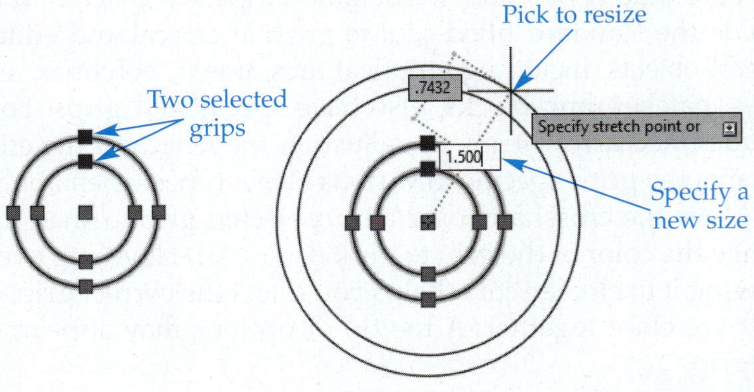

Press [Esc] to deactivate the current grip operation. Press [Esc] again to deselect all objects and hide the grips. You can also right-click and pick **Deselect All** to deselect objects and hide all grips.

NOTE

Use the options in the **Grip size** and **Grips** areas of the **Selection** tab in the **Options** dialog box to control grip size and color.

PROFESSIONAL TIP

By default, you can perform some conventional operations by selecting objects before you access a command. For example, select objects to erase and then activate the **ERASE** command or press [Delete]. The **Noun/verb selection** check box in the **Selection modes** area of the **Selection** tab in the **Options** dialog box and the **PICKFIRST** system variable control this behavior.

noun/verb selection: Performing tasks in AutoCAD by selecting the objects before activating a command.

verb/noun selection: Performing tasks in AutoCAD by activating a command before selecting objects.

Standard Grip Commands

Standard grip boxes provide access to the **STRETCH**, **MOVE**, **ROTATE**, **SCALE**, and **MIRROR** commands. In addition, the **Copy** option of the **MOVE** command imitates the **COPY** command. Select grips to display options at the dynamic input cursor and at the command line. Do not attempt to use conventional means of command access, such as the ribbon. The first command is **STRETCH**, as indicated by the ** STRETCH ** Specify stretch point or [Base point/Copy/Undo/eXit]: prompt. Use the **STRETCH** command, or press [Enter] to cycle through the **MOVE**, **ROTATE**, **SCALE**, and **MIRROR** commands.

An alternative to cycling through commands is to select grips, right-click, and select an option from the shortcut menu. A third method to activate a command is to enter the first two characters of the command name. Type MO for **MOVE**, MI for **MIRROR**, RO for **ROTATE**, SC for **SCALE**, or ST for **STRETCH**.

Stretching

Stretching using grips is similar to stretching using the **STRETCH** command, except that the selected grip acts as the stretch base point. In addition, depending on the selected grip and type of object, stretching using a grip can result in a move, rotate, or scale operation. See Figure 13-3. Stretch individual grips, or select multiple grips as needed depending on the desired result. See Figure 13-4.

Use the **Base point** option to specify a base point instead of using the selected grip as the base point. Select the **Copy** option to perform the stretch without changing the original object. Select the **Undo** option to undo the previous operation. Select the **eXit** option or press [Esc] to exit without completing the stretch. When you finish stretching, the selected grips return to the unselected state. Press [Esc] to hide the grips.

Dynamic input and other drawing aids, such as polar tracking, are very useful for grip editing. Dynamic input is especially effective with the **STRETCH** grip command. Figure 13-5 shows an example of using dimensional input to modify the size of a circle or offset the circle a specific distance. In this example, enter the new radius of the circle in the distance input field, or press [Tab] to enter an offset in the other distance input field. Another example is modifying the length of an ellipse axis by selecting the appropriate quadrant grip and using dimensional input to edit the value. These are just two examples of using dynamic input with grips. You can apply similar processes to edit most objects.

Figure 13-3.
Examples of using the **STRETCH** grip command with dynamic input active. Note the selected grip in each case and the relevant dynamic input fields.

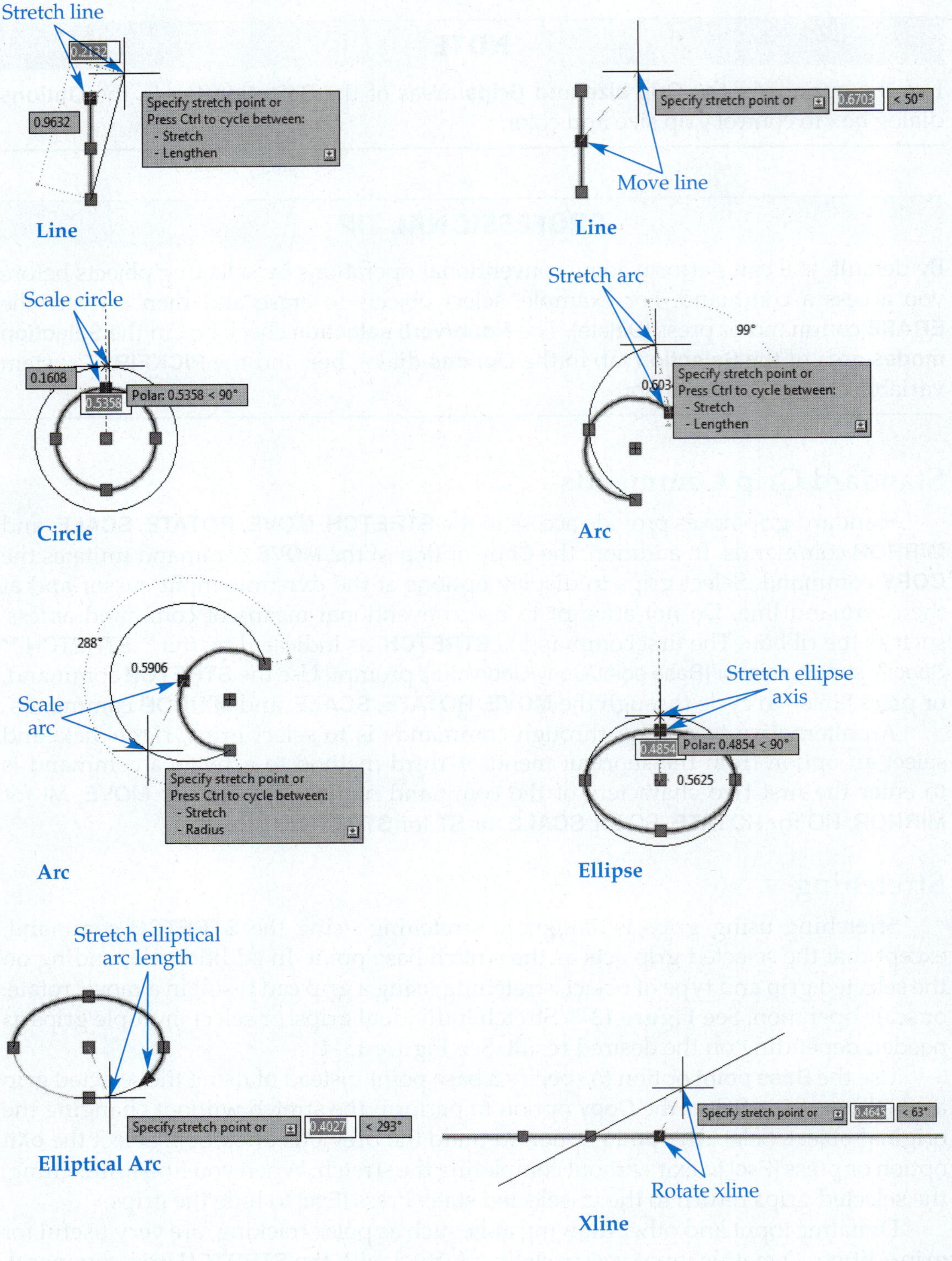

PROFESSIONAL TIP

Use grid snaps, coordinate entry, polar tracking, object snaps, and object snap tracking with any grip editing command to improve accuracy. When dynamic input is active, depending on the selected grip and type of objects, pressing [Tab] activates a different set of inputs. Press [Tab] until you receive the desired input fields.

Figure 13-4.
Stretching a drawing consisting of three lines and an arc. A—Select a corner to stretch. B—Hold down [Shift] to select multiple grips to stretch.

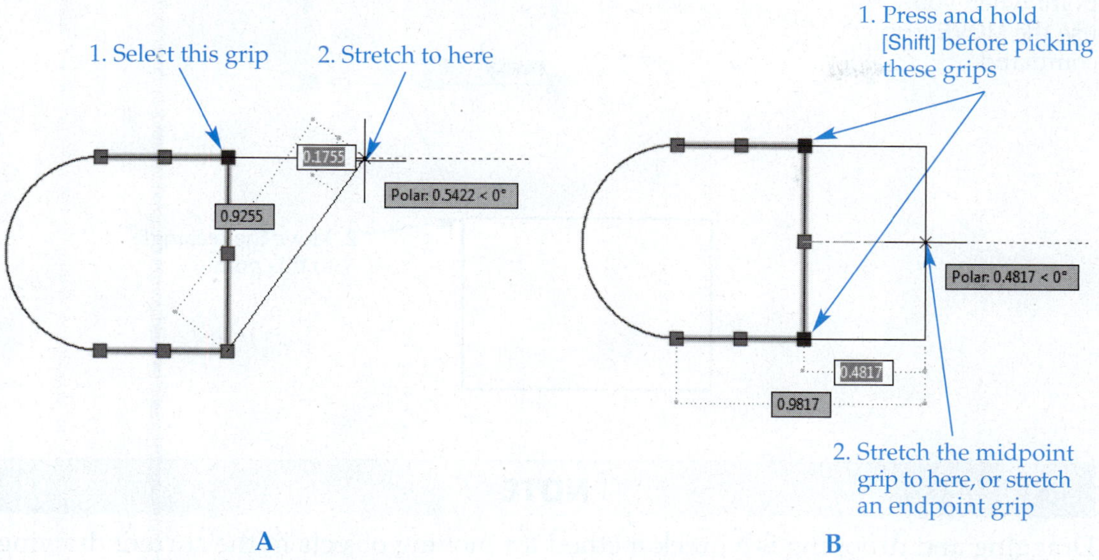

Figure 13-5.
Using the dimensional input feature of dynamic input with the **STRETCH** grip command.

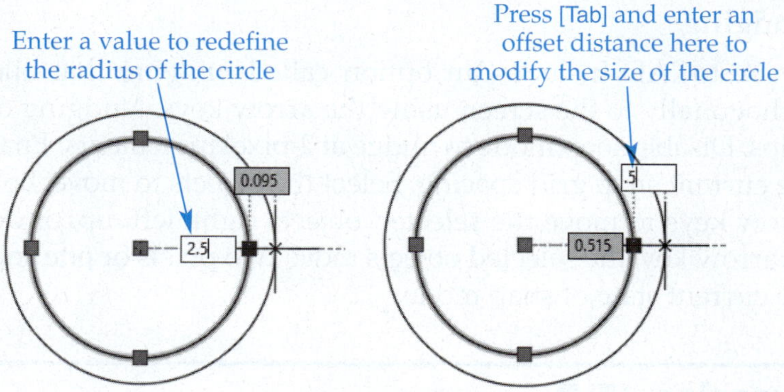

Exercise 13-1

www.g-wlearning.com/CAD/

Complete the exercise on the companion website.

Moving

To move objects using grips, select the objects to move, pick a grip to use as the base point, and activate the **MOVE** grip command. Specify a new location for the base point to move the objects. See Figure 13-6. The **Base point**, **Copy**, **Undo**, and **eXit** options are similar to those for the **STRETCH** grip command.

Dragging and Dropping

You can also use a drag-and-drop operation to move objects. Select the objects to move and then press and hold down the pick button on a portion of any selected object, but do not select a grip. While still holding the pick button, drag the objects to the desired location and release the pick button to complete the move. Preview the operation before moving.

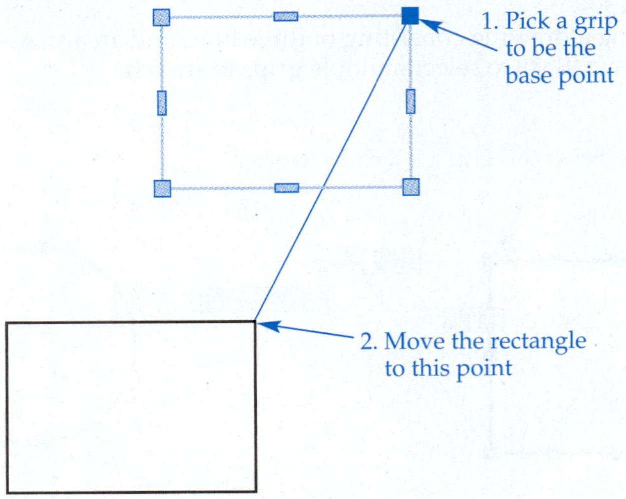

Figure 13-6. The selected grip is the default base point when you use the **MOVE** grip command.

NOTE

Dragging and dropping is a quick method for moving objects in the current drawing or to another open drawing, but is inaccurate because you cannot use drawing aids.

Nudging

nudging: Moving objects orthogonally by selecting the objects and using the arrow keys on the keyboard.

AutoCAD includes an option called ***nudging*** that allows you to move objects orthogonally to the screen using the arrow keys. Nudging does not involve the use of grips. Disable snap mode to nudge at 2-pixel increments. Enable snap mode to nudge at the current snap grid spacing. Select the objects to move, hold down [Ctrl], and use the arrow keys to move the selected objects right, left, up, or down. Each time you press an arrow key, the selected objects move two pixels or one snap distance, depending on the current state of snap mode.

Exercise 13-2

www.g-wlearning.com/CAD/

Complete the exercise on the companion website.

Rotating

To rotate objects using grips, select the objects to rotate, pick a grip to use as the base point, and activate the **ROTATE** grip command. Specify a rotation angle to rotate the objects. The **Base point**, **Copy**, **Undo**, and **eXit** options are similar to those for the **STRETCH** grip command.

Use the **Reference** option to specify a new angle in relation to an existing angle. The reference angle is often the current angle of the objects. If you know the value of the current angle, enter the value at the prompt. Otherwise, pick two points to identify the angle. Enter a value for the new angle or pick a point. Figure 13-7 shows **ROTATE** grip command options.

Exercise 13-3

www.g-wlearning.com/CAD/

Complete the exercise on the companion website.

Figure 13-7.
Using the **ROTATE** grip command with the default rotation angle option and with the **Reference** option.

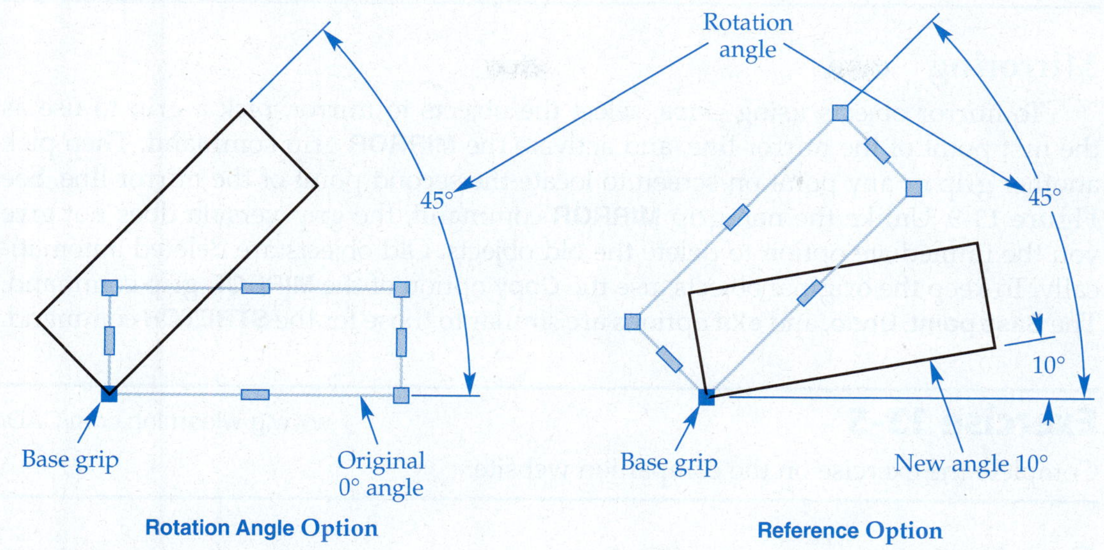

Scaling

To scale objects using grips, select the objects to scale, pick a grip to use as the base point, and activate the **SCALE** grip command. Enter a scale factor or pick a point to increase or decrease the size of the objects. The **Base point**, **Copy**, **Undo**, and **eXit** options are similar to those for the **STRETCH** grip command.

Use the **Reference** option to specify a new size in relation to an existing size. The reference size is often the current length, width, or height of the objects. If you know the current size, enter the value at the prompt. Otherwise, pick two points to identify the size. Enter a value for the new size or pick a point. Figure 13-8 shows **SCALE** grip command options.

Figure 13-8.
When using the **SCALE** grip command, enter a scale factor or use the **Reference** option.

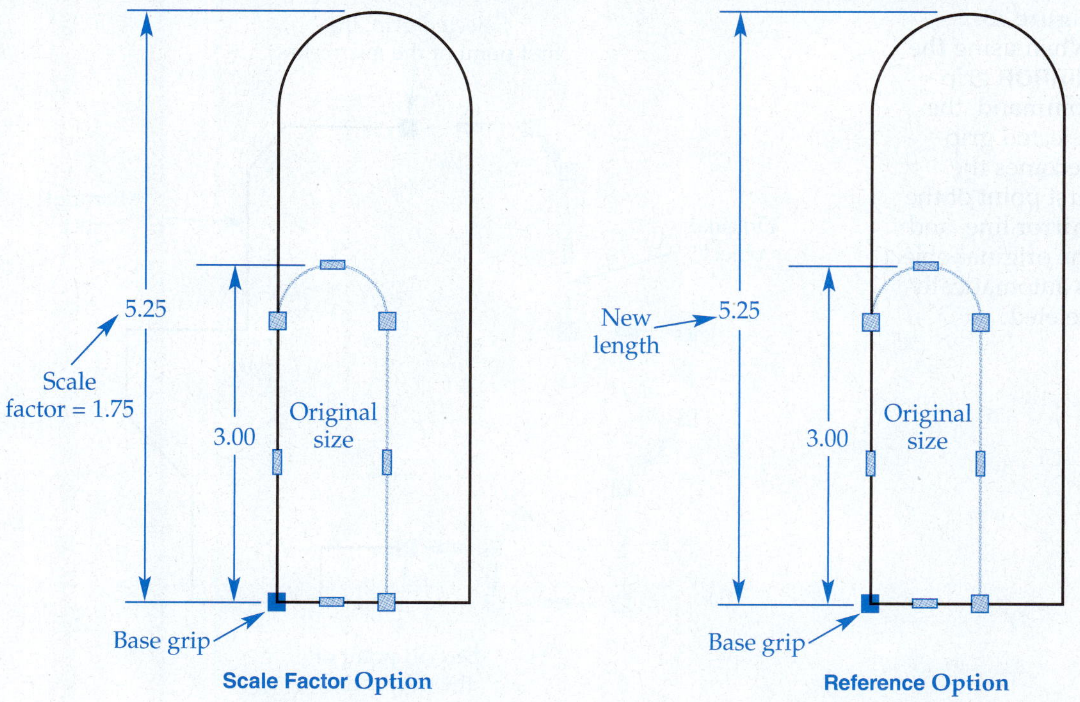

Exercise 13-4

www.g-wlearning.com/CAD/

Complete the exercise on the companion website.

Mirroring

To mirror objects using grips, select the objects to mirror, pick a grip to use as the first point of the mirror line, and activate the **MIRROR** grip command. Then pick another grip or any point on-screen to locate the second point of the mirror line. See Figure 13-9. Unlike the non-grip **MIRROR** command, the grip version does not give you the immediate option to delete the old objects. Old objects are deleted automatically. To keep the original objects, use the **Copy** option of the **MIRROR** grip command. The **Base point**, **Undo**, and **eXit** options are similar to those for the **STRETCH** command.

Exercise 13-5

www.g-wlearning.com/CAD/

Complete the exercise on the companion website.

Copying

Each standard grip editing command includes the **Copy** option. The effect of using the **Copy** option depends on the selected objects, grip, and command. The original selected objects remain unchanged, and the copy stretches when the **STRETCH** grip command is active, rotates when the **ROTATE** grip command is active, or scales when the **SCALE** grip command is active. The **Copy** option of the **MOVE** grip command is the true copy operation, allowing you to copy from the selected grip. The selected grip acts as the copy base point. Create as many copies of the selected object as needed, and then exit the command.

Exercise 13-6

www.g-wlearning.com/CAD/

Complete the exercise on the companion website.

Figure 13-9. When using the **MIRROR** grip command, the selected grip becomes the first point of the mirror line, and the original object is automatically deleted.

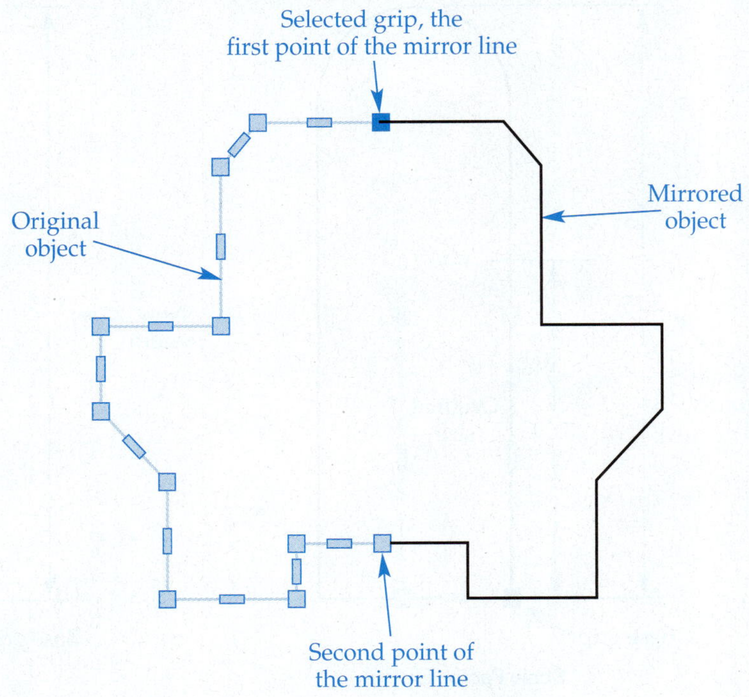

Object-Specific Grip Options

Several objects have specialized grips or additional grip options. Context-sensitive commands are available at the endpoint grips of a line or arc, the midpoint grip of an arc, and the endpoint grip of an elliptical arc when the endpoint is at a quadrant. You can access and apply the same context-sensitive, object-specific grip commands in three different ways. **Figure 13-10** shows each technique. **Figure 13-11** illustrates and explains the process of using object-specific grip commands to edit lines, arcs, and elliptical arcs. Use dynamic input when possible to complete the operation.

Select an mtext object to display a standard grip at the justification point, grips for modifying the mtext boundary width and height, and grips for adjusting columns. See **Figure 13-12**. Using grips is an alternative to re-entering the text editor to make changes to mtext layout, as described in Chapters 9 and 10. This textbook explains specialized grips related to other objects when applicable.

Editing Associative Arrays

AutoCAD creates associative rectangular, polar, and path arrays by default. To create a non-associative array, deselect the **Associative** button in the **Properties** panel of the **Array Creation** contextual ribbon tab, or select **No** at the **ASsociative** option, as described in Chapter 12. An easy way to edit an associative array is to select the array to display grips and the context-sensitive **Array** ribbon tab. See **Figure 13-13**. Use the same techniques shown in **Figure 13-10** to access and apply the same context-sensitive array grip commands. The **Array** ribbon tab includes most of the same functions as the **Array Creation** ribbon tab, as explained in Chapter 12, and offers additional editing tools.

ARRAYEDIT
Ribbon
Home
> Modify
Edit Array
Type
ARRAYEDIT

NOTE

Use the grips that appear when creating an array in the same manner as when editing an array to define array characteristics. However, to access context-sensitive array grip commands during array creation, pick a grip and then press [Ctrl] to cycle through options.

Figure 13-10.
Apply one of the following methods to access context-sensitive grip commands at the endpoint grips of a line or arc, the midpoint grip of an arc, and the endpoint grip of an elliptical arc when the endpoint is at a quadrant. A—Hover over an unselected grip. B—Pick a grip and press [Ctrl] to cycle through options. C—Pick a grip and then right-click to display options.

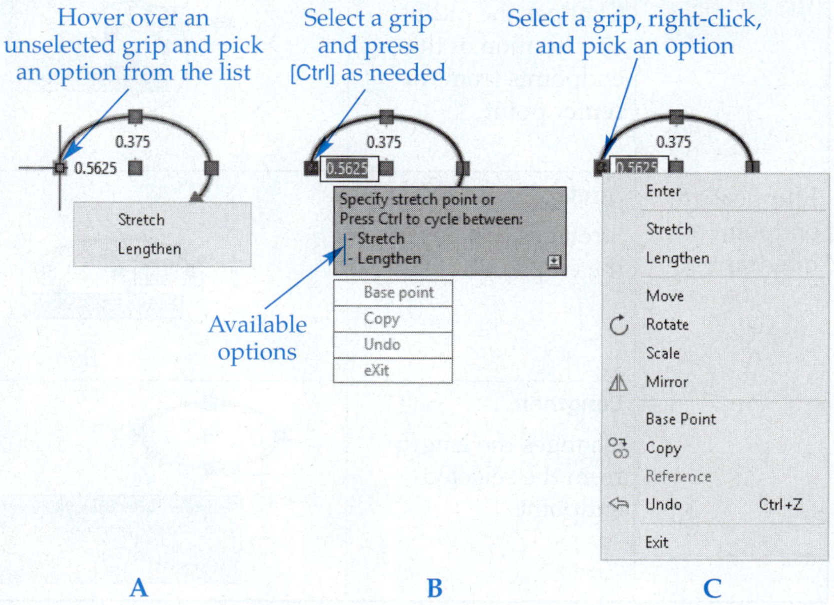

Figure 13-11.
Examples of options available for modifying lines, arcs, and elliptical arcs using grips.

Grip	Option	Process	Result
Line endpoint	**Stretch** Stretches the line at the selected endpoint	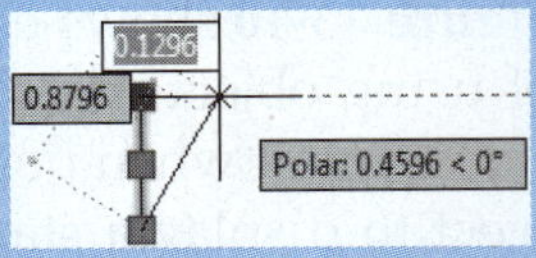	
	Lengthen Changes the length from the selected endpoint	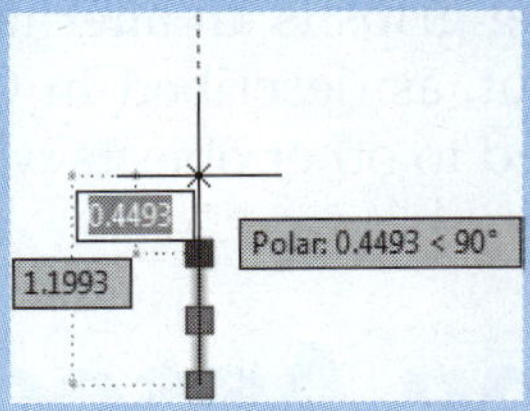	
Arc endpoint	**Stretch** Stretches the arc at the selected endpoint	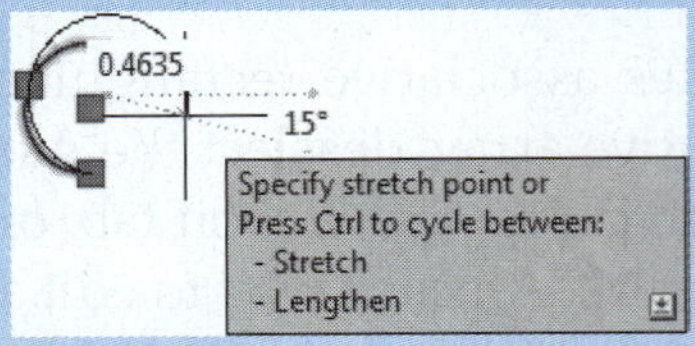	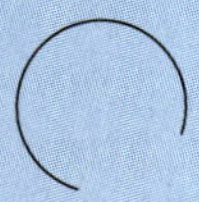
	Lengthen Changes the length from the selected endpoint	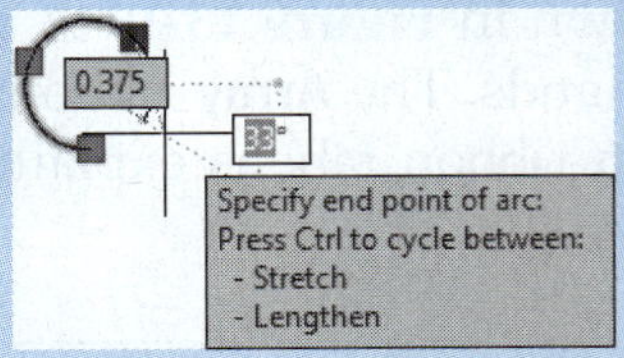	
Arc midpoint	**Stretch** Stretches the radius and center point from the endpoints	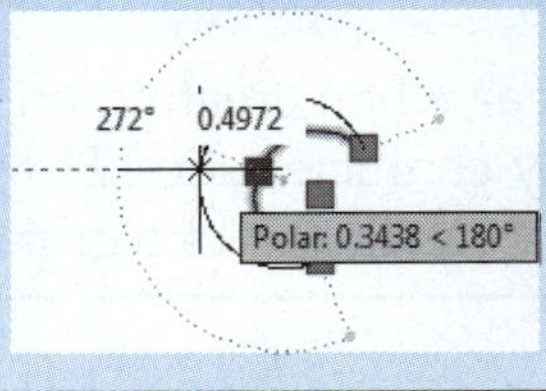	
	Radius Changes the radius and location of the endpoints from the center point	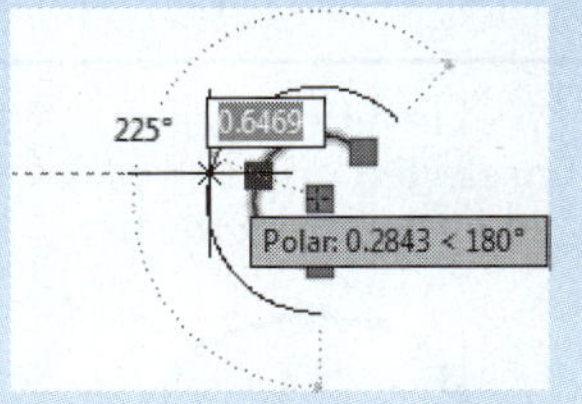	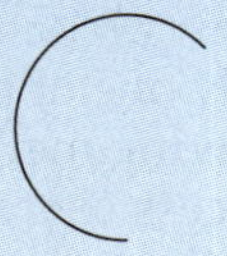
Elliptical arc endpoint at quadrant	**Stretch** Stretches the axis of the ellipse	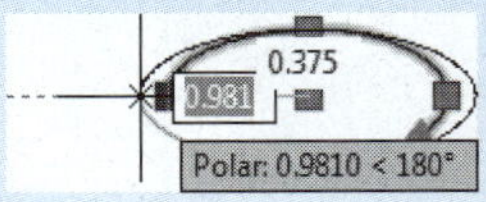	
	Lengthen Changes the length from the selected endpoint	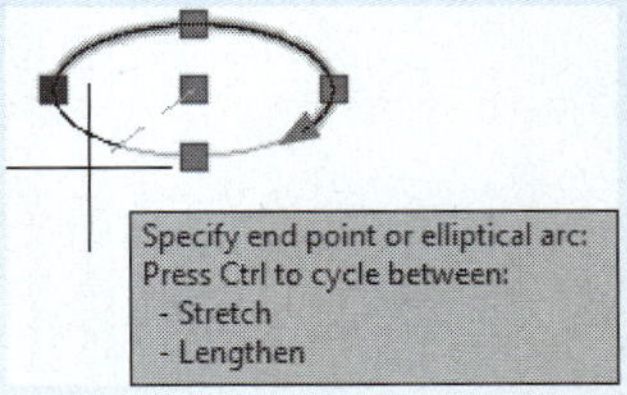	

Figure 13-12.
Use grips to apply standard editing commands to the justification base point of mtext and to adjust the mtext boundary or columns.

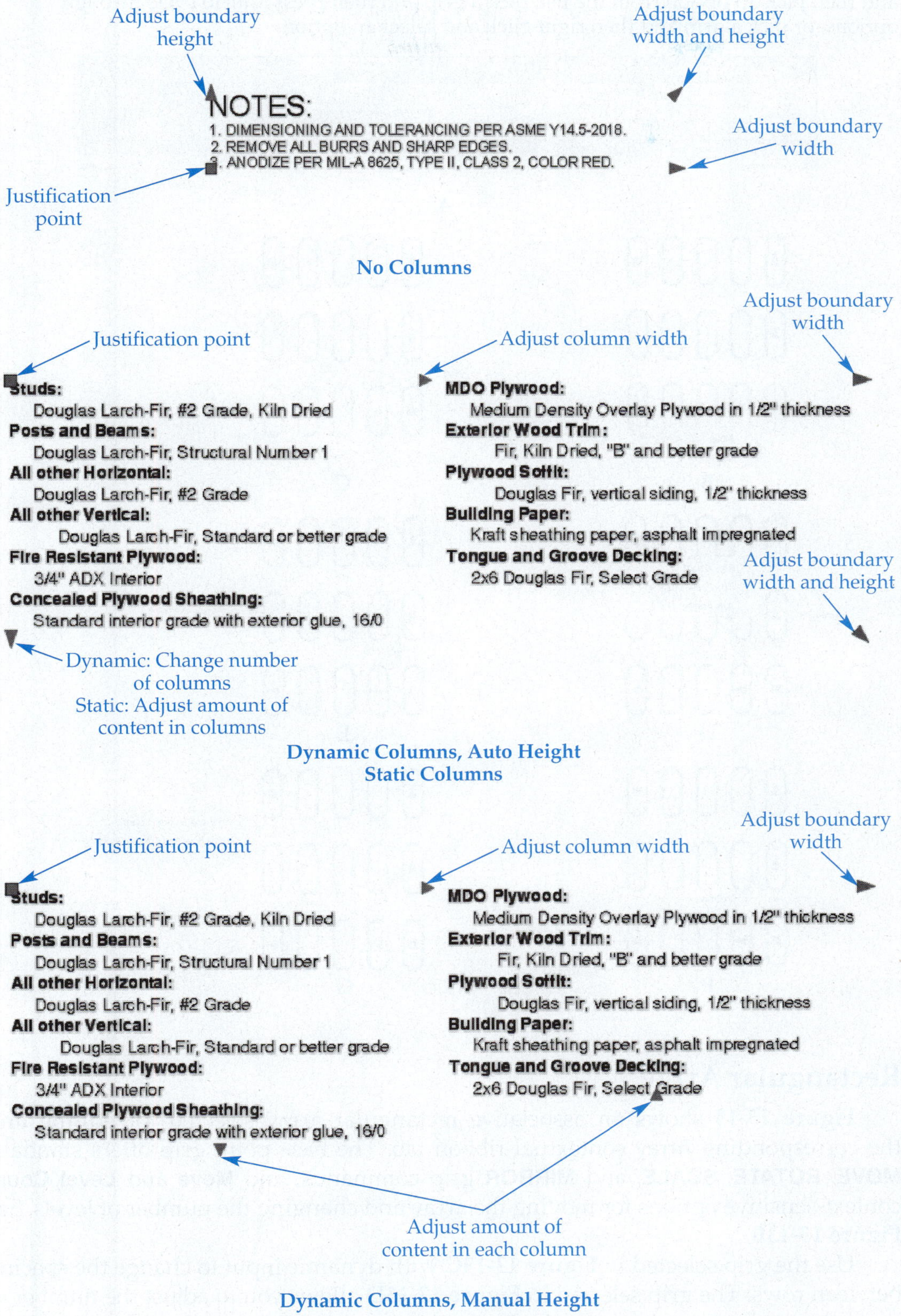

Figure 13-13.
Editing an associative rectangular array used to create a pattern of slots. To access context-sensitive array grip commands, hover over an unselected grip to display a menu of options and then pick an option from the list, pick a grip and then press [Ctrl] to cycle through options, or pick a grip and then right-click and select an option.

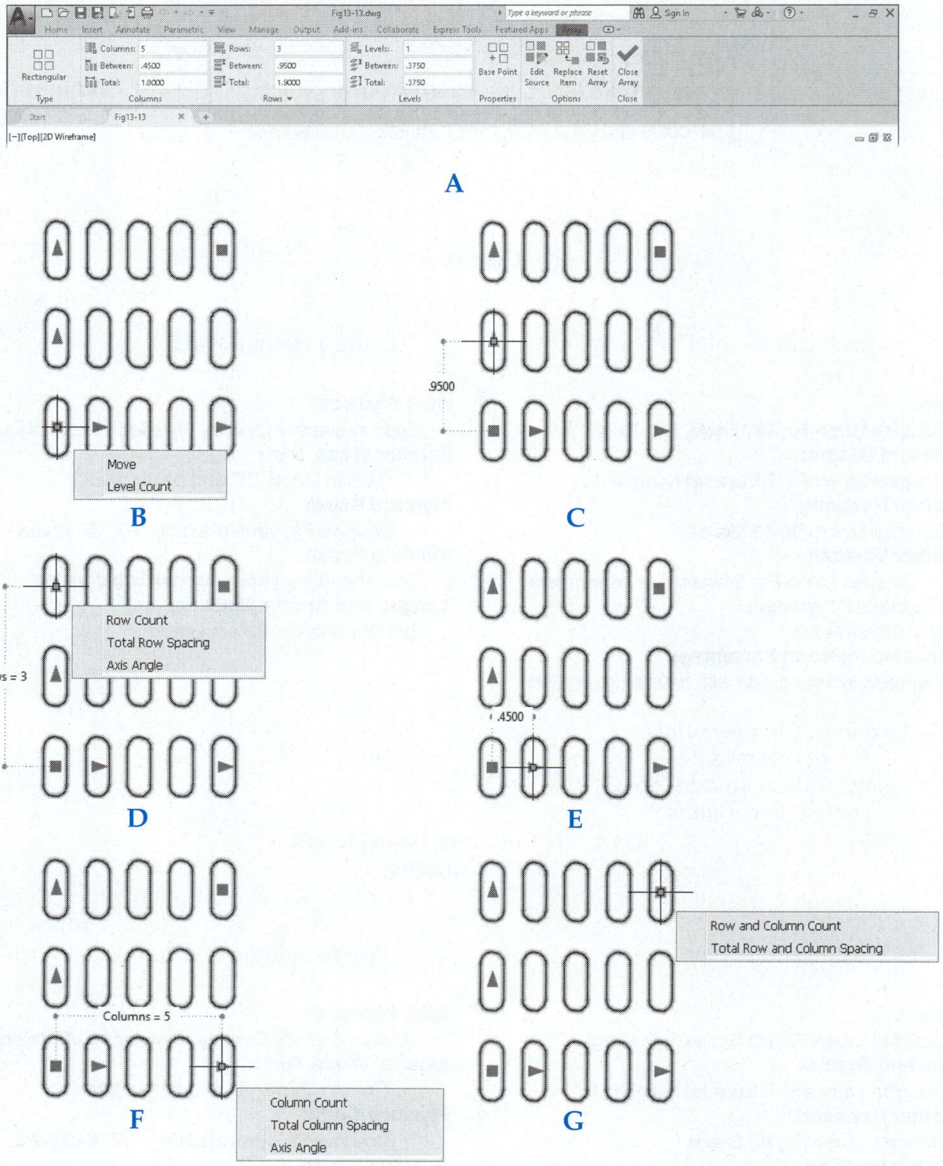

Rectangular Array

Figure 13-13 shows an associative rectangular array selected for editing and the corresponding **Array** contextual ribbon tab. The base point grip offers standard **MOVE**, **ROTATE**, **SCALE**, and **MIRROR** grip commands, and **Move** and **Level Count** context-sensitive options for moving the array and changing the number of levels. See **Figure 13-13B**.

Use the grip selected in **Figure 13-13C** with dynamic input to change the spacing between rows. The grip selected in **Figure 13-13D** allows you to adjust the number of rows, the total spacing between the first and last rows, and the angle of the array from rows. Use the grip selected in **Figure 13-13E** with dynamic input to change the spacing between columns. The grip selected in **Figure 13-13F** allows you to adjust the number of columns, the total spacing between the first and last columns, and the angle of the array from the columns. Use the grip selected in **Figure 13-13G** to edit the number of rows and columns and the spacing between rows and columns dynamically.

The **Axis Angle** option is a unique grip function that allows you to specify the angles of the row and column axes. The alignment of rows and columns rotate, not the objects. See Figure 13-14.

Polar Array

Figure 13-15 shows an associative polar array selected for editing and the corresponding **Array** ribbon tab. The center point grip offers standard **MOVE**, **ROTATE**, **SCALE**, and **MIRROR** grip commands. See Figure 13-15B. Use the base point grip selected in Figure 13-15C with dynamic input to stretch the radius of the array; modify the number of rows, levels, and items; and change the angle to fill. The grip selected in Figure 13-15D allows you to adjust the angle between items.

Path Array

Figure 13-16 shows an associative path array selected for editing and the corresponding **Array** ribbon tab. The base point grip offers standard **MOVE**, **ROTATE**, **SCALE**, and **MIRROR** grip commands, and **Move**, **Row Count**, and **Level Count** context-sensitive options to move the array and change the number of rows and levels. See Figure 13-16B.

Figure 13-14.
Use the **Axis Angle** option with dynamic input to set the angles of the row and column axes in the rectangular array. Select the appropriate grip to specify an angle from 0° horizontal or vertical.

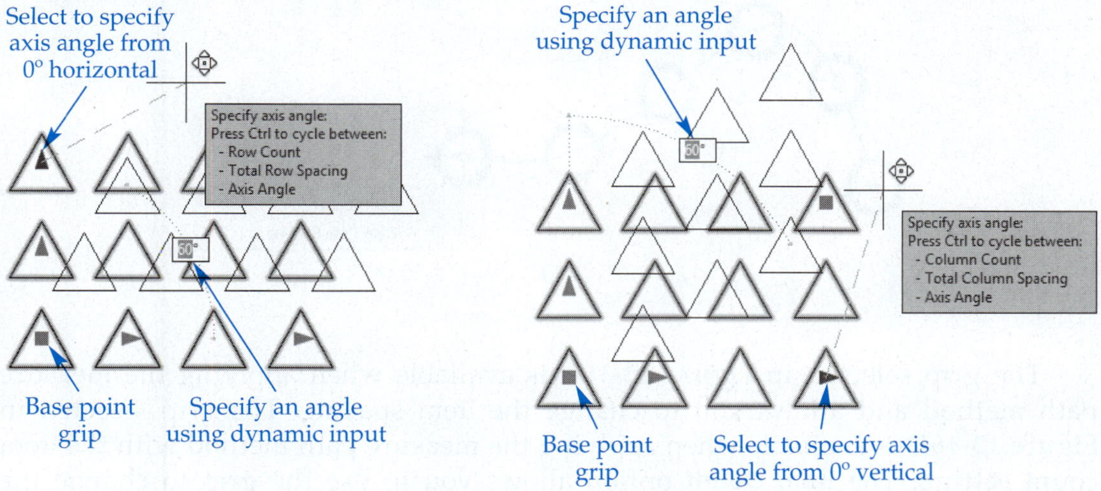

Figure 13-15.
Editing an associative polar array of teeth on a spur gear. To access context-sensitive array grip commands, hover over an unselected grip to display a menu of options and then pick an option from the list, pick a grip and then press [Ctrl] to cycle through options, or pick a grip and then right-click and select an option.

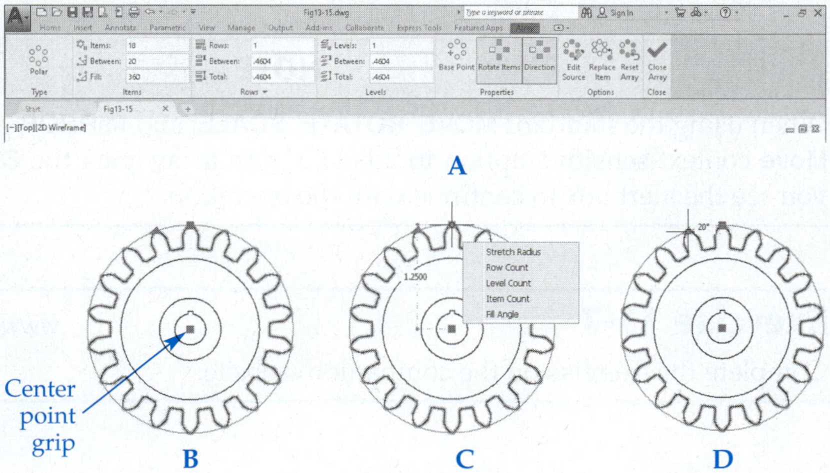

Figure 13-16.
Editing an associative path array of trees along a trail. To access context-sensitive array grip commands, hover over an unselected grip to display a menu of options and then pick an option from the list, pick a grip and then press [Ctrl] to cycle through options, or pick a grip and then right-click and select an option.

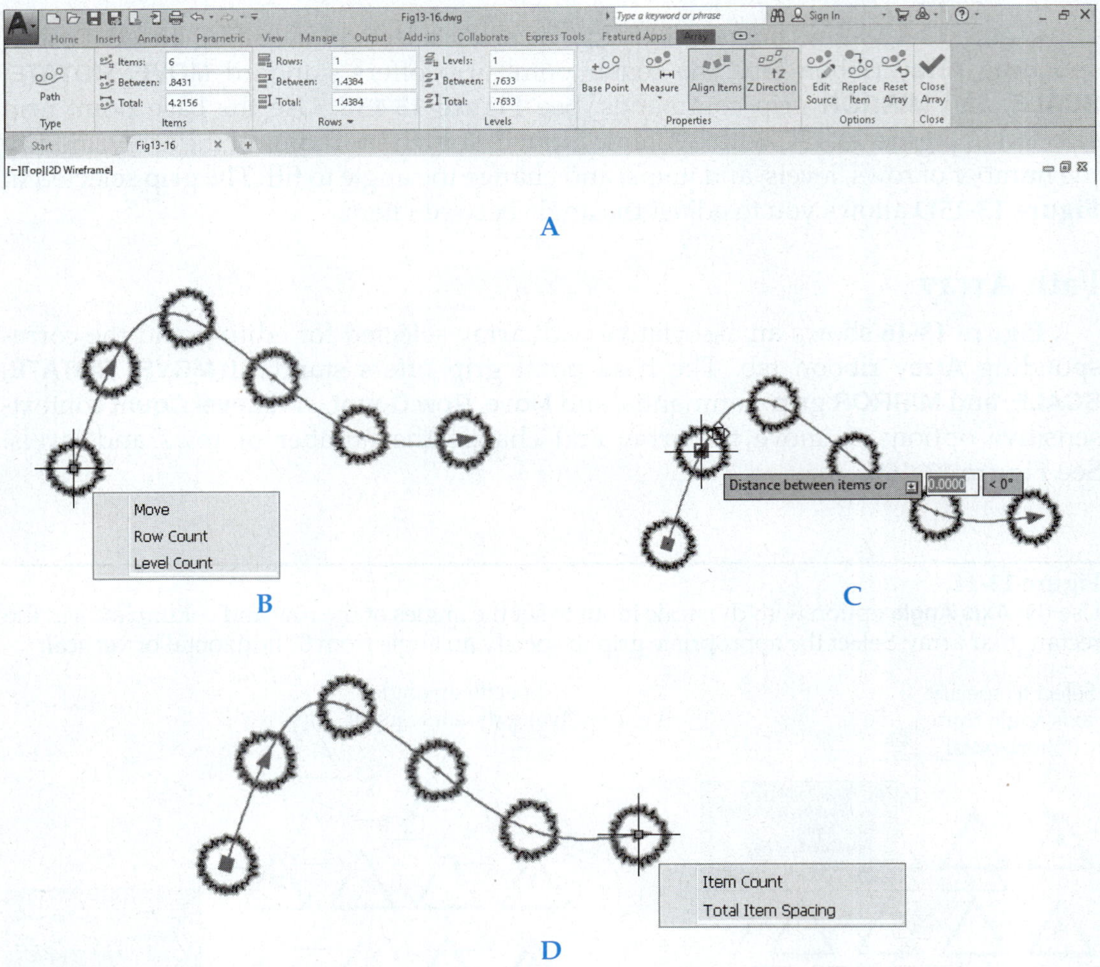

The grip selected in Figure 13-16C is available when applying the measure path method and allows you to change the item spacing. The grip selected in Figure 13-16D is available when applying the measure path method with the item count setting. The **Item Count** option allows you to use the grip to change the total number of items. Use the **Total Item Spacing** option to specify the total item distance with the grip.

NOTE

When using the standard **MOVE**, **ROTATE**, **SCALE**, and **MIRROR** grip commands or the **Move** context-sensitive option to adjust a path array, pick the **Continue** button when you see the alert box to continue with the operation.

Exercise 13-7

www.g-wlearning.com/CAD/

Complete the exercise on the companion website.

Editing Source Objects

Associative rectangular, polar, and path arrays include an option to edit the source objects, such as adding or removing geometry, and apply the changes to the array without exploding or recreating the pattern. An easy way to edit the source objects of an associative array is to select the array to display grips and the context-sensitive **Array** ribbon tab. Then pick the **Edit Source** button from the **Options** panel of the **Array** ribbon tab. Pick any item in the selected array. You do not have to pick the original source objects. Pick the **OK** button when you see the alert box to enter the array editing state.

Ribbon
Array
> Options
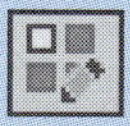
Edit Source

Use drawing and editing commands to modify the selected item. Figure 13-17 shows a basic change made using the **STRETCH** command to modify the items in a rectangular array. When you finish editing, pick the **Save Changes** button in the **Edit Array** panel on the **Home** ribbon tab to save changes and exit the array editing state, or pick the **Discard Changes** button to exit without saving.

Ribbon
Home
> Edit Array
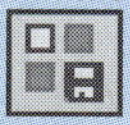
Save Changes

Ribbon
Home
> Edit Array
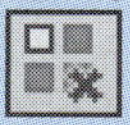
Discard Changes

NOTE

The layer you assign to source objects applies to all items in the array. When working with an associative array, edit and then assign a different layer to the source objects in order to change the layer on which the array is drawn.

Replacing Items

Associative rectangular, polar, and path arrays also offer the option to replace specific items with different objects. The easiest way to replace items in an associative array is to select the array to display grips and the context-sensitive **Array** ribbon tab. Then pick the **Replace Item** button from the **Options** panel of the **Array** ribbon tab. Pick objects not related to the array to use as the replacement for items in the array. Press [Enter] to continue. A rubberband line attaches to the base point of the array for reference. Select an appropriate base point for the replacement objects. Use the **centroid** and **Key Point** options if necessary, as described in Chapter 12.

Ribbon
Array
> Options
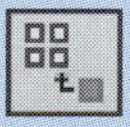
Replace Item

Next, pick items in the array to replace with the selected replacement objects, and then press [Enter] to continue. An alternative is to select the **Source objects** option to replace all items in the array. To finalize the array and exit the command, press [Enter] or select the **eXit** option. AutoCAD updates the array and erases the replacement objects. Figure 13-18 shows an example of replacing specific items in a path array.

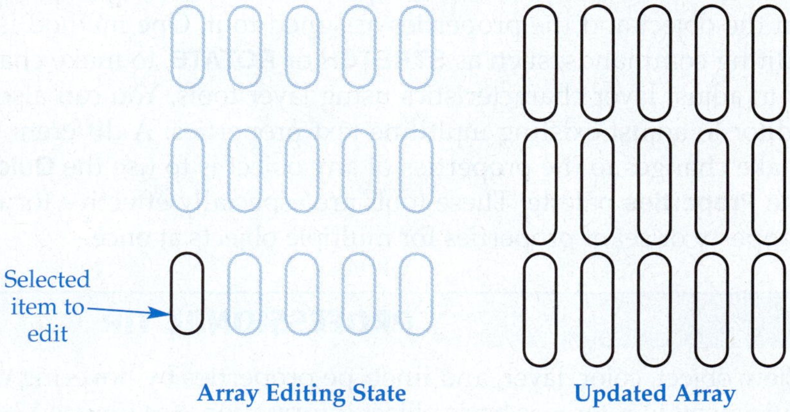

Figure 13-17. Editing source objects to modify an associative array. This example shows changing the size of slots in a rectangular array.

Figure 13-18.
Replacing items in an associative array. This example shows replacing specific trees along a trail in a path array.

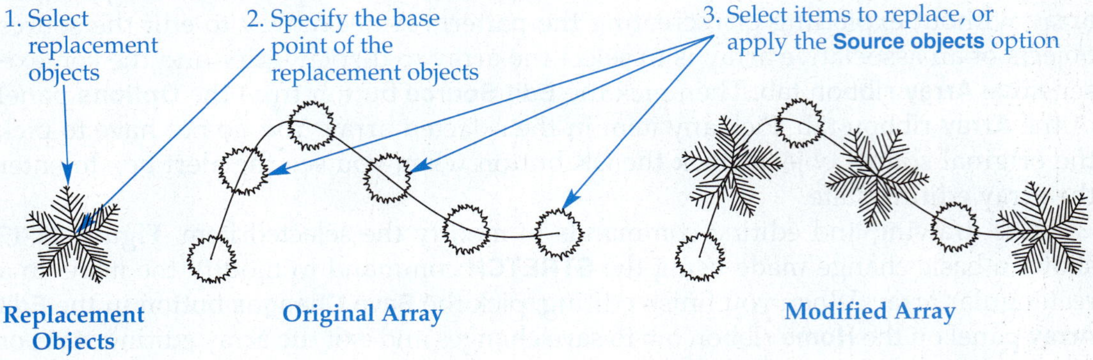

NOTE

The easiest way to return an associative array to its original state, without replacements, is to select the array to display grips and the context-sensitive **Array** ribbon tab. Then pick the **Reset Array** button from the **Options** panel. However, you cannot reset an array to the original state if you use the **Source objects** option and then exit array editing.

Exercise 13-8

www.g-wlearning.com/CAD/

Complete the exercise on the companion website.

Object Properties

Every object has properties. Properties include geometry characteristics, such as the coordinates of the endpoints of a line in X,Y,Z space, the diameter of a circle, or the area of a rectangle. Layer is another property and is associated with all objects. The layer you assign to an object defines other properties, including color, linetype, and lineweight. Most objects also include object-specific properties. For example, multiline text has a variety of text properties, and an associative rectangular array has column, row, and other properties that define the array.

AutoCAD provides many options for adjusting object properties, depending on the object and the properties assigned to it. One method is to use grip editing or editing commands, such as **STRETCH** or **ROTATE**, to make changes. Another method is to adjust layer characteristics using layer tools. You can also use the multiline text editor to adjust existing multiline text properties. A different technique to view and make changes to the properties of any object is to use the **Quick Properties** palette or the **Properties** palette. These tools are especially effective for modifying a particular property or set of properties for multiple objects at once.

PROFESSIONAL TIP

View object, color, layer, and linetype properties by hovering over an object. This is a quick way to reference basic object information. See Figure 13-19.

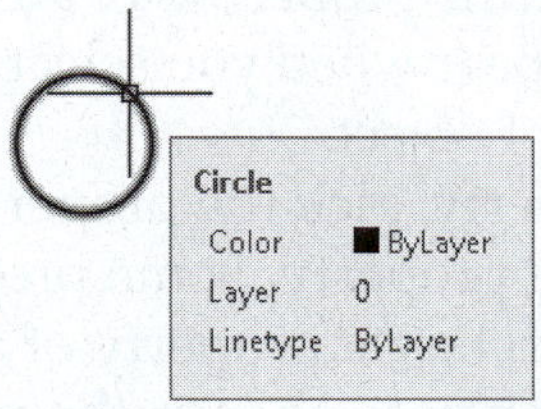

Figure 13-19.
Hover over an object to view its color, layer, and linetype properties.

Using the Quick Properties Palette

The **Quick Properties** palette, shown in **Figure 13-20**, appears by default when you double-click on certain objects. For example, the **Quick Properties** palette appears by default when you double-click on a line, but the text editor opens when you double-click on an mtext or text object. You can also set the **Quick Properties** palette to display with a single-click on an object. A quick way to enable or disable this function is to pick the **Quick Properties** button on the status bar or type [Ctrl]+[Shift]+[P]. The **Quick Properties** button does not appear on the status bar by default. Select the **Quick Properties** option in the status bar **Customization** flyout to add the **Quick Properties** button to the status bar.

The **Quick Properties** palette floats by default above and to the right of where you pick. The drop-down list at the top of the **Quick Properties** palette identifies the selected object. Properties associated with the selected object are displayed below the drop-down list in rows. For example, if you pick a circle, the **Quick Properties** palette lists rows of circle properties.

The **Quick Properties** palette lists common properties associated with the selected objects by default. You should recognize most of the properties included in the **Quick Properties** palette. To pick multiple objects, use window selection and then right-click and select **Quick Properties**, or double-click sequentially on more than one object. When you pick multiple objects, use the **Quick Properties** palette to modify all of the objects, or pick a specific object type from the drop-down list to modify. See **Figure 13-21**.

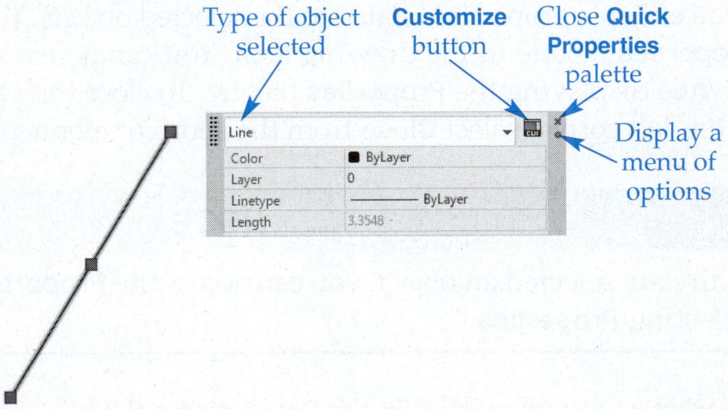

Figure 13-20.
Use the **Quick Properties** palette to display and modify certain object properties, such as the basic properties of the selected line shown.

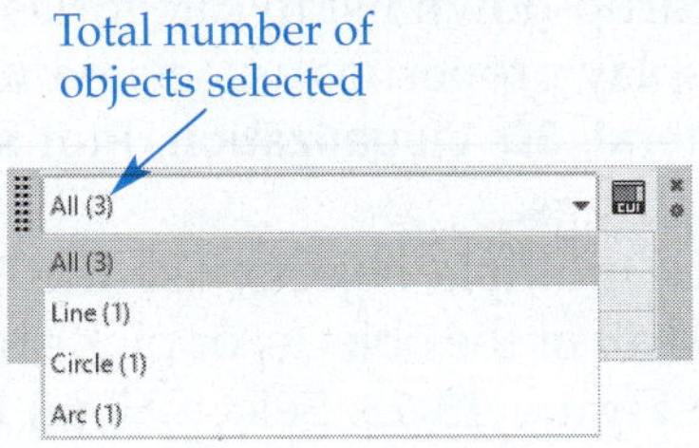

Figure 13-21.
The **Quick Properties** palette with three different types of objects selected. You can edit the objects individually or select **All (3)** to edit the objects together.

Select **All (*n*)** to change the properties of all selected objects. Only properties shared by all selected objects appear when you select **All (*n*)**. Select the appropriate object type to modify a single type of object.

To change a property, pick the property or current value. The way you change a value depends on the property. Some properties, such as the Radius property of an arc or circle or the Text height property of an mtext object, display a text box. Enter a new value in the text box to change the property. Many text boxes display a calculator icon on the right side that opens the **QuickCalc** palette for calculating values. Chapter 15 covers using **QuickCalc**. Other properties, such as the Layer property, display a drop-down list of selections. Hover over a value in the drop-down list to preview the effect before assigning. A pick button is available for geometric properties, such as the Center X and Center Y properties of a circle. Select the pick button to specify a new coordinate. Select a dialog box icon to open a dialog box related to the property. Press [Esc] or pick the **Close** button in the upper-right corner of the palette to hide the **Quick Properties** palette.

NOTE

Right-click on a **Quick Properties** grab bar or pick the **Options** button on the **Quick Properties** palette to access options for adjusting the display and function of the **Quick Properties** palette. The **Quick Properties** tab of the **Drafting Settings** dialog box includes some of the same settings, as well as additional options.

Exercise 13-9

www.g-wlearning.com/CAD/

Complete the exercise on the companion website.

Using the Properties Palette

PROPERTIES

Ribbon
View
> Palettes
Properties Palette
Home
> Properties
Properties
Type
PROPERTIES
PROPS
CH
MO
[Ctrl]+[1]

The **Properties** palette, shown in Figure 13-22, provides the same function as the **Quick Properties** palette, but the **Properties** palette allows you to view *all* properties and adjust all editable properties related to the selected objects. You can dock, lock, and resize the **Properties** palette in the drawing area. You can access commands and continue to work while displaying the **Properties** palette. To close the palette, pick the **Close** button in the top-left corner, select **Close** from the shortcut menu, or press [Ctrl]+[1].

NOTE

If you already selected an object, you can access the **Properties** palette by right-clicking and selecting **Properties**.

Working with properties in the **Properties** palette is similar to working with properties in the **Quick Properties** palette. The drop-down list at the top of the **Properties** palette identifies the selected object. Properties associated with the selected object display below the drop-down list in categories and property rows. The categories and rows update to display properties associated with your selections. If you do not select an object, the **General**, **3D Visualization**, **Plot style**, **View**, and **Misc** categories list the current drawing settings.

When you pick multiple objects, use the **Properties** palette to modify properties that are common to all of the objects, or pick a specific object type from the drop-down list to modify. See Figure 13-23. Select **All (*n*)** to change the properties of all selected objects. Only properties shared by all selected objects appear when you select **All (*n*)**.

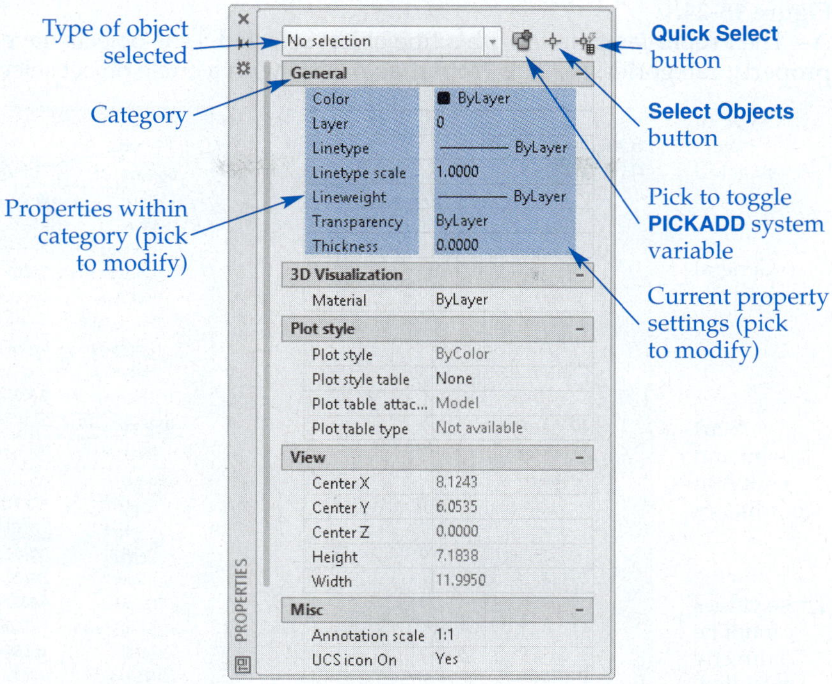

Figure 13-22.
Use the **Properties** palette to modify drawing settings and object properties.

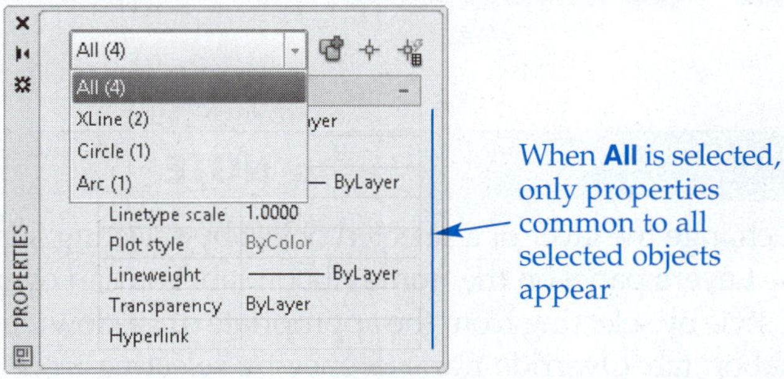

Figure 13-23.
The **Properties** palette with four objects selected. You can edit the objects individually or select **All (4)** to edit properties that are common to all of the objects.

You should recognize most of the properties listed in the **Properties** palette. Do not adjust properties that you do not recognize. For example, the **3D Visualization** category and any properties related to the Z axis are for use in 3D applications.

To change a property, pick the property or current value. The way you change a value depends on the property, just as it does in the **Quick Properties** palette. Use the appropriate text box, drop-down list, or button to modify the value. After you make changes to the objects, press [Esc] to clear grips and remove the objects from the **Properties** palette. Close the **Properties** palette when you are finished.

The upper-right portion of the **Properties** palette contains three buttons. The left button toggles the value of the **PICKADD** system variable, which determines whether you need to hold down [Shift] when adding objects to a selection set. Pick the **Select Objects** button in the middle to deselect the currently selected objects and change the crosshairs to a pick box, allowing you to select other objects. Pick the **Quick Select** button on the right to access the **Quick Select** dialog box from which you can create a selection set, as described later in this chapter.

General Properties

The **General** category of the **Properties** palette allows you to modify general object properties such as color, layer, linetype, linetype scale, plot style, lineweight, transparency, and thickness. See Figure 13-24. The **Quick Properties** palette also lists some general properties.

Figure 13-24.
A—The **Properties** palette with a line object selected. Line objects have options in only three property categories. B—The **Properties** palette with a circle object selected.

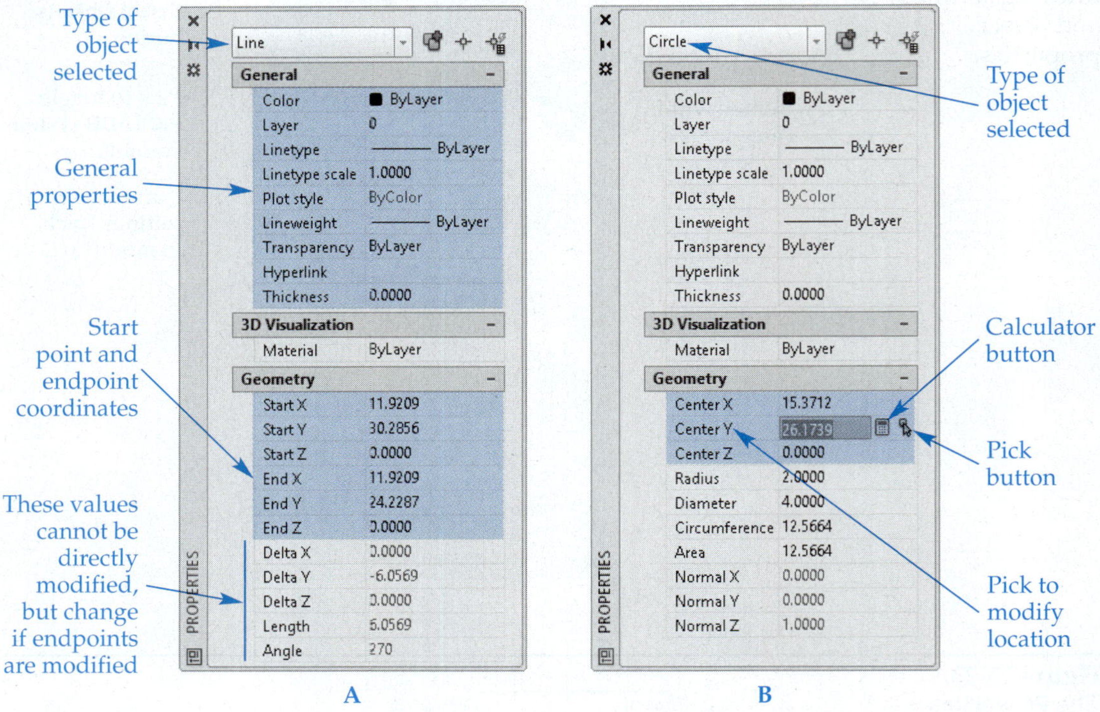

NOTE

You can change the layer of a selected object by selecting a layer from the **Layer** drop-down list in the **Layers** panel on the **Home** ribbon tab. You can override color, linetype, lineweight, and plot style by selecting from the appropriate drop-down list in the **Properties** panel on the **Home** ribbon tab. Override transparency by selecting from the flyout and using the slider, also in the **Properties** panel on the **Home** ribbon tab. Hover over a layer, color, linetype, or lineweight in the drop-down list to preview the effect before assigning. Use the **Property preview** check box of the **Selection** tab in the **Options** dialog box to control property previewing.

CAUTION

Color, linetype, lineweight, and transparency should typically be set as ByLayer. Changing color, linetype, lineweight, or transparency to a value other than ByLayer overrides logical properties, making the property an ***absolute value***. Therefore, if the color of an object is set to red, for example, it appears red regardless of the color assigned to the layer on which you draw the object.

absolute value: In property settings, a value set directly instead of referenced by a layer or a block. An absolute value ignores the current layer settings.

Linetype scale should usually be set globally so the linetype scale of all objects is constant. Adjusting the linetype scale of individual objects can create nonstandard drawings and make it difficult to adjust linetype scale globally. For most applications, you should not override color, linetype, linetype scale, plot style, lineweight, transparency, or thickness.

Geometry Properties

The **Geometry** category of the **Properties** palette allows you to modify object coordinates and dimensions. Refer again to Figure 13-24. The properties in the **Geometry** category vary depending on the selection. Figure 13-24A highlights the X, Y, and Z coordinates that you can use to relocate the start point and endpoint of a line. Figure 13-24B highlights the X, Y, and Z coordinates that you can use to relocate the center of a circle. Enter a value, select the calculator button to calculate a value, or use the pick button to specify a point on-screen. The **Quick Properties** palette also lists some geometry properties.

PROFESSIONAL TIP

The Radius, Diameter, Circumference, and Area properties are especially useful for modifying the size of a circle. Circle properties are an example of the useful information and instinctive adjustments often available using the **Properties** palette. The **Properties** palette provides editable properties for most objects.

Exercise 13-10

www.g-wlearning.com/CAD/

Complete the exercise on the companion website.

Text Properties

The **Text** category appears when you select an mtext or single-line text object. Figure 13-25 shows text properties associated with mtext. The **Properties** palette provides a convenient way to modify a variety of text properties without re-entering the text editor. The **Properties** palette is especially effective to adjust a particular property for multiple selected text objects. For example, change the annotative setting of all text in the drawing using the Annotative property row, or reset the height of multiple mtext or text objects using the Height property row. The **Quick Properties** palette also lists some text-specific properties.

Adding an Mtext Frame

Use the Text frame property of the **Text** category in the **Properties** palette, shown in Figure 13-25, to draw a frame around mtext objects. See Figure 13-26. The distance

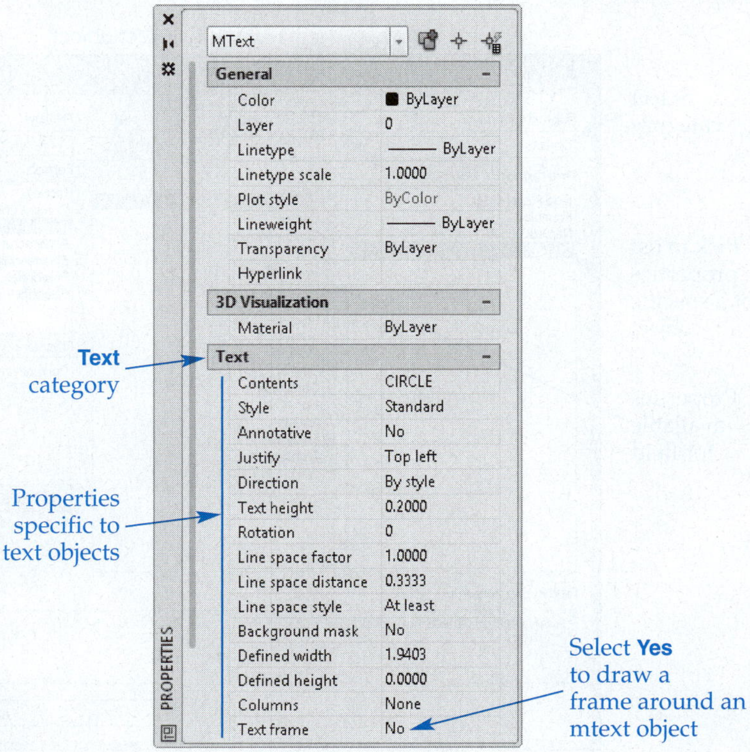

Figure 13-25. The **Properties** palette shows the properties of selected multiline text. The properties of single-line text are slightly different from those of mtext.

Figure 13-26.
An example of using the Text frame property of the **Text** category in the **Properties** palette to draw a frame around mtext.

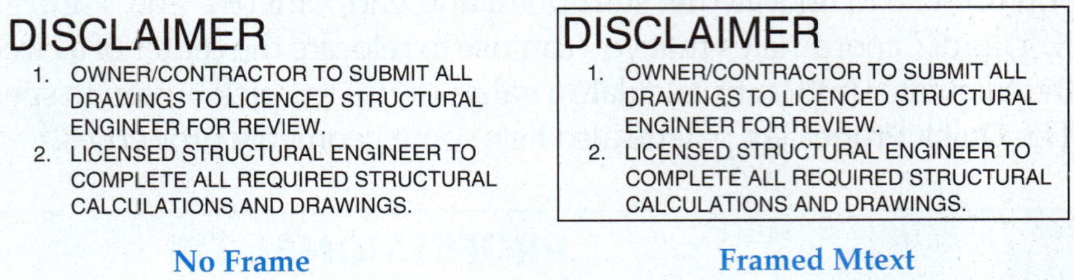

between the mtext and text frame is based on the text height. The default frame offset measured from the bottom of the text is 1.5 times the text height. If a background mask is used, the background mask border offset factor described in Chapter 9 determines the distance between the mtext and the text frame.

NOTE

If you apply a text frame to an mtext object with multiple columns, AutoCAD frames each column separately, and does not frame the gutter.

PROFESSIONAL TIP

Fields provide an effective way to display drawing information within mtext and text objects. You can use fields with many AutoCAD functions and tools, including object properties, drawing properties, attributes, and sheet sets. You can update fields when changes occur to the reference data. Chapter 10 explains using fields. **Figure 13-27** shows acquiring object properties to display in fields.

Figure 13-27.
Pick the **Object** field to add a property for a specific object to a field. Pick the **Select object** button to select the object. Properties specific to the object type appear. Select the property and format for the field.

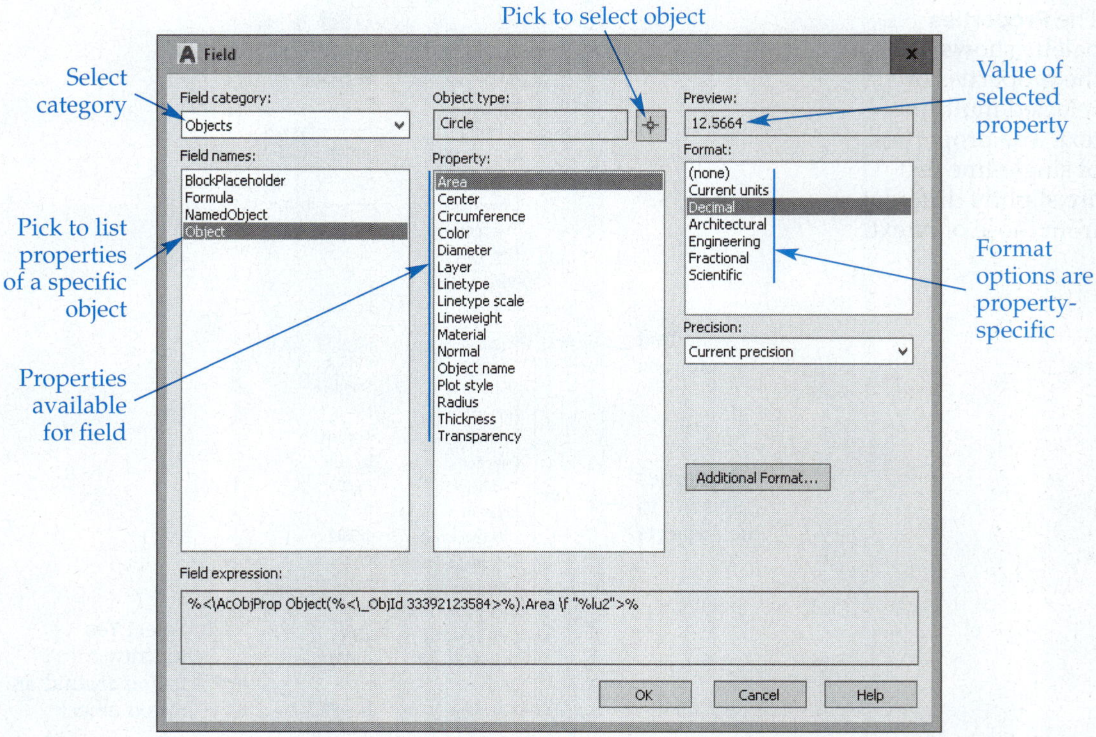

Exercises 13-11 and 13-12

www.g-wlearning.com/CAD/

Complete the exercises on the companion website.

Matching Properties

Use the **MATCHPROP** command to match the properties assigned to one object to other objects. You can match properties in the same drawing or between drawings. Access the **MATCHPROP** command and select the ***source object***. A paintbrush icon appears next to the pick box and AutoCAD lists the properties it will match. Hover over ***destination objects*** to preview the match.

To change the match properties, select the **Settings** option *before* picking the destination objects to display the **Property Settings** dialog box. See Figure 13-28. Properties are replaced in the destination objects if the corresponding **Property Settings** dialog box check boxes are active. For example, to match only the layer property and text style of one text object to another text object, uncheck all boxes except the **Layer** and **Text** property check boxes. Exit the **Property Settings** dialog box when you finish adjusting match properties. Finally, select the destination objects, and then press [Enter] to match properties.

Ribbon
Home
> Properties

Match Properties

Type
MATCHPROP
PAINTER
MA

MATCHPROP

source object: When matching properties, the object with the properties you want to copy to other objects.

destination object: When matching properties, the object that receives the properties of the source object.

Exercise 13-13

www.g-wlearning.com/CAD/

Complete the exercise on the companion website.

Editing between Drawings

You can edit in more than one drawing at a time and edit between open drawings. For example, you can copy objects from one drawing to another. You can also refer to a drawing to obtain information, such as a distance, while working in a different drawing. Figure 13-29 shows two drawings, each of a different section for the same home alteration project, tiled vertically. The Windows ***copy and paste*** function allows you to copy objects from one drawing to another. For example, copy the rafters and exterior studs from the Proposed Entry Section A drawing to paste and reuse them in the

copy and paste: A Windows function that allows you to copy an object from one location to the Windows Clipboard and then paste it to another location.

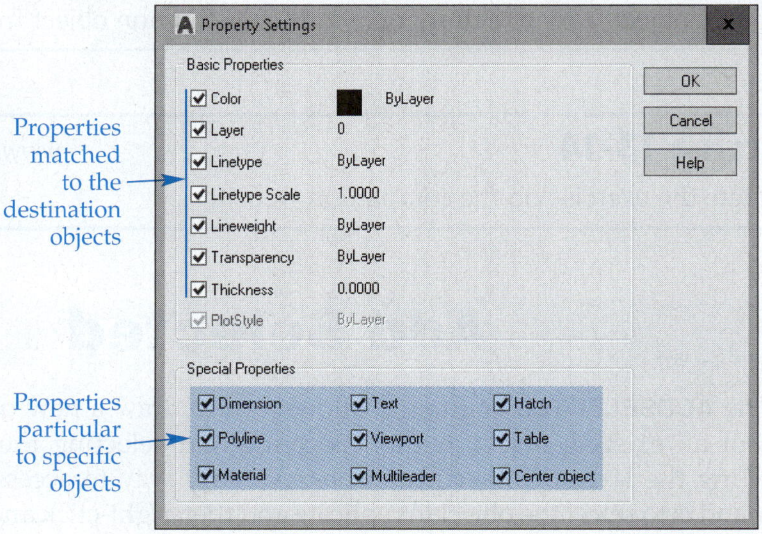

Figure 13-28. In the **Property Settings** dialog box for the **MATCHPROP** command, select the properties to match to the destination objects.

Figure 13-29.
Tile multiple drawings to make editing between drawings easier.

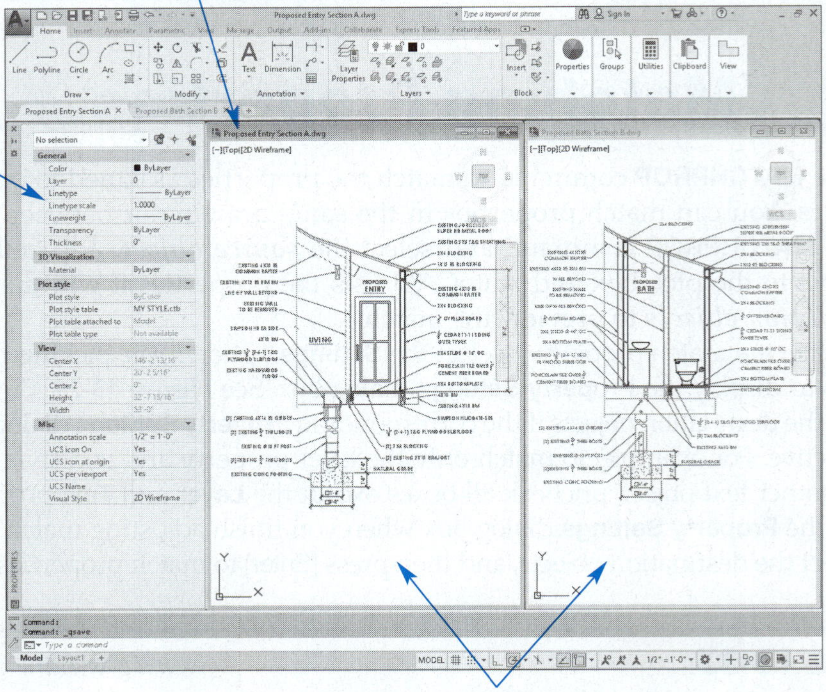

Proposed Bath Section B drawing. Select the objects to copy from the source drawing and select a copy option. Then switch to the destination drawing and select a paste option.

You can cut, copy, and paste between drawings using options from the **Clipboard** panel on the **Home** ribbon tab, the **Clipboard** cascading shortcut menu, or the Windows-standard keyboard shortcuts. **Figure 13-30** explains cut, copy, and paste options available when you right-click after selecting objects or after cutting or copying objects. Many of the same options are available from the **Clipboard** panel on the **Home** ribbon tab and by typing, as shown.

PROFESSIONAL TIP

You may find it more convenient to use the **MATCHPROP** command to match properties between drawings. To use the **MATCHPROP** command between drawings, select the source object from one drawing and the destination object from another.

Exercise 13-14

www.g-wlearning.com/CAD/

Complete the exercise on the companion website.

Add Selected

Type
ADDSELECTED

The **ADDSELECTED** command allows you to draw a new object using the properties of an existing object, without locating and selecting the object command or presetting the layer or other properties. An easy way to access the **ADDSELECTED** command is to select the object to replicate and then right-click and select **Add Selected**.

Figure 13-30.
Right-click options available for cutting, copying, and pasting objects in the active drawing or between drawings.

Cutting or Copying		
Option	**Keyboard Shortcut**	**Description**
Cut	[Ctrl]+[X]	Erases selected objects from the drawing and places the objects on the Windows Clipboard.
Copy	[Ctrl]+[C]	Copies selected objects to the Clipboard.
Copy with Base Point	[Ctrl]+[Shift]+[C]	Copies selected objects to the Clipboard using a specific base point to position the copied objects for pasting. When prompted, select a logical base point, such as a corner or center point of an object.
Pasting		
Option	**Keyboard Shortcut**	**Description**
Paste	[Ctrl]+[V]	Pastes the objects on the Clipboard to the drawing. If you used the **Copy with Base Point** option, the objects attach to the crosshairs at the specified base point.
Paste as Block	[Ctrl]+[Shift]+[V]	Pastes all objects on the Clipboard to the drawing as a block. The pasted objects act as a single object. Blocks are covered later in this textbook. Use the **EXPLODE** command to break up the block.
Paste to Original Coordinates		Pastes the objects on the Clipboard to the same coordinates at which they were located in the original drawing.

AutoCAD initiates the drawing command and assigns properties corresponding to the selected object. For example, pick a circle, right-click and select **Add Selected**, and draw a circle as if you had accessed the **CIRCLE** command. AutoCAD applies the properties of the selected circle to the new circle, regardless of the current settings.

Select Similar

Type
SELECTSIMILAR

The **SELECTSIMILAR** command provides another method of creating a selection set. AutoCAD selects all objects in the drawing that match certain properties of objects you select. An easy way to access the **SELECTSIMILAR** command is to pick the object type you want to select throughout the drawing and then right-click and select **Select Similar**. By default, all objects of the same type and layer are selected. For example, pick a polyline object drawn on a HIDDEN layer, and then right-click and select **Select Similar** to create a selection set of all polylines drawn on the HIDDEN layer.

To specify the properties for selecting similar objects, you must type **SELECTSIMILAR** before picking objects and select the **SEttings** option to display the **Select Similar Settings** dialog box. Use the check boxes to filter the properties that must match the objects you pick in order for other objects to select. The more boxes you check, the more properties must match in order for AutoCAD to select objects.

NOTE

You can pick multiple objects with different properties to select all similar objects. For example, pick a line, arc, and spline, each assigned to a different layer, to select all lines, arcs, and splines with matching layers.

Quick Select

QSELECT

Ribbon
Home
> Utilities
Quick Select

Type
QSELECT
COUNT
NUMBER

The **QSELECT** command is similar to the **SELECTSIMILAR** command, but it provides additional filters. For example, you can use either command to pick all circles in a drawing, but the **QSELECT** command allows you to pick ∅2″ circles. Access the **QSELECT** command to display the **Quick Select** dialog box shown in Figure 13-31. The **Quick Select** dialog box provides options for specifying the exact objects to include in or exclude from the selection set.

Begin the process by selecting the **Entire drawing** option from the **Apply to:** drop-down list to have access to all object types in the drawing for creating a selection set. An alternative is to pick specific objects to create an initial filter of just the selected objects. Pick objects before accessing the **QSELECT** command, or pick the **Select objects** button to return to the drawing window temporarily to pick objects. The **Apply to:** drop-down list then displays **Current selection**. To return to the entire drawing format, select the **Entire drawing** option or check **Append to current selection set** at the bottom of the dialog box.

Pick a specific object type from the **Object type:** drop-down list to create a selection set according to the object type. The **Multiple** option displays properties common to different objects in the entire drawing or the selected objects. Pick a property from the **Properties:** list to narrow the selection set. The items in the **Properties:** list vary depending on the specified object type or **Multiple** option.

relative operators: In math, functions that determine the relationship between data items.

The **Value:** text box or drop-down list allows you to specify a property value to narrow the selection set. Use the **Operator:** drop-down list to assign a ***relative operator*** to control which objects are selected according to the specified property value. For example, to select all ∅2″ circles in the drawing, select the **Entire drawing** option, a **Circle** object type, the **Diameter** property, and the **= Equals** operator, and type a value of 2.

Figure 13-31.
Use the **Quick Select** dialog box to create a specific selection set.

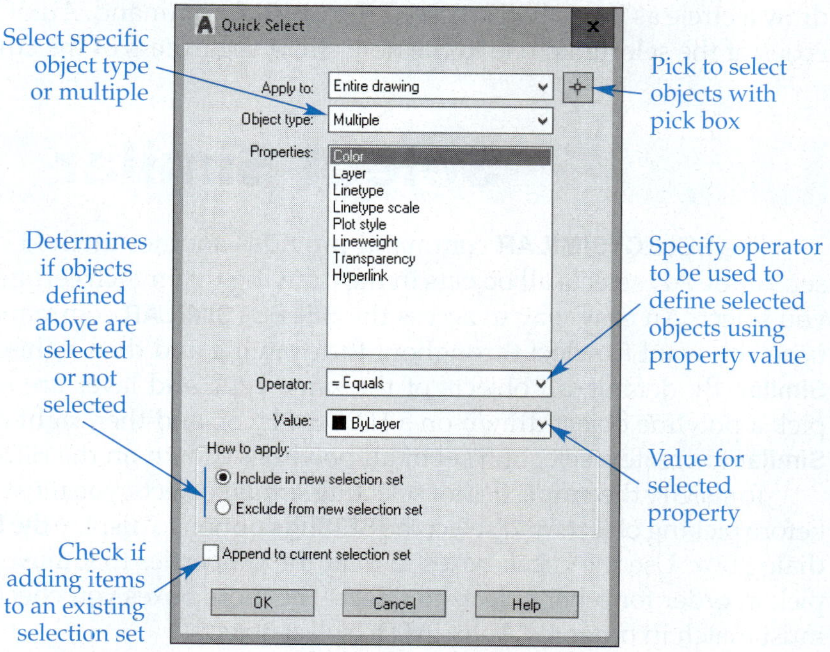

The **Include in new selection set** option creates a selection set according to the quick select settings, as previously described. The **Exclude from new selection set** option reverses the selection set to select all objects except those you specify. Using the previous example, all objects except ∅2″ circles would be selected. Pick the **OK** button to create the selection set.

NOTE

You can also access the **Quick Select** dialog box by right-clicking in the drawing area and selecting **Quick Select...** or by picking the **Quick Select** button on the **Properties** palette.

PROFESSIONAL TIP

Creating a selection set according to specific properties can be very useful. For example, suppose you design a sheet metal part with many different size holes (circle objects) at different locations, and 20 of the holes accept 1/8″ screws. A design change occurs, and the holes must accept 3/16″ screws instead. You could select and modify each circle individually, but it is more efficient to create a selection set of the 20 circles of the same size and modify them at the same time. Use the **Quick Properties** palette or the **Properties** palette to adjust properties of objects selected using the **SELECTSIMILAR** and **QSELECT** commands. You can also use parametric tools, explained in Chapter 22, to make all the circles equal in size. Then, when you change the diameter of one circle, all circles change to the new value.

Exercise 13-15

www.g-wlearning.com/CAD/

Complete the exercise on the companion website.

Supplemental Material

www.g-wlearning.com/CAD/

Object Selection Filters

For detailed information about selecting multiple objects using the **Object Selection Filters** dialog box, go to the companion website, navigate to this chapter in the **Contents** tab, and select **Object Selection Filters**.

Supplemental Material

www.g-wlearning.com/CAD/

Object Groups

For information about creating object groups, go to the companion website, navigate to this chapter in the **Contents** tab, and select **Object Groups**.

Express Tools

www.g-wlearning.com/CAD/

Selection Express Tools

For information about the most useful selection set express tools, go to the companion website, navigate to this chapter in the **Contents** tab, and select **Selection Express Tools**.

Chapter Review

Answer the following questions using the information in this chapter.

1. Name the editing commands that are available using standard grips.
2. How can you select a grip command other than the default **STRETCH**?
3. What is the purpose of the **Base point** option in the grip commands?
4. Which **ROTATE** grip option would you use to rotate an object from an existing 60° angle to a new 25° angle?
5. Explain the function of the **Copy** option in the grip commands.
6. Identify the ribbon tab that provides tools for editing an associative array.
7. Explain how to change the source objects in an array.
8. Explain how to replace one or more items in an associative array.
9. Describe the options available for editing object properties.
10. Describe what happens when you double-click on a line.
11. Identify at least two ways to access the **Properties** palette.
12. How can you change the linetype of an object using the **Properties** palette?
13. For most applications, what value should you use for the color, linetype, and lineweight of objects?
14. Explain how to change the radius of a circle from 1.375 to 1.875 using the **Properties** palette.
15. What command changes the properties of existing objects to match the properties of a different object?
16. Explain how the Windows copy and paste function works to copy an object from one drawing to another.
17. Name the option that pastes a group of objects as a block when pasted.
18. What is the purpose of the **ADDSELECTED** command? Provide an example.
19. What is the purpose of the **SELECTSIMILAR** command? Provide an example.
20. List the information you would specify in the **Quick Select** dialog box to select all Ø6″ circles in a drawing.

Drawing Problems

Start AutoCAD if it is not already started. Start a new drawing for each problem using an appropriate template of your choice. The template should include layers and text styles, when necessary, for drawing the given objects. Add layers and text styles as needed. Draw all objects using appropriate layers. Use appropriate text styles, justification, and format. Follow the specific instructions for each problem. Use only drawing and editing commands and techniques you have already learned. Do not draw dimensions. Use your own judgment and approximate dimensions when necessary.

▼ Basic

1. General Draw the objects labeled A. Then use the **STRETCH** grip command to make the objects look like the objects labeled B. Save the drawing as P13-1.

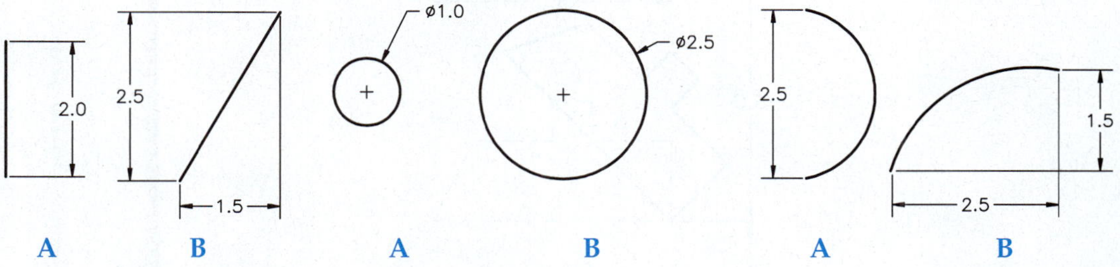

2. General Use the **LINE** command to draw the object labeled A. Use the **Copy** option of the **MOVE** grip command to copy the object to the position labeled B. Edit Object A so it looks like Object C. Edit Object B so it looks like Object D. Save the drawing as P13-2.

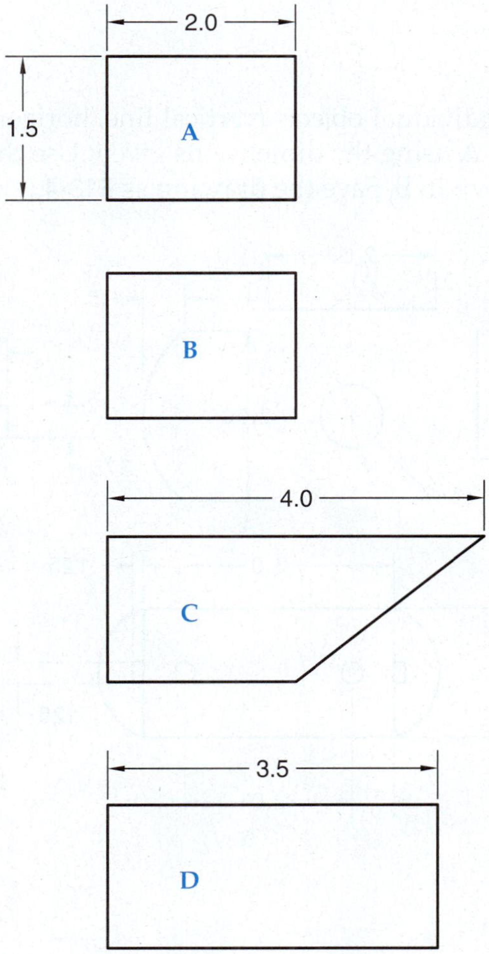

3. **General** Use the **LINE** command to draw the view labeled A. Copy the object, without rotating it, to a position below, as indicated by the dashed lines. Rotate the object 45°. Copy the rotated object labeled B to a position below, as indicated by the dashed lines. Use the **Reference** option to rotate the object 25°, as shown by Object C. Save the drawing as P13-3.

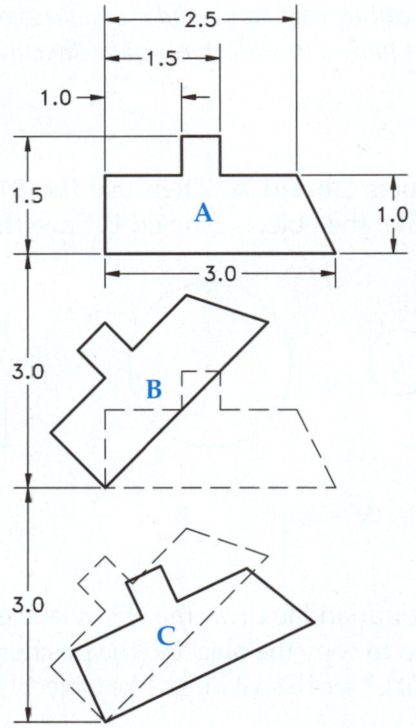

▼ Intermediate

4. **Piping** Draw the individual objects (vertical line, horizontal line, circle, arc, and three-line shape) in A using the dimensions given. Use these objects and grips to create the view shown in B. Save the drawing as P13-4.

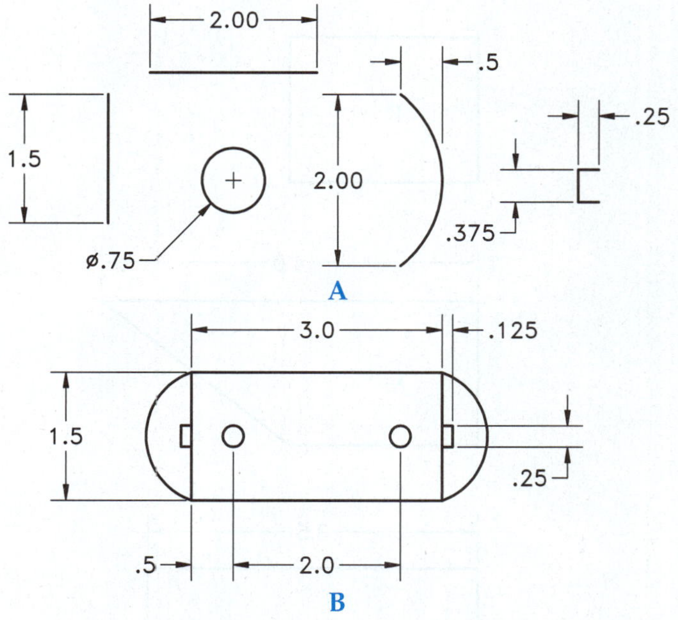

5. **Piping** Save P13-4 as P13-5. In the P13-5 file, copy the view two times to positions B and C. Use the **SCALE** grip command to scale the view in position B to 50% of its original size. Use the **Reference** option of the **SCALE** grip command to enlarge the view in position C from the existing 3.0 length to 4.5, as shown in C. Resave the drawing.

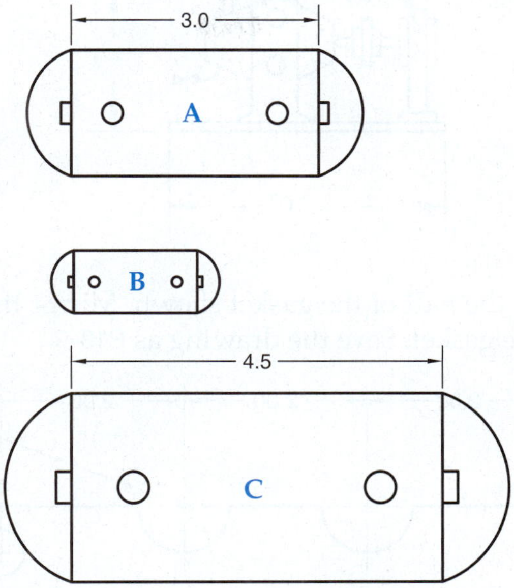

6. **Mechanical** Draw the portion of the gasket shown in A. Use an associative rectangular array to pattern the ∅.25 holes. Use the **MIRROR** grip command to complete the gasket as shown in B. Save the drawing as P13-6.

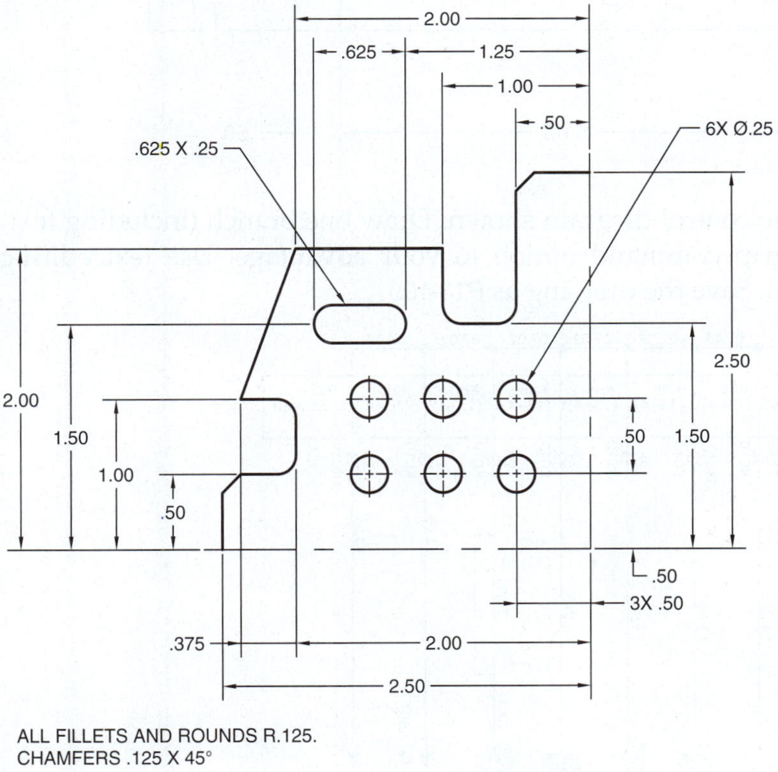

A

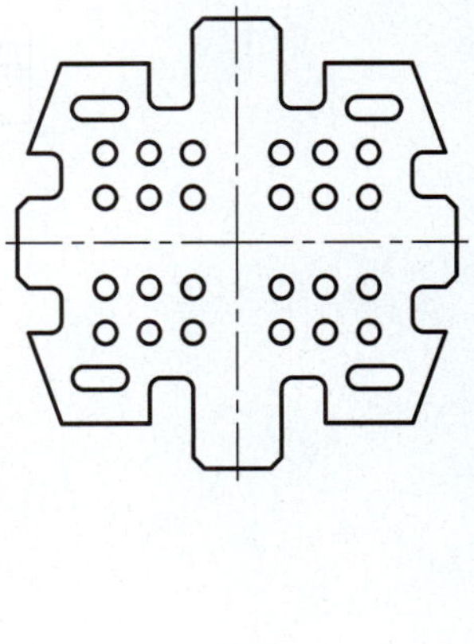

B

7. **Mechanical** Save P13-6 as P13-7. In the P13-7 file, use the **Properties** palette to change the diameters of the circles from .25 to .125. Change the layer assigned to the slots to a layer that uses the PHANTOM linetype. Be sure the linetype scale allows the linetypes to display correctly.

8. **Mechanical** Draw an assembly view similar to the view shown within the boundaries of the given dimensions. All other dimensions are flexible. Save the drawing as P13-8.

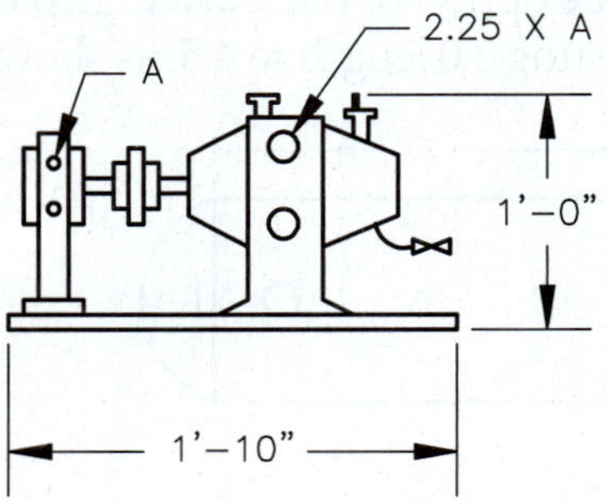

9. **Mechanical** Draw the half of the gasket shown. Mirror the drawing to complete the other half of the gasket. Save the drawing as P13-9.

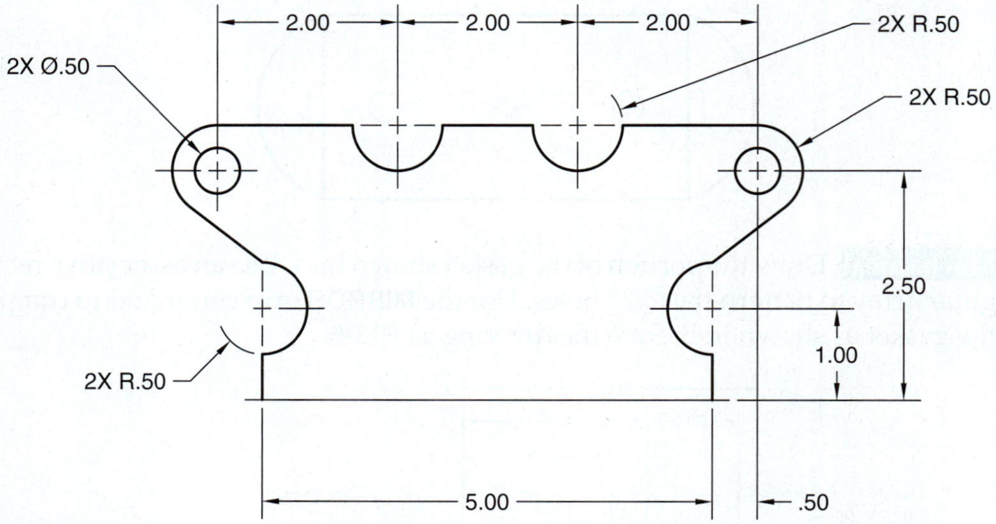

10. **Electronics** Draw the control diagram shown. Draw one branch (including text) and use the **Copy** grip command option to your advantage. Use text editing commands as needed. Save the drawing as P13-10.

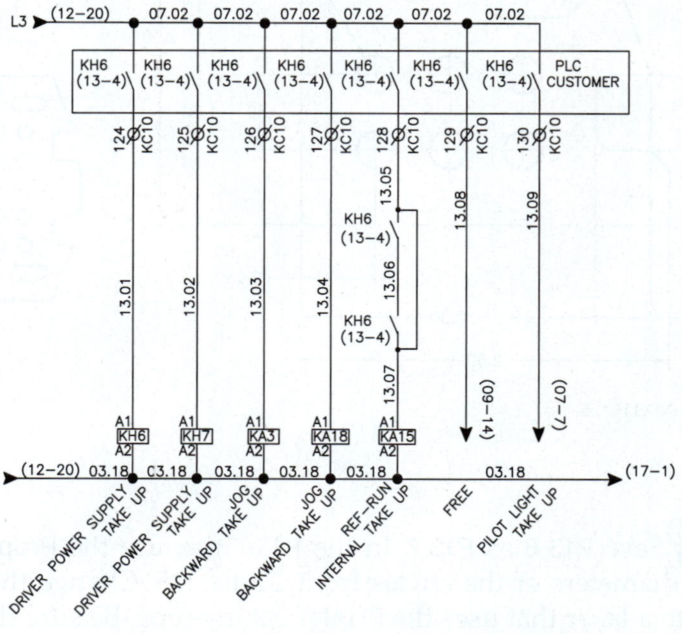

(Design and drawing by EC Company, Portland, Oregon)

▼ Advanced

11. **Mechanical** Draw the folded and flat pattern views of the sheet metal bracket shown. The part material is 18-gage steel. Save the drawing as P13-11.

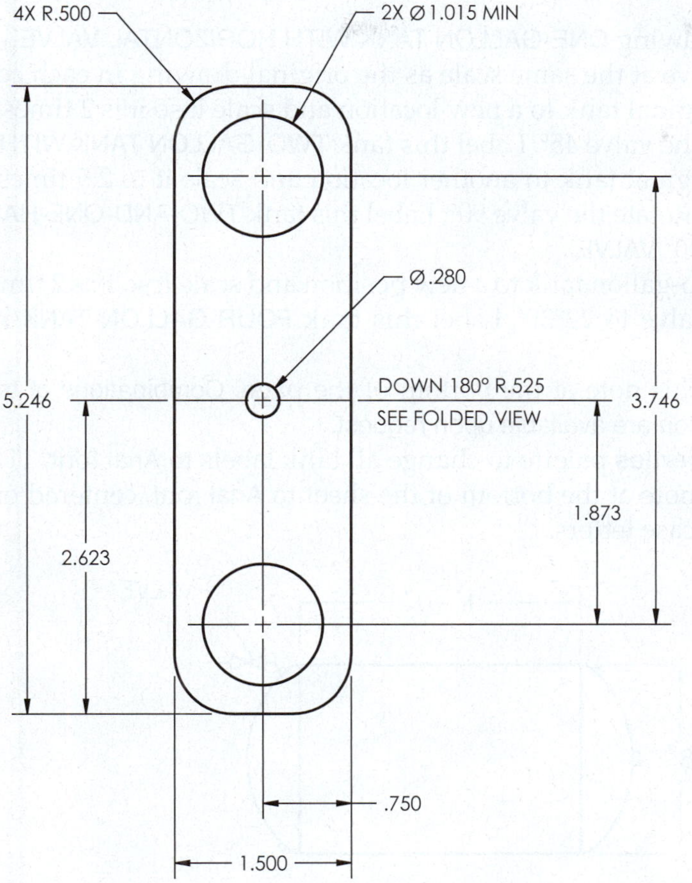

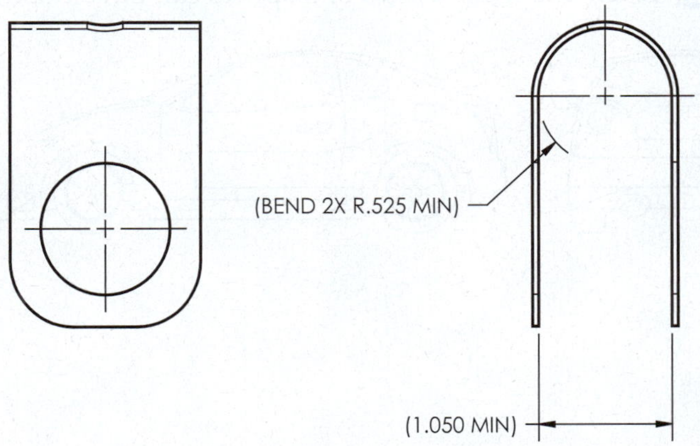

Drawing Problems - Chapter 13

12. **Piping** Draw a tank similar to the one shown within the boundaries of the given dimensions. All other dimensions are flexible. Save the drawing as P13-12. After drawing the tank, create a page for a vendor catalog, as follows:
- All labels should use Romand font, centered directly below the view. Use a text height of .5″.
- Label the drawing ONE-GALLON TANK WITH HORIZONTAL VALVE.
- Keep the valve at the same scale as the original drawing in each copy.
- Copy the original tank to a new location and scale it so it is 2 times its original size. Rotate the valve 45°. Label this tank TWO-GALLON TANK WITH 45° VALVE.
- Copy the original tank to another location and scale it to 2.5 times the size of the original. Rotate the valve 90°. Label this tank TWO-AND-ONE-HALF-GALLON TANK WITH 90° VALVE.
- Copy the two-gallon tank to a new position and scale it so it is 2 times this size. Rotate the valve to 22°30′. Label this tank FOUR-GALLON TANK WITH 22°30' VALVE.
- Left-justify this note at the bottom of the page: Combinations of tank size and valve orientation are available upon request.
- Use the **Properties** palette to change all tank labels to Arial font, 1″ high.
- Change the note at the bottom of the sheet to Arial font, centered on the sheet, using uppercase letters.

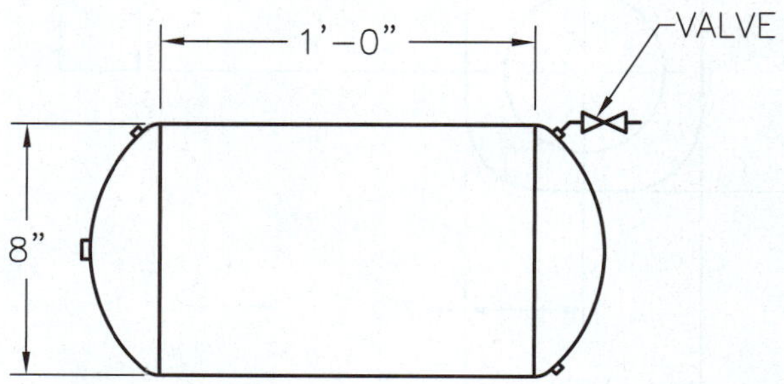

13. **General** Draw the three views of a sports car. Save the drawing as P13-13.

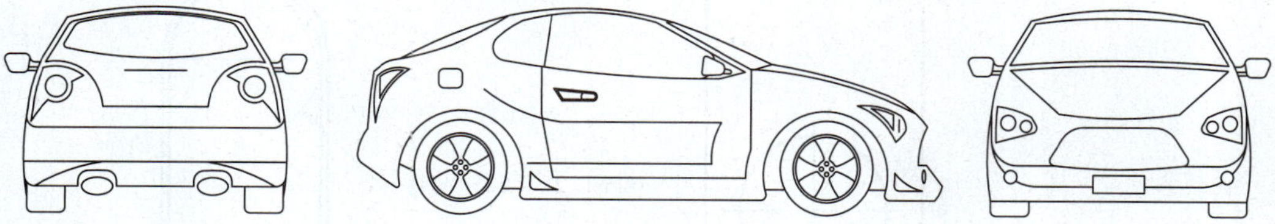

14. **General** Draw the views of the sailboat shown. Save the drawing as P13-14.

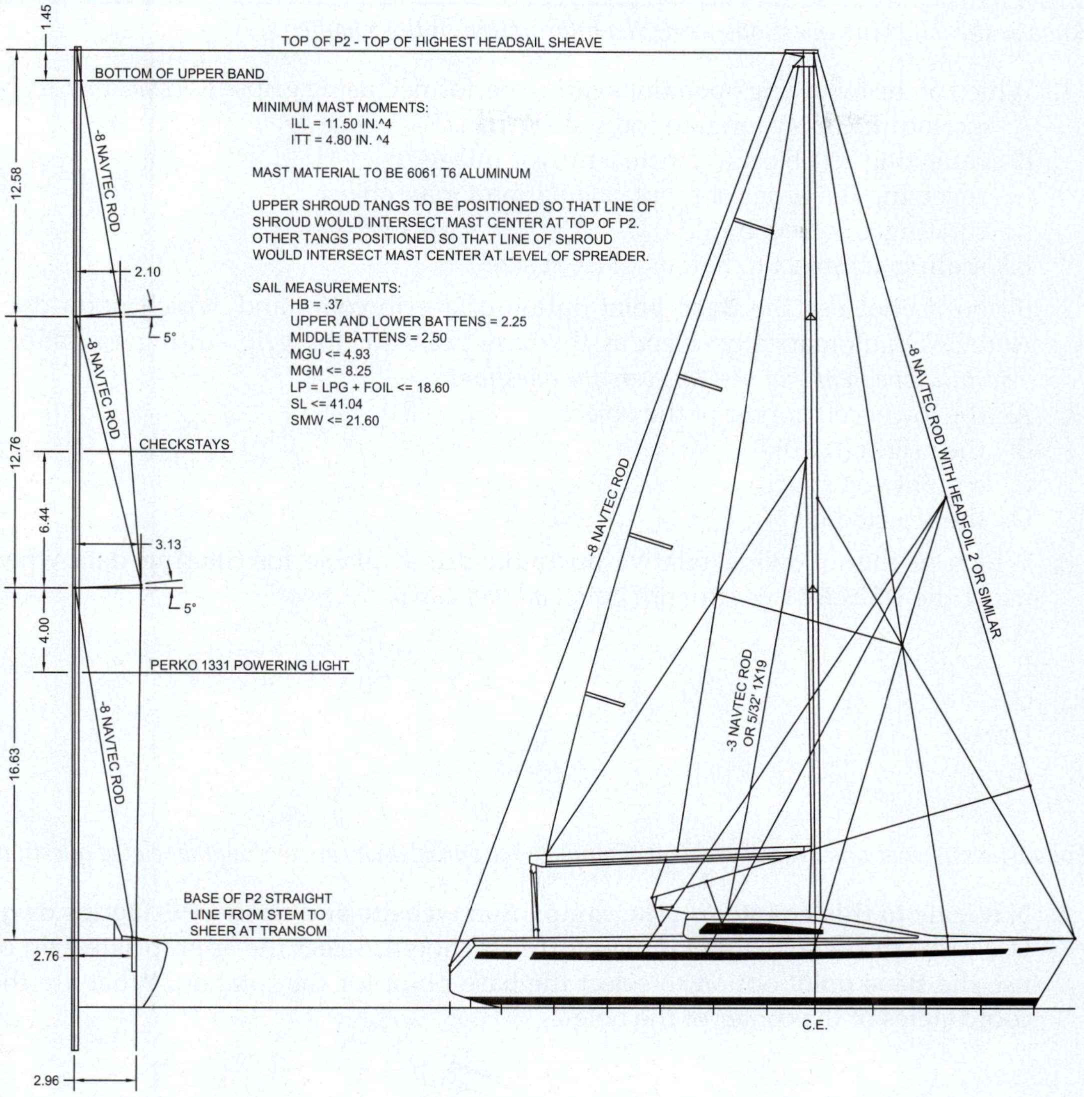

15. **Architectural** Prepare a freehand, dimensioned 2D sketch of a patio or deck plan. Include an outdoor kitchen with a grill, single-burner cooktop, and refrigerator. Add ample seating areas, a table with chairs, and a hot tub. Use dimensions based on your experience, research, and measurements. Start a new drawing from scratch or use an architectural-unit template of your choice. Draw the patio or deck from your sketch. Save the drawing as P13-15.

AutoCAD Certified Professional Exam Practice

Answer the following questions using the information in this chapter.

1. Which of the following operations can be performed using grips? *Select all that apply.*
   A. exploding a polygon into individual lines
   B. reflecting an object to create a mirror image
   C. reversing the order of point calculation for an object
   D. rotating an object by 17°
   E. scaling an object to half its current size
2. If you do not use the **Base point** option of a grip command, which point does AutoCAD automatically select as the base point for the grip editing operation? *Select the one item that best answers the question.*
   A. the lower-left corner of the object
   B. the origin (0,0,0)
   C. a point you specify
   D. the selected grip
3. Which of the following relative operators are available for filtering data when using the **QSELECT** command? *Select all that apply.*
   A. *
   B. /
   C. =
   D. <
   E. >

Follow the instructions in each problem. Complete each problem by answering the related question.

4. **Navigate to this chapter on the companion website and open CPE-13grips.dwg.** Use grips to rotate the metal plate 23.5°, as shown. Select the appropriate grip or use the **Base point** option to select the base point for the rotation. What are the coordinates of the center of the hole?

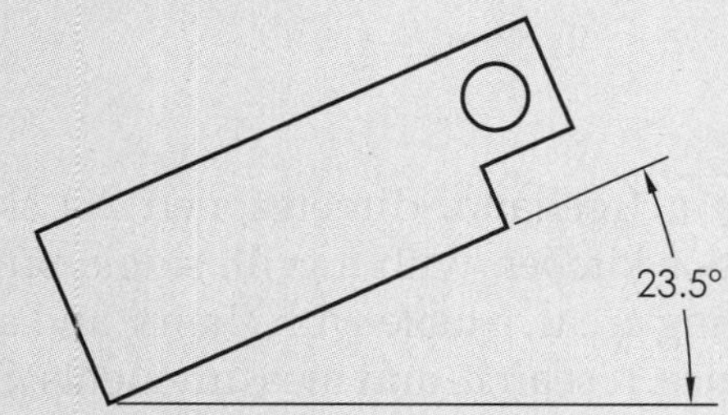

5. **Navigate to this chapter on the companion website and open CPE-13select.dwg.** Enter the **SELECTSIMILAR** command using the command line or dynamic input and specify the **SEttings** option. Disable all of the **Similar Based On** check boxes except **Color**, **Layer**, and **Name**. Select the port in the lower-left corner of the connector system and press [Enter]. How many ports are selected based on your settings?

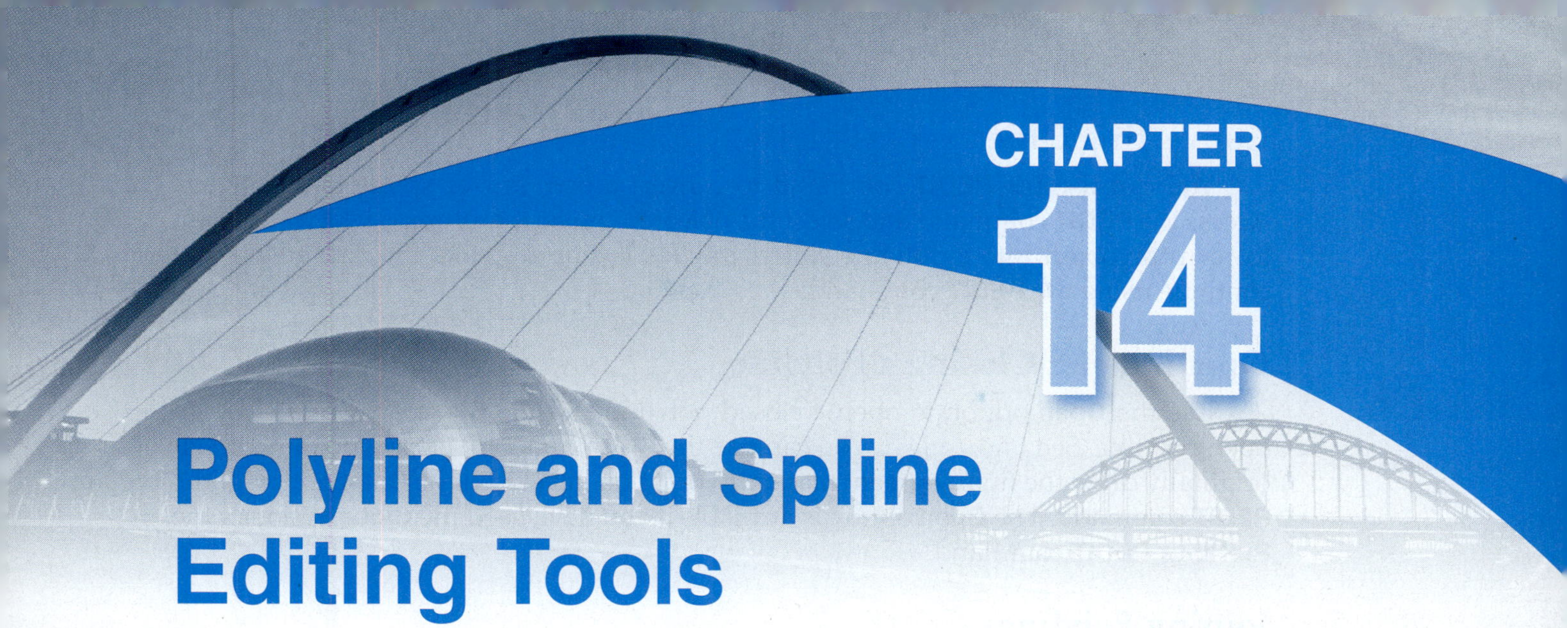

CHAPTER 14

Polyline and Spline Editing Tools

Learning Objectives

After completing this chapter, you will be able to:

✓ Edit polylines with the **PEDIT** command.
✓ Use context-sensitive polyline grip commands.
✓ Create polyline boundaries.
✓ Edit splines with the **SPLINEDIT** command.
✓ Use context-sensitive spline grip commands.

You can modify polyline and spline objects using standard editing commands such as **ERASE**, **STRETCH**, and **SCALE**, but AutoCAD also provides specific commands to edit polylines and splines. Use the **PEDIT** command or polyline grip commands to modify polylines. Use the **SPLINEDIT** command or spline grip commands to modify splines. This chapter also describes how to create polyline boundaries and explores additional options for converting polylines and splines.

Using the PEDIT Command

Access the **PEDIT** command and select a polyline to edit, or select the **Multiple** option to select multiple polylines to edit. To select a wide polyline, pick the edge of a polyline segment rather than the center. Select an option to activate the appropriate editing function.

PEDIT

Ribbon
Home
> Modify
Edit Polyline
Type
PEDIT
EDITPOLYLINE
PE

NOTE

You can also access **PEDIT** command options by double-clicking on a polyline, or by right-clicking on a polyline and selecting **Edit Polyline** from the **Polyline** cascading menu.

PROFESSIONAL TIP

Select the **Undo** option immediately after performing an unwanted edit to restore previous objects without leaving the command. Use the **Undo** option more than once to step back through each operation.

Converting Objects to a Polyline

You can use the **PEDIT** command to convert a line, arc, or spline to a polyline. Access the **PEDIT** command and select the object to convert. A prompt asks if you want to turn the object into a polyline. Select the **Yes** option to make the conversion and continue using the **PEDIT** command.

Opening and Closing a Polyline

Use the **Open** option to open a closed polyline and the **Close** option to close an open polyline. See Figure 14-1. You can use the **Close** option to close a polyline if you originally drew the final segment manually, or if you have used the **Open** option of the **PEDIT** command. The **Open** option is available if you have used the **Close** option of the **PLINE** or **PEDIT** command.

Joining Polylines

The **Join** option of the **PEDIT** command provides the same function as the **JOIN** command, allowing you to create a single polyline object from connected polylines or from a polyline connected to lines and arcs. As when you use the **JOIN** command, the objects to be joined must share a common endpoint and cannot overlap or have gaps between segments. Select the **Join** option and pick the objects to join. You can include the original polyline in the selection set, but it is not necessary. See Figure 14-2.

PROFESSIONAL TIP

After you join objects into a continuous polyline, use the **Close** option to close the polyline if necessary.

Figure 14-1.
Open and closed polylines.

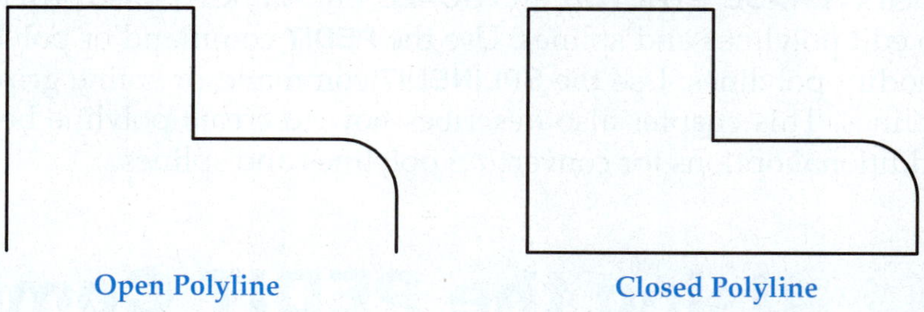

Figure 14-2.
Joining a polyline to other connected lines and an arc.

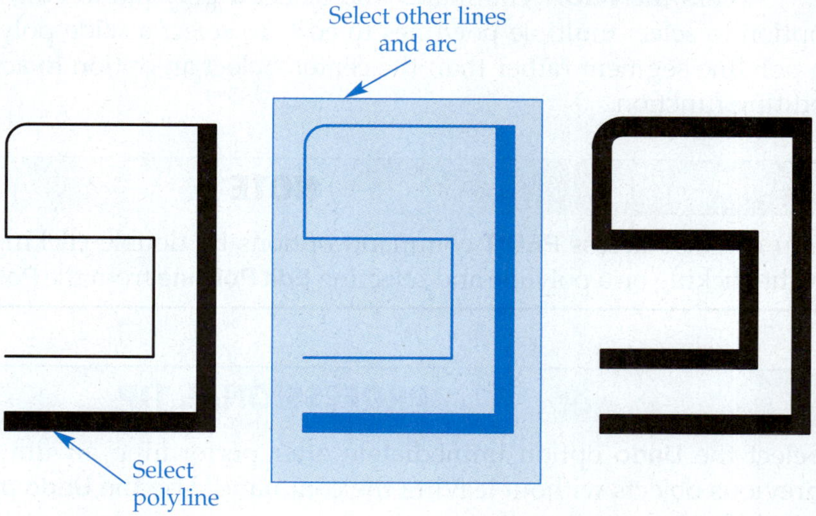

Changing Polyline Width

Select the **Width** option of the **PEDIT** command to specify a new width to assign to the polyline. See Figure 14-3. The width of the original polyline can be constant, or it can vary, but all segments change to the new, constant width.

Exercise 14-1

www.g-wlearning.com/CAD/

Complete the exercise on the companion website.

Editing a Polyline Vertex

The **Edit vertex** option of the **PEDIT** command allows you to perform several operations at a ***polyline vertex***. Select the **Edit vertex** option to display an "X" marker at the first polyline vertex. The options available with the **Edit vertex** option are more difficult to use than other editing tools. The **Break**, **Insert**, **Move**, and **Straighten** options provide functions that you can perform more easily using context-sensitive polyline grips or standard editing commands such as **BREAK** and **STRETCH**. However, the **Tangent** and **Width** options are unique to the **Edit vertex** option and are sometimes useful. The **Tangent** option is most appropriate when used with the **Fit** option, as explained in the next section.

polyline vertex: The point at which two polyline segments meet.

NOTE

If an **Edit vertex** option does not appear to take effect, use the **Regen** option to regenerate the polyline. Use the **eXit** option to return to the **PEDIT** prompt.

Use the **Width** option to change the starting and ending widths of a polyline segment. Enter the **Edit vertex** option and use the **Next** and **Previous** options to move the "X" marker to the first vertex where you want to change the polyline width. Activate the **Width** option and specify the starting and ending width of the polyline segment. See Figure 14-4.

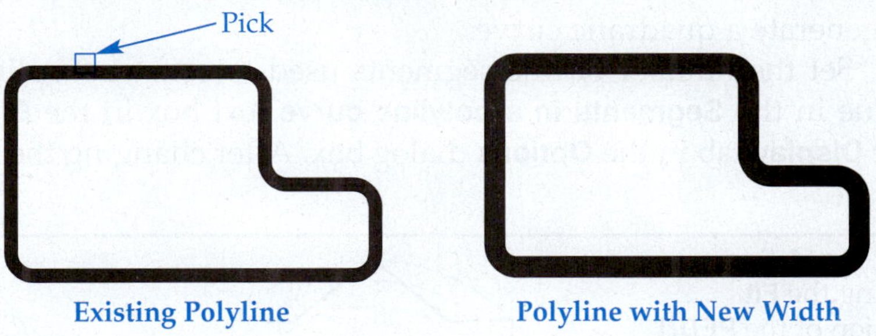

Figure 14-3. Changing the width of a polyline.

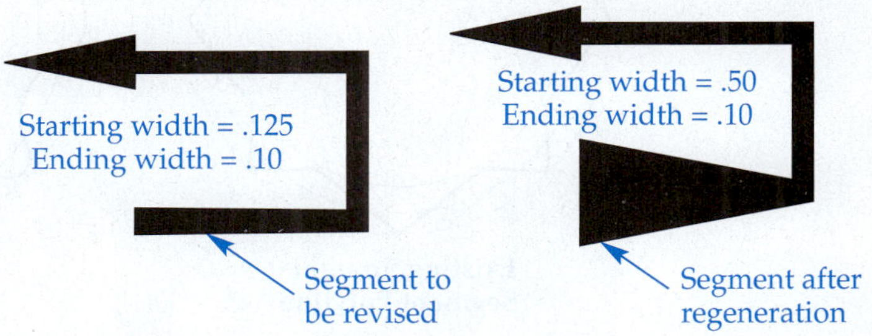

Figure 14-4. Changing the width of a polyline segment with the **Width** vertex editing option. Use the **Regen** option to display the change if the width does not appear to change.

Exercise 14-2

www.g-wlearning.com/CAD/

Complete the exercise on the companion website.

Fit Option

curve fitting: Converting a polyline into a series of smooth curves.

fit curve: A curve that passes through all of its fit points.

The **Fit** option of the **PEDIT** command uses ***curve fitting*** to convert straight polyline segments to a series of smooth curves. For example, use curve fitting to create lines on a graph, add an underlayment symbol to a structural section, or construct a freeform shape. Start by picking accurate points and drawing straight polyline segments. Then use the **Fit** option to smooth the segments. The **Fit** option creates a ***fit curve*** by constructing pairs of arcs that pass through control points. You can use the vertices of the polyline as the control points or specify different control points.

Prior to curve fitting, you can use the **Tangent** option of the **Edit vertex** option to assign each vertex a tangent direction. AutoCAD then fits the curve based on the preset tangent directions. Specifying tangent directions is optional and provides a way to edit vertices when the **Fit** option of the **PEDIT** command does not produce the best results. Enter the **PEDIT** command, select the **Edit vertex** option, and use the **Next** and **Previous** options to move the "X" marker to the first vertex to be changed. Select the **Tangent** option and specify a tangent direction in degrees or pick a point in the expected direction. An arrow at the vertex indicates the direction. Continue to move the marker to other vertices and use the **Tangent** option as necessary.

When all of the vertices you want to change include a specified tangent direction, enter the **Fit** option to create the curve. See Figure 14-5. You can also enter the **PEDIT** command, select a polyline, and enter the **Fit** option without adjusting tangencies. If the resulting curve is not correct, use the **Edit vertex** option to make changes as necessary.

Spline Option

spline curve: A curve that passes through the first and last fit points and is influenced by the other fit points.

cubic curve: A very smooth curve created by the **PEDIT Spline** option with **SPLINETYPE** set at 6.

quadratic curve: A curve created by the **PEDIT Spline** option with **SPLINETYPE** set at 5. The curve is tangent to the polyline segments between the intermediate control points.

When you edit a polyline with the **Fit** option, the resulting curve passes through each polyline vertex. The **Spline** option also smoothes the corners of a straight-segment polyline, but the **Spline** option creates a ***spline curve*** that approximates a true B-spline. See Figure 14-6. Before using the **Spline** option, you can choose to create a spline using ***cubic*** or ***quadratic*** calculations by adjusting the **SPLINETYPE** system variable. The default setting of 6 draws a cubic curve. Change the value to 5 to generate a quadratic curve.

Set the number of line segments used to construct spline curves by entering a value in the **Segments in a polyline curve** text box in the **Display resolution** area of the **Display** tab in the **Options** dialog box. After changing the value, reissue the **Spline**

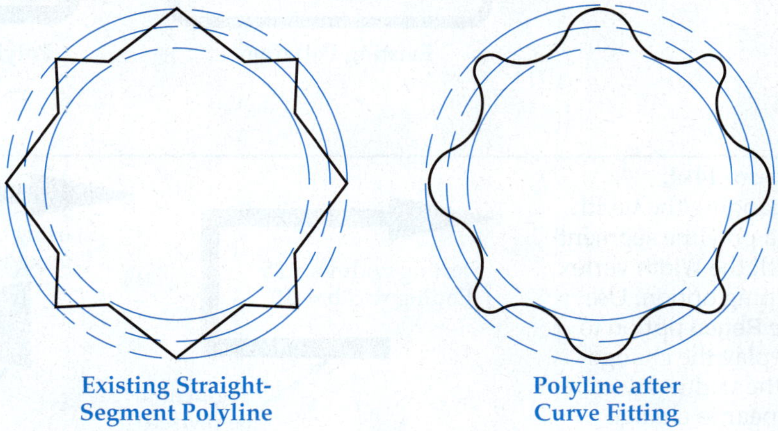

Figure 14-5. Using the **Fit** option of the **PEDIT** command to turn a polyline into a smooth curve.

Figure 14-6.
A comparison of polylines edited with the **Fit** and **Spline** options of the **PEDIT** command. The **SPLINETYPE** system variable controls whether the **Spline** option creates a quadratic or cubic curve.

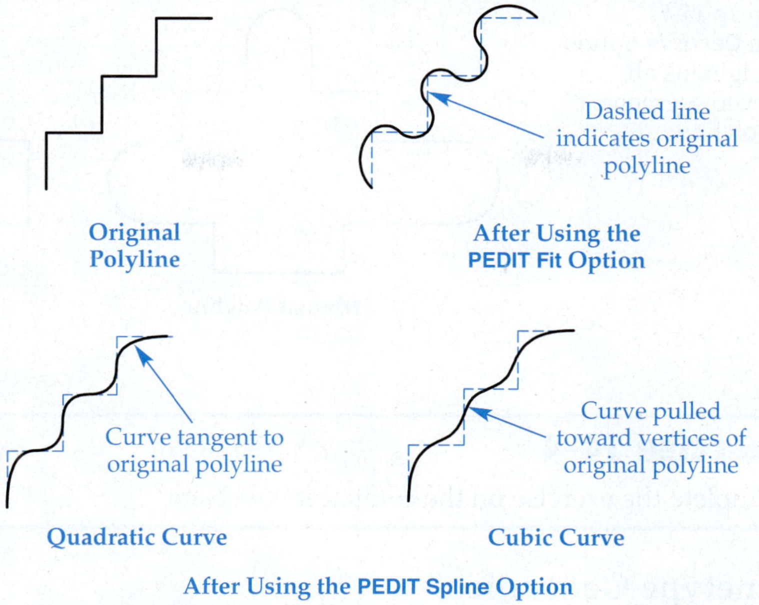

option of the **PEDIT** command to apply the setting. The default value is 8, which creates a smooth spline curve with moderate regeneration time. Increasing the **Segments in a polyline curve** value creates a smoother spline curve, but increases regeneration time and drawing file size. See Figure 14-7.

NOTE

The **Fit** and **Spline** options of the **PEDIT** command create approximations of a B-spline curve. Use the **SPLINE** command to create a true B-spline curve.

Exercise 14-3

www.g-wlearning.com/CAD/

Complete the exercise on the companion website.

Straightening All Polyline Segments

The **Decurve** option of the **PEDIT** command returns a polyline edited with the **Fit** or **Spline** option to its original form. Specified tangent directions remain, however, for future reference. You can also use the **Decurve** option to straighten arc segments of a polyline. See Figure 14-8.

Figure 14-7.
A comparison of curves drawn with different display resolution settings.

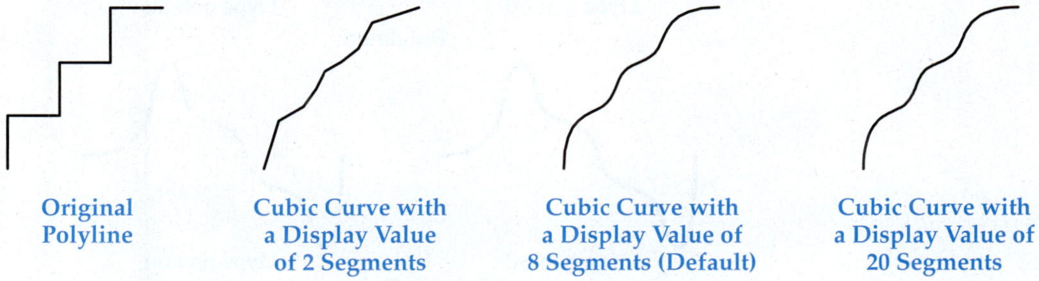

Figure 14-8.
The **Decurve** option straightens all curved segments of a polyline.

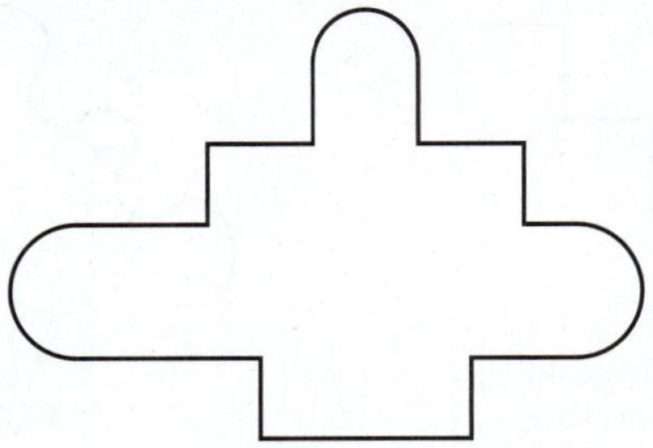

Exercise 14-4

www.g-wlearning.com/CAD/

Complete the exercise on the companion website.

Linetype Generation

The **Ltype gen** option determines how linetypes other than Continuous generate in relation to polyline vertices. For example, if you use a Center linetype and disable the **Ltype gen** option, the polyline has a long dash at each vertex. When you activate the **Ltype gen** option, the polyline has a constant pattern in relation to the polyline as a whole. See Figure 14-9.

NOTE

The **Reverse** option of the **PEDIT** command applies the same operation as the **REVERSE** command explained in Chapter 12. Reversing affects the vertex options of the **PEDIT** command. The **PLINEREVERSEWIDTHS** system variable also applies to the **PEDIT** command.

PROFESSIONAL TIP

Use the **MATCHPROP** command, described in Chapter 13, to match the properties assigned to one polyline to other polylines.

Figure 14-9.
A comparison of polylines and splined polylines with the **Ltype gen** option of the **PEDIT** command on and off.

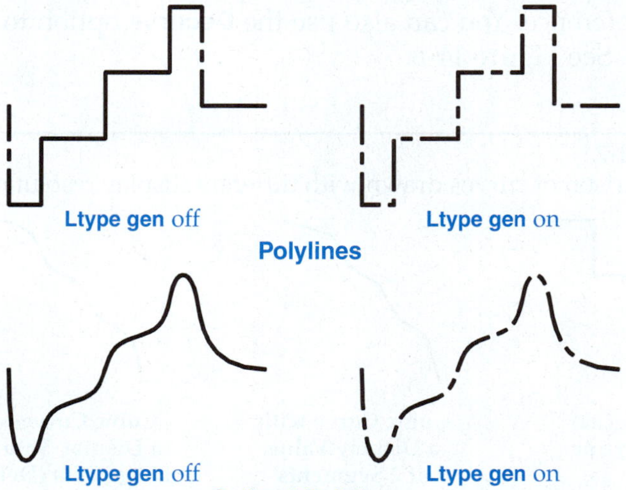

Polyline Grip Commands

Pick a polyline object to display primary grips at each vertex and secondary grips at the midpoint of each segment. See **Figure 14-10**. Vertex grips have the standard filled-square grip appearance, and the midpoint grips appear as filled rectangles. Use any of the grips to apply standard grip editing commands, including **STRETCH**, **MOVE**, **ROTATE**, **SCALE**, **MIRROR**, and **Copy**. Polyline grips also offer commands specific to working with polylines. Using these context-sensitive commands is often the best way to add or remove a vertex, convert a straight segment to an arc, or convert an arc to a straight segment.

You can access and apply the same context-sensitive polyline grip commands in three different ways. See **Figure 14-10**. Different options are available depending on whether you hover over or select a vertex or a midpoint grip, and whether the segment is straight or an arc. **Figure 14-11** illustrates the process of using polyline grip commands to edit polylines. Specify a point if necessary to complete an operation. **Base point**, **Copy**, **Undo**, and **eXit** options are available for stretching a vertex. These options behave the same as when using the standard **STRETCH** grip edit command.

Figure 14-10.
Pick a polyline object to display grips at each vertex and at the midpoint of each segment. Apply one of the following methods to access context-sensitive polyline grip commands. A—Hover over an unselected grip to display a menu of options, and then pick an option from the list. B—Pick a grip and then press [Ctrl] to cycle through options. C—Pick a grip and then right-click and select an option.

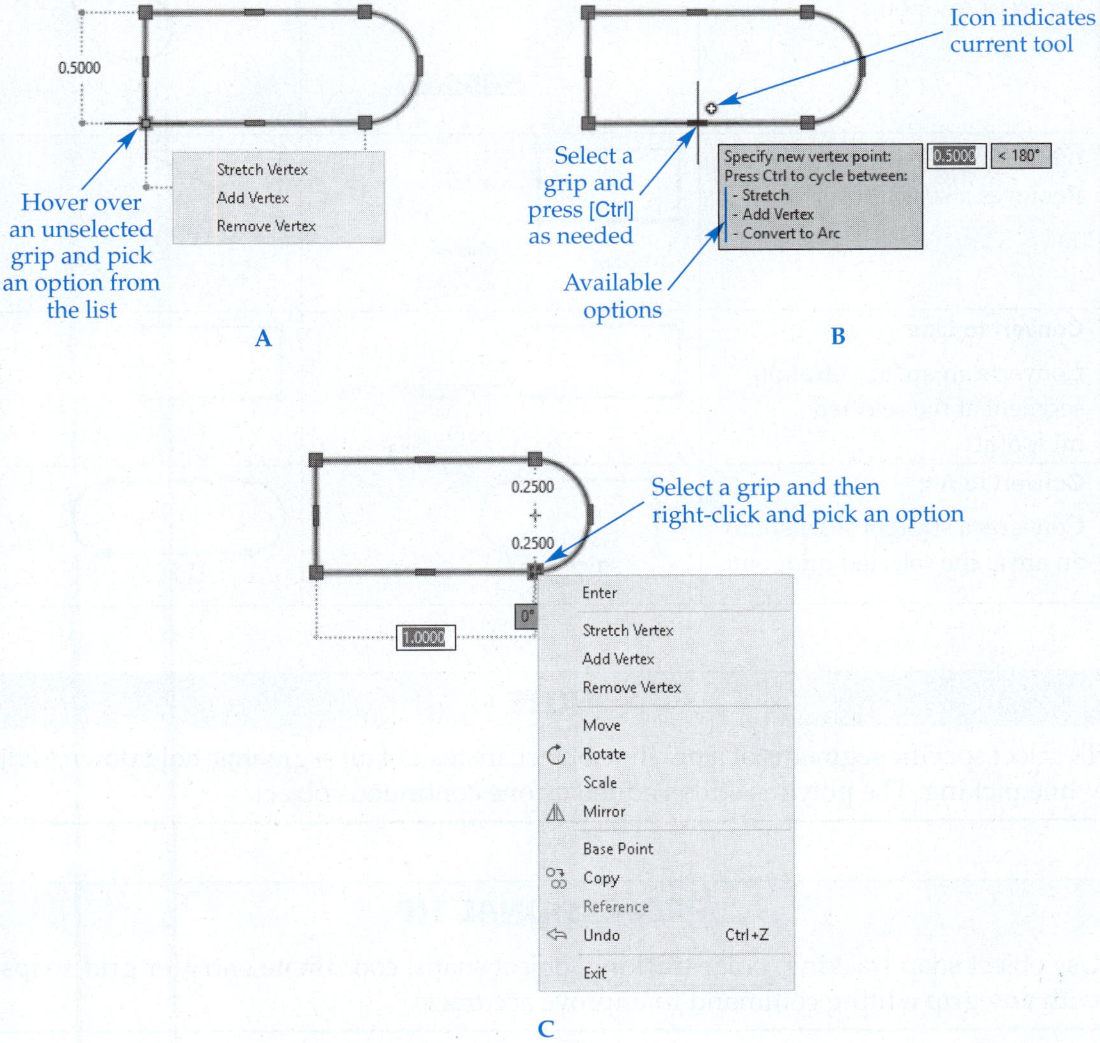

Figure 14-11.
Examples of options available for modifying polylines using grips.

Option	Process	Result
Stretch Vertex Stretches the polyline at the selected vertex	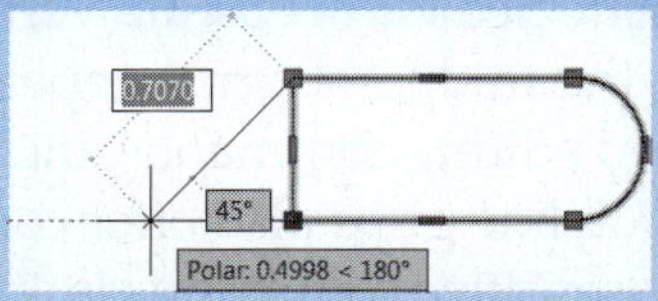	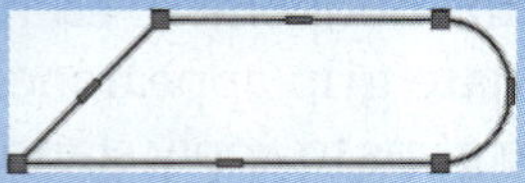
Stretch (straight segment) Stretches the polyline at the selected midpoint	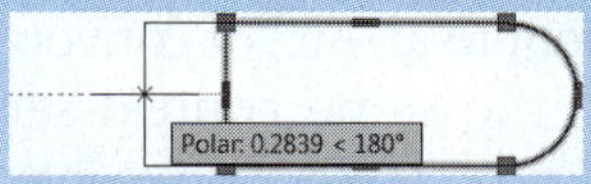	
Stretch (arc) Changes the radius of the arc at the selected midpoint	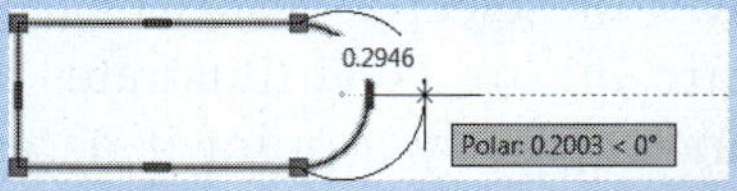	
Add Vertex (straight segment) Adds a vertex using a selected vertex or midpoint	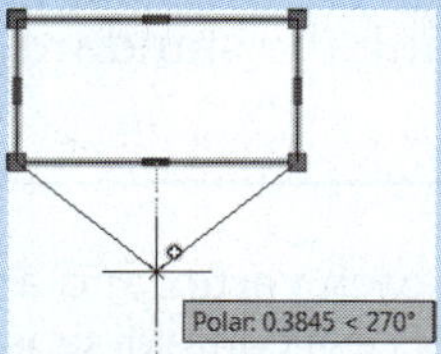	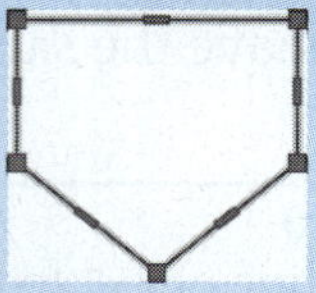
Add Vertex (arc) Adds a vertex using a selected vertex or midpoint	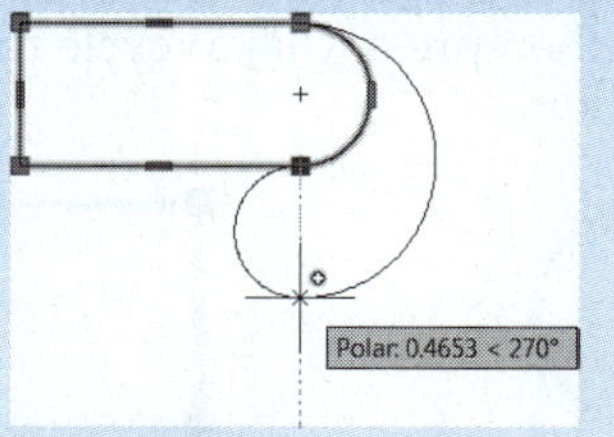	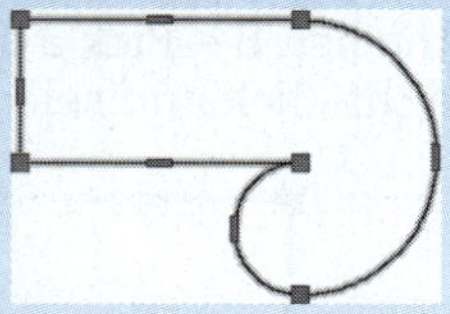
Remove Vertex Removes a selected vertex	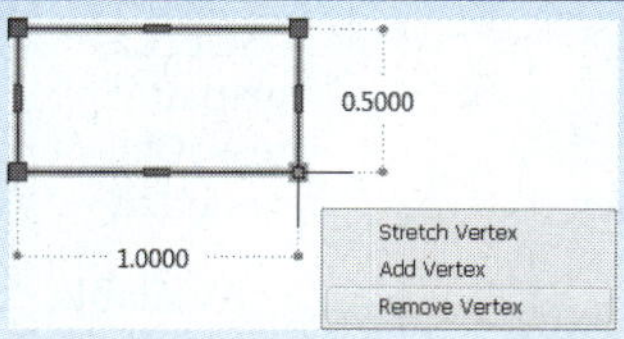	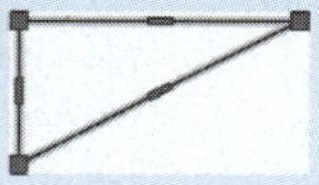
Convert to Line Converts an arc to a straight segment at the selected midpoint	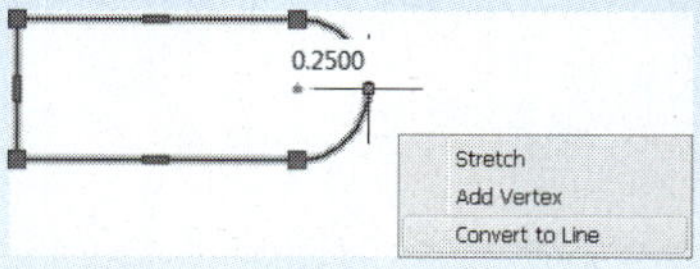	
Convert to Arc Converts a straight segment to an arc at the selected midpoint	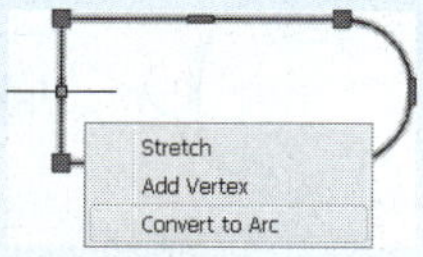	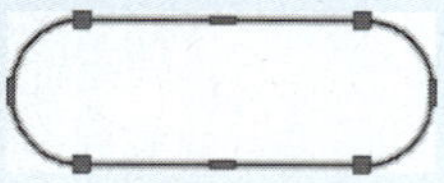

NOTE

To select specific segments of a polyline object, instead of all segments, hold down [Ctrl] while picking. The polyline still is edited as one continuous object.

PROFESSIONAL TIP

Use object snap tracking, polar tracking, object snaps, coordinate entry, or grid snaps with any grip editing command to improve accuracy.

Exercise 14-5

www.g-wlearning.com/CAD/

Complete the exercise on the companion website.

Creating a Polyline Boundary

Ribbon
Home
> Draw
Boundary
Type
BOUNDARY
BO

BOUNDARY

Access the **BOUNDARY** command to create a polyline boundary from linear objects that form a closed area. The **Boundary Creation** dialog box appears as shown in Figure 14-12. Select the default **Polyline** option from the **Object type:** drop-down list to create a polyline around the specified area. Select **Region** to create a ***region*** that you can use for area calculations, shading, extruding a solid model, and other purposes.

The **Current viewport** setting is active by default in the **Boundary set** drop-down list. This setting defines the ***boundary set*** from everything in the current viewport, even if it is not in the current display. An alternative is to pick the **New** button to return to the drawing window and select objects to use to create a boundary set. When you are finished selecting objects, press [Enter]. The **Boundary Creation** dialog box returns with **Existing set** active in the **Boundary set** drop-down list to indicate that the boundary set references the selected objects.

The **Island detection** setting specifies whether ***islands*** within the boundary are treated as boundary objects. See Figure 14-13. Check **Island detection** to form separate boundaries from islands within a boundary.

region: A closed 2D area that has physical properties such as a centroid and product of inertia.

boundary set: The part of the drawing AutoCAD evaluates to define a boundary.

island: A closed area inside a boundary.

Figure 14-12.
The **Boundary Creation** dialog box.

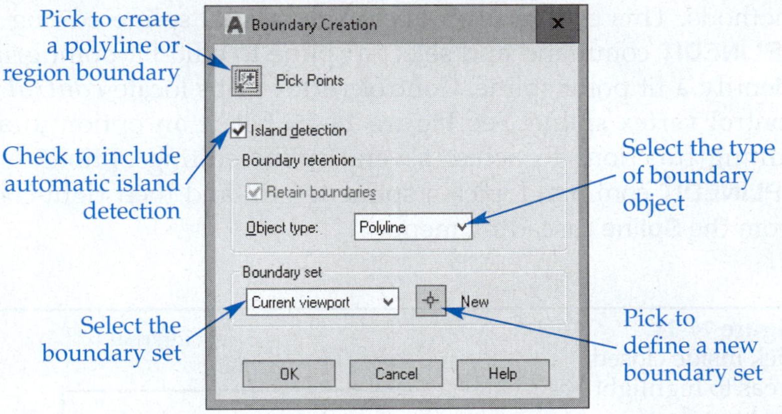

Figure 14-13.
You can include or exclude islands when defining a boundary set.

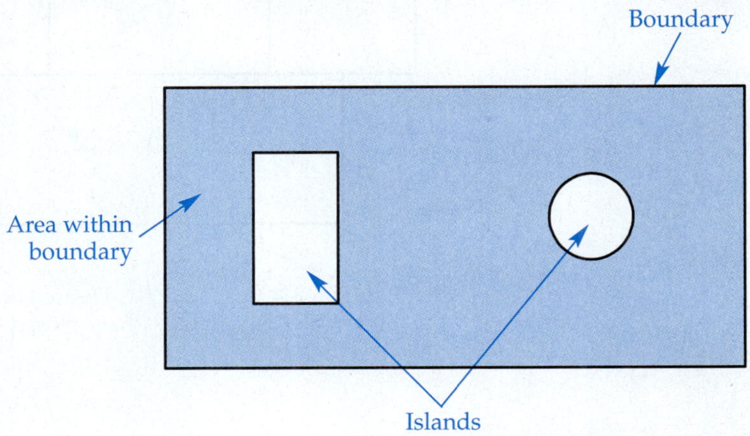

Select the **Pick Points** button, located in the upper-left corner of the **Boundary Creation** dialog box, to create boundaries. Pick a point in each boundary area. If a point you pick is inside a closed area, the boundary becomes highlighted, as shown in Figure 14-14. The **Boundary Definition Error** alert box appears if the point you pick is not within a closed polygon. Pick **Close** and try again.

Unlike an object created with the **JOIN** command or the **Join** option of the **PEDIT** command, a polyline boundary created with the **BOUNDARY** command does not replace the original objects. The polyline traces over the defining objects with a polyline. The separate objects still exist underneath the newly created boundary. To avoid duplicate geometry, move the boundary to another location on-screen, erase the original objects, and then move the boundary back to its original position.

PROFESSIONAL TIP

You can simplify area calculations by using the **BOUNDARY** command or by joining objects with the **JOIN** command or the **Join** option of the **PEDIT** command before using the **MEASUREGEOM** command. To retain the original objects, explode the polyline after the area calculation if you used the **JOIN** command or the **Join** option of the **PEDIT** command. Erase the polyline boundary after the calculation if you used the **BOUNDARY** command. Chapter 15 covers the **MEASUREGEOM** command.

Using the SPLINEDIT Command

SPLINEDIT

Ribbon
Home
> Modify

Edit Spline
Type
SPLINEDIT
SPE

fit points: Points through which a spline passes that determine the shape of the spline.

control vertices: Points that are used to define the curve shape and change the curve design. Adding control points typically increases the complexity of the curve.

Splines are complex curves that you can adjust and calculate using a variety of methods. This chapter introduces specialized spline editing commands. Access the **SPLINEDIT** command and select a spline to edit. Fit point grips locate ***fit points*** and identify a fit point spline. Control vertex grips locate ***control vertices*** and identify a control vertex spline. See Figure 14-15. Select an option to activate the appropriate editing function. To access several spline-editing options without first issuing the **SPLINEDIT** command, pick a spline to edit and then right-click and select an option from the **Spline** cascading menu.

Figure 14-14.
Pick inside closed areas to highlight the boundary.

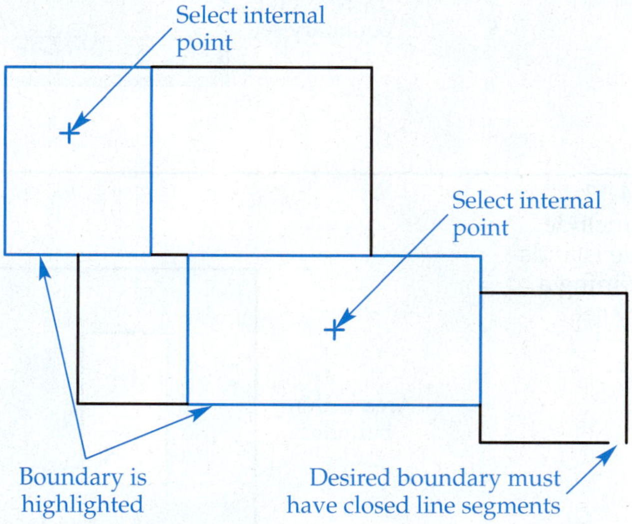

Figure 14-15.
Access the **SPLINEDIT** command to display fit point or control vertex grips and a list of options for editing splines.

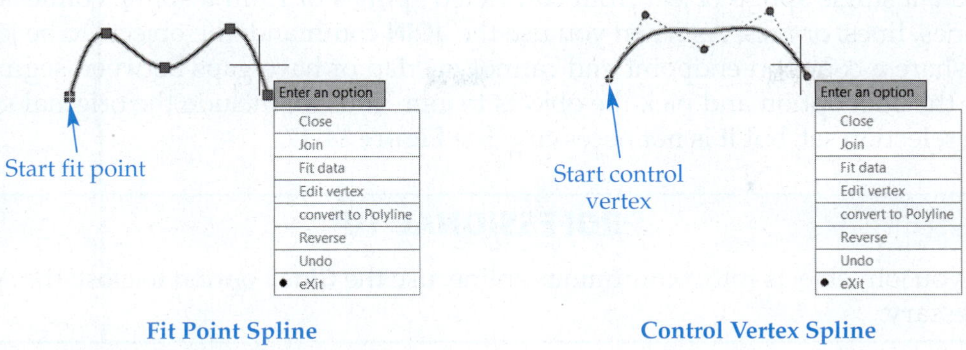

NOTE

You can also access the **SPLINEDIT** command by double-clicking on a spline.

PROFESSIONAL TIP

Select the **Undo** option immediately after performing an unwanted edit to restore the previous spline without leaving the command. Use the **Undo** option more than once to step back through each operation.

Opening and Closing a Spline

Use the **Open** option to open a closed spline and the **Close** option to close an open spline. See Figure 14-16. The **Open** option is available if you originally closed the spline by drawing the final segment manually or if you have used the **Close** option of the **SPLINE** or **SPLINEDIT** command. The effect of closing a spline varies depending on the spline creation method and associated options.

NOTE

The **Fit data** option includes **Open** and **Close** options that behave the same as the **Open** and **Close** options previously described.

Figure 14-16.
Closing an open spline. Note the significant difference between closing a spline created using the **Fit** method and a spline created using the **CV** method, even though the open spline is the same shape.

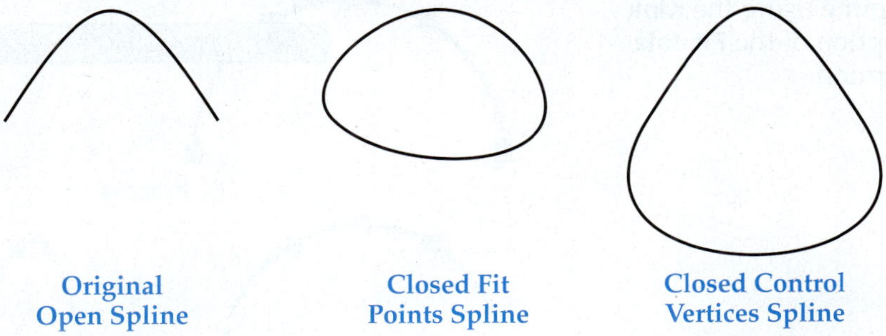

Joining Splines

The **Join** option provides the same function as the **JOIN** command, allowing you to create a single spline object from connected splines or from a spline connected to polylines, lines, or arcs. As when you use the **JOIN** command, the objects to be joined must share a common endpoint and cannot overlap or have gaps between segments. Select the **Join** option and pick the objects to join. You can include the original spline in the selection set, but it is not necessary. See Figure 14-17.

PROFESSIONAL TIP

After you join objects into a continuous spline, use the **Close** option to close the spline if necessary.

Fit Data Option

The **Fit data** option allows you to perform several operations at a spline fit point. If you apply the **Fit data** option to a spline showing control vertices, the spline converts to show fit points. The options available with the **Fit data** option are more difficult to use than other editing tools. The **Add**, **Close**, **Delete**, **Move**, **Purge**, and **Tangents** options provide functions that you can perform more easily using standard editing commands or context-sensitive spline grips. You will learn to grip-edit splines later in this chapter.

The **Kink** option of the **Fit data** option allows you to select a point on the spline to add a sharp point known as a *kink*. See Figure 14-18. It is easier to add a kink to a fit

Figure 14-17.
Joining a spline to other connected objects. The single object is now a complex spline that you can further edit to create a freeform shape.

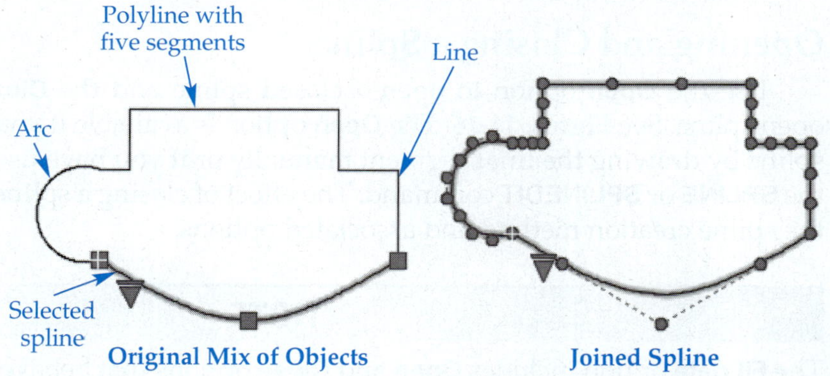

Figure 14-18.
Adding a kink to a spline using the **Kink** option of the **Fit data** option.

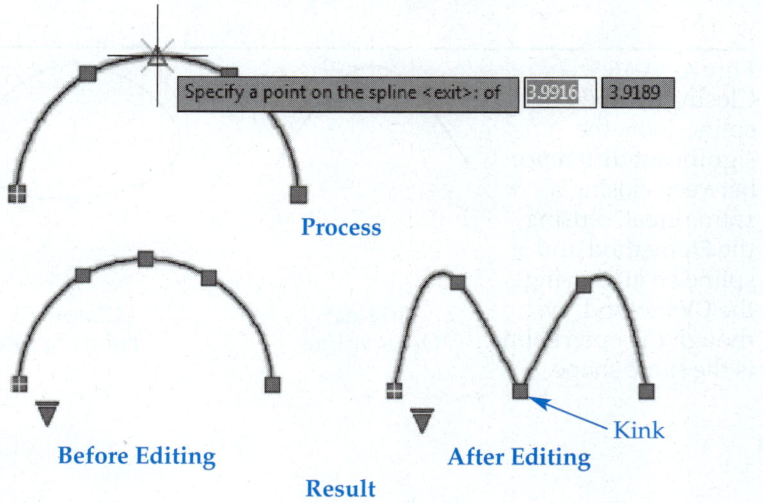

point spline using the **Add Fit Kink** option available from the **Spline** cascading menu that appears when you pick a spline to edit and then right-click. You can add a kink to a control vertex spline using the **Kink** option of the **Fit data** option, but doing so changes the spline to a fit point spline.

The **toLerance** option is unique to the **Fit data** option and allows you to adjust fit tolerance values. The results are immediate, so you can adjust the fit tolerance as necessary to produce different results. Use the **eXit** option to return to the **SPLINEDIT** prompt.

Edit Vertex Option

The **Edit vertex** option allows you to perform several operations at a spline control vertex. If you apply a specific **Edit vertex** option to a spline showing fit points, the spline converts to show control vertices. As with the **Fit data** option, the options associated with the **Edit vertex** option are more difficult to use than other editing tools. The **Add**, **Delete**, and **Move** options provide functions that you can perform more easily using standard editing commands or context-sensitive spline grips.

The **Elevate order** and **Weight** options are unique to the **Edit vertex** option and are sometimes useful. Figure 14-19 explains and illustrates these options. When using the **Weight** option, use the **Next**, **Previous**, and **Select point** options to navigate to different vertices. Use the **eXit** option to return to the **SPLINEDIT** prompt.

Converting Splines and Polylines

The **convert to Polyline** option of the **SPLINEDIT** command allows you to convert a spline to a polyline. Select the spline to convert and enter a value at the Specify a precision <current>: prompt. The higher the specified precision, the more vertices added to the polyline, making the polyline smoother. See Figure 14-20.

order: In a spline, the degree of the spline polynomial + 1.

Figure 14-19.
Spline editing options that are unique to the **Edit vertex** option of the **SPLINEDIT** command.

Option	Process	Result
Elevate order Adds control points for greater control of spline shape and spline *order* refinement. Specify a value between the current number of vertices and 26.	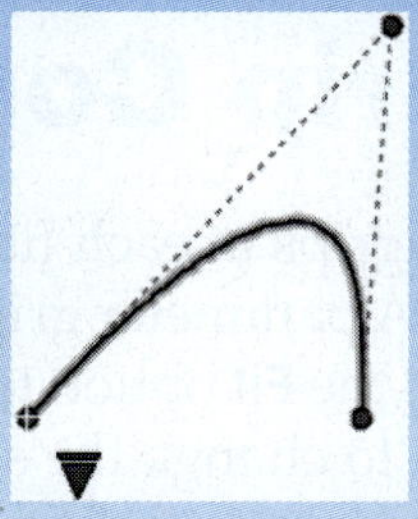 3 Original Control Vertices	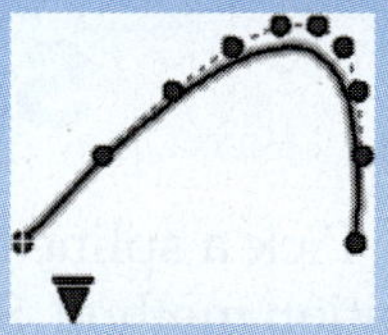 Order Elevated to 10
Weight Adjusts the spline weight at the selected vertex. The larger the weight, the closer the spline pulls toward the vertex. Specify a value greater than or equal to 1.	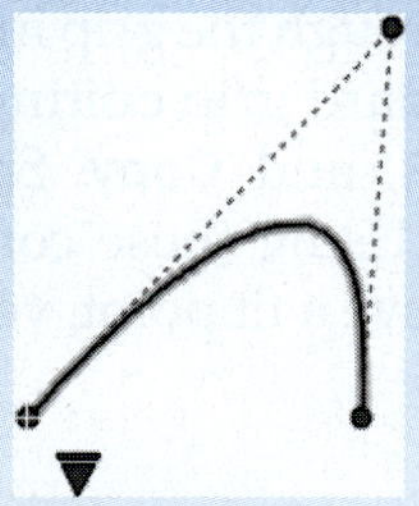 Original Weight of 1 Applied to the Second Vertex	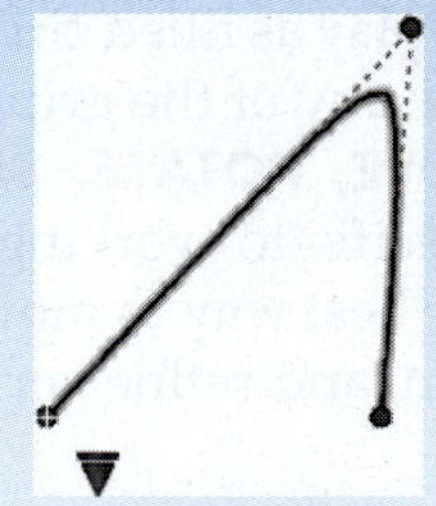 Modified Weight of 5 Applied to the Second Vertex

Figure 14-20.
The effects of changing precision when converting a spline to a polyline.

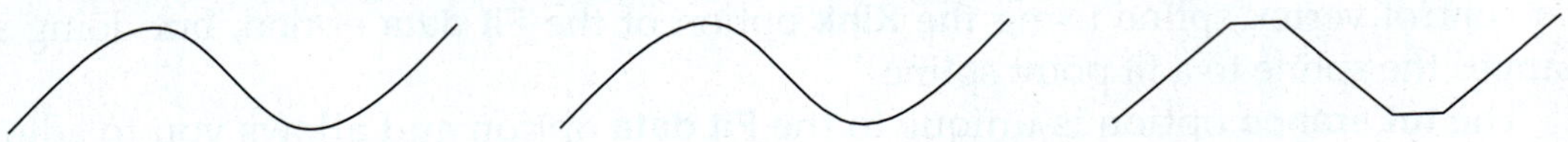

Use the **Object** option of the **SPLINE** command to convert a spline-fitted polyline object, created using the **PEDIT** command, to a spline object. When you access the **SPLINE** command, activate the **Object** option instead of defining points. Then pick a spline-fitted polyline object to convert the polyline to a spline.

NOTE

The **Reverse** option of the **SPLINEDIT** command applies the same operation as the **REVERSE** command. Reversing affects the fit point and vertex options of the **SPLINEDIT** command.

PROFESSIONAL TIP

A spline created by fitting a spline curve to a polyline is a linear approximation of a true spline and is not as accurate as a spline drawn using the **SPLINE** command. An additional advantage of spline objects over smoothed polylines is that splines use less disk space.

Exercise 14-6

www.g-wlearning.com/CAD/

Complete the exercise on the companion website.

Spline Grip Commands

Pick a spline object to display grips at each fit point or vertex, depending on the creation method. See Figure 14-21. A parameter grip appears near the start point of the spline. Pick the grip and then select the **Fit** option to change the spline to use fit points, or pick the **Control Vertices** option to change the spline to use control vertices. These options are also available when you select a spline and right-click.

Fit point grips have the standard filled-square grip appearance, and vertex grips display as filled circles. A + icon through the grip indicates the start point of the spline. Use any of the grips to apply standard grip editing commands, including **STRETCH**, **MOVE**, **ROTATE**, **SCALE**, **MIRROR**, and **Copy**. Spline grips also offer commands specific to working with splines. Using these context-sensitive commands is often the best way to move, add, or remove a fit point, vertex, or kink, adjust tangent direction, and refine vertices.

Figure 14-21.
Pick a spline object to display grips at each fit point or vertex and a parameter grip near the start point. Apply one of the following methods to access context-sensitive spline grip commands. A—Hover over an unselected grip to display a menu of options, and then pick an option from the list. B—Pick a grip and then press [Ctrl] to cycle through options. C—Pick a grip, right-click, and select an option.

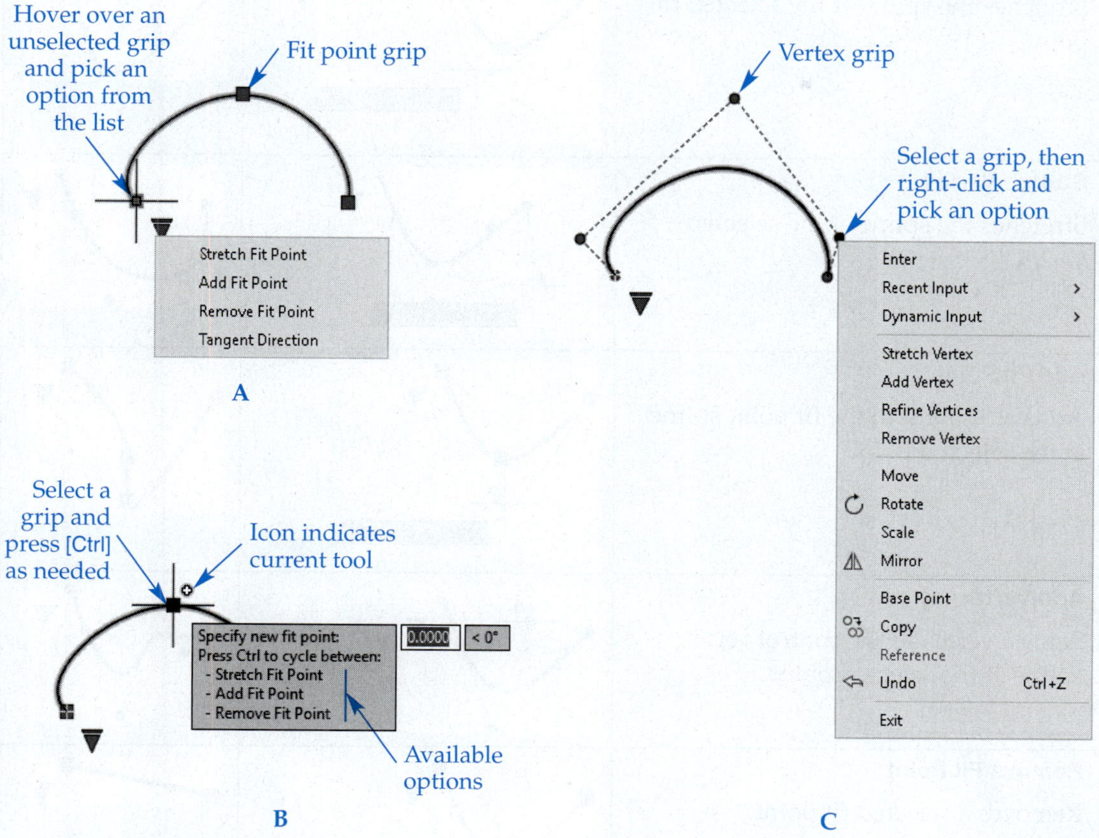

There are three different ways to access and apply the same context-sensitive spline grip commands. Figure 14-21 explains each technique. An icon identifies the current operation. Different options are available depending on whether you hover over or select a fit point or a vertex, and whether the point is at the end of the spline. Figure 14-22 provides examples of using spline grip commands to edit splines. **Base point**, **Copy**, **Undo**, and **eXit** options are available for stretching a fit point or vertex. They behave the same as in the standard **STRETCH** grip edit command.

PROFESSIONAL TIP

Use object snap tracking, polar tracking, object snaps, coordinate entry, or grid snaps with any grip editing command to improve accuracy.

Exercise 14-7

www.g-wlearning.com/CAD/

Complete the exercise on the companion website.

Figure 14-22.
Examples of options available for modifying splines using grips.

Option	Process	Result
Stretch Fit Point Stretches the spline at the selected fit point	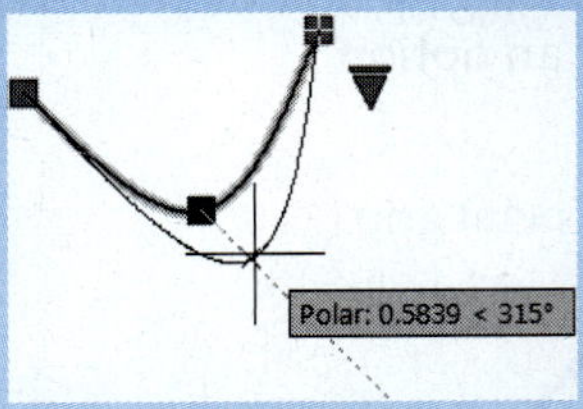	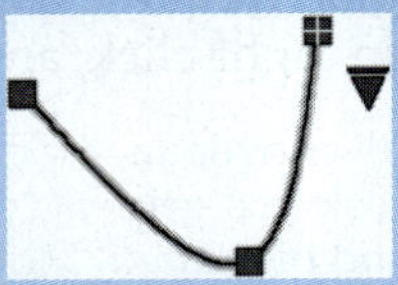
Stretch Vertex Stretches the spline at the selected vertex	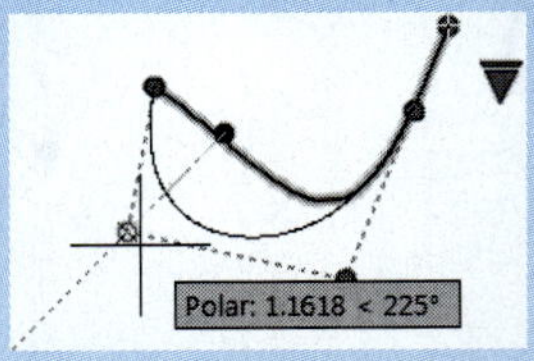	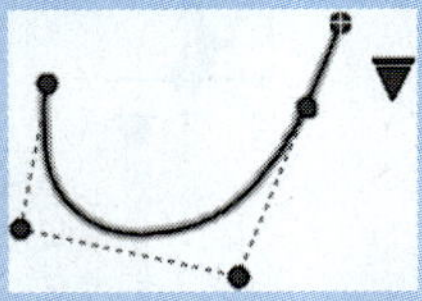
Add Fit Point Adds a fit point to the fit point spline at the selected point	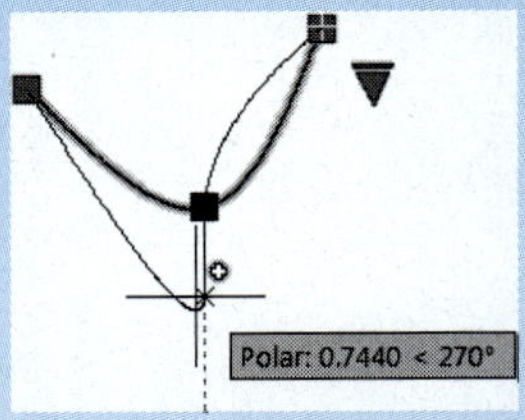	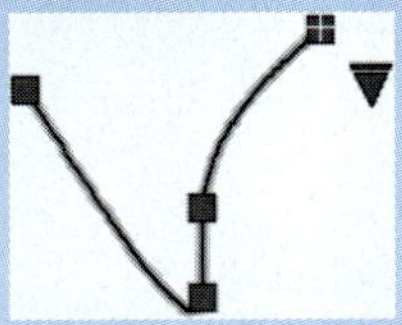
Add Vertex Adds a vertex to the control vertex spline at the selected point	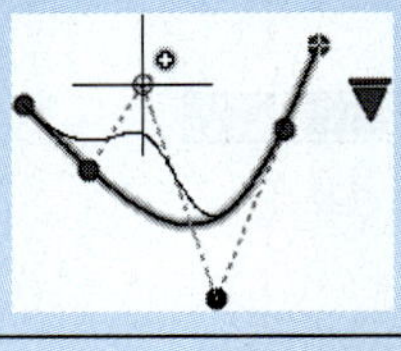	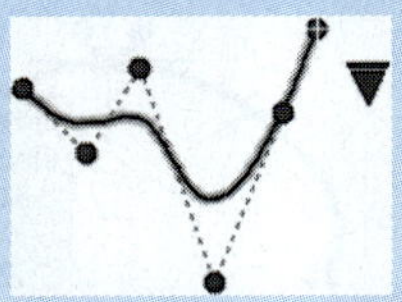
Remove Fit Point Removes a selected fit point	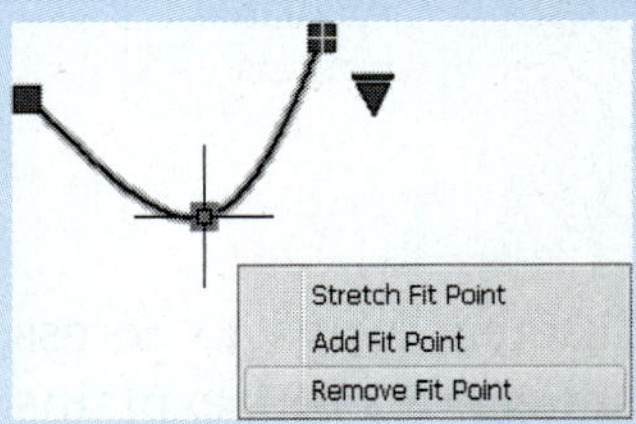	
Remove Vertex Removes a selected vertex	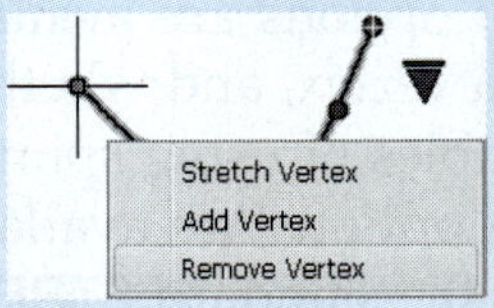	
Tangent Direction Edits the selected start point or endpoint tangent direction for fit point splines	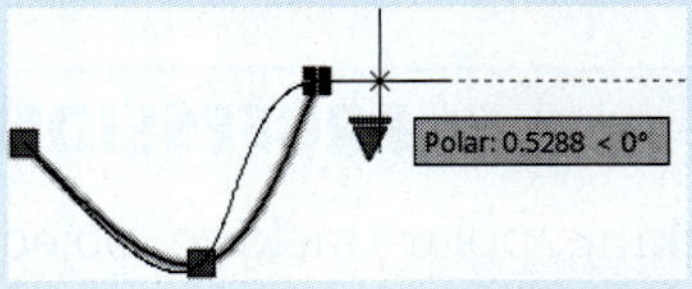	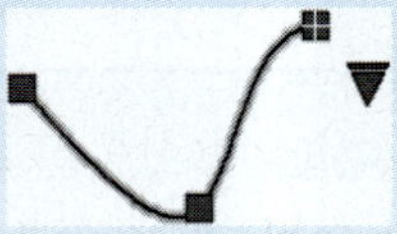
Refine Vertices Adds vertices relative to the selected vertex to fine-tune the spline for editing	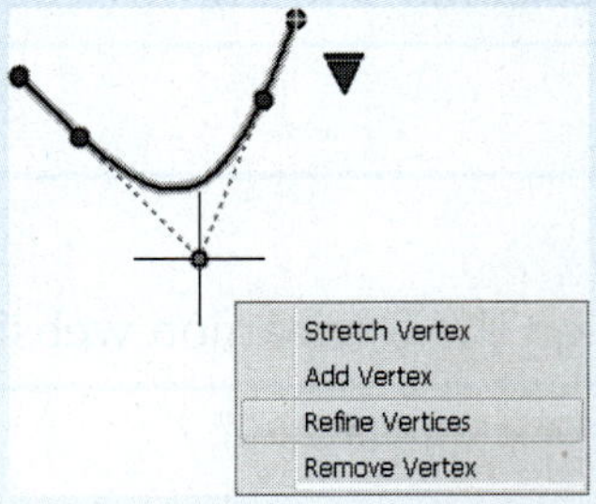	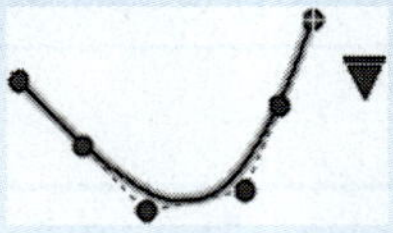

Chapter Review

Answer the following questions using the information in this chapter.

1. Name the command and option required to turn three connected lines into a single polyline.
2. When you enter the **Edit vertex** option of the **PEDIT** command, where does AutoCAD place the "X" marker?
3. If you change the starting and ending widths of a polyline using the **Edit vertex** option of the **PEDIT** command and nothing appears to happen, what should you do?
4. Which **PEDIT** command option allows you to change the starting and ending widths of a polyline?
5. Name the **PEDIT** command option used for curve fitting.
6. How do you move the "X" marker to edit a different polyline vertex?
7. Explain the difference between a fit curve and a spline curve.
8. Compare a quadratic curve, cubic curve, and fit curve.
9. Which **SPLINETYPE** system variable setting allows you to draw a quadratic curve?
10. Explain how you can adjust the way polyline linetypes are generated using the **PEDIT** command.
11. Describe how to use grips to stretch a straight polyline segment at the midpoint.
12. Describe how to use grips to straighten a polyline arc.
13. Name the command used to create a polyline boundary.
14. Which **Fit data** option of the **SPLINEDIT** command allows you to add a sharp point to a spline?
15. Identify the **Edit vertex** option of the **SPLINEDIT** command that lets you increase the number of control points appearing on a spline curve.
16. Name the **Edit vertex** option of the **SPLINEDIT** command that controls the pull exerted by a control point on a spline.
17. Name the **SPLINE** command option that allows you to turn a spline-fitted polyline into a true spline.
18. Describe how to use grips to add a fit point to a control vertex spline.
19. Describe how to use grips to remove a vertex from a control vertex spline.
20. Which spline grip option allows you to add vertices relative to a selected vertex to fine-tune the spline?

Drawing Problems

Start AutoCAD if it is not already started. Start a new drawing for each problem using an appropriate template of your choice. The template should include layers and text styles, when necessary, for drawing the given objects. Add layers and text styles as needed. Draw all objects using appropriate layers. Use appropriate text styles, justification, and format. Follow the specific instructions for each problem. Use only drawing and editing commands and techniques you have already learned. Do not draw dimensions. Use your own judgment and approximate dimensions when necessary.

▼ Basic

1. General Use the **LINE** command to draw two connected lines. Use the **PEDIT** command to convert one of the lines to a polyline, and then use the **Join** option to convert the line and polyline into a single polyline object. Use the **LINE** command to draw a rectangle. Use the **PEDIT** command to convert one of the lines to a polyline, and then use the **Join** option to convert the three remaining lines and polyline into a single polyline object. Save the drawing as P14-1.

2. General Use the **SPLINE** command to draw a spline of your own design with at least four fit points. Use the **convert to Polyline** option of the **SPLINEDIT** command to convert the spline to a polyline. Save the drawing as P14-2.

3. Mechanical Use the **RECTANGLE** command to draw a 50 mm × 25 mm rectangle. Use grip editing to convert the straight 25 mm segments to R12.5 mm arcs, creating a full-round slot. Save the drawing as P14-3.

4. Mechanical Use the **POLYGON** command to draw the hexagon shown in A. Use the **Remove Vertex** and **Stretch** polyline grip edit commands to edit the hexagon as shown in B. Save the drawing as P14-4.

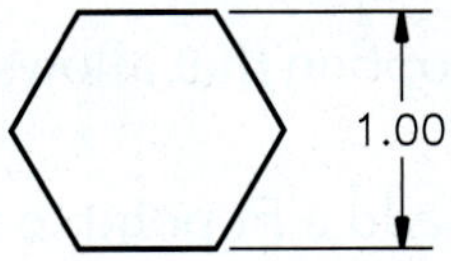

A

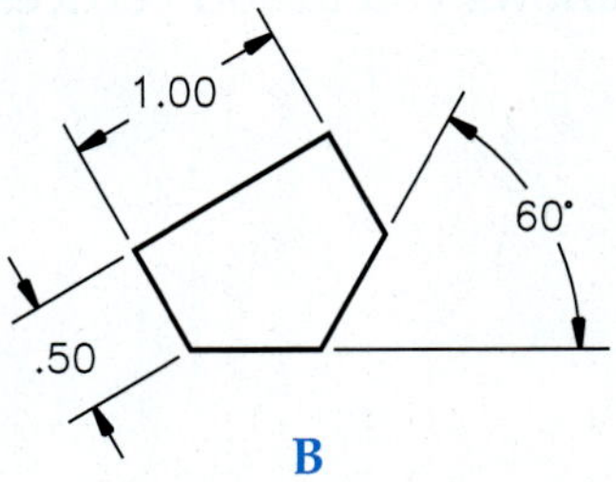

B

▼ Intermediate

5. **General** Save P4-15 as P14-5. If you have not yet completed Problem 4-15, do so now. In the P14-5 file, use polyline editing to change the original objects to a rectangle as shown. Use the **Decurve** and **Width** options of the **PEDIT** command and polyline grip editing.

Original — After Editing

6. **General** Save P4-16 as P14-6. If you have not yet completed Problem 4-16, do so now. Make the following changes in the P14-6 file.
   A. Combine the two polylines using the **Join** option of the **PEDIT** command.
   B. Change the beginning width of the left arrow to 1.0 and the ending width to .2.
   C. Draw a polyline .062 wide, similar to Line A, as shown.

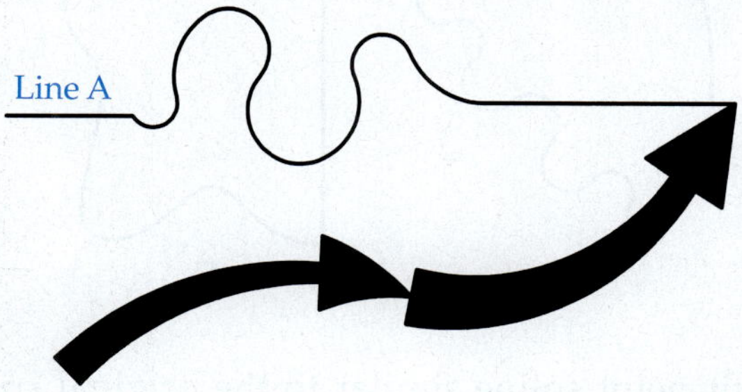

7. **General** Draw five polylines .032 wide, using the following absolute coordinates for all five objects.

Point	Coordinates	Point	Coordinates	Point	Coordinates
1	1,1	5	3,3	9	5,5
2	2,1	6	4,3	10	6,5
3	2,2	7	4,4	11	6,6
4	3,2	8	5,4	12	7,6

Leave the first polyline as drawn. Use the **Fit** option of the **PEDIT** command to smooth the second polyline. Use the **Spline** option of the **PEDIT** command to turn the third polyline into a quadratic curve. Make the fourth polyline into a cubic curve. Make the fifth polyline a cubic curve and then use the **Decurve** option of the **PEDIT** command to return it to its original form. Save the drawing as P14-7.

8. **Architectural** Use the **PLINE** command to draw four copies of a patio plan similar to the plan shown in A. Draw the house walls 6″ wide. Leave the first plan as drawn. Use the **PEDIT** command to create the designs shown. Use the **Fit** option for B, a quadratic spline for C, and a cubic spline for D. Change the **SPLINETYPE** system variable as required. Save the drawing as P14-8.

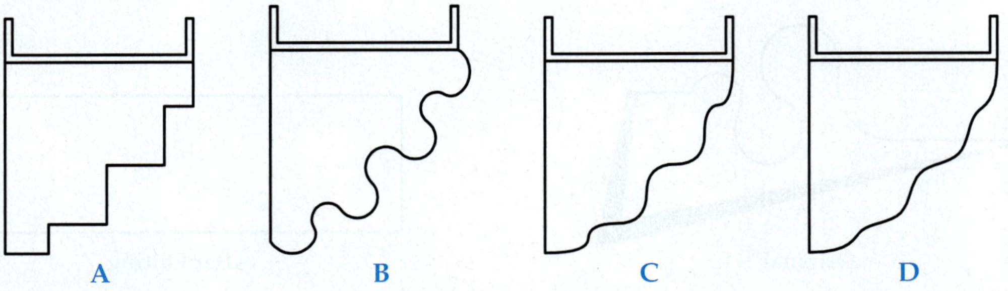

9. **Architectural** Save P14-8 as P14-9. In the P14-9 file, use grip editing to create four new patio designs similar to A, B, C, and D below.

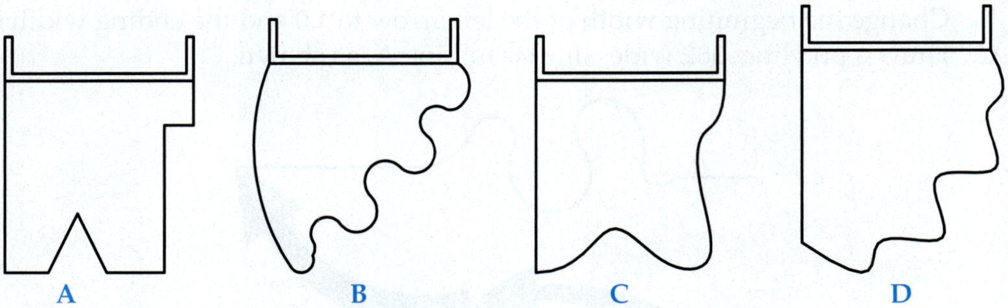

10. **General** Draw a fit point spline similar to the original spline shown. Copy the spline seven times to create a layout similar to the layout shown. Use the **SPLINEDIT** command or grip editing to perform the specified operations. Save the drawing as P14-10.

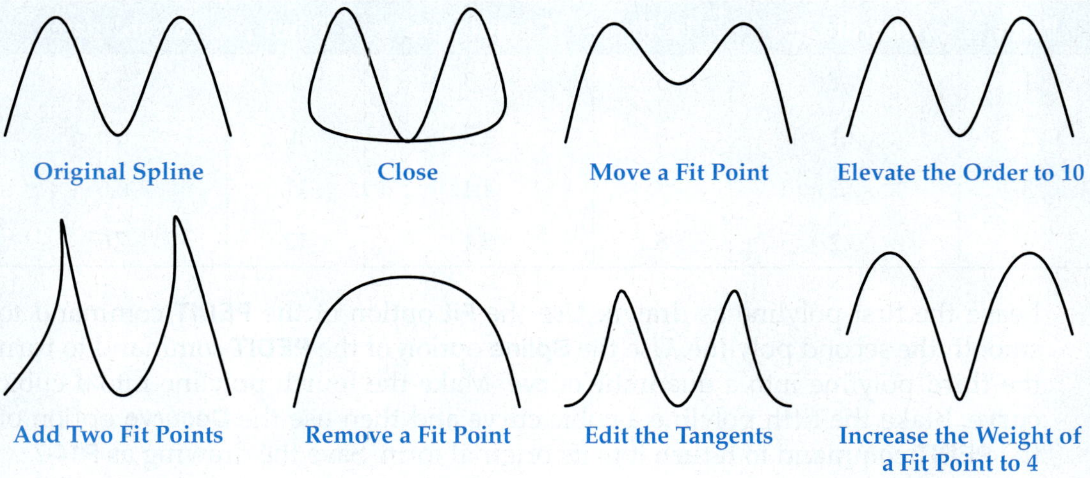

▼ Advanced

11. **Architectural** Draw the exterior door elevation shown. Use polylines and the **OFFSET** command to draw the features. Save the drawing as P14-11.

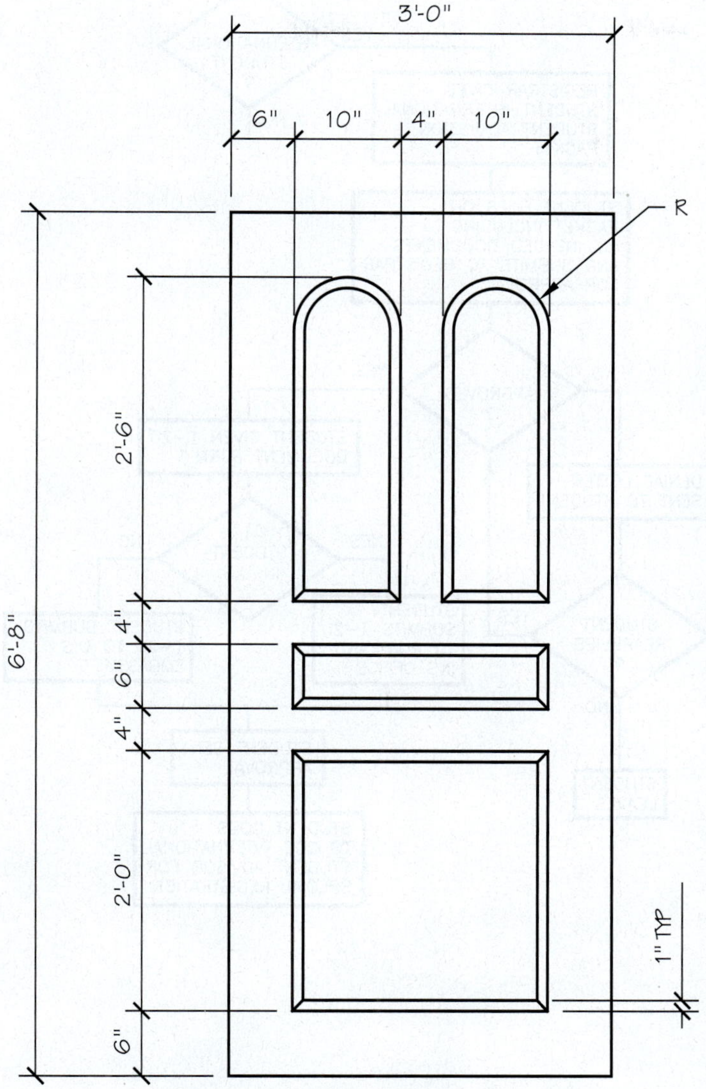

12. Graphic Design Draw the flow chart shown. Use polylines to draw the connecting lines, arrows, and diamonds. Save the drawing as P14-12.

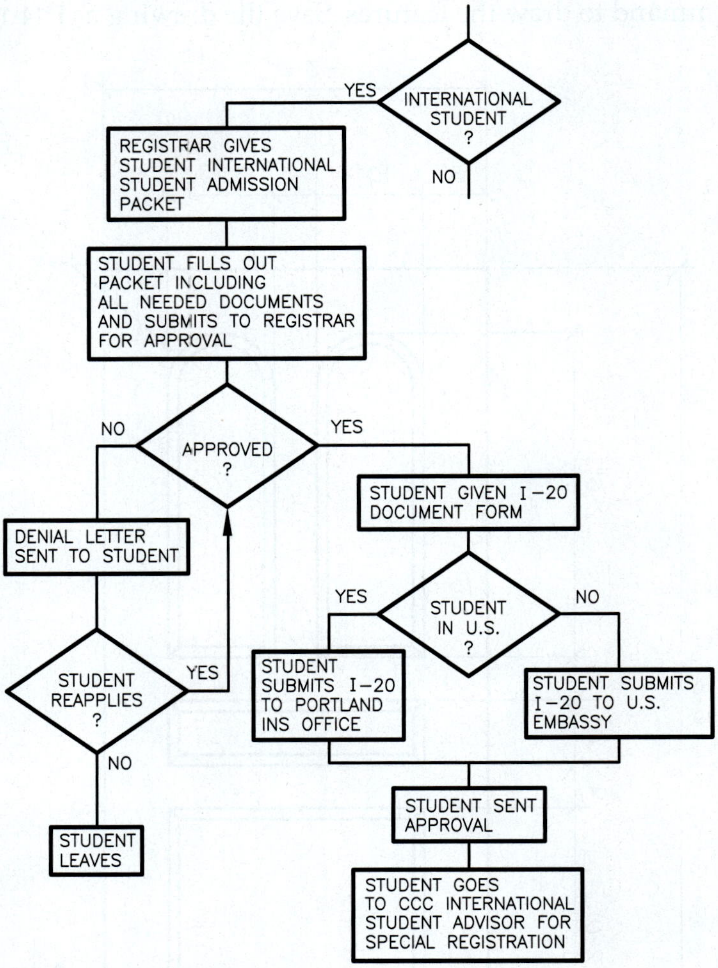

13. **Graphic Design** Draw the flow chart shown. Use polylines to draw the connecting lines, arrows, and diamonds. Save the drawing as P14-13.

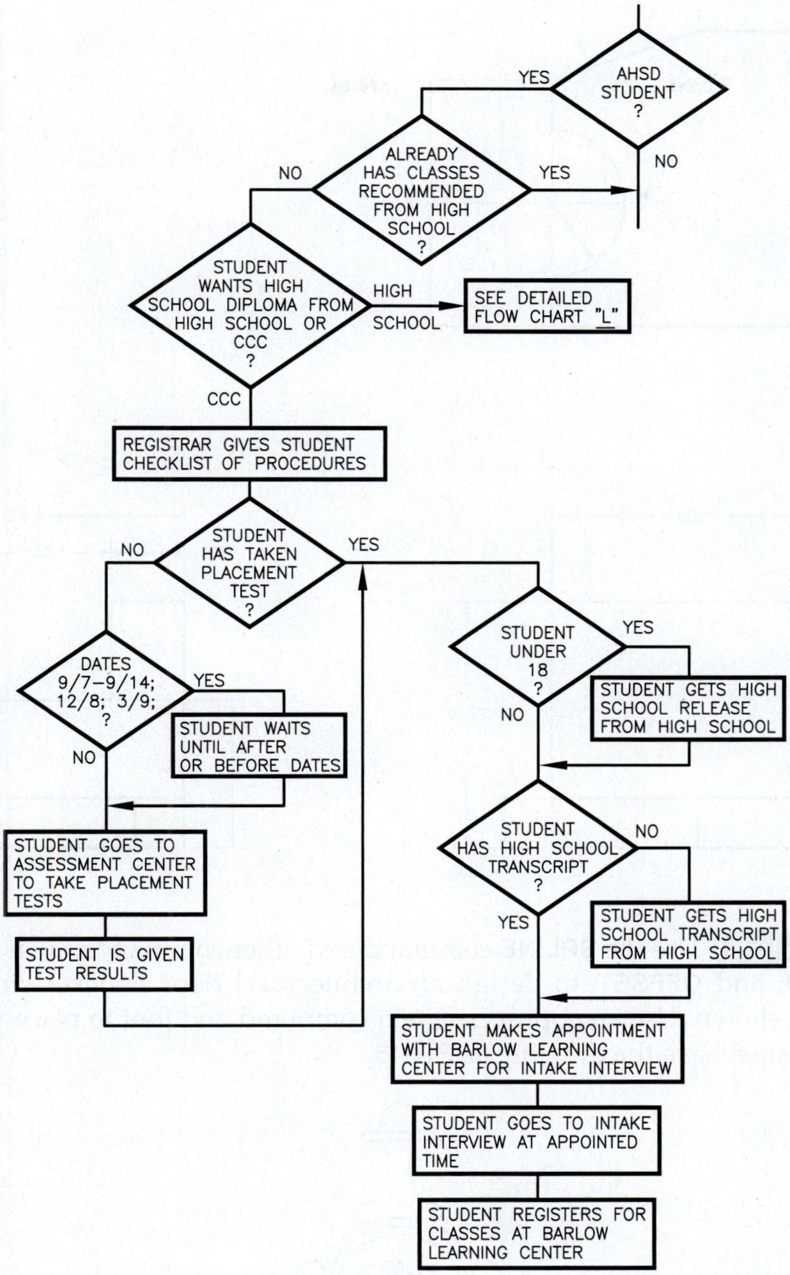

14. **Mechanical** Draw the part views shown. Save the drawing as P14-14.

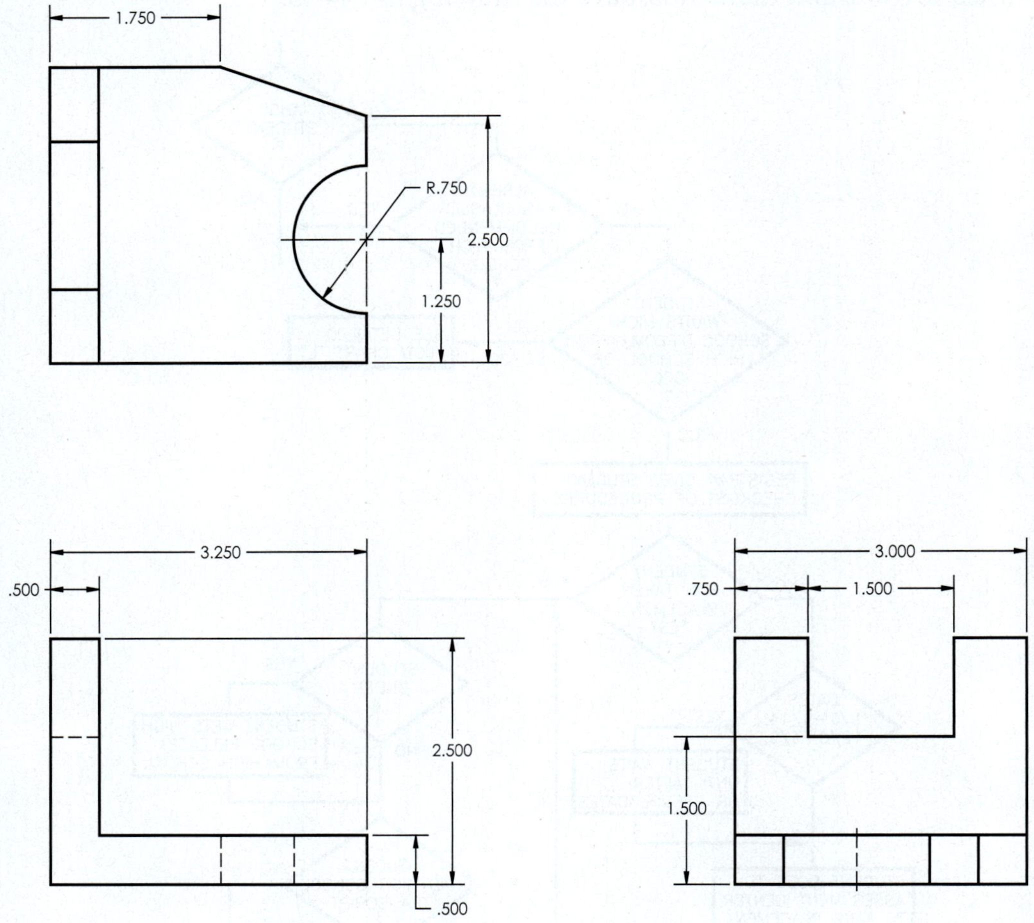

15. **Architectural** Use the **SPLINE** command and other commands, such as **ELLIPSE**, **MIRROR**, and **OFFSET**, to design an architectural door knocker similar to the knocker shown. Use an appropriate text command and font to place your initials in the center. Save the drawing as P14-15.

16. **Architectural** Draw the roof plan shown. Save the drawing as P14-16.

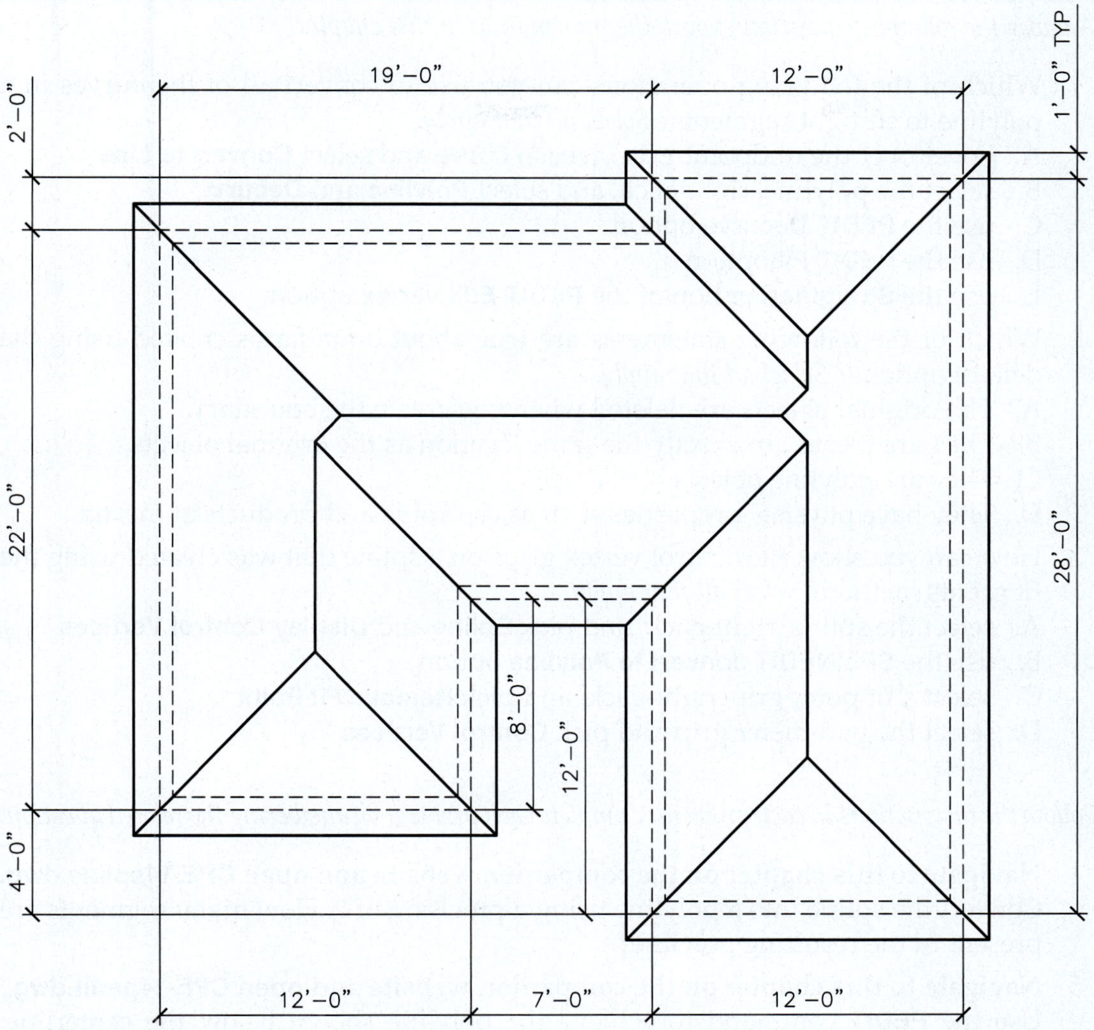

17. **Architectural** Find a tree appropriate for residential landscaping. Prepare a freehand 2D sketch of the tree elevation. Use available measuring devices, such as a tape measure, to dimension the size and location of features accurately. Start a new drawing from scratch or use an architectural template of your choice. Draw the tree elevation from your sketch. Save the drawing as P14-17.

18. **Mechanical** Prepare a freehand, dimensioned 2D sketch of a view of a new design for a cell phone. Include circles, ellipses, polylines, and freeform ergonomic edges in the design. Use available measuring devices, such as a tape measure and caliper, to measure an actual cell phone for reference if necessary. Start a new drawing from scratch or use a decimal-unit template of your choice. Draw all of the views needed to describe the cell phone completely from your sketch. Use the **SPLINE** command to draw the freeform curves. Save the drawing as P14-18. Prepare a slide show presentation of your research, design process, and drawing and present it to your class or office.

AutoCAD Certified Professional Exam Practice

Answer the following questions using the information in this chapter.

1. Which of the following operations can you use to convert all of the curves in a polyline to straight segments? *Select all that apply.*
   A. hover over the midpoint grip on each curve and select **Convert to Line**
   B. select the polyline, right-click, and select **Polyline** and **Decurve**
   C. use the **PEDIT Decurve** option
   D. use the **PEDIT Fit** option
   E. use the **Straighten** option of the **PEDIT Edit vertex** option
2. Which of the following statements are true about boundaries created using the default options? *Select all that apply.*
   A. The original objects are deleted when you create the boundary.
   B. They are created in exactly the same location as the original objects.
   C. They are polyline objects.
   D. They have physical properties such as centroids and products of inertia.
3. How can you show the control vertex grips on a spline that was created using the fit points method? *Select all that apply.*
   A. select the spline, right-click, and pick **Spline** and **Display Control Vertices**
   B. use the **SPLINEDIT convert to Polyline** option
   C. select a fit point grip, right-click, and pick **Remove Fit Point**
   D. select the parameter grip and pick **Control Vertices**

Follow the instructions in each problem. Complete each problem by answering the related question.

4. **Navigate to this chapter on the companion website and open CPE-14spline.dwg.** Convert the spline into a polyline using a precision of 2. How many segments are present in the resulting polyline?
5. **Navigate to this chapter on the companion website and open CPE-14pedit.dwg.** Use the **PEDIT** command to achieve the polyline shown below the centerline. Mirror the result using the centerline as the mirror line. What are the coordinates of Point 1?

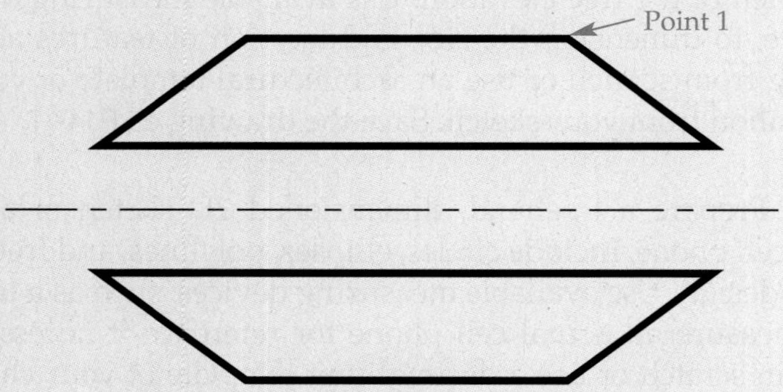

CHAPTER 15

Obtaining Drawing Information

Learning Objectives

After completing this chapter, you will be able to:

- ✓ Measure distance, radius, diameter, angles, and area.
- ✓ List data related to a single point, an object, a group of objects, or an entire drawing.
- ✓ Determine the drawing status.
- ✓ Determine the amount of time spent in a drawing session.
- ✓ Perform calculations using the **QuickCalc** palette.
- ✓ Compare drawings using the **COMPARE** command.

This chapter describes tools that allow you to retrieve geometric values such as point coordinates, distance, angle, and area. You will learn methods for referencing object data, determining drawing status, and calculating time spent working on a drawing. You will also learn how to use the **QuickCalc** palette to calculate values while you work and how to compare drawings using the **COMPARE** command.

Taking Measurements

Taking measurements from a drawing is common during the designing and drafting process. Mechanical drafting examples include measuring a circle to confirm the size of a hole, measuring the angle between two surfaces, and calculating the volume of a part. Architectural drafting examples include checking the dimensions of a room, measuring the location of features within a room, and calculating square footage.

Using Grips

Grips and dynamic input provide one way to view basic object measurements. To identify the location of a point that corresponds to a grip, confirm that the coordinate display field in the status bar is on. Then pick the object to activate grips and hover over a grip. The coordinates of the point appear in the coordinate display field of the status bar. Use grips and dynamic input to view measurements between grips. Pick the object to activate grips and hover over a grip to display measurements. The information that appears varies depending on the object type and the selected grip. See Figure 15-1.

Figure 15-1.
Examples of hovering over grips to display measurements at the dynamic input cursor. Dynamic input must be active to view measurements, but it does not have to be active to list the coordinates of a grip in the status bar.

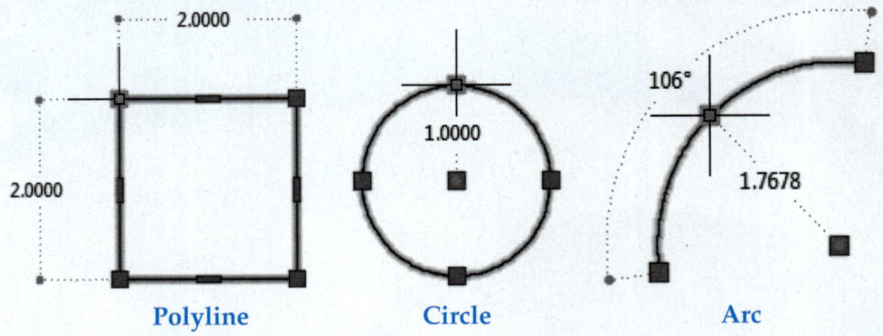

Exercise 15-1

www.g-wlearning.com/CAD/

Complete the exercise on the companion website.

Using the MEASUREGEOM Command

MEASUREGEOM
Ribbon
Home > Utilities
Measure
Type
MEASUREGEOM
MEA

The **MEASUREGEOM** command allows you to measure distance, radius, angle, area, and volume. The ribbon is an effective way to access **MEASUREGEOM** command options directly. See Figure 15-2. When you access the **MEASUREGEOM** command by typing, you must activate a measurement option before you begin. The **MEASUREGEOM** command remains active after you take measurements, allowing you to continue measuring without reselecting the command. Select the **eXit** option or press [Esc] to exit the command.

Quick Option

AutoCAD 2020 NEW

Use the **Quick** option of the **MEASUREGEOM** command to view object length, radius, and angle dynamically as you move the cursor over objects in the drawing. Contact the large crosshairs cursor with objects to display measurements. Figure 15-3 shows using the **Quick** option to measure the lengths of polyline segments, the angle between polyline segments, and the radius of a circle. Right angles (90° angles) between objects are identified by a square icon. The **Quick** option can also be used to measure the distance between parallel objects. The parallel distance appears directly on the large crosshairs cursor.

Use the **Mode** option to make the **Quick** option the command default when accessing the **MEASUREGEOM** command by typing. The **Mode** option is set to **No** by default.

Exercise 15-2

www.g-wlearning.com/CAD/

Complete the exercise on the companion website.

Figure 15-2.
Use the flyout in the **Utilities** panel of the **Home** ribbon tab to access specific **MEASUREGEOM** command options.

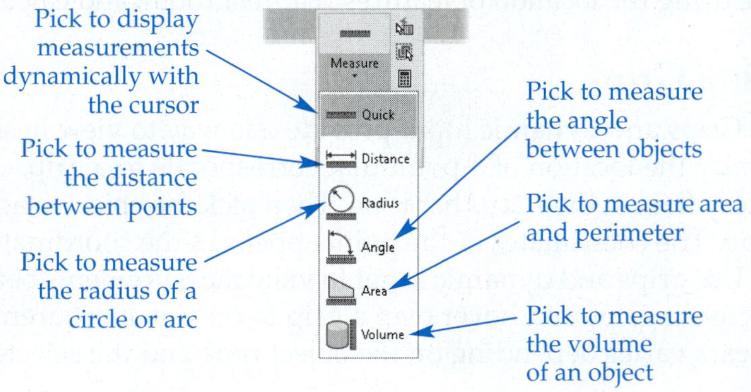

Measuring Distance

Use the **Distance** option of the **MEASUREGEOM** command to find the distance between points. Specify the first point, followed by the second point. The linear distance between the points, angle in the XY plane, and delta (change in) X and Y values appear on-screen, as shown in Figure 15-4, and at the command line. In a 2D drawing, the angle from the XY plane and delta Z values are always 0, as indicated at the command line. The first point you specify defines the vertex of the angular dimension, as shown in Figure 15-4.

After you specify the first point, you can select the **Multiple points** option to measure the distance between multiple points. AutoCAD calculates the distance between each point and displays the value at the command line during selection. Several options are available for picking multiple points, as described in Figure 15-5. When you finish measuring, use the **Total** or **Close** option to display the total distance between all points. Figure 15-6 shows an example of using the **Multiple points** option to calculate the perimeter of a shape.

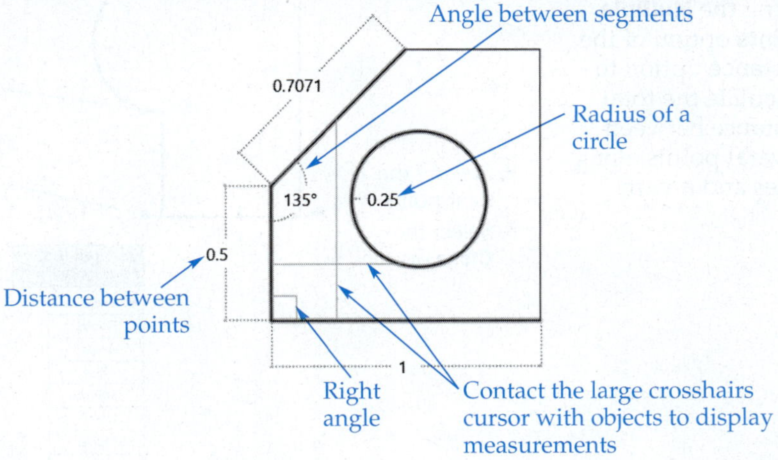

Figure 15-3. Using the **Quick** option to display measurements. Right angles are identified by the square icon.

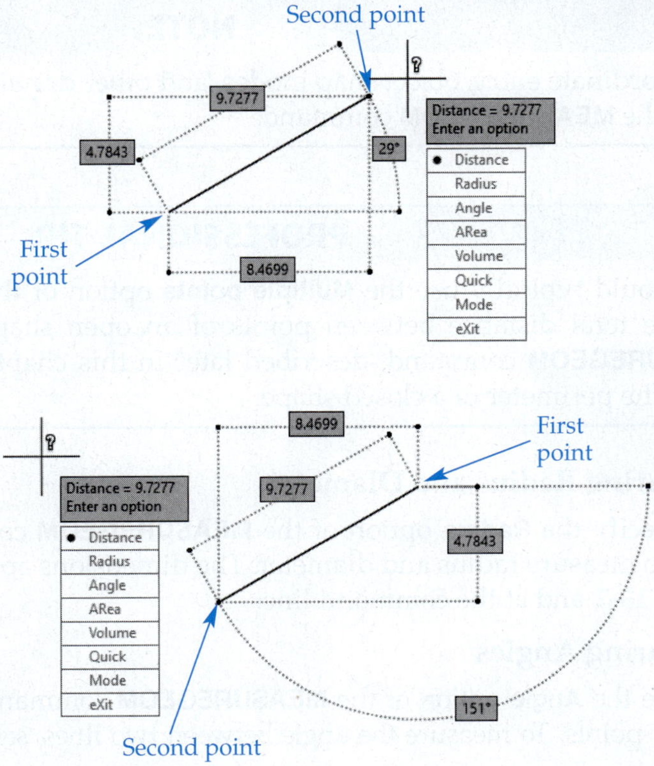

Figure 15-4. The data provided by the **Distance** option. Notice that the first point defines the vertex of the angular value.

Figure 15-5.
Options available for the **Multiple points** option of the **Distance** option.

Option	Description
Arc	Measures the length of an arc; includes the same options available for drawing arcs. Choose the **Line** option to return to measuring the distance between linear points.
Length	Measures the specified length of a line.
Undo	Cancels the effects of an unwanted selection, returning to the previous measurement point.
Total	Finishes multiple point selection and calculates the total distance between points.
Close	Connects the current point to the first point; finishes multiple point selection and calculates the total distance between points.

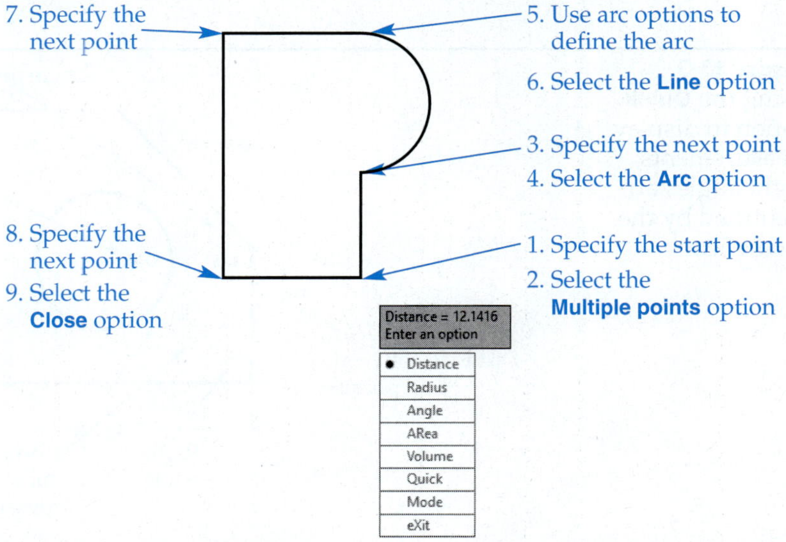

Figure 15-6.
An example of using the **Multiple points** option of the **Distance** option to calculate the total distance between several points along lines and an arc.

NOTE

Use coordinate entry, object snap modes, and other drawing aids to pick points when using the **MEASUREGEOM** command.

PROFESSIONAL TIP

You should typically use the **Multiple points** option of the **Distance** option to calculate the total distance between points of an open shape. The **Area** option of the **MEASUREGEOM** command, described later in this chapter, is more suited to calculating the perimeter of a closed shape.

Measuring Radius and Diameter

Specify the **Radius** option of the **MEASUREGEOM** command and pick an arc or circle to measure radius and diameter. The dimensions appear on-screen, as shown in Figure 15-7, and at the command line.

Measuring Angles

Use the **Angle** option of the **MEASUREGEOM** command to find the angle between lines or points. To measure the angle between two lines, select the first line followed by

Figure 15-7.
The data provided by the **Radius** option when you measure a circle or arc.

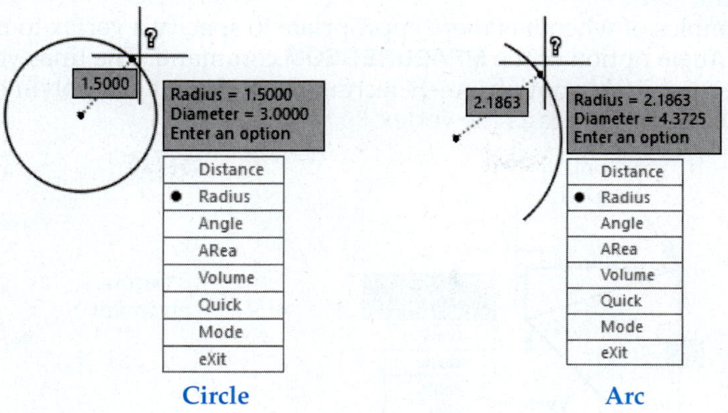

Figure 15-8.
The data provided by the **Angle** option. The lines you see appear temporarily when using this option. A—Selecting two lines. B—Picking an arc.

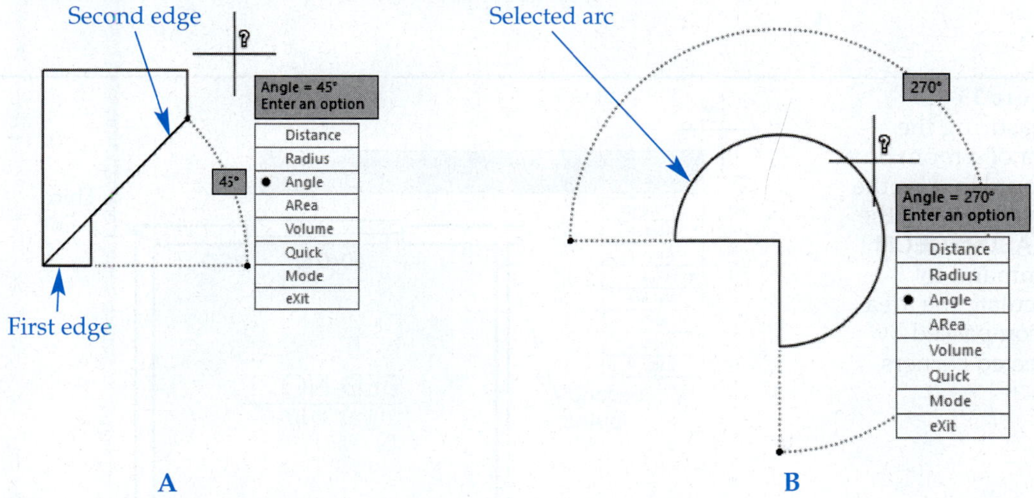

the second line. The dimension appears on-screen, as shown in Figure 15-8A, and at the command line. Select an arc to measure the angle between arc endpoints. See Figure 15-8B.

NOTE

Measuring an angle by selecting a circle and a point is typically most suitable for measuring 3D objects, such as a cylinder.

Select the **Specify vertex** option when it is more appropriate to measure the angle between points, instead of selecting objects. Then select the angle vertex, followed by the first angle endpoint, and finally the second angle endpoint. See Figure 15-9.

Exercise 15-3

www.g-wlearning.com/CAD/

Complete the exercise on the companion website.

Measuring Area

Use the **Area** option of the **MEASUREGEOM** command to find the area between selected points or the area of an object. Specify the first point (corner) of the area to measure, followed by all other points (corners). The **Arc**, **Length**, and **Undo** options are similar to those available with the **Multiple points** option of the **Distance** option. A light

Figure 15-9.
Examples of when it is more appropriate to specify a vertex to measure an angle when using the **Angle** option of the **MEASUREGEOM** command. The lines you see appear temporarily when using this option. A—Selecting the endpoint of a polyline as the vertex. B—Selecting the center of an arc as the vertex.

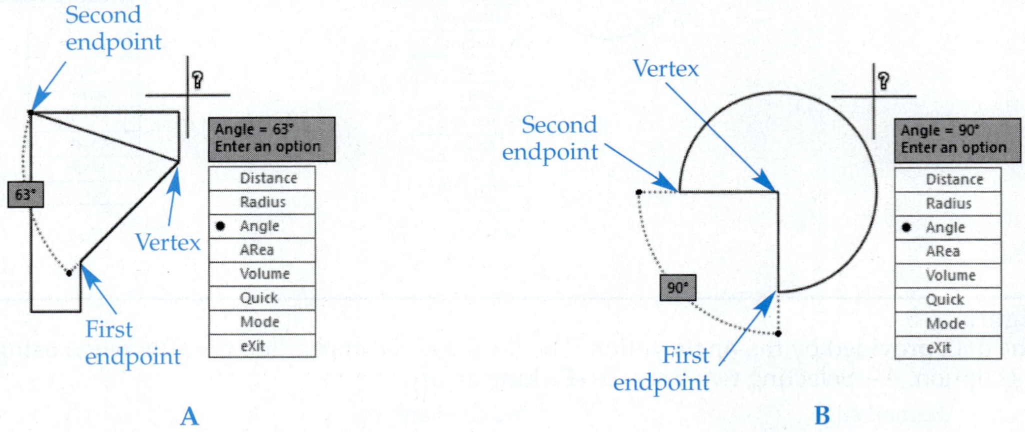

Figure 15-10.
Measuring the area of a room on a floor plan. Use the **Area** option of the **MEASUREGEOM** command to calculate the area encompassed by selected corners.

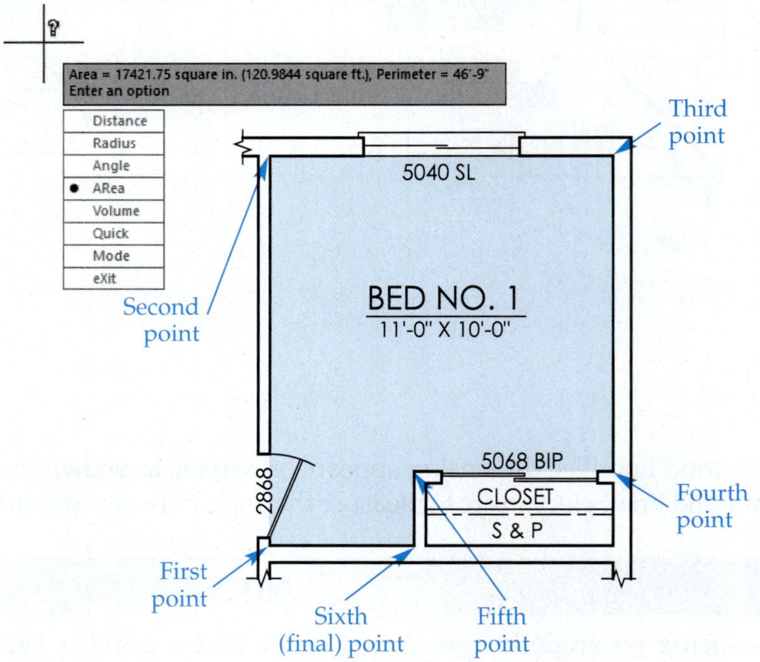

green background fills the area to help you visualize the area. When you finish specifying perimeter corners, use the **Total** option to display the area between the points and the perimeter. The values appear on-screen, as shown in Figure 15-10, and at the command line.

Use the **Object** option of the **Area** option to find the area of a polyline object, circle, or spline without picking corners. Access the **Area** option, activate the **Object** option instead of picking vertices, and then select the object to display values. AutoCAD displays the area of the object and a second value. The second value returned by the **Object** option varies depending on the object type, as shown in Figure 15-11.

NOTE

You can use the **Object** option with an open polyline or spline object. A calculation is made as if a closure exists between the first and last points of the object.

Figure 15-11.
Values returned for common objects when you use the **Object** option.

Object	Values Returned
Polyline	Area and length or perimeter
Circle	Area and circumference
Spline	Area and length or perimeter
Rectangle	Area and perimeter

The **Area** option includes options that allow you to calculate the sum of multiple areas during a single operation. Before selecting corners or an object, activate the **Add area** option and define the first area by picking corners or using the **Object** option. Continue adding areas as needed, or use the **Subtract area** option to remove areas from the selection set. AutoCAD calculates a running total of the area as you add or remove areas.

Figure 15-12 shows an example of using the **Add area** and **Subtract area** options of the **Area** option in the same operation. In this example, select the **Add area** option, then select the **Object** option, and pick the rectangle (polyline). Press [Enter] at the (ADD mode) Select objects: prompt to continue. The area and perimeter of the rectangle and a total area appear.

Then select the **Subtract area** option to enter **SUBTRACT** mode. Select the **Object** option and pick the circles to subtract the area of the circles from the area of the rectangle. The area and circumference of each circle appear, along with the total area of the rectangle minus the total area of the subtracted circles. Press [Enter] at the (SUBTRACT mode) Select objects: prompt to continue. Press [Enter] to display the total area and return to the **MEASUREGEOM** command prompt.

PROFESSIONAL TIP

Calculating the area, circumference, and perimeter of shapes drawn with the **LINE** command can be time-consuming, because you must specify each vertex. Use the **PLINE** command when possible to construct a single object or join existing objects. Use the **Object** option of the **Area** option to add or subtract objects.

Figure 15-12.
Measuring the area of a part view. To calculate the total area of the rectangle minus the areas of the two circles, first select the outer boundary of the rectangle (polyline) using the **Add area** option of the **Area** option. Select the inner circle boundaries using the **Subtract area** option. The total calculation appears.

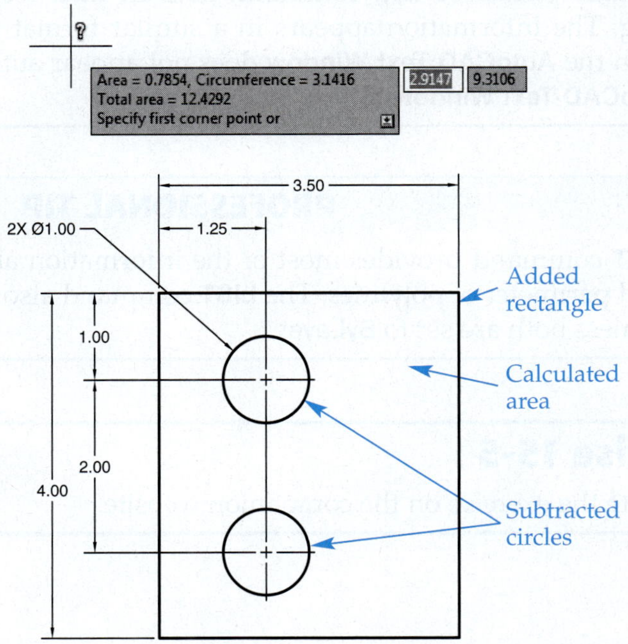

NOTE

You can use the **Volume** option of the **MEASUREGEOM** command to measure the volume of a basic 2D drawing, but the option is most appropriate for measuring 3D objects. *AutoCAD and Its Applications—Advanced* covers 3D modeling applications.

ID
Ribbon
Home
> Utilities
ID Point
Type
ID
LOCATION
COORDINATES

DIST
Type
DIST
LENGTH

AREA
Type
AREA

NOTE

Additional inquiry commands are also available. The **ID** command displays the coordinates of a selected point. The **DIST** command finds the distance between points, and the **AREA** command calculates area. These commands function much like the options available with the **MEASUREGEOM** command, but they are not as interactive.

Exercise 15-4

www.g-wlearning.com/CAD/

Complete the exercise on the companion website.

Listing Drawing Data

LIST
Ribbon
Home
> Properties
List
Type
LIST
LI
LS

The **LIST** command displays data associated with selected objects. Access the **LIST** command, select objects, and press [Enter]. The data for each object is displayed at the command line and in the **AutoCAD Text Window**. If the list data does not fit in the window, AutoCAD prompts you to press [Enter] to display additional data. Figure 15-13A shows the data displayed for a line. The Delta X and Delta Y values indicate the horizontal and vertical distances between the *from point* and *to point* of the line. Figure 15-13B identifies the measurements of a line provided by the **LIST** command.

DBLIST
Type
DBLIST

NOTE

The **DBLIST** (database list) command lists all data for every object in the current drawing. The information appears in a similar format used by the **LIST** command, although the **AutoCAD Text Window** does not appear automatically. Press [F2] to open the **AutoCAD Text Window**.

PROFESSIONAL TIP

The **LIST** command provides most of the information about an object, including the area and perimeter of polylines. The **LIST** command also reports object color and linetype, unless both are set to ByLayer.

Exercise 15-5

www.g-wlearning.com/CAD/

Complete the exercise on the companion website.

Figure 15-13.
A—Using the **LIST** command to list the properties of a line in the **AutoCAD Text Window**.
B—The various data and measurements of a line provided by the **LIST** command.

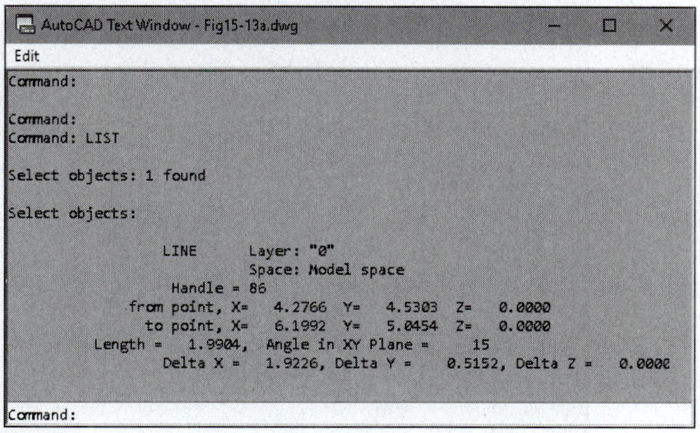

A

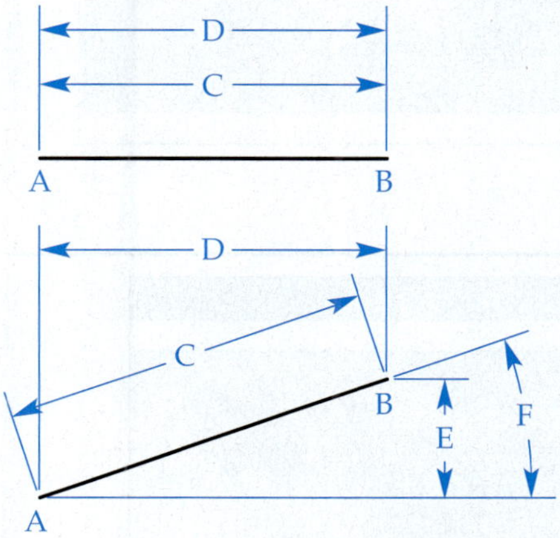

Point	Data
A	XY coordinate
B	XY coordinate
Measurement	**Data**
C	Length
D	Delta X
E	Delta Y
F	Angle

B

Reviewing the Drawing Status

The **STATUS** command provides a method to display a variety of drawing information at the command line and in the **AutoCAD Text Window**. See Figure 15-14. The number of objects in a drawing refers to the total number of objects—erased and existing. Free dwg disk (C:) space: represents the space left on the drive that contains the drawing file. Drawing aid settings appear, along with the current settings for layer, linetype, and color. Press [Enter] if necessary to proceed to additional information.

Type **STATUS**

STATUS

Checking the Time

The **TIME** command displays the current time and time related to the current drawing session at the command line and in the **AutoCAD Text Window**. See Figure 15-15. The drawing creation time starts when you begin a new drawing, not when you first save a new drawing. The **SAVE** command affects the Last updated: time. However, all drawing session time is lost if you exit AutoCAD without saving. Use the **Display** option to update the timer display with current times. Select the **OFF** option to disable the timer and the **ON** option to turn the timer back on. Select the **Reset** option to reset the timer.

Type **TIME**

TIME

Figure 15-14.
Using the **STATUS** command to review drawing information.

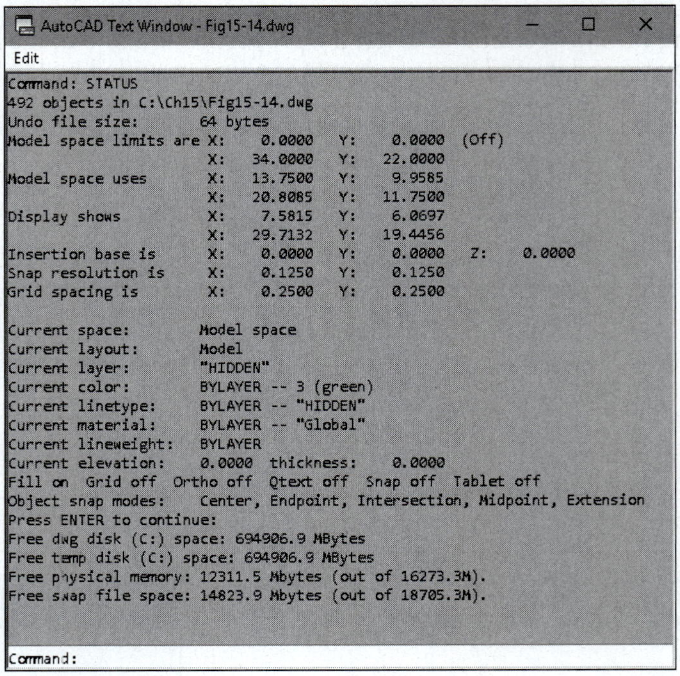

Figure 15-15.
Using the **TIME** command to review the current time and time related to the current drawing session.

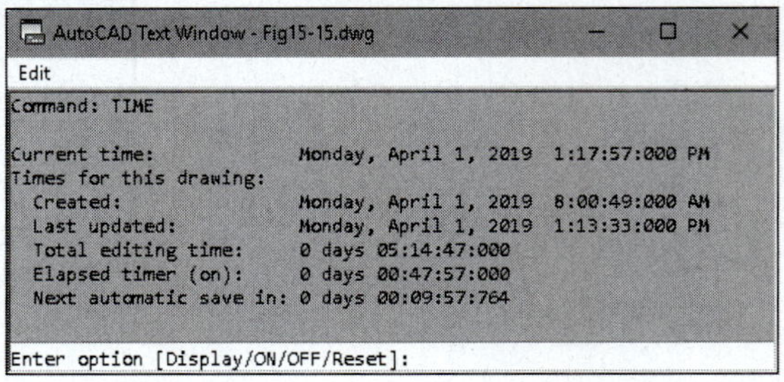

NOTE

The Windows operating system maintains the date and time settings for the computer. You can change these settings in the Windows Control Panel.

NOTE

TEXTSCR

Type
TEXTSCR

If you close the command line, data associated with the **LIST**, **STATUS**, and **TIME** commands appears in the **AutoCAD Text Window**. Use the **TEXTSCR** command to display the **AutoCAD Text Window** at any time. Press [F2] to close the **AutoCAD Text Window**.

PROFESSIONAL TIP

field: A text object that can display a specific property value, setting, or characteristic.

Fields are an effective way to display drawing information within mtext and text objects. You can use fields with many AutoCAD functions and tools, including object properties, drawing properties, attributes, and sheet sets. You can update fields when changes occur to the reference data. Chapter 10 explains using fields.

Exercise 15-6

www.g-wlearning.com/CAD/

Complete the exercise on the companion website.

Using QuickCalc

Most drafting projects require you to make calculations. For example, you may need to double-check dimensions, or you may need to calculate a distance or angle when working from a sketch with missing dimensions. Drafters often make calculations using a handheld calculator. An alternative is to use **QuickCalc**, which is a palette containing a basic calculator, scientific calculator, units converter, and variables feature. See Figure 15-16. Use **QuickCalc** as you would a handheld calculator. You can also use **QuickCalc** while a command is active to paste calculations when a prompt asks for a specific value.

Ribbon
View
> Palettes
Quick Calculator
Type
QUICKCALC
QC
MATH
CALCULATOR
[Ctrl]+[8]

QUICKCALC

NOTE

Access the **QuickCalc** palette quickly by right-clicking in the drawing window and selecting **QuickCalc**. You can also access **QuickCalc** from specific items, such as the button that sometimes appears next to a text box in a dialog box, the **Quick Properties** palette, and the **Properties** palette.

Entering Expressions

The basic math functions used in numeric expressions include addition, subtraction, multiplication, division, and exponential notation. You can enter grouped expressions by using parentheses to break up the expressions to be calculated separately. For example, to calculate 6 + 2 and then multiply the sum by 4, enter (6+2)*4. The result will be wrong if you do not add the parentheses. Figure 15-17 shows the symbols used for basic math operators.

Figure 15-16. The **QuickCalc** palette allows you to perform a variety of calculations.

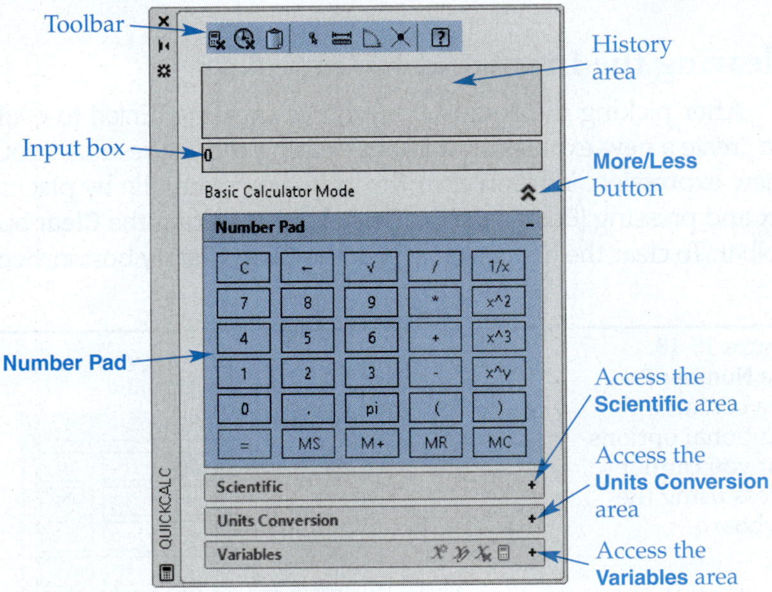

Figure 15-17.
Common math operators.

Symbol	Function	Example
+	Addition	3+26
–	Subtraction	270–15.3
*	Multiplication	4*156
/	Division	265/16
^	Exponent	22.6^3
( )	Grouped expressions	2*(16+2^3)

Add expressions in the input box by picking buttons on the **Number Pad** or by using keyboard keys. After creating the expression, pick the equal (=) button on the number pad or press [Enter] to evaluate the expression. Figure 15-18 shows options found on the **Number Pad** that are not available from the keyboard.

The result of an evaluated expression appears in the input box, and the expression moves to the history area. Figure 15-19 shows the **QuickCalc** palette before and after calculating 96.27 + 23.58. When you are using only the input box in **QuickCalc**, pick the **More/Less** button below the input box to hide the additional sections, saving valuable drawing space.

NOTE

If you move the cursor outside of the **QuickCalc** palette, the drawing area is automatically activated. Pick anywhere inside the **QuickCalc** palette to reactivate **QuickCalc**.

PROFESSIONAL TIP

If you make a mistake in the input box, you do not need to clear the input box and start over again. Use the left and right arrow keys to move through the field. Right-click in the input box to access a shortcut menu of options useful for copying and pasting.

Clearing the Input and History Areas

After picking the equal (=) button or pressing [Enter] to evaluate an expression, you can create a new expression without clearing the last result. AutoCAD automatically starts a new expression. You can clear the input box manually by placing the cursor in the input box and pressing [Backspace] or [Delete] or by picking the **Clear** button from the **QuickCalc** toolbar. To clear the history area, pick the **Clear History** button. See Figure 15-19B.

Figure 15-18.
The **Number Pad** area contains additional options that you cannot access using the keyboard.

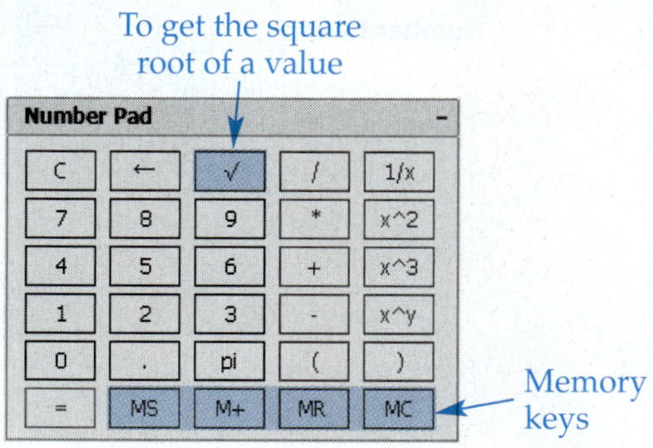

Figure 15-19.
A—Add an expression to the input box. B—After creating the expression, press [Enter] to evaluate the expression. The history area stores the expression and result. Use the buttons on the **QuickCalc** toolbar to clear the input box and history area.

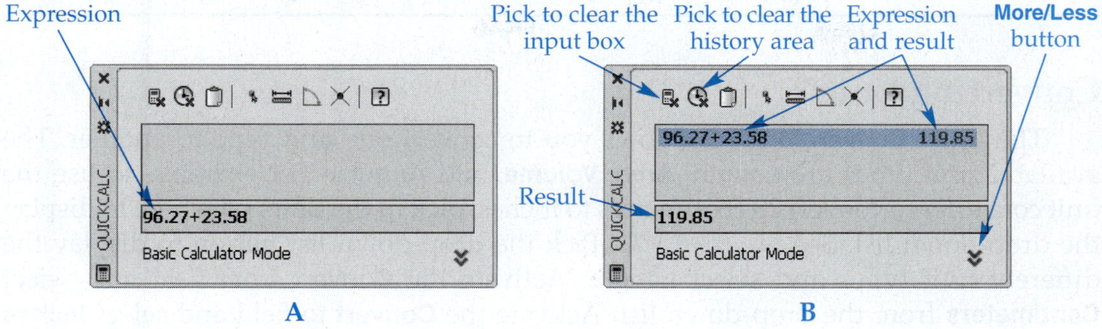

CAUTION

If you enter an expression that **QuickCalc** cannot evaluate, AutoCAD displays an error dialog box. Pick the **OK** button, correct the error, and try again.

Exercise 15-7

www.g-wlearning.com/CAD/

Complete the exercise on the companion website.

Scientific Calculations

The **Scientific** area of **QuickCalc** includes trigonometric, exponential, and other advanced functions. See Figure 15-20. To use one of the functions, add a value to the input box, pick the appropriate function button, and pick the equal (=) button or press [Enter]. When you pick a function button, the input box value appears in parentheses after the expression. For example, to get the sine of 14, clear the input box, type 14 in the input box, and pick the **sin** button. The input box now reads sin(14). Pick the equal (=) button or press [Enter] to view the result.

Figure 15-20.
The scientific functions available in **QuickCalc**.

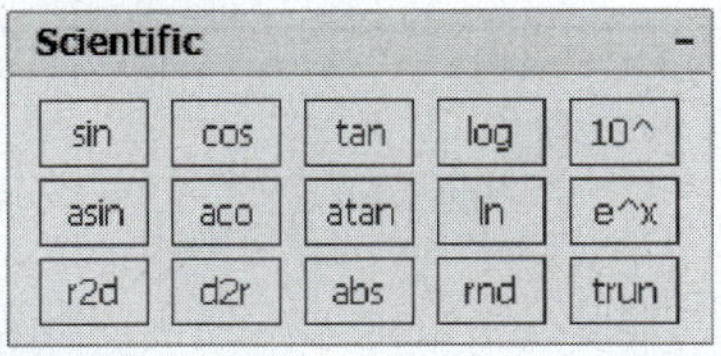

sin Sine	cos Cosine	tan Tangent	log Base-10 Log	10^ Base-10 Exponent
asin Arcsine	aco Arccosine	atan Arctangent	ln Natural Log	e^x Natural Exponent
r2d Convert Radians to Degrees	d2r Convert Degrees to Radians	abs Absolute Value	rnd Round	trun Truncate

NOTE

You can pick a function button first, but doing so puts a default value of 0 in parentheses. Place the cursor in the input box to type a different number in the parentheses if needed.

Converting Units

The **Units Conversion** area allows you to convert one unit type to another. The available unit types are **Length**, **Area**, **Volume**, and **Angular**. For example, to use the unit converter to convert 23 centimeters to inches, pick in the **Units type** field to display the drop-down list. See Figure 15-21. Pick the drop-down list button to display the different unit types and select **Length**. Activate the **Convert from** field and select **Centimeters** from the drop-down list. Activate the **Convert to** field and select **Inches** from the drop-down list. Type 23 in the **Value to convert** field and pick the equal (=) button or press [Enter]. The **Converted value** field displays the converted units.

To pass the converted value to the input box for use in an expression, pick the **Return Conversion to Calculator Input Area** button. See Figure 15-22. If the button is not visible, pick once on the converted units in the **Converted value** field.

Exercise 15-8

www.g-wlearning.com/CAD/

Complete the exercise on the companion website.

Using Variables

variable: A text item that represents another value and is available for future reference.

constant: An expression or value that stays the same.

function: An expression or value that asks for user input to get values to pass to the expression.

If you use an expression or value frequently, you can save it as a ***variable***. Use the **Variables** area, shown in Figure 15-23, to create, edit, and delete variables and to pass variables to the input box. The **Variables** area of **QuickCalc** includes predefined ***constant*** and ***function*** variables.

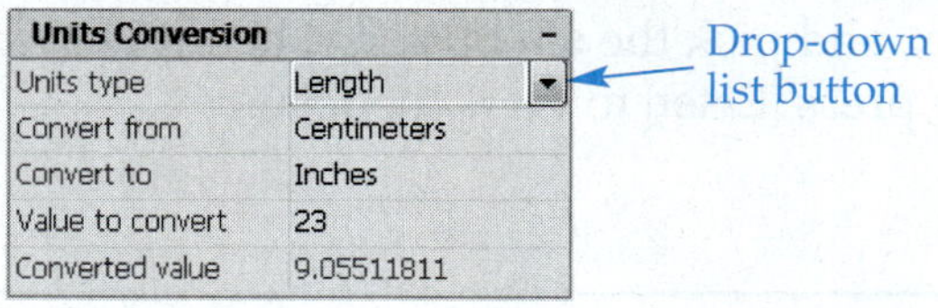

Figure 15-21.
Picking the current unit type activates the field and displays the drop-down list button.

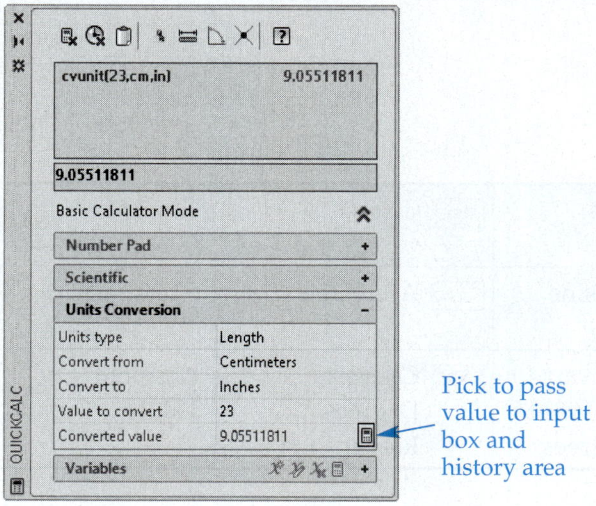

Figure 15-22.
After you convert a value, pass the value to the input box for use in an expression.

Figure 15-23.
The **Variables** area of **QuickCalc** allows you to store values and expressions for later use.

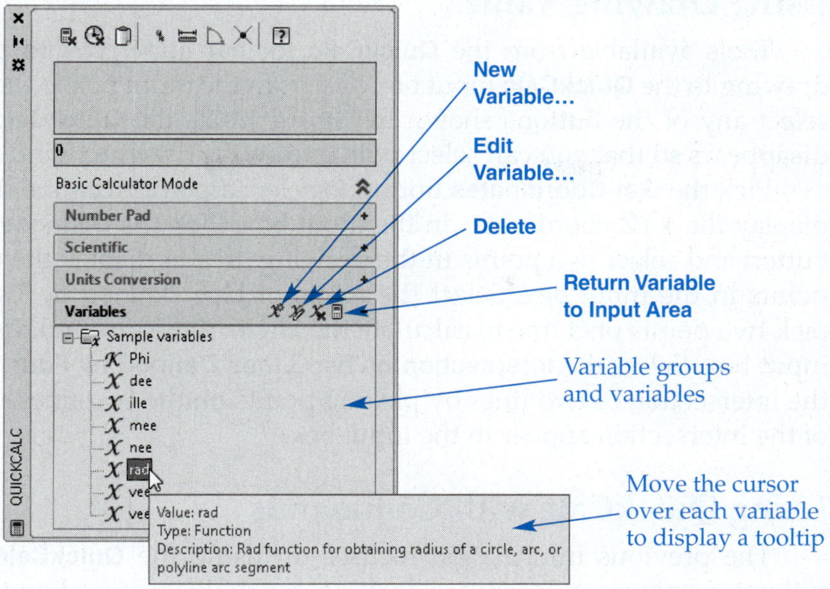

To create a new variable, pick the **New Variable...** button to open the **Variable Definition** dialog box shown in **Figure 15-24**. Select the variable type in the **Variable type** area. Type a name for the variable in the **Name:** field. Select a group to contain the variable using the **Group with:** drop-down list. Type the value or the expression for the variable in the **Value or expression:** field. Type a description for the variable in the **Description:** field. Pick the **OK** button to save the variable and display it in the **Variables** area.

To edit a variable, pick the variable to modify and pick the **Edit Variable...** button to reopen the **Variable Definition** dialog box. Pick the **Delete** button to delete the selected variable. Pick the **Return Variable to Input Area** button or double-click the variable name to pass the selected variable to the input box.

Additional options for variables are accessed by right-clicking in the **Variables** area to display a shortcut menu. Pick the **New Category** option to create a new category for saved variables. To rename a variable, select the **Rename** option and type a new name.

PROFESSIONAL TIP

The AutoCAD help system explains the predefined variables and their functions.

Figure 15-24.
Use the **Variable Definition** dialog box to define new variables.

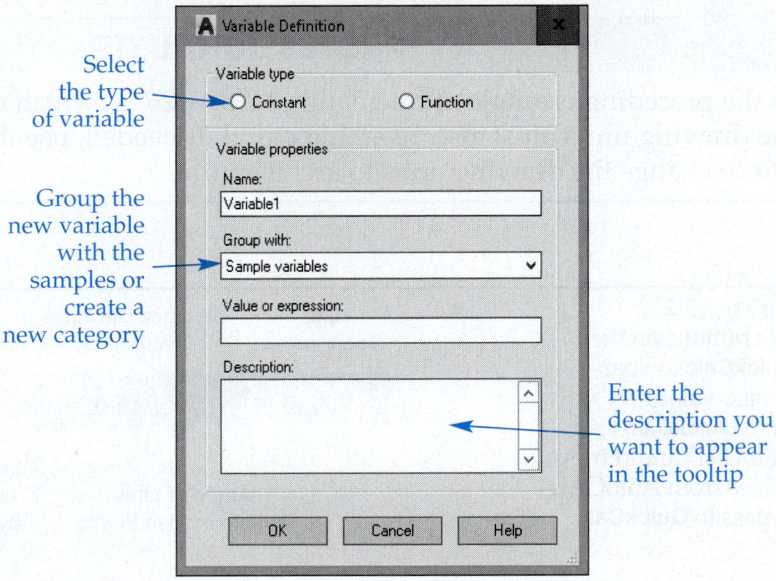

Using Drawing Values

Tools available from the **QuickCalc** toolbar allow you to pass values from the drawing to the **QuickCalc** input box and from the input box to the drawing. When you select any of the buttons shown in Figure 15-25, the **QuickCalc** palette temporarily disappears so that you can select points from the drawing window.

Pick the **Get Coordinates** button to select a point from the drawing window and display the XYZ coordinates in the input box. Pick the **Distance Between Two Points** button and select two points in the drawing area to display the distance between the points in the input box. Select the **Angle of Line Defined by Two Points** button and pick two points on a line to calculate the angle of the line and display the angle in the input box. Select the **Intersection of Two Lines Defined by Four Points** button to find the intersection of two lines by picking points on the two lines. The XYZ coordinates of the intersection appear in the input box.

Using QuickCalc with Commands

The previous information focuses on using the **QuickCalc** palette to calculate unknown values while you are drafting, much like using a handheld calculator or the Windows Calculator. You can also use **QuickCalc** while a command is active to pass a calculated value to the command line as a response to a prompt. AutoCAD provides a few alternatives for using **QuickCalc** while a command is active.

If the **QuickCalc** palette is active when you access a command, when the prompt requesting an unknown value appears, calculate the value using the **QuickCalc** palette and then pick the **Paste value to command line** button to pass the value as a response to the prompt. For example, to draw a line a distance of 14′-8″ + 26′-3″ horizontally from a start point, activate the **QuickCalc** palette, access the **LINE** command, and pick a start point. Then use polar tracking or ortho mode to move the crosshairs to the right or left of the start point so the line is at a 0° or 180° angle. At the Specify next point or [Undo]: prompt, enter 14'8" + 26'3" in the **QuickCalc** palette input box and pick the equal (=) button or press [Enter]. The result in the input box is 40'-11". Pick the **Paste value to command line** button to respond to the prompt with a 40'-11" value. Press [Enter] to draw the 40′-11″ line.

If the **QuickCalc** palette is not active while you are using a command, you can still calculate and use a value. When the prompt requesting an unknown value appears, access **QuickCalc** by right-clicking and selecting **QuickCalc** or by typing 'QC. A **QuickCalc** window, which is not the same as the **QuickCalc** palette, opens in command calculation mode. Use the necessary tools to evaluate an expression. Then pick the **Apply** button to pass the value back to the command line and close the **QuickCalc** window.

PROFESSIONAL TIP

In the preceding example of calculating 14′-8″ + 26′-3″, which uses architectural units, the drawing units must also be architectural. If needed, use the **Drawing Units** dialog box to change the drawing units to architectural.

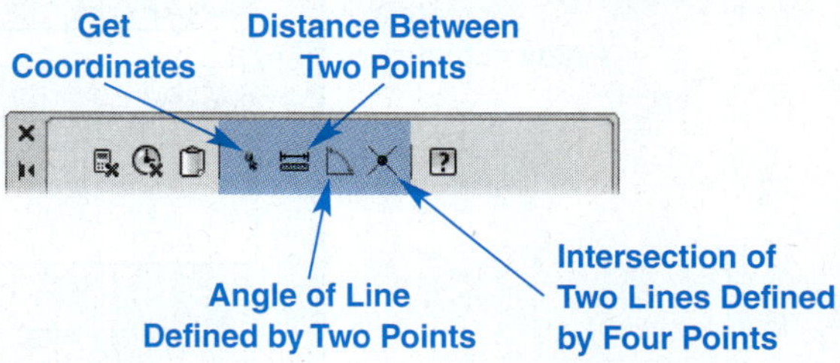

Figure 15-25.
Use buttons on the **QuickCalc** toolbar to pass values from **QuickCalc** to AutoCAD and retrieve values from AutoCAD to pass to **QuickCalc**.

Exercise 15-9

www.g-wlearning.com/CAD/

Complete the exercise on the companion website.

Using QuickCalc with Object Properties

QuickCalc also allows you to calculate expressions for an object while using the **Properties** palette. Pick a field that contains a numeric value to display the calculator icon. Pick the calculator icon to open the **QuickCalc** window, again not the same item as the **QuickCalc** palette, in property calculation mode. Use expressions and values in the same manner as when using **QuickCalc** at any other time. Once you evaluate the expression in the input box, pick the **Apply** button to pass the value to the property field in the **Properties** palette. The object automatically updates based on the new value.

Additional QuickCalc Options

The history area provides access to additional **QuickCalc** options. Right-click in the history area to access the shortcut menu shown in Figure 15-26A. Figure 15-26B explains the menu options.

Picking the **Properties** button on the **QuickCalc** palette displays a shortcut menu. The settings allow you to change the palette appearance, including its ability to dock, hide, or appear transparent.

Comparing Drawings

AutoCAD 2020 NEW

The **COMPARE** command is used to evaluate the differences between two drawing files. Use the **COMPARE** command when it is necessary to observe and document the differences between two drawings, such as when you are working on a design project and other members of the project team make changes to the design.

Figure 15-26.
The options available in the **QuickCalc** history area shortcut menu.

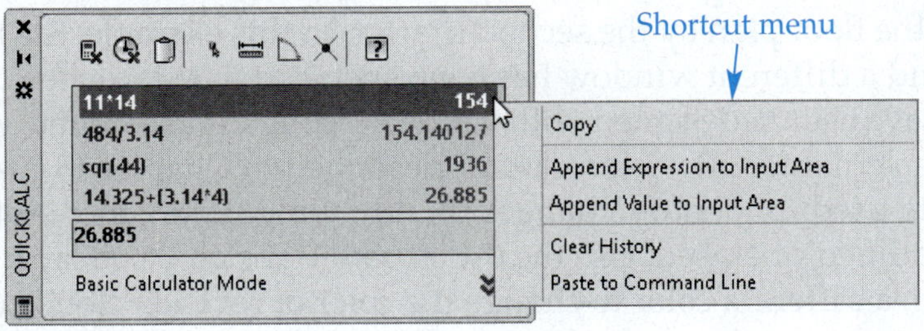

A

Option	Description
Copy	Copies the expression and value to the Windows Clipboard.
Append Expression to Input Area	Passes the expression to the input box.
Append Value to Input Area	Passes the value to the input box.
Clear History	Clears the history area.
Paste to Command Line	Passes the value to the command line.

B

COMPARE

Ribbon
Collaborate
> Compare

DWG Compare
Application Menu
Drawing Utilities
> DWG Compare
Type
COMPARE

The **COMPARE** command displays a comparison between an open drawing called the *base drawing* and a specified drawing called the *reference drawing*. The comparison is displayed in the base drawing. Access the **COMPARE** command and use the **Select a drawing to compare** dialog box to locate the reference drawing. Pick the **Open** button to begin the comparison. See Figure 15-27. During the comparison, the **DWG Compare** toolbar appears near the top of the drawing window.

In the comparison state, any objects that are unique to the base drawing are dark green (color 82). Any objects that are unique to the reference drawing are red (color 1). To select different colors, pick the **Settings** button on the left end of the **DWG Compare** toolbar to display the **Settings** panel. Pick the color swatches corresponding to the base drawing and reference drawing in the **Difference** area of the **Settings** panel to open the **Select Color** dialog box and pick the desired colors.

To activate the **COMPARE** command when no drawing is open, select **DWG Compare** in the **Application Menu**. In the **DWG Compare** dialog box, use the **Base** area to specify the base drawing and the **Reference** area to specify the reference drawing. Pick the **Compare** button to begin the comparison.

NOTE

The general intent of the **COMPARE** command is to help evaluate a design before and after changes have been made. Therefore, the base and reference drawings are typically separate files with unique names. The **DWG Compare** dialog box warns you if you specify the same drawing file for the base and reference drawings.

In Figure 15-27, a residential floor plan saved by one drafter, named A-02-MMM, is compared to the same floor plan modified by a second drafter, named A-02-DPM. The A-02-MMM file is the base drawing. AutoCAD brings the A-02-DPM reference drawing into the base drawing as an external reference and places it on the 0-Compare-A-02-DPM layer. External references are explained later in this textbook.

All objects in the base and reference files that have the same X, Y, and Z position display in the same gray color (color 253). Objects that differ between the two drawings display in the specified base and reference drawing colors. A rectangular revision cloud, drawn on the 0-Compare-Revision Cloud layer and displayed in orange (color 40), appears around each ***change set***. The eight change sets in Figure 15-27 identify revisions made to the floor plan by the second drafter. In this example, the refrigerator has been moved and a different window has been specified in the kitchen, the dining and living rooms have been redesigned with vaulted ceilings, and an entry wall and broom closet have been enlarged. Revision clouds, described in Chapter 18, are cloud-shaped polyline objects used to identify changes in a drawing. Like other AutoCAD objects, they can be modified or erased. Use the 0-Compare-Revision Cloud layer as needed. For example, assign a different color to change the color of revision clouds or use the layer to add revision marks and notes.

change set: A grouping of objects that differ between two drawing files selected for comparison with the **COMPARE** command.

The **DWG Compare** toolbar and the **Settings** panel offer an easy approach to working with drawing comparison options. Figure 15-28 describes each item in the **DWG Compare** toolbar and the **Settings** panel.

Exercise 15-10

www.g-wlearning.com/CAD/

Complete the exercise on the companion website.

Figure 15-27.
Using the **COMPARE** command to analyze a residential floor plan before and after a design revision. A—The **DWG Compare** toolbar appears during the comparison. Pick the **Settings** button in the **DWG Compare** toolbar to display the **Settings** panel. B—Change sets in the comparison indicate differences between the base drawing and the reference drawing. A revision cloud appears around each change set. In this example, there are eight change sets.

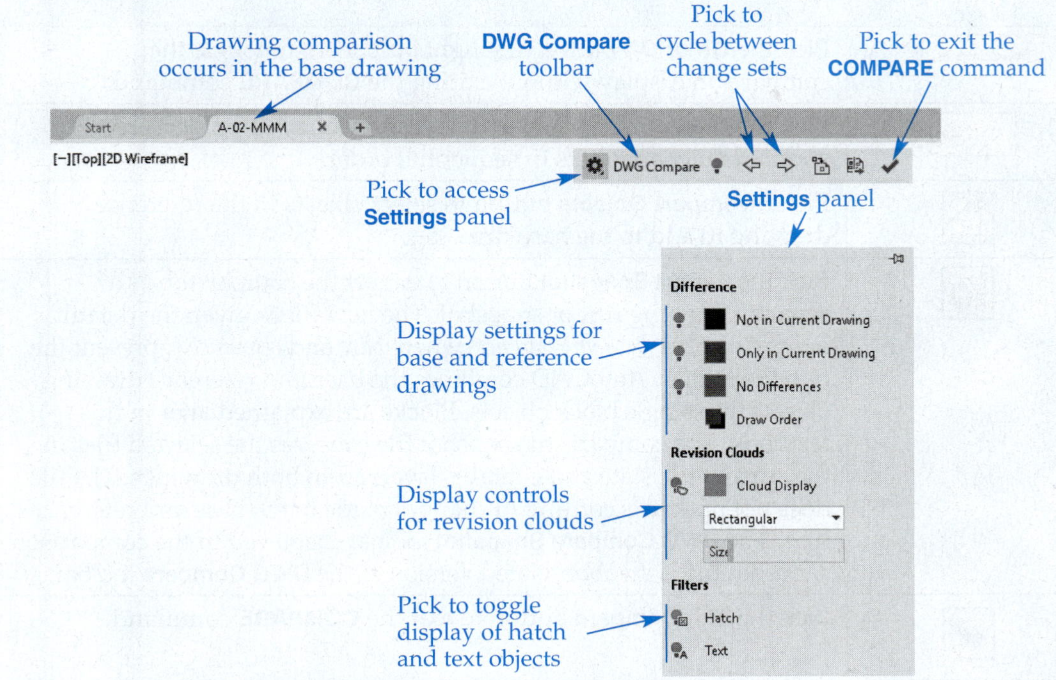

DWG Compare Toolbar and **Settings** Panel

A

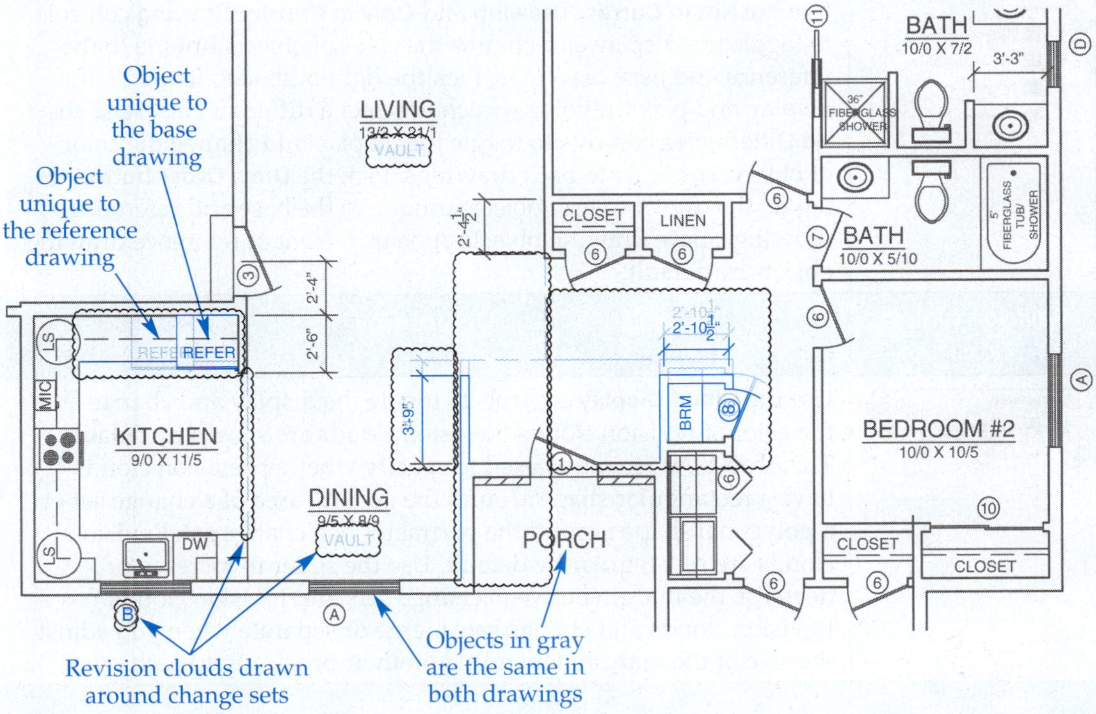

B

Figure 15-28.
The display options and tools available in the **DWG Compare** toolbar and the **Settings** panel.

DWG Compare Toolbar	
Command or Option	**Description**
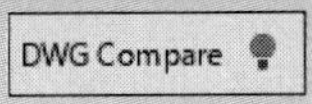	Pick the **On or Off** button (the light bulb icon) to toggle the comparison display without exiting the **COMPARE** command.
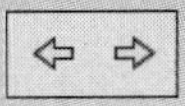	Pick the **Next** button (right arrow) or **Previous** button (left arrow) to zoom in on change sets in sequential order.
	Pick the **Import Objects** button to select objects in the reference drawing to add to the base drawing.
	Pick the **Export Snapshot** button to export the comparison as a separate drawing file, or snapshot. The new file is given the default name Compare_*base* vs *reference*, where *base* and *reference* represent the two file names. AutoCAD combines the base and reference drawing objects into single block objects. Blocks are explained later in this textbook. The comparison snapshot file captures the selected files in the current file state and contains layers from both drawings. The file does not update according to changes made to the base and reference files. The **DWG Compare Snapshot** toolbar displayed in the comparison snapshot file is an abbreviated version of the **DWG Compare** toolbar.
	Pick the **Exit Compare** button to exit the **COMPARE** command.
Settings Panel	
Difference area	**Description**
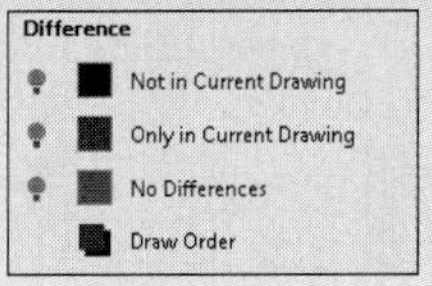	Use the **Not in Current Drawing** and **Only in Current Drawing** controls to toggle the display and change the color of objects unique to the reference and base drawings. Pick the light bulb icon to toggle the display and pick the color swatch to select a different color. Use the **No Differences** controls to toggle the display and change the color of objects common to both drawings. Pick the **Draw Order** button to toggle the draw order of objects unique to the base and reference drawings. Base drawing objects appear in front of reference drawing objects by default.
Revision Clouds area	**Description**
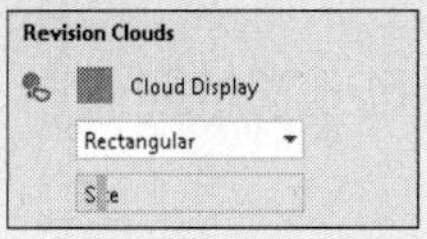	Use the **Cloud Display** controls to toggle the display and change the color of revision clouds. Revision clouds are shown by default. The drop-down menu is used to specify whether revision clouds have a rectangular shape around the general area of a change set or a polygonal shape around the perimeter of a change set. Revision clouds are rectangular by default. Use the slider to increase or decrease the margin between change sets and revision clouds. Revision clouds and change sets merge or separate when you adjust the size of the margin depending on their proximity to each other.
Filters area	**Description**
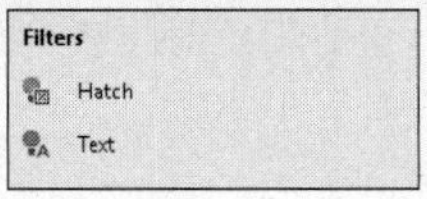	Pick the filter buttons to toggle the display of hatch and text objects common and unique to the base and reference drawings. Hatch objects, explained later in this textbook, are not shown by default. Text appears by default.

Chapter Review

Answer the following questions using the information in this chapter.

1. Explain how to use grips to identify the location of a point and the dimensions of an object.
2. What types of information does the **Distance** option of the **MEASUREGEOM** command provide?
3. What information does the **Area** option of the **MEASUREGEOM** command provide?
4. To add the areas of several objects using the **Area** option of the **MEASUREGEOM** command, when do you select the **Add area** option?
5. Explain how calculating the area of a polyline using the **Area** option of the **MEASUREGEOM** command is different from calculating the area of an object drawn with the **LINE** command.
6. What is the purpose of the **LIST** command?
7. Describe the meaning of *delta X* and *delta Y*.
8. What command, other than **MEASUREGEOM** and **AREA**, provides the area and perimeter of an object?
9. What is the function of the **DBLIST** command?
10. Which command allows you to list drawing aid settings for the current drawing?
11. What information does the **TIME** command provide?
12. When does the drawing creation time start?
13. What term describes a text object that can display a specific property value, setting, or characteristic?
14. List three ways to open the **QuickCalc** palette.
15. Give the proper symbol to use for the following math functions.
    A. Addition
    B. Subtraction
    C. Multiplication
    D. Division
    E. Exponent
    F. Grouped expressions
16. Under which area of the **QuickCalc** palette can you find the square root function?
17. Under which area of the **QuickCalc** palette can you find the arccosine function?
18. When using one of the scientific functions, which should you do first: pick the scientific function button or type in the value to use in the input box? Why?
19. Name the four types of units that you can convert using **QuickCalc**.
20. What term describes a text item that represents another value and can be accessed later as needed?
21. Which **QuickCalc** button passes the value in the **QuickCalc** input box to respond to a prompt?
22. How can you start **QuickCalc** while a command is active?
23. When using the **QuickCalc** window while a command is active, how do you pass the value to respond to a prompt?
24. Explain how to access **QuickCalc** when using the **Properties** palette in order to change an object property.
25. What is a grouping in a drawing comparison that contains objects that differ between the base and reference drawing files?

Drawing Problems

Start AutoCAD if it is not already started. Start a new drawing for each problem using an appropriate template of your choice. The template should include layers and text styles, when necessary, for drawing the given objects. Add layers and text styles as needed. Draw all objects using appropriate layers. Use appropriate text styles, justification, and format. Follow the specific instructions for each problem. Use only drawing and editing commands and techniques you have already learned. Do not draw dimensions. Use your own judgment and approximate dimensions when necessary.

▼ Basic

1. General Use **QuickCalc** to calculate the result of the following equations.
   A. 27.375 + 15.875
   B. 16.0625 – 7.1250
   C. 5 × 17′-8″
   D. 48′-0″ ÷ 16
   E. (12.625 + 3.063) + (18.250 – 4.375) – (2.625 – 1.188)
   F. 7.252^2

2. General Use **QuickCalc** to convert 4.625″ to millimeters.

3. General Use **QuickCalc** to convert 26 mm to inches.

4. General Use **QuickCalc** to convert 65 miles to kilometers.

5. General Use **QuickCalc** to convert 750 cubic feet to cubic meters.

6. General Use **QuickCalc** to find the square root of 360.

7. General Use **QuickCalc** to calculate 3.25 squared.

▼ Intermediate

8. Mechanical Show the calculation and answer used with the **LINE** command to make an 8″ line .006 in./in. longer in a pattern to allow for shrinkage in the final casting. Show only the expression and answer.

9. Architectural Solve for the deflection of a structural member. The formula is written as $PL^3/48EI$, where P = pounds of force, L = length of beam, E = modulus of elasticity, and I = moment of inertia. The values to be used are P = 4000 lbs, L = 240″, and E = 1,000,000 lbs/in^2. The value for I is the result of the beam (width × height^3)/12, where width = 6.75″ and height = 13.5″.

10. General Calculate the coordinate located at 4,4,0 + 3<30.

11. General Calculate the coordinate located at (3 + 5,1 + 1.25,0) + (2.375,1.625,0).

Drawing Problems - Chapter 15

12. **Mechanical** Draw the part view shown. Check the time when you start the drawing. Draw all the features using the **PLINE** and **CIRCLE** commands. Use the **Area** option of the **MEASUREGEOM** command and the **Object**, **Add area**, and **Subtract area** options to calculate the following.
   A. The area and perimeter of Object A.
   B. The area and perimeter of Object B. The slot ends are full radius.
   C. The area and circumference of one of the circles (C).
   D. The area of Object A, minus the area of Object B.
   E. The area of Object A, minus the areas of the other three features.

   Enter the **TIME** command and note the editing time spent on your drawing. Save the drawing as P15-12. Save a copy of P15-12 as P15-12_REV1. In the P15-12_REV1 file, change the diameter of both circles to .75 and the width of the slot to .50. Use the **COMPARE** command to analyze the differences between P15-12 and P15-12_REV1. Export a snapshot of the drawing comparison. Save the snapshot using the default file name.

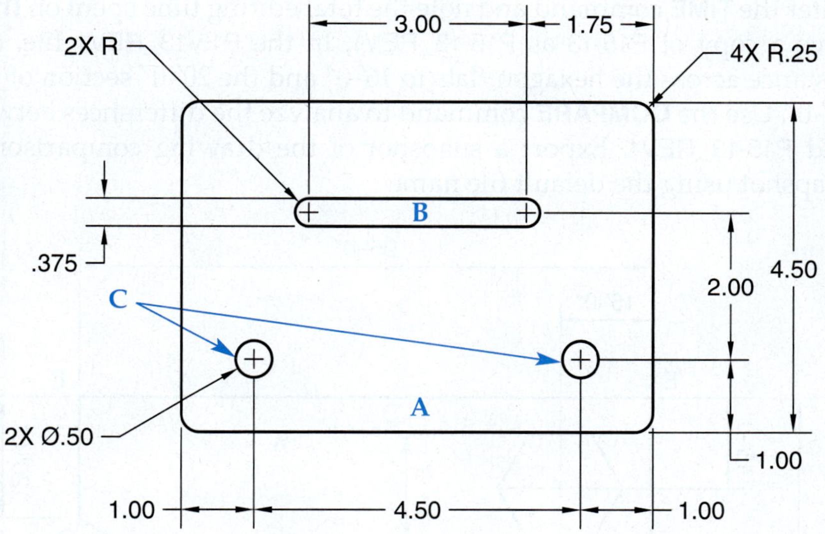

13. Architectural Draw the deck shown using the **PLINE** command. Use the **POLYGON** command to draw the hot tub. Save the drawing as P15-13. Use the following guidelines to complete this problem.
   A. Specify architectural units for the drawing. Use 1/2″ fractions and decimal degrees. Leave the remaining settings for the drawing units at the default values.
   B. Set the limits to 100′,80′ and use the **All** option of the **ZOOM** command.
   C. Set the grid spacing to 2′ and the snap spacing to 1′.
   D. Calculate the following measurements.
      a. The area and perimeter of the deck (A).
      b. The area and perimeter of the hot tub (B).
      c. The area of the deck minus the area of the hot tub.
      d. The distance between Point C and Point D.
      e. The distance between Point E and Point C.
      f. The coordinates of Points C, D, and F.
   E. Enter the **DBLIST** command and check the information listed for the drawing.
   F. Enter the **TIME** command and note the total editing time spent on the drawing. Save a copy of P15-13 as P15-13_REV1. In the P15-13_REV1 file, change the distance across the hexagon flats to 16′-0″ and the 20′-0″ section of the deck to 24′-0″. Use the **COMPARE** command to analyze the differences between P15-13 and P15-13_REV1. Export a snapshot of the drawing comparison. Save the snapshot using the default file name.

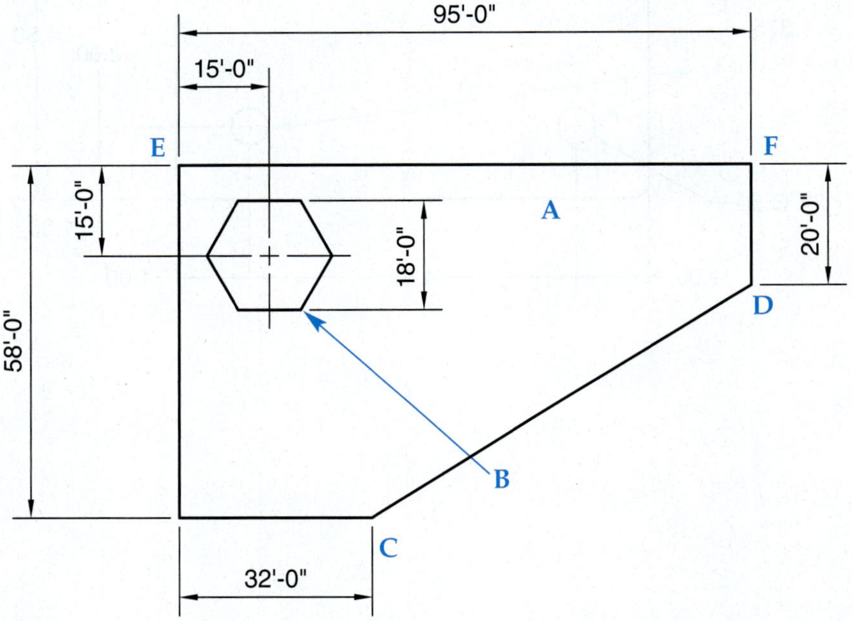

14. General The drawing shown is a side view of a pyramid. The pyramid has four sides. Create an auxiliary view showing the true size of a pyramid face. Save the drawing as P15-14. Calculate the following.
   A. The area of one side.
   B. The perimeter of one side.
   C. The area of all four sides.
   D. The area of the base.
   E. The true length (distance) from the midpoint of the base on one side to the apex.

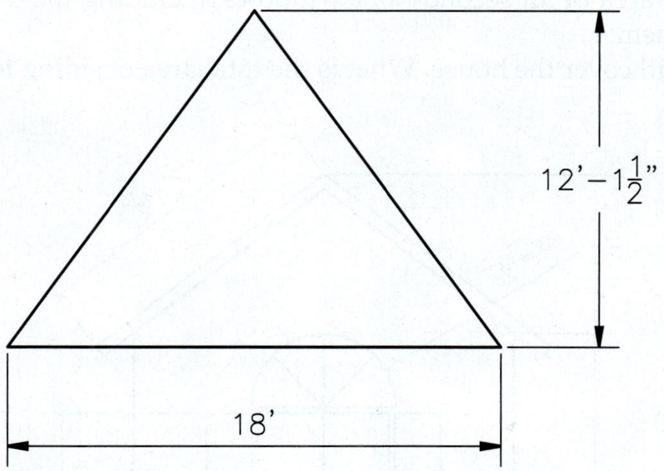

▼ Advanced

15. General Make the following calculations given the right triangle shown.
   A. Length of side c (hypotenuse).
   B. Sine of angle A.
   C. Sine of angle B.
   D. Cosine of angle A.
   E. Tangent of angle A.
   F. Tangent of angle B.

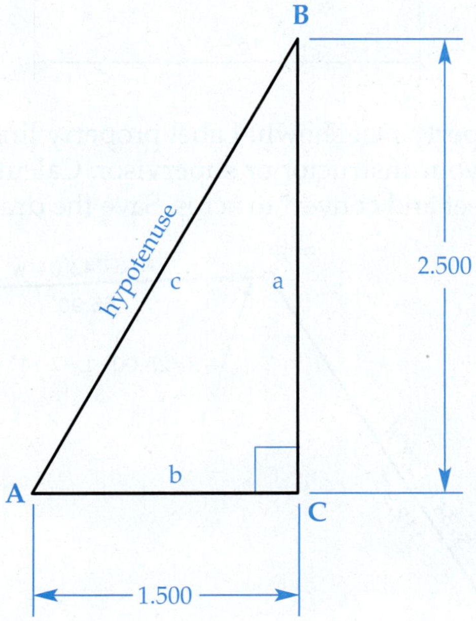

16. Architectural Draw the house elevation shown. Draw the windows as single lines only (the location of the windows is not critical). The spacing between each of the second-floor windows is 3″. The width of this end of the house is 16′-6″. The length of the roof is 40′. Use the **PLINE** command to assist in creating the specific shapes in this drawing, except as previously noted. Save the drawing as P15-16. Calculate the following.

A. The total area of the roof.

B. The diagonal distance from one corner of the roof to the other.

C. The area of the first-floor window.

D. The total area of all second-floor windows, including the 3″ spaces between each of them.

E. Siding will cover the house. What is the total area of siding for this end?

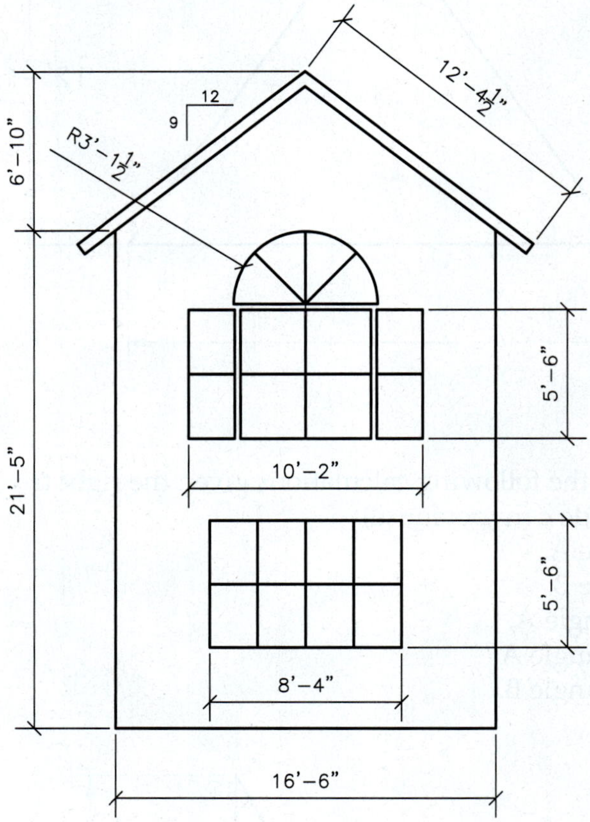

17. Civil Draw the property plat shown. Label property line bearings and distances only if required by your instructor or supervisor. Calculate the area of the property plat in square feet and convert to acres. Save the drawing as P15-17.

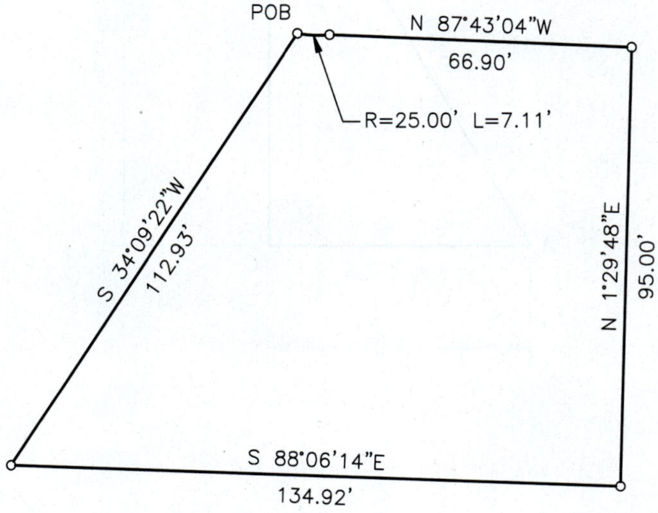

18. **Civil** Draw the subdivision plat. Label the drawing as shown. Calculate the acreage of each lot and record each value as a label inside the corresponding lot (for example, .249 AC). Save the drawing as P15-18.

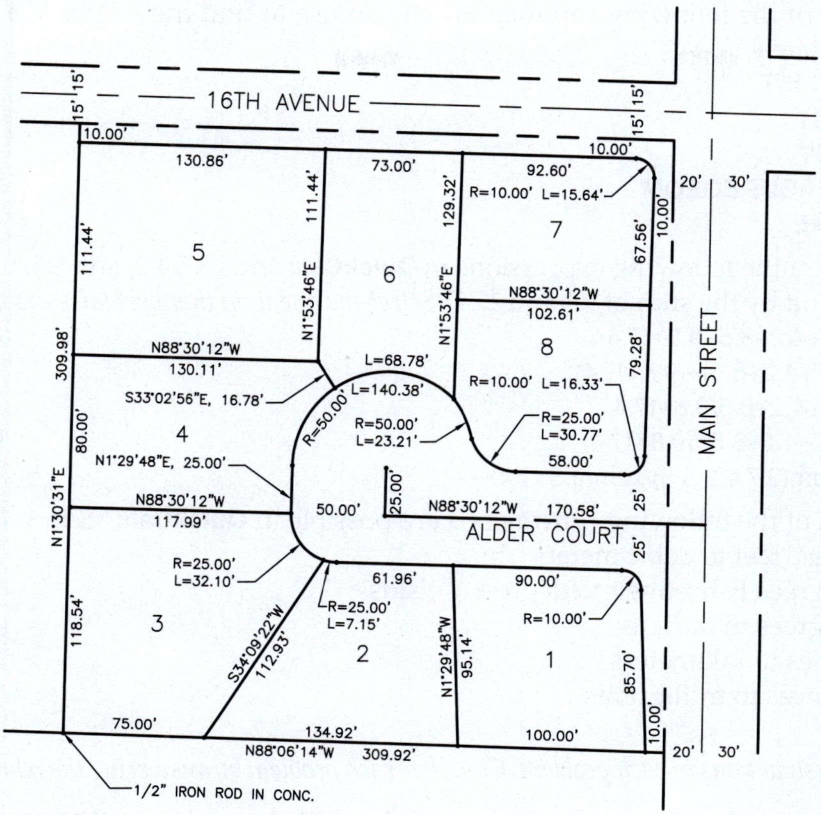

19. **Mechanical** Research the design of an existing ball valve with the following specifications: brass body, chrome-plated brass ball, full port, lever style, non-threaded (solder) female ends to accept a Ø1″ pipe, rated for 500 psi at 100°. Create a freehand, dimensioned 2D sketch of the existing design from manufacturer's specifications or from measurements taken from an actual valve. Start a new drawing from scratch or use a decimal-unit template of your choice. Draw the views needed to describe the valve completely from your sketch. Save the drawing as P15-19. Print an 8.5″ × 11″ copy of the drawing extents using a 1:1 scale and a landscape orientation. Save a copy of P15-19 as P15-19_REV1. In the P15-19_REV1 file, make three changes to the ball valve design. Use the **COMPARE** command to analyze the differences between P15-19 and P15-19_REV1. Export a snapshot of the drawing comparison. Save the snapshot using the default file name.

AutoCAD Certified Professional Exam Practice

Answer the following questions using the information in this chapter.

1. Which of the following commands can you use to find the length of a line? *Select all that apply.*
   A. **DBLIST**
   B. **DIST**
   C. **LIST**
   D. **MEASUREGEOM**
   E. **TIME**
2. Which of the following expressions in **QuickCalc** adds 3.7, 4.2, and 8.5, then divides the result by the sum of 9.8 and 17.4? *Select the one item that best answers the question.*
   A. 3.7+4.2+8.5÷9.8+17.4
   B. (3.7+4.2+8.5)÷(9.8+17.4)
   C. 3.7+4.2+8.5/9.8+17.4
   D. (3.7+4.2+8.5)/(9.8+17.4)
   E. =sum(3.7,4.2,8.5)/sum(9.8,17.4)
3. Which of the following conversions are possible in **QuickCalc**? *Select all that apply.*
   A. cubic feet to cubic meters
   B. degrees Fahrenheit to degrees Celsius
   C. degrees to radians
   D. miles to kilometers
   E. ounces to milligrams

Follow the instructions in each problem. Complete each problem by answering the related question.

4. **Navigate to this chapter on the companion website and open CPE-15circum.dwg.** Use appropriate commands to find the answers to the following questions.
   A. What is the circumference of Circle 1?
   B. What is the total area of Circles 3, 4, 5, 6, 7, and 8?

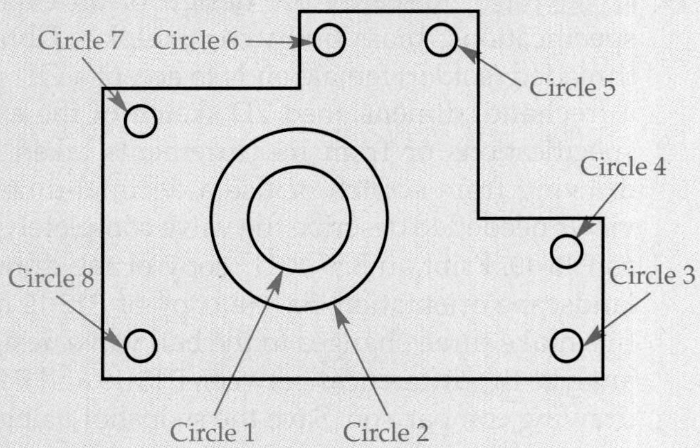

5. **Navigate to this chapter on the companion website and open CPE-15perimeter.dwg.** What is the perimeter of the gear?

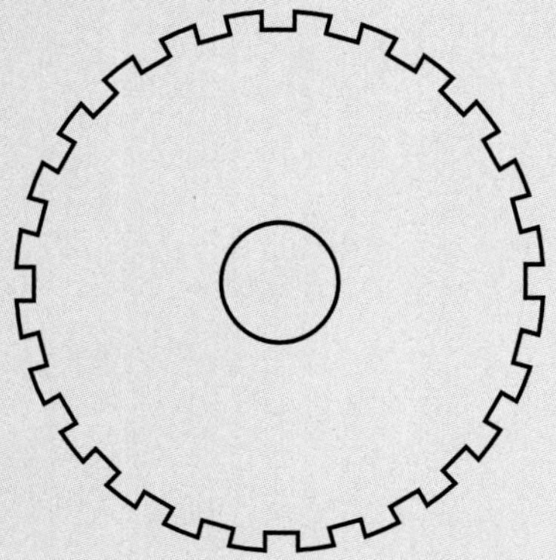

CHAPTER 16

Dimension Standards and Styles

Learning Objectives

After completing this chapter, you will be able to:

- ✓ Describe common dimension standards and practices.
- ✓ Set drawing scale using manual and annotative techniques.
- ✓ Create and manage dimension styles.

A ***dimension*** typically includes numerical values, lines, symbols, and notes. **Figure 16-1** shows typical dimension elements and dimensioning applications. Use dimension commands to dimension the size and location of features and objects. Dimensions you draw are created as ***associative dimensions*** by default. Dimension styles control the initial appearance of dimension elements. Dimensional constraint tools, explained in Chapter 22, allow you to use dimensions to control object size and location.

dimension: A description of the size, shape, or location of features on an object or structure.

associative dimension: A dimension in which all elements are linked to, or associated with, the dimensioned object; updates when the associated object changes.

Dimension Standards and Practices

Dimensions help communicate drawing information. Each drafting field uses different dimensioning practices. Dimensioning practices depend on product requirements, manufacturing or construction accuracy, standards, and tradition. It is important for you to draw dimensions according to industry and company standards. Dimension standards help to ensure that product manufacturing or construction is accurate. In addition, consistent dimension formatting is critical to legibility and drawing clarity. A drawing should use the same general dimension format throughout when possible.

This textbook presents mechanical drafting dimensioning standards according to the ASME Y14.5-2018 standard, *Dimensioning and Tolerancing*, published by the American Society of Mechanical Engineers (ASME). When appropriate, this textbook also references International Standards Organization (ISO) standards and discipline-specific standards, including the United States National CAD Standard® (NCS) and American Welding Society (AWS) standards.

Figure 16-1.
Dimensions describe the size and location of objects and features. Follow accepted conventions when dimensioning. The mechanical drafting views shown in this figure include unidirectional dimensions. All dimension values and notes read horizontally on the drawing.

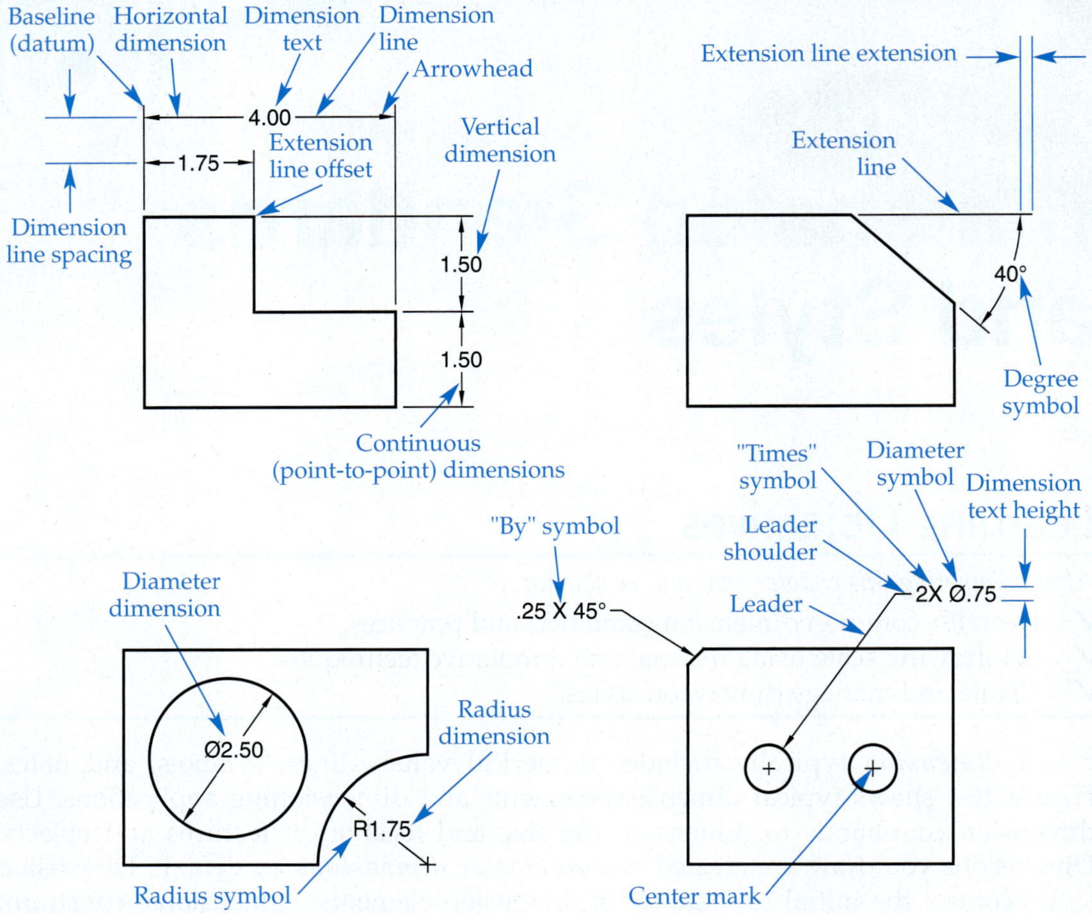

unidirectional dimensioning: A dimensioning system in which all dimension values are displayed horizontally on the drawing.

aligned dimensioning: A dimensioning system in which dimension values align with dimension lines.

size dimensions: Dimensions that provide the size of physical features.

feature: Any physical portion of a part or object, such as a surface, hole, window, or door.

location dimensions: Dimensions that locate features on an object without specifying the size of the feature.

rectangular coordinate system: A system for locating dimensions from surfaces, centerlines, or center planes using linear dimensions.

polar coordinate system: A coordinate system in which angular dimensions locate features from surfaces, centerlines, or center planes.

Unidirectional Dimensioning

Unidirectional dimensioning is common in mechanical drafting. The term *unidirectional* means "in one direction." Unidirectional dimensioning allows you to read all dimensions from the bottom of the sheet. Unidirectional dimensions often have arrowheads at the ends of dimension lines. The dimension value usually appears in a break near the center of the dimension line. Refer to **Figure 16-1**.

Aligned Dimensioning

Aligned dimensioning is most common on architectural and structural drawings. Dimension values read at the same angle as the dimension line. Horizontal values read horizontally, and vertical values are rotated 90° to read from the right side of the sheet. Notes usually read horizontally. Tick marks, dots, or arrowheads commonly terminate aligned dimension lines. In architectural drafting, you generally place the dimension number above the dimension line and use tick mark terminators. See **Figure 16-2**.

Size and Location Dimensions

Size dimensions provide the size of physical geometric ***features***. See **Figure 16-3**. ***Location dimensions*** locate features. See **Figure 16-4**. Dimension to the center of some circular features, such as holes and arcs, in the view in which they appear circular. Dimension to the edges of rectangular features. The ***rectangular coordinate system*** and the ***polar coordinate system*** are the two basic systems for applying location dimensions.

Figure 16-2.
An example of aligned dimensioning on an architectural floor plan. Notice the tick marks used instead of arrowheads and the placement of the dimensions above the dimension line.

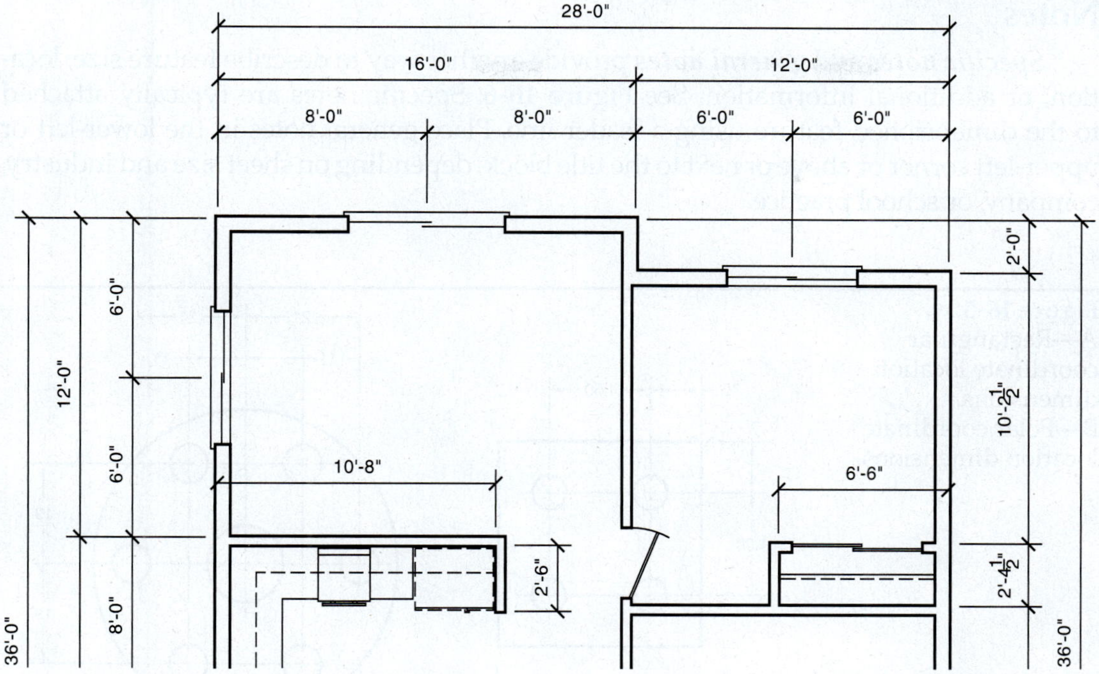

Figure 16-3.
Size dimensions describe the size of features, such as those in this part view.

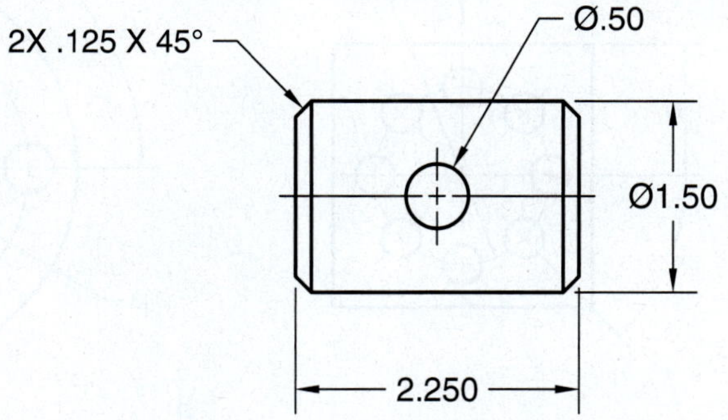

Figure 16-4.
Using location dimensions to locate circular and rectangular features on part drawings.

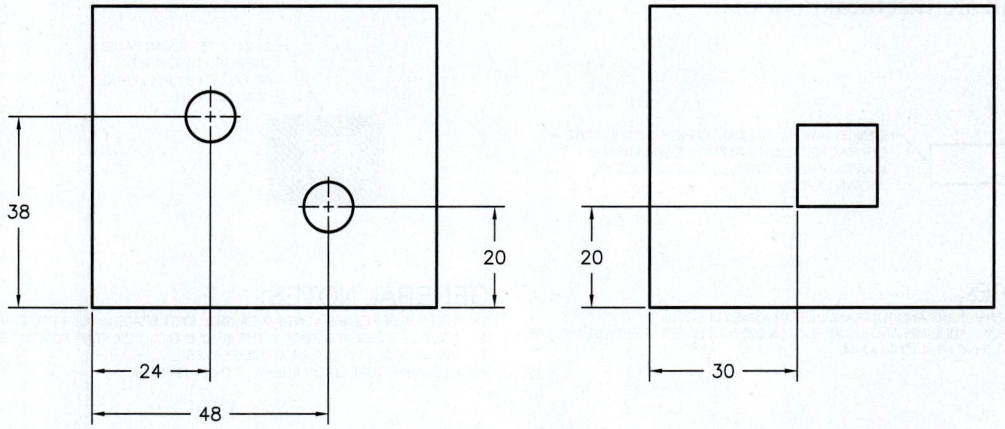

See Figure 16-5. An example of location dimensions used in architectural drafting is dimensioning to the center of windows and doors on a floor plan.

Notes

specific notes: Notes that relate to individual or specific features on the drawing.

general notes: Notes that apply to the entire drawing.

Specific notes and ***general notes*** provide another way to describe feature size, location, or additional information. See Figure 16-6. Specific notes are typically attached to the dimensioned feature using a leader line. Place general notes in the lower-left or upper-left corner or above or next to the title block, depending on sheet size and industry, company, or school practice.

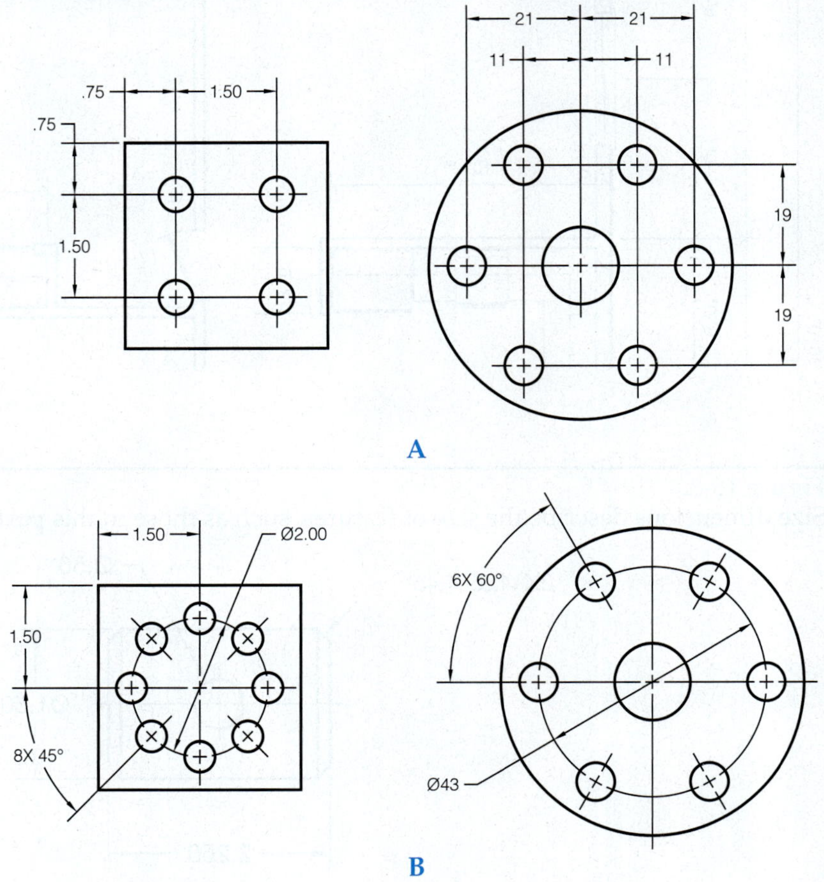

Figure 16-5.
A—Rectangular coordinate location dimensions.
B—Polar coordinate location dimensions.

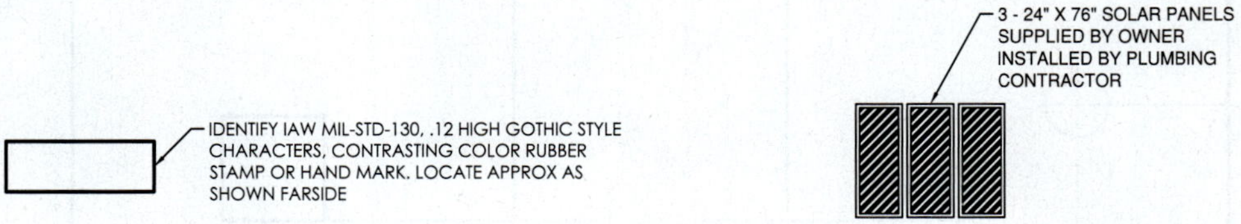

A B

NOTES:
1. DIMENSIONING AND TOLERANCING PER ASME Y14.5-2018.
2. REMOVE ALL BURRS AND SHARP EDGES R.020 MAX UNLESS OTHERWISE SPECIFIED.
3. HEAT TREAT H900 TO Rc 42-44.

C

GENERAL NOTES:
1. PROVIDE SCREENED VENTS @ EA. 3RD. JOIST SPACE @ ALL ATTIC EAVES.
2. PROVIDE SCREENED ROOF VENTS @ 10'-0" O.C. (1/300 VENT TO ATTIC SPACE).
3. USE 1/2" CCX PLY. @ ALL EXPOSED EAVES.
4. USE 300# COMPOSITION SHINGLES OVER 15# FELT.

D

Figure 16-6.
A—A specific note added to a mechanical part drawing. B—A specific note added to an architectural roof plan. C—General notes on a mechanical part drawing. D—General notes on an architectural floor plan.

Dimensioning Features and Objects

In mechanical drafting, you dimension flat surfaces using measurements for each feature. If you provide an overall dimension, you should omit one dimension, because the overall dimension controls the omitted value. See Figure 16-7A. In architectural drafting, it is common to place all dimensions without omitting any when possible to help make construction easier. See Figure 16-7B.

Dimensioning Cylindrical Shapes

You typically dimension the diameter and the length of a cylindrical shape in the view in which the cylinder appears rectangular. See Figure 16-8. The diameter symbol next to the dimension indicates that the feature is cylindrical. This allows you to omit the view in which the cylinder appears as a circle.

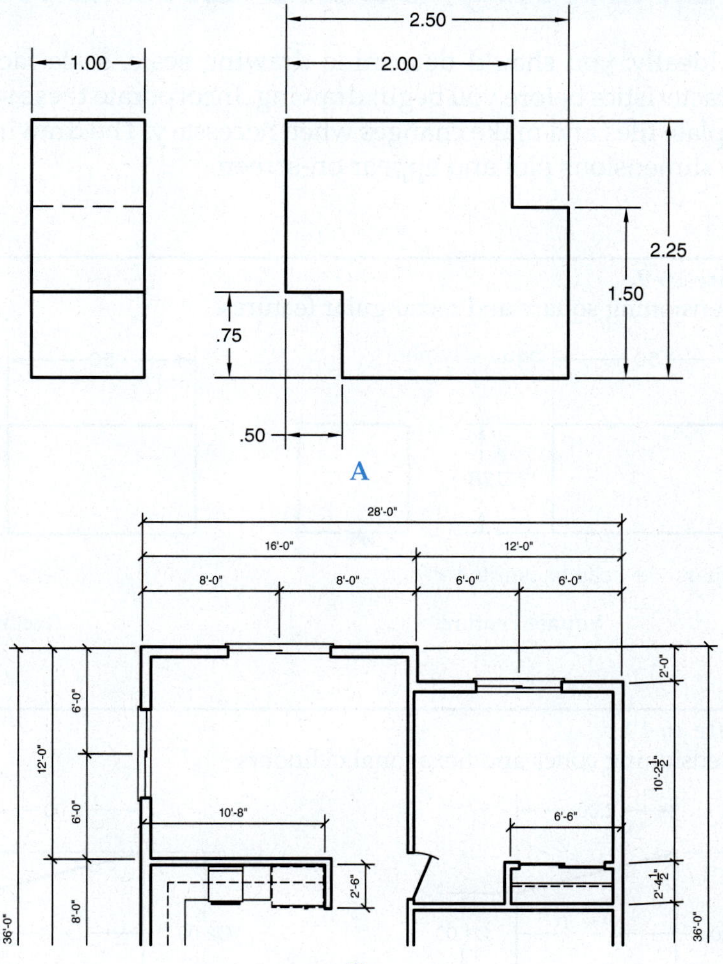

Figure 16-7. A—Dimensioning flat surfaces in mechanical drafting. B—Dimensioning architectural features.

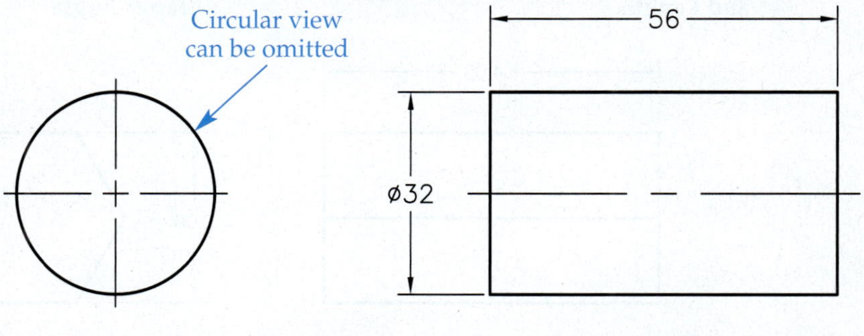

Figure 16-8. Dimensioning cylindrical shapes.

Dimensioning Square and Rectangular Features

You usually dimension square and rectangular features in the views that show the length and height. If appropriate, add a square symbol preceding the dimension for a square feature to eliminate the need for an additional view. See Figure 16-9.

Dimensioning Cones and Regular Polygons

One method to dimension a conical shape is to dimension the length and the diameters at both ends. An alternative is to dimension the taper angle and the length. Regular polygons that have an even number of sides are usually dimensioned by giving the distance across the flats and the length. Figure 16-10 shows examples of dimensioning cones and regular polygons.

Drawing Scale and Dimensions

Ideally, you should determine drawing scale, scale factors, and dimension size characteristics before you begin drawing. Incorporate these settings into your drawing template files and make changes when necessary. The drawing scale factor determines how dimensions plot and appear on-screen.

Figure 16-9.
Dimensioning square and rectangular features.

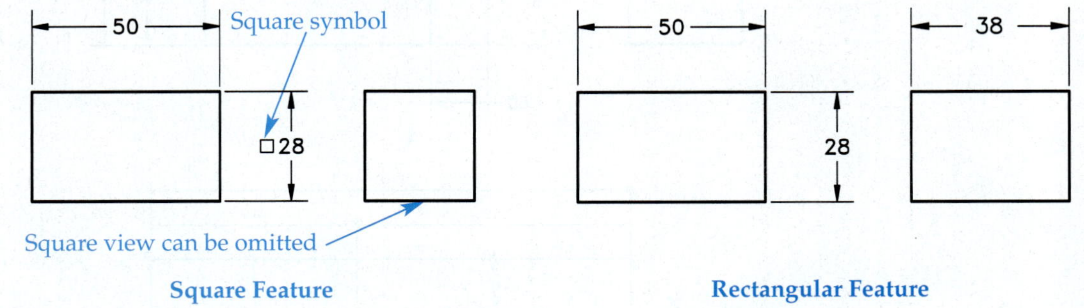

Figure 16-10.
Dimensioning cones and hexagonal cylinders.

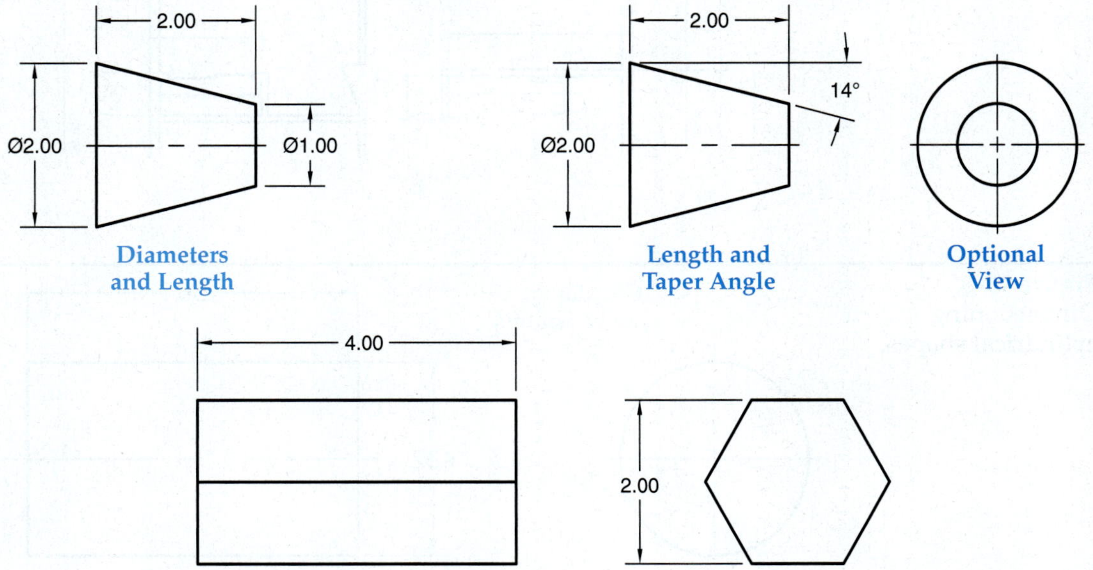

To help understand the concept of drawing scale, look at the portion of a floor plan shown in **Figure 16-11**. You should draw everything in model space at full scale. This means that the bathtub, for example, is actually drawn 5′ long. However, at this scale, dimension appearance becomes an issue, because full-scale dimension elements, such as 1/8″ dimension text, are too small compared to the other full-scale objects. See **Figure 16-11A**. As a result, you must adjust the size of dimensions according to the drawing scale. See **Figure 16-11B**. You can calculate the scale factor manually and apply it to dimensions, or you can allow AutoCAD to calculate the scale factor using annotative dimensions.

Scaling Dimensions Manually

To adjust the size of dimension elements manually according to a specific drawing scale, you must calculate the drawing scale factor. You then multiply the scale factor by the plotted size of dimension elements to get the model space size of dimension elements. Apply the scale factor to all dimension elements by entering the scale factor in the **Fit** tab of the **New** (or **Modify**) **Dimension Style** dialog box, described later in this chapter. For example, if you manually scale dimensions for a drawing with a 1/4″ = 1′-0″ scale, or a scale factor of 48, you must enter 48 in the **Fit** tab of the **New** (or **Modify**) **Dimension Style** dialog box. Refer to Chapter 9 for information on determining the drawing scale factor.

Annotative Dimensions

AutoCAD scales annotative dimensions according to the annotation scale you select, which reduces the need for you to calculate the scale factor. Once you select an annotation scale, AutoCAD applies the corresponding scale factor to annotative dimensions and all other annotative objects. When you place annotative dimensions using the previous example, you set an annotation scale of 1/4″ = 1′-0″. Then when you draw annotative dimensions, AutoCAD scales dimension elements automatically according to the preset 1/4″ = 1′-0″ annotation scale.

Annotative dimensions offer several advantages over manually scaled dimensions, including the ability to control dimension appearance based on the drawing scale and plotted size of dimension elements, while reducing the need to focus on the scale factor. Annotative dimensions are especially effective when the drawing scale changes or when a single sheet includes views at different scales.

Figure 16-11.
An example of a portion of a floor plan drawn at full scale in model space. A—Dimensions drawn at full scale (1/8″ high text) are too small compared to the large full-scale objects. B—Dimensions scaled (6″ high text) to display and plot correctly relative to the size of features on the drawing.

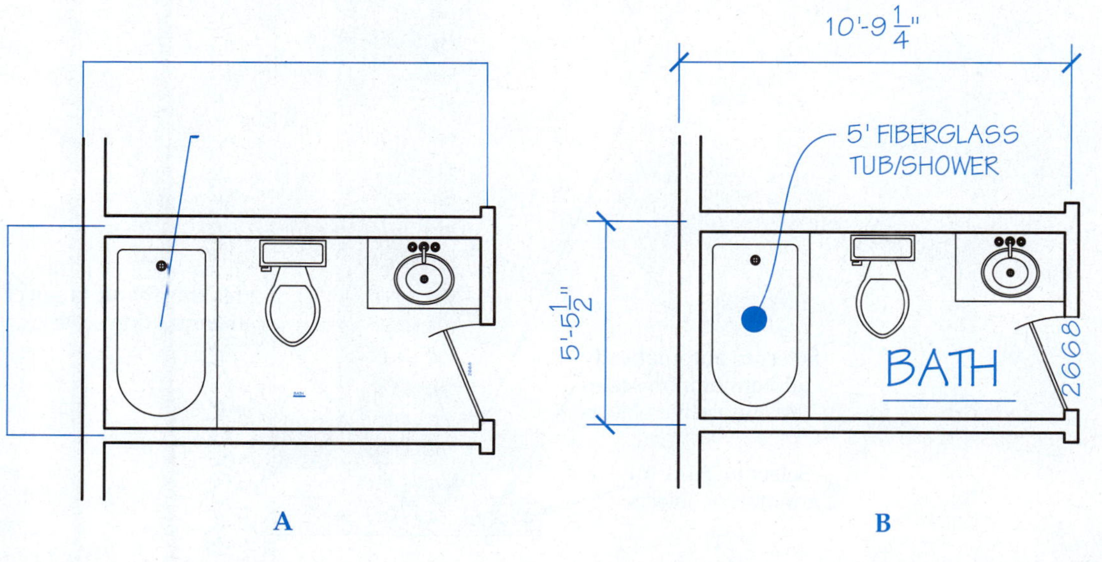

PROFESSIONAL TIP

If you anticipate preparing scaled drawings, you should use annotative dimensions and other annotative objects instead of manual scaling. However, scale factor does influence non-annotative items and is still an important value to identify and use throughout the drawing process.

Setting Annotation Scale

The annotation scale is typically the same as the drawing scale. You should usually set annotation scale before you begin adding dimensions so that dimension characteristics are scaled automatically. However, this is not always possible. It may be necessary to adjust the annotation scale throughout the drawing process, especially if you prepare multiple drawings with different scales on one sheet. This textbook approaches annotation scaling in model space, using the process of selecting the appropriate annotation scale before placing dimensions. To draw dimensions using a different scale, pick the new annotation scale and then draw the dimensions.

The **Select Annotation Scale** dialog box appears when you access a dimension command and an annotative dimension style is current. This dialog box is a convenient way to set annotation scale before adding dimensions. You will learn about dimension styles later in this chapter. You can also select the annotation scale from the **Annotation Scale** flyout on the status bar. See Figure 16-12. See Chapter 9 for information on creating and editing annotation scales.

NOTE

This textbook describes many additional annotative object tools. Some of these tools are more appropriate for working with layouts, as explained later in this textbook.

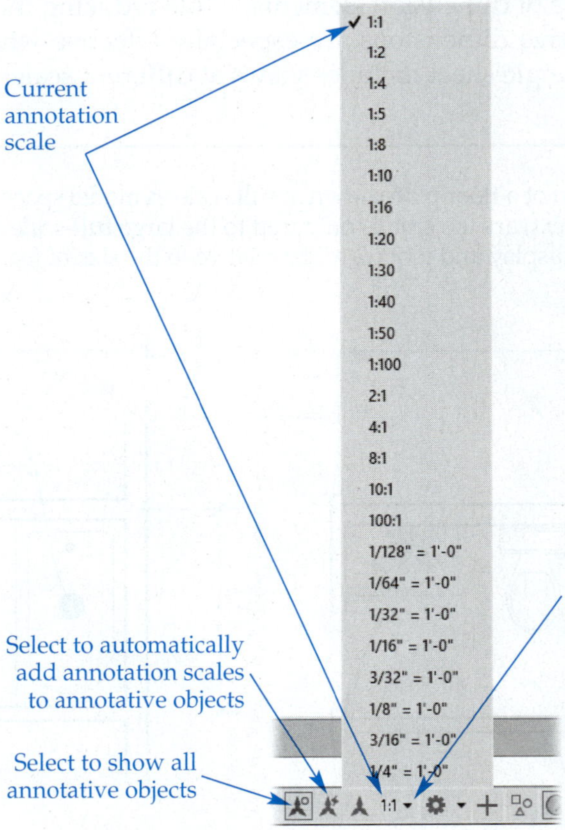

Figure 16-12.
Pick the **Annotation Scale** flyout on the status bar to activate an annotation scale.

Dimension Styles

A ***dimension style*** presets many dimension characteristics. Dimension style settings usually apply to a specific drafting field or dimensioning application and correspond to appropriate drafting standards. For example, a mechanical drafting dimension style may use unidirectional placement; reference a text style assigned the Arial, Romans, or Century Gothic font; center text in a break in the dimension line; and terminate dimension lines with arrowheads. See Figure 16-1. An architectural drafting dimension style may use an aligned dimension format; reference a text style assigned the SansSerif or Stylus BT font; place text above the dimension line; and terminate dimension lines with tick marks. See Figure 16-2.

dimension style: A saved configuration of dimension appearance settings.

Some drawings require a single dimension style. However, you may need multiple dimension styles depending on the variety of dimensions you apply and different dimension characteristics. Add dimension styles to drawing templates for repeated use.

Create a dimension style for each frequently used dimension appearance or function. For example, create a dimension style for unspecified tolerances and a different dimension style for a common specified tolerance. Another example is developing a separate dimension style to add dimensions with a common prefix or suffix.

PROFESSIONAL TIP

AutoCAD provides the flexibility to control the appearance of dimensions for various dimensioning requirements without using a separate dimension style. For example, you can change the precision of or add a prefix to a limited number of special-case dimensions, instead of creating separate dimension styles.

Dimension Style Manager

Create, modify, and delete dimension styles using the **Dimension Style Manager** dialog box. See Figure 16-13. The **Styles** list box displays existing dimension styles.

Ribbon
Home
> Annotation
Dimension Style
Annotate
> Dimensions
Dimension Style
Annotate
> Dimensions
> Dimension Style
Manage Dimension Styles...

Type
DIMSTYLE
DIMSTY
DDIM
DST
D

DIMSTYLE

Figure 16-13.
The **Dimension Style Manager** dialog box.

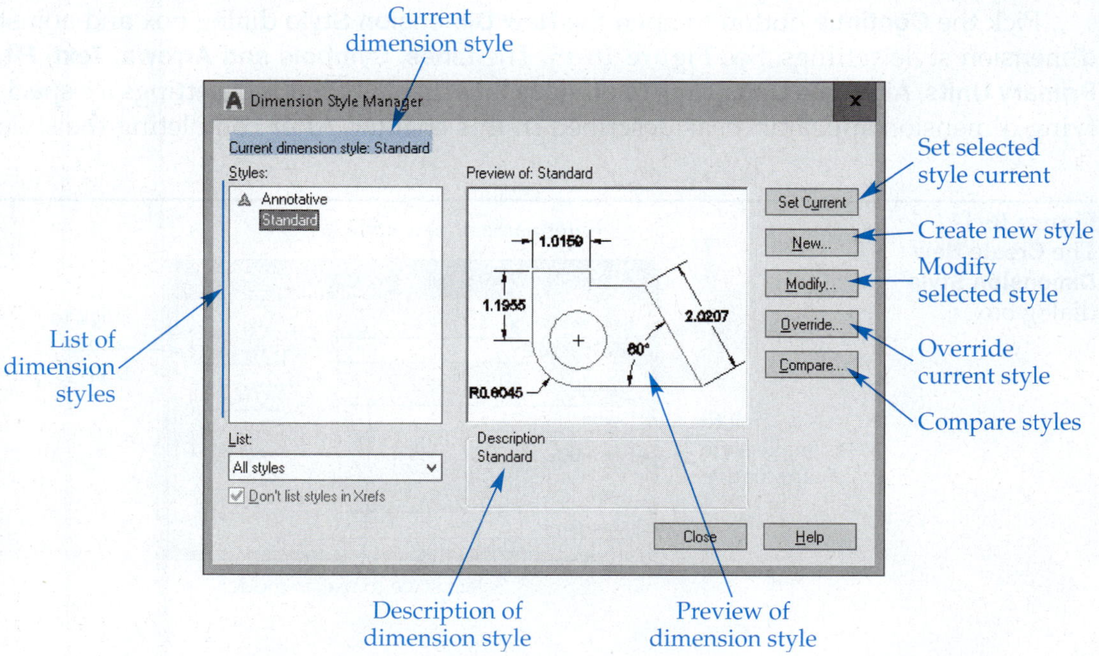

The Annotative dimension style allows you to create annotative dimensions, as indicated by the icon to the left of the style name. The Standard dimension style does not use the annotative function.

To make a dimension style current, double-click the style name, right-click on the name and select **Set current**, or pick the name and select the **Set Current** button. Use the **List** drop-down list to filter the number of dimension styles displayed in the **Styles** list box. Pick the **All styles** option to show all dimension styles in the file, or pick the **Styles in use** option to show only the current style and styles used in the drawing. If the current drawing contains external references (xrefs), use the **Don't list styles in Xrefs** box to eliminate xref-dependent dimension styles from the **Styles** list box. This setting is valuable because you cannot set xref dimension styles current or use them to create new dimensions. You will learn about external references later in this textbook.

The **Description** area provides information about the selected dimension style. The **Preview of** image displays a representation of the dimension style and changes according to the selections you make. If you change dimension settings without creating a new dimension style, the changes are automatically stored as a dimension style override.

Creating New Dimension Styles

To create a new dimension style, select an existing dimension style from the **Styles** list box to use as a base for formatting the new dimension style. Then pick the **New...** button to open the **Create New Dimension Style** dialog box. See **Figure 16-14**. You can base the new dimension style on the formatting of a different dimension style by selecting from the **Start With** drop-down list. Notice that Copy of followed by the name of the existing style appears in the **New Style Name** text box. Replace the default name with a more descriptive name, such as Mechanical or Architectural. Dimension style names can have up to 255 characters, including uppercase or lowercase letters, numbers, dashes (–), underlines (_), and dollar signs ($).

Pick the **Annotative** check box to make the dimension style annotative. You can also make the dimension style annotative by selecting the **Annotative** check box in the **Fit** tab of the **New** (or **Modify**) **Dimension Style** dialog box, described later in this chapter. The **Use for** drop-down list specifies the type of dimensions to which the new style applies. Use the **All dimensions** option to create a new dimension style for all types of dimensions. If you select the **Linear dimensions**, **Angular dimensions**, **Radius dimensions**, **Diameter dimensions**, **Ordinate dimensions**, or **Leaders and Tolerances** option, you create a sub-style of the dimension style specified in the **Start With** text box.

Pick the **Continue** button to open the **New Dimension Style** dialog box and adjust dimension style settings. See **Figure 16-15**. The **Lines**, **Symbols and Arrows**, **Text**, **Fit**, **Primary Units**, **Alternate Units**, and **Tolerances** tabs display groups of settings for specifying dimension appearance, as described in this chapter. After completing the style

Figure 16-14.
The **Create New Dimension Style** dialog box.

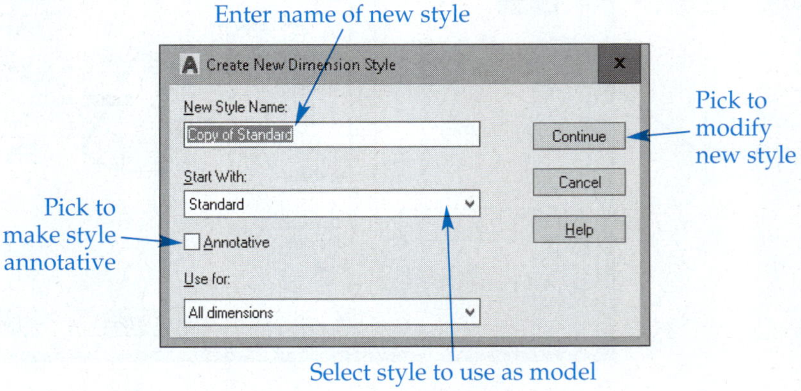

Figure 16-15.
The **Lines** tab of the **New** (or **Modify**) **Dimension Style** dialog box.

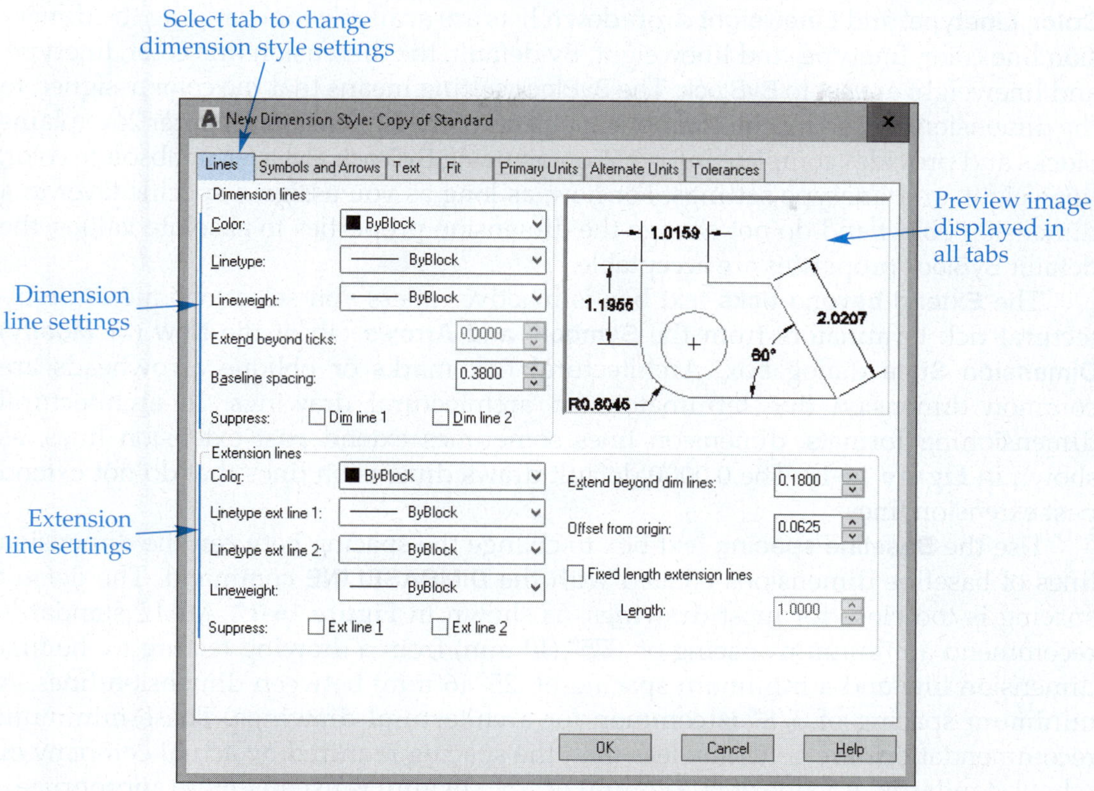

definition, pick the **OK** button to return to the **Dimension Style Manager** dialog box. Pick the **Close** button to exit the **Dimension Style Manager** dialog box.

NOTE

The preview image shown in the upper-right corner of each tab of the **New** (or **Modify**) **Dimension Style** dialog box displays a representation of the dimension style and changes according to the selections you make.

CAUTION

AutoCAD stores dimension style settings as ***dimension variables***. Dimension variables have limited practical uses and are more likely to apply to advanced applications such as scripting and customizing. Changing dimension variables by typing the variable name is not a recommended method for changing dimension style settings. Changes made in this manner can introduce inconsistencies with other dimensions. You should make changes to dimensions by redefining styles or performing style overrides.

dimension variables: System variables that store the values of dimension style settings.

Lines Tab

Figure 16-15 shows the **Lines** tab of the **New** (or **Modify**) **Dimension Style** dialog box. The **Lines** tab controls the display of dimension and extension lines. A dimension style presets the appearance of dimension and extension lines for common applications. You can edit specific dimensions when necessary without using a separate dimension style.

Dimension Line Settings

The **Dimension lines** area of the **Lines** tab allows you to set dimension line format. **Color**, **Linetype**, and **Lineweight** drop-down lists are available for changing the dimension line color, linetype, and lineweight. By default, the dimension line color, linetype, and lineweight are set to ByBlock. The ByBlock setting means that the color assigned to the dimension is used for the component objects of the dimension. Chapter 24 explains blocks and provides complete information on using ByBlock, ByLayer, or absolute color, lineweight, and linetype settings. For now, as long as you assign a specific layer to a dimension object and do not change the dimension properties to absolute values, the default ByBlock properties are acceptable.

The **Extend beyond ticks** text box is inactive unless you select oblique or architectural tick terminators from the **Symbols and Arrows** tab of the **New** (or **Modify**) **Dimension Style** dialog box. Architectural tick marks or oblique arrowheads are common dimension line terminators on architectural drawings. In architectural dimensioning formats, dimension lines sometimes extend past extension lines, as shown in **Figure 16-16**. The 0.0000 default draws dimension lines that do not extend past extension lines.

Use the **Baseline spacing** text box to change the spacing between the dimension lines of baseline dimensions created with the **DIMBASELINE** command. The default spacing is too close for most drawings, as shown in **Figure 16-17**. ASME standards recommend a minimum spacing of .375″ (10 mm) from a drawing feature to the first dimension line and a minimum spacing of .25″ (6 mm) between dimension lines. A minimum spacing of 3/8″ is common for architectural drawings. These minimum recommendations are generally less than the spacing required by actual company or school standards. A value of .5″ (12 mm) or .75″ (19 mm) is usually more appropriate.

Figure 16-16. Using the **Extend beyond ticks** setting to allow the dimension line to extend past the extension line. With the default value of 0, the dimension line ends at the extension line.

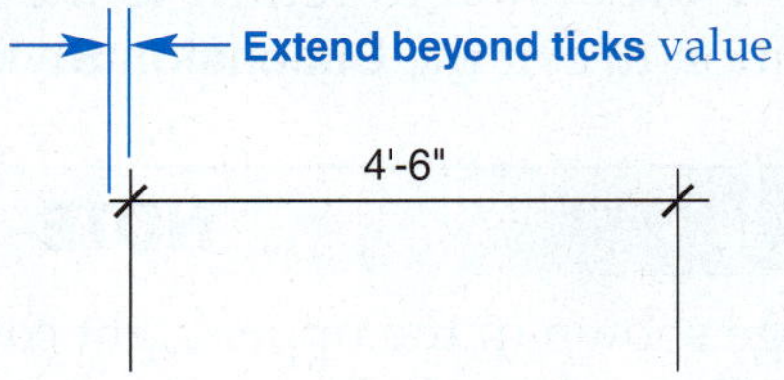

Figure 16-17. The **Baseline spacing** setting controls the spacing between dimension lines when you use the **DIMBASELINE** command.

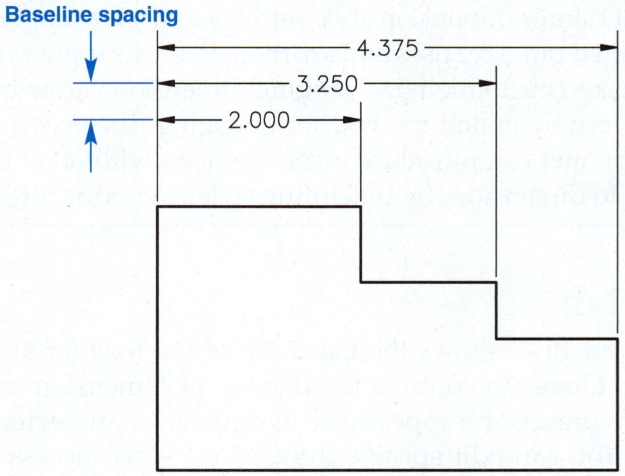

Check the **Suppress** boxes to hide the first, second, or both sides of dimension lines and dimension line terminators for dimension lines broken by a value. The **Dim line 1** and **Dim line 2** check boxes refer to the first and second points you pick when drawing a dimension. Both dimension lines appear by default. Figure 16-18 shows the result of using dimension line suppression options.

Extension Line Settings

The **Extension lines** area of the **Lines** tab allows you to set extension line format. **Color**, **Linetype ext line 1**, **Linetype ext line 2**, and **Lineweight** drop-down lists are available for changing the extension line color, linetype, and lineweight from the default ByBlock setting, if necessary. You can use the **Linetype ext line 1** and **Linetype ext line 2** drop-down lists to override the linetype applied to each extension line. Extension lines 1 and 2 correspond to the first and second points you pick when drawing a dimension.

Use the **Extend beyond dim lines** option to set the distance the extension line runs past the dimension line. See Figure 16-19. A .125″ (3 mm) extension line extension is standard practice. The **Offset from origin** option specifies the distance between the object and the beginning of the extension line. Most applications require this small offset. A .063″ (1.5 mm) extension line offset is standard practice. When an extension line meets a centerline, however, use a setting of 0.0 to prevent a gap.

The **Fixed length extension lines** check box sets a given length for extension lines. Check this box to activate the **Length** text box. The value in the **Length** text box sets a restricted length for extension lines, measured from the dimension line toward the extension line origin.

Figure 16-18.
Using the **Dim line 1** and **Dim line 2** dimensioning settings. "Off" is equivalent to an unchecked **Suppress** check box in the **Lines** tab.

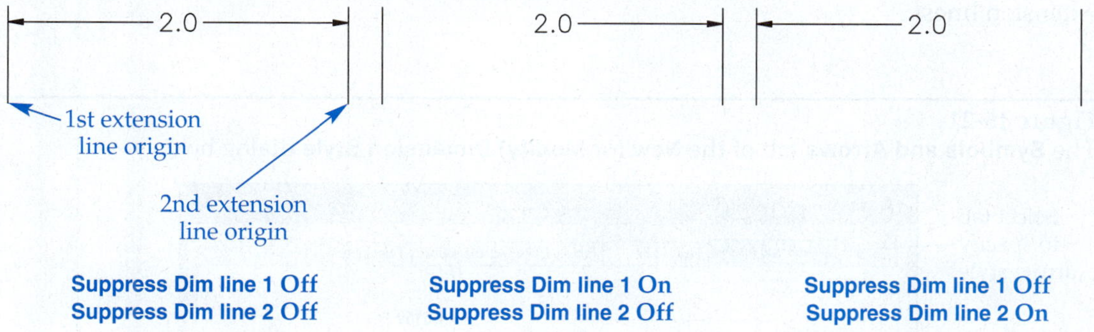

Figure 16-19.
The extension line extension and extension line offset settings.

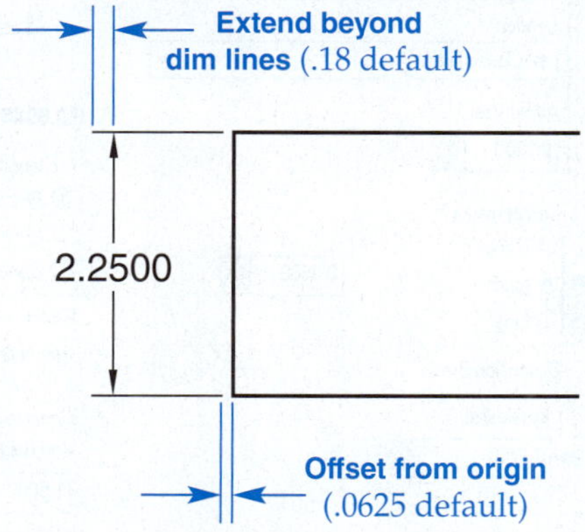

Check the **Suppress** boxes to hide the first, second, or both extension lines. The **Ext line 1** and **Ext line 2** check boxes refer to the first and second points you pick when drawing a dimension. Both extension lines appear by default. Figure 16-20 shows an example of suppressing extension lines when they coincide with object lines.

Symbols and Arrows Tab

Figure 16-21 shows the **Symbols and Arrows** tab of the **New** (or **Modify**) **Dimension Style** dialog box. The **Symbols and Arrows** tab controls the appearance of dimension line and leader terminators, center marks, and other symbol components of dimensions. A dimension style presets the appearance of symbols and arrows for common applications. You can edit specific dimensions when necessary without using a separate dimension style.

Arrowhead Settings

Use the appropriate drop-down list in the **Arrowheads** area to select the arrowhead to use for the first, second, and leader arrowheads. The default arrowhead is **Closed filled**, which is recommended by ASME standards, although **Closed blank**, **Closed**, or **Open** arrowheads are sometimes used. A leader pointing to a surface in a view where the surface appears as a plane terminates with a small dot. Figure 16-22 shows arrowhead styles. If you pick a new arrowhead in the **First** drop-down list, AutoCAD automatically makes the same selection for the **Second** drop-down list.

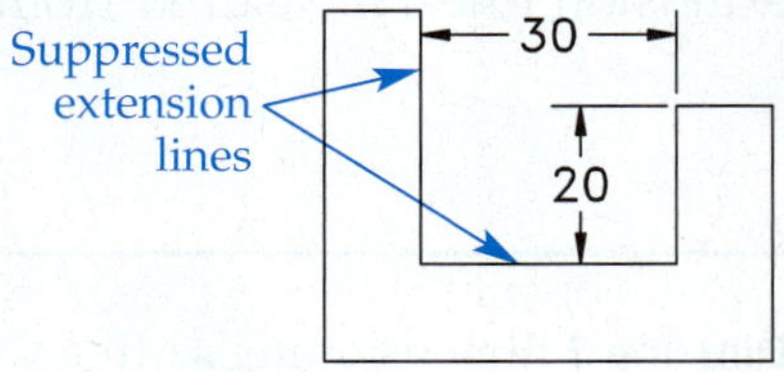

Figure 16-20. Examples of when it is appropriate to suppress extension lines. Place dimension lines away from objects, when possible, to display both extension lines.

Figure 16-21. The **Symbols and Arrows** tab of the **New** (or **Modify**) **Dimension Style** dialog box.

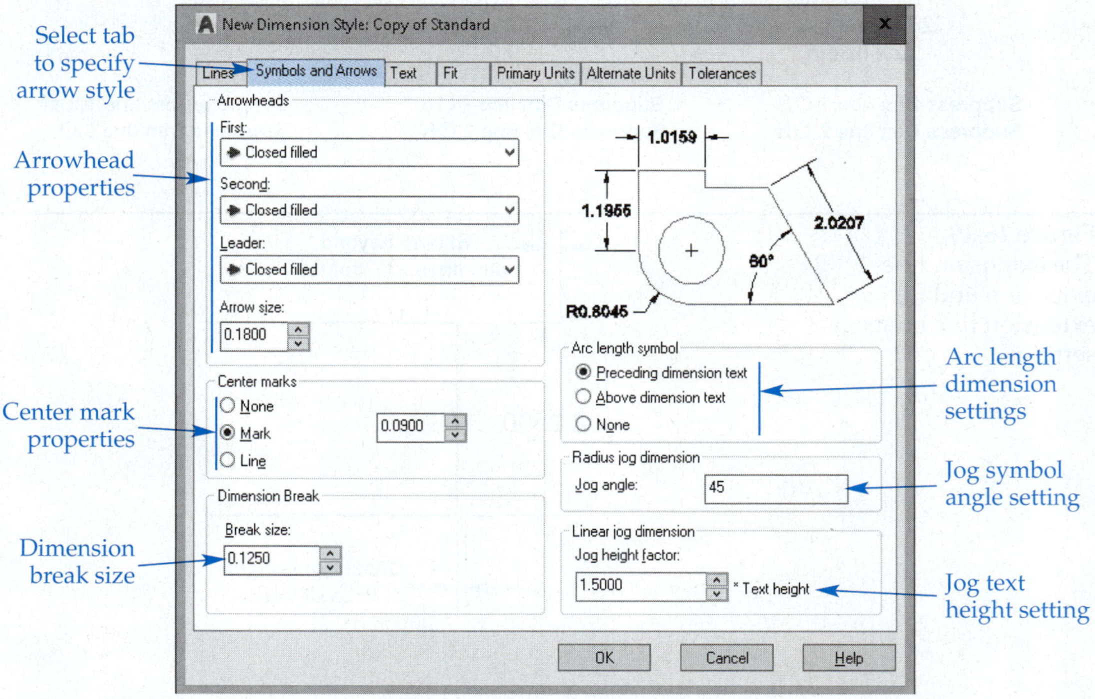

When you select the **Oblique** or **Architectural tick** arrowhead, the **Extend beyond ticks** text box in the **Lines** tab becomes activated.

Notice that Figure 16-22 does not contain an example of a user arrow. The **User Arrow...** option allows you to access a custom arrowhead. You must first design an arrowhead that fits inside a 1-unit square (unit block) with a dimension line "tail" 1 unit in length, and save the arrowhead as a block. Blocks are described later in this textbook. The **Select Custom Arrow Block** dialog box appears when you pick **User Arrow...** from an **Arrowheads** drop-down list. Pick a block from the **Select from Drawing Blocks:** drop-down list and then pick **OK** to apply the arrowhead to the style.

Use the **Arrow size** text box to change arrowhead size. Figure 16-23 shows the measurement you specify to set the arrowhead size. A .125″ (3 mm) arrowhead size is most common, especially on mechanical drawings.

Center Mark Settings

Centerlines are used to show and locate the centers of circular features. Centerlines typically extend past objects, as shown in the centerlines example of Figure 16-24. ASME standards refer to a two-segment cross that does not extend past objects as the center dashes of centerlines. See the center dashes example of Figure 16-24. Center dashes typically apply to circular objects that are too small to receive centerlines. Center dashes are also common for rectangular coordinate dimensioning without dimension lines, as described in Chapter 18, regardless of circular object size.

Figure 16-22.
Examples of dimensions drawn using the options in the **Arrowheads** drop-down lists.

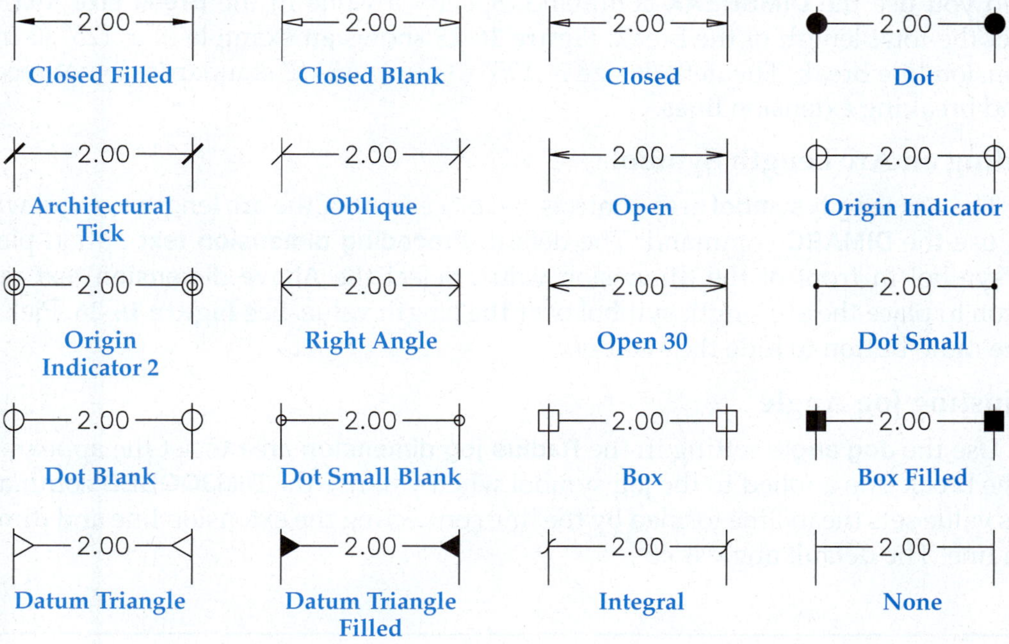

Figure 16-23.
An arrowhead size of .125″ (3 mm) is common practice. The **Closed filled**, **Closed blank**, and **Closed** arrowhead styles adhere to the ASME standard 3:1 ratio of length to width.

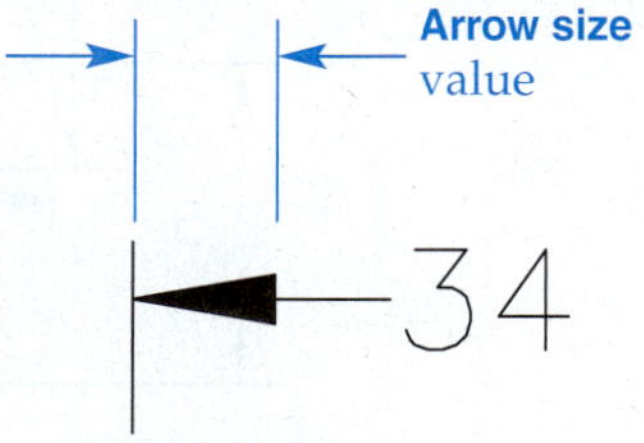

Figure 16-24.
Arcs and circles displayed with centerlines and center dashes. Common sizes are .125″ (3 mm) for the centerline extension past objects and .0625″ (1.5 mm) for the centerline dash half-length.

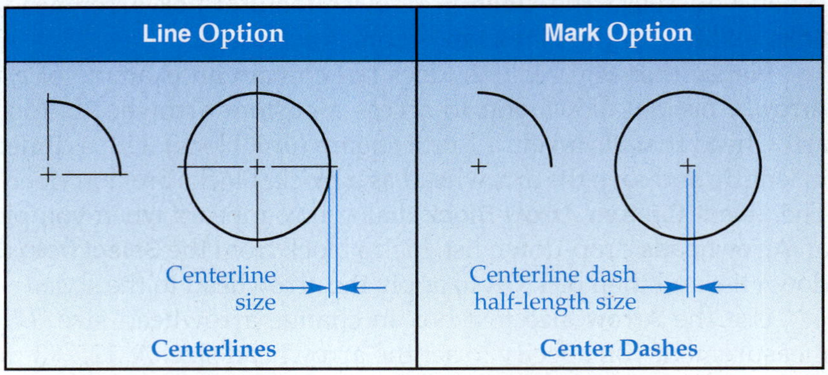

The **Center marks** area controls the appearance of center dashes and centerlines drawn when using circular feature dimensioning commands. The **None** option results in no center dashes or centerlines. Fillets and rounds generally have no center dashes. The **Mark** option places center dashes. The **Line** option places centerlines.

The size you specify in the text box in the **Center marks** area defines half the length of a centerline dash and the distance that the centerline extends past the object. A value of .0625″ (1.5 mm) is appropriate for the centerline dash half-length, but does not provide for the preferred .125″ (3 mm) extension past the object.

Adjusting Break Size

The **Dimension Break** area controls the amount of extension line that is hidden when you use the **DIMBREAK** command. Specify a value in the **Break size** text box to set the total length of the break. Figure 16-25 shows an example of a .125″ (3 mm) extension line break. The default size is .125″ (3 mm). ASME standards do not recommend breaking extension lines.

Adding an Arc Length Symbol

The **Arc length symbol** area controls the placement of the arc length symbol when you use the **DIMARC** command. The default **Preceding dimension text** option places the symbol in front of the dimension value. Select the **Above dimension text** radio button to place the arc length symbol over the length value. See Figure 16-26. Pick the **None** radio button to hide the symbol.

Adjusting Jog Angle

Use the **Jog angle** setting in the **Radius jog dimension** area to set the appearance of the break line applied to the jog symbol when you use the **DIMJOGGED** command. This value sets the incline formed by the line connecting the extension line and dimension line. The default angle is 45°.

Figure 16-25.
Use the **Break size** setting to specify the length of the break created using the **DIMBREAK** command.

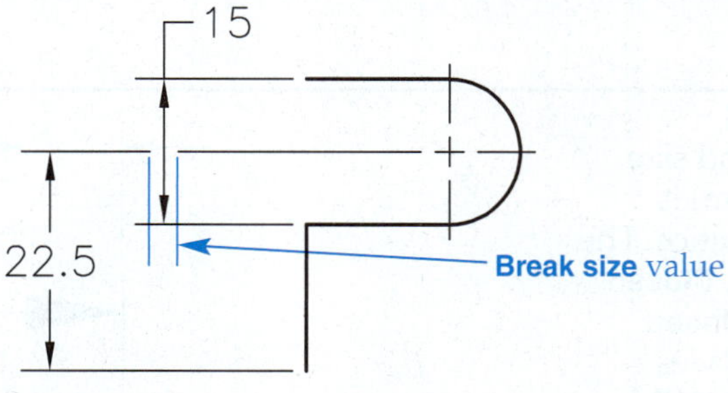

Figure 16-26.
You can place the arc length symbol in front of or above the arc dimension text.

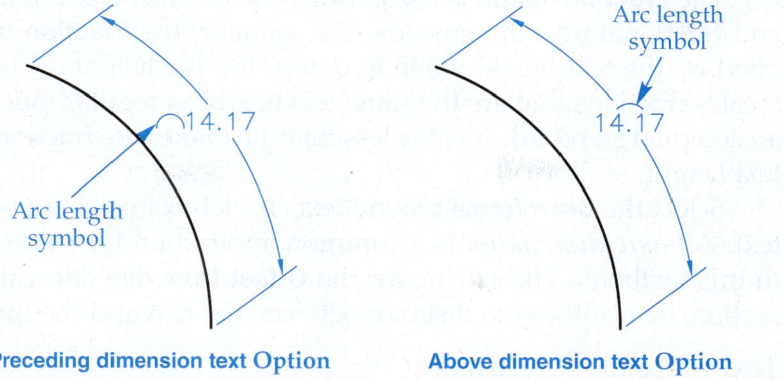

Setting Jog Height

The **Jog height factor** setting in the **Linear jog dimension** area controls the size of the break symbol created using the **DIMJOGLINE** command. This value sets the height of the break symbol based on a multiple of the text height. For example, the default value of 1.5 creates a break symbol that is .18″ tall if the text height is .12″.

NOTE

Chapter 17 describes the **DIMJOGLINE** command in more detail. Chapter 18 covers the **DIMJOGGED** and **DIMARC** commands, and Chapter 20 covers the **DIMBREAK** command.

Exercise 16-1

www.g-wlearning.com/CAD/

Complete the exercise on the companion website.

Text Tab

Figure 16-27 shows the **Text** tab of the **New** (or **Modify**) **Dimension Style** dialog box. Use the **Text** tab to control the display of dimension text. A dimension style presets the appearance of dimension text for common applications. You can edit specific dimensions when necessary without using a separate dimension style.

Text Appearance Settings

Use the **Text appearance** area to set the dimension text style, color, height, and frame. A dimension style references a text style for the appearance of dimension values. Pick an existing text style from the **Text style** drop-down list. To create or modify a text style, pick the ellipsis (**...**) button next to the drop-down list to launch the **Text Style** dialog box. Use the **Text color** drop-down list to specify the appropriate text color, which should be ByBlock for typical applications. The **Fill color** drop-down list is used to set the dimension background color. By default, no background color is used.

Use the **Text height** text box to specify the dimension text height. Dimension text height is commonly the same as the text height used for most other drawing text, except for titles, which often display a taller text height. The default dimension text height of .18″ (4.5 mm) is an acceptable standard. Many companies use a text height of .125″ (3 mm). The ASME standard recommends a .12″ (3 mm) text height. The ASME standard text height for titles and labels is .24″ (6 mm).

The **Fraction height scale** setting controls the height of fractions for architectural- and fractional-unit dimensions. The value in the **Fraction height scale** box is multiplied by the text height value to determine the height of the fraction. A value of 1.0 creates fractions that are the same text height as regular (non-fractional) text, which is an accepted standard. A value less than 1.0 makes the fraction smaller than the regular text height.

Select the **Draw frame around text** check box to create a box around the dimension text. A ***basic dimension*** is a common application for framed text, as described later in this textbook. The setting for the **Offset from dim line** value, explained in the next section, determines the distance between the text and the frame.

basic dimension: A theoretically perfect dimension used to describe the exact size, profile, orientation, and location of a feature.

Text Placement Settings

The **Text placement** area controls text placement relative to the dimension line. See Figure 16-28. The **Vertical** drop-down list provides vertical justification options. Use the default **Centered** option to place dimension text centered in a gap in the dimension line. This is the most common dimensioning practice in mechanical drafting and many other fields.

Select the **Above** option to place the dimension text horizontally above horizontal dimension lines. For vertical and angled dimension lines, the text appears in a gap in the dimension line. This option is common for architectural drafting and building construction. Architectural drafting typically uses aligned dimensioning, in which the dimension text aligns with the dimension lines, and all text reads from either the bottom or the right side of the sheet.

Pick the **Outside** option to place the dimension text outside the dimension line and either above or below a horizontal dimension line or to the right or left of a vertical dimension line. The direction in which you move the cursor determines the above/below and left/right placement. Select the **JIS** option to align the text according to the Japanese Industrial Standards (JIS). Select the **Below** option to place the dimension text horizontally below a horizontal dimension line.

Figure 16-27.
The **Text** tab of the **New** (or **Modify**) **Dimension Style** dialog box.

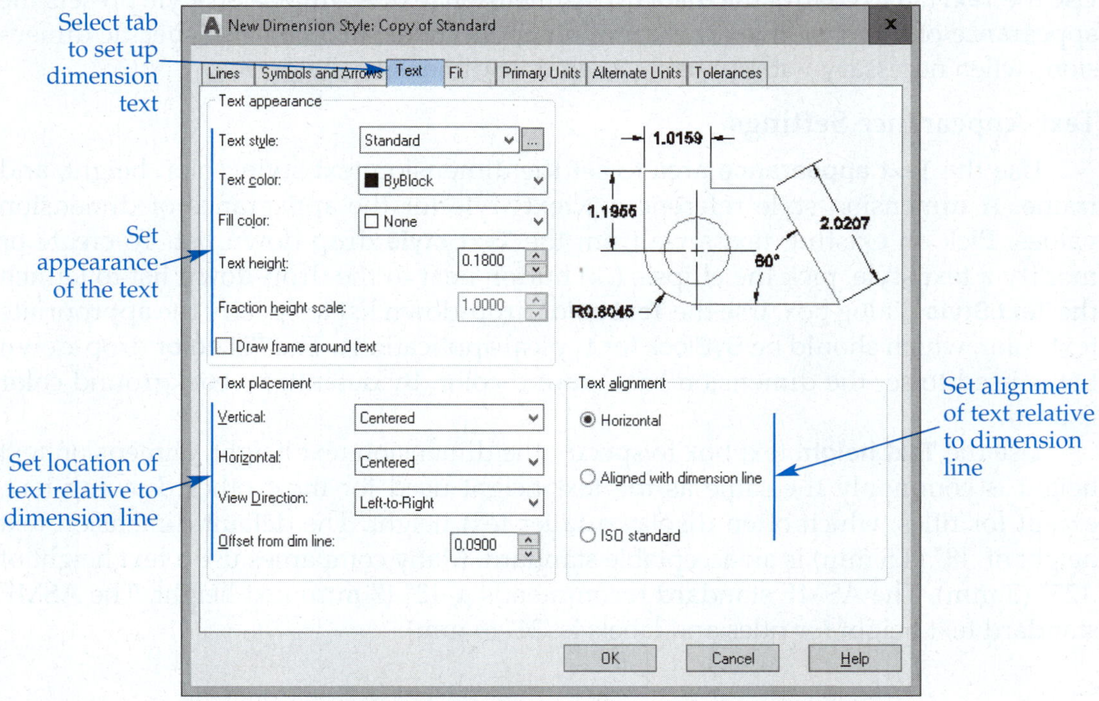

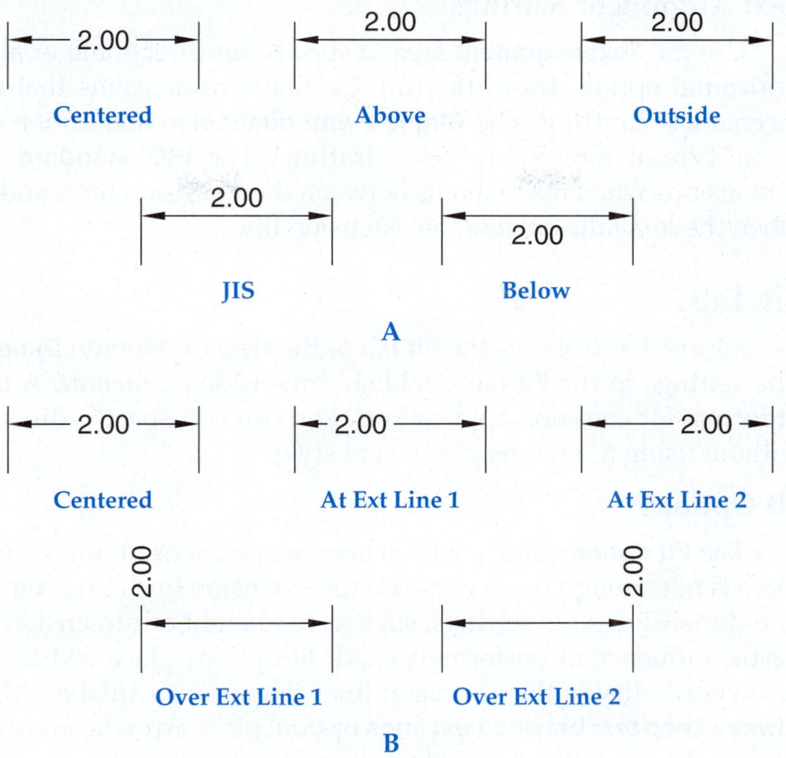

Figure 16-28. Dimension text justification options. A—Vertical justification options, with the horizontal **Centered** justification. B—Horizontal justification options, with the vertical **Centered** justification.

The **Horizontal** drop-down list provides options for controlling the horizontal placement of dimension text. Pick the default **Centered** option to center dimension text between the extension lines. Select the **At Ext Line 1** option to locate the text next to the extension line you place first, or pick **At Ext Line 2** to locate the text next to the extension line you place second. Select **Over Ext Line 1** to place the text aligned with and over the first extension line, or pick **Over Ext Line 2** to place the text aligned with and over the second extension line. Placing text aligned with and over an extension line is not common practice.

The **View Direction** drop-down list determines the reading direction of dimension text. Use the default **Left-to-Right** option to make text readable from left to right or from bottom to top, depending on the text placement and alignment. Select the **Right-to-Left** option to flip dimension text. Text may appear inverted and reads from right to left or from top to bottom, depending on the text placement and alignment. Changing text view direction to right-to-left is not common practice.

The **Offset from dim line** text box sets the gap between the dimension line and dimension text, the distance between the leader shoulder and text, and the space between text and a frame. The gap should be half the text height for most applications. Figure 16-29 shows the gap in linear dimensions and dimensions that use a leader or frame.

Figure 16-29. The **Offset from dim line** value controls the space between dimension text and the dimension line, leader shoulder, and frame.

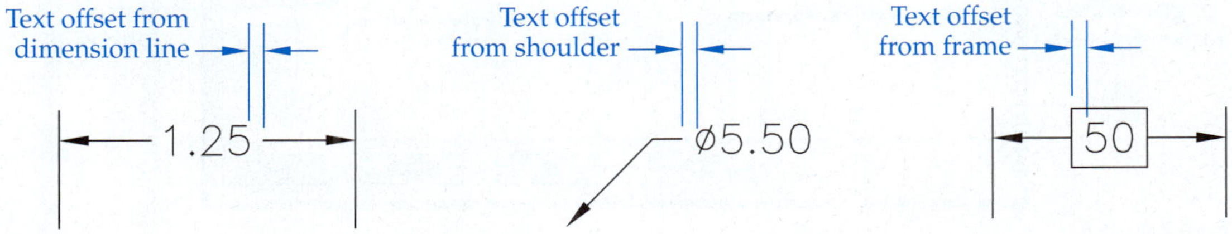

Text Alignment Settings

Use the **Text alignment** area to specify unidirectional or aligned dimensions. The **Horizontal** option draws the unidirectional dimensions that are commonly used in mechanical drafting. The **Aligned with dimension line** option creates aligned dimensions, typical for architectural drafting. The **ISO standard** option creates aligned dimensions when the text falls between the extension lines and horizontal dimensions when the text falls outside the extension lines.

Fit Tab

fit format: The arrangement of dimension text and arrowheads on a drawing.

Figure 16-30 shows the **Fit** tab of the **New** (or **Modify**) **Dimension Style** dialog box. The settings in the **Fit** tab establish dimension *fit format*. A dimension style presets fit format for common applications. You can edit specific dimensions when necessary without using a separate dimension style.

Fit Options

The **Fit options** area controls how text, dimension lines, and arrows behave when there is not enough room between the extension lines to accommodate all of the items. The dimension style settings, such as the height of dimension text, offset, and arrowheads, influence fit performance. All fit options place text and dimension lines with arrowheads inside the extension lines if space is available. All fit options, except the **Always keep text between ext lines** option, place arrowheads, dimension lines, and text outside the extension lines when space is limited.

Select the default **Either text or arrows (best fit)** radio button to move either the dimension value or the arrows outside extension lines first. Pick the **Arrows** radio button to attempt to place arrowheads outside extension lines first, followed by text. Pick the **Text** radio button to attempt to place text outside extension lines first, followed by arrowheads. Select the **Both text and arrows** radio button to move both text and arrowheads outside extension lines.

Figure 16-30.
The **Fit** tab of the **New** (or **Modify**) **Dimension Style** dialog box.

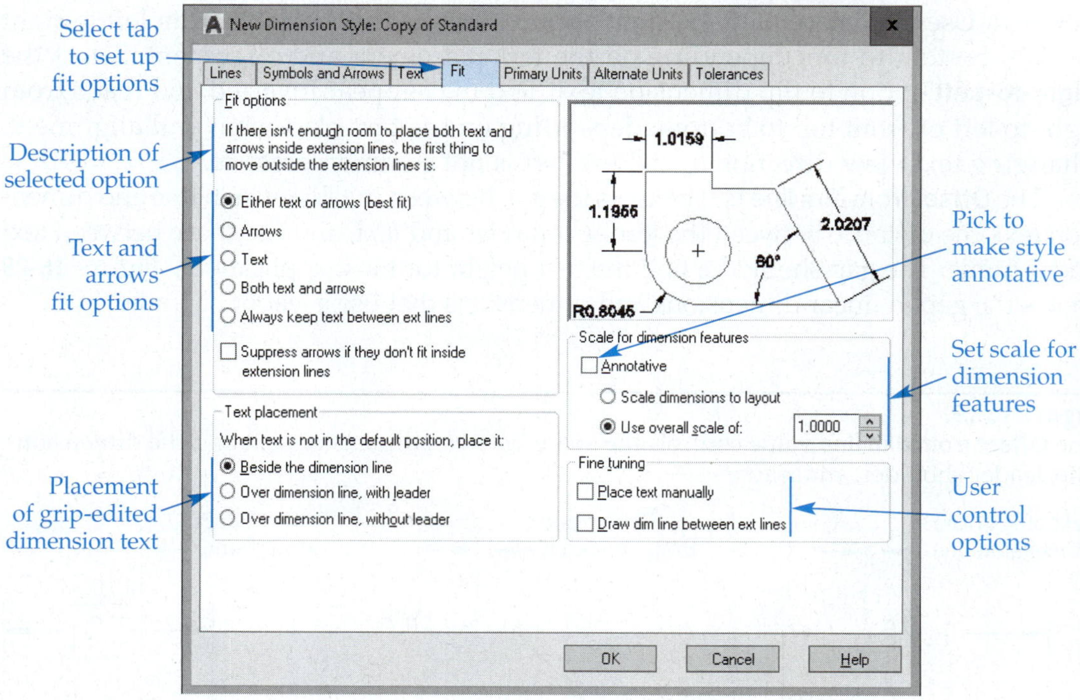

Select the **Always keep text between ext lines** radio button to place the dimension value between extension lines. This option typically causes interference between the dimension value and extension lines when there is limited space between extension lines. Pick the **Suppress arrows if they don't fit inside extension lines** check box to remove the arrowheads if they do not fit between extension lines. Use this option with caution, because it can create dimensions that violate standards.

Text Placement Settings

Sometimes it becomes necessary to move the dimension text from its default position. You can stretch the dimension text independently of the dimension. The selected option in the **Text placement** area presets the effect of stretching dimension text.

Select the **Beside the dimension line** radio button to restrict dimension text movement. You can stretch the text with the dimension line, but only within the same plane as the dimension line. Pick the **Over dimension line, with leader** radio button to stretch the text in any direction away from the dimension line. A leader line forms connecting the text to the dimension line. Select the **Over dimension line, without leader** radio button to stretch the text in any direction away from the dimension line without including a leader.

PROFESSIONAL TIP

To return the dimension text to its default position, select the dimension and use the **Reset Text Position** context-sensitive grip option, or right-click and select **Reset Text Position** from the shortcut menu.

Text Scale Options

Use the **Scale for dimension features** area to set the dimension scale factor. Select the **Annotative** check box to create an annotative dimension style. The **Annotative** check box is already selected when you modify the Annotative dimension style or pick the **Annotative** check box in the **Create New Dimension Style** dialog box.

Select the **Scale dimensions to layout** radio button to dimension in a floating viewport in a paper space layout. You must add dimensions to the model in a floating viewport in order for this option to function. Scaling dimensions to the layout adjusts the overall scale according to the active floating viewport by setting the overall scale equal to the viewport scale factor.

Pick the **Use overall scale of** radio button to scale a drawing manually. Enter the drawing scale factor to be applied to all dimension elements in the corresponding text box. The scale factor is multiplied by the plotted dimension size to get the dimension size in model space. For example, if you manually scale dimensions for a drawing with a 1:2 (half) scale, or a scale factor of 2 ($2 \div 1 = 2$), enter 2 in the **Use overall scale of** text box.

PROFESSIONAL TIP

You can draw dimensions in model space and layout space. Set the model space dimension scale according to the drawing scale factor to achieve the correct dimension appearance. Associative paper space dimensions automatically adjust to model modifications and do not require scaling. In addition, if you dimension in paper space, you can dimension the model differently in two viewports. However, paper space dimensions are not visible when you work in model space, so you must be careful not to move a model space object into a paper space dimension. Avoid using non-associative paper space dimensions.

Fine Tuning Settings

The **Fine tuning** area provides additional options for controlling the placement of dimension text. Select the **Place text manually** check box to increase flexibility when placing dimensions, allowing you to locate text to the side within extension lines or outside of extension lines. However, the **Place text manually** feature can make it more cumbersome to offset dimension lines equally, and it is not necessary for standard dimensioning practices.

The **Draw dim line between ext lines** option forces AutoCAD to place the dimension line inside the extension lines, even when the text and arrowheads are outside. The default application is to place the dimension line and arrowheads outside the extension lines. See Figure 16-31. Although some companies prefer the appearance, forcing the dimension line inside the extension lines is not an ASME standard.

Exercise 16-2

www.g-wlearning.com/CAD/

Complete the exercise on the companion website.

Primary Units Tab

Figure 16-32 shows the **Primary Units** tab of the **New** (or **Modify**) **Dimension Style** dialog box. The **Primary Units** tab controls linear and angular dimension units.

Figure 16-31.
The effect of the **Draw dim line between ext lines** option in the **Fine tuning** area of the **Fit** tab.

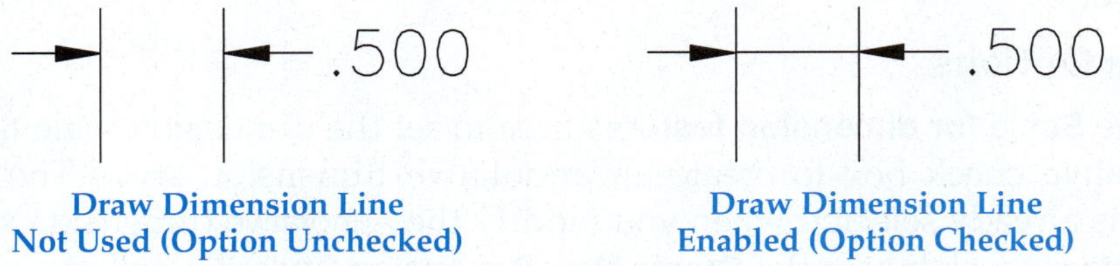

Figure 16-32.
The **Primary Units** tab of the **New** (or **Modify**) **Dimension Style** dialog box.

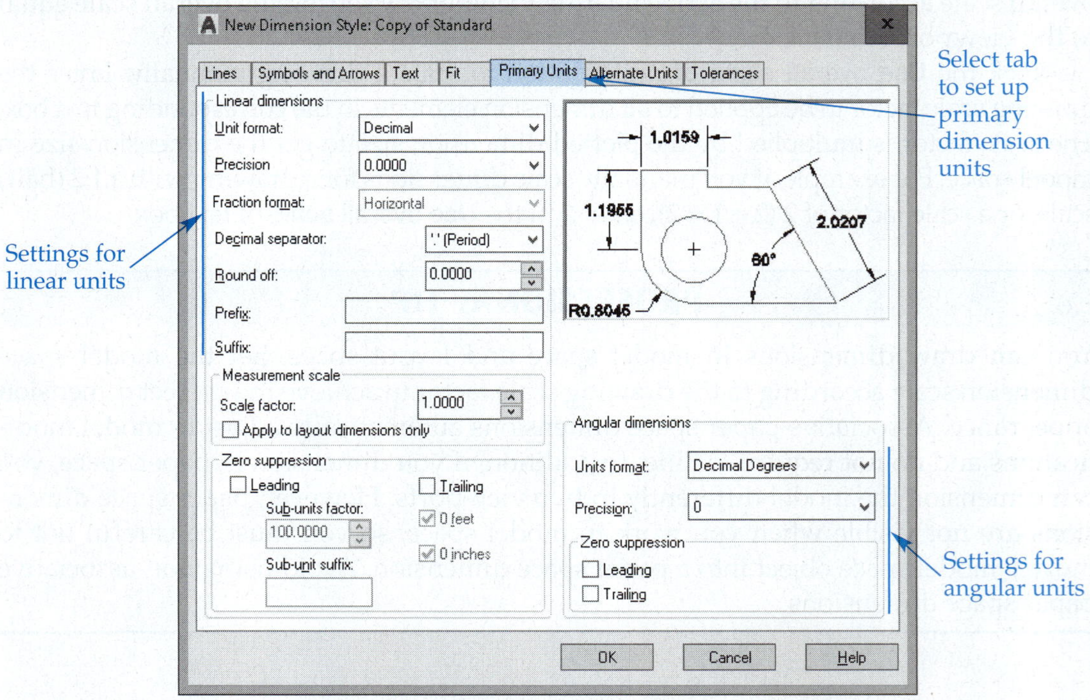

A dimension style presets units for typical dimensioning requirements. You can edit specific dimensions when necessary without using a separate dimension style.

Linear Dimension Settings

The **Linear dimensions** area allows you to specify settings for primary linear dimensions. The options in the **Unit format** drop-down list include the same options available in the **Length** area of the **Drawing Units** dialog box. The **Windows Desktop** option, also available, sets the unit format to use the Windows system settings for unit display. Typically, primary linear dimension unit format is the same as the corresponding drawing units set in the **Drawing Units** dialog box.

The **Precision** drop-down list sets the precision applied to dimensions, which may be the same as the related drawing units precision. A drawing often includes different precisions applied to specific dimensions. With decimal units, precision determines the number of zeros that follow the decimal place.

Precision settings in mechanical drafting depend on the accuracy required to manufacture specific features. Some features require greater precision, generally due to fits between mating parts. For example, a precision setting of 0.00 represents less exactness than a setting of 0.0000. Chapter 19 further explains this concept. Common precisions in mechanical drafting include 0.0000, 0.000, and 0.00.

When you specify fractional units, precision values identify the smallest desired fractional denominator. Precisions of 1/16 to 1/64 are common, but you can select a precision from 1/2 to 1/256. A precision of 0 displays no fractional values.

Use the **Decimal separator** drop-down list to specify commas, periods, or spaces as separators for decimal numbers. The **'.' (Period)** option is the default and is appropriate for typical applications. The **Fraction format** drop-down list is available if the unit format is **Architectural** or **Fractional**. The options for controlling the display of fractions are **Diagonal**, **Horizontal**, and **Not Stacked**.

The **Round off** text box specifies the accuracy of rounding for dimension numbers. The default is 0, which means that no rounding takes place and all associated dimensions specify the value exactly as measured. If you enter a value of .1, all dimensions are rounded to the closest .1 unit. For example, an actual measurement of 1.188 is rounded to 1.2. Rounding is inappropriate for most applications.

Enter a value in the **Prefix** text box to add a ***prefix*** to dimensions. A typical application for a prefix is SR3.5, where SR means "spherical radius." The prefix replaces the ∅ or R symbol when applied to a diameter or radius dimension.

prefix: A special note or application placed before the dimension value.

Enter a value in the **Suffix** text box to add a ***suffix*** to dimensions. A typical application for a suffix is 3.5 MAX, where MAX is the abbreviation for "maximum." Other examples include adding the suffix in (for inch) when you are placing a limited number of inch dimensions on a metric drawing or using the suffix mm (for millimeter) when you are placing a limited number of millimeter dimensions on an inch drawing.

suffix: A special note or application placed after the dimension value.

PROFESSIONAL TIP

A prefix or suffix is normally a special specification, used in only a few cases on a drawing. As a result, it is often most effective to enter a prefix or suffix using the **MText** or **Text** option of the related dimensioning command, or edit an existing dimension value to add a prefix or suffix.

Set the scale factor of linear dimensions in the **Scale factor** text box in the **Measurement scale** area. If you set a scale factor of 1, dimension values display their measured value. If the scale factor is 2, dimension values display two times the measured value. For example, an actual measurement of 2 inches displays as 2 with a scale factor of 1, but the same measurement displays as 4 when the scale factor is 2.

Check **Apply to layout dimensions only** to make the linear scale factor active only for dimensions created in paper space.

Zero Suppression Options

sub-units: Unit formats smaller than the primary unit format. For example, centimeters can be defined as a sub-unit of meters.

The **Zero suppression** area provides options for suppressing primary unit leading and trailing zeros and for controlling function ***sub-units***. Uncheck **Leading** to leave a zero on decimal units less than 1, such as 0.5. This option is suitable for creating metric dimensions as recommended by the ASME standard. Check **Leading** to remove the 0 on decimal units less than 1, such as .5. Apply this option to create inch dimensions as recommended by the ASME standard. The **Leading** check box is not available for architectural units.

Uncheck **Trailing** to leave zeros after the decimal point based on the precision. This setting is usually suitable for decimal-inch dimensioning because trailing zeros control tolerances for manufacturing processes. Check **Trailing** for metric dimensions to conform to the ASME standard. The **Trailing** check box is not available for architectural units.

The **0 feet** and **0 inches** check boxes are available for architectural and engineering units. Check **0 feet** to remove the zero in dimensions given in feet and inches when there are zero feet. For example, check **0 feet** to display a dimension as 11", or uncheck **0 feet** to display the same dimension as 0'-11".

Check **0 inches** to remove the zero when the inch portion of dimensions displayed in feet and inches has no value, such as 12'. If **0 inches** is unchecked, the same dimension reads 12'-0".

The **Sub-units factor** and **Sub-unit suffix** text boxes become available when you use decimal units and select the **Leading** check box. Most drawings use a single format for all dimension values. For example, all dimensions on a decimal-inch drawing are measured in inches or decimals of an inch. Sub-units allow you to apply a different unit format to dimensions that are smaller than the primary unit format, without using decimals. For example, if you use meters to dimension most objects on a metric civil engineering drawing, you can use a **Sub-units factor** value of 100 (100 cm/m) and a **Sub-unit suffix** of cm to dimension objects smaller than one meter using centimeters instead of decimals of a meter. Then, when you dimension an object that is 0.5 meters, the dimension reads 50 cm.

NOTE

For drawings that do not require sub-units but do suppress leading zeros, specify no sub-unit suffix. As long as you do not add a suffix, there is no need to change the sub-unit factor, although a factor of 0 also disables sub-units.

Angular Dimension Settings

The **Angular dimensions** area allows you to specify settings for primary angular dimensions. The options in the **Units format** drop-down list are similar to those in the **Angle** area of the **Drawing Units** dialog box. Typically, the primary angular dimension unit format is the same as the corresponding drawing units. Use the **Precision** drop-down list to set the appropriate angular dimension value precision. The **Zero suppression** area has check boxes for suppressing leading and trailing zeros on angular dimensions. Zero suppression for angular units is usually the same as applied to linear dimensions.

Alternate Units Tab

Figure 16-33 shows the **Alternate Units** tab of the **New** (or **Modify**) **Dimension Style** dialog box. Use the **Alternate Units** tab to set ***alternate units*** or ***dual dimensioning units***. Dual dimensioning practices are no longer a recommended ASME standard. ASME recommends that drawings be dimensioned using inch units or metric units only. However, other applications do use alternate units. Companies that use manufacturers and vendors both in the United States and internationally may require dual dimensioning.

alternate units (dual dimensioning units): Dimensions in which measurements in one system, such as inches, are followed by bracketed measurements in another system, such as millimeters.

Select the **Display alternate units** check box to enable alternate units. The **Alternate Units** tab includes most of the same settings found in the **Primary Units** tab. The value in the **Multiplier for alt units** text box is multiplied by the primary unit to establish the value for the alternate unit. A value of 25.4 allows you to use millimeters as alternate units on an inch-unit drawing. The **Placement** area controls the location of the alternate-unit dimension value. You can place alternate units either after or below the primary value.

NOTE

Chapter 19 describes the **Tolerances** tab found in the **New** (or **Modify**) **Dimension Style** dialog box.

Exercise 16-3

www.g-wlearning.com/CAD/

Complete the exercise on the companion website.

Figure 16-33.
The **Alternate Units** tab of the **New** (or **Modify**) **Dimension Style** dialog box.

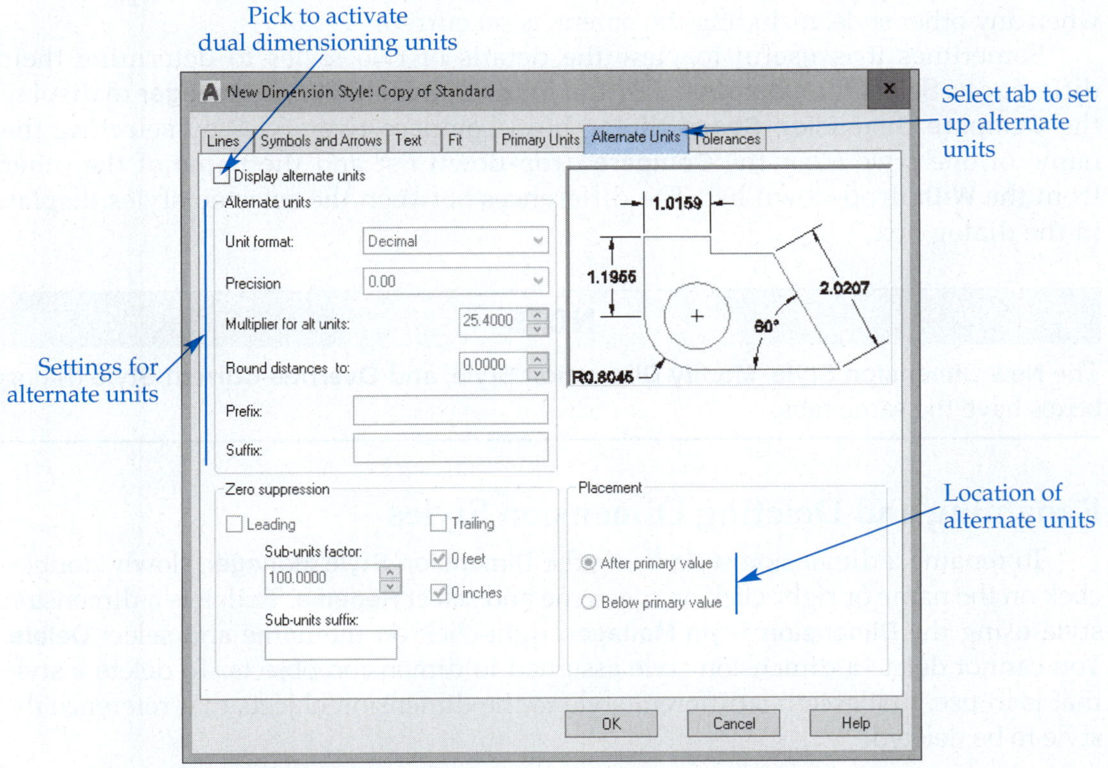

Developing Dimension Styles

Creating and using dimension styles is an important element of drafting with AutoCAD. Carefully evaluate the characteristics of dimensions and check school, company, and national standards to verify the accuracy of dimension settings. Figure 16-34 provides possible settings for three common dimension styles. Use the AutoCAD default values for settings not listed.

PROFESSIONAL TIP

To save valuable drafting time, add dimension styles to template drawings.

Exercise 16-4 www.g-wlearning.com/CAD/

Complete the exercise on the companion website.

Changing Dimension Styles

Use the **Dimension Style Manager** to change the characteristics of an existing dimension style. Pick the **Modify...** button to open the **Modify Dimension Style** dialog box. When you make changes to a dimension style, such as selecting a different text style or linear precision, existing dimensions assigned the modified dimension style are updated. Use a different dimension style with different characteristics when appropriate to prevent updating existing dimensions.

override: A temporary change to the current style settings; the process of changing a current style temporarily.

child: A style override.

parent: The dimension style from which a style override is formed.

To ***override*** a dimension style, pick the **Override...** button in the **Dimension Style Manager** to open the **Override Current Style** dialog box. An example of an override is including a text prefix for a few of the dimensions in a drawing. The **Override...** button is available only for the current style. Once you create an override, it is current and appears as a branch, called the ***child***, of the ***parent*** style. Override settings are lost when any other style, including the parent, is set current.

Sometimes it is useful to view the details of two styles to determine their differences. Select the **Compare...** button in the **Dimension Style Manager** to display the **Compare Dimension Styles** dialog box. Compare two styles by selecting the name of one style from the **Compare** drop-down list and the name of the other from the **With** drop-down list. The differences between the selected styles display in the dialog box.

NOTE

The **New Dimension Style**, **Modify Dimension Style**, and **Override Current Style** dialog boxes have the same tabs.

Renaming and Deleting Dimension Styles

To rename a dimension style using the **Dimension Style Manager**, slowly double-click on the name or right-click on the name and select **Rename**. To delete a dimension style using the **Dimension Style Manager**, right-click on the name and select **Delete**. You cannot delete a dimension style assigned to dimension objects. To delete a style that is in use, first assign a different style to the dimension objects that reference the style to be deleted.

Figure 16-34.
Dimension settings for typical mechanical and architectural drawings.

Setting	Mechanical—Inch	Mechanical—Metric (mm)	Architectural—US Customary
Baseline spacing	.5	12	1/2"
Extend beyond dimension lines	.125	3	1/8"
Offset from origin	.063	1.5	1/16" or 3/32"
Arrowhead options	**Closed filled**, **Closed**, or **Open**	**Closed filled**, **Closed**, or **Open**	**Architectural tick**, **Dot**, **Closed filled**, **Oblique**, or **Right angle**
Arrow size	.125	3	1/8"
Center marks	**Line**	**Line**	**Mark**
Center mark size	.0625	1.5	1/16"
Text style	Arial, Romans, or Century Gothic	Arial, Romans, or Century Gothic	SansSerif, Arial, Century Gothic, or Stylus BT
Text height	.12	3	1/8"
Vertical and horizontal text placement	**Centered**	**Centered**	Vertical: **Above** Horizontal: **Centered**
View direction	**Left-to-right**	**Left-to-right**	**Left-to-right**
Offset from dimension line	.063	1.5	1/16"
Text alignment	**Horizontal**	**Horizontal**	**Aligned with dimension line**
Linear unit format	**Decimal**	**Decimal**	**Architectural**
Linear precision	0.0000	0.00	1/16"
Linear zero suppression	Suppress only the leading zero	Suppress only the trailing zero	Suppress only the 0 feet zero
Sub-units factor	0	Disabled	Disabled
Sub-units suffix	None	Disabled	Disabled
Angular unit format	**Decimal degrees**	**Decimal degrees**	**Decimal degrees**
Angular precision	0	0	0
Angular zero suppression	Suppress only the leading angular dimension zero	Suppress only the trailing angular dimension zero	Suppress only the leading angular dimension zero
Alternate units	Do not display	Do not display	Do not display
Tolerances	By application	By application	None

RENAME

Type
RENAME

NOTE

You can also rename styles using the **Rename** dialog box. Select **Dimension styles** in the **Named Objects** list to rename a dimension style.

Setting a Dimension Style Current

Set a dimension style current using the **Dimension Style Manager** by double-clicking the style in the **Styles** list box, right-clicking on the name and selecting **Set current**, or picking the style and selecting the **Set Current** button. To set a dimension style current without opening the **Dimension Style Manager**, use the **Dimension Style** drop-down list in the expanded **Annotation** panel of the **Home** ribbon tab or on the **Dimensions** panel of the **Annotate** ribbon tab.

Command line content search offers another way to find a dimension style and make a dimension style current without using the **Dimension Style Manager**. Begin typing a dimension style name using the command line. Dimension styles that match the letters you enter appear in the suggestion list as you type. Select a dimension style from the suggestion list to set current.

PROFESSIONAL TIP

You can import dimension styles from existing drawings using **DesignCenter**. See Chapter 5 for more information about using **DesignCenter** to import file content.

Exercise 16-5

www.g-wlearning.com/CAD/

Complete the exercise on the companion website.

NOTE

The **Dimension** panel of the **Express Tools** ribbon tab includes a **DIMEX** command that allows you to export a dimension style as a separate .dim file, and a **DIMIM** command that allows you to import a .dim file as a dimension style into the current drawing. For most applications, use **DesignCenter** to reuse dimension styles from existing drawings or templates, without creating a separate file.

Template Development

www.g-wlearning.com/CAD/

Adding Dimension Styles

For instructions on adding dimension styles to each drawing template, go to the companion website, navigate to this chapter in the **Contents** tab, and select **Adding Dimension Styles**.

Chapter Review

Answer the following questions using the information in this chapter.

1. List at least three factors that influence a company's dimensioning practices.
2. Name the current ASME document that specifies mechanical drafting dimensioning practices.
3. Name two basic coordinate systems used to create location dimensions.
4. Define the term *general notes*.
5. Explain the difference between placing specific and general notes on a drawing.
6. Explain how to dimension a cylinder using a single view.
7. Describe two ways to dimension a cone.
8. When is the best time to determine the drawing scale and scale factors for a drawing?
9. Identify advantages to using annotative dimensions over manually scaled dimensions.
10. Define *dimension style*.
11. Name the dialog box used to create dimension styles.
12. Identify two ways to access the dialog box identified in Question 11.
13. Name the dialog box tab used to control the appearance of dimension lines and extension lines.
14. Name at least four arrowhead types that are available in the **Symbols and Arrows** tab for common use on architectural drawings.
15. Name the dialog box tab used to control the settings that display the dimension text.
16. How do you assign a text style to a dimension style?
17. What is the ASME-recommended height for dimension numbers and notes on drawings?
18. Name the dialog box tab used to control settings that adjust the location of dimension lines, dimension text, arrowheads, and leader lines.
19. How can you delete a dimension style from a drawing?
20. How do you set a dimension style current?

Drawing Problems

Start AutoCAD if it is not already started. Follow the specific instructions for each problem.

▼ Basic

1. **Mechanical** Start a new drawing from a template and create an Arial text style using the Arial font. Create the Mechanical-Inch dimension style shown in Figure 16-34. Use the default AutoCAD settings for the dimension style settings not listed. Save the drawing as P16-1.

2. **Mechanical** Start a new drawing from a template and create an Arial text style using the Arial font. Create the Mechanical-Metric dimension style shown in Figure 16-34. Use the default AutoCAD settings for the dimension style settings not listed. Save the drawing as P16-2.

3. **Architectural** Start a new drawing from a template and create a Stylus BT text style using the Stylus BT font. Create the Architectural-US Customary dimension style shown in Figure 16-34. Use the default AutoCAD settings for the dimension style settings not listed. Save the drawing as P16-3.

4. **General** Write a short report explaining the difference between unidirectional and aligned dimensioning. Use a word processor and include sketches giving examples of each method.

5. **General** Write a short report explaining the difference between size and location dimensions. Use a word processor and include sketches giving examples of each method.

6. **General** Write a short report describing the difference between dimensioning for mechanical drafting (drafting for manufacturing) and architectural drafting. Use a word processor and include sketches giving examples of each method.

7. **Mechanical** Make freehand sketches showing standard practices for applying location dimensions. Make one sketch showing rectangular coordinate dimensioning and another showing polar coordinate dimensioning.

8. **Mechanical** Make freehand sketches showing the standard practice for dimensioning a cylindrical object, a square object, and a conical object.

▼ Intermediate

9. **Mechanical** Find a copy of the ASME Y14.5-2018 standard, *Dimensioning and Tolerancing*, and write a report of approximately 350 words explaining the importance and basic content of this standard.

10. **Mechanical** Interview your drafting instructor or supervisor and determine what dimension standards exist at your school or company. Write them down and keep them with you as you learn AutoCAD. Make notes as you progress through this textbook on how you use these standards. Also, note how you could change the standards to match the capabilities of AutoCAD.

▼ Advanced

11. **Mechanical** Create a freehand sketch of **Figure 16-1**. Label each dimension item. To the side of the sketch, write a short description of each item.

12. **Civil** Research civil drafting and create a template establishing the dimension styles for a civil drawing.

13. **Mechanical** Visit a local manufacturing company at which design drafting work is part of the business. Write a report with sketched examples identifying the standards used at the company.

14. **Architectural** Interview a local architect or architectural designer. Write a report with sketched examples identifying the standards used by the person you interview.

15. **Civil** Visit a local civil engineering company at which design drafting work is part of the business. Write a report with sketched examples identifying the standards used at the company.

AutoCAD Certified Professional Exam Practice

Answer the following questions using the information in this chapter.

1. Which of the following items are common components of a dimension? *Select all that apply.*
   A. boundaries
   B. lines
   C. numerical values
   D. symbols
   E. xlines
2. What is the term for a dimension style in which all text reads from the bottom of the sheet? *Select the one item that best answers the question.*
   A. aligned dimensioning
   B. polar dimensioning
   C. rectangular dimensioning
   D. unidirectional dimensioning
3. What is the smallest number of views that can be used to dimension a cylindrical shape? *Select the one item that best answers the question.*
   A. one
   B. two
   C. three
   D. four

Follow the instructions in each problem. Complete each problem by answering the related question.

4. **Navigate to this chapter on the companion website and open CPE-16annoscale.dwg.** What annotation scale is currently assigned to this drawing?
5. **Navigate to this chapter on the companion website and open CPE-16dimstyle.dwg.** What text style is assigned to the Mechanical dimension style? What font is assigned to the text style?

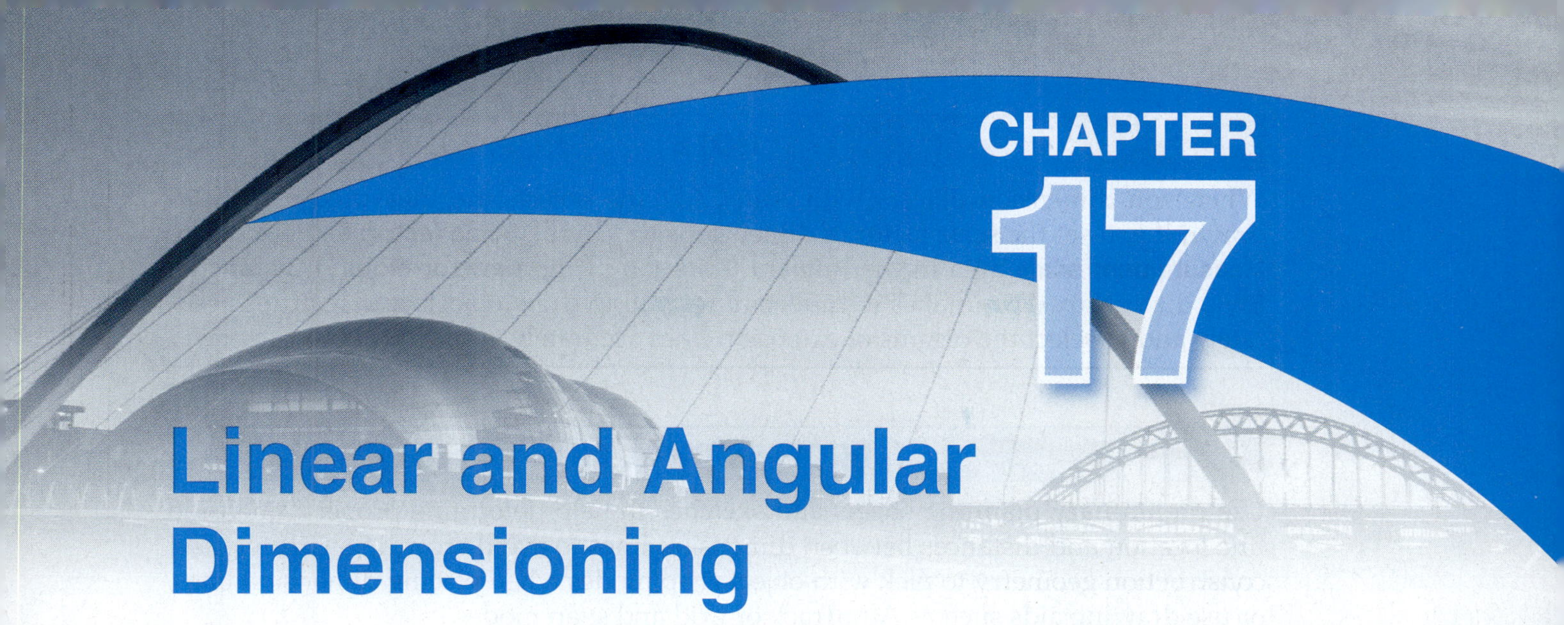

CHAPTER 17

Linear and Angular Dimensioning

Learning Objectives

After completing this chapter, you will be able to:

- ✓ Add linear dimensions to a drawing.
- ✓ Add angular dimensions to a drawing.
- ✓ Draw datum and chain dimensions.
- ✓ Override the layer assigned to new dimensions.
- ✓ Dimension multiple objects using the **QDIM** command.

Drawings often require a variety of dimensions to describe the size and shape of features. Linear and angular dimensions are two of the most common types of dimensions. This chapter covers the process of adding linear and angular dimensions to a drawing using several dimensioning commands. It also covers adding a break symbol to a dimension line and using the **QDIM** command.

Introduction to Placing Dimensions

AutoCAD includes many commands for dimensioning a drawing. The **DIM** command includes a variety of options for drawing and editing many types of dimensions and provides flexibility for different requirements. You can usually use the **DIM** command or another dimension command to accomplish the same task. For example, use the **DIM** or **DIMLINEAR** command to draw a single linear dimension. The **DIM** command can help to automate the dimensioning process. However, you may prefer or need to use other dimension commands for certain applications.

Dimension commands reference the current dimension style and the objects or points you select to create a dimension object. When you use the **DIM** command to create a linear dimension, for example, you create a rotated or aligned dimension object that includes all related dimension style characteristics, dimension and extension lines, arrowheads, and a dimension value that describes the distance between the extension line origins.

NOTE

When you dimension objects with AutoCAD, the objects are measured exactly as drawn if you set the scale factor of dimensions to 1 in the **Scale factor:** text box in the **Measurement scale** area in the **Primary Units** tab of the **New** (or **Modify**) **Dimension Style** dialog box. This makes it important for you to draw objects and features accurately and to select the origins of extension lines accurately.

PROFESSIONAL TIP

Use preliminary planning sheets and sketches to help determine proper dimension line location and distances between dimension lines to avoid crowding. Create offset construction geometry to pick with object snap modes when placing dimension lines, or use drawing aids such as AutoTrack or grid and snap modes.

Linear Dimensions

DIM
Ribbon
Home
> Annotation
Annotate
> Dimensions
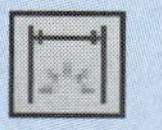
Dimension
Type
DIM

Linear dimensions usually measure straight distances, such as distances between horizontal, vertical, or slanted surfaces. The **DIM** command is one method for drawing a single linear dimension. Select objects or specify points to establish the origin of each extension line. To dimension a line or polyline segment as shown in Figure 17-1, access the **DIM** command and move the pick box over the object to dimension. When the object becomes highlighted and you see a preview of the dimension, pick the object to allow AutoCAD to originate an extension line from each endpoint. Now use the options that appear at the Specify dimension line location or second line for angle [Mtext/Text/text aNgle/Undo]: prompt as needed. Select the **Undo** option to undo the selection without exiting the **DIM** command. To apply the default settings and create a linear dimension, specify a point to locate the dimension line.

Specify points to define the origin of each extension line when selecting a single object is not possible or does not result in the correct dimension. To create the dimension shown in Figure 17-2, pick a point to locate the origin of the first extension line, and

Figure 17-1.
Using the **DIM** command to dimension a line or polyline segment by selecting the object. AutoCAD determines the extension line origins automatically. ASME standards recommend a minimum spacing of .375″ (10 mm) from a drawing feature to the first dimension line and a minimum spacing of .25″ (6 mm) between dimension lines. A value of .5″ (12 mm) or .75″ (19 mm) is usually more appropriate. Maintain consistent spacing throughout a drawing.

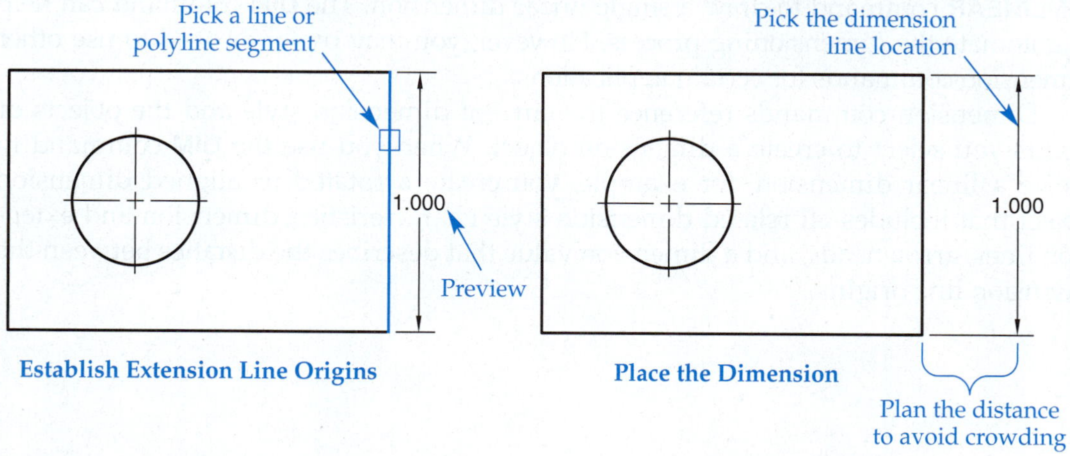

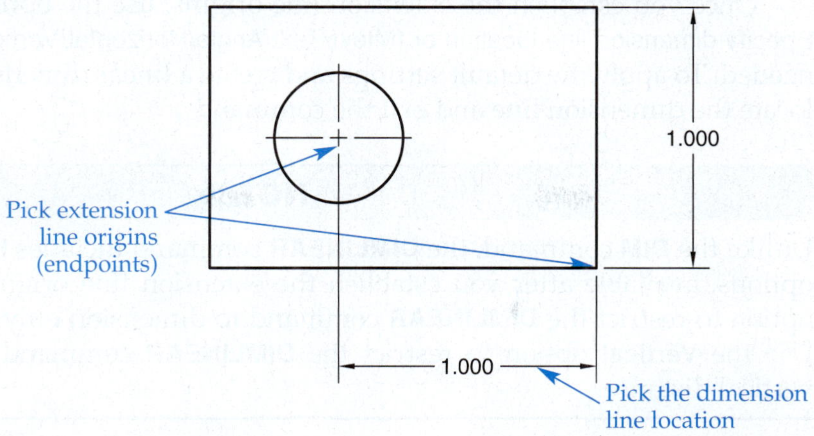

Figure 17-2. Specifying points to establish extension line origins. The **Endpoint** and **Intersection** object snap modes are useful for locating origins accurately.

then pick a point to locate the origin of the second extension line. Use object snap modes and other drawing aids to pick the exact origins. Once you establish the extension line origins, the process of completing the dimension is the same as previously described.

The **DIM** command remains active after you create a dimension, allowing you to continue dimensioning without reselecting the command. Select the **Undo** option as needed to undo dimensions created during the current **DIM** command sequence. Exit the command when finished.

Exercise 17-1

www.g-wlearning.com/CAD/

Complete the exercise on the companion website.

Using the DIMLINEAR Command

DIMLINEAR
Ribbon: Home > Annotation; Annotate > Dimensions — Linear
Type: DIMLINEAR, DLI

The **DIMLINEAR** command is an alternative to the **DIM** command for creating linear dimensions. Access the **DIMLINEAR** command, pick a point to locate the origin of the first extension line, and then pick a point to locate the origin of the second extension line. Use object snap modes and other drawing aids to pick the exact origin of extension lines. Another option is to press [Enter] at the Specify first extension line origin or <select object>: prompt and then pick a line, polyline segment, circle, or arc to dimension. When you select a line, polyline segment, or arc, extension lines originate from the endpoints. When you pick a circle, extension lines originate from opposite quadrants depending on where you place the dimension line. See Figure 17-3.

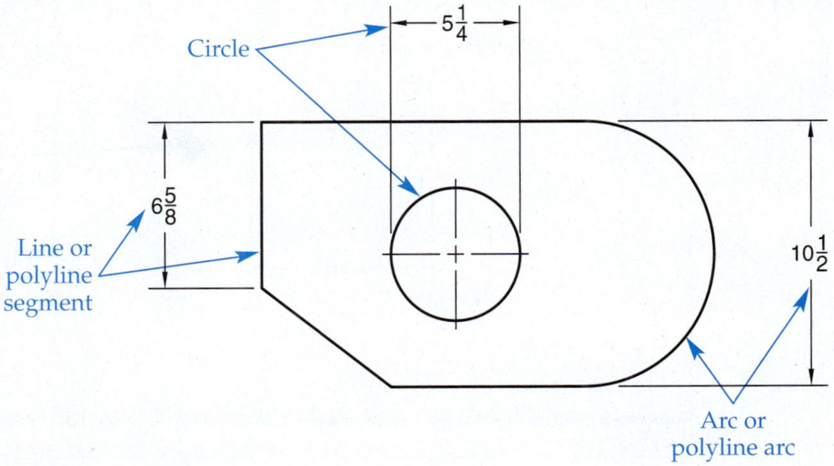

Figure 17-3. Use the **Select object** option of the **DIMLINEAR** command to establish extension line origins by picking a line, polyline segment, arc, or circle.

Once you establish the extension line origins, use the options that appear at the Specify dimension line location or [Mtext/Text/Angle/Horizontal/Vertical/Rotated]: prompt as needed. To apply the default settings and create a linear dimension, specify a point to locate the dimension line and exit the command.

NOTE

Unlike the **DIM** command, the **DIMLINEAR** command includes **Horizontal** and **Vertical** options, available after you establish the extension line origins. Use the **Horizontal** option to restrict the **DIMLINEAR** command to dimension only a horizontal distance. Use the **Vertical** option to restrict the **DIMLINEAR** command to dimension only a vertical distance.

Adjusting Dimension Text

The value attached to the dimension corresponds to the distance between the origins of extension lines. Use the **Mtext** option of the **DIM** or **DIMLINEAR** command after you establish the extension line origins to adjust the dimension value using the multiline text editor. See Figure 17-4. The highlighted value is the current dimension value calculated by AutoCAD. Add to or modify the dimension text and then close the text editor. The command continues, allowing you to pick the dimension line location.

The **Text** option of the **DIM** or **DIMLINEAR** command also allows you to adjust the dimension text, even though the final dimension value is mtext. The current dimension value appears in brackets at the dynamic input cursor and at the command line. Add to or modify the value as necessary and then press [Enter] to exit the option. The command continues, allowing you to pick the dimension line location.

Figure 17-4.
When you use the **Mtext** option, the **Text Editor** ribbon tab appears with the dimension value calculated by AutoCAD in a text box for editing.

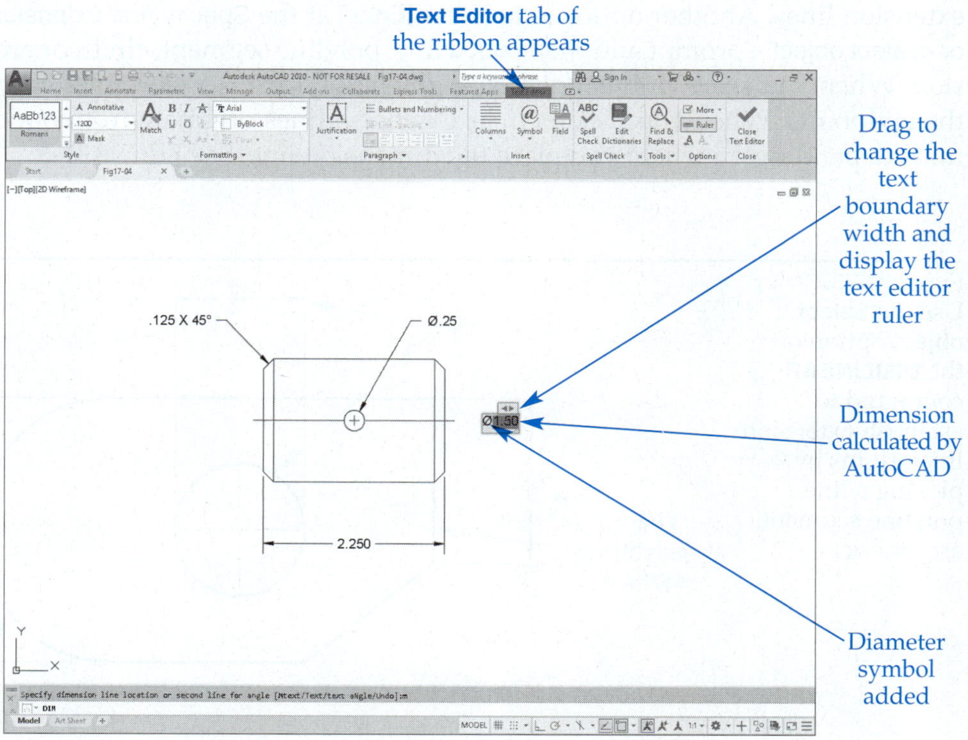

NOTE

Dimension values are horizontal or aligned with the dimension line, according to the current dimension style format. The **text aNgle** option of the **DIM** command and the **Angle** option of the **DIMLINEAR** command have limited applications, but allow you to rotate the dimension text. To use these options, enter the desired angle at the Specify angle of dimension text: prompt.

PROFESSIONAL TIP

If you forget to use the **Mtext** or **Text** option before placing a dimension, or to modify an existing dimension value, double-click on the dimension value to display the multiline text editor. Chapter 20 provides more information on editing dimensions.

Including Symbols with Dimension Text

Some AutoCAD dimension commands automatically place appropriate symbols with the dimension value. For example, when you dimension an arc using the **DIM** or **DIMRADIUS** command, an R appears before the dimension value. When you dimension a circle using the **DIM** or **DIMDIAMETER** command, a ∅ symbol appears before the dimension value. The ASME standard recommends these symbols. However, most dimension commands do not automatically place certain symbols or add necessary characters.

One option is to use the **Mtext** option to activate the multiline text editor. Place the cursor at the location to add a symbol and then use options from the **Symbol** flyout or cascading submenu, or type characters. For example, pick the **Diameter** symbol from the **Symbol** flyout to add ∅ before the dimension value. Another example is typing open and close parentheses around the highlighted value to create a reference dimension.

You can also add content using the **Text** option. Place the cursor at the location to add a symbol and type control codes or characters. For example, type %%C to display ∅ or type parentheses around the value to create a reference dimension.

Another way to place symbols with dimension text is to create a dimension style that references a text style using the gdt.shx font. A text style with the gdt.shx font allows you to place common dimension symbols, including geometric dimensioning and tolerancing (GD&T) symbols, using the lowercase letter keys.

NOTE

Although you can add a prefix and suffix to a dimension style, it is usually more appropriate to adjust the limited number of dimensions that require a prefix or suffix.

PROFESSIONAL TIP

AutoCAD dimensioning should be accurate and neat. You can achieve consistent, professional results by using the following guidelines.

- Never round off decimal values when entering locations, distances, or angles. For example, enter .4375 for 7/16, rather than .44.
- Set the precision to the most common precision level in the drawing before adding dimensions. Adjust the precision as needed for each dimension.
- Use drawing aids, such as object snaps, to ensure the accuracy of dimensions.
- Never type a different dimension value from what appears highlighted or in <> brackets. To change a dimension, revise the drawing or dimension settings. Only adjust dimension text when it is necessary to add prefixes and suffixes or to use a different text format.

Reference Material

www.g-wlearning.com/CAD/

Drafting Symbols

For more information about common drafting symbols and the gdt.shx font, go to the companion website, select the **Resources** tab, and select **Drafting Symbols**.

Exercises 17-2 and 17-3

www.g-wlearning.com/CAD/

Complete the exercises on the companion website.

Dimensioning Angled Surfaces and Auxiliary Views

When you dimension a surface drawn at an angle, such as a feature shown in an auxiliary view, it is often necessary to align the dimension line with the surface, so that the extension lines are perpendicular to the surface. See Figure 17-5.

DIMALIGNED

Ribbon
Home
> Annotation
Annotate
> Dimensions

Aligned

Type
DIMALIGNED
DAL

You can usually use the **DIM** or **DIMALIGNED** command when the length of the extension lines is equal. The **Rotated** option of the **DIMLINEAR** command is often necessary when extension lines are unequal. Notice the difference between the aligned and rotated dimension objects shown in Figure 17-5. To draw the rotated dimension objects, access the **DIMLINEAR** command and establish the extension line origins. Then select the **Rotated** option. At the Specify angle of dimension line <0>: prompt, enter a value or specify two points on the line to define the dimension line angle. Specify a point to locate the dimension line and exit the command. See Figure 17-6.

Exercises 17-4 and 17-5

www.g-wlearning.com/CAD/

Complete the exercises on the companion website.

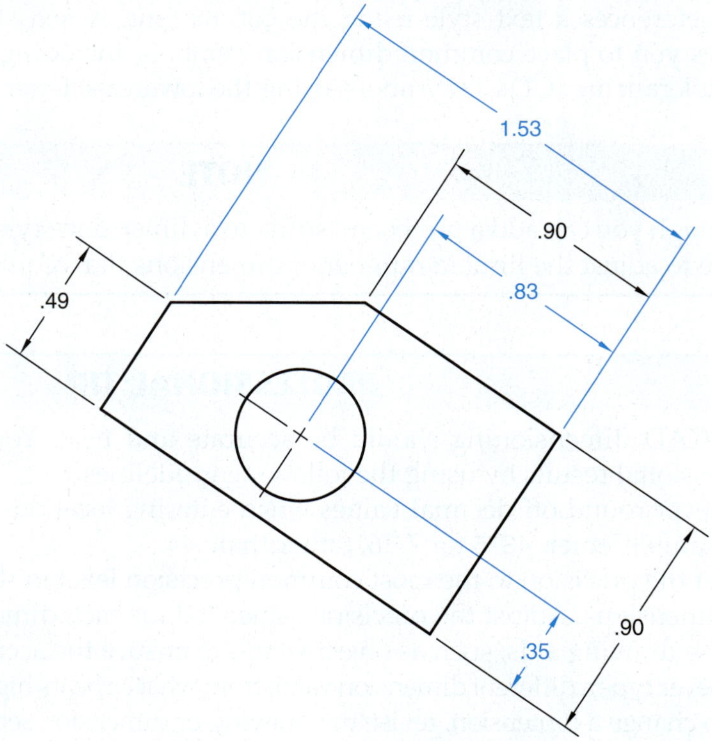

Figure 17-5.
Aligned and rotated dimensions are used to dimension surfaces drawn at an angle. Create aligned dimensions using the **DIM** or **DIMALIGNED** command. The aligned dimension objects are shown in black and the rotated dimension objects are shown in color. Notice the length of the extension lines.

Figure 17-6.
The **Rotated** option of the **DIMLINEAR** command allows you to specify the angle of the dimension line.

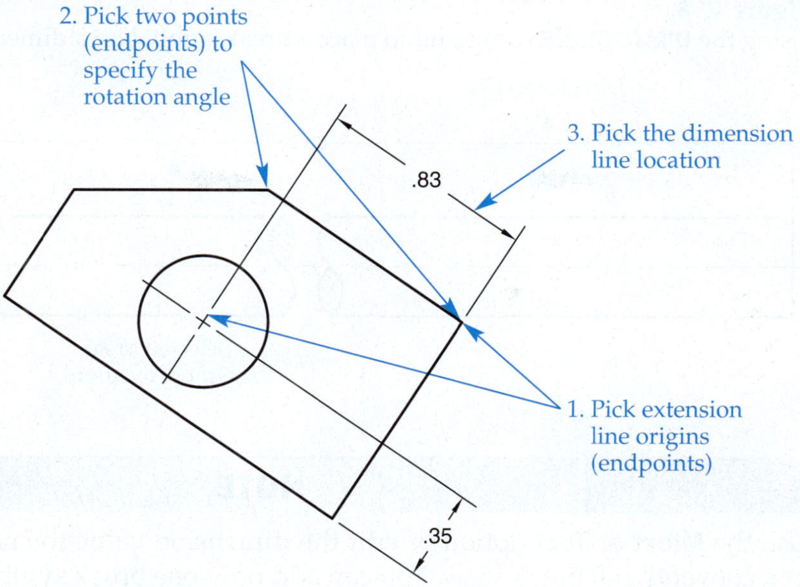

Dimensioning Long Objects

When you create a drawing of a long part that has a constant shape, the view may not fit on the sheet, or it may look strange compared to the rest of the drawing. To overcome this problem, use a ***conventional break*** (or ***break***) to shorten the view. Figure 17-7 shows examples of standard break lines. For many long parts, a conventional break is required to display views or increase view scale without increasing sheet size. Dimensions added to conventional breaks describe the actual length of the product in its unbroken form. The dimension line includes a break symbol to indicate that the drawing view is broken and that the feature is longer than it appears.

conventional break (break): Removal of a portion of a long, constant-shaped object to make the object fit better on the sheet.

Using the DIMJOGLINE Command

Use the **DIMJOGLINE** command to add a break symbol to a dimension line created using the **DIM**, **DIMLINEAR**, or **DIMALIGNED** command. Access the **DIMJOGLINE** command and pick a linear or aligned dimension line. Then pick a location on the dimension line to place the break symbol. See Figure 17-8. An alternative to selecting the location of the break symbol is to press [Enter] to accept the default location. You can move the break later using grip editing or by reusing the **DIMJOGLINE** command to select a different location. To remove the break symbol, access the **DIMJOGLINE** command and select the **Remove** option.

DIMJOGLINE
Ribbon
Annotate > Dimensions
Dimjogline
Type
DIMJOGLINE

Figure 17-7.
Standard break lines.

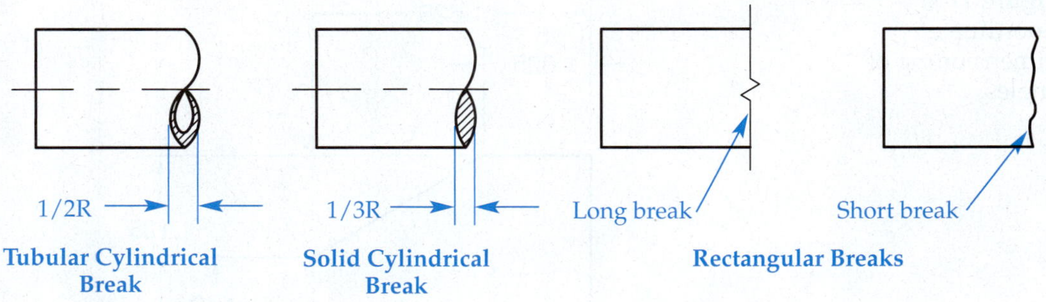

Figure 17-8.
Using the **DIMJOGLINE** command to place a break symbol on a dimension line.

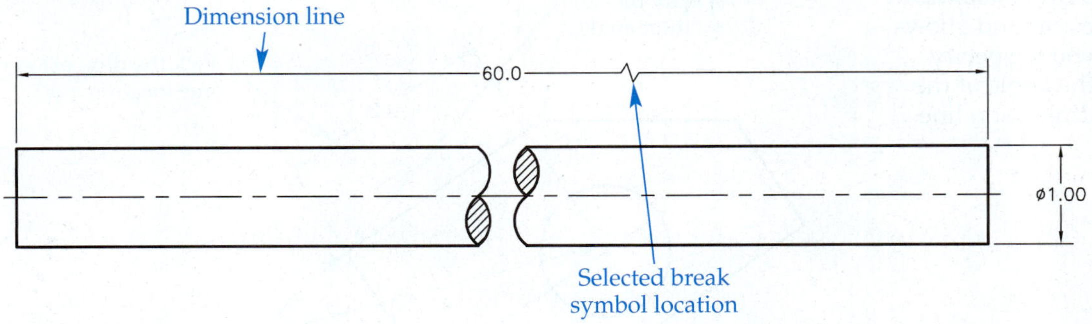

NOTE

Use the **Mtext** or **Text** option or edit the dimension value to specify the actual length in a conventional break view. You can add only one break symbol to a dimension line.

Exercise 17-6

www.g-wlearning.com/CAD/

Complete the exercise on the companion website.

Dimensioning Angles

coordinate dimensioning: A method of dimensioning angles in which dimensions locate the corner of the angle.

DIMANGULAR
Ribbon
Home > Annotation
Annotate > Dimensions
Angular
Type
DIMANGULAR
DAN

angular dimensioning: A method of dimensioning angles in which one corner of an angle is located with a dimension and the value of the angle is provided in degrees.

vertex: The point at which the two lines that form an angle meet.

Coordinate and angular dimensioning are accepted methods for dimensioning angles. Figure 17-9 shows an example of ***coordinate dimensioning*** using the **DIM** or **DIMLINEAR** command.

Figure 17-10 shows examples of ***angular dimensioning*** using the **DIM** or **DIMANGULAR** command. You can dimension the angle between any two nonparallel lines from the ***vertex*** of the angle. AutoCAD automatically draws extension lines if needed.

When using the **DIM** command, pick the first leg of the angle to dimension, then hover over the second leg of the angle. When the Select line to specify second side of angle: prompt appears, pick the second leg. The last prompt asks you to pick the location of the dimension line arc. To use the **DIMANGULAR** command, pick the first leg of the angle to dimension, pick the second leg of the angle, and then pick the location of the dimension line arc. Figure 17-11 shows examples of angular dimensions and the effect that limited space may have on dimension fit and placement. Fit characteristics apply to most dimensions.

Figure 17-9.
Coordinate dimensioning of angles.

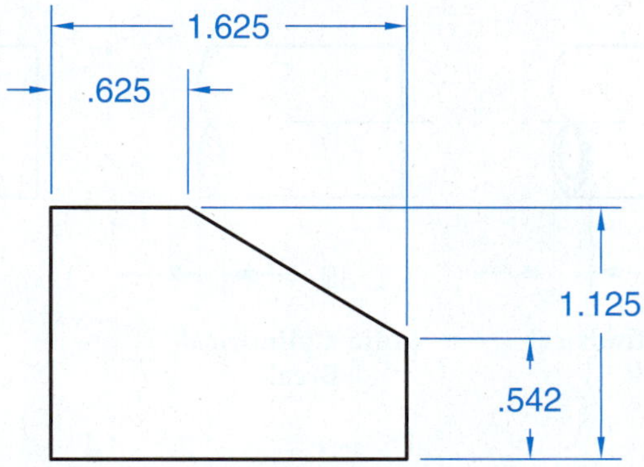

Figure 17-10.
Two examples of drawing angular dimensions.

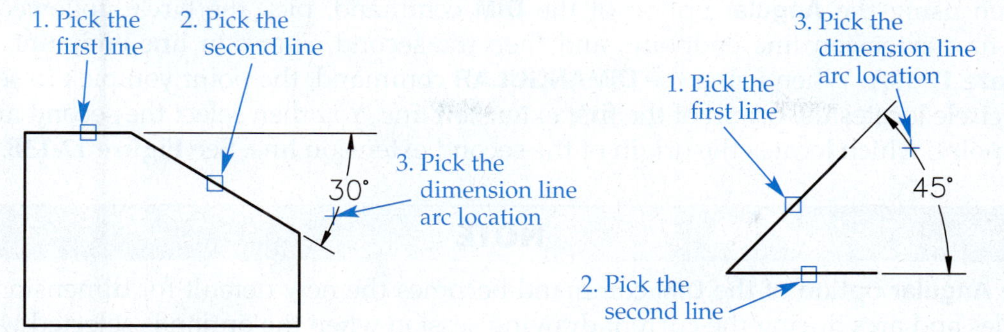

Figure 17-11.
The location of the dimension line determines the arrangement of the dimension line arc, text, and arrows.

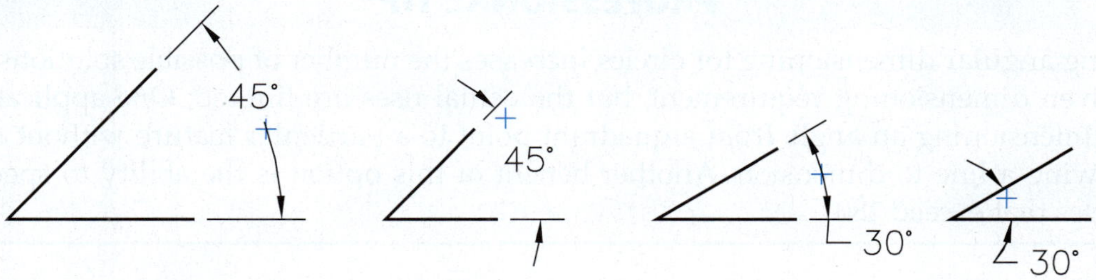

PROFESSIONAL TIP

You can create four different dimensions (two different angles) with an angular dimension. To preview these options before selecting the dimension line location, use the cursor to move the dimension around an imaginary circle. Unlike the **DIM** command, the **DIMANGULAR** command includes a **Quadrant** option that you can use to isolate a specific quadrant of the imaginary circle and force the dimension to produce the value found in the selected quadrant.

Dimensioning Angles on Arcs and Circles

Use the **DIM** or **DIMANGULAR** command to dimension the included angle of an arc or a portion of a circle. When using the **DIM** command, it is helpful to select the **Angular** option before selecting an object or specifying points. When you dimension an arc, the center point becomes the angle vertex, and the two arc endpoints establish the extension line origins. See Figure 17-12.

Figure 17-12.
Placing an angular dimension on an arc.

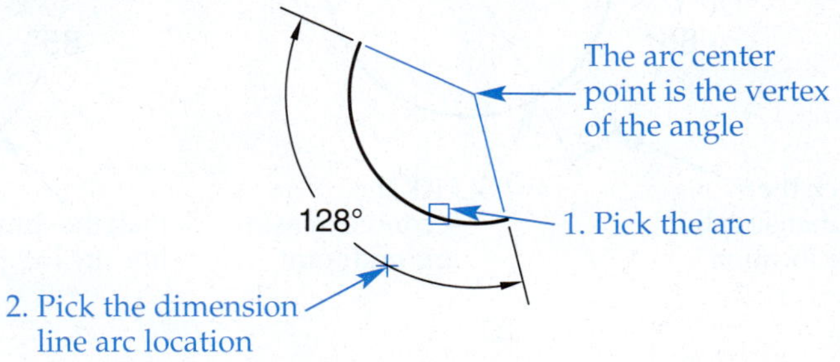

When you dimension the included angle of a portion of a circle, the center point becomes the angle vertex, and two specified points locate the extension line origins. When using the **Angular** option of the **DIM** command, pick the circle, followed by the first extension line endpoint, and then the second extension line endpoint. See Figure 17-13A. When using the **DIMANGULAR** command, the point you pick to select the circle locates the origin of the first extension line. You then select the second angle endpoint, which locates the origin of the second extension line. See Figure 17-13B.

NOTE

The **Angular** option of the **DIM** command becomes the new default for dimensioning circles and arcs during the current drawing session when the option is selected while hovering over an object. Hover over a circle or arc again while the **DIM** command is active and select an option, such as **Diameter** or **Radius**, to change the setting.

PROFESSIONAL TIP

Using angular dimensioning for circles increases the number of possible solutions for a given dimensioning requirement, but the actual uses are limited. One application is dimensioning an angle from a quadrant point to a particular feature without first drawing a line to dimension. Another benefit of this option is the ability to specify angles that exceed 180°.

Angular Dimensioning through Three Points

You can also use the **DIM** or **DIMANGULAR** command to establish an angular dimension according to a specified vertex and two angle line endpoints. See Figure 17-14. To use the **DIM** command, preset the **Angular** option and then select the **Vertex** option. To use the **DIMANGULAR** command, press [Enter] after the first prompt. Now pick the vertex, followed by the two points. You can also use this method to dimension angles over 180°.

Figure 17-13.
A—Using the **DIM** command to place an angular dimension on a circle. B— Using the **DIMANGULAR** command to place an angular dimension on a circle.

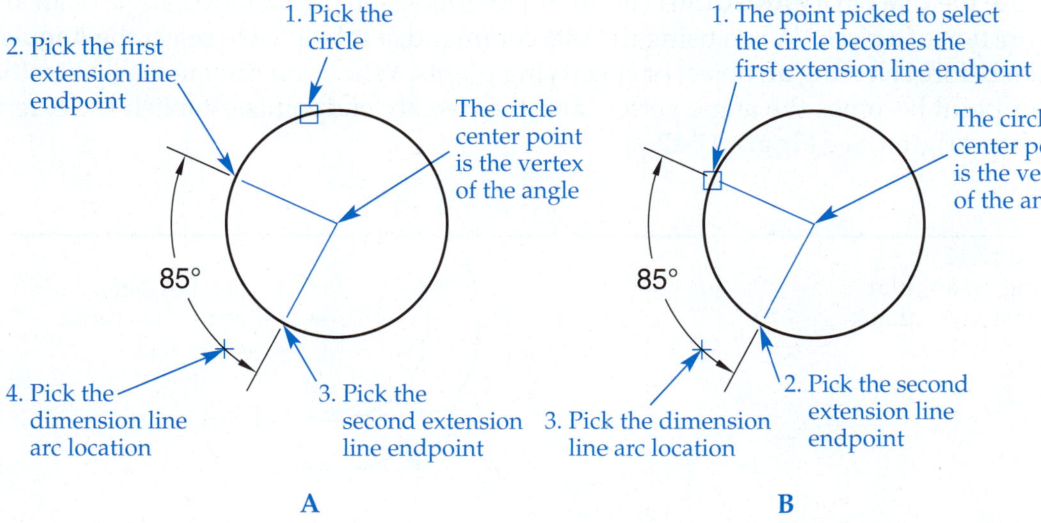

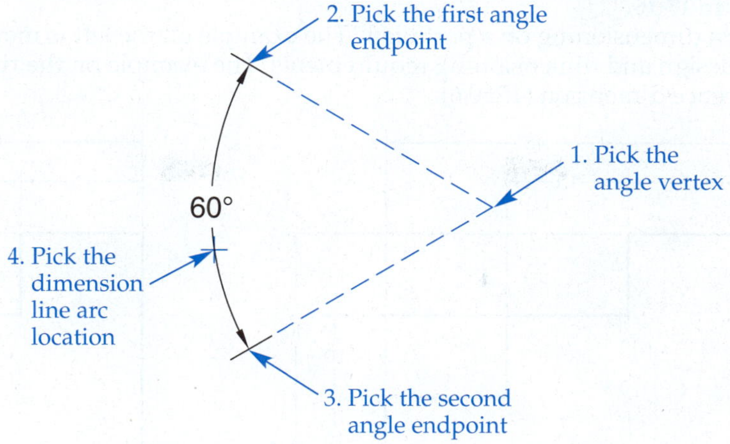

Figure 17-14. Placing angular dimensions using three points.

Exercise 17-7

www.g-wlearning.com/CAD/

Complete the exercise on the companion website.

Baseline, Chain, and Direct Dimensioning

Baseline dimensioning is commonly used in mechanical drafting. In baseline dimensioning, each dimension references a ***datum*** and is independent of the others. This achieves more accuracy in manufacturing. Figure 17-15 shows an object dimensioned with baseline dimensioning and surface datums.

Mechanical drafting sometimes uses ***chain dimensioning***. However, this method provides less accuracy than baseline dimensioning because each dimension is dependent on other dimensions in the chain, resulting in ***tolerance buildup***. In mechanical drafting, it is common to leave one dimension out and provide an overall dimension that controls the missing value. See Figure 17-16.

baseline dimensioning: A method of dimensioning in which several dimensions originate from a common surface, centerline, or center plane.

datum: Theoretically perfect surface, plane, point, or axis from which measurements can be taken.

chain dimensioning: A method of dimensioning in which dimensions appear in a line from one feature to the next.

tolerance buildup: Accumulation that occurs when the tolerance of each individual dimension builds on the next.

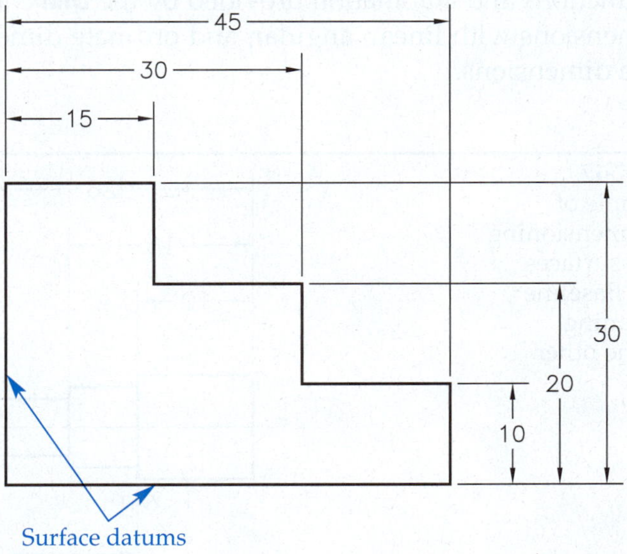

Figure 17-15. Baseline dimensioning on a part view. Maintain consistent spacing throughout the drawing.

Figure 17-16.
Chain dimensioning on a part view. The example on the left is more common, depending on the design and dimensioning requirements. The example on the right includes an optional reference dimension (45±0.6).

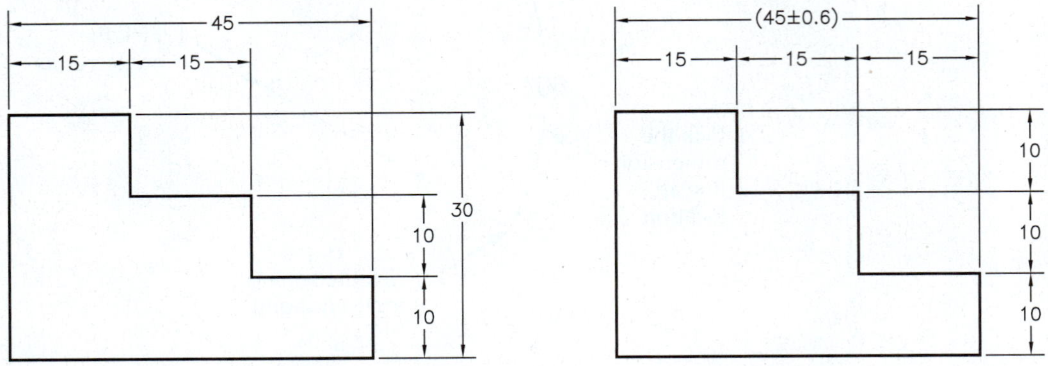

Architectural drafting uses chain dimensioning in most applications to reduce the need to calculate or find dimension values during construction. Architectural drafting practices usually show dimensions for all features, plus an overall dimension.

direct dimensioning: A type of dimensioning applied to control the specific size or location of one or more specific features.

Direct dimensioning controls the specific size or location of one or more specific features, often resulting in the least tolerance buildup. Figure 17-17 shows an example of direct dimensioning on a part view. Baseline or chain dimensioning could be used to dimension the view, but would create more tolerance accumulation.

NOTE

You can use object snaps to snap to elements of a dimension object, except for the endpoint of an extension line near the extension line origin. If you have a need to snap to an extension line endpoint, uncheck the **Ignore dimension extension lines** option in the **Drafting** tab of the **Options** dialog box.

Baseline Dimensioning

DIMBASELINE
Ribbon
Annotate > Dimensions
Baseline
Type
DIMBASELINE
DBA

The **DIM** and **DIMBASELINE** commands automate baseline dimensioning by allowing you to select several points to define a series of baseline dimensions. This textbook focuses on the **DIM** command for baseline dimensioning because of the additional functions and automation provided by the **DIM** command. You can create baseline dimensions with linear, angular, and ordinate dimensions. Chapter 18 describes ordinate dimensions.

Figure 17-17.
An example of direct dimensioning between surfaces X and Y. Baseline dimensioning locates the other surfaces.

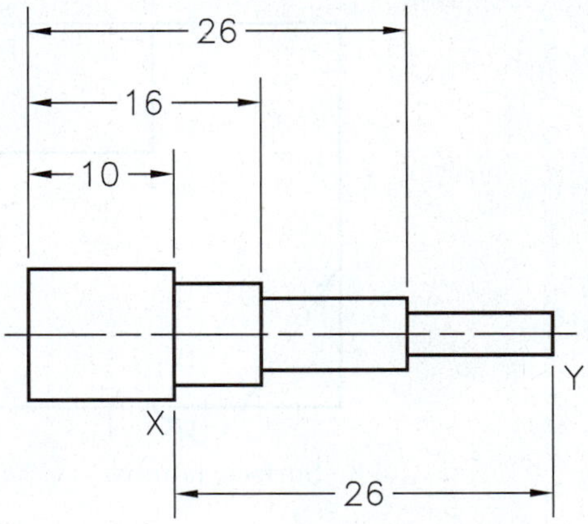

You must create the first dimension in the group of dimensions before using the **DIM** or **DIMBASELINE** command for baseline dimensioning. For example, for linear baseline dimensions, use the **DIM** or **DIMLINEAR** command to dimension the first feature. Then access the **DIM** or **DIMBASELINE** command to place additional dimensions.

When using the **DIM** command, it is helpful to select the **Baseline** option before making any selections. Then at the Specify first extension line origin as baseline or [Offset]: prompt, pick the dimension to serve as the base. The extension line nearest to the point where you select the dimension establishes the datum. Therefore, although you can select anywhere on the dimension object, it is good practice to pick the extension line originating from the datum. Now select the second extension line origin. Continue picking extension line origins until you have dimensioned all of the features. See Figure 17-18. Use the **Undo** option to undo previously drawn dimensions. Use the **Select** option to pick a different dimension to serve as the base for a new set of baseline dimensions.

AutoCAD treats linear dimensions created using the **Baseline** option of the **DIM** command as a set. This is an advantage to using the **DIM** command instead of the **DIMBASELINE** command. Dimensions are organized in correct order, even if you do not select extension line origins in the correct sequence, or if you use the **Baseline** option of the **DIM** command later to add dimensions. For example, the correct arrangement shown in Figure 17-18 can be created by selecting second extension line origins in a different order than shown.

You can also draw angular baseline dimensions. Use the **DIM** or **DIMANGULAR** command to dimension the first feature. Then access the **DIM** or **DIMBASELINE** command to place additional dimensions. Figure 17-19 shows angular baseline dimensions created using the **DIM** command. As with linear dimensions, you can pick an existing angular dimension other than the most recent dimension.

When placing linear or angular baseline dimensions with the **DIM** command, use the **Offset** option before selecting an extension line origin to override the default spacing between dimension lines. The default spacing is specified in the **Baseline spacing:** text box in the **Lines** tab of the **New** (or **Modify**) **Dimension Style** dialog box. The new offset applies to all baseline dimensions.

Figure 17-18.
Using the **Baseline** option of the **DIM** command to add linear baseline dimensions. AutoCAD automatically places, offsets, and organizes dimension objects. The order in which you pick second extension lines is not critical when using the **DIM** command.

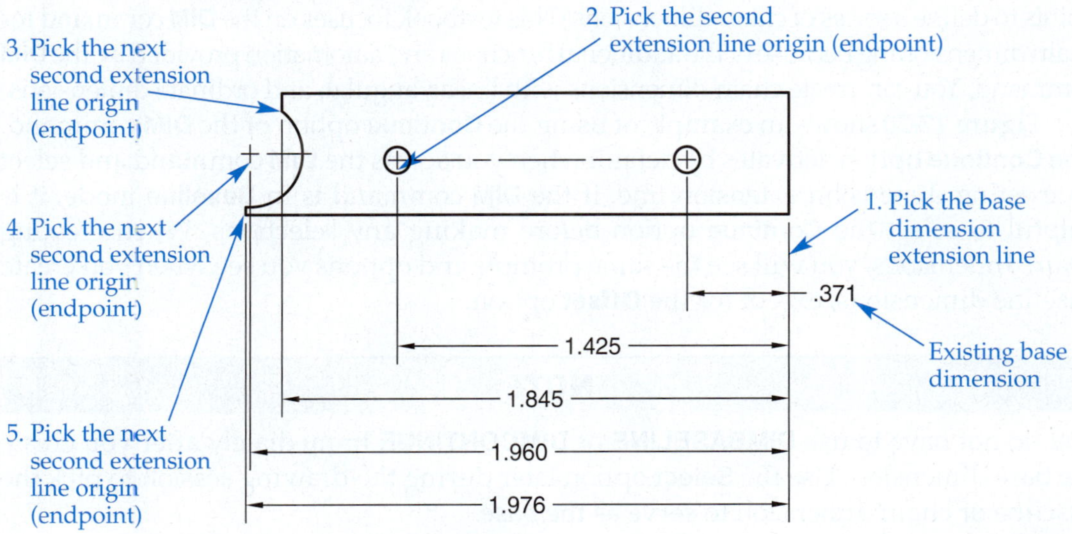

Figure 17-19.
Using the **Baseline** option of the **DIM** command to add angular baseline dimensions.

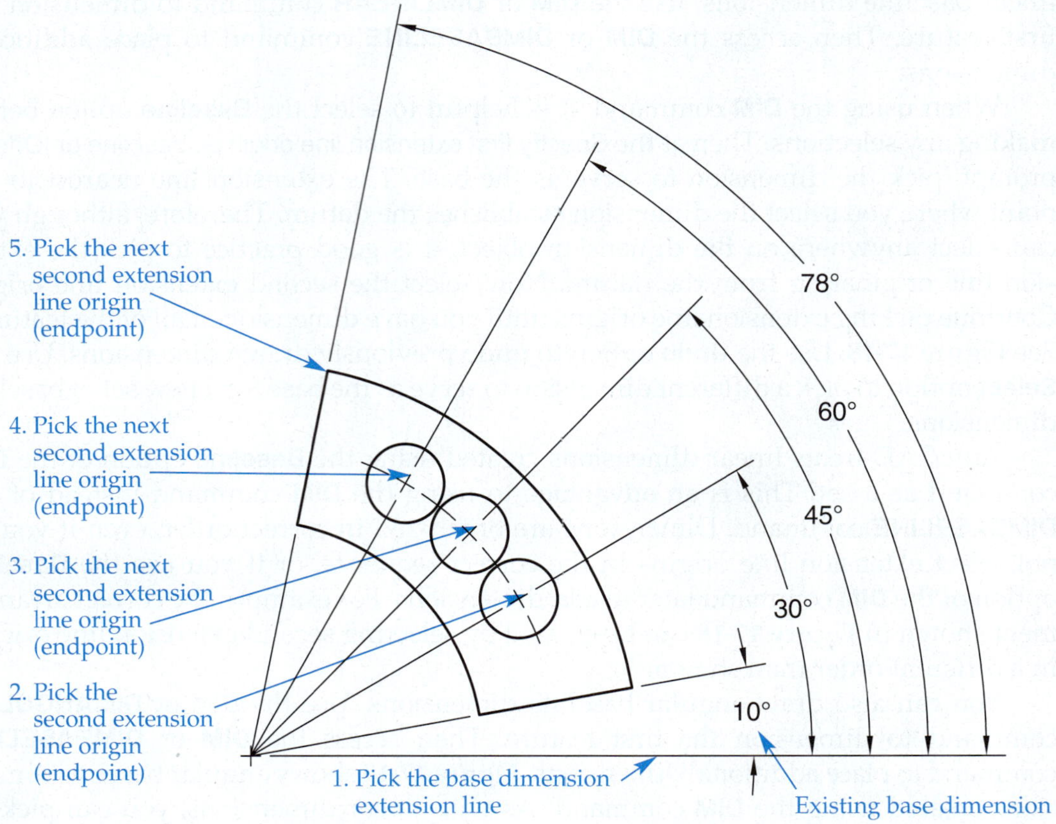

PROFESSIONAL TIP

The ASME standard minimum distance between dimension lines is .25″ (6 mm), but this is usually too close. A distance of .75″ (19 mm) from the object to the first dimension line and .5″ (12 mm) between other dimension lines is often ideal.

continued dimensioning: The AutoCAD term for chain dimensioning.

DIMCONTINUE
Ribbon
Annotate > Dimensions
Continue
Type
DIMCONTINUE
DCO

Chain Dimensioning

AutoCAD refers to chain dimensioning as ***continued dimensioning***. The **DIM** and **DIMCONTINUE** commands automate chain dimensioning by allowing you to select several points to define a series of chain dimensions. This textbook focuses on the **DIM** command for chain dimensioning because of the additional functions and automation provided by the **DIM** command. You can create chain dimensions with linear, angular, and ordinate dimensions.

Figure 17-20 shows an example of using the **Continue** option of the **DIM** command. The **Continue** option activates by default when you access the **DIM** command and select an existing dimension extension line. If the **DIM** command is in **Baseline** mode, it is helpful to select the **Continue** option before making any selections. When creating chain dimensions, you will see the same prompts and options you see when you create baseline dimensions, except for the **Offset** option.

NOTE

You do not have to use **DIMBASELINE** or **DIMCONTINUE** immediately after you create the base dimension. Use the **Select** option later during the drawing session to pick the baseline or chain dimension to serve as the base.

Figure 17-20.
Using the **Continue** option of the **DIM** command to create chain dimensions on a portion of a foundation plan.

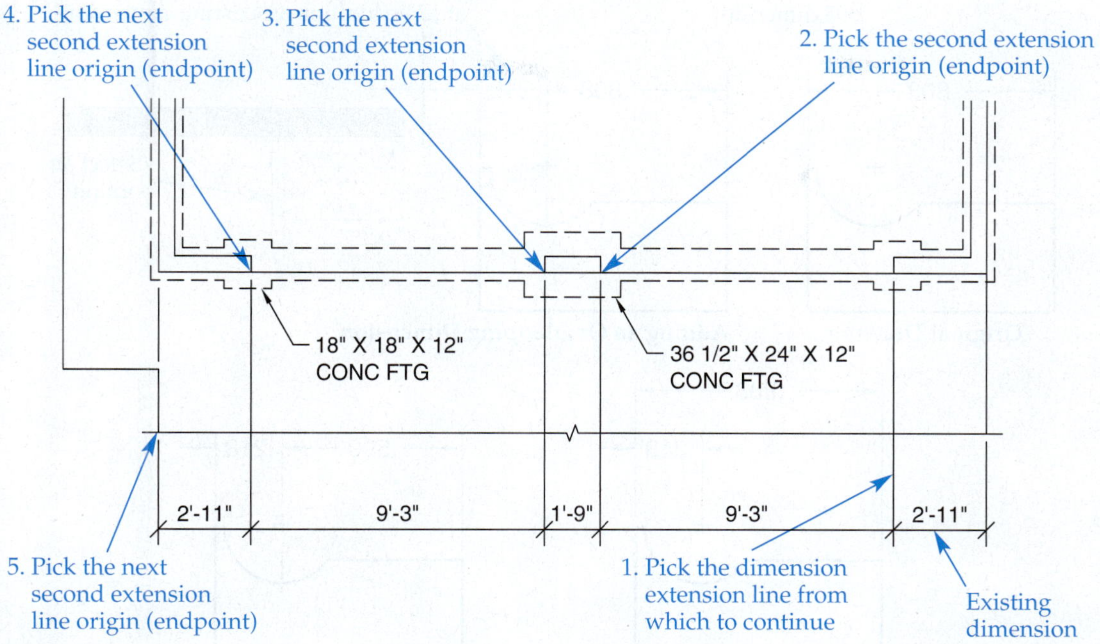

NOTE

By default, new dimensions created using the **DIM**, **DIMBASELINE**, and **DIMCONTINUE** commands are assigned the same layer and dimension style of the selected dimension. Change the **DIMCONTINUEMODE** system variable from 1 to 0 to assign the current layer and dimension style to new dimensions, regardless of the layer and dimension style assigned to the selected dimension.

NOTE

The **Baseline Dimension** option, available for grip-editing linear and angular dimensions, and the **Continue Dimension** option, available for grip-editing linear, angular, and ordinate dimensions, provide effective alternatives to the **DIM**, **DIMBASELINE**, and **DIMCONTINUE** commands. Chapter 20 provides more information on editing dimensions.

Exercise 17-8

www.g-wlearning.com/CAD/

Complete the exercise on the companion website.

Adjusting Overlapping Dimensions

AutoCAD recognizes when a dimension overlaps another dimension while using the **DIM** command and prompts you to take action before placing the dimension. This is another advantage to using the **DIM** command, especially when adding baseline and chain dimensions. Figure 17-21 shows an example of placing an overlapping dimension and the options available for handling the overlap.

Figure 17-21.
Options for handling overlapping dimensions while using the **DIM** command.

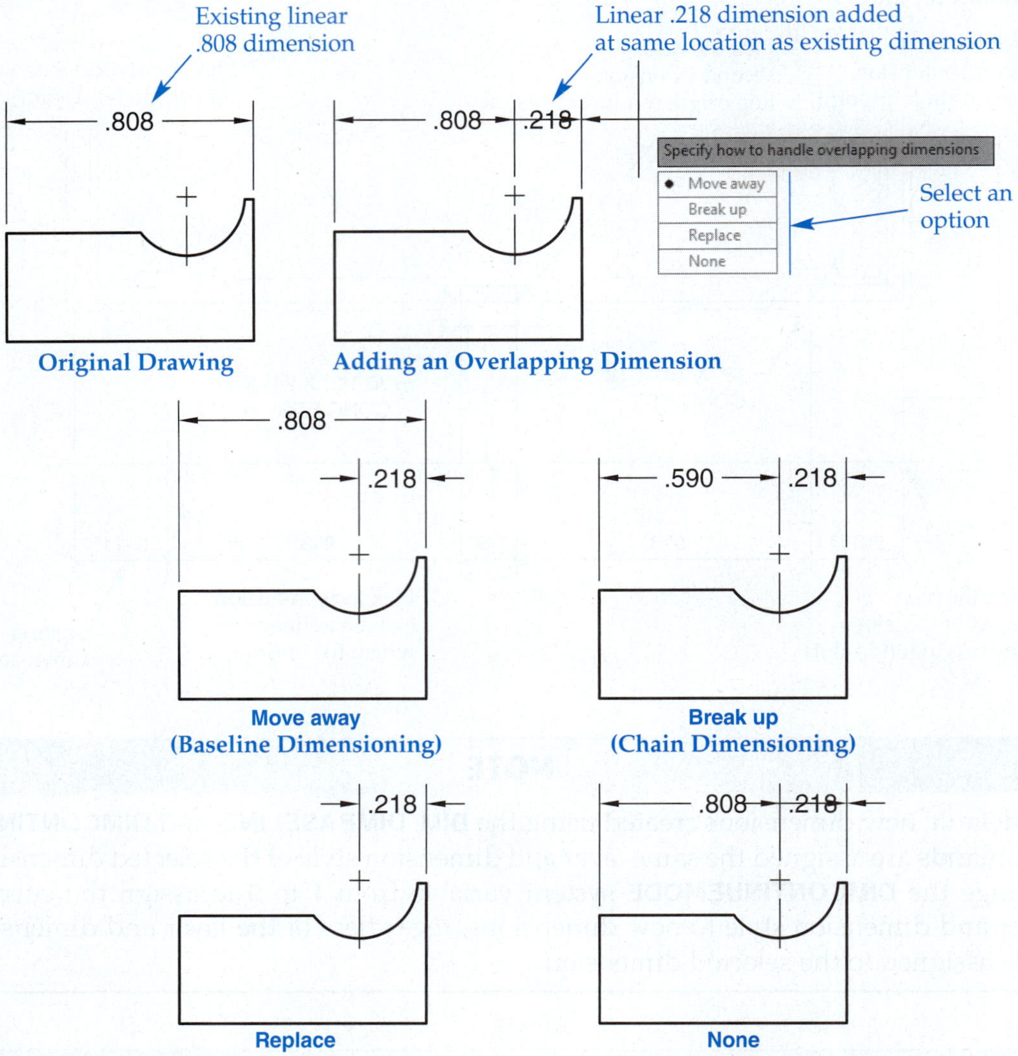

Dimension Layer Override

Though you can always change the layer assigned to an object, you generally make the appropriate layer current before drawing. For example, set a Dimension layer current before dimensioning. An alternative available when using the **DIM** command is to assign a layer other than the current layer to new dimension objects. This is most practical if you plan to assign all dimensions to one layer.

An easy way to override the layer assigned to dimension objects is to pick the layer name from the **Dim Layer Override** drop-down list in the **Dimensions** panel of the **Annotate** ribbon tab. See Figure 17-22. Use the vertical scroll bar to move up and down through a long list. You can also set the dimension layer override using the **Layer** option of the **DIM** command. Enter the layer name or select an object assigned to the layer to use as the override layer. The **DIMLAYER** system variable stores the dimension layer override setting.

NOTE

You can use the **Layer** option of the **DIM** command to enter a name of a new layer that does not currently exist to use for the dimension layer override. However, it is better practice to select an existing layer with appropriate properties.

Figure 17-22.
Use the **Dim Layer Override** drop-down list to select the layer to assign to new dimension objects.

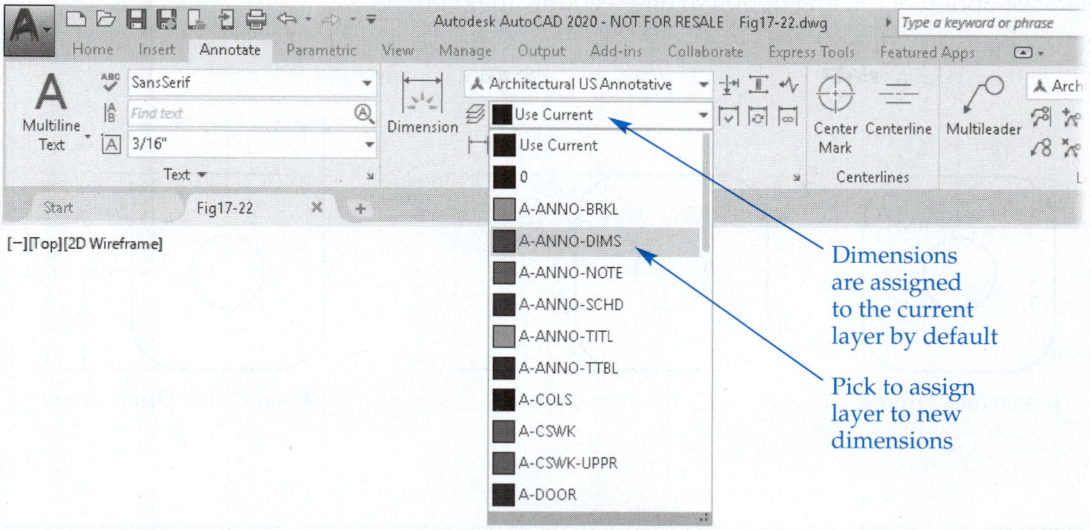

Using QDIM to Dimension

Ribbon
Annotate
> Dimensions
Quick Dimension
Type
QDIM

The **QDIM** command automates the dimensioning process by creating a group of dimensions based on the objects you select. You can also use the **QDIM** command to edit dimensions, as explained in Chapter 20.

Access the **QDIM** command and select the objects to dimension. Press [Enter] to display a preview of the dimensions attached to the cursor. By default, AutoCAD establishes an extension line from every line and polyline endpoint and every circle and arc center point.

To apply chain dimensioning, use the default **Continuous** option and specify the location of the dimension lines. See Figure 17-23A. To apply baseline dimensioning, select the **Baseline** option and specify the location of the first dimension line. See Figure 17-23B. If the **QDIM** command does not reference the appropriate datum, use the **datumPoint** option before locating the dimensions to specify a different datum point.

Use the **Edit** option before locating the dimensions to add dimensions to, or remove dimensions from, the current set. Marks indicate the points acquired by the **QDIM** command. Use the **Add** option to specify a point to add a dimension, or use the **Remove** option to specify a point to remove the corresponding dimension. Press [Enter] to return to the previous prompt and continue using the **QDIM** command. Figure 17-23C shows a view dimensioned using the **QDIM** command twice. The **Remove** option of the **Edit** option was used each time to remove unwanted dimensions.

The **Ordinate** option allows you to apply rectangular coordinate dimensioning without dimension lines, as explained in Chapter 18. The **Diameter** option dimensions all selected circles and arcs with diameter dimensions and no linear dimensions. The **Radius** option dimensions all selected circles and arcs with radius dimensions and no linear dimensions. Chapter 18 describes diameter and radius dimensions.

NOTE

The **Settings** option provides an **Endpoint** or **Intersection** toggle meant to set the object snap mode for locating extension line origins. The **Staggered** option creates a unique grouping of dimensions, similar to baseline dimensioning, but beginning from the center objects and expanding outward.

Figure 17-23.
The **QDIM** command can dimension multiple selected objects, without you having to pick extension line origins. This example shows selecting all objects initially and then using the **Remove** option of the **Edit** option to deselect unwanted points.

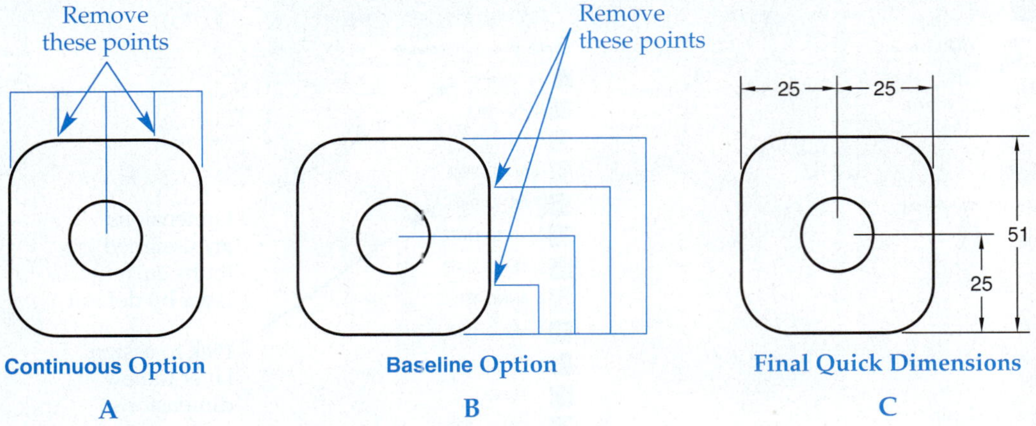

Exercise 17-9

www.g-wlearning.com/CAD/

Complete the exercise on the companion website.

Chapter Review

Answer the following questions using the information in this chapter.

1. Name the two options of the **DIM** and **DIMLINEAR** commands that allow you to change dimension text.
2. Give two examples of symbols that automatically appear with some dimensions.
3. What is one purpose of the AutoCAD gdt.shx font?
4. Name the three dimensioning commands that provide linear dimensions for angled surfaces.
5. Which command allows you to place a break symbol in a dimension line?
6. Name two commands used to dimension angles in degrees.
7. Describe a way to specify an angle in degrees if the angle is greater than 180°.
8. Define *baseline dimensioning.*
9. Define *direct dimensioning.*
10. How do you place a baseline dimension from the origin of the previously drawn dimension?
11. What is the conventional term for the type of dimensioning AutoCAD refers to as continued dimensioning?
12. Explain how to assign a layer other than the current layer to new dimension objects.
13. Which command other than **DIM** and **DIMBASELINE** creates baseline dimensions?
14. Explain how to remove dimensions from the current set when you use the **QDIM** command.
15. Name at least three options for dimensioning that are available through the **QDIM** command.

Drawing Problems

Start AutoCAD if it is not already started. Start a new drawing for each problem using an appropriate template of your choice. The template should include appropriate layers, text styles, and dimension styles for drawing the given objects. Add layers, text styles, and dimension styles as needed. Draw all objects using appropriate layers. Use appropriate text and dimension styles, justification, and format. Follow the specific instructions for each problem. Use only drawing and editing commands and techniques you have already learned. Use your own judgment and approximate dimensions when necessary. Apply dimensions accurately using ASME or appropriate industry standards.

Note: Some of the problems in this chapter are built on problems from previous chapters. If you have not yet completed those problems, complete them now.

▼ Basic

1. **Mechanical** Start a new drawing from scratch or use a fractional-unit template of your choice. Save the file as P17-1. If you have not yet completed Problem 3-7, do so now. Open P3-7 and copy one instance of Object A and Object B to the P17-1 drawing. The P17-1 file should be active. Dimension the views as shown in Problem 3-7.

2. **General** Save P3-5 as P17-2. If you have not yet completed Problem 3-5, do so now. In the P17-2 file, dimension the drawing as shown.

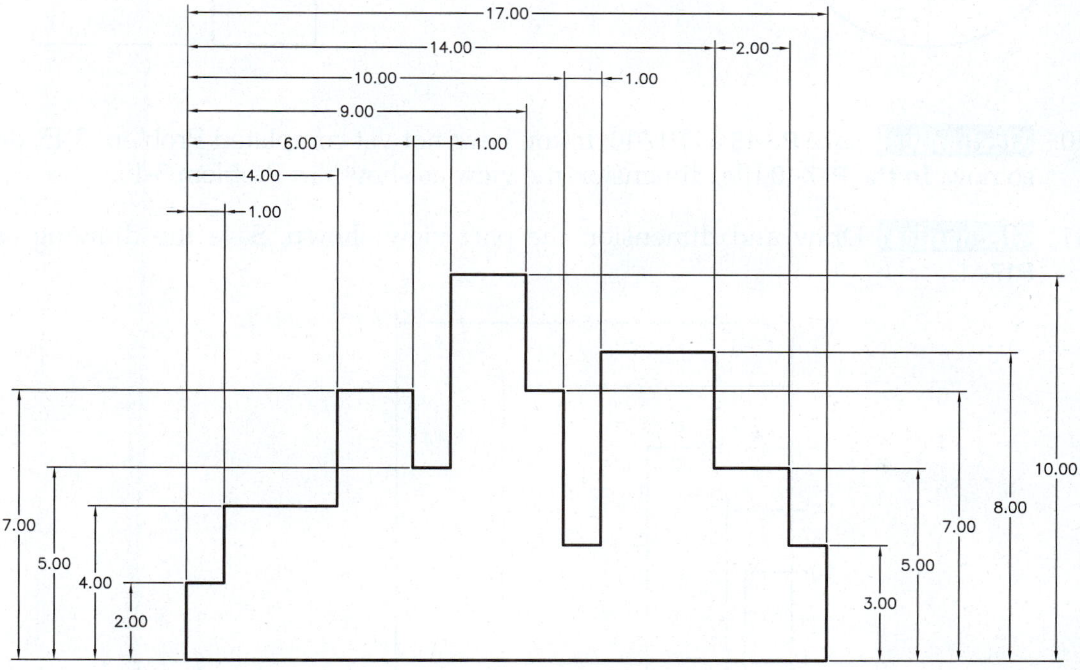

3. **General** Save P3-10 as P17-3. If you have not yet completed Problem 3-10, do so now. In the P17-3 file, dimension the views as shown in Problem 3-10.

4. **Mechanical** Save P3-13 as P17-4. If you have not yet completed Problem 3-13, do so now. In the P17-4 file, dimension the view as shown in Problem 3-13.

5. **Mechanical** Save P3-14 as P17-5. If you have not yet completed Problem 3-14, do so now. In the P17-5 file, dimension the view as shown in Problem 3-14. Note that this is a metric drawing.

6. **Mechanical** Save P3-8 as P17-6. If you have not yet completed Problem 3-8, do so now. In the P17-6 file, dimension the view as shown in Problem 3-8.

7. **Mechanical** Save P3-9 as P17-7. If you have not yet completed Problem 3-9, do so now. In the P17-7 file, dimension the view as shown in Problem 3-9.

8. **General** Write a report explaining the difference between baseline and chain dimensioning. Use a word processor and include sketches giving examples of each method.

▼ Intermediate

9. **Mechanical** Draw and dimension the views of the shaft shown. Save the drawing as P17-9.

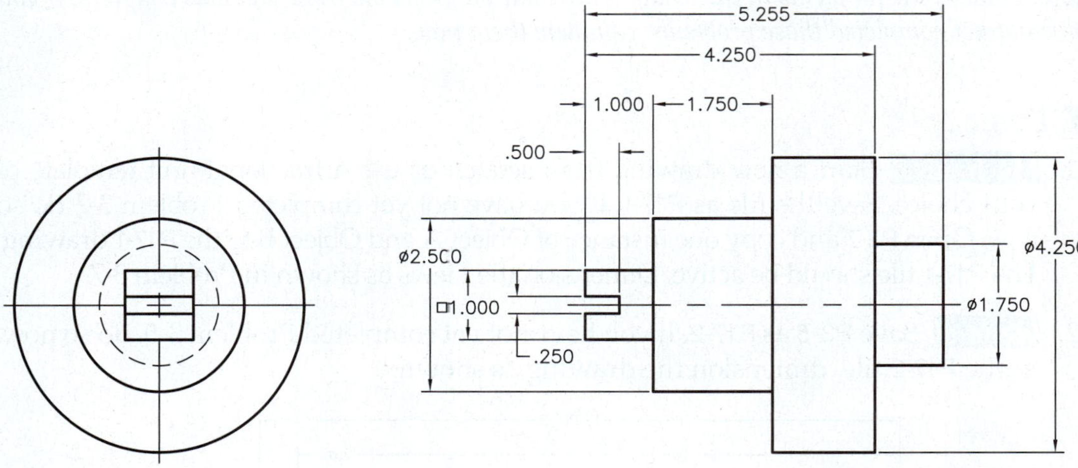

10. **Mechanical** Save P3-15 as P17-10. If you have not yet completed Problem 3-15, do so now. In the P17-10 file, dimension the view as shown in Problem 3-15.

11. **Mechanical** Draw and dimension the part view shown. Save the drawing as P17-11.

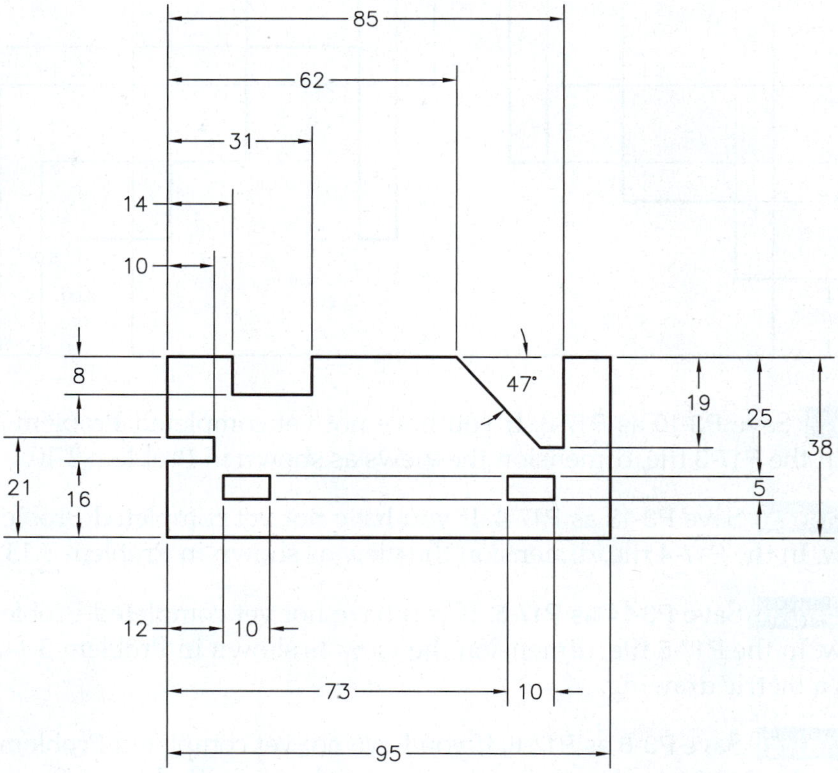

12. **Architectural** Draw and dimension the partial floor plan shown. Save the drawing as P17-12.

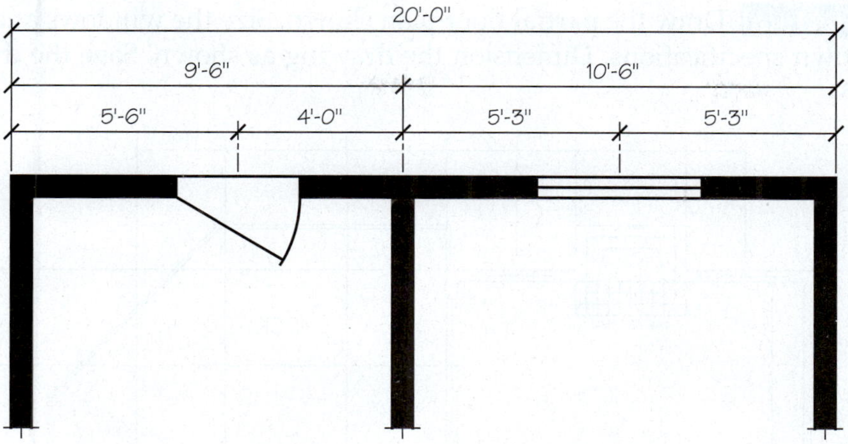

13. **Architectural** Draw and dimension the partial floor plan shown. Save the drawing as P17-13.

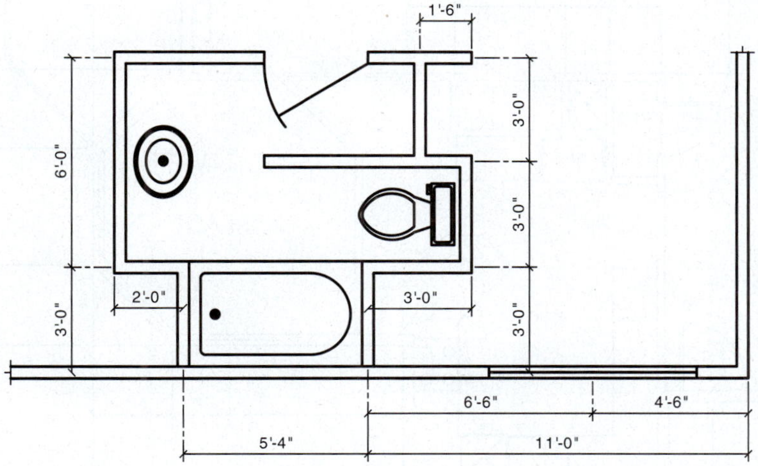

▼ Advanced

14. **Mechanical** Save P3-3 as P17-14. If you have not yet completed Problem 3-3, do so now. In the P17-14 file, make one copy of the view at a new location to the right of the original view. Dimension the view on the left using baseline dimensioning. Dimension the view on the right using chain dimensioning.

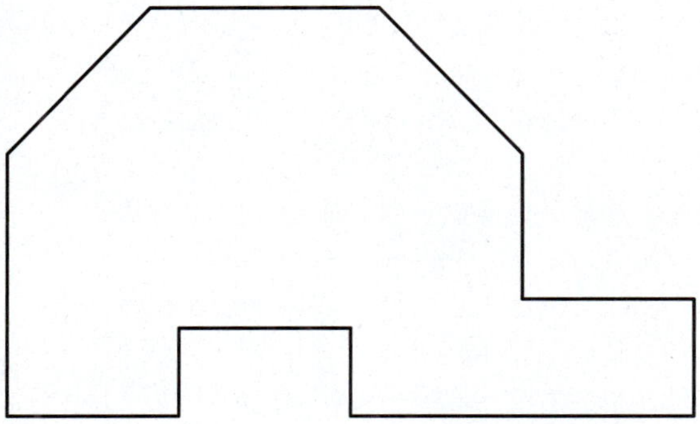

15. **Architectural** Save P8-9 as P17-15. If you have not yet completed Problem 8-9, do so now. In the P17-15 file, dimension the desk as shown in Problem 8-9.

16. **Architectural** Draw the partial floor plan shown. Size the windows and doors to your own specifications. Dimension the drawing as shown. Save the drawing as P17-16.

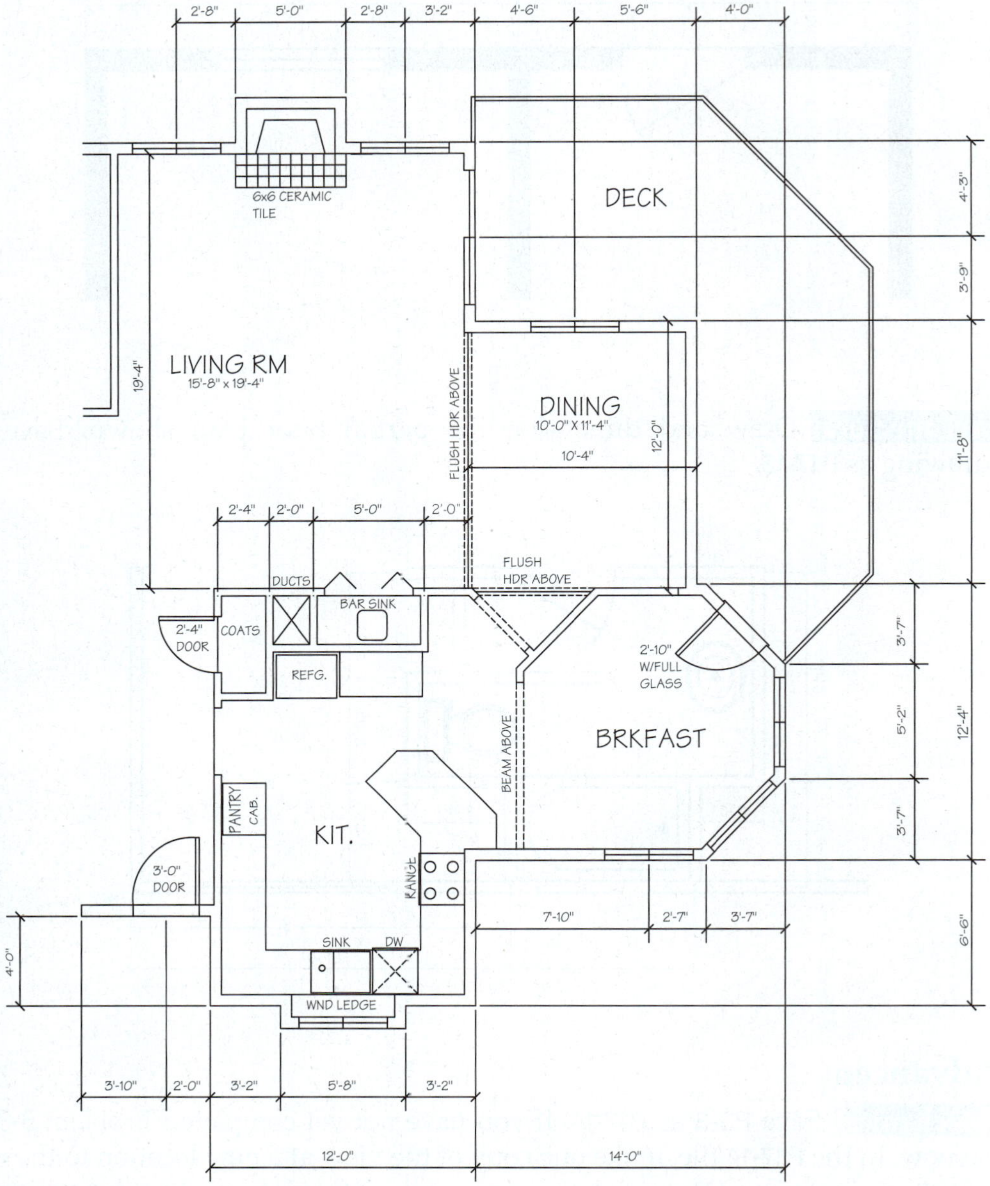

AutoCAD Certified Professional Exam Practice

Answer the following questions using the information in this chapter.

1. Which of the following is the minimum ASME-recommended spacing from the object to the first dimension line? *Select all that apply.*
   A. .375″
   B. .5″
   C. .75″
   D. 10 mm
   E. 19 mm
   F. 12 mm
2. How can you add a symbol to dimension text? *Select all that apply.*
   A. apply control codes or characters
   B. create a dimension style that references the gdt.shx font
   C. hold [Ctrl] and press an appropriate letter key
   D. press the appropriate function key
   E. use the **Mtext** option of the dimensioning command
   F. use the **Text** option of the dimensioning command
3. Which of the following commands continues dimensioning from a previously placed dimension? *Select all that apply.*
   A. **DIMANGULAR**
   B. **DIMBASELINE**
   C. **DIMCONTINUE**
   D. **DIMJOGLINE**
   E. **DIM**

Follow the instructions in each problem. Complete each problem by answering the related question.

4. **Navigate to this chapter on the companion website and open CPE-17angles.dwg.**
   What are the measured values of ANGLE A and ANGLE B? Use the appropriate dimensioning command(s) to find the measurements.

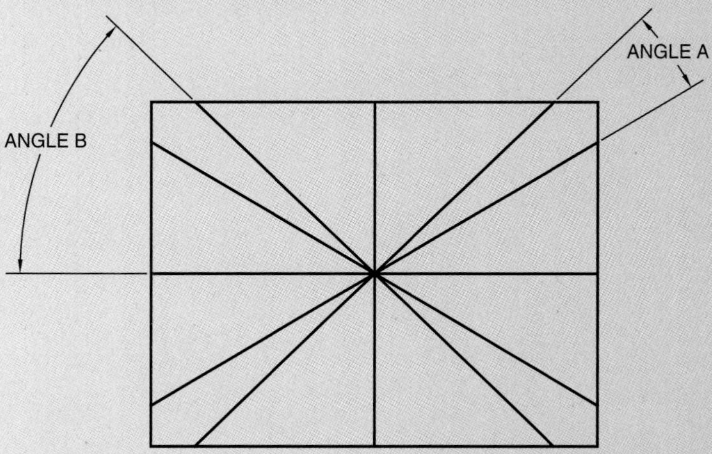

5. **Navigate to this chapter on the companion website and open CPE-17dimstyle.dwg.** Use the **Dimension Style Manager** to change the baseline spacing in the Standard dimension style to .5 and leave all other settings as they are currently set. Use the **DIM** command to finish the dimensions as shown. What are the coordinates of Point 1?

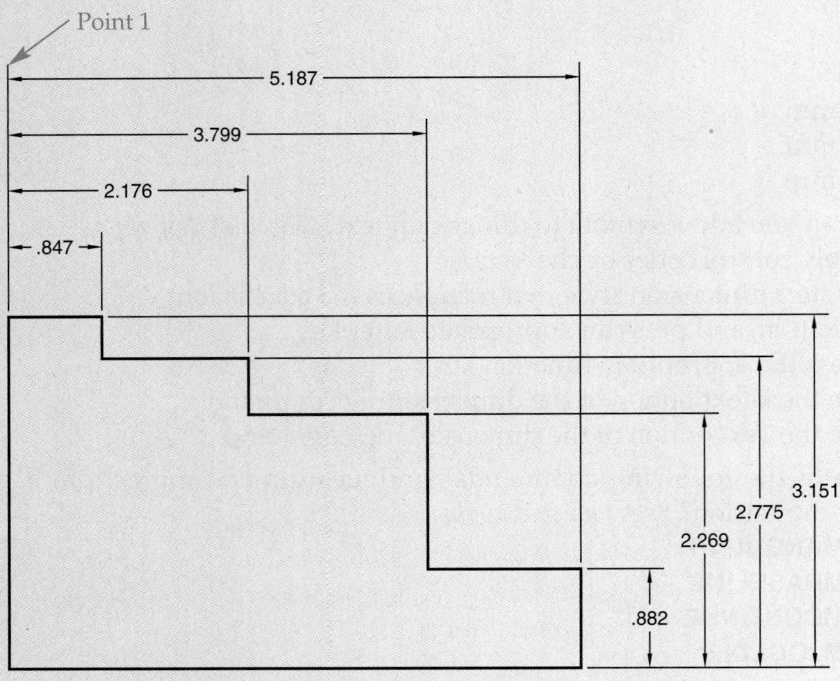

CHAPTER 18

Dimensioning Features and Alternate Practices

Learning Objectives

After completing this chapter, you will be able to:

- ✓ Dimension circles, arcs, and other curves.
- ✓ Create center dashes and centerlines.
- ✓ Create and use multileader styles.
- ✓ Draw leaders using the **MLEADER** command.
- ✓ Apply alternate dimensioning practices.
- ✓ Create ordinate dimension objects.
- ✓ Mark up a drawing using the **REVCLOUD** and **WIPEOUT** commands.

A drawing must describe the size and location of all features for manufacturing or construction. This chapter explains options for dimensioning object features and introduces alternate mechanical drafting dimensioning practices. This chapter also introduces basic redlining techniques using the **REVCLOUD** and **WIPEOUT** commands.

Dimensioning Circles, Arcs, and Other Curves

AutoCAD includes several commands for dimensioning circular and noncircular curves. You can use AutoCAD to dimension a circle or arc with a diameter or radius. However, circles are generally dimensioned by their diameter, and arcs are dimensioned by their radius. This is also the ASME standard for dimensioning circles and arcs. Curves that do not have a constant radius require other dimensioning methods, as described in this chapter.

You can use the **DIM** command to dimension the diameter or radius of a circle or arc. Alternative commands, such as **DIMDIAMETER** and **DIMRADIUS**, are also available for completing the same tasks. Dimension commands reference the current dimension style and the objects or points you select to create a dimension object. When you use the **DIM** command to dimension the diameter of a circle, for example, you create a diametric dimension object that includes all related dimension style characteristics, centerlines, arrowheads, a leader, and a dimension value associated with the diameter. The leader points to the center of the circle (or the center of the arc, when dimensioning an arc) as recommended by the ASME standard.

NOTE

The **DIM** command dimensions a circle with a diameter dimension and an arc with a radius dimension by default for the current drawing session. If you set a different option while hovering over an object, such as the **Angular** option explained in Chapter 17, the setting changes to the new default for the corresponding object type. You must hover over a circle or arc while the **DIM** command is active and select an option, such as **Diameter** or **Radius**, to change the setting.

Dimensioning Circles

DIM

Ribbon
Home
> Annotation
Annotate
> Dimensions
Dimension
Type
DIM

The **DIM** command is one option for drawing a single diametric dimension object. Access the **DIM** command and move the pick box over the circle to dimension. When the circle becomes highlighted and you see a preview of the dimension, pick the circle to attach a leader to the circle and the dimension value to the crosshairs. Now use the options that appear at the Specify diameter dimension location or [Radius/Mtext/Text/text aNgle/Undo]: prompt as needed. Select the **Radius** option to dimension the circle by the radius. Select the **Undo** option to undo the selection without exiting the **DIM** command. To apply the default settings and create a diametric dimension, specify a point to locate the dimension value. See Figure 18-1.

The **DIM** command remains active after you create a dimension, allowing you to continue dimensioning without reselecting the command. Select the **Undo** option as needed to undo dimensions created during the current **DIM** command sequence. Exit the command when finished.

NOTE

DIMDIAMETER

Ribbon
Home
> Annotation
Annotate
> Dimensions
Diameter
Type
DIMDIAMETER
DDI

The **DIMDIAMETER** command is an alternative to the **DIM** command for dimensioning the diameter of a circle or arc. Use the **DIM** and **DIMDIAMETER** commands in a similar manner.

The **DIM** and **DIMDIAMETER** commands and many other dimension commands described in this chapter include the **Mtext**, **Text**, and **Angle** or **text aNgle** options explained in Chapter 17. Use the **Mtext** or **Text** option to add information to or change the dimension value. The **Angle** or **text aNgle** option changes the dimension text angle, although this practice is not common.

Figure 18-1.
Using the **DIM** command to dimension a circle.

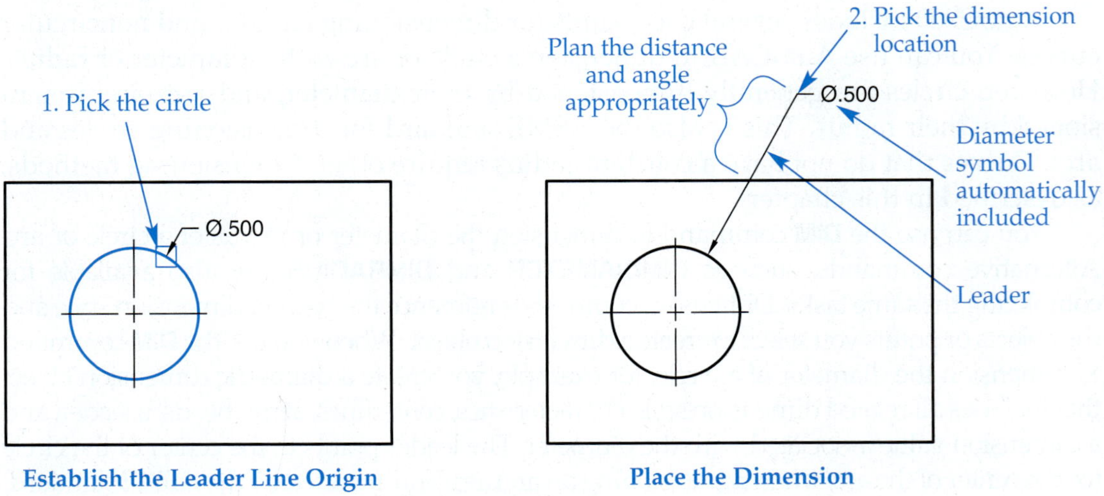

Exercise 18-1

www.g-wlearning.com/CAD/

Complete the exercise on the companion website.

Dimensioning Holes

You generally dimension holes in the view in which they appear as circles. Give location dimensions to the center and a leader showing the diameter. The **DIM** or **DIMDIAMETER** command is effective for dimensioning holes. To note multiple holes of the same size, dimension the size of one hole using the **DIM** or **DIMDIAMETER** command and the **Mtext** or **Text** option. Precede the diameter with the number of holes followed by X and then a space. See the 2X ∅.50 dimension in Figure 18-2.

PROFESSIONAL TIP

The ASME standard recommends a small space between the object and the extension line. To specify the space, adjust the **Offset from origin** setting in the dimension style to an appropriate positive value, such as .063″ (1.5 mm). When dimensioning to centerlines to locate circular features, pick the endpoint of the center mark as shown. Picking the endpoint of a centerline leaves an unacceptable space between the centerline and the beginning of the extension line.

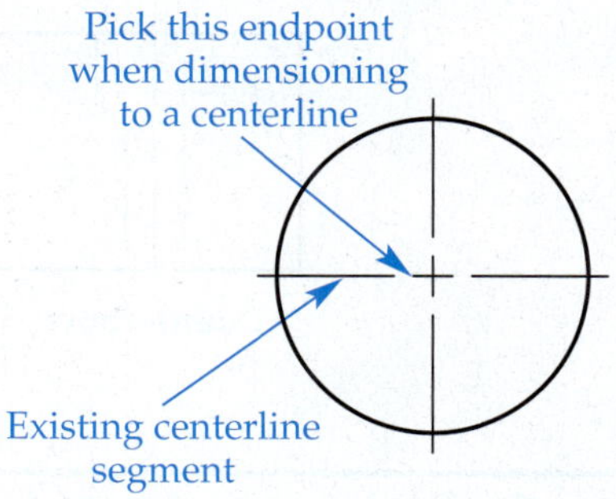

Dimensioning for Manufacturing Processes

Counterbore, ***spotface***, and ***countersink*** manufacturing processes are examples of hole features that are dimensioned using symbols. Dimension these manufacturing processes in the view in which holes appear as circles, with a leader providing machining information in a note. See Figure 18-3. The **Mtext** option of the **DIM** or **DIMDIAMETER** command provides a convenient method for dimensioning manufacturing processes. Many symbols are available in the gdt.shx font. You can also create custom symbols as ***blocks***. Blocks are described later in this textbook.

counterbore: A larger-diameter hole machined at one end of a smaller hole that provides a place for the fastener head.

spotface: A larger-diameter hole machined at one end of a smaller hole that provides a smooth, recessed surface for a washer; similar to a counterbore, but not as deep.

countersink: A cone-shaped recess at one end of a hole that provides a mating surface for a fastener head of the same shape.

block: A symbol previously created and saved for reuse.

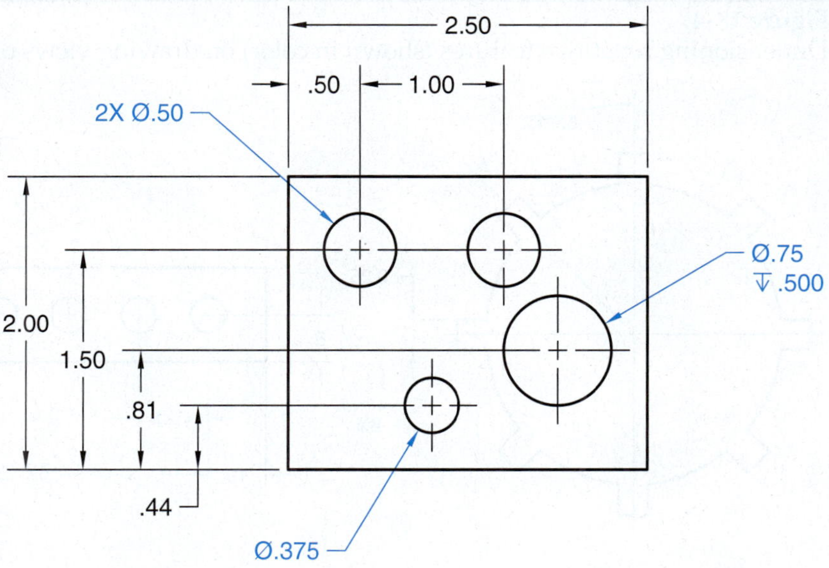

Figure 18-2.
Dimensioning holes on a mechanical part drawing view.

Figure 18-3.
Dimension notes for machining processes. You can insert symbols as blocks or use lowercase letters in the gdt.shx font.

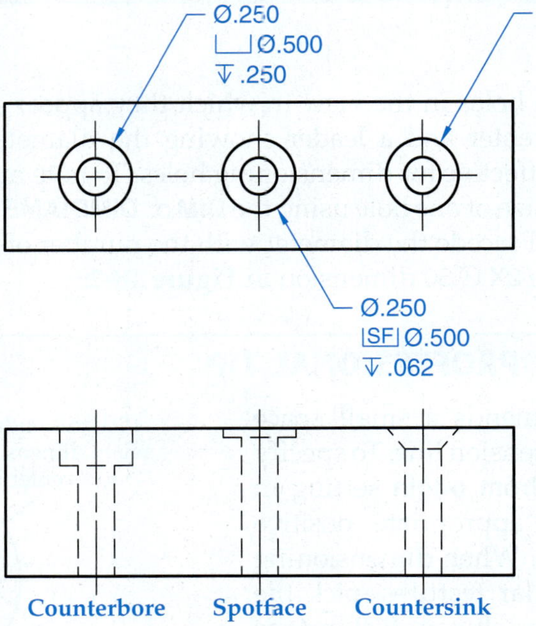

Reference Material

www.g-wlearning.com/CAD/

Drafting Symbols

For more information about common drafting symbols and the gdt.shx font, go to the companion website, select the **Resources** tab, and select **Drafting Symbols**.

Dimensioning Repetitive Features

repetitive features: Multiple features having the same shape and size.

Dimension *repetitive features* with the number of repetitions followed by an X, a space, and the dimension value. See Figure 18-4. Use a dimension command appropriate for the application, and use the **Mtext** or **Text** option to add repetitive information to the dimension value. Use the **MLEADER** command, described later in this chapter, to create the 8X note shown in Figure 18-4.

Figure 18-4.
Dimensioning repetitive features (shown in color) on drawing views of mechanical parts.

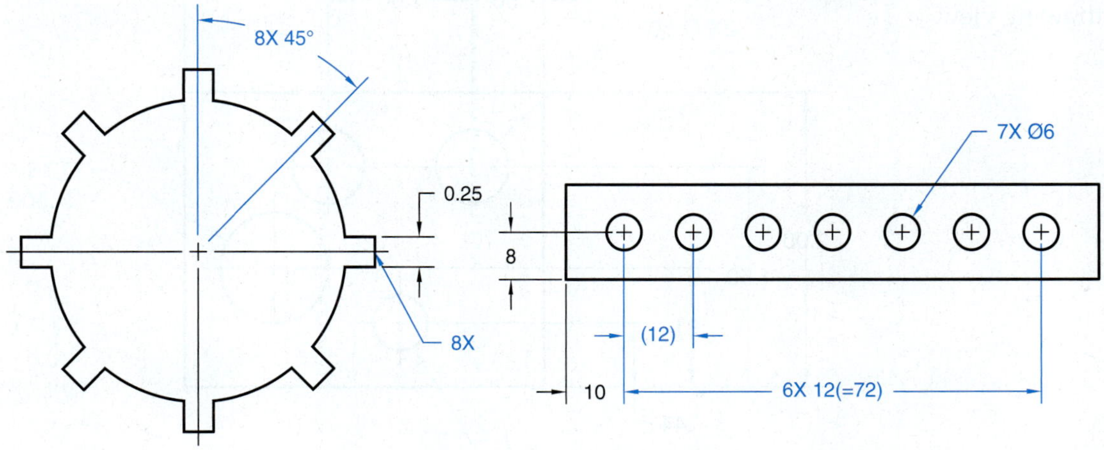

Exercise 18-2

www.g-wlearning.com/CAD/

Complete the exercise on the companion website.

Dimensioning Arcs

The **DIM** command is one option for drawing a single radial dimension object. Access the **DIM** command and move the pick box over the arc or polyline arc to dimension. When the arc becomes highlighted and you see a preview of the dimension, pick the arc to attach a leader to the arc and the dimension value to the crosshairs. Now use the options that appear at the Specify radius dimension location or [Diameter/Angular/Mtext/Text/text aNgle/Undo]: prompt as needed. Select the **Diameter** option to dimension the arc by the diameter. Select the **Undo** option to undo the selection without exiting the **DIM** command. To apply the default settings and create a radial dimension, specify a point to locate the dimension value. See Figure 18-5.

NOTE

The **DIMRADIUS** command is an alternative to the **DIM** command for dimensioning the radius of an arc, polyline arc, or circle. Use the **DIM** and **DIMRADIUS** commands in a similar manner.

DIMRADIUS
Ribbon
Home
> Annotation
Annotate
> Dimensions
Radius
Type
DIMRADIUS
DRA

Dimensioning Arc Length

Use the **DIM** or **DIMARC** command to dimension the length of an arc or polyline arc. When using the **DIM** command, by default, you must hover over an arc while the command is active and select the **arc Length** option. When the arc becomes highlighted and you see a preview of the dimension, pick the arc to originate an extension line from each endpoint and locate the angle vertex at the center point. Pick the location of the dimension line arc to place the dimension. By default, the arc length symbol occurs before the text. The ASME standard recommends placing the symbol over the text, as shown in Figure 18-6. The dimension style controls the symbol placement.

DIMARC
Ribbon
Home
> Annotation
Annotate
> Dimensions
Arc Length
Type
DIMARC
DAR

Figure 18-5.
Using the **DIM** command to dimension an arc.

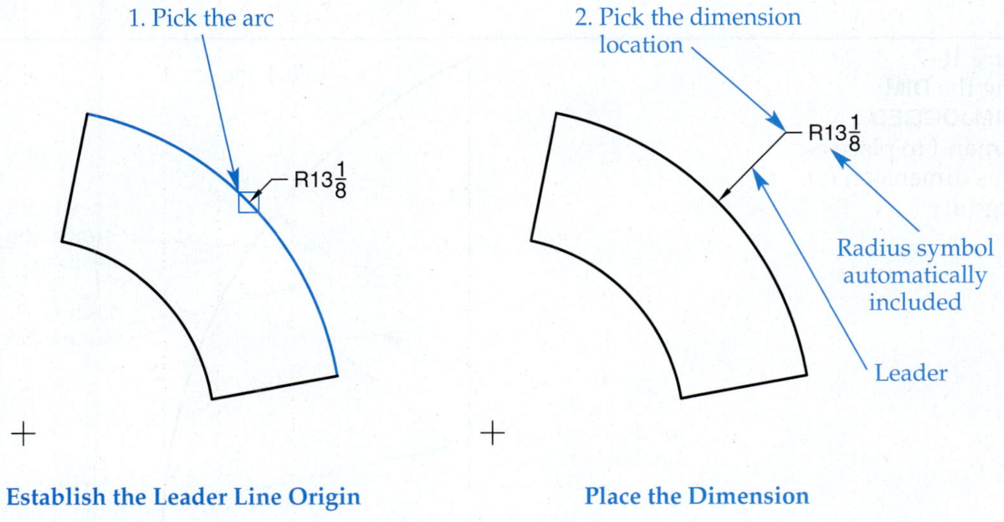

Figure 18-6.
Using the **DIM** or **DIMARC** command to dimension the length of an arc.

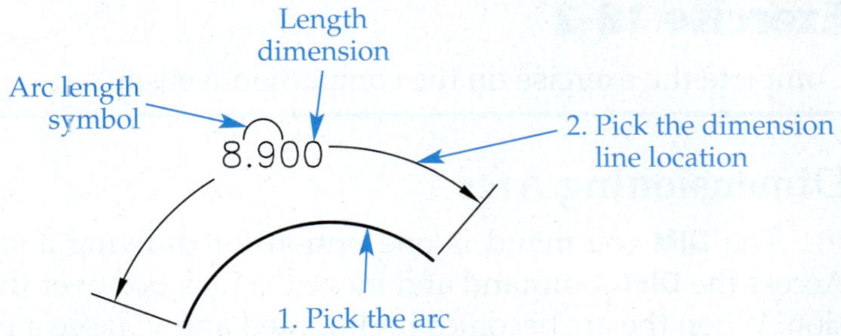

NOTE

Unlike the **DIM** command, the **DIMARC** command includes a **Partial** option and a **Leader** option, which is available when the arc is greater than 90°. Before placing the arc length dimension, select the **Partial** option and then select two points on the arc to dimension the length between the points. The **Leader** option allows you to add a leader pointing to the arc from the dimension value.

Dimensioning Large Arcs

DIMJOGGED

Ribbon
Home
> Annotation
Annotate
> Dimensions
Jogged
Type
DIMJOGGED
JOG

Use the **DIM** or **DIMJOGGED** command to dimension an arc or polyline arc that is so large that the center point cannot appear on the layout. Jogging a dimension line is most appropriate for large arcs, but you can select a circle. When using the **DIM** command, by default, you must hover over an arc or circle while the command is active and select the **Jogged** option. Pick the arc or circle to dimension and then specify a location for the origin of the center. This point represents the center of the arc or circle. The radius associated with the selected object does not change. Select a location for the dimension line and then pick a location for the break symbol. See Figure 18-7. Grip editing is an effective way to move the components of the dimension after you place the dimension.

Dimensioning Fillets and Rounds

fillets: Small arcs on inside corners designed to strengthen inside corners.

rounds: Small arcs on outside corners used to relieve sharp corners.

You can dimension ***fillets*** and ***rounds*** individually as arcs, using the **DIM** or **DIMRADIUS** command, or collectively in a general note. See Figure 18-8. On mechanical drawings, it is common to include a general note such as ALL FILLETS AND ROUNDS R.125 UNLESS OTHERWISE SPECIFIED on the drawing.

Figure 18-7.
Using the **DIM** or **DIMJOGGED** command to place a radius dimension for a large arc.

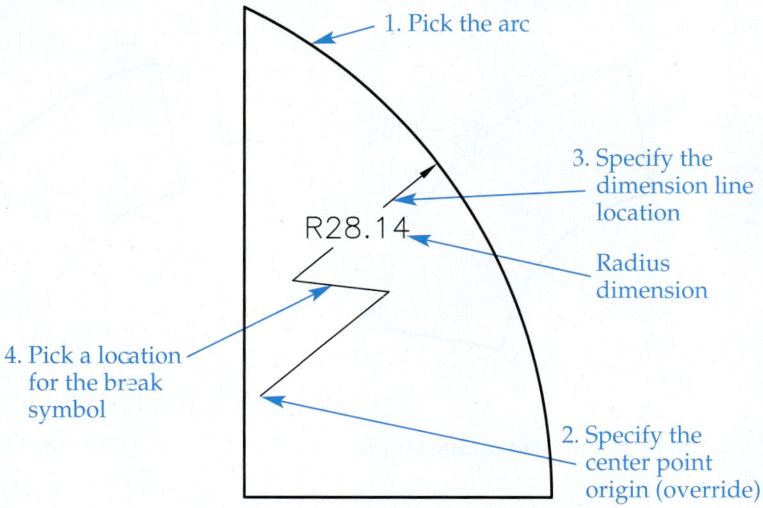

Figure 18-8.
Dimensioning fillets and rounds.

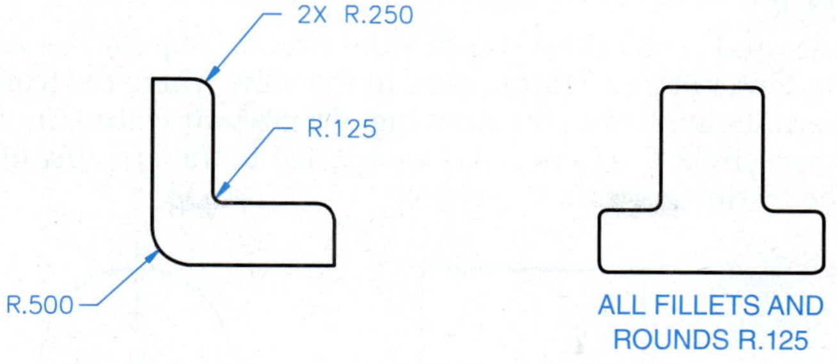

Exercise 18-3

www.g-wlearning.com/CAD/

Complete the exercise on the companion website.

Dimensioning Other Curves

Dimension curves as arcs when possible. When an arc does not have a constant radius, dimension to points along the curve using the **DIM** or **DIMLINEAR** command. See Figure 18-9.

Adding Center Dashes and Centerlines

Centerlines are used to show and locate the centers of circular features and to represent the axes of circular or symmetrical objects. Centerlines typically extend past objects, as shown in Figure 18-10A and Figure 18-10B. ASME standards refer to a two-segment cross that does not extend past objects as the center dashes of centerlines. See Figure 18-10C. Center dashes typically apply to circular objects that are too small to receive centerlines. Center dashes are also common for rectangular coordinate dimensioning without dimension lines, regardless of circular object size, as described later in this chapter.

Figure 18-9.
Dimensioning curves that do not have a constant radius.

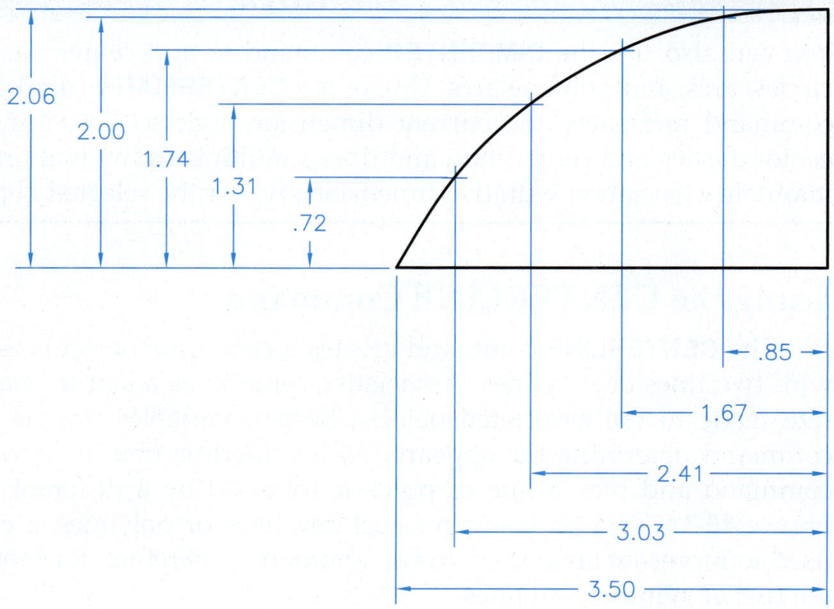

Figure 18-10.
Centerlines and center dashes applied to circular objects. A—A centerline used to show and locate the center of a circular feature in the view where the feature appears as lines, or edges. B—Centerlines applied to the drawing of a circular feature in the view where the feature appears as a circle. C—Center dashes applied to the drawing of a circular feature in the view where the feature appears as a circle.

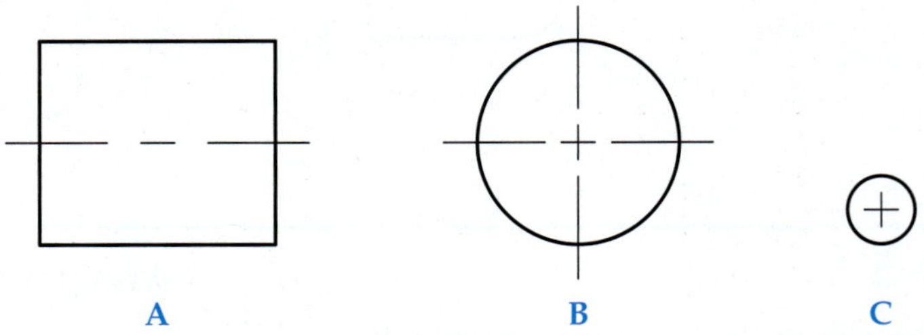

Depending on the current dimension style setting, when you use the **DIM**, **DIMDIAMETER**, and **DIMRADIUS** commands, small circles and arcs receive center dashes; large circles and arcs display center dashes or centerlines; or no symbol appears. Center dashes and centerlines that occur when using these commands are elements of the diametric or radial dimension object. Use the **CENTERMARK** or **DIMCENTER** command to add center dashes or centerlines to circular objects without creating a dimension. AutoCAD also includes a **CENTERLINE** command for drawing a centerline between two lines or polylines.

Using the CENTERMARK Command

The **CENTERMARK** command creates center dashes or centerlines as center mark objects, associated with circles, arcs, or polyline arcs. An associative center mark adapts to changes in location and size made to the associated object. System variables unique to the **CENTERMARK** command determine the appearance of center mark objects. The **CENTERMARKEXE** system variable determines whether center dashes or centerlines are applied with the **CENTERMARK** command. The default ON setting draws centerlines. Change the setting to OFF to draw center dashes without centerline extensions.

CENTERMARK
Ribbon
Annotate
> Centerlines

Center Mark
Type
CENTERMARK

Access the **CENTERMARK** command and pick a circle or arc to draw center dashes or centerlines. Continue picking objects to draw additional center dashes or centerlines. Exit the command when finished.

DIMCENTER
Type
DIMCENTER
DCE

NOTE

You can also use the **DIMCENTER** command to add center dashes or centerlines to circles, arcs, and polyline arcs. Unlike the **CENTERMARK** command, the **DIMCENTER** command references the current dimension style to determine the appearance of center dashes and centerlines, and draws multiple individual line objects that do not maintain association with the dimension style or the selected object.

Using the CENTERLINE Command

CENTERLINE
Ribbon
Annotate
> Centerlines

Centerline
Type
CENTERLINE

The **CENTERLINE** command creates a centerline object bisecting and associated with two lines or polylines. Associative centerlines adapt to changes in location and size made to the associated objects. System variables unique to the **CENTERLINE** command determine the appearance of centerline objects. Access the **CENTERLINE** command and pick a line or polyline followed by a different line or polyline. See Figure 18-11. Though you can select any lines or polylines, a centerline is typically used to represent an axis or line of symmetry. Therefore, for most applications, select parallel or symmetrical lines.

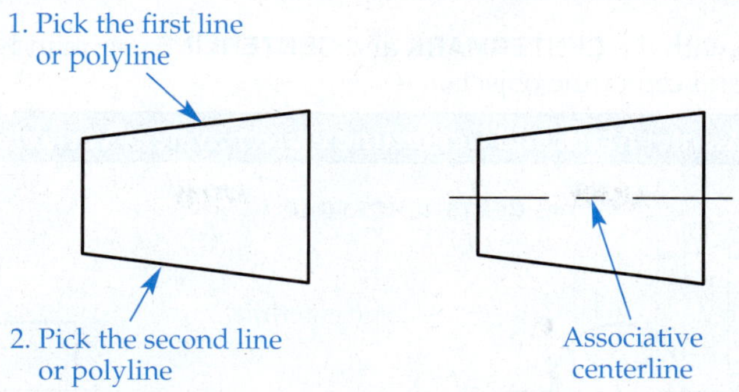

Figure 18-11. Using the **CENTERLINE** command to add a centerline between two lines of a conical feature.

Adjusting Center Mark and Centerline Object Settings

System variables control the initial appearance of associative center marks and centerlines. Adjust the system variables as necessary to create center dashes and centerlines according to drawing requirements and drafting standards. Figure 18-12 describes the system variables to adjust before adding center mark and centerline objects. Changes made to these system variables are stored in the drawing file. Adjust center mark and centerline system variables in drawing templates for repeated use. Chapter 20 explains options for editing existing center mark and centerline objects.

Exercise 18-4

www.g-wlearning.com/CAD/

Complete the exercise on the companion website.

Drawing Leader Lines

The **DIM**, **DIMDIAMETER**, and **DIMRADIUS** commands include ***leader lines*** when you dimension circles and arcs. AutoCAD multileaders created using the **MLEADER** command allow you to add leader lines for other applications, such as specific notes. Multileaders consist of single or multiple lines of ***annotations***, including symbols, and leaders. Multileader styles control multileader characteristics, such as leader format, annotation style, and arrowhead size. You can create multi-segment leaders and align and group separate leaders as described in Chapter 20.

leader line: A line that connects a note or symbol to a specific feature or location on a drawing.

annotation: Textual information presented in notes, specifications, comments, and symbols.

NOTE

The **QLEADER** and **LEADER** commands are available to create leader lines, but they provide fewer options than the **MLEADER** command.

Multileader Styles

A ***multileader style*** presets many multileader characteristics. Multileader style settings correspond to appropriate drafting standards and usually apply to a specific drafting field or dimensioning application. In mechanical drafting, properly drawn leaders have one straight segment extending from the feature to a horizontal ***shoulder*** that is 1/8″–1/4″ (3 mm–6 mm) long. While most other fields also use straight leaders, AutoCAD provides the option of drawing curved leaders, which are common in architectural drafting. See Figure 18-13.

multileader styles: Saved configurations for the appearance of leaders.

shoulder: A short horizontal line usually added to the end of straight leader lines.

Figure 18-12.
System variables associated with the **CENTERMARK** and **CENTERLINE** commands. Adjust these settings before adding center mark and centerline objects.

Center Mark and Centerline Object System Variables

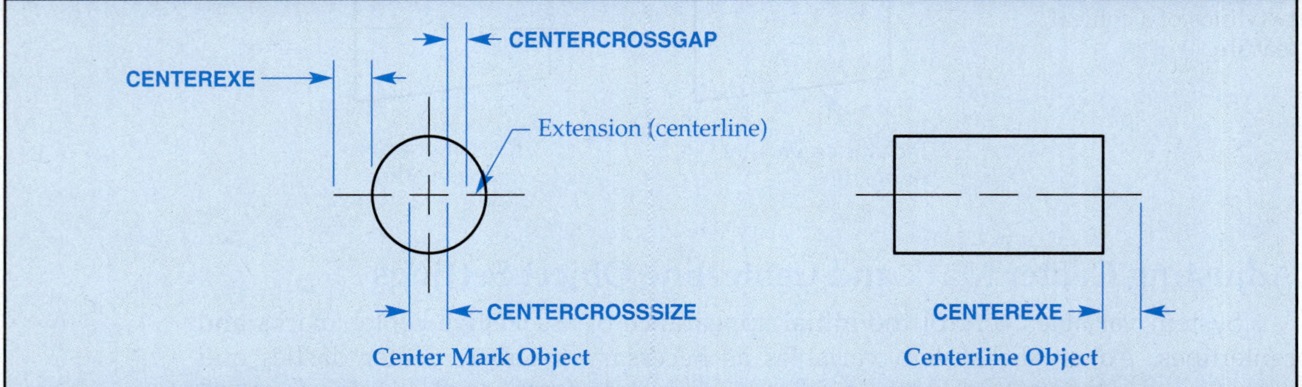

System Variable	Description and Settings	Default	Recommended Value or Setting
CENTERCROSSGAP	An absolute value, such as .0625″ (1.5 mm), specifies the space between the center dash and the extension regardless of the circle diameter. A relative value, such as 0.05x, sets a scale factor multiplied by the diameter of the circle or arc (x) to determine the gap distance. The ByLinetype setting allows the linetype assigned to the center mark object to control the gap distance regardless of the circle diameter.	0.05x	.0625″ (1.5 mm)
CENTERCROSSSIZE	An absolute value, such as .125″ (3 mm), sets the length of the center dash regardless of the circle diameter. A relative value, such as 0.1x, sets a scale factor multiplied by the diameter of the circle or arc (x) to determine the length of the center dash. The ByLinetype setting allows the linetype assigned to the center mark object to control the length of the center dash regardless of the circle diameter.	0.1x	.125″ (3 mm)
CENTEREXE	Specify an absolute value, such as .125″ (3 mm), to set the distance that the centerline extends past the object.	0.1200	.125″ (3 mm)
CENTERLAYER	Assigns a layer to center mark and centerline objects. The use current setting allows the current layer to control the layer assignment. Enter the name of an existing or new layer to override the current layer.	use current	use current
CENTERLTSCALE	Specify a real number other than 0 to set the linetype scale assigned to center mark and centerline objects. Use the default setting to align the scale with the default global LTSCALE setting.	1.0000	1.0000
CENTERLTYPE	Assigns a linetype to center mark and centerline objects. Use the use current or ByLayer setting to allow the current layer or linetype override to control the linetype assignment. Enter the name of a linetype to override the current linetype.	Center2	use current or ByLayer
CENTERLTYPEFILE	Specifies the .lin file associated with the **CENTERLTYPE** setting.	acad.lin	acad.lin
CENTERMARKEXE	Enter ON to draw centerlines or OFF to draw center dashes.	ON	ON or OFF

Figure 18-13.
Multileader styles control the display of leaders created using the **MLEADER** command.

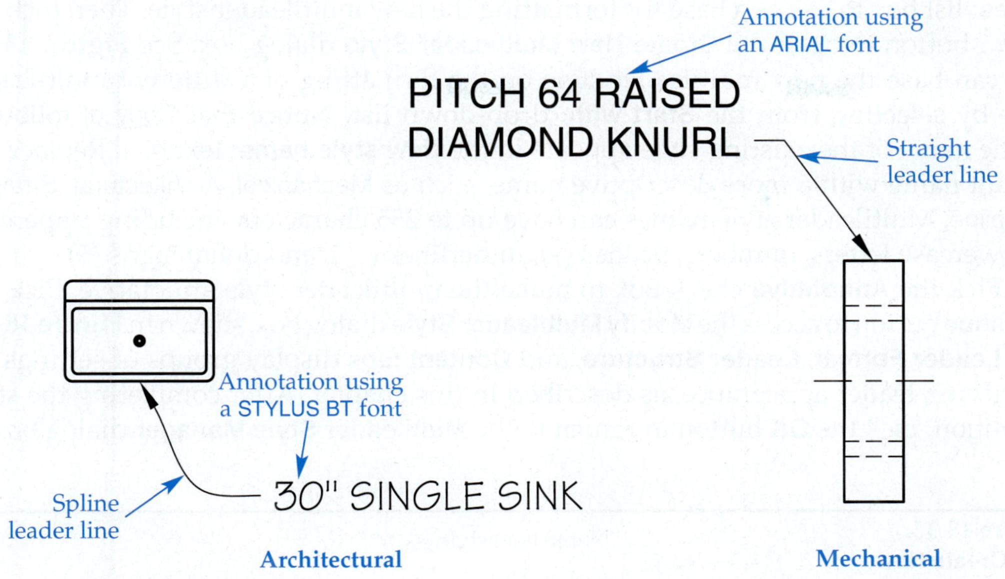

Multileader Style Manager

Create, modify, and delete multileader styles using the **Multileader Style Manager** dialog box. See Figure 18-14. The **Styles:** list box displays existing multileader styles. The Annotative multileader style allows you to create annotative leaders, as indicated by the icon to the left of the style name. The Standard multileader style does not use the annotative function. To make a multileader style current, double-click the style name, right-click on the name and select **Set current**, or pick the name and select the **Set Current** button. Use the **List:** drop-down list to filter the multileader styles displayed in the **Styles:** list box.

Ribbon
Home
> Annotation
Multileader Style
Annotate
> Leaders
Multileader Style
Type
MLEADERSTYLE
MLS

MLEADERSTYLE

NOTE

The **Preview of:** image displays a representation of the multileader style and changes according to the selections you make.

Figure 18-14.
The **Multileader Style Manager** dialog box.

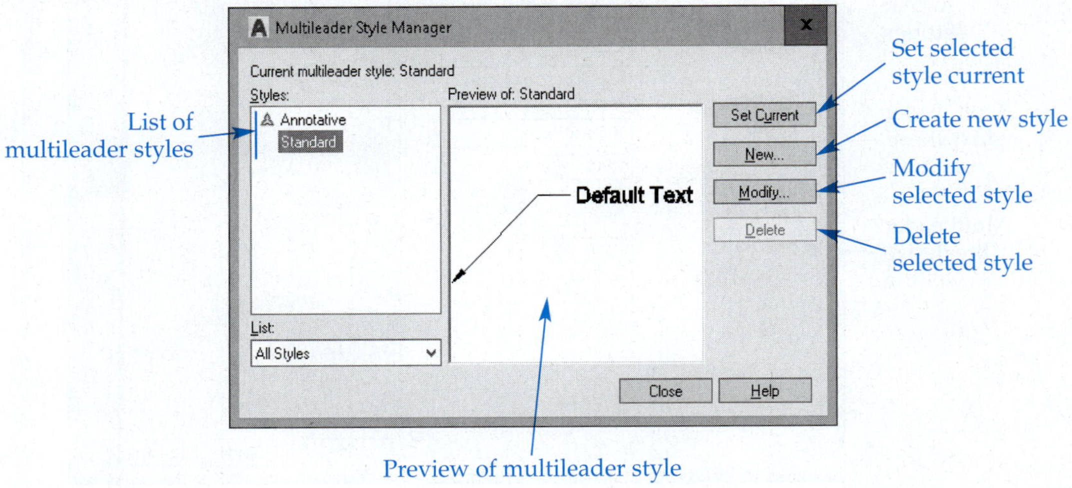

Creating New Multileader Styles

To create a new multileader style, select an existing multileader style from the **Styles:** list box to use as a base for formatting the new multileader style. Then pick the **New...** button to open the **Create New Multileader Style** dialog box. See **Figure 18-15**. You can base the new multileader style on the formatting of a different multileader style by selecting from the **Start with:** drop-down list. Notice that Copy of followed by the name of the existing style appears in the **New style name:** text box. Replace the default name with a more descriptive name, such as Mechanical, Architectural, Straight, or Spline. Multileader style names can have up to 255 characters, including uppercase or lowercase letters, numbers, dashes (–), underlines (_), and dollar signs ($).

Pick the **Annotative** check box to make the multileader style annotative. Pick the **Continue** button to access the **Modify Multileader Style** dialog box, shown in **Figure 18-16**. The **Leader Format**, **Leader Structure**, and **Content** tabs display groups of settings for specifying leader appearance, as described in this chapter. After completing the style definition, pick the **OK** button to return to the **Multileader Style Manager** dialog box.

Figure 18-15.
The **Create New Multileader Style** dialog box.

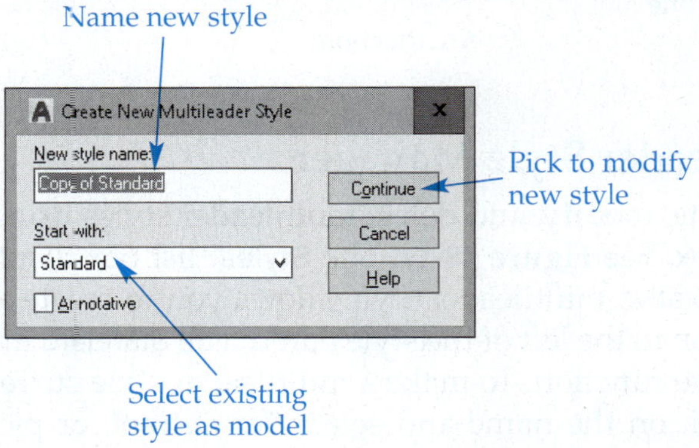

Figure 18-16.
The **Leader Format** tab of the **Modify Multileader Style** dialog box.

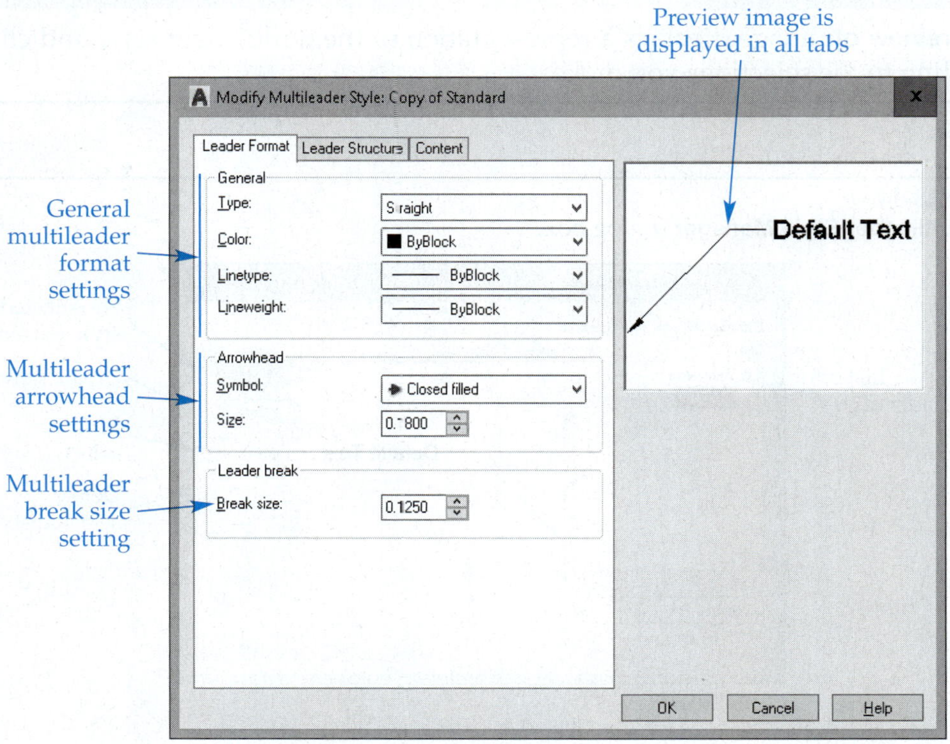

NOTE

The preview image in the upper-right corner of each **Modify Multileader Style** dialog box tab displays a representation of the multileader style and changes according to the selections you make.

Leader Format Tab

Figure 18-16 shows the **Leader Format** tab of the **Modify Multileader Style** dialog box. The **Leader Format** tab presets the appearance of the leader line and arrowhead. You can edit specific leaders when necessary without using a separate multileader style.

General Settings

The **General** area contains a **Type:** drop-down list that you can use to specify the leader line shape. The **Straight** option produces leaders with straight line segments. The **Spline** option produces curved leader lines, which are common in architectural drafting. Pick the **None** option to create a multileader style that does not use a leader line. Use the **None** option to create a leader that you can associate with other leaders using the **MLEADERALIGN** and **MLEADERCOLLECT** commands, described in Chapter 20.

Color:, **Linetype:**, and **Lineweight:** drop-down lists are available for changing the leader line color, linetype, and lineweight. By default, the leader line color, linetype, and lineweight are set to ByBlock. The ByBlock setting means that the color assigned to the multileader is used for the component objects of the multileader. Chapter 24 explains blocks and provides complete information on using ByBlock, ByLayer, or absolute color, lineweight, and linetype settings. For now, as long as you assign a specific layer to a multileader and do not change the multileader properties to absolute values, the default ByBlock properties are acceptable.

Arrowhead Settings

The **Arrowhead** area sets the leader arrowhead style and size. Select the arrowhead style from the **Symbol:** drop-down list. The arrowhead symbol options are the same as those for dimension style arrowheads. Set the arrowhead size using the **Size:** text box. A .125″ (3 mm) arrowhead size is most common, especially on mechanical drawings. Leader arrowheads are typically the same size as dimension arrowheads.

Adjusting Break Size

The **Leader break** area controls the amount of leader line hidden by the **DIMBREAK** command. Specify a value in the **Break size:** text box to set the total length of the break. The default size is .125″ (3 mm). ASME standards do not recommend breaking leader lines.

Leader Structure Tab

Figure 18-17 shows the **Leader Structure** tab of the **Modify Multileader Style** dialog box. Use the **Leader Structure** tab to control leader construction and size. You can edit specific leaders when necessary without using a separate multileader style.

Setting Constraints

The **Constraints** area restricts the leader line angle and the number of points you can select to create a leader. Check the **Maximum leader points** check box to set a maximum number of vertices on the leader line. The multileader automatically forms once you pick the maximum number of points. To use fewer than the maximum number of points, press [Enter] at the Specify next point: prompt. Deselect the **Maximum leader points** check box to allow an unlimited number of vertices.

Figure 18-17.
The **Leader Structure** tab of the **Modify Multileader Style** dialog box.

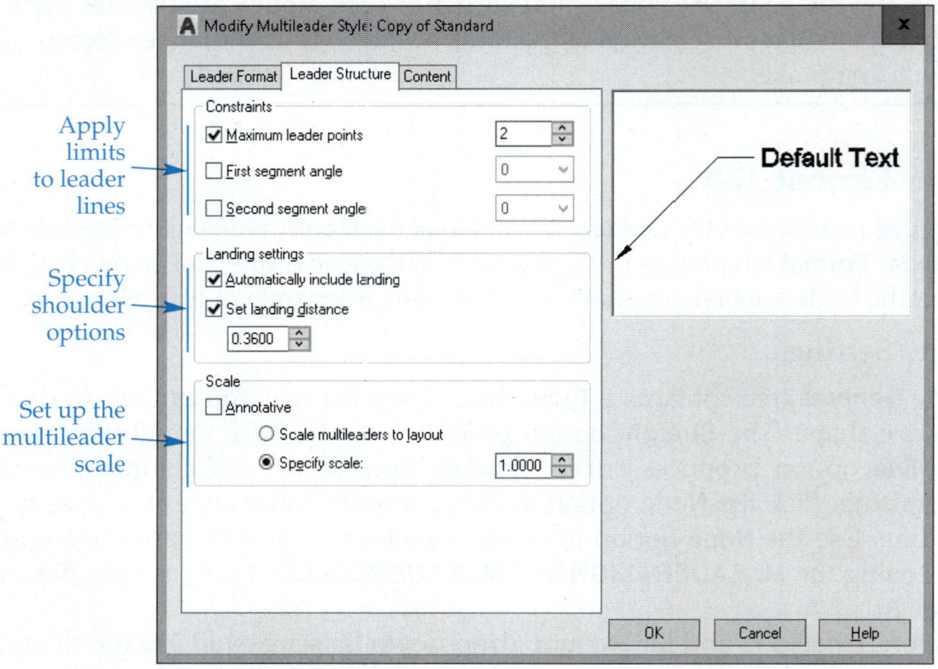

Use the **First segment angle** and **Second segment angle** check boxes to restrict the first two leader line segments to certain angles. Deselect the check boxes to draw leader lines at any angle. Select the appropriate check box and pick a value from the drop-down list to restrict the angle of the leader segment according to the selected value. Ortho mode overrides angle constraints, so it is advisable to turn ortho mode off while you are placing leaders.

PROFESSIONAL TIP

Leader lines should have angles not less than 15° and not greater than 75° from horizontal when possible. Use the **First segment angle** and **Second segment angle** settings to help maintain this standard.

Landing Settings

landing: The AutoCAD term for a leader shoulder.

The **Landing settings** area controls the display and size of the *landing* and is available only for straight-line multileader styles. Select the **Automatically include landing** check box to display a shoulder automatically when you select the second leader line point. This is the preferred method for creating straight leader lines. Deselect the check box to create leaders without shoulders or to pick a third point to draw the shoulder manually. When you check **Automatically include landing**, the **Set landing distance** check box becomes enabled. Check **Set landing distance** to define a specific shoulder length, typically 1/8″–1/4″ (3 mm–6 mm), in the text box. If you deselect the text box, a prompt asks for the shoulder length when you place a leader.

Scale Options

The **Scale** area sets the multileader scale factor. Check the **Annotative** check box to create an annotative multileader style. The **Annotative** check box is already selected when you modify the Annotative multileader style or pick the **Annotative** check box in the **Create New Multileader Style** dialog box.

Select the **Scale multileaders to layout** radio button to add leaders in a floating viewport in a paper space layout. You must add leaders to the model in a floating viewport in order for this option to function. Scaling leaders to the layout allows the overall scale to adjust according to the active floating viewport by setting the overall scale equal to the viewport scale factor.

Pick the **Specify scale:** radio button to scale a drawing manually. Enter the drawing scale factor to apply to all leader elements in the corresponding text box. The scale factor is multiplied by the plotted leader size to get the size of the leader in model space. For example, if you manually scale multileaders for a drawing with a 1:2 (half) scale, or a scale factor of 2 (2 ÷ 1 = 2), enter 2 in the **Specify scale:** text box.

Content Tab

Figure 18-18 shows the **Content** tab of the **Modify Multileader Style** dialog box. The **Content** tab controls the display of text or a block with the leader line. Use the **Multileader type:** drop-down list to select the type of object to attach to the end of the leader line or shoulder. Figure 18-19 shows an example of a leader drawn with each content option. You can edit specific leaders when necessary without using a separate multileader style.

Figure 18-18.
The **Content** tab of the **Modify Multileader Style** dialog box with the **Mtext** multileader type selected.

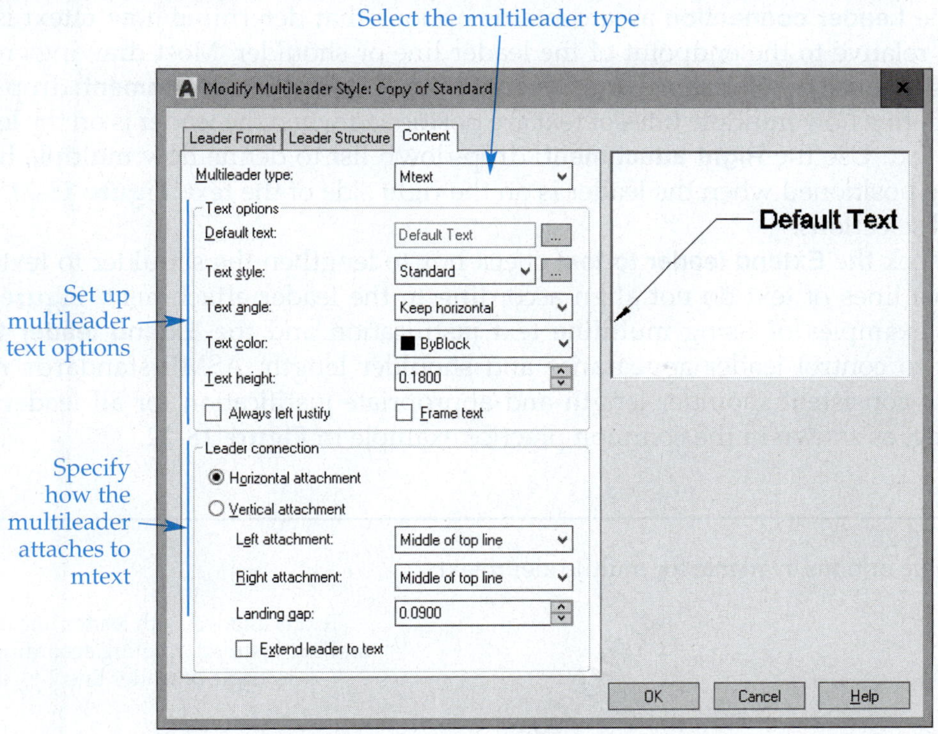

Figure 18-19.
Examples of each multileader content type.

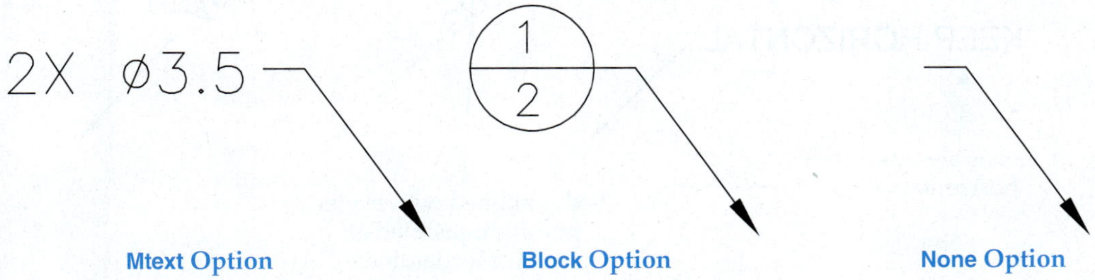

Attaching Mtext

Pick the **Mtext** option from the **Multileader type:** drop-down list, as shown in Figure 18-18, to attach multiline text to the leader. The **Text options** and **Leader connection** areas of the **Content** tab appear when you select the **Mtext** content type. The **Default text:** option allows you to specify a value to attach to leaders during leader placement. This is useful when the same note or symbol is required throughout a drawing. Pick the ellipsis (...) button to return to the drawing window and use the multiline text editor to enter the default text value. Close the text editor to return to the **Modify Multileader Style** dialog box.

A multileader style references a text style for the appearance of leader text. Pick an existing text style from the **Text style:** drop-down list. To create or modify a text style, pick the ellipsis (...) button next to the drop-down list to open the **Text Style** dialog box.

Select an option from the **Text angle:** drop-down list to control the angle at which text appears in reference to the angle of the leader line or shoulder. See Figure 18-20. Use the **Text color:** drop-down list to specify the text color, which should be ByBlock for typical applications. Use the **Text height:** text box to specify the leader text height. Leader text height is usually the same as dimension text height.

The **Always left justify** option forces leader text to left-justify, regardless of the leader line direction. Check the **Frame text** check box to create a box around the leader text. A basic dimension is a common application for framed text, as described in Chapter 19.

The **Leader connection** area contains options that determine how mtext is positioned relative to the endpoint of the leader line or shoulder. Most drawings require leaders that use the **Horizontal attachment** option. Use the **Left attachment:** drop-down list to define how multiple lines of text are positioned when the leader is on the left side of the text. Use the **Right attachment:** drop-down list to define how multiple lines of text are positioned when the leader is on the right side of the text. Figure 18-21 shows typical selections.

Check the **Extend leader to text** check box to lengthen the shoulder to text when multiple lines of text do not align according to the leader attachment. Figure 18-22 shows examples of using multiline text justification and the **Extend leader to text** option to control leader appearance and shoulder length. ASME standards recommend a consistent shoulder length and appropriate justification for all leaders on a drawing, as shown in the common practice example in Figure 18-22.

Figure 18-20.
Text angle options available for multileader mtext.

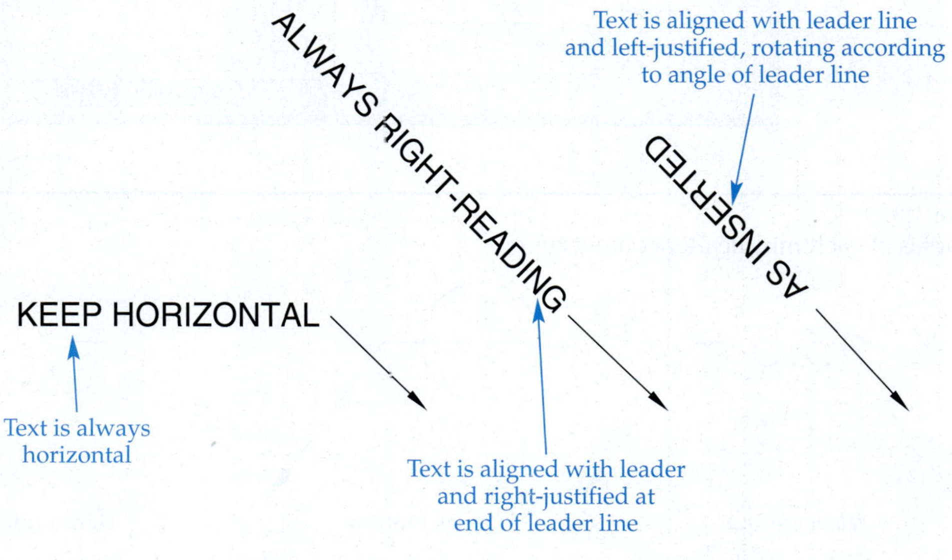

Figure 18-21.
Horizontal multileader text alignment options. The shaded examples are the recommended ASME standards.

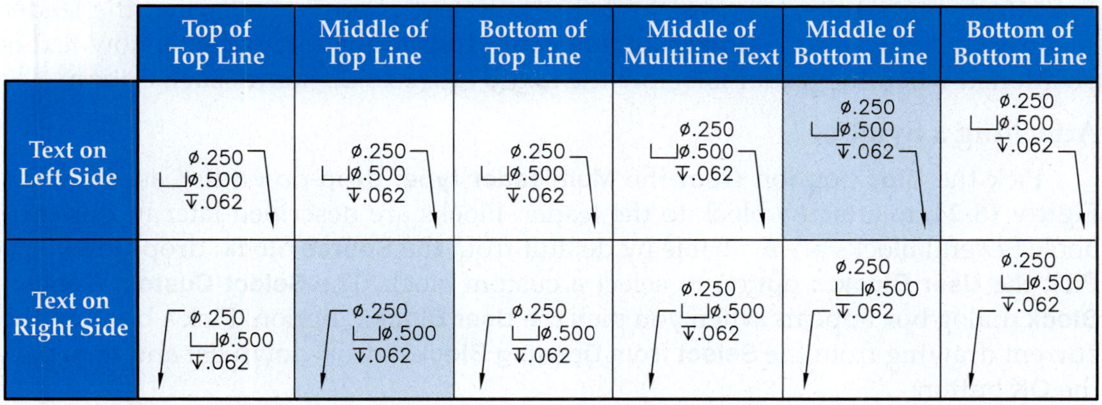

Figure 18-22.
Techniques for controlling multileader shoulder length include modifying justification according to leader attachment to maintain a consistent shoulder length throughout the drawing or extending the leader to text to maintain a consistent justification throughout the drawing.

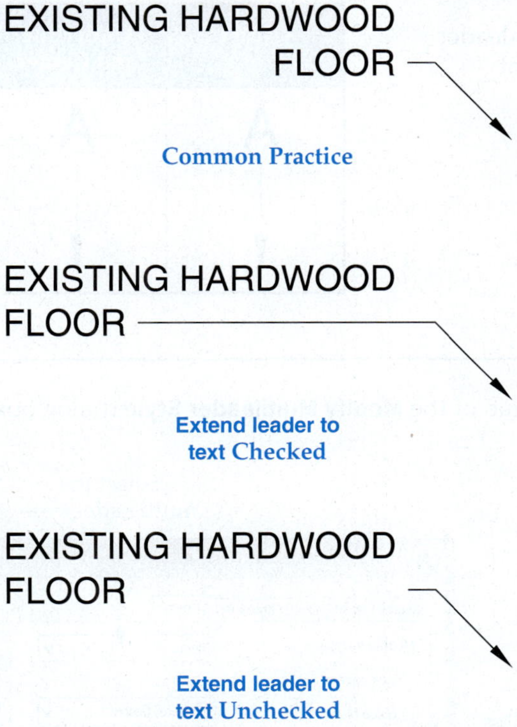

The **Underline bottom line** option places all lines of text above a line that extends from the endpoint of the leader line or shoulder. The **Underline top line** option places the first line of text above a line that extends from the endpoint of the leader line or shoulder. The **Underline all text** option places the first line of text above a line that extends from the endpoint of the leader line or shoulder and underlines the lines of text below. The **Landing gap:** text box specifies the space between the leader line or shoulder and the text, and the space between the text and the frame, when used. The default is .09″, but .063″ (1.5 mm) is standard practice.

PROFESSIONAL TIP

Common drafting practice is to use the **Middle of bottom line** option for left attachment and the **Middle of top line** option for right attachment. The **Middle of text** option is sometimes necessary to display a single line of text that includes a horizontally stacked fraction.

Selecting the **Vertical attachment** radio button is uncommon. This option eliminates the possible use of a shoulder and connects the leader endpoint to the top center or bottom center of the text, depending on the leader line position. Use the **Top attachment:** drop-down list to define how text is positioned when the leader is above the text. Use the **Bottom attachment:** drop-down list to define how text is positioned when the leader is below the text. Figure 18-23 shows each option.

Attaching a Symbol

Pick the **Block** option from the **Multileader type:** drop-down list, as shown in Figure 18-24, to attach a block to the leader. Blocks are described later in this textbook. Several blocks are available by default from the **Source block:** drop-down list. Pick the **User Block...** option to select a custom block. The **Select Custom Content Block** dialog box appears when you pick the **User Block...** option. Pick a block in the current drawing from the **Select from Drawing Blocks:** drop-down list and then pick the **OK** button.

Figure 18-23.
Vertical multileader text alignment options.

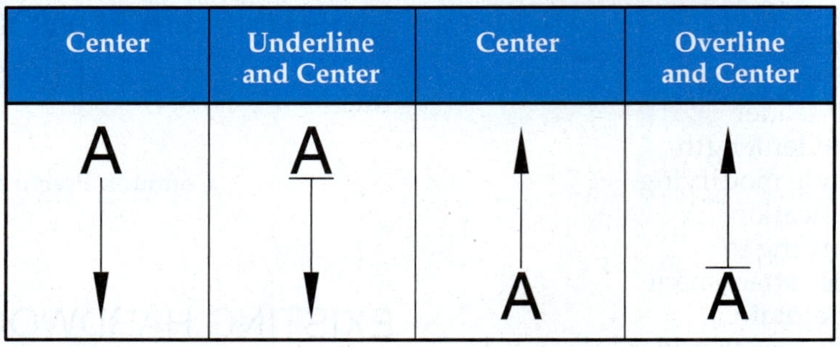

Figure 18-24.
The **Content** tab of the **Modify Multileader Style** dialog box with the **Block** multileader type selected.

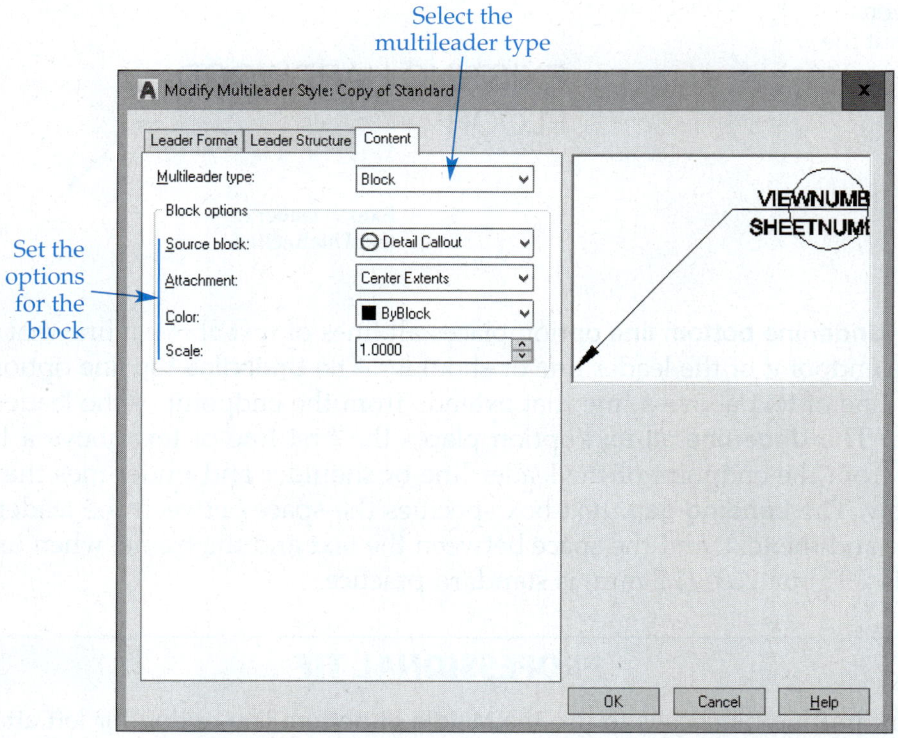

Use the **Attachment:** drop-down list to specify how to attach the block to the leader. Select the **Insertion point** option to attach the block to the leader according to the block insertion point, or base point. Select the **Center Extents** option to attach the block directly to the leader, aligned to the center of the block, even if the block insertion point is not on the block itself. See Figure 18-25.

Use the **Color:** drop-down list to specify the block color, which should be ByBlock for typical applications. Use the **Scale:** text box to proportionately increase or decrease the block size. Scale does not affect the appearance of the leader line, arrowhead, or shoulder, or the scale applied to the multileader object.

Using No Content

Select the **None** option from the **Multileader type:** drop-down list to end the leader without annotation. Use the **None** option when you need to create only a leader, without text or a symbol attached to the leader line or shoulder.

PROFESSIONAL TIP

Add leaders to existing multileaders using the **Add Leader** tool, explained in Chapter 20. This eliminates the need to create a separate multileader style that uses the **None** multileader content type for most applications.

Exercise 18-5

www.g-wlearning.com/CAD/

Complete the exercise on the companion website.

Changing Multileader Styles

Use the **Multileader Style Manager** to change the characteristics of an existing multileader style. Pick the **Modify...** button to open the **Modify Multileader Style** dialog box. When you make changes to a multileader style, such as selecting a different text or arrowhead style, all existing leaders assigned the modified multileader style are updated. Use a different multileader style with different characteristics when appropriate.

Renaming and Deleting Multileader Styles

To rename a multileader style using the **Multileader Style Manager**, slowly double-click on the name or right-click on the name and select **Rename**. To delete a multileader style using the **Multileader Style Manager**, right-click on the name and select **Delete**. You cannot delete a multileader style that is assigned to multileader objects. To delete a style that is in use, assign a different style to the multileaders that reference the style to be deleted.

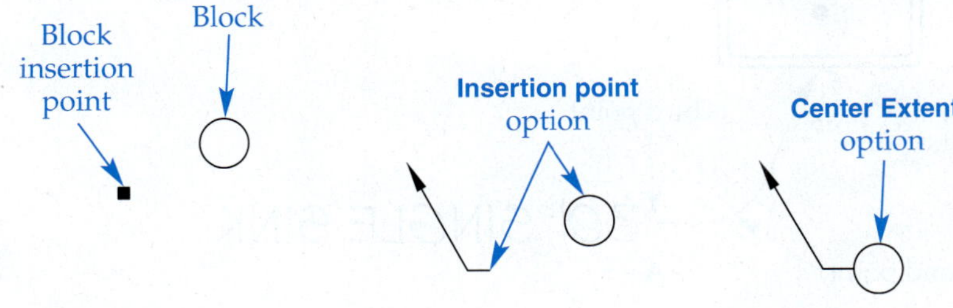

Figure 18-25. Adjusting multileader block attachment.

NOTE

You can also rename styles using the **Rename** dialog box. Select **Multileader styles** in the **Named Objects** list to rename a multileader style.

RENAME
Type
RENAME
REN

Setting a Multileader Style Current

Set a multileader style current using the **Multileader Style Manager** by double-clicking the style in the **Styles:** list box, right-clicking on the name and selecting **Set current**, or picking the style and selecting the **Set Current** button. To set a multileader style current without opening the **Multileader Style Manager**, use the **Multileader Style** drop-down list located in the expanded **Annotation** panel on the **Home** ribbon tab or the **Leaders** panel on the **Annotate** ribbon tab.

PROFESSIONAL TIP

You can import multileader styles from existing drawings using **DesignCenter**. See Chapter 5 for more information about using **DesignCenter** to import file content.

Inserting Multileaders

MLEADER
Ribbon
Home
> Annotation
Annotate
> Leaders
Multileader
Type
MLEADER
MLD

Use the **MLEADER** command to insert a multileader. How you insert a multileader depends on the current multileader style settings and the option you select to construct the leader. There are three general methods for inserting a multileader. The method used determines what portion of the leader you locate first. Review the components of a leader, shown in Figure 18-26, before reading the options for creating a multileader.

Figure 18-26.
Examples of leaders created using the **MLEADER** command. A—An architectural leader created using a spline leader line, the **Specify leader arrowhead location** option, and three leader points. B—A mechanical leader created using a straight leader line, the **leader Landing first** option, and two leader points.

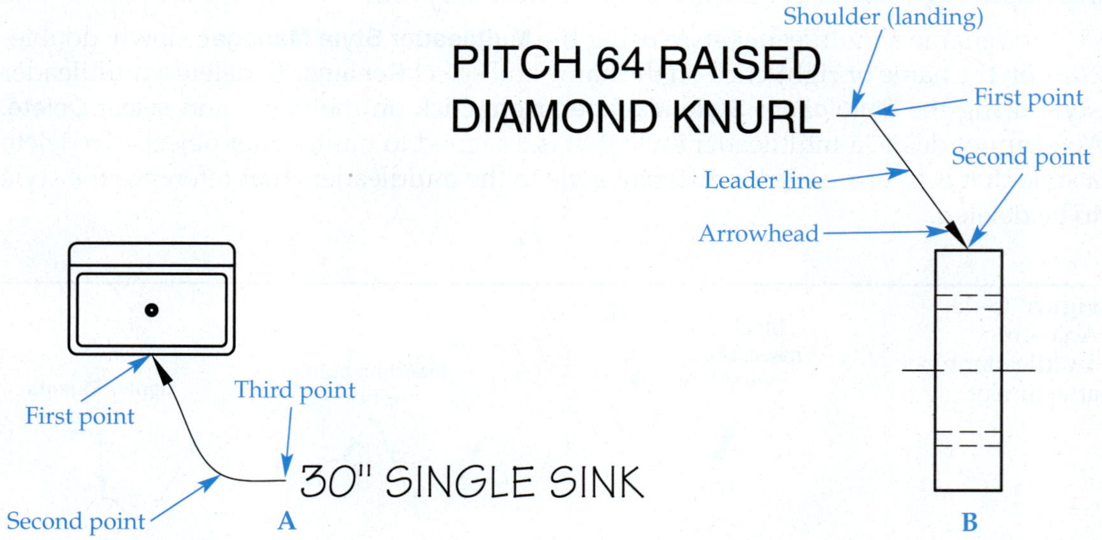

To use the default **Specify leader arrowhead location** option, first specify the point at which the arrowhead touches. Then specify a point to locate where the leader ends and the shoulder begins. If the **Mtext** option is active, enter leader text using the multiline text editor.

Select the **leader Landing first** option to specify a point to locate where the leader ends and the shoulder begins first. Then specify the point at which the arrowhead touches. If the **Mtext** option is active, enter leader text using the multiline text editor.

Use the **Content first** option to define the leader content first. The **Mtext** option allows you to type text using the multiline text editor. Then specify the point at which the arrowhead touches.

Select **Options** to access a list of options to override the current multileader style characteristics. The options are the same as those found in the **Modify Multileader Style** dialog box.

NOTE

Additional tools are available for adding and removing multiple leader lines, arranging multiple leaders, and combining leader content, as explained in Chapter 20.

Exercise 18-6

www.g-wlearning.com/CAD/

Complete the exercise on the companion website.

Dimensioning Chamfers

Dimension ***chamfers*** of 45° with a leader giving the angle and linear dimension or with two linear values. See Figure 18-27. Place the leader using the **MLEADER** command. Chamfers other than 45° include either the angle and a linear dimension or two linear dimensions. See Figure 18-28. Use the **DIM** command or the **DIMLINEAR** and **DIMANGULAR** commands for this purpose.

chamfer: An angled surface used to relieve sharp corners.

Exercise 18-7

www.g-wlearning.com/CAD/

Complete the exercise on the companion website.

Figure 18-27. Dimensioning 45° chamfers.

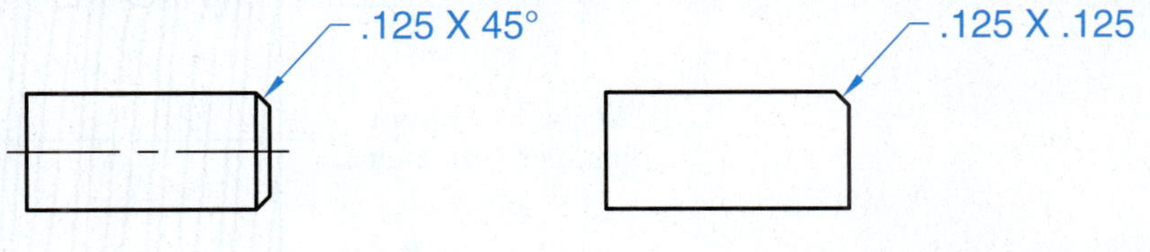

Figure 18-28. Dimensioning chamfers that are not 45°.

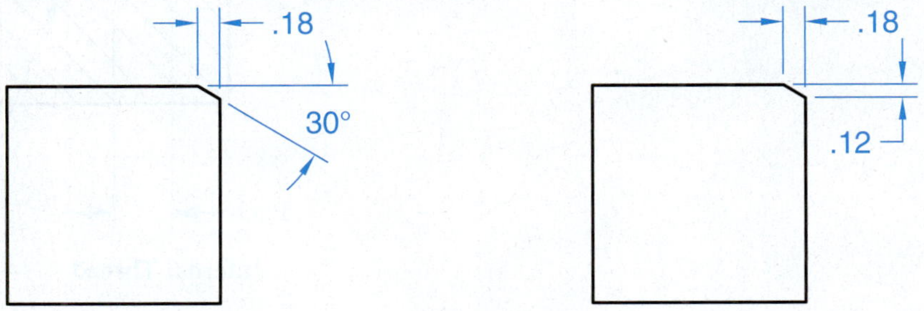

Thread Drawings and Notes

Figure 18-29 shows the elements of external and internal screw threads. However, threads commonly appear on a drawing as a simplified representation in which a hidden line indicates thread depth. See **Figure 18-30**. Threaded parts often include a chamfer to help engage the mating thread.

Thread representations show the reader that a thread exists, but the thread note gives the exact specifications. The thread note typically connects to the thread with a leader. See **Figure 18-31**. The most common thread forms are the Unified and metric screw threads, but a variety of other thread forms are required for specific applications.

Figure 18-29.
Features of external and internal screw threads.

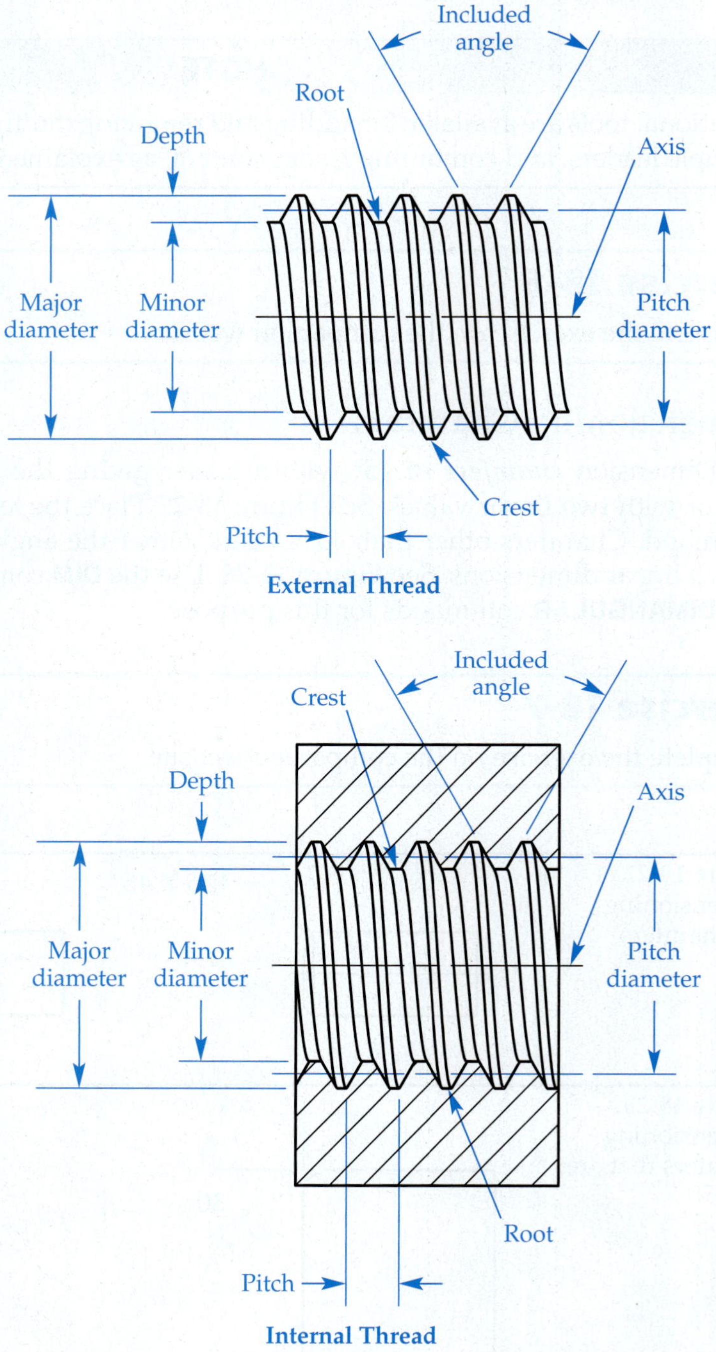

Figure 18-30.
Simplified thread representations.

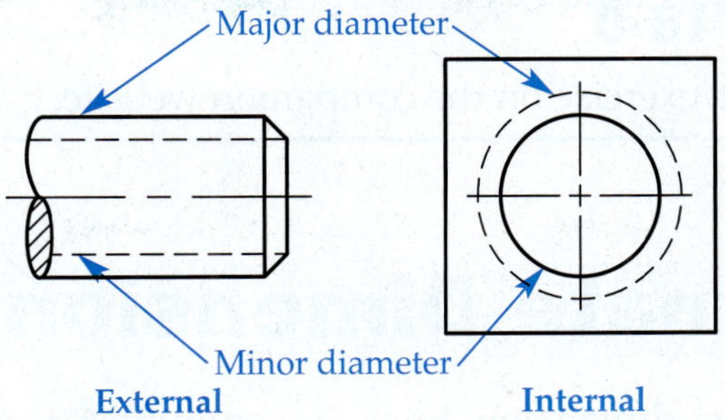

Figure 18-31.
Displaying the thread note with a leader.

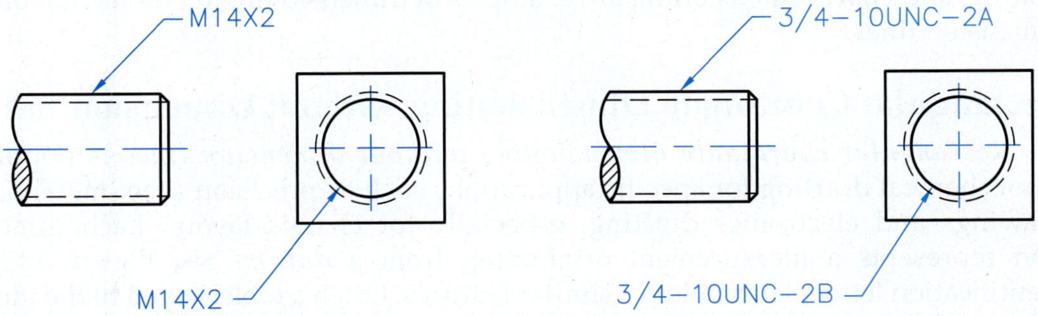

The following format specifies the thread note for Unified screw threads:

3/4-10UNC-2A
(1) (2) (3) (4)(5)

(1) Major diameter of thread, given as a fraction or number
(2) Number of threads per inch
(3) Thread series: UNC = Unified National Coarse, UNF = Unified National Fine
(4) Class of fit: 1 = large tolerance, 2 = general-purpose tolerance, 3 = tight tolerance
(5) Thread type: A = external thread, B = internal thread

The following format specifies the thread note for metric threads:

M14X2
(1) (2) (3)

(1) M = metric thread
(2) Major diameter in millimeters
(3) Pitch in millimeters

There are too many screw threads to describe in detail in this textbook. Refer to the *Machinery's Handbook*, published by Industrial Press, Inc., or a comprehensive mechanical drafting text for more information.

NOTE

The **Dimension** panel of the **Express Tools** ribbon tab includes the **QLATTACH**, **QLATTACHSET**, and **QLDETACHSET** commands. These commands allow you to attach an annotation object to and detach it from leaders created using the **QLEADER** and **LEADER** commands. Use multileaders instead of these leader commands and express tools.

Exercise 18-8

www.g-wlearning.com/CAD/

Complete the exercise on the companion website.

Alternate Dimensioning Practices

Omitting dimension lines is common for drawings in industries that use computer-controlled machining processes and for drawings in which unconventional dimensioning practices are required because of product features. Rectangular coordinate dimensioning without dimension lines, tabular dimensioning, and chart dimensioning are examples of dimensioning methods that omit dimension lines.

Rectangular Coordinate Dimensioning without Dimension Lines

rectangular coordinate dimensioning without dimension lines: A type of dimensioning that includes only extension lines and text aligned with the extension lines.

datum: Theoretically perfect surface, plane, point, or axis from which measurements are made while dimensioning.

Rectangular coordinate dimensioning without dimension lines is popular in mechanical drafting for specific applications such as precision sheet metal part drawings and electronics drafting, especially for chassis layout. Each dimension represents a measurement originating from a ***datum***. See Figure 18-32. Identification letters label holes or similar features. Often a table, keyed to the identification letters, indicates feature size or specifications.

Tabular Dimensioning

tabular dimensioning: A form of rectangular coordinate dimensioning without dimension lines in which dimensions appear in a table.

In ***tabular dimensioning***, each feature receives a label with a letter or number that correlates to a table. See Figure 18-33. Some companies take this practice a step further and display the location and size of features using X and Y coordinates in the table. The depth of features is also provided using Z coordinates where appropriate. Each feature is labeled with a letter or number that correlates to the table, as shown in Figure 18-34. Chapter 21 explains drawing tables.

Chart Dimensioning

chart dimensioning: A type of dimensioning in which the variable dimensions are shown with letters that correlate to a chart in which the possible dimensions are given.

Chart dimensioning may take the form of unidirectional, aligned, or tabular dimensioning or rectangular coordinate dimensioning without dimension lines. Chart dimensioning provides flexibility when dimensions change as the requirements of the product change. See Figure 18-35.

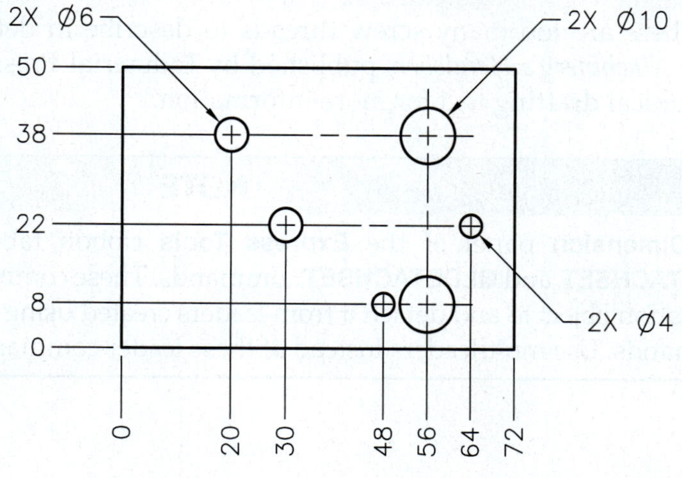

Figure 18-32.
Rectangular coordinate dimensioning without dimension lines.

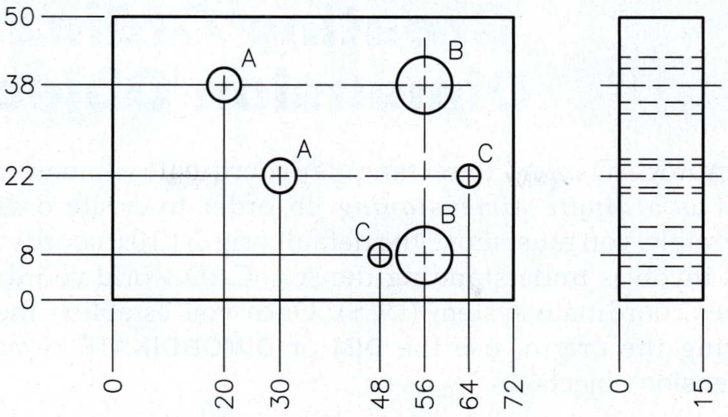

Figure 18-33. Tabular dimensioning. Letters reference holes, and the table presents the related information.

HOLE	A	B	C
DIA.	6	10	4

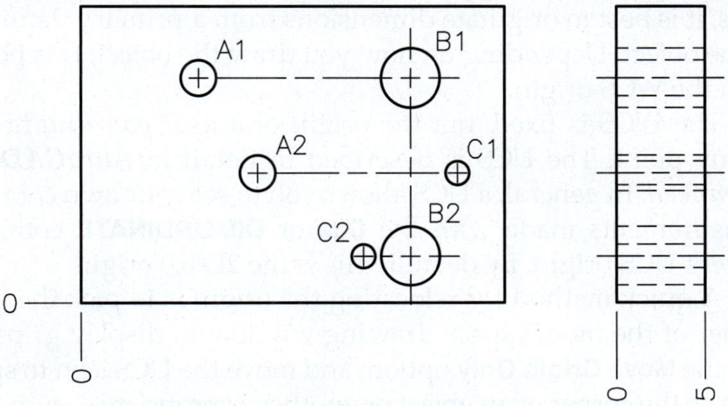

Figure 18-34. Holes can be located with X, Y, and Z references given in a table with tabular dimensioning.

HOLE	QTY.	DIA.	X	Y	Z
A1	1	6	20	38	THRU
A2	1	6	30	22	THRU
B1	1	10	56	38	THRU
B2	1	10	56	8	THRU
C1	1	4	64	22	THRU
C2	1	4	48	8	THRU

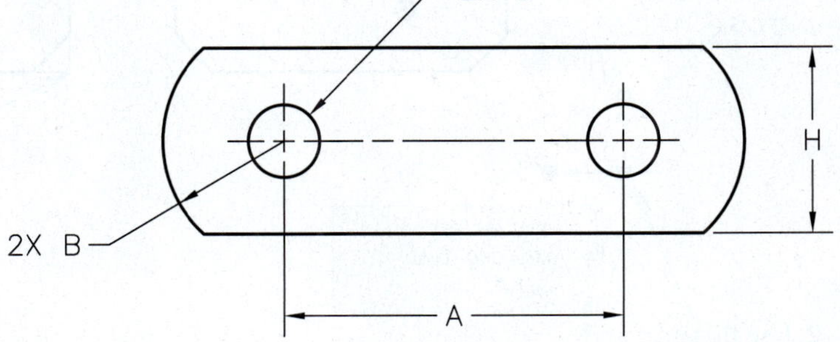

Figure 18-35. Chart dimensioning.

CHAIN NO.	A	B	D	H
SST1000	2.1	.75	.44	1.125
SST1001	3.0	.90	.48	1.525
SST1002	3.0	1.17	.95	2.125

Creating Ordinate Dimension Objects

ordinate dimensioning: The AutoCAD term for rectangular coordinate dimensioning without dimension lines.

AutoCAD refers to rectangular coordinate dimensioning without dimension lines as ***ordinate dimensioning***. In order to create ordinate dimension objects accurately, you must move the default origin (0,0,0 coordinate) to the object datum. This involves understanding the AutoCAD world coordinate system (WCS) and a user coordinate system (UCS). Once you establish the datum by temporarily moving the origin, use the **DIM** or **DIMORDINATE** command to place ordinate dimension objects.

Using the WCS and a UCS

world coordinate system (WCS): The AutoCAD rectangular coordinate system. In 2D drafting, the WCS contains four quadrants, separated by the X and Y axes.

user coordinate system (UCS): A user override of the WCS in which the origin (0,0,0) is moved to a location specified by the user.

The ***world coordinate system (WCS)*** 0,0,0 origin has been used for the drawings you have created throughout this textbook. In most cases, this is appropriate. However, when you apply rectangular coordinate dimensioning without dimension lines, it is best to originate dimensions from a primary datum, which is often a corner of the object. Depending on how you draw the object, this point may or may not align with the WCS origin.

The WCS is fixed, but the origin of a ***user coordinate system (UCS)*** can move to any point. The UCS is described in detail in ***AutoCAD and Its Applications—Advanced***. In general, a UCS allows you to set your own coordinate system and origin. Measurements made with the **DIM** or **DIMORDINATE** command originate from the current UCS origin. By default, this is the 2D 0,0 origin.

A quick method for relocating the origin is to pick the UCS icon in the lower-left corner of the model space drawing window to display grips. Use the origin grip box and the **Move Origin Only** option, and move the UCS icon to specify a new origin point, such as the corner of an object or another appropriate datum. Press [Esc] to deselect the UCS icon and hide grips. See Figure 18-36.

Figure 18-36.
Before you use the **DIM** or **DIMORDINATE** command, move the UCS origin to the appropriate datum location and add center marks to circular features that you plan to dimension.

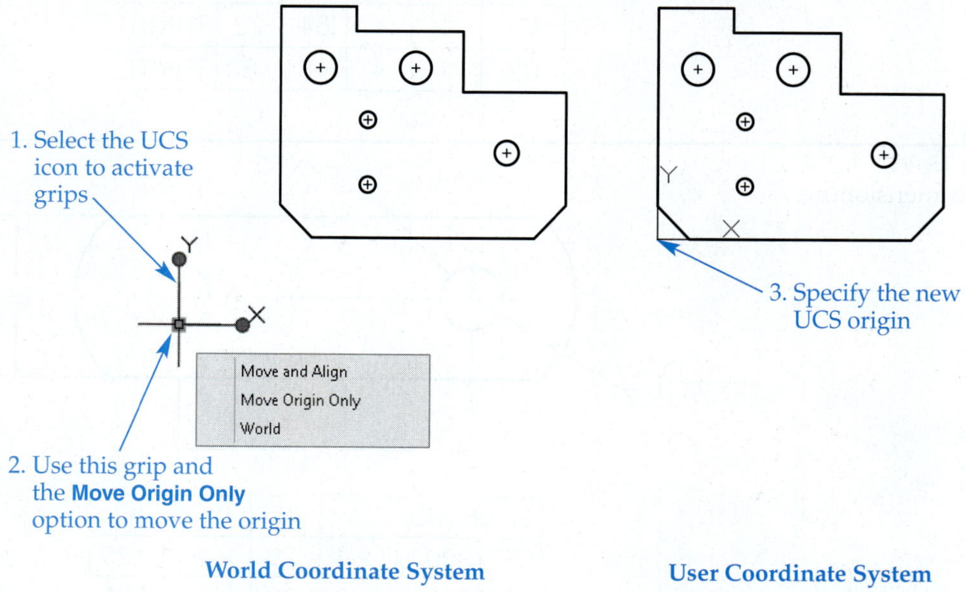

When you finish drawing dimensions from a datum, you can leave the UCS origin at the datum or move it back to the WCS origin. To return to the WCS, grip-edit the UCS icon and apply the **World** option, or select the **WCS** option from the drop-down menu below the view cube.

Placing Ordinate Dimensions

Use the **DIM** or **DIMORDINATE** command to place an extension line and a dimension at the point you specify. The dimension measures an X or Y coordinate distance from the UCS origin. Since you are working in the XY plane, you may want to turn polar tracking or ortho mode on before using the **DIM** or **DIMORDINATE** command. In addition, if the drawing includes circular features, use the **DIMCENTER** command to place center dashes as shown in Figure 18-36. This conforms to ASME standards and provides an endpoint to pick when dimensioning to the center of circular features.

> **DIMORDINATE**
> Ribbon
> **Home**
> **> Annotation**
> **Annotate**
> **> Dimensions**
> **Ordinate**
> Type
> **DIMORDINATE**
> **DOR**

When using the **DIM** command, it is helpful to select the **Ordinate** option before specifying points. When you see the Specify feature location prompt, pick a point to locate the origin of the extension line. If the feature is the corner of the object, pick an endpoint or other appropriate position. If the feature is a circle, pick the end of the center dash, as shown in Figure 18-37, not the center of the object. This leaves the required space between the center mark and the extension line. Zoom in if needed and use object snap modes when necessary. The next prompt asks for the leader endpoint, which refers to the extension line endpoint. Specify the endpoint of the extension line.

If the X axis or Y axis distance between the feature and the extension line endpoint is large, the axis AutoCAD uses for the dimension by default may not be correct. When this happens, use the **Xdatum** or **Ydatum** option to specify the axis from which the dimension originates. The **Mtext**, **Text**, and **Angle** options are identical to the options available with other dimensioning commands. When using the **DIM** command, the **Undo** option is available to undo a selection without exiting the **DIM** command. Pick the extension line endpoint to complete the process.

Figure 18-38A shows ordinate dimensions placed on the object. Notice that the dimension text aligns with the extension lines. Aligned dimensioning is standard with ordinate dimensioning. Add missing lines, such as centerlines or fold lines, to complete the drawing. Identify the holes with letters and create a correlated dimensioning table if appropriate. See Figure 18-38B.

Figure 18-37.
Pick the endpoints of center dashes to establish the correct offset, or develop a specific dimension style with an extension line origin offset of 0 for placing dimensions from centerline endpoints.

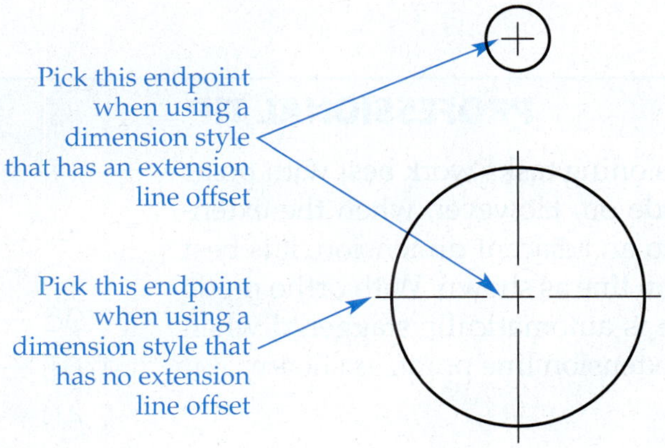

Figure 18-38.
A—Placing ordinate dimensions. B—Completing the drawing.

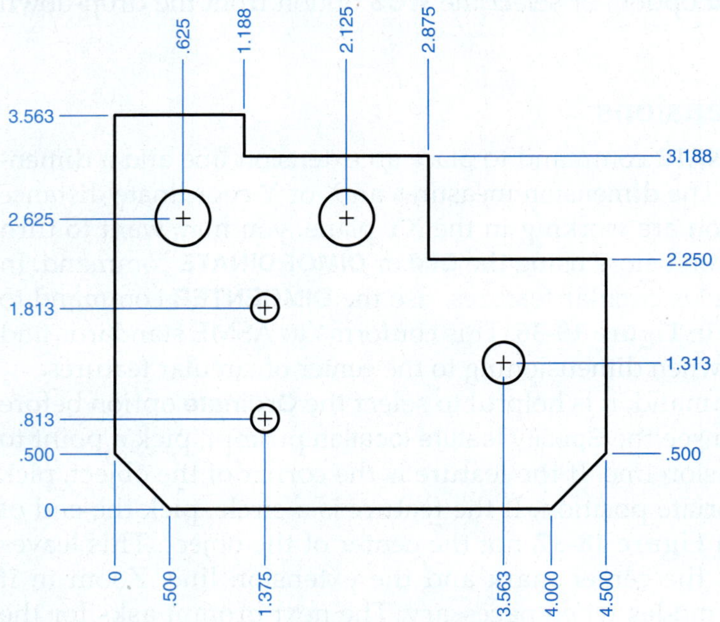

A

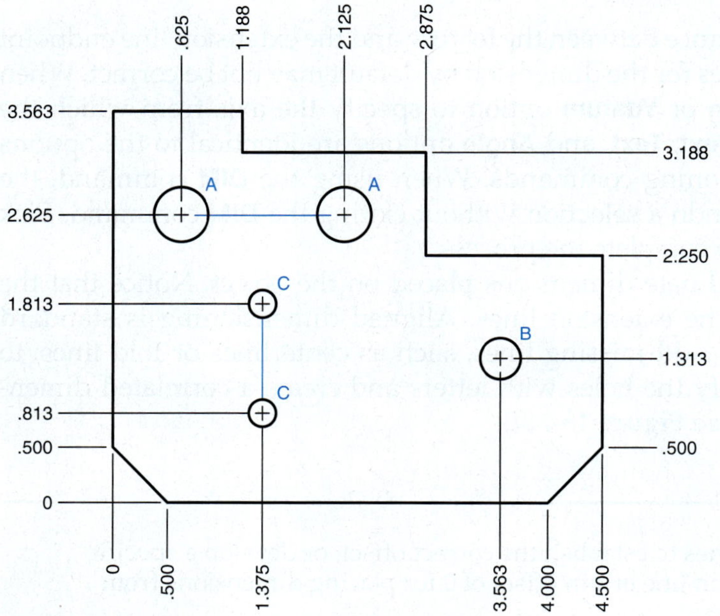

HOLE DATA		
HOLE	QTY.	DIAMETER
A	2	.500
B	1	.375
C	2	.250

B

PROFESSIONAL TIP

Most ordinate dimensioning tasks work best with polar tracking or ortho mode on. However, when the extension line is too close to an adjacent dimension, it is best to stagger the extension line as shown. With ortho mode off, the extension line is automatically staggered when you pick the second extension line point, as shown.

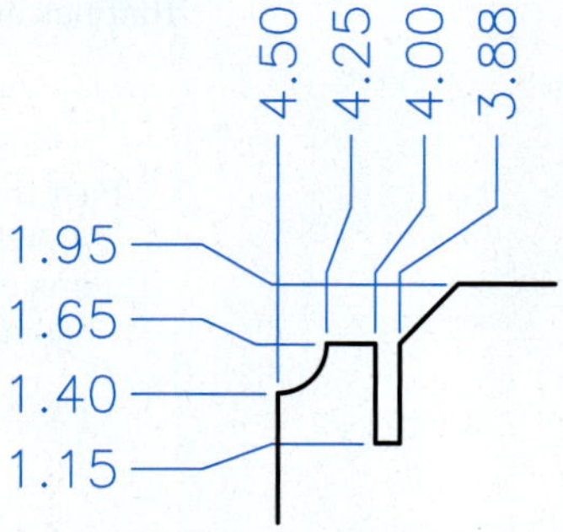

Exercise 18-9

www.g-wlearning.com/CAD/

Complete the exercise on the companion website.

Marking Up Drawings

Marking up, or *redlining*, is not a dimensioning practice, but it is similar to dimensioning in that it helps to explain information to the reader. Redlines are typically added directly to a final drawing by someone who reviews the drawing for accuracy and design changes. You might experience this process with your instructor or supervisor. Common mark-up techniques include redlining a plot with a red pen, using separate mark-up software to review exported drawings, or redlining directly in the drawing file. AutoCAD redlines become part of the drawing to document revision history.

marking up (redlining): The process of reviewing a drawing and marking required changes.

You can redline a drawing with any appropriate AutoCAD command, typically using a separate layer. Redlining often includes basic objects, text, and leaders. In some cases, you may add redline dimensions and even an entire drawing or detail. The **REVCLOUD** and **WIPEOUT** commands are also common mark-up commands.

Creating Revision Clouds

revision cloud: A polyline of sequential arcs used to form a cloud shape around changes in a drawing.

A *revision cloud* is a polyline of sequential arcs forming a cloud-shaped object. See Figure 18-39. A revision cloud points the drafter and other members of a design team to a specific portion of the drawing that may require an edit.

Use the **Rectangular**, **Polygonal**, or **Freehand** option of the **REVCLOUD** command to specify points to draw a revision cloud. See Figure 18-40. To draw a revision cloud using the **Rectangular** option, specify opposite corners of the rectangle. To draw a revision cloud using the **Polygonal** option, specify the location of each vertex. AutoCAD automatically closes the revision cloud with a side attached to the first and last vertex. Use the **Undo** option to remove the last side drawn and continue from the previous vertex. After locating the last vertex, press [Enter] to complete the revision cloud. To draw a revision cloud using the **Freehand** option, pick a start point, and then move the crosshairs around the perimeter of the revision cloud until you return close to the start point. AutoCAD closes the revision cloud automatically and exits the command.

REVCLOUD

Ribbon
Home
> Draw
Annotate
> Markup

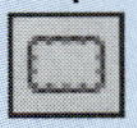
Rectangular Revision Cloud

Polygonal Revision Cloud

Freehand Revision Cloud

Type
REVCLOUD

Figure 18-39.
Examples of revision clouds identifying areas of a drawing to be modified. Notice the leaders added to describe the changes. You can create revision clouds using the Normal style (lower example) or the Calligraphy style (upper example).

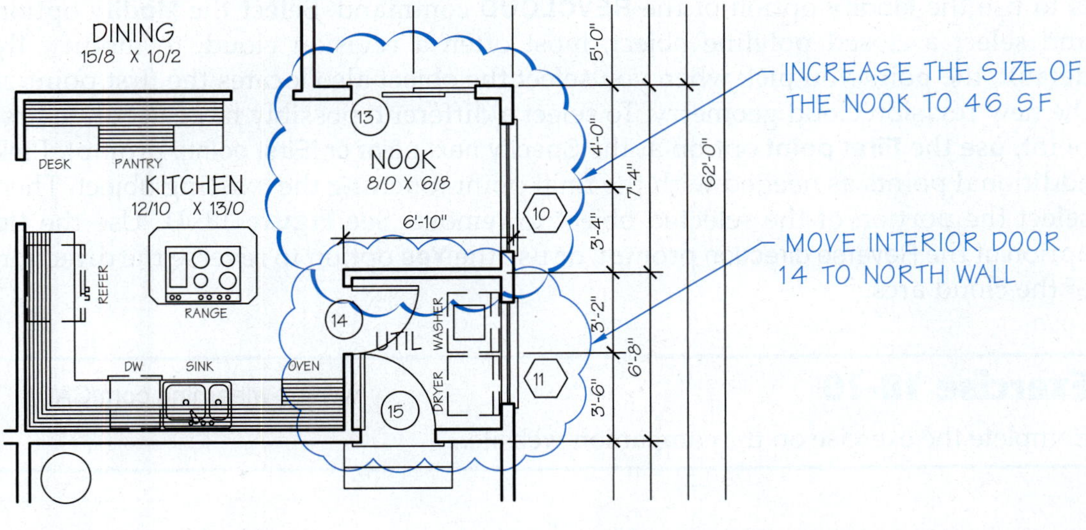

Figure 18-40.
Options for drawing revision clouds with the **REVCLOUD** command.

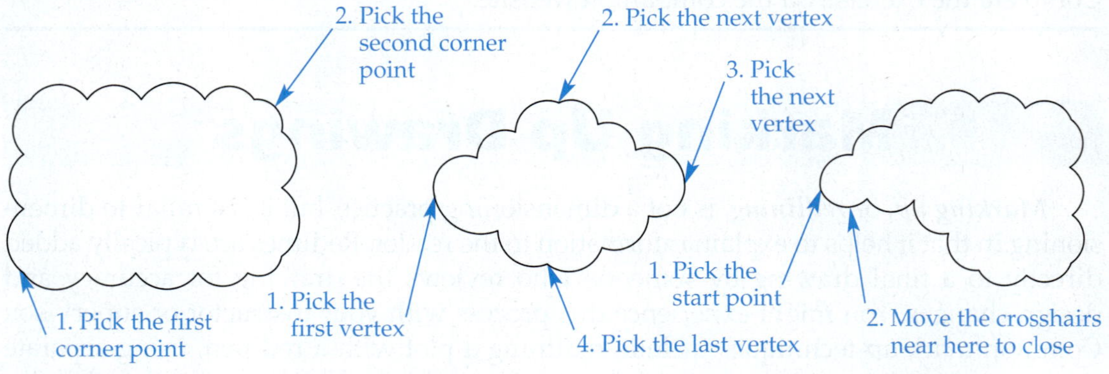

Defining Arc Length

Use the **Arc length** option to specify the size of revision cloud arcs. The value measures the length of an arc from the arc start point to the arc endpoint. AutoCAD prompts for the minimum arc length and then for the maximum arc length. Specifying different minimum and maximum values causes the revision cloud to have an uneven, hand-drawn appearance.

Converting Objects to Revision Clouds

Use the **Object** option to convert an object to a revision cloud. You can convert an object such as a circle, polyline, spline, ellipse, polygon, or rectangle. When prompted, select the object to convert. Use the **No** option at the Reverse direction prompt, or use the **Yes** option to reverse the direction of the cloud arcs.

Changing Revision Cloud Style

Use the **Style** option to change the revision cloud style. The default style is Normal, which displays arcs with a consistent width. When you specify the Calligraphy style, the start and end widths of the individual arcs are different, creating a more stylized revision cloud. See Figure 18-39.

Modifying Revision Clouds

Edit the shape of a revision cloud using techniques similar to those used when editing a closed polyline object. Grip editing is especially effective for editing a revision cloud and can be used to move, add, and remove vertices. Another option is to use the **Modify** option of the **REVCLOUD** command. Select the **Modify** option and select a closed polyline object, most often a revision cloud, to modify. By default, the point you pick when you select the object also locates the first point of the new revision cloud geometry. To select a different, possibly more accurate first point, use the **First point** option at the Specify next point or [First point]: prompt. Pick additional points as needed with the final point touching the existing object. Then select the portion of the selected object to remove. See Figure 18-41. Use the **No** option at the Reverse direction prompt, or use the **Yes** option to reverse the direction of the cloud arcs.

Exercise 18-10 www.g-wlearning.com/CAD/

Complete the exercise on the companion website.

Figure 18-41.
Using the **Modify** option of the **REVCLOUD** command to edit the shape of a revision cloud.

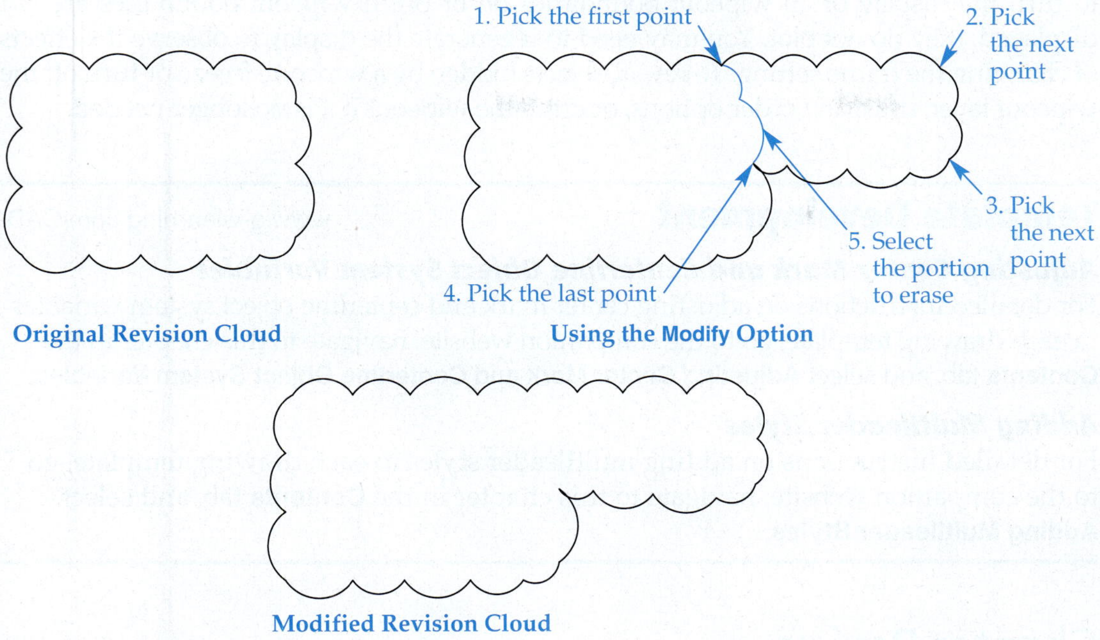

Using the WIPEOUT Command

The **WIPEOUT** command allows you to clear a portion of the drawing without erasing objects. The command is sometimes appropriate for applications similar to those for **REVCLOUD**, most often redlining. Figure 18-42 shows an example of a wipeout used to lay out the location of a proposed building site on a site plan.

Specify the first corner of the wipeout, followed by all other perimeter corners. Use the **Undo** option as needed to reverse the effects of an incorrect selection. Use the **Close** option or press [Enter] to create the wipeout.

WIPEOUT

Ribbon
Home
> Draw
Annotate
> Markup
Wipeout
Type
WIPEOUT

Figure 18-42.
Using the **WIPEOUT** command to clear a proposed building site on a site plan. Objects below the wipeout still exist. Further information has been added to the wipeout in this example.

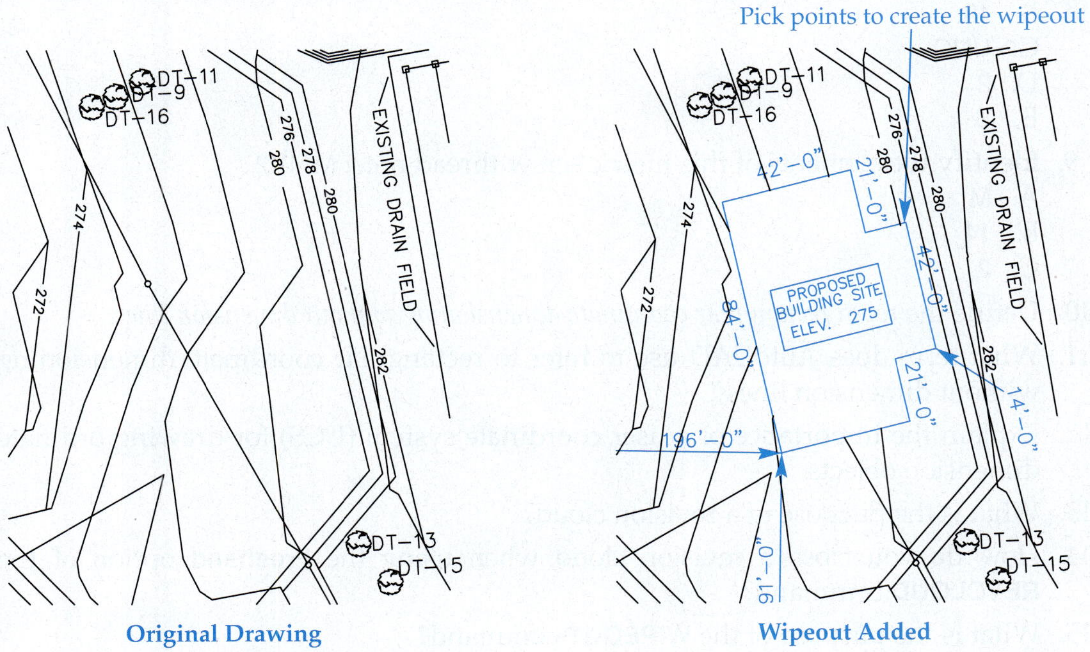

An alternative to picking points is to use the **Polyline** option and select a closed polyline object to convert to a wipeout. Use the **Frames** option and the appropriate selection to turn the display of all wipeout boundaries on or off. If wipeout boundaries are not displayed, they do not plot. You may need to regenerate the display to observe the effects of changing the frame setting. To reveal objects hidden by a wipeout, freeze or turn off the wipeout layer, use draw order options, or erase the wipeout if it is no longer needed.

Template Development

www.g-wlearning.com/CAD/

Adjusting Center Mark and Centerline Object System Variables

For detailed instructions on adjusting center mark and centerline object system variables in each drawing template, go to the companion website, navigate to this chapter in the **Contents** tab, and select **Adjusting Center Mark and Centerline Object System Variables**.

Adding Multileader Styles

For detailed instructions on adding multileader styles to each drawing template, go to the companion website, navigate to this chapter in the **Contents** tab, and select **Adding Multileader Styles**.

Chapter Review

Answer the following questions using the information in this chapter.

1. Which commands provide diameter dimensions for circles?
2. Which commands provide radius dimensions for arcs?
3. What is the purpose of the **CENTERMARK** command?
4. What is the most common size for leader arrowheads?
5. What angle constraints should you use for leaders to maintain standard mechanical drafting practices?
6. What is the usual length for the shoulder (landing) of a leader?
7. Describe two ways to dimension a 45° chamfer.
8. Identify the elements of this Unified screw thread note: 1/2-13UNC-2B.
   A. 1/2
   B. 13
   C. UNC
   D. 2
   E. B
9. Identify the elements of this metric screw thread note: M14X2.
   A. M
   B. 14
   C. 2
10. Define the term *rectangular coordinate dimensioning without dimension lines.*
11. What term does AutoCAD use to refer to rectangular coordinate dimensioning without dimension lines?
12. Explain the importance of a user coordinate system (UCS) for drawing ordinate dimension objects.
13. What is the purpose of a revision cloud?
14. How do you close a revision cloud when using the **Freehand** option of the **REVCLOUD** command?
15. What is the purpose of the **WIPEOUT** command?

Drawing Problems

Start AutoCAD if it is not already started. Start a new drawing for each problem using an appropriate template of your choice. The template should include layers and text, dimension, and multileader styles, when necessary, for drawing the given objects. Add layers and text, dimension, and multileader styles as needed. Draw all objects using appropriate layers. Use appropriate text, dimension, and multileader styles, justification, and format. Follow the specific instructions for each problem. Use only drawing and editing commands and techniques you have already learned. Use your own judgment and approximate dimensions when necessary. Apply dimensions accurately using ASME or appropriate industry standards.

▼ Basic

1. **Mechanical** Save P4-3 as P18-1. If you have not yet completed Problem 4-3, do so now. In the P18-1 file, create or import a new layer for centerlines, and draw the centerlines using the centerline layer.

2. **Mechanical** Save P4-7 as P18-2. If you have not yet completed Problem 4-7, do so now. In the P18-2 file, create or import a new layer for centerlines, and draw the centerlines using the centerline layer.

3. **Mechanical** Draw and dimension the part view shown. Save the drawing as P18-3.

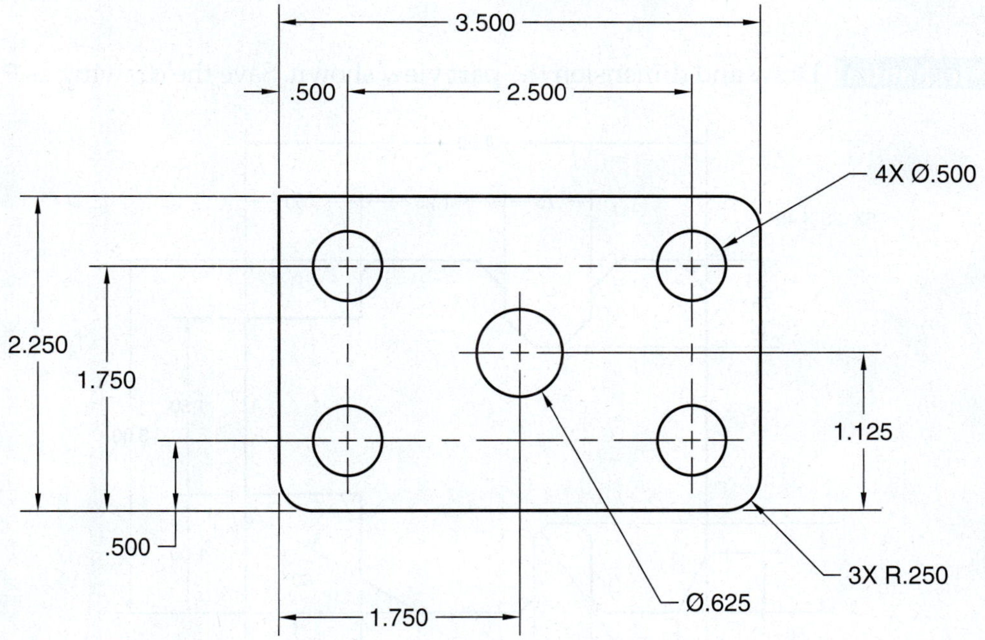

Drawing Problems - Chapter 18

4. **Mechanical** Draw and dimension the part views shown. Save the drawing as P18-4.

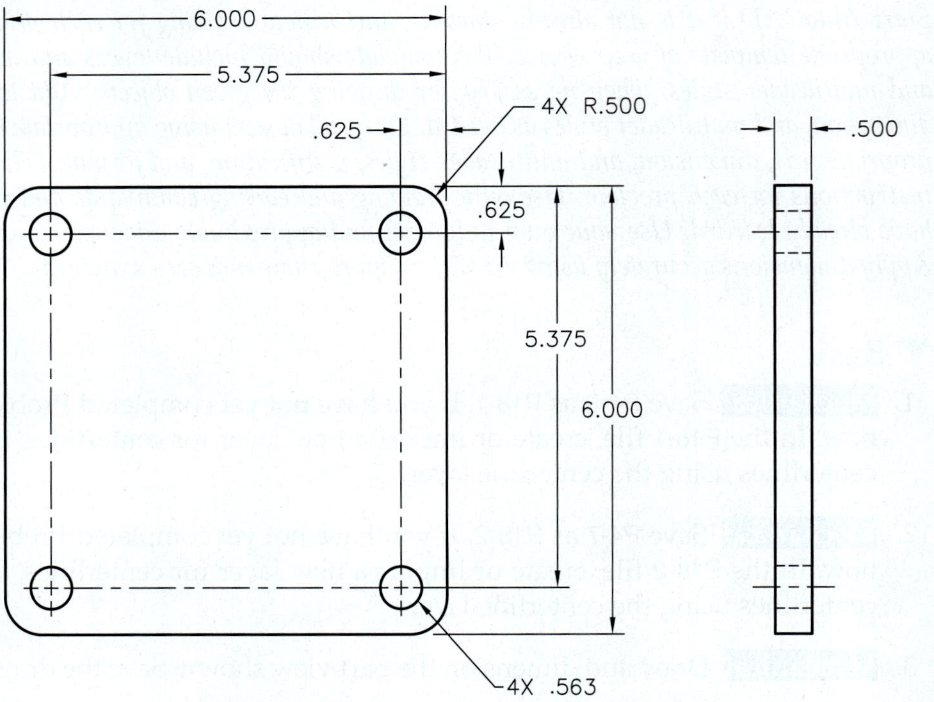

5. **Mechanical** Draw and dimension the part view shown. Save the drawing as P18-5.

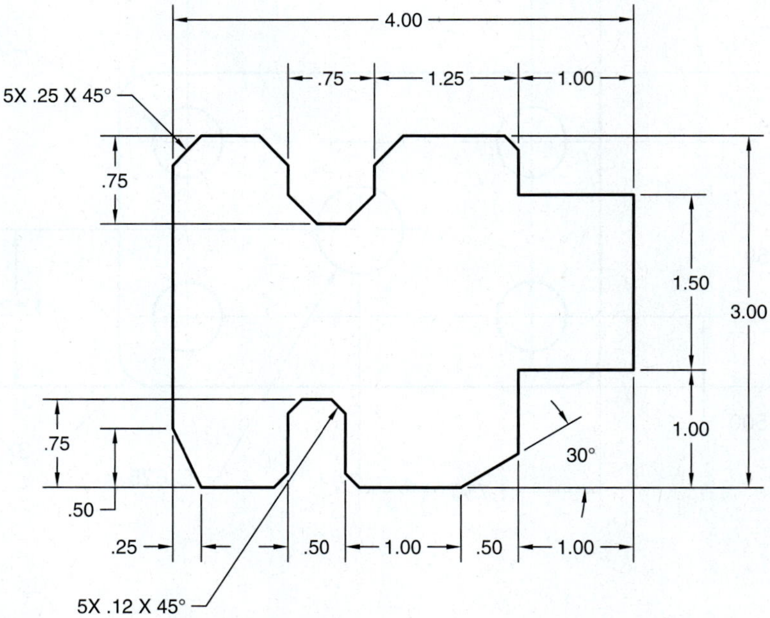

6. **Mechanical** Draw and dimension the part view shown. Save the drawing as P18-6.

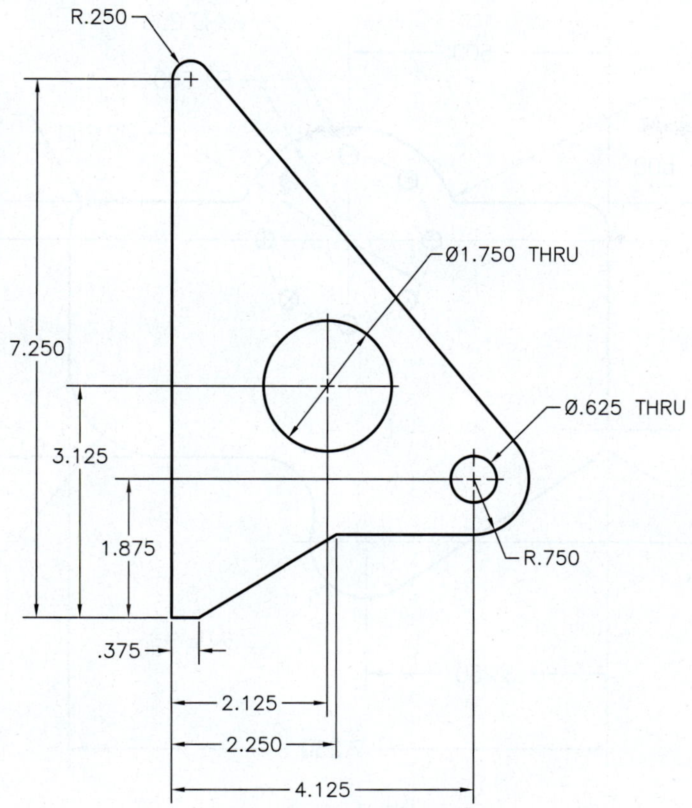

7. **Mechanical** Draw and dimension the pin shown. Save the drawing as P18-7.

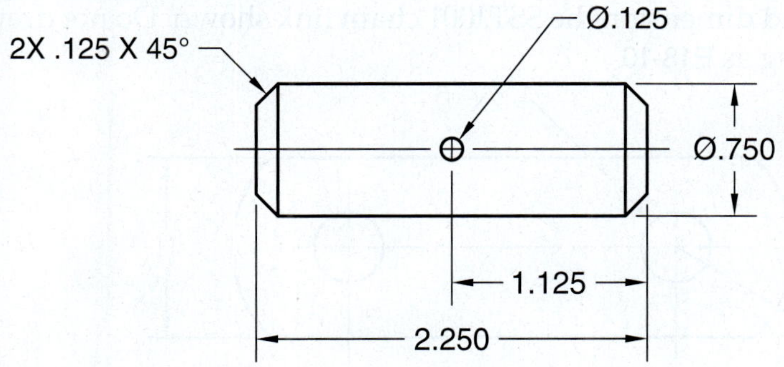

Title: Pin
Material: SAE 4320

8. **Mechanical** Draw and dimension the spline shown. Save the drawing as P18-8.

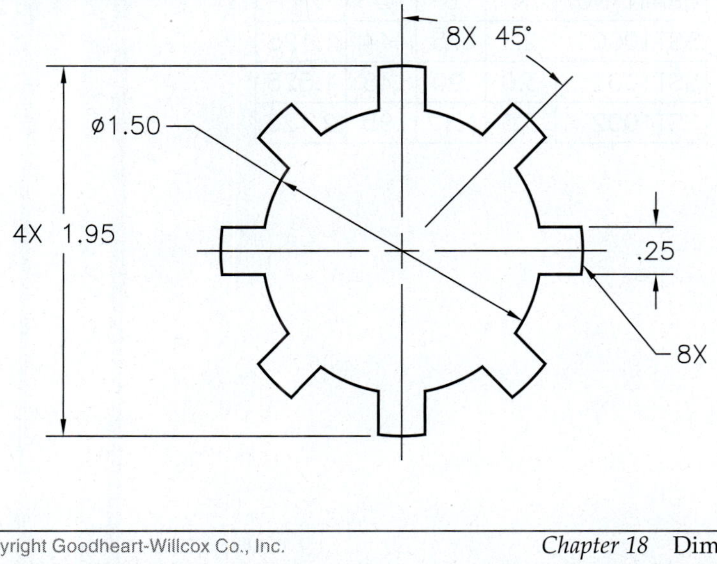

Title: Spline
Material: MS .125 THK

9. **Mechanical** Draw and dimension the gasket shown. Save the drawing as P18-9.

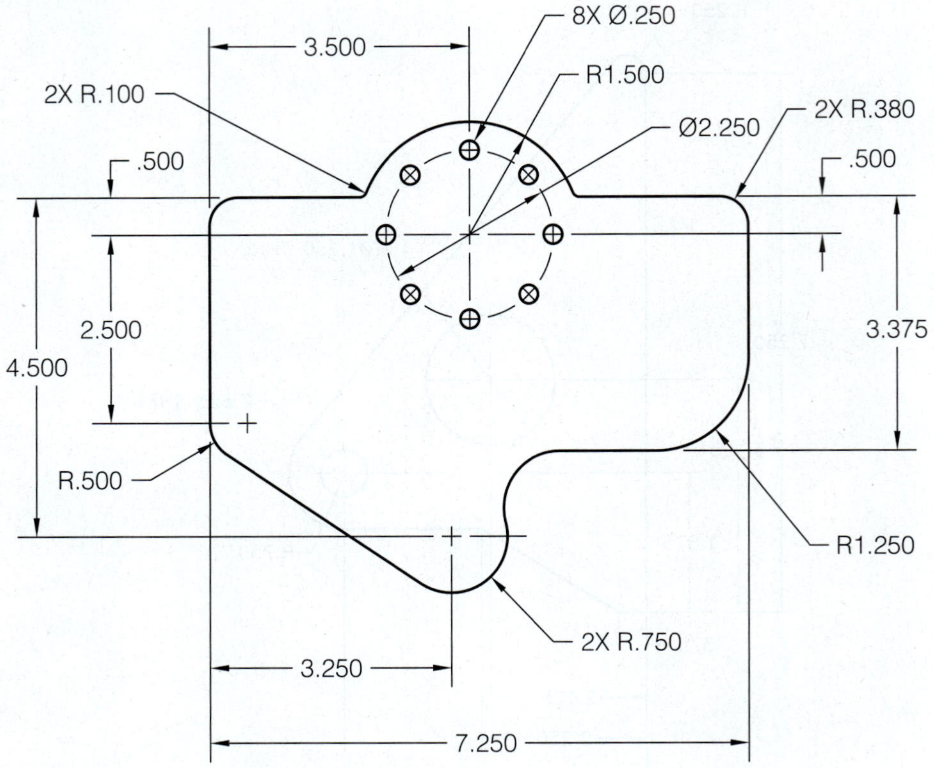

Title: Gasket
Material: 00 Phosphor Bronze

10. **Mechanical** Draw and dimension the SST1001 chain link shown. Do not draw the table. Save the drawing as P18-10.

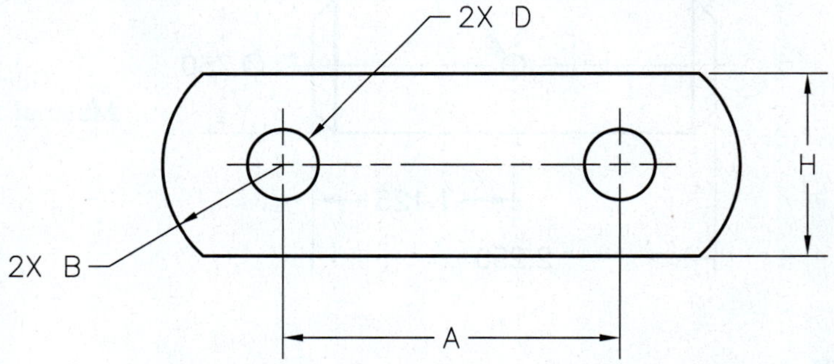

CHAIN NO.	A	B	D	H
SST1000	2.1	.75	.44	1.125
SST1001	3.0	.90	.48	1.525
SST1002	3.0	1.17	.95	2.125

Drawing Problems - Chapter 18

11. **Mechanical** Draw and dimension the part view shown. Save the drawing as P18-11.

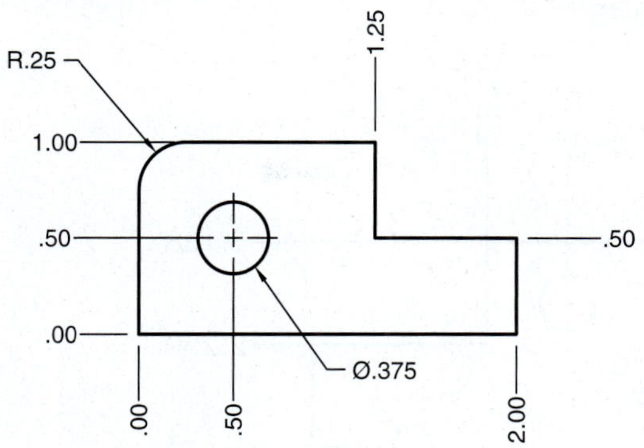

12. **Mechanical** Draw and dimension the chassis spacer shown. Do not draw the table. Save the drawing as P18-12.

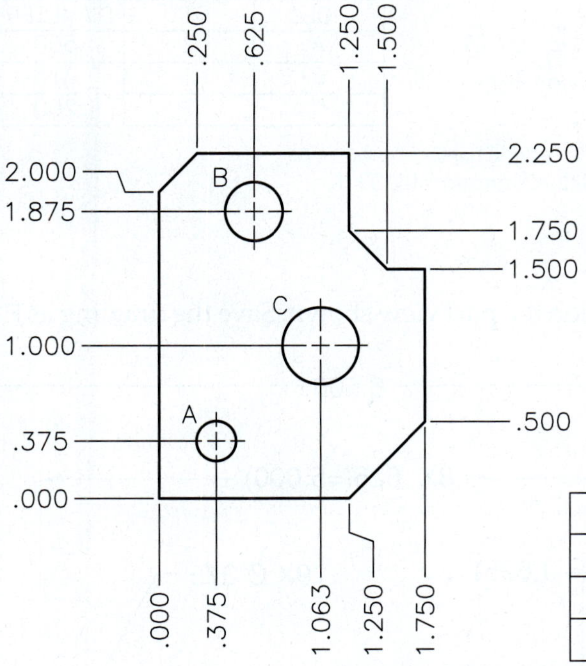

KEY	DIAMETER	DEPTH
A	.25	THRU
B	.38	THRU
C	.50	THRU

Title: Chassis Spacer
Material: .008 Aluminum

13. **Mechanical** Draw and dimension the chassis shown. Do not draw the table. Save the drawing as P18-13.

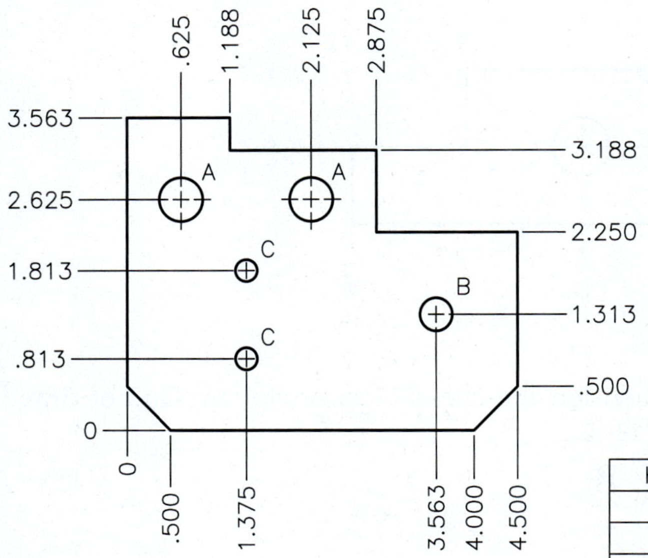

HOLE	QTY	DIAMETER
A	2	.500
B	1	.375
C	2	.250

Title: Chassis
Material: Aluminum .100 THK

14. **Mechanical** Draw and dimension the part view shown. Save the drawing as P18-14.

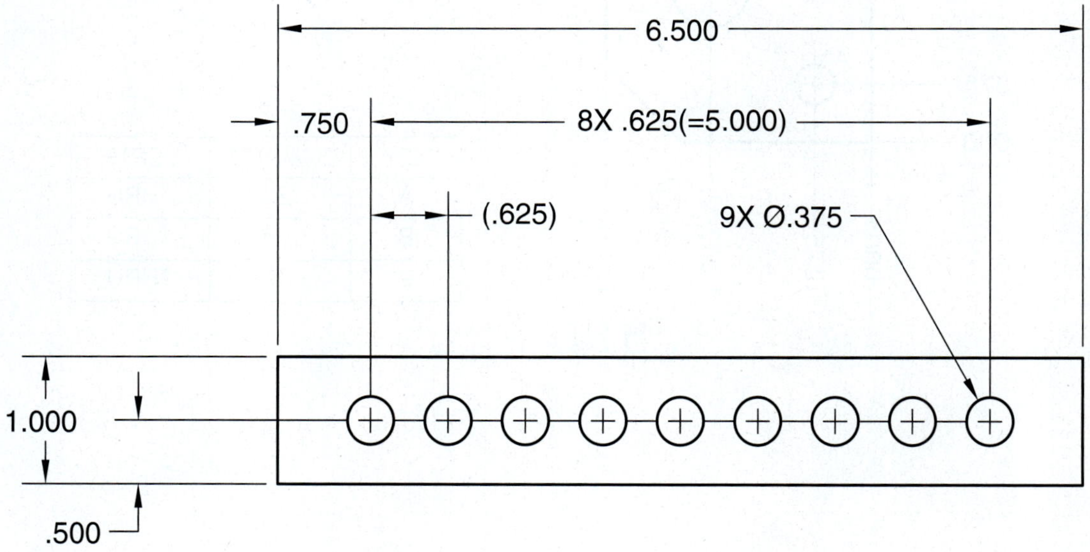

▼ Intermediate

15. **Mechanical** Draw and dimension the thumb screw shown. Save the drawing as P18-15. Print an 8.5″ × 11″ copy of the drawing extents using a 1:1 scale and landscape orientation.

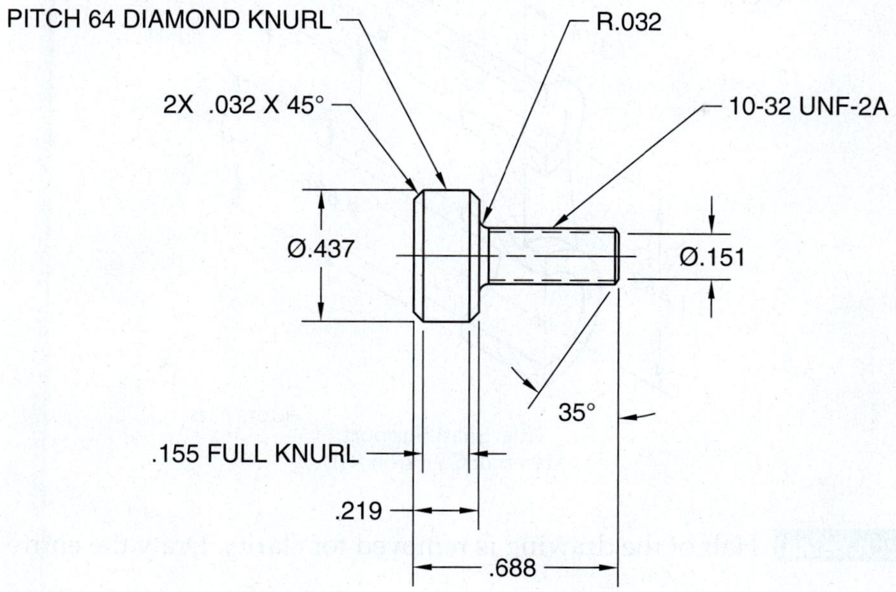

For Problems 16 through 18, draw and dimension the orthographic views needed to describe the part completely. Save the drawings as P18-16, P18-17, *and* P18-18.

16. **Mechanical**

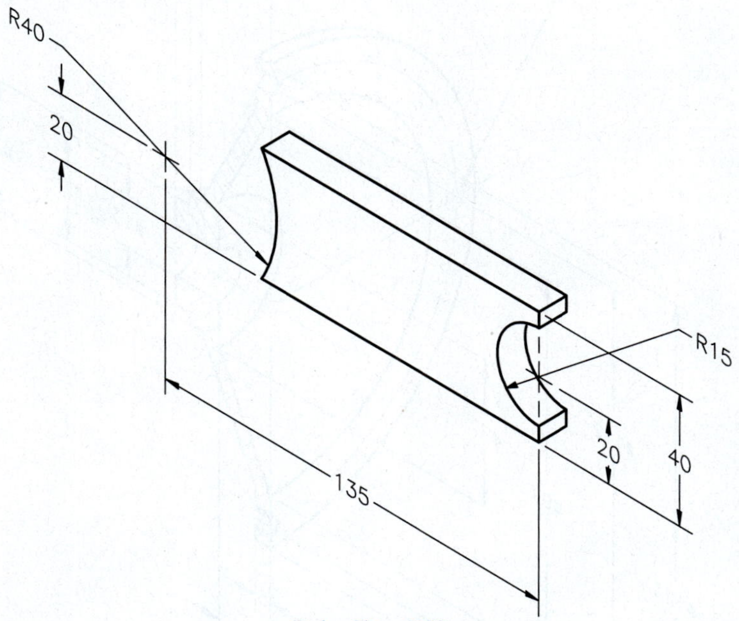

Title: Shim MS
Metric 10 THK

17. **Mechanical** Half of the drawing is removed for clarity. Draw the entire part.

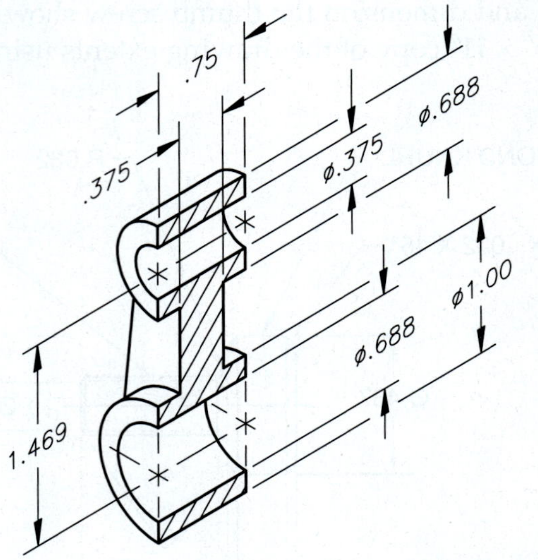

Title: Shaft Support
Material: Cast Iron (CI)

18. **Mechanical** Half of the drawing is removed for clarity. Draw the entire part.

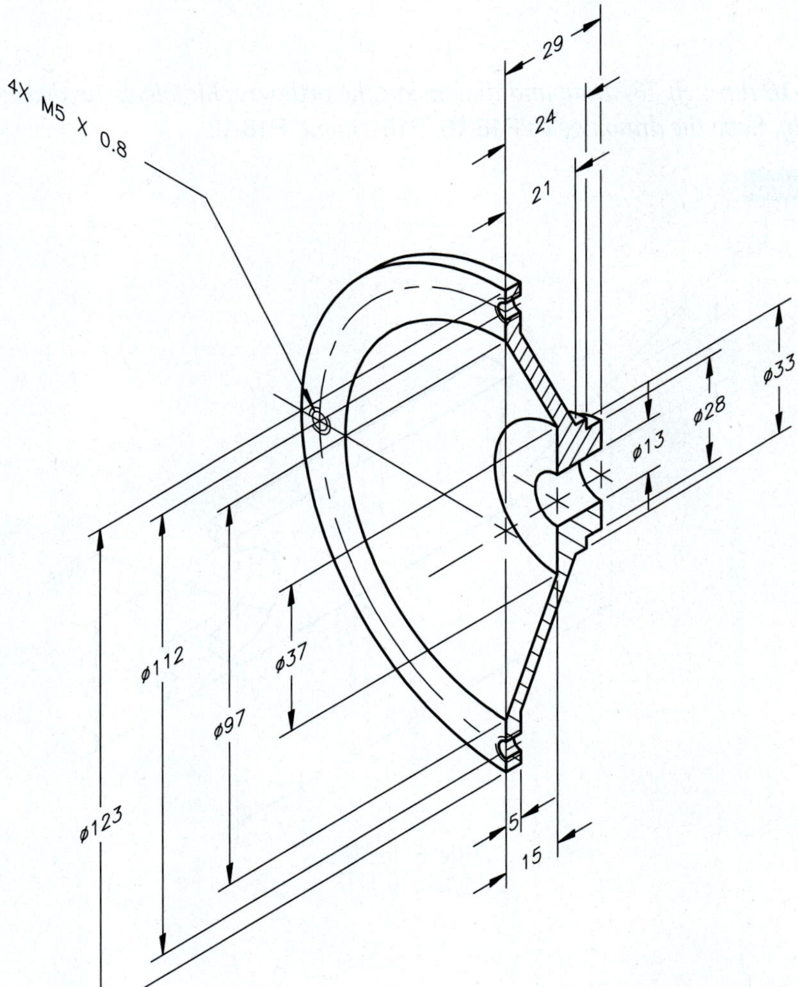

Title: Transmission Cover
Material: Cast Iron (CI)
Metric

19. **Mechanical** Draw and dimension the part views shown. Save the drawing as P18-19.

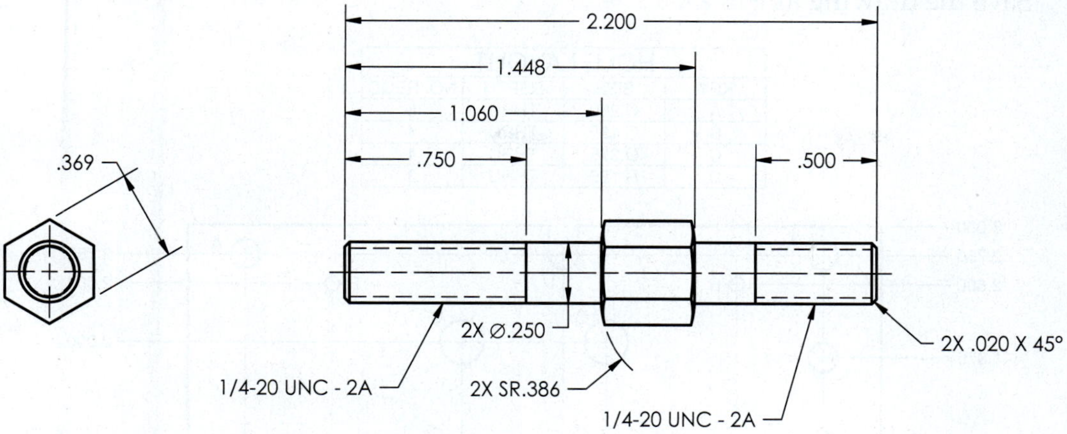

Title: Stud
Material: Stainless Steel

20. **Mechanical** Draw and dimension the part view shown. Do not draw the table. Save the drawing as P18-20.

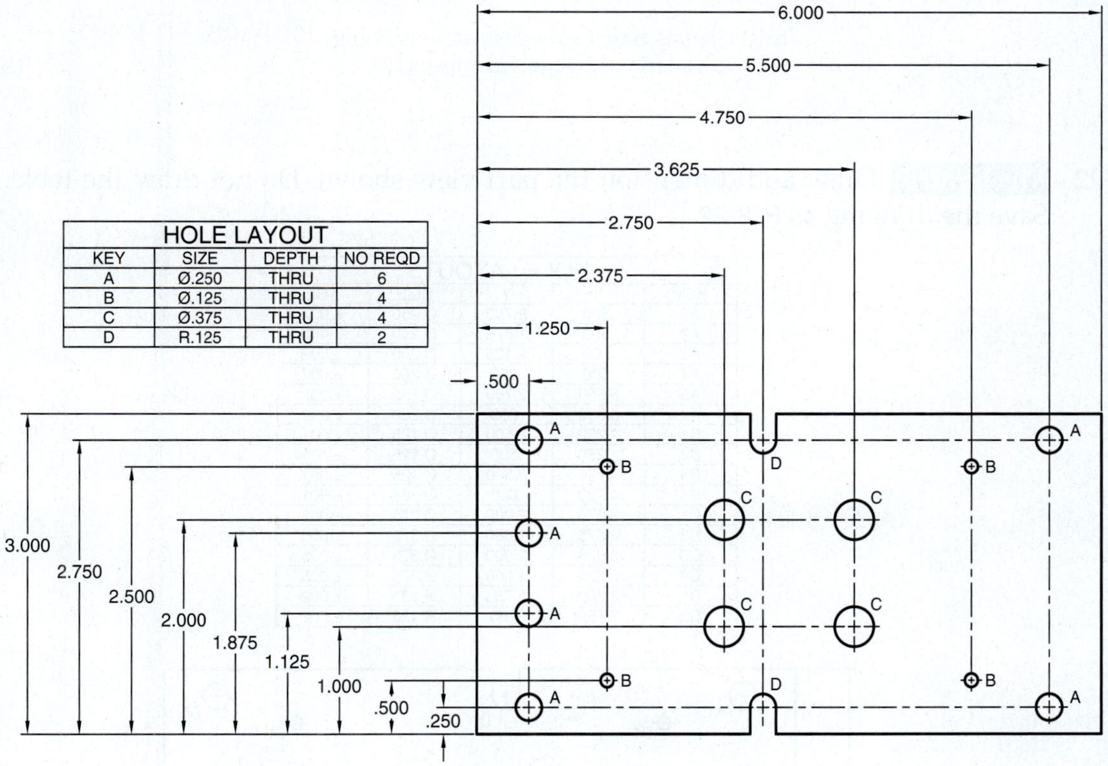

HOLE LAYOUT			
KEY	SIZE	DEPTH	NO REQD
A	Ø.250	THRU	6
B	Ø.125	THRU	4
C	Ø.375	THRU	4
D	R.125	THRU	2

Title: Chassis Base (datum dimensioning)
Material: 12 gage Aluminum

21. **Mechanical** Draw and dimension the part view shown. Do not draw the table. Save the drawing as P18-21.

HOLE LAYOUT			
KEY	SIZE	DEPTH	NO. REQD
A	Ø.250	THRU	6
B	Ø.125	THRU	4
C	Ø.375	THRU	4
D	R.125	THRU	2

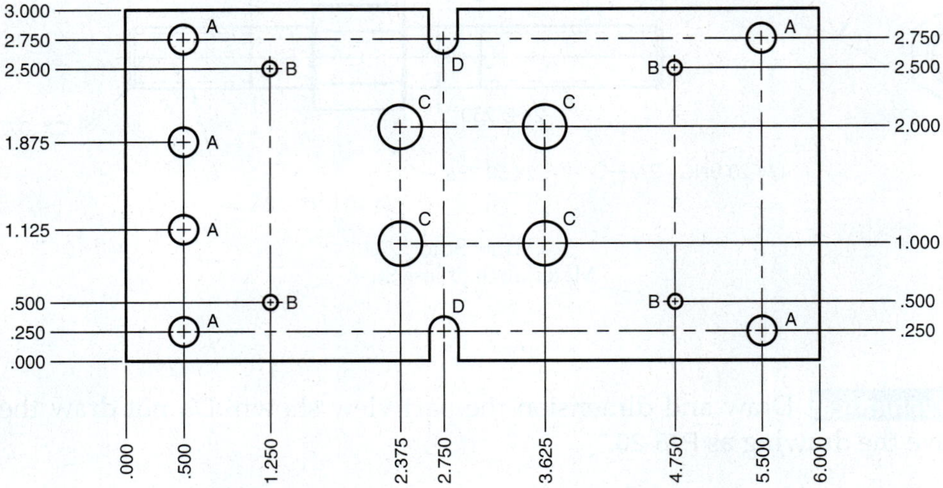

Title: Chassis Base (arrowless dimensioning)
Material: 12 gage Aluminum

22. **Mechanical** Draw and dimension the part view shown. Do not draw the table. Save the drawing as P18-22.

HOLE LAYOUT				
KEY	X	Y	SIZE	TOL
A1	.500	2.750	Ø.250	±.002
A2	.500	1.875	Ø.250	±.002
A3	.500	1.125	Ø.250	±.002
A4	.500	.250	Ø.250	±.002
A5	5.500	2.750	Ø.250	±.002
A6	5.500	.250	Ø.250	±.002
B1	1.250	2.500	Ø.125	±.001
B2	1.250	.500	Ø.125	±.001
B3	4.750	2.500	Ø.125	±.001
B4	4.750	.500	Ø.125	±.001
C1	2.375	2.000	Ø.375	±.005
C2	2.375	1.000	Ø.375	±.005
C3	3.625	2.000	Ø.375	±.005
C4	3.625	1.000	Ø.375	±.005
D1	2.750	2.750	R.125	±.002
D2	2.750	.250	R.125	±.002

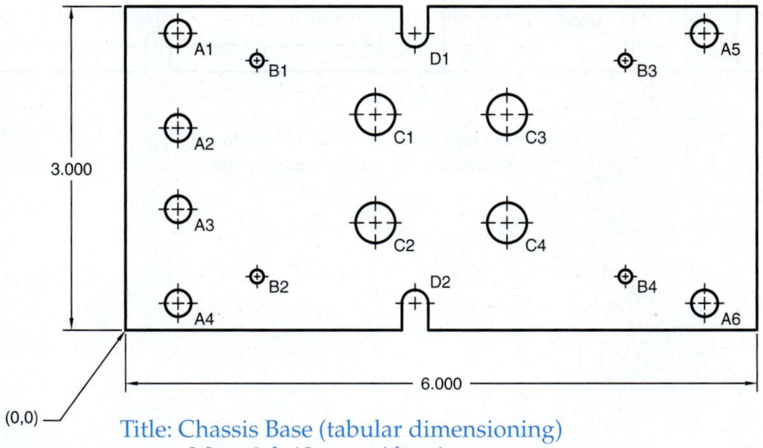

Title: Chassis Base (tabular dimensioning)
Material: 12 gage Aluminum

▼ Advanced

23. **Mechanical** Draw and dimension the part views shown. Do not draw the table. Save the drawing as P18-23.

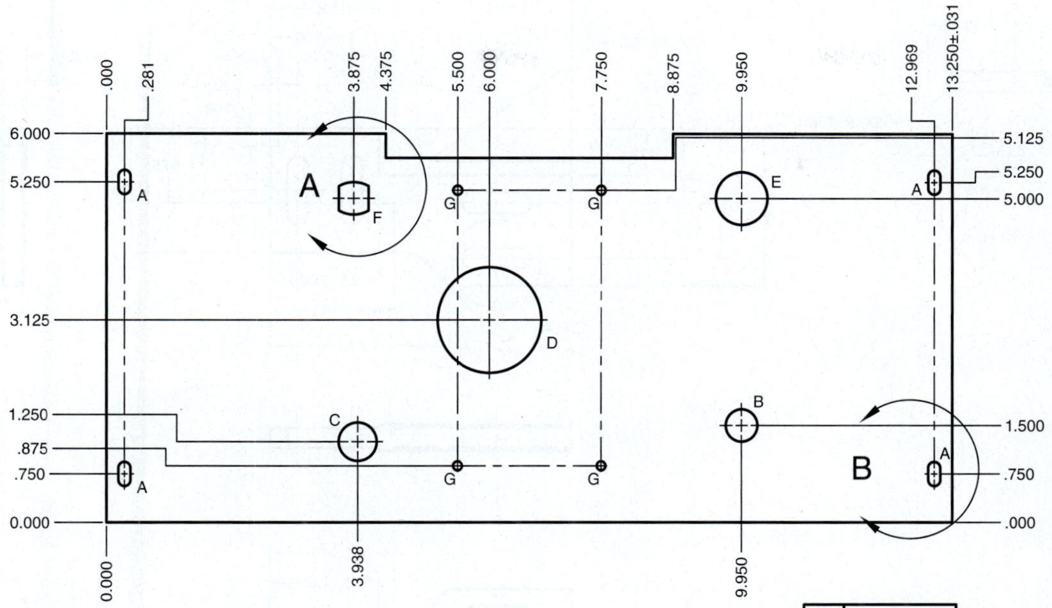

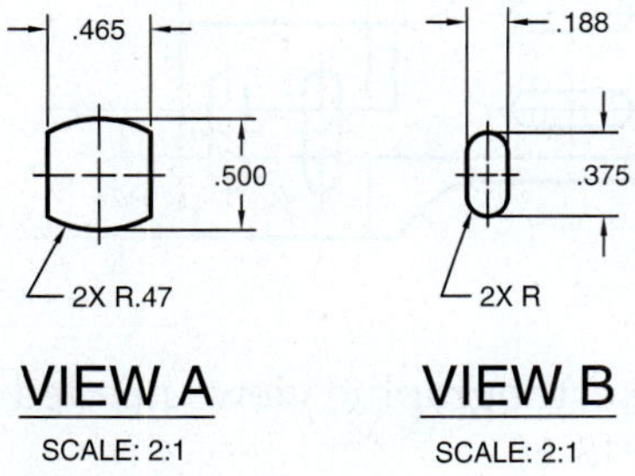

G	Ø.141
F	SEE VIEW A
E	Ø.813
D	Ø1.625
C	Ø.594
B	Ø.500
A	SEE VIEW B
HOLE LEGEND	

24. **Mechanical** Draw and dimension the part views shown. Save the drawing as P18-24.

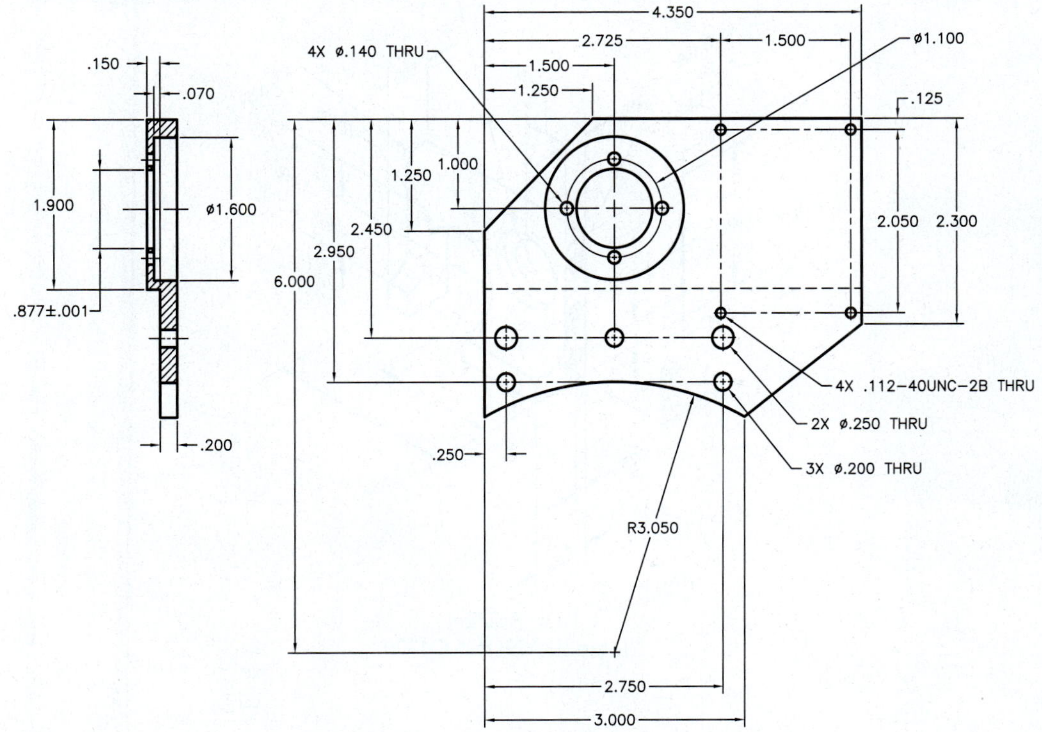

Title: Bracket
Material: SAE 1040

25. **Mechanical** Draw and dimension the part views shown. Save the drawing as P18-25.

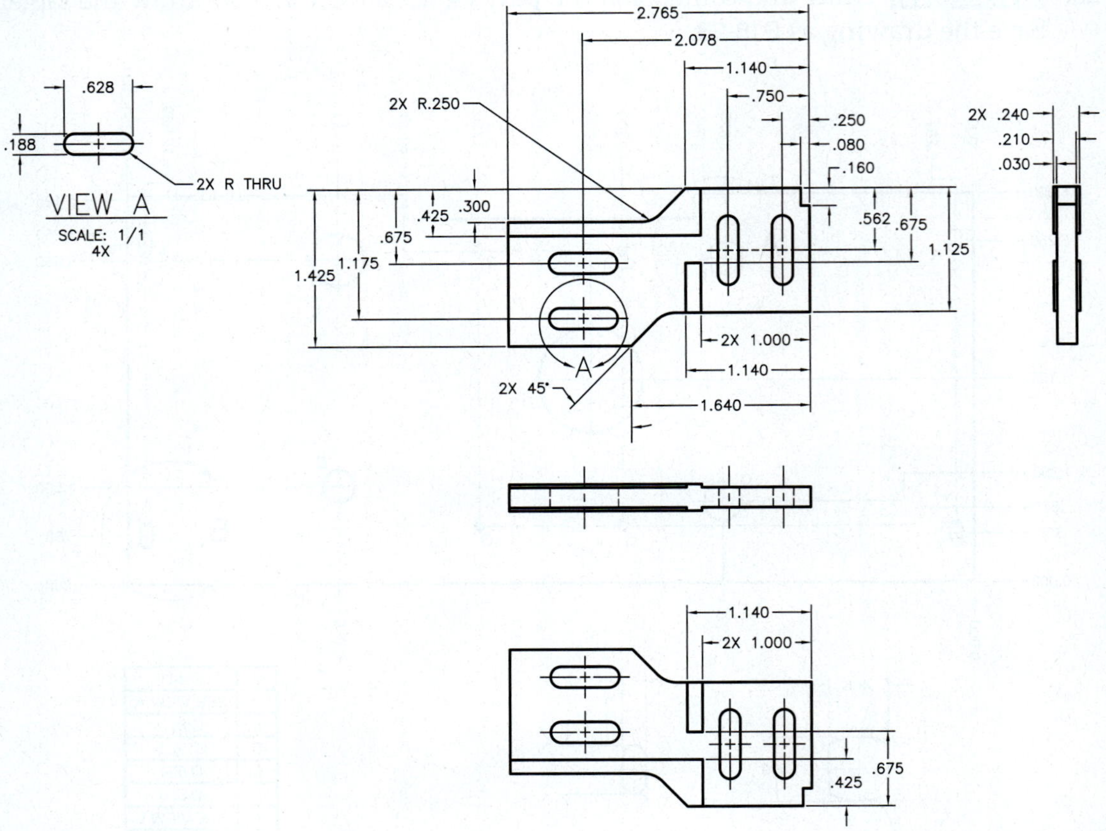

26. **Mechanical** Draw and dimension the orthographic views needed to describe the part completely. Save the drawing as P18-26.

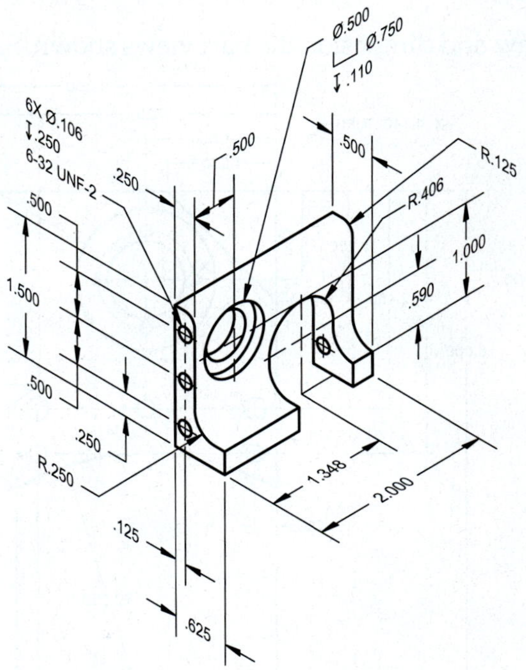

Title: Bracket
Material: AISI/SAE 4320

27. **Architectural** Save P17-16 as P18-27. If you have not yet completed Problem 17-16, do so now. In the P18-27 file, add the client-requested redlines to the floor plan as shown.

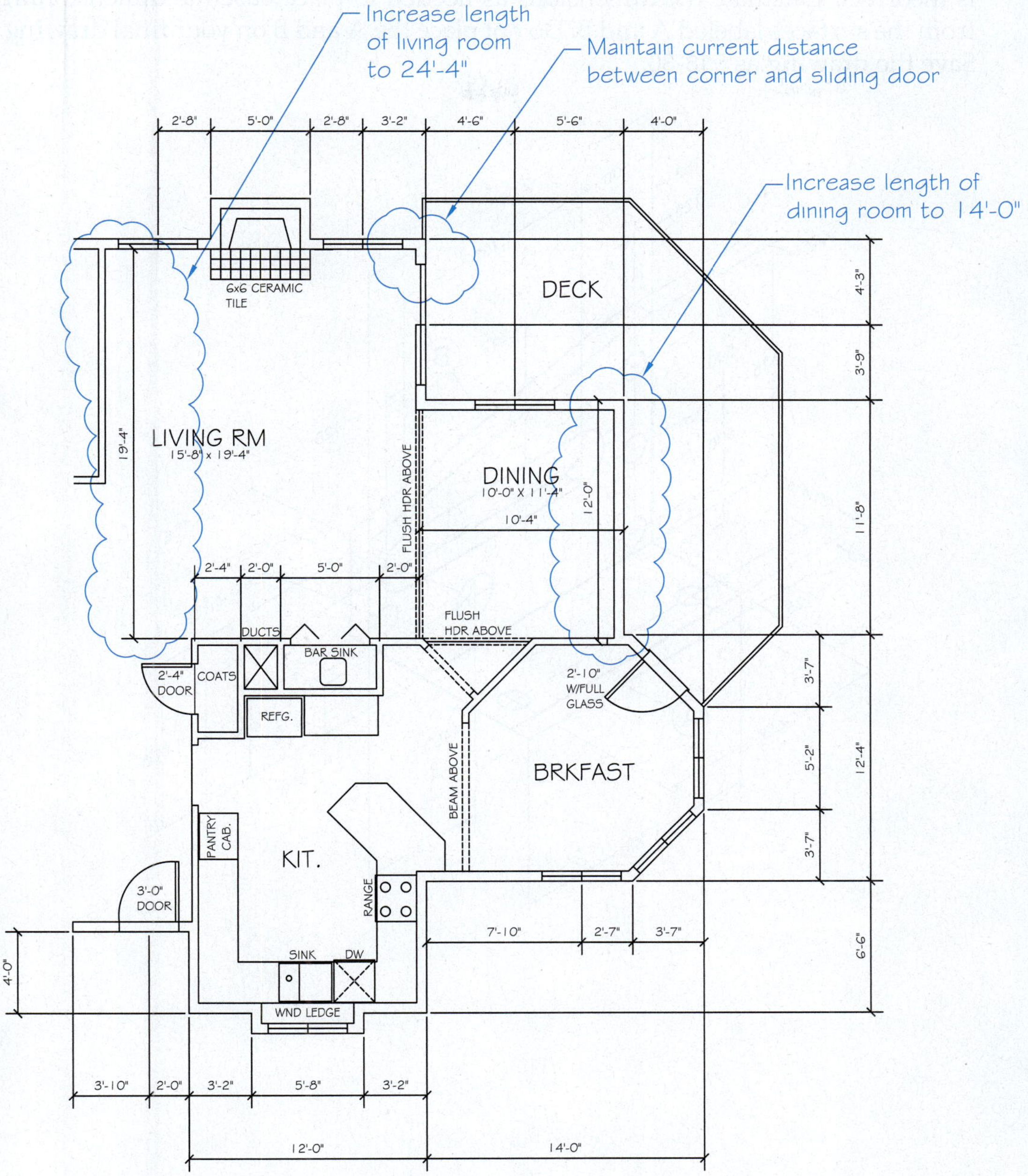

28. **Architectural** Save P5-14 as P18-28. If you have not yet completed Problem 5-14, do so now. In the P18-28 file, dimension the drawing of the kitchen.

29. **Architectural** Save P8-21 as P18-29. If you have not yet completed Problem 8-21, do so now. In the P18-29 file, dimension the most important views of the hanger. Erase the undimensioned views.

30. **Mechanical** Carefully evaluate the problem before beginning. Many of the given dimensions are provided to the inside surfaces of the bracket. This application is incorrect. Calculate the dimensions as needed to place baseline dimensioning from the surfaces labeled A and B. Do not place the A and B on your final drawing. Save the drawing as P18-30.

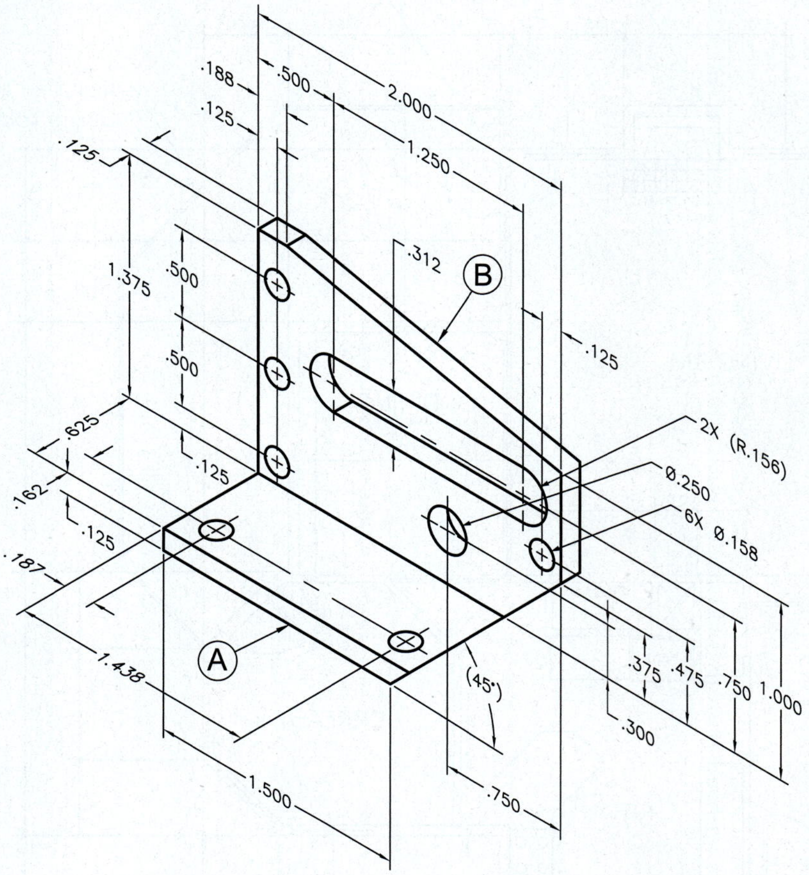

AutoCAD Certified Professional Exam Practice

Answer the following questions using the information in this chapter.

1. What should the **Offset from origin** setting be when you use the **DIM** or **DIMORDINATE** command and dimension to a centerline? *Select all that apply.*
   A. 0
   B. .012
   C. .063
   D. .12 mm
   E. 1 mm
   F. 1.5 mm
2. What is the correct term for the shaded area in the hole shown below? *Select the one item that best answers the question.*
   A. counterbore
   B. countersink
   C. fillet
   D. landing
   E. round
   F. spotface

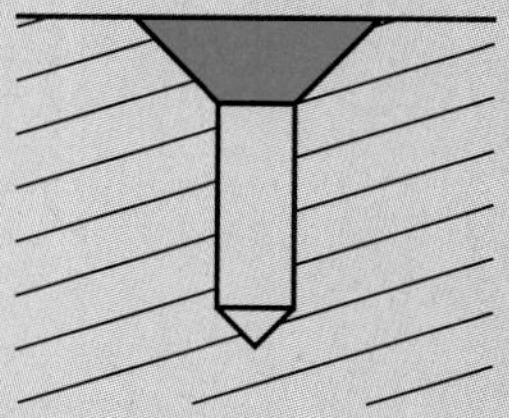

3. Which of the following commands can be used to dimension an arc? *Select all that apply.*
   A. **DIMARC**
   B. **DIMCENTER**
   C. **DIM**
   D. **DIMJOGLINE**
   E. **DIMRADIUS**

Follow the instructions in each problem. Complete each problem by answering the related question.

4. **Navigate to this chapter on the companion website and open CPE-18ordinate.dwg.** Use rectangular coordinate dimensioning without dimension lines to dimension the drawing as shown. What are the values of A, B, and C?

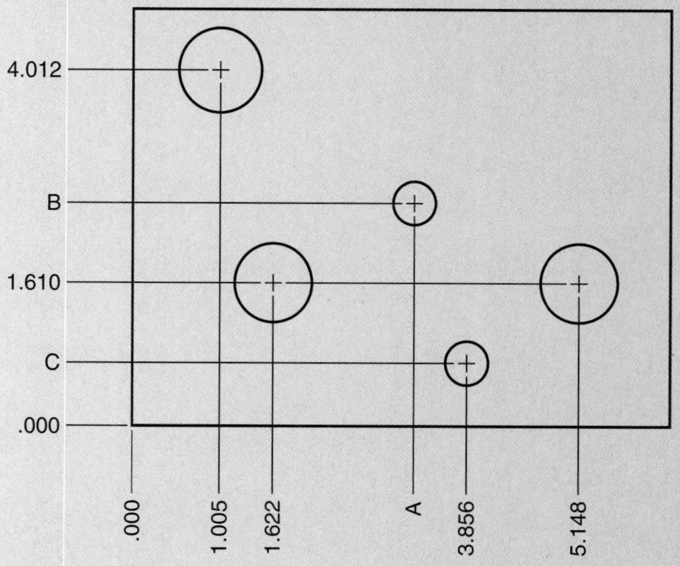

5. **Navigate to this chapter on the companion website and open CPE-18chart.dwg.** Create the drawing shown using the dimensions for item DRI203. Use the **Node** object snap to start the lower-left corner of the object at the point object provided in the drawing file. What are the coordinates of Point 1?

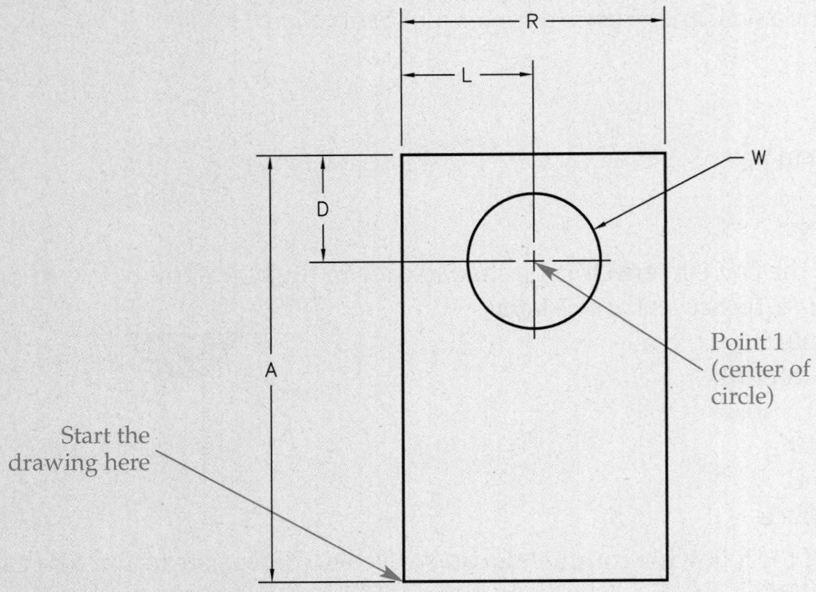

ITEM NO.	A	D	L	R	W
DRI201	4.000	1.000	1.250	2.500	1.261
DRI202	6.000	1.500	1.875	3.750	1.892
DRI203	9.500	2.375	2.969	5.938	2.995

AutoCAD Certified Professional Exam Practice - Chapter 18

CHAPTER 19

Dimensioning with Tolerances

Learning Objectives

After completing this chapter, you will be able to:

- ✓ Define and use dimensioning and tolerancing terminology.
- ✓ Set the precision for dimensions and tolerances.
- ✓ Set up the primary units for use with inch or metric dimensions.
- ✓ Specify an appropriate tolerance method.
- ✓ Create and use specified tolerance dimension styles.
- ✓ Explain the purpose of geometric dimensioning and tolerancing (GD&T).

This chapter introduces general tolerancing as applied to ***conventional dimensioning*** and explains how to create dimensions with specified tolerances for mechanical drawings common in the manufacturing industry. This chapter also introduces geometric dimensioning and tolerancing (GD&T) symbols and offers information on how you can learn more about GD&T.

conventional dimensioning: Dimensioning without the use of geometric tolerancing.

Tolerancing Fundamentals

Every dimension has a ***tolerance***, except dimensions specifically identified as reference, maximum, minimum, or stock. The tolerancing practice depends on specific engineering and manufacturing applications, interrelated features, and industry and company preference. You can apply tolerance to dimensions indirectly by placing the information in the title block or a general note. See Figure 19-1. Any dimension that requires a tolerance that is different from the general tolerances given in the title block or general note must have the specific tolerance applied directly to the dimension on the drawing. See Figure 19-2.

The dimension stated as 12.50±0.25 in Figure 19-2A is in a style known as ***plus-minus dimensioning***. The tolerance of this dimension is the difference between the maximum and minimum ***limits***. In this case, the upper limit is 12.75 (12.50 + 0.25 = 12.75), and the lower limit is 12.25 (12.50 – 0.25 = 12.25). To find the tolerance, subtract the lower limit from the upper limit. The tolerance in this example is 0.50 (12.75 – 12.25 = 0.50). The ***specified dimension*** of the feature shown in Figure 19-2A is 12.50.

tolerance: The total amount by which a specific dimension is permitted to vary.

plus-minus dimensioning: A dimensioning system in which a variance from the dimension applies in both the positive (+) and negative (–) directions or in one direction only.

limits: The largest and smallest numerical values the feature can have.

specified dimension: The part of the dimension from which the limits are calculated.

Figure 19-1.
The title block or a general note provides indirect tolerance specifications. A—An unspecified tolerance on an inch drawing. B—An unspecified tolerance on a metric drawing. Metric tolerancing is generally controlled by the ISO 2768 standard, *General Tolerances*, developed by the International Organization for Standardization (ISO).

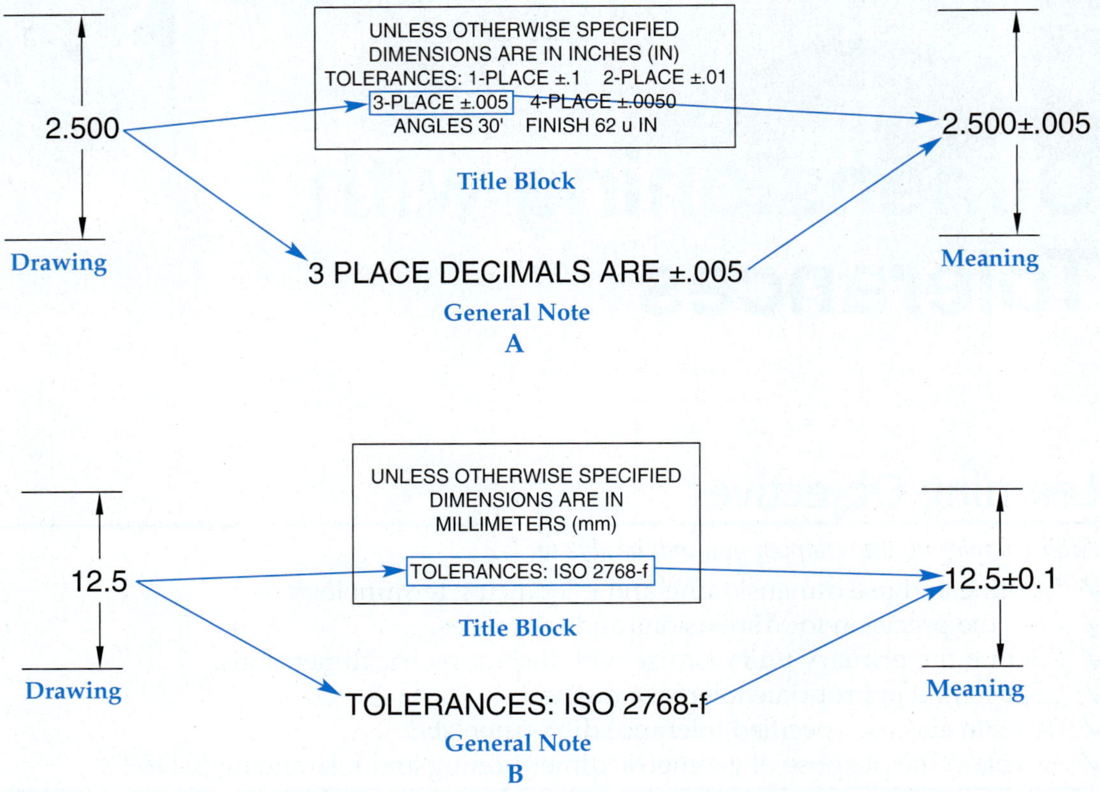

Figure 19-2.
A—Plus-minus dimensioning. B—Limit dimensioning.

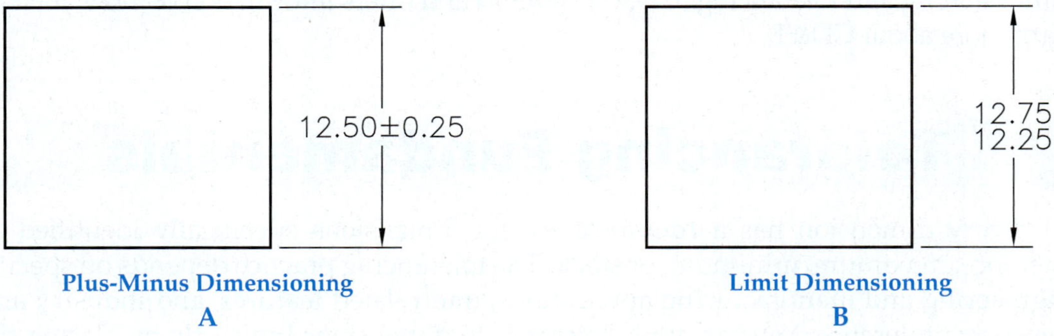

limit dimensioning: Method in which the upper and lower limits are given, instead of the specified dimension and tolerance.

bilateral tolerance: A tolerance style that permits variance in both the positive and negative directions from the specified dimension.

Limit dimensioning, shown in Figure 19-2B, is an alternative method of showing and calculating tolerance. Limit dimensioning is most common for defining fits between mating parts, such as a sliding fit between a hole and shaft or a press fit between a hole and bearing. Some companies, or departments within a company, such as an inspection department, prefer limit dimensioning because it does not require calculating limits. However, the actual dimension of the object in the drawing is unknown.

Plus-minus dimensioning uses a bilateral or unilateral tolerance format, depending on the application. Figure 19-3 shows examples of equal and unequal ***bilateral tolerances***. Bilateral tolerancing is the most common tolerancing method. Manufacturers typically prefer equal bilateral tolerancing because they attempt to

Figure 19-3.
Examples of plus-minus dimensioning values using bilateral tolerances.

Metric	Inch	Angular
24 ± 0.1	$.750 \pm .005$	$45°30' \pm 0°5'$

Equal Bilateral Tolerance

Metric	Inch	Angular
$24^{+0.08}_{-0.20}$	$.750^{+.002}_{-.003}$	$45.5°^{+0.2°}_{-0.5°}$

Unequal Bilateral Tolerance

manufacture features as close to the specified dimension as possible. Figure 19-4 shows examples of ***unilateral tolerances***. Some companies use unilateral tolerances to define fits between mating parts. However, manufacturers who use the drawing to program computer numerical control (CNC) machining equipment often avoid unilateral tolerancing.

Basic dimensions establish true position from datums and between interrelated features, and define true profile. A rectangle, or frame, around the dimension value distinguishes a basic dimension from other types of dimensions. ***Single limits*** are sometimes applied to various features, such as chamfers, fillets, rounds, hole depths, and thread lengths. The abbreviation for minimum (MIN) or maximum (MAX) follows the dimension value to describe a single limit application. The design determines the unspecified limit.

unilateral tolerance: A tolerance style that permits a variation in only one direction from the specified dimension.

basic dimension: A theoretically exact dimension used in geometric dimensioning and tolerancing.

single limits: Limit dimensions used when the specified dimension cannot be any more than the maximum or less than the minimum given value.

Dimensioning Units

The ASME Y14.5-2018 standard, *Dimensioning and Tolerancing*, has separate recommendations for the display of inch, metric, and angular dimensions. Figure 19-5, Figure 19-6, and Figure 19-7 explain the rules for each type of dimension. The US unit of measure commonly used on engineering drawings is the inch. The SI unit of measure commonly used on engineering drawings is the millimeter. Company or school policy and product requirements determine the actual units used on engineering drawings.

Place the general note UNLESS OTHERWISE SPECIFIED, ALL DIMENSIONS ARE IN INCHES (or MILLIMETERS) on the drawing when all dimensions are in inches or millimeters. Follow millimeter dimensions on an inch drawing with the abbreviation mm. Follow inch dimensions on a metric drawing with the abbreviation IN. Figure 19-3 and Figure 19-4 show examples of displaying inch, metric, and angular dimensions with specified tolerances.

Figure 19-4.
A unilateral tolerance allows variation in only one direction from the specified dimension.

Metric	Inch	Angular
$24^{\ 0}_{-0.2}$	$.625^{+.000}_{-.004}$	$25.5°^{\ 0}_{-0.5°}$
$24^{+0.2}_{\ 0}$	$.625^{+.004}_{-.000}$	$25.5°^{+0.5°}_{\ 0}$

Figure 19-5.
Dimensioning rules for inch dimensions.

Rules for Inch Dimensions	Examples
A zero does not precede a decimal inch that is less than one.	.5
Express a specified dimension to the same number of decimal places as its tolerance. Add zeros to the right of the decimal point if needed.	.250±.005 (additional zero added to .25)
Fractional inches generally indicate a larger tolerance, or give nominal sizes, such as in a thread callout.	Dimension value: 2 1/2±1/32 Thread: 1/2-13UNC-2B
Plus and minus values of an inch tolerance have the same number of decimal places.	$.250^{+.005}_{-.010}$ .255 .240
Unilateral tolerances use the + or – symbol, and the 0 value has the same number of decimal places as the value that is greater or less than 0.	$.250^{+.005}_{-.000}$ $.250^{+.000}_{-.005}$
Inch limit tolerance values have the same number of decimal places. When displaying limit tolerance values on one line, the lower value precedes the higher value, and a dash separates the values. When displaying stacked limit tolerance values, place the higher value above the lower value.	One line: 1.000–1.062 Stacked: 1.062 1.000
Basic dimension values have the same number of decimal places as their associated tolerance.	[2.000]

Setting Primary Units

A dimension style controls the appearance of dimensions, including dimension values and tolerance. The initial phase of dimensioning with tolerances involves setting the appropriate values for the primary units of the dimension style. Use the **Primary Units** tab of the **New** (or **Modify**) **Dimension Style** dialog box, shown in Figure 19-8, to set the dimension units and precision.

Use the **Precision** drop-down list in the **Linear dimensions** area to set the number of zeros displayed after the decimal point of the specified dimension. The **Zero suppression** settings control the display of zeros before and after the decimal point. For inch dimensions, the **Leading** options should be on, and the **Trailing** options should be off. For typical metric dimensions, without using subunits, the **Leading** options should be off, and the **Trailing** options should be on.

Figure 19-6.
Dimensioning rules for metric dimensions.

Rules for Metric Dimensions	Examples
Omit the decimal point and zero when the dimension is a whole number.	12
A zero precedes a decimal millimeter that is less than one.	0.5
When the dimension is greater than a whole number by a fraction of a millimeter, the last digit to the right of the decimal point is not followed with a zero. This rule is true unless the dimension displays tolerance values.	12.5
Plus and minus values of a metric tolerance have the same number of decimal places. Add zeros to fill in where needed.	$24^{+0.25}_{-0.10}$ 24.25 24.00
Metric limit tolerance values have the same number of decimal places. When displaying limit tolerance values on one line, the lower value precedes the higher value, and a dash separates the values. When displaying stacked limit tolerance values, place the higher value above the lower value. Examples in ASME Y14.5 show no zeros after the specified dimension to match the tolerance.	One line: 7.5–7.6 Stacked: 7.6 / 7.5 24±0.25 24.5±0.25
When applying unilateral tolerances, use a single 0 without a + or – sign for the 0 part of the value.	$24^{\ 0}_{-0.2}$ $24^{+0.25}_{\ 0}$
Basic dimension values follow the same display rules as stated for other metric numbers.	[24] [24.5]

Figure 19-7.
Dimensioning rules for angular dimensions.

Rules for Angular Dimensions	Examples
Establish angular dimensions in degrees (°) and decimal degrees (30.5°), or in degrees (°), minutes (′), and seconds (″).	24°15′30″
The plus and minus tolerance values and the angle have the same number of decimal places.	30.0°±0.5° (not 30°±0.5°)
Where only specifying minutes or seconds, precede the number of minutes or seconds with 0° or 0°0′, as applicable.	0°45′30″ 0°0′45″

Figure 19-8.
The **Primary Units** tab of the **New** (or **Modify**) **Dimension Style** dialog box sets the unit format and precision of linear dimensions.

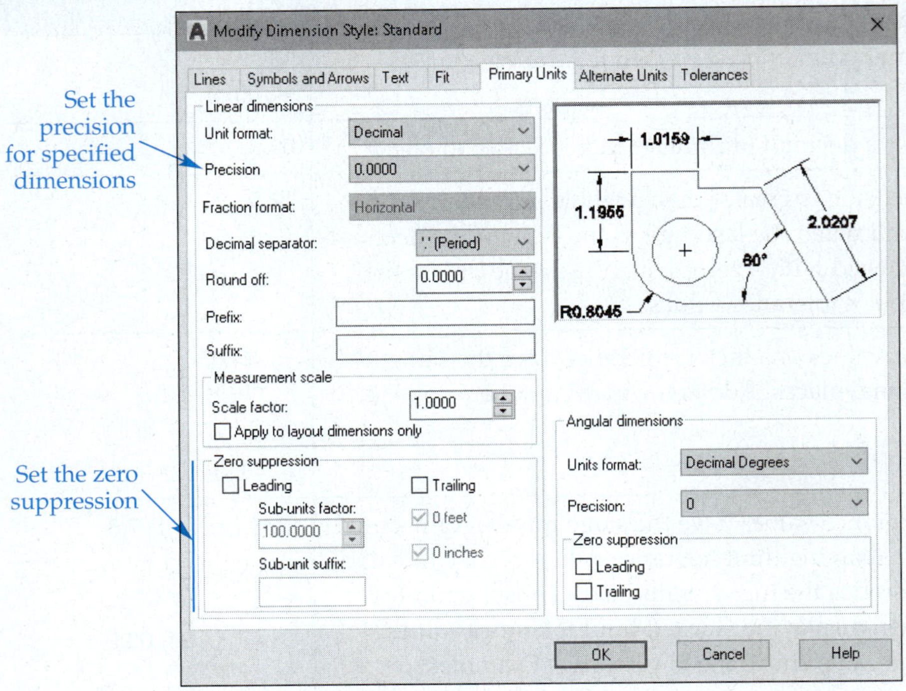

Setting Tolerance Methods

The **Tolerances** tab of the **New** (or **Modify**) **Dimension Style** dialog box, shown in Figure 19-9, allows you to create a specified tolerance dimension style. The default option in the **Method:** drop-down list is **None**. This means dimensions use an unspecified tolerance format. As a result, most of the options in the **Tolerances** tab are disabled. When you pick a different tolerance method from the drop-down list, appropriate options become enabled, and the preview image reflects the selected method. Figure 19-10 shows the drop-down list options.

NOTE

The following information describes tolerance methods and the settings unique to each. You will learn about general settings, including tolerance precision, height, vertical position, alignment, and zero suppression, later in this chapter.

Symmetrical Tolerance Method

symmetrical tolerance: The AutoCAD term for an equal bilateral tolerance.

Select the **Symmetrical** option from the **Method:** drop-down list to create a *symmetrical tolerance*. Use the **Symmetrical** option to draw dimensions that display an equal bilateral tolerance in the plus-minus format. See Figure 19-11. Enter a tolerance value in the **Upper value:** text box. Although the **Lower value:** text box is disabled, you can see that the value in the **Lower value:** text box matches the value in the **Upper value:** text box.

Figure 19-9.
The **Tolerances** tab of the **New** (or **Modify**) **Dimension Style** dialog box contains formatting settings for tolerance dimensions.

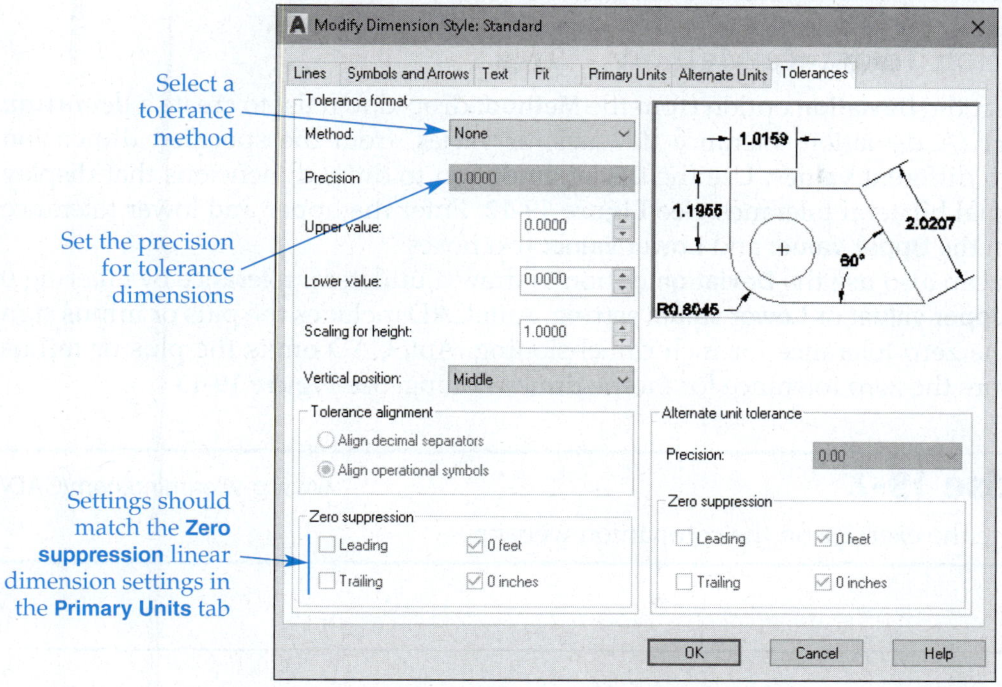

Figure 19-10.
Select a tolerance method from the **Method:** drop-down list in the **Tolerance format** area.

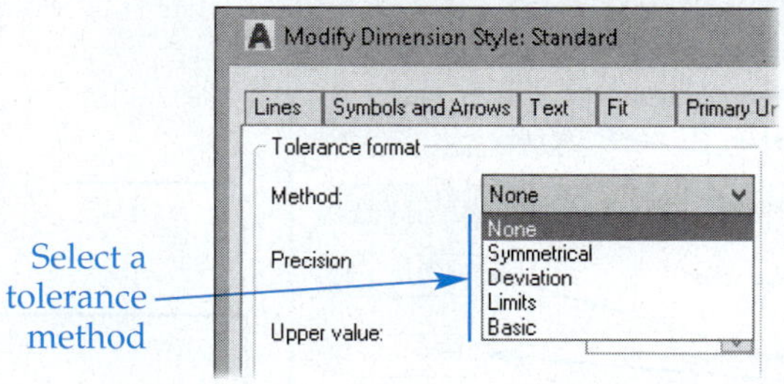

Figure 19-11.
Setting the **Symmetrical** tolerance method option current, with an equal bilateral tolerance value of .005.

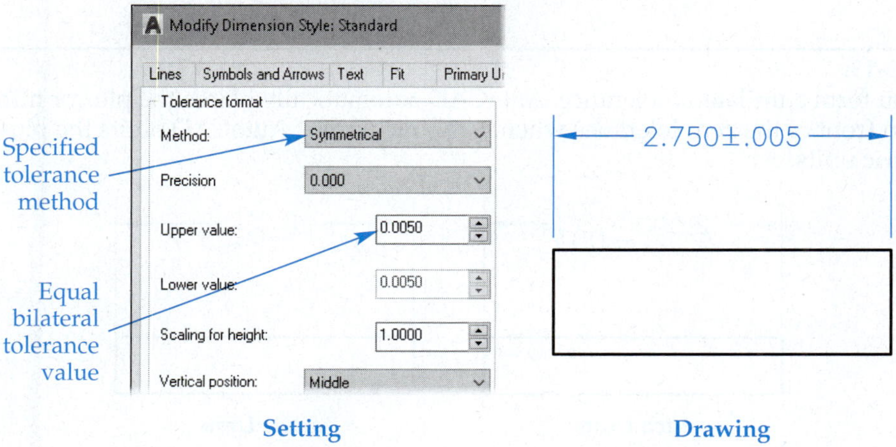

Exercise 19-1

www.g-wlearning.com/CAD/

Complete the exercise on the companion website.

Deviation Tolerance Method

deviation tolerance: The AutoCAD term for an unequal bilateral tolerance.

Select the **Deviation** option from the **Method:** drop-down list to create a ***deviation tolerance***. A deviation tolerance deviates, or varies, from the specified dimension with two different values. Use the **Deviation** option to draw dimensions that display an unequal bilateral tolerance. See Figure 19-12. Enter the upper and lower tolerance values in the **Upper value:** and **Lower value:** text boxes.

You can also use the **Deviation** option to draw a unilateral tolerance by entering 0 for the **Upper value:** or **Lower value:** setting. AutoCAD includes the plus or minus sign before the zero tolerance for inch dimensioning. AutoCAD omits the plus or minus sign before the zero tolerance for metric dimensioning. See Figure 19-13.

Exercise 19-2

www.g-wlearning.com/CAD/

Complete the exercise on the companion website.

Figure 19-12.
Setting the **Deviation** tolerance method option current, with unequal bilateral tolerance values.

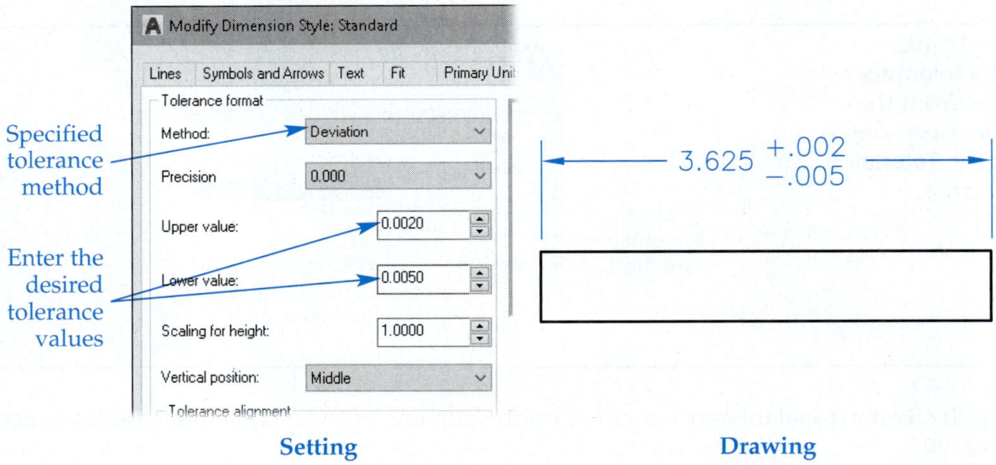

Figure 19-13.
When you form a unilateral tolerance, AutoCAD automatically places the plus or minus symbol in front of the zero tolerance when using inch units. AutoCAD omits the plus symbol with metric units.

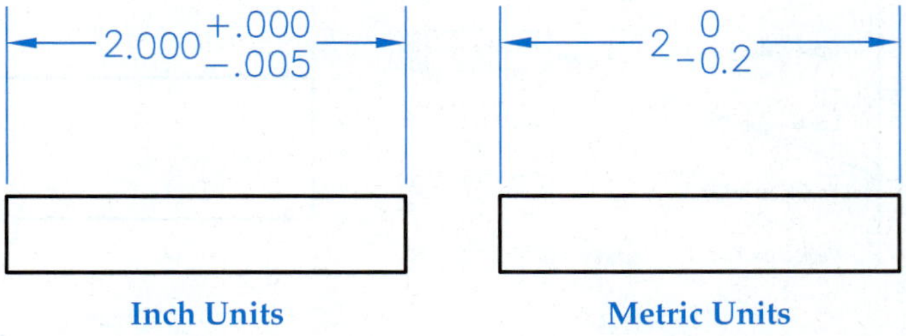

Limits Tolerance Method

Select the **Limits** option from the **Method:** drop-down list to apply limit dimensioning. See Figure 19-14. Use the **Upper value:** and **Lower value:** text boxes to enter the upper and lower tolerance values to add and subtract from the specified dimension. The upper and lower values can be equal or different.

Exercise 19-3

www.g-wlearning.com/CAD/

Complete the exercise on the companion website.

Basic Tolerance Method

Pick the **Basic** option from the **Method:** drop-down list to draw basic dimensions. See Figure 19-15. Few options are enabled in the dialog box because a basic dimension has no tolerance. A rectangle around the dimension value distinguishes a basic dimension from other dimensions.

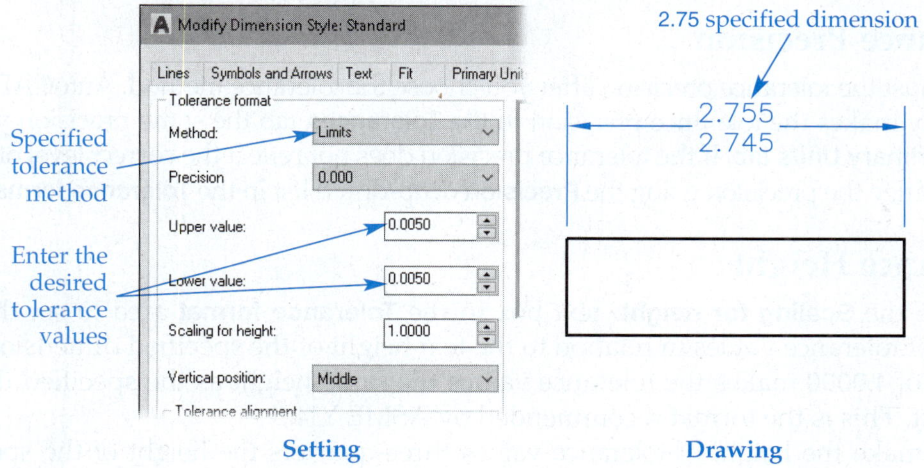

Figure 19-14.
Selecting the **Limits** tolerance method and setting limit values.

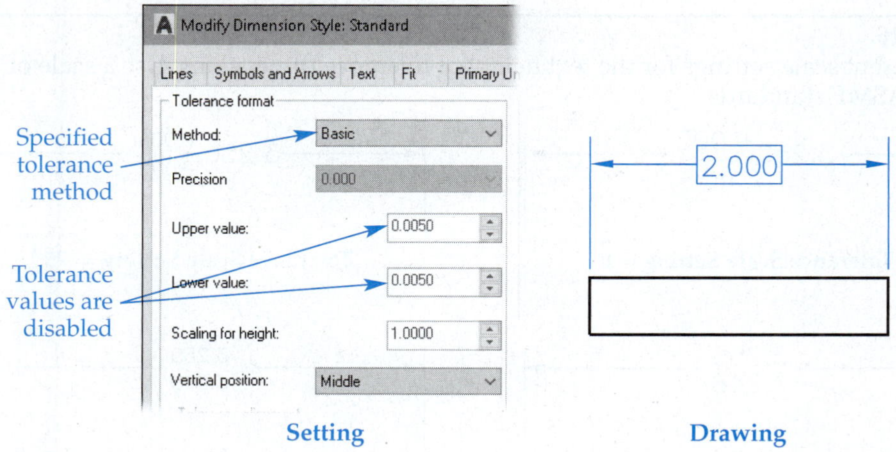

Figure 19-15.
Use the **Basic** tolerance method for basic dimensions. The dimension text for a basic dimension appears inside a frame.

NOTE

Checking **Draw frame around text** in the **Text** tab of the **New** (or **Modify**) **Dimension Style** dialog box also activates the basic tolerance method.

PROFESSIONAL TIP

Choose a tolerance method based on the characteristics of the specified tolerance. If the upper and lower values are equal, choose the **Symmetrical** option to create an equal bilateral tolerance. If the upper and lower values vary, use the **Deviation** option. Use the **Limits** option to show only the minimum and maximum allowed values.

Specifying Tolerance Settings

After you choose the tolerance method, adjust the tolerance settings to achieve the required tolerances. You can set the tolerance precision, height, position, alignment, and zero suppression characteristics.

Tolerance Precision

Adjust the tolerance precision after you choose the tolerance method. AutoCAD automatically makes the tolerance precision in the **Tolerances** tab the same precision you set in the **Primary Units** tab. If the tolerance precision does not reflect the correct level of precision, change the precision using the **Precision** drop-down list in the **Tolerance format** area.

Tolerance Height

Use the **Scaling for height:** text box in the **Tolerance format** area to set the text height of tolerance values in relation to the text height of the specified dimension. The default of 1.0000 makes the tolerance values the same height as the specified dimension text. This is the format recommended by ASME Y14.5.

To make the height of tolerance values three-quarters the height of the specified dimension, type .75 in the **Scaling for height:** text box. Some companies prefer this practice to keep the tolerance portion of the dimension from taking up additional space. Figure 19-16 shows examples of specified tolerance values with different text heights.

Figure 19-16.
Using different scale settings for the text height of tolerance dimensions. Use a scale of 1 to adhere to ASME standards.

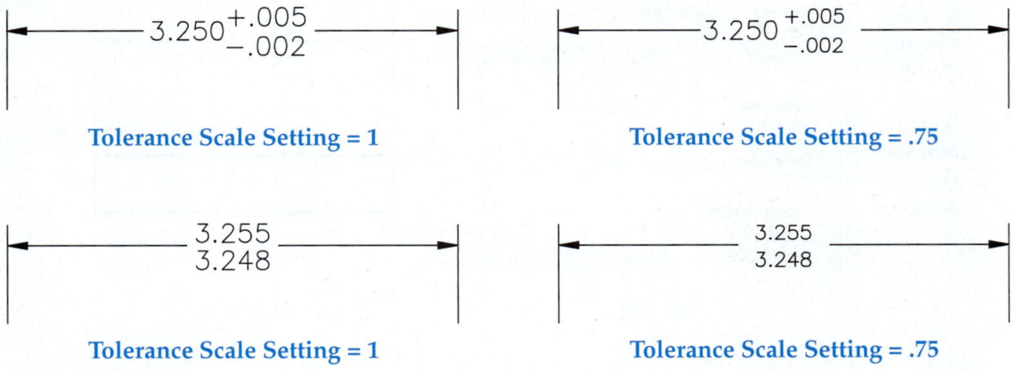

Vertical Position

Use the options in the **Vertical position:** drop-down list in the **Tolerance format** area to control the alignment, or justification, of deviation tolerance dimensions. The default **Middle** option centers the tolerance with the specified dimension. This is the format recommended by ASME Y14.5. The other justification options are **Top** and **Bottom**. Figure 19-17 shows deviation tolerance dimensions with each justification option.

Tolerance Alignment

The options in the **Tolerance alignment** area become enabled when you use a deviation or limits tolerance method. Tolerance alignment controls the left and right tolerance justification. When using a deviation tolerance method, pick the **Align decimal separators** radio button to align the upper and lower tolerance value decimal points vertically. Select the **Align operational symbols** radio button to align the upper and lower tolerance plus and minus symbols vertically. See Figure 19-18. When using the limits tolerance method, pick the **Align decimal separators** radio button to align the upper and lower limit decimal points vertically. Select the **Align operational symbols** radio button to left-justify the upper and lower limits. See Figure 19-19.

Figure 19-17.
Examples of the tolerance justification options for deviation tolerance dimensions. Use the **Middle** option to adhere to ASME standards.

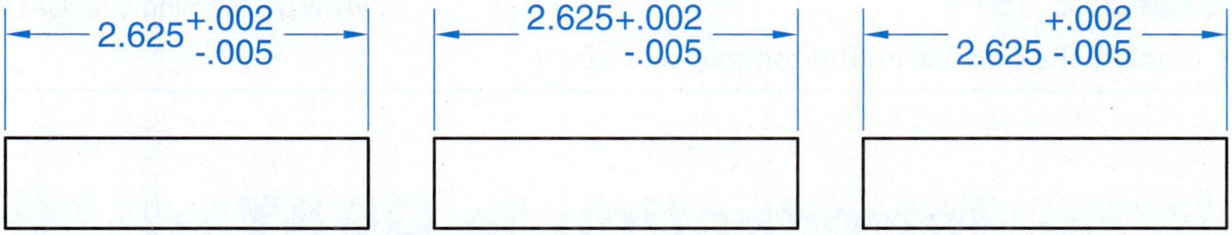

Middle Top Bottom

Figure 19-18.
Changing tolerance alignment for use with a deviation tolerance method.

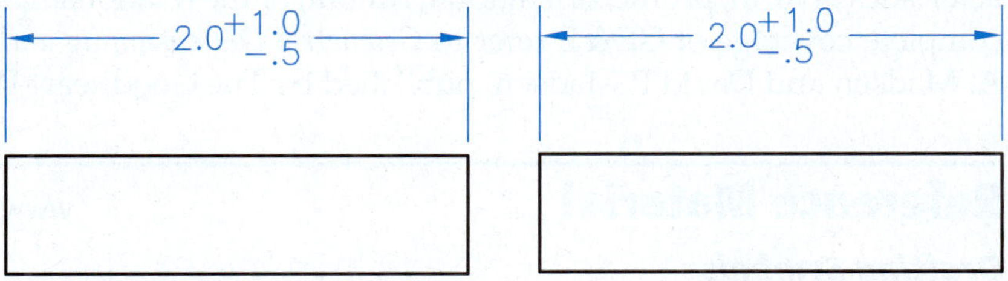

Aligned on Decimal Separator Aligned on Operational Symbols

Figure 19-19.
Changing tolerance alignment for use with a limits tolerance method.

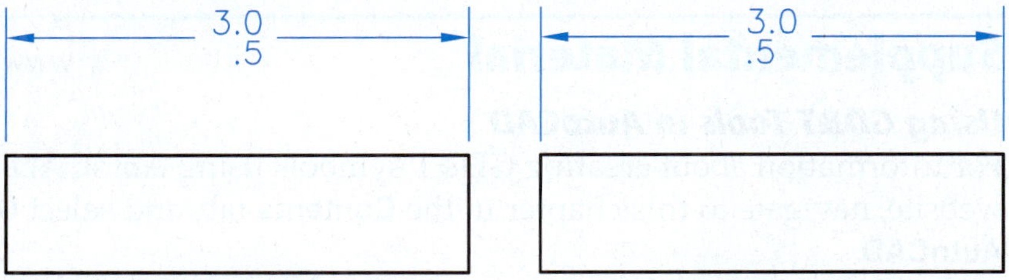

Aligned on Decimal Separator Aligned on Operational Symbols

Zero Suppression

You must select a tolerance method to enable the options in the **Zero suppression** area. The suppression settings for linear dimensions in the **Tolerances** tab should be the same as the zero suppression settings for linear dimensions in the **Primary Units** tab. AutoCAD does not automatically match the tolerance setting to the primary units setting.

Select the **Leading** check box in the **Zero suppression** area of the **Tolerances** tab when you are drawing inch-specified tolerance dimensions. Activate the same option for linear dimensions in the **Primary Units** tab. You can then draw inch-specified tolerance dimensions without placing the zero before the decimal point, as recommended by ASME standards. These settings allow you to create a tolerance dimension value such as .625±.005.

Deselect the **Leading** check box in the **Zero suppression** area of the **Tolerances** tab when you are drawing metric-specified tolerance dimensions. Deactivate the same option for linear dimensions in the **Primary Units** tab. This allows you to place a metric-specified tolerance dimension value with the zero before the decimal point, such as 12±0.2, as recommended by ASME standards.

NOTE

The options in the **Alternate unit tolerance** area become enabled when you pick the **Display alternate units** check box in the **Alternate Units** tab of the **New** (or **Modify**) **Dimension Style** dialog box. Use the **Alternate unit tolerance** area to set specified tolerances for alternate units.

Exercise 19-4

www.g-wlearning.com/CAD/

Complete the exercise on the companion website.

Introduction to GD&T

geometric dimensioning and tolerancing (GD&T): The dimensioning and tolerancing of individual features of a part where the permissible variations relate to characteristics of form, profile, orientation, runout, or the relationship between features.

Geometric dimensioning and tolerancing (GD&T) is the dimensioning and tolerancing of individual features of a part where the permissible variations relate to characteristics of form, profile, orientation, runout, or the relationship between features. For complete coverage of GD&T, refer to *Geometric Dimensioning and Tolerancing* by David A. Madsen and David P. Madsen, published by The Goodheart-Willcox Company, Inc.

Reference Material

www.g-wlearning.com/CAD/

Drafting Symbols

For the names and examples of GD&T symbols and symbol applications, go to the companion website, select the **Resources** tab, and select **Drafting Symbols**.

Supplemental Material

www.g-wlearning.com/CAD/

Using GD&T Tools in AutoCAD

For information about creating GD&T symbols using AutoCAD, go to the companion website, navigate to this chapter in the **Contents** tab, and select **Using GD&T Tools in AutoCAD**.

Chapter Review

Answer the following questions using the information in this chapter.

1. Define the term *tolerance.*
2. What are the limits of the tolerance dimension 3.625±.005?
3. Give an example of an equal bilateral tolerance in inches and in metric units.
4. Give an example of an unequal bilateral tolerance in inches and in metric units.
5. Give an example of a unilateral tolerance in inches and in metric units.
6. What is the purpose of the **Symmetrical** tolerance method option?
7. What is the purpose of the **Deviation** tolerance method option?
8. What is the purpose of the **Limits** tolerance method option?
9. How do you set the number of zeros displayed after the decimal point for a tolerance dimension?
10. Explain the result of setting the **Scaling for height:** option to 1 in the **Tolerances** tab.
11. What setting should you use for the **Scaling for height:** option if you want the tolerance dimension height to be three-quarters of the specified dimension height?
12. Name the tolerance dimension justification option recommended by the ASME standards.
13. Which **Zero suppression** settings should you choose for linear and tolerance dimensions when using inch units?
14. Which **Zero suppression** settings should you choose for linear and tolerance dimensions when using metric units?
15. What is the purpose of geometric dimensioning and tolerancing?

Drawing Problems

Start AutoCAD if it is not already started. Start a new drawing for each problem using an appropriate template of your choice. The template should include layers and text, dimension, and multileader styles, when necessary, for drawing the given objects. Add layers and text, dimension, and multileader styles as needed. Draw all objects using appropriate layers. Use appropriate text, dimension, and multileader styles, justification, and format. Follow the specific instructions for each problem. Use only drawing and editing commands and techniques you have already learned. Use your own judgment and approximate dimensions when necessary. Apply dimensions accurately using ASME or appropriate industry standards.

▼ Basic

1. **Mechanical** Draw and dimension the part view shown. Save the drawing as P19-1.

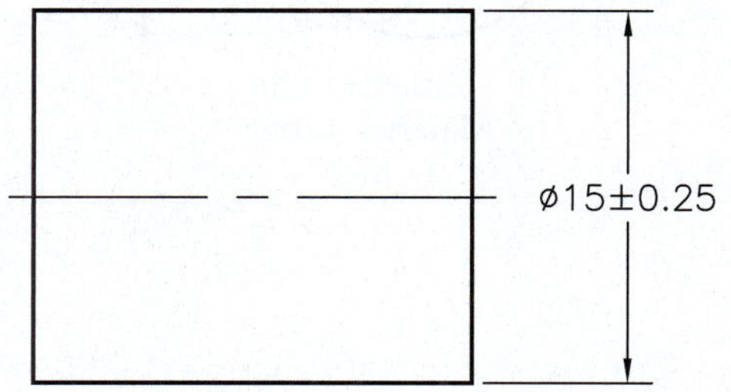

2. **Mechanical** Draw and dimension the part view shown. Save the drawing as P19-2.

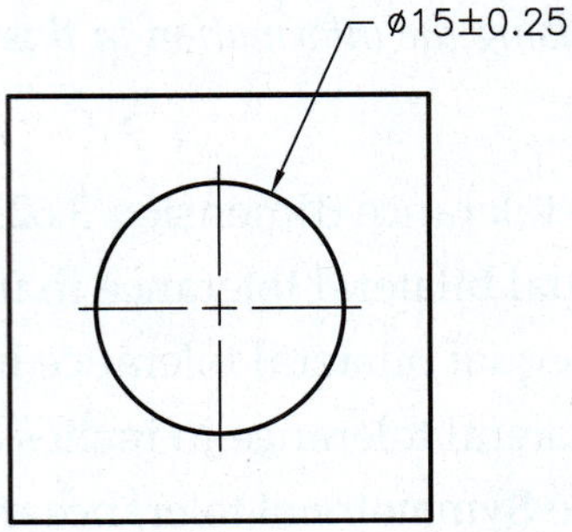

For Problems 3 through 6, draw and dimension the orthographic views needed to describe the part completely. Save the drawings as P19-3, P19-4, P19-5, *and* P19-6.

3. **Mechanical**

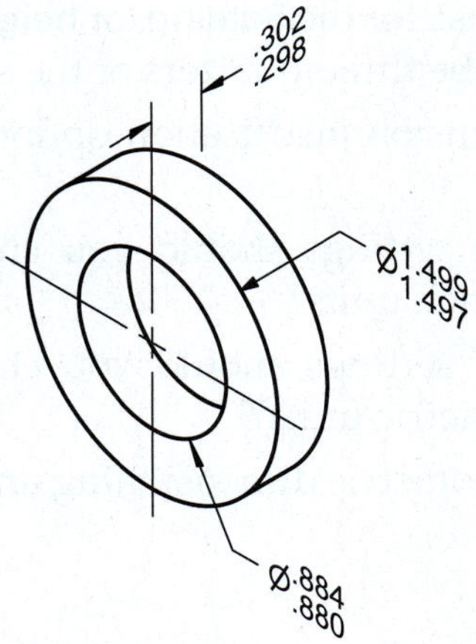

Title: Washer
Material: SAE 1020
Inch

4. **Mechanical**

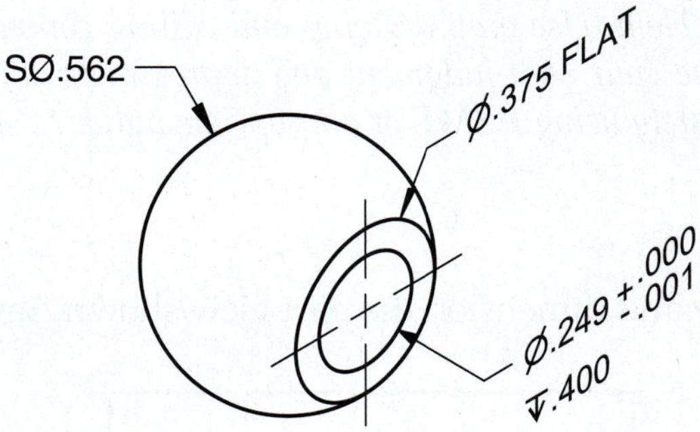

Title: Handle
Material: Bronze
Inch

5. Mechanical

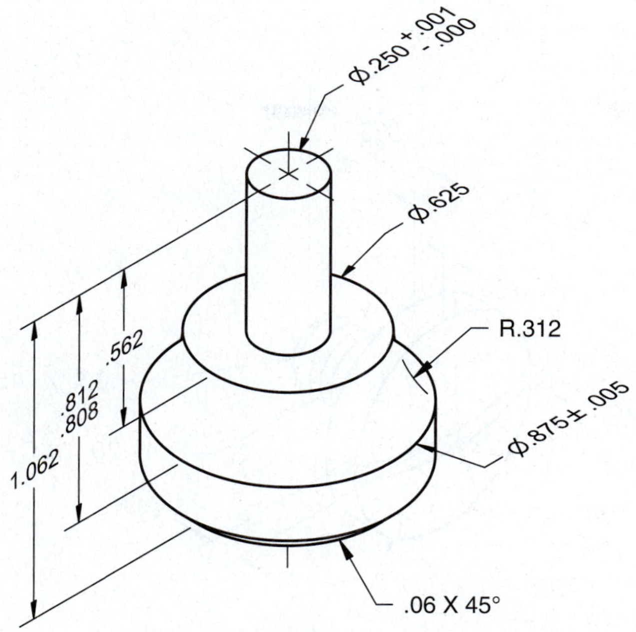

6. Mechanical

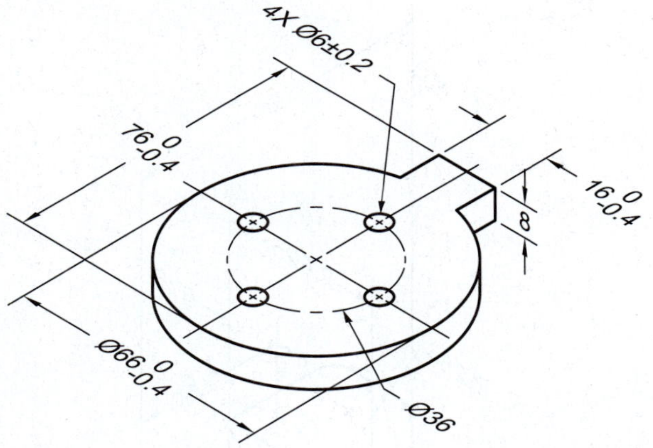

▼ Intermediate

7. Mechanical Draw and dimension the threaded stud part views shown. Save the drawing as P19-7.

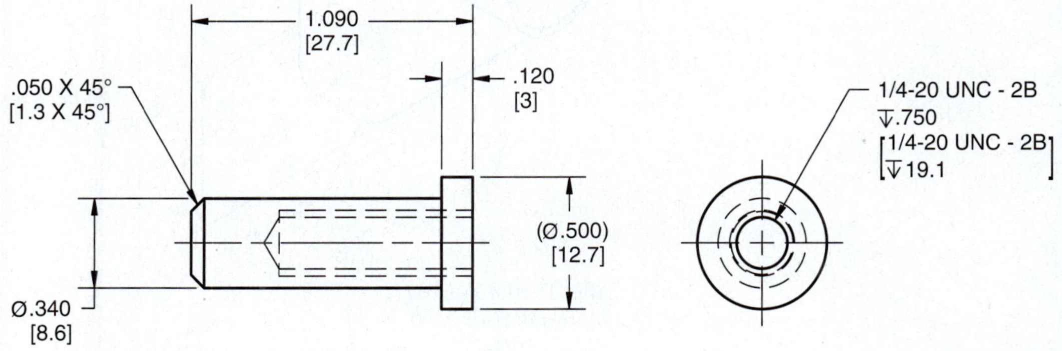

Drawing Problems - Chapter 19

For Problems 8 through 10, draw and dimension the orthographic views needed to describe the part completely. Save the drawings as P19-8, P19-9, *and* P19-10.

8. **Mechanical**

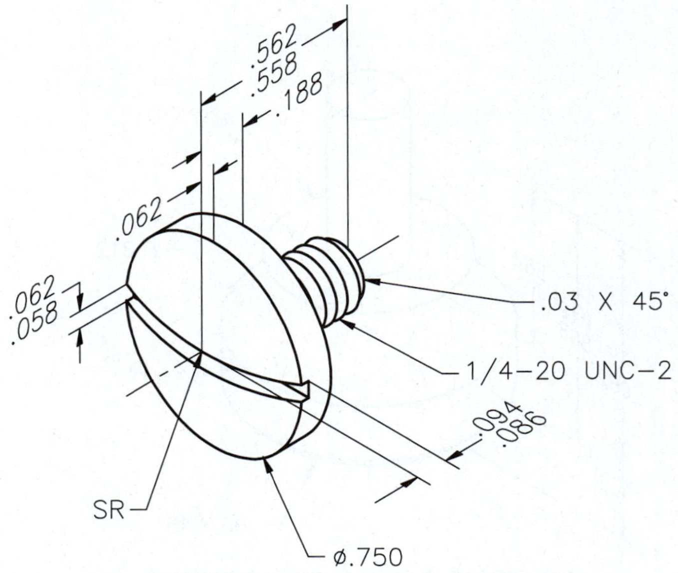

Title: Screw
Material: SAE 4320
Inch

9. **Mechanical** A portion of the drawing is removed for clarity. Draw the entire part.

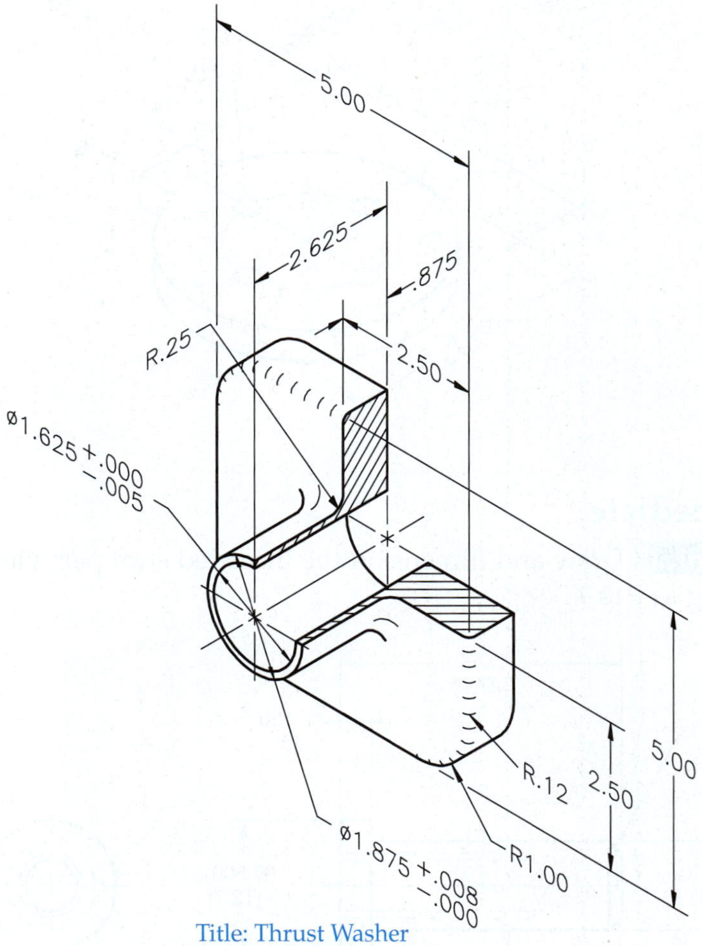

Title: Thrust Washer
Material: SAE 5150
Inch

10. **Mechanical**

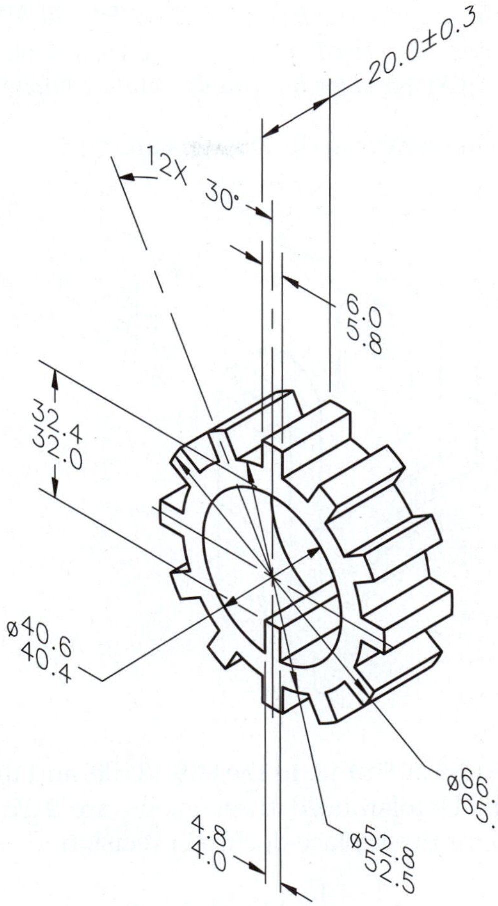

Title: Locking Collar
Material: SAE 1080
Metric

▼ Advanced

11. **Mechanical** Draw and dimension the vise clamp part views shown. Save the drawing as P19-11.

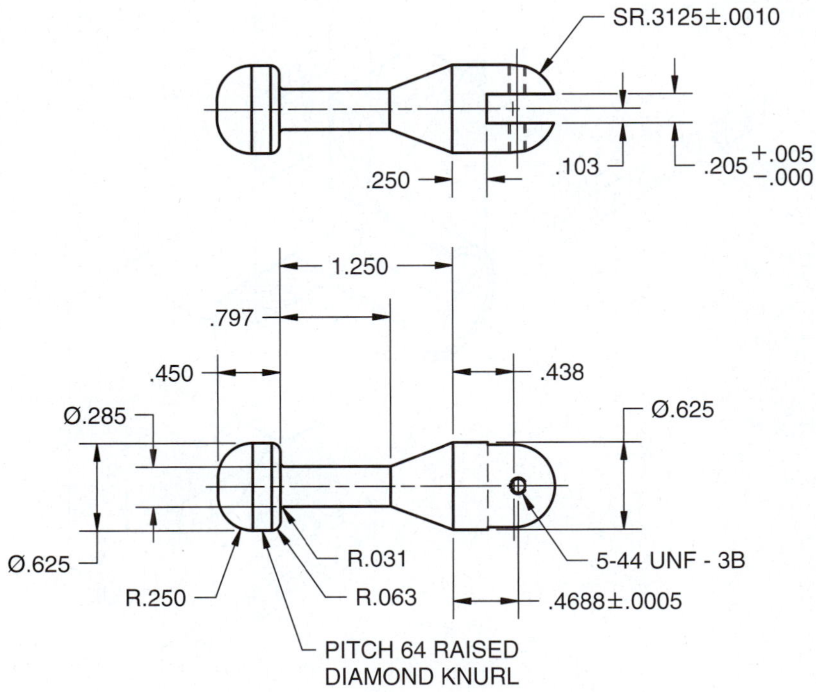

For Problems 12 through 16, draw and dimension the orthographic views needed to describe the part completely. Save the drawings as P19-12, P19-13, P19-14, P19-15, *and* P19-16. *Use the GD&T commands and practices described in the "Using GD&T Tools in AutoCAD" supplement available in the Supplemental Material for this chapter on the companion website.*

12. **Mechanical** *Note:* Untoleranced dimensions are ±0.3.

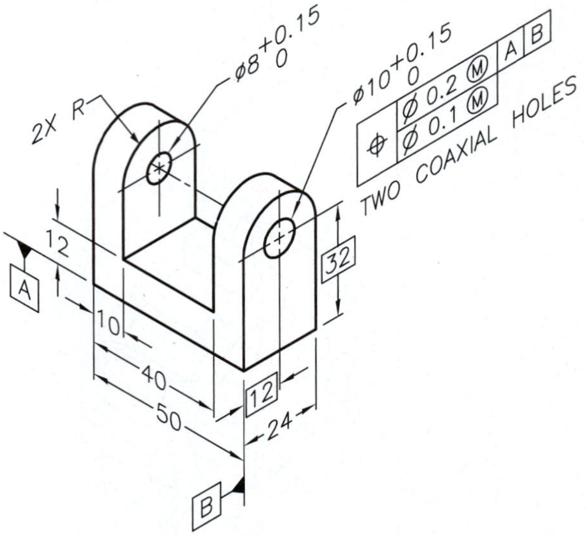

13. **Mechanical** Save P19-9 as P19-13. In the P19-13 file, add the geometric tolerancing applications shown. Untoleranced dimensions are ±.02 for two-place decimal precision and ±.005 for three-place decimal precision.

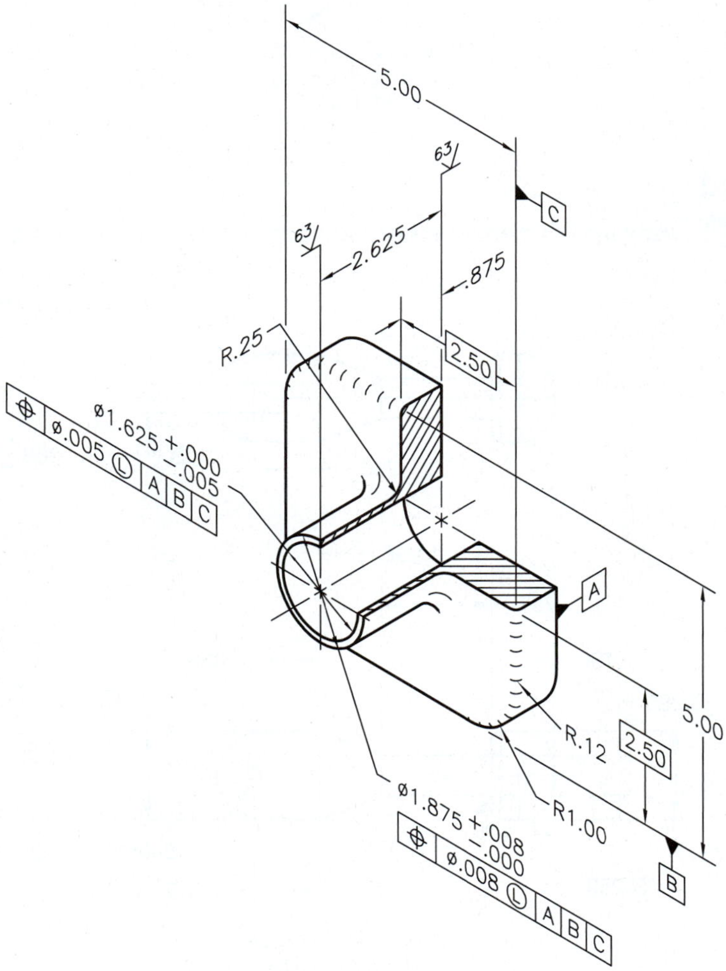

14. **Mechanical** Save P19-6 as P19-14. In the P19-14 file, add the geometric tolerancing applications shown. Untoleranced dimensions are ±0.5.

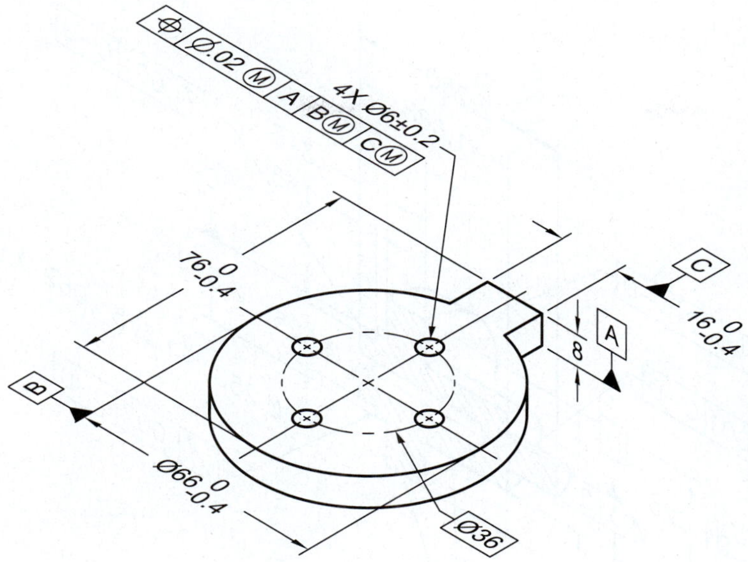

15. **Mechanical** Save P19-10 as P19-15. In the P19-15 file, add the geometric tolerancing applications shown.

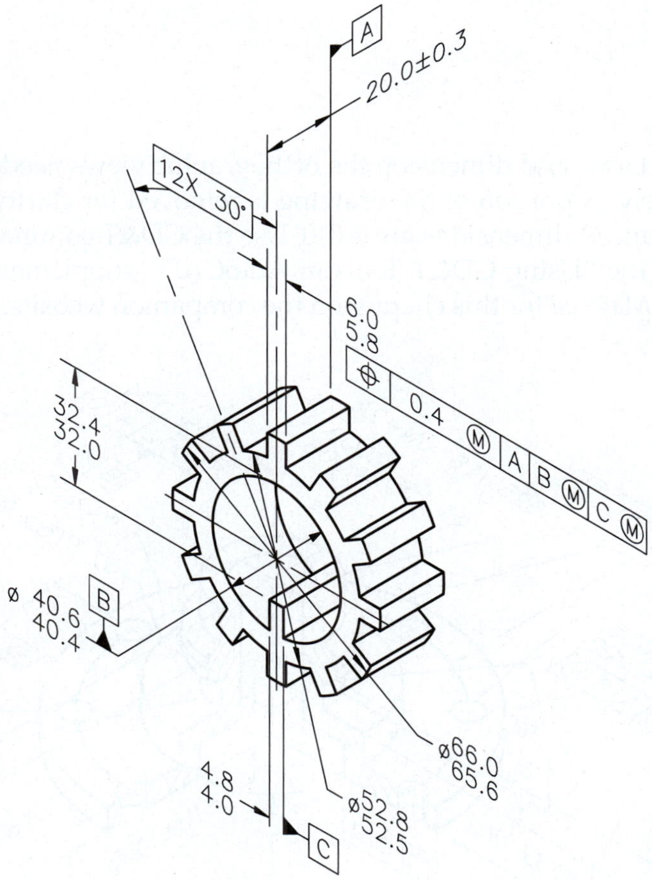

16. **Mechanical** *Note:* Half of the drawing is removed for clarity. Draw the entire part. Untoleranced dimensions are ±.010.

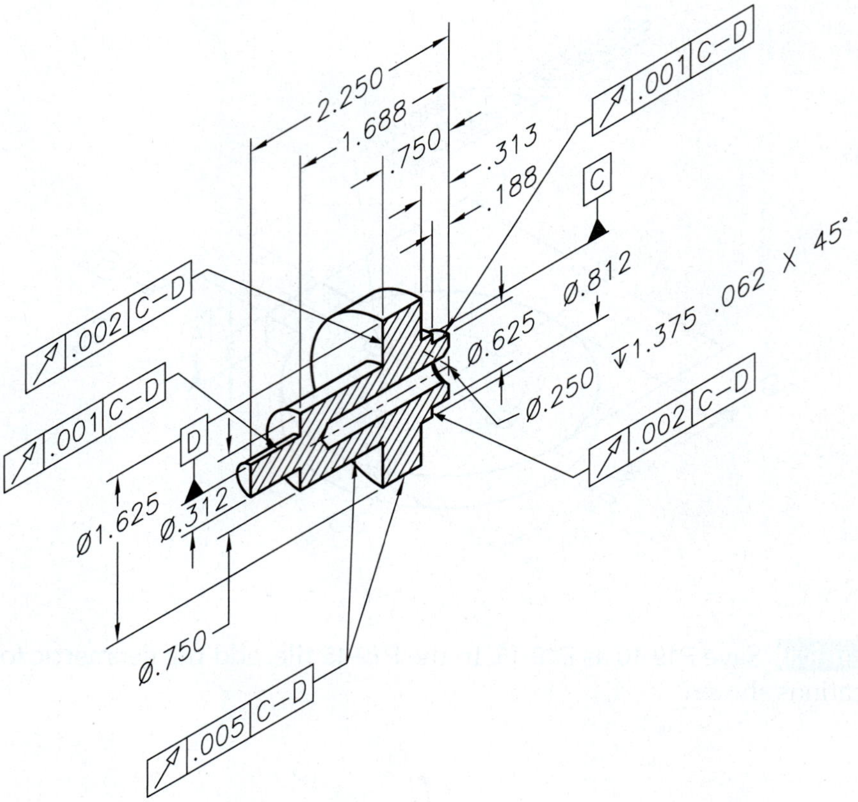

17. **Mechanical** Draw and dimension the orthographic views needed to describe the part completely. A portion of the drawing is removed for clarity. Draw the entire part. Untoleranced dimensions are ±.010. Use the GD&T commands and practices described in the "Using GD&T Tools in AutoCAD" supplement available in the *Supplemental Material* for this chapter on the companion website. Save the drawing as P19-17.

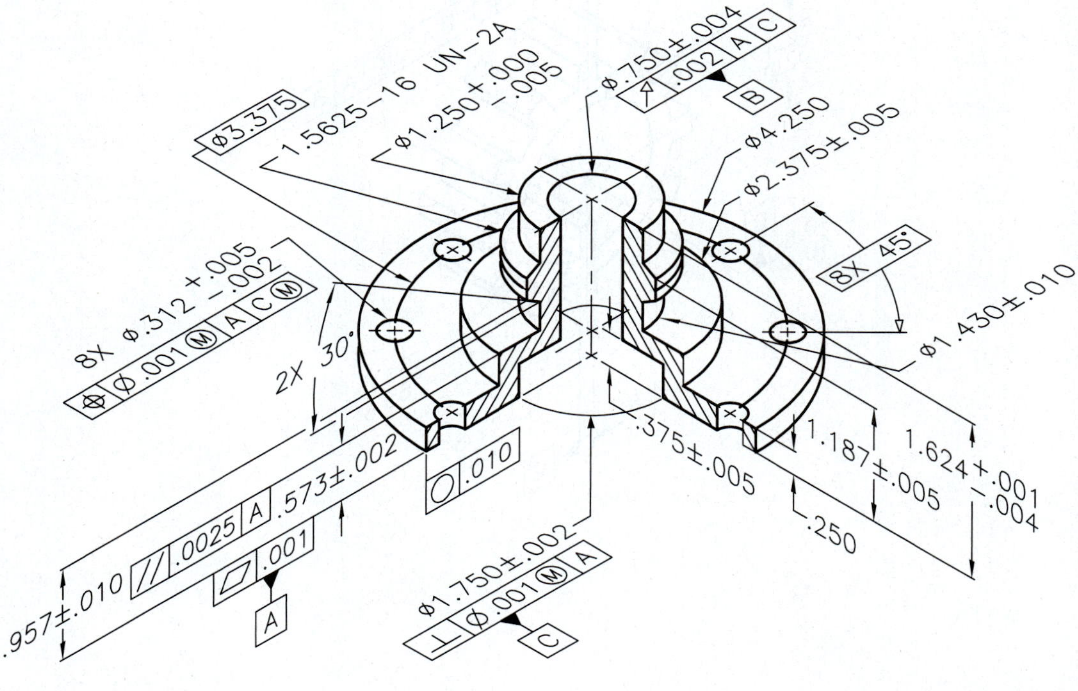

18. **Mechanical** Save P18-24 as P19-18. If you have not yet completed Problem 18-24, do so now. In the P19-18 file, use the GD&T commands and practices described in the "Using GD&T Tools in AutoCAD" supplement available in the *Supplemental Material* for this chapter on the companion website.

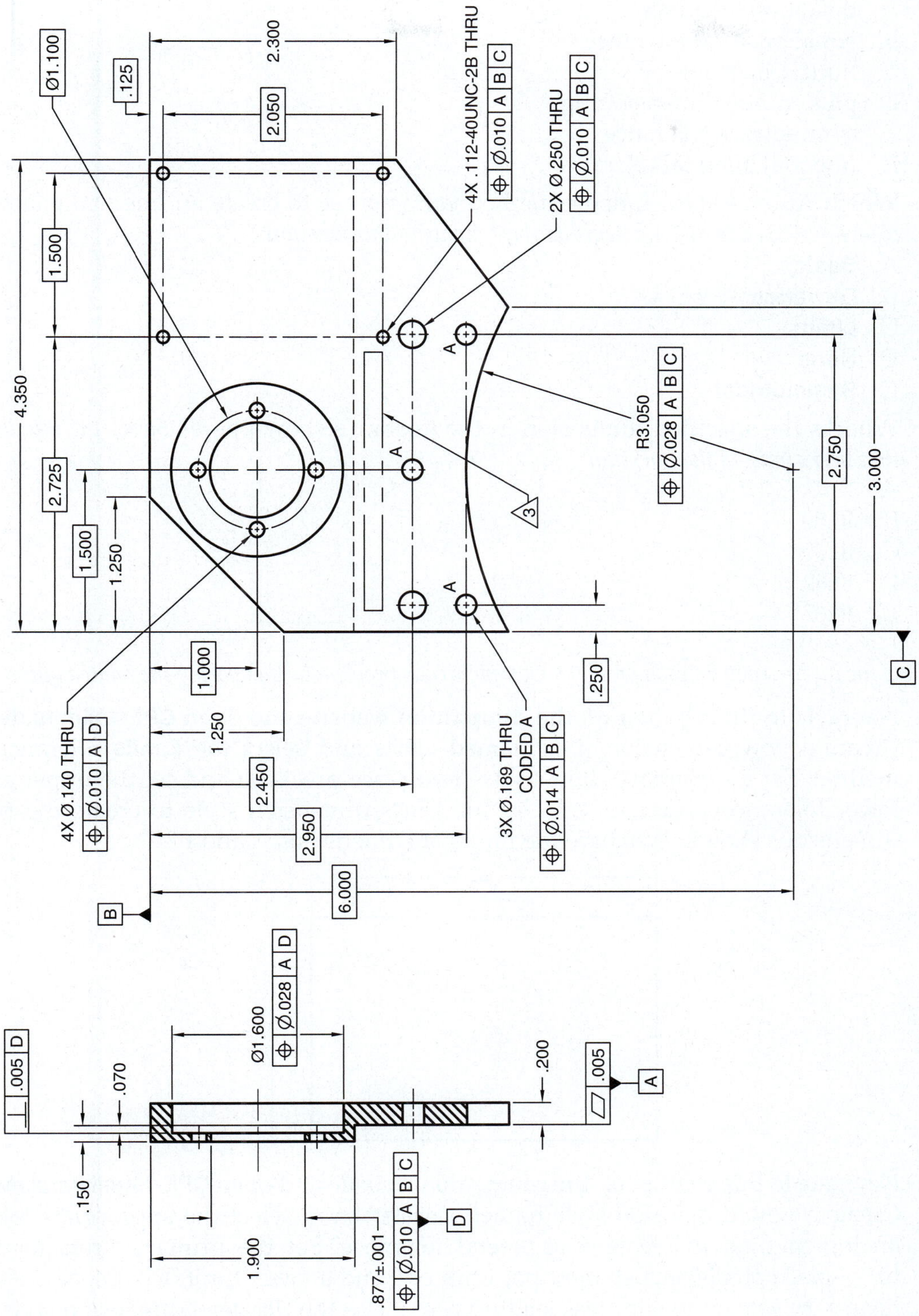

19. **Mechanical** Research the design of an existing shaft collar with the following specifications: two-piece clamp-on, 1" bore, 1/4-28UNF screw threads. Create a freehand, dimensioned 2D sketch of the existing design from manufacturer's specifications, or from measurements taken from an actual shaft collar. Start a new drawing from scratch or use a decimal-unit template of your choice. Draw and dimension each part of the collar from your sketch. Save the drawing as P19-19.

AutoCAD Certified Professional Exam Practice

Answer the following questions using the information in this chapter.

1. Which of the following terms describes the dimension 18.75±.25? *Select all that apply.*
   A. deviation tolerance
   B. equal bilateral tolerance
   C. limit dimensioning
   D. plus-minus dimensioning
   E. symmetrical tolerance
   F. unequal bilateral tolerance
2. Which AutoCAD tolerancing method can you use to create an unequal bilateral tolerance? *Select the one item that best answers the question.*
   A. **Basic**
   B. **Deviation**
   C. **Limits**
   D. **None**
   E. **Symmetrical**
3. What is the specified dimension in the tolerance shown here? *Select the one item that best answers the question.*
   A. 10.17
   B. 10.50
   C. 10.60
   D. 10.62
   E. 10.67

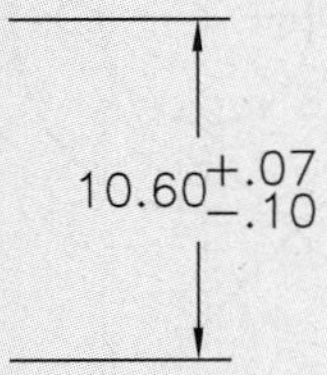

Follow the instructions in each problem. Complete each problem by answering the related question.

4. **Navigate to this chapter on the companion website and open CPE-19limits.dwg.** Create a new dimension style named Limits and select the **Limits** tolerancing method. Set the primary units to use two-place precision and set the upper and lower tolerance values to .25. Use the Limits dimension style to create the two dimensions shown. What are the limits of Dimensions A and B?

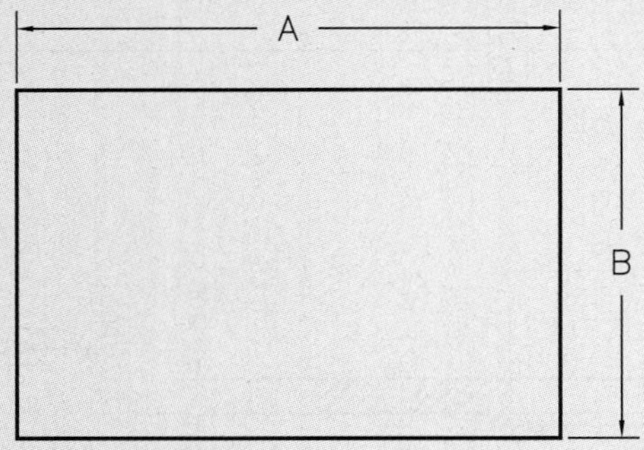

5. **Navigate to this chapter on the companion website and open CPE-19unilateral.dwg.** Create a new dimension style named Unilateral and select the appropriate tolerancing method to create a unilateral tolerance. Set the primary units to use three-place precision. Set an upper limit of 0 and a lower limit of –.021. Make the necessary setting to suppress leading zeros. Use the Unilateral dimension style to create the two dimensions shown. What are the limits of Dimensions C and D?

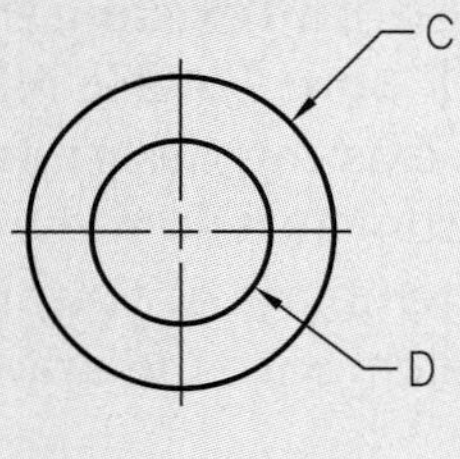

CHAPTER 20

Editing Dimensions

Learning Objectives

After completing this chapter, you will be able to:

- ✓ Describe and control associative dimensions.
- ✓ Control the appearance of existing dimensions and dimension text.
- ✓ Override dimension style settings.
- ✓ Change dimension line spacing and alignment.
- ✓ Break dimension, extension, and leader lines.
- ✓ Create inspection dimensions.
- ✓ Edit associative center mark and centerline objects.
- ✓ Edit existing multileaders.

You can modify dimensions using standard editing commands such as **ERASE** and **STRETCH**. AutoCAD also provides specific commands and options to adjust dimensions. This chapter describes techniques for editing dimension placement, appearance, and values.

Associative Dimensioning

A dimension is a group of elements treated as a single object. For example, you can access the **ERASE** command and pick any portion of a dimension to erase the entire dimension object. Additionally, dimensions reference objects or points. When you edit objects and the corresponding dimensions with commands such as **STRETCH**, **MOVE**, **ROTATE**, and **SCALE**, dimensions change accordingly. See Figure 20-1.

An ***associative dimension*** forms by default when you select objects or pick points using object snaps. For example, if you dimension the ∅1.0 circle in Figure 20-1 using the **DIM** or **DIMDIAMETER** command, and then change the size of the circle to ∅1.50, the diameter dimension adapts to show the correct size of the modified circle. Create associative dimensions when possible and practical by selecting objects or using object snaps. Associative dimensions are related directly to object size and make revisions easier.

associative dimension: A dimension associated with an object. The dimension value updates automatically when the object changes.

A ***non-associative dimension*** forms when you select points without using object snaps. A non-associative dimension is still a single object that updates when you make changes to the dimension, such as stretching the extension line origin.

non-associative dimension: A dimension linked to point locations, not an object; does not update when the object changes.

Figure 20-1.
An example of a revised drawing. Dimensions adjust to the modified geometry, and the dimension values update to reflect the size and location of the modified geometry.

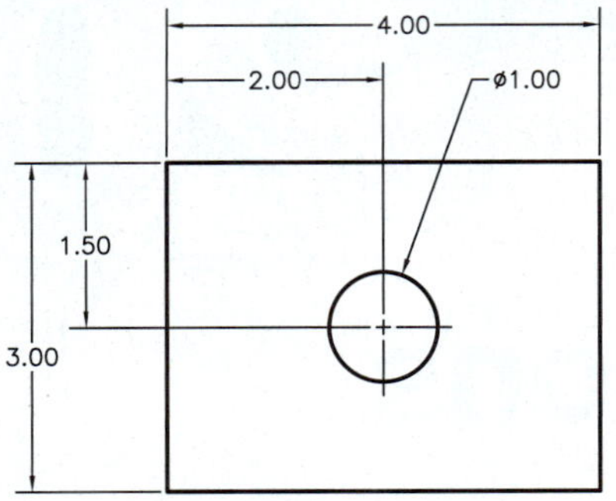

Original Drawing

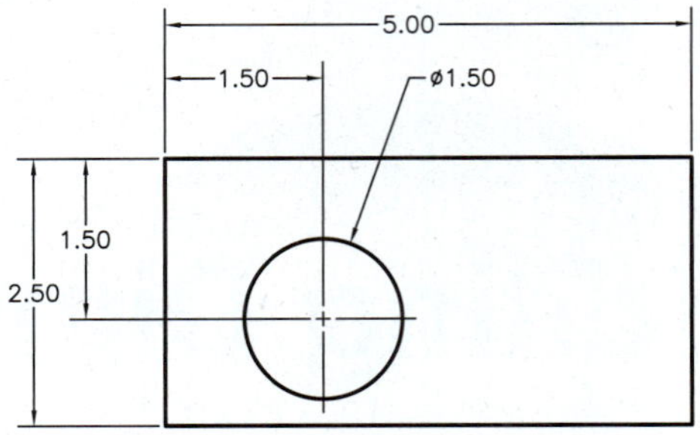

Revised Drawing

Non-associative dimensions are appropriate when associative dimensions would result in dimensioning difficulty or unacceptable standards. When using non-associative dimensions, remember to edit the dimension with the object it dimensions, or adjust the dimension after the object changes.

NOTE

Refer to the Associative property in the **General** area of the **Properties** palette to determine whether a dimension is associative.

PROFESSIONAL TIP

Dimension commands allow you to dimension a drawing, but they do not control object size and location. Chapter 22 explains how to use dimensional constraints to control object size and location. If you anticipate creating a drawing with features that will require significant or constant change, you may want to use dimensional constraints instead of, or in addition to, traditional dimension objects.

Associating Dimensions with Objects

Dimensions are associated with objects by default when you select objects or pick points using object snaps. To deactivate associative dimensioning for new objects, access the **Options** dialog box and deselect the **Make new dimensions associative** check box in the **Associative Dimensioning** area of the **User Preferences** tab.

DIMREASSOCIATE

Ribbon
Annotate
> Dimensions

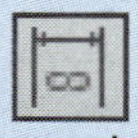
Reassociate

Type
DIMREASSOCIATE
DRE

Use the **DIMREASSOCIATE** command to convert a non-associative dimension to an associative dimension. Access the command, select the dimension to associate with an object, and press [Enter]. An X marker appears at a dimension origin, such as the origin of a linear dimension extension line or the center of a radial dimension. Select a point on an object to associate with the marker location. Repeat the process to locate the second object point, if required. Use the **Next** option to advance to the next definition point. Use the **Select object** option to select an object to associate with the dimension. The extension line endpoints automatically associate with the object endpoints.

NOTE

Type
DIMDISASSOCIATE

To disassociate a dimension from an object, grip-edit the dimension to stretch a grip point away from the associated object, or use the **DIMDISASSOCIATE** command.

NOTE

The **Dimension** panel of the **Express Tools** ribbon tab includes the **DIMREASSOC** command, not to be confused with the **DIMREASSOCIATE** command. The **DIMREASSOC** command allows you to change the overridden value of an associated dimension back to the actual associated dimension value. Access the **DIMREASSOC** command and select associative dimensions to change. The **DIMREASSOCIATE** command creates an associative dimension, but does not change an overridden dimension value.

Definition Points

Definition points, or ***defpoints***, form automatically when you create a dimension. Use the **Node** object snap to snap to a definition point. AutoCAD automatically creates a Defpoints layer and places definition points on the layer. By default, the Defpoints layer does not plot. You can only plot definition points if you rename the Defpoints layer and then set the renamed layer to plot. Definition points are accessible even if you turn off or freeze the Defpoints layer.

definition points (defpoints): The points used to specify the location of the dimension and the dimension text.

Exercise 20-1

www.g-wlearning.com/CAD/

Complete the exercise on the companion website.

CAUTION

A dimension is a single object even though it may include extension lines, a dimension line, arrowheads, and text. You may be tempted to explode a dimension using the **EXPLODE** command to modify individual dimension elements. You should rarely, if ever, explode dimensions. Exploded dimensions lose their association to related features and dimension styles.

PROFESSIONAL TIP

You can edit individual dimension properties without exploding a dimension by using dimension shortcut menu options, editing grips, or using the **Properties** palette to create a ***dimension style override***.

dimension style override: A temporary alteration of dimension style settings that does not actually modify the style.

Dimension Editing Tools

As the drawing process evolves and design changes occur, you will find it necessary to make changes to dimensioned objects and dimensions. AutoCAD includes dimension-specific editing tools and techniques to help you adjust dimensions as necessary.

Assigning a Different Dimension Style

To assign a different dimension style to existing dimensions, pick the dimensions to change and select a different dimension style from the **Dimension Style** drop-down list in the **Annotation** panel on the **Home** ribbon tab. The **Dimension Style** drop-down list is also available in the **Dimensions** panel on the **Annotate** ribbon tab. Another technique is to select the dimensions to change and select a different dimension style from the **Quick Properties** or **Properties** palette.

A third method is to select the dimensions to edit and then right-click to display the shortcut menu shown in Figure 20-2. Use the **Dimension Style** cascading menu to assign a different dimension style to the dimension or to save a new dimension style based on the properties of the selected dimension.

Ribbon
Annotate
> Dimensions

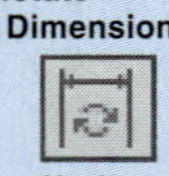

Update

The **Update** dimension tool provides another way to change the dimension style assigned to existing dimensions. Before you access the **Update** tool, set the dimension style to be assigned to existing dimensions current. Then access the **Update** tool and pick the dimensions to change them to the current style.

Editing the Dimension Value

AutoCAD provides the flexibility to control the appearance of dimensions for various dimensioning requirements without using a separate dimension style. For example, add a prefix or suffix to the dimension value or adjust the dimension text format for a limited number of special-case dimensions, instead of creating separate dimension styles.

TEXTEDIT

Type
TEXTEDIT
ED

Use the **TEXTEDIT** command, described in Chapter 10, to modify an existing dimension value. Figure 20-3 shows adding a diameter symbol to a linear diameter dimension. Use this editing method if you forget to use the **Mtext** or **Text** option of the **DIM** or **DIMLINEAR** command. An easy way to activate the **TEXTEDIT** command is to double-click on the dimension value. The multiline text editor appears and the highlighted value represents the current dimension value. Add to or modify the dimension text and then close the text editor.

CAUTION

You can replace the highlighted dimension value, but this action disassociates the dimension value with the object or points that it dimensions. Therefore, leave the default value intact whenever possible.

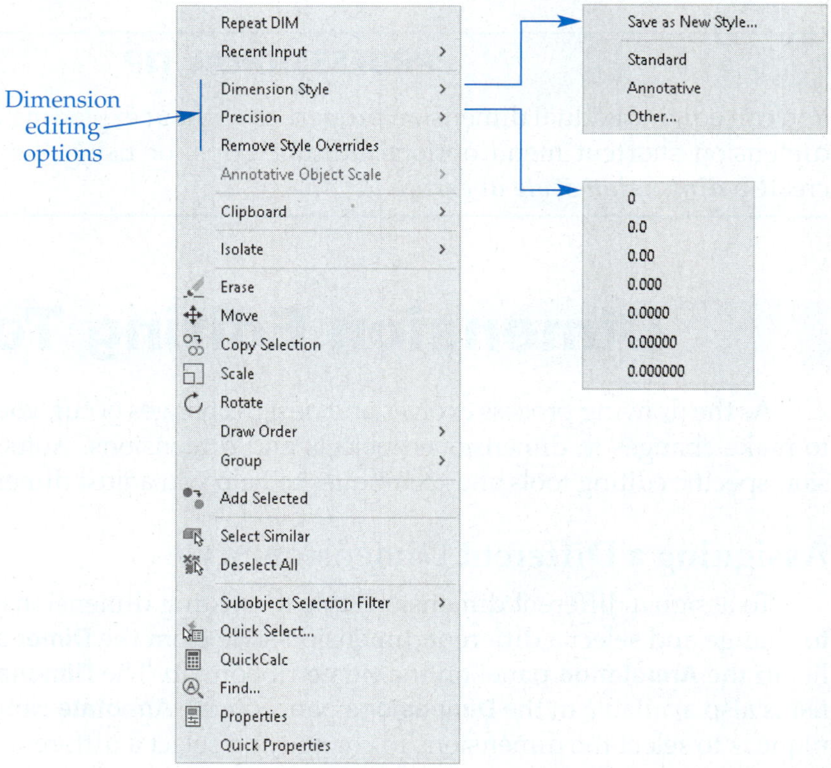

Figure 20-2.
Select a dimension and then right-click to access this shortcut menu with options for adjusting individual dimensions.

Figure 20-3.
When you use the **TEXTEDIT** command to edit a dimension value, the multiline text editor appears and the highlighted value represents the current dimension value calculated by AutoCAD. This example shows adding a diameter symbol to an existing dimension.

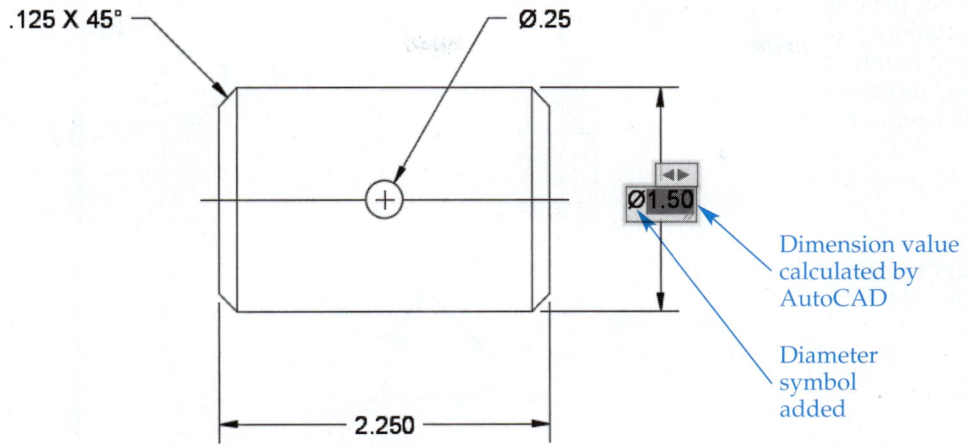

Exercise 20-2

www.g-wlearning.com/CAD/

Complete the exercise on the companion website.

Dimension Grip Commands

A drawing typically includes certain dimensions that require special treatment or modification. For example, proper dimensioning practice requires dimensions that are clear and easy to read. This sometimes involves moving the text of adjacent dimensions to separate the text elements. See Figure 20-4. AutoCAD offers a variety of commands and options for adjusting dimensions, but often a quick method for making basic changes is to use grips. Select a dimension to display grips at editable locations on the dimension.

Most dimension grips offer the standard **STRETCH**, **MOVE**, **ROTATE**, **SCALE**, **MIRROR**, and **Copy** grip commands. Remember when using these commands that the dimension is a single object with variables controlled by the dimension style. Certain dimension grips also provide access to context-sensitive options. You can access and apply context-sensitive grip commands in three different ways. See Figure 20-5.

Figure 20-4.
Staggering dimension text for improved readability.

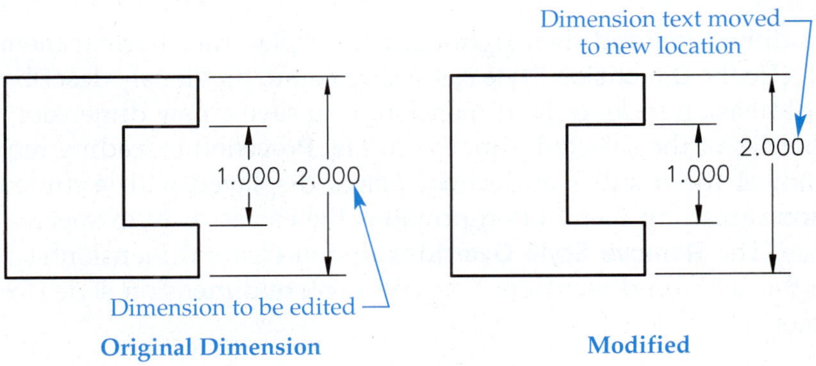

Figure 20-5. Using grips to make basic changes to dimensions. A—Hover over an unselected grip to display a menu of options, and then pick an option from the list. B—Pick a grip and press [Ctrl] to cycle through options. C—Pick a grip, right-click, and select an option.

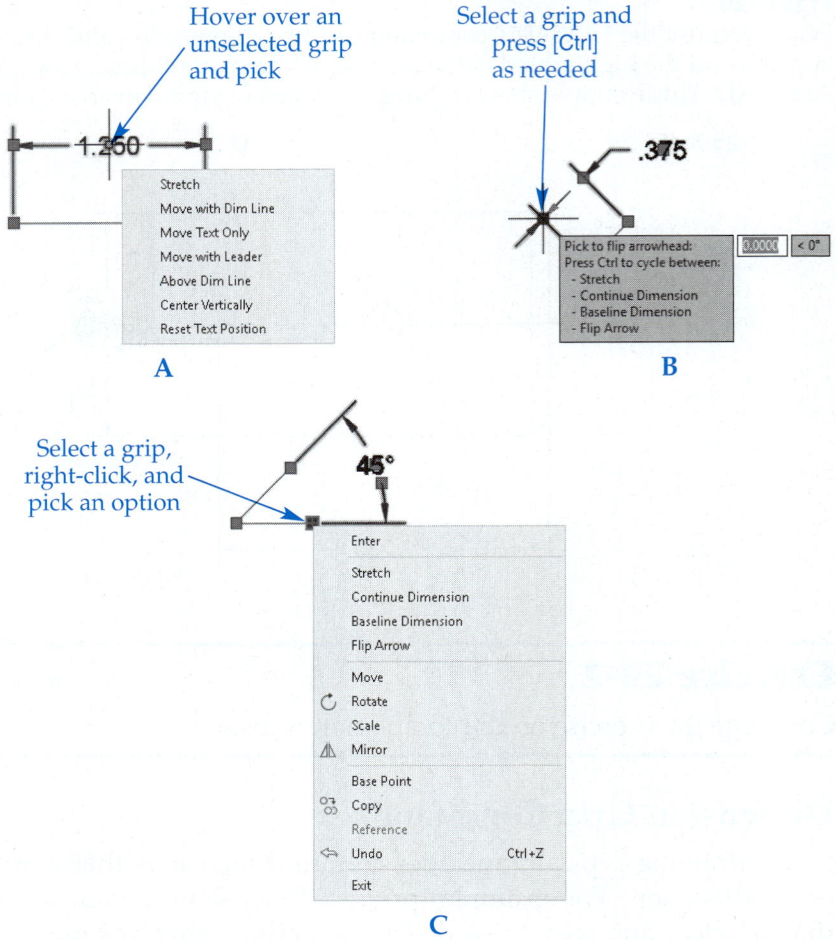

Figure 20-6 shows the process of using the dimension text grip commands to edit the display of a dimension value. **Figure 20-7** shows the process of using the grip commands available at the ends of dimension lines or extension lines, depending on the type of dimension. The options also vary depending on the type of dimension. Use drawing aids and construction geometry when necessary to complete the operation accurately.

PROFESSIONAL TIP

Activate the **Place text manually** check box in the **Fit** tab of the **New** (or **Modify**) **Dimension Style** dialog box to provide greater flexibility for the initial placement of dimensions when necessary.

Select a dimension and then right-click to display the shortcut menu shown in **Figure 20-2**. Use the **Dimension Style** cascading menu, previously described, to assign a different dimension style to the dimension or to save a new dimension style based on the properties of the selected dimension. The **Precision** cascading menu includes options to adjust the number of decimal places displayed with a dimension value. The **Precision** cascading menu often provides the easiest way to specify an alternative tolerance. The **Remove Style Overrides** option clears dimension style overrides assigned to the selected dimension. You will explore dimension style overrides later in this chapter.

Figure 20-6.
Examples of options available for modifying dimension text using grips.

Option	Process	Result
Stretch Stretches the text with or without the dimension line	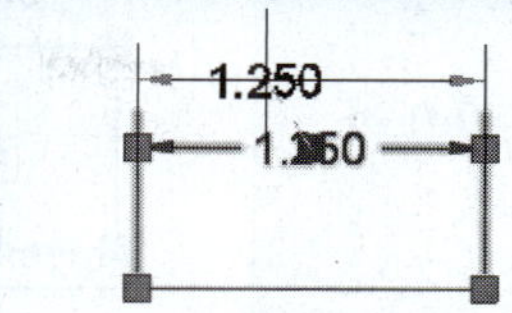	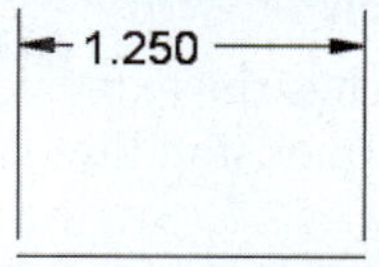
Move with Dim Line Always stretches the text with the dimension line	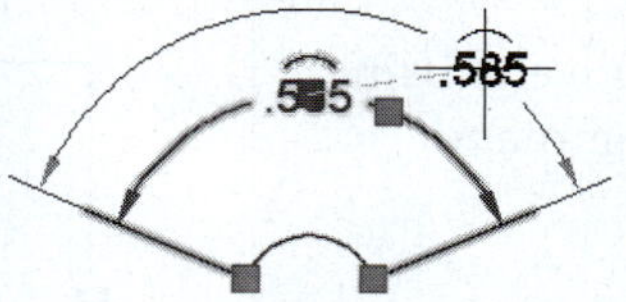	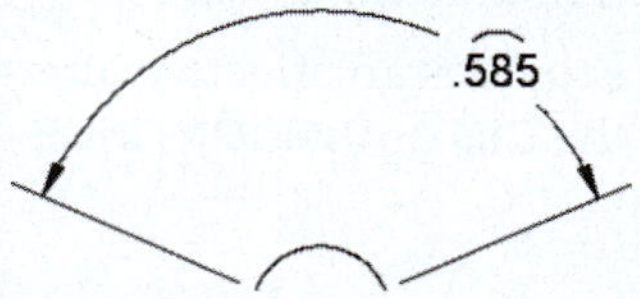
Move Text Only Moves text independently of the dimension, which is usually not appropriate	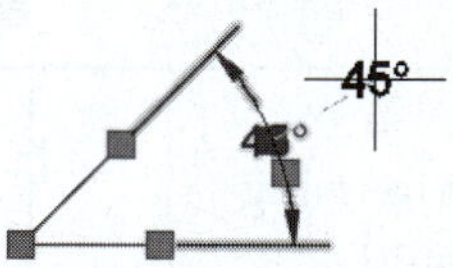	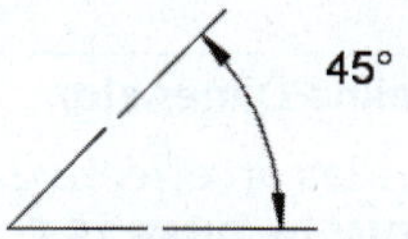
Move with Leader Moves text independently of the dimension and adds a leader line	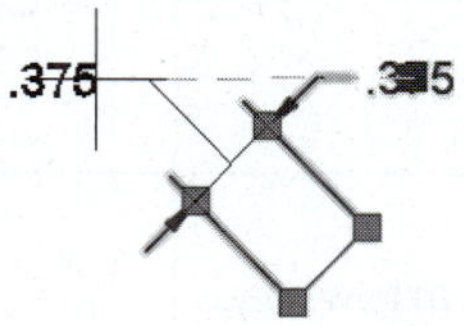	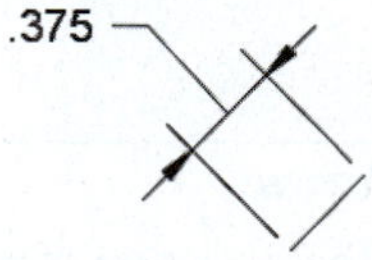
Above Dim Line Places text above the dimension line	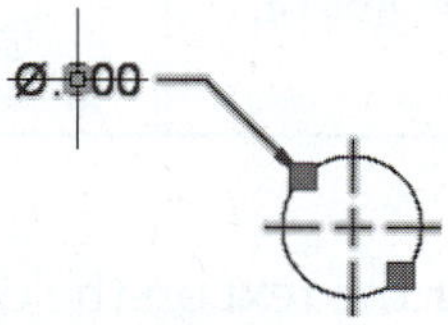	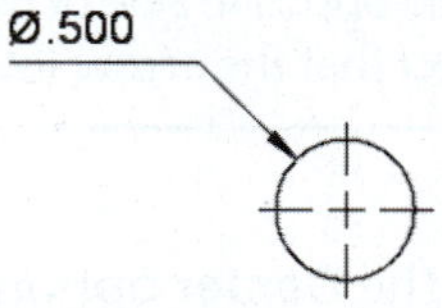
Center Vertically Returns text to centered between the dimension line	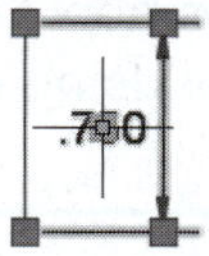	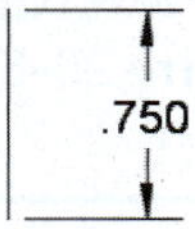
Reset Text Position Returns text to its original position	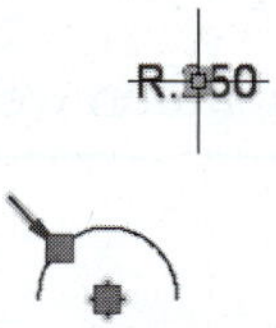	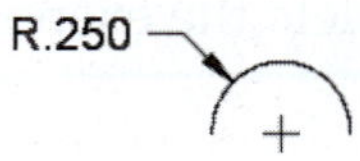

Exercise 20-3

www.g-wlearning.com/CAD/

Complete the exercise on the companion website.

Using the DIMTEDIT Command

Ribbon
Annotate
> Dimensions
DIMTEDIT
Type
DIMTEDIT

DIMTEDIT

The **DIMTEDIT** command provides another method to change the placement and orientation of existing dimension text. Access the **DIMTEDIT** command and select the dimension to alter. Move the cursor to stretch the text and specify a new point. AutoCAD automatically reestablishes the break in the dimension line.

The **DIMTEDIT** command also provides options to relocate dimension text to a specific position and rotate the text. However, it is usually quicker to select the appropriate button from the expanded **Dimensions** panel of the **Annotate** ribbon tab. Use the **Left** option to move horizontal text to the left and vertical text down. Use the **Right** option to move horizontal text to the right and vertical text up.

Figure 20-7.
Examples of options available for modifying or creating dimensions using grips.

Option	Process	Result
Stretch Stretches the extension lines and moves the dimension line	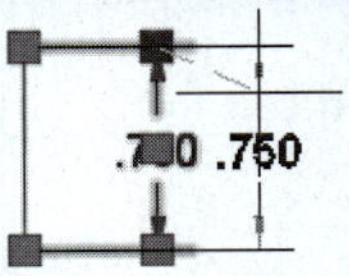	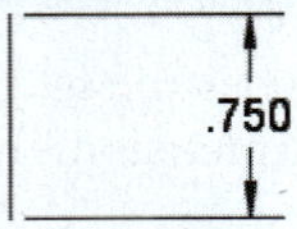
Continue Dimension Provides an effective alternative to using the **DIM** or **DIMCONTINUE** command	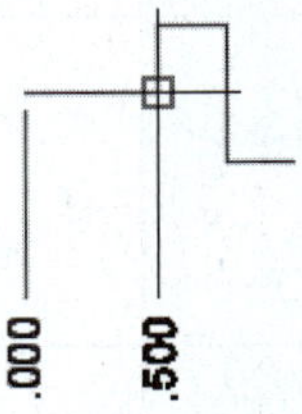	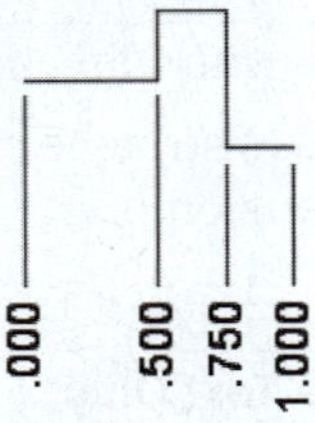
Baseline Dimension Provides an effective alternative to using the **DIM** or **DIMBASELINE** command	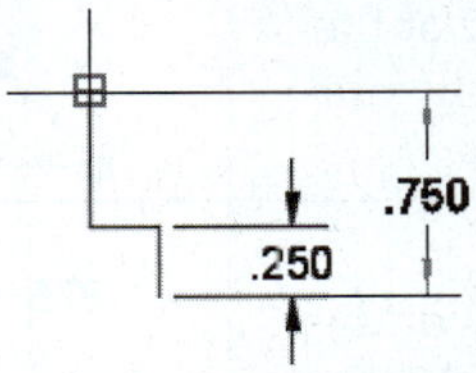	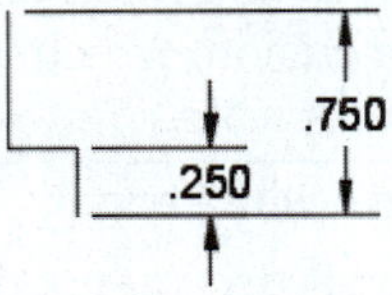
Flip Arrow Flips the direction of a dimension arrowhead to the opposite side of the extension line or object that the arrow touches	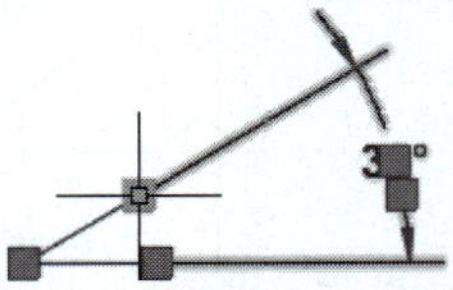	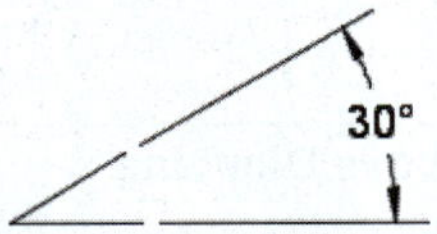

Use the **Center** option to center the text on the dimension line. Use the **Home** option to move the text back to its original position. Use the **Angle** option to rotate the dimension text. Figure 20-8 shows the result of using each **DIMTEDIT** command option.

Exercise 20-4

www.g-wlearning.com/CAD/

Complete the exercise on the companion website.

Figure 20-8.
A comparison of the **DIMTEDIT** command options.

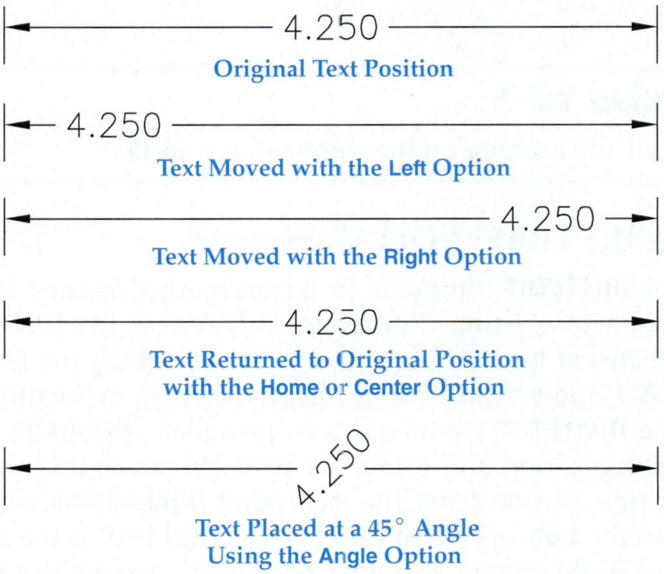

Using the DIMEDIT Command

Ribbon
Annotate
> Dimensions
Oblique
Type
DIMEDIT

DIMEDIT

The **DIMEDIT** command, not to be confused with the **DIMTEDIT** command, provides **Home** and **Rotate** options that function the same as the **Home** and **Angle** options of the **DIMTEDIT** command. The **New** option is similar to the **TEXTEDIT** command. When you activate the **New** option, the multiline text editor appears. Enter new dimension text, close the text editor, and select the dimension to change.

The **Oblique** option is unique to the **DIMEDIT** command and allows you to change the extension line angle without affecting the associated dimension value. Figure 20-9A shows an example of adjusting the placement of dimensions when space is limited by changing existing linear dimensions to use oblique extension lines. Figure 20-9B shows an example of using oblique extension lines to orient extension lines properly with the angle of the stairs in a stair section. Notice that the associated values and orientation of the dimension lines in these examples do not change.

To create oblique extension lines, dimension the object using the **DIM** or **DIMLINEAR** and **DIMALIGNED** commands as appropriate, even if the dimensions are crowded or overlap. Then access the **Oblique** option of the **DIMEDIT** command. A quick way to access the **Oblique** option is to pick the corresponding button from the expanded **Dimensions** panel of the **Annotate** ribbon tab. Then pick the linear and aligned dimensions to be redrawn at an oblique angle and specify the obliquing angle. Plan carefully to make sure you enter the correct obliquing angle. Obliquing angles originate from 0° East and revolve counterclockwise. Enter a specific value or pick two points to define the obliquing angle.

Figure 20-9. Drawing dimensions with oblique extension lines.

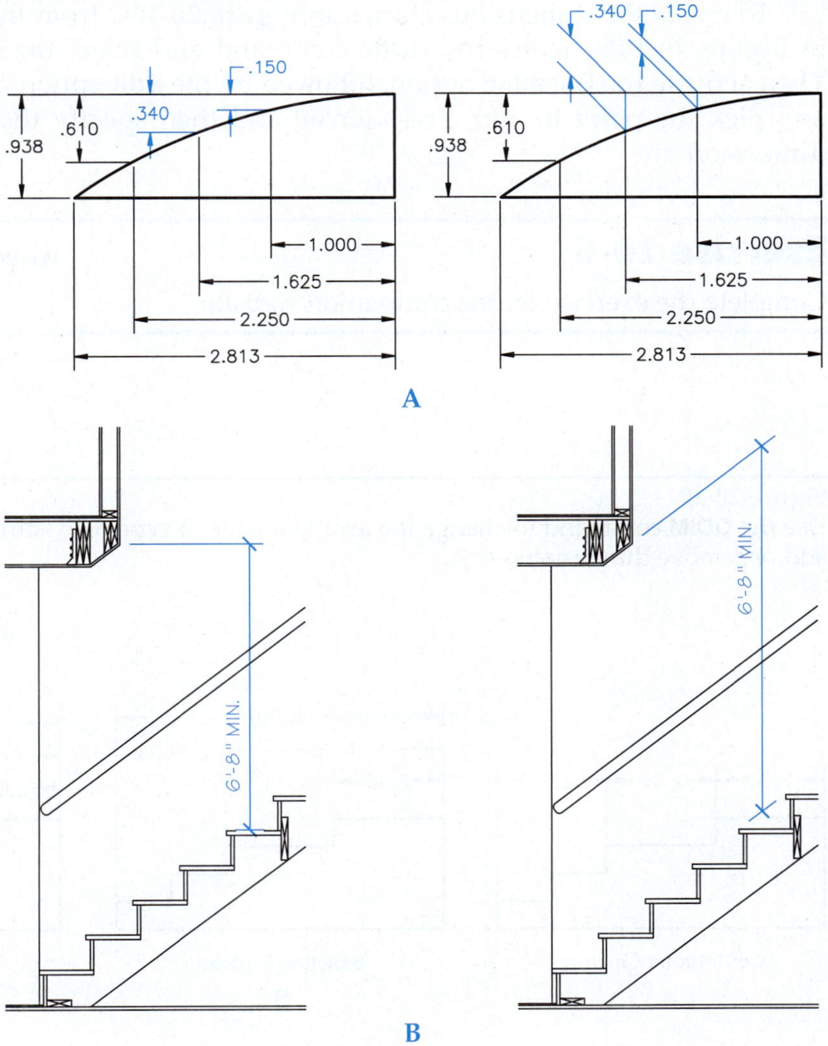

Exercise 20-5

www.g-wlearning.com/CAD/

Complete the exercise on the companion website.

Editing Dimensions with the QDIM Command

QDIM

Ribbon
Annotate
> Dimensions
Quick Dimension
Type
QDIM

The **QDIM** command provides options for replacing, adding, removing, and rearranging linear or ordinate dimensions. The **QDIM** command does not change diameter or radius dimensions. Access the **QDIM** command and select the dimensions to modify, and any other objects to be dimensioned. The **QDIM** command replaces the selected dimensions and adds dimensions to selected objects. Press [Enter] to display a preview of the dimensions attached to the cursor.

The **Continuous** option changes selected linear dimensions to chain dimensions. See Figure 20-10A. The **Baseline** option changes selected linear dimensions to baseline dimensions. See Figure 20-10B. The **Ordinate** option changes selected linear dimensions to rectangular coordinate dimensions without dimension lines. You must reselect the location of the dimensions. If the **QDIM** command does not reference the appropriate datum, use the **datumPoint** option before locating the dimensions to specify a different datum point.

Use the **Edit** option before locating the dimensions to add dimensions to, or remove dimensions from, the current set. Marks indicate the points acquired by the **QDIM** command. Use the **Add** option to specify a point to add a dimension, or use the **Remove** option to specify a point to remove the corresponding dimension. Press [Enter] to return to the previous prompt and continue using the **QDIM** command.

To create the dimensions shown in Figure 20-10C from the dimensions shown in Figure 20-10A, access the **QDIM** command and select the existing dimensions. Then activate the **Baseline** option, followed by the **Edit** option. Select the **Add** option and pick the point to add. Press [Enter] and then specify the location of the first dimension line.

Exercise 20-6

www.g-wlearning.com/CAD/

Complete the exercise on the companion website.

Figure 20-10.
Use the **QDIM** command to change the arrangements or type of existing dimensions and to add or remove dimensions.

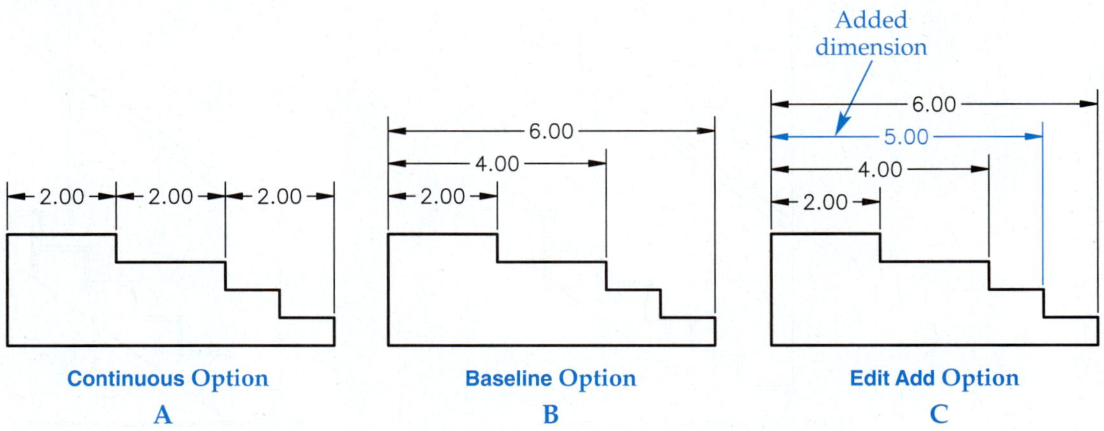

Overriding Dimension Style

Drawings often include dimensions that require settings slightly different from the assigned dimension style. These dimensions may be too few to merit creating a new style. Perform a dimension style override for these situations. For example, use a dimension style with an **Offset from origin** value of .063 for most applications. Apply a dimension style override with an **Offset from origin** value of 0 to three dimensions that should not display an extension line offset.

Existing Dimensions

The **Properties** palette is an effective tool for overriding the dimension style assigned to existing dimensions. The **Properties** palette divides dimension properties into several categories. See Figure 20-11. To change a property, access the proper category, pick the property to highlight, and adjust the corresponding value.

Figure 20-11.
The **Properties** palette allows you to edit dimension properties and create dimension style overrides.

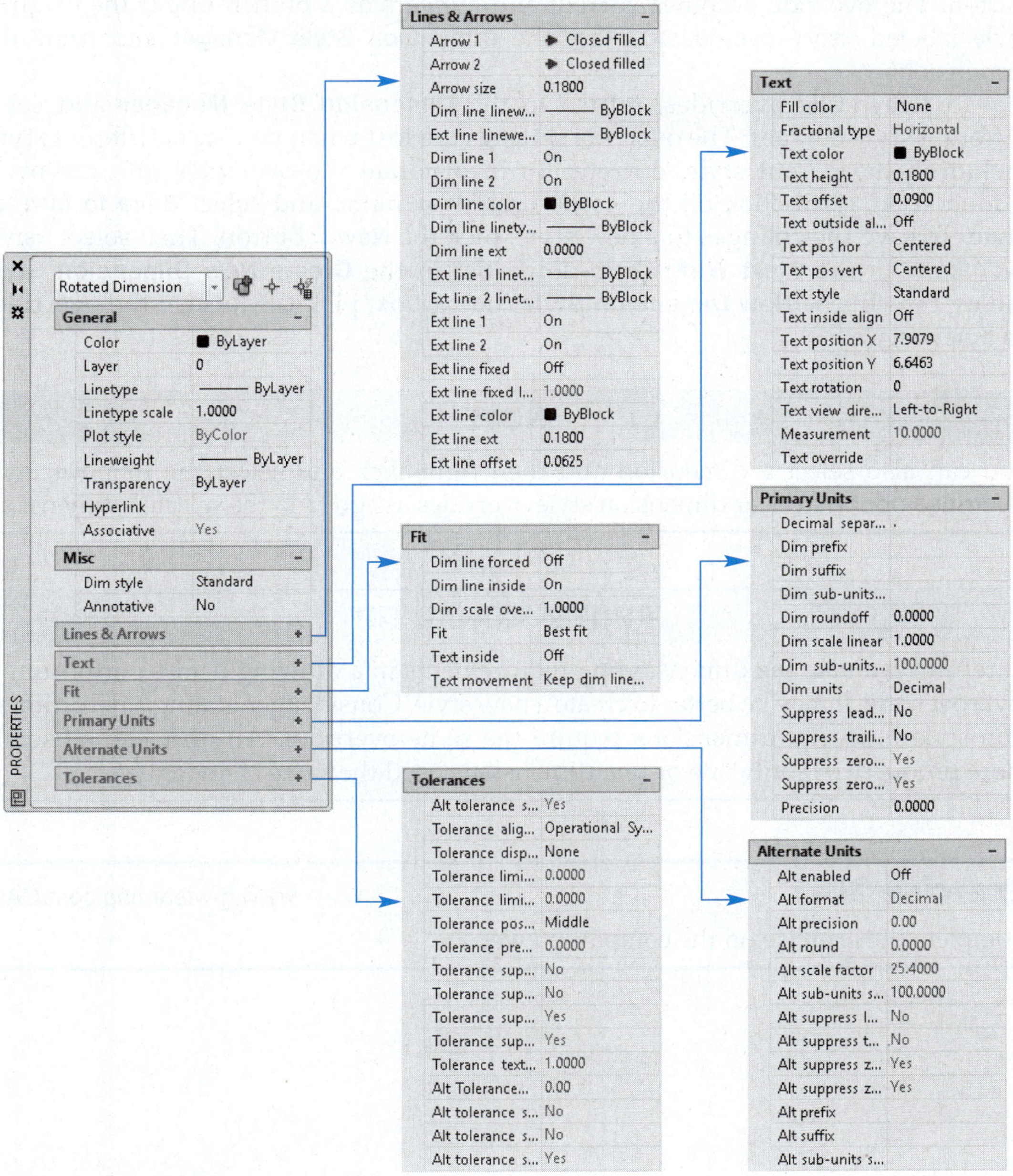

Most changes made using the **Properties** palette override the dimension style assigned to the selected dimension. The changes do not alter the original dimension style and do not apply to new dimensions.

NOTE

The **Quick Properties** palette provides a limited number of dimension properties and style overrides.

New Dimensions

Use the **Dimension Style Manager** to override the dimension style assigned to dimensions you are about to create. An example of an override is including a text prefix for a few dimensions. Select the dimension style to override from the **Styles:** list and then pick the **Override...** button to open the **Override Current Style** dialog box. The **Override...** button is available only for the current style. The **Override Current Style** dialog box includes the same tabs as the **New Dimension Style** and **Modify Dimension Style** dialog boxes. Make the necessary changes and pick the **OK** button. The override becomes current and appears as a branch under the original style labeled <style overrides>. Close the **Dimension Style Manager** and draw the dimensions.

To clear style overrides, return to the **Dimension Style Manager** and set a different style current. The override settings are lost when you set a different style, including the parent style, current. To incorporate the overrides into the overridden style, right-click on the <style overrides> name and select **Save to current style**. To save the changes to a new style, pick the **New...** button. Then select <style overrides> in the **Start With:** drop-down list in the **Create New Dimension Style** dialog box. In the **New Dimension Style** dialog box, pick **OK** to save the overrides as a new style.

NOTE

You can also select a dimension and then right-click and select the **Remove Style Overrides** option to clear dimension style overrides assigned to the selected dimension.

PROFESSIONAL TIP

Carefully evaluate the dimensioning requirements in a drawing before performing a style override. It may be better to create a new style. Consider generating a new dimension style if several dimensions require the same overrides. An override is usually more productive if only one or two dimensions need the same changes.

Exercise 20-7

www.g-wlearning.com/CAD/

Complete the exercise on the companion website.

Spacing and Aligning Dimensions

The amount of space between a drawing view and the first dimension line and the space between dimension lines vary depending on the drawing and industry or company standards. ASME standards recommend a minimum spacing of .375″ (10 mm) from a drawing feature to the first dimension line and a minimum spacing of .25″ (6 mm) between dimension lines. A minimum spacing of 3/8″ is common for architectural drawings. These minimum recommendations are generally less than the spacing required by actual company or school standards. A value of .5″ (12 mm) or .75″ (19 mm) is usually more appropriate.

Typically, the spacing between dimension lines is equal, and chain dimensions align. See **Figure 20-12**. You generally determine the correct location and spacing of dimension lines before and while dimensioning. However, you can adjust dimension line spacing and alignment after you place dimensions. This is a common requirement when there is a need to increase or decrease the space between dimension lines, such as when the drawing scale changes, or when dimensions are unequally spaced or misaligned.

Grips or the **STRETCH**, **DIMTEDIT**, and **QDIM** commands are common methods for adjusting the location and alignment of dimension lines. However, you must determine the exact location or amount of stretch applied to each dimension line before using these commands. An alternative is to use the **DIM** or **DIMSPACE** command to adjust the space equally between dimension lines or to align dimension lines.

Figure 20-12.
Correct drafting practice requires equal space and alignment between dimension lines for readability. The correct example uses a spacing of .75″ (19 mm) from the drawing feature to the first dimension line and a spacing of .5″ (12 mm) between dimension lines.

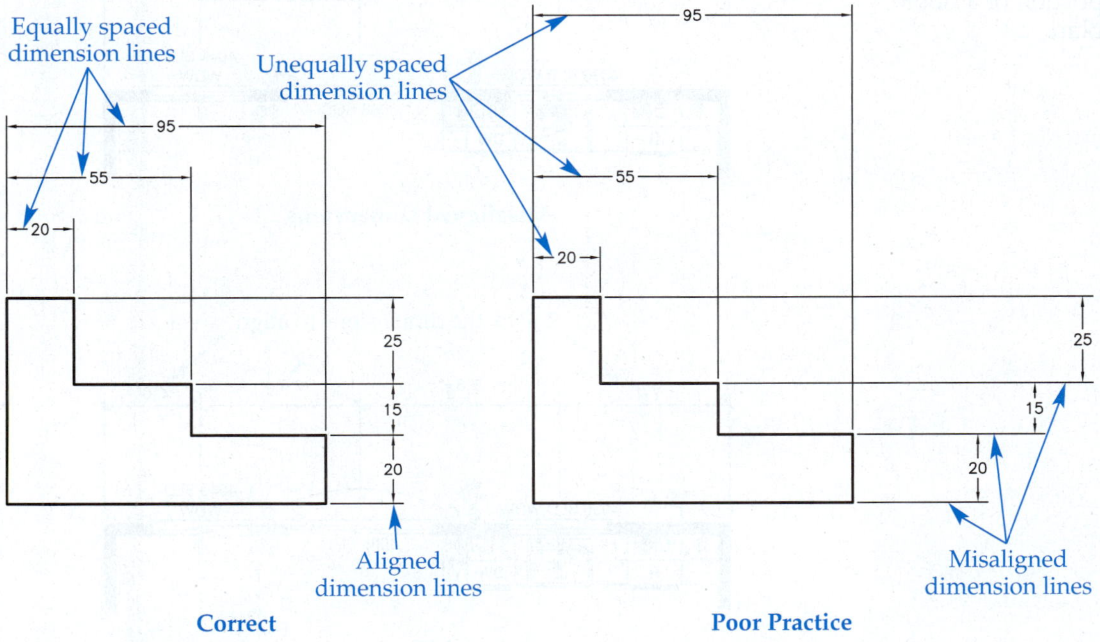

Using the DIM Command

base dimension: The dimension line that remains in the same location, with which other dimension lines align or space.

To align dimensions using the **DIM** command, select the **Align** option and then select the ***base dimension***, followed by the dimensions to align. Press [Enter] to complete the operation. See Figure 20-13. To space dimensions, select the **Distribute** option and then select the **Equal** or **Offset** option. The **Equal** option spaces each selected dimension line equally between the first and last selected dimension. See Figure 20-14A. Pick the dimensions and then press [Enter] to complete the operation. The **Offset** option spaces all dimensions from the base dimension according to the value specified in the **Baseline spacing:** text box in the **Lines** tab of the **New** (or **Modify**) **Dimension Style** dialog box. Select the base dimension, followed by the dimensions to distribute. See Figure 20-14B. Use the **Offset** option before selecting the dimensions to distribute to specify a different offset value. Press [Enter] to complete the operation.

Using the DIMSPACE Command

DIMSPACE
Ribbon
Annotate > Dimensions
Adjust Space
Type
DIMSPACE

To space dimensions using the **DIMSPACE** command, select the base dimension, followed by each dimension to be spaced. Press [Enter] to display the Enter value or [Auto]: prompt. Enter a value to space the dimension lines equally. For example, enter .5 to space the selected dimension lines .5″ apart. Enter a value of 0 to align the dimensions. See Figure 20-15. Use the **Auto** option to space dimension lines using a value that is twice the height of the dimension text.

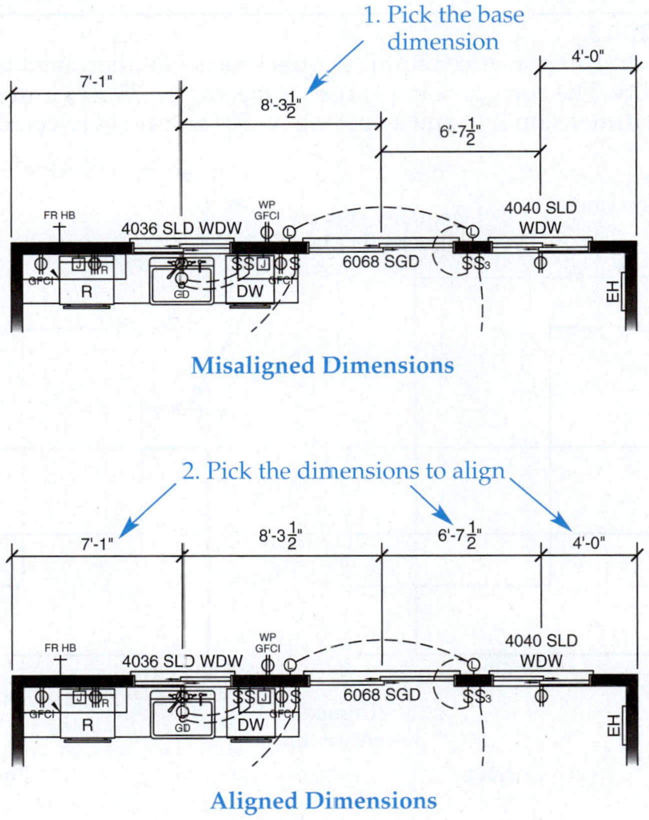

Figure 20-13.
Using the **Align** option of the **DIM** command to align dimensions on a portion of a floor plan.

Figure 20-14.
Using the **Distribute** option of the **DIM** command to space dimension lines correctly. A—Using the **Equal** option and selecting dimensions for equal distribution. B—Using the **Offset** option and selecting dimensions for an offset distribution (shown at top). Shown at bottom is the result of the equal distribution created in A. C—Correctly spaced dimension lines.

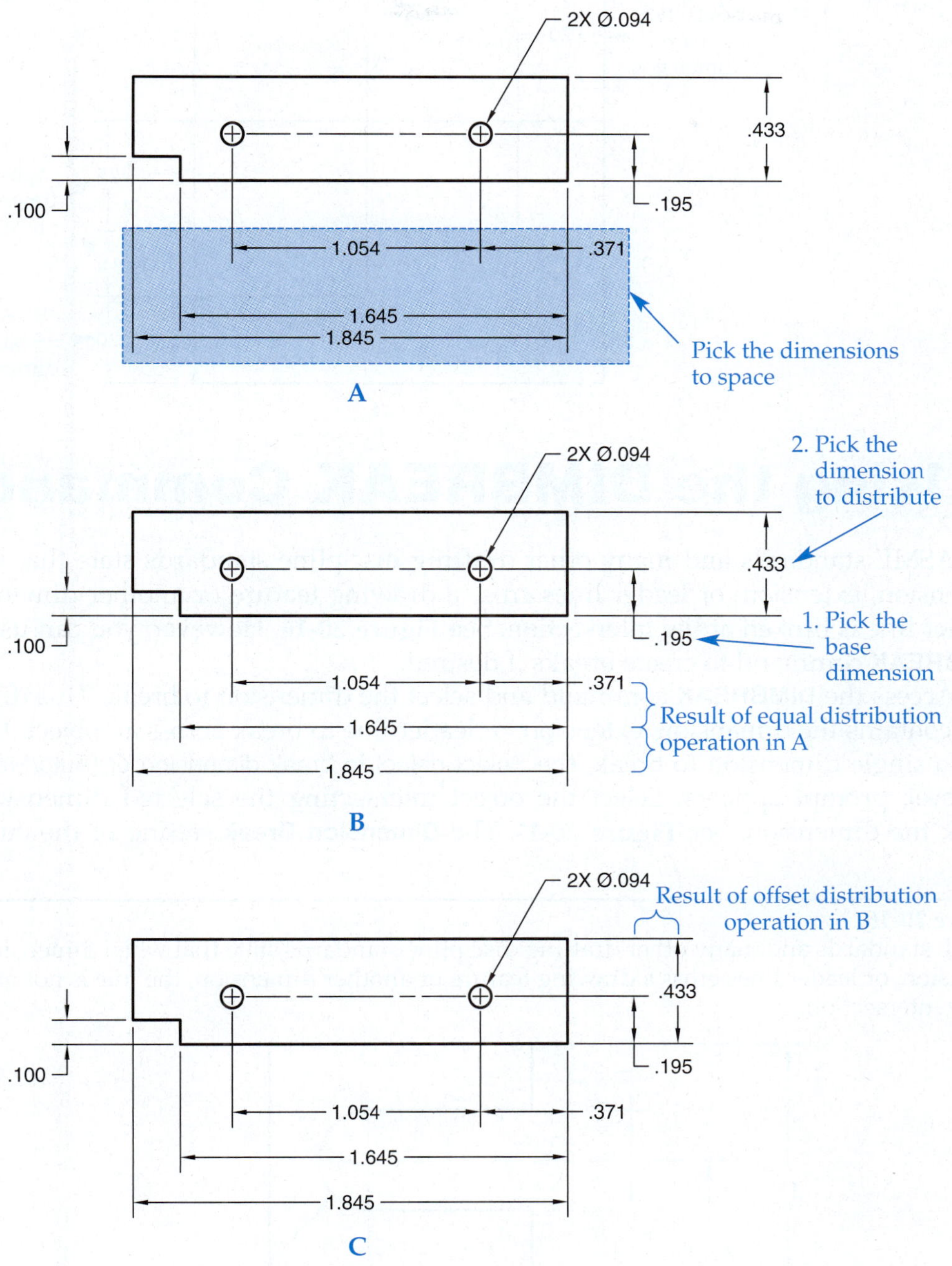

NOTE

Use the **DIM** or **DIMSPACE** command to space and align linear and angular dimensions.

Exercise 20-8

www.g-wlearning.com/CAD/

Complete the exercise on the companion website.

Figure 20-15.
Using the **DIMSPACE** command to space and align dimension lines correctly.

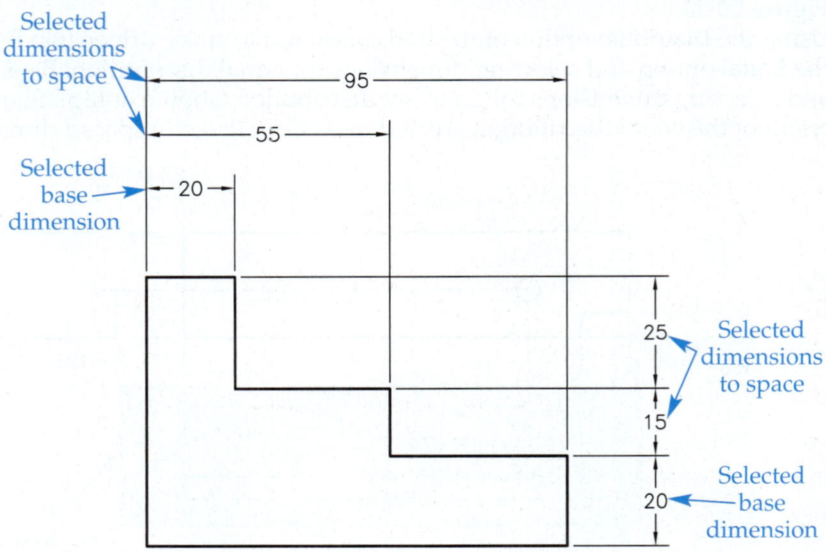

Using the DIMBREAK Command

ASME standards and many other drafting discipline standards state that when dimension, extension, or leader lines cross a drawing feature or another dimension, neither line is broken at the intersection. See **Figure 20-16**. However, you can use the **DIMBREAK** command to create breaks if desired.

DIMBREAK

Ribbon
Annotate
> Dimensions

Break

Type
DIMBREAK

Access the **DIMBREAK** command and select the dimension to break. This dimension contains the dimension, extension, or leader line to break across an object. If you pick a single dimension to break, the Select object to break dimension or [Auto/Manual/Remove]: prompt appears. Select the object intersecting the selected dimension to break the dimension. See **Figure 20-17**. The **Dimension Break** setting of the current

Figure 20-16.
ASME standards and many other drafting discipline standards state that when dimension, extension, or leader lines cross a drawing feature or another dimension, the line is not broken at the intersection.

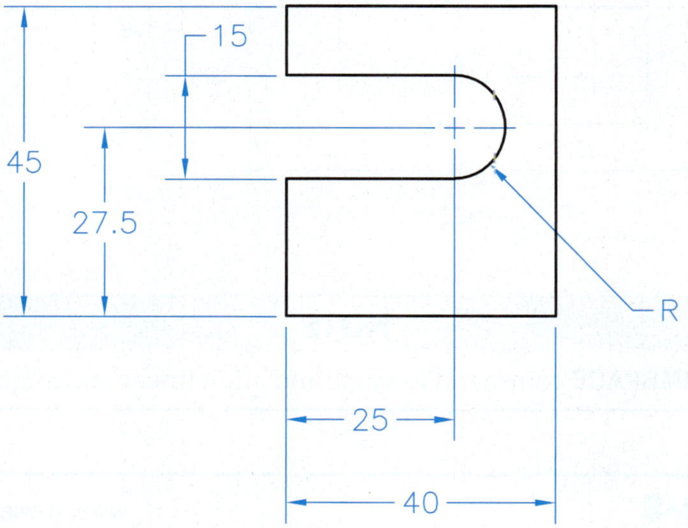

Figure 20-17.
Use the **DIMBREAK** command to break dimension, extension, or leader lines when they cross an object. *Caution:* This example violates ASME standards and is for reference only. Extension and leader lines do not break over object lines, but some drafters prefer to break an extension line when it crosses a dimension line.

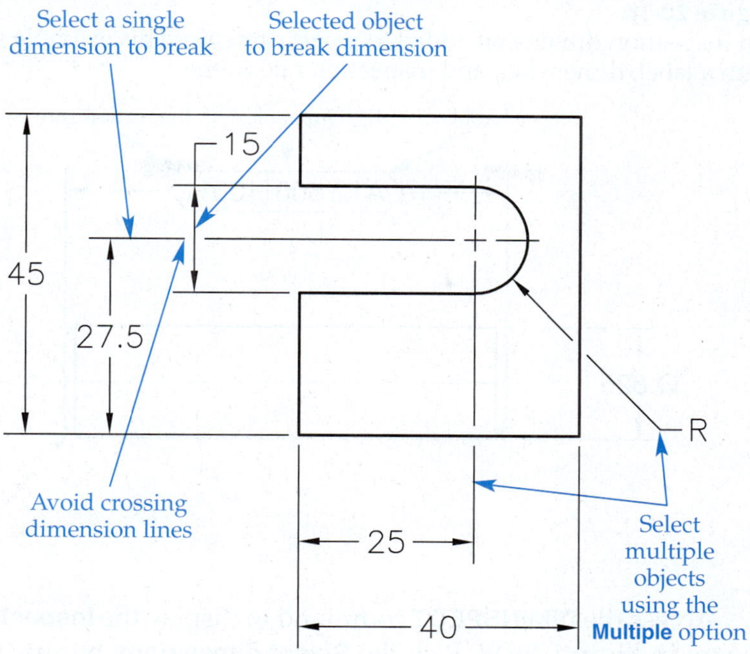

dimension style controls the break size. Select additional objects if necessary to break the dimension at additional locations.

The **Auto** option breaks the dimension, extension, or leader line automatically at all locations where objects intersect the selected object. Use the **Manual** option to define the size of the break by selecting two points along the dimension, extension, or leader line, instead of using the break size set in the current dimension style. Use the **Remove** option to remove a break created using the **DIMBREAK** command.

Another technique is to use the **Multiple** option to select more than one dimension. Press [Enter] after you select dimensions to display the Select object to break dimensions or [Auto/Remove]: prompt. Use the **Auto** option to break the selected dimension, extension, or leader lines everywhere they intersect another object. Use the **Remove** option to remove breaks created using the **DIMBREAK** command.

Exercise 20-9 www.g-wlearning.com/CAD/

Complete the exercise on the companion website.

Creating Inspection Dimensions

Inspections and tests occur throughout the design and manufacturing of a product. Tests help ensure the correct size and location of product features. In some cases, size and location dimensions include information about how frequently a test on the dimension occurs for consistency and tolerance during the manufacturing process. See Figure 20-18. Use the **DIMINSPECT** command to add inspection information to most types of existing dimensions.

Figure 20-18.
An inspection dimension added to a part drawing. This example shows an angular shape with a label, dimension, and inspection rate frame.

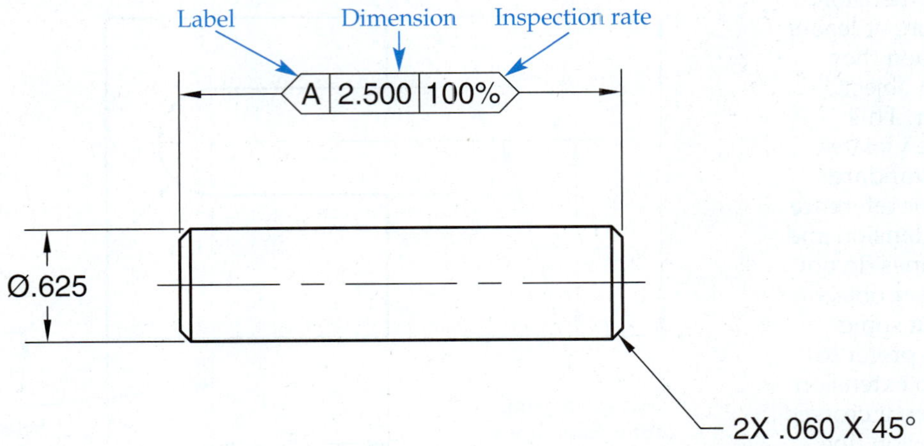

DIMINSPECT

Ribbon
Annotate
> Dimensions
Inspect
Type
DIMINSPECT

Access the **DIMINSPECT** command to display the **Inspection Dimension** dialog box, shown in Figure 20-19. Pick the **Select dimensions** button and select the dimensions to which you want to apply inspection information. You can select multiple dimensions, although the same inspection specifications apply to each. Pick the appropriate radio button in the **Shape** area to define the shape of the inspection dimension frame. The inspection dimension contains the inspection label, the dimension value, and the inspection rate. Select the **None** option to omit frames around values.

Pick the **Label** check box to include a label, and type the label in the text box. The label appears on the left side of the inspection dimension and identifies the dimension. The inspection dimension shown in Figure 20-18 is labeled A. The dimension frame houses the dimension value specified when you created the dimension. The dimensioned length of the part shown in Figure 20-18 is 2.500, as created using the **DIM** or **DIMLINEAR** command. The **Inspection rate** check box is active by default. Enter a value in the text box to indicate how often to test the dimension. The inspection rate for the dimension shown in Figure 20-18 is 100%. This rate has different

Figure 20-19.
The **Inspection Dimension** dialog box allows you to add inspection information to existing dimensions.

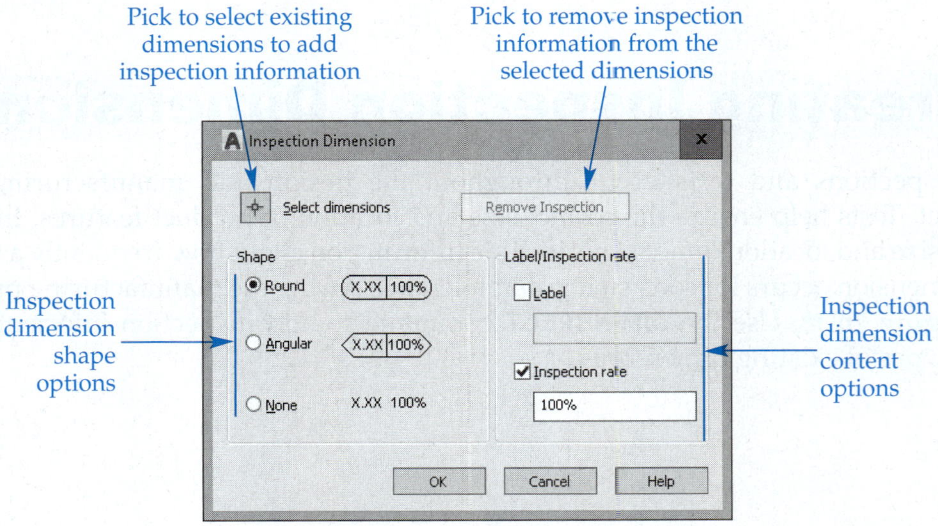

meanings depending on the application. In this example, the inspection rate of 100% means that the manufacturer must check the length of the part for tolerance every time the part is added to an assembly.

To remove an inspection dimension, access the **DIMINSPECT** command, pick the **Select dimensions** button in the **Inspection Dimension** dialog box, and select the dimensions from which you want to remove inspection information. Press [Enter] to return to the **Inspection Dimension** dialog box, and pick the **Remove Inspection** button to return the dimension to its condition prior to adding the inspection content.

Exercise 20-10

www.g-wlearning.com/CAD/

Complete the exercise on the companion website.

Center Mark and Centerline Editing Tools

Edit center mark and centerline objects using methods similar to those you use to edit dimensions. Use the **Properties** palette or **Quick Properties** palette to override specific center mark and centerline properties. In addition to these general editing techniques, specific tools are available to edit center marks and centerlines.

Use the **CENTERDISASSOCIATE** command to remove the association between a center mark or centerline and the associated geometry. Access the command, select the center marks or centerlines to disassociate, and press [Enter]. Moving a center mark or centerline away from the associated geometry also removes association. Use the **CENTERREASSOCIATE** command to form an association between a disassociated center mark or centerline and selected geometry. Access the command and select a disassociated center mark or centerline. Next, pick a circle or arc to associate a center mark, or two lines or polylines to associate a centerline.

Type: CENTER-DISASSOCIATE

CENTERDISASSOCIATE

Center Mark and Centerline Grip Commands

Often the quickest method for making basic changes to center marks and centerlines is to use grips. Select a center mark or centerline object to display grips at editable locations on the object. See **Figure 20-20**. Remember that the center mark or centerline is a single object with an appearance initially controlled by system variables.

Type: CENTER-REASSOCIATE

CENTERREASSOCIATE

The grip at the center of a center mark object with extensions provides access to context-sensitive options. Access and apply context-sensitive grip commands using one of three methods, similar to grip-editing dimensions, as shown in **Figure 20-5**. **Figure 20-20** explains the process of using center mark and centerline object grip commands. Use drawing aids and construction geometry when possible to complete the operation accurately.

NOTE

Use the **CENTERRESET** command to change the distance that all centerline extensions of a selected center mark object extend past the object according to the current **CENTEREXE** system variable. A quick way to access the command is to select a center mark or centerline object and then right-click and select the **Reset Extension Length** option.

Type: CENTERRESET

CENTERRESET

Figure 20-20.
Use grips to make basic changes to center mark and centerline objects.

Grip	Option	Process	Result
Center grip of center dashes and centerlines	Moves and disassociates the center mark or centerline object	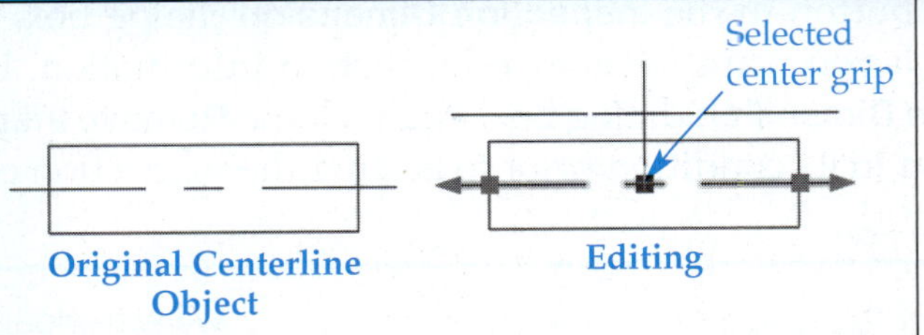	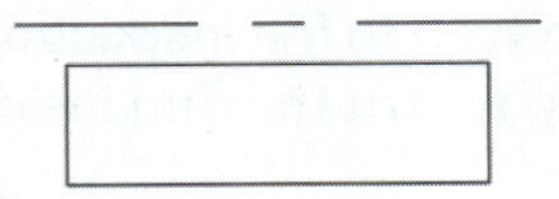
Center grip of center mark objects with centerline extensions	**Stretch** Moves and disassociates the center mark object	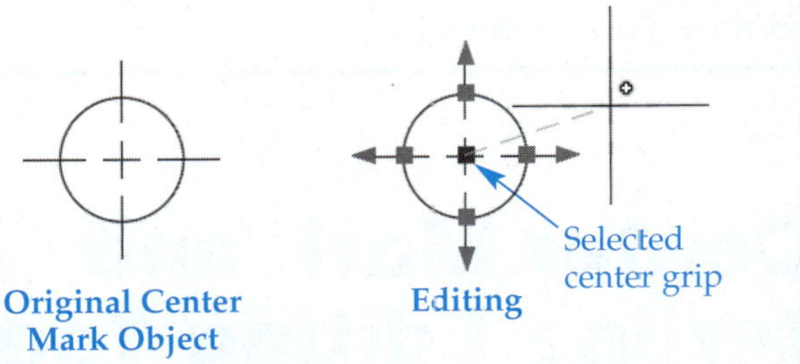	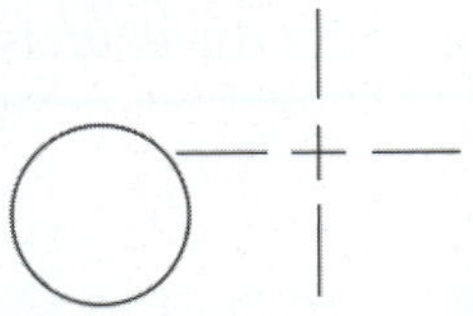
	Change Extension Length Changes the distance that all centerline extensions extend past the object	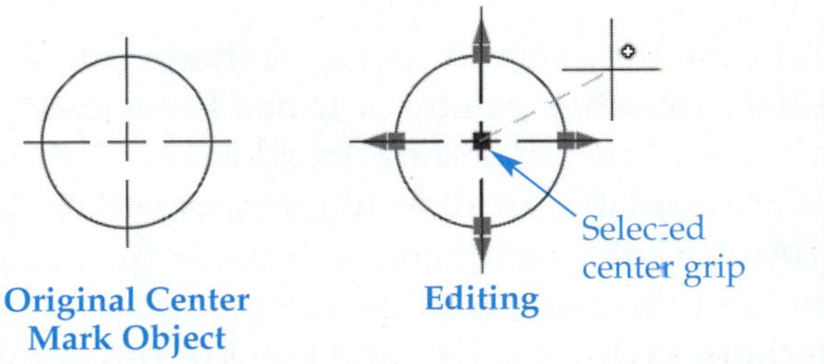	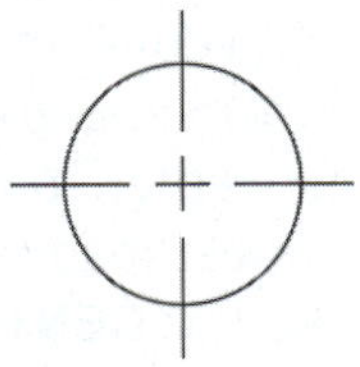
Length grip of center mark or centerline	Redefines the location where a center mark or centerline extends past the object	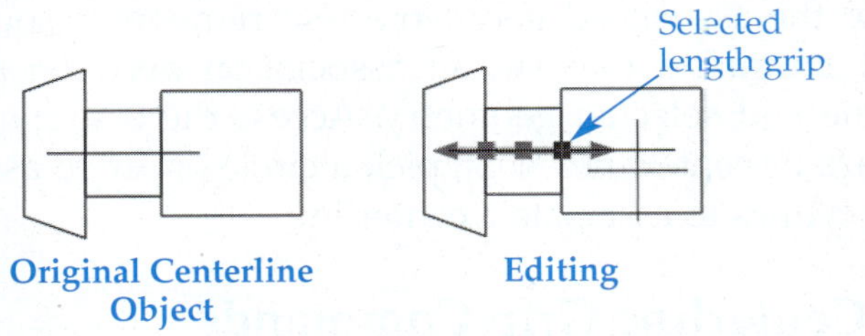	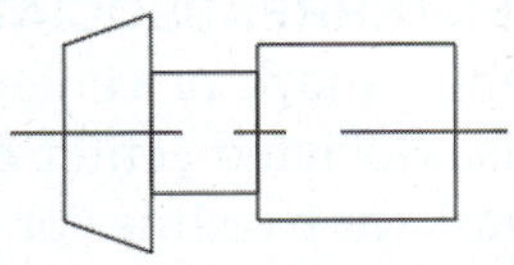 After Two Operations
Extension endpoint grip of center mark or centerline	Changes the distance that a center mark or centerline extends past the object	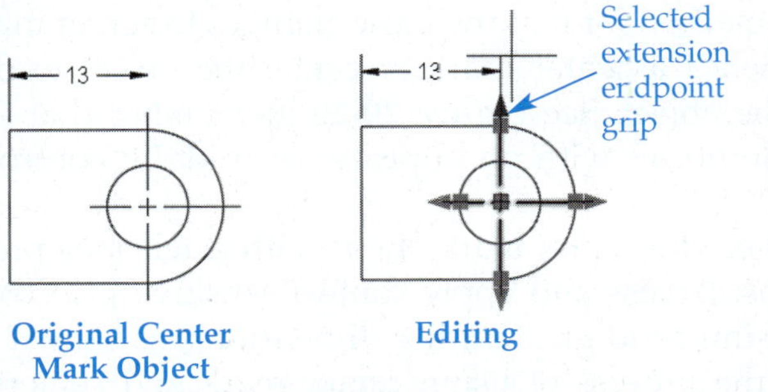	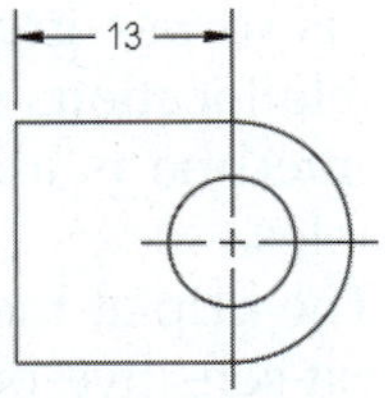

Exercise 20-11

www.g-wlearning.com/CAD/

Complete the exercise on the companion website.

Multileader Editing Tools

Edit multileaders using methods similar to those you use to edit dimensions. Use the **Properties** palette or **Quick Properties** palette to override specific multileader properties. In addition to these general editing techniques, specific tools allow you to add and remove leader lines and space, align, and group multileader objects.

Assigning a Different Multileader Style

To assign a different multileader style to existing multileaders, pick the multileaders to change and select a different multileader style from the **Multileader Style** drop-down list in the **Annotation** panel on the **Home** ribbon tab. The **Multileader Style** drop-down list is also available in the **Leaders** panel on the **Annotate** ribbon tab. Another technique is to select the multileaders to change and select a different multileader style from the **Quick Properties** palette or the **Properties** palette.

A third method is to select a multileader to edit and then right-click to display the shortcut menu shown in Figure 20-21. Use the **Multileader Style** cascading menu to assign a different multileader style to the multileader or to save a new multileader style based on the properties of the selected multileader.

Editing Multileader Mtext

An easy method to modify multileader text created using the **Mtext** content option is to double-click on the multileader text to activate the multiline text editor. Add to or modify the multileader text and then close the text editor. You can also use the **TEXTEDIT** command, described in Chapter 10, to edit multileader mtext.

Figure 20-21.
Select a multileader and then right-click to access this shortcut menu with options for adjusting individual multileaders.

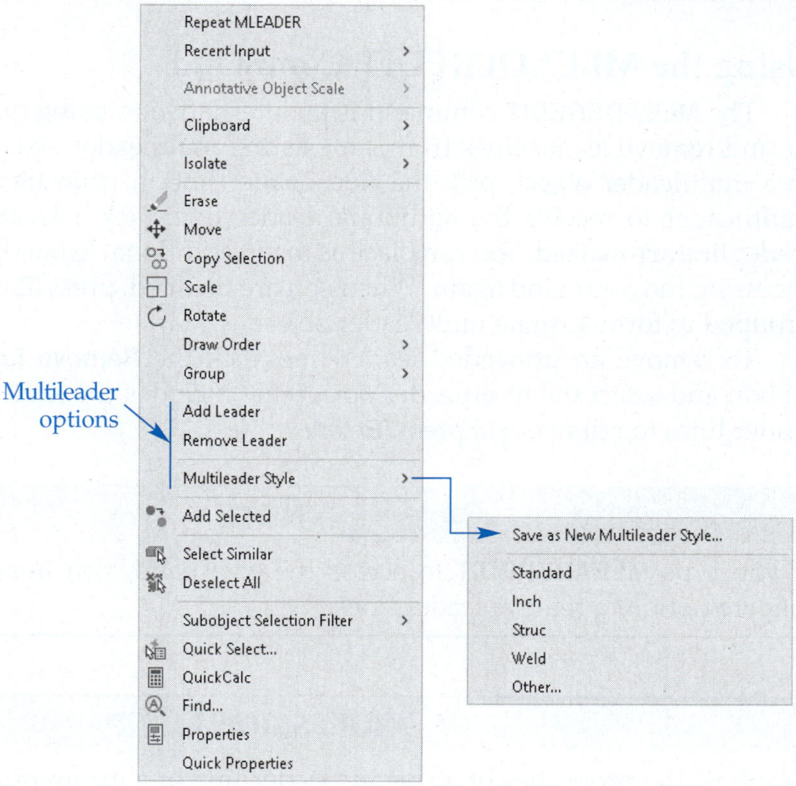

Multileader Grip Commands

Often the quickest method for making basic changes to multileaders is to use grips. For example, use grips to perform stretch operations and add or remove leaders and vertices. Select a multileader to display grips at editable locations on the multileader. See Figure 20-22. Most multileader grips offer the standard **STRETCH**, **MOVE**, **ROTATE**, **SCALE**, **MIRROR**, and **Copy** grip commands. Remember that the multileader is a single object with variables controlled by the multileader style.

Context-Sensitive Options

Certain multileader grips provide access to context-sensitive options. Access and apply context-sensitive grip commands using one of three methods, similar to grip-editing dimensions, as shown in Figure 20-5. Figure 20-22 explains the process of using multileader grip commands. Use drawing aids and construction geometry when possible to complete the operation accurately.

NOTE

Multiple leaders are not a recommended ASME standard. For example, do not use two leaders from the same value to dimension two chamfers of the same size. However, multiple leaders are appropriate for some applications, such as welding symbols. Multiple leaders are also appropriate for some architectural drawings and structural and civil engineering drawings.

Shortcut Menu Options

Select a multileader and then right-click to display the shortcut menu shown in Figure 20-21. Use the **Add Leader** option to add a leader line to the multileader object, or use the **Remove Leader** option to remove a leader line.

Using the MLEADEREDIT Command

MLEADEREDIT

Ribbon
Home
> Annotation
Annotate
> Leaders

Add Leader

Type
MLEADEREDIT

MLEADEREDIT

Ribbon
Home
> Annotation
Annotate
> Leaders

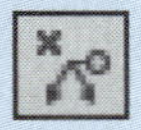

Remove Leader

Type
MLEADEREDIT

The **MLEADEREDIT** command is an alternative to using grips to add leader lines to, and remove leader lines from, an existing multileader object. To add a leader line to a multileader object, pick the **Add Leader** button from the ribbon and select the multileader to receive the additional leader line. Pick a location for the additional leader line arrowhead. You can place as many additional leader lines as needed without accessing the command again. When you are finished, press [Enter]. All leader lines are grouped to form a single multileader object.

To remove an unneeded leader line, pick the **Remove Leader** button from the ribbon and select the multileader object that includes the leader to remove. Select the leader lines to remove and press [Enter].

NOTE

If you type **MLEADEREDIT** to access the command, you must activate the **Remove leaders** option to remove leader lines.

PROFESSIONAL TIP

To adjust the properties of a specific leader line in a group of leaders attached to the same content, hold down [Ctrl] and pick the leader to modify. Then access the **Properties** palette. Properties specific to the selected leader appear, and all other properties are filtered out.

Figure 20-22.
Use grips to make basic changes to multileaders, and add and remove leader lines and vertices.

Grip	Option	Process	Result
Landing endpoint	Lengthens the landing from the endpoint closest to the content	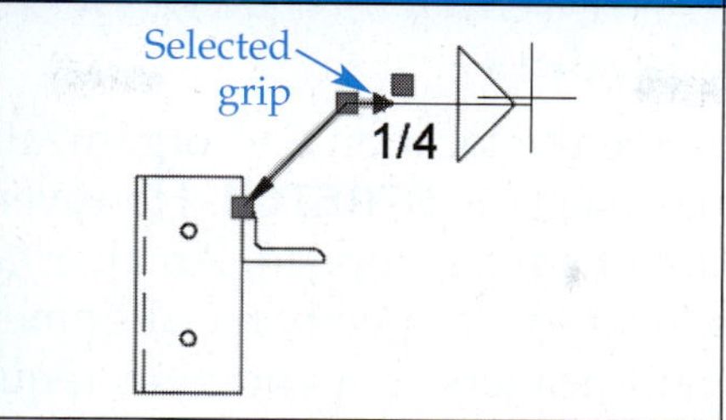	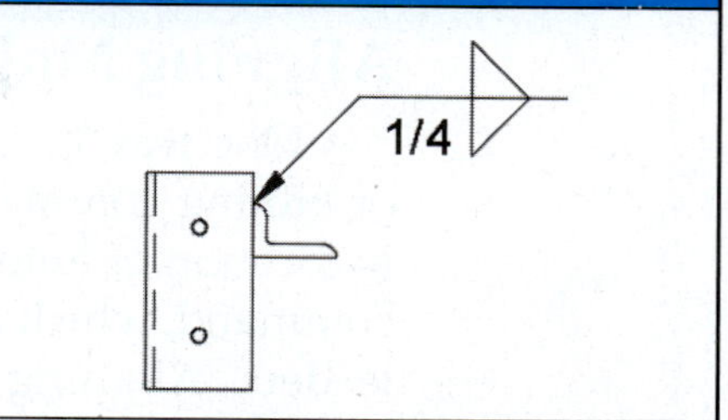
Leader line vertex	**Stretch** Stretches the leader at the selected vertex	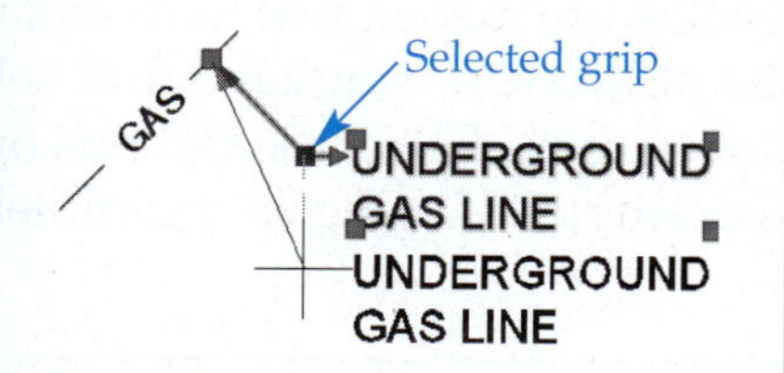	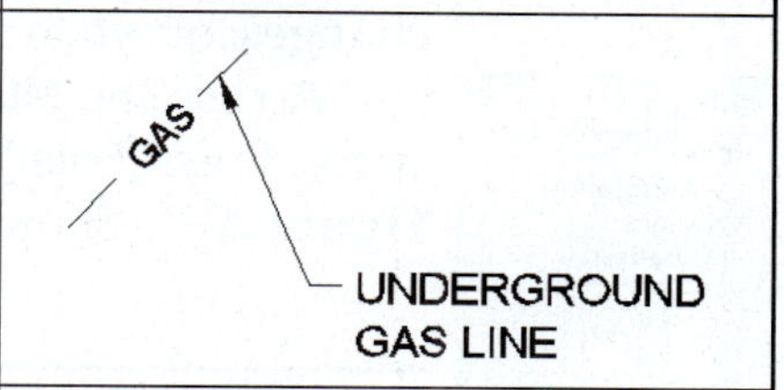
	Lengthen Landing Lengthens the landing from the selected vertex	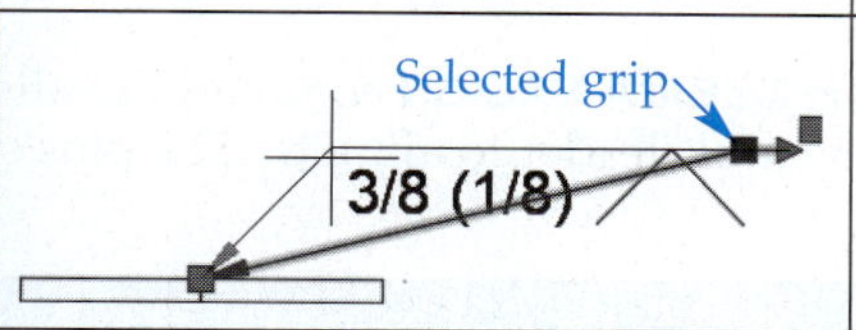	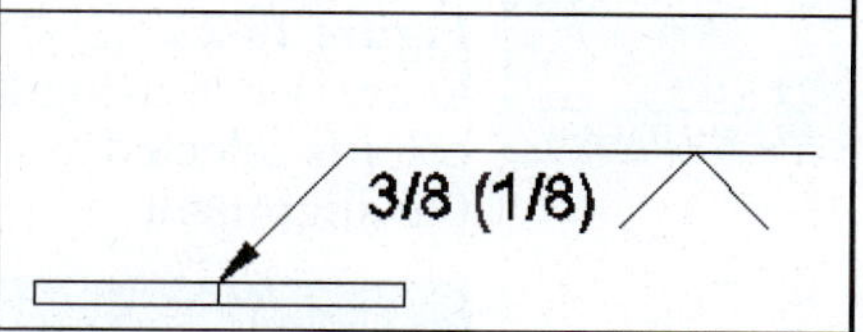
	Add Leader Adds leaders to the content	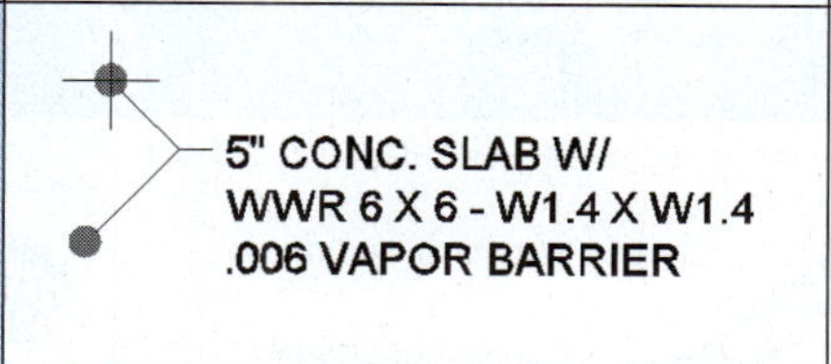	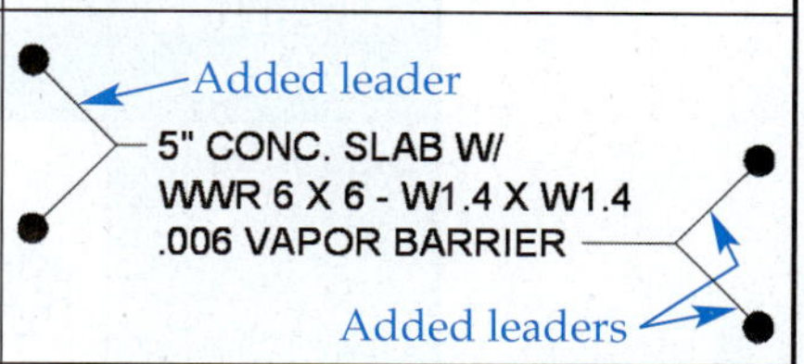
Leader line endpoint or vertex	**Stretch** Stretches the leader line at the selected endpoint or vertex		
	Add Vertex Adds a leader point	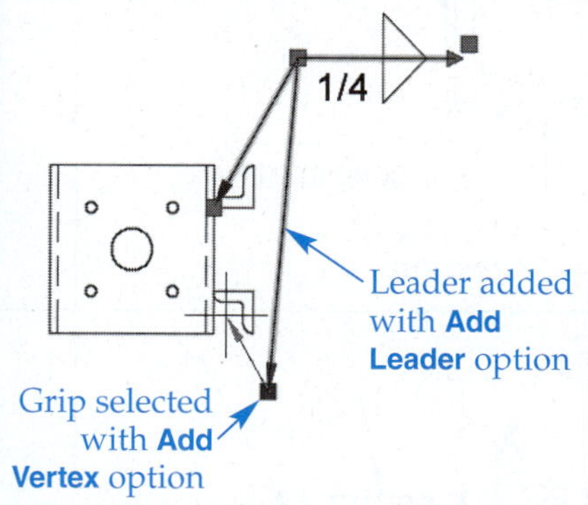	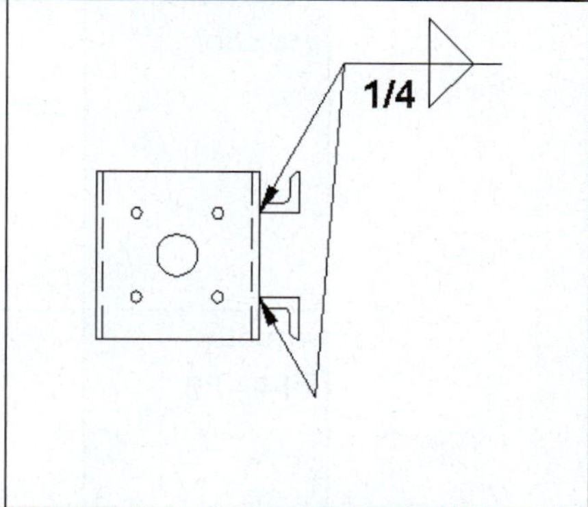
Leader line endpoint	**Remove Leader** Removes a selected leader line	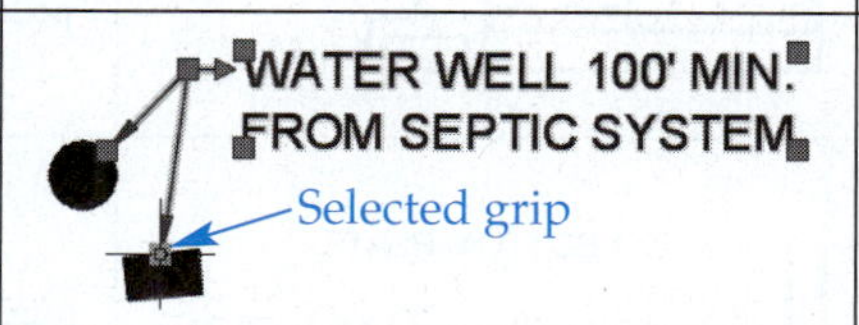	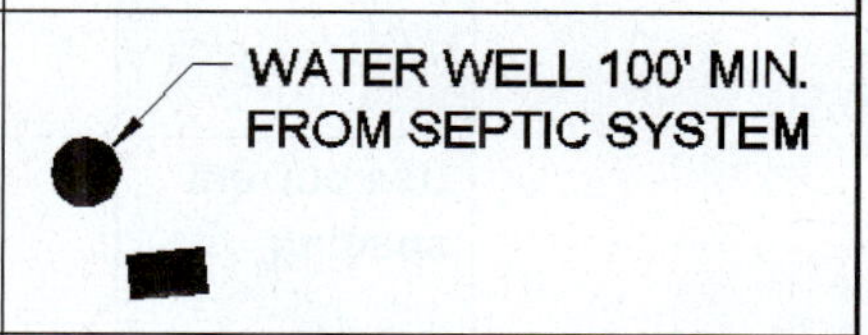
Vertex	**Remove Vertex** Removes a selected vertex	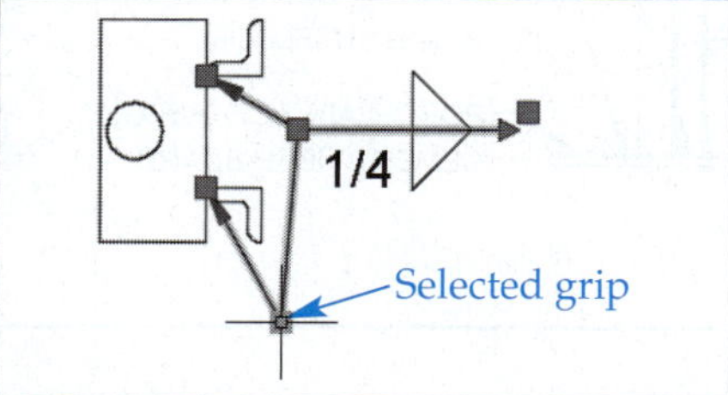	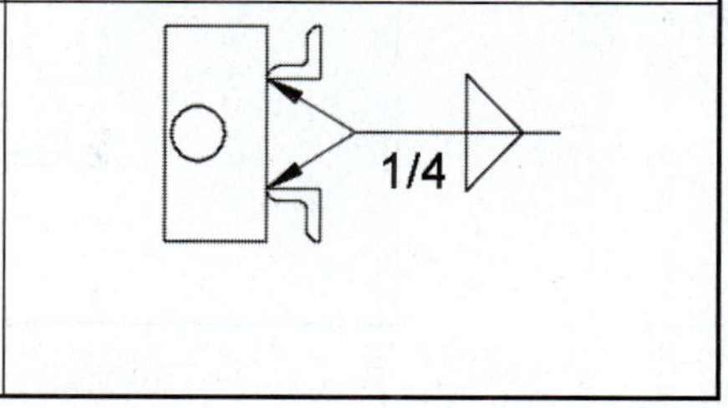

Exercise 20-12

www.g-wlearning.com/CAD/

Complete the exercise on the companion website.

Aligning Multileaders

One way to change the location and organization of multileaders is to use grips or editing commands, such as **STRETCH**. However, you must determine the change in location or amount of stretch applied. An alternative is to use the **MLEADERALIGN** command, which allows you to specify the alignment, spacing, and direction of multileaders. Aligning multileaders is a common requirement when the drawing scale changes, or when leaders are misaligned or unequally spaced.

MLEADERALIGN

Ribbon
Home
> Annotation
Annotate
> Leaders
Align

Type
MLEADERALIGN

Access the **MLEADERALIGN** command and select all of the leaders to align and space. Press [Enter] to continue. Select the **Options** option to set the alignment method. Figure 20-23 shows examples of using each multileader alignment method.

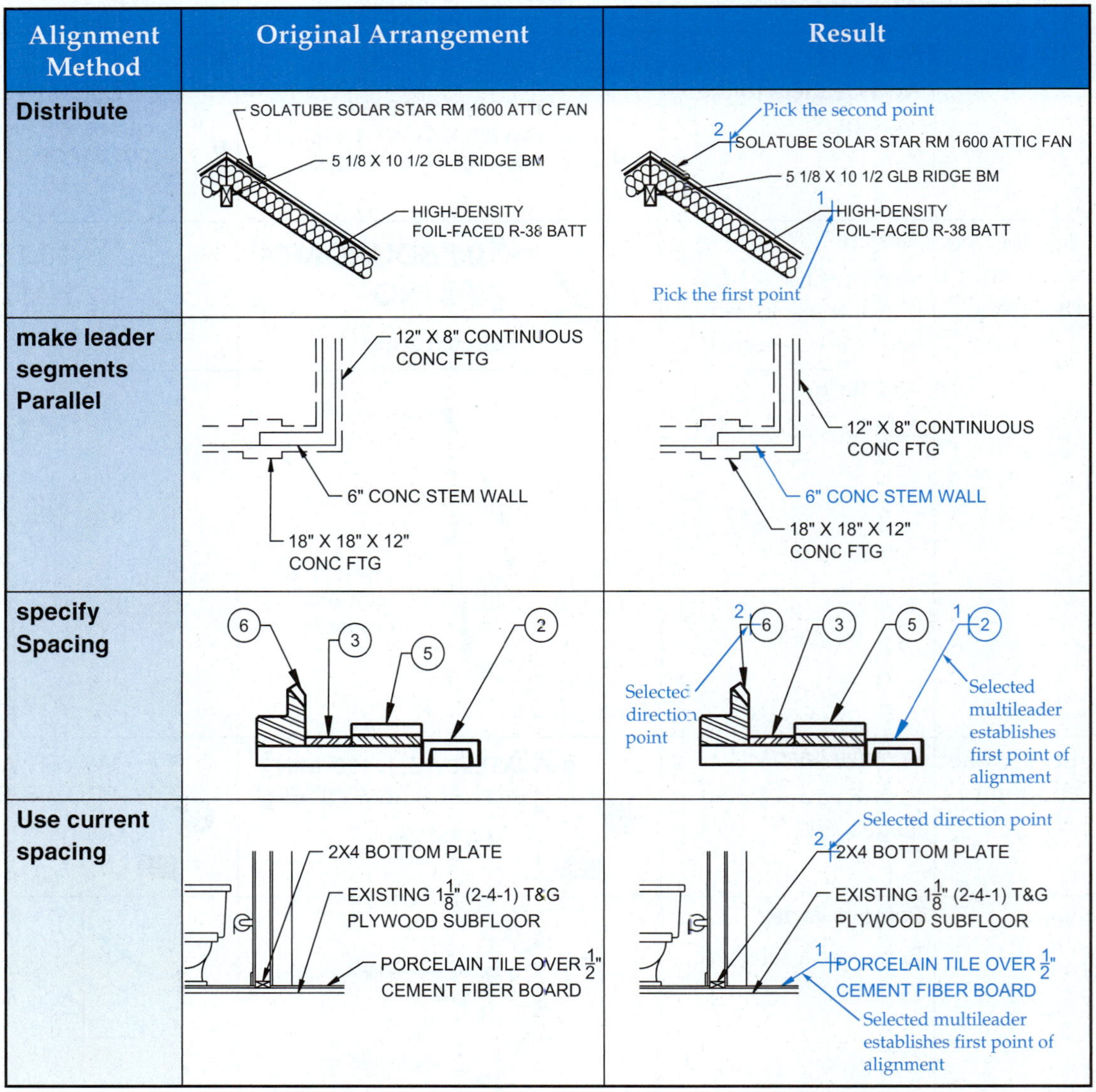

Figure 20-23.
Examples of using the **MLEADERALIGN** command to align multileaders. The multileader in color is selected as the multileader to align to. The plus signs indicate points used to create the alignment.

Use the **Distribute** option to align and space the selected leaders equally between two points. Pick the first point and then the second point to create the alignment. Use the **make leader segments Parallel** option to make all leaders parallel to a selected leader. Select the leader to which other leaders should align. The length of each leader line, except for the selected leader, increases or decreases in order to become parallel with the selected leader. Use the **specify Spacing** option to space the leaders equally according to the distance, or clear space, between the extents of each leader. Enter the spacing, select the leader to which other leaders should align, and then specify the alignment direction by picking or entering a point. Use the **Use current spacing** option to align leaders without changing the space between the leaders. Select the leader to which other leaders should align and then specify the alignment direction by picking or entering a point.

NOTE

The second point you locate when using the **specify Spacing** and **Use current spacing** options only determines alignment direction. The second point you locate when using the **Distribute** option controls alignment direction and spacing.

Exercise 20-13

www.g-wlearning.com/CAD/

Complete the exercise on the companion website.

Grouping Multileaders

balloons: Circles that contain a number or letter to identify the assembly component and correlate the component to a parts list or bill of materials. Balloons connect to a component with a leader line.

grouped balloons: Balloons that share the same leader, which typically connects to the most obviously displayed component.

You can group separate multileaders created using a **Block** multileader content style to use a single leader line. This practice is common when adding ***balloons*** to assembly drawings. ***Grouped balloons*** allow you to identify closely related clusters of assembly components, such as a screw, flat washer, and nut. See Figure 20-24. Use the **MLEADERCOLLECT** command to group multiple existing leaders using a single leader line.

Access the **MLEADERCOLLECT** command and select the leaders to group. The order in which you select leaders determines how they are grouped. Select leaders in a sequential order, ending with the leader line you want to keep.

MLEADERCOLLECT

Ribbon
Home
> Annotation
Annotate
> Leaders
Collect
Type
MLEADERCOLLECT

Figure 20-24.
An example of grouped balloons identifying closely related parts. Some of the parts or features may be hidden.

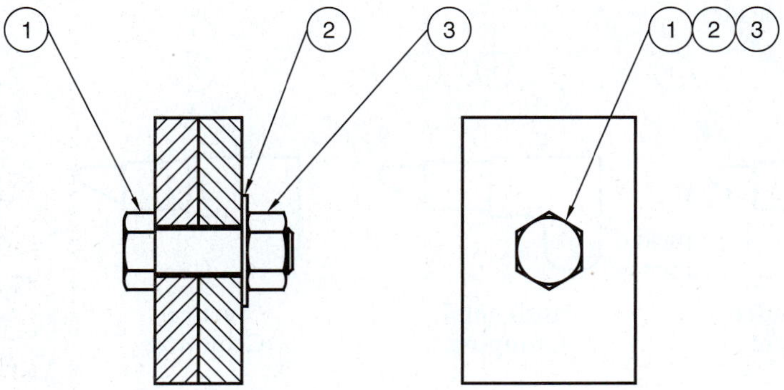

The options illustrated in Figure 20-25 are available after you select the leaders and press [Enter]. Select the **Horizontal** option to align grouped content horizontally, or the **Vertical** option to align grouped content vertically. Pick a point to locate the grouped leader. Select the **Wrap** option to wrap grouped content to additional lines as needed when the number of items exceeds a specified width or quantity. Enter the width at the Specify wrap width or [Number]: prompt, or use the **Number** option to enter a quantity not to exceed before the grouped leaders wrap. Then pick a point to locate the grouped leader.

NOTE

You can only use the **MLEADERCOLLECT** command to group symbols attached to leaders created using the **Block** content style.

PROFESSIONAL TIP

MATCHPROP

Ribbon
Home
> Properties
Match Properties
Type
MATCHPROP
MA

Use the **MATCHPROP** command, described in Chapter 13, to match the properties assigned to one dimension, center mark, centerline, or multileader object to other objects of the same type.

Exercises 20-14 and 20-15

www.g-wlearning.com/CAD/

Complete the exercises on the companion website.

Supplemental Material

www.g-wlearning.com/CAD/

Isometric Dimensions

The **Oblique** option of the **DIMEDIT** command is one option for dimensioning isometric drawings. For information about constructing dimensions for isometric views, go to the companion website, navigate to this chapter in the **Contents** tab, and select **Isometric Dimensions**.

Figure 20-25.
Options for grouping leaders using the **MLEADERCOLLECT** command.

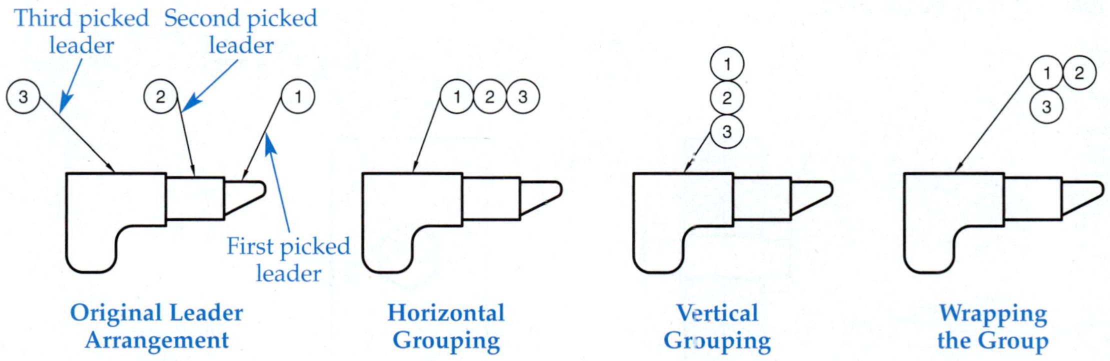

Chapter Review

Answer the following questions using the information in this chapter.

1. Define *associative dimension*.
2. Why is it important to have associative dimensions for editing objects?
3. Which **Options** dialog box setting controls associative dimensioning?
4. Which command allows you to convert non-associative dimensions to associative dimensions?
5. Which command allows you to disassociate a dimension from an object?
6. What are definition points?
7. List the context-sensitive options available for grip-editing dimension text.
8. Name three methods of changing the dimension style of a dimension.
9. How does the **Update** dimension tool affect selected dimensions?
10. Explain how to modify an existing dimension to add a prefix or suffix to the dimension text if you forget to use the **Mtext** or **Text** option of the **DIM** or **DIMLINEAR** command.
11. Name the command that allows you to control the placement and orientation of an existing dimension text value.
12. Name two applications in which you might need to create oblique extension lines.
13. Which option of the **QDIM** command can you use to add a new baseline dimension to an existing set of baseline dimensions?
14. When you use the **Properties** palette to edit a dimension property, what is the effect on the dimension style?
15. Which commands can you use to adjust the space equally between dimension lines or align dimension lines without having to determine the exact location or amount of stretch needed?
16. What two options are available when you use the **Multiple** option of the **DIMBREAK** command?
17. What command allows you to add information about how frequently the manufacturer should test a dimension for consistency and tolerance during the manufacturing of a product?
18. When using grips to edit a center mark, which context-sensitive option allows you to change the distance that all centerline extensions extend past the object?
19. Name two applications in which leaders with multiple leader lines are appropriate.
20. Identify the four options available to change multileader alignment when using the **MLEADERALIGN** command.

Drawing Problems

Start AutoCAD if it is not already started. Start a new drawing for each problem using an appropriate template of your choice. The template should include layers and text, dimension, and multileader styles, when necessary, for drawing the given objects. Add layers and text, dimension, and multileader styles as needed. Draw all objects using appropriate layers. Use appropriate text, dimension, and multileader styles, justification, and format. Follow the specific instructions for each problem. Use only drawing and editing commands and techniques you have already learned. Use your own judgment and approximate dimensions when necessary. Apply dimensions accurately using ASME or appropriate industry standards.

Note: Some of the problems in this chapter are built on problems from previous chapters. If you have not yet completed those problems, complete them now.

▼ Basic

1. **Mechanical** Save P17-9 as P20-1. If you have not yet completed Problem 17-9, do so now. In the P20-1 file, edit the drawing as follows:
   A. Erase the front (circular) view.
   B. Stretch the vertical dimensions to provide more space between dimension lines. Be sure the space you create is the same between all vertical dimensions.
   C. Stagger the existing vertical dimension text numbers if they are not staggered as shown in the original problem.
   D. Erase the 1.750 horizontal dimension and then stretch the 5.255 and 4.250 dimensions to make room for a new baseline dimension from the baseline to where the 1.750 dimension was located. This should result in a new baseline dimension that equals 2.750. Be sure all horizontal dimension lines are equally spaced.

2. **Mechanical** Save P18-3 as P20-2. If you have not yet completed Problem 18-3, do so now. In the P20-2 file, edit the drawing as follows:
   A. Stretch the total length from 3.500 to 4.000, leaving the holes the same distance from the edges.
   B. Fillet the upper-left corner. Modify the 3X R.250 dimension accordingly.
   C. Change the two lower holes from ∅.500 to ∅.375. Modify the 4X ∅.500 dimension accordingly and add a new dimension to dimension the two lower holes.

3. **Architectural** Save P17-13 as P20-3. If you have not yet completed Problem 17-13, do so now. In the P20-3 file, edit the drawing as follows: Make the bathroom 8′-0″ wide by stretching the walls and vanity that are currently 6′-0″ wide to 8′-0″. Do this without increasing the size of the water closet compartment. Provide two equally spaced oval sinks where there is currently one.

4. **Mechanical** Save P18-21 as P20-4. If you have not yet completed Problem 18-21, do so now. In the P20-4 file, edit the drawing as follows:
   A. Lengthen the part .250 on each side for a new overall dimension of 6.500.
   B. Change the width of the part from 3.000 to 3.500 by widening an equal amount on each side.

5. **Mechanical** Save P18-19 as P20-5. If you have not yet completed Problem 18-19, do so now. In the P20-5 file, edit the drawing as follows:
   A. Shorten the .750 thread on the left side to .500.
   B. Shorten the .388 hexagon length to .300.

Drawing Problems - Chapter 20

6. **Mechanical** Draw the shim shown at A. Then edit the .150 and .340 values using oblique dimensions as shown at B. Save the drawing as P20-6.

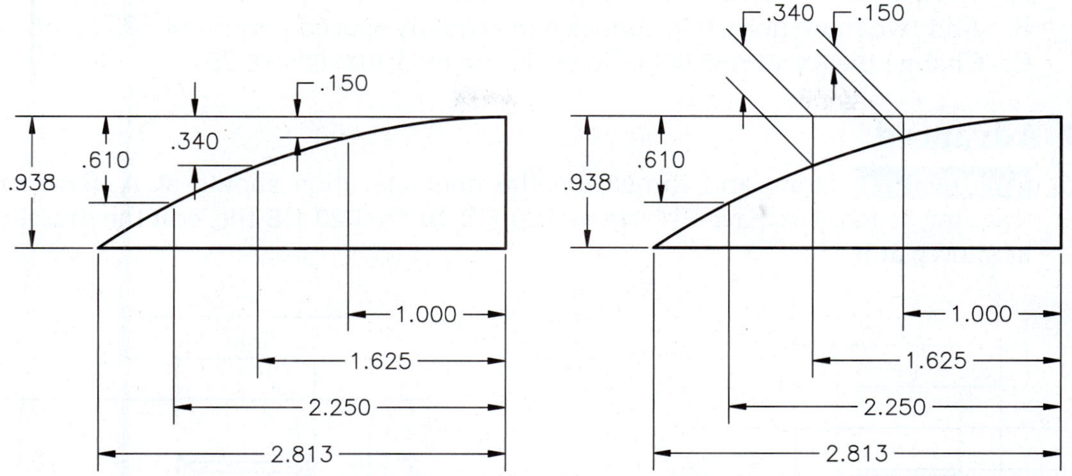

Draw this object.
A

Modify the dimensions as shown here.
B

Title: Shim

▼ Intermediate

7. **Mechanical** Draw and dimension the swivel screw shown. Save the drawing as P20-7.

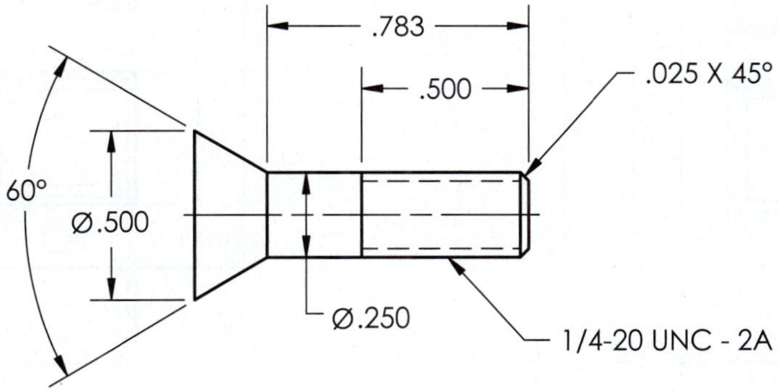

8. **Mechanical** Save P18-7 as P20-8. If you have not yet completed Problem 18-7, do so now. In the P20-8 file, edit the drawing as follows:
   A. Use the existing drawing as the model and make four copies.
   B. Leave the original drawing as it is and edit the other four pins in the following manner, keeping the Ø.125 hole exactly in the center of each pin.
   C. Give one pin a total length of 1.500.
   D. Create the next pin with a total length of 2.000.
   E. Edit the third pin to a length of 2.500.
   F. Change the last pin to a length of 3.000.
   G. Organize the pins on your drawing in a vertical row ranging in length from the smallest to the largest. You may need to change the drawing limits.

9. **Mechanical** Save P18-8 as P20-9. If you have not yet completed Problem 18-8, do so now. In the P20-9 file, edit the drawing as follows:
   A. Modify the spline to have twelve projections, rather than eight.
   B. Change the angular dimension, linear dimension, and 8X dimension to reflect the modification.

10. **Mechanical** Save P18-14 as P20-10. If you have not yet completed Problem 18-14, do so now. In the P20-10 file, edit the drawing as follows:
   A. Stretch the total length from 6.500 to 7.750.
   B. Add two more holes that continue the equally spaced pattern of .625 apart.
   C. Change the 8X .625(=5.000) dimension to read 10X .625(=6.250).

▼ Advanced

11. **Architectural** Draw and dimension the door elevation shown at A. Save the drawing as P20-11A. Save P20-11A as P20-11B. In the P20-11B file, edit the drawing as shown at B.

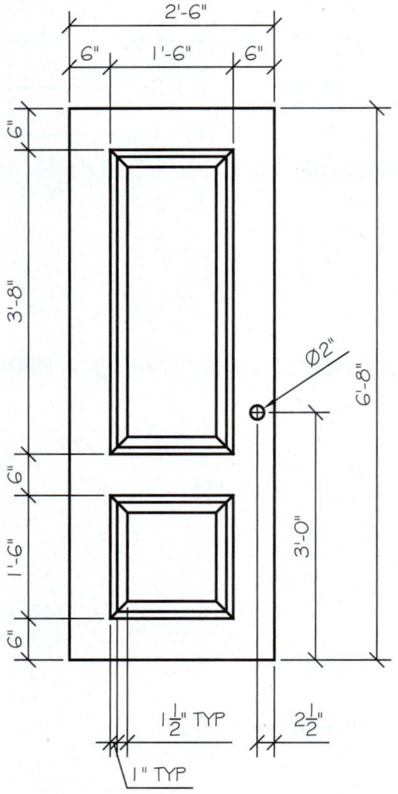

A

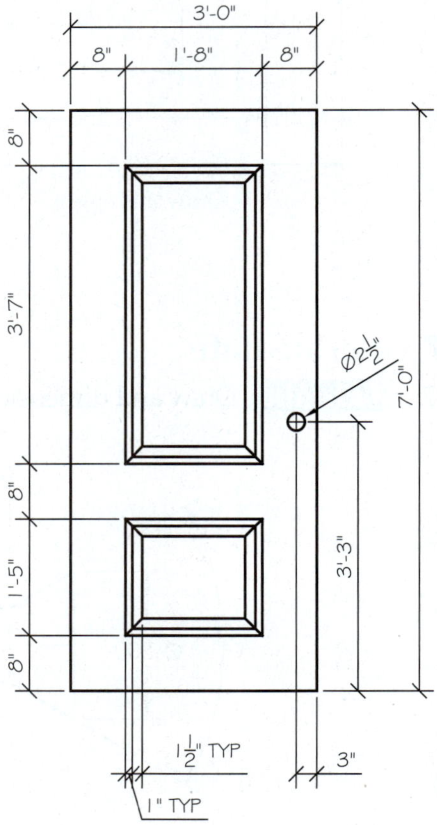

B

12. **Architectural** Save P17-16 as P20-12. If you have not yet completed Problem 17-16, do so now. In the P20-12 file, make the client-requested revisions to the floor plan as shown. Make sure the dimensions reflect the changes.

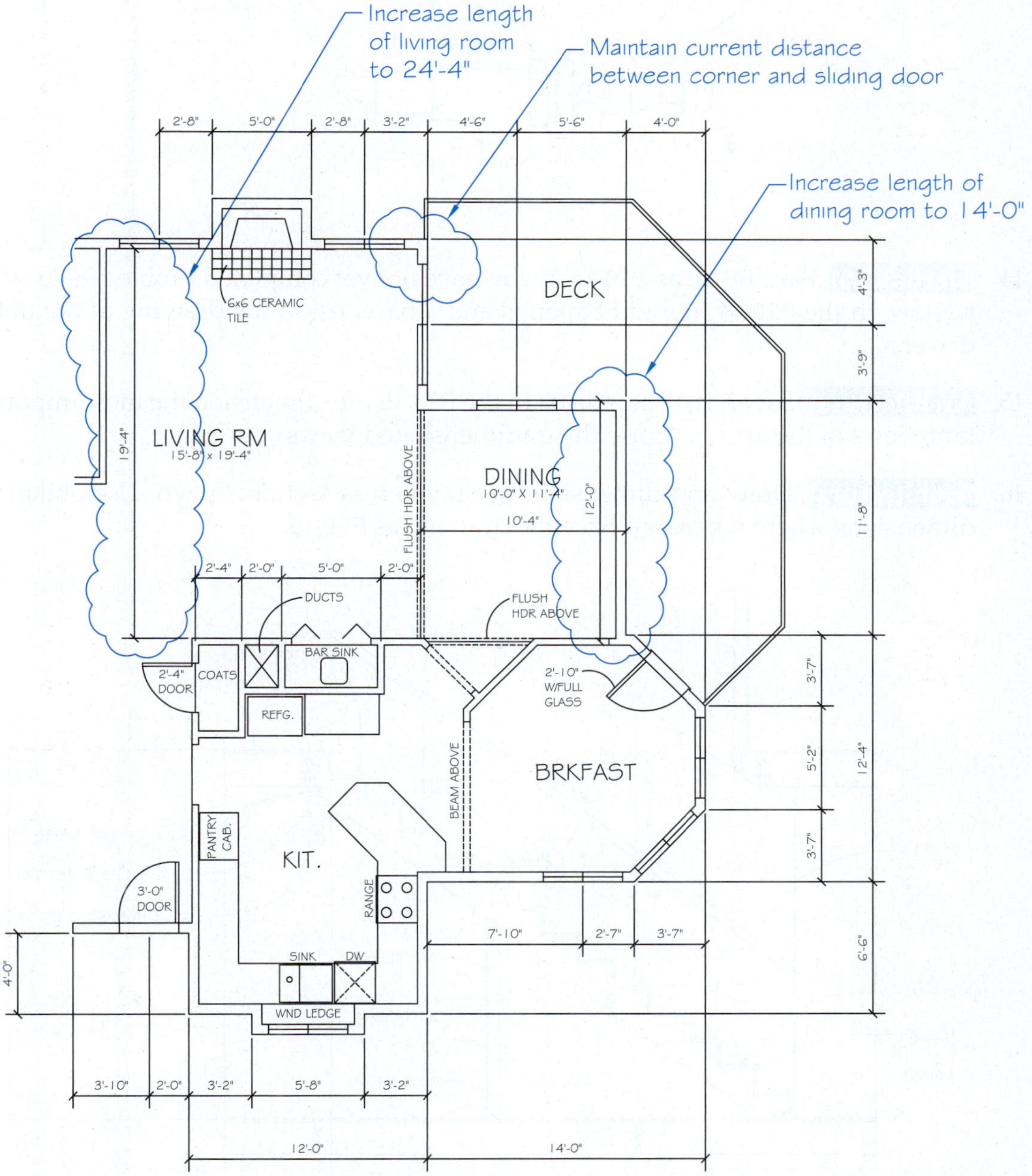

13. Mechanical Design and draw a vise clamp similar to the vise clamp shown. Add balloons and a parts list to the drawing. Save the drawing as P20-13.

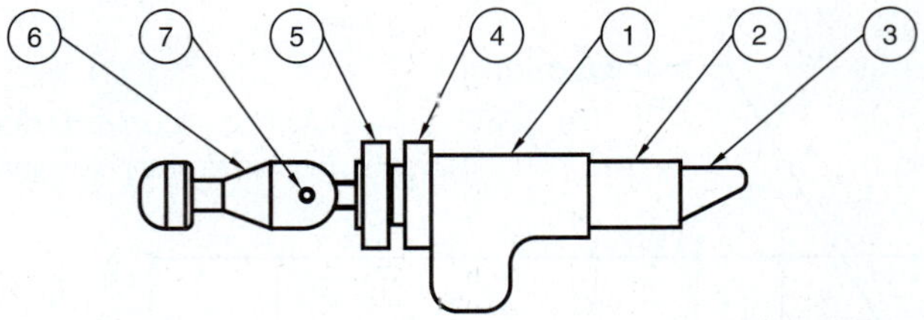

14. Mechanical Save P8-20 as P20-14. If you have not yet completed Problem 8-20, do so now. In the P20-14 file, add balloons and a parts list to the drawing of the nut driver.

15. Architectural Save P11-17 as P20-15. In the P20-15 file, dimension the most important views of the anchor. Erase the undimensioned views.

16. Architectural Draw and dimension the stairs cross section shown. Use oblique dimensions where necessary. Save the drawing as P20-16.

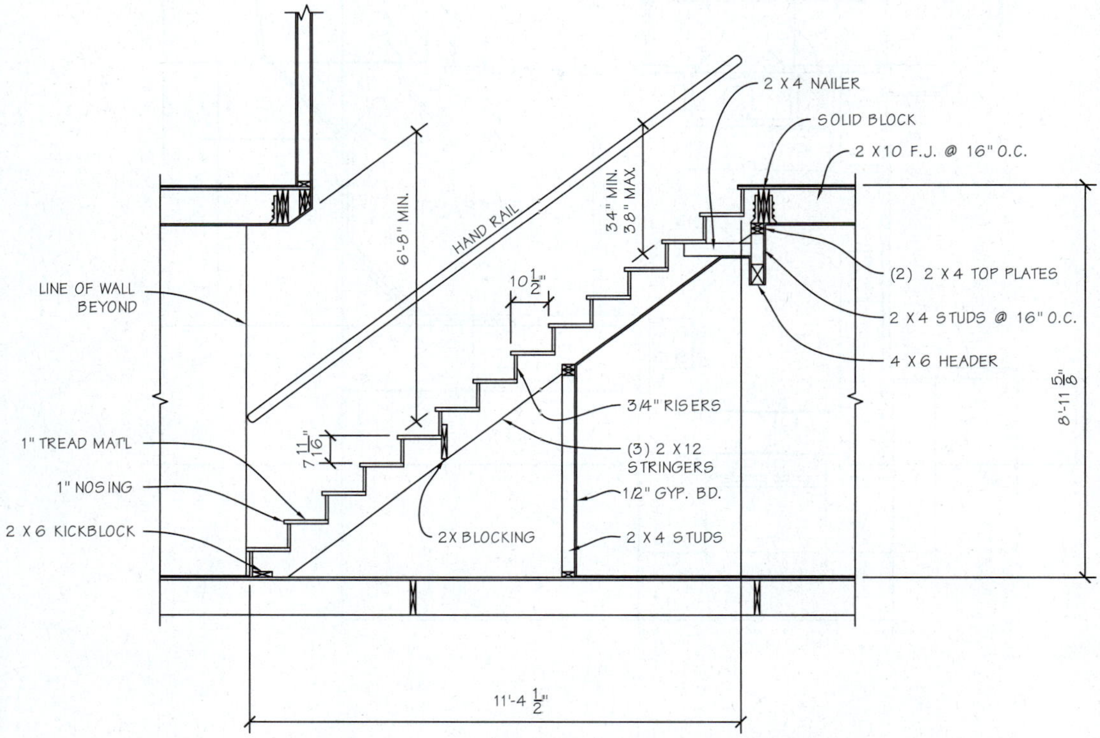

17. General Use a word processor to write a report of at least 250 words explaining the importance of associative dimensioning. Cite at least three examples from actual industry applications. Show at least four drawings illustrating your report.

AutoCAD Certified Professional Exam Practice

Answer the following questions using the information in this chapter.

1. Which of the following commands can you use to add a prefix to an existing linear dimension value? *Select all that apply.*
   A. **TEXTEDIT**
   B. **DIMEDIT New** option
   C. **DIMLINEAR Mtext** option
   D. **DIMTEDIT**
   E. **QDIM Edit** option
2. Which of the following commands allow you to space dimensions equally? *Select all that apply.*
   A. **DIMBASELINE**
   B. **DIMBREAK**
   C. **DIM**
   D. **DIMSPACE**
   E. **MATCHPROP**
3. Which of the following commands allow you to adjust the location and alignment of dimension lines? *Select all that apply.*
   A. **TEXTEDIT**
   B. **DIMSPACE**
   C. **DIMBREAK**
   D. **DIM**
   E. **STRETCH**

Follow the instructions in each problem. Complete each problem by answering the related question.

4. **Navigate to this chapter on the companion website and open CPE-20align.dwg.** Use the appropriate command to align leaders 1 and 3 horizontally with leader 2, as shown. Use ortho mode to ensure that the balloon alignment is exactly horizontal. What are the coordinates of the balloon grip on leader 1?

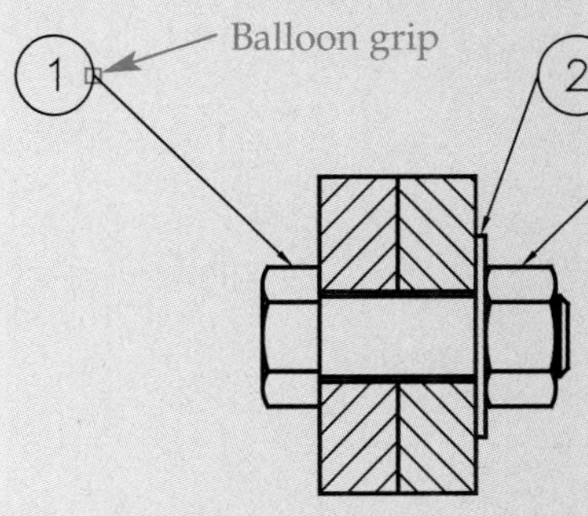

AutoCAD Certified Professional Exam Practice - Chapter 20

5. **Navigate to this chapter on the companion website and open CPE-20distribute.dwg.** Use the appropriate command to distribute the four leaders equally between Point A and Point B. What are the coordinates of the balloon grip of leader 4?

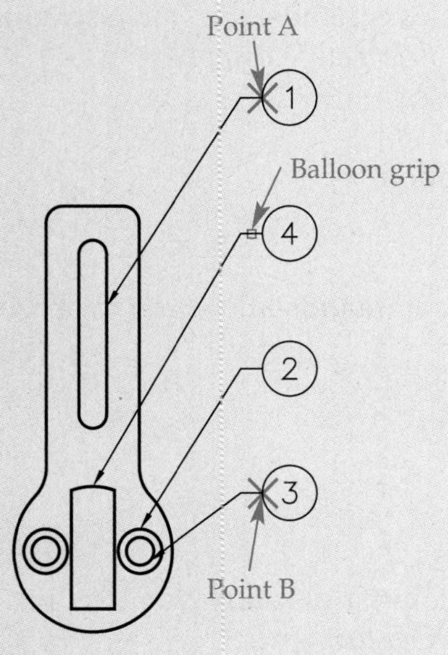

AutoCAD Certified Professional Exam Practice - Chapter 20

CHAPTER 21

Tables

Learning Objectives

After completing this chapter, you will be able to:

- ✓ Create and modify table styles.
- ✓ Insert tables into a drawing.
- ✓ Edit tables.
- ✓ Create formulas in table cells to perform calculations.

This chapter describes how to create ***tables***, which are common elements on technical drawings. Examples of table applications include bills of materials, parts lists, schedules, legends, tabular and chart dimensions, revision history and status blocks, and other ***associated lists***. Figure 21-1 shows an example of a schedule and a parts list and highlights the features of a table.

table: An arrangement of rows and columns that organize data to make it easier to read.

associated list: The ASME term describing tables added or related to engineering drawings.

Table Styles

A ***table style*** presets many table characteristics. Create a table style for each different table appearance or function. For example, to draw a parts list, use a table style preset to the standard format of a parts list. To draw a wire list, use a different table style preset to the standard format of a wire list. Add table styles to drawing templates for repeated use. Avoid adjusting table formats independently of the table style assigned to the table.

table style: A saved collection of table settings, including direction, text appearance, and margin spacing.

Table Style Dialog Box

Create, modify, and delete table styles using the **Table Style** dialog box. See Figure 21-2. The **Styles:** list box displays existing table styles. The Standard table style is the default. To make a table style current, double-click the style name, right-click the name and select **Set current**, or pick the name and select the **Set Current** button. Below the **Styles:** list box is a drop-down list that you can use to filter the number of table styles displayed in the **Table Style** dialog box. Pick the **All styles** option to show all table styles in the file, or pick the **Styles in use** option to show only the current style and styles used in the drawing.

TABLESTYLE

Ribbon
Home
> Annotation
Table Style
Annotate
> Tables
Table Style
Type
TABLESTYLE
TS

Figure 21-1.
A—An example of a window schedule added to an architectural floor plan to identify windows and related information. This table uses a down direction. B—An example of a parts list added to a mechanical assembly drawing to identify assembly components. This table uses an up direction.

Title row; Header row; Column; Data rows; Cell

WINDOW SCHEDULE				
MARK	SIZE	TYPE	ROUGH OPEN	QTY
A	12 X 60	JOB BUILT	VERIFY	2
B	96 X 60	W4N5 CSMT	8'-0 3/4" X 5'-0 7/8"	1
C	48 X 60	W2N5 CSMT	4'-0 3/4" X 5'-0 7/8"	2
D	48 X 36	W2N3 CSMT	4'-0 3/4" X 3'-6 1/2"	2
E	42 X 42	2N3 CSMT	3'-6 1/2" X 3'-6 1/2"	2
F	72 X 48	G64 SLD	6'-0 1/2" X 4'-0 1/2"	1

A

Column; Cell; Data rows; Header row; Title row

5	4		224707-33	3/4 X 1/4-20UNC HEX NUT
4	4		224707-20	3/4 FLAT WASHER
3	4		224707-10	3/4 X 1/4-20UNC HEX BOLT
2	1		087930-02	HOUSING COVER
1	1		087930-01	HOUSING
FIND NO	QTY REQD	DIA	PART OR IDENT NO	NOMENCLATURE OR DESCRIPTION
PARTS LIST				

B

Figure 21-2.
The **Table Style** dialog box.

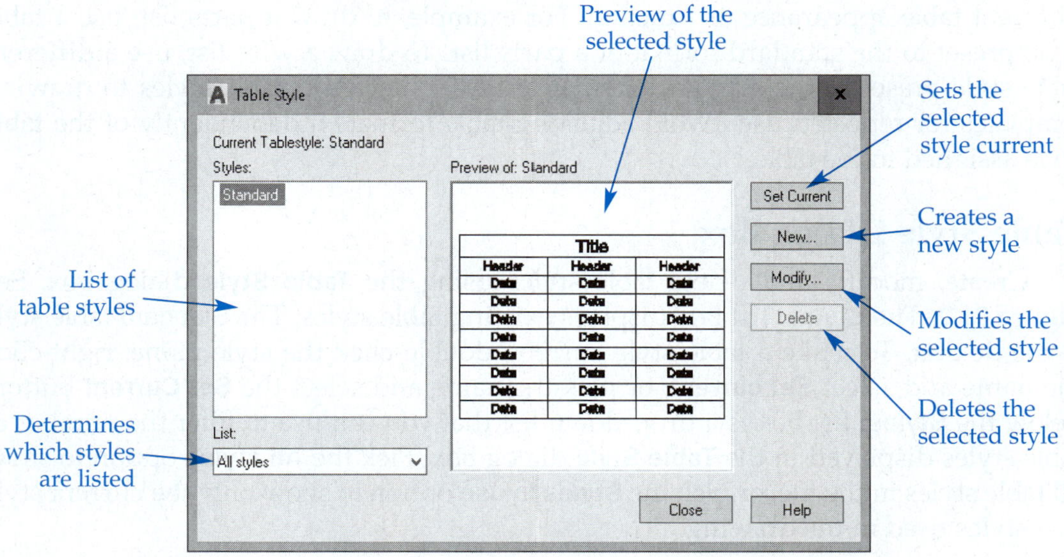

NOTE

You can also open the **Table Style** dialog box from the **Insert Table** dialog box, described later in this chapter, by picking the **Launch the Table Style dialog** button.

Creating New Table Styles

To create a new table style, select an existing table style from the **Styles:** list box to use as a base for formatting the new table style. Then pick the **New...** button to open the **Create New Table Style** dialog box. See Figure 21-3. You can base the new table style on the formatting of a different table style by selecting from the **Start With:** drop-down list. Notice that Copy of followed by the name of the existing style appears in the **New Style Name:** text box. Replace the default name with a more descriptive name, such as Parts List, Parts List No Heading, or Door Schedule.

Table style names can have up to 255 characters, including uppercase and lowercase letters, numbers, dashes (–), underlines (_), and dollar signs ($). After typing the table style name, pick the **Continue** button to open the **New Table Style** dialog box and adjust table style settings. See Figure 21-4. Pick the **OK** button to apply changes and close the **New Table Style** dialog box. Pick the **Close** button to exit the **Table Style** dialog box.

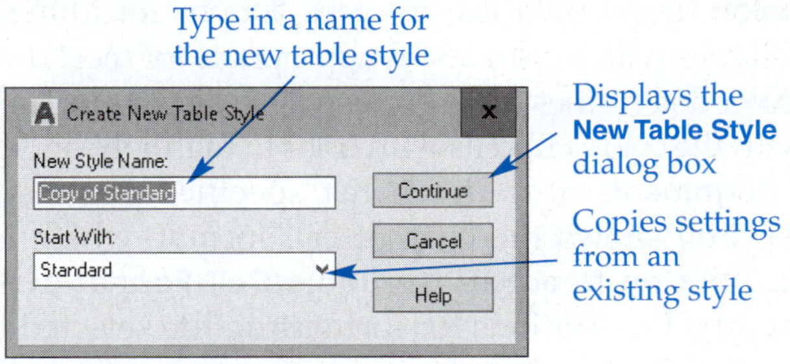

Figure 21-3.
In the **Create New Table Style** dialog box, specify the name of the new table style and the existing style to copy as a basis for the new style.

Figure 21-4.
Use the **New Table Style** dialog box to specify the formatting properties of a new table style. This figure shows adjusting the Data cell style.

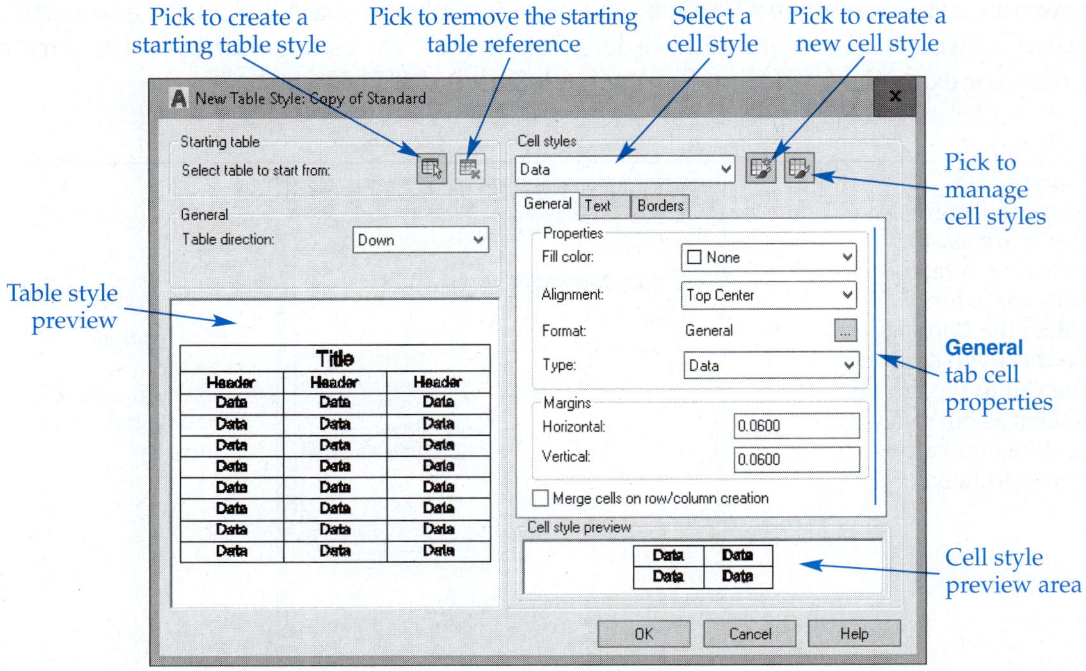

PROFESSIONAL TIP

Record the names and details about the table styles you create and keep this information in a log for future reference.

Table Direction

The **Table direction:** setting in the **General** area of the **New Table Style** dialog box determines the placement of title and header rows and the order of data rows. Select **Down** from the drop-down list to place data rows below the title and header rows. Select **Up** to place data rows above the title and header rows. See Figure 21-1.

Cell Styles

cell styles: Styles that allow you to assign specific formatting to data, header, and title row cells.

Use the **Cell styles** area of the **New Table Style** dialog box to control ***cell styles***. Pick the **Data**, **Header**, or **Title** option from the **Cell styles** drop-down list to display the properties corresponding to the selected style. Figure 21-4 shows the Data cell style selected. Set cell formatting properties using the **General**, **Text**, and **Borders** tabs. The options in these tabs are the same for adjusting data, header, and title cell styles.

General Tab Settings

The **General** tab, shown in Figure 21-4, allows you to set general table characteristics. The **Fill color:** drop-down list provides options for filling cells. The default **None** setting does not fill cells with a color and is appropriate for most drafting applications. The drawing window color determines the on-screen table display. Pick a color from the drop-down list to fill cells with the color. Fill cells with color to highlight or organize table information.

The **Alignment:** drop-down list specifies text justification within the cell. The **Format:** setting shows the current cell format, which is General by default. Pick the ellipsis (**...**) button to access the **Table Cell Format** dialog box. See Figure 21-5. The **Data type:** area lists options for formatting the selected table cell. Pick the appropriate format, such as Text or Currency, to access options for adjusting the format characteristics. Different options are available depending on the selected format.

Use the **Type:** drop-down list of the **New Table Style** dialog box to specify the cell data type. Pick the **Data** option to define a data cell type. Choose the **Label** option if the cell is a label type, such as a column heading or the table title. The **Margins** area provides a **Horizontal:** and **Vertical:** text box for controlling the horizontal and vertical space between cell content and borders. The default varies depending on the current units. For example, for decimal units the default is .06″ (1.5 mm).

Figure 21-5.
Many different data types are available to format a table cell. For example, select the Currency data type to enable the cell to recognize values as currency and format values appropriately.

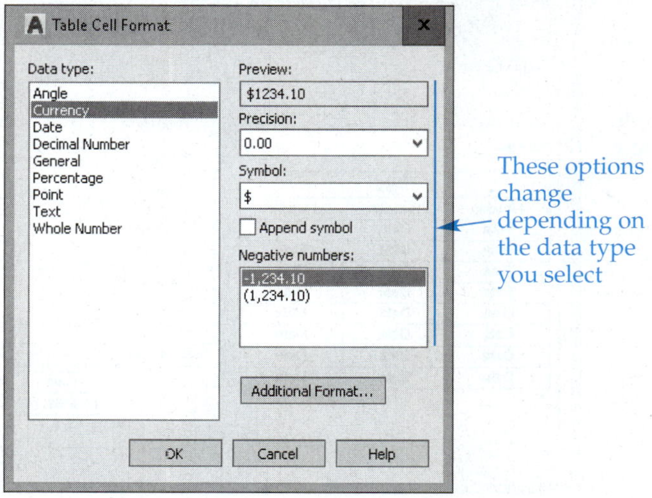

Pick the **Merge cells on row/column creation** check box to merge the row of cells together to form a single cell. This box is checked by default for the Title cell style. The title cell provides an example of when it is suitable to merge cells. The title applies to the entire table, or to each column.

Text Tab Settings

The **Text** tab, shown in Figure 21-6, allows you to set text characteristics for the selected cell style. The **Text style:** drop-down list displays the text styles found in the current drawing. Select a style or pick the ellipsis (**...**) button to the right of the drop-down list to open the **Text Style** dialog box to create or modify a text style.

Use the **Text height:** text box to specify the text height. The default varies depending on the current units and cell style, such as .18″ (4.5 mm) for data and header and .25″ (6.3 mm) for title when the drawing specifies decimal units. The **Text height:** text box is inactive if you assign a text height other than 0 to the text style. Use the **Text color:** drop-down list to set the text color. The **Text angle:** text box controls the rotation angle of text within the table cell. Figure 21-7 shows an example of a 90° text angle applied to the Header cell style.

Borders Tab Settings

The **Borders** tab, shown in Figure 21-8, allows you to control the border display and characteristics for the selected cell style. Use the **Lineweight:** drop-down list to assign a unique lineweight to cell borders. Use the **Linetype:** drop-down list to assign a unique linetype to cell borders. As when creating layers, you must load linetypes before you can apply them to borders. Use the **Color:** drop-down list to set the cell border color.

Pick the **Double line** check box to add another line around the default single-line border style. The **Spacing:** text box is available when you check **Double line**, allowing you to enter the distance between the double lines. The default spacing varies depending on the current units, such as .045″ (1.14 mm) when the drawing uses decimal units.

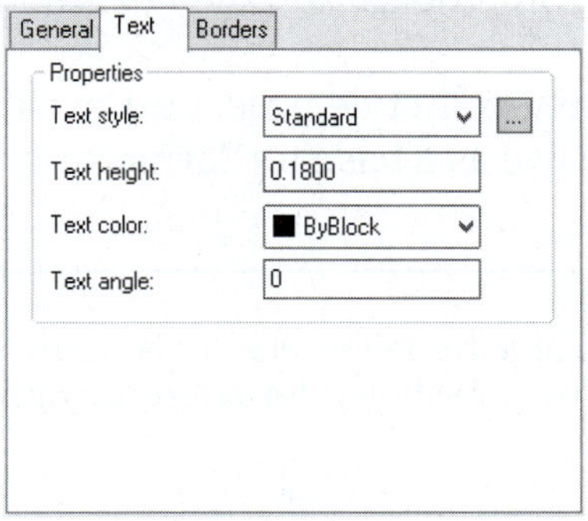

Figure 21-6.
The **Text** tab in the **New Table Style** dialog box allows you to set text properties.

Figure 21-7.
An example of a room schedule added to an architectural floor plan to list the characteristics of floor areas. In this table, a 90° text angle has been applied to the header cell style.

ROOM SCHEDULE					
NUMBER	NAME	LENGTH	WIDTH	HEIGHT	AREA
1	BEDROOM 1	11'-0"	10'-0"	9'-0"	110 SQ. FT.
2	BEDROOM 2	10'-0"	11'-0"	9'-0"	110 SQ. FT.
3	MASTER BEDROOM	12'-0"	14'-0"	9'-0"	168 SQ. FT.
4	LIVING ROOM	12'-0"	16'-0"	9'-0"	192 SQ. FT.
5	DINING ROOM	11'-0"	12'-0"	9'-0"	132 SQ. FT.
6	KITCHEN	11'-0"	10'-0"	9'-0"	110 SQ. FT.

Figure 21-8.
The **Borders** tab in the **New Table Style** dialog box allows you to set cell border properties.

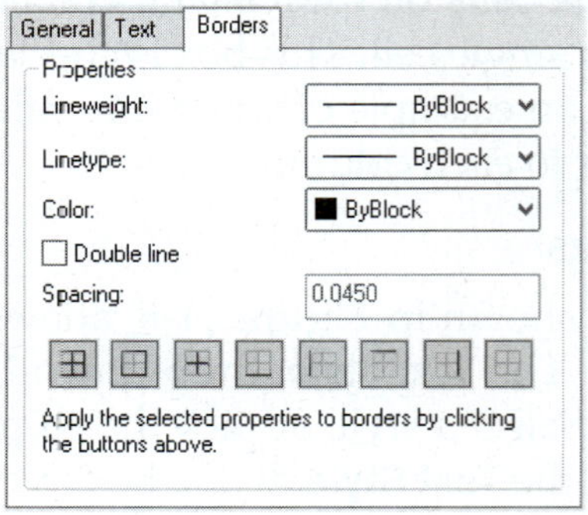

The **Border** buttons control how the **Lineweight:**, **Linetype:**, **Color:**, and **Double line** properties apply to cell borders. From left to right, the options are **All Borders**, **Outside Borders**, **Inside Borders**, **Bottom Border**, **Left Border**, **Top Border**, **Right Border**, and **No Borders**. After you set border properties, select or deselect the border buttons as needed. See Figure 21-9.

NOTE

AutoCAD treats a table like a block. Chapter 24 explains blocks and provides complete information on using ByBlock, ByLayer, or absolute color, lineweight, and linetype with blocks. For now, as long as you assign a specific layer to a table and do not change the table properties to absolute values, the default ByBlock cell properties are acceptable.

Creating Cell Styles

The default Data, Header, and Title cell styles are adequate for typical table applications. However, you can develop additional cell styles to increase the flexibility and options for creating tables. For example, you could create a cell style called Data Yellow that is the same as the Data cell style but fills cells with a yellow color. Then when you draw a table, you can choose the Data or the Data Yellow cell style, depending on the application.

To create a new cell style, select an existing cell style from the **Cell styles** area drop-down list to use as a base for formatting the new cell style. Then pick the **Create**

Figure 21-9.
Border options available for table cells. This figure shows applying border options to data cell borders. The thin lines are shown for reference only.

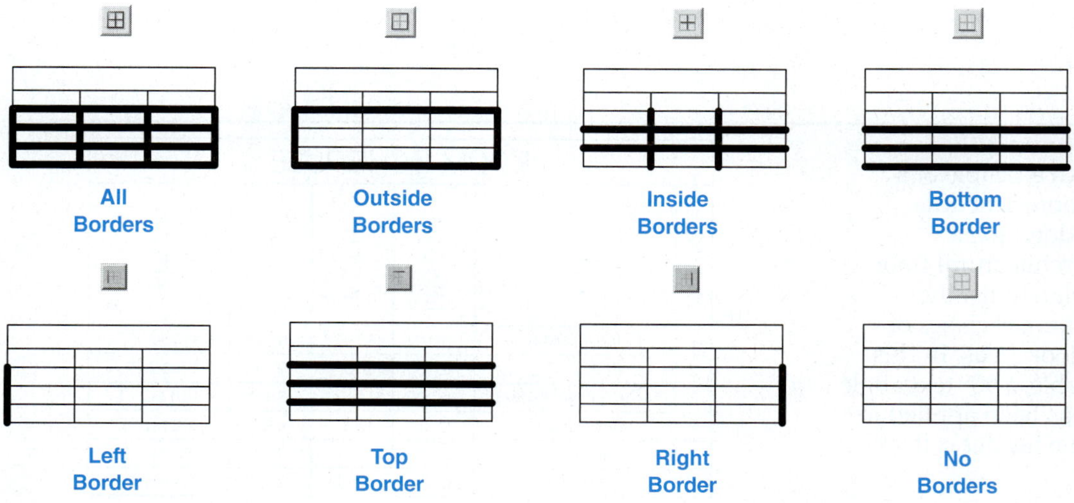

a new cell style button from the **Cell styles** area, or select **Create new cell style...** from the **Cell styles** area drop-down list to display the **Create New Cell Style** dialog box. Type a name for the new cell style in the **New Style Name:** text box. You can base the new cell style on the formatting of a different cell style by selecting from the **Start With:** drop-down list.

Use the **Manage Cell Styles** dialog box, shown in Figure 21-10, to create, rename, and delete cell styles. To access this dialog box, pick the **Manage Cell Styles dialog** button from the **Cell styles** area, or select **Manage cell styles...** from the **Cell styles** area drop-down list.

NOTE

The preview areas of the **New Table Style** dialog box allow you to see how the selected table style characteristics appear in a table. This provides a convenient way to observe changes made to a table style, without creating a table.

Exercise 21-1

www.g-wlearning.com/CAD/

Complete the exercise on the companion website.

Starting Table Styles

One technique for creating a table is to use a starting table style to base a new table on an existing table. You can consider a starting table style to be a table template that includes preset table properties and specific rows, columns, and data entries. Using a starting table style is much like copying a complete table and editing the table as needed. A starting table style can save time if you often prepare similar tables. For example, use a starting table style of a standard parts list to create a parts list quickly. Another example is using a starting table style of a finished door schedule to add a similar door schedule to a plan that contains most of the same doors.

A starting table style references the characteristics of an existing table, including the number of columns and rows and the table direction. Other table style characteristics, such as text style, are set according to the selected base table style. As a result, it is usually most appropriate to create a new starting table style using a base table style that is the same as that used to draw the reference table. For example, if you create a door schedule using a table style named Door Schedule, you should base the new starting table style on the Door Schedule table style.

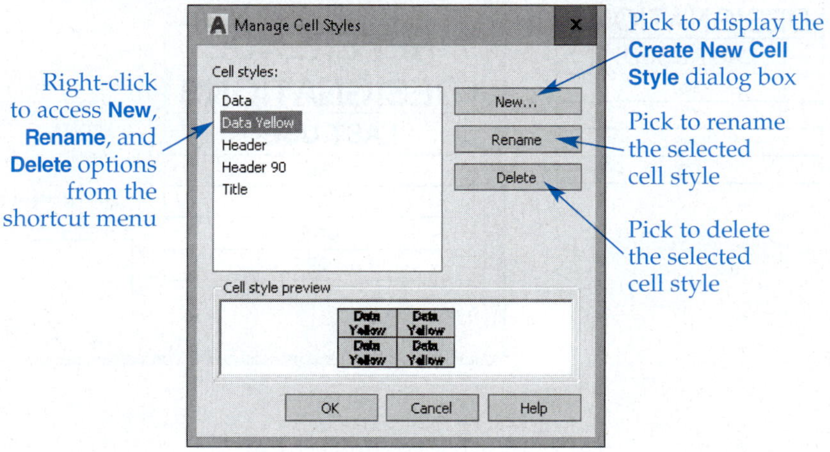

Figure 21-10.
Use the **Manage Cell Styles** dialog box to create new cell styles and to rename and delete existing cell styles.

To create a starting table style, pick the **Select table to start from:** button in the **Starting table** area. Then pick a border line of the existing table to reference. The preview displays the selected table and the table style settings of the base table style. See Figure 21-11. Modify the table direction and cell style options using the **General** and **Cell styles** areas. Pick the **Remove Table** button to remove the table reference from the table style. Pick the **Start from Table Style** insertion option, described later in this chapter, to add a table to a drawing using a starting table style.

Changing, Renaming, and Deleting Table Styles

To edit a table style, select a table style from the **Styles:** list box in the **Table Style** dialog box. Then pick the **Modify...** button to access the **Modify Table Style** dialog box, which is the same as the **New Table Style** dialog box. If you make changes to a table style, such as merging cells, all existing table objects assigned the modified table style are updated. Use a different table style with different characteristics when appropriate.

To rename a table style using the **Table Style** dialog box, slowly double-click on the name or right-click on the name and select **Rename**. To delete a table style using the **Table Style** dialog box, right-click on the name and choose **Delete**, or pick the style and select the **Delete** button. You cannot delete a table style that is assigned to table objects. To delete a style that is in use, assign a different style to the tables that reference the style. You cannot delete or rename the Standard style.

NOTE

RENAME

Type
RENAME

You can also rename styles using the **Rename** dialog box. Select **Table styles** in the **Named Objects** list to rename the style.

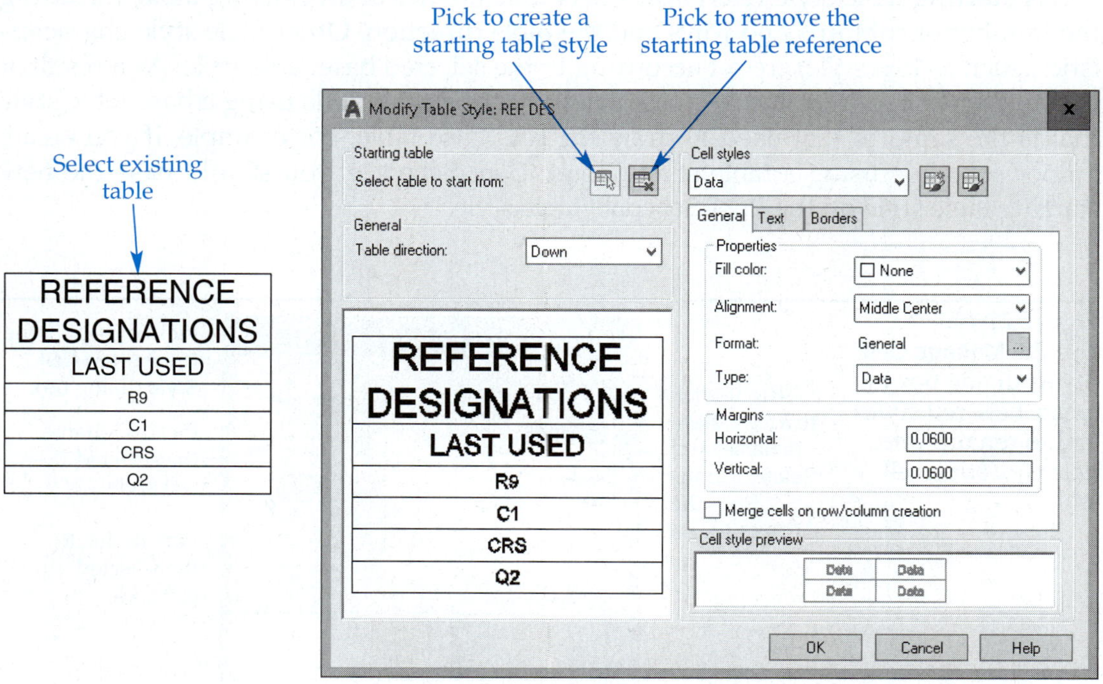

Figure 21-11.
Creating a starting table style that references an existing table. The table in this example is a list of reference designations identifying last-used components on an electrical schematic.

Setting a Table Style Current

Set a table style current using the **Table Style** dialog box by double-clicking the style in the **Styles:** list box, right-clicking on the style and selecting **Set current**, or picking the style and selecting the **Set Current** button. To set a table style current without opening the **Table Style** dialog box, use the **Table Style** flyout on the expanded **Annotation** panel of the **Home** ribbon tab, or the **Tables** panel of the **Annotate** ribbon tab. See Figure 21-12.

PROFESSIONAL TIP

You can import table styles from existing drawings using **DesignCenter**. See Chapter 5 for more information about using **DesignCenter** to reuse drawing content.

Inserting Tables

Use the **TABLE** command to insert an empty table with a specified number of rows and columns. After you insert the table, you can type text and insert content into the table cells. The **TABLE** command also provides other methods for inserting tables, such as beginning a table using a starting table style, forming a table from data in an existing Microsoft® Excel spreadsheet or CSV (comma-separated) file, and creating a table by referencing AutoCAD data. Access the **TABLE** command to display the **Insert Table** dialog box. See Figure 21-13.

TABLE

Ribbon
Home
> Annotation
Annotate
> Tables
Table

Type
TABLE
TB

Placing an Empty Table

To place an empty table, select a table style from the **Table style** drop-down list, or pick the **Launch the Table Style dialog** button to create or modify a table style. Next, pick the **Start from empty table** radio button in the **Insert options** area to create an empty table. The preview area shows a representation of a table using the current table style, but it does not adjust to column and row settings. Pick the **Specify insertion point** radio button in the **Insertion behavior** area to create a table by using the values in the **Column & row settings** area and then selecting a single point to place the table in the drawing.

Figure 21-12.
A fast way to set a table style current is to use one of the drop-down lists on the ribbon.

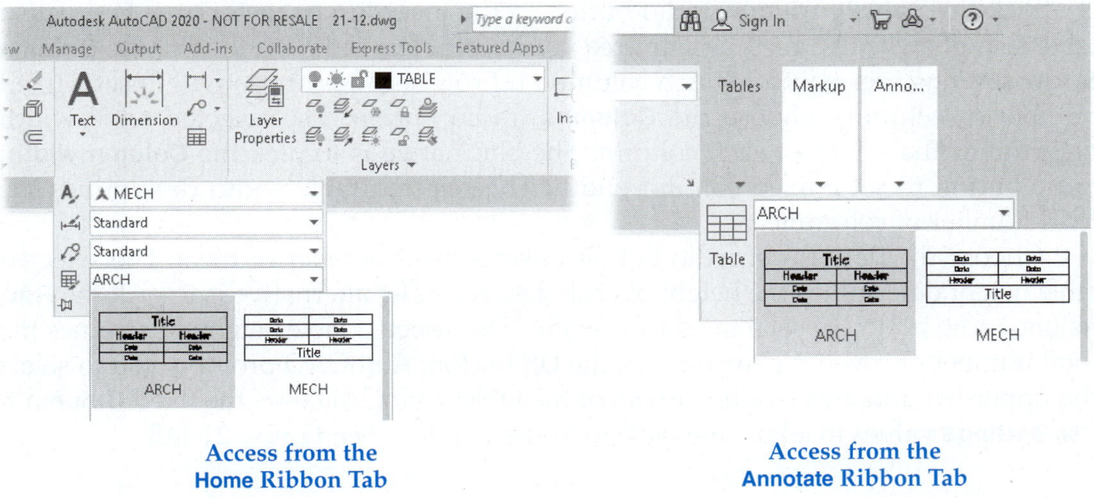

Figure 21-13.
The **Insert Table** dialog box, shown with the **Start from empty table** insert option selected.

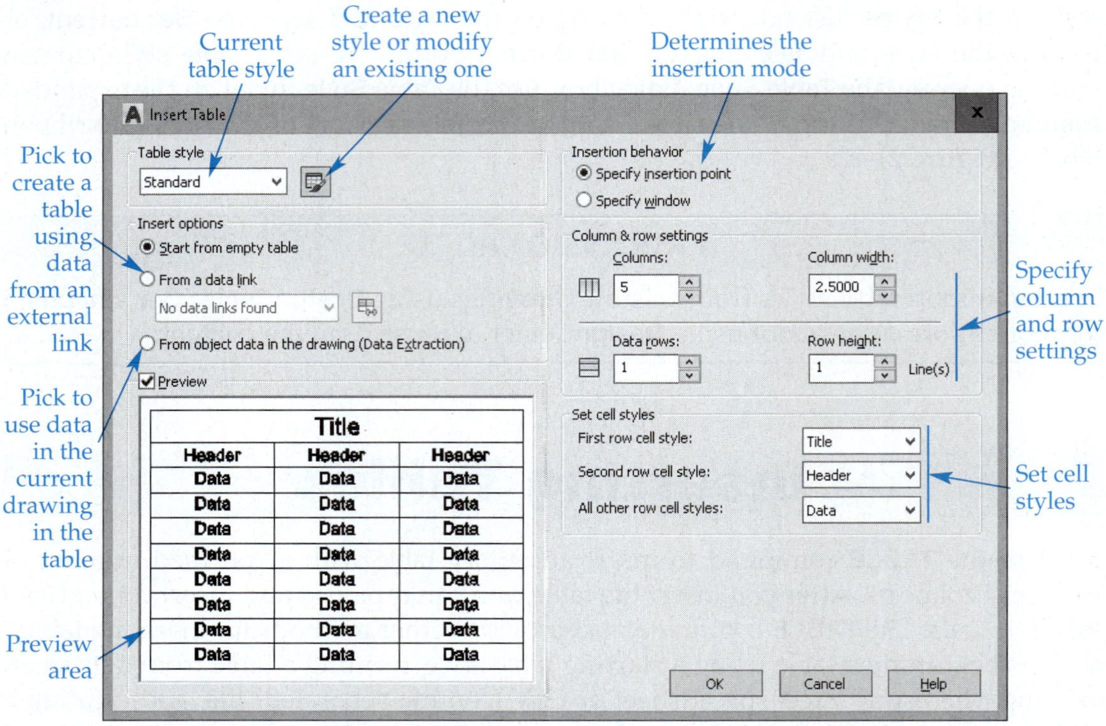

Use the **Columns:** text box to specify the total number of table columns. Choose a **Column width:** value to establish the initial width of each column. To help avoid initial crowding, enter a column width larger than necessary and then resize the columns later. Use the **Data rows:** text box to specify the total number of data rows. Choose a **Row height:** value to establish the initial height of each row based on the number of lines typed and the margin settings assigned to the table style. When you pick the **OK** button, AutoCAD prompts you to specify the table insertion point. See **Figure 21-14A**.

NOTE

When you use the **Specify insertion point** option to place a table, the cursor attaches to the table based on the table style direction.

Pick the **Specify window** radio button in the **Insertion behavior** area to create a table that fits within a rectangular area you create. The radio buttons in the **Column & row settings** area control which column and row settings are active. To set a fixed number of columns, choose the **Columns:** radio button. The selected table width determines the width of each column. The alternative is to pick the **Column width:** radio button to set a fixed column width. The selected table width determines the total number of columns.

Choose the **Data rows:** radio button to set a fixed number of rows. The selected table height determines the height of each data row. The alternative is to pick the **Row height:** radio button to set a fixed row height. The selected table height determines the total number of rows. When you pick the **OK** button, AutoCAD prompts you to select the upper-left and lower-right corners of the table. AutoCAD uses the fixed **Column & row settings** values to adjust the table to fit the window. See **Figure 21-14B**.

Figure 21-14.
Two ways to insert an empty table. A—Using the **Specify insertion point** radio button to select a single insertion point to create a table with three fixed columns and five fixed data rows. B—Using the **Specify window** radio button to specify an area using two pick points to create a table with three columns and five data rows that adjust according to the window size.

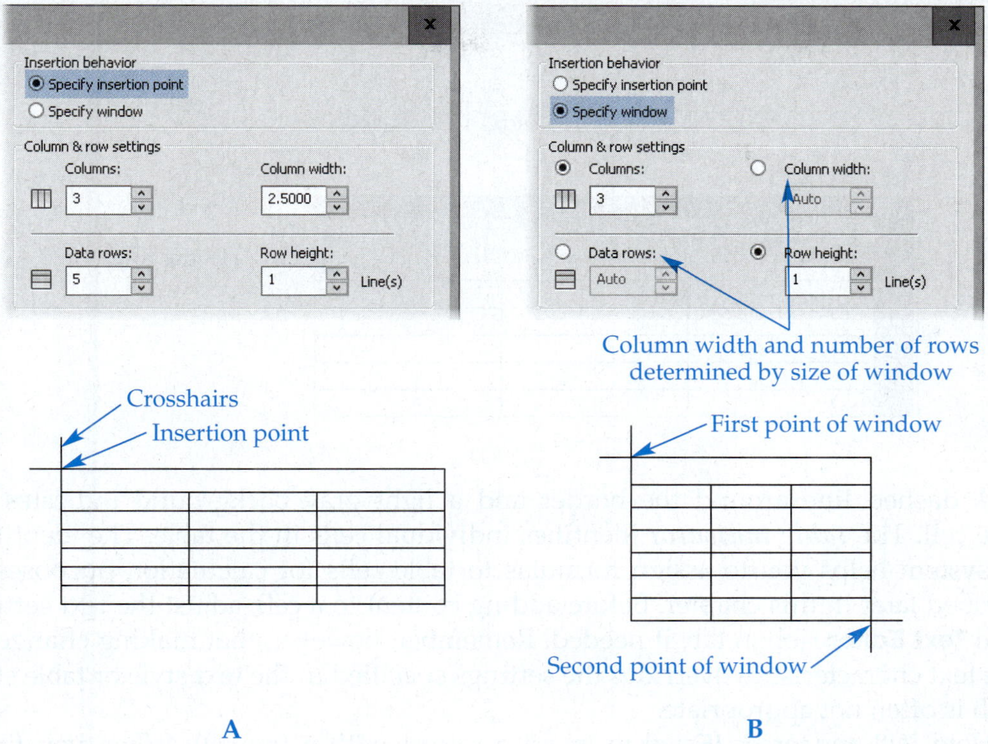

NOTE

You specify the total number of data rows when constructing a table. Depending on cell style settings, the table may also include a header row and title row. The current table style text height and cell margin settings determine the default value for row height. For example, enter a row height of 1 if you plan to have a single line of text in each cell.

PROFESSIONAL TIP

You can add, delete, and fully adjust rows and columns as needed. Therefore, it is not critical that you enter the exact number and size of columns and rows before inserting a table.

Exercise 21-2

www.g-wlearning.com/CAD/

Complete the exercise on the companion website.

Adding Cell Content

When you insert a table, the **Text Editor** contextual ribbon tab appears, with the text editor cursor in the title cell ready for typing. See Figure 21-15. Typing in a cell is like typing mtext. The options and settings available in the **Text Editor** ribbon tab and shortcut menu function the same in table cells as in editing mtext.

Figure 21-15.
Use the **Text Editor** ribbon tab to add and modify table cell text. A blinking cursor, dashed border, and light gray background indicate the active cell.

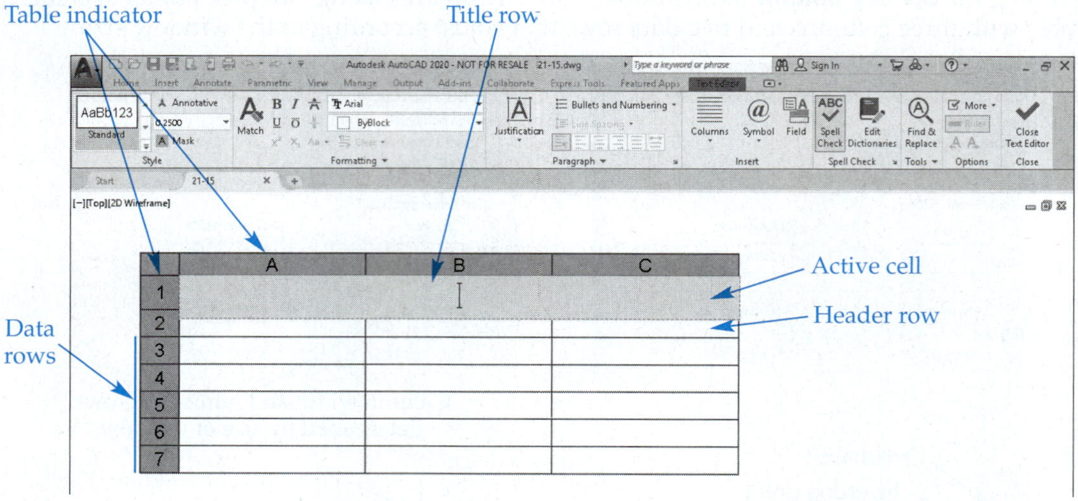

A dashed line around the border and a light gray background indicates the active cell. The ***table indicator*** identifies individual cells in the table. The identification system helps you to assign formulas to table cells for calculation purposes, as described later in this chapter. Before adding content to a cell, adjust the text settings in the **Text Editor** ribbon tab, if needed. Remember, however, that making changes to some text characteristics overrides the settings specified in the text style or table style, which is often not appropriate.

table indicator: The grid of letters and numbers that identify individual cells in a table.

Hold [Alt] and press [Enter] to insert a return within the cell. When you finish entering text in the active cell, press [Tab] to move to the next cell. Hold [Shift] and press [Tab] to move the cursor backward (to the left or up) and make the previous cell active. Press [Enter] to make the cell directly below the current cell active, or exit the text editor if the cursor is at the last cell. You can also use the arrow keys to navigate table cells.

When you finish typing, exit the text editor system using the **Close Text Editor** option, or pick outside of the table editor. You can also press [Esc] twice. An easy way to reopen the text editor to make changes to text in a cell is to double-click in the cell. Figure 21-16 shows a finished table.

CLOSE
Ribbon
Text Editor > Close
Close Text Editor

Exercise 21-3

www.g-wlearning.com/CAD/

Complete the exercise on the companion website.

Using a Starting Table Style

To place a table using a starting table style, select a starting table style from the **Table style** drop-down list or select the **Launch the Table Style dialog** button to create or modify a starting table style. The **Start from Table Style** radio button becomes activated in the **Insert options** area. See Figure 21-17. The preview area shows a preview of the parent table with the current table style settings and table options.

The **Specify insertion point** option is the only method for inserting a table using a starting table style. However, you can add columns and rows to the table using the **Additional columns:** and **Additional rows:** text boxes. You can also select the items from the parent table to include in the new table using the check boxes in the **Table options** area. For example, pick the **Data cell text** check box to create a new table that contains all the text added to the data cells of the parent table.

Figure 21-16.
A completed parts list table added to a mechanical assembly drawing to identify assembly components.

PARTS LIST				
FIND NO	QTY REQD	DIA	PART OR IDENT NO	NOMENCLATURE OR DESCRIPTION
1	1		100-TBL-001	TABLE
2	2		100-CLJA-45	CLAMP JAWS
3	2		202-PIV-32	PIVOT ARM
4	1		340-HAND-06	LOCKING HANDLE
5	3		38009561	1/4-28UNF NYLOCK NUT
6	2		567-ADJR-98	ADJUSTING ROD HANDLE
7	2		786-SPG-64	1/8 SPRING PIN

Figure 21-17.
You can use the **Insert Table** dialog box to create a new table using a starting table style.

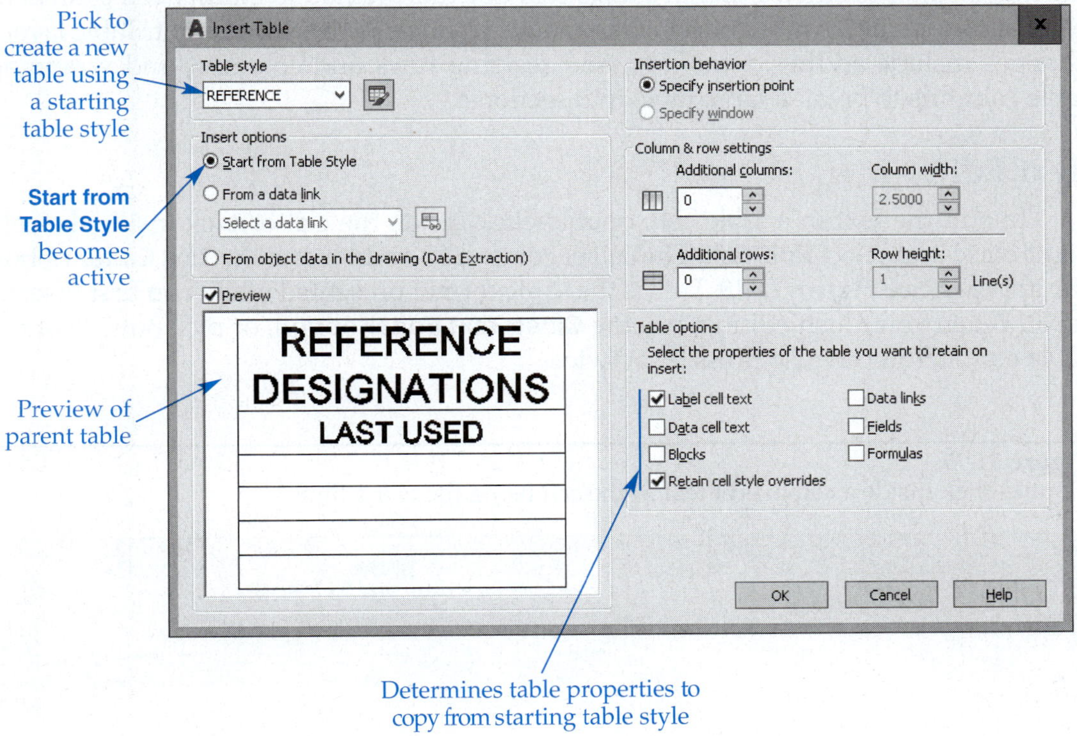

Pick the **OK** button and specify the insertion point of the table. The **Text Editor** ribbon tab appears with the text editor cursor in the title cell, ready for adding new content or editing existing values. Exit the text editor when you are finished. Figure 21-18 shows a table created by referencing an existing starting table style, with two additional rows.

Exercise 21-4

www.g-wlearning.com/CAD/

Complete the exercise on the companion website.

Figure 21-18.
Use a starting table style to create a new or modified table quickly. This example shows creating a new table of electronic reference designations using the starting table style created in Figure 21-11.

Existing table style used to form a starting table style

REFERENCE DESIGNATIONS
LAST USED
R9
C1
CRS
Q2

In the new table, all table options are retained and two rows are added

REFERENCE DESIGNATIONS
LAST USED
R9
C1
CRS
Q2
T4
R6

Editing Tables

AutoCAD provides several options to edit existing tables. One option is to re-enter the mtext editor to edit the text in a table cell. Use this method to modify cell content or change text format. Another option is to make changes to the table layout. Table layout changes include adding, removing, and resizing rows and columns, and wrapping table columns to break a large table into sections.

Text Editor

To edit the text in a table cell, double-click inside the cell or pick inside the cell, right-click, and select **Edit Text**. The cell becomes a text editor and the **Text Editor** ribbon tab appears. See **Figure 21-19**. This is the same format presented when you first insert a table. When you finish editing, use the **Close Text Editor** option, or pick outside of the table editor. You can also press [Esc] twice.

Figure 21-19.
Double-click inside a cell to edit text in the cell using the text editor.

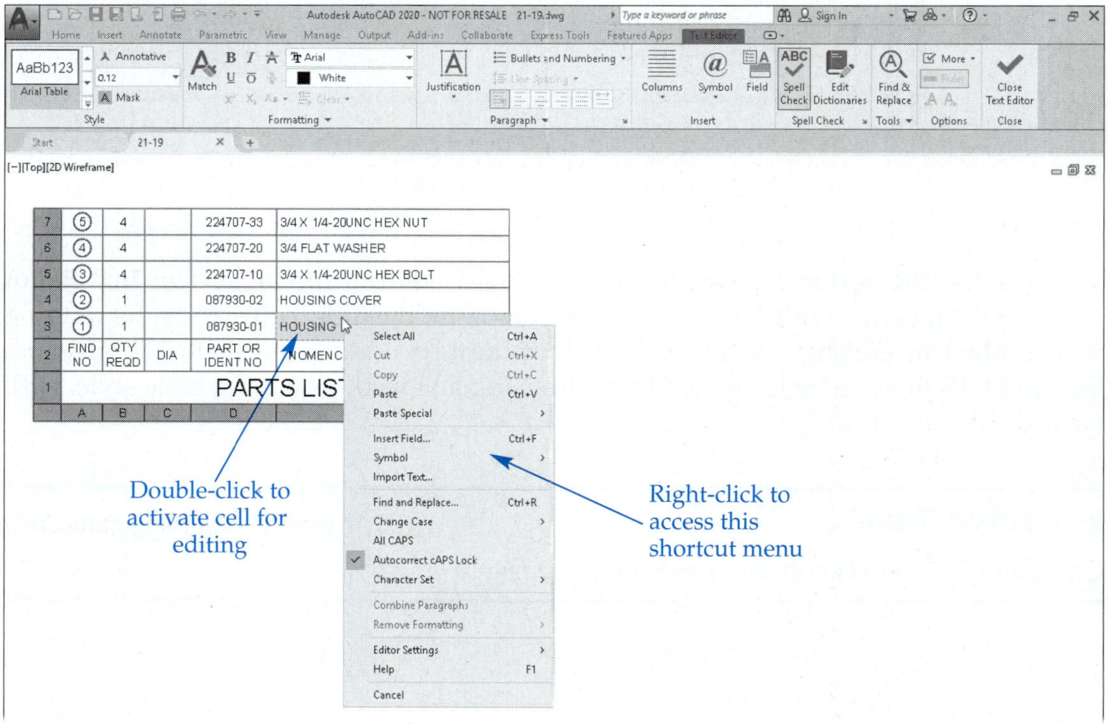

Exercise 21-5

www.g-wlearning.com/CAD/

Complete the exercise on the companion website.

Table Cell Editor

You can access several table layout settings by picking (single-clicking) inside a cell to make the cell active and display the **Table Cell** contextual ribbon tab. The highlighted cell includes grips. The **Table Cell** ribbon tab contains options for adjusting table and individual cell layout. You can access many of the same options in the **Table Cell** ribbon tab and Windows Clipboard functions from the shortcut menu that appears when you right-click away from the ribbon. See Figure 21-20. Use an option shown in Figure 21-21 to select multiple cells and apply changes to all the cells at once.

NOTE

Make sure you pick completely inside of the cell. If you accidentally select one of the cell borders, the entire table becomes the selected object. Editing table layout by selecting a cell border is described later in this chapter.

Auto-Fill

An effective method for copying the content of one cell to multiple cells is to use the ***auto-fill*** function. Figure 21-22A shows the basic steps to use auto-fill. Pick inside the cell that contains the content to copy. Select the diamond-shaped auto-fill grip, and then right-click to choose an auto-fill option, if necessary. Finally, move the cursor to the last cell to fill and pick inside the cell.

auto-fill: A table function that fills selected cells based on the contents of another cell.

Figure 21-20.
Pick inside a cell to access several options for modifying the table layout.

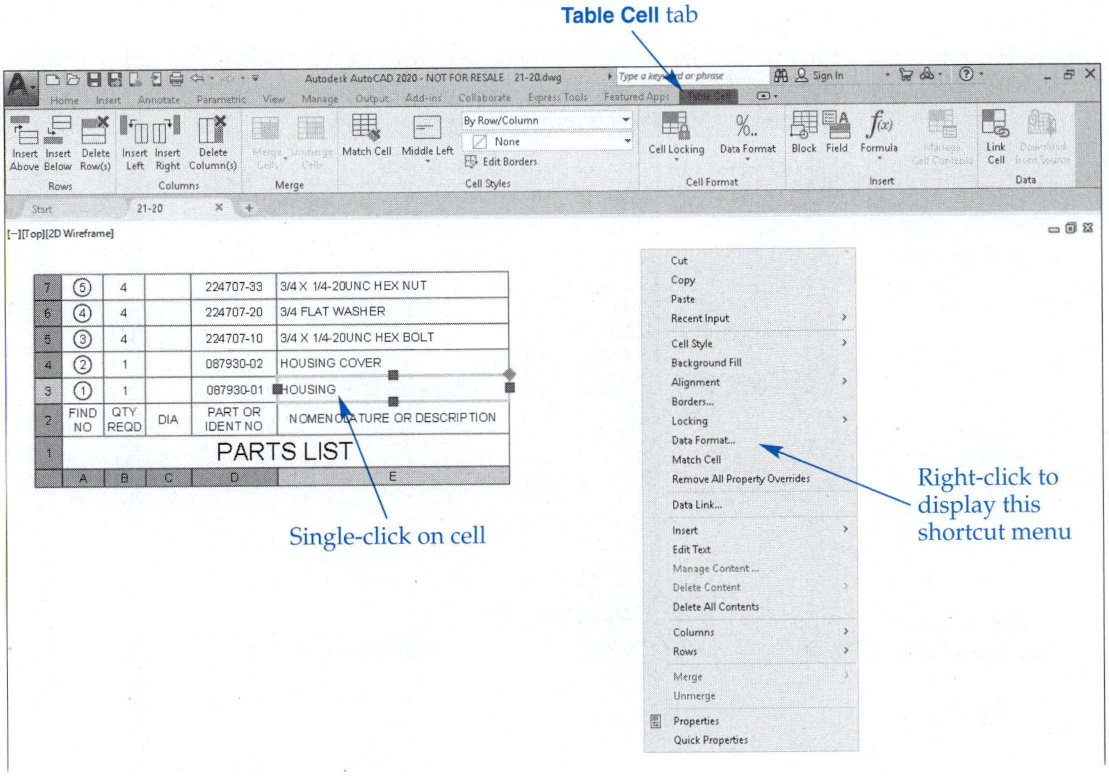

Figure 21-21.
Selecting multiple cells in a table to edit. A—Using the pick-and-drag method. B—Picking a range of cells using [Shift]. C—Selecting a row, column, or the entire table.

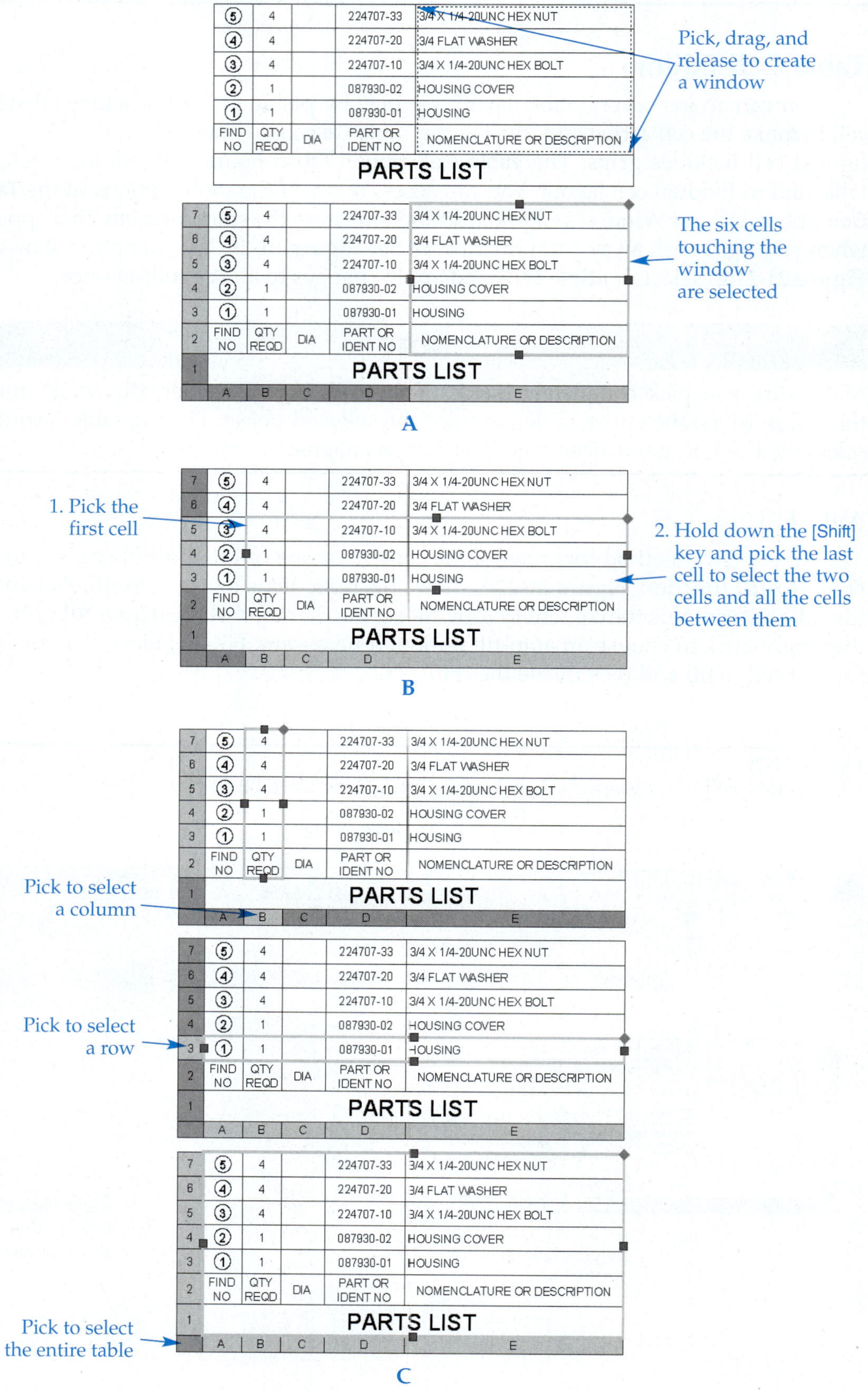

The **Fill Series** option fills cells with the content of the selected cell and applies format overrides. This option also automatically increases or decreases values of certain data types, such as dates and whole numbers, as the fill occurs. See Figure 21-22B. The **Fill Series Without Formatting** option fills cells with the content of the selected cell, but does not include format overrides.

Figure 21-22.
A—Using the auto-fill function to copy cell content to multiple cells. B—Using the **Fill Series** option to fill sequential whole number data. C—Using the **Copy Cells** option to copy data that would fill sequentially when using the **Fill Series** option. D—The completed door schedule uses auto-fill multiple times to fill cells quickly. A door schedule typically appears on an architectural floor plan to identify doors and related information.

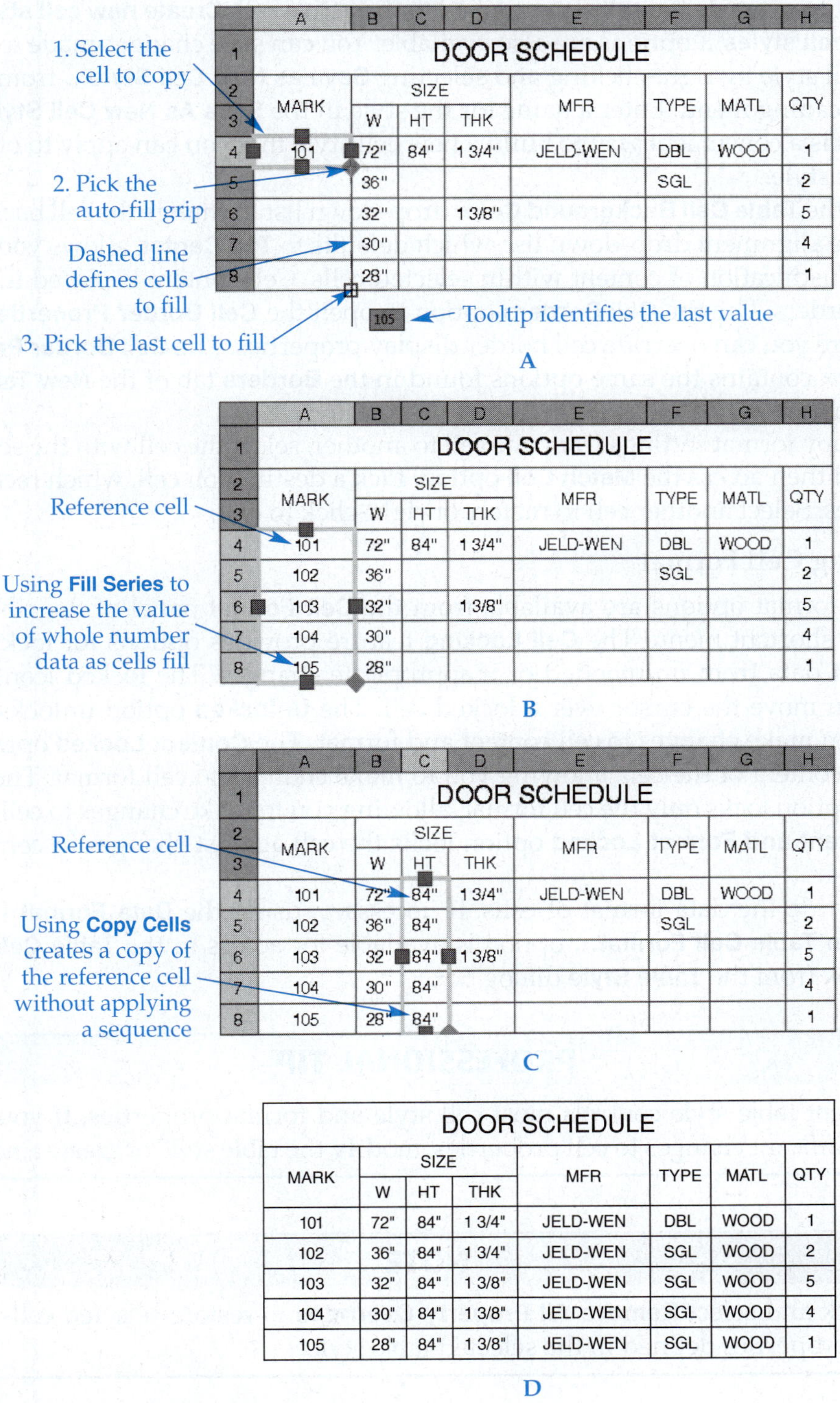

DOOR SCHEDULE							
MARK	SIZE			MFR	TYPE	MATL	QTY
	W	HT	THK				
101	72"	84"	1 3/4"	JELD-WEN	DBL	WOOD	1
102	36"	84"	1 3/4"	JELD-WEN	SGL	WOOD	2
103	32"	84"	1 3/8"	JELD-WEN	SGL	WOOD	5
104	30"	84"	1 3/8"	JELD-WEN	SGL	WOOD	4
105	28"	84"	1 3/8"	JELD-WEN	SGL	WOOD	1

The **Copy Cells** option copies the content of the selected cell and applies format overrides, but creates a static cell copy that does not adjust data values. See Figure 21-22C. The **Copy Cells Without Formatting** option copies the content of the selected cell, but does not include any format overrides. The **Fill Formatting Only** option fills the cells only with format overrides applied to the selected cell, allowing you to enter cell content manually. Figure 21-22D shows a table finished using multiple auto-fills.

Modifying Cell Style

Ribbon
Table Cell
> Cell Styles
> Table Cell Styles

The table style and the text style assigned to the table style determine the appearance of most cell properties. However, you can override the cell style assigned to specific cells if necessary. Cell style options are available from the **Cell Styles** panel of the **Table Cell** ribbon tab and from the shortcut menu. Use the **Table Cell Styles** drop-down list to override the cell style applied to the active cell. **Create new cell style...** and **Manage cell styles...** options are also available. You can save changes made to a cell as a new cell style by right-clicking and selecting **Save as New Cell Style...** from the **Cell Style** cascading menu. Enter a name for the style in the **Save As New Cell Style** dialog box. This is a convenient way to build a new cell style that you can apply to other cells or a table style.

Ribbon
Table Cell
> Cell Styles
> Table Cell Background Color

Ribbon
Table Cell
> Cell Styles
> Alignment

Ribbon
Table Cell
> Cell Styles
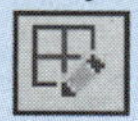
Cell Borders

Ribbon
Table Cell
> Cell Styles
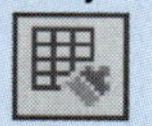
Match Cell

Use the **Table Cell Background Color** drop-down list to change the cell background color. The alignment drop-down list, which defaults to **Top Center**, allows you to override the justification of content within selected cells. Cell content is placed in relation to cell borders. Use the **Cell Borders** option to open the **Cell Border Properties** dialog box, where you can override cell border display properties. The **Cell Border Properties** dialog box contains the same options found in the **Borders** tab of the **New Table Style** dialog box.

To copy format settings from one cell to another, select the cell with the settings to copy, and then access the **Match Cell** option. Pick a destination cell, which receives the properties. Select another cell to match, or right-click to exit.

Adjusting Cell Format

Ribbon
Table Cell
> Cell Format
> Cell Locking

Ribbon
Table Cell
> Cell Format
> Data Format

Cell format options are available from the **Cell Format** panel of the ribbon and from the shortcut menu. The **Cell Locking** feature provides options for locking cells to protect data from unintended or inappropriate changes. The locked icon appears when you move the cursor over a locked cell. The **Unlocked** option unlocks the cell so you can make changes to cell content and format. The **Content Locked** option locks only the content of the cell, allowing you to make changes to cell format. The **Format Locked** option locks only the cell format, allowing you to make changes to cell content. The **Content and Format Locked** option locks the cell against changes in content and format.

Override the data format of cells, if necessary, using the **Data Format** function. A **Custom Table Cell Format...** option is available for access to the **Table Cell Format** dialog box from the **Table Style** dialog box.

PROFESSIONAL TIP

The current table style controls most cell style and format properties. If you plan to make significant changes to cell properties, modify the table style or create a new style.

NOTE

Right-click and select **Remove All Property Overrides** to restore selected cells to their original properties defined in the selected table style.

Inserting Fields and Blocks

In addition to text, table cells can contain fields, formulas, and blocks. To insert these items, use the options available from the **Insert** panel of the **Table Cell** ribbon tab, or from the **Insert** cascading menu of the shortcut menu. You will learn about formulas later in this chapter. Choose the **Field...** option to insert a field into a table cell using the same **Field** dialog box available for creating mtext and text objects. You can also insert fields into a cell using the text editor after double-clicking inside the cell to activate it.

Ribbon
Table Cell
> Insert

Insert Field

Blocks are AutoCAD symbols that are useful in tables for applications such as creating a legend, displaying a view or flag note symbol in a parts list, and adding tags to a schedule. This chapter describes options for inserting a block in a table. You will learn about blocks later in this textbook. Select the **Block...** option to insert a block into a table cell using the **Insert a Block in a Table Cell** dialog box. Figure 21-23 describes the options available in this dialog box. Double-click on a block to open the **Edit a Block in a Table Cell** dialog box to make changes. A cell can contain both text and blocks.

Ribbon
Table Cell
> Insert

Insert Block

Adding and Resizing Columns and Rows

You can add, delete, and resize columns and rows as needed. Select a single cell or a group of cells to add, depending on the requirement. To delete and resize columns and rows, you do not need to select entire columns and rows. The following options are available from the **Columns** panel of the **Table Cell** ribbon tab or the **Columns** cascading menu of the shortcut menu:

- **Insert Left.** Add a new column to the left of the selection.
- **Insert Right.** Add a new column to the right of the selection.
- **Delete.** Eliminate entire columns.
- **Size Columns Equally.** Size multiple columns to the same width.

The following options are available from the **Rows** panel of the **Table Cell** ribbon tab or the **Rows** cascading menu of the shortcut menu:

- **Insert Above.** Add a new row above the selection.
- **Insert Below.** Add a new row below the selection.
- **Delete.** Delete entire rows.
- **Size Rows Equally.** Size multiple rows to the height of the tallest row.

Figure 21-23.
Options in the **Insert a Block in a Table Cell** dialog box.

Feature	Description
Name	Used to choose the block from a drop-down list of the blocks stored in the current drawing.
Browse...	Displays the **Select Drawing File** dialog box, where a drawing file can be selected and inserted into the table cell as a block.
AutoFit	Scales the block automatically to fit inside the cell.
Scale	Sets the block insertion scale. For example, a value of 2 inserts the block at twice its original size. A value of .5 inserts the block at half its created size. The **Scale** option is not available if the **AutoFit** check box is checked.
Rotation angle	Rotates the block to the specified angle.
Overall cell alignment	Determines the justification of the block in the cell and overrides the current cell alignment setting.

NOTE

To insert a new row at the bottom of a table that uses a **Down** direction, position the cursor in the lower-right cell and press [Tab]. To insert a new row at the top of a table that uses an **Up** direction, position the cursor in the upper-right cell and press [Tab].

PROFESSIONAL TIP

You can also adjust column and row size using grips. Grip boxes appear in the middle of cell border lines. To resize a column or row, select a grip, move the crosshairs, and pick.

Merging Cells

Ribbon
Table Cell
> Merge

Merge Cells

Merging allows you to combine adjacent cells. The default Title cell style is an example of merged cells. Merge options are available from the **Merge** panel of the ribbon and from the **Merge** cascading menu in the shortcut menu. Select the cells to merge and select the appropriate **Merge Cells** option. Select the **All** option to merge all cells into one cell. The **By Row** and **By Column** options allow you to merge cells in multiple rows or columns without removing the horizontal or vertical borders. Use the **Unmerge Cells** option to separate merged cells back to individual cells.

NOTE

The **Delete All Contents** option deletes the contents in the selected cell. You can accomplish the same task by picking a cell and pressing [Delete].

Exercise 21-6

www.g-wlearning.com/CAD/

Complete the exercise on the companion website.

Picking a Cell Edge to Edit Table Layout

Pick the edge, or border, of a cell to access additional methods for adjusting table layout. The display includes the table indicator grid, grips you can use to adjust row height and column width, and the table break function. Once you pick a cell border, right-click to display the shortcut menu shown in Figure 21-24.

Adjusting Table Style

Select a table style from the **Table Style** cascading menu to apply a different table style to the selected table. Pick the **Set As Table in Current Table Style** option to create a starting table style based on the selected table and the current table style. This is a convenient technique for creating a starting table style without opening the **Table Style** dialog box. If the selected table was drawn using a starting table style, selecting **Set As Table in Current Table Style** redefines the starting table. To save modifications made to the table as a new table style, pick the **Save as New Table Style...** option and enter a name for the style in the **Save As New Table Style** dialog box.

Resizing Columns and Rows

Use the grip boxes or arrowheads at the corners of columns and rows to adjust column and row size. To resize a column or row, select a grip box, move the crosshairs, and pick. Use the arrowhead grips to increase or decrease row height and/or column width uniformly.

Figure 21-24.
Pick a cell border to access several additional options for modifying table layout.

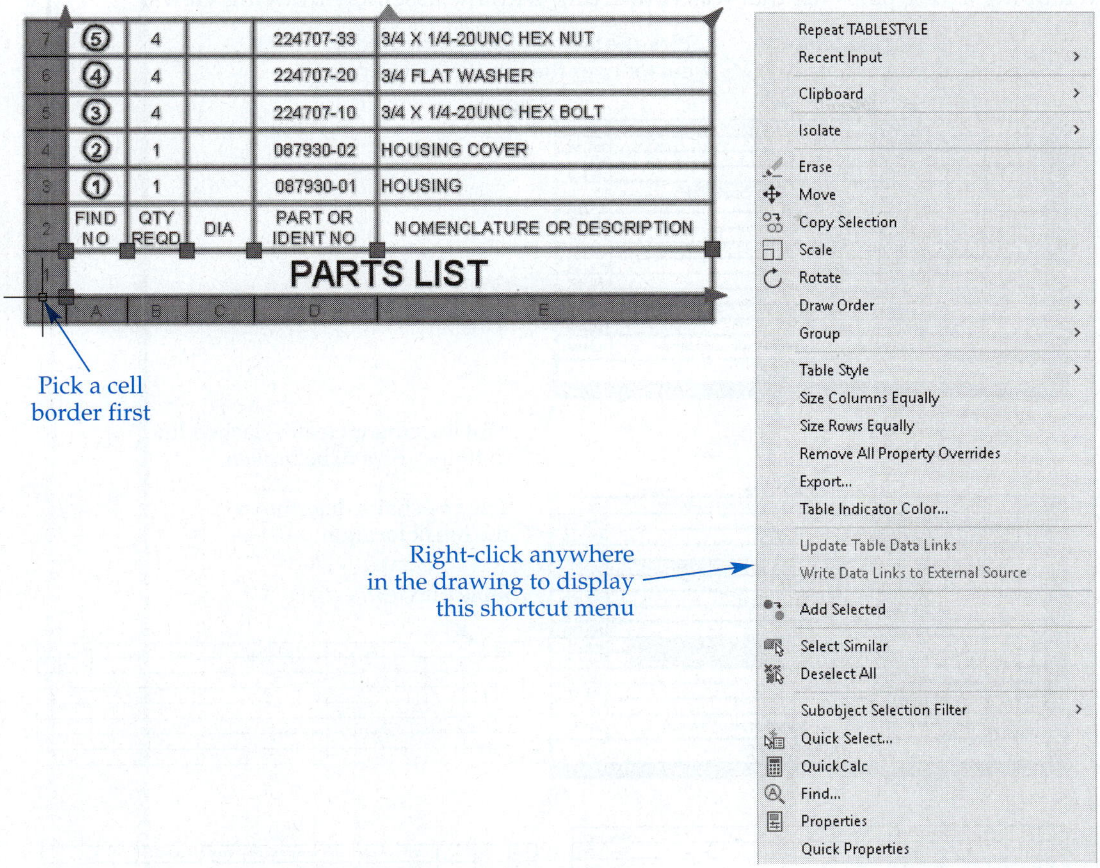

The **Size Columns Equally** option sizes all columns to the same width. AutoCAD divides the total width of the table evenly among the columns. The **Size Rows Equally** option sizes all rows to match the height of the tallest row in the table.

NOTE

The default grip box stretches the column or row without changing the size of the table. Hold [Ctrl] while stretching to increase or decrease the size of the table with the column or row resize.

Table Breaks

Use the table break function to break a table into separate sections while maintaining a single table object. Breaking a table is common when it is necessary to fit a long table in a specific area or on a certain size sheet. The table breaking grip is located midway between the sides of the table at the top or the bottom of the table, depending on the table direction. See **Figure 21-25A**.

To break a table, select the table breaking grip as shown in **Figure 21-25A** and move the crosshairs into the table to display a preview of the table sections and a vector line as shown in **Figure 21-25B**. The crosshairs determines the location of the break. The closer to the table title and headers you move the crosshairs, the more sections you create, as shown in the table preview. When the preview of the table looks correct, pick the location to form the table breaks. See **Figure 21-25C**.

Figure 21-25.
The procedure for breaking, or wrapping, a table into sections. This example shows wrapping a long parts list that was conflicting with the assembly drawing views.

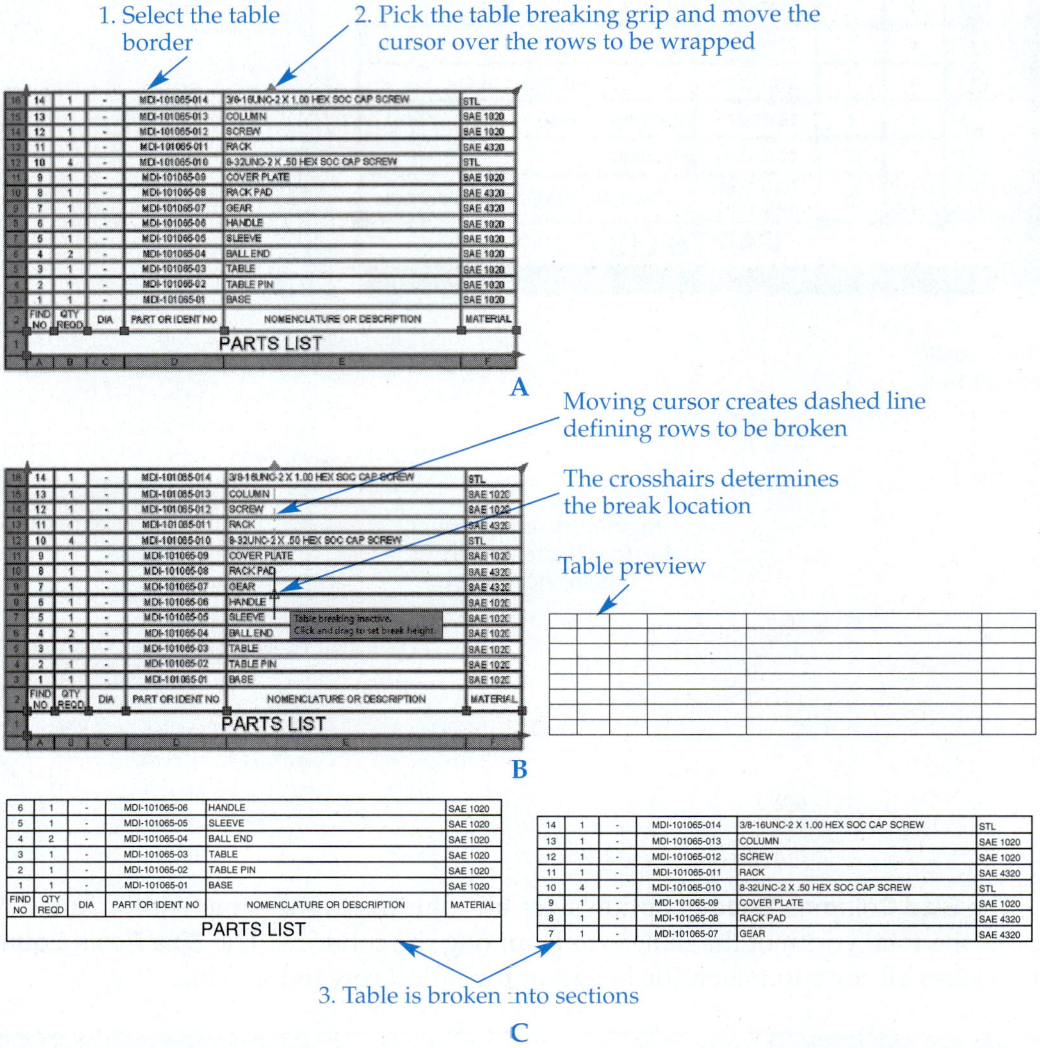

NOTE

After you add table breaks, several options become available from the **Properties** palette for adjusting the table sections, as explained later in this chapter.

Additional Table Layout Options

The following additional table options are available from the shortcut menu:

- **Remove All Property Overrides.** Restores the original properties to the table, defined according to the selected table style.
- **Export....** Exports the table as a CSV file.
- **Table Indicator Color....** Allows you to change the color of the table indicator shown when you pick inside a cell.

Exercise 21-7

www.g-wlearning.com/CAD/

Complete the exercise on the companion website.

Table Properties

The **Properties** palette displays certain table properties depending on whether you select inside a cell or pick a cell edge. See Figure 21-26. The **Cell** and **Content** categories appear when you select inside a cell and allow you to adjust the properties of the selected cell. The **Table** and **Table Breaks** categories appear when you select a cell edge and include common table settings and table break properties. The **Quick Properties** palette also lists certain table-specific properties.

PROFESSIONAL TIP

If the height of table rows is taller than desired, or if rows become unequal in height, enter a very small value in the **Table height** row of the **Table** category to return all rows to the smallest height possible based on the margin spacing between cell content and cell borders.

Several useful options are available for adjusting table breaks in the **Table Breaks** category of the **Properties** palette. The **Enabled** option toggles between the broken and unbroken display. The **Yes** value appears when you create table breaks and enable breaking. Pick **No** to return the table to an unbroken display. The **Direction** option defines direction of broken table flow, or wrap. The default **Right** option wraps the table to the right. Select **Left** to wrap the table to the left, or pick **Up** to wrap the table above.

The **Repeat top labels** option repeats cells that use a Label cell type at the beginning of each table section. Typically, the title cell and header cells use a Label cell type. Choose **Yes** to add the title and header cells to the wrapped table sections. The **Repeat bottom labels** option repeats cells that use a Label cell type at the end of each section. The **Manual positions** option allows you to move table sections independently while maintaining the table as a single object. When you select **No**, table sections move as a group.

Figure 21-26.
Table properties in the **Properties** palette. A—The properties displayed when you select inside a cell. B—The properties displayed when you pick a cell edge.

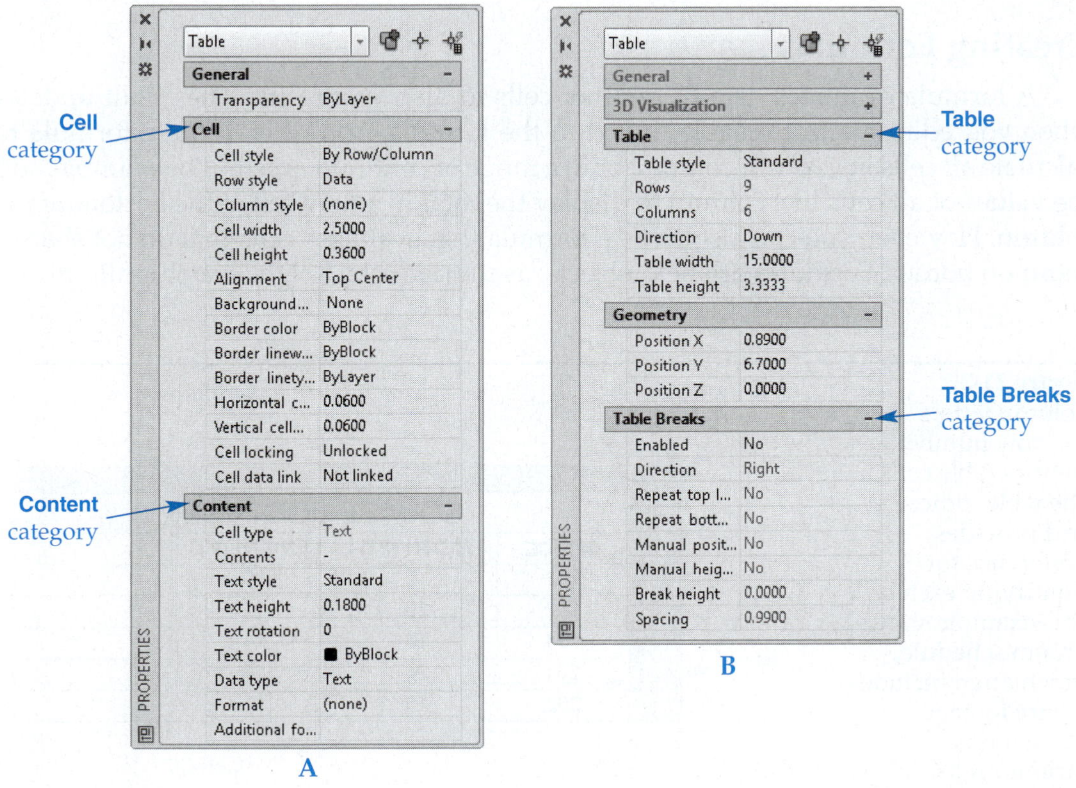

The **Manual heights** option adds a table-breaking grip to each section, which allows you to adjust the number of rows in each section independently and add additional breaks. When you select **No**, the table-breaking grip appears at the original section only and controls the number of breaks. The **Break height** text box allows you to define the height of each table section. The selected height determines the number of sections. The **Spacing** text box allows you to define the spacing between table sections. A value of 0 places the sections together.

PROFESSIONAL TIP

Use the **MATCHPROP** command, described in Chapter 13, to match the properties assigned to one table to other tables.

Exercise 21-8

www.g-wlearning.com/CAD/

Complete the exercise on the companion website.

Calculating Values in Tables

formulas: Mathematical expressions that allow you to perform calculations within table cells.

Performing calculations on table data using ***formulas*** is a common requirement. Examples include calculating and showing the total number of parts in a parts list, the total cost of items in a bill of materials, or the total glazing area in a window schedule. Formulas calculate operations based on numeric data in table cells. AutoCAD allows you to write formulas for sums, averages, counts, and other mathematical functions.

The table indicator grid that appears when you edit a table cell provides an identification system for cells. Letters identify columns, and numbers identify rows. Use the combination of column letter and row number to describe cells in formulas. For example, C6 identifies the cell located in Column C, Row 6. See Figure 21-27.

Creating Formulas

A formula evaluates data from other cells to display a result. The result updates when you edit data in the table linked to the formula. You often write a formula to calculate all cells in a row or column, or in a range of continuous cells. For example, add the values of all cells in a column to display the total in a new cell at the bottom of the column. However, you can also write a formula that evaluates cells that do not share a common border. Formulas are field objects, as indicated by a light gray highlight.

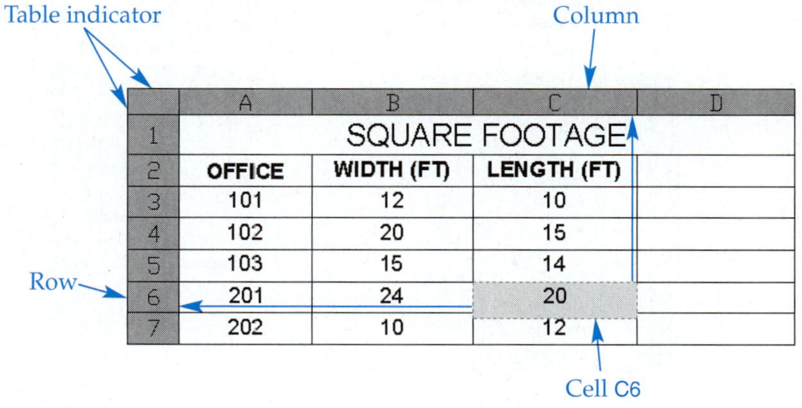

Figure 21-27. Column letters and row numbers identify table cells. The table indicator grid provides a reference for identifying each cell. This example shows a room schedule, which often includes square footage calculations for various areas.

You must enter the proper syntax in the table cell to create an accurate formula. An example of a complete expression using the standard syntax is =(C3+D4). The equal sign (=) tells AutoCAD to perform a calculation. The open parenthesis marks the beginning of the expression. Type C3+D4 to tell AutoCAD to add the value of cell C3 to the value of cell D4. The closing parenthesis marks the end of the expression.

When identifying a cell in an expression, you must enter the letter before the number. For example, you cannot enter 3C to designate the cell C3. Common operators used for mathematical functions include + for addition, – for subtraction, * for multiplication, / for division, and ^ for exponentiation. If you enter an incorrect expression or an expression evaluating cells without numeric data, AutoCAD displays the pound character (#) to indicate an error.

NOTE

Parentheses are unnecessary in some expressions, but other expressions cannot be calculated without them. It is good practice to use parentheses in all expressions.

Input a formula using the text editor. Figure 21-28 shows an example of typing the multiplication formula =(B3*C3) in cell D3. The result appears when you close the text editor of the cell, such as when you move to another cell.

You can use grouped expressions in formulas by enclosing expression sets in parentheses. For example, the expression =(E1+F1)*E2 multiplies the sum of E1 and F1 by E2. Another example: =(E1+F1)*(E2+F2)/G6 multiplies the sum of E1 and F1 by the sum of E2 and F2 and divides the product by G6.

Sum, Average, and Count Formulas

An alternative to entering formulas manually is to select a formula from AutoCAD to calculate the sum, average, or count of a range of cells. Pick inside a cell where the calculation is to occur, and then choose a formula using the **Formula** drop-down list, which is also available by right-clicking and selecting from the **Formula** cascading menu on the **Insert** cascading menu.

Ribbon
Table Cell
> Insert
> Formula

Expression

	A	B	C	D
1	SQUARE FOOTAGE			
2	OFFICE	WIDTH (FT)	LENGTH (FT)	
3	101	12	10	=(B3*C3)
4	102	20	15	
5	103	15	14	
6	201	24	20	
7	202	10	12	

A

Result

SQUARE FOOTAGE			
OFFICE	WIDTH (FT)	LENGTH (FT)	
101	12	10	120
102	20	15	
103	15	14	
201	24	20	
202	10	12	

B

Figure 21-28. Entering a multiplication formula to calculate the square footage of a room. A—Type the expression in the table cell using the correct syntax. B—The calculation occurs when you close the text editor of the cell.

Select the **Sum** option and then use window selection to add the values of selected cells. See Figure 21-29A. The range, or window, can include cells from several columns and rows. Be sure to select all of the cells to be included in the calculation. When you select the second point, the expression appears in the cell. Notice in Figure 21-29B that the resulting expression is =Sum(D3:D7). This formula specifies that the selected cell is equal to the sum of cells D3 through D7. The colon (:) indicates the range of cells for the calculation. Figure 21-29 shows an example of calculating the square footage for each office, and then calculating the total square footage.

The **Average** option creates a formula that calculates the average value of selected cells. The average is the sum of the selected cells divided by the number of cells selected. The **Count** option creates a formula that counts the number of selected cells. The count only includes cells that contain a value.

You can type sum, average, and count formulas directly into a cell without using the supplied formulas. If you calculate a value over a range of cells, use the colon symbol (:) to designate the range. You can also write an expression that evaluates individual cells instead of a range. The cells do not have to share a common border. To write an expression using nonadjacent cells, use a comma to separate the cell names. For example, to average cells D1, D3, and D6, type =Average(D1,D3,D6).

You can include a range of cells and individual cells in the same expression. For example, to count cells A1 through B10 in addition to cells C4 and C6, enter =Count(A1:B10,C4,C6). Figure 21-30 shows examples of sum, average, and count formulas.

NOTE

When using architectural units in a drawing, you can type the foot (′) and inch (″) symbols in table cells for use in values and formulas. When you use the foot symbol for a cell value, a formula in another cell automatically converts the resulting value to inches and feet.

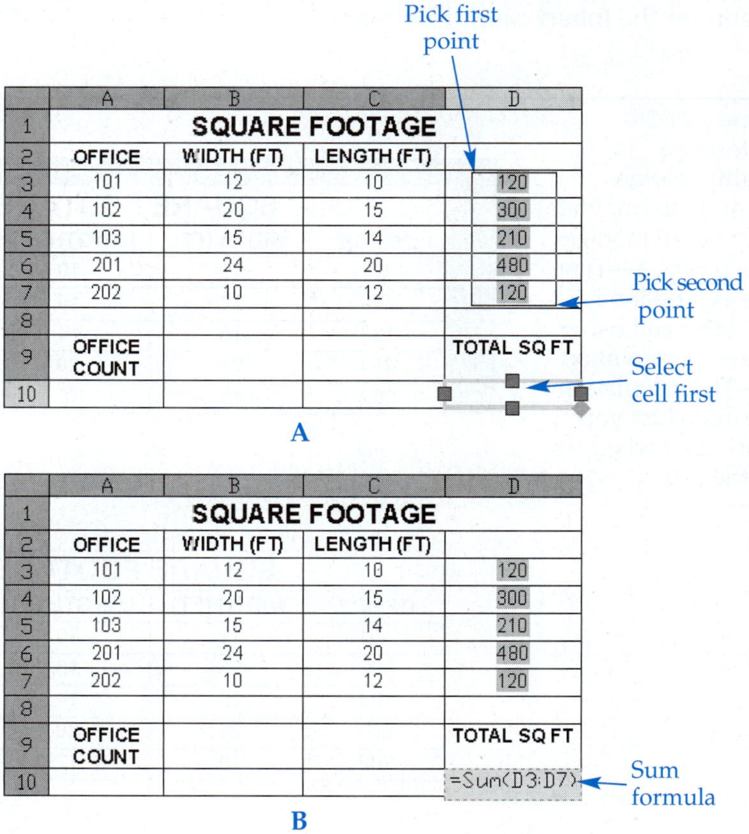

Figure 21-29. Creating a sum formula in a table cell. A—Pick a cell to hold the formula and select a range of cells for the formula by windowing around the cells. B—After you pick the second point of the window, the formula displays in the cell.

Figure 21-30.
Examples of sum, average, and count formulas and their resulting values.

	A	B	C	D
1	SQUARE FOOTAGE			
2	OFFICE	WIDTH (FT)	LENGTH (FT)	SQ FT
3	101	12	10	120
4	102	20	15	300
5	103	15	14	210
6	201	24	20	480
7	202	10	12	120
8				
9	OFFICE COUNT	AVERAGE SQ FT PER ROOM		TOTAL SQ FT
10	5	246		1230

=Count(A3:A7) =Average(D3:D7) =Sum(D3:D7)

Other Formula Options

The **Formula** drop-down list and **Insert** cascading menu contain additional options for writing table formulas. The **Cell** option allows you to select a cell from a different table in the drawing to insert the contents in the current cell. You can then use the cell value in a formula. Select the **Equation** option to place an equal sign (=) in the current cell. You can then type the expression manually.

NOTE

You can use the **FIELD** command to insert and edit table cell formulas. Select **Formula** from the **Field names:** list in the **Field** dialog box to display buttons for creating sum, average, and count formulas. You can also select a cell value from a different table as a starting point. Select table cells in the drawing area to define the formula. You can use the **Formula:** text box in the **Field** dialog box to add to or edit the formula. Unit format options are also available.

Exercise 21-9

www.g-wlearning.com/CAD/

Complete the exercise on the companion website.

Supplemental Material

www.g-wlearning.com/CAD/

Linking a Table to Excel Data

For information about using existing data entered in a Microsoft® Excel spreadsheet or a CSV file to create an AutoCAD table, go to the companion website, navigate to this chapter in the **Contents** tab, and select **Linking a Table to Excel Data**.

Supplemental Material

www.g-wlearning.com/CAD/

Extracting Table Data

For information about using existing AutoCAD text to create a table, go to the companion website, navigate to this chapter in the **Contents** tab, and select **Extracting Table Data**.

Template Development

www.g-wlearning.com/CAD/

Adding Table Styles

For detailed instructions on adding table styles to each drawing template, go to the companion website, navigate to this chapter in the **Contents** tab, and select **Adding Table Styles**.

Chapter Review

Answer the following questions using the information in this chapter.

1. What is the purpose of creating a table style?
2. Describe the procedure for creating a table style based on an existing table style.
3. What is the purpose of the **Alignment:** setting in the **New Table Style** dialog box?
4. Which setting would you adjust in the **New Table Style** dialog box to increase the spacing between the text and the top of the cell?
5. How can creating a new table using a starting table style save time?
6. How can you make a table style current without opening the **Table Style** dialog box?
7. List two ways to open the **Insert Table** dialog box.
8. Describe the two ways to insert an empty table and explain how the methods differ.
9. By default, what two types of rows are at the top of a table?
10. What ribbon tab opens when you insert a table?
11. If you finish typing in a cell and want to move to the next cell in the same row, what two keyboard keys can you use?
12. List two ways to make a cell active for editing.
13. Explain how to insert a field into a table cell.
14. How can you insert a new row at the bottom of a table?
15. How are table cells identified in formulas?
16. Write the table cell formula that adds the value of C3 and the value of D4.
17. What is the function of the colon (:) in the formula =Sum(D3:D7)?
18. What is the difference between a sum formula and a count formula?
19. Write the table cell formula that averages the values of cells D1, D3, and D6.
20. Explain how to write a formula that calculates a function for cells that do not share common borders.

Drawing Problems

Start AutoCAD if it is not already started. Start a new drawing using an appropriate template of your choice. The template should include layers, text styles, and table styles for drawing the given objects. Add layers, text styles, and table styles as needed. Draw all objects using appropriate layers. Use appropriate text styles, table styles, justification, and format. Follow the specific instructions for each problem. Use only drawing commands and techniques you have already learned. Use your own judgment and approximate dimensions when necessary.

Note: Some of the problems in this chapter are built on problems from previous chapters. If you have not yet completed those problems, complete them now.

▼ Basic

1. **Electronics** Create the electronic schematic reference designations list shown. Save the drawing as P21-1.

REFERENCE DESIGNATIONS	
LAST USED	DATE
R9	3/28
C1	3/28
CRS	3/28
Q2	3/28

2. **Mechanical** Save P18-10 as P21-2. If you have not yet completed Problem 18-10, do so now. In the P21-2 file, complete the drawing of the chain link by adding the table shown in Problem 18-10.

3. **Mechanical** Save P18-12 as P21-3. If you have not yet completed Problem 18-12, do so now. In the P21-3 file, complete the drawing of the chassis spacer by adding the table shown in Problem 18-12.

4. **Mechanical** Save P18-13 as P21-4. If you have not yet completed Problem 18-13, do so now. In the P21-4 file, complete the drawing of the chassis by adding the table shown in Problem 18-13.

5. **Mechanical** Save P18-20 as P21-5. If you have not yet completed Problem 18-20, do so now. In the P21-5 file, complete the drawing of the part view by adding the table shown in Problem 18-20.

6. **Mechanical** Save P18-21 as P21-6. If you have not yet completed Problem 18-21, do so now. In the P21-6 file, complete the drawing of the part view by adding the table shown in Problem 18-21.

7. **Mechanical** Save P18-23 as P21-7. If you have not yet completed Problem 18-23, do so now. In the P21-7 file, complete the drawing of the part views by adding the table shown in Problem 18-23.

8. **Mechanical** Create the parts list shown in Figure 21-1B. Do not draw the circles around the find numbers. Save the drawing as P21-8.

9. **Mechanical** Create the parts list shown. Save the drawing as P21-9.

5	1	–	210014–29	1/2–12UNC HEX NUT
4	1	–	320014–33	1/2 FLAT WASHER
3	2	–	632043–43	7/16 EXTERNAL SNAP RING
2	2	–	255010–41	1/4–20UNC WING NUT
1	2	–	803010–11	3/4 X 1/4–20UNC BOLT
FIND NO	QTY REQD	DIA	PART OR IDENT NO	NOMENCLATURE OR DESCRIPTION
PARTS LIST				

10. **Architectural** Create the window schedule shown in Figure 21-1A. Do not draw the polygons around the marks. Save the drawing as P21-10.

▼ Intermediate

11. **Mechanical** Save P18-22 as P21-11. If you have not yet completed Problem 18-22, do so now. In the P21-11 file, complete the drawing of the part view by adding the table shown in Problem 18-22.

12. **Mechanical** Convert the drawing shown to a drawing with the holes located using the **DIMORDINATE** command based on the X and Y coordinates given in the table. Create a table with columns for Hole (identification), Quantity, Description, and Depth (Z axis). Save the drawing as P21-12.

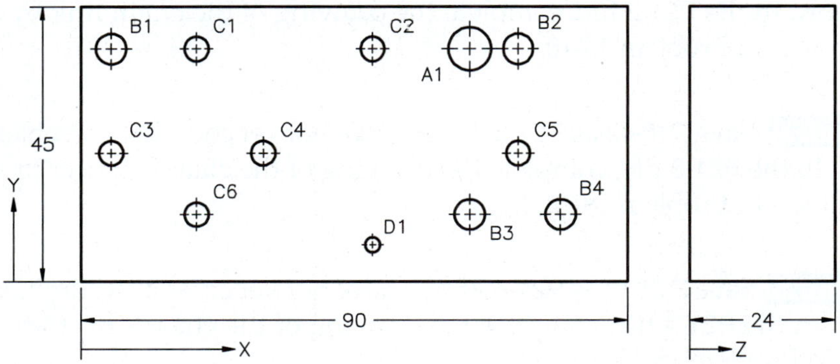

HOLE	QTY	DESC	X	Y	Z
A1	1	ø7	64	38	18
B1	1	ø5	5	38	THRU
B2	1	ø5	72	38	THRU
B3	1	ø5	64	11	THRU
B4	1	ø5	79	11	THRU
C1	1	ø4	19	38	THRU
C2	1	ø4	48	38	THRU
C3	1	ø4	5	21	THRU
C4	1	ø4	30	21	THRU
C5	1	ø4	72	21	THRU
C6	1	ø4	19	11	THRU
D1	1	ø2.5	48	6	THRU

Title: Base
Material: Bronze

13. **Architectural** Draw the finish schedule shown. Use the **Symbol** option of the text editor to locate and insert bullet symbols as shown. Save the drawing as P21-13.

INTERIOR FINISH SCHEDULE											
ROOM	FLOOR				WALLS				CEIL		
	CARPET	VINYL	TILE	HARDWOOD	PAINT	PAPER	TEXTURE	SPRAY	SMOOTH	BROCADE	PAINT
FOYER			●		●		●			●	●
KITCHEN			●			●		●	●		●
DINING				●	●		●			●	●
FAMILY	●				●		●			●	●
LIVING	●				●		●			●	●
MASTER BED	●				●		●			●	●
MASTER BATH			●			●		●	●		●
BATH 2		●				●		●	●		●
BED 2	●				●		●			●	●
BED 3	●				●		●			●	●
UTILITY		●				●		●	●		●

14. **Architectural** Create the door schedule from the architect's sketch exactly as shown. Use a text style assigned the CountryBlueprint font. Save the drawing as P21-14. Then make the following changes:

- Edit the text style to use the SansSerif font.
- Replace the abbreviation SYM. with the word MARK.
- Remove the period after the abbreviation QTY.
- Remove the periods from the abbreviation S.C.
- Replace the abbreviation RP. with the words RAISED PANEL.
- Remove the period after the abbreviation SLDG.

DOOR SCHEDULE			
SYM.	SIZE	TYPE	QTY.
1	36x80	S.C. RP. METAL INSULATED	1
2	36x80	S.C. FLUSH METAL INSULATED	2
3	32x80	S.C. SELF CLOSING	2
4	32x80	HOLLOW CORE	5
5	30x80	HOLLOW CORE	5
6	30x80	POCKET SLDG.	2

15. **Architectural** Create the window schedule from the architect's sketch exactly as shown. Use a text style assigned the CityBlueprint font. Save the drawing as P21-15. Then make the following changes:
- Edit the text style to use the SansSerif font.
- Replace the abbreviation SYM. with the word MARK.
- Remove the period after the abbreviations QTY, SLDG, and AWN.
- Replace the abbreviation CSM. with the abbreviation CSMT.

WINDOW SCHEDULE				
SYM.	SIZE	MODEL	ROUGH OPEN	QTY.
A	12x60	JOB BUILT	VERIFY	2
B	96x60	W4N5 CSM.	8'–0 3/4" x 5'–0 7/8"	1
C	48x60	W2N5 CSM.	4'–0 3/4" x 5'–0 7/8"	2
D	48x36	W2N3 CSM.	4'–0 3/4" x 3'–6 1/2"	2
E	42x42	2N3 CSM.	3'–6 1/2" x 3'–6 1/2"	2
F	72x48	G64 SLDG.	6'–0 1/2" x 4'–0 1/2"	1
G	60x42	G536 SLDG.	5'–0 1/2" x 3'–6 1/2"	4
H	48x42	G436 SLDG	4'–0 1/2" x 3'–6 1/2"	1
J	48x24	A41 AWN.	4'–0 1/2" x 2'–0 7/8"	3

16. **Architectural** Create the door schedule shown using a table break. Save the drawing as P21-16. Make the following changes:
- Change the schedule so items 11, 12, 13, and 14 continue directly below SYMBOL items 1 through 10, with only one DOOR SCHEDULE heading at the top.
- Replace the word SYMBOL with the word MARK.
- Replace the word QUANTITY with the abbreviation QTY.
- Replace the abbreviation S.C.R.P. with the words SOLID CORE RAISED PANEL (no period).
- Replace the abbreviation S.C.- with the words SOLID CORE (no hyphen, but include a space).
- Replace the abbreviation H.C. with the words HOLLOW CORE.
- Replace WOOD FRAME-TEMP. SLDG GL. with the abbreviations TMPD SGD.
- Make all data rows equal in height. Set the vertical spacing between table sections to one-half the text height.

DOOR SCHEDULE			
SYMBOL	SIZE	MODEL	QUANTITY
1	3'-0" X 6'-8"	S.C. R.P. METAL INSULATED	1
2	3'-0" X 6'-8"	S.C.-FLUSH-METAL INSULATED	2
3	2'-8" X 6'-8"	S.C.-SELF CLOSING	2
4	2'-8" X 6'-8"	H.C.	5
5	2'-6" X 6'-8"	H.C.	3
6	2'-6" X 6'-8"	POCKET	2
7	2'-4" X 6'-8"	POCKET	1
9	5'-0" X 6'-0"	BI-PASS	2
10	3'-0" X 6'-8"	BI-FOLD	1

DOOR SCHEDULE			
SYMBOL	SIZE	MODEL	QUANTITY
11	4'-0" X 6'-8"	BI-FOLD	1
12	2'-0" X 6'-0"	SHATTER PROOF	1
13	6'-0" X 6'-8"	WOOD FRAME-TEMP. SLDG GL.	1
14	9'-0" X 7-0"	OVERHEAD GARAGE	2

▼ Advanced

17. **Mechanical** Save P21-9 as P21-17. In the P21-17 file, make the following changes:
    - Change PARTS LIST to PURCHASE PARTS LIST.
    - Change Part Number 803010-11 as follows:
      Find No: 7, Description: $\frac{1}{4}$-20UNC-2 X $\frac{3}{4}$ BOLT HEX HD.
    - Change Part Number 255010-41 as follows:
      Find No: 11, Description: $\frac{1}{4}$-20UNC WING NUT.
    - Change Part Number 632043-43 as follows:
      Find No: 15, Description: Ø $\frac{7}{16}$ EXTERNAL SNAP RING.
    - Change Part Number 320014-33 as follows:
      Find No: 19, Description: Ø $\frac{1}{2}$ FLAT WASHER.
    - Change Part Number 210014-29 as follows:
      Find No: 21, Description: $\frac{1}{2}$-12UNC-2 HEX NUT.

18. **General** Create a table of your own design and use at least six of the applications of calculating values in tables described in this chapter. Save the drawing as P21-18.

19. **General** Access the companion website content for this chapter and review the **Linking a Table to Excel Data** supplement available in the Supplemental Material. Use this information to create a table by linking to Microsoft® Excel data. To do this, create your own Excel spreadsheet or find an existing Excel spreadsheet containing suitable data. Link a table to the Excel data. Save the drawing as P21-19.

20. **Electronics** Access the companion website content for this chapter and review the **Extracting Table Data** supplement available in the Supplemental Material. Use the description to create a table similar to the given examples and then extract the table data into an AutoCAD drawing. Save the drawing as P21-20.

21. **Mechanical** Research the requirements for a set of working drawings for a mechanical assembly. Identify a mechanical assembly consisting of several parts and possibly subassemblies. Use the **TABLE** command to create a parts list identifying each component. Save the drawing as P21-21.

22. **Architectural** Research an example of a residential home design. Use the **TABLE** command to create a door and window schedule identifying all of the doors and windows in the home. Save the drawing as P21-22.

23. **Electronics** Research an example of an electronic wiring harness schematic. Use the **TABLE** command to create a wire list identifying all of the required wires. Save the drawing as P21-23.

24. **Civil** Obtain a copy of leveling field notes prepared by a surveyor. Use the **TABLE** command to reproduce the notes and calculate unspecified values. Save the drawing as P21-24.

AutoCAD Certified Professional Exam Practice

Answer the following questions using the information in this chapter.

1. What is the meaning of the formula =((C3*D2)/4)? *Select the one item that best answers the question.*
   A. add the values of C3 and D2 and divide the result by 4
   B. multiply the values of both C3 and D2 by 4 and add the results
   C. multiply the value of C3 by the value of D2 and divide the result by 4
   D. divide the value of D2 by 4 and multiply the result by the value of C3
2. How can you remove existing breaks in a table? *Select all that apply.*
   A. select the table, right-click, and select **Remove All Property Overrides**
   B. move the table breaking grip down to the last row
   C. select the table breaking grip and press [Delete]
   D. use the **Properties** palette to set **Table Breaks Enabled** to **No**
3. In a table that uses the **Down** direction, what is the cell identification for a cell in the fourth row from the top, in the fifth column from the left? *Select the one item that best answers the question.*
   A. 4E
   B. 5D
   C. D4
   D. D5
   E. E4

Follow the instructions in each problem. Complete each problem by answering the related question.

4. **Navigate to this chapter on the companion website and open CPE-21formula.dwg.** Perform the following tasks on the existing table:
   A. In the first data row of the AREA column, create a formula to calculate the area (length × width) of each room listed in the table.
   B. Use auto-fill to copy the formula to the remaining rows in the AREA column.
   C. Add a row at the bottom of the table.
   D. In the new bottom row, merge cells A11 through E11, set the alignment to **Top Right**, and enter the text TOTAL AREA.
   E. In the bottom-right cell, create a formula to calculate the total area of all the rooms.

   What is the total area of the house?

CHAPTER 22

Parametric Drafting

Learning Objectives

After completing this chapter, you will be able to:

- ✓ Explain parametric drafting processes and applications.
- ✓ Add and manage geometric constraints.
- ✓ Add and manage dimensional constraints.
- ✓ Adjust the form of dimensional constraints.

Parametric drafting tools allow you to assign ***parameters***, or ***constraints***, to objects. The parametric concept, also known as *intelligence*, provides a way to associate objects and limit design changes. You cannot change a constraint so that it conflicts with other parametric geometry. A database stores and allows you to manage all parameters. You typically use parametric tools with standard drafting practices to create a more interactive drawing.

parametric drafting: A form of drafting in which parameters and constraints drive object size and location to produce drawings with features that adapt to changes made to other features.

parameters (constraints): Geometric characteristics and dimensions that control the size, shape, and position of drawing geometry.

Parametric Fundamentals

Parametric drafting can increase your ability to control every aspect of a drawing during and after the design and documentation process. Parametric tools can change the way you construct and edit geometry. However, in general, use parametric tools as a supplement to standard drafting practices and drawing aids. When used correctly, this technique allows you to produce accurate parametric drawings efficiently.

Understanding Constraints

Add parameters using ***geometric constraints*** and ***dimensional constraints***. Well-defined constraints allow you to incorporate and preserve specific design intentions and increase revision efficiency. For example, if two holes through a part, drawn as circles, must always be the same size, use a geometric constraint to make the circles equal and add a dimensional constraint to size one of the circles. The size of both circles changes when you modify the dimensional constraint value. See **Figure 22-1**.

geometric constraints: Geometric characteristics applied to restrict the size or location of geometry.

dimensional constraints: Measurements that numerically control the size or location of geometry.

Figure 22-1.
An example of a parametric relationship. The dimensional constraint controls the size of both circles with the aid of an equal geometric constraint.

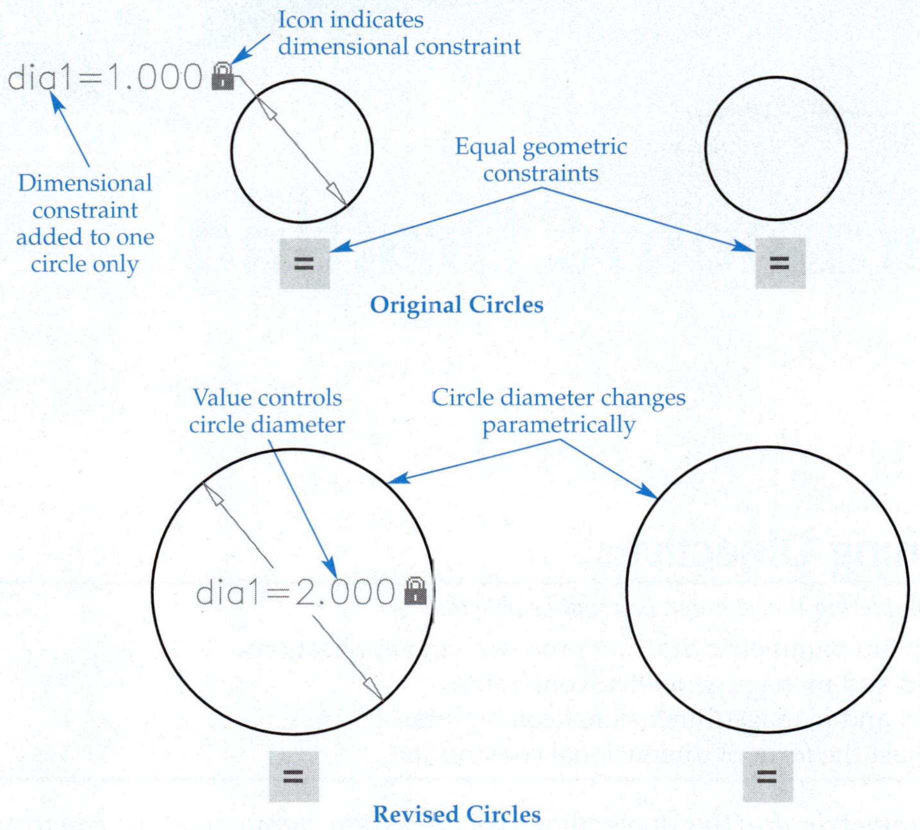

under-constrained: Describes a drawing that includes constraints, but not enough to size and locate all geometry.

fully constrained: Describes a drawing in which objects have no freedom of movement.

over-constrained: Describes a drawing that contains too many constraints.

reference dimension: A dimension used for reference purposes only. Parentheses enclose reference dimensions to differentiate them from other dimensions.

You must add constraints to make an object parametric. Dimensional constraints create parameters that direct object size and location. In contrast, an associative dimension is associated with an object, but it does not control object size or location. Figure 22-2 shows an example of a drawing that is ***under-constrained***, ***fully constrained***, and ***over-constrained***. As you progress through the design process, you will often fully or almost fully constrain the drawing to ensure that the design is accurate. However, a message appears if you attempt to over-constrain the drawing. See Figure 22-3. AutoCAD does not allow you to over-constrain a drawing, as shown by the ***reference dimension*** in Figure 22-2.

Figure 22-4 shows an extreme example of constraining, for reference only. Study the figure to understand how constraints work, and how applying constraints differs from and complements traditional drafting with AutoCAD. Typically, you should prepare initial objects as accurately as possible using standard commands and drawing aids. Add geometric constraints while you are drawing, or add them later to existing objects. Apply dimensional constraints after creating the geometry.

Parametric Applications

Parametric tools aid the design and revision process, place limits on geometry to preserve design intent, and help form geometric constructions. Consider using constraints to help maintain relationships between objects in a drawing, especially during the design process, when changes are often frequent. However, you must decide if the additional steps required to make a drawing parametric are appropriate and necessary for the application.

Figure 22-2.
Levels of parametric constraint.

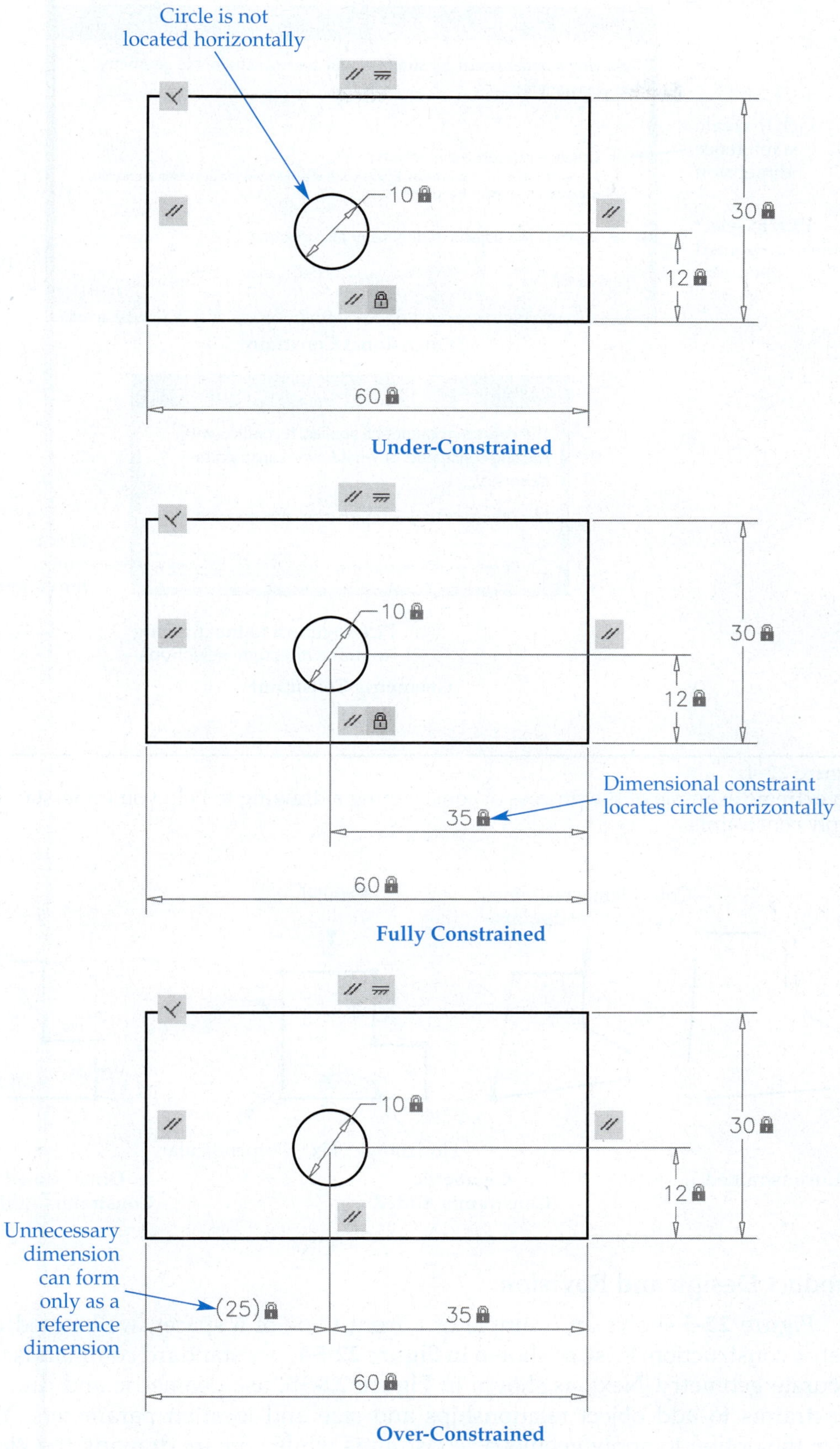

Figure 22-3.
Error messages appear when you attempt to over-constrain objects.

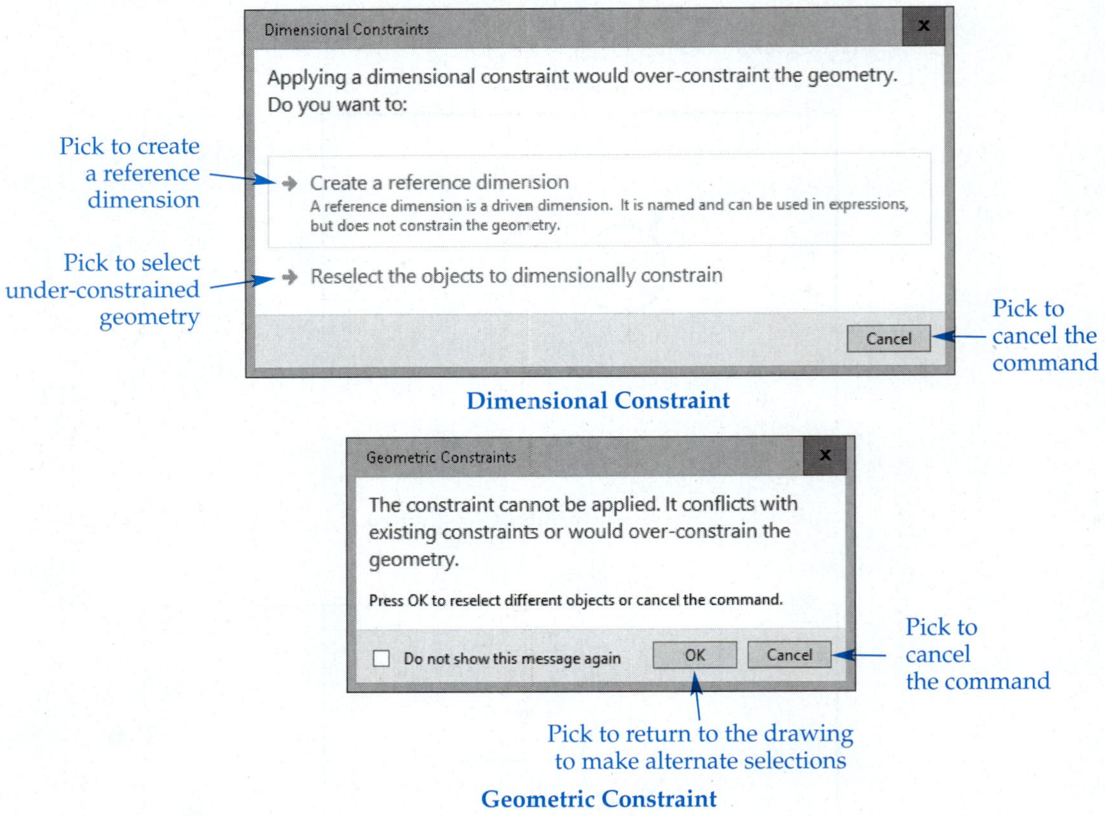

Figure 22-4.
An extreme example of the process of constraining a drawing to help you understand how to apply constraints.

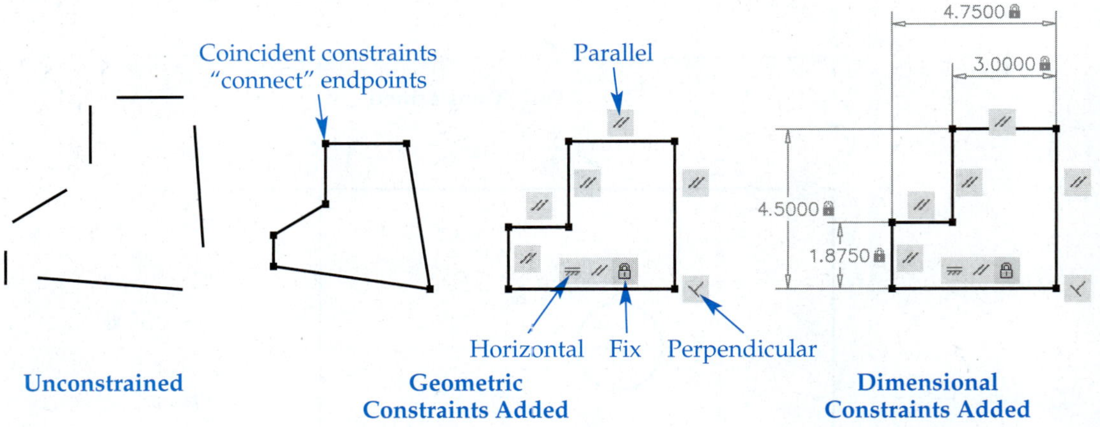

Product Design and Revision

Figure 22-5 shows an example of a front view of a spacer, well suited to parametric construction. First, as shown in **Figure 22-5A**, use standard commands to create accurate geometry. Next, as shown in **Figure 22-5B**, use geometric and dimensional constraints to add object relationships and size and location parameters. You also have the option to apply geometric constraints while you are drawing the view. This example uses centerlines, drawn on a separate construction layer, to apply the correct constraints. Then, as shown in **Figure 22-5C**, use the constraints to explore design alternatives and make changes to the drawing efficiently.

Figure 22-5D shows converting the dimensional constraint format to a formal appearance, to which you can assign a dimension style. You can still use converted dimensions to adjust geometry parametrically. This example shows converting all dimensions except the .250 radius and .200 diameter dimensions. Using standard associative .250 radius and .200 diameter dimensions allows you to add the 6X prefix and relocate the dimensions. You can then hide the constraint information to view the finished drawing.

Geometric Construction

Use constraints to form geometric constructions when standard commands are inefficient or ineffective. For example, suppose you know that the angle of a line is 30°, and you know the line is tangent to a circle. However, you do not know the length of the line or the location of the line endpoints. One option is to position a 30° construction line, using the **Ang** option of the **XLINE** command, anywhere in the drawing. See Figure 22-6A. Use a tangent geometric constraint to form a tangent relationship between the xline and circle. See Figure 22-6B. Then hide or delete the constraint if necessary. See Figure 22-6C.

Unsuitable Applications

You may find that parametric drafting is unsuitable or ineffective for some applications. For example, it may be unsuitable to add parameters to a drawing if the drawing is of a finalized product that will not require revision, or if you can easily modify drawing geometry without associating objects.

In addition, if your drawing includes a large number of objects, you may find it cumbersome to add the constraints required to form a fully intelligent drawing. For instance, you can use constraints to form all necessary relationships between objects in a floor plan. See Figure 22-7. In this example, constraints connect walls, specify walls as perpendicular or parallel, control wall thickness, position windows between walls, locate sinks on vanities, and form many other parametric relationships. You can then adjust dimensional constraints as needed to update the drawing.

If you effectively constrain all objects shown in Figure 22-7, you have the option, for example, to change the 11'-10 1/2" dimensional constraint to increase the width of the master bedroom. The entire floor plan adjusts to the modified room size. Consider, however, what this process requires. You must constrain all wall endpoints; constrain the points where doors and windows meet walls; constrain the distances between walls and objects, such as cabinets, sinks, and fixtures; and form all other geometric and dimensional constraints.

NOTE

You can also use parametric tools when constructing blocks, as described later in this textbook.

PROFESSIONAL TIP

Constraints can also be used in multiview layout, to help maintain alignment between views.

Exercise 22-1

www.g-wlearning.com/CAD/

Complete the exercise on the companion website.

Figure 22-5.
The front view of this spacer is a good candidate for parametric drafting. A—Accurate view geometry constructed using standard AutoCAD practices. B—Adding geometric and dimensional constraints to constrain the drawing. C—Changing the values of a few dimensional constraints to update the entire drawing. D—Reusing dimensional constraints to help prepare a formal drawing.

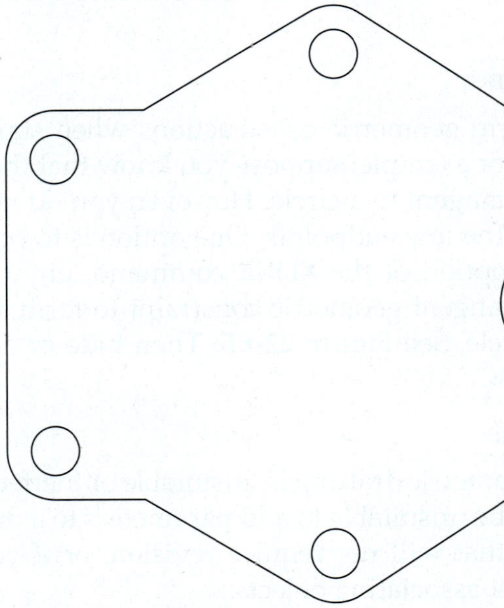

A

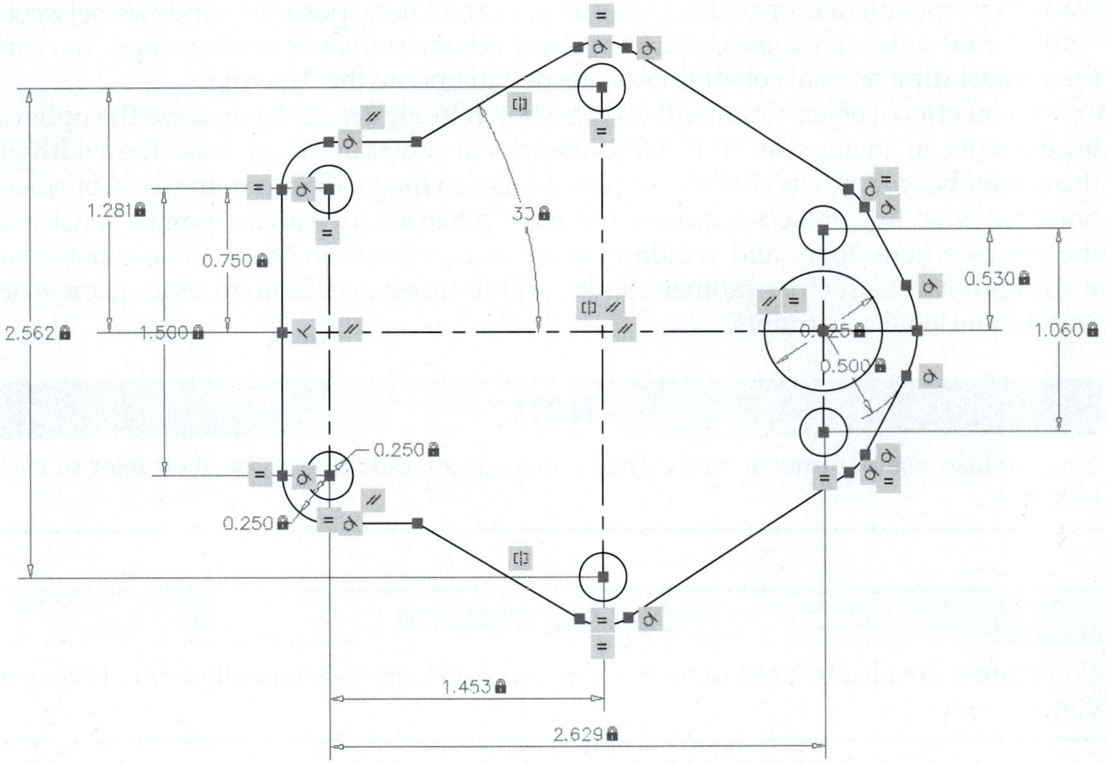

B

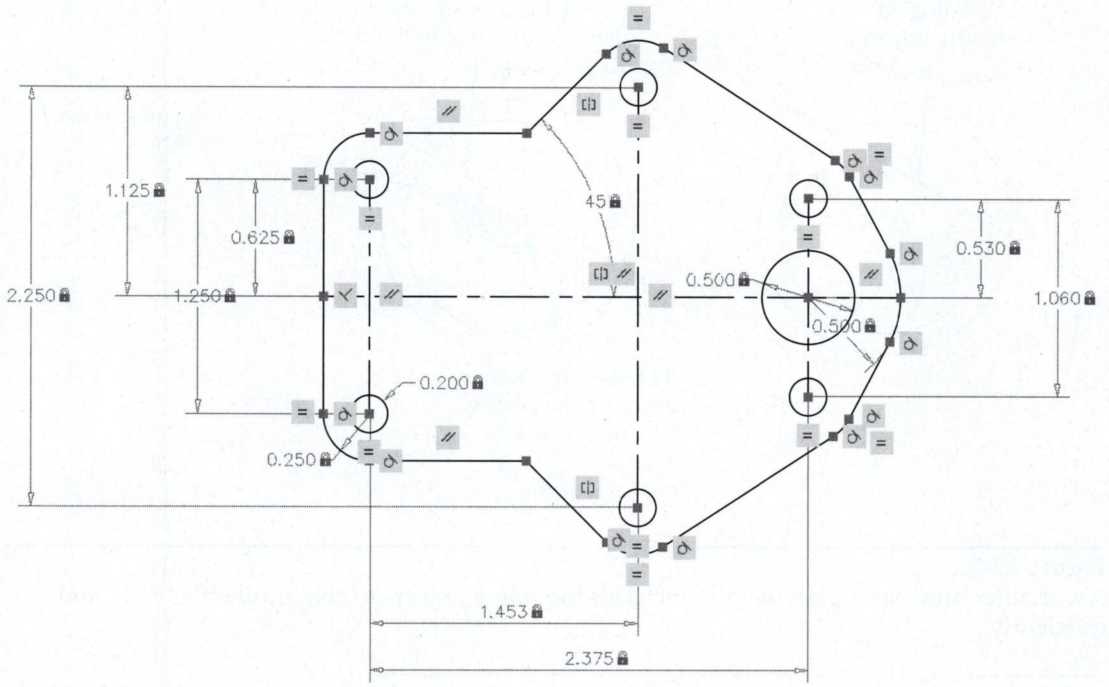
1.125
0.625
2.250
1.250
45
0.500
0.530
1.060
0.500
0.200
0.250
1.453
2.375

C

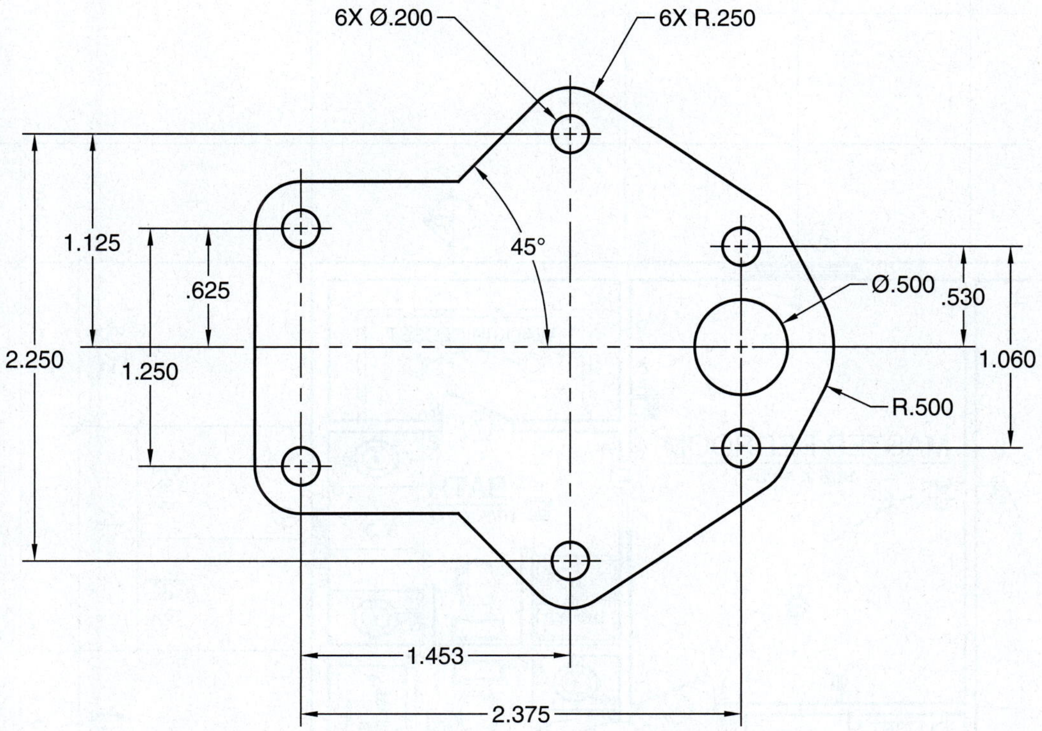
6X Ø.200
6X R.250
1.125
.625
2.250
1.250
45°
Ø.500
.530
1.060
R.500
1.453
2.375

D

Figure 22-6.
A—A 30° xline placed near an associated circle. B—Using a tangent geometric constraint to form a tangent construction. Notice the appropriate selection process. C—The final drawing.

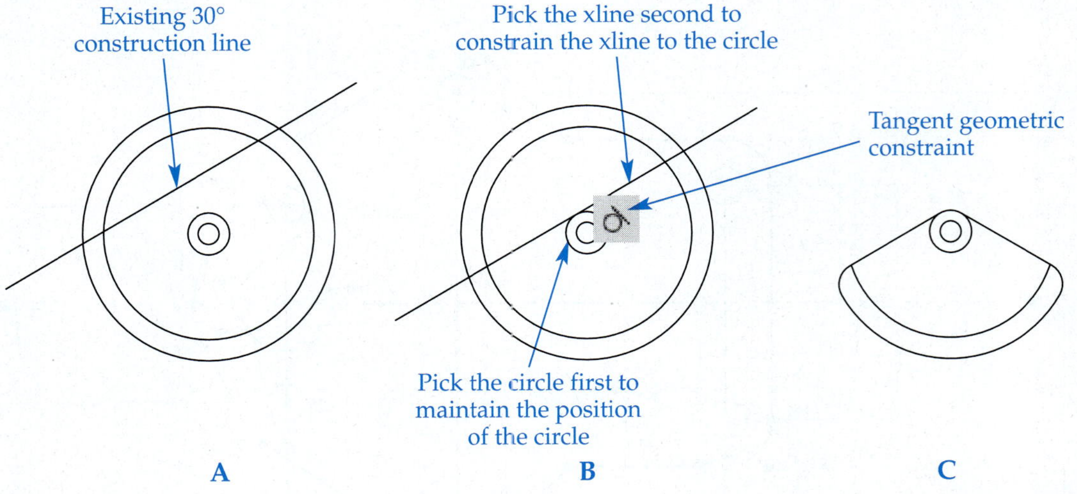

Figure 22-7.
An architectural floor plan usually includes too many objects to constrain effectively and efficiently.

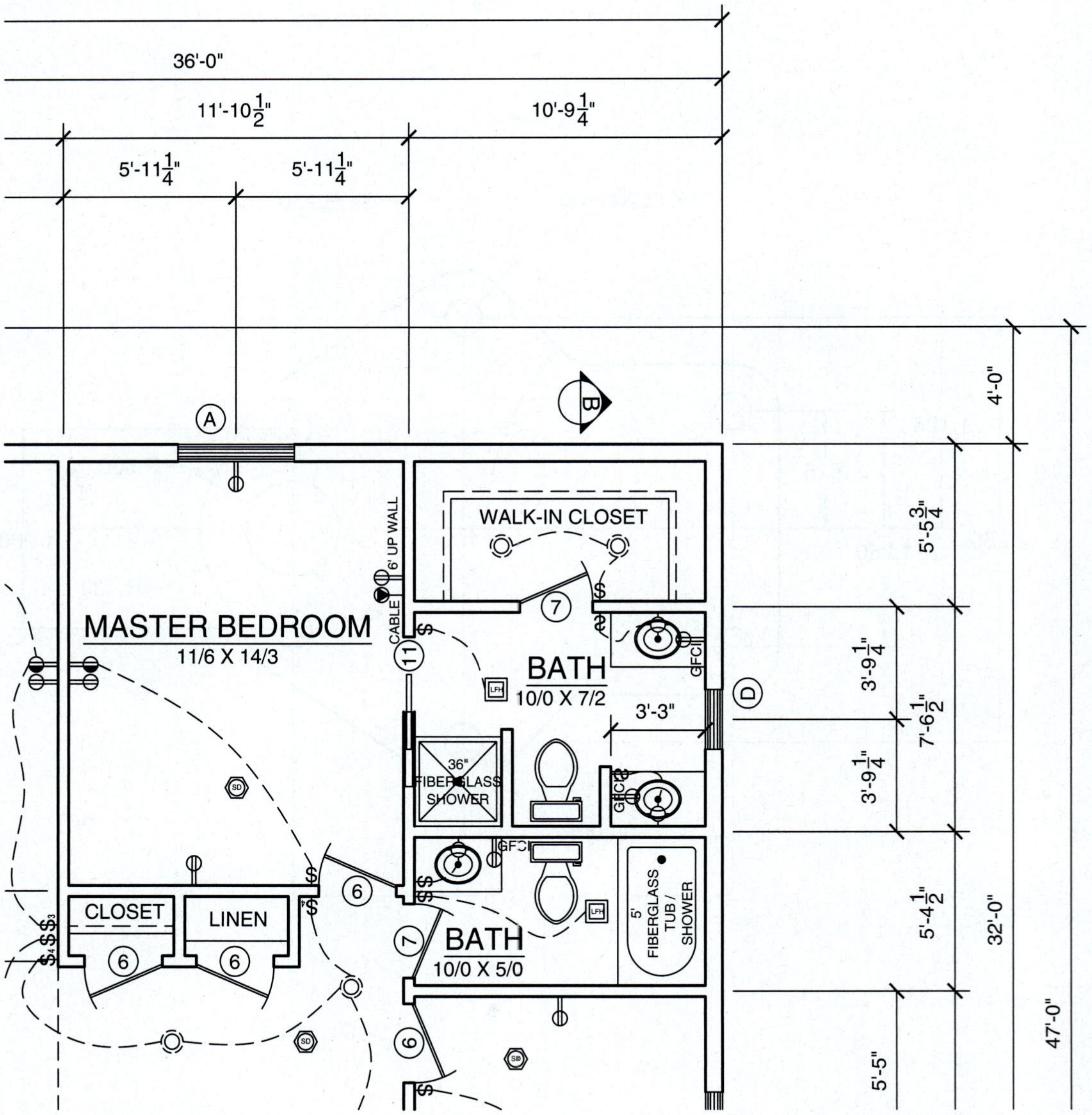

Geometric Constraints

Geometric constraint tools allow you to add the geometric relationships required to build a parametric drawing. You typically add geometric constraints, or at least a portion of the necessary geometric constraints, before dimensional constraints to help preserve design intent. You can ***infer*** certain geometric constraints while drawing and editing, or manually apply geometric constraints to existing unconstrained geometry.

infer: Automatically detect and apply using logic.

Inferring constraints is a fast way to add geometric relationships. However, often a combination of inferred and manually added geometric constraints is necessary. By default, a constraint-specific icon is visible when you infer constraints to indicate the presence of a geometric constraint. View, adjust, and remove geometric constraints as needed. Figure 22-8 describes each geometric constraint.

PROFESSIONAL TIP

Placing too many geometric constraints can cause problems as you progress through the design process. Apply only the geometric constraints necessary to generate the required geometric constructions.

Inferring Geometric Constraints

The **CONSTRAINTINFER** system variable controls whether AutoCAD infers constraints as you create new geometry. The **Infer Constraints** button on the status bar provides a quick way to toggle **Infer Constraints** on and off. The **Infer Constraints** button does not appear on the status bar by default. Select the **Infer Constraints** option in the status bar **Customization** flyout to add the **Infer Constraints** button to the status bar. When **Infer Constraints** is on, constraints are inferred when you draw a new object. Appropriate object snaps, other drawing aids, and **Infer Constraints** must be active in order to infer constraints, except when you use the **RECTANGLE** command. For example, use the **LINE** command and a horizontal (0° or 180°) or vertical (90° or 270°) polar tracking angle, or activate **Ortho** mode, to constrain a horizontal or vertical line. See Figure 22-9A.

Constraints are also inferred when you edit an object. Some commands, such as **FILLET** and **CHAMFER**, infer constraints automatically. However, you must use grip editing or appropriate object snaps to infer constraints when stretching, moving, or copying. For example, use grip editing and the center point and midpoint grips, or use the **MOVE** command and **Center** and **Midpoint** object snap modes, to move the center point of a circle coincident to the midpoint of a line. See Figure 22-9B.

NOTE

The **Quadrant**, **Extension**, and **Apparent Intersect** object snaps do not infer constraints. The **OFFSET**, **BREAK**, **TRIM**, **EXTEND**, **SCALE**, **MIRROR**, **ARRAY**, and **MATCHPROP** commands do not infer constraints. Exploding a polyline removes all inferred constraints.

CAUTION

The **Infer Constraints** tool can save drafting time by placing geometric constraints while you draw or edit. However, use caution to ensure that appropriate geometric constructions occur. You may have to replace certain constraints manually.

Figure 22-8.
Geometric constraints form geometric relationships between points and/or objects. Only some of the geometric constraints can be inferred.

Constraint	Icon	Inferable	Description
Coincident	Inactive Active	Yes	Constrains two points to the same location, or constrains a point along a curve. A point can be coincident to a curve without touching the curve.
Horizontal	Object Point-to-point	Yes	Horizontally aligns two points or a line, polyline, ellipse axis, mtext, or text object; positions geometry along the X axis on the default XY plane.
Vertical	Object Point-to-point	Yes	Vertically aligns two points or a line, polyline, ellipse axis, mtext, or text object; positions geometry along the Y axis on the default XY plane.
Parallel	Object-to-object	Yes	Creates a parallel constraint between a line, polyline, ellipse axis, mtext, or text object with another line, polyline, ellipse axis, mtext, or text object.
Perpendicular	Object-to-object	Yes	Forms a perpendicular constraint between a line, polyline, ellipse axis, mtext, or text object with another line, polyline, ellipse axis, mtext, or text object.
Tangent	Object-to-object	Yes	Forms a tangent constraint between a circle, arc, or ellipse and a line, polyline, circle, arc, or ellipse.
Collinear	Object-to-object	No	Aligns a line, polyline, ellipse axis, mtext, or text object with another line, polyline, ellipse axis, mtext, or text object.
Concentric	Object-to-object	No	Constrains the center of a circle, arc, or ellipse to the center of another circle, arc, or ellipse.
Equal	Object-to-object	No	Sizes and locates an object in reference to another object.
Symmetric	Object-to-object Point-to-point Line of symmetry	No	Establishes symmetry between objects or points and a line of symmetry.
Fix	Object Point	No	Secures an object to its current location in space.
Smooth	Object-to-object	No	Connects and creates a curvature-continuous situation, or G2 curve, between a spline and a line, polyline, spline, or arc.

Figure 22-9.
A—Inferring a horizontal and vertical constraint while drawing a line. B—Editing a drawing to infer a coincident constraint.

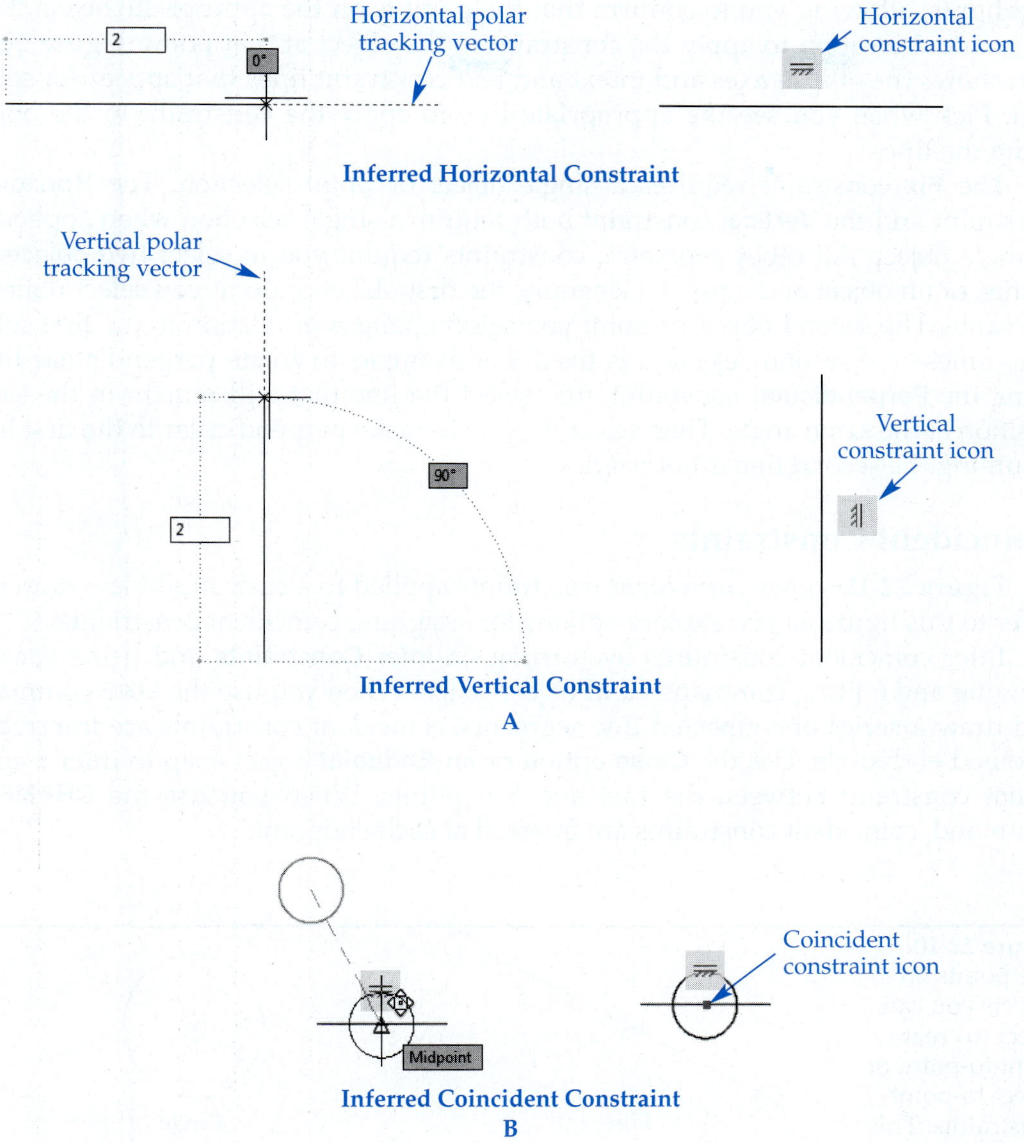

Exercise 22-2

www.g-wlearning.com/CAD/

Complete the exercise on the companion website.

Manual Geometric Constraints

Type
GEOMCONSTRAINT GCON

Infer constraints when possible and appropriate. You can also add geometric constraints manually to apply geometric constraints that are not inferred automatically, as indicated in Figure 22-8. Add geometric constraints manually to include additional geometric constraints as needed, or to constrain a nonparametric drawing. A quick way to place geometric constraints manually is to pick the appropriate button from the **Geometric** panel of the **Parametric** ribbon tab. You can also type GC followed by the name of the constraint, such as GCTANGENT, or select an option from the **GEOMCONSTRAINT** command. Follow the prompts to make the required selection(s), form the constraint, and exit the command.

Select objects, points, or an object and a point, depending on the objects and geometric constraint. Figure 22-10 shows the point markers that appear as you move the crosshairs on an object to select a point. The object associated with a specific point highlights, allowing you to confirm that the point is on the appropriate object. Pick the marked location to apply the constraint to the object at that point. Figure 22-10 also shows the ellipse axes and mtext and text constraint lines that appear for selection. Pick when you see the appropriate line to apply the constraint to the object using the line.

The **Fix** constraint requires a single object or point selection. The **Horizontal** constraint and the **Vertical** constraint both require a single selection when applied to a single object. All other geometric constraints require you to select two objects or points, or an object and a point. Generally, the first object or point you select remains the same. The second object or point you select changes in relation to the first selection, unless the second selection is fixed. For example, to create perpendicular lines using the **Perpendicular** constraint, first select the line that will remain in the same position at the same angle. Then select the line to make perpendicular to the first line, assuming the second line is not fixed.

Coincident Constraints

coincident: A geometric construction that specifies two points sharing the same position.

Figure 22-11 shows ***coincident*** constraints applied to a basic multiview drawing. Refer to this figure as you explore options for assigning coincident constraints.

Infer coincident constraints by turning on **Infer Constraints** and using specific drawing and editing commands and object snaps. When you use the **LINE** command and draw a series of connected line segments, coincident constraints are inferred at adjoined endpoints. Use the **Close** option or an **Endpoint** object snap to infer a coincident constraint between the first and last points. When you use the **CHAMFER** command, coincident constraints are inferred at each endpoint.

Figure 22-10.
The points on objects you can select to create point-to-point or object-to-point constraints. This figure also shows selectable constraint lines at ellipse axes and through mtext and text objects. Use the lines to establish horizontal, vertical, parallel, perpendicular, or collinear constraints.

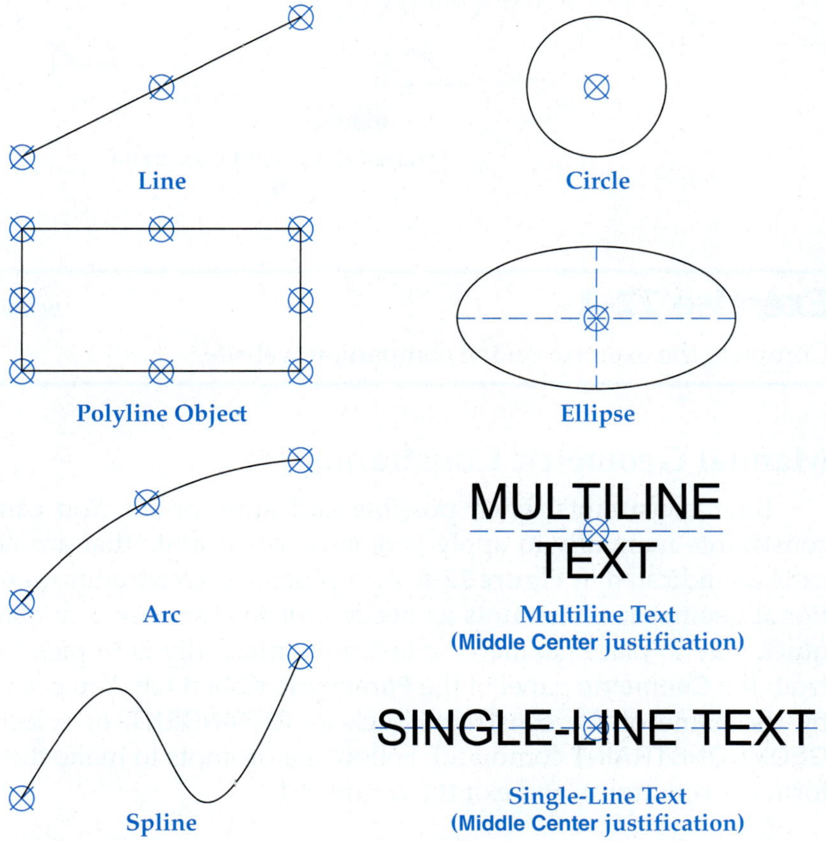

Figure 22-11.
Coincident constraints required for a basic multiview drawing. For clarity, labels do not indicate all required coincident constraints. This drawing is used as an example throughout this chapter.

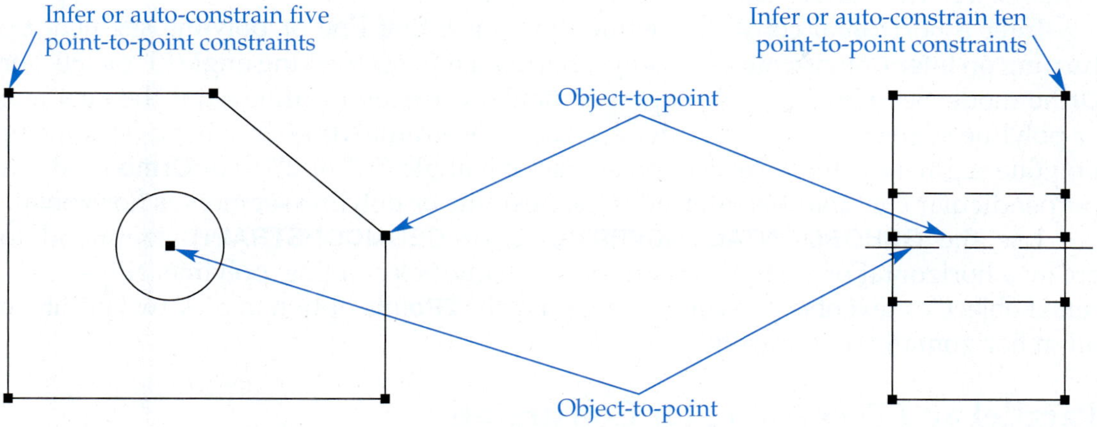

The **Endpoint**, **Midpoint**, **Center**, **Insertion**, and **Node** object snaps infer a point-to-point coincident constraint. For example, snap to the midpoint of a line to constrain the center of a donut to the midpoint of the line. The **Nearest** object snap infers an object-to-point coincident constraint. For example, snap to a nearest location on a line to constrain the center of a circle to the line. The center of the circle is free to move anywhere aligned with the line, and does not have to touch the line.

Use the **GCCOINCIDENT** or **GEOMCONSTRAINT** command to create a coincident constraint manually. Move the pick box near a point on an existing object to display a point marker. Pick the marked location, and then pick a point on another object to make the two points coincide.

Use the **Object** option to select an object and a point, as required to constrain a point along a curve. The point does not have to contact the curve, and you can select the object first or second. Refer to Figure 22-11. The **Autoconstrain** option allows you to select multiple objects to form coincident constraints at every possible coincident intersection in a single operation. Right-click or press [Enter] to apply the **Autoconstrain** option and end the command.

Ribbon
Parametric > Geometric
Coincident
Type
GEOMCONSTRAINT > Coincident
GCCOINCIDENT

NOTE

Connected polyline segments act as if they are constrained, but they do not use or accept coincident constraints at adjoined endpoints.

PROFESSIONAL TIP

When you select points, be sure to select the points corresponding to the surface to be constrained.

Exercise 22-3

www.g-wlearning.com/CAD/

Complete the exercise on the companion website.

Horizontal and Vertical Constraints

Figure 22-12 shows examples of horizontally and vertically constrained objects and points. Horizontal and vertical constraints are commonly used to define a horizontal or vertical surface datum or to align points.

Infer a horizontal constraint while drawing a first line or polyline segment by turning on **Infer Constraints** and using a horizontal polar tracking angle (0° or 180°) or **Ortho** mode. See Figure 22-9A. A perpendicular constraint is inferred if the next line or polyline segment is vertical. Infer a vertical constraint while drawing a first line or polyline segment using a vertical polar tracking angle (90° or 270°) or **Ortho** mode. A perpendicular constraint is inferred if the next line or polyline segment is horizontal.

Use the **GCHORIZONTAL**, **GCVERTICAL**, or **GEOMCONSTRAINT** command to create a horizontal or vertical constraint manually. Select a line, polyline, ellipse axis, mtext object, or text object to constrain, or use the **2Points** option to pick two points to align horizontally or vertically.

Ribbon
Parametric > Geometric
Horizontal
Type
GEOMCONSTRAINT > Horizontal
GCHORIZONTAL

Ribbon
Parametric > Geometric
Vertical
Type
GEOMCONSTRAINT > Vertical
GCVERTICAL

parallel: A geometric construction that specifies that objects such as lines remain parallel and will never intersect, no matter how long they become.

perpendicular: A geometric construction that defines a 90° angle between objects such as lines.

Parallel and Perpendicular Constraints

Figure 22-13 shows ***parallel*** and ***perpendicular*** constraints applied to a basic multiview drawing. Parallelism and perpendicularity are common geometric characteristics that provide specific controls related to the orientation of features.

Figure 22-12.
Examples of vertically and horizontally constrained objects and points.

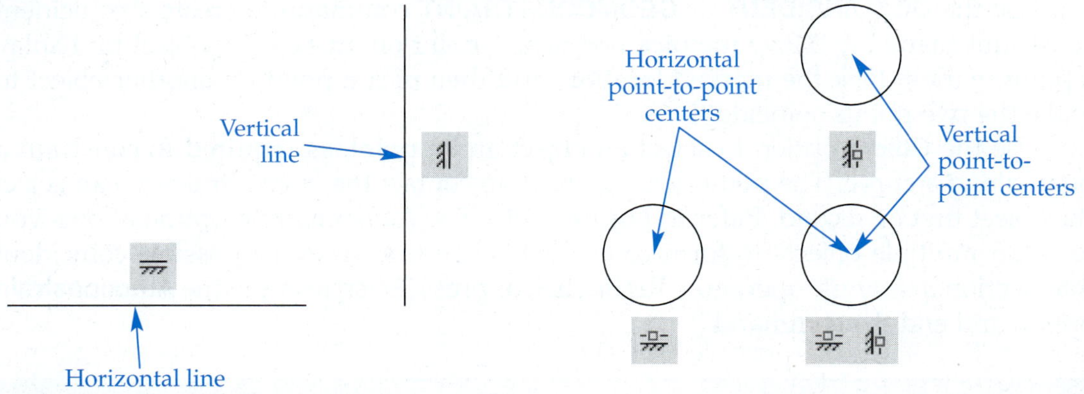

Figure 22-13.
Parallel and perpendicular constraints required for the example multiview drawing.

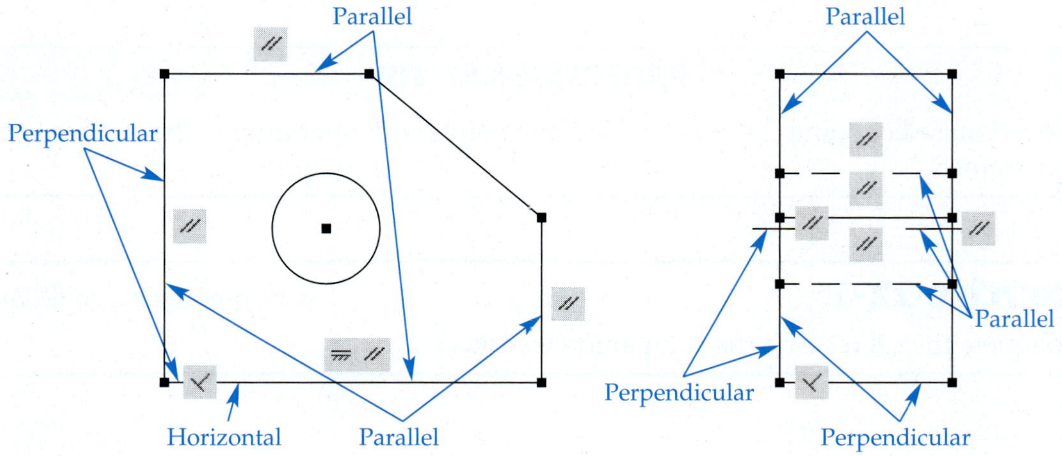

Infer parallel or perpendicular constraints by turning on **Infer Constraints** and using drawing and editing commands and object snaps. A perpendicular constraint is inferred if the next segment of a connected line or polyline is horizontal or vertical. When you use the **RECTANGLE** command to draw a polyline object, parallel and perpendicular constraints are inferred to create a true rectangle. Additional parallel and perpendicular constraints are inferred when you use the **Chamfer** option of the **RECTANGLE** command, depending on the specified distances. An additional perpendicular constraint is inferred when you use the **Fillet** option.

The **Parallel** object snap infers a parallel constraint between objects. The **Perpendicular** object snap infers a perpendicular and coincident constraint between objects.

Use the **GCPARALLEL**, **GCPERPENDICULAR**, or **GEOMCONSTRAINT** command to create a parallel or perpendicular constraint manually. Select a line, polyline, ellipse axis, mtext object, or text object and a line, polyline, ellipse axis, mtext object, or text object.

Ribbon
Parametric
> Geometric

Parallel
Type
GEOMCONSTRAINT
> Parallel
GCPARALLEL

Ribbon
Parametric
> Geometric

Perpendicular
Type
GEOMCONSTRAINT
> Perpendicular
GCPERPENDICULAR

Exercise 22-4

www.g-wlearning.com/CAD/

Complete the exercise on the companion website.

Tangent and Collinear Constraints

Figure 22-14 shows tangent and collinear constraints applied to a basic multiview drawing. Collinear constraints cannot be inferred.

Infer tangent constraints by turning on **Infer Constraints** and using the **FILLET** command with a radius greater than 0. You can also infer tangent constraints on arc segments of polylines, and by using the **Tangent** object snap.

NOTE

The **Continue** option of the **ARC** command does not infer a coincident or tangent constraint between the existing object and the new arc.

Use the **GCTANGENT** or **GEOMCONSTRAINT** command to create a tangent constraint manually. Select a circle, arc, or ellipse and then a line, polyline, circle, arc, or ellipse. Use the **GCCOLLINEAR** or **GEOMCONSTRAINT** command to create a collinear constraint manually. A collinear constraint is commonly used for applications such as aligning multiview surfaces. Select a line, polyline, ellipse axis, mtext object, or text object and then another line, polyline, ellipse axis, mtext object, or text object. Activate the **Multiple** option to select multiple objects to align in a single operation. Right-click or press [Enter] to complete a multiple constrain operation.

Ribbon
Parametric
> Geometric

Tangent
Type
GEOMCONSTRAINT
> Tangent
GCTANGENT

Ribbon
Parametric
> Geometric

Collinear
Type
GEOMCONSTRAINT
> Collinear
GCCOLLINEAR

Exercises 22-5 and 22-6

www.g-wlearning.com/CAD/

Complete the exercises on the companion website.

Concentric and Equal Constraints

Ribbon
Parametric
> Geometric

Concentric

Type
GEOMCONSTRAINT
> Concentric
GCCONCENTRIC

Use the **GCCONCENTRIC** or **GEOMCONSTRAINT** command to assign a ***concentric*** constraint. Select a circle, arc, or ellipse and a second circle, arc, or ellipse. Use the **GCEQUAL** or **GEOMCONSTRAINT** command to size objects equally and, in some cases, to locate objects. Select two objects, or activate the **Multiple** option to select multiple objects to equalize in a single operation. Right-click or press [Enter] to complete a multiple constrain operation. Figure 22-15 shows examples of concentric and equal constraints.

Ribbon
Parametric
> Geometric

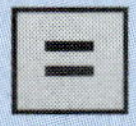

Equal

Type
GEOMCONSTRAINT
> Equal
GCEQUAL

Symmetric, Fix, and Smooth Constraints

Use the **GCSYMMETRIC** or **GEOMCONSTRAINT** command to establish symmetry between objects or points and a line of symmetry. Select one object, followed by another, and finally, a line of symmetry. Select a line or a single polyline segment to specify the line of symmetry. Use the **2Points** option to pick points followed by the line of symmetry to constrain symmetrical points.

Ribbon
Parametric
> Geometric

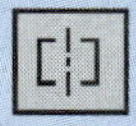

Symmetric

Type
GEOMCONSTRAINT
> Symmetric
GCSYMMETRIC

concentric: Describes arcs, circles, and/or ellipses sharing the same center point.

Figure 22-14.
A—Collinear and tangent constraints required for the example multiview drawing.
B—Common tangent constraint applications.

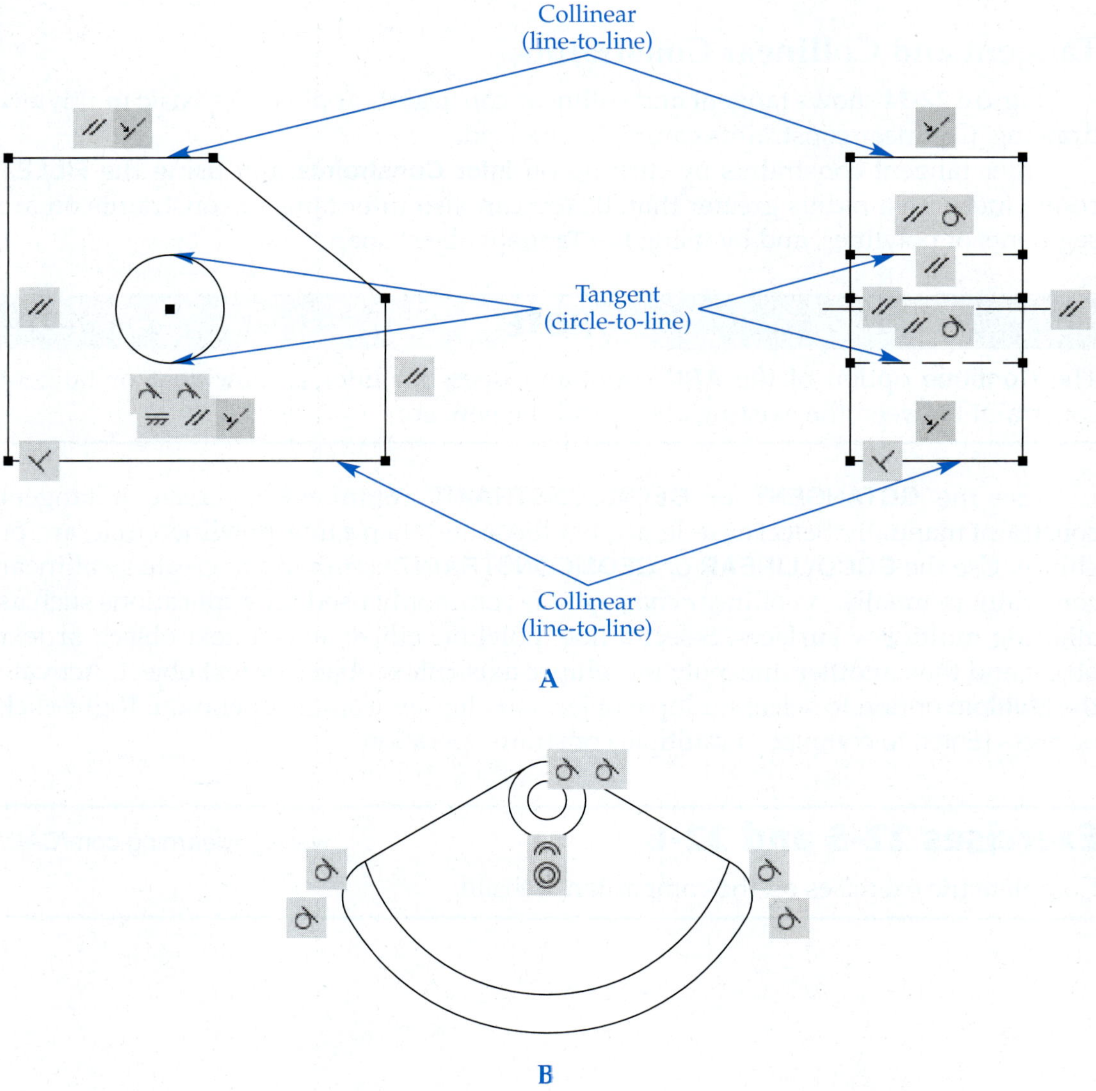

Figure 22-15.
Examples of concentric and equal constraints. All circles are concentric to an arc. All small arcs are equal, and all small circles are equal. Use the **Multiple** option to help make multiple objects equal in a single operation.

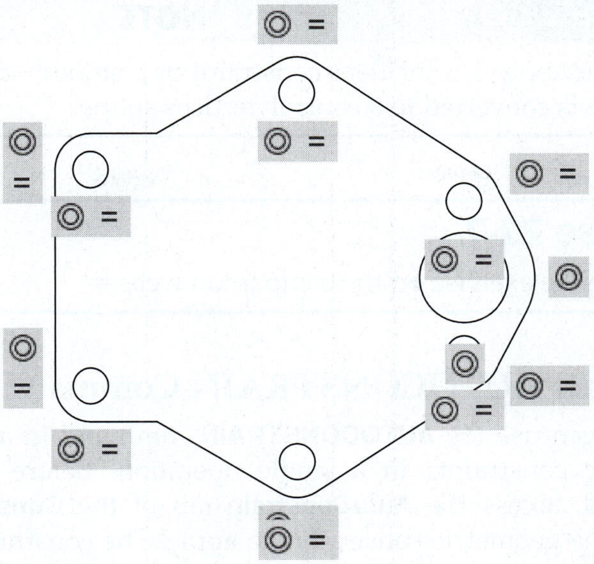

Use the **GCFIX** or **GEOMCONSTRAINT** command to secure a point or object to its current location in space to help preserve design intent. A single fix constraint is often required to fully constrain a drawing. Use the default method to fix a point, or activate the **Object** option to select an object to fix. Figure 22-16 shows examples of symmetric and fix constraints. Use the **GCSMOOTH** or **GEOMCONSTRAINT** command to connect and create a curvature-continuous situation, or G2 curve, between a selected spline and a line, polyline, spline, or arc.

> **Ribbon**
> Parametric
> > Geometric
> Fix
> **Type**
> GEOMCONSTRAINT
> > Fix
> GCFIX

> **Ribbon**
> Parametric
> > Geometric
> Smooth
> **Type**
> GEOMCONSTRAINT
> > Smooth
> GCSMOOTH

Figure 22-16.
A symmetrical parametric drawing created by adding symmetric constraints to circles and arcs. A fix constraint secures the drawing in space.

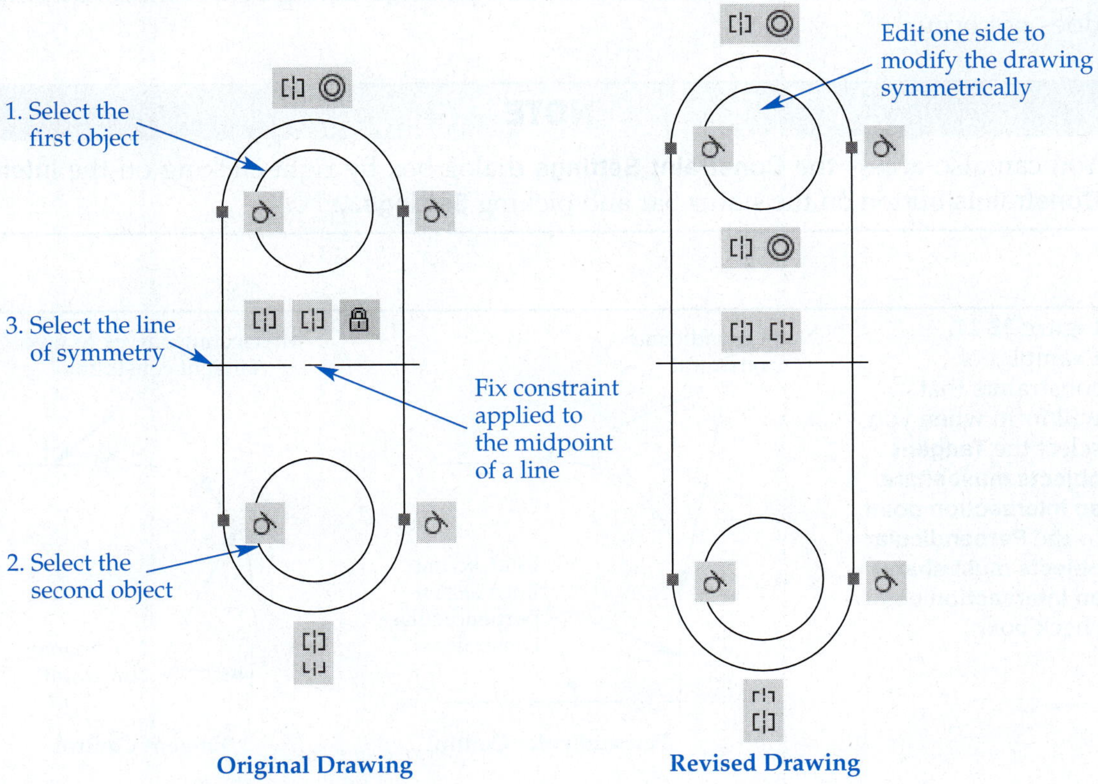

NOTE

When you apply a coincident constraint or a smooth constraint to a fit points spline, the spline is converted to a control vertices spline.

Exercise 22-7

www.g-wlearning.com/CAD/

Complete the exercise on the companion website.

Using the AUTOCONSTRAIN Command

Ribbon
Parametric > Geometric

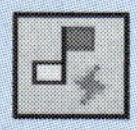

AutoConstrain

Type
AUTOCONSTRAIN

Ribbon
Parametric > Geometric

Constraint Settings

Type
CONSTRAINT-SETTINGS

You can use the **AUTOCONSTRAIN** command in an attempt to add all required geometric constraints in a single operation. Before using the **AUTOCONSTRAIN** command, access the **AutoConstrain** tab of the **Constraint Settings** dialog box to specify the geometric constraints to apply. The constraint priority determines which constraints are applied first. The higher the priority, the more likely and often the constraint will form if appropriate geometry is available. Select a constraint and use the **Move Up** and **Move Down** buttons to change its priority. Use the corresponding **Apply** check marks and the **Select All** and **Clear All** buttons to omit specific constraints during the constraining procedure.

The check boxes determine whether tangent and perpendicular constraints can form if objects do not intersect. See Figure 22-17. The **Tolerances** area controls how specific constraints form based on the distance between and angle of objects. A distance less than or equal to the value specified in the **Distance** text box receives constraints. An angle less than or equal to the value specified in the **Angle** text box receives constraints. See Figure 22-18.

Once you specify the settings, access the **AUTOCONSTRAIN** command and select the objects to constrain. The **Settings** option is available before selection to access the **AutoConstrain** tab of the **Constraint Settings** dialog box. A fix constraint does not occur.

NOTE

You can also access the **Constraint Settings** dialog box by right-clicking on the **Infer Constraints** button on the status bar and picking **Settings...**.

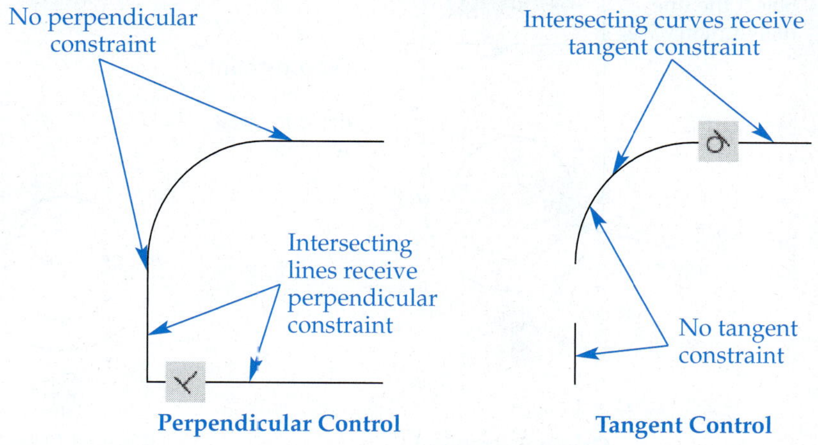

Figure 22-17. Examples of constraints that will form when you select the **Tangent objects must share an intersection point** or the **Perpendicular objects must share an intersection point** check box.

Figure 22-18.
Examples of constraints that form based on **Distance** and **Angle** tolerance values.

	Distance between Objects Greater Than Distance Tolerance	Distance between Objects Less Than or Equal to Distance Tolerance	Angle Greater Than Angle Tolerance	Angle Less Than or Equal to Angle Tolerance
Before				
After	No Effect	Coincident Endpoints	No Effect	Horizontal

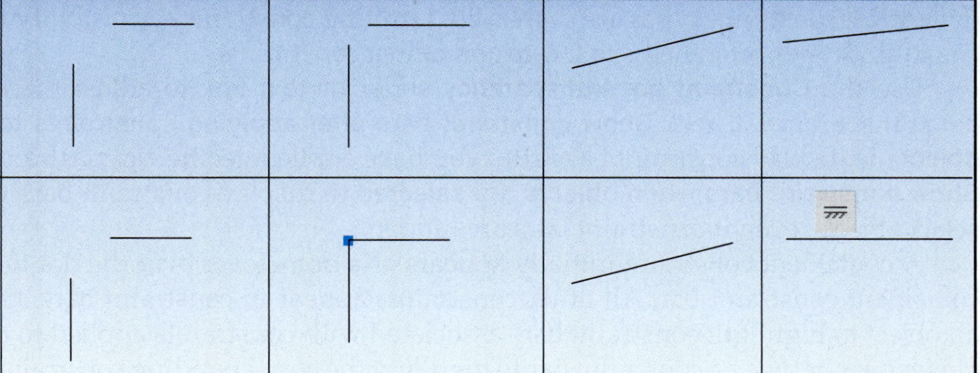

CAUTION

The **AUTOCONSTRAIN** command can save drafting time by placing geometric constraints in a single operation. However, use caution to ensure that appropriate geometric constructions occur and geometry does not shift. You may have to replace certain constraints.

Managing Geometric Constraints

Geometric constraint bars appear by default to indicate geometric constraints. See Figure 22-19. Refer to Figure 22-8 for help recognizing constraint icons. The **CONSTRAINTBAR** command includes options to show or hide constraint bars. A quick way to access these options is to pick the appropriate button from the **Geometric** panel of the **Parametric** ribbon tab.

Pick the **Show/Hide** button, pick objects to manage, and then right-click or press [Enter] to access options. Choose the **Show** option to display hidden constraint bars, the **Hide** option to hide visible constraint bars, or the **Reset** option to display hidden

Ribbon
Parametric
> Geometric

Show/Hide Geometric Constraints

Type
CONSTRAINTBAR

geometric constraint bars: Toolbars that allow you to view and remove geometric constraints.

Figure 22-19.
Use geometric constraint bars to view and delete geometric constraints. Horizontal, vertical, symmetric, and fix constraints use different icons depending on whether the constraint is a point-to-point or object-to-point constraint.

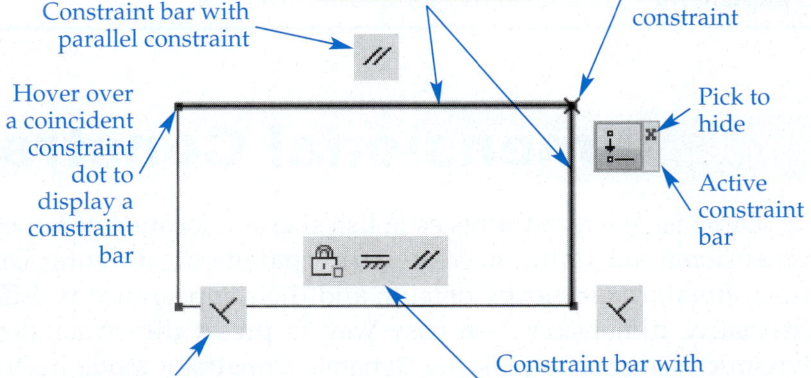

constraint bars and move constraint bars back to the default positions. Select the **Show All** button to display all constraint bars. Pick the **Hide All** button to hide all constraint bars. Hiding constraint bars does not remove geometric constraints.

Access the **Geometric** tab of the **Constraint Settings** dialog box to specify the geometric bars that appear and other geometric bar characteristics. Select the check boxes for individual constraints to display the constraints in constraint bars. Use the **Select All** and **Clear All** buttons to select or deselect all constraint-type check boxes. By default, all constraint types are displayed. Limiting constraint bar visibility to specific constraint types often helps to locate and adjust constraints.

Use the **Constraint bar transparency** slider or text box to adjust the constraint bar transparency. Check **Show constraint bars after applying constraints to selected objects** to display constraint bars after you have applied geometric constraints. Check **Show constraint bars when objects are selected** to display constraint bars when you select objects, even if constraint bars are hidden.

A coincident constraint initially appears as a dot. Hover over the dot to show the coincident constraint bar. All other constraints appear in constraint bars. Hover over an object to highlight constraint bars associated with constraints applied to the object. Hover over or select a constraint bar to highlight the corresponding constrained objects and display markers that identify constrained points. See Figure 22-19. The visual effect of hovering over an object or geometric constraint bar helps you recognize the objects and points associated with the constraint.

If a constraint bar blocks your view, drag it to a new location. To hide a specific constraint bar, pick the **Hide Constraint Bar** button, or right-click and choose **Hide**. The right-click shortcut menu also includes options for hiding all constraint bars and for accessing the **Geometric** tab of the **Constraint Settings** dialog box.

Design changes sometimes require deleting existing constraints. To delete geometric constraints, hover over an icon in the constraint bar and press [Delete], or right-click and select **Delete**.

PROFESSIONAL TIP

To confirm that constraints, especially geometric constraints, are present and appropriate, select an object to display grips and attempt to stretch a grip. As an object becomes constrained, you should observe less freedom of movement. Stretching or attempting to stretch grips is one of the fastest ways to assess design options and to analyze where a constraint is still required. You know the drawing is constrained when you are no longer able to stretch the geometry.

Exercise 22-8

www.g-wlearning.com/CAD/

Complete the exercise on the companion website.

dynamic format: A dimensional constraint format used specifically for controlling the size or location of geometry.

Dimensional Constraints

Ribbon
Parametric > Dimensional > Dynamic Constraint Mode
Type
CCONSTRAINT-FORM DCFORM

Dimensional constraints establish size and location parameters. You must include dimensional constraints to create a truly parametric drawing. Dimensional constraints use a ***dynamic format*** by default, and their appearance is different from traditional associative dimensions. An easy way to preset dimensional constraints to use the dynamic format is to pick the **Dynamic Constraint Mode** button from the expanded **Dimensional** panel of the **Parametric** ribbon tab.

You cannot modify how dynamic dimensional constraints appear, but you can change them to an ***annotational format***. To preset new dimensional constraints to use an annotational format, pick the **Annotational Constraint Mode** button from the expanded **Dimensional** panel of the **Parametric** ribbon tab. The annotational format uses the current dimension style. This textbook focuses on using the **Dynamic Constraint Mode** function and then assigning annotational format later. View, adjust, and remove dimensional constraints as needed.

Ribbon
Parametric
> Dimensional
> Annotational Constraint Mode
Type
CCONSTRAINTFORM
DCFORM

annotational format: A dimensional constraint format in which the constraints look like traditional dimensions, using a dimension style. Annotational dimensional constraints can still control the size or location of geometry.

Adding Dimensional Constraints

A quick way to create dimensional constraints is to pick the appropriate button on the **Dimensional** panel of the **Parametric** ribbon tab. You can also type DC followed by the name of the constraint, such as DCLINEAR, or select an option from the **DIMCONSTRAINT** command. To create a dimensional constraint, follow the prompts to make the required selections, pick a location for the dimension line, enter a value to form the constraint, and exit the command.

The process of selecting points or objects to locate dimensional constraint extension lines is the same as that for adding geometric constraints. When a dimensional constraint command requires you to pick two points or objects, the first point or object you select generally remains the same. In some cases, you can consider the first point or object the datum. The second point or object you select changes in relation to the first selection, unless the second selection is fixed. When selecting points, be sure to select the points corresponding to the surface to constrain.

A text editor appears after you select the location for the dimension line, allowing you to specify the dimension value. See Figure 22-20A. Each dimensional constraint is a parameter with a specific name, expression, and value. By default, linear dimensions receive d names, angular dimensions receive ang names, diameter dimensions receive dia names, and radial dimensions receive rad names. Every parameter must have a unique name. The name of the first of each type of dimension includes a 1, such as d1. The next dimension includes a 2, such as d2, and so on. Use the text editor to enter a more descriptive name, such as Length, Width, or Diameter. Follow the name with the = symbol and then the dimension value. Changing the current dimension value modifies object size or location. Press [Enter] or pick outside of the text editor to form the constraint. See Figure 22-20B.

The most basic way to specify the value of a dimensional constraint is to type a value in the text editor. Dimensional constraint units reflect the current work environment and unit settings, including length, angle type, and precision. Accept the current value if the drawing is accurate. Enter a different value if the drawing is inaccurate, or to change object size.

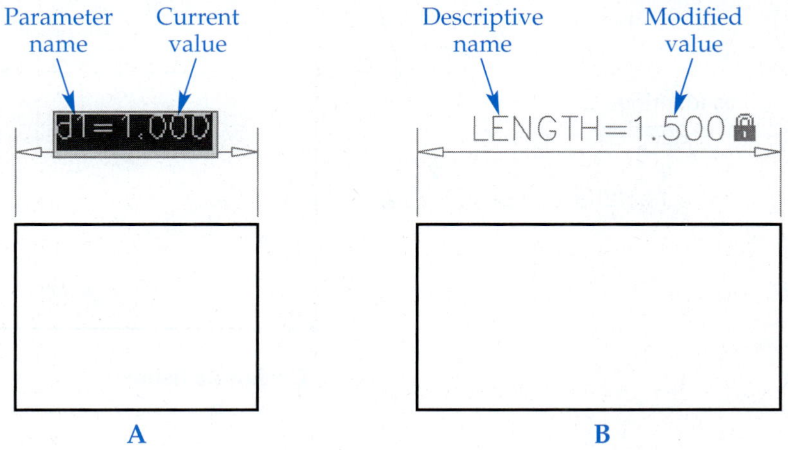

Figure 22-20. A—Use the text editor that appears after you establish the location of the dimension line to specify a parametric dimension. B—An example of modifying a parameter name and value. Notice that the dimension controls the object size.

Another option is to enter an expression in the text editor. You can use an expression if you do not know an exact value, much like using a calculator. Usually, however, expressions include parameters to associate dimensional constraints. This enables the drawing to adapt according to parameter changes. In the active text editor, move the text cursor to the location to add the parameter, and then pick another existing dimensional constraint to copy its name to the expression. An alternative is to type the parameter name in the expression. An fx: in front of the dimension text indicates an expression. **Figure 22-21** shows an example of using an existing parameter in an expression to control the size of an object. In this example, the design requires that the height of the object always be half the value of the length.

NOTE

If you add an existing parameter to the text of a different dimensional constraint, without making a calculation, such as d1 = d2, the dimensional constraint uses the same value as the reference parameter. This is necessary for many applications, but use an equal geometric constraint when possible.

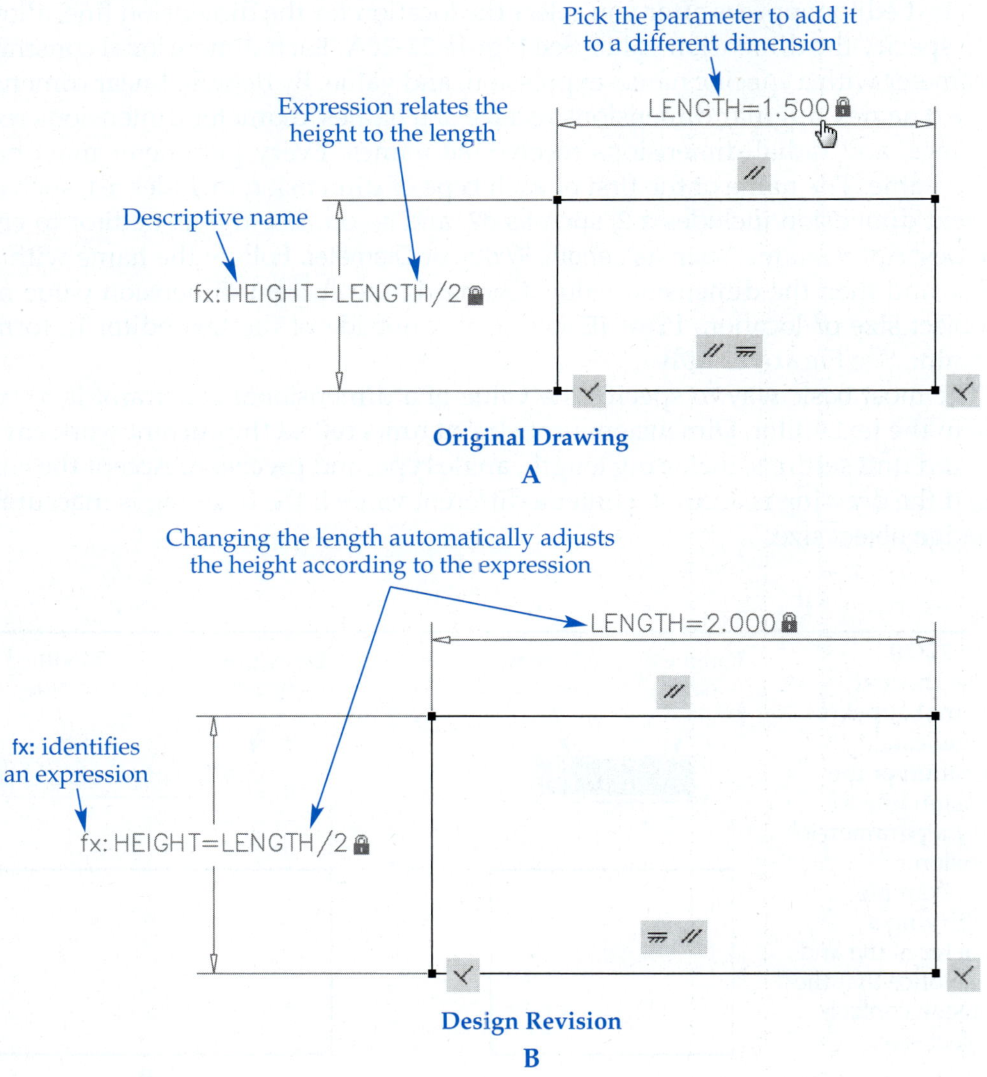

Figure 22-21.
A—A modified parameter name and expression that references another parameter. B—The dimension that includes a parameter references the parameter when changes occur.

Linear Dimensional Constraints

Use the **DCLINEAR** or **DIMCONSTRAINT** command to place a horizontal or vertical linear dimensional constraint. The **Horizontal** option sets the command to constrain only a horizontal distance. The **Vertical** option sets the command to constrain only a vertical distance. The **Horizontal** and **Vertical** options are helpful when it is difficult to produce the appropriate linear dimensional constraint, such as when you are dimensioning the horizontal or vertical distance of an angled surface.

Pick two points to specify the origin of the dimensional constraint, or use the **Object** option to select a line, polyline, or arc to constrain. Move the dimension line to an appropriate location and pick. Specify the dimension value and adjust the parameter name if desired. Press [Enter] or pick outside of the text editor to form the constraint. Figure 22-20 and Figure 22-21 show examples of linear dimensions.

> Ribbon
> **Parametric > Dimensional**
> **Linear**
> Type
> **DIMCONSTRAINT > Linear**
> **DCLINEAR**

> Ribbon
> **Parametric > Dimensional**
> **Horizontal**
> Type
> **DIMCONSTRAINT > Horizontal**
> **DCHORIZONTAL**

> Ribbon
> **Parametric > Dimensional**
> **Vertical**
> Type
> **DIMCONSTRAINT > Vertical**
> **DCVERTICAL**

Exercise 22-9

www.g-wlearning.com/CAD/

Complete the exercise on the companion website.

Aligned Dimensional Constraints

Use the **DCALIGNED** or **DIMCONSTRAINT** command to place a linear dimensional constraint with a dimension line that is aligned with an angled surface and extension lines perpendicular to the surface. Pick two points to specify the origin of the dimensional constraint, or use the **Object** option to select a line, polyline, or arc to constrain.

Often when you apply aligned dimensions, such as when you dimension an auxiliary view, it is necessary to pick a point and an aligned surface or two aligned surfaces. Use the **Point & line** option to select a point and an alignment line. Use the **2Lines** option to select two alignment lines. Move the dimension line to an appropriate location and pick. Specify the dimension value and adjust the parameter name if desired. Press [Enter] or pick outside of the text editor to form the constraint. Figure 22-22 shows examples of aligned dimensional constraints created using each method.

> Ribbon
> **Parametric > Dimensional**
> **Aligned**
> Type
> **DIMCONSTRAINT > Aligned**
> **DCALIGNED**

Angular Dimensional Constraints

Use the **DCANGULAR** or **DIMCONSTRAINT** command to place an angular dimensional constraint between two objects or three points. Pick two lines, two polyline segments, or an arc, or use the **3Point** option to select the angle vertex, followed by two points to locate each side of the angle. Move the dimension line to an appropriate location and pick. Specify the dimension value and adjust the parameter name if desired. Press [Enter] or pick outside of the text editor to form the constraint. Figure 22-23 shows examples of angular dimensional constraints created using each method.

> Ribbon
> **Parametric > Dimensional**
> **Angular**
> Type
> **DIMCONSTRAINT > Angular**
> **DCANGULAR**

Exercise 22-10

www.g-wlearning.com/CAD/

Complete the exercise on the companion website.

Diameter and Radius Dimensional Constraints

Ribbon
Parametric
> Dimensional
Diameter
Type
DIMCONSTRAINT
> Diameter
DCDIAMETER

Ribbon
Parametric
> Dimensional
Radius
Type
DIMCONSTRAINT
> Radius
DCRADIUS

Access the **DCDIAMETER** or **DIMCONSTRAINT** command to form a diameter dimensional constraint. Access the **DCRADIUS** or **DIMCONSTRAINT** command to form a radius dimensional constraint. You can select a circle or arc when using either command. In formal drafting, diameter constraints are applied to circles and radius constraints are applied to arcs. See Figure 22-24. In some parametric applications, however, most often when you are dimensioning construction geometry, it is appropriate to constrain arcs using a diameter and circles using a radius.

Figure 22-22.
A fully constrained auxiliary view drawing with dimensional constraints created using the **Aligned** option. Follow the extension lines to the objects to identify the selections.

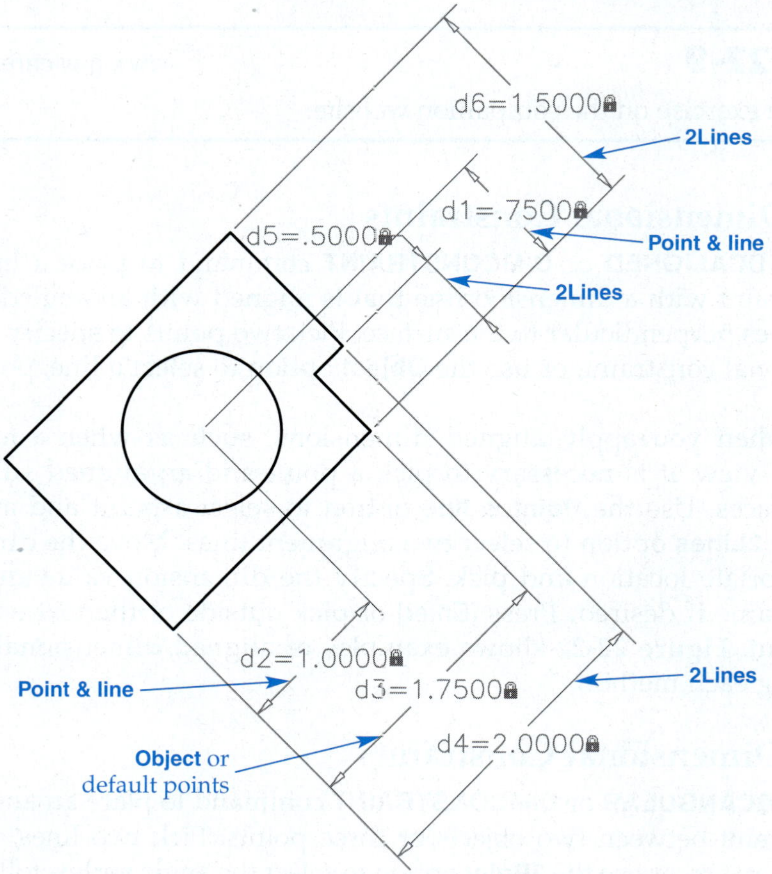

Figure 22-23.
Forming angular dimensional constraints.

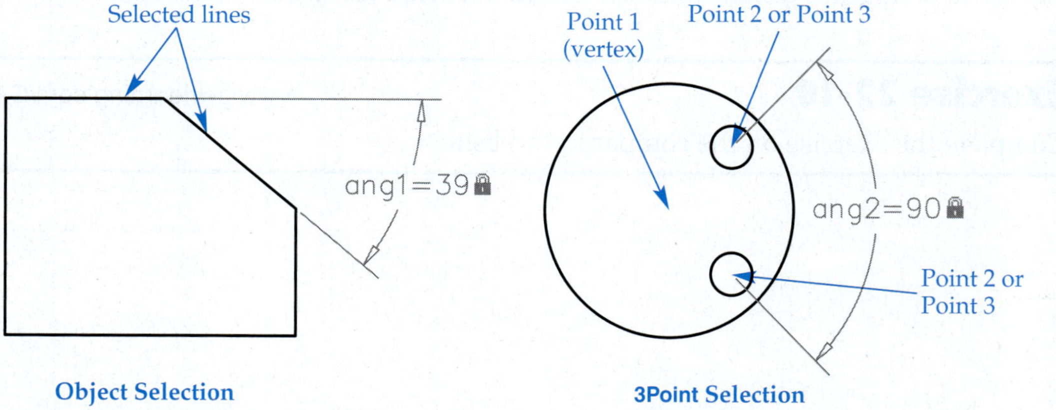

Figure 22-24.
Forming diameter and radius dimensional constraints. In this example, equal geometric constraints control the size of the undimensioned arcs.

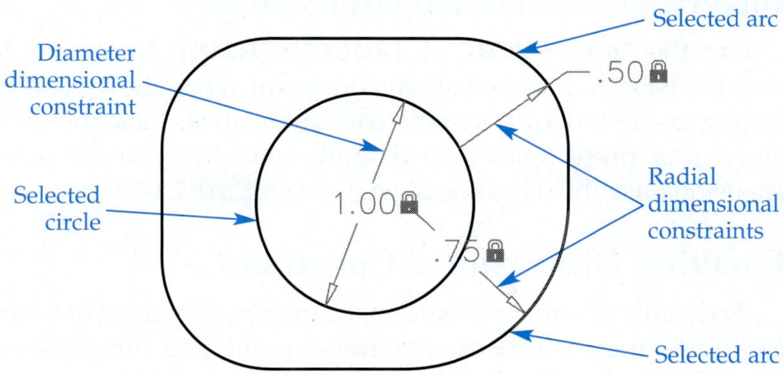

Reference Dimensional Constraints

When you try to constrain a fully constrained object, the drawing should become over-constrained. However, AutoCAD does not allow over-constraining to occur. You can either cancel the command without accepting the dimension, or accept the dimension and allow it to become a reference dimension. See Figure 22-25. Reference dimensions are sometimes required to form specific parameters or expressions. You cannot edit a reference dimension to change the size of an object, but a reference dimension changes when you modify related dimensions.

NOTE

Dimensional constraints define and constrain drawing geometry. They use a unique AutoCAD style and do not comply with ASME standards. Do not be overly concerned about the placement or display characteristics of dimensional constraints, but if possible, apply dimensional constraints just as you would add dimensions to a drawing, using correct drafting practices. In addition, move and manipulate dimensional constraints so the drawing is as uncluttered as possible. Use grips to make basic adjustments to dimensional constraint position.

Figure 22-25.
Parentheses identify a reference dimension. You cannot directly modify reference dimensions.

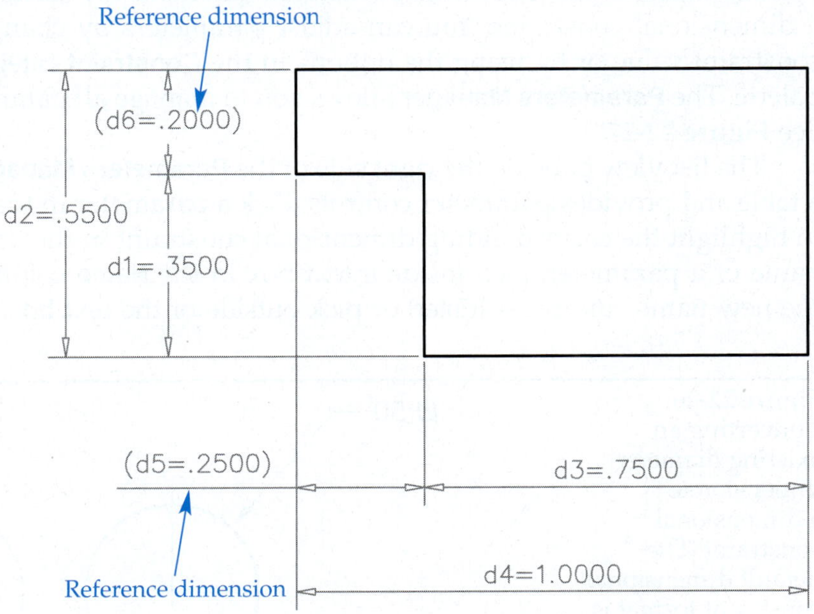

Converting Associative Dimensions

Ribbon
Parametric
> Dimensional
Convert
Type
DIMCONSTRAINT
> Convert
DCCONVERT

Use the **DCCONVERT** or **DIMCONSTRAINT** command to convert an associative dimension to a dimensional constraint. This allows you to prepare a parametric drawing using existing associative dimensions. Pick the associative dimensions to convert and press [Enter]. By default, new dimensional constraints and converted dimensions use the dynamic format. See Figure 22-26.

Managing Dimensional Constraints

Ribbon
Parametric
> Dimensional
Show/Hide Dynamic Constraints
Type
DCDISPLAY

By default, all dimensional constraints are displayed in the dynamic format, include a lock icon, and include the parameter name and dimension value. The **DCDISPLAY** command provides methods to show or hide dimensional constraints. A quick way to access these options is to pick the appropriate button from the **Dimensional** panel of the **Parametric** ribbon tab. Pick the **Show/Hide** button, pick dimensional constraints to manage, and then right-click or press [Enter] to access options. Choose the **Show** option to display hidden dimensional constraints, or the **Hide** option to hide visible dimensional constraints.

Select the **Show All** button to display all dimensional constraints. Pick the **Hide All** button to hide all dimensional constraints. Hiding dimensional constraints does not remove them. You typically hide dimensional constraints to prepare a formal drawing, or when you no longer need to see dimensional constraints for the current design phase.

Ribbon
Parametric
> Dimensional
Constraint Settings
Type
CONSTRAINT-SETTINGS

Use the **Dimensional** tab of the **Constraint Settings** dialog box to adjust additional dimensional constraint settings. The **Dimension name format:** drop-down list allows you to display dimensional constraints with the parameter name and value, parameter name, or value. Use the **Show lock icon for annotational constraints** check box to toggle the lock icon on or off for annotational constraints. If you hide dimensional constraints and select the **Show hidden dynamic constraints for selected objects** check box, you can pick an object to show associated dimensional constraints temporarily.

Design changes sometimes require deleting existing constraints. Use the **ERASE** command to eliminate specific dimensional constraints.

Working with Parameters

dimensional constraint parameters: Parameters that form when you insert a dimensional constraint.

Ribbon
Parametric
> Manage
Parameters Manager
Type
PARAMETERS

A ***dimensional constraint parameter*** automatically forms every time you add a dimensional constraint. You can adjust parameters by changing the dimensional constraint value or by using the options in the **Constraint** category of the **Properties** palette. The **Parameters Manager** allows you to manage all parameters in the drawing. See Figure 22-27.

The list view pane on the right side of the **Parameters Manager** lists parameters in a table and provides parameter controls. Pick a parameter in the **Parameters Manager** to highlight the corresponding dimensional constraint in the drawing. To change the name of a parameter, pick inside a text box in the **Name** column to activate it, type the new name, and press [Enter] or pick outside of the text box. Enter a new value or

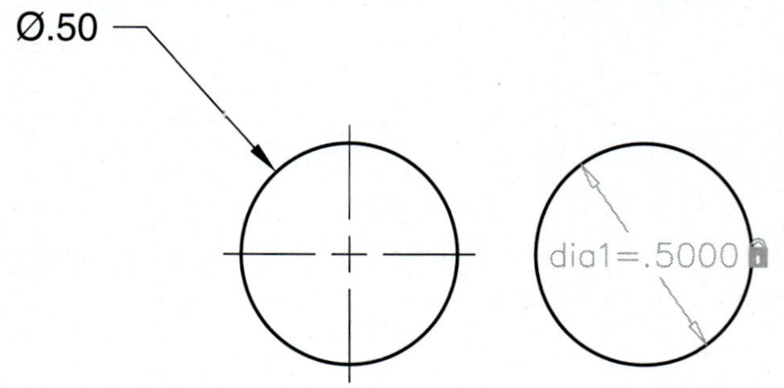

Figure 22-26. Converting an existing diameter dimension to a dimensional constraint. The default dimensional constraint format is dynamic.

Figure 22-27.
The **Parameters Manager** is a good resource for reviewing and editing parameters and for creating user parameters. This figure shows the expanded parameters filter tree.

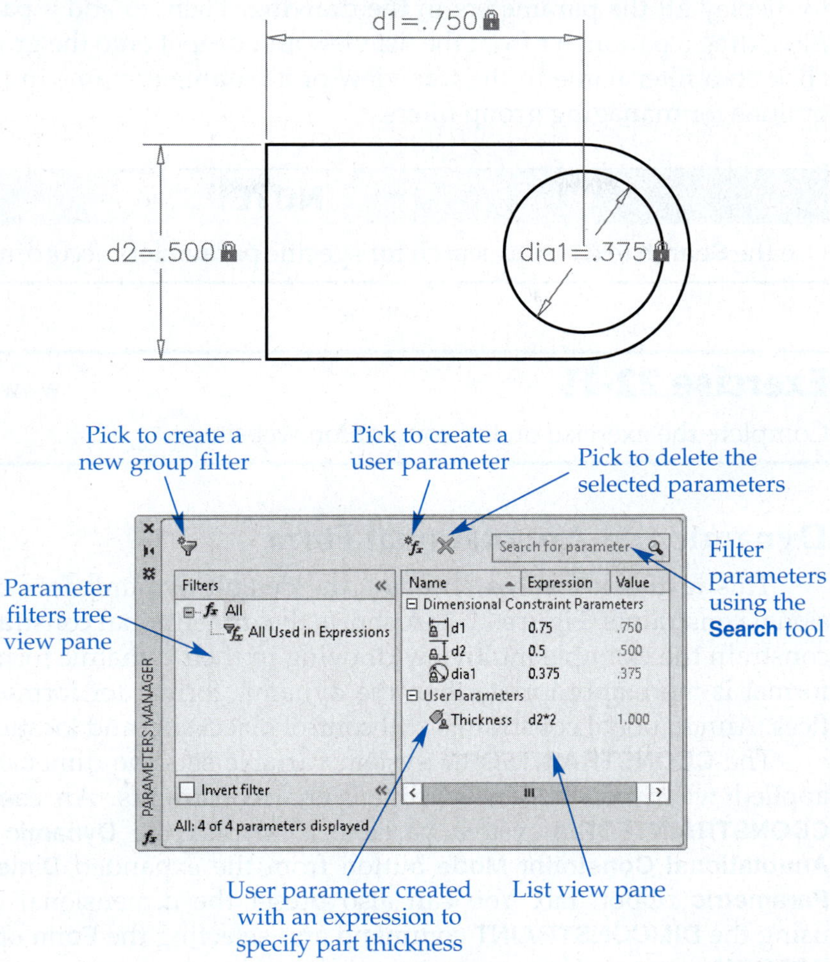

expression for the parameter in an **Expression** column text box. The value appears in the **Value** column display box for reference. Use the **Delete** button or right-click on a parameter and select **Delete Parameter** to remove the parameter and the corresponding dimensional constraint.

To specify ***user parameters,*** pick the **User Parameters** button to display a **User Parameters** node. User parameters function like dimensional constraints in the **Parameters Manager**. Create user parameters in order to access specific parameters throughout the design process. For example, if you know the thickness of a part will always be twice a certain dimensional constraint, create a user parameter similar to the parameter shown in Figure 22-27 to define the thickness. You can then use the custom parameter for reference and in expressions when you place additional dimensional constraints.

user parameters: Additional parameters you define.

Pick the **Expand Parameters** filter tree button or right-click in the list pane and select **Show Filter Tree** to display the tree view pane on the left side of the **Parameters Manager**. ***Parameter filters*** are typically appropriate when it becomes difficult to manage a very large number of parameters. Filter a large list of parameters to make it easier to work with the parameters needed for a specific drawing task.

parameter filters: Settings that screen out, or filter, parameters you do not want to display in the list view pane of the **Parameters Manager.**

Parameter filters are listed in alphabetical order in the **All** node. Select the **All** node to display all parameters in the drawing. Pick the **All Used in Expressions** filter to display only parameters used in or referenced by expressions. Pick the **Invert filter** check box to invert, or reverse, a parameter filter. Select the **Creates a new parameter group** button or right-click on a filter name and select **New Group Filter** to create a custom parameter group filter. A new group filter appears in the filter tree view ready to accept a custom name, if desired. Select the **All** node at the top of the filter tree area to display all the parameters in the drawing. Then, to add a parameter to the group filter, drag a parameter from the list view and drop it onto the group filter name. Right-click on a filter name in the tree view or a parameter name in the list view to access options for managing group filters.

NOTE

Use the **Search** text box to search for specific parameters according to parameter name.

Exercise 22-11

www.g-wlearning.com/CAD/

Complete the exercise on the companion website.

Dynamic and Annotational Form

This textbook focuses on using the default dynamic format for placing dimensional constraints. Figure 22-28A shows the dimensional constraints required to fully constrain the example multiview drawing in their dynamic format. The annotational format is more appropriate than the dynamic format for formal dimensioning practices. Annotational constraints still control object size and location.

The **CCONSTRAINTFORM** system variable sets the dimensional constraint form applied when you add new dimensional constraints. An easy way to adjust the **CCONSTRAINTFORM** system variable is to pick the **Dynamic Constraint Mode** or **Annotational Constraint Mode** button from the expanded **Dimensional** panel of the **Parametric** ribbon tab. You can also preset the dimensional constraint format by using the **DIMCONSTRAINT** command and selecting the **Form** option, or by using the **DCFORM** command before creating a dimensional constraint.

NOTE

The specified annotational or dynamic dimensional constraint format is a system variable and applies to new dimensional constraints until you change it to the alternate setting.

Use the Constraint Form drop-down list in the **Constraint** category of the **Properties** palette to change the dimensional constraint form assigned to existing dimensions. Figure 22-28B shows the dimensional constraints in dynamic format that require changing to the annotational format for the example multiview drawing. Annotational dimensions, especially those converted from dynamic dimensions, often require that you make format and organizational changes to prepare the final drawing. You may also have to add nonparametric dimensions. Make the following changes to create the final drawing shown in Figure 22-28C:

- Hide all geometric constraints and dynamic dimensions.
- Disable the lock icon display.
- Change the dimension name format to **Value**.
- Make basic dimension style overrides and dimension location adjustments if necessary.

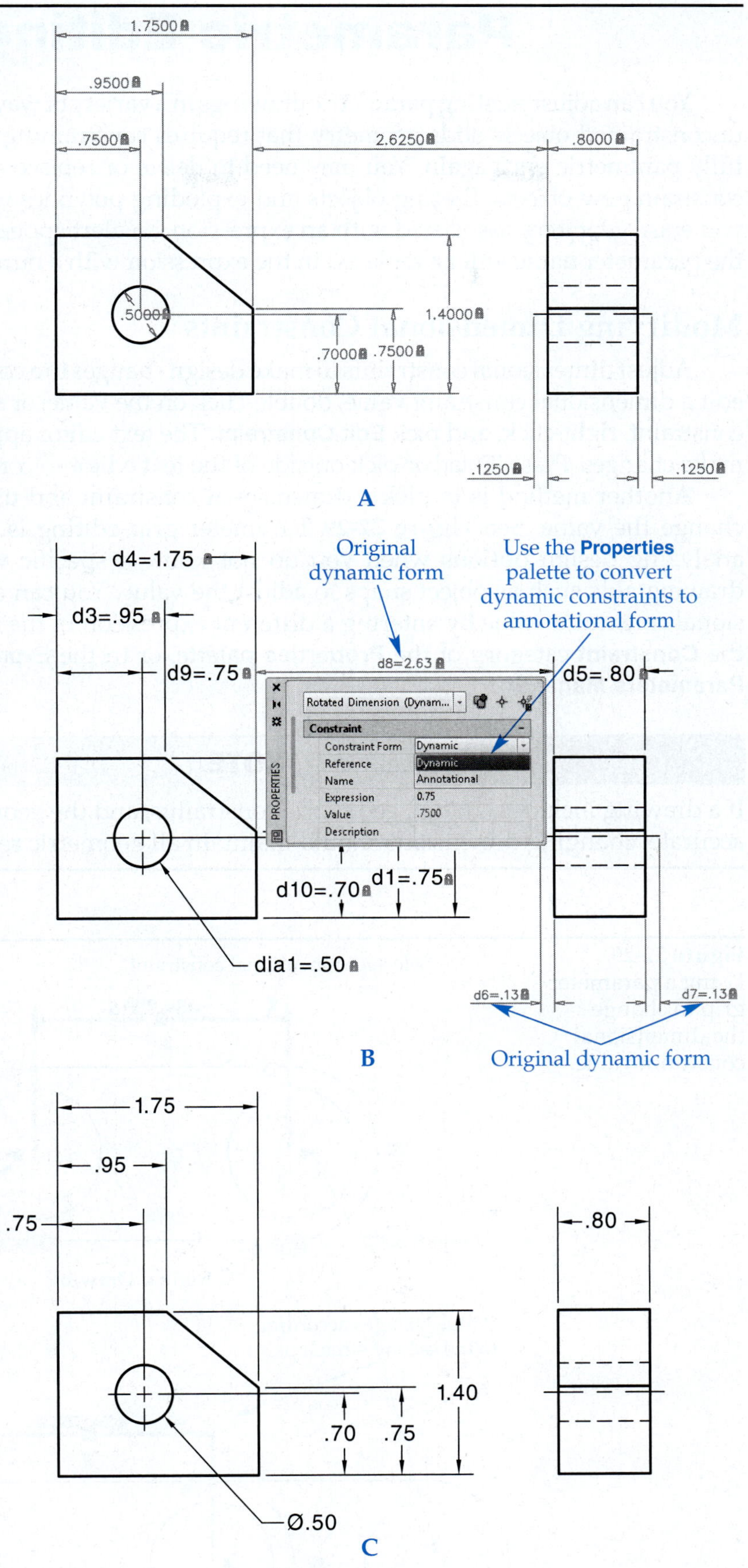

Figure 22-28. The typical process for changing a parametric drawing to a final, formal appearance. A—A fully constrained drawing with dynamic dimensional constraints. Notice that dimensions apply to all items, including centerline extensions and the distance between views. B—Changing dynamic dimensions to annotational dimensions. Change only those dimensions required for formal dimensioning. C—The appearance of the final drawing.

Parametric Editing

You can adjust existing parametric drawings in a variety of ways. Drawing additional unconstrained objects adds geometry that requires constraining to make the drawing fully parametric once again. You may need to delete or replace existing constraints to constrain new objects. Erasing objects and exploding polylines removes constraints. If you erase geometry associated with an expression, an alert appears informing you that the parameter name will be replaced in the expression with a numerical value.

Modifying Dimensional Constraints

Adjust dimensional constraints to make design changes to a constrained drawing. To edit a dimensional constraint value, double-click on the value, or select the dimensional constraint, right-click, and pick **Edit Constraint**. The text editor appears, allowing you to make changes. Press [Enter] or pick outside of the text editor to complete the operation.

Another method is to pick a dimensional constraint and use parameter grips to change the value. See Figure 22-29. Parameter grip-editing is most appropriate for analyzing design options when you do not know a specific value, or when using drawing aids such as object snaps to adjust the value. You can also modify a dimensional constraint value by entering a different expression in the Expression text box in the **Constraint** category of the **Properties** palette, or in the Expression text box of the **Parameters Manager**.

NOTE

If a drawing includes enough geometric constraints, and the geometric constraints are accurate, changing dimensions should maintain all geometric relationships.

Figure 22-29.
Using a parameter grip to change the dimensional constraint value.

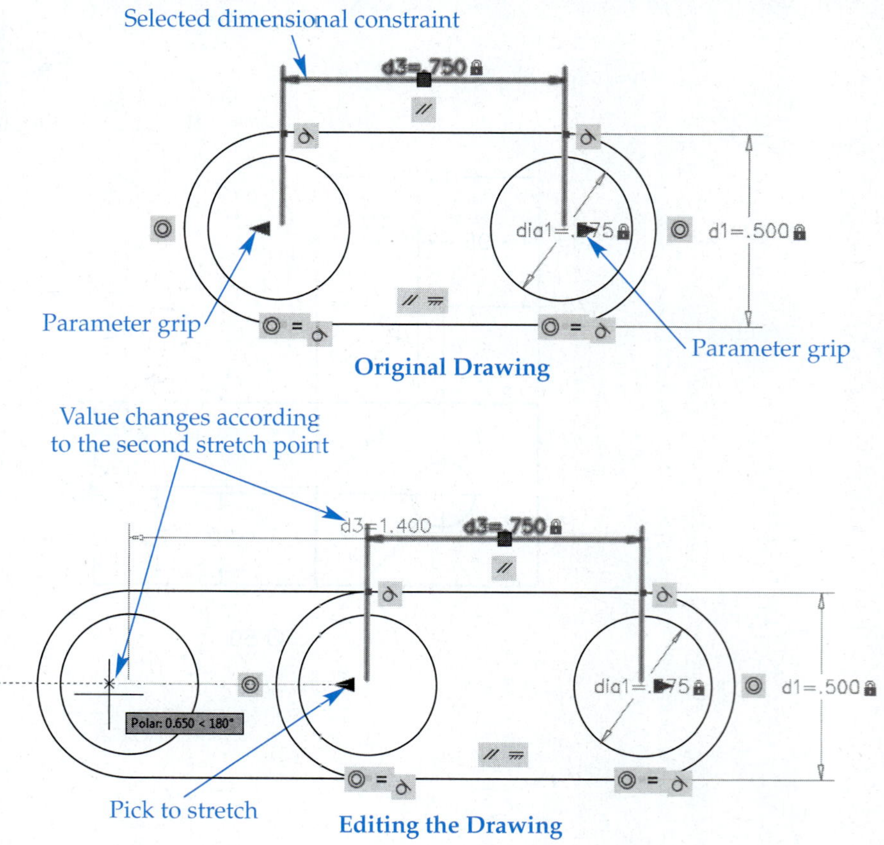

Exercise 22-12

www.g-wlearning.com/CAD/

Complete the exercise on the companion website.

Removing Constraints

Constraints limit your ability to make changes to a drawing. For example, you cannot rotate a horizontally or vertically constrained line. You may have to relax or delete constraints to make significant changes to a drawing. Relax constraints using standard editing practices. When you use a basic editing command such as **ROTATE**, a message appears asking if you want to relax constraints. To relax constraints while stretching, moving, or scaling with grips, you may need to press [Shift] to toggle relaxing on and off. See Figure 22-30A. Other edits automatically remove constraints. See Figure 22-30B. The only constraints, if any, that you remove are those required to achieve the edit.

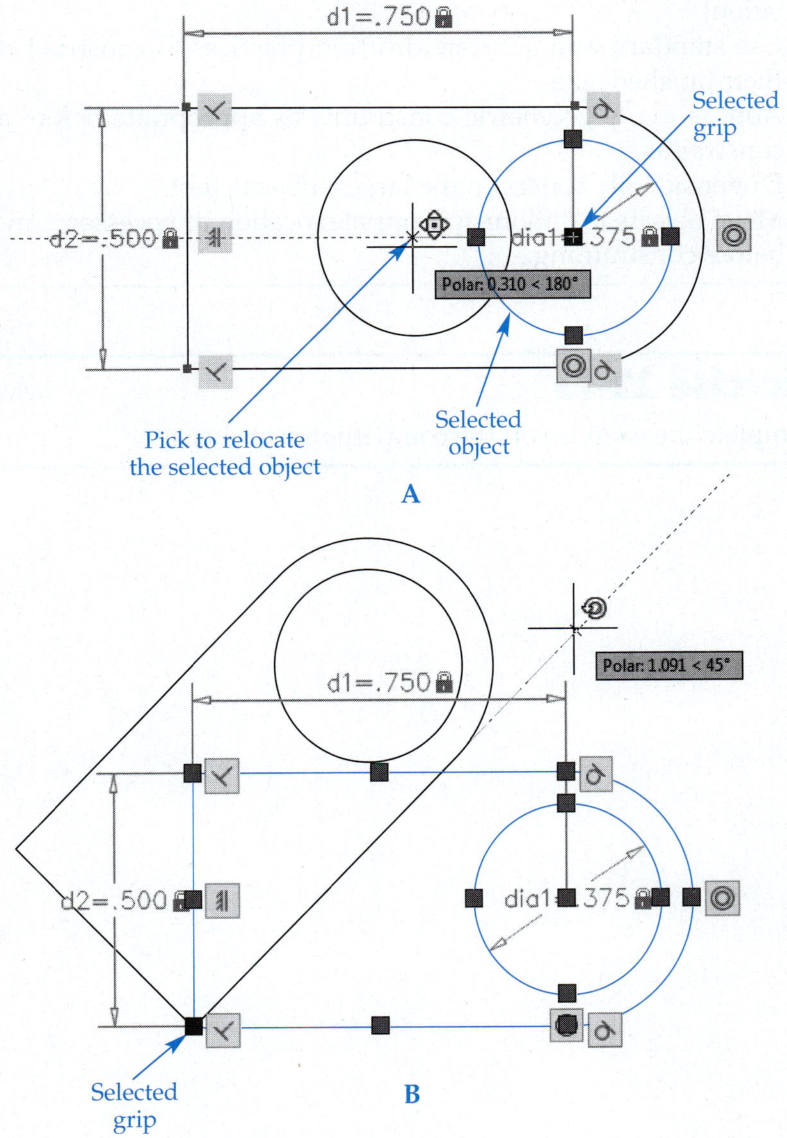

Figure 22-30. Examples of relaxing constraints. A—Access the **MOVE** command and then press [Shift] to remove the concentric constraint. The process is similar for stretching and scaling. B—Rotating removes a vertical constraint in this example.

Ribbon
Parametric > Manage
Delete Constraints
Type
DELCONSTRAINT

The **DELCONSTRAINT** command is available in addition to the methods previously described for deleting individual constraints. Access the **DELCONSTRAINT** command and select the geometric and dimensional constraints to remove. The **DELCONSTRAINT** command is especially effective for removing a significant number of constraints. Use the **All** option to remove all constraints from the drawing.

NOTE

Some editing techniques, such as mirroring or moving, may only require removal of fix constraints, especially when you are editing an entire drawing. Exploding a polyline removes all constraints.

PROFESSIONAL TIP

Occasionally, constraining geometry causes objects to twist out of shape, making it difficult to control the size and position of the drawing. Use the **UNDO** command to return to the previous design. Consider the following suggestions to help avoid this situation:

- Use standard and accurate drafting practices to construct objects at or close to their finished size.
- Add as many geometric constraints as appropriate before adding dimensional constraints.
- Dimensionally constrain the largest objects first.
- Move objects to a more appropriate location, if necessary, and change object size before constraining.

Exercise 22-13

www.g-wlearning.com/CAD/

Complete the exercise on the companion website.

Chapter Review

Answer the following questions using the information in this chapter.

1. Give an example demonstrating how to use constraints to form a geometric construction when standard AutoCAD commands are inefficient or ineffective.
2. Describe two applications in which parametric drafting may be unsuitable or ineffective.
3. Describe the purpose of geometric constraint tools, and identify what you see on-screen that indicates the presence of a geometric constraint.
4. Name the tool that forms geometric constraints while you draw or edit.
5. Name the commands that allow you to assign geometric constraints manually.
6. When you use geometric constraint commands that allow you to pick two objects or points, describe what generally happens to the first and second objects you select.
7. List the object snaps that infer coincident constraints.
8. Identify common uses for horizontal and vertical constraints.
9. Describe how to infer horizontal and vertical constraints.
10. List the object types that can form parallel or perpendicular constraints.
11. Name the types of objects you can constrain with the tangent constraint.
12. What does the collinear constraint allow you to do?
13. Explain the basic function of the equal constraint.
14. Describe the default function of the symmetric constraint.
15. Name the command you can use to attempt to add all required geometric constraints in a single operation.
16. Describe how to specify the appearance and characteristics of geometric constraint bars.
17. Compare the appearance of a coincident constraint with the display of other constraints.
18. Explain how to determine which objects and points are associated with a constraint.
19. What should you do if constraint bars block your view or if you want to hide constraint bars?
20. Name the command that allows you to assign linear, diameter, radius, and angular dimensional constraints.
21. What is the most basic method to specify dimension values when you create a dimensional constraint?
22. Name the commands that allow you to place a linear dimensional constraint with a dimension line aligned with an angled surface and extension lines perpendicular to the surface.
23. Which commands allow you to place an angular dimensional constraint between two objects or three points?
24. Explain the options AutoCAD provides when you try to over-constrain a drawing.
25. Explain how to convert an associative dimension to a dimensional constraint, and give the advantage of using this option.
26. What happens every time you add a dimensional constraint?
27. How do you adjust parameters?
28. Explain how to edit a dimensional constraint value.
29. Describe how to relax constraints.
30. Which command provides an efficient method of removing a significant number of constraints in a single operation?

Drawing Problems

Start AutoCAD if it is not already started. Start a new drawing for each problem using an appropriate template of your choice. The template should include layers and text, dimension, multileader, and table styles, when necessary, for drawing the given objects. Add layers and text, dimension, multileader, and table styles as needed. Draw all objects using appropriate layers. Use appropriate text, dimension, multileader, and table styles, justification, and format. Follow the specific instructions for each problem. Use only drawing and editing commands and techniques you have already learned. Use your own judgment and approximate dimensions when necessary. Apply formal dimensions accurately using ASME or appropriate industry standards.

Note: Dimensional constraints shown for reference are created using AutoCAD and may not comply with ASME standards.

▼ Basic

1. General Use the **POLYGON** command to draw the hexagon as shown. Fully constrain the hexagon as shown. All sides are equal. Fix the midpoint of the construction line. Edit the d1 parameter to change the distance across the flats to 4.000. Save the drawing as P22-1.

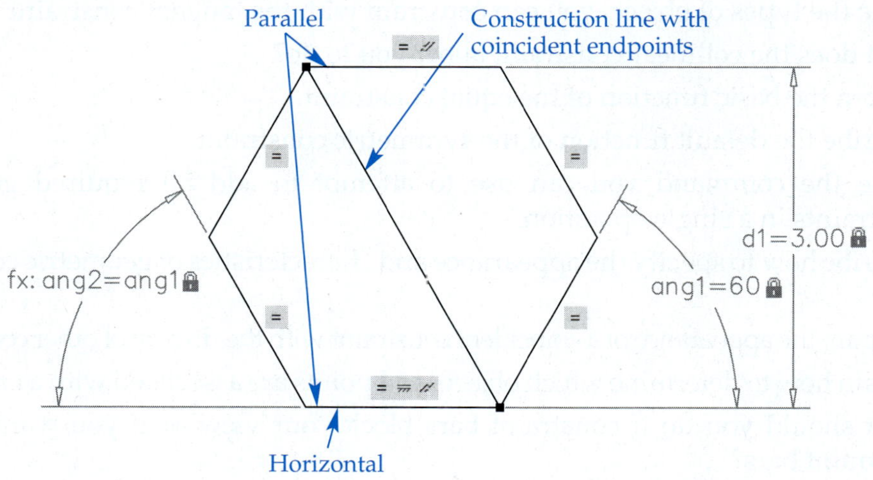

2. General Use the **RECTANGLE** and **CIRCLE** commands to draw the view shown. Fully constrain the view as shown. The circle is tangent to the rectangle in two locations. Edit the d2 parameter to change the distance to 4.500. Edit the d3 parameter to change the distance to 2.000. Save the drawing as P22-2.

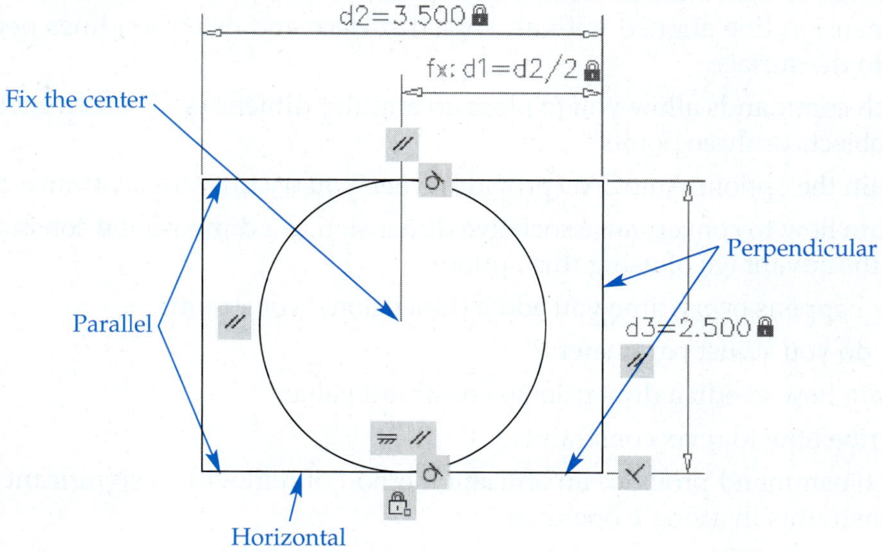

3. **General** Draw and fully constrain the circles shown. Both of the smaller circles are tangent to the larger circle. Edit the dia1 parameter to change the diameter to 2.000. Save the drawing as P22-3.

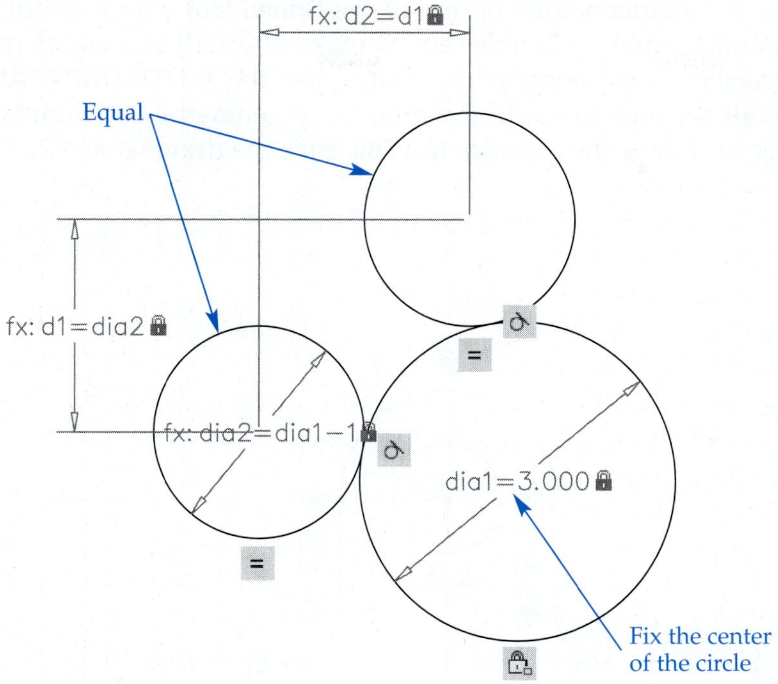

4. **Mechanical** Draw and fully constrain the pipe spacer shown. Infer as many constraints as possible and appropriate. Use the **AUTOCONSTRAIN** command, with default settings, to apply additional geometric constraints. Make all circles equal in size. Edit the d1 parameter to change the distance to 2.000. Edit the rad1 parameter to change the radius to 1.250. Save the drawing as P22-4.

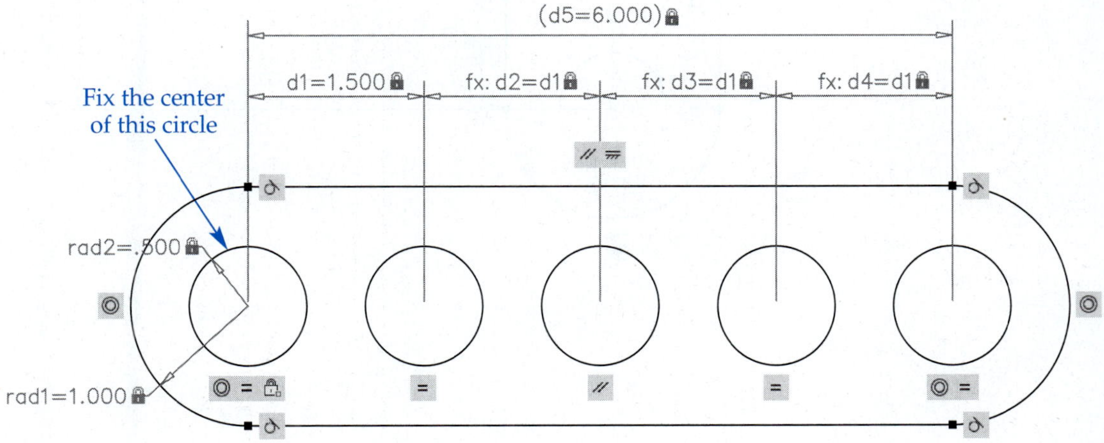

▼ Intermediate

5. Mechanical Draw and fully constrain the view shown. Do not infer constraints. Apply geometric constraints in the following order: fix the center of the center circle; use the **Autoconstrain** option of the **Coincident** option to apply all coincident constraints; make all circles equal; make all small arcs equal; make all large arcs concentric to the appropriate circles; use the **AUTOCONSTRAIN** command with default settings to apply the remaining geometric constraints. Edit the d1 parameter to change the distance to 1.750. Save the drawing as P22-5.

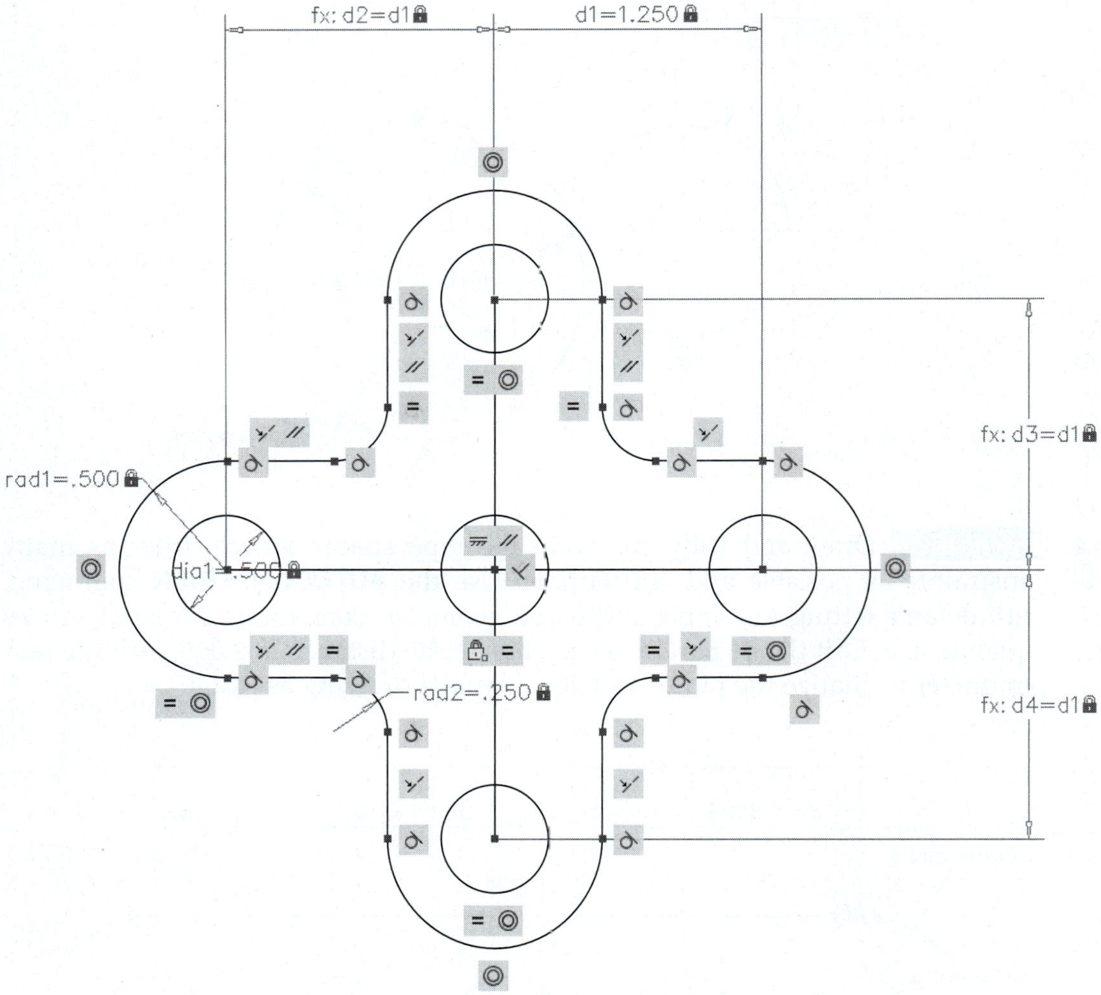

6. Mechanical Draw and fully constrain the view shown. Follow the guidelines below.
   - Infer constraints when possible and appropriate.
   - Use the **Fillet** option of the **RECTANGLE** command to create the 4.6250 by 6.3750 rectangle with .6250 rounded corners.
   - Use the **RECTANGLE** command to create the construction rectangle and then add the circles. Notice that the circles are not concentric to the arcs.
   - Apply geometric constraints in the following order: fix the center of the lower-left circle; make all circles equal; make all arcs equal.
   - Use the **AUTOCONSTRAIN** command with default settings to apply the remaining geometric constraints.
   - Edit the d1 parameter to change the distance to 1.0000.
   - Edit the d5 parameter to change the distance to 5.1250.
   - Save the drawing as P22-6.

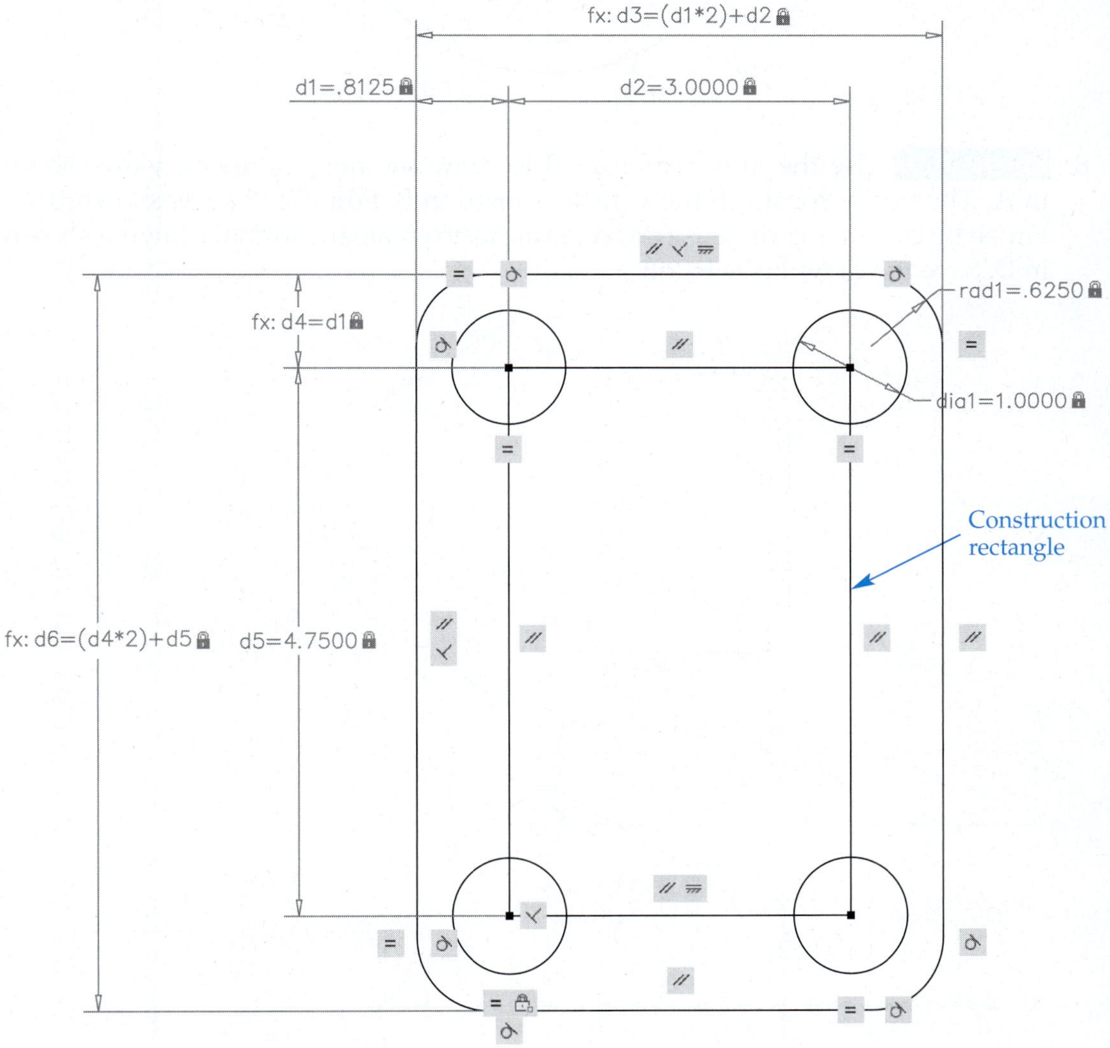

▼ Advanced

7. **Mechanical** Draw and fully constrain the view shown. Save the drawing as P22-7.

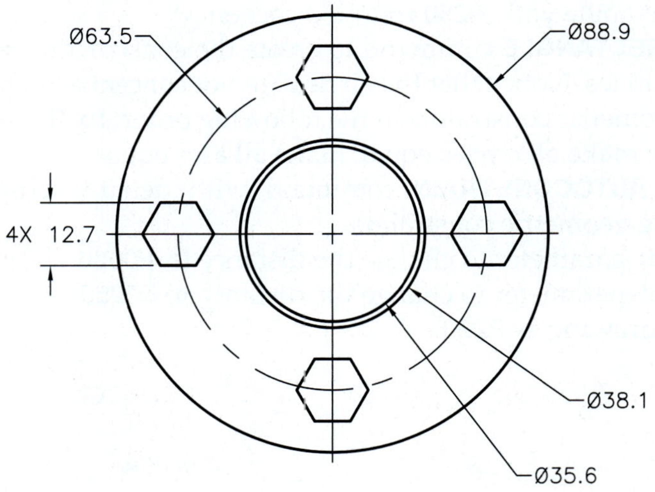

8. **Mechanical** Use the information in B to draw the nonparametric view shown in A. Then fully constrain the view as shown in B. Edit the view as shown in C. Finish by converting dimensional constraints to create the formal drawing shown in D. Save the drawing as P22-8.

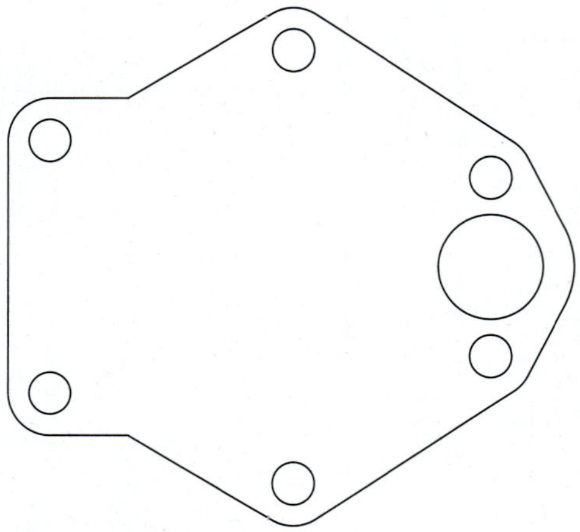

A

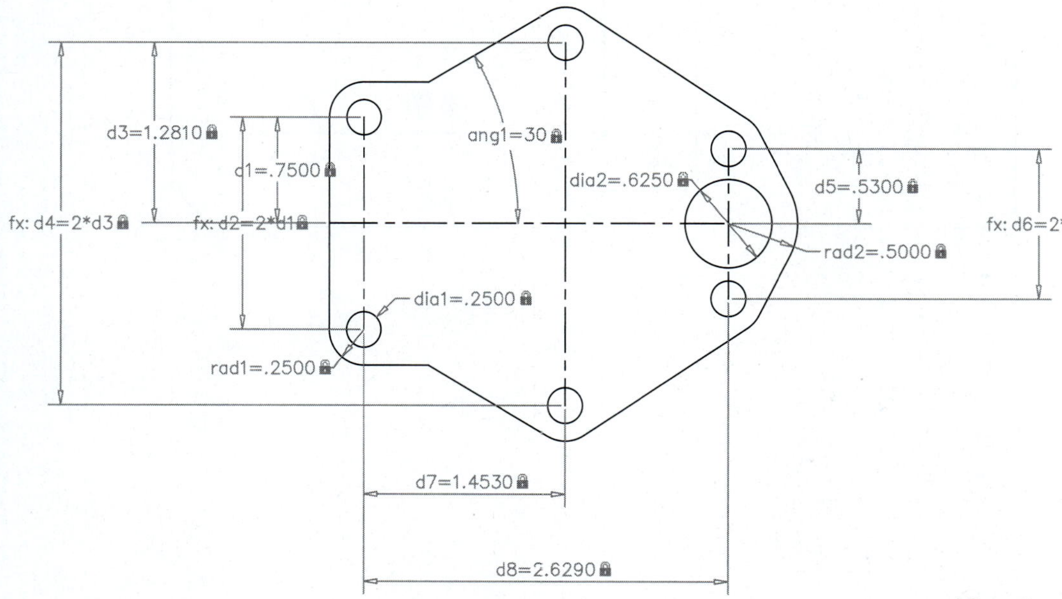

B

(Continued)

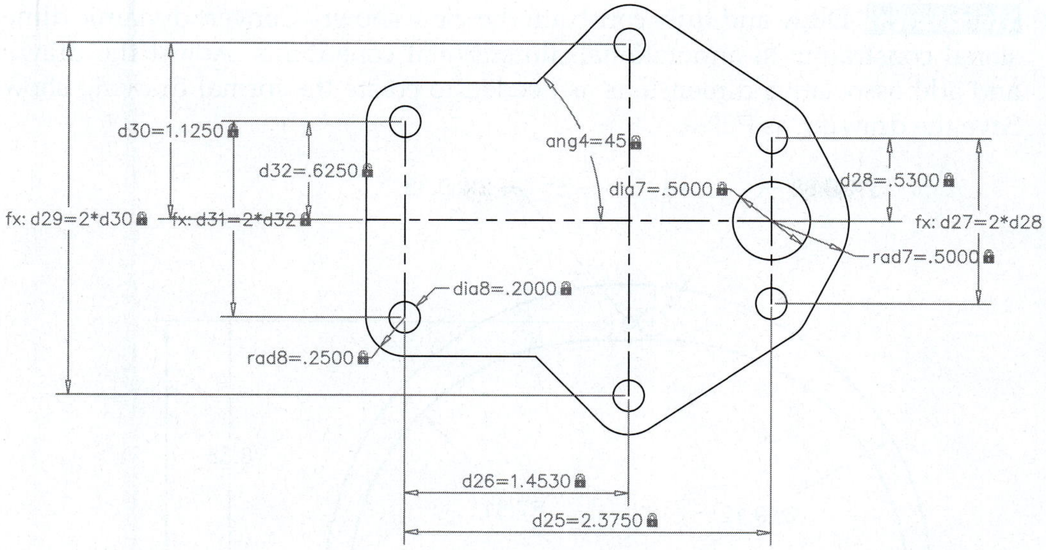
d30=1.1250
d32=.6250
fx: d29=2*d30
fx: d31=2*d32
ang4=45
dia7=.5000
d28=.5300
fx: d27=2*d28
rad7=.5000
dia8=.2000
rad8=.2500
d26=1.4530
d25=2.3750

C

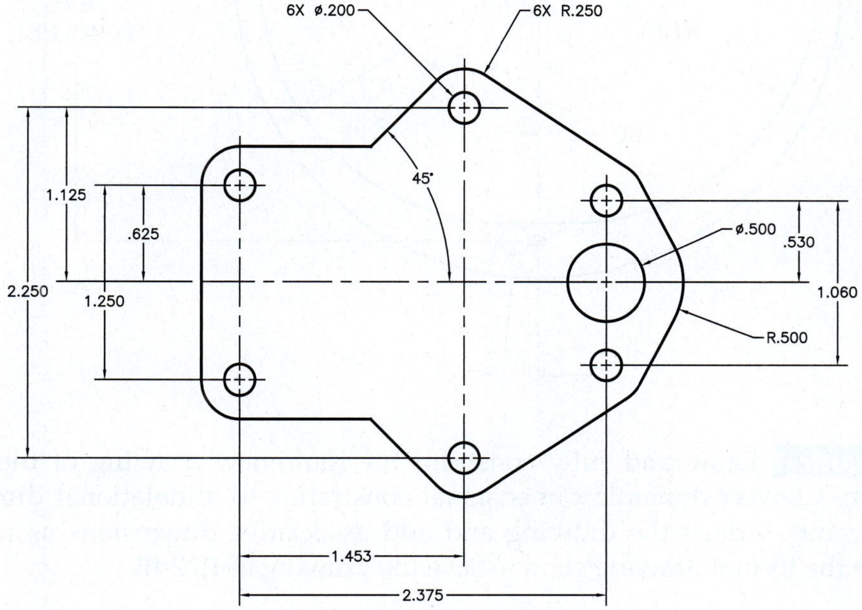
6X ⌀.200
6X R.250
45°
1.125
.625
2.250
1.250
⌀.500
.530
1.060
R.500
1.453
2.375

D

9. **Mechanical** Draw and fully constrain the view shown. Convert dynamic dimensional constraints to annotational dimensional constraints. Adjust the drawing and add associative dimensions as needed to create the formal drawing shown. Save the drawing as P22-9.

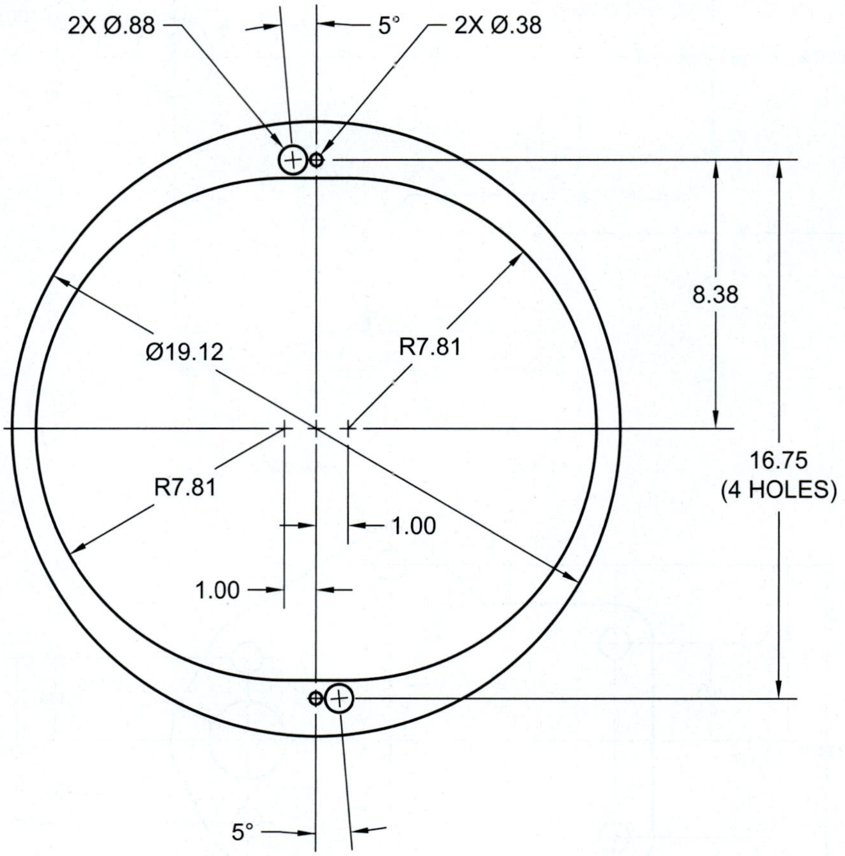

10. **Mechanical** Draw and fully constrain the multiview drawing of the support shown. Convert dynamic dimensional constraints to annotational dimensional constraints. Adjust the drawing and add associative dimensions as needed to create the formal drawing shown. Save the drawing as P22-10.

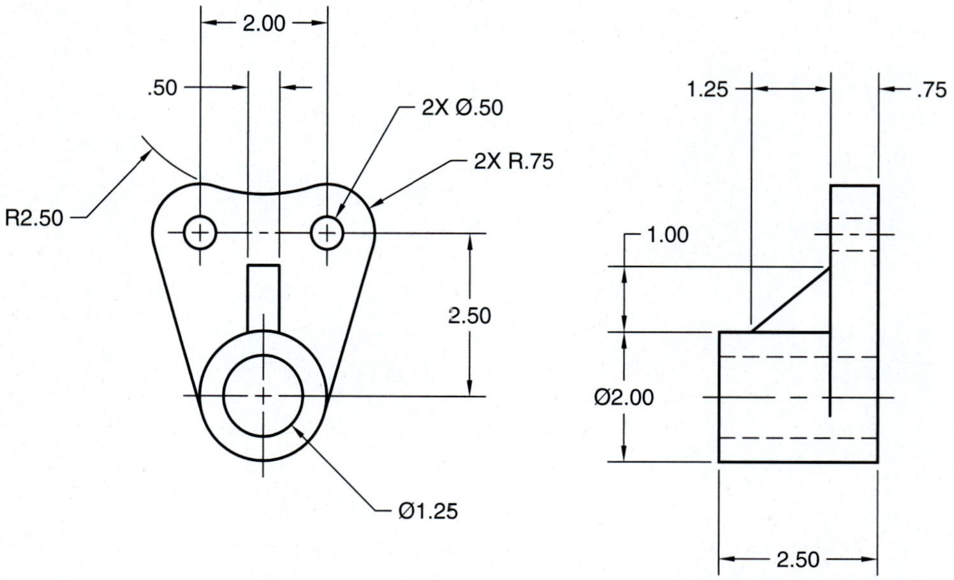

11. **Mechanical** Use the isometric drawing provided to create a multiview orthographic drawing for the part. Include only the views necessary to fully describe the object. Draw and fully constrain the drawing. Convert dynamic dimensional constraints to annotational dimensional constraints. Adjust the drawing and add associative dimensions as needed to create a formal drawing. Save the drawing as P22-11.

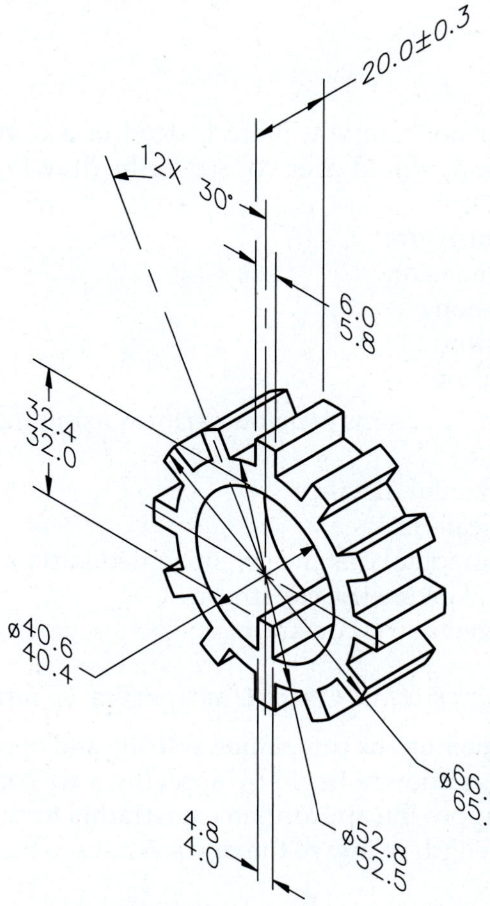

Title: Locking Collar
Material: SAE 1080
Metric

12. **Mechanical** Research the design of an existing manifold with the following specifications: one-piece; nickel-plated 6061 aluminum, NPT female connections; two 1/2″ inlets; eight 3/8″ outlets on one side, 1″ center-to-center; four mounting holes; for use with air, water, or hydraulic oil. Create a freehand, dimensioned 2D sketch of the existing design from manufacturer's specifications, or from measurements taken from an actual manifold. Start a new drawing from scratch or use a decimal-unit template of your choice. Draw and fully constrain a multiview orthographic drawing of the manifold. Include only the views necessary to fully describe the part. Convert dynamic dimensional constraints to annotational dimensional constraints. Adjust the drawing and add associative dimensions as needed to create a formal drawing. Save the drawing as P22-12.

AutoCAD Certified Professional Exam Practice

Answer the following questions using the information in this chapter.

1. Which of the following are forms of dimensional constraints? *Select all that apply.*
   A. annotational
   B. construction
   C. dynamic
   D. geometric
   E. static
2. What type of dimension can you place instead of a dimensional constraint if a dimensional constraint would over-constrain the drawing? *Select the one item that best answers the question.*
   A. annotational dimension
   B. construction dimension
   C. dynamic dimension
   D. inferred dimension
   E. reference dimension
3. Which of the following actions can you perform using the **DCLINEAR** command? *Select all that apply.*
   A. constrain a horizontal distance
   B. constrain a vertical distance
   C. constrain the horizontal distance of an angled surface
   D. place a horizontal geometric constraint
   E. place a vertical geometric constraint

Follow the instructions in each problem. Complete each problem by answering the related question.

4. **Navigate to this chapter on the companion website and open CPE-22parameter.dwg.** Constrain the lines as shown. Begin by applying a fix constraint to the lower-left intersection. Then apply the appropriate constraints to achieve the figure shown. Do not change the length of any of the lines. What are the coordinates of Point A?

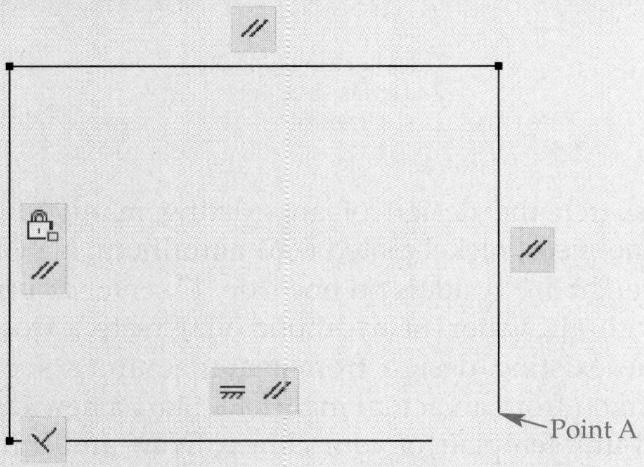

5. **Navigate to this chapter on the companion website and open CPE-22edit.dwg.** Show the dimensional constraints and edit the LENGTH dimension to 7.500. Do not change any other settings. What are the coordinates of Point A (the center of the right arc)?

CHAPTER 23

Section Views and Graphic Patterns

Learning Objectives

After completing this chapter, you will be able to:

✓ Identify sectioning techniques.
✓ Add graphic patterns using the **HATCH** command.
✓ Insert hatch patterns using **DesignCenter**.
✓ Insert hatch patterns using tool palettes.
✓ Edit existing hatch patterns.

Drawings often require ***graphic patterns*** to describe specific information. For example, the front elevation of the house shown in Figure 23-1 contains patterns and fills that graphically represent building materials and shading. One of the most common graphic patterns is a group of section lines added to a section view. This chapter focuses on using the **HATCH** command to draw graphic patterns. You will also learn other methods of adding hatches and how to edit hatches.

graphic pattern: A patterned arrangement of objects or symbols.

Section Views

It is poor practice, and often not possible, to dimension internal, hidden features. ***Section views***, also called ***sectional views*** or ***sections***, clarify hidden features. Typically, you add section views to a multiview drawing to describe the exterior and interior features of a product. Often, as shown in Figure 23-2, a primary view is a section or includes a section.

section view (sectional view, section): A view that shows internal features as if a portion of the object is cut away.

When a drawing includes a section, one of the other views contains a ***cutting-plane line*** to show the location of the cut. The cutting-plane line is a thick dashed or phantom line in accordance with the ASME Y14.2 *Line Conventions and Lettering* standard. The standard cutting-plane line terminates with arrowheads that point toward the cutting plane, indicating the line of sight when you are looking at the section view.

cutting-plane line: The line that cuts through the object to expose internal features.

Each end of the cutting-plane line is labeled with a letter. The letters correlate to the section view title below the section view, such as SECTION A-A, to key the cutting plane with the section view. When you section more than one view, labels continue with B-B through Z-Z, if necessary. Do not use letters *I*, *O*, *Q*, *S*, *X*, and *Z*, because they can be confused with numbers. Labeling section views is necessary for drawings with multiple sections. You can often omit the label when only one section view is present and its location is obvious.

Figure 23-1.
Graphic patterns describe repetitive drawing information, such as the siding, brick, roofing, concrete, and shading added to the front elevation of a house. In this illustration, all of the items shown in color are graphic patterns.

Figure 23-2.
A two-view mechanical part drawing with a full section view (front) and a broken-out section (left-side).

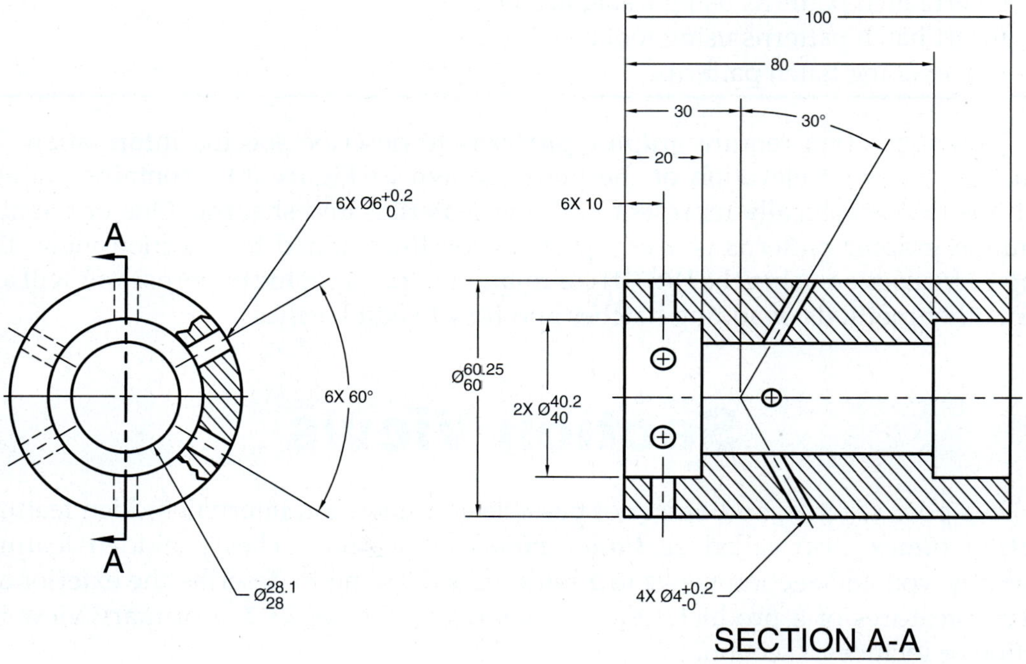

Other drafting fields, including architectural, structural, and civil drafting, also use sectioning. Cross sections through buildings show construction methods and materials. See Figure 23-3. Profiles and cross sections on a civil engineering drawing show the contour and construction of land for utility, transportation, and other civil engineering projects. Cutting-plane lines used in nonmechanical fields are often composed of letter and number symbols to help coordinate the sections in a set of drawings.

Figure 23-3.
An architectural section view showing the construction of a proposed bathroom for a residential renovation project.

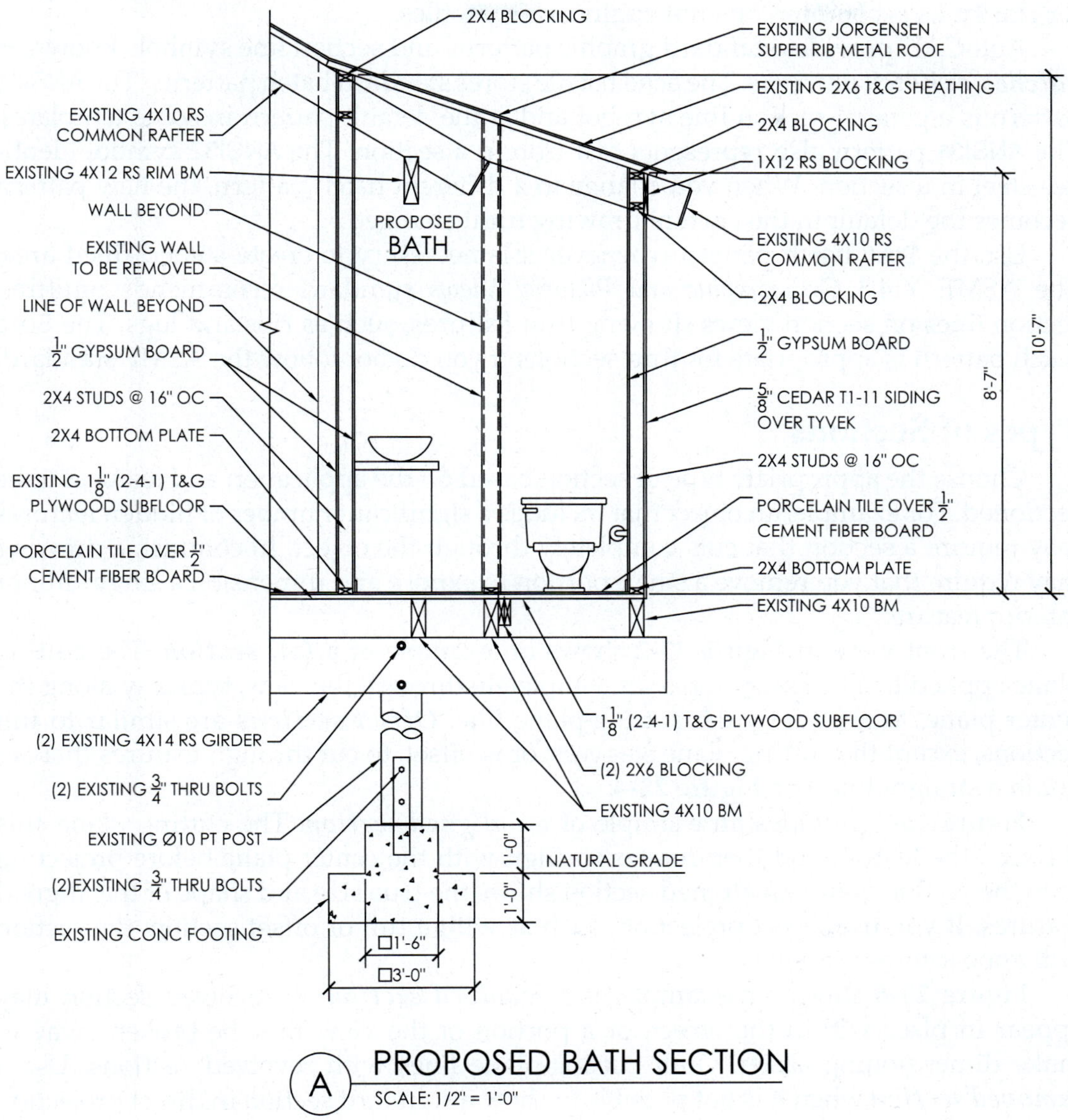

Section Lines

Section lines are the graphic pattern used in a section view. Section lines distinguish hidden features from exterior objects. Section views include section line symbols to show where material is cut to reveal hidden features. The standard on mechanical drawings is 45° section lines, unless another angle is required to satisfy other section line rules. Avoid drawing section lines at angles greater than 75° or less than 15° from horizontal. Section lines should never be parallel or perpendicular to adjacent lines on the drawing. In addition, section lines should not cross object lines.

section lines: Lines that show where material is cut away.

Equally spaced section lines, with a minimum .063″ (1.5 mm) spacing, are standard on mechanical drawings, and are adequate for basic applications such as the front view in Figure 23-2. Depending on the drawing and discipline, you may use different patterns to clarify the drawing or to represent the specific material cut by the section. A specific material pattern is not necessary on a part drawing if the title block or a note clearly indicates the material. However, use different or coded section lines on an assembly drawing to represent each component or different material.

Section line symbols on a nonmechanical drawing can be lines or can consist of graphic patterns that represent specific materials, such as insulation, earth, and concrete on an architectural or structural section. Section lines may also be omitted for clarity, as is common on civil engineering profiles.

hatches (hatch patterns): AutoCAD section line symbols and graphic patterns.

AutoCAD provides standard graphic patterns and section line symbols known as ***hatches***, or ***hatch patterns***. The acad.pat file stores standard hatch patterns. The ANSI31 pattern is a general section line symbol and is the default pattern in some templates. The ANSI31 pattern also represents cast iron in a section. The ANSI32 symbol identifies steel in a section. When you change to a different hatch pattern, the new pattern becomes the default in the current drawing until changed.

Use the Solid hatch pattern whenever it is necessary to create a solid filled area. The ASME Y14.3 *Orthographic and Pictorial Views* standard recommends omitting section lines on section views showing thin features, such as ribs and lugs. The Solid hatch pattern is appropriate for thin sections if you do not follow the ASME standard.

Types of Sections

Choose the appropriate type of section based on the application and features to be sectioned. For example, an object that includes a significant number of hidden features may require a section that cuts completely through the object. In contrast, a drawing may require that you remove a small portion to expose and dimension a single, minor interior feature.

full section: Section that shows half the object removed.

offset section: Section that has a staggered cutting plane.

The front view in Figure 23-2 shows an example of a ***full section***. The cutting plane applied to a full section passes completely through the view, typically along the center plane, as shown by the cutting-plane line. ***Offset sections*** are similar to full sections, except the cutting plane staggers, or is offset, to cut through features that are not in a straight line. See Figure 23-4.

aligned section: Section used when a feature is out of alignment with the center plane.

Figure 23-5 provides an example of an ***aligned section***. The cutting plane cuts through the feature and then rotates to align with the center plane before projecting onto the section view. An aligned section shows the true size and shape of the aligned features. If you use direct projection, such as with a full or offset section, the section will appear foreshortened.

revolved section: Section that clarifies the contour of an object that has the same shape throughout its length.

removed section: Standard section view, but removed from direct projection from the cutting plane.

Figure 23-6 shows an example of a ***revolved section***. A revolved section may appear in place within the object, or a portion of the view may be broken away to make dimensioning easier. Omit cutting-plane lines with revolved sections. Use a ***removed section*** when it is not possible to show a standard section in direct projection from the cutting plane, or when a revolved section is inappropriate. See Figure 23-7. A cutting-plane line identifies the location of the section. Multiple removed sections require labeled cutting-plane lines and related views. Drawing only the ends of the cutting-plane lines simplifies the views.

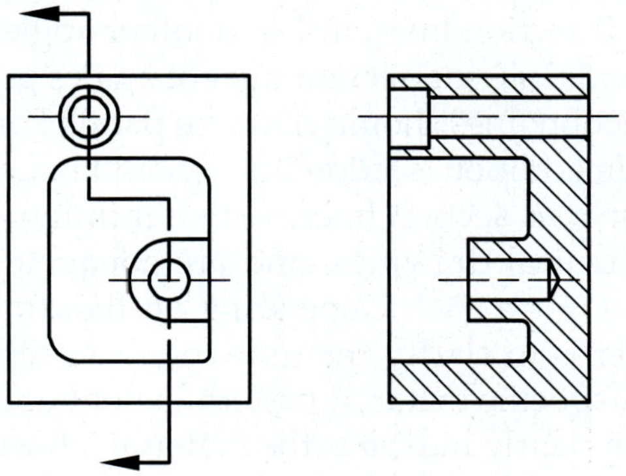

Figure 23-4. A two-view mechanical part drawing with an offset section view.

Figure 23-5.
A two-view mechanical part drawing with an aligned section view.

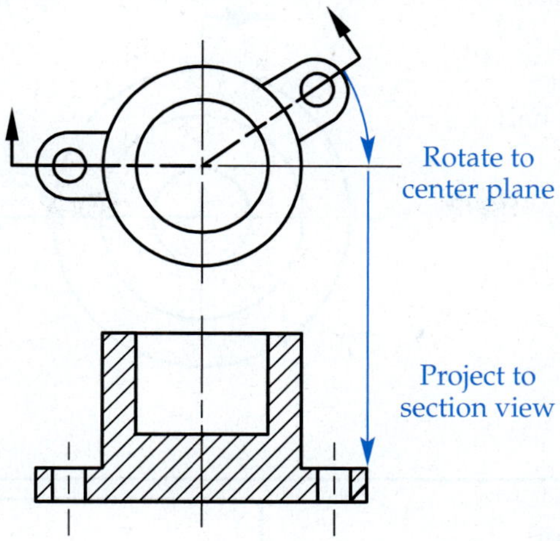

Figure 23-6.
The same revolved section drawn in place and with broken object lines. Breaking the view is appropriate for this part to clarify the profile. Revolving in place is most common when the section shows a cylindrical or similar feature.

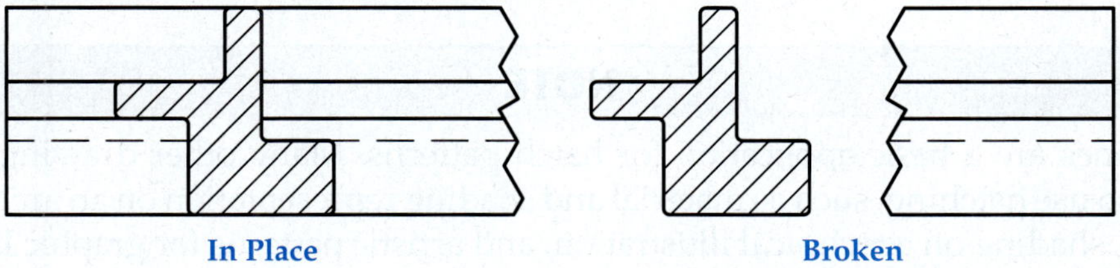

Figure 23-7.
A drawing with two removed sections (Section A-A and Section C-C) and a full section (Section B-B).

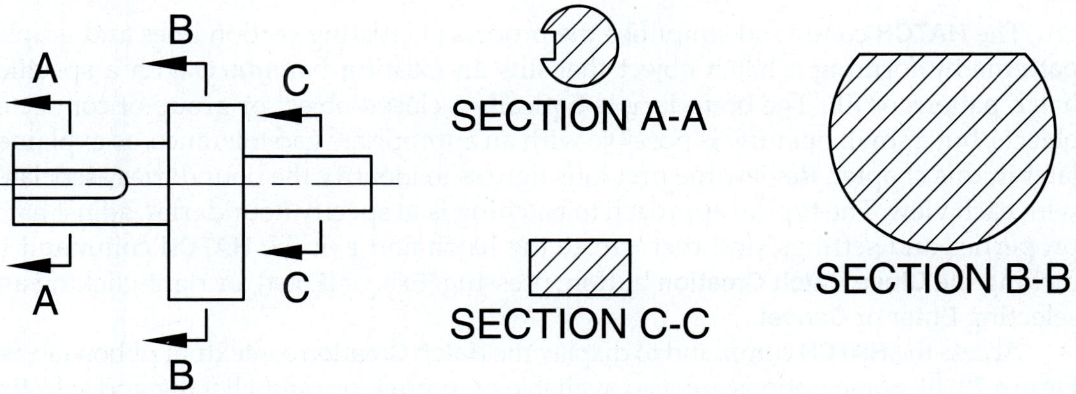

Figure 23-8 shows an example of a ***half section***. The term *half* describes how half of the view appears in section, while the other half remains as an exterior view. Half sections are commonly used to draw symmetrical objects. A centerline separates the sectioned portion of the view from the unsectioned portion. You normally omit hidden lines from the unsectioned half.

half section: Section that shows one-quarter of the object removed.

Broken-out sections clarify specific hidden features. See Figure 23-9. The left-side view in Figure 23-2 also shows a broken-out section.

broken-out section: Section that shows a small portion of the object removed.

Figure 23-8.
A two-view mechanical part drawing with a half section.

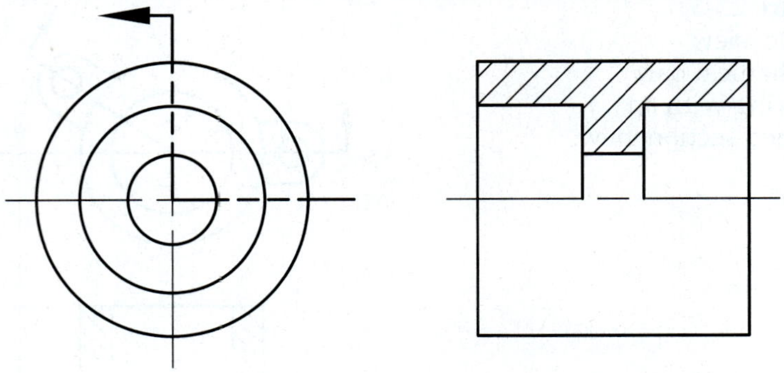

Figure 23-9.
Use a broken-out section to display specific hidden features.

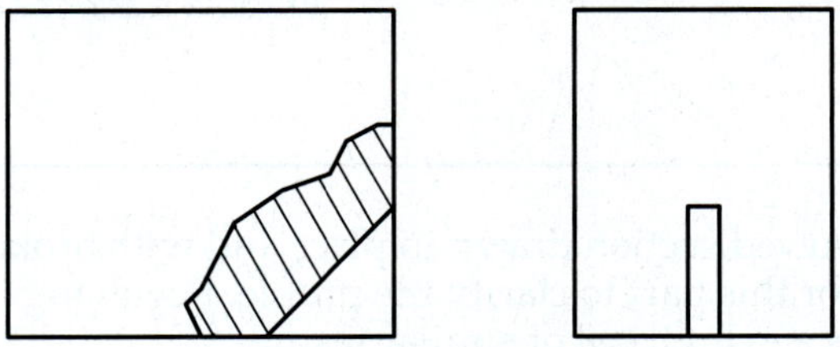

NOTE

Section lines are a basic application for hatch patterns. Many other drawing requirements also use hatching, such as material and shading representation on an architectural elevation, shading on a technical illustration, and artistic patterns for graphic layouts.

Using the HATCH Command

boundary: The area filled by a hatch.

HATCH

Ribbon
Home
> Draw
Hatch

Type
HATCH
BHATCH
KNURL
POCHE'
BH
H

Ribbon
Hatch Creation
> Close
Close Hatch Creation

The **HATCH** command simplifies the process of creating section lines and graphic patterns by forming a hatch object that fills an existing ***boundary*** with a specified hatch pattern or fill. The boundary is typically a closed object or group of connected objects, but a small opening is possible with an appropriate gap tolerance, as explained later in this chapter. Review the previous figures to identify the boundaries associated with each view. The typical approach to hatching is to specify boundaries, adjust hatch properties and settings, and then create the hatch and exit the **HATCH** command by picking the **Close Hatch Creation** button, pressing [Esc] or [Enter], or right-clicking and selecting **Enter** or **Cancel**.

Access the **HATCH** command to display the **Hatch Creation** contextual ribbon tab. See Figure 23-10. Some options are also available by typing, or right-clicking and selecting from the shortcut menu. The default method for placing a hatch is to pick a point within a boundary and allow AutoCAD to identify the boundary. The **Select objects** option allows you to select objects in addition to or instead of internal points, as explained later in this chapter. The **SeTtings** option displays the **Hatch and Gradient** dialog box, which is an alternative to the **Hatch Creation** ribbon tab for creating hatch patterns. The **Select objects** and **SeTtings** options are also available from the **Hatch Creation** ribbon tab.

NOTE

The **Hatch Creation** ribbon tab is contextual. As a result, pressing [Esc] or [Enter], or right-clicking and selecting **Enter** or **Cancel** exits the **HATCH** command and creates a hatch if you have specified boundaries.

NOTE

Command line content search offers another way to activate the **HATCH** command by selecting a specific hatch or fill. Begin typing a pattern or fill name using the command line. Patterns and fills that match the letters you enter appear in the suggestion list as you type. Choose a pattern or fill from the suggestion list to activate the **HATCH** command with the selected pattern or fill preset.

Specifying Boundaries

The default and typically easy method for placing a hatch is to pick a point within a boundary and allow AutoCAD to detect the boundary. Move the crosshairs inside a boundary to preview the hatch, as shown in Figure 23-10. Use the preview to help identify the boundary. The initial preview may not use the correct hatch pattern or boundary properties, but should allow you to determine if selecting the point will fill the acceptable boundary. You can adjust the pattern and boundary properties interactively after you specify boundaries.

If the boundary looks correct, pick the point to apply the hatch and highlight the boundary. Continue picking points to specify additional boundaries to include with the hatch object. If a preview does not appear, AutoCAD cannot detect the boundary or the current hatch scale is too large for the size of the boundary. The most common reason AutoCAD cannot detect a boundary is because the boundary contains gaps. Exit the **HATCH** command and edit the drawing to create a closed boundary.

Island Detection

When picking points to specify hatch boundaries, you may need to adjust how AutoCAD treats ***islands***, as shown in Figure 23-11. Use the flyout in the expanded **Options** panel to specify island detection. Pick the **Normal Island Detection** option to hatch every other boundary, stepping inward from the outer boundary. Select the **Outer Island Detection** option to hatch only the outermost boundary. Choose the **Ignore Island Detection** option to ignore all islands and hatch everything within the outer boundary.

islands: Boundaries inside another boundary.

Ribbon

Hatch Creation > Options

Normal Island Detection

Outer Island Detection

Ignore Island Detection

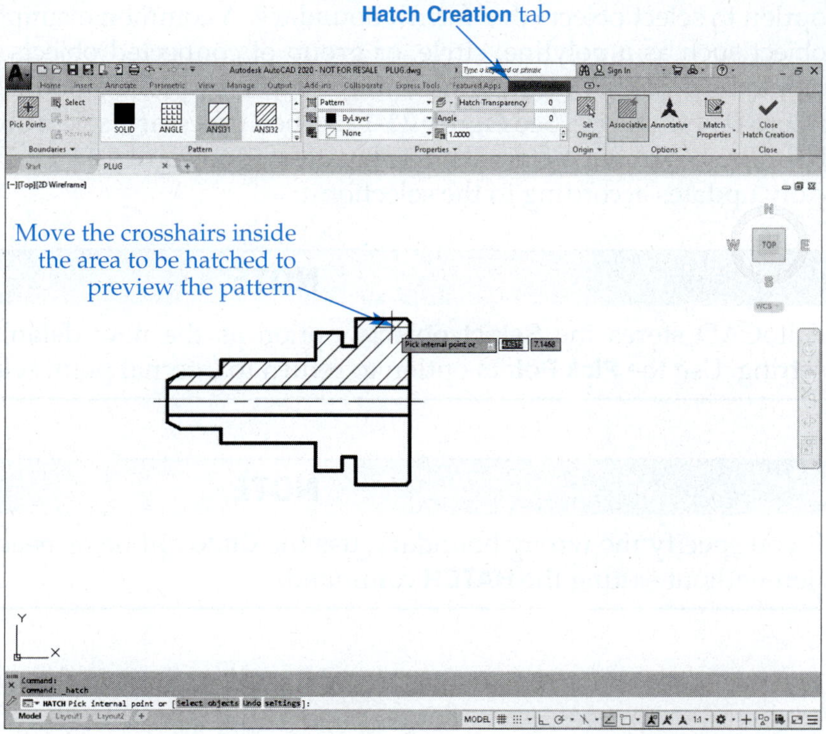

Figure 23-10.
The **Hatch Creation** contextual ribbon tab. Use the default **Pick Points** option and the crosshairs to preview the boundary to hatch.

Figure 23-11.
Adjusting island detection when picking a single internal point. The **Outer Island Detection** option is appropriate for this example and for most applications.

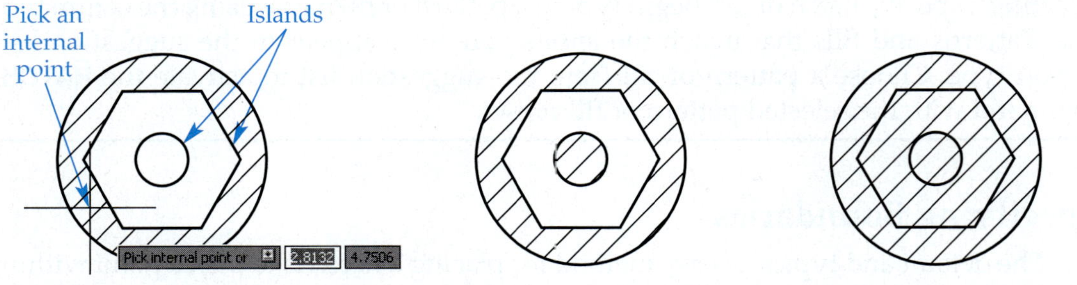

Outer Island Detection | Normal Island Detection | Ignore Island Detection

NOTE

The **No Island Detection** option applies to hatches created with versions of AutoCAD prior to AutoCAD 2011.

PROFESSIONAL TIP

Pick the **Outer** island display style to ensure that inner islands are not hatched unintentionally.

Exercise 23-1

www.g-wlearning.com/CAD/

Complete the exercise on the companion website.

Selecting Objects

Ribbon
Hatch Creation > Boundaries
Select Boundary Objects

Ribbon
Hatch Creation > Boundaries
Pick Points

An alternative method to specify a boundary is to use the **Select Boundary Objects** option to select objects that form a boundary. A common example is selecting a closed object such as a polyline, circle, or group of connected objects to hatch an area that would be difficult or time-consuming to hatch by picking points because of numerous internal boundaries. See Figure 23-12. When necessary, select objects within the hatch boundary to exclude from the hatch pattern. See Figure 23-13. The preview automatically updates according to the selections.

NOTE

AutoCAD stores the **Select objects** option as the new default boundary selection setting. Use the **Pick Points** option to return to internal point selection mode.

NOTE

If you specify the wrong boundary, use the **Undo** option as needed to undo the selection without exiting the **HATCH** command.

Removing Boundaries

The **Remove Boundary Objects** option is available after you specify a boundary. Use the **Remove Boundary Objects** option to select unwanted boundaries or objects to remove from the selection set.

Ribbon
Hatch Creation
> Boundaries

Remove Boundary Objects

Exercise 23-2

www.g-wlearning.com/CAD/

Complete the exercise on the companion website.

Boundary Set

By default, the **HATCH** command evaluates the entire current viewport to detect boundaries. To limit the evaluation area, possibly increasing hatch performance, pick the **Select New Boundary Set** option and use a window to define the area to evaluate. See Figure 23-14A. Right-click or press [Enter] to create the boundary set.

Ribbon
Hatch Creation
> Boundaries

Select New Boundary Set

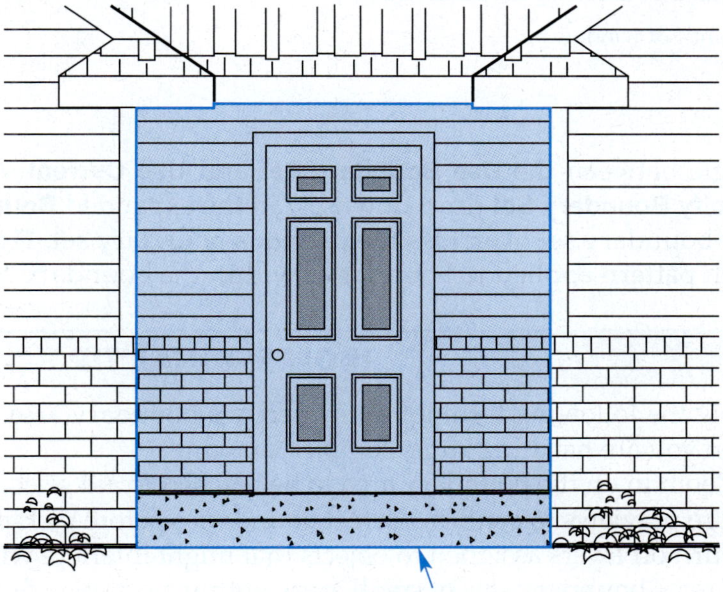

Figure 23-12. An example of an architectural elevation on which it is easier to use the **Select Boundary Objects** option to specify a hatch boundary. The closed polyline around the area to be hatched is for construction purposes only.

Figure 23-13. Using the **Select Boundary Objects** option to exclude a polyline hexagon and text object from the hatch pattern.

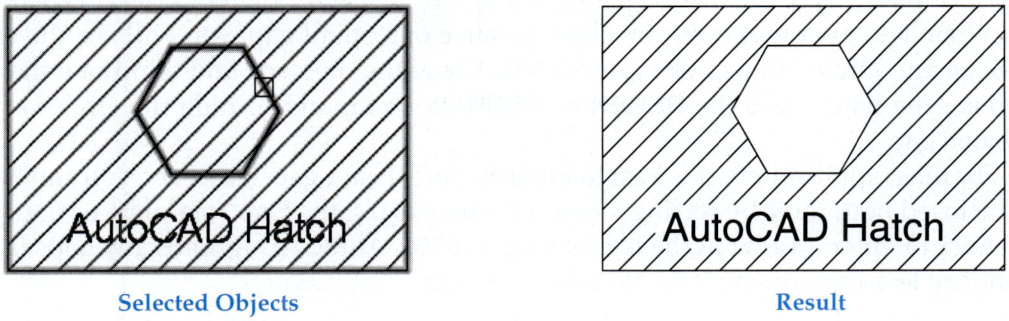

Figure 23-14.
A—The boundary set limits the area that AutoCAD evaluates during a hatching operation.
B—The results after hatching is applied.

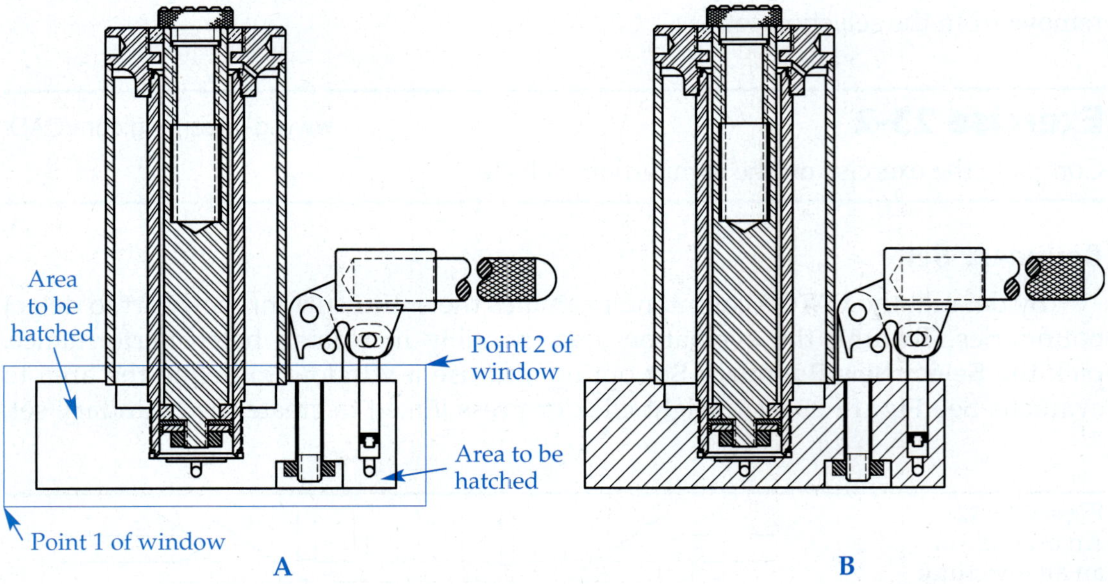

Toggle between the **Use Boundary Set** and **Use Current Viewport** option from the **Specify Boundary Set** drop-down list in the expanded **Boundaries** panel. Creating a new boundary set overrides the previous boundary set. Figure 23-14B shows a new hatch pattern applied to boundaries within the boundary set.

PROFESSIONAL TIP

Apply the following techniques to specify a boundary area and save time, especially when you are hatching large, complex drawings:

- Zoom in on the boundary area to be hatched to aid selection.
- Use previews to confirm correct boundaries before picking.
- Turn off layers assigned to objects that might interfere with boundary definition.
- Create boundary sets of small areas within a complex drawing.

Hatching Unclosed Areas and Correcting Boundary Errors

The **HATCH** command works well unless there is a gap in a boundary, or you pick a point or select objects outside a likely boundary. When you select a point or objects where no boundary can form, an error message states that a valid boundary cannot be determined. Close the message and try again to specify the boundary. When you try to hatch an area that does not close because of a small gap, you will see the error message and circles shown in Figure 23-15. Close the message and eliminate the gap to create the hatch. Use the **REGEN** or **REDRAW** command to hide the circles at the probable gap.

gap tolerance: The amount of gap allowed between segments of a boundary to be hatched.

Ribbon
Hatch Creation
> Options
> Gap Tolerance

For most applications, it is best to identify and close a gap. However, you can hatch an unclosed boundary by setting a ***gap tolerance*** using the **Gap Tolerance** option. Use the slider or enter a value in the text box up to 5000. AutoCAD ignores any gaps in the boundary less than or equal to the tolerance value.

Figure 23-15.
Close a boundary to create a hatch pattern.

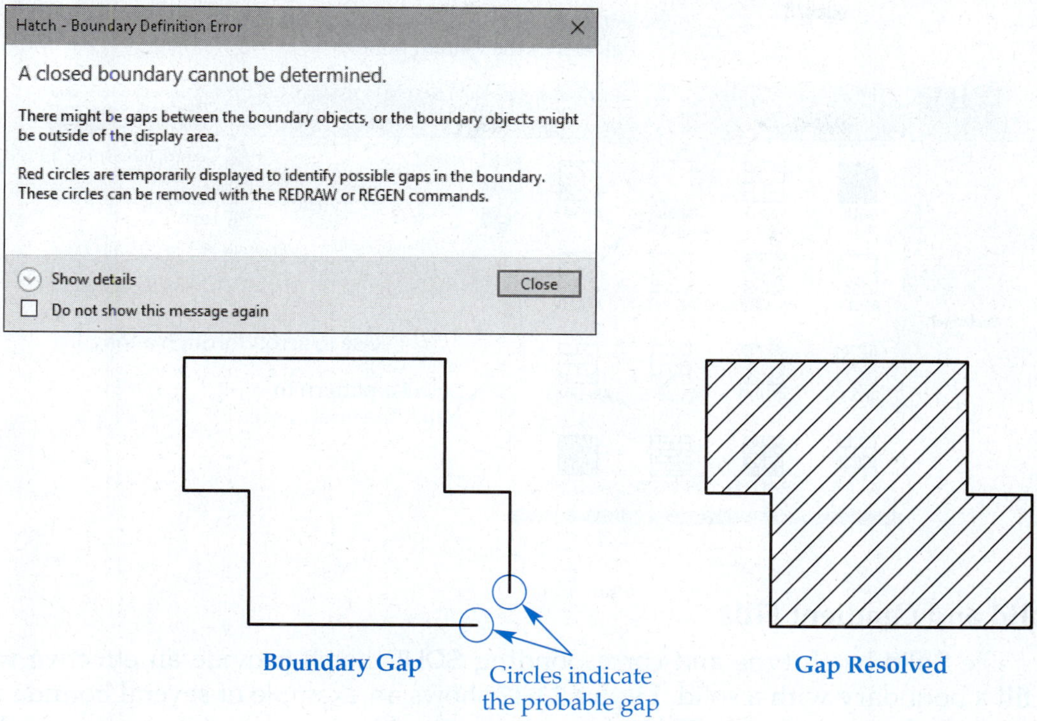

Retaining Boundaries

By default, a hatch pattern boundary forms according to objects in the drawing. The boundary is temporary, which is appropriate for most applications. Select the **Retain Boundaries - Polyline** option to form a separate polyline object overlapping the boundary. Choose the **Retain Boundaries - Region** option to form a separate ***region*** object overlapping the boundary. Retaining a boundary forms additional boundary objects that you can use even if you remove or edit the original objects.

region: A closed two-dimensional area.

Ribbon
Hatch Creation > Boundaries
Retain Boundaries - Polyline
Retain Boundaries - Region

Selecting a Hatch Pattern

The **Pattern** panel includes all of the hatch patterns and fills supplied with AutoCAD. Use the scroll buttons to the right of the patterns to locate a pattern, or pick the expansion arrow to display patterns in a temporary window. Scroll through a long list as necessary to find the appropriate pattern. See Figure 23-16. The **Hatch Type** drop-down list in the **Properties** panel includes Solid, Gradient, Pattern, and User defined categories to help you locate specific hatches in the **Pattern** panel. Select a category to display related hatches in the **Pattern** panel. Select a hatch pattern from the **Pattern** panel to apply to the specified boundaries.

Ribbon
Hatch Creation > Properties
Pattern

Patterns

The **Pattern** hatch type provides patterns stored in the acad.pat and acadiso.pat files. AutoCAD includes many different patterns to accommodate most hatch requirements. For example, use the default ANSI31 style for section lines, or use the BRICK or one of the AR-B patterns to represent brick on an architectural elevation. The **Properties** panel provides settings specific to the selected hatch pattern, as explained later in this chapter.

Exercise 23-3

www.g-wlearning.com/CAD/

Complete the exercise on the companion website.

Figure 23-16.

Select a hatch pattern from the **Pattern** panel of the **Hatch Creation** ribbon tab.

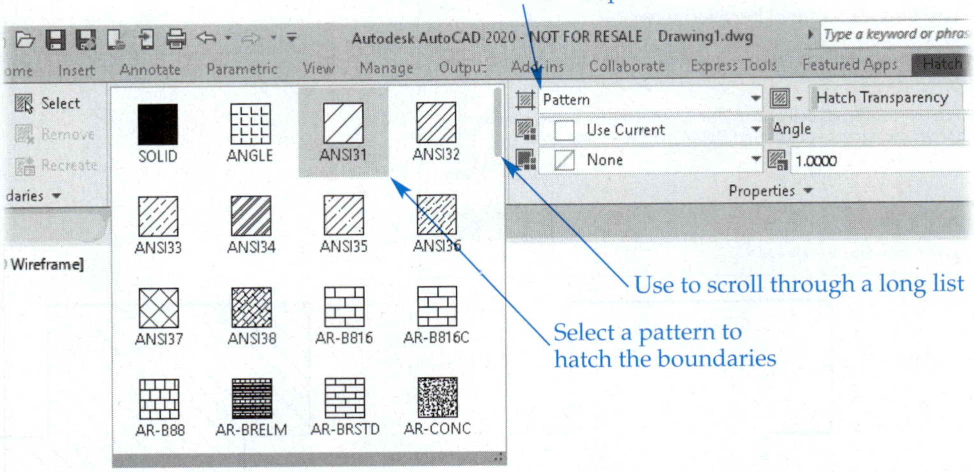

Solid and Gradient Fills

The **Solid** hatch type and corresponding SOLID hatch provide an effective way to fill a boundary with a solid. Figure 23-17 shows an example of several boundaries filled with the SOLID hatch. This example uses specific transparent layers assigned to different hatch objects. Use an option associated with the **Gradient** hatch type to create a gradient fill, as explained later in this chapter.

Exercise 23-4

www.g-wlearning.com/CAD/

Complete the exercise on the companion website.

Figure 23-17.

A portion of a storm water pollution control plan with transparent SOLID hatching. The solid fills represent different impervious and non-impervious surfaces.

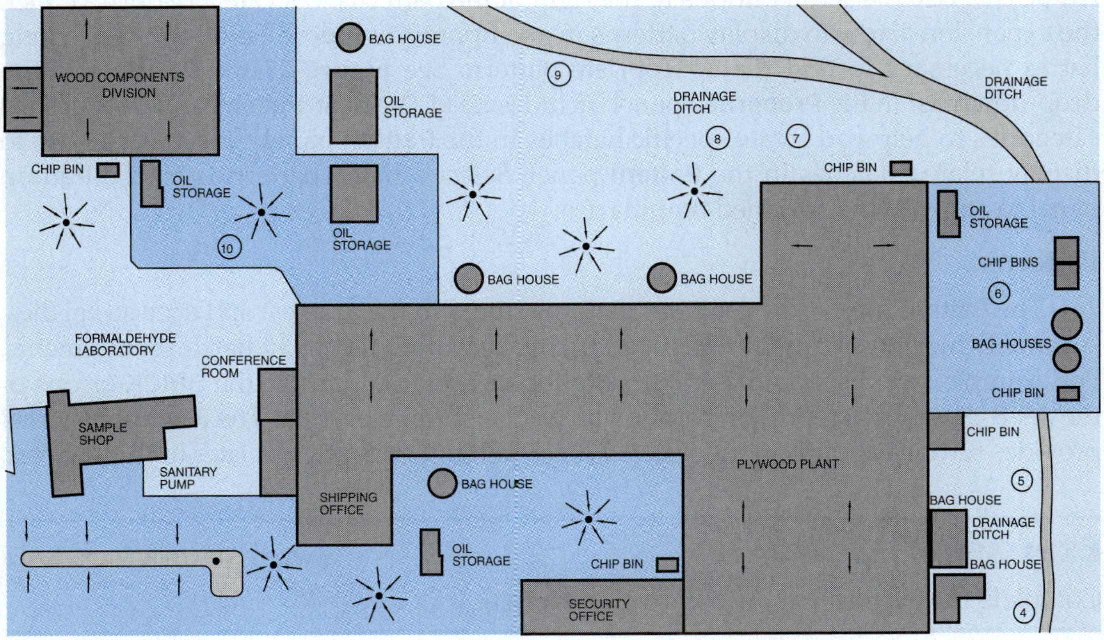

NOTE

The **SOLID** command allows you to draw basic solid shapes without creating a boundary. Access the **SOLID** command and pick points in a specific sequence to form different shapes. The SOLID hatch applied using the **HATCH** command is typically a better method of creating solid fills.

Type
SOLID
SO

SOLID

User Defined Hatch

The **User defined** hatch type and corresponding USER hatch create a pattern of equally spaced lines for basic hatching applications. The lines use the linetype assigned to the current layer. Use options in the **Properties** panel of the **Hatch Creation** ribbon tab to adjust the USER hatch appearance, as explained later in this chapter.

NOTE

Create and save custom hatch patterns in PAT files. Add the files to the AutoCAD search path to have access to custom hatch patterns from the **Pattern** panel.

Pattern Size

The **Properties** panel provides hatch pattern appearance control. Scale, or spacing, is a primary hatch property. Use the **Hatch Pattern Scale** option to adjust hatches of the **Pattern** type. The default scale is 1. If the pattern appears too small or large, specify a different scale. See Figure 23-18. For example, by default, the ANSI31 hatch is a pattern of lines spaced .125″ (3 mm) apart. If you change the scale to 2, the pattern of lines is spaced .25″ (6 mm) apart.

Ribbon
Hatch Creation
> Properties
Hatch Pattern Scale

The **Hatch Spacing** option replaces the **Hatch Pattern Scale** option when you select the USER hatch. Specify the exact distance between lines. Changes to scale or spacing update automatically in the preview of the specified boundaries.

PROFESSIONAL TIP

Use a smaller hatch scale for small objects and a larger hatch scale for larger objects. This makes section lines look appropriate for the drawing scale. Often you must use your best judgment when selecting a hatch scale.

Use an appropriate hatch scale or spacing to display the hatch pattern on-screen and plot correctly according to the drawing scale. To understand the concept of hatch size, look at the section view shown in Figure 23-19. In this example, which uses the ANSI31 hatch pattern, the section line spacing should be the same distance apart regardless of drawing scale. The section lines on the full-scale (1:1) drawing display correctly. However, the section lines are too close on the half-scale (1:2) drawing, and they are too far apart on the double-scale (2:1) drawing. To obtain the correct results, you must adjust the hatch size according to the drawing scale. You can calculate scale factor manually and apply it to hatch scale or spacing, or you can allow AutoCAD to calculate the scale factor using annotative hatch patterns.

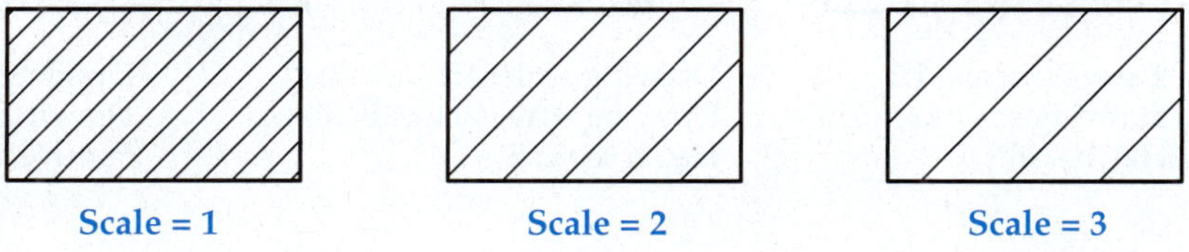

Figure 23-18. Hatch pattern scale applies to hatches of the **Pattern** type and to custom hatch patterns.

Figure 23-19.
The hatch pattern may appear incorrect if the drawing scale changes.

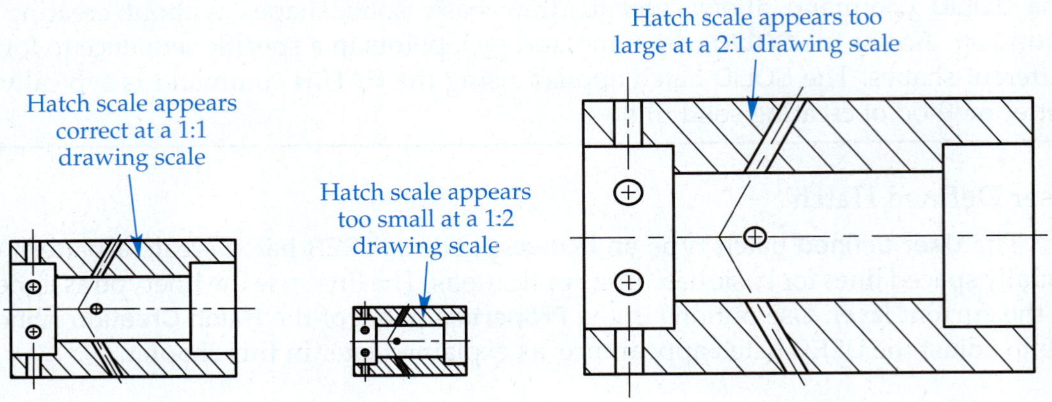

Scaling Hatch Patterns Manually

To adjust hatch size manually according to a specific drawing scale, you must first calculate the drawing scale factor. Then multiply the scale factor by the plotted hatch scale or spacing to get the model space hatch scale or spacing. Specify the scale of predefined or custom hatch patterns using the **Hatch Pattern Scale** option. Specify the spacing of the USER hatch using the **Hatch Spacing** option. **Figure 23-20** shows examples of adjusting hatch scale according to drawing scale. Refer to Chapter 9 for information on determining the drawing scale factor.

Annotative Hatch Patterns

Ribbon
Hatch Creation > Options
Annotative

Pick the **Annotative** button in the **Options** panel to make the hatch pattern annotative. Once you choose an annotation scale, AutoCAD applies the corresponding scale factor to annotative hatches and all other annotative objects.

The hatch pattern is displayed at the proper size regardless of the drawing scale, much like the example shown in **Figure 23-20**, but without requiring you to change the hatch scale or spacing. For example, if you specify a value using the **Hatch Pattern Scale** or **Hatch Spacing** option appropriate for an annotation scale of 1/4″ = 1′-0″, and then change the annotation scale to 1″ = 1′-0″, the appearance of the hatch pattern relative to the drawing scale does not change. It looks the same on the 1/4″ = 1′-0″ scale drawing as it does on the 1″ = 1′-0″ scale drawing. Refer to Chapter 9 for information on setting annotation scale.

Figure 23-20.
The hatch pattern scale may require adjusting, depending on the drawing scale.

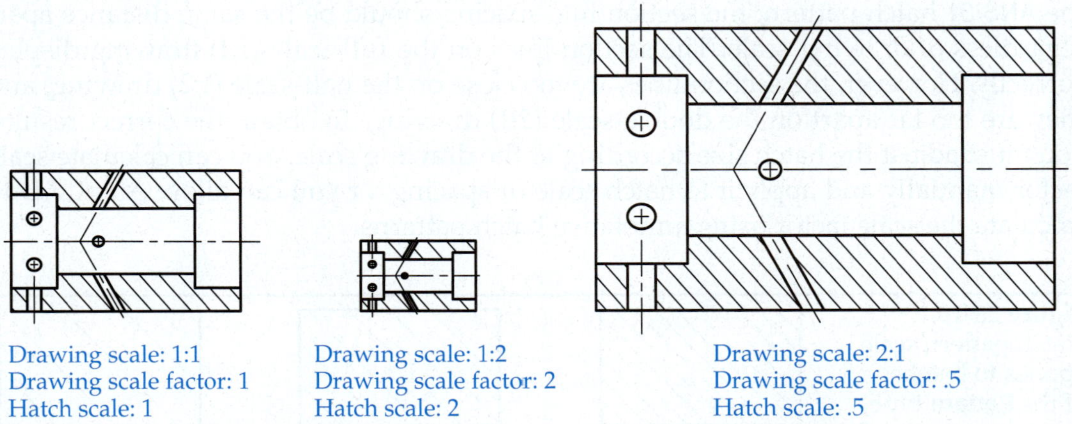

Relative to Paper Space Option

The **Relative to Paper Space** option in the expanded **Properties** panel allows you to scale the hatch pattern relative to the scale of the active layout viewport. You must enter a floating layout viewport in order to select the **Relative to Paper Space** option. The hatch scale automatically adjusts according to the viewport scale. For example, a floating viewport scale set to 4:1 uses a scale factor of .25 (1 ÷ 4 = .25). If you enter a hatch scale of 1, the hatch automatically appears at a scale of .25 (1 × .25 = .25).

> Ribbon
> **Hatch Creation**
> **> Properties**
> **Relative to Paper Space**

NOTE

You can control the ISO pen width for predefined ISO patterns using the **ISO Pen Width:** drop-down list in the expanded **Properties** panel.

> Ribbon
> **Hatch Creation**
> **> Properties**
> **> ISO Pen Width**

Additional Pattern Properties

The **Properties** panel provides other settings that you can preview interactively as you make changes. The **Hatch Angle** option controls pattern rotation. Use the slider or enter a value up to 359 in the text box to rotate the pattern relative to the X axis. For example, the ANSI31 hatch is a pattern of 45° lines. Change the angle to 15° to form a pattern of 60° (45 + 15 = 60) lines. The **Double** option is available with the USER hatch and allows you to create a pattern of double lines. Figure 23-21 shows examples of user-defined hatch patterns.

> Ribbon
> **Hatch Creation**
> **> Properties**
> **> Hatch Angle**

> Ribbon
> **Hatch Creation**
> **> Properties**
> **Double**

The **Background Color** option allows you to fill the specified boundaries with the hatch pattern and a background color similar to a solid fill. See Figure 23-22. Hatching with a solid fill and pattern is not a common drafting practice, but is appropriate in some applications. Choose a color from the drop-down list or pick the **More Colors...** option to pick a color from the **Select Color** dialog box. The color must be different from the color assigned to the pattern layer.

> Ribbon
> **Hatch Creation**
> **> Properties**
> **Background Color**

The **Hatch Layer Override** option corresponds to the **HPLAYER** system variable, which assigns a layer to new hatch objects. Select the current layer or choose a different layer if appropriate. When you access the **HPLAYER** system variable from outside the **HATCH** command by typing the variable name, you have the options to type the name of an existing layer or the name of a new layer to create and assign to new hatch objects. The **Hatch Color** and **Hatch Transparency** options override the color and transparency assigned to the current layer. For most applications, you should not override color or transparency.

> Ribbon
> **Hatch Creation**
> **> Properties**
> **Hatch Layer Override**

Figure 23-21.
Examples of user-defined hatch patterns with different hatch angles and spacing.

Angle	0°	45°	0°	45°
Spacing	.125	.125	.250	.250
Single Hatch	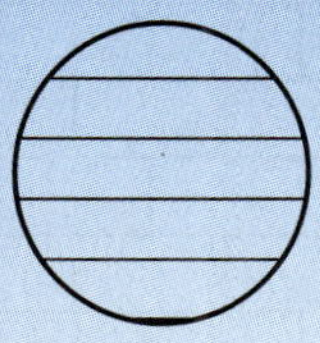	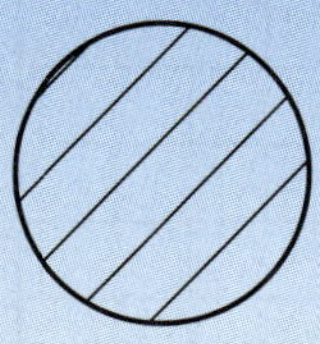	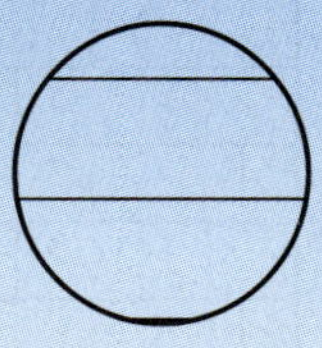	
Double Hatch	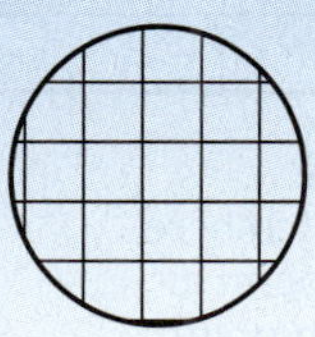	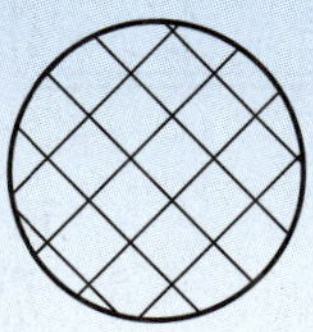	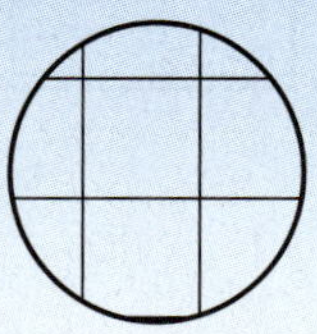	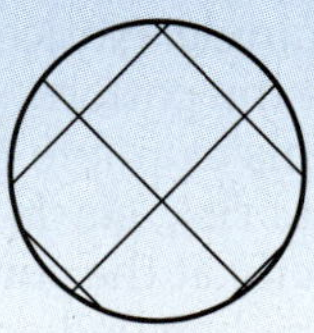

Figure 23-22.
Use the **Background Color** option on the **Properties** panel to fill boundaries with a color and a hatch pattern. This example shows an architectural window elevation with a Color 9 background and the ANSI34 pattern.

Setting the Hatch Origin Point

Ribbon
Hatch Creation
> Origin
Set Origin

The **Origin** panel includes options that control the position of hatch patterns. Select an option from the expanded **Origin** panel, or pick the **Set Origin** button. The default setting is **Use Current Origin**, which refers to the current UCS origin, to define the point from which the hatch pattern forms and how the pattern repeats. In some cases, it is important that a hatch pattern align with, or originate from, a specific point. A common example is hatching a representation of bricks. To specify a different origin point, pick the **Set Origin** button and select an origin point. See Figure 23-23.

The expanded **Origin** panel includes **Bottom Left**, **Bottom Right**, **Top Left**, **Top Right**, and **Center** options. Select an option to align the hatch origin point with the corresponding point on the hatch boundary. For example, use the **Bottom Left** option to create the pattern shown in Figure 23-23B. Select the **Store as Default Origin** option to save the custom origin point.

Gradient Fill

gradient fill: A shading transition between the tones of one color or two separate colors.

Select the **Gradient** hatch type and choose a gradient style from the **Pattern** panel to create a ***gradient fill***. See Figure 23-24. Gradient fills are commonly used to simulate color-shaded objects and to create the appearance of a lit surface with a gradual transition from an area of highlight to a darker area. Use two colors to simulate a transition from light to dark between the colors. Several different gradient fill patterns are available to create linear sweep, spherical, radial, or curved shading.

Figure 23-23.
A—The default **Use Current Origin** setting. B—Pick the **Set Origin** button and select the lower-left corner (endpoint) of the rectangle. Notice that the pattern, or in this example the first brick, starts exactly at the corner of the hatched area.

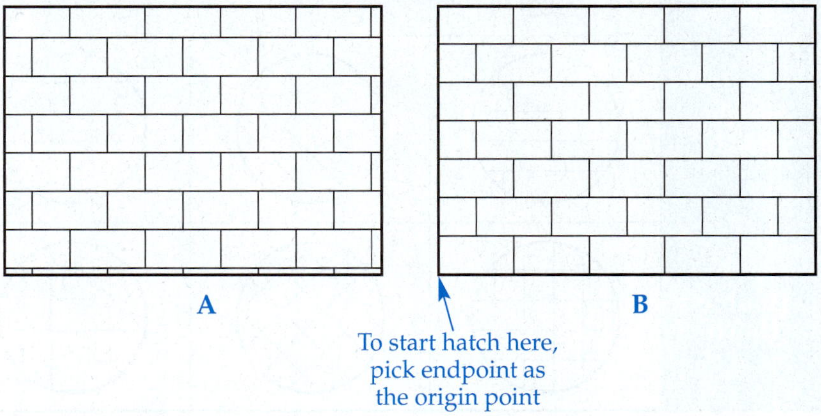

Figure 23-24.
Select a gradient style from the **Pattern** panel to access options for creating gradient fills, such as this two-color representation of a tropical lagoon during sunset.

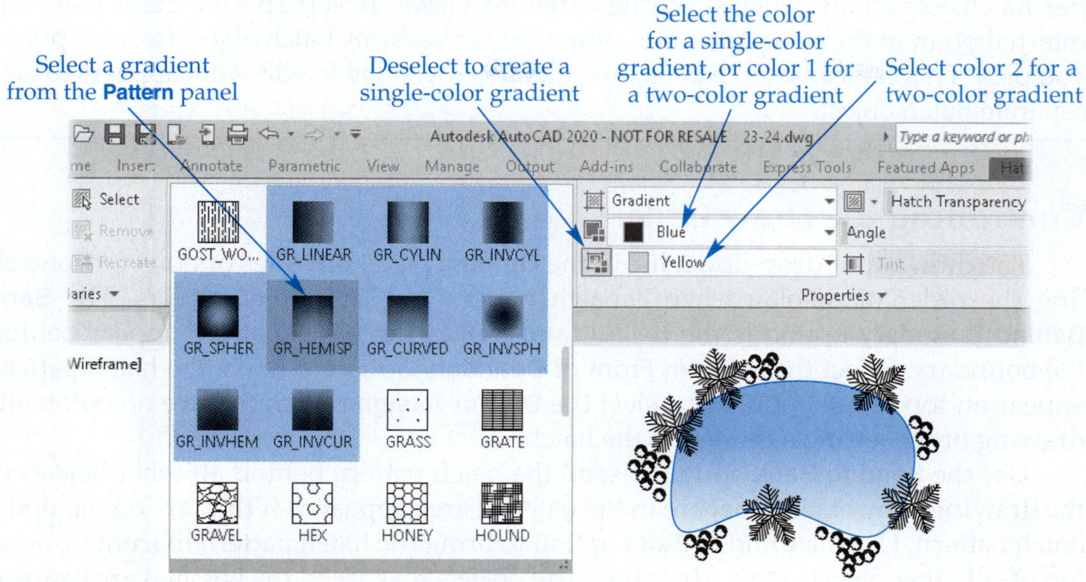

The **Gradient Colors** button is active by default to specify a fill using a smooth transition between two colors. Use the **Gradient Color 1** and **Gradient Color 2** drop-down lists to specify gradient colors. Deselect the **Gradient Colors** button to create a fill that has a smooth transition between the darker ***shades*** and lighter ***tints*** of one color. The **Gradient Tint and Shade** option becomes enabled when you use the single-color option. Use the slider or enter a value up to 100 in the text box to specify the tint or shade of a color used for a one-color gradient fill.

The **Centered** button is active by default and applies a symmetrical configuration. If you deselect the **Centered** button, the gradient fill shifts to simulate the projection of a light source from the left of the boundary. Use the **Angle** option to specify the gradient fill angle relative to the current UCS. The default angle is 0°.

Ribbon
Hatch Creation > Properties
Gradient Colors

Ribbon
Hatch Creation > Properties
Gradient Tint and Shade

shade: A specific color mixed with black.

tint: A specific color mixed with white.

Exercise 23-5

www.g-wlearning.com/CAD/

Complete the exercise on the companion website.

Hatch Composition Options

The **HATCH** command creates an ***associative hatch pattern*** by default. To create a ***non-associative hatch pattern***, deselect the **Associative** button in the **Options** panel. An associative hatch is appropriate for most applications. If you stretch, scale, or otherwise edit the objects that define the boundary of an associative hatch, the pattern automatically adjusts to fill the modified boundary. A non-associative hatch pattern does not respond to changes made to the original boundary. Instead, non-associative hatch boundary grips are available for changing the extents of the hatch, separate from the original boundary objects.

associative hatch pattern: A hatch pattern that updates automatically when you edit associated objects.

non-associative hatch pattern: A hatch that is independent of objects and updates when the boundary changes, but not when you make changes to objects.

You can select multiple points and objects during a single hatch operation. By default, multiple boundaries form a single hatch object. Selecting and editing one of the hatch patterns selects and edits all patterns created during the same operation. If this is not the preferred result, select the **Create Separate Hatches** option before applying the hatch. Individual hatch patterns form for each boundary.

Ribbon
Hatch Creation > Options
Create Separate Hatches

PROFESSIONAL TIP

When you create an associative hatch, it is often best to specify a single internal point per hatch, especially when hatching different views. If you specify more than one internal point in the same operation, AutoCAD creates one hatch object from all points you pick. This can cause unexpected results when you try to edit what appears to be a separate hatch object.

Controlling the Draw Order

Ribbon
Hatch Creation > Options
Send Behind Boundary

The draw order drop-down list in the **Options** panel provides options for controlling the order of display when a hatch pattern overlaps other objects. The **Send Behind Boundary** option is the default and makes the hatch pattern appear behind the boundary. Select the **Bring in Front of Boundary** option to make the hatch pattern appear on top of the boundary. Select the **Do Not Assign** option to have no automatic drawing order setting assigned to the hatch.

Use the **Send to Back** option to send the hatch pattern behind all other objects in the drawing. Any objects that are in the hatched area appear as if they are on top of the hatch pattern. Use the **Bring to Front** option to bring the hatch pattern in front of, or on top of, all other objects in the drawing. Any objects that are in the hatched area appear as if they are behind the hatch pattern.

NOTE

Use the **DRAWORDER** command to change the draw order setting after creating a hatch pattern.

Reusing Existing Hatch Properties

Ribbon
Hatch Creation > Options
Use the Current Origin Point
Use the Source Origin Point

You can specify hatch pattern characteristics by referencing an identical hatch pattern from the drawing. Pick the **Match Properties - Use the Current Origin Point** option to match the properties of a selected hatch, except use the origin point specified in the **Origin** panel. Pick the **Match Properties - Use the Source Origin Point** option to match the properties of a selected hatch, including the hatch origin point. Select an existing hatch pattern to match, and then specify the boundaries for the new hatch object.

PROFESSIONAL TIP

Use the **MATCHPROP** command, described in Chapter 13, to match the properties assigned to one existing hatch to other existing hatches.

Exercise 23-6

www.g-wlearning.com/CAD/

Complete the exercise on the companion website.

Hatching Using DesignCenter

Ribbon
View
> Palettes
DesignCenter
Type
ADCENTER
ADC

ADCENTER

To pattern a boundary using **DesignCenter**, use the **Tree View** pane to locate and select a PAT file to display the patterns in the **Content** pane. See Figure 23-25A. The most effective technique to transfer a hatch pattern from **DesignCenter** to the active drawing is to use a drag-and-drop operation. Press and hold down the pick button on the pattern to import, and then drag the cursor to the drawing window. A hatch pattern symbol appears with the cursor. See Figure 23-25B. Release the pick button in a boundary to apply the hatch pattern. See Figure 23-25C.

An alternative to the drag-and-drop method is copy and paste. Right-click on a hatch pattern in **DesignCenter** and pick **Copy**. Move the cursor into the active drawing, right-click, and select **Paste** from the **Clipboard** cascading menu. A hatch pattern symbol appears with the crosshairs. Pick in a boundary to apply the hatch pattern. You can also use **DesignCenter** with the **HATCH** command. Right-click on a hatch pattern in **DesignCenter** and select **BHATCH...** to access the **Hatch Creation** ribbon tab with the selected hatch pattern active.

The same rules that apply to the **HATCH** command also apply to dragging and dropping or copying and pasting hatches. When you insert hatch patterns using **DesignCenter**, the angle, scale, and island detection settings match the settings of the previous hatch pattern. Edit the hatch pattern, as described later in this chapter, to change the settings after you insert the hatch pattern.

Figure 23-25.
A—Pick a PAT file in **DesignCenter** to display the available hatch patterns in the **Content** pane. B—The hatch pattern symbol appears under the cursor during the drag-and-drop and paste operations. C—Pick a point to apply the hatch pattern.

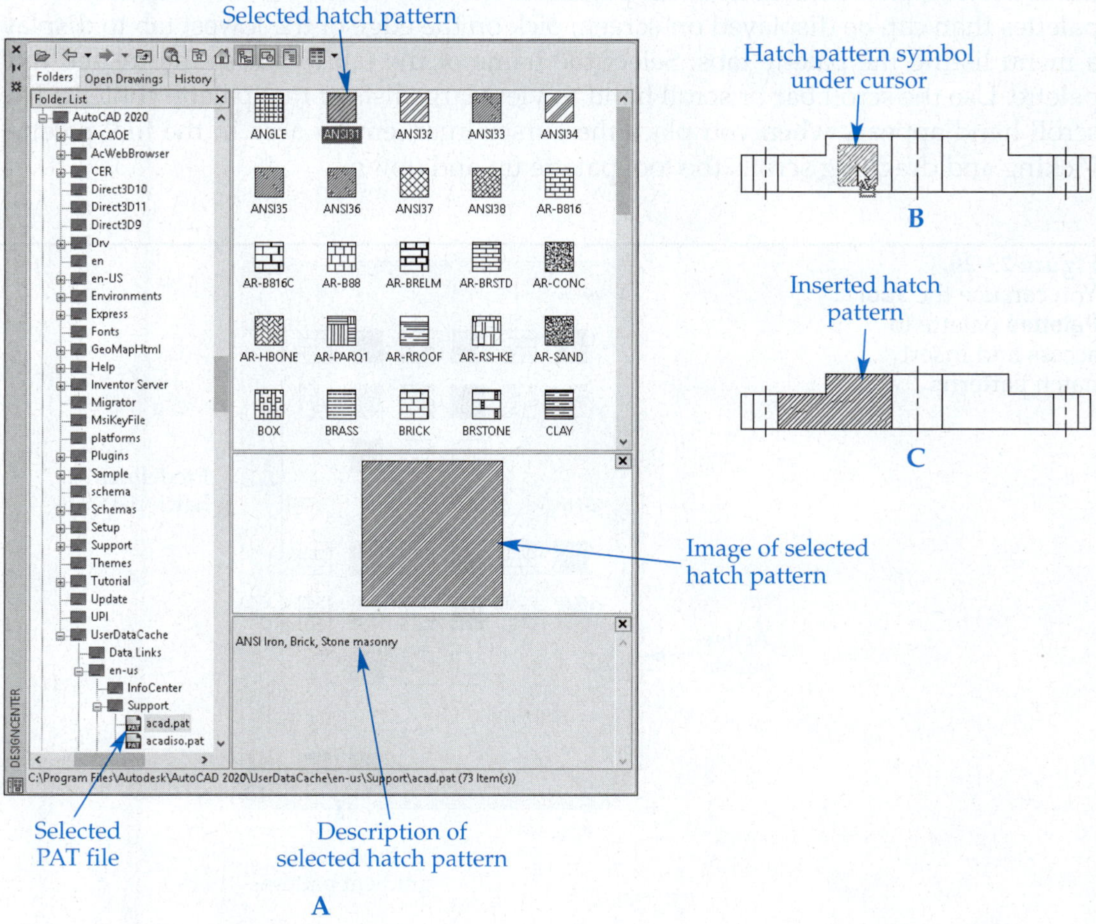

NOTE

AutoCAD includes two PAT files: acad.pat and acadiso.pat. To verify the location of AutoCAD support files, access the **Files** tab in the **Options** dialog box and check the path listed under the **Support File Search Path**.

Exercise 23-7

www.g-wlearning.com/CAD/

Complete the exercise on the companion website.

Hatching Using Tool Palettes

TOOLPALETTES

Ribbon
View
> Palettes
Tool Palettes

Type
TOOLPALETTES
TP

tool palette: A palette that contains tabs to help organize commands and other features.

The **Tool Palettes** palette, shown in Figure 23-26, provides an alternative means of storing and inserting hatch patterns. ***Tool palettes*** can also store and activate other drawing content and tools, such as blocks, images, tables, external reference files, drawing and editing commands, user-defined macros, script files, and AutoLISP expressions. The **Command Tool Samples** tool palette contains examples of custom commands. The **Hatches and Fills** palette contains several commonly used hatch patterns and gradient fills. You can add tool palettes to the **Tool Palettes** palette and add tools to tool palettes.

Locating and Viewing Content

Tabs along the side of the **Tool Palettes** palette divide each tool palette. Pick a tab to view related content in a tool palette. If the **Tool Palettes** palette contains more palettes than can be displayed on-screen, pick on the edge of the lowest tab to display a menu listing the palette tabs. Select the name of the tab to access the related tool palette. Use the scroll bar or scroll hand to view large lists of tool palette content. The scroll hand appears when you place the cursor in an empty area in the tool palette. Picking and dragging scrolls the tool palette up and down.

Figure 23-26.
You can use the **Tool Palettes** palette to access and insert hatch patterns.

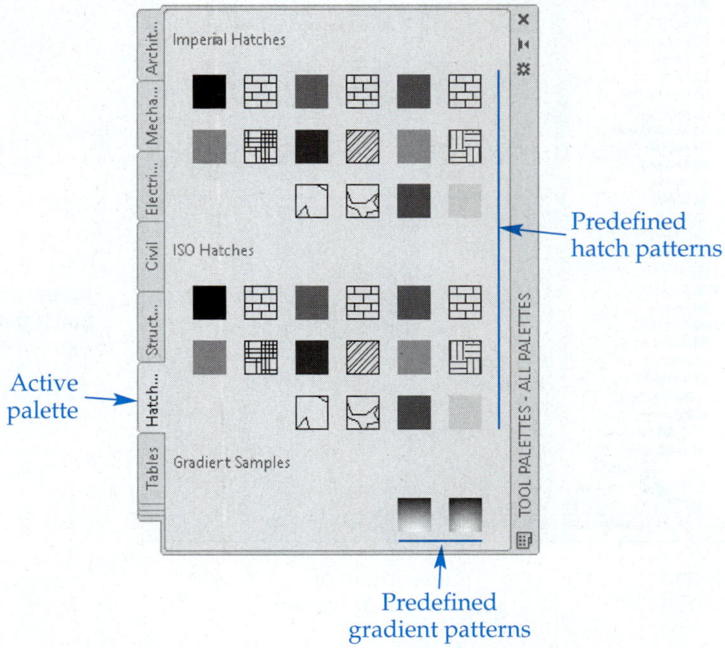

NOTE

By default, icons represent the tools in each tool palette. Several tool palette view options are available, including displaying a tool as an image of your choice.

Inserting Hatch Patterns

To insert a hatch pattern from the **Tool Palettes** palette, access a tool palette containing hatches and fills. To drag and drop a pattern, press and hold down the pick button on the hatch to be inserted, and then drag the cursor into the drawing. A hatch pattern symbol appears with the cursor. Release the pick button in a boundary to apply the hatch pattern. An alternative to the drag-and-drop method is to pick once on the hatch image to attach the hatch to the crosshairs, and then pick a boundary in the drawing to apply the hatch pattern. Edit the hatch pattern, as described later in this chapter, to change the settings after you insert the hatch pattern.

Exercise 23-8

www.g-wlearning.com/CAD/

Complete the exercise on the companion website.

Editing Hatch Patterns

A hatch pattern is a single object that you can edit using the **Properties** palette or standard editing commands such as **ERASE**, **COPY**, and **MOVE**. AutoCAD also provides specific tools to edit hatches. Edit a hatch object to apply a different pattern or pattern properties, or when the drawing changes, such as when you add or remove objects that change existing hatch boundaries.

Click on a hatch object to display the **Hatch Editor** contextual ribbon tab, or double-click to display both the **Quick Properties** palette and the **Hatch Editor** ribbon tab. The **Hatch Editor** ribbon tab is similar to the **Hatch Creation** ribbon tab, but focuses on options for adjusting the existing hatch. You can select multiple hatch objects to apply the same changes to each hatch.

The **Recreate Boundary** and **Display Boundary Objects** options are specific to the **Hatch Editor** ribbon tab. The most practical application for the **Recreate Boundary** option is to recreate erased boundary geometry. See Figure 23-27. Pick the **Recreate Boundary** button and follow the prompts to recreate the boundary as a region or polyline. You can also specify whether to associate the hatch with the objects. The **Display Boundary Objects** option, available for associative hatches, selects the associative boundary. Use grips to change the size and shape of the boundary.

Ribbon
Hatch Editor
> Boundaries

Recreate Boundary

Ribbon
Hatch Editor
> Boundaries

Display Boundary Objects

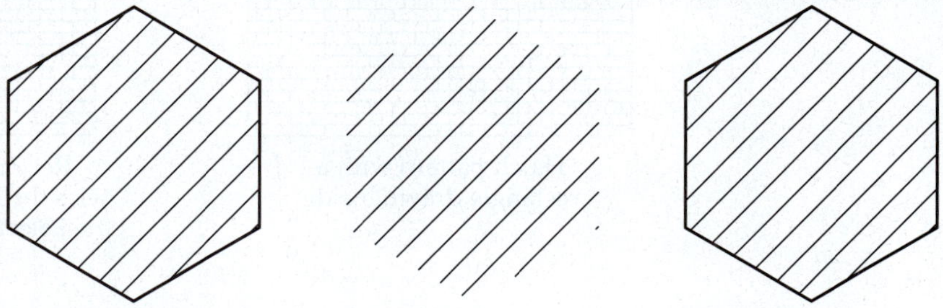

Figure 23-27. Using the **Recreate Boundary** option to recreate a lost object associated with the hatch boundary.

Original Hatched Polygon — Polygon Erased — Recreated Polygon Boundary

HATCHEDIT

Ribbon
Home
> Modify
Edit Hatch
Type
HATCHEDIT
HE

NOTE

The **HATCHEDIT** command allows you to edit a single hatch object using the **Hatch Edit** dialog box. The **-HATCHEDIT** command allows you to edit one or more selected hatches using the command line.

Exercise 23-9

www.g-wlearning.com/CAD/

Complete the exercise on the companion website.

Adding and Removing Boundaries

Use the **Pick Points**, **Select Boundary Objects**, and **Remove Boundary Objects** options of the **Hatch Editor** ribbon tab to add boundaries to and remove boundaries from existing hatch objects. For example, Figure 23-28 shows drawing a rectangle to create a window on a portion of an architectural elevation. To add the window as an island in the boundary, double-click the hatch pattern to display the **Hatch Editor** ribbon tab, and then use the **Select Boundary Objects** option to pick the rectangle. Close the **Hatch Editor** ribbon tab to complete the operation.

Editing Associative Hatch Patterns

When you edit an object associated with a hatch pattern, the hatch pattern changes to adapt to the edit. Figure 23-29 shows examples of stretching an associated object and removing an island from an associative boundary. As long as you edit the objects associated with the boundary, the hatch pattern updates.

Exercise 23-10

www.g-wlearning.com/CAD/

Complete the exercise on the companion website.

Center Grip Editing

The circle center grip that appears when you select a hatch object with the crosshairs provides convenient access to hatch editing options. There are three different ways to access and apply the same context-sensitive hatch grip commands.

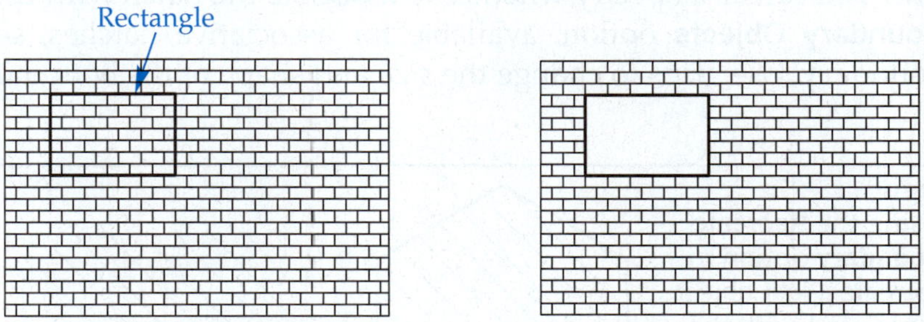

Figure 23-28. Adding an object to an existing hatch pattern boundary.

Hatch pattern with a rectangle drawn inside

Rectangle is now associated with the hatch pattern after using the **Select Boundary Objects** option

Figure 23-29.
Editing objects with associative hatch patterns. A—The hatch pattern stretches with the object. B—The hatch pattern revises to fill the area of an erased island.

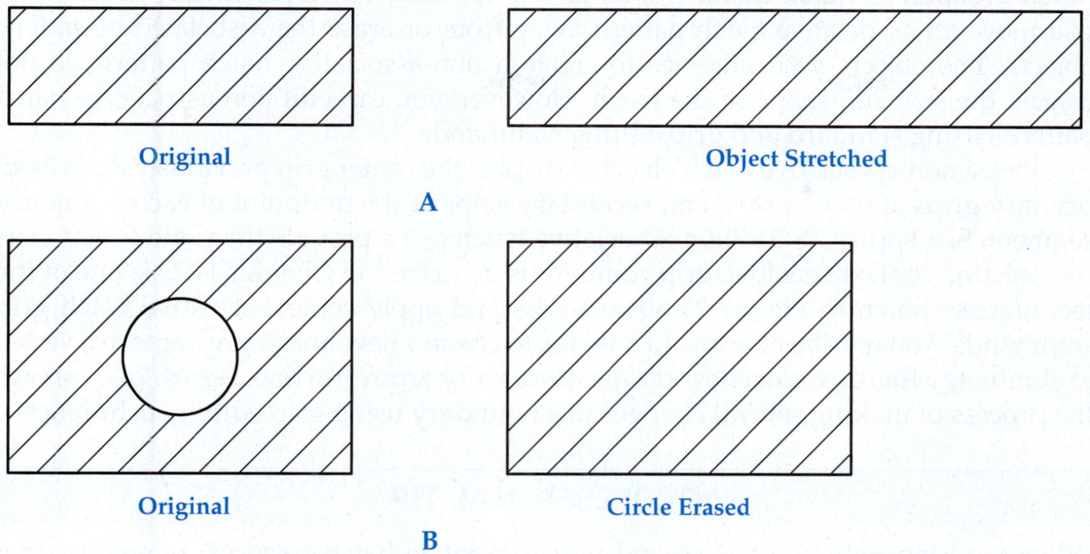

Figure 23-30 explains and illustrates each technique. The **Stretch** option functions as a primary grip option, providing access to the standard **STRETCH**, **MOVE**, **ROTATE**, **SCALE**, and **MIRROR** commands. Use the **Origin Point** option to specify a new origin point for the hatch pattern. Use the **Hatch Angle** option to rotate the hatch, and use the **Hatch Scale** option to edit the pattern size.

Figure 23-30.
Use the center hatch object grip to edit a hatch pattern. Apply one of the following methods to access context-sensitive hatch grip commands. A—Hover over the unselected grip to display a menu of options, and then pick an option from the list. B—Pick the grip and then press [Ctrl] to cycle through options. C—Pick the grip and then right-click and select an option.

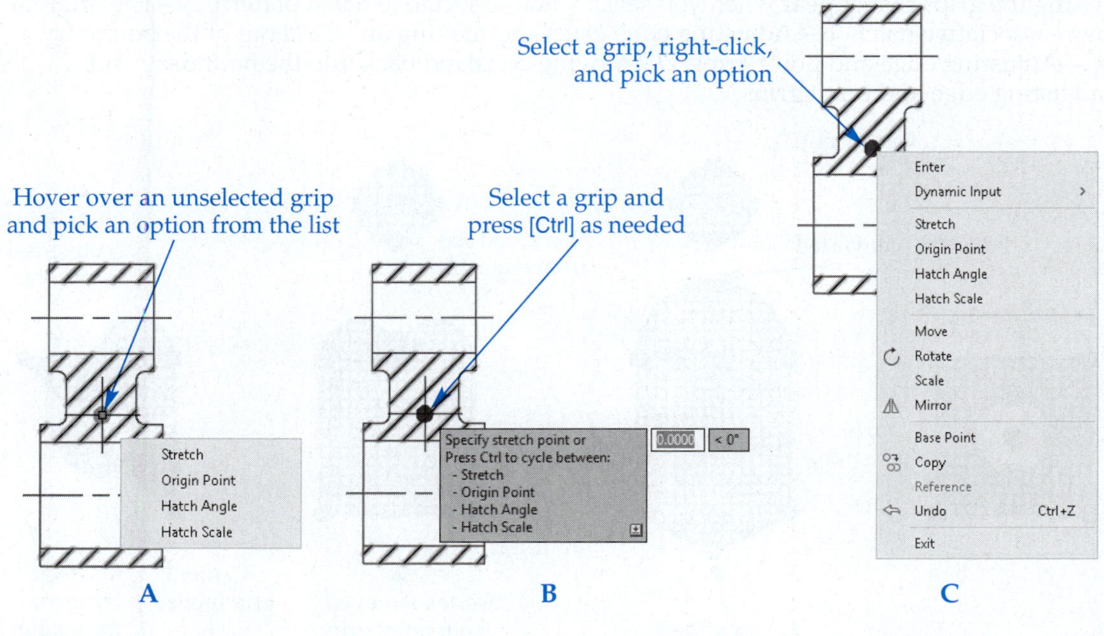

Editing Non-Associative Hatch Patterns

Create a non-associative hatch pattern by deselecting the **Associative** option in the **Hatch Creation** or **Hatch Editor** ribbon tab. A non-associative hatch also forms when you move an associative hatch pattern away from or erase the associated boundary objects. The objects you reference to create a non-associative hatch pattern do not control the size and shape of the hatch. However, you can edit non-associative hatch patterns using standard and grip editing commands.

Pick a non-associative hatch object to display the center grip previously described, primary grips at each vertex, and secondary grips at the midpoint of each boundary segment. See Figure 23-31. Non-associative hatch grips provide the same function as the polyline context-sensitive grip commands described in Chapter 14. Use one of the techniques shown in Figure 23-30 to access and apply context-sensitive hatch grip commands. You may be able to add a vertex to create a new line or arc, remove a vertex to eliminate a line or arc, or convert a line to an arc or an arc to a line. Figure 23-31 shows the process of making several changes to a boundary using grip editing techniques.

PROFESSIONAL TIP

When working with associative and non-associative hatch patterns, remember that associative hatch patterns are associated with objects. The objects define the hatch boundary. Non-associative hatch patterns are not associated with objects, but they do show association with the hatch boundary.

NOTE

The **MIRRHATCH** system variable is set to 0 by default. This setting prevents a hatch from reversing during a mirror operation. Change the **MIRRHATCH** value to 1 to mirror a hatch in relation to the original object. See Figure 23-32.

Figure 23-31.
Using the grips that appear when you select a non-associative hatch pattern. A—The original non-associative hatch. B—Adjusting edge grips and moving an island out of the boundary. C—Adjusting edge and point grips. D—Moving an island back into the boundary and adjusting edge and point grips.

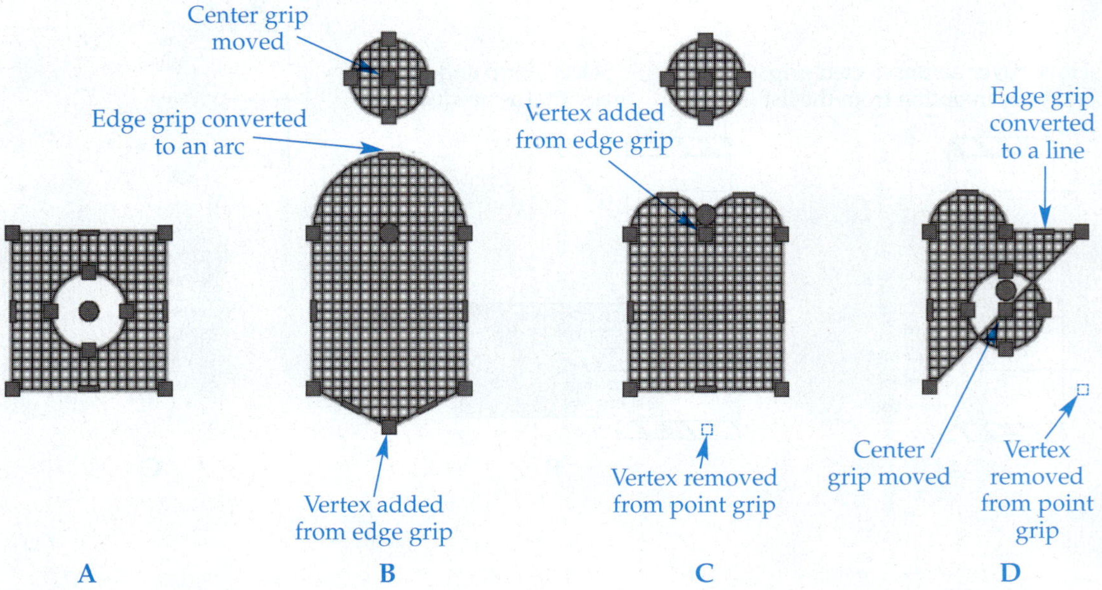

Figure 23-32.
The **MIRRHATCH** system variable options.

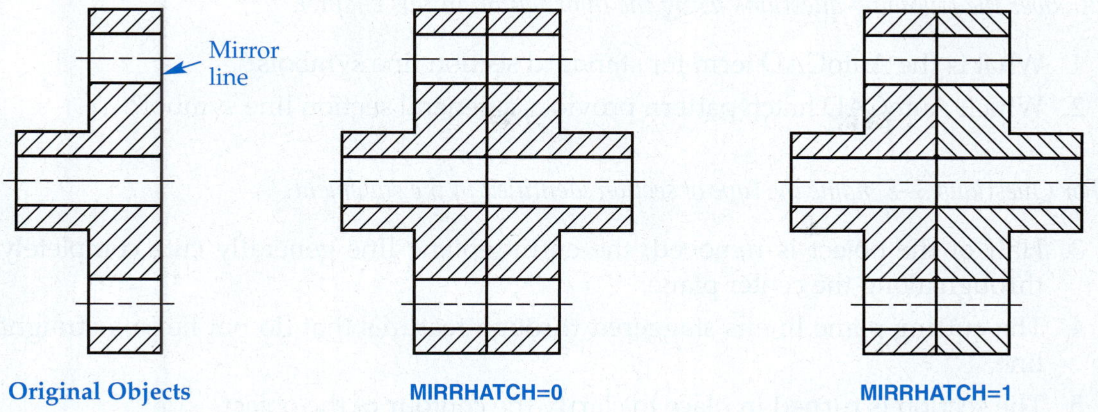

Exercise 23-11

www.g-wlearning.com/CAD/

Complete the exercise on the companion website.

Express Tools

www.g-wlearning.com/CAD/

The SUPERHATCH Command

The **Draw** panel of the **Express Tools** ribbon tab includes a **SUPERHATCH** command. For information about the **SUPERHATCH** command, go to the companion website, navigate to this chapter in the **Contents** tab, and select **The SUPERHATCH Command**.

Chapter Review

Answer the following questions using the information in this chapter.

1. What is the AutoCAD term for standard section line symbols?
2. Which AutoCAD hatch pattern provides a general section line symbol?

For Questions 3–8, name the type of section identified in the statement.

3. Half of the object is removed; the cutting-plane line generally cuts completely through along the center plane.
4. The cutting-plane line is staggered through features that do not lie in a straight line.
5. The section is turned in place to clarify the contour of the object.
6. The section is rotated and is located away from the object. A cutting-plane line normally identifies the location of the section.
7. The cutting-plane line cuts through one-quarter of the object; used primarily on symmetrical objects.
8. A small portion of the view is removed to clarify an internal feature.

9. Explain the three island detection style options.
10. Describe the basic difference between using the **Pick Points** and **Select Boundary Objects** options in the **Hatch Creation** ribbon tab.
11. How do you limit AutoCAD hatch evaluation to a specific area of the drawing?
12. What is the purpose of the **Gap Tolerance** setting in the **Hatch Creation** ribbon tab?
13. Explain how to select a hatch pattern or fill using the **Hatch Creation** ribbon tab.
14. Name the two files supplied with AutoCAD that contain hatch patterns.
15. What considerations should you take into account when choosing a hatch scale?
16. How do you change the hatch angle in the **Hatch Creation** ribbon tab?
17. What is a gradient fill, and how do you create a gradient fill with the **HATCH** command?
18. Define *associative hatch pattern*.
19. What is the result of stretching an object associated with an associative hatch pattern?
20. Explain how to use an existing hatch pattern on a drawing as the pattern for another hatch.
21. Explain how to use drag and drop to insert a hatch pattern from **DesignCenter** into an active drawing.
22. Explain two ways to insert a hatch pattern from a tool palette into a drawing.
23. What is typically the easiest way to access the **Hatch Editor** ribbon tab?
24. How does the **Hatch Editor** ribbon tab compare to the **Hatch Creation** ribbon tab?
25. What happens if you erase an island inside an associative hatch pattern?

Drawing Problems

Start AutoCAD if it is not already started. Start a new drawing for each problem using an appropriate template of your choice. The template should include layers and text, dimension, multileader, and table styles, when necessary, for drawing the given objects. Add layers as needed and add text, dimension, multileader, and table styles as needed. Draw all objects using appropriate layers. Use appropriate text, dimension, multileader, and table styles, justification, and format. Follow the specific instructions for each problem. Use only drawing and editing commands and techniques you have already learned. Use your own judgment and approximate dimensions when necessary. Apply formal dimensions accurately using ASME or appropriate industry standards.

▼ Basic

1. **Graphic Design** Draw the game board shown. Save the drawing as P23-1.

2. **Graphic Design** Draw the bar graph shown. Save the drawing as P23-2.

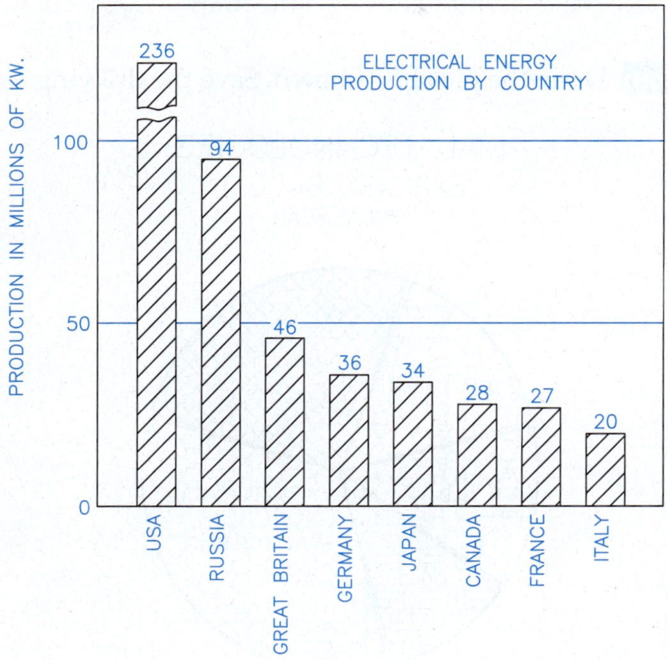

Drawing Problems - Chapter 23

3. **Electronics** Draw the component layout shown. Save the drawing as P23-3.

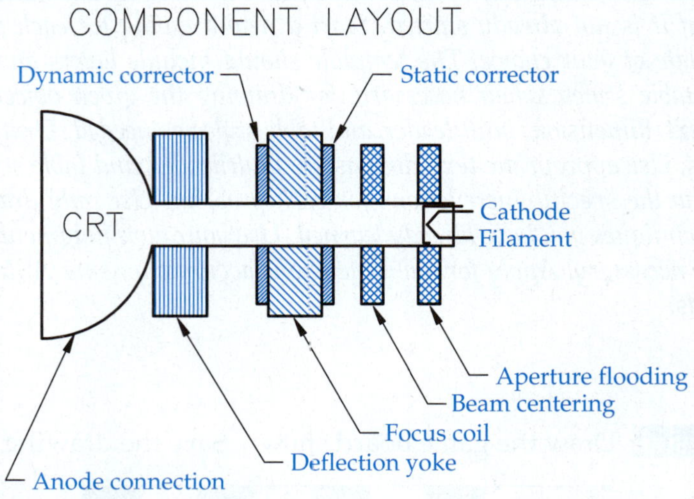

4. **Graphic Design** Draw the bar graphs shown. Save the drawing as P23-4.

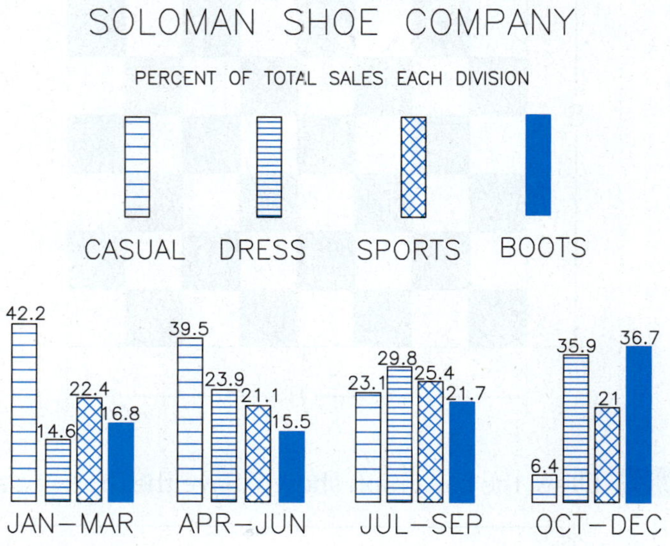

5. **Graphic Design** Draw the pie chart shown. Save the drawing as P23-5.

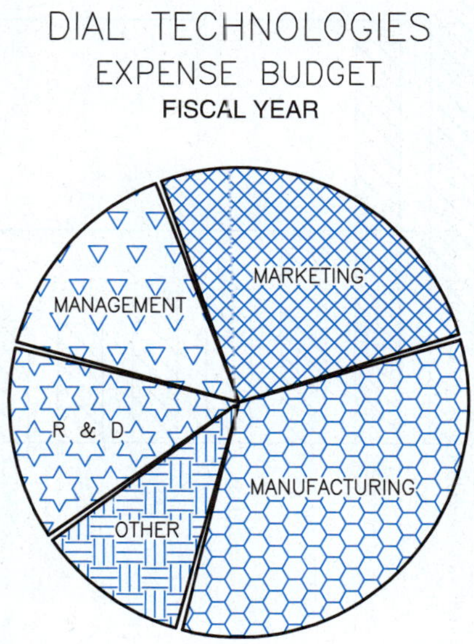

Drawing Problems - Chapter 23

6. **Graphic Design** Draw the bar graph shown. Save the drawing as P23-6.

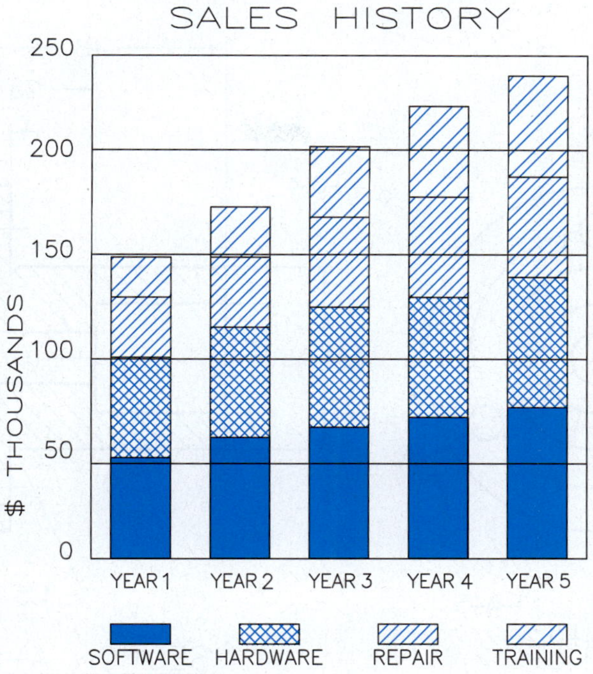

▼ Intermediate

7. **Mechanical** Draw and dimension the views of the hydraulic valve cylinder shown in **Figure 23-2**. Save the drawing as P23-7.

8. **Mechanical** Draw and dimension the views of the hub shown. Save the drawing as P23-8.

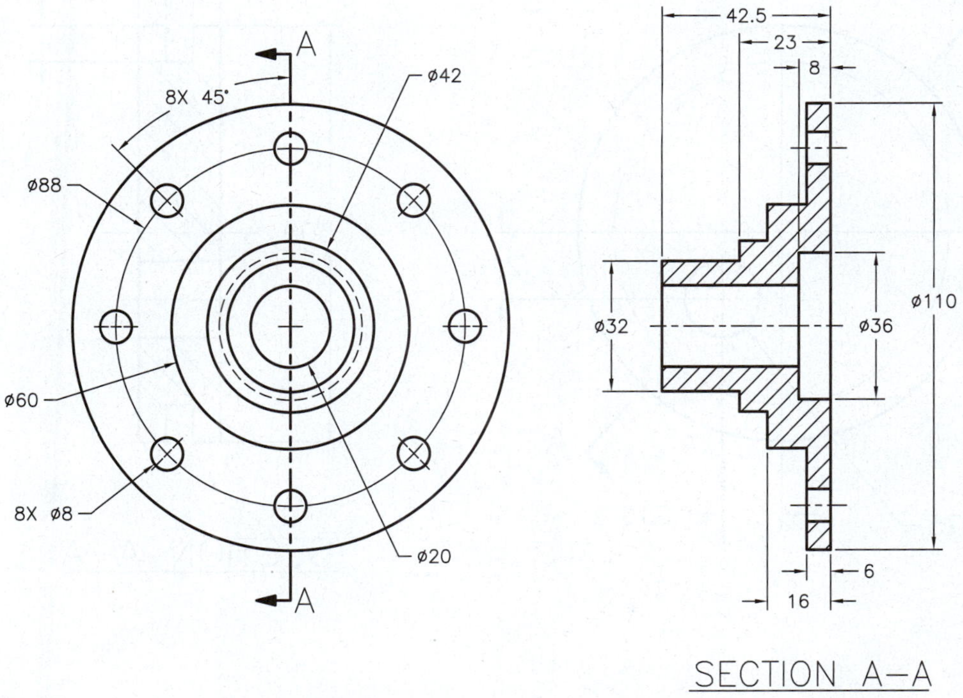

9. Mechanical Draw and dimension the views of the hub shown. Save the drawing as P23-9.

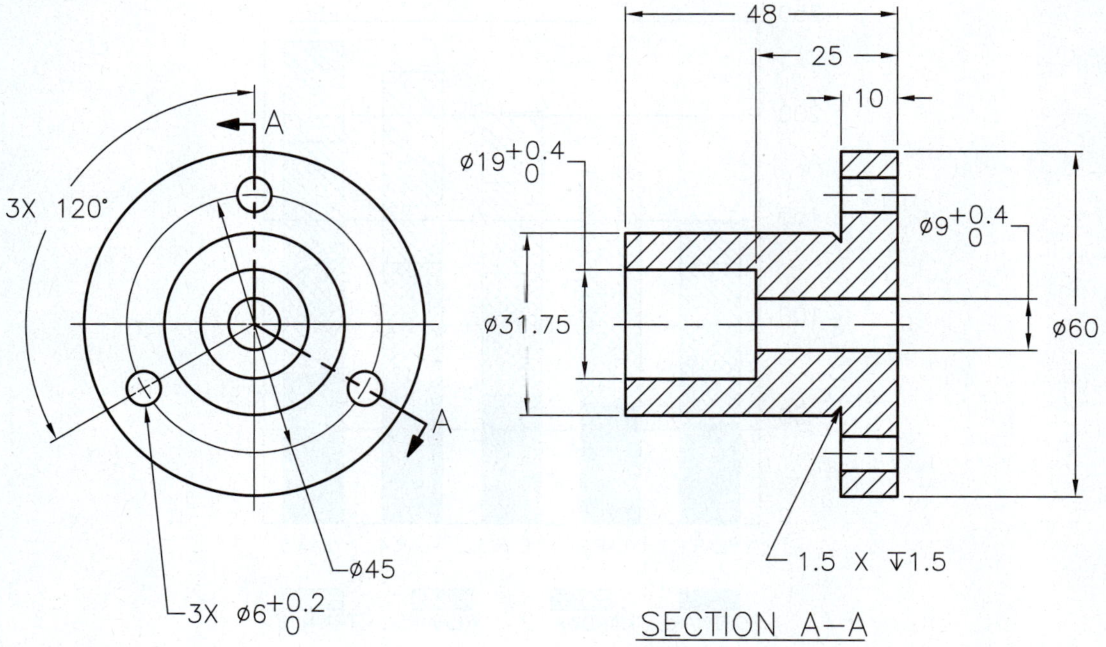

10. Mechanical Draw and dimension the views of the chain guide as shown. Save the drawing as P23-10.

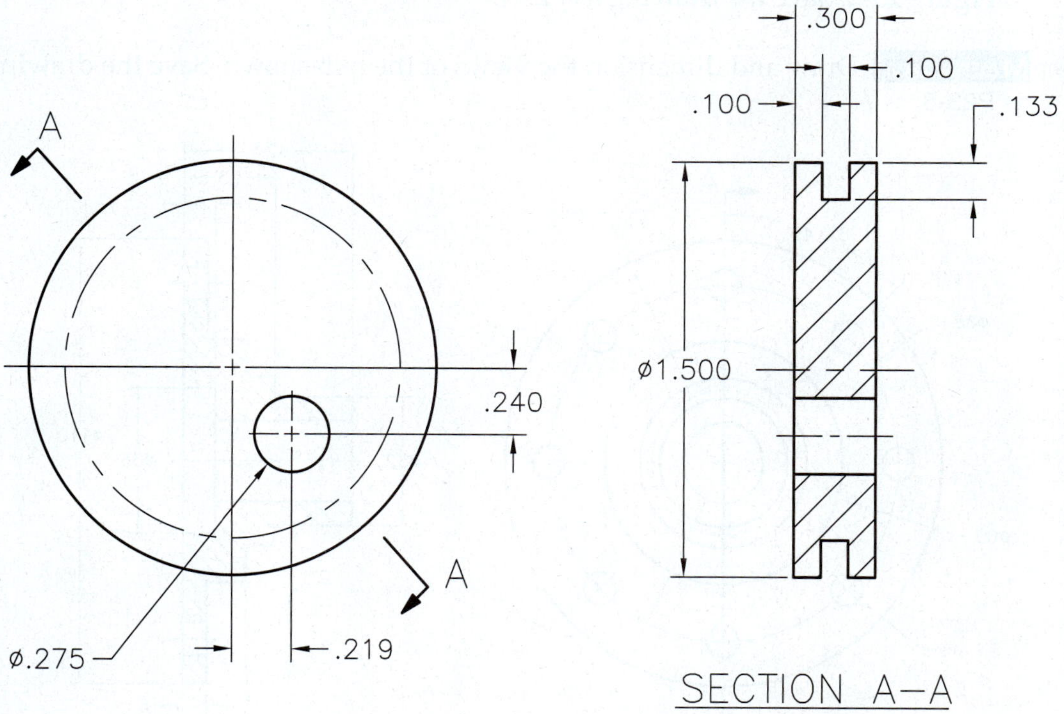

11. **Mechanical** Draw and dimension the views of the sleeve and add the notes as shown. Save the drawing as P23-11.

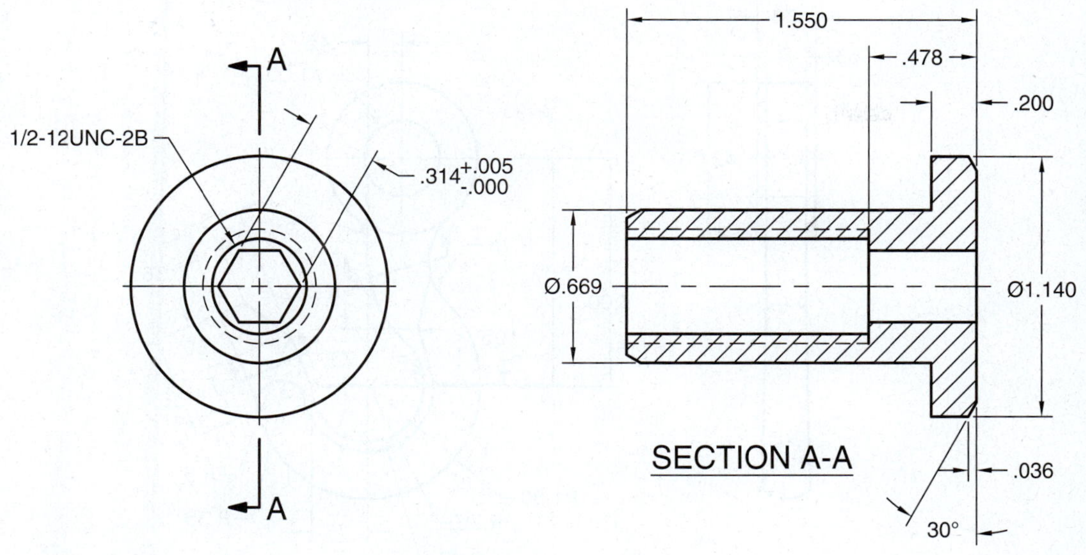

1. DIMENSIONING AND TOLERANCING PER ASME Y14.5-2018.
2. REMOVE ALL BURRS AND SHARP EDGES.
3. CASE HARDEN 45-50 ROCKWELL.
4. PAINT ACE GLOSS BLACK ALL OVER.

▼ Advanced

12. **Mechanical** Draw and dimension the views shown. Add the following notes: OIL QUENCH 40-45C, CASE HARDEN .020 DEEP, and 59-60 ROCKWELL C SCALE. Save the drawing as P23-12.

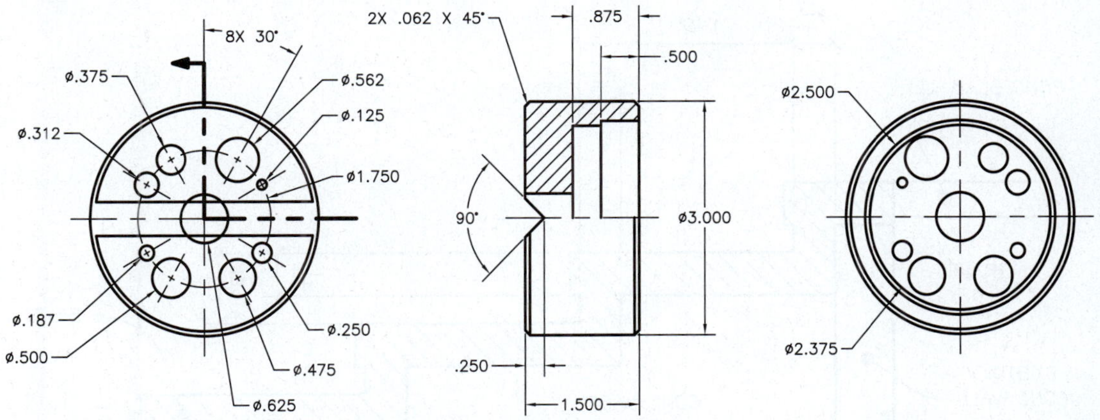

Name: Diffuser
Material: AISI 1018

13. **Mechanical** Draw and dimension the views of the tow hook as shown. Save the drawing as P23-13.

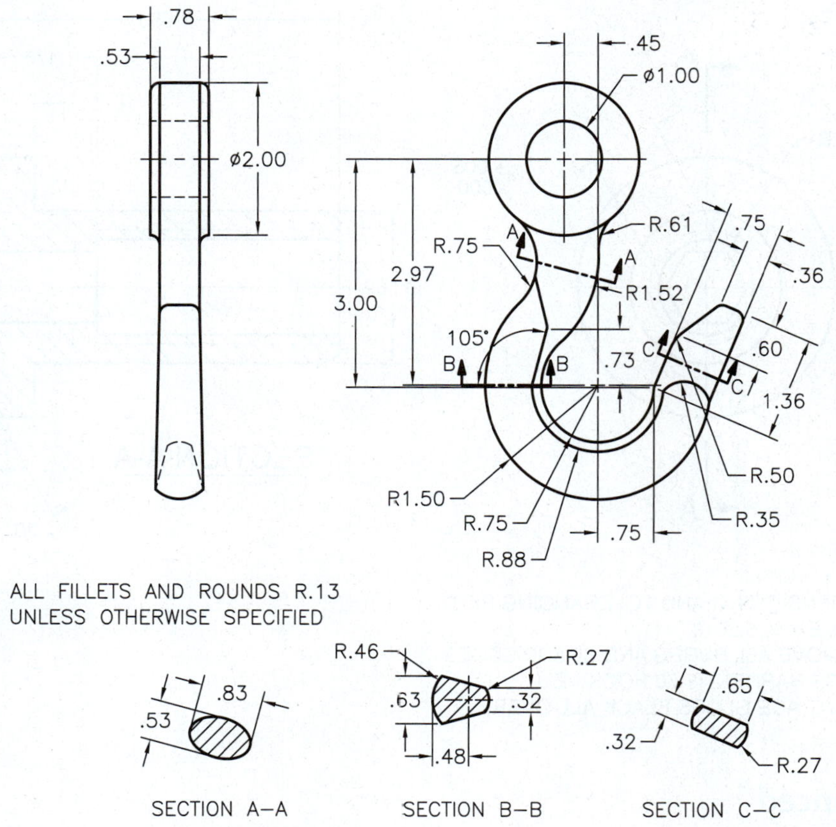

14. **Architectural** Draw the stair detail as shown. Save the drawing as P23-14.

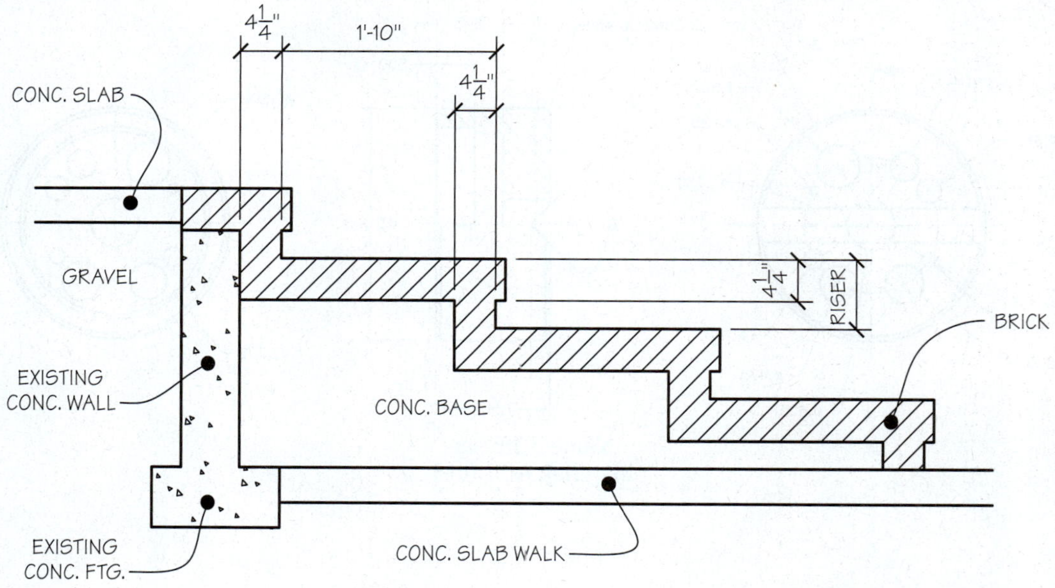

15. **Architectural** Draw the foundation detail shown, but make the following changes:
   A. Use the SansSerif font
   B. Use .125″ leader shoulders
   C. When text is on the left side of a leader, position text at the middle of the bottom line
   D. When text is on the right side of a leader, position text at the middle of the top line

   Save the drawing as P23-15.

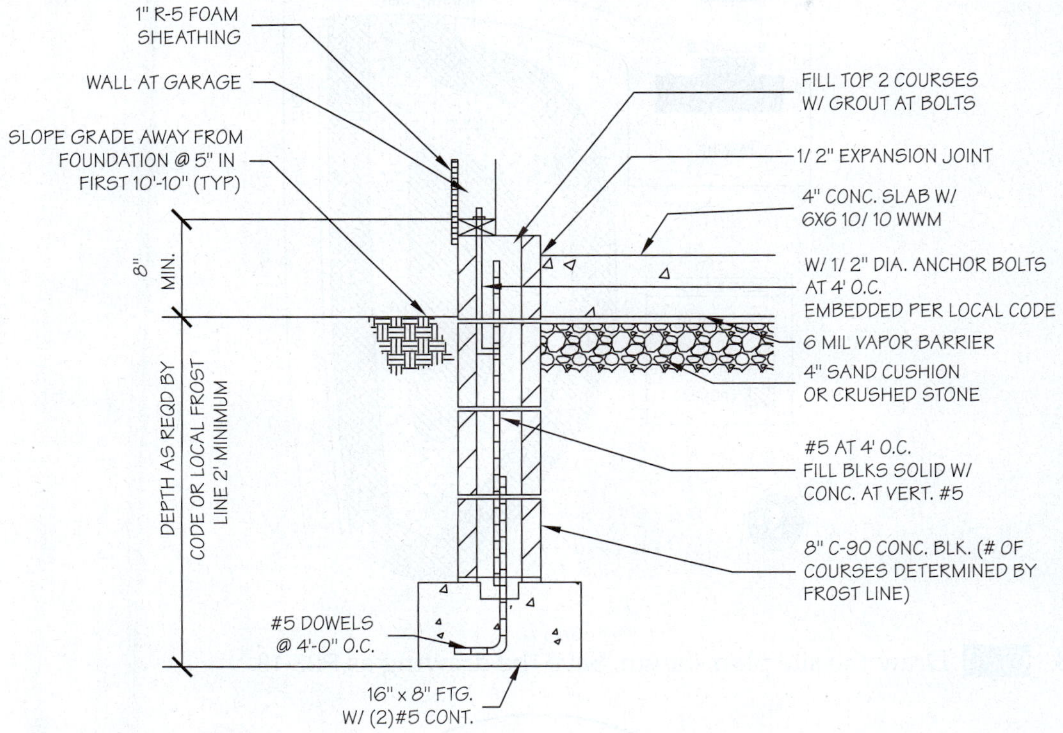

16. **Architectural** Draw the front elevation shown. Save the drawing as P23-16.

Drawing Problems - Chapter 23

17. **Civil** Draw the site plan shown. Save the drawing as P23-17.

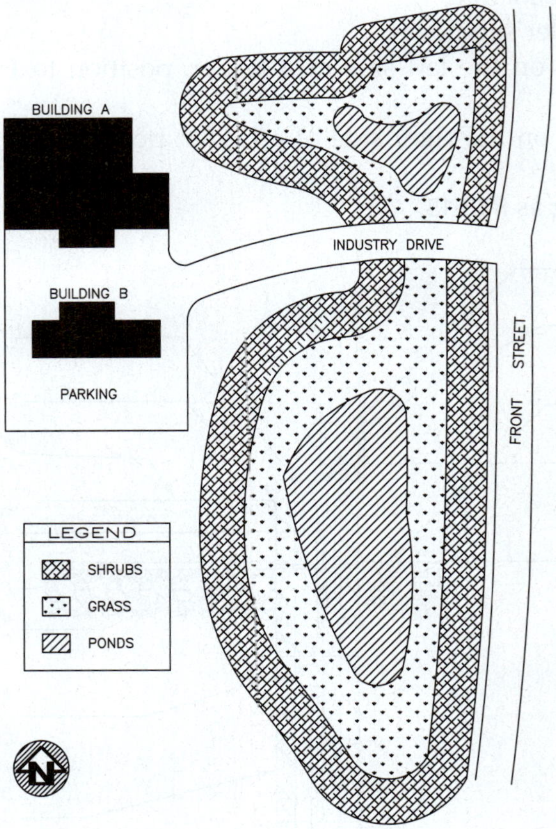

18. **Civil** Draw the site plan shown. Save the drawing as P23-18.

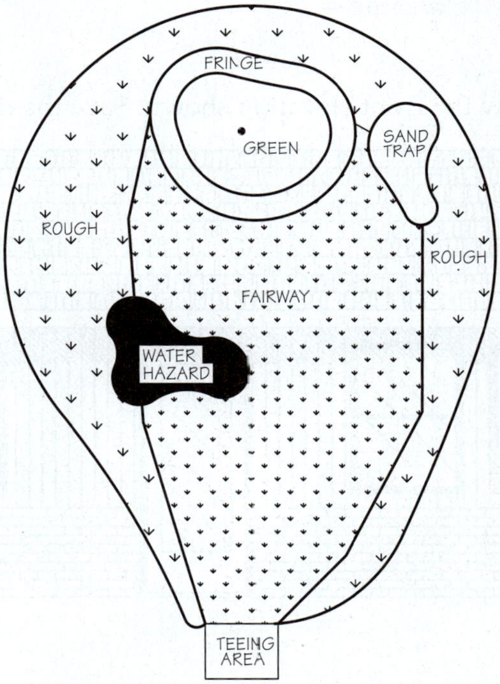

19. **Civil** Draw the site plan shown. Save the drawing as P23-19.

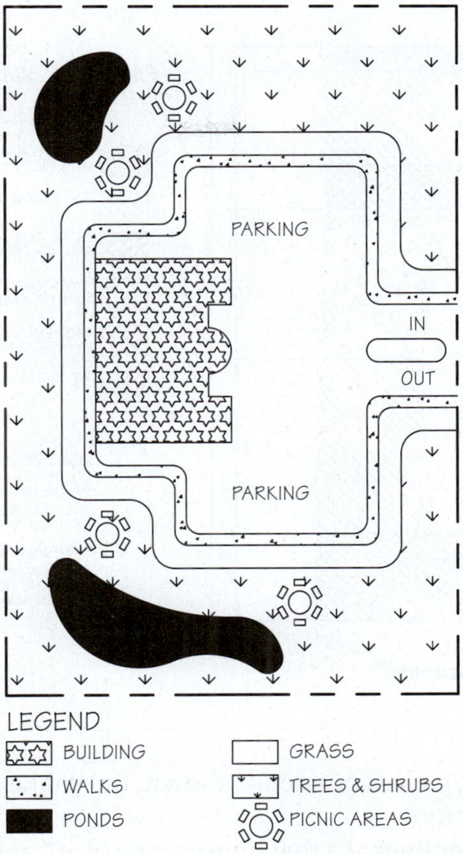

20. **Architectural** Draw the structural detail shown. Save the drawing as P23-20.

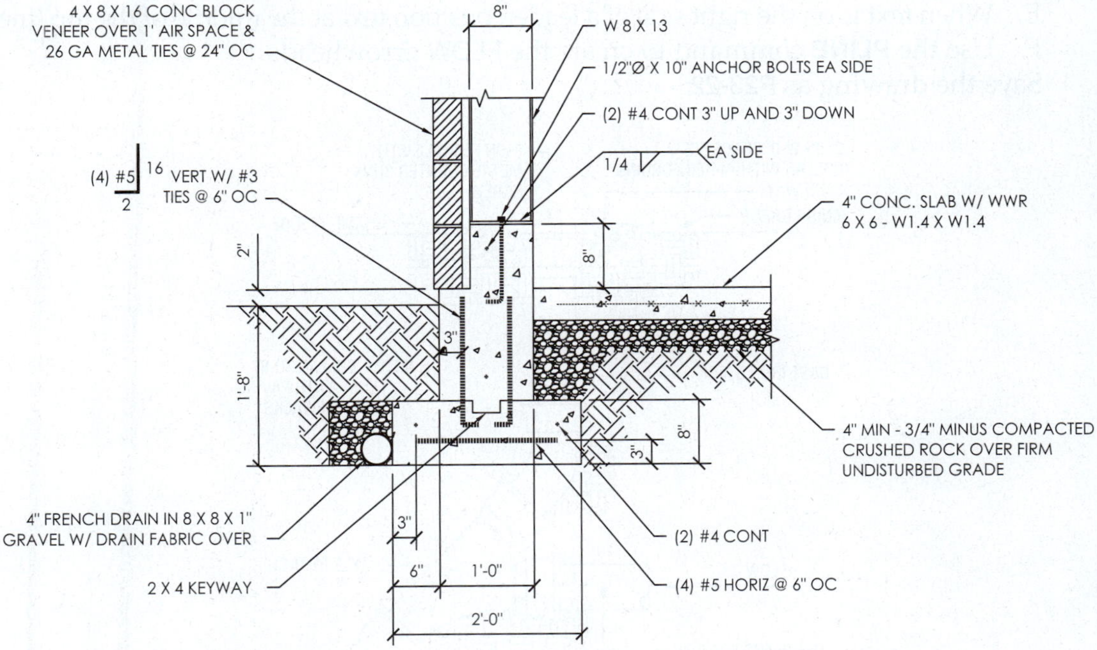

21. **Civil** Draw the profile shown. Save the drawing as P23-21.

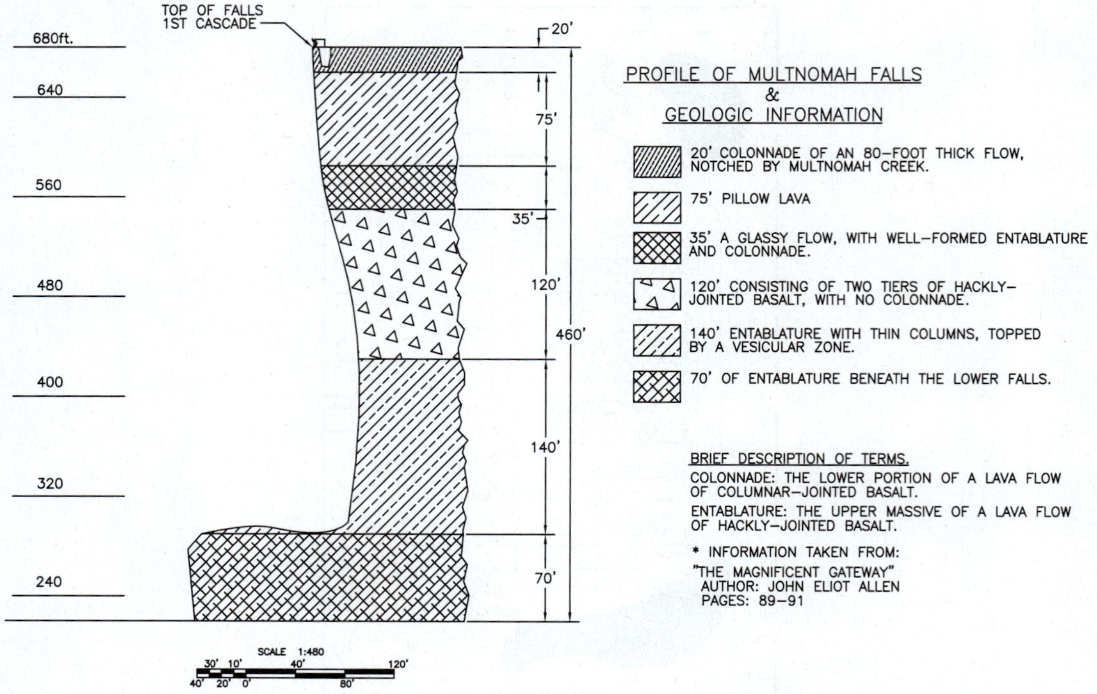

22. **Piping** Draw the drop cleanout detail shown, but make the following changes:

A. Use the SansSerif font
B. Use diagonal leader lines at a maximum angle of 60° and a minimum angle of 30°
C. Use .125″ leader shoulders
D. When text is on the left side of a leader, position text at the middle of the bottom line
E. When text is on the right side of a leader, position text at the middle of the top line
F. Use the **PLINE** command to create the FLOW arrowheads.

Save the drawing as P23-22.

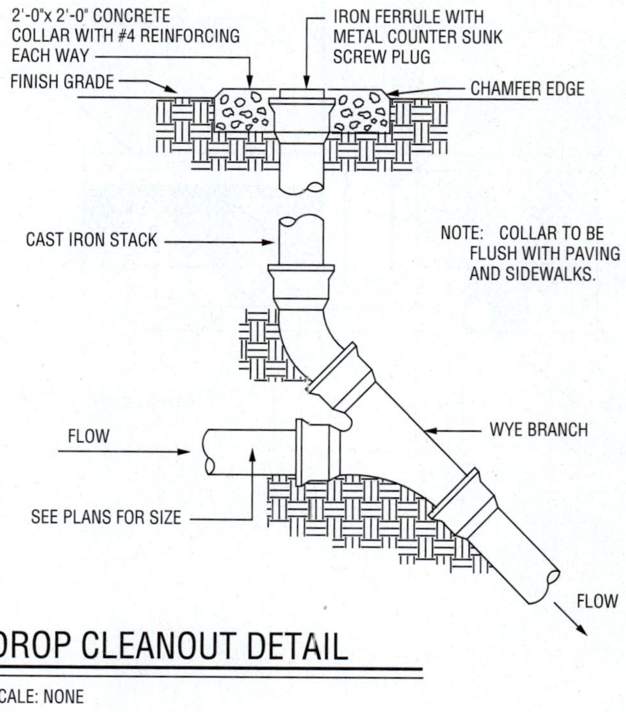

23. General Draw the map shown, using the **SPLINE** command to create the curved shapes. Draw the map at full scale, following the map shown as closely as possible. Save the drawing as P23-23.

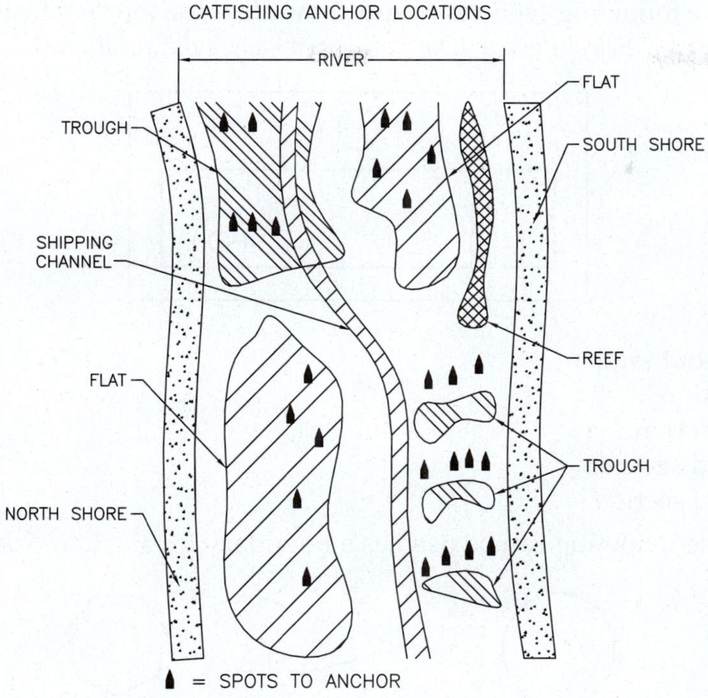

Drawing Problems - Chapter 23

AutoCAD Certified Professional Exam Practice

Answer the following questions using the information in this chapter.

1. Which of the following sections would show all of the internal features of the part most efficiently? *Select the one item that best answers the question.*

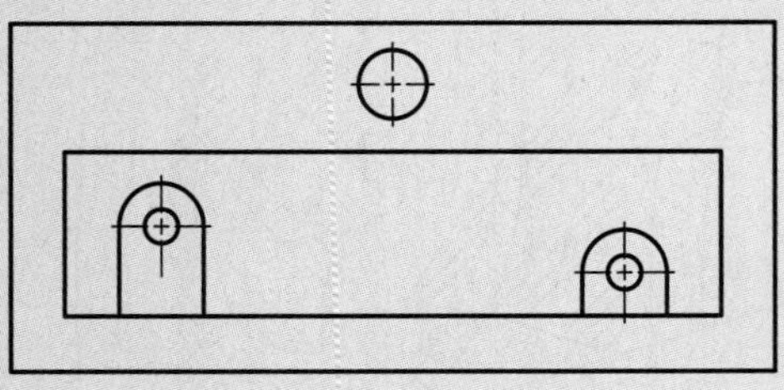

   A. broken-out section
   B. full section
   C. offset section
   D. removed section
   E. revolved section

2. Which of the following can be used as a boundary for a hatch? *Select all that apply.*

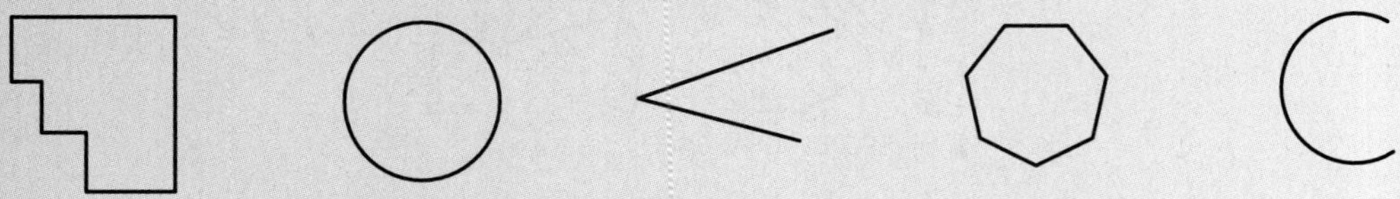

A B C D E

3. Which of the following options would you choose to create the hatch shown? *Select the one item that best answers the question.*
   A. **Ignore Island Detection**
   B. **No Island Detection**
   C. **Normal Island Detection**
   D. **Outer Island Detection**

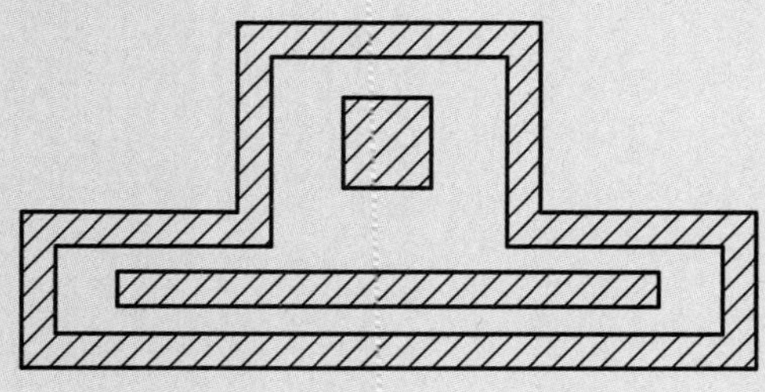

Follow the instructions in each problem. Complete each problem by answering the related question.

4. **Navigate to this chapter on the companion website and open CPE-23scale.dwg.** What is the scale of the hatch in this drawing?
5. **Navigate to this chapter on the companion website and open CPE-23boundary.dwg.** Recreate the boundary for the hatch. Use the boundary and the appropriate measurement command to answer the following question: What is the total area of the hatched object?

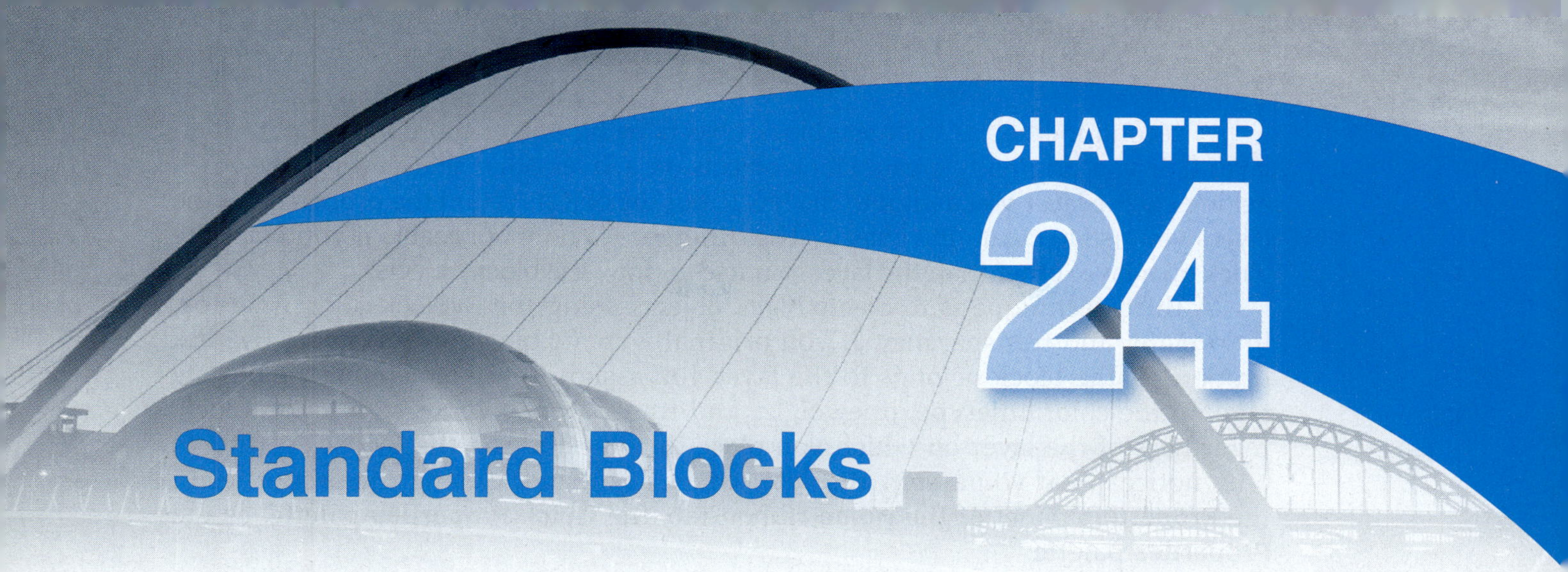

CHAPTER 24

Standard Blocks

Learning Objectives

After completing this chapter, you will be able to:

- ✓ Create and save blocks.
- ✓ Insert blocks into a drawing.
- ✓ Edit a block and update the block in a drawing.
- ✓ Create blocks as drawing files.
- ✓ Construct and use a symbol library.
- ✓ Purge unused content from a drawing.

The ability to create and use ***blocks*** is a major benefit of drawing with AutoCAD. The **BLOCK** command stores a block within a drawing as a ***block definition***. The **WBLOCK** command saves a ***wblock*** as a separate drawing file. You can insert blocks as often as needed and share blocks between drawings. You also have the option to scale, rotate, and adjust blocks to meet specific drawing requirements.

block: An object, such as a symbol, saved and stored in a drawing for future use.

block definition: Information about a block stored within the drawing file.

wblock: A block definition saved as a separate drawing file.

Constructing Blocks

A block can consist of any object or group of objects, including annotation, or can be an entire drawing. Review each drawing and project to identify items you can use more than once. Fasteners, punches, subassemblies, fixtures, and appliances are examples of items to consider converting to blocks. Draw the objects once and then save them as a block for multiple use.

Selecting a Layer

Identify the appropriate layer on which to create block elements before drawing the objects. To do this, you must understand how layers and object properties apply to blocks. The 0 layer is the preferred layer on which to draw block objects. If you originally create block objects on the 0 layer, the block inherits the properties of any layer you assign to the block. Draw the objects for all blocks on the 0 layer and then assign the appropriate layer to each block when you insert the block. If you draw block objects on a layer other than layer 0, place all the objects on layer 0 before creating the block.

A second method is to create block objects using one or more layers other than layer 0. If you originally create block objects on a layer other than layer 0, the block belongs to the layer you assign to the block, but the objects retain the properties of the layers you use to create the objects. The difference is only noticeable if you place the block on a layer other than the layer you use to draw the block objects.

A third technique is to create block objects using the ByBlock color, linetype, lineweight, and transparency. If you originally create block objects using ByBlock properties, the block belongs to the layer you assign to the block, but the objects take on the color, linetype, lineweight, and transparency you assign to the block, regardless of the layer on which you place the block. Using the ByBlock setting is only noticeable if you assign absolute values to the block using the properties in the **Properties** panel of the **Home** ribbon tab, the **Quick Properties** palette, or the **Properties** palette.

Another option is to create block objects using an absolute color, linetype, lineweight, and transparency. If you originally create block objects using absolute values, such as a Blue color, a Continuous linetype, a 0.05 mm lineweight, and a transparency value of 50, the block belongs to the layer you assign to the block. However, the objects display the specified absolute values regardless of the properties assigned to the drawing or the layer on which you place the block.

CAUTION

Drawing block objects on a layer other than layer 0, or using ByBlock or absolute properties, can cause significant confusion. The result is often a situation in which a block belongs to a layer, but the block objects display properties of a different layer, or absolute values. In most cases, draw block objects on layer 0, and then assign a specific layer to each block.

Drawing Block Elements

Draw the elements of a block as you would any other geometry. If you plan to define a block as annotative, to scale the block according to the drawing scale, you can include annotative or non-annotative objects, such as text, with the block. As long as you specify the block as annotative, all objects act annotative, even if some objects are non-annotative. However, you must use non-annotative objects when preparing a non-annotative block.

insertion base point: The point on a block that defines where the block is positioned during insertion.

When you finish drawing the objects, determine the best location for the ***insertion base point***. When you insert the block into a drawing, the insertion base point positions the block. Figure 24-1 shows examples of common blocks and a possible insertion base point for each.

PROFESSIONAL TIP

A single block allows you to create multiple features that are identical except for scale. In these cases, draw the base block to fit inside a one-unit square. This makes it easy to scale the block when you insert it into a drawing to create variations of the block. Drawing a block in this manner is explained in more detail later in this chapter.

BLOCK

Ribbon
Home
> Block
Insert
> Block Definition

Create Block

Type
BLOCK
BMAKE
B

Creating Blocks

Once you draw objects and identify an appropriate insertion base point, you are ready to define the objects as a block. Use the **BLOCK** command and the corresponding **Block Definition** dialog box to create a block. See Figure 24-2.

Figure 24-1.
Common drafting symbols and their insertion points for placement on drawings. Colored grips indicate the insertion points.

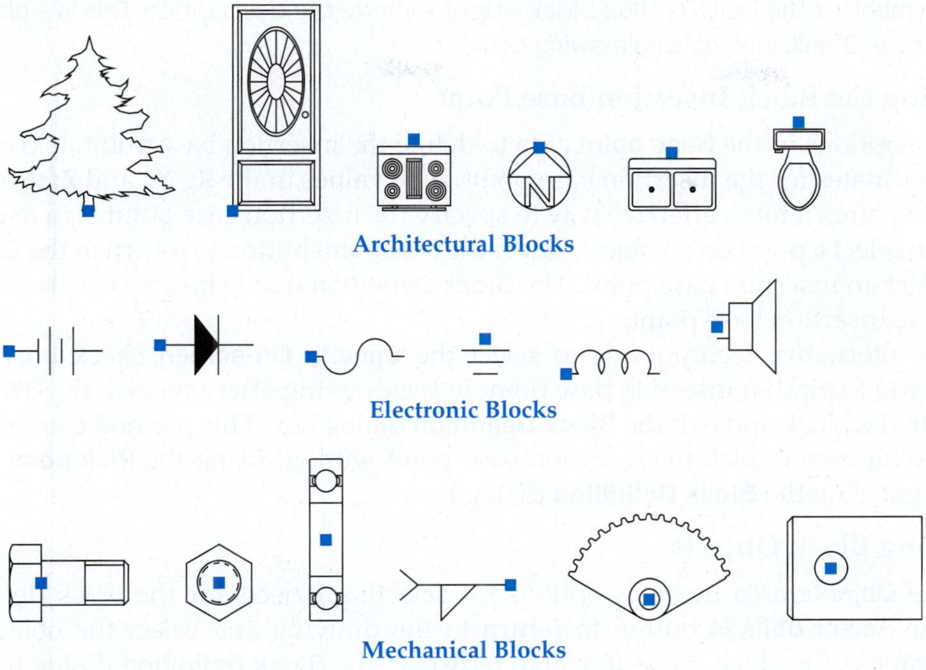

Figure 24-2.
Use the **Block Definition** dialog box to create a block.

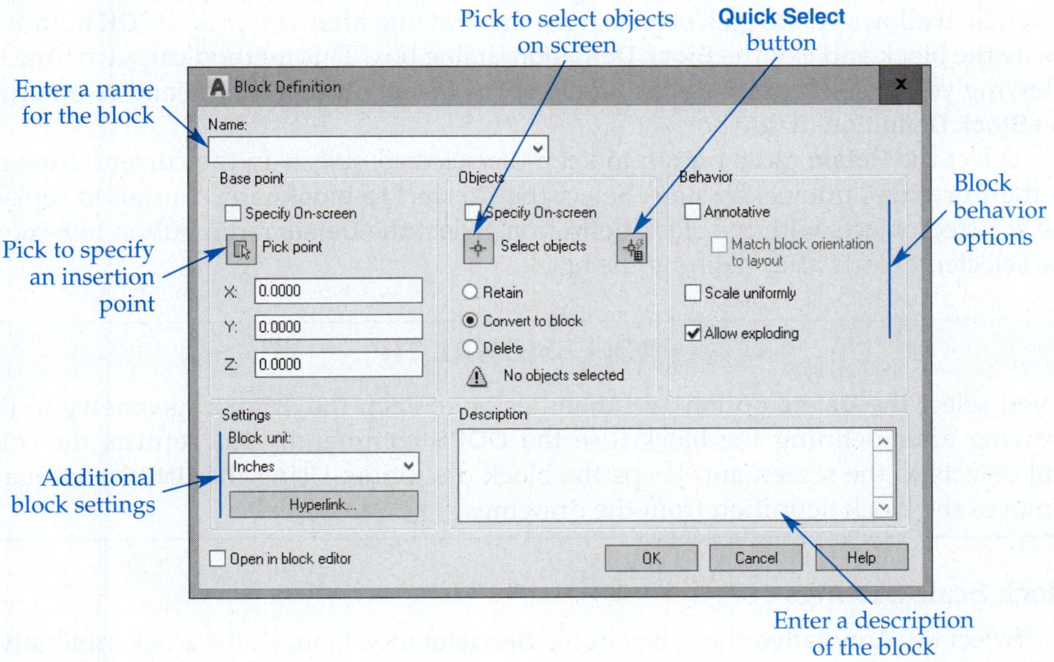

Naming and Describing the Block

Enter a descriptive name for the block in the **Name:** text box. For example, name a vacuum pump PUMP or a 3′-0″ × 6′-8″ door DOOR_3068. The block name cannot exceed 255 characters. It can include numbers, letters, spaces, the dollar sign ($), hyphen (-), and underscore (_). Use the drop-down list to access an existing name to recreate a block or to use a block name as reference when naming a new block with a similar name.

A block name is often descriptive enough to identify the block. However, you can enter a description of the block in the **Description** text box to help identify the block. For example, the PUMP block might include the description This is a vacuum pump symbol, or the DOOR_3068 block might include the description This is a plan view, 3' wide by 6'-8" tall, interior, single-swing door.

Defining the Block Insertion Base Point

Use options in the **Base point** area to define the insertion base point. If you know the coordinates for the insertion base point, type values in the **X:**, **Y:**, and **Z:** text boxes. However, often a more effective way to specify the insertion base point is to use object snaps to select a point on an object. Select the **Pick point** button to return to the drawing and select an insertion base point. The **Block Definition** dialog box reappears after you select the insertion base point.

An alternative technique is to select the **Specify On-screen** check box, which allows you to pick an insertion base point in the drawing after you pick the **OK** button to create the block and exit the **Block Definition** dialog box. This method can save time by allowing you to pick the insertion base point without using the **Pick point** button and re-entering the **Block Definition** dialog box.

Selecting Block Objects

The **Objects** area includes options for selecting objects for the block definition. Pick the **Select objects** button to return to the drawing and select the objects that will compose the block. Press [Enter] to redisplay the **Block Definition** dialog box. The number of selected objects appears in the **Objects** area, and an image of the selection appears next to the **Name:** drop-down list. Use the **QuickSelect** button and **Quick Select** dialog box to define a selection set filter.

An alternative method for selecting objects is to select the **Specify On-screen** check box, which allows you to pick objects from the drawing after you pick the **OK** button to create the block and exit the **Block Definition** dialog box. This method can save time by allowing you to select objects without using the **Select objects** button and re-entering the **Block Definition** dialog box.

Select the **Retain** radio button to keep the selected objects in the current drawing in their original, unblocked state. Select the **Convert to block** radio button to replace the selected objects with the block definition. Select the **Delete** radio button to remove the selected objects after defining the block.

PROFESSIONAL TIP

If you select the **Delete** option and then decide to keep the original geometry in the drawing after defining the block, use the **OOPS** command. This returns the original objects to the screen and keeps the block definition. Using the **UNDO** command removes the block definition from the drawing.

Block Scale Settings

Select the **Annotative** check box in the **Behavior** area to make the block annotative. AutoCAD scales annotative blocks according to the annotation scale. When you select the **Annotative** check box, the **Match block orientation to layout** check box becomes available. Select this check box to match the orientation of a block in layout viewports with the layout orientation, even if the drawing view rotates.

If you check **Scale uniformly** in the **Behavior** area, you do not have the option of specifying different X and Y scale factors when you insert the block. You will learn options for scaling blocks later in this chapter.

Additional Block Definition Settings

If you check **Allow exploding** in the **Behavior** area, you have the option of exploding the block. If you do not check **Allow exploding**, you cannot explode the block after inserting it. Select a unit type from the **Block unit:** drop-down list in the **Settings** area to specify the insertion units of the block. Pick the **Hyperlink...** button to access the **Insert Hyperlink** dialog box to insert a hyperlink in the block. If you check **Open in block editor**, the new block immediately opens in the **Block Editor** when you create the block and exit the **Block Definition** dialog box. The **Block Editor** is described later in this chapter.

NOTE

To verify that a block has been saved, reopen the **Block Definition** dialog box. Pick the **Name:** drop-down list to display a list of blocks in the current drawing.

PROFESSIONAL TIP

You can use blocks to create other blocks. Insert existing blocks into a view and then save all of the objects as a block. This is a process known as ***nesting***. You must give the top-level block a name that is different from any nested block. Proper planning and knowledge of all existing blocks can speed up the drawing process and the creation of complex views.

nesting: Creating a block that includes other blocks.

Exercise 24-1 www.g-wlearning.com/CAD/

Complete the exercise on the companion website.

Inserting Blocks

AutoCAD provides several options for inserting a block into a drawing. Remember to make the layer you want to assign to the block current before inserting the block. You should also determine the proper size and rotation angle for the block before insertion. The term ***block reference*** describes an inserted block. ***Dependent objects*** are any ***named objects***, such as blocks and layers. AutoCAD automatically updates dependent objects in a drawing the next time you open the drawing.

block reference: A specific instance of a block inserted into a drawing.

dependent objects: Named objects in a drawing that have been inserted or referenced into another drawing.

named objects: Blocks, dimension styles, groups, layers, linetypes, materials, multileader styles, multiline styles, plot styles, shapes, table styles, text styles, and visual styles that have specific names.

Inserting Blocks Using the Ribbon

A quick way to insert a block defined in the current file is to use the **Insert** drop-down list in the **Block** panel of the **Home** ribbon tab or the **Block** panel of the **Insert** ribbon tab. See Figure 24-3. Resize the preview area or scroll through the list as necessary to find and select the desired block. The selected block attaches to the cursor and the Specify insertion point or [Basepoint/Scale/X/Y/Z/Rotate]: prompt, or a similar prompt, appears. Specify a point to insert the block, or select an option to make adjustments before inserting. You will learn more about base point, scale, and rotate options in the following sections.

The **Recent Blocks...** and **Blocks from Other Drawings...** options at the bottom of the **Insert** drop-down list are used to access the **Recent** and **Other Drawing** tabs of the **Blocks** palette, described in the next section.

-INSERT

Ribbon
Home
 > Block
Insert
 > Block
 > Insert
Type
-INSERT

Figure 24-3.
Inserting a block using the **Insert** drop-down list in the **Block** panel of the **Home** ribbon tab or the **Block** panel of the **Insert** ribbon tab.

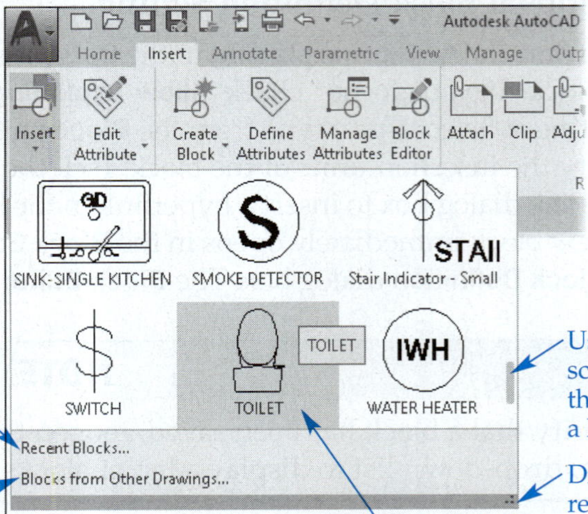

AutoCAD 2020 NEW

Inserting Blocks Using the Blocks Palette

Ribbon
View
> Palettes
Blocks
Type
BLOCKSPALETTE

Ribbon
Home
> Block
Insert
> Block
Insert Block
> Recent Blocks...
Type
INSERT
SYMBOL
I

Access the **BLOCKSPALETTE** or **INSERT** command to display the **Blocks** palette. See Figure 24-4. The **Current Drawing** tab lists blocks defined in the current file, the same as those that appear in the **Insert** drop-down list of the ribbon. The **Recent** tab lists recently created or inserted blocks from the current and previous drawing sessions. This provides a convenient way to reuse blocks that were recently defined or inserted. The **Other Drawing** tab lists block definitions found in another file. To select a block from another file, pick the ellipsis (**...**) button shown in Figure 24-4 and select a file from the **Select Drawing File...** dialog box. The file appears in the **Other Drawing** tab along with any blocks defined in the file. You can insert the entire drawing (wblock) as a block, as explained later in this chapter.

To insert a block from the **Blocks** palette, hover over the block to display the name and other information, including the annotative and dynamic status. Dynamic blocks

Figure 24-4.
The **Blocks** palette allows you to select and prepare a block for insertion. Select a block from the current drawing, a recently created or inserted block, or a block from a different drawing.

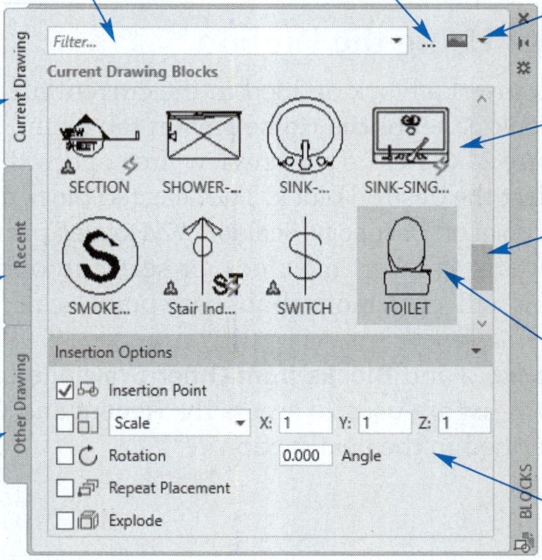

are described later in this textbook. Pick once on the block to attach the block to the crosshairs at the insertion base point. Specify a point in the drawing to insert the block.

An alternative method for inserting a block is to use drag-and-drop. Press and hold down the pick button on the block and drag the cursor into the drawing. The block attaches to the cursor at the insertion base point. Release the pick button to insert the block at the location of the cursor. When using this method, options are not presented for adjusting the insertion base point, scale, and rotation of the block.

A third technique for inserting a block is to right-click on the block and select an insert option from the shortcut menu. The settings in the **Insertion Options** area, explained in the following sections, apply when using this technique and when picking on a block to insert. These settings do not apply when using drag-and-drop.

Additional features in the **Blocks** palette allow you to manage how blocks are displayed. Refer to Figure 24-4. Use the filter drop-down list to filter for specific blocks by entering a name or partial name. Use the * wildcard character to narrow the filter criteria. By default, blocks are represented by large icons. Select an option from the drop-down list in the upper-right corner to change the icon size or to display blocks in detail or list format.

NOTE

To remove a block from the **Recent** tab, right-click on the block and pick **Remove from Recent List**.

Specifying the Block Insertion Point

When you pick a block to insert or use drag-and-drop, you specify where to locate the block during insertion. When you right-click on a block and select **Insert**, the **Insertion Point** check box in the **Insertion Options** area determines the insertion behavior. Check the **Insertion Point** check box to receive a prompt for locating the block during insertion. To insert the block using absolute coordinates, deselect the **Insertion Point** check box and enter coordinates in the **X:**, **Y:**, and **Z:** text boxes.

Scaling Blocks

The scale options in the **Insertion Options** area allow you to specify scale values for the block in relation to the X, Y, and Z axes. Deselect the **Scale** check box to enter scale values in the **X:**, **Y:**, and **Z:** text boxes. Select the **Uniform Scale** option to specify a scale value for the X axis that also applies to the scale of the Y and Z axes. Check the **Scale** check box to receive prompts for scaling the block during insertion. If you created the block with **Scale uniformly** checked in the **Block Definition** dialog box, the scale factor you enter applies to the X, Y, and Z axes.

Blocks are classified as ***real blocks***, ***schematic blocks***, or ***unit blocks***, depending on how you scale the block during insertion. Examples of real blocks include a bolt, a bathtub, a pipe fitting, and the car shown in Figure 24-5A. Examples of schematic blocks include notes, detail bubbles, tags, and section symbols. See Figure 24-5B. Annotative blocks are typically classified as schematic blocks. When you insert an annotative schematic block, AutoCAD automatically determines the block scale based on the annotation scale. When you insert a non-annotative schematic block, you must specify the scale factor.

There are three general types of unit blocks. An example of a ***1D unit block*** is a 1″ line. An example of a ***2D unit block*** is a 1″ × 1″ square. An example of a ***3D unit block*** is a 1″ × 1″ × 1″ cube. To use a unit block, determine the individual scale factors for each axis that apply when you insert the block. This allows you to stretch or compress the block to create modified versions of the block. For example, when inserting a 2D unit block, assign different scale factors for the X and Y axes to change the block dimensions.

real block: A block originally drawn at a 1:1 scale and then inserted using 1 for both the X and Y scale factors.

schematic block: A block originally drawn at a 1:1 scale and then inserted using the drawing scale factor for both the X and Y scale values.

unit block: A 1D, 2D, or 3D unit block drawn to fit in a 1-unit, 1-unit-square, or 1-unit-cubed area so that it can be scaled easily.

1D unit block: A 1-unit, one-dimensional object, such as a straight line segment, saved as a block.

2D unit block: A 2D object that fits into a 1-unit × 1-unit square, saved as a block.

3D unit block: A 3D object that fits into a 1-unit × 1-unit × 1-unit cube, saved as a block.

Figure 24-5.
A—Real blocks, such as this car, are drawn at full scale and inserted using a scale factor of 1 for both the X and Y axes. B—A schematic block is inserted using the scale factor of the drawing for the X and Y axes. C—A 2D unit block is often inserted at different scales for the X and Y axes.

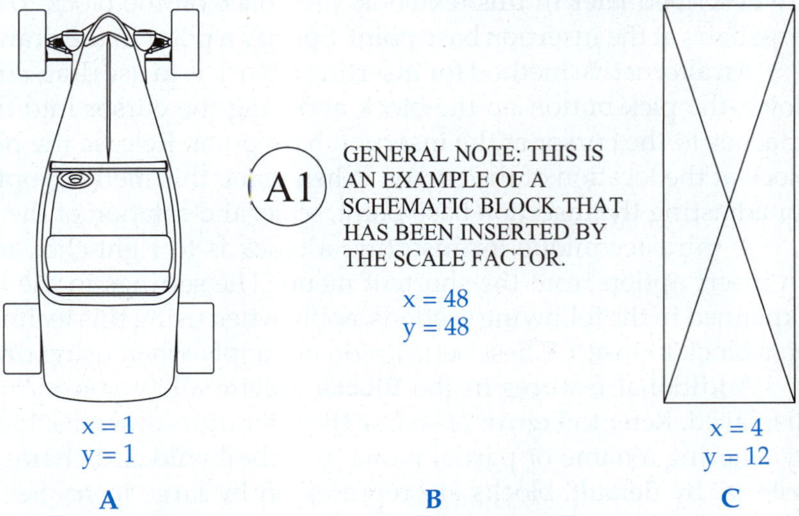

Figure 24-5C shows an example of specifying 4 for the X axis and 12 for the Y axis to create a 4″ × 12″ beam. **Figure 24-6** shows another example of using a 2D unit block. Enter negative scale factors to create a mirror image of a block.

NOTE

For most applications, insert annotative blocks at a scale of 1 to apply the annotation scale correctly. Entering a scale other than 1 adjusts the scale of the block by multiplying the scale value by the annotative scale factor.

Rotating Blocks

The rotation options in the **Insertion Options** area allow you to insert the block at a specific angle. Deselect the **Rotation** check box to enter a value in the **Angle** text box. The default angle of 0° inserts the block as defined using the **Block Definition** dialog box. Check the **Rotation** check box to receive a prompt for rotating the block during insertion. You cannot rotate a block defined using the **Match block orientation to layout** option in the **Block Definition** dialog box.

PROFESSIONAL TIP

You can rotate a block based on the current UCS. Be sure the proper UCS is active, and then insert the block using a rotation angle of 0°. If you decide to change the UCS later, any inserted blocks retain their original angle.

Figure 24-6.
A comparison of different X and Y scale factors used for inserting a 2D unit block. This example shows a block of a plan view window symbol inserted into a 6″ wall.

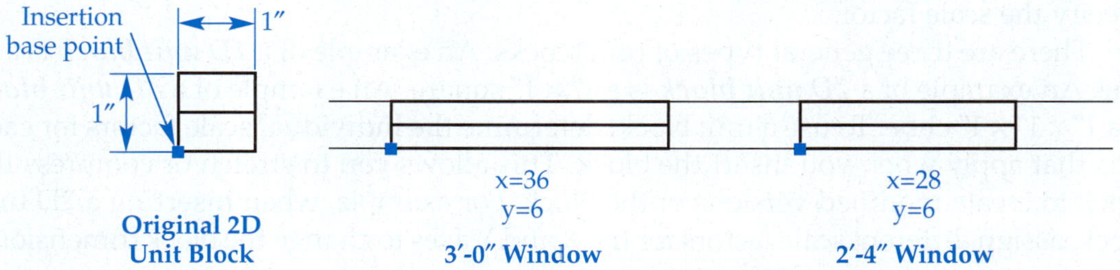

Repeat Block Placement

The **Repeat Placement** check box in the **Insertion Options** area is deselected by default. As a result, when you pick a block to insert or use the **Insert** shortcut menu option, you add one block reference to the drawing. Check **Repeat Placement** to insert additional references of the same block in the same command sequence. To exit the command, press [Esc].

Exploding Blocks

A block is saved as a single object, no matter how many objects the block includes. Check **Explode** in the **Insertion Options** area to explode the block into the original objects during insertion. Alternatively, right-click on the block and select **Insert and Explode** without checking **Explode**. Exploding a block is most common for editing or creating a block definition. If you explode the block on insertion, it assumes its original properties, including its original layer, color, and linetype. The **Explode** check box is disabled if you unchecked **Allow exploding** in the **Block Definition** dialog box.

Working with On-Screen Prompts

When you pick a block to insert or use the **Insert** shortcut menu option, prompts appear for any check boxes that are checked in the **Insertion Options** area of the **Blocks** palette. When you see a Specify insertion point or [Basepoint/Scale/X/Y/Z/Rotate]: prompt, or similar prompt, specify a point to insert the block, or select an option to make adjustments before inserting. The options allow you to specify a different base point; enter a value for the overall scale; enter independent scale factors for the X, Y, and Z axes; and enter a rotation angle.

The insertion base point specified when the block was created may not always be the best point for inserting the block. Instead of inserting and then moving the block, use the **Basepoint** option to specify a different base point before locating the block. Select the **Basepoint** option when prompted. The block temporarily appears on-screen, allowing you to select an alternate insertion base point. The block reattaches to the crosshairs at the new point and the command resumes, allowing you to select the insertion point in the drawing.

NOTE

When specifying an insertion point for a block, additional options are available when the Specify insertion point or [Basepoint/Scale/X/Y/Z/Rotate] prompt, or similar prompt, appears. These options, which do not appear on the command line, allow you to preview the scale of the X, Y, and Z axes and the rotation angle before entering actual values. Enter PX, PY, PZ, or PR on the command line and specify a value to preview the scale of an axis or the rotation angle.

NOTE

Command line content search offers another way to insert a block. Begin typing the letters in a block name on the command line. Blocks that match the letters you enter appear in the suggestion list as you type. Selecting a block from the suggestion list activates the **-INSERT** command with the selected block ready to insert.

Exercise 24-2

www.g-wlearning.com/CAD/

Complete the exercise on the companion website.

Using the MINSERT Command

Type
MINSERT

The **MINSERT** command allows you to insert multiple arranged copies of a block. Figure 24-7 shows an example of a **MINSERT** command application. To follow this example, set architectural units, draw a 4′ × 3′ rectangle, and save the rectangle as a block named DESK. Then access the **MINSERT** command and enter DESK. Pick a point as the insertion point and then accept the X scale factor of 1, the Y scale factor of use X scale factor, and the rotation angle of 0. The arrangement is to be three rows and four columns. In order to make the horizontal spacing between desks 2′ and the vertical spacing 4′, you must consider the size of the desk when entering the distance between rows and columns. Enter 7' (3′ desk depth + 4′ space between desks) at the Enter distance between rows or specify unit cell prompt. Enter 6' (4′ desk width + 2′ space between desks) at the Specify distance between columns prompt.

The pattern created using the **MINSERT** command is a single minsert block object that you cannot explode. Do not confuse this object with a rectangular array created using the **ARRAY** command. Use the **Properties** palette to modify the number of rows and columns, change the spacing between objects, or change other properties. If you rotate the initial block, all objects in the pattern rotate about their insertion points. If you rotate the patterned objects about the insertion point while using the **MINSERT** command, all objects align on that point.

Exercise 24-3

www.g-wlearning.com/CAD/

Complete the exercise on the companion website.

Inserting Entire Drawings

To insert an entire drawing into the current drawing as a block, access the **INSERT** command and pick the ellipsis (...) button in the **Blocks** palette. Use the **Select Drawing File...** dialog box to select a file to insert. The file appears in the **Other Drawing** tab along with any blocks defined in the file. Insert the file as you would a block.

When you insert a drawing into another drawing, the inserted drawing becomes a block reference and functions as a single object. The drawing is inserted on the current layer, but only objects drawn on the 0 layer inherit the color, linetype, lineweight, and transparency properties of the current layer. Explode the inserted drawing if necessary. When the block is exploded, the objects revert to their original layers. Inserting a drawing can bring any existing block definitions and other drawing content, such as layers and dimension styles, into the current drawing.

Figure 24-7.
Creating an arrangement of desks using the **MINSERT** command. An alternative is to use the **ARRAY** command after you position one block.

By default, a drawing has an insertion base point of 0,0,0 when you insert it into another drawing. To change the insertion base point of the drawing, access the **BASE** command and select a new insertion base point. Save the drawing before inserting it into another drawing.

BASE
Ribbon
Home
> Block
Set Base Point
Type
BASE

Exercise 24-4

www.g-wlearning.com/CAD/

Complete the exercise on the companion website.

Inserting Blocks Using DesignCenter

ADCENTER
Ribbon
View
> Palettes
DesignCenter
Type
ADCENTER
ADC

DesignCenter provides a method to insert blocks or entire drawings as blocks in the current drawing. To insert a block using **DesignCenter**, use the **Tree View** pane to locate and select a file containing the block to be inserted. Select the **Blocks** branch in the **Tree View** pane or double-click on the **Blocks** icon in the **Content** pane to display blocks defined in the file. See Figure 24-8. Drag-and-drop is an effective technique to transfer a block from **DesignCenter** to the active drawing.

An alternative to drag-and-drop is to right-click on a block in **DesignCenter** and use one of the shortcut menu options. Select **Insert Block...** to access the **Insert** dialog box with the selected block active. The insertion options in the **Insert** dialog box are similar to those in the **Blocks** palette and allow you to scale, rotate, or explode the block during insertion. Once you pick **OK** to close the dialog box, specify a point to insert the block or select an option to make adjustments before inserting.

Select **Copy** in the shortcut menu to insert the block using a copy and paste operation. Move the cursor into the active drawing, right-click, and select **Paste** from the

Figure 24-8.
Use **DesignCenter** to insert blocks from files or drawings from folders. Several example blocks are available in the AutoCAD Sample folder shown.

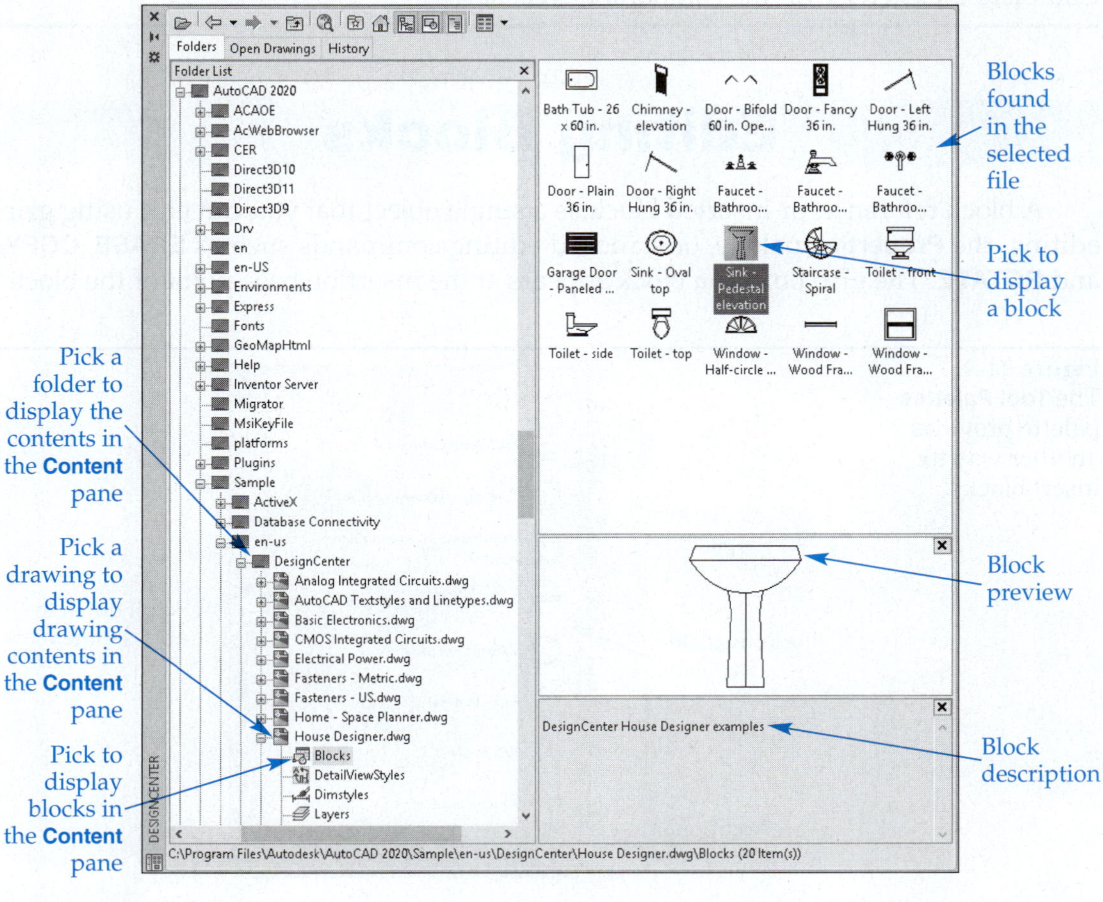

Clipboard cascading menu. The block attaches to the cursor at the insertion base point. Specify a point to insert the block.

To insert a drawing or DXF file using **DesignCenter**, use the **Tree View** pane to locate and select a folder to display the contents of the folder in the **Content** pane. Drag and drop or copy and paste the file from the **Content** pane into the current drawing. You can also right-click a file icon in the **Content** pane and select **Insert as Block...**.

NOTE

Blocks are inserted from **DesignCenter** based on the type of block units you specify when you create the block. For example, if the original block was a 1 × 1 square and you specified the block units as feet when you created the block, then the block inserts as a 12″ × 12″ square.

Inserting Blocks Using Tool Palettes

block insertion tools: Blocks located on a tool palette.

The **Tool Palettes** palette, shown in Figure 24-9, provides another means of storing and inserting blocks. AutoCAD refers to blocks in a tool palette as ***block insertion tools***. Tool palettes can also store and activate other drawing content and tools, such as hatch patterns and AutoCAD commands.

Inserting a block from the **Tool Palettes** palette is similar to using the **Blocks** palette. Access the tool palette (tab) containing the block and hover over the block icon to display the name and description. Pick on the block or use a drag-and-drop operation to insert the block into the drawing. The pick method offers an advantage over drag-and-drop by presenting options for adjusting the insertion base point, scale, and rotation.

Exercise 24-5 www.g-wlearning.com/CAD/

Complete the exercise on the companion website.

Editing Blocks

A block reference, or inserted block, is a single object that you can edit using grip editing, the **Properties** palette, or standard editing commands such as **ERASE**, **COPY**, and **ROTATE**. The grip box for a block appears at the insertion base point of the block.

Figure 24-9.
The **Tool Palettes** palette provides another way to insert blocks.

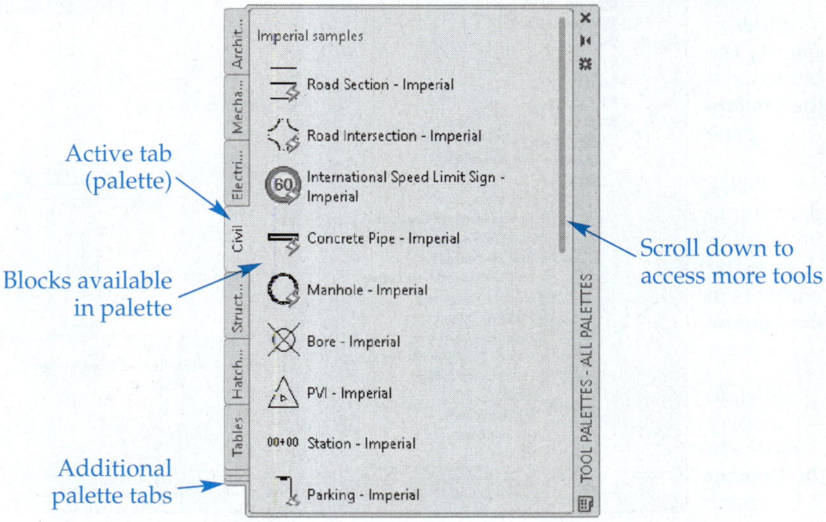

A different form of block editing involves redefining the block by modifying the block definition or changing the objects that compose the block. You can redefine a block using the **Block Editor** or by exploding and then recreating the block.

Changing Block Properties to ByLayer

If you originally set block element properties such as color, linetype, lineweight, and transparency to absolute values, and you want to change the properties to ByLayer, you can either edit the block definition or use the **SETBYLAYER** command. Access the **SETBYLAYER** command and use the **Settings** option to display the **SetByLayer Settings** dialog box. Select the check boxes that correspond to the object properties you want to convert to ByLayer. Pick the **OK** button to exit the **SetByLayer Settings** dialog box.

Ribbon
Home
> Modify
Set to ByLayer
Type
SETBYLAYER

SETBYLAYER

Next, select the block with the properties you want to set to ByLayer and press [Enter]. The Change ByBlock to ByLayer? prompt appears next. Select **Yes** if you wish to change all object properties currently set to ByBlock to ByLayer. The next prompt asks if you want to include blocks in the conversion. Select **Yes** to convert the properties of the selected block and all references of the same block in the drawing to ByLayer.

PROFESSIONAL TIP

To change the properties of several blocks at once, use the **SELECTSIMILAR** or **QSELECT** command to create a selection set of block references.

Using the Block Editor

Use the **BEDIT** command to make changes to a block definition using the **Block Editor**. Access the **BEDIT** command to display the **Edit Block Definition** dialog box. See Figure 24-10. Select the name of an existing block to edit from the list box. Pick the <Current Drawing> option to edit a block saved as the current drawing, such as a wblock. A preview and description of the selected block appear. You can create a new block by typing a unique name in the **Block to create or edit** field. Pick the **OK** button to open the selected block in the **Block Editor**. See Figure 24-11. If you typed a new block name, the drawing area is empty, allowing you to create a new block.

Ribbon
Home
> Block
Insert
> Block Definition
Block Editor
Type
BEDIT

BEDIT

Modifying a Block

Use drawing and editing commands to modify or create the block definition. Specify the block insertion base point at the UCS origin, or 0,0,0 point. The tools in the panels of the **Block Editor** ribbon tab are specifically for modifying and creating block

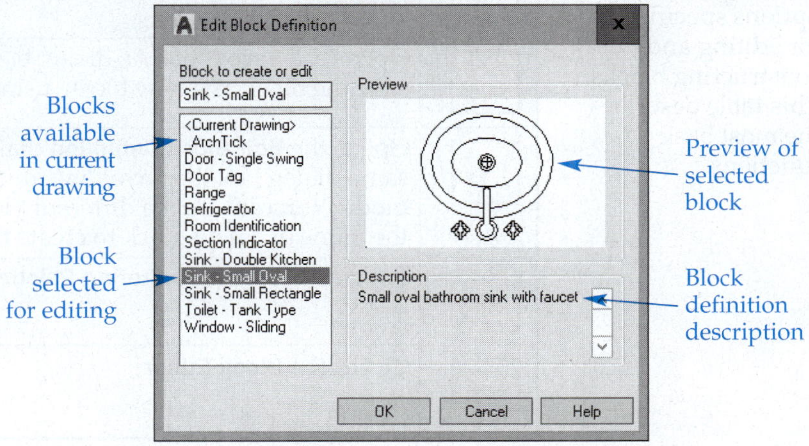

Figure 24-10. The **Edit Block Definition** dialog box.

Figure 24-11.
The context-sensitive **Block Editor** ribbon tab and the **Block Authoring Palettes** palette are available in block editing mode. Only the block geometry appears in the **Block Editor**.

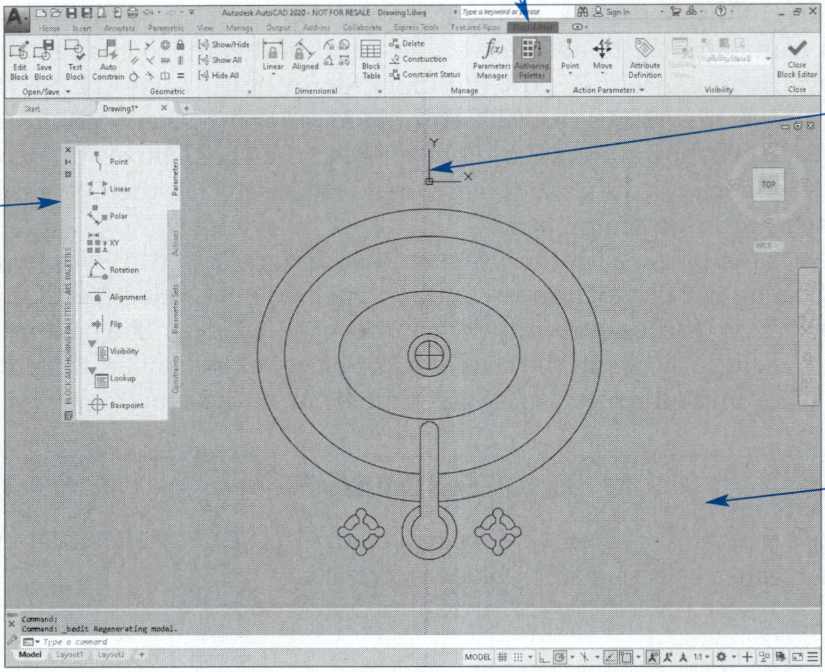

geometry. Figure 24-12 describes some of the basic commands available in the **Block Editor** ribbon tab. Parametric tools allow you to constrain block geometry and form block tables. Many of the commands and options found on the **Block Editor** ribbon tab relate to dynamic blocks. This textbook explains dynamic blocks, block tables, and other block editing tools in later chapters.

When you finish editing, close the **Block Editor** to return to the drawing. If you have not saved your changes, a dialog box appears asking if you want to save changes. Pick the appropriate option to save or discard changes, or pick the **Cancel** button to return to the **Block Editor**.

Figure 24-12.
The **Block Editor** ribbon tab contains several tools and options specifically for editing and constructing blocks. This table describes the most basic functions.

Button	Description
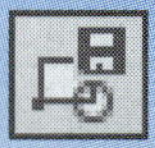	Saves changes to the block and updates the block definition.
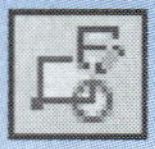	Opens the **Save Block As** dialog box, allowing you to save the block as a new block, using a different name.
	Opens the **Edit Block Definition** dialog box, which is the same dialog box displayed when you enter block editing mode. You can select a different block to edit or specify the name of a new block to create from scratch.
	Toggles the **Block Authoring Palettes** palette off and on.
	Closes the **Block Editor**.

NOTE

Double-click on a block to display the **Edit Block Definition** dialog box with the block selected. Open a block directly in the **Block Editor** by selecting the block and then right-clicking and selecting **Block Editor**. Another option is to open a block directly in the **Block Editor** when you create the block by checking the **Open in block editor** check box in the **Block Definition** dialog box.

Adding a Block Description

To change the description assigned to the original block definition, open the block in the **Block Editor** and display the **Properties** palette with no objects selected. Make changes to the description using the Description property in the **Block** category. Pick the **Save Block** button and the **Close Block Editor** button to return to the drawing.

NOTE

You can also edit blocks in place using the **REFEDIT** command. Chapter 31 describes in-place editing using the **REFEDIT** command as it applies to external references. Use the same techniques to edit blocks.

Exercise 24-6

www.g-wlearning.com/CAD/

Complete the exercise on the companion website.

Exploding and Redefining a Block

Ribbon
Home
> Modify
Explode
Type
EXPLODE
BREAKUP
DISASSEMBLE
X

EXPLODE

To explode a block during insertion, check **Explode** in the **Insertion Options** area of the **Blocks** palette or right-click on the block and select **Insert and Explode**. This is useful when you want to edit the individual objects of the block directly after insertion. You can also use the **EXPLODE** command after inserting the block to explode the block, but only if you check **Allow exploding** in the **Block Definition** dialog box. Access the **EXPLODE** command, select the objects to be exploded, and press [Enter] to complete the operation.

Follow this procedure to redefine an existing block using the **EXPLODE** and **BLOCK** commands:

1. Insert the block to redefine.
2. Make sure you know the exact location of the insertion base point, which is lost during the explosion.
3. Use the **EXPLODE** command to explode the block.
4. Edit the elements of the block as needed.
5. Recreate the block definition using the **BLOCK** command.
6. Assign the block the same original name and, if appropriate, the same insertion point.
7. Select the objects to include in the block.
8. Pick the **OK** button in the **Block Definition** dialog box to save the block. When a message appears asking if you want to redefine the block, pick **Redefine block**.

CAUTION

A common mistake is to forget to use the **EXPLODE** command before redefining the block. When you try to recreate the block using the same name, an alert box indicates that the block references itself. This means you are trying to create a block that already exists. Cancel the command, explode the block, and try again to redefine the block.

NOTE

Once a block is modified, whether from changes made using the **BEDIT** command or from redefinition using the **EXPLODE** and **BLOCK** commands, all instances of that block in the drawing update according to the changes.

Exercise 24-7

www.g-wlearning.com/CAD/

Complete the exercise on the companion website.

Understanding the Circular Reference Error

circular reference error: An error that occurs when a block definition references itself.

A ***circular reference error*** occurs when you try to redefine a block that already exists using the same name. AutoCAD informs you that the block references itself. A block can be composed of many objects, including other blocks. When you use the **BLOCK** command to incorporate an existing block into a new block, AutoCAD detects all objects that compose the new block, including existing block definitions. A problem occurs if you select an instance of the redefined block as an element of the new definition. The new block refers to a block of the same name, or references itself. Figure 24-13A illustrates the process of correctly redefining a block named BOX to avoid a circular reference error. Figure 24-13B shows an incorrect redefinition resulting in a circular reference error.

NOTE

Additional shortcut menu options are available in the **Recent** and **Other Drawing** tabs in the **Blocks** palette when inserting a block with the same name as an existing block reference. The **Insert and Redefine** and **Redefine Only** shortcut menu options are available when you right-click on a block from a different drawing file and the block has the same name as an existing block reference in the current drawing. The **Insert** option is also available. If you select the **Insert** option, the block reference found in the current file is inserted. The **Insert and Redefine** option inserts the block reference from the source, and redefines all blocks with the same name in the current file to match the source. The **Redefine Only** option redefines all blocks with the same name in the current file without inserting a block object. The **Insert and Redefine** and **Redefine Only** shortcut menu options are also available when using **DesignCenter**.

Renaming Blocks

Type
RENAME
REN

Access the **RENAME** command to rename a block using the **Rename** dialog box, without editing the block definition. See Figure 24-14. Select **Blocks** from the **Named Objects** list, and then pick the block to be renamed in the **Items** list. The current name appears in the **Old Name:** text box. Type the new block name in the **Rename To:** text box. Pick the **Rename To:** button to display the new name in the **Items** list. Pick the **OK** button to exit the **Rename** dialog box.

Updating Block Icons

A block icon forms when you define a block. The icon appears when you insert and edit blocks to help you recognize the block. Block icons require updating when an icon does not appear, as is often the case when you store a block in a drawing created with an older version of AutoCAD, or when the icon does not reflect changes made to

Figure 24-13.
A—The correct procedure for redefining a block. B—Redefining a block without first exploding it creates an invalid circular reference.

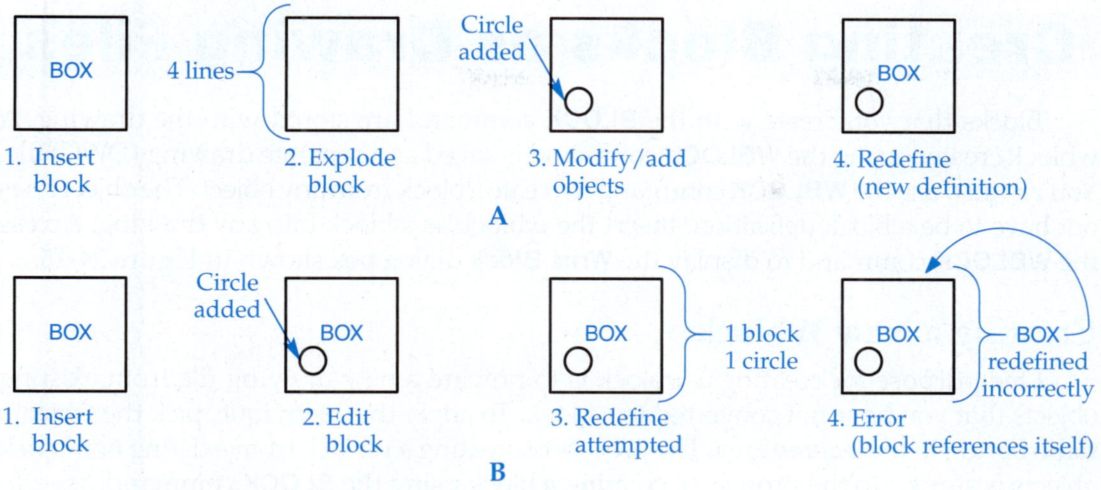

Figure 24-14.
The **Rename** dialog box allows you to change the name of blocks and other named objects.

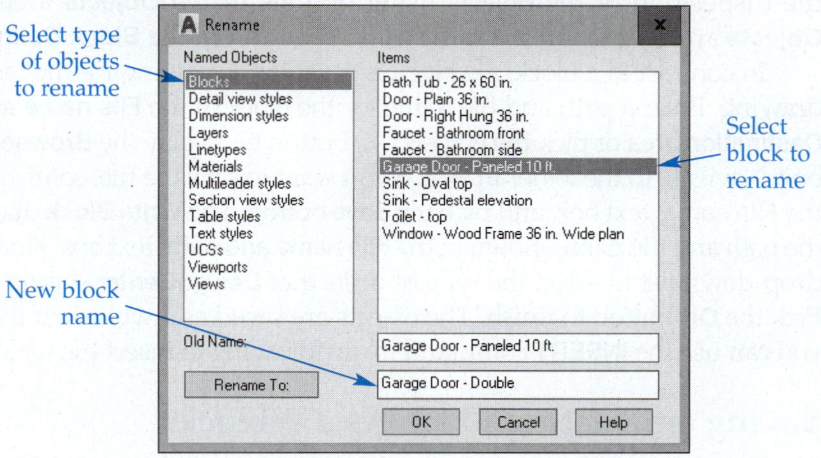

the block. To create or update a block icon, open the drawing that contains the block, access the **BLOCKICON** command, enter the name of the block, and press [Enter].

Type
BLOCKICON

Copying Nested Objects

Some modify commands, such as **TRIM** and **EXTEND**, allow you to pick individual objects nested within a block reference or external reference (xref). However, most other commands recognize a block reference as a single block reference object and an xref as a single external reference object. For example, use the **COPY** command to create a copy of a block reference or xref. In contrast, the **NCOPY** command allows you to copy individual objects nested within a block reference or xref, which is more efficient than exploding a block and copying objects, for example.

Type
NCOPY

Access the **NCOPY** command and use the **Settings** option to specify the method for copying nested objects. The difference between the default **Insert** option and the **Bind** option is most apparent when you are copying named objects nested within an xref. Chapter 31 explains xrefs. Use the **Insert** option to copy objects nested within a block. Individually select objects nested within a block reference or xref to copy, specify a base point, and pick a location to place the copy. The **Multiple** option is available before you specify a base point and allows you to create several copies of the same object using a single **NCOPY** operation. The **NCOPY** command provides the same options as the standard **COPY** command, allowing you to specify a base point and a second point, select a displacement using the **Displacement** option, or define the first point as the displacement.

Creating Blocks as Drawing Files

Type
WBLOCK

Blocks that you create with the **BLOCK** command are stored with the drawing. A wblock created using the **WBLOCK** command is saved as a separate drawing (DWG) file. You can also use the **WBLOCK** command to create a block from any object. The object does not have to be a block definition. Insert the wblock as a block into any drawing. Access the **WBLOCK** command to display the **Write Block** dialog box shown in Figure 24-15.

Creating a New Wblock

One purpose for creating a wblock is to prepare a new drawing file from existing objects that you have not converted to a block. To apply this technique, pick the **Objects** radio button in the **Source** area. The process of creating a wblock from existing non-block objects is similar to the process of creating a block using the **BLOCK** command. Specify an insertion base point using options in the **Base point** area, and select the objects and the disposition of the objects using options in the **Objects** area. The **Base point** and **Objects** areas function the same as those found in the **Block Definition** dialog box.

In contrast to a block, a wblock is saved as a drawing file, not as a block in the current drawing. Enter a path and file name for the block in the **File name and path:** text box in the **Destination** area or pick the ellipsis (**...**) button to display the **Browse for Drawing File** dialog box. Navigate to the folder in which you want to save the file, confirm the name of the file in the **File name:** text box, and pick the **Save** button. The **Write Block** dialog box redisplays with the path and file name shown in the **File name and path:** text box. Finally, use the **Insert units:** drop-down list to select the type of units that **DesignCenter** should use to insert the block. Pick the **OK** button to finish. The objects are saved as a wblock in the specified folder. Now you can use the **INSERT** command in any drawing to insert the wblock.

Saving an Existing Block As a Wblock

To create a wblock from an existing block, pick the **Block** radio button in the **Source** area. Select the block to save as a wblock from the drop-down list. Use the options in the **Destination** area to locate the wblock, and pick the **OK** button to finish.

Figure 24-15.
Using the **Write Block** dialog box to create a wblock from selected objects without first defining a block. You can also create a wblock from an existing block definition (using the **Block** radio button) or a drawing (using the **Entire drawing** radio button).

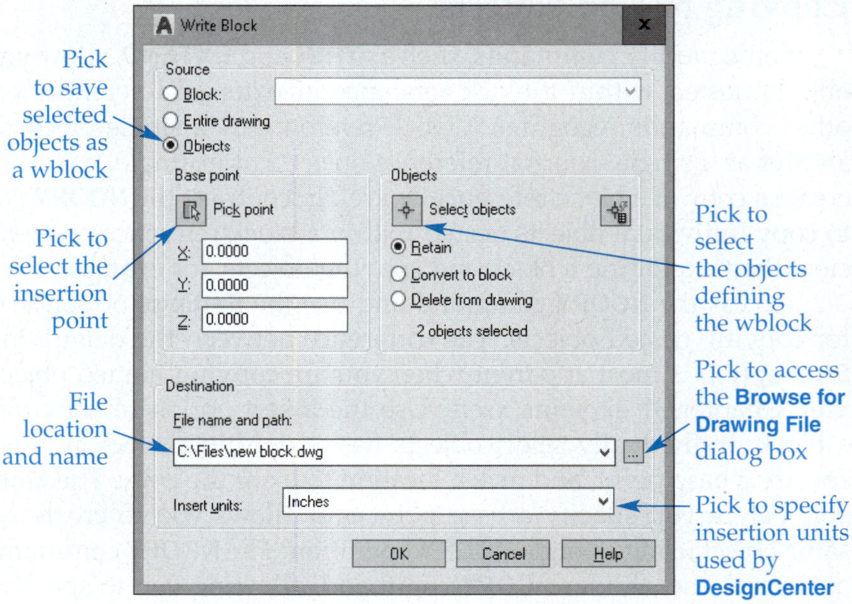

Storing a Drawing As a Wblock

To store an entire drawing as a wblock, pick the **Entire drawing** radio button in the **Source** area. Use the options in the **Destination** area to locate the wblock. In this case, the whole drawing is saved as if you are using the **SAVE** command. However, all uninserted, or unused, blocks in the drawing are deleted. If the drawing contains any unused blocks, the **Entire drawing** method may reduce the size of a drawing considerably. Pick the **OK** button to finish.

Exercise 24-8

www.g-wlearning.com/CAD/

Complete the exercise on the companion website.

Revising an Inserted Drawing

If you insert a wblock into multiple drawings and then need to make changes to the wblock, use **DesignCenter** to access the original drawing file. Right-click on the file icon in the **Content** area and select **Insert as Block...**. Then, in the **Insert** dialog box, check **Specify On-screen** in the **Insertion point** area and pick the **OK** button. When a message asks to redefine the block, pick the **Redefine block** option. All of the wblock references update. Press [Esc] to cancel the command so that you do not insert a new block.

PROFESSIONAL TIP

If you work on projects that use inserted drawings that require revision, use reference drawings (xrefs) instead of inserting drawing files. Chapter 31 explains reference drawings placed using the **XREF** command. All xrefs automatically update when you open a drawing file that contains the xref content.

Symbol Libraries

Store frequently used symbols in ***symbol libraries*** to increase productivity. Continue to add new symbols to libraries to increase the number of symbols available for reuse. Establish whether you will store symbols as blocks or drawing files, and identify a storage location and system.

symbol library: A collection of related blocks, shapes, views, symbols, or other content.

Creating Symbol Libraries

There are two general methods to create a symbol library. One option is to save multiple blocks in a single drawing. See Figure 24-16. The other option is to save each block to a separate wblock file. Consider the following guidelines when developing a symbol library:

- Follow industry and company or school standards for blocks and symbols.
- Identify each block with a descriptive name and insertion point location.
- When saving multiple blocks in a drawing file, save one group of symbols per drawing file and use folders to organize files.
- When using wblocks, give each file a meaningful name and use folders to organize files.
- Provide all users with a hard copy of the symbol library showing each symbol, insertion points, storage locations, and any other necessary information.
- If a network is not in use, place symbol library files on each workstation in the classroom or office.

Figure 24-16.
An example of an architectural symbol library created by saving multiple blocks in a single drawing. The "X" symbols indicate insertion points and are not part of the blocks. *(Courtesy of Ron Palma, 3D-DZYN)*

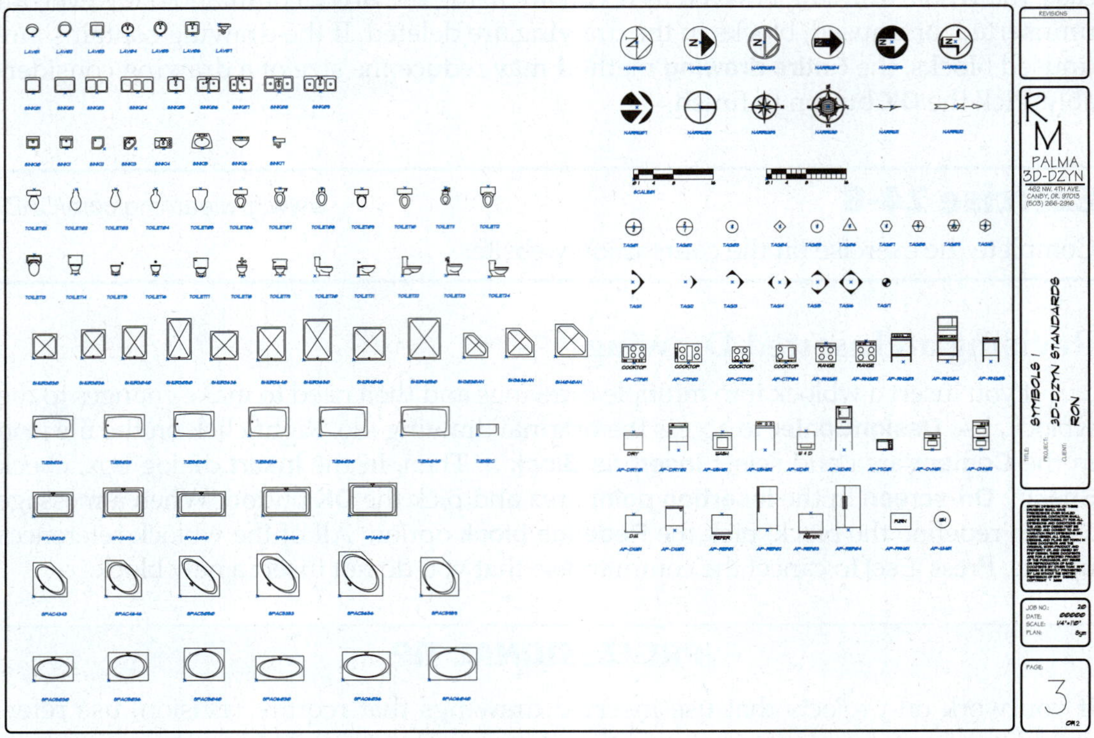

- Keep backup copies of all files in a secure location.
- When you revise symbols, update all files containing the edited symbols as appropriate.
- Inform all users of any changes to saved symbols.

Storing Symbol Libraries

Store symbol libraries on the local or network hard drive. This location is easy to access, quick, and more convenient to use than portable media. Removable media, such as a removable hard drive, USB flash drive, or CD, are appropriate for backup purposes if a network drive with an automatic backup function is not available. In the absence of a network, use removable media to transport files from one workstation to another.

Use a system of folders and files to store and organize symbol libraries. Store content outside of the AutoCAD system folders to keep the system folders uncluttered, and to differentiate your folders and files from AutoCAD system folders and files. A good method is to create a separate folder with a name such as \Blocks to store all blocks. Then use folders with names such as \Architectural, \Electrical, \Plumbing, and \Structural to group blocks associated with specific disciplines or applications.

If you save multiple symbols within a single drawing, use the **Other Drawing** tab of the **Blocks** palette, **DesignCenter**, or the **Tool Palettes** palette to insert the symbols as needed. This system often works best when you use several drawing files to group similar or discipline-specific symbols. For example, create different symbol library drawing files with names such as Electronics, Fasteners, Piping, and Structural Shapes. Limit the symbols in a drawing to a reasonable number so you can easily find and load the symbols.

Arrange drawing files saved on the hard drive in a logical manner. All workstations in a non-networked classroom or office should have folders with the same names.

Assign one person to update and copy symbol libraries to all workstations. Copy drawing files to each workstation as necessary. Keep the master and backup versions of the symbol libraries in separate locations.

Purging Named Objects

AutoCAD 2020 NEW

Ribbon
Manage
> Cleanup
Purge
Type
PURGE
REMOVE
PU

purge: Delete unused named objects from a drawing file.

A drawing typically accumulates unused named objects that may be unnecessary. Unused named objects increase drawing file size and may make it more difficult to locate and use content that is common and necessary. Access the **PURGE** command to *purge* unused items from the drawing using the **Purge** dialog box. See Figure 24-17.

The **Purgeable Items** button at the top of the dialog box is active when you enter the **PURGE** command. Pick the **Find Non-Purgeable Items** button to view content that cannot be purged. Use the **Named Items Not Used** tree view to locate and highlight items to purge. Pick the item name to display a preview in the **Preview** window. Check the **All items** check box to purge all unused named objects, or check specific categories and individual named items as necessary.

Before purging, check the **Confirm each item to be purged** check box in the **Options** area to have an opportunity to review each item before deleting. Check **Purge nested items** to purge nested items. Checking **Zero-length geometry** in the **Purge Unnamed Objects** area is an effective way to remove all zero-length objects, such as a line or arc drawn as a dot. These objects are often mistakes or unintended results of drawing and editing processes. Check **Empty text objects** to remove all text objects that include only spaces. Check **Orphaned data** to remove any data remaining from referencing a design (DGN) file, such as a MicroStation® DGN file.

Pick the **Purge Checked Items** button to purge the selected named objects, or pick the **Purge All** button to purge all unused items. Picking the **Purge All** button activates **All items** in the tree view. Purging may cause other named objects to become unreferenced.

Figure 24-17.
The **Purge** dialog box.

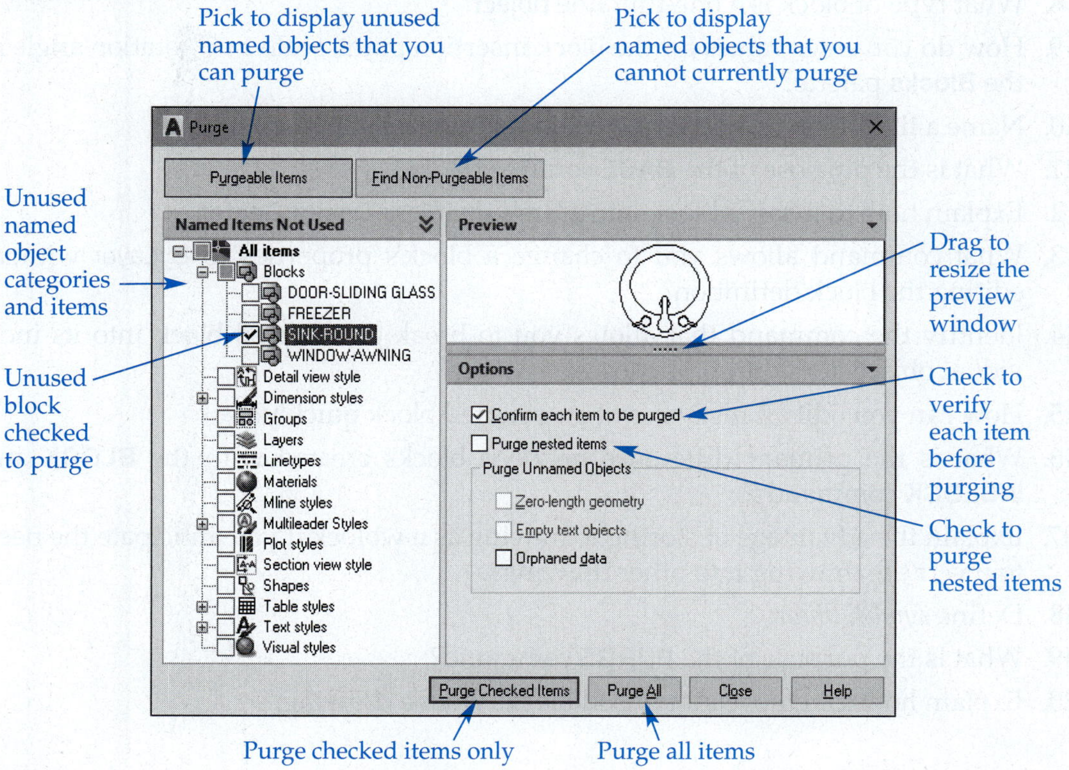

As a result, you may need to purge more than once to purge the drawing of all unused named objects. Messages appear to guide you through the purge operation.

NOTE

Selecting **Find Non-Purgeable Items** in the **Cleanup** panel of the **Manage** ribbon tab displays the **PURGE** dialog box with the **Find Non-Purgeable Items** button active. Select a non-purgeable item from the tree view to display a preview of the item, a possible reason for why the item cannot be purged, and additional details such as the amount of file space the item is using and how many occurrences of the item the file contains. Use the information to help determine if purging an item is beneficial and what changes you may be able to make in order to purge the item. When attempting to locate a non-purgeable item such as a block, pick the **Select Objects** button associated with the item in the **Purge** dialog box. This closes the dialog box, zooms to the item in the drawing, and selects the item.

Chapter Review

Answer the following questions using the information in this chapter.

1. Why would you draw blocks on the 0 layer?
2. What properties do blocks drawn on a layer other than layer 0 assume when they are inserted?
3. What characters can be used in a block name?
4. Define the term *nesting* in relation to blocks.
5. What is a block reference?
6. Explain three ways to insert a block from the **Current Drawing** tab of the **Blocks** palette.
7. Describe the effect of entering negative scale factors when inserting a block.
8. What type of block is a one-unit line object?
9. How do you enter values for the block insertion point, scale, and rotation angle in the **Blocks** palette?
10. Name a limitation of a pattern created with the **MINSERT** command.
11. What is the purpose of the **BASE** command?
12. Explain how to insert a block into a drawing from **DesignCenter.**
13. What command allows you to change a block's properties to ByLayer without editing the block definition?
14. Identify the command that allows you to break an inserted block into its individual objects for editing purposes.
15. How can you edit all instances of an inserted block quickly?
16. What is the primary difference between blocks created with the **BLOCK** and **WBLOCK** commands?
17. Explain the advantage of storing a drawing as a wblock if you anticipate the need to insert the drawing into other drawings.
18. Define *symbol library*.
19. What is the purpose of the **PURGE** command?
20. Explain how to remove all unused blocks from a drawing.

Drawing Problems

Start AutoCAD if it is not already started. Start a new drawing for each problem using an appropriate template of your choice. The template should include layers and text, dimension, multileader, and table styles, when necessary, for drawing the given objects. Add layers and text, dimension, multileader, and table styles as needed. Draw all objects using appropriate layers. Use appropriate text, dimension, multileader, and table styles, justification, and format. Follow the specific instructions for each problem. Use only drawing and editing commands and techniques you have already learned. Use your own judgment and approximate dimensions when necessary. Apply formal dimensions accurately using ASME or appropriate industry standards.

Note: *Some of the problems in this chapter are built on problems from previous chapters. If you have not yet completed those problems, complete them now.*

▼ Basic

1. Architectural Save P12-17 as P24-1. If you have not yet completed Problem 12-17, do so now. In the P24-1 file, erase all copies of the symbols, leaving the original objects intact. These include the steel column symbols and the bay and column line tags. Then do the following. Reference the sketch shown in Problem 12-17 if necessary.
   A. Make blocks of the steel column symbol and the tag symbols.
   B. Use the **ARRAYRECT** or **MINSERT** command to place the symbols in the drawing.
   C. Dimension the drawing as shown.

2. Architectural Save P12-18 as P24-2. If you have not yet completed Problem 12-18, do so now. In the P24-2 file, erase all of the desk workstations except one. Then do the following. Reference the sketch shown in Problem 12-18 if necessary.
   A. Create a block of the workstation.
   B. Insert the block into the drawing using the **MINSERT** command.
   C. Dimension one of the workstations as shown.

3. Electronics Complete this problem after completing Problem 24-7. Save P24-7 as P24-3. In the P24-3 file, modify the NAND gates to become XNOR gates, as shown, by modifying the block definition.

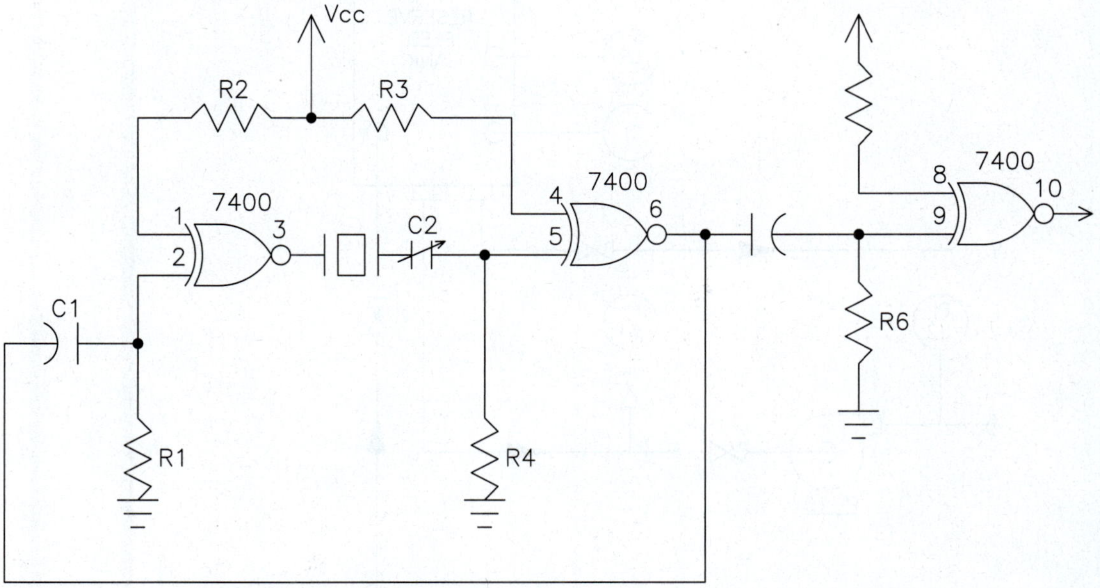

Drawing Problems - Chapter 24

▼ Intermediate

Problems 4–7 represent a variety of diagrams created using symbols as blocks. Create each drawing as shown. The drawings are not to scale. Create the symbols first as blocks or wblocks and then save the symbols in a symbol library using one of the methods described in this chapter.

4. **Electronics** Draw the integrated circuit schematic for a clock. Save the drawing as P24-4.

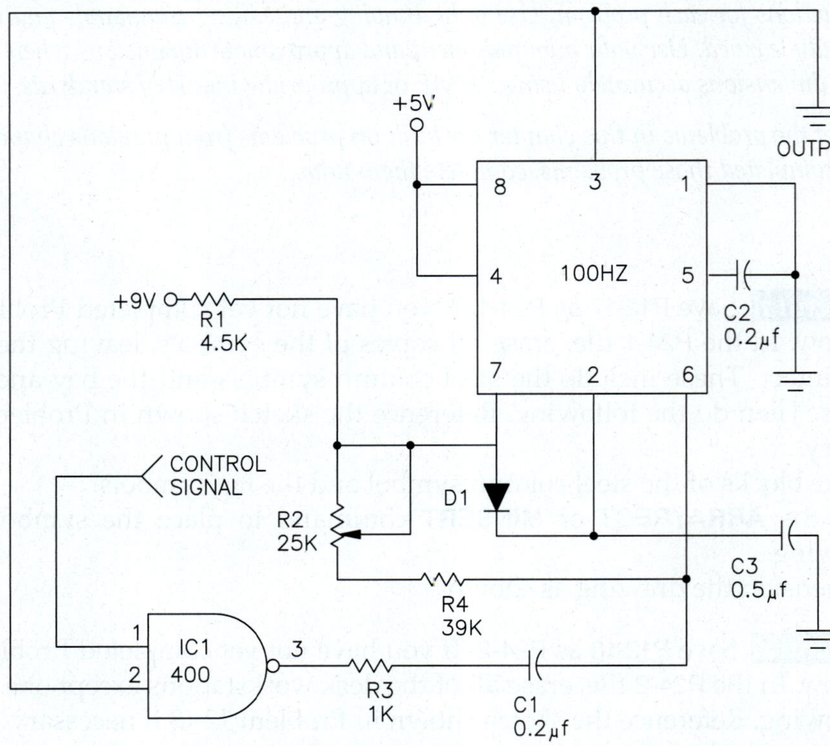

Integrated Circuit for Clock

5. **Piping** Draw the piping flow diagram. Save the drawing as P24-5.

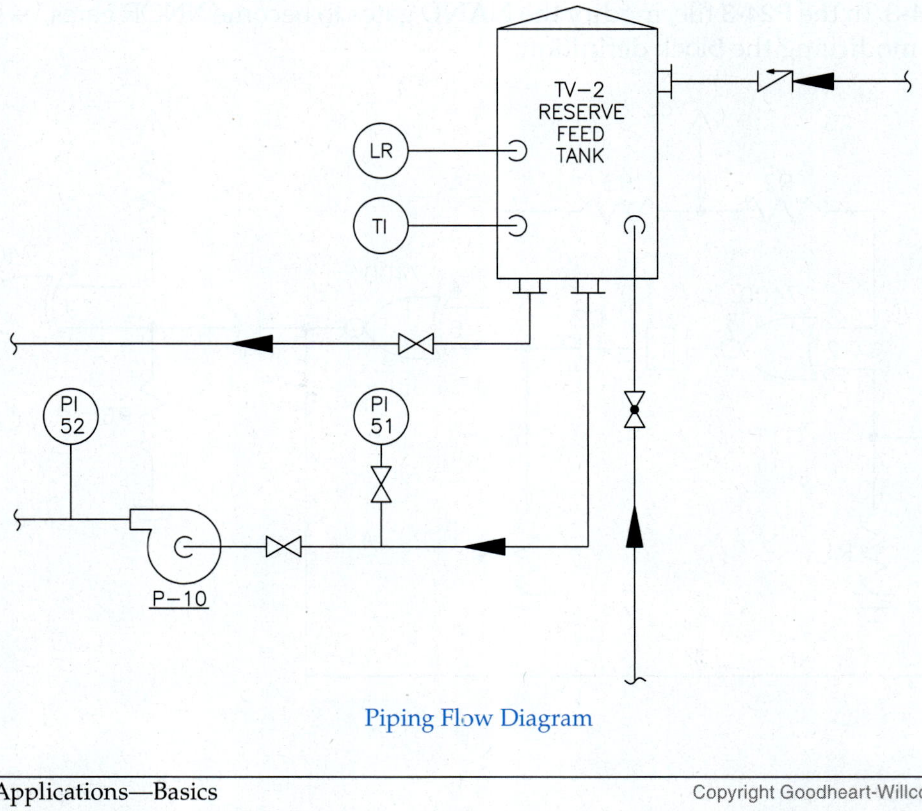

Piping Flow Diagram

6. **Electronics** Draw the logic diagram of a marking system. Save the drawing as P24-6.

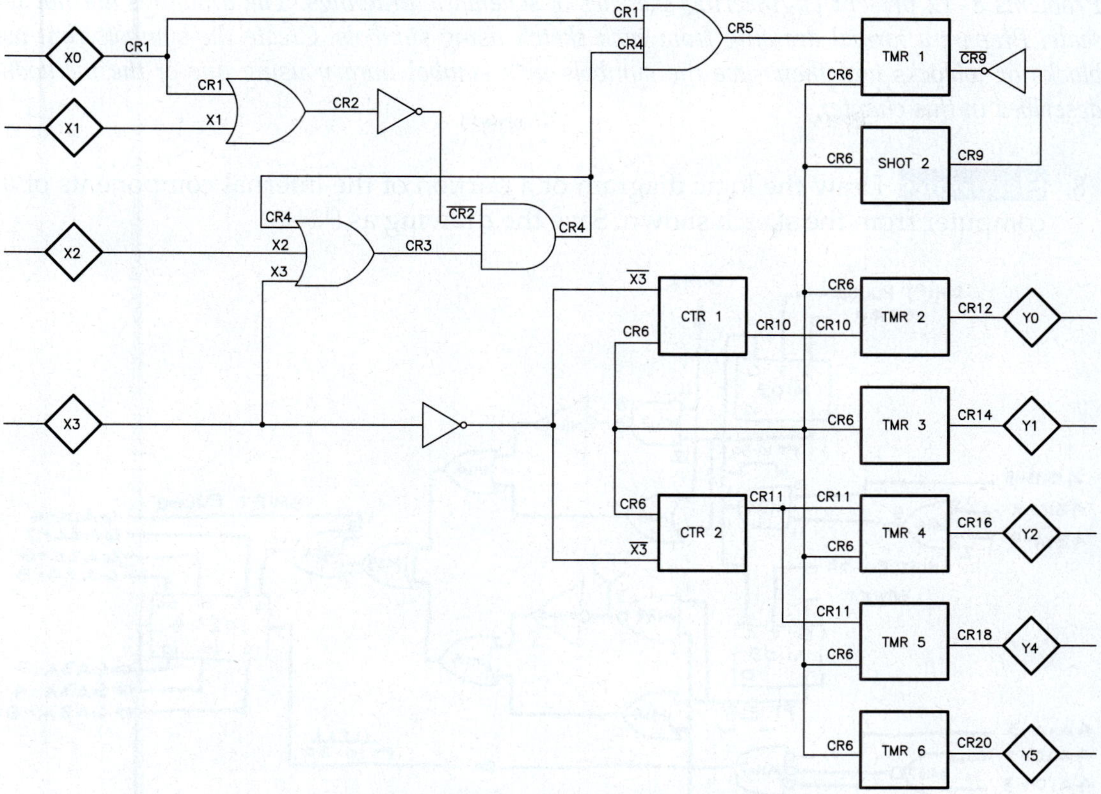

Logic Diagram of Marking System

7. **Electronics** Draw the digital logic circuit. Create each type of component in the circuit as a block. Save the drawing as P24-7.

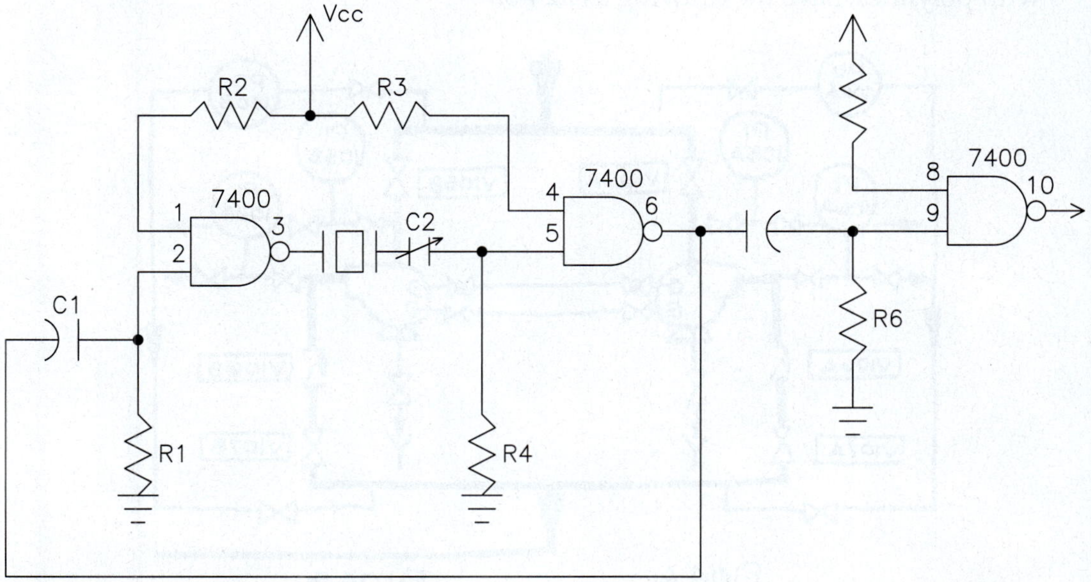

▼ Advanced

Problems 8–12 present engineering sketches of schematic drawings. The drawings are not to scale. Prepare a formal drawing from each sketch using symbols. Create the symbols first as blocks or wblocks and then save the symbols in a symbol library using one of the methods described in this chapter.

8. **Electronics** Draw the logic diagram of a portion of the internal components of a computer from the sketch shown. Save the drawing as P24-8.

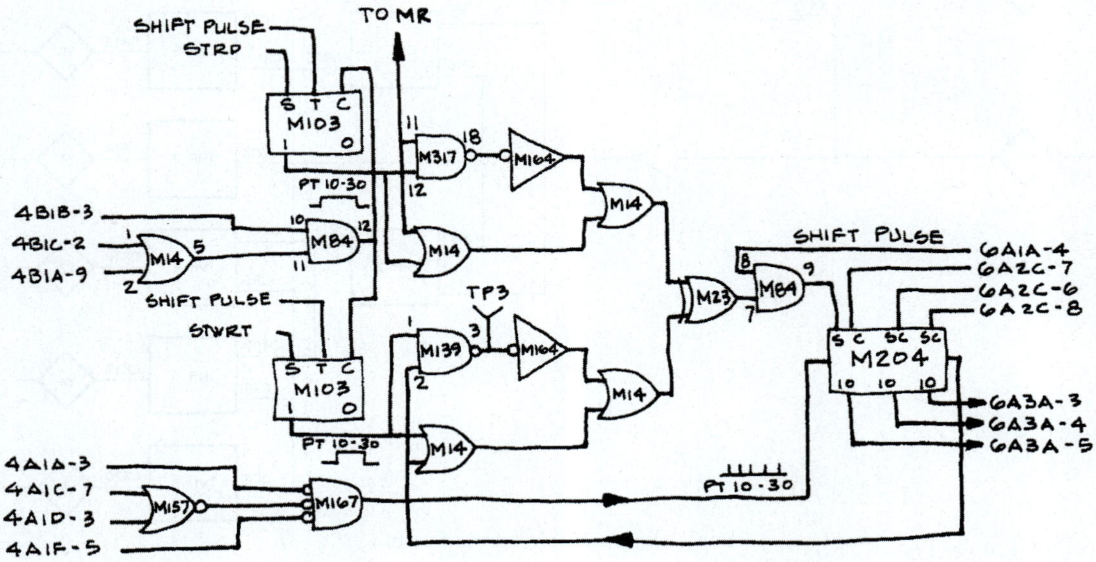

9. **Piping** Draw the piping flow diagram of a cooling water system from the sketch shown. Look closely at this drawing before you begin. Draw the thick flow lines with polylines. Save the drawing as P24-9.

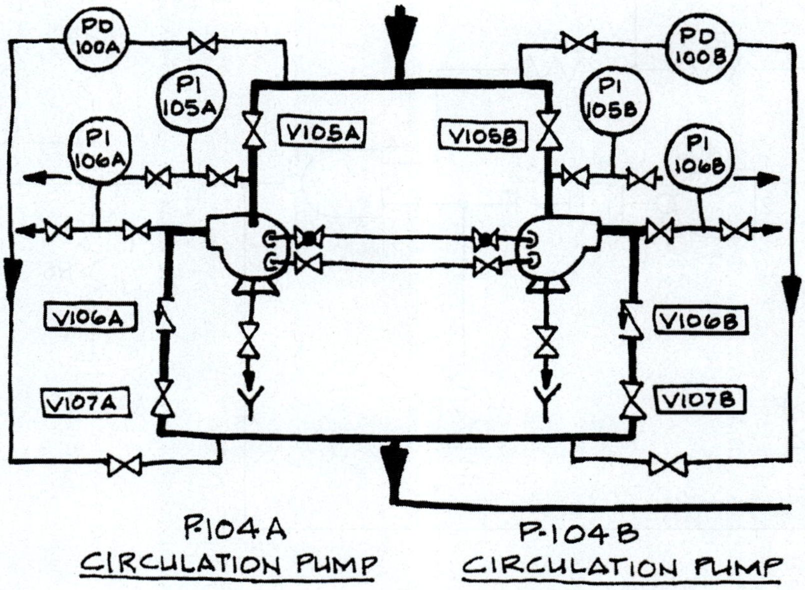

10. **Piping** Draw the general arrangement of a basement floor plan for a new building from the sketch shown. The engineer has shown one example of each type of equipment. Use the following instructions to complete the drawing.

A. All text should be 1/8″ high, except the text for the bay and column line tags, which should be 3/16″ high. The diameter of the line balloons for the bay and column lines should be twice the diameter of the text height.

B. The column and bay steel symbols represent wide-flange structural shapes and should be 8″ wide × 12″ high.

C. The PUMP and CHILLER installations (except PUMP #4 and PUMP #5) should be drawn per the dimensions given for PUMP #1 and CHILLER #1. Use the dimensions shown for the other PUMP units.

D. TANK #2 and PUMP #5 (P-5) should be drawn per the dimensions given for TANK #1 and PUMP #4.

E. Tanks T-3, T-4, T-5, and T-6 are all the same size and are aligned 12′ from column line A.

F. Plan this drawing carefully and create as many blocks as necessary to increase your productivity. Dimension the drawing as shown, and provide location dimensions for all equipment not shown in the engineer's sketch.

G. Save the drawing as P24-10.

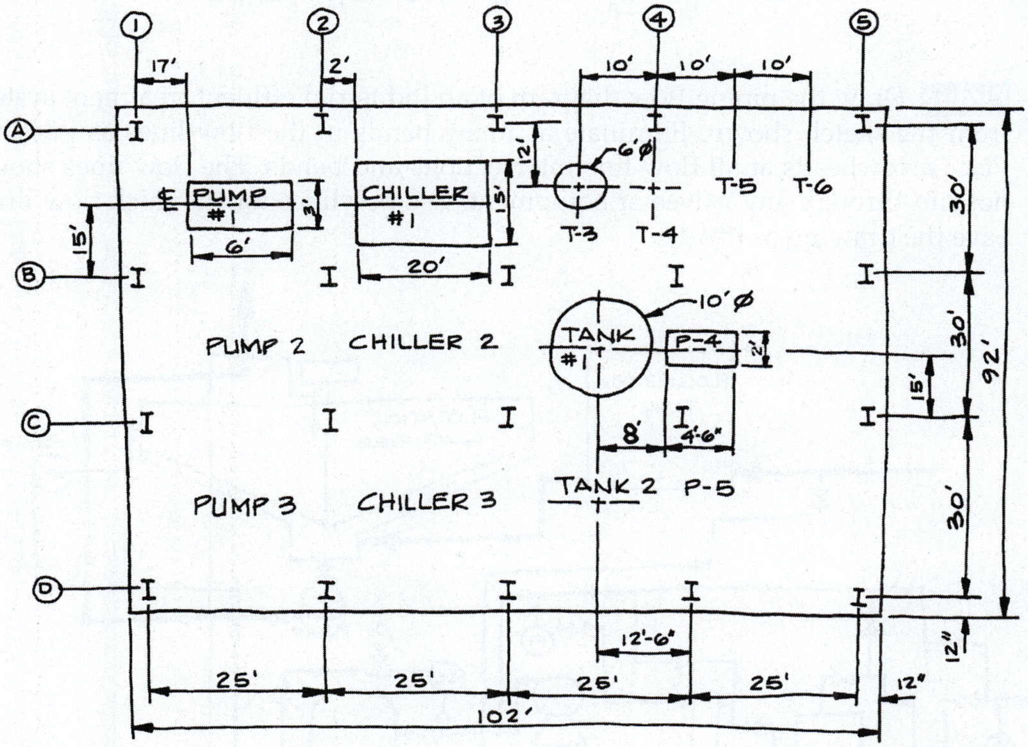

11. **Piping** Save P24-10 as P24-11. Using the P24-11 file, revise the drawing. The engineer has provided you with a sketch of the necessary revisions. It is up to you to alter the drawing as quickly and efficiently as possible. Do not add the dimensions shown on the sketch; the dimensions are provided for construction purposes only. Revise the drawing so all chillers and the four tanks reflect the changes.

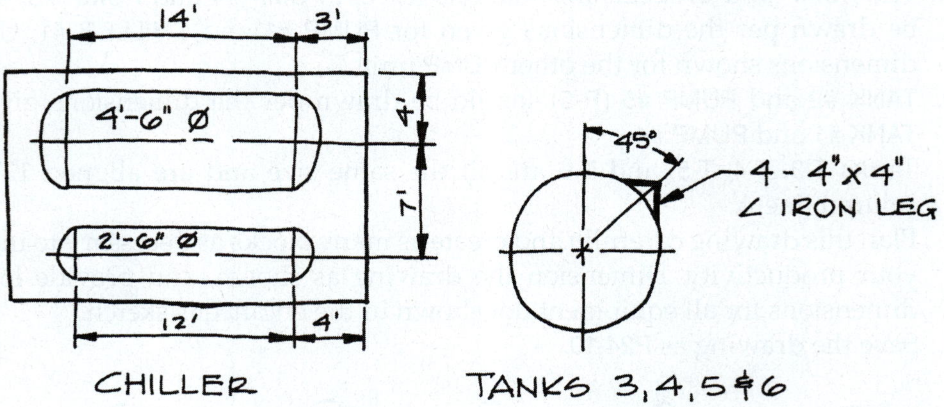

12. **Piping** Draw the piping flow diagram of an industrial effluent treatment system from the sketch shown. Eliminate as many bends in the flow lines as possible. Place arrowheads at all flow line intersections and bends. The flow lines should not run through any valves or equipment. Use polylines for the thick flow lines. Save the drawing as P24-12.

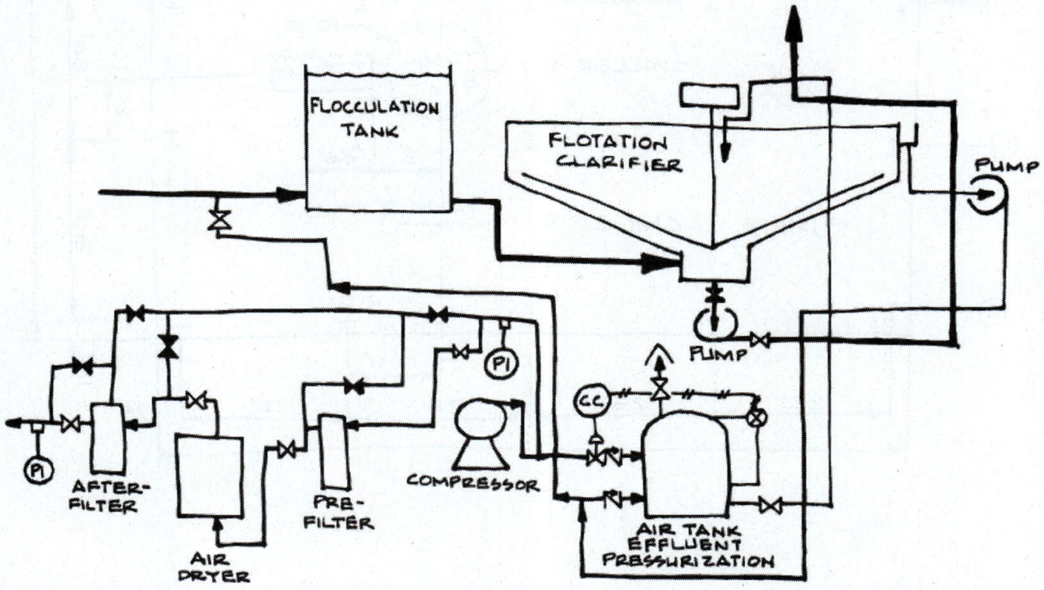

13. **Electronics** Create computer, plotter, and printer/copier blocks and then draw the network diagram shown. Save the drawing as P24-13.

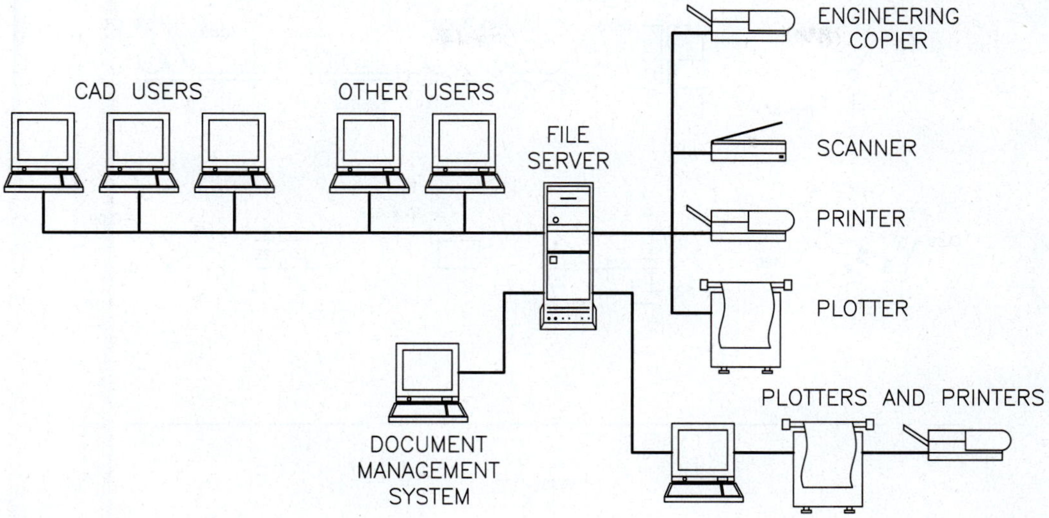

14. **Piping** Draw the piping diagram, creating blocks for each type of fitting. Save the drawing as P24-14.

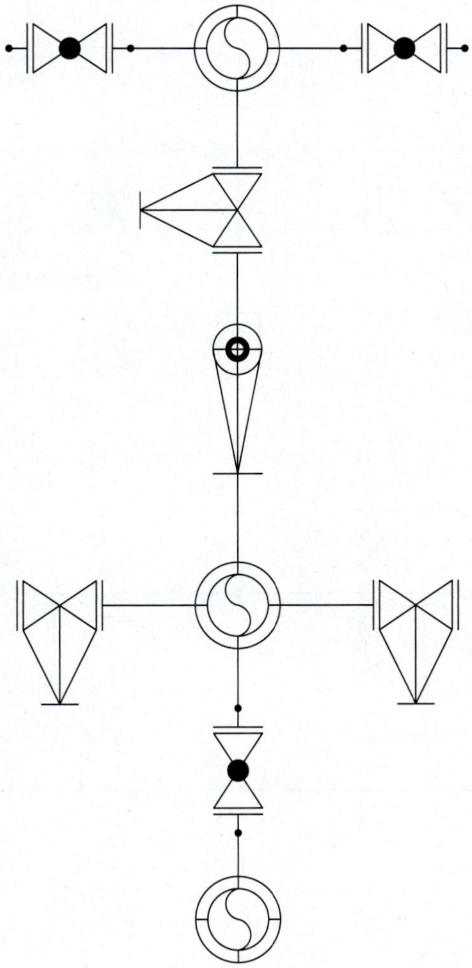

Drawing Problems - Chapter 24

15. **Electronics** Create component blocks based on the dimensions shown. Then use the blocks to draw the schematic below. Save the drawing as P24-15.

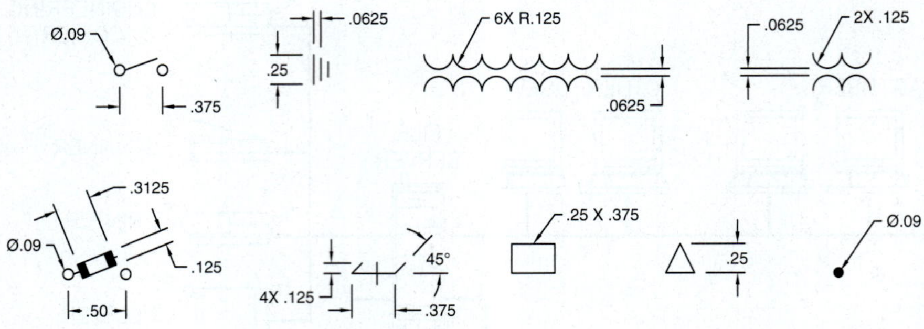

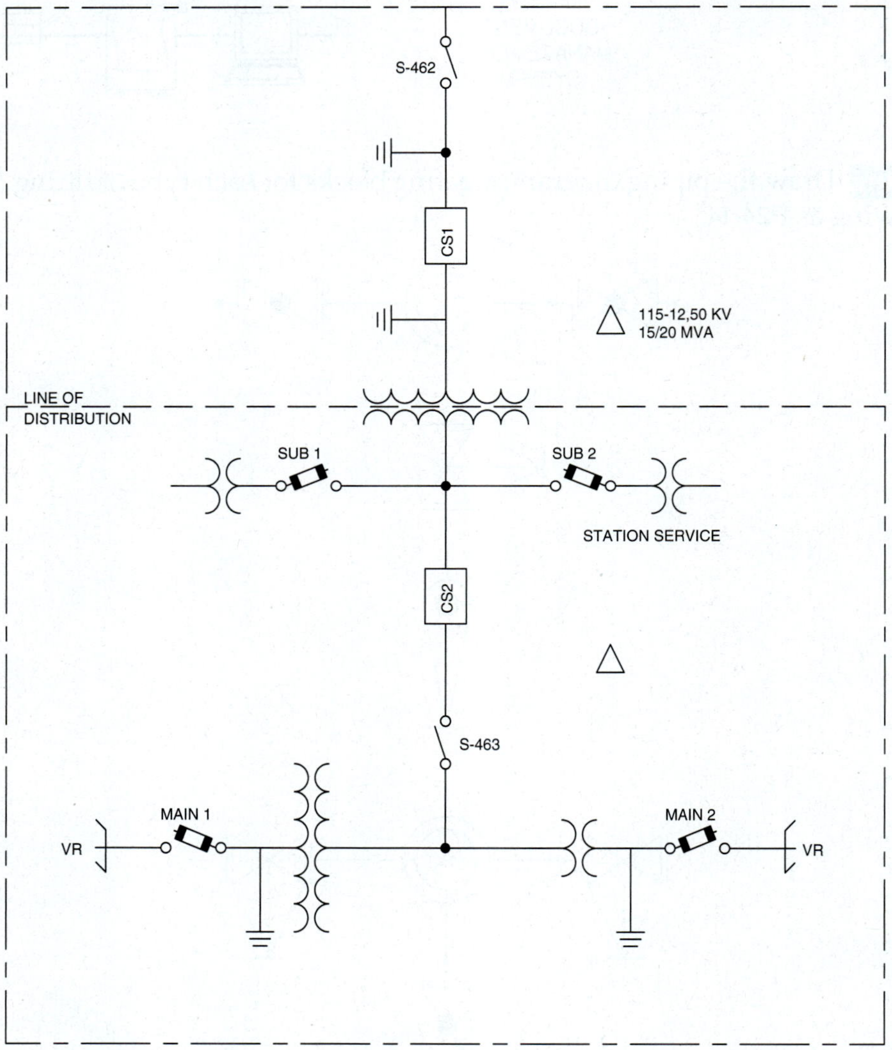

16. General Create a symbol library for one of the drafting disciplines listed and save the symbol library as a template or drawing file. Then, after checking with your supervisor or instructor, draw a problem using the library. If you save the symbol library as a template, start the problem with the template. If you save the symbol library as a drawing file, start a new drawing and insert the symbol library into the new file. Specialty areas you might create symbols for include:
    - Mechanical (machine features, fasteners, tolerance symbols)
    - Architectural (doors, windows, fixtures)
    - Structural (steel shapes, bolts, standard footings)
    - Civil (mapping symbols, survey markers, utilities)
    - Industrial piping (fittings, valves)
    - Piping flow diagrams (tanks, valves, pumps)
    - Electrical schematics (resistors, capacitors, switches)
    - Electrical one-line (transformers, switches)
    - Electronics (IC chips, test points, components)
    - Logic diagrams (AND gates, NAND gates, buffers)
    - GD&T (GD&T symbols)

    Save the drawing as P24-16 or choose an appropriate file name, such as ARCH-PRO or ELEC-PRO. Display the symbol library created in this problem and print a hard copy. Put the printed copy in your notebook for reference.

AutoCAD Certified Professional Exam Practice

Answer the following questions using the information in this chapter.

1. A drawing has five layers: 0 (white), Objects (blue), Dimensions (red), Center (yellow), and Hidden (green). If you create objects on the 0 layer, assign the objects an absolute color of green, create a block of the objects, and then insert the block on the Objects layer, what color will the block be? *Select the one item that best answers the question.*
   A. blue
   B. green
   C. red
   D. white
   E. yellow
2. You have a 1″ × 1″ unit block of a window. To use the block to represent a 3′ window in a 4″ wall, which of the following scale factors would you use? *Select the one item that best answers the question.*
   A. x = 1, y = 1
   B. x = 3, y = 4
   C. x = 4, y = 30
   D. x = 30, y = 4
   E. x = 36, y = 4
3. Which of the following can you use to insert a block into a drawing? *Select all that apply.*
   A. **BLOCK** command
   B. **DesignCenter**
   C. **INSERT** command
   D. **Tool Palettes** palette
   E. **WBLOCK** command

Follow the instructions in each problem. Complete each problem by answering the related question.

4. **Navigate to this chapter on the companion website and open CPE-24block.dwg.** This drawing contains a block named Hole. Insert the block using the default settings. Select the lower-left corner of the existing object as the insertion point. What are the coordinates of the center of the hole?
5. **Navigate to this chapter on the companion website and open CPE-24insert.dwg.** This drawing contains a 2D unit block named Door. Use the appropriate scale factors and rotation to place a 2′-4″ door 6″ above the intersection of the two walls. Edit the wall lines to finish the doorway as shown. What are the coordinates of Point A?

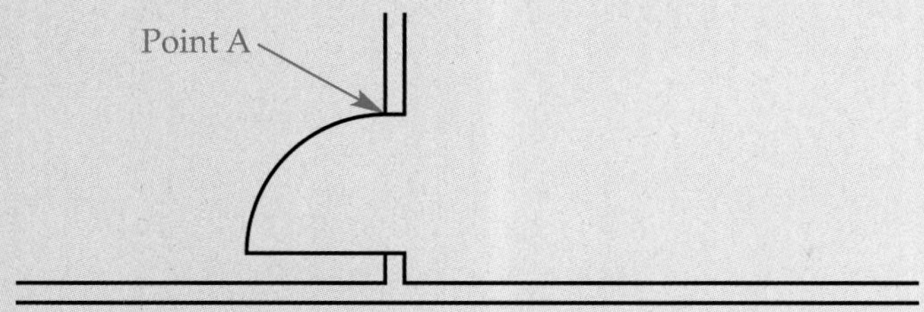

CHAPTER 25

Block Attributes

Learning Objectives

After completing this chapter, you will be able to:

- ✓ Define attributes.
- ✓ Create blocks that contain attributes.
- ✓ Insert blocks with attributes into a drawing.
- ✓ Edit attribute values in existing blocks.
- ✓ Edit single and multiple attribute references.
- ✓ Create title blocks, revision history blocks, and parts lists with attributes.
- ✓ Display attribute values in fields.

Attributes enhance blocks that require text or numerical information. For example, a door tag block contains a letter or number that links the door to a door schedule. Adding an attribute to the door tag block allows you to include a unique letter or number with the symbol, without adding block definitions for each door tag. You can also extract attribute data to automate drawing requirements, such as preparing schedules, parts lists, and bills of materials.

attributes: Text-based data assigned to a specific object. Attributes turn a drawing into a graphical database.

Defining Attributes

Attributes and geometry are often used together to create a block. See Figure 25-1. However, you can prepare blocks that include only attributes. Create attributes with other objects during the initial phase of block development. Add as many attributes as needed to describe the symbol or product, such as the name, number, manufacturer, type, size, price, and weight of an item. Access the **ATTDEF** command to assign attributes using the **Attribute Definition** dialog box. See Figure 25-2.

ATTDEF

Ribbon
Home
> Block
Insert
> Block Definition
Block Editor
> Action Parameters
Define Attributes
Type
ATTDEF
ATT

Attribute Modes

Use the **Mode** area to set attribute modes. Symbols often require attributes to appear with the block. An alternative is to check **Invisible** to hide attributes, but still include attribute data in the drawing that you can reference and *extract*. The geranium symbol in Figure 25-1 is an example of a block with attributes that might be invisible, depending on the application. The other symbols in Figure 25-1 are examples of blocks with visible attributes. Blocks can include both visible and invisible attributes.

extract: Gather content from the drawing file database to display in the drawing or in an external document.

Figure 25-1.
Examples of blocks with defined attributes.

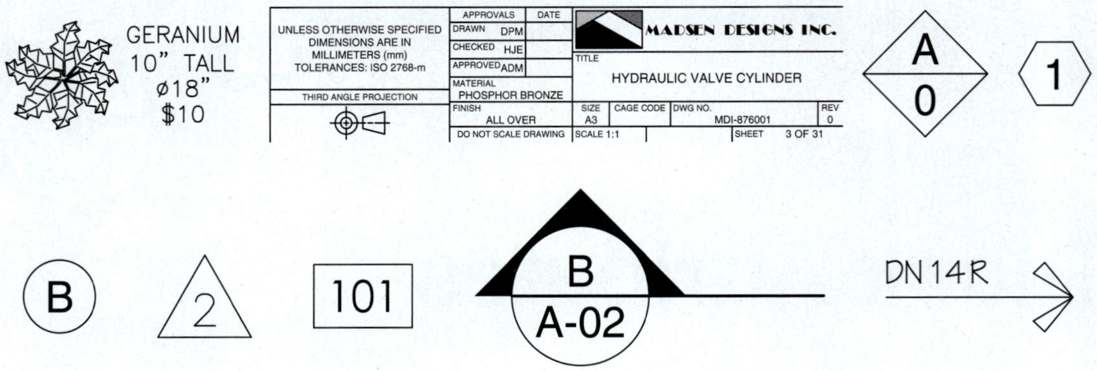

Figure 25-2.
Use the **Attribute Definition** dialog box to assign attributes to blocks.

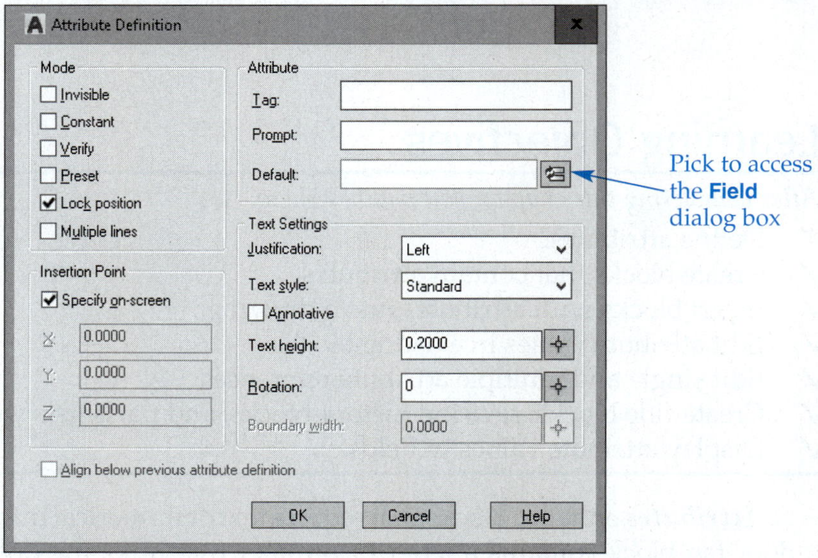

Pick the **Constant** check box if the value of the attribute should always be the same. All insertions of the block display the same value for the attribute, without prompting for a new value when you insert the block. Uncheck the **Constant** check box to use different attribute values for multiple insertions of the block. Check the **Verify** check box to display a prompt that asks if the attribute value is correct when you insert the block. Check the **Preset** check box to have the attribute assume preset values during block insertion. The **Preset** option disables the command line attribute prompt when you insert the block. Uncheck **Preset** to display the normal prompt. Uncheck the **Lock position** check box to have the ability to move the attribute independently of the block after insertion.

You can create single-line or multiple-line attributes. Check the **Multiple lines** check box to activate options for creating a multiple-line attribute. Uncheck the **Multiple lines** check box to create a single-line attribute.

Tag, Prompt, and Value

The **Attribute** area provides text boxes for assigning a tag, prompt, and default value to the attribute. Use the **Tag:** text box to enter the attribute name, or tag. For example, the tag for a size attribute for a valve block could be SIZE. You must enter a tag in order to create an attribute. The tag cannot include spaces. The attribute definition applies uppercase characters to the tag, even if you type lowercase characters in the text box.

Type a statement in the **Prompt:** text box that will display when you insert or edit the block. For example, if you specify SIZE as the attribute tag, you might specify What is the valve size? or Enter valve size: as the prompt. You can also leave the prompt blank. The **Prompt:** text box is disabled when you select the **Constant** attribute mode.

Use the **Default:** text box to specify a default attribute value or a description of an acceptable value for reference. For example, you might type the most common size for the SIZE attribute, or a message regarding the type of information needed, such as 10 SPACES MAX or NUMBERS ONLY. If you deselect the **Multiple lines** attribute mode, enter the default value directly in the text box. Single-line attribute values can include up to 255 characters.

If you select the **Multiple lines** attribute mode, pick the ellipsis (**...**) button to enter the drawing area and place multiline text using the **Text Formatting** toolbar and text editor. See Figure 25-3. Enter the default text, and then pick the **OK** button on the toolbar to return to the **Attribute Definition** dialog box. Use the **Insert Field** button to include a field in the default value. You can leave the default value blank.

NOTE

The abbreviated **Text Formatting** toolbar shown in Figure 25-3 appears by default. Set the **ATTIPE** system variable to 1 to display the complete **Text Formatting** toolbar. The **ATTIPE** system variable is set to 0 by default.

Text Settings

Use options in the **Text Settings** area to specify attribute text format. Use the **Justification:** drop-down list to select a justification for the attribute text. The default justification is **Left**. In single-line attributes, the text itself is justified. In the **Multiple lines** attribute mode, the text boundary is justified.

Use the **Text style:** drop-down list to select a text style available in the current drawing to assign to the attribute. Check the **Annotative** check box to make the attribute text height annotative. AutoCAD scales annotative attributes according to the specified annotation scale, which reduces the need for you to calculate the scale factor. If you plan to define a block as annotative, you can include annotative or non-annotative objects, such as attributes, with the block. However, you must use non-annotative objects when preparing a non-annotative block.

Specify the height of the attribute text in the **Text height:** text box, or pick the **Text Height** button next to the text box to specify two points in the drawing to set the text height. Specify the rotation angle for the attribute text in the **Rotation:** text box, or pick the **Rotation** button next to the text box to pick two points in the drawing to set the text rotation.

Figure 25-3.
Define multiple-line attributes directly on-screen. The abbreviated **Text Formatting** toolbar appears instead of the **Text Editor** ribbon tab.

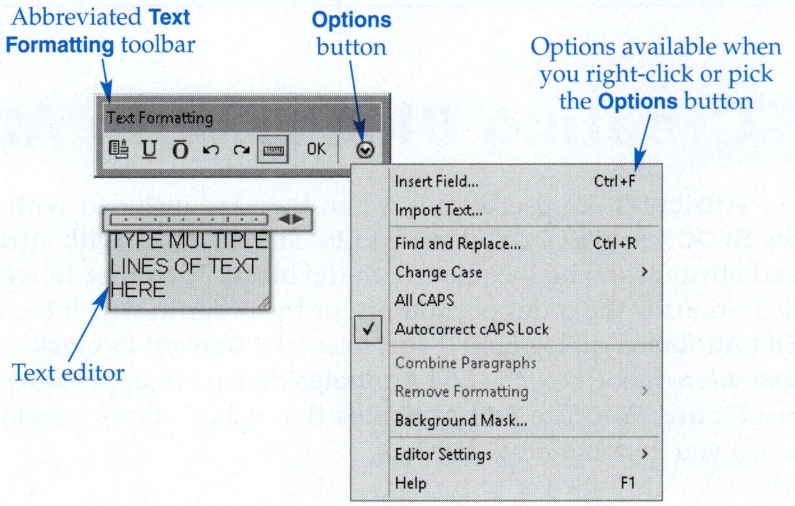

The **Boundary width:** option is available in the **Multiple lines** attribute mode. Type a width in the **Boundary width:** text box, or pick the **Boundary width** button next to the text box to specify two points in the drawing to set a text boundary width.

Defining the Attribute Insertion Point

Use options in the **Insertion Point** area to define how and where to position the attribute during insertion. Check the **Specify on-screen** check box to pick an insertion point in the drawing after you pick the **OK** button to create the attribute and exit the **Attribute Definition** dialog box. An alternative is to type values in the **X:**, **Y:**, and **Z:** text boxes if you know the coordinates for the insertion point.

The **Align below previous attribute definition** check box is enabled if the drawing already contains at least one attribute. Check the box to place the new attribute directly below the most recently created attribute using the justification of that attribute. This is an effective technique for placing a group of different attributes in the same block. The **Text Settings** and **Insertion Point** areas are deactivated.

Placing the Attribute

After defining all elements of the attribute, pick the **OK** button to close the **Attribute Definition** dialog box. The attribute tag appears on-screen if you specified coordinates for the insertion point, or if you used the **Align below previous attribute definition** option. Otherwise, AutoCAD prompts you to specify an insertion point. If the attribute mode is set to **Invisible**, do not be concerned that the tag is visible. The tag disappears when you include the attribute with a block definition.

Editing Attribute Properties

The **Properties** palette provides options for editing attributes before you include them in a block. See Figure 25-4. Change the color, linetype, or layer of the selected attribute in the **General** category. Use options in the **Text** category to adjust the tag, prompt, or value. If the value contains a field, the value appears as normal text in the **Properties** palette. Modified field text is automatically converted to text and is disassociated from the field. The **Text** category also contains options to change the attribute text settings. Additional text and attribute options are available in the **Misc** category.

PROFESSIONAL TIP

A powerful feature of the **Properties** palette for editing attributes is the ability to change the original attribute modes. The Invisible, Constant, Verify, and Preset mode settings are available in the **Misc** category.

Creating Blocks with Attributes

Attributes are useful only when they are included with a block definition. Use the **BLOCK** or **WBLOCK** command to define a block with attributes. Select all objects and attributes to be included with the block. The order in which you select attribute definitions is the order of prompts, or the order in which the attributes appear in the **Edit Attributes** dialog box. If you select the **Convert to block** radio button in the **Block Definition** dialog box, the **Edit Attributes** dialog box appears when you create the block. See Figure 25-5. The **Edit Attributes** dialog box allows you to adjust attribute values when you insert or edit the block.

Figure 25-4.
The **Properties** palette allows you to modify attributes.

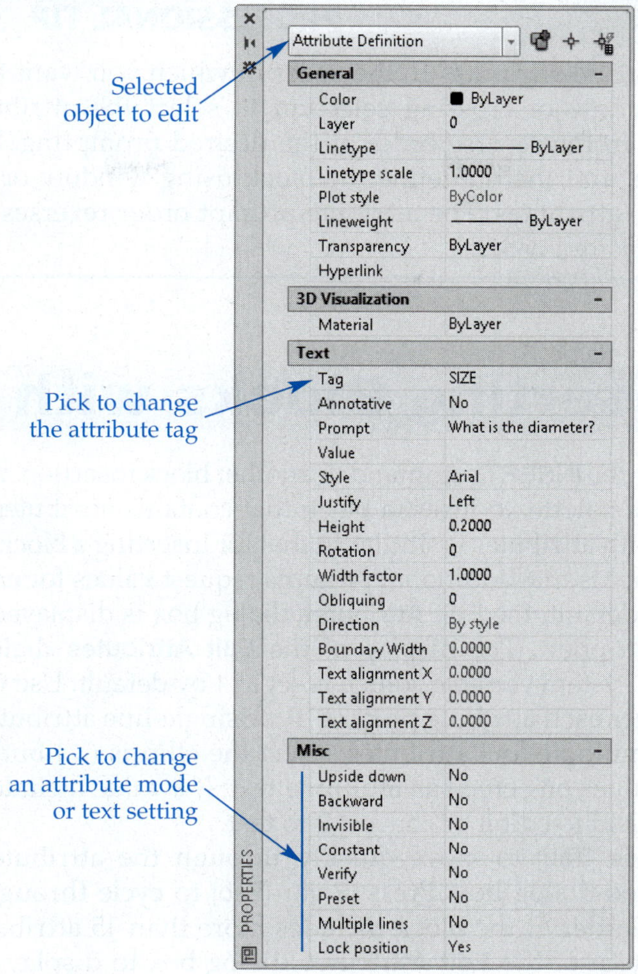

Figure 25-5.
The **Edit Attributes** dialog box allows you to enter attribute definitions when you insert or edit a block.

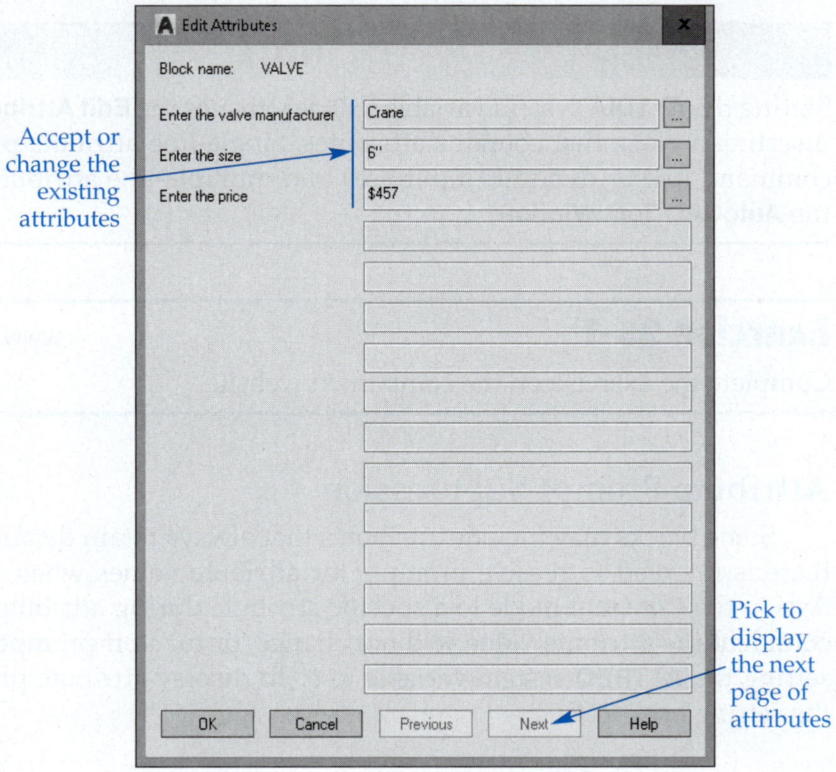

PROFESSIONAL TIP

If you create attributes in the order in which you want to receive prompts and then use window or crossing selection to select the attributes, the attribute prompts display in the reverse order of the desired prompting. To change the order, insert, explode, and then redefine the block using window or crossing selection again to pick the attributes. The attribute prompt order reverses again, placing the prompts in the desired order.

Inserting Blocks with Attributes

Use the **INSERT** command or another block insertion method, such as **DesignCenter** or a tool palette, to insert a block that contains attributes. The process of inserting a block with attributes is similar to that for inserting a block without attributes. The only difference is that additional prompts request values for each attribute.

By default, the **Edit Attributes** dialog box is displayed after you specify the block insertion point. The display of the **Edit Attributes** dialog box is controlled by the **ATTDIA** system variable, which is set to 1 by default. Use the **Edit Attributes** dialog box to answer each attribute prompt. Type single-line attribute values in the text boxes. To define multiple-line attributes, select the ellipsis (**...**) button next to the text boxes to enter values on-screen as multiline text. If a value includes a field, you can right-click on the field to edit it or convert it to text.

Press [Tab] to move quickly through the attributes and buttons in the **Edit Attributes** dialog box. Press [Shift]+[Tab] to cycle through attributes and buttons in reverse order. If the block includes more than 15 attributes, pick the **Next** button at the bottom of the **Edit Attributes** dialog box to display the next page of attributes. When you finish entering values, pick the **OK** button to close the dialog box and create the block.

NOTE

Setting the **ATTDIA** system variable to 0 deactivates the **Edit Attributes** dialog box when inserting a block that contains attributes. Single-line attribute prompts appear at the command line or dynamic input area and multiple-line attribute prompts display in the **AutoCAD Text Window**.

Exercise 25-1

www.g-wlearning.com/CAD/

Complete the exercise on the companion website.

Attribute Prompt Suppression

Some blocks may include attributes that always retain default values. In this case, there is no need to receive prompts for attribute values when you insert the block. Assign the **Constant** mode to a specific attribute during attribute definition if you are confident the attribute value will not change, or turn off prompts for all attributes by setting the **ATTREQ** system variable to 0. To display attribute prompts again, change the setting back to 1.

PROFESSIONAL TIP

Project and drawing planning should involve setting system variables such as **ATTREQ**. Setting **ATTREQ** to 0 before using blocks can save time in the drawing process. Always remember to set **ATTREQ** back to 1 to receive prompts instead of accepting defaults. If anticipated attribute prompts do not appear, check the current **ATTREQ** setting and adjust it if necessary.

Controlling Attribute Display

Some attributes only provide content to generate parts lists or bills of materials and to speed accounting. These types of attributes usually do not display on-screen or plot. Use the **ATTDISP** command to control the display of attributes on-screen. An easy way to activate the **ATTDISP** command option is to pick the corresponding button from the ribbon.

Use the default **Retain Attribute Display** (**Normal**) option to display attributes exactly as created. Use the **Display All Attributes** (**ON**) option to display all attributes, both visible and invisible. Use the **Hide All Attributes** (**OFF**) option to suppress the display of all attributes, including visible attributes.

ATTDISP
Ribbon
Insert
> Block
Home
> Block
Retain Attribute Display
Display All Attributes
Hide All Attributes
Type
ATTDISP

PROFESSIONAL TIP

Use the **ATTDISP** command to display invisible attributes when necessary or to hide all attributes if they interfere with the current drawing session. In a drawing in which attributes should be visible but are not, check the current setting of **ATTDISP** and adjust it if necessary.

Editing Attribute References

Once you insert a block with attributes, tools are available for editing attribute values and settings. Access the **ATTEDIT** command and select a block with attributes to redisplay the **Edit Attributes** dialog box. Another option is to modify the attributes of a single selected block using the **EATTEDIT** command and corresponding **Enhanced Attribute Editor**. See Figure 25-6. Use the **Select block** button to return to the drawing to select a different block to modify.

ATTEDIT
Type
ATTEDIT

EATTEDIT
Ribbon
Home
> Block
Insert
> Block
Edit Attribute (Single)
Type
EATTEDIT

Figure 25-6.
Select the attribute to modify and change the value in the **Attribute** tab of the **Enhanced Attribute Editor**.

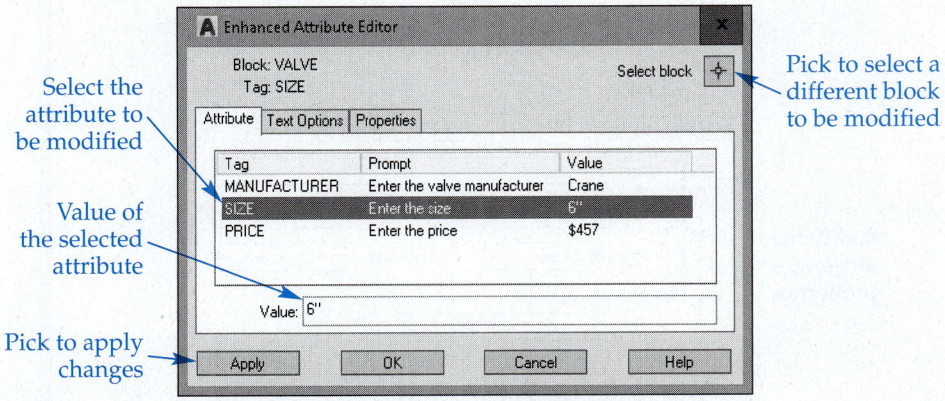

The **Attribute** tab, shown in Figure 25-6, displays all of the attributes assigned to the selected block. Pick the attribute you want to modify and enter a new value in the **Value:** text box. If the attribute is a multiple-line attribute, pick the ellipsis (**...**) button to modify the text on-screen. Pick the **Apply** button after adjusting the value.

NOTE

Type
ATTIPEDIT

A quick way to access the **EATTEDIT** command is to double-click on a block containing attributes. You can also edit multiple-line attribute values without accessing the **Enhanced Attribute Editor** using the **ATTIPEDIT** command.

Use the **Text Options** tab, shown in Figure 25-7A, to modify the text options of an attribute. The **Properties** tab, shown in Figure 25-7B, provides general property adjustments for an attribute. Each attribute in a block is a separate item. The settings you apply in the **Text Options** and **Properties** tabs affect the active attribute in the **Attribute** tab. Pick the **Apply** button to apply changes. Pick the **OK** button to close the dialog box.

Exercise 25-2

www.g-wlearning.com/CAD/

Complete the exercise on the companion website.

Figure 25-7.
A—The **Text Options** tab provides options in addition to those set in the **Attribute Definition** dialog box. B—The **Properties** tab allows you to modify the properties of an attribute.

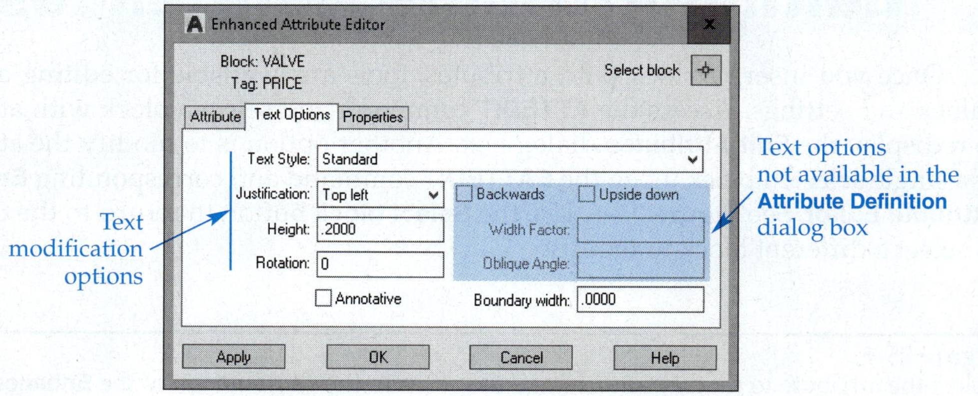

A

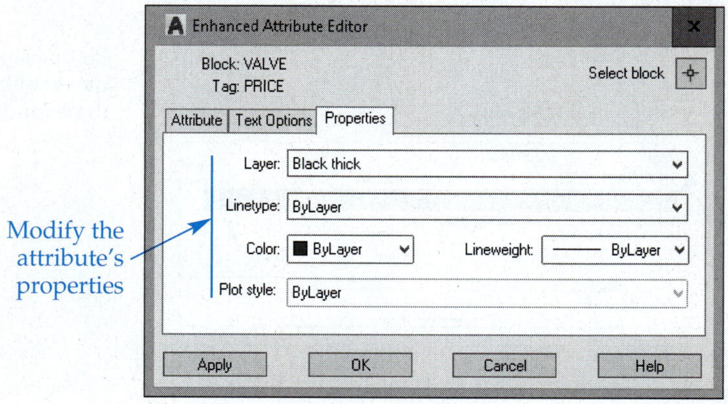

B

Using the FIND Command to Edit Attributes

The **FIND** command, which displays the **Find and Replace** dialog box described in Chapter 10, provides a quick way to edit attributes. You can also access the **FIND** command when no command is active by right-clicking in the drawing area and selecting **Find...**, or by entering the text in the **Find text** text box in the **Text** panel of the **Annotate** ribbon tab and pressing [Enter]. Specify the portion of the drawing to search, the attribute value to find, and the replacement value.

FIND

Ribbon
Annotate
> Text

Find
Type
FIND

Editing Multiple Attribute References

The **Edit Attributes** dialog box and the **Enhanced Attribute Editor** allow you to edit attribute values and settings by selecting blocks one at a time. The **-ATTEDIT** command edits the attributes of multiple blocks. When you access the **-ATTEDIT** command, a prompt asks if you want to edit attributes individually. Use the default **Yes** option to select specific blocks with attributes to edit. Use the **No** option to apply ***global attribute editing***.

-ATTEDIT

Ribbon
Home
> Block
Insert
> Block

Edit Attribute (Multiple)
Type
-ATTEDIT
-ATE

global attribute editing: Editing or changing all insertions, or instances, of the same block in a single operation.

If you select the **Yes** option, prompts appear to specify the block name, attribute tag, and attribute value. To edit attribute values selectively, respond to each prompt with the correct name or value, and then select one or more attributes. If you see the message 0 found after selecting attributes, you picked an incorrectly specified attribute. It is often quicker to press [Enter] at each of the three specification prompts and then pick the attribute to edit. Select an option and follow the prompts to edit the attribute.

If you select the **No** option, the Edit only attributes visible on screen? prompt appears. Select the **Yes** option to edit all visible attributes, or select the **No** option to edit all attributes, including invisible attributes. The same three prompts previously described for individual block editing appear.

Figure 25-8A shows a VALVE block inserted three times with the manufacturer specified as CRANE. In this example, the manufacturer was supposed to be POWELL. To change the attribute for each insertion, access the **-ATTEDIT** command and specify global editing. Press [Enter] at each of the three specification prompts. When the Select Attributes: prompt appears, pick CRANE on each of the VALVE blocks and press [Enter]. At the Enter string to change: prompt, enter CRANE, and at the Enter new string: prompt, enter POWELL. See the result in **Figure 25-8B**.

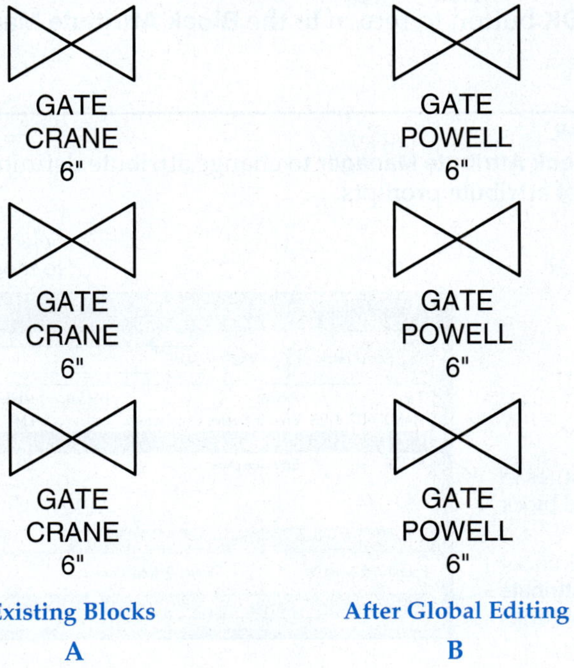

Figure 25-8. Using the global editing technique with the **-ATTEDIT** command allows you to change the same attribute on several block insertions.

PROFESSIONAL TIP

Use care when assigning the **Constant** mode to attribute definitions. The **-ATTEDIT** command displays 0 found if you attempt to edit a block attribute that has a **Constant** mode setting. Assign the **Constant** mode only to attributes you know will not change.

NOTE

You can also use the **-ATTEDIT** command to edit individual attribute values and properties. However, it is more efficient to use the **Edit Attributes** dialog box or the **Enhanced Attribute Editor** to change individual attributes.

Exercise 25-3

www.g-wlearning.com/CAD/

Complete the exercise on the companion website.

Editing Attribute Definitions

BATTMAN

Ribbon
Home
> Block
Insert
> Block Definition
Block Attribute Manager...
Type
BATTMAN

Once you create a block with attributes, tools are available for modifying attribute definitions. One option is to edit attribute definitions using the **BATTMAN** command, which displays the **Block Attribute Manager**. See Figure 25-9. To manage the attributes in a block, select the block name from the **Block:** drop-down list or pick the **Select block** button to return to the drawing and pick a block.

The tag, prompt, default value, and modes for each attribute are listed by default. To select the attribute properties listed in the **Block Attribute Manager**, pick the **Settings...** button to open the **Block Attribute Settings** dialog box. See Figure 25-10. Check the properties to list in the **Display in list** area. Check the **Emphasize duplicate tags** check box to highlight attributes with identical tags in red. Check **Apply changes to existing references** to apply changes made in the **Block Attribute Manager** to existing blocks. Pick the **OK** button to return to the **Block Attribute Manager**.

Figure 25-9.
Use the **Block Attribute Manager** to change attribute definitions, delete attributes, and change the order of attribute prompts.

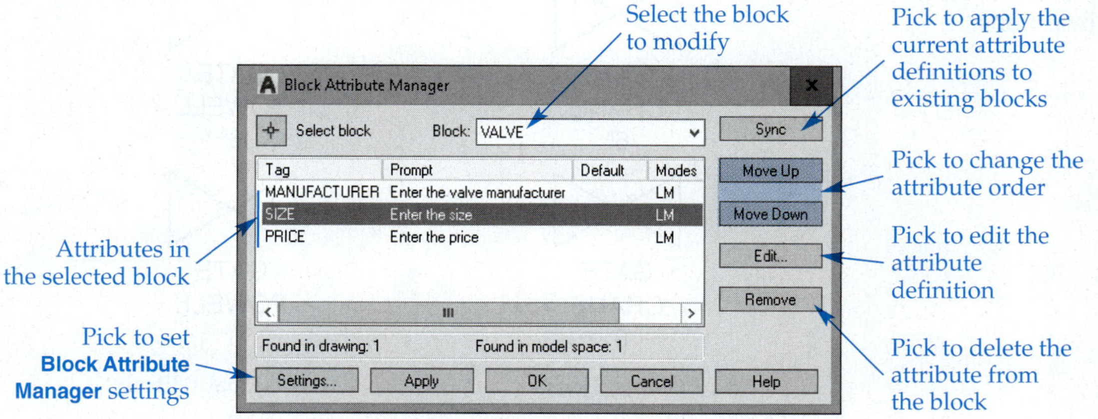

Figure 25-10.
The **Block Attribute Settings** dialog box controls the attribute properties displayed in the **Block Attribute Manager** and attribute editing functions.

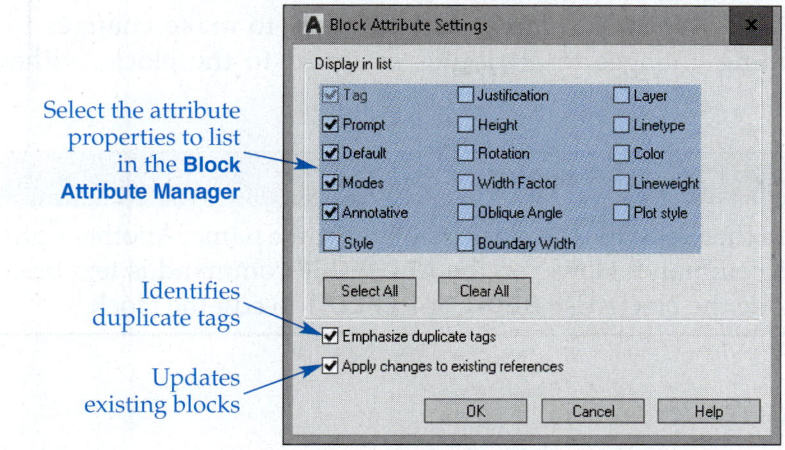

The attribute list in the **Block Attribute Manager** reflects the order in which prompts appear when you insert a block. Use the **Move Up** and **Move Down** buttons to change the order of the selected attribute within the list, modifying the prompt order. To delete an attribute, pick the **Remove** button. To modify an attribute, select the attribute and pick the **Edit...** button to display the **Edit Attribute** dialog box. See Figure 25-11. Use the **Attribute** tab to modify the modes, tag, prompt, and default value. The settings in the **Text Options** and **Properties** tabs of the **Edit Attribute** dialog box are identical to those in the tabs found in the **Enhanced Attribute Editor**. Check **Auto preview changes** at the bottom of the dialog box to display changes to attributes immediately in the drawing area.

After modifying the attribute definition in the **Edit Attribute** dialog box, pick the **OK** button to return to the **Block Attribute Manager**. Then pick the **OK** button to return to the drawing. When you modify attributes within a block, future insertions of the block reflect the changes. Existing blocks update only if you select the **Apply changes to existing references** check box in the **Block Attribute Settings** dialog box.

NOTE

The **Block Attribute Manager** modifies attribute definitions, not attribute values. Modify attribute values using the **Edit Attributes** dialog box or the **Enhanced Attribute Editor**.

Figure 25-11.
Use the **Edit Attribute** dialog box to modify attribute definitions and properties.

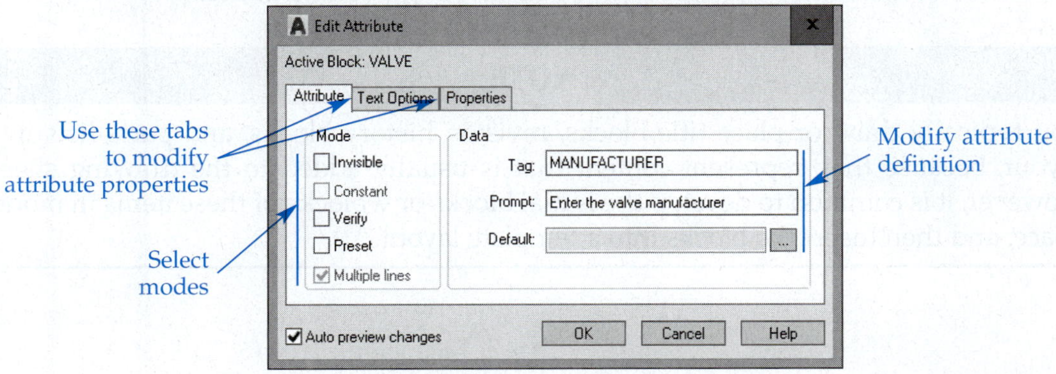

Redefining a Block with Attributes

To add attributes to or revise the geometry of a block, edit the block definition using the **BEDIT** or **REFEDIT** commands. Chapter 31 explains the **REFEDIT** command. The **BEDIT** and **REFEDIT** commands allow you to make changes to a block definition, including changes to attributes assigned to the block, without exploding the block.

NOTE

You can also explode and then redefine the block using the same name. Another option is to use the **ATTREDEF** command. However, the **ATTREDEF** command is text based, and you must first explode the block. Use **BEDIT** or **REFEDIT** to edit the block.

Synchronizing Attributes

Redefining a block automatically updates the properties of all of the same blocks in the drawing, but does not apply changes made to attributes. For example, if you add an object to a block, all existing blocks of the same name update to display the new object. However, if you add an attribute to a block, all existing blocks of the same name continue to display the original attributes, without the new attribute. Synchronize the blocks to update the attribute redefinition.

ATTSYNC

Ribbon
Home
> Block
Insert
> Block Definition

Synchronize Attributes

Type
ATTSYNC

Synchronize blocks in the **Block Attribute Manager** by picking the **Sync** button. This method is convenient because it allows you to make changes to and remove attributes using the **Block Attribute Manager**. To synchronize attributes without using the **Block Attribute Manager**, use the **ATTSYNC** command. Access the **ATTSYNC** command and use the default **Select** option to pick a block that contains the attributes you want to synchronize. An alternative is to use the **Name** option to type the block name, or use the **?** option to list the names of all blocks in the drawing. Then select the **Yes** option to synchronize attributes, or the **No** option to select a different block.

Automating Drafting Documentation

Attributes automate the process of placing symbols that require textual information. Attributes are especially useful for automating common detailing or documentation tasks such as preparing title block information, revision history block data, schedules, or a parts list or bill of materials. Filling out these items is usually one of the more time-consuming tasks associated with drafting documentation.

NOTE

You typically draw or place title blocks, revision history blocks, and parts lists in a layout, because they represent content that is usually added to the drawing sheet. However, it is common to develop the initial blocks or wblocks of these items in model space, and then insert the blocks into a template layout.

Title Blocks

To create an automated title block, first use the correct layer, typically layer 0, to draw title block objects and add text that does not change, such as the titles of compartments. Format the title block in accordance with industry and company or school standards. Include your company or school logo if appropriate. If you work in an industry that produces items for the federal government, also include the applicable ***Commercial and Government Entity Code (CAGE Code)***. Figure 25-12 shows a title block, dimensioning and tolerancing block, and angle of projection block drawn in accordance with the ASME Y14.1 *Decimal Inch Drawing Sheet Size and Format* standard.

Commercial and Government Entity Code (CAGE Code): A five-digit numerical code identifier applicable to any organization that produces items used by the federal government.

Next, define attributes for each area of the title block. As you create attributes, determine the appropriate text height and justification for each definition. Common title block attributes include drawing title, drawing number, drafter, checker, dates, drawing scale, sheet size, material, finish, revision letter, and tolerance information. See Figure 25-13. Create approval attributes with a prompt such as ENTER INITIALS OR SEEK SIGNATURE, providing the flexibility to type initials or leave the cell blank for written initials. Apply the same practice to date attributes. Include any other information that may be specific to the organization or drawing application. Assign default values to the attributes wherever possible. For example, if you consistently specify the same tolerances for dimensions, assign default values to the tolerance attributes.

Figure 25-12.
Sheet blocks must comply with applicable standards. This title block, dimensioning and tolerancing block, and angle of projection block comply with the ASME Y14.1 *Decimal Inch Drawing Sheet Size and Format* standard.

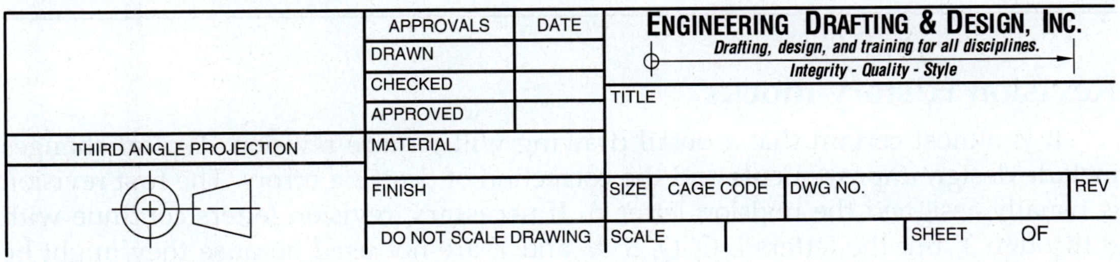

Figure 25-13.
Define attributes for each area of the title block. Attributes should define all information that might change, including general tolerances. The dimensioning and tolerance block in this example uses a multiple line attribute. This figure shows the attributes in color for illustrative purposes only.

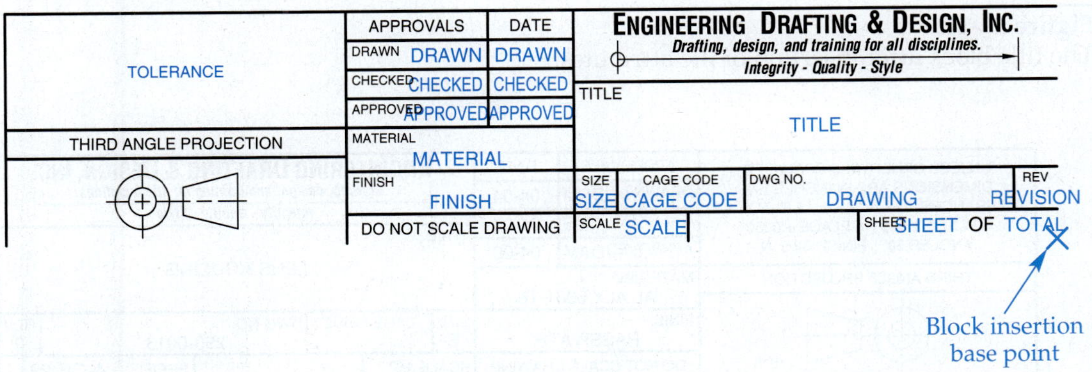

PROFESSIONAL TIP

The size of each area within the title block limits the number of characters displayed in a line of text. Include a reminder about the maximum number of characters in the attribute prompt, such as Enter drawing name (15 characters max). Each time you insert a block or drawing containing the attribute, the prompt displays the reminder.

After you define each attribute in the title block, you are ready to create the block. One option is to use the **BLOCK** command to create a block of the title block within the current file. When specifying the insertion base point, pick a corner of the title block that is convenient to use each time you insert the block. Figure 25-13 shows the insertion base point that is appropriate for that particular title block. Use the **Delete** option in the **Block Definition** dialog box to remove the selected objects from the drawing.

Another option is to use the **WBLOCK** command to save the drawing as a file. Give the file a descriptive name, such as TITLE_B or FORMAT_B for a B-size title block. Figure 25-14 shows the attribute block created in Figure 25-13, inserted and filled in using attributes.

NOTE

If you are creating a template, insert the block at the appropriate location and save the file as a drawing template. Edit the values in an existing title block using the **Enhanced Attribute Editor**.

Revision History Blocks

It is almost certain that a detail drawing will require revision. Typical changes include design improvements and the correction of drafting errors. The first revision is usually assigned the revision letter *A*. If necessary, revision letters continue with *B* through *Y*, but the letters *I*, *O*, *Q*, *S*, *X*, and *Z* are not used because they might be confused with numbers.

Drawing layout formats include an area with columns specifically designated to record drawing changes. This area, commonly called the ***revision history block***, is normally located at the upper-right corner of the drawing sheet. A column for ***zones*** is included only if applicable.

revision history block: A block that provides space for the revision letter, a description of the change, the date, and approvals.

zones: A system of letters and numbers used on large drawings to help direct the attention of the person reading the print to a location on the drawing.

Figure 25-14.
The title block after insertion of the attributes.

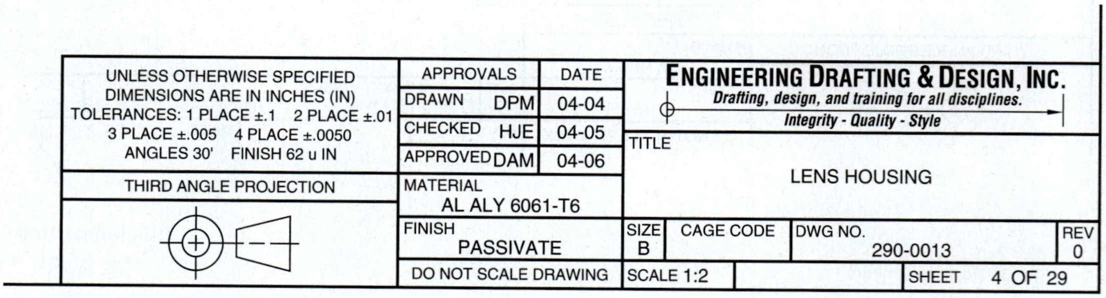

Figure 25-15.
Creating a revision history block using two separate blocks. A—The first block forms the title and heading rows. B—The second block includes attributes and is added each time an engineering change is employed. The revision history block shown complies with the ASME Y14.1 *Decimal Inch Drawing Sheet Size and Format* standard. This figure shows the attributes in color for illustrative purposes only.

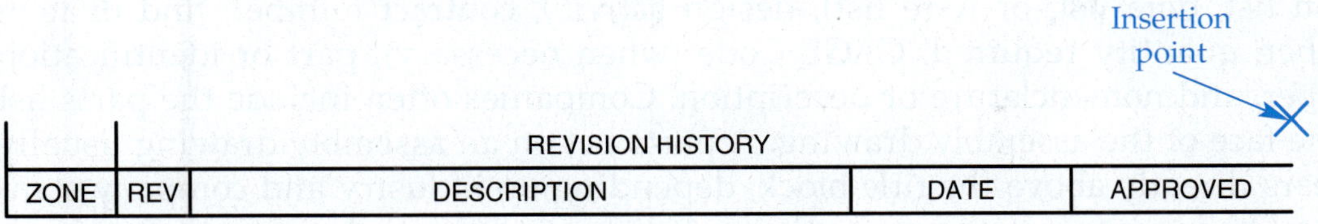

A

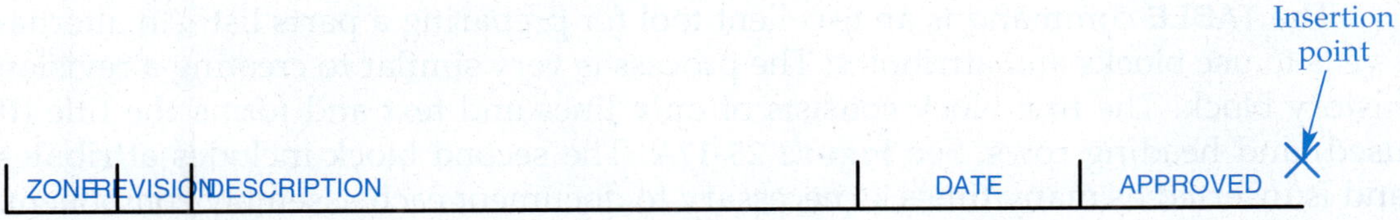

B

The **TABLE** command is an excellent tool for preparing a revision history block. An alternative is to use blocks and attributes to document revisions. The process is similar to creating a title block, but a revision history block requires two separate blocks. The first block consists of only lines and text and forms the title and heading rows. See **Figure 25-15A**. Insert the second block, which includes attributes, whenever a revision is required. See **Figure 25-15B**.

Format the revision history block according to industry and company or school standards, and use the correct layer, typically layer 0. As you create attributes, determine the appropriate text height and justification for each definition. Define attributes for the zone (if necessary), revision letter, description, date, and approval. Assign the APPROVED attribute a prompt such as ENTER INITIALS OR SEEK SIGNATURE, providing the flexibility to type initials or leave the cell blank for written initials. Apply the same practice to the date attribute.

Use the **BLOCK** or **WBLOCK** command to create the blocks. If you create wblocks, use descriptive file names such as REVBLK or REV. **Figure 25-16** shows an example of revision information added by inserting the two blocks created in **Figure 25-15** in the inside corner of the border at the upper-right corner of the drawing sheet.

Figure 25-16.
The completed revision history block after inserting one title and heading block and two change blocks.

REVISION HISTORY				
ZONE	REV	DESCRIPTION	DATE	APPROVED
C3	A	ADDED .125 CHAMFER	03-16	ADM
B3	B	REDUCED SHAFT LENGTH FROM 4.580	04-06	BAE

Parts Lists

Assembly drawings require a parts list, or bill of materials, that provides information about each component of the assembly or subassembly. Columns can identify a variety of information depending on the organization and application. Common elements include the list type (such as a parts list, index list, application list, data list, or wire list), design activity, contract number, find or item number, quantity required, CAGE Code (when necessary), part or identification number, and nomenclature or description. Companies often include the parts list on the face of the assembly drawing. A parts list on an assembly drawing usually appears directly above the title block, depending on industry and company standards. Other organizations create the parts list as a separate document, often in an 8-1/2″ × 11″ format.

The **TABLE** command is an excellent tool for preparing a parts list. An alternative is to use blocks and attributes. The process is very similar to creating a revision history block. The first block consists of only lines and text and forms the title (if used) and heading rows. See Figure 25-17A. The second block includes attributes and is inserted as many times as necessary to document each assembly component. See Figure 25-17B.

Format the parts list according to industry and company or school standards, and use the correct layer, typically layer 0. As you create attributes, select the appropriate text height and justification for each definition. Define attributes for the column or cell in the parts list.

Use the **BLOCK** or **WBLOCK** command to create the blocks. If you save wblocks, use descriptive file names, such as PL for parts list or BOM for bill of materials. Figure 25-18 shows an example of the beginning of a parts list developed by inserting the blocks created in Figure 25-17.

Figure 25-17.
Creating a parts list using two separate blocks. A—The first block forms the title (if used) and heading rows. B—The second block includes attributes and is inserted as many times as necessary to define each assembly component. The parts list shown complies with the ASME Y14.1 *Decimal Inch Drawing Sheet Size and Format* standard.

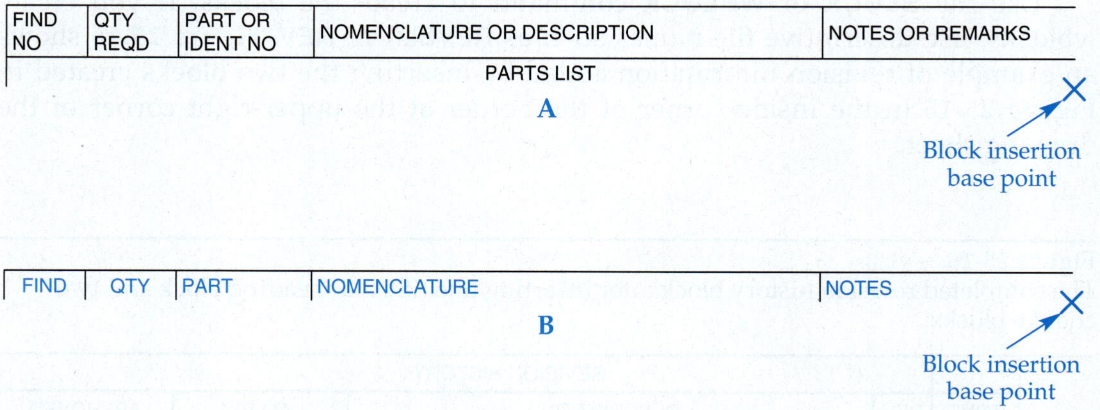

Figure 25-18.
A parts list after inserting blocks and editing attribute values.

4	4	30-004579-04	ULTRA THIN RETAINER SLEEVE	SAE 1020
3	12	85741K4R	8-32UNC-2 X .500 HEX SOCKET HEAD CAP SCREW	SAE 4320
2	2	30-004579-02	HIGH PRESSURE RACK PAD	BLACK UHMW
1	1	30-004579-01	MAIN MOUNTING PLATE	AL ALY 6061-T6
FIND NO	QTY REQD	PART OR IDENT NO	NOMENCLATURE OR DESCRIPTION	NOTES OR REMARKS
PARTS LIST				

UNLESS OTHERWISE SPECIFIED DIMENSIONS ARE IN INCHES (IN) TOLERANCES: 1 PLACE ±.1 2 PLACE ±.01 3 PLACE ±.005 4 PLACE ±.0050 ANGLES 30' FINISH 62 u IN	APPROVALS	DATE	ENGINEERING DRAFTING & DESIGN, INC. Drafting, design, and training for all disciplines. Integrity - Quality - Style
	DRAWN DPM	04-06	
	CHECKED ADM	04-09	TITLE
	APPROVED DAM	04-12	CRYOGENIC PLATE SUBASSEMBLY
THIRD ANGLE PROJECTION	MATERIAL NOTED		
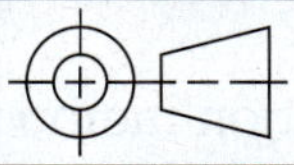	FINISH ALL OVER		SIZE C / CAGE CODE / DWG NO. 30-004579 / REV 0
	DO NOT SCALE DRAWING		SCALE 1:1 / SHEET 1 OF 5

Using Fields to Reference Attributes

FIELD

Ribbon
Insert
> Data
Field
Type
FIELD

Use fields to link text to attribute values. To display a field as an attribute value, access the **Field** dialog box from within the **MTEXT** or **TEXT** command, or use the **FIELD** command. In the **Field** dialog box, select **Objects** from the **Field category:** drop-down list, and select **Object** in the **Field names:** list box. Then pick the **Select object** button to return to the drawing window and select the block containing the attribute.

When you select the block, the **Field** dialog box reappears with the available properties (attributes) listed. Pick an attribute tag to display the corresponding value in the **Preview:** box. Select the format and pick the **OK** button to insert the attribute value as a field in the text object.

Supplemental Material

www.g-wlearning.com/CAD/

Extracting Attribute Data

For information about using attributes to create a table and exporting attribute data to an external file, go to the companion website, navigate to this chapter in the **Contents** tab, and select **Extracting Attribute Data**.

Chapter Review

Answer the following questions using the information in this chapter.

1. What is an attribute?
2. Explain the purpose of the **ATTDEF** command.
3. Describe the function of the following attribute modes:
   A. **Invisible**
   B. **Constant**
   C. **Verify**
   D. **Preset**
4. What is the purpose of the **Default:** text box in the **Attribute Definition** dialog box?
5. How can you edit attributes before including the attributes with a block?
6. How can you change the visibility of an existing attribute?
7. If you select attributes using the window or crossing selection method to define a block, in what order will attribute prompts appear?
8. What purpose does the **ATTREQ** system variable serve?
9. List the three options for attribute display.
10. Explain how to change the value of an inserted attribute.
11. What does *global attribute editing* mean?
12. After you save a block with attributes, what method can you use to change the order of prompts when you insert the block?
13. List examples of detailing or documentation tasks that attributes can help automate.
14. What element of an assembly drawing provides information about each component of the assembly or subassembly?
15. How can you link text to an attribute value?

Drawing Problems

Start AutoCAD if it is not already started. Start a new drawing for each problem using an appropriate template of your choice. The template should include layers and text, dimension, multileader, and table styles, when necessary, for drawing the given objects. Add layers and text, dimension, multileader, and table styles as needed. Draw all objects using appropriate layers. Use appropriate text, dimension, multileader, and table styles, justification, and format. Follow the specific instructions for each problem. Use only drawing and editing commands and techniques you have already learned. Use your own judgment and approximate dimensions when necessary. Apply dimensions accurately using ASME or appropriate industry standards.

Note: Some of the problems in this chapter are built on problems from previous chapters. If you have not yet completed those problems, complete them now.

▼ Basic

1. **General** Use a word processor to list each attribute mode. Provide a description of each.

2. **Architectural** Draw the structural steel wide flange shapes shown. Do not dimension the shapes. Create attributes using the information given. Make a block of each shape. Name the first block W12 X 40. Name the second block W12 X 30. Insert each block once to test the attributes. Save the drawing as P25-2.

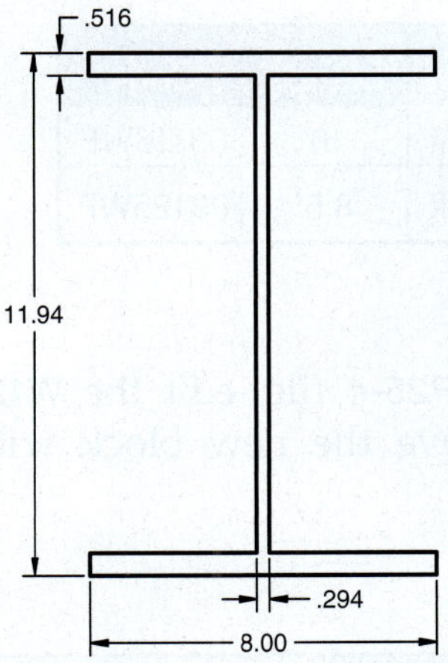

Attributes			
	Steel	W12 × 40	Constant, Visible
	Mfr.	Ryerson	Invisible
	Price	Quote	Invisible
	Weight	40 lbs/ft	Constant, Invisible
	Length	10′	Invisible
	Code	03116WF	Invisible

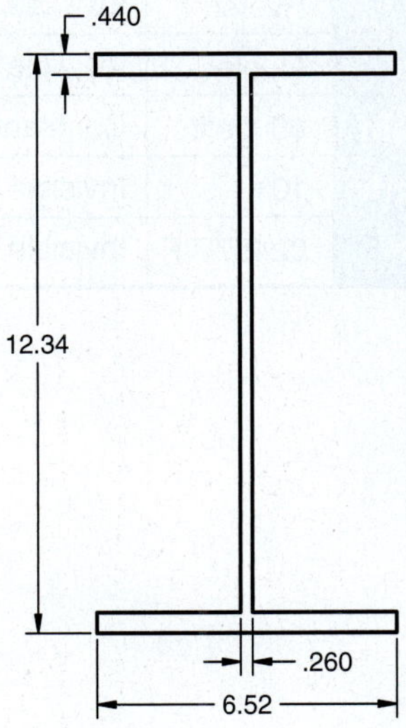

Attributes			
	Steel	W12 × 30	Constant, Visible
	Mfr.	Ryerson	Invisible
	Price	Quote	Invisible
	Weight	30 lbs/ft	Constant, Invisible
	Length	10′	Invisible
	Code	03125WF	Invisible

▼ Intermediate

3. **Architectural** Save P25-2 as P25-3. In the P25-3 file, construct the floor plan shown. Insert the W12 X 40 and W12 X 30 blocks as shown. The chart below the drawing provides the required attribute data. Enter the appropriate information for the attributes as prompted. You can speed the drawing process by using **ARRAY** or **COPY**. Dimension the drawing.

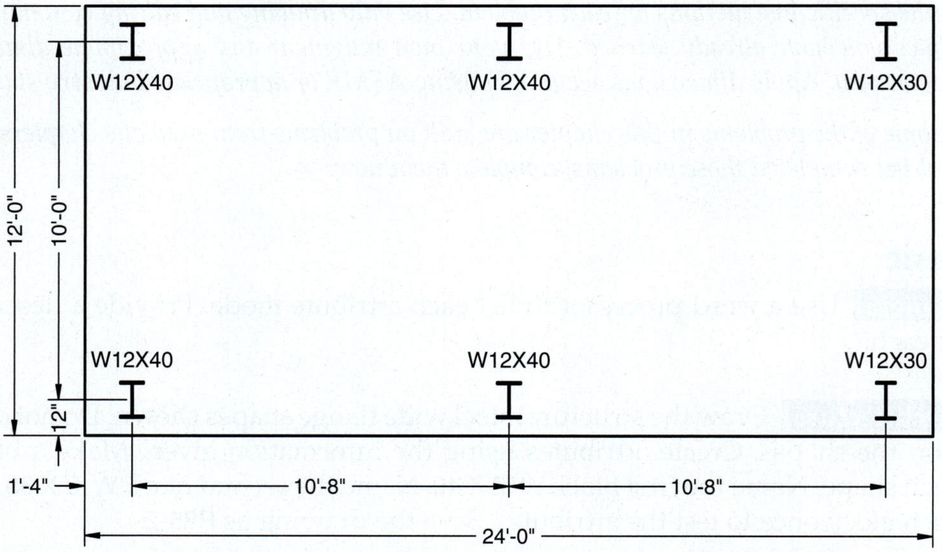

Steel	Mfr.	Price	Weight	Length	Code
W12 × 40	Ryerson	Quote	40 lbs/ft	10′	03116WF
W12 × 30	Ryerson	Quote	30 lbs/ft	8.5′	03125WF

4. **Architectural** Save P25-2 as P25-4. In the P25-4 file, edit the W12 X 40 block according to the following information. Save the new block with the name W10 X 60.

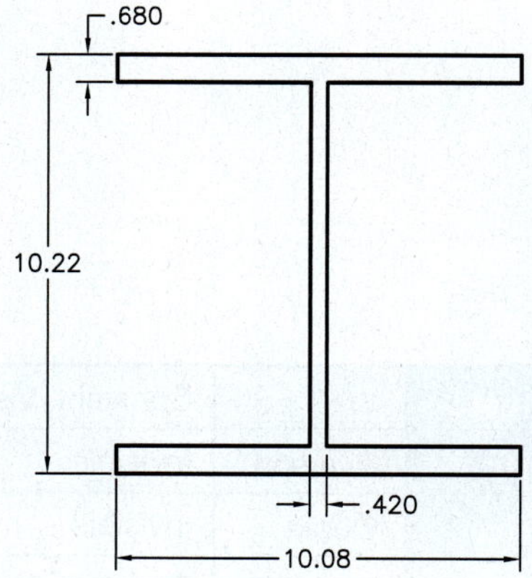

Attributes			
	Steel	W10 × 60	Constant, Visible
	Mfr.	Ryerson	Invisible
	Price	Quote	Invisible
	Weight	60 lbs/ft	Constant, Invisible
	Length	10′	Invisible
	Code	02457WF	Invisible

▼ Advanced

5. **Architectural** Save P8-21 as P25-5. If you have not yet completed Problem 8-21, do so now. In the P25-5 file, add constant, invisible attributes to each view of the glulam (glued laminated) beam hanger to identify all necessary specifications. Create a block of each view with the attributes. Insert the block of each view.

6. **Architectural** Save P11-17 as P25-6. If you have not yet completed Problem 11-17, do so now. In the P25-6 file, add constant, invisible attributes to each view of the mudsill anchor to identify all necessary specifications. Create a block of each view with the attributes. Insert the block of each view.

7. **Architectural** Save P25-3 as P25-7. In the P25-7 file, create a tab-separated extraction file for the blocks in the drawing. Extract the following information for each block:
   - Block name
   - Steel
   - Manufacturer
   - Price
   - Weight
   - Length
   - Code

   Save the tab-separated extraction file as P25-7.

8. **Architectural** Save P25-2 as P25-8. In the P25-8 file, create a table from the block attribute data and insert the table into the drawing.

9. **General** Select a drawing from Chapter 24 and create a bill of materials for the drawing using the **Data Extraction** wizard. Use the comma-separated format to display the file. Display the file in Windows Notepad. Save the drawing and the comma-separated extraction file as P25-9.

10. **General** Create a drawing of the computer workstation layout in the classroom or office in which you are working. Provide attribute definitions for all of the items listed here.
    - Workstation ID number
    - Computer brand name
    - Model number
    - Processor chip
    - Amount of RAM
    - Hard disk capacity
    - Video graphics card brand and model
    - CD-ROM/DVD-ROM speed
    - Date purchased
    - Price
    - Vendor phone number
    - Other data as you see fit

    Generate and extract a file for all of the computers in the drawing. Save the drawing and extracted file as P25-10. Prepare a slide show presentation of your research, design process, and drawing and present it to your class or office.

AutoCAD Certified Professional Exam Practice

Answer the following questions using the information in this chapter.

1. Which group of **ATTDEF** settings is most efficient for an attribute that is not expected to change and that will be used only to tabulate data? *Select the one item that best answers the question.*
   A. **Invisible, Constant, Lock position**
   B. **Invisible, Verify, Preset**
   C. **Visible, Constant, Preset**
   D. **Visible, Constant, Verify**
   E. **Visible, Verify, Lock position**
2. Which of the following tools can you use to change attribute modes after an attribute has been created? *Select all that apply.*
   A. **-ATTEDIT**
   B. **ATTREDEF**
   C. **BATTMAN**
   D. **EATTEDIT**
   E. **Properties** palette
3. Which of the following commands allow you to add attributes to a block without exploding the block? *Select all that apply.*
   A. **ATTREDEF**
   B. **BEDIT**
   C. **EATTEDIT**
   D. **FIELD**
   E. **REFEDIT**

Follow the instructions in the problem. Complete the problem by answering the related question.

4. **Navigate to this chapter on the companion website and open CPE-25attribute.dwg.** Insert the existing BUSH block into the drawing. Without exploding the block, change the attribute mode from **Invisible** to **Visible**, and change the text height to 12. What type of plant is this, and how much does it cost?

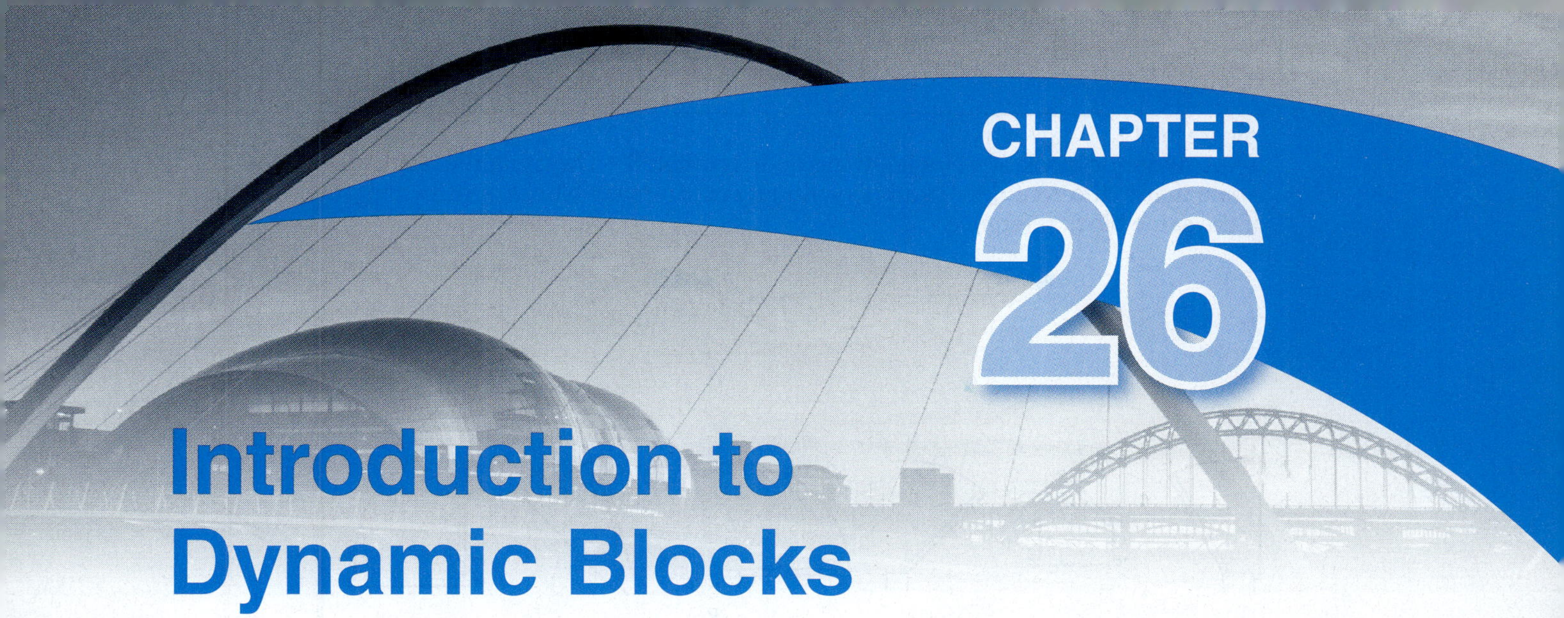

CHAPTER 26

Introduction to Dynamic Blocks

Learning Objectives

After completing this chapter, you will be able to:

- ✓ Explain the function of dynamic blocks.
- ✓ Assign action parameters and actions to blocks.
- ✓ Modify parameters and actions.
- ✓ Assign value sets to parameters.
- ✓ Create and use chain actions.

A standard block typically represents a very specific item, such as a hex head bolt that is 1″ long. In this example, if the same hex head bolt is available in three other lengths, you must create three additional standard blocks. An alternative is to create a single ***dynamic block*** that can be adjusted to show each different bolt length. Dynamic blocks increase productivity and reduce the size of symbol libraries, making symbol libraries more manageable.

dynamic block: An adjustable block to which you can assign parameters, actions, and geometric constraints and constraint parameters.

Dynamic Block Fundamentals

A dynamic block is a parametric symbol that you can adjust to change size, shape, and geometry without drawing additional blocks, and without affecting other instances of the block reference. Figure 26-1 shows an example of a dynamic block of a plan view single-swing door symbol. In this example, the dynamic properties of the block allow you to create many different single-swing door symbols according to specific parameters, such as door size, wall thickness, swing location, swing angle representation, wall angle, and exterior or interior usage.

The process of constructing and using dynamic blocks is identical to the process of creating standard blocks, except for the addition of ***action parameters*** and (usually) ***actions*** that control block geometry. AutoCAD refers to action parameters as ***parameters*** in the context of dynamic blocks. A dynamic block can contain multiple parameters, and a single parameter can include multiple actions. You can use geometric constraints and ***constraint parameters*** as an alternative or in addition to parameters and actions.

action parameter (parameter): A specification for block construction that controls block characteristics such as positions, distances, and angles of dynamic block geometry.

action: A definition that controls how dynamic block parameters behave.

constraint parameters: Dimensional constraints that control the size or location of block geometry numerically.

Figure 26-1.
A—A dynamic block of a single-swing door symbol. B—The dynamic block allows you to create many different door symbols without creating new blocks or affecting other instances of the same block reference.

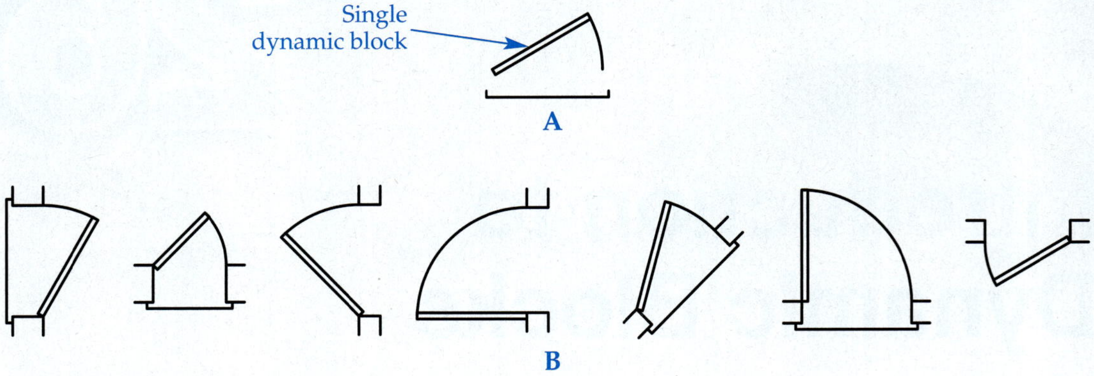

parameter grips: Special grips that allow you to change the parameters of a dynamic block.

Figure 26-2A shows an example of a bolt symbol created as a dynamic block and selected for grip editing. The bolt shaft objects include a linear parameter with a stretch action, as indicated by the ***parameter grips***. The length of the bolt increases when you stretch the right-hand linear parameter grip to the right. See **Figure 26-2B**.

NOTE

As you learn to create and use dynamic blocks, you will notice that many actions function like editing commands with which you are already familiar, allowing operations such as stretch, move, scale, array, and rotate.

PROFESSIONAL TIP

AutoCAD includes several files of dynamic block symbols in the Program Files/Autodesk/AutoCAD 2020/Sample/en-us/Dynamic Blocks folder. The **Tool Palettes** palette also includes sample dynamic blocks. Explore the sample symbols as you learn to create and use dynamic blocks.

Figure 26-2.
A linear parameter with a stretch action assigned to the shaft objects in the block of a bolt. A—Selecting the block displays the linear parameter grips. B—Selecting and moving a linear parameter grip stretches the bolt shaft.

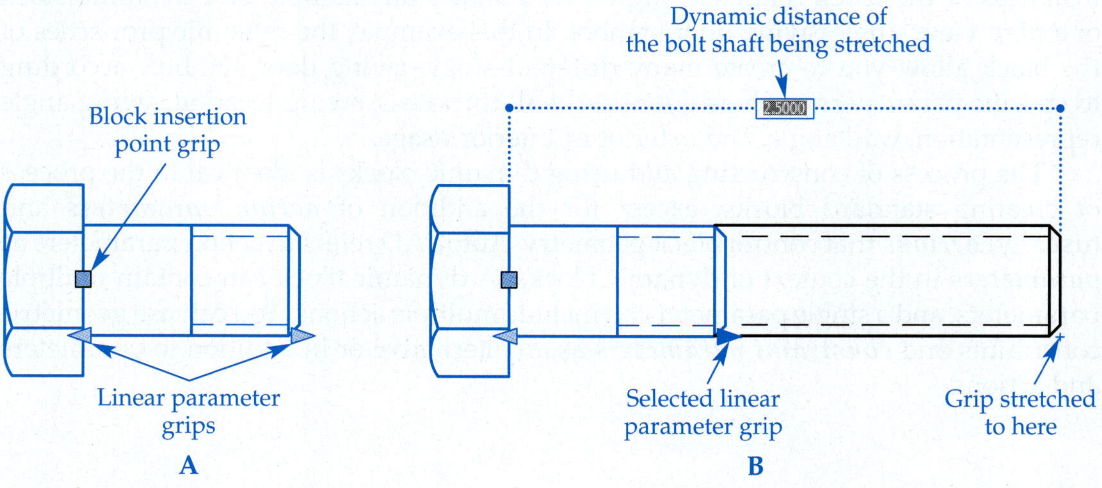

Assigning Dynamic Properties

Edit an existing block in the **Block Editor** to assign dynamic properties. Access the **BEDIT** command to display the **Edit Block Definition** dialog box shown in **Figure 26-3** and select the block from the list box. A preview and the description of the selected block appear. Pick the **OK** button to open the selection in the **Block Editor**. See **Figure 26-4**.

To create a dynamic block from scratch within the **Block Editor**, type a name for the new block in the **Block to create or edit** text box. To edit a block saved as the current drawing, such as a wblock, pick the **<Current Drawing>** option.

Ribbon
Home
> Block
Insert
> Block Definition

Block Editor

Type
BEDIT

BEDIT

Figure 26-3.
The **Edit Block Definition** dialog box.

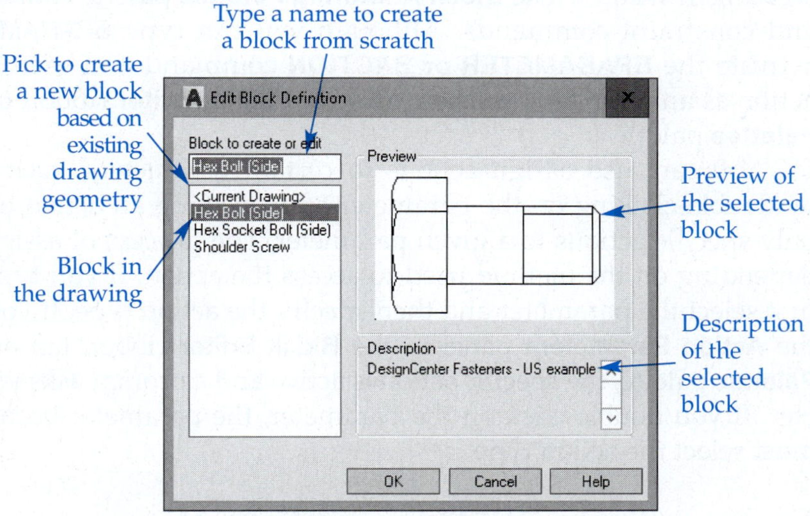

Figure 26-4.
The **Block Editor** ribbon tab and the **Block Authoring Palettes** palette are available in block editing mode. Notice the drawing window background, which has changed color to indicate block editing mode, and the location of the block relative to the origin.

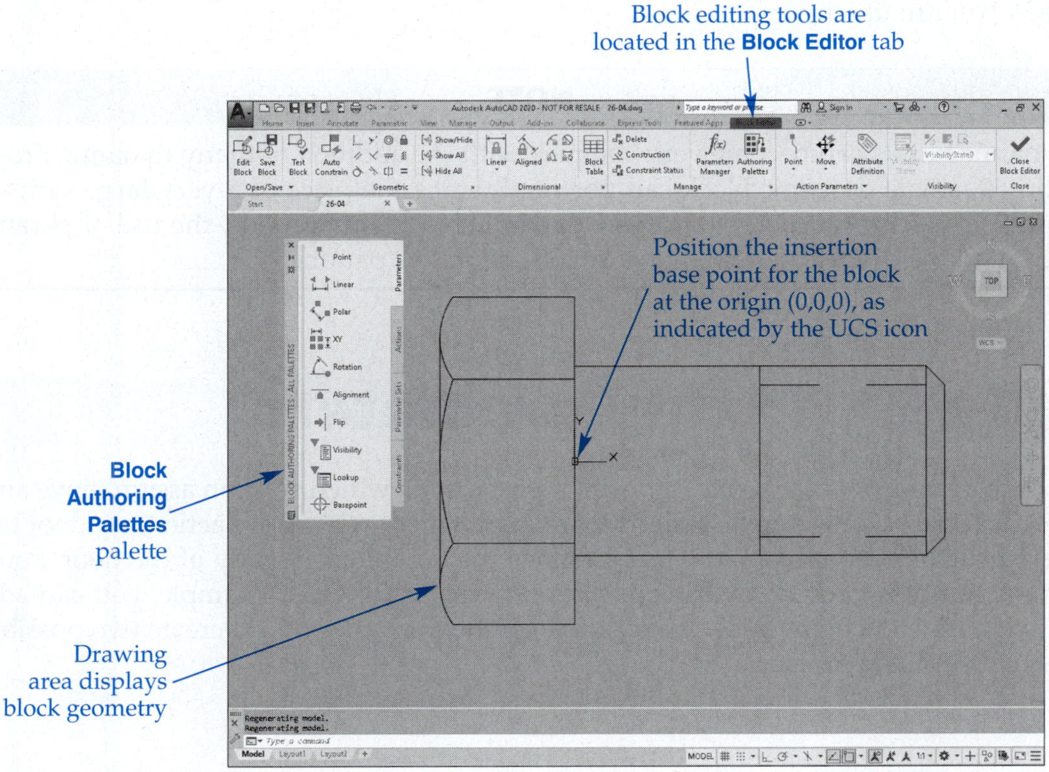

NOTE

Double-click on a block to display the **Edit Block Definition** dialog box with the block selected. You can open a block directly in the **Block Editor** by selecting the block, right-clicking, and selecting **Block Editor**. Another option is to open a block directly in the **Block Editor** during block creation by selecting the **Open in block editor** check box in the **Block Definition** dialog box.

The **Block Editor** ribbon tab and **Block Authoring Palettes** palette provide easy access to commands and options for assigning dynamic block properties and creating attributes. The **Block Authoring Palettes** palette contains parameter, action, and constraint commands. Although you can type BPARAMETER or BACTION to activate the **BPARAMETER** or **BACTION** command and then select a parameter or action as an option, it is easier to use the **Block Editor** ribbon tab or **Block Authoring Palettes** palette.

You can also assign actions to certain parameters, such as point parameters, by double-clicking on the parameter and selecting an action option. You can assign only specific actions to a given parameter. The process of assigning an action varies, depending on the method used to access the action. If you type BACTION, you must first select the parameter and then specify the action type. If you pick the action from the **Action Parameters** panel in the **Block Editor** ribbon tab or the **Block Authoring Palettes** palette, the specific action is active and a prompt asks you to select the parameter. If you double-click on the parameter, the parameter becomes selected, but you must select the action type.

Saving a Block with Dynamic Properties

BSAVE
Ribbon
Block Editor > Open/Save
Save Block
Type
BSAVE

BSAVEAS
Ribbon
Block Editor > Open/Save
Save Block As
Type
BSAVEAS

Once you add one or more parameters to a block and assign actions to the parameters, you are ready to save and use the dynamic block. Use the **BSAVE** command to save the block, or use the **BSAVEAS** command to save the block using a different name. Remember that saving changes to a block updates all blocks of the same name in the drawing. Use the **BCLOSE** command to exit the **Block Editor** when you are finished.

NOTE

Dynamic blocks can become very complex with the addition of many dynamic properties. A single dynamic block can potentially take the place of a very large symbol library. This chapter focuses on basic dynamic block applications, the use of parameters, and the process of assigning a single action to a parameter.

Point Parameters

point parameter: A parameter that defines an XY coordinate location in the drawing.

A ***point parameter*** creates a position property to which you can assign move and stretch actions. For example, assign a point parameter with a move action to a door tag that is part of a door block so you can move the tag independently of the door. Point parameters also provide multiple insertion point options. For example, you can add point parameters to the ends of a welding symbol reference line to create two possible insertion point options.

Figure 26-5.
A point parameter consists of the grip location and a label. This example shows adding a point parameter to the tag of a plan view door symbol to allow the tag to be positioned independently of the door.

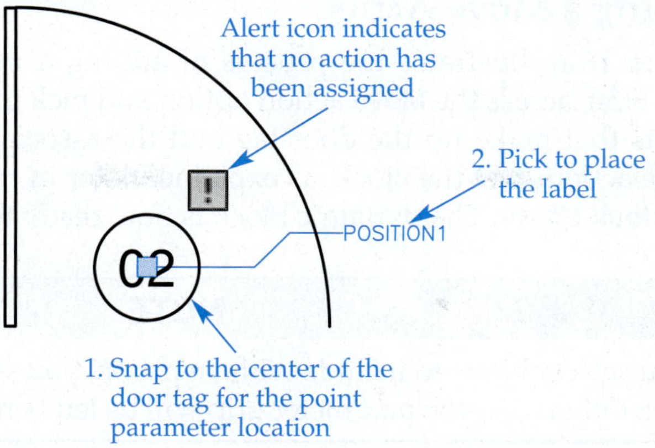

Figure 26-5 provides an example of adding a point parameter. Access the **Point** parameter option and specify a location for the parameter. The parameter location determines the base point from which dynamic actions occur. **Figure 26-5** shows picking the center of the door tag circle to identify the base point of a move action. Adjust the parameter location after initial placement if necessary. The yellow alert icon indicates that no action is assigned to the parameter.

Ribbon
Block Editor
> Action
Parameters

Point

Type
BPARAMETER
> Point

After you specify the parameter location, pick a location for the ***parameter label.*** The label appears only in block editing mode. By default, the label for the first point parameter is Position1. Move the label as needed after initial placement.

parameter label: A label that indicates the purpose of a parameter.

Next, enter the number of grips to associate with the parameter. The default **1** option creates a single grip at the parameter location that allows you to use grip editing to carry out the assigned action. If you select the **0** option, you can only use the **Properties** palette to adjust the block.

Parameter options are available before you specify the parameter location. Most of the options are also available from the **Properties** palette if you have already created the parameter. Use the **Label** option to enter a more descriptive label name. The **Name** option allows you to specify a name for the parameter that displays as the Parameter type property in the **Properties** palette. The **Chain** option specifies whether a chain action can affect the parameter. Chain actions are described later in this chapter. The **Description** option allows you to type a description, such as the purpose of or application for the parameter. The description displays in the drawing area as a tooltip. The **Palette** option determines whether the label appears in the **Properties** palette when you select the block.

PROFESSIONAL TIP

Change the parameter label name to something more descriptive, and add a description. This helps you identify each parameter when you are controlling a block dynamically. It is especially important to change the parameter label name and add a description when you add multiple parameters to a block. Consider keeping the default parameter type as part of the name. For example, change the name of the door symbol point parameter from Point to Point – Tag Center.

Exercise 26-1

www.g-wlearning.com/CAD/

Complete the exercise on the companion website.

Assigning a Move Action

Ribbon
Block Editor
> Action
Parameters

Move

Type
BACTIONTOOL
> Move

move action: An action used to move a block object independently of other objects in the same block.

Figure 26-6 illustrates the process of adding a ***move action*** to the door block example. First, access the **Move** action option and pick the point parameter. Then select the objects that make up the door tag and the associated parameter. Press [Enter] to assign the action. Test the block, as explained later in this chapter. Save the block and exit the **Block Editor**. The dynamic block is now ready to use.

NOTE

When you select objects to include with an action, you should also select the associated parameter. Otherwise, the parameter grip will be left behind when you apply the action.

PROFESSIONAL TIP

After you create an action, use the **Properties** palette to change the action name to something more descriptive, but keep the default action type with the name. For example, change the name of the door symbol move action from Move to Move – Tag.

Using a Move Action Dynamically

Figure 26-7A shows the door block reference, selected for editing. The point parameter grip appears as a light blue square in the center of the door tag. The insertion base point specified when the block was created appears as a standard unselected grip. Select the point parameter grip and move the door tag as shown in Figure 26-7B. Pick a point to specify a new location for the door tag. See Figure 26-7C.

NOTE

During block insertion, press [Ctrl] to cycle through the positions of any parameters added to the dynamic block. This is one method of selecting a different insertion base point, corresponding to the position of a parameter, to use when inserting the block.

PROFESSIONAL TIP

After block insertion, use the **Properties** palette to adjust dynamic properties. If you do not include grips with a parameter, the options in the **Custom** category are the only way to adjust the block.

Figure 26-6.
Assigning a move action to a point parameter.

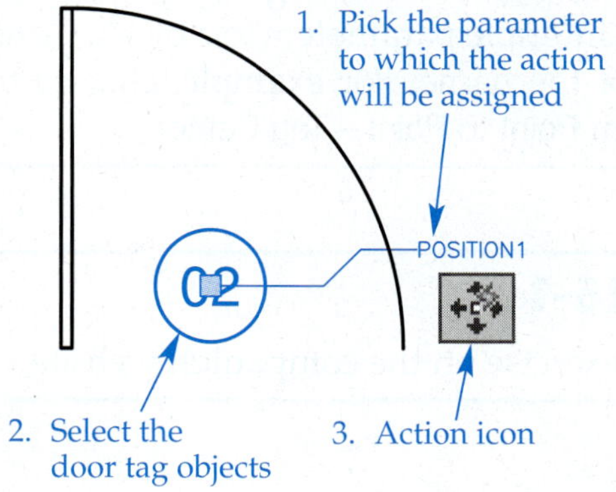

Figure 26-7.
Dynamically moving an action assigned to a point parameter. A—Select the block to display grips. The point parameter grip is shown as a light blue square. B—Select and move the point parameter grip. C—The door tag is at a new location, but it is still part of the block.

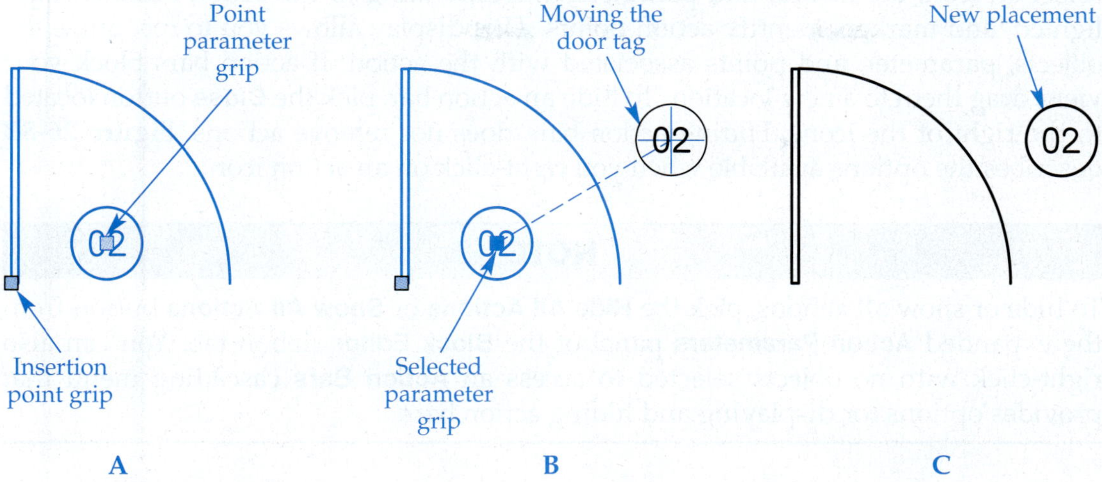

Testing and Adjusting Dynamic Properties

After you exit the **Block Editor**, you can insert a block and use grip editing or the **Properties** palette to confirm appropriate dynamic function. If the block does not respond as desired, however, you must re-enter the **Block Editor** to make changes. A more convenient option is to access the **BTESTBLOCK** command from inside the **Block Editor** to enter the **Test Block Window**. The **Test Block Window** provides standard AutoCAD commands and options, allowing you to test dynamic function without exiting block editing mode. Pick the **Close Test Block** button to re-enter the **Block Editor**. AutoCAD discards changes made while testing so that you can adjust the original block as needed. Block testing is especially important when a block includes multiple dynamic properties.

> **BTESTBLOCK**
> Ribbon
> **Block Editor > Open/Save**
> **Test Block**
> Type
> **BTESTBLOCK**

Adjusting Parameters

Use standard editing commands such as **MOVE** to make changes to existing parameter labels or grips. You can move parameter grips independently of the parameter label, which is often required if multiple grips are stacked or are near the same location. Use the **ERASE** command to remove a parameter or parameter grips.

Grip editing is especially effective for adjusting parameters. When you select a parameter, grips appear at the parameter location and label. Use the **Properties** palette to adjust the properties of the selected parameter. The settings in the **Properties** palette change depending on the type of parameter. Limited property options are also available by selecting a parameter and right-clicking. Use the **Grip Display** cascading menu to redefine the number of grips or to relocate grips with the parameter location. Use the **Rename Parameter** option to change the name of the label.

Adjusting Actions

action bars: Toolbars that allow you to view, remove, and adjust actions.

Action bars appear by default when you assign actions. See Figure 26-8A. Each action displays an icon to identify and control the action. When you hover over or select an icon, the objects and parameter corresponding to the action become highlighted, and markers identify action points. This display allows you to recognize the objects, parameter, and points associated with the action. If action bars block your view, drag them to a new location. To hide an action bar, pick the **Close** button located to the right of the icons. Hiding action bars does not remove actions. Figure 26-8B describes the options available when you right-click on an action icon.

NOTE

To hide or show all actions, pick the **Hide All Actions** or **Show All Actions** button from the expanded **Action Parameters** panel of the **Block Editor** ribbon tab. You can also right-click with no objects selected to access an **Action Bars** cascading menu that provides options for displaying and hiding action bars.

Figure 26-8.
A—Action bars appear by default when you add actions. B—Options for adjusting actions when you right-click on an action icon.

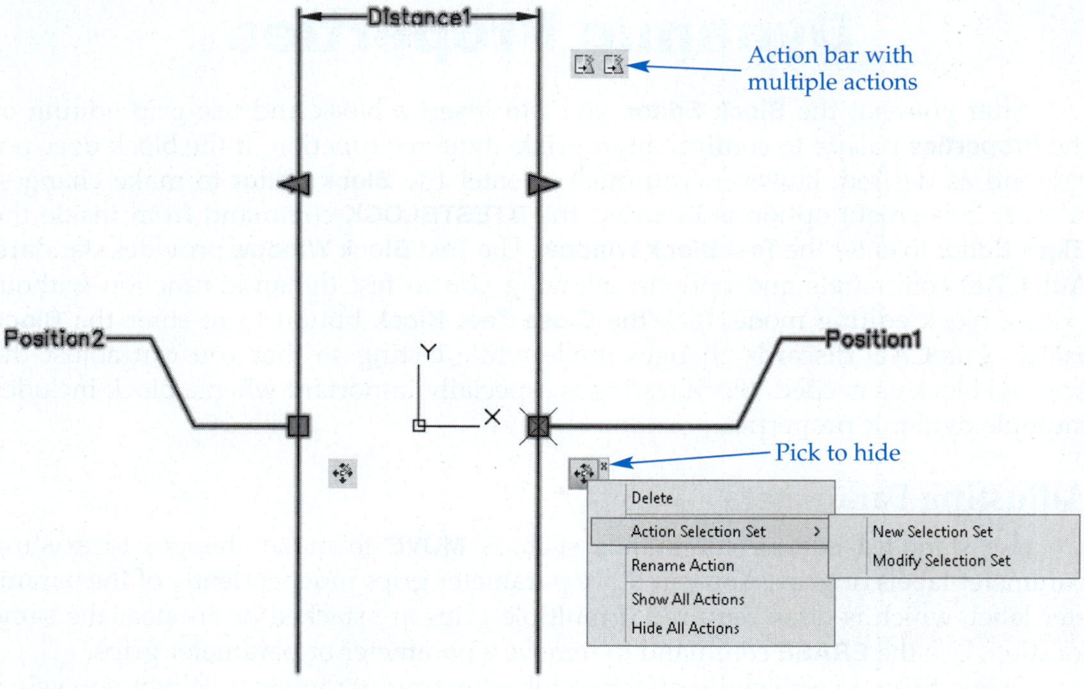

A

Option	Description
Delete	Deletes the action.
New Selection Set	Allows you to select objects to associate with the action; eliminates the original selection set.
Modify Selection Set	Adds or removes objects associated with the action.
Rename Action	Provides a text box for renaming the action.
Show All Actions	Displays all actions.
Hide All Actions	Hides all actions.

B

Exercise 26-2

www.g-wlearning.com/CAD/

Complete the exercise on the companion website.

Linear Parameters

A ***linear parameter*** creates a distance property to which you can assign move, scale, stretch, and array actions. For example, assign a linear parameter with a stretch action to a block of a bolt symbol to make the bolt shaft longer or shorter. Assign a second linear parameter and stretch action to the bolt head to control the bolt head diameter.

linear parameter: A parameter that creates a measurement reference between two points.

Ribbon
Block Editor
> Action
Parameters
Linear
Type
BPARAMETER
> Linear

Figure 26-9 provides an example of adding a linear parameter. For this example, activate the **Linear** parameter option and use the **Label** option to name the linear parameter Shaft Length. Next, pick the start point and endpoint of the linear parameter to determine the locations from which dynamic actions occur. If you plan to assign a single action to the parameter, select the point associated with the action second. Figure 26-9 shows picking the endpoint of the lower edge of the shaft and then using polar tracking or the **Extension** object snap to pick the point where the edge of the shaft would meet the end if extended. You must select points that are horizontal or vertical to each other to create a horizontal or vertical linear parameter.

Once you select the start point and endpoint, pick a location for the parameter label. Next, enter the number of grips to associate with the parameter. The default **2** option creates grips at the start point and endpoint, allowing you to use grip editing to carry out the action assigned to either point. Select the **1** option to assign a grip at the endpoint only, as shown in Figure 26-9. You will be able to grip-edit the block only if an action is associated with the endpoint. If you select the **0** option, you can adjust the block only by using the **Properties** palette.

Name, **Label**, **Chain**, **Description**, **Base**, **Palette**, and **Value set** options are available before you specify points. Most of these options are also available from the **Properties** palette if you have already created the parameter. The **Base** option allows you to assign the start point or midpoint of the linear parameter as the action base point. The **Value set** option allows you to specify values for the action. The **Base** and **Value set** options are described later in this chapter.

Figure 26-9.
Defining a linear parameter.

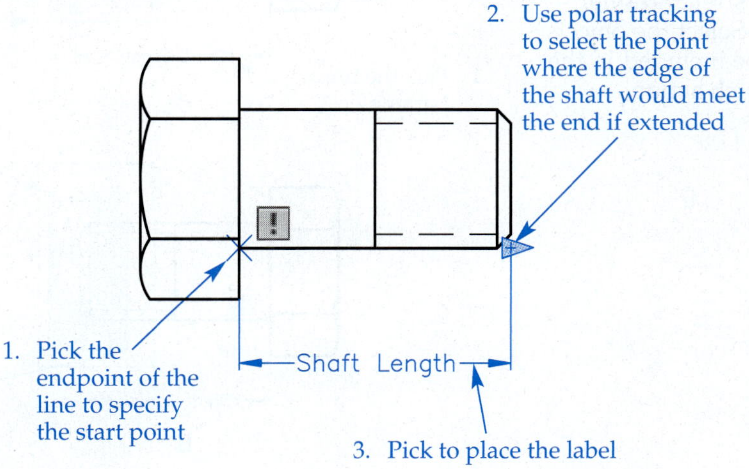

Assigning a Stretch Action

Ribbon
Block Editor
> Action Parameters
Stretch
Type
BACTIONTOOL
> Stretch

stretch action: An action used to change the size and shape of block objects with a stretch operation.

Figure 26-10 illustrates the process of adding a ***stretch action*** to the bolt symbol example. Access the **Stretch** action option and pick the Shaft Length parameter. Then specify a parameter point to associate with the action. Move the crosshairs near the appropriate parameter point to display the red snap marker, and pick to select. Alternatives include selecting the **sTart point** option to pick the start point of the linear parameter or the **Second point** option to select the endpoint. If you plan to use grip editing to control the block, and added a single grip, specify the point with the grip.

Next, create a window to define the stretch frame. This is the same technique you apply when using the **STRETCH** command. See Figure 26-10A. Then pick the objects to stretch, including the associated parameter. You do not need to use a crossing window, because the previous operation defines the stretch. However, crossing selection is often quicker. See Figure 26-10B. Press [Enter] to place the action icon. Test and save the block, and exit the **Block Editor**. The dynamic block is now ready to use.

Using a Stretch Action Dynamically

Figure 26-11 shows the bolt block reference selected for editing. The linear parameter grip is a light blue arrow at the far end of the bolt shaft. The insertion base point specified when the block was created appears as a standard unselected grip. Select the parameter grip and stretch the shaft to the new length. Use dynamic input to view the stretch dimension, and enter an exact length value in the distance field. You can also use the **Properties** palette to define the distance.

PROFESSIONAL TIP

The dynamic input distance field is a property of the linear parameter, allowing you to enter an exact distance. To get the best results when using a linear parameter, it is important that you locate the first and second points correctly.

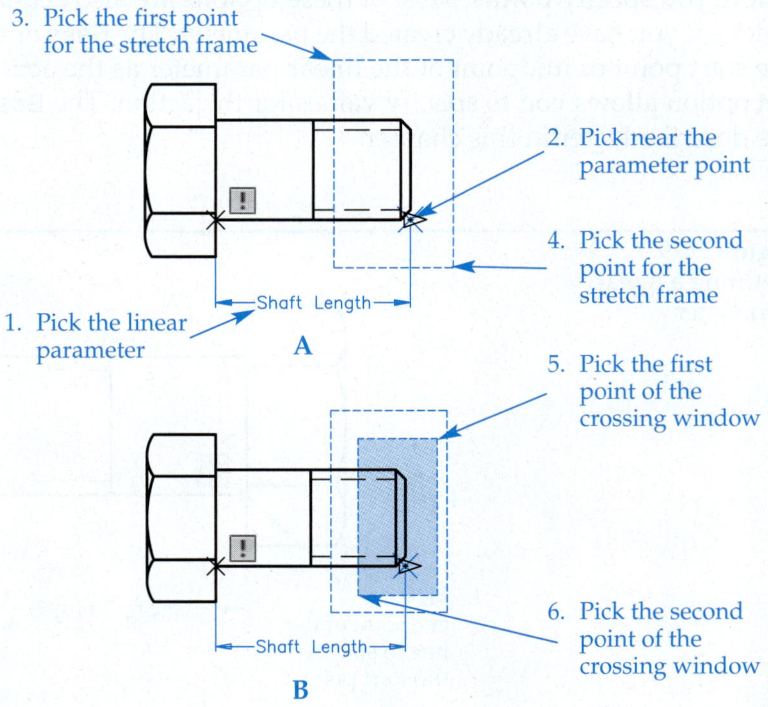

Figure 26-10. Assigning a stretch action to a linear parameter. A—Specify the parameter and parameter grip, and create a window for the stretch frame. B—Select the objects to be included in the stretch action.

Figure 26-11.
Selecting the inserted block in the drawing displays the parameter grip. Use the parameter grip to stretch the bolt shaft to the new length.

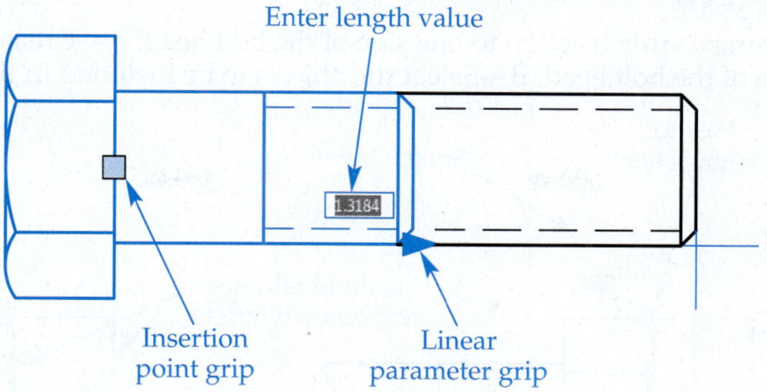

Exercise 26-3

www.g-wlearning.com/CAD/

Complete the exercise on the companion website.

Stretching Objects Symmetrically

The **Linear** parameter option includes a **Base** option that allows you to assign the start point or midpoint of the linear parameter as the action base point. Use the **Midpoint** setting to specify the midpoint as the action base point. The midpoint base point maintains symmetry when you adjust the block. You can set the base point preference before picking the first point or later using the **Properties** palette.

Figure 26-12 shows an example of a linear parameter with a stretch action assigned to the objects composing the bolt head. For this example, activate the **Linear** parameter option and use the **Base** option to select the **Midpoint** setting. Next, use the **Label** option to change the label name to Head Diameter. Select the start point and endpoint of the linear parameter to define the parameter and automatically calculate the midpoint. This example uses the upper-left and lower-left corners of the bolt head. After you select the start point and endpoint, pick a location for the parameter label. Enter the number of grips to associate with the parameter. The Figure 26-12 example uses the default **2** option to create grips at the start point and endpoint.

Figure 26-13 demonstrates the process of assigning a stretch action to one side of the bolt head. First, access the **Stretch** action option and pick the Head Diameter parameter. Then pick the upper linear parameter point to associate with the action. Create a crossing window to define the stretch frame, as shown in Figure 26-13A. Then select the objects to stretch, including the associated parameter, as shown in Figure 26-13B. Press [Enter] to place the action icon.

Figure 26-12.
The base point of a linear parameter appears as an X. Use the **Midpoint** option to locate the base point halfway between the start point and endpoint.

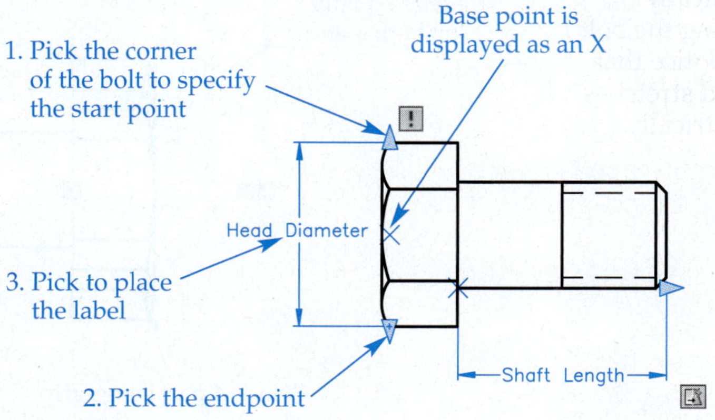

Figure 26-13.
Assigning a stretch action to one side of the bolt head. A—Create a crossing window around the top of the bolt head. B—Select the objects to be included in the stretch action.

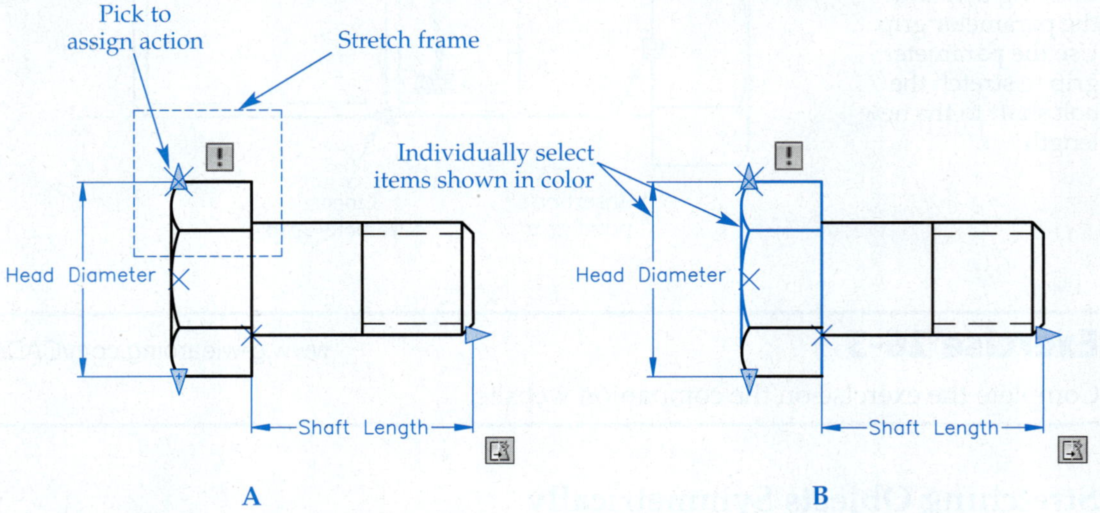

Repeat the previous sequence to assign a second stretch action to the opposite side of the bolt head. Test and save the block, and exit the **Block Editor**. The dynamic block is now ready to use. **Figure 26-14** illustrates using the lower grip point or dynamic input to stretch the bolt block reference. Use either grip to stretch the bolt head. You can also use the **Properties** palette to define the distance.

Assigning a Scale Action

scale action: An action used to scale some of the objects within a block independently of the other objects.

Figure 26-15 shows a block of a vanity symbol that includes a sink. In this example, a ***scale action*** is assigned to a linear parameter to adjust the size of the sink while maintaining the dimensions of the vanity. Activate the **Linear** parameter option and use the **Base** option to select the **Midpoint** setting. Next, use the **Label** option to change the label name to SINK LENGTH. Select the start point and endpoint of the linear parameter to define the parameter and automatically calculate the midpoint. This example uses two quadrants of the sink. After you select the start point and endpoint, pick a location for the parameter label. Next, select the number of grips to associate with the parameter. The **Figure 26-15** example uses the default **2** option to create grips at the start point and endpoint.

Figure 26-14.
Dynamically stretching the bolt head. Notice that the head stretches symmetrically.

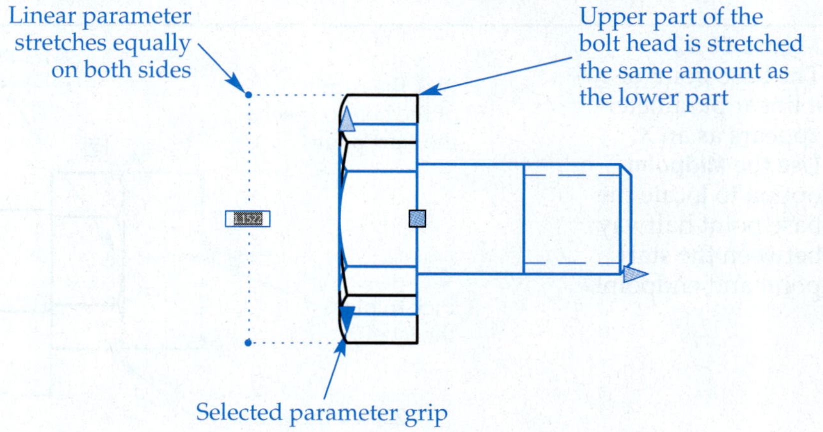

Figure 26-15. A linear parameter assigned to the sink objects in a block of a vanity with a sink. Locate the parameter base point and independent base type at the center of the sink to scale the sink symmetrically.

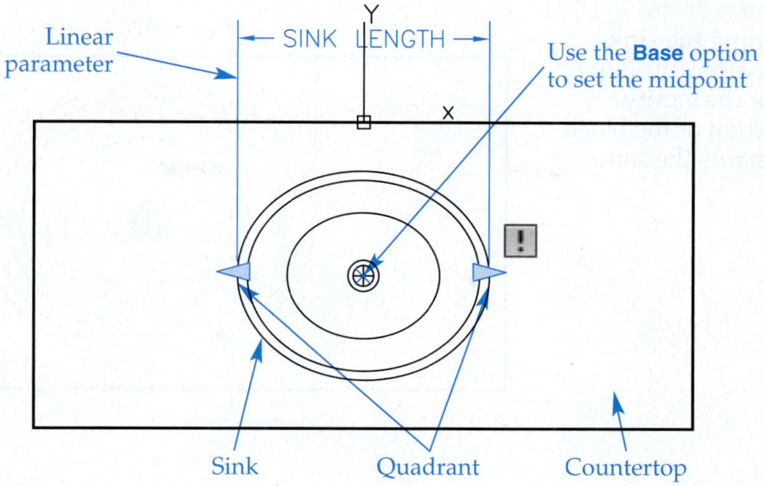

Ribbon
Block Editor > Action Parameters
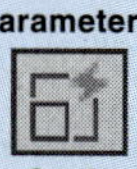
Scale
Type
BACTIONTOOL > Scale

Now assign a scale action to the parameter. First, access the **Scale** action option and pick the SINK LENGTH linear parameter. Then select the objects to include in the scale action, including the associated parameter. Press [Enter] to assign the action.

When using a scale action, it is critical to scale objects relative to the correct base point. Access the **Properties** palette and display the properties of the scale action. The **Overrides** category includes options for adjusting the base point. The default **Base type** option is **Dependent**, which scales the objects relative to the base point of the associated parameter. Select the **Independent** option to specify a different location. The **Base X** and **Base Y** values default to the parameter start point. Enter the coordinates relative to the block insertion base point, or use the pick button that appears when you select the **Base X** and **Base Y** values to select points on-screen. For the sink example, it is important that the objects be scaled relative to the exact center of the sink so that the sink will be centered within the vanity as the scale changes. Test and save the block, and exit the **Block Editor**. The dynamic block is now ready to use.

PROFESSIONAL TIP

In the previous example, the **Independent** option allows you to set the center of the sink as the base point for the scale action. This is necessary because the base point of the linear parameter is not the specified midpoint. The parameter uses a midpoint base to scale the parameter and parameter grips from the parameter midpoint. The **Independent** option of the scale action controls the point from which the geometry, not the parameter, is scaled.

Using a Scale Action Dynamically

Figure 26-16 shows using the right grip or dynamic input to scale the sink in the vanity block reference. You can use either grip to scale the sink. You can also use the **Properties** palette to define the distance.

Exercise 26-4

www.g-wlearning.com/CAD/

Complete the exercise on the companion website.

Figure 26-16.
Scaling the sink dynamically. Notice that the vanity portion of the block remains the same.

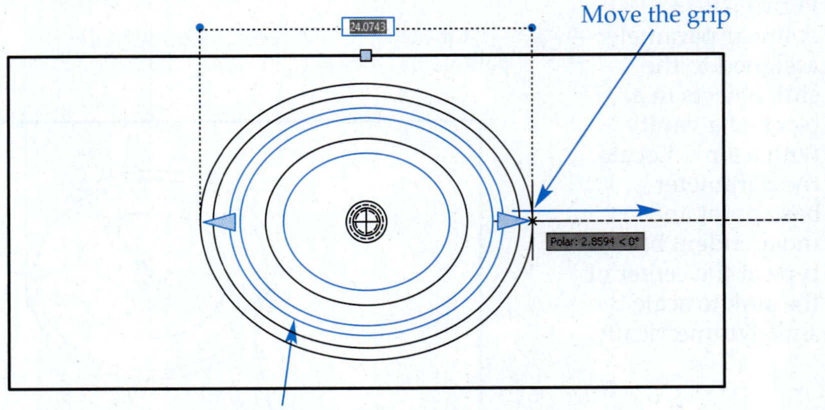

Polar Parameters

Ribbon
Block Editor
> Action Parameters
Polar
Type
BPARAMETER
> Polar

polar parameter: A parameter that includes a distance property and an angle property.

A ***polar parameter*** provides the same function as a linear parameter, but creates an angle, or rotation, property in addition to a distance property. You can assign move, scale, stretch, polar stretch, and array actions to a polar parameter. For example, assign a polar parameter with a polar stretch action to the steel channel shape shown in Figure 26-17 to adjust the depth to create different size channels, and rotate the block at the same time if necessary. See Figure 26-18.

To insert the polar parameter, access the **Polar** parameter option. Use the **Label** option to change the label name to Depth, but keep the default angle name. Then specify the base point, such as the left endpoint shown. Pick an endpoint aligned with the base point, such as the opposite endpoint shown. After you select the base point and endpoint, pick a location for the parameter label. Next, enter the number of grips to associate with the parameter. Select the **1** option to assign a grip at the endpoint only, as shown in Figure 26-17. You will be able to grip-edit the block only if an action is associated with the endpoint.

NOTE

Name, **Label**, **Chain**, **Description**, **Palette**, and **Value set** options are available before you specify the parameter. Most of these options are also available from the **Properties** palette if you have already created the parameter.

Figure 26-17.
Adding a polar parameter.

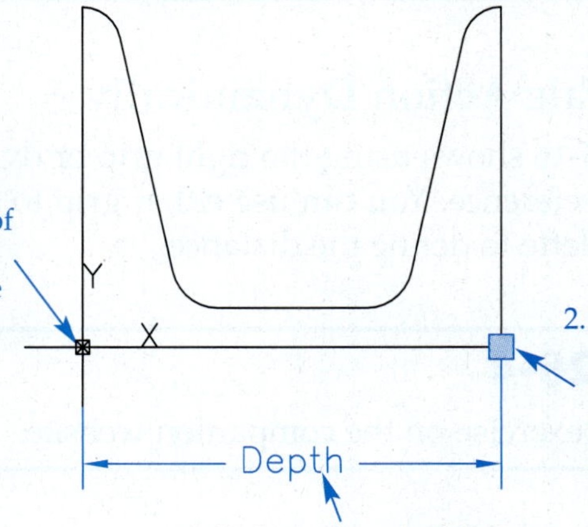

Figure 26-18.
An example of a portion of a motor mount with side views of different steel channel shapes that were created and rotated using a single block with a polar parameter assigned a polar stretch action.

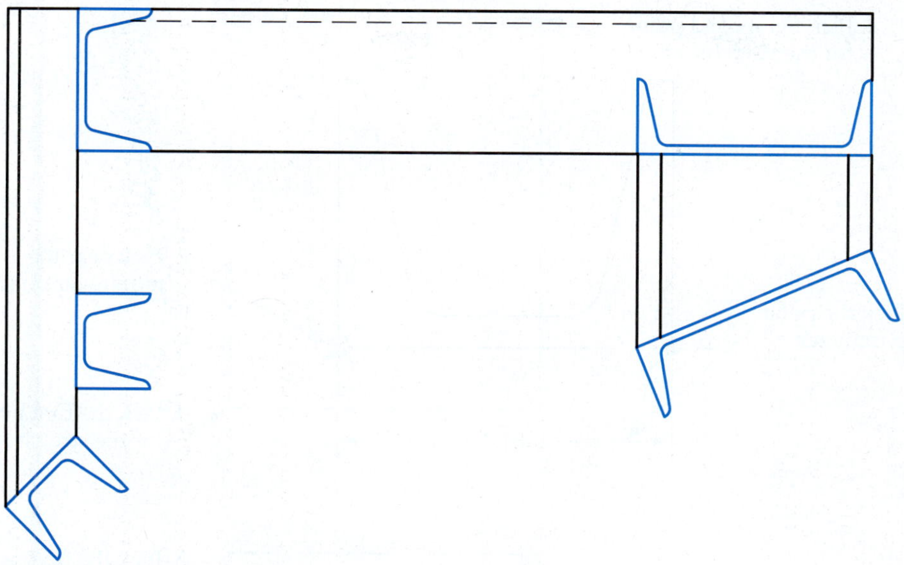

Assigning a Polar Stretch Action

Figure 26-19 illustrates the process of adding a ***polar stretch action*** to the steel channel shape example. Access the **Polar Stretch** action option and pick the Depth parameter. Then specify a parameter point to associate with the action. Move the crosshairs near the appropriate parameter point to display the red snap marker, and pick to select. Alternatives include selecting the **sTart point** option to pick the start point of the polar parameter, or the **Second point** option to select the endpoint. If you plan to use grip editing to control the block, and added a single grip, specify the point with the grip.

Ribbon
Block Editor
> Action Parameters
Polar Stretch
Type
BACTIONTOOL
> Polar Stretch

polar stretch action: An action used to change the size, shape, and rotation of block objects with a stretch operation.

Next, create a window to define the stretch frame. This is the same technique you apply when using the **STRETCH** command. See Figure 26-19A. Pick the objects to stretch, including the associated parameter. You do not need to use a crossing window, because the previous operation defines the stretch. However, crossing selection is often quicker. See Figure 26-19B. Then select the objects to rotate, which are often the objects of a block. See Figure 26-19C. Press [Enter] to place the action icon. Test and save the block, and exit the **Block Editor**. The dynamic block is now ready to use.

Using a Polar Stretch Action Dynamically

Figure 26-20 shows the steel channel shape block reference selected for editing. The polar parameter grip appears as a light blue square at the far end of the channel depth. The insertion base point specified when the block was created appears as a standard unselected grip. Select the parameter grip and stretch the channel depth to the new length and angle. Use dynamic input to view the stretch dimension and rotation angle, and enter an exact length value in the distance field and rotation angle in the angle field. You can also use the **Properties** palette to define the distance and rotation angle.

Figure 26-19.
Assigning a polar stretch action to a polar parameter. A—Specify the parameter and parameter grip, and create a window for the stretch frame. B—Select the objects to be included in the stretch action. C—Select the objects to be included in the rotation action.

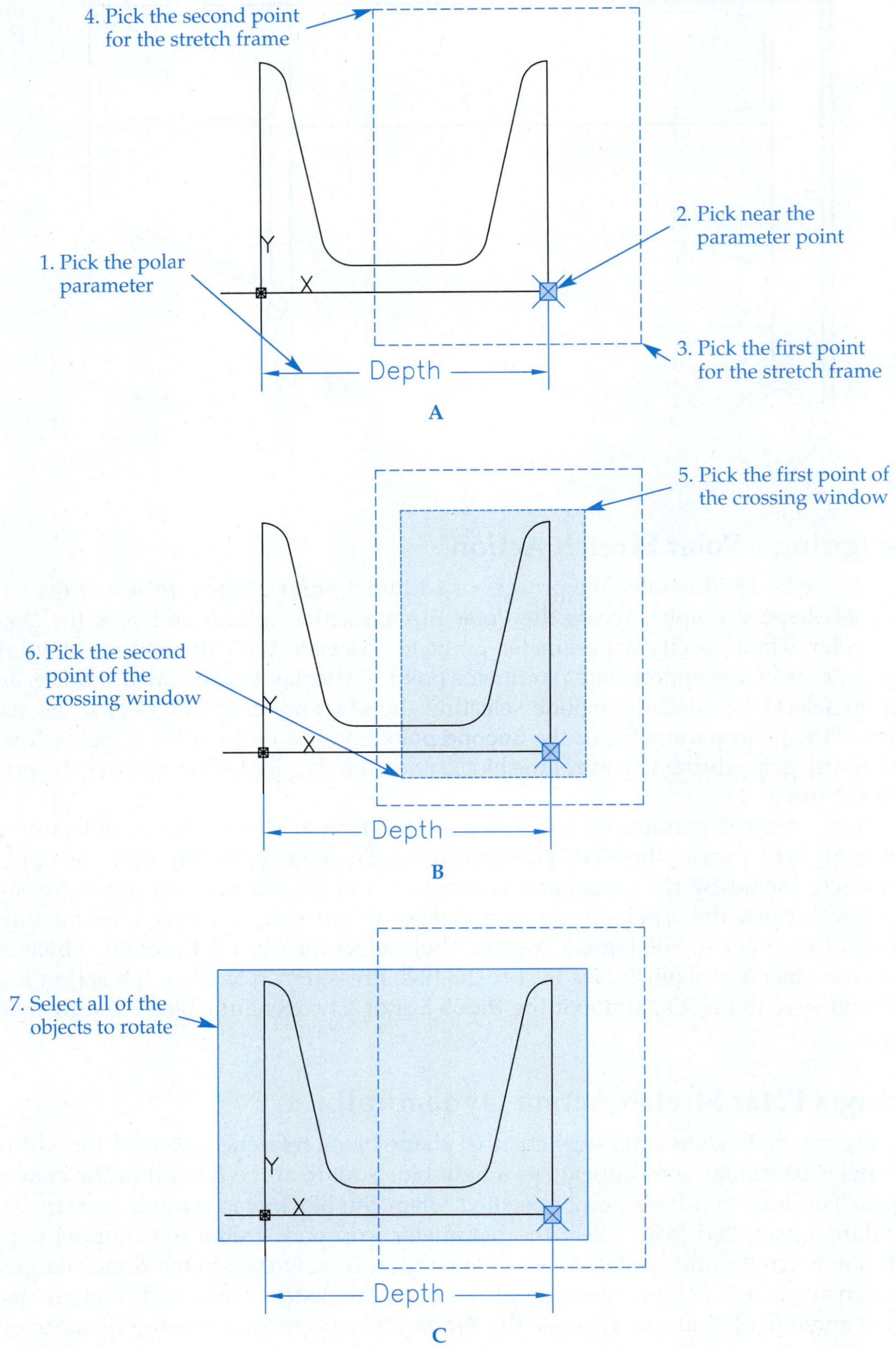

Figure 26-20.
Applying the polar stretch action.

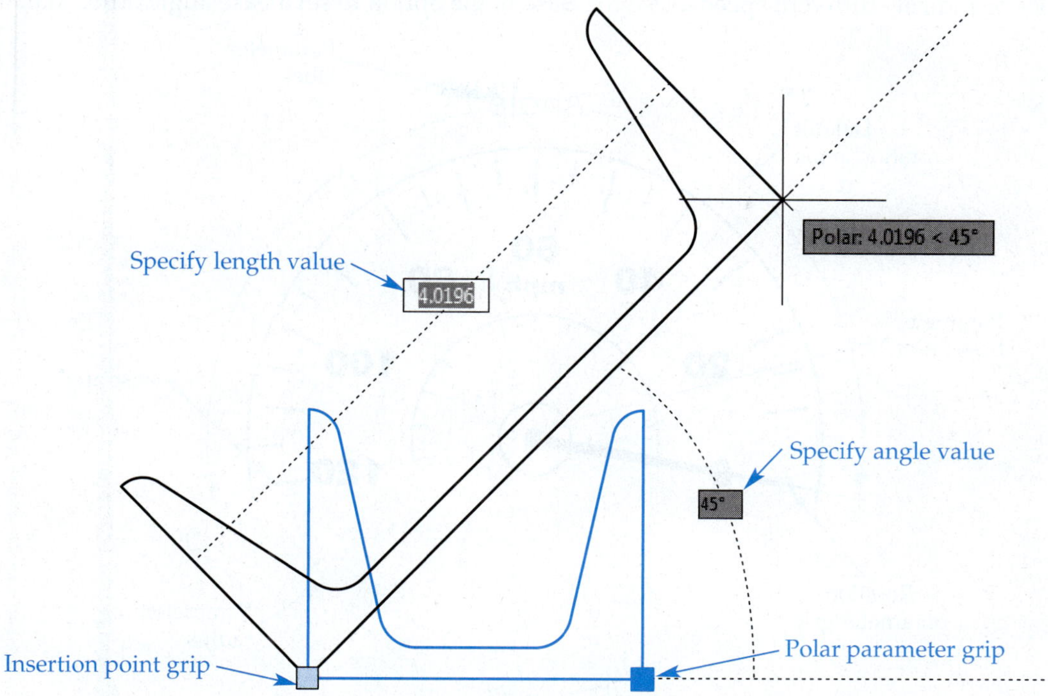

NOTE

The **Properties** palette includes **Distance multiplier** and **Angle offset** options for move, stretch, and polar stretch actions. Enter a value in the **Distance multiplier** text box to multiply by the parameter value when adjusting the block. For example, if you assign a distance multiplier of 2 to a move action and move an object 4 units, the object actually moves 8 units. Enter an angle in the **Angle offset** text box to change the parameter grip angle. For example, if you assign an offset angle of 45 to a move action and move an object 10°, the object actually moves 55°.

Exercise 26-5

www.g-wlearning.com/CAD/

Complete the exercise on the companion website.

Rotation Parameters

rotation parameter: A parameter that allows objects in a block to rotate independently of the block.

rotate action: An action used to rotate objects within a block without affecting the other objects in the block.

A ***rotation parameter*** creates a rotation property to which you can assign a ***rotate action***. For example, assign a rotation parameter with a rotate action to the needle in the speedometer block shown in Figure 26-21 to rotate the needle around the circumference of the dial. To insert the rotation parameter, access the **Rotation** parameter option and pick the center of the circular base of the needle as the rotation base point. Then pick a point, such as the needle endpoint shown, to specify the parameter radius. Set the default rotation angle from 0° east, or if rotation should originate from an angle other than 0°, use the **Base angle** option. Figure 26-21 shows using the **Base angle** option to base the rotation at –10° and then specifying a default rotation angle of 200° to align the rotation with the 120 and 0 mph marks.

Ribbon
Block Editor
> Action
Parameters

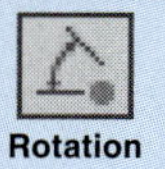

Rotation

Type
BPARAMETER
> Rotation

Figure 26-21.
A rotation parameter with a rotate action allows you to rotate the needle in a speedometer block to indicate different speeds. Use the **Base angle** option to set a base angle other than 0°.

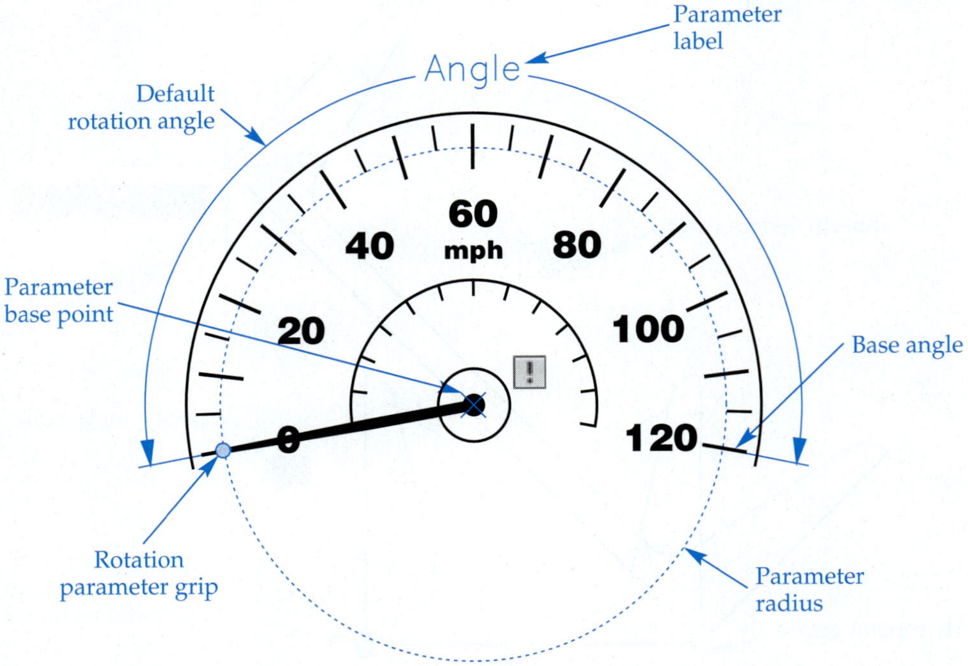

After you define the rotation parameter, pick a location for the parameter label. Next, enter the number of grips to associate with the parameter. The default **1** option creates a single grip at the parameter radius that allows you to use grip editing to carry out the rotate action.

NOTE

Name, **Label**, **Chain**, **Description**, **Palette**, and **Value set** options are available before you specify the parameter. Most of these options and the **Base angle** option are also available from the **Properties** palette if you have already created the parameter.

Assigning a Rotate Action

Ribbon
Block Editor > Action Parameters
Rotate
Type
BACTIONTOOL > Rotate

To assign a rotate action to the speedometer example, access the **Rotate** action option and pick the rotation parameter. Then select the objects that make up the needle and the rotation parameter. Press [Enter] to place the action. If necessary, access the **Properties** palette and adjust the **Base type** option. The default **Dependent** option sets the rotation point as the base point of the rotation parameter, which is appropriate for the speedometer example. Test and save the block, and exit the **Block Editor**. The dynamic block is now ready to use.

Using a Rotate Action Dynamically

Figure 26-22 shows using the rotation parameter grip or dynamic input to rotate the needle inside the speedometer block reference. The **Endpoint** object snap is most appropriate to select a specific speed for this example. You can also use the **Properties** palette to define the angle.

Figure 26-22.
Dynamically rotating the needle in a speedometer block using a rotate action assigned to a rotation parameter.

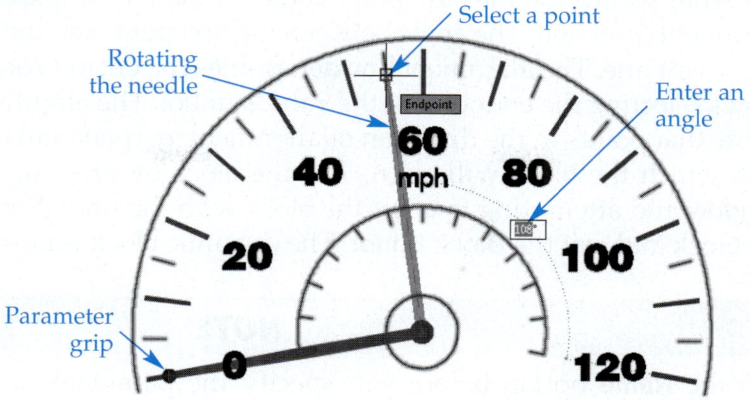

Exercise 26-6

www.g-wlearning.com/CAD/

Complete the exercise on the companion website.

Alignment Parameters

An ***alignment parameter*** creates an alignment property. When you move a block with an alignment parameter near another object, the block rotates to align with the object based on the angle and alignment line defined in the block. An alignment parameter saves time by eliminating the need to rotate a block or assign a rotation parameter. An alignment parameter affects the entire block, and therefore requires no action.

alignment parameter: A parameter that aligns a block with another object in the drawing.

Creating an Alignment Parameter

Figure 26-23 provides an example of adding an alignment parameter to the block of a gate valve symbol to align the gate valve with pipes. Access the **Alignment** parameter option and pick the point in the center of the valve to locate the parameter grip and define the first point of the alignment line. Next, specify the alignment direction, or use the **Type** option to specify the alignment type. Alignment type does not affect how the block aligns; it determines the direction of the alignment grip. Select the **Perpendicular** option to point the grip perpendicular to the alignment line, or select the **Tangent** option to point the grip tangent to the alignment line. Set the **Perpendicular** option for the gate valve example.

Ribbon
Block Editor
> Action
Parameters
Alignment
Type
BPARAMETER
> Alignment

Figure 26-23.
Adding an alignment parameter to a gate valve block.

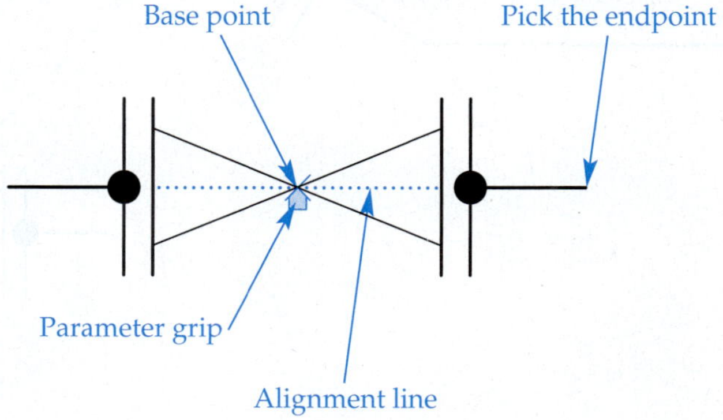

After specifying the base point and alignment type, pick a second point to set the alignment direction. The angle between the first point and the second point defines the alignment line. The alignment line determines the default rotation angle. Figure 26-23 shows selecting the endpoint of the valve symbol. The alignment parameter grip is an arrow that points in the direction of alignment, perpendicular or tangent to the object with which the block will align. Test the block by drawing a line in the **Test Block Window** and attempting to align the block with the line. When you are finished, save the block and exit the **Block Editor**. The dynamic block is now ready to use.

NOTE

Use the **Name** option before you specify the parameter to rename the parameter. Alignment parameters do not include labels. The alignment type can be adjusted from the **Properties** palette if you have already created the parameter.

Using an Alignment Parameter Dynamically

Figure 26-24 shows using the alignment parameter grip to align the gate valve block with a pipeline. Select the block to display grips and then pick the parameter grip. Move the block near another object to align the block with the object. The rotation depends on the alignment path and type and the angle of the other object.

NOTE

When you manipulate a block with an alignment parameter, the **Nearest** object snap temporarily turns on, if it is not already on.

Exercise 26-7

www.g-wlearning.com/CAD/

Complete the exercise on the companion website.

Figure 26-24.
Move the gate valve block near a line to align the block with the line.

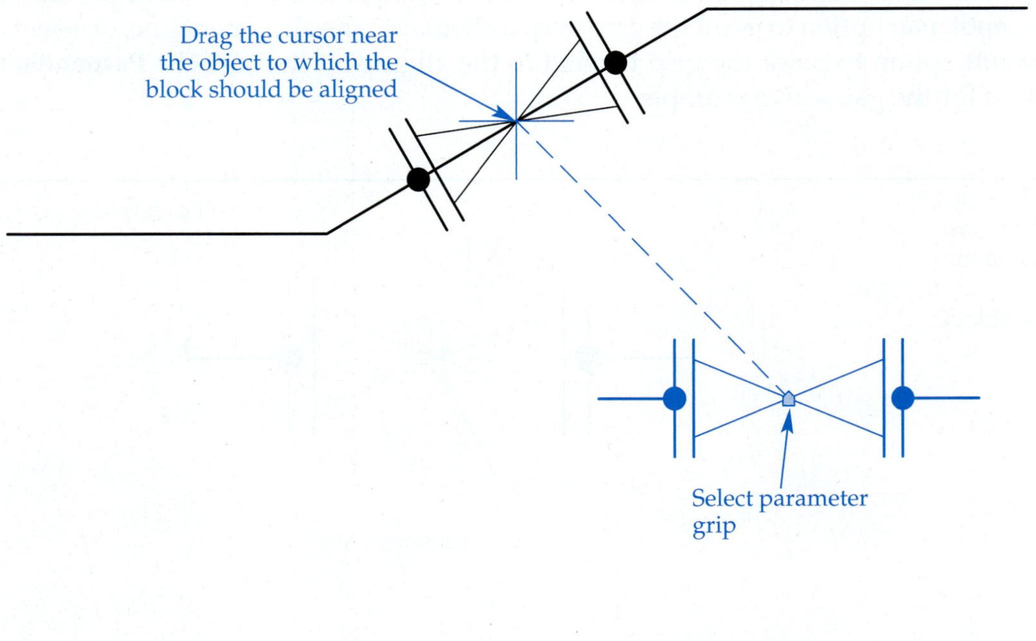

Flip Parameters

A ***flip parameter*** creates a flip property to which you can assign a ***flip action***. For example, assign a flip parameter with a flip action to a door symbol to provide the option to place the door on either side of a wall. Another example is using a flip parameter to control the side of a reference line on which a weld symbol appears for arrow side or other side applications.

Ribbon
Block Editor
> Action Parameters
Flip

Type
BPARAMETER
> Flip

flip parameter: A parameter that mirrors selected objects within a block.

flip action: An action used to flip the entire block or selected objects within the block.

Figure 26-25 shows an example of adding a flip parameter. Access the **Flip** parameter option and pick the base point, followed by the endpoint of the reflection line. See Figure 26-25A. Pick a location for the parameter label, and then enter the number of grips to associate with the parameter. The default **1** option creates a single flip grip that allows you to use grip editing to carry out the flip action.

Flipping a block mirrors the block over the reflection line. However, for the door symbol, with the line in the position shown in Figure 26-25A, an incorrect flip will result when you flip the block to the other side of a wall. To mirror the block properly, you must locate the reflection line to account for wall thickness. To place the door on a 4″ wall, for example, use the **MOVE** command to move the reflection line 2″ lower than the door. The label and parameter grip also move. In addition, move the parameter grip horizontally to the middle of the door opening to help place and flip the block properly. See Figure 26-25B.

NOTE

Name, **Label**, **Description**, and **Palette** options are available before you specify the parameter. Most of these options are also available from the **Properties** palette if you have already created the parameter.

PROFESSIONAL TIP

A block reference with a flip parameter mirrors about the reflection line. You must place the reflection line in the correct location so the flip creates a symmetrical, or mirrored, copy. This typically requires the reflection line to be coincident with the block insertion point.

Figure 26-25.
A—Adding a flip parameter to an architectural floor plan door block. B—Moving the parameter so the block will flip correctly about the centerline of a wall.

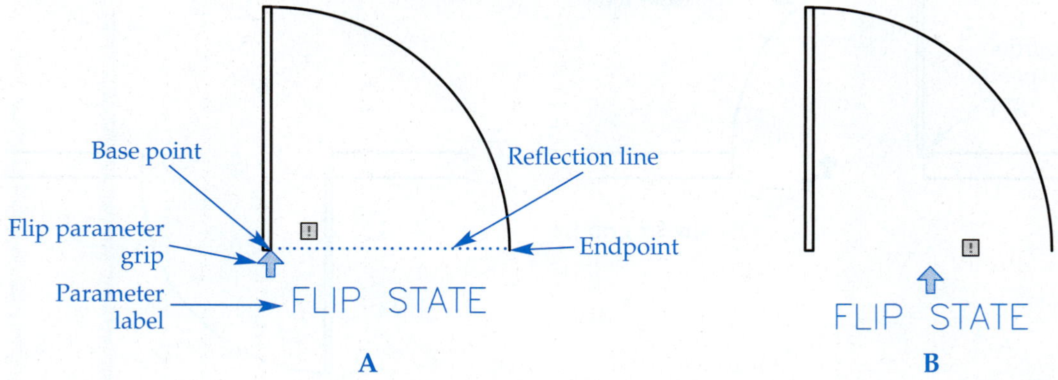

Assigning a Flip Action

Ribbon
Block Editor
> Action
Parameters

Flip

Type
BACTIONTOOL
> Flip

To assign a flip action to the door example, access the **Flip** action option and pick the flip parameter. Then select the objects that make up the door and the flip parameter. Press [Enter] to place the action. Test and save the block, and exit the **Block Editor**. The dynamic block is now ready to use.

Using a Flip Action Dynamically

Figure 26-26A shows a reference of the door block selected for editing. Pick the flip parameter grip to flip the block to the other side of the reflection line, as shown in Figure 26-26B. Unlike other parameters and actions that require stretching, moving, or rotating, a single pick initiates a flip action.

PROFESSIONAL TIP

Add another flip parameter with a flip action to a door symbol to flip the door from side to side. This one block takes the place of four blocks to accommodate different door positions.

Exercise 26-8

www.g-wlearning.com/CAD/

Complete the exercise on the companion website.

XY Parameters

XY parameter: A parameter that specifies distance properties in the X and Y directions.

An *XY parameter* creates horizontal and vertical distance properties. You can assign move, scale, stretch, and array actions to XY parameters. The XY parameter can include up to four parameter grips—one at each corner of a rectangle defined by the parameter. You can use the XY parameter for a variety of applications, depending on the assigned actions.

Figure 26-26.
A—Select the block to display the flip parameter grip. B—Pick the flip parameter grip to flip the block about the reflection line. The entire block flips because all of the objects within the block are included in the selection set for the action.

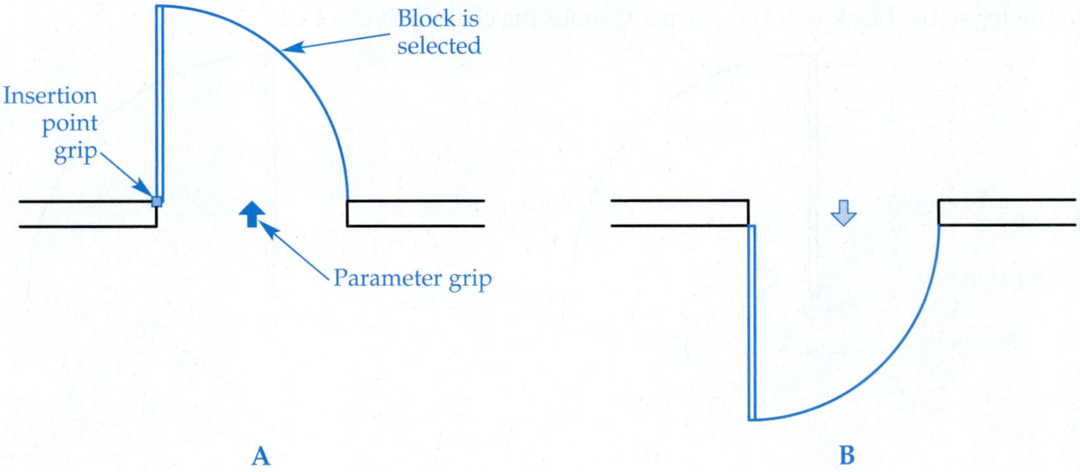

Figure 26-27 provides an example of inserting an XY parameter. Access the **XY** parameter option and pick the base point, which is the origin of the X and Y distances. Next, pick a point to specify the XY point, which is the corner opposite the base point. Finally, enter the number of grips to associate with the parameter. The default **1** option assigns a grip at the XY point that allows you to use grip editing to carry out the assigned action. Select the **4** option, as shown in Figure 26-27, to assign a grip at each XY corner to maximize flexibility, or select a smaller number to limit dynamic options. If you select the **0** option, you can only adjust the block using the **Properties** palette.

Ribbon
Block Editor > Action Parameters
XY
Type
BPARAMETER > XY

NOTE

Name, **Label**, **Chain**, **Description**, **Palette**, and **Value set** options are available before you specify the parameter. Most of these options are also available from the **Properties** palette if you have already created the parameter.

Assigning an Array Action

Figure 26-28 illustrates using an ***array action*** assigned to an XY parameter. This example shows dynamically arraying the block of an architectural glass block to create an architectural feature of glass blocks, such as a wall, without using a separate array operation. Access the **Array** action option and pick the XY parameter. Then select the objects to be included in the array, and press [Enter] to accept the selection.

Ribbon
Block Editor > Action Parameters
Array
Type
BACTIONTOOL > Array

array action: An action used to array objects within the block based on preset specifications.

At the Enter the distance between rows or specify unit cell: prompt, enter a value for the distance between rows or pick two points to set the row and column values. At the Enter the distance between columns: prompt, specify a value for the distance between columns. The second prompt does not appear if you select two points to define the row and column values. In the glass block example, be sure to allow for a mortar joint when setting the row and column distance. Before assigning the action, you may want to draw a construction point offset from the block by the width of the mortar joint. Then you can pick two points to define the row and column values. Be sure to erase the construction point before saving the block. Test and save the block, and exit the **Block Editor**. The dynamic block is now ready to use.

Using an Array Action Dynamically

Figure 26-28 illustrates using the upper-right grip or dynamic input to array a reference of the architectural glass block. You can use any available grip to apply an array, depending on where you want the array to occur. You can also use the **Properties** palette to define the array. Notice that proper action definition produces mortar joints. The resulting array remains a single block.

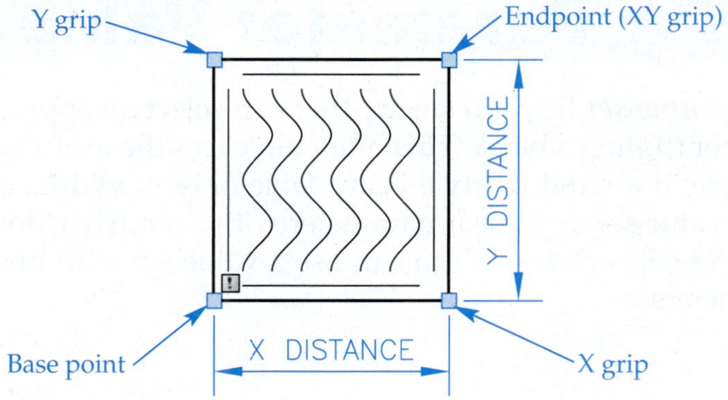

Figure 26-27. Adding an XY parameter to a block of an architectural glass block. The XY parameter consists of X and Y distance properties and four grips.

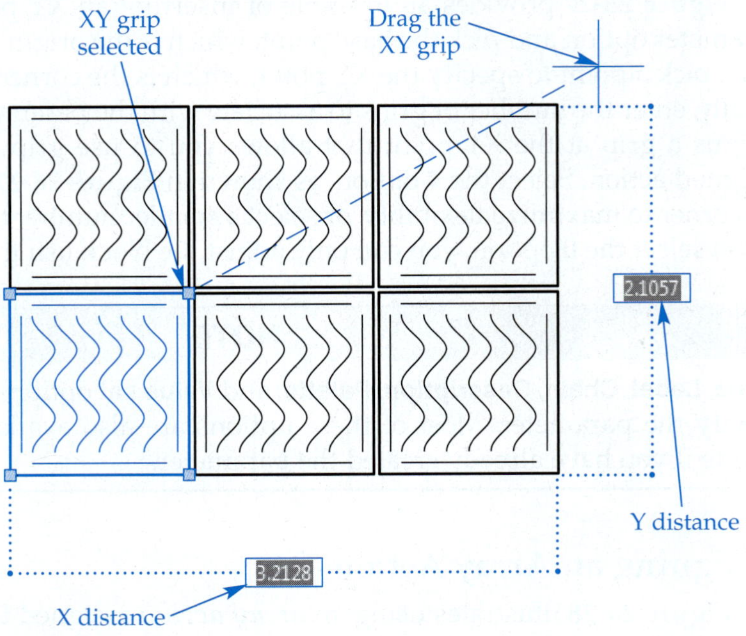

Figure 26-28. Dynamically creating an array of architectural glass blocks using a block with an XY parameter and an array action. The pattern of rows and columns forms as you move the XY parameter. Notice the mortar joints that form between the glass blocks because of proper action definition.

Exercise 26-9

www.g-wlearning.com/CAD/

Complete the exercise on the companion website.

Base Point Parameters

base point parameter: A parameter that defines an alternate base point for a block.

Ribbon
Block Editor
> Action Parameters
Basepoint
Type
BPARAMETER
> Basepoint

The **Block Editor** origin (0,0,0 point) determines the default location of the block insertion base point. Typically, you construct blocks in the **Block Editor** in reference to the origin, using the origin as the location of the insertion base point. The base point you choose when creating a block using the **BLOCK** command attaches to the origin when you open the block in the **Block Editor**. Add a ***base point parameter*** to override the default origin base point.

Access the **Basepoint** parameter option and pick a point to place the base point parameter. The parameter appears as a circle with crosshairs. After you save the block, the location of the base point parameter becomes the new base point for the block. You cannot assign actions to a base point parameter, but you can include a base point parameter in the selection set for actions.

Parameter Value Sets

value set: A set of allowed values for a parameter.

A ***value set*** helps to ensure that you select an appropriate value when dynamically controlling a block. This often increases the usefulness of a dynamic block. For example, if a window style is available only in widths of 36″, 42″, 48″, 54″, and 60″, add a value set to a linear parameter with a stretch action to limit selection to these sizes. See **Figure 26-29**. You can use a value set with linear, polar, XY, and rotation parameters.

Figure 26-29.
When you adjust a block that includes a value set, tick marks appear at locations corresponding to the values in the value set. You can only adjust the block to one of the tick marks.

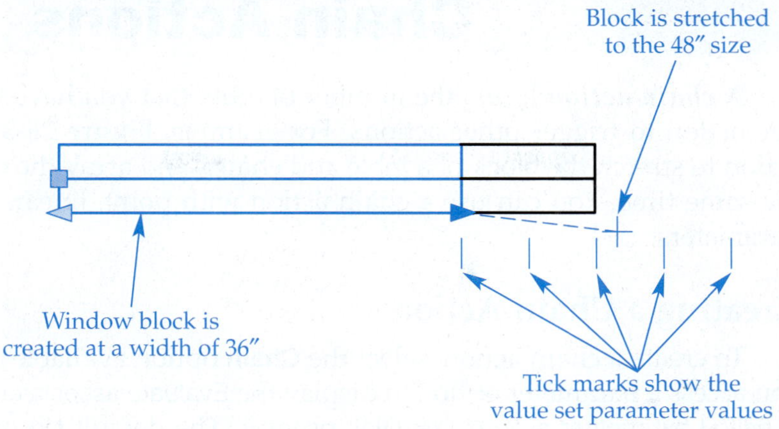

Creating a Value Set

To create a value set, select the **Value set** option available at the first prompt after you access a parameter option, and then select a value set type. Select the **List** option to create a list of possible sizes. Type all of the valid values for the parameter separated by commas. For the window block example, type 36,42,48,54,60. Then press [Enter] to return to the initial parameter prompt, and add the parameter. After you insert the parameter, the valid values appear as tick marks.

Select the **Increment** option to specify an incremental value. With this option, you can set minimum and maximum values to provide a limit for the increments. For the window block example, use the **Value set** option again to set 6″ width increments. This time, select the **Increment** option and type 6 for the distance increment, 36 for the minimum distance, and 60 for the maximum distance. The initial parameter prompt returns after you enter the maximum distance.

After you add a parameter with a value set, assign an action to the parameter. For the window block in Figure 26-29, assign a stretch action to the linear parameter. This allows the window to stretch to the valid widths specified in the value set. Test and save the block, and exit the **Block Editor**. The dynamic block is now ready to use.

NOTE

You can also use the options in the **Value Set** category of the **Properties** palette to specify value sets during block definition.

Using a Value Set with a Parameter

Figure 26-29 shows using a linear parameter grip with a stretch action to specify the width of a window block reference. Tick marks appear at the positions of valid values. As you stretch the grip, the modified block snaps to the nearest tick mark. When using dynamic input, you can also enter a value in the input field. If you type a value that is not included in the value set, AutoCAD applies the nearest valid value. You can also use the **Properties** palette to select a value.

Exercise 26-10

www.g-wlearning.com/CAD/

Complete the exercise on the companion website.

Chain Actions

chain action: An action that triggers another action when you modify a parameter.

A ***chain action*** limits the number of edits that you have to perform by allowing one action to trigger other actions. For example, Figure 26-30 shows using a chain action to stretch the block of a table and chairs, and array the chairs along the table at the same time. You can use a chain action with point, linear, polar, XY, and rotation parameters.

Creating a Chain Action

To create a chain action, select the **Chain** option available at the first prompt after you access a parameter option to display the Evaluate associated actions when parameter is edited by another action? [Yes/No]: prompt. The default **No** option does not create a chain action. Select the **Yes** option to create a chain action.

Figure 26-31 shows the default arrangement of the table and chairs block example. For this example, access the **Linear** parameter option and use the **Label** option to change the label name to CHAIR ARRAY. Next, select the **Chain** option and select **Yes**. To complete the parameter, select the start point and endpoint shown in Figure 26-31A, and assign a single grip to the endpoint. Then assign an array action to the parameter, selecting only the upper-right and lower-right chairs as the objects to array. At the Enter the distance between columns: prompt, use object snaps to snap to the endpoint of one of the chairs and then snap to the equivalent endpoint on the chair next to the first chair.

Add another single-grip linear parameter, labeled TABLE STRETCH, as shown in Figure 26-31B. Assign a stretch action to the parameter associated with the TABLE STRETCH parameter grip. Use a crossing window around the right end of the table and the CHAIR ARRAY parameter grip. See Figure 26-31C. Select the table, the chair at the right end of the table, and the CHAIR ARRAY parameter as the objects to stretch. Test and save the block, and exit the **Block Editor**. The dynamic block is now ready to use.

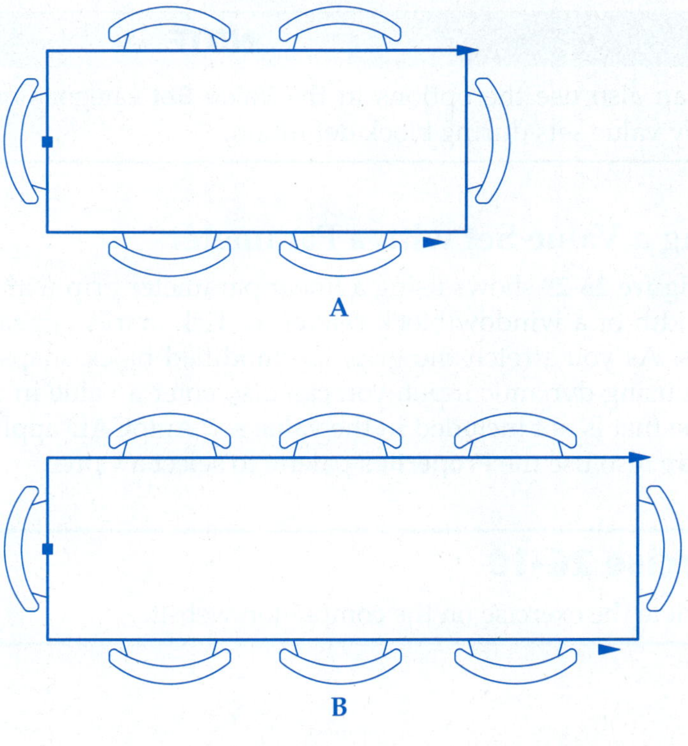

Figure 26-30.
A—A block of a table with six chairs. B—Use a chain action with a linear parameter to array the chairs automatically when you stretch the table.

Figure 26-31.
A—Inserting a linear parameter to use with an array action for the chairs. B—Inserting a linear parameter to stretch the table. C—Assigning a stretch action to the linear parameter. When you specify the crossing window, be sure to include the CHAIR ARRAY parameter grip in the frame.

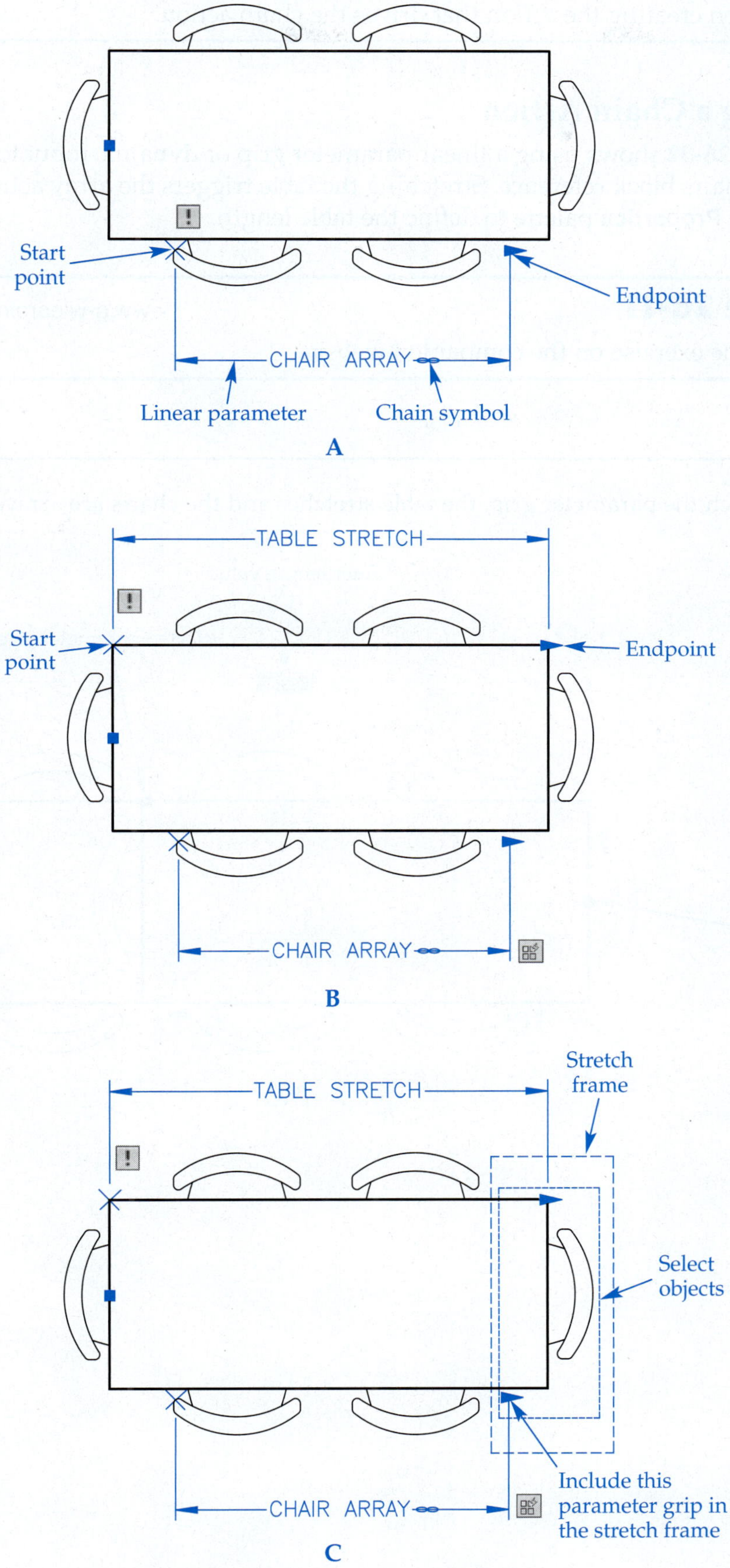

PROFESSIONAL TIP

The keys to successfully creating a chain action are to set the **Chain** option to **Yes** for the parameter affected automatically and to include the parameter in the object selection set when creating the action that drives the chain action.

Applying a Chain Action

Figure 26-32 shows using a linear parameter grip or dynamic input to stretch the table and chairs block reference. Stretching the table triggers the array action. You can also use the **Properties** palette to define the table length.

Exercise 26-11

www.g-wlearning.com/CAD/

Complete the exercise on the companion website.

Figure 26-32.
As you stretch the parameter grip, the table stretches and the chairs are arrayed at the same time.

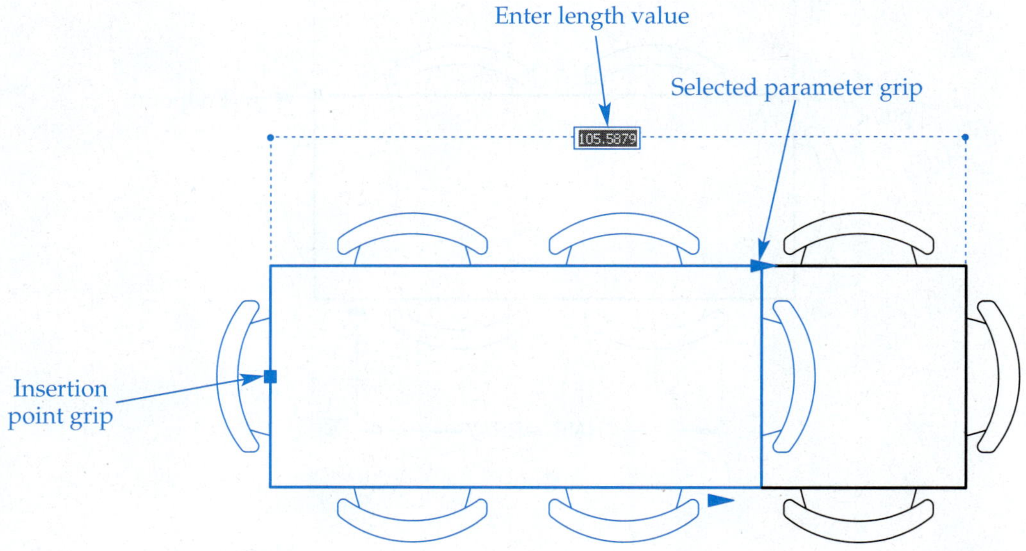

Chapter Review

Answer the following questions using the information in this chapter.

1. Define *dynamic block*.
2. What is the function of a dynamic block?
3. How are standard blocks and dynamic blocks the same? How are they different?
4. Identify the property that forms when you create a point parameter and list the actions you can assign.
5. What is the purpose of a move action?
6. When do action bars appear?
7. What information can you get by hovering over or selecting an action bar?
8. What is the basic function of a linear parameter?
9. Explain the function of a stretch action.
10. Describe how to use a stretch action symmetrically.
11. What is the basic function of a scale action?
12. Describe the polar parameter type and list the actions that you can assign to the parameter.
13. Identify the property that forms when you create a rotation parameter and list the actions you can assign.
14. Describe what happens when you move a block with an alignment parameter near another object in the drawing. How does this save drawing time?
15. Give at least one practical example of using a flip parameter.
16. Describe the properties an XY parameter creates and list the actions that you can assign to the parameter.
17. Give an example of using an array action assigned to an XY parameter.
18. When would you add a base point parameter?
19. Describe the basic use of a value set.
20. Explain the function of a chain action.

Drawing Problems

Start AutoCAD if it is not already started. Start a new drawing for each problem using an appropriate template of your choice. The template should include layers and text, dimension, multileader, and table styles, when necessary, for drawing the given objects. Add layers and text, dimension, multileader, and table styles as needed. Draw all objects using appropriate layers. Use appropriate text, dimension, multileader, and table styles, justification, and format. Follow the specific instructions for each problem. Use only drawing and editing commands and techniques you have already learned. Use your own judgment and approximate dimensions when necessary. Apply dimensions accurately using ASME or appropriate industry standards.

Note: Some of the problems in this chapter are built on problems from previous chapters. If you have not yet completed those problems, complete them now.

▼ Basic

1. **Architectural** Save P24-1 as P26-1. If you have not yet completed Problem 24-1, do so now. In the P26-1 file, erase all copies of the steel column symbols except for the one in the lower-left corner. Insert an XY parameter into the steel column block and associate an array action with the parameter. Use the proper values for the array action to array the block dynamically to match the drawing. Use the dynamic block to create the rest of the steel columns in the drawing.

2. **Mechanical** Create a block named WIRE ROLL as shown. Do not include the dimensions. Insert a linear parameter on the entire length of the roll. Use a value set with the following values: 36″, 42″, 48″, and 54″. Assign a stretch action to the parameter and associate the action with either parameter grip. Create a crossing window that will allow the length of the roll to stretch. Select all of the objects on one end and the length lines as the objects to stretch. Insert the WIRE ROLL block four times into a drawing and stretch each block to use a different value set length. Save the drawing as P26-2.

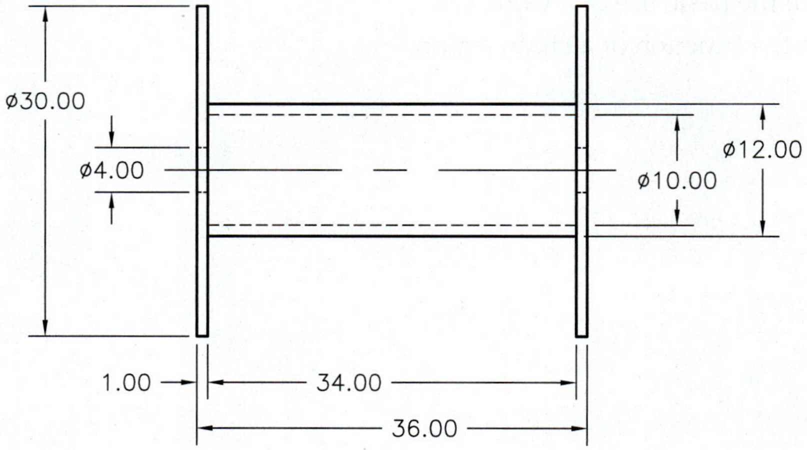

3. **Piping** Create a block named 90D ELBOW as shown on the left. Do not include the dimensions. Insert two flip parameters and two flip actions. The purpose of one of the flip parameter/action combinations is to flip the elbow horizontally. The second flip parameter/action combination flips the elbow vertically. Use the dynamic block to create the drawing shown on the right. Save the drawing as P26-3.

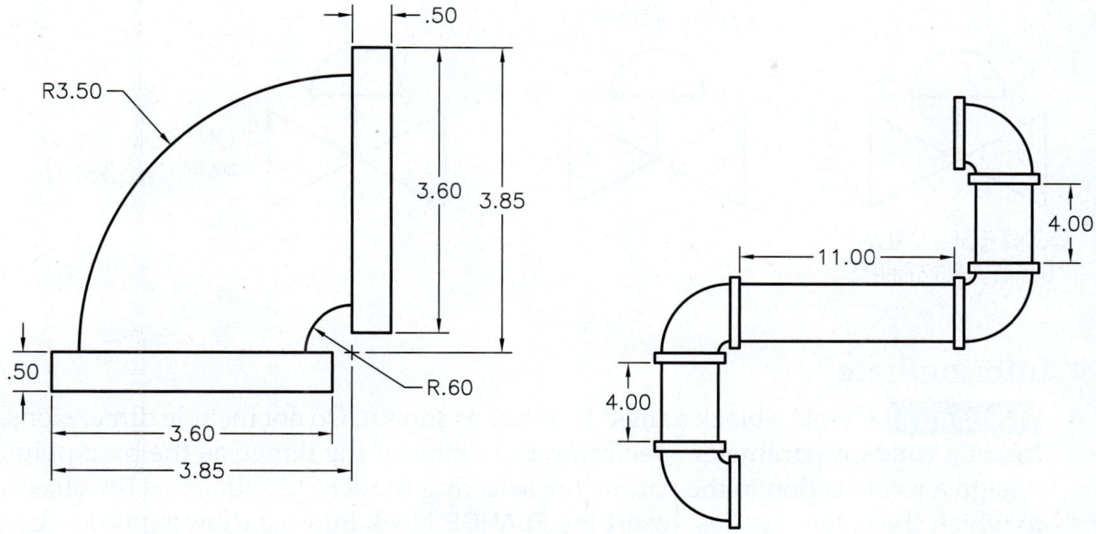

4. **Architectural** Create a block of the 48″ window as shown on the left. Do not include the dimensions. Insert an alignment parameter so the length of the window can align with a wall. Then draw the walls shown on the right. Insert the window block as needed. Use the alignment parameter to align the window to the walls. Center the windows on wall segments unless dimensioned. Save the drawing as P26-4.

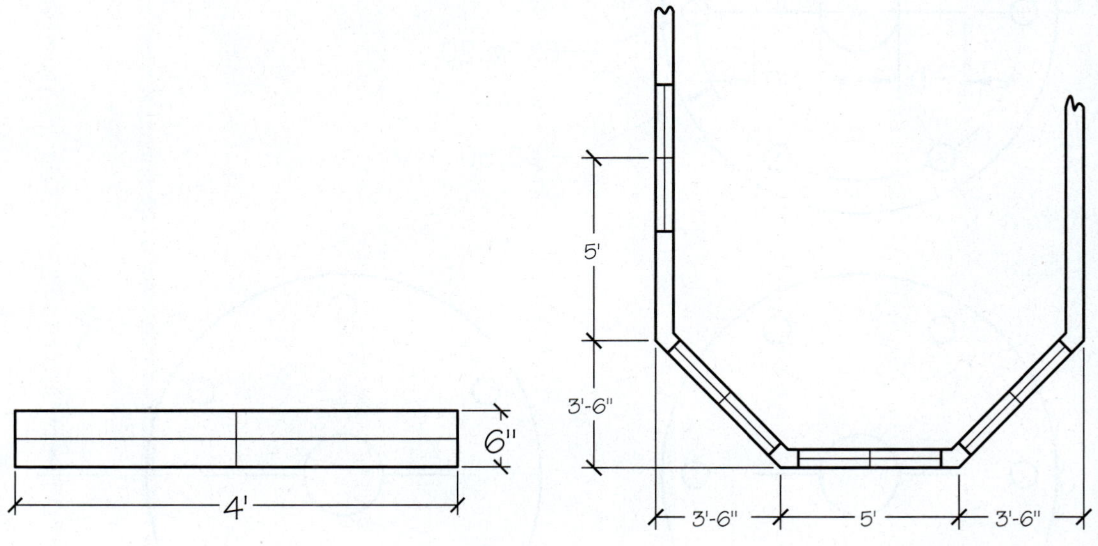

5. **Piping** Create a block named CONTROL VALVE as shown on the left. Include the label in the block. Insert a point parameter and assign a move action to the parameter. Select the two lines of text as the objects to which the action applies. Insert the CONTROL VALVE block into the drawing three times. Use the point parameter to move the text to match the three positions shown. Save the drawing as P26-5.

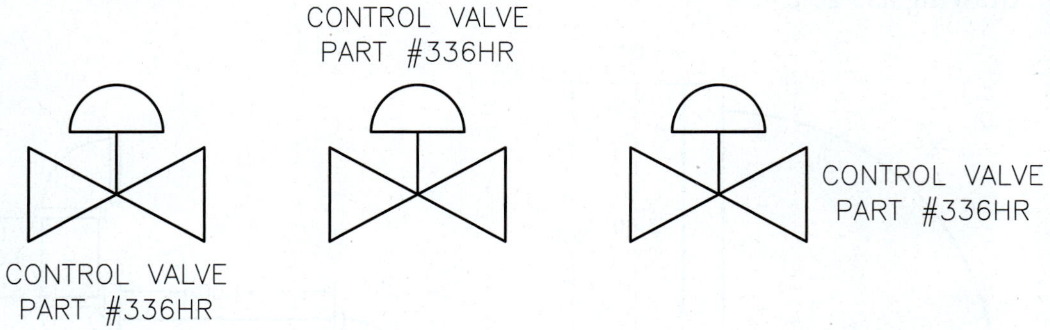

▼ Intermediate

6. **Mechanical** Create a block named FLANGE as shown. Do not include dimensions. Insert a rotation parameter specifying the center of the flange as the base point. Assign a rotate action to the parameter, selecting the six ∅.20 circles as the objects to which the action applies. Insert the FLANGE block into the drawing twice. Use the rotation parameter to create the two configurations shown. Save the drawing as P26-6.

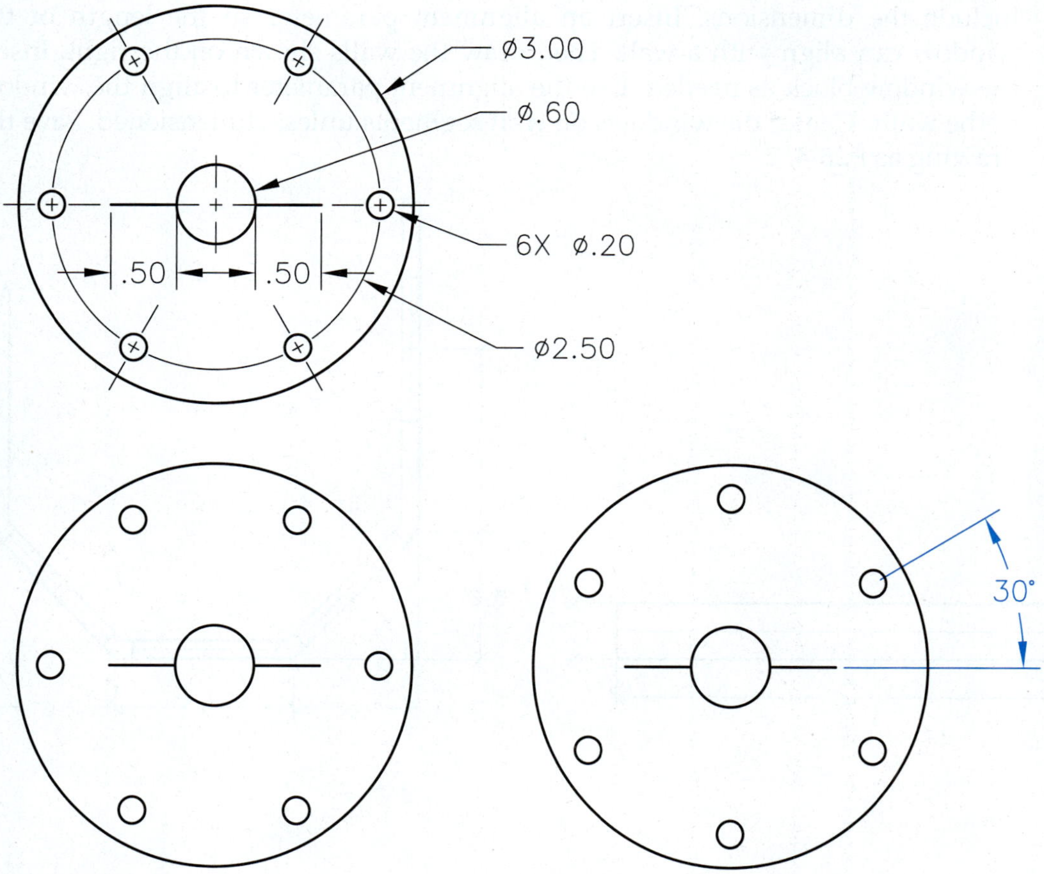

7. Mechanical Save P26-6 as P26-7. In the P26-7 file, open the FLANGE block in the **Block Editor** and use the **Properties** palette to apply the following settings to the rotation parameter:
   A. **Angle name**—BOLT HOLES
   B. **Angle description**—ROTATION OF BOLT HOLE PATTERN
   C. **Ang type**—INCREMENT
   D. **Ang increment**—30
   E. Save the changes and exit the **Block Editor**.

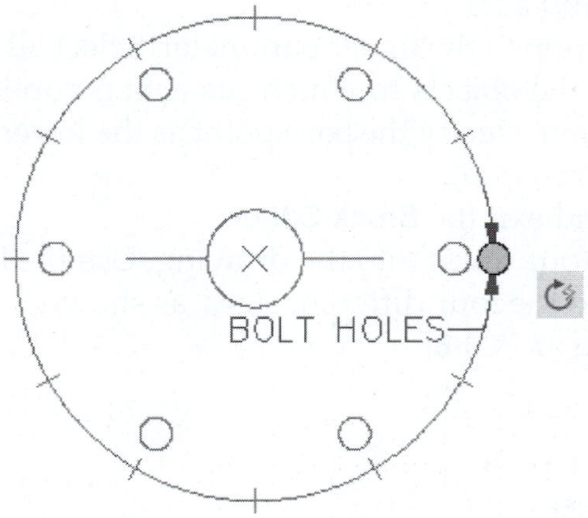

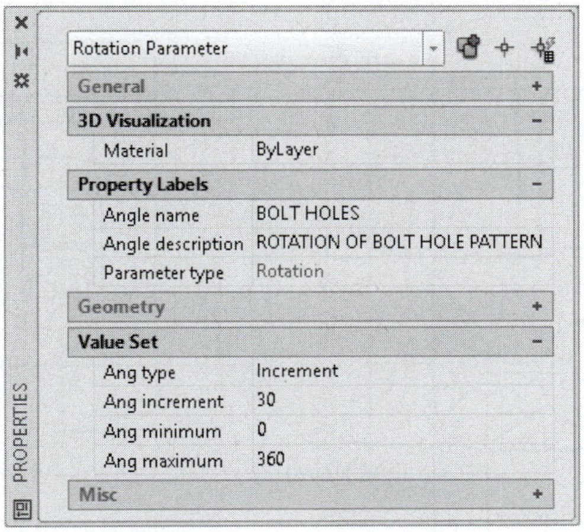

Drawing Problems - Chapter 26

▼ Advanced

8. **Mechanical** The drawing shown is a fan with an enlarged view of the motor. This fan can have one of four motors of different sizes. Create the fan as a dynamic block.
   A. Draw all the objects. Do not dimension the drawing or draw the enlarged view.
   B. Create a block named FAN consisting of the objects shown in the dimensioned view of the fan.
   C. Open the block in the **Block Editor** and insert a linear parameter along the top of the motor (the 1.50″ dimension). Use a value set with the following values: 1.125, 1.5, 1.875, and 2.25.
   D. Assign a scale action to the linear parameter. Select all of the objects that make up the motor as the objects to which the action applies. Use an independent base point type and specify the base point as the lower-left corner of the motor (the implied intersection).
   E. Save the block and exit the **Block Editor**.
   F. Insert the block four times into the drawing. Use the linear parameter grip to scale the motor to the four different sizes, as shown.
   G. Save the drawing as P26-8.

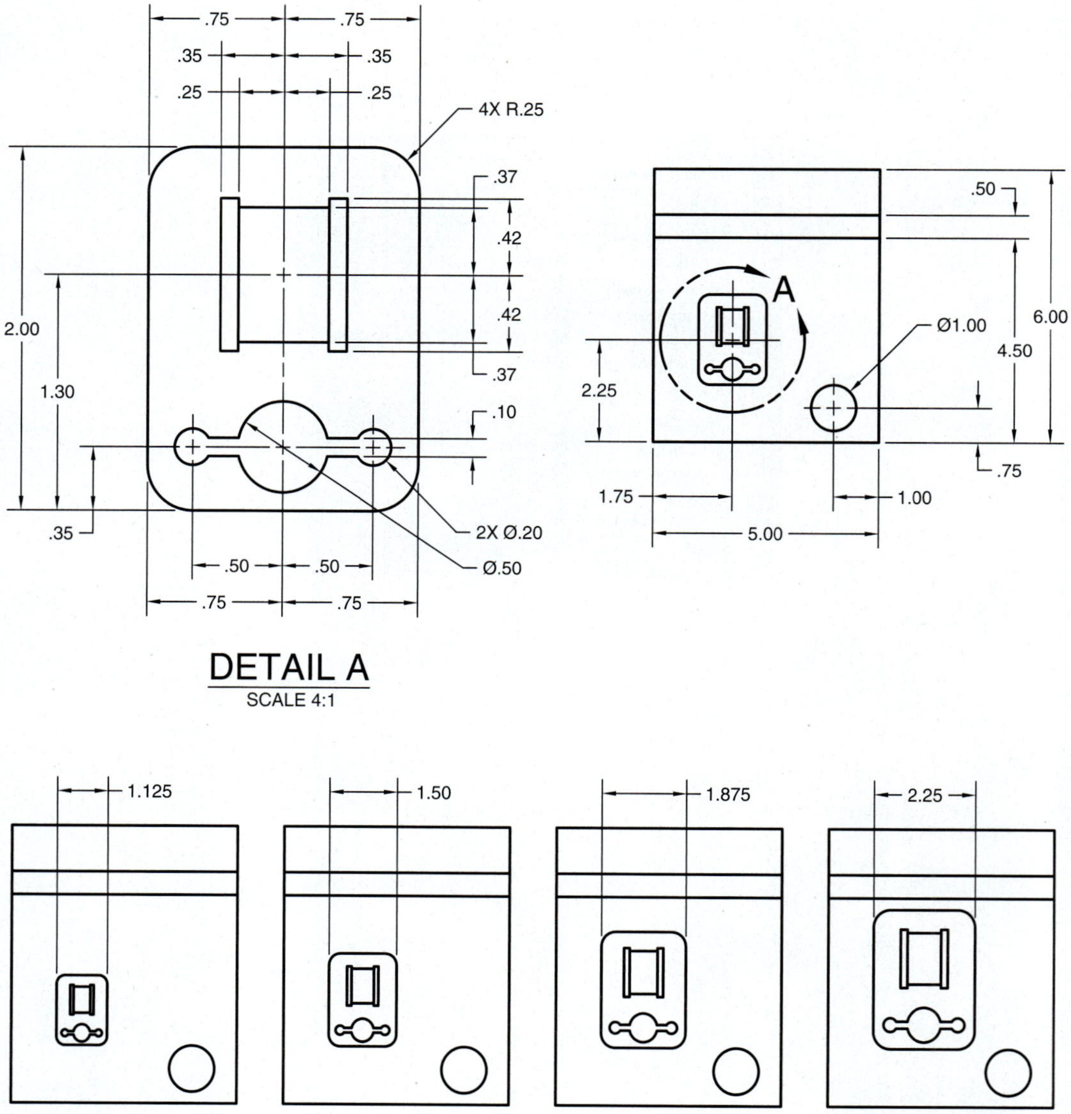

9. **Architectural** Construct a block of an architectural glass block similar to the block shown. Use an XY parameter with an array action to create a glass block wall of 10 rows and 15 columns. Construct the block to include an appropriate mortar joint. Save the drawing as P26-9.

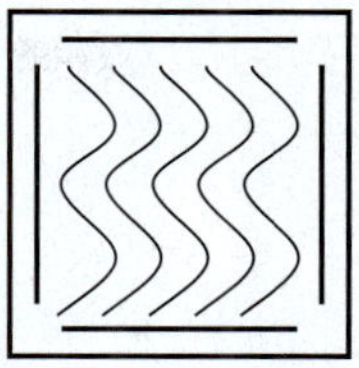

10. **Mechanical** Construct a block of a steel C4×5.4 shape. Reference the American Institute of Steel Construction (AISC) manual for dimensions. Add a polar parameter with a polar stretch action to adjust the depth and angle of block references. Insert and adjust the block as needed to create a portion of a motor mount similar to the drawing shown in **Figure 26-18**. Save the drawing as P26-10.

11. **Mechanical** Construct a block of a speedometer similar to the speedometer shown in **Figure 26-21**. Add a rotation parameter with a rotate action to adjust the needle reading for block references. Insert the block four times and use the dynamic needle to create a speedometer that reads 5 mph, 25 mph, 55 mph, and 110 mph. Save the drawing as P26-11.

12. **Architectural** Construct a block of a table with six chairs similar to the block shown in **Figure 26-30A**. Use linear parameters with array, stretch, and chain actions as needed to stretch and add chairs to block references. Insert the block twice and use the dynamic stretch and array to create a table with eight chairs and a table with ten chairs. Save the drawing as P26-12.

AutoCAD Certified Professional Exam Practice

Answer the following questions using the information in this chapter.

1. Which of the following commands can be used to save a dynamic block? *Select all that apply.*
   A. **BACTION**
   B. **BTESTBLOCK**
   C. **BPARAMETER**
   D. **BSAVE**
   E. **BSAVEAS**
2. Which of the following actions can be assigned to a point parameter? *Select all that apply.*
   A. flip
   B. move
   C. rotate
   D. scale
   E. stretch
3. Which of the following block properties can be changed using a polar stretch action? *Select all that apply.*
   A. color
   B. layer
   C. shape
   D. size
   E. rotation

Follow the instructions in the problem. Complete the problem by answering the related question.

4. **Navigate to this chapter on the companion website and open CPE-26align.dwg.** Edit the existing SOFA block to include an alignment parameter that will align the middle of the back of the sofa with a wall. Insert the block into the drawing. Use the alignment parameter and other appropriate tools to center the sofa on the window as shown, exactly 4″ from the wall. What are the coordinates of the middle of the sofa back? (Use the coordinates of the quadrant point on the sofa back.)

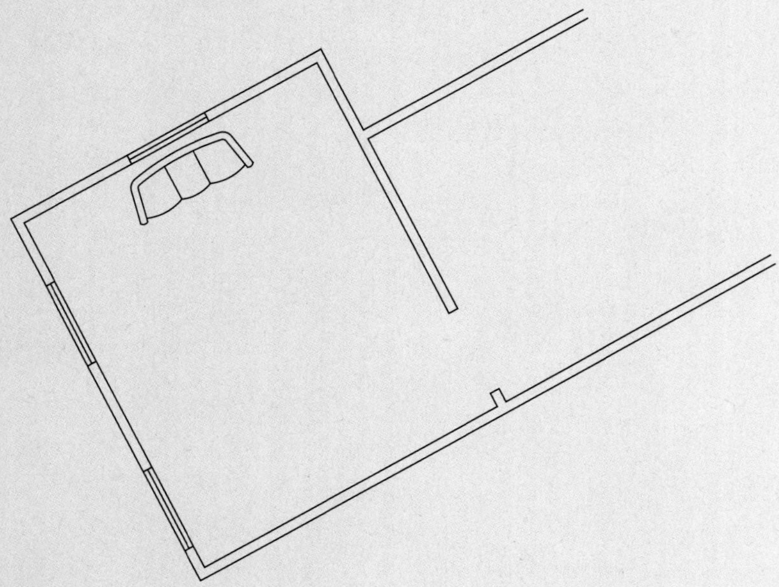

AutoCAD Certified Professional Exam Practice - Chapter 26

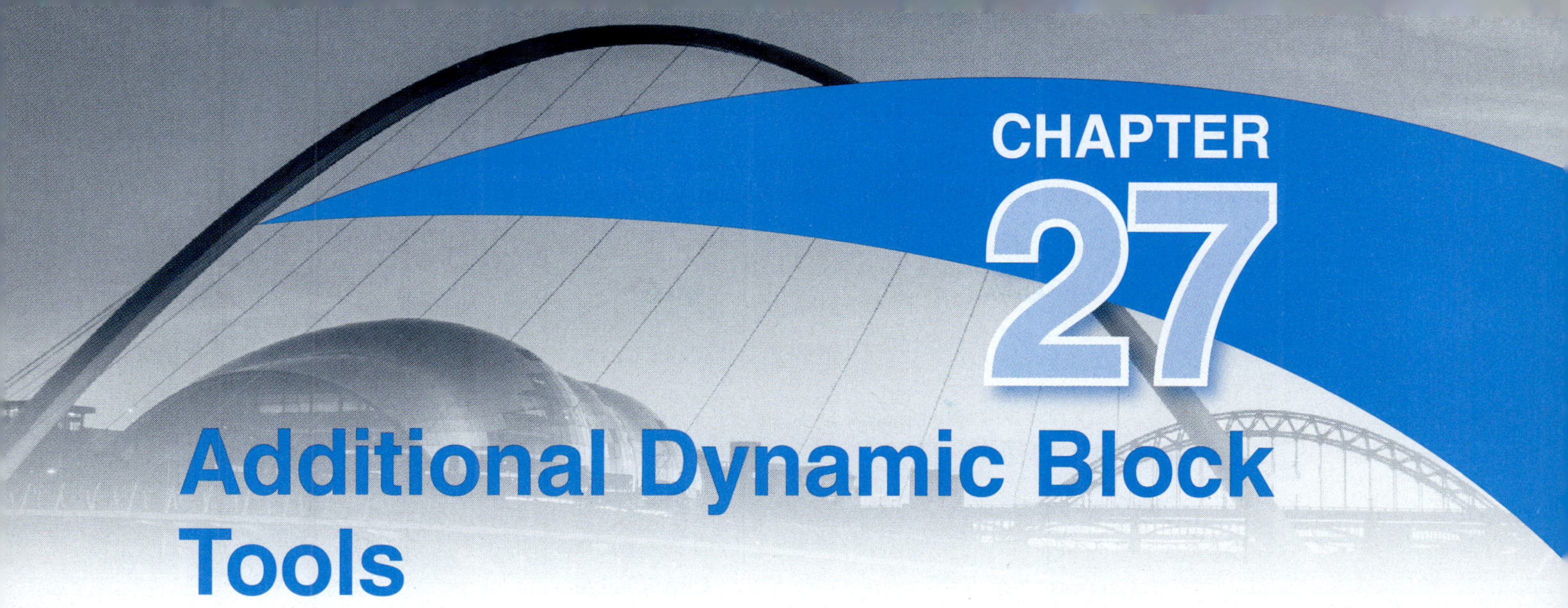

CHAPTER 27

Additional Dynamic Block Tools

Learning Objectives

After completing this chapter, you will be able to:

- ✓ Apply visibility parameters.
- ✓ Create and use lookup parameters.
- ✓ Use parameter sets.
- ✓ Constrain block geometry.
- ✓ Use a block properties table.

This chapter describes adding visibility and lookup parameters to enhance the usefulness of blocks. It also explains how to apply geometric constraints and constraint parameters to blocks as an alternative or in addition to using parameters and actions. Finally, this chapter explores the process of using a block properties table.

Visibility Parameters

A ***visibility parameter*** allows you to assign ***visibility states*** to objects within a block. Selecting a visibility state displays only the objects in the block that are associated with the visibility state. Visibility states expand the capacity of blocks in a symbol library by allowing you to hide or make visible specific objects. A block can include only one visibility parameter. Visibility parameters do not require an action.

visibility parameter: A parameter that allows you to assign multiple views to objects within a block.

visibility states: Views created by selecting block objects to display or hide.

Figure 27-1 provides an example of using a visibility parameter to create four different valve symbols from a single block. To create the block, draw all of the objects representing the different variations, as shown in Figure 27-1A. Then assign a visibility parameter and add visibility states that identify the objects that are visible in each variation. Insert the block and select a visibility state to display the corresponding objects. See Figure 27-1B.

To add a visibility parameter, access the **Visibility** parameter option and pick a location for the parameter label. The parameter automatically includes a single grip by default. When you insert the block and select the grip, a shortcut menu appears listing visibility states. There is no prompt to select objects because the visibility parameter is associated with the entire block.

Ribbon
Block Editor > Action Parameters
Visibility
Type
BPARAMETER > Visibility

BPARAMETER

Figure 27-1.
A—All of the objects composing each unique valve symbol shown together. B—Create each different valve from a single block using a visibility parameter with different visibility states.

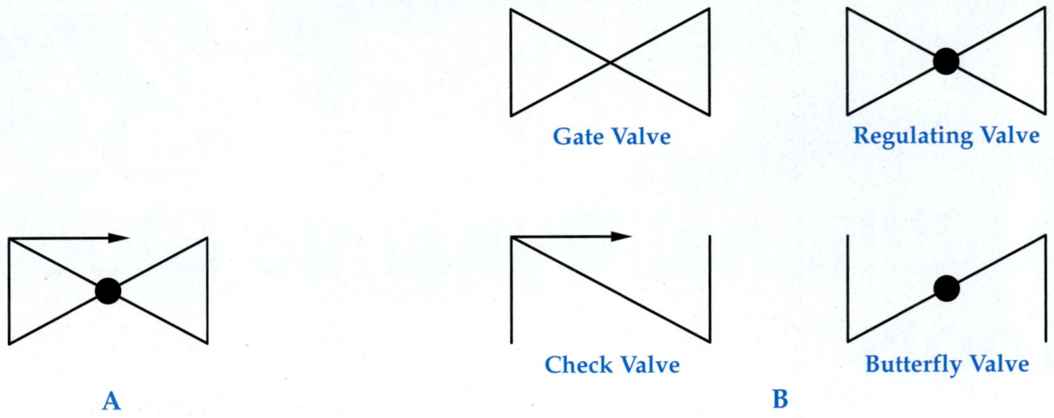

NOTE

Name, **Label**, **Description**, and **Palette** options are available before you specify the parameter. Most of the options are also available from the **Properties** palette if you have already created the parameter.

Creating Visibility States

BVSTATE

Ribbon
Block Editor > Visibility
Visibility States
Type
BVSTATE

The tools in the **Visibility** panel of the **Block Editor** ribbon tab become enabled when you add a visibility parameter. See Figure 27-2. To create a visibility state, access the **BVSTATE** command to display the **Visibility States** dialog box. See Figure 27-3A. Pick the **New...** button to open the **New Visibility State** dialog box shown in Figure 27-3B. Type the name of the new visibility state in the **Visibility state name:** text box. For the valve block example shown in Figure 27-1, an appropriate name could be GATE VALVE, REGULATING VALVE, CHECK VALVE, or BUTTERFLY VALVE, depending on which valve the visibility state represents.

Pick the **Hide all existing objects in new state** radio button to make all of the objects in the block invisible when you create the new visibility state. This allows you to select only the objects that should be visible for the visibility state. Pick the **Show all existing objects in new state** radio button to make all of the objects in the block visible when you create the new visibility state. This allows you to hide objects that should be invisible for the visibility state. Pick the **Leave visibility of existing objects unchanged in new state** radio button to display the objects that are currently visible when you create the new visibility state.

Figure 27-2.
The visibility tools in the **Visibility** panel of the **Block Editor** ribbon tab.

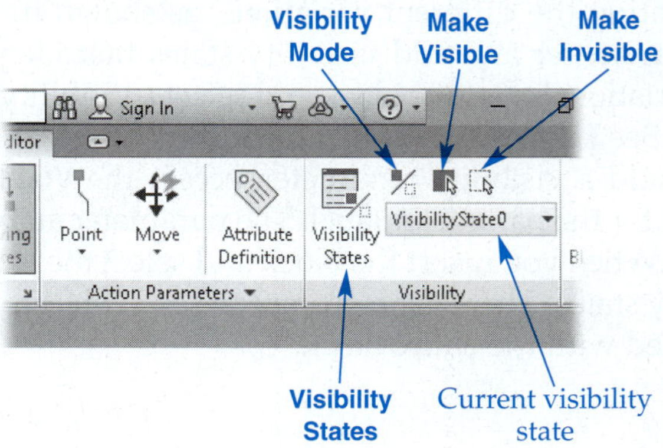

Figure 27-3.
A—Manage visibility states using the **Visibility States** dialog box. B—Create new visibility states using the **New Visibility State** dialog box.

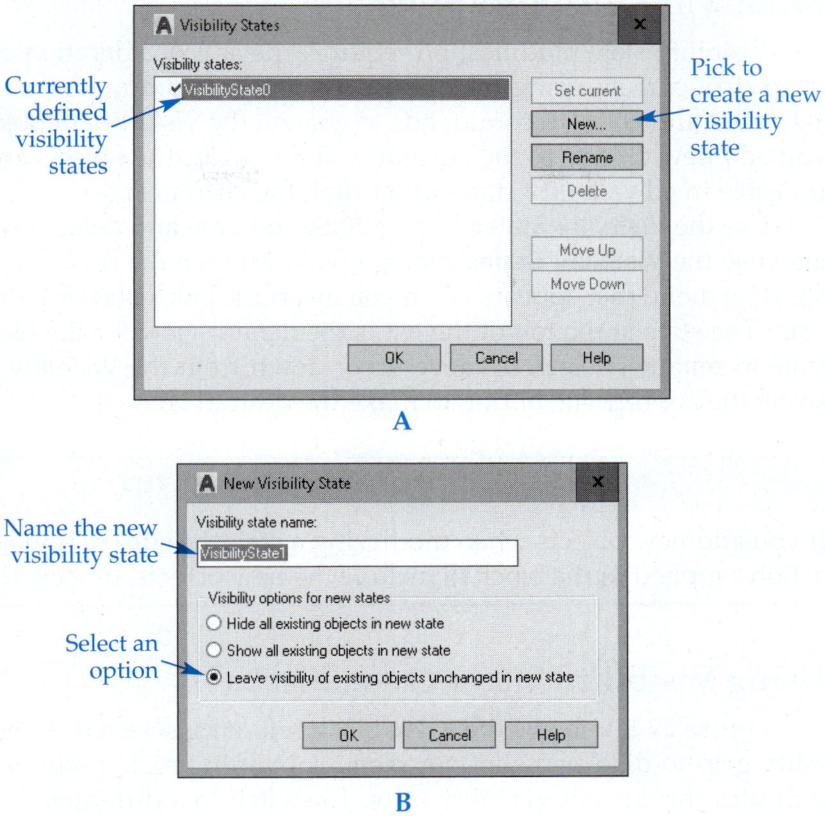

Pick the **OK** button to create the new visibility state. The new state is added to the list in the **Visibility States** dialog box and becomes the current state, as indicated by the check mark next to the name. Pick the **OK** button to return to block editing mode.

Next, use the **BVSHOW** and **BVHIDE** commands to display only the objects that should be visible in the current state. Pick the **Make Visible** button to select objects to make visible. Invisible objects are temporarily displayed semi-transparently for selection. Pick the **Make Invisible** button to select objects to make invisible. For example, to make a visibility state to depict the gate valve shown in Figure 27-4B from the valve block shown in Figure 27-4A, pick the **Make Invisible** button to turn off the filled circle and the arrow. The changes are saved to the visibility state automatically. Pick the **Visibility Mode** button to toggle the visibility mode on and off. Turn on visibility mode to display invisible objects as semi-transparent. Turn off visibility mode to display only visible objects.

Repeat the process to create additional visibility states for the block. The valve block example requires four visibility states. The **Current visibility state** drop-down list displays the current visibility state. Select a state from the drop-down list to make the state current. After you create all visibility states, test and save the block and exit the **Block Editor**. The dynamic block is now ready to use.

BVSHOW
Ribbon
Block Editor > Visibility
Make Visible
Type
BVSHOW

BVHIDE
Ribbon
Block Editor > Visibility
Make Invisible
Type
BVHIDE

Figure 27-4.
A—The VALVE block with all objects visible. B—The VALVE block after making the arrow and filled circle invisible to display the GATE VALVE visibility state.

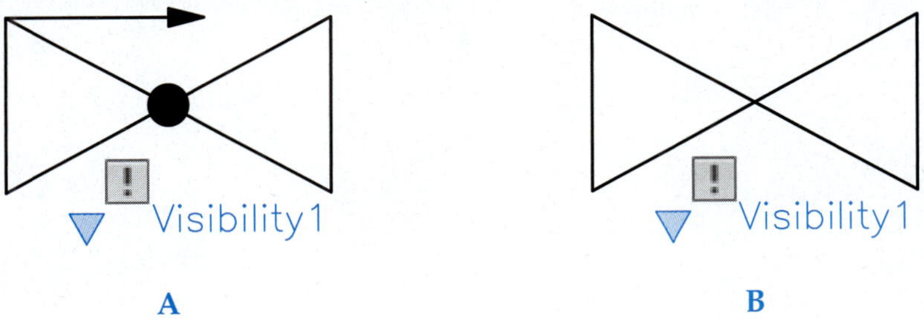

Modifying Visibility States

Visibility state modification requires special consideration. Set the state you want to modify current using the **Current visibility state** drop-down list, and then use the **BVSHOW** and **BVHIDE** commands to change the visibility of objects as needed. When you add new objects to the current visibility state, the objects are automatically set as invisible in all visibility states other than the current state.

Use the **Visibility States** dialog box to rename and delete visibility states. You can also use the **Visibility States** dialog box to arrange the order of visibility states in the shortcut menu that appears when you insert the block and pick the visibility parameter grip. The state at the top of the list is the default view for the block. Pick the visibility state to rename, delete, or move up or down from the **Visibility states:** list box. Then select the appropriate button to make the desired change.

PROFESSIONAL TIP

If you add new objects when modifying a state, be sure to update the parameters and actions applied to the block to include the new objects, if needed.

Using Visibility States Dynamically

Figure 27-5A shows the valve block reference selected for editing. Select the visibility grip to display a shortcut menu containing each visibility state. A check mark indicates the current visibility state. To switch to a different view of the block, select the name of the visibility state from the list. See Figure 27-5B. You can also use the **Properties** palette to select a visibility state.

Exercise 27-1

www.g-wlearning.com/CAD/

Complete the exercise on the companion website.

Figure 27-5.
A—Pick the visibility parameter grip to display a shortcut menu with the available visibility states. The current state is checked. B—Select a different visibility state from the shortcut menu to display a different view of the block.

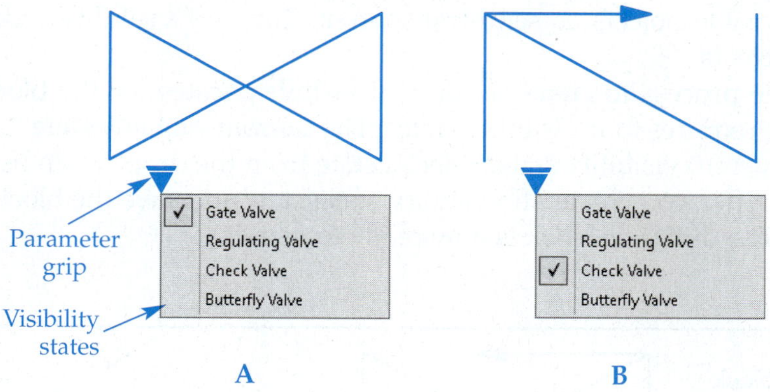

Lookup Parameters

A ***lookup parameter*** creates a lookup property to which you can assign a ***lookup action***. For example, Figure 27-6 shows three valve symbols created from a single block by adjusting the rotation parameter of the middle line. The lookup action allows the middle line rotation to control the length of the start and end lines.

lookup parameter: A parameter that allows tabular properties to be used with existing parameter values.

lookup action: An action used to select a preset group of parameter values to carry out actions with stored values.

To create the valve block shown in Figure 27-7, first draw the geometry of the 0° symbol. Then add a linear parameter and label it Start Line. Select the start point as the bottom of the start line and the endpoint as the top of the start line. Assign a stretch action to the parameter, associated with the top parameter grip. Draw the crossing window around the top of the start line and select the start line as the object to stretch.

Add another linear parameter, labeled End Line. Select the bottom of the end line as the start point and the top of the end line as the endpoint. Assign a stretch action to the parameter, associated with the top parameter grip. Draw the crossing window around the top of the end line and select the end line as the object to stretch.

Next, add a rotation parameter labeled Middle Line. Specify the center of the circle as the base point. Select the right endpoint of the middle line to set the radius, and specify the default rotation angle as 0. Assign a rotation action to the parameter. Pick the center of the circle as the rotation base point, and select the middle line as the object to rotate.

Figure 27-6.
A lookup parameter allows you to create these three valve symbols using the same block. Notice how the geometry changes.

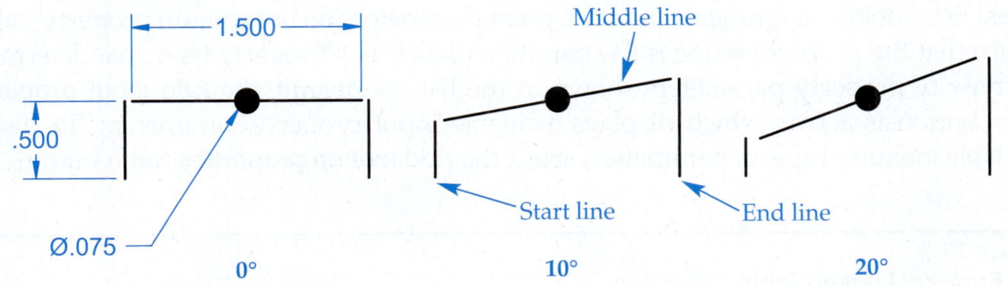

Figure 27-7.
The block of the valve symbol example with linear parameters and stretch actions assigned to the start and end lines and a rotation parameter and rotate action assigned to the middle line.

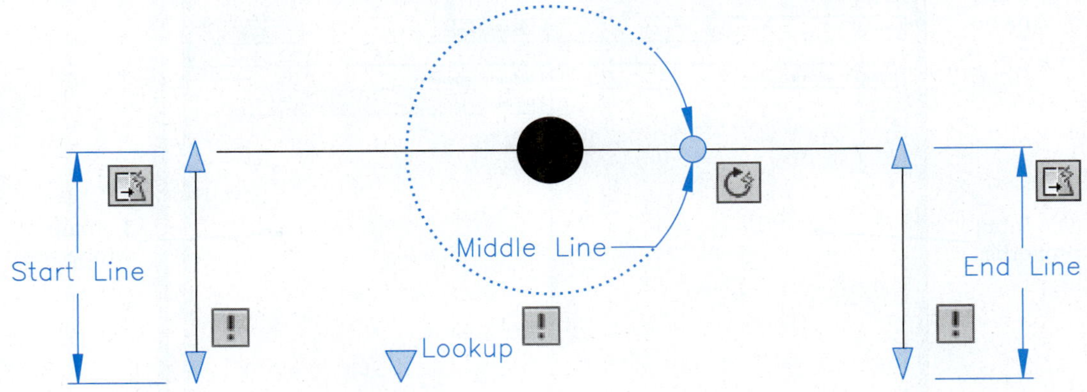

BPARAMETER

Ribbon
Block Editor
> Action Parameters
Lookup
Type
BPARAMETER
> Lookup

To add a lookup parameter, access the **Lookup** parameter option and pick a location for the parameter label. Then enter the number of grips to associate with the parameter. The default **1** option creates a single lookup grip that allows you to use grip editing to carry out the lookup action. When you insert the block and select the grip, a shortcut menu appears listing rotation options. There is no prompt to select objects because a lookup parameter is associated with the entire block.

NOTE

Name, **Label**, **Description**, and **Palette** options are available before you specify the parameter. Most of the options are also available from the **Properties** palette if you have already created the parameter.

Assigning a Lookup Action

BACTIONTOOL

Ribbon
Block Editor
> Action Parameters
Lookup
Type
BACTIONTOOL
> Lookup

To assign a lookup action, access the **Lookup** action option and select a lookup parameter. The **Property Lookup Table** dialog box appears, allowing you to create a lookup table. See Figure 27-8.

Creating a Lookup Table

lookup table: A table that groups the properties of parameters into custom-named lookup records.

A ***lookup table*** groups parameter properties into custom-named lookup records. The **Action name:** display box indicates the name of the lookup action associated with the table. The table is initially blank. To add a parameter property, pick the **Add Properties...** button to open the **Add Parameter Properties** dialog box. See Figure 27-9.

All parameters in the block that contain property values appear in the **Parameter properties:** list. Lookup, alignment, and base point parameters do not contain property values. Notice that the property name is the parameter label. The **Property type** area determines the type of property parameters shown in the list. By default, the **Add input properties** radio button is active, which displays available input property parameters. To display available lookup property parameters, select the **Add lookup properties** radio button.

Figure 27-8.
The **Property Lookup Table** dialog box.

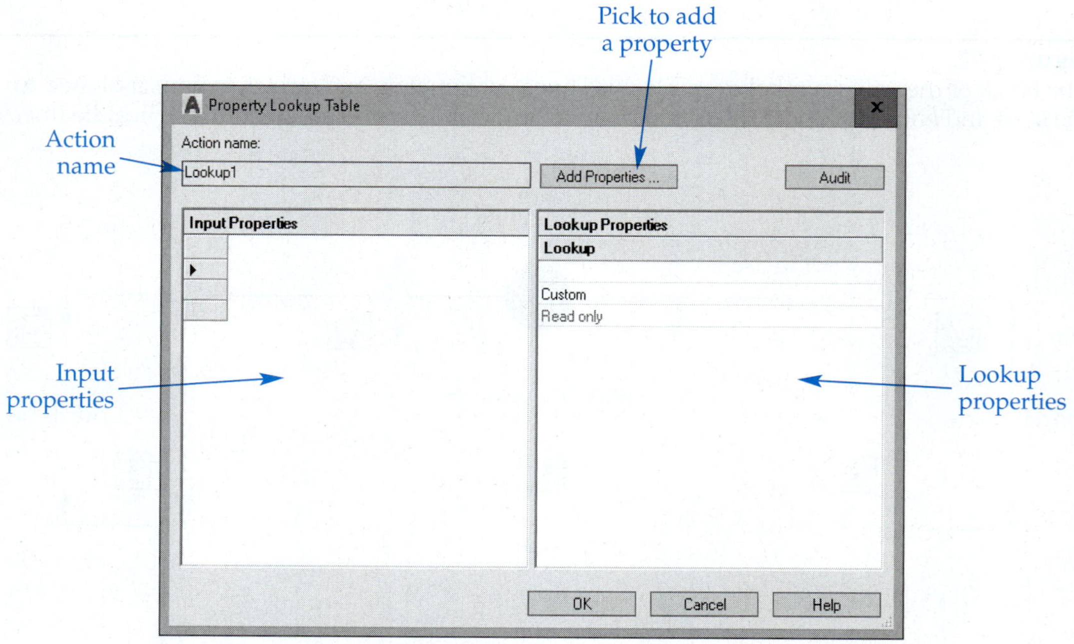

Figure 27-9.
Parameter properties are listed in the **Add Parameter Properties** dialog box.

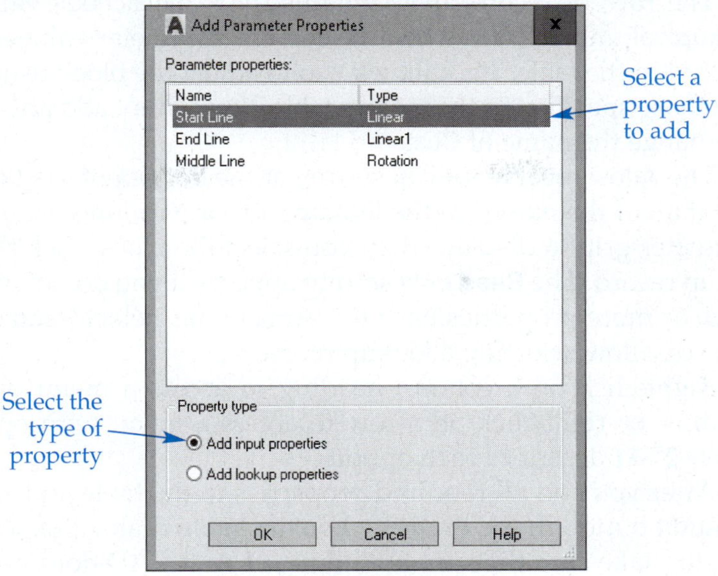

To add parameter properties to the lookup table, select the properties in the **Parameter properties:** list and pick the **OK** button. A new column, named as the parameter property, forms for each parameter in the **Input Properties** area of the **Property Lookup Table** dialog box. See **Figure 27-10**. Use the **Input Properties** area to specify a value for parameters added to the table. Type a value in each cell in the column. Add a custom name for each row, or record, in the **Lookup** column in the **Lookup Properties** area. This area displays the name that appears in the shortcut menu when you insert the block and select the lookup parameter grip.

For the valve symbol example, add the Middle Line, Start Line, and End Line parameter properties to the table. Then complete the lookup table as shown in **Figure 27-10**. Start with the Middle Line values. Press [Enter] after typing the value to add a new blank row and then type the remaining values in each cell. Use the [Enter], [Tab], or arrow keys, or pick in a different cell to navigate through the table.

Figure 27-10.
A lookup table with multiple parameters and values added.

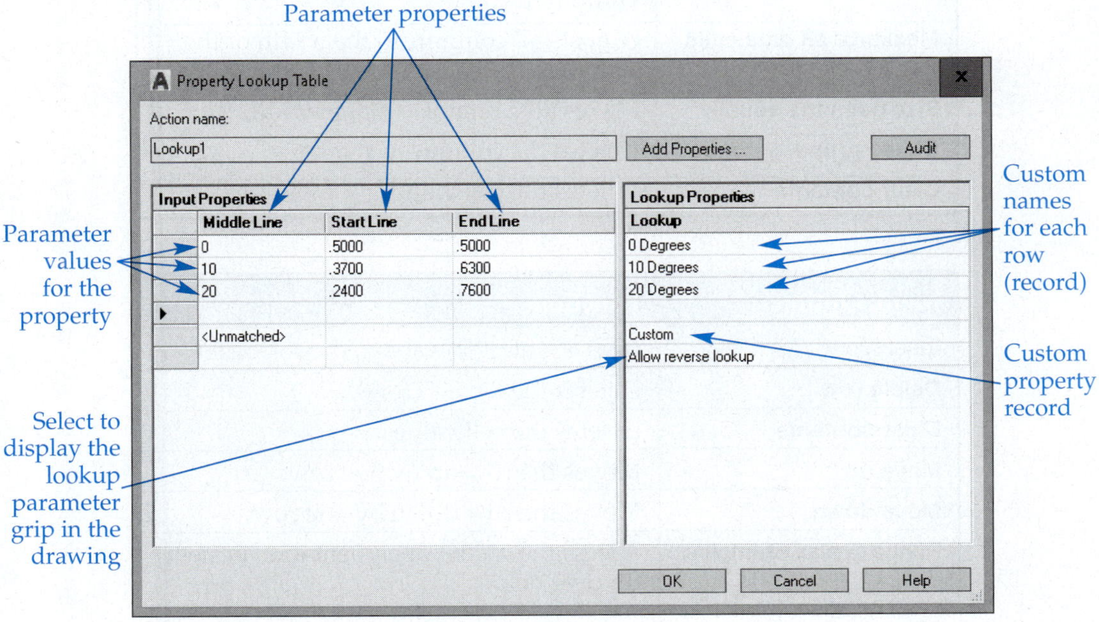

The row, or record, that contains the <Unmatched> value, named Custom in the **Lookup** column, applies when the current parameter values of the block do not match a record in the table. This allows you to adjust the block using parameter values other than those specified in the lookup table. You cannot add any values to the row, but you can change the name of Custom.

The **Allow reverse lookup** setting at the bottom of the **Lookup** column is available only if all of the names in the lookup table are unique. This option allows the lookup parameter grip to display when you select the block. Pick the grip to select a specific lookup record. The **Read only** setting appears if you do not name a lookup property, or if two or more properties have the same name. Select **Read only** from the drop-down list to disallow selecting a lookup record.

Right-click on a column heading to access a menu with options for adjusting columns, or right-click on a row to access a menu with options for adjusting rows. Figure 27-11 describes each option.

After you add all required properties to the table and assign values to each, pick the **Audit** button in the **Property Lookup Table** dialog box to check each record in the table to make sure they are all unique. If AutoCAD does not find errors, pick the **OK** button to return to the **Block Editor**. Test and save the block, and exit the **Block Editor**. The dynamic block is now ready to use.

NOTE

To redisplay the **Property Lookup Table** dialog box, right-click on a lookup action and pick **Display Lookup Table**.

Figure 27-11.
A—Options available when you right-click on a column. B—Options available when you right-click on a row.

Menu Option	Function
Sort	Sorts the records (rows) in ascending or descending order. Pick again to reverse the sort order.
Maximize all headings	Adjusts all columns to the width of the column headings.
Maximize all data cells	Adjusts all columns to the width of the values in the cells.
Size columns equally	Makes all columns equal in width.
Delete property column	Deletes the column.
Clear contents	Deletes the cell values.

A

Menu Option	Function
Insert row	Inserts a new row above the selected row.
Delete row	Deletes the record (row).
Clear contents	Deletes the cell values.
Move up	Moves the row up by one row.
Move down	Moves the row down by one row.
Range syntax examples...	Access the online documentation system for user help.

B

Using a Lookup Action Dynamically

Figure 27-12 shows a valve block reference selected for editing. The figure shows the **Property Lookup Table** dialog box for reference only. Since **Allow reverse lookup** is set in the lookup table, the lookup parameter grip appears along with the other parameter grips. Pick the lookup parameter grip to display a shortcut menu containing each lookup record. The entries in the menu match the entries in the **Lookup** column of the **Property Lookup Table** dialog box. A check mark indicates the current record. To switch to a different view of the block, select the name of the record from the list.

You can change other parameters assigned to the block, such as the linear and rotation parameters of the example block, independently of the named records. When you change any of the parameters, the lookup parameter becomes Custom, because the current parameter values do not match one of the records in the lookup table.

Exercise 27-2

www.g-wlearning.com/CAD/

Complete the exercise on the companion website.

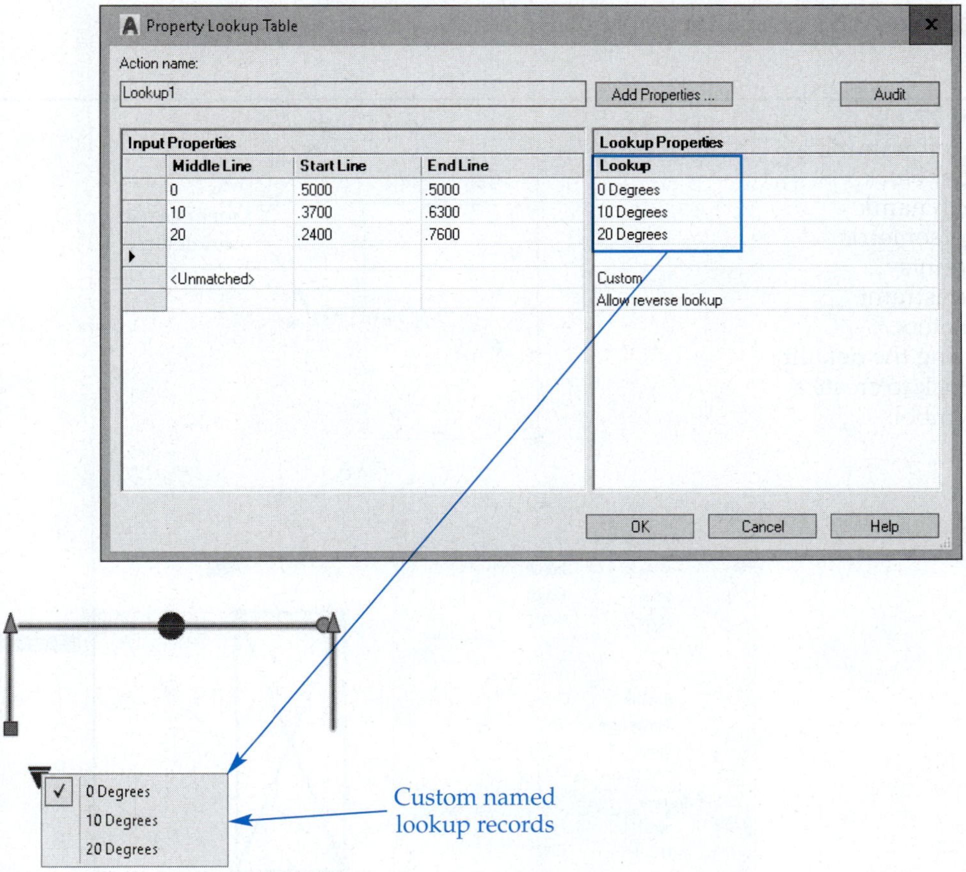

Figure 27-12.
The lookup parameter grip appears when you select the block. The list of available lookup records is displayed when you pick the lookup parameter grip. Notice the correlation between the available options and the lookup property names in the **Property Lookup Table** dialog box.

Parameter Sets

The **Parameter Sets** tab of the **Block Authoring Palettes** palette contains common parameters and actions grouped to enhance productivity. Follow the prompts to create a parameter and automatically associate an action with the parameter. The action forms without any selected objects, as is indicated by the yellow alert icon. If the parameter set contains an action that must include associated objects, as most do, right-click on the action icon and select the appropriate option from the **Action Selection Set** cascading menu. The prompts may differ depending on the type of action.

Constraining Block Geometry

constraint parameters: Dimensional constraints available for block construction to control the size or location of block geometry numerically.

Geometric constraints and ***constraint parameters*** can directly replace action parameters and actions. For example, the block of the cut framing member shown in Figure 27-13A uses geometric constraints to maintain geometric relationships and two linear constraint parameters to specify the member size. When you insert and select the block to edit, use the constraint parameter grips or options in the **Properties** palette to adjust the block. See Figure 27-13B. An alternative is to create the block using two linear parameters.

You may find that geometric constraints and constraint parameters are easier to use than action parameters and actions for certain tasks. However, for some blocks, you will discover that action parameters and actions require less effort than adding geometric constraints and constraint parameters. Decide which dynamic block commands and options are appropriate for the blocks you create.

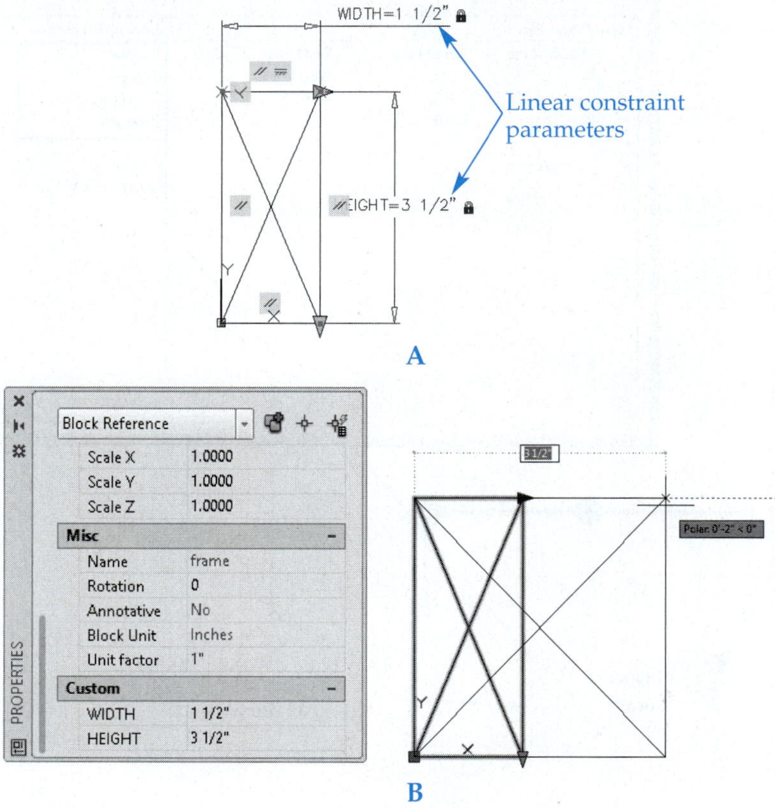

Figure 27-13.
A—A cut framing member block made dynamic using geometric constraints and constraint parameters.
B—Using the default 2×4 block to create a 4×4 symbol.

A combination of dynamic properties is also effective. For example, parameters and actions such as alignment, array, and flip offer dynamic controls that are often not possible using geometric constraints and constraint parameters. Figure 27-14 shows how adding an alignment parameter to the cut framing member block allows you to size and align instances of the block.

Using Geometric Constraints

The geometric constraint commands and options available in the **Block Editor** are identical to those you use to constrain a parametric drawing geometrically. The tools in the **Geometric** panel of the **Parametric** ribbon tab are duplicates of the tools in the **Geometric** panel of the **Block Editor** ribbon tab. Use the geometric constraints in the **Block Editor** as you would in the drawing environment, including the options for relaxing and deleting constraints. The same shortcut menu, **Constraint Settings** dialog box, and **Properties** palette functions apply. Review Chapter 22 for information on adding geometric constraints.

Assign constraints to block objects before you define the block or during block editing to create a dynamic block. See Figure 27-15A. Once you define and insert the block, only constraint parameters, action parameters, or actions influence geometric constraints. This allows you to use blocks as objects in parametric drawings. For example, you can insert and rotate the block, as shown in Figure 27-15B, even though the block definition includes a horizontal constraint. Use constraints in the drawing to locate blocks and establish geometric relationships between blocks and other objects. See Figure 27-15C.

NOTE

Use geometric constraints in the block environment to form geometric constructions in specific situations when standard AutoCAD commands are inefficient or ineffective.

Exercise 27-3

www.g-wlearning.com/CAD/

Complete the exercise on the companion website.

Figure 27-14.
Using geometric constraints and constraint parameters to adjust the size of a cut framing member symbol. An alignment parameter aligns each member for specific applications.

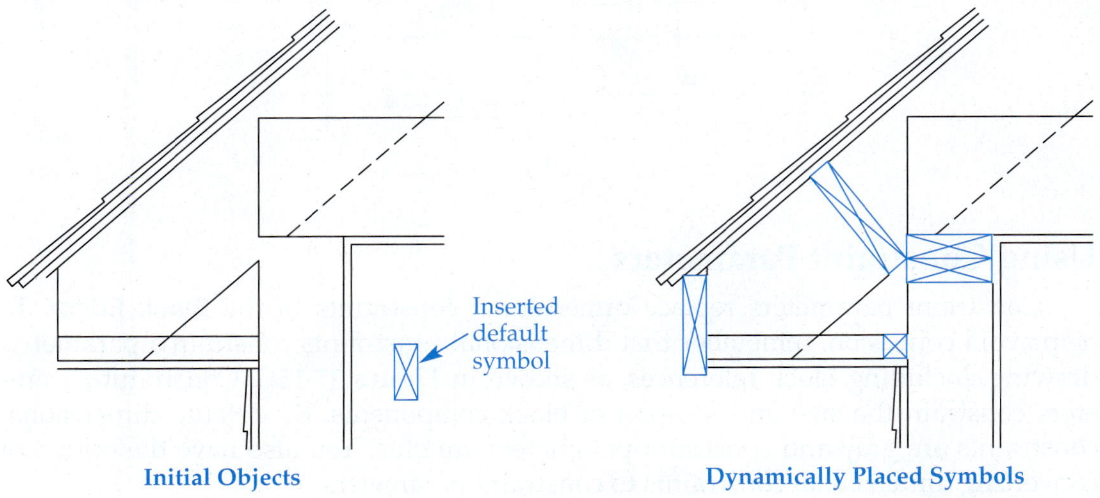

Figure 27-15.
A—A wide flange block made dynamic using geometric constraints and constraint parameters. B—You can rotate the block reference in the drawing because the constraints define the size and shape of block geometry during definition and when the block is adjusted dynamically. C—Constrain blocks in a drawing as you would any other geometry.

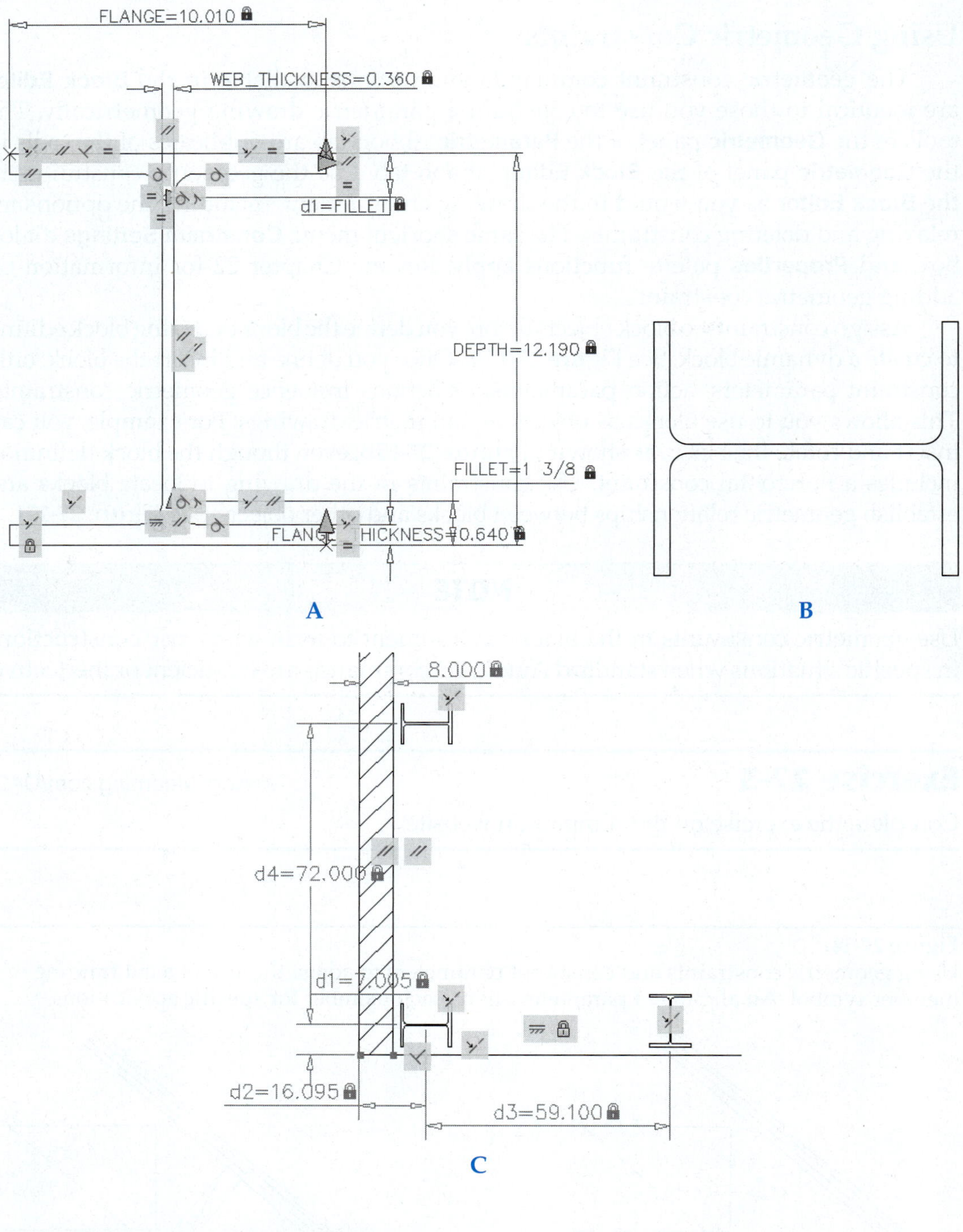

Using Constraint Parameters

Constraint parameters replace dimensional constraints in the **Block Editor**. To help avoid confusion, remember that dimensional constraints constrain a parametric drawing, including block references, as shown in Figure 27-15C. Constraint parameters constrain the size and location of block components. By default, dimensional constraints are gray and constraint parameters are blue. You also have the option of converting dimensional constraints to constraint parameters.

You can often use constraint parameters instead of action parameters and actions. If you do not use action parameters, you must include constraint parameters to create a dynamic block. The constraint parameter commands and options available in the **Block Editor** function much like the commands and options you use to constrain a parametric drawing dimensionally. Review Chapter 22 for information on adding dimensional constraints.

The **BCPARAMETER** command replaces the **DIMCONSTRAINT** command in the **Block Editor** and provides **Linear**, **Horizontal**, **Vertical**, **Aligned**, **Angular**, **Radius**, and **Diameter** options. The **Linear** option is the default in the **Block Editor** ribbon tab. You can also use the **BCPARAMETER** command to convert dimensional constraints to constraint parameters. Each constraint parameter is a separate **BCPARAMETER** command option. A quick way to add constraint parameters using the **BCPARAMETER** command is to pick the appropriate button from the **Dimensional** panel of the **Block Editor** ribbon tab.

Ribbon
Block Editor
> Dimensional
Linear
Type
BCPARAMETER
> (option)
BCPARAMETER

The process of adding constraint parameters is identical to that for adding dimensional constraints, except that the number of grips displayed for constraint parameters can be changed. Constraint parameters are essentially a combination of dimensional constraints and action parameters. The constraint parameters given custom names in Figure 27-16 are those that can be adjusted for specific block references. As when creating a parametric drawing, the other constraint parameters are required to define the block and define specific geometric relationships. Notice the expressions applied to these values.

To create a constraint parameter, follow the prompts to make the required selections, pick a location for the dimension line, and enter a value to form the constraint. If you plan to assign a single grip to a constraint parameter, select the point associated with the grip second. The number of grips displayed can be set using the **Number of Grips** option in the **Properties** palette.

NOTE

If you attempt to over-constrain a block, a message appears indicating that adding the geometric constraint or constraint parameter is not allowed. You cannot create reference constraint parameters.

Figure 27-16.
Using constraint parameters to form a dynamic block of a spacer. A single grip is all that is required for each constraint parameter. Do not assign or rename constraint parameters that do not control geometry.

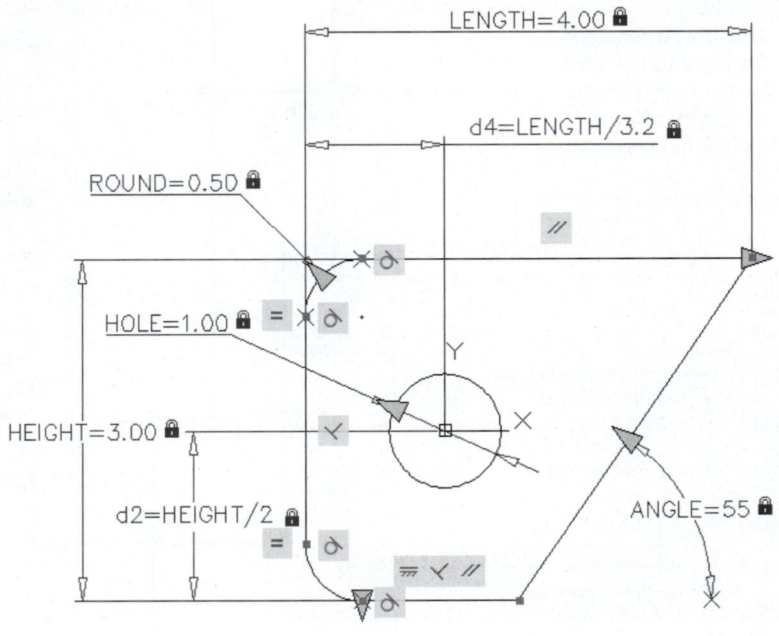

PROFESSIONAL TIP

As when adding dimensional constraints or action parameters, change the constraint parameter name to a custom, more descriptive name. Naming labels helps organize parameters and helps identify each parameter when you control the block dynamically. Custom parameters also appear in the **Custom** category of the **Properties** palette.

BCPARAMETER

Ribbon
Block Editor > Dimensional
Convert
Type
BCPARAMETER > Convert

Use the **Convert** option of the **BCPARAMETER** command to convert a dimensional constraint to a constraint parameter. This allows you to prepare a dynamic block using existing dimensional constraints. Access the **Convert** option and pick the dimensional constraint to convert. The dimensional constraint becomes the corresponding constraint parameter and includes the default number of grips.

Controlling Constraint Parameters

Control and adjust constraint parameters using a combination of the same techniques you use to manage dimensional constraints and action parameters. Many of the options from shortcut menus, the **Constraint Settings** dialog box, and the **Properties** palette apply. Right-click with no objects selected to access options for displaying and hiding parametric constraints and for accessing the **Constraint Settings** dialog box. Select a constraint parameter and then right-click to display a shortcut menu with options for editing the constraint, changing the name format, and redefining the grips.

As with dimensional constraints and the action parameters, the **Properties** palette provides an effective way to control and enhance constraint parameters. You can also use the **Parameters Manager**. Figure 27-17 shows a foundation detail block with linear constraint parameters. Notice the multiple options available in the **Properties** palette for adjusting the selected constraint parameter.

Figure 27-17.
Adjust constraint parameters as you would dimensional constraints and action parameters. Use the **Properties** palette to add value sets.

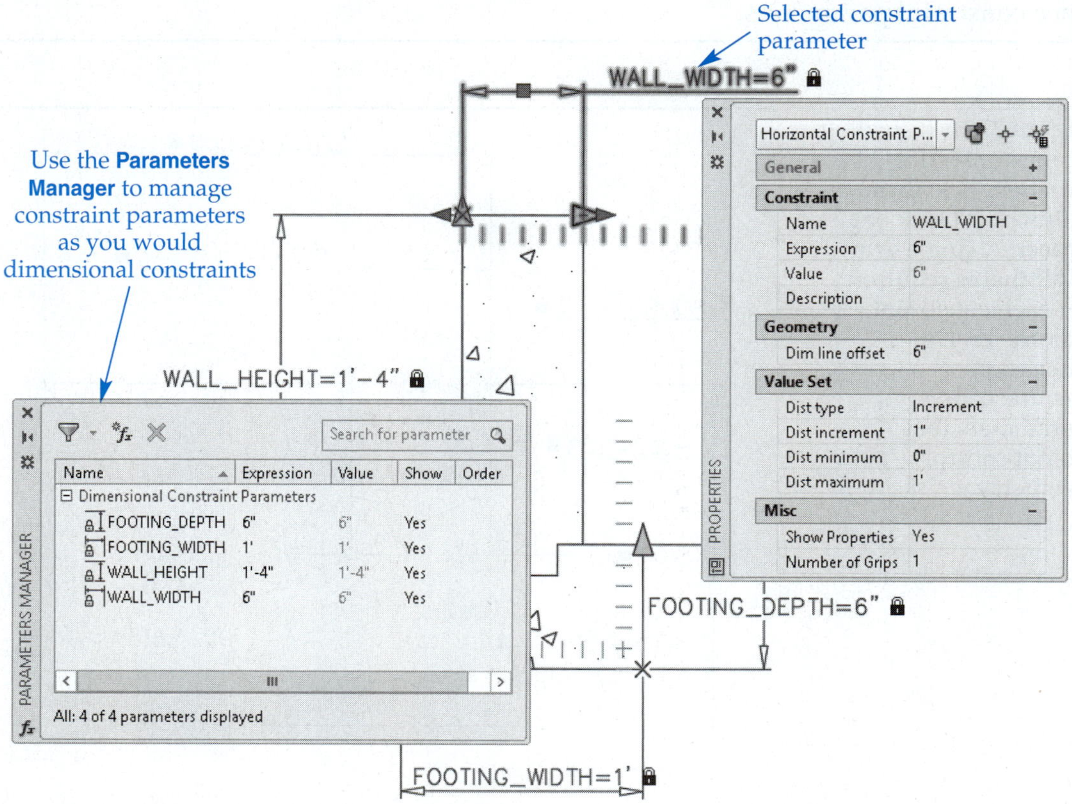

Use the options in the **Value Set** category of the **Properties** palette to assign value sets to a constraint parameter. Each constraint parameter in the Figure 27-17 example uses an incremental value to help ensure that you select an appropriate value when adjusting a block reference. You can also create a list of possible sizes. The processes of creating a value set in the **Properties** palette and using value sets are identical for constraint parameters and action parameters.

Exercise 27-4

www.g-wlearning.com/CAD/

Complete the exercise on the companion website.

Additional Parametric Tools

The **Block Editor** offers additional options for adding constraints to blocks. Many of the tools, such as the **DELCONSTRAINT** command, function the same in block editing mode as in drawing mode. However, the **Block Editor** does offer some unique parametric construction commands.

The **BCONSTRUCTION** command allows you to create construction geometry to aid geometric construction and constraining. Construction geometry appears only in the block definition. See Figure 27-18. Access the **BCONSTRUCTION** command and select the objects to convert to or revert from construction geometry. Then press [Enter]. Next, select the **Convert** option to convert non-construction objects to the construction format, or select **Revert** to return construction geometry to the standard format. You can also use the **Hide all** option to hide all existing construction geometry before selecting objects, or use the **Show all** option to display all construction geometry.

Ribbon
Block Editor > Manage
Construction Geometry
Type
BCONSTRUCTION

BCONSTRUCTION

Use the **BCONSTATUSMODE** command to toggle constraint status identification on and off. When you turn constraint status mode on, objects with no constraints appear white (black) by default, objects assigned some form of constraints are blue, and fully constrained geometry is magenta. If the block contains a constraint error, objects associated with the error are red. Using constraint status is helpful, especially if you want to constrain objects in a certain order or confirm that geometry has been fully constrained.

Ribbon
Block Editor > Manage
Constraint Display Status
Type
BCONSTATUSMODE

BCONSTATUSMODE

NOTE

Use the **BESETTINGS** command to access the **Block Editor Settings** dialog box. There you can adjust parameter and parameter grip color and appearance, constraint status colors, and other **Block Editor** settings.

Figure 27-18.
A weld nut block in which construction geometry aids geometric construction and constraining.

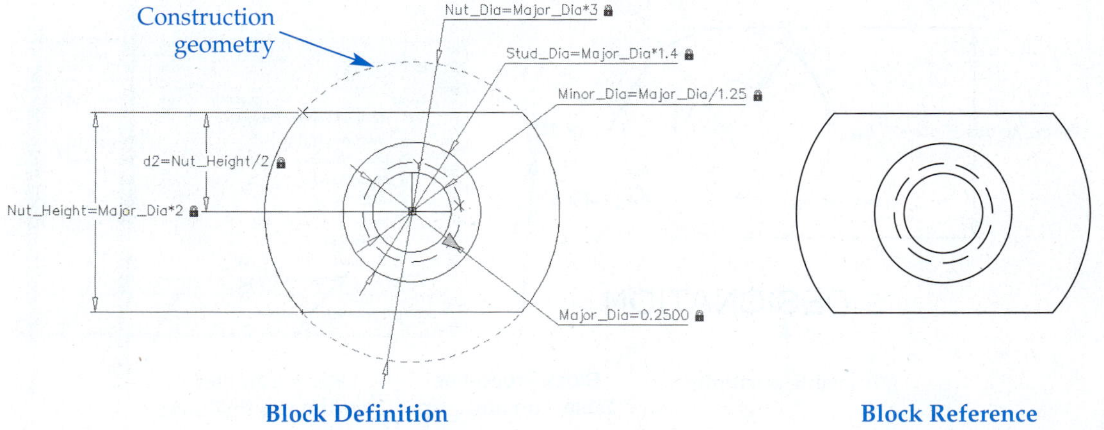

Block Properties Tables

block properties table: A table of action parameters and/or constraint parameters that allows you to create multiple block properties and then select them to create block references.

A ***block properties table*** allows you to assign specific values to multiple block properties, and then select a specific group, or row, of properties to create block references. The concept is similar to using a lookup action parameter. A block properties table can include action parameters, constraint parameters, or both. You can also add attributes to the table, which is often appropriate for naming each record, or row.

Figure 27-19 shows the block of the front view of a heavy hex nut in the **Block Editor**. The block includes an appropriate level of constraints and includes constraint parameters to direct dynamic changes. The block also includes an invisible and preset attribute for defining the designation of each different nut and, as shown in the **Parameters Manager**, a user-defined parameter for the nut thickness.

PROFESSIONAL TIP

It is critical that you assign the **Preset** mode to attributes that you include in a block properties table. This allows the attribute value to adjust to the selected block record. The **Preset** mode requires no default value, and you will not receive a prompt to adjust the value.

BTABLE
Ribbon
Block Editor
> Dimensional
Block Table
Type
BTABLE

After you create parameters and attributes, access the **BTABLE** command and select the parameter location. Then enter the number of grips to associate with the parameter. The default **1** option creates a single grip that allows you to select a table record from the grip shortcut menu. If you select the **0** option, you can only use the **Properties** palette to select a record. The **Palette** option, available before you specify the parameter location or from the **Properties** palette, determines whether the label appears in the **Properties** palette when you select the block reference. The **Block Properties Table** dialog box appears, allowing you to create a block properties table. See Figure 27-20.

Creating a Block Properties Table

A block properties table groups the properties of parameters into custom records, or rows. To add parameter properties, pick the **Add Properties** button to open the **Add Parameter Properties** dialog box. See Figure 27-21. All parameters in the block that contain property values appear in the **Parameter properties:** list. Lookup, alignment, and base point parameters do not contain property values. Notice that the property name is the parameter label.

Figure 27-19.
A heavy hex nut block definition ready to use to create a block properties table.

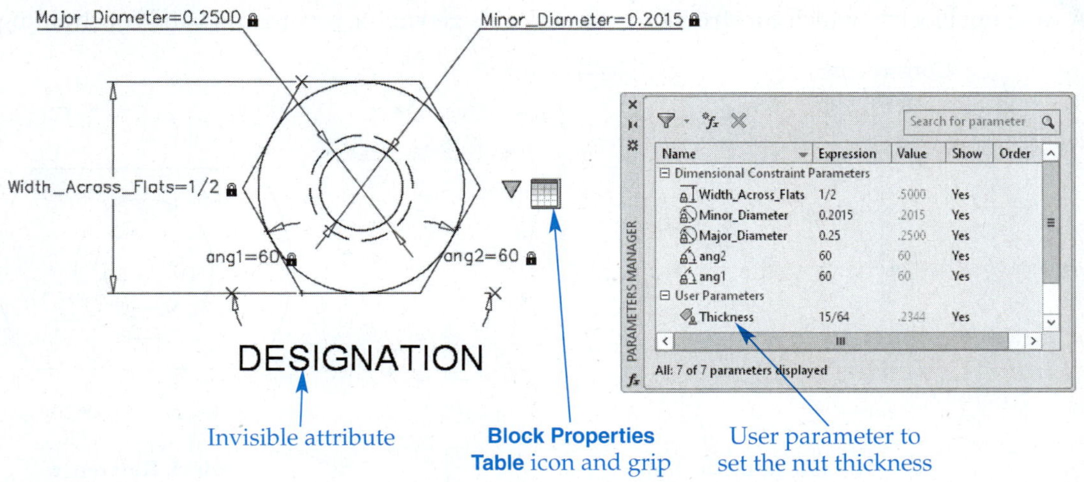

Figure 27-20.
The **Block Properties Table** dialog box.

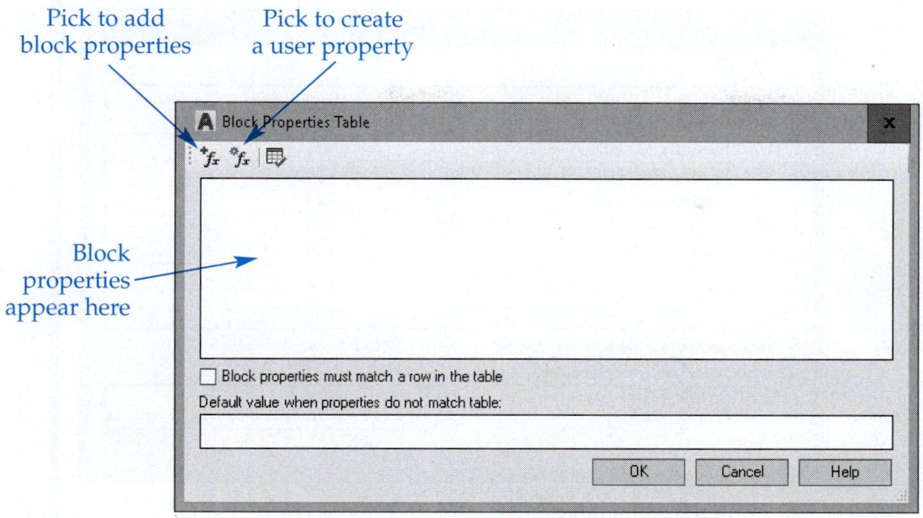

Figure 27-21.
Parameter properties are listed in the **Add Parameter Properties** dialog box.

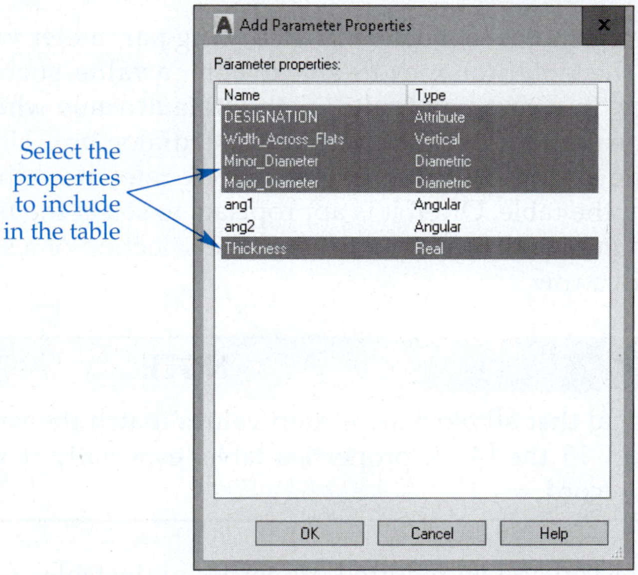

To add parameter properties to the table, select the properties in the **Parameter properties:** list and pick the **OK** button. A column appears in the table for each parameter property. Type a value in each cell in the column. A new row forms automatically when you enter a value in a cell. See Figure 27-22. Press [Enter], [Tab], [Shift]+[Enter], or the arrow keys, or pick in a different cell to navigate through the table.

For the nut block example, complete the table as shown in Figure 27-22. The **DESIGNATION** column references the attribute property. The value you enter in the **DESIGNATION** text box in each row specifies the record name. This value appears in the shortcut menu when you insert the block and select the block properties table parameter grip.

NOTE

Right-click on a column heading to access a menu with options for adjusting columns. Right-click on a row to access a menu with options for adjusting rows. The options are similar to those for adjusting lookup table columns and rows.

Figure 27-22.
A block properties table with multiple parameters and values added.

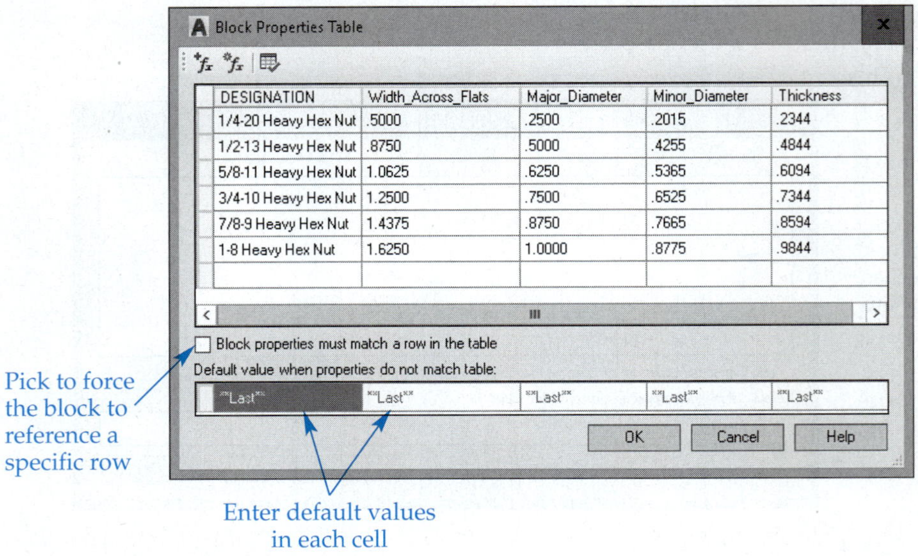

You can adjust a block reference using parameter values other than those specified in the table. You may be able to enter a value, such as the value of an attribute property, in a text box found in the **Default value when properties do not match table:** area of the **Block Properties Table** dialog box. Use the ****Last**** option to use the value assigned to the previous block reference when you specify a value not found in the table. Often it is appropriate to select the **Block properties must match a row in the table** check box to force the selection of a specific record, matching all values in a row.

NOTE

It is critical that all block definition values match the values specified in the default block row in the block properties table, especially if you force the selection of a specific record.

After you add all required properties to the table and assign values to each, pick the **Audit** button in the **Block Properties Table** dialog box to check each record in the table. Make sure the records are unique and that there are no discrepancies between the block definition and the table values. If AutoCAD does not find errors, pick the **OK** button to return to the **Block Editor**. Test and save the block, and exit the **Block Editor**. The dynamic block is now ready to use.

NOTE

To redisplay the **Block Properties Table** dialog box, double-click on the parameter, or access the **BTABLE** command.

Using a Block Properties Table Dynamically

Figure 27-23 shows the inserted nut block example selected for editing. Since the block table parameter includes a grip, a grip appears that you can use to select a specific block style. The entries in the shortcut menu match the rows in the **Block Properties Table** dialog box. A check mark indicates the current record. To switch to a different view of the block, select the name of the record from the list. You can also pick the **Properties Table...** option to display the **Block Properties Table** dialog box in drawing mode. Double-click a row to activate it. In this example, no other grips were assigned to blocks. This makes the table and the **Properties** palette the only two methods to select a block reference format.

PROFESSIONAL TIP

The options for developing dynamic blocks and creating parametric drawings can become confusing. Keep the following concepts in mind as you proceed:

- Use constraints as an alternative or in addition to action parameters and actions.
- Constraints allow you to create a parametric drawing or a dynamic block.
- Assign constraints to create a dynamic block during block definition or while editing the block.
- Treat inserted blocks like any other object when preparing a parametric drawing.

Exercise 27-5

www.g-wlearning.com/CAD/

Complete the exercise on the companion website.

Express Tools

www.g-wlearning.com/CAD/

Block Express Tools

The **Blocks** and **Draw** panels of the **Express Tools** ribbon tab include additional commands related to blocks. For information about the most useful block express tools, go to the companion website, navigate to this chapter in the **Contents** tab, and select **Block Express Tools**.

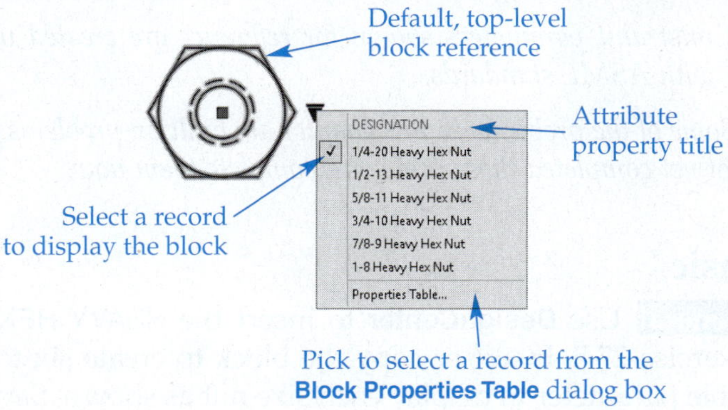

Figure 27-23. The block table parameter grip displays when you select the block. The list of available records appears when you pick the parameter grip. Notice the correlation between the available options and the names in the **Block Properties Table** dialog box.

Chapter Review

Answer the following questions using the information in this chapter.

1. Define *visibility parameter.*
2. What are visibility states?
3. How do you display the shortcut menu that allows you to select from the existing visibility states of a block?
4. When you select a block and display the visibility parameter shortcut menu, what indicates the current visibility state?
5. Define *lookup parameter.*
6. Explain the basic function of a lookup action.
7. Identify the basic function of a lookup table.
8. What is a parameter set?
9. What ribbon tab, in addition to the **Parametric** ribbon tab, contains geometric constraint tools?
10. When can you assign constraints to block objects to create a dynamic block?
11. What takes the place of dimensional constraints in the **Block Editor**?
12. What command and option allow you to convert a dimensional constraint to a constraint parameter?
13. How can you create construction geometry to aid geometric construction in the **Block Editor**?
14. How can you toggle constraint status identification on and off in the **Block Editor**?
15. What does a block properties table allow you to do?

Drawing Problems

Start AutoCAD if it is not already started. Start a new drawing for each problem using an appropriate template of your choice. The template should include layers and text, dimension, multileader, and table styles, when necessary, for drawing the given objects. Add layers and text, dimension, multileader, and table styles as needed. Draw all objects using appropriate layers. Use appropriate text, dimension, multileader, and table styles, justification, and format. Follow the specific instructions for each problem. Use only drawing and editing commands and techniques you have already learned. Use your own judgment and approximate dimensions when necessary. Apply dimensions accurately using ASME or appropriate industry standards.

Note: Constraint parameters shown for reference are created using AutoCAD and may not comply with ASME standards.

Note: Some of the problems in this chapter are built on problems from previous chapters. If you have not yet completed those problems, complete them now.

▼ Basic

1. General Use **DesignCenter** to insert the HEAVY HEX NUT block you created in Exercise 27-5. Insert or copy the block to create six total symbols. Use the block table parameter to display each size nut as shown. Save the drawing as P27-1.

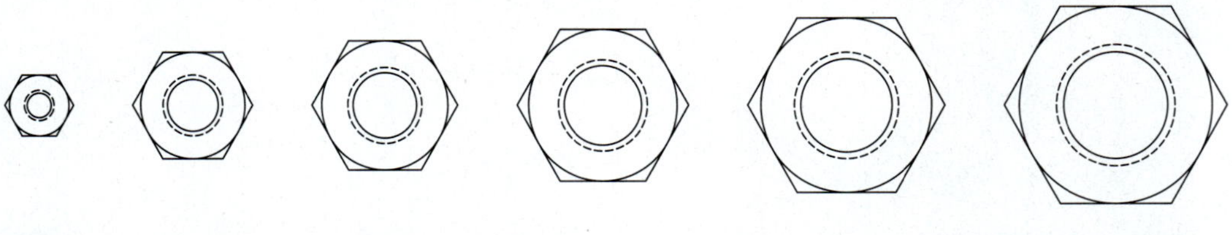

2. **Architectural** Save P22-2 as P27-2. If you have not yet completed Problem 22-2, do so now. In the P27-2 file, create a block named FIXTURE, select all objects, and pick the center of the circle as the insertion base point. Open the block in the **Block Editor** and convert the dimensional constraints to constraint parameters. Insert the FIXTURE block into the drawing three times to create the 4×4, 6×6, and 4×5 symbols as shown.

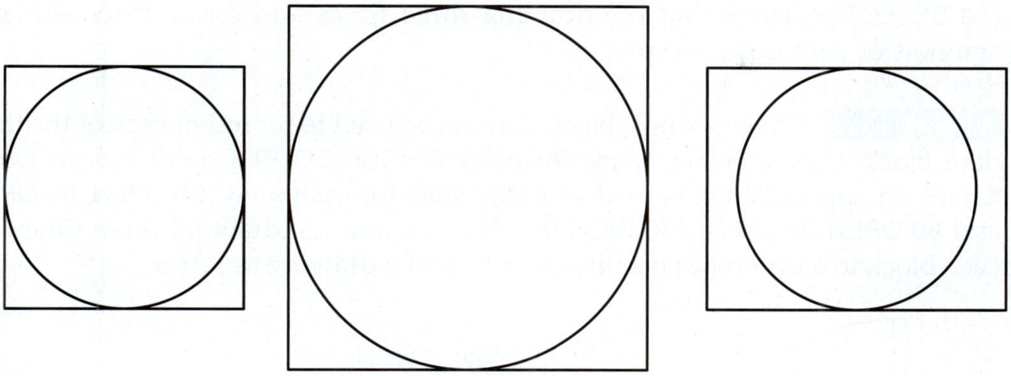

3. **Mechanical** Save P22-6 as P27-3. If you have not yet completed Problem 22-6, do so now. In the P27-3 file, create a block named PLATE, select all objects, and pick the center of the plate as the insertion base point. Open the block in the **Block Editor** and convert the construction rectangle to construction geometry. Convert the dimensional constraints to constraint parameters. Insert the PLATE block into the drawing and create the drawing shown. Do not add dimensions.

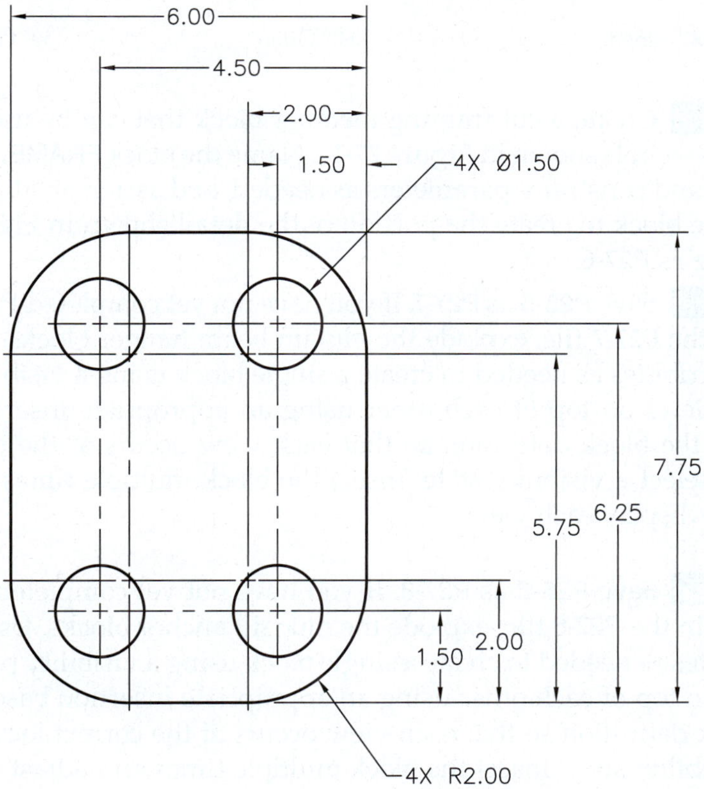

Drawing Problems - Chapter 27

▼ Intermediate

4. **Mechanical** Save P22-5 as P27-4. If you have not yet completed Problem 22-5, do so now. In the P27-4 file, create a block named SELECTOR, select all objects, and pick the center of the center circle as the insertion base point. Open the block in the **Block Editor** and convert the construction lines to construction geometry. Convert the dimensional constraints to constraint parameters. Insert the SELECTOR block into the drawing three times and create three different symbols of your own design.

5. **Architectural** Create a single block that can be used to represent each of the three door blocks shown below. Name the block 30 INCH DOOR. Do not include labels. Create an appropriately named visibility state for each view: 90 OPEN, 60 OPEN, and 30 OPEN. Insert the 30 INCH DOOR block into the drawing three times. Set each block to a different visibility state. Save the drawing as P27-5.

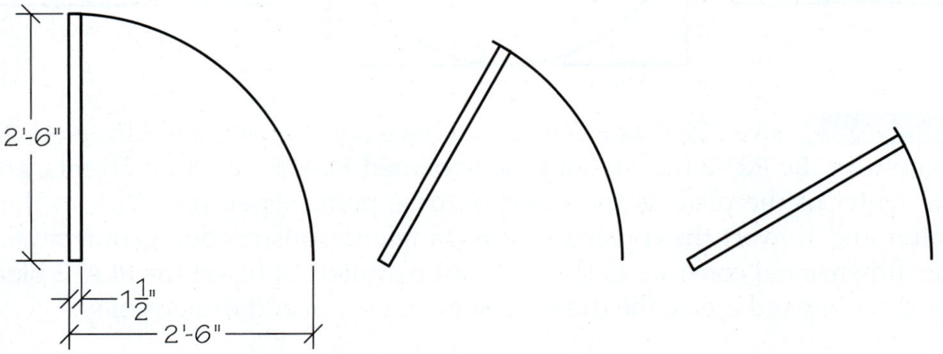

90° Open 60° Open 30° Open

6. **Architectural** Create a cut framing member block that can be used to represent each of the symbols shown in Figure 27-14. Name the block FRAME. Add geometric constraints and constraint parameters as needed, and assign an alignment parameter. Use the block to create the portion of the detail shown in Figure 27-14. Save the drawing as P27-6.

7. **Architectural** Save P25-5 as P27-7. If you have not yet completed Problem 25-5, do so now. In the P27-7 file, explode the glulam beam hanger blocks. Use the geometry and attributes as needed to create a single block using a visibility parameter. Stack the views on top of each other using an appropriate insertion base point location in the block definition so that each view occurs at the correct location when you select a visibility state. Insert the block multiple times and adjust the visibility to display each view.

8. **Architectural** Save P25-6 as P27-8. If you have not yet completed Problem 25-6, do so now. In the P27-8 file, explode the mudsill anchor blocks. Use the geometry and attributes as needed to create a single block using a visibility parameter. Stack the views on top of each other using an appropriate insertion base point location in the block definition so that each view occurs at the correct location when you select a visibility state. Insert the block multiple times and adjust the visibility to display each view.

▼ Advanced

9. **Architectural** Create a foundation footing and wall block that can be used to represent various construction requirements. Name the block Foundation. Add the geometric constraints and constraint parameters shown. Include 1″ increment value sets for each constraint parameter. The block should stretch symmetrically as shown. Save the drawing as P27-9.

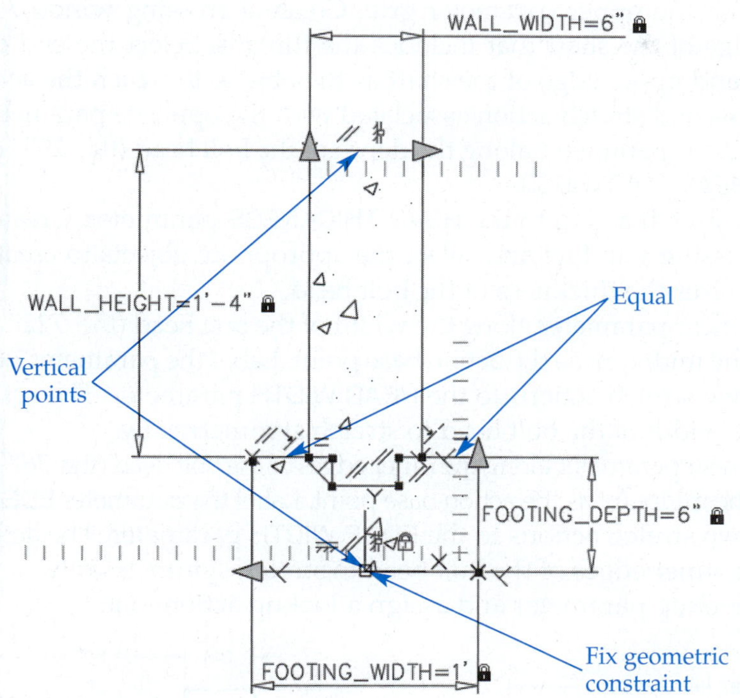

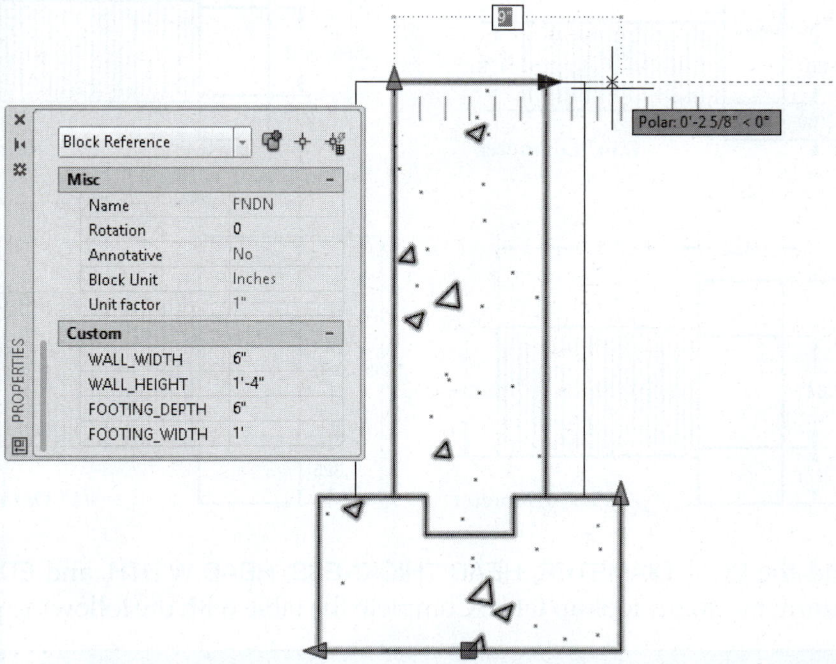

Drawing Problems - Chapter 27

10. **Mechanical** The bolt shown is available in four different sizes. Create a dynamic block that will allow the bolt diameter and size of the bolt head to be changed in a single operation.

A. Draw the objects that make up the 7/16″ diameter bolt and create a block named BOLT. Do not include dimensions.

B. Insert a linear parameter along the bolt diameter (the .438″ dimension). Specify the midpoint as the action base point. Label the parameter BOLT DIAMETER.

C. Assign two stretch actions to the BOLT DIAMETER parameter. Associate the first action with the upper parameter grip. Create a crossing window around the upper edge of the shaft that includes the threads. Select the end of the shaft, threads, and upper edge of the shaft as the objects to which the action applies. Create a second stretch action associated with the opposite parameter grip.

D. Insert a linear parameter along the depth of the bolt head (the .297″ dimension). Label it HEAD THICKNESS.

E. Assign a stretch action to the HEAD THICKNESS parameter. Create an appropriate crossing window and select the appropriate objects to create an action that increases the thickness of the bolt head.

F. Insert a linear parameter along the width of the bolt head (the .722″ dimension). Specify the midpoint as the action base point. Label the parameter HEAD WIDTH.

G. Assign two stretch actions to the HEAD WIDTH parameter. The actions should allow the width of the bolt head to stretch symmetrically.

H. Insert a linear parameter along the inner edges of the bolt head (the .361″ dimension). Specify the midpoint as the action base point. Label the parameter EDGE WIDTH.

I. Assign two stretch actions to the EDGE WIDTH parameter. The actions should allow the inner edges of the bolt head to stretch symmetrically.

J. Insert a lookup parameter and assign a lookup action to it.

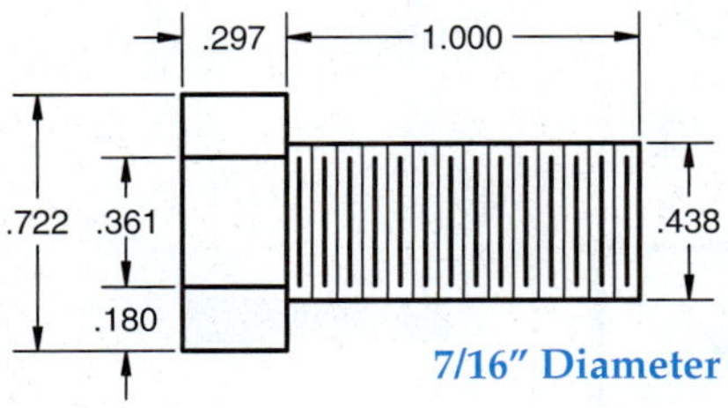

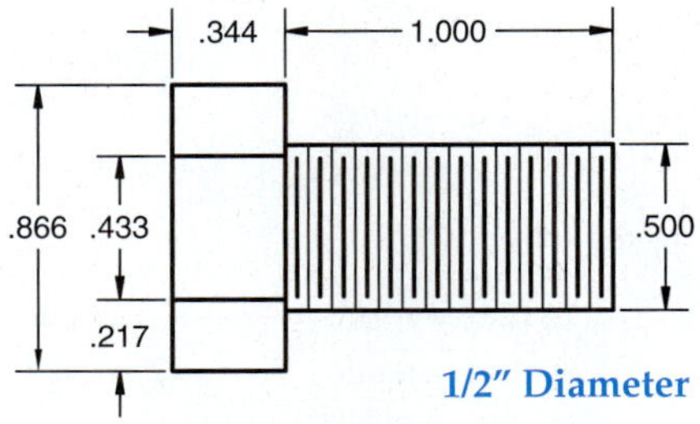

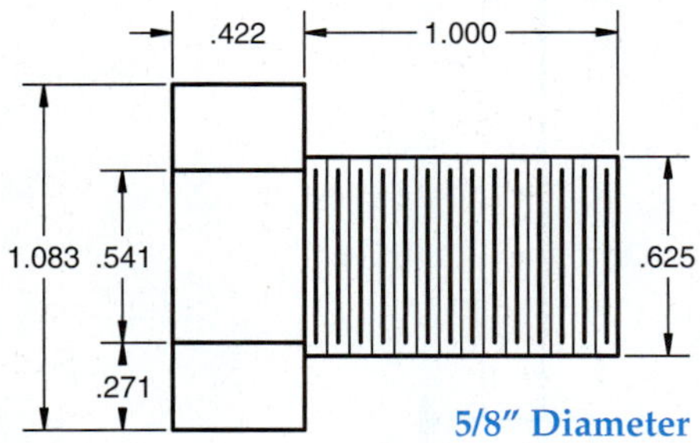

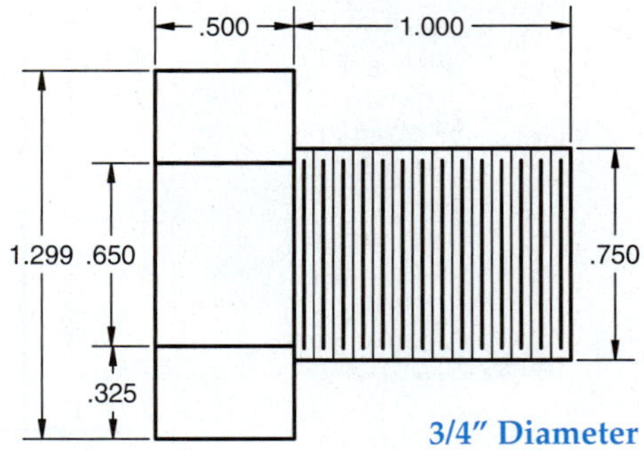

K. Add the BOLT DIAMETER, HEAD THICKNESS, HEAD WIDTH, and EDGE WIDTH parameters to the lookup table. Complete the table with the following properties:

Bolt Diameter	Head Thickness	Head Width	Edge Width	Lookup
.438	.297	.722	.361	7/16″ Diameter
.500	.344	.866	.433	1/2″ Diameter
.625	.422	1.083	.541	5/8″ Diameter
.750	.500	1.299	.650	3/4″ Diameter

L. Set the table to allow reverse lookup, save the block, and exit the **Block Editor**.

M. Insert the block four times into the drawing. Specify a different lookup property for each block.

N. Save the drawing as P27-10.

11. Mechanical Repeat Problem 27-10, but this time use geometric constraints, constraint parameters, and a block properties table instead of action parameters. Save the drawing as P27-11.

AutoCAD Certified Professional Exam Practice

Answer the following questions using the information in this chapter.

1. How many visibility parameters can you assign to a block? *Select the one item that best answers the question.*
   A. 1
   B. 2
   C. 4
   D. 16
   E. unlimited number
2. Study the property lookup table for a pipeline valve. Which option would you select from the lookup parameter grip to display the valve in the 45° position? *Select the one item that best answers the question.*

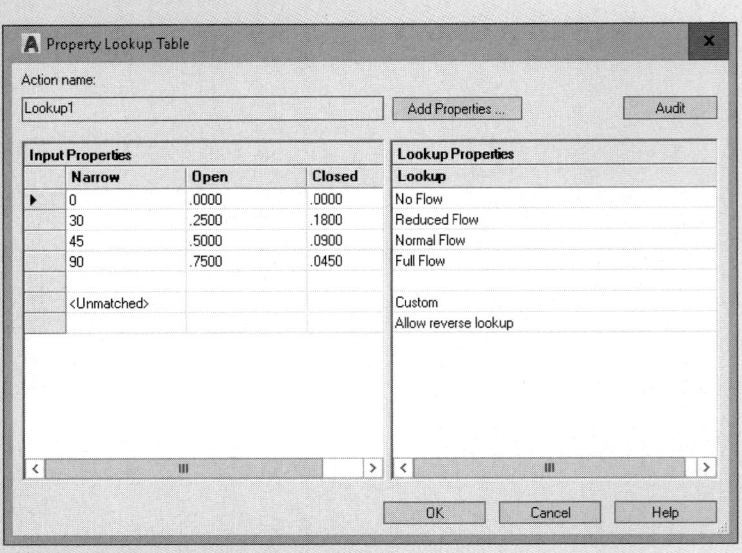

   A. Full Flow
   B. No Flow
   C. Normal Flow
   D. Reduced Flow
3. Which of the following actions can you perform using the **BCPARAMETER** command? *Select all that apply.*
   A. convert a dimensional constraint to a constraint parameter
   B. create a perpendicular geometric constraint
   C. create a radius constraint parameter
   D. place a linear constraint parameter
   E. place a parallel geometric constraint

Follow the instructions in the problem. Complete the problem by answering the related question.

4. **Navigate to this chapter on the companion website and open CPE-27constraint.dwg.** Use the existing constraint parameter to make the outer circle tangent to the line, as shown. What is the diameter of the inner circle?

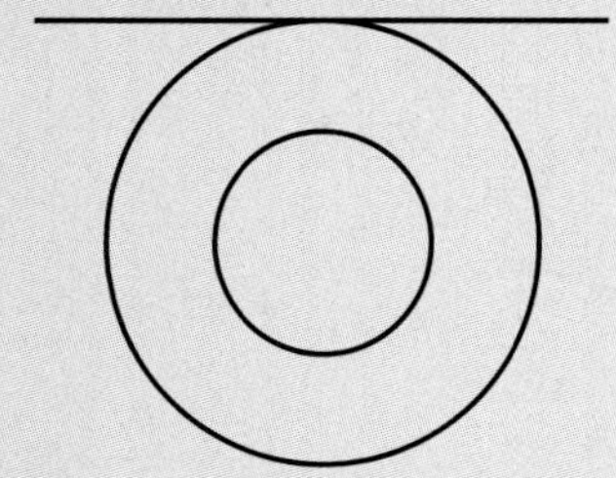

CHAPTER 28

Layout Setup

Learning Objectives

After completing this chapter, you will be able to:

✓ Describe the purpose for and proper use of layouts.
✓ Begin to prepare layouts for plotting.
✓ Manage layouts.
✓ Use the **Page Setup Manager** to define plot settings.
✓ Use plot styles and plot style tables.

You may often print or plot a drawing in model space to make a quick hard copy for check or reference purposes. Ordinarily, however, you create a drawing in model space and then lay out the drawing for plotting in paper space. A hard copy is useful on the shop floor or at a construction site. A design team can check and redline a hard copy without a computer or CADD software. Most of the same steps you follow to prepare a drawing for plotting apply to exporting a drawing to an electronic format.

Introduction to Layouts

The first step to make an AutoCAD drawing is to create a ***model*** in ***model space***. See Figure 28-1A. Model space is usually active by default. You have been using model space throughout this textbook to create objects and dimensioned drawing views. Once you complete a model, use a ***layout*** in ***paper space*** to prepare the final drawing for plotting. See Figure 28-1B. A layout represents the sheet of paper used to lay out and plot a drawing. A layout often includes the following items, depending on the drawing:

- Floating viewports to display content from model space
- Border
- Sheet blocks such as a title block and revision history block
- General notes
- Bill of materials, parts list, schedules, legend, and other associated lists
- Sheet annotation and symbols such as titles, north arrow, and graphic scale
- Page setup information

model: A 2D or 3D drawing composed of various objects, such as lines, circles, and text, usually created at full size.

model space: The environment in AutoCAD in which you create drawings and designs.

layout: A specific arrangement of views or drawings for plotting or printing on paper.

paper space: The environment in AutoCAD in which you create layouts.

Figure 28-1.
A—Design and draft objects in model space. B—Use paper space to finalize and lay out drawings and designs on paper for plotting or exporting.

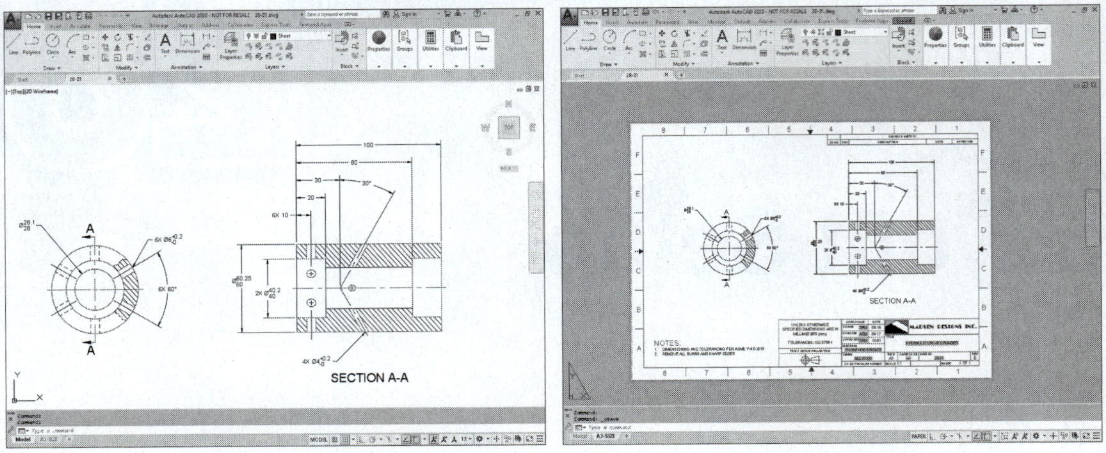

A B

floating viewport: A viewport added to a layout in paper space to display objects drawn in model space.

A major element of the layout system is the ***floating viewport***. Consider a layout to be a virtual sheet of paper and a floating viewport as a window cut into the paper to show objects drawn in model space. In Figure 28-1B, a single viewport exposes objects drawn in model space. Draw the floating viewport on a layer that you can turn off or freeze so the viewport does not plot and is not displayed on-screen. Chapter 29 explains using floating viewports.

Layouts with floating viewports offer the ability to construct properly scaled drawings. A single drawing can have multiple layouts, each representing a different paper space, or plot, definition. Each layout can include multiple floating viewports to provide additional or alternate drawing views, prepared at different scales if necessary. You can use a single drawing file to prepare several different final drawings and drawing views. For example, an architectural drawing file might include several details that are too large to place on a single sheet of paper. You can use multiple layouts, and, if necessary, differently scaled floating viewports to prepare as many sheets as needed to plot all of the details in the drawing.

Working with Layouts

Learn to use tools and options for displaying and managing layouts before you prepare a layout for plotting. The layout and model tabs and the model space and paper space tools in the status bar are available by default. These are effective tools for navigating between model space and layouts, and for managing layouts. You can also type MODEL and press [Enter] to return to model space from a layout.

Model and Layout Tabs

The model and layout tabs appear left of the status bar. See Figure 28-2A. A quick way to toggle the display of model and layout tabs on and off is to pick the **Layout Tabs** button from the **Interface** panel of the **View** ribbon tab. The model space tab is to the left of the layout tabs. Layout tabs are arranged in the order they were created, from left to right. If there are so many layouts that the tabs would spread into the status bar, pick the overflow button that appears to the right of the last layout tab to access a menu of layouts.

Hover over an inactive tab to display a preview of the contents. Pick a layout tab to enter paper space with the selected layout current, or pick the **Model** tab to re-enter

Figure 28-2.
A—Using the layout and model tabs to activate model space and paper space. B—Options available when you right-click on a tab.

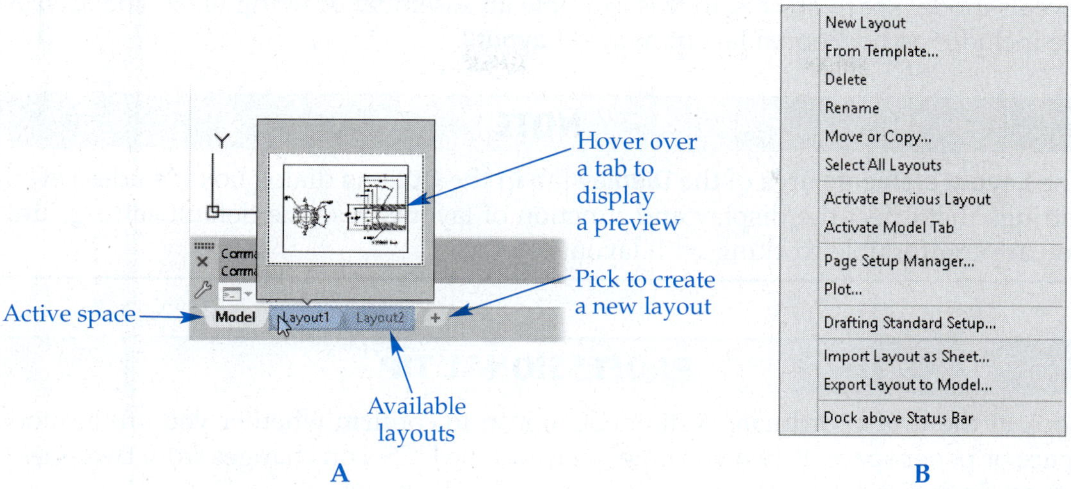

model space. Right-click on a tab to access a shortcut menu with options for controlling layouts and the layout and model tabs. See Figure 28-2B. Pick **Activate Model Tab** to enter model space. Select **Activate Previous Layout** to make the previously current layout current. Pick **Select All Layouts** to select all layouts in the drawing for purposes such as publishing or deleting. The shortcut menu is also a resource for adding, moving, renaming, and deleting layouts.

Figure 28-3 shows an example of using the supplied acad.dwt drawing template and picking the **Layout1** tab to display the default **Layout1** layout. The layout uses

Figure 28-3.
Pick the **Layout1** tab to display **Layout1** provided in the default acad.dwt template.

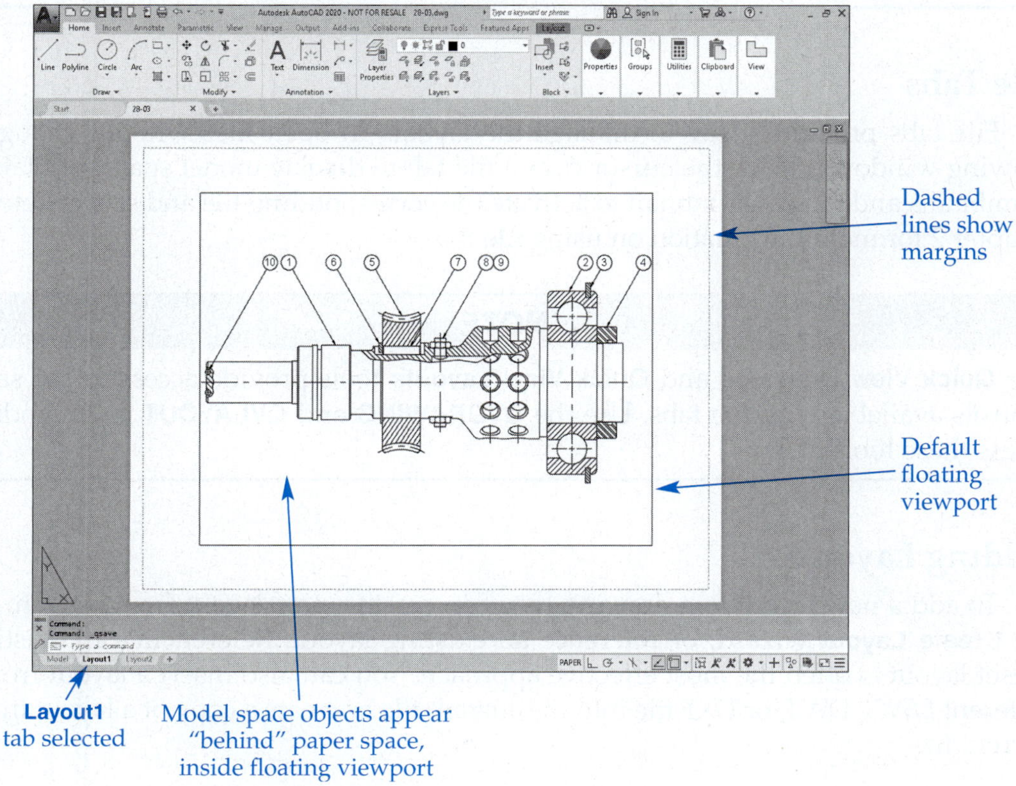

margin: The extent of the printable area; objects drawn past the margin (dashed lines) do not print.

default settings based on an 8.5″ × 11″ sheet of paper in a landscape (horizontal) orientation. The white rectangle you see on the gray background is a representation of the sheet. Dashed lines mark the sheet ***margin***. A large, rectangular floating viewport reveals model space objects, in this example an assembly drawing view. The acad.dwt file includes an additional layout named **Layout2**.

NOTE

The **Layout elements** area of the **Display** tab in the **Options** dialog box includes several settings that affect the display and function of layouts. Use the default settings until you are comfortable working with layouts.

PROFESSIONAL TIP

Look at the user coordinate system (UCS) icon to confirm whether you are in model space or paper space. When you enter a layout, the UCS icon changes from two lines to a triangle that indicates the X and Y coordinate directions.

Model and Paper Buttons

Type MSPACE

Type PSPACE

Pick the **MODEL** button on the status bar to exit model space and enter paper space. If the file contains multiple layouts, the top-level layout is displayed unless you previously accessed a different layout. While a layout is active, pick the **PAPER** button to use the **MSPACE** command, which activates a floating viewport. The **MSPACE** command does not return you to model space. Select the **MODEL** button to use the **PSPACE** command and deactivate a floating viewport.

Exercise 28-1

www.g-wlearning.com/CAD/

Complete the exercise on the companion website.

File Tabs

File tabs provide a way to manage the layouts in open files without changing drawing windows. Move the cursor over a file tab to display model space and layout thumbnails and pick a thumbnail to activate the corresponding file and space. Refer to Chapter 2 for more information on using file tabs.

NOTE

Type QVDRAWING

Type QVLAYOUT

The **Quick View Drawings** and **Quick View Layouts** tools provide access to the same controls available with file tabs. Use the **QVDRAWING** and **QVLAYOUT** commands to access these tools.

Adding Layouts

To add a new layout to a drawing, you can create a new layout from scratch, use the **Create Layout** wizard, or reference an existing layout. Referencing an existing, preset layout is often the most effective approach. You can also insert a layout from a different DWG, DWT, or DXF file into the current file or create a copy of a layout in the current file.

Starting from Scratch

Ribbon
Layout
> Layout
New Layout

To create a new layout from scratch, pick the **New Layout** button from the **Layout** panel of the **Layout** ribbon tab, or pick the plus (+) button to the right of the last layout tab. You can also right-click on the model tab or a layout tab and pick **New Layout**. A new layout appears on the far right of the layout list. The settings applied to the new layout depend on the template used to create the original file. The name of the layout is set according to the names of other existing layouts. For example, when you add a new layout to a default drawing started from the acad.dwt template, a new layout named **Layout3** appears and includes an 8.5″ × 11″ sheet of paper, a landscape (horizontal) orientation, and a large floating viewport.

Using the Create Layout Wizard

Type
LAYOUTWIZARD

Use the **Create Layout** wizard to build a layout from scratch using values and options you enter in the wizard. The wizard provides options for naming the new layout and selecting the printer, paper size, drawing units, paper orientation, title block, and viewport configuration. The pages of the wizard guide you through the process of developing the layout.

Using a Template

Ribbon
Layout
> Layout
From Template

To create a new layout from a layout stored in an existing DWG, DWT, or DXF file, pick the **From Template** button from the **Layout** panel of the **Layout** ribbon tab. You can also right-click on the model tab or a layout tab and pick **From Template...**. The **Select Template From File** dialog box appears. See Figure 28-4A. The Template folder in the path set by the AutoCAD Drawing Template File Location appears by default. Select the file containing the layout you want to add to the current drawing and pick the **Open** button. The **Insert Layout(s)** dialog box appears, listing all layouts in the selected file. See Figure 28-4B. Highlight the layout(s) to copy and pick the **OK** button.

Using DesignCenter

Ribbon
View
> Palettes
DesignCenter
Type
ADCENTER
ADC

ADCENTER

DesignCenter provides an effective way to add existing layouts to the current drawing. Use the folder list to locate and select a drawing or template file. Pick the Layouts branch in the folder list, or double-click on the **Layouts** icon in the content pane. See Figure 28-5. Select one or more layouts to copy from the content pane and then drag and drop, or use the **Add Layout(s)** or **Copy** option from the shortcut menu to insert the layouts into the current drawing.

Moving and Copying Layouts

Organize layouts in a logical order to reduce confusion and aid in the publishing process, described in Chapter 33. An easy way to rearrange layouts is to drag and drop a tab left or right. Another option to move a layout is to right-click on a layout tab and pick **Move or Copy...** to display the **Move or Copy** dialog box. See Figure 28-6. Be sure the **Create a copy** check box is unchecked and pick the layout that will appear to the right of the selected layout, or pick **(move to end)** to move the layout to the right of all other layouts.

An easy way to create a copy of a layout is to hold down [Ctrl] while dragging and dropping the tab to copy left or right. To create a copy of a layout using the **Move or Copy...** dialog box, select the **Create a copy** check box. The default name of the new layout is the name of the current or selected layout plus a number in parentheses. When you add and rename layouts, the layouts do not automatically rearrange into a predetermined order.

Figure 28-4.
Adding a layout using an existing layout stored in a different DWG, DWT, or DXF file. A—Select the file containing the layout. B—Highlight the layout to add to the current drawing.

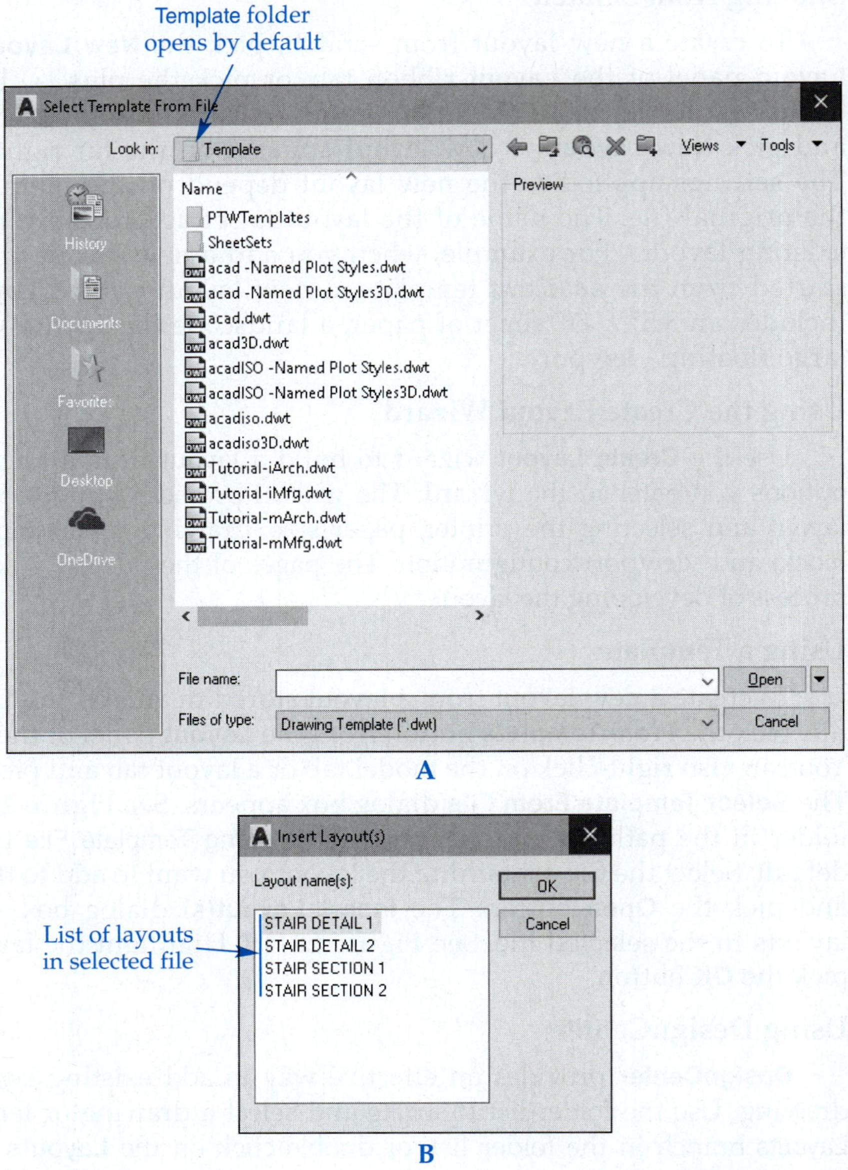

Figure 28-5.
Using **DesignCenter** to share layouts between drawings.

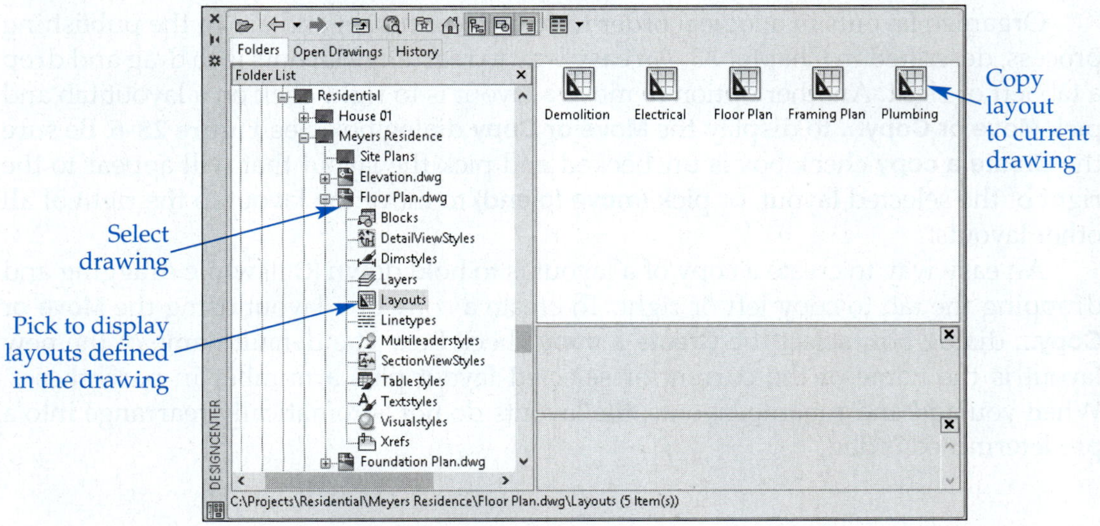

Figure 28-6.
The **Move or Copy** dialog box offers one way to reorganize and copy layouts within a drawing.

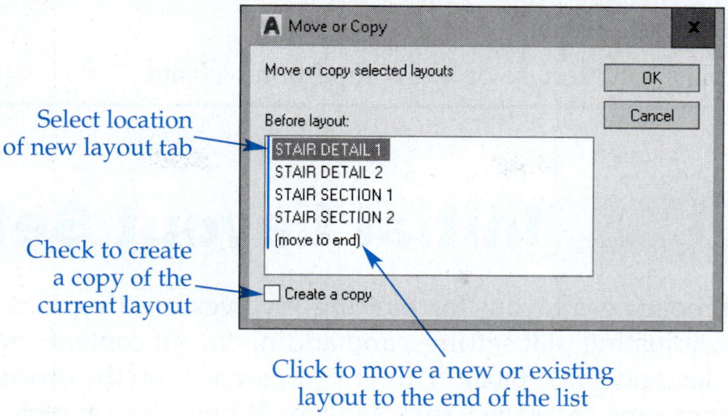

Renaming Layouts

Layouts are easier to recognize and use when they have descriptive names. To rename a layout, right-click on a layout tab and pick **Rename**. You can also slowly double-click on the current name to activate it for editing. When the layout name is highlighted, type a new name and press [Enter].

Deleting Layouts

To delete an unused layout from the drawing, right-click on the layout tab and pick **Delete**. An alert message warns that the layout will be deleted permanently. Pick the **OK** button to remove the layout.

NOTE

You can use the **LAYOUT** command to create and manage layouts, but for most applications, options available from items such as the ribbon and the model and layout tabs are much easier and more convenient to use than the **LAYOUT** command.

Type
LAYOUT

Exporting a Layout to Model Space

Application Menu
Save As
Save Layout as a Drawing
Type
EXPORTLAYOUT

The **EXPORTLAYOUT** command allows you to save the entire layout display as model space objects in a separate DWG file. The **EXPORTLAYOUT** command produces a "snapshot" of the current layout display that you can use for applications in which it is necessary to combine model space and paper space objects, such as when exporting a file as an image. Model space and paper space are not exported together as an image.

A quick way to access the **EXPORTLAYOUT** command is to right-click on a layout tab and pick **Export Layout to Model...**. The **Export Layout to Model Space Drawing** dialog box appears and functions much like the **Save As** dialog box. Pick a location for the file, use the default file name or enter a different name, and pick the **Save** button. Everything shown in the layout, including objects drawn in model space, is converted to model space and saved as a new file.

CAUTION

The **EXPORTLAYOUT** command eliminates the relationship between model space and paper space. Export a layout only when it is necessary to export model space and paper space together as a single unit. Use a descriptive name to differentiate between the exported layout file and the original file.

Exercise 28-2

www.g-wlearning.com/CAD/

Complete the exercise on the companion website.

Initial Layout Setup

Preparing a layout for plotting involves creating and modifying floating viewports, adjusting plot settings, and adding layout content such as annotation, symbols, a border, and a title block. This chapter focuses on the process of preparing layouts for plotting using the **Page Setup Manager**. When you complete this initial phase, you will be better prepared to add content to layouts and create and manage floating viewports, as described in Chapter 29.

page setup: A saved collection of settings required to create a finished plot of a drawing.

A ***page setup*** establishes most of the settings that determine how a drawing plots. Plot settings include plot device selection, paper size and orientation, plot area and offset, plot scale, and plot style. The **Page Setup Manager** and related **Page Setup** dialog box allow you to create and modify saved page setups that control how layouts appear on-screen and plot. This is where initial layout setup occurs. You then use the **Plot** dialog box to create the actual plot using the saved page setup. The **Page Setup** and **Plot** dialog boxes include most of the same settings.

PROFESSIONAL TIP

Layout setup usually involves several steps. A well-defined page setup decreases the amount of time required to prepare a drawing for plotting. Once a layout is set up, only a few steps are required to produce a plot. Add fully defined layouts to drawing templates for convenient future use.

Page Setups

PAGESETUP

Application Menu
Print

Page Setup

Ribbon
Output
> Plot
Layout
> Layout

Page Setup Manager

Type
PAGESETUP

Access the **PAGESETUP** command to create page setups using the **Page Setup Manager**. See Figure 28-7. The **Page setups** area of the **Page Setup Manager** contains a list box that lists available page setups, and includes buttons to add and modify page setups. The **Selected page setup details** area provides information about the highlighted page setup.

NOTE

You can also access the **Page Setup Manager** by right-clicking on the model tab or a layout tab and selecting **Page Setup Manager...**.

Page setups available for the current model tab or layout appear in the **Page setups** list box. For example, when you access the **Page Setup Manager** in paper space from the default **Layout1** tab, *Layout1* appears in the **Page setups** list box. This indicates that the default, unnamed page setup is set current for **Layout1**. Another example is a custom page setup listed as *B-SIZE (Mechanical-Inch PDF)*. This indicates that a page setup named Mechanical-Inch PDF is set current for the active **B-SIZE** layout tab. You have the option to create or use other page setups instead of the current page setup associated with the layout.

Figure 28-7.
Use the **Page Setup Manager** to modify existing page setups and to create and import page setups.

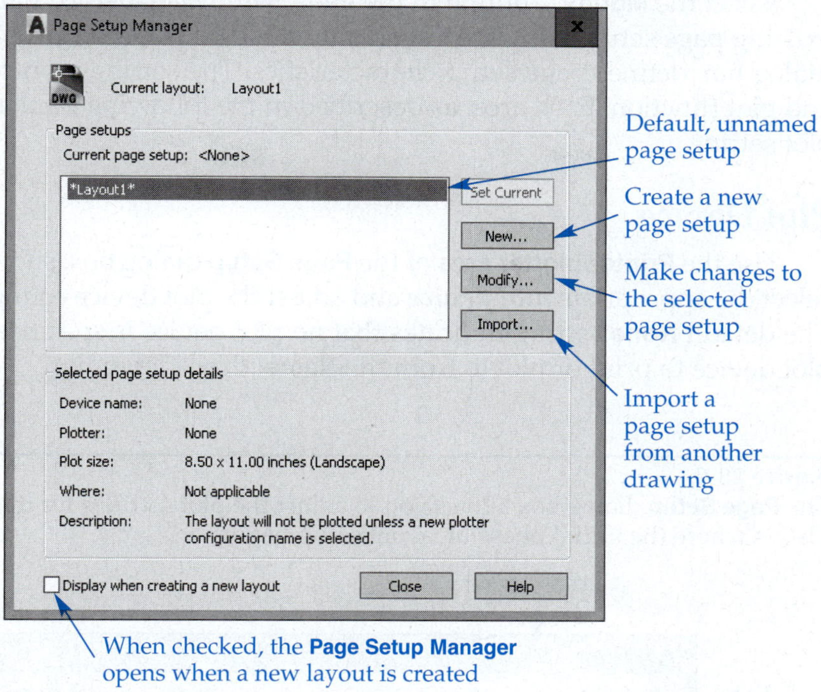

NOTE

When preparing a layout for plotting, check to be sure that you are in paper space and that the appropriate layout is current.

Create a new page setup to use different plot characteristics without overriding plot settings or spending time making page setup changes. For example, create two page setups to plot to two different printers or plotters. Pick the **New...** button to create a new page setup using the **New Page Setup** dialog box. See Figure 28-8. Type a name for the new page setup and choose an option from the **Start with:** list box. Pick the **OK** button to create the page setup and display the **Page Setup** dialog box. Pick the **Import...** button in the **Page Setup Manager** to use existing page setups from a DWG, DWT, or DXF file.

To attach a different page setup to the current layout, select a page setup from the list in the **Page Setup Manager** and pick the **Set Current** button, or right-click on a page setup and choose **Set Current**. The layout will now plot according to the selected page setup. When you make a different page setup current, the selected page setup overrides the layout page setup. The page setup name appears in parentheses next to the layout name. To rename or delete an existing page setup, right-click on the page setup and pick **Rename** or **Delete**.

Figure 28-8.
The **New Page Setup** dialog box appears when you create a new page setup. Selecting **None** in the **Start with:** list box does not select a printer. Selecting **Default output device** selects the default printer assigned to the computer.

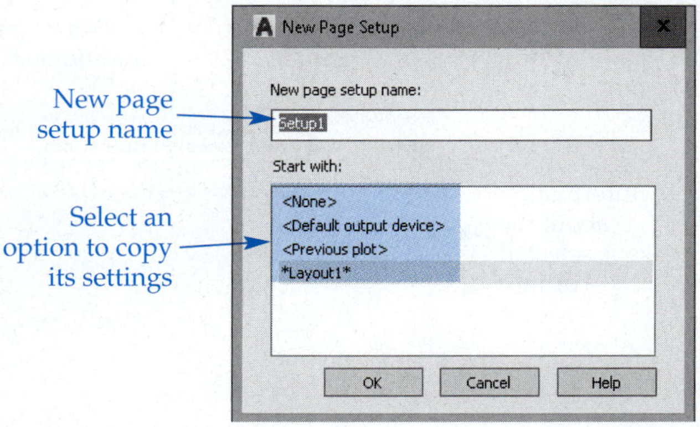

Select the **Modify...** button in the **Page Setup Manager** to change the settings of an existing page setup using the **Page Setup** dialog box. See Figure 28-9. The **Page Setup** dialog box defines page setup characteristics. The settings control layout appearance and plot function. Each area, as described in the following sections, controls a specific plot setting.

Plot Device

plot device: The printer, plotter, or alternative plotting system to which the drawing is sent.

configured: Installed and ready to use.

Use the **Printer/plotter** area of the **Page Setup** dialog box, shown in Figure 28-10, to select the appropriate ***plot device*** and adjust the plot device configuration if necessary. The default **None** setting indicates that no plot device is specified. Select a ***configured*** plot device to print or plot to from the **Name:** drop-down list.

Figure 28-9.
The **Page Setup** dialog box allows you to adjust the plot settings for the selected page setup. This is where the initial phase of layout setup begins.

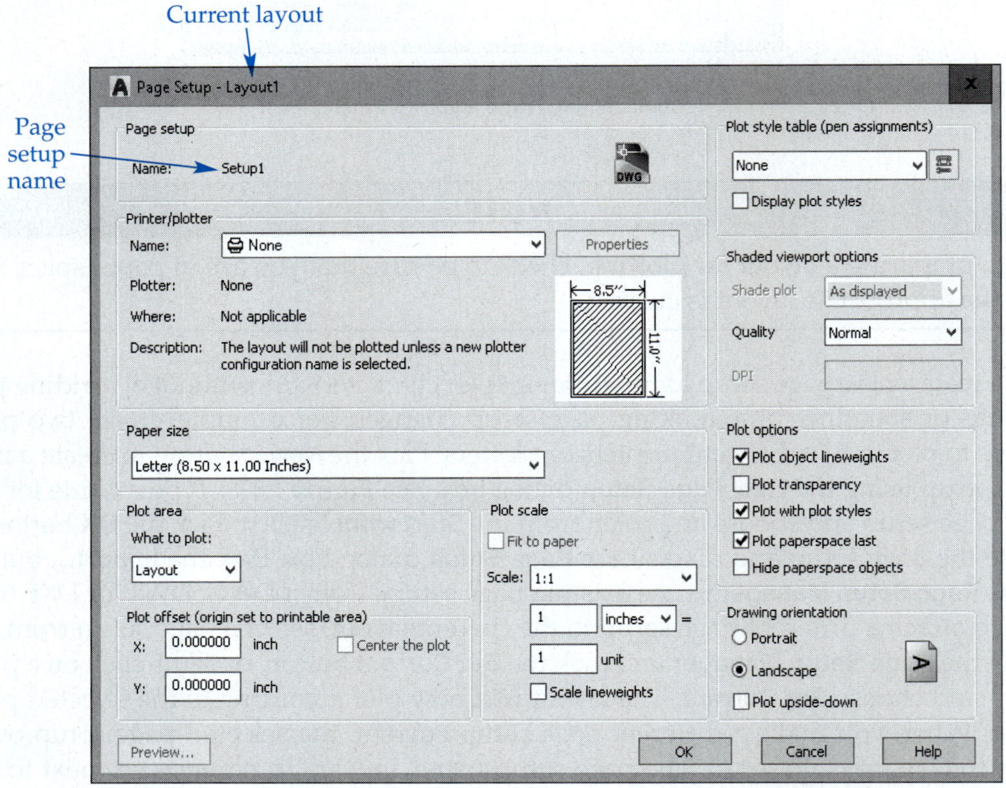

Figure 28-10.
The **Printer/plotter** area of the **Page Setup** dialog box.

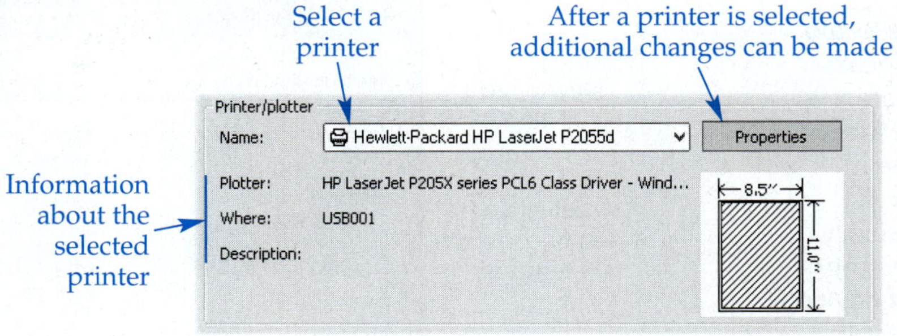

Electronic Plots

Exporting a drawing is an effective way to display and share a drawing for some applications. One way to export a drawing is to plot a layout to a different file type. Electronic plotting uses the same general process as hard-copy plotting, but the plot exists electronically instead of on an actual sheet of paper.

exporting: Transferring electronic data from a database, such as a drawing file, to a different format used by another program.

A common electronic plotting method is to create a portable document format (PDF) file. For example, send a PDF file of a layout to a manufacturer, vendor, contractor, agency, or plotting service. The recipient uses the free Adobe® Reader software to view the plot electronically, and to plot the drawing to scale without having AutoCAD software. This method also helps avoid inconsistencies that sometimes occur when sharing AutoCAD files. Use the **DWG To PDF.pc3**, **AutoCAD PDF (General Documentation).pc3**, **AutoCAD PDF (High Quality Print).pc3**, **AutoCAD PDF (Smallest File).pc3**, or **AutoCAD PDF (Web and Mobile).pc3** option to plot a PDF file. Select the **DWF6 ePlot.pc3** or **DWFx ePlot (XPS Compatible).pc3** option to plot to the appropriate design web format (DWF) file. The recipient of a DWF file uses a viewer such as the Autodesk® Design Review software to view and mark up the plot.

NOTE

You will explore additional information on creating electronic plots later in this textbook.

PROFESSIONAL TIP

If you use a plotting service or blueprint shop to plot large sheets, ask if the printer accepts PDF files. Plotting a layout to a PDF file and then using the PDF file to create a hard copy is often the most convenient way to share a drawing and help ensure that the hard copy plots as expected.

CAUTION

You can use the **PLOTTERMANAGER** command to add printers or plotters to the **Printer/plotter** list. However, avoid editing and saving modified plot configuration (PC3) files unless you know doing so is appropriate. These files are critical to the proper functioning of your plotter.

PLOTTERMANAGER

Application Menu
Print
Manage Plotters
Ribbon
Output > Plot
Plotter Manager
Type
PLOTTERMANAGER

Sheet Size

The **Paper size** area, shown in Figure 28-11, controls the ***sheet size***. The sheet size determines the size of the virtual sheet of paper displayed in a layout and corresponds to the actual sheet size you plan to use when plotting. To determine sheet size, take into account the size of the drawing. Evaluate additional space for dimensions, notes,

sheet size: The size of the paper used to lay out and plot drawings.

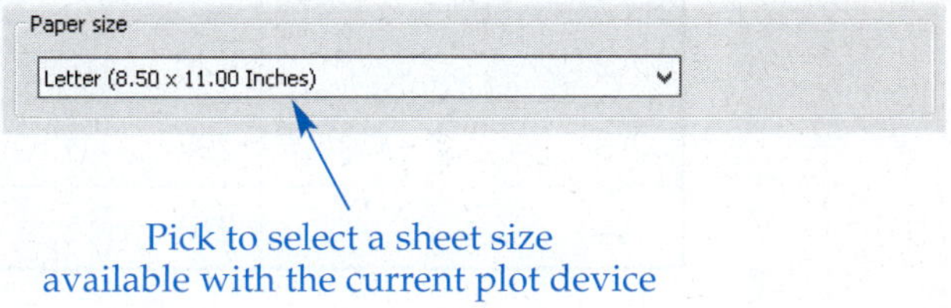

Figure 28-11. The **Paper size** area allows you to define the sheet size applied to the layout, which corresponds to the sheet size on which you plan to plot.

a border, clear space between the drawing and border, sheet blocks (such as the title block and revision history block), zoning, and an area for general notes, other annotations, and symbols. Select the appropriate sheet size from the drop-down list in the **Paper size** area.

Standard Sheet Sizes

The ASME Y14.1 *Decimal Inch Drawing Sheet Size and Format* standard and the ASME Y14.1M *Metric Drawing Sheet Size and Format* standard specify the American Society of Mechanical Engineers (ASME) standard sheet sizes and formats. Figure 28-12 lists the ASME Y14.1 standard US customary (inch) sheet size specifications. Figure 28-13 lists the ASME Y14.1M standard metric sheet size specifications.

Figure 28-14 shows standard ASME sheet sizes and layout. To describe sheet size values verbally, state the horizontal measurement and then the vertical measurement. For example, describe a C-size sheet as 22 (horizontal) × 17 (vertical). Longer lengths are known as *elongated* and *extra-elongated* drawing sizes. These are available in multiples of the short side of the sheet size.

Some companies that prepare architectural or related drawings prefer inch-unit architectural sheet sizes, which vary slightly from the ASME standard. AutoCAD includes architectural and modified architectural sheet sizes depending on the specified plot device. Standard architectural sheet sizes in inches are listed in Figure 28-15.

Figure 28-12. ASME Y14.1 standard sheet sizes for US customary (inch) drawings.

Size Designation	Size (in inches)	
	Vertical	Horizontal
A	8 1/2 11	11 (horizontal format) 8 1/2 (vertical format)
B	11	17
C	17	22
D	22	34
E	34	44
F	28	40
Sizes G, H, J, and K are roll sizes.		

Figure 28-13. ASME Y14.1M standard sheet sizes for metric drawings.

Size Designation	Size (in millimeters)	
	Vertical	Horizontal
A0	841	1189
A1	594	841
A2	420	594
A3	297	420
A4	210	297

Figure 28-14.
A—Standard decimal inch drawing sheet sizes and layout based on the ASME Y14.1 standard.
B—Standard metric drawing sheet sizes and layout based on the ASME Y14.1M standard.

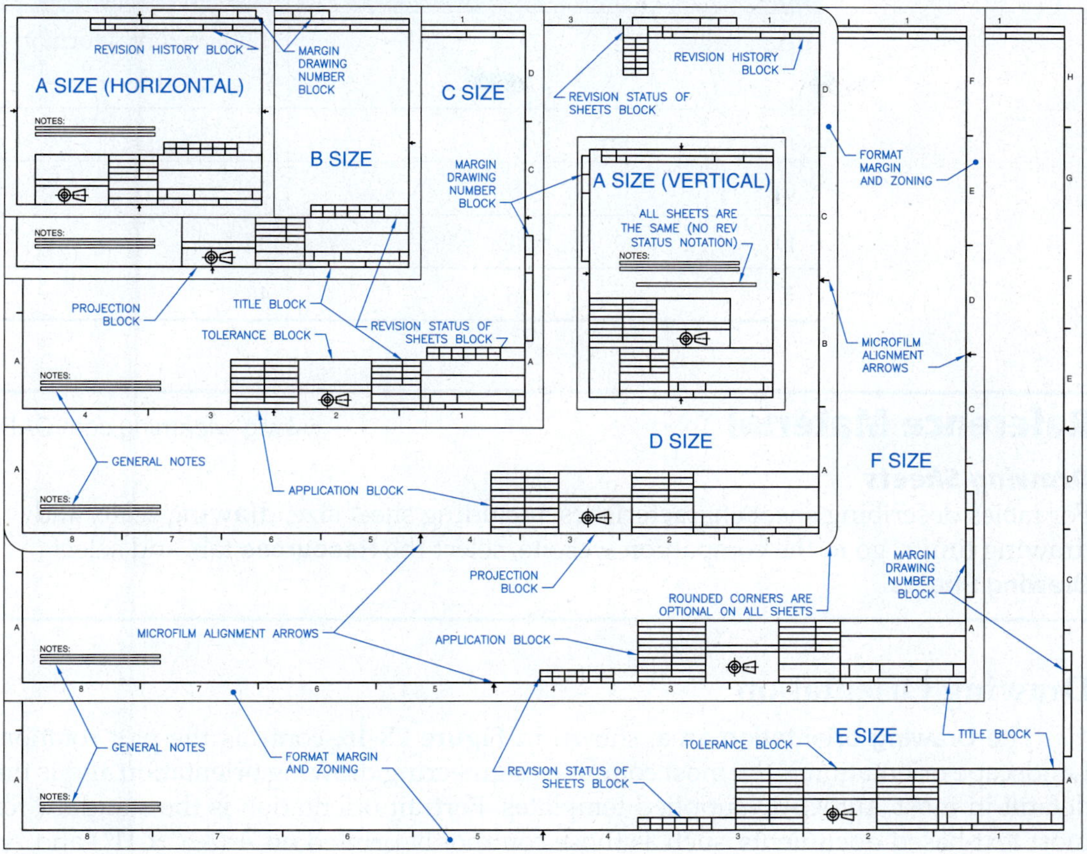

A

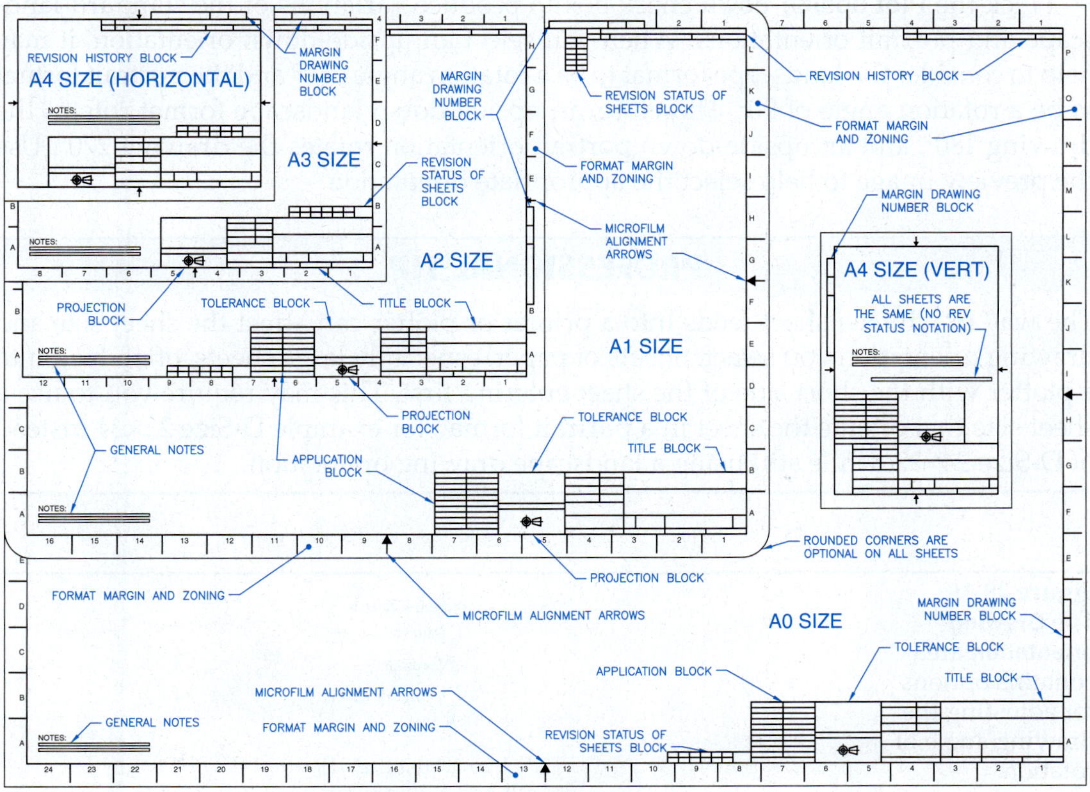

B

Figure 28-15.
Architectural sheet sizes.

Size Designation	Size (in inches)	
	Vertical	Horizontal
A	9 12	12 (horizontal format) 9 (vertical format)
B	12	18
C	18	24
D	24	36
E	36	48

Reference Material

www.g-wlearning.com/CAD/

Drawing Sheets

For tables describing sheet characteristics, including sheet size, drawing scale, and drawing limits, go to the companion website, select the **Resources** tab, and select **Drawing Sheets**.

Drawing Orientation

The **Drawing orientation** area, shown in Figure 28-16, controls the plot rotation. Landscape orientation is the most common engineering drawing orientation and is the default in most AutoCAD-supplied templates. Portrait orientation is the standard for most text-based documents, such as those commonly printed on A (8.5″ × 11″) and A4 (210 mm × 297 mm) sheets.

Pick the **Plot upside-down** check box to produce variations of the standard landscape and portrait orientations. When you select an upside-down orientation, it may help to consider the landscape format to be a rotation angle of 0° and the portrait format to be a rotation angle of 90°. Therefore, an upside-down landscape format rotates the drawing 180°, and an upside-down portrait orientation rotates the drawing 270°. Use the preview image to help select the appropriate orientation.

PROFESSIONAL TIP

The way in which a sheet feeds into a printer or plotter can affect the sheet size and drawing orientation you select. Sheets of paper, especially large sheets, often feed into a plotter with the short side of the sheet entering first. This may require you to use a sheet size that orients the sheet in a portrait format, for example D-Size 22×34 instead of D-Size 34×22, while still using a landscape drawing orientation.

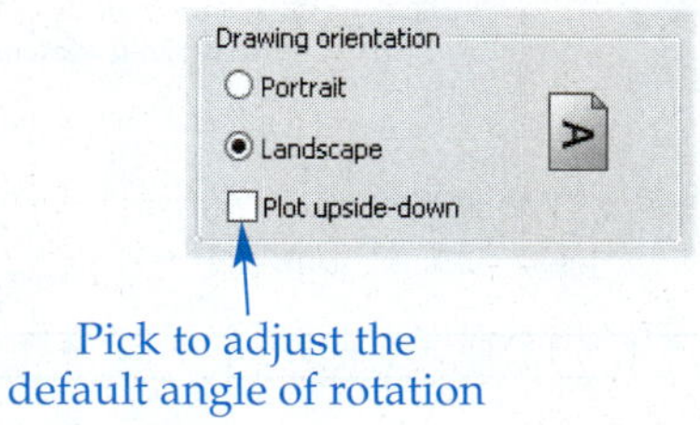

Figure 28-16.
The **Drawing orientation** area contains options for adjusting the drawing angle of rotation.

Exercise 28-3

www.g-wlearning.com/CAD/

Complete the exercise on the companion website.

Plot Area

The **Plot area** section, shown in Figure 28-17, allows you to choose the portion of the drawing to plot. Select an option from the **What to plot:** drop-down list. The **Layout** option is available when you plot a layout. When you select this option, everything inside the margins of the layout plots. The **Layout** option is the default and is a common setting for plotting a layout. Use other plot area options primarily for plotting in model space, or to adjust the area to plot in paper space.

The **View** option is available when named views exist in the drawing, and if the model space or paper space environment associated with the current page setup includes a named view. When you pick the **View** option, an additional drop-down list appears in the **Plot area** section, allowing you to select a specific view to define as the plot area. Refer to Chapter 6 for information about other **Plot area** options.

Plot Offset

The **Plot offset** area, shown in Figure 28-18, controls the distance the drawing is offset from the plot origin. You can specify the plot origin as the lower-left corner of the printable area or the lower-left corner of the sheet by selecting the appropriate radio button in the **Specify plot offset relative to** area on the **Plot and Publish** tab of the **Options** dialog box.

The **Plot offset (origin set to printable area)** title appears when you use the default **Printable area** option in the **Options** dialog box. The values you enter in the **X:** and **Y:** text boxes define the offset from the printable area. Use the default values of 0 to locate the plot origin at the lower-left corner of the printable area, which corresponds to the lower-left corner of the layout margin (dashed rectangle). To move the drawing away from the default printable area origin, enter positive or negative values in the text boxes. For example, to move the drawing one unit to the right and two units above the lower-left corner of the margin, enter 1 in the **X:** text box and 2 in the **Y:** text box.

Select the **Edge of paper** radio button in the **Options** dialog box to display the **Plot offset (origin set to layout border)** title. The values you enter in the **X:** and **Y:** text boxes

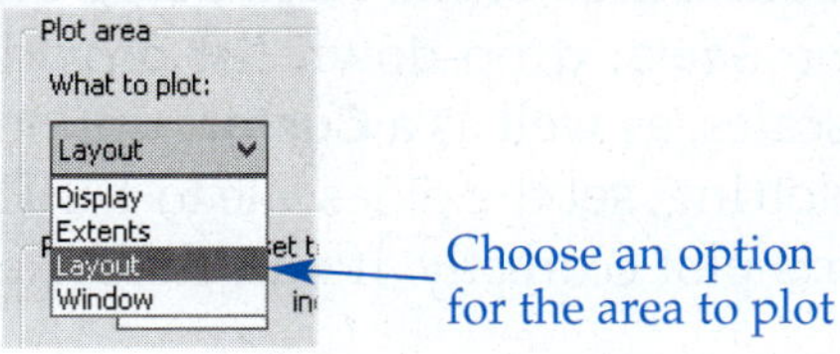

Figure 28-17. Use the **Plot area** section to define the portion of the drawing to plot.

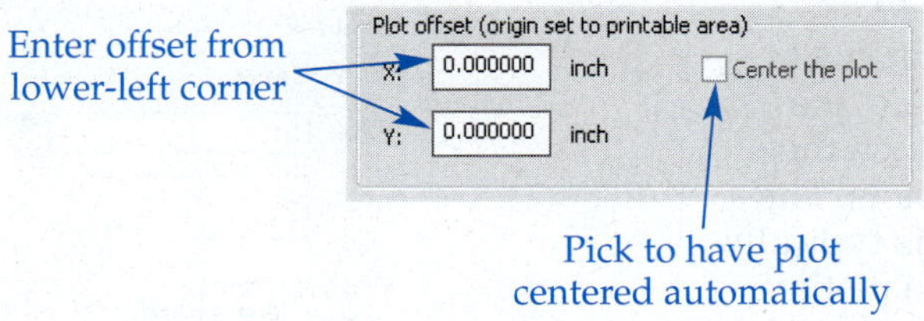

Figure 28-18. The **Plot offset** area controls how far the drawing is offset from the lower-left corner of the printable area or sheet edge.

define the offset from the edge of the sheet. Use the default values of 0 to locate the plot origin at the lower-left corner of the sheet. Change the values in the text boxes to move the drawing away from the sheet origin.

When you select any option other than the **Layout** option from the **What to plot:** drop-down list in the **Plot area** section, the **Center the plot** check box becomes active. Check **Center the plot** to shift the plot origin automatically as needed to center the selected plot area in the printable area.

PROFESSIONAL TIP

If a drawing does not center on the sheet when you plot using the **Layout** option, open the **Options** dialog box and select the **Plot and Publish** tab. Pick the **Edge of paper** radio button in the **Specify plot offset relative to** area. Then use values of 0 in the X and Y offset text boxes in the **Page Setup** dialog box to locate the plot origin at the lower-left corner of the sheet. Depending on the specific plot configuration, you may also need to pick the **Plot upside-down** check box in the **Drawing orientation** area of the **Page Setup** dialog box to locate the origin exactly at the lower-left corner of the sheet. Alternatively, resolve the issue by changing the plot offset to the X and Y values of the lower-left corner of the printable area.

Plot Scale

Always draw objects at their actual size, or full scale, in model space, regardless of the size of the objects. For example, if you draw a small machine part and the length of a line in the drawing is 2 mm, draw the line 2 mm long in model space. If you draw a building and the length of a line in the drawing is 80′, draw the line 80′ long in model space. For layout and printing purposes, you must then scale drawings with features of these sizes to fit properly on a sheet, according to a specific ***drawing scale***.

drawing scale: The ratio between the actual size of objects in the drawing and the size at which the objects plot on a sheet of paper.

When you scale a drawing, you increase or decrease the displayed size of model space objects. A properly scaled floating viewport in a layout allows for this process, as described in Chapter 29. When setting up a layout for plotting, remember that the layout is also at full scale. One difference between model space and paper space is that objects in model space can be very large or very small, while objects in paper space always correspond to sheet size. In order for objects on the layout and in model space to appear correct when plotted, you must plot a layout at full scale, or 1:1.

Use the **Plot scale** area of the **Page Setup** dialog box to specify the plot scale. See Figure 28-19. The **Scale:** drop-down list provides predefined decimal and architectural or related scales, as well as a **Custom** option. For most applications, when setting up a layout for plotting, set the plot scale to 1:1. This ensures that the layout and scaled floating viewports plot correctly. If you choose a scale other than 1:1, the layout does not plot to scale.

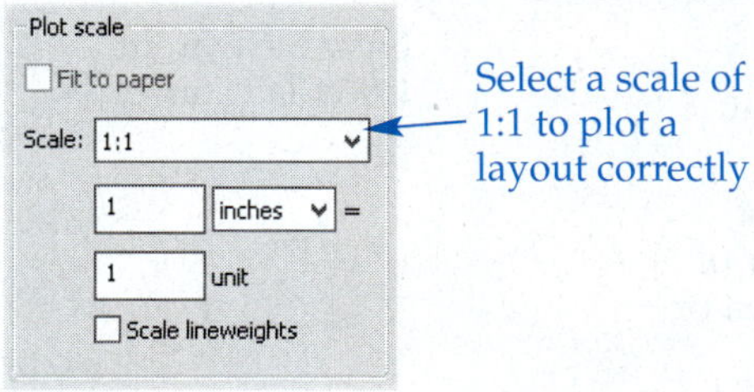

Figure 28-19. Use the **Plot scale** area to adjust the scale at which the drawing plots. The plot scale is typically set to 1:1 to plot a layout even though the drawing scale may not be 1:1.

NOTE

When preparing to plot in model space, you may choose to select a scale other than 1:1. For example, to plot an architectural floor plan, you might set the plot scale to 1/4″ = 1′-0″. If the desired scale is not available, enter values in the text boxes below the drop-down list of predefined scales and select the correct unit of measure from the drop-down list. The **Custom** option automatically displays when you enter values. For example, 1 inch = 600 units is a custom scale entry used to plot at a scale of 1″ = 50′ (50 × 12″ = 600″). Refer to Chapter 9 for more information about drawing scale and scale factors.

Pick the **Fit to paper** check box to adjust the plot scale automatically to fit on the selected sheet. Fitting to paper is useful if you are not concerned about plotting to scale, such as when you are creating a check copy on a sheet that is too small to plot at the appropriate scale. Select the **Scale lineweights** check box to scale (increase or decrease the weight of) lines when the plot scale changes.

PROFESSIONAL TIP

If you need to plot an inch drawing on a metric sheet, use a custom scale of 1 inch = 25.4 units. Conversely, use a custom scale of 25.4 mm = 1 unit to plot a metric drawing on an inch sheet.

Exercise 28-4

www.g-wlearning.com/CAD/

Complete the exercise on the companion website.

Plot Styles

Object properties control the appearance of objects on-screen. By default, what you see on-screen is what plots. For example, if you draw objects on a layer that uses a Red color, Continuous linetype, and 0.60 mm lineweight, the objects display and plot red, continuous, and thick, assuming you show lineweights on-screen and use a color plotter. If you want this result, you are ready to continue with the page setup and plotting process.

To define exactly how objects plot regardless of what displays on-screen, you must assign ***plot styles*** to objects. Use plot styles to maintain object properties in the drawing, but plot objects according to specific plotting properties. For example, to plot all objects in a drawing as dark as possible, you should plot them using the color black. In this example, plot styles allow you to plot all objects black without making them black on-screen. Plot styles also allow you to plot objects using shades of gray instead of color, or to plot objects lighter or darker than they display on-screen.

plot styles: Configurations of properties, including color, linetype, lineweight, line end treatment, and fill style, that are applied to objects for plotting purposes only.

Plot Style Tables

Plot style tables contain plot styles. Choose to use either a ***color-dependent plot style table*** or a ***named plot style table***. A color-dependent plot style table forces objects to plot according to object color. Color-dependent plot style tables contain 255 preset plot styles—one for each AutoCAD Color Index (ACI) color. Each color-dependent plot style is linked to an index color. Plot style properties control how to treat objects of a certain color when the objects plot. For example, the plot style Color 1, which is Red, defines how to plot all objects that are red on-screen. If you assign the Black plot color to plot style Color 1, all red objects are plotted black, even though the objects are red on-screen.

plot style table: A configuration, saved as a separate file, that groups plot styles and provides complete control over plot style settings.

color-dependent plot style table: A file that contains plot style settings used to assign plot values to object colors.

named plot style table: A file that contains plot style settings used to assign plot values to objects or layers.

A named plot style table forces objects to plot according to named plot style values, which you can assign to a layer or object. Any layer or object assigned a named plot style plots using the settings specified for that plot style. For example, create a layer named OBJECT that uses a Red color and a plot style named BLACK that uses a Black color. Then assign the BLACK plot style to the OBJECT layer. Objects drawn on the OBJECT layer plot using the BLACK plot style and plot black in color, even though the objects are red on-screen.

Ideally, decide which plot style table is appropriate before you begin drawing. The templates created in the Template Development feature of this textbook, for example, assume that drawings plot so that all objects appear dark, or black, with different object linetypes and lineweights. You can use a color-dependent plot style table or a named plot style table to create this effect. Default AutoCAD plot style behavior uses color-dependent plot style tables. For most applications, it is usually best to use color-dependent plot style tables, because they are the default and do not require you to assign named plot styles to layers or objects.

Configuring Plot Style Table Type

To configure the plot style type used by default when you create new drawings, pick the **Plot Style Table Settings...** button on the **Plot and Publish** tab of the **Options** dialog box. This opens the **Plot Style Table Settings** dialog box shown in Figure 28-20.

To use a named plot style table, pick the **Use named plot styles** radio button from the **Default plot style behavior for new drawings** area before you start a new drawing file. Select a specific plot style table to use as the default for new drawings from the **Default plot style table:** drop-down list in the **Current plot style table settings** area. When you select the **Use named plot styles** radio button, the **Default plot style for layer 0:** and **Default plot style for objects:** options become active.

NOTE

When you use a template to create a new drawing, the plot style settings defined in the template override the settings you specify in the **Plot Style Table Settings** dialog box. For example, if you configure the template to use named plot style tables, you can only select a named plot style table to apply to the plot, even if you select the **Use color dependent plot styles** radio button in the **Plot Style Table Settings** dialog box.

STYLESMANAGER

Application Menu
Print
Manage Plot Styles
Type
STYLESMANAGER

Pick the **Add or Edit Plot Style Tables...** button in the **Plot Style Table Settings** dialog box to open the **Plot Styles** window. See Figure 28-21. The **Plot Styles** window lists available color-dependent and named plot style tables saved in the Plot Style Table

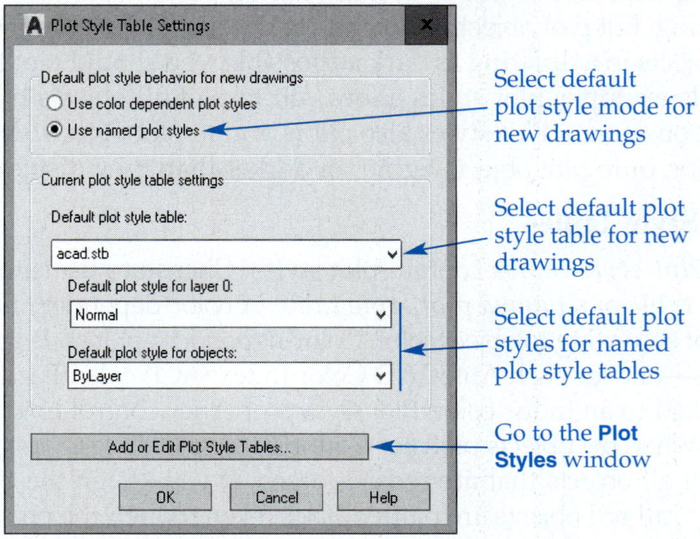

Figure 28-20.
Use the **Plot Style Table Settings** dialog box to set up the default plot style types and tables for new drawings. This example shows the options available when a named plot style table is specified.

Figure 28-21.
The **Plot Styles** window lists available plot style files and allows you to create and edit plot style tables.

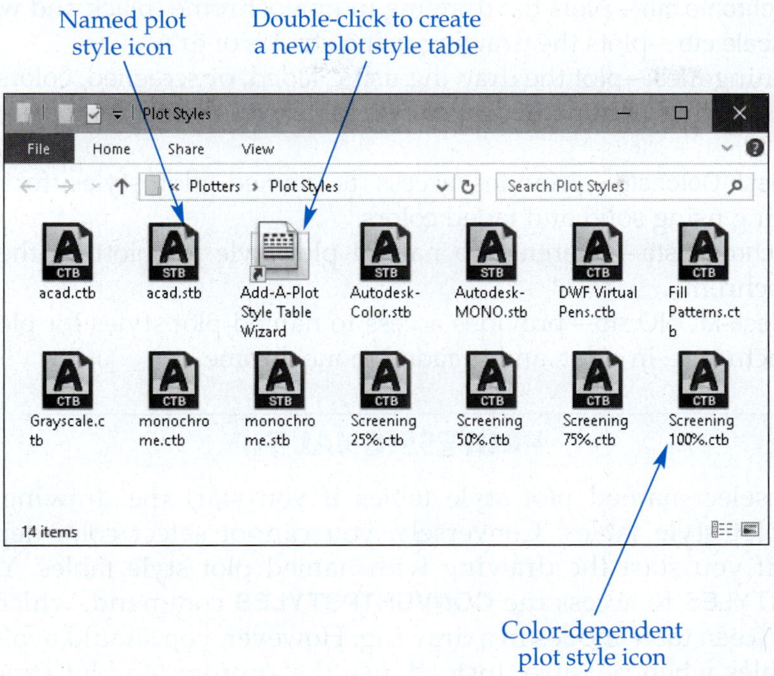

Search Path, as defined in the expanded **Printer Support File Path** option in the **Files** tab of the **Options** dialog box. Color-dependent plot style table files (CTB) use the .ctb extension. Named plot style table files (STB) use the .stb extension. Double-click on Add-A-Plot Style Table Wizard to create a new plot style table, or double-click on an existing plot style table file to edit the file.

Supplemental Material

www.g-wlearning.com/CAD/

Creating and Editing Plot Style Tables

For detailed information about creating and editing plot style tables, go to the companion website, navigate to this chapter in the **Contents** tab, and select **Creating and Editing Plot Style Tables.**

Selecting a Plot Style Table

Use the **Plot style table (pen assignments)** area of the **Page Setup** dialog box, shown in Figure 28-22, to activate and manage plot style tables. Select a plot style table from the drop-down list. Use the default **None** option instead of selecting a specific color-dependent or named plot style table. The **None** option plots exactly what appears on-screen without using plot styles, assuming you use a color plotter to plot objects with color.

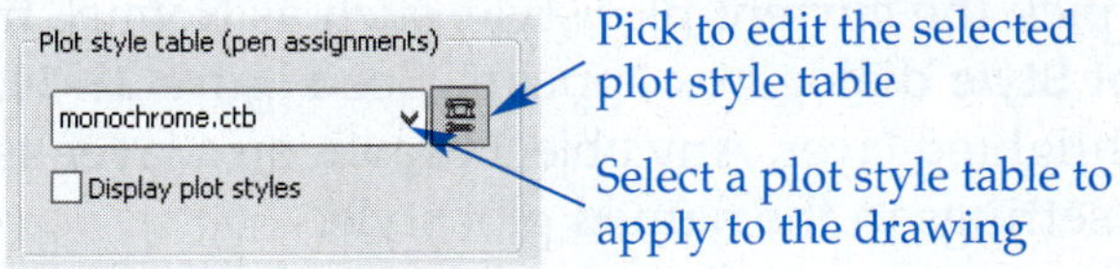

Figure 28-22.
Use the **Plot style table (pen assignments)** area to select, create, and edit plot style tables.

Only color-dependent or named plot style tables appear, depending on the type of plot style table assigned to the current drawing. The most often used color-dependent plot style tables are:

- Monochrome.ctb—plots the drawing in monochrome (black and white).
- Grayscale.ctb—plots the drawing using shades of gray.
- Screening files—plot the drawing using faded, or screened, colors.

When a drawing uses named plot style tables, it is common to select one of the following:

- Autodesk-Color.stb—provides access to named plot styles for plotting the drawing using solid and faded colors.
- Monochrome.stb—references a named plot style for plotting the drawing in monochrome.
- Autodesk-MONO.stb—provides access to named plot styles for plotting objects monochrome, in color, and in faded monochrome.

PROFESSIONAL TIP

You cannot select named plot style tables if you start the drawing with color-dependent plot style tables. Conversely, you cannot select color-dependent plot style tables if you start the drawing with named plot style tables. You can type CONVERTPSTYLES to access the **CONVERTPSTYLES** command, which allows you to switch between table modes in a drawing. However, you should avoid converting plot style tables when possible. Instead, use the appropriate plot style type when beginning a new drawing.

Exercise 28-5

www.g-wlearning.com/CAD/

Complete the exercise on the companion website.

Applying Plot Styles

After you select a plot style table from the drop-down list in the **Plot style table (pen assignments)** area of the **Page Setup** dialog box, the plot styles contained in the selected plot style table are ready to assign to objects in the drawing. When you select a color-dependent plot style table, plot styles are automatically applied to objects according to object color. No additional steps are required to apply color-dependent plot styles to objects.

When you select a named plot style table, the named plot styles contained in the table are ready to assign to layers or individual objects. Any layer or object assigned a named plot style plots using the settings specified for that style. Named plot style tables contain as many plot styles as have been created. For example, the monochrome.stb plot style table contains the default Normal plot style and a style named Style 1. When you apply the Normal plot style to layers or objects, objects plot exactly as they appear on-screen. Style 1 assigns the color Black to all layers or objects that use Style 1 for plotting. In order to plot all objects in the drawing black, even if the objects have different colors on-screen, you must apply Style 1 to all layers.

Assign named plot styles to layers in the **Layer Properties Manager**. See Figure 28-23. The **Plot Style** column appears when the drawing uses named plot styles. Select a layer and then pick the current plot style, such as Normal, from the **Plot Style** column. The **Select Plot Style** dialog box appears. See Figure 28-24. Select a named style to assign to the highlighted layer. Any object drawn on a layer assigned a named plot style plots using the settings in the named plot style.

Figure 28-23.
Assign named plot styles to layers using the **Layer Properties Manager**. This example uses the monochrome.stb plot style table and the Style 1 plot style assigned to all layers to plot all objects black.

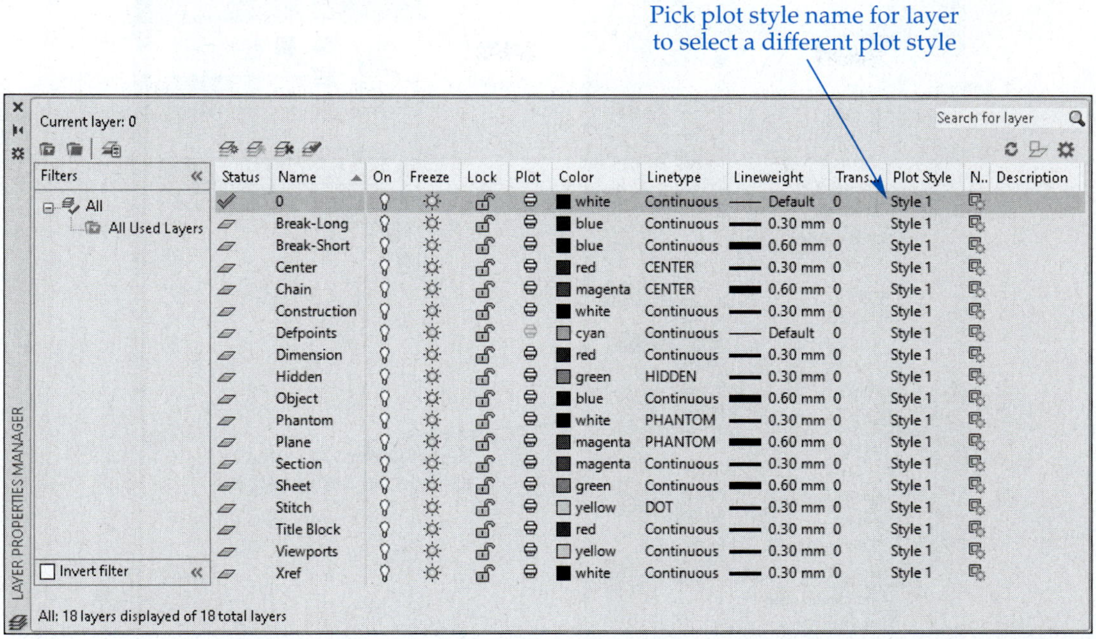

Figure 28-24.
Use the **Select Plot Style** dialog box to assign a plot style to a layer.

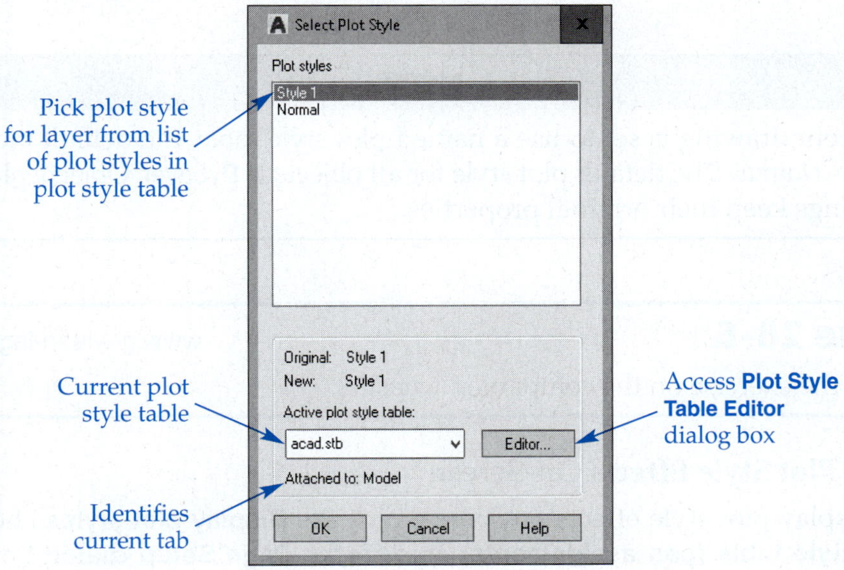

You can also assign named plot styles to individual objects. If you assign a plot style to an object drawn on a layer that has already been assigned a named plot style, the plot style assigned to the object overrides the plot style settings for the layer. Assign plot styles to objects using the **Properties** palette or the **Plot Style Control** drop-down list in the **Properties** panel on the **Home** ribbon tab. See **Figure 28-25**.

Figure 28-25.
You can assign a plot style to individual objects using A—the **Properties** palette or B—the **Plot Style Control** drop-down list in the ribbon.

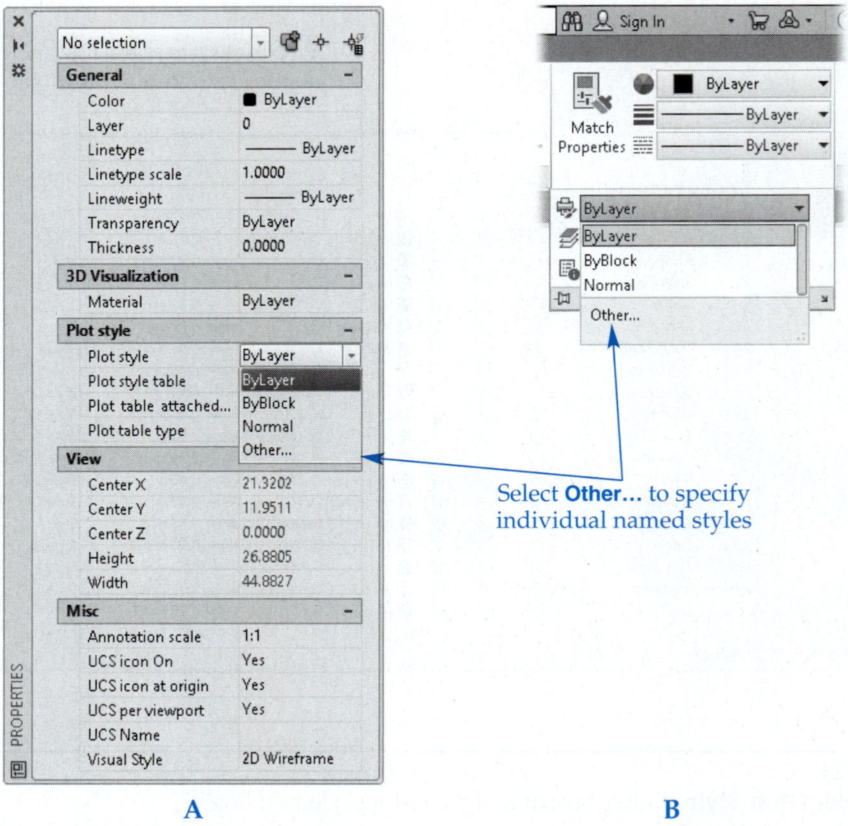

NOTE

If the current drawing is set to use a named plot style table, the default plot style for all layers is Normal. The default plot style for all objects is ByLayer. Objects plotted with these settings keep their original properties.

Exercise 28-6

www.g-wlearning.com/CAD/

Complete the exercise on the companion website.

Viewing Plot Style Effects On-Screen

To display plot style effects on-screen, pick the **Display plot styles** check box in the **Plot style table (pen assignments)** area in the **Page Setup** dialog box. Objects appear on-screen as they will plot. Typically, it is not appropriate to work with objects displayed as they will plot, especially if you print in monochrome. In this example, the color assigned to a layer appears black, which defeats the purpose of assigning colors to layers. A better practice is to use the preview feature of the **Page Setup** or **Plot** dialog box, as described later in this chapter, to preview the effects of plot styles before plotting.

Other Plotting Options

The **Plot options** area of the **Page Setup** dialog box contains additional options that affect how specific items plot. If you plan to plot objects using a plot style table, be sure to select the **Plot with plot styles** check box. This check box toggles the use of plot styles on and off, allowing you to create a plot quickly with or without using plot styles.

Checking the **Plot object lineweights** check box plots all objects with a lineweight greater than 0 using the assigned lineweight. Deselect the check box to plot all objects using a 0, or thin, lineweight.

The **Plot transparency** check box provides the flexibility to plot transparent objects transparent or opaque. Check **Plot transparency** for objects that are assigned to a transparent layer and objects in which the transparency has been overridden to appear transparent. For many drawings, transparency should be reproduced on the plot, as shown in Figure 28-26, especially if you are using transparent fills. However, sometimes transparency is only for on-screen display purposes. For this application, deselect the **Plot transparency** check box.

Pick the **Plot paperspace last** check box to plot paper space objects after model space objects. This option ensures that objects in paper space that overlap objects in model space plot over the model space objects. The **Plot paperspace last** check box is disabled when you plot in model space because model space does not include paper space objects. Select the **Hide paperspace objects** check box to remove hidden lines from 3D objects created in paper space. This option is only available when you plot from a layout tab and affects only objects drawn in paper space. It does not affect any 3D objects in a viewport.

NOTE

The **Shaded viewport options** area of the **Page Setup** dialog box provides settings that control viewport shading. These options set the type and quality of shading for plotting 3D models from a shaded or rendered viewport. ***AutoCAD and Its Applications—Advanced*** provides detailed instruction on 3D modeling and rendering.

Completing Page Setup

After you select all appropriate page setup options, pick the **Preview...** button in the lower-left corner of the **Page Setup** dialog box to preview the effects of the page setup. What you see on-screen is the exact plot appearance, assuming you use a color plotter to make color prints. The **Realtime Zoom** command is automatically activated in the preview window. Additional view commands are available from the toolbar near the top of the window or from a shortcut menu. Use view commands to help confirm that the plot settings are correct. When you finish previewing the plot, pick the **Close** button on the toolbar, press [Esc] or [Enter], or right-click and select **Exit** to return to the **Page Setup** dialog box. Pick the **OK** button to exit the **Page Setup** dialog box, and pick the **Close** button to exit the **Page Setup Manager**.

Exercise 28-7

www.g-wlearning.com/CAD/

Complete the exercise on the companion website.

Figure 28-26.
An example of an architectural elevation with transparent layers plotted to show the existing structure. Opaque layers identify the proposed structure.

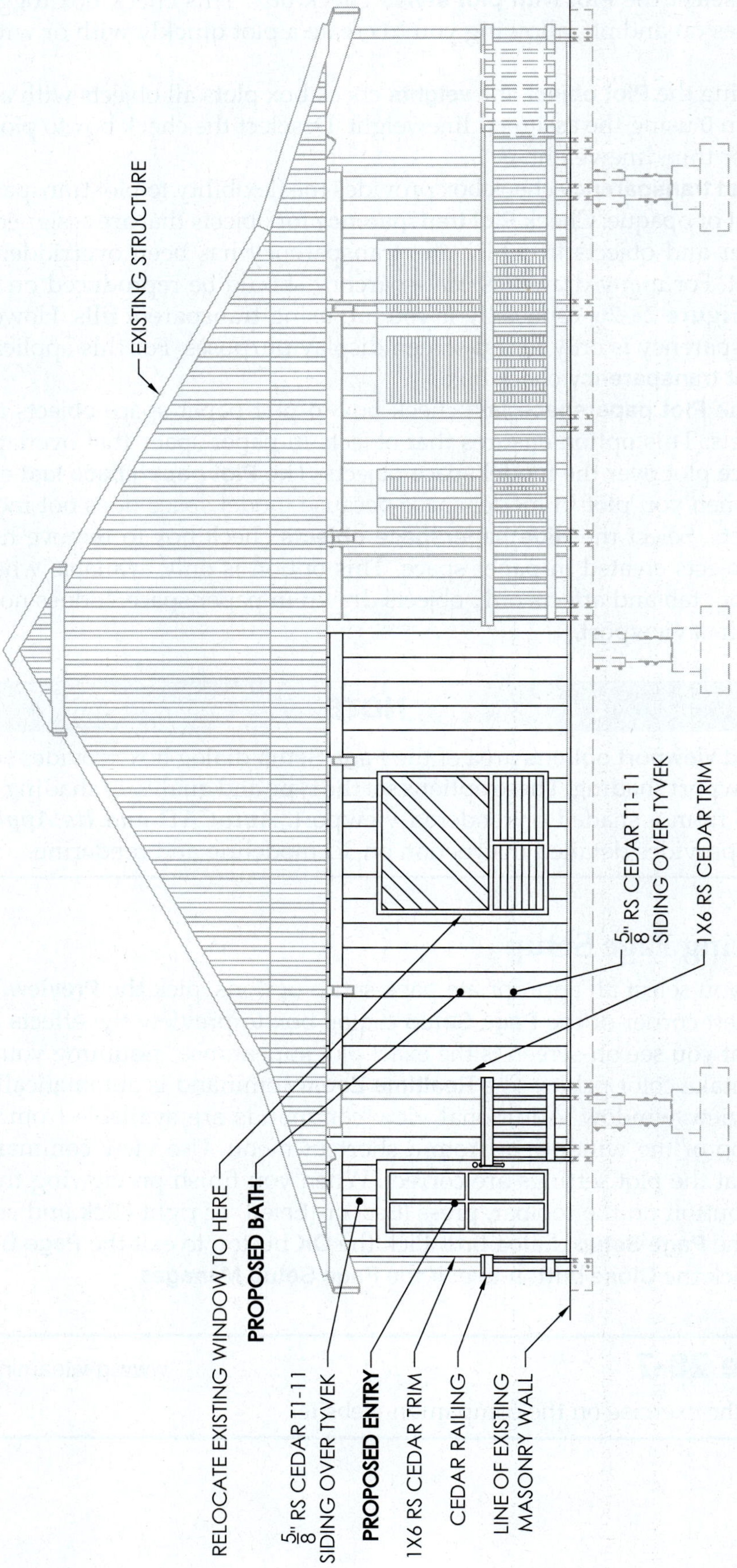

Chapter Review

Answer the following questions using the information in this chapter.

1. What is a model?
2. Define *model space.*
3. What is the purpose of a layout?
4. What is paper space?
5. What is the purpose of a floating viewport?
6. What is the purpose of the dashed rectangle that appears on the default layout?
7. If you pick the **MODEL** button to enter paper space and the file contains multiple layouts, none of which you previously accessed, which layout is displayed by default?
8. What is a page setup?
9. Describe the basic function of the **Page Setup Manager** and the related **Page Setup** dialog box.
10. Describe the function of the **Plot** dialog box and explain when it is used in relation to the **Page Setup Manager** and **Page Setup** dialog box.
11. Explain the importance of a well-defined page setup.
12. What is a plot device?
13. Define *sheet size.*
14. What factors should you consider when you select a sheet size for a drawing?
15. Which ASME standards specify sheet sizes?
16. What is the plot offset of a layout?
17. Define *drawing scale.*
18. What are plot styles?
19. Explain the purpose of a plot style table.
20. How do you assign plot styles in a color-dependent plot style?
21. Explain the function of a named plot style table.
22. How can you be certain that your page setup options will produce the desired plot?

Drawing Problems

Start AutoCAD if it is not already started. Follow the specific instructions for each problem.

Note: Some of the problems in this chapter are built on problems from previous chapters. If you have not yet completed those problems, complete them now.

▼ Basic

1. General Use a word processor to describe five items commonly found in a layout.

2. General Start a new drawing using the acad.dwt template. Create a plot style table named Black35mm.cbt that will plot all colors in the AutoCAD drawing in black ink on the paper, with a lineweight of 0.35 mm. (*Hint:* To make the same change to a property of all the plot styles, select the first plot style in the list, in this case Color 1, then scroll to the end of the list, hold down the [Shift] key, and select the last plot style in the list, in this case Color 255.) Save the drawing as P28-2.

3. General Start a new drawing using the acad -Named Plot Styles.dwt template. Create a plot style table named BlackShades.stb. Create the following plot styles:
   - Black100% with color set to black, and all other properties set to their default values.
   - Black50% with color set to black, screening set to 50, and all other properties set to their default values.
   - Black25% with color set to black, screening set to 25, and all other properties set to their default values.

   Save the drawing as P28-3.

▼ Intermediate

4. Mechanical Save P11-15 as P28-4. If you have not yet completed Problem 11-15, do so now. In the P28-4 file, delete the default **Layout2**. Create a new B-size sheet layout by following these steps:
   A. Rename the default **Layout1** as **B-SIZE**.
   B. Select the **B-SIZE** layout and access the **Page Setup Manager**.
   C. Modify the **B-SIZE** page setup according to the following settings:
   - **Printer/plotter**: Select a printer or plotter that can plot a B-size sheet
   - **Paper size**: Select the appropriate B-size sheet (varies with printer or plotter)
   - **Plot area**: Layout
   - **Plot offset**: 0,0
   - **Plot scale**: 1:1 (1 inch = 1 unit)
   - **Plot style table**: monochrome.ctb
   - **Plot with plot styles**
   - **Plot paperspace last**
   - Do not check **Hide paperspace objects**
   - **Drawing orientation**: Select the appropriate orientation (varies with printer or plotter)

5. **Mechanical** Save P11-16 as P28-5. If you have not yet completed Problem 11-16, do so now. In the P28-5 file, delete the default **Layout2**. Create a new A2-size sheet layout according to the following steps:
   A. Rename the default **Layout1** as **A2-SIZE**.
   B. Select the **A2-SIZE** layout and access the **Page Setup Manager**.
   C. Modify the **A2-SIZE** page setup according to the following settings:
      - **Printer/plotter**: Select a printer or plotter that can plot an A2-size sheet
      - **Paper size**: Select the appropriate A2-size sheet (varies with printer or plotter)
      - **Plot area**: Layout
      - **Plot offset**: 0,0
      - **Plot scale**: 1:1 (1 mm = 1 unit)
      - **Plot style table**: monochrome.ctb
      - **Plot with plot styles**
      - **Plot paperspace last**
      - Do not check **Hide paperspace objects**
      - **Drawing orientation**: Select the appropriate orientation (varies with printer or plotter)

6. **Mechanical** Save P8-18 as P28-6. If you have not yet completed Problem 8-18, do so now. In the P28-6 file, delete the default **Layout2**. Create a new B-size sheet layout according to the following steps:
   A. Rename the default **Layout1** as **B-SIZE**.
   B. Select the **B-SIZE** layout and access the **Page Setup Manager**.
   C. Modify the **B-SIZE** page setup according to the following settings:
      - **Printer/plotter**: Select a printer or plotter that can plot a B-size sheet
      - **Paper size**: Select the appropriate B-size sheet (varies with printer or plotter)
      - **Plot area**: Layout
      - **Plot offset**: 0,0
      - **Plot scale**: 1:1 (1 inch = 1 unit)
      - **Plot style table**: monochrome.ctb
      - **Plot with plot styles**
      - **Plot paperspace last**
      - Do not check **Hide paperspace objects**
      - **Drawing orientation**: Select the appropriate orientation (varies with printer or plotter)

▼ Advanced

7. **General** Use a word processor to write a report of approximately 250 words explaining the difference between model space and paper space and describing the importance of using layouts. Cite at least three examples from actual industry applications of using layouts to prepare a multi-sheet drawing. Use at least four drawings to illustrate your report.

For Problems 8 through 11: Start a new drawing for each problem using an appropriate template of your choice. The template should include layers and text, dimension, multileader, and table styles, when necessary, for drawing the given objects. Add layers and text, dimension, multileader, and table styles as needed. Draw all objects using appropriate layers. Use appropriate text, dimension, multileader, and table styles, justification, and format. Follow the specific instructions for each problem. Use only drawing and editing commands and techniques you have already learned. Use your own judgment and approximate dimensions when necessary. Apply dimensions accurately using ASME or appropriate industry standards.

8. **Architectural** Draw and dimension the wood beam details shown. Establish the missing information using your own specifications, or determine the correct size of items not dimensioned. Prepare a single layout for plotting on a C-size sheet using the monochrome.ctb plot style table. Delete all other layouts. Save the drawing as P28-8.

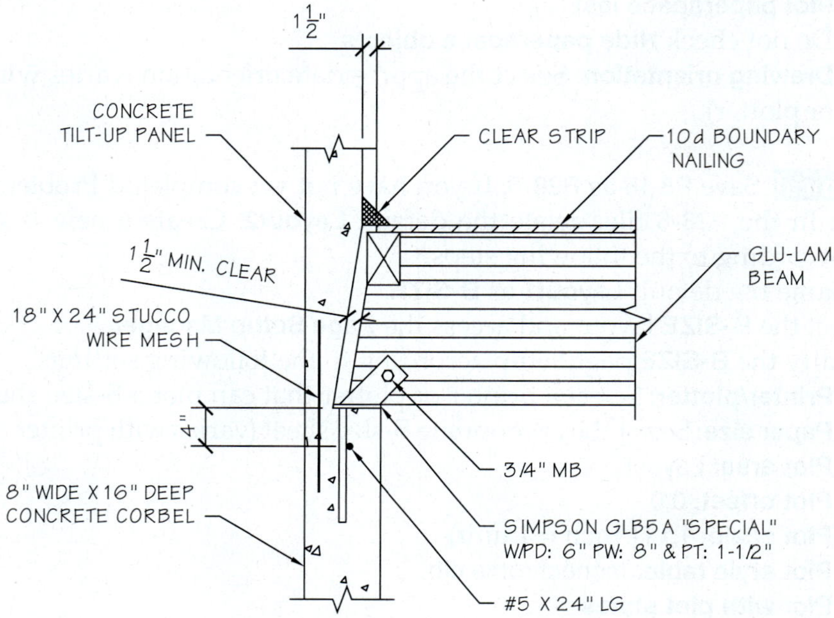

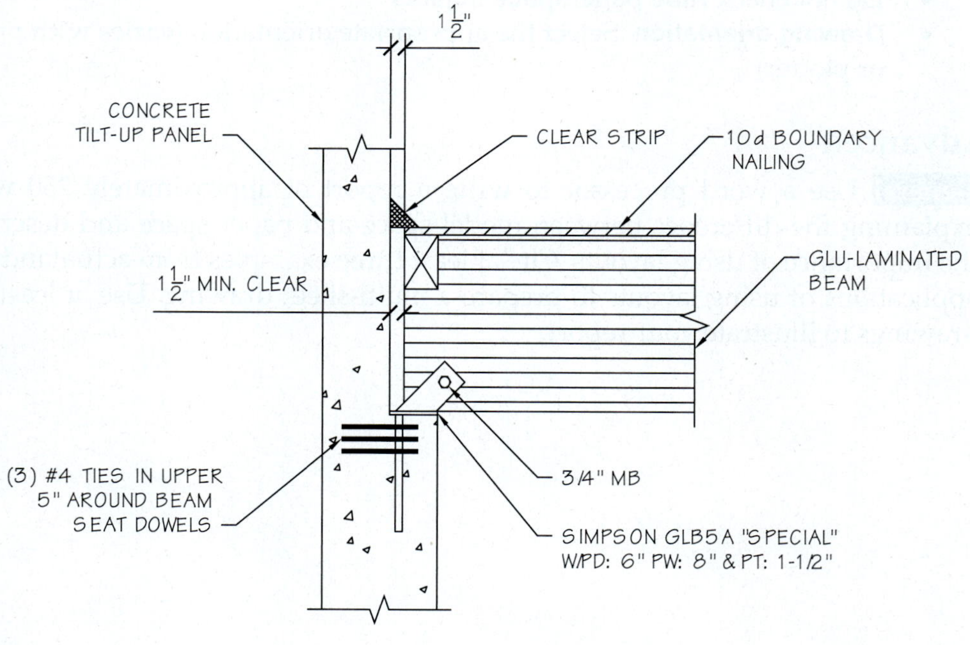

9. **Mechanical** Draw and dimension the views of the support shown. Prepare a single layout for plotting on an A3-size sheet using the monochrome.ctb plot style table. Delete all other layouts. Save the drawing as P28-9.

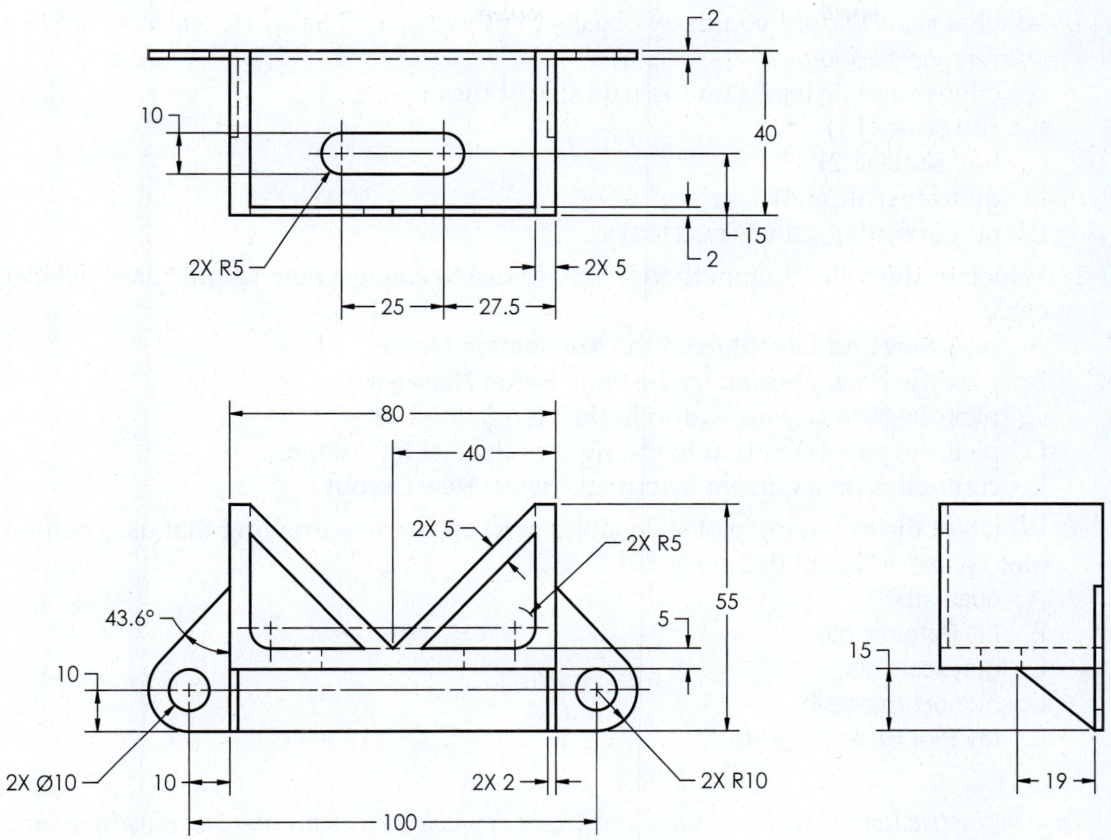

10. **Mechanical** Research the design of an existing drive sprocket with the following specifications: single strand with hub, accepts ANSI 50 chain, 28 teeth, 5/8″ pitch, finished 1″ bore with 1/4″ × 1/8″ ANSI keyway and two set screws, steel material. Create a freehand, dimensioned 2D sketch of the existing design from manufacturer's specifications or from measurements taken from an actual sprocket. Start a new drawing from scratch or use a decimal-unit template of your choice. Draw and dimension the views required to document the design. Prepare a single layout for plotting on an appropriate sheet using the monochrome.ctb plot style table. Delete all other layouts. Save the drawing as P28-10. Prepare a slide show presentation of your research, design process, and drawing and present it to your class or office.

11. **Architectural** Create a freehand, dimensioned 2D sketch of the floor plan of a precast concrete restroom for a public park. Design the restroom to fit a 16′-0″ × 16′-0″ concrete slab foundation. Divide the restroom to provide separate stalls for men and women. Use dimensions based on your experience, research, and measurements. Start a new drawing from scratch or use an architectural template of your choice. Draw and dimension the floor plan from your sketch. Prepare a single layout for plotting on an appropriate sheet using the monochrome.ctb plot style table. Delete all other layouts. Save the drawing as P28-11. Prepare a slide show presentation of your research, design process, and drawing and present it to your class or office.

Drawing Problems - Chapter 28

AutoCAD Certified Professional Exam Practice

Answer the following questions using the information in this chapter.

1. At what scale should you draw objects in model space? *Select the one item that best answers the question.*
   A. choose a scale based on the true size of the objects
   B. full scale (1:1)
   C. half scale (1:2)
   D. quarter scale (1:4)
   E. no particular scale is necessary
2. Which of the following methods can you use to create a new layout? *Select all that apply.*
   A. pick **New** and **Layout** from the **Application Menu**
   B. pick the **New...** button in the **Page Setup Manager**
   C. pick the **New Layout** button in the navigation bar
   D. pick the plus (**+**) button to the right of the last layout tab
   E. right-click on a current layout and select **New Layout**
3. Which of the following plot style tables can you use in a drawing that uses named plot styles? *Select all that apply.*
   A. acad.stb
   B. Fill Patterns.ctb
   C. grayscale.ctb
   D. monochrome.stb
   E. My Plot Style Table.stb

Follow the instructions in each problem. Complete each problem by answering the related question.

4. **Navigate to this chapter on the companion website and open CPE-28pagesetup.dwg.** Assuming that you print this drawing using a color printer, in what color will objects drawn on the Objects layer plot?
5. **Navigate to this chapter on the companion website and open CPE-28sheetsize.dwg.** Use appropriate inquiry commands to determine the answer to this question. What sheet size is **Layout1** most likely designed to plot?

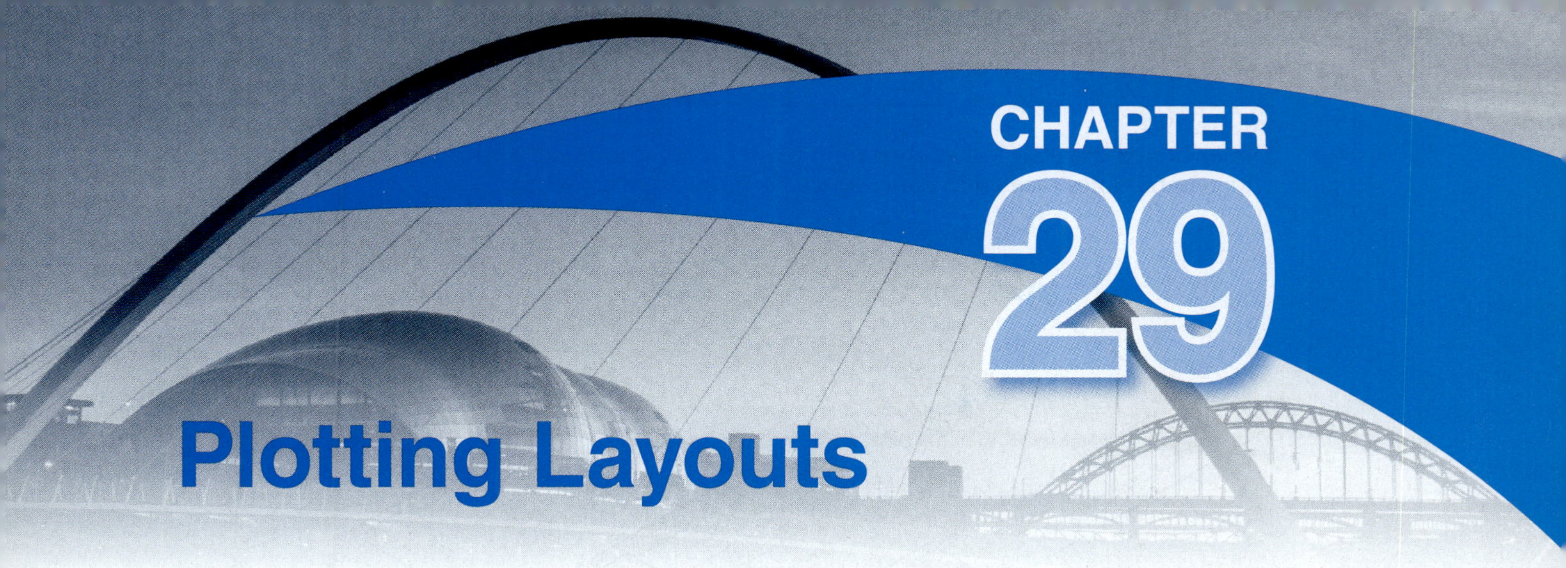

CHAPTER 29

Plotting Layouts

Learning Objectives

After completing this chapter, you will be able to:

✓ Add layout content.
✓ Use floating viewports to create properly scaled final drawings.
✓ Preview and plot layouts.

This chapter explores the additional steps needed to complete layout setup. You will learn to add content to a layout and to place and use floating viewports. You will also use the **PLOT** command to preview the plot and send the layout to a printer, plotter, or alternate electronic format.

Layout Content

Model space provides an environment to create drawing views and add dimensions and annotations directly to views. Layouts provide an effective method to display model space content using floating viewports and to add items such as:

- Border
- Sheet blocks, such as title blocks and revision history blocks
- General notes
- Bill of materials, parts list, schedules, legends, and other associated lists
- Sheet annotation and symbols such as titles, north arrow, and graphic scale

Layouts provide flexibility to lay out, scale, and prepare a final drawing. Consider the objects placed in model space to be model content and the objects you add to a layout to be sheet content. A complete drawing forms when you bring model and sheet content together. See **Figure 29-1**.

Drawing in Paper Space

Most 2D drawing and editing commands and options described throughout this textbook function the same in paper space as in model space. However, some tools are specific to or most commonly used in either model space or paper space. For example, floating viewports are specific to paper space.

Figure 29-1.
Dimensioned drawing views created in model space, combined with a border, sheet blocks, and general notes created in paper space, form the final drawing.

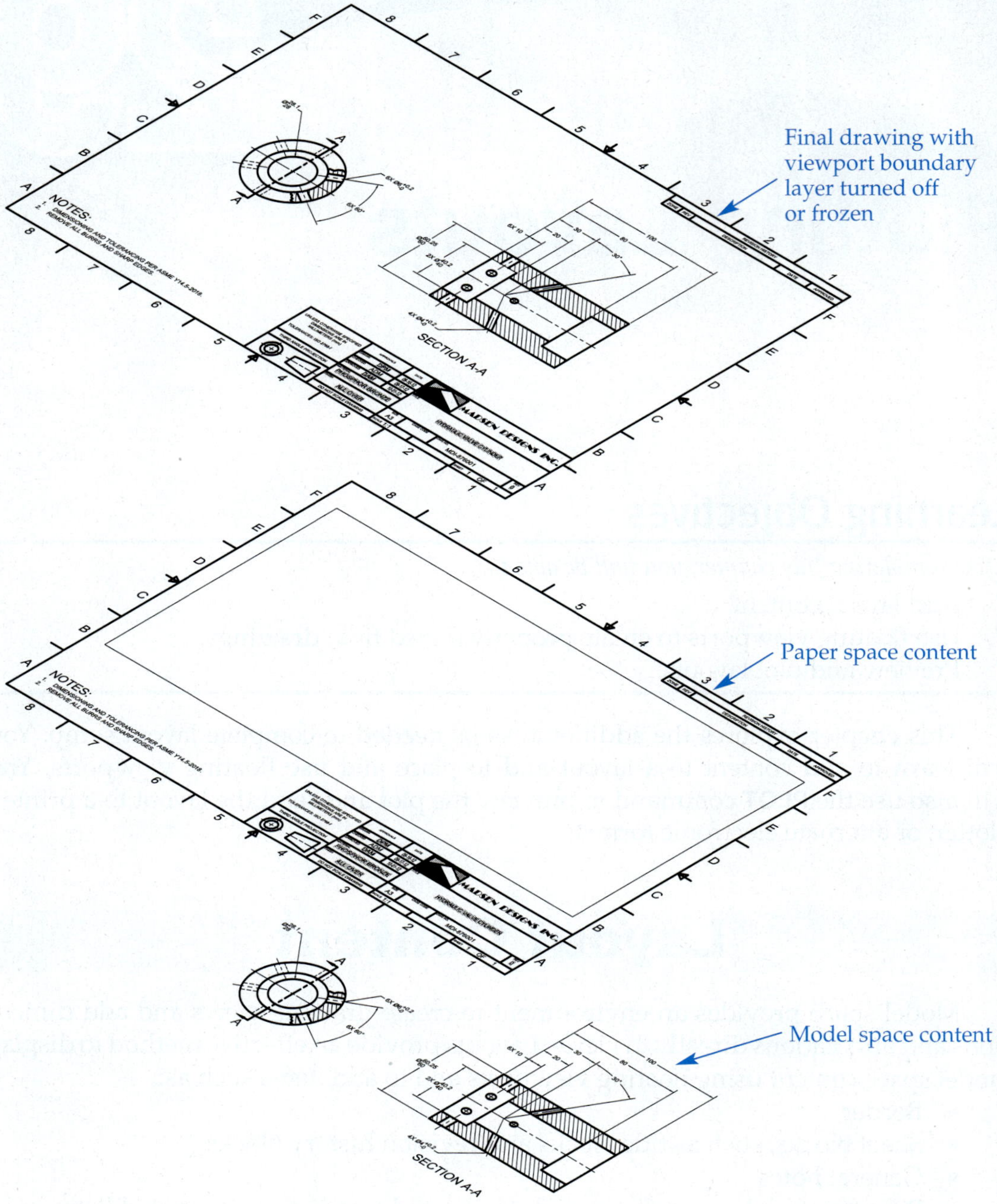

NOTE

Although paper space is a 2D environment, you can display 3D models created in model space in paper space floating viewports.

As in model space, you typically draw geometry on a layout at full scale. One difference between model space and paper space is that objects in model space may be very large or very small, while paper space objects always correspond to the sheet size. Draw all layout content using the actual size you want the objects to appear on the plotted sheet. Use multiline or single-line text or blocks with attributes to add content such as general notes, view titles, and similar annotations to a layout. Place a bill of materials, a parts list, schedules, or similar tabular information using the **TABLE** command or blocks with attributes. Typically, you will create items such as a border, title block, and revision history block as blocks, often with attributes, and then insert these items into the layout.

Use layers appropriate for the layout and the objects added to the layout. Consider using a single layer named SHEET, for example, on which you draw all layout content. Another option is to use layers specific to layout items, such as a BORDER layer for the border and a TITLE or A-ANNO-TTBL layer for the title block. Assign a layer named Viewport, VPORT, or A-ANNO-NPLT, for example, to floating viewports so you can turn off, freeze, or set the viewport boundary to "no plot" before plotting.

NOTE

The default floating viewport boundary assumes the layer that is current when you first access a layout. If you use the default viewport, it may be necessary to change the layer on which it is drawn.

The Layout Origin

Most drawing and editing in paper space occurs on the sheet, which is the white rectangle you see on the gray background. However, it is possible, and necessary in some applications, to create objects off the sheet. When drawing and editing on a layout, remember that the origin (0,0) is controlled by the X and Y values you enter in the **Plot offset** area of the **Page Setup** dialog box. For example, if you draw a line with a start point of 1,1, the line begins 1 unit to the right and 1 unit up from the plot origin. The default origin position is at the lower-left corner of the printable area. See Figure 29-2. Refer to Chapter 28 for more information about plot origin and the **Plot offset** area of the **Page Setup** dialog box.

Layout content often references the edge of the sheet. For example, you might position a border 1/2" inside of the sheet edge. In this situation, you usually want the layout origin to be the lower-left corner of the sheet. The best option is to select the **Edge of paper** radio button in the **Specify plot offset relative to** area in the **Plot and Publish** tab of the **Options** dialog box. In the **Plot offset** area of the **Page Setup** dialog box, use values of 0.000 in the X and Y offset text boxes to locate the plot origin at the lower-left corner of the sheet. This is an excellent way to center the drawing on the sheet.

NOTE

Depending on the specific plot configuration, you may need to pick the **Plot upside-down** check box in the **Drawing orientation** area of the **Page Setup** dialog box in addition to using the **Edge of paper** offset option. Draw an object to see if 0,0 is actually located exactly at the lower-left corner of the sheet. If not, pick the **Plot upside-down** check box to solve the problem.

Figure 29-2.
The default location of the plot origin is often not appropriate, because all point entry is in reference to the lower-left corner of the printable area. Change the location of the plot origin to define the lower-left corner of the sheet at coordinates 0,0.

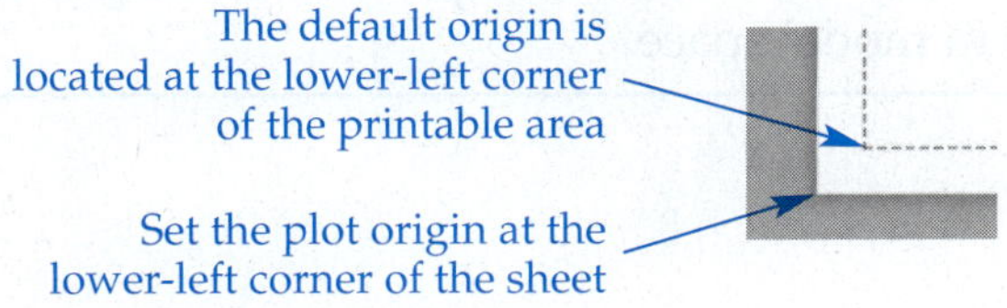

PROFESSIONAL TIP

Defining the plot offset relative to the edge of the paper maintains the plot offset location from the sheet edge even when you use a different plotter or plot configuration, as long as you use the same sheet size.

A second option to relate content to the lower-left corner of the sheet is to select the **Printable area** radio button in the **Specify plot offset relative to** area in the **Plot and Publish** tab of the **Options** dialog box. Then identify the values of the lower-right corner of the printable area. Information about the printable area is available in the **Device and Document Settings** tab of the **Plotter Configuration Editor**. Next, in the **Plot offset** area of the **Page Setup** dialog box, change the X and Y plot offset values to the values of the lower-left corner of the printable area. The values you enter must be negative. The origin offsets from the printable area, so if you use a different plotter or plot configuration, the printable area may change, causing the offset to shift.

Exercise 29-1

www.g-wlearning.com/CAD/

Complete the exercise on the companion website.

Floating Viewports

The primary advantage of using floating viewports in a paper space layout is the ability to prepare scaled drawings without increasing or decreasing the actual size of drawing views or sheet content. You can also create multiple viewports on a single layout to show differently scaled or alternate drawing views. For example, a single sheet might contain a floor plan drawn at a 1/4″ = 1′-0″ scale, an eave detail drawn at a 3/4″ = 1′-0″ scale, and a foundation detail drawn at a 3/4″ = 1′-0″ scale. See Figure 29-3.

A floating viewport boundary is the portion of the viewport that you see. Everything inside the viewport shows through from model space. Use commands such as **MOVE**, **ERASE**, **STRETCH**, and **COPY** to modify the viewport boundary in paper space. Use display commands such as **VIEW**, **PAN**, and **ZOOM** to modify the display of model space objects in the floating viewport. Additional options are available for adjusting how objects appear, according to the layers you assign to objects. Layer control allows you to define how the drawing appears within the viewport. Be sure a layout tab is current as you work through the following sections describing floating viewports.

PROFESSIONAL TIP

When you first select a layout and enter paper space, a rectangular floating viewport appears, showing the objects in model space. As part of layout setup, consider using the **ERASE** command to erase the default viewport and draw a new viewport (or several viewports). Erasing the default viewport, or erasing everything in the layout, does not erase objects in model space.

Figure 29-3.
An example of an architectural layout, ready to plot, that includes three floating viewports used to display drawing views at different scales. The layer on which the viewports are drawn is turned off or frozen for plotting.

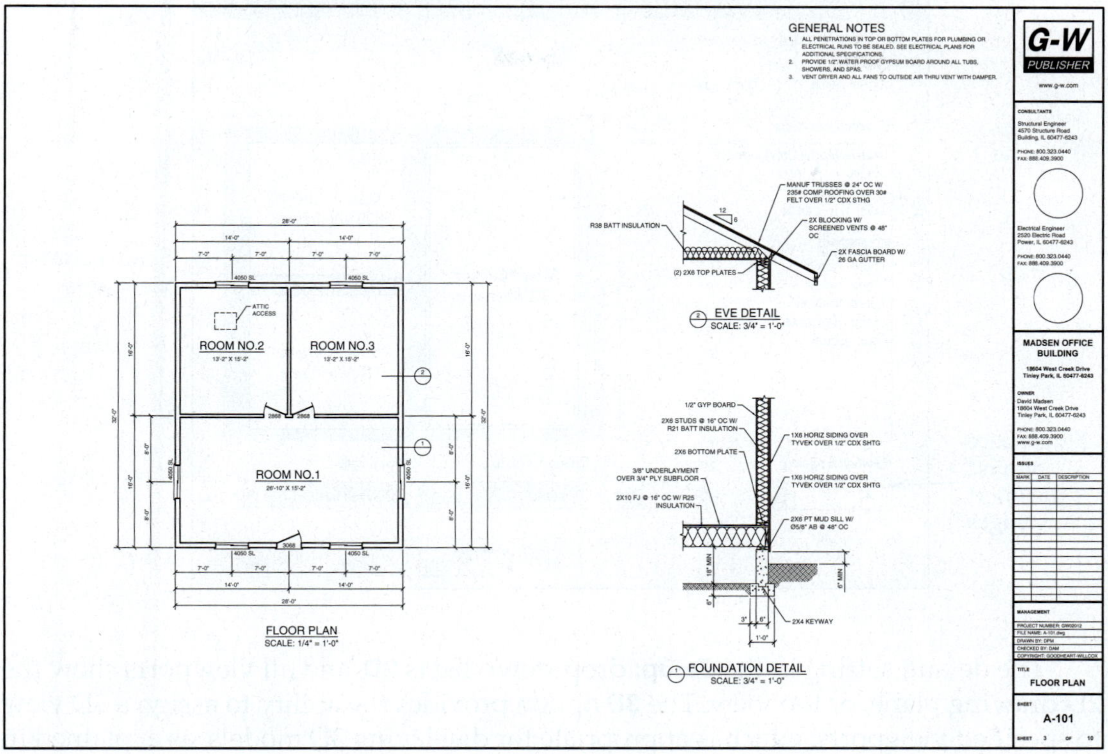

Specifying a Single Rectangular Viewport

An easy way to place a single rectangular floating viewport is to pick the **Rectangular** button from the **Layout Viewports** panel of the **Layout** ribbon tab. Specify a first and second corner to create the floating viewport and exit the command. Figure 29-1 shows an example of a single rectangular floating viewport used to display dimensioned views drawn in model space.

> **Ribbon**
> Layout
> > Layout Viewports
> Rectangular

Using the Viewports Dialog Box

The **Rectangular** button previously described offers a shortcut for creating a single rectangular viewport. Access the **VIEWPORTS** command to display the **Viewports** dialog box with multiple viewport configuration options. See Figure 29-4. The **Viewports** dialog box is similar in paper space and model space.

> **Ribbon**
> Layout
> > Layout Viewports
> New Viewports...
> **Type**
> VIEWPORTS
> VPORTS

VIEWPORTS

The **Standard viewports:** list contains preset viewport configurations. Select the ***Active Model Configuration*** option to convert a model space viewport configuration into individual floating viewports. For example, if model space displays two tiled viewports, use the ***Active Model Configuration*** option to create two floating viewports.

The remaining items in the **Standard viewports:** list are preset viewport configurations. The configuration name identifies the number of viewports and the arrangement or location of the largest viewport. Select a configuration to see a preview of the floating viewports in the **Preview** area. The **Viewport Spacing:** text box is available when you select a standard viewport configuration that includes two or more viewports, as shown in Figure 29-4. Enter a value to define the space between multiple viewports.

Figure 29-4.
The **Viewports** dialog box with the **New Viewports** tab selected. The **Viewport Spacing:** setting is available when paper space is active.

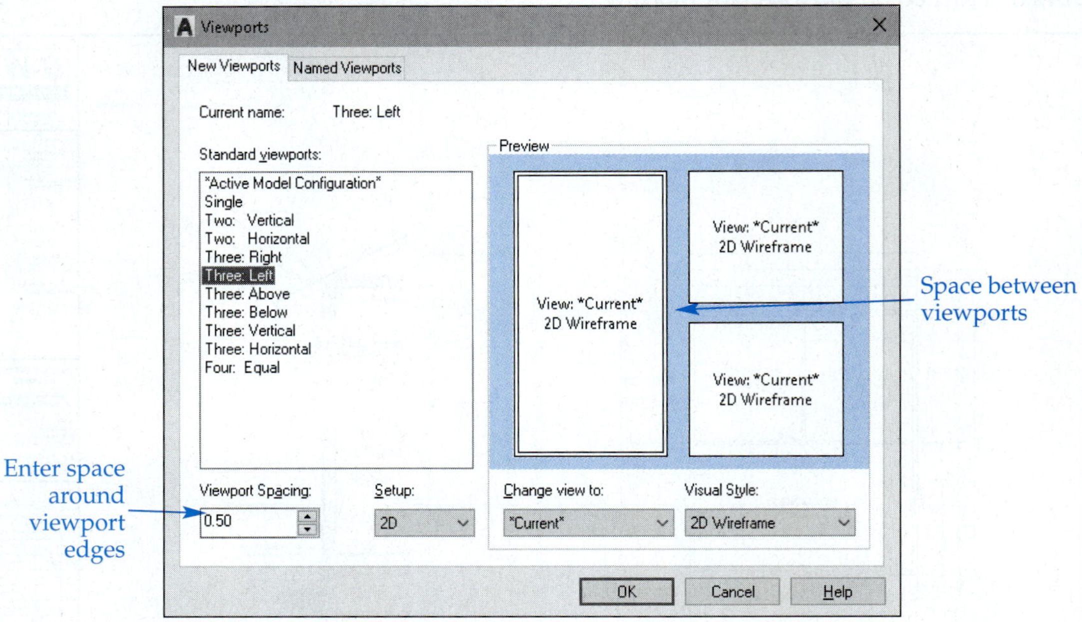

The default setting in the **Setup:** drop-down list is **2D**, and all viewports show the 2D drawing plane, or top view. The **3D** option provides the ability to assign a 3D view to specified viewports, which is appropriate for displaying 3D models, as explained in *AutoCAD and Its Applications—Advanced*.

The viewport configuration is displayed in the **Preview** image. To change a view in a viewport, pick the viewport in the **Preview** image and select from the **Change view to:** and **Visual Style:** drop-down lists. The **Change view to:** drop-down list includes any named model views saved with the file. Pick a named view from the list to apply to the selected viewport. Use the **2D Wireframe** option of the **Visual Style:** drop-down list for standard 2D drafting applications. *AutoCAD and Its Applications—Advanced* explains the visual style settings appropriate for 3D modeling.

Pick the **OK** button to specify a first and second corner to define the area occupied by the viewport configuration. See Figure 29-5. An alternative is to choose the **Fit** option before specifying the first point to fill the printable area with the viewport configuration.

NOTE

The text-based **–VPORTS** and **MVIEW** commands offer the same floating viewport options that you can more easily access from the **Layout Viewports** panel of the **Layout** ribbon tab, the **Viewports** dialog box, or other methods such as a shortcut menu and the **Properties** palette. These techniques are described throughout this chapter.

Exercise 29-2

www.g-wlearning.com/CAD/

Complete the exercise on the companion website.

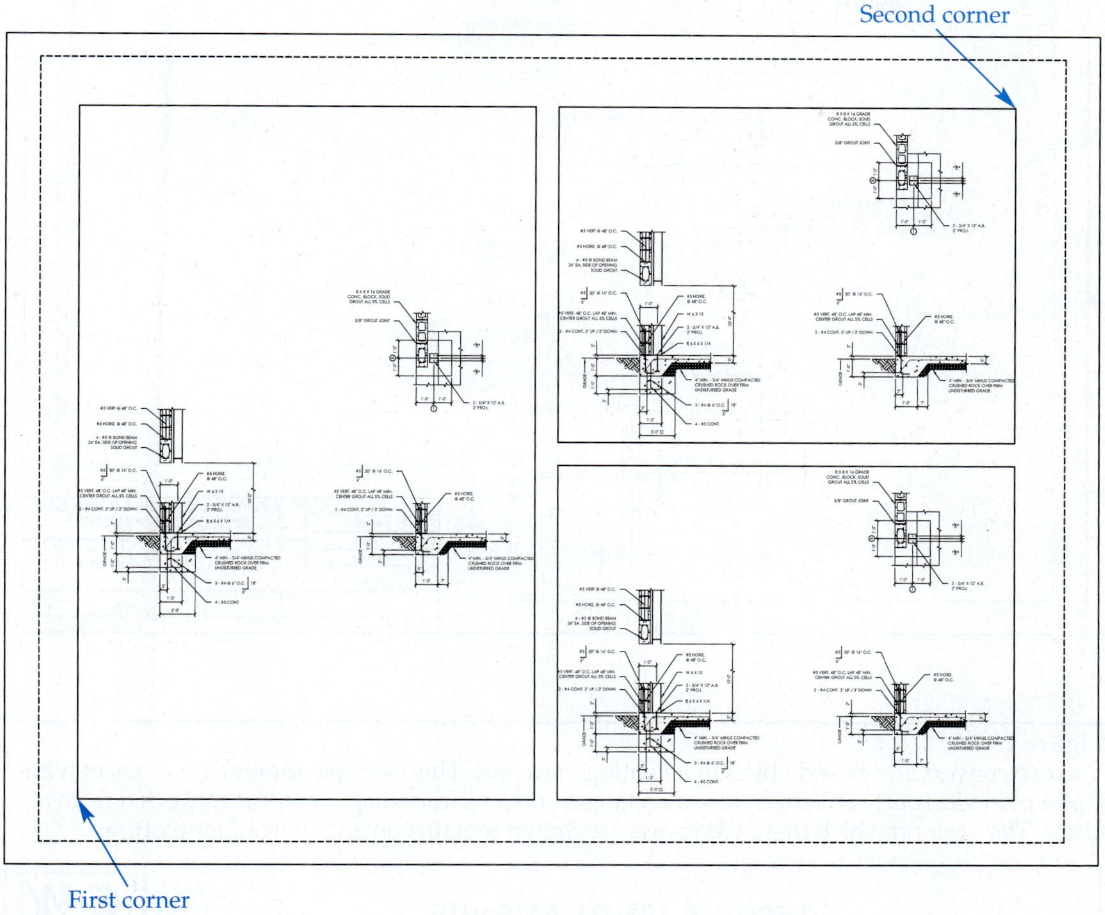

Figure 29-5.
Select two points on the layout to specify the area to be filled by the viewport configuration. Notice that the viewport spacing forms as specified in the **Viewports** dialog box. This example shows the initial placement of three structural details using a three-viewport configuration. Each viewport will eventually display a single detail.

Forming Polygonal Floating Viewports

An easy way to construct a polygonal floating viewport is to pick the **Polygonal** button from the **Layout Viewports** panel of the **Layout** ribbon tab. Construct a polygonal viewport using the same techniques you use to draw a closed polyline object. The viewport can be any closed shape composed of lines and arcs. Figure 29-6 shows a polygonal floating viewport used to define the maximum drawing view area 1/2″ in from the border and title block.

Ribbon
Layout
> Layout Viewports
Polygonal

Converting Objects to Floating Viewports

An easy way to convert a closed object drawn in paper space to a floating viewport is to pick the **Object** button from the **Layout Viewports** panel of the **Layout** ribbon tab. Select a closed shape, such as a circle, ellipse, or closed polyline shape, to convert the object to a viewport. Figure 29-7 shows an example of a circle and a closed polyline converted to floating viewports.

Ribbon
Layout
> Layout Viewports
Object

Exercise 29-3

www.g-wlearning.com/CAD/

Complete the exercise on the companion website.

Figure 29-6.
A structural detail displayed in a polygonal floating viewport. The layer on which the viewport is drawn is turned off or frozen for plotting.

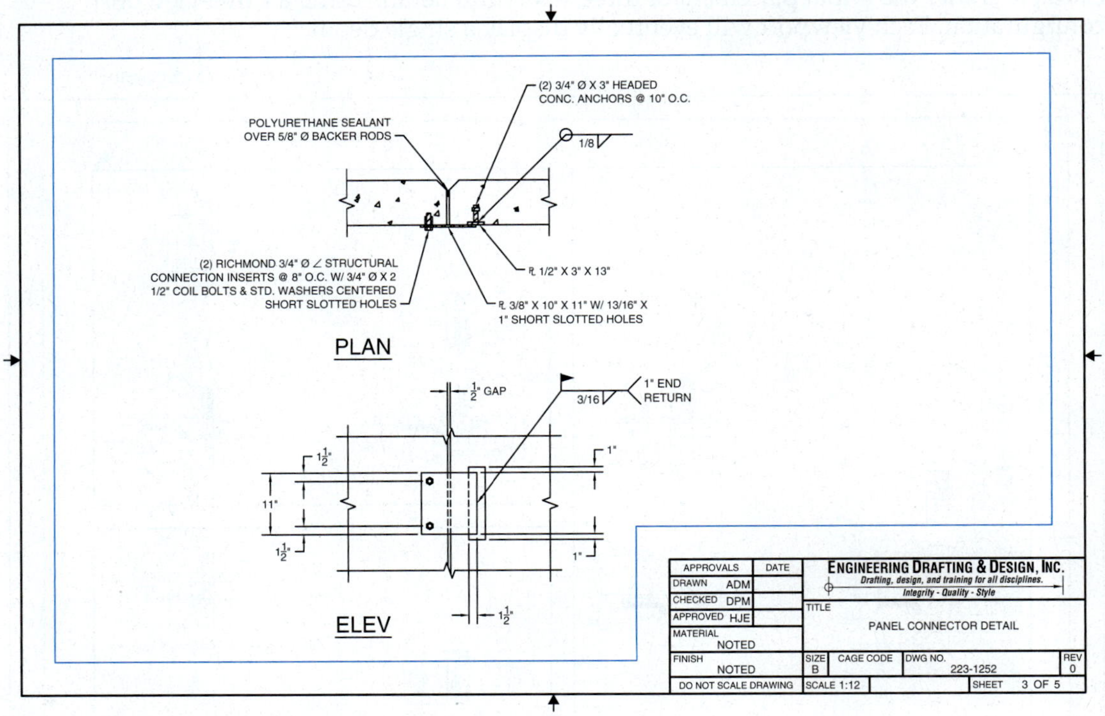

Figure 29-7.
You can convert any closed object to a floating viewport. This example shows a cover sheet with a plot plan viewport converted from a rectangle and a vicinity map viewport converted from a circle. The layer on which these viewports are drawn remains on and thawed for plotting.

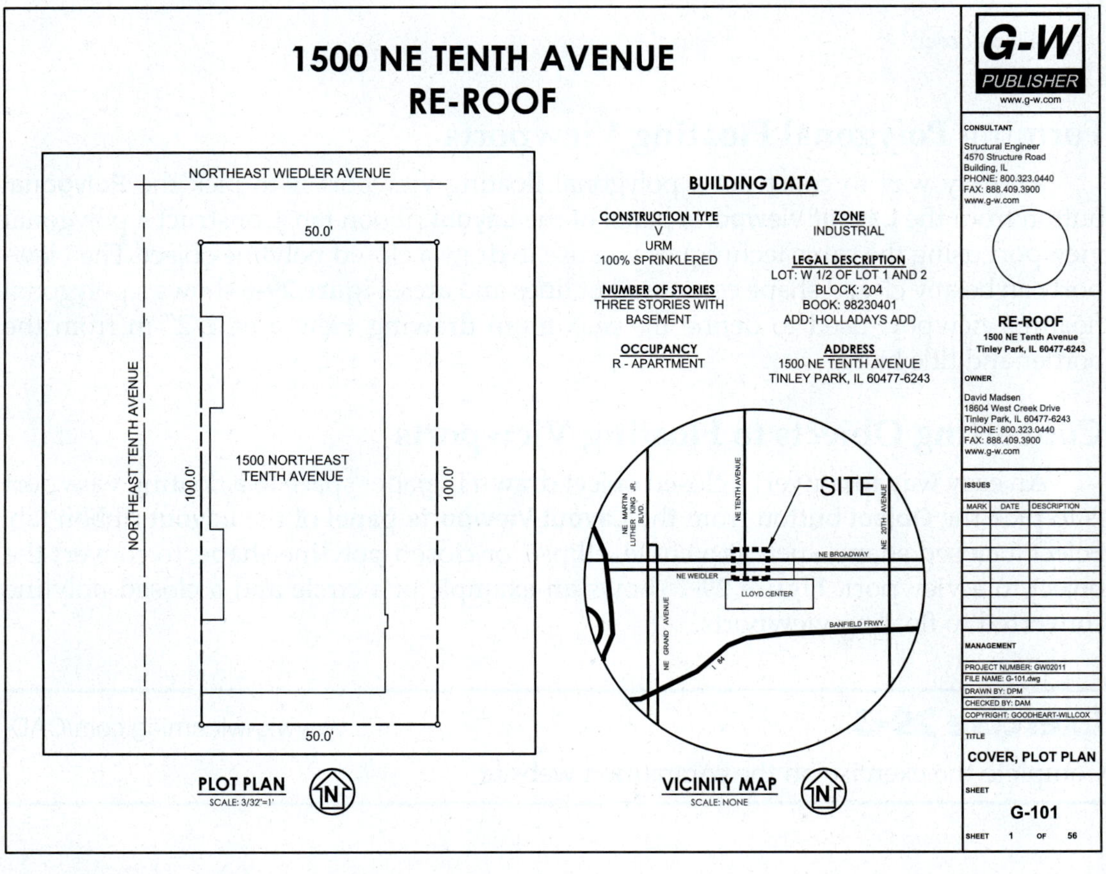

Inserting Named Views as Floating Viewports

Ribbon
Layout
> Layout Viewports
>Insert View

AutoCAD allows you to insert a named model view saved with the file as a floating viewport. A quick way to apply this option is to use the **Insert View** drop-down list in the **Layout Viewports** panel of the **Layout** ribbon tab. Scroll through a long list as needed to find the appropriate named view. The boundary of the selected view attaches to the cursor. You have the option to right-click and select the viewport scale from a shortcut menu before placing the view. You will learn more about scaling floating viewports later in this chapter. Specify a point to insert the floating viewport.

Inserting New Views as Floating Viewports

Ribbon
Layout
> Layout Viewports
> Insert View
> New View

The **New View** option provides a method for placing a floating viewport based on a specified area in model space. A quick way to apply this option is to use the **Insert View** drop-down list in the **Layout Viewports** panel of the **Layout** ribbon tab. If there are no named views in the drawing, pick the **Insert View** button. If the drawing contains named views, select the **New View** option from the **Insert View** drop-down list. AutoCAD temporarily displays model space, allowing you to specify the corners of a view to capture. Press [Enter] to return to paper space with the boundary of the selected view attached to the cursor. Right-click and select the viewport scale from the shortcut menu if desired, and then specify a point to insert the floating viewport.

Exercise 29-4 www.g-wlearning.com/CAD/

Complete the exercise on the companion website.

Adjusting the Floating Viewport Boundary

For purposes of adjusting a floating viewport boundary, you should consider the boundary a closed object. For example, treat rectangular viewports like rectangles, polygonal viewports like closed polyline objects, circular viewports like circles, and elliptical viewports like ellipses. Use grip editing and editing commands to modify the size, shape, and location of floating viewports. Floating viewports have a filled-square center grip that you can use to move the viewport and a filled-triangle grip for adjusting the viewport scale, as explained later in this chapter. When you adjust a floating viewport, the "hole" cut through the sheet changes.

Exercise 29-5 www.g-wlearning.com/CAD/

Complete the exercise on the companion website.

Clipping Viewports

Ribbon
Layout
> Layout Viewports
Clip
Type
VPCLIP

VPCLIP

The **VPCLIP** command allows you to redefine the boundary of an existing viewport. To access the command from a shortcut menu, select a viewport and then right-click and pick **Viewport Clip**. You can clip a floating viewport to an existing closed object that you draw before accessing the **VPCLIP** command, or you can clip the viewport to a polygonal shape that you create while using the command.

After you access the **VPCLIP** command, select the viewport to clip. Then select an existing closed shape, such as a circle, ellipse, or closed polyline, to recreate the viewport in the shape of the selected object. See Figure 29-8. An alternative, after you select the existing viewport, is to use the **Polygonal** option to redefine the viewport to a polygonal shape. This option works the same as the **Polygonal** option previously described, except the existing viewport transforms into the new shape.

Figure 29-8.
Clipping a viewport showing a wall section to a closed polyline to create an eave detail. AutoCAD removes the original viewport and converts the rectangle to a viewport.

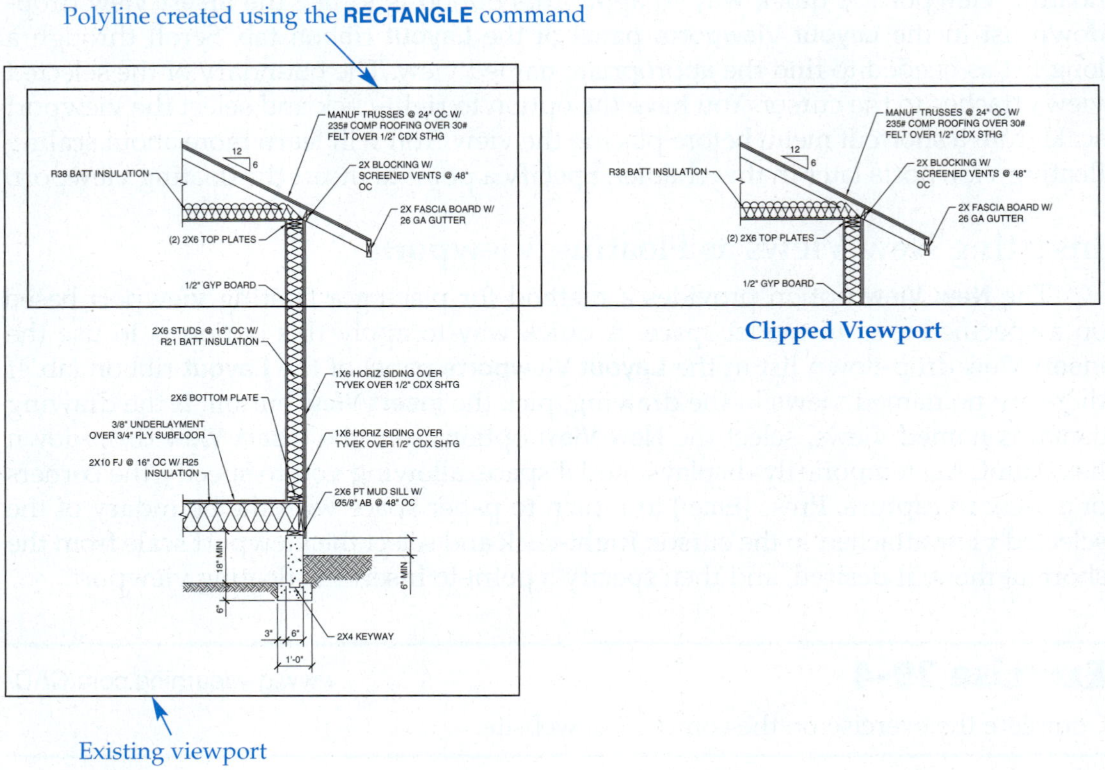

AutoCAD recognizes a clipped floating viewport as clipped. The **VPCLIP** command offers a **Delete** option when you select a clipped viewport. Use the **Delete** option to remove the clipped definition and convert the shape into a viewport sized to fit the extents of the original clipping object or polygonal shape.

NOTE

A clipping object or polygonal shape does not need to be on or overlap the viewport to be clipped.

PROFESSIONAL TIP

Create floating viewports and adjust the size and shape of viewport boundaries after you insert the border, title block, and other layout content. This will allow you to position viewports so they do not interfere with layout information.

Rotating Model Space Content

The **VPROTATEASSOC** system variable setting determines what happens to model space content when you rotate a floating viewport. The variable has a value of 1 by default. As a result, when you rotate a viewport, the display of objects in model space rotates to align with the viewport. See Figure 29-9A. The orientation of objects in model space does not change. In order to maintain the original alignment of model space content in a floating viewport, as shown in Figure 29-9B, access the **VPROTATEASSOC** system variable before rotating the viewport and enter a value of 0.

Figure 29-9.
A highway cut and fill plan shown through a viewport. A—The default **VPROTATEASSOC** setting of 1 rotates model space content on a floating viewport to align with viewport rotation. B—Change the value to 0 to maintain the original model space orientation when you rotate a viewport.

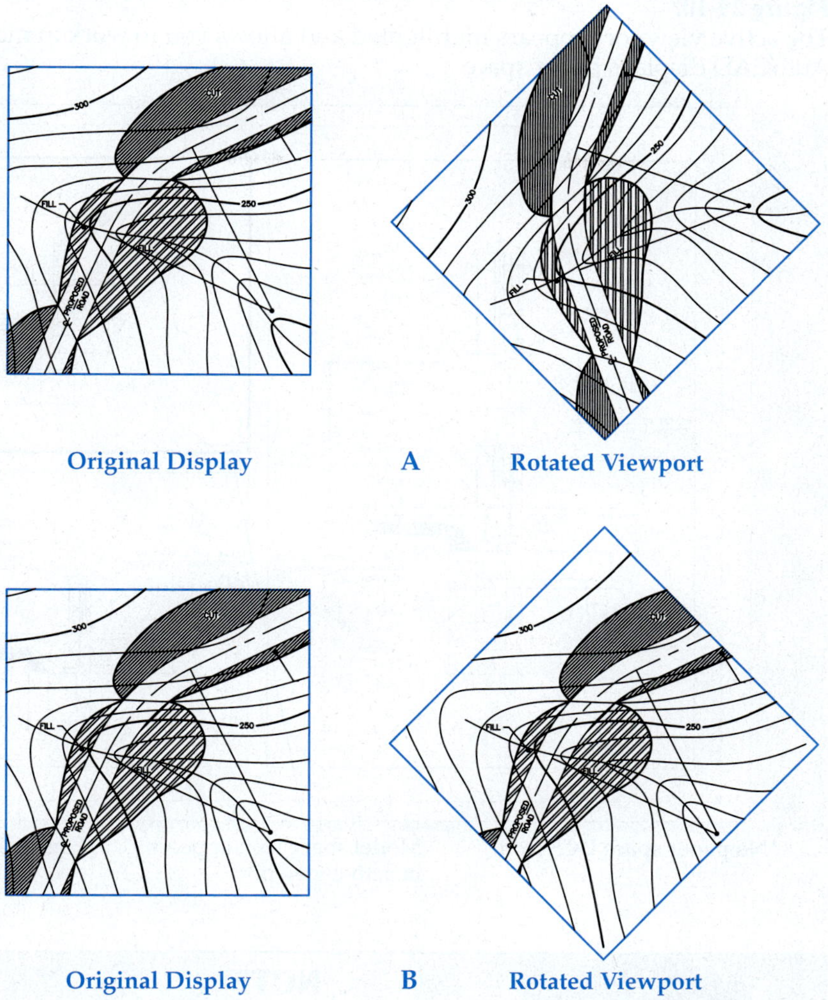

NOTE

Other commands, such as **UCS** and **MVSETUP**, include options for rotating items shown through a viewport. The **VPROTATEASSOC** system variable automates the process. The entire display of model space rotates with the rotation of the viewport. Drawing characteristics such as unidirectional dimensions do not update according to the rotation.

Activating and Deactivating Floating Viewports

Activate a floating viewport to work with model space objects while in paper space. This allows you to adjust the display of the model space drawing shown in the viewport. Repeat the process of activating and adjusting a viewport for every floating viewport in the layout to achieve the final drawing.

To activate a floating viewport, double-click inside the viewport boundary, select the **PAPER** button on the status bar, or type MSPACE or MS. If the layout contains a single viewport, the viewport appears highlighted, indicating that it is current. On the layout, the UCS icon disappears, and the model space UCS icon is displayed in the corner of the active layout viewport. You are now working directly in model space, through the paper space viewport. The active and highlighted viewport is the viewport you double-click on or the newest viewport, depending on how you access the **MSPACE** command. By default, the active floating viewport includes the viewport controls, the navigation bar, and the view cube. See Figure 29-10. To make a different viewport active, pick once inside the viewport.

Figure 29-10.
The active viewport appears highlighted and allows you to work in model space while AutoCAD displays paper space.

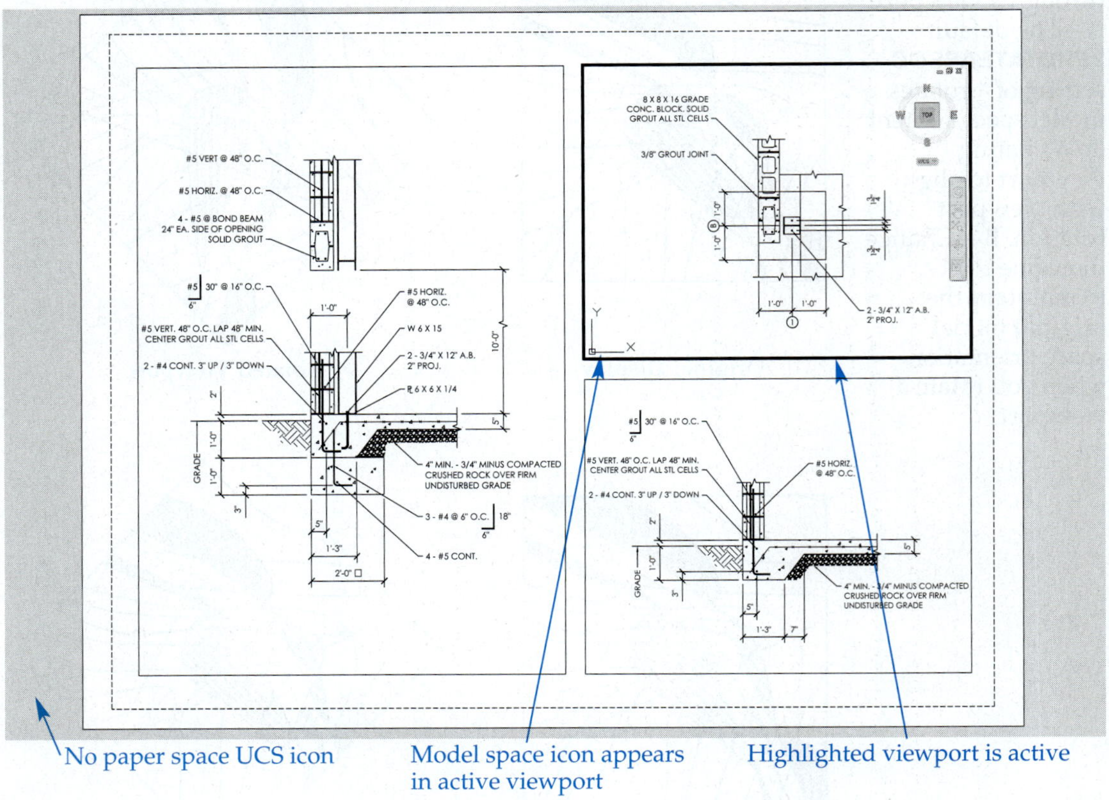

NOTE

The viewport controls are initially difficult to see, but they are located in the upper-left corner of the active floating viewport.

After you adjust the display of all floating viewports, you must re-enter paper space to plot and continue working with the layout. To activate paper space, double-click outside a viewport, select the **MODEL** button on the status bar, or type PSPACE or PS. The layout space UCS icon reappears, and the model space UCS icon disappears from the corners of the viewports.

Scaling a Floating Viewport

The scale you assign to a floating viewport is usually the same as the drawing scale. A quick way to set viewport scale is to activate a viewport, or pick a viewport boundary in paper space without activating the viewport. Then select the appropriate scale from the **Viewport Scale** flyout on the status bar or the shortcut menu that appears when you select the viewport boundary and pick the filled-triangle scale grip. See Figure 29-11. An alternative is to pick the viewport to be scaled in paper space and access the **Properties** palette. Then choose a viewport scale from the Standard scale drop-down list.

NOTE

AutoCAD adjusts the size of the floating viewport boundary when you use the filled-triangle scale grip to specify the viewport scale. The other methods of scaling a viewport do not change the size of the viewport boundary.

Figure 29-11.
Using the **Viewport Scale** flyout button on the status bar to set the viewport scale. This example shows scaling a dimensioned multiview drawing in a single viewport to finalize a part drawing.

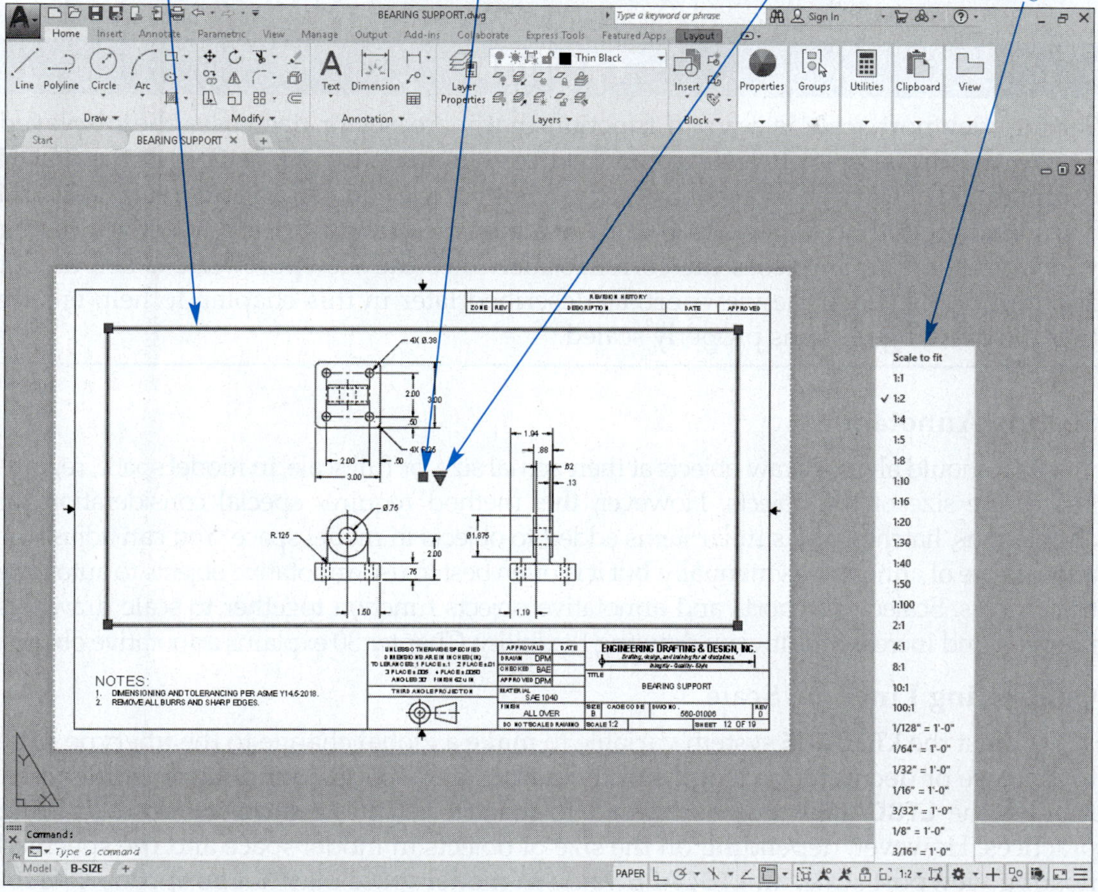

If a scale is unavailable, or to change an existing scale, pick the **Viewport Scale** flyout on the status bar and choose **Custom...** to access the **Edit Drawing Scales** dialog box. Move the highlighted scale up or down in the list using the **Move Up** or **Move Down** button. To remove the highlighted scale from the list, pick the **Delete** button.

Select the **Edit...** button to open the **Edit Scale** dialog box, where you can change the name of the scale and adjust the scale by entering the paper and drawing units. For example, a scale of 1/4″ = 1′-0″ uses a paper units value of .25 or 1 and a drawing units value of 12 or 48. To create a new annotation scale, pick the **Add...** button to display the **Add Scale** dialog box, which provides the same options as the **Edit Scale** dialog box. Pick the **Reset** button to restore the list to display the default scales.

NOTE

Changes you make in the **Edit Drawing Scales** dialog box are stored with the drawing and are specific to the drawing. To make changes to the default scale list saved to the system registry, pick the **Default Scale List...** button on the **User Preferences** tab of the **Options** dialog box to access the **Default Scale List** dialog box. The options are the same as those in the **Edit Drawing Scales** dialog box, but changes are saved as the default for new drawings, resetting the **Edit Drawing Scales** dialog box.

PROFESSIONAL TIP

Use the **MATCHPROP** command, described in Chapter 13, to match the properties assigned to one floating viewport to other floating viewports.

CAUTION

Setting viewport scale is a zoom function that increases or decreases the displayed size of the drawing in the viewport. You can also use the **XP** option of the **ZOOM** command to specify the scale of the active viewport. If you use an option of the **ZOOM** command other than a specific XP value to adjust the drawing inside an active floating viewport, the drawing loses the correct scale. Once the viewport scale is set, do not zoom in or out. Lock the viewport, as described later in this chapter, to help ensure that the drawing remains properly scaled.

Scaling Annotations

You should always draw objects at their actual size, or full scale, in model space, regardless of the size of the objects. However, this method requires special consideration for annotations, hatches, and similar items added to objects in model space. You can adjust the appearance of annotations manually, but it is often best to use annotative objects to automate the process. Scaled viewports and annotative objects function together to scale drawings properly and increase multiview drawing flexibility. Chapter 30 explains annotative objects.

Controlling Linetype Scale

Adjust the **LTSCALE** system variable to make a global change to the linetype scale to increase or decrease the lengths of the dashes and spaces found in some linetypes. Modify the **LTSCALE** value as needed to make linetypes match standard drafting practices. However, depending on the size of objects in model space and the specified floating viewport scale, an **LTSCALE** value in model space may not be appropriate for paper space.

For example, an **LTSCALE** value of .5 is appropriate for an inch-unit mechanical part drawing plotted at full scale. In this example, apply an **LTSCALE** value of .5 to model space and paper space because both environments function at full scale. If you scale the drawing to 2:1, a linetype scale of .25 (scale factor of 1/2 × **LTSCALE** value of .5 = .25) is needed in model space and paper space in order for lines to appear correct in both environments. By default, AutoCAD calculates the appropriate linetype scale display in model space and paper space according to the **LTSCALE** setting.

The **CELTSCALE**, **PSLTSCALE**, and **MSLTSCALE** system variables control how the **LTSCALE** system variable applies, or does not apply, to linetypes in model space and paper space. The **CELTSCALE**, **PSLTSCALE**, and **MSLTSCALE** system variables are set to **1** by default, and should be set to **1** to apply the **LTSCALE** value correctly in model space and paper space. All linetypes will then appear with the same lengths of dashes and dots regardless of the floating viewport scale, and no matter whether you are in paper space or model space.

Using the previous example, lines will appear correctly in model space and at a scale of 1:1 and 2:1 in paper space. However, when you scale a floating viewport or change the annotation scale in model space, remember to use the **REGEN** command to regenerate the display. Otherwise, the linetype scale will not update according to the new scale. The **MSLTSCALE** system variable is associated with the selected annotation scale, as described in Chapter 30.

Adjusting a View

When you first create a floating viewport, AutoCAD performs a **ZOOM Extents** to display everything in model space through the viewport. The **Scale to fit** viewport scale option accomplishes the same task. When you scale a viewport, AutoCAD adjusts the view from the center of the viewport, which often results in the appropriate display. However, you must adjust the view when you change the size or shape of the viewport, when a centered view is not appropriate, or when you need to display a specific portion of the drawing. Use the **PAN** command in an active viewport to redefine the displayed location of the view.

Boundary Adjustment

The viewport boundary can "cut off" a scaled model space drawing. This may be acceptable to display a portion of a view. However, to display the entire view, you can either increase the size of the viewport boundary or select a different scale to reduce the displayed size of the view to fit the viewport. If it is not appropriate to increase the size of the viewport or decrease the scale, use a larger sheet size. Figure 29-12 shows a drawing with two viewports. The rectangular viewport shows everything in model space at full scale. The circular viewport cuts off model space objects and displays objects at a 2:1 scale to create a detail.

Precision Adjustment and Alignment

The **PAN** command is effective for adjusting model space content in a floating viewport, but it offers limited precision, especially when you are attempting to align views in different viewports. One way to adjust and align views precisely is to use the **MOVE** command to move viewports, because objects shown in a viewport move with

Figure 29-12.
An example of a multiview part drawing in which it is appropriate to show all model space objects, but also necessary to display only a portion of model space to create a view enlargement.

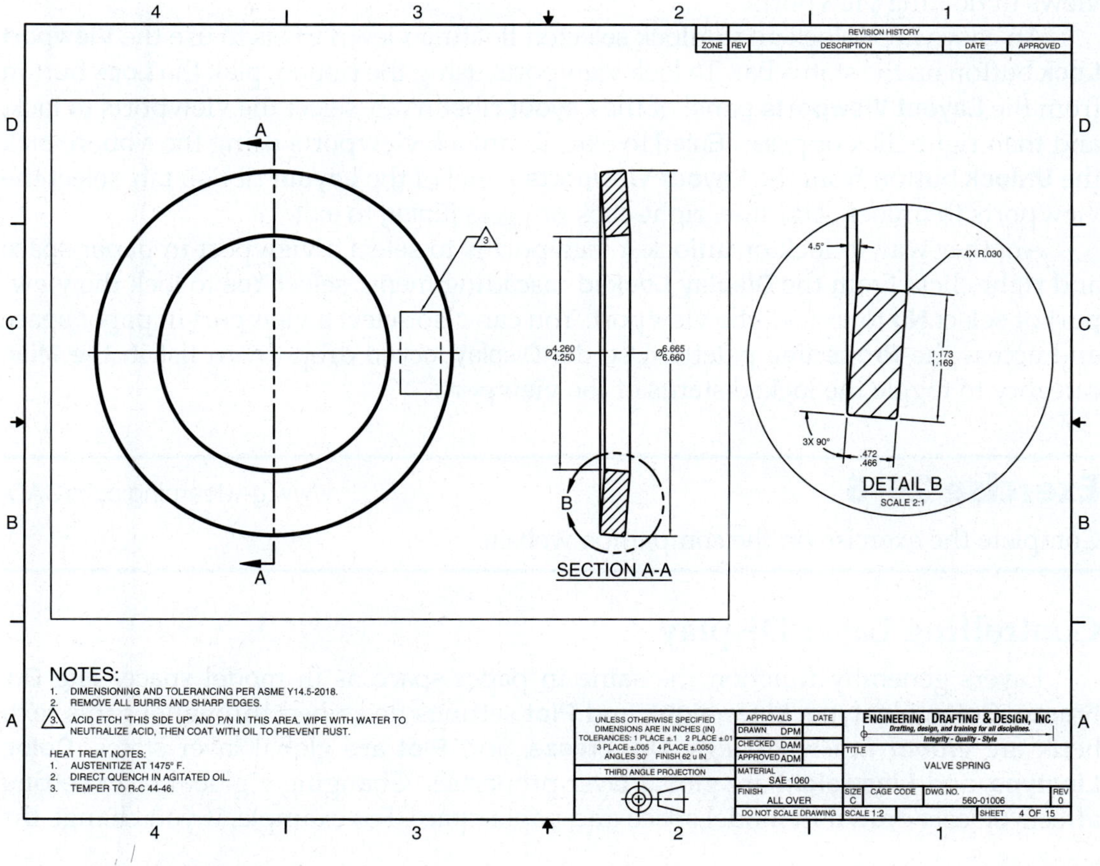

the viewport. Draw construction geometry, such as a line, on the layout or use object snaps and AutoTrack to reference specific points between views. Access the **MOVE** command and select the viewport as the object to move, but specify the base point and second point on objects in model space or on construction geometry on the layout. Use object snaps to aid point selection. See Figure 29-13.

MVSETUP

The **MVSETUP** command provides another way to adjust and align views precisely. Choose the **Align** option and then select the **Horizontal** option to align views horizontally or the **Vertical alignment** option to align views vertically. A viewport becomes activated when you select an alignment option. Pick inside the viewport that contains the model space view with which to align. Then specify a base point. Use object snaps, AutoTrack, or construction geometry to aid point selection. Activate the viewport that contains the model space view to pan. Then specify the point from which model space pans horizontally or vertically. See Figure 29-14. Continue using the **MVSETUP** command or cancel to exit.

NOTE

The **MVSETUP** command includes several additional options that are outdated, perform operations that you can accomplish more easily using other commands, or have limited application.

Locking and Unlocking Floating Viewports

After you adjust the drawing in the viewport to reflect the proper scale and view, lock the viewport to prevent changing the scale and view orientation in the viewport. Locking the viewport allows you to use display commands such as **ZOOM** and **PAN** to aid in working with objects in *paper space* without changing the scale or position of views in floating viewports.

An easy way to lock and unlock selected floating viewports is to use the **Viewport Lock** button on the status bar. To lock viewports using the ribbon, pick the **Lock** button from the **Layout Viewports** panel of the **Layout** ribbon tab, select the viewports to lock, and then right-click or press [Enter] to exit. To unlock viewports using the ribbon, pick the **Unlock** button from the **Layout Viewports** panel of the **Layout** ribbon tab, select the viewports to unlock, and then right-click or press [Enter] to exit.

Another way to lock or unlock a viewport is to select a viewport in paper space and right-click. From the **Display Locked** cascading menu, select **Yes** to lock the viewport or select **No** to unlock the viewport. You can also select a viewport in paper space and access the **Properties** palette. Use the Display locked drop-down list in the **Misc** category to toggle the locked status of the viewport.

Exercise 29-6

www.g-wlearning.com/CAD/

Complete the exercise on the companion website.

Controlling Layer Display

Layers generally function the same in paper space as in model space. The **On**, **Freeze**, **Color**, **Linetype**, **Lineweight**, and **Plot** settings described throughout this textbook are *global layer settings*. **On**, **Freeze**, and **Plot** are global layer states. **Color**, **Linetype**, and **Lineweight** are global layer properties. Changing a global layer setting affects objects drawn in model space and paper space. For example, if you change the

global layer settings: Layer settings applied to both model space and paper space.

Figure 29-13.
Moving a floating viewport to position the center of the front elevation of this home in the center of a sheet. A—Select the viewport to be moved and a base point associated with model space objects. The X identifies the specified point for illustrative purposes only. B—Move the viewport to a point on the layout. This example uses construction lines to aid selection. Object snap and AutoTrack are also effective.

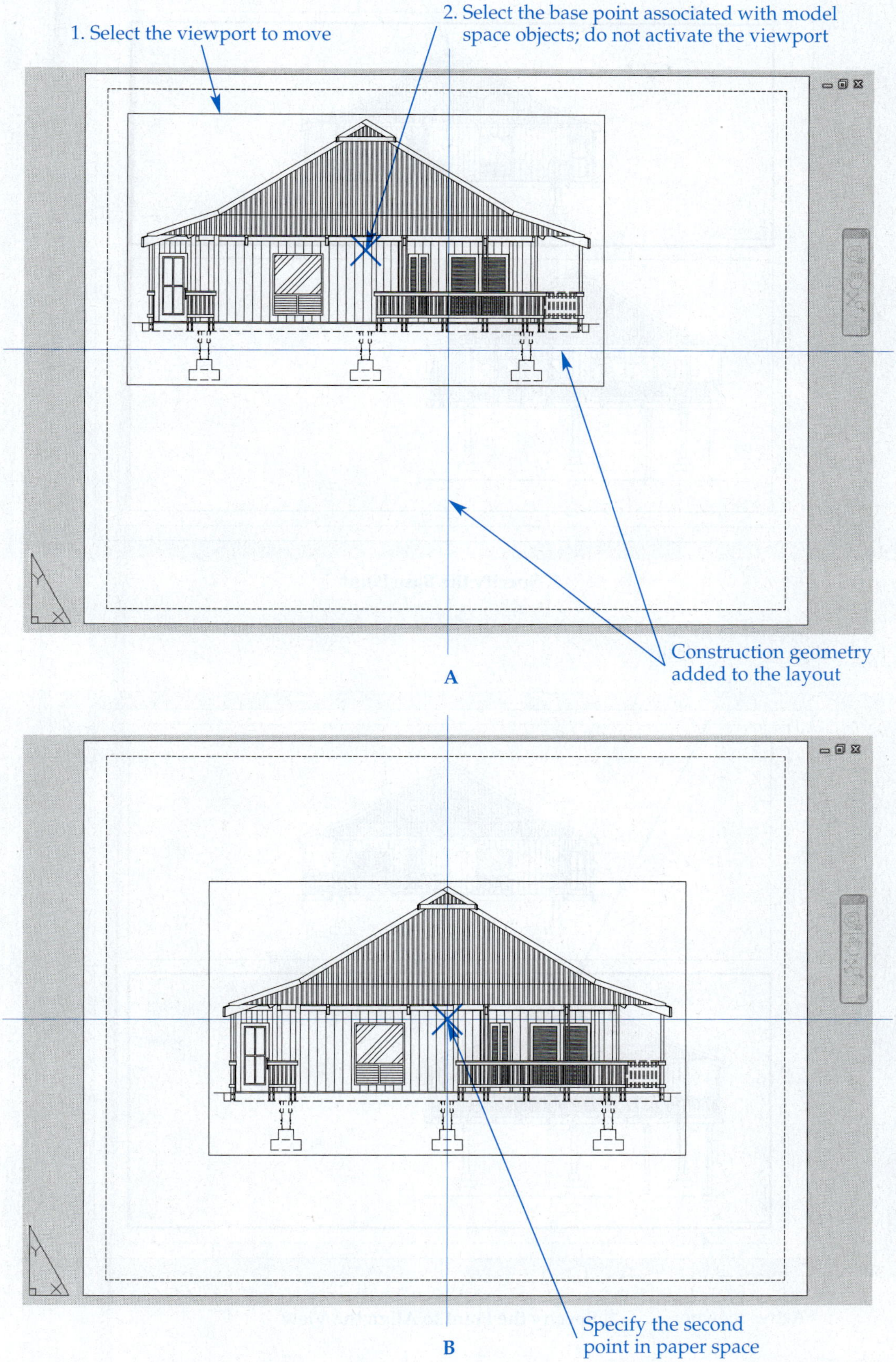

Figure 29-14.
Using the **MVSETUP** command, **Align** option, and **Vertical alignment** option to align views vertically in different floating viewports. The same process applies to aligning views horizontally.

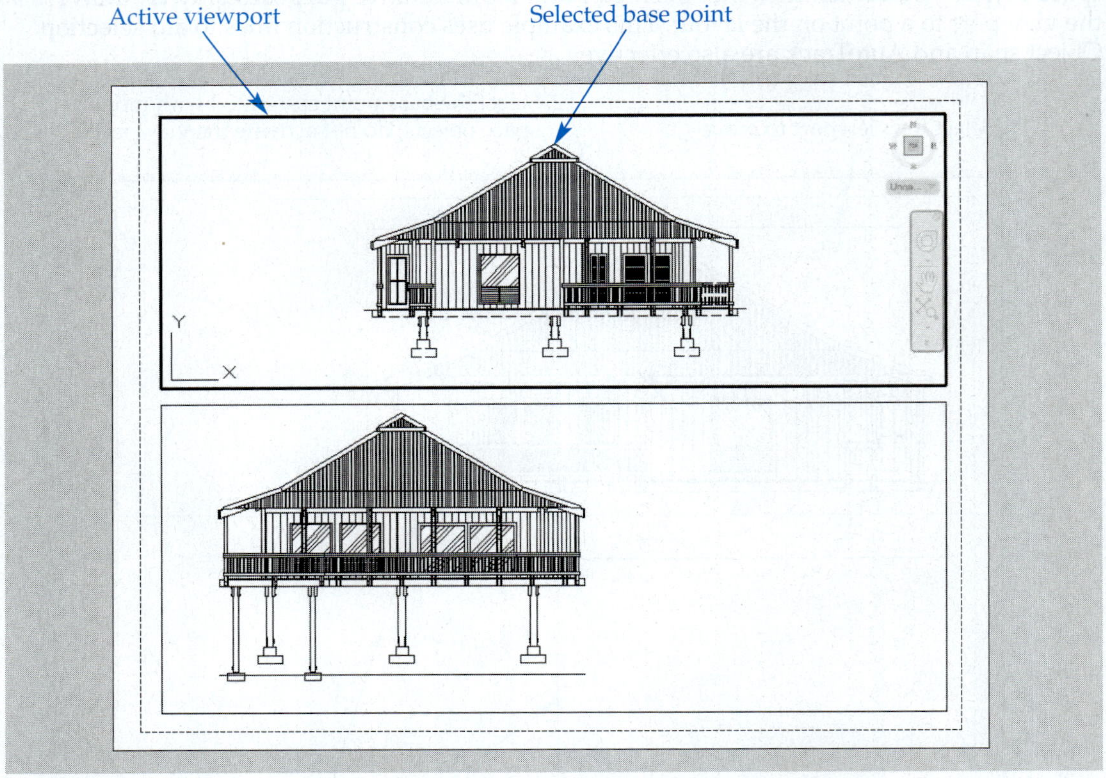

1. Specify the Base Point

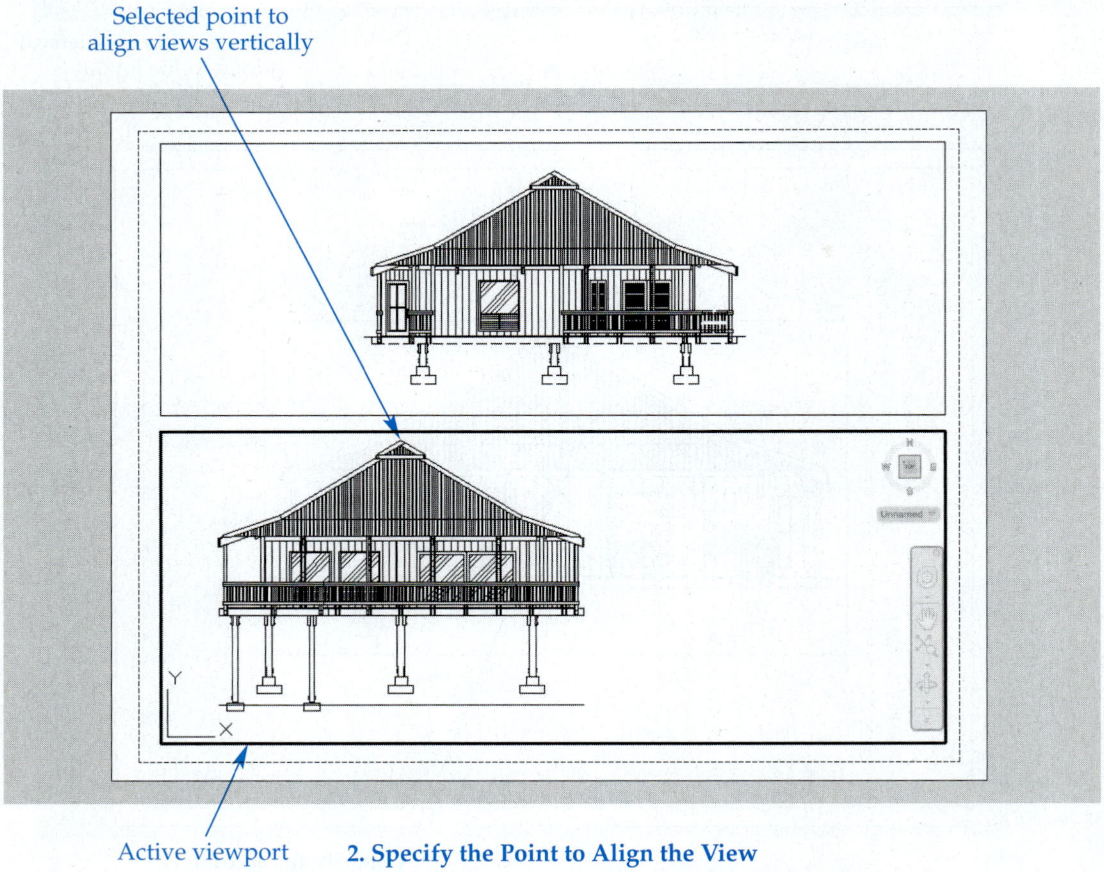

2. Specify the Point to Align the View

(Continued)

Figure 29-14.
Continued.

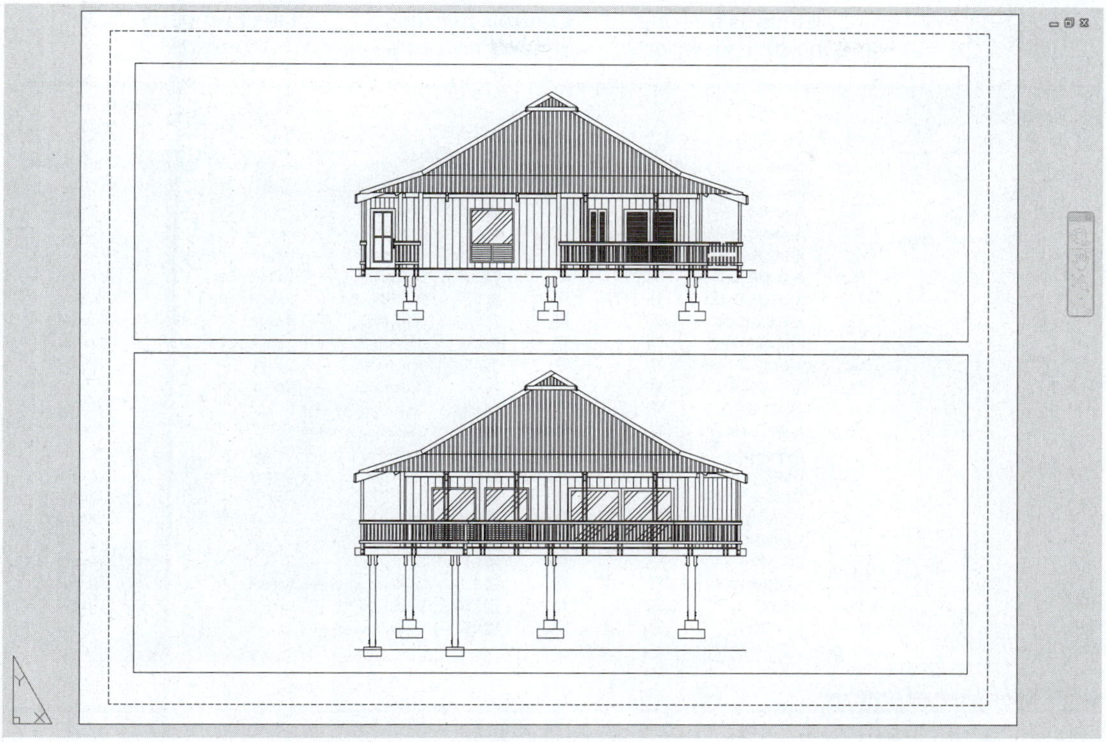

Aligned View

color of a layer in model space and lock the layer, all objects drawn on that layer in paper space also change color and become locked.

AutoCAD provides the option to freeze layers in a floating viewport and apply ***layer property overrides***. These features expand the function of the layer system and improve your ability to reuse drawing content.

Use the **LAYER** command and the corresponding **Layer Properties Manager** to control layer display in floating viewports. See Figure 29-15. This is the same **Layer Properties Manager** used to manage layers throughout this textbook. The **VP Freeze**, **New VP Freeze**, **VP Color**, **VP Linetype**, **VP Lineweight**, **VP Transparency**, and **VP Plot Style** columns control layer display options for floating viewports. Except for the **New VP Freeze** column, these columns appear only in layout mode. You probably need to use the scroll bar at the bottom of the palette to see the columns. The options can apply to layout content, such as the viewport boundary. However, layer settings typically apply to an active floating viewport. Be sure the floating viewport to which you want to apply layer control settings is active as you work through the following sections.

LAYER
Ribbon
Home
> Layers
Layer Properties
Type
LAYER
LA

layer property overrides: Color, linetype, lineweight, transparency, and plot style properties applied to specific viewports in paper space.

NOTE

The text-based **VPLAYER** command also controls layer display in floating viewports. The **Layer Properties Manager** is faster and easier to use than the **VPLAYER** command.

Freezing and Thawing

Freeze layers in the active viewport to create different views using a single drawing. For example, Figure 29-16A shows the model space display of a floor plan with electrical plan content added directly to the floor plan using electrical plan layers. Figure 29-16B shows two layouts from the same drawing file. One layout displays a

Figure 29-15.
Use the **Layer Properties Manager** to control the display of layers in floating viewports.

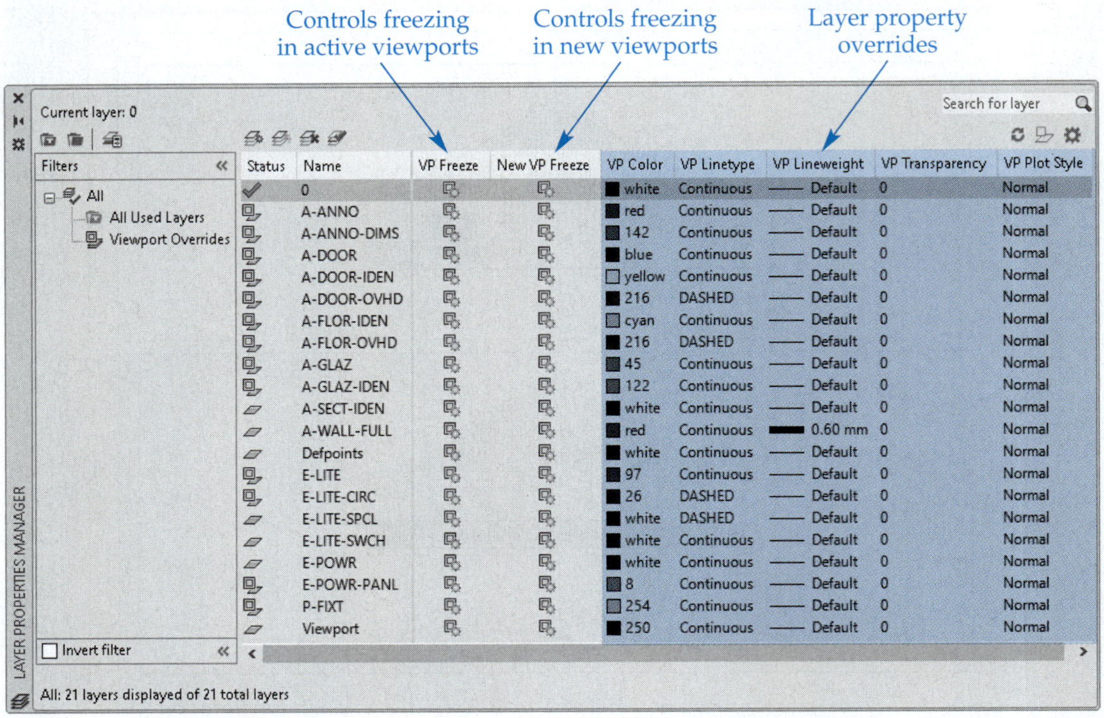

floor plan with no electrical information, and the other layout displays an electrical plan without specific floor plan content.

In the Figure 29-16 example, you draw many objects, such as doors, walls, and windows, on layers that maintain the global **Thaw** setting. As a result, these objects appear in model space and in both floating viewports. Freeze layers in specific viewports (**VP Freeze**) to create two different drawings. This example shows viewport layer freezing in two different viewports, each viewport in a different layout, but you can apply the same concept to multiple viewports in the same layout.

VP Freeze

VP Thaw

The **VP Freeze** column of the **Layer Properties Manager** controls freezing and thawing layers in the current viewport. Pick the **VP Thaw** icon or the **VP Freeze** icon to toggle freezing and thawing in the current viewport. The **VP Freeze** icon freezes layers only in the selected floating viewport, while the **Freeze** icon freezes layers globally in all floating viewports. To freeze a layer in all layout viewports, including those created before picking the **VP Freeze** icon or **VP Thaw** icon, right-click and pick **In All Viewports** from the **VP Freeze Layer** cascading menu. To freeze a layer in all layout viewports except the current viewport, right-click and select **In All Viewports Except Current** from the **VP Freeze Layer** cascading menu. To thaw a layer in all layout viewports, right-click and choose **VP Thaw Layer in All Viewports**.

PROFESSIONAL TIP

The **VP Freeze** function is also available in the **Layer Control** drop-down list in the **Layers** panel on the **Home** ribbon tab. This provides a quick way to freeze and thaw layers in a viewport without accessing the **Layer Properties Manager**.

New VP Freeze

New VP Thaw

The **New VP Freeze** column of the **Layer Properties Manager** controls freezing and thawing of layers in newly created floating viewports. Pick the **New VP Thaw** icon or the **New VP Freeze** icon to toggle freezing and thawing in any new floating viewport. This feature has no effect on the active viewport.

Figure 29-16.
Creating a floor plan and separate electrical plan using the same drawing file. A—An example of "overlapping" layers in model space. B—Layers frozen in separate layouts to create two different drawing views.

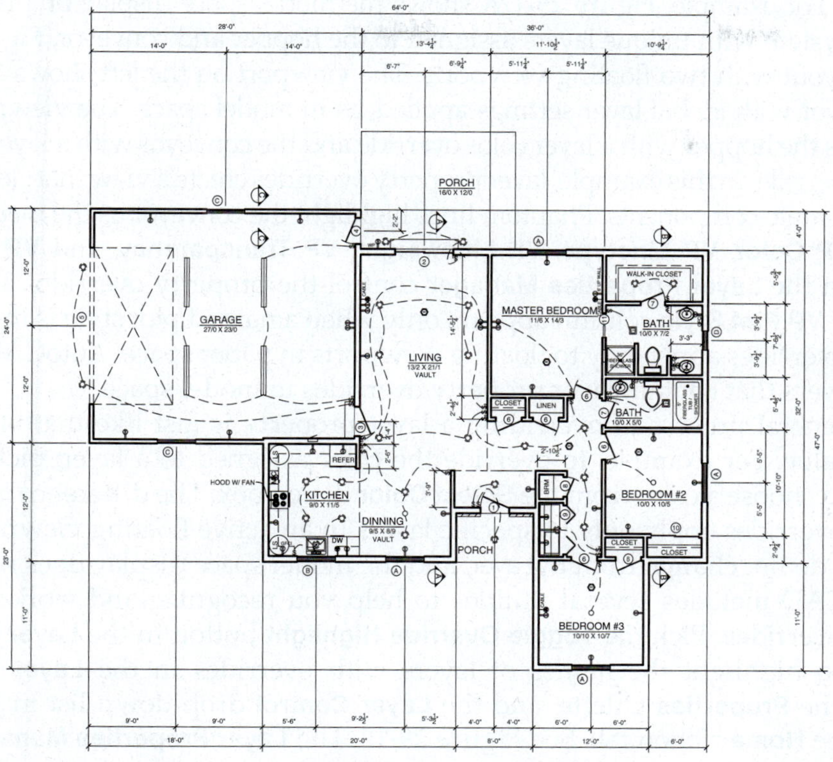

A

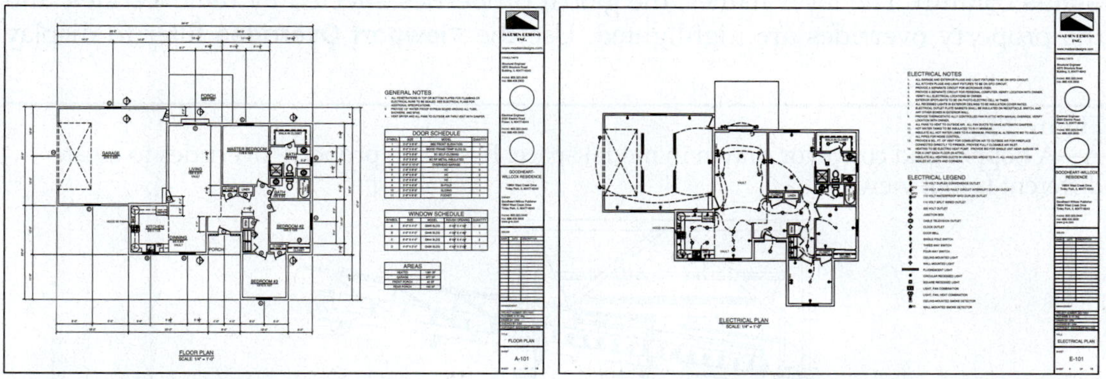

Floor Plan Layout
Created by freezing electrical layers in the floating viewport

Electrical Plan Layout
Created by freezing floor plan layers in the floating viewport

B

NOTE

Right-click on a layer in the **Layer Properties Manager** and select **New Layer VP Frozen in All Viewports** to create a new layer preset with the **VP Freeze** and **New VP Freeze** icons selected.

Exercise 29-7

www.g-wlearning.com/CAD/

Complete the exercise on the companion website.

Layer Property Overrides

You can use layer property overrides to create different views without changing individual object properties, creating separate drawing files, or readjusting global layer properties. For example, **Figure 29-17A** shows the model space display of a hopper and conveyor system with unique layers assigned to the hopper and conveyor. **Figure 29-17B** shows a layout with two floating viewports. The viewport on the left shows the hopper and conveyor with global layer settings applied, as in model space. The viewport on the right shows the hopper with a layer color override and the conveyor with a layer color and linetype override. In this example, layer property overrides create a view that clearly shows the two separate components. Phantom lines highlight the conveyor as the mechanism.

The **VP Color**, **VP Linetype**, **VP Lineweight**, **VP Transparency**, and **VP Plot Style** columns in the **Layer Properties Manager** control the property overrides assigned to layers. The **VP Plot Style** column appears only when a named plot style is in use. Layer property overrides apply only to floating viewports in paper space. AutoCAD does not identify layers that contain layer property overrides in model space.

The general process of overriding a layer property is just like that of changing a global value. For example, to override the color assigned to a layer, pick the color swatch and choose a color from the **Select Color** dialog box. The difference is that layer property overrides apply only to specific layers in an active floating viewport. Object properties do not change from ByLayer, and the model space display does not change.

AutoCAD includes several utilities to help you recognize and work with layer property overrides. Pick the **Toggle Override Highlight** button in the **Layer Properties Manager** to highlight the names of layers with overrides in the **Layer Properties Manager**, the **Properties** palette, and the **Layer Control** drop-down list in the **Layers** panel of the **Home** ribbon tab. See **Figure 29-18**. The **Layer Properties Manager** identifies layers that contain layer property overrides with a sheet and viewport icon in the **Status** column. The layer names, the global properties affected by the overrides, and the property overrides are highlighted. Use the **Viewport Overrides** filter to display

Figure 29-17.
A—A hopper and conveyor drawn in model space. B—Using property overrides to create different layout views.

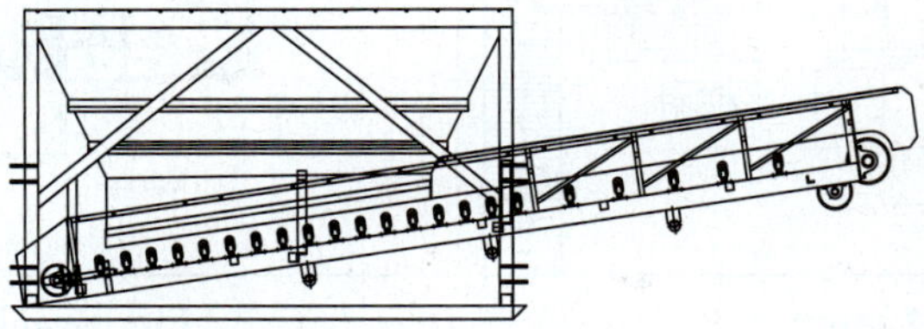

Model Space
Displays exactly as drawn
A

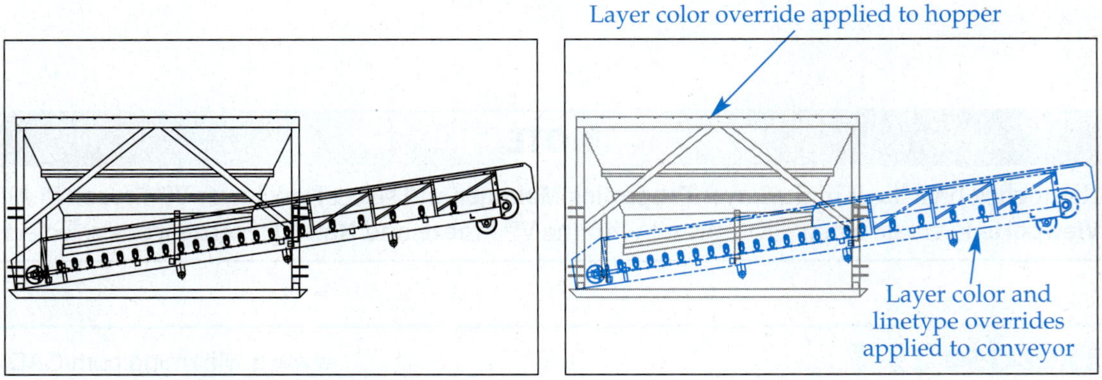

Left Viewport
Displays model space exactly as drawn

Right Viewport
Layer property overrides applied
B

Figure 29-18.
When the **Toggle Override Highlight** button is selected in the **Layer Properties Manager**, layers with property overrides are highlighted in the **Layer Properties Manager**, the **Properties** palette, and the **Layer Control** drop-down list on the ribbon.

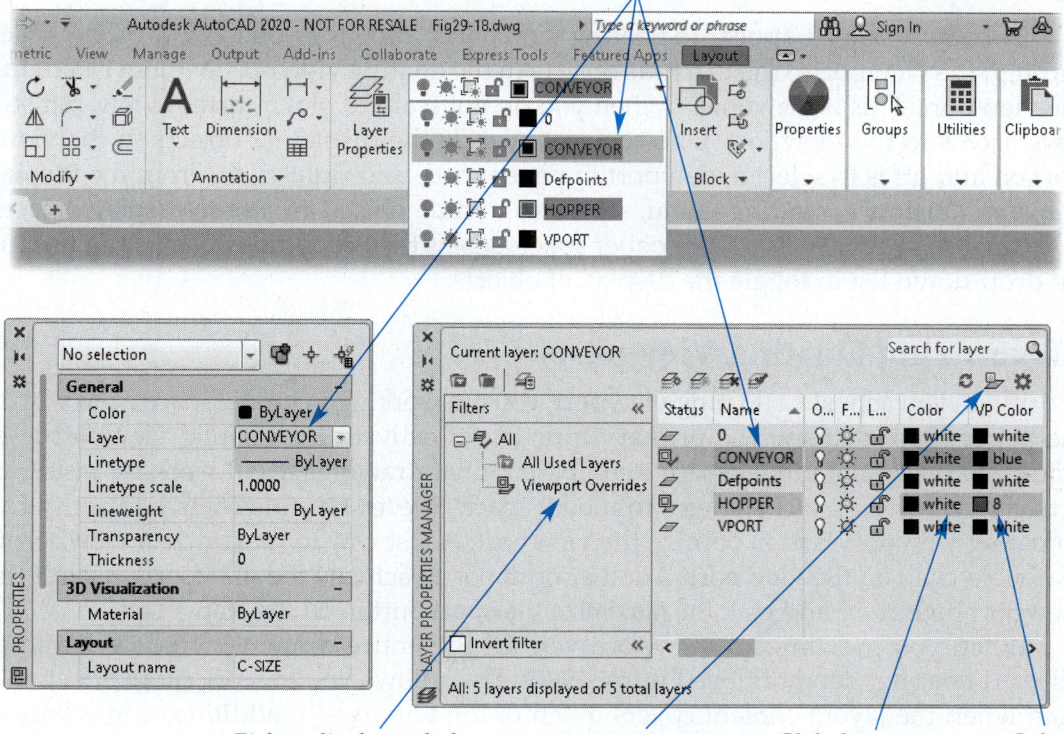

and manage only those layers that include layer property overrides. You can also save layer property overrides in a layer state.

The **Properties** palette highlights the names of layers with property overrides and the overridden properties. The **Layer Control** drop-down list in the **Layers** panel of the **Home** ribbon tab also highlights the names of layers with property overrides.

NOTE

The **Viewport Overrides** icon appears in the status bar when you activate a floating viewport or assign layer property overrides to the active viewport.

If layer property overrides are no longer necessary, you should remove the overrides from the layer. Changing a property back to the original, or global, value does not remove the override. Right-click on a layer that contains layer property overrides in the **Layer Properties Manager** and pick **Remove Viewport Overrides for** to access a cascading menu of options for removing layer property overrides. Pick **Selected Layers** and then **In Current Viewport only** or **In All Viewports** to remove layer property overrides from the current viewport or from all viewports that include overrides.

NOTE

You can also use the **Layer** option of the **MVIEW** command to remove layer property overrides.

Exercise 29-8

www.g-wlearning.com/CAD/

Complete the exercise on the companion website.

Turning Off Floating Viewport Objects

By default, objects appear in floating viewports, allowing you to view model space through the viewports. You can hide objects in the floating viewport without removing the viewport. This is convenient when you do not want to plot a certain view, but still have access to the viewport. One option to toggle the display of objects in the viewport on and off is to select a viewport in paper space and right-click. From the **Display Viewport Objects** cascading menu, select **No** to hide objects or **Yes** to display objects. You can also select a viewport in paper space, access the **Properties** palette, and use the On drop-down list to toggle the display of objects.

Maximizing Floating Viewports

VPMAX

Status Bar

Maximize Viewport

Type

VPMAX

When you activate a floating viewport, you are working in model space from within the paper space display. The primary function of activating a floating viewport is to adjust the display of model space to prepare a final drawing. Avoid working inside an active viewport to make changes to model space objects. One alternative to activating a floating viewport is to maximize the viewport. A fast way to maximize a viewport is to double-click on the viewport. Another option is to activate the viewport or select the viewport boundary and pick the **Maximize Viewport** button on the status bar.

When you maximize a viewport, you fill the entire drawing window with the selected floating viewport. See Figure 29-19. This allows you to work more effectively than when the layout content covers much of the window. In addition, a maximized

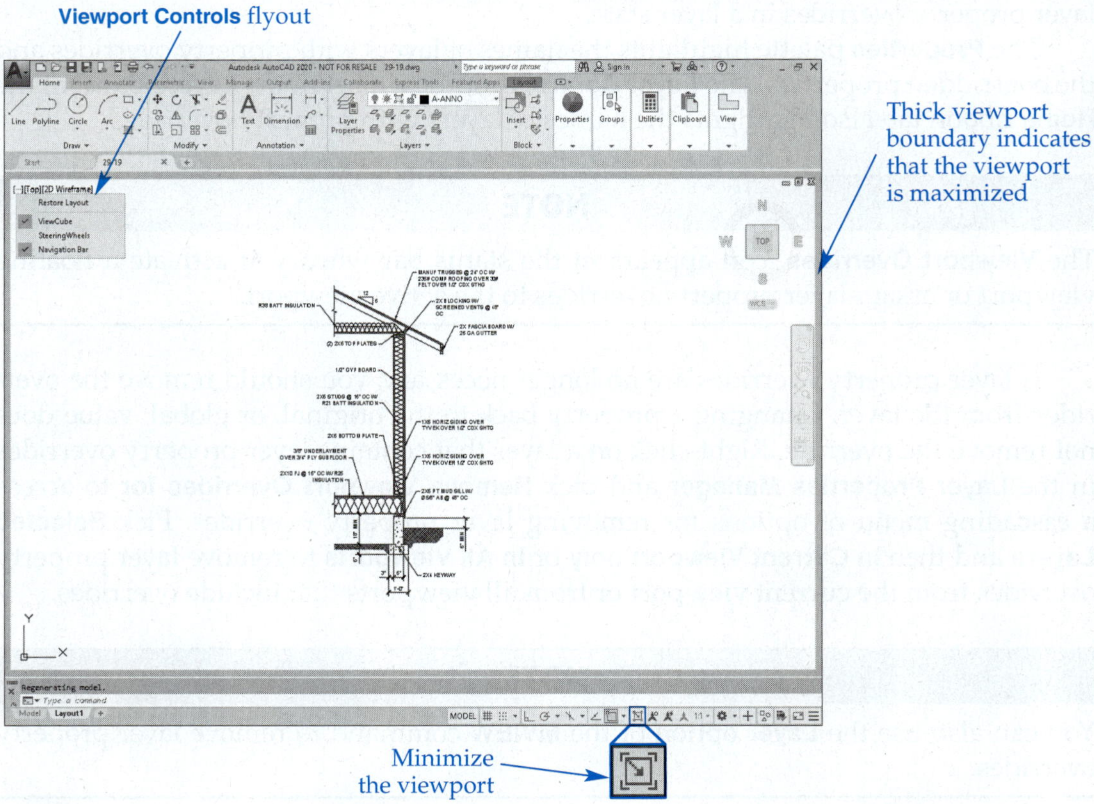

Figure 29-19.
Maximize a floating viewport to work in an environment similar to model space, but with layout characteristics, such as layers frozen in the viewport and layer property overrides.

viewport displays objects exactly as they appear in the floating viewport, including frozen layers and layer overrides. Typically, you should maximize a floating viewport to use view commands such as **ZOOM** and **PAN** and make changes to objects in model space while remaining in paper space.

To redisplay the entire layout, double-click on the thick viewport boundary, pick the **Minimize Viewport** button on the status bar, right-click and choose **Minimize Viewport**, select **Restore Layout** from the **Viewport Controls** flyout, or type VPMIN.

VPMIN
Status Bar
Minimize Viewport
Type
VPMIN

NOTE

If you maximize a viewport, use the **ZOOM** or **PAN** command, and then return to the entire layout, the previous view and viewport scale are restored by AutoCAD. You can maximize a floating viewport even if the viewport is not active.

PROFESSIONAL TIP

If you do not want to see floating viewport boundaries on the plot, remember to freeze or turn off the layer assigned to the viewport before plotting.

Exercise 29-9

www.g-wlearning.com/CAD/

Complete the exercise on the companion website.

Using the CHSPACE Command

Type
CHSPACE

The **CHSPACE** command provides a convenient way to move objects between model space and paper space without having to copy and paste between the different environments. The **CHSPACE** command is only available in paper space. Activate a floating viewport to move objects from model space to paper space, or deactivate floating viewports to move objects from paper space to model space. Then access the **CHSPACE** command, select the objects to transfer, and right-click or press [Enter]. To move objects to model space, select a viewport to activate and then right-click or press [Enter] to complete the operation. To move objects to paper space, right-click or press [Enter] twice to complete the operation.

Plotting

PLOT
Quick Access
Plot
Ribbon
Output
> Plot
Plot
Application Menu
Print
> Plot
Type
PLOT
[Ctrl]+[P]

After you prepare a layout for plotting, you are ready to plot. If you develop an appropriate page setup and layout, the process of creating the actual print should be almost automatic. Select the layout to plot and access the **PLOT** command. The **Plot** dialog box appears with the name of the layout displayed on the title bar. See Figure 29-20.

NOTE

You can also access the **Plot** dialog box by selecting from the shortcut menu available from the model or layout tab, or by picking the **Plot...** button in a model or layout thumbnail image displayed using file tabs.

The **Page Setup** and **Plot** dialog boxes are very similar, except the **Plot** dialog box provides additional options specific to creating a plot. All the settings in the **Plot** dialog box correspond to those in the **Page Setup** dialog box. Pick the **>**, or **More Options**, button in the lower-right corner of the **Plot** dialog box to toggle the display of additional dialog box areas, as shown in Figure 29-20. Specify a number in the **Number of copies** text box to indicate how many copies of the layout to plot. The **Plot options** area provides additional plot settings. Pick the **Plot in background** check box to continue working while the plot processes.

Changing plot settings in the **Plot** dialog box overrides the page setup for a specific plotting requirement. This is a convenient way to make a plot using slightly modified plot settings without creating a new page setup. For example, you can make a "check print" by selecting a printer, using an A- or B-size sheet, and scaling the plot to fit the paper. After the drawing prints, the settings return to those originally assigned in the page setup, allowing you to plot the final drawing using the appropriate printer, sheet size, and 1:1 scale.

Supplemental Material

www.g-wlearning.com/CAD/

Adding a Plot Stamp

plot stamp: Text added only to the hard copy that includes information such as the drawing name or the date and time the drawing was printed.

For detailed information about adding a ***plot stamp*** to the plot using options in the **Plot** dialog box, go to the companion website, navigate to this chapter in the **Contents** tab, and select **Adding a Plot Stamp**.

Saving Changes to the Layout

If you make changes in the **Plot** dialog box and want to save changes to the layout page setup for future plots, pick the **Save changes to layout** check box in the **Plot**

Figure 29-20.
Use the **Plot** dialog box to finalize the layout and send the drawing to a printer, plotter, or file.

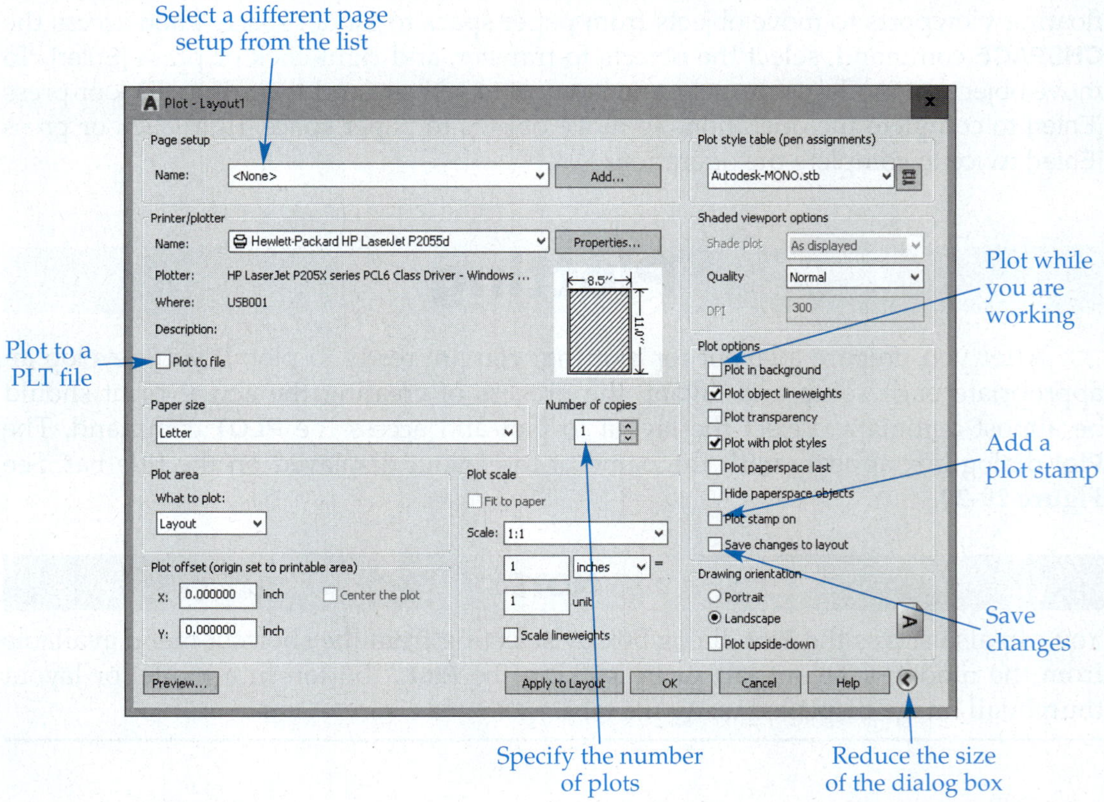

options area. You can also save changes by picking the **Apply to Layout** button. If you do not save the changes or apply them to the layout, changes made in the **Plot** dialog box are discarded, and the original page setup appears the next time you open the **Plot** dialog box.

Page Setup Options

The **Plot** dialog box provides an alternate means of creating a page setup. To apply this technique, access the **Plot** dialog box and make changes to plot settings, just as you would in the **Page Setup** dialog box. Then select the **Add...** button in the **Page setup** area to display the **Add Page Setup** dialog box. Enter a name for the page setup in the **New page setup name:** text box. All current settings in the **Plot** dialog box are saved with the new page setup. Select a page setup from the **Name:** drop-down list to restore the settings in the **Plot** dialog box. Pick the **<Previous plot>** option to reference the settings used to create the last plot, or pick the **Import...** option to import a page setup from a DWG, DWT, or DXF file.

NOTE

When using the **Plot** dialog box to define settings for a page setup, name the page setup after you make changes to settings. If you name the page setup and want to make changes later, such as changes to a plot style, use the **Page Setup Manager** instead.

Previewing the Plot

The final step before plotting is to preview the plot. The plot preview shows what the plot should look like based on plot and layout settings. Always preview the plot to check the drawing for errors and view the effects of plot settings before sending the information to the plot device. This will help you eliminate unnecessary plots. To preview the plot, pick the **Preview...** button in the lower-left corner of the **Plot** dialog box to enter preview mode. See Figure 29-21. What you see on-screen is what will plot, assuming you use a color plotter to make color prints and load the correct sheet size in the plot device.

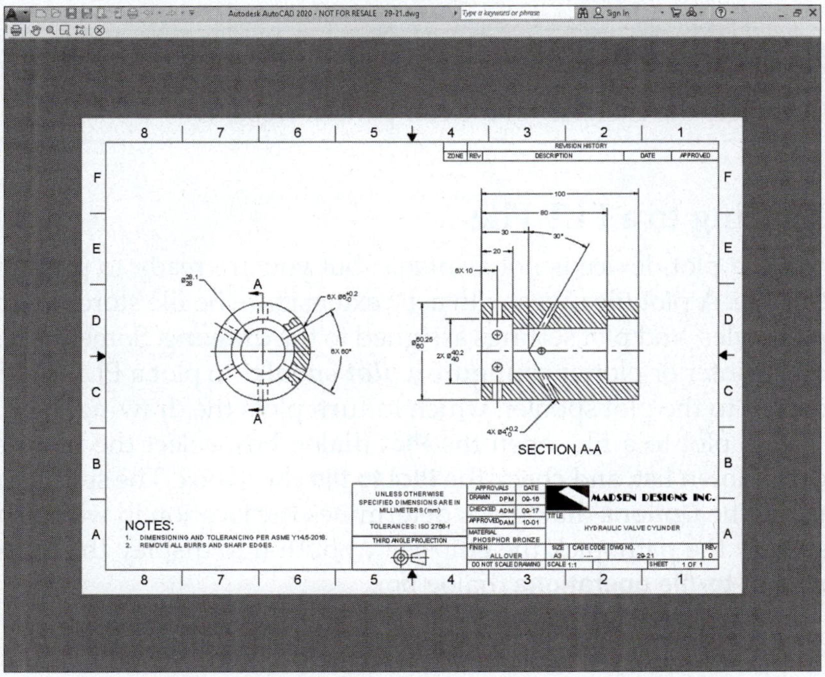

Figure 29-21. Previewing a plot is an excellent way to help confirm that the plot will be correct before sending the information to the plot device.

The **Realtime Zoom** command is automatically active in the preview window. Additional view commands are available from the toolbar near the top of the window or from a shortcut menu. Use view commands to help confirm that the plot settings are correct. When you finish previewing the plot and are ready to plot, pick the **Plot** button on the toolbar, or right-click and select **Plot**. When you finish previewing the plot, pick the **Close** button on the toolbar, press [Esc] or [Enter], or right-click and select **Exit** to return to the **Plot** dialog box.

NOTE

You must select a plot device other than **None** in order to activate the **Preview...** button.

Output

When plotting to a plotter or printer, pick the **OK** button to send the plot to the plot device and close the **Plot** dialog box. When plotting to a PDF file, pick the **OK** button to display the **Browse for Plot File** dialog box. Specify a location and name for the file and then pick the **Save** button to close the **Plot** dialog box and open the PDF file automatically using installed Adobe® software. When plotting to a DWF or DWFx file, pick the **OK** button to display the **Browse for Plot File** dialog box. Specify a location and name for the file and then pick the **Save** button to close the **Plot** dialog box. View the DWF or DWFx file using Autodesk® Design Review software.

AutoCAD notifies you of the success or failure of a hard copy or electronic plot with a message from the status bar tray. To view additional details about the plot, pick **Click to view plot and publish details...** to display the **Plot and Publish Details** dialog box. You can also access the **Plot and Publish Details** dialog box using the **VIEWPLOTDETAILS** command.

NOTE

AutoCAD provides additional tools for exporting drawings, sharing drawings online, automating the process of transmitting drawings electronically (eTransmit), and ***publishing***. Publishing a set of drawing sheets is described later in this textbook.

publishing: Preparing a sequential set of multiple drawings for hard copy or electronic plotting of the set.

Exercise 29-10

www.g-wlearning.com/CAD/

Complete the exercise on the companion website.

Plotting to a PLT File

If a plot device is not available, but you are ready to plot, an alternative is to plot to a file. A plot file saves with a .plt extension. The file stores all the drawing geometry, plot styles, and plot settings assigned to the drawing. Some offices or schools with only one printer or plotter configure a ***plot spooler*** to plot a PLT file. The PLT file is usually copied to the plot spooler, which in turn plots the drawing.

plot spooler: A disk drive with memory that allows you to plot files.

To plot to a file, open the **Plot** dialog box, select the plot device from the **Name:** drop-down list, and check the **Plot to file** check box. The setting in the **Plot and Publish** tab of the **Options** dialog box determines the location in which the plot file is saved. To specify the path, pick the ellipsis (**...**) button to display the **Select default location for all plot-to-file operations** dialog box.

NOTE

The **Plot and Publish** tab of the **Options** dialog box contains other general plot and publish settings in addition to those described in this textbook. Most plot and publish settings seldom require adjustment.

Express Tools

www.g-wlearning.com/CAD/

Layout Express Tools

The **Layout** panel of the **Express Tools** ribbon tab includes additional layout commands. For information about the most useful layout express tools, go to the companion website, navigate to this chapter in the **Contents** tab, and select **Layout Express Tools**.

Template Development

www.g-wlearning.com/CAD/

Adding Layouts

For detailed instructions on adding layouts to each drawing template, go to the companion website, navigate to this chapter in the **Contents** tab, and select **Adding Layouts**.

Chapter Review

Answer the following questions using the information in this chapter.

1. Name the two types of content that are brought together to create a complete drawing.
2. Explain how to create a polygonal viewport.
3. How can you convert an object created in paper space into a floating viewport?
4. Describe how to insert a named model view as a floating viewport.
5. How do you activate a floating viewport?
6. How can you tell that a viewport is active in paper space?
7. How do you reactivate paper space after activating a floating viewport for editing?
8. How does the scale you assign to a floating viewport compare with the drawing scale?
9. To what value should the **CELTSCALE**, **PSLTSCALE**, and **MSLTSCALE** system variables be set so that the **LTSCALE** value will be applied correctly in both model space and paper space?
10. Viewport edges may cut off the drawing when the viewport is correctly scaled. List three options to display the entire view.
11. Why should you lock a viewport after you adjust the drawing in the viewport to reflect the proper scale and view?
12. Give an example of why you would hide objects in a floating viewport without removing the viewport.
13. What is a plot stamp?
14. If you make changes to the page setup using the **Plot** dialog box, how can you save these changes to the page setup so that the changes apply to future plots?
15. Give at least two reasons why you should always preview a plot before sending the information to the plot device.

Drawing Problems

Start AutoCAD if it is not already started. Start a new drawing for each problem using an appropriate template of your choice. The template should include layers and text, dimension, multileader, and table styles, when necessary, for drawing the given objects. Add layers and text, dimension, multileader, and table styles as needed. Draw all objects using appropriate layers. Use appropriate text, dimension, multileader, and table styles, justification, and format. Follow the specific instructions for each problem. Use only drawing and editing commands and techniques you have already learned. Use your own judgment and approximate dimensions when necessary. Apply dimensions accurately using ASME or appropriate industry standards.

Note: Some of the problems in this chapter are built on problems from previous chapters. If you have not yet completed those problems, complete them now.

▼ Basic

1. **Mechanical** Follow the instructions in the Template Development portion of the companion website for this chapter to add and set up layouts for the Mechanical-Inch template file.

2. **Mechanical** Follow the instructions in the Template Development portion of the companion website for this chapter to add and set up layouts for the Mechanical-Metric template file.

3. **Architectural** Follow the instructions in the Template Development portion of the companion website for this chapter to add and set up layouts for the Architectural-US template file.

4. **Architectural** Follow the instructions in the Template Development portion of the companion website for this chapter to add and set up layouts for the Architectural-Metric template file.

5. **Civil** Follow the instructions in the Template Development portion of the companion website for this chapter to add and set up layouts for the Civil-US template file.

6. **Civil** Follow the instructions in the Template Development portion of the companion website for this chapter to add and set up layouts for the Civil-Metric template file.

7. **Mechanical** Save P28-4 as P29-7. If you have not yet completed Problem 28-4, do so now. In the P29-7 file, make the **B-SIZE** layout current. Create a new layer named VPORT. Delete the default floating viewport and create a single floating viewport .5″ from the edges of the sheet on the VPORT layer. Scale model space in the viewport to 1:1. Plot the layout, leaving the VPORT layer on and thawed.

8. **Mechanical** Save P28-5 as P29-8. If you have not yet completed Problem 28-5, do so now. In the P29-8 file, activate the **A2-SIZE** layout. Create a new layer named VPORT. Delete the default floating viewport and create a single floating viewport 10 mm from the edges of the sheet on the VPORT layer. Scale model space in the viewport to 1:1. Plot the layout, leaving the VPORT layer on and thawed.

9. **Mechanical** Save P28-6 as P29-9. If you have not yet completed Problem 28-6, do so now. In the P29-9 file, activate the **B-SIZE** layout. Create a new layer named VPORT. Delete the default floating viewport and create a single floating viewport .5″ from the edges of the sheet on the VPORT layer. Scale model space in the viewport to 1:1. Plot the layout, leaving the VPORT layer on and thawed.

▼ Intermediate

10. **Architectural** Save P28-8 as P29-10. If you have not yet completed Problem 28-8, do so now. In the P29-10 file, create a floating viewport on a layer named VPORT and scale model space in the viewport using an appropriate scale. Plot the layout, leaving the VPORT layer on and thawed.

11. **Mechanical** Save P8-1 as P29-11. If you have not yet completed Problem 8-1, do so now. In the P29-11 file, delete the default **Layout2**. Create a new A-size sheet layout according to the following steps:
    A. Rename the default **Layout1** to **A-SIZE**.
    B. Select the **A-SIZE** layout and access the **Page Setup Manager**.
    C. Modify the **A-SIZE** page setup according to the following settings:
    - **Printer/plotter**: Select a printer or plotter that can plot an A-size sheet.
    - **Paper size**: Select the appropriate A-size sheet (varies with printer or plotter).
    - **Plot area**: Layout
    - **Plot offset**: 0,0
    - **Plot scale**: 1:1 (1 in = 1 unit)
    - **Plot style table**: monochrome.ctb
    - **Plot with plot styles**
    - **Plot paperspace last**
    - Do not check **Hide paperspace objects**
    - **Drawing orientation**: Select the appropriate orientation (varies with printer or plotter)

    D. Create a new layer named VPORT.
    E. Delete the default floating viewport and create a single floating viewport on the VPORT layer .5″ from the edges of the sheet.
    F. Scale model space in the viewport to 1:2. Plot the layout, leaving the VPORT layer on and thawed.

12. **Mechanical** Save P8-11 as P29-12. If you have not yet completed Problem 8-11, do so now. In the P29-12 file, create a layout and floating viewport as needed to plot the drawing at an appropriate scale.

▼ Advanced

13. **Electrical** Open P29-13 from the companion website. Create a layout, plot style, and page setup so the layout can be plotted as follows. Using color-dependent plot styles, have the equipment (shown in color in the diagram) plot with a lineweight of 0.8 mm and 80% screening on an A-size sheet oriented horizontally. Plotted text height should be 1/8″. Plot in paper space at a scale of 1:1. Save the drawing as P29-13.

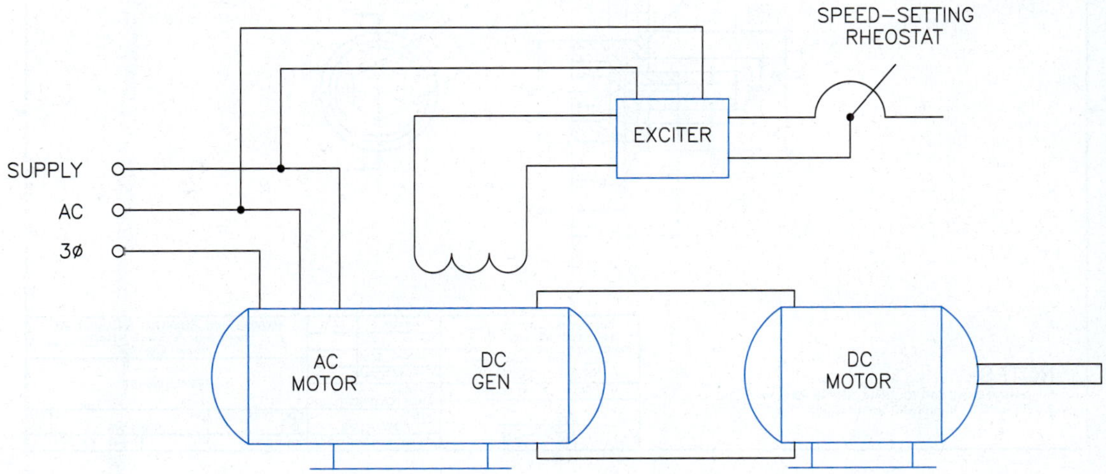

Drawing Problems - Chapter 29

14. **Electrical** Open P29-14 from the companion website. Create four layouts with names and displays as follows:
    - The **Entire Schematic** layout plots the entire schematic on a B-size sheet.
    - The **3 Wire Control** layout plots only the 3 Wire Control diagram on an A-size sheet, oriented horizontally.
    - The **Motor** layout plots the motor symbol and connections in the lower center of the schematic on an A-size sheet, oriented vertically.
    - The **Schematic** layout plots the schematic without the 3 Wire Control and motor components on an A-size sheet, oriented horizontally.

    Set up the layouts so they will plot with a text height of 1/8″. Plot in paper space at a scale of 1:1. Save the drawing as P29-14.

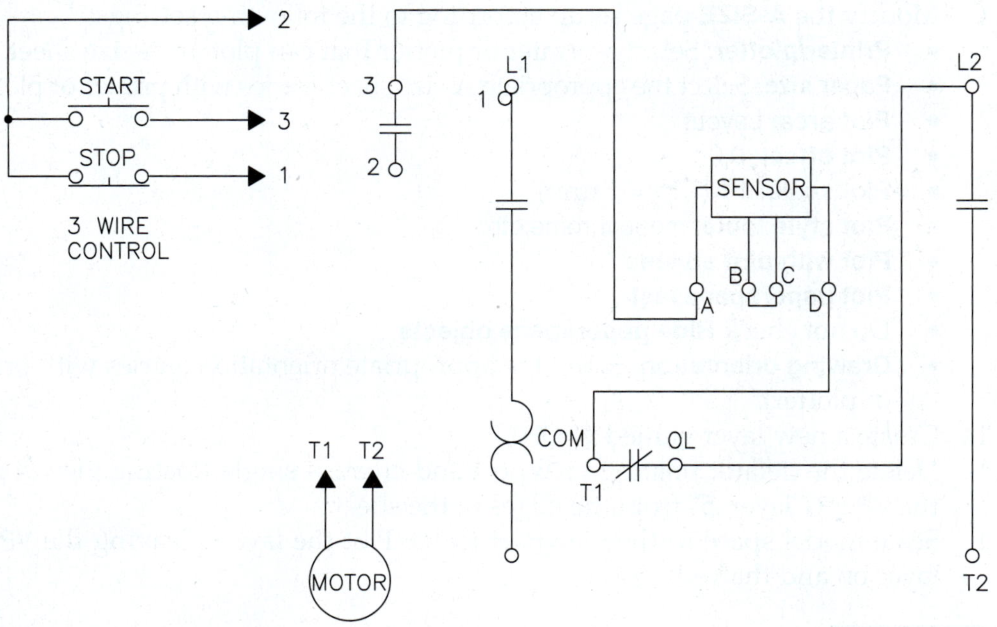

15. **Mechanical** Draw and dimension the female insert shown. Use a layout to plot the drawing. Save the drawing as P29-15.

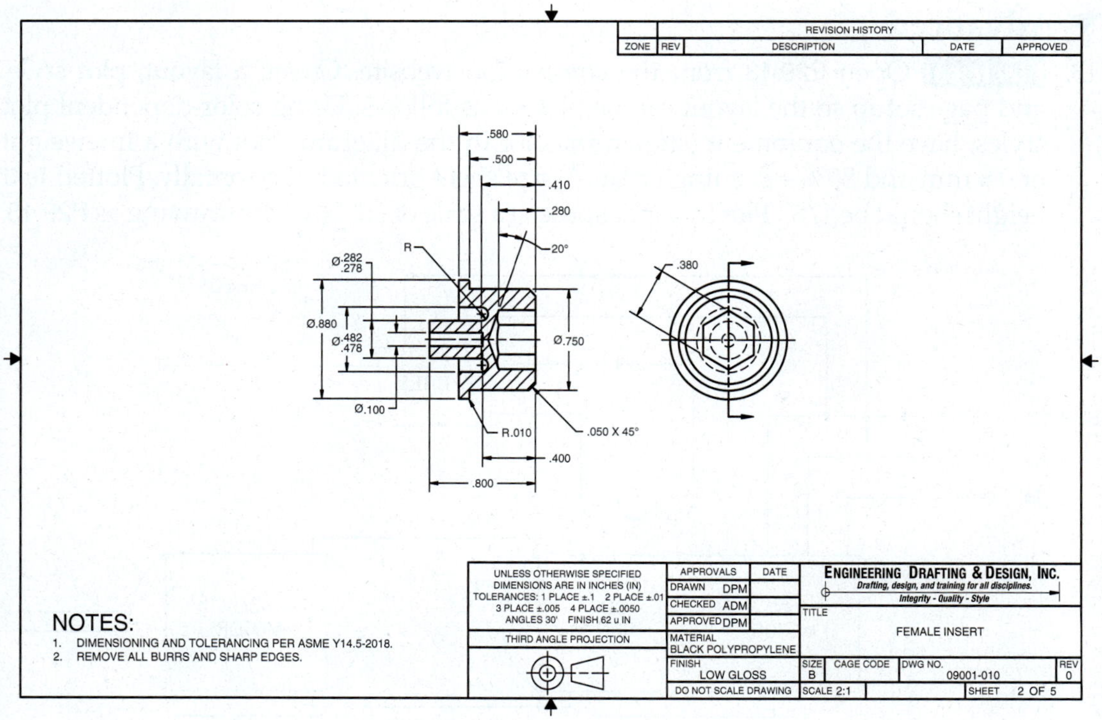

AutoCAD Certified Professional Exam Practice

Answer the following questions using the information in this chapter.

1. Which of the following drawing elements should you typically create in paper space? *Select all that apply.*
   A. border
   B. drawing views
   C. general drawing notes
   D. thread notes
   E. title block
2. Which of the following methods can you use to create a floating viewport? *Select all that apply.*
   A. **EXPORTLAYOUT** command
   B. **PLOT** command
   C. **Viewports** dialog box
   D. **Layout Viewports** panel on the **Layout** ribbon tab
   E. **VPCLIP** command
3. Which of the following features of the **Layer Properties Manager** apply layer property overrides? *Select all that apply.*
   A. **Freeze**
   B. **New VP Freeze**
   C. **New VP Thaw**
   D. **Thaw**
   E. **VP Plot Style**

Follow the instructions in each problem. Complete each problem by answering the related question.

4. **Navigate to this chapter on the companion website and open CPE-29adjust.dwg.** Move the polygonal viewport up so its top edge aligns with the top edge of the rectangular viewport, and the left edge of the upper portion of the polygonal viewport is .25″ from the rectangular viewport, as shown. What are the 2D coordinates of Point A?

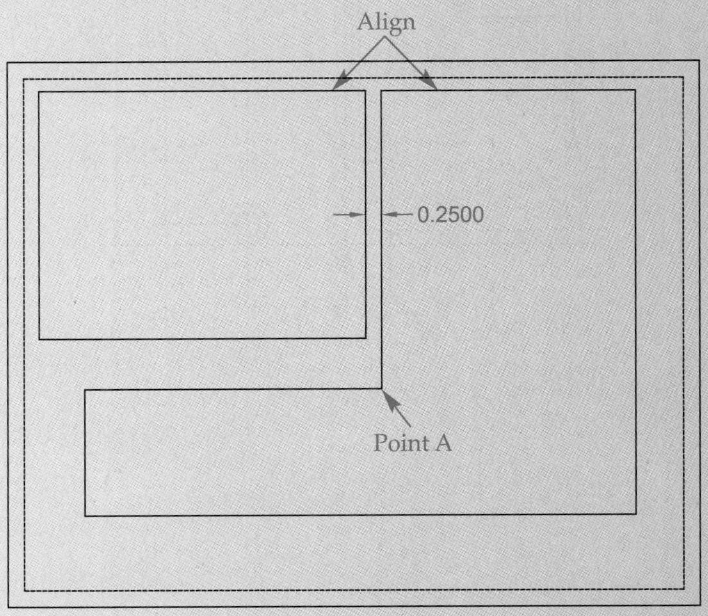

AutoCAD Certified Professional Exam Practice - Chapter 29

5. **Navigate to this chapter on the companion website and open CPE-29scale.dwg.** Make the **Stair Detail** layout current. Set the contents of the existing viewport to display at a scale of 1/4″ = 1′-0″ and activate the viewport. Do not make any other changes. What are the 2D coordinates of Point B?

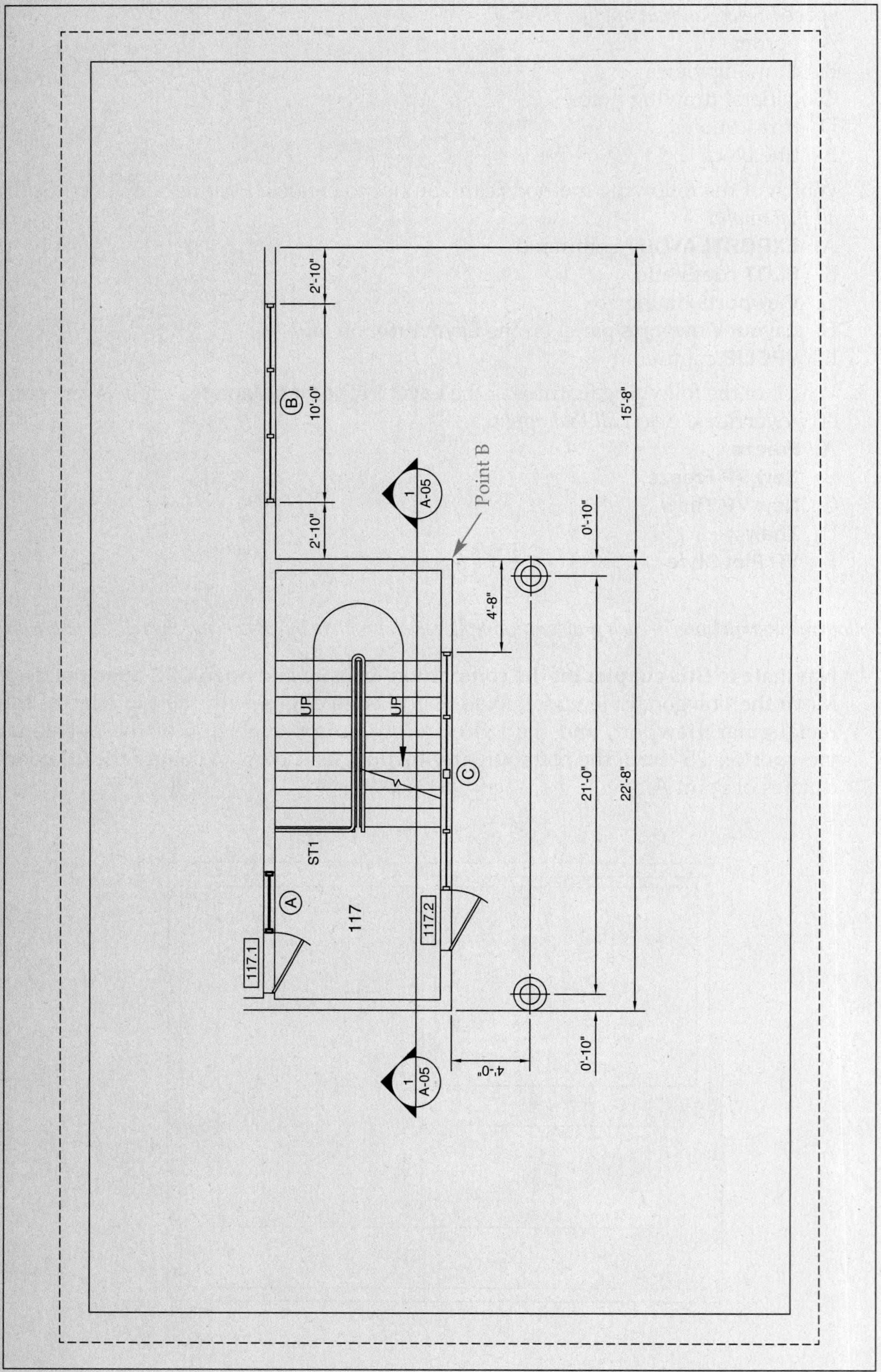

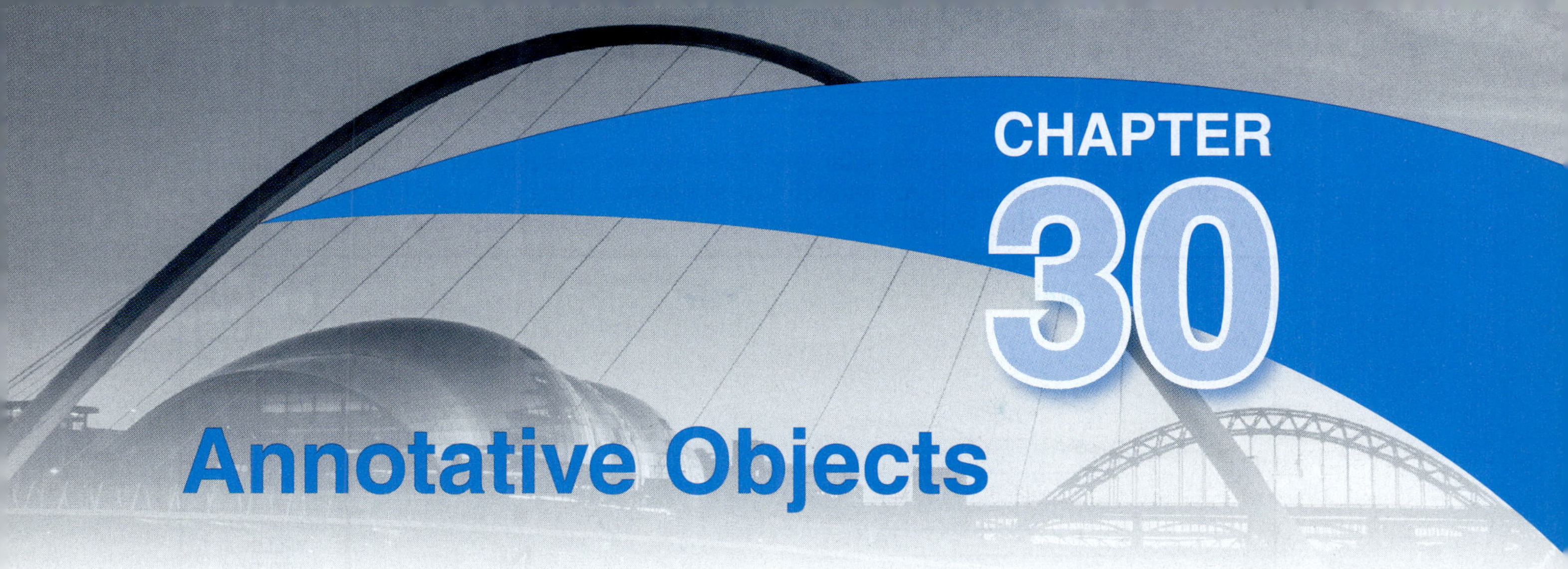

CHAPTER 30

Annotative Objects

Learning Objectives

After completing this chapter, you will be able to:

- ✓ Explain the differences between manual and annotative object scaling.
- ✓ Specify objects as annotative.
- ✓ Create and use annotative objects in model space.
- ✓ Display annotative objects in scaled layout viewports.
- ✓ Adjust the scale of annotations according to a new drawing scale.
- ✓ Use annotative objects to help prepare multiview drawings.

Annotations and related items, such as dimension and hatch objects, must be scaled so that information appears on-screen and plots correctly relative to scaled objects. AutoCAD provides annotative tools to automate the process of scaling ***annotative objects***. Annotative tools also provide flexibility for working with layouts to create multiview drawings.

annotations: Letters, numbers, words, and notes used to describe information on a drawing.

annotative objects: AutoCAD objects that can adapt automatically to the current drawing scale.

Introduction to Annotative Objects

Always draw objects at their actual size, or full scale, in model space, regardless of the size of the objects. For example, if you draw a small machine part and the length of a line in the drawing is 2 mm, draw the line 2 mm long in model space. If you draw a building and the length of a line in the drawing is 80′, draw the line 80′ long in model space. These examples describe features on a drawing that are too small or too large for layout and printing purposes. To fit these objects properly on a sheet, you ***scale*** them to a specific drawing scale.

When you scale a drawing, you increase or decrease the displayed size of model space objects. A properly scaled floating viewport in a layout allows for this process. Scaling a drawing greatly affects the display of items added to objects in model space, such as annotations, because these items should be the same size on a plotted sheet, regardless of the displayed size, or scale, of the rest of the drawing. See **Figure 30-1**.

scale: (verb) The process of enlarging or reducing objects to fit properly on a sheet of paper. (noun) The ratio between the actual size of drawing objects and the size at which objects plot on a sheet of paper.

Traditional manual scaling of annotations, hatches, and other objects requires determining the drawing scale factor and then multiplying the scale factor by the plotted size of the objects. In contrast, annotative objects are scaled automatically according to the selected annotation scale, which is the same as the drawing scale. This reduces the need to focus on the scale factor and manually adjust the size of objects according to the drawing scale.

Figure 30-1.
The large drawing features in this example of a residential site plan require scaling in order to fit on a standard size sheet. Annotations are scaled according to the plotted size of the drawing; otherwise, they would be too small to see.

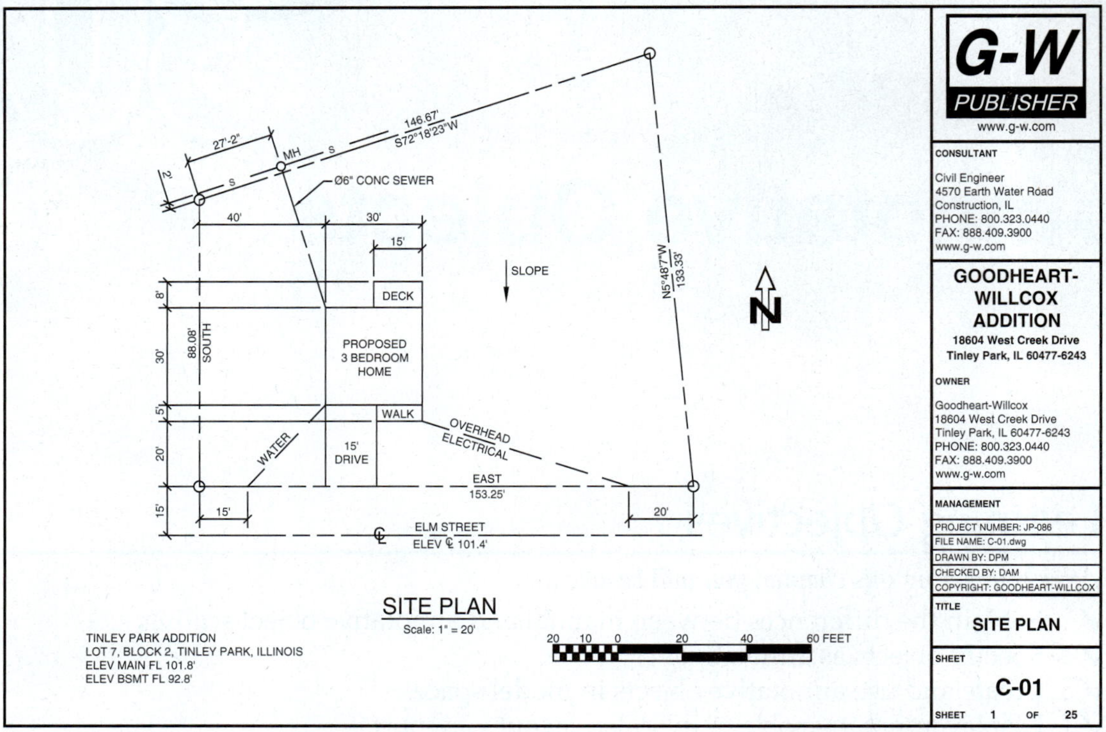

PROFESSIONAL TIP

Use annotative objects instead of traditional manual scaling even if you do not anticipate using a drawing scale other than 1:1.

Defining Annotative Objects

Objects that can be defined as annotative include single-line and multiline text, dimensions, leaders and multileaders, GD&T symbols created using the **TOLERANCE** command, hatch patterns, blocks, and attributes. The method you use to define objects as annotative varies depending on the object type. You can make objects annotative when you first draw them or convert non-annotative objects to annotative status as needed.

Creating New Annotative Objects

Single-line and multiline text is annotative when it is drawn using an annotative text style. To make a text style annotative, pick the **Annotative** check box in the **Size** area of the **Text Style** dialog box. See Figure 30-2. A drawing may include a combination of annotative and non-annotative text, dimension, and multileader styles. An example of text that is typically not annotative is text added directly to a layout, which is usually printed at a scale of 1:1.

Dimensions, standard leaders, and GD&T symbols created using the **TOLERANCE** command are annotative when they are drawn using an annotative dimension style. To make a dimension style annotative, pick the **Annotative** check box in the **Fit** tab of the **New** (or **Modify**) **Dimension Style** dialog box. See Figure 30-3.

Figure 30-2.
Single-line and multiline text objects are annotative when you draw them using an annotative text style.

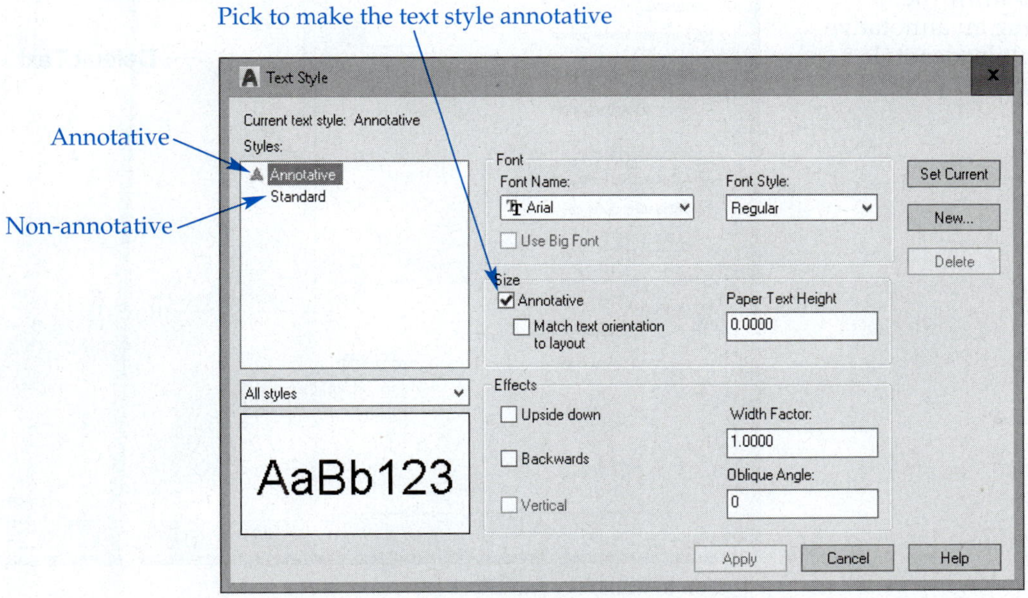

Figure 30-3.
Dimensions, leaders, and GD&T symbols created using the **TOLERANCE** command are annotative when you draw them using an annotative dimension style.

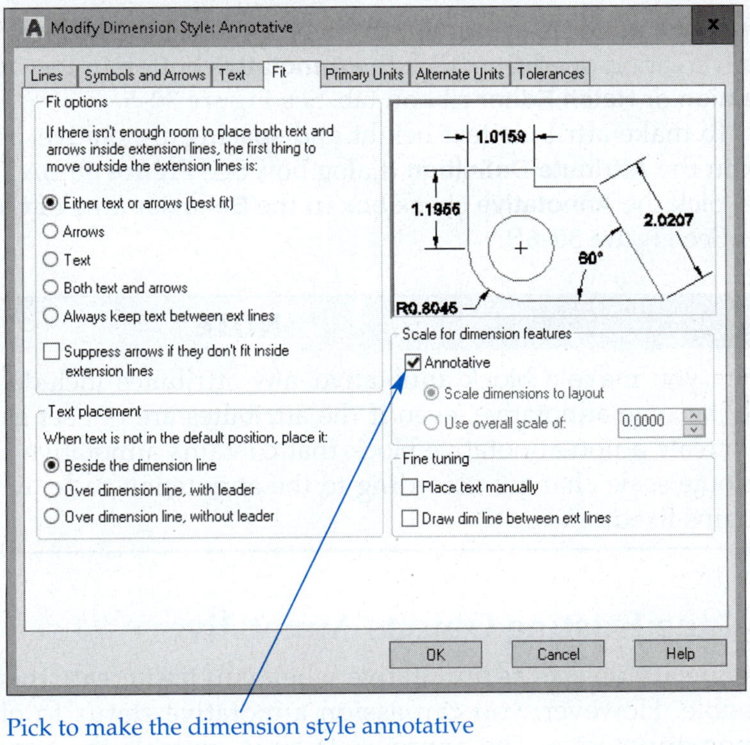

Multileaders are annotative when they are drawn using an annotative multileader style. To make a multileader style annotative, pick the **Annotative** check box in the **Leader Structure** tab of the **Modify Multileader Style** dialog box. See Figure 30-4.

NOTE

When you create an annotative multileader using the **Block** multileader type, the block automatically becomes annotative, even if the block is not set as annotative.

Figure 30-4.
Multileaders are annotative when you draw them using an annotative multileader style.

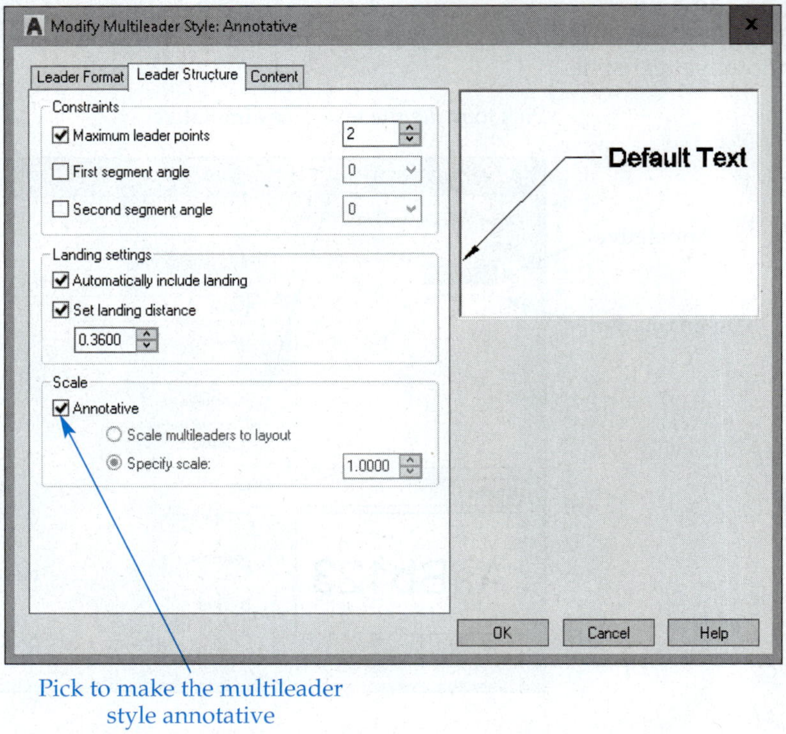

Hatch patterns are annotative when you set the hatch scale as annotative during hatch creation or editing. Pick the **Annotative** button in the **Options** panel of the **Hatch Creation** or **Hatch Editor** ribbon tab. See Figure 30-5.

To make attribute text height and spacing annotative, pick the **Annotative** check box in the **Attribute Definition** dialog box. See Figure 30-6A. To make a block annotative, pick the **Annotative** check box in the **Behavior** area of the **Block Definition** dialog box. See Figure 30-6B.

NOTE

When you make a block annotative, any attributes included in the block automatically become annotative, even if the attributes are not set as annotative. However, if you create a non-annotative block that contains annotative attributes, the annotative attribute scale changes according to the annotation scale, while the size of the block remains fixed.

Making Existing Objects Annotative

Specify objects as annotative when you first create them in model space when possible. However, you can assign annotative status to objects originally drawn as non-annotative. The appropriate style controls the annotative status of single-line and multiline text, dimensions, standard leaders and multileaders, and GD&T symbols created using the **TOLERANCE** command. Change the style assigned to the object to an annotative style to make the object annotative. You must edit or recreate existing hatch patterns, blocks, and attributes in order to make the objects annotative.

Figure 30-5.
Set the hatch pattern scale to **Annotative** when you create or edit the hatch pattern.

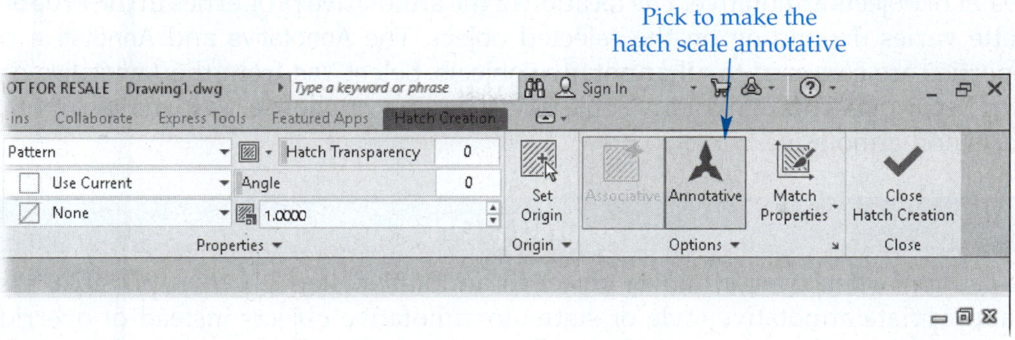

Figure 30-6.
Set attributes (A) and blocks (B) as annotative during definition.

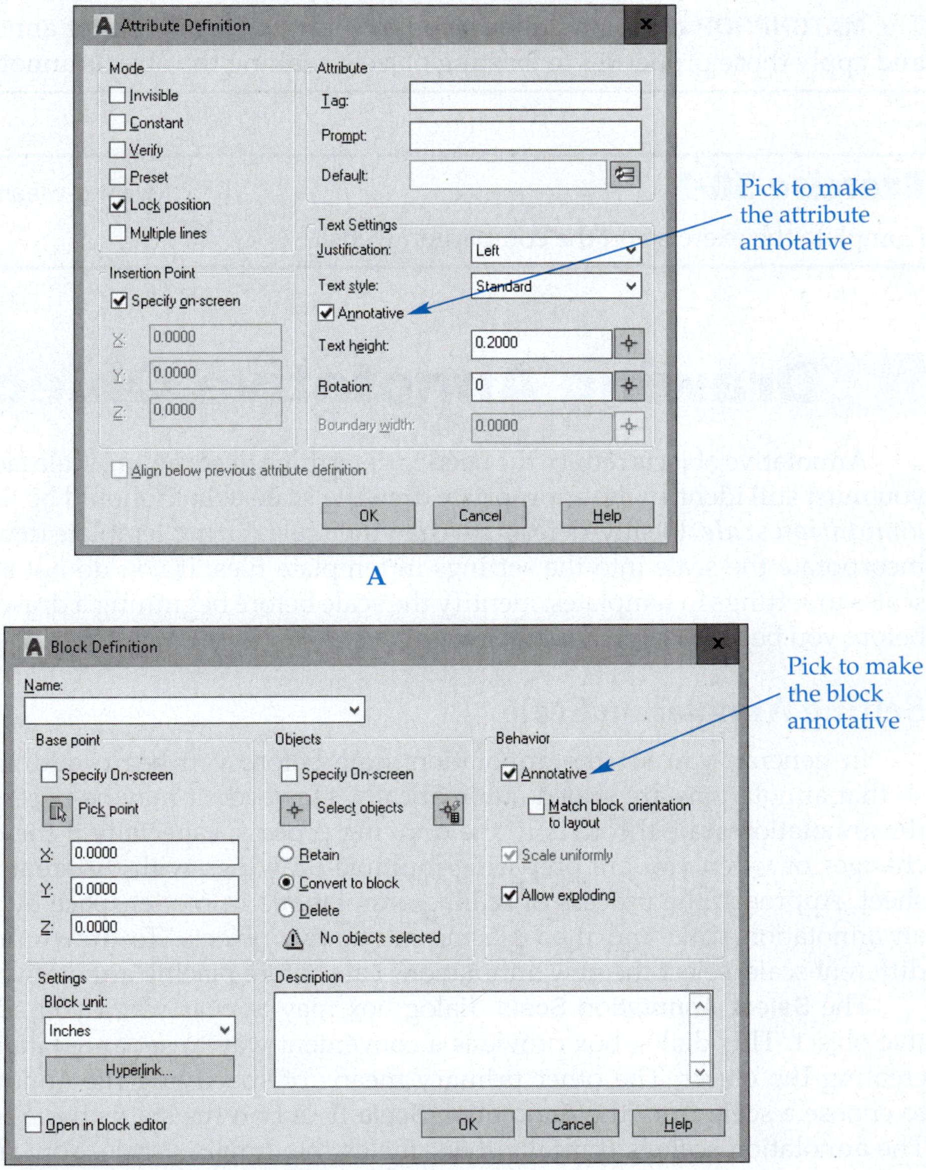

B

One way to make existing objects annotative is to override the non-annotative status using the **Properties** palette. This technique is most effective to make a limited number of objects annotative. The location of the annotative properties in the **Properties** palette varies depending on the selected object. The Annotative and Annotative scale properties are common to all annotative objects. Select Yes from the Annotative drop-down list to make non-annotative objects annotative, or choose No to make annotative objects non-annotative.

CAUTION

Use caution when overriding an object to annotative status. When possible, assign an appropriate annotative style or status to annotative objects instead of overriding specific objects.

NOTE

The **MATCHPROP** command allows you to select the properties of annotative objects and apply those properties to existing objects, making the objects annotative.

Exercise 30-1

www.g-wlearning.com/CAD/

Complete the exercise on the companion website.

Drawing Annotative Objects

annotation scale: The scale AutoCAD uses to calculate the scale factor applied to annotative objects.

Annotative objects reduce the need to determine the drawing scale factor. However, you must still identify the appropriate drawing scale, which should be the same as the ***annotation scale***. Ideally, determine drawing scale during template development and incorporate the scale into the settings in template files. If you do not apply drawing scales to settings in templates, identify the scale before beginning a drawing, or at least before you begin placing annotations.

Setting Annotation Scale

In general, you set the annotation scale before you begin adding annotations so that annotations are scaled automatically. However, it may be necessary to adjust the annotation scale throughout the drawing process, especially if the drawing scale changes or when you are preparing multiple drawings with different scales on one sheet. Approach the process of scaling annotations in model space by first selecting an annotation scale and then placing annotative objects. To draw annotations at a different scale, select the new annotation scale before placing annotative objects.

The **Select Annotation Scale** dialog box may appear when you add an annotative object. This dialog box provides a convenient way to set annotation scale before creating the object. The other primary means of specifying the annotation scale is to choose a scale from the **Annotation Scale** flyout on the status bar. See Figure 30-7. The annotation scale is typically the same as the drawing scale. You can also set the annotation scale in the **Properties** palette when no objects are selected by choosing the annotation scale from the **Annotation scale** drop-down list in the **Misc** category. See Chapter 9 for information on creating and editing annotation scales.

Figure 30-7.
The status bar includes several annotation scale options.

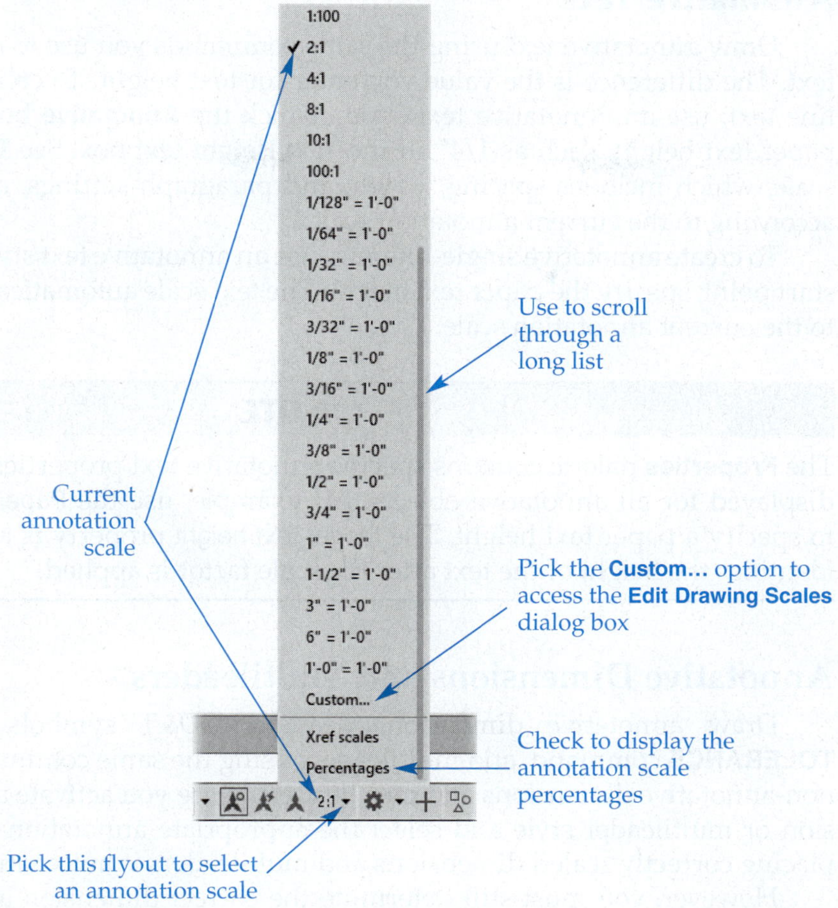

NOTE

Annotation scale sets the drawing scale in model space for controlling annotative objects. Viewport scale sets the drawing scale in a layout floating viewport to define the drawing scale. Both scales should be the same and should match the drawing scale.

Controlling Model Space Linetype Scale

The **CELTSCALE**, **PSLTSCALE**, and **MSLTSCALE** system variables control how the **LTSCALE** system variable applies to linetypes in model space and paper space. Leave the **CELTSCALE**, **PSLTSCALE**, and **MSLTSCALE** system variables at their default setting of 1 to apply the **LTSCALE** value correctly according to the current annotation scale. However, when you change the annotation scale, remember to use the **REGEN** command to regenerate the display. Otherwise, the linetype scale will not update according to the new scale.

PROFESSIONAL TIP

When you open a drawing in AutoCAD 2020 that was created in an AutoCAD version earlier than AutoCAD 2008, the **MSLTSCALE** system variable is set to 0. Change the value to 1 to take advantage of annotative linetype scaling.

Annotative Text

Draw annotative text using the same commands you use to draw non-annotative text. The difference is the value you enter for text height. To create annotative multiline text, use an annotative text style or pick the **Annotative** button. Then enter the paper text height, such as 1/4″, in the **Text Height** text box. See Figure 30-8. The text scale, which includes spacing, width, and paragraph settings, automatically adjusts according to the current annotation scale.

To create annotative single-line text, use an annotative text style. After you pick the start point, specify the paper text height. The text scale automatically adjusts according to the current annotation scale.

NOTE

The **Properties** palette contains specific annotative text properties in addition to those displayed for all annotative objects. For example, use the Paper text height property to specify a paper text height. The Model text height property is a reference value that identifies the height of the text after the scale factor is applied.

Annotative Dimensions and Multileaders

Draw annotative dimensions, leaders, GD&T symbols created using the **TOLERANCE** command, and multileaders using the same commands you use to draw non-annotative dimensions and multileaders. Once you activate an annotative dimension or multileader style and select the appropriate annotation scale, the process of placing correctly scaled dimensions and multileaders is automatic.

However, you must still determine the correct dimension and text location and spacing from objects when you add dimensions and text to scaled drawings. This involves multiplying the scale factor by the plotted spacing. For example, if the first dimension line should be 3/4″ from an object when plotted at a 1/4″ = 1′-0″ scale (scale factor = 48), the correct spacing in model space is 36″ (48 × 3/4″ = 36″) from the object.

Annotative Hatch Patterns

The difference between annotative and non-annotative hatch patterns is the way in which the drawing scale affects the hatch scale. When you create annotative hatch patterns, the scale you enter in the **Hatch Pattern Scale** text box produces the same results regardless of the specified annotation scale. For example, if you enter a value in the **Hatch Pattern Scale** text box that is appropriate for an annotation scale of 1:1, and then change the annotation scale to 4:1, the hatch pattern scale does not change relative to the drawing display. It looks the same on the 1:1 scaled drawing as on the 4:1 scaled drawing.

In contrast, when you create non-annotative hatch patterns, if you enter a value in the **Hatch Pattern Scale** text box that is appropriate for a drawing scaled to 1:1, and then change the drawing scale to 4:1, the displayed scale of the hatch pattern increases. The hatch looks four times as large on the 4:1 drawing as on the 1:1 drawing.

Annotative Blocks and Attributes

schematic block: A block originally drawn at a 1:1 scale.

Annotative blocks, often classified as ***schematic blocks***, are commonly used for annotation purposes. When you insert an annotative schematic block, AutoCAD determines the block scale based on the current annotation scale, eliminating the need to enter a scale factor. For most applications, insert annotative blocks at a scale of 1 to apply the annotation scale correctly. Entering a scale other than 1 adjusts the scale of the block by multiplying the block scale by the annotation scale.

Figure 30-8.
Create annotative multiline text using an annotative text style, or pick the **Annotative** button.

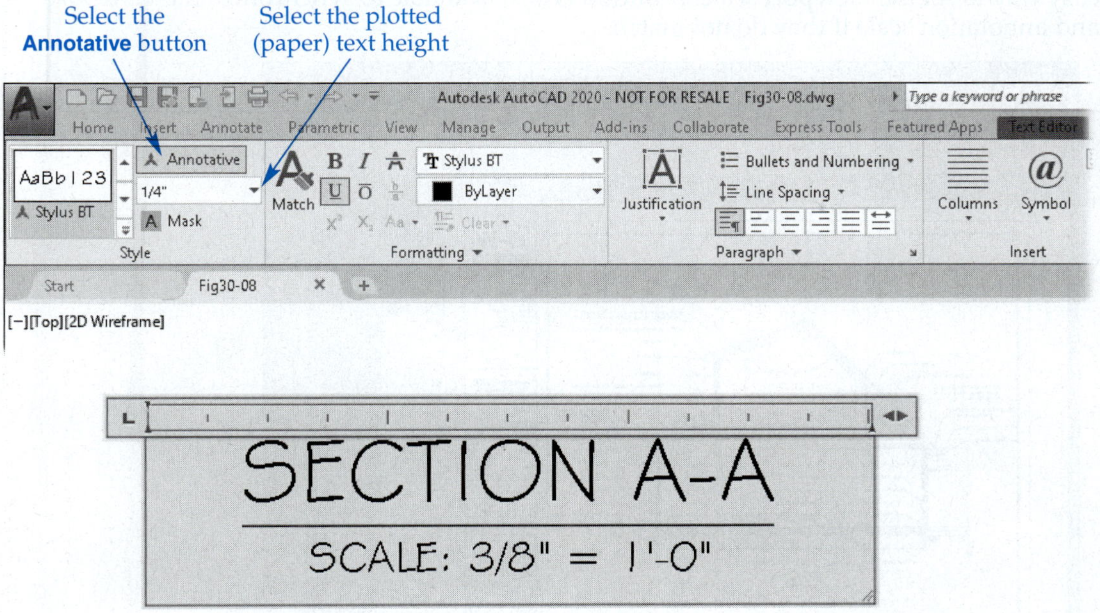

PROFESSIONAL TIP

When you create unit and schematic blocks that contain text and attributes, you should usually not make the text and attributes annotative. The text height you specify is set according to the full-scale size of the block, not necessarily the paper height. Any non-annotative text and attributes you select when you make a block annotative also automatically become annotative.

Exercise 30-2

www.g-wlearning.com/CAD/

Complete the exercise on the companion website.

Displaying Annotative Objects in Layouts

Once you create drawing features and symbols and add annotative objects according to the appropriate annotation scale, you are ready to display and plot the drawing using a paper space layout. Refer to Chapter 29 to review the process of using and scaling floating viewports. Figure 30-9 shows a drawing scaled to 3/8″ = 1′-0″. In this example, drawing features are drawn at full scale in model space. The annotation scale in model space was set to 3/8″ = 1′-0″, and annotative text, dimensions, multileaders, hatch patterns, and blocks were added. The annotative objects in paper space are automatically scaled according to the 3/8″ = 1′-0″ annotation scale.

Figure 30-9.
Scaling a drawing in a floating paper space viewport. The **Viewport Scale** flyout provides an easy way to set the viewport scale. A button is also available to synchronize the viewport and annotation scale if they do not match.

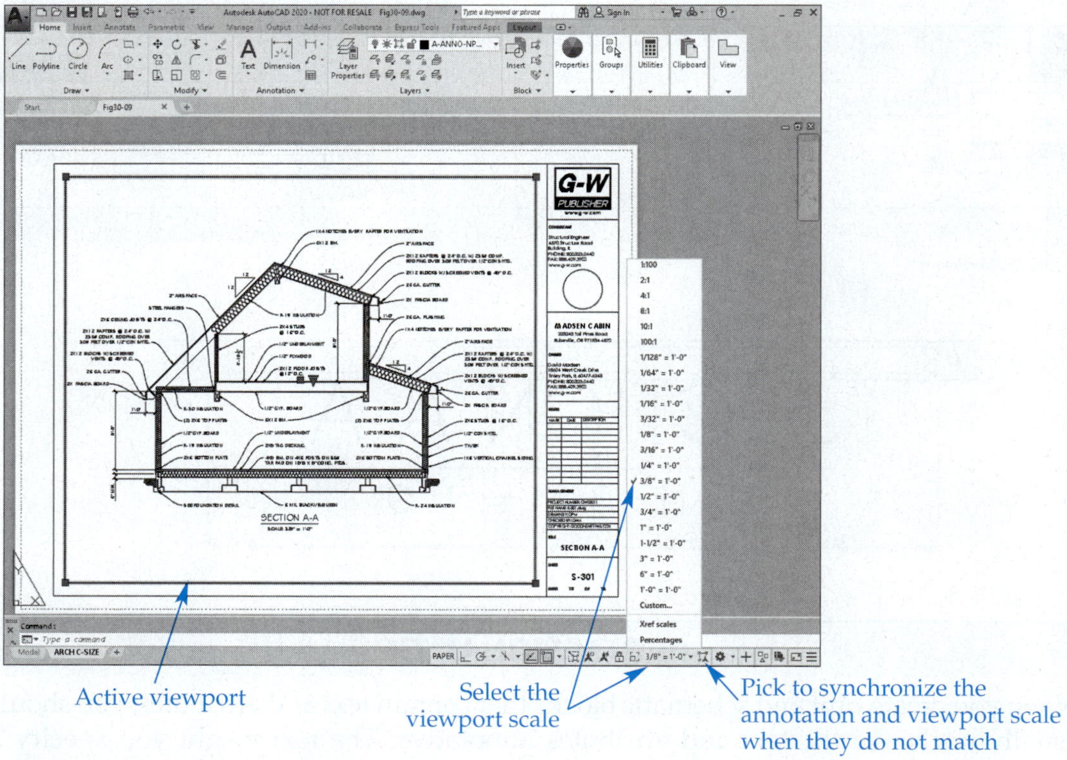

Synchronizing Annotation and Viewport Scale

Viewport Scale Sync

Status Bar

Viewport Scale Sync

In the Figure 30-9 example, the viewport scale and the annotation scale are the same, which is typical when scaling annotative objects. If you select a different viewport scale from the **Viewport Scale** flyout, the annotation scale automatically adjusts according to the viewport scale. However, if you adjust the viewport scale using a view tool such as zooming, the annotation scale does not change. The viewport scale and the annotation scale must match in order for the drawing and annotative objects to be scaled correctly. Pick the **Viewport Scale Sync** button in the status bar to synchronize the viewport and annotation scales.

The **Properties** palette also provides viewport and annotation scale controls. You must be in paper space and pick a floating viewport to access viewport properties. Choose a viewport scale from the **Standard scale** drop-down list. Adjust the annotation scale using the **Annotation scale** option. See Figure 30-10.

PROFESSIONAL TIP

Lock the viewport display to avoid zooming and disassociating the viewport scale from the annotation scale. Refer to Chapter 29 for more information on locking floating viewports.

Exercise 30-3

www.g-wlearning.com/CAD/

Complete the exercise on the companion website.

Figure 30-10.
The **Properties** palette also allows you to set the viewport and annotation scale.

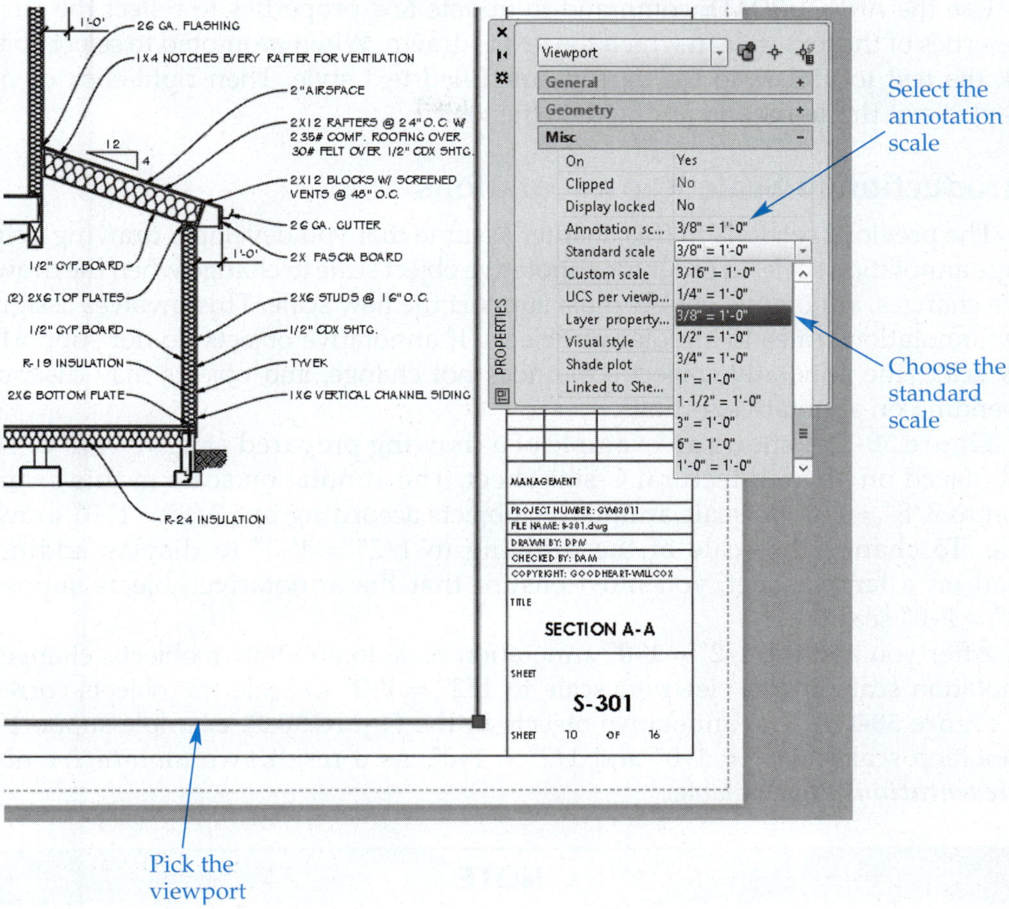

Changing Drawing Scale

No matter how much you plan a drawing, drawing scale can change throughout the drawing process. Reduce the drawing scale if it is necessary to use a smaller sheet. Increase the drawing scale if drawing features are redesigned and become larger, or if additional drawing detail is required.

Changing the drawing scale affects the size and position of annotations. If you change the drawing scale, remember that the annotation scale is the same as the drawing scale.

To change the annotation scale in model space, select a new annotation scale from the **Annotation Scale** flyout. To change the annotation scale in an active viewport in a layout, adjust the viewport scale by selecting the drawing scale from the **Viewport Scale** flyout. Again, the viewport and annotation scales should be the same for most applications.

Using the ANNOUPDATE Command

If you create single-line text using a non-annotative text style and then change the text style to annotative using the **Text Style** dialog box, text drawn using the style does not automatically become annotative. The properties of the text remain set according to the non-annotative text style. Likewise, if you create annotative text using an annotative text style and then change the text style to non-annotative using the **Text Style** dialog box, text drawn in the style does not automatically become non-annotative. The properties of the text remain set according to the annotative style.

Type
ANNOUPDATE

Use the **ANNOUPDATE** command to update text properties to reflect the current properties of the text style in which the text is drawn. When prompted to select objects, pick the text to update to the current, modified text style. Then right-click or press [Enter] to exit the command and update the text.

Introduction to Scale Representations

The previous sections in this chapter assume that you develop a drawing using a single annotation scale. In order for annotative object scale to change when the drawing scale changes, annotative objects must support the new scale. This involves assigning new annotation scales to annotative objects. If annotative objects do not support the new scale, the annotative object scale does not change, and objects may disappear, depending on annotative settings.

Figure 30-11A shows an example of a drawing prepared at a 3/8″ = 1′-0″ scale and placed on an architectural C-size sheet. The annotation scale in this example is set to 3/8″ = 1′-0″, to scale annotative objects according to a 3/8″ = 1′-0″ drawing scale. To change the scale of the drawing to 1/2″ = 1′-0″ to display additional detail on a larger sheet, you must ensure that the annotative objects support a 1/2″ = 1′-0″ scale.

After you add the 1/2″ = 1′-0″ annotation scale to annotative objects, change the annotation scale or the viewport scale to 1/2″ = 1′-0″ to scale the objects correctly. See Figure 30-11B. The annotative objects in the Figure 30-11 example support two annotation scales: 3/8″ = 1′-0″ and 1/2″ = 1′-0″. As a result, two ***annotative object representations*** are available.

annotative object representation: Display of an annotative object at an annotation scale that the object supports.

NOTE

Annotative objects display an icon when you hover the crosshairs over the objects. Objects that support a single annotation scale display the annotative icon shown in Figure 30-12A. Annotative objects that support more than one annotation scale display the annotative icon shown in Figure 30-12B.

Understanding Annotation Visibility

Before changing the current annotation scale, you should understand how the annotation scale affects annotative object visibility. The annotative object scale does not change if annotative objects do not support the selected annotation scale. In addition, annotative objects disappear when an annotation scale that the objects do not support is current. For example, if annotative objects only support an annotation scale of 3/8″ = 1′-0″, and you set an annotation scale of 1/2″ = 1′-0″, the annotative object scale remains set at 3/8″ = 1′-0″, and the objects disappear.

An easy way to turn annotative object visibility on and off according to the current annotation scale is to pick the **Annotation Visibility** button on the status bar. See Figure 30-13. Picking the **Annotation Visibility** button sets the value of the **ANNOALLVISIBLE** system variable. Turning on annotation visibility is most effective when you are adding annotation scales to or deleting them from annotative objects. If you add multiple annotation scales to annotative objects, the annotative object representation is based on the current scale.

Figure 30-11.
A—A drawing created using an annotation scale of 3/8″ = 1′-0″ on an architectural C-size sheet. Annotative objects automatically appear at the correct scale. B—The same drawing shown in A, modified to an annotation scale of 1/2″ = 1′-0″ and placed on an architectural D-size sheet. An annotation scale of 1/2″ = 1′-0″ is added to all of the annotative objects, allowing the objects to adapt to the new scale automatically.

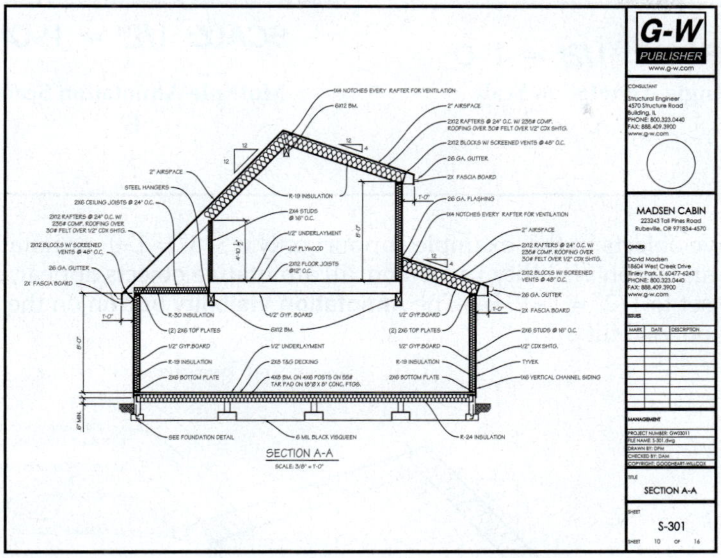

A

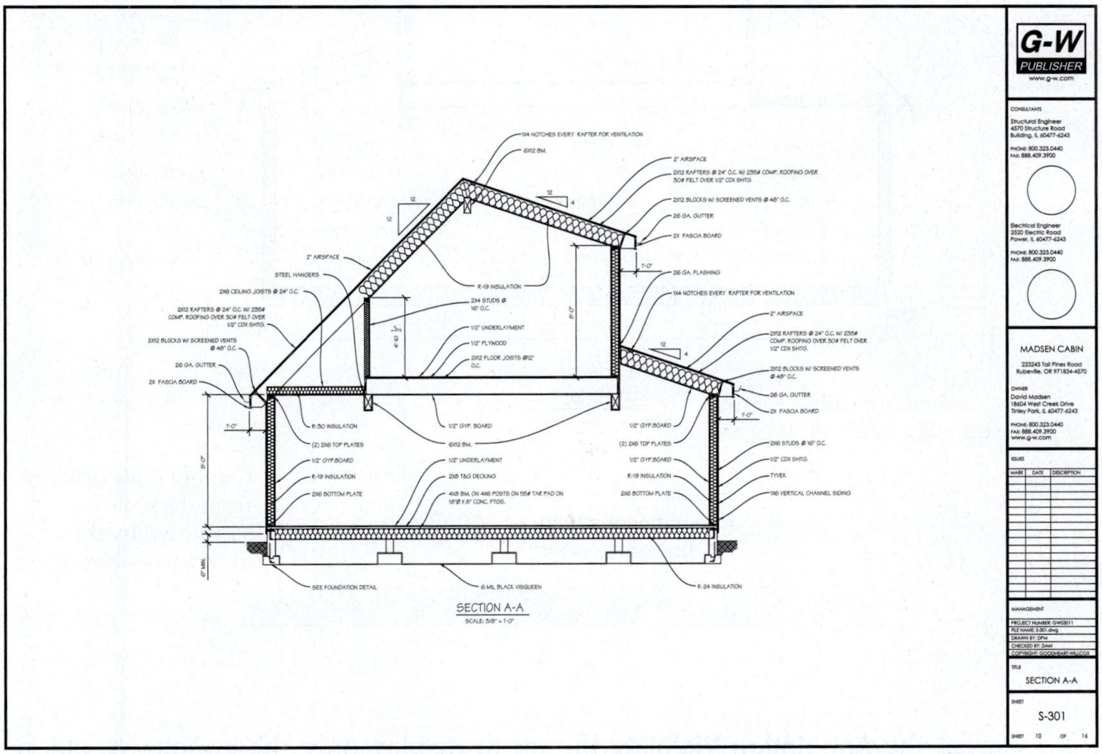

B

Figure 30-12.
The icons displayed when annotative objects support single or multiple annotation scales.

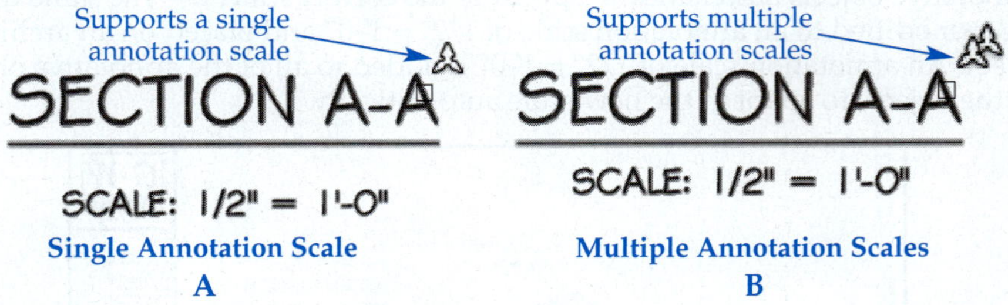

Figure 30-13.
A—The annotative objects in this example support only a 3/8″ = 1′-0″ annotation scale. However, with annotation visibility turned on, all annotative objects appear, even with the annotation scale set to 1/2″ = 1′-0″. B—The **Annotation Visibility** button on the status bar controls annotation visibility.

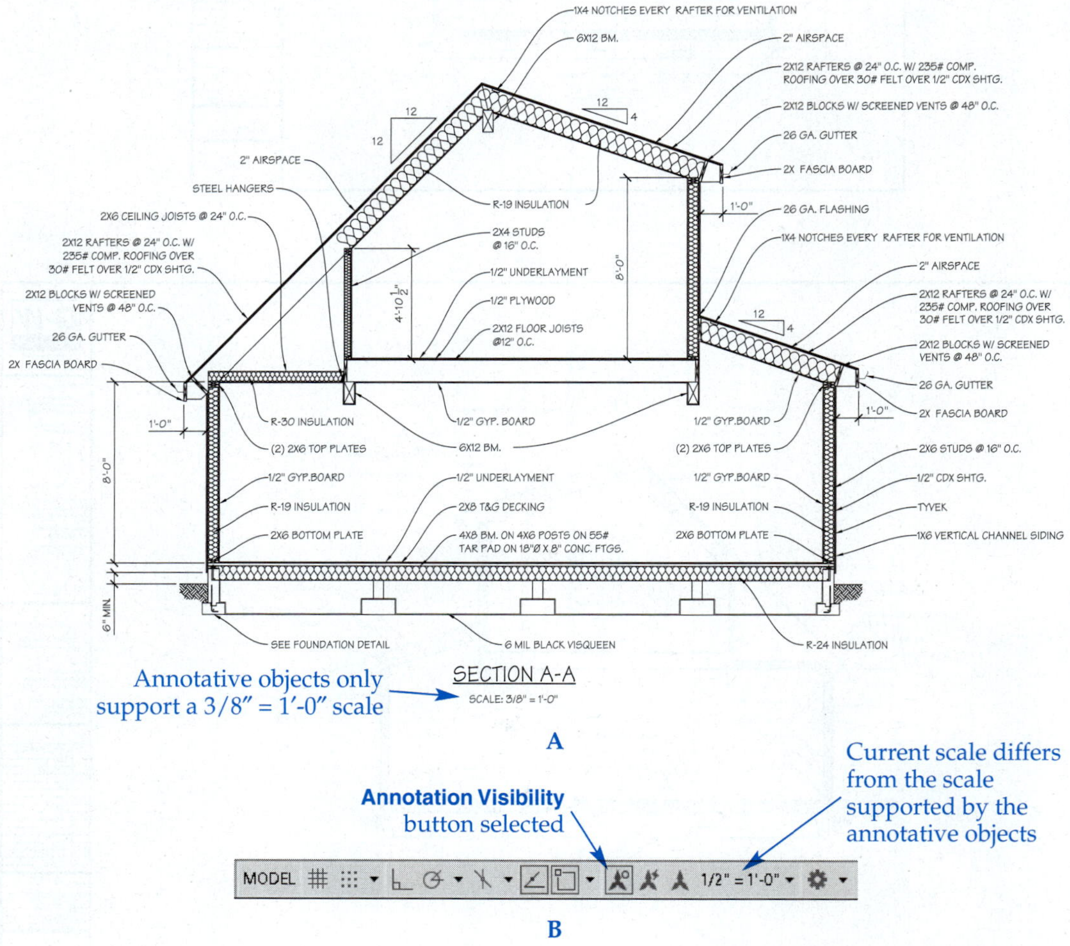

Deselect the **Annotation Visibility** button to display only the annotative objects that support the current annotation scale. Any annotative objects unsupported by the current annotation scale disappear. See **Figure 30-14**. Turning annotation visibility off is most effective when you are annotating a drawing, or a portion of a drawing, using a different annotation scale without showing annotative object representations specific to a different annotation scale. Turning off visibility of annotative objects that do not support the current annotation scale is also effective for preparing multiview drawings because it eliminates the need to create separate layers for objects displayed at different scales. This practice is described later in this chapter.

Figure 30-14.
Deselect the **Annotation Visibility** button to display only the annotative objects that support the current annotation scale. The annotative objects in this example do not appear because they support only a 3/8″ = 1′-0″ annotation scale, and the current annotation scale is 1/2″ = 1′-0″.

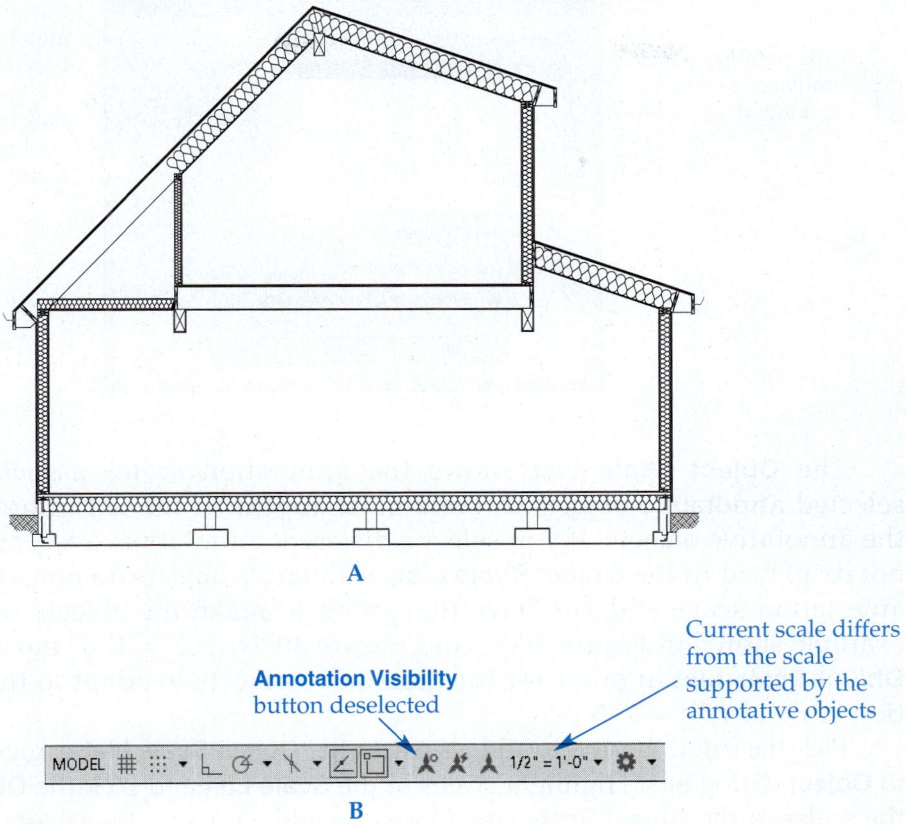

Adding and Deleting Annotation Scales

One method for assigning additional annotation scales to annotative objects is to add the scales to selected objects. This method is appropriate whenever the drawing scale changes, but it is especially effective for adding annotation scales only to specific objects, such as when you are creating multiview drawings.

Delete an annotation scale from annotative objects if the annotation scale is no longer in use, should not be displayed in a specific view, or makes it difficult to work with annotative objects. When you delete an annotation scale from annotative objects, the scale can no longer be applied to the objects. Add annotation scales to or delete annotation scales from selected objects using annotation scaling tools or the **Properties** palette.

Using the OBJECTSCALE Command

> **Ribbon**
> Annotate
> > Annotation Scaling
> Add/Delete Scales
> **Type**
> OBJECTSCALE
>
> **OBJECTSCALE**

The **OBJECTSCALE** command provides one method of adding and deleting annotation scales supported by annotative objects. A quick way to access the **OBJECTSCALE** command is to select an annotative object and then right-click and pick **Add/Delete Scales...** from the **Annotative Object Scale** cascading menu. If you activate the **OBJECTSCALE** command by right-clicking on objects, the **Annotation Object Scale** dialog box appears, allowing you to add or remove annotation scales supported by the selected objects. See Figure 30-15. If you access the **OBJECTSCALE** command before selecting objects, all annotative objects are displayed, even those objects that do not support the current annotation scale. Select the annotative objects to modify and right-click or press [Enter] to display the **Annotation Object Scale** dialog box.

Figure 30-15.
Use the **Annotation Object Scale** dialog box to add annotation scales to and delete them from annotative objects.

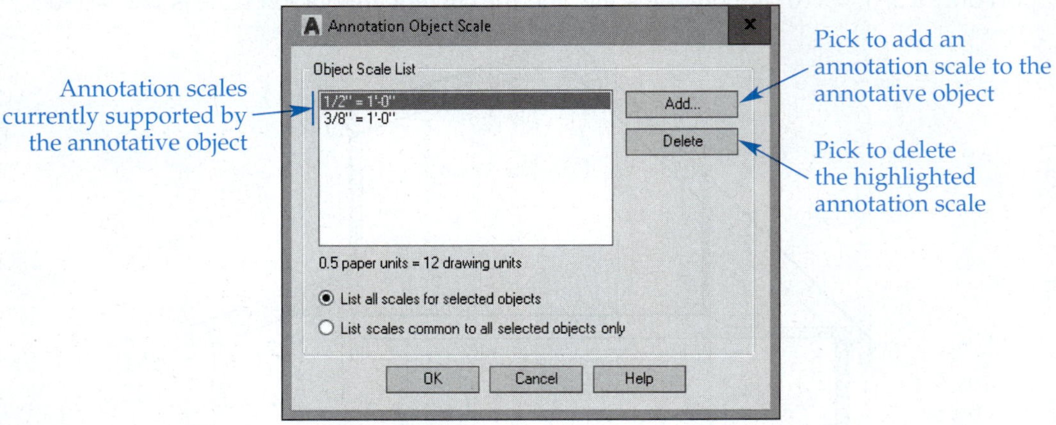

The **Object Scale List** shows the annotation scales associated with the selected annotative objects. A scale must appear in the list in order to apply to the annotative objects. If you select a different annotation scale, and that scale is not displayed in the **Object Scale List**, annotative objects do not adapt to the new annotation scale, and you have the option to make the objects invisible. In the example shown in Figure 30-13 and Figure 30-14, 1/2″ = 1′-0″ must appear in the **Object Scale List** in order for the annotative objects to adapt to the new annotation scale of 1/2″ = 1′-0″.

Pick the **Add...** button to add a scale to the **Object Scale List** using the **Add Scales to Object** dialog box. Highlight scales in the **Scale List** and pick the **OK** button to add the scales to the **Object Scale List**. Once you add a scale to the **Object Scale List**, you can pick an annotation scale that corresponds to a listed scale to scale the selected annotative objects. To remove a scale from the **Object Scale List**, highlight the scale and pick the **Delete** button.

If you select multiple annotative objects, pick the **List scales common to all selected objects only** radio button to display only the annotative scales common to the selected objects. Pick the **List all scales for selected objects** radio button to show all annotation scales associated with any of the selected objects, even if some of the objects do not support the listed scales. Listing all scales for selected objects is helpful when you want to delete a scale that applies only to certain objects.

NOTE

If a desired scale is not available in the **Add Scales to Object** dialog box, close the **Annotation Object Scale** dialog box and access the **Edit Drawing Scales** dialog box to add a new scale to the list of available scales.

Using the Properties Palette

The **Properties** palette also allows you to add annotation scales to selected annotative objects. See Figure 30-16. The location of the annotative properties in the **Properties** palette varies depending on the selected object. The Annotative scale property displays the annotation scale currently applied to the selected annotative objects and contains an ellipsis button (**...**) that you can pick to open the **Annotation Object Scale** dialog box.

Figure 30-16.
The Annotative scale property in the **Properties** palette provides another way to access the **Annotation Object Scale** dialog box.

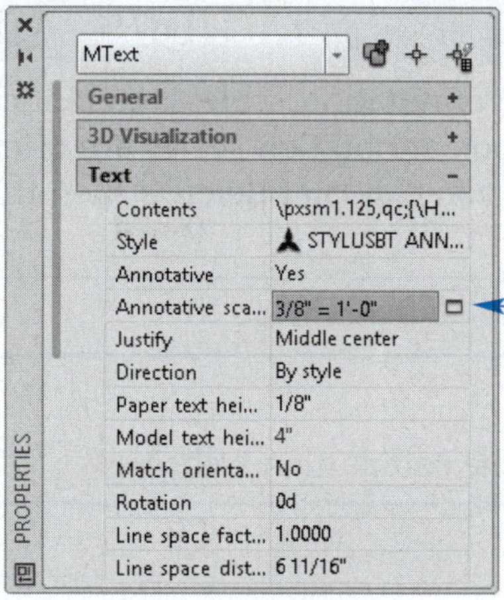

Automatically Adding Annotation Scales

Another technique for assigning additional annotation scales to annotative objects is to add a selected annotation scale automatically to all annotative objects in the drawing. This eliminates the need to add annotation scales to individual annotative objects and quickly produces newly scaled drawings.

The **ANNOAUTOSCALE** system variable controls the ability to add an annotation scale to all existing annotative objects. Enter 1, –1, 2, –2, 3, –3, 4, or –4, depending on the desired effect. Figure 30-17A describes each option. After you enter the initial value, an easy way to toggle the **ANNOAUTOSCALE** system variable on and off is to pick the button on the status bar. See Figure 30-17B.

Figure 30-17.
A—**ANNOAUTOSCALE** system variable options. B—After you enter the initial **ANNOAUTOSCALE** system variable setting, use the button on the status bar to toggle **ANNOAUTOSCALE** on and off.

Value	Mode	Description
1	On	Adds the selected annotation scale to annotative objects, not including those drawn on a layer that is turned off, frozen, locked, or frozen in a viewport.
–1	Off	1 behavior is used when **ANNOAUTOSCALE** is turned back on.
2	On	Adds the selected annotation scale to annotative objects, not including those drawn on a layer that is turned off, frozen, or frozen in a viewport.
–2	Off	2 behavior is used when **ANNOAUTOSCALE** is turned back on.
3	On	Adds the selected annotation scale to annotative objects, not including those drawn on a layer that is locked.
–3	Off	3 behavior is used when **ANNOAUTOSCALE** is turned back on.
4	On	Adds the selected annotation scale to all annotative objects regardless of the status of the layer on which the annotative object is drawn. 4 is the AutoCAD default setting when toggled on.
–4	Off	4 behavior is used when **ANNOAUTOSCALE** is turned back on. –4 is the AutoCAD default setting when toggled off.

A

B

CAUTION

Use caution when adding annotation scales automatically. Due to the effectiveness and transparency of the tool, annotation scales are often added to annotative objects unintentionally. Although you can later delete scales, this causes additional work and confusion.

Exercise 30-4 www.g-wlearning.com/CAD/

Complete the exercise on the companion website.

Preparing Multiview Drawings

Drawings for many different engineering fields often contain views, sections, and details drawn at different scales. Annotative objects offer several advantages for these drawings, especially when views in model space appear at different scales in layouts. Use scaled viewports to display multiple views using a single file. You can assign a different annotation scale to each drawing view that contains annotative objects, reducing the need to calculate multiple drawing scale factors, while maintaining the appropriate scale of previously drawn annotative objects. Additionally, by adjusting annotative object scale representation visibility and position, you can prepare differently scaled multiview drawings, while eliminating the need to use separate, scale-specific layers and annotations.

Creating Differently Scaled Drawings

Figure 30-18A shows an example of two different drawing views, a full section (SECTION A-A) and a stair section, both drawn at full scale in model space. The full section in Figure 30-18A uses a 3/8″ = 1′-0″ scale. To prepare the full section, set the annotation scale in model space to 3/8″ = 1′-0″, and then add annotative objects. The annotative objects are automatically scaled according to the 3/8″ = 1′-0″ annotation scale. The stair section in Figure 30-18A uses a 1/2″ = 1′-0″ scale. To prepare the stair section, change the annotation scale in model space from 3/8″ = 1′-0″ to 1/2″ = 1′-0″, and then add annotative objects. These annotative objects are automatically scaled according to the 1/2″ = 1′-0″ annotation scale. If you look closely, you can see the different scales applied to the drawing views.

Figure 30-18A shows annotation visibility on, allowing you to see all annotative objects and observe the effects of using different scales. Figure 30-18B shows annotation visibility off to show only annotative objects that support the current annotation scale, which is 1/2″ = 1′-0″ in this example.

The next step is to display and plot the drawing using multiple paper space viewports. Figure 30-19 shows an architectural D-size sheet layout with two floating viewports. One viewport displays the full section at a viewport scale of 3/8″ = 1′-0″. The other viewport displays the stair section at a viewport scale of 1/2″ = 1′-0″. Notice that the annotative objects are the same size in both views.

Exercise 30-5 www.g-wlearning.com/CAD/

Complete the exercise on the companion website.

Figure 30-18.
Two different drawing views drawn at full scale in model space. The full section (**SECTION A-A**) uses an annotation scale of 3/8″ = 1′-0″, and the stair section uses an annotation scale of 1/2″ = 1′-0″. A—Annotation visibility is on. B—Annotation visibility is off with the current annotation scale set to 1/2″ = 1′-0″ (the view scale of the stair section).

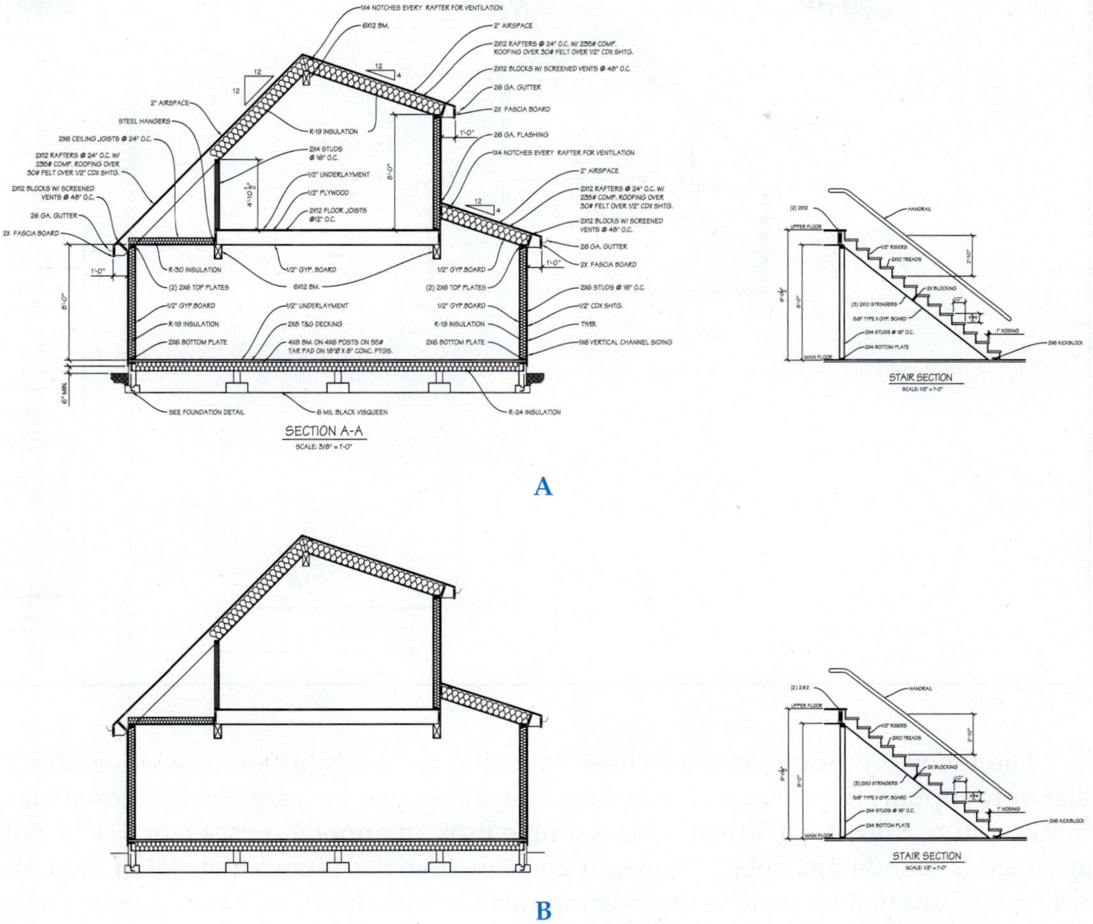

A

B

Reusing Annotative Objects

Often the same drawing features appear in different views at different scales. For example, you may plot a drawing on a large sheet using a large scale, and plot the same drawing on a smaller sheet using a smaller scale. Another example is preparing a view enlargement or detail.

Annotative objects significantly improve the ability to reuse existing drawing features. Use annotation visibility to hide annotative objects not supported by the current annotation scale. You can also adjust the position of scale representations according to the appropriate annotation scale. These options allow you to include differently scaled annotative objects on the same sheet without creating copies of the objects and without using scale-specific layers.

Using Invisible Scale Representations

If annotative objects do not support an annotation scale, the annotative objects disappear when the annotation scale that the objects do not support is current. This is a valuable technique for displaying certain items at a specific scale. Pick the **Annotation Visibility** button on the status bar to turn on and off annotative object visibility.

Figure 30-19.
Using viewports with different scales to create a multiview drawing. Notice that the annotative objects are the same size in both views.

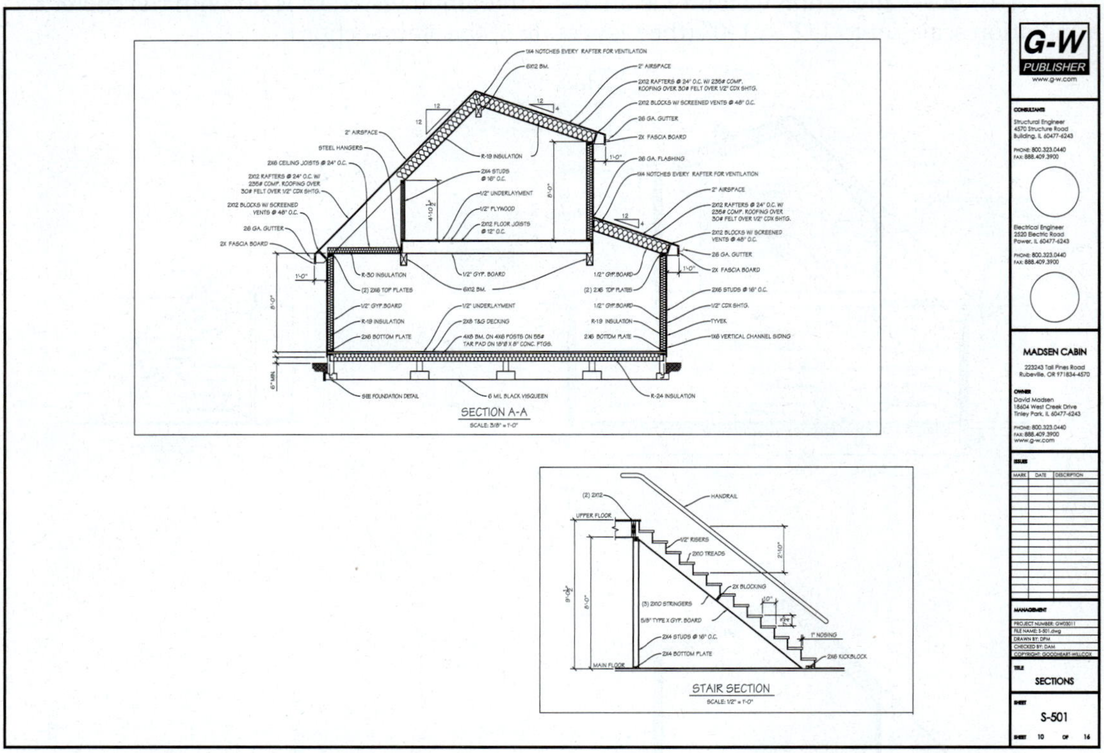

The following example shows how adjusting the visibility of annotative objects that only support the current annotation scale allows you to create an additional view from existing drawing features. This example uses an annotation scale of 3/4″ = 1′-0″ to create a foundation detail. To begin constructing the foundation detail, add the 3/4″ = 1′-0″ annotation scale to the existing earth hatch pattern so it will appear on the full section and the foundation detail. See **Figure 30-20**. Next, with the current annotation scale set to 3/4″ = 1′-0″, add annotative objects specific to the foundation detail. See **Figure 30-21**. These objects support only the 3/4″ = 1′-0″ annotation scale, hiding the objects on the full section, which uses a 3/8″ = 1′-0″ scale.

PROFESSIONAL TIP

If objects already support an annotation scale, but you do not want to display those annotations at the current scale, delete the annotation scale from the objects.

Adjusting Scale Representation Position

When you reuse annotative objects, the location and spacing of annotative objects on one scale are often not appropriate for another scale. Reposition each scale representation to overcome this issue.

In the foundation detail example in **Figure 30-21**, some of the existing 3/8″ = 1′-0″ scaled dimensions and multileaders from the full section are reused in the foundation detail. See **Figure 30-22A**. The first step is to add a 3/4″ = 1′-0″ annotation scale to the objects. Next, with **Annotation Visibility** turned off, as shown in **Figure 30-22B**, you can see the resulting position of the selected objects, which is initially the same as the position of the 3/8″ = 1′-0″ objects. The only difference is that now the 3/8″ = 1′-0″ objects also support a 3/4″ = 1′-0″ scale.

Figure 30-20.
Reuse the earth hatch pattern by adding the 3/4″ = 1′-0″ foundation detail scale to the annotative hatch pattern.

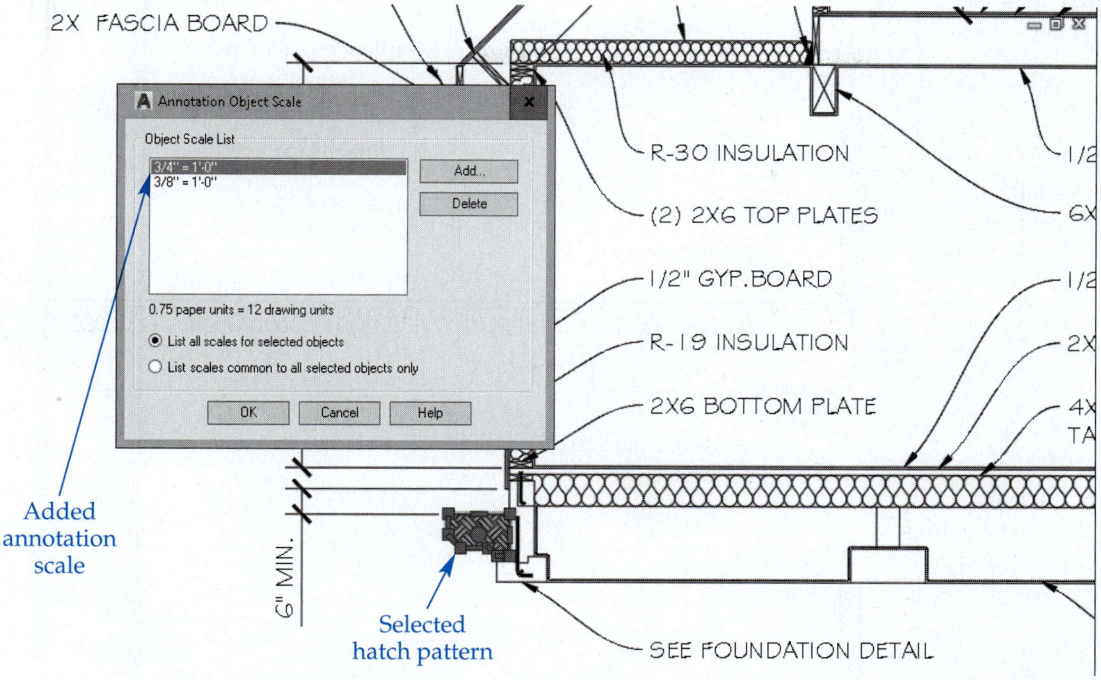

Figure 30-21.
Adding annotative text, dimensions, multileaders, and hatch patterns specific to the foundation detail using a 3/4″ = 1′-0″ annotation scale.

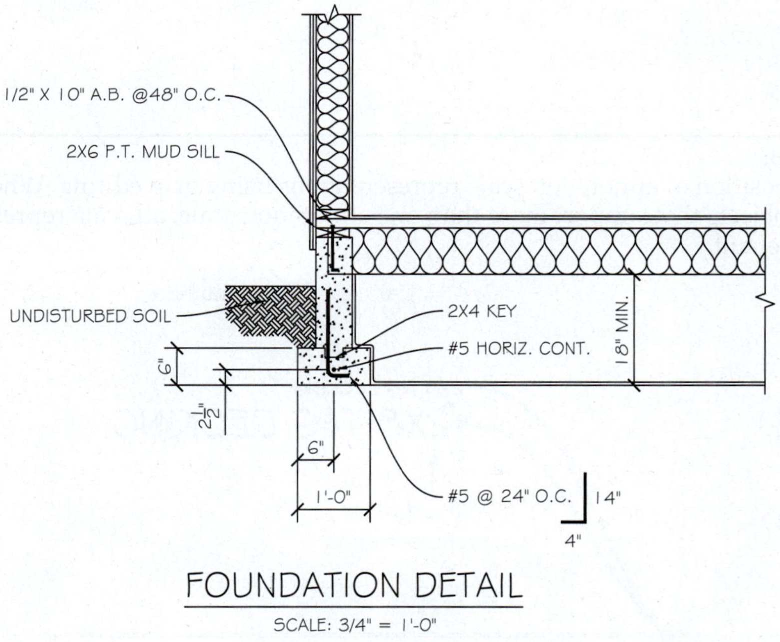

Use grip editing to adjust the position of annotation scale representations. When you select annotative objects that support more than one annotation scale, all scale representations appear by default. See **Figure 30-23**. An annotative object is a single object, but it can contain several scale representations. Grips are displayed on the scale representation that corresponds to the current annotation scale.

Figure 30-22.
A—Reusing some of the existing 3/8″ = 1′-0″ scaled objects to create another drawing view.
B—Adding a 3/4″ = 1′-0″ annotation scale to existing objects and setting the annotation scale to 3/4″ = 1′-0″.

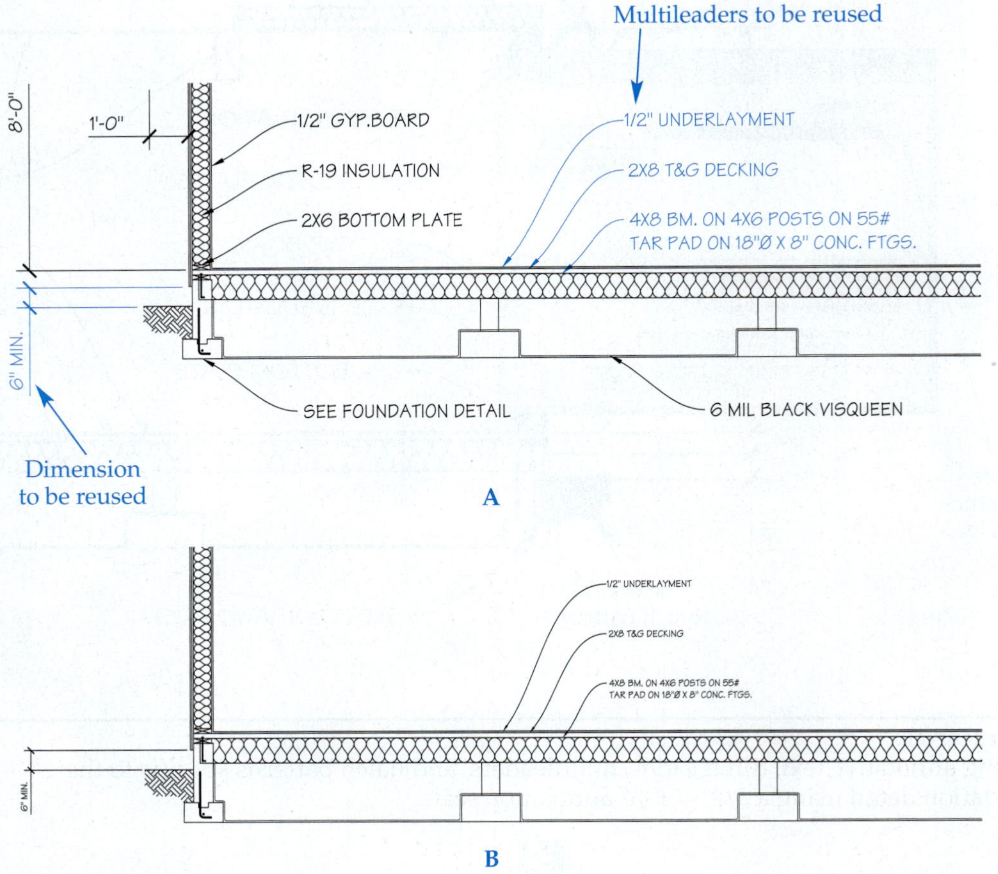

Figure 30-23.
Adjust the position of annotation scale representations using grip editing. When you select annotative objects that support more than one annotation scale, all scale representations appear by default.

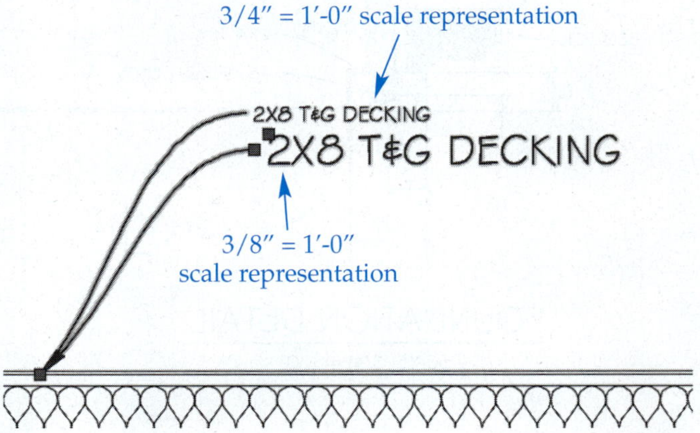

Using grips to edit scale representations is similar to editing the object used to create the scale representation. The difference when editing a scale representation is that you adjust a scaled copy of the object. Figure 30-24 shows the effects of editing the position of dimension and multileader scale representations on the foundation detail. The figure shows selecting the representations to help demonstrate the effects of editing scale representation position. Notice that you can edit all elements of the scale representation to produce the desired annotations at the appropriate locations.

PROFESSIONAL TIP

Use the **TEXTALIGN**, **DIMSPACE**, and **MLEADERALIGN** commands to adjust text, dimension, and multileader spacing and alignment after changing the drawing scale.

The **SELECTIONANNODISPLAY** system variable controls the display of selected scale representations and is set to 1 by default. As a result, all scale representations display and appear dimmed when you pick an annotative object that supports multiple annotation scales. See Figure 30-24. The display can be confusing if the selected object supports several annotation scales. Set the **SELECTIONANNODISPLAY** system variable to 0 to display only the scale representation that corresponds to the current annotation scale.

NOTE

You can edit scale representations individually only by using grip editing. When you use modify commands to edit an annotative object, all of the scale representations change at once.

Figure 30-24.
Editing the position of scale representations is much like creating scaled copies of existing annotations.

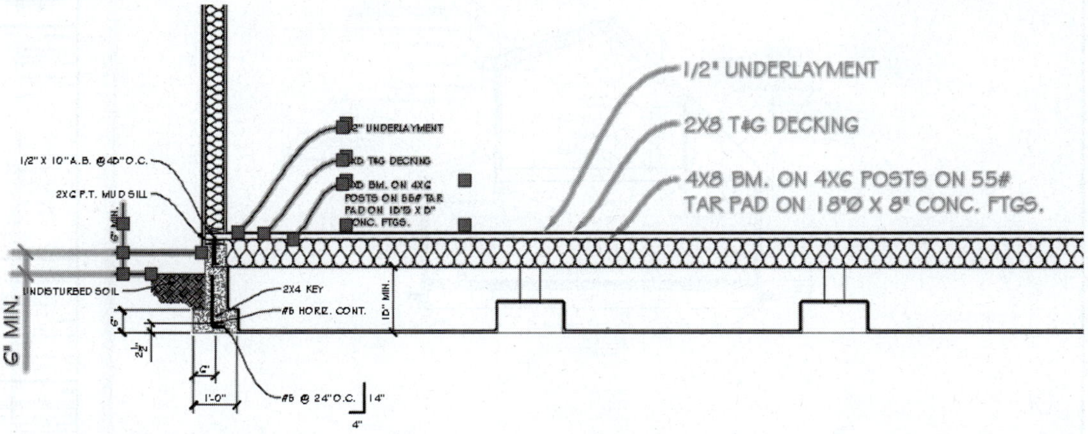

Resetting Scale Representation Position

ANNORESET

Ribbon
Annotate
> Annotation Scaling
Sync Scale Positions
Type
ANNORESET

The **ANNORESET** command removes multiple scale representation positions, allowing you to change the position of all selected scale representations to the position of the scale representation that is set for the current annotation scale. A quick way to access the **ANNORESET** command is to select annotative objects, right-click, and pick the option from the **Annotative Object Scale** cascading menu. If you activate the **ANNORESET** command by right-clicking on objects, the position of the selected objects resets. If you access the command before selecting objects, pick the annotative objects. Then right-click or press [Enter] to exit the command and reset the scale representation positions.

Completing a Multiview Drawing

The last step in creating a multiview drawing is to display and plot the drawing using multiple paper space viewports. Figure 30-25 shows an architectural D-size sheet layout with three floating viewports. One viewport displays the full section at a 3/8″ = 1′-0″ viewport scale. A second viewport displays the stair section at a 1/2″ = 1′-0″ viewport scale. A third viewport displays the foundation detail at a 3/4″ = 1′-0″ viewport scale.

PROFESSIONAL TIP

When you save drawings using annotative objects to earlier versions of AutoCAD that do not support annotative objects, scale representations may convert to non-annotative objects, but automatically become assigned to unique layers. To use this function, select the **Maintain visual fidelity for annotative objects** check box in the **Open and Save** tab of the **Options** dialog box.

Figure 30-25.
A complete multiview drawing created using annotative objects.

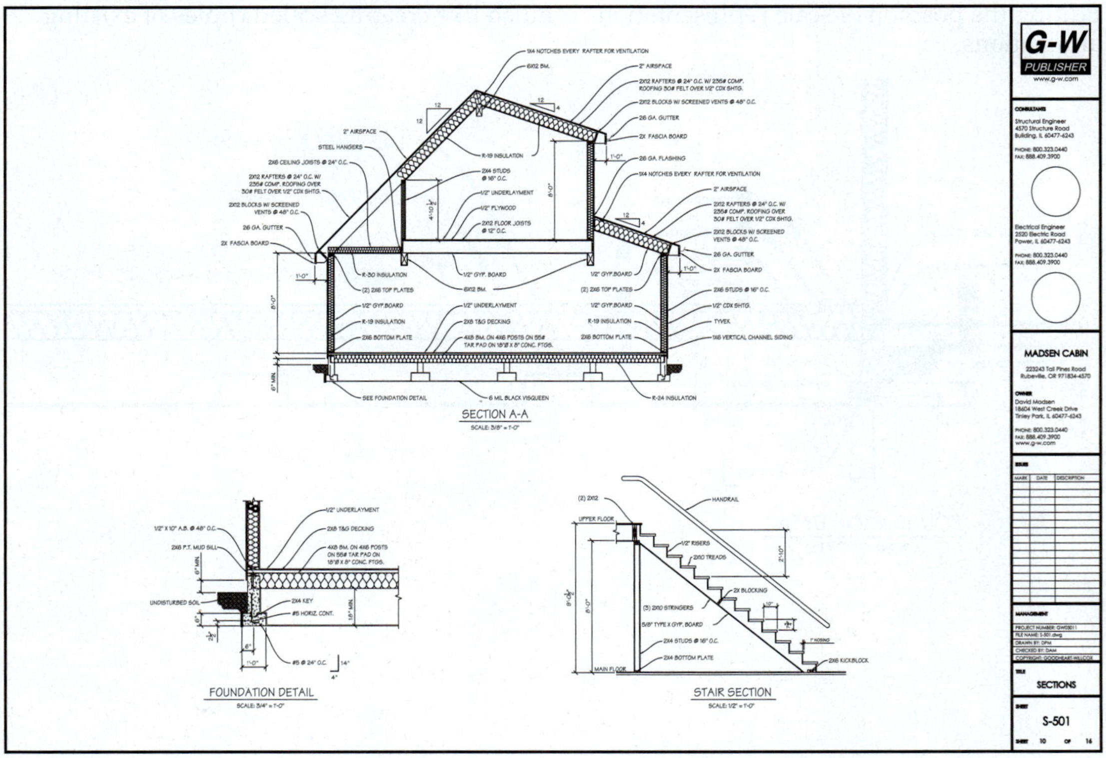

Exercise 30-6

www.g-wlearning.com/CAD/

Complete the exercise on the companion website.

Chapter Review

Answer the following questions using the information in this chapter.

1. What are annotative objects?
2. Explain the practical differences between manual and annotative object scaling.
3. Identify at least four types of objects that you can make annotative.
4. How do you set text scale, including spacing, width, and paragraph settings, to adjust automatically according to the current annotation scale?
5. Explain why it is important for the viewport scale to match the annotation scale.
6. Which **MSLTSCALE** system variable setting should you use so you do not have to calculate the drawing scale factor when entering an **LTSCALE** value?
7. Name the command used to update text properties according to the current properties of the text style on which the text is drawn.
8. What is an annotative object representation?
9. Describe the result of setting the **ANNOAUTOSCALE** system variable to a value of 4.
10. Explain the effect of turning annotation visibility on and off.

Drawing Problems

Start AutoCAD if it is not already started. Start a new drawing for each problem using an appropriate template of your choice. The template should include layers and text, dimension, multileader, and table styles, when necessary, for drawing the given objects. Add layers and text, dimension, multileader, and table styles as needed. Draw all objects using appropriate layers. Use appropriate text, dimension, multileader, and table styles, justification, and format. Follow the specific instructions for each problem. Use only drawing and editing commands and techniques you have already learned. Use your own judgment and approximate dimensions when necessary. Apply dimensions accurately using ASME or appropriate industry standards.

Note: Some of the problems in this chapter are built on problems from previous chapters. If you have not yet completed those problems, complete them now.

▼ Basic

1. **General** Save P23-9 as P30-1. If you have not yet completed Problem 23-9, do so now. In the P30-1 file, convert all the non-annotative objects to annotative objects.

2. **General** Save P24-10 as P30-2. If you have not yet completed Problem 24-10, do so now. In the P30-2 file, convert all of the non-annotative objects to annotative objects.

▼ Intermediate

3. **Mechanical** Draw the section view and side view shown. Use annotative objects to prepare a full-scale drawing of the part. Change the annotation scale to 2:1 and adjust the scale representations as needed according to the new scale. Save the drawing as P30-3.

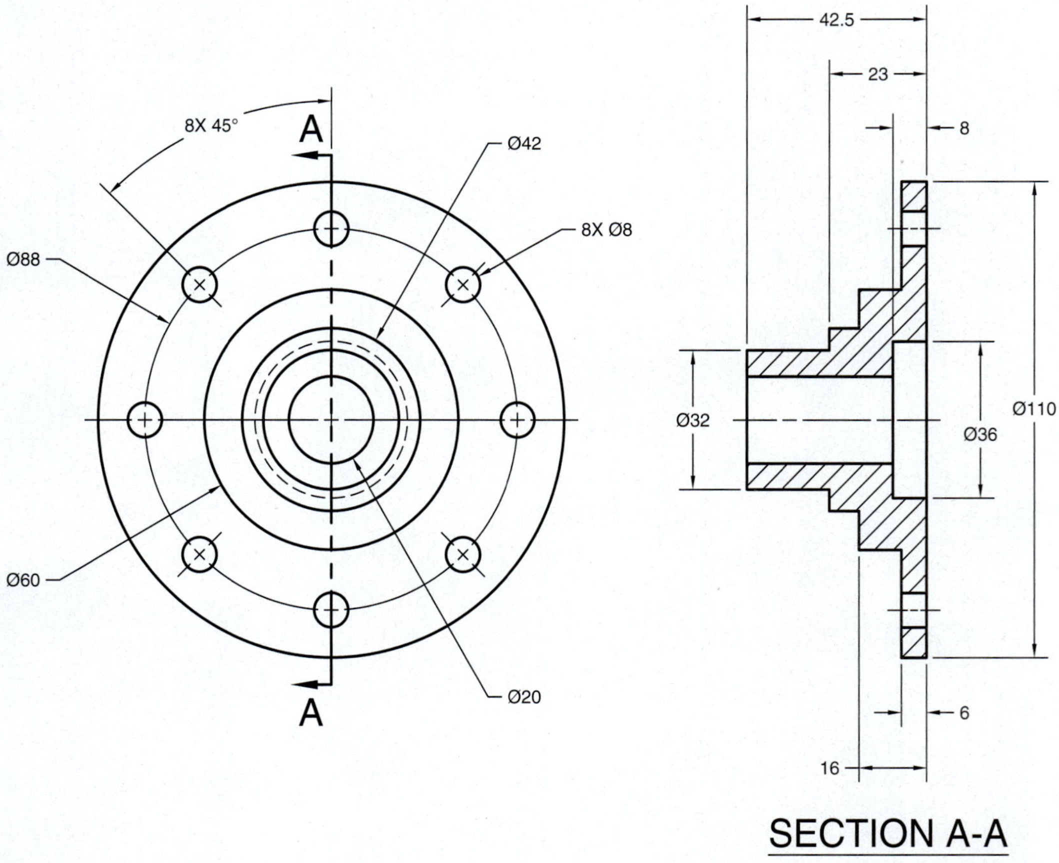

4. **Mechanical** Draw the section view and side views shown. Use annotative objects to prepare a full-scale drawing of the part. Change the annotation scale to 2:1 and adjust the scale representations as needed according to the new scale. Save the drawing as P30-4.

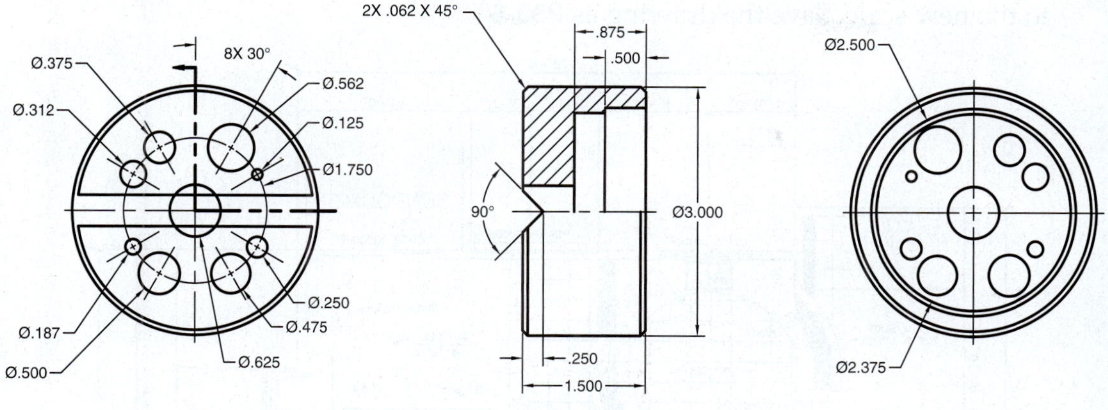

Name: Diffuser
Material: AISI 1018

5. **Mechanical** Draw the fan shown at full scale in model space. Use annotative objects to prepare a full-scale view of the fan as shown and a view enlargement of the motor. You should not have to create a copy of the motor or develop scale-specific layers. Save the drawing as P30-5.

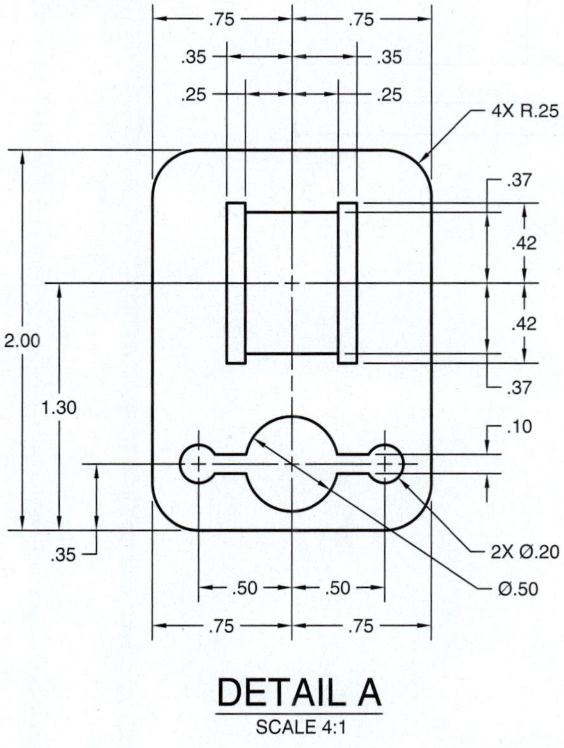

Motor Detail

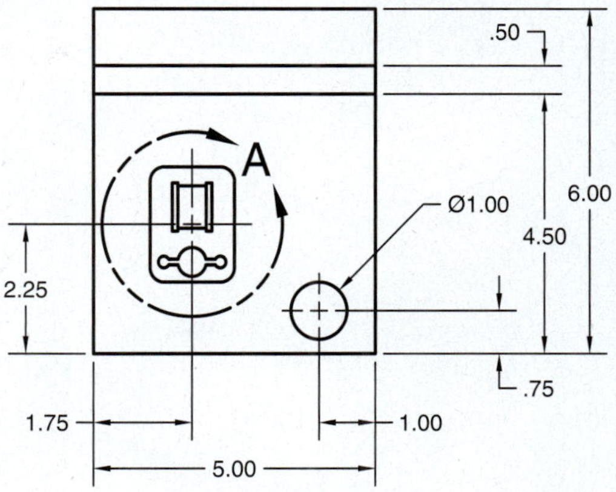

Fan

▼ Advanced

6. **Architectural** Draw the floor plan shown at full scale in model space. Use annotative objects to prepare a 1/4″ = 1′-0″ view. Change the annotation scale to 1/8″ = 1′-0″ and adjust the annotative object representations as needed according to the new scale. Save the drawing as P30-6.

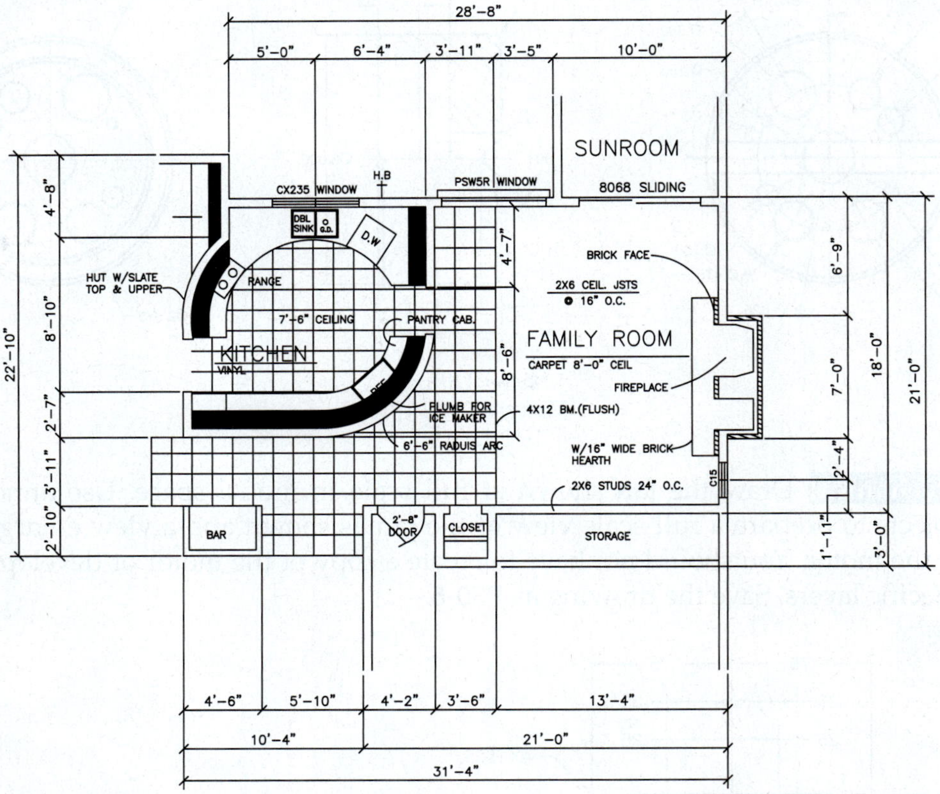

7. **Mechanical** Draw the part shown at full scale in model space. Use annotative objects to prepare the full-scale view and the view enlargement shown. You should not have to create a copy of the part or develop scale-specific layers. Plot the layout. Save the drawing as P30-7.

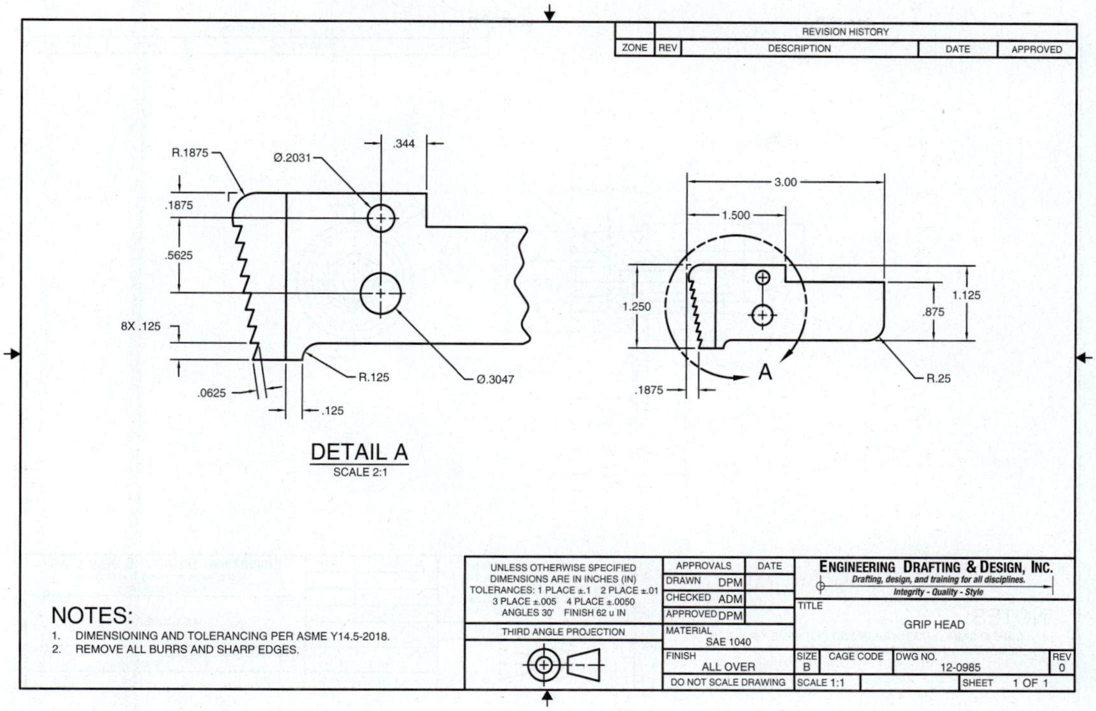

8. Mechanical Draw, dimension, and plot the male insert shown using a layout. Use annotative objects. Use a larger sheet and annotative object scaling to plot a 4:1 scale drawing of the insert. You should not have to create a copy of the part or develop scale-specific layers. Save the drawing as P30-8.

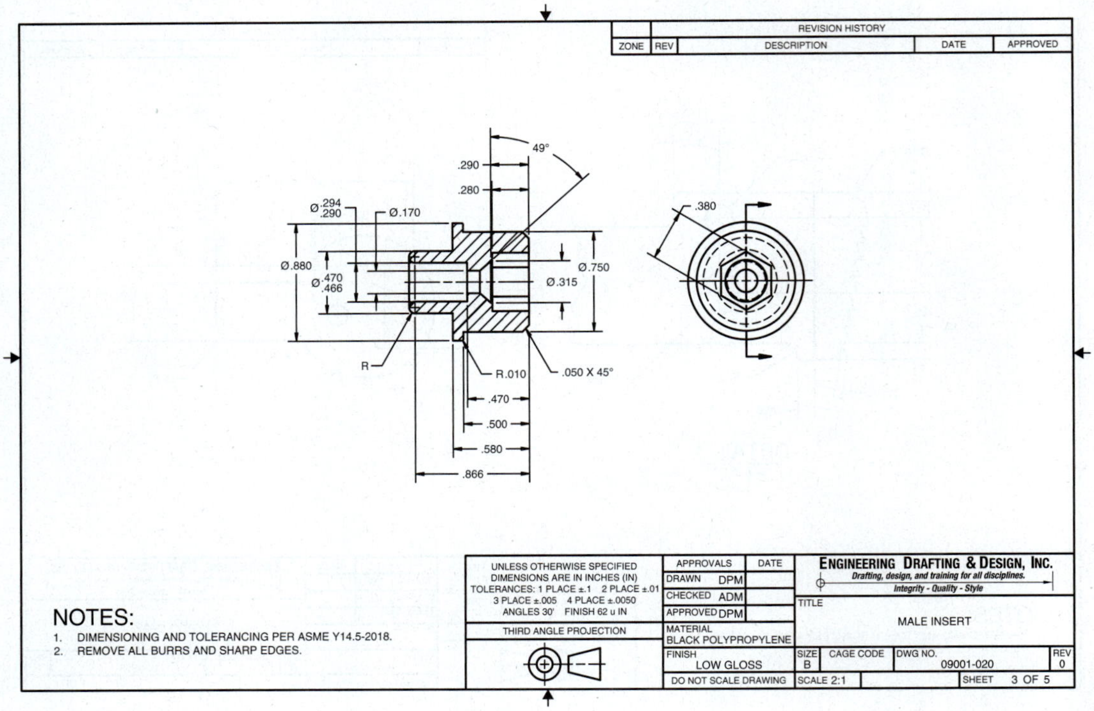

9. Mechanical Draw the sheet metal flat pattern shown. Design the shape of punch A using your own judgment and dimensions proportionate to other part features. Use annotative objects to prepare the full-scale view of the flat pattern and the view enlargement of the punch. Dimension the punch. You should not have to create a copy of the part or develop scale-specific layers. Plot the layout. Save the drawing as P30-9.

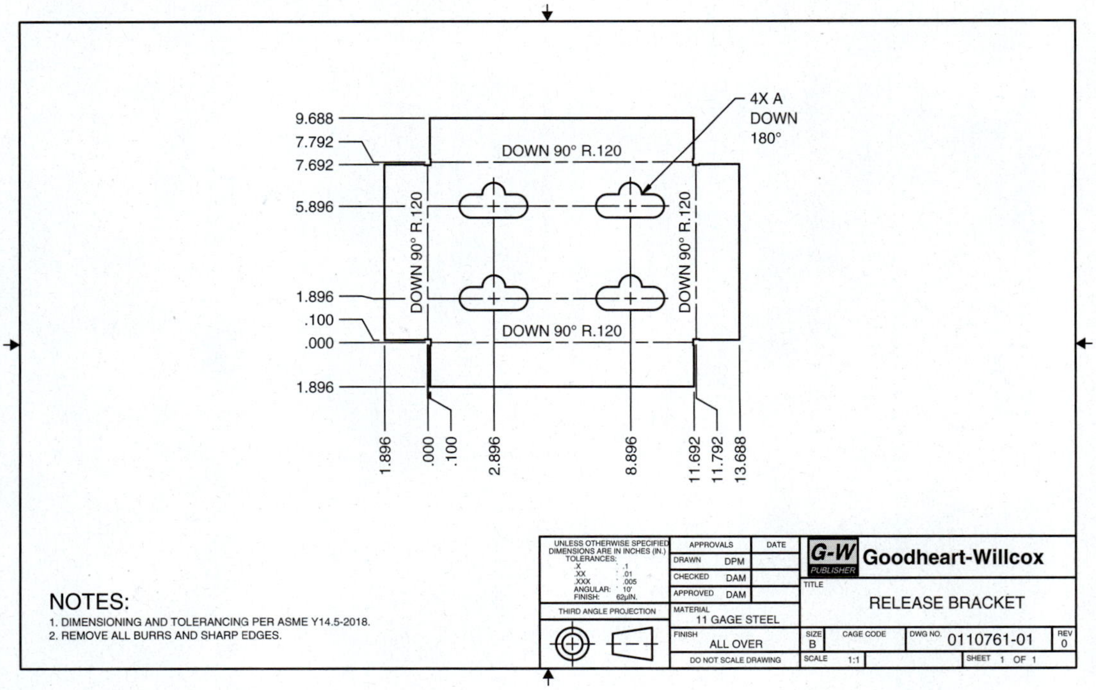

10. **Mechanical** Research the design of an existing pair of paper-cutting scissors consisting of at least three separate parts. Create freehand, dimensioned 2D sketches of the existing design from manufacturer's specifications, or from measurements taken from actual scissors. Start a new drawing from scratch or use a decimal-unit template of your choice. Prepare detail drawings of each part and an assembly drawing with balloons and a parts list. Use annotative objects and separate layouts for each drawing to prepare a set of drawings in one file. Plot each drawing. Save the drawing as P30-10. Prepare a slide show presentation of your research, design process, and drawing and present it to your class or office.

11. **Architectural** Create a freehand, dimensioned 2D sketch of a complete floor plan for a home with three bedrooms and two bathrooms. Start a new drawing from scratch or use an architectural template of your choice. Draw, dimension, and plot the floor plan from your sketch using a layout. Use annotative objects. Save the drawing as P30-11. Prepare a slide show presentation of your research, design process, and drawing and present it to your class or office.

12. **Civil** Obtain a hard copy of a plot plan of a small residential subdivision, or a portion of a subdivision. Draw, dimension, and plot the subdivision using a layout. Use annotative objects. Save the drawing as P30-12.

AutoCAD Certified Professional Exam Practice

Answer the following questions using the information in this chapter.

1. A drawing of a mechanical part is set up to be printed at a scale of 2:1 in a layout in paper space. The length of one feature on the part is 35.125″. How long should that feature be drawn in model space? *Select the one item that best answers the question.*
   A. 3.513″
   B. 17.563″
   C. 35.125″
   D. 70.250″
2. Which of the following statements are true about blocks? *Select all that apply.*
   A. Attributes included in an annotative block automatically become annotative even if they are not set to be annotative.
   B. Attributes included in an annotative block change according to the annotation scale, but the block remains fixed.
   C. You can make a block annotative when you originally create the block.
   D. You can make a block annotative after its creation by using the **Properties** palette.
3. How can you change the overall annotation scale of a layout? *Select all that apply.*
   A. right-click and select an annotation scale from the cascading list
   B. select **Scale List** in the **Annotation Scaling** panel of the **Annotate** ribbon tab
   C. use the **Annotation Scale** flyout
   D. use the **Properties** palette
   E. use the **Viewport Scale** flyout

Follow the instructions in the problem. Complete the problem by answering the related question.

4. **Navigate to this chapter on the companion website and open CPE-30annoscale.dwg.** Change the annotation scale to 1/2″ = 1′-0″ to display annotative hatch patterns created at that scale. What hatch pattern is used for the undisturbed soil in this drawing?

CHAPTER 31

External References

Learning Objectives

After completing this chapter, you will be able to:

✓ Explain the function of external references.
✓ Prepare drawings for xref insertion.
✓ Attach existing drawings to the current drawing.
✓ Work with xrefs and dependent objects.
✓ Manage multiple xrefs in a drawing.
✓ Bind external references and selected dependent objects to a drawing.
✓ Edit external references in the current drawing.

External references (xrefs) expand on the concept of reusing existing content in AutoCAD. Xrefs provide an effective way to relate existing base drawings, complex symbols, images, and details to other drawings. Xrefs also help multiple users share content. This chapter focuses on using xref drawings and provides common xref applications.

external reference (xref): A DWG, raster image, DWF, DWFx, DGN, PDF, Navisworks (NWD or NWC), or point cloud file incorporated into a drawing for reference only.

Introduction to Xrefs

An xref is a drawing (DWG), raster image, design web format (DWF or DWFx), design (DGN), portable document format (PDF), point cloud (RCP or RCS), or Navisworks® (NWD or NWC) file that you reference into a ***host drawing***. Inserting an xref is similar to inserting an entire drawing as a block. However, unlike a block, which is actually stored in the file in which you insert the block, the file geometry in a ***reference file*** is not added to the host drawing. File data appears on-screen for reference only. The result is usable information, but the host file remains much smaller than if you insert a block or copy and paste objects. Xrefs are also easier to manage in a host drawing than blocks or pasted objects.

host drawing: The drawing into which xrefs are incorporated.

reference file: An xref; a file referenced by the host drawing.

Another benefit of using xrefs is the link between reference and host files. Any changes you make to reference files are reflected in host drawings, so the host drawings display the most recent reference content. AutoCAD reloads each xref whenever the host drawing loads. This allows you or a design drafting team to work on a multi-file project, with the assurance that any revisions to reference files are displayed in host drawings.

NOTE

Add as many xrefs as needed for a drawing at any time during the drawing process. You can also include xrefs within referenced files. The host drawing updates to recognize each new xref.

Xref Files

Files that you can reference into a current drawing include DWG, raster image, DWF, DWFx, DGN, PDF, RCP, RCS, NWD, and NWC files. DWF and DWFx files are drawings compressed for publication, viewing, and mark-up using a viewer, such as the Autodesk® Design Review software. DWF and DWFx files are commonly used to share drawings with members of a design drafting team who do not use AutoCAD. The high compression also makes DWF and DWFx files easy to transmit electronically.

Reference a raster image file for any application that requires adding an image to a drawing, such as for a company logo in a title block. A DGN file is native to software such as MicroStation® and is referenced to work with DGN file data. PDF file reference allows you to work with PDF file content. An NWD or NWC file is native to Navisworks software and is referenced to work with NWD or NWC file data. Reference RCP or RCS files to work with point clouds, as described in ***AutoCAD and Its Applications—Advanced***.

NOTE

IMPORT

Application Menu
Import
> Other Formats
Ribbon
Insert
> Import
Import
Type
IMPORT

AutoCAD includes an **IMPORT** command that allows you to import data from any of several different file formats, including DGN and PDF files. Unlike using an xref, importing typically converts data from the selected file into AutoCAD objects. If you import data that includes raster images, such as raster images imported with a PDF file, the raster images import as xrefs.

Xref Applications

DWG files are the most common xref files and are the focus of this chapter. The term *xref* often applies specifically to referenced DWG files. In general, use xrefs to reuse existing drawing information and help develop other drawings. There are countless applications for xref drawings in every drafting field. The following sections provide typical xref drawing applications. As you work with AutoCAD, you will discover a variety of uses for xref drawings.

Reference Existing Geometry

One of the most common applications for xref drawings is to reference existing geometry to use as a pattern or source of needed information in the host drawing. For example, a floor plan includes size and shape information required to prepare additional plans, elevations, sections, and details. Figure 31-1 shows an example of referencing a floor plan file into a new drawing to use as an outline for creating a roof plan file.

Figure 31-2 shows an example of the roof plan file created in Figure 31-1 attached to a new drawing as an xref and then used to project an elevation. The roof plan xref includes a ***nested xref*** of the floor plan. In this example, the elevation file references the roof plan. The roof plan in turn references the floor plan.

nested xrefs: Xrefs contained within other xrefs.

Create a Multiview Drawing

Another xref application is to create commonly used drawings, such as sections and details, as separate drawing files and then attach each drawing as an xref to a host drawing, known as the ***master drawing***. Use floating viewports and layer viewport

master drawing: A host drawing created by attaching several frequently used xrefs.

Figure 31-1.
Using a floor plan xref drawing as a pattern, or outline, to draw a roof plan.

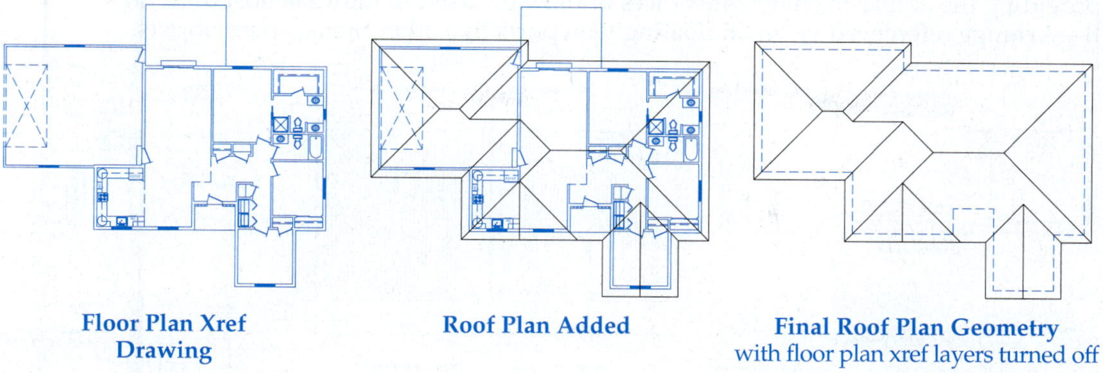

Figure 31-2.
Using a roof plan xref drawing that contains a nested floor plan xref drawing as a pattern for projecting geometry needed to create an elevation. The projection lines, drawn using the **XLINE** command, are for reference.

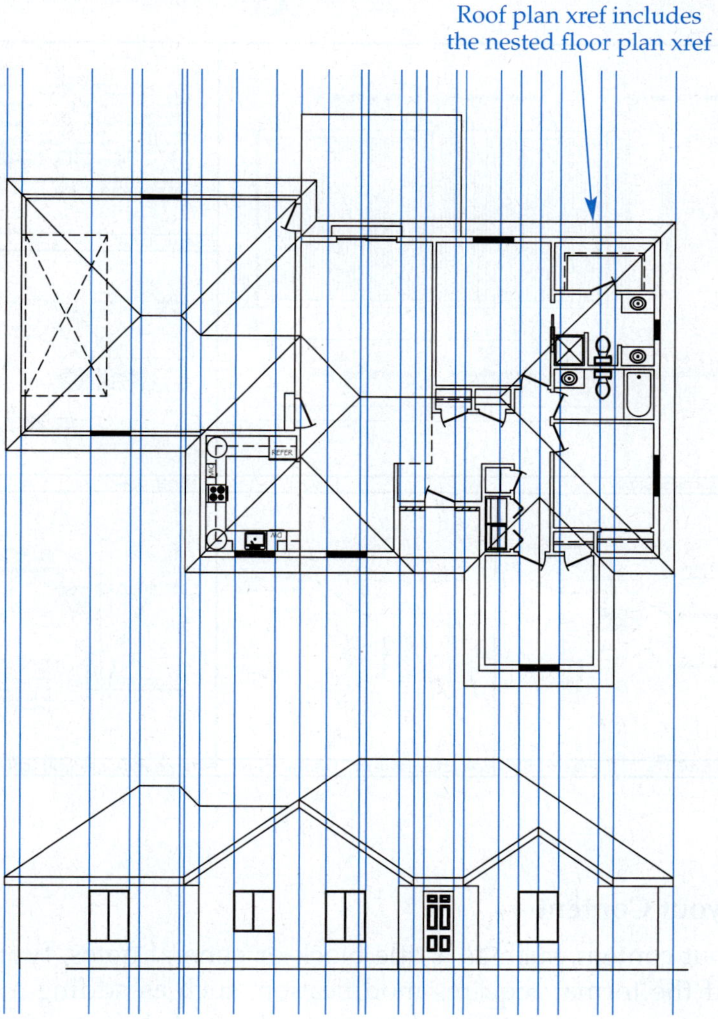

freezing to create a multiview layout. You can prepare a multiview drawing entirely from existing xref drawings or from a combination of objects created "in place" in the master drawing and attached xrefs.

Figure 31-3A shows an example of five stock details referenced into the model space environment of a new drawing. Floating viewports arrange the details in a layout, as shown in **Figure 31-3B**. When you make changes to details in the referenced detail files, the files are updated in the host file.

Figure 31-3.
A—Xref frequently used drawing views into model space, reducing the size of the file and providing the ability to change instances of the view used in multiple host drawings.
B—Arrange referenced views in floating viewports like other model space objects.

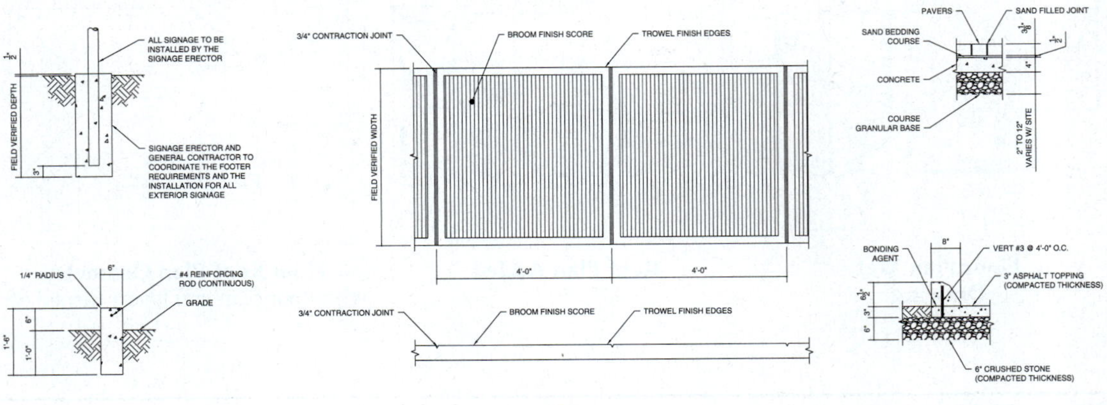

A

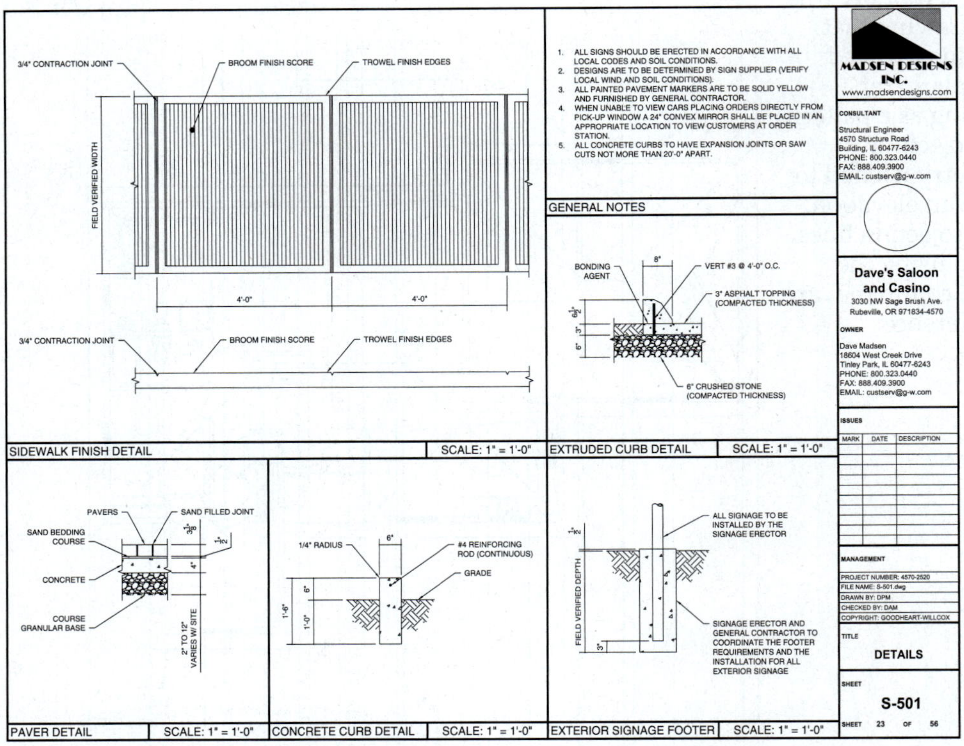

B

Add Layout Content

Layout content, such as a title block or general notes, typically has a standard format. If the format requires modification, such as adding a new note to a list of general notes, you can make changes to the xref drawing and easily update each host file that references the xref. This is the same concept as using xref drawings to build a multiview drawing. **Figure 31-3B** shows general notes added to the layout as an xref.

Arrange Sheet Views

You can use external references to arrange sheet views in layouts when you are working with sheet sets. Chapters 32 and 33 describe sheet sets.

PROFESSIONAL TIP

You can use blocks, wblocks, and non-block objects in a manner similar to xref drawings. However, xrefs are much easier to manage, provide greater flexibility, and update automatically in host drawings. Always use xref drawings if you plan to use the drawing in multiple host drawings, and if the design might change.

Preparing Xref and Host Drawings

Before you place an xref drawing, you should prepare the xref and host drawing files for xref insertion. When you place an xref drawing, everything you see in model space is inserted into the host file as a single item on a specified layer. Layout content is not included. The default insertion base point for an xref file is the model space origin, or 0,0,0. If it is critical that xref objects coincide with the 0,0,0 point for insertion, move all objects in model space as needed in the xref file.

An alternative to moving objects to the origin is to use the **BASE** command to change the insertion base point of the drawing. Access the **BASE** command and select a new insertion base point. Save the drawing before using it as an xref.

BASE
Ribbon
Home
> Block
Insert
> Block Definition
Set Base Point
Type
BASE

The host file should include a unique layer (named XREF or A-ANNO-REFR, for example) assigned to xrefs. By default, the **XREFLAYER** system variable is set to assign the current layer to xref drawings. Make the appropriate layer current and proceed to place the xref drawing. To place xref drawings on a specified layer instead of the current layer, access the **XREFLAYER** system variable and type a layer name. AutoCAD adds the layer to the host drawing if it is not found. Type a period (.) to return to using the current layer.

Placing Xref Drawings

To place an xref, access the **ATTACH** command to display the **Select Reference File** dialog box. The **Drawing (*.dwg)** option in the **Files of type:** drop-down list limits the files displayed in the dialog box to drawings. Use the **Select Reference File** dialog box to locate the drawing file to add to the host file as an xref. Then pick the **Open** button to display the **Attach External Reference** dialog box. See Figure 31-4.

ATTACH
Ribbon
Insert
> Reference
Attach
Type
ATTACH

The **Attach External Reference** dialog box includes options for specifying how and where to place the selected file in the host drawing as an xref. If an external reference already exists in the current drawing, place another copy by choosing the file from the **Name:** drop-down list. To place a different xref drawing, pick the **Browse...** button and select the new file in the **Select Reference File** dialog box.

You can also place an xref using the **External References** palette shown in Figure 31-5. The **External References** palette is a complete external reference management tool. To place an xref drawing using the **External References** palette, pick the **Attach DWG** button from the **Attach** flyout, or right-click on the **File References** pane and select **Attach DWG...**. The **Select Reference File** dialog box appears, displaying only drawing files. Locate and select a file to add as an xref and pick the **Open** button to display the **Attach External Reference** dialog box.

EXTERNALREFERENCES
Ribbon
Insert
> Reference
External References
Type
EXTERNAL-REFERENCES
XREF
XR

NOTE

The **XATTACH** command is identical to the **ATTACH** command, but it initially displays only drawing files in the **Select Reference File** dialog box.

XATTACH
Type
XATTACH
XA

Figure 31-4.
Use the **Attach External Reference** dialog box to specify how to place an xref in the host drawing. Pick the **Show Details** button to display additional file details, as shown.

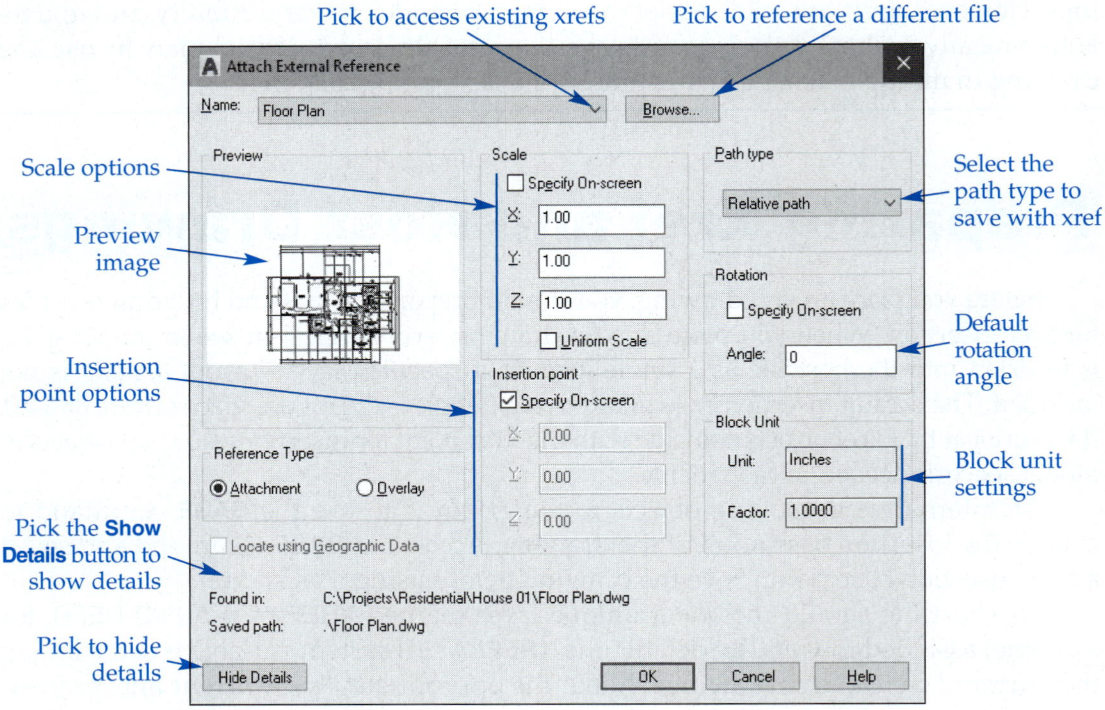

Figure 31-5.
The **External References** palette provides access to all options for externally referenced files.

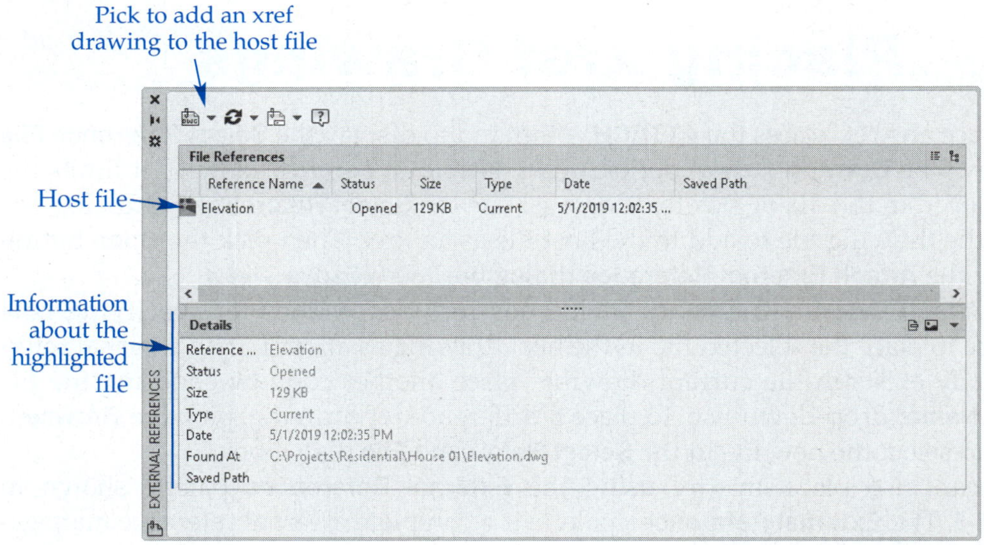

Attachment vs. Overlay

attachment: An xref linked with or referenced into the current drawing.

overlay: An xref displayed in the host drawing, but not attached to it.

You can choose to insert an xref drawing as an ***attachment*** or an ***overlay*** by selecting the **Attachment** or **Overlay** radio button in the **Reference Type** area. Attach xrefs for most applications. An xref overlay allows you to share content with others in a design drafting team, typically while working in a networked environment. You can overlay drawings without referencing nested xrefs.

Nesting occurs when an xref file references another xref file. An attached xref that has nested xrefs is the ***parent xref***. When you attach an xref, the host drawing receives any nested xrefs that the xref contains. The difference between an attached xref and an overlaid xref is related to the way in which nested xrefs are handled. If you overlay an xref in a host drawing and then attach the host drawing to the current drawing, the overlaid xref does not appear in the current drawing.

parent xref: An xref that contains one or more other xrefs.

For example, suppose you attach a floor plan xref to a host file to create a foundation plan, and then overlay the foundation plan xref to a host file to draw a section. Overlaying the foundation plan brings the foundation and floor plan geometry into the section file for reference. If a member of your design drafting team uses your section, or is working on a drawing that already has the floor plan, foundation plan, or both plans attached, she or he can attach or overlay the section xref into a drawing without bringing in the floor plan and foundation plan.

NOTE

You can change an overlay to an attachment or an attachment to an overlay after insertion. Managing xrefs is described later in this chapter.

Selecting the Path Type

Use the **Path type** drop-down list in the **Path type** area to set how AutoCAD stores the path to the xref file. The path locates the xref file when you open the host file. The path appears in the **Attach External Reference** dialog box when you pick the **Show Details** button, and later appears in the **External References** palette. See **Figure 31-6**.

The **Full path** option saves an ***absolute path***. When using the **Full path** option, you must locate xref drawings in the drive and folder specified in the saved path. You can move the host drawing to any location, but the xref drawings must remain in the saved path. This option is acceptable if it is unlikely that you will move or copy the host and xref drawings to another computer, drive, or folder.

absolute path: A path to a file defined by the location of the file on the computer system.

The default **Relative path** option is often appropriate if you share drawings with a client or eventually archive drawings. The **Relative path** option saves a ***relative path***. If the host drawing and xref files are located in a single folder and subfolders, you can copy the

relative path: A path to a file defined according to the location of the file relative to the host drawing.

Figure 31-6.
You can reference a file using a full path, a relative path, or no path. The path type is displayed in the **Saved Path** column in the **External References** palette.

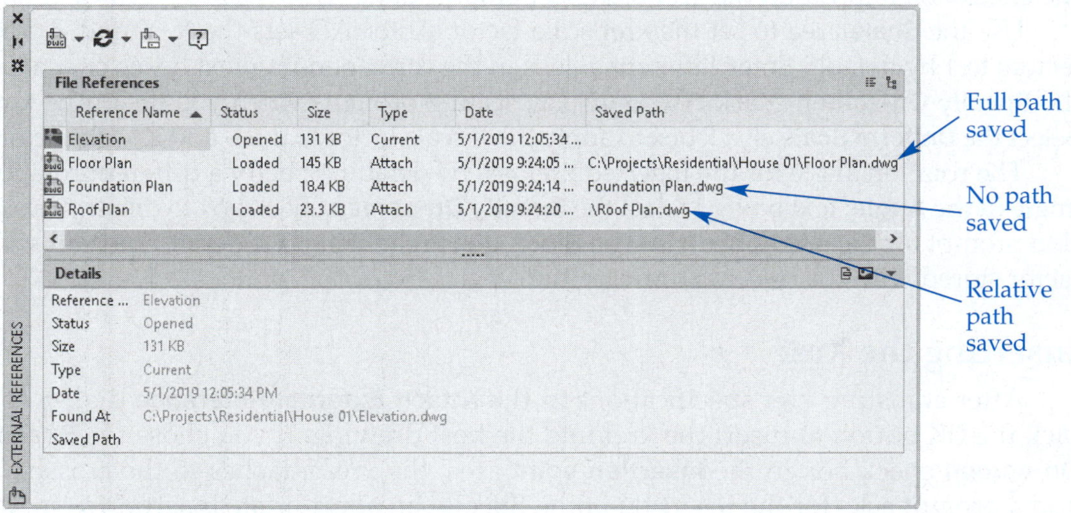

folder to any location without losing the connection between files. For example, copy the folder from the C: drive of one computer to the D: drive of another computer, to a folder on a USB drive, or to an archive server. If you perform these types of transfers with the path type set to **Full path**, you need to open the host drawing after copying and redefine the saved paths for all xref files. You cannot use a relative path if the xref file is on a drive other than the drive on which the host file is stored. A full path preceded by an asterisk in the **Saved Path** column of the **External References** palette indicates that the xref path type is set to relative, but the host file has not previously been saved. Once you save the host file, the asterisk disappears and the path is shown as relative. If you save a copy of the host file to a different location, AutoCAD prompts you to update relative paths.

Select the **No path** option if you do not want to save the path to the xref file. If you choose the **No path** option, the xref file loads only if you include the path to the file in one of the Support File Search Path locations or if the xref file is in the same folder as the host file. Specify the Support File Search Path locations in the **Files** tab of the **Options** dialog box.

NOTE

AutoCAD searches for xref files in all paths of the current project name. Project search paths are listed under Project Files Search Path in the **Files** tab of the **Options** dialog box. Create a new project as follows:

1. Pick Project Files Search Path to highlight it, and then pick the **Add...** button.
2. Enter a project name.
3. Pick the plus sign icon (+), and then pick the word **Empty**.
4. Pick the **Browse...** button and locate the folder that is to become part of the project search path. Then pick **OK**.
5. Complete the project search path definition by entering the **PROJECTNAME** system variable and specifying the same name you specified in the **Options** dialog box.

Additional Xref Placement Options

The remaining items in the **Attach External Reference** dialog box allow you to control or identify xref insertion location, scaling, rotation angle, and block unit settings. Deselect the **Specify On-screen** check box in the **Insertion point** area to enter 2D or 3D coordinates in the text boxes for insertion of the xref. Activate the **Specify On-screen** check box to specify the insertion location on-screen. The **Locate using Geographic Data** check box is active if the xref and host drawings include ***geographic data***. Pick the check box to position the xref using geographic data.

geographic data: Information added to a drawing to describe specific locations and directions on Earth.

Use the **Scale** area to set the xref scale factor. AutoCAD sets the X, Y, and Z scale factors to 1 by default. Enter different values in the corresponding text boxes or activate the **Specify On-screen** check box to display scaling prompts when you insert the xref. Select the **Uniform Scale** check box to apply the X scale factor to the Y and Z scale factors.

The rotation angle for the inserted xref is 0 by default. Specify a different rotation angle in the **Angle:** text box, or select the **Specify On-screen** check box to display a rotation prompt for the rotation angle. The **Block Unit** area displays the unit type and scale factor stored with the selected drawing file.

Inserting the Xref

After adjusting xref specifications in the **Attach External Reference** dialog box, pick the **OK** button to insert the xref into the host drawing. If you chose the **Specify On-screen** check box in the **Insertion point** area, the xref attaches to the crosshairs and a prompt asks for the insertion point. Specify an appropriate insertion point for the xref.

The options for attaching an xref are essentially the same as those for inserting a block. However, remember that xref geometry is not added to the database of the host file, as are inserted blocks. Therefore, using external references helps keep your drawing file size to a minimum.

Exercise 31-1

www.g-wlearning.com/CAD/

Complete the exercise on the companion website.

Placing Xrefs with DesignCenter and Tool Palettes

To place an xref into the current drawing using **DesignCenter**, first use the **Tree View** pane to locate the folder containing the drawing you want to attach. Then display the drawing files located in the selected folder in the **Content** pane. Right-click on the drawing file in the **Content** pane and select **Attach as Xref...**. Another method is to drag and drop the drawing into the current drawing area using the right mouse button. When you release the button, select the **Create Xref** option.

You must add an xref or drawing file to a tool palette in order to use the **Tool Palettes** window to place the file as an xref. To add an xref to a tool palette, drag an existing xref from the current drawing or an xref from the **Content** pane of **DesignCenter** into the **Tool Palettes** window. Use drag and drop to attach the xref to the current drawing from the palette. Xref files in tool palettes display an external reference icon.

A drawing file (not an xref) added to a tool palette from the current drawing or using **DesignCenter** is a block tool. To convert the block tool to an xref tool, right-click on the image in the **Tool Palettes** window and select **Properties...** to display the **Tool Properties** dialog box. Then change the **Insert as** field status from **Block** to **Xref** using the **Insert as** drop-down list. The **Reference type** row controls whether the xref is inserted as an attachment or an overlay.

Working with Xref Objects

An xref is inserted as a single external reference object. Xref drawings appear faded 50% by default to help differentiate the xref from the host drawing. Xref fading is an on-screen display function only and does not apply to plots. The **XDWGFADECTL** system variable controls fading of xref drawings on-screen. An easy way to adjust fading is to use the options in the expanded **Reference** panel of the **Insert** ribbon tab. See Figure 31-7A. Pick the **Xref Fading** button to activate or deactivate xref fading, and use the slider or text box to increase or decrease fading. See Figure 31-7B.

Select an xref in the **External References** palette to highlight all visible instances of the xref in the drawing. Use editing commands such as **MOVE** and **COPY** to modify the xref as needed. However, there are some significant differences between xrefs and other objects. For example, if you erase an xref, the xref definition remains in the file, similar to an erased block. You must detach an xref, as explained later in this chapter, to remove it from the file completely.

NOTE

To select an xref, pick an object displayed on-screen that is part of the xref.

Figure 31-7.
A—Use options in the expanded **Reference** panel of the **Insert** ribbon tab to control xref fading. B—Default fading applied to xref objects.

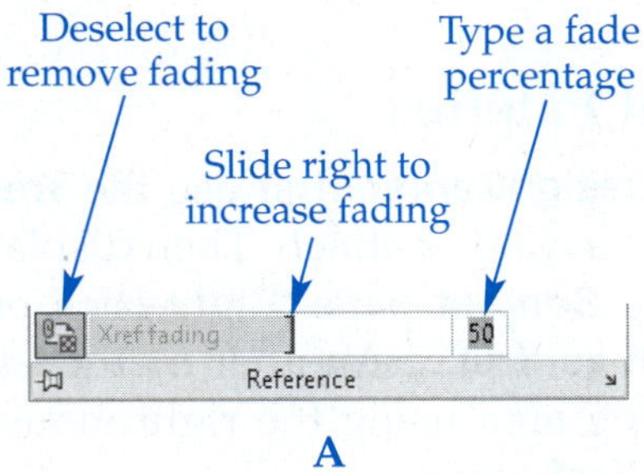

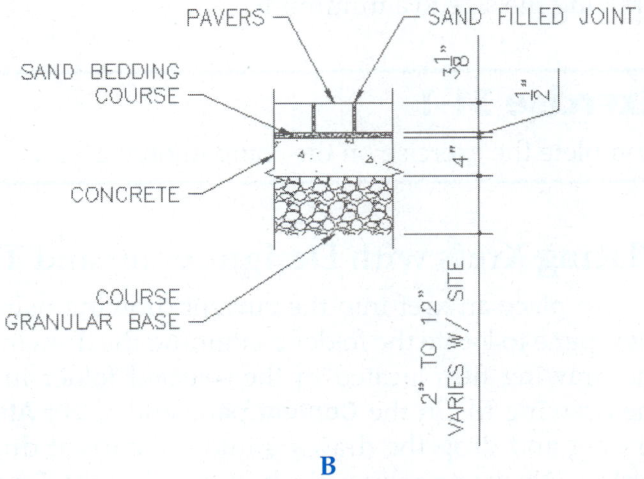

Dependent Objects

dependent objects: Objects displayed in the host drawing, but defined in the xref drawing.

When you place an xref, the host file receives all named objects in the xref file, such as layers and blocks, as ***dependent objects***. Dependent objects are displayed in the host drawing for reference only. The xref drawing stores the actual object definitions. Binding, explained later in this chapter, allows you to add dependent objects to the host drawing for use.

Working with Dependent Layers

When you attach an xref, dependent objects are assigned unique names that consist of the xref file name followed by the actual object name, separated by a vertical bar symbol (|). For example, a layer named A-DOOR in a reference drawing named Floor Plan comes into the host drawing as Floor Plan|A-DOOR. See Figure 31-8. This name, and the light gray color used to display the name, distinguish xref-dependent layers from layers in the host drawing. The names also make it easier to manage layers when several xrefs are attached to the host drawing, because the layers from each reference file are preceded by their file names. You cannot rename xref-dependent objects.

NOTE

Use the Xref filter in the **Layer Properties Manager** to display and manage dependent layers. You can also save dependent layers in a layer state.

You cannot set xref-dependent layers current. However, you can override all xref-dependent layer properties as necessary, except for the name. For example, if you reference a drawing to check the relationship of objects between two drawings, consider changing the xref-dependent colors to make it easier to differentiate between the content of the host drawing and the xref drawing. Changing xref-dependent properties affects the display only in the host drawing and does not alter the original reference file.

AutoCAD includes several utilities to help you recognize and work with xref-dependent layer overrides. Figure 31-8 shows options available in the **Layer Properties Manager**. Use the Xref Overrides filter to display and manage only dependent layers with property overrides. A sheet and external reference status icon in the **Status** column identifies a dependent layer with property overrides. Hover over the status icon to display a tooltip that lists the overrides. Pick the **Toggle Override Highlight** button to highlight the names of layers with overrides and the overridden properties.

Figure 31-8.
The xref drawing name and a vertical bar symbol (|) precede xref-dependent layer names in the host drawing. Notice how xref-dependent layers are also light gray in color.

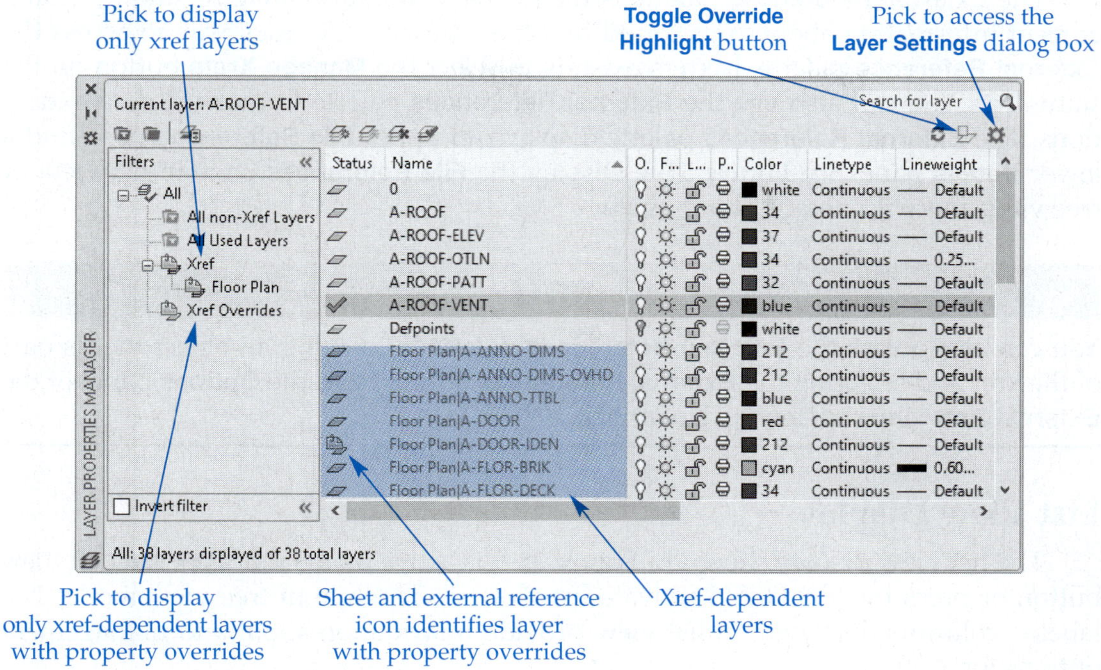

One option to remove dependent layer property overrides from all layers is to right-click on the Xref Overrides filter and select the properties to reset from the **Reset Xref Layer Properties** cascading menu. The menu is also available when you right-click on a layer name and pick **All layers** from the **Reset Xref Layer Properties for** cascading menu. To remove dependent layer property overrides from specific layers, select the layer names to reset and right-click and pick **Selected layers** from the **Reset Xref Layer Properties for** cascading menu.

The **Xref Layer Settings** area in the **Layer Settings** dialog box provides options for controlling xref-dependent layers. The **Retain overrides to xref layer properties** radio button is selected by default and maintains changes to xref-dependent layer properties. Property overrides remain even after you reload the xref. Check the **Xref layer properties to reload (except overrides)** check box and use the property check boxes to select specific properties to reset, except overridden properties, when you reload the xref. Pick the **Don't retain overrides to xref layer properties** radio button to discard xref-dependent overrides when you reload the xref. If you assign absolute properties, instead of ByLayer, to an object in the xref file, check the **Treat xref object properties as ByLayer** check box to force the object to appear as ByLayer.

NOTE

The Layer property overrides property in the **Misc** category of the **Properties** palette indicates whether a selected xref object includes xref-dependent property overrides. To determine whether a floating viewport contains an xref object with xref-dependent property overrides, hover over the viewport boundary to display the object properties.

Exercise 31-2

www.g-wlearning.com/CAD/

Complete the exercise on the companion website.

Managing Xrefs

Status Bar

Manage Xrefs

The **External References** palette is the primary tool for managing and accessing current information about xrefs found in a host drawing. A quick way to access the **External Reference** palette, if xrefs exist, is to select the **Manage Xrefs** button on the status bar. You can also use the **External References** palette to manage data extractions. The **External References** palette displays an upper **File References** pane and a lower **Details** pane. See Figure 31-9. Display the **File References** pane in list view or tree view and with details or a preview.

NOTE

You can also access the **External References** palette by picking an object that is part of the xref and selecting the **External References** button from the **Options** panel of the **External Reference** contextual ribbon tab.

List View Display

The list view display shown in Figure 31-9 is active by default. Pick the **List View** button or press the [F3] key to activate list view mode while in tree view mode. The labeled columns displayed in list view provide information about and management options for xrefs.

The **Reference Name** column displays the current drawing file name followed by the names of all existing xrefs in alphabetical or chronological order. The standard AutoCAD drawing file icon identifies the host drawing, and a sheet of paper with a paper clip icon identifies xref drawings. Each xref type has a different icon. The **Status** column describes the status of each xref, which can be:

- **Loaded.** The xref is attached to the drawing.
- **Unloaded.** The xref is attached but cannot be displayed or regenerated.
- **Unreferenced.** The xref has nested xrefs that are not found or are unresolved. An unreferenced xref is not displayed.

Figure 31-9.
The **External References** palette allows you to view and manage referenced files. The **File References** pane appears in the **List View** mode and the **Details** pane appears in the **Details** mode.

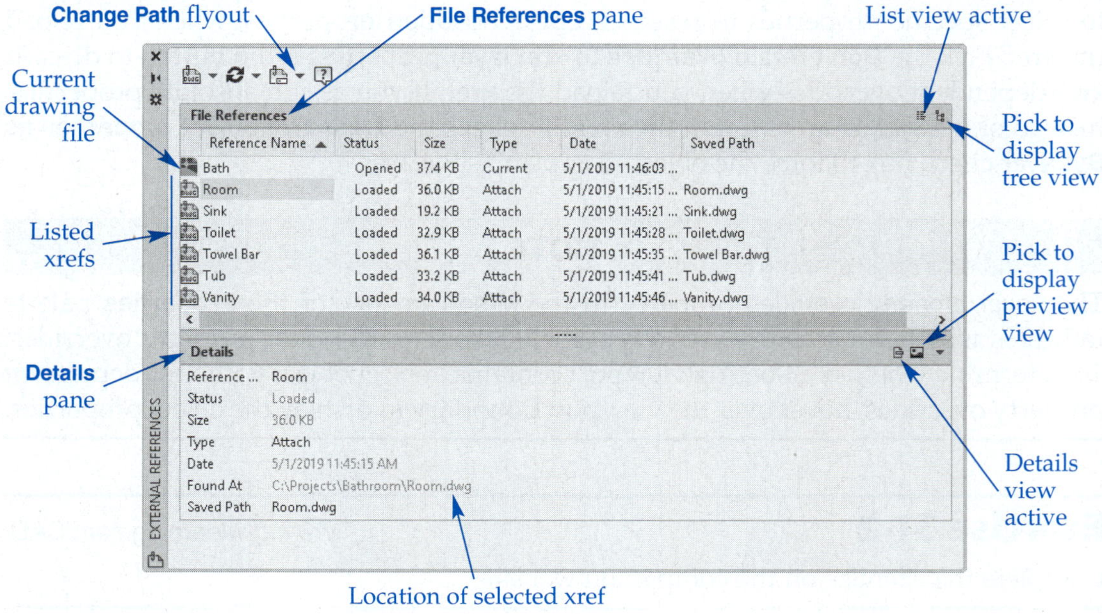

- **Not Found.** The xref file is not found in the specified search paths.
- **Unresolved.** The xref file is missing or cannot be found.
- **Orphaned.** The parent of the nested xref cannot be found.

The **Size** column lists the file size for each xref. The **Type** column indicates whether the xref is attached or referenced as an overlay. To change the reference type from an attachment to an overlay or from an overlay to an attachment, double-click on the current type, or right-click on the reference name or current type and make a selection from the **Xref Type** cascading menu. The **Date** column indicates the date the xref was last modified.

The **Saved Path** column lists the path name saved with the xref. If only a file name appears, the path was not saved. Prefixes describe the relative paths to xref files. In Figure 31-6, the characters .\ precede the Roof Plan reference file. The period (.) represents the folder containing the host drawing. From that folder, AutoCAD looks in the House 01 folder that contains the Roof Plan drawing. The Elevation reference file in Figure 31-10 uses a similar specification. In this example, the same folder contains the Elevation xref and the host drawing. The characters ..\ precede the specification for the Wall xref. The double period instructs AutoCAD to move up one folder level from the current location. The double period repeats to move up multiple folder levels. For example, AutoCAD locates the Panel xref in Figure 31-10 by moving up two folder levels from the folder of the host drawing and opening the Symbols folder.

To change the path status of an xref, select the reference name in the **File References** pane of the **External References** palette and then choose from the **Change Path** flyout shown in Figure 31-9, or right-click on the reference name and make a selection from the **Change Path Type** cascading menu. Pick **Make Absolute** to assign the **Full path** setting, **Make Relative** to specify the **Relative path** setting, or **Remove Path** to specify the **No path** setting.

Figure 31-10.
Relationship between the symbols in the **Saved Path** list and the file locations within the folder structure.

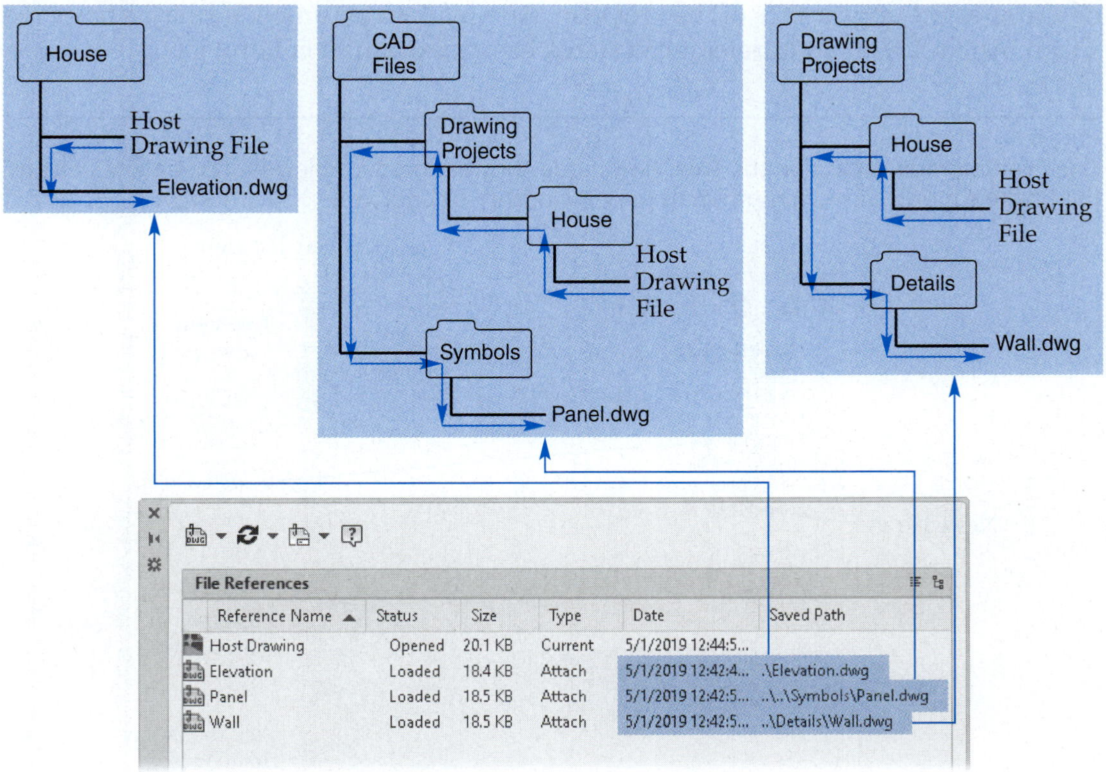

NOTE

The path saved to the xref is one of several locations AutoCAD searches when you open a host drawing and an xref requires loading. AutoCAD searches path locations to load xref files in the following order:

1. The full or relative path associated with the xref
2. The current folder of the host drawing
3. The project paths specified in the Project Files Search Path
4. The support paths specified in the Support File Search Path
5. The Start in: folder path specified for the AutoCAD application shortcut associated using the **Properties** option in the desktop icon shortcut menu

PROFESSIONAL TIP

Adjust the column widths in list view mode as necessary to view complete information. To adjust the width of a column, move the cursor to the edge of the button at the top of the column until the cursor changes to a horizontal resizing cursor. Press and hold the left mouse button and drag the column to the desired width. If columns extend beyond the width of the palette, a horizontal scroll bar appears at the bottom of the list.

Tree View Display

Pick the **Tree View** button or press the [F4] key to see a list of xrefs in the **File References** pane, and show nesting levels. See Figure 31-11. Nesting levels are displayed in a format similar to the arrangement of folders. The status of the xref determines the appearance of the xref icon. An xref with an unloaded status has a red x icon. An xref with an unreferenced status has an exclamation point icon.

Figure 31-11.
The **File References** pane in the **Tree View** mode shows nested xref levels. The **Details** pane in the **Preview** mode shows a thumbnail preview of the selected xref.

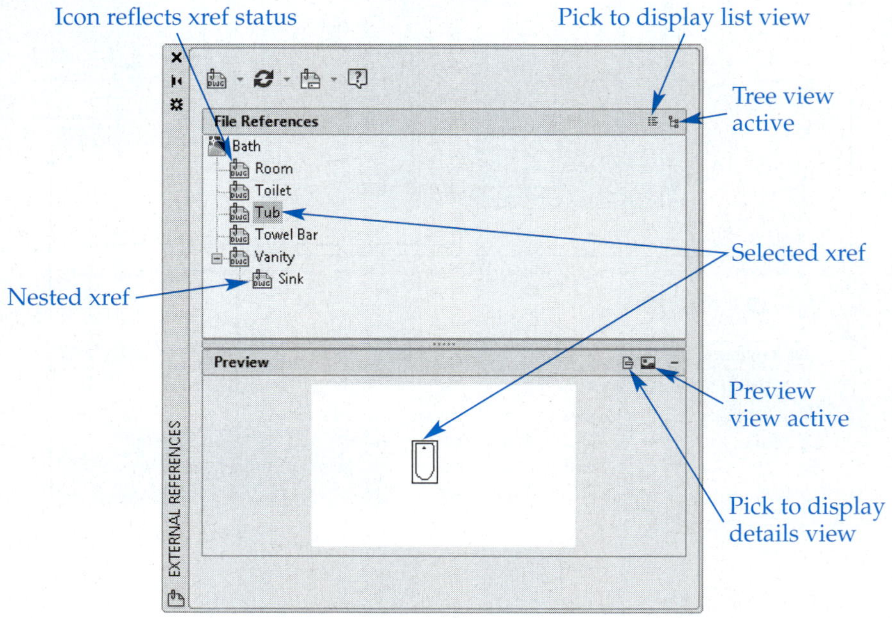

NOTE

Use standard selection practices as necessary to select multiple xrefs, allowing you to make the same changes to several xrefs at the same time.

Viewing Details or a Preview

The **Details** mode, shown in Figure 31-9, is active by default. Pick the **Details** button to display details while in **Preview** mode. The information listed in the **Details** pane corresponds to the host file or xref selected in the **File References** pane. The rows in the **Details** mode display the same type of information found in the **List View** mode of the **File References** pane. However, in the **Details** pane, you can modify the reference name by entering a new name in the **Reference Name** text box. You can also change the reference type from an attachment to an overlay or from an overlay to an attachment by picking the appropriate option from the **Type** drop-down list. In addition, you can use the **Saved Path** row to update the location of an xref path, as explained later in this chapter. All other details are for reference only. Pick the **Preview** button on the **Details** pane to display an image of the xref selected in the **File References** pane. See Figure 31-11.

Detaching, Reloading, and Unloading Xrefs

Each time you open a host drawing containing an attached xref, the xref loads and appears on-screen. This association remains permanent until you ***detach*** the xref. Erasing an xref does not remove the xref from the host drawing. To detach an xref, right-click on the reference name in the **File References** pane of the **External References** palette and pick **Detach**. All instances of the xref and all of its nested xrefs are detached from the current drawing, along with all referenced data.

detach: Remove an xref from a host drawing.

In some situations, you may need to update, or ***reload***, an xref file in the host drawing. For example, if you edit an xref while the host drawing is open, the updated version may be different from the version you see. To update the xref, right-click on the reference name in the **File References** pane of the **External References** palette and pick **Reload**, or pick the **Reload All References** button from the flyout at the top of the palette to reload all unloaded xrefs. Reloading xrefs forces AutoCAD to read and display the most recently saved version of each xref.

reload: Update an xref in the host drawing.

To ***unload*** an xref, right-click on the reference name in the **File References** pane of the **External References** palette and pick **Unload**. An unloaded xref is not displayed or regenerated, so performance increases. Reload the xref to redisplay it.

unload: Suppress the display of an xref without removing the xref from the host drawing.

NOTE

If AutoCAD cannot find an xref, an alert appears when you open the host drawing. Choose the appropriate option to ignore the problem or fix the problem using the **External References** palette.

Updating the Xref Path

A file path saved with an xref is displayed in the **Saved Path** column of the **File References** pane and the **Saved Path** row of the **Details** pane in the **External References** palette. If the **Saved Path** location does not include an xref file, when you open the host drawing, AutoCAD searches the ***library path***. A link to the xref forms if AutoCAD finds a file with a matching name. In such a case, the **Saved Path** location differs from where AutoCAD actually found the file.

library path: The path AutoCAD searches by default to find an xref file, including the current folder and locations set in the **Options** dialog box.

Check for matching paths in the **External References** palette by comparing the path listed in the **Saved Path** column of the **File References** pane and **Saved Path** row of the **Details** pane with the listing in the **Found At** row of the **Details** pane. When you move an xref and the new location is not in the library path, the xref status is **Not Found**.

To find or replace the **Saved Path** location, right-click on the reference name in the **File References** pane and pick **Select New Path...**; select the reference name in the **File References** pane and pick **Select New Path...** from the **Change Path** flyout; or select the path in the **Saved Path** edit box of the **Details** pane and pick the ellipsis (**...**) button to the right of the edit box. The **Select new path** dialog box appears, allowing you to locate the new folder and select the desired file. Pick the **Open** button to update the path. If there are other missing xrefs, AutoCAD will prompt you to apply the same selected path to the other missing references.

Another option for assigning a different search path to multiple xrefs found in the same existing path is to right-click on the selected reference names in the **File References** pane and pick **Find and Replace...** or select the reference names in the **File References** pane and pick **Find and Replace...** from the **Change Path** flyout. The **Find and Replace Selected Paths** dialog box appears, allowing you to specify the appropriate paths. Pick the ellipsis (**...**) buttons to locate the folder specified in the saved path and the folder located in the replacement path. Pick the **Replace All** button to replace the existing path for all selected xrefs with the new path.

The Manage Xrefs Button

By default, when you edit, save, and close an xref, and then open the host drawing, changes made to the xref automatically appear without any notification. If you make changes to an xref and save the file while the host drawing is open, a notification appears in the status bar. Changes are indicated by the appearance of the **Manage Xrefs** button, a balloon message, or both.

TRAYSETTINGS

Type
TRAYSETTINGS

The **Tray Settings** dialog box controls notifications in the status bar tray for xref changes and other system updates. Access the **TRAYSETTINGS** command to display the **Tray Settings** dialog box. Select the **Display icons from services** check box to display the **Manage Xrefs** button in the status bar tray when you attach an xref to the current drawing. If you modified an xref and saved changes in the current file since opening the file, the **Manage Xrefs** button appears with an exclamation sign. Pick the **Manage Xrefs** button or right-click on the **Manage Xrefs** button and select **External References...** to open the **External References** palette to reload the xref.

Select the **Display notifications from services** check box in the **Tray Settings** dialog box to display a balloon message notification with the name of the modified xref file. See Figure 31-12A. You can then pick the xref name in the balloon message to reload the file. The example in Figure 31-12 shows adding a Towel Bar xref to the Room parent xref drawing. The xref reloads in the host drawing named Bath. See Figure 31-12B. You can also reload xrefs by right-clicking on the **Manage Xrefs** button and selecting **Reload DWG Xrefs**.

Exercise 31-3

www.g-wlearning.com/CAD/

Complete the exercise on the companion website.

Figure 31-12.
The **Manage Xrefs** button in the status bar tray provides a notification when you modify and save an xref file. A—A balloon message and an exclamation point appear at the icon. B—Reloading the xref file updates the current drawing and changes the appearance of the icon.

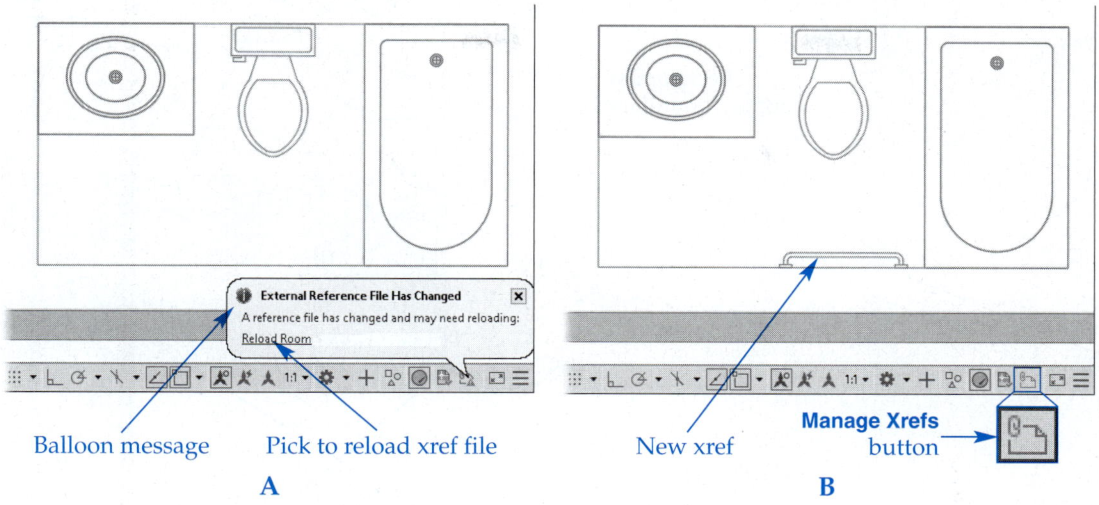

Clipping Xrefs

Clip, or crop, an xref to display only a specific portion, or an xref ***subregion***. All geometry that falls outside the clipping boundary is invisible, and objects that are partially within the subregion appear trimmed at the boundary. Although clipped objects appear trimmed, the xref file does not change. Clipping applies to a selected instance of an xref, not to the actual xref definition.

subregion: The displayed portion of a clipped xref.

Use the **XCLIP** command to create and modify clipping boundaries. Quick ways to access the **XCLIP** command include picking an object that is part of the xref and selecting the **Create Clipping Boundary** button from the **Clipping** panel of the **External Reference** contextual ribbon tab, or right-clicking and selecting **Clip Xref**. If you access the **XCLIP** command before selecting an xref, pick an object associated with the xref to clip. Then press [Enter] to accept the default **New boundary** option and select an option to create the clipping boundary.

XCLIP
Ribbon
External Reference > Clipping
Create Clipping Boundary
Type
XCLIP
XC

When you select the **New boundary** option, a prompt asks you to specify the clipping boundary. Use the default **Rectangular** option to create a rectangular boundary. Then pick opposite corners of the rectangular boundary. See **Figure 31-13**. Note that the geometry outside the clipping boundary no longer appears after clipping. The **New boundary** option includes additional methods for specifying the clip boundary and area to clip, as described in **Figure 31-14**.

Edit a clipped xref as you would an unclipped xref. The clipping boundary moves with the xref. Note that nested xrefs are clipped according to the clipping boundary for the parent xref. Use the **XCLIPFRAME** system variable to toggle the display and plotting behavior of the clipping boundary frame. The default value of 2 displays the frame but does not plot it. Set the value to 0 to hide and not plot the frame. Set the value to 1 to display and plot the frame. These settings can also be accessed and controlled from the ribbon using the frame settings drop-down list in the **Reference** panel of the **Insert** ribbon tab. The **FRAMESELECTION** system variable controls the ability to select a hidden frame (**XCLIPFRAME** system variable set to 0). The default value of 1 allows you to select hidden frames. The frame appears for reference when you select the xref. Change the value to 0 to disable selection of hidden frames. You can still pick an object that is part of the xref to select the xref.

FRAME
Ribbon
Insert > Reference
Frames settings vary
Type
FRAME

Figure 31-13.
Clipping a large site plan xref to display a specific area. A—Using the **Rectangular** boundary selection option. B—The clipped xref.

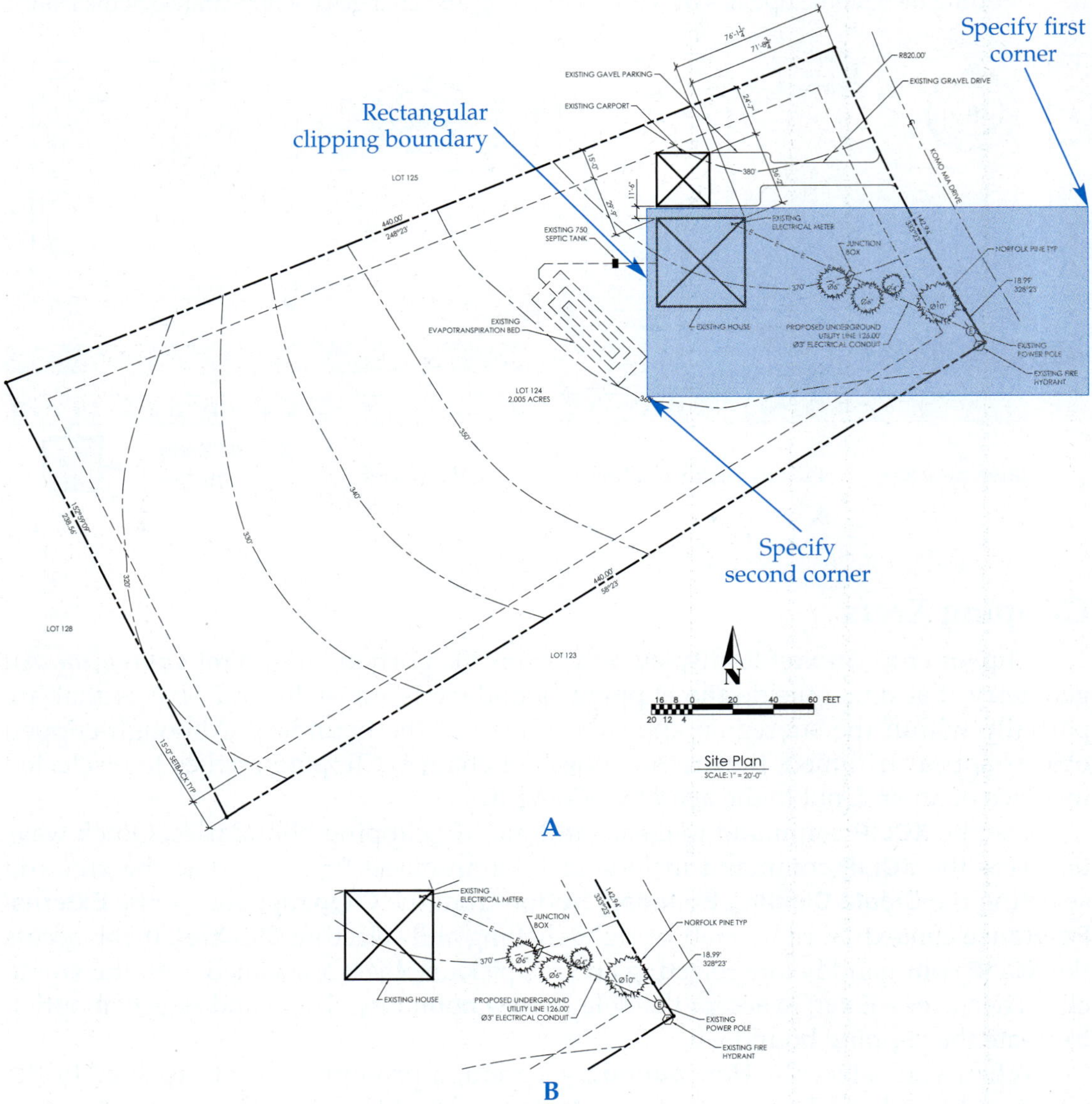

Figure 31-14.
Additional options available for the **New boundary** option of the **XCLIP** command.

Option	Description
Select polyline	Select an existing polyline object as the clip boundary. If the polyline does not close, the start and endpoints of the boundary connect.
Polygonal	Draw an irregular polygon as a boundary.
Invert clip	Inverts the selection so that the portion of the xref that lies outside of the clipping boundary is clipped. Only the portion of the xref outside of the boundary is displayed.

The other options of the **XCLIP** command apply after you define a clip boundary. The **ON** and **OFF** options turn the clipping feature on or off. The **Clipdepth** option allows you to define front and back clipping planes to control the portion of a 3D drawing that displays. Clipping 3D models is described in *AutoCAD and Its Applications—Advanced*.

Use the **Delete** option to remove an existing clipping boundary, returning the xref to its unclipped display. A quick way to remove clipping is to pick an object that is part of the xref and select the **Remove Clipping** button from the **Clipping** panel of the **External Reference** contextual ribbon tab. Use the **generate Polyline** option to create and display a polyline object at the clip boundary to frame the clipped portion.

Exercise 31-4

www.g-wlearning.com/CAD/

Complete the exercise on the companion website.

Demand Loading and Xref Editing Controls

Demand loading controls how much of an xref loads when you attach the xref to the host drawing. It improves performance and saves disk space because only a portion of the xref file loads into the host drawing. For example, data on frozen layers is not loaded, and data outside of clipping regions does not load.

demand loading: Loading only the portion of an xref file necessary to regenerate the host drawing.

Demand loading occurs by default. Use the **Open and Save** tab of the **Options** dialog box to check or change the setting. The **Demand load Xrefs:** drop-down list in the **External References (Xrefs)** area contains each demand loading option. Select the **Enabled with copy** option to turn on demand loading. Other users can edit the original drawing because AutoCAD uses a copy of the referenced drawing. Alternatively, pick the **Enabled** option to turn on demand loading. If you use this option, the xref file is considered "in use" while you are referencing the drawing, preventing other users from editing the file. Select the **Disabled** option to turn off demand loading.

NOTE

The **Allow other users to Refedit current drawing** check box in the **Open and Save** tab of the **Options** dialog box is checked by default. It controls whether the current drawing can be edited in place by others while it is open and when it is referenced by another file.

PROFESSIONAL TIP

If you plan to use a drawing as an external reference, save the file with *spatial indexes* and *layer indexes*. These lists help improve performance when you reference drawings with frozen layers and clipping boundaries. Use the following procedure to create spatial and layer indexes:

1. Access the **Save Drawing As** dialog box.
2. Pick **Options...** from the **Tools** flyout button and select the **DWG Options** tab of the **Saveas Options** dialog box.
3. Select the type of index required from the **Index type:** drop-down list.
4. Pick the **OK** button and save the drawing.

spatial index: A list of objects ordered according to their location in 3D space.

layer index: A list of objects ordered according to the layers to which they are assigned.

Binding an Xref

bind: Convert an xref to a permanently inserted block in the host drawing.

Bind an xref to make the xref a permanent part of the host drawing, as if you were inserting the file using the **INSERT** command. Binding is useful when you need to send the full drawing file to another location or user, such as a plotting service or client. To bind an xref using the **External References** palette, right-click on the reference name and pick **Bind...**. The **Bind Xrefs/DGN underlays** dialog box that appears contains **Insert** and **Bind** radio buttons.

Using the Insert and Bind Options

The **Insert** option converts the xref into a normal block, as if you had used the **INSERT** command to place the file. In addition, the drawing is added to the block definition table, and all named objects, such as layers, blocks, and styles, are incorporated into the host drawing as named in the xref. For example, if you bind an xref file named PLATE that contains a layer named OBJECT, the xref-dependent layer PLATE|OBJECT becomes the locally defined layer OBJECT. All other xref-dependent objects lose the xref name and assume the properties of the locally defined objects with the same name. The **Insert** binding option provides the best results for most purposes.

The **Bind** option also converts the xref into a normal block. However, the xref name remains with all dependent objects. Two dollar signs with a number between them replace the vertical line in each name. For example, an xref-dependent layer named Title|Notes becomes Title0Notes. The number inside the dollar signs is automatically incremented if a local object definition with the same name exists. For example, if Title0Notes already exists in the drawing, the newly bound layer becomes Title1Notes. In this manner, all xref-dependent object definitions that are bound receive unique names. Use the **RENAME** command or other appropriate method to rename bound objects.

Binding Specific Dependent Objects

Binding an xref allows you to make all dependent objects in the xref file a permanent part of the host drawing. Dependent objects include named items such as blocks, dimension styles, layers, linetypes, and text styles. Before binding, you cannot directly use any dependent objects from a referenced drawing in the host drawing. For example, you cannot make an xref-dependent layer or text style current in the host drawing.

In some cases, you may only need to incorporate one or more specific named objects, such as a layer or block, from an xref into the host drawing, instead of binding the entire xref. If you only need selected items, it can be counterproductive to bind an entire drawing. Instead, use the **XBIND** command and corresponding **Xbind** dialog box, shown in Figure 31-15, to select specific named objects to bind.

XBIND
Type
XBIND
XB

Xrefs have AutoCAD drawing file icons. Expand a group to select an individually named object. To select an object for binding, highlight the object and pick the **Add** button. The names of all objects selected and added are displayed in the **Definitions to Bind** list. Pick the **OK** button to complete the operation. A message displayed on the command line indicates how many objects of each type are bound.

Individual objects bound using the **XBIND** command are renamed in the same manner as objects bound using the **Bind** option in the **Bind Xrefs/DGN underlays** dialog box. An automatic linetype bind performs so that a layer that includes a linetype not loaded in the host drawing can reference the required linetype definition. The linetype includes a new linetype name, such as xref1$0$hidden. In a similar manner, a previously undefined block may automatically bind to the host drawing as a result of binding nested blocks. Use the **RENAME** command or other appropriate method to rename objects.

Figure 31-15.
Use the **Xbind** dialog box to bind xref-dependent objects individually to the host drawing.

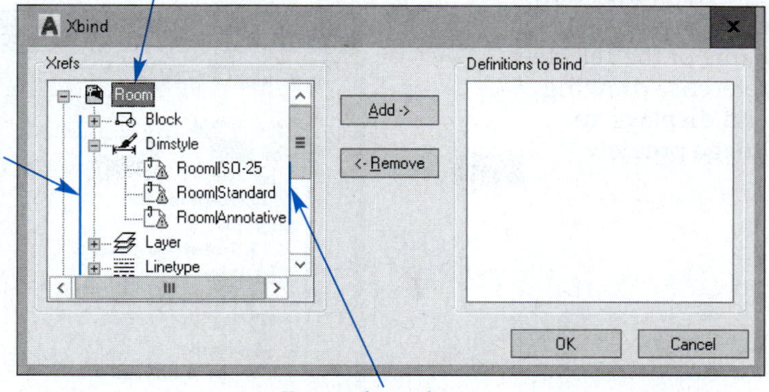

Exercise 31-5

www.g-wlearning.com/CAD/

Complete the exercise on the companion website.

Editing Xref Drawings

One option for editing an xref drawing is to use in-place editing, or ***reference editing***, within the host drawing. You can save any changes made to the xref to the original xref drawing from within the host drawing. Alternatively, you can edit the xref in a separate drawing window as you would any other drawing file.

reference editing: Editing reference drawings from within the host file.

Reference Editing

Ribbon
External Reference > Edit

Edit Reference In-Place
Type
REFEDIT

REFEDIT

The **REFEDIT** command allows you to edit xref drawings in place. Quick ways to initiate reference editing include double-clicking on an xref, picking an object that is part of the xref and selecting the **Edit Reference In-Place** button from the **Edit** panel of the **External Reference** contextual ribbon tab, or selecting an xref and then right-clicking and selecting **Edit Xref In-place**. If you access the **REFEDIT** command without first selecting an xref, you must then pick the xref to edit. The **Reference Edit** dialog box opens with the **Identify Reference** tab active. See Figure 31-16. The example shows the Room reference drawing selected for editing. Notice that nested blocks, such as the Bath Tub - 26 x 60 in. block found in the Tub reference, are listed under their parent xrefs.

The **Automatically select all nested objects** radio button in the **Path:** area is active by default. Use this option to make all xref objects available for editing. To edit specific xref objects, pick the **Prompt to select nested objects** radio button. The Select nested objects: prompt displays after you pick the **OK** button, allowing you to pick objects that belong to the selected xref. Pick all the geometry you want to edit and press [Enter]. The nested objects you select make up the ***working set***. If multiple instances of the same xref appear, be sure to pick objects from the original xref you select.

working set: Nested objects selected for editing during a **REFEDIT** operation.

Additional options for reference editing are available in the **Settings** tab of the **Reference Edit** dialog box. The **Create unique layer, style, and block names** option controls the naming of selected layers and ***extracted*** objects. Check the box to assign the prefix $*n*$, with *n* representing an incremental number, to object names. This is similar to the renaming method used when you bind an xref.

extracted: Temporarily removed from the drawing for editing purposes.

The **Display attribute definitions for editing** option is available if you select a block object in the **Identify Reference** tab of the **Reference Edit** dialog box. Check the box to edit any attribute definitions included in the reference. To prevent accidental changes

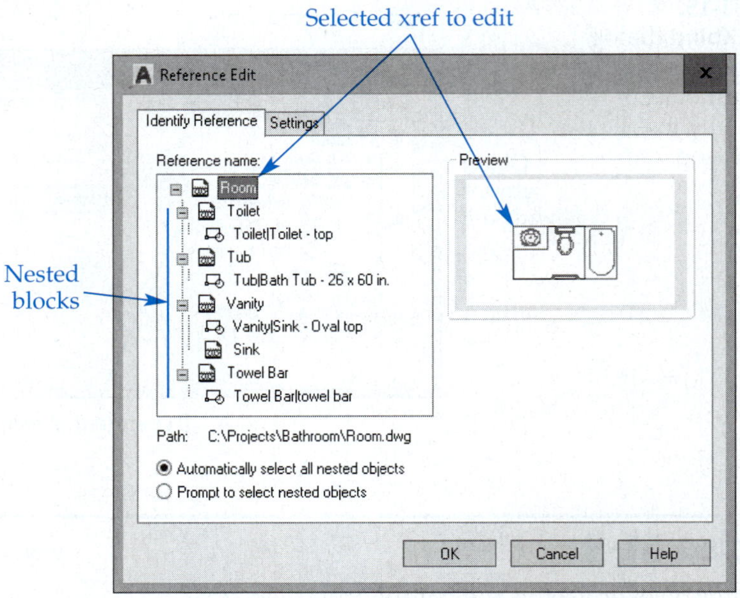

Figure 31-16.
The **Reference Edit** dialog box lists the name of the selected reference drawing and displays an image preview.

to objects that do not belong to the working set, check the **Lock objects not in working set** option. This makes all objects outside of the working set unavailable for selection in reference editing mode.

If the selected xref file contains other references, the **Reference name:** area lists all nested xrefs and blocks in tree view. In the example given, Toilet, Tub, Vanity, and Towel Bar are nested xrefs in the Room xref. If you pick the drawing file icon next to Vanity in the tree view, for example, an image preview appears and the selected xref is highlighted in the drawing window.

When you finish adjusting settings, pick the **OK** button to begin editing the xref. The primary difference between the drawing and reference editing environments is the **Edit Reference** panel that appears in each ribbon tab, including the expanded **External Reference** contextual ribbon tab that appears when you pick an xref. See Figure 31-17. Use the tools in the **Edit Reference** panel to add objects to the working set, remove objects from the working set, and save or discard changes to the original xref file.

Any object you draw during the in-place edit is automatically added to the working set. Use the **Add to Working Set** button to add existing objects to the working set. When you add an object to the working set, the object is extracted, or removed, from the host drawing. The **Remove from Working Set** button allows you to remove selected objects from the working set. Removing a previously extracted object adds the object back to the host drawing.

Figure 31-18A shows reference-editing the Vanity xref nested in the Room xref. Notice that all objects not in the working set become faded. The objects in the working set appear in normal display mode. Once you define the working set, use drawing and editing commands to alter the xref. Pick the **Save Changes** button to save the changes.

Figure 31-17.
The **External Reference** contextual ribbon tab. The **Edit Reference** panel appears in each ribbon tab during reference editing. Most other commands and options function the same in the drawing and reference editing environments.

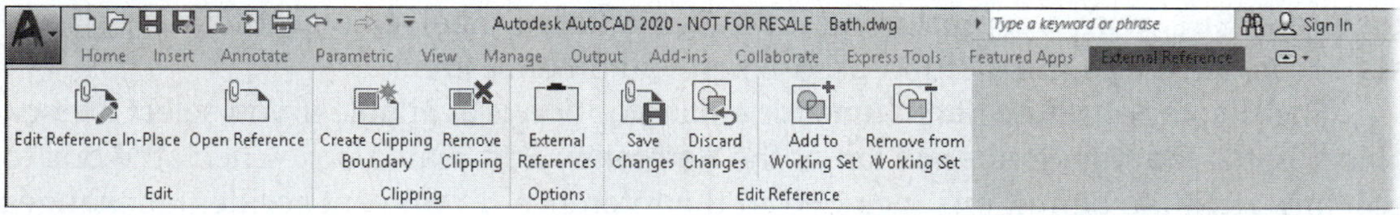

Figure 31-18. Reference editing. A—Objects in the drawing that are not a part of the working set appear faded during the reference editing session. B—All instances of the xref update immediately after reference editing.

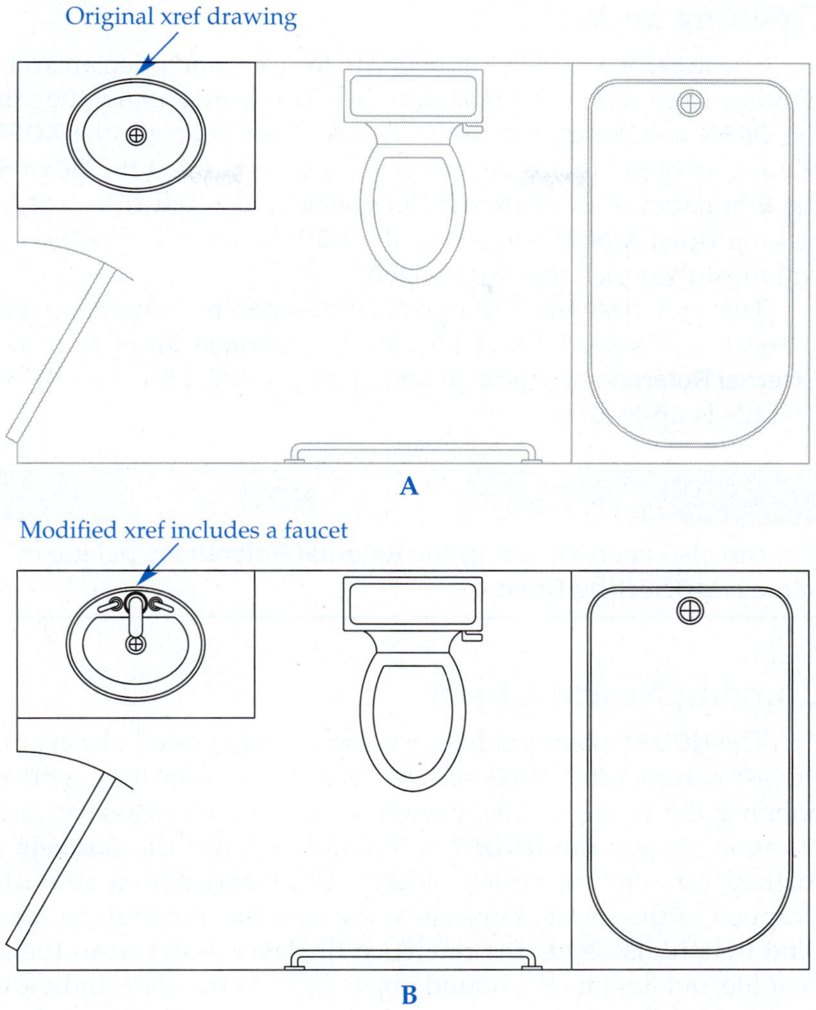

Pick the **OK** button when AutoCAD asks if you want to continue with the save and redefine the xref. All instances of the xref are updated. **Figure 31-18B** shows the xref after editing to redesign the sink and add a faucet. Pick the **Discard Changes** button to exit reference editing without saving changes.

NOTE

In-place reference editing is best suited for minor revisions. Conduct major xref revisions in the reference drawing file.

CAUTION

All edits made using reference editing are saved back to the reference drawing file and affect any host drawing that references the file. For this reason, it is critically important that you edit external references only with the permission of your supervisor or instructor.

Opening an Xref File

Type
XOPEN

The **XOPEN** command allows you to select and open an xref in a separate drawing window from within the host drawing. This is essentially the same procedure as using the **OPEN** command, but faster. Quick ways to access the **XOPEN** command include picking an object that is part of the xref and selecting the **Open Reference** button from the **Edit** panel of the **External Reference** contextual ribbon tab, or right-clicking and picking **Open Xref**. If you access the **XOPEN** command without first selecting an xref, you must then pick the xref to open.

The xref drawing file opens in a separate drawing window. After you make changes and save the xref file, use the **Manage Xrefs** icon in the status bar or the **External References** palette to reload the modified xref file. Reloading ensures that the host file is up-to-date.

NOTE

You can also open an xref in the **External References** palette by right-clicking the xref name and selecting **Open**.

Copying Nested Objects

NCOPY

Ribbon
Express Tools > Blocks

Copy Nested Objects

Type
NCOPY

The **NCOPY** command allows you to copy nested objects in an xref drawing. The **NCOPY** command extracts selected objects from the nest, which is more efficient than opening the reference file, exploding a block if necessary, and copying objects, for example. Access the **NCOPY** command and use the **Settings** option to specify the method for copying nested objects. The **Insert** option adds the reference file layer assigned to the selected objects to the host file and assigns the layer to the copy. The **Bind** option also adds the reference file layer assigned to the selected objects to the host file, but assigns the bound object layer to the copy. Individually select the objects to copy and then follow the prompts as you would when using the **COPY** command, described in Chapter 12, to complete the operation.

Exercise 31-6

www.g-wlearning.com/CAD/

Complete the exercise on the companion website.

Express Tools

www.g-wlearning.com/CAD/

Xref Express Tools

The **Blocks** and **Modify** panels of the **Express Tools** ribbon tab include additional commands related to xrefs and blocks. For information about the most useful xref express tools, go to the companion website, navigate to this chapter in the **Contents** tab, and select **Xref Express Tools**.

Chapter Review

Answer the following questions using the information in this chapter.

1. What types of files can you reference into an AutoCAD drawing?
2. What effect does the use of referenced drawings have on drawing file size?
3. What is a nested xref?
4. List at least three common applications for xrefs.
5. On what layer should you consider inserting xrefs into a host drawing?
6. Which command allows you to attach an xref drawing to the current file?
7. What is the difference between an overlaid xref and an attached xref?
8. What is the difference between an absolute path and a relative path?
9. Describe the process of placing an xref using **DesignCenter**.
10. What must you do before you can use a tool palette to place an xref?
11. If you attach an xref file named FPLAN to the current drawing, and FPLAN contains a layer called ELECTRICAL, what name will appear for this layer in the **Layer Properties Manager**?
12. What is the purpose of the **Detach** option in the **External References** palette?
13. When are xrefs updated in the host drawing?
14. What could you do to suppress an xref temporarily without detaching it from the host drawing?
15. Which command allows you to display only a specific portion of an externally referenced drawing?
16. What are spatial and layer indexes, and what function do they perform?
17. Why would you want to bind a dependent object to a master drawing?
18. What does the layer name WALL0NOTES mean?
19. What command allows you to edit external references in place?
20. What command allows you to open a parent xref drawing in a new AutoCAD drawing window by selecting the xref in the host drawing?

Drawing Problems

Start AutoCAD if it is not already started. Start a new drawing for each problem using an appropriate template of your choice. The template should include layers and text, dimension, multileader, and table styles, when necessary, for drawing the given objects. Add layers and text, dimension, multileader, and table styles as needed. Draw all objects using appropriate layers. Use appropriate text, dimension, multileader, and table styles, justification, and format. Follow the specific instructions for each problem. Use only drawing and editing commands and techniques you have already learned. Use your own judgment and approximate dimensions when necessary. Apply dimensions accurately using ASME or appropriate industry standards.

Note: Some of the problems in this chapter are built on problems from previous chapters. If you have not yet completed those problems, complete them now.

▼ Basic

1. General Attach a dimensioned problem from Chapter 17 into a new drawing as an xref. Save the drawing as P31-1.

2. General Attach a dimensioned problem from Chapter 18 into a new drawing as an xref. Save the drawing as P31-2.

3. General Attach a dimensioned problem from Chapter 19 into a new drawing as an xref. Save the drawing as P31-3.

▼ Intermediate

4. Architectural Attach the EX29-10.dwg file used in Exercise 29-10 into a new drawing as an xref. Copy the xref two times. Use the **XCLIP** command to create a clipping boundary on each view. Apply an inverted rectangular clip to the original xref, a polyline boundary on the first copy, and a polygonal boundary on the second copy. Save the drawing as P31-4.

5. Architectural Attach the EX29-10.dwg file used in Exercise 29-10 into a new drawing as an xref. Bind the xref to the new drawing. Rename the layers to the names assigned to the original EX29-10 (xref) file. Explode the block created by binding the xref. Save the drawing as P31-5.

6. Mechanical Create the multi-detail drawing shown according to the following information:
   - Use the Mechanical-Inch.dwt drawing template file available on the companion website.
   - Set drawing units to Fractional.
   - Xref the following files into model space: Detail-Item 1.dwg, Detail-Item 2.dwg, Detail-Item 3.dwg, Detail-Item 4.dwg, Detail-Item 5.dwg, and Detail-Item 6.dwg. These files are available on the companion website.
   - Use six floating viewports on the **C-SIZE** layout to arrange and scale the details. Use a 1:2 scale.
   - Adjust the title block information using the drawing property fields and attributes.
   - Plot the drawing.

   Save the drawing as P31-6.

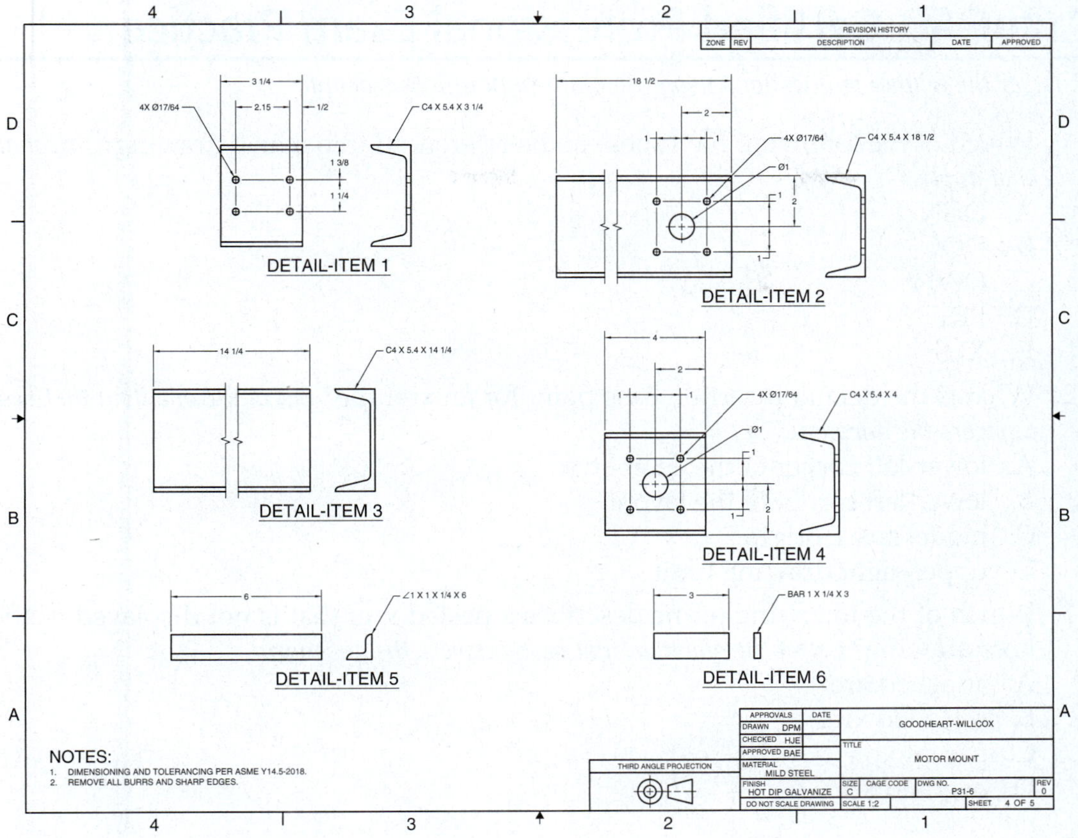

▼ Advanced

7. **Architectural** Design and draw a basic residential floor plan using an appropriate template. Save the file as P31-7FLOOR. Attach the P31-7FLOOR file into a new file as an xref. Use the xref to help draw a roof plan. Save the roof plan file as P31-7ROOF.

8. **Architectural** Attach the P31-7ROOF file into a new file as an xref. Use the xref to help draw front and rear elevations. Save the elevation file as P31-8.

9. **General** Use a word processor to write a report of approximately 250 words explaining the purpose of external references. Include a description of the types of files that you can reference. Cite at least three examples from actual industry applications of using external references to help prepare drawings. Use at least four sketches to illustrate your report.

10. **Architectural** Attach the P28-11 file into a new file as an xref. Use the xref to help draw a roof plan. Save the file as P31-10ROOF. Attach the P28-11 file into a new file as an xref. Use the xref to help draw a slab foundation plan. Save the file as P31-10FDTN. Attach the P31-10ROOF file and the P31-10FDTN file into a new file as xrefs. Use corresponding insertion base points to overlap the plans exactly. Use the xrefs to help draw front, rear, right-side, and rear exterior elevations and an interior elevation for each wall. Save the elevation file as P31-10.

11. **Architectural** Draw the details shown in Figure 31-3A using a separate file for each detail and name the files according to the detail names shown in Figure 31-3B. Xref the files to create the layout shown in Figure 31-3B. Plot the layout. Save the complete drawing as P31-11.

AutoCAD Certified Professional Exam Practice

Answer the following questions using the information in this chapter.

1. Which of the following file types can be referenced into a host drawing? *Select all that apply.*
   A. DGN
   B. DOC
   C. DWFx
   D. PDF
   E. XLS
2. What is the default insertion base point for an xref file? *Select the one item that best answers the question.*
   A. lower-left corner of the geometry
   B. lower-left corner of the layout
   C. model space origin
   D. upper-right drawing limit
3. Which of the following terms describes a nested xref that is not displayed in the host drawing? *Select the one item that best answers the question.*
   A. attached xref
   B. reloaded xref
   C. parent xref
   D. overlaid xref

Follow the instructions in the problem. Complete the problem by answering the related question.

4. **Navigate to this chapter on the companion website and open CPE-31backyard.dwg.** What layers in this drawing are former xref-dependent layers that have been bound to the drawing?

CHAPTER 32

Introduction to Sheet Sets

Learning Objectives

After completing this chapter, you will be able to:

- ✓ Describe the functions of an AutoCAD sheet set.
- ✓ Create sheet sets from examples and existing drawings.
- ✓ Manage sheet sets.
- ✓ Create and manage subsets.
- ✓ Add sheets to a sheet set.

A design project typically requires a set of drawings and documents that completely specify the design. Preparing accurate drawings and making revisions in a timely manner involves significant organization, especially when a project includes multiple related ***sheets*** and ***views***. ***Sheet sets*** help organize a set of drawings and simplify project management. This chapter explains how to organize and manage sheet sets, subsets, and sheets. In order to work effectively with sheet sets, you must understand the elements of a sheet set such as drawing templates, blocks, and fields, as explained throughout this textbook.

sheet: A printed drawing or electronic layout that displays project design requirements.

view: 2D representation of an object.

sheet set: A collection of drawing sheets for a project; the AutoCAD tool that aids project organization.

Sheet Set Fundamentals

A sheet set is an electronic database of information about a project and the set of drawings required to document the design. A sheet set provides a way to organize files related to a set of drawings, similar to using Windows Explorer and a folder with subfolders to contain files. A basic application of a sheet set is to group drawings in the proper order for easy access and quick opening. A sheet set also provides functions that automate managing and creating a set of drawings, which improves speed and accuracy. **Figure 32-1** shows an example of using a sheet set to manage the drawings for a small architectural project of a Cappuccino Express drive-through coffee stand. The sheet set organizes the four required layouts in the appropriate order for viewing, editing, ***publishing*** (also known as batch plotting), and ***archiving***.

publishing: Preparing a sequential set of multiple drawings for hard copy or electronic plotting of the set.

archiving: Gathering and storing all drawings and associated files related to a project.

Every sheet set has a sheet set data (DST) file specific to the project. A DST file is essentially an electronic version of a design project. It is similar to a folder containing subfolders and each sheet in a set of drawings. The purpose of a DST file is to manage the paths to files related to a set of drawings and the procedures required for the

Figure 32-1.
Using a sheet set to manage a set of architectural drawings. The **Sheet Set Manager** is a palette for creating and working with sheet sets.

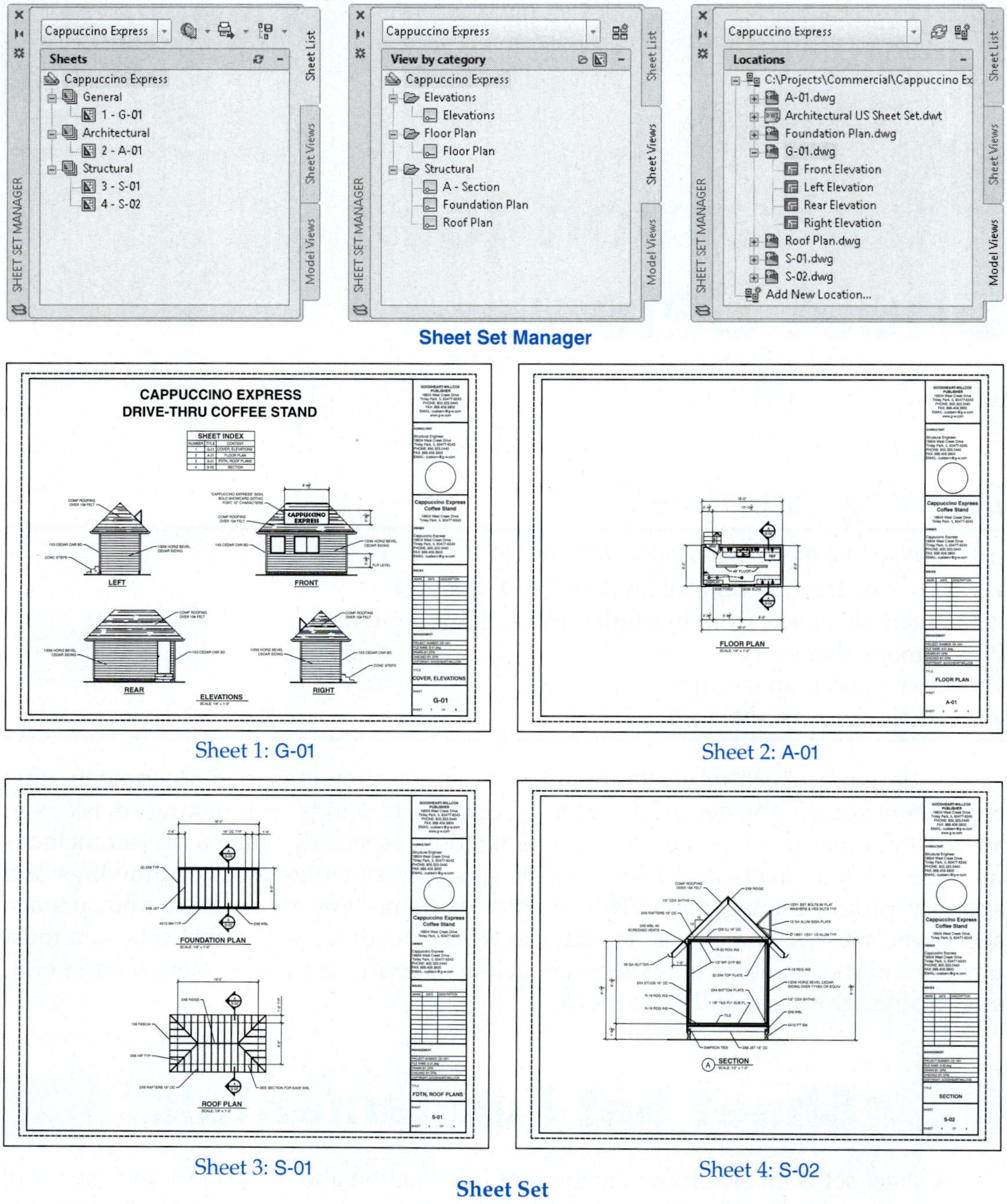

sheet set to function properly. A sheet set stores and displays all information about the project, including:

- Paths to required drawing layouts, or sheets.
- The structure and organization of sheets in the sheet set.
- Project properties common to all sheets and usually found in sheet blocks, such as the name and number of the project.
- A path to a template file and storage location for creating new sheets.
- Paths to drawing and model space views.
- Paths to blocks with attributes containing fields that link information from one drawing or view to another drawing or view on different sheets of the sheet set.
- A path to a page setup used to plot all sheets using the same settings, if appropriate.

Introduction to Sheet Set Fields

One way to enhance the usefulness of a sheet set is to use text and blocks with attributes that include ***fields*** linked to elements of the sheet set. Figure 32-2 highlights some of the fields used in the Cappuccino Express sheet set shown in Figure 32-1. Common applications for sheet set fields include titles, title blocks, callout blocks, view label blocks, and sheet list tables. You will learn about these items in Chapter 33.

field: A text object that can display a specific property value, setting, or characteristic.

Fields display properties and allow values to change during the course of the project. For example, the SHEET_NUMBER attribute in the title block on each sheet uses a **CurrentSheetNumber** field that increments the sheet number when you add a sheet to the set. Fields are also associated with sheets and views. For example, the title under each view is a block that includes a VIEW attribute with a **SheetView** field that displays the name of the view. The title updates when you change the name of the view.

Figure 32-2.
Examples of fields added to text and attributes linked to sheet set properties. Use the same tools and options to add fields that reference sheet set data as you would for fields associated with other items.

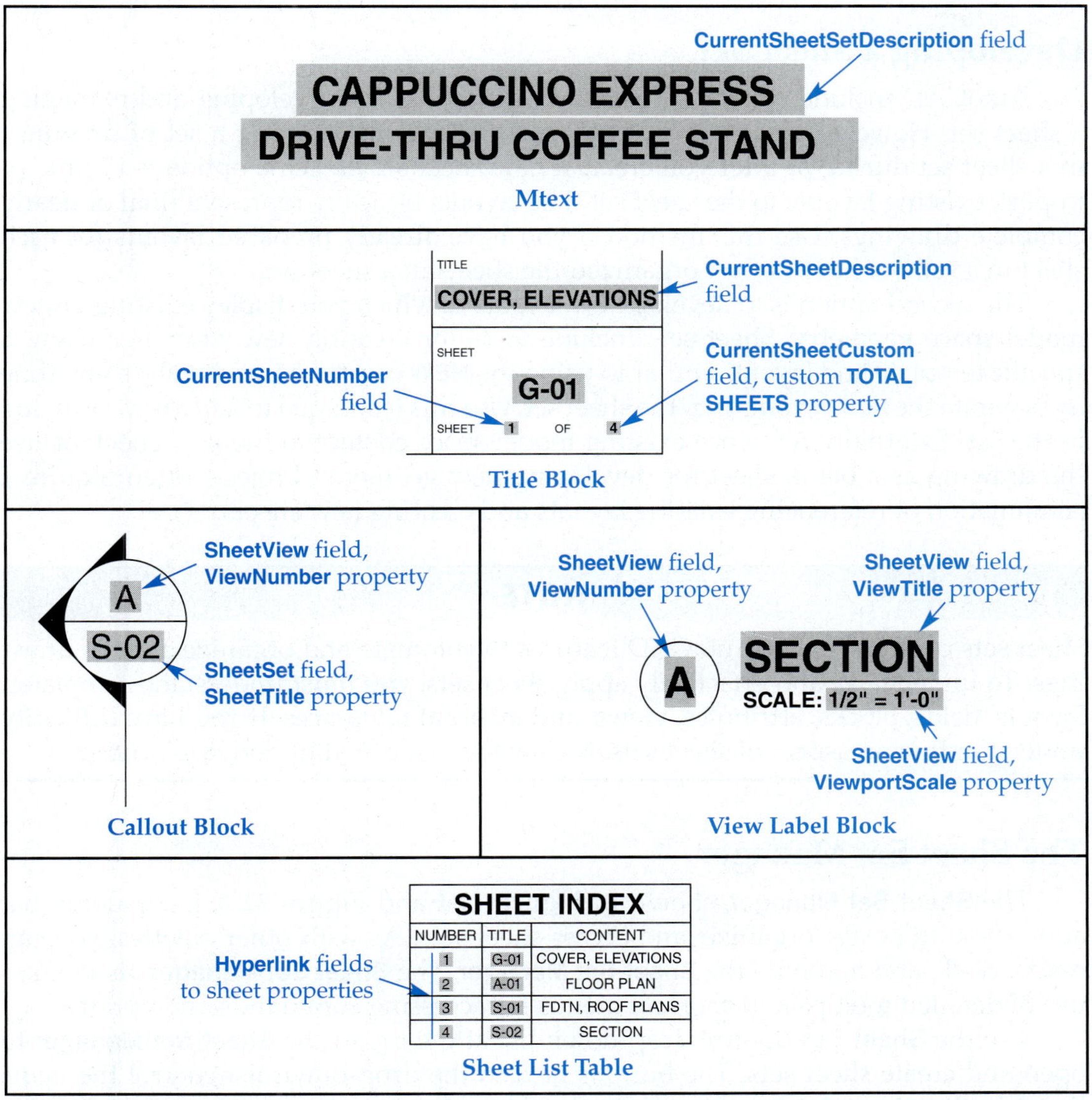

Preparing for a Sheet Set

Preparing to create a sheet set involves several processes. You can adjust every aspect of a sheet set whenever necessary, but for best efficiency, prepare for a sheet set in advance by identifying and creating all of the required elements. Preparing for a sheet set includes considering basic properties such as the project name and number, and identifying where you will store the DST file. Preparation also includes developing a drawing template (DWT) file for creating new sheets, drawing blocks with attributes containing sheet set fields, and defining a page setup to plot all sheets using the same settings.

A sheet in a sheet set is a layout in paper space. Although a drawing file can have multiple layouts, it is recommended to use only one layout per drawing when assigned to a sheet set. When drawings are accessed by multiple users, only one user at a time can work on a drawing with multiple layouts. Therefore, prepare a single layout per drawing if you plan to use a sheet set. Delete all other layouts found in existing drawings and in the template you assign to the sheet set. Use an external reference to reuse model space content in other layouts, and in different drawing files when necessary.

If you plan to organize existing drawings in a sheet set, store all files related to the project in a designated folder when possible. Use a limited number of subfolders to group files, such as an Architectural folder to group architectural drawings and a Structural folder to group structural drawings for a building project. AutoCAD allows you to reference existing subfolders to structure a sheet set.

Developing a Sheet Set

AutoCAD includes multiple tools and techniques for developing and managing a sheet set. However, there are two primary options for building a set of drawings in a sheet set during or after you create a new sheet set file. One option is to link, or import, existing layouts to the sheet set. The layouts typically represent final or nearly complete drawings. Use this method if you have already prepared layouts for each sheet in a set, but have not yet organized the sheets in a sheet set.

The second option is to develop new layouts in which you display existing or new model space geometry. Sheet sets include tools for creating new drawings using a specific template and layout, similar to using the **NEW** command, and at the same time incorporate the new layouts into the sheet set. Use this technique to add a new drawing to the set. Externally reference existing model space content to the new sheet, or use the drawing as a blank sheet for developing new geometry. Projects often require a combination of referencing existing layouts and creating new sheets.

NOTE

Sheet sets combine many AutoCAD features to automate and organize a set of drawings. To understand and effectively apply sheet sets, you must understand templates, layouts, fields, blocks, attributes, views, and external references. If you have difficulty understanding an aspect of sheet sets, review the associated underlying concept.

SHEETSET

Quick Access
Sheet Set Manager
Ribbon
View
> Palettes
Sheet Set Manager
Type
SHEETSET
SSM

The Sheet Set Manager

The **Sheet Set Manager**, shown in Figure 32-1 and Figure 32-3, is a palette that allows you to create, organize, and access sheet sets. As with other palettes, you can resize, dock, and autohide the **Sheet Set Manager**. The **Sheet Set Manager** also makes use of detailed tooltips and shortcut menus for accessing commands and options.

Use the **Sheet Set Control** drop-down list at the top of the **Sheet Set Manager** to open and create sheet sets. The buttons next to the drop-down list control the items listed in the **Sheet Set Manager** and vary depending on the current tab. Right-click

Figure 32-3.
The **Sheet Set Manager** contains **Sheet List**, **Sheet Views**, and **Model Views** tabs. Pick the **Sheet Set Control** drop-down list to access options to create or open a sheet set.

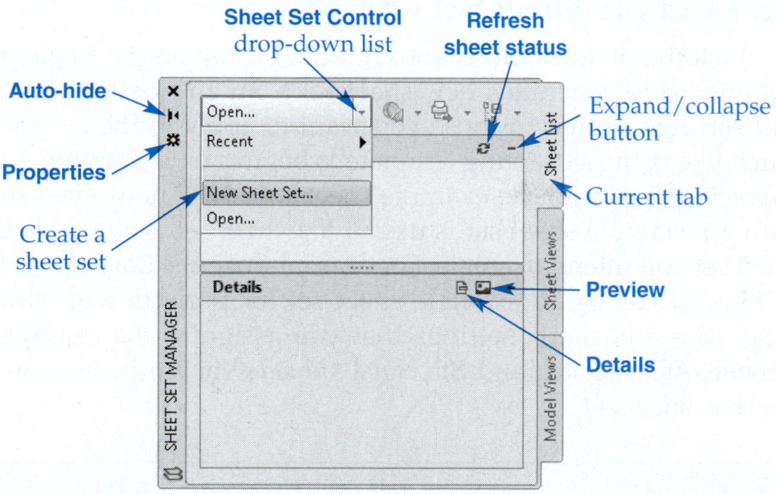

above the **Sheets** pane and select **Preview/Details Pane** to display the **Preview** or **Details** pane. Pick the **Details** or **Preview** button to toggle the corresponding display. The **Details** pane, shown in Figure 32-3, lists properties associated with the item selected in the **Sheet Set Manager**. The **Preview** pane displays an image of a selected sheet or view.

The **Sheet List** tab is the primary resource for managing sheets within a sheet set. Some projects require using only the **Sheet List** tab. The **Sheet Views** tab allows you to organize and place named views on layouts. Sheet views also provide a method to insert blocks that relate information on one sheet to a view on another sheet, such as the section bubbles shown in Figure 32-1, and sheet blocks associated with views, such as the view titles shown in Figure 32-1. The **Model Views** tab allows you to open files related to the current sheet set, and includes an option to create new layout views using existing model space content.

NOTE

When no drawing is open, the **Sheet Set Manager** button appears in the **Quick Access** toolbar. Pick the button to access the **Sheet Set Manager** without opening a drawing.

Creating Sheet Sets

To create a new sheet set, first begin a new file or open an existing file, even if the file does not relate to the sheet set. Then access the **NEWSHEETSET** command to display the **Create Sheet Set** wizard. See Figure 32-4. The **NEWSHEETSET** command is also available from the **Sheet Set Manager** by picking the **New Sheet Set...** option from the **Sheet Set Control** drop-down list. Use the **Create Sheet Set** wizard to create a sheet set from an *example sheet set*, from scratch without selecting existing layouts to include as sheets, or from scratch but with the option to select existing layouts to include as sheets.

Type
NEWSHEETSET

example sheet set: An existing sheet set used as a template for developing a new sheet set.

Figure 32-4.
On the **Begin** page, choose whether to start from an example sheet set or an existing drawing.

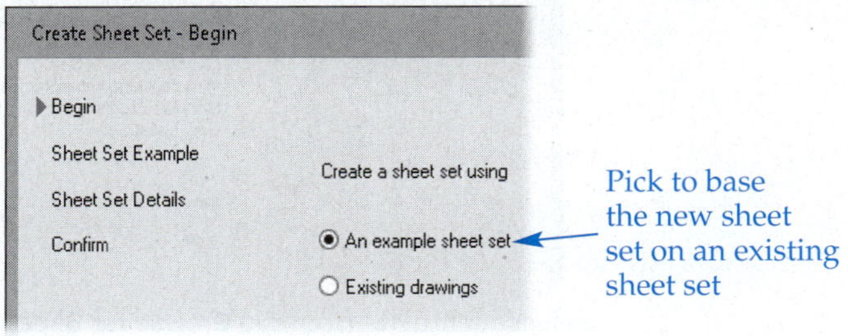

An Example Sheet Set Option

Pick the **An example sheet set** radio button on the **Begin** page of the **Create Sheet Set** wizard to generate a new sheet set from an existing sheet set that has properties and settings similar to those you want to apply to the new sheet set. This concept is much like using a drawing template to begin a new drawing. Later, you can modify the characteristics from the example sheet set in the new sheet set as needed. Beginning with an example sheet set is useful if a sheet set is available that closely matches the sheet set you intend to create. For example, use the Cappuccino Express sheet set shown in Figure 32-1 as an example sheet set for projects with similar characteristics. All properties and many settings that you assign to the example sheet set, such as the General, Architectural, and Structural ***subsets*** shown in Figure 32-1, are reproduced in the new sheet set.

subsets: Groups of similar layouts, such as those in the same discipline, sometimes based on folder hierarchy.

PROFESSIONAL TIP

Avoid using an example sheet set if the new sheet set is significantly different from an available example sheet set. It may take more time to modify the sheet set than to create a new sheet set that is specific to the new project.

Sheet Set Example Page

Pick the **Next** button to display the **Sheet Set Example** page. See Figure 32-5. When you create a sheet set from an example sheet set, you start from an existing DST file. The **Select a sheet set to use as an example** radio button is active by default, and a list box displays all DST files in the default Template folder. Select an example sheet set from the list, or pick the **New Sheet Set** option in the **Select a sheet set to use as an example** list box to begin a new sheet set from scratch.

You can use any existing sheet set as an example sheet set. To use an example sheet set not saved in the Template folder, pick the **Browse to another sheet set to use as an example** radio button and then select the ellipsis (...) button. Use the **Browse for Sheet Set** dialog box to locate and select a DST file in another folder.

Sheet Set Details Page

Select the **Next** button to display the **Sheet Set Details** page. See Figure 32-6. Use the **Sheet Set Details** page to modify the existing sheet set data and create settings for the new project. Most sheet set details are specific to a project. Enter the name, or title, of the sheet set in the **Name of new sheet set:** text box. The name is typically the project number or a short description of the project. Type a description for the sheet set in the **Description (optional):** area. The **Store sheet set data file (.dst) here:** text box

Figure 32-5.
Use the **Sheet Set Example** page to select an example sheet set.

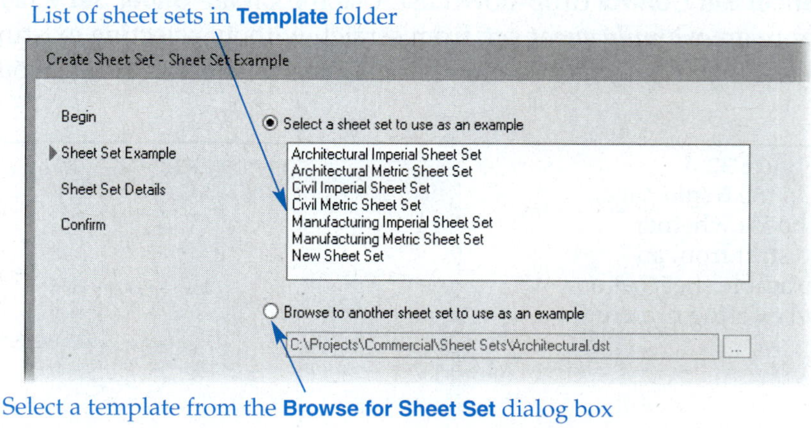

Figure 32-6.
On the **Sheet Set Details** page, type a name and description for the sheet set and specify the location of the DST file. You can also access additional sheet set properties.

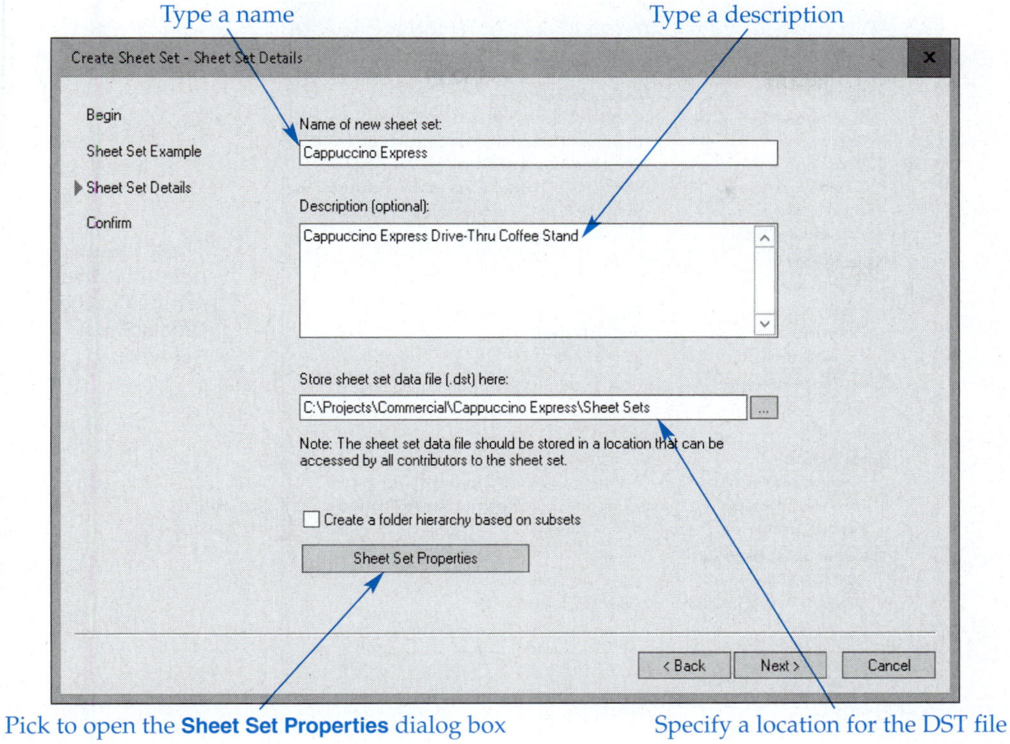

determines where the sheet set file is saved. Pick the ellipsis (**...**) button and select a folder in the **Browse for Sheet Set Folder** dialog box to change the default DST file location. Pick the **Create a folder hierarchy based on subsets** check box to allow AutoCAD to create subfolders that match the subsets in the example sheet set. Layouts in each folder are listed under each subset.

Pick the **Sheet Set Properties** button to adjust and create additional sheet set properties using the **Sheet Set Properties** dialog box. See **Figure 32-7**. Use the **Sheet Set Properties** dialog box to edit all properties except the **Sheet set data file** property. Methods of adjusting properties depend on the property, but include typing in a text box, selecting from a flyout, or picking the ellipsis (**...**) button to navigate to a specific location.

The **Sheet Set** category includes **Name**, **Sheet set data file**, and **Description** properties that correspond to the values you define at the **Sheet Set Details** page. The **Model view** property specifies the folder(s) containing drawing files with model space content that you intend to use for constructing new views on layouts in the sheet set. The **Label block for views** property specifies the block used to label views. The **Callout blocks** property specifies blocks available for use as callout blocks. The **Page setup overrides file** property determines the location of a DWT file containing a page setup that applies to all sheets in the sheet set. The **Project Control** category lists properties common to most projects, including **Project number**, **Project name**, **Project phase**, and **Project milestone**.

The properties in the **Sheet Creation** category determine the location of the drawing files for new sheets and the template and layout used to create them. You also have the option to create a new sheet and drawing file without referencing an existing layout. Use the **Sheet storage location** property in the **Sheet Creation** category to specify where to store new sheets. Remember the specified location so that you can locate the new drawings. Use the **Sheet creation template** property to define the drawing template and layout for creating new sheets. The **Select Layout as Sheet Template** dialog box appears when you select the ellipsis (**...**) button. See **Figure 32-8**.

Figure 32-7.
The **Sheet Set Properties** dialog box stores properties related to the sheet set, which are typically properties and settings that apply to the entire project.

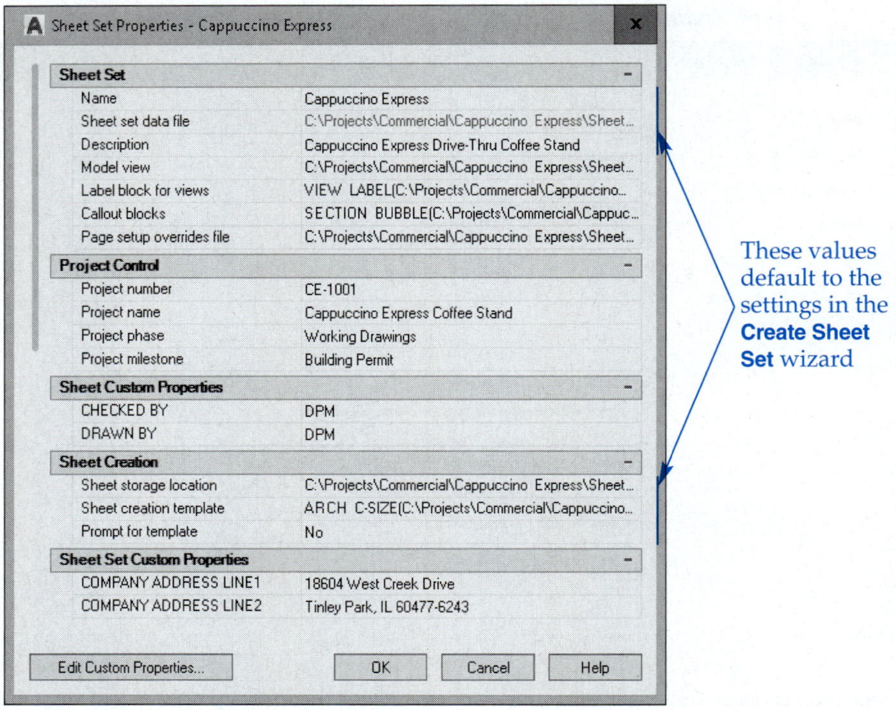

Figure 32-8.
Selecting an existing layout as a template for new sheets in a sheet set.

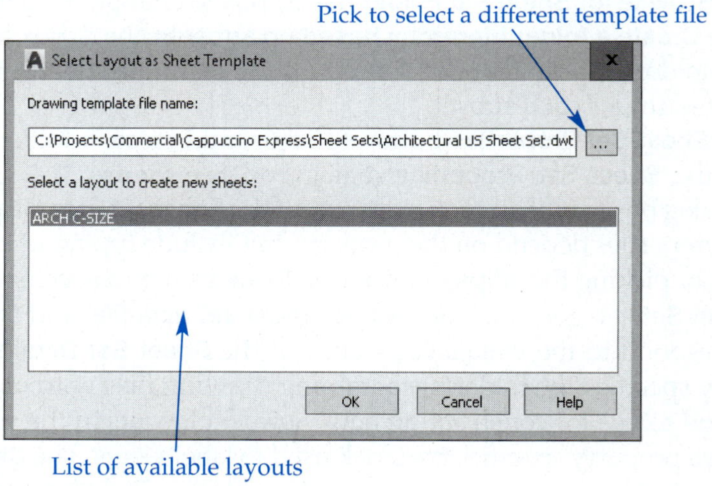

All layouts in the selected template appear in the list box. Select the appropriate layout and pick the **OK** button. Set the **Prompt for template** property to **No** to use the specified **Sheet creation template** property to create all new sheets. Select **Yes** to have the option to choose a different template to create a new sheet.

The **Sheet Set Custom Properties** category lists custom properties that you create by picking the **Edit Custom Properties...** button to access the **Custom Properties** dialog box. Custom properties are necessary to specify additional information on sheets, such as the properties shown in **Figure 32-7**, which are linked to fields in attributes of the title block. Use the **Add...** and **Delete** buttons in the **Custom Properties** dialog box to add and delete custom properties. After you set all values in the **Sheet Set Properties** dialog box, pick the **OK** button to return to the **Sheet Set Details** page.

NOTE

You are required to specify all relevant sheet set details if you choose the **New Sheet Set** option on the **Sheet Set Example** page.

Confirm Page

Pick the **Next** button to display the **Confirm** page. See Figure 32-9. The **Sheet Set Preview:** area displays all information associated with the sheet set. Use the **Back** button to return to previous pages to make changes. Pick the **Finish** button to create the sheet set.

PROFESSIONAL TIP

Copy and paste the information in the **Sheet Set Preview:** area of the **Confirm** page into a word processing program to save and print for reference.

Exercise 32-1

www.g-wlearning.com/CAD/

Complete the exercise on the companion website.

Existing Drawings Option

To add existing layouts to a new sheet set during the sheet set creation process, select the **Existing drawings** radio button on the **Begin** page of the **Create Sheet Set** wizard. See Figure 32-4. The **Existing drawings** option generates a new sheet set from scratch, without the option to use an example sheet set, but with the option to add existing layouts as sheets.

The **Existing drawings** option is useful if an example sheet set is unavailable or inappropriate, and if you have already developed multiple final or nearly final layouts.

Figure 32-9.
Use the **Confirm** page to preview settings before creating the sheet set.

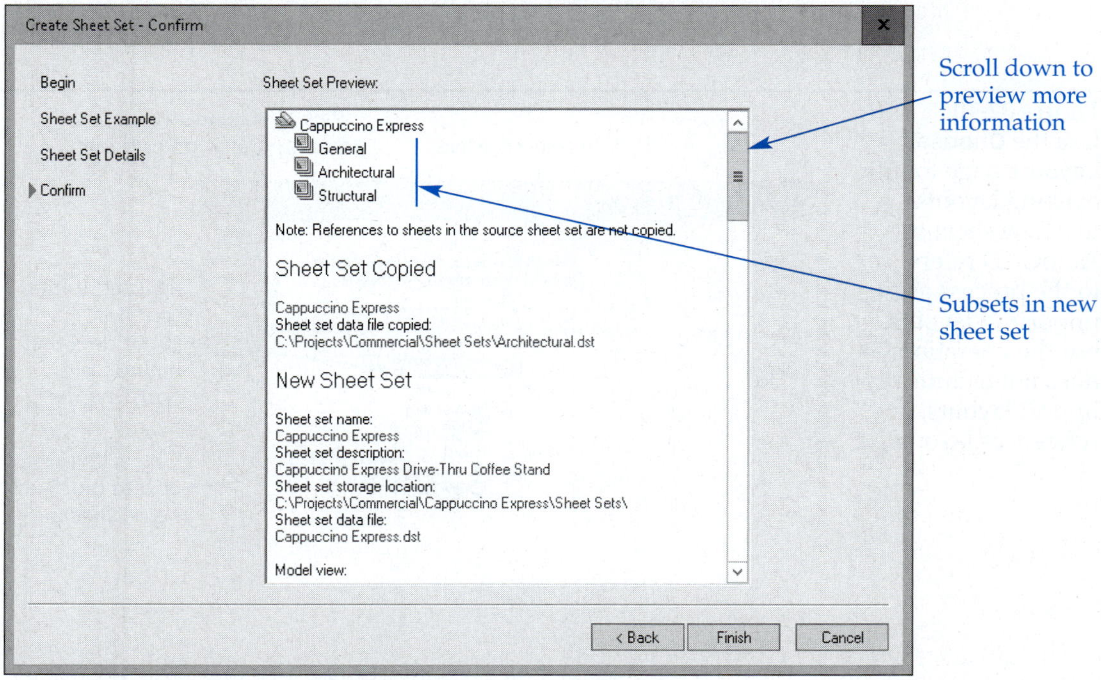

If you choose to apply the **Existing drawings** option, it is recommended that all drawings only include a single layout. In addition, organize files in a structured hierarchy of folders before creating the sheet set. This saves time because you can reproduce the subfolders as subsets in the new sheet set.

Sheet Set Details Page

Select the **Next** button to display the **Sheet Set Details** page. This is nearly the same page that appears when you create a new sheet set from an example sheet set, as shown in Figure 32-6. Use the **Sheet Set Details** page to define sheet set data and create properties for the new project. Because the **Existing drawings** option does not use an example sheet set, you must specify all relevant details, including custom properties.

Choose Layouts Page

Pick the **Next** button to display the **Choose Layouts** page. See Figure 32-10. The **Choose Layouts** page allows you to specify layouts in existing drawings to add to the sheet set as sheets. Pick the **Browse...** button to display the **Browse for Folder** dialog box, and select the folder containing the drawing files with the desired layouts. The list box displays the selected folder, subfolders, drawing files, and all layouts within the drawings. Repeat the process to add other folders to the list box as needed.

Check the boxes corresponding to the layouts you want to add to the new sheet set. Deselect layouts that you do not want to add to the sheet set, such as layouts in unneeded reference files or other drawings not directly associated with the set of drawings. Subfolders, files, and layouts are selected as one item. Uncheck a folder to uncheck all related drawing files and layouts. Uncheck a drawing file to uncheck all related layouts.

To adjust sheet naming and subset options, pick the **Import Options...** button to display the **Import Options** dialog box. See Figure 32-11. The name of a drawing with one layout, the name of the layout, and the name of the sheet in the sheet set are often the same. However, it is possible to adjust sheet naming when necessary. The **Prefix sheet titles with file name** check box is used to include the drawing file name with the name of the layouts that become sheets. For example, the sheet name Electrical Plan – First Floor Electrical forms from a layout named First Floor Electrical imported from the drawing file Electrical Plan.dwg.

Figure 32-10.
Use the **Choose Layouts** page to link existing layouts to a new sheet set. AutoCAD refers to the process as importing layouts, but the operation does not technically import layouts; it references layouts.

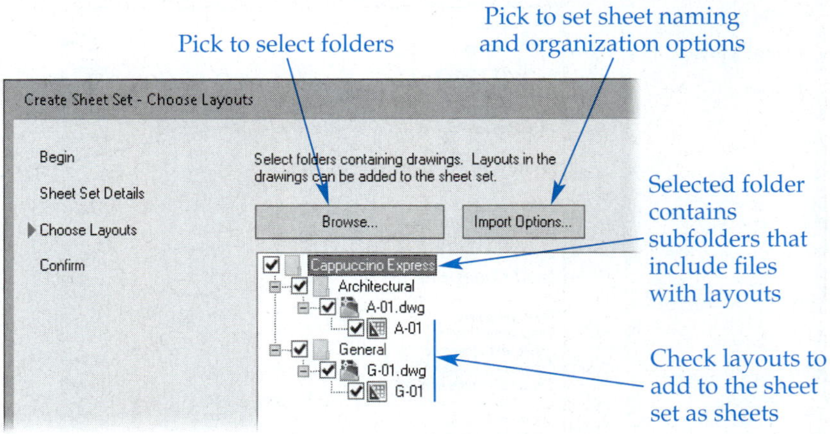

Figure 32-11.
The **Import Options** dialog box allows you to specify sheet naming conventions and folder structuring options.

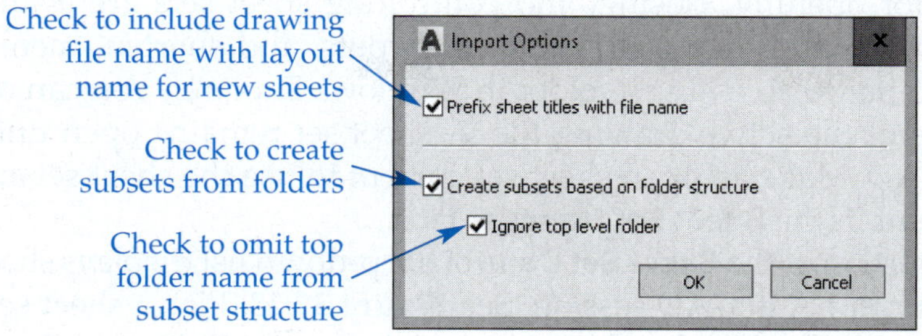

Select the **Create subsets based on folder structure** check box to organize a sheet set so that subfolders are grouped into subsets. Layouts in each folder are listed in each subset. The **Ignore top level folder** option determines whether the top level folder creates a subset. Figure 32-12A shows the **Choose Layouts** page with layouts imported from the Cappuccino Express folder for the Cappuccino Express sheet set. This sheet set uses the **Create subsets based on folder structure** and **Ignore top level folder** options, but not the **Prefix sheet titles with file name** option. Figure 32-12B shows the result of the configuration in the **Sheet Set Manager**. As shown in Figure 32-12B, each sheet can include a number preceding the sheet name, separated by a dash.

Confirm Page

Pick the **Next** button to display the **Confirm** page. This is the same page that appears when you create a new sheet set from an example sheet set. The **Sheet Set Preview:** area displays all information associated with the sheet set. Use the **Back** button to return to previous pages to make changes. Pick the **Finish** button to create the sheet set.

Exercise 32-2

www.g-wlearning.com/CAD/

Complete the exercise on the companion website.

Figure 32-12.
Creating a sheet set named Cappuccino Express with subsets. A—Layouts imported from the Architectural and General subfolders in the Cappuccino Express folder. Subfolders are designated as subsets for the new sheet set. B—Subsets shown in the **Sheet Set Manager**.

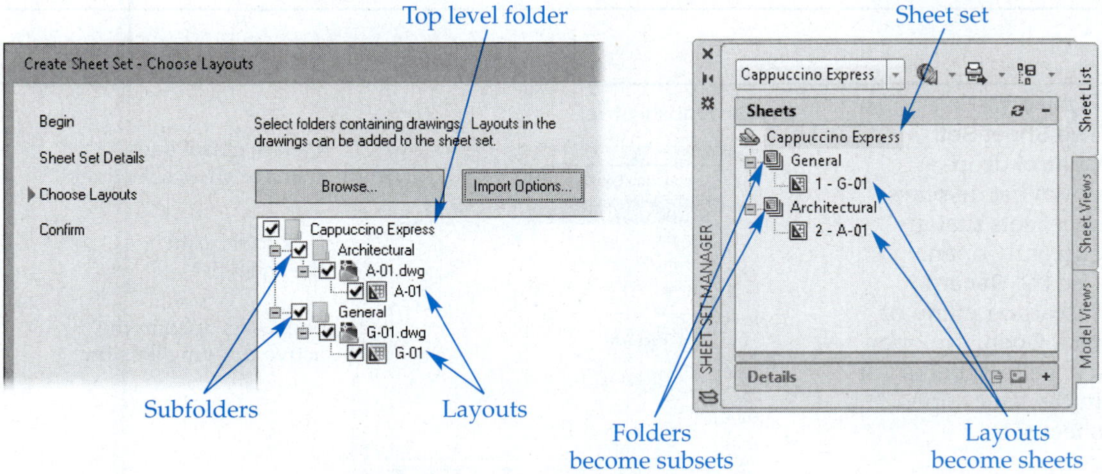

A B

Managing Sheet Sets

Options for opening, closing, and controlling sheet sets are available from the **Sheet Set Control** drop-down list, buttons, flyouts, and shortcut menus. Opening a sheet set is similar to accessing a folder in Windows Explorer. You can open any sheet set, regardless of the active drawing file. A sheet set remains open until you close it or exit AutoCAD. Make an open sheet set current to use the sheet set and display the sheet set content in the **Sheet Set Manager** tabs.

OPENSHEETSET

Type
OPENSHEETSET

The top portion of the **Sheet Set Control** drop-down list displays sheet sets that are open in the current AutoCAD session. See Figure 32-13. Pick a sheet set from the list to make it current. The list clears when you exit AutoCAD. Use the **Recent** cascading menu to open a recently opened sheet set. Select **Open...** to display the **Open Sheet Set** dialog box. Then navigate to a DST file and open the file in the **Sheet Set Manager**. You can also use the **OPENSHEETSET** command from outside of the **Sheet Set Manager** to open a sheet set.

NOTE

The **OPENSHEETSET** command is also available by picking the **Open a Sheet Set...** option from the **Get Started** column on the **Create** page of the **Start Tab**.

To close a sheet set, right-click on an open sheet set in the **Sheet Set Control** drop-down list or on the sheet set title in the **Sheet List** tab, and select **Close Sheet Set**. Sheet sets automatically close when you exit AutoCAD. Closing a sheet set removes the sheet set from the **Sheet Set Manager** and allows you to delete the DST file if necessary. To save all of the drawing files with layouts associated with the current sheet set at the same time, right-click on a sheet set title in the **Sheet List** tab and select **Resave All Sheets**.

Right-click on the sheet set and pick **Properties...** to display the **Sheet Set Properties** dialog box. This is the same dialog box that is available from the **Sheet Set Details** page when you create a new sheet set, as shown in Figure 32-7.

NOTE

Close drawing files with layouts associated with the current sheet set to update the sheet set according to changes made to the drawings. To update changes to the sheet list manually, pick the **Refresh sheet status** button. See Figure 32-13.

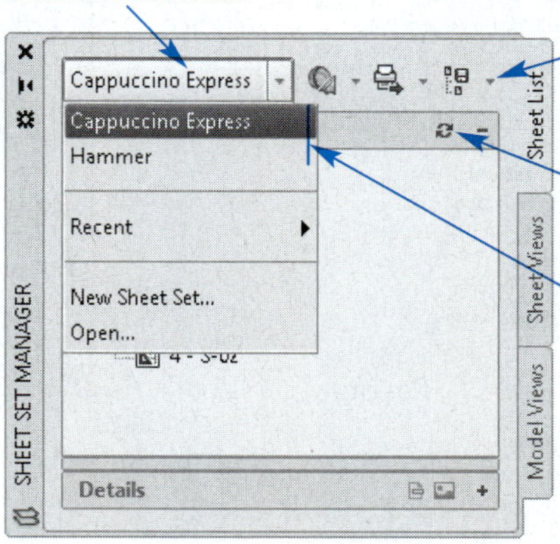

Figure 32-13.
The **Sheet Set Control** drop-down list displays sheet sets that are currently open. Use the **Recent** cascading menu or pick **Open...** to open a sheet set that is not in the list of open sheet sets.

Subsets

Subsets provide a way to organize sheets in a sheet set, similar to using subfolders to organize files with Windows Explorer. The example in Figure 32-10 uses a General subset to group all general sheets for a building project and an Architectural subset to group all architectural sheets. Another example is using subsets to organize assemblies, subassemblies, and parts for a mechanical design project. Although subsets are not required, they can provide significant help in managing a sheet set, especially when the sheet set includes multiple sheets. For example, you can set a subset not to publish instead of taking the time to set each individual sheet not to publish.

Creating Subsets

The **Existing drawings** option for creating a new sheet set includes the **Create subsets based on folder structure** option that allows you to form subsets based on an existing folder hierarchy during the process of creating a sheet set. For existing sheet sets, use the **Sheet List** tab of the **Sheet Set Manager** to create new subsets. Add a subset to a sheet set or to an existing subset to create the desired subset hierarchy. Right-click on the sheet set or subset name in the **Sheet List** tab and select **New Subset...** to open the **Subset Properties** dialog box. See Figure 32-14. Type the name of the subset in the **Subset Name** text box. For example, name a subset Structural to contain all structural sheets in a sheet set for a building project. Select **Yes** from the **Create Folder Hierarchy** drop-down list to create a new folder that corresponds to the subset.

The **Publish Sheets in Subset** property determines whether sheets in the subset are published. The **Do Not Publish Sheets** setting prevents sheets in the subset from being published. An icon identifies subsets set not to publish. The **New Sheet Location** property determines the path to which new sheets are saved when you add a new sheet from the subset. The default location is the folder associated with the sheet set or subset in which you create the new subset, or the folder you specified when you created the sheet set.

Use the **Sheet Creation Template** property to specify a drawing template and layout to use when creating new sheets in the subset. For example, assign a specific drawing template and layout to an Architectural subset to create new architectural discipline sheets, and assign a different drawing template and layout to a Structural subset to create new structural discipline sheets. The process for specifying a sheet creation template and layout for a subset is identical to the process for selecting the sheet creation template and layout for a sheet set. Use the **Prompt for Template** drop-down list to indicate whether a prompt should ask for a sheet template instead of using the specified sheet creation template.

Figure 32-14.
The **Subset Properties** dialog box allows you to define a new subset.

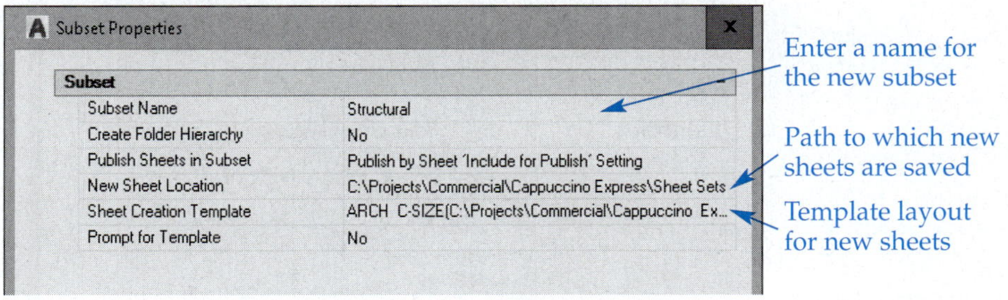

Managing Subsets

Drag and drop subsets as needed to restructure the sheet set. Right-click on a subset in the **Sheet List** tab and pick **Collapse** to collapse the nodes of a subset. To make changes to an existing subset, right-click on the subset and select **Properties...** to redisplay the **Subset Properties** dialog box. The **Rename Subset...** shortcut menu option also opens the **Subset Properties** dialog box. To delete a subset, right-click on the subset and select **Remove Subset**. The **Remove Subset** option is not available if the subset contains sheets.

Exercise 32-3

www.g-wlearning.com/CAD/

Complete the exercise on the companion website.

Sheets

A sheet set or subset contains sheets, just as a folder or subfolder contains drawing files in Windows Explorer. A sheet is a path to a single layout in a drawing file. See Figure 32-15. To open a sheet, double-click on the sheet or right-click on the sheet and select **Open**. The drawing file that contains the referenced layout tab opens and displays the layout corresponding to the sheet. You can also open a sheet as ***read-only***.

read-only: Describes a drawing file opened for viewing only. You can make changes to the drawing, but you cannot save changes without using the **SAVEAS** command.

NOTE

When you use any technique to open a file, including opening files from the **Sheet Set Manager**, the files are added to the open files list. Use file tabs or the window control commands on the ribbon to view open files. Save and close any files that you do not need open.

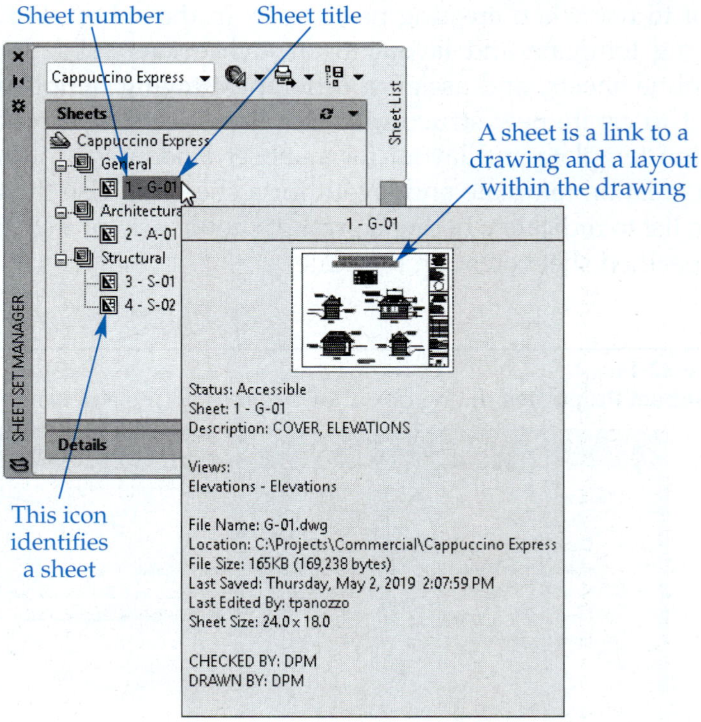

Figure 32-15. Sheets in the set of drawings for the example Cappuccino Express building project. Hover over a sheet to display a detailed tooltip.

Adding a Layout as a Sheet

The **Existing drawings** option for creating a new sheet set includes the **Choose Layouts** page that allows you to add existing layouts to a sheet set as sheets. You can also use the **Sheet Set Manager** to add a layout to an existing sheet set. Another method to add a layout to a sheet set is to add the layout directly from an open drawing. To add an existing layout to a sheet set using the **Sheet Set Manager**, right-click on the sheet set or the subset that will contain the sheet, and select **Import Layout as Sheet...**. The **Import Layouts as Sheets** dialog box appears. See Figure 32-16.

Pick the **Browse for Drawings...** button to select a drawing file. All layouts in the drawing file appear in the list box and are checked for import by default. The **Status** column indicates whether the layout can be imported into the sheet set. You can only use a layout in one sheet set. In other words, you cannot import a layout that is already part of a sheet set. Each checked layout is imported as a separate sheet. Importing layouts only links layouts to the sheet set. The original file remains unchanged, and no new drawing files or layouts form. Uncheck a box to exclude a layout from importing. Select the **Prefix sheet titles with file name** check box to include the name of the file in the sheet title. Pick the **Import Checked** button to import the selected layouts as sheets.

A layout tab provides a way to add existing layouts to a sheet set directly from an open drawing. Right-click on the layout tab or thumbnail image you want to import and select **Import Layout as Sheet...**. The **Import Layouts as Sheets** dialog box appears with the selected layout listed. You must save the drawing and set up the layout to make the **Import Layout as Sheet...** menu option available.

NOTE

Right-click on a subset in the **Sheet Set Manager** and select **Import Layout as Sheet...** to import layouts into a subset. Importing a layout from a layout tab does not provide an initial option to add layouts as sheets in a subset. You must drag and drop the sheets into the appropriate subset after importing.

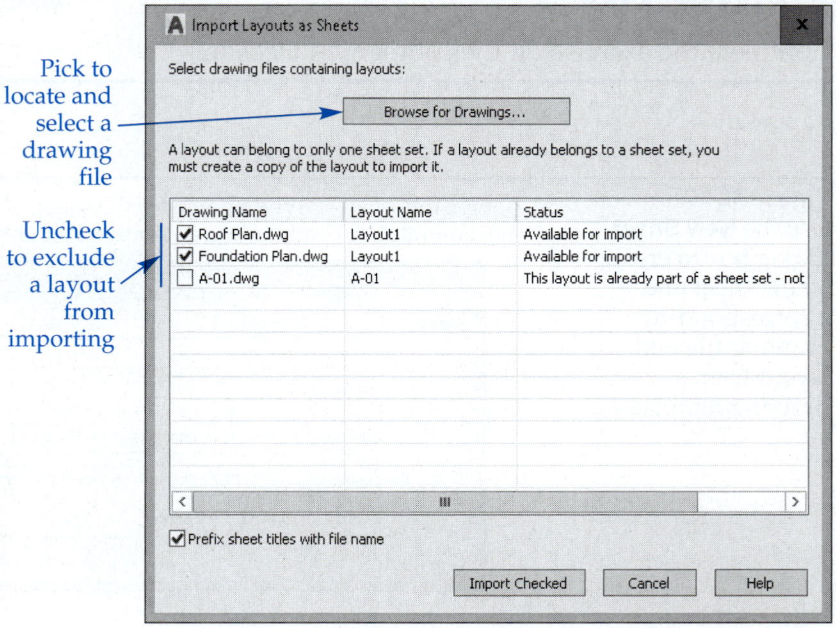

Figure 32-16. Use the **Import Layouts as Sheets** dialog box to link existing layouts to a sheet set as sheets. AutoCAD refers to the process as importing layouts, but the operation does not technically import layouts; it references layouts.

PROFESSIONAL TIP

To use the same layout as a sheet in a different sheet set, create a new file and attach the existing drawing file as an xref. Create a layout of the model space drawing in the new file and access the layout through the **Import Layouts as Sheets** dialog box.

Adding a Sheet Using a Template

Another option for adding a sheet to a sheet set is to create a new drawing using a specific template and layout, and at the same time incorporate the new layout into the sheet set as a sheet. Externally reference existing model space content to the new sheet, or use the drawing as a blank sheet for developing new geometry.

Before adding a sheet from a template, consider assigning a specific drawing template to use for new sheet creation in the **Sheet creation template** setting in the **Sheet Set Properties** and **Subset Properties** dialog boxes. Then set the **Prompt for template** property to **No** to force AutoCAD to use only the specified template, or choose **Yes** to have the option to select a different template when you create a new sheet.

Right-click on the sheet set or subset that will contain the new sheet, and select **New Sheet...**. If you do not assign a drawing template to use for new sheet creation, or if you selected **Yes** for the **Prompt for template** option, AutoCAD displays the **Select Layout as Sheet Template** dialog box. The **New Sheet** dialog box appears if a specified template layout exists, or after you select the template layout. See Figure 32-17.

Type the sheet number in the **Number:** text box and the sheet name in the **Sheet title:** text box. A new sheet creates a new drawing file, and the sheet title becomes the name of the layout in the drawing file. Enter the file name in the **File name:** text box. The file name is the sheet number and title by default. Edit the drawing file name, if necessary. For example, you might remove the sheet number from the name. The **Folder path:** display box shows where the drawing file will be saved, as specified in the **Sheet Set Properties** or **Subset Properties** dialog box. Check **Open in drawing editor** to open the file and display the layout.

Exercise 32-4

www.g-wlearning.com/CAD/

Complete the exercise on the companion website.

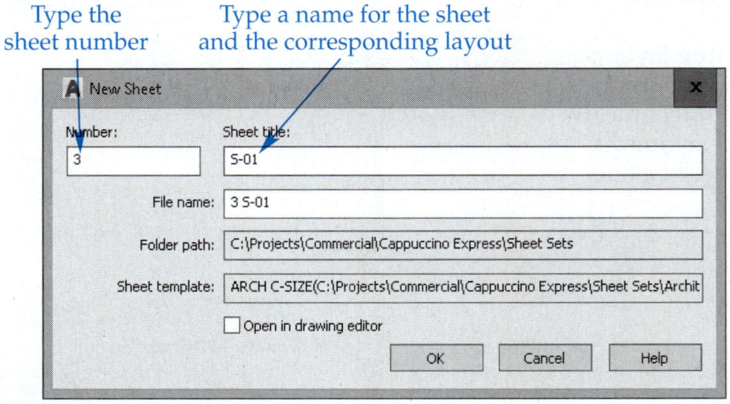

Figure 32-17. Use the **New Sheet** dialog box to create a new sheet and a corresponding drawing file and layout from a drawing template.

Managing Sheets

Drag and drop sheets as needed to restructure the sheet set and move sheets from the sheet set to subsets or between subsets. Right-click on a sheet in the **Sheet Set Manager** and pick **Rename & Renumber...** to access the **Rename & Renumber Sheet** dialog box. Use the **Number:** text box to renumber the sheet and the **Sheet title:** text box to rename the sheet. Check **Rename layout to match: Sheet title** to name the layout to match the modified sheet name. The **Layout name:** text box is available if you deselect the **Rename layout to match: Sheet title** check box, and allows you to change the layout name independently of the sheet name. The **Prefix with sheet number** check box is also available.

Options for renaming the file are available only if the drawing file is closed. Check **Rename drawing file to match: Sheet title** to name the file to match the modified sheet name. The **File name:** text box is available if you deselect the **Rename drawing file to match: Sheet title** check box, and allows you to change the file name independently of the sheet name. The **Prefix with sheet number** check box is also available. If the sheet is one of several in a subset, pick the **Next** and **Previous** buttons to access different sheets in the subset.

Right-click on a sheet and pick **Properties...** to display the **Sheet Properties** dialog box. See Figure 32-18. The **Sheet Properties** dialog box offers another way to change the sheet name and number, and provides additional properties. Use the text boxes to type the appropriate information. It is important that you specify all relevant information to define the project completely, so that text and blocks containing attributes with fields display the correct content. The **Include for Publish** option determines whether the sheet is published or plotted with the sheet set. The default value is **Yes**.

The **Expected layout** and **Found layout** text boxes display the path to the file where you initially saved the sheet and the path to the file where AutoCAD found the sheet. If the paths are different, pick the ellipsis (**...**) button to update the **Expected layout** setting using the **Import Layout as Sheet** dialog box. The **Sheet Properties** dialog box includes the same **Rename options** area found in the **Rename & Renumber Sheet** dialog box.

Right-click on a sheet and select **Remove Sheet** to remove the sheet from the sheet set. Removing a sheet does not delete the drawing file.

Figure 32-18.
The **Sheet Properties** dialog box allows you to modify sheet properties. Add all relevant sheet details that will appear in title blocks and similar content.

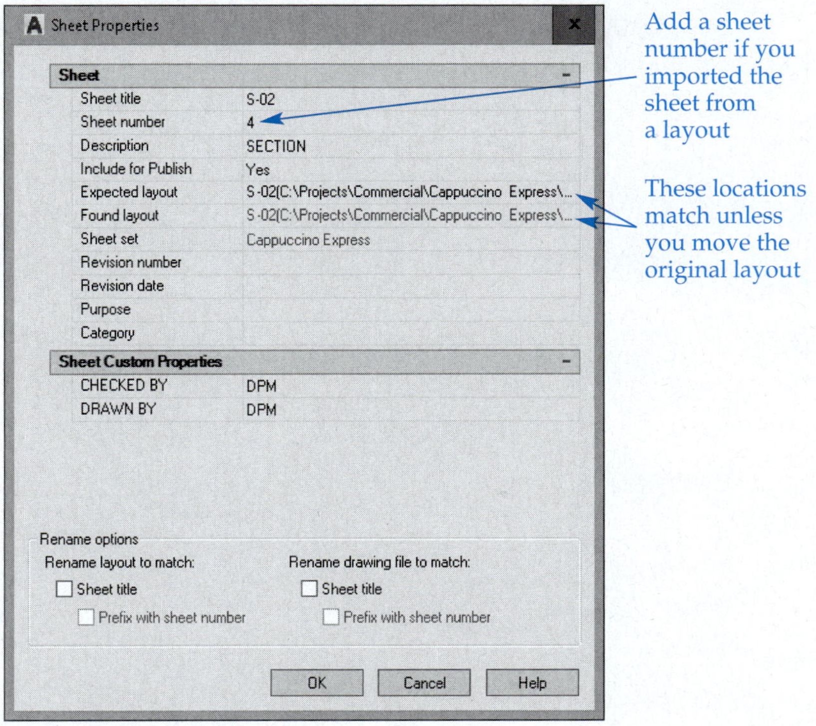

NOTE

When you change the location of a drawing file that has layouts associated with a sheet set, the association with the sheet set is broken. Update the specified path to the drawing file in the **Sheet Properties** dialog box or re-import the layouts into the sheet set.

Exercise 32-5

www.g-wlearning.com/CAD/

Complete the exercise on the companion website.

Chapter Review

Answer the following questions using the information in this chapter.

1. What does the term *sheet* refer to in relation to a sheet set and a drawing file?
2. What is a sheet set?
3. What is the purpose of a sheet set data file, and what extension does the file have?
4. What is the purpose of fields in a sheet set?
5. Explain when you would add existing layouts to a sheet set and when you should add a new layout.
6. Describe the general options for creating a new sheet set.
7. What are subsets in relation to a sheet set?
8. How do you add a custom property to a sheet set?
9. Explain how to create a subset in an existing sheet set.
10. How do you modify a sheet name or number?

Drawing Problems

Start AutoCAD if it is not already started. Follow the specific instructions for each problem.

▼ Basic

1. General Create a new sheet set using the **Create Sheet Set** wizard and the New Sheet Set example sheet set. Name the new sheet set My Sheet Set.

2. Civil Create a new sheet set using the **Create Sheet Set** wizard and the Civil Imperial Sheet Set example sheet set. Name the new sheet set Civil Sheet Set.

▼ Intermediate

3. General Use a word processor to write a report of approximately 250 words explaining the purpose of sheet sets. Cite at least three examples from actual industry applications of using sheet sets to help manage a set of drawings. Use at least two sketches to illustrate your report.

▼ Advanced

4. Mechanical Design and draw an arbor press. Prepare layouts for the assembly and each component. Then create a new sheet set to organize the drawings and layouts.

AutoCAD Certified Professional Exam Practice

Answer the following question using the information in this chapter.

1. Which of the following actions can you perform using the **Sheet Set Manager**? *Select all that apply.*
   A. add sheets to a sheet set
   B. change the name of a sheet set data file
   C. organize files related to a project
   D. open drawing files that contain layouts referenced in a sheet set
   E. purge unused blocks from a sheet set

CHAPTER

33

Additional Sheet Set Tools

Learning Objectives

After completing this chapter, you will be able to:

- ✓ Create and use sheet views with callout and view label blocks.
- ✓ Add model views to a sheet set.
- ✓ Create sheet list tables.
- ✓ Manage sheet set fields.
- ✓ Publish a sheet set.
- ✓ Archive a sheet set.

After you prepare a sheet set with subsets and sheets, you are ready to continue sheet set development using sheet views and objects specifically related to sheet set applications. This chapter explains this process and describes options for publishing and archiving a sheet set. In order to work effectively with sheet sets, you must understand the elements of a sheet set such as drawing templates, blocks, and fields, as explained throughout this textbook.

Sheet Views

Use the **Sheet Views** tab of the **Sheet Set Manager** to group views by category, open views for viewing and editing, and add ***callout blocks*** and ***view label blocks***. ***Sheet views*** provide a way to access specific views and to link drawing views to sheets in the sheet set. You must create sheet views to add callout and view label blocks from the **Sheet Set Manager**. Callout and view label blocks use attributes containing fields that link data between the sheet set and drawing views. The fields update automatically to reflect changes in sheet numbering and organization. See Figure 33-1.

One option to create a sheet view is to use the **NEWVIEW** command in paper space to prepare a named view of a layout. This technique essentially forms a copy of a sheet as a sheet view that you can reference to add callout and view label blocks. The second option to create a sheet view is to add a ***model view*** to a layout. You will learn about model views later in this chapter.

callout block: A block that uses attributes containing fields that link the view number and sheet title between the sheet set and drawing (sheet) views.

view label block: A block that uses attributes containing fields that link the view name, number, and scale to drawing (sheet) views.

sheet view: A layout or model view saved for use in a sheet set; allows you to add views to layouts and insert callout and view label blocks.

model view: A drawing file or named model space view added to a layout to create a sheet view.

Figure 33-1.
Sheet views allow you to access specific views and add callout and view label blocks that link drawing views to sheets throughout the sheet set.

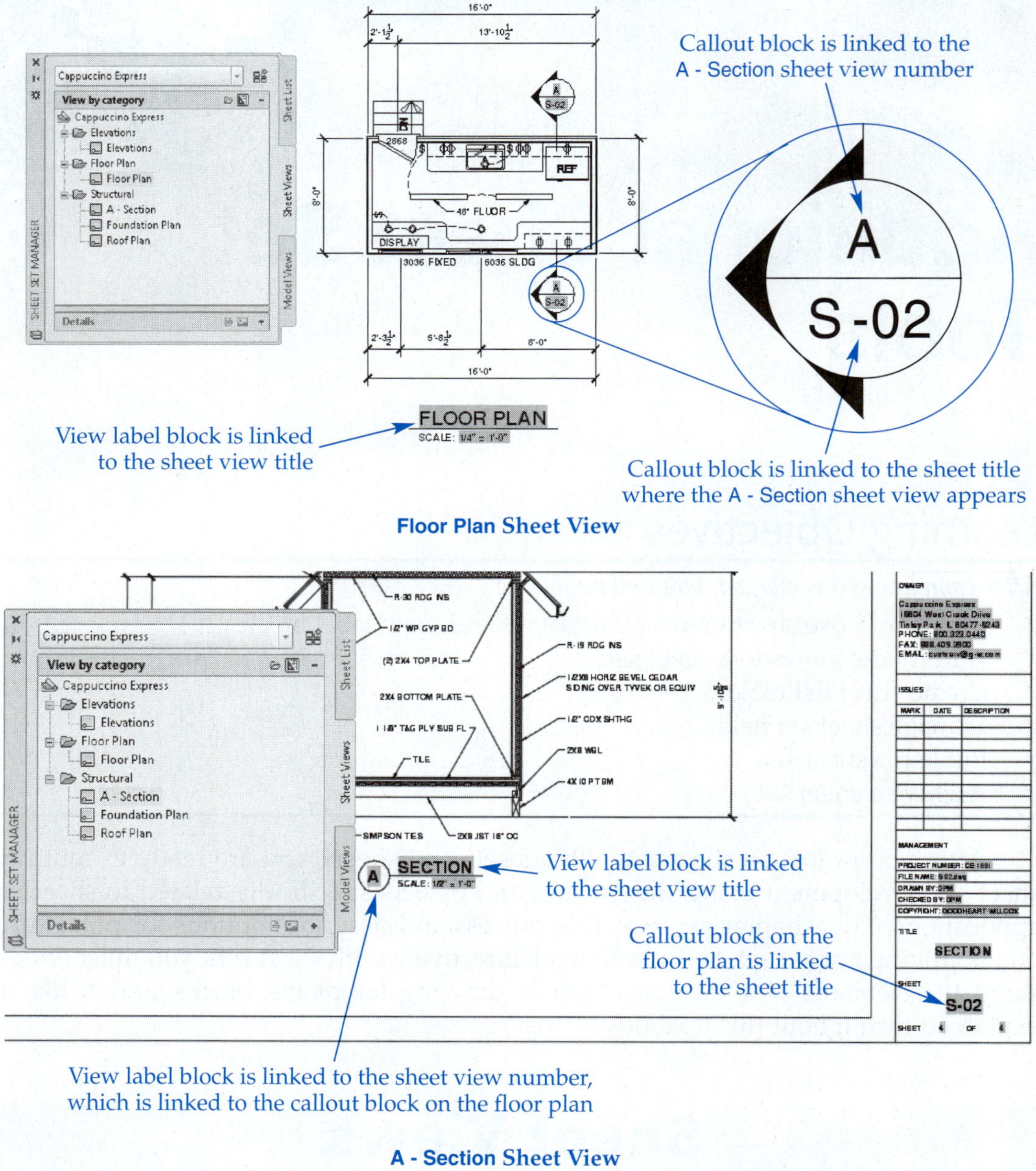

View Categories

View categories organize views in the **Sheet Views** tab, similar to subsets in the **Sheet List** tab. However, view categories group views, while subsets group sheets that might include several sheet views. For example, an Architectural subset includes an Elevations sheet and other architectural sheets. The Elevations sheet includes Front Elevation, Left Elevation, Rear Elevation, and Right Elevation views, added to the sheet using sheet views of the same name and grouped in an Elevations sheet view category.

One method to add a view category is to type a value in the **View category:** text box in the **New View/Shot Properties** dialog box accessed with the **NEWVIEW** command. To add a view category without using the **NEWVIEW** command, pick the **Sheet Views** tab and select the **View by category** button. See Figure 33-2. Pick the **New View Category** button or right-click on the sheet set name and select **New View Category...** to open the **View Category** dialog box. See Figure 33-3. Type a name for the category in the **Category name:** text box.

NEWVIEW
Ribbon
View
> Named Views
New View
Type
NEWVIEW
NVIEW

Figure 33-2.
Create view categories in the **Sheet Views** tab of the **Sheet Set Manager**.

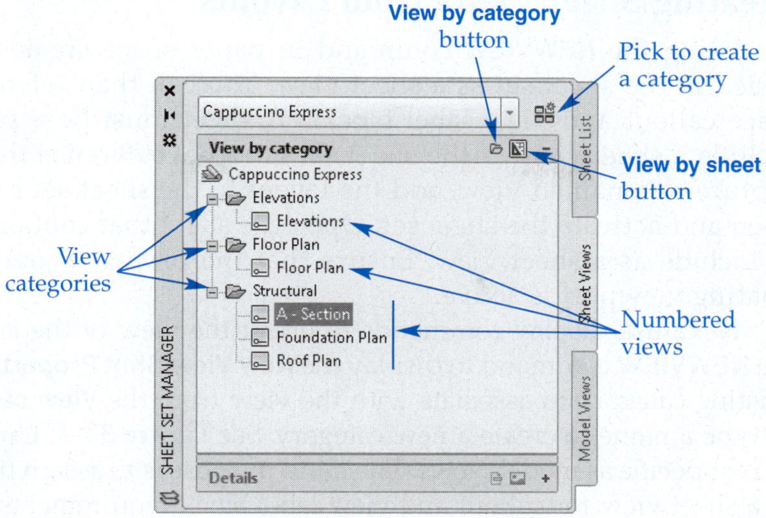

Figure 33-3.
The **View Category** dialog box allows you to name the category and select callout blocks for use with views.

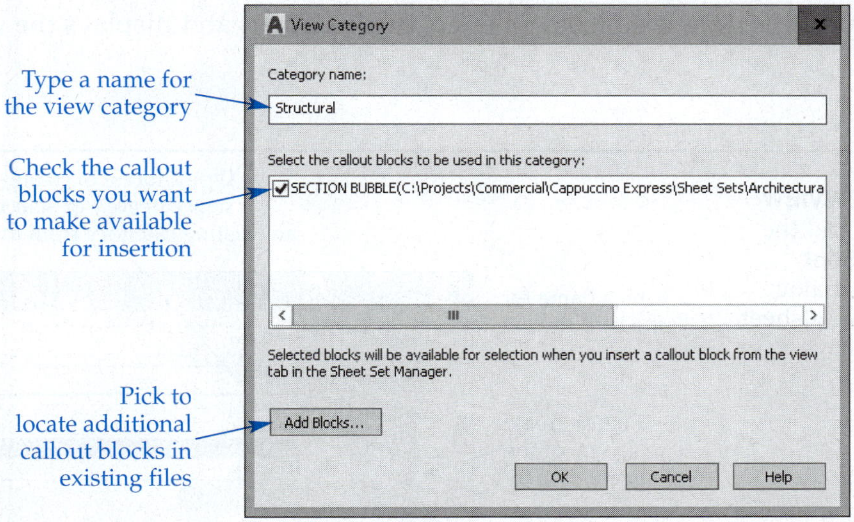

The **View Category** dialog box lists all available callout blocks for the current view category. Check the box next to the callout block to make the block available for all views added to the category. If a block is not in the list, pick the **Add Blocks...** button to access the **List of Blocks** dialog box, from which you can locate a drawing file with blocks to add. After you select the necessary callout blocks, pick the **OK** button to create the new category.

View categories are listed alphanumerically. To change the category name and add callout blocks, right-click on a view category in the **Sheet Views** tab and pick **Rename...** or **Properties...** to reopen the **View Category** dialog box. To delete a category, right-click on the category and select **Remove Category**. The **Remove Category** option is not available if the category includes sheet views. You must remove all sheet views from the category before deleting the category.

Exercise 33-1

www.g-wlearning.com/CAD/

Complete the exercise on the companion website.

Creating Sheet Views from Layouts

Using the **NEWVIEW** command in paper space creates a layout view that is added to the sheet set as a sheet view. You can then reference the sheet view to place callout and view label blocks. A sheet must be a part of the sheet set to include a sheet view. If the sheet set does not reference the layout you intend to capture as a named view, add the layout to the sheet set before continuing. Then open and activate the sheet set. Open the sheet that contains the layout you want to include as a sheet view. Ensure that the proper layout is active, and that no floating viewport is active.

Next, use display commands to orient the view of the layout as needed. Access the **NEWVIEW** command to display the **New View/Shot Properties** dialog box. Select an existing category to associate with the view from the **View category:** drop-down list, or type a name to create a new category. See **Figure 33-4**. Layout views are typically not as specific as model views. The main purpose is to assign the layout to the sheet set as a sheet view for callout and view label block requirements. Therefore, the **Current display** boundary option is often acceptable. Specify the remaining view settings and pick the **OK** button as needed to save the view.

The new layout view appears in the **Sheet Set Manager** in the specified view category. To display the view from the **Sheet Set Manager**, double-click on the view or right-click on the view and select **Display**. If the drawing file is open, the view is set current. If the drawing file is not open, the file opens and displays the view.

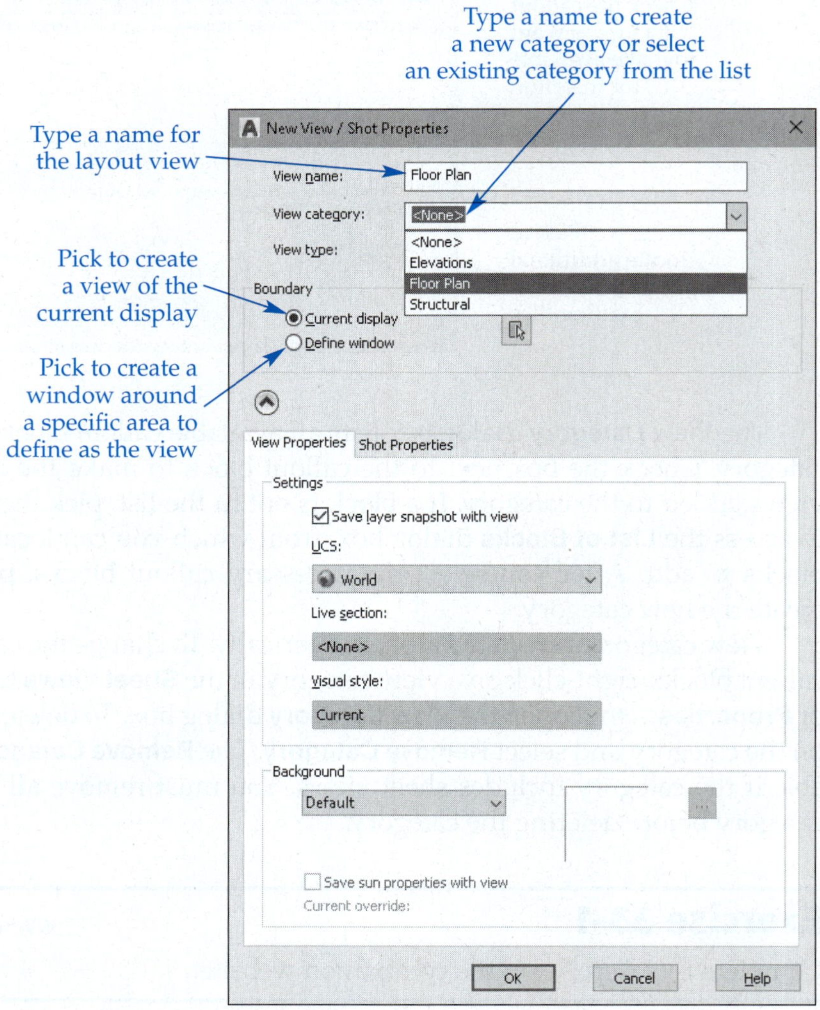

Figure 33-4. Use the **NEWVIEW** command and the **New View/Shot Properties** dialog box to create a sheet view.

Managing Sheet Views

Display sheet views by category or sheet using the appropriate button shown in Figure 33-2. Sheet views are listed alphanumerically in both formats. Pick the **View by category** button to display sheet views within view categories. Pick the **View by sheet** button to display sheet views in the sheets where they are located. The **View by sheet** display also provides an option to change the category assigned to a sheet view. Right-click on a sheet view and select a different category from the **Set category** cascading menu.

To change the name or number of a sheet view, right-click on the sheet view in the **Sheet Views** tab and select **Rename & Renumber...** to display the **Rename & Renumber View** dialog box. See Figure 33-5. Enter a number or letter for the view in the **Number:** text box. A view number is typically not required for views that are not linked to other sheets, such as the Floor Plan view shown in Figure 33-1. However, the view number is critical for views referenced on other sheets, such as the Section view shown in Figure 33-1, which uses the letter A as the view number. The view number is displayed in front of the view name in the **Sheet Set Manager**. Modify the view name in the **View title:** text box. Use the **Next** and **Previous** buttons to renumber or rename different views in the sheet or view category. Pick the **OK** button when you are finished.

Exercise 33-2

www.g-wlearning.com/CAD/

Complete the exercise on the companion website.

Callout Blocks

A common example of a callout block is the section view information at the end of a cutting-plane line, as shown on the floor plan in Figure 33-6. In this example, the upper portion of the callout block includes an attribute with a **SheetView** field that references the **ViewNumber** property of the A - Section sheet view. The lower portion of the callout block includes an attribute with a **SheetSet** field that references the **SheetTitle** property of the 4 - S-02 sheet. The result is a link from the cutting-plane line on the floor plan to the sheet where the section is located and the view identification.

Callout blocks automate references between drawing content and views by updating to reflect changes in sheet numbering and organization. Other examples of callout blocks include viewing-plane line symbols, such as exterior and interior elevation bubbles, and detail identification. You can also use callout blocks as balloons on an assembly drawing to associate components with a parts list, as found in a set of drawings. A callout block typically appears on a different sheet from the sheet view it references. Callout blocks can also include hyperlink fields, or ***hyperlinks***, that provide convenient access to the reference sheet view.

hyperlinks: Links in a document that connect it to related information in other documents or on the Internet.

PROFESSIONAL TIP

Use dynamic blocks with suitable parameters, such as rotation, flip, and visibility parameters, to reduce the number of blocks needed for a sheet set.

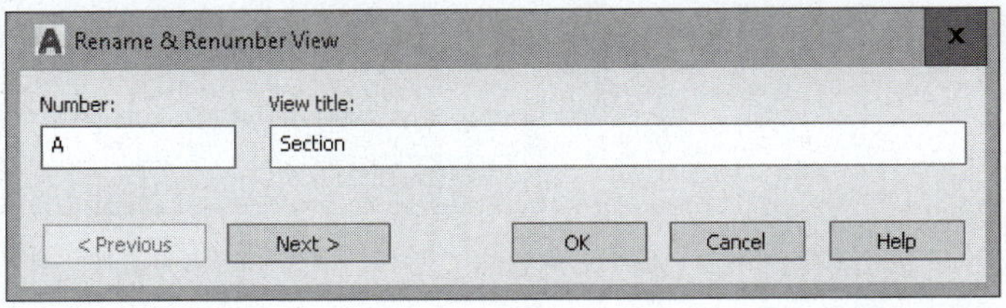

Figure 33-5.
Use the **Rename & Renumber View** dialog box to renumber or rename a sheet view.

Figure 33-6.
Using a callout block assigned to a sheet view to link a cutting plane on a floor plan to a section view on a separate sheet.

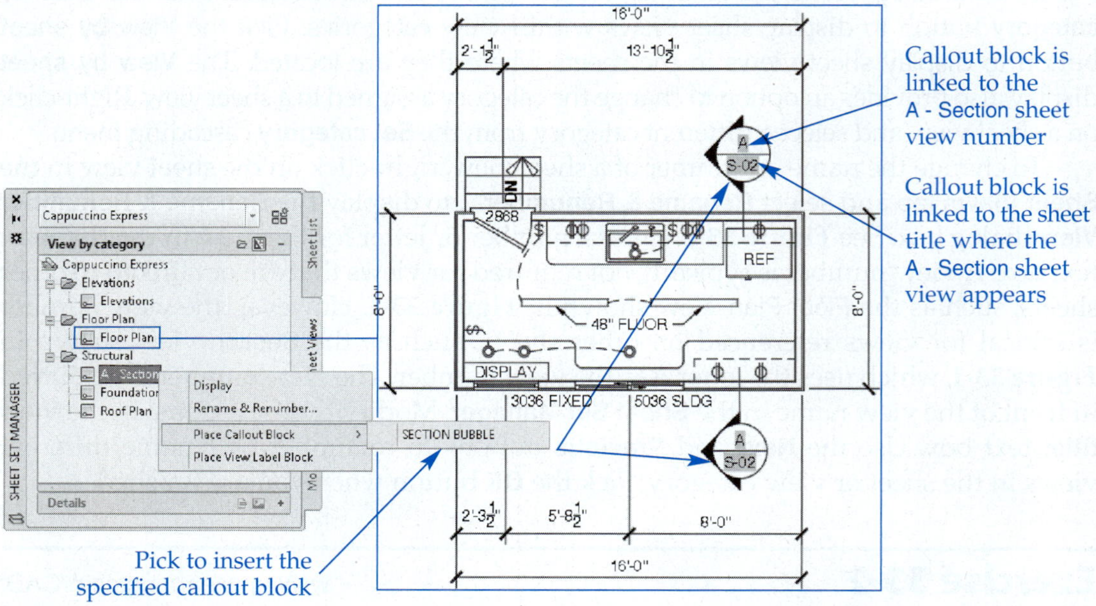

Before you can insert a callout block from the **Sheet Set Manager**, the block must be available to the sheet set and assigned to the view category. A sheet set or view category can reference multiple callout blocks. Callout blocks are often added to a sheet set during sheet set creation. To assign callout blocks to an existing sheet set, right-click on the sheet set in the **Sheet Set Manager** and pick **Properties...** to display the **Sheet Set Properties** dialog box. Specify the available blocks in the **Callout blocks** text box. Pick the ellipsis (...) button to access the **List of Blocks** dialog box. See **Figure 33-7**. Pick the **Add...** button to access the **Select Block** dialog box, and then pick the ellipsis button to navigate to and select the drawing or template file that contains callout blocks.

Figure 33-7.
All of the callout blocks available to a sheet set appear in the **List of Blocks** dialog box.

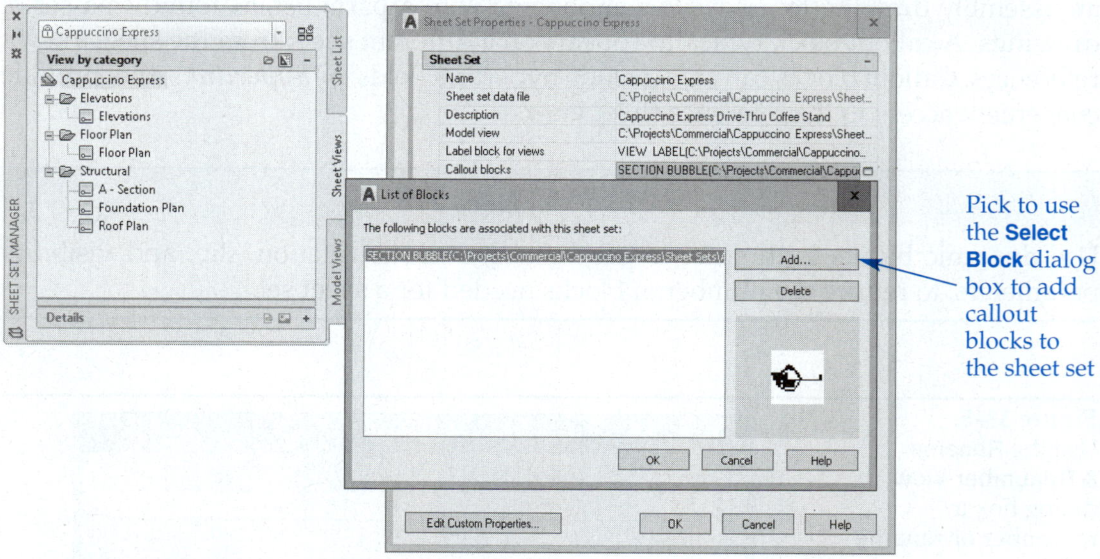

If the file consists of only the objects that make up the block, choose the **Select the drawing file as a block** radio button to use the file as a callout block. If the file includes blocks saved as blocks, pick the **Choose blocks in the drawing file:** radio button and select the blocks to use as callout blocks. Use the **Delete** button to remove a block from the **List of Blocks** dialog box.

After you add callout blocks to a sheet set, you must assign the appropriate callout blocks to each view category. To specify the callout blocks available for a view category, right-click on the category and select **Properties...** to open the **View Category** dialog box. See Figure 33-3. Use the check boxes to select the callout blocks to assign to the view category. Only check those callout blocks that apply to the specific category. Pick the **Add Blocks...** button to add new blocks to the view category using the **Select Block** dialog box.

To insert a callout block, open the sheet on which the reference is to appear and activate model space to add the symbol to model space. For example, open the floor plan shown in Figure 33-6 to add the cutting-plane line callouts referencing the section view. Then pick the **Sheet Views** tab of the **Sheet Set Manager**, right-click on the sheet view to reference, and select the block from the **Place Callout Block** cascading menu. For example, right-click on the A - Section sheet view to create the cutting-plane line reference shown in Figure 33-6. Specify an insertion point for the block and follow the prompts to modify the block as needed.

Exercise 33-3

www.g-wlearning.com/CAD/

Complete the exercise on the companion website.

View Label Blocks

View label blocks automate reference between drawing views and view identification by updating to reflect changes in view number, title, and scale. A common example of a view label block is the view title and scale information displayed under the section view shown in Figure 33-8. In this example, the bubble portion of the view label block

Figure 33-8.
Using a view label block assigned to a sheet view to title a section view on the sheet where the sheet view is located.

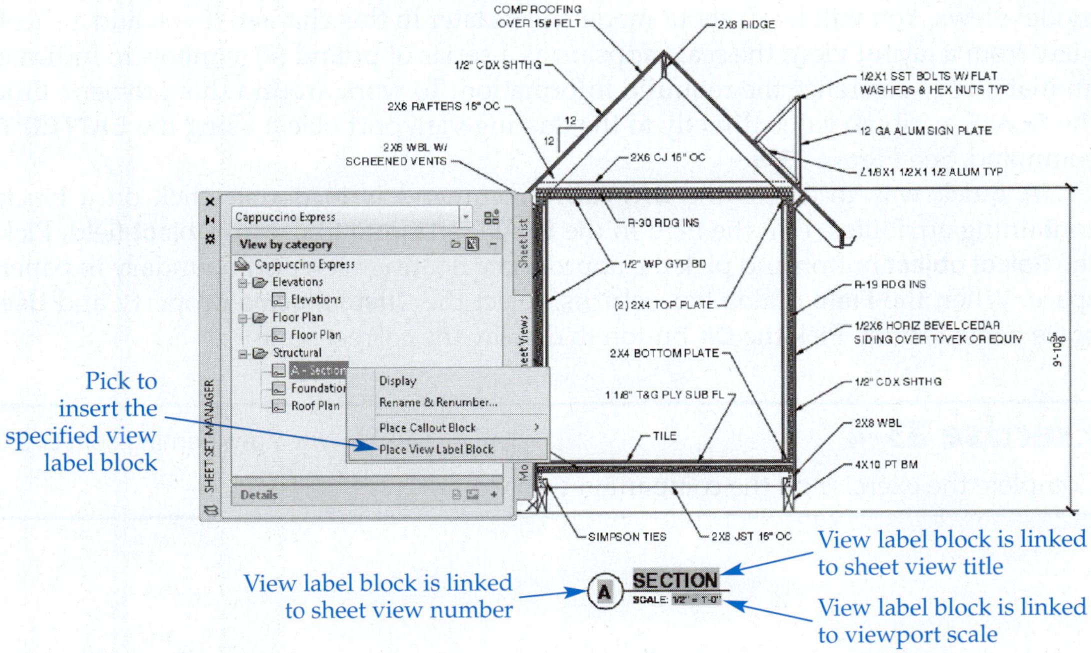

includes an attribute with a **SheetView** field that references the **ViewNumber** property of the A - Section sheet view. The title includes an attribute with a **SheetView** field that references the **ViewTitle** property of the A - Section sheet view. The scale includes an attribute with a **SheetView** field that references the **ViewportScale** property of the A - Section sheet view.

NOTE

View label blocks can also include hyperlinks, but hyperlinks are usually not useful in view label blocks because a view label block typically appears on the same sheet as the sheet view it references.

Before you can insert a view label block from the **Sheet Set Manager**, the block must be available to the sheet set. A sheet set can only reference a single view label block, and you cannot assign different view label blocks to each view category. View label blocks are often added to a sheet set during sheet set creation. To assign a view label block to an existing sheet set, right-click on the sheet set in the **Sheet Set Manager** and pick **Properties...** to display the **Sheet Set Properties** dialog box.

Specify the view label block in the **Label block for views** text box. Pick the ellipsis (**...**) button to access the **Select Block** dialog box, and then pick the ellipsis button to navigate to and select the drawing or template file that contains the view label block. If the file consists of only the objects that make up the block, choose the **Select the drawing file as a block** radio button to use the file as a view label block. If the file includes blocks saved as blocks, pick the **Choose blocks in the drawing file:** radio button and select the block to use as the view label block.

To insert a view label block, open the sheet containing the view to be labeled. For example, open the section view layout shown in Figure 33-8 to add the view label referencing the section view. Items such as labels are often created in paper space, but you can insert a view label into model space if necessary. Pick the **Sheet Views** tab of the **Sheet Set Manager**, right-click on the sheet view to reference, and select **Place View Label Block**. For example, right-click on the A - Section sheet view to create the view label shown in Figure 33-8. Specify an insertion point for the block and follow the prompts to modify the block as needed.

If the view label block includes a **SheetView** field that references the **ViewportScale** property, AutoCAD only recognizes the viewport scale of sheet views created from model views. You will learn about model views later in this chapter. If you add a sheet view from a layout view, the scale appears as a series of pound (#) symbols to indicate an inability to reference the required information. To work around this problem, link the SCALE attribute value directly to the floating viewport object using the **EATTEDIT** command. See Figure 33-9.

A quick way to access the **EATTEDIT** command is to double-click on a block containing attributes. Edit the field in the SCALE attribute to use an **Object** field. Pick the **Select object** button and pick the appropriate floating viewport boundary in paper space. When the **Field** dialog box returns, select the **Custom scale** property and **Use scale name** format. Pick the **OK** button to display the correct scale.

Exercise 33-4

www.g-wlearning.com/CAD/

Complete the exercise on the companion website.

Figure 33-9.
Using a field to display the correct viewport scale when referencing a layout view to place a view label.

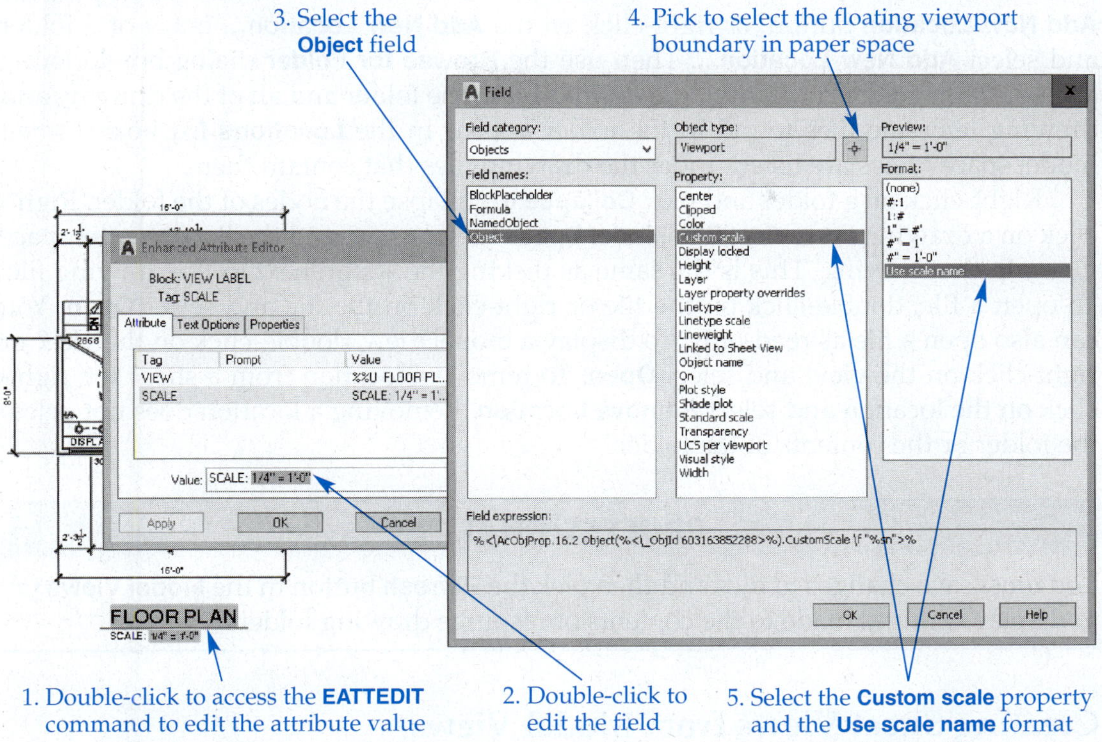

Model Views

The **Model Views** tab of the **Sheet Set Manager** allows you to access *resource drawings*. See **Figure 33-10**. Add folders to the **Model Views** tab to list drawing files, drawing template files, and named model views found in the files. One purpose of resource drawings is to access files associated with the project, but not included as sheets. This function is similar to locating files in a different folder using Windows Explorer. Another purpose of resource drawings is to create sheet views from model space content or named model views. You can use this function in addition or as an alternative to creating sheet views from layout views.

resource drawings: Drawing files that include named model space views referenced for use as sheet views.

Figure 33-10.
Use the **Model Views** tab to access resource drawings and create sheet views from model space.

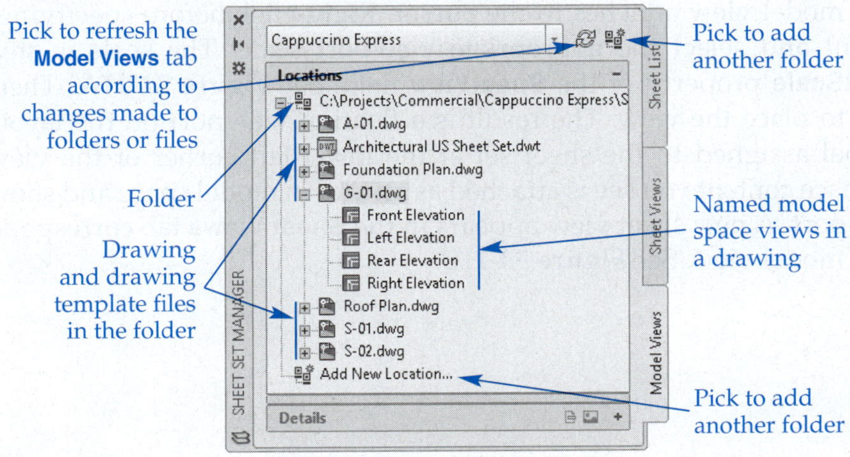

Adding Folders

Folders are often added to the **Model Views** tab during sheet set creation. To add a folder to an existing sheet set, double-click on the **Add New Location...** entry, pick the **Add New Location** button, or right-click on the **Add New Location...** entry or a folder and select **Add New Location...**. Then use the **Browse for Folder** dialog box to locate and select the folder containing the desired files. The folder and all of the drawing and drawing template files found in the folder appear in the **Locations** list box. Named model space views are listed under the drawing files that contain them.

Right-click on a folder and pick **Collapse** to collapse the nodes of the folder. Right-click on a drawing and select **See Model Space Views** to expand the list of model space views in the drawing. This is the same as picking the + sign next to the drawing file. To open a file, double-click on the file or right-click on the file and select **Open**. You can also open a file as read-only. To display a model view, double-click on the view or right-click on the view and select **Open**. To remove a location from a sheet set, right-click on the location and select **Remove Location**. Removing a location does not delete the folder or the contents of the folder.

PROFESSIONAL TIP

You must save changes to files and then pick the **Refresh** button in the **Model Views** tab to display changes made to the contents of resource drawing folders.

Creating Sheet Views from Model Views

Create a sheet view from an entire model space drawing or from a named model space view listed in the **Model Views** tab. When you create a sheet view from a drawing or model view, AutoCAD combines and automates the traditional process of adding an xref to a new drawing, creating a floating viewport, scaling the viewport, and adding a view label block.

To create a sheet view from the entire drawing in model space, ensure that the resource folder includes the drawing file. To create a sheet view from a named model view, open the file containing the model space content to reference. Then use the **NEWVIEW** command in model space to create a named model view. Select an existing view category, or type a name to create a new category.

To create a sheet view from model space content, activate the sheet in the sheet set that will receive the sheet view, and erase any existing floating viewports. Create a new sheet if necessary, and assign the sheet to the sheet set. Set an appropriate viewport layer current. Drag and drop the drawing or view from the **Model Views** tab or right-click on the drawing or view and select **Place on Sheet**. An alert appears if the sheet set does not include the current layout.

The model view attaches to the cursor. Right-click before specifying the insertion point and select the appropriate viewport scale. The scale is stored as the **ViewportScale** property of the **SheetView** field. See Figure 33-11A. Then specify a location to place the view. The result is a floating viewport on the layout with the view label assigned to the sheet set at the lower-left corner of the viewport. The model space content you see is attached as an xref in model space and shows through the viewport. A new sheet view appears in the **Sheet Views** tab corresponding to the inserted model view. See Figure 33-11B.

Figure 33-11.
Referencing a drawing as a model view to create a sheet view. The same basic process applies to forming sheet views from named model views. A—Drag and drop the drawing and select the viewport scale. B—Specify the location for the view.

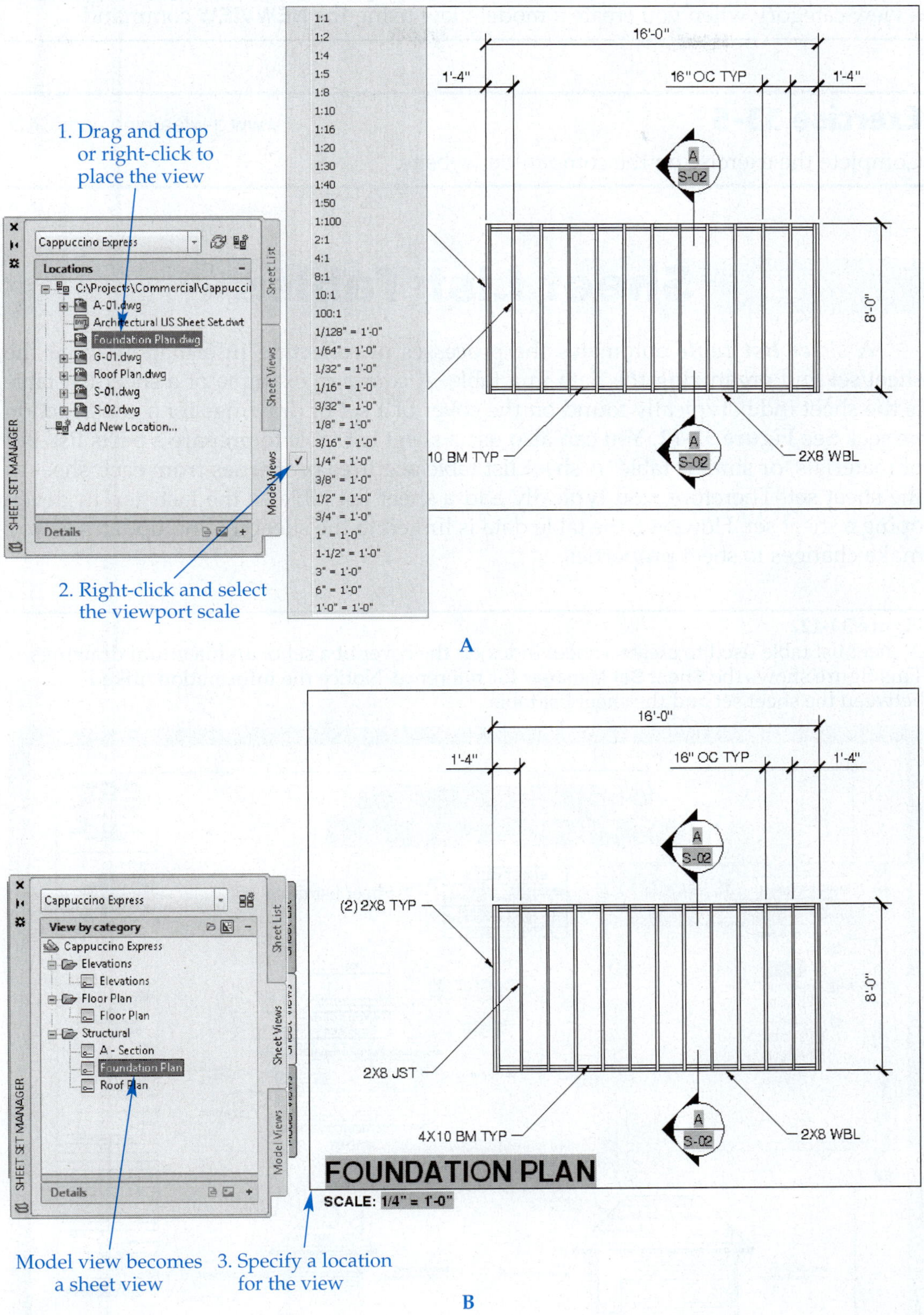

NOTE

Drag and drop a sheet view created from a model view into the appropriate view category. A view automatically appears in the correct category if you type the name of a view category when you create a model view using the **NEWVIEW** command.

Exercise 33-5

www.g-wlearning.com/CAD/

Complete the exercise on the companion website.

Sheet List Tables

sheet list table: An AutoCAD table that references a table style and selected items in a sheet set to create a list of sheets in the sheet set and related information.

A ***sheet list table*** automates the processes of collecting information about the sheet set and organizing the data in a table. A common example of a sheet list table is the sheet index typically found on the cover of a set of drawings for a construction project. See Figure 33-12. You can also use a sheet list table to prepare a parts list, bill of materials, or similar table. A sheet list table acquires properties from each sheet in the sheet set. Therefore, you typically add a sheet list table as the last step in developing a sheet set. However, the table data is linked to the sheet set and updates as you make changes to sheet properties.

Figure 33-12.
A sheet list table used to create a sheet index on the cover of a set of architectural drawings. This figure shows the **Sheet Set Manager** for reference. Notice the information linked between the sheet set and the sheet list table.

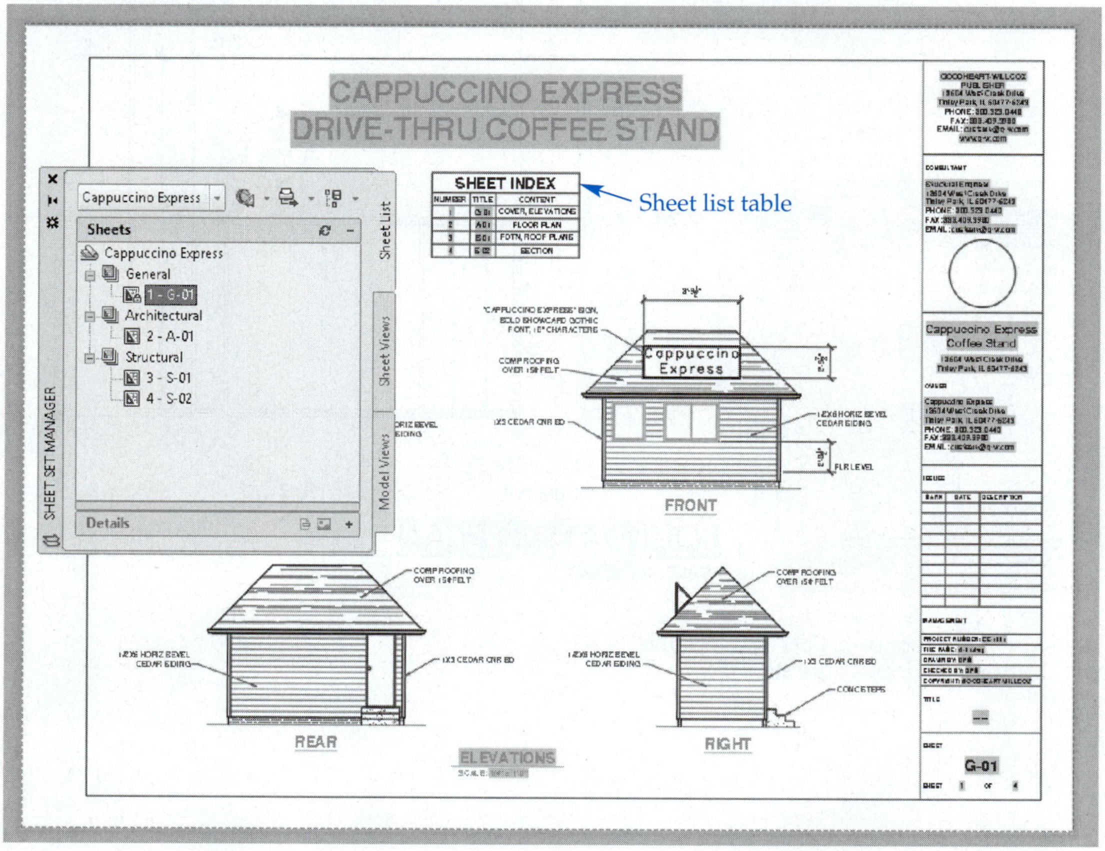

Inserting a Sheet List Table

To insert a sheet list table, access the **Sheet Set Manager** and open the sheet on which you want to place the table. Right-click on the sheet set, a subset, or a sheet in the **Sheet Set Manager** and select **Insert Sheet List Table...** to open the **Sheet List Table** dialog box. See Figure 33-13. For most applications, right-click on the sheet set to access the **Sheet List Table** dialog box to create a sheet list table that is initially set to include all subsets and sheets. Right-clicking on a subset or sheet to access the **Sheet List Table** dialog box initially limits the sheets in the table, although you can add and remove sheets from the table as needed.

Select a table style to apply to the sheet list table from the **Table Style name:** drop-down list, or pick the ellipsis (...) button to access the **Table Style** dialog box. The preview area displays a representation of the table. Select the **Show Subheader** check box to include subheader rows based on selected subsets. Figure 33-12 shows a table without subheader rows. Figure 33-14 shows the same table, but with subheader rows acquired from the General, Architectural, and Structural subsets.

Figure 33-13.
A—Use the **Sheet List Table** dialog box to set up a sheet list table. B—Specify the sheets to include in the table.

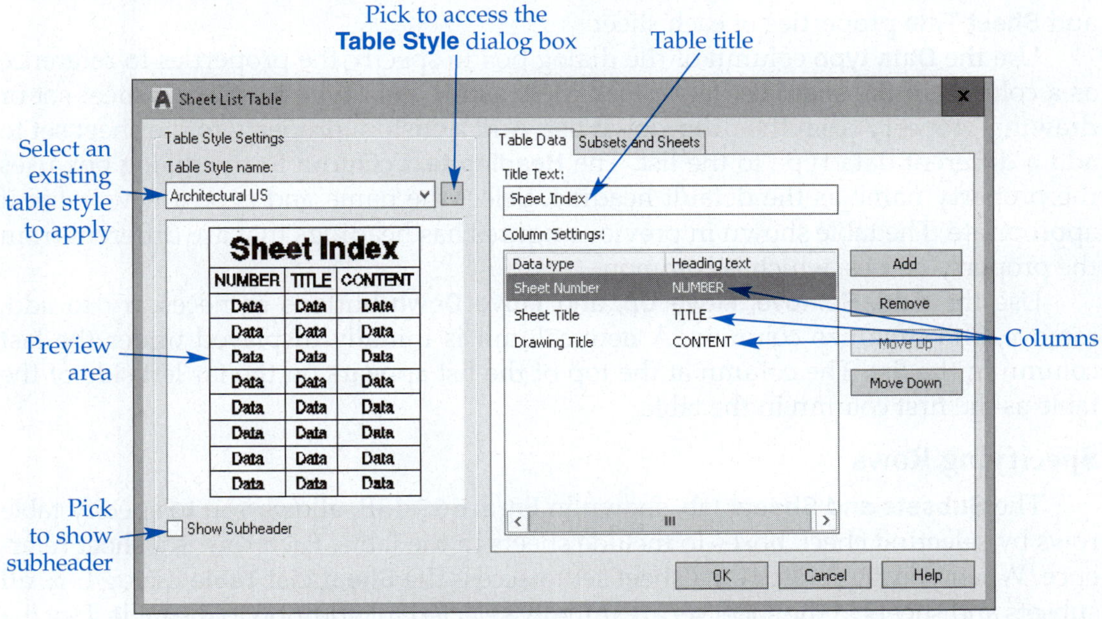

A

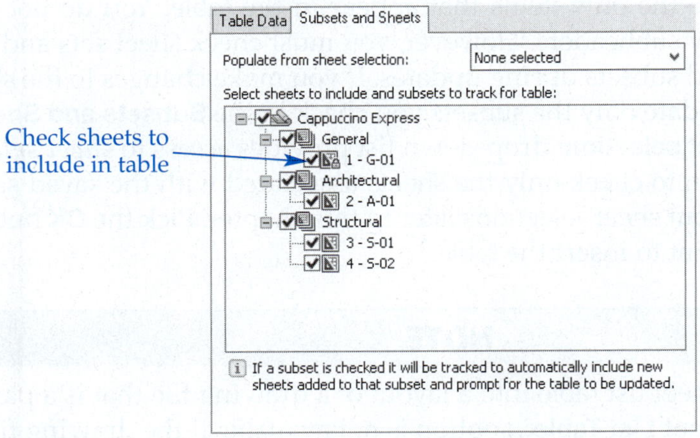

B

Figure 33-14.
A sheet list table with subheaders added by referencing subsets.

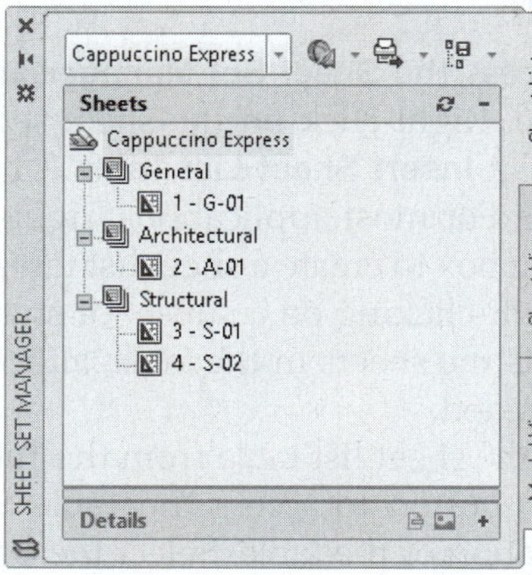

SHEET INDEX		
NUMBER	TITLE	CONTENT
General		
1	G-01	COVER, ELEVATIONS
Architectural		
2	A-01	FLOOR PLAN
Structural		
3	S-01	FDTN, ROOF PLANS
4	S-02	SECTION

Specifying the Title and Columns

Use the **Table Data** tab, shown in Figure 33-13A, to specify the title of the table, the title of columns, and column organization. Type the title in the **Title Text:** text box or accept the default Sheet List Table title. A sheet list table can include various types of information from the drawing file and the sheet set. The default sheet list table includes **Sheet Number** and **Sheet Title** columns corresponding to the **Sheet Number** and **Sheet Title** properties of each sheet.

Use the **Data type** column of the dialog box to specify the properties to reference as a column in the sheet list table. Pick the existing data type to select a sheet set or drawing property from the drop-down list. Add a custom property to the sheet set to add a different data type to the list. The **Heading text** column in the dialog box uses the property name as the default heading. Select the name and type a new value if appropriate. The table shown in previous figures has headings that are different from the property names, which is common.

Use the **Add**, **Remove**, **Move Up**, and **Move Down** buttons as necessary to add, remove, and organize columns. A new column is initially displayed under the last column in the list. The column at the top of the list appears on the far left side of the table as the first column in the table.

Specifying Rows

The **Subsets and Sheets** tab, shown in Figure 33-13B, allows you to specify table rows by selecting check boxes to include sheets in the table. Each row is a sheet reference. When you right-click on a sheet set to access the **Sheet List Table** dialog box, all subsets and sheets in the sheet set are initially selected for addition to the table. Use the appropriate check boxes to add and remove sheets from the table.

Checked sheets are the only items that appear in the table. You do not need to check subsets to display subheaders. However, you must check sheet sets and subsets to include sheet sets and subsets during updates. If you make changes to the sheet set, a prompt appears to update only the subsets you check in the **Subsets and Sheets** tab. The **Populate from sheet selection:** drop-down list provides access to *sheet selections*. Choose a sheet selection to check only the sheets associated with the saved selection. You will learn more about sheet selections later in this chapter. Pick the **OK** button and specify an insertion point to insert the table.

sheet selections: Groups of subsets and sheets that are often used to publish the same group of sheets.

NOTE

You can only insert a sheet list table into a layout of a drawing file that is a part of the sheet set. The **Insert Sheet List Table...** option is not available if the drawing file is not part of the sheet set or if model space is current.

Editing a Sheet List Table

The information in a sheet list table is linked directly to the data source, or property, field. Update a sheet list table to display changes made to the sheet set. For example, update a sheet list table after changing the sheet numbers in the **Sheet Set Manager** or after adding a sheet to the sheet set. Use the **DATALINKUPDATE** command to update a sheet list table. A quick way to access the command is to select the table to update, right-click, and select **Update Table Data Links**. You can also select inside a table header or data cell and then right-click and pick **Update Sheet List Table** from the **Sheet List Table** cascading menu.

Ribbon
Insert
> Linking & Extraction
Download from Source
Type
DATALINKUPDATE
DATALINKUPDATE

To modify the properties for the table, select inside a table header or data cell and then right-click. Choose **Edit Sheet List Table Settings...** from the **Sheet List Table** cascading menu to reopen the **Sheet List Table** dialog box. Pick the **OK** button to update the sheet list table.

NOTE

Use the same techniques to modify a sheet list table that you use to modify any other table object. For example, you can change text or add columns and rows. However, avoid unlocking cells and overriding values linked to the data source. When you use the **DATALINKUPDATE** command on the modified table, an alert indicates that manual modifications will be discarded.

Sheet List Table Hyperlinks

The **Sheet Number** and **Sheet Title** properties contain hyperlinks that you can use to access sheets. To open a sheet using a hyperlink, hover the crosshairs over a sheet number or sheet title until the hyperlink icon and tooltip appear. Hold down the [Ctrl] key and pick the hyperlink to open the selected sheet.

Exercise 33-6

www.g-wlearning.com/CAD/

Complete the exercise on the companion website.

Sheet Set Fields

Throughout this chapter, you have experienced how text and blocks with attributes containing fields link text information to elements of the sheet set. Text and multiline text objects, title blocks, callout blocks, view label blocks, and sheet list tables use sheet set fields to automate documentation and the process of changing values during the course of the project. The following sections provide additional information on applying sheet set fields.

AutoCAD provides specific field types for use with sheet sets. In the **Field** dialog box, pick **SheetSet** from the **Field category:** drop-down list to filter fields specific to sheet sets in the **Field names:** list box. See Figure 33-15. Use fields in the **SheetSet** category to display values defined in the sheet set, subset, sheet, or sheet view. Some fields have several property and format options.

Figure 33-15.
An example of a link between a current sheet set field and the corresponding sheet set property.

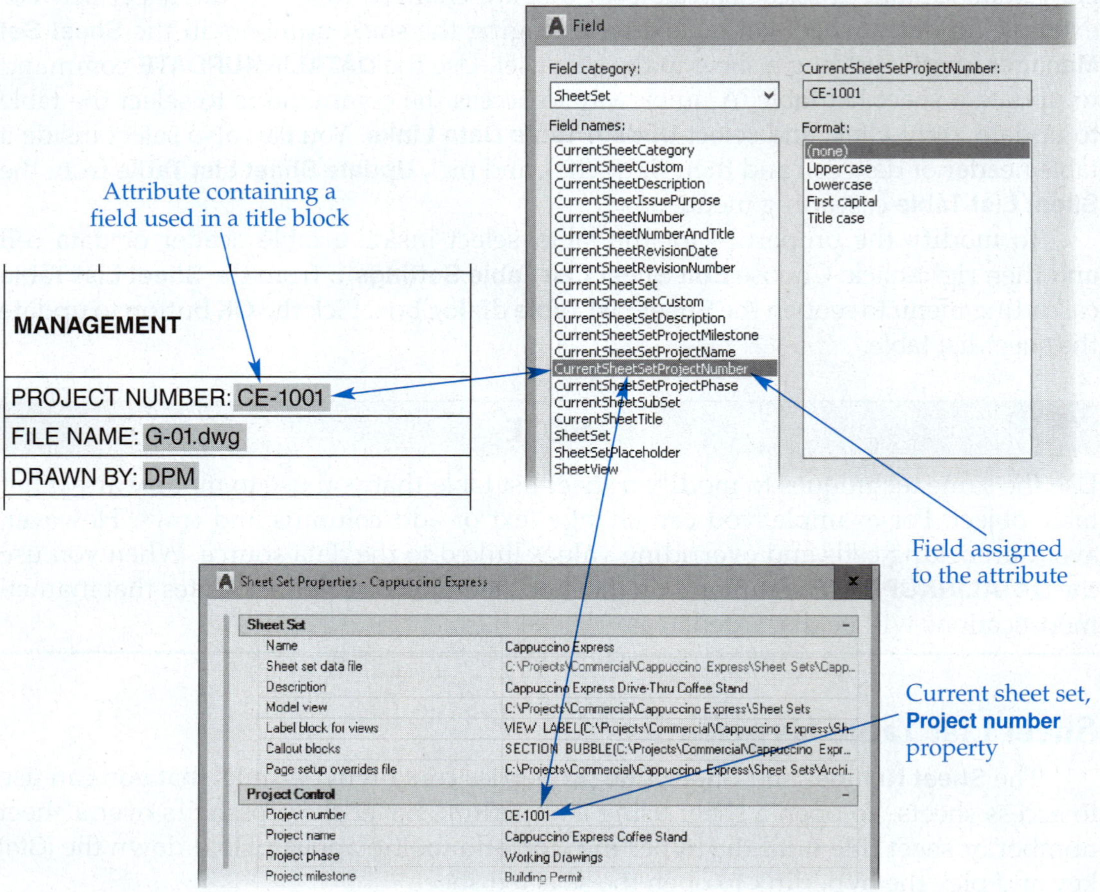

PROFESSIONAL TIP

When you prepare attributes with fields, avoid using the **Multiple lines** option, because multiple-line attributes with fields often display an inappropriate format. Use several single-line attributes and custom properties as needed. In addition, choose the **Preset** option in the **Attribute Definition** dialog box or **Properties** palette to have the attribute assume preset sheet set property values during block insertion.

Current Sheet Set Fields

Current sheet set fields, such as the **CurrentSheetSetProjectNumber** field shown in Figure 33-15, are linked to properties assigned in the **Sheet Set Properties** dialog box. The field value is specific to the current sheet set, which is where the field is located. For example, if you insert a title block onto a new sheet in a sheet set named Big House, the **CurrentSheetSetProjectName** field will display Big House. If you insert the same title block onto a new sheet in a sheet set named Little House, the **CurrentSheetSetProjectName** field displays Little House.

Figure 33-16.
An example of a link between a current sheet field and the corresponding sheet property.

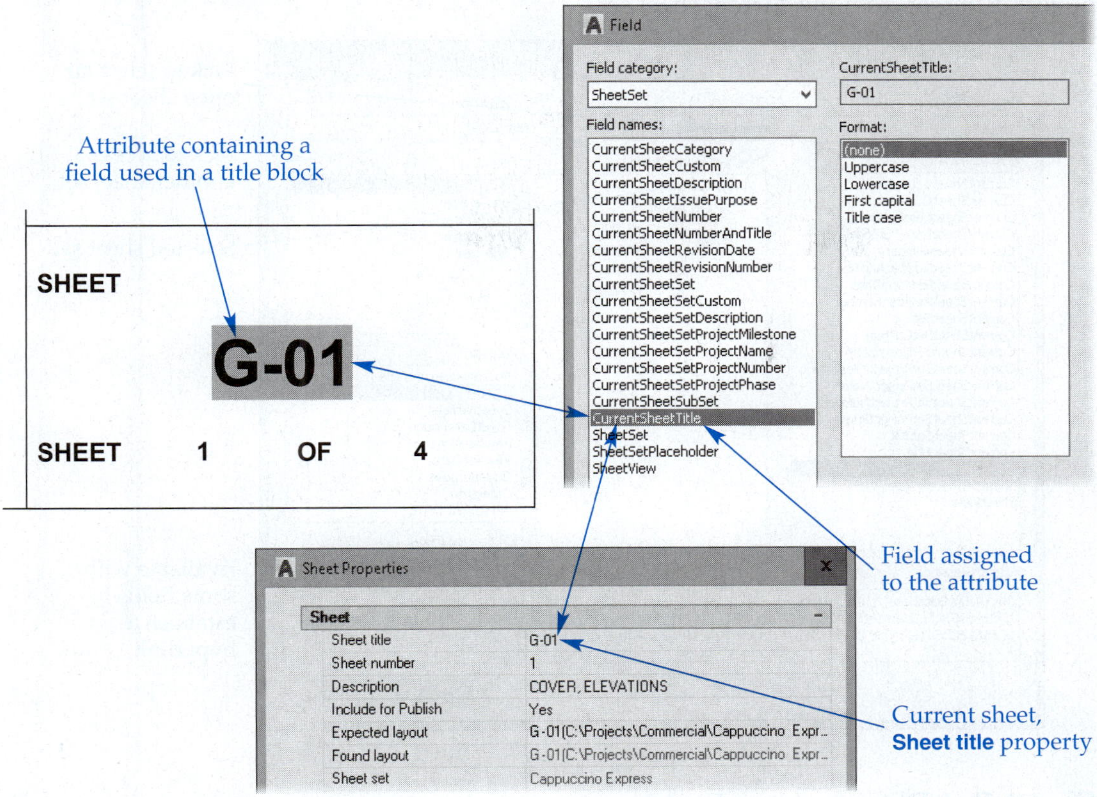

Current Sheet Fields

Current sheet fields, such as the **CurrentSheetTitle** field shown in Figure 33-16, are linked to properties assigned in the **Sheet Properties** or **Sheet Set Properties** dialog boxes. The field value is specific to the current sheet, which is where the field is located. For example, a title block with an attribute containing the **CurrentSheetNumber** field displays 1 on sheet 1 and 2 on sheet 2 in a sheet set.

Sheet Set Fields

The various **CurrentSheetSet** and **CurrentSheet** fields are appropriate for most applications, but they link values only to the current project. Pick the **SheetSet** field to display options for linking values to any existing sheet set. See Figure 33-17. Use the **Sheet set:** drop-down list to display the contents of an open sheet set in the **Sheet navigation tree:**, or pick the ellipsis (**...**) button to open another sheet set. Select the sheet set, a subset, or a sheet to specify the field value.

The **Property:** list box, which appears when you pick the sheet set or a sheet, allows you to select a property to associate with the sheet set or sheet. Use the **Format:** list box to override the format used in the **Sheet Set Properties**, **Subset Properties**, or **Sheet Properties** dialog box, depending on the selection. For example, if you type Assembly in the **Subset Properties** dialog box, pick the **Uppercase** format to force AutoCAD to display ASSEMBLY. Pick the **Associate hyperlink** check box to include a hyperlink to the property with the field.

NOTE

Custom sheet set and sheet fields are available from the **Property:** list box.

Figure 33-17.
The **SheetSet** field allows you to associate a value to a specific sheet set, typically related to a project different from the current sheet set.

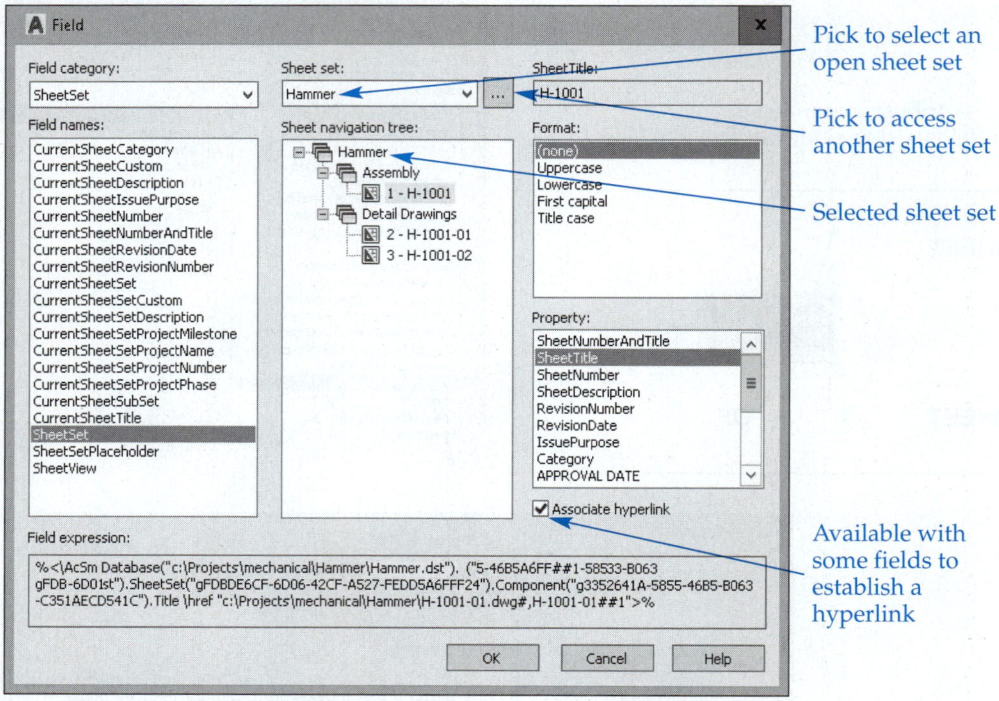

Sheet Set Placeholder Fields

sheet set placeholder: A temporary value for a field that later references specific properties for values.

The **SheetSetPlaceholder** field inserts a ***sheet set placeholder*** that allows you to prepare text or a block without knowing the specific location of the property that the field will reference. Callout and view label blocks use **SheetSetPlaceholder** fields. See Figure 33-18. When you add a callout or view label block to a drawing in a sheet set, AutoCAD locates the information corresponding to the placeholders. The placeholders automatically become **CurrentSheet**, **CurrentSheetSet**, or **SheetView** fields, depending on the type of placeholder, and display the correct values.

Sheet View Fields

SheetView fields form when you use callout and view label blocks with attributes containing sheet set placeholder fields. Select the **SheetView** field to display options for linking values to any existing sheet view, or to place values without using callout or view label blocks. The process for assigning a **SheetView** field is similar to that for assigning a **SheetSet** field, except that you select a sheet set, view category, or sheet view from the **Sheet navigation tree:** list box.

PROFESSIONAL TIP

UPDATEFIELD
Ribbon
Insert
> Data
Update Fields
Type
UPDATEFIELD

If a field does not display the expected value, update the field manually using the **UPDATEFIELD** command, or use the **REGEN** command to apply an automatic update. In some cases, you may need to delete and then reinsert a block to display the correct field values.

Custom Properties

Custom properties are necessary to include additional information about a sheet set and sheets. For example, many of the properties referenced by fields in a title block are

Figure 33-18.
A view label block in the **Block Editor**. Callout and view label blocks use **SheetSetPlaceholder** fields to assign a field with the necessary characteristics.

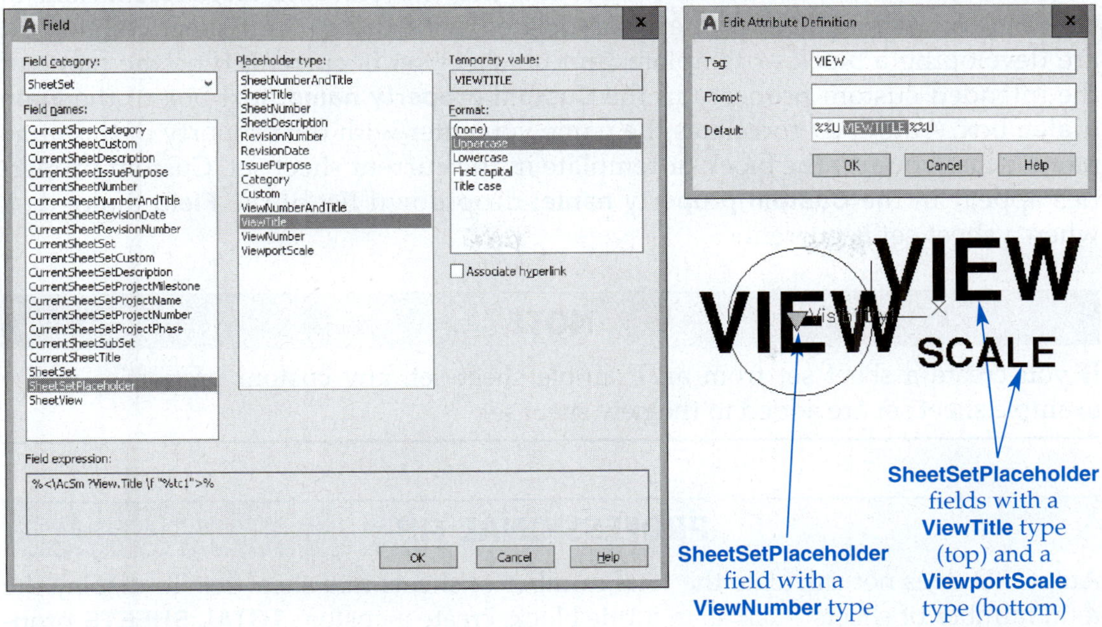

custom properties. To add or modify custom properties, access the **Sheet Set Properties** dialog box and pick the **Edit Custom Properties...** button to display the **Custom Properties** dialog box. See **Figure 33-19A**.

Use the **Add...** button to add custom properties using the **Add Custom Property** dialog box. See **Figure 33-19B**. Specify the name, default value, and owner of the property. The **Sheet Set** owner creates a custom sheet set property that applies to all sheets in the set. The **Sheet** owner creates a custom sheet property that might vary from sheet to sheet. Use the **Delete** button to delete custom properties. Custom sheet set and sheet properties appear in the **Sheet Set Properties** dialog box. Custom sheet properties also appear in the **Sheet Properties** dialog box.

Figure 33-19.
A—Use the **Custom Properties** dialog box to add and delete custom properties. Study the examples of custom properties and default values added to this sheet set. B—Define a custom property using the **Add Custom Property** dialog box.

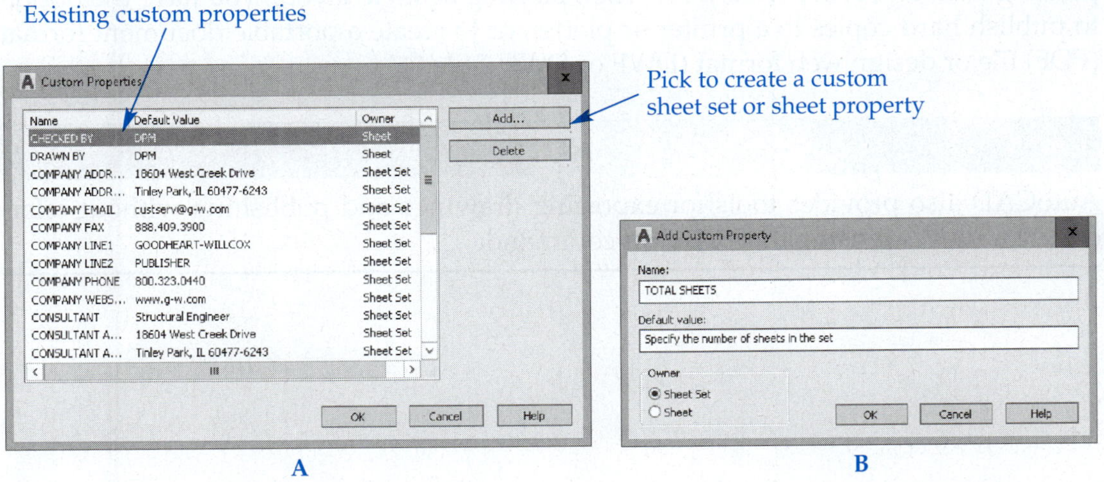

Custom Property Fields

Choose the **CurrentSheetSetCustom** field to reference a custom sheet set property associated with the current sheet set. Use the **CurrentSheetCustom** field to reference a custom sheet property associated with the current sheet. When you are developing a block or template, and no sheet set is current, type the name of the intended custom property in the **Custom property name:** text box of the **Field** dialog box. AutoCAD correlates the name you enter with the property of the same name when you use the block or template in the current sheet set. Custom properties appear in the **Custom property name:** drop-down list of the **Field** dialog box when a sheet set is current.

NOTE

If you create a sheet set from an example sheet set, any custom properties in the example sheet set are added to the new sheet set.

PROFESSIONAL TIP

AutoCAD does not calculate the total number of sheets in a sheet set. To display the total number of sheets, such as in a title block, create a custom **TOTAL SHEETS** property owned by the sheet set. At the end of the project, view the number assigned to the last sheet and type the value in the **TOTAL SHEETS** property in the **Sheet Set Properties** dialog box.

Exercise 33-7

www.g-wlearning.com/CAD/

Complete the exercise on the companion website.

Publishing a Sheet Set

A sheet set provides options to publish the sheet set, subsets, or individual sheets in the sheet set. Publishing is also referred to as *batch plotting*. Publishing a sheet set is a convenient way to create a hard copy or electronic version of a set of drawings, without plotting each layout separately. As when plotting a single layout, you have the option to publish hard copies to a printer or plotter, or to create a portable document format (PDF) file or design web format (DWF or DWFx) file.

NOTE

AutoCAD also provides tools for exporting drawings and publishing without referencing a sheet set using the **PUBLISH** command.

Publish Options

Pick the sheet set to publish, or press [Shift] or [Ctrl] to select specific items to publish, such as individual subsets or sheets. Access publish options from the **Publish** flyout, shown in Figure 33-20, or right-click on the sheet set, subsets, or sheets and select an option from the **Publish** cascading menu. Pick the **Publish to DWF**, **Publish to DWFx**, or **Publish to PDF** option to create a DWF, DWFx, or PDF file from the selected items. A dialog box appears, allowing you to specify a name and location for the file. The resulting file contains multiple pages with each sheet on a separate page. Use the **Publish to Plotter** option to plot the selected items to the default printer or plotter using the plot settings from each layout.

The **Publish using Page Setup Override** option is available if you assigned a named page setup from a file to the **Page setup overrides file** property in the **Sheet Set Properties** dialog box. Pick the option to publish all sheets using the settings specified in the saved page setup. Use the **Include for Publish** option to identify whether the selected items will be published. Pick the **Edit Subset and Sheet Publish Settings...** option to display the **Publish Sheets** dialog box. Check specific sheets to include for publishing.

The **Publish in Reverse Order** option publishes sheets in the opposite order from the order displayed in the **Sheet Set Manager**. With certain printers, publishing a sheet set in reverse order is helpful so that the last sheet is on the bottom of the stack, at the end of the set. Select the **Include Plot Stamp** option to plot the plot stamp assigned to the layout or page setup override with the selected sheets. Select the **Plot Stamp Settings...** option to open the **Plot Stamp** dialog box to specify the plot stamp settings. The **Manage Page Setups...** option opens the **Page Setup Manager**, allowing you to create or modify a page setup. The **Sheet Set DWF Publish Options...** selection displays the **Sheet Set DWF Publish Options** dialog box, which includes settings for creating a DWF or DWFx file. The **Sheet Set PDF Publish Options...** selection displays the **Sheet Set PDF Publish Options** dialog box, which includes settings for creating a PDF file. The **Publish Dialog Box...** option opens the **Publish** dialog box, which lists the sheets in the current sheet set or the sheet selection.

Figure 33-20.
The **Publish** flyout provides options for preparing a selected sheet set, or subsets or sheets within the sheet set, for publishing.

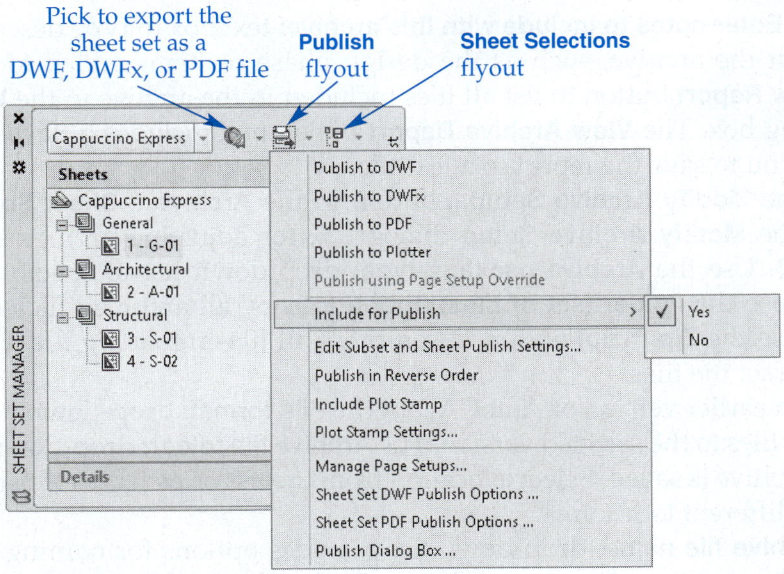

Sheet Selections

Sheet selections provide a way to group specific subsets or sheets for publishing or for populating rows in a sheet list table. The most common application for a sheet selection is to create a selection set of specific subsets or sheets to publish. For example, create a sheet selection named Plans and Elevations from the Plans and Elevations subsets to publish only the plan and elevation sheets in the sheet set.

To save a sheet selection, pick the subsets or sheets to be included in the selection set. Then pick the **Sheet Selections** flyout on the **Sheet Set Manager** and select **Create...** to access the **New Sheet Selection** dialog box. Enter a name for the selection set and pick the **OK** button. The new selection set appears when you pick the **Sheet Selections** button. Pick a sheet selection from the **Sheet Selections** flyout to highlight the associated sheets in the **Sheet Set Manager**. To rename or delete a sheet selection, pick **Manage...** from the **Sheet Selections** flyout to display the **Sheet Selections** dialog box. Select the sheet selection and pick the **Rename** or **Delete** button.

Archiving a Sheet Set

Archive sets of drawings as necessary throughout a project. For example, when you present a set of drawings to a client for the first time, the client may want to make design changes. Archive the files at this phase for future reference before making modifications. Archive copies of all drawings and related files to a single location. Related files include external references, font files, plot style table files, template files, and other documents associated with the project.

Type
ARCHIVE

Access the **ARCHIVE** command to archive a sheet set using the **Archive a Sheet Set** dialog box. See Figure 33-21. A quick way to activate the **ARCHIVE** command is to right-click on the sheet set in the **Sheet Set Manager** and select **Archive...**. The **Sheets** tab, shown in Figure 33-21A, displays all subsets and sheets in the sheet set. Check the sheets you want to archive. The **Files Tree** tab shown in Figure 33-21B and the alternative **Files Table** tab shown in Figure 33-21C list drawing files and related files. Pick the **Add File...** button in the **Files Tree** or **Files Table** tab to access the **Add File To Archive** dialog box. Use the dialog box to include files that are not part of the sheet set in the archive. You can include any type of file with the archive—the archive is not limited to AutoCAD files.

Use the **Enter notes to include with this archive:** text box to type descriptive information about the archive, such as the design phase or items added to the archive. Pick the **View Report** button to list all files included in the archive in the **View Archive Report** dialog box. The **View Archive Report** dialog box includes a **Save As...** button that allows you to save the report to a text file.

Select the **Modify Archive Setup...** button in the **Archive a Sheet Set** dialog box to display the **Modify Archive Setup** dialog box for adjusting archive settings. See Figure 33-22. Use the **Archive package type:** drop-down list to specify the archive format. Choose the **Folder (set of files)** option to copy all archived files into a single folder. Choose the **Zip (*.zip)** option to compress all files into a ***zip file***. Use a zip file utility to extract the files.

zip file: A file that contains one or more folders and/or files compressed using the ZIP file format.

Select an earlier version of AutoCAD in the **File format:** drop-down list to convert the archived files to the selected version. The **Archive file folder:** drop-down list defines where the archive is saved. Select a location from the list or pick the ellipsis (...) button to choose a different location.

The **Archive file name:** drop-down list provides options for naming the archive. Choose the **Prompt for a filename** option to display the **Specify Zip File** dialog box so that you can specify a name for the archive package. Pick the **Overwrite if necessary**

Figure 33-21.
The **Archive a Sheet Set** dialog box allows you to specify archive settings and files to archive. A—Use the **Sheets** tab to select sheets in the set to archive. B—Use the **Files Tree** tab to include other documents that relate to a project in the archive. C—The **Files Table** tab provides an alternative display for selecting documents to archive.

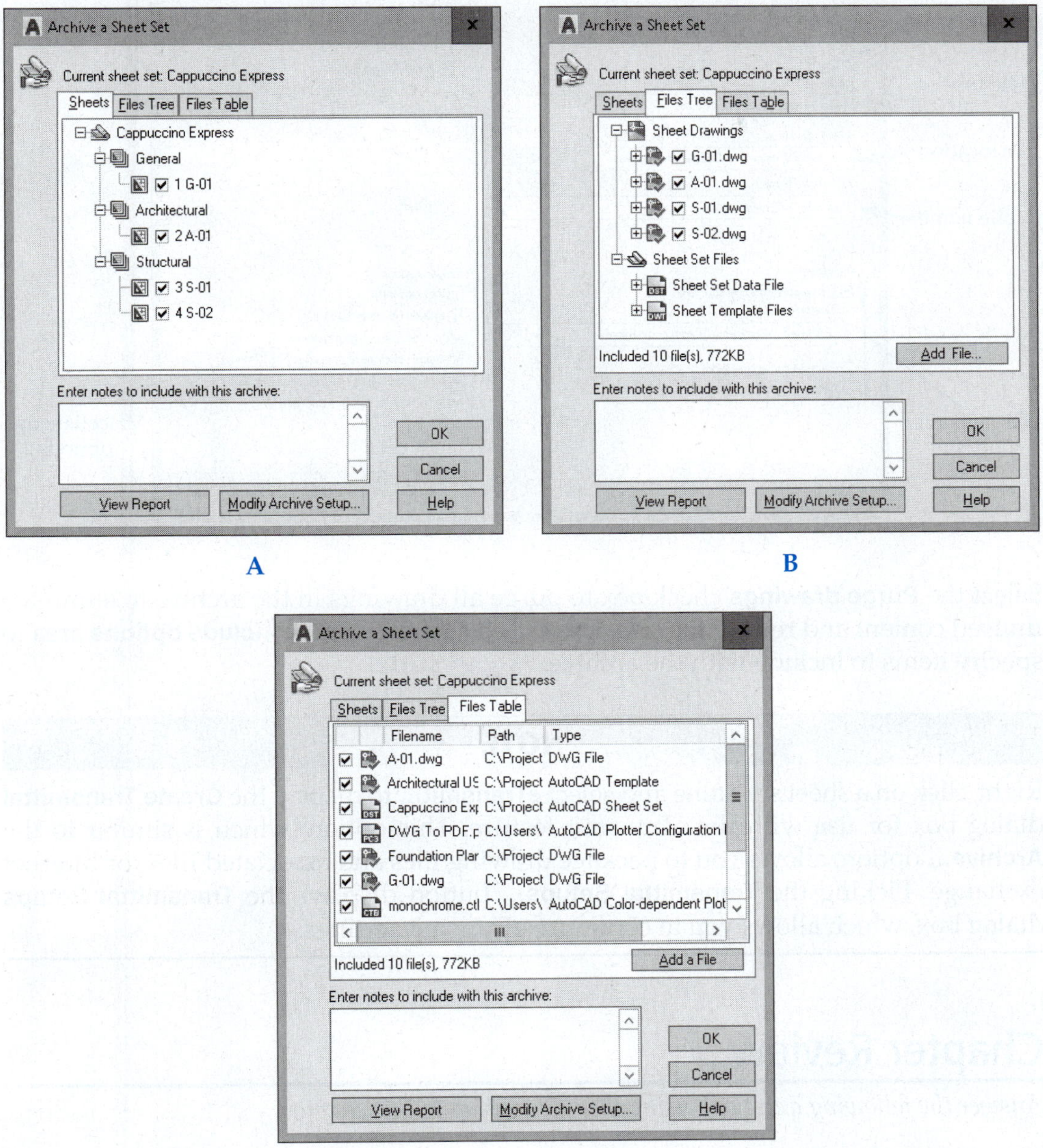

option to overwrite the file name if a file with the same name already exists. Select the **Increment file name if necessary** option to create a new file with an incremental number added to the file name if a file with the same name already exists. This option allows you to save multiple versions of the archive.

Select the **Use organized folder structure** radio button to allow the archive file to duplicate the folder structure for the files. The **Source root folder:** setting determines the root folder for files that use relative paths, such as xrefs. Pick the **Place all files in one folder** radio button to archive all files into a single folder. Select the **Keep files and folders as is** radio button to use the same folder structure for all the files in the sheet set.

The **Set default plotter to 'none'** check box disassociates the plotter name from the drawing files, which is useful if you send files to someone using a different plotter.

Figure 33-22.
Specify archive file settings using the **Modify Archive Setup** dialog box.

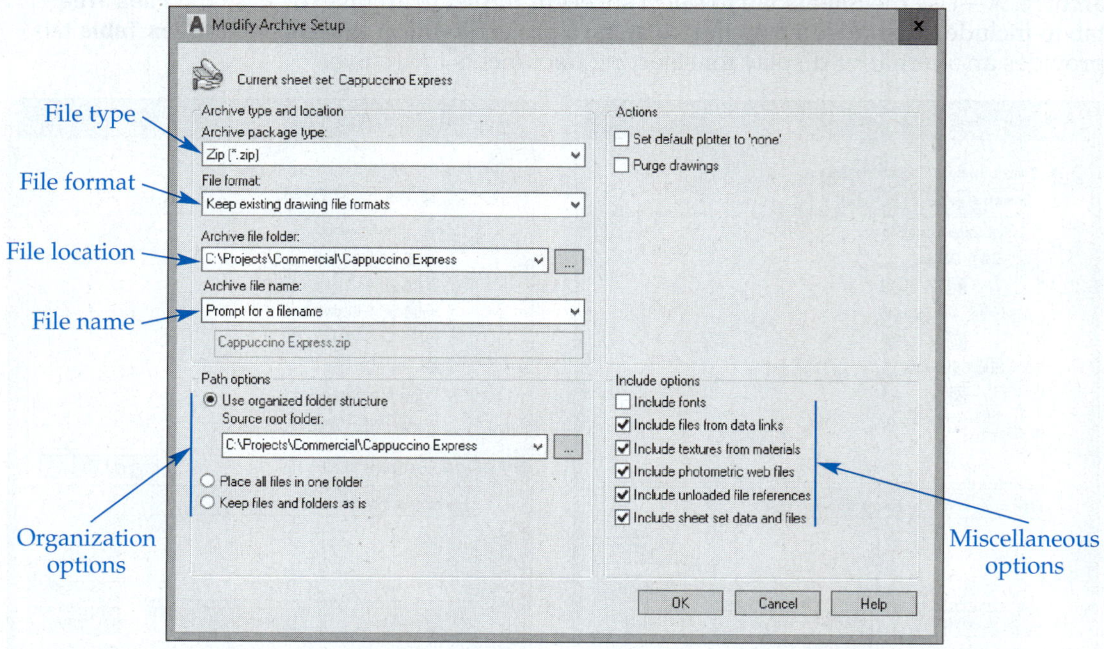

Select the **Purge drawings** check box to purge all drawings in the archive to eliminate unused content and reduce file size. Use the check boxes in the **Include options** area to specify items to include with the archive.

NOTE

Right-click on a sheet set name and select **eTransmit...** to display the **Create Transmittal** dialog box for use with the eTransmit feature. This option, which is similar to the **Archive...** option, allows you to package drawing files and associated files for Internet exchange. Picking the **Transmittal Setups...** button displays the **Transmittal Setups** dialog box, which allows you to configure eTransmit settings.

Chapter Review

Answer the following questions using the information in this chapter.

1. What is a sheet view?
2. Explain why AutoCAD callout blocks and view labels update automatically when you make changes to the related sheet set.
3. Explain how to create a view category from the **Sheet Set Manager** and associate callout blocks to the category.
4. What information do the upper and lower values in a callout block typically provide?
5. What are *resource drawings*?
6. Explain how to add a column heading to a sheet list table.
7. How can you update a sheet list table to reflect changes made in the **Sheet Set Manager**?
8. Explain how to publish a sheet set to a DWF, DWFx, or PDF file.
9. How do you create a sheet selection set?
10. What is the purpose of archiving a sheet set?

Drawing Problems

Start AutoCAD if it is not already started. Follow the specific instructions for each problem.

Note: Some of the problems in this chapter are built on problems from previous chapters. If you have not yet completed those problems, complete them now.

▼ Basic

1. **Mechanical** Open the arbor press sheet set you created in Problem 32-4. Publish and archive the final sheet set.

▼ Intermediate

2. **Mechanical** Complete the Hammer sheet set you started in Exercise 32-1. Save a copy of the drawing files H-1001.dwg, H-1001-01.dwg, and H-1001-02.dwg from the companion website to the Hammer folder you created during Exercise 32-1. Open the Hammer sheet set created in Exercise 32-1. Continue creating the sheet set as follows:
   A. Access the **Sheet Set Properties** dialog box and adjust the properties as shown in A, but use APPROVED BY, CHECKED BY, DRAWN BY, and COMPANY values appropriate to your drawings.

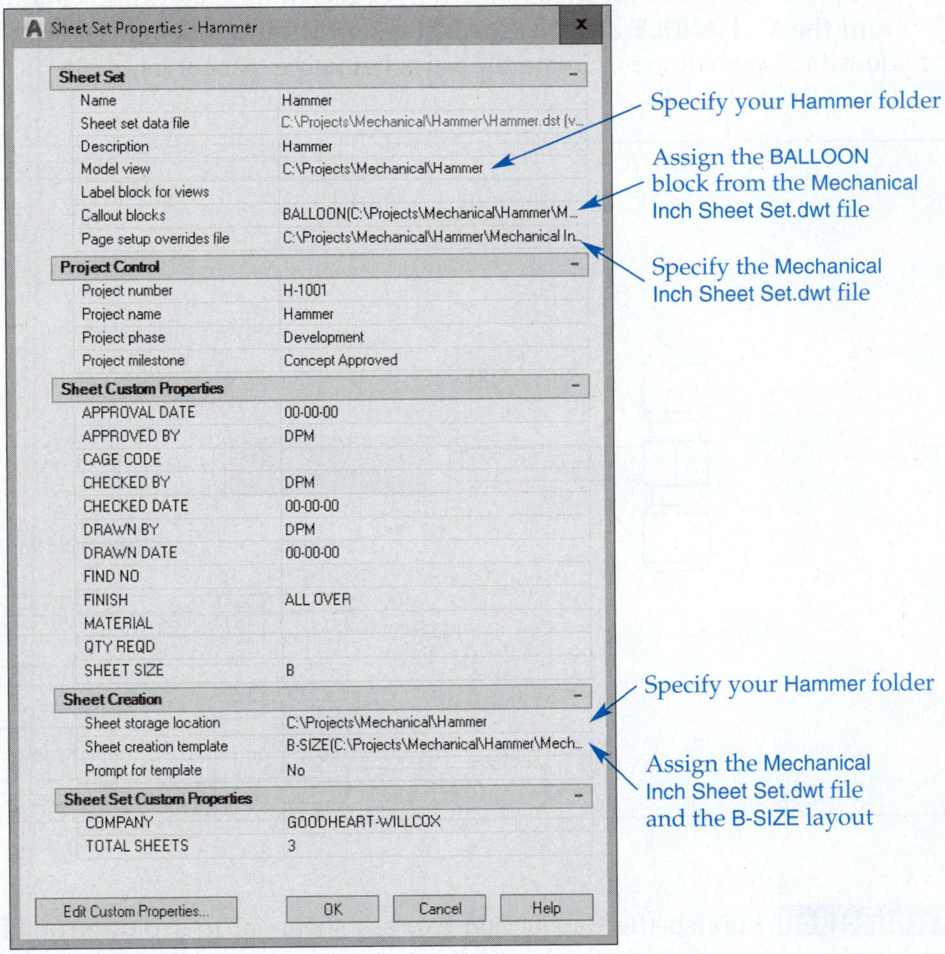

A

   B. Import the H-1001 layout into the Assembly subset and the H-1001-01 and H-1001-02 layouts into the Detail Drawings subset. Access the **Sheet Properties** dialog box for each sheet and adjust the properties as shown in B.

Drawing Problems - Chapter 33

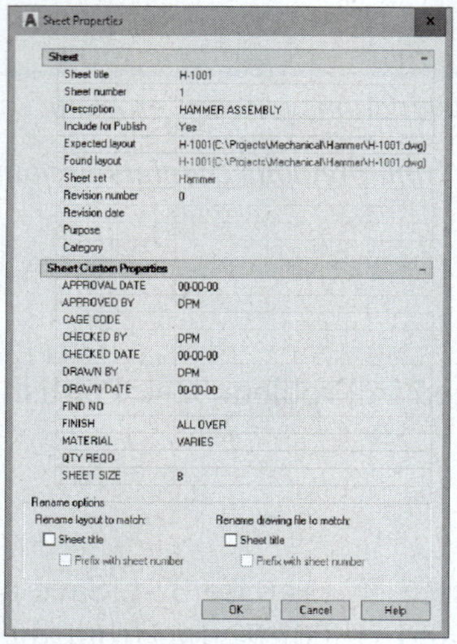

1 - H-1001

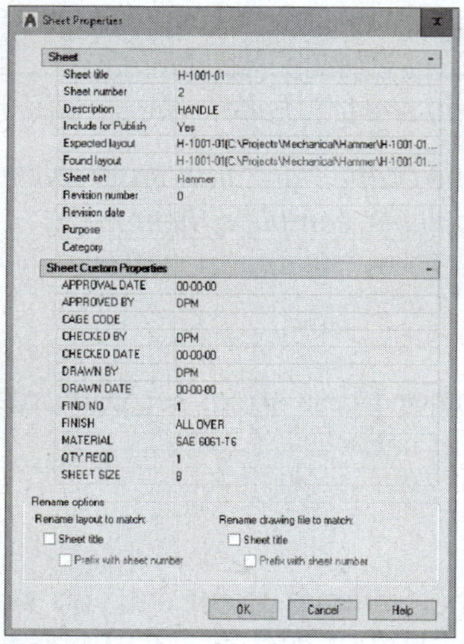

2 - H-1001-01

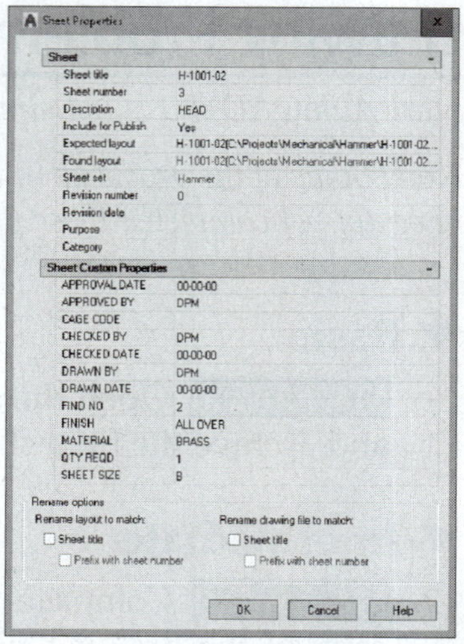

3 - H-1001-02

B

C. Create a sheet view of the 2 - H-1001-01 layout. Number the view 1 and name it HANDLE. Create a sheet view of the 3 - H-1001-02 layout. Number the view 2 and name it HEAD. Assign the BALLOON callout block to both views.

D. Complete the H-1001 layout as shown in C. Use multileaders and callout blocks from the 1 - HANDLE and 2 - HEAD sheet views in model space to create the identification balloons. Create the parts list using a sheet list table.

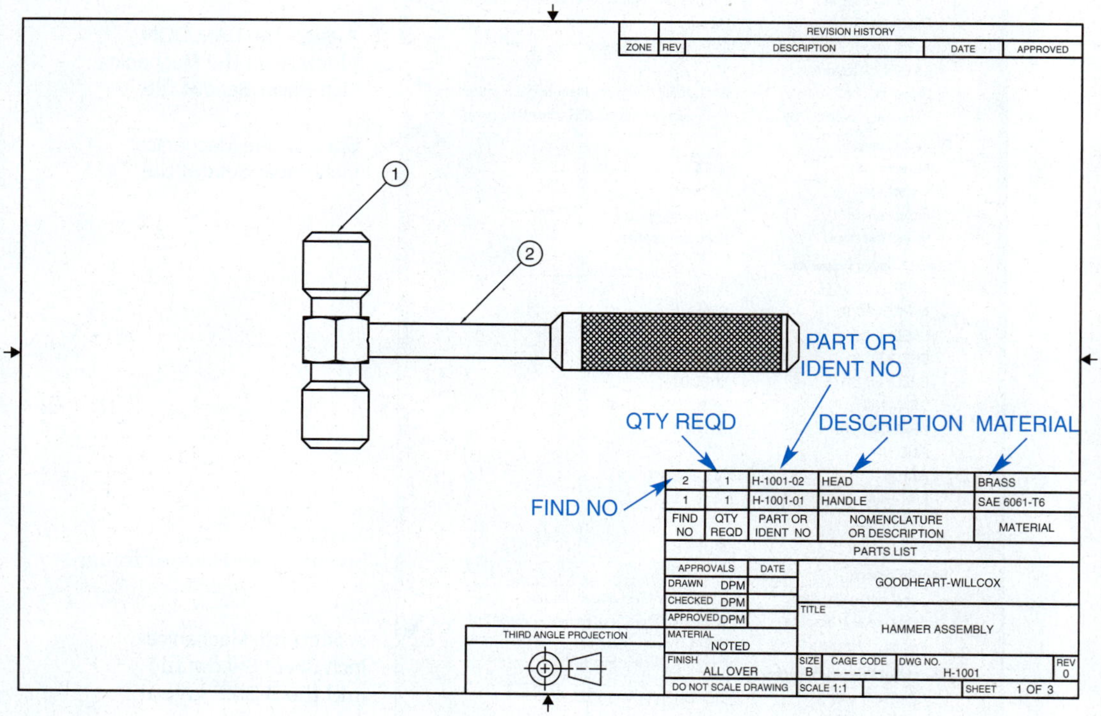

C

3. **Architectural** Publish the Cappuccino Express sheet set to a plotter or PDF file. Archive the sheet set using the zip file format.

4. **Mechanical** Publish the Hammer sheet set to a plotter or PDF file. Archive the sheet set using the zip file format.

5. **Electronics** Create a new sheet set using the **Create Sheet Set** wizard and the **Existing drawings** option. Name the new sheet set Schematic Drawings. On the **Choose Layouts** page, pick the **Browse...** button and browse to the folder where you saved the P29-14.dwg file from Chapter 29. Import all of the layouts from the file into the new sheet set. Continue creating the sheet set as follows:
   A. In the **Sheet Set Properties** dialog box, assign the layout named **ISO A1 Layout** from the Tutorial-mMfg.dwt template file in the AutoCAD 2020 Template folder as the sheet creation template.
   B. Open a new drawing file using a template of your choice and create a block for a view label. Save the drawing file and then assign the block to the sheet set using the **Label block for views** setting in the **Sheet Set Properties** dialog box.
   C. Create a new view category and name it Schematics.
   D. Open the **3 Wire Control** layout, create a new view, and add it to the Schematics view category. Double-click on the new view name in the **Sheet Views** tab and insert the view label block you previously created. Renumber the view and save the drawing.
   E. Add a custom property named **Checked by** to the sheet set and set the owner type to **Sheet**. Add another custom property named **Client** and set the owner type to **Sheet Set**.

6. **Architectural** Create a new sheet set using the **Create Sheet Set** wizard and the Architectural Imperial Sheet Set example sheet set. Name the new sheet set Floor Plan Drawings. Under the Architectural subset, create a new sheet named Floor Plan. Number the sheet A1. In the **Model Views** tab, add a new location by browsing to the folder where you saved the P17-16.dwg file from Chapter 17. Open the P17-16.dwg file and continue as follows:
   A. Create three model space views named Kitchen, Living Room, and Dining Room. Orient each display as needed to describe the area of the floor plan. Save and close the drawing.
   B. Open the A1 - Floor Plan sheet. Create a new layer named Viewport and set it current.
   C. In the **Model Views** tab, expand the listing under the P17-16.dwg file. Right-click on each view name and select **Place on Sheet**. Insert each view into the layout. Delete the default view labels inserted with the views. Double-click inside each viewport and set the viewport scale as desired.
   D. In the **Sheet Views** tab, renumber the views. Insert a new view label block under each view.
   E. Save and close the drawing.

▼ Advanced

7. **Architectural** Plan a new shopping center for your area. Determine how many stores to include. If possible, obtain a copy of a survey for vacant land in your area suitable for building the shopping center. Determine the components of a complete set of plans for the shopping center, including a site plan, floor plans, foundation plans, roof plans, elevations, and any needed sections. Establish the components for a new sheet set to organize the drawings and layouts.

8. **Architectural** Plan a new residence with approximately 3500–4000 square feet, four bedrooms, three baths, and a den/office, kitchen, dining room, nook, family room, and three-car garage. Create a complete set of plans for the residence, including a site plan, floor plans, foundation plans, roof plans, elevations, and any needed sections. Prepare layouts for each drawing. Then create a new sheet set to organize the drawings and layouts. Archive the final sheet set.

AutoCAD Certified Professional Exam Practice

Answer the following questions using the information in this chapter.

1. Which of the following methods can you use to create a sheet view? *Select all that apply.*
   A. add a model view to a layout
   B. double-click on an existing sheet
   C. right-click on an existing sheet
   D. use the **NEWVIEW** command
2. To which of the following file types can you publish a sheet set? *Select all that apply.*
   A. DST
   B. DWF
   C. EXE
   D. PDF
   E. ZIP

INDEX

1D unit block, 735
2D Cartesian coordinate system, 63
2D unit block, 735
2D wheel, 167
2-Point option, 96
3D Modeling workspace, 9
3D models, 2–3
3D unit block, 735
3-Point option, 97

A

absolute coordinates, 65
absolute path, 947
absolute value, 137, 384
acquired point, 200
action, 783
action bars, 790
action parameter, 783
Add Custom Property dialog box, 1007
Add Parameter Properties dialog box, 825
Add Scale dialog box, 254
ADDSELECTED command, 388
alert, 5
ALIGN command, 337
aligned dimensioning, 456
aligned section, 694
aligning objects, 337
alignment parameters, 801–802
  creating, 801–802
  using dynamically, 802
alignment paths, 206
alternate units, 479
American Institute of Architects (AIA) *CAD Layer Guidelines*, 134
angular dimension settings, 478
angular dimensioning, 494
  dimensioning rules, 563
  through three points, 496–497
ANNOALLVISIBLE system variable, 920
ANNOAUTOSCALE system variable, 925
Annotation Object Scale dialog box, 924
annotation scale,
  adding and deleting, 923–924
  automatically adding, 924
  setting, 253–254, 462, 914–915
  using **OBJECTSCALE** command, 923
  using **Properties** palette, 924
Annotation Scale dialog box, 253
annotation visibility, 920–923
annotational format, 669
annotations, 249, 519, 909
annotative blocks and attributes, 916–917
annotative dimensions, 461–462
  and multileaders, 916
annotative hatch patterns, 916
annotative object representation, 920
annotative objects, 909–940
  changing drawing scale, 919–926
  displaying in layouts, 917–919
  making existing objects annotative, 912–914
  preparing multiview drawings, 926–932
annotative text, 252, 916
ANNOUPDATE command, 919–920
Application Menu, 6, 11–13, 44
  searching for commands, 12
ARC command, 99, 107

ARCHIVE command, 1010
Archive a Sheet Set dialog box, 1009, 1010
archiving, 969
arcs, 99–102
  Continue option, 101–102
AREA command, 434
array, 341
array action, 805
ARRAY command, 342
arraying objects, 341–351
  base point, 350
  divide path method, 349–350
  measure path method, 349
  path array, 348
  polar array, 345
  rectangular array, 342
ARRAYPATH command, 349
ARRAYRECT command, 342
ASME Y14.1 standard sheet sizes, 856–858
ASME Y14.2 standard, 127–129, 249
associative arrays, 373–380
  replacing items, 379–380
associative dimension, 455, 581–583
  definition points, 583
associative hatch pattern, 707
ATTACH command, 945
Attach External Reference dialog box, 946
attachment, 946
ATTDEF command, 761
ATTDIA system variable, 766
ATTDISP command, 767
ATTEDIT command, 767
-ATTEDIT command, 769
ATTREDEF command, 772
Attribute Definition dialog box, 762, 768, 1004
attribute definitions, 770–772
attribute insertion point, defining, 764
attributes,
  attribute insertion point, 764
  attribute modes, 761–762
  controlling attribute display, 767
  creating blocks, 764–766
  defining, 761–764
  editing attribute properties, 764, 767–770
  tag, prompt, and value, 762–763
  text settings, 763–764
ATTSYNC command, 772
AutoCAD Color Index (ACI), 136
AutoCAD interface, 6–19
AUTOCONSTRAIN command, 666–667
Autodesk App Store, 29
Auto-fill, 629
automatic save, 39
automatic windowing, 80
automating drafting documentation, 772–777
AutoSnap feature, 76
AutoSnap marker, 193
AutoStack Properties dialog box, 262, 263
AutoStacking, using to create stacked text, 262
AutoTrack, 206–210
  object snap tracking, 206–208
  polar tracking, 208–209
  with polar snaps, 209–210
auxiliary view, 230
Axis, End option, 102–103

B

background mask, 265
Background Mask dialog box, 265
backup files, 39–40
BACTION command, 786
balloons, 605
BASE command, 945
base dimension, 594
base point, 317
base point parameter, 806
baseline dimensioning, 497–499
basic dimension, 472, 561
basic plotting, 184–186
  introduction to printing and plotting, 184
BATTMAN command, 770
BCLOSE command, 786
BCONSTATUSMODE command, 833
BCONSTRUCTION command, 833
BCPARAMETER command, 831
BEDIT command, 741, 772, 785
BESETTINGS command, 833
big font, 256
bilateral tolerance, 560
binding, 960–961
  dependent objects, 960
BLEND command, 310
block, 224, 513, 729
  block scale settings, 732
Block Attribute Settings dialog box, 771
block attributes, 761–782
  automating drafting documentation, 772–777
  editing attribute definitions, 770–772
  extracting attribute data, 777
  inserting blocks with attributes, 766–770
  using fields to reference attributes, 777

Index - Basics

Block Authoring Palettes palette, 785
BLOCK command, 764, 806
block definition, 729
Block Definition dialog box, 731, 736
Block Editor, 741–743
Block Editor Settings dialog box, 833
block insertion tools, 740
block properties table, 834–837
  Block Properties Table dialog box, 835
  creating, 834–836
block properties to ByLayer, changing, 741
block reference, 733
BLOCKICON command, 745
blocks,
  constructing, 729–733
  creating as drawing files, 746–747
  editing, 740–745
  inserting, 733–740
  inserting using **DesignCenter**, 739–740
Blocks palette, 734
BLOCKSPALETTE command, 734
boundary, 696
BOUNDARY command, 409
Boundary Creation dialog box, 409
boundary edge, 315
boundary set, 409
BPARAMETER command, 786
break, 493
BREAK command, 224, 310–311
broken-out section, 695
BSAVE command, 786
BSAVEAS command, 786
BSI, 5
B-spline, 112
BTABLE command, 834
BTESTBLOCK command, 789
button, 5
BVHIDE command, 821
BVSHOW command, 821
BVSTATE command, 820

C

CAD. *See* computer-aided drafting
CADD. *See* computer-aided design and drafting
CAGE Code. *See* Commercial and Government Entity Code
callout blocks, 989, 993–995
Cartesian (rectangular) coordinate system, 63
cascading menu, 10
cascading submenu, 10
cell format, editing tables, 632
cell styles, 618
cells, merging, 634
center dashes and centerlines, 517–519
center mark and centerline editing tools, 599–600
Center object snap, 198
Center, Diameter option, 96
Center, Radius option, 95
CENTERDISASSOCIATE command, 599
CENTERLINE command, 518
CENTERMARK command, 518
CENTERMARKEXE system variable, 518
CENTERREASSOCIATE command, 599
CENTERRESET command, 599
centroid, 198, 344
chain action, 808–810
chain dimensioning, 497, 500–501
chamfer, 110, 308, 531
CHAMFER command, 110, 308–309, 660
Character Map dialog box, 263, 264
chart dimensioning, 534
check box, 27
Check Spelling dialog box, 292
Check Spelling Settings dialog box, 293
child, 480
chord length, 99
chord length table, 101
CHSPACE command, 899
CIRCLE command, 23, 95, 208
  2-Point option, 96
  3-Point option, 97
  Center, Diameter option, 96
  Center, Radius option, 95
  Tan, Tan, Radius option, 97–98
  Tan, Tan, Tan option, 98
circular array, 341
circular reference error, 744
circumscribed polygon, 108
Clean Screen command, 172
click, 5
CLOSE command, 40
CLOSEALLOTHER command, 40
closing files, 40
coincident, 660
Color Books tab, 137
color-dependent plot style table, 861
Column Settings dialog box, 273–275
command alias, 19–20
command line, 17, 24, 70
commands,
  ADDSELECTED, 388
  ALIGN, 337
  ANNOUPDATE, 919–920

ARC, 99, 107
ARCHIVE, 1010
AREA, 434
ARRAY, 342
ARRAYPATH, 348
ARRAYRECT, 342
ATTACH, 945
ATTDEF, 761
ATTDISP, 767
ATTEDIT, 767, 769
-ATTEDIT, 769
ATTREDEF, 772
ATTSYNC, 772
AUTOCONSTRAIN, 666–667
BACTION, 786
BASE, 945
BATTMAN, 770
BCLOSE, 786
BCONSTATUSMODE, 833
BCONSTRUCTION, 833
BCPARAMETER, 831
BEDIT, 741, 772, 785
BESETTINGS, 833
BLEND, 310
BLOCK, 764, 806
BLOCKICON, 745
BLOCKSPALETTE, 734
BOUNDARY, 409
BPARAMETER, 786
BREAK, 224, 310–311
BSAVE, 786
BSAVEAS, 786
BTESTBLOCK, 789
BVHIDE, 821
BVSHOW, 821
BVSTATE, 820
CENTERDISASSOCIATE, 599
CENTERLINE, 518
CENTERMARK, 518
CENTERREASSOCIATE, 599
CENTERRESET, 599
CHAMFER, 110, 308–309, 660
CHSPACE, 899
CIRCLE, 23, 95, 208
Clean Screen, 172
CLOSE, 40
CLOSEALLOTHER, 40
COMPARE, 443
CONSTRAINTBAR, 667
CONVERTPSTYLES, 864
COPY, 338, 365
CUI, 7
DATALINKUPDATE, 1003
DBLIST, 434
DCALIGNED, 671
DCANGULAR, 671
DCCONVERT, 674
DCDIAMETER, 672
DCFORM, 676
DCRADIUS, 672
DELCONSTRAINT, 680, 833
DIM, 487, 512, 584
DIMANGULAR, 495
DIMARC, 515
DIMBASELINE, 466, 498
DIMBREAK, 470, 596–597
DIMCONSTRAINT, 671, 672, 831
DIMCONTINUE, 500
DIMDIAMETER, 491, 512
DIMEDIT, 589
DIMEX, 482
DIMIM, 482
DIMINSPECT, 597
DIMJOGLINE, 471, 493–494
DIMLINEAR, 487, 489–490, 584
DIMRADIUS, 491
DIMREASSOC, 583
DIMREASSOCIATE, 582
DIMSPACE, 593–594, 931
DIMTEDIT, 587–588, 593
DIST, 434
DIVIDE, 204, 224
DONUT, 112
DRAWORDER, 175
EATTEDIT, 767, 996
ELLIPSE, 105
ERASE, 77, 78, 144, 365, 581, 789
EXIT, 6
EXPLODE, 322, 345, 583, 743
EXPORTLAYOUT, 851
EXTEND, 315
FIELD, 641, 777
FILETAB, 47
FILETABCLOSE, 47
FILLET, 110, 305–307, 365
FIND, 295, 769
GCCOINCIDENT, 661
GCCOLLINEAR, 663
GCCONCENTRIC, 664
GCEQUAL, 664
GCFIX, 665
GCPARALLEL, 663
GCPERPENDICULAR, 663
GCSMOOTH, 665
GCTANGENT, 663
GEOMCONSTRAINT, 659, 663, 664
GOTOSTART, 34, 48
HATCH, 14, 696–708

HIDEOBJECTS, 173
ID, 434
IMPORT, 942
INPUTSEARCHOPTIONS, 25
INSERT, 18, 734, 738, 766
ISOLATEOBJECTS, 173
JOIN, 312–313
JUSTIFYTEXT, 298
LAYER, 131, 893
LAYOUT, 851
LEADER, 519
LENGTHEN, 310–320
LIMITS, 55
LINE, 15, 19, 64, 68, 70, 71, 74, 75, 95, 136, 175, 209, 210, 433, 657, 660
LINETYPE, 139
LINEWEIGHT, 140
LIST, 434
MATCHPROP, 387, 638, 708
MEASURE, 204, 225
MEASUREGEOM, 410, 428–434
MINSERT, 738
MIRROR, 339, 367, 585
MLEADER, 519, 530
MLEADERALIGN, 604, 931
MLEADERCOLLECT, 606
MLEADEREDIT, 602–603
MOVE, 333, 365, 585, 803
MTEDIT, 294
MTEXT, 249, 253, 259, 285, 286, 777
MVSETUP, 890
NAVVCUBE, 170
NCOPY, 745, 964
NEW, 12
NEWSHEETSET, 973
NEWVIEW, 178, 990
OFFSET, 221, 228
OOPS, 77, 79, 732
OPEN, 40, 964
OPENSHEETSET, 980
PAGESETUP, 12, 852
PAN, 163, 166, 167, 175, 899
PEDIT, 341, 401–406
PLINE, 105, 402
PLOT, 54, 899
PLOTTERMANAGER, 855
POINT, 204, 223
POLYGON, 108, 109, 222
PUBLISH, 1008
PURGE, 749
QDIM, 503, 590, 593
QLATTACH, 533
QLATTACHSET, 533
QLEADER, 519
QNEW, 33, 48
QSAVE, 37
QSELECT, 390
RAY, 228
RECTANGLE, 20, 106, 109, 223, 234, 314, 657, 663
REDO, 13, 77, 79
REFEDIT, 772, 961
REGEN, 106, 171, 1006
REGENALL, 171
RENAME, 744, 960
REVCLOUD, 539, 541
REVERSE, 340
REWIND, 167
RIBBON, 16
ROTATE, 170, 335, 585
SAVE, 12, 435, 747
SAVEAS, 38, 45, 56, 57, 982
SCALE, 321–322, 585
SCALETEXT, 297
searching for, 12
SELECTSIMILAR, 389
SNAP, 73
SPELL, 292
SPLINE, 25
SPLINEDIT, 341
STATUS, 435
STRETCH, 317, 367, 581, 585, 797
SUPERHATCH, 715
TABLE, 623, 775
TEXT, 249, 285, 286, 777
TEXTALIGN, 296, 297, 931
TEXTEDIT, 294, 584, 601
TEXTSCR, 436
TIME, 435
TOLERANCE, 910
TRAYSETTINGS, 956
U, 77, 78
UNDO, 77, 79, 167
UNISOLATEOBJECTS, 173
UPDATEFIELD, 290, 1006
using transparently, 173
VIEW, 989, 990, 992, 998
VIEWBACK, 167
VIEWFORWARD, 167
VPLAYER, 893
-VPORTS, 182
WBLOCK, 746, 774
WIPEOUT, 539, 541
WORKSPACE, 9
XATTACH, 945
XBIND, 960
XCLIP, 957
XLINE, 226, 227, 234, 653, 942

XOPEN, 964
XREF, 747
ZOOM, 54, 164, 166, 167, 175, 183, 899
Commercial and Government Entity Code (CAGE Code), 773
COMPARE command, 443
comparing drawings, 443–446
composition, 251
computer-aided design and drafting (CADD), 1
computer-aided drafting (CAD), 1
concentric, 664
constraining block geometry, 828–833
constraint parameters,
controlling, 832–833
using, 830–832
Constraint Settings dialog box, 666
CONSTRAINTBAR command, 667
CONSTRAINTINFER system variable, 657
constraints, 649
construction lines, 226–228
RAY command, 228
XLINE command, 226, 227, 234
context-oriented help, 27
context-sensitive, 10
continued dimensioning, 500
Continuous linetype, 138
control code sequence, 287
control vertices, 410
conventional break, 493
conventional dimensioning, 559
CONVERTPSTYLES command, 864
coordinate dimensioning, 494
coordinate display, 67
coordinate entry methods, 65–67
absolute coordinates, 65
dynamic input, 67–69
polar coordinates, 66
relative coordinates, 65
coordinates, 20
copy and paste, 387
COPY command, 338, 365
copying nested objects, 745
copying objects, 338–339
Array option, 339
counterbore, 513
countersink, 513
Create Layout wizard, 849
Create New Dimension Style dialog box, 464, 475
Create New Multileader Style dialog box, 522
Create New Table Style dialog box, 617
Create page, 7
Create Sheet Set wizard, 976
Create Transmittal dialog box, 1012
crosshairs and cursor, 9
cubic curve, 404
CUI command, 7
current layer, 135
cursor menu, 10
curve fitting, 404
Custom Properties dialog box, 1007
Customization flyout, 17
Customize User Interface (CUIx) file, 12
cutting edge, 313
cutting-plane line, 691

D

DATALINKUPDATE command, 1003
datum, 497, 534
DBLIST command, 434
DCALIGNED command, 671
DCANGULAR command, 671
DCCONVERT command, 674
DCDIAMETER command, 672
DCFORM command, 676
DCRADIUS command, 672
default, 7
Default Scale List dialog box, 254
deferred perpendicular, 202
deferred tangency, 202
definition point, 204, 583
defpoints, 583
DELCONSTRAINT command, 680, 833
demand loading, 959
dependent objects, 733, 950
DesignCenter,
introduction, 149
hatching, 709–710
layouts, 849–850
detach, 955
deviation tolerance, 566
dialog box, 18
diameter, 96
Dictionaries dialog box, 293
DIM command, 487, 512, 584
DIMANGULAR command, 495
DIMARC command, 515
DIMBASELINE command, 466, 498
DIMBREAK command, 470, 596–597
DIMCONSTRAINT command, 671, 672, 831
DIMCONTINUE command, 500
DIMCONTINUEMODE system variable, 501
DIMDIAMETER command, 491, 512
DIMEDIT command, 589

dimension, 455
dimension constraints, 668–677
dimension editing tools, 583–590
    assigning different dimension style, 583–584
    DIMEDIT command, 589
    dimension grip commands, 585–587
    DIMTEDIT command, 587–588
    editing dimension value, 584–585
    QDIM command, 590
dimension layer override, 502–503
dimension standards and styles, 455–486
dimension styles, 463–482
    adding, 482
    changing, 480
    creating, 464–465
    Dimension Style Manager dialog box, 463
    overriding, 583, 591–592
    renaming and deleting, 480–482
    setting current, 482
dimension variables, 465
dimensional constraint parameters, 674
dimensional constraints, 668–677
    converting associative dimensions, 674
    dynamic and annotational form, 676–677
    managing, 674
    working with parameters, 674–676
dimensional input, 68
dimensioning,
    aligned dimensioning, 456
    alternate practices, 534–535
    arcs, 515
    circles, 512
    cones and regular polygons, 460
    cylindrical shapes, 459
    features and objects, 459
    fillets and rounds, 516
    holes, 513
    manufacturing processes, 513–514
    repetitive features, 514
    size and location, 456
    square and rectangular features, 460
    unidirectional dimensioning, 456
    with tolerances, 559–580
dimensioning units, 561–562
dimensions, spacing and aligning, 593–596
DIMEX command, 482
DIMIM command, 482
DIMINSPECT command, 597
DIMJOGLINE command, 471, 493–494
DIMLINEAR command, 487, 489–490, 584
DIMRADIUS command, 491
DIMREASSOC command, 583
DIMREASSOCIATE command, 582
DIMSPACE command, 593–594, 931
DIMTEDIT command, 587–588, 593
DIN, 5
direct dimensioning, 498
direct distance entry, 74
Direction Control dialog box, 54
displacement, 317
DIST command, 434
DIVIDE command, 204, 224
dock, 10
documents,
    controlling drawing windows, 45–46
    file tabs, 46–49
    managing multiple, 45–49
DONUT command, 112
donuts and filled circles, drawing, 112
double-click, 5
Drafting & Annotation workspace, 164
drafting documentation, automating, 772–777
Drafting Settings dialog box, 86, 196, 208, 210
drafting standards, 5
drafting symbols, 492, 514, 570
draw order, controlling, 175–176
drawing aids, 70–77
    grid mode, 72–73
    object snap, 76
    object snap tracking, 76–77
    ortho mode, 75
    polar tracking, 74
    snap mode, 73
drawing and editing, 61–94
    introduction to drawing aids, 70–77
    object selection, 79–86
drawing data, listing, 434
drawing exchange file (DXF), 37
drawing files, 33
drawing information,
    checking the time, 435–436
    listing drawing data, 434
    obtaining, 427–454
    reviewing drawing status, 435
    taking measurements, 427–434
    using **QuickCalc**, 437–443
drawing lines, 64–65
drawing points, 223–226
    setting point style, 224
drawing properties, 49–51
Drawing Recovery Manager, 39
drawing scale,
    and dimensions, 460–462
    and text height, 252–254
    annotation visibility, 920–923

Index - Basics

changing, 919–926
definition, 860
introduction to scale representations, 920
drawing sessions, 3
drawing settings, 51–55
drawing units, 53–54
introduction to model space, 51
drawing sheets, 56, 858
drawing standards, 4
drawing standards file (DWS), 34
drawing status, reviewing, 435
drawing template, 4, 33, 36–40
drawing units, 53
Drawing Units dialog box, 53–55
drawing window, 8
drawings and templates, 33–60
basic drawing settings, 51–55
closing files, 40
drawing properties, 49–51
introduction to templates, 56–58
managing multiple documents, 45–49
opening saved file, 40–45
starting a new drawing, 33–36
DRAWORDER command, 175
drop-down list, 7
dual dimensioning units, 479
duplicate objects, deleting, 323
DWG Compare toolbar, 444
dynamic blocks,
alignment parameters, 801–802
base point parameters, 806
chain actions, 808–810
definition, 783
flip parameters, 803–804
linear parameters, 791–796
parameter value sets, 806–807
point parameters, 786–789
polar parameters, 796–799
rotation parameters, 799–801
XY parameters, 804–806
dynamic columns, 272
dynamic input, 19, 67–69

E

EATTEDIT command, 767, 996
Edit a Block in a Table Cell dialog box, 633
Edit Attribute dialog box, 771
Edit Attributes dialog box, 764–765
Edit Block Definition dialog box, 741, 785
Edit Drawing Scales dialog box, 254
Edit Layer State dialog box, 148
editing,
attribute references, 767–770
between drawings, 387–388
blocks, 740–745
dimensions, 572–614
ERASE command, 77–78
hatch patterns, 711–715
OOPS command, 79
REDO command, 79
tables, 628–638
U command, 78
UNDO command, 79
ELLIPSE command, 102–105
Axis, End option, 102–103
Center option, 102
drawing elliptical arcs, 103–105
Rotation option, 103
Endpoint object snap, 198
ERASE command, 77, 78, 144, 365, 581, 789
escape key, 25
EXIT command, 6
EXPLODE command, 322, 345, 583, 743
exploding and redefining a block, 743
Export layer state dialog box, 148
exporting, 855
EXPORTLAYOUT command, 851
EXTEND command, 315
extending, 315–316
Extension object snap, 200
extension path, 200
external reference (xref), definition, 941
external references. *See* Xrefs

F

fence selection, 83–84
field, 971
FIELD command, 641, 777
Field dialog box, 289, 641, 1008
fields, 289–291, 436
editing and converting, 290–291
inserting, 289
updating, 290
fields and blocks, inserting, 633
file properties, 36
file tabs, 46–49
files,
backup files, 39–40
closing, 40
finding, 41–43
opening, 40–45
recovering damaged file, 40

FILETAB command, 47
FILETABCLOSE command, 47
fillet, 110, 305, 516
FILLET command, 110, 305–307, 365
 creating full rounds, 306
 setting fillet radius, 305–306
 trim settings, 307
Find and Replace dialog box, 295, 296
FIND command, 295, 769
finding files, 41–43
fit curve, 404
fit format, 474
fit points, 410
flip action, 803
flip parameters, 803–804
float, 10
floating viewports, 846, 878–899
 activating and deactivating, 885–886
 boundary adjustment, 889
 clipping viewports, 883–884
 controlling layer display, 890–897
 controlling linetype scale, 888
 converting objects to floating viewports, 881–882
 forming polygonal floating viewports, 881
 inserting views, 883
 locking and unlocking, 890
 maximizing, 898–899
 scaling annotations, 888
flyout, 13
flythrough, 2
folders, adding, 998
font, 249
foreshortened, 230
formulas, 638–641
 creating, 638–639
 sum, average, and count formulas, 639–640
fractional text, 251
FRAMESELECTION system variable, 957
Full Navigation wheel, 167
full section, 694
fully constrained, 650
function, 440
function keys, 27

G

gap tolerance, 700
GB, 5
GCCOINCIDENT command, 661
GCCOLLINEAR command, 663
GCCONCENTRIC command, 664
GCEQUAL command, 664
GCFIX command, 665
GCPARALLEL command, 663
GCPERPENDICULAR command, 663
GCSMOOTH command, 665
GCTANGENT command, 663
GD&T, 570
general notes, 458
geographic data, 948
GEOMCONSTRAINT command, 659, 663, 664
Geometric Center object snap, 198
geometric constraint bars, 667
geometric constraints,
 AUTOCONSTRAIN command, 666–667
 coincident, 660–661
 concentric and equal, 664
 horizontal and vertical, 662
 inferring, 657–659
 managing, 667–668
 parallel and perpendicular, 662–663
 symmetric, fix, and smooth, 664–666
geometric dimensioning and tolerancing (GD&T), 570
global attribute editing, 769
global layer settings, 890
global linetype scale, 139
GOST, 5
GOTOSTART command, 34, 48
grab bar, 11
gradient fill, 706–707
graphic pattern, 691
graphical user interface (GUI), 6
graphics window, 8
grid, 72
grips, 77, 365–373
 copying, 372
 editing associative arrays, 373–380
 mirroring, 372
 moving, 369
 nudging, 370
 object properties, 380–386
 rotating, 370
 scaling, 371
 stretching, 367–369
grouped balloons, 605
gutter, 275

H

half section, 695
hard copy, 184

HATCH command, 14, 696–708
annotative hatch patterns, 704
boundary set, 699–700
controlling draw order, 708
correcting boundary errors, 700
gradient fill, 706–707
hatch composition options, 707–708
island detection, 697–698
removing boundaries, 699
retaining boundaries, 701
reusing existing hatch properties, 708
scaling hatch patterns manually, 704
selecting objects, 698
setting hatch origin point, 706
solid and gradient fills, 702
specifying boundaries, 697–701
user defined hatch, 703
hatch patterns, 694, 701
editing, 711–715
Help window, 27, 29
hexagon, 109
HIDDEN linetype, 138
HIDEOBJECTS command, 173
host drawing, 941
hover, 9
HPLAYER system variable, 705
hyperlinks, 993

I

icon, 5
ID command, 434
implied intersection, 306
implied windowing, 80
IMPORT command, 942
Import layer state dialog box, 148
Import Layouts as Sheets dialog box, 983
Import Options dialog box, 979
inch dimensions, dimensioning rules, 562
included angle, 99
infer, 657
InfoCenter, 29
initial template setup, 58
input search tools, 24–25
INPUTSEARCHOPTIONS command, 25
inscribed polygon, 108
INSERT command, 18, 734, 738, 766
Insert dialog box, 739, 747
Insert Table dialog box, 617, 624, 627
insertion base point, 730
Inspection Dimension dialog box, 598
inspection dimensions, creating, 597–599
interface, 6
Intersection object snap, 200
islands, 409, 697
ISOLATEOBJECTS command, 173
isometric circles and arcs, 105
isometric dimensions, 606
isometric drawings, 74
isometric text, 298

J

JIS, 5
JOIN command, 312–313
joining different object types, 313
joining same object type, 312–313
justification point, 285
justify, 267
JUSTIFYTEXT command, 298

K

key point, 344
keyboard shortcut, definition, 25
keyboard shortcuts, 25–27

L

landing, 524
landscape, 185
lasso selection, 81–82
last selection, 84
LAYER command, 131, 893
layer display,
controlling, 890–897
freezing and thawing, 893–895
layer property overrides, 893–897
Layer Filter Properties dialog box, 133
layer filters, 146
layer index, 959
layer property overrides, 893
layer settings, 149
Layer Settings dialog box, 133
layer state, 147
Layer States Manager dialog box, 133
Layer Transparency dialog box, 142

Index - Basics

layers,
adding, 133, 151
deleting, 145
freezing and thawing, 143–144
layer color, 136–137
layer description, 143
layer filters, 146
layer linetype, 137–138
layer lineweight, 140
layer names, 133
layer plotting properties, 143
layer status, 135
linetype scale, 139
loading linetypes, 138–139
locking and unlocking, 144
renaming, 134
selecting multiple, 134
setting current layer, 135–136
transparency, 141–143
used in drafting fields, 131
LAYOUT command, 851
layout setup, 845–874
initial setup, 852
page setups, 852–868
working with layouts, 846–852
layouts,
definition, 51, 845
drawing in paper space, 875–877
layout origin, 877–878
moving and copying, 849
LEADER command, 519
leader line, 519
LENGTHEN command, 319–320
library path, 955
limit dimensioning, 560
limits, 54, 559
LIMITS command, 55
LINE command, 15, 19, 64, 68, 70, 71, 74, 75, 95, 136, 175, 209, 210, 433, 657, 660
line conventions, 127
line spacing, 269
line standards and layers, 127–162
layers used in drafting fields, 131
linear and angular dimensioning, 487–510
adjusting overlapping dimensions, 501–502
baseline, chain, and direct dimensioning, 497–501
dimension layer override, 502–503
dimensioning angles, 494–497
linear dimensions, 487–494
adjusting dimension text, 490–491
dimensioning angled surfaces and auxiliary views, 492
DIMJOGLINE command, 493–494
DIMLINEAR command, 489–490
long objects, 493
symbols with dimension text, 491
linear parameters, 791–796
assigning scale action, 794–795
assigning stretch action, 792
stretching objects symmetrically, 793–794
using scale action dynamically, 795
using stretch action dynamically, 792–793
LINETYPE command, 139
linetype scale, 139
linetypes, loading, 138
lineweight, 140
LINEWEIGHT command, 140
Lineweight dialog box, 140
list box, 16
LIST command, 434
List of Blocks dialog box, 994–995
Load or Reload Linetypes dialog box, 138
location dimensions, 456
Lock UI option, 11
Lock User Interface flyout, 11
locked layer fading, 144
locking and unlocking layers, 144
LOCKUI system variable, 11
lookup action, 823
lookup parameters, 823–827
assigning lookup action, 824
creating lookup table, 824–826
using dynamically, 827
lookup table, 824

M

major axis, 102
Manage Cell Styles dialog box, 621
margin, 848
markers, definition, 76, 193
marking up, 539–542
creating revision clouds, 539
modifying revision clouds, 540
WIPEOUT command, 541
master drawing, 942
MATCHPROP command, 387, 638, 708
MEASURE command, 204, 225
MEASUREGEOM command, 410, 428–434
angles, 430–431
area, 431–433

distance, 429–430
radius and diameter, 430
measurements, taking from a drawing, 427–434
metric dimensions, dimensioning rules, 563
Microsoft® Excel spreadsheet, 641
Midpoint object snap, 198
minor axis, 102
MINSERT command, 738
MIRRHATCH system variable, 714
MIRROR command, 339, 367, 585
mirror line, 339
mirroring objects, 339–340
MIRRTEXT system variable, 339
MLEADER command, 519, 530
MLEADERALIGN command, 604, 931
MLEADERCOLLECT command, 606
MLEADEREDIT command, 602–603
model, 51, 845
model space, 51, 845
model viewports, 180–184
adjusting, 183
configuring, 181
joining, 183
working in, 183
model views, 997–1000
adding folders, 998
creating sheet views from model views, 998–1000
modeless dialog boxes, 16
Modify Archive Setup dialog box, 1010
Modify Dimension Style dialog box, 480
Modify Multileader Style dialog box, 522
modifying objects, 305–332
BLEND command, 310
BREAK command, 310–311
CHAMFER command, 308–309
exploding objects, 322–323
extending, 315–316
FILLET command, 305–307
JOIN command, 312–313
LENGTHEN command, 319–320
SCALE command, 321–322
stretching, 317–318
trimming, 313–314
move action, 788
MOVE command, 333, 365, 585, 803
moving objects, 333–335
Displacement option, 333
using first point as displacement, 334
MSLTSCALE system variable, 915
MTEDIT command, 294
MTEXT command, 249, 253, 259, 285, 286, 777
multileader editing tools, 601–606
aligning multileaders, 604–605
assigning different multileader style, 601
context-sensitive options, 602
editing multileader mtext, 601
grouping multileaders, 605–606
multileader grip commands, 602
shortcut menu options, 602
Multileader Style Manager dialog box, 521
multileader styles, 519–530
attaching a symbol, 528–529
attaching mtext, 526–528
changing, 529
creating new, 522–523
renaming and deleting, 529–530
scale options, 524–525
setting constraints, 523–524
setting current, 530
using no content, 529
multiline, 108
multiline text, 259–275
adding symbols, 263
automatic bullets and numbering, 270–271
character formatting, 266
columns, 272
dynamic columns, 273
importing text, 275
lists, 270
manual stacking, 263
matching and removing formatting, 271–272
paragraph formatting, 267
stacking text, 262
static columns, 273
style settings, 264–265
superscript and subscript text, 263
using background mask, 265
using text editor, 260–261
multiview drawing construction, 232–235
auxiliary views, 234–235
orthographic views, 233
multiview drawings, 228–232, 926–932
auxiliary view, 230
creating differently scaled drawings, 926–927
preparing, 926–932
removed views, 230–232
reusing annotative objects, 927
selecting front view, 228–229
showing hidden features, 232
symmetry and circle centers, 232
MVSETUP command, 890

N

named objects, 733
named plot style table, 861
named views, 176–180
  activating a view, 179
  managing views, 180
navigation bar, 163–164
navigation wheel, 167
NAVVCUBE command, 170
NCOPY command, 745, 964
nested xrefs, 942
nesting, 733
NEW command, 12
New Page Setup dialog box, 852
New Sheet dialog box, 984
New Table Style dialog box, 256, 617
NEWVIEW command, 178, 990
New View/Shot Properties dialog box, 178, 990
New Visibility State dialog box, 821
New VP Freeze icon, 894
New VP Thaw icon, 894
NEWSHEETSET command, 973
non-associative array, 345
non-associative dimension, 581
non-associative hatch pattern, 707
non-breaking space, 263
non-uniform rational basis spline (NURBS), 112
notes, 458
noun/verb selection, 367
nudging, 370
NURBS, 112

O

object commands,
  arcs, 99–102
  circles, 95–99
  donuts and filled circles, 112
  ellipses, 102–105
  polylines, 105–108
  rectangles, 109–111
  regular polygons, 108–109
  true splines, 112–113
object groups, 391
object properties, 380–387
  adding mtext frame, 385–387
  general properties, 383–384
  geometry properties, 384–385
  Properties palette, 382–383
  text properties, 385–387
  using **Quick Properties** palette, 381–383
object selection, 79–86
  changing selection set, 84–85
  cycling through stacked objects, 85–86
  fence selection, 83–84
  lasso selection, 81–82
  last selection, 84
  previous selection, 84
  selecting all objects, 84
  window and crossing polygon selection, 82–83
  window and crossing selection, 79–81
object selection filters, 391
object snap, 76, 193–220
  and AutoTrack, 193–220
  center object snap, 198
  endpoint object snap, 198
  extension and extended intersection object snaps, 200–201
  geometric center object snap, 198–199
  intersection object snap, 200
  mid between 2 points snap, 205–206
  midpoint object snap, 198
  nearest object snap, 204
  node object snap, 204
  object snap overrides, 197
  parallel object snap, 204
  perpendicular object snap, 202
  quadrant object snap, 199–200
  running object snaps, 194–197
  snap from, 205
  tangent object snap, 202–203
  temporary track point snap, 205
  tracking, 206
object snap override, 197
Object Snap shortcut menu, 197
object snap tracking, 76–77
OFFSET command, 221, 228
offset section, 694
OOPS command, 77, 79, 732
OPEN command, 40, 964
OPENSHEETSET command, 980
option, 1
Options dialog box, 7, 9, 12, 19, 166, 196, 290, 848, 859
ordinate dimension objects, 536–539
  using WCS and a UCS, 536–537
ordinate dimensioning, 536
origin, 63
ortho, 75

Index - Basics

orthographic projection, 228
over-constrained, 650
overlay, 946
overlay system, 131
override, 480

P

Page Setup dialog box, 854
page setups, 852–868
  drawing orientation, 858
  electronic plots, 855
  plot area, 859
  plot device, 854–855
  plot offset, 859–860
  plot scale, 860–861
  sheet size, 855–856
  standard sheet sizes, 856–858
PAGESETUP command, 12, 852
palette, 16
pan, 63, 166
PAN command, 163, 166, 167, 175, 899
panels, 14
paper (layout) space, 51, 845
paper text height, 252
paragraph alignment, 267
Paragraph dialog box, 260, 267, 268
parallel, 662
parallel alignment path, 204
parallel offsets, 221–223
  erasing source object, 222
  Layer option, 223
  specifying offset distance, 221–222
  Through option, 222
parameter filters, 675
parameter grips, 784
parameter label, 787
parameter sets, 828
parameter value sets, 806–807
parameters, 649, 783
parametric drafting, 649–690
  dimensional constraints, 668–677
  geometric constraints, 657–668
  parametric editing, 678–680
  parametric fundamentals, 649–656
parent, 480
parent xref, 947
partial auxiliary view, 230
partial open, 45
parts list, 776
path array, 341
PEDIT command, 341, 401–406
  changing polyline width, 402
  converting objects to polyline, 402
  editing polyline vertex, 403
  Fit option, 404
  joining polylines, 402
  linetype generation, 406
  opening and closing polyline, 402
  straightening all polyline segments, 405
PEDIT Spline option, 404
perpendicular, 662
pick, 5
pick box, 77
PICKADD system variable, 383
PICKFIRST system variable, 367
planning sheet, 4
PLINE command, 105, 402
PLINEREVERSEWIDTHS system variable, 341, 406
PLOT command, 54, 899
plot device, 854
Plot dialog box, 185, 899
plot spooler, 902
plot stamp, 900
plot style table, 861
Plot Style Table Settings dialog box, 862
plot styles, 861–866
  applying, 864–866
  completing page setup, 867
  configuring plot style table type, 862–863
  creating and editing plot style tables, 863
  plot style tables, 861–862
  viewing effects on-screen, 866
PLOTTERMANAGER command, 855
plotting, 899–902
  adding a plot stamp, 900
  adding layouts, 903
  layout express tools, 903
  output, 902
  page setup options, 900
  plotting to PLT file, 902
  previewing, 901
  saving changes to layout, 900
plotting layouts, 875–908
  floating viewports, 878–899
  layout content, 875–878
plus-minus dimensioning, 559
POINT command, 204, 223
point entry, 63
point of tangency, 97

point parameters, 786–789
assigning move action, 788
using move action dynamically, 788–789
Point Style dialog box, 224
pointer input, 68
polar array, 341, 377
polar coordinate, 66
polar coordinate system, 456
polar parameters, 796–799
assigning polar stretch action, 797
using polar stretch action dynamically, 797–799
polar stretch action, 797
polar tracking, 74, 208–209
polygon, 64
POLYGON command, 108, 109, 222
polyline and spline editing tools, 401–426
creating polyline boundary, 409–410
PEDIT command, 401–406
polyline grip commands, 407–408
spline grip commands, 414–416
SPLINEDIT command, 410–414
polyline boundary, creating, 409–410
polyline grip commands, 407–409
polyline vertex, 403
polylines, 105–108
drawing polyline arcs, 107–108
halfwidth, 106
Length option, 106
setting polyline width, 105
pop-up menu, 10
portrait, 185
prefix, 477
preview box, 18
previous selection, 84
primary units, setting, 562–564
printing and plotting, 184–186
making a plot, 184–186
plotting in model space, 184
projection plane, 228
prompts, responding to, 61–62
Properties dialog box, 49, 50
property columns, adjusting, 145
Property Lookup Table dialog box, 824, 827
PUBLISH command, 1008
publishing, 186, 902, 969
publishing a sheet set, 1008–1010
publish options, 1009
sheet selections, 1010
purge, 132, 749
PURGE command, 749
Purge dialog box, 749
purging named objects, 749–750

Q

QDIM command, 503, 590, 593
QLATTACH command, 533
QLATTACHSET command, 533
QLDETACHSET command, 533
QLEADER command, 519
QNEW command, 33, 48
QSAVE command, 37
QSELECT command, 390
quadrant, 96, 199
quadrant object snap, 199–200
quadratic curve, 404
Quick Access toolbar, 8, 13, 14
Quick Properties palette, 381–383
Quick Select dialog box, 390
Quick Start template, 34
QuickCalc,
clearing input and history areas, 438
converting units, 440
drawing values, 442
entering expressions, 437
QuickCalc palette, 437
scientific calculations, 439
using, 437–443
variables, 440–441
with commands, 442
with object properties, 443

R

radio button, 54
radius, 95
ray, 228
RAY command, 228
read-only, 45, 982
real block, 735
realtime panning, 166
realtime zoom, 164–165
RECTANGLE command, 20, 106, 109, 223, 234, 314, 657, 663
rectangles, 109–111
adding rounded corners, 110
drawing rotated rectangle, 111–112
rectangle dimensions, 111
setting the width, 110–111
specifying area, 111
rectangular array, 341, 376
rectangular coordinate dimensioning without dimension lines, 534
rectangular coordinate system, 456

rectangular coordinates, 63
redlining, 539
REDO command, 13, 77, 79
redrawing, 171
REFEDIT command, 772, 961
reference dimension, 650
Reference Edit dialog box, 961
reference editing, 961
reference file, 941
REGEN command, 106, 171, 1006
REGENALL command, 171
regenerating, 171
region, 409, 701
regular polygon, 108
relative coordinates, 65
relative operators, 390
relative path, 947
reload, 955
removed section, 694
removed view, 230
Rename & Renumber View dialog box, 993
RENAME command, 744, 960
Rename dialog box, 258, 482, 622, 745
repetitive features, 514
resource drawings, 997
REVCLOUD command, 539, 541
REVERSE command, 340
reversing object's point calculation, 340–341
revision cloud, 539
revision history block, 774
revolved section, 694
REWIND command, 167
RIBBON command, 16
ribbon elements, displaying and organizing, 16
ribbon panels, 14–15
right-click, 5
right-click menu, 10
root point, 226
rotate action, 799
ROTATE command, 170, 335, 585
rotating objects, 335–337
  copying while rotating, 337
  Reference option, 336
rotation parameters, 799–801
  assigning rotate action, 800
  using rotate action dynamically, 800–801
round, 110, 305, 516
rubberband line, 64
running object snaps, 76, 194

S

SAVE command, 12, 435, 747
SAVEAS command, 38, 45, 56, 57, 982
scalable fonts, 256
scale, 909
scale action, 794
SCALE command, 321–322, 585
  copying while scaling, 322
  Reference option, 321
scale factor, 252
SCALETEXT command, 297
schematic block, 735, 916
scroll bar, 16
search text box, 12
section, 691
section lines, 693–694
section view, 691–696
  and graphic patterns, 691–728
  section lines, 693–694
  types of sections, 694–696
Select Annotation Scale dialog box, 253, 462
Select Color dialog box, 137
Select Drawing File... dialog box, 734
Select File dialog box, 42
Select Layout as Sheet Template dialog box, 975
Select Linetype dialog box, 138
Select Plot Style dialog box, 865
Select Reference File dialog box, 945
selected grip, 366
selection express tools, 391
selection set, 79
SELECTIONANNODISPLAY system variable, 931
SELECTSIMILAR command, 389
Set Mtext Height dialog box, 261, 274
Set Mtext Width dialog box, 261, 274
shade, 707
sheet, 51, 969
Sheet List Table dialog box, 1001
sheet list tables, 1000–1003
  editing, 1003
  inserting, 1001
  sheet list table hyperlinks, 1003
  specifying rows, 1002
  specifying title and columns, 1002
Sheet Properties dialog box, 945, 1005
sheet selections, 1002
sheet set fields, 1003–1008
  current sheet set fields, 1005
  custom properties, 1006–1007
  custom property fields, 1008
  sheet set placeholder fields, 1006

sheet set placeholder, 1006
Sheet Set Properties dialog box, 975, 1004
sheet set tools,
  archiving a sheet set, 1010–1012
  model views, 997–1000
  publishing a sheet set, 1008–1010
  sheet list tables, 1000–1003
  sheet set fields, 1003–1008
  sheet views, 989–997
sheet sets, 969–988
  fundamentals, 969–973
  managing, 980
  preparing, 972
  Sheet Set Manager, 972–973
  sheets, 982–986
  subsets, 981–982
sheet size, 56
sheet views, 989–997
  callout blocks, 993–995
  creating from layouts, 992
  managing, 993
  view categories, 990–991
  view label blocks, 995–997
sheets, 982–986
  adding a layout as a sheet, 983–984
  adding a sheet using a template, 984
shortcut key, 25, 27, 261
shortcut menu, 10
shoulder, 519
single limits, 561
single-line text, 285–288
  adding symbols, 287–288
  align and fit justification, 287
  checking spelling, 291–294
  fields, 289–291
  revising text, 294–298
  underscoring or overscoring text, 288
size dimensions, 456
slider, 9
snap, 73
SNAP command, 73
snap grid, 73
snap resolution, 73
snapping, 76, 193
soft copy, 184
solid model, 2
source object, 387
source points, 337
spacing and aligning dimensions, 593–596
  DIM command, 594
  DIMSPACE command, 594–596
spatial index, 959
specific notes, 458
specified dimension, 559
Specify Zip File dialog box, 1009
SPELL command, 292
spelling,
  aligning text, 296
  changing dictionaries, 293–294
  changing justification, 298
  checking, 291–294
  cutting, copying, and pasting text, 294
  finding and replacing text, 294
  scaling text, 297
  using **SPELL** command, 292–293
spline, 112
SPLINE command, 25, 112
spline curve, 404
spline grip commands, 414–416
Spline option, 404
SPLINEDIT command, 341, 410–414
  converting splines and polylines, 413–414
  Edit vertex option, 413
  Fit Data option, 412
  joining splines, 412
  opening and closing a spline, 411
SPLINETYPE system variable, 405
spotface, 513
Stack Properties dialog box, 262
stacked objects, 85
standard blocks, 729–760
  constructing blocks, 729–733
  creating blocks as drawing files, 746–747
  editing blocks, 740–745
  inserting blocks, 733–740
  purging named objects, 749–750
  symbol libraries, 747–749
standards, 4
Start Tab, 7–8
starting AutoCAD, 5
static columns, 273
STATUS command, 435
status toggle buttons, 17
sticky panel, 11
stretch action, 792
STRETCH command, 317, 367, 581, 585, 797
stretching, 317–318
  Displacement option, 317
  using first point as displacement, 318
subregion, 957
Subset Properties dialog box, 981, 1005
subsets, 974
sub-units, 478
suffix, 477
SUPERHATCH command, 715
surface model, 2
symbol libraries, 747–749
  creating, 747

Index - Basics

storing, 748–749
symmetrical tolerance, 564
system variables,
ANNOALLVISIBLE, 920
ANNOAUTOSCALE, 925
ATTDIA, 766
CENTERMARKEXE, 518
CONSTRAINTINFER, 657
DIMCONTINUEMODE, 501
FRAMESELECTION, 957
HPLAYER, 705
LOCKUI, 11
LTSCALE, 139
MIRRHATCH, 714
MIRRTEXT, 339
MSLTSCALE, 915
PICKADD, 383
PICKFIRST, 367
PLINEREVERSEWIDTHS , 341, 406
REGENMODE, 171
SELECTIONANNODISPLAY, 931
SPLINETYPE, 405
XCLIPFRAME, 957
XDWGFADECTL, 949
XREFLAYER, 945

T

tab, 14
table cell editor, editing tables, 629
TABLE command, 623, 775
table indicator, 626
table layout,
adjusting table style, 634
picking a cell edge to edit, 634
resizing columns and rows, 634–635
table breaks, 635–636
table properties, 637–638
Table Style dialog box, 615–616
table styles, 615–623
cell styles, 618
changing, renaming, and deleting table styles, 622
creating cell styles, 620–621
creating new table styles, 616–617
setting current, 623
starting table styles, 621–622
tab settings, 618–619
table direction, 618
Table Style dialog box, 615–616
using a starting table style, 626
tables, 615–648
adding table styles, 641
calculating values, 638–641
editing, 628–638
extracting table data, 641
inserting, 623–628
linking table to Excel data, 641
table styles, 615–623
tabular dimensioning, 534
Tan, Tan, Radius option, 97–98
Tan, Tan, Tan option, 98
tangent, 97
Tangent object snap, 203
template development, 56–58
initial template setup, 58
Template Options dialog box, 57
templates,
definition, 4, 33
introduction to, 56–58
template development, 56–58
using, 849
text, 249
scaling text manually, 252
text boundary, 259
text box, 9
TEXT command, 249, 285, 286, 777
text editor, 260
text express tools, 298
text height, 252
text presentation, 251
text standards and composition, 249–251
Text Style dialog box, 255, 258
text styles, 255–259
adding, 275
and multiline text, 249–284
changing, renaming, and deleting text styles, 258
creating new text styles, 255–256
effects, 257–258
font options, 256
setting current, 259
size options, 256
Text Style dialog box, 255
text typefaces and formats, 250
TEXTALIGN command, 296, 297, 931
TEXTEDIT command, 294, 584, 601
TEXTSCR command, 436
thread drawings and notes, 532–534
tiled viewport, 180
TIME command, 435
tint, 707
title blocks, 773–774
tolerance, 559

Index - Basics

tolerance buildup, 497
TOLERANCE command, 910
tolerance methods,
basic tolerance method, 567–568
deviation tolerance method, 566
limits tolerance method, 567
setting, 564–568
symmetrical tolerance method, 564–566
tolerance settings,
specifying, 568–570
tolerance alignment, 569
tolerance height, 568
tolerance precision, 568
vertical position, 569
zero suppression, 570
tolerance text, 262
tool buttons, 13
tool palettes, 710–711
inserting hatch patterns, 711
locating and viewing content, 710
toolbars, 11
tooltip, 9
tracking vectors, 205
Transmittal Setups dialog box, 1012
transparency, 144
transparently, 173
Tray Settings dialog box, 956
TRAYSETTINGS command, 956
tree view display, 954
trimming, 313–314
to an implied intersection, 314
without selecting cutting edge, 314
True Color tab, 137
TrueType fonts, 266
Try page, 8

U

U command, 77, 78
under-constrained, 650
UNDO command, 77, 79, 167
UNDO options, 79
unidirectional dimensioning, 456
unilateral tolerance, 561
UNISOLATEOBJECTS command, 173
unit block, 735
unload, 955
unselected grips, 365
update, 290
UPDATEFIELD command, 290, 1006
US National CAD Standard (NCS), 36, 127, 130, 249, 455
user coordinate system (UCS), 536
user parameters, 675
using commands transparently, 175

V

value set, 806
variable, 440
Variable Definition dialog box, 441
verb/noun selection, 367
vertex, 494
view, 969
View Category dialog box, 991
VIEW command, 989, 990, 992, 998
view cube, 170
view label block, definition, 989
view label blocks, 995–997
view tools, 163–174
additional zoom options, 165
Center navigation command, 168
cleaning the screen, 172–173
controlling draw order, 175–176
isolating or hiding objects, 173–174
model viewports, 180–184
named views, 176–180
navigation bar, 163–164
navigation wheel, 167
panning, 166
realtime zooming, 164–165
regenerating the screen, 171
rewinding, 168–170
steering wheels, 167
view back and forward, 167
zooming, 164
view transitions and resolution, 173
VIEWBACK command, 167
VIEWFORWARD command, 167
viewing-plane line, 232
viewport controls, 170–171
Viewports dialog box, 181
visibility parameters, 819–822
visibility state modification, 822
visibility states, 819
Visibility States dialog box, 821–822
VP Freeze icon, 894
VP Thaw icon, 894
VPLAYER command, 893
-VPORTS command, 182

W

walkthrough, 2
wblock,
  creating, 746–747
  definition, 729
  revising inserted drawing, 747
  saving existing block as a wblock, 746
  storing drawing as a wblock, 747
WBLOCK command, 746, 774
wedges, 168
window and crossing polygon selection, 82–83
window and crossing selection, 79–81
Windows Clipboard, 48
windows, controlling, 10–11
Windows Explorer, 44
WIPEOUT command, 539, 541
wireframe model, 2
wireframe visual style, 3
work,
  file storage and naming, 36–37
  planning, 3–5
  saving, 5, 36–40
working procedures checklist, 5
working set, 961
working with layouts, 846–852
  adding, 848–851
  deleting, 851
  exporting layout to model space, 851
  file tabs, 848
  model and layout tabs, 846–848
  model and paper buttons, 848
  renaming layouts, 851
workspace, 8
WORKSPACE command, 9
world coordinate system (WCS), 536
Write Block dialog box, 746

X

XATTACH command, 945
XBIND command, 960
Xbind dialog box, 960
XCLIP command, 957
XCLIPFRAME system variable, 957
XDWGFADECTL system variable, 949
xline, 226
XLINE command, 226, 227, 653, 942
XOPEN command, 964
XREF command, 747
XREFLAYER system variable, 945
xrefs, 941–968
  adding layout content, 944
  arranging sheet views, 944–945
  attachment vs. overlay, 946–947
  binding, 960–961
  clipping, 957
  copying nested objects, 964
  creating multiview drawing, 942–944
  demand loading and editing controls, 958
  dependent objects, 950
  detaching, reloading, and unloading xrefs, 955
  editing drawings, 961–963
  inserting, 948–949
  managing, 952–959
  opening an xref file, 964
  placing with **DesignCenter** and tool palettes, 949
  placing xref drawings, 945–949
  preparing xref and host drawings, 945
  reference editing, 961
  referencing existing geometry, 942
  selecting path type, 947–948
  updating xref path, 955
  working with xref objects, 949–951
  xref express tools, 964
XY parameters, 804–806
  assigning array action, 805
  using dynamically, 805–806

Z

zero suppression, 570
zero suppression options, 478
zip file, 1010
zones, 774
zoom, 63, 164
ZOOM command, 54, 164, 166, 167, 175, 193, 899
zoom in, 64, 164
zoom options, 166
zoom out, 64, 164

2020

AutoCAD®

and Its Applications

ADVANCED

by

Terence M. Shumaker
Faculty Emeritus
Former Chairperson
Drafting Technology
Autodesk Premier Training Center
Clackamas Community College
Oregon City, OR

David A. Madsen
President
Madsen Designs Inc.
Faculty Emeritus
Former Department Chairperson
Drafting Technology
Autodesk Premier Training Center
Clackamas Community College
Oregon City, OR
Director Emeritus
American Design Drafting Association

Jeffrey A. Laurich
Instructor, Mechanical Design Technology
Fox Valley Technical College
Appleton, WI

J.C. Malitzke
President
Digital JC CAD Services Inc.
Former Department Chair
Computer Integrated Technologies
Former Manager and Instructor
Autodesk Authorized Training Center
Moraine Valley Community College
Palos Hills, IL

Craig P. Black
Instructor, Mechanical Design Technology
Former Manager and Instructor
Autodesk Premier Training Center
Fox Valley Technical College
Appleton, WI

Publisher
The Goodheart-Willcox Company, Inc.
Tinley Park, IL
www.g-w.com

Manufactured in the United States of America.

ISBN 978-1-63563-866-0

1 2 3 4 5 6 7 8 9 – 20 – 24 23 22 21 20 19

Cataloging Data

Names: Shumaker, Terence M., author. | Madsen, David A., author. | Laurich, Jeffrey A., author. | Malitzke, J.C., author. | Black, Craig P., author.

Title: AutoCAD and Its Applications: Advanced 2020 / by Terence M. Shumaker, David A. Madsen, Jeffrey A. Laurich, J.C. Malitzke, Craig P. Black.

Description: 27th edition. | Tinley Park, IL : The Goodheart-Willcox Company, Inc., [2020] | Includes index.

Identifiers: ISBN 9781635638660

Subjects: LCSH: Computer graphics. | AutoCAD.

Classification: LCC T385.S46123 2020 | DDC 620'.0042028553--dc23

INTRODUCTION

AutoCAD and Its Applications—Advanced provides complete instruction in mastering three-dimensional design and modeling using AutoCAD. Topics are covered in an easy-to-understand sequence and progress in a way that allows you to become comfortable with the commands as your knowledge builds from one chapter to the next. In addition, ***AutoCAD and Its Applications—Advanced*** offers:

- Examples and discussions of industrial practices and standards.
- Professional tips explaining how to effectively and efficiently use AutoCAD.
- Exercises to reinforce the chapter topics. These exercises should be completed where indicated in the text as they build on previously learned material.
- Review questions at the end of each chapter for testing knowledge of commands and key AutoCAD concepts.
- A large selection of drawing problems supplements each chapter. Problems are presented as 3D illustrations, actual plotted drawings, and engineering sketches.

This text focuses on advanced AutoCAD applications. Basic AutoCAD applications are covered in ***AutoCAD and Its Applications—Basics,*** which can be ordered directly from Goodheart-Willcox Publisher. ***AutoCAD and Its Applications*** texts are also available for previous releases of AutoCAD.

Fonts Used in This Text

Different typefaces are used throughout each chapter to define terms and identify AutoCAD commands. Important terms appear in ***bold-italic face, serif*** type. AutoCAD menus, commands, system variables, dialog box names, and toolbar button names are printed in **bold-face, sans serif** type. File names, folder names, and paths appear in the body of the text in Roman, sans serif type. Keyboard keys are shown inside of square brackets [ ] and appear in Roman, sans serif type. For example, [Enter] means to press the enter (return) key.

Introducing AutoCAD Commands

There are several ways to select AutoCAD drawing and editing commands. Selecting commands from the ribbon is slightly different from entering them at the keyboard. When a command is introduced, the command-entry methods are illustrated in the margin next to the text reference.

CONE

Ribbon
Home
> Modeling

Cone
Type
CONE

The example in the margin next to this paragraph illustrates the different methods of initiating the **CONE** command to draw a solid cone primitive while the 3D environment is active. The 3D environment consists of a drawing file based on the acad3D.dwt template and the 3D Modeling workspace current, as described in Chapter 1. This book assumes the 3D environment is current for all procedures and discussions.

Flexibility in Design

Flexibility is the keyword when using ***AutoCAD and Its Applications—Advanced***. This text is an excellent training aid for both individual and classroom instruction. It is also an invaluable resource for any professional using AutoCAD. ***AutoCAD and Its Applications—Advanced*** teaches you how to apply AutoCAD to common 3D design and modeling tasks.

When working through the text, you will see a variety of notices. These include Professional Tips, Notes, and Cautions that help you develop your AutoCAD skills.

PROFESSIONAL TIP

These ideas and suggestions are aimed at increasing your productivity and enhancing your use of AutoCAD commands and techniques.

NOTE

A note alerts you to important aspects of a command, function, or activity that is being discussed. These aspects should be kept in mind while you are working through the text.

CAUTION

A caution alerts you to potential problems if instructions or commands are incorrectly used or if an action can corrupt or alter files, folders, or storage media. If you are in doubt after reading a caution, always consult your instructor or supervisor.

AutoCAD and Its Applications—Advanced provides several ways for you to evaluate your performance. Included are:

- **Exercises.** The companion website contains exercises for each chapter. These exercises allow you to perform tasks that reinforce the material just presented. You can work through the exercises at your own pace. However, the exercises are intended to be completed when called out in the text.
- **Chapter reviews.** Each chapter includes review questions at the end of the chapter. Questions require you to give the proper definition, command, option, or response to perform a certain task. You may also be asked to explain a topic or list appropriate procedures.
- **Drawing problems.** There are a variety of drawing and design problems at the ends of chapters. These are presented as real-world CAD drawings, 3D illustrations, and engineering sketches. The problems are designed to make you think, solve problems, use design techniques, research and use proper drawing standards, and correct errors in the drawings or engineering sketches. Labels are used to represent the discipline to which a drawing problem applies.

  Mechanical These problems address mechanical drafting and design applications, such as manufactured part designs.

Architectural These problems address architectural and structural drafting and design applications, such as floor plans, furniture, and presentation drawings.

Piping These problems address piping drafting and design applications, such as tank drawings and pipe layout.

General These problems address a variety of general drafting and design applications. These problems should be attempted by everyone learning advanced AutoCAD techniques for the first time.

NOTE

Some problems presented in this text are given as engineering sketches. These sketches are intended to represent the kind of material from which a drafter is expected to work in a real-world situation. As such, engineering sketches often contain errors or slight inaccuracies and are most often not drawn according to proper drafting conventions and applicable standards. Additionally, other drawings may contain errors or inaccuracies. Errors in these problems are *intentional* to encourage you to apply appropriate techniques and standards in order to solve the problem. As in real-world applications, sketches should be considered preliminary layouts. Always question inaccuracies in sketches and designs and consult the applicable standards or other resources.

Companion Website

The companion website is located at www.g-wlearning.com/CAD/. Select the entry for ***AutoCAD and Its Applications—Advanced 2020*** to access the material for this book. The companion website contains the exercises for each chapter and the appendix material for the textbook.

As you work through each chapter, exercises on the companion website are referenced. The exercises are intended to be completed as the references are encountered in the text. The solid modeling tutorial in Appendix A should be completed after Chapter 14. The surface modeling tutorial in Appendix B should be completed after Chapter 10. The remaining appendix material is intended as reference material.

Also included on the companion website are the exercise activities. The exercise activities are intended to supplement the exercises on the companion website. These activities are referenced within the appropriate exercises and can be completed as additional practice.

Additionally, full-color images of figures illustrating key topics in Chapters 17, 18, and 19 are provided in electronic format on the companion website. Figures with a corresponding color image on the companion website are identified with a reference above the figure caption. The full-color images are intended to enhance visualization and understanding of important topics relating to materials, lighting, and rendering and presentation.

Online Instructor Resources

The Online Instructor Resources provide support materials designed to help instructors effectively plan and teach AutoCAD courses. Support materials are available in one accessible location and include tips for the instructor, answer keys, exercise and drawing problem solution files, course syllabi, and a sample course exam. Also included is assessment software with question banks featuring hundreds of questions corresponding to each chapter in the text. The Online Instructor Resources are available as a subscription and can be accessed at school or at home.

About the Authors

Terence M. Shumaker is Faculty Emeritus, the former Chairperson of the Drafting Technology Department, and former Director of the Autodesk Premier Training Center at Clackamas Community College in Oregon City, Oregon. Terence taught at the community college level for over 28 years. He worked as a training consultant for Autodesk, Inc., and conducted CAD program development workshops around the country. He has professional experience in surveying, civil drafting, industrial piping, and technical illustration. He is the author of Goodheart-Willcox's ***Process Pipe Drafting*** and coauthor of the ***AutoCAD and Its Applications*** series.

David A. Madsen is the president of Madsen Designs Inc. (www.madsendesigns.com). David is Faculty Emeritus and the former Chairperson of Drafting Technology and the Autodesk Premier Training Center at Clackamas Community College in Oregon City, Oregon. David was an instructor and a department chairperson at Clackamas Community College for nearly 30 years. In addition to teaching at the community college level, David was a Drafting Technology instructor at Centennial High School in Gresham, Oregon. David is a former member of the American Design Drafting Association (ADDA) Board of Directors. He was honored with Director Emeritus status by the ADDA in 2005. David has extensive experience in mechanical drafting, architectural design and drafting, and building construction. He holds a Master of Education degree in Vocational Administration and a Bachelor of Science degree in Industrial Education. David is coauthor of the ***AutoCAD and Its Applications*** series, ***Architectural Drafting Using AutoCAD, Geometric Dimensioning and Tolerancing,*** and other drafting and design textbooks.

Jeffrey A. Laurich has been an instructor in Mechanical Design Technology at Fox Valley Technical College in Appleton, Wisconsin, since 1991. He has also taught business and industry professionals in the Autodesk Premier Training Center at FVTC. Jeff teaches drafting, AutoCAD, Design of Tooling, GD&T, and 3ds Max. He created a certificate program at FVTC titled Computer Rendering and Animation that integrates the 3D capabilities of AutoCAD and 3ds Max. Jeff has professional experience in furniture design, surveying, and cartography. He holds a diploma in Mechanical Drafting, an Associate of Applied Science degree in Natural Resources Technology, and a Bachelor of Science degree in Career and Technical Education and Training. Together with other instructors at FVTC, Jeff has created courses in electric guitar building and technical illustration.

J.C. Malitzke is president of Digital JC CAD Services Inc. (digitaljccad.com), a training and consulting company using Autodesk products. He is the former department chair of Computer Integrated Technologies at Moraine Valley Community College in Palos Hills, Illinois. J.C. was a professor of Mechanical Design and Drafting/CAD and taught for the Autodesk Authorized Training Center at Moraine Valley. J.C. has been actively using and teaching Autodesk products for more than 30 years. He is a founding member and past chair of the Autodesk Training Center Executive Committee and the Autodesk Training Center Leadership Council. J.C. has been the coauthor and principal investigator on two National Science Foundation grants. He has won numerous awards, including: Drafting Educator of the Year by the Illinois Drafting Educators Association; the Instructor Quality Award by Autodesk; Autodesk University Instructor Award; Professor of the Year, Co-Innovator of the Year, and Co-Master Teacher awards at Moraine Valley Community College; and the Illinois Outstanding Faculty Member of the Year awarded by the Illinois Community College Trustees Association. J.C. was also one of the recipients of the Top 10 Most Popular Autodesk University Classes Ever award for his class Compelling 3D Features in AutoCAD. J.C. is an Autodesk Certified Instructor (ACI) and an Autodesk Certification Evaluator (ACE), and holds a Bachelor's degree in Education and a Master's degree in Industrial Technology from Illinois State University.

Craig P. Black is an instructor in the Mechanical Design Technology department and the former instructor and manager of the Autodesk Premier Training Center at Fox Valley Technical College in Appleton, Wisconsin. Craig has been teaching at FVTC since 1990. In 2009, he created the CAD Management certificate program at FVTC. In 2013, he created the Mechanical CAD Drafting Technical Diploma at FVTC. Craig has served two terms on the Autodesk Training Center Executive Committee (now known as the Autodesk Training Center Leadership Council) and chaired the committee in 2001. He has been working with Autodesk software products for more than 25 years. He has presented on various topics at a number of Autodesk University annual training sessions and has been contracted to teach training sessions on Autodesk products across the United States. In addition to teaching, Craig also does AutoCAD customization and AutoLISP and DCL programming for area businesses and industries. Prior to his current position, Craig worked in the civil, architectural, electrical, and mechanical drafting and design disciplines.

Acknowledgments

The authors and publisher would like to thank the following individuals and companies for their assistance and contributions.

Autodesk, Inc.
City & County of Denver
Fitzgerald, Hagan, & Hackathorn
Fort Mason Center for Arts & Culture
Laser Design and GKS Services
Bill Strenth, Pittsburg State University School of Construction

Trademarks

ADDA Technical Publication

The content of this text is considered a fundamental component to the design drafting profession by the American Design Drafting Association. This publication covers topics and related material relevant to the delivery of the design drafting process. Although this publication is not conclusive, it should be considered a key reference tool in furthering the knowledge, abilities, and skills of a properly trained designer or drafter in the pursuit of a professional career.

BRIEF CONTENTS

Three-Dimensional Design and Modeling

1 Introduction to Three-Dimensional Modeling 1
2 Creating Primitives and Composites 21
3 Viewing and Displaying Three-Dimensional Models 49
4 Understanding Three-Dimensional Coordinates and User Coordinate Systems 79
5 Using Model Space Viewports 115
6 Model Extrusions and Revolutions. 131
7 Sweeps and Lofts 153
8 Creating and Working with Solid Model Features 173
9 Mesh Modeling. 213
10 Advanced Surface Modeling 237
11 Subobject Editing. 271
12 Solid Model Editing. 307
13 Text and Dimensions in 3D 331
14 Model Documentation, Analysis, and File Exchange 343
15 Point Clouds 399

Model Visualization and Presentation

16 Visual Style Settings and Basic Rendering. 417
17 Materials in AutoCAD 441
18 Lighting 481
19 Advanced Rendering. 511
20 Using Show Motion to View a Model 529
21 Cameras, Walkthroughs, and Flybys 549

EXPANDED CONTENTS

Three-Dimensional Design and Modeling

Chapter 1
Introduction to Three-Dimensional Modeling 1
Using Rectangular 3D Coordinates 1
Right-Hand Rule of 3D Drawing. 3
Basic Overview of the Interface. 4
Displaying 3D Views. 7
Introduction to 3D Model Display Using Visual Styles 12
Rendering a Model 14
3D Construction Techniques 15
3D Object Snaps. 16
Sample 3D Construction. 17
Guidelines for Working with 3D Drawings. 19

Chapter 2
Creating Primitives and Composites .21
Overview of Solid Modeling 21
Constructing Solid Primitives 22
Constructing a Planar Surface. 34
Creating Composite Solids. 35
Creating a Helix. 40
Working with Regions 42

Chapter 3
Viewing and Displaying Three-Dimensional Models 49
Using the Navigation Bar to Display Models 50
Creating and Displaying Named Views. 51
Dynamically Displaying Models with the View Cube. 53
Creating a Continuous 3D Orbit 62
Plan Command Options. 63
Displaying Models with Steering Wheels 63
Controlling the Display of the Navigation Bar 75

Chapter 4
Understanding Three-Dimensional Coordinates and User Coordinate Systems 79
Introduction to Spherical Coordinates 79
Using Spherical Coordinates 80
Introduction to Cylindrical Coordinates 82
Using Cylindrical Coordinates 83
3D Polylines 84
Introduction to Working with User Coordinate Systems. 84
Working with User Coordinate Systems 89
Working with the UCS Icon 94
Using a Dynamic UCS 96
Additional Ways to Change the UCS 97
Managing User Coordinate Systems and Displays. 105

Chapter 5
Using Model Space Viewports 115
Understanding Viewports 115
Creating Viewports 116
Drawing in Multiple Viewports 125

Chapter 6
Model Extrusions and Revolutions. 131
Creating Extruded Models 131
Creating Revolved Models 136
Using Extrude and Revolve to Create Surfaces 138
Using Extrude and Revolve as Construction Tools. 141

Chapter 7
Sweeps and Lofts 153
Creating Swept Surfaces and Solids 153
Creating Lofted Objects 158

Chapter 8
Creating and Working with Solid Model Features 173
Changing Properties Using the Properties Palette 173
Solid Model History 174
Selection Cycling 176
Aligning Objects in 3D 177
Move, Rotate, and Scale Gizmos 181
3D Moving 182
3D Rotating. 182
3D Mirroring 184
Creating 3D Arrays 185
Filleting Solid Objects. 194
Chamfering Solid Objects 196
Converting to Solids 198
Slicing a Solid. 199

Removing Features 204
Constructing Features on Solid Models 204

Chapter 9
Mesh Modeling 213
Overview of Mesh Modeling 213
Tessellation Division Values 214
Drawing Mesh Primitives 216
Converting Between Mesh and Surface or Solid Objects 217
Smoothing and Refining a Mesh Object 221
Editing Meshes 223

Chapter 10
Advanced Surface Modeling 237
Overview 237
Understanding Surface Model Types 238
Creating Network Surfaces 242
Creating Surfaces from Existing Surfaces 245
Trimming Surfaces 253
NURBS Surfaces 256
Additional Surface Modeling Methods 259

Chapter 11
Subobject Editing 271
Grip Editing 271
Using Grips with Surfaces 275
Overview of Subobject Editing 275
Subobject Editing Fillets and Chamfers 278
Face Subobject Editing 279
Edge Subobject Editing 287
Vertex Subobject Editing 293
Using Subobject Editing as a Construction Tool 296
Other Solid Editing Tools 299

Chapter 12
Solid Model Editing 307
Overview of the SOLIDEDIT Command 307
Face Editing 309
Edge Editing 317
Extracting a Wireframe 318
Body Editing 319
Using SOLIDEDIT as a Construction Tool 322

Chapter 13
Text and Dimensions in 3D 331
Creating Text with Thickness 331
Text and the UCS 332
Dimensioning in 3D 334

Chapter 14
Model Documentation, Analysis, and File Exchange 343
Introduction to Model Documentation. .343
Drawing Views .344
Section Views .352
Detail Views .362
Auxiliary Views .365
Dimensioning Drawing Views .366
Using the EXPORTLAYOUT Command .369
Creating Section Planes .369
Using and Editing Section Planes .372
Model Analysis .384
Solid Model File Exchange .389

Chapter 15
Point Clouds .399
Laser Scanning .400
Autodesk ReCap Pro .400
Inserting a Point Cloud File .407
Point Cloud Display Options .408
Working with Point Clouds .410

Model Visualization and Presentation

Chapter 16
Visual Style Settings and Basic Rendering .417
Controlling Solid Model Display .417
Controlling Line Display .420
Overview of the Visual Styles Manager .421
Visual Style Settings .422
Creating Your Own Visual Style .431
Plotting Visual Styles .433
Introduction to Rendering .433

Chapter 17
Materials in AutoCAD .441
Materials Browser .442
Applying and Removing Materials .445
Material Display Options .446
Materials Editor .446
Creating and Modifying Materials .448
Maps .460
Adjusting Material Maps .473

Chapter 18
Lighting 481
Types of Lights 481
Properties of Lights 482
AutoCAD Lights 485
Determining Proper Light Intensity 499
Shadows 500
Adding a Background 500

Chapter 19
Advanced Rendering 511
Render Window 511
Render Presets 513
Rendering Fundamentals 518
Render Environment and Exposure 520

Chapter 20
Using Show Motion to View a Model 529
Understanding Show Motion 530
Show Motion Toolbar 530
New View/Shot Properties Dialog Box 531
Creating a Still Shot 534
Creating a Walk Shot 535
Creating a Cinematic Shot 537
Displaying or Replaying a Shot 543
Thumbnail Shortcut Menu 544
Creating and Using View Categories 545

Chapter 21
Cameras, Walkthroughs, and Flybys 549
Creating Cameras 549
Animation Preparation 553
Walking and Flying 554
Motion Path Animation 561

Index 567

Companion Website Contents

www.g-wlearning.com/CAD/

Contents Tab

Exercises

Drawing Files

- Exercise Files
- Exercise Activity Files

Color Images

Resources Tab

Appendices

- Appendix A Solid Modeling Tutorial
- Appendix B Surface Modeling Tutorial
- Appendix C Common File Extensions
- Appendix D AutoCAD Command Aliases
- Appendix E Advanced Application Commands

Reference Material

- Drafting Symbols
- Drawing Sheet Sizes, Settings, and Scale Parameters
- Standards and Related Documents
- Project and Drawing Problem Planning Sheet
- Standard Tables

Support Tab

Support and Information

CHAPTER 1

Introduction to Three-Dimensional Modeling

Learning Objectives

After completing this chapter, you will be able to:

- ✓ Describe how to locate points in 3D space.
- ✓ Describe the right-hand rule of 3D visualization.
- ✓ Explain the function of the ribbon.
- ✓ Identify the functions of the viewport controls and the view cube.
- ✓ Display 3D objects from preset isometric viewpoints.
- ✓ Display 3D objects from any desired viewpoint.
- ✓ Set a visual style current.

Three-dimensional (3D) design and modeling is a powerful tool for use in design, visualization, testing, analysis, manufacturing, assembly, and marketing. Three-dimensional models also form the basis of computer animations, architectural walk-throughs, parametric models used in building information modeling (BIM), and virtual worlds used in the entertainment industry and for gaming platforms. Drafters who can design objects, buildings, and "worlds" in 3D are in demand for a wide variety of positions, both inside and outside of the traditional drafting and design disciplines.

The first 15 chapters of this book present a variety of modeling techniques for drawing and designing in 3D. In these chapters, you will learn how to construct solid models, surface models, and mesh models, and work with point clouds. The skills you learn will provide you with the ability to construct any object in 3D and prepare you for entry into an exciting aspect of graphic communication.

To be effective in creating and using 3D objects, you must first have good 3D visualization skills. These skills include the ability to see an object in three dimensions and to visualize it rotating in space. Visualization skills can be obtained by using 3D techniques to construct objects and by trying to see two-dimensional sketches and drawings as 3D models. This chapter provides an introduction to several aspects of 3D drawing and visualization. Subsequent chapters expand on these aspects and provide a detailed examination of 3D drawing, editing, visualization, and display techniques.

Using Rectangular 3D Coordinates

In two-dimensional drawing, you see one plane defined by two dimensions. These dimensions are usually located on the X and Y axes and what you see is the XY plane. However, in 3D drawing, another coordinate axis—the Z axis—is added. This results in two additional planes—the XZ plane and the YZ plane. If you are looking at a standard AutoCAD screen after starting a new drawing using the default acad.dwt template, the positive Z axis comes directly out of the screen toward you. AutoCAD can only draw lines in 3D if it knows the X, Y, and Z coordinate values of each point on the object. For 2D drawing, only two of the three coordinates (X and Y) are needed.

Compare the 2D and 3D coordinate systems shown in **Figure 1-1**. Notice that the positive values of Z in the 3D coordinate system come up from the XY plane. In a new drawing based on the acad.dwt template, consider the surface of your computer screen as the XY plane. Anything behind the screen is negative Z and anything in front of the screen is positive Z.

The object in **Figure 1-2A** is a 2D drawing showing the top view of an object. The XY coordinate values of the origin and each point are shown. Think of the object as being drawn directly on the surface of your computer screen. However, this is actually a 3D object. When displayed in a pictorial view, the Z coordinates can be seen. Notice in **Figure 1-2B** that the first two values of each coordinate match the X and Y values of the 2D view. Three-dimensional coordinates are always expressed as (X,Y,Z). The 3D object was drawn using positive Z coordinates. Therefore, the object comes out of your computer screen when it is viewed from directly above. The object can also be drawn using negative Z coordinates. In this case, the object would extend behind, or into, the screen.

Study the nature of the rectangular 3D coordinate system. Be sure you understand Z values before you begin constructing 3D objects. It is especially important that you carefully visualize and plan your design when working with 3D constructions.

PROFESSIONAL TIP

All points in three-dimensional space can be drawn using one of three coordinate entry methods—rectangular, spherical, or cylindrical. This chapter uses the rectangular coordinate entry method. Complete discussions on the spherical and cylindrical coordinate entry methods are provided in Chapter 4.

Exercise 1-1

www.g-wlearning.com/CAD/

Complete the exercise on the companion website.

Figure 1-1.
A comparison of 2D and 3D coordinate systems.

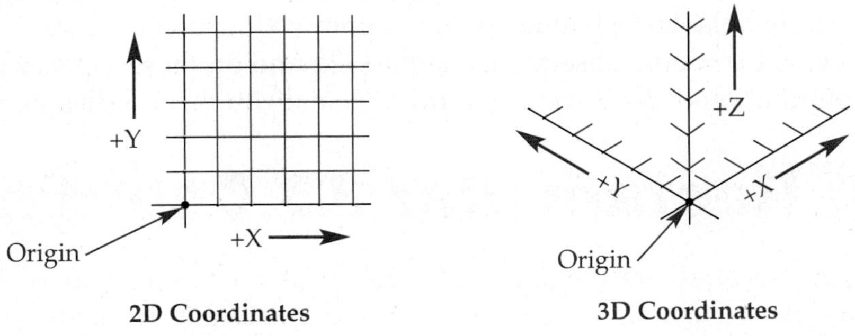

Figure 1-2.
A—The points making up a 2D object require only two coordinates. B—Each point of a 3D object must have an X, Y, and Z value. Notice that the first two coordinates (X and Y) are the same for each endpoint of a vertical line.

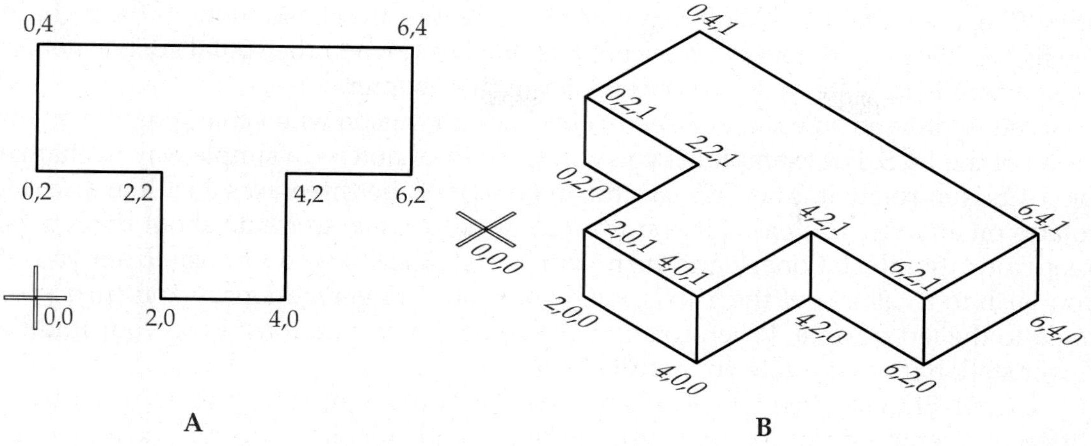

Right-Hand Rule of 3D Drawing

In order to effectively draw in 3D, you must be able to visualize objects in 3D space. The ***right-hand rule*** is a simple method for visualizing the 3D coordinate system. It is a representation of the positive coordinate values in the three axis directions. The AutoCAD world coordinate system (WCS) and a user coordinate system (UCS) are based on this concept of visualization.

To use the right-hand rule, position the thumb, index finger, and middle finger of your right hand as shown in **Figure 1-3**. Imagine that your thumb is the X axis, your index finger is the Y axis, and your middle finger is the Z axis. Hold your hand in front of you so that your middle finger is pointing directly at you, as shown in **Figure 1-3**. This is the plan view of the XY plane. The positive X axis is pointing to the right and the positive Y axis is pointing up. The positive Z axis comes toward you and the origin of this system is the palm of your hand.

The concept behind the right-hand rule can be visualized even better if you are sitting at a computer and the AutoCAD drawing window is displayed. Make sure the current drawing is based on the acad.dwt template. If the UCS icon is not displayed in the lower-left corner of the screen, turn it on by using the **UCSICON** command, or

UCSICON
Ribbon
View
> Viewport Tools
UCS Icon
Type
UCSICON

Figure 1-3.
Positioning your hand to use the right-hand rule to understand the relationship of the X, Y, and Z axes.

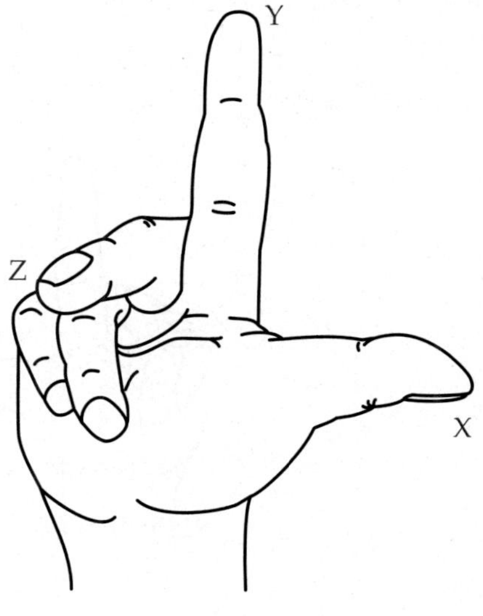

select the **UCS Icon** button in the **Viewport Tools** panel of the **View** ribbon tab. Now, orient your right hand as shown in **Figure 1-3** and position it next to the UCS (or WCS) icon. Your index finger and thumb should point in the same directions as the Y and X axes, respectively. Your middle finger will be pointing out of the screen directly at you, representing the Z axis. See **Figure 1-4**. Notice the illustration on the right in the figure. This is the shaded UCS icon. It is displayed when the visual style is not 2D Wireframe. Visual styles are introduced later in this chapter.

The right-hand rule can be used to eliminate confusion when changing the orientation of the UCS. For example, as you will learn in Chapter 4, a simple way to change the UCS is to rotate it. The UCS can rotate on any of the three axes, just like a wheel rotates on an axle. Therefore, if you want to visualize how to rotate about the X axis, keep your thumb stationary and turn your hand either toward or away from you. If you wish to rotate about the Y axis, keep your index finger stationary and turn your hand to the left or right. When rotating about the Z axis, you must keep your middle finger stationary and rotate your entire arm.

If your 3D visualization skills are weak or you are having trouble visualizing different orientations of the UCS, use the right-hand rule. It is a useful technique for improving your 3D visualization skills. Rotating the UCS around one or more of the axes becomes easier once you begin drawing 3D objects. A complete discussion of UCSs is provided in Chapter 4.

Basic Overview of the Interface

AutoCAD provides different working environments tailored to either 2D drawing or 3D drawing or annotating a drawing. These environments are called *workspaces* and can be quickly restored. There are three default workspaces available in AutoCAD. The Drafting & Annotation workspace is designed for drawing in 2D and annotating a drawing. The workspace for basic 3D modeling and editing is called 3D Basics. The full range of 3D modeling features is found in the 3D Modeling workspace.

Figure 1-4.
Using the right-hand rule to visualize the X, Y, and Z axis directions.

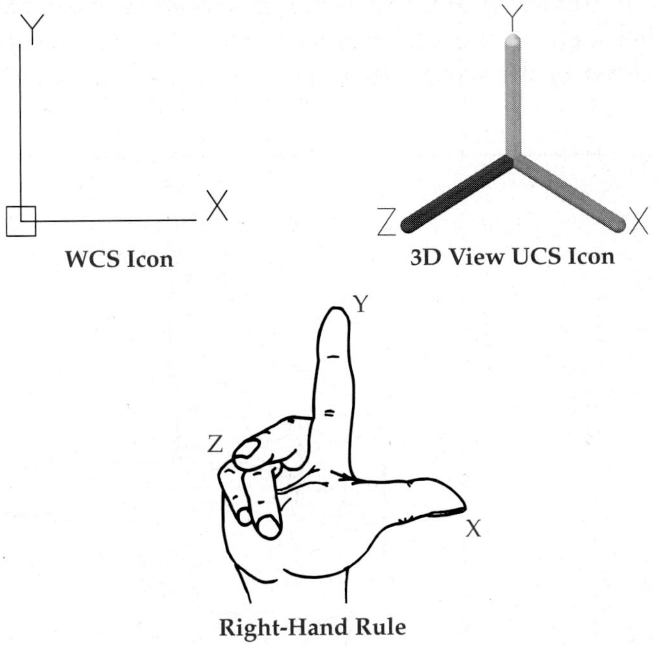

Workspaces can be created, customized, and saved to allow a variety of graphical user interface configurations. The 3D Modeling workspace provides quick access to the full range of tools required to construct, edit, view, and visualize 3D models. It is the workspace used throughout this book. This section provides an overview of the 3D Modeling workspace and the layout of the ribbon and its panels.

NOTE

In order to use the default 3D "environment," you must start a new drawing file based on the acad3D.dwt template and set the 3D Modeling workspace current. All discussions in the remainder of this book assume that AutoCAD is in this default 3D environment.

Workspaces

A ***workspace*** is a drawing environment in which menus, toolbars, palettes, and ribbon panels are displayed for a specific task. A workspace stores not only which of these tools are visible, but also their on-screen locations. You can quickly change workspaces using the **Workspace Switching** button on the status bar or the **WSCURRENT** command. See **Figure 1-5**. The default 3D Modeling workspace is shown in **Figure 1-6**.

Ribbon Panels

Type
RIBBON

RIBBON

Within the 3D Modeling workspace, the **Home** tab on the ribbon displays 12 panels. The tools in these panels provide all of the functions needed to design and view your 3D model. See **Figure 1-7A**. The panel title appears at the bottom of the panel. The up arrow button at the right-hand end of the tab names controls the display of the ribbon tabs, panels, and panel titles. Picking it cycles through the display of the full ribbon, tab titles and panel buttons, tab and panel titles only, or tab titles only. Picking the down arrow displays a drop-down menu containing the minimize options. See **Figure 1-7B**.

Many panels contain more tools than those displayed in the default view. These panels can be expanded by picking the flyout arrow to the right of the panel title. See **Figure 1-8A**. The flyout portion is displayed as long as the cursor remains over the panel. It retracts when the cursor moves off the panel. To retain the flyout display, pick the pushpin icon in the lower-left corner of the expanded panel. See **Figure 1-8B**. The panel continues to be displayed until you pick the pushpin icon again to release it.

Multiple panel flyouts can be displayed using the pushpin, but they may overlap in the process. This may obscure menu tools in adjacent panels. Simply pick on the panel you wish to use and its tools are displayed in full.

You can display only those panels that you need. Right-click anywhere on the ribbon to display the shortcut menu. Select **Show Panels** to display a cascading menu that contains a list of the panels available for that tab. See **Figure 1-9**. The panels that

Figure 1-5. Switching workspaces using the **Workspace Switching** button on the status bar.

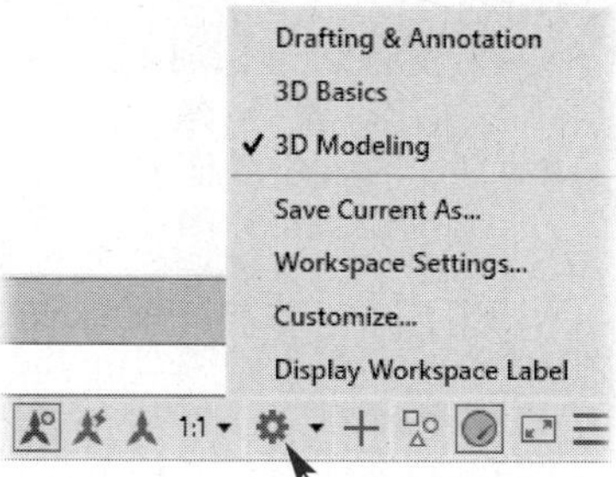

Figure 1-6.
The 3D Modeling workspace with a drawing file based on the acad3D.dwt template.

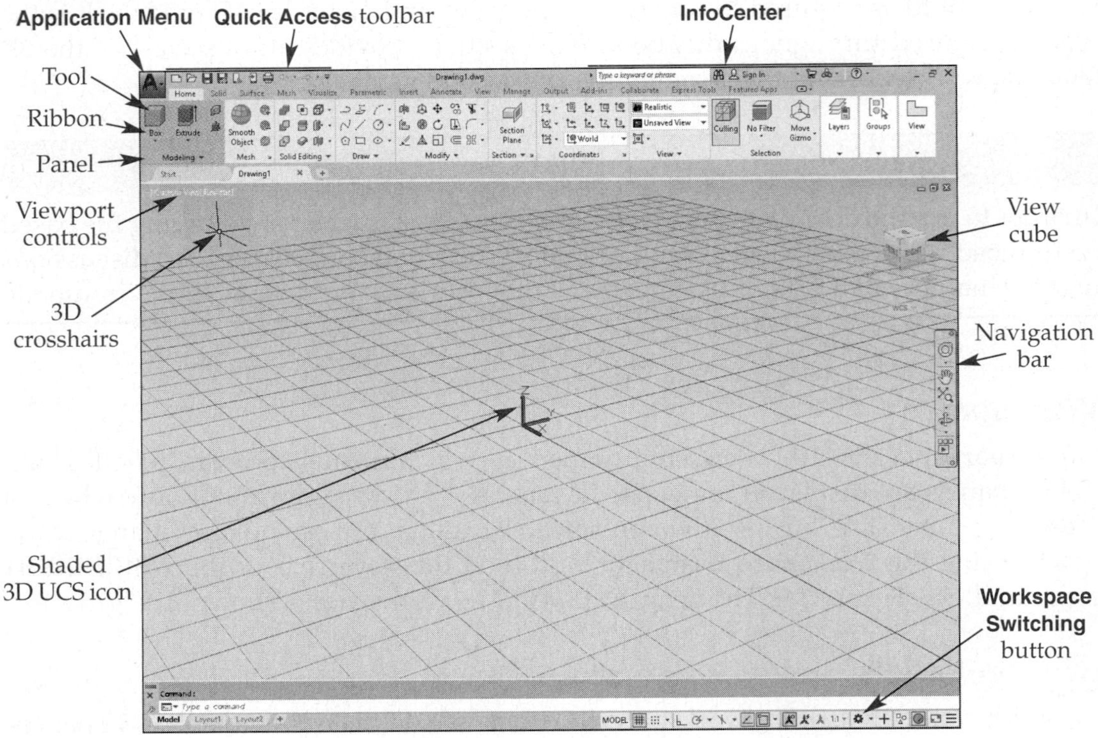

Figure 1-7.
A—The ribbon is composed of tabs and panels that provide access to 3D modeling and viewing commands without using toolbars and menus. B—The drop-down menu allows you to select a specific ribbon display.

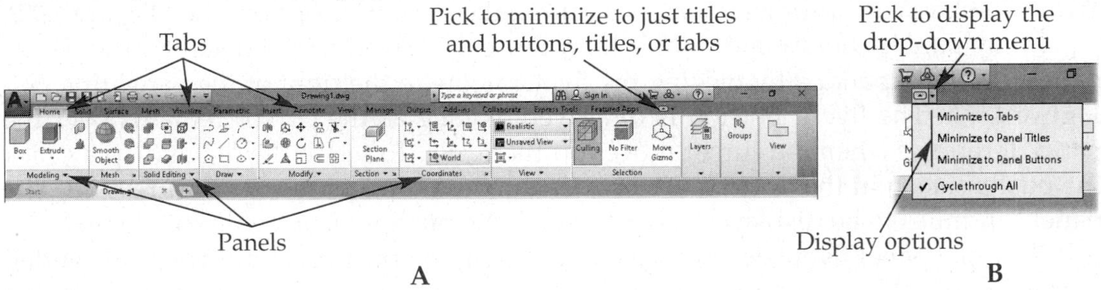

Figure 1-8.
A—Picking the flyout arrow expands the panel. The flyout is displayed as long as the cursor remains over the panel. It retracts when the cursor moves off the panel. B—Multiple panel flyouts can be displayed using the pushpin, but may overlap in the process. Pick the panel you wish to use to display its tools in full.

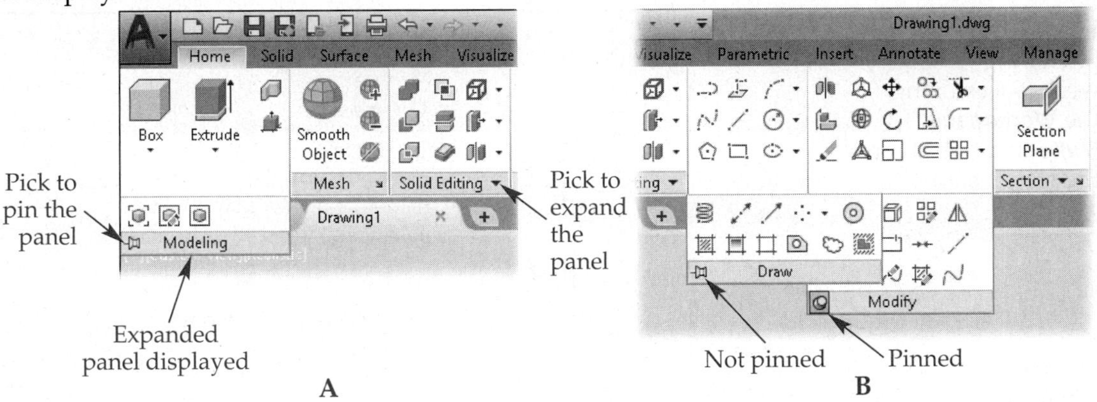

Figure 1-9.
Panels in a tab can be displayed or hidden using the shortcut menu.

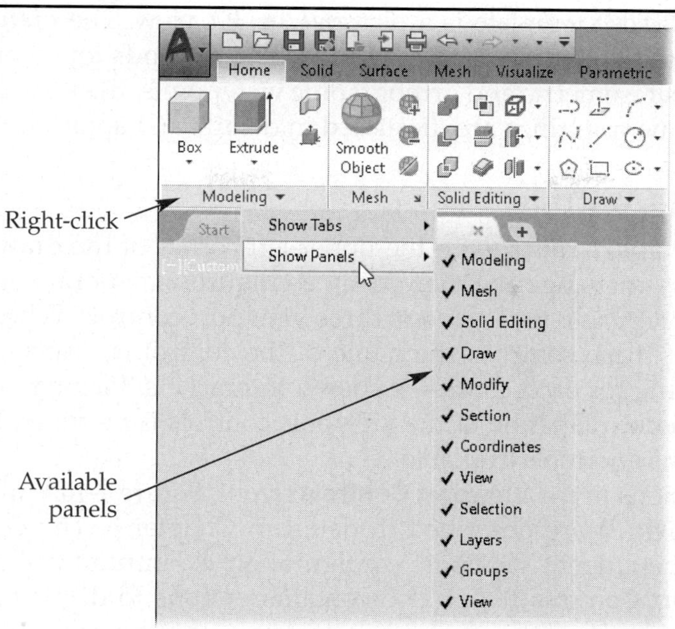

are currently displayed in the ribbon have a check mark next to their name. Select any of the checked panels that you wish to remove from the current display. Unchecked panels can be displayed by selecting their name. Each tab has its own set of available panels.

Each panel contains command tools. These are discussed in detail in the appropriate chapters of this book.

NOTE

The Drafting & Annotation, 3D Basics, and 3D Modeling workspaces display the ribbon by default. In each workspace, the ribbon displays a set of panels specifically related to that workspace. As soon as you select a different workspace or reload the current workspace, the panels associated with the workspace are displayed in the ribbon.

PROFESSIONAL TIP

All elements of the AutoCAD interface, including ribbon panels, toolbars, menus, and the drawing window appearance, can be customized using the customization functions in AutoCAD. In addition, custom commands can be created and assigned to interface elements. The **Options** dialog box, accessed with the **OPTIONS** command, and the **Customize User Interface** dialog box, accessed with the **CUI** command, are used to make customizations in AutoCAD. The AutoCAD help system provides detailed documentation and examples of AutoCAD customization.

OPTIONS

Application Menu
Options
Type
OPTIONS
OP

CUI

Ribbon
Manage > Customization
User Interface
Type
CUI

Displaying 3D Views

AutoCAD provides several methods of changing your viewpoint to produce different pictorial views. The default view in the 2D environment based on the acad.dwt template is a plan, or top, view of the XY plane. The default view in the 3D environment based on the acad3D.dwt template is a pictorial, or 3D, view. The ***viewpoint*** is the location in space from which the object is viewed. The methods for changing your viewpoint include preset isometric and orthographic viewpoints, the view cube, and camera lens settings. Camera settings are discussed in detail in Chapter 21.

Viewport Controls

The ***viewport controls*** in the upper-left corner of the drawing window provide a quick way to change the viewpoint, configure viewports, and manage the model display. See **Figure 1-10**. There are three viewport controls. When the cursor is moved over each of these controls, the name of the control is displayed in the tooltip. The current setting for each control is shown in brackets. Picking on a control displays a flyout. A brief explanation of the viewport controls is provided here to introduce you to the options and tools available.

The options in the **Viewport Controls** flyout, **Figure 1-10A**, allow you to work with viewports, which are described in detail in Chapter 5. The default setting (-) indicates that the current viewport configuration is "minimized" to a single viewport. The **Viewport Controls** flyout also contains options to display AutoCAD navigation tools, including the view cube, steering wheels, and the navigation bar. The view cube is introduced in this chapter, and other navigation tools are introduced in Chapter 3.

The **View Controls** flyout, **Figure 1-10B**, enables you to change the viewpoint to one of the preset orthographic or isometric viewpoints. This is discussed in more detail in the next section. The default setting when you start a new drawing with the acad3D.dwt template is Custom View. The other options in the **View Controls** flyout can be used to display a perspective or parallel projection or to access the **View Manager** dialog box, which is used to create named views. Creating named views is discussed in Chapter 3.

Figure 1-10.
Using the AutoCAD viewport controls. Picking on a control displays a flyout. A—The **Viewport Controls** flyout. B—The **View Controls** flyout. C—The **Visual Style Controls** flyout.

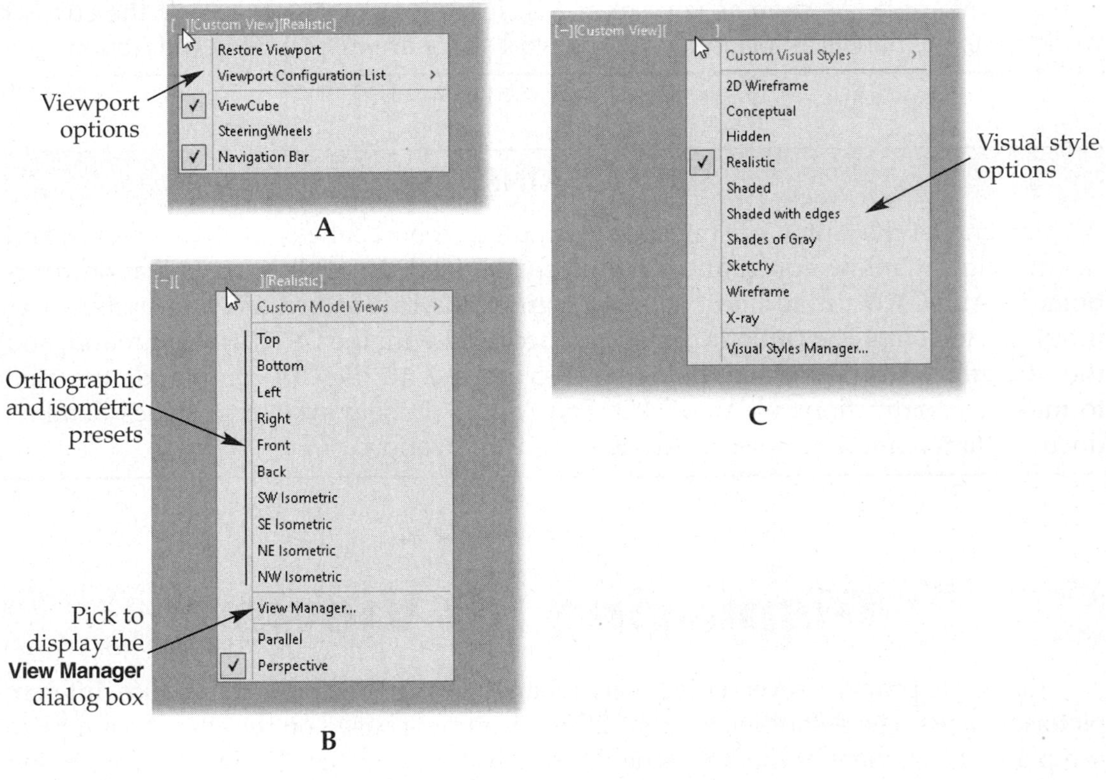

The **Visual Style Controls** flyout, **Figure 1-10C**, allows you to quickly display the drawing in one of 10 different visual styles. The default setting when you start a new drawing with the acad3D.dwt template is Realistic. Use of visual styles is introduced in this chapter and discussed completely in Chapter 16.

Isometric and Orthographic Viewpoint Presets

A 2D isometric drawing is based on angles of 120° between the three axes. AutoCAD provides preset viewpoints that allow you to view a 3D object from one of four isometric locations. See **Figure 1-11**. Each of these viewpoints produces an isometric view of the object. In addition, AutoCAD has presets for the six standard orthographic views of an object. The isometric and orthographic viewpoint presets are based on the world coordinate system (WCS).

Ribbon
Home
> View
> SW Isometric
> SE Isometric
> NE Isometric
> NW Isometric

VIEW

The four preset isometric views are southwest, southeast, northeast, and northwest. The six orthographic presets are top, bottom, left, right, front, and back. To switch your viewpoint to one of these presets, pick the view name in the **View Controls** flyout in the viewport controls, as previously discussed. Refer to **Figure 1-10B**. You can also select one of the presets in the **View** panel in the **Home** tab of the ribbon. See **Figure 1-12**. The presets are also available in the **Named Views** panel in the **Visualize** tab.

Ribbon
Home
> View
> Top
> Bottom
> Left
> Right
> Front
> Back

VIEW

NOTE

You can also change the viewpoint to one of the orthographic or isometric presets by using the **VIEW** command to access the **View Manager** dialog box.

Once you select a view, the viewpoint in the current viewport is automatically changed to display an appropriate isometric or orthographic view. Since these presets are based on the WCS, selecting a preset produces the same view of the object regardless of the current user coordinate system (UCS).

A view that looks straight down on the current drawing plane is called a ***plan view***. An important aspect of the orthographic presets is that selecting one not only changes the viewpoint, but, by default, it also changes the UCS to be plan to the orthographic view. All new objects are created on that UCS instead of the WCS (or previous UCS). Working with UCSs is explained in detail in Chapter 4. However, in order to

Figure 1-11.
There are four preset isometric viewpoints in AutoCAD. This illustration shows the direction from which the cube will be viewed for each of the presets. The grid represents the XY plane of the WCS.

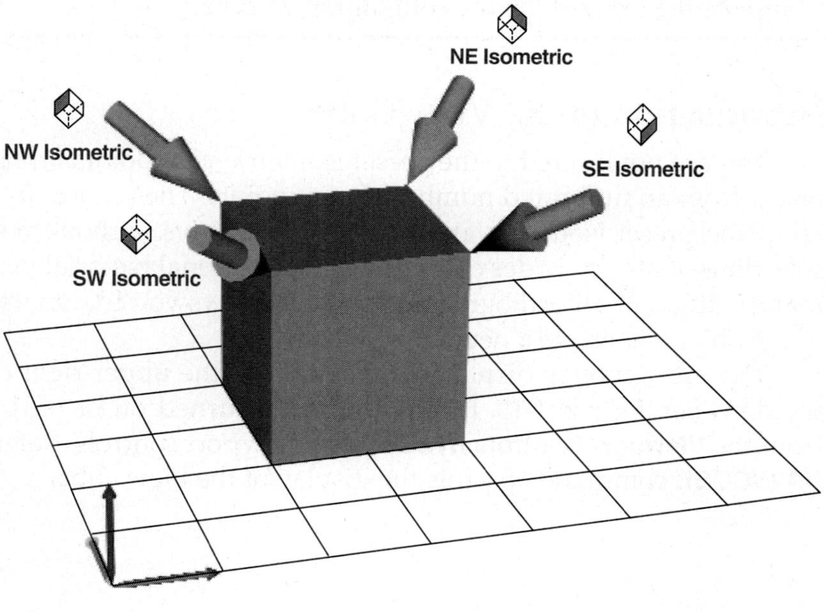

Figure 1-12.
Selecting preset views using the **View** panel in the **Home** tab of the ribbon.

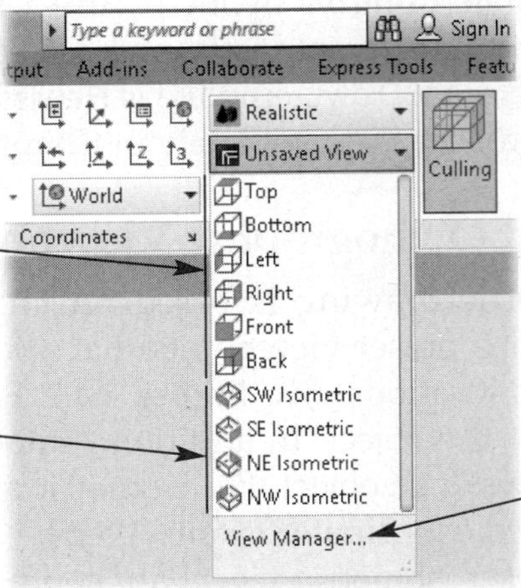

change the UCS to the WCS, type UCS to access the **UCS** command and then type W for the **World** option.

When an isometric or other 3D view is displayed, you can easily switch to a plan view of the current UCS by typing the **PLAN** command and selecting the **Current UCS** option. The **PLAN** command is discussed in more detail in Chapter 3.

NOTE

Selecting an orthographic view of a model using one of the methods described previously produces a plan view, but it may not achieve the results you desire. Three-dimensional models can be displayed in AutoCAD using either parallel or perspective projection. Displaying a plan view in either projection is possible. However, a true plan view, as used in 2D orthographic projections, can only be created when the model is displayed as a parallel projection. You can quickly change the display from perspective to parallel, or vice versa, by picking either **Parallel** or **Perspective** in the **View Controls** flyout in the viewport controls. The current projection is checked in the flyout.

Exercise 1-2

www.g-wlearning.com/CAD/

Complete the exercise on the companion website.

Introduction to the View Cube

You are not limited to the preset isometric viewpoints. In fact, you can view a 3D object from an unlimited number of viewpoints. The ***view cube*** allows you to display all of the preset isometric and orthographic views without making menu or ribbon selections. It also provides quick access to additional pictorial views and easy dynamic manipulation of all orthographic views. It allows you to dynamically rotate the view of the objects to create a new viewpoint.

NAVVCUBE
Ribbon
View
> Viewport Tools

View Cube
Type
NAVVCUBE
CUBE

The view cube is displayed by default in the upper-right corner of the drawing window. See **Figure 1-13**. It can be quickly turned on or off by selecting **ViewCube** from the **Viewport Controls** flyout in the viewport controls. Refer to **Figure 1-10A**. The **NAVVCUBE** command controls the display of the view cube.

Figure 1-13.
Using the view cube to change the viewpoint.

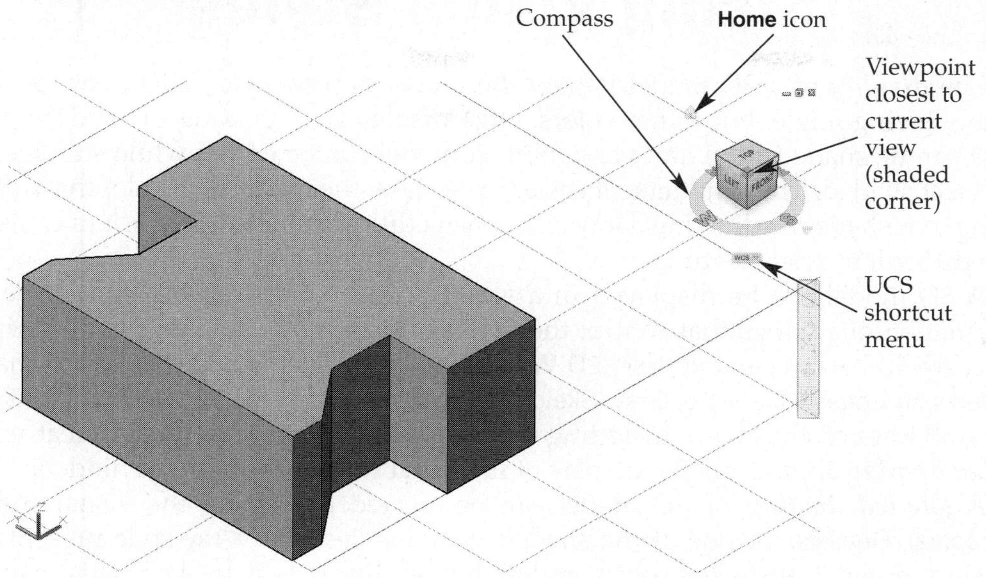

As the cursor is moved over the view cube, individual faces, edges, and corners are highlighted. If you pick one of the named faces on the view cube, that orthographic plan view is displayed. However, the UCS is not changed. If you pick one of the corners, a standard isometric view is displayed. Picking one of the four corners on the top face of the cube displays one of the preset isometric views. If you pick an edge, an isometric view is displayed that looks at the edge you selected. If you pick on the cube and drag the mouse, the view changes dynamically.

If you get lost while changing the viewpoint, just pick the **Home** icon (the small house shown in **Figure 1-13**) on the view cube to display the view defined as the home view. By default, the southwest isometric view is the home view. You can test this by selecting the **Home** icon and looking at the compass at the base of the view cube. Note that the left face of the cube aligns with the west point on the compass. The front surface aligns with the south point on the compass. Therefore, the top-front corner points to the southwest.

One edge, corner, or face of the view cube is always shaded or highlighted. Refer to **Figure 1-13**. This shading indicates the viewpoint on the cube that is closest to the current view.

The **NAVVCUBE** command has many options. This discussion is merely an introduction to the command. The command options and view cube features are covered in detail in Chapter 3.

PROFESSIONAL TIP

The **UNDO** command reverses the effects of the **NAVVCUBE** command.

Exercise 1-3

www.g-wlearning.com/CAD/

Complete the exercise on the companion website.

Introduction to 3D Model Display Using Visual Styles

The *display* of a 3D model is how the model is presented. This does not refer to the viewing angle, but rather colors, edge display, and shading or rendering. An object can be shaded from any viewpoint. A model can be edited while still keeping the object shaded. This can make it easier to see how the model is developing without having to reshade the drawing. However, when editing a shaded object, it may also be more difficult to select features.

A 3D model can be displayed in a variety of visual styles. A ***visual style*** is a combination of settings that control the display of edges and shading in a viewport. There are 10 basic visual styles—2D Wireframe, Conceptual, Hidden, Realistic, Shaded, Shaded with Edges, Shades of Gray, Sketchy, Wireframe, and X-ray. A ***wireframe display*** shows all lines on the object, including those representing back or internal features. A ***hidden display*** suppresses the display of lines that would normally be hidden.

A ***shaded display*** of the model can be created by setting the visual style to Conceptual, Realistic, or one of the shaded visual styles. The X-ray style presents the model in muted, translucent colors, and all hidden lines are displayed. This may be a good choice to use in the design of a model. The Realistic visual style is considered the most realistic *shaded* display. A more detailed shaded model, a ***rendered display*** of the model, can be created with the **RENDER** command. A rendering is the most realistic presentation.

VSCURRENT

Ribbon
Home
> View
Visualize
> Visual Styles
Type
VSCURRENT
VS

Examples of the basic visual styles are shown in **Figure 1-14**. To change styles, select one from the **Visual Style Controls** flyout in the viewport controls. Refer to **Figure 1-10C**. You can also use the **VSCURRENT** command or the drop-down list in the **View** panel on the **Home** tab or the **Visual Styles** panel on the **Visualize** tab in the ribbon. See **Figure 1-15**.

In the default 3D environment based on the acad3D.dwt template, the default display mode, or visual style, is Realistic. In this visual style, all objects appear as solids and are displayed in their assigned layer colors. Other display options are available. These options are discussed in detail in Chapter 16, but are given here as an introduction.

- **2D Wireframe.** Displays all lines of the model using assigned linetypes and lineweights. The 2D UCS icon and 2D grid are displayed, if turned on. If the **HIDE** command is used to display a hidden-line view, use the **REGEN** command to redisplay the wireframe view.
- **Wireframe.** Displays all lines of the model. The 3D grid and the 3D UCS icon are displayed, if turned on.
- **Hidden.** Displays all visible lines of the model from the current viewpoint and hides all lines not visible. Objects are not shaded or colored.
- **Sketchy.** Edges appear hand-sketched.
- **Shades of Gray.** Gray shades are shown with highlighted edges.
- **Conceptual.** The object is smoothed and shaded with transitional colors to help highlight details.
- **Realistic.** The model is shaded and smoothed using assigned layer colors and materials.
- **Shaded.** A smooth-shaded model is displayed, but edges are not shown.
- **Shaded with Edges.** Edges are displayed on the smooth-shaded model.
- **X-Ray.** The model appears transparent.

A variety of options can be used to change individual components of each of these visual styles. A complete discussion of visual styles is given in Chapter 16. Detailed discussions on rendering, materials, lights, and animations appear in Chapters 16 through 21.

Figure 1-14.
The default AutoCAD visual styles.

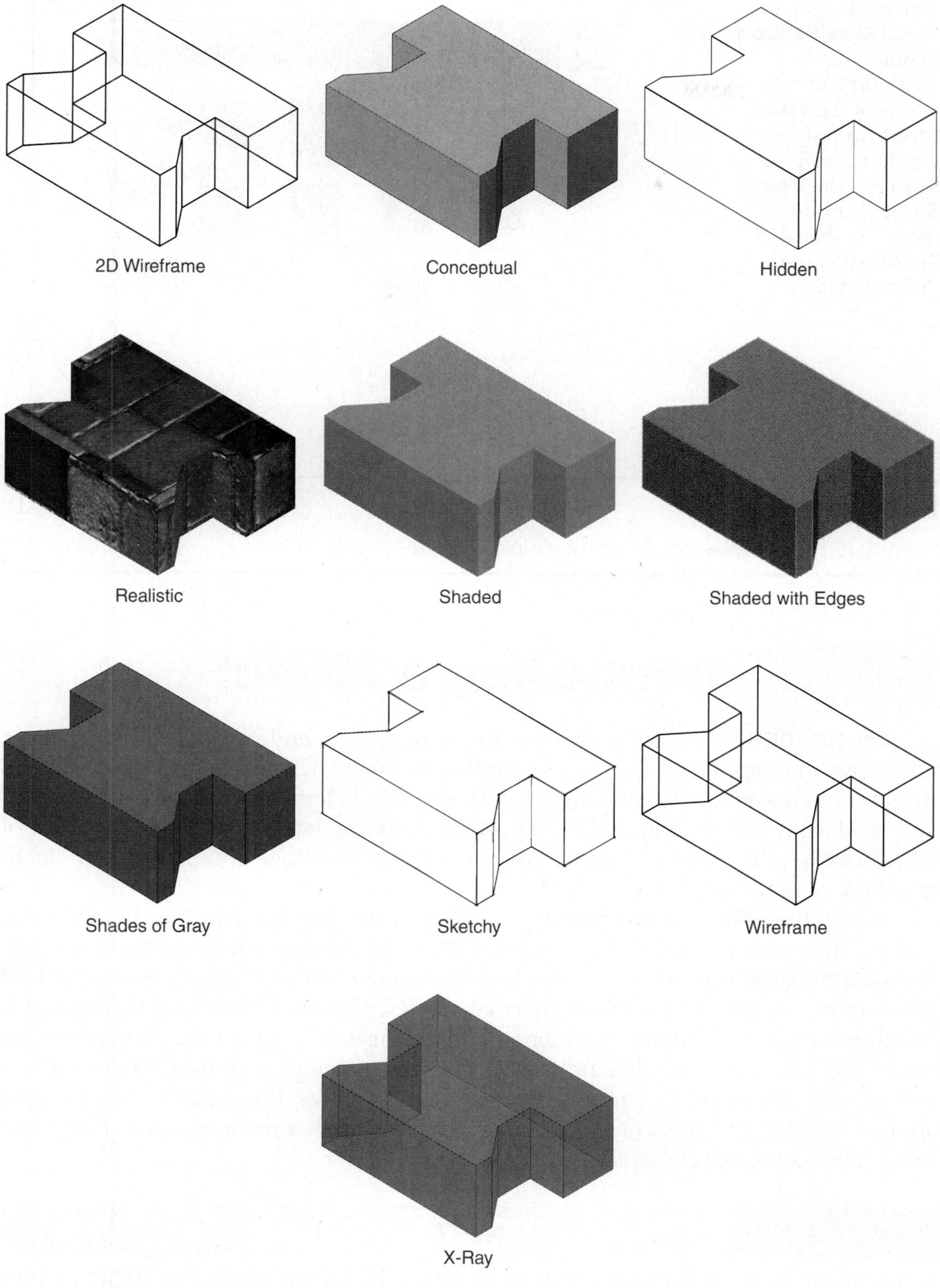

NOTE

When the visual style is 2D Wireframe, you can quickly view the model with hidden lines removed by typing HIDE. The **HIDE** command can be used at any time to remove hidden lines from a wireframe display. If **HIDE** is used when any other visual style is current, the Hidden visual style is set current.

Figure 1-15.
Visual styles can be managed using the **Visual Style Controls** flyout in the viewport controls or one of the **Visual Styles** drop-down lists in the ribbon. Shown is the **Visual Styles** drop-down list in the **Visual Styles** panel in the **Visualize** tab.

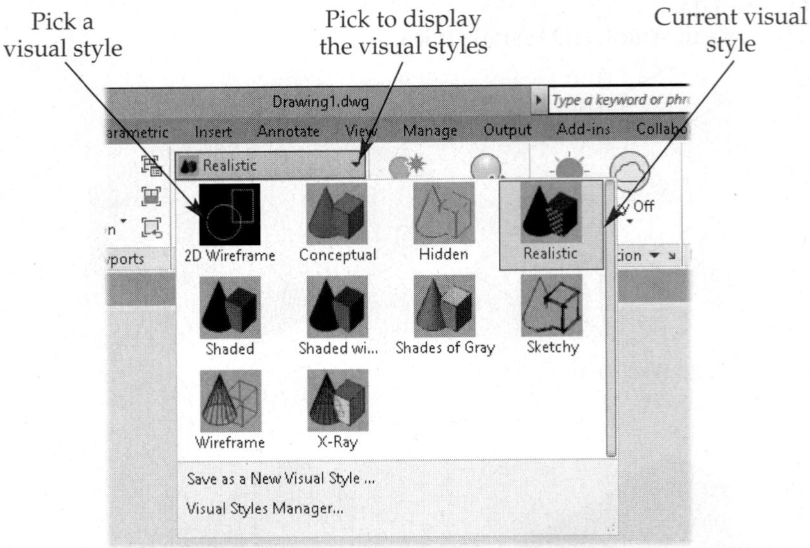

Exercise 1-4

www.g-wlearning.com/CAD/

Complete the exercise on the companion website.

Rendering a Model

RENDER
Ribbon
Visualize > Render
Render to Size
Type
RENDER
RR

RENDERWINDOW
Ribbon
Visualize > Render
Render Window
Type
RENDERWINDOW
RW

The **RENDER** command creates a realistic image of a model, **Figure 1-16**. However, rendering an image takes longer than shading an image. There are a variety of settings that you can use to fine-tune renderings. These include lights, materials, backgrounds, and rendering preferences. AutoCAD provides six rendering presets, the highest of which can require up to 12 hours to perform. Render settings are discussed in detail in Chapters 16 through 19.

When the **RENDER** command is initiated, the **Render** window is displayed and the image is rendered. See **Figure 1-17**. If the **Render** window closes, use the **RENDERWINDOW** command to redisplay the window. The rendering that is produced is based on a variety of render settings that are discussed in Chapters 16 through 19. If no lights have been created in the scene, AutoCAD uses default lighting to generate the rendering. The AutoCAD default lighting consists of one or two distant light sources that evenly illuminate all surfaces and generate shadows. If no materials have been applied to objects in the scene, objects are rendered with a matte material that is the same color as the object display color.

NOTE

If the image is rendered in the viewport, clean the screen using the **ZOOM** or **PAN** command. Setting the rendering destination as the viewport is discussed in Chapter 19.

Exercise 1-5

www.g-wlearning.com/CAD/

Complete the exercise on the companion website.

Figure 1-16.
Rendering produces the most realistic display of a scene.

Figure 1-17.
The rendered model is displayed in the **Render** window.

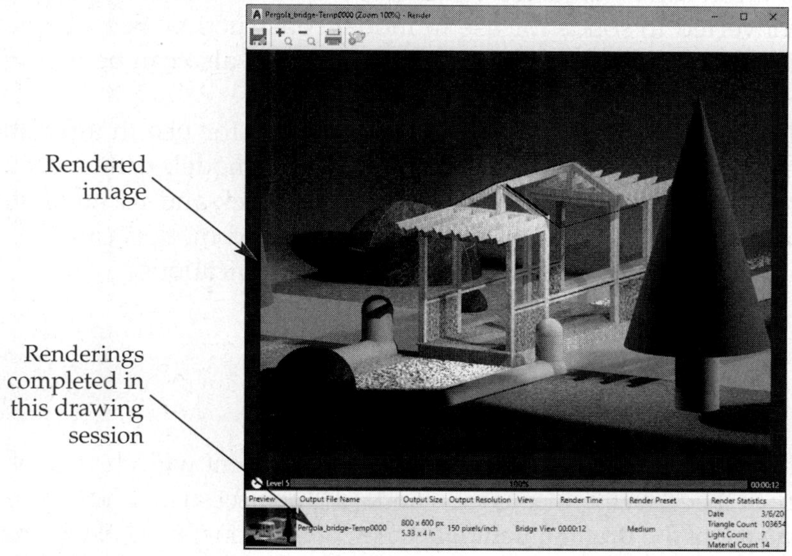

3D Construction Techniques

Before constructing a 3D model, you should determine the purpose of your design. What will the model be used for—manufacturing, analysis, or presentation? This helps you determine which tools you should use to construct and display the model. Three-dimensional objects can be drawn as solids, meshes, or surfaces and displayed in wireframe, hidden-line removed, and shaded views.

A ***wireframe object***, or model, is an object constructed of lines in 3D space. Wireframe models are hard to visualize because it is difficult to determine the angle of view and the nature of the surfaces represented by the lines. The **HIDE** command has no effect on a true wireframe model because there is nothing to hide. All lines are always visible because there are no surfaces or faces between the lines. True wireframe models have very limited applications.

Surface modeling represents solid objects by creating a skin in the shape of the object. However, there is nothing inside of the object. Think of a surface model as a balloon filled with air. A surface model looks more like the real object than a wireframe and can be used for rendering. Surface models are often constructed for applications such as civil engineering terrain modeling, automobile body design, sheet metal design and fabrication, and animation. Surface modeling techniques are discussed in Chapter 10.

Like surface modeling, ***solid modeling*** represents the shape of objects, but it also provides data related to the physical properties of the objects. Solid models can be analyzed to determine mass, volume, moments of inertia, and centroid. A solid model is not just a skin, it represents a solid object. Some third-party programs allow you to perform finite element analysis (FEA) on the model. Solid model files can also be exported for use in stereolithography, rapid prototyping, and 3D printing. These processes can produce a plastic or polymer prototype for analysis and testing. This is discussed in Chapter 14. In addition, solid models can be rendered. Most 3D objects are created as solid models.

In AutoCAD, solid models can be created from primitives. ***Primitives*** are basic shapes used as the foundation to create complex shapes. Some of these basic shapes include boxes, cylinders, spheres, and cones. Detailed shapes and primitives can be created using 3D mesh primitives, mesh modeling techniques, and surface modeling. A 3D ***mesh object***, which is a type of surface model, can have a free-flowing shape because the size of the mesh can be adjusted to achieve various levels of smoothness. Mesh objects can be converted to solids for use in model construction. See Chapter 9 for a detailed discussion on 3D mesh modeling. Solid primitives also can be modified to create a finished product. See **Figure 1-18**.

Surface and solid models can be exported from AutoCAD for use in animation and rendering software, such as Autodesk 3ds Max®. Rendered models can be used in any number of presentation formats, including slide shows, black-and-white or color prints, and animations recorded to video files. Surface and solid models can also be used to create virtual worlds for entertainment and gaming applications.

3D Object Snaps

DSETTINGS

Type
DSETTINGS
DS
SE
DDRMODES

The construction and editing of a 3D model can be more efficient with the use of 3D object snaps. These work in the same manner as the standard 2D object snaps and can be set using the **3D Object Snap** tab of the **Drafting Settings** dialog box. If you use 3D object snaps, turn on only those options you need to construct the object. See **Figure 1-19**.

The **Vertex**, **Midpoint on edge**, **Center of face**, **Perpendicular**, and **Nearest to face** 3D object snap modes should be familiar to you from your work in 2D. The **Knot** 3D object

Figure 1-18.
A—The two cylinders and the box are solid primitives. B—With a couple of quick modifications, the large cylinder becomes a shaft with a machined keyway.

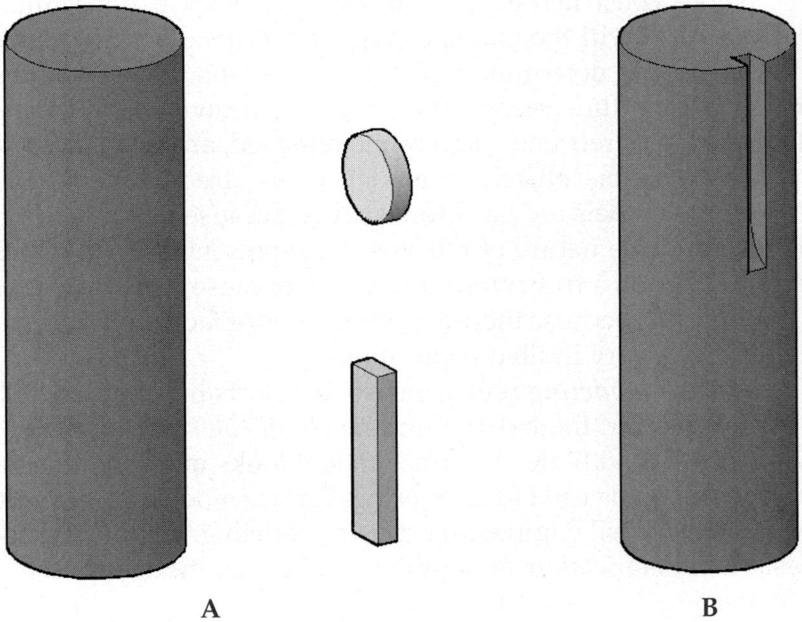

Figure 1-19.
The 3D object snaps can be set in the **Drafting Settings** dialog box.

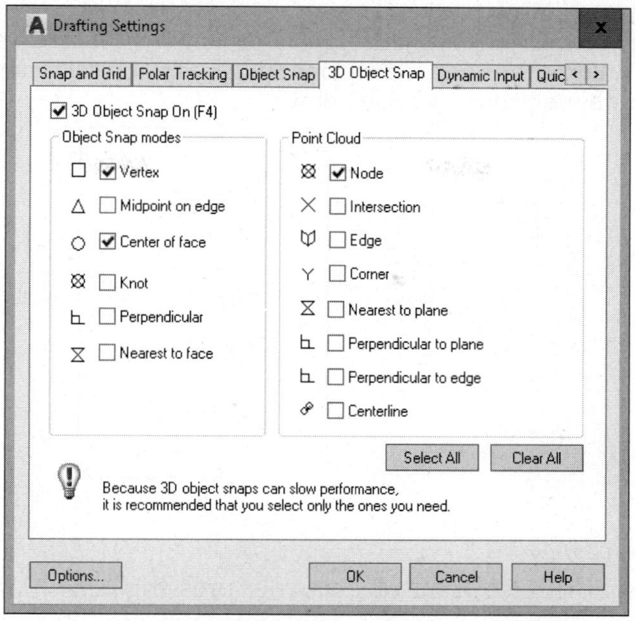

snap mode refers to a spline and the point at which one curve ends and the next curve begins. The point cloud object snap modes can be used when working with point clouds. Point clouds are discussed in Chapter 15.

The 3D AutoSnap markers that appear when drawing can be displayed in a different color than 2D AutoSnap markers. Set the color for 3D AutoSnap markers by accessing the **Display** tab in the **Options** dialog box. Pick the **Colors...** button in the **Window Elements** area to access the **Drawing Window Colors** dialog box. Select the desired context in the **Context:** list and then select a color for the 3d Autosnap marker element in the **Interface element:** list.

Application Menu
Options
Type
OPTIONS
OP
OPTIONS

Sample 3D Construction

The example provided here illustrates how you can move from a layout sketch to an exact 2D drawing with parametric constraints, then quickly extrude the object into a 3D solid model. Techniques illustrated here are discussed throughout the book. The object shown is the plastic housing for a 24/15-pin cable connector. The final rendition of this object would include internal cutouts, rounded edges, and holes, but is not shown in this example.

Creating a 2D Sketch

A 2D layout can be drawn quickly using the Drafting & Annotation workspace. In doing so, you can create a quick sketch without close attention to exact dimensions or you can use direct distance entry and polar tracking for accuracy. If appropriate, constraints can then be used to apply accurate dimensions and geometric relationships, such as perpendicularity and parallelism, to the object. See **Figure 1-20A**. In this example, only half of the object is drawn, since it is symmetrical. Then the **MIRROR** command is used to complete the layout. See **Figure 1-20B**.

Ribbon
Home
> Draw
Region
Type
REGION
REG
REGION

Converting to a Region

The object is still composed of just 2D lines. Therefore, it must be changed to an object that can be converted into a solid model. The **REGION** command is used for this purpose. After selecting this command, you are prompted to select objects. Be sure to

Figure 1-20.
A—Half of the 2D object is drawn using geometric constraints. B—The object is mirrored to create a full profile of the connector body. C—The object is now a "region" and is displayed in the southwest isometric viewpoint.

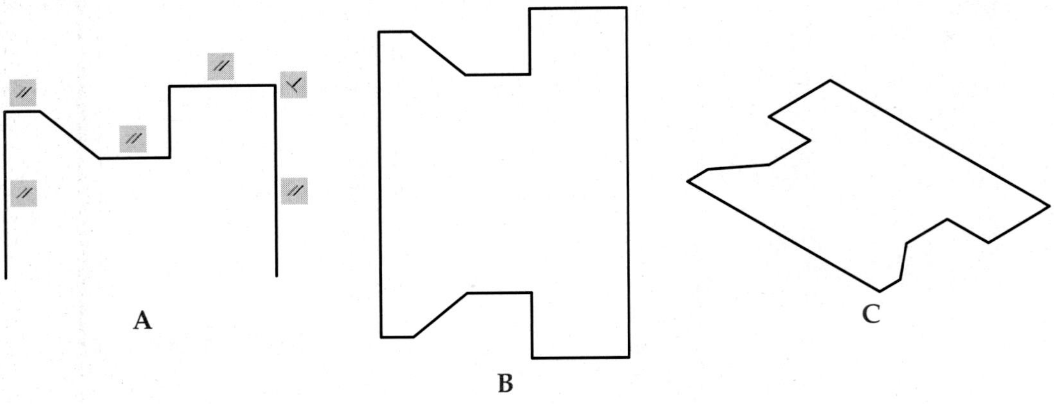

pick all of the lines of the object. After pressing [Enter], you are prompted that a loop is extracted, a region is created, and, if constraints were applied, that they are removed. The appearance of the object will not change.

Converting to a Solid

The object is now a "region" that can be quickly converted to a solid. First change to the 3D Modeling workspace, and then use the viewport controls or the view cube to display the object from the southwest isometric viewpoint. Working with models is more intuitive when a 3D display is used. See **Figure 1-20C**.

EXTRUDE

Ribbon
Home
> Modeling
Extrude
Type
EXTRUDE
EXT

The region can be converted to a solid using either the **EXTRUDE** or **PRESSPULL** command. For example, pick the **Extrude** button in the **Modeling** panel on the **Home** tab of the ribbon and select the region. Using direct distance entry, enter a height of .55, as shown in **Figure 1-21**, and press [Enter].

Figure 1-21.
The object is extruded into a solid and displayed as a wireframe.

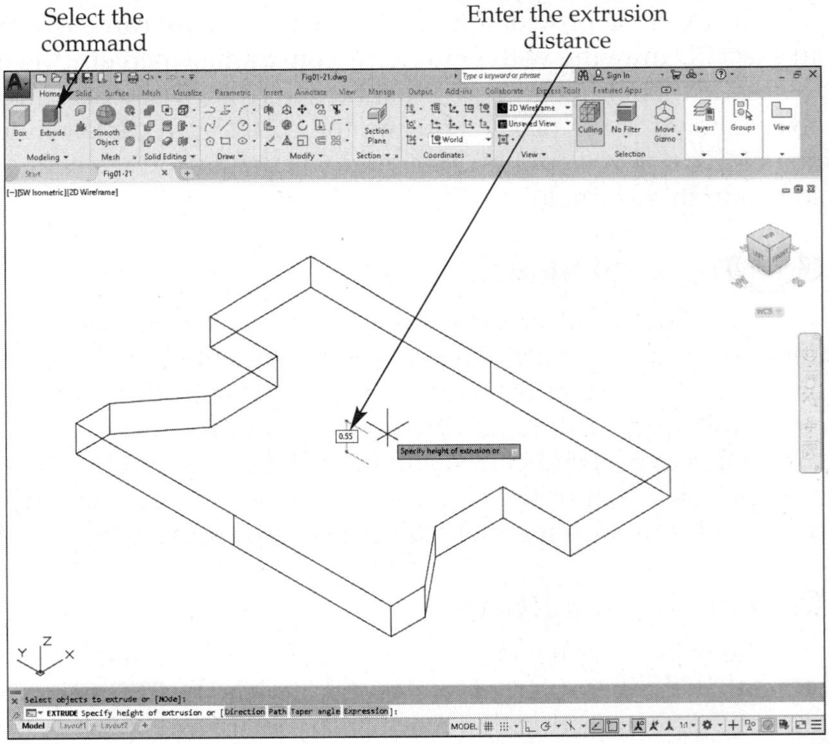

Using a Visual Style

The object you have created is now a 3D solid. If the drawing was based on the acad3D.dwt template, the model is displayed in the Realistic visual style. If the drawing was based on the acad.dwt template, the model is displayed in the 2D Wireframe visual style.

As mentioned previously, there are 10 different visual styles from which to choose. Select an option from the **Visual Style Controls** flyout in the viewport controls or the **View** panel on the **Home** tab of the ribbon. The Conceptual visual style provides a quick display of the model using shaded tones of the object color, as shown in **Figure 1-22**.

The techniques used in this example are just a brief introduction to the creation of 3D models. Detailed descriptions of modeling, display, and editing techniques are included in the following chapters.

Guidelines for Working with 3D Drawings

Working in 3D, like working with 2D drawings, requires careful planning to efficiently produce the desired results. The following guidelines can be used when working in 3D.

Planning

- Determine the type of final drawing you need and the manner in which it will be displayed. Then, choose the method of 3D construction that best suits your needs—wireframe, surface, mesh, or solid.
- If appropriate for the project, use 2D constraints to create a 2D layout or sketch.
- For an object requiring only one pictorial view, it actually may be quicker to draw an object in 3D rather than in AutoCAD's isometric mode. AutoCAD's 3D solid modeling tools enable you to quickly create an accurate model, and then display it in the required isometric format using preset views. The **VIEWBASE** command can then be used to create a 2D drawing of the model.
- It is best to use AutoCAD's 3D commands to construct objects and layouts that need to be viewed from different angles for design purposes.
- Construct only the features needed for the function of the drawing. This saves space and time, and makes visualization much easier.
- Use 2D or 3D object snap modes in a pictorial view in conjunction with UCS icon manipulation to save having to create new UCSs.
- Keep in mind that when the grid is displayed, the pattern appears at the current elevation and parallel to the XY plane of the current UCS.

Figure 1-22.
The Conceptual visual style provides a quick display of the model using shaded tones of the object colors.

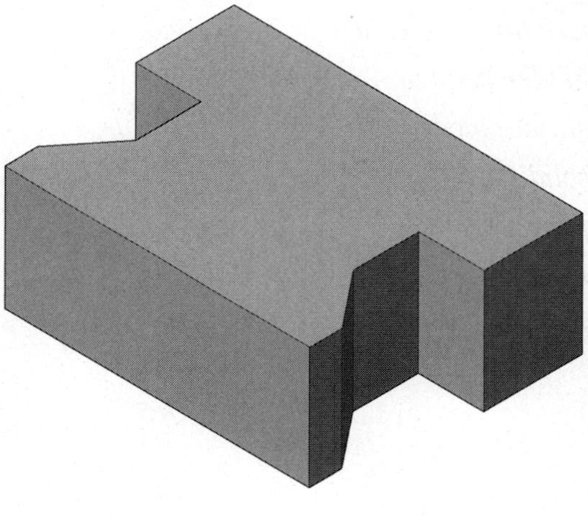

Editing

- Use the **Properties** palette to change the color, layer, or linetype of 3D objects.
- Use grips to edit a solid-modeled object (see Chapter 11).
- Do as much editing as possible from a 3D viewpoint. It is quicker and the results are immediately seen.

Displaying

- Use the **HIDE** command and visual styles to help visualize complex drawings.
- To quickly change views, use the preset isometric views, view cube, and **PLAN** command.
- Create and save 3D views for quicker pictorial views. This avoids having to repeatedly use the view cube or other methods to change the viewing angle.
- Freeze unwanted layers before displaying objects in 3D and especially before using **HIDE**. AutoCAD regenerates layers that are turned off, which may cause an inaccurate hidden display to be created. Frozen layers are not regenerated.
- You may have to slightly move objects that touch or intersect if the display removes a line you need to see or plot. However, be sure to move the objects back to maintain accuracy in the model.

Chapter Review

Answer the following questions using the information in this chapter.

1. What are the three coordinates needed to locate any point in 3D space?
2. In a 2D drawing, what is the value for the Z coordinate?
3. What purpose does the right-hand rule serve?
4. Which three fingers are used in the right-hand rule?
5. What is the definition of a *viewpoint*?
6. What is the function of the *ribbon* and its panels?
7. How do you turn the display of individual panels on or off in the ribbon?
8. How can you quickly change the display from perspective projection to parallel projection, or vice versa?
9. How many preset isometric viewpoints does AutoCAD have? List them.
10. How does changing the UCS impact using one of the preset isometric viewpoints?
11. List the six preset orthographic viewpoints.
12. When selecting a preset orthographic viewpoint, what happens to the UCS?
13. Which AutoCAD tool allows you to dynamically change your viewpoint using an on-screen cube icon?
14. Define *wireframe display.*
15. Define *hidden display.*
16. Define *surface model.*
17. Define *solid model.*
18. Define *primitive.*
19. Define *mesh object.*
20. What is a *visual style*?

CHAPTER 2

Creating Primitives and Composites

Learning Objectives

After completing this chapter, you will be able to:

✓ Construct 3D solid primitives.
✓ Explain the dynamic feedback presented when constructing solid primitives.
✓ Create complex solids using the **UNION** command.
✓ Remove portions of a solid using the **SUBTRACT** command.
✓ Create a new solid from the common volume between two solids.
✓ Create regions.

Overview of Solid Modeling

In Chapter 1, you were introduced to the three basic types of 3D models—wireframe objects, solid models, and surface models. A solid model is probably the most useful and, hence, most common type of 3D model. A solid model accurately and realistically represents the shape and form of a final object. In addition, a solid model contains data related to the object's volume, mass, and centroid.

Solid modeling is very flexible. A model can start with solid primitives, such as a box, cone, or cylinder, and a variety of editing functions can then be performed. Think of creating a solid model as working with modeling clay. Starting with a basic block of clay, you can add more clay, remove clay, cut holes, round edges, etc., until you have arrived at the final shape and form of the object.

PROFESSIONAL TIP

Snaps can be used on solid objects. For example, using 2D object snaps, you can snap to the center of a solid sphere using the **Center** object snap. The **Endpoint** object snap can be used to select the corners of a box, apex of a cone, corners of a wedge, etc. Using 3D object snaps, you can snap to the vertex or midpoint of an edge or to the center of a face.

Constructing Solid Primitives

As you learned in Chapter 1, a primitive is a basic building block. The eight ***solid primitives*** in AutoCAD are a box, cone, cylinder, polysolid, pyramid, sphere, torus, and wedge. These primitives can also be used as building blocks for complex solid models. This section provides detailed information on drawing all of the solid primitives. All of the 3D modeling primitive commands can be accessed using one of several methods. You can use the **Primitive** panel in the **Solid** tab of the ribbon or the **Modeling** panel in the **Home** tab of the ribbon. You can also type the name of the solid primitive. Using the **Modeling** panel in the **Home** tab of the ribbon is shown in **Figure 2-1**.

The information required to construct a solid primitive depends on the type of primitive being drawn. For example, to draw a solid cylinder, you must provide a center point for the base, a radius or diameter of the base, and the height of the cylinder. A variety of command options are available when creating primitives, but each primitive is constructed using just a few basic dimensions. These are shown in **Figure 2-2**.

Certain familiar editing commands can be used on solid primitives. For example, you can fillet or chamfer the edges of a solid primitive. In addition, there are other editing commands that are specifically for use on solids. You can also perform Boolean operations on solids. These operations allow you to add one solid to another, subtract one solid from another, or create a new solid based on how two solids overlap.

PROFESSIONAL TIP

Solid objects of a more free-form nature can be created by first constructing a mesh primitive, editing it, and then converting it to a solid. This process is covered in detail in Chapter 9.

Using Dynamic Input and Dynamic Feedback

Dynamic input allows you to construct models in a "heads up" fashion with minimal eye movement around the screen. When a command is initiated, the command prompts are then displayed in the dynamic input area, which is at the lower-right corner of the crosshairs. As the pointer is moved, the dynamic input area follows it. The dynamic input area displays values of the cursor location, dimensions, command

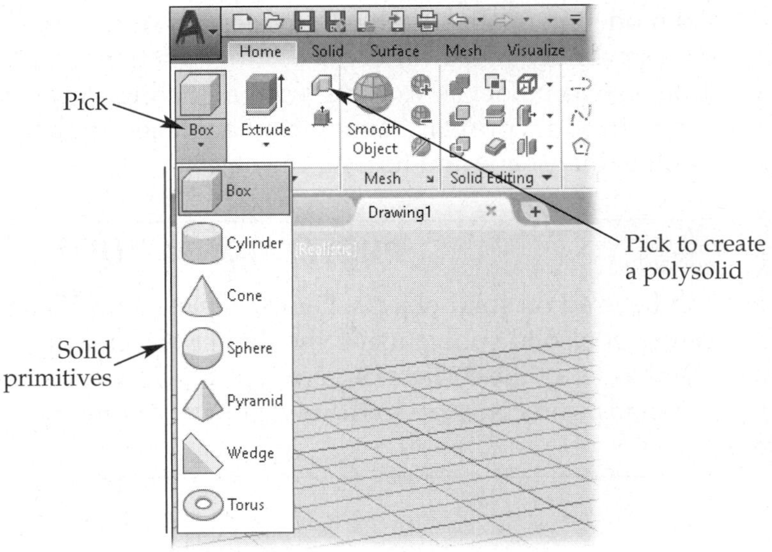

Figure 2-1.
The **Modeling** panel in the **Home** tab of the ribbon.

Figure 2-2.
An overview of AutoCAD's solid primitives and the dimensions required to draw them.

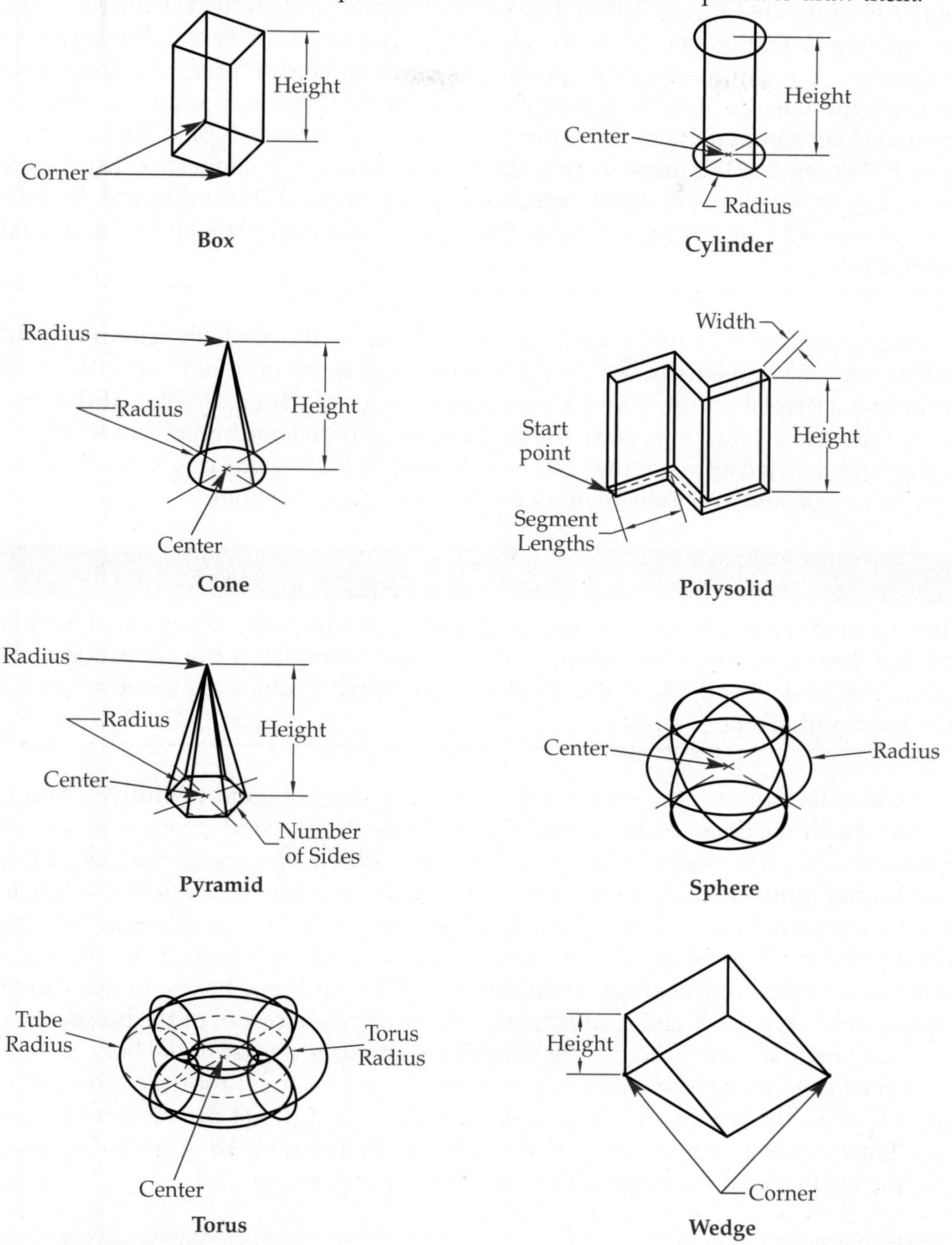

prompts, and command options (in a drop-down list). Coordinates and dimensions are displayed in boxes called *input fields.* When command options are available, a drop-down list arrow appears. Press the down arrow key on the keyboard to display the list. You can use the pointer to select the option or press the down arrow key until a dot appears by the desired option and press [Enter].

The command line provides an alternative to dynamic input. When a command is active, the available options are highlighted and can be selected directly at the command line. As the pointer is moved over an option, the pointer changes to a hand cursor and the option is selectable. This method requires no keyboard input.

NOTE

AutoCAD provides a suggestion list of commands and system variables at the dynamic input area as you type. As additional characters are typed, the suggestion list changes. You can then use the pointer to select an item or press the down arrow key until the desired item is highlighted, and press [Enter]. In addition, AutoCAD recognizes certain synonyms of command names when you type a command. For example, typing SPIRAL displays the **HELIX** command in the suggestion list. Options controlling AutoComplete, AutoCorrect, and other AutoCAD input search functions can be accessed by right-clicking inside the command line window and selecting **Input Search Options...**.

As an example of using dynamic input, if you select the modeling command **BOX**, the first item that appears in the dynamic input area is the prompt to specify the first corner and a display of the X and Y coordinate values of the crosshairs. At this point, you can use the pointer to specify the first corner or type coordinate values. Type the X value and then a comma or the [Tab] key to move to the Y value input box. This locks the typed value and any movement of the pointer will not change it.

CAUTION

When using dynamic input to enter coordinate values from the keyboard, it is important that you avoid pressing [Enter] until you have completed the coordinate entry. When you press [Enter], all of the displayed coordinate values are accepted and the next command prompt appears.

In addition to entering coordinate values for sizes of solid primitives, you can provide direct distance dimensions. For example, the second prompt of the **BOX** command is for the second corner of the base. When you move the pointer, two dimensional input fields appear. Also, notice that a preview of the base is shown in the drawing area. This is the ***dynamic feedback*** that AutoCAD provides as you create a solid primitive. See **Figure 2-3A**. Once you enter or pick the first corner of the base, the dynamic input area changes to display X and Y coordinate boxes. In this case, the values entered are the X and Y coordinates of the opposite corner of the box base. If X, Y, *and* Z coordinates are entered, the point is the opposite corner of the box.

After establishing the location and size of the box base, the next prompt asks you to specify the height. Again, you can either enter a direct dimension value and press [Enter] or select the height with the pointer. See **Figure 2-3B**. AutoCAD provides dynamic feedback on the height of the box as the pointer is moved.

NOTE

The techniques previously described can be used with any form of dynamic input. The current input field is always highlighted. You can always enter a value and use the [Tab] key to lock the input and move to the next field.

BOX

Ribbon
Home
> Modeling
Solid
> Primitive
Box
Type
BOX

Box

A ***box*** has six flat sides forming square corners. It can be constructed starting from an initial corner or the center. See **Figure 2-4**. A cube can be constructed, **Figure 2-4A**, as well as a box with unequal-length sides, **Figure 2-4B**.

Figure 2-3.
A—Specifying the base of a box with dynamic input on. Notice the preview of the base.
B—Setting the height of a box with dynamic input on. Notice the preview of the height.

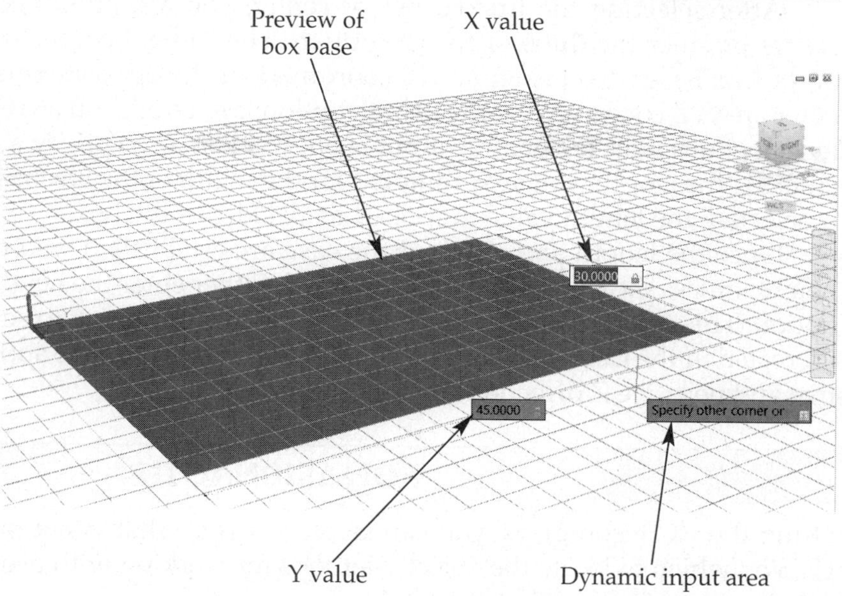

A

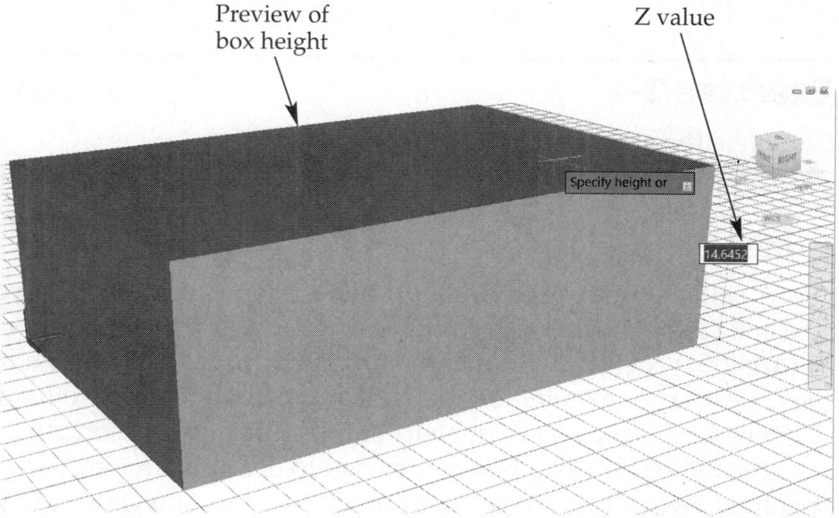

B

Figure 2-4.
A—A box created using the **Cube** option. B—A box created by selecting the center point.

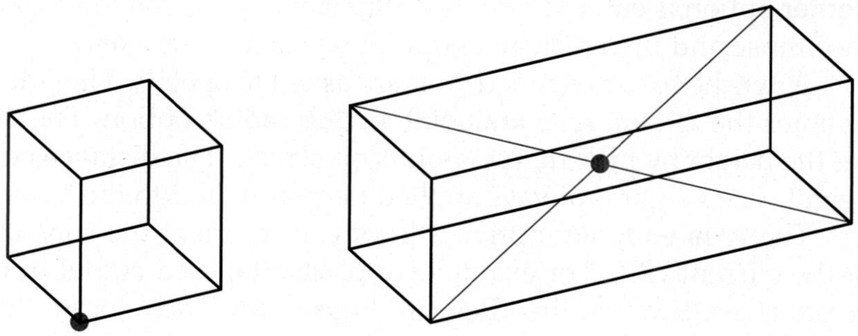

When the command is initiated, you are prompted to select the first corner or enter the **Center** option. The first corner is one corner on the base of the box. The center is the geometric center of the box, as shown in **Figure 2-4B**. If you select the **Center** option, you are next prompted to select the center point.

After selecting the first corner or center, you are prompted to select the other corner or enter the **Cube** or **Length** option. The "other" corner is the opposite corner of the box base if you enter an XY coordinate or the opposite corner of the box if you enter an XYZ coordinate. If the **Length** option is entered, you are first prompted for the length of one side. If dynamic input is on, you can also specify a rotation angle. After entering the length, you are prompted for the width of the box base. If the **Cube** option is selected, the length value is applied to all sides of the box.

Once the length and width of the base are established, you are prompted for the height, unless the **Cube** option was selected. Either enter the height or select the **2Point** option. This option allows you to pick two points on screen to set the height. The box is created. If the **Center** option was selected, the height will be applied in both the positive and negative Z directions.

PROFESSIONAL TIP

Using the UCS icon grips, you can select a surface that is not parallel to the current UCS on which to locate the object. Methods for working with user coordinate systems are discussed in detail in Chapter 4.

Exercise 2-1

www.g-wlearning.com/CAD/

Complete the exercise on the companion website.

Cone

CONE

Ribbon
Home
> Modeling Solid
> Primitive

Cone

Type
CONE

A ***cone*** has a circular or elliptical base with edges that converge at a single point. The cone may be *truncated* so the top is flat and the cone does not have an apex. See **Figure 2-5**. When the command is initiated, you are prompted for the center point of the cone base, or to enter an option. If you pick the center, you must then set the radius of the base. To specify a diameter, enter the **Diameter** option after specifying the center.

The **3P**, **2P**, and **Ttr** options are used to define a circular base using three points on the circle, two points on the circle, or two points of tangency on the circle and a radius. The **Elliptical** option is used to create an elliptical base.

If the **Elliptical** option is entered, you are prompted to pick both endpoints of one axis and then one endpoint of the other axis of an ellipse that defines the base. If the **Center** option is entered after the **Elliptical** option, you are asked to select the center of the ellipse and then pick an endpoint on each of the axes.

After the base is defined, you are asked to specify a height. You can enter a height or enter the **2Point**, **Axis endpoint**, or **Top radius** option. The **2Point** option is used to set the height by picking two points on screen. The distance between the points is the height. The height is always applied perpendicular to the base.

The **Axis endpoint** option allows you to orient the cone at any angle, regardless of the current UCS. For example, to place a tapered cutout in the end of a block, first create a construction line. Refer to **Figure 2-6**. Then, locate the cone base and give a coordinate location of the apex, or axis endpoint. You can then subtract the cone from the box to create the tapered hole.

The **Top radius** option allows you to specify the radius of the top of the cone. If this option is not used, the radius is zero, which creates a pointed cone. Setting the radius to a value other than zero produces a ***frustum cone***, or a cone where the top is truncated, and does not come to a point.

Figure 2-5.
A—A circular cone. B—A frustum cone. C—An elliptical cone.

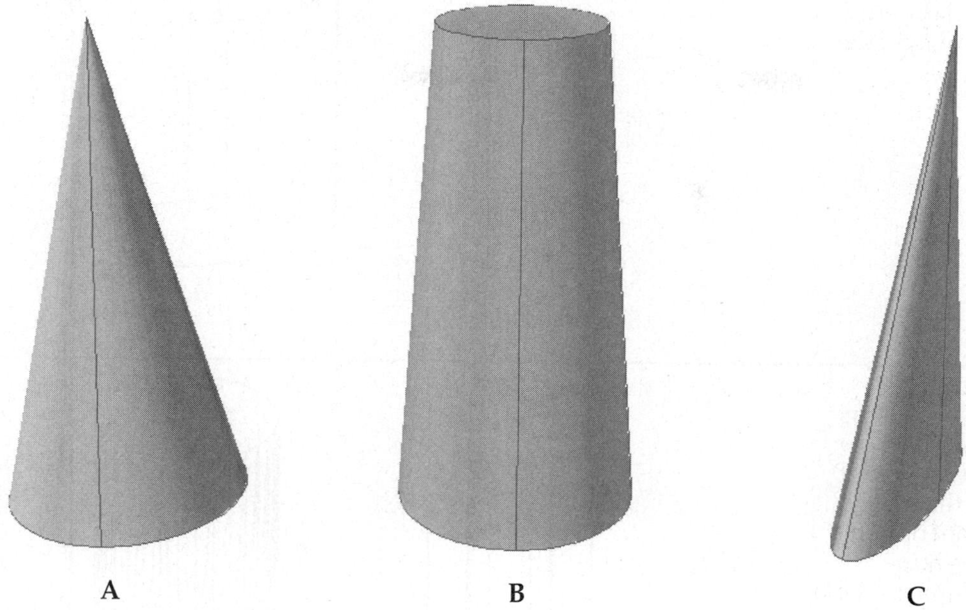

Figure 2-6.
A—Cones can be positioned relative to other objects using the **Axis endpoint** option.
B—The cone is subtracted from the box.

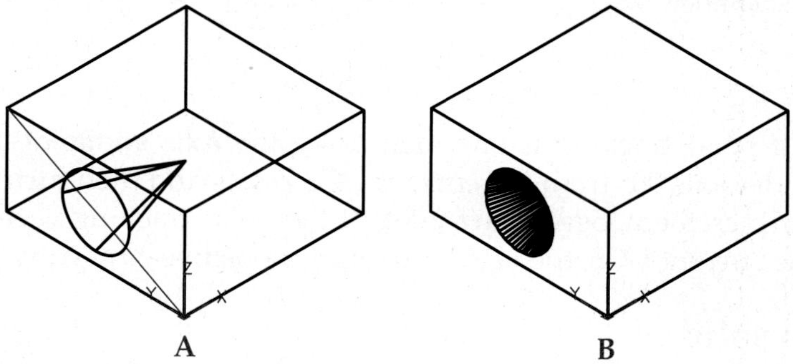

Cylinder

A ***cylinder*** has a circular or elliptical base and edges that extend perpendicular to the base. See **Figure 2-7**. When the command is initiated, you are prompted for the center point of the cylinder base, or to enter an option. If you pick the center, you must then set the radius of the base. To specify a diameter, enter the **Diameter** option after specifying the center.

Ribbon
Home
> Modeling
Solid
> Primitive
Cylinder
Type
CYLINDER
CYL

CYLINDER

The **3P**, **2P**, and **Ttr** options are used to define a circular base using three points on the circle, two points on the circle, or two points of tangency on the circle and a radius. The **Elliptical** option is used to create an elliptical base.

If the **Elliptical** option is entered, you are prompted to pick both endpoints of one axis and then one endpoint of the other axis of an ellipse defining the base. If the **Center** option is entered, you are asked to select the center of the ellipse and then pick an endpoint on each of the axes.

After the base is defined, you are asked to specify a height or to enter the **2Point** or **Axis endpoint** option. The **2Point** option is used to set the height by picking two points on screen. The distance between the points is the height. The **Axis endpoint** option allows you to orient the cylinder at any angle, regardless of the current UCS, just as with a cone.

The **Axis endpoint** option is useful for placing a cylinder inside of another object to create a hole. The cylinder can then be subtracted from the other object to create a hole. Refer to **Figure 2-8**. If the axis endpoint does not have the same X and Y coordinates as the center of the base, the cylinder is tilted from the XY plane.

Figure 2-7.
A—A circular cylinder. B—An elliptical cylinder.

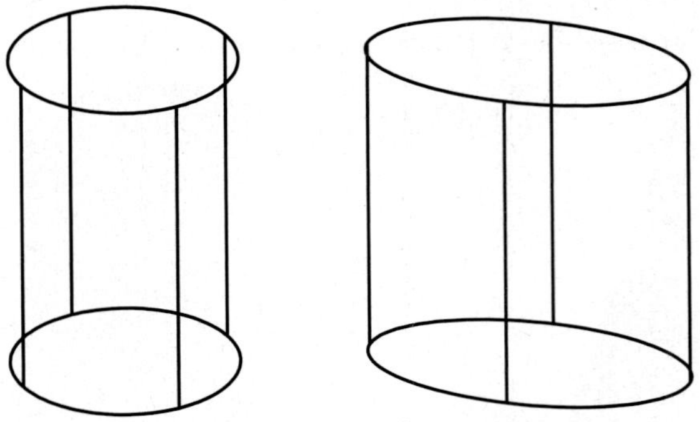

Figure 2-8.
A—A cylinder is drawn inside of another cylinder using the **Axis endpoint** option. B—The large cylinder has a hole after **SUBTRACT** is used to remove the small cylinder.

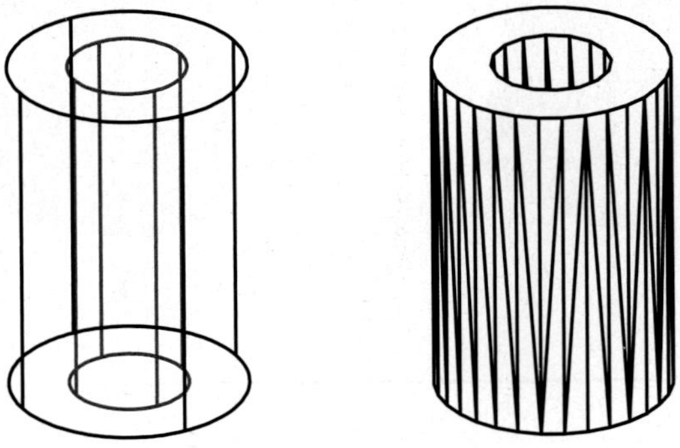

If polar tracking is on when using the **Axis endpoint** option, you can rotate the cylinder axis 90° from the current UCS Z axis and then turn the cylinder to any preset polar increment. See **Figure 2-9A**. If the polar tracking vector is parallel to the Z axis of the current UCS, the tooltip displays a positive or negative Z value. See **Figure 2-9B**.

Polysolid

POLYSOLID

Ribbon
Home
> Modeling
Solid
> Primitive

Polysolid

Type
POLYSOLID
PSOLID

The ***polysolid*** primitive is simply a polyline constructed as a solid object by applying a width and height to the polyline. Many of the options used to create polylines are used with the **POLYSOLID** command. The principal difference is that a solid object is constructed using **POLYSOLID**.

When the command is initiated, you are prompted to select the start point or enter an option. By default, the width of the polysolid is equally applied to each side of the line you draw. This is center justification. Using the **Justify** option, you can set the justification to center, left, or right. The justification applies to all segments created in the command session. See **Figure 2-10**. If you select the wrong justification option, you must exit the command and begin again.

The default width is .25 units and the default height is 4 units. These values can be changed using the **Height** and **Width** options of the command. The height value is saved in the **PSOLHEIGHT** system variable. The width value is saved in the **PSOLWIDTH** system variable. Using these system variables, the default width and height can be set outside of the command.

The **Object** option allows you to convert an existing 2D object into a polysolid. AutoCAD entities such as lines, circles, arcs, polylines, polygons, and rectangles can be converted. The 2D object cannot be self intersecting. Some objects, such as 3D polylines, cannot be converted.

Once you have set the first point on the polysolid, pick the endpoint of the first segment. Continue adding segments as needed. The **Undo** option allows you to remove the last segment drawn and continue from the previous point without exiting the command. To complete the command, press [Enter].

Figure 2-9.
A—If polar tracking is on, you can rotate the cylinder axis 90° from the current UCS Z axis and then move the cylinder to any angle in the XY plane. B—If the polar tracking vector is moved parallel to the current Z axis of the UCS, the tooltip displays a positive or negative Z dimension.

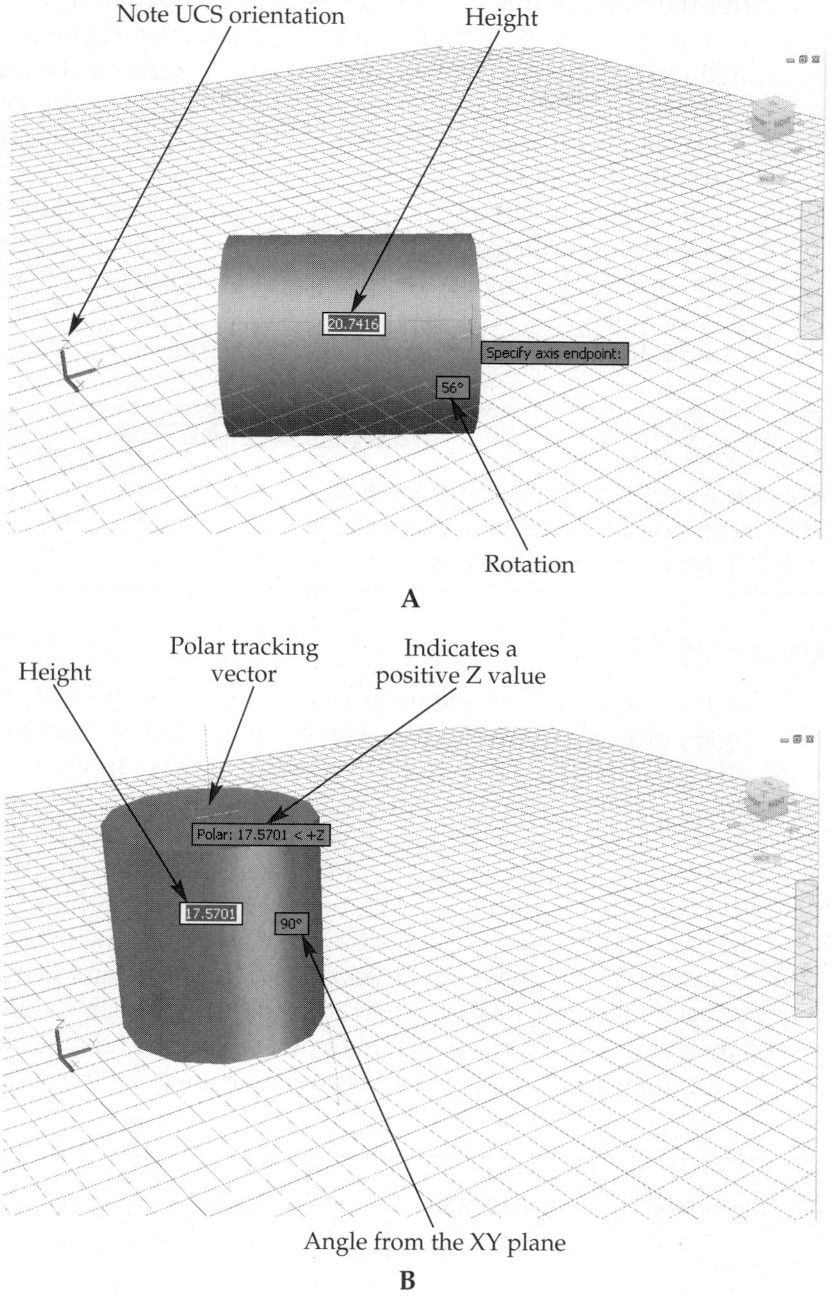

Figure 2-10.
When you begin the **POLYSOLID** command, use the **Justify** option to select the alignment.

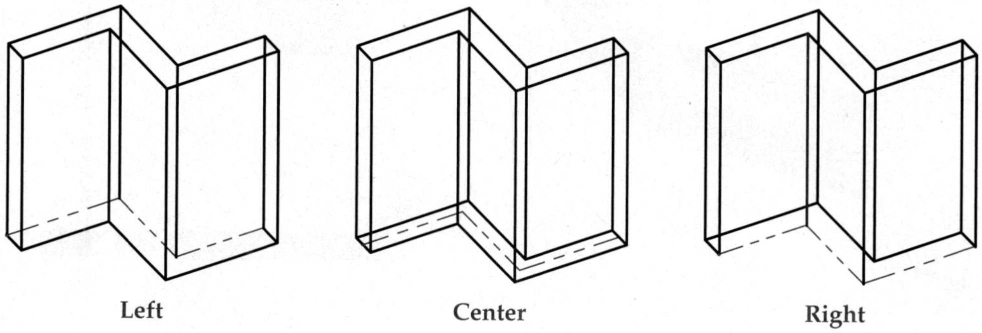

After the first point is set, you can enter the **Arc** option. The current segment will then be created as an arc instead of a straight line. See **Figure 2-11**. Arc segments will be created until you enter the **Line** option. The suboptions for the **Arc** option are:

- **Close.** If there are two or more segments, this option creates an arc segment between the active point and the first point of the polysolid. You can also close straight-line segments.
- **Direction.** Specifies the tangent direction for the start of the arc.
- **Line.** Returns the command to creating straight-line segments.
- **Second point.** Locates the second point of the arc. This is not the endpoint of the segment.
- **Undo.** Removes the previous segment of the polysolid arc.

PROFESSIONAL TIP

The **Object** option of the **POLYSOLID** command is a powerful tool for converting 2D objects to 3D solids. For example, you can create a single-line wall plan using a polyline and then quickly convert it to a 3D model.

Pyramid

A ***pyramid*** has a base composed of straight-line segments and edges that converge at a single point. The pyramid base can be composed of three to 32 sides and is drawn much like a 2D polygon. A pyramid primitive may be drawn with a pointed apex or as a ***frustum pyramid***, which has a truncated, or flat, apex. See **Figure 2-12**.

PYRAMID

Ribbon
Home
> Modeling
Solid
> Primitive
Pyramid
Type
PYRAMID
PYR

Once the command is initiated, you are prompted for the center of the base or to enter an option. To set the number of sides on the base, select the **Sides** option. Then, enter the number of sides. You are returned to the first prompt.

The base of the pyramid can be drawn by either picking the center and the radius of a base circle or by picking the endpoints of one side. The default method is to pick the center. Simply specify the center and then set the radius. To pick the endpoints of one side, select the **Edge** option. Then, pick the first endpoint of one side followed by the second endpoint. If dynamic input is on, you can also set a rotation angle for the pyramid.

Figure 2-11.
The **Arc** option of the **POLYSOLID** command is used to create curved segments.

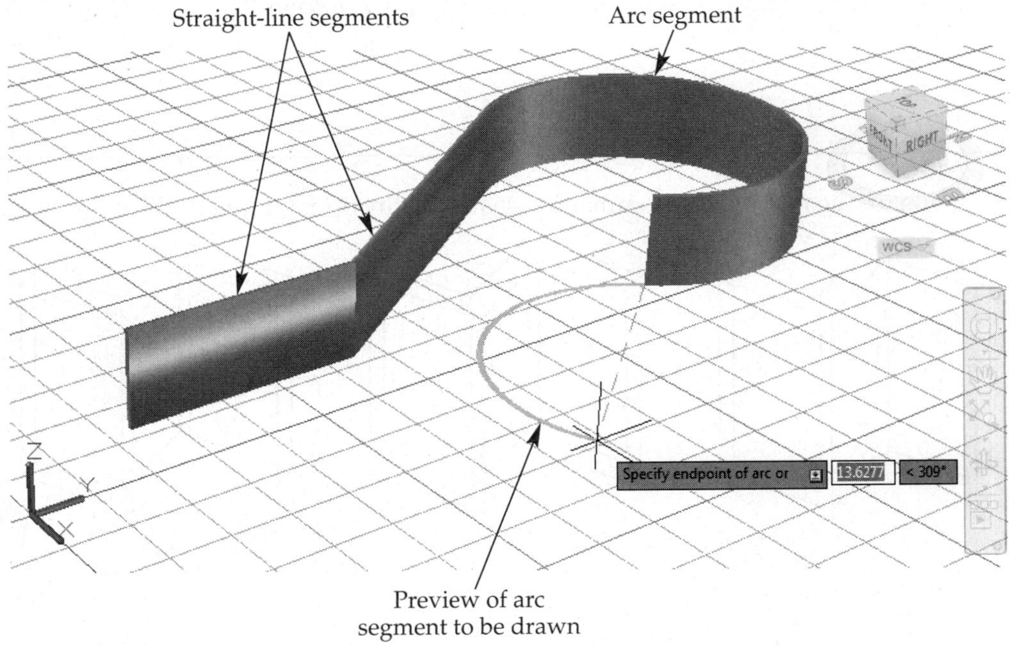

Figure 2-12.
A sampling of pyramids that can be constructed with the **PYRAMID** command.

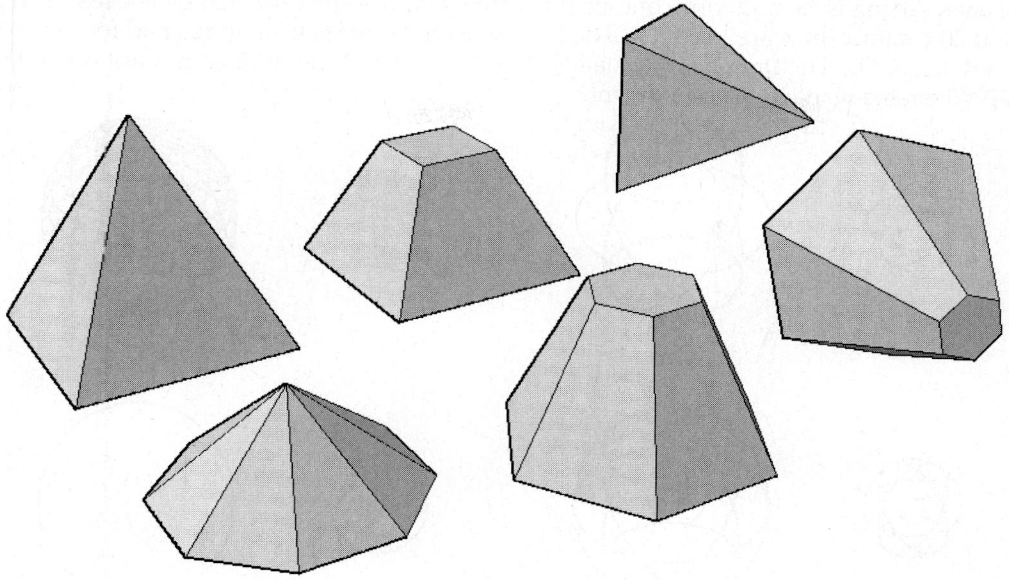

If drawing the base from the center point, the polygon is circumscribed about the base circle by default. To inscribe the polygon on the base circle, enter the **Inscribed** option before setting the radius. To change back to a circumscribed polygon, select the **Circumscribed** option before setting the radius.

After locating and sizing the base, you are prompted for the height. To create a frustum pyramid, enter the **Top radius** option. Then, set the radius of the top circle. The top will be either inscribed or circumscribed based on the base circle. You are then returned to the height prompt.

The height value can be set by entering a direct distance. You can also use the **2Point** option to set the height. With this option, pick two points on screen. The distance between the two points is the height value. The **Axis endpoint** option can also be used to specify the center of the top in the same manner as for a cone or cylinder.

PROFESSIONAL TIP

Six-sided frustum pyramids can be used for bolt heads and as "blanks" for creating nuts.

Sphere

SPHERE

Ribbon
Home
> Modeling
Solid
> Primitive

Sphere
Type
SPHERE

A *sphere* is a round, smooth object like a baseball or globe. Once the command is initiated, you are prompted for the center of the sphere or to enter an option. If you pick the center, you must then set the radius of the sphere. To specify a diameter, enter the **Diameter** option after specifying the center. The **3P**, **2P**, and **Ttr** options are used to define the sphere using three points on the surface of the sphere, two points on the surface of the sphere, or two points of tangency on the surface of the sphere and a radius.

Spheres and other curved objects can be displayed in a number of different ways. The manner in which you choose to display these objects should be governed by the display requirements of your work. Notice in **Figure 2-13** the lines that define the shape of the spheres in a wireframe display. These lines are called ***contour lines***, also known as *tessellation lines* or *isolines*. The **Visual Styles Manager** can be used to set the

Figure 2-13.
Different displays of spheres established with visual style settings. A—The Draw true silhouettes setting is No and four contour lines are used. B—The Draw true silhouettes setting is No and 20 contour lines are used. C—The Draw true silhouettes setting is Yes and four contour lines are used. D—The Draw true silhouettes setting is Yes and the **HIDE** command is used with the 2D Wireframe visual style set current.

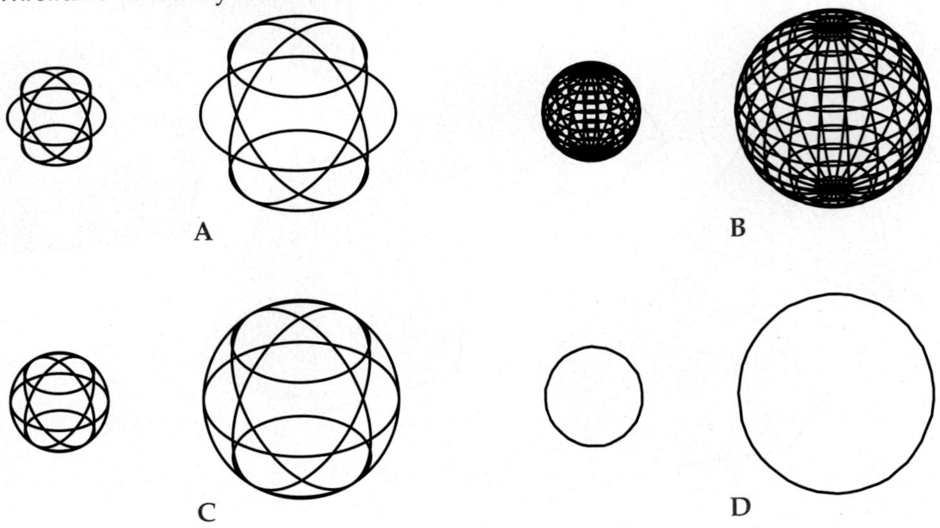

display of contour lines and silhouettes on spheres and other curved 3D surfaces for a given visual style. See **Figure 2-14**.

With the **Visual Styles Manager** displayed, select the 2D Wireframe image tile. The Contour lines setting in the **2D Wireframe options** area establishes the number of lines used to show the shape of curved objects. A similar setting appears in all of the other visual styles if their Show property in the **Edge Settings** area is set to Isolines. The default value is four, as shown in **Figure 2-13A**, but it can be set to a value from zero

Figure 2-14.
The **2D Wireframe options** area of the **Visual Styles Manager** is used to control the display of contour lines and silhouettes on spheres and other curved 3D surfaces in a given visual style.

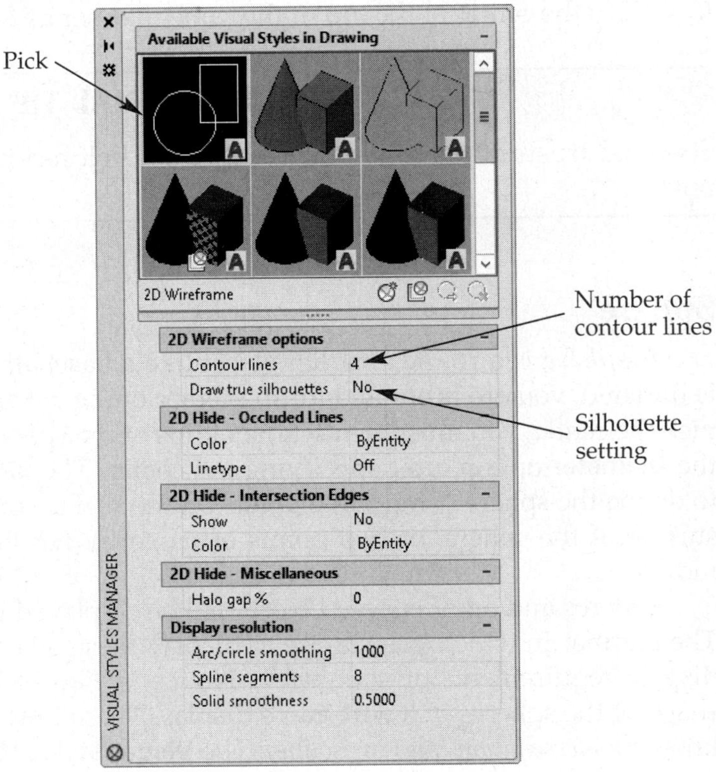

to 2047. **Figure 2-13B** displays spheres with 20 contour lines. It is best to use a lower number during construction and preliminary displays of the model and, if needed, higher settings for more realistic visualization. The contour lines setting is also available in the **Display** tab of the **Options** dialog box or by typing ISOLINES.

The Draw true silhouettes setting in the **2D Wireframe options** area of the **Visual Styles Manager** controls the display of silhouettes on 3D solid curved surfaces. The setting is either Yes or No. Notice the sphere silhouette in **Figures 2-13C** and **2-13D**. The Draw true silhouettes setting is stored in the **DISPSILH** system variable.

Torus

TORUS

Ribbon
Home
> Modeling Solid
> Primitive
Torus

Type
TORUS
TOR

A basic ***torus*** is a cylinder bent into a circle, similar to a doughnut or inner tube. There are three types of tori. See **Figure 2-15**. A torus with a tube diameter that touches itself is called ***self intersecting*** and has no center hole. To create a self-intersecting torus, the tube radius must be greater than the torus radius. The third type of torus looks like a football. It is drawn by entering a negative torus radius and a positive tube radius of greater absolute value, such as –1 and 1.1.

Once the command is initiated, you are prompted for the center of the torus or to enter an option. If you pick the center, you must then set the radius of the torus. To specify a diameter, select the **Diameter** option after specifying the center. This defines a base circle that is the centerline of the tube. The **3P**, **2P**, and **Ttr** options are used to define the base circle of the torus using either three points, two points, or two points of tangency and a radius.

Once the base circle of the torus is defined, you are prompted for the tube radius, or to enter an option. The tube radius defines the cross-sectional circle of the tube. To specify a diameter of the cross-sectional circle, enter the **Diameter** option. You can also use the **2Point** option to pick two points on screen that define the diameter of the cross-sectional circle. If you are working conceptually, you can move the pointer and see the tube size change dynamically.

Wedge

WEDGE

Ribbon
Home
> Modeling Solid
> Primitive
Wedge

Type
WEDGE
WE

A ***wedge*** has five sides, four of which are at right angles and the fifth at an angle other than 90°. See **Figure 2-16**. Once the command is initiated, you are prompted to select the first corner of the base or to enter an option. By default, a wedge is constructed by picking diagonal corners of the base and setting a height. To pick the center point,

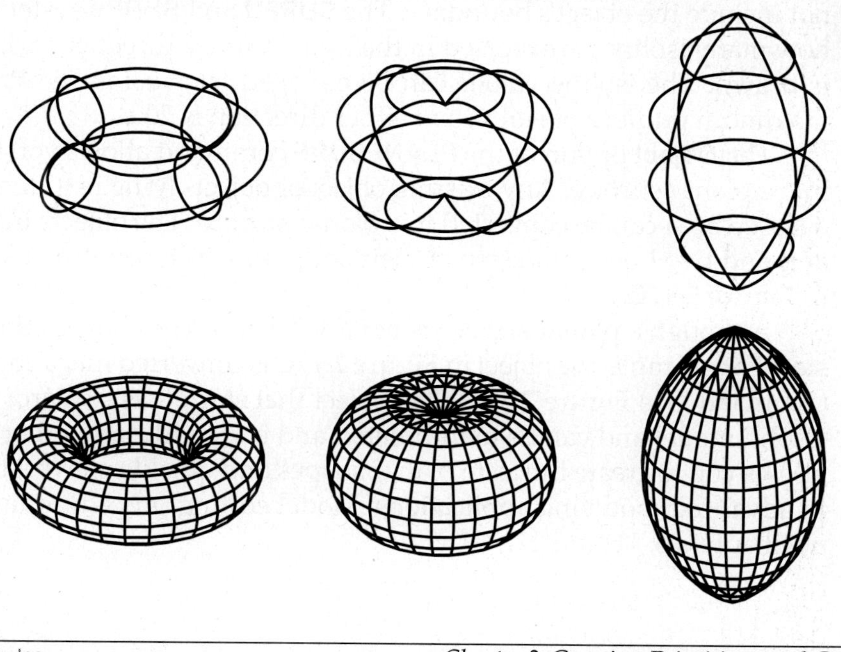

Figure 2-15.
The three types of tori are shown as wireframes and with hidden lines removed.

Figure 2-16.
A—A wedge drawn by picking corners and specifying a height. B—A wedge drawn using the **Center** option. Notice the location of the center.

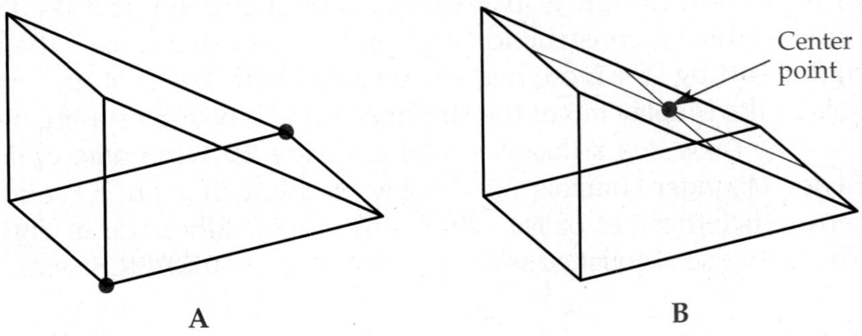

select the **Center** option. The center point of a wedge is the middle of the angled surface. You must then pick a point to set the width and length before entering a height.

After specifying the first corner or the center, you can enter the length, width, and height instead of picking a second corner. When prompted for the second corner, select the **Length** option and specify the length. You are then prompted for the width. After the width is entered, you are prompted for the height. Either enter the height or select the **2Point** option to pick two points on screen.

To create a wedge with equal length, width, and height, select the **Cube** option when prompted for the second corner. Then, enter a length. The same value is automatically used for the width and height.

Exercise 2-2

www.g-wlearning.com/CAD/

Complete the exercise on the companion website.

Constructing a Planar Surface

PLANESURF

Ribbon
Surface
> Create

Planar Surface
Type
PLANESURF

A ***planar surface*** primitive is an object consisting of a single plane and is created parallel to the current XY plane. The surface that is created has zero thickness and is composed of a mesh of lines. It is created with the **PLANESURF** command. The command prompts you to specify the first corner and then the second corner of a rectangle. Once drawn, the surface is displayed as a mesh with lines in the local directions, similar to the X and Y directions. See **Figure 2-17A**. These lines, called isolines, do not include the object's boundary. The **SURFU** and **SURFV** system variables determine how many isolines are created in the local U and V directions when the planar surface is drawn. The isoline values can be changed later using the **Properties** palette. The maximum number of isolines in either direction is 200.

The **Object** option of the **PLANESURF** command allows you to convert a 2D object into a planar surface. Any existing object or objects lying in a single plane and forming a closed area can be converted to a planar surface. The objects in **Figure 2-17B** are two arcs and two lines connected at their endpoints. The resulting planar surface is shown in **Figure 2-17C**.

Although a planar surface is not a solid, it can be converted into a solid in a single step. For example, the object in **Figure 2-17C** is converted into a solid using the **THICKEN** command. See **Figure 2-17D**. The object that started as two arcs and two lines is now a solid model and can be manipulated and edited like any other solid. This capability allows you to create intricate planar shapes and quickly convert them to a solid for use in advanced modeling applications. Model editing procedures are discussed in detail in Chapters 8, 11, and 12.

Figure 2-17.
A—A rectangular planar surface with four isolines in the U direction and eight isolines in the V direction. B—These two arcs and two lines form a closed area and lie on a single plane. C—The arcs and lines are converted into a planar surface. D—The planar surface is converted into a solid.

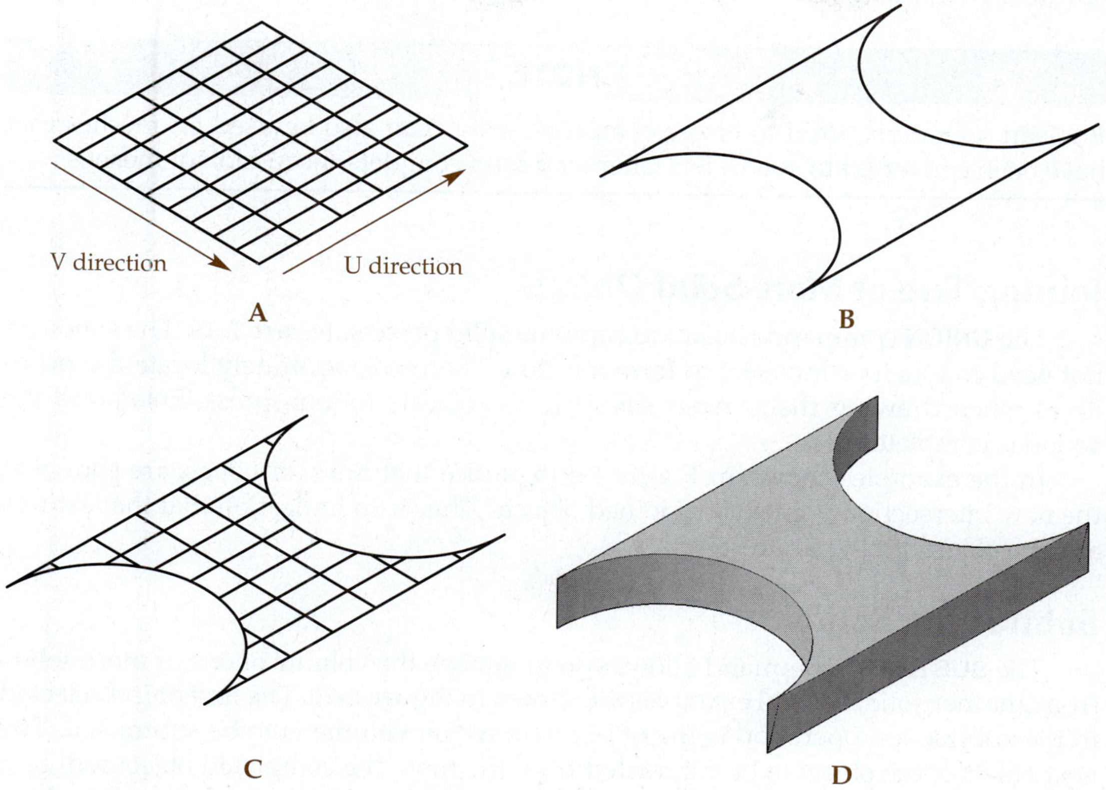

Creating Composite Solids

A ***composite solid*** is a solid model constructed of two or more solids, often primitives. Solids can be subtracted from each other, joined to form a new solid, or overlapped to create an intersection or interference. The commands used to create composite solids are found in the **Solid Editing** panel of the **Home** tab or the **Boolean** panel of the **Solid** tab in the ribbon. See **Figure 2-18**.

Introduction to Booleans

Three operations form the basis of constructing many complex solid models. Joining two or more solids is called a ***union*** operation. Subtracting one solid from another is called a ***subtraction*** operation. Forming a solid based on the volume of overlapping solids is called an ***intersection*** operation. Unions, subtractions, and intersections as a

Figure 2-18.
Selecting a Boolean command in the **Boolean** panel on the **Solid** tab of the ribbon. These commands are also located in the **Solid Editing** panel on the **Home** tab.

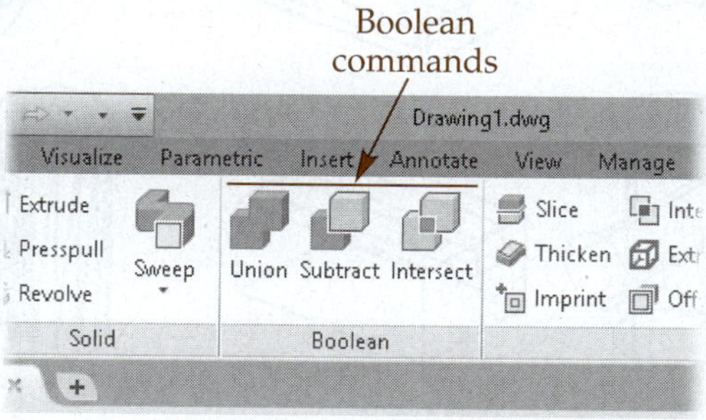

group are called ***Boolean operations***. George Boole (1815–1864) was an English mathematician who developed a system of mathematical logic where all variables have the value of either one or zero. Boole's two-value logic, or *binary algebra*, is the basis for the mathematical calculations used by computers and, with respect to modeling, for those required in the construction of composite solids.

NOTE

Boolean operations used to create composite solids can also be used on meshes that have been converted to solids. See Chapter 9 for a complete discussion of meshes.

Joining Two or More Solid Objects

UNION
Ribbon
Home
> Solid Editing
Solid
> Boolean

Solid, Union
Type
UNION
UNI

The **UNION** command is used to combine solid objects, **Figure 2-19**. The solids do not need to touch or intersect to form a union. Therefore, accurately locate the primitives when drawing them. After selecting the objects to join, press [Enter] and the action is completed.

In the examples shown in **Figure 2-19B**, notice that lines, or edges, are shown at the new intersection points of the joined objects. This is an indication that the features are one object, not separate objects.

Subtracting Solids

SUBTRACT
Ribbon
Home
> Solid Editing
Solid
> Boolean

Solid, Subtract
Type
SUBTRACT
SU

The **SUBTRACT** command allows you to remove the volume of one or more solids from another solid. Several examples are shown in **Figure 2-20**. The first object selected in the subtraction operation is the object *from* which volume is to be subtracted. The next object is the object to be subtracted from the first. The completed object will be a new solid. If the result is the opposite of what you intended, you may have selected the objects in the wrong order. Just undo the operation and try again.

Creating New Solids from the Intersection of Solids

INTERSECT
Ribbon
Home
> Solid Editing
Solid
> Boolean

Solid, Intersect
Type
INTERSECT
IN

When solid objects intersect, the overlap forms a common volume, a space that both objects share. This shared space is called an ***intersection***. An intersection (common volume) can be made into a composite solid using the **INTERSECT** command. **Figure 2-21** shows several examples. A solid is formed from the common volume. The original objects are removed.

Figure 2-19.
A—The solid primitives shown here have areas of intersection and overlap. B—Composite solids after using the **UNION** command. Notice the lines displayed where the previous objects intersected.

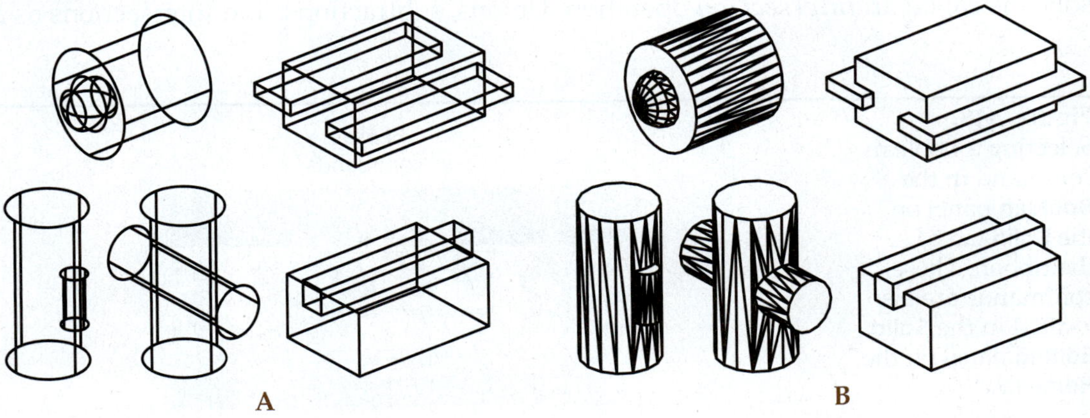

Figure 2-20.
A—The solid primitives shown here have areas of intersection and overlap. B—Composite solids after using the **SUBTRACT** command.

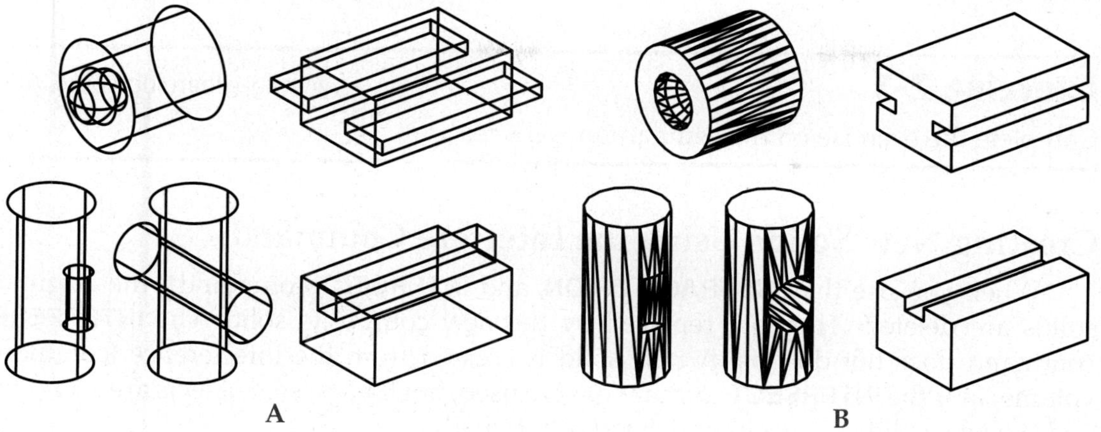

Figure 2-21.
A—The solid primitives shown here have areas of intersection and overlap.
B—Composite solids after using the **INTERSECT** command.

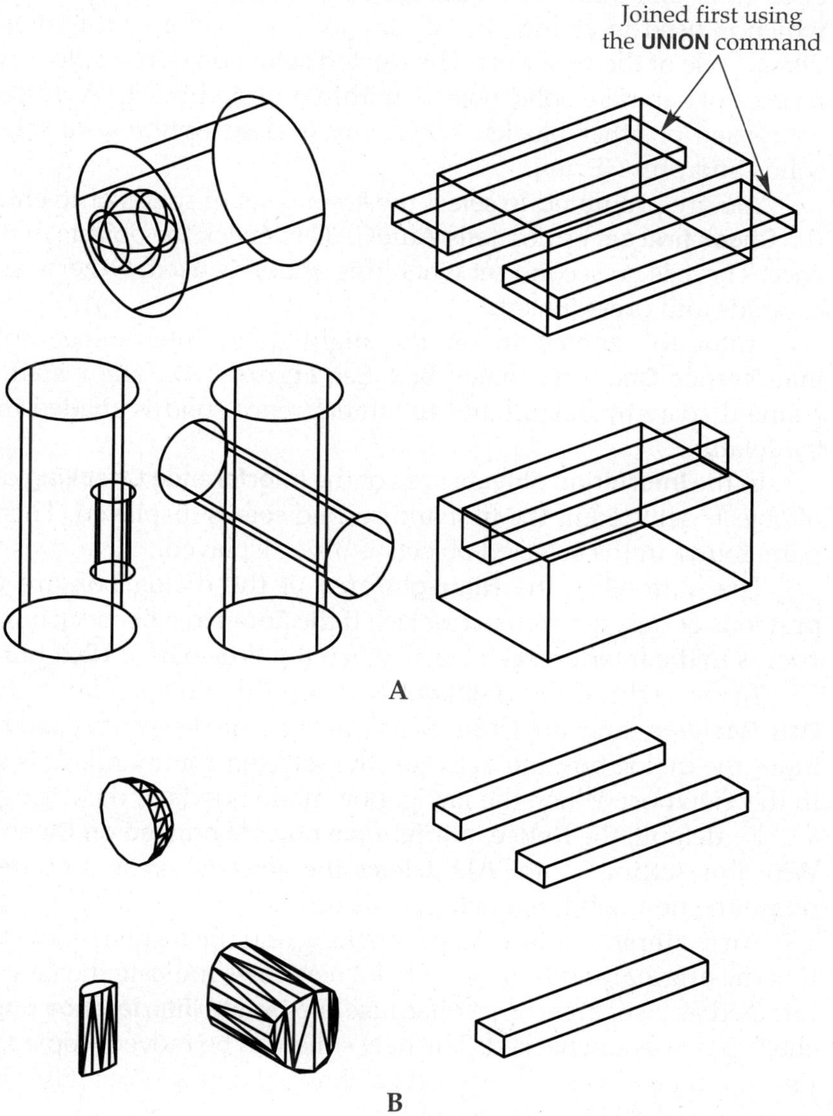

The **INTERSECT** command is also useful in 2D drawing. For example, if you need to create a complex shape that must later be used for inquiry calculations or hatching, draw the main object first. Then, draw all intersecting or overlapping objects. Next, create regions of the shapes. Finally, use **INTERSECT** to create the final shape. The resulting shape is a region and has solid properties. Regions are discussed later in this chapter.

Exercise 2-3

www.g-wlearning.com/CAD/

Complete the exercise on the companion website.

Creating New Solids Using the Interfere Command

INTERFERE

Ribbon
Home
> Solid Editing
Solid
> Solid Editing
Interfere
Type
INTERFERE
INF

When you use the **SUBTRACT**, **UNION**, and **INTERSECT** commands, the original solids are deleted. They are replaced by the new composite solid. The **INTERFERE** command does not do this. A new solid is created from the interference (common volume) as if the **INTERSECT** command were used, but the original objects are retained and the new solid can be either deleted or retained.

Once the command is initiated, you are prompted to select the first set of solids or to enter an option. The **Settings** option opens the **Interference Settings** dialog box, which is used to change the visual style and color of the interference solid and the visual style of the viewport. The **Nested selection** option allows you to check the interference of separate solid objects within a nested block. A ***nested block*** is one that is composed of other blocks. When any needed options are set, select the first set of solids and press [Enter].

You are prompted to select the second set of solids or to enter an option. Entering the **Check first set** option tells AutoCAD to check the objects in the first set for interference. There is no second set when this option is used. Otherwise, select the second set of solids and press [Enter].

AutoCAD zooms in on the highlighted interference solid and displays the **Interference Checking** dialog box. See Figure 2-22. The visual style is set to a wireframe display by default and the interference solid is shaded in a color, which is red by default.

In the **Interfering objects** area of the **Interference Checking** dialog box, the number of objects selected in the first and second sets is displayed. The number of interfering pairs found in the selected objects is also displayed.

The buttons in the **Highlight** area of the dialog box are used to highlight the previous or next interference object. If the **Zoom to pair** check box is checked, AutoCAD zooms to the interference objects when the **Previous** or **Next** button is selected.

To the right of the **Highlight** area are three navigation buttons—**Zoom Realtime**, **Pan Realtime**, and **3D Orbit**. Selecting one of these navigation options temporarily hides the dialog box and activates the selected command. This allows you to navigate in the viewport. When the navigation mode is exited, the dialog box is redisplayed.

By default, the **Delete interference objects created on Close** check box is checked. With this setting, AutoCAD deletes the object(s) created by interference. In order to retain the new solid(s), uncheck this box.

An example of interference checking and the result is shown in Figure 2-23. Notice that the original solids are intact, but new lines indicate the new solid. The new solid is retained as a separate object because the **Delete interference objects created on Close** check box was unchecked. The new solid can be moved, copied, and manipulated just like any other object. Figure 2-23C shows the new object after it has been moved and a conceptual display generated.

When the **INTERFERE** command is used, AutoCAD compares the first set of solids to the second set. Any solids that are selected for both the first and second sets are automatically included as part of the first selection set and eliminated from the second. If

Figure 2-22.
The **Interference Checking** dialog box is used to check for interference between solids. To retain the interference solid, uncheck the **Delete interference objects created on Close** check box.

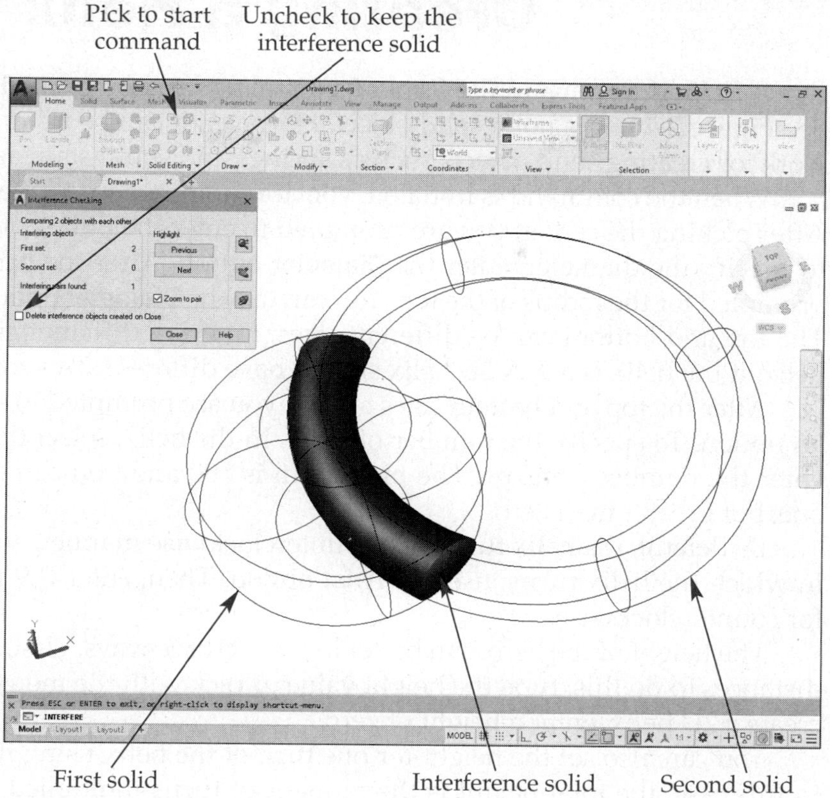

Figure 2-23.
A—Two solids form an area of intersection. B—After using **INTERFERE**, a new solid is defined (shown here in color) and the original solids remain. C—The new solid can be moved or copied.

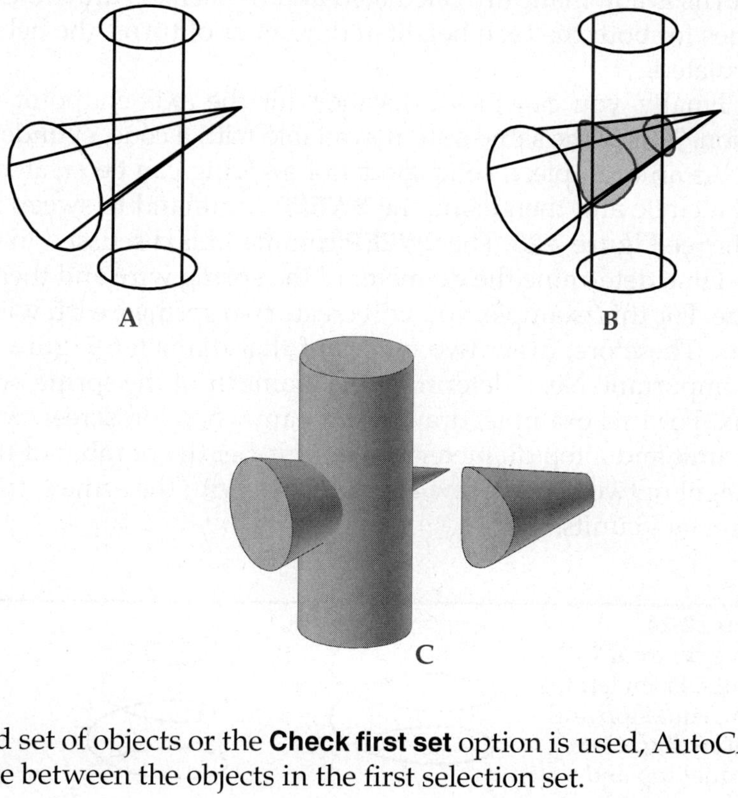

you do not select a second set of objects or the **Check first set** option is used, AutoCAD calculates the interference between the objects in the first selection set.

Exercise 2-4

www.g-wlearning.com/CAD/

Complete the exercise on the companion website.

Creating a Helix

HELIX

Ribbon
Home
> Draw

Helix

Type
HELIX

A *helix* is a spline in the form of a spiral and can be created as a 2D or 3D object. See **Figure 2-24**. It is not a solid object. However, it can be used as the path or framework for creating solid objects such as springs and spiral staircases.

When the command is initiated, you are prompted for the center of the helix base. After picking the center, you are prompted to enter the radius of the base. If you want to specify the diameter, enter the **Diameter** option. After the base is defined, you are prompted for the radius of the top. You can use the **Diameter** option to enter a diameter. The top and bottom can be different sizes. Entering different sizes creates a tapered helix, if the helix is 3D. A 2D helix should have different sizes for the top and bottom.

After the top and bottom sizes are set, you are prompted to set the height or select an option. To specify the number of turns in the helix, select the **Turns** option. Then, enter the number of turns. The maximum is 500 and you can enter values less than one, but greater than zero.

By default, the helix turns in a counterclockwise manner. To change the direction in which the helix turns, use the **Twist** option. Then, enter CW for clockwise or CCW for counterclockwise.

The height of the helix can be set in one of three ways. First, you can enter a direct distance. To do this, type the height value or pick with the mouse to set the height. To create a 2D helix, enter a height of zero.

You can also set the height for one turn of the helix using the **Turn height** option. In this case, the total height is the number of turns multiplied by the turn height. If you provide a value for the turn height and then specify the helix height, the number of turns is automatically calculated and the helix is drawn. Conversely, if you provide values for both the turn height and number of turns, the helix height is automatically calculated.

Finally, you can pick a location for the axis endpoint using the **Axis endpoint** option. This is the same option available with a cone, cylinder, or pyramid.

As an example, a solid model of a spring can be created by constructing a helix and a circle and then using the **SWEEP** command to sweep the circle along the helix path. See **Figure 2-25**. The **SWEEP** command is discussed in detail in Chapter 7.

First, determine the diameter of the spring wire and then draw a circle using that value. For this example, you will create two springs, each with a wire diameter of .125 units. Therefore, draw two circles of that diameter, **Figure 2-26**. Their locations are not important. Next, determine the diameter of the spring and draw a corresponding helix. For this example, draw a helix anywhere on screen with a bottom diameter of one unit and a top diameter of one unit. Set the number of turns to eight and specify a height of two units. Draw another helix with the same settings, except make the top diameter .5 units.

Figure 2-24.
Three types of helices. From left to right, equal top and bottom diameters, unequal top and bottom diameters, and unequal top and bottom diameters with the height set to zero.

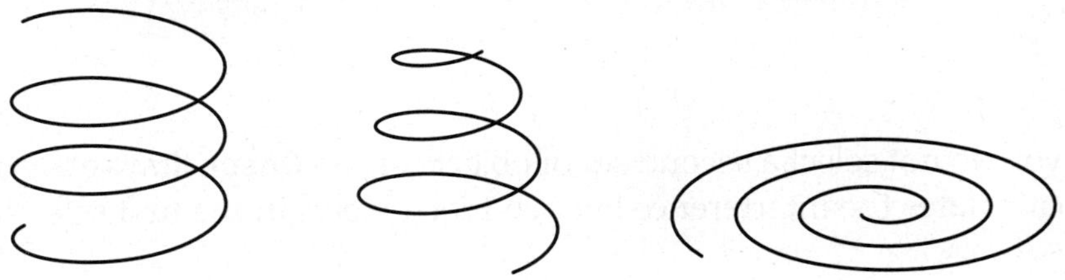

Figure 2-25.
A helix can be used as a path to create a spring.

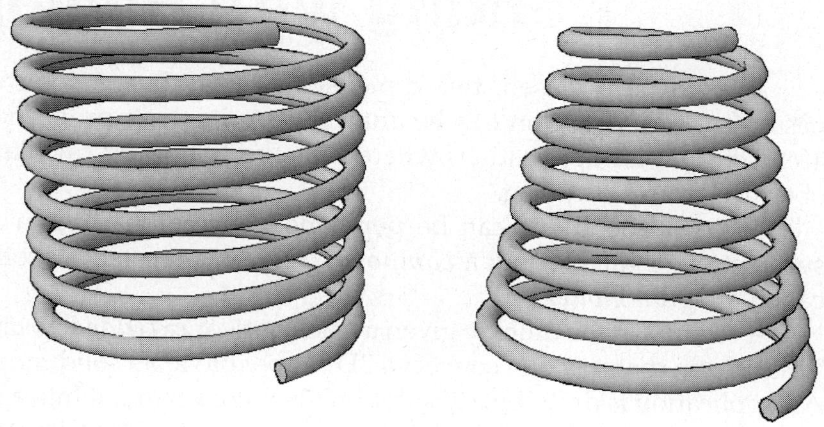

Figure 2-26.
To create a spring, first draw a circle the same diameter as the spring wire. Then, draw the helix and sweep the circle along the helix. Shown here are the two helices used to create the springs in Figure 2-25.

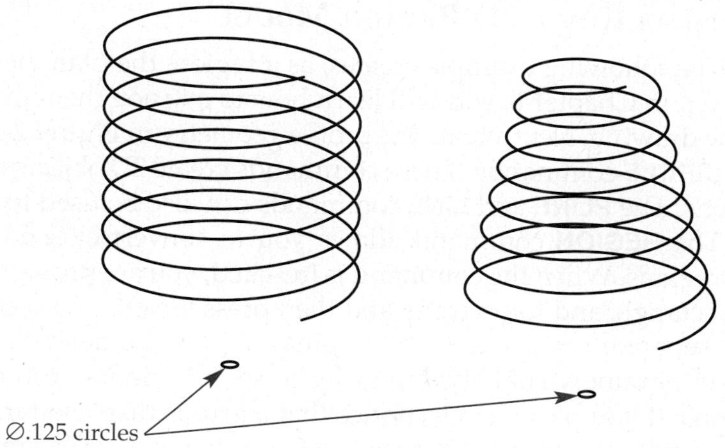

Initiate the **SWEEP** command. You are first prompted to select the objects to sweep; pick one circle and press [Enter]. Next, you are prompted to select the sweep path. Select one of the helices. The first sweep, or spring, is completed. Repeat the procedure for the other circle and helix. The drawing is now composed of two swept solids. The two helices and the two circles are consumed by the **SWEEP** command.

Ribbon
Solid
> Solid
Sweep
Type
SWEEP

SWEEP

NOTE

The **DELOBJ** system variable determines whether profile and path objects used to create 3D models are retained or deleted from the drawing. The default setting is 3. At this setting, all profile-defining and path-defining geometry is deleted. To retain the original objects used to create the model, set the **DELOBJ** system variable to 0.

Exercise 2-5

www.g-wlearning.com/CAD/

Complete the exercise on the companion website.

Working with Regions

A ***region*** is a closed, two-dimensional solid. It is a solid model without thickness (Z value). A region can be analyzed for its mass properties. Therefore, regions are useful for 2D applications where area and boundary calculations must be quickly obtained from a drawing.

Boolean operations can be performed on regions. When regions are unioned, subtracted, or intersected, a ***composite region*** is created. A composite region is also called a ***region model***.

A region can be quickly given a thickness, or ***extruded***, to create a 3D solid object. This means that you can convert a 2D shape into a 3D solid model in just a few steps. An application is drawing a 2D section view, converting it into a region, and extruding the region into a 3D solid model. Extruding is covered in Chapter 6.

Constructing a 2D Region Model

The following example creates, as a region, the plan view of a base for a support bracket. In Chapter 6, you will learn how to extrude the region into a solid. First, start a new drawing. Next, create the profile geometry in Figure 2-27 using the **RECTANGLE** and **CIRCLE** commands. These commands create 2D objects that can be converted into regions. The **PLINE** and **LINE** commands can also be used to create closed 2D objects.

REGION
Ribbon
Home
> Draw
Region
Type
REGION
REG

The **REGION** command allows you to convert closed, two-dimensional objects into regions. When the command is initiated, you are prompted to select objects. Select the rectangle and four circles and then press [Enter]. The rectangle and each circle are now separate regions and the original objects are deleted. You may need to switch to a wireframe visual style in order to see the circles. You can individually pick the regions. If you pick a circle, notice that a grip is displayed in the center, but not at the four quadrants. This is because the object is not a circle anymore. However, you can still snap to the quadrants.

In order to create the proper solid, the circular regions must be subtracted from the rectangular region. Using the **SUBTRACT** command, select the rectangle as the object to be subtracted *from*, and then select all of the circles as the objects to subtract. Now, if you select the rectangle or any of the circles, you can see that a single region has been created from the five separate regions. If you set the Conceptual or Realistic visual style current, you can see that the circles are now holes in the region. See Figure 2-28. Solids created in this manner can be given a thickness using either the **EXTRUDE** or **PRESSPULL** command.

Calculating the Area of a Region

A region is not a polyline. It is an enclosed area called a ***loop***. Certain properties of the region, such as area, are stored as a value of the region. The **MEASUREGEOM** command can be used to determine the length of all sides and the area of the loop. This can be a useful advantage of a region.

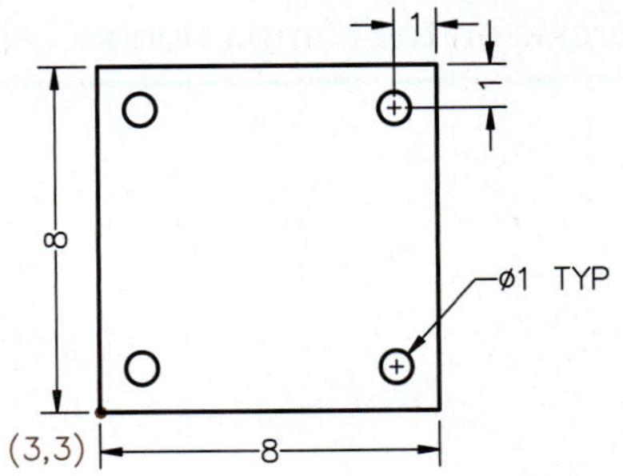

Figure 2-27.
These 2D objects can be converted to regions in order to create a region model. The region model can then be made into a 3D solid.

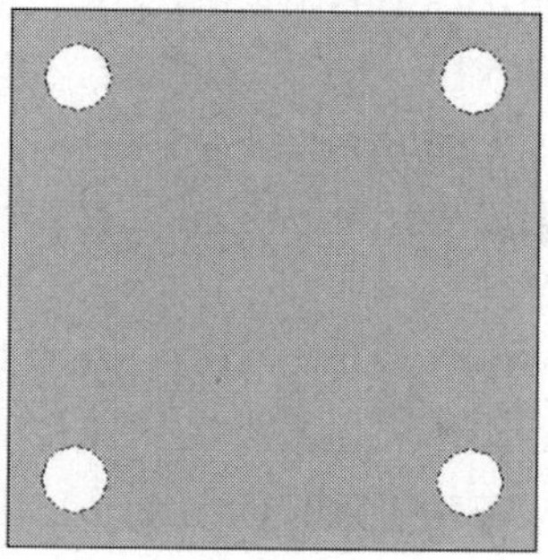

Figure 2-28.
Once the circular regions are subtracted from the rectangular region, they appear as holes. This is clear when a shaded visual style is set current.

For example, suppose a parking lot is being repaved. You need to calculate the surface area of the parking lot to determine the amount of material needed. This total surface area excludes the space taken up by planting dividers, sidewalks, and lampposts because you will not be paving under these items. If the parking lot and all objects inside of it are drawn as a region model, the **MEASUREGEOM** command can give you this figure in one step using the **Object** option of the **Area** option. When the **Area** option is used, the calculated area is highlighted light green. Depending on the number of smaller regions inside the large region, the highlighting may appear incorrect, but the resulting calculation is correct. If a polyline is used to draw the parking lot, all internal features must be subtracted when the **MEASUREGEOM** command is used.

PROFESSIONAL TIP

Regions can prove valuable when working with many items:
- Roof areas excluding chimneys, vents, and fans.
- Bodies of water, such as lakes, excluding islands.
- Lawns and areas of grass excluding flower beds, trees, and shrubs.
- Landscaping areas excluding lawns, sidewalks, and parking lots.
- Concrete surfaces, such as sidewalks, excluding openings for landscaping, drains, and utility covers.

You can find many other applications for regions that can help in your daily tasks.

Exercise 2-6

www.g-wlearning.com/CAD/

Complete the exercise on the companion website.

Chapter Review

Answer the following questions using the information in this chapter.

1. What is a *solid primitive*?
2. How is a solid cube created?
3. How is an elliptical cylinder created?
4. Where is the center of a wedge located?
5. What is a *frustum pyramid*?
6. What is a *polysolid*?
7. Name at least four AutoCAD 2D entities that can be converted to a polysolid.

8. What type of entity does the **HELIX** command create and how can it be used to create a solid model?
9. What is a *composite solid*?
10. Which types of mathematical calculations are used in the construction of composite solids?
11. How are two or more solids combined to make a composite solid?
12. What is the function of the **INTERSECT** command?
13. How does the **INTERFERE** command differ from **INTERSECT** and **UNION**?
14. What is a *region*?
15. How can a 2D section view be converted to a 3D solid model?

Drawing Problems

Draw the objects in the following problems using the appropriate solid primitive commands and Boolean operations. When necessary, determine thickness dimensions for parts or use dimensions specified by your instructor. If appropriate, apply geometric constraints to the base 2D drawing. Do not add dimensions to the models. Save the drawings as P2-*(problem number). Display and plot the problems as indicated by your instructor.*

1. General

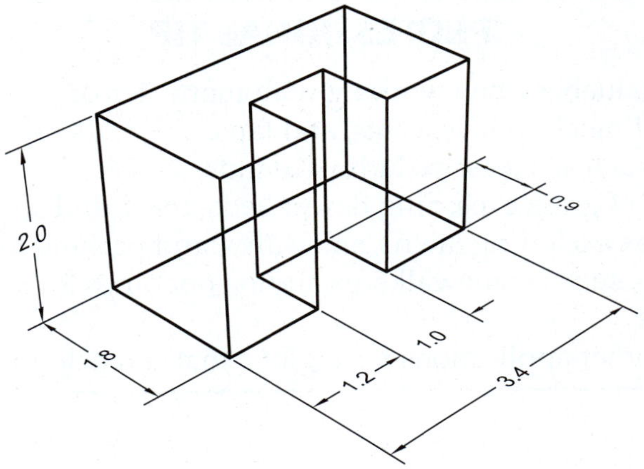

2. General

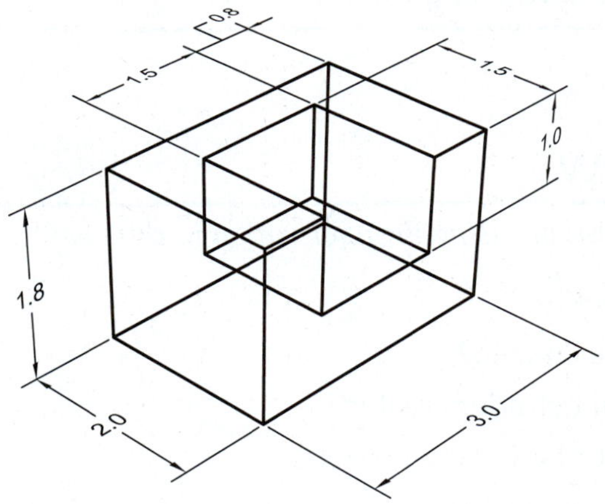

Drawing Problems - Chapter 2

3. **General**

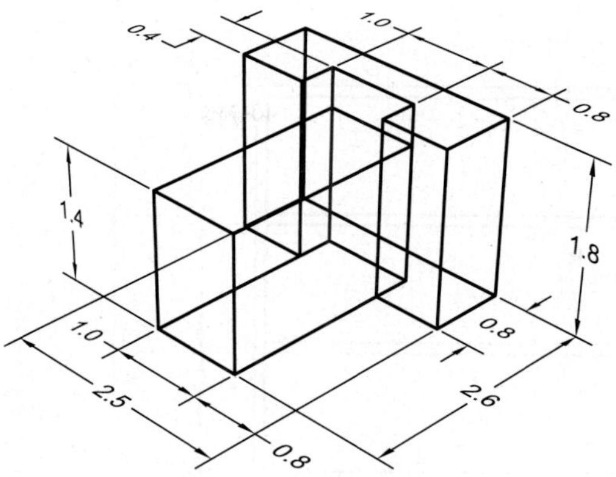

4. **Mechanical**

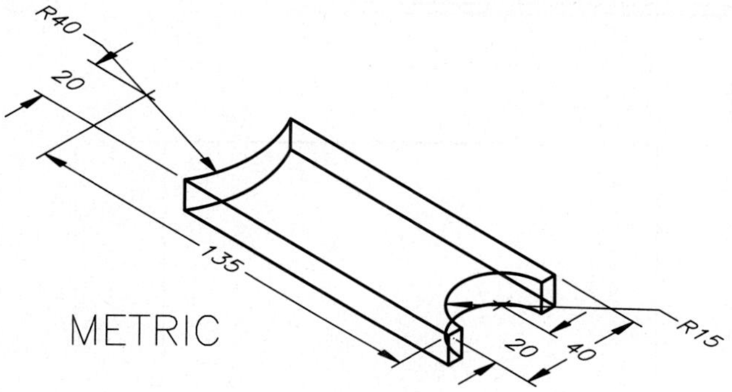

5. **Mechanical**

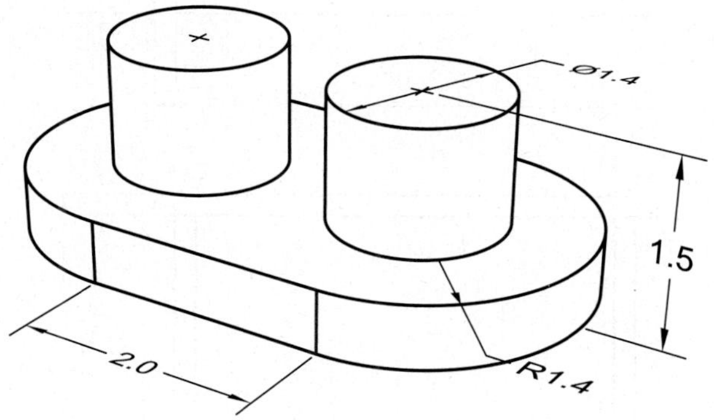

6. **Mechanical**

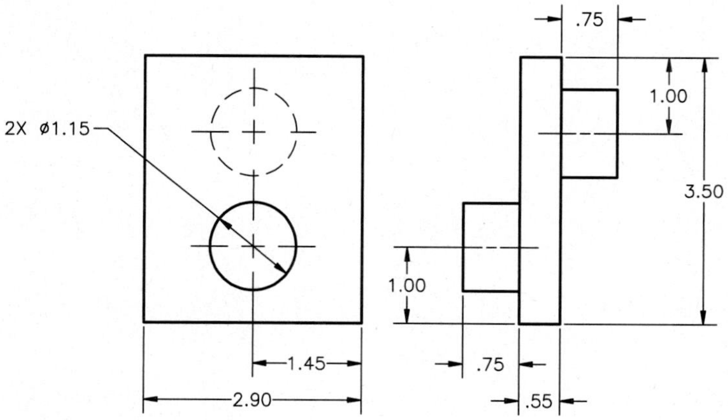

7. **Architectural**

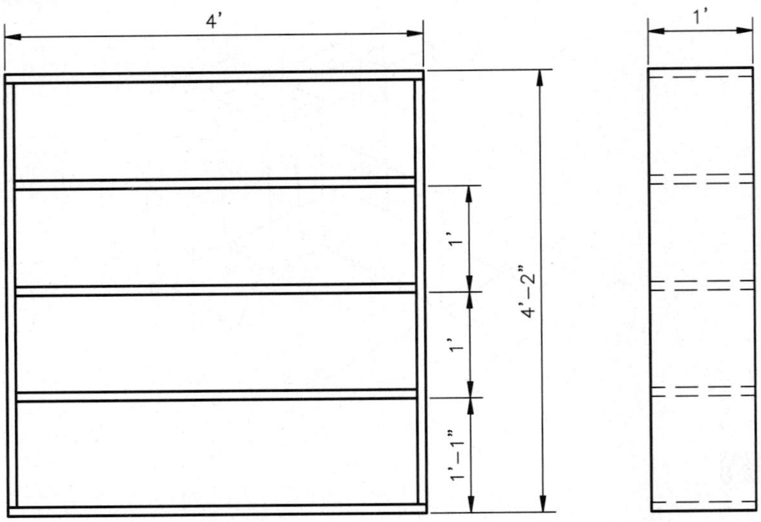

8. **Architectural**

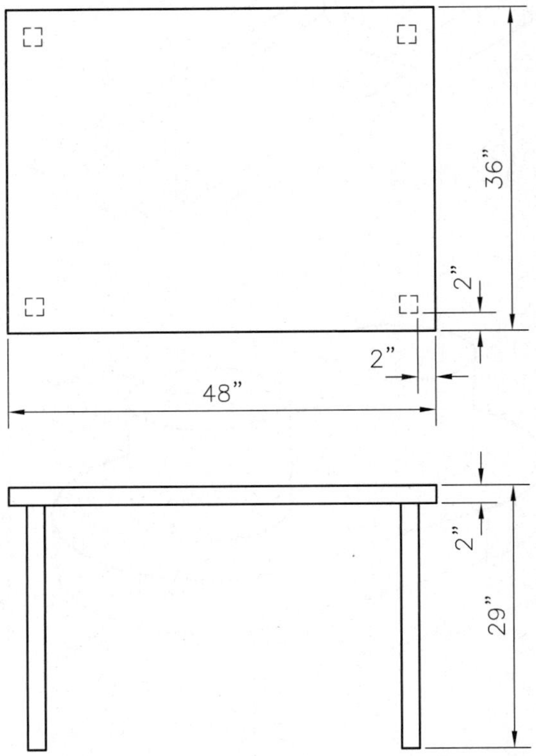

9. **Architectural**

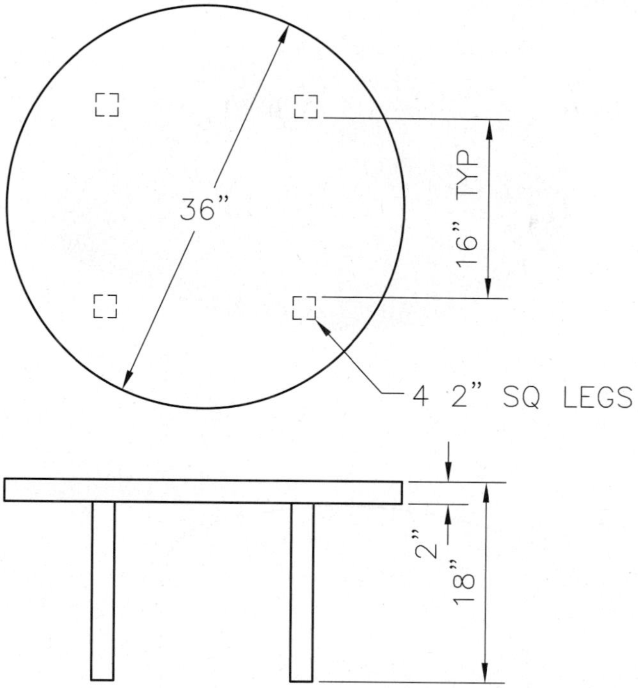

10. **Architectural**

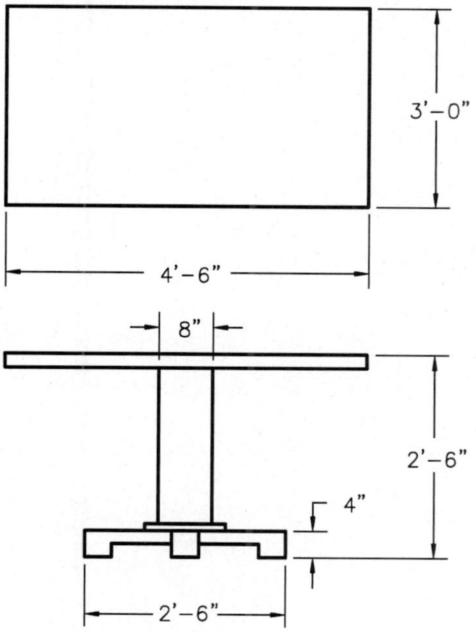

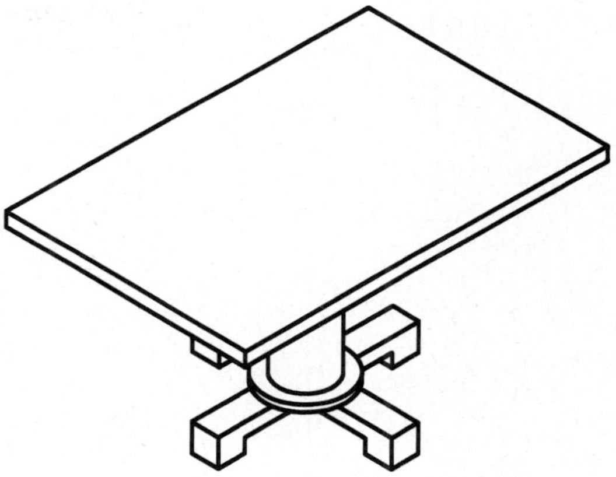

11. **Architectural** Draw the seat for a kitchen chair shown below. In later chapters, you will complete this chair. The seat is 1″ thick.

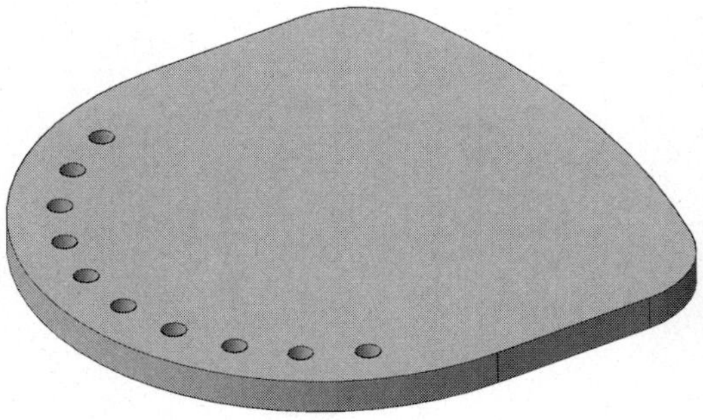

A

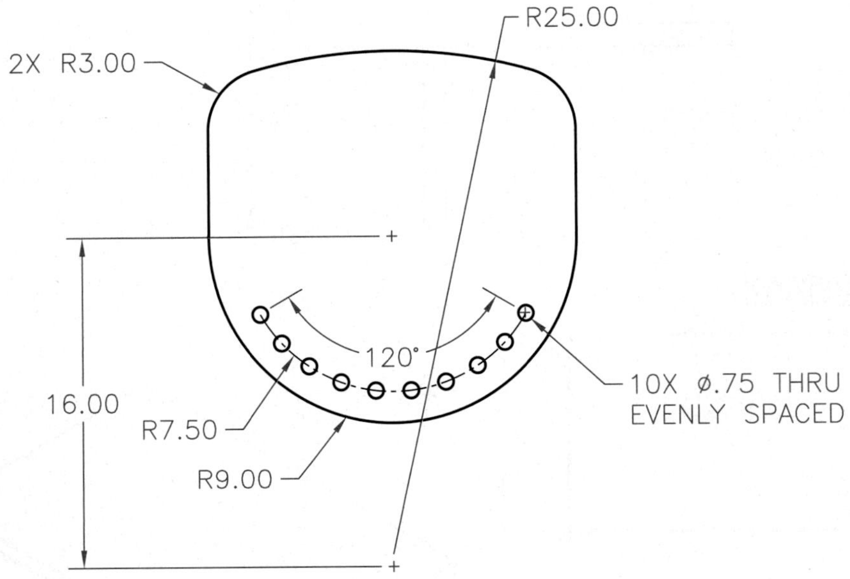

B

CHAPTER 3

Viewing and Displaying Three-Dimensional Models

Learning Objectives

After completing this chapter, you will be able to:

✓ Use the viewport controls to display views and control the display of view-navigation tools.
✓ Use the navigation bar to perform a variety of display manipulation functions.
✓ Create and save named views.
✓ Use the view cube to rotate the view of the model dynamically in 3D space.
✓ Use the view cube to display orthographic plan views of all sides on the model.
✓ Use steering wheels to display a 3D model from any angle.

AutoCAD provides several tools with which you can display and present 3D models in pictorial and orthographic views:

- The viewport controls, discussed in Chapter 1.
- Preset isometric viewpoints, discussed in Chapter 1.
- The view cube. This on-screen tool provides access to preset and dynamic display options.
- Steering wheels.
- The **3DORBIT**, **3DFORBIT**, and **3DCORBIT** commands. These commands provide dynamic display and continuous orbiting functions for demonstrations and presentations.

Once a viewpoint has been selected, you can enhance the display by applying visual styles. The **View** panel in the **Home** tab of the ribbon provides a variety of ways to display a model, including wireframe representation, hidden line removal, and simple shading. An introduction to visual styles is provided in Chapter 1 and complete coverage is provided in Chapter 16.

A more advanced representation can be produced by creating a rendering with the **RENDER** command. A rendering produces the most realistic image with highlights, shading, and materials, if applied. **Figure 3-1** shows a 3D model of a cast iron plumbing cleanout after using **HIDE**, setting the Conceptual visual style current, and using **RENDER**. Notice the difference in the three displays.

Figure 3-1.
A—Hidden display (hidden lines removed). B—The **Conceptual** visual style set current.
C—Rendered with lights and materials.

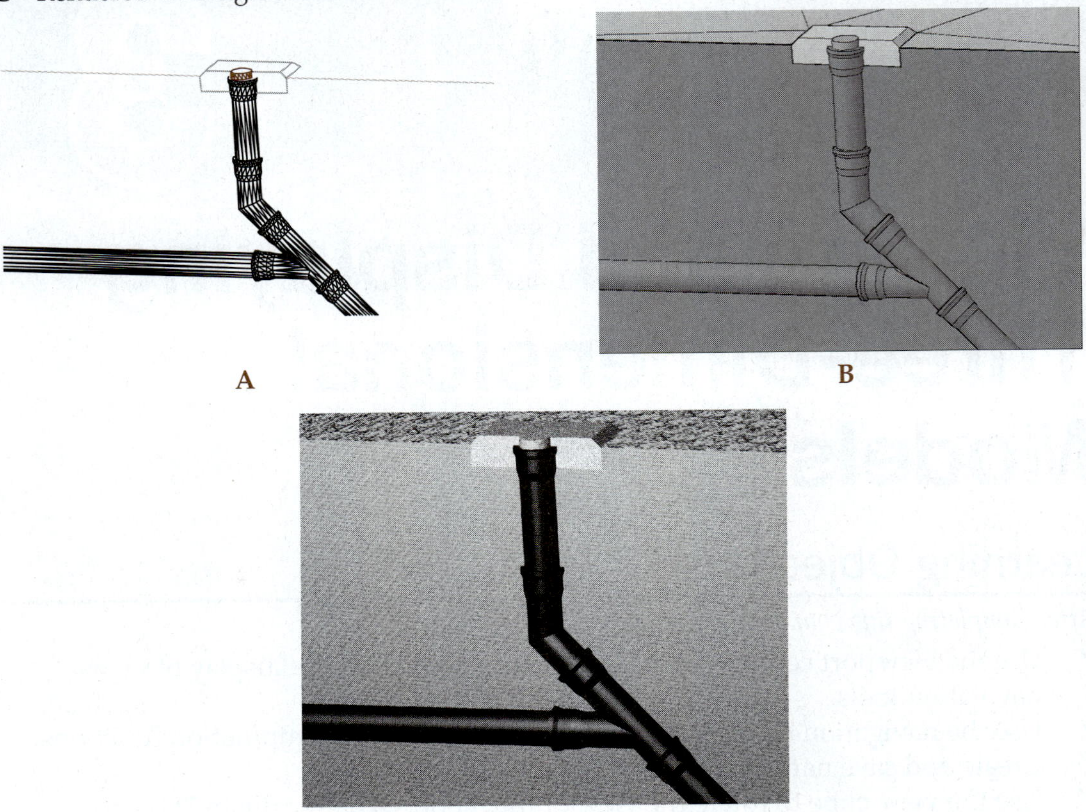

A B C

PROFESSIONAL TIP

The viewport controls, displayed in the upper-left corner of the AutoCAD drawing window, are discussed in Chapter 1. These controls provide quick access to preset isometric and orthographic views and settings for the view cube, navigation bar, and steering wheels. Use the viewport controls whenever possible to increase your drawing efficiency.

Keep in mind that undocked palettes may hide the viewport controls. If this happens, either dock the palette or move it out of the way.

Using the Navigation Bar to Display Models

By default, the navigation bar appears below the view cube on the right side of the screen. See **Figure 3-2**. It allows you to use the navigation tools described in this chapter, as well as the **ZOOM** and **PAN** commands and three orbit options. The **NAVBAR** command is used to turn on or off the display of the navigation bar. The navigation bar can be controlled using the viewport controls by selecting **Navigation Bar** in the **Viewport Controls** flyout. You can also select the **Navigation Bar** button in the **Viewport Tools** panel of the **View** ribbon tab. The appearance and location of the navigation bar can be customized, as discussed later in this chapter.

Figure 3-2.
The navigation bar appears below the view cube by default. This bar allows you to use the navigation tools quickly for drawing-display purposes.

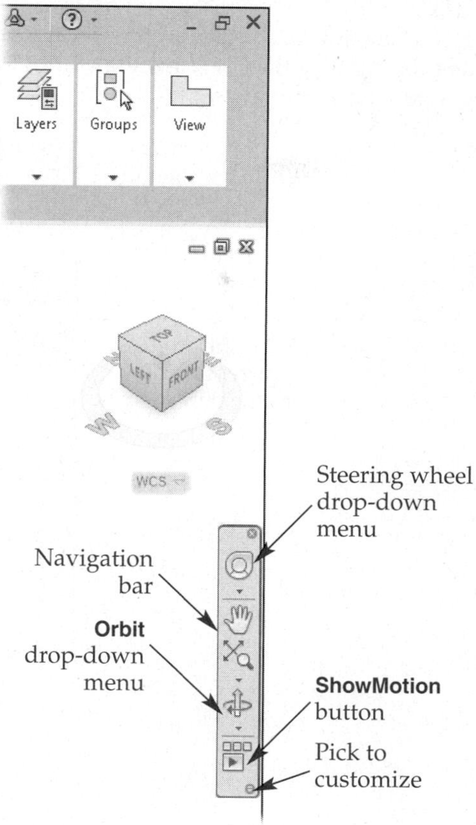

Creating and Displaying Named Views

A ***named view*** is a single-frame display of a model from a given viewpoint. Named views are easy to create and have a wide variety of uses in AutoCAD. The **New View/Shot Properties** dialog box is used for creating named views. To access this dialog box, first pick the **ShowMotion** button in the navigation bar. Refer to **Figure 3-2**. When the **ShowMotion** toolbar appears at the bottom-center of the drawing window, pick the **New Shot...** button. See **Figure 3-3A**. The **New View/Shot Properties** dialog box appears, **Figure 3-3B**. The **NEWVIEW** command also provides access to the **New View/Shot Properties** dialog box.

Ribbon
View
> Viewport Tools
ShowMotion
Type
NAVSMOTION
MOTION
Toolbar
Navigation Bar
ShowMotion
NAVSMOTION

To create a named view, pick the **View Properties** tab. Enter a view name and select **Still** for the view type. Pick the **Current display** radio button to accept the model display as is. You can also pick the **Define window** radio button to return to the model and refine the display. Pick **OK** to complete the named view creation. Two images now appear above the **ShowMotion** toolbar. See **Figure 3-4A**. The large image is the view category and displays the default name of <None>. Above it is the thumbnail of the view just created with its name shown below. Regardless of the view that is displayed on the screen, you can always display a named view by picking the **Play** button in the view thumbnail. See **Figure 3-4B**.

Ribbon
Visualize
> Named Views
View
> Named Views
New View
Type
NEWVIEW
NVIEW
NEWVIEW

The new view name will appear under the **Model Views** branch in the **View Manager** dialog box when using the **VIEW** command. Named views are saved with the drawing and can be displayed at any time by selecting from the drop-down list in the **Named Views** panel of the **Visualize** or **View** ribbon tab.

A named view can be used for any show motion view. The functions of the view category and view thumbnails are much more obvious when working with a variety of different views. A complete discussion of show motion is provided in Chapter 20.

Figure 3-3.
A—Pick the **New Shot...** button in the **ShowMotion** toolbar to display the **New View/Shot Properties** dialog box. B—The **New View/Shot Properties** dialog box provides options for creating a new named view.

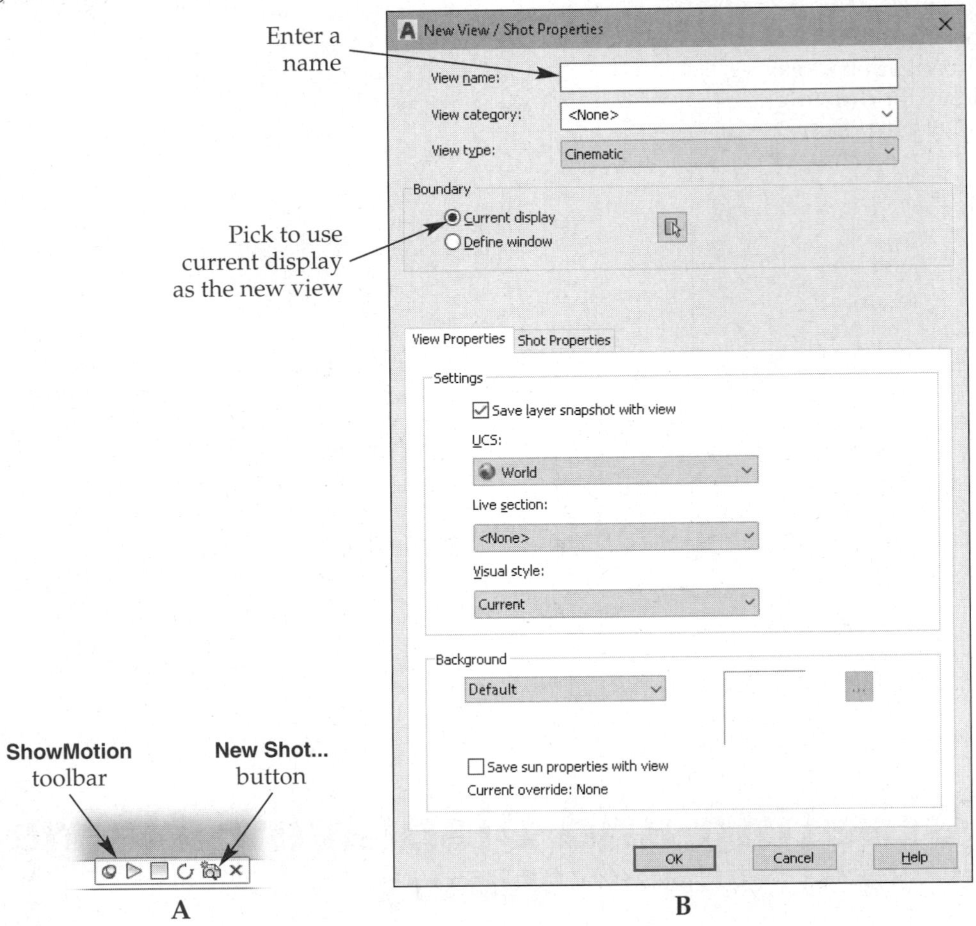

Figure 3-4.
A—Thumbnail images for the view category and named view are displayed above the **ShowMotion** toolbar. B— If the view has been changed, pick the **Play** button in the thumbnail of the desired view to redisplay the view.

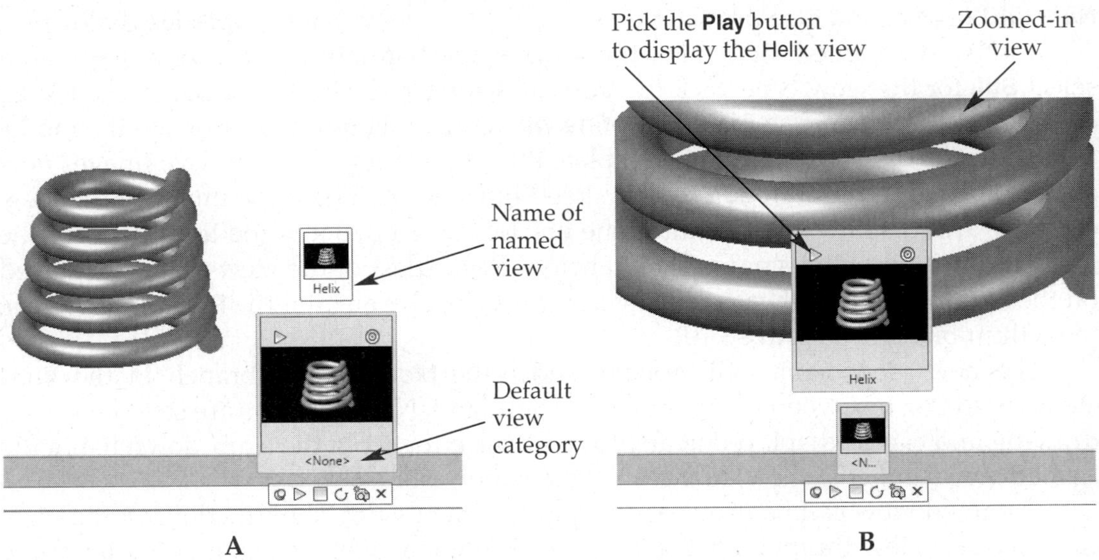

Dynamically Displaying Models with the View Cube

The *view cube* navigation tool allows you to quickly change the current view of the model to a preset pictorial or orthographic view or any number of dynamically user-defined 3D views. The tool is displayed, by default, in the upper-right corner of the drawing area. If the view cube is not displayed, select **View Cube** in the **Viewport Controls** flyout in the viewport controls. You can also pick the **View Cube** button in the **Viewport Tools** panel of the **View** ribbon tab.

Ribbon
View
> Viewport Tools
View Cube
Type
NAVVCUBE
CUBE

NAVVCUBE

The **View Cube Settings** dialog box offers many settings for expanding and enhancing the view cube. This dialog box is discussed in detail later in this chapter.

PROFESSIONAL TIP

The location of the view cube can be quickly changed by picking the customize button on the navigation bar. Then, select **Docking positions** in the shortcut menu and select the location of your choice in the cascading menu.

Understanding the View Cube

The view cube tool is composed of a cube labeled with the names of all six orthographic faces. See **Figure 3-5**. A compass labeled with the four compass points (N, S, E, and W) rests at the base of the cube. The compass can be used to change the view. It also provides a visual cue to the orientation of the model in relation to the current user coordinate system (UCS). The ***user coordinate system*** describes the orientation of the X, Y, and Z axes. A complete discussion of user coordinate systems is given in Chapter 4.

Below the cube and compass is a button labeled WCS, which displays a shortcut menu. The label on this button displays the name of the current user coordinate system, which is, by default, the world coordinate system (WCS). This shortcut menu gives you the ability to switch from one user coordinate system to another. A user coordinate system is any coordinate system that is not the world coordinate system. The benefit of creating new user coordinate systems is the increase in efficiency and productivity when working on 3D models.

When the cursor is not over the view cube, the tool is displayed in its dimmed or ***inactive state***. When the cursor is moved onto the view cube, the tool is displayed in its ***active state*** and a little house appears to the upper-left of the cube. This house is the **Home** icon. Picking this icon always restores the same view, called the ***home view***. Later in this chapter, you will learn how to change the home view.

Figure 3-5.
The cube in the view cube tool is labeled with the names of all six orthographic faces. The view cube also contains a **Home** icon, compass, UCS shortcut menu, and shortcut menu button.

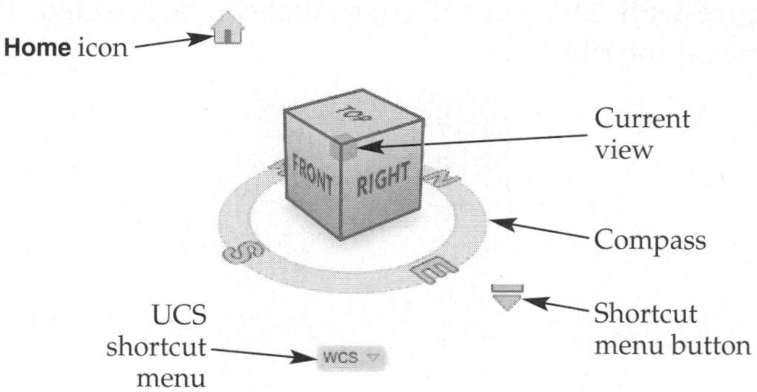

Notice that a corner, edge, or face on the cube is highlighted. In the default view of a drawing based on the acad3D.dwt template, the corner between the top, right, and front faces is shaded. The shaded part of the cube represents the standard view that is closest to the current view. Move the cursor over the cube and notice that edges, corners, and faces are highlighted as you move the pointer. If you pick a highlighted edge, corner, or face, a view is displayed that looks directly at the selected feature. By selecting standard views on the view cube, you can quickly move between pictorial and orthographic views.

PROFESSIONAL TIP

Picking the top face of the view cube displays a plan view of the current UCS XY plane. This is often quicker than using the **PLAN** command, which is discussed later in this chapter.

Dynamic Displays

The easiest way to change the current view using the view cube is to pick and drag the cube. Simply move the cursor over the cube, press and hold the left mouse button, and drag the mouse to change the view. The view cube and model view dynamically change with the mouse movement. When you have found the view you want, release the mouse button. This is similar to the **Orbit** tool in steering wheels, which are discussed later in the chapter.

NOTE

By default, the view cube does not move freely without snapping to the closest view. This action is reset by right-clicking on the view cube and picking **View Cube Settings...** in the cursor menu. In the **View Cube Settings** dialog box, uncheck **Snap to closest view** in the **When dragging on the View Cube** area.

When dragging the view cube, you are not restricted to the current XY plane, as you are when using the compass. The compass is discussed later in this chapter.

Pictorial Displays

The view cube provides immediate access to 20 different preset pictorial views. When one of the corners of the cube is selected, a standard isometric pictorial view is displayed. See **Figure 3-6A**. Eight corners can be selected. Picking one of the four corners on the top face of the cube restores one of the preset isometric views. You can select a preset isometric view using the viewport controls or the **Home** or **Visualize** tab of the ribbon. However, using the view cube may be a quicker method.

Selecting one of the edges of the cube sets the view perpendicular to that edge. See **Figure 3-6B**. There are 12 edges that can be selected. These standard views are not available on the ribbon.

Figure 3-6.
A—When one of the corners of the cube is selected, a standard isometric view is displayed.
B—Selecting one of the edges of the cube produces the same rotation in the XY plane as an isometric view, but a zero elevation view in the Z plane.

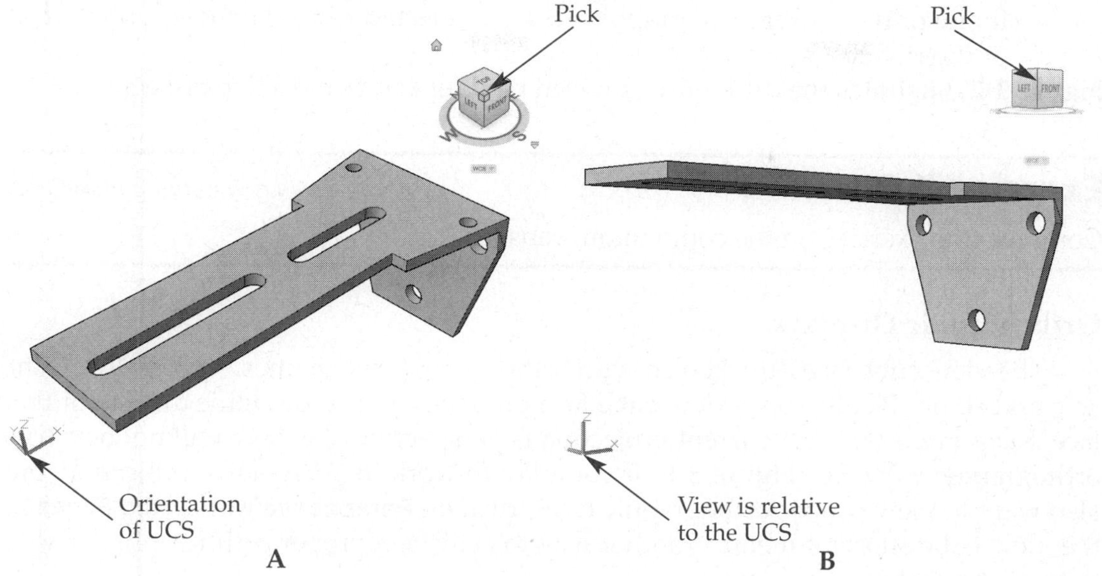

PROFESSIONAL TIP

The quickest method of dynamically rotating a 3D model is achieved by using the mouse wheel button. Simply press and hold the [Shift] key while pressing down and holding the mouse wheel. Now move the mouse in any direction and the view rotates accordingly. This is a transparent function that executes the **3DORBIT** command and can be used at any time. Since it is transparent, it can even be used while you are in the middle of a command. This is an excellent technique to use because it does not require selecting another tool or executing a command.

Projection

The view displayed in the drawing window can be in one of two projections. The ***projection*** refers to how lines are applied to the viewing plane. In a pictorial view, lines in a ***perspective projection*** appear to converge as they recede into the background. The points at which the lines converge are called ***vanishing points***. In 2D drafting, it is common to represent an object in pictorial as a one- or two-point perspective, especially in architectural drafting. In a ***parallel projection***, lines remain parallel as they recede. This is how an orthographic or axonometric (isometric, dimetric, or trimetric) view is created.

To quickly change the projection using the viewport controls, pick **Parallel** or **Perspective** from the **View Controls** flyout. To change the projection using the view cube, right-click on the view cube to display the shortcut menu, or pick the shortcut menu button at the lower-right of the view cube. Refer to **Figure 3-5**.

Three display options are given:

- **Parallel.** Displays the model as a parallel projection. This creates an orthographic or axonometric view.
- **Perspective.** Displays the model in the more realistic, perspective projection. Lines recede into the background toward invisible vanishing points.

- **Perspective with Ortho Faces.** Displays the model in perspective projection when a pictorial view is displayed and parallel projection when an orthographic view is displayed. The parallel projection is only set current if a face on the view cube is selected to display the orthographic view. It is not set current if one of the preset orthographic views is selected from the viewport controls or the ribbon.

Figure 3-7 illustrates the difference between parallel and perspective projection.

Exercise 3-1

www.g-wlearning.com/CAD/

Complete the exercise on the companion website.

Orthographic Displays

The view cube faces are labeled with orthographic view names, such as Top, Front, Left, and so on. Picking on a view cube face produces an orthographic display of that face. Keep in mind, if the current projection is perspective, the view will not be a true orthographic view. See **Figure 3-8**. If you plan to work in perspective projection, but also want to view proper orthographic faces, turn on **Perspective with Ortho Faces** in the view cube shortcut menu. Another way to achieve a proper orthographic view is to turn on parallel projection.

When a face is selected on the view cube, the cube rotates to orthographically display the named face. The view rotates accordingly. In addition, notice that a series of triangles point to the four sides of the cube (when the cursor is over the tool). See **Figure 3-9A**. Picking one of these triangles displays the orthographic view corresponding to the face to which the triangle is pointing, **Figure 3-9B**. This is a quick and efficient method to precisely rotate the display between orthographic views.

When an orthographic view is displayed, two ***roll arrows*** appear to the upper-right of the view cube. Picking either of these arrows rotates the current view 90° in the selected direction and about an axis perpendicular to the view. See **Figure 3-9C**. Using the triangles and roll arrows on the view cube provides the greatest flexibility in manipulating the model between orthographic views.

Figure 3-7.
In a parallel projection, parallel lines remain parallel. In a perspective projection, parallel lines converge to a vanishing point. Notice the three receding lines on the box in the perspective projection. If these lines are extended, they will intersect.

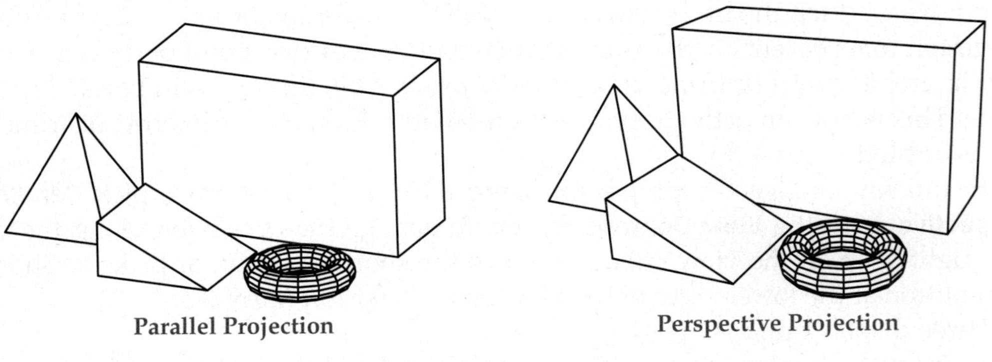

Figure 3-8.
A—To properly view orthographic faces using the view cube, turn on **Perspective with Ortho Faces** in the shortcut menu. B—When an orthographic view is set current with perspective projection on, the view is not a true orthographic view. Notice how you can see the receding surfaces.

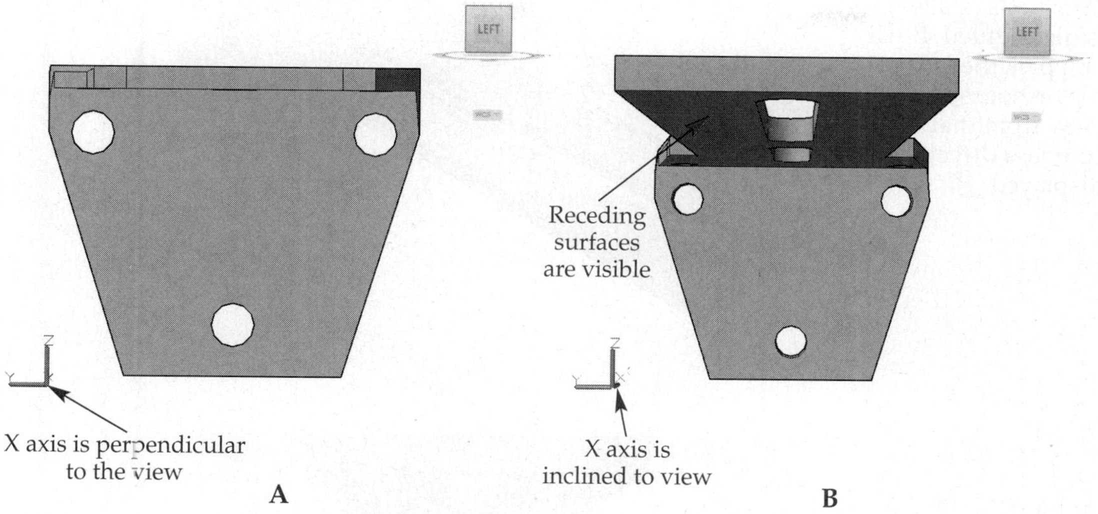

Figure 3-9.
A—The selected orthographic face is surrounded by triangles. Pick one of the triangles to display that orthographic face. B—The orthographic face corresponding to the picked triangle is displayed. Picking a roll arrow rotates the current view 90° in the selected direction. C—The rotated view.

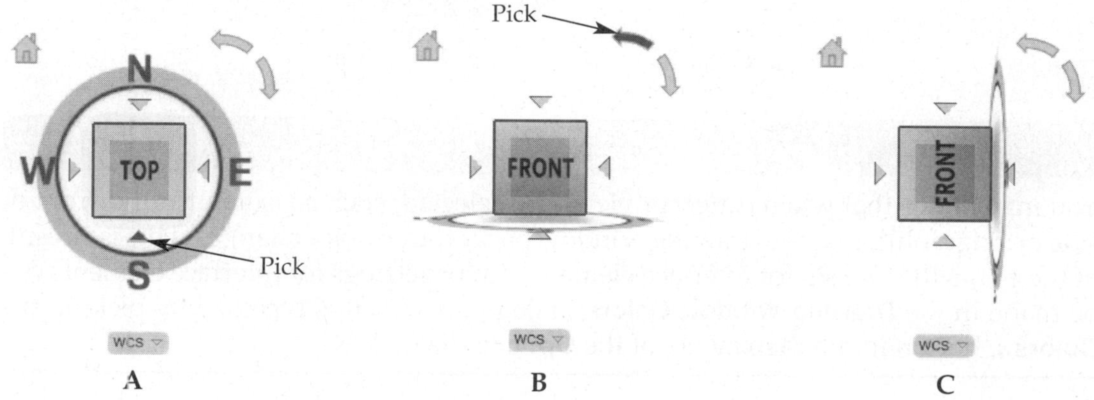

Setting Views with the Compass

The compass allows you to dynamically rotate the model in the XY plane. To do so, pick and hold on one of the four labels (N, S, E, or W). Then, drag the mouse to rotate the view. The view pivots about the Z axis of the current UCS. Try this a few times and notice that you can completely rotate the model by continuously moving the cursor off the screen. For example, pick the letter W and move the pointer either right or left (or up or down). Notice as you continue to move the mouse in one direction, the model continues to rotate in that direction.

You can use the compass to display the model in a view plan to the right, left, front, or back face of the cube. When the pointer is moved to one of the four compass directions, the letter is highlighted, **Figure 3-10A**. Simply single pick on the compass label that is next to the face you wish to view. See **Figure 3-10B**.

Figure 3-10.
A—When the pointer is moved over one of the four compass directions, the letter is highlighted. B—If you pick the letter, the orthographic view from that compass direction is displayed.

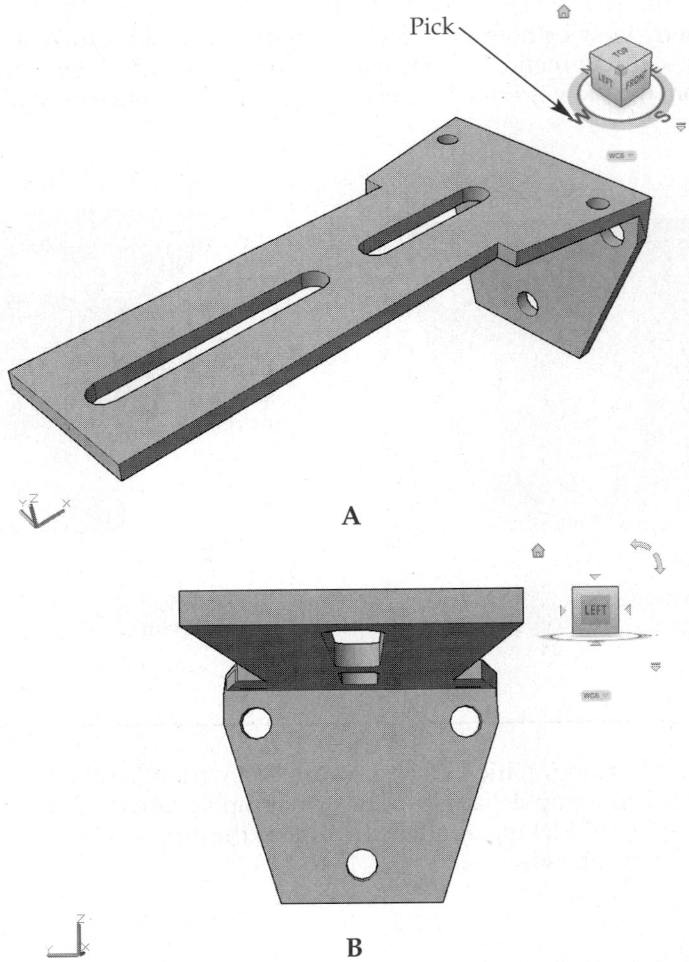

NOTE

You may notice that when different views are selected, such as isometric and front or side orthographic, that the drawing window background color changes. This is a result of the colors that are set for interface elements. Color settings for interface elements can be made in the **Drawing Window Colors** dialog box, which is accessed by picking the **Colors...** button in the **Display** tab of the **Options** dialog box.

Home View

The default home view is the southwest isometric view of the WCS. Picking the **Home** icon in the view cube always displays the view defined as the home view, regardless of the current UCS. You can easily set the home view to any display you wish. First, using any navigation method, display the model as required. Next, right-click on the view cube and pick **Set Current View as Home** from the shortcut menu. Now, when you pick the **Home** icon in the view cube, this view is set current. Remember, the **Home** icon does not appear until the cursor is over the view cube.

PROFESSIONAL TIP

Should you become disoriented after repeated use of the view cube, it is far more efficient to pick the **Home** icon than it is to use **UNDO** or try to select an appropriate location on the view cube.

UCS Settings

The UCS shortcut menu in the view cube lists the WCS and the names of all named UCSs in the current drawing. If there are no named UCSs, the listing is **WCS** and **New UCS**. See **Figure 3-11A**. To create a new UCS, pick **New UCS** from the shortcut menu. Next, use the appropriate **UCS** command options to create the new UCS. See Chapter 4 for complete coverage of the **UCS** command.

As new UCSs are created and saved, their names are added to the UCS shortcut menu. See **Figure 3-11B**. Now, if you wish to work on the model using a specific UCS, simply select it from the list. The UCS is then restored.

Exercise 3-2 www.g-wlearning.com/CAD/

Complete the exercise on the companion website.

View Cube Settings Dialog Box

The appearance and function of the view cube can be changed using the **View Cube Settings** dialog box, **Figure 3-12**. This dialog box is displayed by selecting **View Cube Settings...** from the view cube shortcut menu or by using the **Settings** option of the **NAVVCUBE** command. The next sections discuss the options found in the dialog box.

Type
NAVVCUBE
CUBE

NAVVCUBE

Figure 3-11.
The UCS shortcut menu below the view cube lists all named UCSs in the current drawing. A—The menu entries are **WCS** and **New UCS**. In this example, there are no saved UCSs in the drawing. B—The names of new UCSs are added to the UCS shortcut menu. The current UCS is indicated with a check mark.

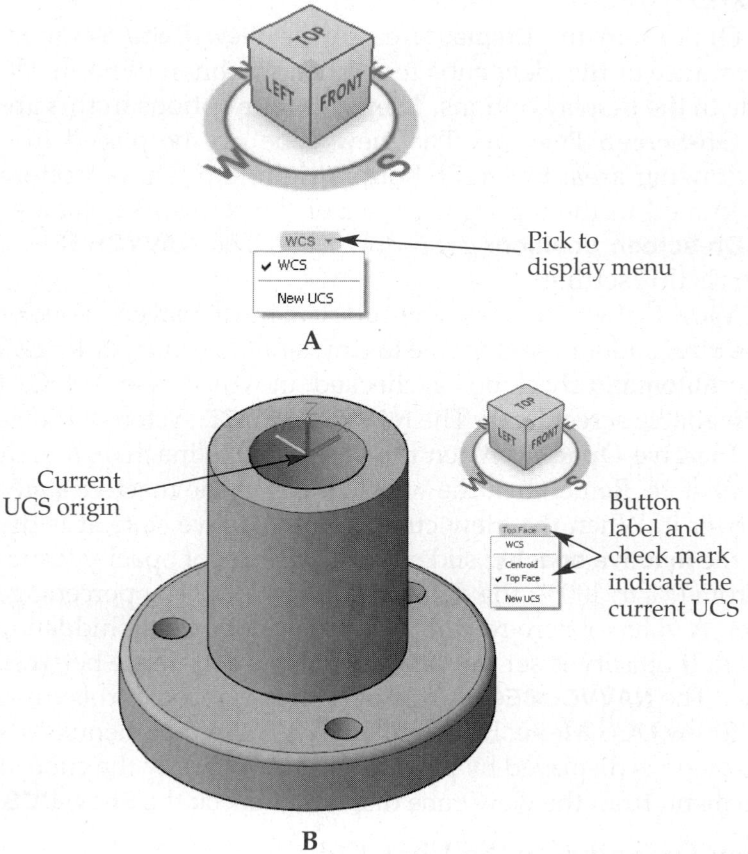

Figure 3-12.
The **View Cube Settings** dialog box.

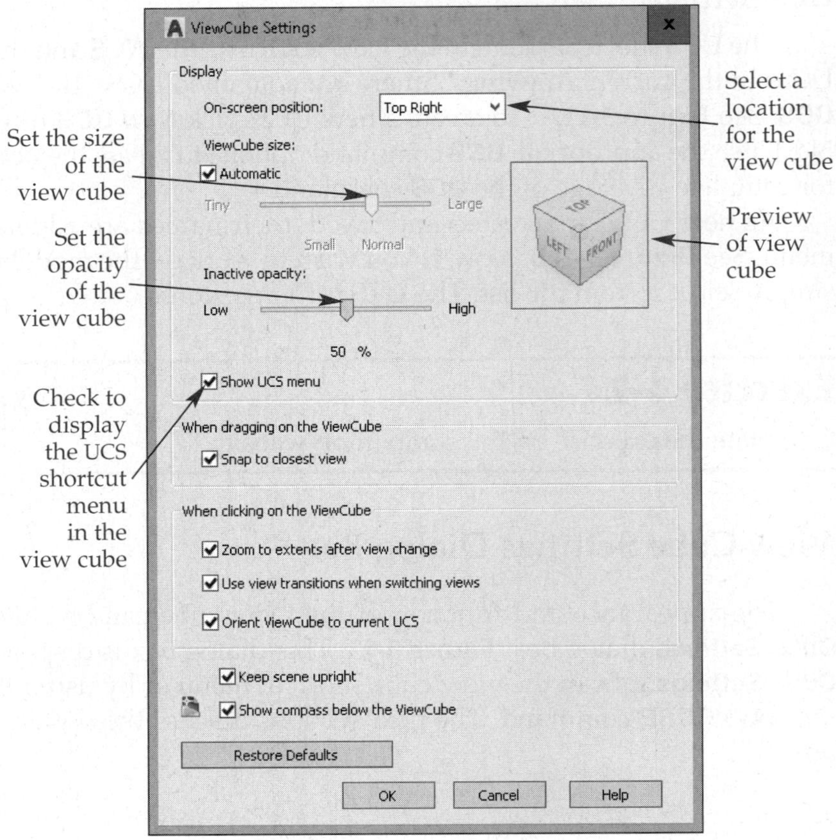

Display

Options in the **Display** area of the **View Cube Settings** dialog box control the appearance of the view cube tool. The thumbnail dynamically previews any changes made to the display options. There are four options in this area of the dialog box.

On-Screen Position. The view cube can be placed in one of four locations in the drawing area: top-right, bottom-right, top-left, or bottom-left corner. By default, it is located in the top-right corner of the screen. To change the location, select it in the **On-screen position:** drop-down list. The **NAVVCUBELOCATION** system variable controls this setting.

View Cube Size. The view cube can be displayed in one of four sizes. Use the **View Cube size:** slider to set the size to tiny, small, normal, or large. The slider is unavailable if the **Automatic** check box is checked, in which case AutoCAD sets the size based on the available screen area. The **NAVVCUBESIZE** system variable controls this setting.

Inactive Opacity. When the view cube is inactive, it is displayed in a semitransparent state. Remember, the view cube is in the inactive state whenever the pointer is not over it. When the view cube is in the active state, it is displayed at 100% opacity. Use the **Inactive opacity:** slider to set the level of opacity (transparency). The value can be from 0% to 100%; the default value is 50%. The percentage is displayed below the slider. A value of zero results in the view cube being hidden until the cursor is moved over it. If opacity is set to 100%, there is no difference between the inactive and active states. The **NAVVCUBEOPACITY** system variable controls this setting.

Show UCS Menu. By default, the UCS shortcut menu is displayed in the view cube. This menu is displayed by picking the button below the cube. If you wish to remove the UCS menu from the view cube display, uncheck the **Show UCS menu** check box.

When Dragging on the View Cube

By default, when you drag the view cube, the view "snaps" to the closest standard view that can be displayed by the view cube. This is because the **Snap to closest view** check box is checked by default. Uncheck this check box if you want the view to rotate freely without snapping to a preset standard view as you drag the view cube.

When Clicking on the View Cube

Options in the **When clicking on the View Cube** area control how the final view is displayed and how the labels on the view cube can be related to the UCS. There are three options in this area.

Zoom to Extents After View Change. If the **Zoom to extents after view change** check box is checked, the model is zoomed to the extents of the drawing whenever the view cube is used to change the view. Uncheck this if you want to use the view cube to change the display without fitting the model to the current viewport.

Use View Transitions When Switching Views. The default transition from one view to the next is a smooth rotation of the view. Although this transition may look nice, if you are working on a large model it may require more time and computer resources than you are willing to use. Therefore, if you are switching views a lot using the view cube, it may be more efficient to uncheck the **Use view transitions when switching views** check box. When this is unchecked, a view change just cuts to the new view without a smooth transition. This option does not affect the view when dragging the view cube or its compass.

Orient View Cube to Current UCS. As you have seen, the view cube is aligned to the current UCS by default. In other words, the top face of the cube is always perpendicular to the Z axis of the UCS. However, this can be turned off by unchecking the **Orient View Cube to current UCS** check box. The **NAVVCUBEORIENT** system variable controls this setting. When unchecked, the faces of the view cube are not reoriented when the UCS is changed. It may be easier to visualize view changes if the view cube faces are oriented to the current UCS.

When the **Orient View Cube to current UCS** check box is unchecked, WCS is displayed above the UCS shortcut menu in the view cube (unless the WCS is current). See **Figure 3-13**. This is a reminder that the labels on the view cube relate to the WCS and not to the current UCS.

Additional Options

Three additional options are available near the bottom of the **View Cube Settings** dialog box. These options are located below the **When clicking on the View Cube** area.

Keep Scene Upright. If the **Keep scene upright** check box is checked, the view of the model cannot be turned upside down when selecting a face, edge, or corner of the view cube. When unchecked, the view may be rotated so it is upside down. This can be confusing, so it is best to leave this box checked.

Show Compass below the View Cube. The compass is displayed by default in the view cube. However, if you do not find the compass useful, you can turn it off. To hide the compass, uncheck the **Show compass below the View Cube** check box.

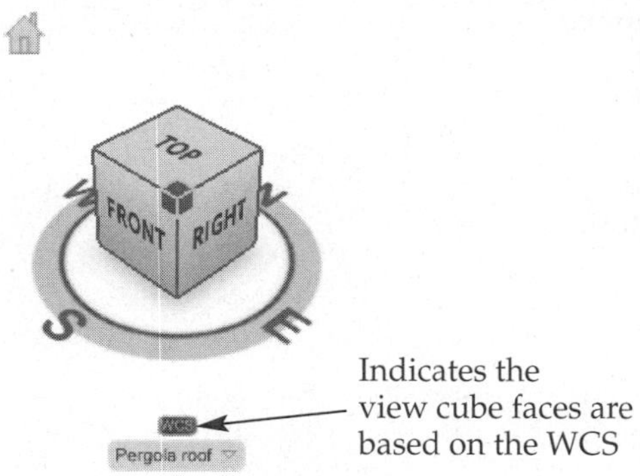

Figure 3-13. When the **Orient View Cube to current UCS** option is off in the **View Cube Settings** dialog box, the UCS shortcut menu has WCS displayed above it.

Restore Defaults. After making several changes in the **View Cube Settings** dialog box, you may get confused about the effect of different option settings on the model display. In this case, it is best to pick the **Restore Defaults** button to return all settings to their original values. Then change one setting at a time and test it to be sure the view cube functions as you intended.

Exercise 3-3

www.g-wlearning.com/CAD/

Complete the exercise on the companion website.

Creating a Continuous 3D Orbit

3DCORBIT

Ribbon
View
> **Navigate**
Continuous Orbit
Type
3DCORBIT

The **3DCORBIT** command provides the ability to create a continuous orbit of a model. By moving your pointing device, you can set the model in motion in any direction and at any speed, depending on the power of your computer. An impressive display can be achieved using this command. The command is located in the **Orbit** drop-down list in the **Navigate** panel of the **View** tab on the ribbon. See **Figure 3-14A**. If the **Navigate** panel is not displayed, right-click on the ribbon to access the shortcut menu and select **Navigate** from the **Show Panels** cascading menu. Additionally, the **3DCORBIT** command can be selected on the navigation bar.

Once the command is initiated, the continuous orbit cursor is displayed. See **Figure 3-14B**. Press and hold the pick button and move the pointer in the direction that you want the model to rotate and at the desired speed of rotation. Release the button when the pointer is moving at the appropriate speed. The model will continue to rotate until you pick the left mouse button, press [Enter] or [Esc], or right-click and pick **Exit** or another option in the shortcut menu. At any time while the model is orbiting, you can left-click and adjust the rotation angle and speed by repeating the process for starting a continuous orbit.

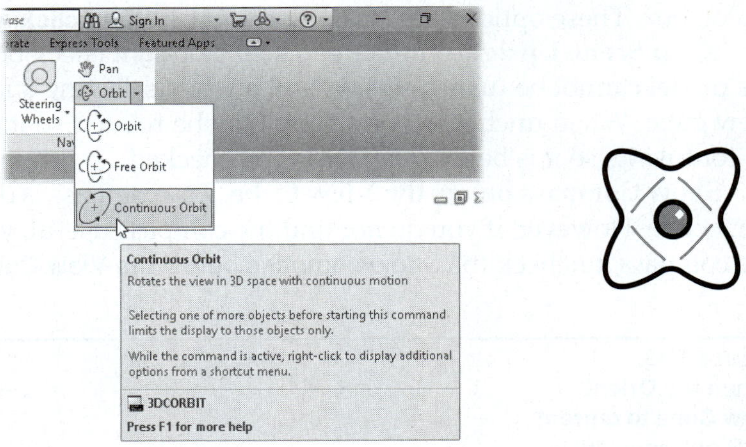

A

B

Figure 3-14. A—Selecting the **3DCORBIT** command. B—This is the continuous orbit cursor in the **3DCORBIT** command (or **Continuous** option of the **3DORBIT** command). Pick and hold the left mouse button. Then, move the cursor in the direction in which you want the view to rotate and release the mouse button.

Plan Command Options

The **PLAN** command, introduced in Chapter 1, allows you to create a plan view of any user coordinate system (UCS) or the world coordinate system (WCS). The **PLAN** command automatically performs a **ZOOM Extents**. This fills the drawing window with the plan view. The command options are:

Type
Plan
PLAN

- **Current UCS.** This creates a view of the object that is plan to the current UCS.
- **UCS.** This displays a view plan to a named UCS. The preset UCSs are not considered named UCSs.
- **World.** This creates a view of the object that is plan to the WCS. If the WCS is the current UCS, this option and the **Current UCS** option produce the same results.

This command may have limited usefulness when working with 3D models. The dynamic capabilities of the view cube are much more intuitive and may be quicker to use. However, you may find instances where the **PLAN** command is easier to use, such as when the view cube is not currently displayed.

Displaying Models with Steering Wheels

Steering wheels, or *wheels*, are dynamic menus that provide quick access to view-navigation tools. A steering wheel follows the cursor as it is moved around the drawing. Each wheel is divided into wedges and each wedge contains a tool. See **Figure 3-15**. The **NAVSWHEEL** command is used to display a steering wheel, but a quicker method is to pick **Steering Wheels** from the **Viewport Controls** flyout in the viewport controls. The steering wheel drop-down menu on the navigation bar can also be used. All of the steering wheel formats discussed in this chapter can be selected in the drop-down menu on the navigation bar.

Ribbon
View
> Navigate

Steering Wheels
Type
NAVSWHEEL
WHEEL
Toolbar
Navigation Bar

Steering Wheels
NAVSWHEEL

AutoCAD provides two basic types of wheels: View Object and Tour Building. Most of the options contained in these two wheels are included in the Full Navigation wheel. Each of these three types can be used in a full-wheel display or a minimized format. This section discusses the Full Navigation wheel and the two basic wheels in their full and mini formats. There is also a 2D Wheel that is displayed in paper (layout) space or when the **NAVSWHEELMODE** system variable is set to 3. The 2D Wheel provides basic view-navigation tools for 2D drafting.

Figure 3-15.
A—Full Navigation wheel. B—View Object wheel. C—Tour Building wheel.

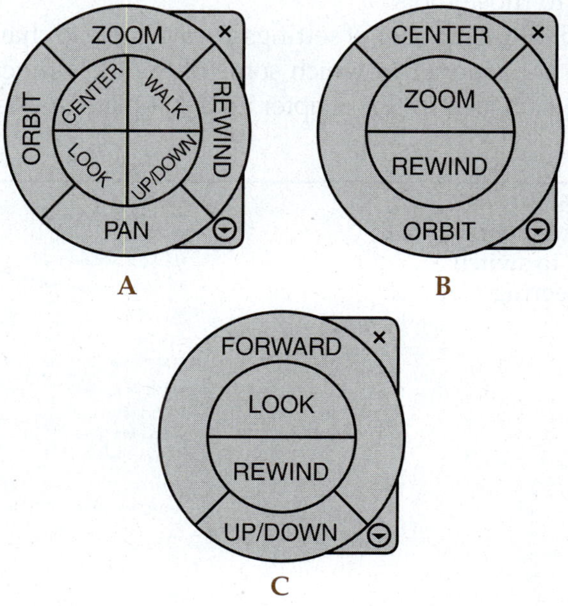

Using a Steering Wheel

Once a steering wheel is displayed, move the cursor around the screen. Notice as you move the cursor, the wheel follows it. When you stop the cursor, the wheel stops. If you move the cursor anywhere inside of the wheel, the wheel remains stationary. Note also that as you move the cursor inside of the wheel, a wedge (tool) is highlighted. If you pause the cursor over a tool, a tooltip is displayed that describes the tool.

To use a specific tool, simply pick and hold on the highlighted wedge, then move the cursor as needed to change the view of the model. Once you release the pick button, the tool ends and the wheel is redisplayed. Some options display a *center point* about which the display will move. The **Center** tool is used to set the center point. These features are all discussed in the next sections.

To change wheels, right-click to display the shortcut menu. Then, select **Full Navigation Wheel** from the menu to display that wheel. Or, select **Basic Wheels** to display a cascading menu. Select either **View Object Wheel** or **Tour Building Wheel** from the cascading menu to display that wheel. Refer to **Figure 3-16**.

PROFESSIONAL TIP

The view cube is a valuable visualization tool, especially when used in conjunction with a steering wheel. As you use the wheel options to manipulate the model, watch the view cube move. This provides you with a dynamic "map" of where your viewpoint is at all times in relation to the model, the UCS, and the view cube compass.

Exercise 3-4

www.g-wlearning.com/CAD/

Complete the exercise on the companion website.

Full Navigation Wheel

The Full Navigation wheel contains many of the tools available in the View Object and Tour Building wheels. Refer to **Figure 3-15A**. These tools are discussed in the following sections. The Full Navigation wheel also contains the **Pan** and **Walk** tools, which are not available in either of the other two wheels. These tools are discussed next. The tools shared with the View Object and Tour Building tools are discussed in the sections corresponding to those tools.

In addition, a number of settings are available to change the appearance of the steering wheel and the manner in which some of the tools function. Refer to the Steering Wheel Settings section later in this chapter for a complete discussion of these settings.

Figure 3-16.
The shortcut menu allows you to switch between steering wheels.

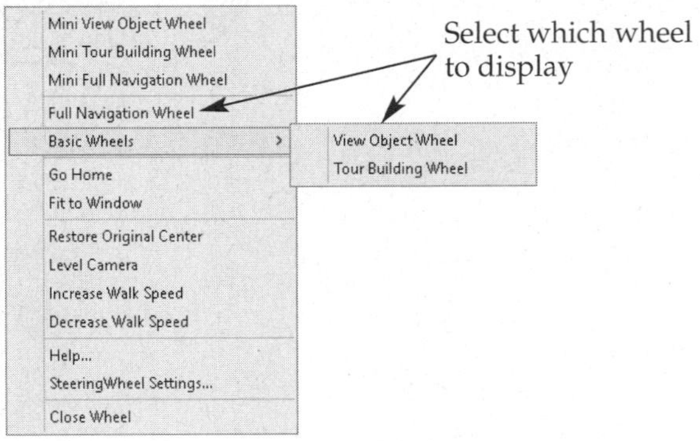

Pan

The **Pan** tool allows you to move the model in the direction that you drag the cursor. This tool functions the same as the AutoCAD **RTPAN** command. When you pick and hold on the tool, the cursor changes to four arrows with the label Pan Tool below the cursor.

If you use the **Pan** tool with the perspective projection current, it may appear that the model is slowly rotating about a point. This is not the case. What you are seeing is merely the effect of the vanishing points. As you pan, the relationship between the viewpoint and the vanishing points changes. You can quickly test this by closing the wheel, right-clicking on the view cube, and picking **Parallel** from the shortcut menu. Now display the wheel again and use the **Pan** tool. Notice the difference. The model pans without appearing to rotate.

Walk

The **Walk** tool is used to simulate walking toward, through, or away from the model. When you pick and hold on the tool, the center circle icon is displayed at the bottom-center of the drawing area. See **Figure 3-17**. The cursor changes to an arrow pointing away from the center of the circle as you move it off the circle. The arrow indicates the direction in which the view will move as you move the mouse. This gives the illusion of walking in that direction in relation to the model.

The farther away from the center tool you move the pointer, the faster the walk. Move the pointer away from the center tool until you have achieved the walk speed desired, then hold the pointer. Release the pick button when the desired view is displayed.

NOTE

If you begin moving the cursor at a point far away from the center circle icon, the walk speed will be substantial and the model may quickly disappear from view. If this occurs, close the steering wheel and pick the **Home** icon in the view cube to restore the home view. Display the steering wheel again, move it to the bottom-center of the drawing area, and then select the **Walk** tool. The cursor will be closer to the center circle icon, and the walk speed will be slower.

If you hold down the [Shift] key while clicking the **Walk** tool, the **Up/Down** slider is displayed. This allows you to change the screen Y axis orientation of the view. This equates to elevating the camera view relative to the object. Releasing the [Shift] key returns you to the standard walk mode. The up and down arrow keys can also be used to change the "height" of the view. The speed of walking can be increased with the plus key (+).

The **Up/Down** slider is also used with the **Up/Down** tool on the Tour Building wheel. It is explained in the Tour Building Wheel section.

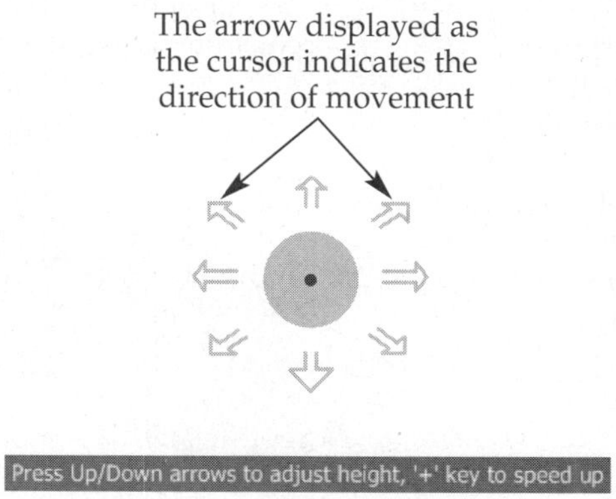

Figure 3-17.
The **Walk** tool displays the center circle icon. As you move the cursor around the icon, an arrow is displayed to indicate the direction in which the view is being moved.

View Object Wheel

If you think of your model as a building, the tools in the View Object wheel are used to view the outside of the building. This wheel contains four navigation tools: **Center**, **Zoom**, **Rewind**, and **Orbit**. Refer to **Figure 3-15B**.

Center

The **Center** tool is used to set the center point for the current view. Many tools, such as **Zoom** and **Orbit**, are applied in relation to the center point. Pick and hold the **Center** tool, then move the cursor to a point on the model and release. The display immediately changes to center the model on that point. See **Figure 3-18**. The selected point must be on an object. Notice that the center point icon resembles a globe with three orbital axes. These axes relate to the three axes of the model shown on the UCS icon.

Zoom

The **Zoom** tool is used to dynamically zoom the view in and out, just as with the AutoCAD **RTZOOM** command. The tool uses the center point set with the **Center** tool. When you pick and hold the **Zoom** tool in the wheel, the center point is displayed at its current location. The cursor changes to a magnifying glass with the label **Zoom Tool** displayed below it. See **Figure 3-19**. There are three different ways to use the **Zoom** tool, as discussed in this section.

When the **Zoom** tool is accessed in the Full Navigation wheel, the center point is relocated to the position of the steering wheel. If you wish to zoom on the existing center point when using the Full Navigation wheel, first press the [Ctrl] key and then access the **Zoom** tool. This prevents the tool from relocating the center point. You can also move the center point using the **Center** tool, then switch to the View Object wheel and access the **Zoom** tool from that wheel. In either case, the zoom is relative to the location of the center point.

Figure 3-18.
The **Center** tool allows you to select a new center point for the current view. This becomes the point about which many steering wheel tools operate.

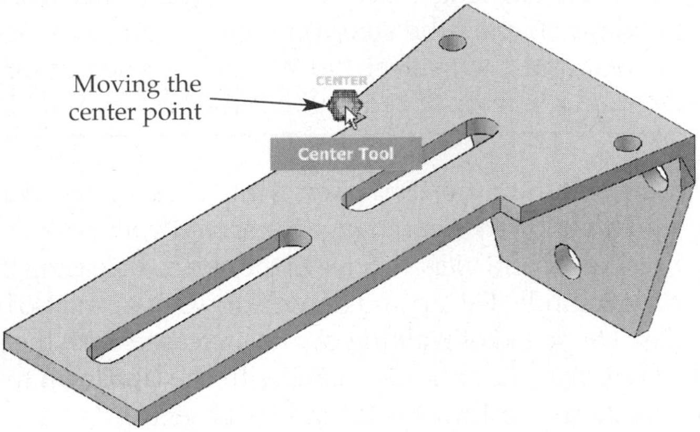

Figure 3-19.
The **Zoom** tool operates in relation to the center point. If the tool is selected from the Full Navigation wheel, the center point is automatically relocated to the cursor location.

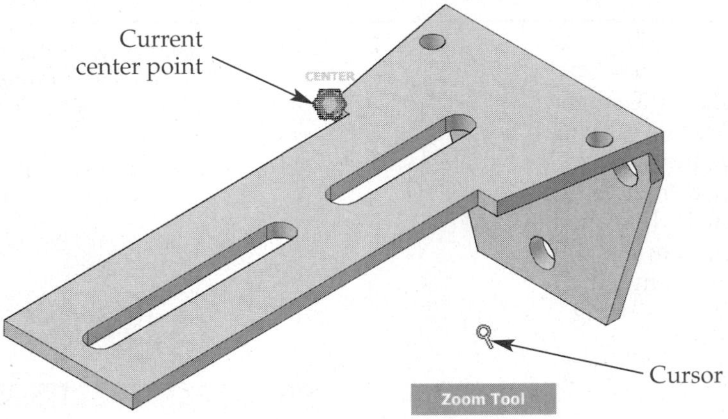

NOTE

Once you close the steering wheel, the center point is reset. The next time a steering wheel is displayed, the center point will be in the middle of the current view.

Pick and Drag. Pick and drag the cursor to dynamically zoom. This is similar to performing a realtime zoom. As the cursor is moved up or to the right, the view-point moves closer (zooms in). Move the pointer to the left or down and the viewpoint moves farther away (zooms out). The zooming is based on the current center point. When you have achieved the appropriate zoom location, release the pointer button.

Single Click. If you select the **Zoom** tool with a single click (using the View Object wheel), the view of the model zooms in by an incremental percentage. Each time you single click on the tool, the view is zoomed in by 25%. The zoom is in relation to the center point.

In order for this function to work when the **Zoom** tool is accessed from the Full Navigation wheel, you must check the **Enable single click incremental zoom** check box in the **Steering Wheels Settings** dialog box. This dialog box is discussed in detail later in the chapter.

Shift and Click. If you press and hold the [Shift] key and then pick the **Zoom** tool (using the View Object wheel), the view is zoomed out by 25%. As with the single-click method, the **Enable single click incremental zoom** check box must be checked in order to use this with the Full Navigation wheel.

Rewind

The **Rewind** tool allows you to step back through previous views. A single pick on this tool displays the previous view. If you pick and hold the tool, a "slide show" of previous views is displayed as thumbnail images. See **Figure 3-20**. The most recent view is displayed on the right-hand side. The oldest view is displayed on the left-hand side. The slide representing the current view is highlighted with an orange frame and a set of brackets. While holding the pick button, move the cursor to the left. Notice that the set of brackets moves with the cursor. As a slide is highlighted, the corresponding view is restored in the viewport. Release the pick button when you find the view you want and it is set current. When the brackets are hovered over two adjacent views, the orange highlighting is dimmed around each view. Selecting a view at this point enables you to display a view that is "between" two slides.

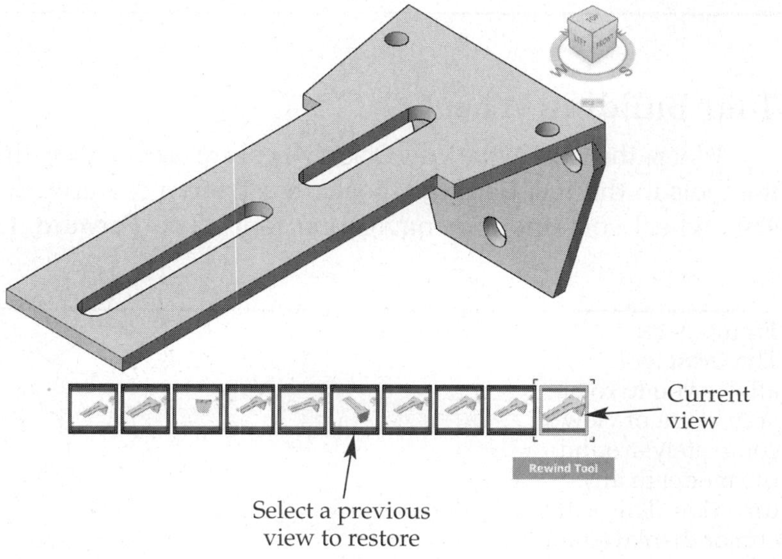

Figure 3-20.
A single pick on the **Rewind** tool displays the previous view. A "slide show" of previous views is displayed as thumbnail images if you press and hold the pick button.

By default, thumbnail images of previous views outside of a steering wheel are not immediately displayed in the navigation history. The **Steering Wheels Settings** dialog box allows you to control when thumbnail images are generated in the navigation history. This dialog box is discussed later in the chapter.

The navigation history is maintained in the drawing file and is different for each open drawing. However, it is not saved when a drawing is closed.

Orbit

The **Orbit** tool allows you to rotate your point of view completely around the model in any direction. The view pivots about the center point set with the **Center** tool. When using the Full Navigation wheel, the center point can be quickly set by pressing and holding the [Ctrl] key, then picking and holding the **Orbit** tool. Next, drag the center point to the desired pivot point on the model and release. Now you can use the **Orbit** tool.

To use the **Orbit** tool, pick and hold on the tool in the steering wheel. The current center point is displayed with the label Pivot. Also, the cursor changes to a point surrounded by two circular arrows. See Figure 3-21. Move the cursor around the screen and the view of the model pivots about the center point. If this is not the result you wanted, just reset the pivot point using the **Center** tool.

PROFESSIONAL TIP

Since the **Orbit** tool is most often used to move your viewpoint quickly to another side of the model, it is more intuitive to locate the pivot point somewhere on the model. First, use the **Center** tool to establish the pivot point. Then, use the **Orbit** tool to rotate the view around the pivot point to the desired viewpoint. If the center point is not set, it defaults to the center of the screen.

CAUTION

When the **Zoom** tool is selected from the Full Navigation wheel, the center point is changed to the steering wheel location. Therefore, that center point is used as the pivot point for the **Orbit** tool. When the **Zoom** tool is selected from the View Object wheel, the center point is not changed. Therefore, it is best to switch to the View Object wheel to use the **Zoom** tool or to press the [Ctrl] key before accessing the tool in the Full Navigation wheel.

Tour Building Wheel

Where the tools in the View Object wheel are used to view the outside of the "building," the tools in the Tour Building wheel are used to move around inside of the "building." This wheel contains four navigation tools. The **Forward**, **Look**, and **Up/Down** tools

Figure 3-21.
The **Orbit** tool allows you to rotate your point of view completely around the model in any direction. This is the cursor displayed for the tool.

are discussed here. The **Rewind** tool is addressed in the View Object Wheel section, discussed earlier.

Forward

The principal tool in the Tour Building wheel is the **Forward** tool. It is similar to the **Walk** tool in the Full Navigation wheel. However, it only allows forward movement from the current viewpoint. This tool requires a center point to be set on the model from within the tool. The existing center point cannot be used.

First, move the cursor and steering wheel to the point on the model that will be the target (center point). Next, pick and hold the **Forward** tool. The pick point becomes the center point and a drag distance indicator is displayed. See **Figure 3-22**. This indicator shows the starting viewpoint, the center point, and the surface of the model that you selected. Hold the mouse button down while moving the pointer up. The orange location slider moves to show the current viewpoint relative to the center point. As you move closer to the model, the green center point icon increases in size, which also provides a visual cue to the zoom level.

Look

The **Look** tool is used to rotate the view about the center of the view. When the tool is activated, the cursor appears as a half circle with arrows. See **Figure 3-23**. As you move the cursor down, the model moves up in the view as if you are actually tilting your head down. Similarly, as you "look" away from the model to the right or left, the model appears to move away from your line of sight. If the vertical movement does not seem intuitive to you, it is easy to change. Open the **Steering Wheels Settings** dialog box and check the **Invert vertical axis for Look tool** check box. Now, when using the **Look** tool, moving the cursor up or down also moves the object in the same direction.

The distance between you and the model remains the same and the orientation of the model does not change. Therefore, you would not want to use this tool if you wanted to see another side of the model.

Up/Down

As the name indicates, the **Up/Down** tool moves the view up or down along the Y axis of the screen, regardless of the orientation of the current UCS. Pick and hold on the tool and the vertical distance indicator appears. See **Figure 3-24**. Two marks on this indicator show the upper and lower limits within which the view can be moved. The

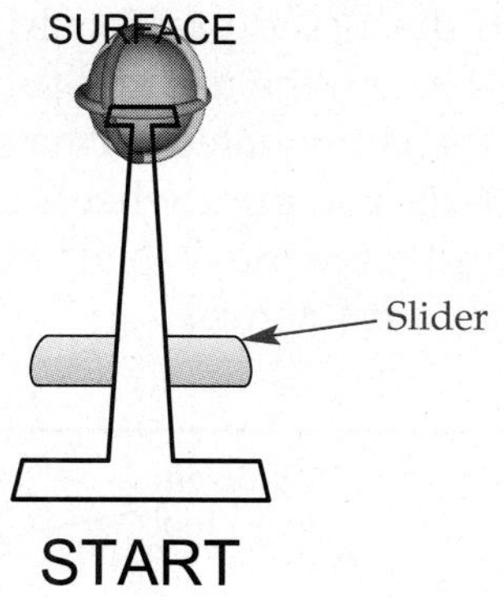

Figure 3-22. The drag distance indicator is displayed when using the **Forward** tool. This indicator shows the start point of the view and the selected surface of the model. The slider indicates the current view position relative to the starting point.

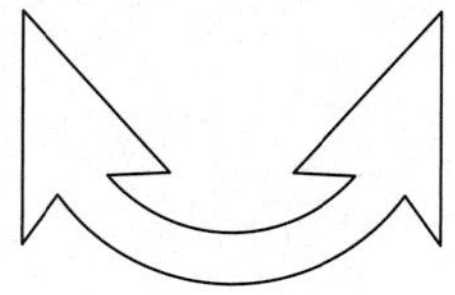

Figure 3-23. The **Look** tool cursor appears as a half circle with arrows.

Figure 3-24.
The indicator displayed when using the **Up/Down** tool shows the upper and lower limits within which the view can be moved. The top position is the location of the view when the tool is selected.

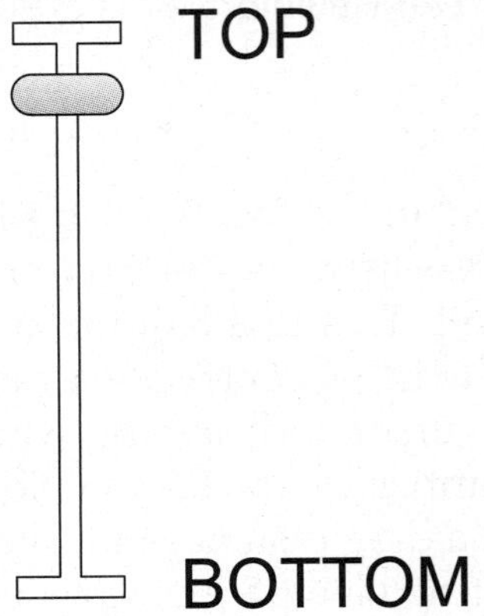

orange slider shows the position of the view as you move the cursor. When the tool is first activated, the view is at the top position. The **Up/Down** tool has limited value. The **Pan** tool is far more versatile.

PROFESSIONAL TIP

Use steering wheel tools such as **Forward**, **Look**, **Orbit**, and **Walk** to manipulate a view and fully explore the model. The **Rewind** tool can then be used to replay all of the previous views saved in the navigation history. This process reveals model views that can be saved as named views for later use in model construction or for shots created with show motion. Using show motion is discussed in detail in Chapter 20. Remember, views created using steering wheels are not saved with the drawing file. Therefore, if there is a possibility they will be needed later, it may be a time-saver to create and save named views as previously discussed in this chapter.

Exercise 3-5

www.g-wlearning.com/CAD/

Complete the exercise on the companion website.

Mini Wheels

The three wheels discussed previously were presented in their *full wheel* formats. As you gain familiarity with the use and function of each wheel and its tools, you may wish to begin using the abbreviated formats. The abbreviated formats are called ***mini wheels***. See **Figure 3-25**. The mini wheels can be selected from the navigation bar or the steering wheel shortcut menu. When you select a mini wheel, it follows the pointer in the same manner as a full wheel.

Figure 3-25.
In addition to full-size steering wheels, mini wheels can be used. A—Mini Full Navigation wheel. B—Mini View Object wheel. C—Mini Tour Building wheel.

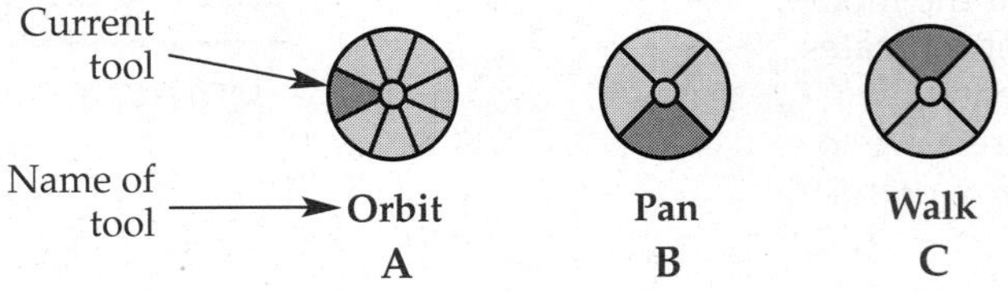

Slowly move the mouse in a small circle and notice that each wedge of the wheel is highlighted. The name of the currently highlighted tool appears below the mini wheel. When the tool you need is highlighted, simply click and hold to activate the tool. All tools in the mini wheels function the same as those in the full-size wheels. The only difference is the appearance of the wheels.

Steering Wheel Shortcut Menu

A quick method for switching between the different wheel formats is to use the steering wheel shortcut menu. Pick the menu arrow at the lower-right corner of a full wheel to display the menu. See **Figure 3-26**. To select a wheel or change wheel formats, simply select the appropriate entry in the menu. Note that a check mark is *not* placed by the current wheel.

In addition to wheel formats, the shortcut menu provides options for viewing the model. These additional options are described as follows:

- **Go Home.** Returns the display to the home view. This is the same as picking the **Home** icon in the view cube.
- **Fit to Window.** Resizes the current view to fit all objects in the drawing inside of the window. This is essentially a zoom extents operation.
- **Restore Original Center.** Restores the original center point of the drawing using the current drawing extents. This does not change the current zoom factor. Therefore, if you are zoomed close into an object and pick this option, the object may disappear from view by moving off the screen.
- **Level Camera.** The camera (your viewpoint) is rotated to be level with the XY ground plane.
- **Increase Walk Speed.** The speed used by the **Walk** tool is increased by 100%.
- **Decrease Walk Speed.** The speed used by the **Walk** tool is decreased by 50%.
- **Help....** Displays the help system documentation for steering wheels.
- **Steering Wheel Settings....** Displays the **Steering Wheels Settings** dialog box, which is discussed in the next section.
- **Close Wheel.** Closes the steering wheel. This is the same as pressing [Esc] to close the wheel.

Exercise 3-6

www.g-wlearning.com/CAD/

Complete the exercise on the companion website.

Figure 3-26. Select the down arrow to display the steering wheel shortcut menu. Both full-size wheels and mini wheels can be displayed using this menu.

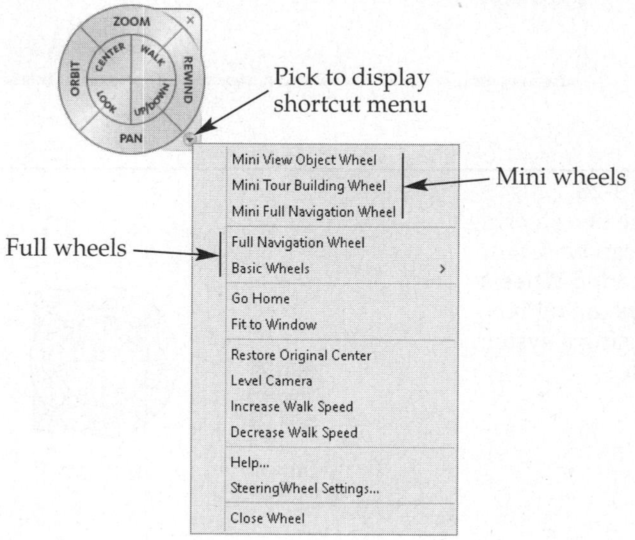

Steering Wheel Settings

The **Steering Wheels Settings** dialog box provides options for wheel appearance. See **Figure 3-27**. It also contains settings for the display and operation of several tools. It is displayed by selecting **Steering Wheel Settings...** in the steering wheel shortcut menu.

Changing Wheel Appearance

The two areas at the top of the **Steering Wheels Settings** dialog box allow you to change the size and opacity of all wheels. The settings in the **Big Wheels** area are for the full-size wheels. The settings in the **Mini Wheels** area are for the mini wheels. The **Wheel size:** slider in each area is used to display the wheels in small, normal, or large size. See **Figure 3-28**. The mini wheel has a fourth, extra large size. These sliders set the **NAVSWHEELSIZEBIG** and **NAVSWHEELSIZEMINI** system variables.

The **Wheel opacity:** slider in each area controls the transparency of the wheels. These sliders can be set to a value from 25% to 90% opacity. The sliders control the **NAVSWHEELOPACITYBIG** and **NAVSWHEELOPACITYMINI** system variables. The appearance of the full wheel in three different opacity settings is shown in **Figure 3-29**.

Figure 3-27.
The **Steering Wheels Settings** dialog box provides options for wheel appearance and the display and operation of several tools.

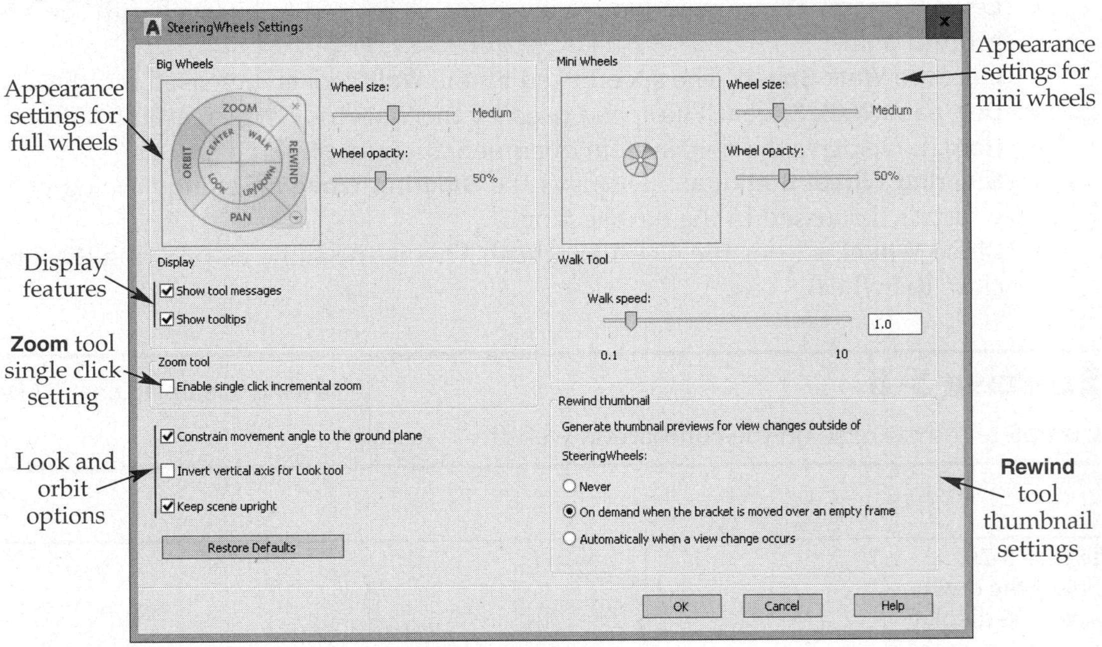

Figure 3-28.
The size of a steering wheel can be set in the **Steering Wheels Settings** dialog box or by using a system variable.

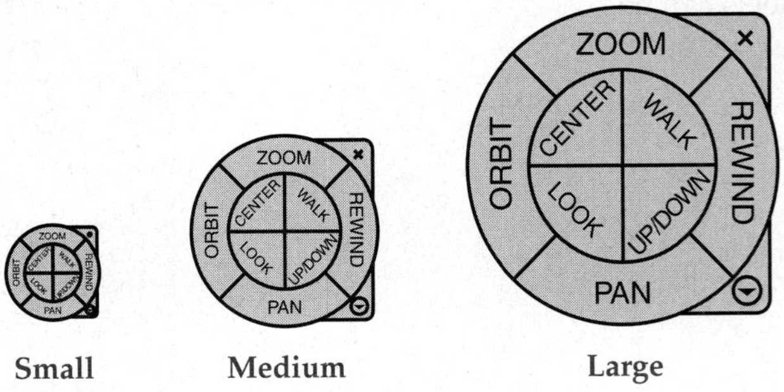

Figure 3-29.
The opacity of a steering wheel can be changed.
A—Opacity of 25%.
B—Opacity of 50%.
C—Opacity of 90%.

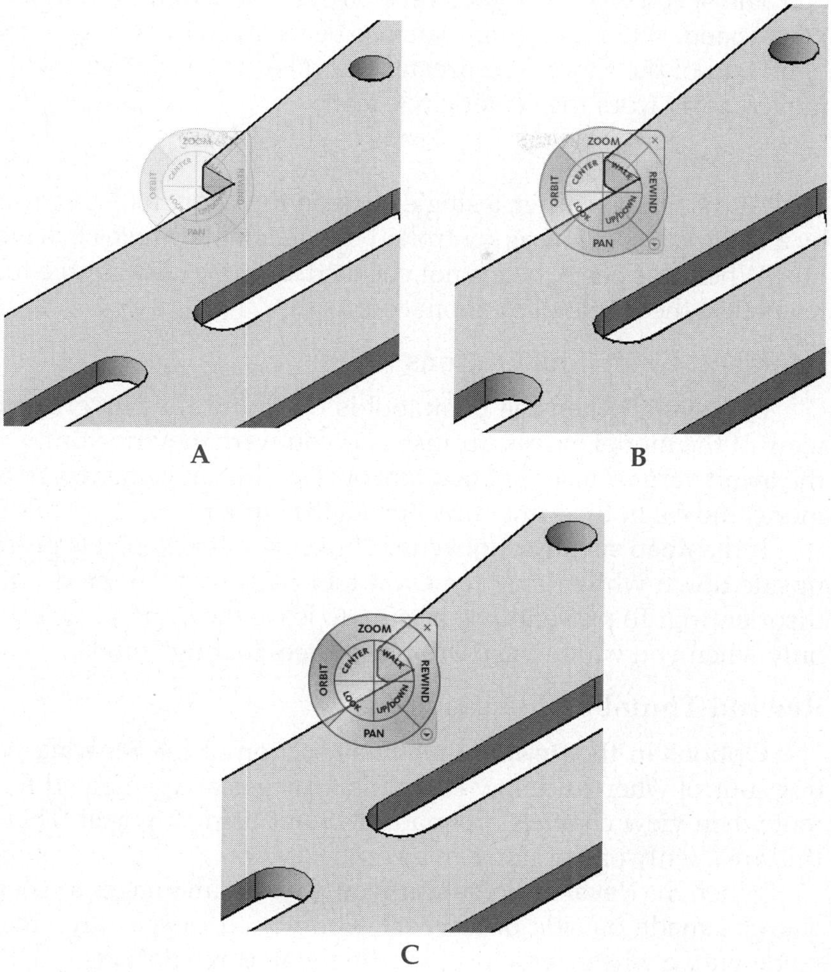

Display Features

The **Display** area of the **Steering Wheels Settings** dialog box controls two features of the wheel display. These settings determine whether messages and tooltips are displayed.

When a tool is selected, its name is displayed below the cursor. In addition, some tools have features or restrictions that can be indicated in a tool message. To see these messages, the **Show tool messages** check box must be checked. Otherwise, the messages are not displayed, but the restrictions remain in effect.

A tooltip is a short message that appears below the wheel when the cursor is hovered over the wheel. When the **Show tooltips** check box is checked, you can hold the cursor stationary over a tool for approximately three seconds and the tooltip will appear. Then, as you move the cursor over tools in the wheel, the appropriate tooltip is immediately displayed.

Walk Tool

By default, when the **Walk** tool is used, you move parallel to the ground plane. This is because the **Constrain movement angle to the ground plane** check box is checked. Test this by selecting the **Walk** tool and then move the cursor toward the top of the screen. You appear to be "walking" over the top of the model. Now, open the **Steering Wheels Settings** dialog box and uncheck the **Constrain movement angle to the ground plane** check box. This allows you to "fly" in the direction the cursor is moved when using the **Walk** tool. Exit the dialog box and again select the **Walk** tool. Move the cursor toward the top of the screen. This time it appears that you are flying directly toward or into the model.

The speed at which you walk through or around the model is controlled by the **Walk speed:** slider. The value can also be changed by typing in the text box at the right-hand end of the slider. The greater the value, the faster you will move as the cursor is moved away from the center circle icon.

Zoom Tool

As discussed earlier, a single click on the **Zoom** tool zooms in on the current view by a factor of 25%. This is controlled by the **Enable single click incremental zoom** check box. When this check box is not checked, a single click on the tool has no effect. Some users find the single-click zoom confusing.

Look and Orbit Tool Options

By default, when the **Look** tool is used and the cursor is moved downward, the view of the model moves up, just as if you were moving your eyes down. If you check the **Invert vertical axis for Look tool** option, this movement is reversed. In this case, the model moves in the same direction as the cursor.

If the **Keep scene upright** check box is not checked, it is possible to turn the model upside down while using the **Orbit** tool. This may not be desirable because it can be disorienting. To prevent this, be sure to leave the option checked. Uncheck this option only when you want to use **Orbit** in a "free-floating" mode.

Rewind Thumbnail Options

Options in the **Rewind thumbnail** section of the **Steering Wheels Settings** dialog box control when and how thumbnail images are generated for use with the **Rewind** tool when view changes are made without using a wheel. There are three options in this area. Only one option can be on.

When the **Never** radio button is on, thumbnail images are never generated for view changes made outside of a wheel. Thumbnail images are created for view changes made with a wheel, which is the case for all three options.

When the **On demand when the bracket is moved over an empty frame** radio button is on, thumbnail images are not automatically generated for view changes made outside of a wheel. The frames for these views display a double arrow icon when the **Rewind** tool in a wheel is used. However, as the brackets are moved over these frames, thumbnail images are generated. This is the default setting. See **Figure 3-30**.

When the **Automatically when a view change occurs** option is selected, a thumbnail image is generated any time a view change is made outside of a wheel. When the **Rewind** tool in a wheel is used, the frames for these views automatically display thumbnail images.

Figure 3-30.
A—By default, when the **Rewind** tool is selected, frames representing view changes made outside of a steering wheel display double-arrow icons.
B—As the brackets are moved over the blank frames, thumbnail images are generated of the views created outside of a wheel.

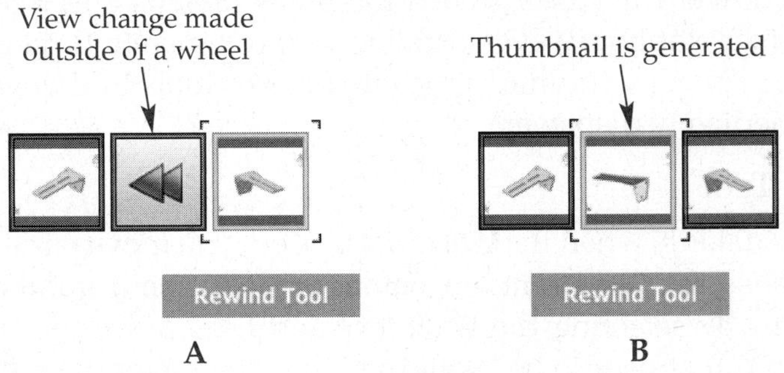

Restore Defaults

Picking the **Restore Defaults** button in the **Steering Wheels Settings** dialog box returns all of the settings in the dialog box to their default values. Select this when at any time you are not sure how the settings are affecting the appearance and function of the steering wheel tools. Then, make changes one at a time as needed.

Exercise 3-7 www.g-wlearning.com/CAD/

Complete the exercise on the companion website.

Controlling the Display of the Navigation Bar

The navigation bar is a flexible tool that can be customized or positioned to suit the needs of your drawing project. The three principal navigation tools are for the view cube, steering wheels, and show motion (discussed in Chapter 20). The view cube is displayed on the navigation bar when the normal display of the view cube is turned off.

The navigation bar can be placed at any location around the edge of the screen. In addition, the bar can be linked to the location of the view cube. The navigation bar can be quickly turned on or off by picking **Navigation Bar** in the **Viewport Controls** flyout in the viewport controls, or by picking the **Navigation Bar** button in the **Viewport Tools** panel of the **View** ribbon tab.

Repositioning the Navigation Bar

The navigation bar can be moved around the screen by picking the small customize button at the lower-right corner of the bar, and then selecting **Docking positions** to display the cascading menu. See Figure 3-31. The default setting is for the bar to be linked to the location of the view cube. This means that if the position of the view cube is changed, the navigation bar will follow. Test this by first confirming that **Link to View Cube** is checked in the menu. Next, move the view cube to one of the other locations, such as top left, using the **View Cube Settings** dialog box. Notice that the view cube and navigation bar move to the new location. Picking the **Undo** button on the **Quick Access** toolbar can quickly restore the previous location.

You can also move the navigation bar to a position of your choosing without moving the view cube. Simply uncheck the **Link to View Cube** option. Notice that a band appears at the top of the bar. When the pointer is moved into this band and you click, it becomes a move cursor. See Figure 3-32. Drag the bar to any edge location on the screen. The bar will dock at the closest screen edge. You cannot place it in a floating position in the middle of the screen.

Customizing the Navigation Bar

The tools that are displayed in the navigation bar can be changed by picking the customize button at the lower-right corner of the bar. By default, all of the tools are checked. If you wish to hide a tool, just select it to uncheck it in the list. All of the tools in the navigation bar can be removed or redisplayed in this manner. If the view cube is displayed on the screen, it will be grayed out in the list. If the view cube is currently not displayed, an option for it appears in the menu. In addition, a button for the view cube appears on the navigation bar. See Figure 3-33. The view cube can then be turned on using the navigation bar.

Figure 3-31.
The navigation bar can be moved around the screen by selecting the small customize button at the lower-right corner of the bar, and then selecting **Docking positions** in the menu.

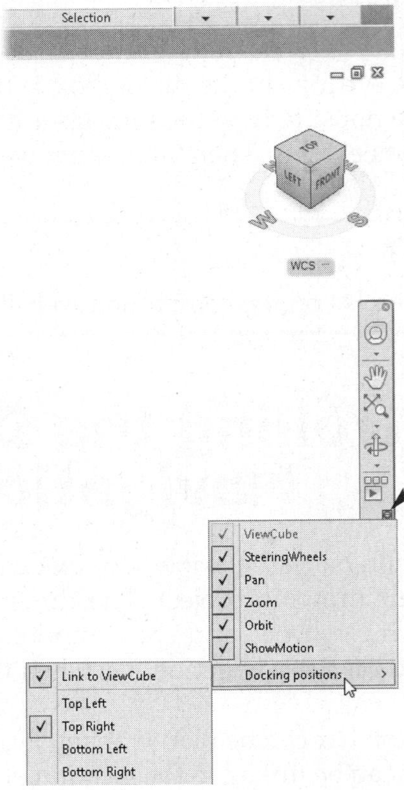

Figure 3-32.
Click and hold to move the navigation bar to any position along the edge of the screen.

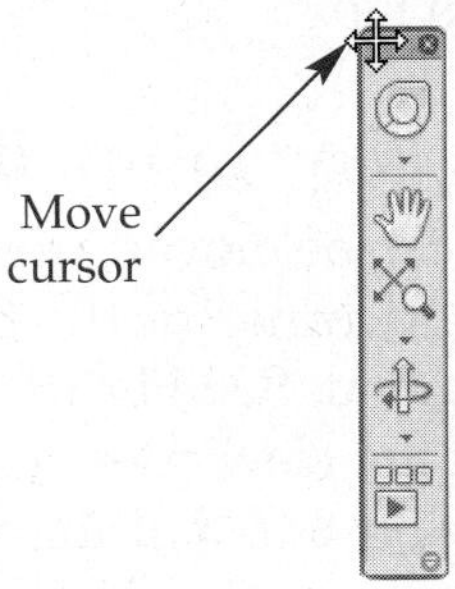

Figure 3-33.
If the view cube is not currently displayed, a button for displaying it appears on the navigation bar.

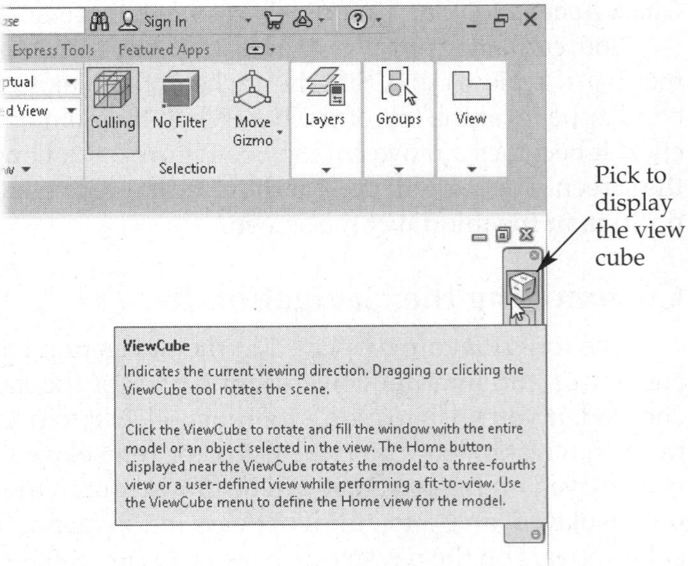

Chapter Review

Answer the following questions using the information in this chapter.

1. New named views are created in which dialog box?
2. Where is a thumbnail image of a new view displayed?
3. How do you select a standard isometric preset view using the view cube?
4. How is a standard orthographic view displayed using the view cube?
5. What is the difference between *parallel projection* and *perspective projection*?
6. What happens when one of the four view cube compass letters is picked?
7. Which command generates a continuous 3D orbit?
8. Which command can be used to produce a view that is parallel to the XY plane of the current UCS?
9. What is a *steering wheel*?
10. Briefly describe how to use a steering wheel.
11. The principal tool in the Tour Building wheel is the **Forward** tool. What is the purpose of this tool?
12. What are the three principal navigation tools found in the navigation bar?

Drawing Problems

1. General Open one of your 3D drawings from Chapter 2. Do the following.
   A. Use the view cube to create a pictorial view of the drawing.
   B. Set the Conceptual visual style current.
   C. Using a steering wheel, display different views of the model.
   D. Use the **Rewind** tool in the steering wheel to restore a previous view.
   E. Save the drawing as P3-1.

2. General Open one of your 3D drawings from Chapter 2. Do the following.
   A. Toggle the projection from parallel to perspective and turn off the view cube compass.
   B. Pick an edge on the view cube. Set the parallel projection current.
   C. Display three additional views based on view cube edges. Alternate between parallel and perspective projection.
   D. Put the model into a continuous orbit.
   E. Save the drawing as P3-2.

3. General Open one of your 3D drawings from Chapter 2. Do the following.
   A. Use a steering wheel to change the view of the model. Do this at least four times.
   B. Use the **Rewind** tool to restore a previous view.
   C. Repeat this for each of the previous views recorded with the steering wheel.
   D. Save the drawing as P3-3.

A user coordinate system (UCS) can be located on any surface and at any orientation for modeling purposes. In this example, a UCS is located on the inclined surface to draw the profile for the slot. The profile is extruded and then subtracted to create the final bracket model.

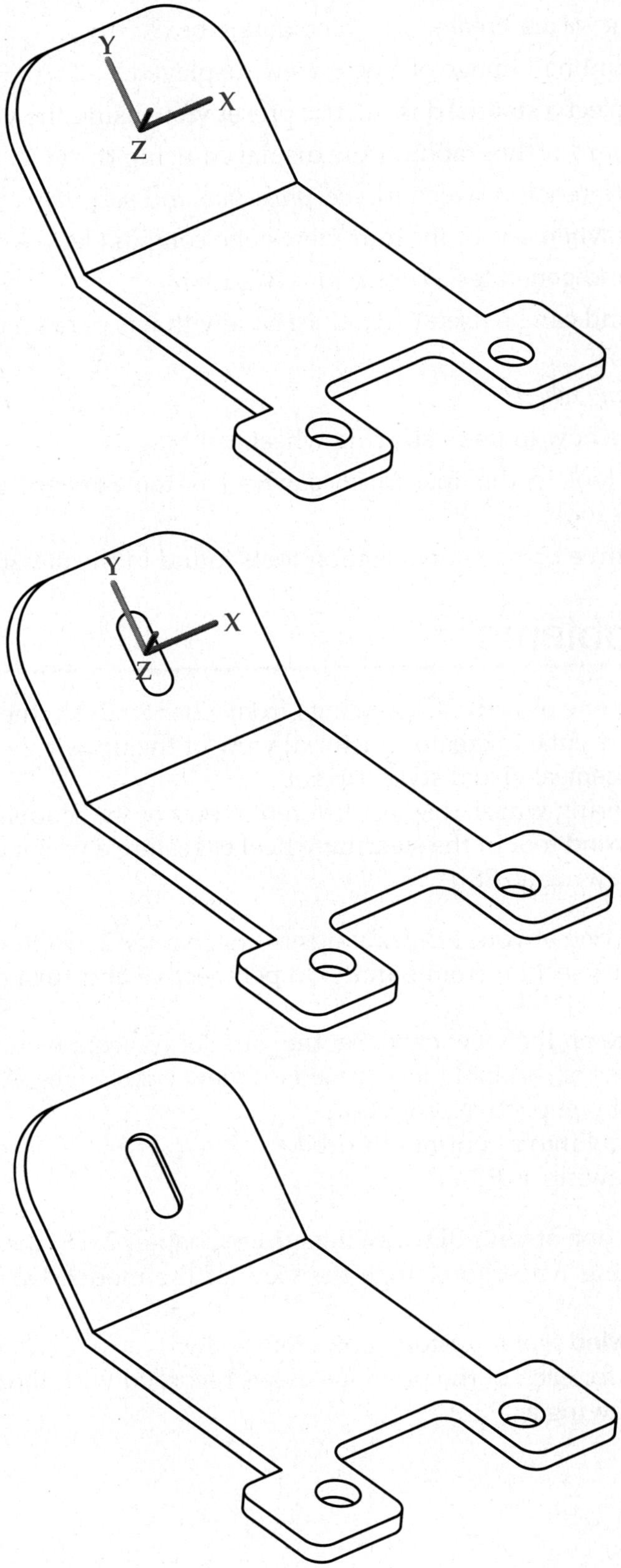

CHAPTER 4

Understanding Three-Dimensional Coordinates and User Coordinate Systems

Learning Objectives

After completing this chapter, you will be able to:

- ✓ Describe rectangular, spherical, and cylindrical methods of coordinate entry.
- ✓ Draw 3D polylines.
- ✓ Describe the function of the world and user coordinate systems.
- ✓ Move the user coordinate system to any surface.
- ✓ Rotate the user coordinate system to any angle.
- ✓ Use the UCS icon grips to move and rotate the UCS.
- ✓ Use a dynamic UCS.
- ✓ Save and manage user coordinate systems.
- ✓ Restore and use named user coordinate systems.
- ✓ Control UCS icon visibility in viewports.

As you learned in Chapter 1, any point in space can be located using X, Y, and Z coordinates. This type of coordinate entry is called *rectangular coordinate entry*. Rectangular coordinate entry is the most common form of coordinate entry. However, there are actually three ways in which to locate a point in space. The other two methods of coordinate entry are spherical coordinate entry and cylindrical coordinate entry. These two coordinate entry methods are discussed in the following sections. In addition, this chapter introduces working with user coordinate systems (UCSs).

Introduction to Spherical Coordinates

Locating a point in 3D space with ***spherical coordinates*** is similar to locating a point on Earth using longitudinal and latitudinal values, with the center of Earth representing the origin. Lines of longitude connect the North and South Poles and provide an east-west measurement on Earth's surface. Lines of latitude horizontally extend around Earth and provide a north-south measurement. The origin (Earth's center) can be that of the default world coordinate system (WCS) or the current user coordinate system (UCS). See **Figure 4-1A**.

Figure 4-1.
A—Lines of longitude, representing the highlighted latitudinal segments in the illustration, run from north to south. Lines of latitude, representing the highlighted longitudinal segments, run from east to west. B—Spherical coordinates require a distance, an angle in the XY plane, and an angle from the XY plane.

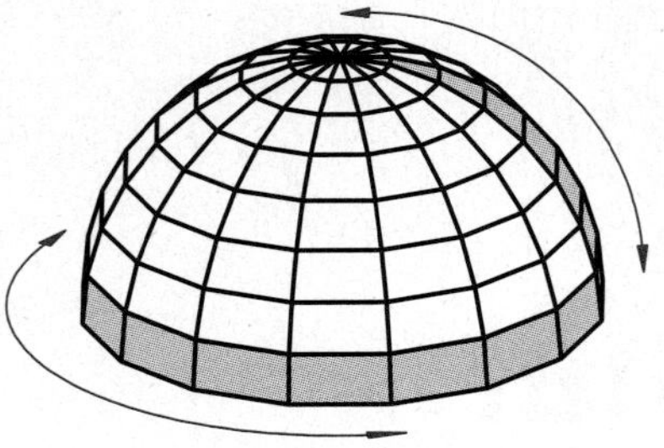

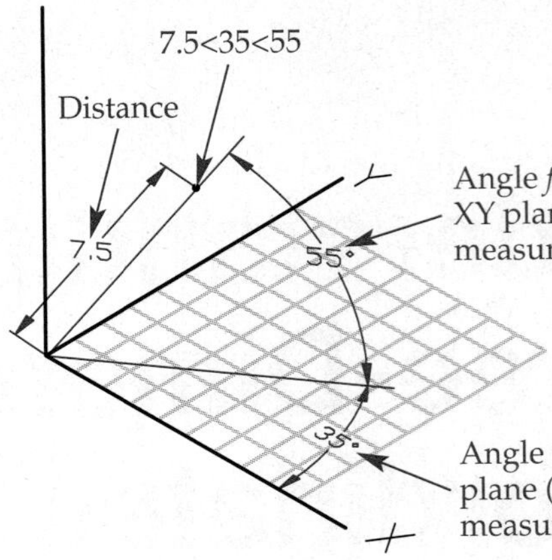

When entering spherical coordinates, the longitude measurement is expressed as the angle *in* the XY plane and the latitude measurement is expressed as the angle *from* the XY plane. See **Figure 4-1B**. A distance from the origin is also provided. The coordinates represent a measurement from the equator toward either the North or South Pole on Earth's surface. The following spherical coordinate entry is shown in **Figure 4-1B**.

7.5<35<55

This coordinate represents an ***absolute*** spherical coordinate, which is measured from the origin of the current UCS. Spherical coordinates can also be entered as ***relative*** coordinates. For example, a point drawn with the relative spherical coordinate @2<35<45 is located two units from the last point, at an angle of 35° *in* the XY plane, and at a 45° angle *from* the XY plane.

If dynamic input is turned on, the "second" or "next" coordinate entry is automatically a *relative* entry (by default). The @ symbol is not entered. To enter *absolute* coordinates with dynamic input turned on, enter an asterisk (*) before the first coordinate.

PROFESSIONAL TIP

Spherical coordinates are useful for locating features on a spherical surface. For example, they can be used to specify the location of a hole drilled into a sphere or a feature located from a specific point on a sphere. If you are working on such a spherical object, you might consider locating a UCS at the center of the sphere, then creating several different user coordinate systems rotated at different angles on the surface of the sphere. Any time a location is required, spherical coordinates can be used. Working with UCSs is introduced later in this chapter.

Using Spherical Coordinates

Spherical coordinates are well suited for locating points on the surface of a sphere. In this section, you will draw a solid sphere and then locate a second solid sphere with its center on the surface of the first sphere.

To draw the first sphere, select the **SPHERE** command. Specify the center point as 7,5 and the radius as 1.5 units. Display a southeast isometric pictorial view of the sphere. Alternately, you can use the view cube to create a different pictorial view. Also, set the Wireframe visual style current and switch to a parallel projection. Your drawing should look similar to **Figure 4-2A**.

Since you know the radius of the sphere, but the center of the sphere is not at the origin of the current UCS (the WCS), a relative spherical coordinate will be used to draw the second sphere. The sphere you drew is a solid and, as such, you can snap to its center using object snap. Set the center running object snap and then enter the **SPHERE** command again to draw the second sphere:

Specify center point or [3P/2P/Ttr]: **FROM**↵
Base point: *(use the* **Center** *object snap to select the center of the existing sphere)*
<Offset>: **@1.5<30<60**↵ *(1.5 is the radius of the first sphere)*
Specify radius or [Diameter]: **.4**↵

The objects should now appear as shown in **Figure 4-2B**. The center of the new sphere is located on the surface of the original sphere. This is clear after setting the Conceptual visual style current, **Figure 4-2C**. If you want the surfaces of the spheres to be tangent, add the radius value of each sphere (1.5 + .4) and enter this value when prompted for the offset from the center of the first sphere:

<Offset>: **@1.9<30<60**↵

Notice in **Figure 4-2B** that the polar axes of the two spheres are parallel. This is because both objects were drawn using the same UCS, which can be misleading unless you understand how objects are constructed based on the current UCS. Test this by locating a cone on the surface of the large sphere, just below the small sphere. First, display a 3D wireframe view of the objects. Then, select the **CONE** command and continue as follows:

Specify center point of base or [3P/2P/Ttr/Elliptical]: **FROM**↵
Base point: **CEN**↵
of *(pick the large sphere)*
<Offset>: **@1.5<30<30**↵
Specify base radius or [Diameter]: **.25**↵
Specify height or [2Point/Axis endpoint/Top radius]: **1**↵ *(before entering this value, make sure to move the crosshairs above the base of the cone to apply the value in the positive Z direction)*

The result of this construction with the Conceptual visual style set current is shown in **Figure 4-3**. Notice how the axis of the cone is parallel to the polar axis of the sphere.

Figure 4-2.
A—A three-unit diameter sphere shown from the southeast isometric viewpoint. B—A .8-unit diameter sphere is drawn with its center located on the surface of the original sphere. Also, lines have been drawn between the poles of the spheres. Notice how the polar axes are parallel. C—The objects after the Conceptual visual style is set current.

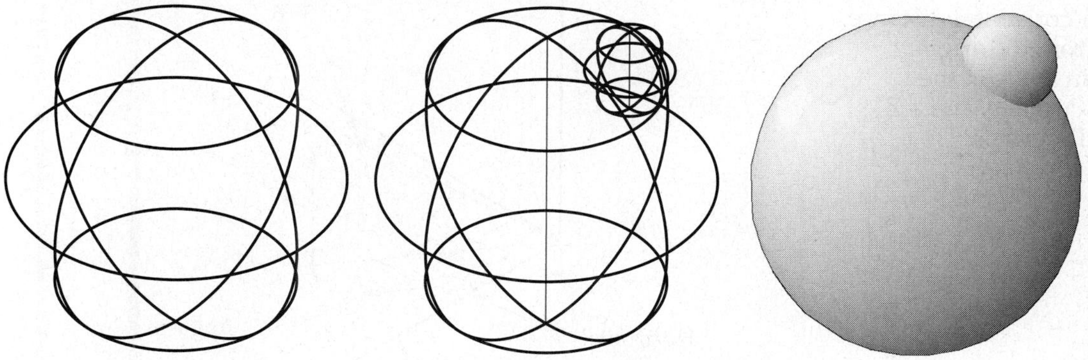

A B C

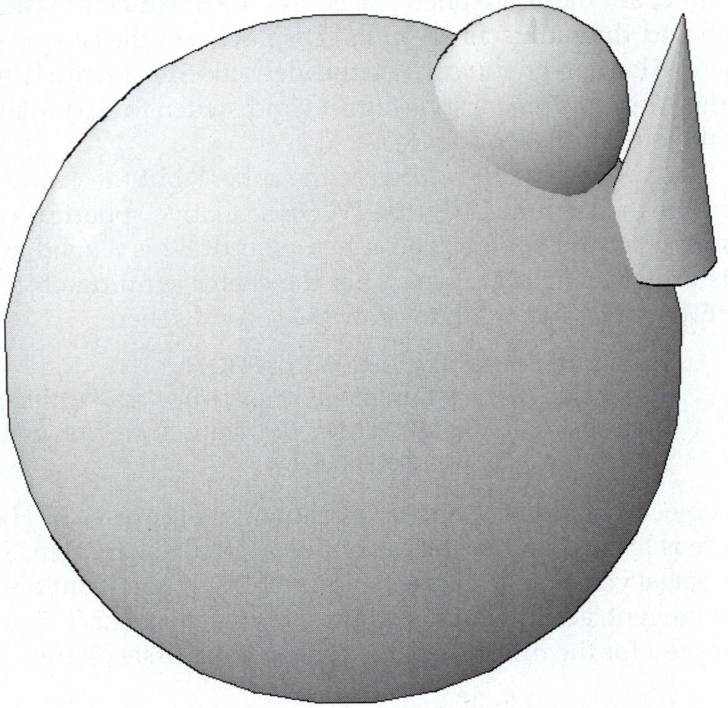

Figure 4-3.
The axis lines of objects drawn in the same user coordinate system are parallel. Notice that the cone does not project from the center of the large sphere.

To draw the cone so that its axis projects from the center of the sphere, you will need to change the UCS. This is discussed later in the chapter.

Introduction to Cylindrical Coordinates

Locating a point in space with ***cylindrical coordinates*** is similar to locating a point on an imaginary cylinder. Cylindrical coordinates have three values. The first value represents the horizontal distance from the origin, which can be thought of as the radius of a cylinder. The second value represents the angle in the XY plane, or the rotation of the cylinder. The third value represents a vertical dimension measured up from the polar coordinate in the XY plane, or the height of the cylinder. See **Figure 4-4**. The absolute cylindrical coordinate shown in the figure is:

7.5<35,6

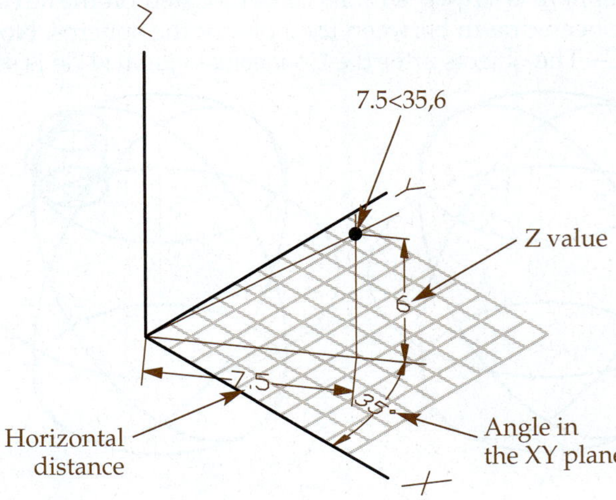

Figure 4-4.
Cylindrical coordinates require a horizontal distance from the origin, an angle in the XY plane, and a Z dimension.

Like spherical coordinates, cylindrical coordinates can also be entered as relative coordinates. For example, a point drawn with the relative cylindrical coordinate @1.5<30,4 is located 1.5 units from the last point, at an angle of 30° in the XY plane of the previous point, and at a distance of four units up from the XY plane of the previous point.

NOTE

Turn off dynamic input before entering cylindrical coordinates. Dynamic input does not correctly interpret the entry when cylindrical coordinates are typed.

Using Cylindrical Coordinates

Cylindrical coordinates work well for attaching new objects to a cylindrical shape. An example of this is specifying coordinates for a pipe that must be attached to an existing pipe, tank, or vessel. In **Figure 4-5**, a pipe must be attached to a 12′ diameter tank at a 30° angle from horizontal and 2′-6″ above the floor. In order to draw the pipe properly as a cylinder, you will have to change the UCS, which you will learn how to do later in this chapter. An attachment point for the pipe can be drawn using the **POINT** command and cylindrical coordinates. First, set the drawing units to architectural. This can be done by selecting **Drawing Utilities**>**Units** from the **Application Menu** and specifying the length units as Architectural in the **Drawing Units** dialog box. Next, set the **PDMODE** system variable to 3. Enter the **POINT** command and continue as follows:

Current point modes: PDMODE=3 PDSIZE=0'-0"
Specify a point: **FROM**↵
Base point: **CEN**↵
of *(pick the base of the cylinder)*
<Offset>: **@6'<30,2'6"**↵ *(The radius of the tank is 6'.)*

The point can now be used as the center of the pipe (cylinder), **Figure 4-5B**. However, if you draw the pipe now, it will be parallel to the tank (large cylinder). By changing the UCS, as shown in **Figure 4-5C**, the pipe can be correctly drawn. You will learn how to do this later in this chapter.

Figure 4-5.
A—A plan view of a tank shows the angle of the pipe attachment. B—A 3D view from the southeast quadrant shows the pipe attachment point located with cylindrical coordinates. C—By creating a new UCS, the pipe can be drawn as a cylinder and correctly located without editing.

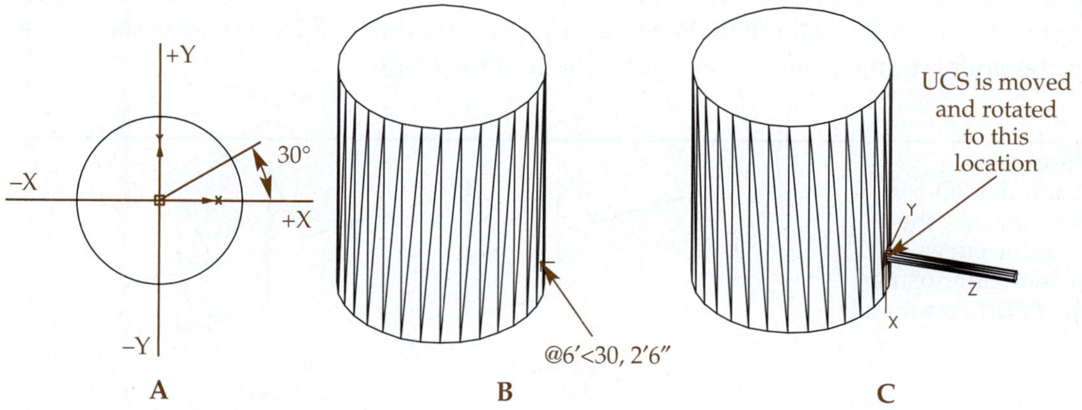

Exercise 4-1

www.g-wlearning.com/CAD/

Complete the exercise on the companion website.

3D Polylines

A polyline drawn with the **PLINE** command is a 2D object. All segments of the polyline must be drawn parallel to the XY plane of the current UCS. A ***3D polyline***, on the other hand, can be drawn in 3D space. The Z coordinate value can vary from point to point in the polyline.

3DPOLY

Ribbon
Home
> Draw

3D Polyline

Type
3DPOLY
3P

The **3DPOLY** command is used to draw 3D polylines. Any form of coordinate entry is valid for drawing 3D polylines. If polar tracking is on when using the **3DPOLY** command, you can pick points in the Z direction if the polar tracking alignment path is parallel to the Z axis.

The **Close** option can be used to draw the final segment and create a closed shape. There must be at least two segments in the polyline to use the **Close** option. The **Undo** option removes the last segment without canceling the command.

The **PEDIT** command can be used to edit 3D polylines. The **Spline curve** option of the **PEDIT** command is used to turn the 3D polyline into a B-spline curve based on the vertices of the polyline. A regular 3D polyline and the same polyline turned into a B-spline curve are shown in **Figure 4-6**.

Exercise 4-2

www.g-wlearning.com/CAD/

Complete the exercise on the companion website.

Introduction to Working with User Coordinate Systems

All points in a drawing or on an object are defined with XYZ coordinate values (rectangular coordinates) measured from the 0,0,0 origin. Since this system of coordinates is fixed and universal, AutoCAD refers to it as the ***world coordinate system (WCS)***. A ***user coordinate system (UCS)***, on the other hand, can be defined with its origin at any location and with its three axes in any orientation desired, while remaining at 90° to each other. The **UCS** command is used to change the origin, position, and rotation of the coordinate system to match the surfaces and features of an object under construction. The symbol that identifies the orientation of the coordinate system is called the ***UCS icon***. When set up to do so, the UCS icon reflects the changes in the orientation of the UCS and placement of the origin.

Figure 4-6.
A regular 3D polyline and the B-spline curve version after using the **PEDIT** command.

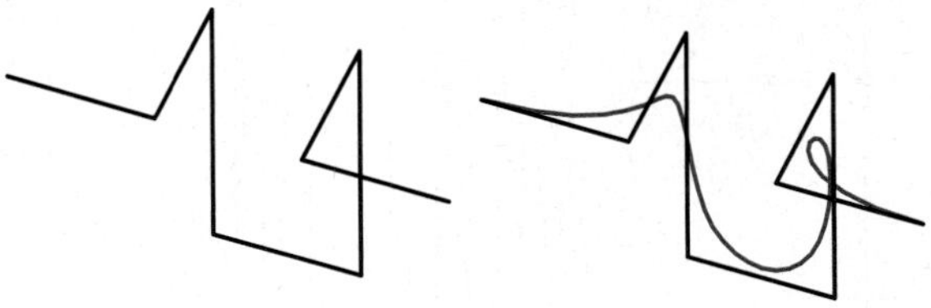

Regular 3D
Polyline

B-spline
Curve

The **UCS** command is used to create and manage UCSs. The quickest method of working with the UCS icon and the **UCS** command is to use the UCS icon shortcut menu and grip menus. Using this method will save you several steps and is described in this section. The **UCS** command and its options can be accessed on the **Coordinates** panel of the **Home** and **Visualize** tabs on the ribbon or by typing the command. Three selections in the ribbon provide access to all UCS options. These selections are introduced here and discussed in detail later in this chapter.

Ribbon
Home
> **Coordinates**
Visualize
> **Coordinates**
UCS
Type
UCS

UCS

- **UCS.** Picking this button executes the **UCS** command. This command allows you to create a named UCS and provides access to all of the command options. The **UCS** command can also be selected by picking on the UCS drop-down list below the view cube. All of the command options are covered later in this chapter.
- **Named UCS Combo Control.** This drop-down list contains the six orthographic UCS options, which are covered later in this chapter. Any saved UCSs will be listed here, too.
- **UCS, Named UCS.** Picking this button displays the **UCS** dialog box. The three tabs in the dialog box contain a variety of UCS and UCS icon options and settings. These options and settings are described as you progress through the chapter.

The UCS Icon Shortcut Menu and Grip Menus

Creating and managing user coordinate systems is described in detail later in this chapter, but this is an introduction to a quick and dynamic method of working with the UCS and the UCS icon.

Many of the functions associated with managing the UCS and the UCS icon can be accomplished using the UCS icon shortcut menu. Grip menus are also available when the UCS icon is selected. The origin grip menu is useful for moving the UCS origin or realigning its orientation. For example, to move the UCS to the base of the wedge in **Figure 4-7A**, first select the UCS icon, then place the pointer over the UCS icon origin grip. Hover over the grip, but do not select it. The grip menu displays three options. Select **Move Origin Only**, then use an object snap to move the UCS origin to the base of the wedge. See **Figure 4-7B**.

To align a new UCS with the face of an object, select the icon and pick **Move and Align**. See **Figure 4-8A**. Then drag the icon to the face of the object. The alignment of the UCS icon depends on which edge of the face the pointer is moved over. In **Figure 4-8B**, the pointer is first moved over the pointed end of the wedge.

Figure 4-7.
A—The **Move Origin Only** option can be selected from the UCS icon origin grip menu. B—Use an object snap to relocate the UCS origin.

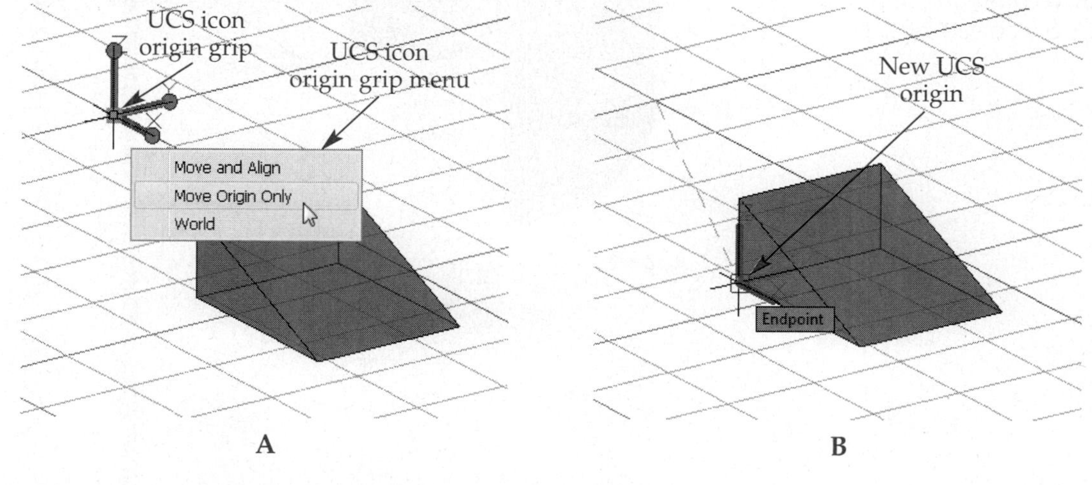

Figure 4-8.
A—To align the UCS with the face of an object, select the UCS icon and pick **Move and Align** from the UCS icon origin grip menu. B—The UCS icon is dragged over an edge and a point is selected using object snaps.

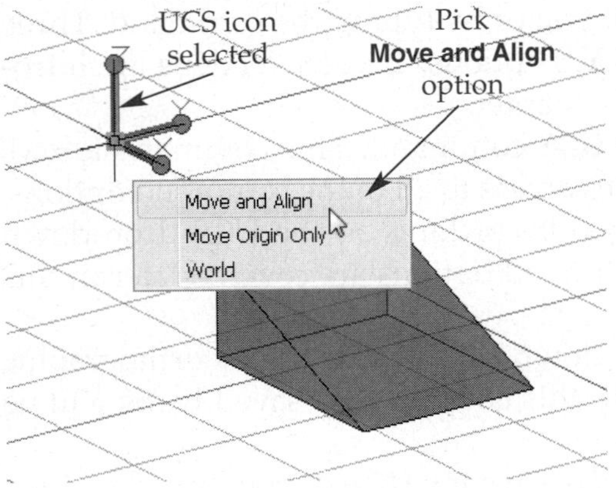

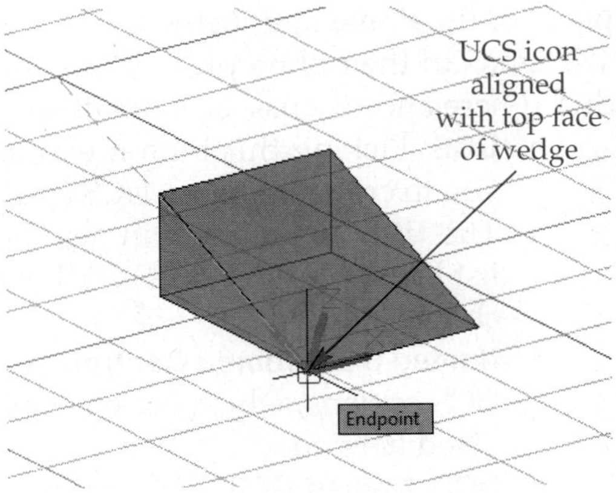

All of the options of the **UCS** command can be selected from the UCS icon shortcut menu, which is accessed by right-clicking on the UCS icon. See **Figure 4-9**. The options are discussed in this chapter as they apply.

Moving and Rotating the UCS

Earlier in this chapter, you used spherical coordinates to locate a small sphere on the surface of a larger sphere. You also drew a cone with the center of its base on the surface of the large sphere. However, the axis of the cone, which is a line from the center of the base to the tip of the cone, is not pointing to the center of the sphere. Refer to **Figure 4-3**. This is because the Z axes of the large sphere and cone are parallel to the world coordinate system (WCS) Z axis. The WCS is the default coordinate system of AutoCAD.

In order for the axis of the cone to project from the sphere's center point, the UCS must be changed. Working with different UCSs is discussed in more detail in the next section. However, the following is a quick overview and describes how to draw a cone with its axis projecting from the center of the sphere.

Figure 4-9.
The UCS icon shortcut menu provides access to the options of the **UCS** command and UCS icon settings.

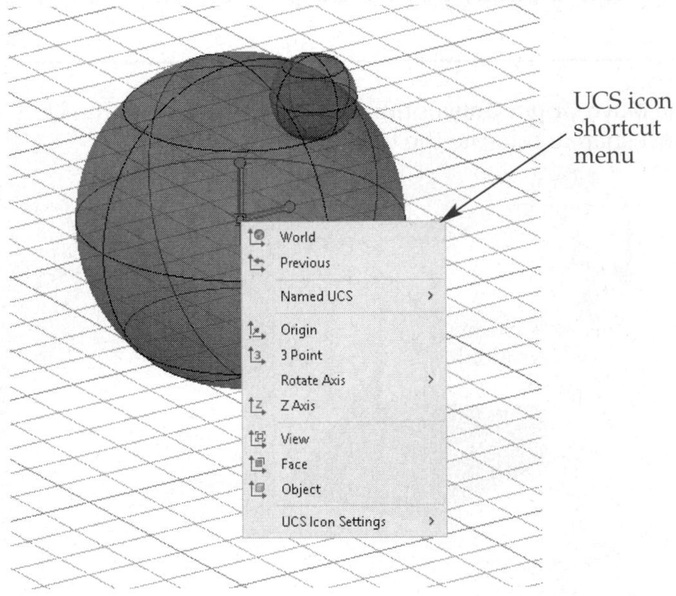

First, draw a three-unit diameter sphere with its center at 7,5. Display the drawing from the southeast isometric preset. To help see how the UCS is changing, make sure the UCS icon is displayed at the origin of the current UCS. The UCS icon drop-down list is found in the **Coordinates** panel of the **Home** and **Visualize** tabs of the ribbon and provides access to three options of the **UCSICON** command. See **Figure 4-10**. Pick the **Show UCS Icon at Origin** button to ensure the icon is displayed at the origin. This option can also be selected from the UCS icon shortcut menu. Also, set the X-ray visual style current, and set the opacity to 25%. In the **Visual Styles** panel of the **Visualize** tab, use the **Opacity** slider to adjust the value. You can also type a numerical value to the right of the slider.

Now, the sphere is drawn and the UCS icon is displayed at the origin of the current UCS (or at the lower-left corner of the screen, depending on the zoom level). However, the WCS is still the current user coordinate system. You are ready to start changing the UCS to meet your needs.

Pick the UCS icon and move the pointer over the origin grip. Select **Move Origin Only**. Now move the UCS origin to the center of the sphere using the center object snap. Notice that the UCS icon is now displayed at the center of the sphere, and its orientation has remained the same, **Figure 4-11**. Also, if the grid is displayed, the red X and green Y axes of the grid intersect at the center of the sphere.

Ribbon
Home
> Coordinates
Visualize
> Coordinates
Origin
Type
UCS

UCS

Study **Figure 4-12** and continue as follows. Keep in mind that the point you are locating—the center of the cone on the sphere's surface—is 30° from the X axis and 30° from the XY plane. For ease of visualization, zoom in so the sphere fills the screen.

Right-click on the UCS icon to display the UCS icon shortcut menu. Then, select **Rotate Axis>Z**. Enter 30 for the rotation angle or select it dynamically. Watch the position of the UCS icon change when you press [Enter]. See **Figure 4-12B**.

Ribbon
Home
> Coordinates
Visualize
> Coordinates
Z
Type
UCS

UCS

Figure 4-10.
The UCS icon drop-down list provides access to the **Origin**, **Off**, and **On** options of the **UCSICON** command.

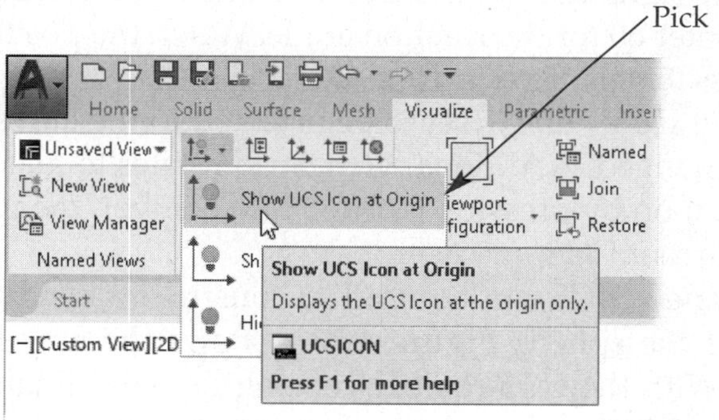

Figure 4-11.
The UCS origin is moved to the center of the sphere.

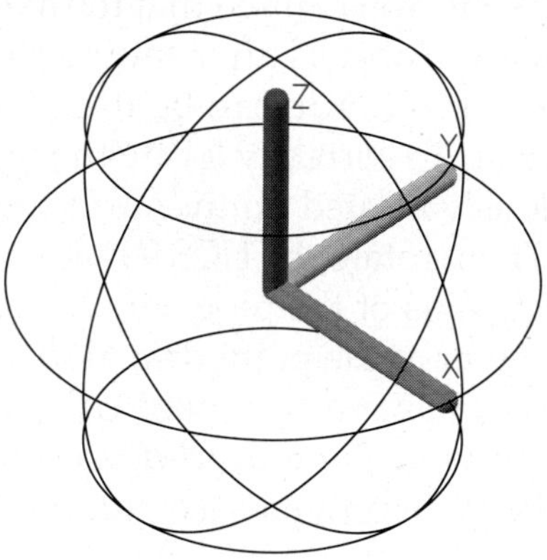

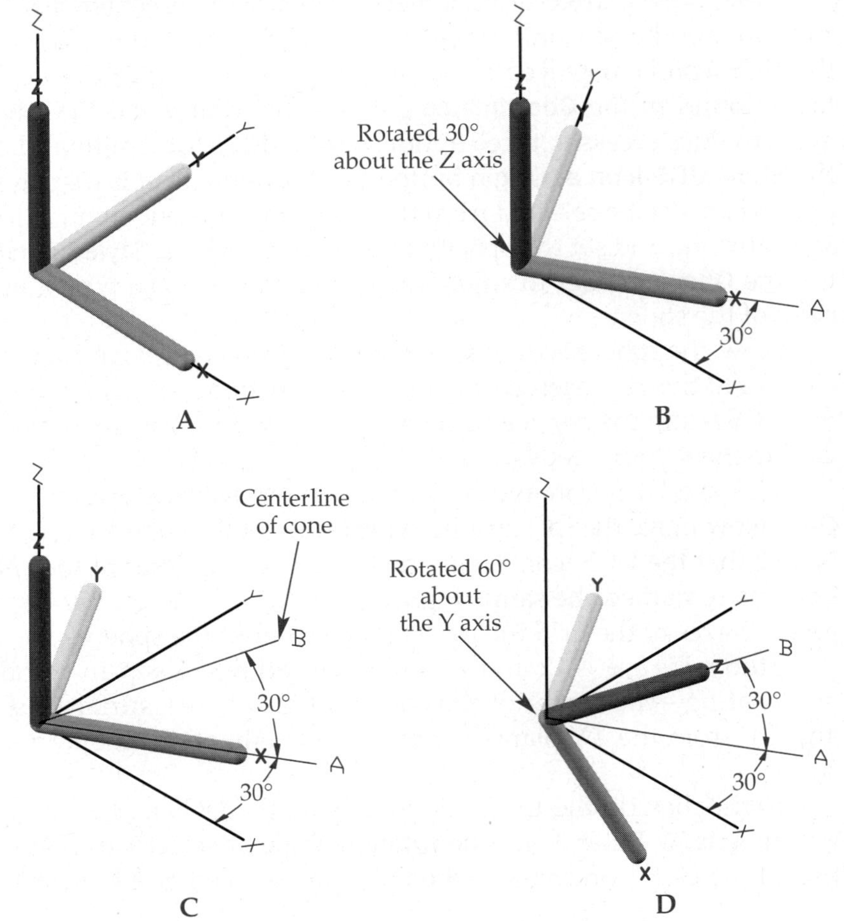

Figure 4-12.
A—The world coordinate system. B—The new UCS is rotated 30° in the XY plane about the Z axis. C—A line rotated up 30° from the XY plane represents the axis of the cone. D—The UCS is rotated 60° about the Y axis. The centerline of the cone coincides with the Z axis of this UCS.

UCS

Ribbon
Home
> Coordinates
Visualize
> Coordinates
Y
Type
UCS

Next, right-click on the UCS icon and select **Rotate Axis>Y** in the UCS icon shortcut menu. Enter 60 for the rotation angle. Watch the position of the UCS icon change when you press [Enter]. See **Figure 4-12D**.

If the view cube is set to be oriented to the current UCS, then it rotates when the UCS is changed. Additionally, the grid rotates to match the new UCS. If the grid is not on, turn it on by pressing [Ctrl]+[G]. Remember, the grid is displayed on the XY plane of the current UCS.

The new UCS can be used to construct a cone with its axis projecting from the center of the sphere. **Figure 4-13A** shows the new UCS located at the center of the sphere. With the UCS rotated, rectangular coordinates can be used to draw the cone. Enter the **CONE** command and specify the center of the base as 0,0,1.5 (the radius of the sphere is 1.5 units). Enter a radius of .25 and a height of 1. The completed cone is shown in **Figure 4-13B**. You can see that the axis projects from the center of the sphere. **Figure 4-13C** shows the objects after setting the Conceptual visual style current.

This same basic procedure can be used in the tank and pipe example presented earlier in this chapter. To correctly locate the pipe (cylinder), first move the UCS origin to the point previously located on the surface of the cylinder. Next, rotate the UCS 30° about the Z axis. Then, rotate the UCS 90° about the Y axis. The Z axis of this new UCS aligns with the long axis of the pipe. Finally, use rectangular coordinates to draw the cylinder with its center at the point drawn in **Figure 4-5B**.

Once you have changed to a new UCS, you can quickly return to the WCS by picking WCS in the view cube drop-down list or by picking **World** in the UCS icon shortcut menu. The WCS provides a common "starting place" for creating new UCSs.

Figure 4-13.
A—A new UCS is created with the Z axis projecting from the center of the sphere. B—A cone is drawn using the new UCS. The axis of the cone projects from the center of the sphere. C—The objects after the Conceptual visual style is set current.

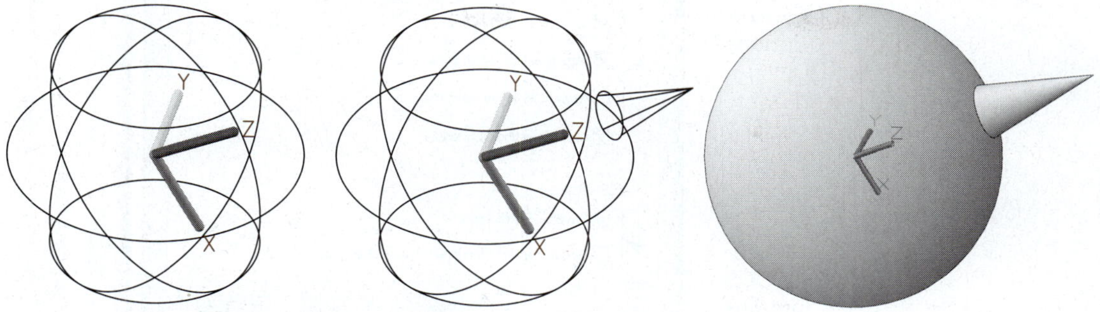

Exercise 4-3

www.g-wlearning.com/CAD/

Complete the exercise on the companion website.

Working with User Coordinate Systems

Once you understand a few of the basic options of user coordinate systems, creating 3D models becomes an easy and quick process. The following sections show how to display the UCS icon, change the UCS in order to work on different surfaces of a model, and name and save a UCS. As you saw in the previous section, working with UCSs is easy.

Displaying the UCS Icon

When a new drawing is started based on the acad3D.dwt template, the UCS icon is located at the WCS origin in the middle of the viewport. The display of the UCS icon is controlled by the **UCSICON** command. Turn the UCS icon on and off using the **Show UCS Icon** and **Hide UCS Icon** buttons on the **Coordinates** panel in the **Home** or **Visualize** tab of the ribbon. Refer to Figure 4-10.

If your drawing does not require viewports and altered coordinate systems, you may want to turn the icon off. The icon disappears until you turn it on again. If you redisplay the UCS icon and it does not appear at the origin, simply pick the **Show UCS Icon at Origin** button on the **Coordinates** panel in the ribbon.

You can also turn the icon on or off and set the icon to display at the UCS origin point using the options in the **Settings** tab of the **UCS** dialog box. See Figure 4-14. This dialog box is displayed by picking the **UCS, Named UCS...** button on the **Coordinates** panel in the ribbon or typing the **UCSMAN** command.

Notice that the **Allow Selecting UCS icon** option is checked by default in the **UCS Icon settings** area of the **Settings** tab. When this option is checked, it enables the display of the UCS icon grips. The UCS icon grips provide for dynamic and intuitive moving and rotating of the UCS icon. The **UCSSELECTMODE** system variable controls the **Allow Selecting UCS icon** setting and is set to 1 (on) by default.

Figure 4-14.
Setting UCS and UCS icon options in the **UCS** dialog box.

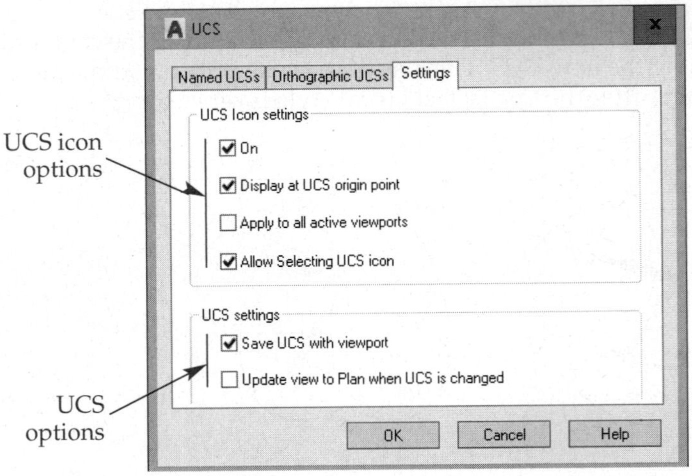

PROFESSIONAL TIP

It is recommended that you have the UCS icon turned on at all times when working in 3D drawings. It provides a quick indication of the current UCS.

Changing the Coordinate System

To construct a three-dimensional object, you must draw shapes at many different angles. Different planes are needed to draw features on angled surfaces. To construct these features, it is easiest to rotate the UCS to match any surface on an object. The following example illustrates this process.

The object in **Figure 4-15** has a cylinder on the angled surface. The **EXTRUDE** command, which is discussed in Chapter 6, is used to create the base of the object. The cylinder is then drawn on the angled feature. In Chapter 13, you will learn how to dimension the object as shown in **Figure 4-15**.

Figure 4-15.
This object can be constructed by changing the orientation of the coordinate system.

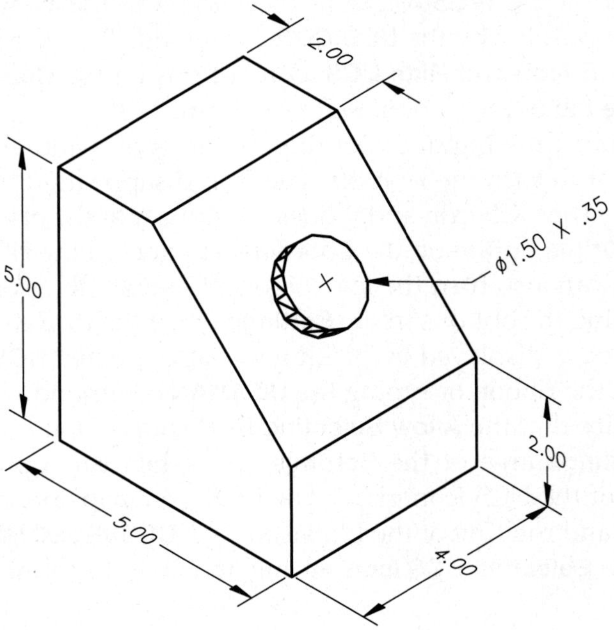

The first step in creating this model is to draw the side view of the base as a wireframe. You could determine the X, Y, and Z coordinates of each point on the side view and enter the coordinates. However, a lot of typing can be saved if all points share a Z value of 0. By rotating the UCS, you can draw the side view entering only X and Y coordinates. Start a new drawing and display the southeast isometric view. If the UCS icon is off, turn it on and display it at the origin.

Now, rotate the UCS 90° around the X axis. The quickest way to do this is to click directly on the UCS icon. Notice that as the pointer is moved over the UCS icon, the color of the icon changes to a light yellow. See **Figure 4-16A**. When you click on the icon, it displays grips. A square origin grip is displayed at the origin, and circular axis grips are displayed at the axis endpoints. See **Figure 4-16B**.

Once the UCS icon grips are displayed, move the pointer over the Y axis grip. The Y grip menu is displayed, **Figure 4-16C**. Select the **Rotate Around X Axis** option. The default value is 90. This is the desired rotation angle, so press [Enter]. The new UCS is now oriented parallel to the side of the new object. The UCS icon is displayed at the origin of the UCS. If needed, pan the view so the UCS icon is near the center of the view.

UCS

Ribbon
Home
> Coordinates
Visualize
> Coordinates

X

Type
UCS

Figure 4-16.
Using the UCS icon to rotate the UCS. A—When the pointer is moved over the UCS icon, the color of the icon changes to light yellow. B—Selecting the UCS icon displays a grip at the origin and grips at all three axis endpoints. C—The menu displayed when the pointer is moved over the Y axis grip to highlight the grip. To rotate the UCS 90° about the X axis, select the **Rotate Around X Axis** option.

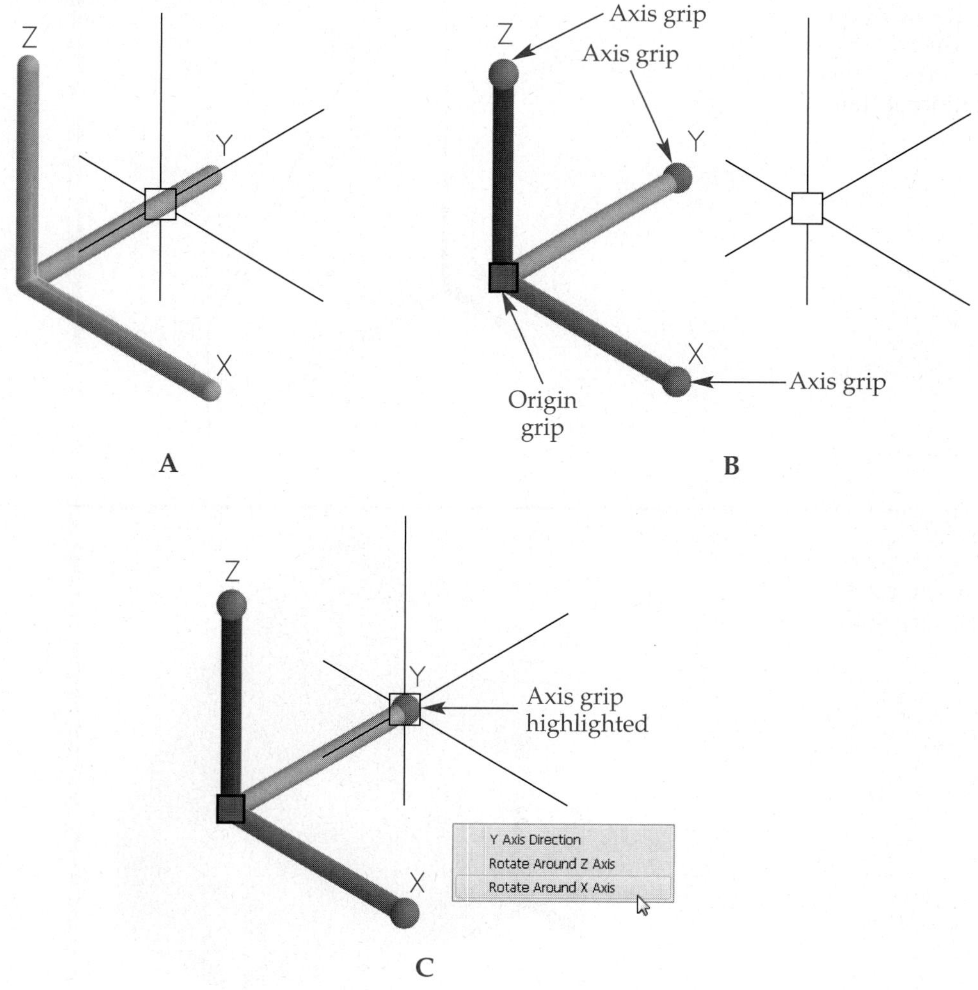

Next, use the **PLINE** command to draw the outline of the side view. Refer to the coordinates shown in **Figure 4-17**. When entering coordinates, you may want to turn off dynamic input. As an alternative, type a pound sign (#) before each coordinate with dynamic input turned on. This temporarily overrides the dynamic input of relative coordinates, which is controlled by the **DYNPICOORDS** system variable. The **PLINE** command is used instead of the **LINE** command for this model because a closed polyline can be extruded into a solid. Be sure to use the **Close** option to draw the final segment. A wireframe of one side of the object is created. Notice the orientation of the UCS icon.

Now, the **EXTRUDE** command is used to create the base as a solid. This command is covered in detail in Chapter 6. Make sure the same UCS used to create the wireframe side is current. Then, select the **EXTRUDE** command by picking the **Extrude** button on the **Modeling** panel of the **Home** tab in the ribbon. When prompted to select objects, pick the polyline and then press [Enter]. Next, move the mouse so the preview extends to the right and enter an extrusion height of 4. If dynamic input is off, you must enter –4.

The base is created as a solid. See **Figure 4-18**. You may want to switch to a parallel projection, as shown in the figure.

Saving a Named UCS

Once you have created a new UCS that may be used again, it is best to save it for future use. For example, you just created a UCS used to draw the wireframe of one side

Figure 4-17.
A wireframe of one side of the base is created. Notice the orientation of the UCS.

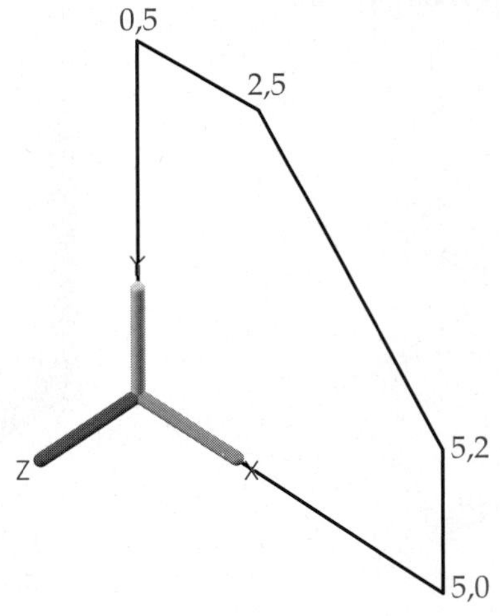

Figure 4-18.
The wireframe is extruded to create the base as a solid.

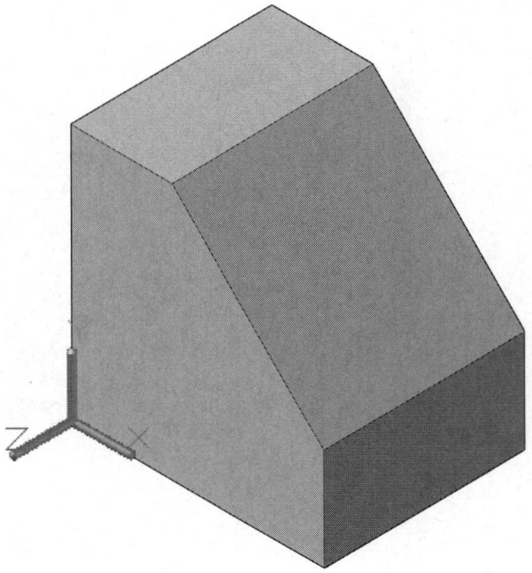

of the object. To save this UCS, use the UCS icon shortcut menu. Move the pointer over the UCS icon and right-click. See **Figure 4-19**. Select **Named UCS>Save** in the shortcut menu, type a name at the prompt, and press [Enter]. The UCS can also be saved by using the **Named>Save** option of the **UCS** command or the **UCS** dialog box.

Ribbon
Home
> Coordinates
Visualize
> Coordinates

UCS, Named
UCS...

Type
UCS

UCS

If using the dialog box, right-click on the entry Unnamed and pick **Rename** in the shortcut menu. See **Figure 4-20**. You can also pick once or double-click on the highlighted name. Then, type the new name in place of Unnamed and press [Enter]. A name can have up to 255 characters. Numbers, letters, spaces, dollar signs ($), hyphens (–), and underscores (_) are valid. Use this method to save a new UCS or to rename an existing one. Now, the coordinate system is saved and can be easily recalled for future use.

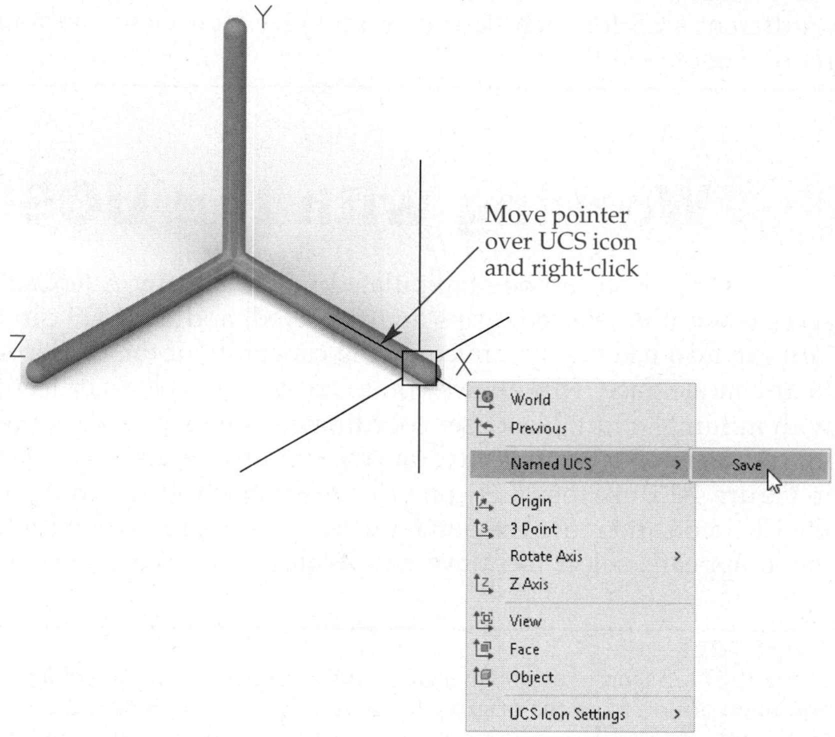

Figure 4-19.
To save a new UCS, move the pointer over the UCS icon and right-click to display the UCS icon shortcut menu. Pick **Named UCS>Save**.

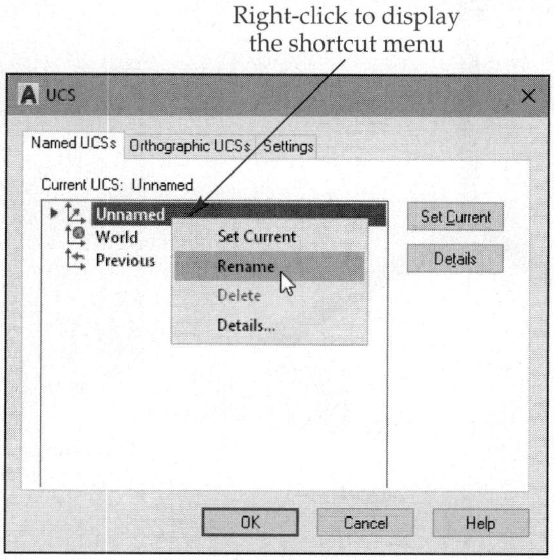

Figure 4-20.
Select **Rename** in the **UCS** dialog box to enter a name and save the Unnamed UCS.

NOTE

If Unnamed does not appear in the **UCS** dialog box, AutoCAD is having a problem. Exit the dialog box and enter the **UCS** command on the command line. Then, use the **Named>Save** option to save the unnamed UCS. The saved UCS will then appear in the **UCS** dialog box.

PROFESSIONAL TIP

Most drawings can be created by rotating the UCS as needed without saving it. If the drawing is complex with several planes, each containing a large amount of detail, you may wish to save a UCS for each detailed face. Then, restore the proper UCS as needed. For example, when working with architectural drawings, you may wish to establish a different UCS for each floor plan and elevation view and for roofs and walls that require detail work.

Working with the UCS Icon

The UCS icon can be manipulated much like any AutoCAD entity. As you have seen, when it is selected, grips are displayed, and the UCS can be rotated with ease. You can also use the dynamic moving capability of the UCS icon to relocate the UCS to any model face. This allows you to create new objects or features on existing faces with minimal effort. New user coordinate systems you create can be saved as needed for future use. For example, to draw the cylinder on the angled face of the object shown in **Figure 4-15**, use the UCS icon grips to establish a UCS on the angled face. First, pick the UCS icon and move the pointer to the origin grip to highlight it. When the shortcut menu appears, select the **Move and Align** option. See **Figure 4-21A**. Now move the

Figure 4-21.
Using the UCS icon to establish a UCS on the angled face of the object. A—Select the **Move and Align** option from the origin grip menu. B—Use the **Endpoint** object snap to relocate the UCS on the angled face. The X-ray visual style is displayed for visualization purposes.

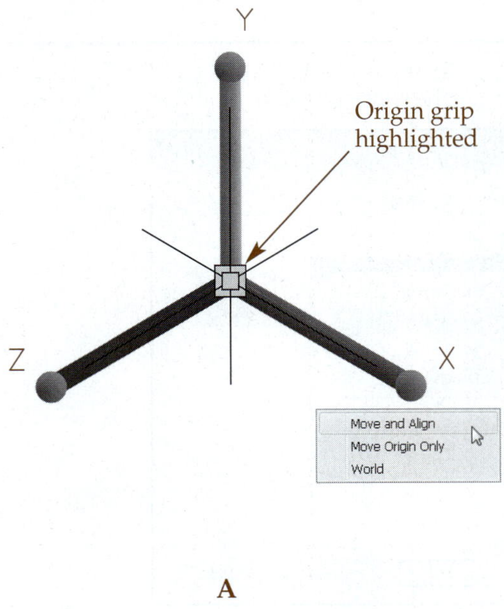

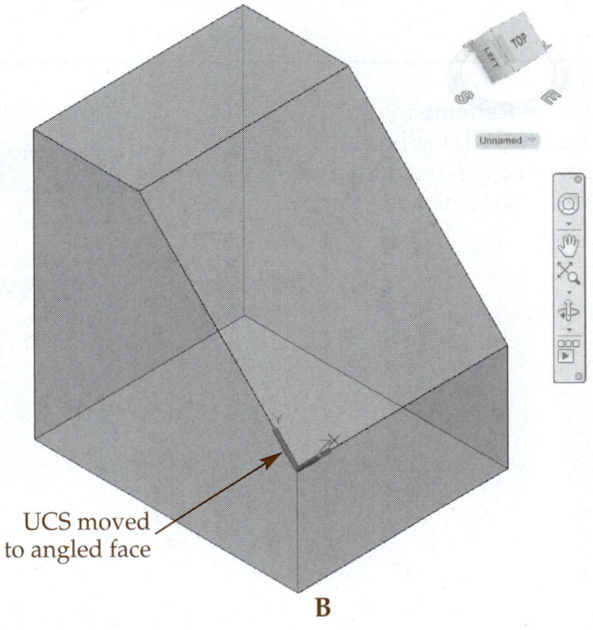

UCS icon onto the angled face of the object. Notice how the icon immediately rotates to match the orientation of the highlighted face it is resting on. Test this by moving the icon to each visible face. As you move the pointer over the object faces, note that hidden faces are not highlighted. Therefore, you cannot work on those faces. If you wish to work on a hidden face, you must first change the viewpoint to make that face visible. You can use the view cube or press and hold the [Shift] key while pressing the mouse wheel button to change the viewpoint dynamically without interrupting the current command.

Now return the UCS icon to the angled face, being careful to ensure that the icon is oriented in the manner shown in **Figure 4-21B**. The X-ray visual style is shown in **Figure 4-21B** to help show the proper UCS orientation. You may have to first move the icon to the top face to achieve the proper UCS orientation, and then move to the angled face. Use the **Endpoint** object snap to select the lower-left corner of the angled face as the new origin.

Next, select the **CYLINDER** command. You are prompted to select a center point of the base. To locate a 1.5″ diameter cylinder in the center of the angled face, use the following procedure.

1. When prompted to specify the center point of the base, press [Shift] and right-click in the drawing area and pick **3D Osnap** from the shortcut menu. Then, pick **Center of face** from the cascading menu. See **Figure 4-22A**.
2. Move the pointer over the angled face. Notice the crosshairs are aligned with the angled face. A circular 3D snap pick point appears at the center of the face. Select this point.
3. Specify the 1.5 unit diameter for the base. See **Figure 4-22B**.
4. Specify a cylinder height of .35 units. See **Figure 4-22C**.
5. The cylinder is properly located on the angled face. See **Figure 4-22D**.

If you do not wish to retain the location of the current UCS after drawing a feature, change it as needed using the UCS icon. If necessary, you can also quickly restore the previous UCS. To do so, simply right-click on the UCS icon and pick **Previous** from the UCS icon shortcut menu. Refer to **Figure 4-19**. Do this after drawing the cylinder so that the UCS is oriented parallel to the side of the object.

PROFESSIONAL TIP

Selecting 3D object snaps can be automated for solid model construction. Simply right-click on the **3D Object Snap** button on the status bar and select the snap mode you wish to use. The **3D Object Snap** button does not appear on the status bar by default. Select the **3D Object Snap** option in the status bar **Customization** flyout to add the **3D Object Snap** button to the status bar. To set multiple snap modes, pick the **Object Snap Settings...** option, and then select the appropriate modes in the **Drafting Settings** dialog box. Be sure to turn on the **3D Object Snap** button before using a modeling command.

Exercise 4-4

www.g-wlearning.com/CAD/

Complete the exercise on the companion website.

Figure 4-22.
Moving the UCS allows you to draw a cylinder on the angled face of the object. The **X-ray** visual style is displayed for visualization purposes. A—To set the center point of the base, use the **Center of face** 3D object snap and select the center of the face. B—Set the radius or diameter of the base. C—Set the height of the cylinder. D—The completed cylinder.

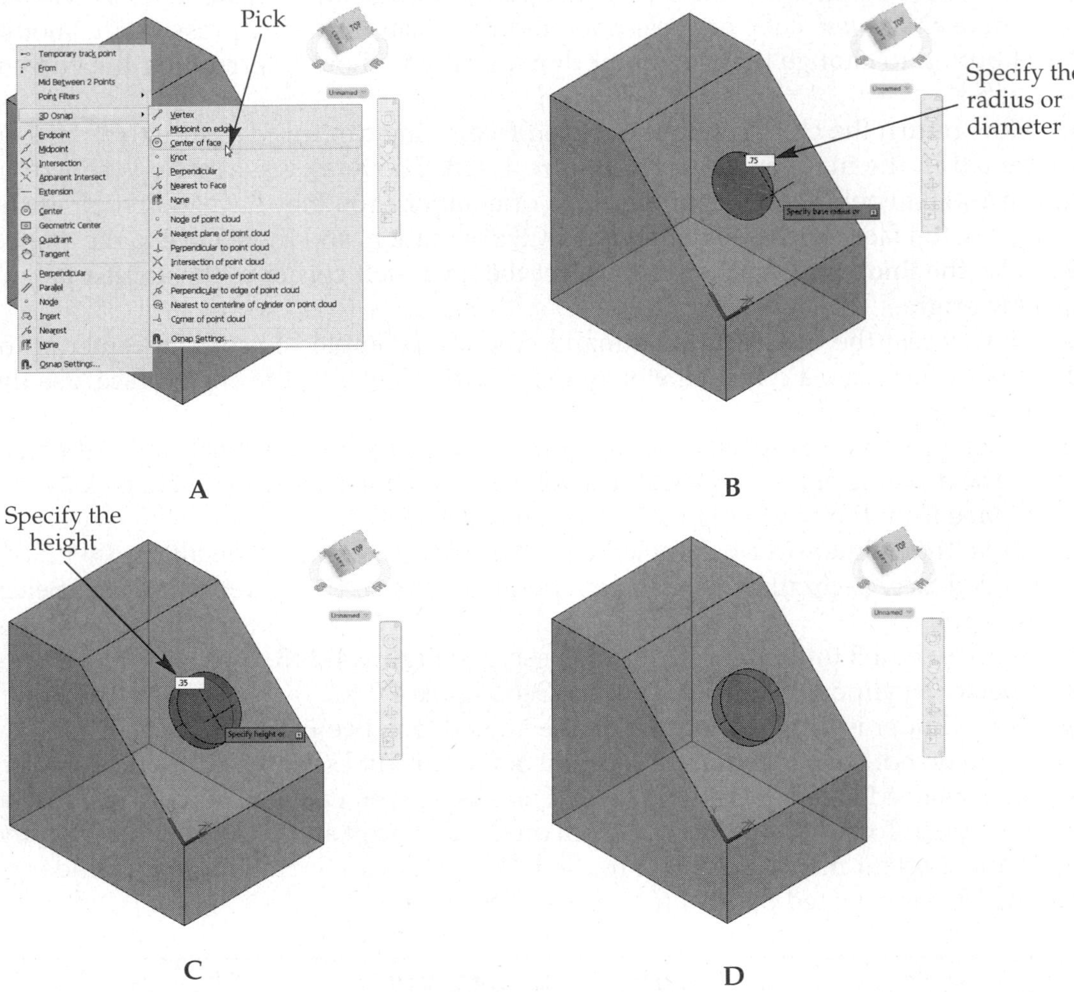

Using a Dynamic UCS

UCSDETECT

Type
UCSDETECT
[Ctrl]+[D]
[F6]
Toolbar
Status Bar
Dynamic UCS

A powerful tool for 3D modeling is the ***dynamic UCS function***. A dynamic UCS is a UCS temporarily located on any existing face of a 3D model. The function is activated by picking the **Dynamic UCS** button on the status bar, pressing the [Ctrl]+[D] key combination, pressing [F6], or setting the **UCSDETECT** system variable to 1. The **Dynamic UCS** button does not appear on the status bar by default. Select the **Dynamic UCS** option in the status bar **Customization** flyout to add the **Dynamic UCS** button to the status bar.

When the pointer is moved over a model surface when using a drawing or modeling command, the XY plane of the UCS is aligned with that surface. This is especially useful when adding primitives or shapes to model surfaces. In addition, dynamic UCSs are useful when inserting blocks and xrefs, locating text, editing 3D geometry, editing with grips, and making area calculations.

A limitation of using a dynamic UCS is that it operates as a temporary override to the current UCS. Once the operation is complete, the UCS reverts to its previous location. A more flexible option is to use the full range of capabilities inherent in the UCS icon as previously discussed.

To see how to work with a dynamic UCS, recreate the cylinder on the angled face of the model completed in the previous discussion. First, delete the existing cylinder by picking on the cylinder and pressing the [Delete] key. Make sure that the UCS is oriented parallel to the side of the object. Refer to Figure 4-18. Next, select the **CYLINDER** command. Make sure the dynamic UCS function is on. Then, move the pointer over one of the surfaces of the object. Notice that the 3D crosshairs change when they are moved over a new surface. The red (X) and green (Y) crosshairs are flat on the face. For ease of visualizing the 3D crosshairs as they are moved across different surfaces, you can display the XYZ labels on the crosshairs. Open the **Options** dialog box, select the **3D Modeling** tab, and activate the **Show labels for dynamic UCS** option in the **3D Crosshairs** area.

The **CYLINDER** command is currently prompting to select a center point of the base. If you pick a point, this sets the center of the cylinder base and temporarily relocates the UCS so its XY plane lies on the selected face. When the command is completed, the UCS reverts to its previous orientation. Using the **CYLINDER** command options and the steps presented earlier, draw a 1.5″ diameter cylinder in the center of the angled face.

When using the dynamic UCS function, experiment with the behavior of the crosshairs as they are moved over different surfaces. The orientation of the crosshairs is related to the edge of the face that they are moved over. Can you determine the pattern by which the crosshairs are turned? The X axis of the crosshairs is always aligned with the edge that is crossed.

If you want to temporarily turn off the dynamic UCS function while working in a command, press and hold the [Shift]+[Z] key combination while moving the pointer over a face. As soon as you release the keys, the dynamic UCS function is reinstated.

Additional Ways to Change the UCS

As you have seen, the UCS can be moved to any location and rotated to any angle desired using the UCS icon grips. Location and alignment are controlled using the origin grip, and exact rotation of the three different UCS axes is controlled using the axis grips. The same UCS options are available when using the **UCS** command, but the UCS icon provides a quick alternative to entering the command or making ribbon or menu selections. The following sections discuss the UCS options available when using the UCS icon grips or the **UCS** command.

Moving and Aligning the UCS

The UCS icon origin grip is used to quickly move and align the UCS. Once the UCS icon is selected to display grips, you can pick on the origin grip and drag the icon dynamically. For example, drag the UCS icon to a model surface and orient the UCS parallel to that surface. If you are using dynamic input, you can enter a distance and angle defining a new location. You can also move the pointer over the grip to display the grip menu, as previously discussed. Pick **Move Origin Only** to relocate the UCS in its current orientation to a different origin point. Pick **Move and Align** to relocate the UCS to a new location and align it with an object face. Use object snaps for accuracy.

When the UCS icon is moved along a curved surface or solid, it will always remain normal to that object. In other words, the Z axis of the UCS will project from the center point of the radius curve. See Figure 4-23. In the example shown, the UCS is moved to a point on the outer surface of the model using the **Move and Align** option. Notice the direction of the Z axis in Figure 4-23C. If dynamic input is on, a tooltip is displayed when using the **Move and Align** option. You can press the [Ctrl] key to cycle through the options. The options are displayed on the command line to indicate the current mode.

Figure 4-23.
Using the UCS icon to move the UCS to a curved surface. In this example, the UCS icon origin grip is used to access the **Move and Align** option. A—The current UCS is located at the center point of the left curve of the model. This is apparent in the top view. B—A top view of the model. C—A new UCS is established on the outer surface by moving the UCS icon to the surface and picking. Note that the Z axis is normal to the surface.

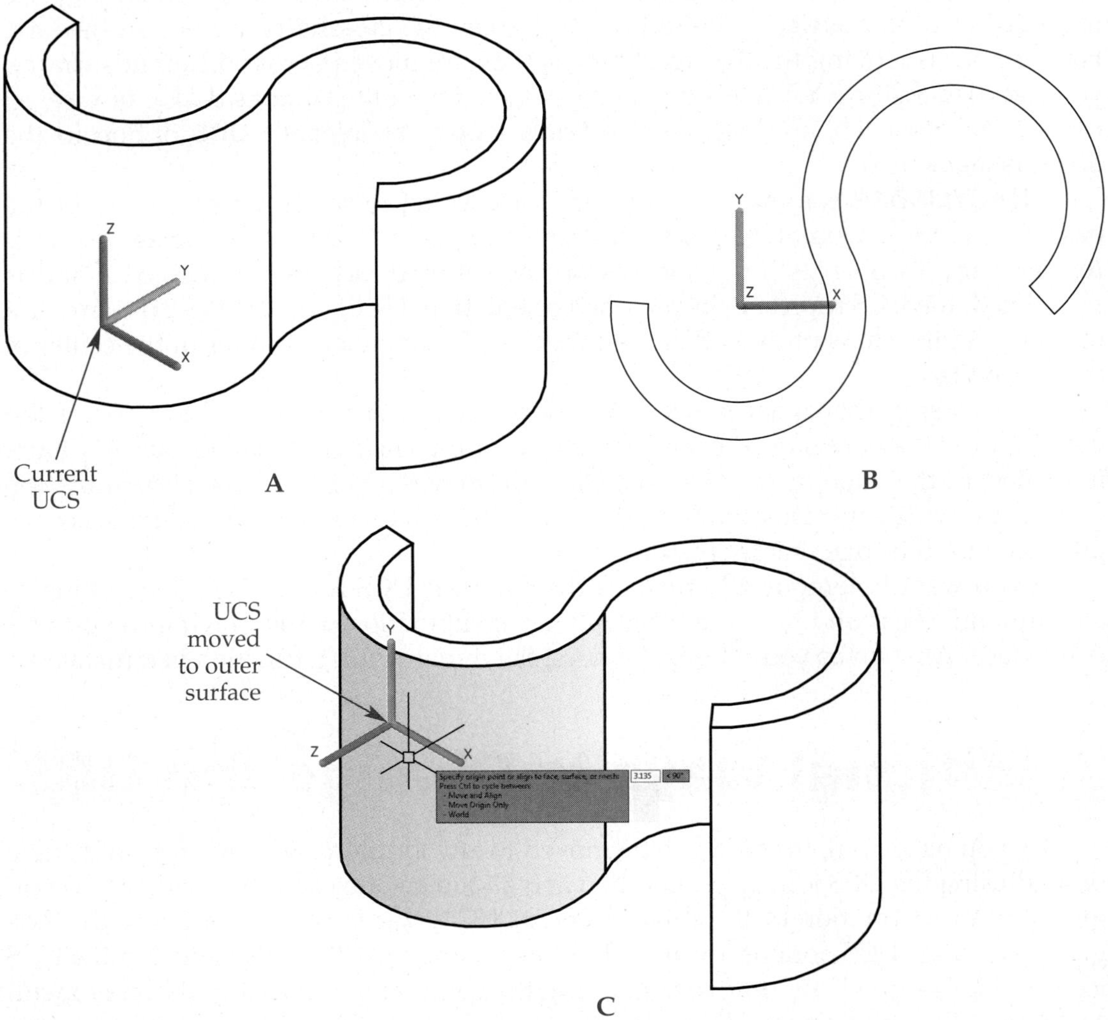

Dynamically moving and aligning the UCS to an object surface can also be achieved by using the **Object** option of the **UCS** command. This option is also available in the UCS icon shortcut menu and is discussed later in this chapter.

Rotating the UCS on an Axis

The axis grips of the UCS icon enable you to rotate the icon in any direction, and to cycle between rotation options. For example, pick the UCS icon to display the grips, and then move the pointer over the Z axis grip (but do not select it). See **Figure 4-24A**. The first item in the Z grip menu is **Z Axis Direction**. Picking this option is the same as picking the Z axis grip. Select this option and notice that it enables you to select the direction of the Z axis. Use object snaps to assist, or turn ortho on to limit selections to 90°. In addition, if dynamic input is on, you can press the [Ctrl] key to cycle through the three options in the grip menu. See **Figure 4-24B**.

The other two options in the grip menu allow the Z axis to be moved in relation to one of the other two axes. Test each of these options to see how they function. The **Rotate Around X Axis** option limits rotation of the Z axis to around the X axis, and the **Rotate Around Y Axis** option limits rotation to around the Y axis.

Figure 4-24.
Using the UCS icon axis rotation options. A—The Z axis options appear after moving the pointer over the Z axis grip. Pick the **Z Axis Direction** option to set the direction of the Z axis. B—The new UCS after rotating the Z axis. Note that the other Z axis options can be accessed by pressing the [Ctrl] key.

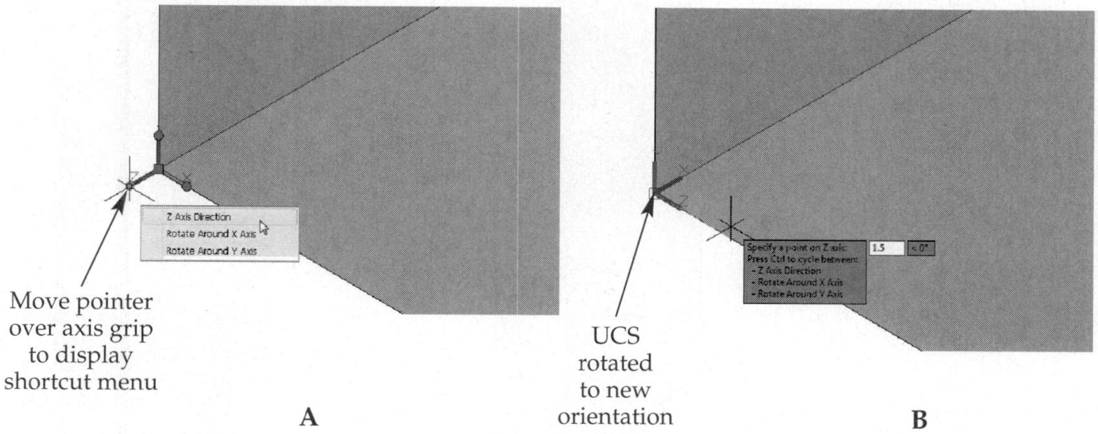

Move the pointer to each of the other two UCS icon axis grips and note that a similar menu is displayed. Each axis of the UCS can be rotated in the same manner using the appropriate axis grip.

Additional UCS Options

Additional options for changing the UCS are discussed in the following sections. Some of these options can be selected from the UCS icon shortcut menu displayed by right-clicking on the UCS icon, shown in Figure 4-19, or by accessing the **Coordinates** panel in the **Home** or **Visualize** tab on the ribbon. All of the options discussed are available when entering the **UCS** command on the command line.

NOTE

Keep in mind that many of the procedures discussed in the following sections can be accomplished using the dynamic moving and rotating functions of the UCS icon grips.

Selecting Three Points to Create a New UCS

The **3 Point** option of the **UCS** command can be used to change the UCS to any flat surface. This option requires that you first locate a new origin, then a point on the positive X axis, and finally a point on the XY plane that has a positive Y value. See Figure 4-25. Use object snaps to select points that are not on the current XY plane. After you pick the third point—the point on the XY plane—the UCS icon changes its orientation to align with the plane defined by the three points. The **3 Point** option is available in the UCS icon shortcut menu.

Ribbon
Home
> **Coordinates**
Visualize
> **Coordinates**
3 Point
Type
UCS

UCS

PROFESSIONAL TIP

When typing the **UCS** command, enter 3 at the Specify origin of UCS or [Face/NAmed/OBject/Previous/View/World/X/Y/Z/ZAxis] <World>: prompt. Notice that the option is not listed in the prompt.

Figure 4-25.
A—A new UCS can be established by picking three points. P1 is the origin, P2 is on the positive X axis, and P3 is on the XY plane and has a positive Y value.
B—The new UCS is created.

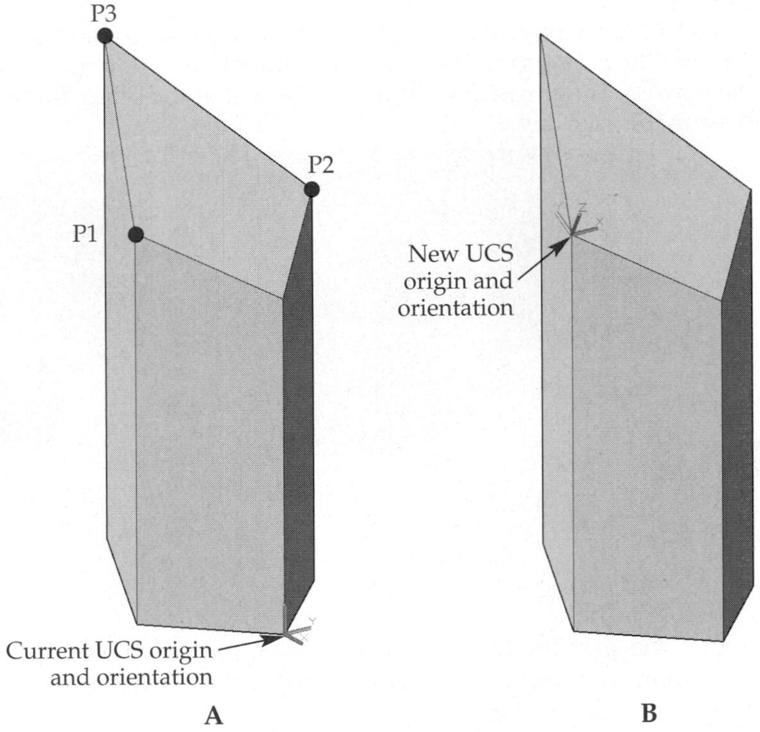

Selecting a New Z Axis

UCS

Ribbon
Home
> Coordinates
Visualize
> Coordinates
Z-Axis Vector
Type
UCS

The **ZAxis** option of the UCS command allows you to select the origin point and a point on the positive Z axis. Once the new Z axis is defined, AutoCAD sets the new X and Y axes. The **ZAxis** option is available in the UCS icon shortcut menu.

You will now add a cylinder to the lower face of the base created earlier. The cylinder extends into the base. Earlier, you moved the UCS to the angled face of the object, drew the cylinder on the angled face, and then restored the UCS to its previous position on the side face. If this is the current orientation, the Z axis does not project perpendicular to the lower face. Therefore, a new UCS must be created on the lower-right face. Change the UCS after entering the **ZAxis** option as follows.

1. Pick the origin of the new UCS. See **Figure 4-26A**. You may have to use an object snap to select the origin.
2. Pick a point on the positive portion of the new Z axis.
3. The new UCS is established and it can be saved if necessary.

Now, use 3D object snaps, AutoTrack, or 2D object snaps to draw a ∅.5″ cylinder centered on the lower face and extending 3″ into the base. Then, subtract the cylinder from the base part to create the hole, as shown in **Figure 4-26B**.

Setting the UCS to an Existing Object

UCS

Ribbon
Home
> Coordinates
Visualize
> Coordinates
Object
Type
UCS

The **Object** option of the **UCS** command can be used to define a new UCS on an object. This option cannot be used with 3D polylines or xlines. There are also certain rules that control the orientation of the UCS. For example, if you select a circle, the center point becomes the origin of the new UCS. The pick point on the circle determines the direction of the X axis. The Y axis is relative to X and the UCS Z axis may or may not be the same as the Z axis of the selected object. The **Object** option is available in the UCS icon shortcut menu.

Look at **Figure 4-27A**. The circle is rotated an unknown number of degrees from the XY plane of the WCS. However, you need to create a UCS in which the circle is lying on the XY plane. Select the **Object** option of the **UCS** command and then pick the circle. The UCS icon may look like the one shown in **Figure 4-27B**. Notice how the

Figure 4-26.
A—Using the **ZAxis** option to establish a new UCS. B—The new UCS is used to create a cylinder, which is then subtracted from the base to create a hole.

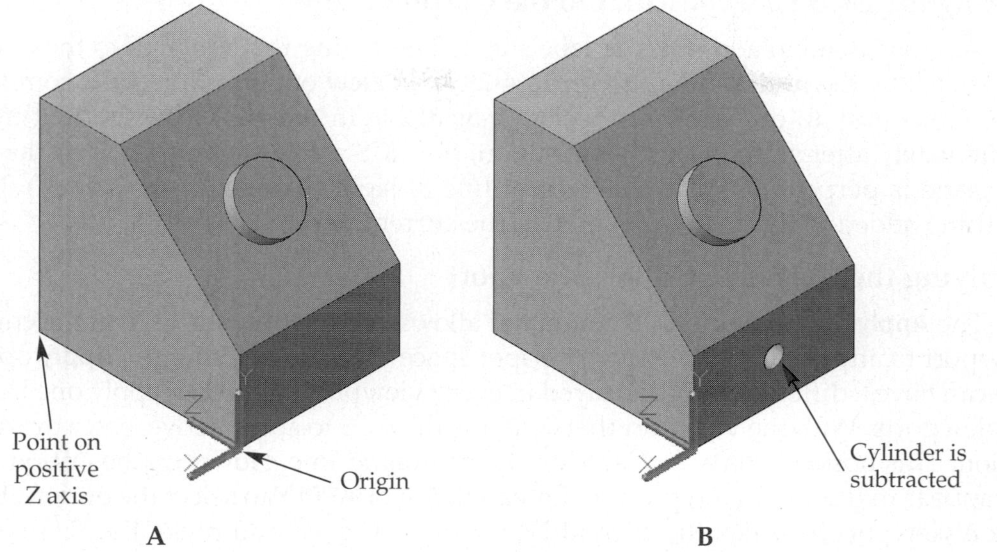

X and Y axes are not aligned with the quadrants of the circle, as indicated by the grip locations. This may not be what you expected. The X axis orientation is determined by the pick point on the circle. Notice how the X axis is pointing at the pick point.

To rotate the UCS in the current plane so the X and Y axes of the UCS are aligned with the quadrants of the circle, use the **ZAxis** option of the **UCS** command. Select the center of the circle as the origin and then enter the absolute coordinate 0,0,1. This uses the current Z axis location, which also forces the X and Y axes to align with the object. Refer to **Figure 4-27C**. This method may not work with all objects.

Setting the UCS to the Face of a 3D Solid

The **Face** option of the **UCS** command allows you to orient the UCS to any face on a 3D solid, surface, or mesh object. The **Face** option is available in the UCS icon shortcut menu. Select the option and then pick a face on the model. After you have selected a face, you have the options of moving the UCS to the adjacent face or flipping the UCS 180° on the X axis, Y axis, or both axes. Use the **Next**, **Xflip**, or **Yflip** option to move or rotate the UCS as needed. Once you achieve the UCS orientation you want,

Ribbon
Home
> Coordinates
Visualize
> Coordinates
Face
Type
UCS

UCS

Figure 4-27.
A—This circle is rotated off of the WCS XY plane by an unknown number of degrees. It will be used to establish a new UCS. B—The circle is on the XY plane of the new UCS. However, the X and Y axes do not align with the circle's quadrants. C—The **ZAxis** option of the **UCS** command is used to align the UCS with the quadrants of the circle.

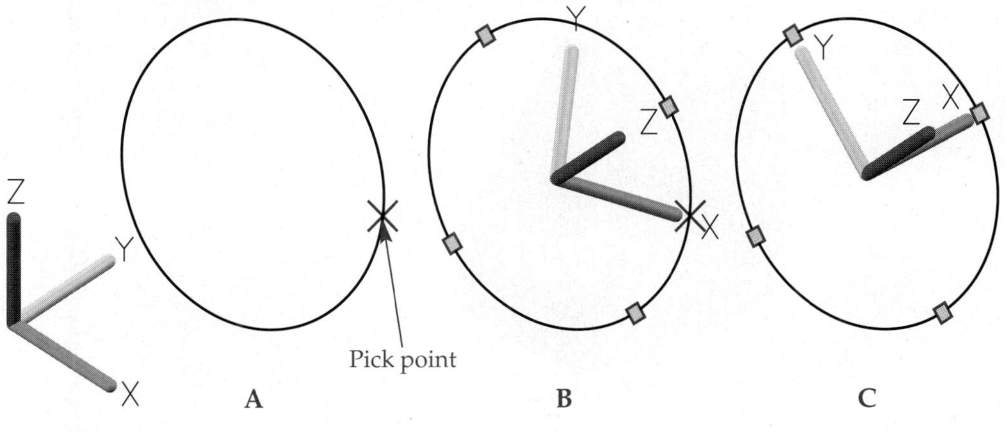

press [Enter] to accept. Notice in **Figure 4-28** how many different UCS orientations can be selected for a single face.

Setting the UCS Perpendicular to the Current View

UCS

Ribbon
Home
> Coordinates
Visualize
> Coordinates
View
Type
UCS

You may need to add notes or labels to a 3D drawing that are plan to the current view, such as the note shown in **Figure 4-29**. The **View** option of the **UCS** command makes this easy to do. The **View** option is available in the UCS icon shortcut menu. Immediately after selecting the **View** option, the UCS rotates to a position so the new XY plane is perpendicular to the current line of sight (parallel to the screen). Now, anything added to the drawing is plan to the current view.

Applying the Current UCS to a Viewport

UCS

Type
UCS

The **Apply** option of the **UCS** command allows you to apply the UCS in the current viewport to any or all model space or paper space viewports. Using the **Apply** option, you can have a different UCS displayed in every viewport or you can apply one UCS to all viewports. With the viewport that contains the UCS to apply active, enter the **Apply** option. This option is only available on the command line. However, the option does not appear in the command prompt. Enter either A or APPLY to select the option. Then, pick a viewport to which the current UCS will be applied and press [Enter]. To apply

Figure 4-28.
Several different UCSs can be selected from a single pick point using the **Face** option of the **UCS** command. Given the pick point, five of the eight possibilities are shown here.

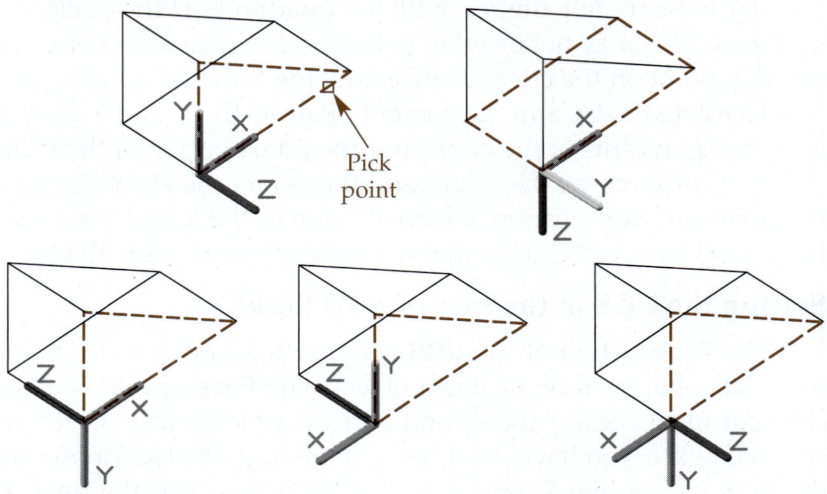

Figure 4-29.
The **View** option of the **UCS** command allows you to place text plan to the current view.

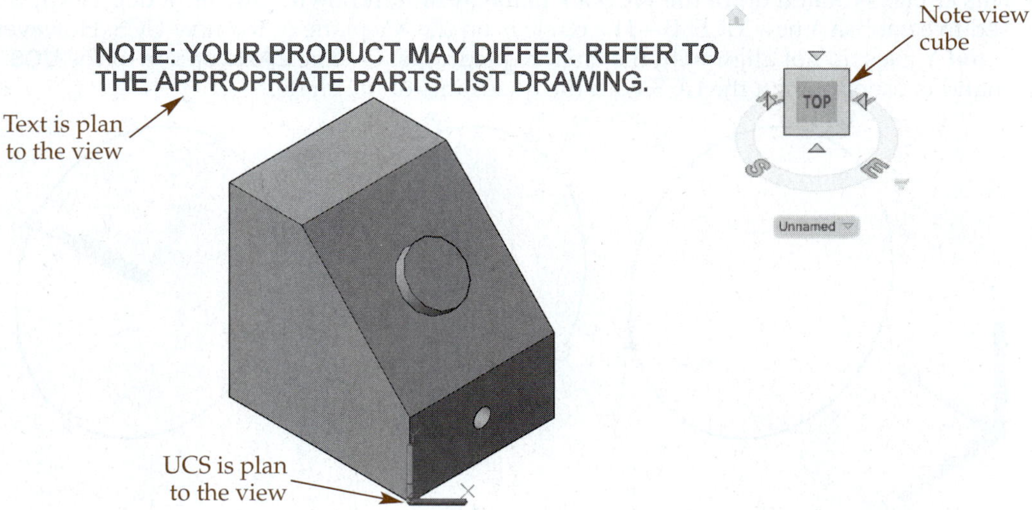

the current UCS to all viewports, enter the **All** option. See Chapter 5 for a complete discussion of model space viewports.

Preset UCS Orientations

AutoCAD has six preset orthographic UCSs that match the six standard orthographic views. With the current UCS as the top view (plan), all other views are arranged as shown in **Figure 4-30**. These orientations can be selected by using the **Named UCS Combo Control** drop-down list in the **Coordinates** panel of the **Home** or **Visualize** tab in the ribbon, entering the command on the command line, or using the **Orthographic UCSs** tab of the **UCS** dialog box. When using the command line, type the name of the UCS (FRONT, BACK, RIGHT, etc.) at the first prompt:

Type
UCS UCSMAN

UCS

Specify origin of UCS or [Face/NAmed/OBject/Previous/View/World/X/Y/Z/ZAxis]
<World>: **FRONT** *or* **FR**↵

Note that there is no orthographic option listed for the command.

The **Relative to:** drop-down list at the bottom of the **Orthographic UCSs** tab of the **UCS** dialog box specifies whether each orthographic UCS is relative to a named UCS or absolute to the WCS. For example, suppose you have a saved UCS named Front Corner that is rotated 30° about the Y axis of the WCS. If you set current the top UCS relative to the WCS, the new UCS is perpendicular to the WCS, **Figure 4-31A**. However, if the top UCS is set current relative to the named UCS Front Corner, the new UCS is also rotated from the WCS, **Figure 4-31B**.

The Z value, or depth, of a preset UCS can be changed in the **Orthographic UCSs** tab of the **UCS** dialog box. First, right-click on the name of the UCS you wish to change. Then, pick **Depth** from the shortcut menu, **Figure 4-32A**. This displays the **Orthographic**

Figure 4-30.
The standard orthographic UCSs coincide with the six basic orthographic views.

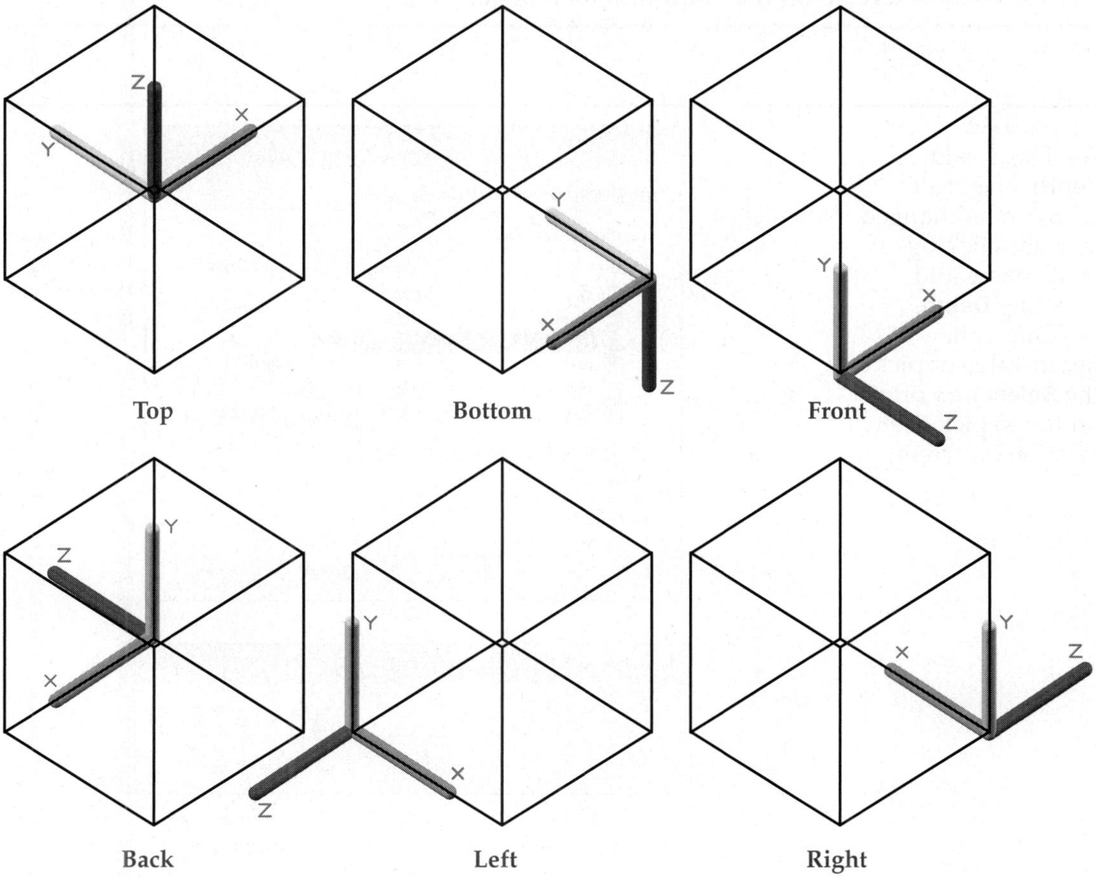

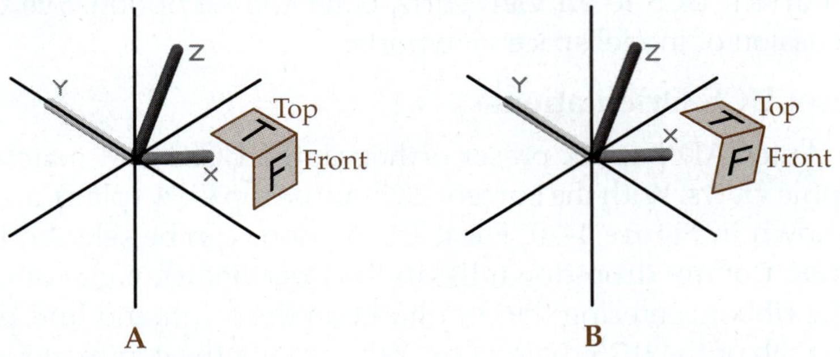

Figure 4-31. The **Relative to:** drop-down list entry in the **Orthographic UCSs** tab of the **UCS** dialog box determines whether the orthographic UCS is based on a named UCS or the WCS. The UCS icon here represents the named UCS. A—Relative to the WCS. B—Relative to the named UCS.

UCS depth dialog box. See **Figure 4-32B**. You can either enter a new depth value or specify the new location on screen by picking the **Select new origin** button. Once the new depth has been selected, it is reflected in the preset UCS list.

CAUTION

Changing the **Relative to:** setting affects *all* preset UCSs and *all* preset viewpoints! Therefore, leave this set to World unless absolutely necessary to change it.

Exercise 4-5

www.g-wlearning.com/CAD/

Complete the exercise on the companion website.

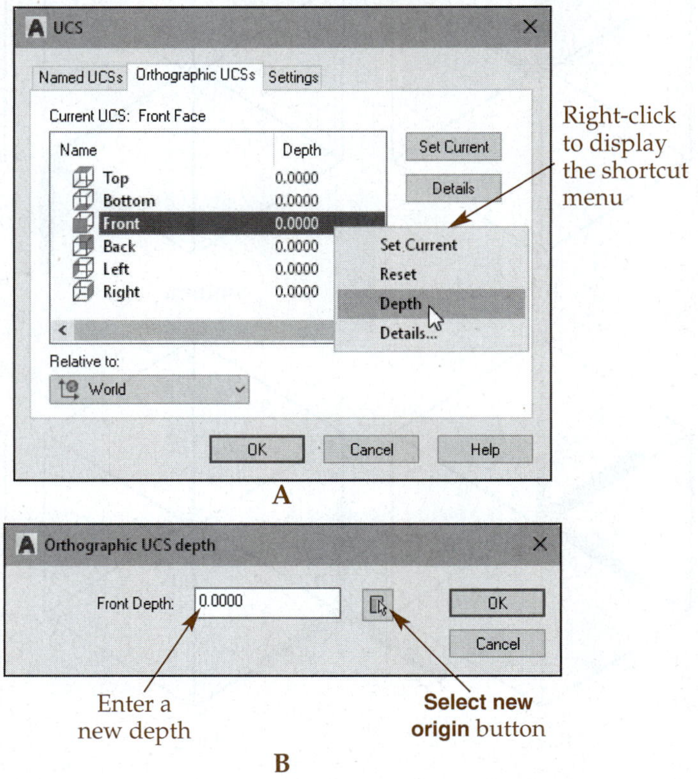

Figure 4-32. A—The Z value, or depth, of a preset UCS can be changed by right-clicking on its name and selecting **Depth**. B—Enter a new depth value or pick the **Select new origin** button to pick a new location on screen.

Managing User Coordinate Systems and Displays

You can create, name, and use as many user coordinate systems as needed to construct your model or drawing. As you saw earlier, AutoCAD allows you to name (save) coordinate systems for future use. User coordinate systems can be saved, renamed, set current, and deleted by using the **Named UCSs** tab of the **UCS** dialog box, **Figure 4-33**.

The **Named UCSs** tab contains the **Current UCS:** list box. This list box contains the names of all saved coordinate systems plus World. If other coordinate systems have been used in the current drawing session, Previous appears in the list. Unnamed appears if the current coordinate system has not been named (saved). The current UCS is indicated by a small triangle to the left of its name in the list and by the label at the top of the list. To make any of the listed coordinate systems active, highlight the name and pick the **Set Current** button.

A list of coordinate and axis values of the highlighted UCS can be displayed by picking the **Details** button. This displays the **UCS Details** dialog box shown in **Figure 4-34**.

If you right-click on the name of a UCS in the list in the **Named UCSs** tab, a shortcut menu is displayed. Using this menu, you can rename the UCS. You can also set the UCS current or delete it using the shortcut menu. The Unnamed UCS cannot be deleted, nor can the World UCS.

PROFESSIONAL TIP

You can also manage UCSs on the command line using the **UCS** command. In addition, by right-clicking on the UCS icon, you can select a UCS from the UCS icon shortcut menu to make it current.

Figure 4-33.
The **UCS** dialog box allows you to rename, list, delete, and set current an existing UCS.

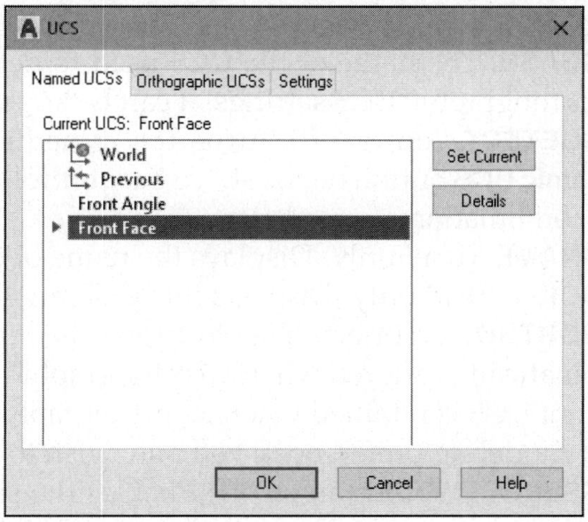

Figure 4-34.
The **UCS Details** dialog box displays the coordinate values of the selected UCS.

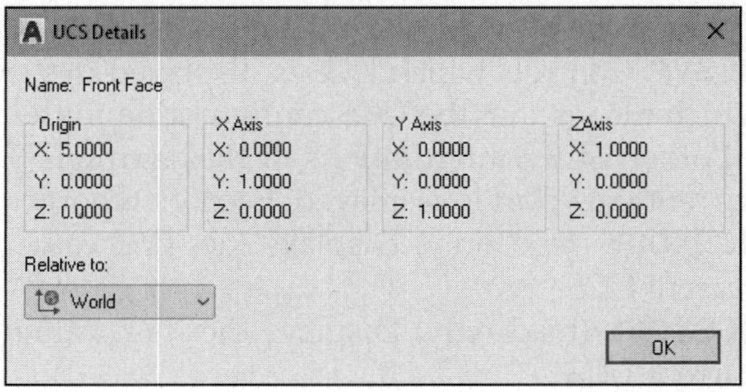

Setting an Automatic Plan Display

After changing the UCS, a plan view is often needed to give you a better feel for the XYZ directions. While you should try to draw in a pictorial view when possible as you construct a 3D object, some constructions may be much easier in a plan view. AutoCAD can be set to automatically make your view of the drawing plan to the current UCS. This is especially useful if you will be changing the UCS often, but want to work in a plan view.

The **UCSFOLLOW** system variable is used to automatically display a plan view of the current UCS. When it is set to 1, a plan view is automatically created in the current viewport when the UCS is changed. Viewports are discussed in Chapter 5. The default setting of **UCSFOLLOW** is 0 (off). After setting the variable to 1, a plan view will be automatically generated the next time the UCS is changed. The **UCSFOLLOW** variable generates the plan view only after the UCS is changed, not immediately after the variable is changed. However, if you select a different viewport, the previous viewport is set plan to the UCS if **UCSFOLLOW** has been set to 1 in that viewport. The **UCSFOLLOW** variable can be individually set for each viewport.

PROFESSIONAL TIP

To get the plan view displayed without changing the UCS, use the **PLAN** command, which is discussed in Chapter 3.

UCS Settings and Variables

As discussed in the previous section, the **UCSFOLLOW** system variable allows you to change how an object is displayed in relation to the UCS. There are also system variables that display information about the current UCS and other system variables that control UCS functions. These variables include:

- **UCSAXISANG.** (stored value) The default rotation angle for the **X**, **Y**, or **Z** option of the **UCS** command.
- **UCSBASE.** The name of the UCS used to define the origin and orientation of the orthographic UCS settings. It can be any named UCS.
- **UCSDETECT.** (on or off) Turns the dynamic UCS function on and off. The **Dynamic UCS** button on the status bar controls this variable, as does the [Ctrl]+[D] key combination.
- **UCSNAME.** (read only) Displays the name of the current UCS.
- **UCSORG.** (read only) Displays the XYZ origin value of the current UCS.
- **UCSORTHO.** (on or off) If set to 1 (on), the related orthographic UCS setting is automatically restored when an orthographic view is restored. If turned off, the current UCS is retained when an orthographic view is restored. Depending on your modeling preferences, you may wish to set this variable to 0.
- **UCSSELECTMODE.** (on or off) Enables the selection and manipulation of the UCS icon with grips. The default setting is 1 (on).
- **UCSVIEW.** (on or off) If this variable is set to 1 (on), the current UCS is saved with the view when a view is saved. Otherwise, the UCS is not saved with the view.
- **UCSVP.** Controls which UCS is displayed in viewports. The default value is 1, which means that the UCS configuration in the viewport is independent from all other UCS configurations. If the setting is 0, the UCS configuration in the current viewport is displayed. Each viewport can be set to either 0 or 1.
- **UCSXDIR.** (read only) Displays the XYZ value of the X axis direction of the current UCS.
- **UCSYDIR.** (read only) Displays the XYZ value of the Y axis direction of the current UCS.

UCS options and variables can also be managed in the **Settings** tab of the **UCS** dialog box. Refer to **Figure 4-14**. The options in the **UCS settings** area are:

- **Save UCS with viewport.** If checked, the current UCS settings are saved with the viewport and the **UCSVP** system variable is set to 1. This variable can be set for each viewport in the drawing. Viewports in which this setting is turned off, or unchecked, will always display the UCS settings of the current active viewport.
- **Update view to Plan when UCS is changed.** This setting controls the **UCSFOLLOW** system variable. When checked, the variable is set to 1. When unchecked, the variable is set to 0.

Chapter Review

Answer the following questions using the information in this chapter.

1. Explain *spherical coordinate entry.*
2. Explain *cylindrical coordinate entry.*
3. A new point is to be drawn 4.5″ from the last point. It is to be located at a 63° angle in the XY plane, and at a 35° angle from the XY plane. Write the proper spherical coordinate notation.
4. Write the proper cylindrical coordinate notation for locating a point 4.5″ in the horizontal direction from the origin, 3.6″ along the Z axis, and at a 63° angle in the XY plane.
5. Name the command that is used to draw 3D polylines.
6. Why is the command in question 5 needed?
7. Which command option is used to change a 3D polyline into a B-spline curve?
8. What is the *world coordinate system (WCS)*?
9. What is a *user coordinate system (UCS)*?
10. What effect does the **Show UCS Icon at Origin** option have on the UCS icon display?
11. Describe how to rotate the UCS so that the Z axis is tilted 30° toward the WCS X axis.
12. How do you return to the WCS from any UCS?
13. Which command controls the display of the user coordinate system icon?
14. What system variable controls the display of grips on the UCS icon? In which dialog box can it be set?
15. Briefly describe how to access the **Move and Align** option using the UCS icon grips.
16. How can you return to a previous UCS configuration using the UCS icon?
17. What is a *dynamic UCS* and how is one activated?
18. What is the function of the **3 Point** option of the **UCS** command?
19. How do you automatically create a display that is plan to a new UCS?
20. What is the function of the **Object** option of the **UCS** command?
21. What is the function of the **Apply** option of the **UCS** command?
22. In which dialog box is the **Orthographic UCSs** tab located?
23. Which command displays the **UCS** dialog box?
24. What appears in the **Named UCSs** tab of the **UCS** dialog box if the current UCS has not been saved?

Drawing Problems

For Problems 1–4, draw each object using solid primitives and Boolean commands to create composite solids. Measure the objects directly to obtain the necessary dimensions. If appropriate, apply geometric constraints to the base 2D drawing. Plot the drawings at a 3:1 scale using display methods specified by your instructor. Save the drawings as P4-*(problem number).*

1. General

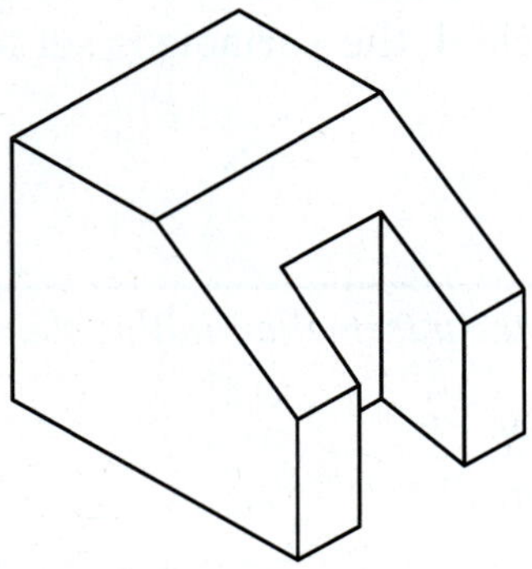

2. General

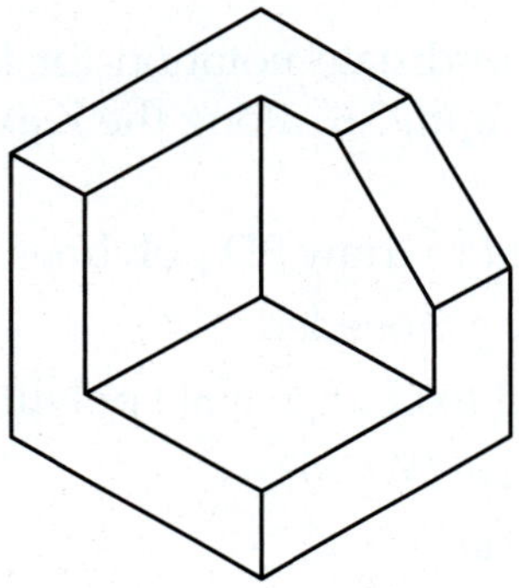

3. General

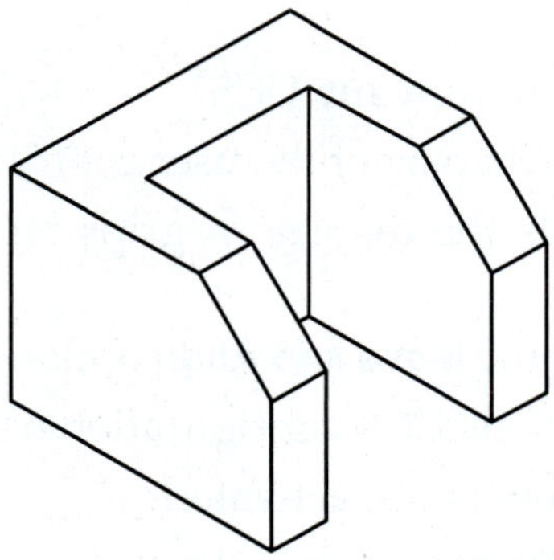

4. General

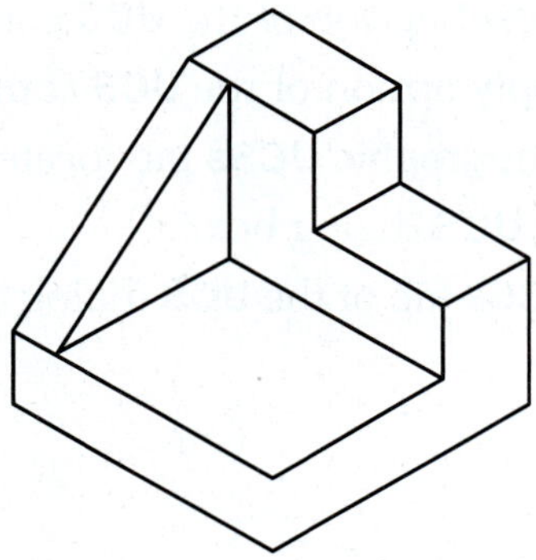

5. **Mechanical** Create the mounting bracket shown. Save the file as P4-5.

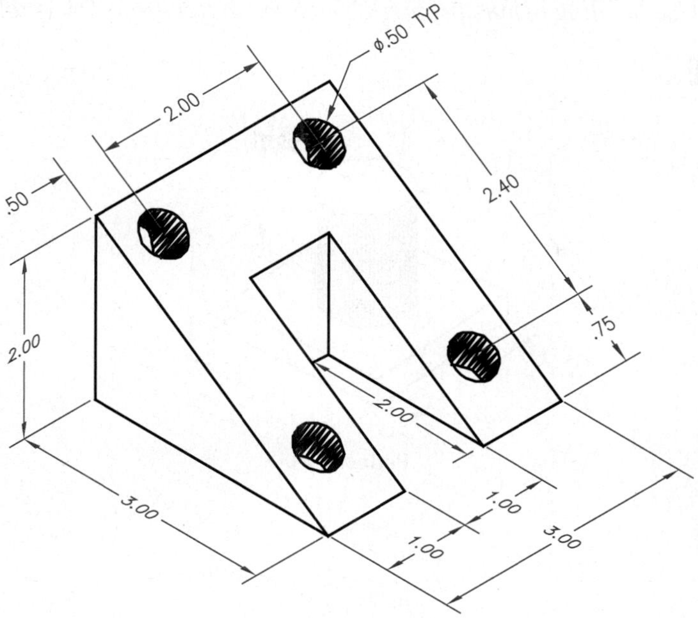

6. **Mechanical** Create the computer speaker as shown below. The large-radius, arched surface is created by drawing a three-point arc. The second point of the arc passes through the point located by the .26 and 2.30 dimensions. Save the file as P4-6.

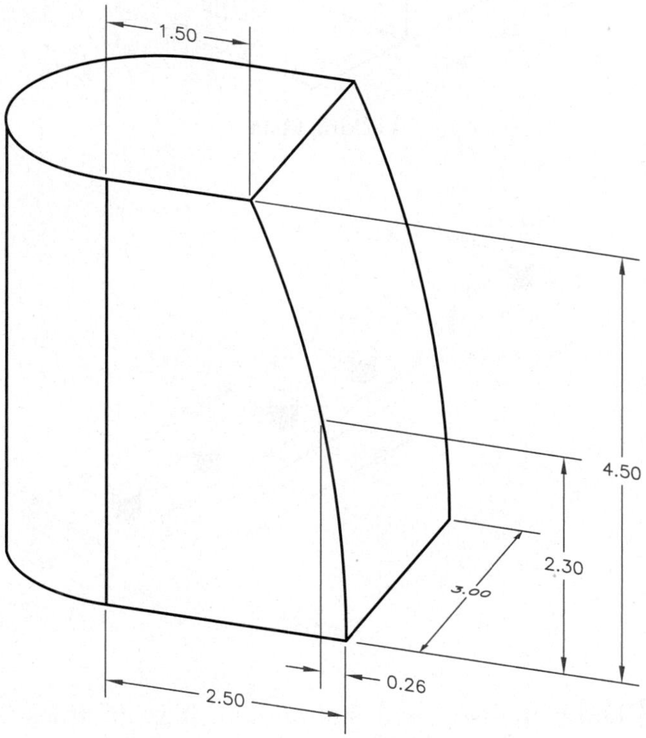

Drawing Problems - Chapter 4

For Problems 7–9, draw each object using solid primitives and Boolean commands to create composite solids. Use the dimensions provided. Save the drawings as P4-*(problem number).*

7. **Mechanical**

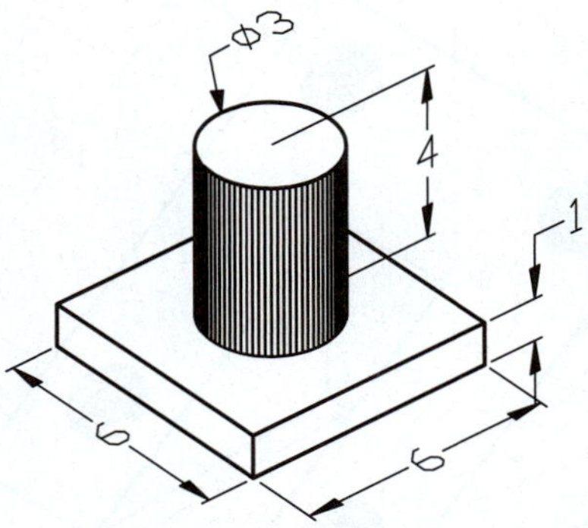

Pedestal #1

8. **Mechanical**

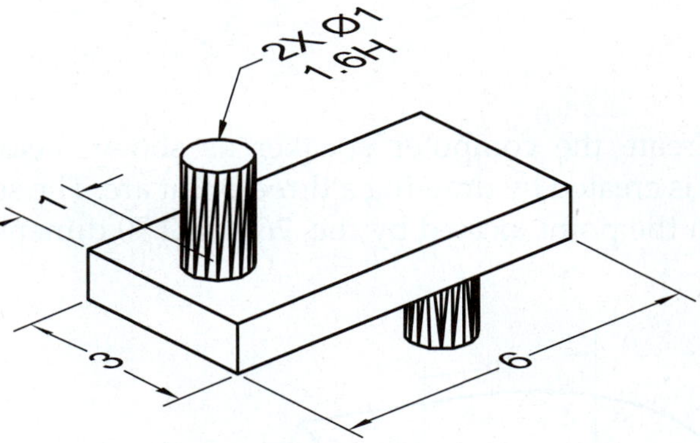

Locking Plate

9. **Mechanical**

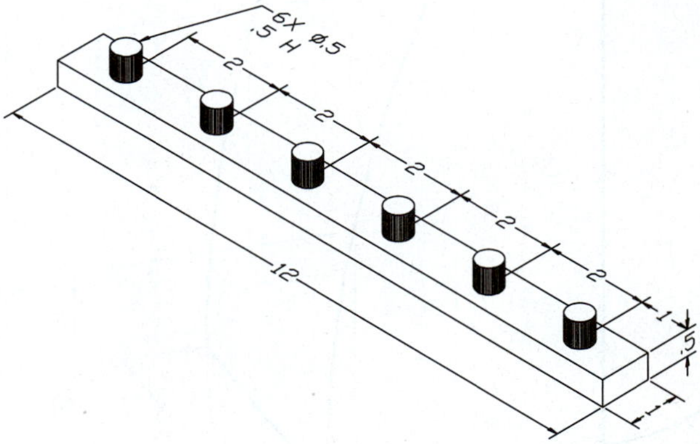

Pin Bar

10. **Architectural** Draw the Ø8″ pedestal shown. It is .5″ thick. The four feet are centered on a Ø7″ circle and are .5″ high. Save the drawing as P4-10.

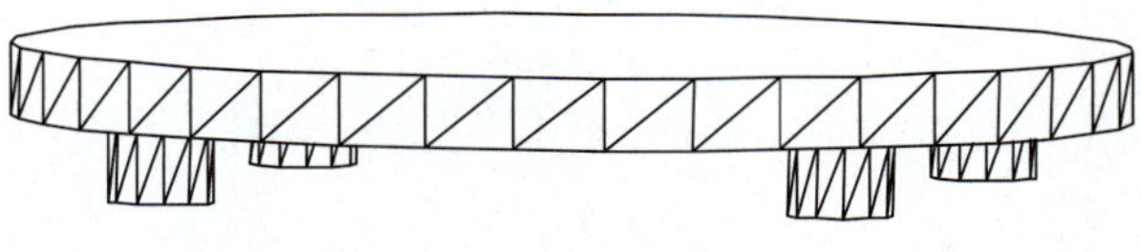

Pedestal #2

11. **Architectural** Four legs (cones), each 3″ high with a Ø1″ base, support this Ø10″ globe. Each leg tilts at an angle of 15° from vertical. The base is Ø12″ and .5″ thick. The bottom surface of the base is 8″ below the center of the globe. Save the drawing as P4-11.

Globe

12. **Architectural** The table legs (A) are 2″ square and 17″ tall. They are 2″ in from each edge. The tabletop (B) is 24″ × 36″ × 1″. Save the drawing as P4-12.

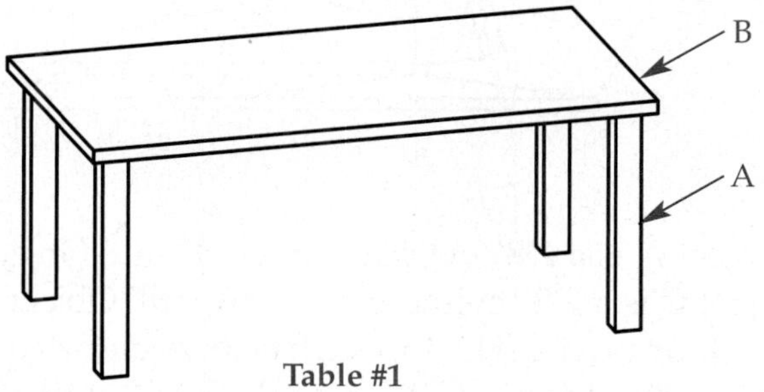

Table #1

13. **Architectural** The table legs (A) for the large table are Ø2″ and 17″ tall. The tabletop (B) is 24″ × 36″ × 1″. The table legs (C) for the small table are Ø2″ and 11″ tall. The tabletop (D) is 24″ × 14″ × 1″. All legs are 1″ in from the edges of the table. Save the drawing as P4-13.

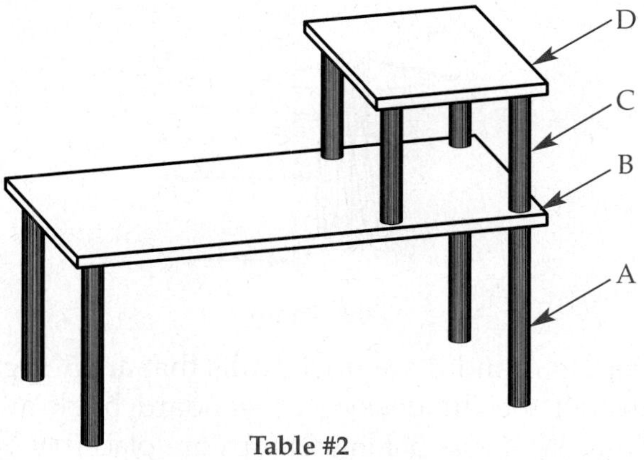

Table #2

14. **General** The spherical objects (A) are Ø4″. Object B is 6″ long and Ø1.5″. Save the drawing as P4-14.

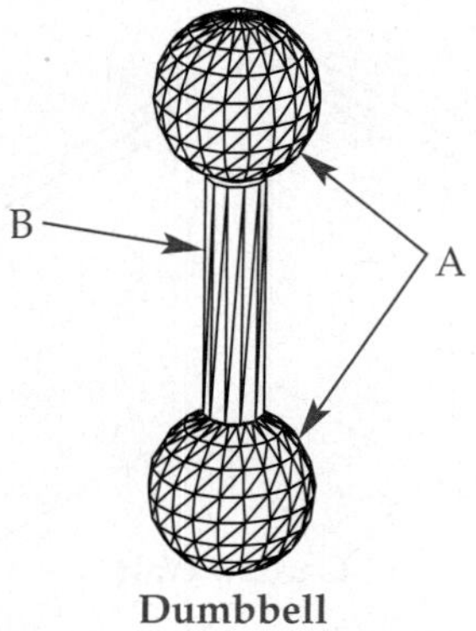

Dumbbell

15. **Architectural** Create the model of the globe using the dimensions shown. Save the file as P4-15.

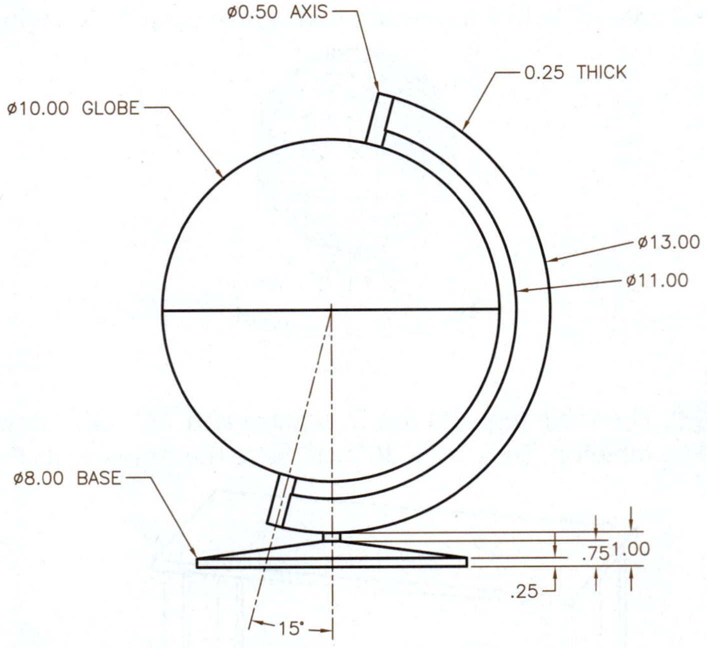

16. **Architectural** Object A is a ∅8″ cylinder that is 1″ tall. Object B is a ∅5″ cylinder that is 7″ tall. Object C is a ∅2″ cylinder that is 6″ tall. Object D is a .5″ × 8″ × .125″ box, and there are four pieces. The top surface of each piece is flush with the top surface of Object C. Object E is a ∅18″ cone that is 12″ tall. Create a smaller cone and hollow out Object E. Save the drawing as P4-16.

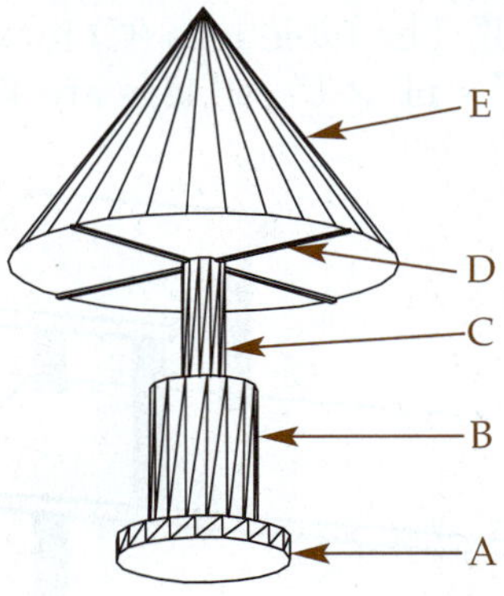

Table Lamp

17. **Architectural** Objects A and B are brick walls that are 5′ high. The walls are two bricks thick. Research the dimensions of standard brick and draw accordingly. Wall B is 7′ long and Wall A is 5′ long. Lamps are placed at each end of the walls. Object C is ∅2″ and 8″ tall. The center is offset from the end of the wall by a distance equal to the width of one brick. Object D is ∅10″. Save the drawing as P4-17.

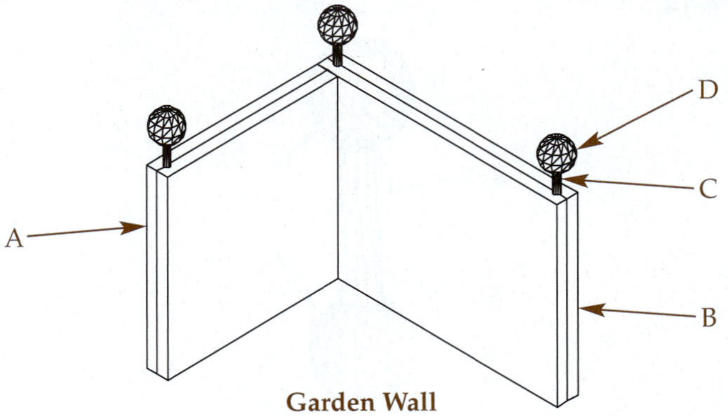

Garden Wall

18. **Architectural** Object A is ∅18″ and 1″ tall. Object B is ∅1.5″ and 6′ tall. Object C is ∅6″ and .5″ tall. Object D is a ∅10″ sphere. Object E is a U-shaped bracket to support the shade (Object F). There are two items; draw them an appropriate size. Object F has a ∅22″ base and is 12″ tall. Save the drawing as P4-18.

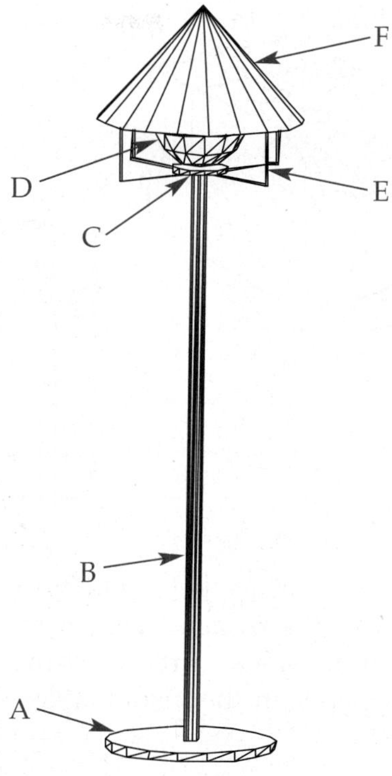

Floor Lamp

19. **General** This is a concept sketch of a desk organizer. Create a solid model using the dimensions given. Create and save new UCSs as needed. Inside dimensions of compartments can vary, but the thickness between compartments should be consistent. Do not add dimensions to the drawing. Plot your drawing on a B-size sheet of paper in a visual style specified by your instructor. Save the drawing as P4-19.

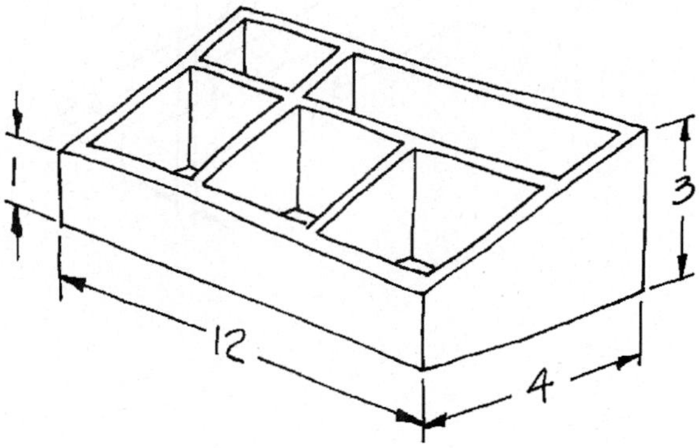

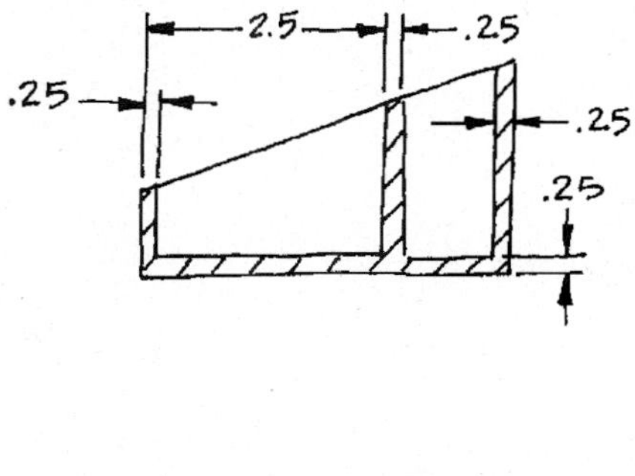

20. **General** This is a concept sketch of a pencil holder. Create a solid model using the dimensions given. Create and save new UCSs as needed. Do not add dimensions to the drawing. Plot your drawing on a B-size sheet of paper in a visual style specified by your instructor. Save the drawing as P4-20.

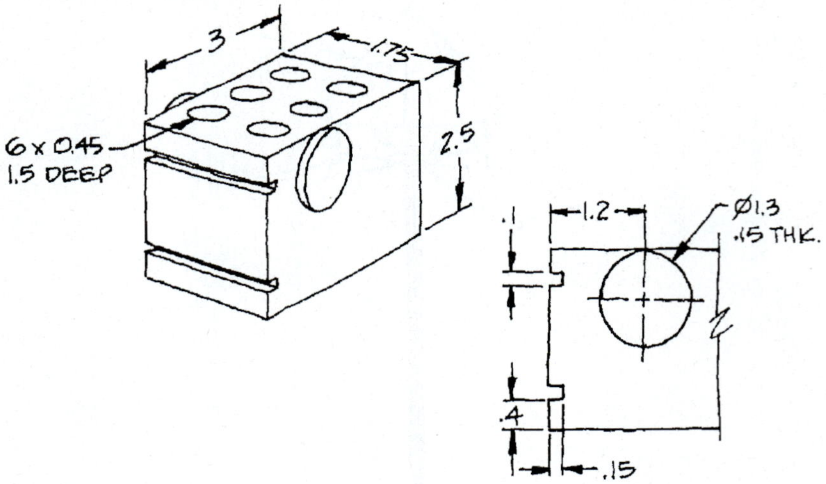

21. **Mechanical** This is an engineering sketch of a window blind mounting bracket. Create a solid model using the dimensions given. Create and save new UCSs as needed. Do not add dimensions to the drawing. Create two plots, each of a different view, on B-size paper in the visual styles specified by your instructor. Save the drawing as P4-21.

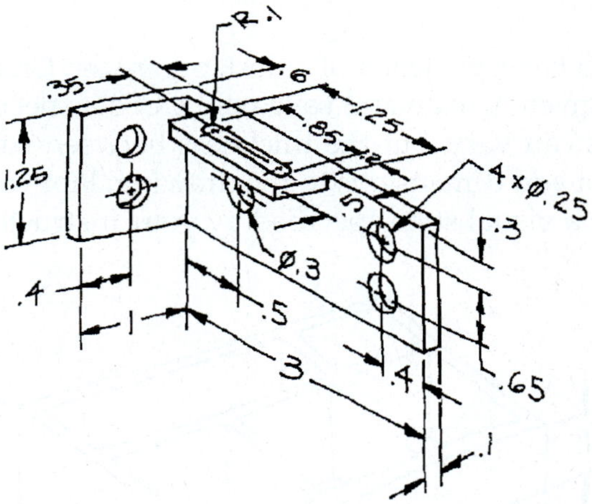

CHAPTER 5

Using Model Space Viewports

Learning Objectives

After completing this chapter, you will be able to:

- ✓ Describe the function of model space viewports.
- ✓ Create and save viewport configurations.
- ✓ Alter the current viewport configuration.
- ✓ Use multiple viewports to construct a drawing.

A variety of views can be displayed in a drawing at one time using model space viewports. This is especially useful when constructing 3D models. Using the **VPORTS** command, you can divide the drawing area into two or more smaller areas. These areas are called ***viewports***. Each viewport can be configured to display a different 2D or 3D view of the model.

The ***active viewport*** is the viewport in which a command will be applied. Any viewport can be made active, but only one can be active at a time. As objects are added or edited, the results are shown in all viewports. A variety of viewport configurations can be saved and recalled as needed. This chapter discusses the use of viewports and shows how they can be used for 3D constructions.

Understanding Viewports

The AutoCAD drawing area can be divided into a maximum of 64 viewports. However, this is impractical due to the small size of each viewport. Usually, the maximum number of viewports practical to display at one time is four. The number of viewports you need depends on the model you are creating. Each viewport can show a different view of an object. This makes it easier to construct 3D objects.

NOTE

The **MAXACTVP** (maximum active viewports) system variable sets the number of viewports that can be used at one time. The initial value is 64, which is the highest setting.

There are two types of viewports used in AutoCAD. The type of viewport created depends on whether it is defined in model space or paper space. ***Model space*** is the space, or mode, where the model or drawing is constructed. ***Paper space***, or layout space, is the space where a drawing is laid out to be plotted. Viewports created in model space are called ***model space viewports*** or simply ***model viewports***, and are also known as ***tiled viewports***. Viewports created in paper space are called ***layout viewports***, also known as ***floating viewports***.

Model space is active by default when you start a new drawing. Model space viewports are created with the **VPORTS** command. Model space viewport configurations are for display purposes only and cannot be plotted. If you plot from model space, the content of the active viewport is plotted.

Model space viewports are described as *tiled viewports*. They are referred to as *tiled* because the edges of each viewport are placed side to side, as with floor tile, and they cannot overlap. Model space viewports are not AutoCAD objects and cannot be edited.

Floating (paper space) viewports are used to lay out the views of a drawing before plotting. They are described as *floating* because they can be moved around and overlapped. Paper space viewports are objects and can be edited. These viewports can be thought of as "windows" cut into a sheet of paper to "see into" model space. You can then display different scaled drawings (views) in these windows. For example, architectural details or sections and details of complex mechanical parts may be displayed in paper space viewports at different scales. Detailed discussions of paper space viewports are provided in ***AutoCAD and Its Applications—Basics***.

The **VPORTS** command can be used to create viewports in a paper space layout. The process is very similar to that used to create model space viewports, which is discussed next. You can also use the **MVIEW** command to create paper space viewports.

Creating Viewports

Creating model space viewports allows you to work with multiple views of the same model. To work on a different view, simply pick with your pointing device in the viewport in which you wish to work. The picked viewport becomes active. Using viewports is a good way to construct 3D models because all views are updated as you draw. However, viewports are also useful when creating 2D drawings.

The project on which you are working determines the number of viewports needed. Keep in mind that the more viewports you display on your screen, the smaller the view in each viewport. Small viewports may not be useful to you. Four different viewport configurations are shown in **Figure 5-1**. As you can see, when 16 viewports are displayed, the viewports are very small. Normally, two to four viewports are used.

Quick Viewport Layout

VPORTS
Ribbon
Home
> View

Multiple viewports

A four-view layout can be instantly displayed by picking the **Multiple viewports** button on the **View** panel in the **Home** tab of the ribbon. See **Figure 5-2**. This selection automatically creates a top, front, side, and pictorial view based on the current UCS. See **Figure 5-3**. This layout may be a good one to start from when working on models that require two or more views for construction purposes.

VPORTS
Ribbon
Home
> View

Single viewport

The screen display can be quickly returned to a single view by picking the **Single viewport** button on the **View** panel in the **Home** tab of the ribbon. Keep in mind, the resulting view will be the viewport that is current, which is the one surrounded by a highlighted frame.

Figure 5-1.
A—Two vertical viewports. B—Two horizontal viewports. C—Three viewports, with the largest viewport positioned at the right. D—Sixteen viewports.

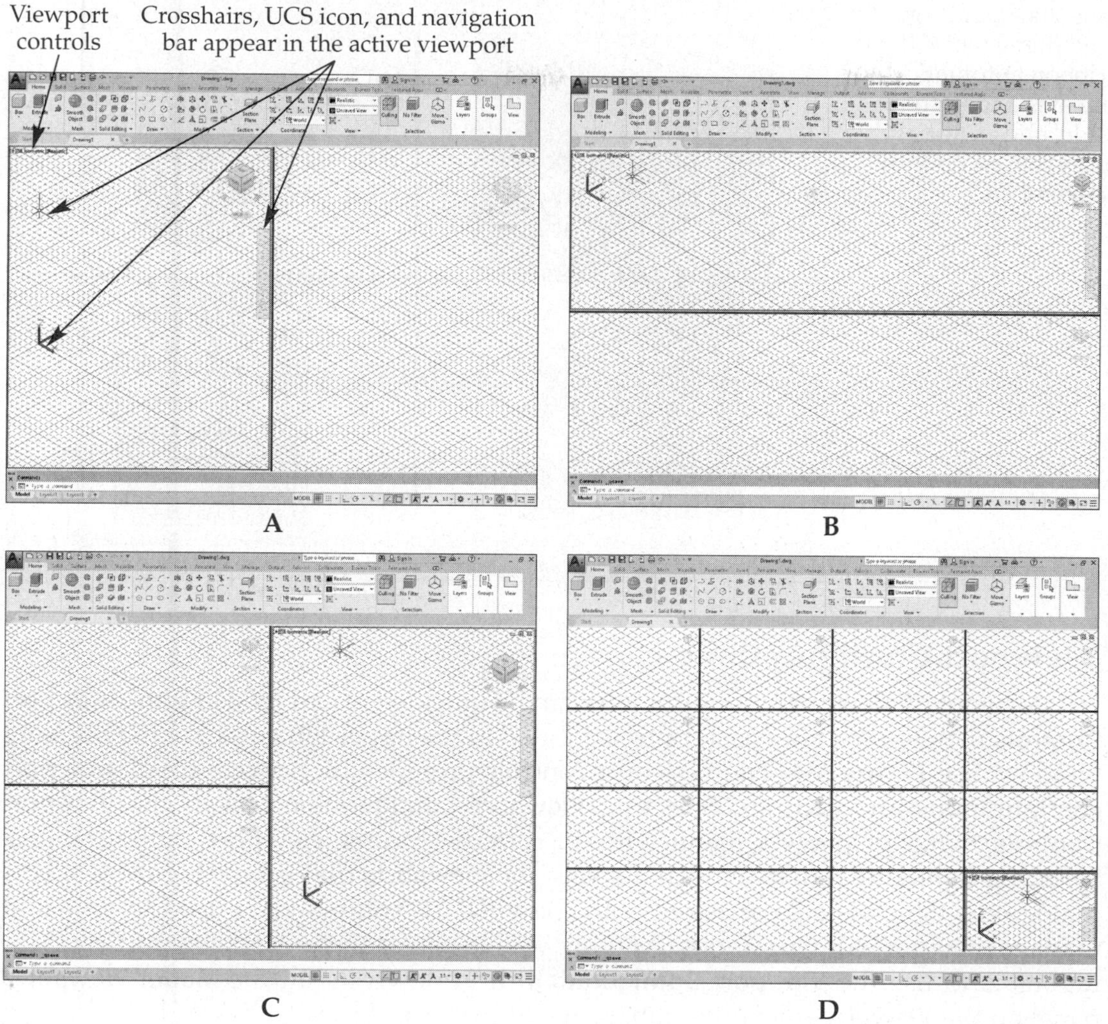

Figure 5-2.
A four-viewport layout can be instantly displayed by picking the **Multiple viewports** button in the **View** panel on the **Home** tab of the ribbon.

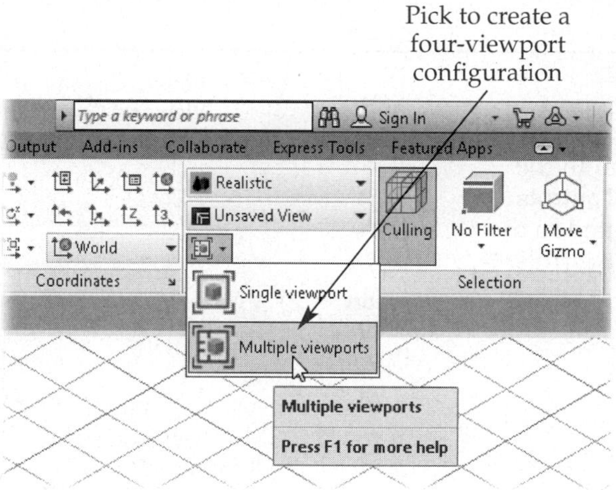

Figure 5-3.
Picking the **Multiple viewports** button in the **View** panel on the **Home** tab of the ribbon automatically creates a top, front, side, and pictorial view based on the current UCS.

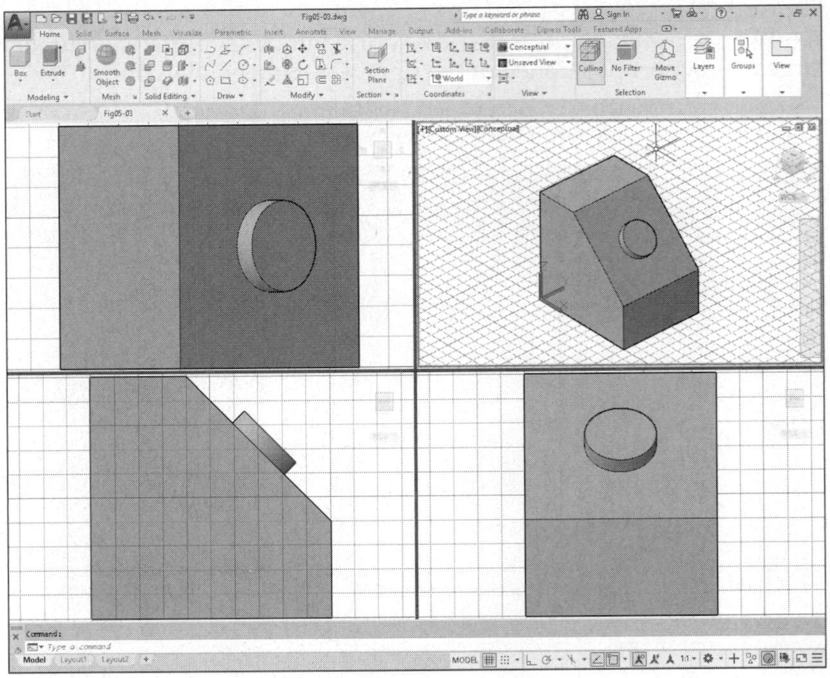

Viewport Configurations

A layout of viewport configurations can be quickly created by using the **Viewport Controls** flyout in the viewport controls located in the upper-left corner of the drawing window. See **Figure 5-4**. The viewport controls provide quick access to all viewport configurations and settings. Selecting **Viewport Configuration List** in the **Viewport Controls** flyout displays a menu with 12 different preset viewport configurations. Selecting **Configure...** in this menu displays the **Viewports** dialog box, **Figure 5-5**. This dialog box, which is also accessed with the **VPORTS** command, is used to save and manage viewport configurations. Another way to access the preset viewport configurations is to use the **Viewport Configuration** drop-down list in the **Model Viewports** panel on the **Visualize** tab of the ribbon. See **Figure 5-6**.

VPORTS

Type
VPORTS

As shown in **Figure 5-5**, there are two tabs in the **Viewports** dialog box. The preset viewport configurations are available in the **New Viewports** tab. The preset viewport configurations include six different options for three-viewport configurations. See

Figure 5-4.
Preset viewport configurations can be accessed in the **Viewport Controls** flyout located in the viewport controls.

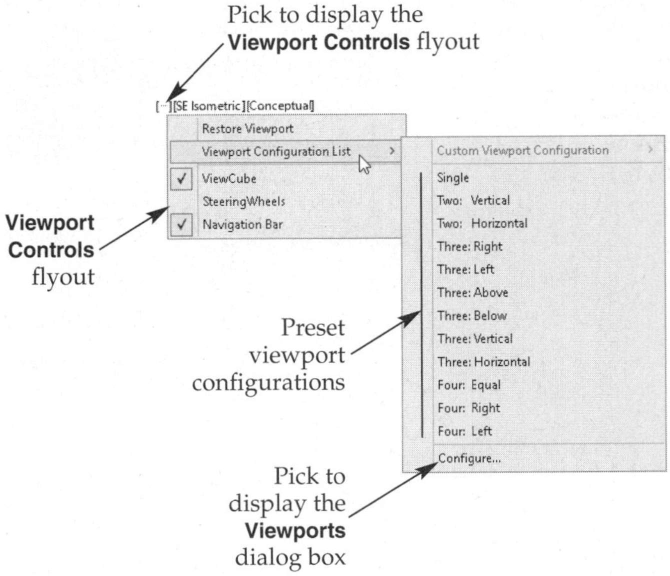

Figure 5-5.
Viewport configurations are created using the **New Viewports** tab of the **Viewports** dialog box.

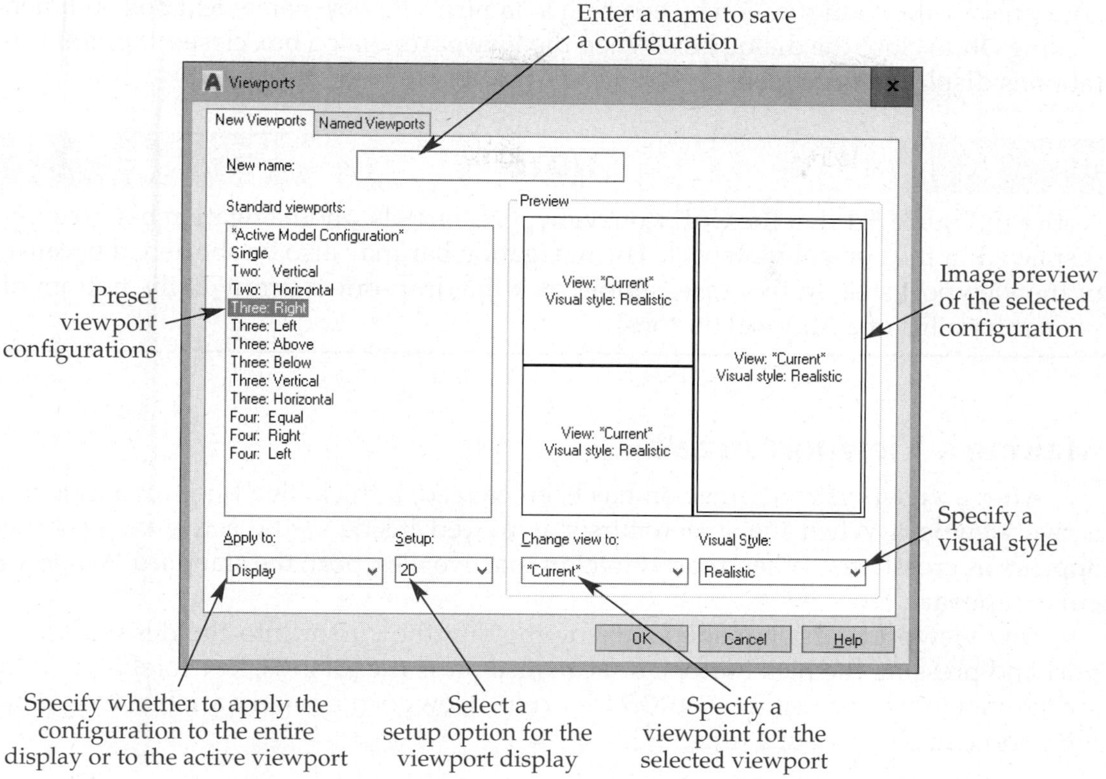

Figure 5-6.
The **Viewport Configuration** drop-down list in the **Model Viewports** panel on the **Visualize** tab of the ribbon offers a quick way to recall a preset viewport configuration.

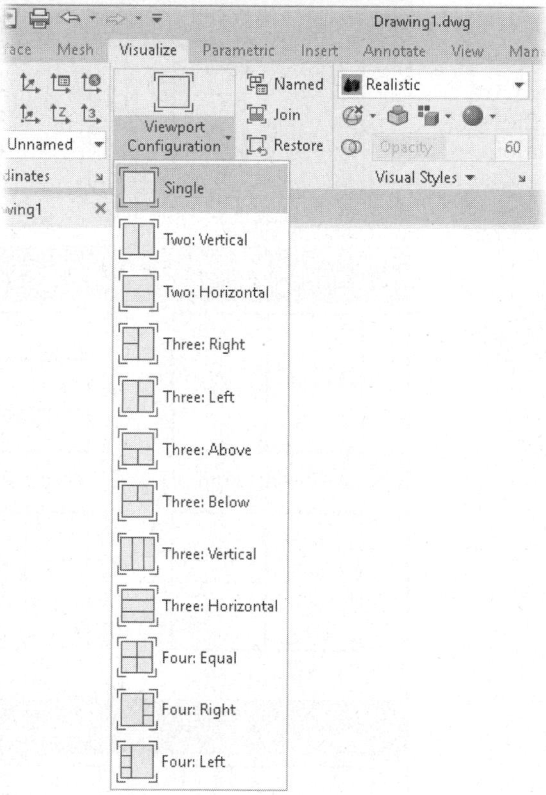

Figure 5-7. When you pick the name of a configuration in the **Standard viewports:** list, the viewport arrangement is displayed in the **Preview** area. After you have selected, you can save the configuration by entering a name in the **New name:** text box and then picking **OK** to close the dialog box. When the **Viewports** dialog box closes, the configuration is displayed on screen.

NOTE

Notice in **Figure 5-1** that the UCS icon, viewport controls, and navigation bar are only displayed in the current viewport. The navigation bar may also be shortened because of the viewport size. In this case, simply pick the drop-down arrow at the bottom of the bar for a list of additional buttons.

Making a Viewport Active

After a viewport configuration has been created, a thick blue line surrounds the active viewport. When the screen cursor is moved inside of the active viewport, it appears as crosshairs. When moved into an inactive viewport, the standard Windows cursor appears.

Any viewport can be made active by moving the cursor into the desired viewport and pressing the pick button. You can also press the [Ctrl]+[R] key combination to switch viewports, or use the **CVPORT** (current viewport) system variable. Only one viewport can be active at a time.

Figure 5-7.
Twelve preset tiled viewport configurations are provided in the **Viewports** dialog box.

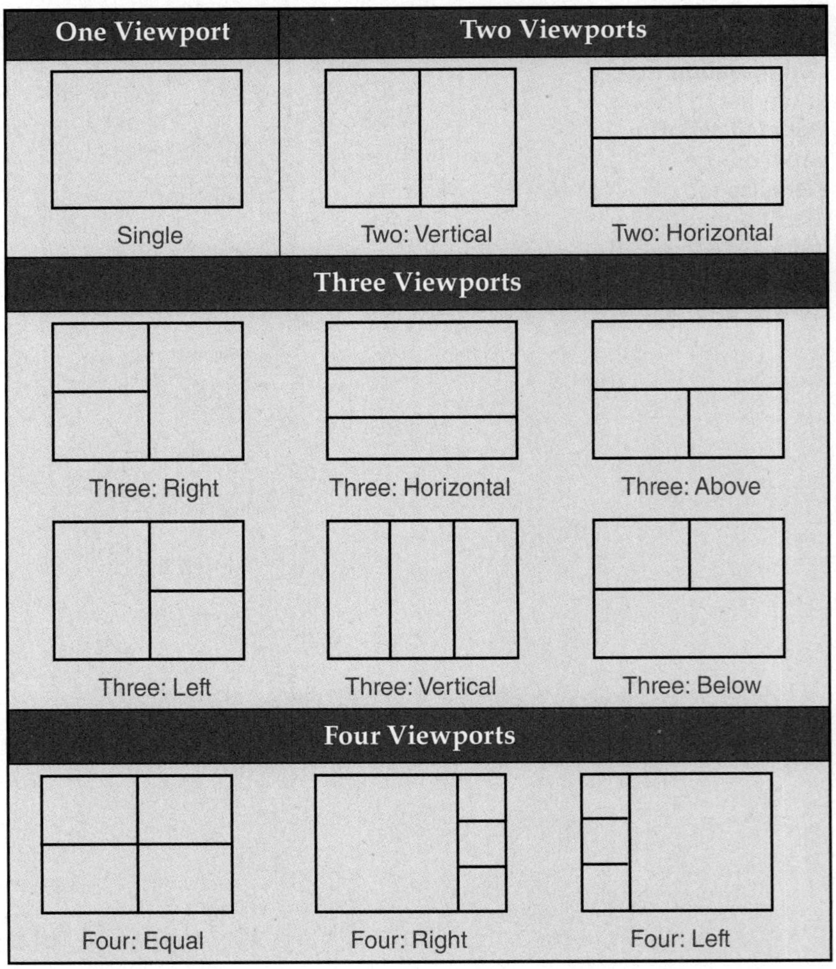

Command: **CVPORT**↵
Enter new value for CVPORT <*current*>: **3**↵

The current value given is the ID number of the active viewport. The ID number is automatically assigned by AutoCAD, starting with 2. To change viewports with the **CVPORT** system variable, simply enter a different ID number. This technique may be used in custom programming for AutoCAD. Using the **CVPORT** system variable is also a good way to determine the ID number of a viewport. The number 1 is not a valid viewport ID number.

PROFESSIONAL TIP

Each viewport can have its own view, viewpoint, UCS, zoom scale, limits, grid spacing, and snap setting. Specify the drawing aids in all viewports before saving the configuration. When a viewport is restored, all settings are restored as well.

Managing Defined Viewports

Ribbon
Visualize
> Model Viewports
Named Viewports...
Type
VPORTS

VPORTS

If you are working with several different viewport configurations, it is easy to restore, rename, or delete existing viewports. You can do so using the **Viewports** dialog box. To access a list of named viewports, open the dialog box and select the **Named Viewports** tab. See **Figure 5-8**. To display a viewport configuration, select its name from the **Named viewports:** list. The selected configuration is displayed in the **Preview** area. If this is the arrangement you want, pick the **OK** button.

Assume you have saved the current viewport configuration. Now, you want to work in a specific viewport, but do not need other viewports displayed on screen. First, pick the viewport you wish to work in to make it active. Next, select **Viewport Configuration List** in the **Viewport Controls** flyout in the viewport controls and select **Single**. You can also use the **Viewport Configuration** drop-down list in the **Model Viewports** panel on the **Visualize** tab of the ribbon. The active viewport is displayed as the only viewport. To restore the original viewport configuration, select **Viewport Configuration List** in the **Viewport Controls** flyout in the viewport controls, and then

Figure 5-8.
The **Named Viewports** tab of the **Viewports** dialog box lists all named viewports and displays the selected configuration in the **Preview** area.

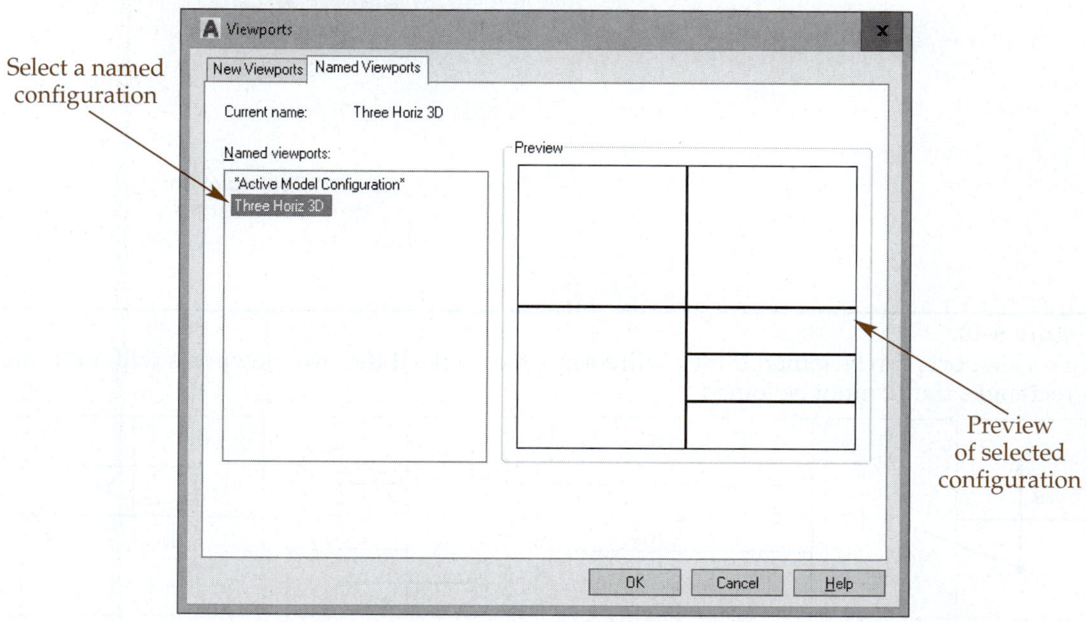

select **Custom Viewport Configuration**. This displays a list of named viewport configurations. Select the desired name and that configuration is displayed.

Viewport configurations can be renamed and deleted using the **Named Viewports** tab of the **Viewports** dialog box. To rename a viewport configuration, right-click on the name and pick **Rename** from the shortcut menu. To delete a viewport configuration, pick **Delete** from the shortcut menu. You can also press the [Delete] key to delete the highlighted viewport configuration. Pick **OK** to exit the dialog box.

PROFESSIONAL TIP

The **-VPORTS** command can be used to manage viewports on the command line. This may be required for some LISP programs where the dialog box cannot be used.

Using the Model Viewports Panel

The **Model Viewports** panel in the **Visualize** tab of the ribbon is shown in **Figure 5-9**. Picking the **Named Viewports...** button on the panel displays the **Viewports** dialog box with the **Named Viewports** tab active. The **Join** button is used to combine two viewports into a single viewport, as described in the next section. Picking the **Restore Viewports** button switches the display between a single viewport and the last multiple viewport configuration displayed. If no previous multiple viewport configuration has been displayed, picking this button displays a configuration of four viewports. Expanding the **Viewport Configuration** drop-down list displays the 12 preset viewport configurations. Selecting a configuration sets it current.

Joining Two Viewports

You can join two adjacent viewports in an existing configuration to form a single viewport. This process is often quicker than creating an entirely new configuration. However, the two viewports must form a rectangle when joined, **Figure 5-10**.

Figure 5-9.
The **Model Viewports** panel in the **Visualize** tab of the ribbon.

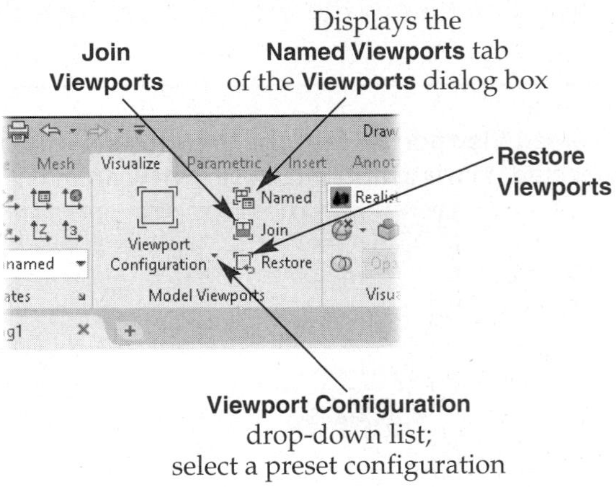

Figure 5-10.
Two viewports can be joined if they will form a rectangle. If the two viewports will not form a rectangle, they cannot be joined.

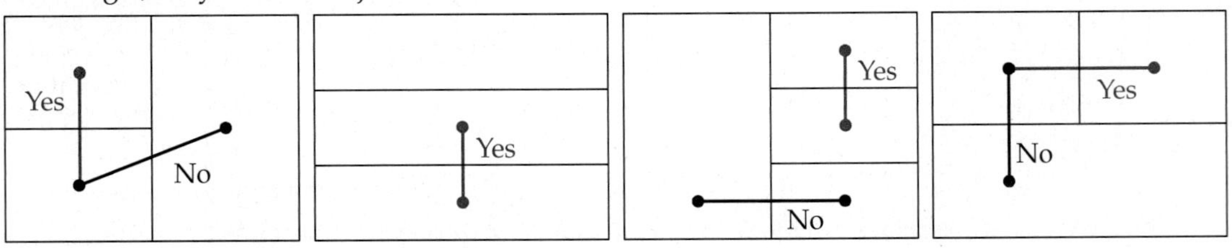

When you enter the **Join** option, AutoCAD first prompts you for the ***dominant viewport***. All aspects of the dominant viewport are used in the new (joined) viewport. These aspects include the limits, grid, UCS, and snap settings.

Ribbon
Visualize
> Model Viewports
Join Viewports
Type
-VPORTS

-VPORTS

Select dominant viewport <current viewport>: *(select the viewport by picking in it or press* [Enter] *to set the current viewport as the dominant viewport)*
Select viewport to join: *(select the other viewport)*

The two selected viewports are joined into a single viewport. If you select two viewports that do not form a rectangle, AutoCAD returns the message:

The selected viewports do not form a rectangle.

Adjusting Viewports

You can adjust a viewport configuration dynamically by picking directly on a viewport boundary and dragging the cursor. Viewports can be resized, added, and joined in this manner. To resize a viewport, pick a viewport boundary and drag it to the desired location. To add a new viewport, pick the small plus sign (+) on a viewport boundary and then drag the green splitter bar to locate the new boundary. See **Figure 5-11**. To join two viewports, pick a viewport boundary and drag it to another viewport boundary.

PROFESSIONAL TIP

Create only the number of viewports and viewport configurations needed to construct your drawing. Using too many viewports reduces the size of the image in each viewport and may confuse you. Also, it helps to zoom each view so that the objects fill the viewport.

Exercise 5-1

www.g-wlearning.com/CAD/

Complete the exercise on the companion website.

Applying Viewports to Existing Configurations and Displaying Different Views

You have total control over what is displayed in model space viewports. In addition to displaying various viewport configurations, you can divide an existing view-

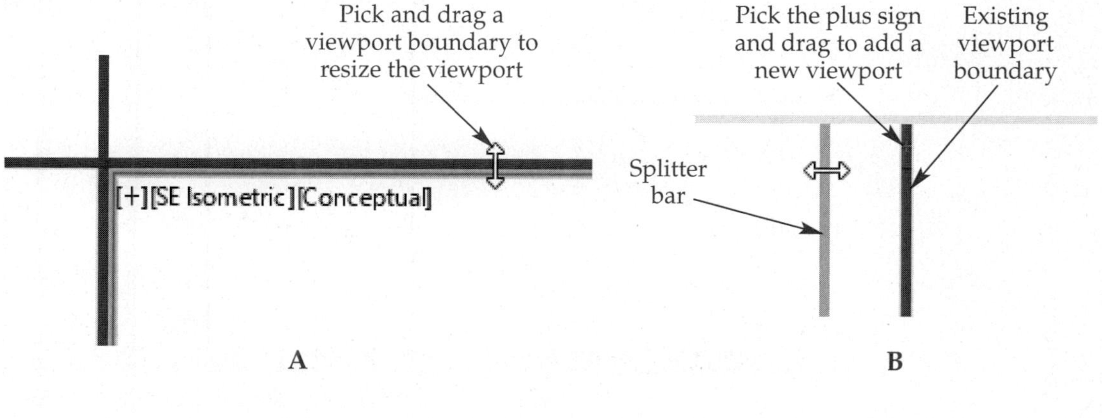

Figure 5-11.
Adjusting viewports dynamically. A—To resize a viewport, pick on a viewport boundary and drag it to a new location. B—To add a new viewport, pick the plus sign on a viewport boundary and then drag the green splitter bar to create a new boundary.

port into additional viewports or assign a different viewpoint to each viewport. The options for these functions are provided in the **New Viewports** tab of the **Viewports** dialog box. These options are located along the bottom of the **Viewports** dialog box, as shown in **Figure 5-5**:

- **Apply to.**
- **Setup.**
- **Change view to.**
- **Visual style.**

Apply To

When a preset viewport configuration is selected from the **Standard viewports:** list, it can be applied to either the entire display or the current viewport. The previous examples have shown how to create viewports that replace the entire display. Applying a configuration to the active viewport rather than the entire display can be useful when you need to display additional viewports.

For example, first create a configuration of three viewports using the Three: Right configuration option. Then, with the right (large) viewport active, open the **Viewports** dialog box again. Notice that the drop-down list under **Apply to:** is grayed out. Now, pick one of the standard configurations. This enables the **Apply to:** drop-down list. The default option is Display, which means the selected viewport configuration will replace the current display. Pick the drop-down list arrow to reveal the second option, Current Viewport. Pick this option and then pick the **OK** button. Notice that the selected viewport configuration has been applied to only the active (right) viewport. See **Figure 5-12**.

Setup

Viewports can be set up to display views in 2D or 3D. The 2D and 3D options are provided in the **Setup:** drop-down list. Displaying different views while working on a drawing allows you to see the results of your work on each view, since changes are reflected in each viewport as you draw. The selected viewport **Setup:** option controls the types of views available in the **Change view to:** drop-down list.

Figure 5-12.
The selected viewport configuration has been applied to the active viewport within the original configuration.

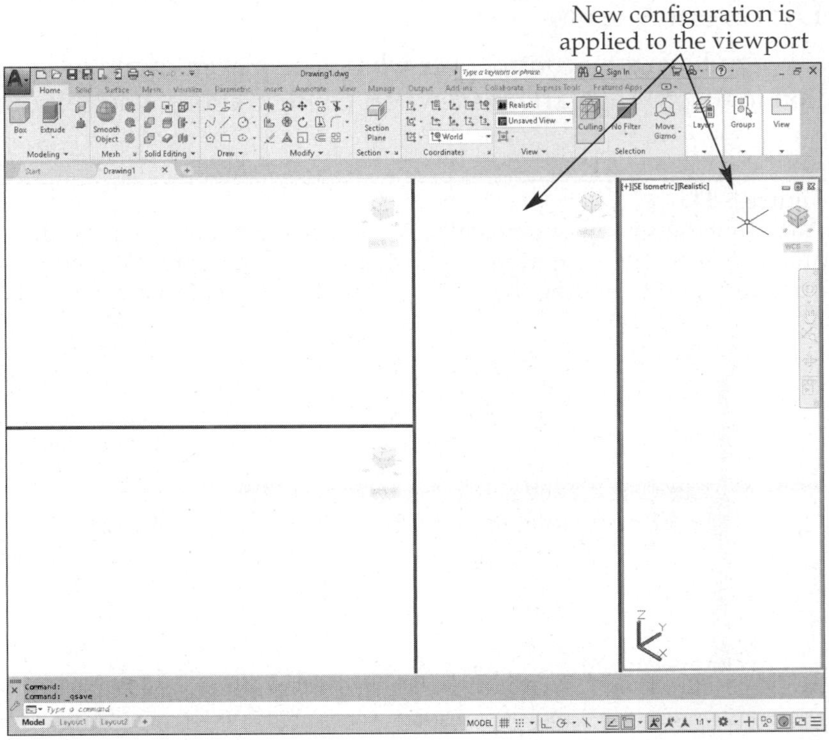

Change View To

The views that can be displayed in a selected viewport are listed in the **Change view to:** drop-down list. If the **Setup:** drop-down list is set to 2D, the views available to be displayed are limited to the current view and any named views. If 3D is active, the options include all of the standard orthographic and isometric views along with named views. When an orthographic or isometric view is selected for a viewport, the resulting orientation is shown in the **Preview** area. To assign a different viewpoint to a viewport, simply pick within a viewport in the **Preview** area to make it active and then pick a viewpoint from the **Change view to:** drop-down list. Important: note that if you set a viewport to one of the orthographic preset views, the UCS is also changed (by default) in that viewport to the corresponding preset.

Visual Style

A visual style can be specified for a viewport. Pick within a viewport in the **Preview** area to make it active and then select a visual style from the **Visual Style:** drop-down list. All preset and saved visual styles are available in the drop-down list.

PROFESSIONAL TIP

If a standard naming convention is used to create viewport configurations, these named configurations can be saved with template drawings. This creates a consistent platform that all students or employees can use.

Exercise 5-2

www.g-wlearning.com/CAD/

Complete the exercise on the companion website.

Drawing in Multiple Viewports

When used with 2D drawings, viewports allow you to display a view of the entire drawing, plus views showing portions of the drawing. This is similar to using the **VIEW** command, except you can have several views on screen at once. You can also adjust the zoom magnification in each viewport to suit different areas of the drawing.

Viewports are also a powerful aid when constructing 3D models. You can specify different viewpoints in each viewport and see the model take shape as you draw. A model can be quickly constructed because you can switch from one viewport to another while drawing and editing. For example, you can draw a line from a point in one viewport to a point in another viewport simply by changing viewports while inside of the **LINE** command. The result is shown in each viewport.

In Chapter 4, you constructed a solid object. It was a base that had an angled surface from which a cylinder projected. Now, you will construct the same object, but using two viewports. First, create a vertical configuration of two viewports. In the **Viewports** dialog box, set the right-hand viewport to display the southeast isometric view. Also, select the Conceptual visual style. Set the left-hand viewport to display the front view. Remember, this will also set the UCS to the front preset orthographic UCS in that viewport. Also, select the Wireframe visual style for the left-hand viewport. Close the dialog box and make the left-hand viewport active. Next, set the parallel projection current in both viewports.

Next, draw a polyline using the coordinates shown in **Figure 5-13**. Remember, to enter absolute coordinates with dynamic input turned on, type a pound sign (#) before the coordinate. This temporarily overrides the dynamic input of relative coordinate values. Be sure to use the **Close** option for the last segment. As you construct the side view, you can clearly see its true size and shape in the left-hand viewport. At the same time, you can see the construction in 3D in the right-hand viewport.

Notice in **Figure 5-13** that each view has a different UCS, as indicated by the UCS icon. Pick in each viewport in your drawing to view the different UCS orientations.

The next step is to extrude the shape to create the base. The **EXTRUDE** command is used the same way as in Chapter 4. With the left-hand viewport current, enter the **EXTRUDE** command, pick the polyline, and enter an extrusion height of –4 units. The base of the object is now complete, **Figure 5-14**.

Now, the cylinder needs to be created on the angled face. First, split the left-hand viewport into two horizontal viewports (top and bottom) using the **New Viewports** tab of the **Viewports** dialog box. Set both of the new viewports to display the current view. Pick the **OK** button to close the dialog box. Then, make the upper-left viewport current and set it up to always display a plan view of the current UCS by setting the **UCSFOLLOW** system variable to 1. In addition, set the **UCSVP** system variable to 0 so that the UCS configuration in the current viewport is displayed.

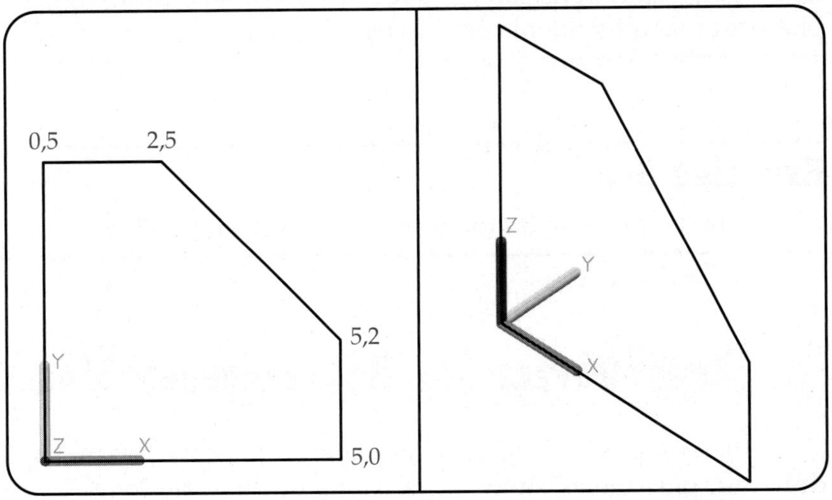

Figure 5-13. The screen is divided into two viewports. A front view of the object appears in the left-hand viewport and a 3D view appears in the right-hand viewport. Activate each viewport to view the UCS icon orientation as shown here.

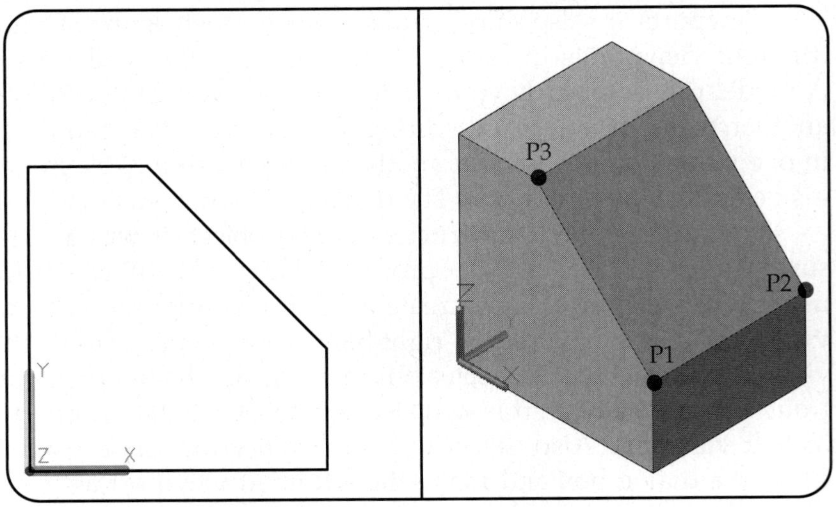

Figure 5-14. The base of the object is now complete. A new UCS will be created on the angled face. If you are using the **3 Point** option of the UCS command, use the pick points indicated.

Using the UCS icon grips, as discussed in Chapter 4, create a new UCS on the angled face in the right-hand viewport. You can move the UCS icon dynamically using the origin grip, or you can use the **3 Point** option of the **UCS** command. If you are using the **3 Point** option, use the pick points shown in **Figure 5-14**. To quickly access the **3 Point** option, right-click on the UCS icon and select **3 Point** from the UCS icon shortcut menu. Notice in your drawing that after making the right-hand viewport active, the view in the upper-left viewport automatically changes to a plan view of the new current UCS. The view changes again after creating the new UCS on the angled face in the right-hand viewport.

With the right-hand viewport current, turn on 3D object snap. Turn off the dynamic UCS function, if it is on. Then, enter the **CYLINDER** command. When locating the center of the cylinder base, use the **Center of face** 3D object snap and select the center of the angled face. Next, enter a diameter of 1.5 units and a height of .35 units.

The object is now complete, **Figure 5-15**. Notice how the viewports have different UCSs. Each viewport can have its own UCS. The view in the upper-left viewport is the plan view of the current UCS. This is because of the **UCSFOLLOW** system variable setting of 1 in the upper-left viewport. In addition, the UCS orientation in the upper-left viewport is the same as the orientation in the right-hand viewport. The UCS orientation in one viewport is not affected by a change to the UCS in another viewport unless the **UCSVP** system variable is set to 0 in a viewport. If a viewport arrangement is saved with several different UCS configurations, then every named UCS remains intact and is displayed when the viewport configuration is restored.

To see the effect of the **UCSVP** system variable in the upper-left viewport, make that viewport current after completing the model in the right-hand viewport. The resulting display is shown in **Figure 5-15**. Next, make the lower-left viewport current, and then make the upper-left viewport current again. Now, notice that the front view is displayed in the upper-left viewport, and the UCS orientation is the same as the orientation in the lower-left viewport.

Figure 5-15.
The cylinder is drawn to complete the object. Notice the plan view in the upper-left viewport. Activate the upper-left viewport after completing the model in the lower-right viewport to view the UCS icon orientation as shown.

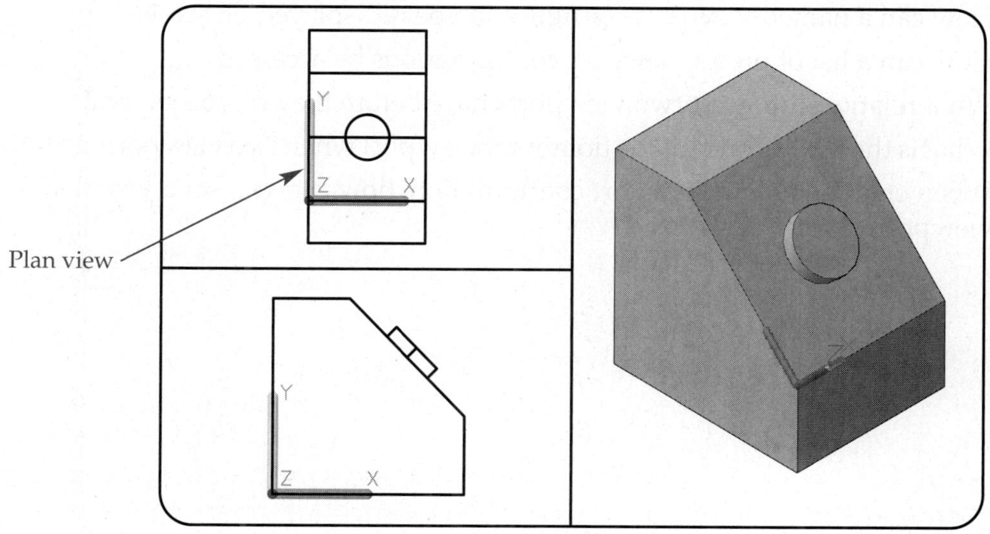

The **REGEN** command affects only the current viewport. To regenerate all viewports at the same time, use the **REGENALL** command. This command can be entered by typing REGENALL.

The Quick Text mode is controlled by the **REGEN** command. Therefore, if you are working with text displayed with the Quick Text mode in viewports, be sure to use the **REGENALL** command in order for the text to be regenerated in all viewports.

NOTE

The **UCSFOLLOW** system variable and the **UCSVP** system variable can be set independently for each viewport. As discussed in the previous example, the UCS configuration in each viewport is controlled by the **UCSVP** system variable. When **UCSVP** is set to 1 in a viewport, the UCS is independent from all other UCSs, which is the default. If **UCSVP** is set to 0 in a viewport, its UCS will change to reflect any changes to the UCS in the current viewport.

Exercise 5-3

www.g-wlearning.com/CAD/

Complete the exercise on the companion website.

Chapter Review

Answer the following questions using the information in this chapter.

1. What is the purpose of *viewports*?
2. How do you name a configuration of viewports?
3. What is the purpose of saving a configuration of viewports?
4. Explain the difference between *tiled* and *floating* viewports.
5. Name the system variable controlling the maximum number of viewports that can be displayed at one time.
6. How can a named viewport configuration be redisplayed on screen?
7. How can a list of preset viewport configurations be accessed?
8. What relationship must two viewports have before they can be joined?
9. What is the significance of the dominant viewport when two viewports are joined?
10. When creating a new viewport configuration, how can you set a visual style in a viewport?

Drawing Problems

1. **General** Construct seven template drawings, each with a preset viewport configuration. Use the following configurations and names. Save each template under the same name as the viewport configuration.

Number of Viewports	Configuration	Name
2	Horizontal	TWO-H
2	Vertical	TWO-V
3	Right	THREE-R
3	Left	THREE-L
3	Above	THREE-A
3	Below	THREE-B
3	Vertical	THREE-V

2. **General** Construct one of the problems from Chapter 4 using viewports. Use one of your template drawings from Problem 5-1. Save the drawing as P5-2.

3. **Mechanical** This is an orthographic drawing of a light fixture bracket. Create it as a solid model. If appropriate, use geometric constraints on the base 2D shapes. Use solid primitives and Boolean commands as needed. Use the dimensions given. Similar holes have the same offset dimensions. Use multiple viewports to construct the drawing. Begin with a four-viewport layout. Then switch viewports as needed to work on specific areas of the model. Create new UCSs as needed. Display an appropriate pictorial view of the drawing in the upper-right viewport. Plot the 3D view of the drawing to scale on a B- or C-size sheet of paper. Save the drawing as P5-3.

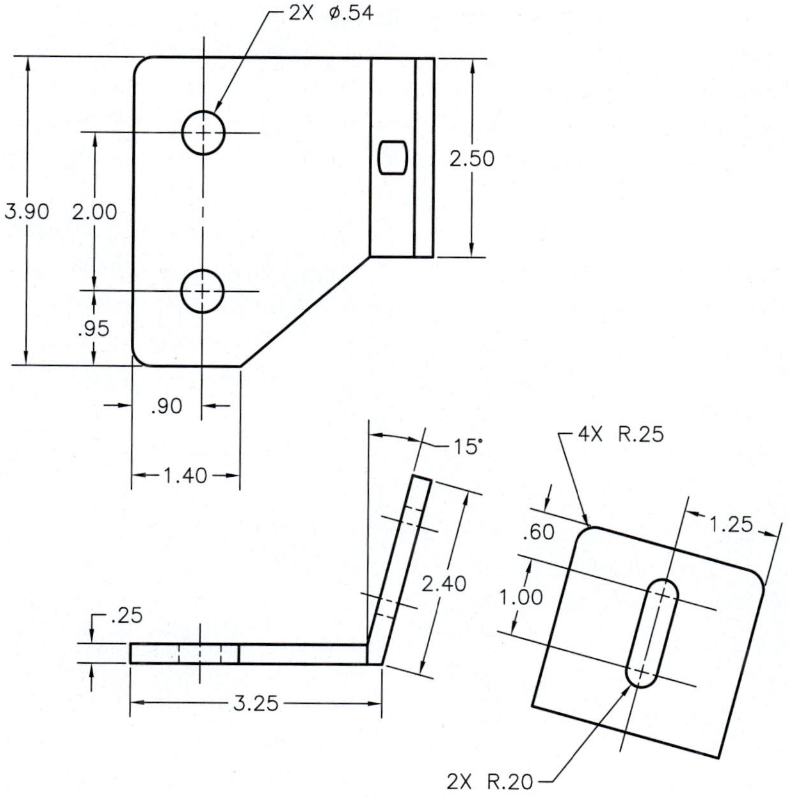

Light Fixture Bracket

4. Mechanical This is an orthographic drawing of an angle bracket. Create it as a solid model. If appropriate, use geometric constraints on the base 2D shapes. Use solid primitives and Boolean commands as needed. Use the dimensions given. Similar holes have the same offset dimensions. Use multiple viewports to construct the drawing. Begin with a four-viewport layout. Then switch viewports as needed to work on specific areas of the model. Create new UCSs as needed. Display an appropriate pictorial view of the drawing in the upper-right viewport. Plot the 3D view of the drawing to scale on a B- or C-size sheet of paper. Save the drawing as P5-4.

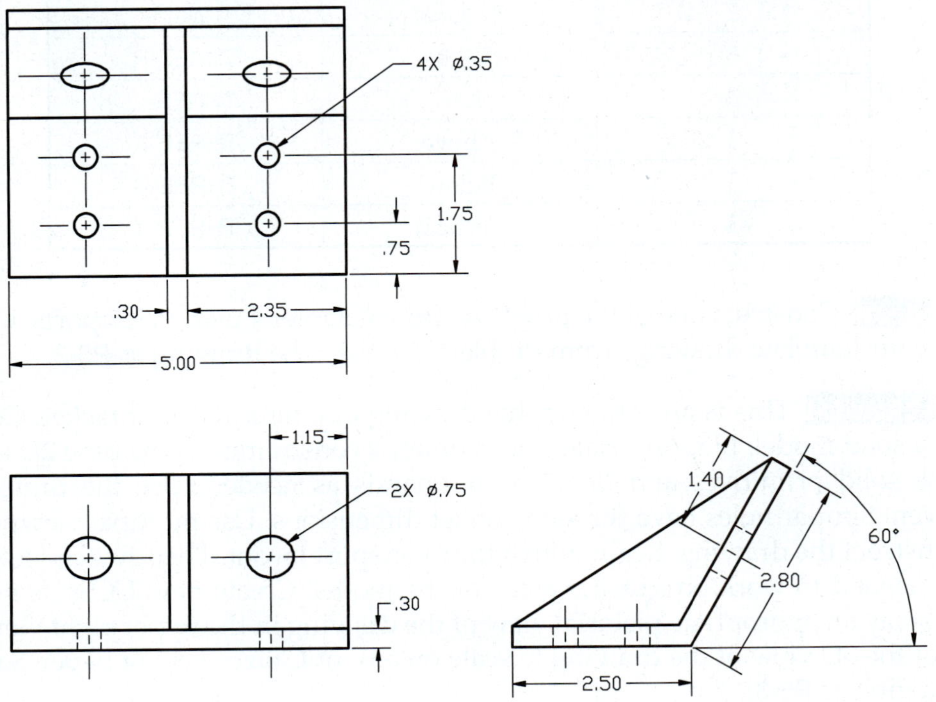

Angle Bracket

CHAPTER 6

Model Extrusions and Revolutions

Learning Objectives

After completing this chapter, you will be able to:

- ✓ Create solids and surfaces by extruding profiles.
- ✓ Extrude regions.
- ✓ Extrude surfaces.
- ✓ Create symmetrical 3D solids and surfaces by revolving profiles.
- ✓ Revolve regions.
- ✓ Revolve surfaces.
- ✓ Extrude and revolve objects using mathematical expressions and constraints.
- ✓ Use solid extrusions and revolutions as construction tools.

A complex shape can be created by *extruding* a profile into a 3D solid or surface model. You have been introduced to the operation in previous chapters. Symmetrical objects can be created by revolving a profile about an axis to create a new shape. This chapter discusses the **EXTRUDE** and **REVOLVE** commands and using these commands as construction tools.

Creating Extruded Models

An extrusion is a profile that has been "extruded" into a 3D solid or surface. The **EXTRUDE** command allows you to create extrusions from lines, arcs, elliptical arcs, 2D and 3D polylines, 2D and 3D splines, circles, ellipses, 2D solids, 3D solid faces and edges, regions, planar surfaces, surface edges, and donuts.

EXTRUDE

Ribbon
Home
> Modeling
Solid
> Solid
Extrude
Surface
> Create
Extrude
Type
EXTRUDE
EXT

By default, closed objects, such as circles, polygons, closed polylines, and donuts, are converted to a ***solid extrusion*** when they are extruded. A solid extrusion represents a solid object and has mass properties. Open-ended objects, such as lines, arcs, polylines, elliptical arcs, and splines, are converted to a ***surface extrusion*** when they are extruded. Surface extrusions have no mass properties.

Extrusions can be created along a straight line or along a path curve. A taper angle can also be applied as you extrude an object. **Figure 6-1** illustrates a polygon extruded into a solid.

When the **EXTRUDE** command is selected, you are prompted to select the objects to extrude. Select the objects and press [Enter]. You are then prompted for the extrusion

Figure 6-1.
The **EXTRUDE** command creates a 3D solid or surface from a profile object. A—The initial, closed profile. B—The extruded solid object shown with hidden lines removed.

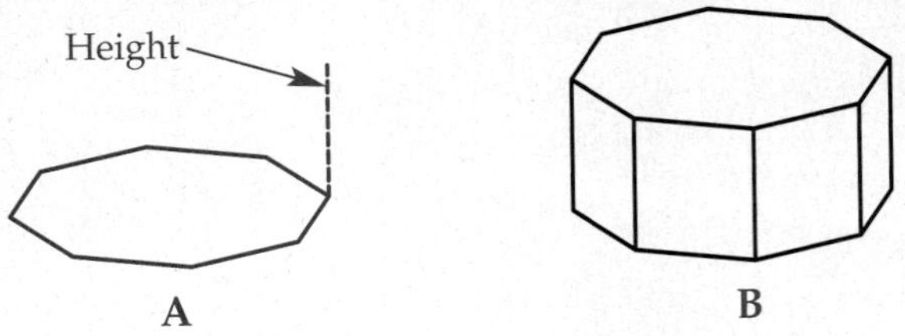

height. If a pictorial view is displayed, you can move the cursor to establish a specific height. The height is always applied along the Z axis of the selected object, not the current UCS. A positive value extrudes the height above the XY plane of the object. A negative height value extrudes the height below the XY plane. If a pictorial view is displayed, you can drag the mouse to set the extrusion above or below the XY plane and then enter the height value.

The **Direction** option provides an alternate way to create the extrusion. After selecting this option, specify a start and end point to define the extrusion direction and height. This is similar to extruding along a path, discussed in the next section.

Before entering a height, you can specify a taper angle. The taper angle can be any value *between* +90° and –90°. A positive angle tapers to the inside of the object from the base. A negative angle tapers to the outside of the object from the base. See **Figure 6-2**. However, the taper angle cannot result in edges that "fold into" the extruded object.

If the command is selected from the **Solid** tab in the ribbon, the mode is automatically set to solid. However, if an open profile is selected, a surface is still created. If the command is selected from the **Surface** tab in the ribbon, the mode is automatically set to surface. In either case, you can use the **Mode** option to change the output type to solid or surface.

Keep in mind, objects that are not closed, such as lines and polylines, *always* result in extruded surfaces. A solid can only be created by the **EXTRUDE** command if the original object is closed.

NOTE

If a surface is created using the **EXTRUDE** command, the type of surface created is controlled by the **SURFACEMODELINGMODE** system variable. The settings available are discussed later in this chapter.

PROFESSIONAL TIP

Objects such as polylines, lines, and arcs that have a thickness can be converted to surfaces using the **CONVTOSURFACE** command. Circles and closed polylines with a thickness can be converted to solids using the **CONVTOSOLID** command. The thickness property for an object is controlled by the Thickness setting in the **Properties** palette.

Figure 6-2.
A—A positive angle tapers to the inside of the object from the base. B—A negative angle tapers to the outside of the object.

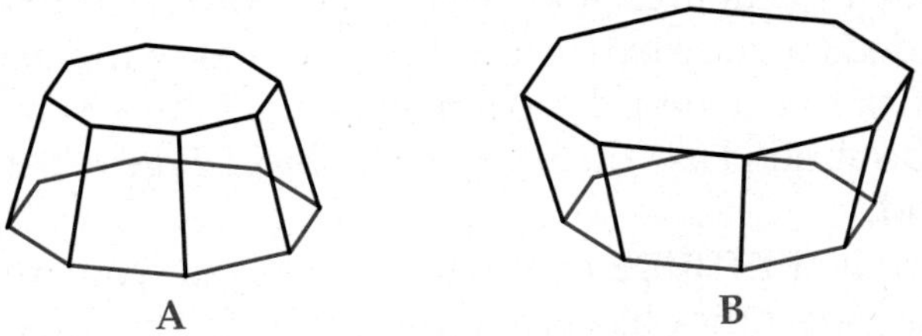

Extrusions along a Path

A profile shape can be extruded along a path to create a 3D solid or surface. The path can be a line, circle, arc, ellipse, elliptical arc, 2D or 3D polyline, helix, 3D solid edge, surface edge, or 2D or 3D spline. Multiple line segments and other objects can be first joined to form a polyline path. The corners of angled segments on the extruded object are mitered, while curved segments are smooth. See **Figure 6-3**.

When open objects, such as lines, arcs, polylines, elliptical arcs, and splines, are used as the profile, they are converted to a swept surface when extruded along a path. A *sweep* is a solid or surface that is created when an open or closed curve is pulled, or swept, along a 2D or 3D path. An extrusion is really a form of a sweep. Sweeps are discussed in detail in Chapter 7.

To extrude along a path, enter the **EXTRUDE** command and select the objects to extrude. When prompted for the height of the extrusion, select the **Path** option. If needed, first enter a taper angle. Then, pick the object to be used as the extrusion path.

Objects can also be extruded along a line at an angle to the base object, **Figure 6-4**. Notice that the plane at the end of the extruded object is parallel to the original object. Also notice that the length of the extrusion is the same as that of the path. The path does not need to be perpendicular to the object.

If the path begins perpendicular to the profile, the cross section of the resulting extrusion is perpendicular to the path, regardless if the path is a straight line, curve, or spline. See **Figure 6-5**. If the path is a spline or curve that does not begin perpendicular to the profile, the profile may not remain perpendicular to the path as it is extruded.

If one of the endpoints of the path is not on the plane of the object to be extruded, the path is temporarily moved to the center of the profile. The extrusion is then created as if the path were connected to the original object, as shown in **Figure 6-4**.

Figure 6-3. A—Angled segments are mitered when an object is extruded. B—Curved segments are smoothed when an object is extruded.

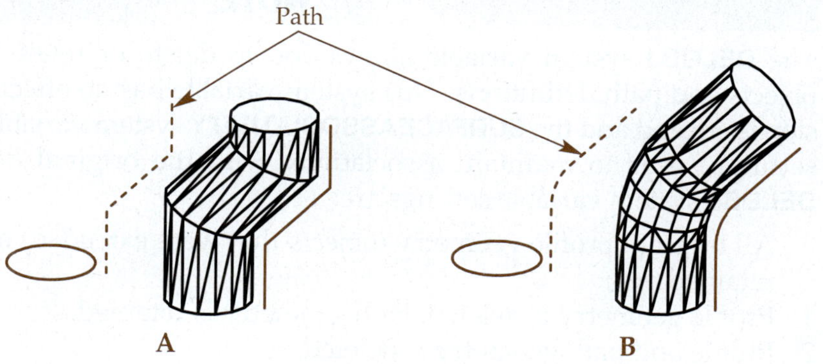

Figure 6-4. A—An object extruded along a path. B—The end of an object extruded along an angled path is parallel to the original object.

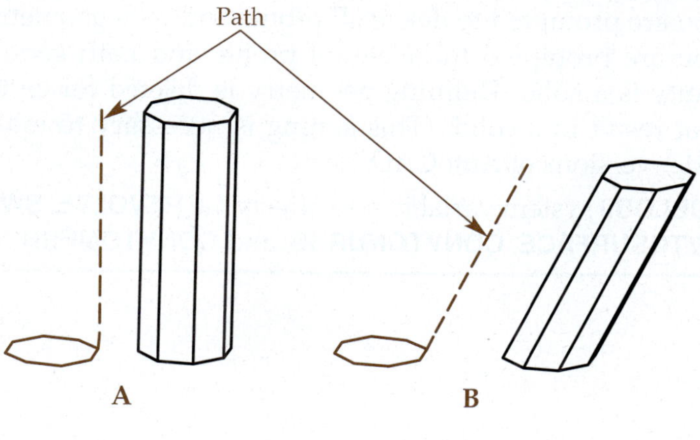

Figure 6-5.
A—Splines can be used as extrusion paths. Notice that the profile on the right is not perpendicular to the start of the path.
B—The resulting extrusions.

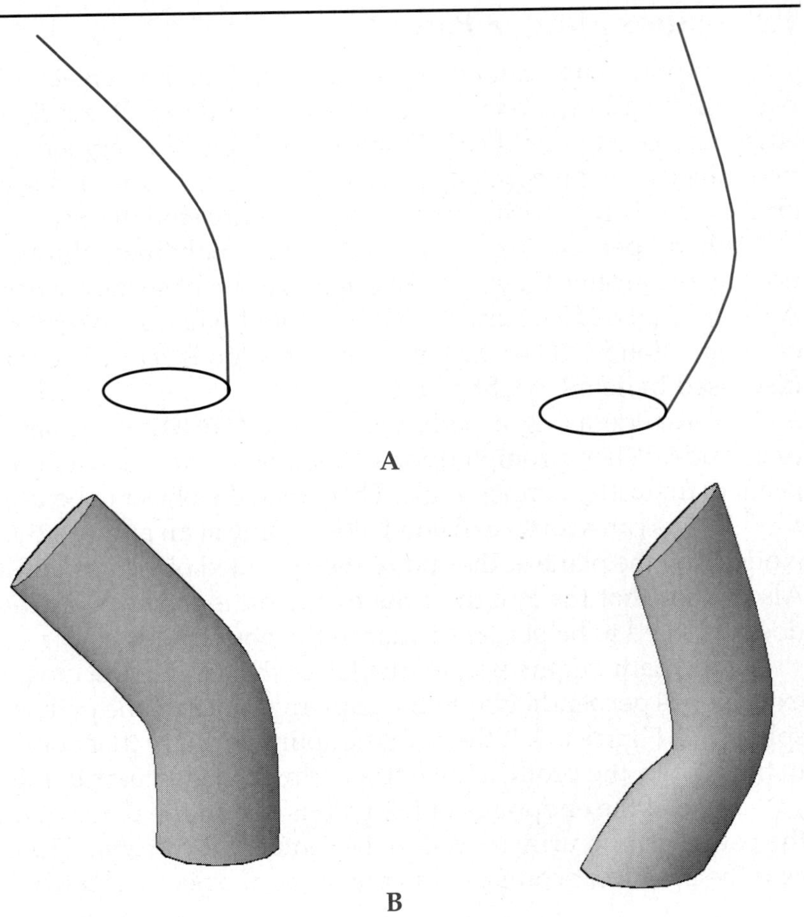

NOTE

The **DELOBJ** system variable allows you to delete or retain the original extruded objects and path definitions. This system variable has no effect when a surface extrusion is created and the **SURFACEASSOCIATIVITY** system variable is set to 1 (the default setting), so as to maintain associativity with the original defining geometry. The **DELOBJ** system variable settings are:

0 All original profile geometry (objects that were extruded) and path geometry is retained.
1 Profile geometry is deleted. Path geometry is retained.
2 Profile and path geometry is deleted.
3 Profile and path geometry is deleted when the extruded entity is a solid. (This is the default setting.)
–1 You are prompted to delete profile geometry. Path geometry is retained.
–2 You are prompted to delete all profile and path geometry.
–3 You are prompted to delete all profile and path geometry when the extruded entity is a solid. Defining geometry is deleted for certain modeling operations that result in a solid. (This setting is necessary to maintain compatibility with older versions of AutoCAD.)

The **DELOBJ** system variable also affects the **REVOLVE, SWEEP, LOFT, CONVTOSOLID, CONVTOSURFACE, CONVTONURBS**, and **CONVTOMESH** commands.

PROFESSIONAL TIP

You can use the **SWEEP** command to sweep objects along a path. In most cases, the **SWEEP** command is preferable to extruding along a path with the **EXTRUDE** command.

Extruding Regions

In Chapter 2, you learned how to create 2D regions. As an example, you created the top view of the base shown in **Figure 6-6A** as a region. Regions can be extruded to create 3D solids. The base you created in Chapter 2 can be extruded to create the final solid shown in **Figure 6-6B**. Any features of the region, such as holes, are extruded the same height as the rest of the object. If the profile was created as polylines, the holes must be separately extruded and then subtracted from the solid. Using this method, you can construct a fairly complex 2D region that includes curved profiles, holes, slots, etc. See **Figure 6-6C**. Then, a complex 3D solid can be quickly created. See **Figure 6-6D**. Additional features can be added using editing commands or Boolean operations.

Exercise 6-1

www.g-wlearning.com/CAD/

Complete the exercise on the companion website.

Figure 6-6.
A—The 2D region that will be extruded. B—The solid object created by extruding the region, shown in a 3D wireframe display. C—A 2D region created from a variety of objects. The objects forming the outer area include polylines, splines, arcs, and lines. The internal circles form two separate regions and are subtracted using a Boolean operation to create a single region. D—The solid object after extruding the region with the Conceptual visual style set current.

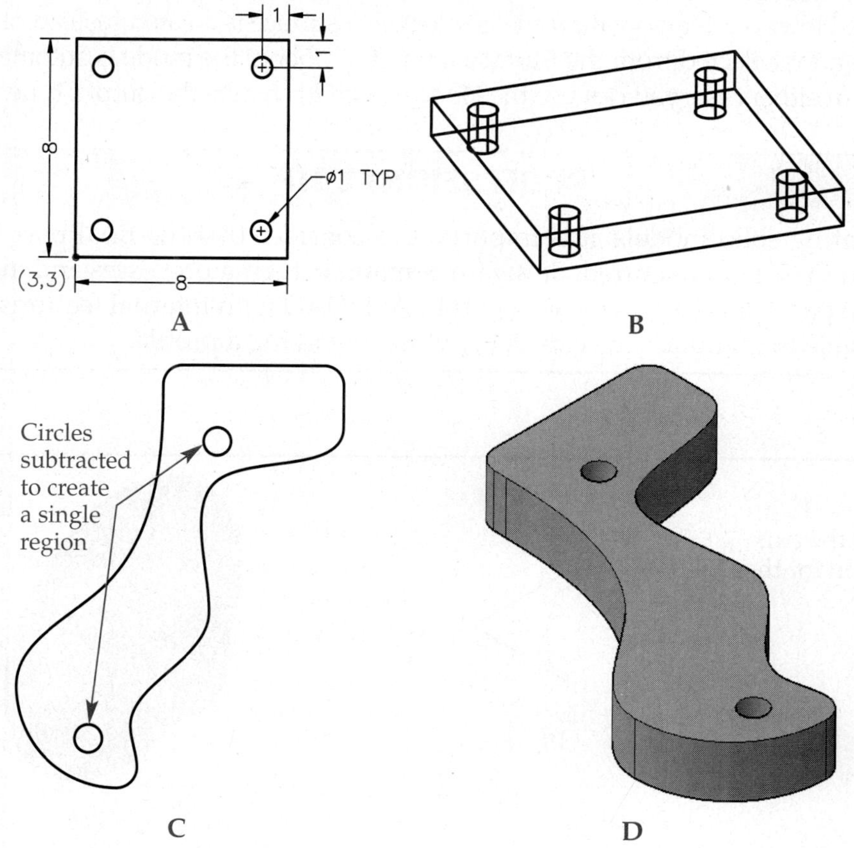

Extruding a Surface

A planar surface can be extruded into a solid object in the same manner as a region. Nonplanar (curved) surfaces can be extruded, but only into extruded surfaces. A planar surface can be quickly converted to a solid using the **EXTRUDE** command. Simply select the surface when prompted to select objects. The surface can be extruded in a specific direction, along a path, or at a taper angle.

Any closed object, such as a circle, rectangle, polygon, or polyline, can be converted into a surface with the **Object** option of the **PLANESURF** command. After accessing the **PLANESURF** command, enter the **Object** option and select the closed object. The resulting surface can then be extruded into a 3D solid. As previously discussed, the **Mode** option of the **EXTRUDE** command allows you to set the output type to solid or surface.

PLANESURF
Ribbon
Surface
> Create
Planar Surface
Type
PLANESURF

Creating Revolved Models

REVOLVE
Ribbon
Home
> Modeling
Solid
> Solid
Revolve
Surface
> Create
Revolve
Type
REVOLVE
REV

The **REVOLVE** command allows you to create solids and surfaces by revolving a shape about an axis. Shapes that can be revolved include lines, arcs, circles, ellipses, polygons, 2D and 3D polylines, 2D and 3D splines, regions, planar surfaces, and donuts. The selected object can be revolved at any angle up to 360°. Open curves and line segments create revolved surfaces. Closed curves can be used to create either solids or surfaces. Surface revolutions have no mass properties.

When the command is selected, you are prompted to pick the objects to revolve. Then, you must define the axis of revolution. The default option is to pick the two endpoints of an axis of revolution. This is shown in **Figure 6-7**. You can also revolve about an object or the X, Y, or Z axis of the current UCS. Once the axis is defined, you are prompted to enter the angle through which the profile will be revolved. When the angle is specified by either moving the cursor or entering a value, the revolution is created.

If the command is selected from the **Solid** tab in the ribbon, the mode is automatically set to solid. However, if an open profile is selected, a surface is created instead of a solid. If the command is selected from the **Surface** tab in the ribbon, the mode is automatically set to surface. In either case, you can use the **Mode** option to change the output type.

PROFESSIONAL TIP

When creating solid models, it is important to consider that the final part will most likely need to be manufactured. Be aware of manufacturing processes and methods as you design parts. It is easy to create a part in AutoCAD with internal features that may be impossible to manufacture, especially when revolving a profile.

Figure 6-7.
Points P1 and P2 are selected as the axis of revolution for the profile.

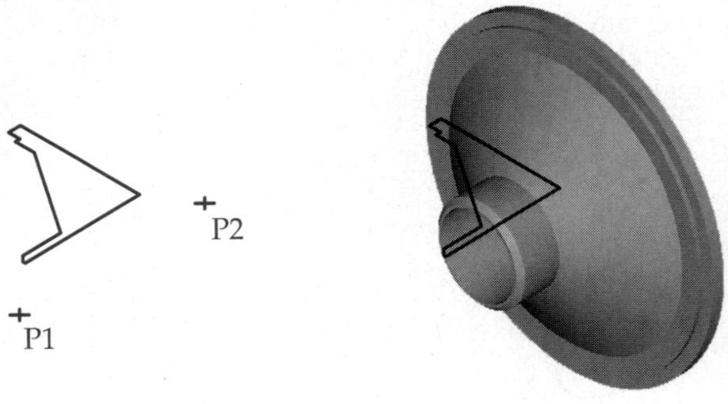

Revolving about an Object

You can select an object, such as a line, as the axis of revolution. Figure 6-8 shows a solid created using the **Object** option of the **REVOLVE** command. Both a full circle (360°) revolution and a 270° revolution are shown. Enter the **Object** option when prompted for the axis of revolution. Then, pick the axis object and enter the angle through which the profile will be rotated. You can use the **Start Angle** option before entering an angle of revolution. This allows you to specify the point at which the revolution starts and then the angle of revolution. The **Reverse** option allows you to reverse the initial direction in which the profile is revolved.

Revolving about the X, Y, or Z Axis

The X axis of the current UCS can be used as the axis of revolution by selecting the **X** option of the **REVOLVE** command. The origin of the current UCS is used as one end of the X axis line. Notice in Figure 6-9 that two different shapes can be created from the same profile by changing the UCS origin. No hole appears in the object in Figure 6-9B because the profile was revolved about an edge that coincides with the X axis. The Y or Z axis can also be used as the axis of revolution. See Figure 6-10.

Exercise 6-2

www.g-wlearning.com/CAD/

Complete the exercise on the companion website.

Figure 6-8.
An axis of revolution can be selected using the **Object** option of the **REVOLVE** command. Here, the line is selected as the axis.

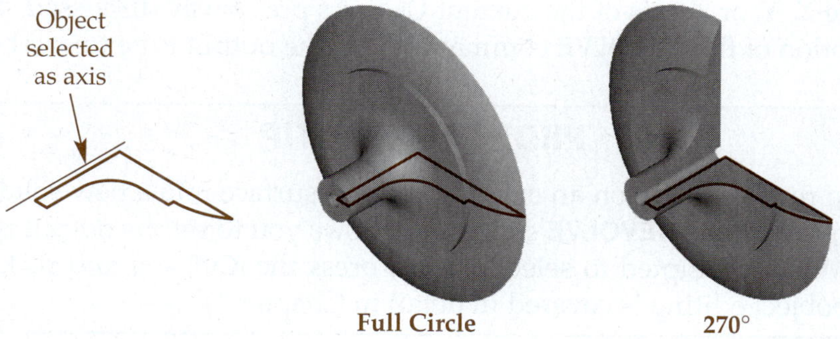

Figure 6-9.
A—A solid is created using the X axis as the axis of revolution. B—A different object is created with the same profile by changing the UCS origin.

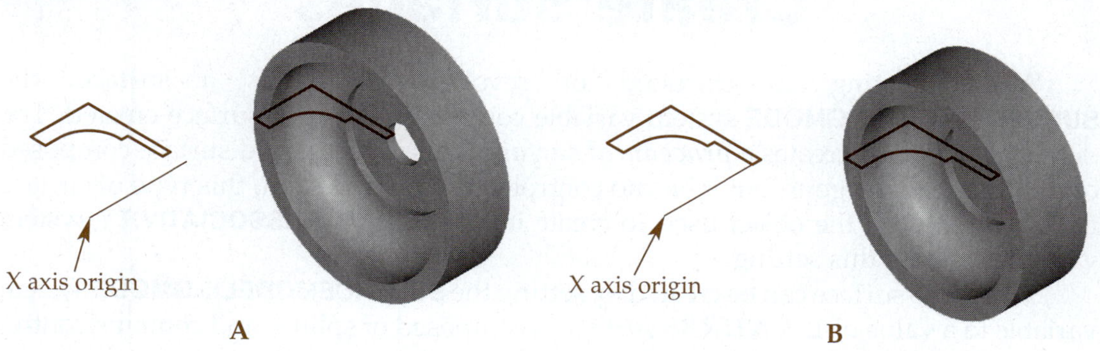

Figure 6-10.
A—A solid is created using the Y axis as the axis of revolution. B—A different object is created by changing the UCS origin.

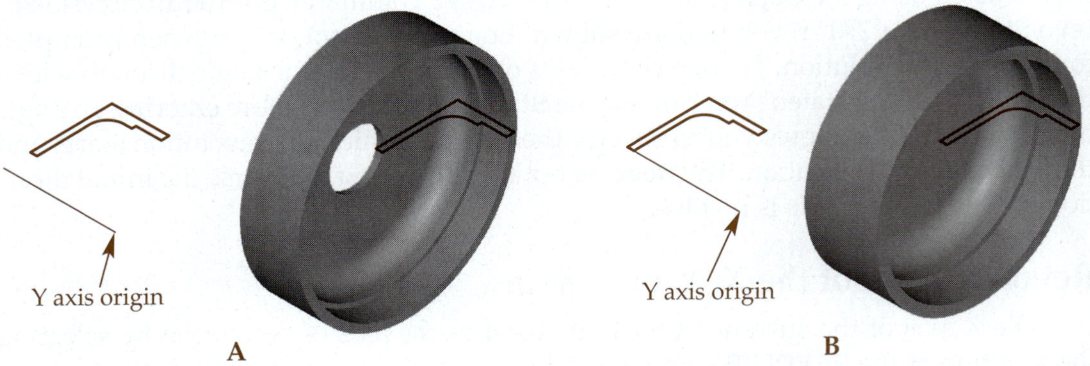

Revolving Regions

Earlier in this chapter, you learned that regions can be extruded. In this manner, holes, slots, keyways, etc., can be created. Regions can also be revolved. A complex 2D shape can be created using Boolean operations on regions. Then, the region can be revolved. One advantage of this method is it may be easier to create a region than trying to create a complex 2D profile as a single, closed polyline. As previously discussed, you can use the **Mode** option of the **REVOLVE** command to set the output type to solid or surface.

Revolving Surfaces

Just as planar and nonplanar surfaces can be extruded, they can also be revolved. When the **REVOLVE** command is selected, simply pick the surface when prompted to select objects. The surface can be revolved about an axis defined by two pick points, an object, or the X, Y, or Z axis of the current UCS. As previously discussed, you can use the **Mode** option of the **REVOLVE** command to set the output type to solid or surface.

PROFESSIONAL TIP

You can also revolve a face on an existing solid or surface into a new solid or surface. The **Mode** option of the **REVOLVE** command allows you to set the output type to solid or surface. When prompted to select objects, press the [Ctrl] key and pick the face to revolve. Subobject editing is covered in detail in Chapter 11.

Using Extrude and Revolve to Create Surfaces

When creating an extruded or revolved model as a surface, the **SURFACEMODELINGMODE** system variable controls the type of surface created. The default setting of 0 creates a ***procedural surface***. This is a standard surface composed of multiple flat polygons, but it has no control vertices. By default, this type of surface is associated with the object used to create it. The **SURFACEASSOCIATIVITY** system variable controls this setting.

A NURBS surface can be created by setting the **SURFACEMODELINGMODE** system variable to a value of 1. A ***NURBS surface*** is composed of splines and contains control vertices that enable the control of the curve shape with great precision.

Surface modeling variables can be set on the **Create** panel in the **Surface** tab of the ribbon. See Figure 6-11. When the **Surface Associativity** button is on, the **SURFACEASSOCIATIVITY** system variable is set to 1. When the **NURBS Creation** button is on, the **SURFACEMODELINGMODE** system variable is set to 1.

Extruding Surfaces from Objects with Dimensional Constraints

When the **Mode** option is set to create a surface, you can use the **Expression** option to create a parametric relationship between a dimensional constraint and the extruded surface. For example, the circle in Figure 6-12 has been given a dimensional constraint named dia1. To create an extruded surface cylinder with a height that is always one-half of the diameter, enter the **Expression** option. Then, at the Enter expression: prompt, enter:

dia1/2

In this manner, using appropriate formulaic expressions, you can create an extruded surface that has a mathematical relationship to a 2D object that has a dimensional constraint.

CAUTION

When you enter an expression while creating a solid, the result is calculated, but the dimensional constraint is removed. Therefore, the parametric aspect of using an expression is lost.

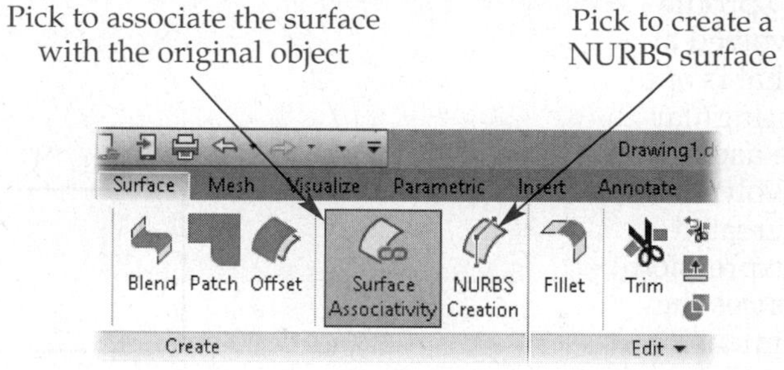

Figure 6-11. Surface modeling variables can be set on the **Create** panel in the **Surface** tab of the ribbon.

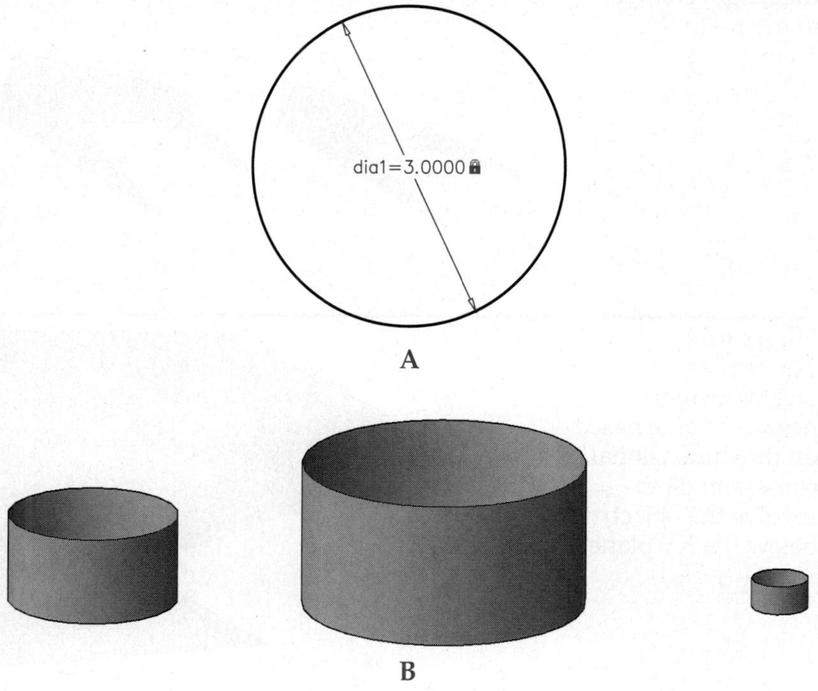

Figure 6-12. A—The circle has been given a dimensional constraint named dia1. B—The circle is extruded using an expression to create a cylinder surface with a height that is one-half the distance of dia1. When the diameter is changed, the height changes to maintain the relationship.

Revolving Surfaces from Objects with Dimensional Constraints

When constructing a revolved surface, the **Expression** option of the **REVOLVE** command can be used with dimensional constraints in the same manner as when using the **EXTRUDE** command. Examples of mathematical expressions used with constraints are shown in **Figures 6-13** and **6-14**. These examples illustrate how linear and angular dimensional constraints can be used to calculate an angular value for a revolved surface.

The profile in **Figure 6-13A** is to be revolved at an angle that is one-half that of the angular dimension (ang2). This is achieved by selecting the **Expression** option and then entering:

ang2/2

As shown in **Figure 6-13B**, the height of the model changes when the angular dimension is changed.

In **Figure 6-14**, the same 2D geometry is used. However, the aligned dimension d2 is used to create a negative value to revolve the object below the XY plane. The following formula is entered:

d2–31

The result of subtracting 31 from the value of d2 is processed as the angular value for the revolution. The result is shown in **Figure 6-14**.

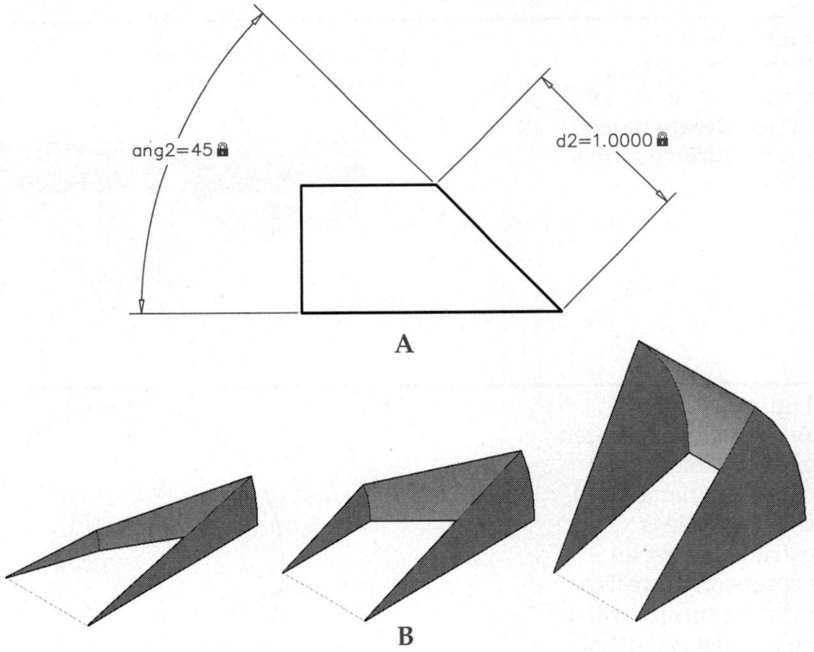

Figure 6-13.
A—The 2D profile is to be revolved at an angle that is one-half of the angular dimension ang2.
B—The revolved surface is created using an expression that references the angular dimension ang2. When the angle is changed, the height changes to maintain the relationship.

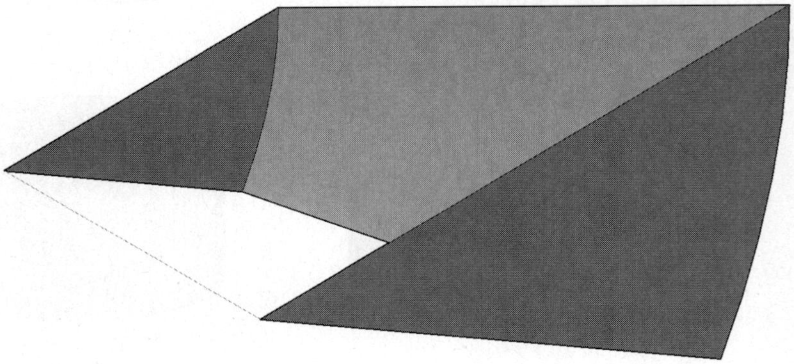

Figure 6-14.
The surface is created using a negative value based on the dimensional constraint d2 to revolve the object below the XY plane.

Using Extrude and Revolve as Construction Tools

It is unlikely that an extrusion or revolution will result in a finished object. Rather, these operations will be used with other model construction methods, such as Boolean operations in solid modeling, to create the final composite 3D object. The next sections discuss how to use **EXTRUDE** and **REVOLVE** with other construction methods to create a finished solid object.

Creating Features with Extrude

You can create a wide variety of features with the **EXTRUDE** command. Study the shapes shown in **Figure 6-15**. These detailed solid objects were created by drawing a profile and then using the **EXTRUDE** command. The objects in **Figures 6-15C** and **6-15D** must be constructed as regions before they are extruded. For example, the five holes (circles) in **Figure 6-15D** must be removed from the base region using the **SUBTRACT** command.

Look at **Figure 6-16**. This is part of a clamping device used to hold parts on a mill table. There is a T-slot milled through the block to receive a T-bolt and one side is stair-stepped, under which parts are clamped. If you look closely at the end of the object, most of the detail can be drawn as a 2D region and then extruded. However, there are also two holes in the top of the block to allow for bolting the clamp to the mill table. These features must be added to the extruded solid.

First, change the UCS to the front preset orthographic UCS. Display a plan view of the UCS. Then, draw the profile shown in **Figure 6-17** using the **PLINE** command. You can draw it in stages, if you like, and then use the **PEDIT** command to join all segments into a single polyline.

Next, use the **EXTRUDE** command to create the 3D solid. Extrude the profile a distance of –6 units with a 0° taper. This will extrude the object away from you. Display the object from the southeast isometric preset viewpoint or use the view cube to display a pictorial view. The object should look similar to **Figure 6-16** without the holes in the top. Set the Conceptual visual style current, if you like.

The two holes are ∅.5 units and evenly spaced on the surface through which they pass. Change to the WCS and draw a construction line from midpoint to midpoint, as shown in **Figure 6-18**. Then, set **PDMODE** to an appropriate value, such as 3, and use the **DIVIDE** command to divide the construction line into three parts. The two points created by the **DIVIDE** command are equally spaced on the surface and can be used to locate the two holes.

Figure 6-15.
Detailed solids can be created by extruding the profile of an object. The profiles are shown here in color.

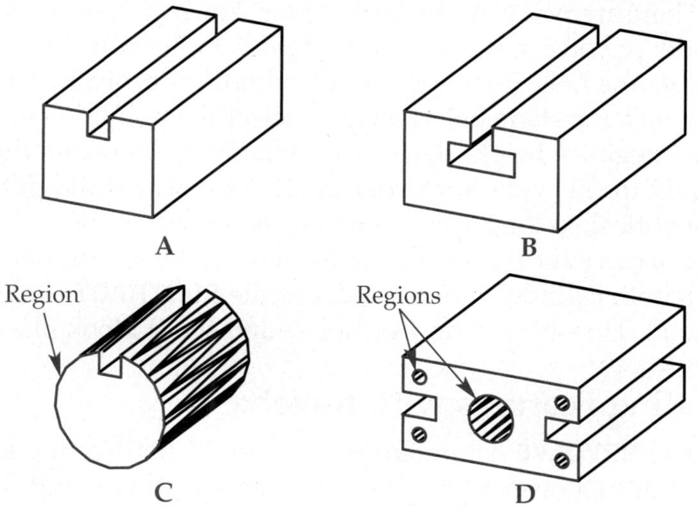

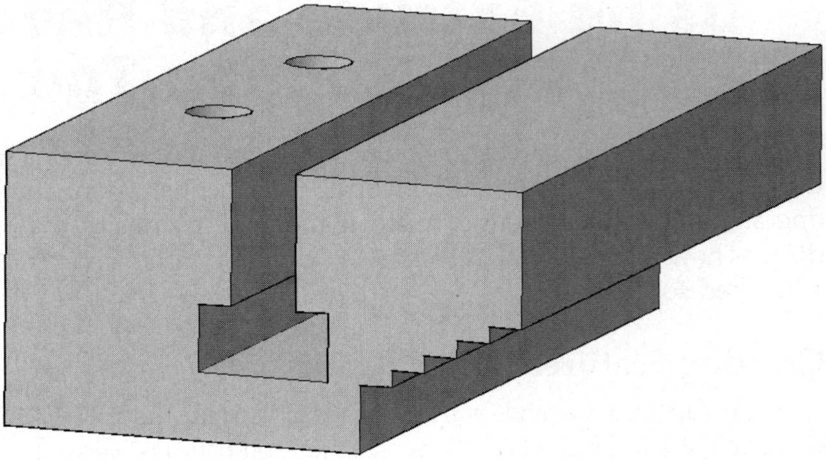

Figure 6-16.
Most of this object can be created by extruding a profile. However, the holes must be added after the extruded solid is created.

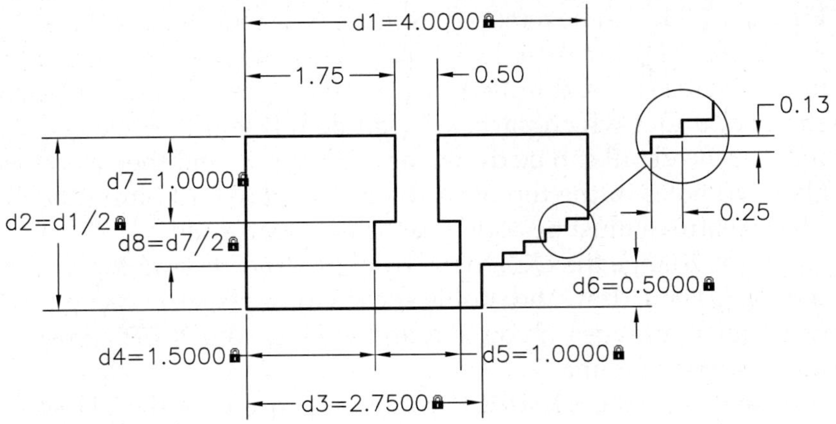

Figure 6-17.
This is the profile that will be extruded for the clamping block. Notice the dimensional constraints that have been applied.

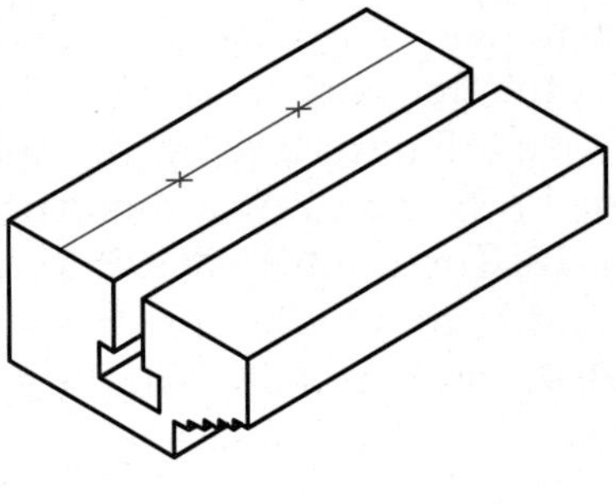

Figure 6-18.
Draw a construction line (shown here in color) and divide it into three parts.

There are two ways to create a hole. You can draw a circle and extrude it to create a cylinder or you can draw a solid cylinder. Either way, you need to subtract the cylinder to create the hole. Drawing a solid cylinder is probably easiest. When prompted for a center, use the **Node** object snap to select the point. Then, enter the diameter. Finally, enter a negative height so that the cylinder extends into the solid or drag the cylinder down in the 3D view so it extends all of the way through the block. The actual height is not critical, as long as it extends through the block.

You can either copy the first cylinder to the second point or draw another cylinder. When both cylinders are located, use the **SUBTRACT** command to remove them from the solid. The object is now complete and should look like **Figure 6-16**.

Creating Features with Revolve

The **REVOLVE** command is very useful for creating symmetrical, round objects. Many times, however, the object you are creating is not completely symmetrical. For example, look at the camshaft in **Figure 6-19**. For the most part, this is a symmetrical, round object. However, the cam lobes are not symmetrical in relation to the shaft and bearings. The **REVOLVE** command can be used to create the shaft and bearings. Then, the cam lobes can be created and added.

Start a new drawing and make sure the WCS is the current UCS. Using the **PLINE** command, draw the profile shown in **Figure 6-20A**. This profile will be revolved through 360°, so you only need to draw half of the true plan view of the cam profile. The profile represents the shaft and three bearings.

Next, display the drawing from the southwest isometric preset viewpoint. Then, use the **REVOLVE** command to create the base camshaft as a 3D solid. Pick the endpoints shown in **Figure 6-20A** as the axis of revolution. Revolve the profile through 360°. Perform a zoom extents and set the Conceptual visual style current to clearly see the object.

Now, you need to create one cam lobe. Change the UCS to the left orthographic preset. Then, draw a construction point in the center of the left end of the camshaft. Use the **Center** object snap and an appropriate **PDMODE** setting. Next, draw the profile shown in **Figure 6-20B**. Use the construction point as the center of the large radius. You may want to create a new layer and turn off the display of the base camshaft.

Once the cam lobe profile is created, use the **REGION** command to create a region. Then, use the **EXTRUDE** command to extrude the region to a height of –.5 units (into the camshaft). The extrusion should have a 0° taper. If you turned off the display of the base camshaft, turn it back on.

Figure 6-19.
For the most part, this object is symmetrical about its center axis. However, the cam lobes are not symmetrical about the axis.

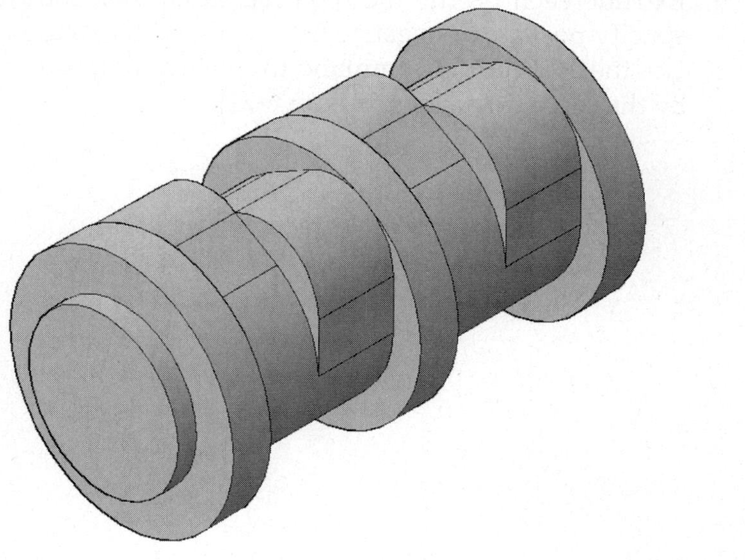

Figure 6-20.
A—This profile will be revolved to create the shaft and bearings. Notice the parallel constraints that have been applied. B—This is the profile of one cam lobe, which will be extruded.

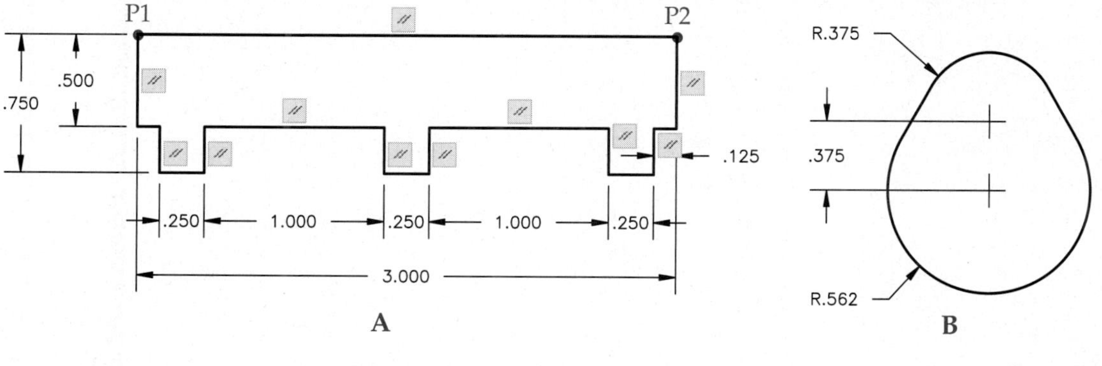

One cam lobe is created, but it is not in the proper position. With the left UCS current, move the cam lobe –.375 units on the Z axis. If a different UCS is current, the axis of movement will be different. This places the front surface of the cam lobe on the back surface of the first bearing. Now, make a copy of the lobe that is located –.5 units on the Z axis. Finally, copy the first two cam lobes –1.25 units on the Z axis.

You now need to rotate the four cam lobes to their correct orientations. Make sure the left UCS is still current. Then, rotate the first and third cam lobes 30°. If a different UCS is current, you can use the **3DROTATE** command. The center of rotation should be the center of the shaft. There are many points on the shaft to which the **Center** object snap can snap; they are all acceptable. You can also use the construction point as the center of rotation. Rotate the second and fourth cam lobes –30° about the same center.

Finally, use the **UNION** command to join all objects. The final object should appear as shown in **Figure 6-19**. Use the view cube to see all sides of the object. You can also create a rotating display using the **3DORBIT** command.

Multiple Intersecting Extrusions

Many solid objects have complex curves and profiles. These can often be constructed from the intersection of two or more extrusions. The resulting solid is a combination of only the intersecting volumes of the extrusions. The following example shows the construction of a coat hook.

1. Construct the first profile, **Figure 6-21A**.
2. Construct the second profile located on a common point with the first, **Figure 6-21B**.
3. Construct the third profile located on the common point, **Figure 6-21C**.
4. Extrude each profile the required dimension into the same area. Be careful to specify positive or negative heights for each extrusion, **Figures 6-21D** and **6-21E**.
5. Use the **INTERSECT** command to create a composite solid from the volume shared by the three extrusions, **Figure 6-21F**.

INTERSECT

Ribbon
Home
> Solid Editing
Solid
> Boolean

Solid, Intersect

Type
INTERSECT
IN

Figure 6-21.
Constructing a coat hook. A—Draw the first profile. B—Draw the second profile. C—Draw the third profile. All three profiles should have a common origin. D—Extrude each profile so that the extruded objects intersect. E—The extruded objects after the Conceptual visual style is set current. F—Use the **INTERSECT** command to create the composite solid. The final solid is shown here with the Conceptual visual style set current.

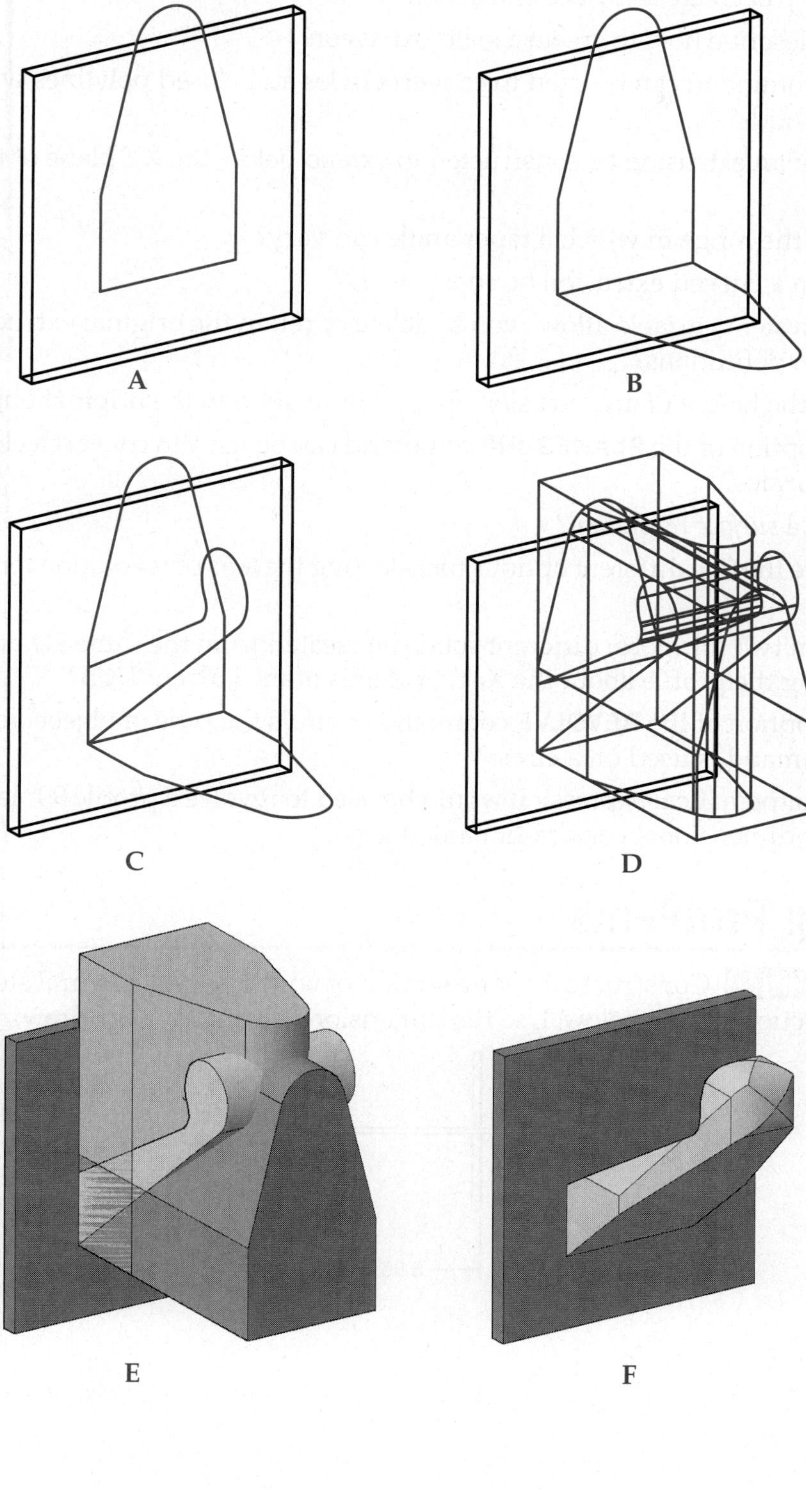

Chapter Review

Answer the following questions using the information in this chapter.

1. What is an *extrusion*?
2. How do you create a surface extrusion?
3. Briefly describe how to create a solid extrusion.
4. Which command can be used to convert circles and closed polylines with a thickness to solids?
5. How can an extrusion be constructed to extend below the XY plane of the current UCS?
6. What is the range in which a taper angle can vary?
7. How can a curved extrusion be constructed?
8. Which system variable allows you to delete or retain the original extruded objects and path definitions?
9. How is the height of an extrusion applied in relation to the original object?
10. Which option of the **PLANESURF** command can be used to convert a closed object into a surface?
11. What is a *surface revolution*?
12. What are the five different options for selecting the axis of revolution for a revolved solid?
13. How can two (or more) different solids be created from the same 2D profile when revolving the profile about the X, Y, or Z axis of the current UCS?
14. Which option of the **REVOLVE** command controls the type of object created when the command is used on a circle?
15. What mathematical expression would be used to revolve a profile 90° less than the angular dimensional constraint named ang4?

Drawing Problems

Drawing Problems - Chapter 6

1. **Architectural** Construct a 12′ long section of wide flange structural steel with the cross section shown below. Use the dimensions given. Save the drawing as P6-1.

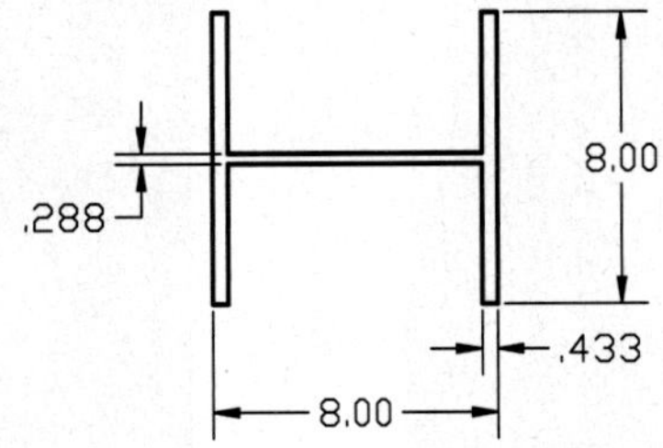

Problems 2–7. These problems require you to use a variety of solid modeling methods to construct the objects. Use **EXTRUDE**, **REVOLVE**, *solid primitives, new UCSs, Boolean commands, and editing tools such as extrusions and revolutions to assist in construction. If appropriate, apply geometric constraints to the base 2D drawing. Do not create section views. Do not add dimensions to the models. For objects that have dimensions with tolerances, use the base dimension from which the upper and lower limits are calculated to create the model. Save each drawing as* P6-*(problem number).*

2. **Mechanical**

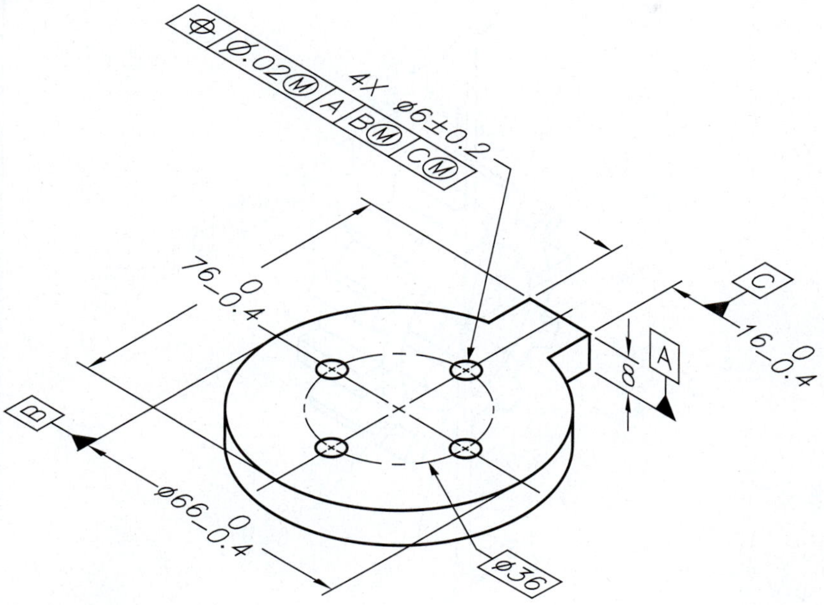

Spring Clip

3. **Mechanical**

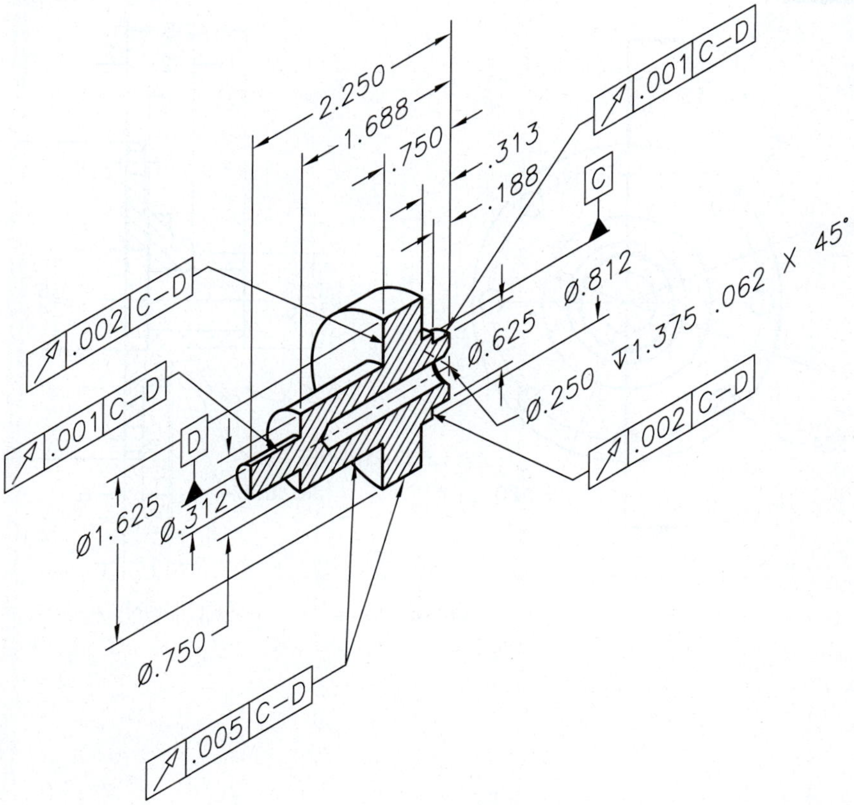

Valve Pin

4. **Mechanical**

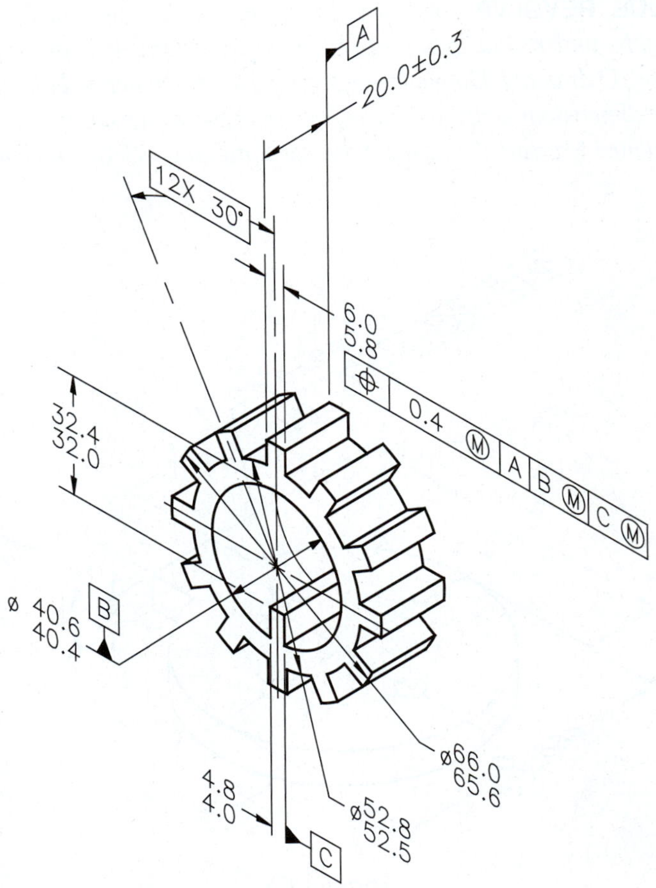

Spline

5. **Mechanical**

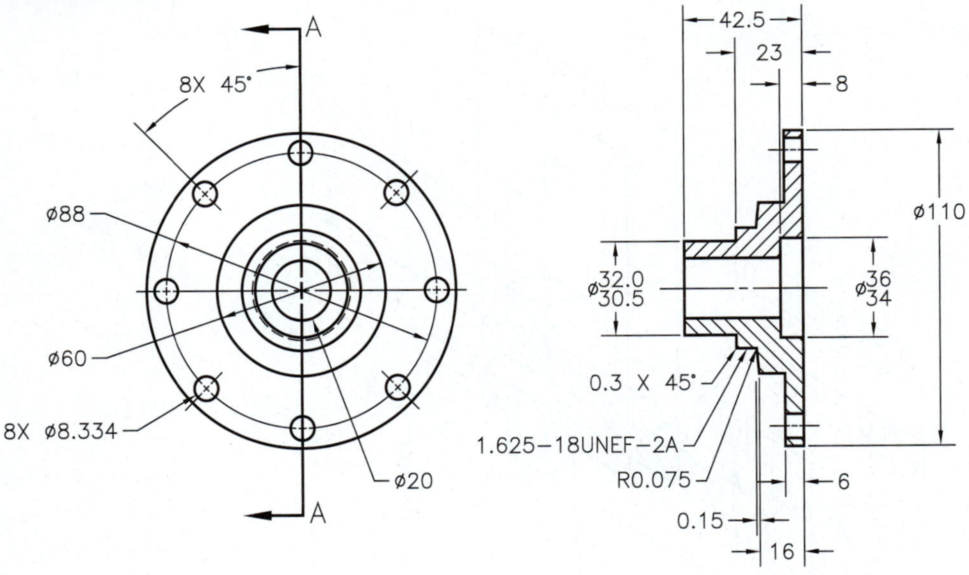

Flange

SECTION A–A

6. **Mechanical**

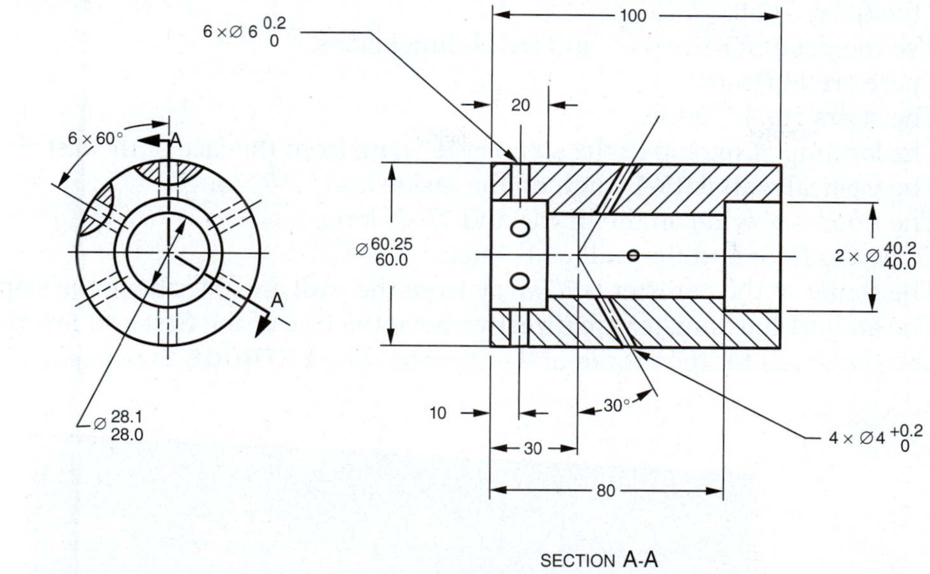

Nozzle

7. **Mechanical**

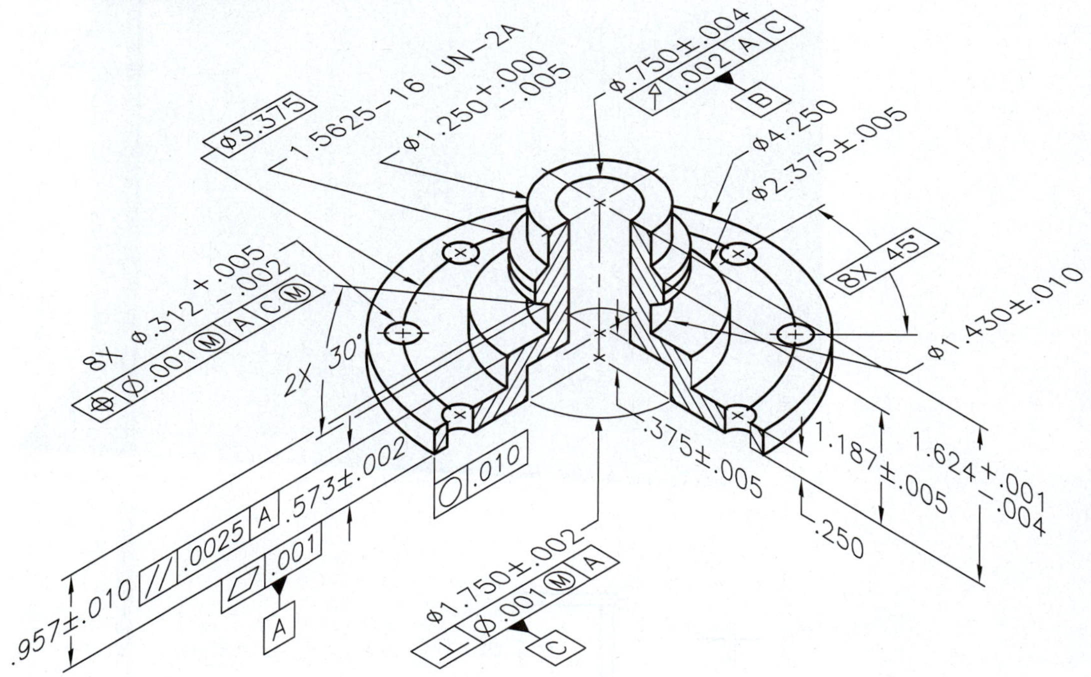

Hub

8. **Architectural** Create the stairway shown below using the following parameters. Save the file as P6-8.
   A. Use the detail for the riser and tread dimensions.
   B. There are 13 risers.
   C. The stairs are 42″ wide.
   D. The landing at the top of the stairs is 48″ long from the face of the last riser.
   E. The vertical wall is 15′-1″ high on the inside and 17′-7″ long.
   F. The floor is 8′ wide on the inside and 17′-7″ long.
   G. Draw the floor and the wall as 1″ thick.
   H. The center of the banister is 3″ away from the wall and 34″ above the steps.
   I. The ends of the banister are directly above the face of the first and last riser.
   J. Use the detail for the profile of the banister. Use **EXTRUDE** as needed.

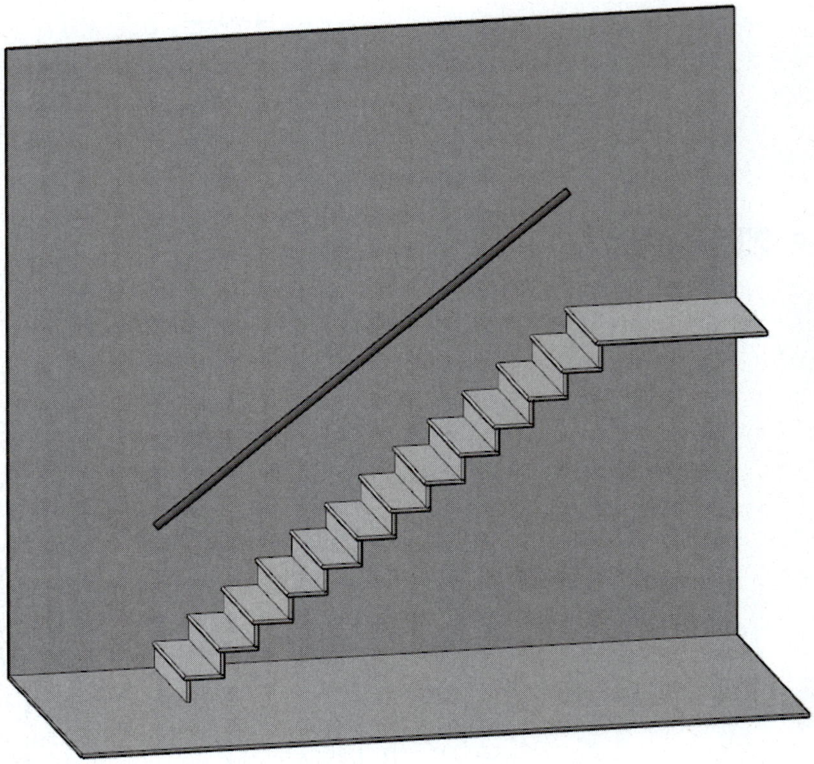

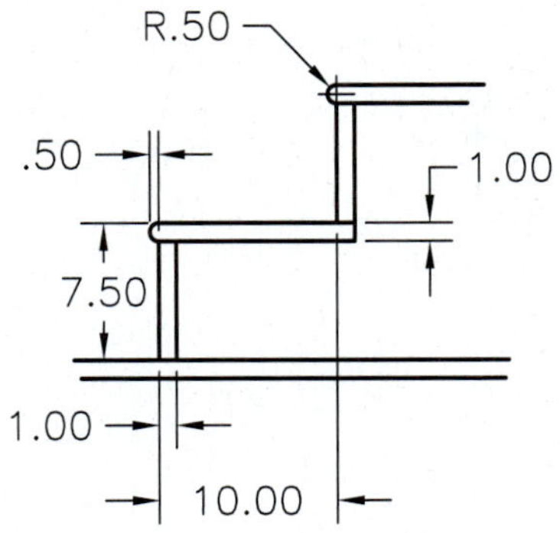

STAIR DETAIL

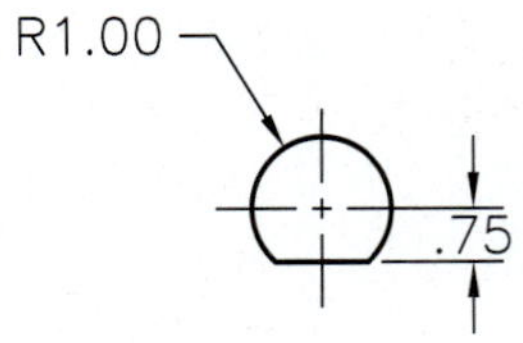

BANISTER DETAIL

9. **Architectural** In this problem, you will refine the seat of the kitchen chair that you started in Chapter 2. You will use an extrusion following a path to create a curved, receding edge under the seat.

A. Open P2-11 from Chapter 2. If you have not yet completed this model, do so now.

B. Create a path for the extrusion by drawing a polyline that exactly matches either the upper or lower edge of the seat. Refer to the drawing shown.

C. Change the view and UCS as needed to display a plan view of the edge of the seat. Draw the profile shown.

D. Extrude the profile along the path and then subtract the extrusion from the seat.

E. Save the drawing as P6-9.

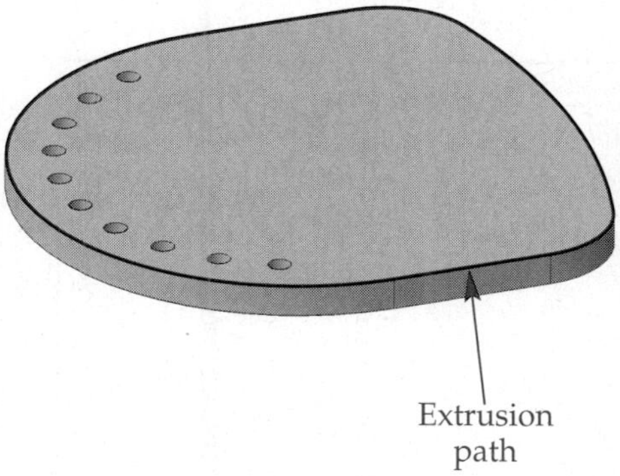

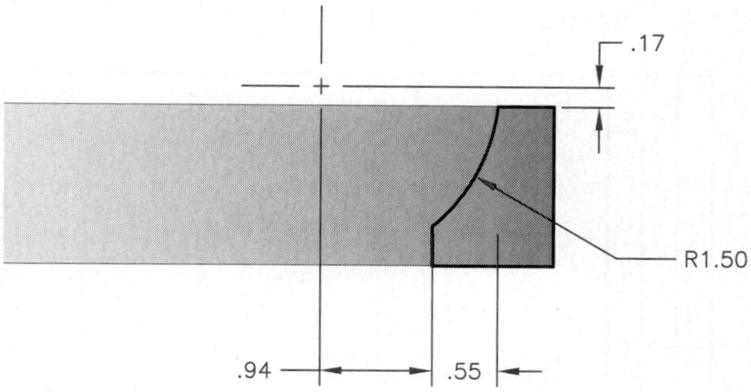

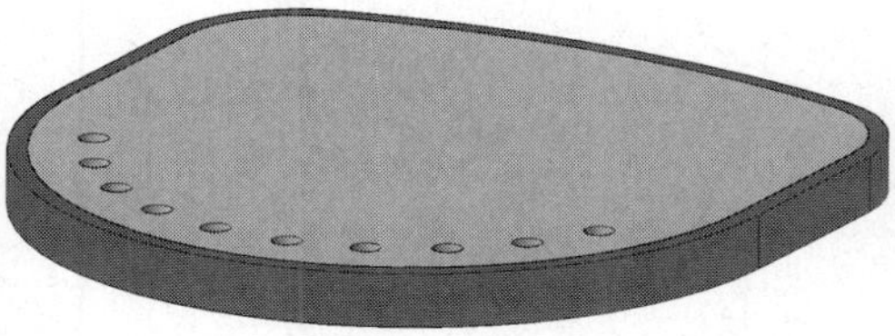

Seat with Extrusion in Place

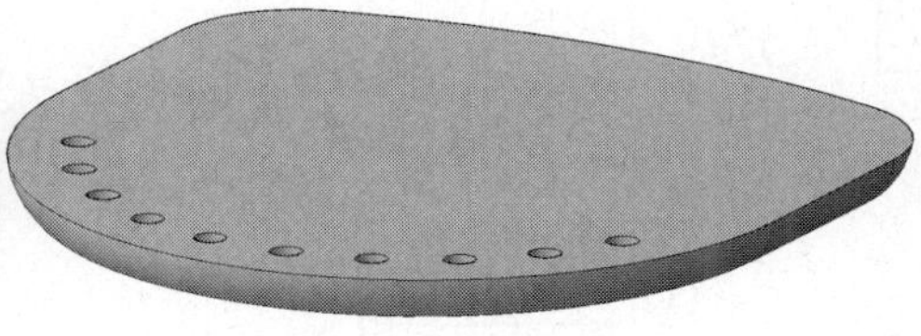

Completed Seat

10. **Architectural** In this problem, you will be taking a manually drawn layout from an archive. You are to create a 3D model of the garage to update the archive.
    A. Review the manually drawn layout. Make note of the construction details shown.
    B. Research any additional details needed to construct the model. For example, the thickness of the doors is not listed as these are purchased items. However, you need to know these dimensions to draw the 3D model.
    C. Using what you have learned, create the garage as a solid model. Be sure to create all components, including the studs in the walls, the footings, and the anchor bolts.
    D. Create layers as needed. For example, you may wish to place the wall sheathing on a layer so it can be hidden to show the studs.
    E. Save the drawing as P6-10.

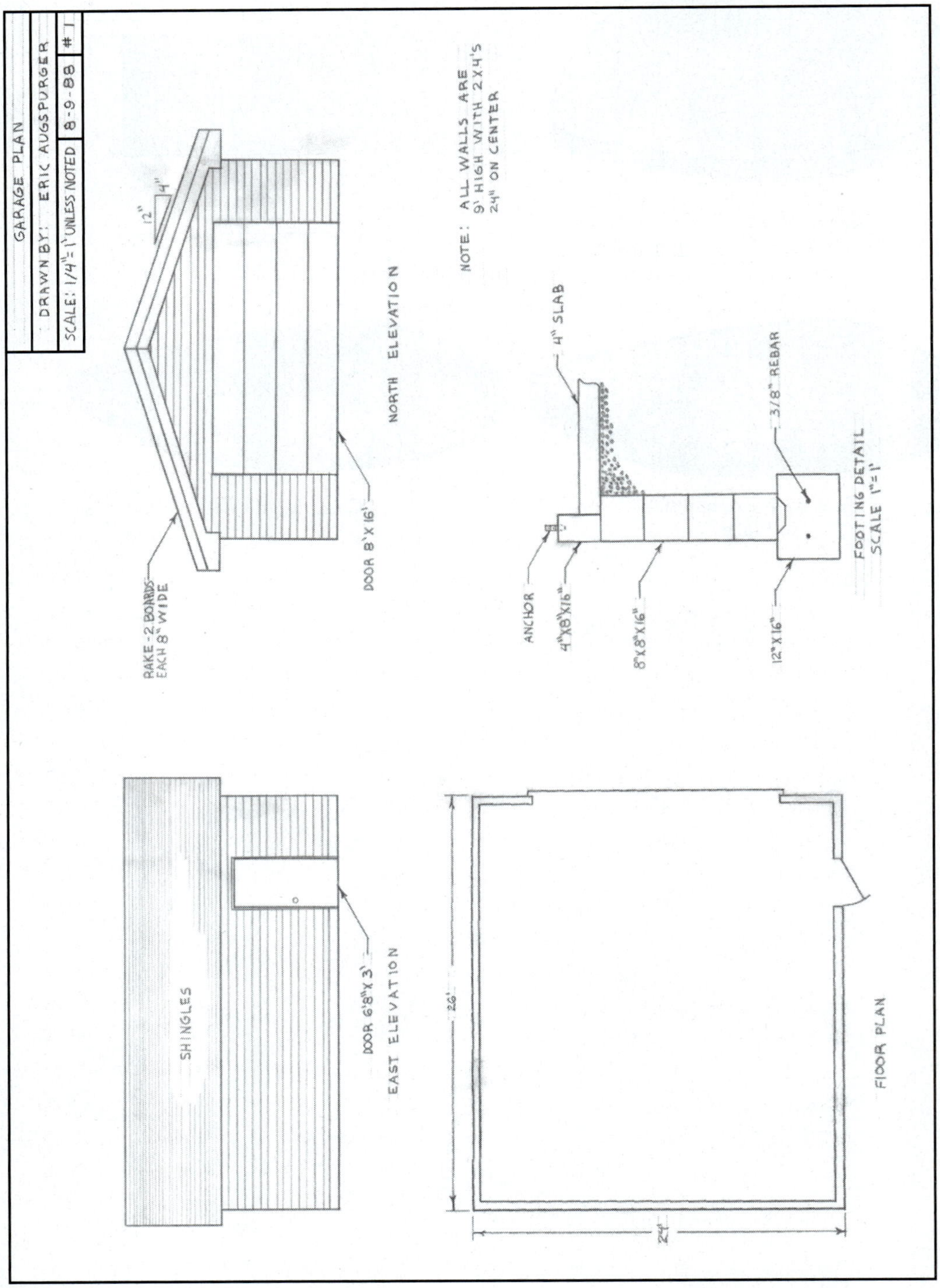

CHAPTER 7

Sweeps and Lofts

Learning Objectives

After completing this chapter, you will be able to:

✓ Sweep shapes along a 2D or 3D path to create a solid or surface object.
✓ Create 3D solid or surface objects by lofting a series of cross sections.

In the previous chapter, you learned about extruded solids and surfaces. Sweeps and lofts are similar to extrusions. In fact, an extrusion is really just a type of sweep. A ***sweep*** is an object created by extruding a single profile along a path object. Sweeping an open shape along the path results in a surface object. If a closed shape is swept, a solid or surface object can be created. A ***loft*** is an object created by extruding between two or more 2D profiles. The shape of the loft object blends from one cross-sectional profile to the next. The profiles can control the loft, or the loft can be controlled by one path or multiple guide curves. As with a sweep, open shapes result in surfaces and closed shapes give you solids. Open and closed shapes cannot be used together in the same loft.

Creating Swept Surfaces and Solids

Ribbon
Home
> Modeling
Solid
> Solid
Surface
> Create
Sweep
Type
SWEEP

SWEEP

The **SWEEP** command is used to create swept surfaces and solids. The command requires at least two objects:

- Shape to be swept.
- 2D or 3D shape to be used as the sweep path.

The profile can be aligned with the path, you can specify the base point, a scale factor can be applied, and the profile can be twisted as it is swept. The command procedure and options are the same for both swept solids and surfaces.

Sweeping an open shape creates a surface. See **Figure 7-1**. The objects that can be swept to create surfaces include lines, arcs, elliptical arcs, 2D polylines, 2D or 3D splines, and 3D edge subobjects. Sweeping a closed shape creates a solid by default, but surfaces may be created as well. See **Figure 7-2**. Closed shapes that can be swept include circles, ellipses, closed 2D polylines, closed 2D splines, regions, and 3D solid face subobjects. The sweep path can be a line, arc, circle, ellipse, elliptical arc, 2D or 3D polyline, 2D or 3D spline, helix, or 3D edge subobject.

Figure 7-1.
A—This open shape will be swept along the path (shown in color). B—The resulting surface.

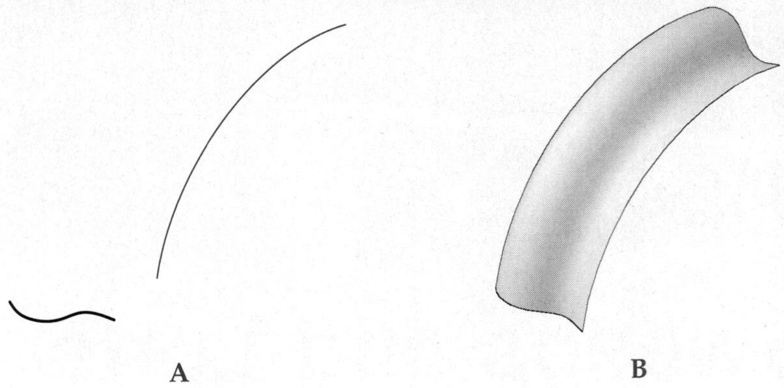

Figure 7-2.
A—This closed shape will be swept along the path (shown in color). B—The resulting solid.

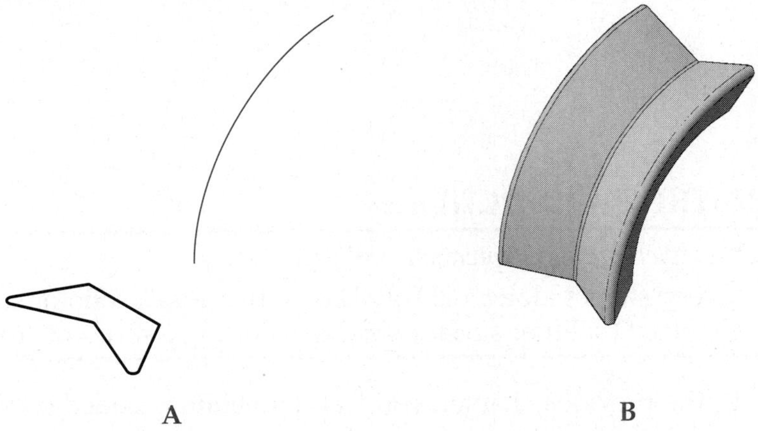

When the command is initiated, you are prompted to select the objects to sweep. Select the profile(s) and press [Enter]. Planar faces of solids may be selected by holding the [Ctrl] key as you select. Multiple profiles can be selected. They are swept along the same path, but separate objects are created.

Next, you are prompted to select the path. The path and profile can lie on the same plane. Select the object to be used as the sweep path. To select the edge of a surface or solid as the path, press the [Ctrl] key and then select the edge. Once you select the path, the profile is moved to be perpendicular to the path and extruded along it. The sweep starts at the endpoint of the path nearest to where you select it.

The **Mode** option that is available when the **SWEEP** command is initiated controls the closed profiles creation mode. This allows you to change the way in which closed shapes are handled. Normally, the **SWEEP** command produces a solid object when a closed shape is swept, but surfaces may be created by changing this mode. For example, a circle swept with a line as the path will normally create a solid cylinder. Setting the **Mode** option to **Surface** will result in a tube created as a surface model.

Exercise 7-1

www.g-wlearning.com/CAD/

Complete the exercise on the companion website.

Changing the Alignment of the Profile

By default, the profile is aligned perpendicular to the sweep path. However, you can create a sweep where the profile is not perpendicular to the path. See **Figure 7-3**. After the **SWEEP** command is initiated, select the profile and press [Enter]. Before selecting the path, enter the **Alignment** option. The default setting of **Yes** means that

Figure 7-3.
A—The profile and path for the sweep. B—By default, the profile is aligned perpendicular to the path when swept. C—Using the **Alignment** option, the profile can be swept so it is not perpendicular to the path.

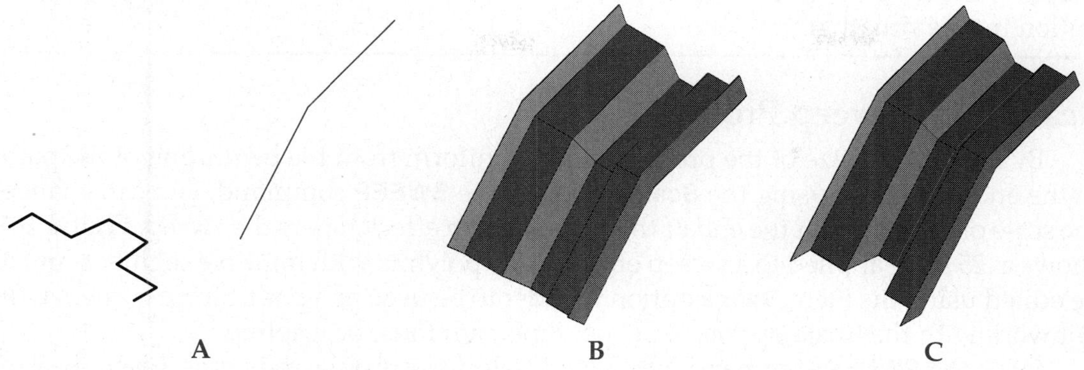

the profile will be moved so it is perpendicular to the path. If you select **No**, the profile is kept in the same position relative to the path as it is swept. The position of the shape determines the alignment. This will not work if the profile and path are coplanar.

Changing the Base Point

The base point is the location on the shape that will be moved along the path to create the sweep. By default, if the shape intersects the path, the profile is swept along the path at the point of intersection. If the shape does not intersect the path, the default base point depends on the type of object being swept. When lines and arcs are swept, the default base point is their midpoint. Open polylines have a default base point at the midpoint of their total length.

The base point can be any point on the shape or anywhere in the drawing. See **Figure 7-4**. To change the base point, use the **Base point** option of the **SWEEP** command. When the command is initiated, select the profile and press [Enter]. Before selecting the path, enter the **Base point** option. Next, pick the new base point. It does not have to be on an existing object. Once the new base point is selected, pick the path to create the sweep.

Figure 7-4.
A—The profile and path for the sweep. B—The sweep is created with the default base point. C—The end of the path is selected as the base point. Notice the difference in this sweep and the one shown in B.

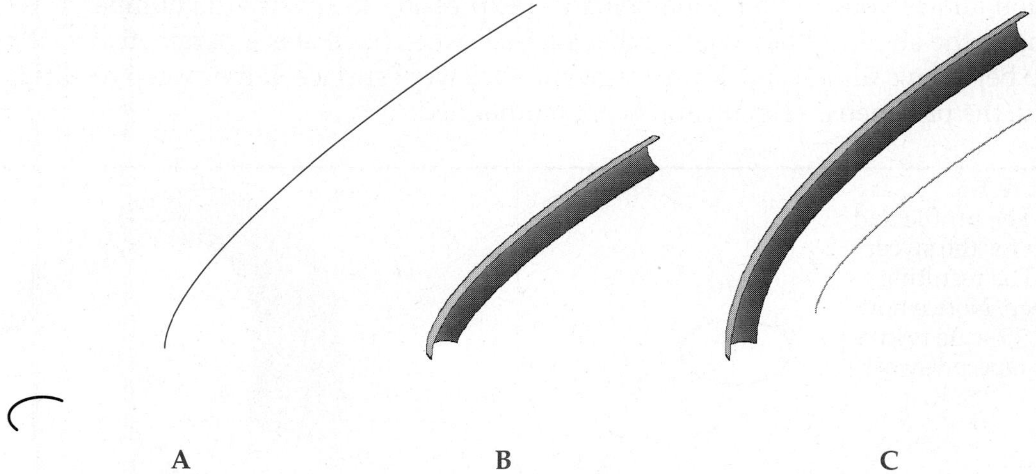

PROFESSIONAL TIP

If the location of the base point in relationship to the path is important, line up the shape with the path before starting the **SWEEP** command. Turn off the **Alignment** option in this situation.

Scaling the Sweep Profile

By default, the size of the profile remains uniform from the beginning of the path to the end. However, using the **Scale** option of the **SWEEP** command, you can change the scale of the profile at the end of the path. This, in effect, tapers the sweep. Figure 7-5 shows a .25 scale applied to a sweep object. A 2D polyline with multiple segments must be edited using the **Fit** or **Spline** option in order to be used as a path. Sharp corners will not work with the **Scale** option. A 3D polyline path must be a spline.

Once the **SWEEP** command is initiated, select the profile and press [Enter]. Before selecting the path, enter the **Scale** option. You are prompted for the scale. Enter the scale value and press [Enter]. The scale value must be greater than zero. You can also enter the **Reference** option. With this option, pick two points for the first reference line and then two points for the second reference line. The difference in scale between the two distances is the scale value. Once the scale is set, pick the path to create the sweep. The **Expression** option allows you to enter a mathematical expression to specify the scaling of the sweep. When used with a surface sweep, it creates a parametric relationship between a dimensional constraint and the swept surface. If the sweep results in a solid, the parametric relationship is not maintained.

Twisting the Sweep

The profile can be rotated as it is swept along the length of the path by using the **Twist** option of the **SWEEP** command. The angle that you enter indicates the rotation of the shape along the path of the sweep. The higher the number, the more twists in the sweep. Figure 7-6 shows how a simple, closed profile and a straight line can be used to create a milling tool. The profile was swept with a 270° twist.

Once the **SWEEP** command is initiated, select the profile and press [Enter]. Then, before selecting the path, enter the **Twist** option. You are prompted for the twist angle or to enter the **Bank** option.

Banking is the natural rotation of the profile on a 3D sweep path, similar to a banked curve on a racetrack. See Figure 7-7. The path must be 3D (nonplanar) to set banking. The banking option is disabled for a 2D path, although you can go through the process of turning it on when creating the sweep. Once you use the **Bank** option to turn banking on, it is on by default the next time the **SWEEP** command is used. To turn it off, enter a twist angle of zero (or the twist angle you wish to use). The **Expression** option allows you to enter a mathematical expression to specify the number of rotations in the sweep. When used with a surface sweep, it creates a parametric relationship between a dimensional constraint and the swept surface. If the sweep results in a solid, the parametric relationship is not maintained.

Figure 7-5.
A—The profile and path for the sweep. B—The resulting sweep. Notice how the .25 scale results in a tapered sweep.

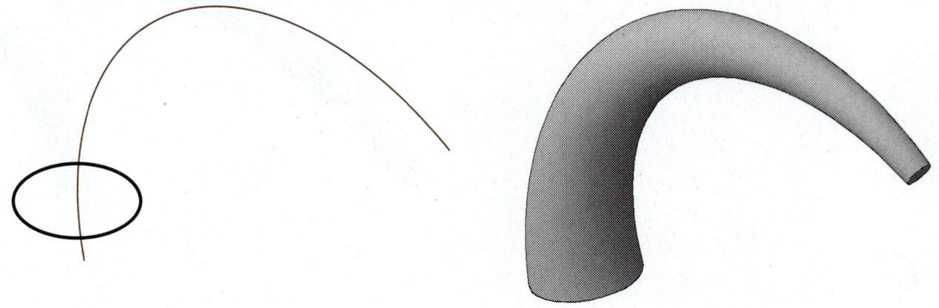

A B

Figure 7-6.
A—The profile and path for creating the end mill. B—The resulting end mill model. Notice how the profile is twisted (rotated) as it is swept.

Figure 7-7.
A—The profile and path for the sweep are shown in color. B—Banking is off for this sweep. When viewed from the side, you can see that the profile does not bank through the curve. Look at the upper-right corner. C—Banking is on for this sweep. Notice how the profile banks or leans through the curve. Compare this to B.

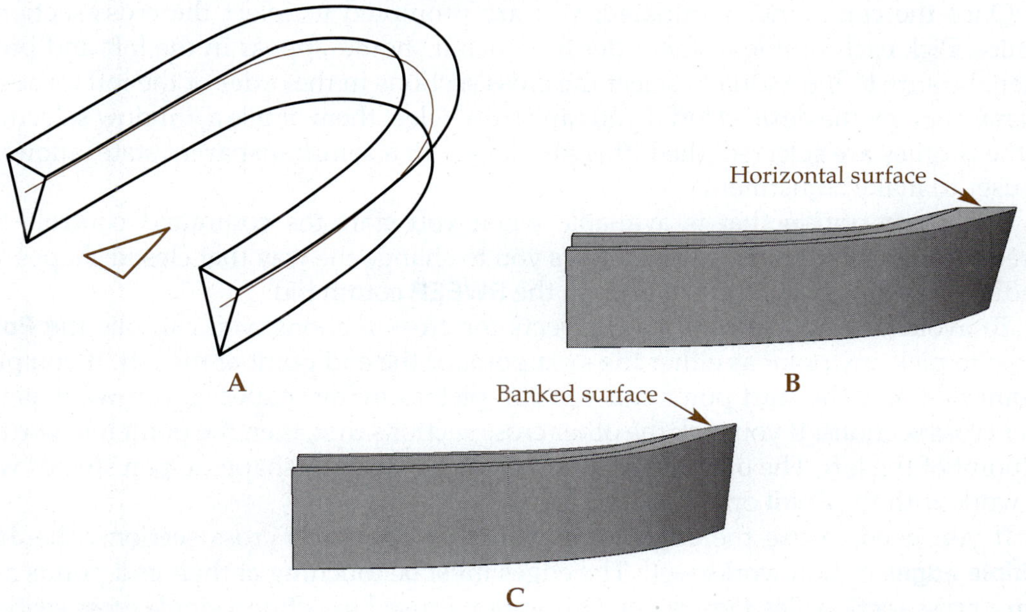

PROFESSIONAL TIP

The **Expression** option available for scaling and twisting a sweep can be used as a simple calculation tool for both swept solids and surfaces. For example, if you want to twist the sweep to an angle one and one half times 45°, select the **Expression** option and enter 1.5*45. The sweep will twist 67.5°.

NOTE

The sweep options can be changed after the sweep is created using the **Properties** palette. In the **Geometry** section, you will find Profile rotation (alignment), Bank (banking), Twist along path (twist angle), and Scale along path (scale) settings.

Exercise 7-2

www.g-wlearning.com/CAD/

Complete the exercise on the companion website.

Creating Lofted Objects

LOFT

Ribbon
Home
> Modeling
Solid
> Solid
Surface
> Create

Loft

Type
LOFT

The **LOFT** command is used to create lofted surfaces and solids based on a series of cross-sectional profiles. **Figure 7-8** shows an example of a loft formed from a rectangle, circle, and polygon. The loft may be guided by only the cross sections, as shown in the figure, by a path, or by guide curves. Lofting open shapes results in a surface object, while lofting closed shapes creates a solid. Open and closed shapes cannot be combined in the same loft.

Objects that can be used as cross sections include lines, circles, arcs, points, ellipses, elliptical arcs, 2D polylines, 2D splines, regions, edge subobjects, surfaces, face subobjects, and helices. Points may be used for the first and last cross sections only. The loft path may be a line, circle, arc, ellipse, elliptical arc, spline, helix, 2D or 3D polyline, or edge subobject. Guide curves may be composed of lines, arcs, elliptical arcs, 2D or 3D splines, 2D or 3D polylines, and edge subobjects. However, 2D polylines are limited to only one segment.

Once the command is initiated, you are prompted to select the cross-sectional profiles. Pick each profile in the order in which it should appear in the loft and press [Enter]. Be sure to individually select the cross sections in the order of the loft creation. You may not get the desired loft if you randomly select them or use a window selection. As the profiles are selected, the loft is previewed in a semitransparent state, allowing the user to make adjustments.

The **Mode** option that is available when you start the command controls the closed profiles creation mode and allows you to change the way that closed shapes are handled. It behaves exactly as it does in the **SWEEP** command.

To avoid having to draw point objects for cross sections, you can use the **Point** option to pick any point as either the start point or the end point of the loft. If you pick a point first, it is the start point and you can pick as many shapes as you want as the other cross sections. If you pick the other cross sections first, then the point must be the endpoint of the loft. The other cross sections must be closed shapes. Open shapes will not work with the **Point** option.

If you need to use the edges of existing 3D objects as cross sections, the **Join multiple edges** option works well. The edges must be touching at their end points and form a cross section. See **Figure 7-9**. This option is used to define a single cross section. When you press [Enter], you are again prompted to select cross sections. The option can be used again to select additional cross sections.

Figure 7-8.
A—The three profiles will be lofted to create a solid. B—The resulting loft with the default settings.

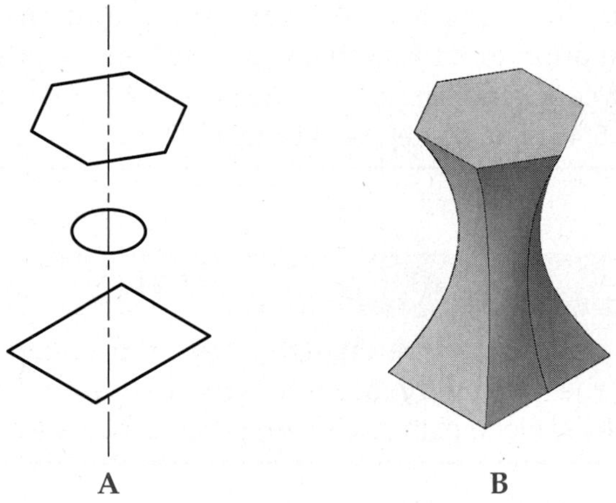

Figure 7-9.
The **Join multiple edges** option allows you to make a quick transition from the edges of one 3D object to another. A—Use the option twice to select two cross sections. B—The resulting loft is a transition between the two objects.

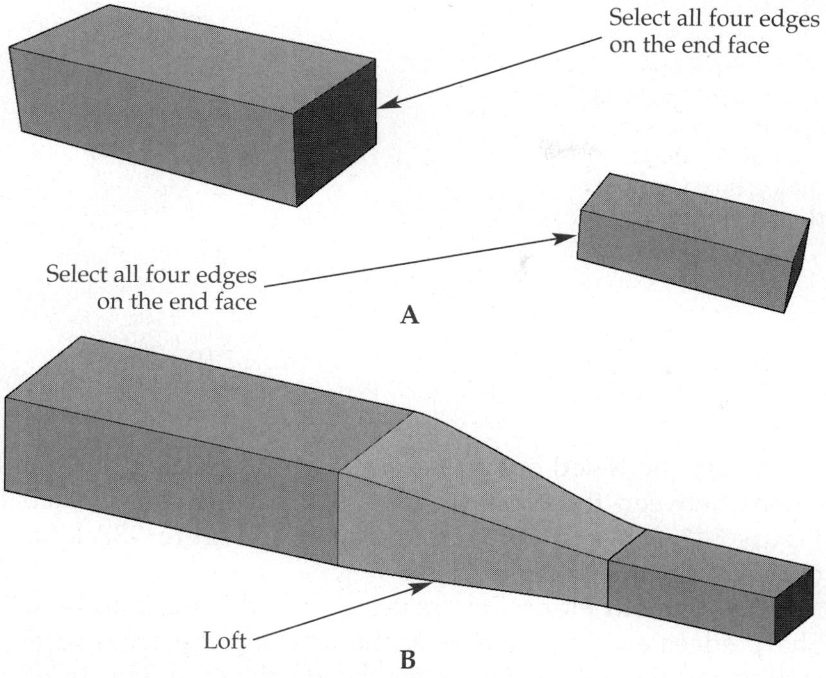

After selecting cross sections, you are prompted to select how the loft is to be controlled. As previously discussed, you can control the loft by the cross sections, a path, or guide curves. These options are discussed in the next sections.

Controlling the Loft with Cross Sections

The **Cross sections only** option of the **LOFT** command is useful when the 2D cross sections are drawn in their proper locations in space. The command determines the transition from one cross section to the next. The cross sections are not moved by the command.

When you select the **Settings** option, the **Loft Settings** dialog box appears, **Figure 7-10**. The settings in this dialog box control the transition or contour between cross sections. As settings are changed, the preview is updated. When all settings have been made, pick the **OK** button to close the dialog box and create the loft.

Figure 7-10.
The **Loft Settings** dialog box is used to control the transition between profiles.

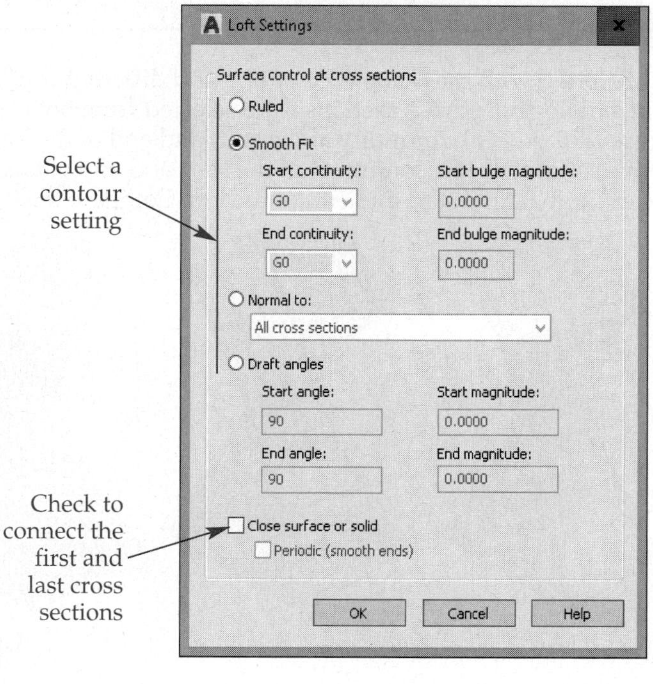

Figure 7-11.
The profiles in Figure 7-8A are lofted with the **Ruled** option selected in the **Loft Settings** dialog box. Compare this to Figure 7-8B.

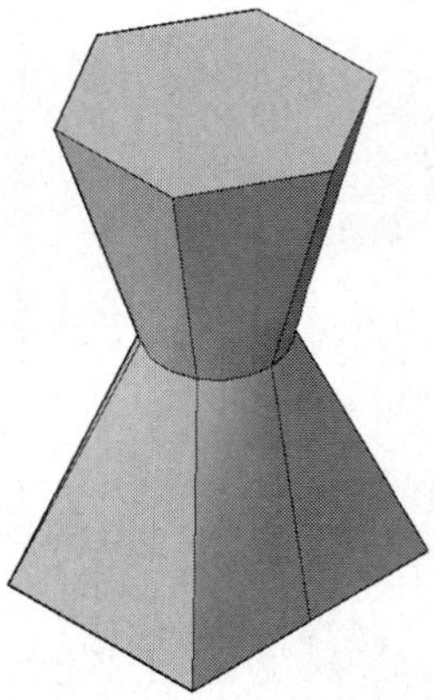

When the **Ruled** option is selected in the dialog box, the loft has straight transitions between the cross sections. Sharp edges are created at each cross section. **Figure 7-11** shows the same cross sections in **Figure 7-8A** lofted with the **Ruled** option on. Compare this to **Figure 7-8B**.

The **Smooth Fit** option creates a smooth transition between the cross sections. Sharp edges are only created at the first and last cross sections. This is the default setting and the one used to create the loft shown in **Figure 7-8B**. The **Start continuity:** and **End continuity:** settings control the tangency and curvature of the first and last cross sections. The **Start bulge magnitude:** and **End bulge magnitude:** settings control the size of the curve at the first and last cross sections. These options only apply if the first and/or last cross sections are regions. See **Figure 7-12**. The start and end continuity can be set to G0, G1, or G2. The G0 (positional continuity) setting creates a sharp edge. The loft transitions precisely at the location of the profile. The G1 (tangential continuity) setting is the default. It creates the largest bulge radius. The G2 (curvature or continuous continuity) setting creates a smaller bulge radius. For all three settings, bulge magnitude is initially set to 0.5. Altering the value will make the bulge either larger or smaller but has no effect when continuity is set to G0.

When the **Normal to:** option is selected in the **Loft Settings** dialog box, you can choose how the normal of the transition is treated at the cross sections. A ***normal*** is a vector extending perpendicular to the cross section. When the transition is normal to a cross section, it is perpendicular to the cross section. You can set the transition normal

Figure 7-12.
Region profiles are lofted with the **Smooth Fit** option and different continuity settings selected in the **Loft Settings** dialog box. Cross sections were selected from bottom to top. Compare these results with Figure 7-8B. A—G0 continuity at the start and end of the loft. Bulge magnitude has no effect at this setting. B—G1 continuity at the start and end of the loft. This is the default setting. C—G2 continuity at the start and end of the loft. Compare the bulge radius to B.

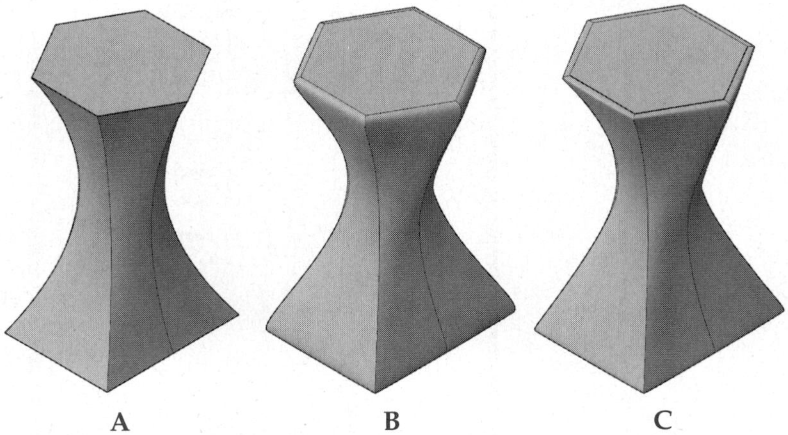

Figure 7-13.
The profiles in Figure 7-8A are lofted with the **Normal to:** option on in the **Loft Settings** dialog box. Cross sections were selected from bottom to top. Compare these results with Figure 7-8B and Figure 7-12. A—Start cross section. B—End cross section. C—Start and End cross sections. D—All cross sections.

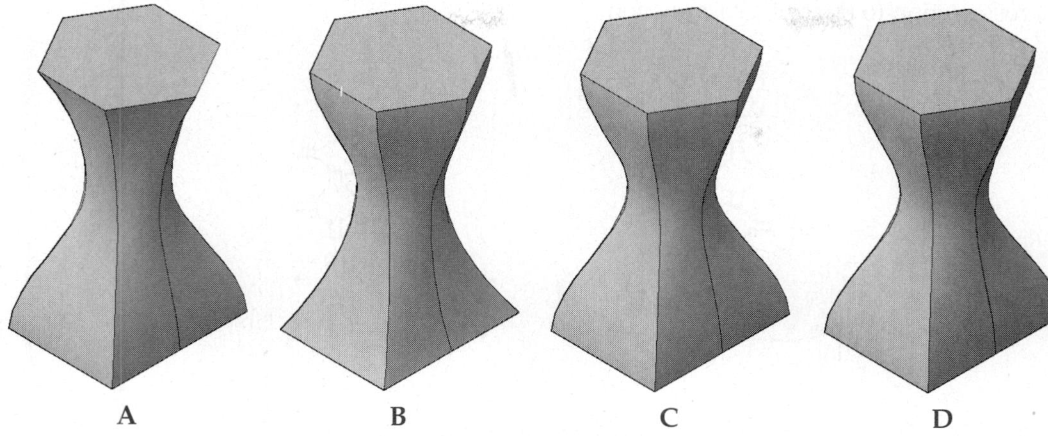

to the first cross section, last cross section, both first and last cross sections, or all cross sections. See **Figure 7-13**. Select the normal setting in the drop-down list. You will have to experiment with these settings to get the desired loft shape.

In manufacturing, plastic or metal parts are sometimes formed in a two-part mold. A slight angle is designed into the parts on the inside and outside surfaces to make removing the part from the mold easier. This taper is called a ***draft angle***. The **Draft angles** option allows you to add a taper to the beginning and end of the loft.

When setting the draft angle, you can set the angle and magnitude. See **Figure 7-14**. The default draft angle is 90°, which means the transition is perpendicular to the cross section. The magnitude represents the relative distance from the cross section, in the same direction as the draft angle, before the transition starts to curve toward the next cross section. Magnitude settings depend on the size of the cross sections, the draft angle values, and the distance between the cross sections. You may have to experiment with different magnitude and angle settings to get the desired loft shape.

The **Close surface or solid** option is used to connect the last cross section to the first cross section. See **Figure 7-15**. This option "closes" the loft, similar to the **Close** option of the **LINE** or **PLINE** command. The shapes from **Figure 7-8** are shown in **Figure 7-16** with the **Close surface or solid** option on. This option is only available when the **Ruled** or **Smooth Fit** option is selected.

Figure 7-14.
When setting the draft angle, you can set the angle and the magnitude. A—Draft angle of 90° and a magnitude of zero. B—Draft angle of 30° and a magnitude of 180. C—Draft angle of 60° and a magnitude of 180.

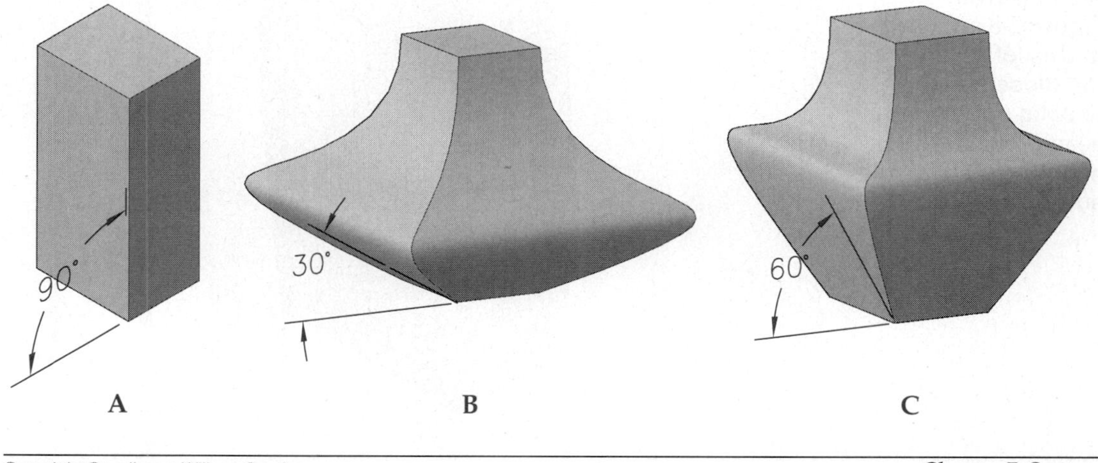

Figure 7-15.
A—These profiles will be used to create a sealing ring. They should be selected in a counterclockwise direction starting with the first cross section. B—The resulting loft with the default settings. Notice the gap between the first and last cross sections. C—By checking the **Close surface or solid** check box in the **Loft Settings** dialog box, the loft continues from the last cross section to the first cross section.

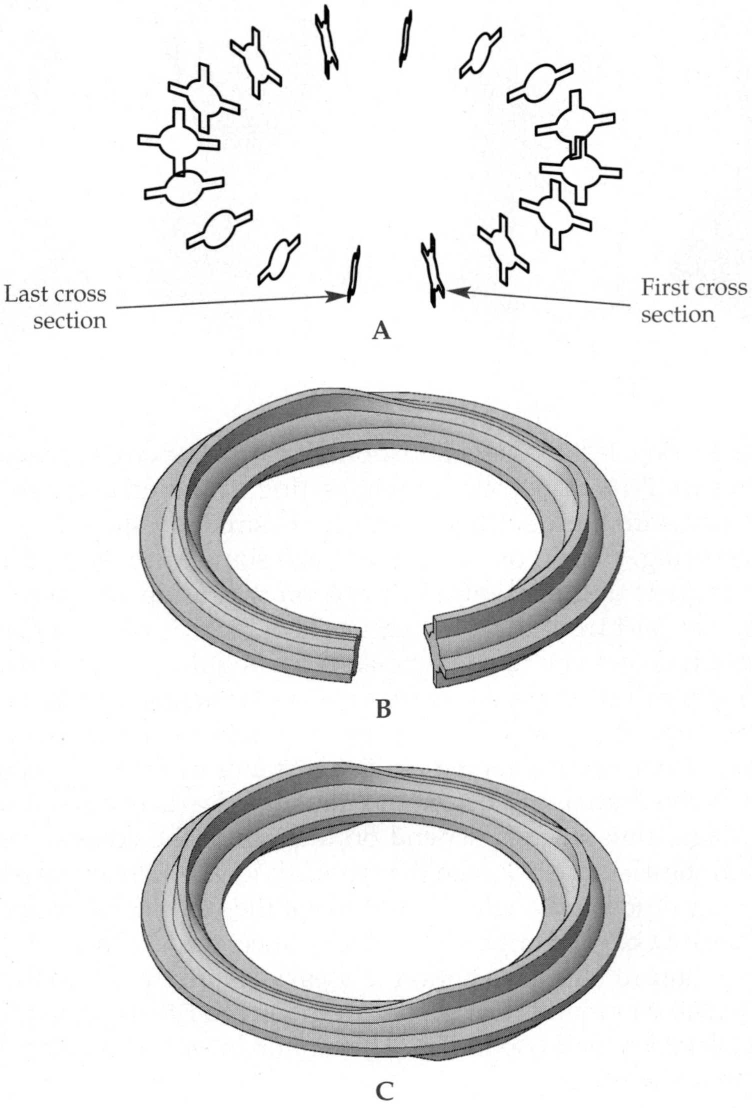

Figure 7-16.
The same cross sections from Figure 7-8 are used in this loft. However, the **Close surface or solid** option has been applied. Notice how the loft is inside out.

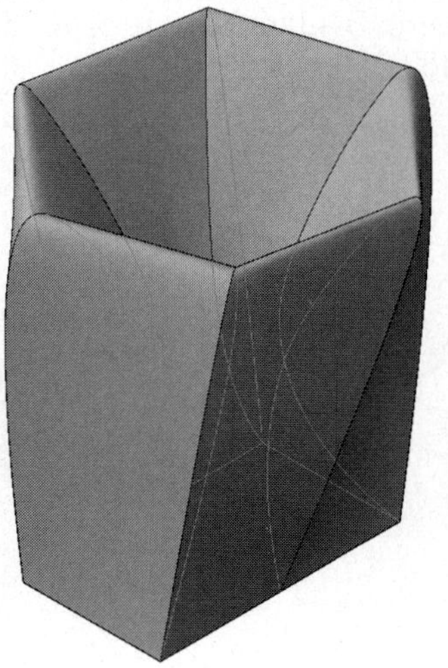

If the **Smooth Fit** option is selected and the **Close surface or solid** check box is checked, the **Periodic (smooth ends)** check box is available. If a closed-loop loft is created similar to the loft in Figure 7-15, the seam may kink if the loft is reshaped in some way. Checking the **Periodic (smooth ends)** check box will help alleviate this problem.

PROFESSIONAL TIP

It may be easier to experiment with the loft settings after the loft is created. Select the loft and change the settings in the **Properties** palette. In addition, special grips and handles appear at the locations of the profiles. Picking on a grip will open a shortcut menu and dragging a handle will alter the loft dynamically.

Exercise 7-3

www.g-wlearning.com/CAD/

Complete the exercise on the companion website.

Controlling the Loft with Guide Curves

Guide curves are lines that control the shape of the transition between cross sections. They do not have to be *curves*. They can be lines, arcs, elliptical arcs, splines (2D or 3D), polylines (2D or 3D), or edge subobjects. There are four rules to follow when using guide curves:

- The guide curve should start on the first cross section.
- The guide curve should end on the last cross section.
- The guide curve should intersect all other cross sections.
- The surface control in the **Loft Settings** dialog box must be set to **Smooth Fit** (**LOFTNORMALS** = 1).

When the **Guides** option of the **LOFT** command is entered, you are prompted to select the guide curves. Select all of the guide curves and press [Enter]. The loft is created. The order in which guide curves are selected is not important.

For example, Figure 7-17A shows two circles that will be lofted. If the **Cross sections only** option is used, a cylinder is created, Figure 7-17B. However, if the **Guides** option is used and the two guide curves shown in Figure 7-17A are selected, one side of the cylinder is deformed similar to a handle or grip. See Figure 7-17C.

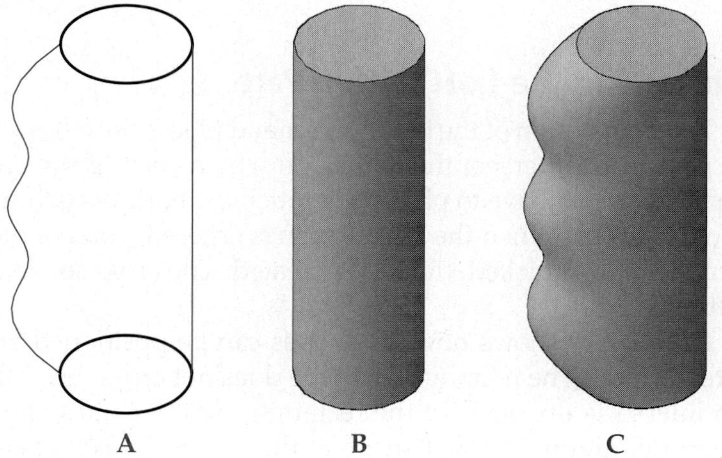

Figure 7-17.
A—These two circles will be lofted. The objects shown in color will be used as guide curves. B—When the circles are lofted using the **Cross sections only** option, a cylinder is created. C—When the **Guides** option is used and the guide curves shown in A are selected, the resulting loft is shaped like a handle or grip.

Figure 7-18.
A—The open profiles shown in black and the guide curve shown in color will be used to create a fabric covering for the three solid objects. B—The resulting fabric covering. This is a surface because the profiles were open.

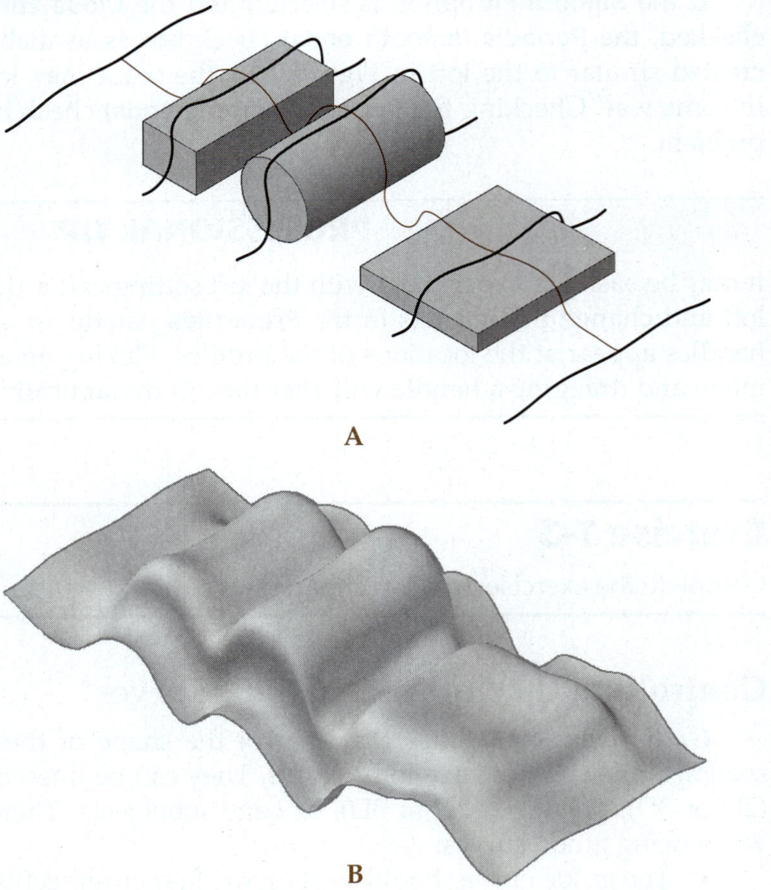

Lofting is used to create open-contour shapes such as fenders, automobile interior parts, fabrics, and other ergonomic consumer products. Figure 7-18 shows the use of open 2D splines in the construction of a fabric covering. Notice how each cross section is intersected by the guide curve. There is a cross section at the beginning of the guide curve and one at the end. These conditions fulfill the rules outlined earlier.

CAUTION

Guide curves only work well when the surface control is set to **Smooth Fit** (**LOFTNORMALS** = 1). If you get an error message when using guide curves or the curves are not reflected in the end result, make sure **LOFTNORMALS** is set to 1 and try it again.

Controlling the Loft with a Path

The **Path** option of the **LOFT** command places the cross sections along a single path. The path must intersect the planes on which each of the cross sections lie. However, the path does *not* have to physically touch the edge of each cross section, as is required of guide curves. When the **Path** option is entered, you are prompted to select the path. Once the path is picked, the loft is created. The cross sections remain in their original positions.

Figure 7-19 shows how 2D shapes can be positioned at various points on a path to create a loft. The rectangular shape does not cross the path. However, as long as the path intersects the plane of the rectangle, which it does, the shape will be included in the loft definition. The last shape at the top of the helix is a point object, causing the loft to taper.

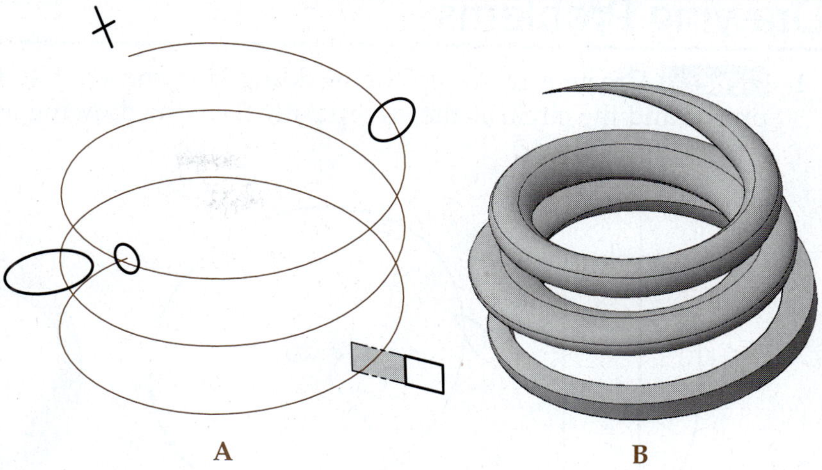

Figure 7-19.
A—The profiles shown in black will be lofted along the path shown in color. Notice how the rectangular profile is not intersected by the path, but the path does intersect the plane on which the rectangle lies. B—The resulting loft.

Exercise 7-4

www.g-wlearning.com/CAD/

Complete the exercise on the companion website.

Chapter Review

Answer the following questions using the information in this chapter.

1. What is a *loft*?
2. What option of the **SWEEP** command determines whether the sweep will be a solid or a surface?
3. When using the **SWEEP** command, on which endpoint of the path does the sweep start?
4. What is the purpose of the **Base point** option of the **SWEEP** command?
5. After the sweep or loft is created, how may the creation options be changed?
6. Which objects may be used as a sweep path?
7. How is the alignment of a sweep set to be perpendicular to the start of the path?
8. Which **SWEEP** command option is used to taper the sweep?
9. What does the **Bank** option of the **SWEEP** command do?
10. What is the difference between the **Ruled** and **Smooth Fit** options in the **LOFT** command?
11. How can you close a loft?
12. List five objects that may be used as guide curves in a loft.
13. What are the four rules that must be followed when using guide curves?
14. When using the **Path** option of the **LOFT** command, what must the path intersect?
15. How can a loft be created so it tapers to a point at its end?

Drawing Problems

1. General Create as a sweep the wedding ring shown. Use the half ellipse as the profile and the circle as the sweep path. Save the drawing as P7-1.

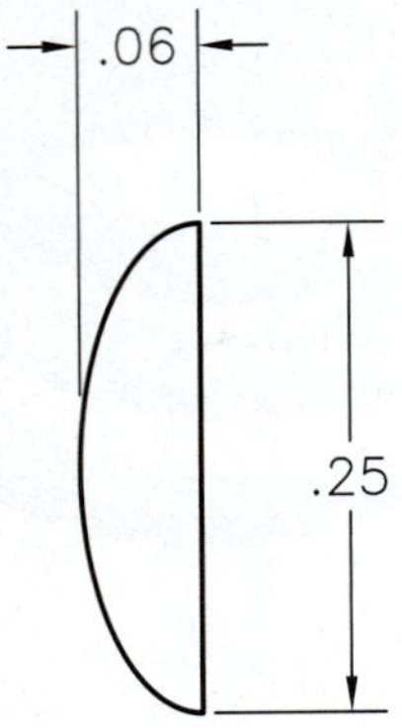

A

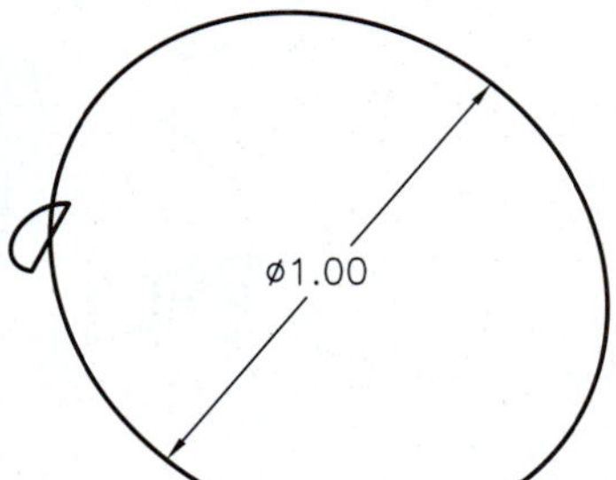

B

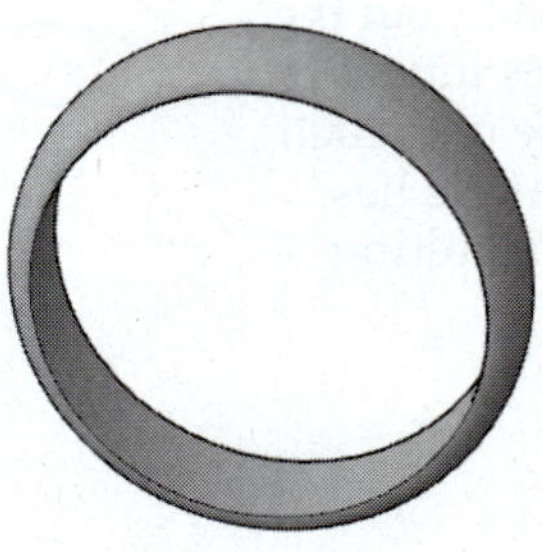

C

2. General Create the aluminum can shown by lofting the circular cross sections. Create two separate loft objects and use one to make a cutout for the recess at the top. Select the six ∅2 and ∅2.50 cross sections from the bottom to the top to create the first loft object. Select the three ∅1.94 and ∅1.75 cross sections from the top down to create the second loft object. Then, subtract the second loft object from the first. Save the drawing as P7-2.

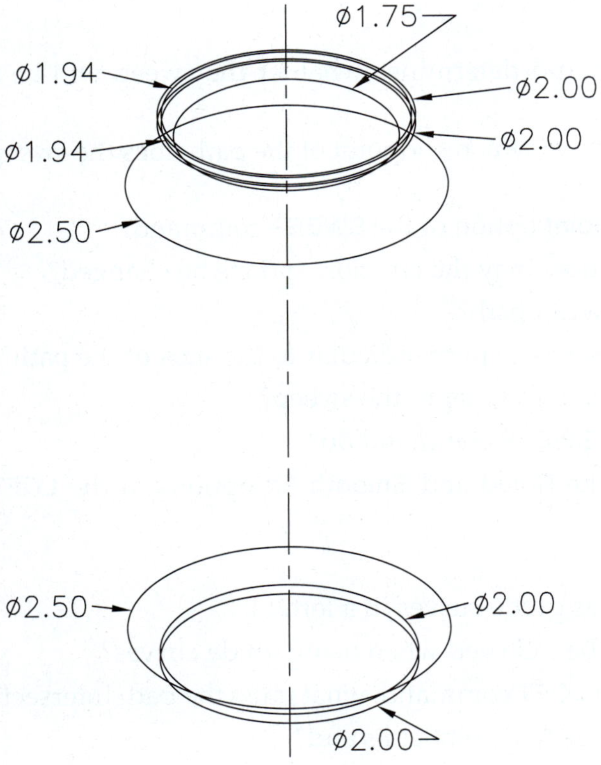

A

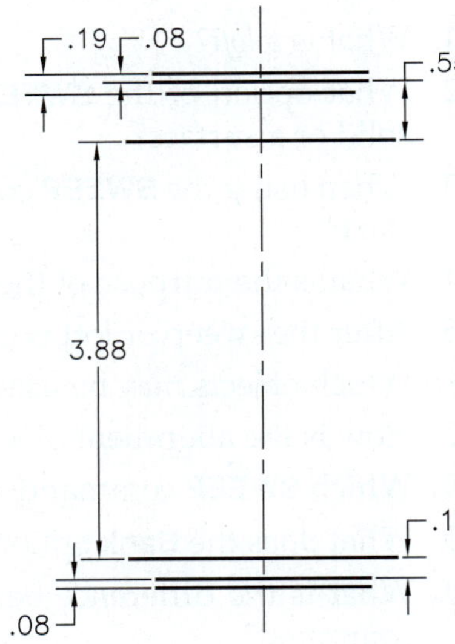

B

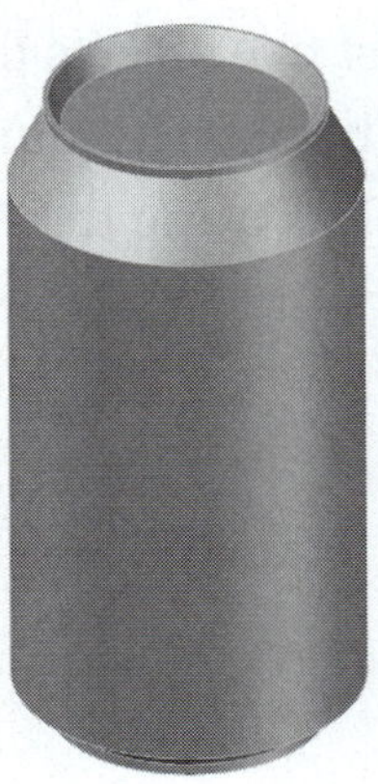

C

3. **General** Create the two shampoo bottles shown. One design uses cross sections only and the other uses a guide curve. Each bottle is made up of two loft objects. Join the pieces so each bottle is one solid. Save the drawing as P7-3.

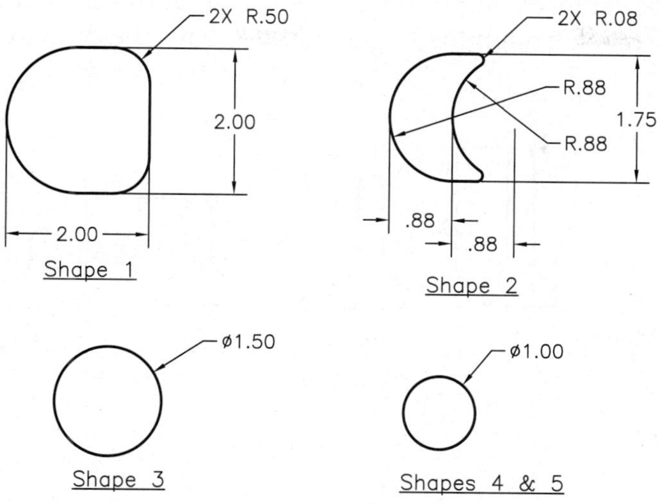

A

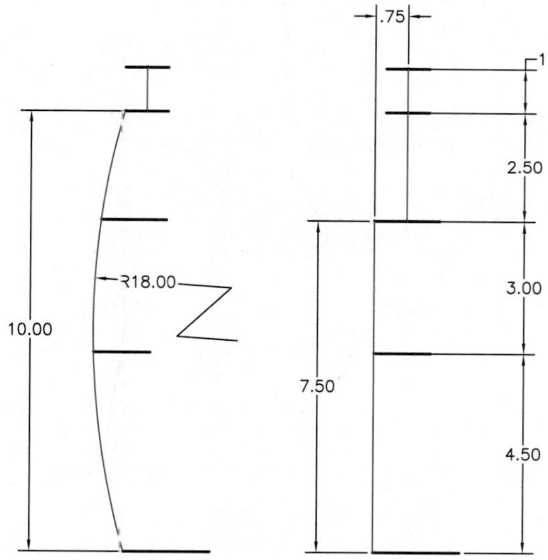

B

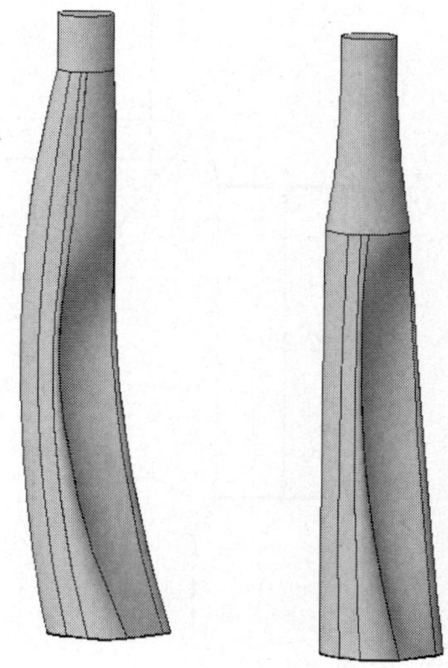

C

4. **Mechanical** Create as a loft the automobile fender shown below. Use the object shown in color as a guide curve or a path. Save the drawing as P7-4.

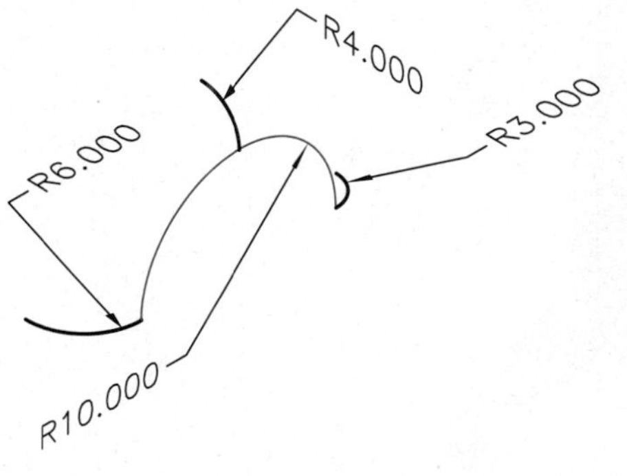

A

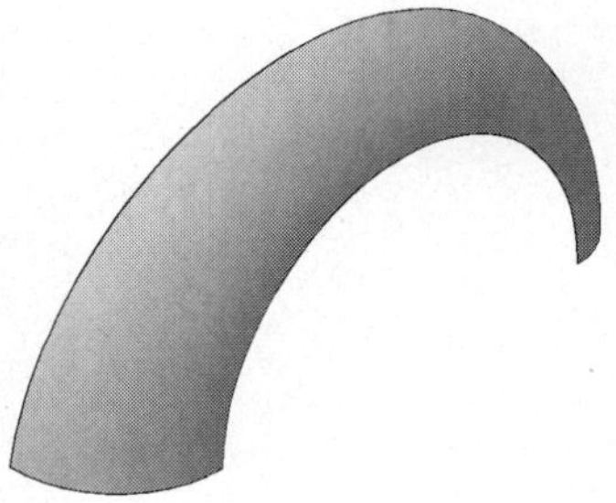

B

5. **Mechanical** Draw as a loft the C-clamp shown. Use the shapes (A, B, C, and D) as the cross sections and the polyline (in color) as the guide curve. Add Ø1 unit cylinders to the ends. Make one cylinder .125H and the other 1.125H. The cylinders should be centered on profile D and located at the ends of the loft as shown. Make a Ø.625 hole through the larger cylinder. Save the drawing as P7-5.

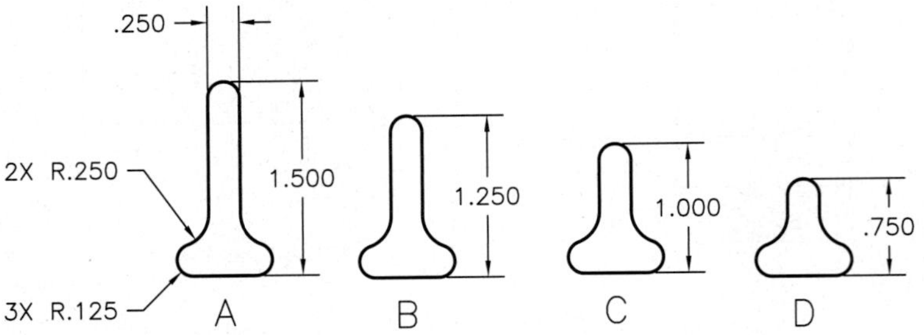

Cross Sections

A

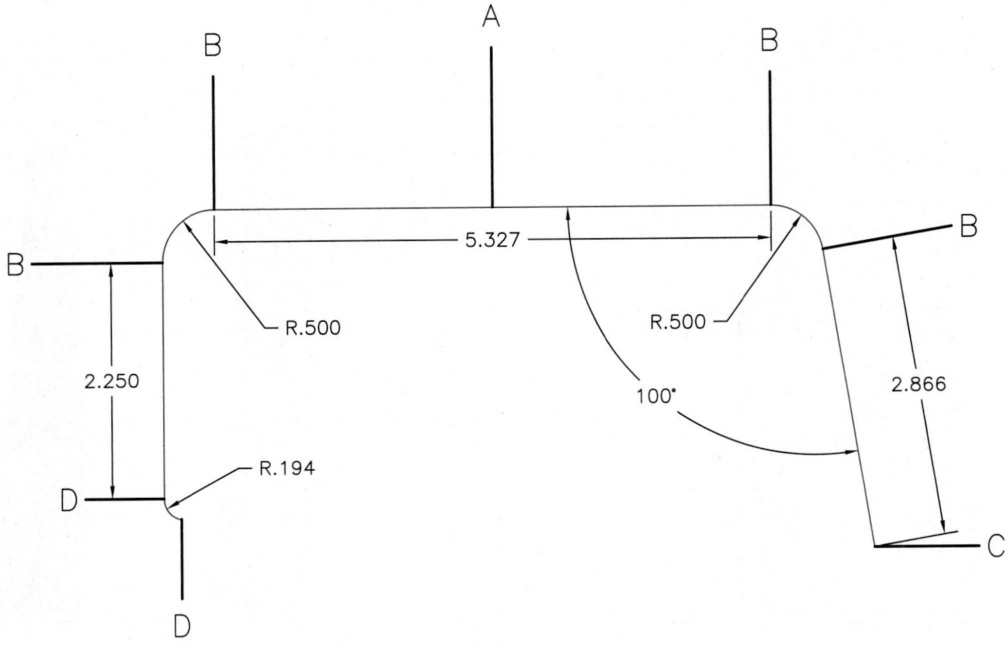

Layout

B

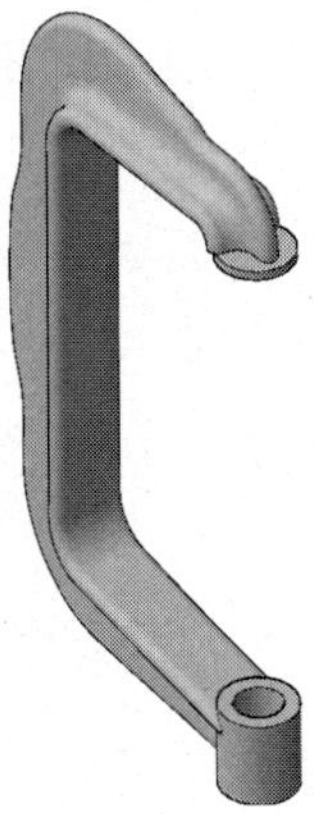

C

6. **Architectural** Create the lamp shade shown. Create two separate loft objects for the top and the bottom. Then, union the two pieces. Finally, scale a copy and hollow out the lamp shade. Save the drawing as P7-6.

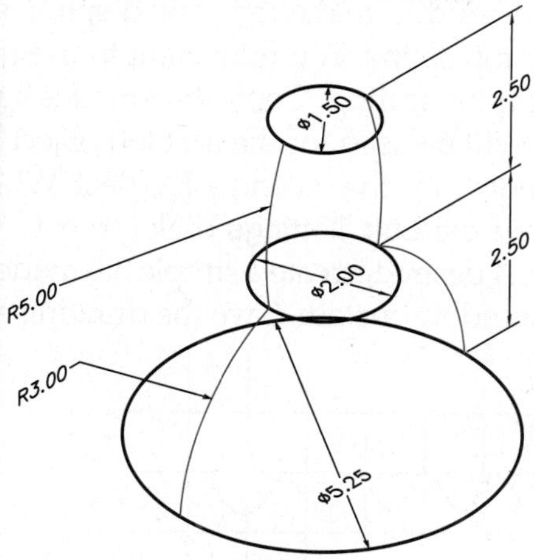

A

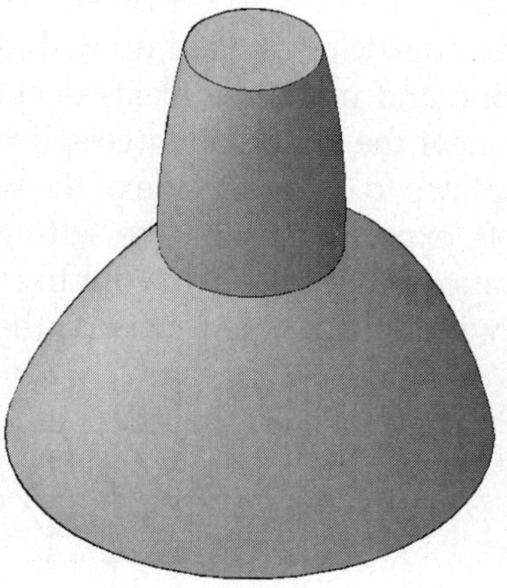

B

7. **General** In this problem, you will draw a racetrack for toy cars by sweeping a 2D shape along a polyline path.
   A. Draw the polyline path shown with the coordinates given. Turn it into a spline.
   B. Draw the 2D profile shown using the dimensions given. Turn it into a region or a polyline.
   C. Use the **SWEEP** command to create the racetrack, as shown in the shaded view.
   D. You may have to use the **Properties** palette to adjust the sweep after it is drawn.
   E. Save the drawing as P7-7.

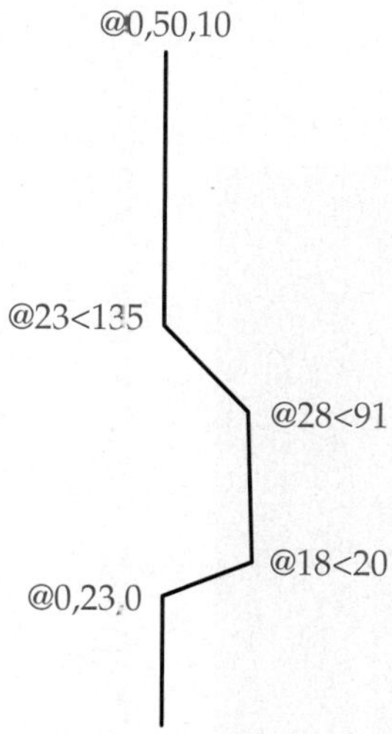

Polyline Path

A

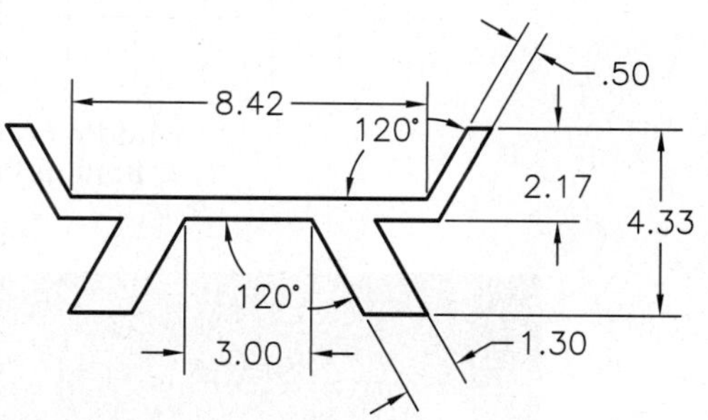

2D Profile

B

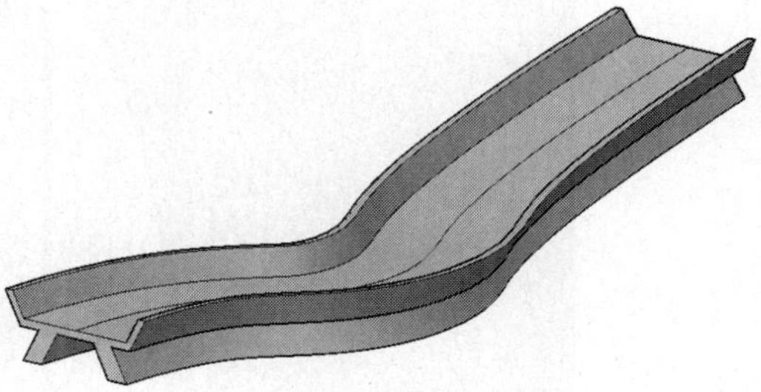

C

8. **Architectural** Use the **LOFT** command to create a window curtain. Create a spline or a polyline converted into a spline similar to the top profile object (A). Position the spline at the top of the window on the left side. Copy the spline and position it about halfway down the window. Use the **SCALE** command to make the copied spline smaller but keep the outside edge lined up with the top spline. You may want to use grips to edit the spline and make the pleats deeper after scaling. Copy the smaller spline to the bottom near the floor. The three splines will be used for the first loft object. Mirror the three splines to the other side of the window for the second loft object. When creating each loft, experiment with the settings in the **Loft Settings** dialog box to change the appearance of the curtain to your liking. If desired, create a simple 3D model of a room with a window opening including the window curtain. Save the drawing as P7-8.

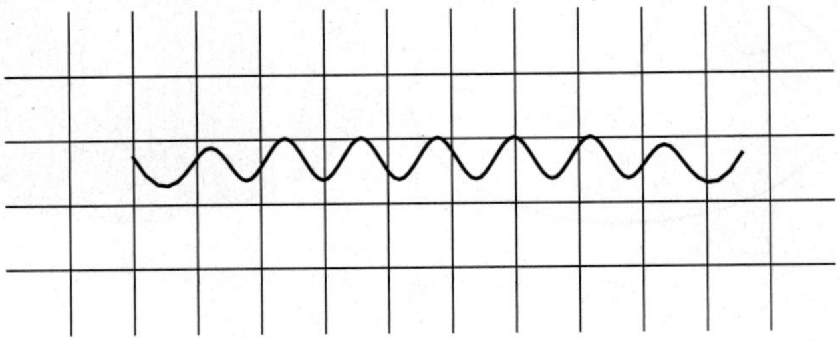

Top Profile
Original Spline
A

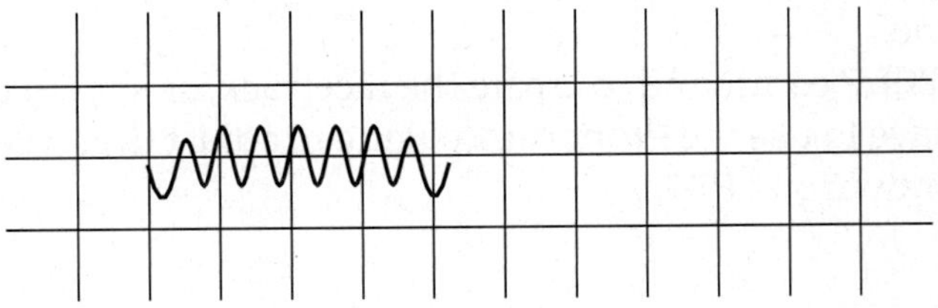

Middle Profile and
Bottom Profile
B

Completed Room
with Window Curtain
C

9. **Architectural** Create the furniture leg shown. Either the **LOFT** command or the **SWEEP** command may be used. However, one command may work better than the other. The profile dimensions refer to the bottom end (small end) of the leg. The top is twice the size of the bottom. Note that the top and the bottom of the leg are parallel. Save the drawing as P7-9.

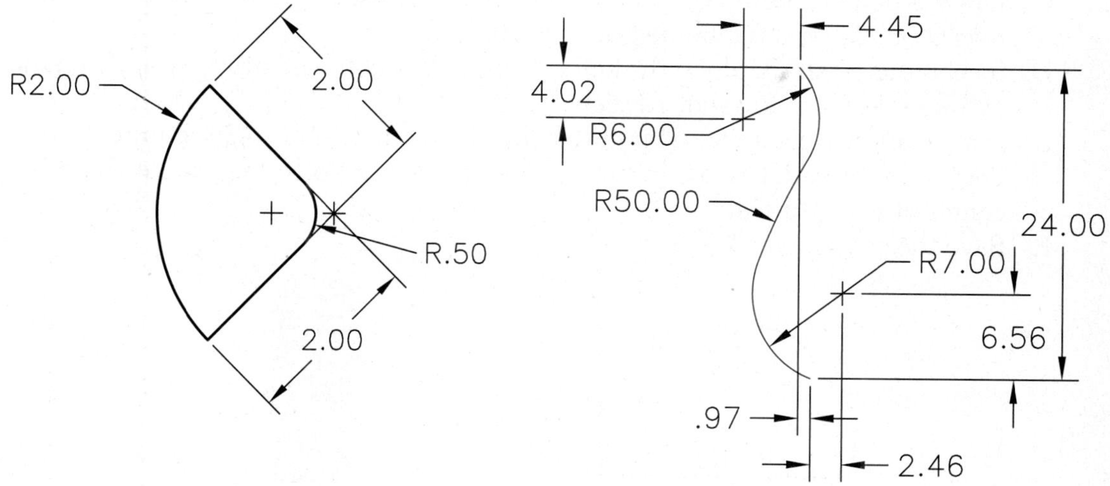

Profile

Path or Guide

A

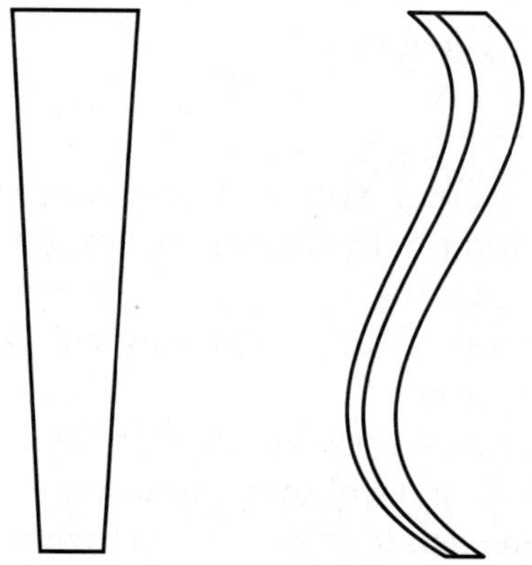

Front View

Side View

B

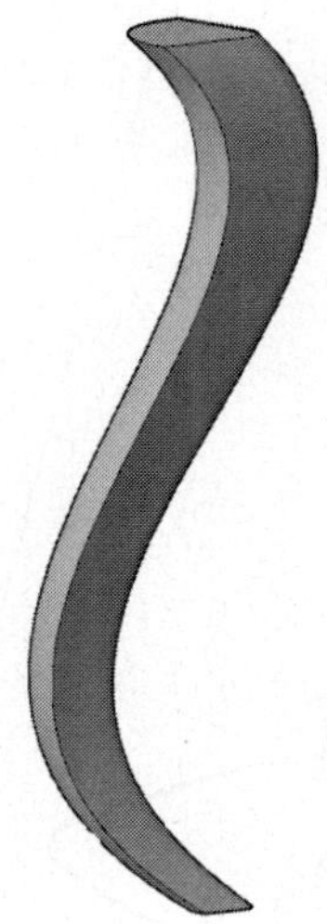

C

10. **Mechanical** In this problem, you will cut a UNC thread in a cylinder by sweeping a 2D shape around a helix and subtracting it.
   A. Draw a Ø.25 cylinder that is 1.00 in height.
   B. Draw the thread cutter profile shown below. The long edge of the cutter should be aligned with the vertical edge of the cylinder.
   C. Draw a helix centered on the cylinder with base and top radii of .125, a turn height of .050, and a total height of 1.000.
   D. Sweep the 2D shape along the helix. Then, subtract the resulting solid from the cylinder. Refer to the shaded view.
   E. If time allows, create another cutter profile to cut a .0313 × 45° chamfer on the end of the thread. Use a circle as a sweep path or revolve the profile about the center of the cylinder.
   F. Save the drawing as P7-10.

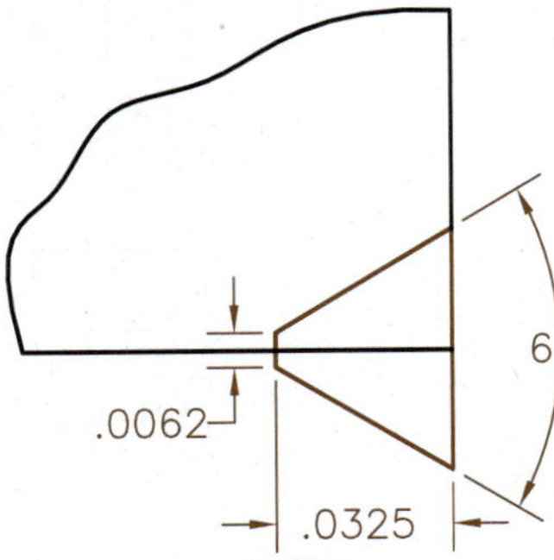

Thread Cutter

A

B

11. **Architectural** In this problem, you will add a seatback to the kitchen chair you started modeling in Chapter 2. In Chapter 6, you refined the seat.
   A. Open P6-9 from Chapter 6.
   B. Draw an arc for the top of the bow. Using a 14.25″ length of the arc, divide it into seven equal segments.
   C. Position eight Ø.50 circles at the division points.
   D. Draw circles at the top of each hole in the seat.
   E. Create a loft between each lower circle and each upper circle.
   F. Using the information in the drawings, create the outer bow for the seatback.
   G. Save the drawing as P7-11.

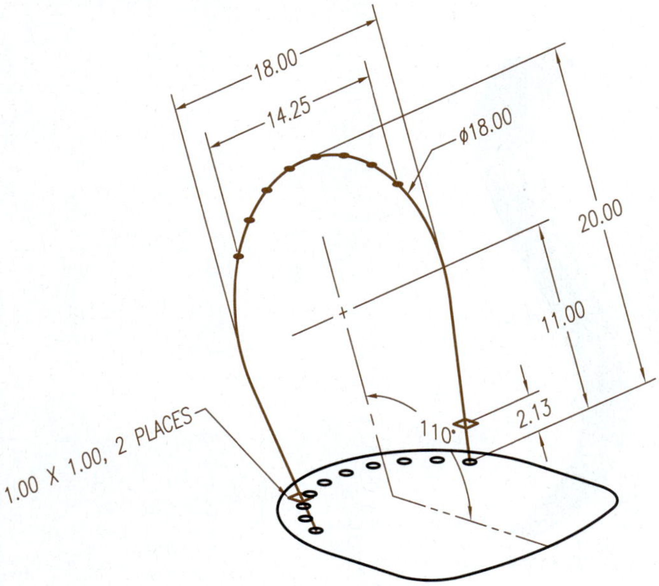

Completed Seatback

CHAPTER 8

Creating and Working with Solid Model Features

Learning Objectives

After completing this chapter, you will be able to:

- ✓ Change properties on solids.
- ✓ Align objects.
- ✓ Rotate objects in three dimensions.
- ✓ Mirror objects in three dimensions.
- ✓ Create 3D arrays.
- ✓ Fillet solid objects.
- ✓ Chamfer solid objects.
- ✓ Slice a solid using various methods.
- ✓ Construct features on solid models.
- ✓ Remove features from solid models.

Changing Properties Using the Properties Palette

PROPERTIES

Ribbon
View
> Palettes
Properties Palette
Type
PROPERTIES
PR
[Ctrl]+[1]

Properties of 3D objects can be modified using the **Properties** palette, which is thoroughly discussed in ***AutoCAD and Its Applications—Basics***. This palette is displayed using the **PROPERTIES** command. You can also select a solid object, right-click, and select **Properties** from the shortcut menu.

The **Properties** palette lists the properties of the currently selected object. For example, **Figure 8-1** lists the properties of a selected solid sphere. Some of its properties are parameters, which is why solid modeling in AutoCAD can be considered parametric. You can change the sphere's radius; diameter; X, Y, Z coordinates; linetype; linetype scale; color; layer; lineweight; and visual settings. The categories and properties available in the **Properties** palette depend on the selected object.

To modify an object property, select the property. Then, enter a new value in the right-hand column. The drawing is updated to reflect the changes. You can leave the **Properties** palette open as you continue with your work.

Figure 8-1.
The **Properties** palette can be used to change many of the properties of a solid.

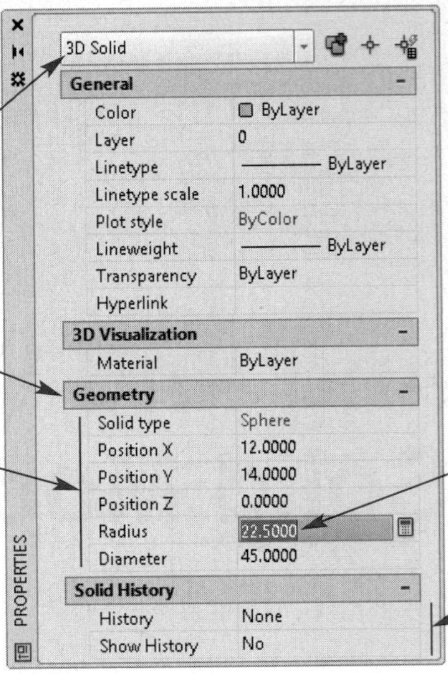

Solid Model History

AutoCAD can automatically record a history of a composite solid model's construction. A ***composite solid*** is created by a Boolean operation or by using the **SOLIDEDIT** command. The retention of a composite solid's history is controlled by the History property setting in the **Solid History** category of the **Properties** palette. By default, the History property is set to None. Refer to **Figure 8-1**. This means that the history is not saved and that edges, vertices, and faces of solids can be directly edited. If the History property in the **Properties** palette is set to Record, the solid history is "recorded," or retained. This allows you to work with the geometry used in modeling operations to create the composite solid. Depending on your modeling preferences, it is generally a good idea to have the history recorded. Then, at any time, you can graphically display all of the geometry that was used to create the model.

The retention of solid model history for newly created composite solids is controlled by the **SOLIDHIST** system variable. By default, the **SOLIDHIST** system variable is set to 0. This means that all new solids have their History property set to None and solid history is not recorded. If the **SOLIDHIST** system variable is set to a value of 1, all new solids have their History property set to Record and solid history is preserved. With either setting of the system variable, 0 (None) or 1 (Record), the **Properties** palette can be used to change the setting for individual solids. The current setting of the **SOLIDHIST** system variable is indicated by the **Solid History** button located in the **Primitive** panel of the **Solid** tab on the ribbon. See **Figure 8-2**. With a default setting of 0 (None), the **Solid History** button has a white background. Picking this button when it has a white background changes the background color to blue and sets the **SOLIDHIST** system variable to 1 (Record).

To view the graphic history of a composite solid, set the Show History property in the **Properties** palette to Yes. All of the geometry used to construct the model is displayed. If the **SHOWHIST** system variable is set to 0, the Show History property is set to No for all solids and cannot be changed. If this system variable is set to 2, the Show History property is set to Yes for all solids and cannot be changed. A **SHOWHIST** setting of 1 allows the Show History property to be individually set for each solid. This is the default setting.

An example of showing the history on a composite solid is provided in **Figure 8-3**. In **Figure 8-3A**, the model appears in its current edited format. The History property

Figure 8-2.
The **Solid History** button, located in the **Primitive** panel of the **Solid** tab on the ribbon, is used to control the **SOLIDHIST** system variable setting. This button has a white background when the **SOLIDHIST** system variable is set to 0 (the default value). If picked, the button has a blue background.

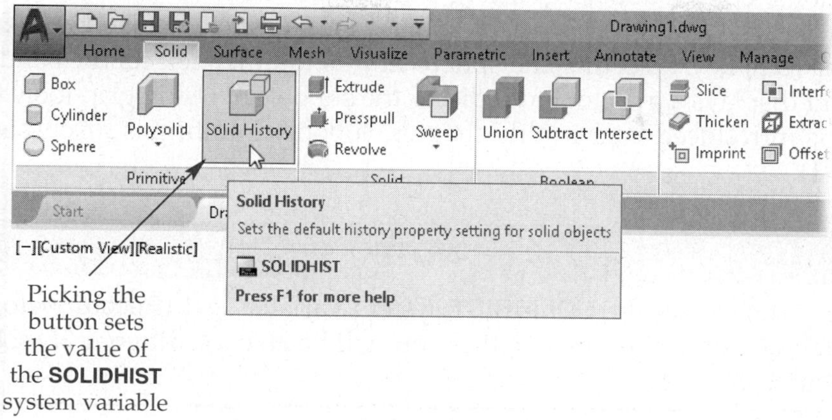

Figure 8-3.
A—The object appears in its current state with the Show History property turned off. The History property was initially set to Record before subtracting two solid primitives. B—The Show History property is set to Yes and the display of isolines has been turned on. C—When the sphere is selected while pressing the [Ctrl] key, grips are displayed to indicate the sphere primitive subobject is selected. The move grip tool can be used to edit the original sphere primitive.

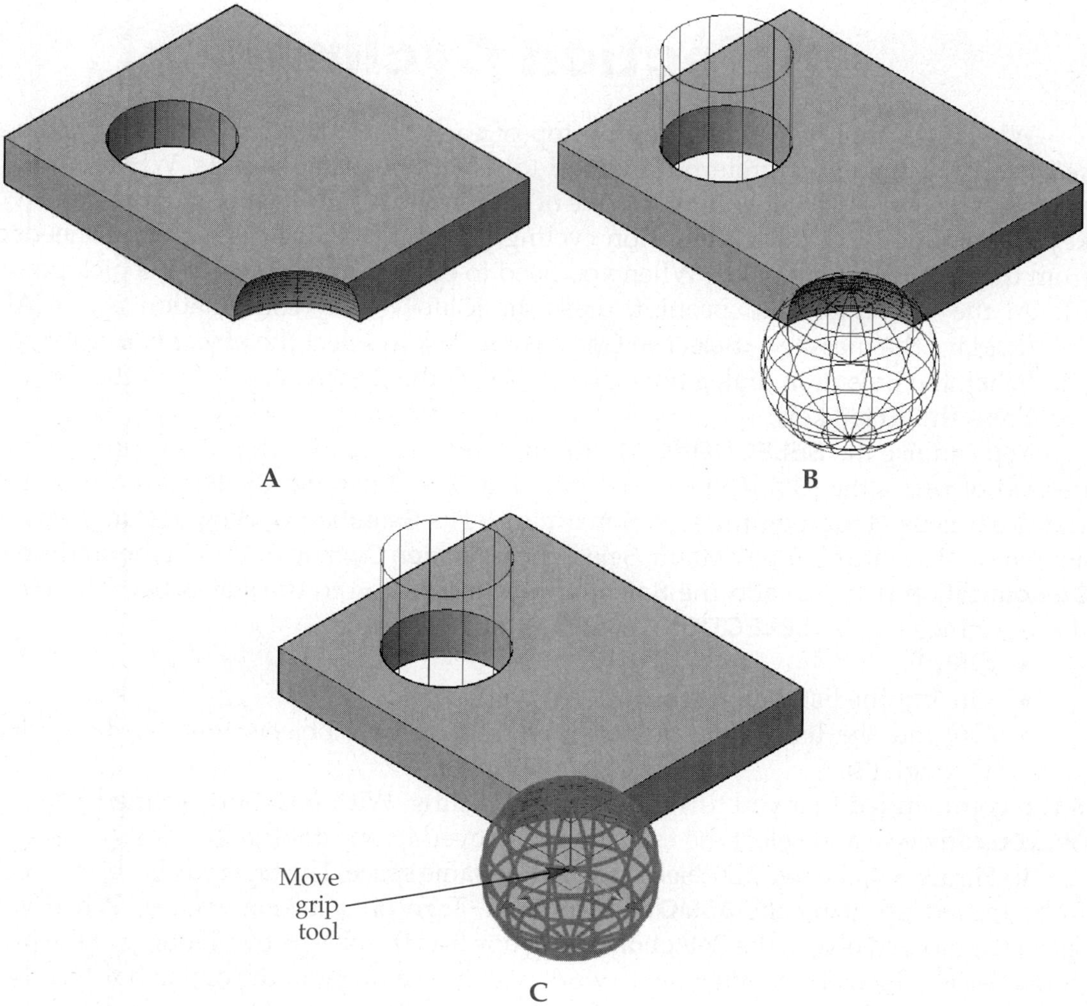

setting was initially set to Record before performing two Boolean subtraction operations. The Show History property is currently set to No and the Conceptual visual style is set current. In **Figure 8-3B**, the Show History property is set to Yes. Isolines have also been turned on. You can see the geometry that was used in the Boolean subtraction operations. Using subobject editing techniques, the individual geometry can be selected and edited. In **Figure 8-3C**, the sphere (the subtracted object) is selected while pressing the [Ctrl] key. Selecting the sphere in this manner selects the solid primitive *subobject* and displays a grip tool for editing the subobject. Using grip tools for editing 3D objects is introduced later in this chapter. Subobject editing is discussed in detail in Chapter 11.

NOTE

It is good practice to set the **SOLIDHIST** system variable to 1 (Record) before starting all 3D modeling work. This ensures that you will be able to edit your 3D solid models if the need arises.

PROFESSIONAL TIP

If the Show History property is set to Yes to display the components of the composite solid, as seen in **Figure 8-3**, the components will appear when the drawing is plotted. Be sure to set the Show History property to No before you print or plot.

Selection Cycling

When selecting objects that are on top of each other or occupy the same space, ***selection cycling*** is the preferred method to select one of the objects. When editing, you may need to erase, move, or copy one of the objects that overlap. Press the [Ctrl]+[W] key combination to turn on selection cycling. You can then select the object needed from the **Selection** dialog box. When you need to cycle through objects at a pick point:

1. At the "select objects" prompt, press the [Ctrl]+[W] key combination. AutoCAD displays the message <Selection Cycling on>. Pick to select the object you want.
2. When the **Selection** dialog box appears, select the desired object from the list.
3. Press the [Enter] key.

SELECTIONCYCLING

Type
SELECTIONCYCLING
Toolbar
Status Bar

Selection Cycling

You can use the **SELECTIONCYCLING** system variable to turn on selection cycling instead of using the [Ctrl]+[W] key combination. The **Selection Cycling** button on the status bar is used for toggling selection cycling. The **Selection Cycling** button does not appear on the status bar by default. Select the **Selection Cycling** option in the status bar **Customization** flyout to add the **Selection Cycling** button to the status bar. There are three settings for the **SELECTIONCYCLING** system variable:

- Off (0).
- On, but the list dialog box does not display (1).
- On and the list dialog box displays the selected objects that can be cycled through (2).

It is recommended that you turn on selection cycling. With selection cycling turned on, you can cycle and select the faces that may overlap one another.

In **Figure 8-4**, the two 3D objects occupy the same space. The tapered 3D object needs to be moved up using the **3DMOVE** command. Turn on selection cycling. When you select the tapered object, the **Selection** dialog box lists the objects overlapping at the pick point. Select which object you want to work with by picking it in the dialog box. As your cursor is over an object in the list, the object is highlighted in the drawing area.

Figure 8-4.
Using selection cycling to choose which object to select.

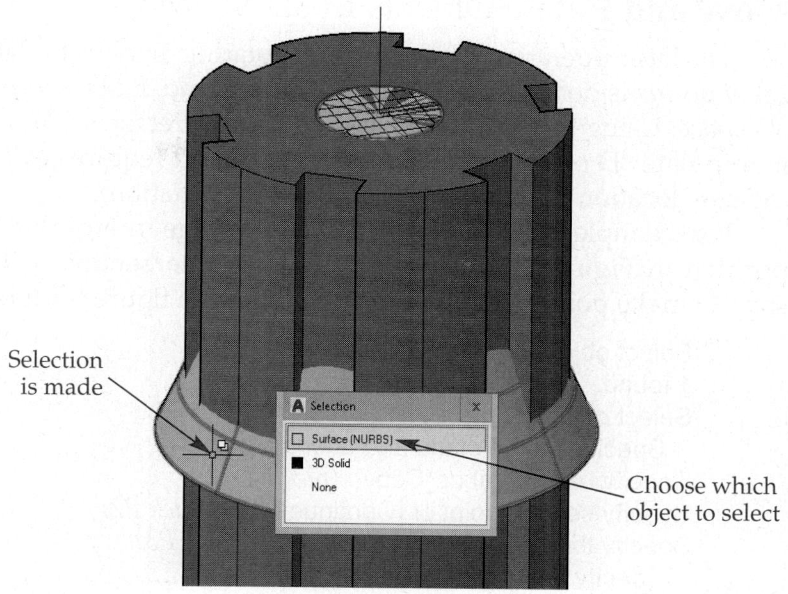

NOTE

You can also press the [Shift] key and space bar at the same time to cycle through overlapping objects.

PROFESSIONAL TIP

If multiple 3D objects occupy the same space, selection cycling gives you the ability to select the correct object needed to complete the edit.

Aligning Objects in 3D

AutoCAD provides two different methods with which to move and rotate objects in a single command. This is called ***aligning*** objects. The simplest method is to align 3D objects by picking source points on the first object and then picking destination points on the object to which the first one is to be aligned. This is accomplished with the **3DALIGN** command, which allows you to both relocate and rotate the object. The second method is possible with the **ALIGN** command. The **ALIGN** command aligns 2D or 3D objects according to three sets of alignment pairs. You first select the source object, then the first source point, and finally the first destination point on the destination object. This technique is repeated two more times to align one object to another. This method of aligning allows you to not only move and rotate an object, but scale the object being aligned.

PROFESSIONAL TIP

Use the **3DALIGN** command to create assemblies.

Move and Rotate Objects in 3D Space

3DALIGN

Ribbon
Home
> Modify
3D Align
Type
3DALIGN
3AL

The basic operation of moving and rotating an object relative to a second object or set of points is done with the **3DALIGN** command. It allows you to reorient an object in 3D space. Using this command, you can correct errors of 3D construction and quickly manipulate 3D objects. The **3DALIGN** command requires existing points (source) and the new location of those existing points (destination).

For example, refer to **Figure 8-5**. The wedge in **Figure 8-5A** is aligned in its new position in **Figure 8-5B** as follows. Set the **Intersection** or **Endpoint** running object snap to make point selection easier. Refer to the figure for the pick points.

Select objects: *(pick the wedge)*
1 found
Select objects: ↵
  Specify source plane and orientation...
Specify base point or [Copy]: *(pick P1)*
Specify second point or [Continue] <C>: *(pick P2)*
Specify third point or [Continue] <C>: *(pick P3)*
  Specify destination plane and orientation...
Specify first destination point: *(pick P4)*
Specify second destination point or [eXit] <X>: *(pick P5)*
Specify third destination point or [eXit] <X>: *(pick P6)*

Figure 8-5.
The **3DALIGN** command can be used to properly orient 3D objects.
A—Before aligning. Note the pick points.
B—After aligning.

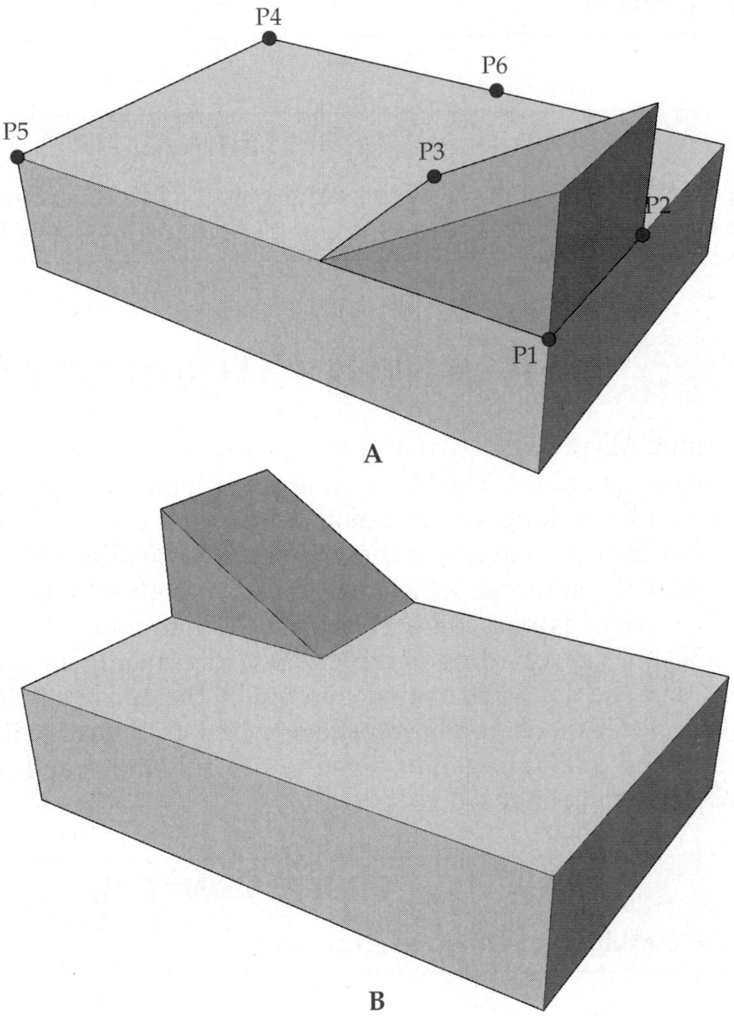

You can also use the **3DALIGN** command to align cylindrical 3D objects. The procedure is similar to aligning planar objects, which require three pick points. However, you only need two pick points per object.

For example, the socket head cap screw in **Figure 8-6** is aligned to a new position using the **3DALIGN** command. Set the **Center** running object snap to make point selection easier. Refer to the figure for the pick points.

Select objects: *(pick the socket head cap screw)*
1 found
Select objects: ↵
  Specify source plane and orientation…
Specify base point or [Copy]: *(pick P1)*
Specify second point or [Continue] <C>: *(pick P2)*
Specify third point or [Continue] <C>: ↵
  Specify destination plane and orientation…
Specify first destination point: *(pick P3)*
Specify second destination point or [eXit] <X>: *(pick P4)*
Specify third destination point or [eXit] <X>: ↵

The idea with this command is you are aligning a plane defined by three points with another plane defined by three points. The planes do not need to correspond to actual planar faces. In the example shown in **Figure 8-5**, the alignment planes coincide with planar faces. However, in the example shown in **Figure 8-6**, the alignment planes define planes on which axes lie.

PROFESSIONAL TIP

You can also use the **3DALIGN** command to copy an object, rather than move it, and realign it at the same time. Just select the **Copy** option at the Specify base point or [Copy]: prompt. Then, continue selecting the points as previously discussed.

Figure 8-6.
Using the **3DALIGN** command to align cylindrical objects. A—Before aligning. Note the pick points. B—After aligning.

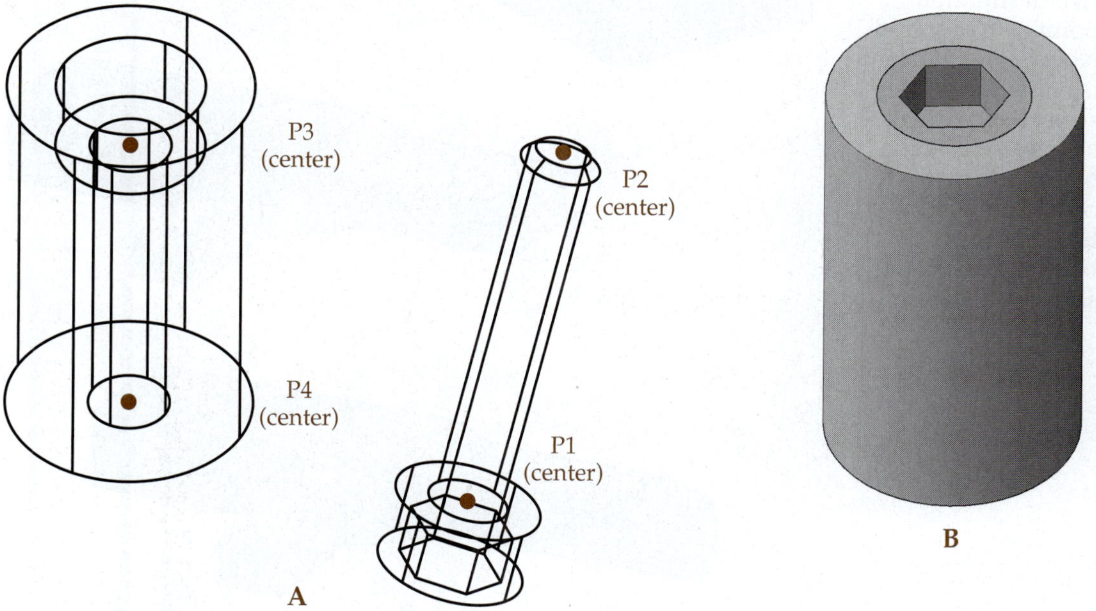

Exercise 8-1

www.g-wlearning.com/CAD/

Complete the exercise on the companion website.

Move, Rotate, and Scale Objects in 3D Space

ALIGN

Ribbon
Home
> Modify

Align

Type
ALIGN
AL

The **ALIGN** command has the same functions of the **3DALIGN** command, but adds the ability to scale an object. See **Figure 8-7**. The 90° bend must be rotated and scaled to fit onto the end of the HVAC assembly. Two source points and two destination points are required, **Figure 8-7A**. Then, you can choose to scale the object.

Select objects: *(pick the 90° bend)*
1 found
Select objects: ↵
Specify first source point: *(pick P1)*
Specify first destination point: *(pick P2; a line is drawn between the two points)*
Specify second source point: *(pick P3)*
Specify second destination point: *(pick P4; a line is drawn between the two points)*
Specify third source point or <continue>: ↵
Scale objects based on alignment points? [Yes/No] <N>: Y↵

The 90° bend is aligned and uniformly scaled to meet the existing ductwork object. See **Figure 8-7B**. You can also align using three source and three destination points. However, when doing so, you cannot scale the object.

PROFESSIONAL TIP

Before using 3D editing commands, set running object snaps to enhance your accuracy and speed.

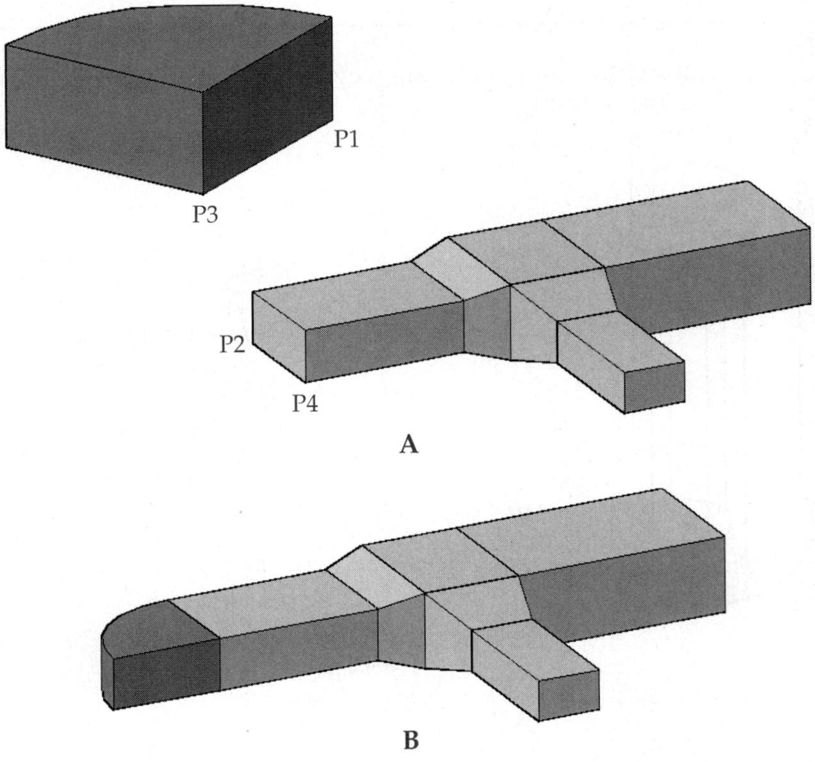

Figure 8-7.
Using the **ALIGN** command. A—Two source points and two destination points are required. Notice how the bend is not at the proper scale. B—You can choose to scale the object during the operation. Notice how the aligned bend is also properly scaled.

Exercise 8-2

www.g-wlearning.com/CAD/

Complete the exercise on the companion website.

Move, Rotate, and Scale Gizmos

A ***gizmo***, also called a *grip tool*, appears when an object or a subobject is selected. This tool is used to specify how the transformation (movement, rotation, or scaling) is applied. Using gizmos (grip tools) is discussed in detail in Chapter 11.

The move gizmo allows movement of an object or a subobject along the X, Y, or Z axis or on the XY, XZ, or YZ plane. The rotate gizmo allows rotation about the X, Y, or Z axis. The scale gizmo allows scaling along the X, Y, or Z axis or XY, XZ, or YZ plane. You can set the default gizmo displayed when an object is selected by using the gizmo flyout in the **Selection** panel of the **Solid** ribbon tab. You can also use the **Show gizmos** flyout on the status bar. The **Show gizmos** flyout does not appear on the status bar by default. Select the **Gizmo** option in the status bar **Customization** flyout to add the **Show gizmos** flyout to the status bar.

The move gizmo is displayed by default. To use the move gizmo, pick any axis to move along that axis. See **Figure 8-8.** To move along a plane, pick the rectangular area at the intersection of the axes. To use the rotate gizmo, pick the circle with the center about which you wish to rotate the selection. To use the scale gizmo, pick an axis to scale along that axis. Pick the triangular area at the intersection of the axes to uniformly scale the selection. Pick the area between the inner and outer triangular areas to nonuniformly scale along that plane. Using gizmos to move and rotate objects is discussed in the following sections.

Ribbon
Solid
> Selection
Move Gizmo
Rotate Gizmo
Scale Gizmo
No Gizmo
Type
DEFAULTGIZMO

DEFAULTGIZMO

NOTE

Mesh objects, discussed in Chapter 9, can be nonuniformly scaled along an axis or plane. The scale gizmo cannot be used to nonuniformly scale solid or surface objects.

Figure 8-8.
Gizmos are used to transform (move, rotate, or scale) objects or subobjects. A—Move gizmo. B—Rotate gizmo. C—Scale gizmo.

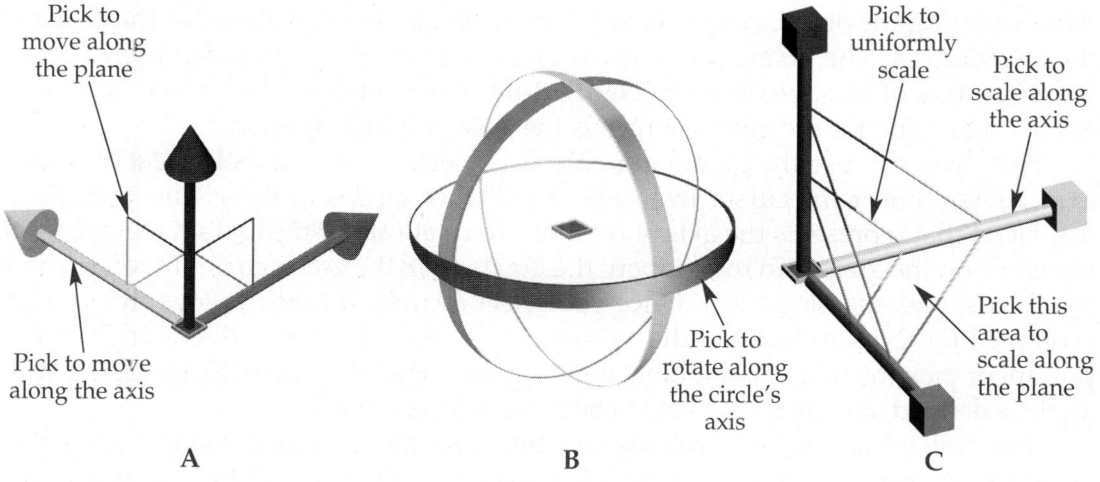

3D Moving

3DMOVE

Ribbon
Home
> Modify
3D Move
Type
3DMOVE
3M

The **3DMOVE** command allows you to quickly move an object along any axis or plane of the current UCS. When the command is initiated, you are prompted to select the objects to move. If the 2D Wireframe visual style is current, the visual style is temporarily changed to the Wireframe visual style because the gizmo is not displayed in 2D mode. After selecting the objects, press [Enter]. The move gizmo is displayed in the center of the selection set. By default, the move gizmo is also displayed when a solid is selected with no command active.

The move gizmo is a tripod that appears similar to the shaded UCS icon. Refer to **Figure 8-8A**. You can relocate the tool by right-clicking on the gizmo and selecting **Relocate Gizmo** from the shortcut menu. Then, move the gizmo to a new location and pick. You can also realign the gizmo using the shortcut menu.

If you move the pointer over the X, Y, or Z axis of the gizmo, the axis changes to yellow. To restrict movement along that axis, pick the axis. If you move the pointer over one of the right angles at the origin of the gizmo, the corresponding two axes turn yellow. Pick to restrict the movement to that plane. You can complete the movement by either picking a new point or by direct distance entry.

NOTE

If the **GTAUTO** system variable is set to 1, the move gizmo is displayed when a solid is selected with no command active. If the **GTLOCATION** system variable is set to 0, the gizmo is placed at the same location as the UCS icon. Both variables are set to 1 by default.

3D Rotating

3DROTATE

Ribbon
Home
> Modify
3D Rotate
Type
3DROTATE
3R

As you have seen in earlier chapters, the **ROTATE** command can be used to rotate 3D objects. However, the command can only rotate objects in the XY plane of the current UCS. This is why you had to change UCSs to properly rotate objects. The **3DROTATE** command, on the other hand, can rotate objects on any axis regardless of the current UCS. This is an extremely powerful editing and design tool.

When the command is initiated, you are prompted to select the object(s) to rotate. After selecting the objects, press [Enter]. The rotate gizmo is displayed in the center of the selection set. The gizmo provides you with a dynamic, graphic representation of the three axes of rotation. Refer to **Figure 8-8B**. After selecting the objects, you must specify a location for the gizmo, which is the base point for rotation.

Now, you can use the gizmo to rotate the objects about the tool's local X, Y, or Z axis. As you hover the cursor over one of the three circles in the gizmo, a vector is displayed that represents the axis of rotation. To rotate about the tool's X axis, pick the red circle on the gizmo. To rotate about the Y axis, pick the green circle. To rotate about the Z axis, pick the blue circle. Once you select a circle, it turns yellow and you are prompted for the start point of the rotation angle. You can enter a direct angle at this prompt or pick the first of two points defining the angle of rotation. When the rotation angle is defined, the object is rotated about the selected axis.

The following example rotates the bend in the HVAC assembly shown in **Figure 8-9A**. Set the **Center of face** 3D object snap. Then, select the command and continue:

Current positive angle in UCS: ANGDIR=*(current)* ANGBASE=*(current)*
Select objects: *(pick the bend)*
1 found
Select objects: ↵
Specify base point: *(acquire the center of the face and pick)*
Pick a rotation axis: *(pick the circle indicated in Figure 8-9A)*
Specify angle start point or type an angle: **180**↵

Note that the rotate gizmo remains visible through the base point and the angle of rotation selections. The rotated object is shown in **Figure 8-9B**.

If you need to rotate an object on an axis that is not parallel to the current X, Y, or Z axes, use a dynamic UCS with the **3DROTATE** command. Chapter 4 discussed the benefits of using a dynamic UCS when creating objects that need to be parallel to a surface other than the XY plane. With the object selected for rotation and the dynamic UCS option active (press the [Ctrl]+[D] key combination), right-click and select **Relocate Gizmo**. Make sure the **Respect Dynamic UCS** option is checked. Then, move the rotate gizmo over a face of the object. The gizmo aligns itself with the surface so that the Z axis is perpendicular to the face. Carefully place the gizmo over the point of rotation using object snaps. Make sure the tool is correctly positioned before picking to locate it. Then, enter an angle or use polar tracking to rotate the object about the appropriate axis on the gizmo.

PROFESSIONAL TIP

By default, the move gizmo is displayed when an object is selected with no command active. To toggle between the move gizmo and the rotate gizmo, select the grip at the tool's origin and press the space bar. Then, pick a location for the tool's origin. You can toggle back to the move gizmo using the same procedure.

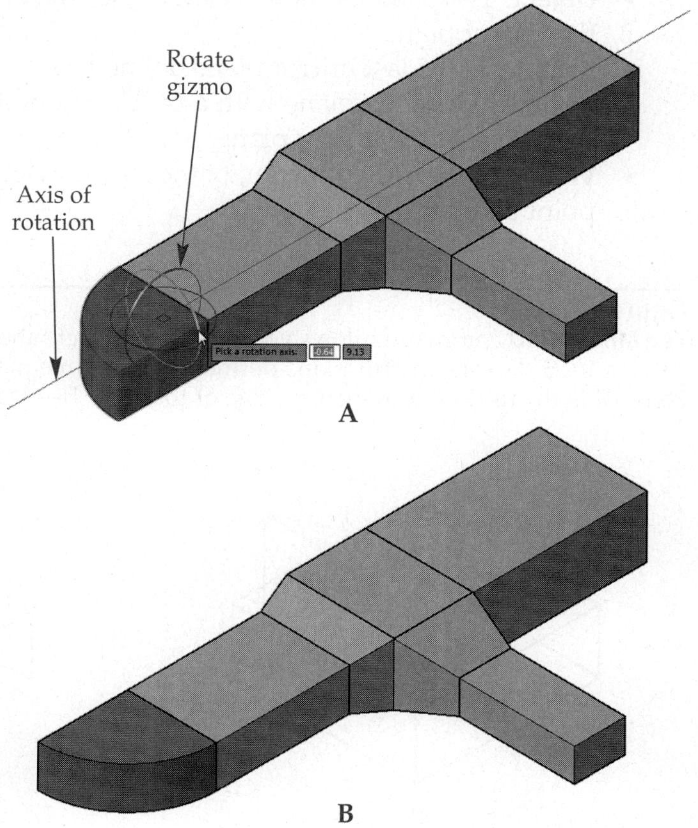

Figure 8-9.
A—Use object snap tracking or object snaps to place the rotate gizmo in the middle of the rectangular face. Then, select the axis of rotation. B—The completed rotation.

Exercise 8-3

www.g-wlearning.com/CAD/

Complete the exercise on the companion website.

3D Mirroring

MIRROR3D

Ribbon
Home
> Modify
3D Mirror
Type
MIRROR3D

The **MIRROR** command can be used to rotate 3D objects. However, like the **ROTATE** command, the **MIRROR** command can only work in the XY plane of the current UCS. Often, to properly mirror objects with this command, you have to change UCSs. The **MIRROR3D** command, on the other hand, allows you to mirror objects about any plane regardless of the current UCS.

The default option of the command is to define a mirror plane by picking three points on that plane, **Figure 8-10A**. Object snaps should be used to accurately define the mirror plane. To mirror the wedge in **Figure 8-10A**, set the **Midpoint** object snap, select the command, and use the following sequence. The resulting drawing is shown in **Figure 8-10B**.

Select objects: *(pick the wedge)*
1 found
Select objects: ↵
Specify first point of mirror plane (3 points) or
[Object/Last/Zaxis/View/XY/YZ/ZX/3points] <3points>: *(pick P1, which is the midpoint of the box's top edge)*
Specify second point on mirror plane: *(pick P2)*
Specify third point on mirror plane: *(pick P3)*
Delete source objects? [Yes/No] <N>: ↵

There are several different ways to define a mirror plane with the **MIRROR3D** command. The options include the following:

- **Object.** The plane of the selected circle, arc, or 2D polyline segment is used as the mirror plane.
- **Last.** Uses the last mirror plane defined.
- **Zaxis.** Defines the plane with a pick point on the mirror plane and a point on the Z axis of the mirror plane.
- **View.** The viewing direction of the current viewpoint is aligned with a selected point to define the plane.

Figure 8-10.
The **MIRROR3D** command allows you to mirror objects about any plane regardless of the current UCS. A—The mirror plane defined by the three pick points is shown here in color. Point P1 is the midpoint of the top edge of the base. B—A copy of the original is mirrored.

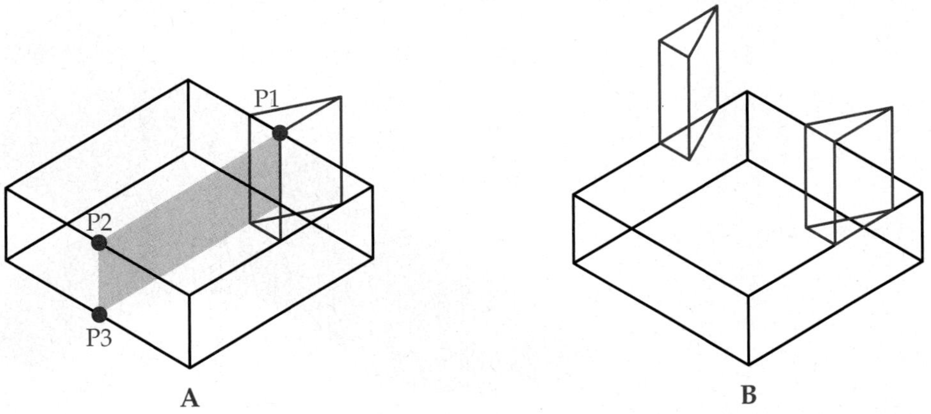

- **XY, YZ, ZX.** The mirror plane is placed parallel to one of the three basic planes of the current UCS and passes through a selected point.
- **3points.** Allows you to pick three points to define the mirror plane, as shown in **Figure 8-10**.

Exercise 8-4

www.g-wlearning.com/CAD/

Complete the exercise on the companion website.

Creating 3D Arrays

An ***array*** is an arrangement of objects in a 2D or 3D pattern. An array can be created as a rectangular, polar, or path array. You probably used arrays to complete some of the problems in previous chapters. A 2D array is created on the XY plane of the current UCS. A 3D array is an arrangement of objects in 3D space. The **ARRAYRECT**, **ARRAYPOLAR**, and **ARRAYPATH** commands can be used to create both 2D and 3D arrays. These commands provide the same functions as the **Rectangular**, **Polar**, and **Path** options of the **ARRAY** command and are available in the drop-down menu located in the **Modify** panel of the **Home** tab on the ribbon. See **Figure 8-11**.

An array can be created as an associative or non-associative array. Creating an associative array creates an ***array object,*** which can be modified as a single entity. For example, you can edit the source object to change all of the items in the array at once. You can also perform other modifications, such as deleting one or more items in the array, while maintaining the associativity of the arrayed items.

When creating a 3D array, the information you specify depends on the type of array being created. Many of the options are similar to those used when creating a 2D array. The following sections discuss 3D rectangular, polar, and path arrays.

NOTE

The legacy **3DARRAY** command can also be used to create 3D arrays, but it is limited to creating rectangular and polar arrays and cannot create associative arrays. Using the **ARRAYRECT**, **ARRAYPOLAR**, or **ARRAYPATH** command is the preferred method to create a 3D array.

3D Rectangular Arrays

In a ***3D rectangular array,*** as with a 2D rectangular array, you must enter the number of rows and columns. However, you must also specify the number of ***levels,*** which represents the third (Z) dimension. The command sequence is similar to that used when creating a 2D array.

Figure 8-11.
The **ARRAYRECT**, **ARRAYPOLAR**, and **ARRAYPATH** commands can be accessed from the drop-down menu located in the **Modify** panel of the **Home** tab on the ribbon.

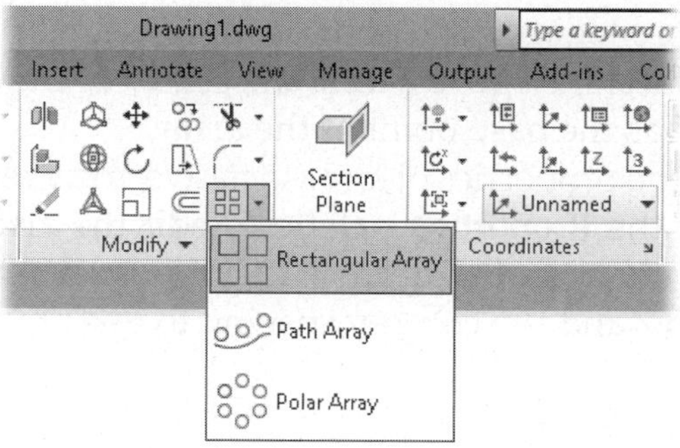

An example of where a 3D rectangular array may be created is the layout of structural columns on multiple floors of a commercial building. In **Figure 8-12A**, you can see two concrete floor slabs of a building and a single structural column. It is now a simple matter of arraying the column in rows, columns, and levels.

ARRAYRECT

Ribbon
Home
> Modify
Rectangular Array
Type
ARRAYRECT

To draw a 3D rectangular array, select the **ARRAYRECT** command. Select the object to array and press [Enter]. An initial pattern of three rows, four columns, and one level forms. You can select one of the array grips and drag to increase or decrease the number of rows, number of columns, or spacing dynamically. See **Figure 8-12B**. Without exiting the command, you can then adjust the row, column, level, and spacing values by using the **Array Creation** ribbon tab. See **Figure 8-12C**. In **Figure 8-12D**, there are three rows, five columns, and two levels. To set the number of levels and spacing between levels, use the **Array Creation** ribbon tab, right-click and select **Levels** from the shortcut menu, or use the prompts on the command line. Note that for the following command sequence, the drawing units have been set to architectural.

Select grip to edit array or [ASsociative/Base point/COUnt/Spacing/COLumns/Rows/Levels/eXit] <eXit>: **R**↵
Enter the number of rows or [Expression] <*current*>: **3**↵
Specify the distance between rows or [Total/Expression] <*current*>: **10'**↵
Specify the incrementing elevation between rows or [Expression] <0">: ↵
Select grip to edit array or [ASsociative/Base point/COUnt/Spacing/COLumns/Rows/Levels/eXit] <eXit>: **COL**↵
Enter the number of columns or [Expression] <*current*>: **5**↵
Specify the distance between columns or [Total/Expression] <*current*>: **10'**↵
Select grip to edit array or [ASsociative/Base point/COUnt/Spacing/COLumns/Rows/Levels/eXit] <eXit>: **L**↵
Enter the number of levels or [Expression] <1>: **2**↵
Specify the distance between levels or [Total/Expression] <*current*>: **12'8"**↵
Select grip to edit array or [ASsociative/Base point/COUnt/Spacing/COLumns/Rows/Levels/eXit] <eXit>: ↵

The result is shown in **Figure 8-12D**. By default, an associative array is created. You can create a non-associative array by selecting the **Associative** option. The **Associative** option setting is maintained by AutoCAD the next time an array command is accessed. You can verify the current setting of this option after entering the command and selecting objects. The Associative = Yes or Associative = No prompt appears. If the Associative = No prompt appears and you want to create an associative array, select the **Associative** option.

As shown in the previous sequence, when setting the distance between rows, you can also define an elevation increment between rows when the Specify the incrementing elevation between rows or [Expression] <0">: prompt appears. The elevation increment is different from the distance between levels. The elevation increment sets the spacing between rows along the Z axis so that each successive row is drawn on a higher or lower plane. This option can also be used when creating a 3D polar array and is discussed in the next section.

The **Base point** option is used to define the base point for the array. By default, the base point defined by AutoCAD is the centroid of the object(s) selected. You may want to select a more logical base point, such as an endpoint of an edge or the center point of a circular face. In **Figure 8-12B**, the center of the cylindrical base of the column has been selected as the base point of the array.

If any of the properties of an associative array require changes, use the **ARRAYEDIT** command or the **Properties** palette to edit the array. The arrayed items will update based on the changes. In **Figure 8-12E**, the array is shown after changing the number of columns to 7 and the column spacing to 6′-8″.

Figure 8-12.
A—Two floors and one structural column are drawn. Creating a 3D rectangular array will place all of the required columns on both floors at the same time. B—After entering the **ARRAYRECT** command, you can pick a grip and drag to set the number of rows or columns or the spacing between rows or columns dynamically. The floor objects are removed from the view for illustration purposes only. C—The **Array Creation** ribbon tab can be used to set values for rows, columns, and levels. D—An associative 3D rectangular array made up of three rows and five columns on two levels. E—An arrangement of three rows and seven columns after editing the properties of the associative array.

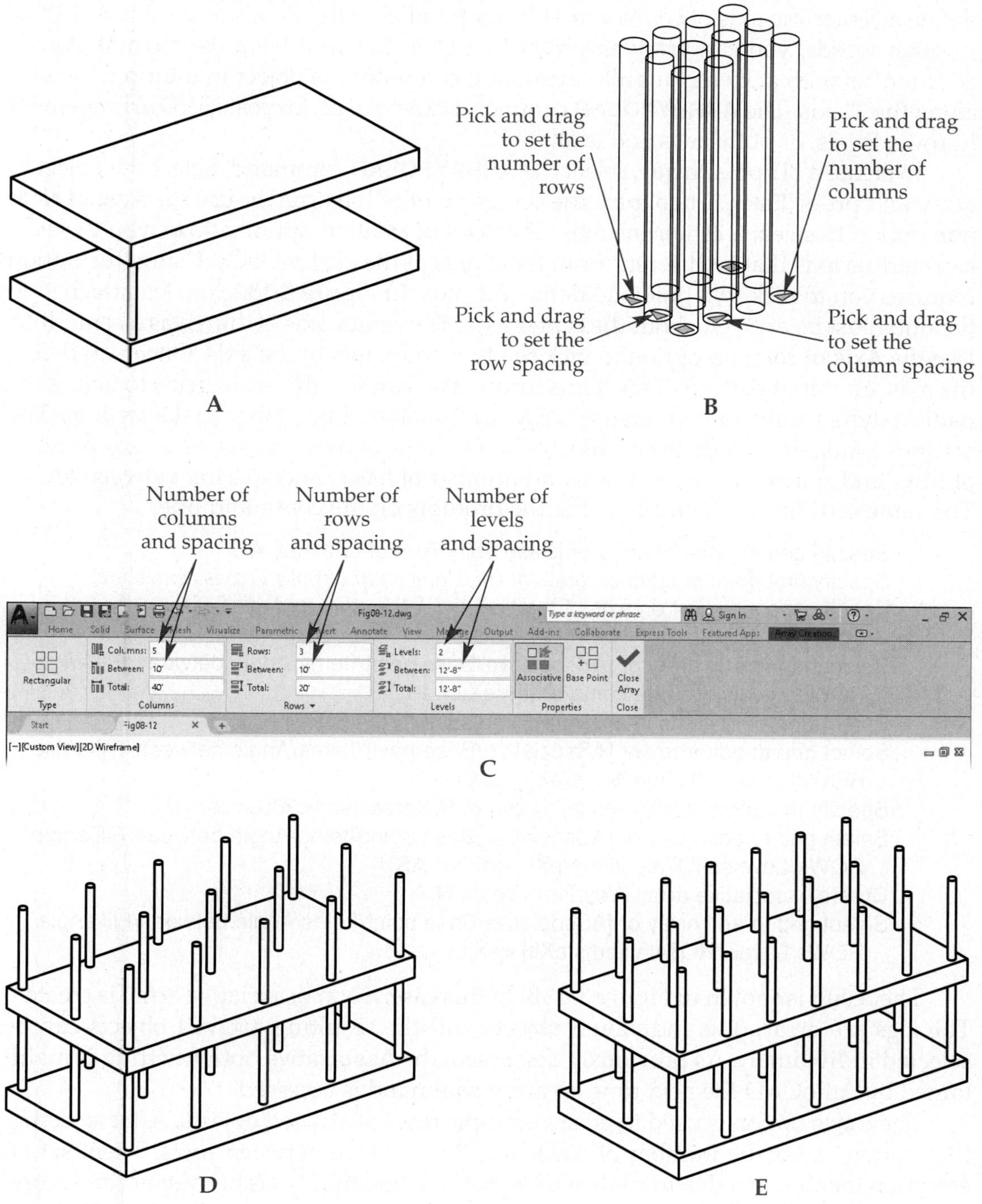

Exercise 8-5

www.g-wlearning.com/CAD/

Complete the exercise on the companion website.

3D Polar Arrays

A ***3D polar array*** is similar to a 2D polar array. However, the axis of rotation in a 2D polar array is parallel to the Z axis of the current UCS. In a 3D polar array, you can define a centerline axis of rotation that is not parallel to the Z axis of the current UCS. In other words, you can array an object in a UCS different from the current one. In addition, as with a 3D rectangular array, you can array the object in multiple "levels" along the Z axis. The **ARRAYPOLAR** command can be used to create a 3D arrangement in rows, levels, or both rows and levels.

ARRAYPOLAR

Ribbon
Home
> Modify
Polar Array
Type
ARRAYPOLAR

To create a 3D polar array, select the **ARRAYPOLAR** command. Select the object to array and press [Enter]. Then, pick the center point of the array or use the **Axis of rotation** option to select a centerline axis. The **Axis of rotation** option allows you to select a centerline axis that is different from the Z axis of the current UCS. Using this option requires you to pick two points to define the axis. In Figure 8-13A, the leg attached to the hub must be arrayed about the center axis. The center axis is drawn as a centerline. Use the **Axis of rotation** option to pick the two endpoints of the axis. Once you define the axis, an initial pattern of six items forms. You can use the array grips to adjust the pattern dynamically or you can use the **Array Creation** ribbon tab to make settings. The settings available include the number of items, angle between items, fill angle, number of rows and spacing between rows, and number of levels and spacing between levels. The same settings can be made using the prompts on the command line.

Specify center point of array or [Base point/Axis of rotation]: **A**↲
Specify first point on axis of rotation: *(pick one endpoint of the axis centerline)*
Specify second point on axis of rotation: *(pick the other endpoint of the axis centerline)*
Select grip to edit array or [ASsociative/Base point/Items/Angle between/Fill angle/ROWs/Levels/ROTate items/eXit] <eXit>: **I**↲
Enter number of items in array or [Expression] <6>: **6**↲
Select grip to edit array or [ASsociative/Base point/Items/Angle between/Fill angle/ROWs/Levels/ROTate items/eXit] <eXit>: **F**↲
Specify the angle to fill (+=ccw, −=cw) or [EXpression] <360>: ↲
Select grip to edit array or [ASsociative/Base point/Items/Angle between/Fill angle/ROWs/Levels/ROTate items/eXit] <eXit>: **AS**↲
Create associative array [Yes/No] <Yes>: **N**↲
Select grip to edit array or [ASsociative/Base point/Items/Angle between/Fill angle/ROWs/Levels/ROTate items/eXit] <eXit>: ↲

The result is shown in Figure 8-13B. In this case, a non-associative array is created. The legs are arrayed as individual objects and the resulting arrayed objects can be selected individually. As previously discussed, the **Associative** option setting is maintained by AutoCAD the next time an array command is accessed.

The **Rows** option is used to create multiple rows of arrayed objects. After selecting this option, enter the number of rows and the distance between rows. Then, set an elevation increment value to control the spacing along the Z axis between each successive row. An array of seats in a theater can be created in this manner. Figure 8-14 shows an example of arraying a single seat to create multiple rows of seats. To create this array, select the **ARRAYPOLAR** command, select the first seat, and use the following command sequence. If needed, use the **Associative** option to create an associative array. For this example, the default **Center point** option is used to set the center point of the array. Using this option is sufficient because the Z axis of the current UCS is

Figure 8-13.
A— Six new legs need to be arrayed about the center axis of the existing part. B—The model after using the **ARRAYPOLAR** command.

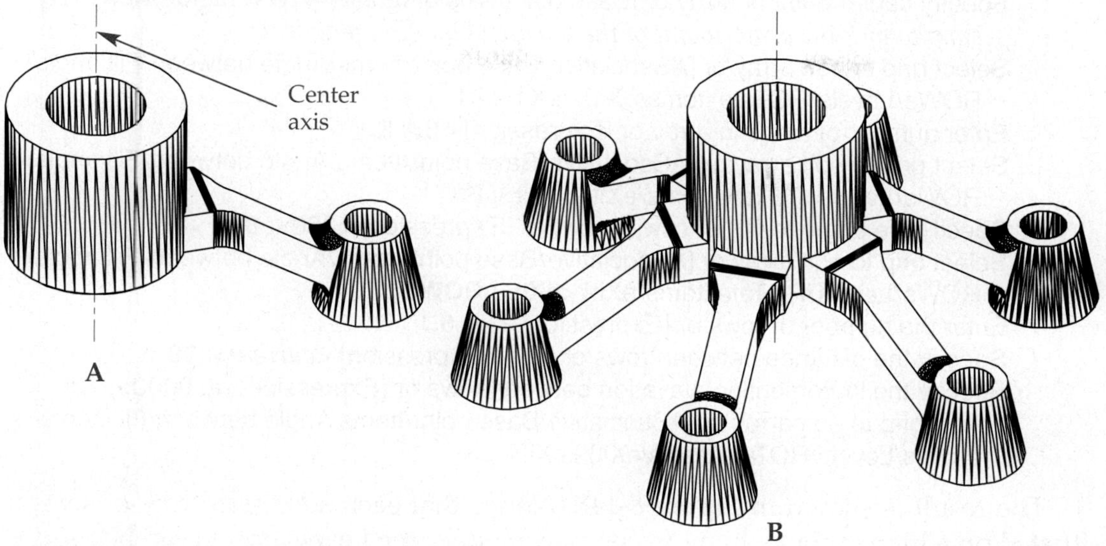

Figure 8-14.
Using the **Rows** option of the **ARRAYPOLAR** command to create a polar array of seats arranged in multiple rows. A—The first seat is modeled on the bottom platform. B—The result after using the **ARRAYPOLAR** command.

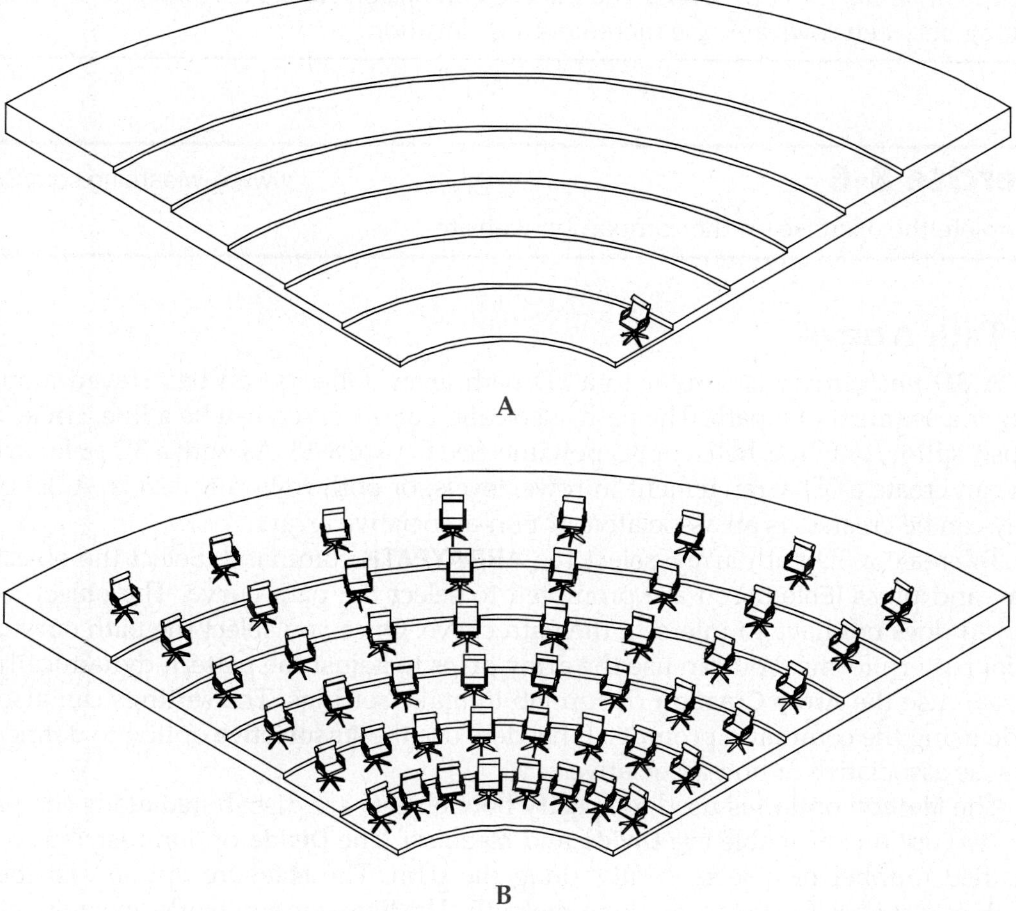

parallel to the required axis of rotation. Once you define the center axis, you can adjust the array using grips, the **Array Creation** ribbon tab, or the command line prompts.

> Specify center point of array or [Base point/Axis of rotation]: *(Use the* **Center** *object snap to pick the center point of the top arc of the first platform)*
> Select grip to edit array or [ASsociative/Base point/Items/Angle between/Fill angle/ROWs/Levels/ROTate items/eXit] <eXit>: **I**↵
> Enter number of items in array or [Expression] <6>: **8**↵
> Select grip to edit array or [ASsociative/Base point/Items/Angle between/Fill angle/ROWs/Levels/ROTate items/eXit] <eXit>: **F**↵
> Specify the angle to fill (+=ccw, −=cw) or [EXpression] <360>: **80**↵
> Select grip to edit array or [ASsociative/Base point/Items/Angle between/Fill angle/ROWs/Levels/ROTate items/eXit] <eXit>: **ROWS**↵
> Enter the number of rows or [Expression] <1>: **6**↵
> Specify the distance between rows or [Total/Expression] <*current*>: **96**↵
> Specify the incrementing elevation between rows or [Expression] <0.0000>: **7**↵
> Select grip to edit array or [ASsociative/Base point/Items/Angle between/Fill angle/ROWs/Levels/ROTate items/eXit] <eXit>: ↵

The result is shown in **Figure 8-14B**. Notice that each successive row of seats is situated on a higher plane. In more complex models, the **Levels** option can be used to create multiple levels of rows.

NOTE

The **Expression** option allows you to enter a mathematical expression to calculate one of the array parameters. For example, you can enter an expression to calculate the number of items, rows, or levels. The **Expression** option can also be used to define the spacing between rows and the incrementing elevation.

Exercise 8-6

www.g-wlearning.com/CAD/

Complete the exercise on the companion website.

3D Path Arrays

A ***3D path array*** is similar to a 2D path array. Objects can be arrayed along a path or a segment of a path. The path, also called a *path curve*, can be a line, circle, arc, ellipse, spline, polyline, helix, or 3D polyline. See **Figure 8-15**. As with a 3D polar array, you can create a 3D arrangement in rows, levels, or both rows and levels. A 3D path array can be created as an associative or non-associative array.

ARRAYPATH
Ribbon
Home
> Modify
Path Array
Type
ARRAYPATH

To create a 3D path array, select the **ARRAYPATH** command. Select the object to array and press [Enter]. You are prompted to select the path curve. The object to be arrayed does not have to intersect the path curve. Once you select the path curve, an initial pattern forms. You can use the array grips to adjust the pattern dynamically or you can use the **Array Creation** ribbon tab to make settings. The settings can also be made using the command prompts. If needed, use the **Associative** option to define the array as associative or non-associative.

The **Method** option is used to specify how objects are distributed along the path. The two options available are **Divide** and **Measure**. The **Divide** option distributes the specified number of objects evenly along the path. The **Measure** option distributes the objects at specific distances along the path. The **Items** option works in conjunction with the **Method** option and allows you to specify the number of objects. When using the **Measure** option, you can specify the spacing between objects or the total distance

Figure 8-15.
Examples of 3D path arrays used to array posts.

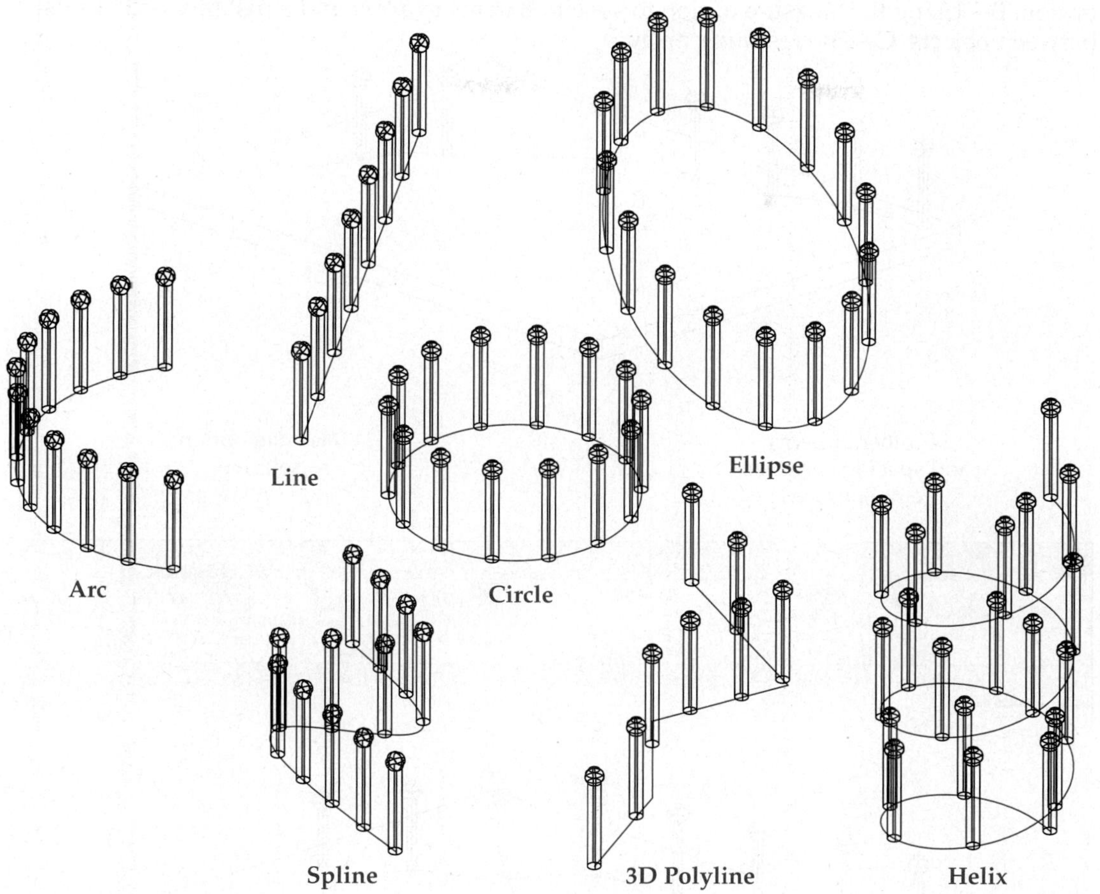

between the first and last objects. If the path curve changes in length, the number of objects will increase or decrease automatically.

The **Base point** option is used to set the array's base point. The default base point of the array is the endpoint of the path curve closest to where you select it. This point serves as the start point of the array. Depending on the result you want, you can select a different base point (start point), such as a point on the object.

The default orientation of the array is the current orientation of the object. The **Tangent direction** option allows you to pick two points to define a different orientation. The **Normal** option is used to align the object "normal" to the path. Using this option aligns the Z axis of the object perpendicular to the path.

After specifying the base point and orientation of the array or using the defaults, continue as follows. When using the **Divide** option, specify the number of items. When using the **Measure** option, specify the distance between each object or the total distance between the first and last objects. Distances can be set using the **Array Creation** ribbon tab. In **Figure 8-16A**, the **Divide** option is used:

Select grip to edit array or [ASsociative/Method/Base point/Tangent direction/Items/
Rows/Levels/Align items/Z direction/eXit] <eXit>: **M**↵
Enter path method [Divide/Measure] <*current*>: **D**↵
Select grip to edit array or [ASsociative/Method/Base point/Tangent direction/Items/
Rows/Levels/Align items/Z direction/eXit] <eXit>: **I**↵
Enter number of items along path or [Expression] <*current*>: **10**↵
Select grip to edit array or [ASsociative/Method/Base point/Tangent direction/Items/
Rows/Levels/Align items/Z direction/eXit] <eXit>: ↵

Figure 8-16.
Using the **ARRAYPATH** command. A—Specifying 10 items to array and using the **Divide** option. B—Using the **Measure** option to specify 8 items to array and a distance of 25 units between objects. C—The resulting array.

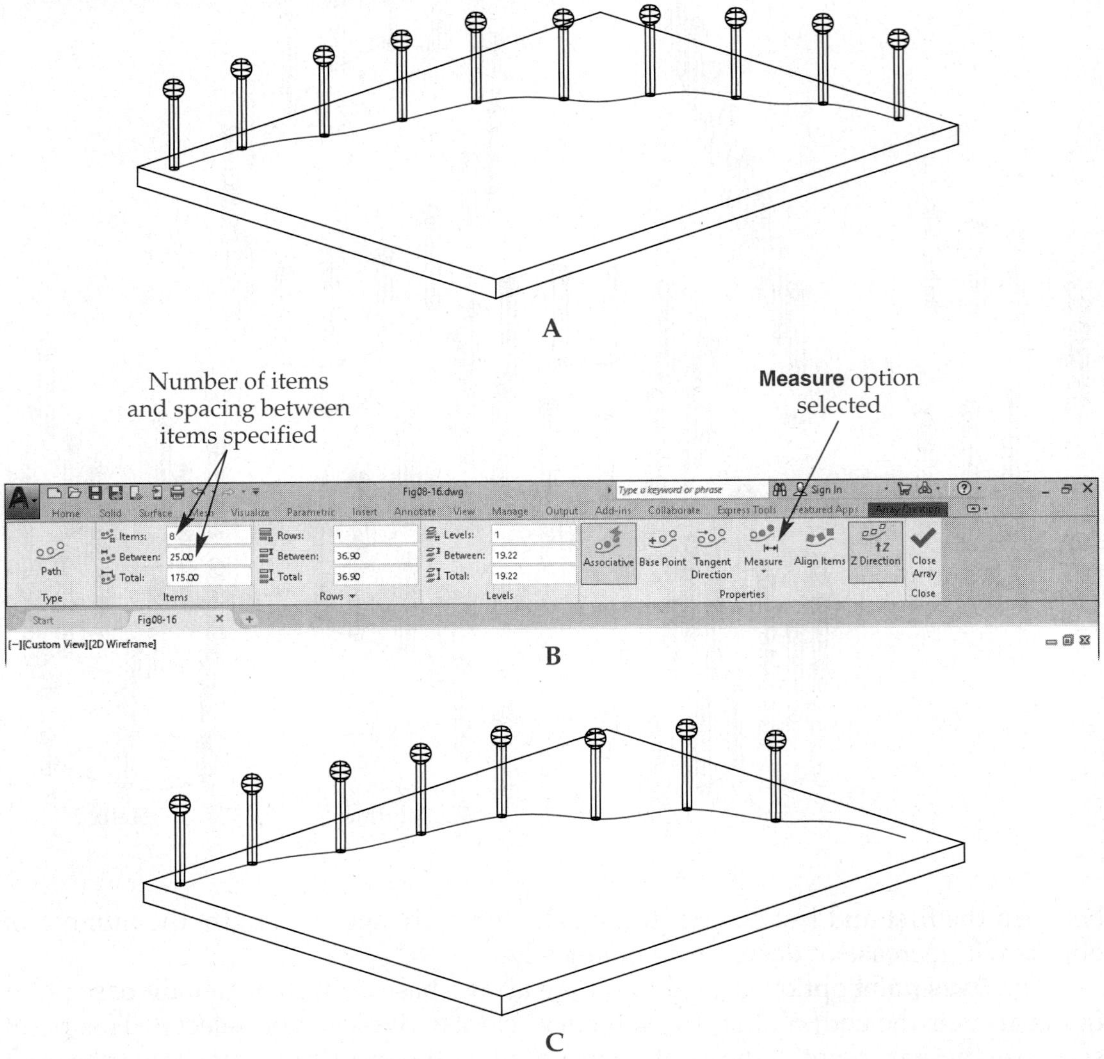

In **Figure 8-16B**, the **Array Creation** ribbon tab is shown after selecting the **Measure** option. The number of items is set to 8 and the distance between items is set to 25 using the settings in the **Items** panel. The result is shown in **Figure 8-16C**.

The **Align items** option is used to align the arrayed objects tangent to the direction of the array path. The **Z direction** option is used to change the Z axis direction of the arrayed objects. By default, the Z axis of each object is aligned in the same orientation used by the original object. If the **Z direction** option is set to **No**, the Z axis of the object changes direction to follow the path as the path changes direction.

The **Rows** and **Levels** options allow you to create an arrangement in rows and levels. The options are similar to those used with the **ARRAYPOLAR** command. In **Figure 8-17**, an oval table is created using the **Levels** option. The table feet and middle supports are created as a 3D path array. One of the feet is arrayed along an elliptical path to create an array consisting of two levels. In this example, a non-associative array is created. This allows the middle supports to be edited as individual objects after creating the array. The middle supports have a smaller diameter than the feet and a longer length. To complete the table, the bottom shelf is copied along the Z axis to create the top.

In **Figure 8-17A**, the bottom shelf has been created as an extruded solid from an ellipse. The shelf is .375″ thick and is centered on the UCS origin. The underside of the shelf rests on the XY plane. A second ellipse is drawn on the XY plane and rests on the underside of the shelf. This ellipse serves as the path for the 3D path array. A construction cylinder has also been created with its base resting on the XY plane. The construction cylinder is 2″ in diameter and 2″ in height, the dimensions of the table feet. The construction cylinder is the item to be arrayed. During the operation, the base point of the array is specified as the top center point of the construction cylinder. This aligns the top of the cylinder with the start of the path. Since the path is an ellipse, AutoCAD defines the start of the path as one of the quadrant points on the major axis. When the base point is selected, the cylinder is then aligned with the quadrant point. This is a sufficient start point for the array.

After selecting the **ARRAYPATH** command, the construction cylinder is selected as the item to array and the construction ellipse is selected as the path. The command sequence continues as follows.

Select grip to edit array or [ASsociative/Method/Base point/Tangent direction/Items/Rows/Levels/Align items/Z direction/eXit] <eXit>: **AS↵**
Create associative array [Yes/No] <Yes>: **N↵**
Select grip to edit array or [ASsociative/Method/Base point/Tangent direction/Items/Rows/Levels/Align items/Z direction/eXit] <eXit>: **M↵**
Enter path method [Divide/Measure] <Measure>: **D↵**
Select grip to edit array or [ASsociative/Method/Base point/Tangent direction/Items/Rows/Levels/Align items/Z direction/eXit] <eXit>: **I↵**
Enter number of items along path or [Expression] <*current*>: **4↵**
Select grip to edit array or [ASsociative/Method/Base point/Tangent direction/Items/Rows/Levels/Align items/Z direction/eXit] <eXit>: **B↵**
Specify base point or [Key point] <end of path curve>: *(Use the* **Center** *object snap to pick the center of the top of the construction cylinder)*
Select grip to edit array or [ASsociative/Method/Base point/Tangent direction/Items/Rows/Levels/Align items/Z direction/eXit] <eXit>: **L↵**
Enter the number of levels or [Expression] <1>: **2↵**
Specify the distance between levels or [Total/Expression] <*current*>: **2-3/8"↵**
Select grip to edit array or [ASsociative/Method/Base point/Tangent direction/Items/Rows/Levels/Align items/Z direction/eXit] <eXit>: ↵

The resulting array is shown in **Figure 8-17B**. The four cylinders on each level of the array are evenly divided along the elliptical path. Because a non-associative array was created, the cylinders can be selected and edited as independent objects. Notice that the four cylinders in the second level of the array rest on the top of the shelf at the correct position along the Z axis. In **Figure 8-17C**, the four cylinders in the second level of the array have been changed to 1″ in diameter and 14″ in height using the **Properties** palette. The edited cylinders form the table supports. A shaded version of the completed table is shown in **Figure 8-17D** after copying the shelf along the Z axis to create the top.

NOTE

Grips are available for editing an associative array after the array is created. Hover over a grip or pick on a grip to edit the array. If you pick a grip and more than one option is available, use the [Ctrl] key to cycle through the options.

Figure 8-17.
Using the **ARRAYPATH** command to create the cylindrical feet and supports for an oval table. A—The bottom shelf is created as an extruded ellipse. A construction ellipse is created to serve as the path for the 3D path array. A construction cylinder is the object to be arrayed. B—A 3D path array consisting of two levels with four items on each level is created. The array is created as a non-associative array. C—The supports are edited to the correct diameter and height. D—The bottom shelf is copied along the Z axis to create the top. Shown is a shaded version of the completed model.

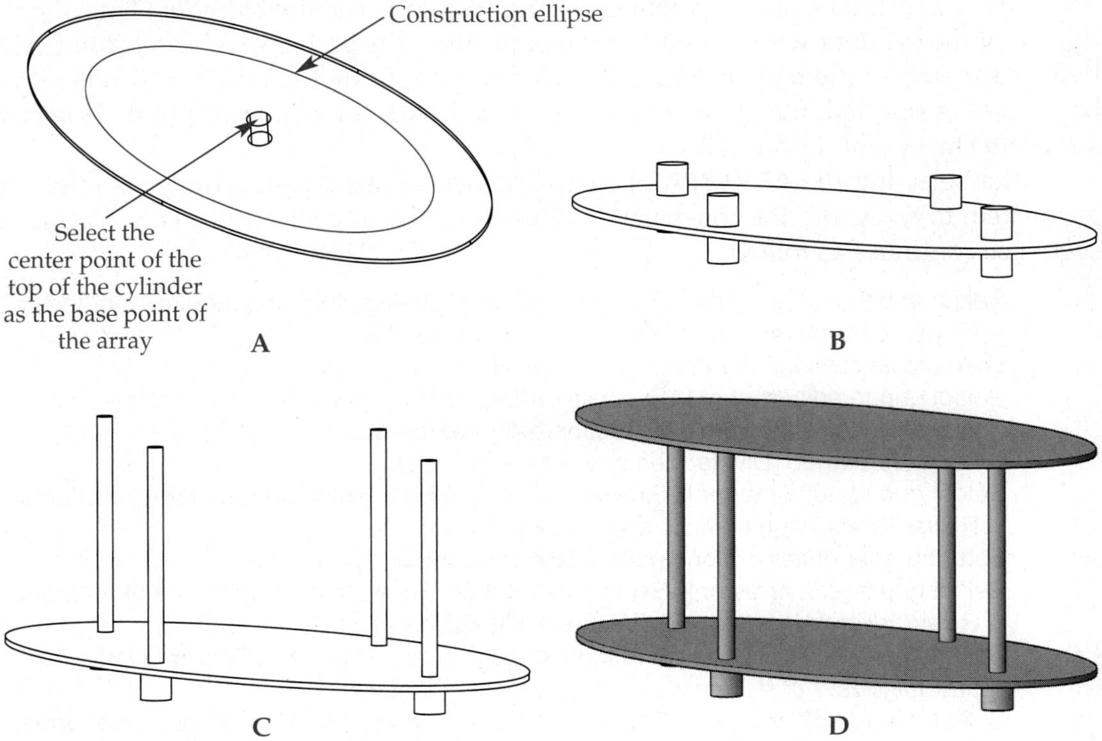

PROFESSIONAL TIP

If you edit the path shape used for an associative 3D path array, the objects follow the new edited path.

Exercise 8-7

www.g-wlearning.com/CAD/

Complete the exercise on the companion website.

Filleting Solid Objects

FILLETEDGE
Ribbon
Solid
> Solid Editing
Fillet Edge
Type
FILLETEDGE

A ***fillet*** is a rounded interior edge on an object. A ***round*** is a rounded exterior edge. See **Figure 8-18**. The **FILLETEDGE** command is used to create fillets and rounds on 3D objects. Additionally, the **FILLET** command can be used to fillet 3D objects.

Before a fillet or round is created at an intersection, the solid objects that intersect need to be joined using the **UNION** command. Then, use the **FILLETEDGE** command. See **Figure 8-19**. Since the object being filleted is actually a single solid and not two objects, only one edge is selected. In the following sequence, the fillet radius is first set at .25, then the fillet is created.

Figure 8-18.
Examples of fillets and rounds. The wireframe displays show the objects before the **FILLET** command is used.

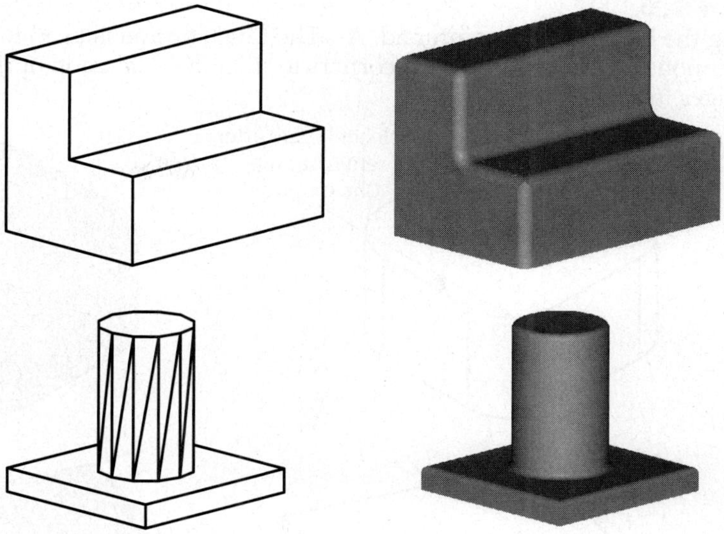

Radius = 1.0000
Select an edge or [Chain/Loop/Radius]: **R.**↵
Enter fillet radius or [Expression] <1.0000>: **.25**↵
Select an edge or [Chain/Loop/Radius]: *(pick the edge to be filleted or rounded)*
Select an edge or [Chain/Loop/Radius]: ↵
Press Enter to accept the fillet or [Radius]: ↵ *(this fillets the selected edge, but you can enter the* **Radius** *option or use the linear stretch grip that appears in the preview to adjust the fillet radius at this point)*

After selecting the first edge, you can continue to select edges to fillet or enter the **Chain**, **Loop**, or **Radius** option. The **Chain** option is used to select a chain of continuous edges that have rounded corners, **Figure 8-20A**. The **Loop** option is similar to the **Chain** option and is used to select a loop of edges, **Figure 8-20B**. When using the **Loop** option, the **Next** option can be used to select the adjacent loop of edges. After using the **Chain** or **Loop** option, you can select individual edges by entering the **Edge** option. Once all edges are selected, press [Enter]. Use the preview to confirm that the current radius is acceptable and press [Enter] to complete the command.

The **FILLET** command works in a similar manner. Once the command is entered, specify the fillet radius and then select the edges to fillet. You can continue to select edges or enter the **Chain**, **Loop**, or **Radius** option. The primary advantages of the **FILLETEDGE** command compared to the **FILLET** command are 1) a preview is shown and 2) a linear stretch grip is associated with the fillet. The linear stretch grip allows for subobject editing, which is discussed in Chapter 11.

Ribbon
Home
> Modify
Fillet
Type
FILLET
F

FILLET

Figure 8-19.
A—Pick the edge where two unioned solids intersect to create a fillet. B—The fillet after rendering.

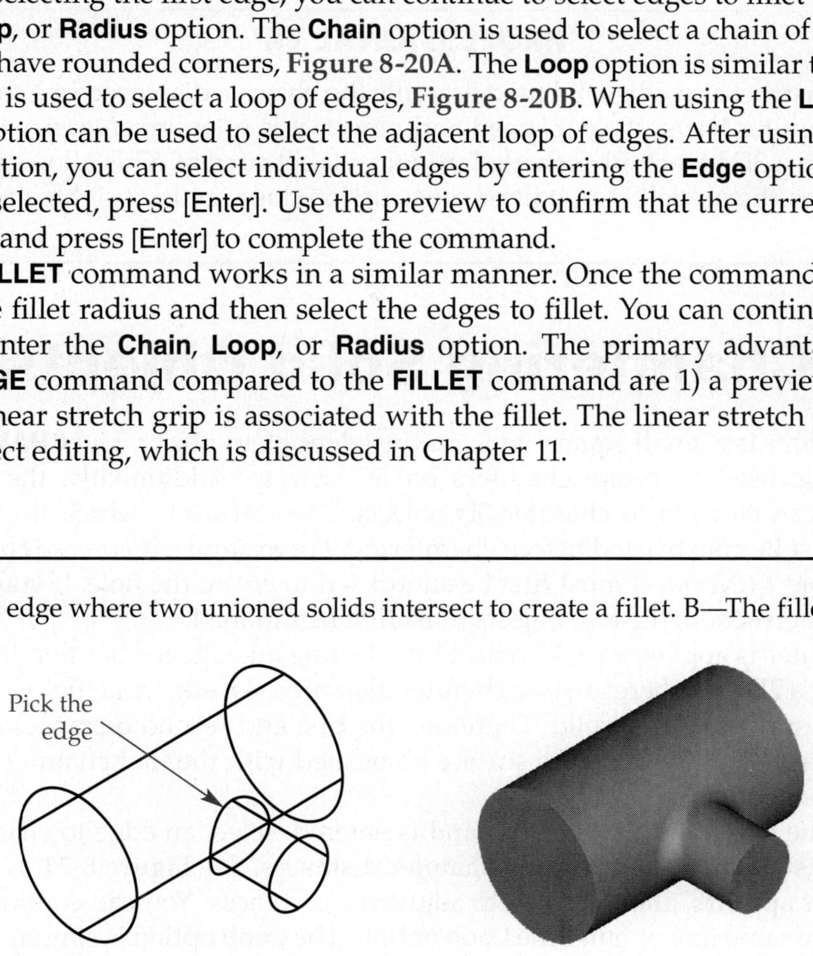

Figure 8-20.
Using the **FILLETEDGE** command. A—The **Chain** option is used to select a chain of continuous edges with rounded corners to fillet. B—The **Loop** option is used to select a loop of edges to fillet.

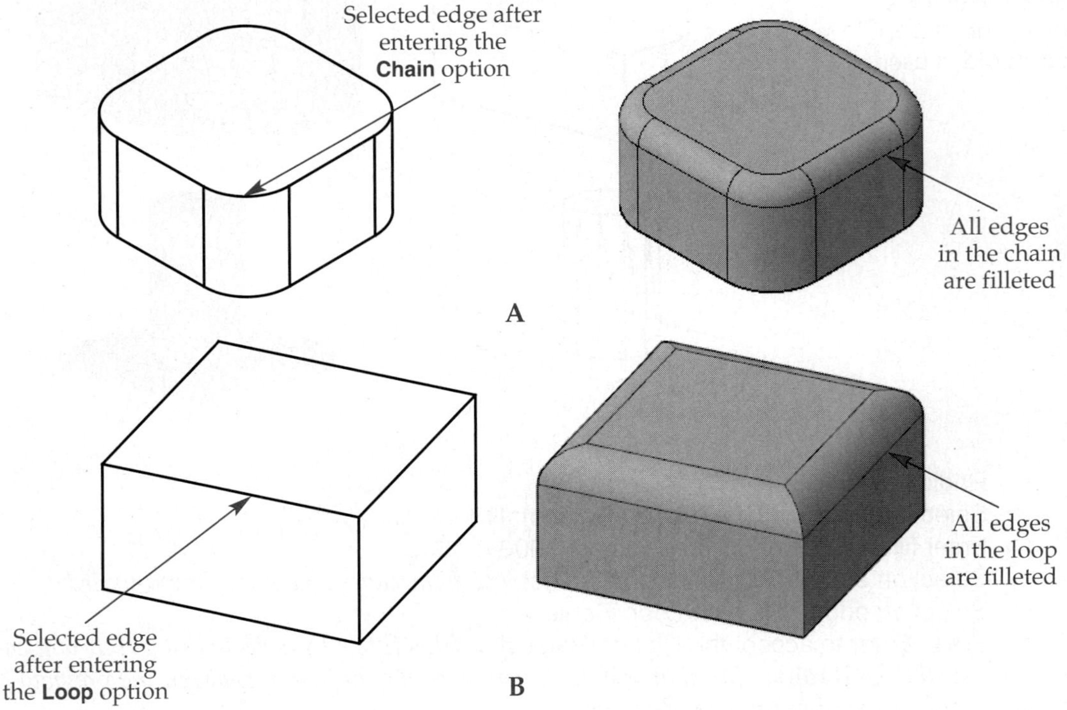

PROFESSIONAL TIP

You can construct and edit solid models while the object is displayed in a shaded view. If your computer has sufficient speed and power, it is often much easier to visualize the model in a 3D view with a shaded visual style set current. This allows you to realistically view the model. If an edit or construction does not look right, just undo and try again.

Chamfering Solid Objects

A ***chamfer*** is a small square edge on the edges of an object. The **CHAMFEREDGE** command is used to create chamfers on 3D objects. Additionally, the **CHAMFER** command can be used to chamfer 3D objects. The feature to which the chamfer is applied must be constructed before chamfering. For example, if you are chamfering a hole, the object (cylinder) must first be subtracted to create the hole. If you are chamfering an intersection, the two objects must first be unioned.

A chamfer is applied to a 3D object by selecting an edge to chamfer. Just as when chamfering a 2D line, there are two chamfer distances. An edge is actually the intersection of two surfaces of the solid. Therefore, the first and second distances correspond to two surfaces on the solid. The surface associated with the first chamfer distance is called the ***base surface***.

CHAMFEREDGE

Ribbon
Solid
> Solid Editing
Chamfer Edge
Type
CHAMFEREDGE

Once the **CHAMFEREDGE** command is entered, select an edge to chamfer. Then, enter the **Distance** option to set the chamfer distances. See **Figure 8-21**. A preview of the chamfer appears, allowing you to adjust the distances. You can continue to select edges on the same face or enter the **Loop** option. The **Loop** option is similar to the **Loop** option used with the **FILLETEDGE** command and allows you to select a loop of edges.

Figure 8-21.
Using the **CHAMFEREDGE** command. A—Select each edge to chamfer. B—The resulting model.

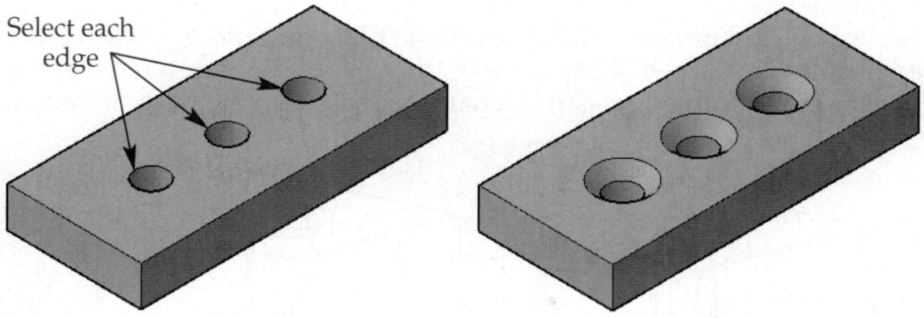

A B

Once all edges are selected, press [Enter]. A preview with linear stretch grips appears and you are prompted to accept the chamfer. If the current chamfer distances are acceptable, press [Enter]. If not, enter the **Distance** option or use the linear stretch grips to set new distance values.

The **CHAMFER** command works in a similar manner. After you enter the command, you must select the edge you want to chamfer. When you make the selection, one of the two surfaces corresponding to the edge is highlighted. The highlighted surface is the base surface and is associated with the first chamfer distance. If the highlighted surface is not the one you want as the base surface, enter N at the [Next/OK] prompt and press [Enter]. This highlights the next surface. An edge is created by two surfaces. Therefore, when you enter N for the next surface, AutoCAD cycles through only two surfaces. When the proper base surface is highlighted, press [Enter].

Ribbon
Home > Modify
Chamfer
Type
CHAMFER
CHA

CHAMFER

Using the **CHAMFER** command to chamfer a hole is shown in **Figure 8-22A**. In **Figure 8-22B**, the cylinder is unioned to the base in order to create the chamfer at the intersection. The ends of the cylinder in **Figure 8-22B** are chamfered by first picking a vertical isoline on the cylindrical face to define the base surface. Then, the top edge is selected, followed by the intersection edge. The following command sequence is illustrated in **Figure 8-22A**.

(TRIM mode) Current chamfer Dist1 = 1.0000, Dist2 = 1.0000
Select first line or [Undo/Polyline/Distance/Angle/Trim/mEthod/Multiple]: *(select a top edge)*
Base surface selection...
Enter surface selection option [Next/OK (current)] <OK>: *(select* **Next** *if the top surface is not selected or press* [Enter]*)*
Specify base surface chamfer distance or [Expression] <1.0000>: **.125**↵
Specify other surface chamfer distance or [Expression] <1.0000>: **.125**↵
Select an edge or [Loop]: *(select the hole diameter edge)*
Select an edge or [Loop]: ↵

NOTE

Editing a fillet or chamfer is discussed in Chapter 11. The **SOLIDEDIT** command discussed in Chapter 12 can also be used to edit a fillet or chamfer.

Exercise 8-8

www.g-wlearning.com/CAD/

Complete the exercise on the companion website.

Figure 8-22.
A—A hole is chamfered with the **CHAMFER** command by picking the top surface, then the edge of the hole. B—The top edge of the cylinder is chamfered by first picking the side, then the top edge. Both edges can be chamfered at the same time, as shown here.

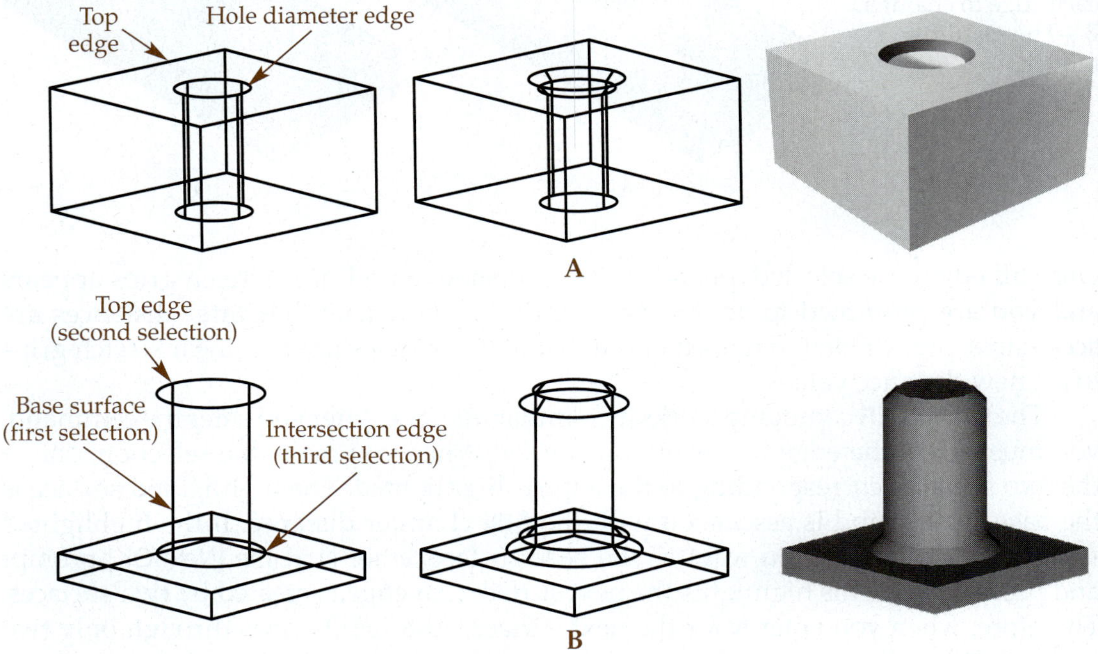

Converting to Solids

Additional flexibility in creating solids is provided by the **CONVTOSOLID** command. This command allows you to directly convert certain closed objects into solids. You can convert:

- Circles with thickness.
- Wide, uniform-width polylines with thickness. This includes polygons and rectangles.
- Closed, zero-width polylines with thickness. This includes polygons, rectangles, and closed revision clouds.
- Mesh primitives and other watertight mesh objects. Working with mesh objects is discussed in Chapter 9.
- Watertight surface models.

CONVTOSOLID

Ribbon
Home
> Solid Editing
Convert to Solid
Type
CONVTOSOLID

First, select the command. Then, select the objects to convert and press [Enter]. The objects are instantly converted with no additional input required. **Figure 8-23** shows two different objects before and after conversion to a solid.

If an object that appears to be a closed polyline with a thickness does not convert to a solid and AutoCAD displays the message Cannot convert an open curve, the polyline was not closed using the **Close** option of the **PLINE** command. Use the **PEDIT** or **PROPERTIES** command to close the polyline and use the **CONVTOSOLID** command again.

PROFESSIONAL TIP

To quickly create a straight section of pipe, draw a donut with the correct ID and OD of the pipe. Then, use the **Properties** palette to give the donut a thickness equal to the length of the section you are creating. Finally, use the **CONVTOSOLID** command to turn the donut into a solid.

Figure 8-23.
A—Two polylines with thickness that will be converted into solids. B—The resulting solids.

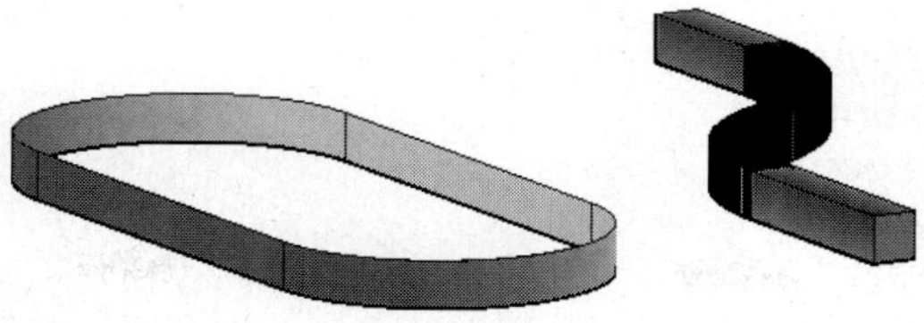

A

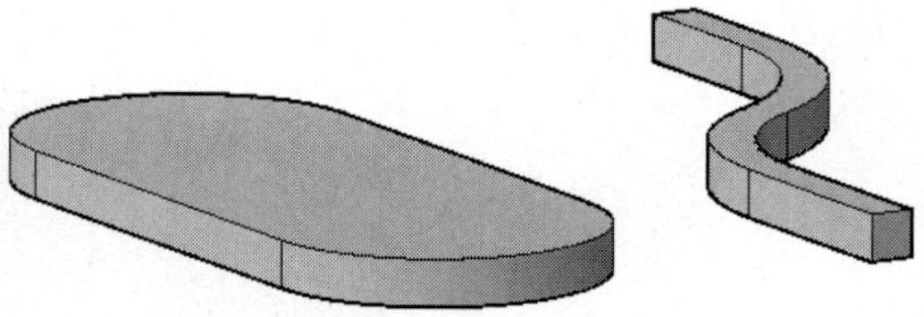

B

Slicing a Solid

A 3D solid can be sliced at any location by using existing objects such as circles, arcs, ellipses, 2D polylines, 2D splines, or surfaces. Why slice a solid? To create complex, angular, contoured, or organic shapes that traditionally cannot be created by just using solids. Good modeling techniques incorporate slicing of solids. For example, think of a computer mouse. This would be difficult to create with solids. But, you can create a basic solid shape, then slice the solid using contoured surfaces as slicing tools. After slicing the solid, you can choose to retain either or both sides of the model. The slices can then be used for model construction or display and presentation purposes.

SLICE

Ribbon
Solid
> Solid Editing
Slice
Type
SLICE

The **SLICE** command is used to slice solids. When the command is initiated, you are asked to select the solids to be sliced. Select the objects and press [Enter]. Next, you must define the slicing path. The default method of defining a path requires you to specify two points on a slicing plane. The plane passes through the two points and is perpendicular to the XY plane of the current UCS. Refer to **Figure 8-24** as you follow this sequence:

1. Select the command and pick the object to be sliced.
2. Pick the start point of the slicing plane. See **Figure 8-24A**.
3. Pick the second point on the slicing plane.
4. You are prompted to specify a point on the desired side to keep. Select anywhere on the back half of the object. The point does not have to be *on* the object. It must simply be on the side of the cutting plane that you want to keep.
5. The object is sliced and the front half is deleted. See **Figure 8-24B**.

When prompted to select the side to keep, you can press [Enter] to keep both sides. If both sides are retained, two separate 3D solids are created. Each solid can then be used for additional modeling purposes.

There are several additional options for specifying a slicing path. These options are listed here and described in the following sections.

- **Planar object**
- **Surface**
- **Zaxis**
- **View**
- **XY**
- **YZ**
- **ZX**
- **3points**

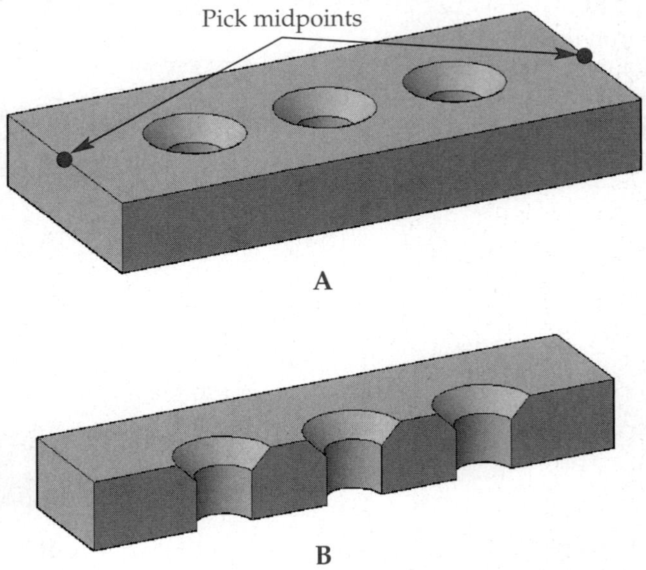

Figure 8-24.
Slicing a solid by picking two points. A—Select two points on the cutting plane. The plane passes through these points and is perpendicular to the XY plane of the current UCS. B—The sliced solid.

NOTE

Once the **SLICE** command has been used, the history of the solid to that point is removed. If a history of the work is important, then save a copy of the file or place a copy of the object on a frozen layer prior to performing the slice.

Planar Object

A second method to create a slice through a 3D solid is to use an existing planar object. Planar objects include circles, arcs, ellipses, 2D polylines, and 2D splines. See **Figure 8-25A**. The plane on which the planar object lies must intersect the object to be sliced. The current UCS has no effect on this option.

Be sure that the planar object has been moved to the location of the slice. Then, select the **SLICE** command, pick the object to slice, and press [Enter]. Next, enter the **Planar Object** option and select the slicing path object (the circle, in this case). Finally, specify which side is to be retained. See **Figure 8-25B**. Again, if both sides are kept, they are separate objects and can be individually manipulated.

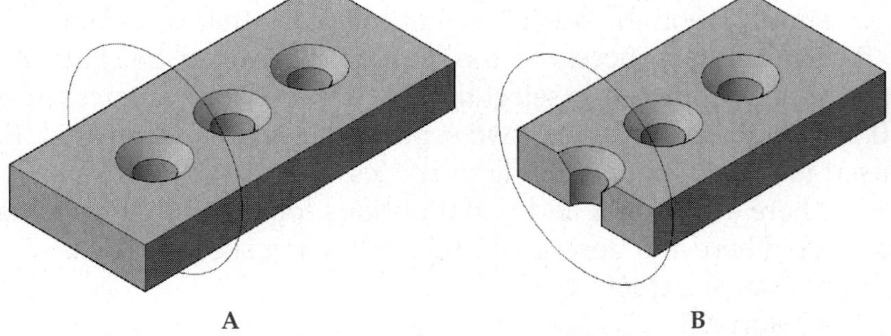

Figure 8-25.
Slicing a solid with a planar object. A—The circle is drawn at the proper orientation and in the correct location. B—The completed slice.

Surface

A surface object can be used as the slicing path. The surface can be planar or nonplanar (curved). Surface modeling is discussed in detail in Chapter 10. Using a surface as a slicing path is a technique that you can use to quickly create a mating die. For example, refer to **Figure 8-26**. First, draw the required surface. The surface should exactly match the stamped part that will be manufactured, **Figure 8-26A**. Then, draw a box that encompasses the surface. Next, select the **SLICE** command, pick the box, and press [Enter]. Then, enter the **Surface** option and select the surface. You may need to turn on selection cycling or work in a wireframe display. Finally, when prompted to select the side to keep, press [Enter] to keep both sides. The two halves of the die can now be moved and rotated as needed, **Figure 8-26B**.

PROFESSIONAL TIP

You can use extruded surfaces, revolved surfaces, lofted surfaces, and swept surfaces as slicing objects.

Exercise 8-9

www.g-wlearning.com/CAD/

Complete the exercise on the companion website.

Z Axis

You can specify one point on the cutting plane and one point on the Z axis of the plane. See **Figure 8-27**. This allows you to have a cutting plane that is not parallel to the current UCS XY plane. First, select the **SLICE** command, pick the object to slice, and press [Enter]. Next, enter the **Zaxis** option. Then, pick a point on the XY plane of the cutting plane followed by a point on the Z axis of the cutting plane. Finally, pick the side of the object to keep.

Figure 8-26.
Slicing a solid with a surface. A—Draw the surface and locate it within the solid to be sliced. The solid is represented here by the wireframe. B—The completed slice with both sides retained. The top can now be moved and rotated as shown here.

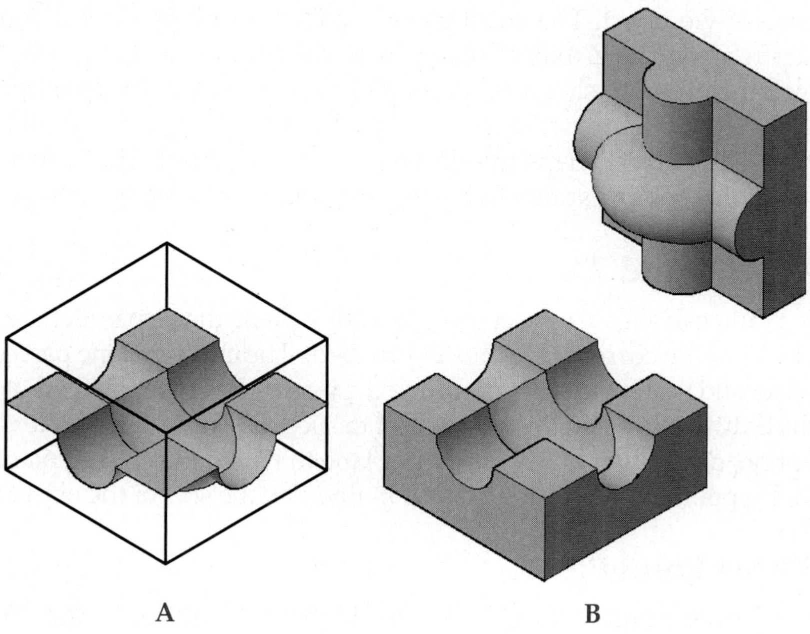

Figure 8-27.
Slicing a solid using the **Zaxis** option. A—Pick one point on the cutting plane and a second point on the Z axis of the cutting plane. B—The resulting slice.

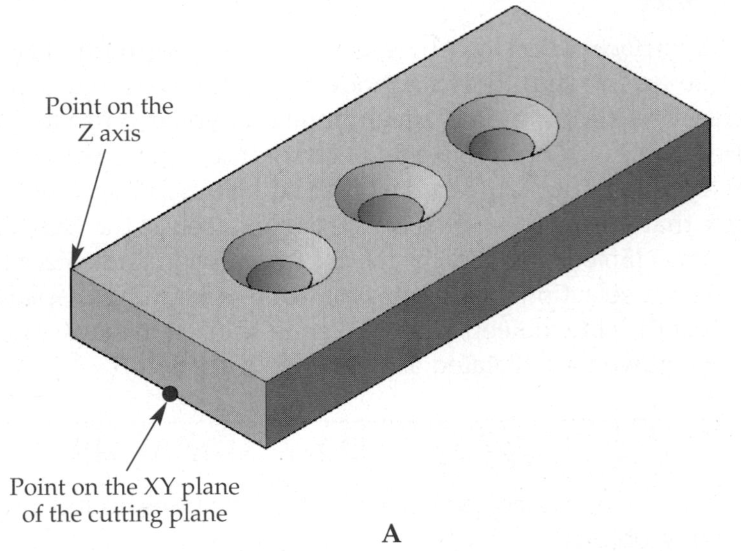

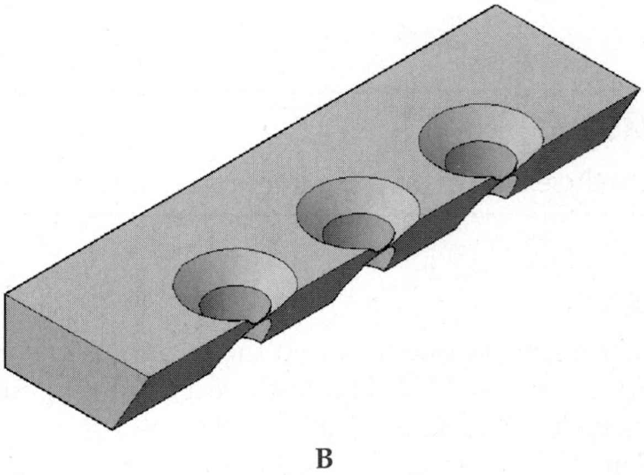

View

A cutting plane can be established that is aligned with the viewing plane of the current viewport. The cutting plane passes through a point you select, which sets the depth along the Z axis of the current viewing plane. First, select the **SLICE** command, pick the object to slice, and press [Enter]. Next, enter the **View** option. Then, pick a point in the viewport to define the location of the cutting plane on the Z axis of the viewing plane. Use object snaps to select a point on an object. The cutting plane passes through this point and is parallel to the viewing plane. Finally, pick the side of the object to keep.

XY, YZ, and ZX

You can slice an object using a cutting plane that is parallel to any of the three primary planes of the current UCS. See **Figure 8-28**. The cutting plane passes through the point you select and is aligned with the primary plane of the current UCS that you specify. First, select the **SLICE** command, pick the object to slice, and press [Enter]. Next, enter the **XY**, **YZ**, or **ZX** option, depending on the primary plane to which the cutting plane will be parallel. Then, pick a point on the cutting plane. Finally, pick the side of the object to keep.

Three Points

Three points can be used to define the cutting plane. This allows the cutting plane to be aligned at any angle, similar to using the **Zaxis** option. See **Figure 8-29**. First, select the **SLICE** command, pick the object to slice, and press [Enter]. Then, enter

Figure 8-28.
Slicing a solid using the **XY**, **YZ**, and **ZX** options. A—The object before slicing. The UCS origin is in the center of the first hole and at the midpoint of the height. B—The resulting slice using the **XY** option. C—The resulting slice using the **YZ** option. D—The resulting slice using the **ZX** option.

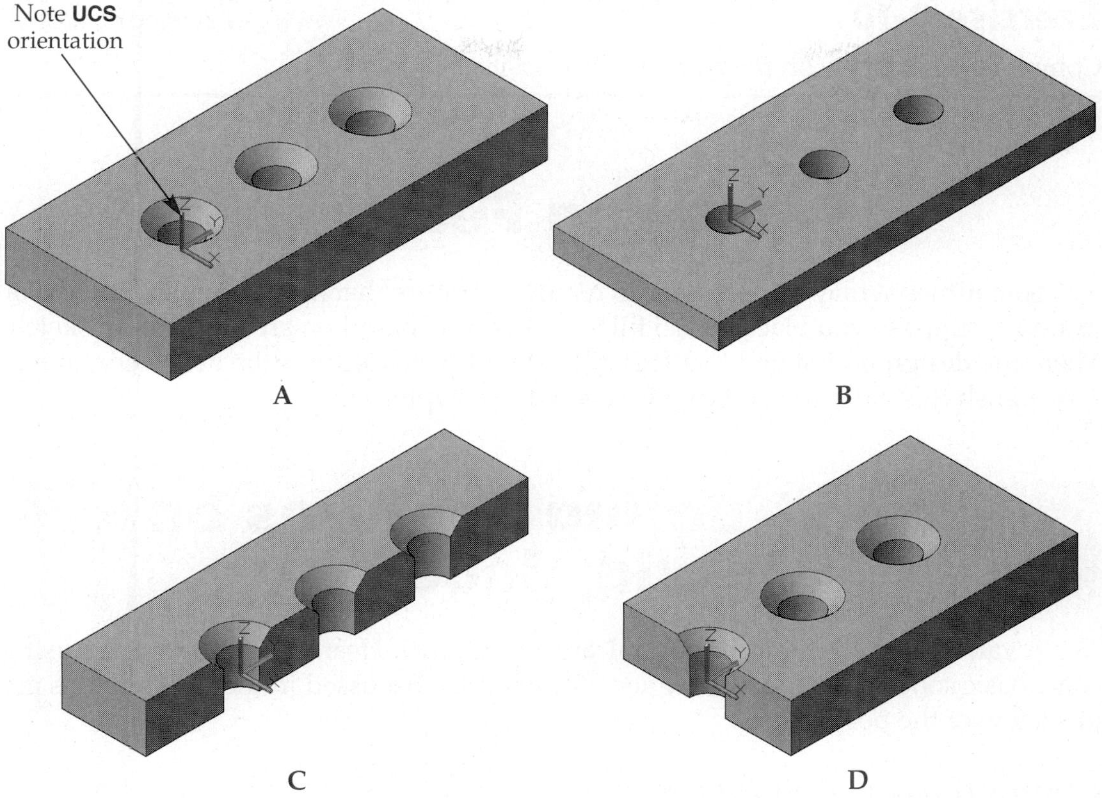

Figure 8-29.
Slicing a solid using the **3points** option. A—Specify three points to define the cutting plane. B—The resulting slice.

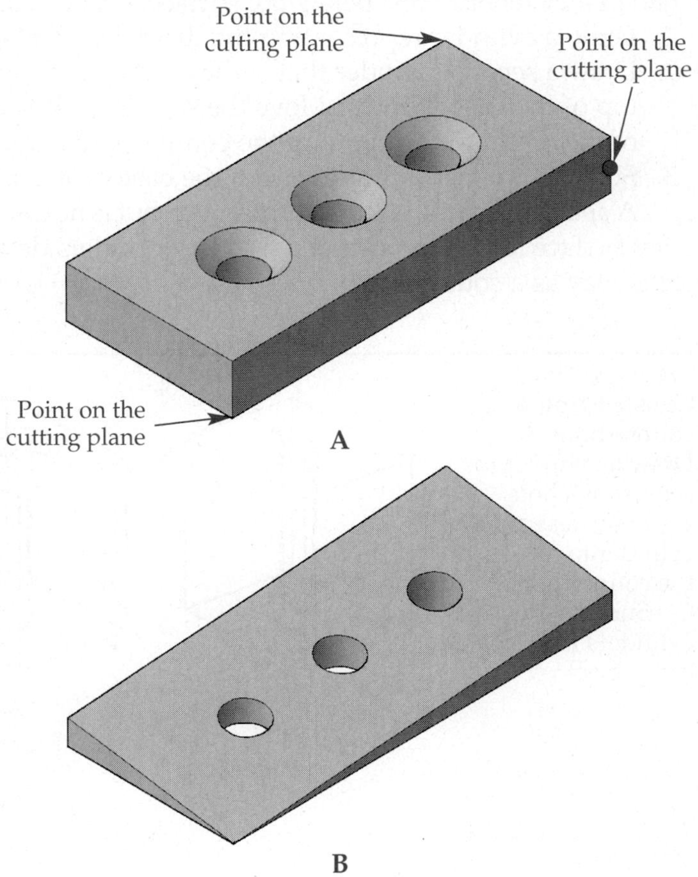

the **3points** option. Pick three points on the cutting plane and then select the side of the object to keep.

Exercise 8-10

www.g-wlearning.com/CAD/

Complete the exercise on the companion website.

Removing Features

Sometimes, it may be necessary to remove a feature that has been constructed. For example, suppose you placed a R.5 fillet on an object based on an engineering sketch. Then, the design is changed to a R.25 fillet. Subobject editing is the best technique to accomplish this. Subobject editing is covered in Chapter 11.

Constructing Features on Solid Models

A variety of machining, structural, and architectural features can be created using some basic solid modeling techniques. The features discussed in the next sections are just a few of the possibilities.

Counterbore and Spotface

A ***counterbore*** is a recess machined into a part, centered on a hole, that allows the head of a fastener to rest below the surface. Create a counterbore as follows.

1. Draw a cylinder representing the diameter of the hole, **Figure 8-30A**.
2. Draw a second cylinder that is the diameter of the counterbore and center it at the top of the first cylinder. Move the second cylinder so it extends below the surface of the object to the depth of the counterbore, **Figure 8-30B**.
3. Subtract the two cylinders from the base object, **Figure 8-30C**.

A ***spotface*** is similar to a counterbore, but is not as deep. See **Figure 8-31**. It provides a flat surface for full contact of a washer or underside of a bolt head. Construct it in the same way as a counterbore.

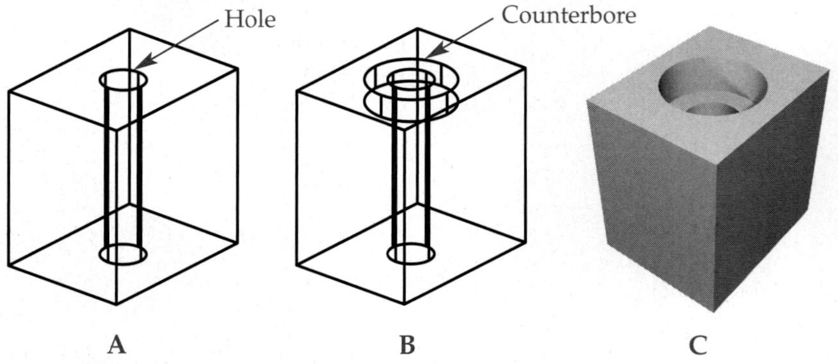

Figure 8-30. Constructing a counterbore. A—Draw a cylinder to represent a hole. B—Draw a second cylinder to represent the counterbore. C—Subtract the two cylinders from the base object.

Figure 8-31.
Constructing a spotface. A—The bottom of the second, larger-diameter cylinder should be located at the exact depth of the spotface. However, the height may extend above the surface of the base. Then, subtract the two cylinders from the base. B—The finished solid.

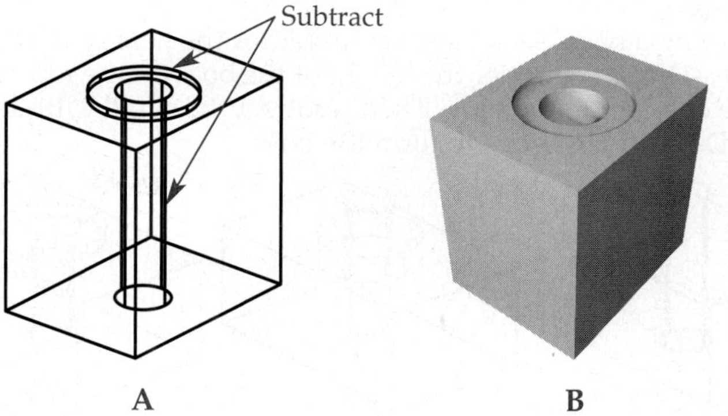

Countersink

A ***countersink*** is like a counterbore with angled sides. The sides allow a flat-head machine screw or wood screw to sit flush with the surface of an object. A countersink can be drawn in one of two ways. You can draw an inverted cone centered on a hole and subtract it from the base or you can chamfer the top edge of a hole. Chamfering is the quickest method.

1. Draw a cylinder representing the diameter of the hole, **Figure 8-32A**.
2. Subtract the cylinder from the base object.
3. Select the **CHAMFEREDGE** or **CHAMFER** command.
4. If using the **CHAMFEREDGE** command, pick the top edge of the hole and enter the chamfer distance(s). Preview the result before completing the operation, **Figure 8-32B**.
5. If using the **CHAMFER** command, select the top edge of the base object, enter the chamfer distance(s), and pick the top edge of the hole.

Boss

A ***boss*** serves the same function as a spotface. However, it is an area raised above the surface of an object. Draw a boss as follows.

1. Draw a cylinder representing the diameter of the hole. Extend it above the base object higher than the boss is to be, **Figure 8-33A**.
2. Draw a second cylinder the diameter of the boss. Place the base of this cylinder above the top surface of the base object a distance equal to the height of the boss. Give the cylinder a negative height value so that it extends inside of the base object, **Figure 8-33B**.
3. Union the base object and the second cylinder (boss). Subtract the hole from the unioned object, **Figure 8-33C**.
4. Fillet the intersection of the boss with the base object, **Figure 8-33D**.

Figure 8-32.
Constructing a countersink. A—Subtract the cylinder from the base to create the hole. B—Chamfer the top of the hole to create a countersink.

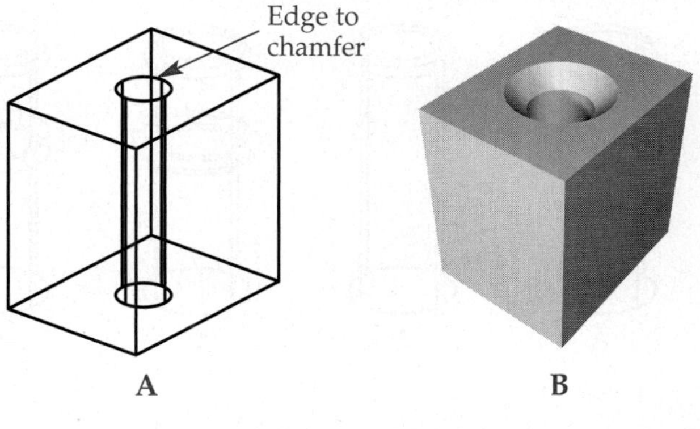

Figure 8-33.
Constructing a boss. A—Draw a cylinder for the hole so it extends above the surface of the object. B—Draw a cylinder the height of the boss on the top surface of the object. C—Union the large cylinder to the base. Then, subtract the small cylinder (hole) from the unioned objects. D—Fillet the edge to form the boss.

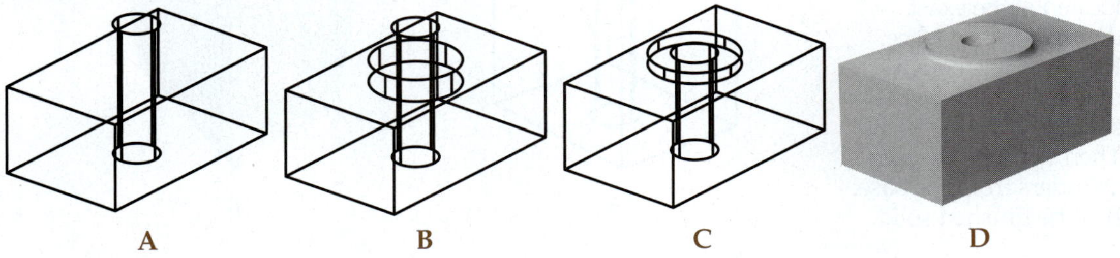

O-Ring Groove

An ***O-ring*** is a circular seal that resembles a torus. It sits inside of a groove constructed so that part of the O-ring is above the surface. An ***O-ring groove*** can be constructed by placing the center of a circle on the outside surface of a cylinder. Then, revolve the circle around the cylinder. Finally, subtract the revolved solid from the cylinder.

1. Construct the cylinder to the required dimensions, **Figure 8-34A**.
2. Rotate the UCS on the X axis (or appropriate axis).
3. Draw a circle with a center point on the surface of the cylinder, **Figure 8-34B**.
4. Revolve the circle 360° about the center of the cylinder, **Figure 8-34C**.
5. Subtract the revolved object from the cylinder, **Figure 8-34D**.

PROFESSIONAL TIP

In many cases, you will draw the O-ring as a torus. A copy of the torus can be used to create the O-ring groove instead of revolving a circle as described in the previous section.

Architectural Molding

Architectural molding features can be quickly constructed using extrusions. First, construct the profile of the molding as a closed shape, **Figure 8-35A**. Then, extrude the profile the desired length, **Figure 8-35B**.

Figure 8-34.
Constructing an O-ring groove. A—Construct a cylinder; this one has a round placed on one end. B—Draw a circle centered on the surface of the cylinder. C—Revolve the circle 360° about the center of the cylinder. D—Subtract the revolved object from the cylinder. E—The completed O-ring groove.

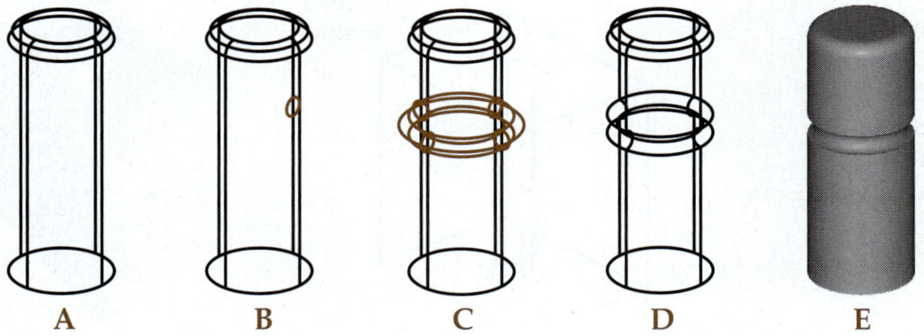

Figure 8-35.
A—The molding profile. B—The profile extruded to the desired length.

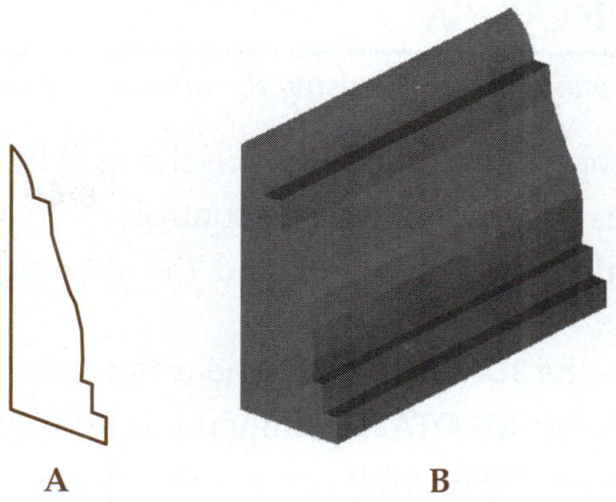

Corner intersections of molding can be quickly created by extruding the same shape in two different directions and then joining the two objects. First, draw the molding profile. Then, copy and rotate the profile to orient the local Z axis in the desired direction, **Figure 8-36A**. Next, extrude the two profiles the desired lengths, **Figure 8-36B**. Finally, union the two extrusions to create the mitered corner molding, **Figure 8-36C**.

Exercise 8-11

www.g-wlearning.com/CAD/

Complete the exercise on the companion website.

Figure 8-36.
Constructing corner molding. A—Copy and rotate the molding profile. B—Extrude the profiles to the desired lengths. C—Union the two extrusions to create the mitered corner. Note: The view has been rotated. D—The completed corner.

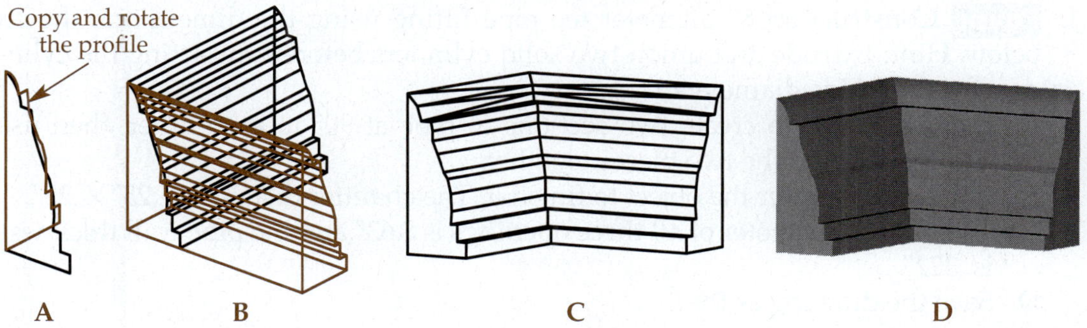

Chapter Review

Answer the following questions using the information in this chapter.

1. Which properties of a solid can be changed in the **Properties** palette?
2. What does the History property control?
3. What is the preferred command for aligning objects to create an assembly of parts?
4. How does the **3DALIGN** command differ from the **ALIGN** command?
5. How does the **3DROTATE** command differ from the **ROTATE** command?
6. How does the **MIRROR3D** command differ from the **MIRROR** command?
7. Which command allows you to create a rectangular array by defining rows, columns, and levels?
8. How does a 3D polar array differ from a 2D polar array?
9. Which option of the **ARRAYPATH** command can be used to evenly distribute the arrayed object along the path?
10. Which command is used to fillet a 3D object?
11. Which command is used to chamfer a 3D object?
12. Name four types of surfaces that can be used to slice objects.
13. Briefly describe the function of the **SLICE** command.
14. Which **SLICE** command option would be used to create a contoured solid object like a computer mouse?
15. Which two commands can be used to create a countersink?

Drawing Problems

1. Piping Construct an 8″ diameter tee pipe fitting using the dimensions shown below. Hint: Extrude and union two solid cylinders before subtracting the cylinders for the inside diameters.
   A. Use **EXTRUDE** to create two sections of pipe at 90° to each other, then use **UNION** to union the two pieces together.
   B. Fillet and chamfer the object to finish it. The chamfer distance is .25″ × .25″.
   C. The outside diameter of all three openings is 8.63″ and the pipe wall thickness is .322″.
   D. Save the drawing as P8-1.

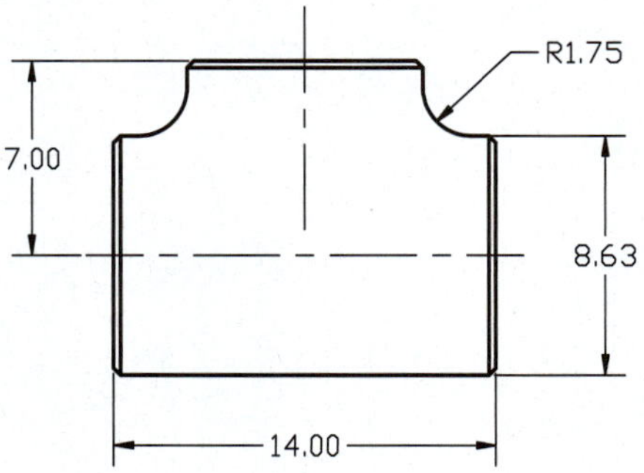

2. **Piping** Construct an 8″ diameter, 90° elbow pipe fitting using the dimensions shown below.
   A. Use **EXTRUDE** or **SWEEP** to create the elbow.
   B. Chamfer the object. The chamfer distance is .25″ × .25″. Note: You cannot use the **CHAMFER** command.
   C. The outside diameter is 8.63″ and the pipe wall thickness is .322″.
   D. Save the drawing as P8-2.

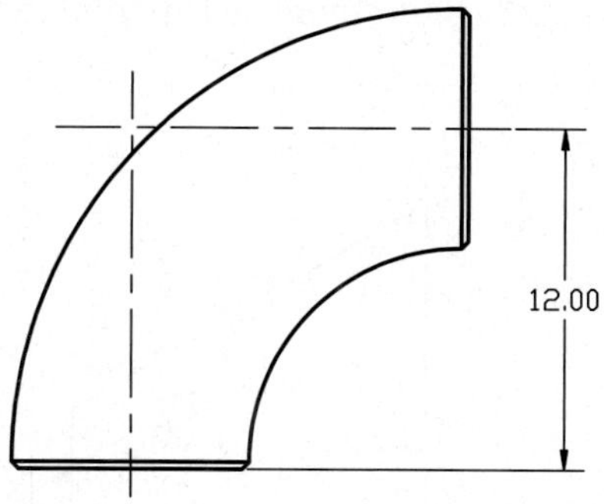

3. **Architectural** Construct picture frame moldings using the profiles shown below.
   A. Draw each of the closed profiles shown. Use your own dimensions for the details of the moldings.
   B. The length and width of A and B should be no larger than 1.5″ × 1″.
   C. The length and width of C and D should be no larger than 3″ × 1.5″.
   D. Construct an 8″ × 12″ picture frame using moldings A and B.
   E. Construct a 12″ × 24″ picture frame using moldings C and D.
   F. Save the drawing as P8-3.

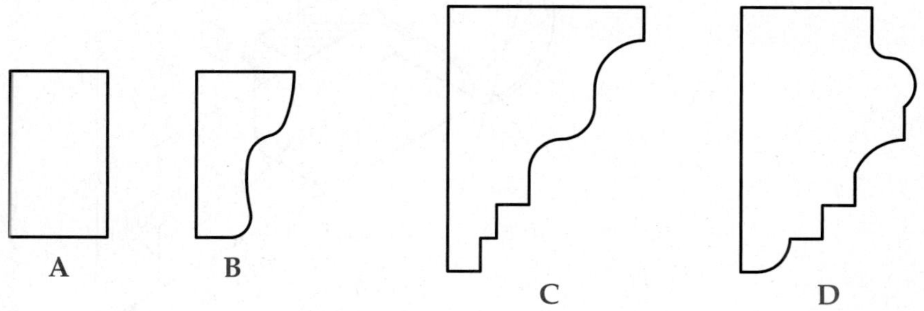

Problems 4–7. These problems require you to use a variety of solid modeling functions to construct the objects. Use all of the solid modeling and editing commands you have learned so far to assist in construction. Create new UCSs as needed and use a dynamic UCS when practical. Use **SOLIDHIST** *and* **SHOWHIST** *to record and view the steps used to create the solid models. Create copies of the completed models and split them as required to show the internal features visible in the section views. Do not add dimensions to the models. For objects that have dimensions with tolerances, use the base dimension from which the upper and lower limits are calculated to create the model. Save each drawing as* P8-*(problem number).*

4. **Mechanical**

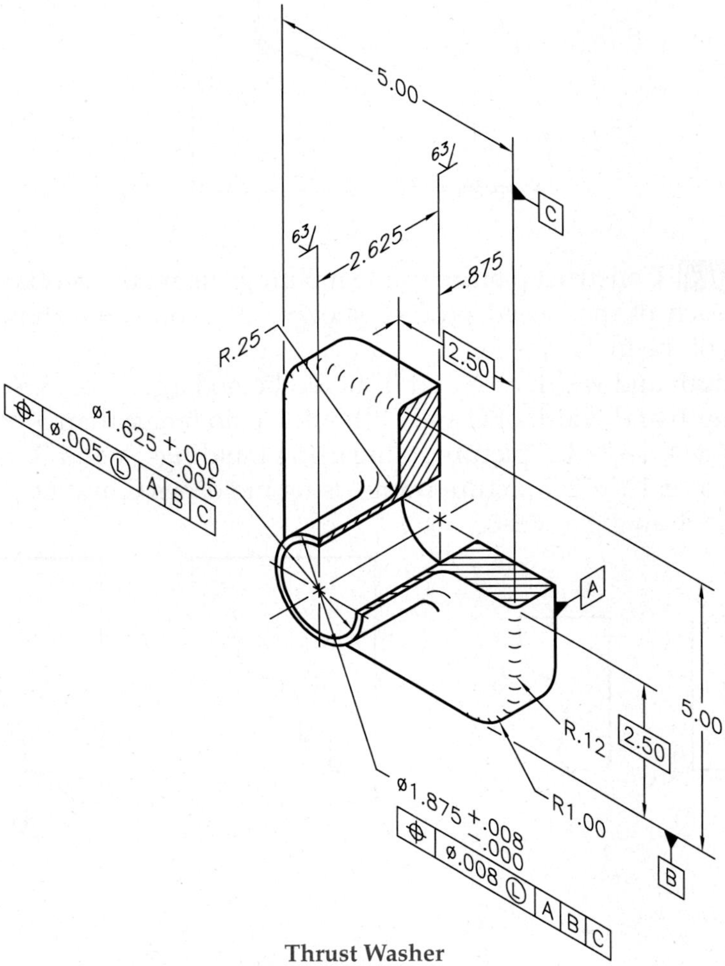

Thrust Washer

Drawing Problems - Chapter 8

5. **Mechanical**

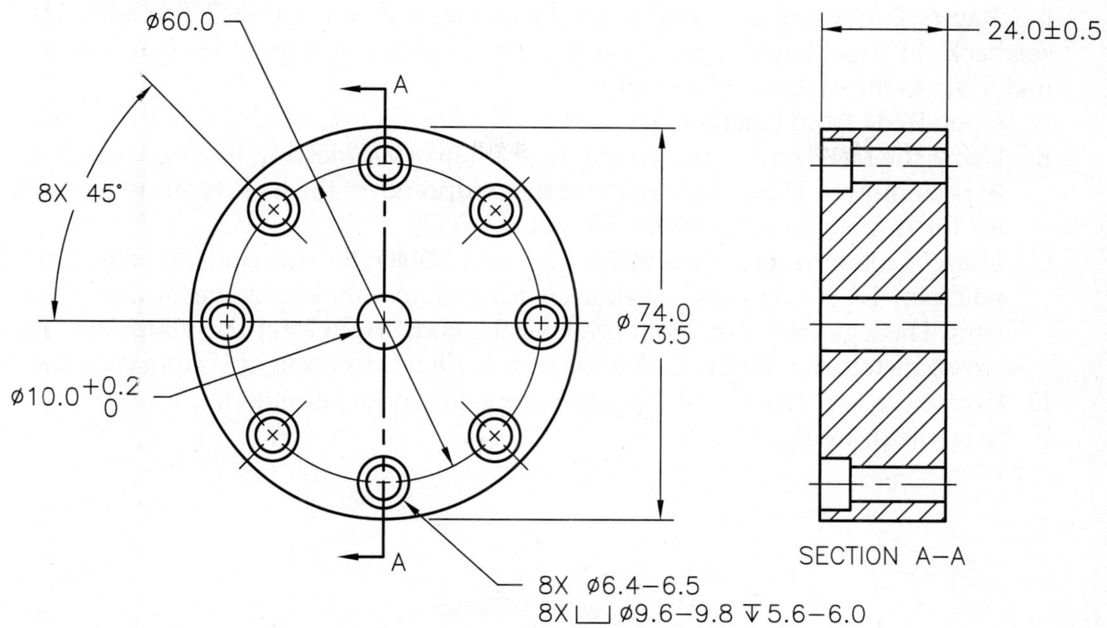

Collar

6. **Mechanical**

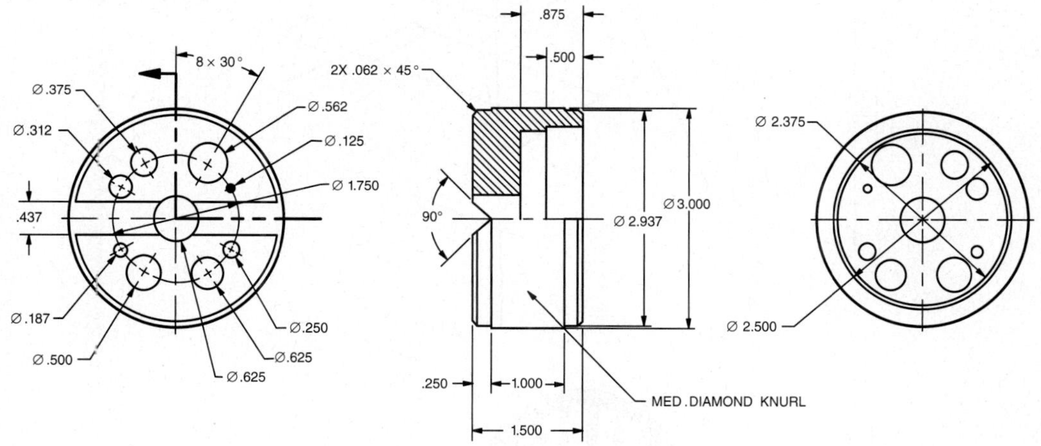

Diffuser

7. **Mechanical**

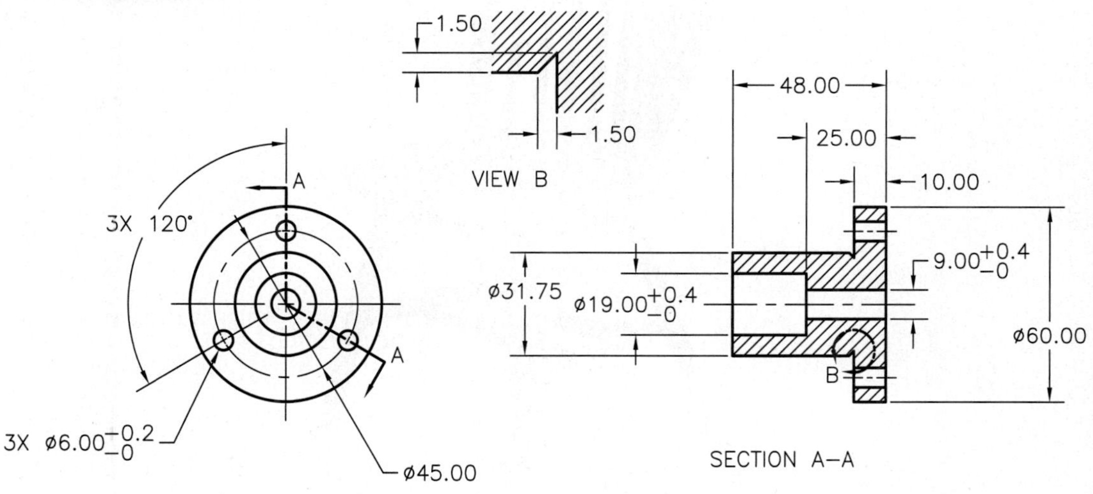

Bushing

8. **Architectural** In this problem, you will add legs to the kitchen chair you started in Chapter 2. In Chapter 6, you refined the seat, and in Chapter 7, you added the seatback. In this chapter, you complete the model. In Chapter 18, you will add materials to the model and render it.
   A. Open P7-11 from Chapter 7.
   B. Using the **LINE** command, create the framework for lofting the legs and crossbars as shown. The crossbars are at the midpoints of the legs. Position the lines for the double crossbar about 4.5″ apart.
   C. Using a combination of the **3DROTATE** and **3DMOVE** commands in conjunction with new UCSs, draw and position circles as shapes for lofting the legs and crossbars. The legs transition from ∅1.00 at the ends to ∅1.25 at the midpoints. The crossbars transition from ∅.75 at the ends to ∅1.00 at a position 2″ from each end.
   D. Create the legs and crossbars. The completed model is shown.
   E. Save the drawing as P8-8.

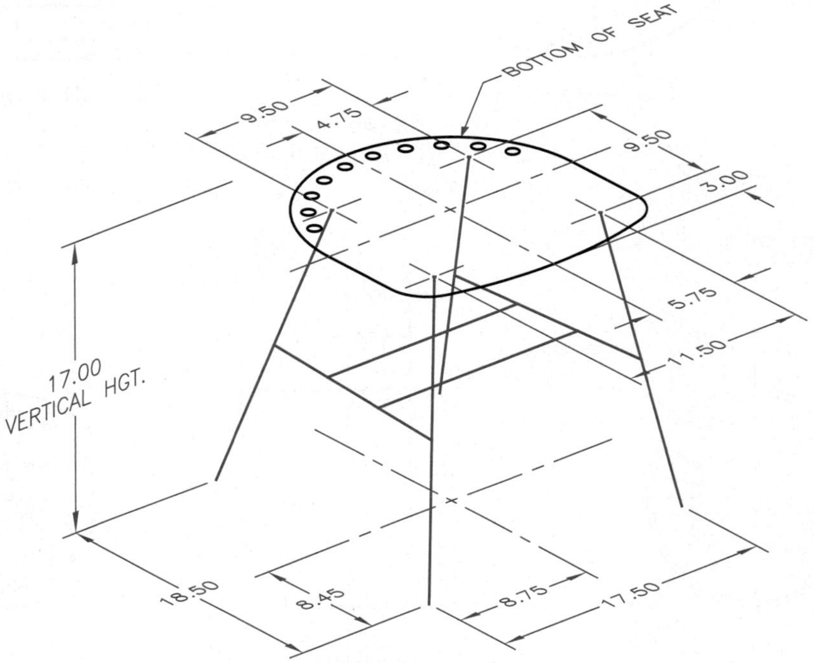

CHAPTER 9

Mesh Modeling

Learning Objectives

After completing this chapter, you will be able to:

✓ Explain tessellation divisions and values.
✓ Create mesh primitives.
✓ Create a smoothed mesh object.
✓ Create a refined mesh object.
✓ Generate a mesh by converting a solid or surface.
✓ Generate a solid or surface by converting a mesh.
✓ Execute editing techniques on mesh objects.
✓ Create a split face on a mesh.
✓ Produce an extruded mesh face.
✓ Apply a crease to mesh subobjects.
✓ Create and close mesh object gaps.
✓ Create a new mesh face by collapsing a mesh face or edge.
✓ Merge mesh faces to form a single mesh face.
✓ Construct a new mesh face by spinning a triangular mesh face.

Overview of Mesh Modeling

Mesh primitives can be used to create freeform designs. See **Figure 9-1**. The tools for creating and editing meshes extend the capability of AutoCAD's 3D modeling tools. There are two key workflows that the designer considers:

- The creation of 3D models, which can be solids, surfaces, or meshes.
- Editing the 3D models to create unique shapes.

Mesh models can be created as mesh primitives, mesh forms, or freeform mesh shapes. A ***mesh model*** consists of vertices, edges, and faces. You can modify a mesh by adjusting the smoothness or by adding creases, extrusions, splits, and gaps. You can also distort a mesh to create unique freeform shapes.

A mesh model is a type of surface model. ***Subdivision surfaces*** is another term for mesh models. Mesh models do *not* have volume or mass. Rather, mesh models only define the shape of the design.

Figure 9-1.
Constructing an ergonomic mouse as a mesh model. A—The basic mesh primitive. B—Using editing tools, the mesh is reformed. C—The completed mesh model.

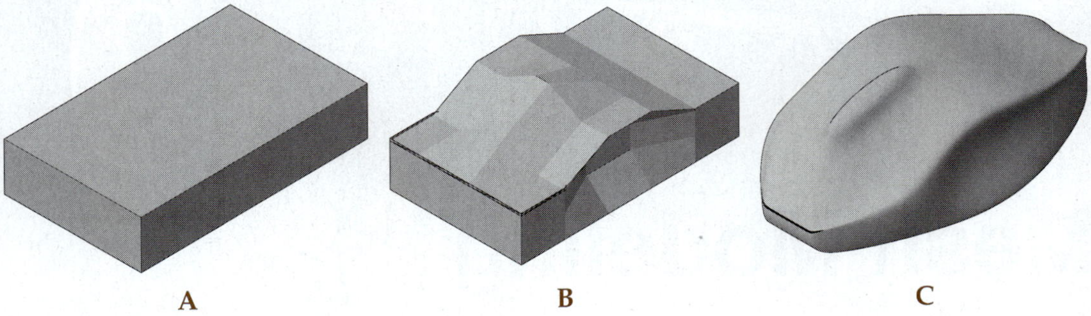

Mesh objects can be created using one of these methods:

- Construct mesh primitives (**MESH** command).
- Convert an existing solid or surface into a mesh object (**MESHSMOOTH** command).
- Construct mesh forms that are ruled, revolved, tabulated, or edge-defined objects (**RULESURF**, **REVSURF**, **TABSURF**, and **EDGESURF** commands).
- Convert legacy surface objects into mesh objects using commands such as **3DFACE**, **3DMESH**, and **PFACE**.

The tools used to create and modify meshes are found on the **Mesh** tab in the ribbon. The commands used to create mesh primitives are similar to those used to create solid primitives, which are discussed in Chapter 2. However, mesh primitives have face mesh objects that are divided into smaller faces. These divisions are based on tessellation division values (smoothness), as discussed in this chapter.

NOTE

A mesh model is a unique type of object in AutoCAD. A mesh model is a type of surface model, but it is different from a procedural surface model or a NURBS surface model. Unique construction and editing methods are used with mesh models. AutoCAD procedural surfaces and NURBS surfaces are precision surfaces, which makes them relatives to solids. Procedural surfaces and NURBS surfaces are discussed in Chapter 10.

PROFESSIONAL TIP

The **RULESURF**, **REVSURF**, **TABSURF**, and **EDGESURF** commands are available for creating mesh forms. These commands are not discussed in this chapter. Other surface modeling methods are typically more useful and provide more flexibility in editing. Refer to Chapter 10 for detailed coverage on surface modeling.

Tessellation Division Values

Tessellation divisions are the basic foundation for the smoothness of a mesh object. Tessellation divisions on a mesh object consist of planar shapes (faces) that fit together to form the surface. See **Figure 9-2A**. These divisions display the edges of a mesh face that can then be edited.

Figure 9-2.
A—Tessellation divisions are key to mesh modeling. They define the smoothness of the mesh model. B—Facets on a mesh face.

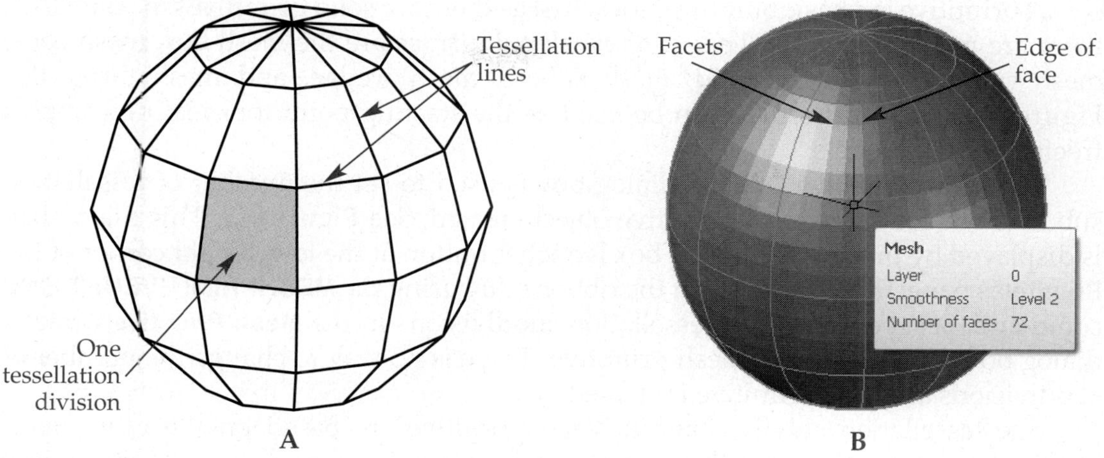

When creating mesh primitives, set the mesh tessellation divisions *before* creating a mesh primitive shape. Setting the proper values for the mesh tessellation divisions ensures the model has enough faces, edges, and vertices for editing. The default tessellation divisions are listed in the **Mesh Primitive Options** dialog box, which is discussed in the next section.

Within the tessellation division, the face consists of structures known as ***facets***. See **Figure 9-2B**. The facets create a grid pattern that is related to the smoothness of the mesh. As the smoothness level of the mesh object increases, the facets increase in number, resulting in a smoother, more rounded surface. As the smoothness level decreases, the facets decrease in number, resulting in a rougher, less rounded surface.

You can change the default smoothness of a mesh primitive by entering the **Settings** option of the **MESH** command, which is discussed in the next section. **Figure 9-3** shows an example of a box mesh primitive created using the default tessellation divisions. The default settings create a box with no smoothness, length divisions of three, width divisions of three, and height divisions of three.

NOTE

Any mesh object or subobject that has a level of smoothness of 1 or higher can be refined by converting facets to editable faces. Refining a mesh is discussed later in this chapter. You can also adjust the appearance of the facets by setting the **VSLIGHTINGQUALITY** system variable. Graphic card and monitor display *may* affect visibility.

Figure 9-3.
This box mesh primitive is drawn with the default settings. A—Displayed with the 2D Wireframe visual style current. B—After the **HIDE** command is used.

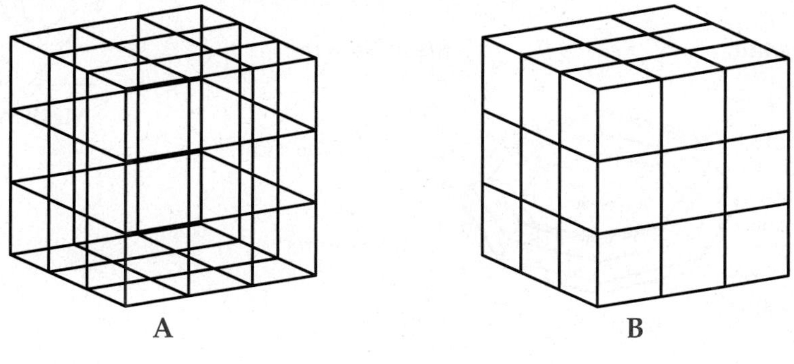

Drawing Mesh Primitives

MESH

Ribbon
Mesh
> Primitives
Mesh Box
Mesh Cone
Mesh Cylinder
Mesh Pyramid
Mesh Sphere
Mesh Wedge
Mesh Torus
Type
MESH

A primitive is a basic building block. Just as there are solid primitives in AutoCAD, there are mesh primitives. The seven ***mesh primitives*** are the mesh box, mesh cone, mesh cylinder, mesh pyramid, mesh sphere, mesh wedge, and mesh torus. See **Figure 9-4**. These primitives can be used as the starting point for creating complex freeform mesh models.

The **Mesh Primitive Options** dialog box is used to set the number of tessellation subdivisions for each mesh primitive object created. See **Figure 9-5**. This dialog box is displayed by picking the dialog box launcher button at the lower-right corner of the **Primitives** panel in the **Mesh** tab of the ribbon or by using the **MESHPRIMITIVEOPTIONS** command. Set the number of tessellation subdivisions in the **Mesh Primitive Options** dialog box *before* creating a mesh primitive. There is no way to change the number of subdivisions after the primitive is created.

The tessellation subdivisions for each primitive are based on the dimensions required to create the primitive. For example, a box has length, width, and height subdivisions. On the other hand, a mesh cylinder has axis, height, and base subdivisions. To set the subdivisions, select the primitive in the tree on the left-hand side of the **Mesh Primitive Options** dialog box. The subdivision properties are then displayed below the tree. Enter the number of subdivisions as required and then close the dialog box. All new primitives of that type will have this number of subdivisions until the setting is changed in the dialog box. Existing primitives are *not* affected.

Figure 9-4.
An overview of AutoCAD's mesh primitives and the dimensions required to draw them.

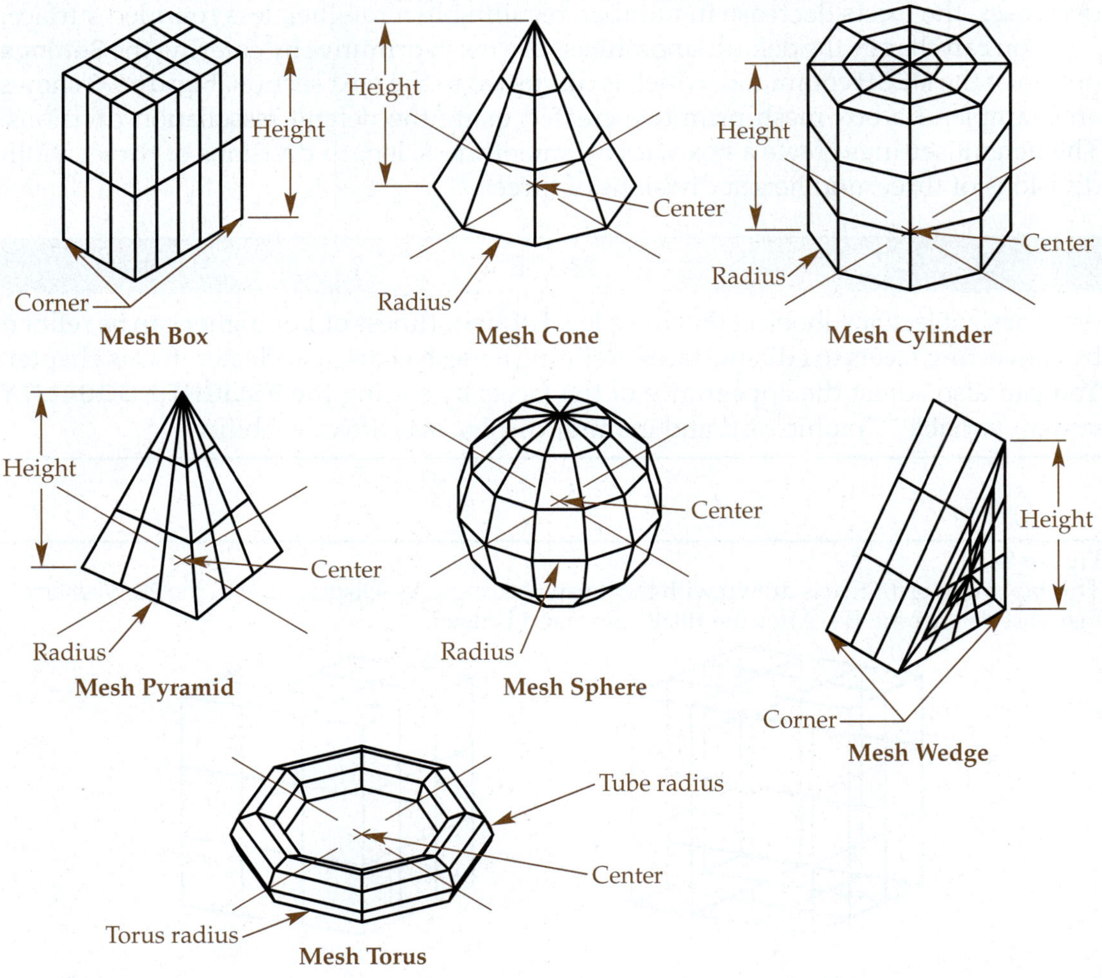

Figure 9-5.
The **Mesh Primitive Options** dialog box.

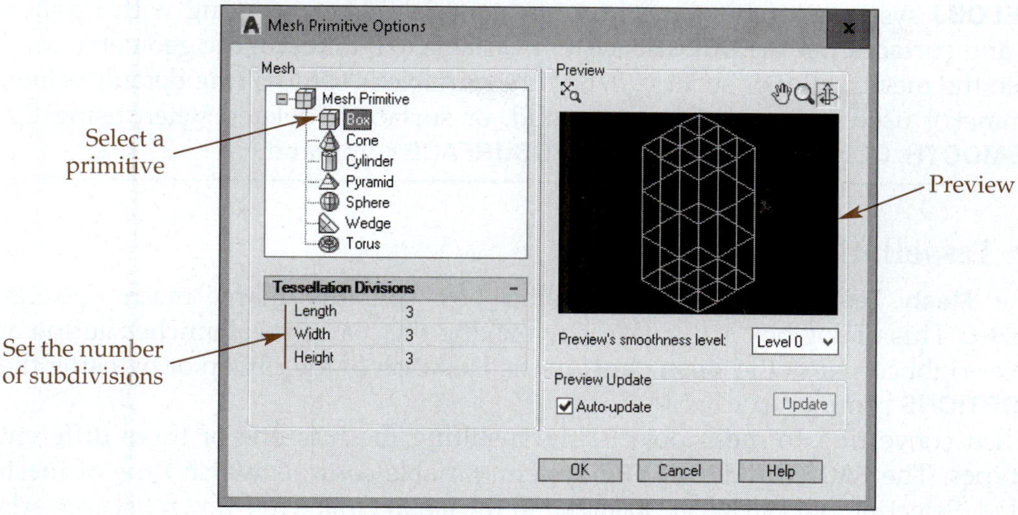

The following is an example of how to create a mesh box primitive. The mesh box is the mesh primitive used for most base shapes. Refer to **Figure 9-4** for the information required to draw mesh primitives.

1. Open the **Mesh Primitive Options** dialog box.
2. Select the box primitive in the tree.
3. Enter 3 in each of the Length, Width, and Height property boxes to set the subdivisions.
4. Close the dialog box.
5. Select the command for drawing a mesh box.
6. Specify the first corner of the base of the mesh box.
7. Specify the opposite corner of the base of the mesh box.
8. Specify the height of the mesh box.

PROFESSIONAL TIP

Creating mesh primitives is similar to creating solid primitives. Creating solid primitives is discussed in Chapter 2.

Exercise 9-1

www.g-wlearning.com/CAD/

Complete the exercise on the companion website.

Converting Between Mesh and Surface or Solid Objects

AutoCAD offers the flexibility of converting between solid, surface, and mesh objects. This allows you to select the type of modeling that offers the best tools for the task at hand, then convert the model into a form appropriate for the end result. The next sections discuss converting between solid, surface, and mesh objects and the settings that control the conversion.

PROFESSIONAL TIP

The **DELOBJ** system variable plays an important role when working with meshes, solids, and surfaces. Set the **DELOBJ** system variable to 0 to retain the geometry used to create the mesh, solid, or surface. When the variable is set to 3 (the default value), the geometry used to create the mesh, solid, or surface is deleted when using the **MESHSMOOTH**, **CONVTOSOLID**, or **CONVTOSURFACE** command.

Mesh Tessellation Options

The **Mesh Tessellation Options** dialog box contains many mesh options, **Figure 9-6**. This dialog box is displayed by picking the dialog box launcher button at the lower-right corner of the **Mesh** panel in the **Mesh** tab of the ribbon or by using the **MESHOPTIONS** command.

When converting to mesh objects, the resulting mesh is one of three different mesh types. The **FACETERMESHTYPE** system variable controls which type of mesh is created. Selecting Smooth Mesh Optimized in the **Mesh type:** drop-down list converts objects to the optimized mesh type (**FACETERMESHTYPE** = 0). This is the default and recommended setting. At this setting, the smoothness level can be set for the resulting mesh using the options in the **Smooth Mesh After Tessellation** area. Selecting Mostly Quads in the **Mesh type:** drop-down list creates faces that are mostly quadrilateral (**FACETERMESHTYPE** = 1). Selecting Triangle in the **Mesh type:** drop-down list creates faces that are mostly triangular (**FACETERMESHTYPE** = 2).

The **Mesh distance from original faces:** setting is the maximum deviation of the mesh faces (**FACETERDEVSURFACE**). Simply put, this setting determines how closely a converted mesh shape matches the original solid or surface shape.

The **Maximum angle between new faces:** setting is the maximum angle of a surface normal of two adjoining faces (**FACETERDEVNORMAL** system variable). The higher the value, the more faces created in very curved areas and the less faces created in flat areas. Increasing the value is good for objects that have curved areas, such as fillets, rounds, holes, or other tightly curved areas.

The **Maximum aspect ratio for new faces:** setting is the upper limit for the ratio of height to width for new faces (**FACETERGRIDRATIO** system variable). By adjusting this value, long faces can be avoided, such as those that would be created from a cylindrical

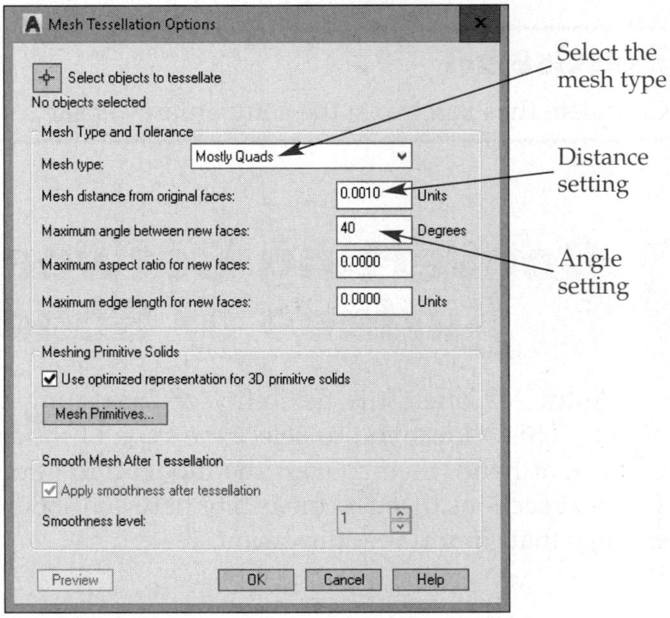

Figure 9-6. The **Mesh Tessellation Options** dialog box.

object during the conversion process. A low value will create a cleaner look in the formed faces. The default setting is 0, which specifies that no limitation is applied to the aspect ratio. A setting of 1 specifies that the height must be equal to the width. A setting greater than 1 specifies that the height may exceed the width. A setting between 0 and 1 specifies that the width may exceed the height.

The **Maximum edge length for new faces:** setting is the maximum length any edge can be (**FACETERMAXEDGELENGTH** system variable). The default setting is 0, which allows the size of the mesh to be determined by the size of the 3D model. A setting of 1 or higher results in a reduced number of faces and less accuracy compared to the original model.

When the **Use optimized representation for 3D primitive solids** check box is checked, the mesh settings in the **Mesh Primitive Options** dialog box are applied when converting a solid primitive. This is the default setting. If the check box is unchecked, the settings in the **Mesh Tessellation Options** dialog box are applied.

NOTE

Converting swept solids and surfaces, regions, closed polylines, 3D face objects, and legacy polygon and polyface mesh objects may produce unexpected results. If this happens, undo the operation and try making setting adjustments in the **Mesh Tessellation Options** dialog box for better results.

Converting from a Solid or Surface to a Mesh

The **MESHSMOOTH** command is used to convert a solid or surface object into a mesh object. See **Figure 9-7**. The command is easy to use. First, enter the command. Then, select the solids or surfaces to be converted. Finally, press [Enter] and the objects are converted into mesh objects. If you select an object that is not a primitive, you may receive a message indicating that the command works best on primitives. If you receive this message, choose to create the mesh.

MESHSMOOTH

Ribbon
Home
> Mesh
Mesh
> Mesh

Smooth Object
Type
MESHSMOOTH
SMOOTH
CONVTOMESH

PROFESSIONAL TIP

With a good understanding of solid modeling, create a solid model first. Then, convert it to a mesh object using the **MESHSMOOTH** command and edit the mesh to create a freeform design.

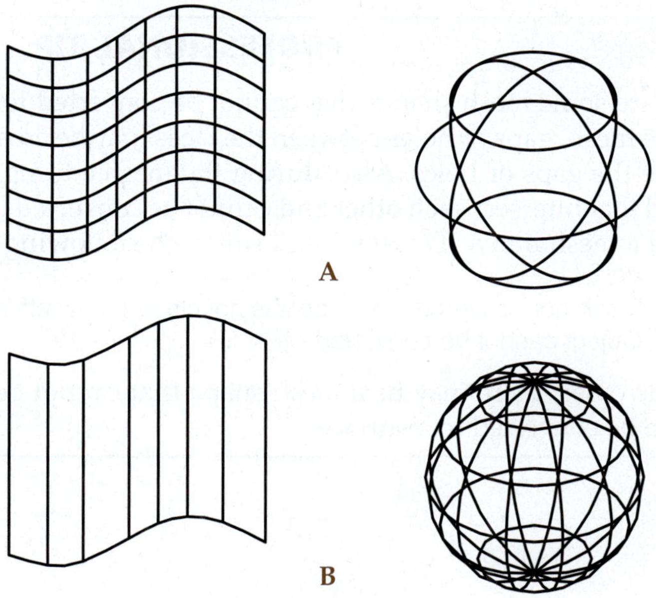

Figure 9-7. Converting an existing solid or surface into a mesh object. A—The existing objects shown are a surface (left) and solid (right). B—After converting each object into a mesh using the **MESHSMOOTH** command.

Exercise 9-2

www.g-wlearning.com/CAD/

Complete the exercise on the companion website.

Converting From a Mesh to a Solid or Surface

SMOOTHMESHCONVERT

Ribbon
Mesh
> Convert Mesh
Smooth, Optimized
Smooth, Not Optimized
Faceted, Optimized
Faceted, Not Optimized
Type
SMOOTHMESH-CONVERT

Mesh objects can be converted into solids or surfaces using the **CONVTOSOLID** and **CONVTOSURFACE** commands. The faces on the resulting solid or surface can be smoothed or faceted and optimized or not. This is controlled by the **SMOOTHMESHCONVERT** system variable. Select one of the four possible settings *before* converting a mesh to a solid or surface. The settings are described as follows:

- **Smoothed and optimized.** Coplanar faces are merged into a single face. The overall shape of some faces can change. Edges of faces that are not coplanar are rounded. (**SMOOTHMESHCONVERT** = 0)
- **Smoothed and not optimized.** Each original mesh face is retained in the converted object. Edges of faces that are not coplanar are rounded. (**SMOOTHMESHCONVERT** = 1)
- **Faceted and optimized.** Coplanar faces are merged into a single, flat face. The overall shape of some faces can change. Edges of faces that are not coplanar are creased or angular. (**SMOOTHMESHCONVERT** = 2)
- **Faceted and not optimized.** Each original mesh face is converted to a flat face. Edges of faces that are not coplanar are creased or angular. (**SMOOTHMESHCONVERT** = 3)

CONVTOSOLID

Ribbon
Mesh
> Convert Mesh
Convert to Solid
Type
CONVTOSOLID

CONVTOSURFACE

Ribbon
Mesh
> Convert Mesh
Convert to Surface
Type
CONVTOSURFACE

To convert a mesh to a solid, first set the smoothing option, as described above. Then, select the **CONVTOSOLID** command. Next, select the mesh objects to convert and press [Enter]. The objects are converted from mesh objects to solid objects based on the selected smoothing option.

To convert a mesh to a surface, first set the smoothing option, as described above. Then, select the **CONVTOSURFACE** command. Next, select the mesh objects to convert and press [Enter]. The objects are converted from mesh objects to surface objects based on the selected smoothing option.

The examples shown in **Figure 9-8A** are simple mesh objects. In **Figure 9-8B**, the mesh objects have been converted into solid objects with the **Faceted, Optimized** button selected in the **Convert Mesh** panel on the **Mesh** tab in the ribbon. In **Figure 9-8C**, the mesh objects have been converted into surface objects with the **Smooth, Optimized** button selected.

PROFESSIONAL TIP

There are some mesh shapes that cannot be converted to a 3D solid. If using grips to edit the mesh, gaps or holes between the faces may be created. Smooth the mesh object to close the gaps or holes. Also, during the mesh editing process, mesh faces may be created that intersect each other and cannot be converted to a 3D solid. Converting this type of a mesh into a 3D solid will result in the following error message:

```
Mesh not converted because it is not closed or it self-intersects.
Object cannot be converted.
```

In some cases, there may be a mesh shape that cannot be converted to a solid object, but can be converted to a surface.

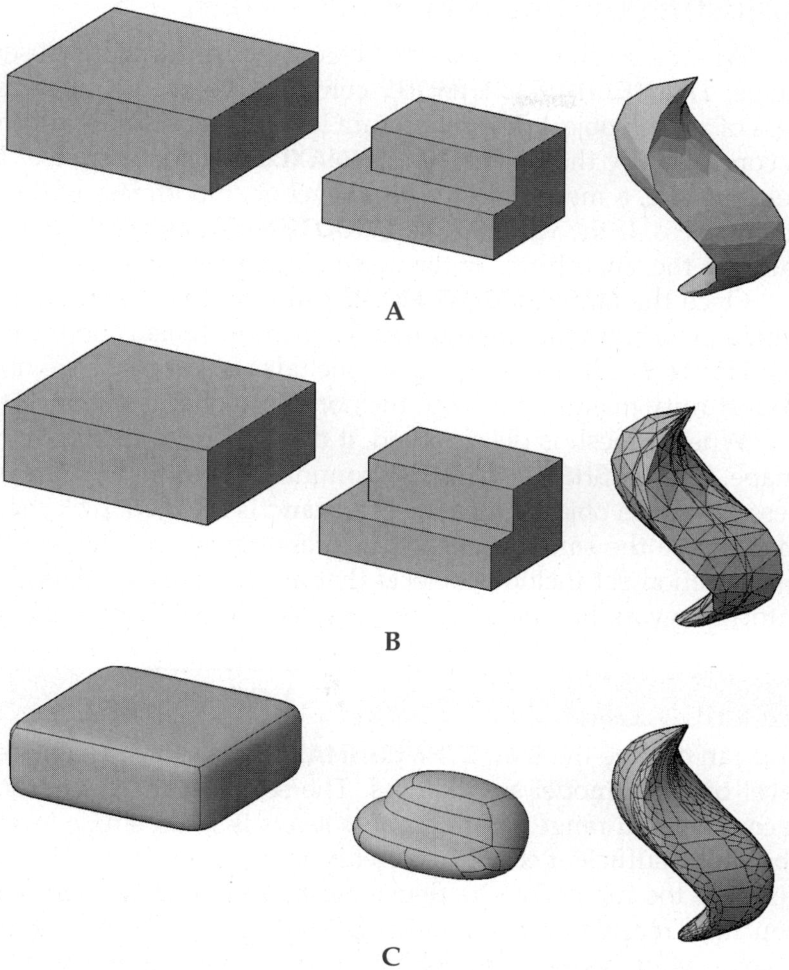

Figure 9-8.
A—Three basic mesh objects.
B—The mesh objects converted into faceted solids.
C—The mesh objects converted into smoothed surfaces.

Exercise 9-3

www.g-wlearning.com/CAD/

Complete the exercise on the companion website.

Smoothing and Refining a Mesh Object

The roundness of a mesh object is increased by increasing the smoothness level. The smoothness can be set before creating the mesh, as described earlier in this chapter. The mesh object can also be refined or have its smoothness increased after it is created. This is described in the following sections. When smoothing or refining a mesh object, the number of tessellation subdivisions is either increased or decreased. The lowest smoothness level is 0 and the highest smoothness level is 4. The default smoothness level is 0.

PROFESSIONAL TIP

When creating primitives, begin with the least amount of faces possible. You can always refine the mesh model to create more faces.

Adjusting the Smoothness of a Mesh

When a mesh is smoothed, it changes form to more closely represent a rounded shape. The **MESHSMOOTHMORE** command is used to increase the level of smoothness of a mesh object. The maximum level of smoothness attained with this command is controlled by the **SMOOTHMESHMAXLEV** system variable. The default setting is 4. For example, a mesh object with a level of smoothness of 0 is considered to have no smoothness. If the value of the **SMOOTHMESHMAXLEV** system variable is 4, you can increase the smoothness of the mesh object up to level 4.

MESHSMOOTHMORE

Ribbon
Home
> Mesh
> Smooth More
Mesh
> Mesh

Smooth More

Type
MESHSMOOTHMORE
MORE

Once the **MESHSMOOTHMORE** command is selected, pick the mesh objects for which to increase the smoothness. Then, press [Enter] and the level is increased by one. See **Figure 9-9**. If the selection set includes objects that are not meshes, you have the opportunity to either filter out the non-mesh objects or convert them to meshes.

When a mesh is desmoothed, it changes form to more closely represent a boxed shape. The **MESHSMOOTHLESS** command is used to decrease the level of smoothness of a mesh object. Once the command is selected, pick the mesh objects for which to decrease the smoothness. Then, press [Enter] and the level is decreased by one. If the selection set includes objects that are not meshes, AutoCAD displays a message informing you that the non-mesh objects will be filtered out.

MESHSMOOTHLESS

Ribbon
Home
> Mesh
> Smooth Less
Mesh
> Mesh

Smooth Less

Type
MESHSMOOTHLESS
LESS

NOTE

You can change the **SMOOTHMESHMAXLEV** system variable setting to create a higher level of mesh model smoothness. The setting can range from 1 to 255. However, a recommended range of smoothness levels is 1–5. Using a level of 5 as the upper limit should be sufficient enough to create a very smooth model. Using a higher level may generate too many faces (a dense mesh) and affect system performance. In this case, you may receive an error message stating that the operation cannot be completed because your system does not have enough physical memory.

PROFESSIONAL TIP

The **MESHSMOOTHMORE** and **MESHSMOOTHLESS** commands change the smoothing one level at a time. You can use the **Properties** palette to change the smoothness to any level in one step.

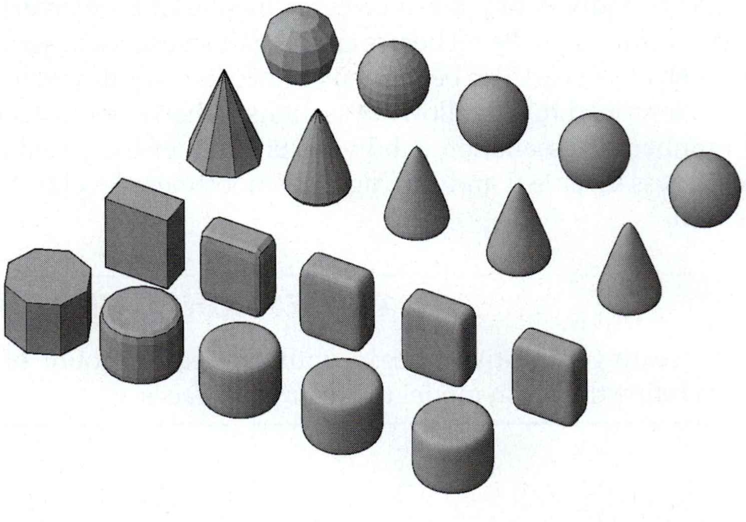

Figure 9-9. Using the **MESHSMOOTHMORE** command to increase the level of smoothness from level 0 to level 4. From left to right, smoothness levels 0, 1, 2, 3, 4.

Exercise 9-4

www.g-wlearning.com/CAD/

Complete the exercise on the companion website.

Refining a Mesh

Refining a mesh increases the number of subdivisions in a mesh object. Refining adds detail to the mesh to make it more realistic with smooth, flowing lines. This also gives a greater selection for editing the mesh faces, vertices, or edges. The **MESHREFINE** command is used to refine a mesh. When the command is used, the number of subdivisions is quadrupled. See **Figure 9-10**. The smoothness level of the original object must be level 1 or higher.

The entire mesh can be refined (all faces at once) or a selected face can be refined. When you refine a mesh *model*, the increased amount of face subdivisions becomes a new smoothness level of 0. However, if you refine an individual *face*, the level of smoothness is not reset.

Be careful not to create too dense of a mesh. Creating a mesh that is too dense may result in face, edge, and vertex subobjects that are very small. This may make it difficult to select subobjects and edit the mesh.

MESHREFINE

Ribbon
Home > Mesh
Mesh > Mesh
Refine Mesh

Type
MESHREFINE
REFINE

Exercise 9-5

www.g-wlearning.com/CAD/

Complete the exercise on the companion website.

Editing Meshes

As discussed earlier, the second key workflow of mesh modeling is the ability to edit the mesh. The tools for editing a mesh are found in the **Mesh Edit** and **Selection** panels on the **Mesh** tab of the ribbon. See **Figure 9-11**. The face, edge, and vertex subobjects can be edited to change the shape of the mesh. These subobjects can be moved, rotated, or scaled. A face on the mesh can be split or extruded.

Subobject filters are used to assist in the selection of a mesh face, edge, or vertex *before* it is moved, rotated, or scaled. This is especially true for a very dense mesh. First, right-click in the drawing window and select **Subobject Selection Filter** to display the cascading menu, **Figure 9-12A**. Then, select the filter you wish to use. You can also select the filter in the **Selection** panel of the **Mesh** tab in the ribbon, **Figure 9-12B**. A subobject filter can also be selected from the **Filters object selection** flyout on the status bar. This flyout does not appear on the status bar by default. Select the **Selection Filtering** option in the status bar **Customization** flyout to add the **Filters object selection** flyout to the status bar.

Figure 9-10.
Refining a mesh.
A—The original mesh with a smoothness level of 1.
B—The refined mesh, which now has more faces and a smoothness level of 0.

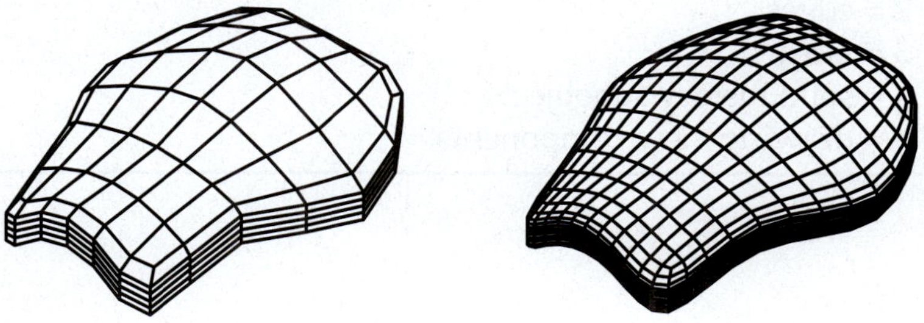

Figure 9-11.
The tools for editing a mesh are found in the **Mesh Edit** and **Selection** panels of the **Mesh** tab in the ribbon.

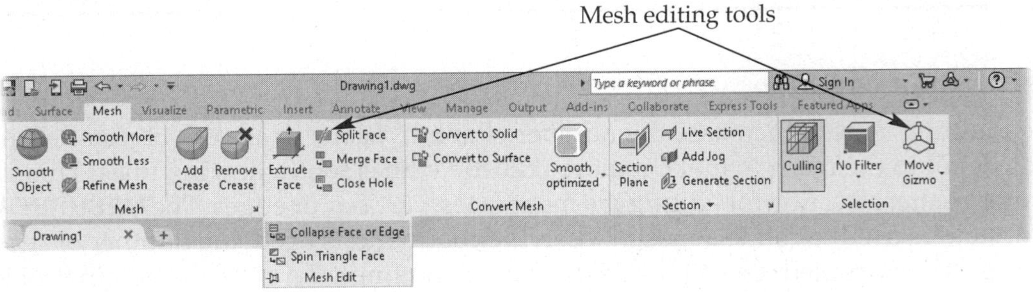

Figure 9-12.
Selecting a subobject filter. A—Using the shortcut menu. B—Using the drop-down list in the **Selection** panel of the **Mesh** ribbon tab.

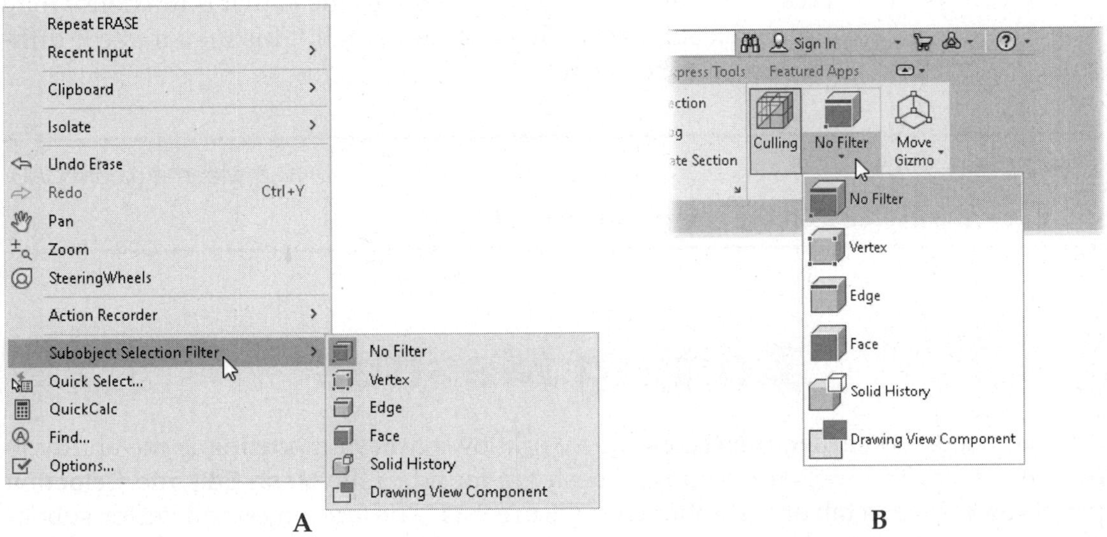

Subobject editing is discussed in detail in Chapter 11. The same procedures discussed in that chapter for solids can be applied to mesh models.

NOTE

Use the [Ctrl] key and left mouse button to select the subobjects or use the **SUBOBJSELECTIONMODE** system variable for subobject filtering. These settings apply:

- 0 = off
- 1 = vertices
- 2 = edges
- 3 = faces
- 4 = solid history subobjects
- 5 = drawing view components

Using Gizmos

A *gizmo*, also called a grip tool, appears when a face, edge, or vertex subobject on a mesh model is selected. This tool is used to specify how a movement, rotation, or scaling transformation is applied. A visual style other than 2D Wireframe must be current in order for the gizmo to appear. Using gizmos (grip tools) is introduced in Chapter 8 and covered in detail in Chapter 11.

You can switch between the move, rotate, and scale gizmos by using the drop-down list in the **Selection** panel on the **Mesh** tab of the ribbon. You can also use the **Show gizmos** flyout on the status bar. The **Show gizmos** flyout does not appear on the status bar by default. Select the **Gizmo** option in the status bar **Customization** flyout to add the **Show gizmos** flyout to the status bar.

When you make a subobject selection on a mesh model, the ribbon displays ***context-sensitive panels*** based on the selection. For example, if a face subobject is selected, the ribbon displays the **Crease** and **Edit Face** panels, **Figure 9-13**. Context-sensitive panels allow improved editing efficiency by displaying tools commonly used for the selected item. Once the selection is canceled, the context-sensitive panels are no longer displayed.

PROFESSIONAL TIP

Before selecting a face, edge, or vertex to perform a move, rotate, or scale operation, set the appropriate selection filter and gizmo.

Extrude a Face

You can select a face subobject on a mesh and extrude it. Extruding a mesh face adds new features to the mesh. This creates new faces that can be edited. The **MESHEXTRUDE** command is used to extrude mesh faces. The command sequence to extrude a mesh face is similar to extruding to create a solid shape. The **EXTRUDE** command is covered in detail in Chapter 6.

Ribbon
Mesh
> Mesh Edit
Extrude Face
Type
MESHEXTRUDE

MESHEXTRUDE

To extrude a face, enter the command. The command automatically turns on the face subobject filter. This is indicated by the **Face** button on the **Selection** panel in the **Mesh** tab on the ribbon. Next, pick the face subobject(s) to extrude. Finally, enter an extrusion height. In **Figure 9-14**, a camera is being developed as a mesh model. The **MESHEXTRUDE** command is used to extrude an individual face as the first step in creating the lens tube.

PROFESSIONAL TIP

Extrude a face using the **MESHEXTRUDE** command instead of moving a face. This will give greater editing control over an individual face.

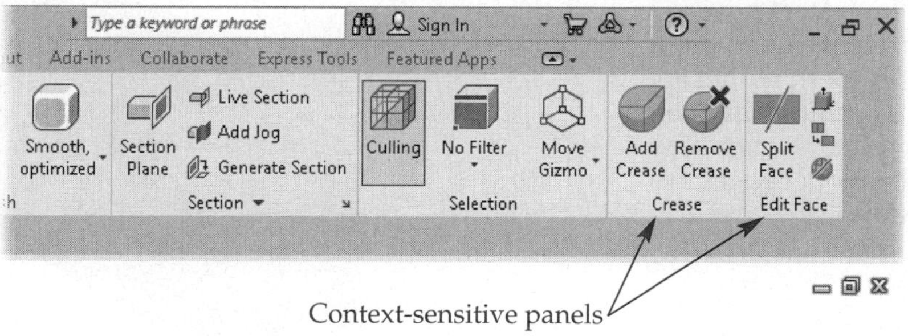

Figure 9-13.
A context-sensitive panel is only displayed when certain objects or subobjects are selected. In this case, the **Crease** and **Edit Face** panels are displayed when a face subobject is selected.

Figure 9-14.
Extruding a mesh face. A—Select the face to extrude. B—The process adds faces to the model.

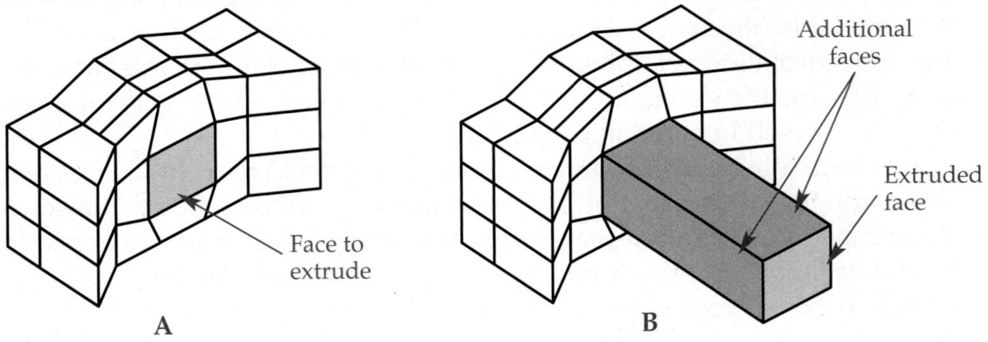

Split a Face

Splitting a mesh face is used to increase the number of faces on the model without refining a mesh. Splitting one face creates two faces. This is easier than refining the mesh. Think of these new split faces as subdivisions of an existing face.

MESHSPLIT
Ribbon
Mesh
> Mesh Edit
Split Face
Type
MESHSPLIT
SPLIT

The **MESHSPLIT** command is used to split a face. Once the command is entered, select the face to split. Then, specify the starting point and the ending point of a line defining the split. You can specify any two points on the mesh face. **Figure 9-15** shows a mesh model with three faces that have been split.

Once the faces are split, mesh editing options, such as extruding, moving, rotating, or scaling, can be used on the new faces. In **Figure 9-16**, the split faces from **Figure 9-15** have been extruded, and the model has been smoothed to level 3.

Exercise 9-6

www.g-wlearning.com/CAD/

Complete the exercise on the companion website.

Figure 9-15.
Three faces have been split on this mesh model. Each face has been split into two faces.

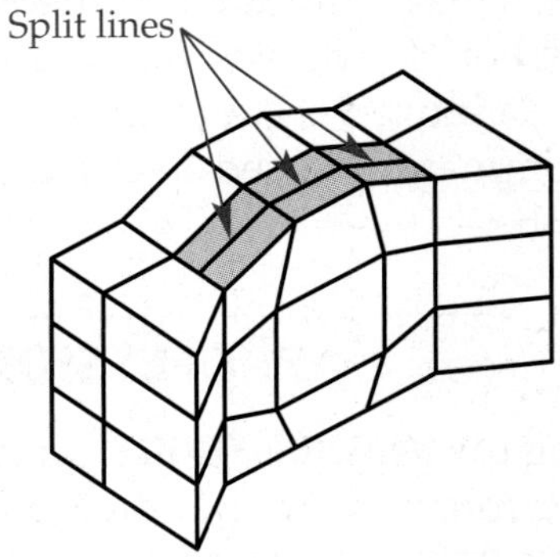

Figure 9-16.
Three faces have been extruded on this mesh model and smoothing has been applied. Notice how the model edges are rounded.

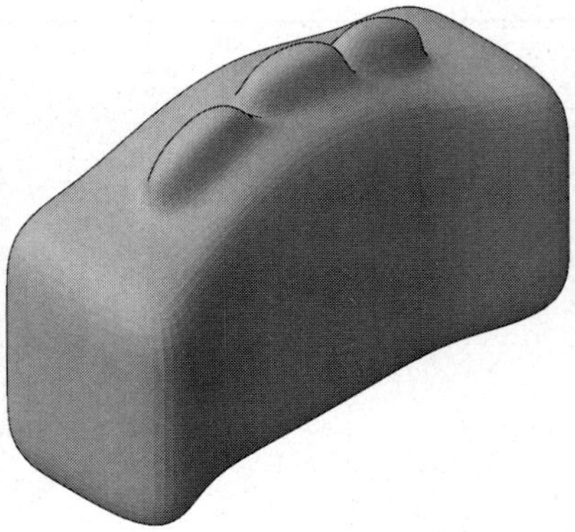

Applying a Crease to a Mesh Model

Ribbon
Mesh
> Mesh
Add Crease
Type
MESHCREASE
CREASE

MESHCREASE

A *crease* is a sharpening of a mesh subobject, much like a crease in folded paper. A crease sharpens or squares off an edge or flattens a face. This prevents the subobject from being smoothed. Creases can be applied to faces, edges, and vertices. A smoothness level of 1 or higher must be assigned to the mesh object for the creases to have an effect, but they can be applied at any smoothness level. Once a crease is applied, any existing smoothing is removed from the subobject. If the mesh smoothness is increased, any creased subobjects are not smoothed. Also, creasing an edge *before* an object is smoothed will limit the mesh editing capabilities.

An example of where a crease might be applied is the bottom of a computer mouse. As the design of the computer mouse begins, the base of the mesh model is rounded when the model is smoothed. See **Figure 9-17A**. By creasing all bottom edges of the mouse, the bottom is squared off. See **Figure 9-17B**. A flat bottom is a requirement of the design intent because the mouse must sit flat on a desk.

In another example, the trigger pads on the game controller shown in **Figure 9-18** have been creased to create a flat surface. When the conceptual design is finished, it can be converted to a solid. This will give the designer a unique shape to which buttons and joysticks can be added. This unique shape would be difficult, if not impossible, to create from scratch as a solid model.

Once the command is entered, select the subobjects to crease. You do not need to press the [Ctrl] key to select subobjects, but filters should be used to make selection easier. Once the subobjects are selected, press [Enter]. This prompt appears:

Specify crease value [Always] <Always>:

If you enter a crease value, this is the highest smoothing level for which the crease is retained. If the smoothness level is set higher than the crease value, the creased subobject is smoothed. The Always option forces the crease to be retained for all smoothness levels. In most cases, this is the recommended option.

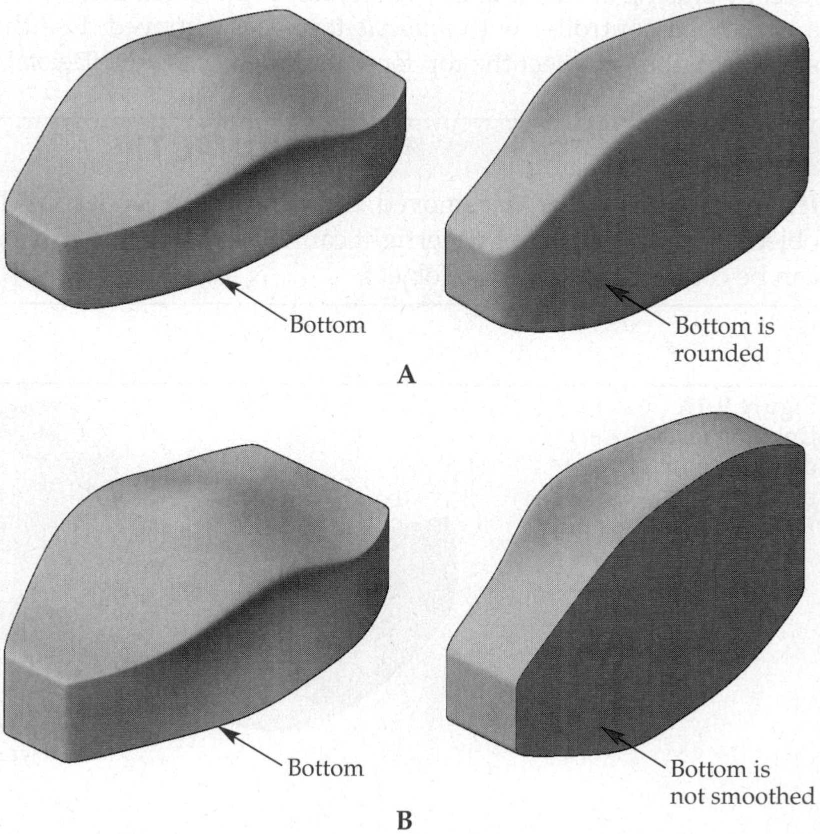

Figure 9-17. Adding creases to a mesh. A—Before the creases are applied, the bottom of the mouse is rounded. This is obvious if the object is rotated (right). B—After the creases are added, the bottom is flat.

Figure 9-18.
In this example, creases have been added to the mesh to create flat areas on the game pad for buttons.

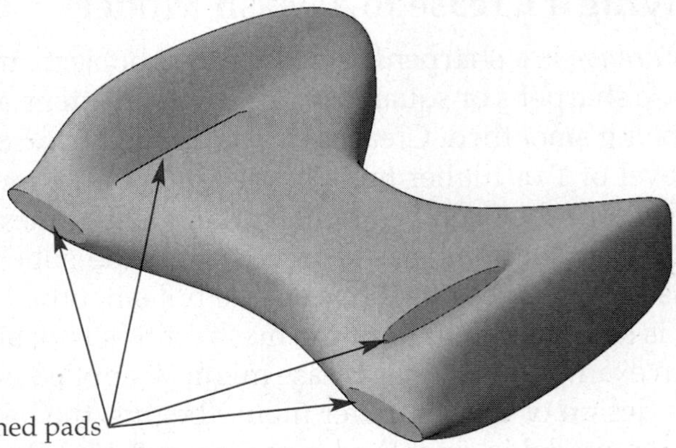

Exercise 9-7

www.g-wlearning.com/CAD/

Complete the exercise on the companion website.

Removing a Crease

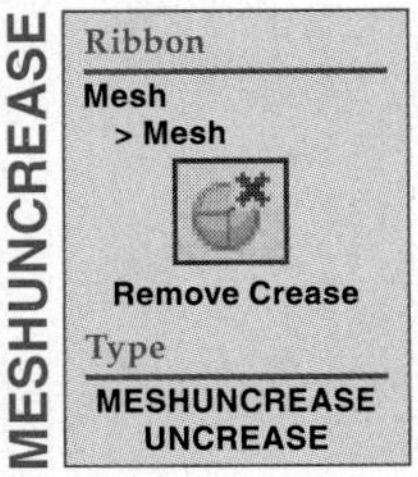

Removing a crease is a simple process. The **MESHUNCREASE** command is used to do this. Enter the command and then select the subobjects from which a crease is to be removed. Then, press [Enter]. The crease is removed and smoothing is applied as appropriate.

Creating a Mesh Gap

As the designer creates meshes, editing the mesh faces is the next step toward final design. There will be times when the designer needs to modify a mesh by deleting faces. The techniques of erasing or deleting a mesh face aid in the design. To delete a mesh face, use the **ERASE** command or press the [Delete] key after selecting the mesh face. By erasing or deleting the mesh face, a gap in the mesh occurs. **Figure 9-19** shows a game pad controller with *multiple* top faces removed. Use the [Ctrl] key or the face subobject filter to select the top faces and then press the [Delete] key.

PROFESSIONAL TIP

If a mesh face is erased or removed, the mesh object is not considered a *watertight* mesh object. A mesh that is not watertight cannot be converted to a solid object. However, it can be converted to a surface object.

Figure 9-19.
Deleting faces creates a gap on the mesh model.

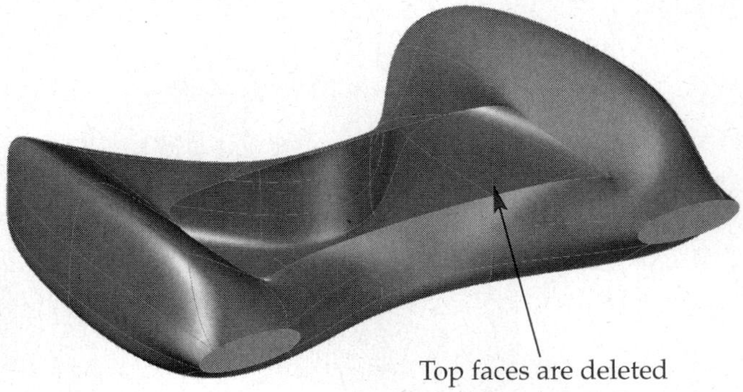

Close Gaps in Mesh Objects

After a mesh face has been erased or deleted, the designer has the ability to close the mesh gap. The **MESHCAP** command is used to close the gap, Figure 9-20. This is done by creating a new face between selected, continuous edges. Once the command is entered, select the edges of the surrounding mesh object faces. The selected edges do not need to form a closed loop. If a closed loop is not formed, AutoCAD automatically closes the loop. Then, press [Enter] to complete the command and close the mesh gap.

Ribbon
Mesh
> Mesh Edit
Close Hole
Type
MESHCAP

MESHCAP

To select multiple edges at once, use the **Chain** option of the **MESHCAP** command. A quick way to access this option is to right-click and select **Chain** from the shortcut menu after entering the **MESHCAP** command. The **Chain** option allows you to select multiple continuous edges with a single pick.

PROFESSIONAL TIP

When using the **MESHCAP** command, the selected mesh face edges should be on the same plane whenever possible. Also, the **MESHCAP** command cannot be used with edges that are shared by adjacent faces, such as edges that form the boundary of a hole feature.

Exercise 9-8

www.g-wlearning.com/CAD/

Complete the exercise on the companion website.

Collapse a Mesh Face or Edge

After a mesh is created, the surrounding mesh faces may be converged to create a different and unique mesh face shape. This is called collapsing a mesh. The designer can collapse surrounding mesh faces at the *center* of a selected mesh edge or face. New mesh faces are created. This helps the designer create new mesh face shapes for further editing.

The **MESHCOLLAPSE** command is used to collapse a mesh. Once the command is selected, pick the face or edge to collapse. The command is immediately applied and then ends.

Ribbon
Mesh
> Mesh Edit
Collapse Face or Edge
Type
MESHCOLLAPSE

MESHCOLLAPSE

Figure 9-21 shows an in-progress conceptual design for a computer mouse. The side must be collapsed to a point and then moved inward. Enter the command and select the middle edge. Be sure to use filters to help select the edge. Once the edge is collapsed, the new vertex can be selected for subobject editing and moved inward.

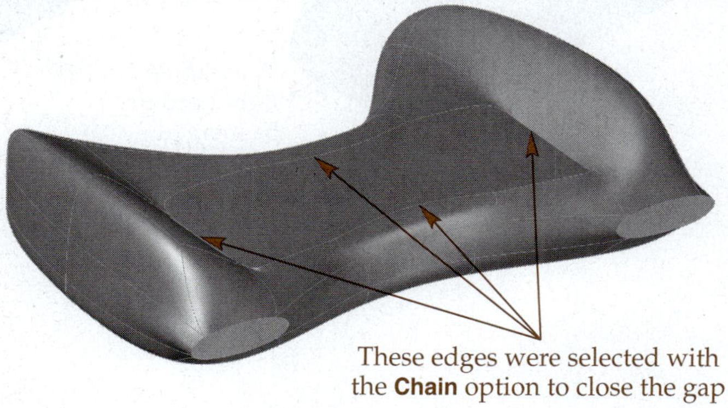

Figure 9-20. A gap on a mesh model can be closed with the **MESHCAP** command.

Figure 9-21.
Collapsing a mesh. A—The edge to be selected. B—The mesh is collapsed and the new vertex is edited.

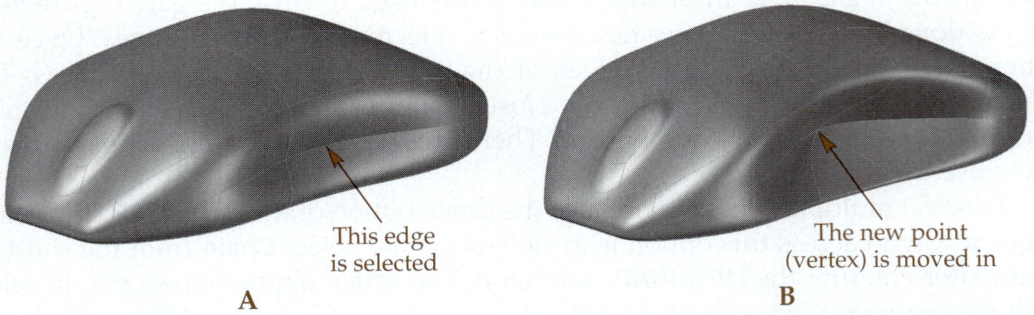

Exercise 9-9

www.g-wlearning.com/CAD/

Complete the exercise on the companion website.

Merge Mesh Faces

MESHMERGE

Ribbon
Mesh
> Mesh Edit
Merge Face
Type
MESHMERGE

The designer can merge adjacent mesh faces into a new, single mesh face. The **MESHMERGE** command is used to do this. Two or more adjacent faces can be selected to merge to create a different or unique face. For best results, the faces should be on the same plane. Do *not* try to merge faces that are not adjacent or are on corners.

Figure 9-22 shows a tape dispenser. The three faces on the front need to be merged to create a new mesh shape. Enter the command and set the face subobject filter. Then, select the three adjoining faces to merge. Press [Enter] to complete the command.

Exercise 9-10

www.g-wlearning.com/CAD/

Complete the exercise on the companion website.

Figure 9-22.
Merging faces. A—The faces to be merged. B—The resulting model.

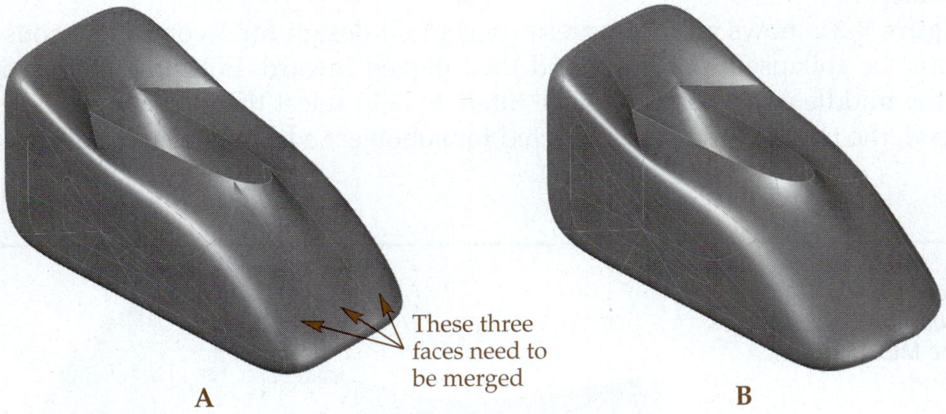

Spin Mesh Faces

Mesh faces have different and unique shapes when created. Additional unique shapes can be created by spinning a triangular mesh face. Spinning a mesh face spins the adjoining edge of *two* triangle mesh faces. Modifying a mesh face by spinning rotates the newly shared edge. The **MESHSPIN** command is used to spin a face.

MESHSPIN

Ribbon
Mesh
> Mesh Edit
Spin Triangle Face
Type
MESHSPIN

PROFESSIONAL TIP

Use the **Vertex** option of the **MESHSPLIT** command to create triangular mesh faces.

Figure 9-23 shows a pocket camera. The camera's top button has been split to create two triangular faces. Before the button top can be extruded and smoothed, the top adjoining faces must be spun.

Select the command, then pick the two triangular faces. Notice how the diagonal line of the triangular faces is now in a different orientation. You can then extrude one of the faces up or down to create a new button. Smooth the camera to create its final design shape.

Figure 9-23.
Spinning a face. A—The face has been split, but the orientation is wrong. B—The face is spun and the orientation is correct.

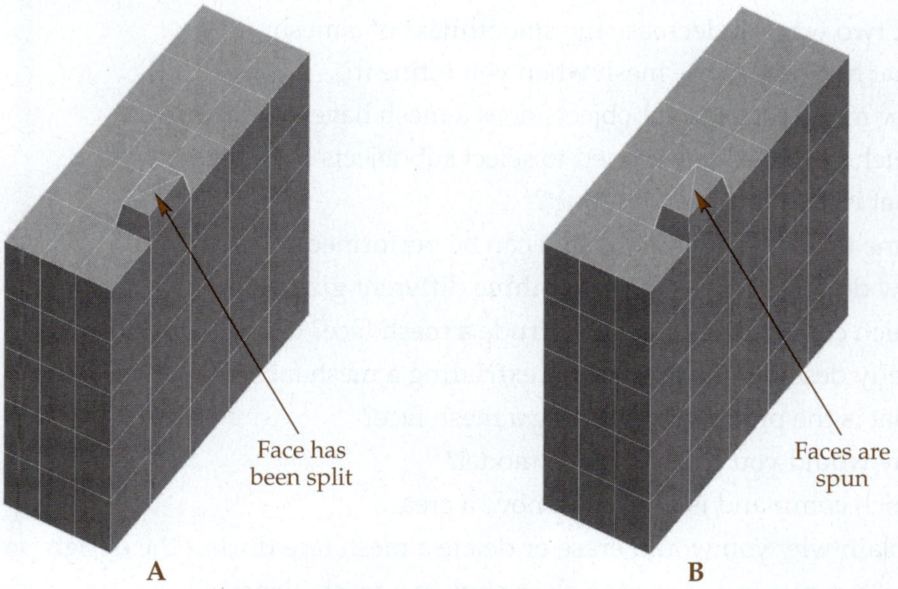

Chapter Review

Answer the following questions using the information in this chapter.

1. Of what does a mesh model consist?
2. What is another term for *mesh models*?
3. What are *tessellation divisions*?
4. When creating a mesh primitive, when should mesh tessellation divisions be set?
5. What are *facets*?
6. For what is the **Mesh Primitive Options** dialog box used?
7. How is a mesh box created?
8. How is a mesh sphere created?
9. How is a mesh torus created?
10. What is the purpose of the **DELOBJ** system variable?
11. Which command converts a mesh object to a surface object?
12. Which command converts a mesh object to a solid object?
13. How is the roundness of a mesh object increased?
14. Which command is used to convert an existing solid or surface to a mesh object?
15. Name the system variable that controls the maximum level of smoothness attained with the **MESHSMOOTHMORE** command.
16. List two ways to decrease the smoothness of a mesh.
17. What happens to the mesh when you refine it?
18. How many types of subobjects does a mesh have? List them.
19. Which keyboard key is used to select subobjects for editing?
20. What is a *context-sensitive panel*?
21. Name the three operations that can be performed with a gizmo.
22. How do you cycle through the three different gizmos?
23. Which command is used to extrude a mesh face?
24. Briefly describe the process for extruding a mesh face.
25. What is the process for splitting a mesh face?
26. Why would you crease a mesh model?
27. Which command is used to remove a crease?
28. Explain why you would erase or delete a mesh face during the design process.
29. Which command is used to close gaps in a mesh object?
30. What is the purpose of collapsing a mesh face or edge?

Drawing Problems

1. **Mechanical** In this problem, you will create an ergonomic computer mouse as a mesh shape. Change visual styles as needed throughout your work.
   A. Set the tessellation divisions for the box primitive to length = 5, width = 3, and height = 3.
   B. Create a 5 × 3 × 1.04 mesh box primitive.
   C. Move the middle faces inward .25 units on both sides of the mesh box, as shown in A.
   D. Move edges on the mesh model to form the mouse shape shown in B.
   E. Smooth the model.
   F. Continue editing faces, edges, and vertices to create a mouse shape similar to the one shown in C.
   G. Crease the model to create a flat bottom.
   H. Save the model as P9-1.

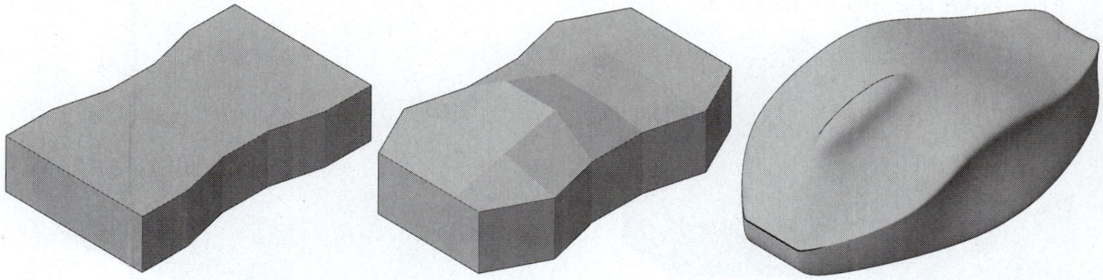

A B C

2. **Mechanical** In this problem, you will create a wireless phone charger cradle as a mesh shape. Change visual styles as needed throughout your work.
   A. Set the tessellation divisions for the box primitive to length = 5, width = 3, and height = 2.
   B. Create a 3 × 2 × 1.5 mesh box primitive.
   C. Move the middle-back edges up 1 unit and the front edges down .75 units, as shown in A.
   D. Crease the middle face on the top of the model, then move it down 1 unit.
   E. Crease all of the faces on the bottom.
   F. Smooth the mesh to a level 3 mesh.
   G. On the top of the model, nonuniformly scale the middle face (which is now round) to create an elliptical shape. Then, move the face toward the back of the model.
   H. Move the middle three edges in the front of the model .5 units toward the back of the model, as shown in B.
   I. Save the model as P9-2.

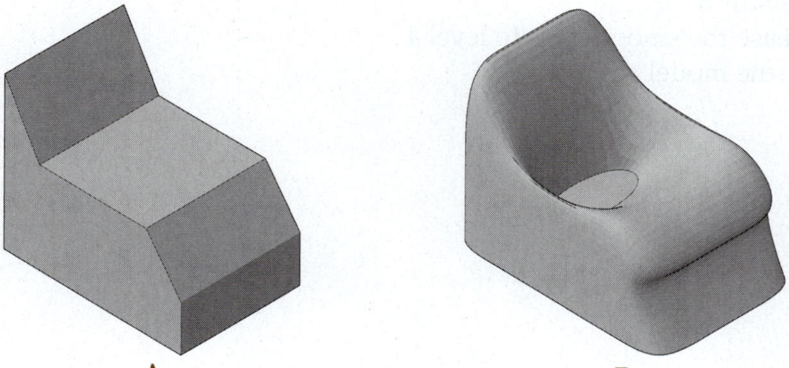

A B

3. **Mechanical** In this problem, you will create a game pad controller as a mesh shape. Change visual styles as needed throughout your work.
   A. Set the tessellation divisions for the box primitive to length = 5, width = 3, and height = 2.
   B. Create a 6 × 4 × 1 mesh box primitive.
   C. Move the middle faces on the back side (long side) inward 1.25 units.
   D. Move the top edges on the right and left side up 1 unit, as shown in A.
   E. Move the front middle face inward .5 units.
   F. Crease the front two corner faces. The top corner of these faces is formed by the edges moved up earlier.
   G. Smooth the model to level 3.
   H. Move the two creased areas outward .25 units.
   I. Split the top face to the left of the indention into three unique faces, as shown in B.
   J. Extrude the three split faces upward .375 units.
   K. Save the model as P9-3.

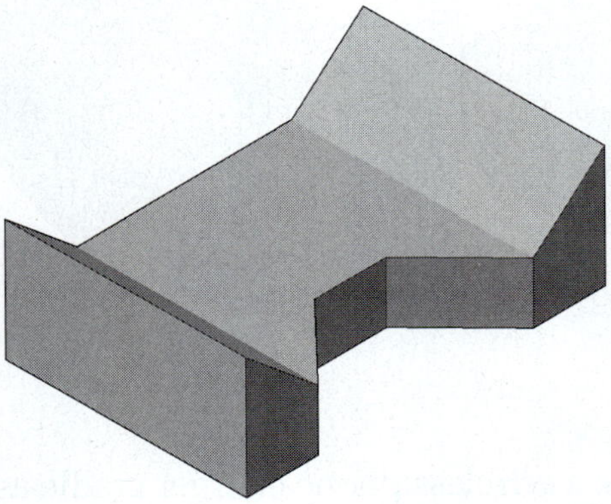

A

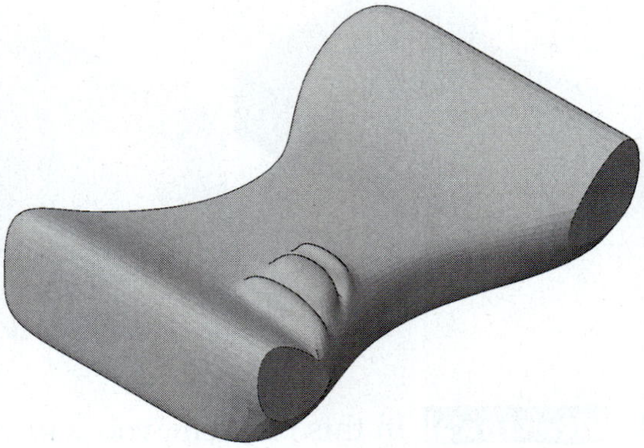

B

4. **General** In this problem, you will create a starship for a gaming application as a mesh shape. Change visual styles as needed throughout your work.
   A. Set the tessellation divisions for the sphere primitive to axis = 12 and height = 6.
   B. Create a Ø4 mesh sphere primitive.
   C. Move the vertex on the top of the sphere up 4 units.
   D. Move the vertex on the right and left sides outward 2 units, as shown in A.
   E. Move the vertex on the bottom of the sphere into the sphere by 1 unit.
   F. Smooth the model to level 2.
   G. On the bottom of the ship, as indicated in A, crease the four middle faces, then move them into the sphere by 1 unit.
   H. Extrude the four middle faces a distance of .75 with a taper of 30°. This forms the landing gear.
   I. Select two faces on the front of the ship and crease them to create windows, as shown in B.
   J. Increase the smoothness to level 4.
   K. Save the model as P9-4.

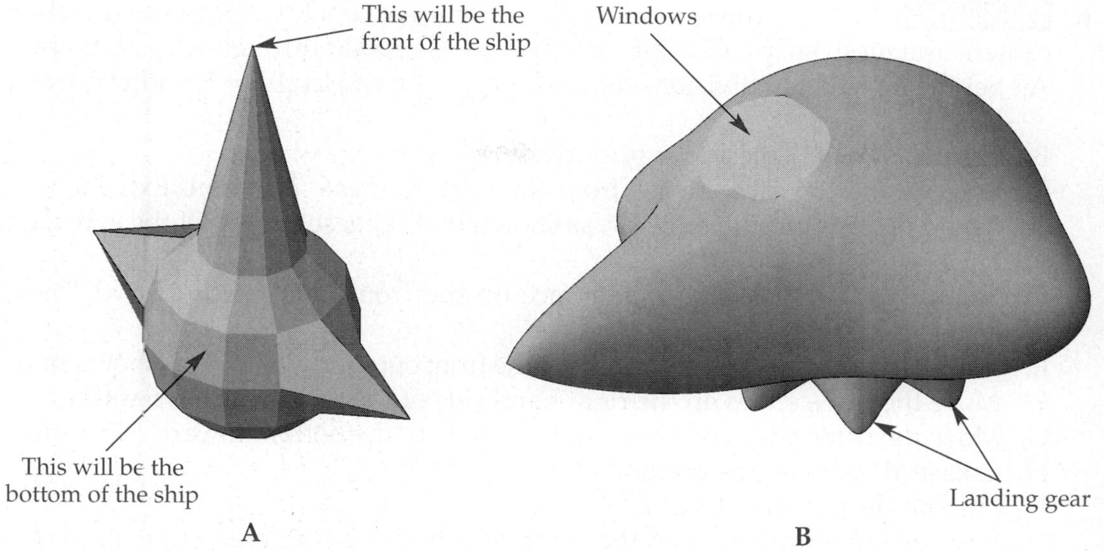

5. **Architectural** In this problem, you will create a home theater chair as a mesh shape. Change visual styles as needed throughout your work.
   A. Set the tessellation divisions for the box primitive to length = 5, width = 3, and height = 2.
   B. Create a 3 × 2 × 1.5 mesh box primitive.
   C. Move the top-back edges (short side) up 1 unit and the top-front edges down .75 units.
   D. Crease the one face in the middle of the top, then move it down 1 unit, as shown in A.
   E. Crease all faces on the bottom.
   F. Smooth the model to level 3.
   G. Nonuniformly scale the middle face on the top, which is now circular, to create an elliptical shape. Then, move the face .5 units toward the chairback.
   H. Move the three top edges on the front .5 units toward the chairback.
   I. Move the middle face in the front top down 1 unit, as shown in B.
   J. Rotate the top three back edges 45° to create a sloped chairback.
   K. Move the single edge in the middle of the back inward .5 units, as shown in C.
   L. Increase the smoothness to level 4.
   M. Save the model as P9-5.

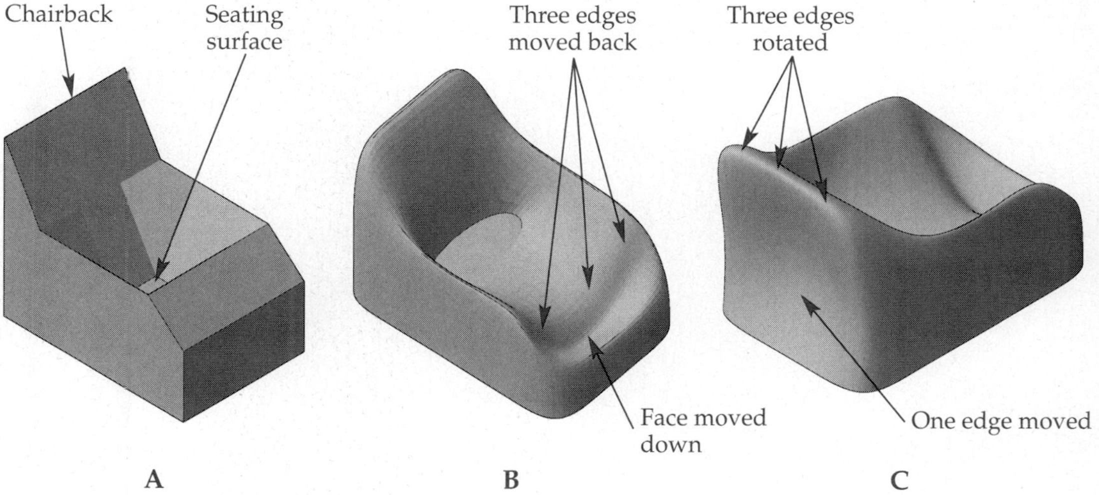

6. **Mechanical** In this problem, you will create a conceptual design for a pocket camera as a mesh shape. Change visual styles as needed throughout your work.
   A. Set the tessellation divisions for the box primitive to length = 5, width = 3, and height = 5.
   B. Create a 3 × 1 × 3 mesh box primitive.
   C. Extrude the top face second from the edge to create a button. Extrude to a height of .25 with a taper of 10°, as shown in A. Crease the top of the extruded face.
   D. Nonuniformly scale the middle face on the front by a factor of two. Then, crease the face.
   E. Move the five edges on the left side of the front outward .25 units, as shown in B.
   F. Move the three edges on the right-hand side of the top inward .375 units.
   G. Move the three edges on the right-hand side of the bottom inward .375 units.
   H. Crease all faces on the bottom.
   I. Smooth the model to level 4.
   J. Extrude the middle face on the front (which was earlier creased) inward .125 units, as shown in C.
   K. Save the model as P9-6.

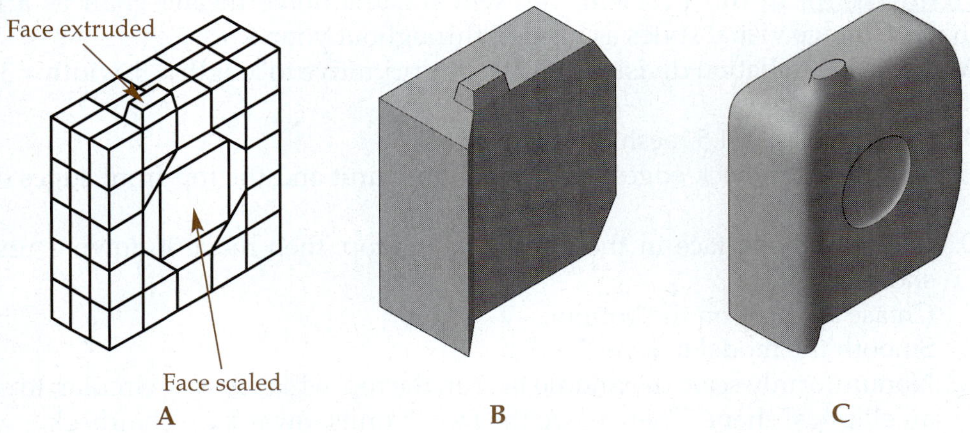

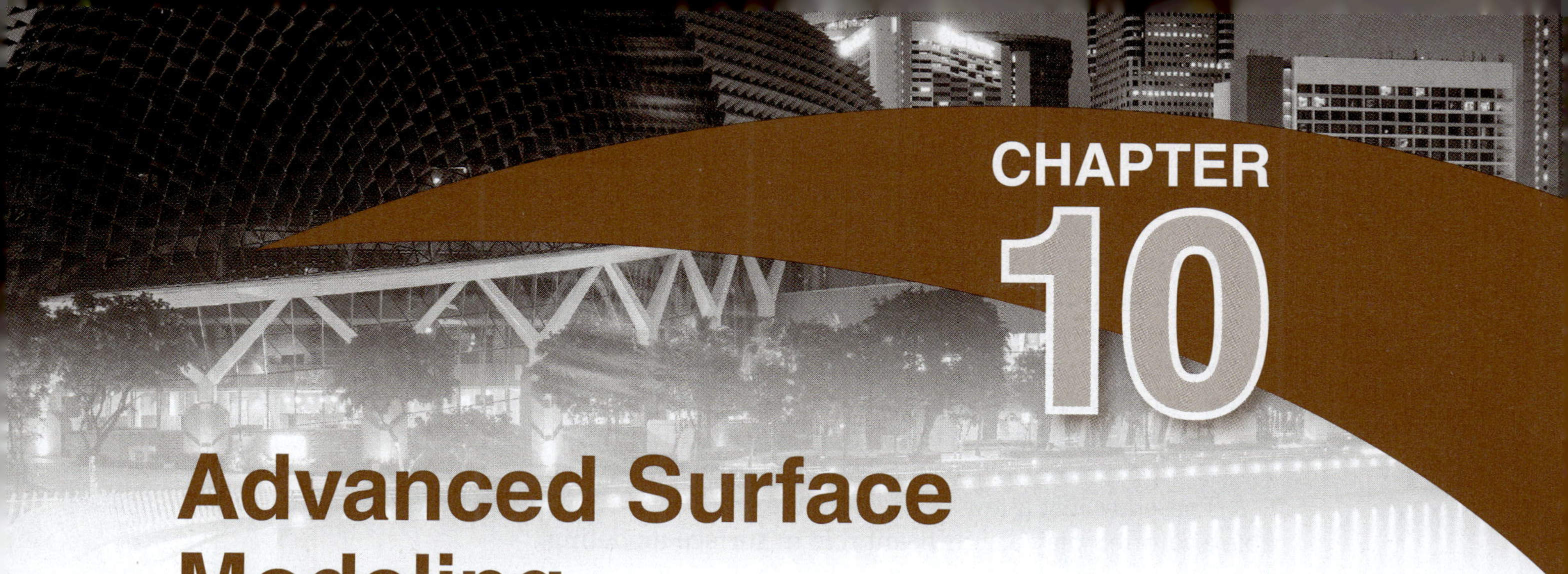

CHAPTER 10

Advanced Surface Modeling

Learning Objectives

After completing this chapter, you will be able to:

- ✓ Understand and work with different types of surface models.
- ✓ Create procedural surfaces.
- ✓ Create NURBS surfaces.
- ✓ Create network surfaces.
- ✓ Create surface models from existing surfaces.
- ✓ Blend and patch surfaces.
- ✓ Offset, fillet, extend, and trim surfaces.
- ✓ Convert existing models to NURBS surfaces.
- ✓ Edit NURBS surface control vertices.
- ✓ Convert 2D objects to surfaces.
- ✓ Thicken a surface into a solid.
- ✓ Sculpt watertight surfaces into solids.
- ✓ Extract curves from existing surfaces.

Overview

This chapter describes advanced surface modeling techniques and workflows used in AutoCAD. Surface modeling provides the ability to create a more freeform shape with tools that solid modeling cannot provide. You have been introduced to basic surface modeling techniques in previous chapters. As you have learned, one way to create surface models is to extrude, revolve, sweep, or loft profiles. This chapter builds on those techniques. In this chapter, you will develop an understanding of surface modeling techniques that can stand alone in the design process or work in combination with other modeling techniques.

As discussed in previous chapters, a ***solid model*** is created with a closed and bounded profile and has mass and volume properties. A ***mesh*** consists of vertices, edges, and faces that define the 3D mesh shape. A mesh does not have mass or volume. A ***surface model*** can be thought of as a thin-walled object with no "Z" depth. A surface model does not have mass or volume.

There are a number of workflows in AutoCAD available to the 3D designer. The following approaches can be considered depending on the nature of the work or the requirements of a specific application:

- Creating 3D models as solids, meshes, procedural surfaces, or NURBS surfaces (procedural surfaces and NURBS surfaces are discussed in the next section).
- Using Boolean operations on solids to create composite solids.
- Slicing composite solids using surfaces.
- Converting solids to mesh models.
- Converting solids to surface models.
- Converting surface models to NURBS surfaces.

These are just a few of the possible workflows. Editing techniques are also available and often play a significant role in surface modeling.

Understanding Surface Model Types

There are two basic types of surface models in AutoCAD: procedural surfaces and NURBS surfaces. A ***procedural surface*** is a standard surface object without control vertices. By default, a procedural surface, when created, is an ***associative surface***. This means that the surface maintains associativity to the defining geometry or to other surrounding surfaces. Editing the defining geometry of an associative surface, or an adjacent surface in a "chain" of associative surfaces, modifies the surface.

A ***NURBS surface*** is based on splines or curves. The acronym *NURBS* stands for *non-uniform rational B-spline*. NURBS surfaces are based on a mathematical model and are used to create organic, freeform shapes. NURBS surfaces have control vertices that can be manipulated to edit the shape of the surface with great precision. Unlike a procedural surface, a NURBS surface cannot be created as an associative surface.

A third type of surface in AutoCAD is a ***generic surface***. A generic surface has no associative history and no control vertices.

The type of surface model created is controlled by the **SURFACEMODELINGMODE** system variable. The default setting, 0, creates procedural surfaces. If the **SURFACEMODELINGMODE** system variable is set to 1, NURBS surfaces are created.

When creating a procedural surface, the **SURFACEASSOCIATIVITY** system variable setting determines whether an associative surface is created. The default setting, 1, creates associative surfaces. This system variable has no effect when creating NURBS surfaces.

Surface models can be created from either closed and bounded geometry or open profile geometry. When using modeling commands such as **EXTRUDE**, **REVOLVE**, **SWEEP**, and **LOFT**, the **Mode** option determines whether a surface model or solid model is created. Open profile curves always create surfaces, regardless of the **Mode** option setting.

The advantage to a procedural surface is the ease with which the surface can be created based on common shapes. In addition, working with procedural surfaces allows the designer to take advantage of associative modeling. Based on the design intent, the designer can use profile curves such as lines, circles, arcs, ellipses, helices, points, polylines, 3D polylines, and splines as the basis for the model. A procedural surface model can then be created using commands such as **EXTRUDE**, **REVOLVE**, **SWEEP**, **LOFT**, or **PLANESURF** (as discussed in previous chapters). If created as an associative surface, the model is linked to the defining geometry and can be modified by editing the geometry.

The commands used to create surface models are located in the **Surface** tab of the ribbon. See **Figure 10-1**. As discussed in Chapter 6, selecting a command from the **Create** panel in the **Surface** tab automatically sets the model creation mode to

Figure 10-1.
The **Surface** tab of the ribbon.

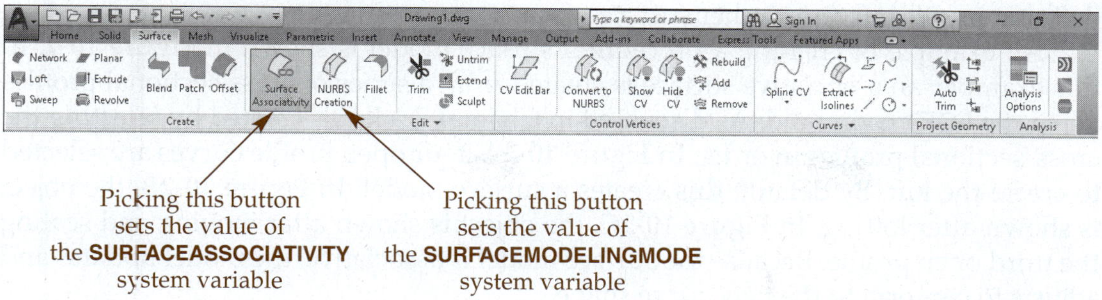

surface. Referring to **Figure 10-1**, notice the **Surface Associativity** and **NURBS Creation** buttons in the **Create** panel. The status of the **Surface Associativity** button indicates the **SURFACEASSOCIATIVITY** system variable setting. By default, the button has a blue background to indicate the system variable is turned on. The status of the **NURBS Creation** button indicates the **SURFACEMODELINGMODE** system variable setting. By default, this button does not have a blue background. This indicates that procedural surfaces are created by default.

PROFESSIONAL TIP

Procedural surfaces are also referred to as *explicit surfaces*. The terms *procedural surface* and *explicit surface* are interchangeable. This text uses procedural surfaces to illustrate various methods of surface creation. NURBS surfaces, as discussed later in this chapter, are used in creating models that represent sophisticated freeform shapes. Procedural surfaces can be converted to NURBS surfaces in order to model more sophisticated surface model shapes.

NOTE

Planar surface creation is discussed in Chapter 2. Extruded and revolved surfaces are discussed in Chapter 6. Swept and lofted surfaces are discussed in Chapter 7.

Exercise 10-1

www.g-wlearning.com/CAD/

Complete the exercise on the companion website.

Working with Associative Surfaces

Procedural surfaces, when created, are associative by default. An associative surface adjusts to the modifications made to the defining profile geometry (or other adjoining surfaces). This provides flexibility in the design. However, it is important to remember that when modifying the shape of an associative surface, you modify the profile geometry, *not* the surface. Modifying the profile geometry maintains the associative relationship. If you pick on the surface and then attempt to modify it, AutoCAD issues a warning that the surface will lose its associativity with the defining curve, surface, or parametric equation. If you choose to continue with the operation, the associativity is lost. You can cancel the operation to preserve the associativity.

Set the **SURFACEASSOCIATIVITY** system variable to 1 to create associative surfaces. If the system variable is set to 0, surfaces that are created have no associativity to defining profile curves or other surfaces.

An example of creating a procedural surface model is shown in **Figure 10-2**. In this example, an associative surface is created from a series of cross-sectional profiles using the **LOFT** command. As discussed in Chapter 7, a loft is created by selecting the cross-sectional profiles in order. In **Figure 10-2A**, four open profile curves are selected to create the loft. By default, this creates a surface model. In **Figure 10-2B**, the object is shown after lofting. In **Figure 10-2C**, the object is shown after moving and scaling the third open profile. Because the surface model is associative, the model updates and adjusts to conform to the new curve shape.

In **Figure 10-3**, a loft surface is created from three closed cross-sectional profiles (circles). The surface is created by using the **Mode** option or by selecting the command from the **Surface** tab in the ribbon. In **Figure 10-3B**, the object is shown after moving and scaling the third closed profile at the top. Because the surface is associative, the model updates and conforms to the new curve shape.

PROFESSIONAL TIP

The **DELOBJ** system variable is ignored if the **SURFACEASSOCIATIVITY** system variable is set to 1. Also, when creating a NURBS surface, the surface is a NURBS surface and is not associative.

Figure 10-2.
Creating a loft surface as an associative surface model. A—The original cross-sectional profiles. B—The model after using the **LOFT** command. C—The model after editing one of the cross-sectional profiles.

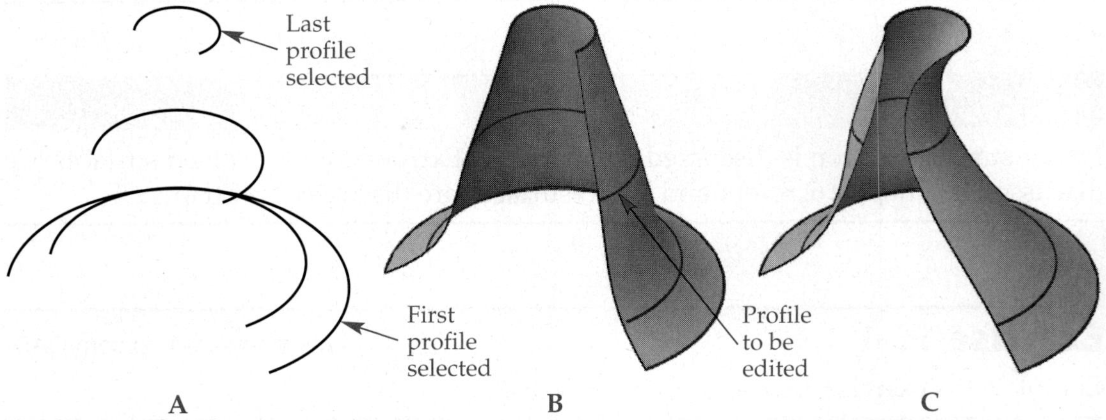

Figure 10-3.
Modifying a loft surface created from closed profiles (circles). A—The original loft surface. B—The model after editing the closed profile at the top.

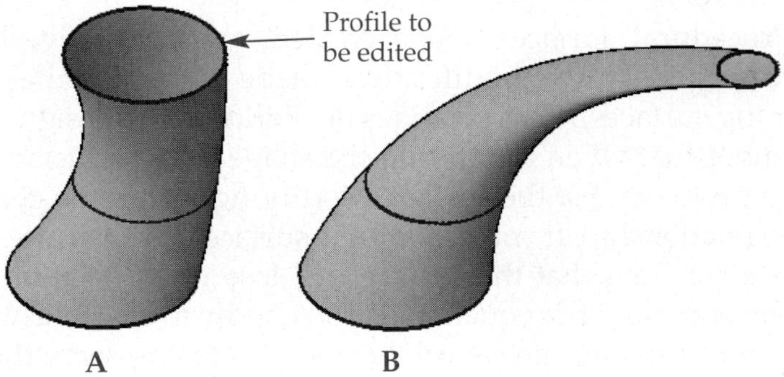

Removing Surface Associativity

The surface associativity can be removed from a surface once the surface has been created. This can be done by selecting the surface and opening the **Properties** palette. The Maintain Associativity property in the **Surface Associativity** category controls the associativity of the surface. See **Figure 10-4**. By default, the property is set to Yes. Selecting **Remove** from the drop-down list removes the associativity and changes the property to None. This converts the surface to a generic surface.

The Show Associativity property in the **Surface Associativity** category controls whether adjoining associative surfaces are highlighted when a surface is selected in order to indicate dependency. When this property is set to Yes and a surface is selected, AutoCAD highlights other surfaces to which the surface is dependent. This can be useful for identifying associative relationships in a chain of surfaces.

PROFESSIONAL TIP

When moving, scaling, or rotating an associative surface, be sure to select the underlying curve geometry defining the surface. Failure to select the underlying geometry will result in the loss of the associativity.

Determining Modeling Workflows for Procedural Surfaces and NURBS Surfaces

When the design of a 3D model requires a freeform shape that would be difficult to create using solids, start by creating a procedural surface. You can convert the surface as required. A practical application is creating a surface model of a car fender. Start with a lofted surface based on four guide curves. Finish creating the fender by creating several procedural surfaces or patches, as discussed later in this chapter. Then, convert the fender surfaces to NURBS surfaces as needed and add further editing techniques for a more freeform sculpted shape.

Different factors determine when to use procedural surface modeling and NURBS surface modeling. For example, create procedural surfaces when it is important to maintain associativity and you plan to edit the original geometry. On the other hand, NURBS surfaces have control vertices that typically permit greater flexibility when editing. NURBS surfaces are often very useful for modeling organic shapes. The extent to which the design will require further editing can serve as a guideline for determining the best modeling approach.

The following sections discuss surface modeling commands and techniques available in AutoCAD. Procedural surfaces are shown in examples as the default creation method. NURBS surface creation and editing techniques are covered later in this chapter.

Figure 10-4.
Associativity settings for a surface are accessed in the **Properties** palette.

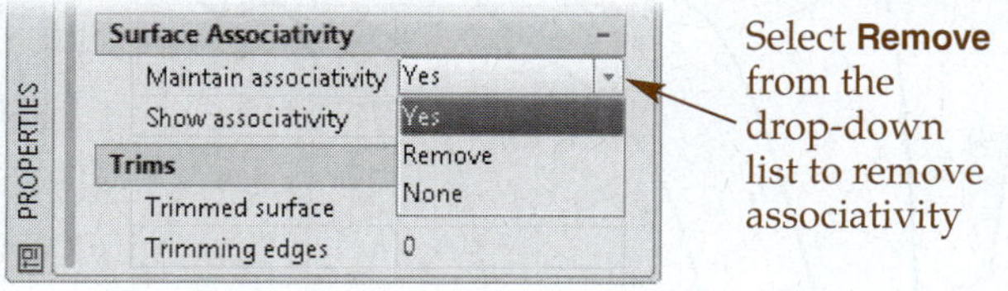

Creating Network Surfaces

A *network surface* is a surface model created by a group or "network" of profile curves or edges. A network surface is similar to a loft surface. As with a loft, the defining profiles can be open or closed curves, such as splines. The defining profiles can also be the edges of existing objects, including region edges, surface edge subobjects, and solid edge subobjects. The curves or edges selected can intersect at coincident points, but do not have to intersect.

SURFNETWORK

Ribbon
Surface
> Create

Network Surface
Type
SURFNETWORK

The **SURFNETWORK** command is used to create network surfaces. After selecting this command, select the curves or edges defining the first direction of the surface. Make sure to select the curves in the order of surface creation. Then, press [Enter]. Next, select the curves or edges defining the second direction of the surface. Press [Enter] when you are done selecting the profiles. This creates the surface and ends the command. See **Figure 10-5**. A network surface is created as an associative surface by default.

The curves selected for the two directions define the U and V directions of the surface. The U and V directions can be thought of as the local directions of the surface and can be defined in either order. The U and V directions define the "flow" of the surface.

Figure 10-6 shows examples of creating network surfaces from similar sets of profile curves. In **Figure 10-6A**, a series of connected profile curves defines the network surface. In **Figure 10-6B**, two of the curves do not intersect with other profiles. Notice the differences between the resulting surface models.

Creating a network surface from region edges, surface subobject edges, and solid subobject edges is shown in **Figure 10-7**. Creating a network surface in this manner may result in some unexpected surface shapes. To select the profile edges, press and hold the [Ctrl] key. You can also use the edge subobject filter by selecting **Edge** from the **Selection** panel in the **Home** tab of the ribbon. In **Figure 10-7**, the four objects are located at different "Z" heights. The network surface is created from the four nonintersecting edges.

Referring to **Figure 10-6**, when working in a wireframe display, isolines appear to represent the curved surfaces of the surface model. The **SURFU** and **SURFV** system variables control the number of isolines displayed in the U and V directions of the surface. The number of isolines does not include the lines defining the object's

Figure 10-5.
Creating a network surface. Splines are used as the profiles in this example. A—The profiles used to define the surface are selected in the numbered order shown. Profiles 1–3 define the first direction and are shown in color. Profiles 4–8 define the second direction. B—The resulting surface model.

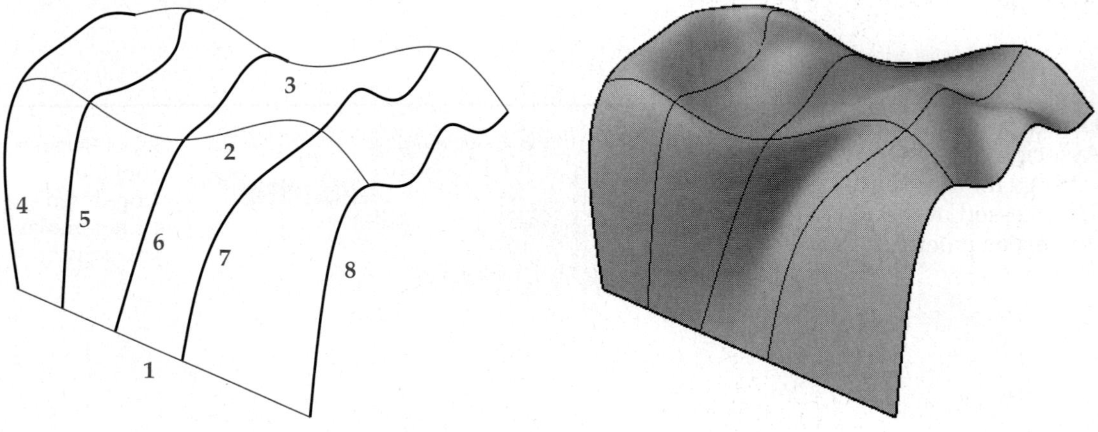

A B

Figure 10-6.
Network surfaces. A—Profiles 1–3 define the first direction of the surface and are shown in color. Profiles 4 and 5 define the second direction of the surface. The resulting surface model is shown in wireframe and shaded form. B—Profiles 1–4 define the first direction of the surface and are shown in color. Profiles 5 and 6 define the second direction of the surface. The resulting surface model is shown in wireframe and shaded form.

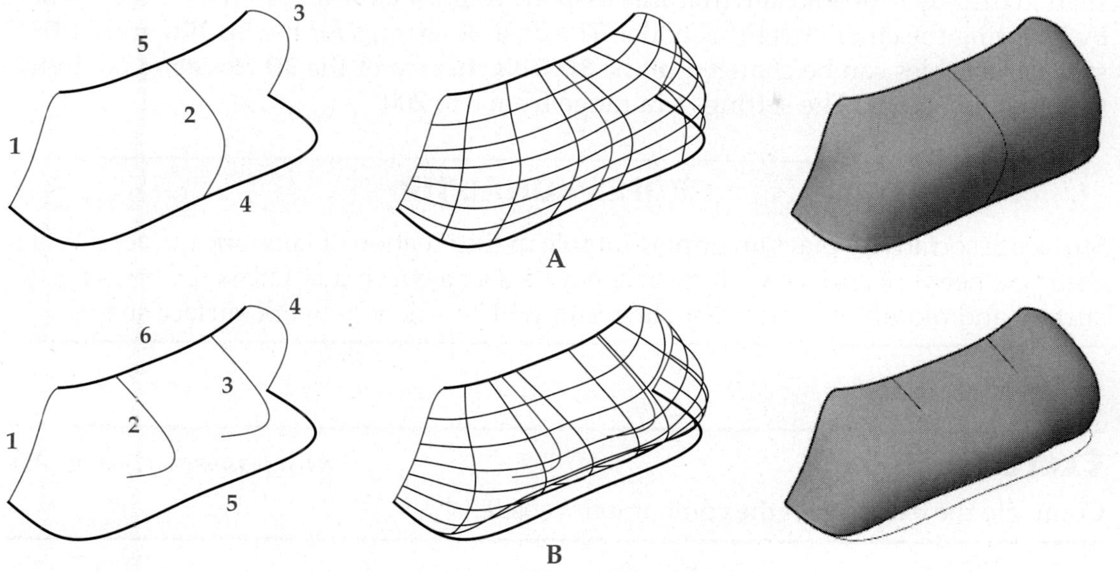

Figure 10-7.
Creating a network surface from the edges of existing objects. A—Two region edges (1 and 2) are selected to define the first direction of the surface. A surface subobject edge (3) and solid subobject edge (4) are selected to define the second direction of the surface. B—The resulting surface model.

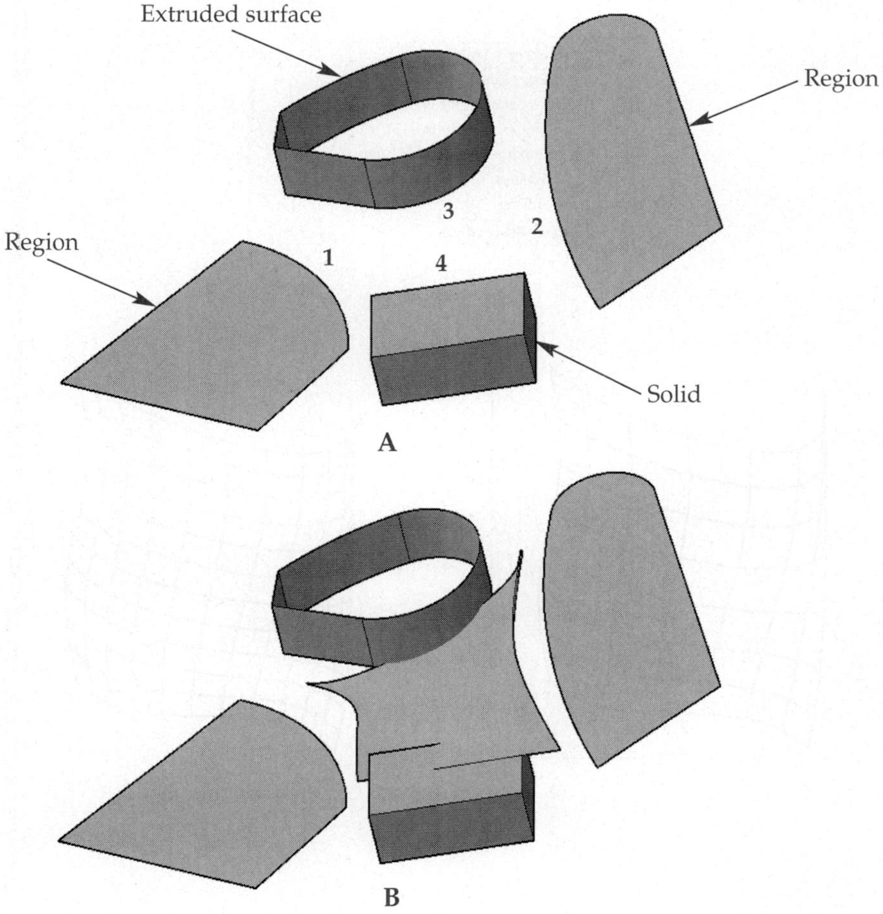

boundary. The default setting for both system variables is 6. These settings can be changed for a surface by selecting the surface and opening the **Properties** palette. The U isolines and V isolines properties in the **Geometry** category control the number of isolines displayed in the U and V directions. See **Figure 10-8**. Setting a higher value can give you a better understanding of the curvature of the model. When working in a shaded display instead of a wireframe display, you can view the isoline representation by hovering the cursor over the model. The default settings for the **SURFU** and **SURFV** system variables can be changed in the **3D Objects** area of the **3D Modeling** tab in the **Options** dialog box. The settings can range from 0 to 200.

PROFESSIONAL TIP

Surface associativity plays an important role in the creation of network surfaces. When editing a network surface with associativity, select a curve that forms the basis for the surface and modify it as needed. The result will be a new network surface shape.

Exercise 10-2

www.g-wlearning.com/CAD/

Complete the exercise on the companion website.

Figure 10-8.
Isolines defining the curvature of the surface model appear when working in a wireframe display. A—The **Properties** palette is used to set the number of isolines in the U and V directions of the surface. B—A network surface with the default number of isolines in the U and V directions. C—The surface after increasing the values of the U isolines and V isolines properties from 6 to 10.

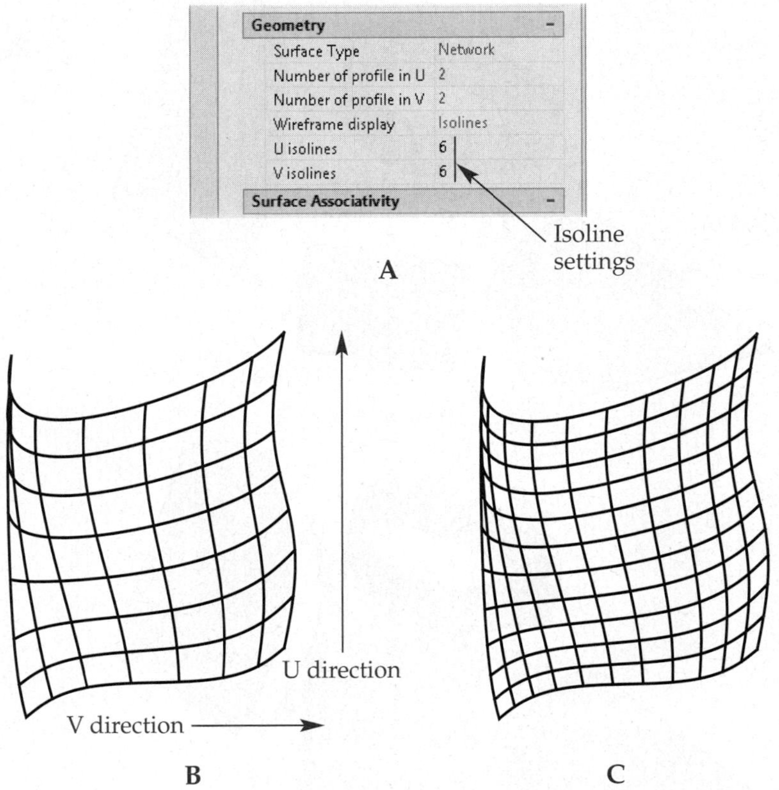

Creating Surfaces from Existing Surfaces

In addition to creating surfaces from profile curves, you can create surfaces from existing surfaces using the **SURFBLEND**, **SURFPATCH**, **SURFOFFSET**, **SURFFILLET**, and **SURFEXTEND** commands. These commands are discussed in the following sections. Surface models created with these commands are created as associative surfaces by default. This maintains associativity between the surfaces used to create the surface and the resulting surface.

Blend Surfaces

When working with surface models, there are situations when you need to "blend" together surfaces that do not meet or touch. The **SURFBLEND** command is used to create a ***blend surface*** between two surface edges or two solid edges. When blending surfaces, you select the surface edges to blend, *not* the surfaces.

Select the **SURFBLEND** command and then select the first edge to blend. Press and hold the [Ctrl] key to select the first edge or use the edge subobject filter. You can select multiple edges or use the **Chain** option to select a chain of continuous edges. Press [Enter] after defining the first edge. Next, select the second edge. Select a single edge or multiple edges, or use the **Chain** option to select a chain of continuous edges. When you press [Enter], a preview of the surface appears. You can press [Enter] to accept the default settings, as shown in Figure 10-9, or you can use the **Continuity** and **Bulge magnitude** options to specify the continuity and bulge magnitude settings at the edges. Different settings can be applied at each edge. The settings are similar to those when creating a loft, as discussed in Chapter 7.

Ribbon
Surface
> Create

Surface Blend
Type
SURFBLEND

SURFBLEND

Continuity defines how the surfaces blend together at the starting and ending edges. The following options are available:

- **G0 (positional continuity).** This option creates a sharp transition between surfaces. The position of the surfaces is maintained continuously at the surface edges. This option is used for creating flat surfaces.
- **G1 (tangential continuity).** This option forms surfaces so that the end tangents match at the edges. The two surfaces blend together tangentially. This is the default option, as shown in Figure 10-9B.
- **G2 (curvature).** This option creates a curvature blend between the surfaces. The surfaces share the same curvature.

Examples of blend surfaces with different continuity settings are shown in Figure 10-10.

Figure 10-9.
A blend surface is created between two existing surfaces by selecting a starting and ending edge. A—Two existing loft surfaces. B—The model after creating a blend surface with the default settings of the **SURFBLEND** command.

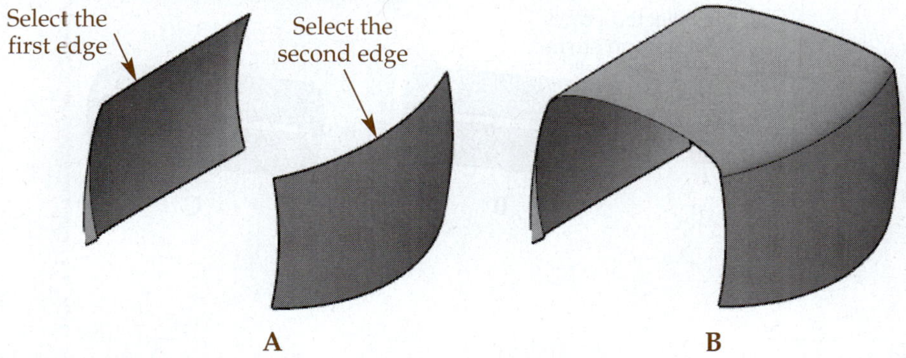

Bulge magnitude defines the size or "bulge" of the radial transition where the surfaces meet. See **Figure 10-11**. The default setting is 0.5. Valid values range from 0 to 1. A greater value is valid, but results in a larger roundness to the blend. Using different surface modeling techniques instead of entering a value greater than 1 is recommended. If the surface is set to G0 (positional continuity), changing the default bulge magnitude value has no effect.

Using different continuity and bulge magnitude settings modifies the surface and provides a way to create different blend surface shapes. You can change the continuity by using the grips that appear when creating the blend surface. Picking on a grip displays a menu with the continuity options. The same grips appear when selecting a blend surface after it has been created. You can also use the **Properties** palette to edit the settings of a blend surface.

Blend surfaces can also be created between the edges of regions or solid objects. An example of creating a blend surface between two solid subobject edges to form a cap is shown in **Figure 10-12**.

PROFESSIONAL TIP

Continuity settings are retained when exporting a 3D model to other 3D CAD modeling applications.

Figure 10-10.
Blend surfaces created with different continuity settings. A—G0 continuity. B—G1 continuity. C—G2 continuity.

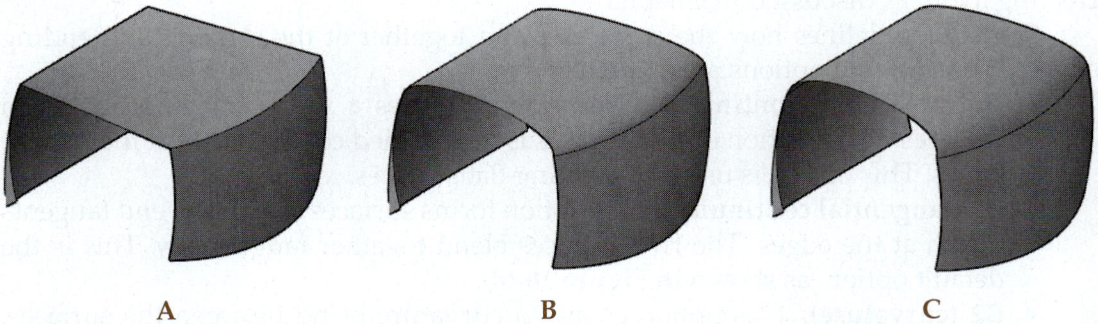

Figure 10-11.
Bulge magnitude determines the size of the radial transition at the edges where surfaces meet. A—The original model consists of two surfaces created from extruded arcs. B—The model after capping the ends with blend surfaces. For each surface, the continuity is set to G1 and the bulge magnitude is set to 0. C—For each surface, the continuity is set to G1. The bulge magnitude at the left edge is set to 0.5. The bulge magnitude at the right edge is set to 1.

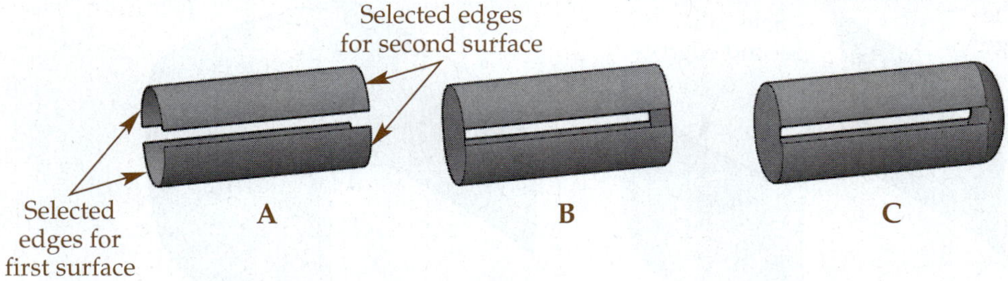

Figure 10-12.
Creating a blend surface between two solid subobject edges. A—The original model consists of two solid boxes. B—A blend surface is created between two subobject edges. The continuity is set to G2 and the bulge magnitude is set to 1.

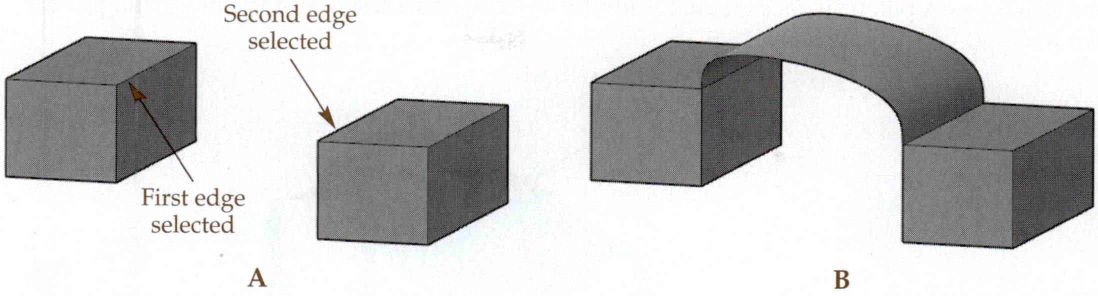

Exercise 10-3

www.g-wlearning.com/CAD/

Complete the exercise on the companion website.

Patch Surfaces

A ***patch surface*** is used to create a "patch" over an opening in an existing surface. A patch surface is used when it is necessary to close an opening or gap in the model. You can think of a patch surface as one of the many squares making up a quilt.

The **SURFPATCH** command is used to create a surface patch based on one or more edges forming a closed loop. You can select one or more surface edges or a series of curves. As when using the **SURFBLEND** command, you can specify the continuity and bulge magnitude to define the curvature of the surface.

Ribbon
Surface
> Create

Surface Patch
Type
SURFPATCH

SURFPATCH

Select the **SURFPATCH** command and then select one or more surface edges defining a closed loop. You can use the **Chain** option to select a chain of continuous surface edges. You can also use the **Curves** option to select multiple curves forming a closed loop. After selecting the edges or curves, press [Enter]. A preview appears and you can press [Enter] to create the surface using the default settings. The **Continuity** and **Bulge magnitude** options can be used to change the default settings as previously discussed. The default continuity setting is G0. The default bulge magnitude setting is 0.5.

The **Guides** option allows you to use a guide curve to constrain the shape of the surface patch. You can select one or more curves to define the guide curve. You can also select points to define the guide curve. When selecting points, use object snaps as needed.

Examples of creating patch surfaces are shown in **Figure 10-13**. In **Figure 10-13A**, the top of the tent requires a patch. To create the patch, the single edge representing the opening in the model is selected with the default **Surface edges** option. In **Figure 10-13B**, the patch surface is created with the continuity set to G1. This is the appropriate setting for the patch surface. In this case, you would want the patch to be tangent to the existing surface. The default bulge magnitude (0.5) is used. In **Figure 10-13C**, the patch surface is created with the continuity set to G1, but the bulge magnitude is set to 1. Notice the different result.

When using the **Guides** option, draw a curve to serve as the guide curve prior to selecting the **SURFPATCH** command. In **Figure 10-14A**, the top of the tent requires a patch. A new middle post has been added to the model, and a spline is drawn to serve as a guide curve. After selecting the **SURFPATCH** command, select the surface edge. Then, select the **Guides** option, select the spline, and press [Enter]. Press [Enter] to create the patch surface, or adjust the continuity and bulge magnitude settings as needed. See **Figure 10-14B**.

Figure 10-13.
Creating a patch surface to close an opening in a surface model. A—The original model. The single surface edge indicated forms a closed loop and is selected to generate the patch surface. B—A patch surface created with the continuity set to G1 and the bulge magnitude set to 0.5. C—A patch surface created with the continuity set to G1 and the bulge magnitude set to 1.

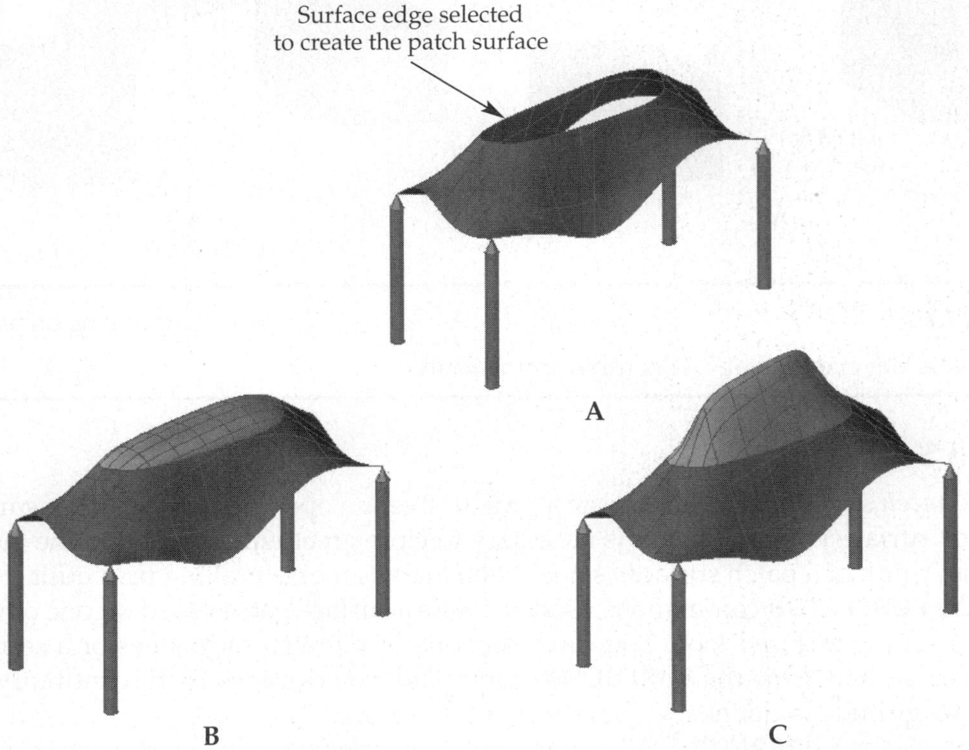

Figure 10-14.
Using a guide curve to constrain the shape of a patch surface. A—The original model. The middle post has been added. The spline is drawn for use with the **Guides** option. B—The patch surface created after selecting the spline as the guide curve. Notice the resulting shape.

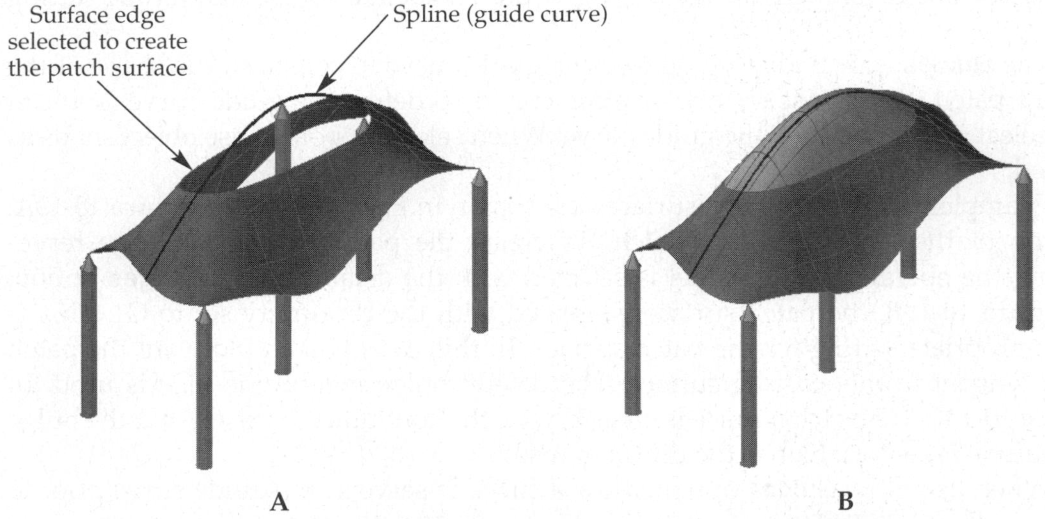

In **Figure 10-15**, the game controller is to be redesigned with a new top shape. The model has been converted from a mesh model to a surface. In the original model (the mesh model), the top faces were deleted. The surface opening has eight continuous

Figure 10-15.
Using the **Chain** option to select multiple surface edges to define the patch surface. A—The original model is a surface converted from a mesh model. The top opening includes eight surface edges. B—The patch surface created after using the **Chain** option to select the edges. The continuity is set to G2 and the bulge magnitude is set to 0.7.

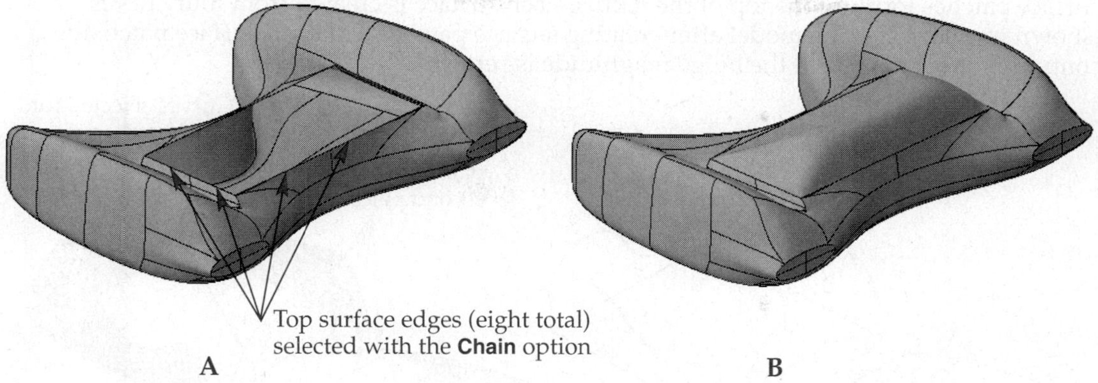

surface edges. The **Chain** option is used to assist in selecting edges to create the surface patch. After selecting the **SURFPATCH** command, select the **Chain** option and select one of the edges. The remaining edges are automatically selected. Next, press [Enter]. You can adjust the continuity and bulge magnitude settings or press [Enter] to create the patch surface. See **Figure 10-15B**.

Using the **Curves** option of the **SURFPATCH** command is shown in **Figure 10-16**. In **Figure 10-16A**, multiple curves have been created to design a sophisticated free-form shape. First, the **LOFT** command is used to create six loft surfaces defining the sides. Refer to the shaded surfaces shown in **Figure 10-16B**. Then, the **Curves** option of the **SURFPATCH** command is used to create three surface patches forming the top of the model. Refer to **Figure 10-16C**. To use the **Curves** option, enter it after selecting the **SURFPATCH** command. Then, select the curves defining the patch surface.

NOTE

Surfaces created using the **SURFPATCH** command may differ from those created with the **LOFT** command. In addition, you may come across design situations where curves used with the **LOFT** command will not create a surface, but will when used with the **SURFPATCH** command.

PROFESSIONAL TIP

The **PREVIEWCREATIONTRANSPARENCY** system variable controls the transparency of surface previews when using the **SURFBLEND**, **SURFPATCH**, and **SURFFILLET** commands. The default setting is 60. Setting a higher value increases the transparency of the surface preview.

Exercise 10-4

www.g-wlearning.com/CAD/

Complete the exercise on the companion website.

Figure 10-16.
Using the **LOFT** and **SURFPATCH** commands to create a freeform design from a series of curves. A—The original model. The **LOFT** command is used to create six separate loft surfaces to form the sides. Each loft surface consists of two cross-sectional curves (shown in color). B—The **Curves** option of the **SURFPATCH** command is used to create three separate surface patches forming the top of the model. Each surface is created from four curves (shown in color). C—The model after creating surface patches. For each surface patch, the continuity is set to G1 and the bulge magnitude is set to 0.

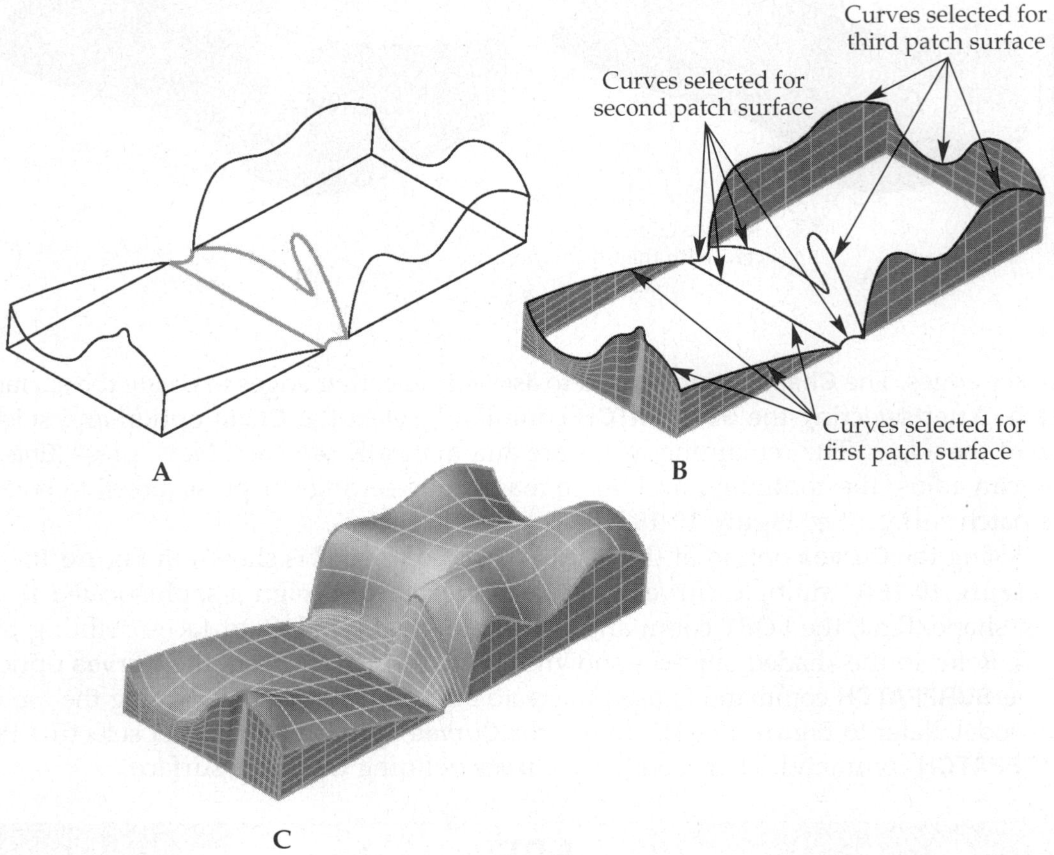

Offsetting Surfaces

The **SURFOFFSET** command allows you to offset a surface to create a new, parallel surface at a specified distance. You can offset a surface in one direction or in both directions from an existing surface. You can also offset a region to create a new surface.

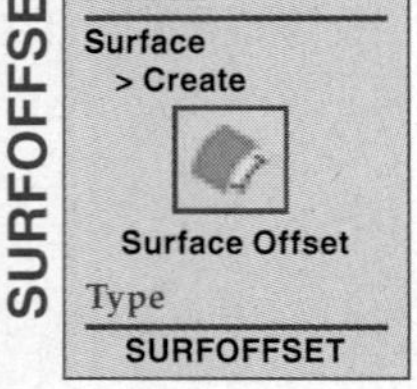

Select the **SURFOFFSET** command and select the surface or region to offset. Then, press [Enter]. A preview of the offset surface appears with offset arrows indicating the direction of the offset. See **Figure 10-17A**. Next, specify the offset distance. If the design calls for the offset surface to be located on the opposite side, select the **Flip direction** option. You can offset the surface to both sides by selecting the **Both sides** option. After specifying the offset distance, press [Enter] to create the offset surface. In **Figure 10-17B**, the hair dryer housing is offset to the inside of the existing surface at a distance of .125.

The **Solid** option allows you to create a new solid based on the specified offset distance. In **Figure 10-17C**, the **Solid** option is used to create a solid. This is similar to using the **THICKEN** command, as discussed later in this chapter.

The **Connect** option can be used when you have more than one surface to offset and you need to maintain connection between the surfaces. When using this option, the original surfaces must be connected. See **Figure 10-18**.

Figure 10-17.
Using the **SURFOFFSET** command. A—After selecting the surface to offset, offset arrows are displayed to indicate the direction of the offset. B—The surface is offset to the inside of the hair dryer housing at a distance of .125. C—The **Solid** option is used to create a new solid using the specified offset distance from the base surface.

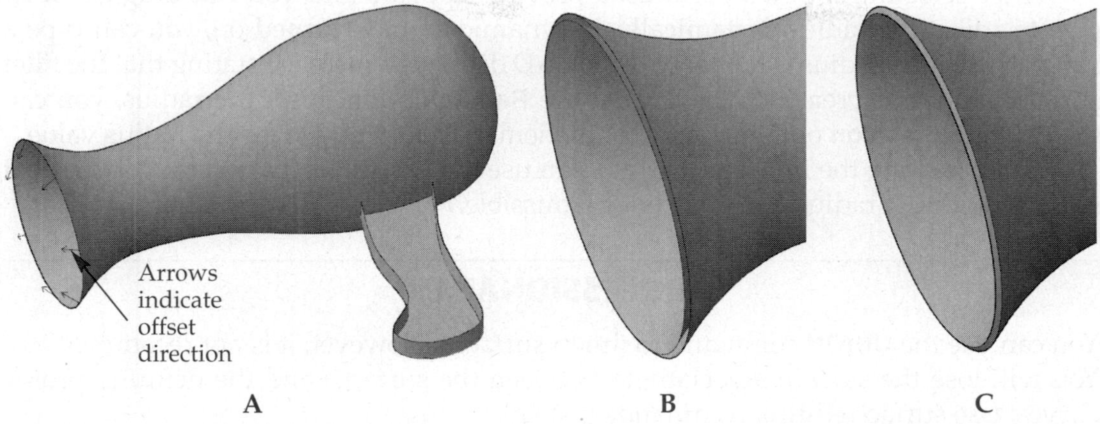

Figure 10-18.
The **Connect** option of the **SURFOFFSET** command is used to maintain the connection between surfaces when offsetting multiple surfaces. A—The original model. B—The vertical surfaces are offset using the **Connect** option.

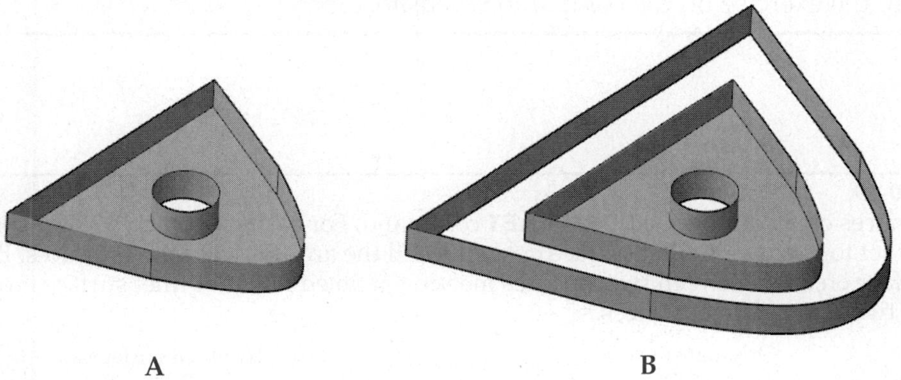

The **Expression** option allows you to enter an expression to constrain the offset distance. Surface associativity must be enabled in order to use this option.

Exercise 10-5

www.g-wlearning.com/CAD/

Complete the exercise on the companion website.

Creating Fillet Surfaces

The **SURFFILLET** command is used to create a fillet between two existing surfaces. The fillet created is a rounded surface that is tangent to the existing surfaces. You can create fillet surfaces from existing surfaces or regions. The existing objects do not have to intersect. Using the **SURFFILLET** command is similar to using other fillet commands in AutoCAD. Commands used to fillet solids are introduced in Chapter 8.

To create a fillet surface, select the **SURFFILLET** command. First, set a radius using the **Radius** option. Then, select two surfaces. AutoCAD stores the radius you specify as the setting for the **FILLETRAD3D** system variable. If you do not specify a radius, the current **FILLETRAD3D** system variable setting is used.

Ribbon
Surface
> Edit
Surface Fillet
Type
SURFFILLET

SURFFILLET

Creating a fillet surface between surfaces that do not meet is shown in **Figure 10-19A**. Creating a fillet surface between surfaces that share an edge or intersect is shown in **Figure 10-19B** and **Figure 10-19C**. By default, the existing surfaces are trimmed to form the new surface. The surface trimming mode can be set by using the **Trim surface** option.

After selecting the two surfaces, a preview appears and you can drag the fillet grip to adjust the radius dynamically. If dynamic input is turned on, you can type a value. If the fillet radius is too large, AutoCAD displays a message stating that the fillet surface cannot be created. When using the **Radius** option to set the radius, you can select the **Expression** option to enter a mathematical expression for the radius value.

After creating the fillet surface, you can use the **Properties** palette to edit the fillet surface radius. A radius of zero is not permissible.

PROFESSIONAL TIP

You can use the **UNION** command to union surfaces. However, it is not recommended. You will lose the surface associativity between the surfaces and the defining profile curves. Use surface editing commands instead.

Exercise 10-6

www.g-wlearning.com/CAD/

Complete the exercise on the companion website.

Figure 10-19.
Fillet surfaces created with the **SURFFILLET** command. For each example, the **Trim surface** option is set to **Yes**. A—A fillet surface created to fill the area between two surfaces. B—A fillet surface created between two surfaces meeting at an edge. C—A fillet surface created between two intersecting surfaces.

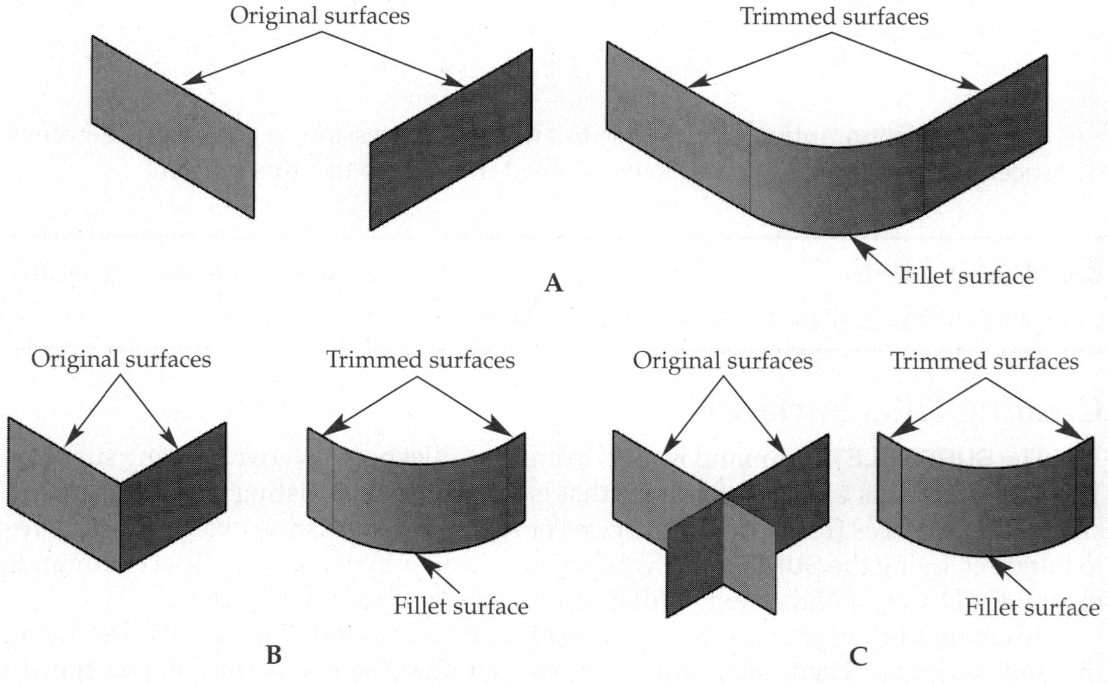

Extending Surfaces

You can add length to an existing surface using the **SURFEXTEND** command. When extending a surface, you can specify whether the new surface is created as a continuation of the existing surface or as a new surface. The surface extends to a new length using the specified distance.

Select the **SURFEXTEND** command and select one or more surface edges to extend. After selecting an edge, press [Enter]. A preview of the extended surface appears and you can drag the cursor dynamically to set the distance. You can also enter a distance by typing a value. The **Expression** option allows you to enter a mathematical expression for the extension distance. If you specify a distance and press [Enter], the surface is extended using the default settings. See **Figure 10-20.**

Ribbon
Surface > Edit
Surface Extend
Type
SURFEXTEND

SURFEXTEND

Before specifying the extension distance, you can use the **Modes** option to specify the extension mode. The two options are **Extend** and **Stretch**. The default **Extend** option is used to extend the surface in the same direction as the existing surface and attempt to maintain the surface shape based on the surface contour. The **Stretch** option is also used to extend the surface in the same direction as the existing surface. However, the resulting extension may not have the same surface contour.

After specifying the extension mode, you can use the **Creation type** option to set the type of surface created. The two options are **Append** and **Merge**. The default **Append** option is used to create a new surface extending from the original surface. This option results in two surfaces instead of one surface. The **Merge** option is used to extend the surface as one surface. After extending a surface, you can use the **Properties** palette to edit the extension distance.

Exercise 10-7

www.g-wlearning.com/CAD/

Complete the exercise on the companion website.

Trimming Surfaces

The **SURFTRIM** command can be used to trim surfaces or regions using other existing surfaces. You can trim any part of a surface where the surface intersects with another surface, region, or curve. In addition, you can project an existing object onto a surface to serve as a trimming boundary. The object to be trimmed and the cutting object do not have to intersect. When an associative surface is trimmed, it remains associative and retains the ability to be modified by editing the cutting object.

Figure 10-20. Extending an existing surface using the **SURFEXTEND** command. A—The original model. B—The extended surface.

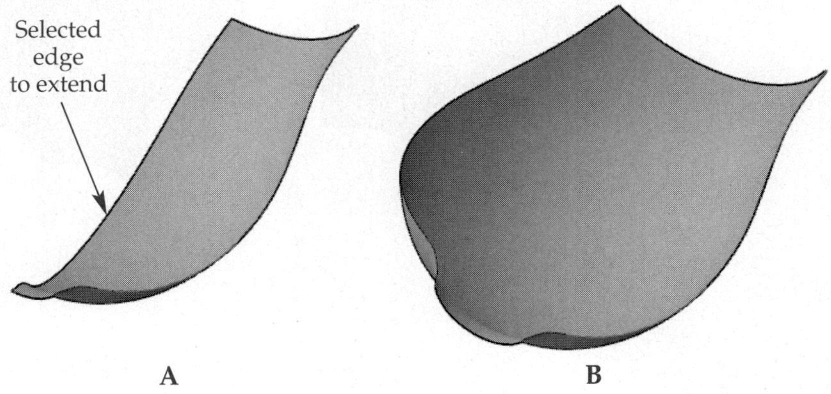

Ribbon
Surface
> Edit

Surface Trim
Type
SURFTRIM

Select the **SURFTRIM** command, select one or more surfaces or regions to trim, and press [Enter]. Next, you are prompted to select the cutting objects. Select one or more curves, surfaces, or regions. Then, press [Enter]. You are then prompted to select the surface areas to be trimmed. Select one or more areas. As you select each area, it is trimmed by the cutting object(s). If you trim an area that you wish to restore, use the **Undo** option. When you are finished trimming, press [Enter] to end the command. See **Figure 10-21.**

The **Extend** and **Projection direction** options are available after selecting the **SURFTRIM** command. The **Extend** option determines whether a surface used as a cutting edge is extended to meet the surface to be trimmed. By default, this option is set to **Yes**. The **Projection direction** option specifies the projection method used for projected geometry, as discussed in the next section.

Exercise 10-8

www.g-wlearning.com/CAD/

Complete the exercise on the companion website.

Using Projected Geometry to Trim Surfaces

With the **SURFTRIM** command, you can trim surfaces using cutting objects other than existing surfaces. Objects used in this manner are referred to by AutoCAD as *curves*. Selecting a curve, such as an arc or circle, allows you to project the geometry onto the surface and use it as the cutting edge. See **Figure 10-22**. In the example shown, the cell phone case is selected as the surface to trim. The arcs located above the cell phone case are selected as the cutting edges. When selected, the arcs are projected onto the surface by AutoCAD. See **Figure 10-22B**. The areas to be trimmed are then selected to complete the sides. See **Figure 10-22C**.

To use a curve instead of an existing surface for trimming, select the curve after selecting the object to trim. You can select lines, arcs, circles, ellipses, polylines, splines, and helices. The **Projection direction** option of the **SURFTRIM** command can be used

Figure 10-21.
Trimming surfaces with the **SURFTRIM** command. A—The original model consists of three loft surfaces. Two of the loft surfaces are to be trimmed. B—The model after trimming.

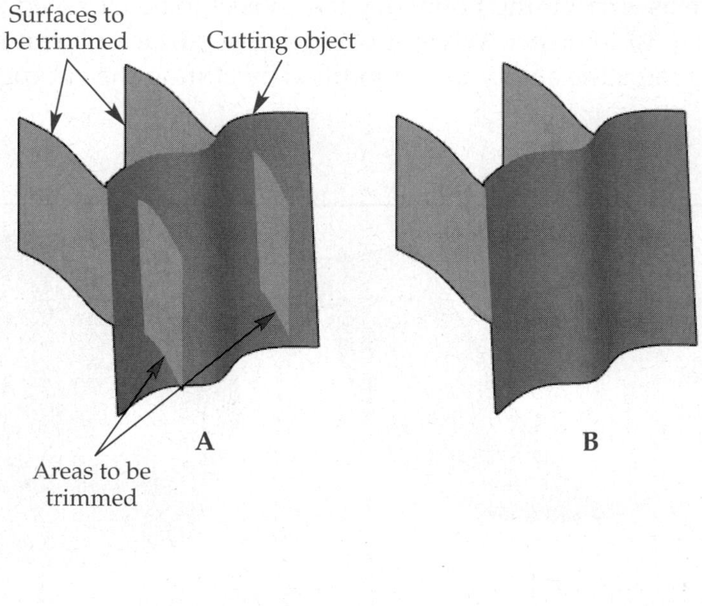

Figure 10-22.
Using projected geometry with the **SURFTRIM** command. A—The cell phone case model is a surface created from a mesh. The arcs drawn above the model are used for trimming the sides. B—Selecting the two curves and pressing [Enter] projects the curves to the model surface. The areas within the projected geometry on each side are selected as the areas to trim. C—The model after trimming.

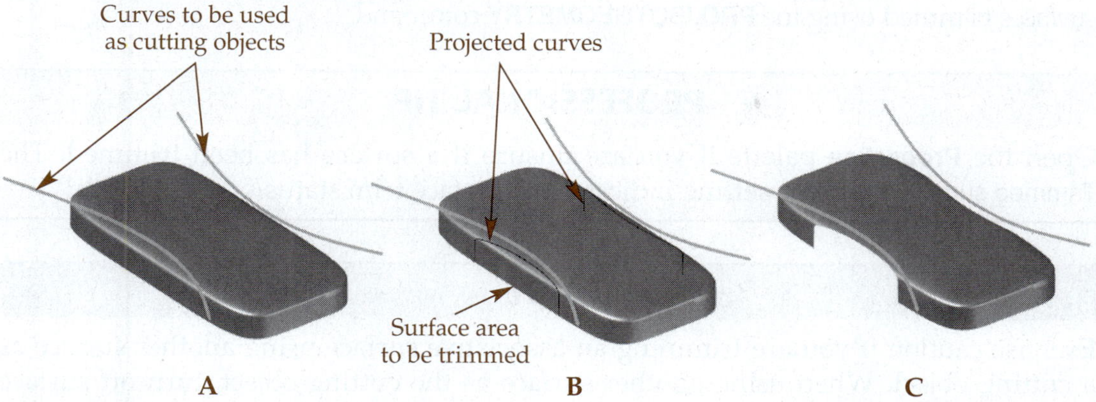

to set the projection method used by AutoCAD when projecting curves onto a surface. The following settings are available.

- **Automatic.** The cutting object is projected onto the surface to be trimmed. The projection is based on the current viewing direction. In a plan view, the projection of the cutting object is in the viewing direction. In a 3D view, the projection of a planar curve is normal to the curve, and the projection of a 3D curve is parallel to the direction of the Z axis of the current UCS. The **Automatic** option is set by default.
- **View.** The cutting object is projected in a direction based on the current view.
- **UCS.** The cutting object is projected in the positive or negative direction of the Z axis of the current UCS.
- **None.** The cutting object is not projected and must lie on the surface in order to perform the trim.

PROFESSIONAL TIP

The **SURFTRIM** command defaults to the **Automatic** option. Automatically projected geometry is used in most trim situations.

NOTE

When using the **PROJECTGEOMETRY** command, the **SURFACEAUTOTRIM** system variable controls automatic trimming. The default setting is 0. A setting of 1 enables automatic trimming of surfaces. The current setting is indicated by the **Auto Trim** button in the **Project Geometry** panel of the **Surface** tab on the ribbon.

Exercise 10-9

www.g-wlearning.com/CAD/

Complete the exercise on the companion website.

Untrimming Surfaces

SURFUNTRIM

Ribbon
Surface
> Edit
Surface Untrim
Type
SURFUNTRIM

If you need to restore a trimmed surface back to its original shape, use the **SURFUNTRIM** command. After selecting this command, select the edge of the surface area to untrim. If the surface has multiple trimmed edges, you can use the **Surface** option. The **SURFUNTRIM** command untrims surfaces trimmed by the **SURFTRIM** command. It does not untrim surfaces trimmed using the **PROJECTGEOMETRY** command.

PROFESSIONAL TIP

Open the **Properties** palette if you are unsure if a surface has been trimmed. The Trimmed surface property setting indicates the surface trim status.

NOTE

Exercise caution if you are trimming an associative surface using another surface as a cutting object. When using another surface as the cutting object, turn off surface associativity (set the **SURFACEASSOCIATIVITY** system variable to 0) before trimming to avoid potential problems in future edits.

NURBS Surfaces

When creating a NURBS surface, you use many of the same commands that you would use to create a procedural surface. You can use splines and various curve shapes to create NURBS surfaces. In addition, you can convert procedural surfaces into NURBS surfaces.

As discussed earlier in this chapter, a NURBS surface is created when the **SURFACEMODELINGMODE** system variable is set to 1. In addition, NURBS surfaces are non-associative. The setting of the **SURFACEASSOCIATIVITY** system variable has no effect when creating a NURBS surface.

The advantage of working with NURBS surfaces is that you use control vertices to control or influence the shape of the surface. The ability to edit control vertices provides significant flexibility in creating and sculpting freeform, organic shapes. For example, in computer animation work, NURBS surface modeling techniques are commonly used for modeling characters to produce the organic shape desired. In AutoCAD, you can create highly sophisticated, freeform shapes using NURBS surface models. This chapter introduces the NURBS surface modeling tools available in AutoCAD.

NURBS Surface Modeling Workflows

There are two common workflows used in NURBS surface modeling. You can begin by creating procedural surfaces and then convert them to NURBS surfaces, or you can create the initial surfaces as NURBS surfaces.

When you start the modeling process by working from procedural surfaces, the following workflow is common:

- Create procedural surfaces using commands such as **EXTRUDE**, **REVOLVE**, **SWEEP**, **LOFT**, **PLANESURF**, and **SURFNETWORK**. The **SURFACEMODELINGMODE** system variable should be set to 0.
- Create other surfaces, such as blend surfaces, patches, fillets, and offset surfaces. Use the commands presented in this chapter.
- Convert the surfaces into NURBS surfaces.
- Edit the NURBS surfaces as needed to create the desired sculpted shape.

When you start the modeling process by creating NURBS surfaces, the following workflow is common:

- Set the **SURFACEMODELINGMODE** system variable to 1 (on). With this setting, NURBS surfaces are created.
- Create the surfaces needed to create the desired model shape. When using this approach, you use splines or curves to define the surface profile. Splines are created using the **SPLINE** command. Splines used for NURBS surface models are typically created with the **Method** option of the **SPLINE** command set to **CV**. This creates splines with control vertices (CVs), also known as ***CV splines***. See **Figure 10-23**. Control vertices play a major role in editing NURBS surfaces.
- Edit the NURBS surfaces as needed to create the desired shape.

There are several important points to keep in mind when you are working with NURBS surfaces. Once a procedural surface is converted to a NURBS surface, the NURBS surface *cannot* be converted back to a procedural surface. In addition, once a NURBS surface is created, it *cannot* be converted to a procedural surface. The design workflow you use is important. Make sure to plan ahead so that your modeling process is suitable for the design.

Creating and Editing NURBS Surfaces

If the surface model you are creating will have only slight modifications, a simple approach is to create a procedural surface and then convert it to a NURBS surface. The **CONVTONURBS** command is used to convert a procedural surface to a NURBS surface. After entering this command, select the surface to convert. When you press [Enter], the surface is converted.

Ribbon
Surface
> Control Vertices
Convert to NURBS
Type
CONVTONURBS

CONVTONURBS

Figure 10-23.
Splines used to create NURBS surfaces are typically created as CV splines. A—A CV spline used for creating an extruded NURBS surface. B—The spline shown selected with control vertices displayed. C—A NURBS surface model created by using the **EXTRUDE** command.

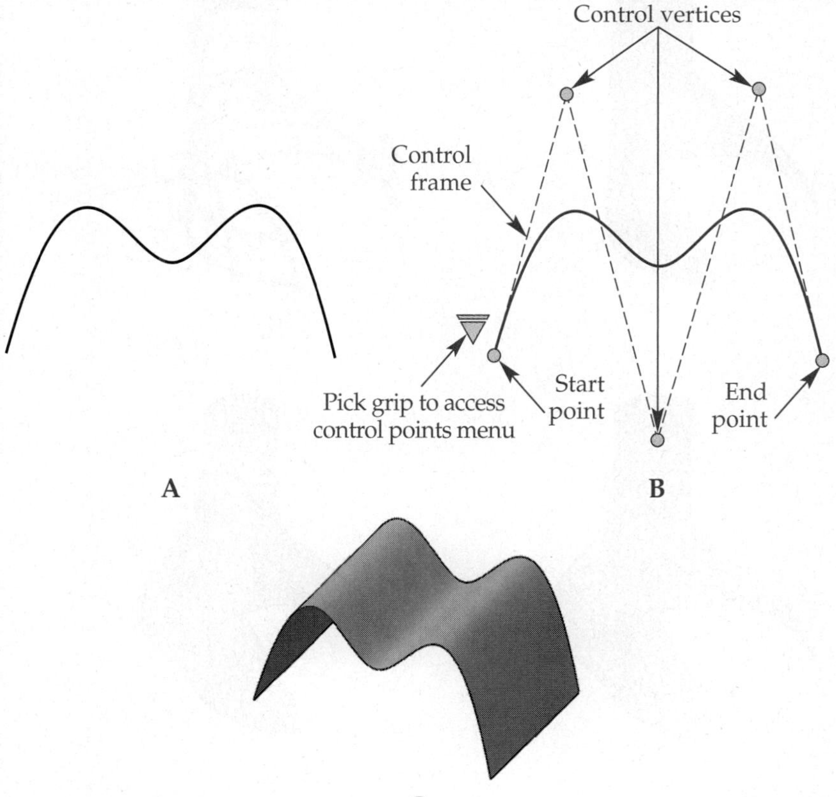

If the intention is to create more complex, freeform shapes, you can create NURBS surfaces without first converting procedural surfaces. In this case, set the **SURFACEMODELINGMODE** system variable to 1. Create profile geometry using splines and use the appropriate surface modeling commands.

The model in **Figure 10-24** shows an example of editing a NURBS surface converted from a procedural surface. The original model shown in **Figure 10-24A** is a loft surface created from open profiles. In **Figure 10-24B**, the model is shown after using the **CONVTONURBS** command. The model is shown selected with control vertices displayed (the wireframe view is shown for reference only). When you select a NURBS surface, control vertices do not appear by default. The display of control vertices is controlled by the **CVSHOW** command. To display control vertices, select the **CVSHOW** command and then select the NURBS surface. Then, press [Enter]. You can also select the command after initially selecting the surface. To remove the display of control vertices, select the **CVHIDE** command. Using this command removes the display of control vertices from all objects in the drawing.

CVSHOW

Ribbon
Surface
> Control Vertices

Surface CV-Show

Type
CVSHOW

CVHIDE

Ribbon
Surface
> Control Vertices

Surface CV-Hide

Type
CVHIDE

Editing control vertices is similar to editing grips. Pick on the control vertex grip and pull or drag. You can press and hold the [Shift] key to select multiple control vertices. You can also use the gizmo that appears. As you pull or drag, the shape of the surface is modified. In **Figure 10-24C**, the model is shown after editing a control vertex on the back end of the surface. A shaded view of the model after editing and using the **CVHIDE** command is shown in **Figure 10-24D**.

Figure 10-24.
Editing a NURBS surface converted from a procedural surface. A—The original model is a loft surface created as a procedural surface. B—The model after using the **CONVTONURBS** and **CVSHOW** commands. A wireframe view is shown for reference only. C—The model after editing one of the control vertices on the back end. D—The model after using the **CVHIDE** command.

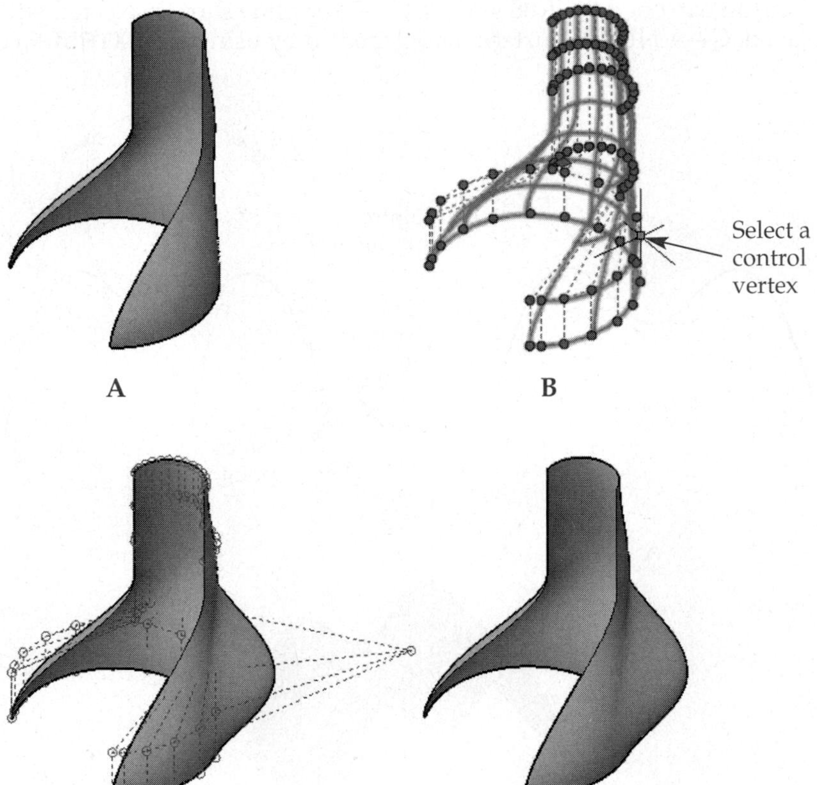

For greater control when editing the control vertices of a NURBS surface, you can use the **3DEDITBAR** command. Select this command and select the NURBS surface to edit. You are then prompted to select a point on the surface. When you select a point, the ***3D edit bar gizmo*** appears, **Figure 10-25A**. This gizmo is similar to the move gizmo that appears when working with 3D objects, as discussed in Chapter 8. However, it contains additional grips for setting the tool options and modifying the tangencies of the surface. See **Figure 10-25B**. The grips include a square grip, triangle grip, and tangent arrow grip. The square grip represents the initial base point of the edit. Picking on the grip and dragging reshapes the surface from the base point. The triangle grip is used to specify the method for reshaping the surface. Picking on the grip displays a shortcut menu. The **Move Point** option is used to reshape the surface by moving the base point. The **Tangent Direction** option is used to adjust the magnitude or bulge of the tangency at the base point. The tangent arrow grip is used to dynamically modify the tangency. The **U Tangent Direction**, **V Tangent Direction**, and **Normal Tangent Direction** options, available in the shortcut menu displayed by right-clicking on the tool, are used to set the direction of adjustment. The shortcut menu contains additional options for relocating the base point and aligning the gizmo.

3DEDITBAR

Ribbon
Surface
> Control Vertices
Surface CV-Edit Bar
Type
3DEDITBAR

Use the 3D edit bar gizmo in the same manner as other gizmo tools. Pull or drag on a grip and then pick to set a distance or use direct distance entry. The movement can be restricted to the X, Y, or Z axis or the XY, XZ, or YZ plane by picking the appropriate axis or plane on the tool. The surface updates dynamically as you drag a grip.

Exercise 10-10

www.g-wlearning.com/CAD/

Complete the exercise on the companion website.

Additional Surface Modeling Methods

Additional methods are available for working with surface models. These methods include converting existing objects to surfaces, using the **THICKEN** and **SURFSCULPT** commands to create solid models from surfaces, and using the **SURFEXTRACTCURVE** command to extract curves from existing surfaces. These methods are discussed in the following sections.

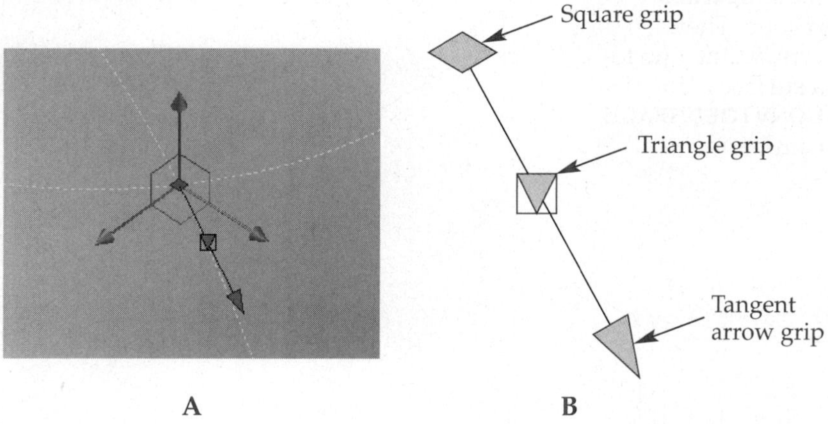

Figure 10-25. Using the **3DEDITBAR** command to edit a NURBS surface. A—The 3D edit bar gizmo appears after selecting a surface and a point on the surface. B—The grips available on the 3D edit bar gizmo.

Converting Objects to Surfaces

AutoCAD provides a great deal of flexibility in converting and transforming objects. For example, a simple line can be quickly turned into a surface. Refer to **Figure 10-26**.

1. Use the Thickness property in the **Properties** palette to give the line a thickness. Notice that the object is still a line object, as indicated in the drop-down list at the top of the **Properties** palette.
2. Select the **CONVTOSURFACE** command.
3. Pick the line. Its property type is now listed in the **Properties** palette as a surface extrusion.

CONVTOSURFACE

Ribbon
Home
> Solid Editing
Convert to Surface
Type
CONVTOSURFACE

Other objects that can be converted to surfaces using the **CONVTOSURFACE** command are 2D solids, 3D solids, arcs with thickness, open polylines with a thickness, regions, and planar 3D faces.

NOTE

The **THICKNESS** system variable can be used to assign a default Thickness property to new 2D objects that you create, such as lines, polylines, polygons, and circles. The value of the **THICKNESS** system variable does not affect the thickness of a planar surface or 3D surfaces.

Thickening a Surface into a Solid

A surface has no thickness. But, a surface can be quickly converted to a 3D solid using the **THICKEN** command.

THICKEN

Ribbon
Solid
> Solid Editing
Thicken
Type
THICKEN

To add thickness to a surface, enter the **THICKEN** command. Then, pick the surface(s) to thicken and press [Enter]. Next, you are prompted for the thickness. Enter a thickness value or pick two points on screen to specify the thickness. See **Figure 10-27**.

The **THICKEN** command can be used in conjunction with the **CONVTOSURFACE** command to quickly convert a 2D line into a solid. For example, create the line and then use the **Properties** palette to give the line a thickness. Convert the line into a surface using the **CONVTOSURFACE** command, and then use the **THICKEN** command to create a solid from the surface.

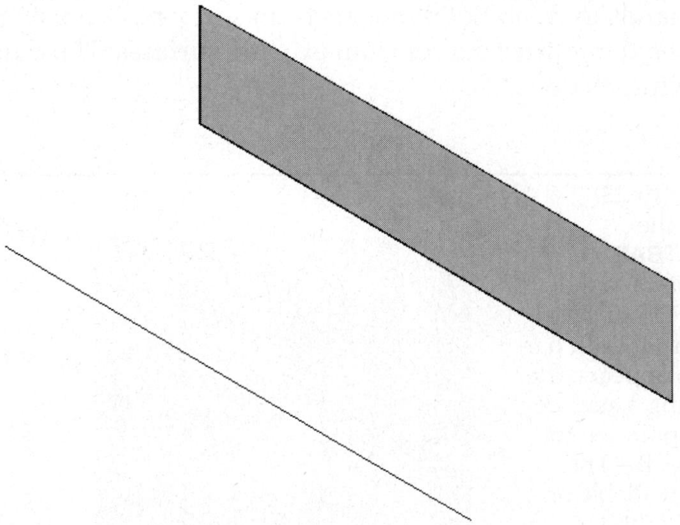

Figure 10-26. Converting a line into a surface. First, draw the line. Next, give the line a thickness using the **Properties** palette. Then, convert the line to a surface using the **CONVTOSURFACE** command.

Figure 10-27.
A—This surface will be thickened into a solid. B—The thickened surface is a 3D solid.

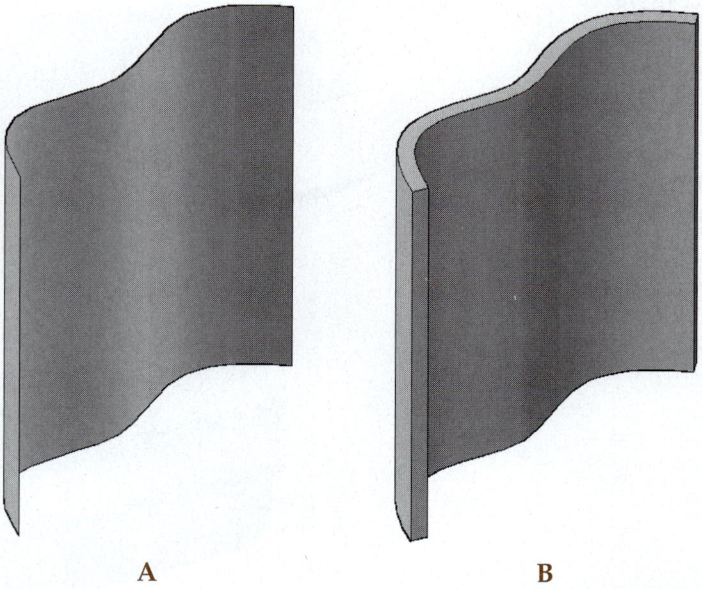

By default, the original surface object is deleted when the 3D solid is created with **THICKEN**. This is controlled by the **DELOBJ** system variable. To preserve the original surface, change the **DELOBJ** value to 0.

Exercise 10-11

www.g-wlearning.com/CAD/

Complete the exercise on the companion website.

Sculpting: Surface to Solid

Ribbon
Surface
> Edit
Surface Sculpt
Type
SURFSCULPT

SURFSCULPT

When designing using surfaces, you can sculpt a surface into a solid using the **SURFSCULPT** command. This is similar to using the **CONVTOSOLID** command as discussed in Chapter 8. The main use for sculpting is to create a solid from a watertight area by trimming and combining multiple surfaces. The command can also be used on solid and mesh objects.

Notice the surfaces in Figure 10-28 create a watertight volume where trimming of the surface can occur. A single-surface enclosed area or an enclosed area formed by multiple surfaces is considered to be a watertight object. Watertight objects can be converted to a solid.

Which command should you use, **CONVTOSOLID** or **SURFSCULPT**? The difference is very subtle.

- **CONVTOSOLID** works the best when you have a watertight mesh and want to convert it to a solid. Also, it works well on polylines and circles with thickness.
- **SURFSCULPT** works the best when you have watertight surfaces or solids that completely enclose a space (no gaps). This is shown in Figure 10-28.
- As a best practice, use **CONVTOSOLID** for converting watertight meshes to solids. Use **SURFSCULPT** for converting surfaces or solids.

NOTE

The watertight surface area must have a G0 continuity (positional continuity) for the **SURFSCULPT** command to work properly.

Figure 10-28.
A—These six surfaces form a watertight volume.
B—The solid is sculpted using the **SURFSCULPT** command.

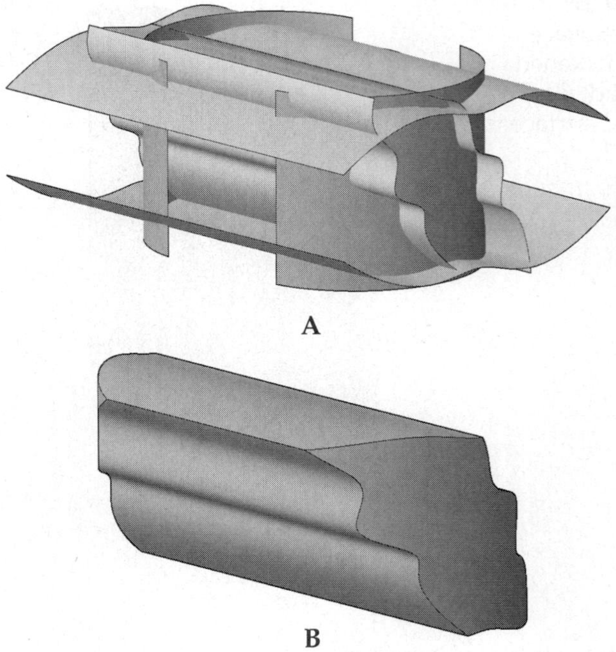

Exercise 10-12

www.g-wlearning.com/CAD/

Complete the exercise on the companion website.

Extracting Curves from Surfaces

Creating the underlying wireframe geometry required in surface or solid modeling can sometimes be difficult. When working with curved surfaces, it may help to create a profile based on the contours of an existing model. The **SURFEXTRACTCURVE** command allows you to extract isoline curves from an existing surface in the local U and V directions of the surface. Surface curves can be extracted from a surface model, solid model, or face of a solid. This provides a way to experiment with wireframe geometry and get a better idea of what the resulting model will look like. The **SURFEXTRACTCURVE** command creates objects such as lines, circles, arcs, polylines, and splines, depending on the existing model. Once the new curves are created, the designer can use the geometry for model edits or conceptual design purposes.

SURFEXTRACTCURVE

Ribbon
Surface
> Curves

Extract Isolines

Type
SURFEXTRACTCURVE

To extract isoline curves from an existing model, enter the **SURFEXTRACTCURVE** command and select a surface. An isoline curve matching the surface contour appears attached to the crosshairs. You can move the crosshairs to other surfaces to preview different curves. The default direction for the curve extraction is in the U direction. You can change the direction to the V direction by selecting the **Direction** option. Move the crosshairs until you locate the desired curve and pick to create the new curve.

In **Figure 10-29A**, a curve is extracted along the U direction of the surface. In **Figure 10-29B**, the direction is changed to the V direction.

The **Chain** option is used to extract curves from a "chain" of adjacent surfaces. In **Figure 10-30A**, the **Direction** option is used to change the default direction and the **Chain** option is used to extract a chain of three curves from the adjacent surfaces of the cell phone case surface model. Notice the curves run the total length of the cell phone case. In **Figure 10-30B**, the same direction is used and the **Chain** option is set to **No** to extract a curve from the top surface only.

Figure 10-29.
Using the **SURFEXTRACTCURVE** command to extract an isoline curve from an existing surface model. A—A curve extracted in the default U direction. The extracted curve is a circle. B—A curve extracted in the V direction. The extracted curve is a polyline.

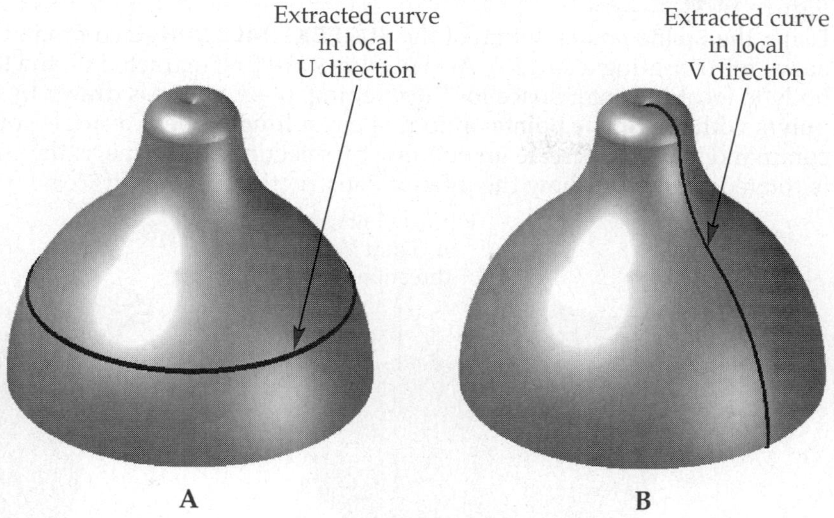

Figure 10-30.
Using the **Chain** option of the **SURFEXTRACTCURVE** command.
A—A chain of curves extracted from three adjacent model surfaces. The extracted curves are splines.
B—The **Chain** option is set to **No** and a curve is extracted from the top surface only. The extracted curve is a spline.

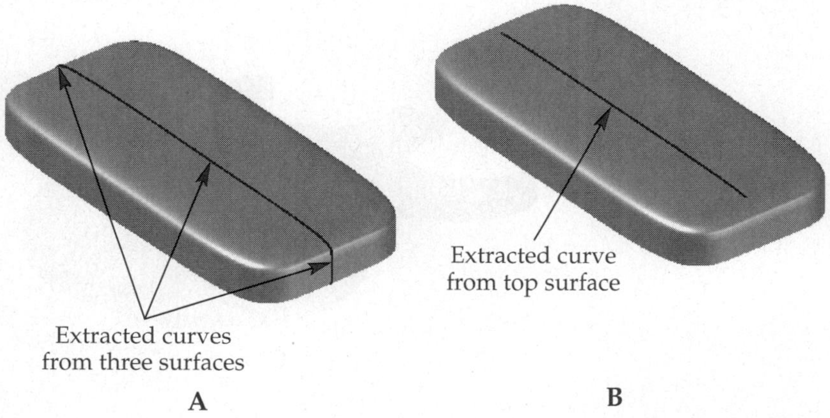

The **Spline points** option allows you to extract a spline by selecting points on a model surface. The spline passes through all specified points. If needed, use the **Close** option to create a closed spline. In **Figure 10-31**, the **Spline points** option is used to construct a spline on the top curved surface of the guitar body. The spline provides a trimming boundary for creating a sound hole or creating an opening to install electrical pickups. In **Figure 10-31A**, four isolines are extracted to locate the space for the opening. In **Figure 10-31B**, a spline is created by selecting a series of points with the **Spline points** option. In **Figure 10-31C**, the **SURFTRIM** command is used to form the opening by trimming the center area. The spline is selected as the cutting curve.

Exercise 10-13

www.g-wlearning.com/CAD/

Complete the exercise on the companion website.

Figure 10-31.
Using the **Spline points** option of the **SURFEXTRACTCURVE** command to extract a spline curve for trimming a surface. A—Four lines are first extracted on the top surface of the guitar body to locate a center space for the opening. B—A spline is drawn by selecting a series of points with the **Spline points** option. The resulting curve is a single spline. C—The **SURFTRIM** command is used to create an opening by selecting the spline as the cutting curve. The view is rotated slightly to show the interior construction.

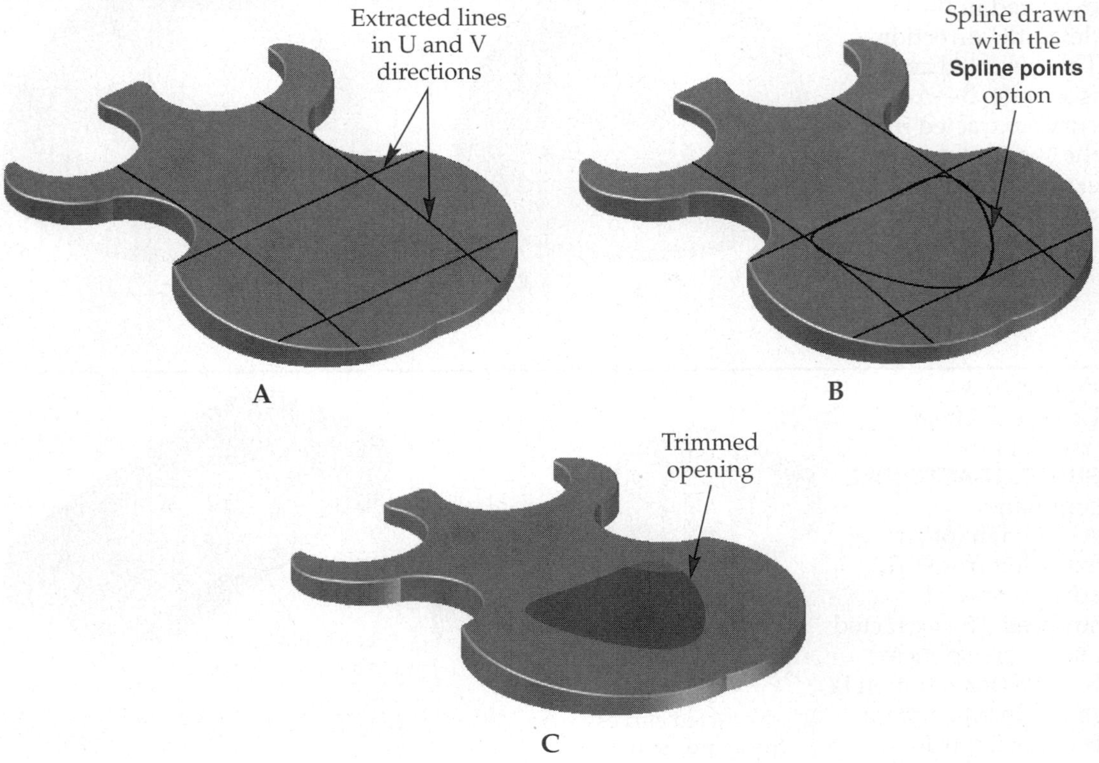

Chapter Review

Answer the following questions using the information in this chapter.

1. Name the two basic types of surface models in AutoCAD.
2. Which type of surface is created when the **SURFACEMODELINGMODE** system variable is set to 0?
3. What is an *associative surface*?
4. Which system variable determines whether a surface model is associative when created?
5. When editing the shape of an associative surface, what should be selected to maintain the surface associativity?
6. What is a *network surface*?
7. What two system variables set the number of isolines displayed in the U and V directions of a surface model?
8. What is the purpose of the **SURFBLEND** command?
9. What are the three options used to define surface continuity? What is the result of using each option?
10. Define *bulge magnitude*.
11. What is the purpose of the **SURFPATCH** command?
12. What is the purpose of the **SURFOFFSET** command?

13. How do you create a new solid when using the **SURFOFFSET** command?
14. What are the two creation type options available when using the **SURFEXTEND** command? What is the purpose of each option?
15. What are the three object types that can be used as cutting objects when trimming a surface?
16. What are the two basic ways to create a NURBS surface?
17. What command is used to display control vertices on a NURBS surface?
18. What command can be used to convert a 2D line or polyline into a surface model?
19. What is the purpose of the **THICKEN** command and which type of object does it create?
20. What is the preferred command to convert a watertight series of surfaces into a solid?
21. Name three objects you can use to extract isoline curves when using the **SURFEXTRACTCURVE** command.
22. What is the default direction used for extracting isoline curves with the **SURFEXTRACTCURVE** command?

Drawing Problems

1. **General** Create a loft surface from the three cross sections shown. The spacing between cross sections is 5 units. The circle cross section is a ∅6 circle. Use your own coordinates for the circle center point and the ellipse center points. To draw the two ellipse cross sections, refer to the dimensions given. The ellipses are centered on the same center point along the Z axis (refer to the top view). The major axis of the top ellipse is parallel to the minor axis of the lower ellipse. Use your own orientation for the ellipse axes relative to the circle (the exact orientation is not important). When creating the loft, create a procedural surface with associativity. After creating the loft, edit it by changing the dimensions of the ellipse cross sections. Refer to the dimensions given. Save the drawing as P10-1.

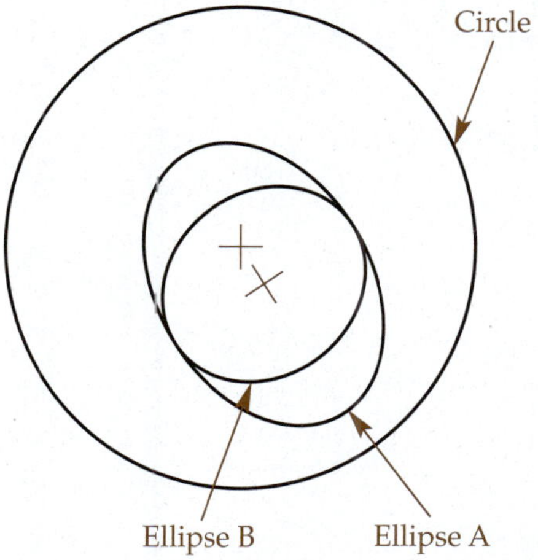

Top View

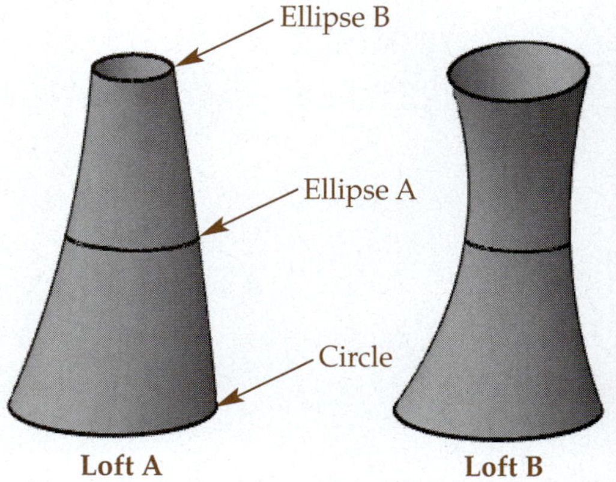

Loft A **Loft B**

Cross Section	Dimensions (Loft A)	Edited Dimensions (Loft B)
Ellipse A	Major diameter = 3.80 Minor diameter = 2.70	Major diameter = 3.00 Minor diameter = 2.00
Ellipse B	Major diameter = 2.70 Minor diameter = 2.30	Major diameter = 6.00 Minor diameter = 4.00

Drawing Problems - Chapter 10

2. General Create the cell phone case shown. Create the case as an extruded surface. Create a patch surface for the top surface. Then, use the arcs to trim the sides.
   A. Draw the base profile in the top view using the dimensions given.
   B. Draw the arcs in the top view. The arcs should extend past the perimeter of the cell phone case. Draw the first arc on the right side of the profile and then mirror it to the other side. Move the arcs along the Z axis so they are located .75 units above the bottom of the cell phone case.
   C. Extrude the base profile to a height of .5 units.
   D. Create the top of the cell phone case by creating a patch surface with G2 continuity and a bulge magnitude of .125.
   E. Trim the sides by projecting the arcs and selecting areas to trim.
   F. Save the drawing as P10-2.

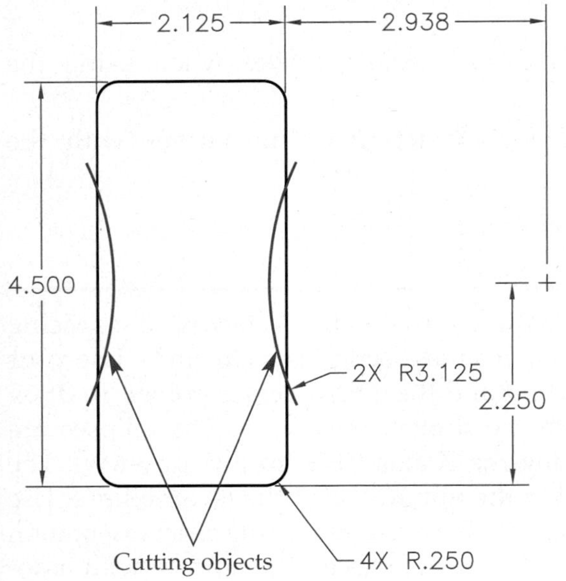

A

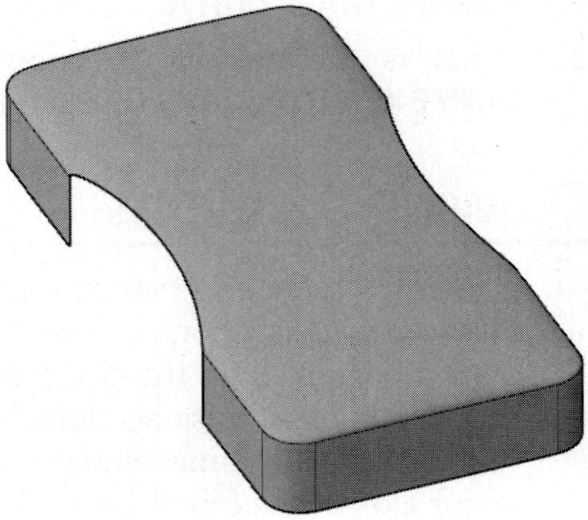
B

3. **General** Create the hair dryer handle shown. Create the base profile using a spline and a straight line segment. Use the profile to create a planar surface. Then, extrude the planar surface and fillet the bottom end. Finally, convert the model to a NURBS surface and edit the top surface.

   A. Draw a spline and a line to create a profile similar to the profile shown. Use the **CV** option of the **SPLINE** command to create a CV spline. Dimensions are not important. Create the general shape by picking points to define the control vertices as indicated. Draw a line to form the straight segment at the top of the handle. Then, use the **JOIN** command to join the line to the spline. The resulting object should be a single, closed spline.
   B. Create a planar surface from the profile.
   C. Extrude the planar surface to a height of .85 units.
   D. Use the **SURFFILLET** command to fillet the bottom end of the handle. Use a radius of .375 units. Set the trimming mode to **Yes**.
   E. Using the **CONVTONURBS** command, convert the model to a NURBS surface. Edit the control vertices of the top surface to create a different shape.
   F. Save the drawing as P10-3.

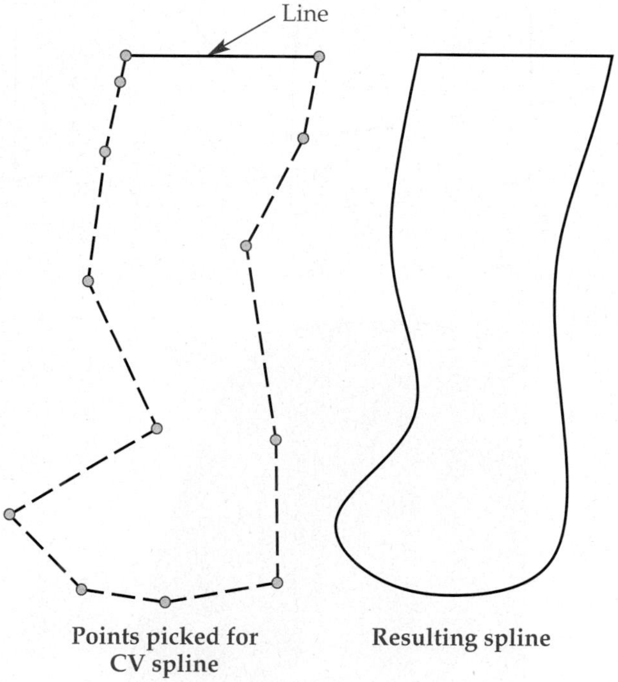

A

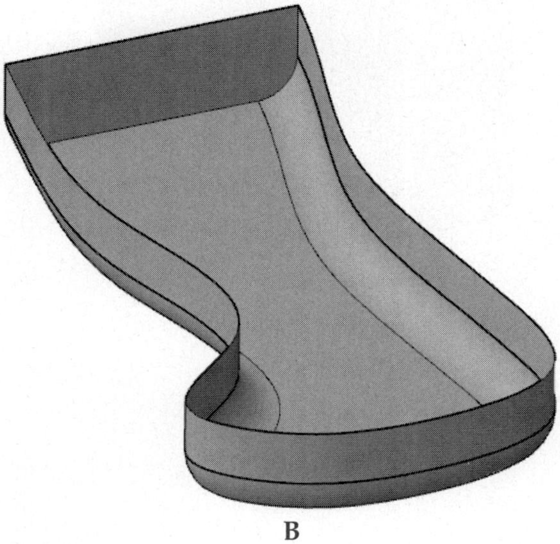

B

4. Mechanical Create the computer speaker using surface modeling commands. Use associative surfaces and edit the height from 4.5 units to 6 units, as shown. Save the drawing as P10-4.

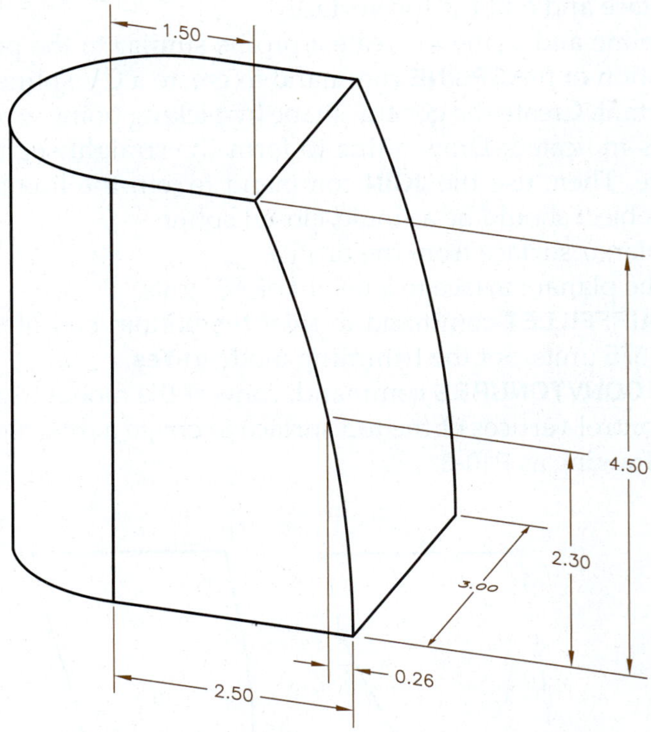

A

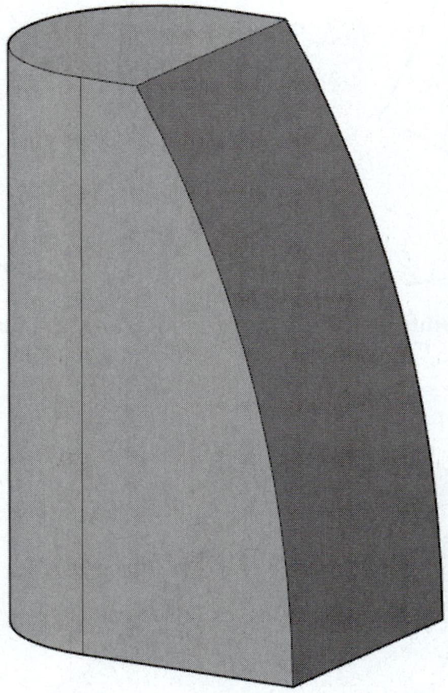

B

5. **Architectural** Create the kitchen chair shown. In earlier chapters, you created this model as a solid model. In this problem, use surface modeling commands to create the model. Use the drawings shown to create the seat, seatback, legs, and crossbars.
   A. Create the top of the seat using the profile shown. Use the edge profile to create the curved, receding edge extending to the bottom of the seat. The bottom of the seat is 1″ below the top.
   B. Create the profile geometry for the seatback. Loft the circular cross sections to create the supports. Loft the square and circular cross sections to create the bow.
   C. Create the framework for lofting the legs and crossbars as shown on the next page. The crossbars are at the midpoints of the legs. Position the lines for the double crossbar about 4.5″ apart. The legs transition from Ø1.00 at the ends to Ø1.25 at the midpoints. The crossbars transition from Ø.75 at the ends to Ø1.00 at a position 2″ from each end.
   D. Save the drawing as P10-5.

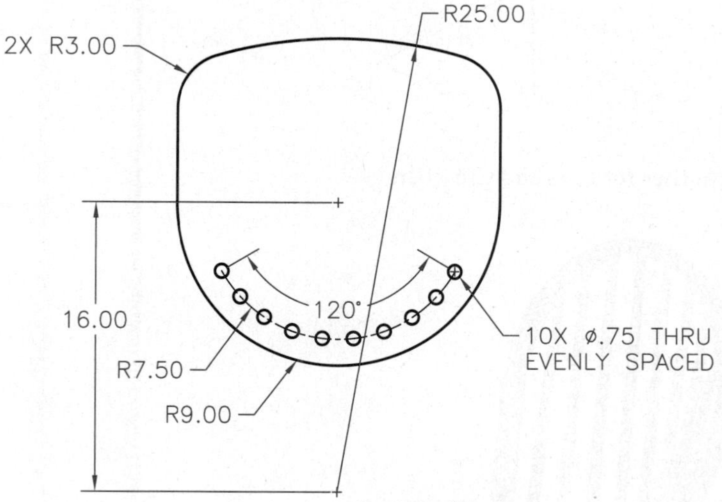

Seat Profile

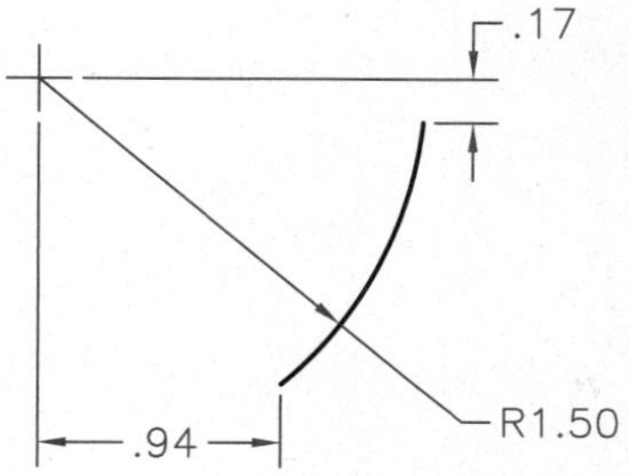

Edge Profile

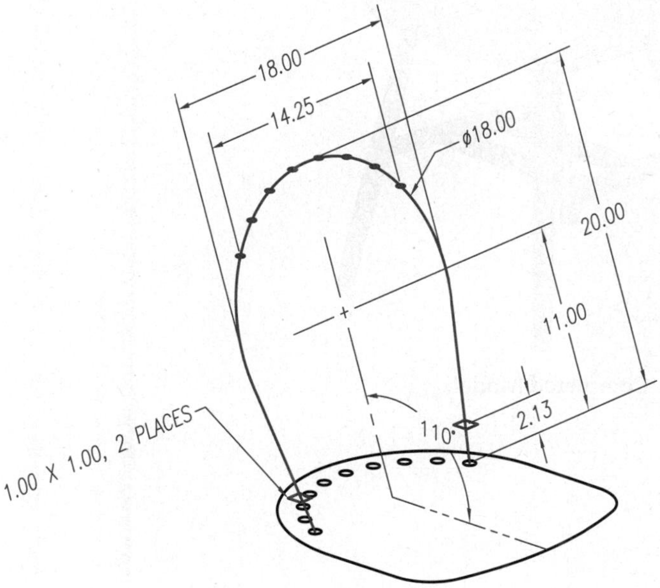

Seatback Profiles

Completed Seatback

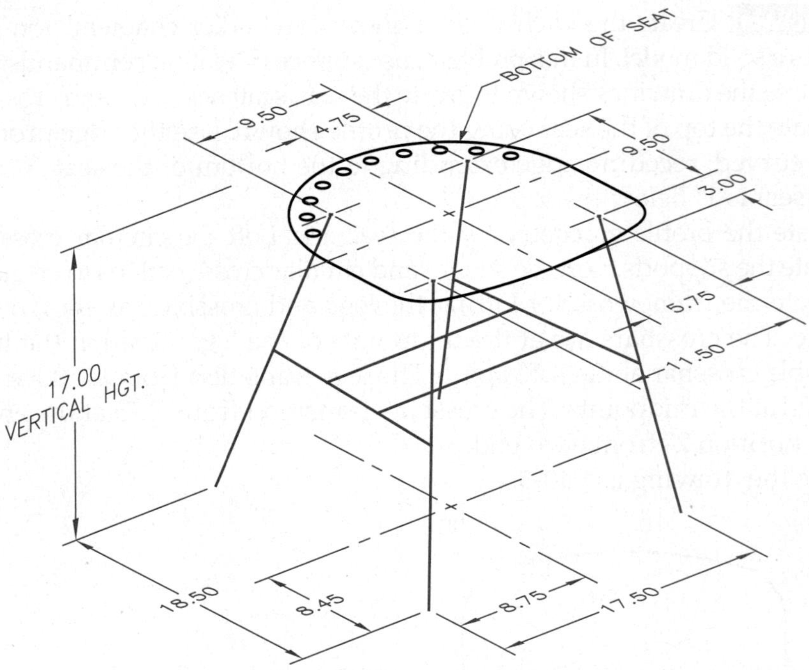

Profiles for Legs and Crossbars

Completed Model

CHAPTER 11

Subobject Editing

Learning Objectives

After completing this chapter, you will be able to:

- ✓ Select subobjects (faces, edges, and vertices).
- ✓ Edit solids using grips.
- ✓ Edit fillet and chamfer subobjects.
- ✓ Edit composite solid model subobjects.
- ✓ Edit face subobjects.
- ✓ Edit edge subobjects.
- ✓ Edit vertex subobjects.
- ✓ Extrude a 2D or 3D object using the **PRESSPULL** command.
- ✓ Offset planar face edges using the **OFFSETEDGE** command.

Grip Editing

There are three basic types of 3D solids in AutoCAD. The commands **BOX**, **WEDGE**, **PYRAMID**, **CYLINDER**, **CONE**, **SPHERE**, **TORUS**, and **POLYSOLID** create 3D solid *primitives*. *Swept objects* are 3D solids created from closed profiles using the **EXTRUDE**, **REVOLVE**, **SWEEP**, and **LOFT** commands. Finally, 3D solid *composites* are created by a Boolean operation (**UNION**, **SUBTRACT**, or **INTERSECT**) or by using the **SOLIDEDIT** command. The **SOLIDEDIT** command is discussed in Chapter 12. Smooth-edged solid primitives are achieved by converting mesh objects to solids.

There are two basic types of grips—base and parameter—that may be associated with a solid object. These grips provide an intuitive means of modifying solids. Base grips are square and parameter grips are typically arrows. The editing techniques that can be performed with these grips are discussed in the next sections.

3D Solid Primitives

The 3D solid primitives all have basically the same types of grips (base and parameter). However, not all grips are available on all primitives. All primitives have a base grip at the centroid of the base. This grip functions like a standard grip in 2D work. It can be used to stretch, move, rotate, scale, or mirror the solid.

Boxes, wedges, and pyramids have square base grips at the corners that allow the size of the base to be changed. See **Figure 11-1**. The object dynamically changes in the viewport as you select and move the grip or after you type the new coordinate location for the grip and press [Enter]. If ortho is off, the length and width can be changed at the same time by dragging the grip, except in the case of a pyramid. The triangular parameter grips on the base allow the length or width to be changed. The height can be changed using the top parameter grip. Each object has one parameter grip for changing the height of the apex and one for changing the height of the plane on which the base sits. A pyramid also has a parameter grip at the apex for changing the radius of the top.

Cylinders, cones, and spheres have four parameter grips for changing the radius of the base, or the cross section in the case of a sphere. See **Figure 11-2**. Cylinders and cones also have parameter grips for changing the height of the apex and the height of the plane on which the base sits. Additionally, a cone has a parameter grip at the apex for changing the radius of the top.

A torus has a parameter grip located at the center of the tube. See **Figure 11-3**. This grip is used to change the radius of the torus. Parameter grips at each quadrant of the tube are used to change the radius of the tube.

Figure 11-1.
Boxes, wedges, and pyramids have square base grips at the corners of the base, a parameter grip at the center of the base or bottom edge, and parameter grips on the sides of the base and center of the top face, edge, or vertex.

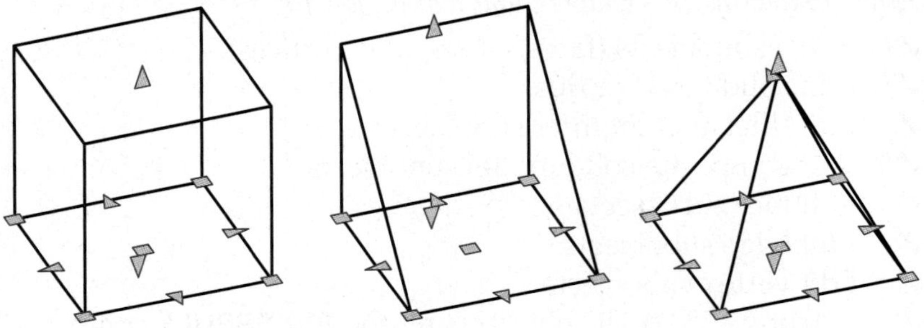

Figure 11-2.
Cylinders, cones, and spheres have four parameter grips for changing their radius. Cylinders and cones have parameter grips for changing their height. Cones also have a parameter grip for changing the radius of the top.

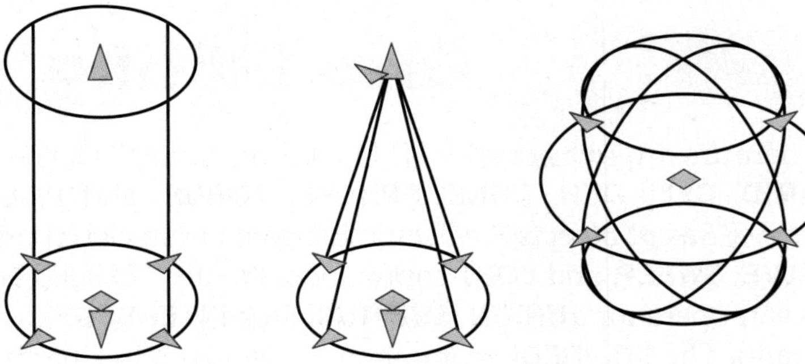

Figure 11-3.
A torus has a parameter grip located at the center of the tube for changing the radius of the torus. There are also parameter grips for changing the radius of the tube.

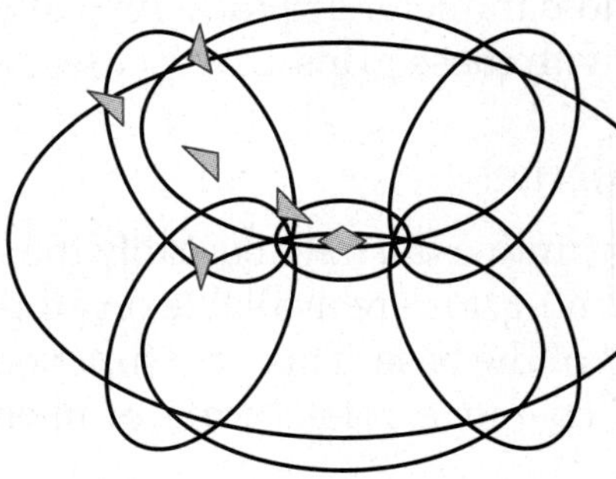

A polysolid does not have parameter grips. Instead, a base grip appears at each corner and edge midpoint of the starting face of the solid. See **Figure 11-4**. Use these grips to change the cross-sectional shape or height of the polysolid. The corners do not need to remain square. Base grips also appear at the endpoint and midpoint of each segment centerline. Use these grips to change the location of each segment's midpoint or endpoints.

Grips in AutoCAD are multifunctional. For example, if dynamic input is turned on, you can hover over the parameter grips of an object to display various design parameters. In **Figure 11-5A**, the box primitive is selected to display grips. Hovering over the parameter grip at the side of the base displays the length dimension. Hovering over parameter grips on other types of primitives, such as a cylinder or sphere, displays the corresponding dimension of the shape. See **Figure 11-5B** and **Figure 11-5C**. You can quickly subobject edit a primitive by selecting a parameter grip and typing a delta value to add or remove geometry. For example, suppose a cylinder has a height of 10 units. The new height requirement is 15. With dynamic input turned on, select the cylinder, pick the parameter grip at the apex, and drag it so the dynamic input changes to delta entry. In the dynamic input box, type 5 for the new height and press [Enter].

Swept Solids

Extrusions, revolutions, sweeps, and lofts are considered swept solids and typically have base grips located at the vertices of their profiles. These can be used to change the size of the profile and, thus, the solid. Other grips that appear include:

- A parameter grip appears on the upper face of an extrusion for changing the height.
- A base grip appears on the axis of a revolved solid for changing the location of the axis in relation to the profile.
- Base grips appear on the vertices of a sweep path for modifying the swept object.

Composite Solids

Composite solids are created by using one of the Boolean commands (**UNION**, **SUBTRACT**, or **INTERSECT**) on a solid. Solids that have been modified using any of the options of the **SOLIDEDIT** command also become composite solids, as do meshes converted into solids. The solid may still look like a primitive, sweep, loft, etc., but it is a composite. The grips available with the previous objects are no longer available, unless performing subobject editing on a composite created with a Boolean command (as discussed later in this chapter). Composite solids have a base grip located at the centroid of the base surface. This grip can be used to stretch, move, rotate, scale, or mirror the solid when the 2D Wireframe visual style is current. See **Figure 11-6A**. This grip also appears if no gizmo is selected to display when a 3D visual style is current. The gizmo drop-down list selection in the **Selection** panel on the **Home** tab of the

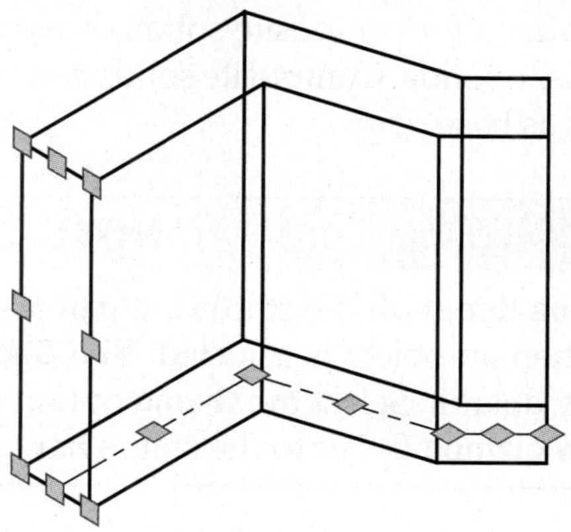

Figure 11-4.
A polysolid has a base grip at each corner and edge midpoint of the starting face of the solid and one at the endpoint and midpoint of each segment.

Figure 11-5.
Using parameter grips to display design data for a box, cylinder, and sphere. A—Hovering over the parameter grip on the side of the base displays the box length. B—Hovering over the parameter grip at the apex of the cylinder displays the cylinder height. C—Hovering over the parameter grip at the cross section of the sphere displays the sphere radius.

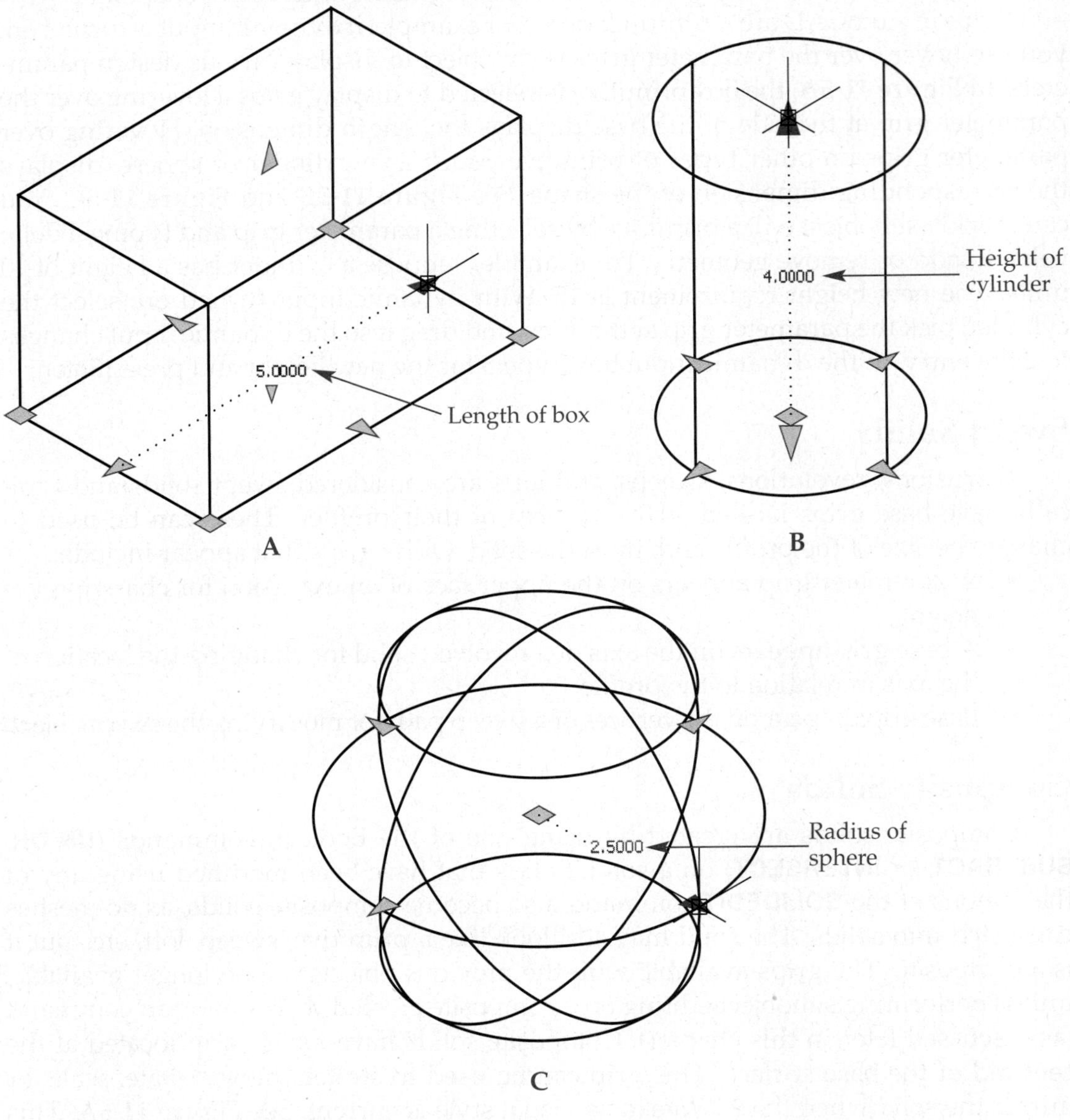

ribbon controls the display of the default gizmo in a 3D visual style. If the move gizmo is set to appear in a 3D visual style (the default option), it appears when you select a composite solid. See **Figure 11-6B**. You can use the move, rotate, or scale gizmo to make modifications.

The grip options for composite solids converted from meshes are similar to those for other composite solids. Composite solids converted from meshes have several grips, but these all act as base grips.

NOTE

The **Show gizmos** flyout on the status bar can also be used to select the default gizmo that appears when an object is selected. The **Show gizmos** flyout does not appear on the status bar by default. Select the **Gizmo** option in the status bar **Customization** flyout to add the **Show gizmos** flyout to the status bar.

Figure 11-6.
Grip editing options available for composite solids. A—A base grip appears on the composite solid when the object is selected if the **2D Wireframe** visual style is current. Right-clicking displays a shortcut menu with the grip editing options. B—The move gizmo appears when the object is selected if a 3D visual style is current and the move gizmo is selected in the **Selection** panel in the **Home** tab of the ribbon. Right-clicking on the gizmo displays the shortcut menu.

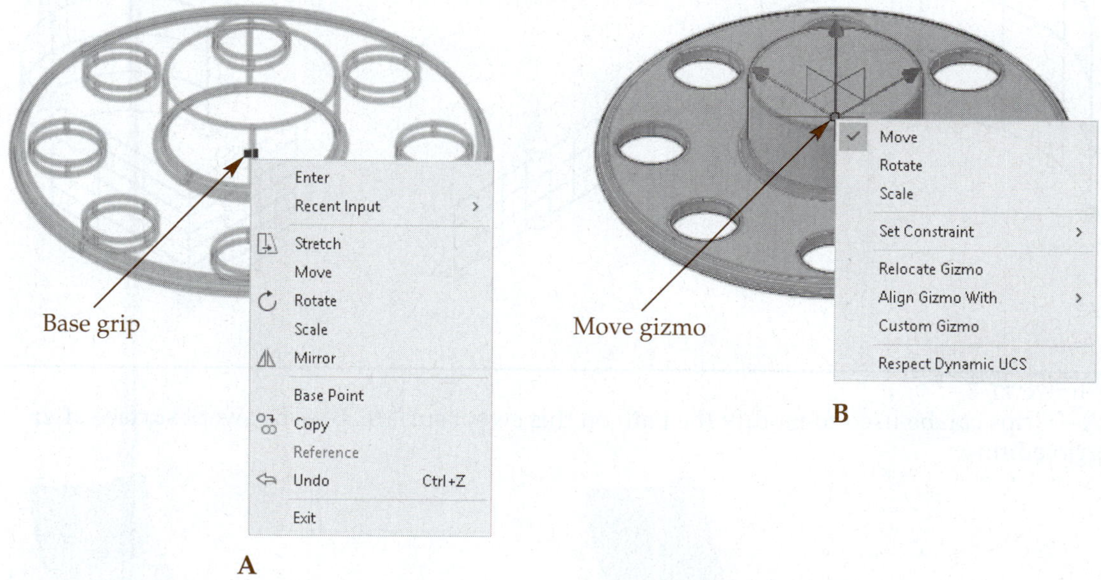

Exercise 11-1

www.g-wlearning.com/CAD/

Complete the exercise on the companion website.

Using Grips with Surfaces

Surfaces can be edited using grips in the same manner previously discussed with solids. A planar surface created with **PLANESURF** can be moved, rotated, scaled, and mirrored, but not stretched. Base grips are located at each corner.

As you learned in Chapter 6, a variety of AutoCAD objects can be extruded to create a surface. Three of these objects—arc, line, and polyline—are shown extruded into surfaces in **Figure 11-7**. Notice the location and type of grips on the surface extrusions. Base grips are located on the original profile that was extruded to make the surface. These grips enable you to alter the shape of the surface. You can use the parameter grips on the top of the surface to change the extrusion height, the extrusion direction, or the radius of the top.

Surfaces that have been extruded, or swept, along a path can be edited with grips. Also, the grips located on the path allow you to change the shape of the surface extrusion. See **Figure 11-8**.

Overview of Subobject Editing

AutoCAD solid primitives, such as cylinders, wedges, and boxes, are composed of three types of subobjects: faces, edges, and vertices. As discussed in Chapter 9, mesh models are also made up of subobjects. In addition, the objects that are used with Boolean commands to create a composite solid are considered subobjects if the solid

Figure 11-7.
Surfaces extruded from an arc, line, and polyline. Notice the grips.

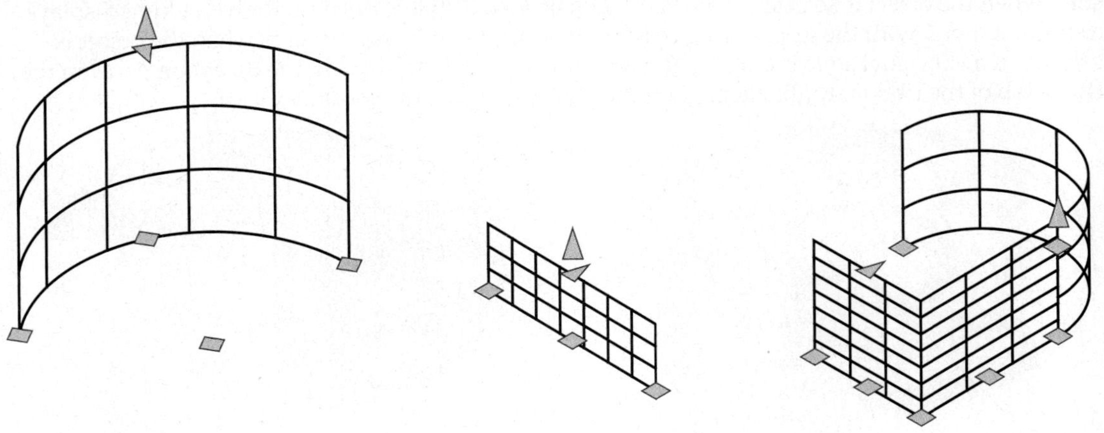

Figure 11-8.
A—Grips can be used to modify the path on this swept surface. B—The swept surface after grip editing.

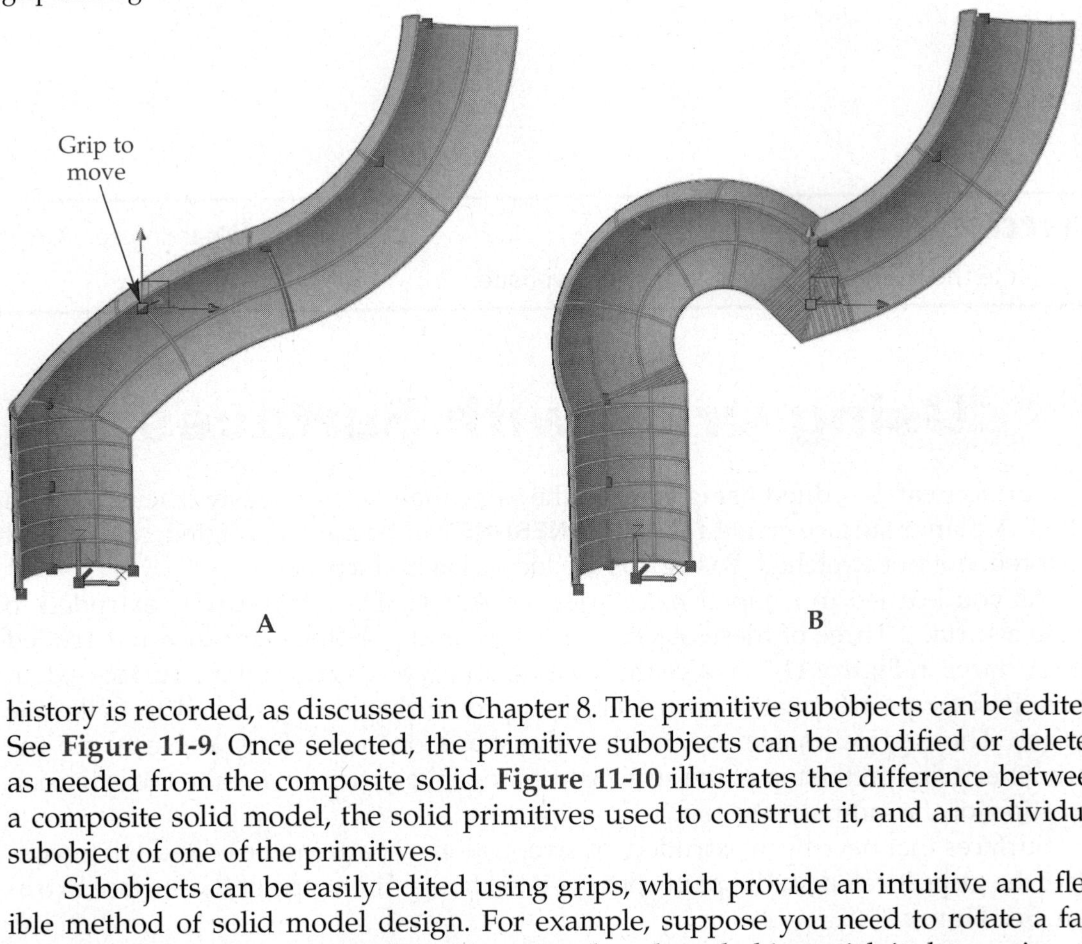

history is recorded, as discussed in Chapter 8. The primitive subobjects can be edited. See **Figure 11-9**. Once selected, the primitive subobjects can be modified or deleted as needed from the composite solid. **Figure 11-10** illustrates the difference between a composite solid model, the solid primitives used to construct it, and an individual subobject of one of the primitives.

Subobjects can be easily edited using grips, which provide an intuitive and flexible method of solid model design. For example, suppose you need to rotate a face subobject in the current XY plane. You can select the subobject, pick its base grip, and then cycle through the editing functions to the rotate function. You can also use the **ROTATE** command on the selected subobject.

To select a subobject, press the [Ctrl] key and pick the subobject. You can select multiple subobjects and subobjects on multiple objects. To select a subobject that is hidden in the current view, first display the model as a wireframe. After creating a selection set, select a grip and edit the subobject as needed. Multiple objects can be

Figure 11-9.
A—If the solid history is recorded, selecting a subobject solid primitive within a composite solid displays its grips. B—The grips on the primitive can be used to edit the primitive.

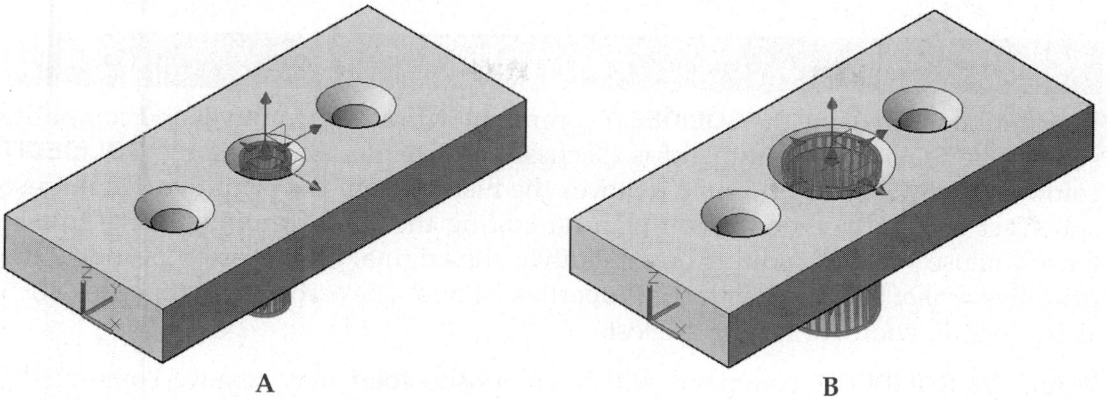

Figure 11-10.
A—The composite solid model is selected. Notice the single base grip. B—The wedge primitive subobject has been selected. Notice the grips associated with the primitive. C—An edge subobject within the primitive subobject is selected for editing.

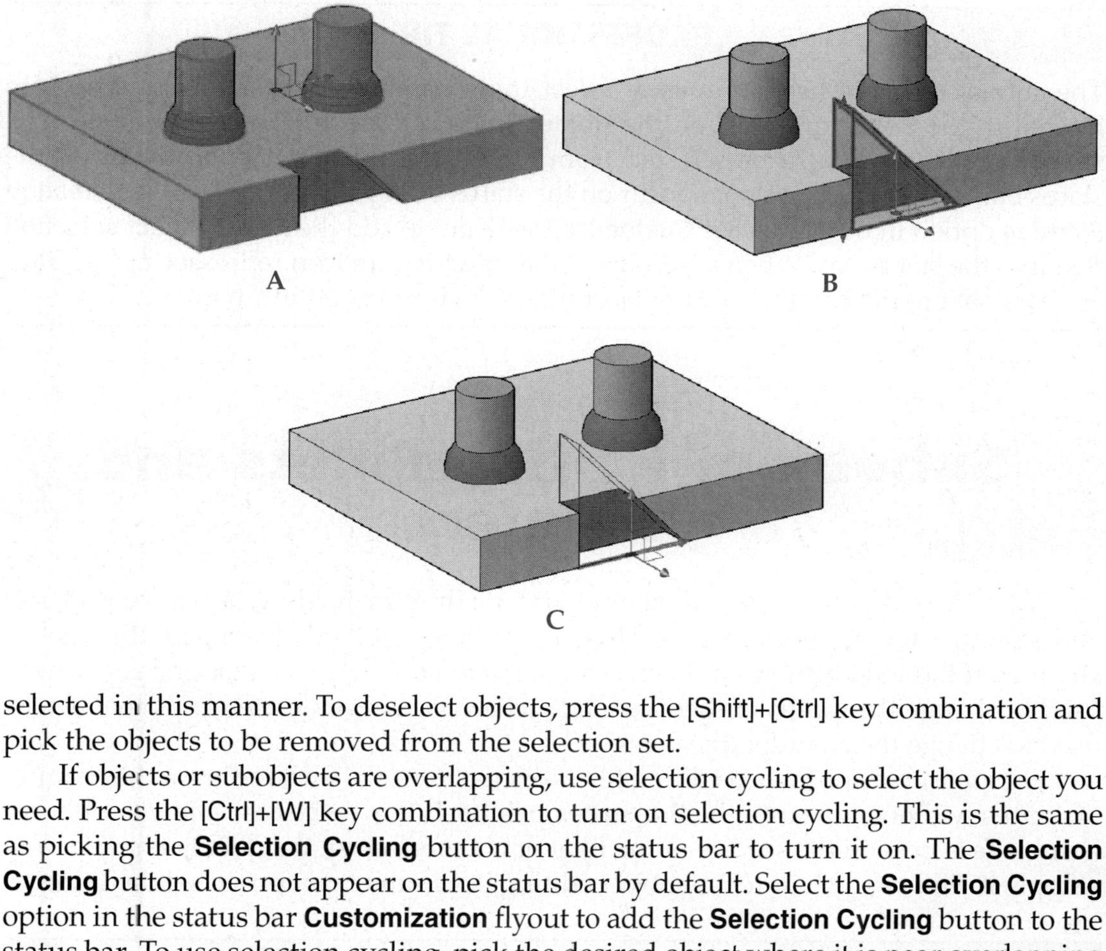

selected in this manner. To deselect objects, press the [Shift]+[Ctrl] key combination and pick the objects to be removed from the selection set.

If objects or subobjects are overlapping, use selection cycling to select the object you need. Press the [Ctrl]+[W] key combination to turn on selection cycling. This is the same as picking the **Selection Cycling** button on the status bar to turn it on. The **Selection Cycling** button does not appear on the status bar by default. Select the **Selection Cycling** option in the status bar **Customization** flyout to add the **Selection Cycling** button to the status bar. To use selection cycling, pick the desired object where it is near overlapping objects and use the **Selection** dialog box to select the object. Selection cycling is introduced in Chapter 8.

> Type
> SELECTIONCYCLING
> Toolbar
> **Status Bar**
> **Selection Cycling**

SELECTIONCYCLING

The [Ctrl] key method can be used to select subobjects for use with editing commands such as **MOVE**, **COPY**, **ROTATE**, **SCALE**, and **ERASE**. Some commands, like **STRETCH** and **MIRROR**, are applied to the entire solid. Other operations may not be applied at all, depending on which type of subobject is selected. You can also use the **Properties** palette to change the color of edge and face subobjects or the material assigned to a face. The color of a subobject primitive can also be changed, but not the color of its subobjects.

CAUTION

Use caution when using the **SOLIDEDIT** command with solid primitives and composite solids. The **SOLIDEDIT** command is discussed in Chapter 12. Using the **SOLIDEDIT** command with a solid primitive removes the history from the primitive. Do not use subobject editing methods if you plan on editing the solid primitive in the future. Once you use subobject editing on a primitive, the original geometric properties of the primitive *cannot* be edited with the **Properties** palette. The associated properties, such as the length, width, or height, are lost.

Using the **SOLIDEDIT** command with a composite solid may remove some of the history from the solid. Therefore, the original objects—some of the subobjects—are no longer available for subobject editing. However, you may still be able to perform some subobject edits that will accomplish your design intent.

PROFESSIONAL TIP

The subobject filters help to quickly select the specific type of subobject. They are located in the **Selection** panel on the **Home**, **Solid**, and **Mesh** tabs of the ribbon. The subobject filters can also be selected from the **Filters object selection** flyout on the status bar. This flyout does not appear on the status bar by default. Select the **Selection Filtering** option in the status bar **Customization** flyout to add the **Filters object selection** flyout to the status bar. When a subobject filter is active, an icon representing the filter appears next to the crosshairs. Subobject filters are introduced in Chapter 9.

Subobject Editing Fillets and Chamfers

If you improperly create a fillet or chamfer, edit it by holding down the [Ctrl] key and selecting the fillet or chamfer. Then, use editing methods to change the fillet or chamfer. If the solid history has been recorded, a fillet has a parameter grip that can be used to change the radius of the fillet and a chamfer has parameter grips that can be used to change the chamfer distances.

On some occasions when editing fillets or chamfers, the fillet radius or the chamfer distances will not appear in the **Properties** palette. If this is the case and you only see a single face grip on the surface of the fillet or chamfer, erase the fillet or chamfer and reapply it. Fillet and chamfer subobject editing is a form of composite solid editing.

PROFESSIONAL TIP

The **Delete** option of the **SOLIDEDIT** command deletes selected faces. This is a quick way to remove features such as chamfers, fillets, holes, and slots. This technique is discussed in Chapter 12.

Exercise 11-2

www.g-wlearning.com/CAD/

Complete the exercise on the companion website.

Face Subobject Editing

Faces of 3D solids can be modified using commands such as **MOVE**, **ROTATE**, and **SCALE** or by using grips and gizmos (grip tools). To select a face on a 3D solid, press the [Ctrl] key and pick within the boundary of the face. Do not pick the edge of the face. The face subobject filter can also be used to limit the selection to a face.

Face grips are circular and located in the center of the face, as shown in **Figure 11-11**. In the case of a sphere, the grip is located in the center of the sphere since there is only one face. The same is true of the curved face on a cylinder or cone.

If you select a solid primitive, all of the grips associated with that primitive are displayed. See **Figure 11-12A**. If you edit a solid primitive face, the history of the primitive is deleted and the object becomes a composite solid. Then, when the object is selected, a single base grip is displayed. See **Figure 11-12B**.

The same holds true for composite solids, if the solid history has been recorded. Solid model history is discussed in Chapter 8. If you select a primitive subobject in a composite solid with a recorded solid history, all of the grips associated with the primitive are displayed. If you edit a solid primitive face of the composite solid, the history of the primitive is deleted and the object becomes a new composite solid. Then, when the subobject is selected, a single base grip is displayed.

While pressing the [Ctrl] key and selecting a face, it may be difficult to select the face you want or to deselect faces you do not need. Use the view cube to transparently change the viewpoint. You can also press and hold the [Shift] key and press and hold the mouse wheel button at the same time to activate the transparent **3DORBIT** command.

NOTE

The recorded history of a composite solid can be displayed by selecting the solid, opening the **Properties** palette, and changing the Show History property in the **Solid History** category to Yes.

Figure 11-11.
Face grips are located in the center of face subobjects.

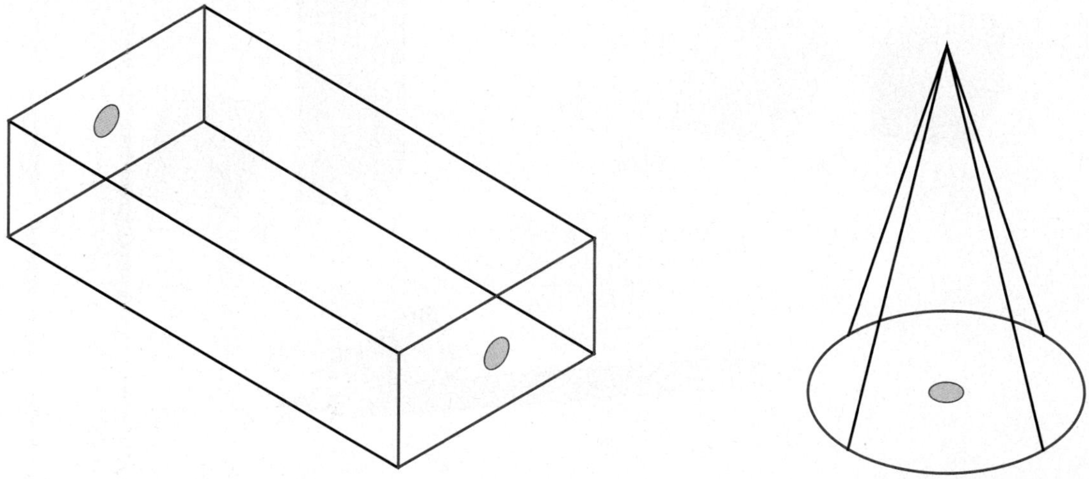

Figure 11-12.
A—This primitive is selected for editing. Notice the grips associated with the primitive. B—If the primitive is edited, the history of that primitive is deleted and a single base grip is displayed.

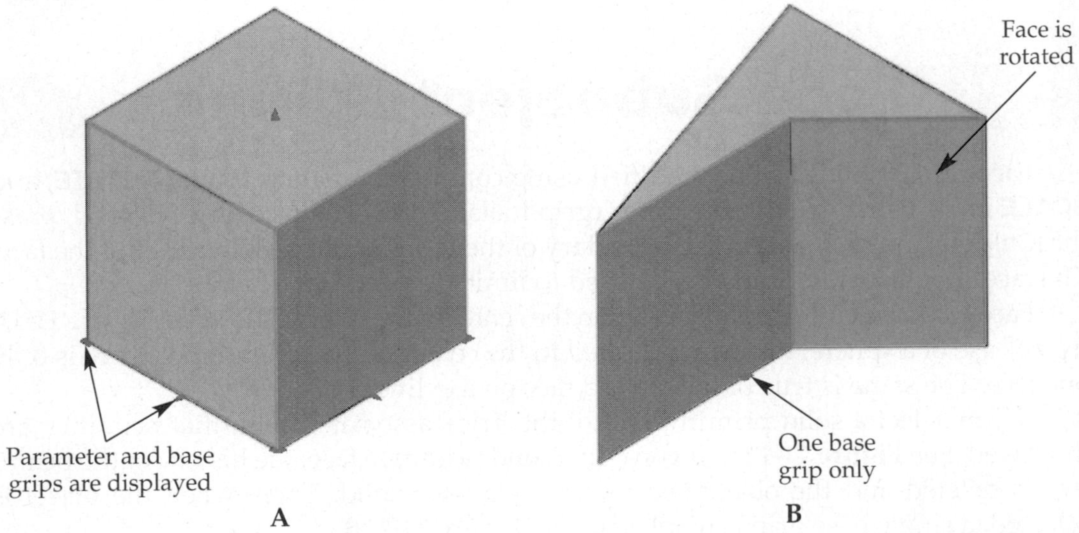

Moving Faces

When a face of a 3D solid is moved, all adjacent faces are dragged and stretched with it. The shape of the original 3D primitive or solid determines the manner in which the face can be moved and how adjacent faces react. A face can be moved using the **MOVE** command, **3DMOVE** command, or move gizmo, or by dragging the face's base grip. When moving a face, use the gizmo, polar tracking, or direct distance entry. Otherwise, the results may appear correct in the view in which the edit is made, but, when the view is changed, the actual result may not be what you wanted. See **Figure 11-13**.

Figure 11-13.
A—The original solid primitives. B—The box is dynamically edited without using exact coordinates or distances. C—When the viewpoint is changed, you can see that dynamic editing has produced unexpected results.

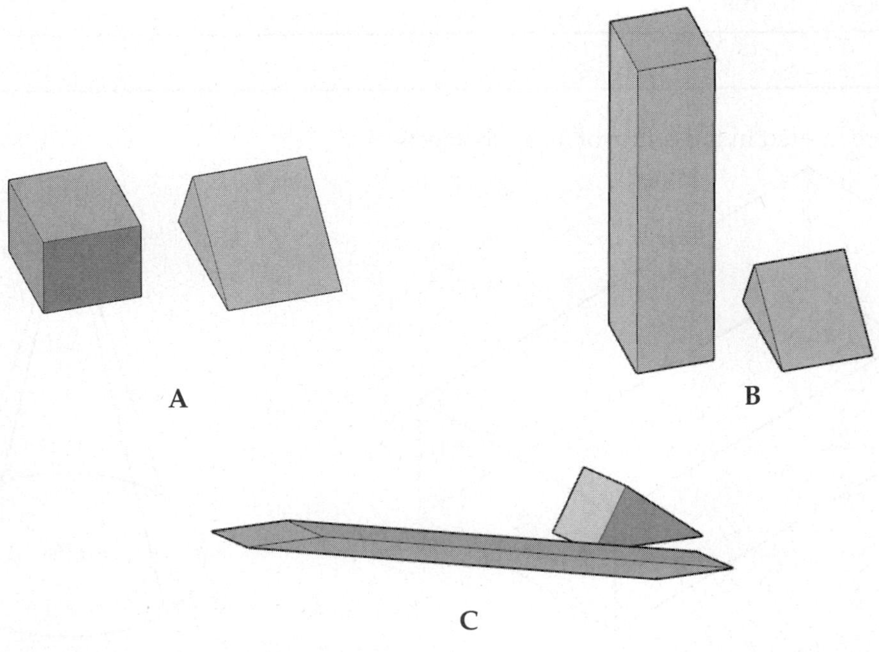

The move gizmo, as discussed earlier in this chapter, is displayed by default when the face is selected. To use this gizmo, move the pointer over the X, Y, or Z axis of the gizmo; the axis changes to yellow. To restrict movement along that axis, pick the axis. If you move the pointer over one of the right angles at the origin of the gizmo, the corresponding two axes turn yellow. Pick to restrict the movement to that plane. You can complete the movement by either picking a new point or by using direct distance entry.

The face base grip is used to access options when dynamically moving a face. First, select the face. Then, hover over the face base grip to display the base grip shortcut menu. See **Figure 11-14A**. The options in this menu determine the effects of the edit on the selected face and adjacent faces. When using the **Extend Adjacent Faces** option, the moved face maintains its shape and orientation. However, its size is modified because the planes of adjacent faces are maintained. See **Figure 11-14B**. If you select **Move Face**, the moved face maintains its size, shape, and orientation. The shape and plane of adjacent faces are changed. See **Figure 11-14C**. If you select **Allow Triangulation**, the moved face maintains its size, shape, and orientation. However, adjacent faces are subdivided into triangular faces, if needed.

During the edit, you can press the [Ctrl] key to cycle through the move options available in the base grip shortcut menu. For example, after selecting the face, pick the face grip or gizmo. Then, press and release the [Ctrl] key to cycle through the options.

Figure 11-14.
A—The original solid primitive. B—Using the **Extend Adjacent Faces** option keeps the adjacent faces in their original planes, but alters the modified face. C—When using the **Move Face** option, the face maintains its shape, size, and orientation.

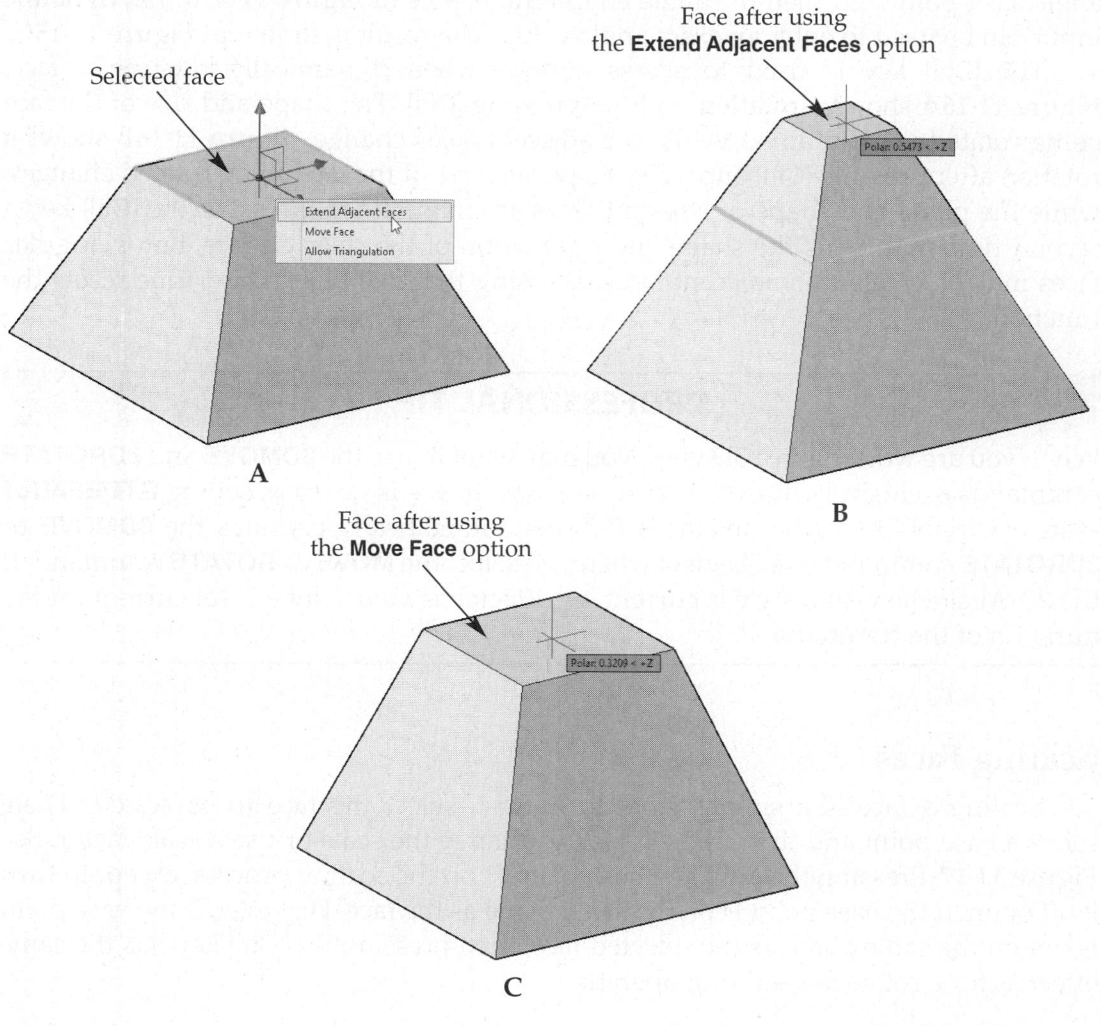

PROFESSIONAL TIP

It is always important to keep the design intent of your solid model in mind. If you are creating a conceptual design, you may be able to use subobject grip editing and gizmos without entering precise coordinates. But, if you are working on a design for manufacturing or production, it is usually critical to use tools such as direct distance entry, gizmos with exact values, and polar tracking for greater accuracy.

Rotating Faces

Before rotating any primitive or subobject, you must know in which plane the rotation is to occur. The **ROTATE** command permits a rotation in the current XY plane. But, you can get around this limitation by using the **3DROTATE** command. This command allows you to select a rotation plane by means of the rotate gizmo discussed in Chapter 8.

The rotate gizmo provides a dynamic, graphic representation of the three axes of rotation. To rotate about the X axis of the tool, pick the red circle on the gizmo. To rotate about the Y axis, pick the green circle. To rotate about the Z axis, pick the blue circle. Once you select a circle, it turns yellow and you are prompted for the start point of the rotation angle. You can enter a direct angle at this prompt or pick a point to define the angle of rotation. When the rotation angle is defined, the face is rotated about the selected axis.

For example, in **Figure 11-15A**, the top face is selected and the rotate gizmo is placed on a corner of the face. After picking the rotation axis on the gizmo, specify the angle start point and then the angle end point. Notice in **Figure 11-15B** that dynamic input can be used to enter an exact angle value. The result is shown in **Figure 11-15C**.

The [Ctrl] key is used to access options when dynamically rotating a face. **Figure 11-16A** shows a rotation without pressing [Ctrl]. The shape and size of the face being rotated is maintained, while the adjacent faces change. **Figure 11-16B** shows a rotation after pressing [Ctrl] once. The shape and size of the face being rotated changes, while the plane and shape of adjacent faces are maintained. Pressing the [Ctrl] key a second time maintains the shape and orientation of the selected face, but triangular faces may be created on adjacent faces. Pressing the [Ctrl] key a third time resets the function.

PROFESSIONAL TIP

When you are working in a 3D view, you may want to use the **3DMOVE** and **3DROTATE** commands exclusively during editing sessions. If this is the case, set the **GTDEFAULT** system variable to 1 (the default is 0). This automatically executes the **3DMOVE** or **3DROTATE** command in a 3D view when you select the **MOVE** or **ROTATE** command. If the 2D Wireframe visual style is current, the Wireframe visual style is set current for the duration of the command.

Scaling Faces

Scaling a face is a simple procedure. First, select the face to be scaled. Then, select a base point and dynamically pick to change the scale or use a scale factor. See **Figure 11-17**. Pressing the [Ctrl] key has no effect on the scaling process, except to turn it off or on, if the base point is on the same plane as the face. However, if the base point is not on the same plane as the selected face, then pressing the [Ctrl] key has the same effect as for a rotate face-editing operation.

Figure 11-15.
Using the rotate gizmo to rotate a face. A—The top face is selected and the gizmo is placed on a base point of the face. B—The axis of rotation is specified and a starting point for the angle is selected. C—The completed rotation.

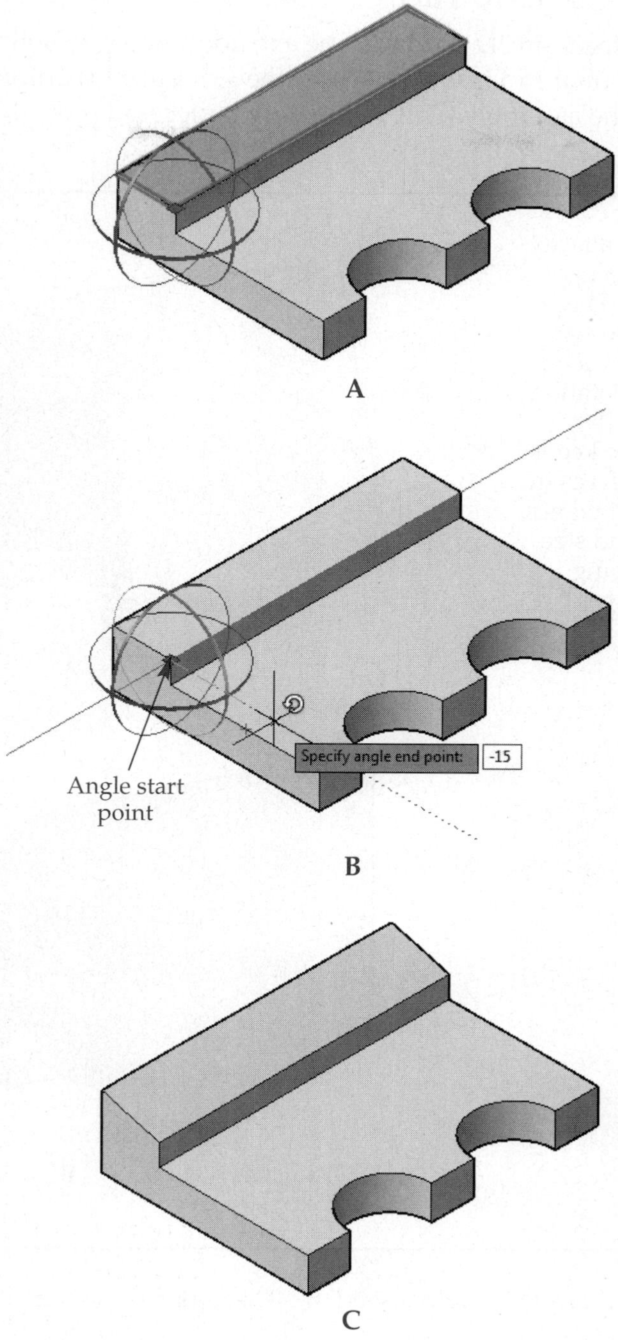

Coloring Faces

To change the color of a face, use the [Ctrl] key selection method to select the face. Next, open the **Properties** palette. See **Figure 11-18**. In the **General** category, pick the drop-down list for the Color property. Select the desired color or pick Select Color... and select a color from the **Select Color** dialog box. To change the material applied to the face, pick the drop-down list for the Material property in the **3D Visualization** category. Select a material from the list. A material must be loaded into the drawing to be available in this drop-down list. Materials are discussed in detail in Chapter 17.

Extruding a Solid Face

Planar faces on 3D solids can be extruded into new solids. Refer to the HVAC duct assembly shown in **Figure 11-19A**. A new, reduced trunk needs to be created on the left end of the assembly. This requires two pieces: a reducer and the trunk.

Figure 11-16.
A—Rotating a face without pressing the [Ctrl] key. The large, top face on the object has been selected for rotation.
B—Pressing the [Ctrl] key once keeps the adjacent faces in their original planes. The shape and size of the face being rotated changes.

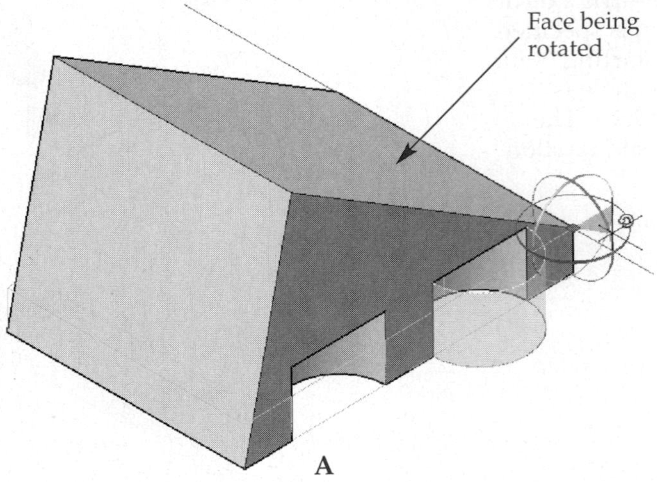

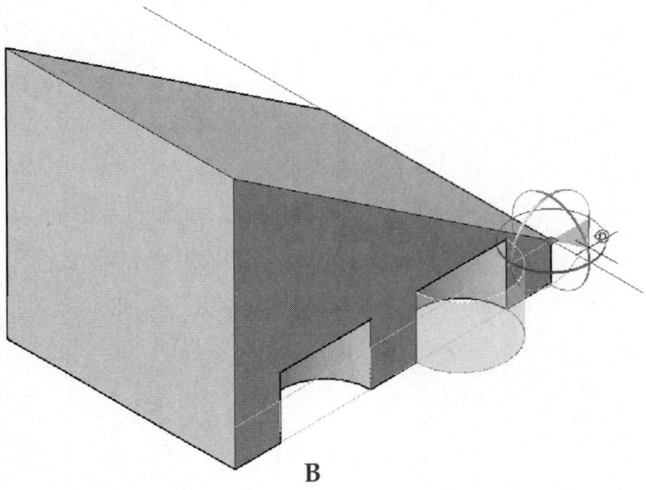

Figure 11-17.
Scaling a face. A—The original solid. B—The dark face is scaled down. C—The dark face is scaled up.

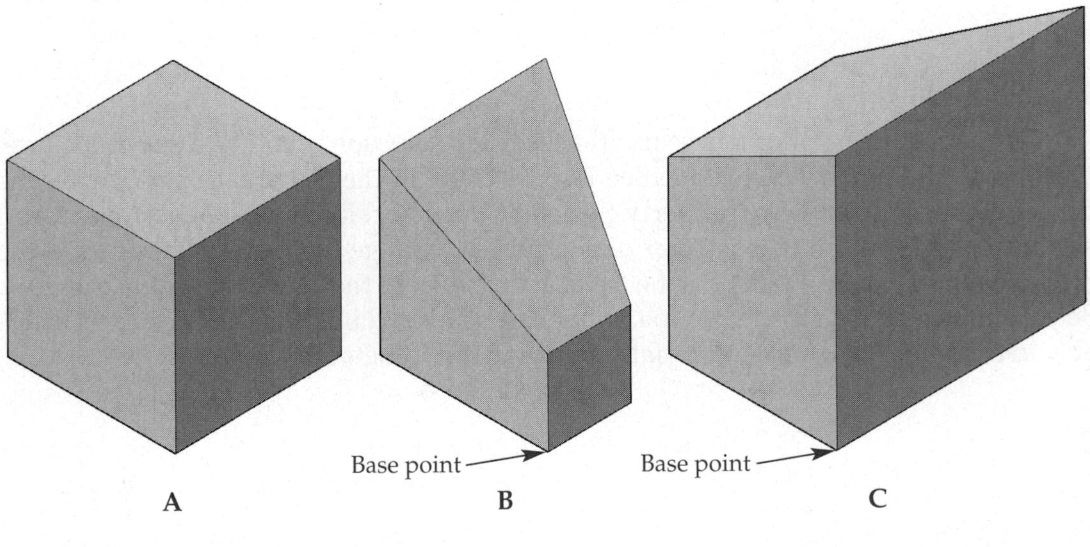

First, select the **EXTRUDE** command. At the "select objects" prompt, press the [Ctrl] key and pick the face subobject to be extruded. Next, since this is a reduced trunk, specify a taper angle. Enter the **Taper angle** option and specify the angle. In this case, a 15° angle is used. Finally, specify the extrusion height. The height of the reducer is 12". See **Figure 11-19B**.

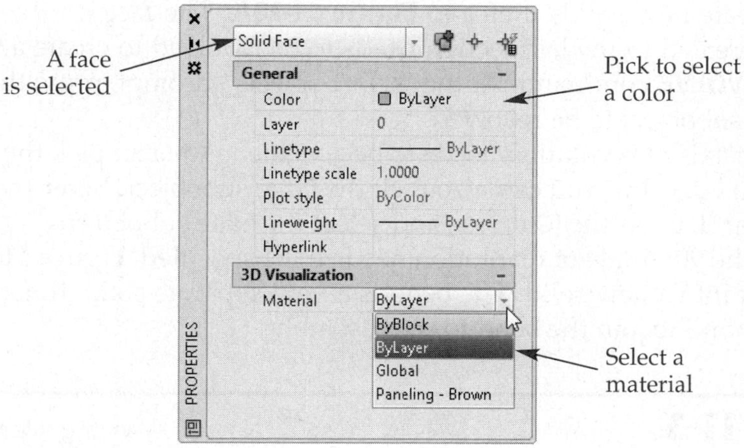

Figure 11-18.
Changing the color of a face or the material assigned to it.

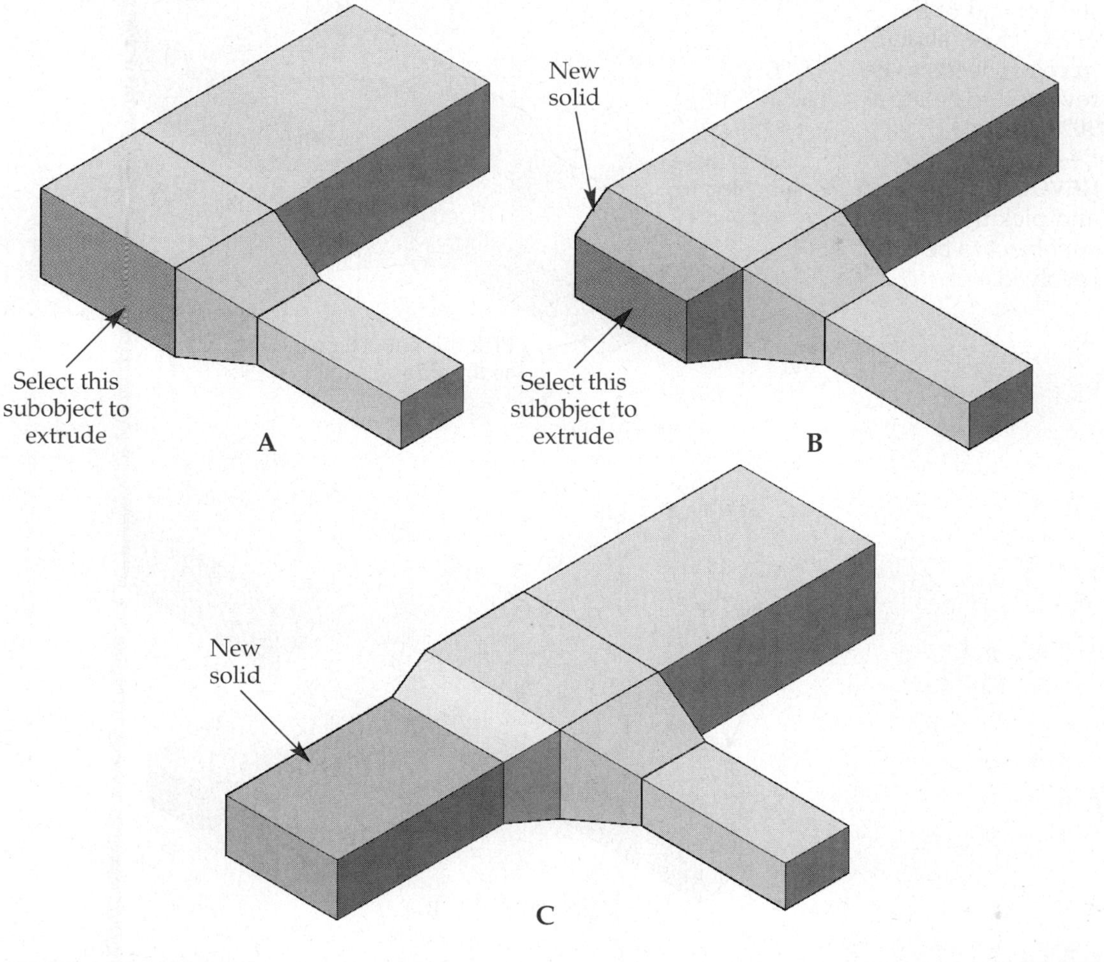

Figure 11-19.
A—A new, reduced trunk needs to be created on the left end of the HVAC assembly. The face shown in color will be extruded. B—The **Taper angle** option of the **EXTRUDE** command is used to create the reducer. The face shown in color is extruded to create the extension. C—The **EXTRUDE** command is used to create an extension from the reducer.

Now, the new trunk needs to be created. Select the **EXTRUDE** command. Press the [Ctrl] key and pick the face to extrude. Since this piece is not tapered, enter the extrusion height, which in this case is 44″. See **Figure 11-19C**. The two new pieces are separate solid objects. If the assembly is to be one solid, use the **UNION** command and join the two new solids to the assembly.

Revolving a Solid Face

Planar faces on 3D solids can be revolved in the same manner as other AutoCAD objects to create new solids. Refer to **Figure 11-20A**. The face on the left end of the HVAC duct created in the last section needs to be revolved to create a 90° bend. First, select the **REVOLVE** command. At the "select objects" prompt, press the [Ctrl] key and pick the face subobject to be revolved.

Next, the axis of revolution needs to be specified. You can pick the two endpoints of the vertical edge, but you can also pick the edge subobject. Enter the **Object** option of the command, press the [Ctrl] key, and select the edge subobject.

Finally, the 90° angle of revolution needs to be specified. **Figure 11-20B** shows the face revolved into a new solid. The bend is a new, separate solid. If necessary, use the **UNION** command to join the bend to the assembly.

Exercise 11-3

www.g-wlearning.com/CAD/

Complete the exercise on the companion website.

Figure 11-20.
A—The face on the left end of the HVAC duct (shown in color) needs to be revolved to create a 90° bend.
B—Use the **REVOLVE** command and pick the face subobject to be revolved.

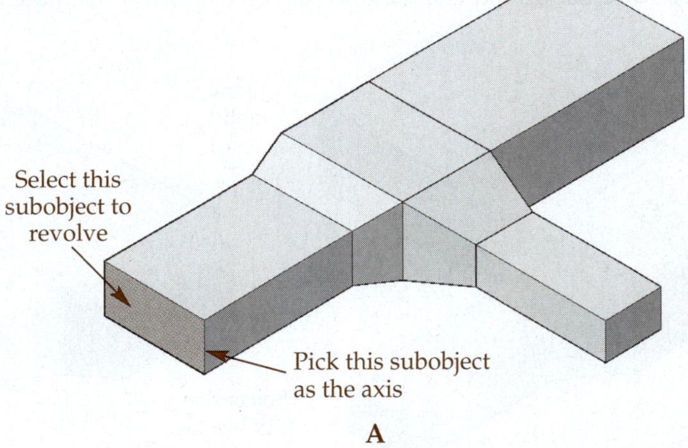

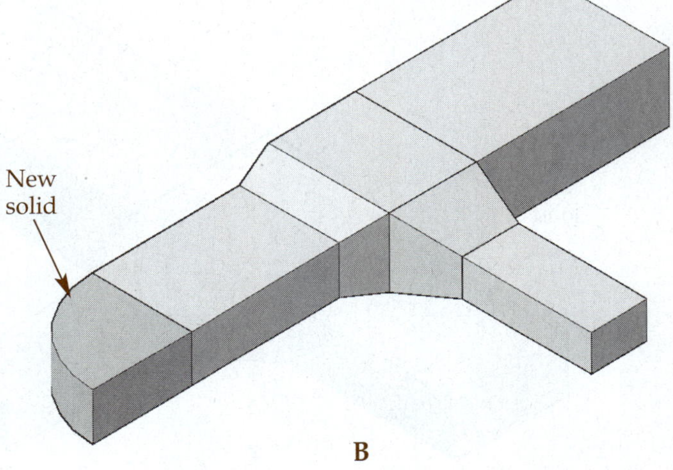

Edge Subobject Editing

Individual edges of a solid can be edited using grips and gizmos in the same manner as faces. To select an edge subobject, press the [Ctrl] key and pick the edge. The edge subobject filter can also be used to limit the selection to an edge.

Grips on linear edges are rectangular and appear in the middle of the edge, Figure 11-21. In addition to solid edges, the edges of regions can be altered using **MOVE**, **ROTATE**, and **SCALE**, but grips are not displayed on regions as they are on solid subobjects.

Remember, editing subobjects of a primitive removes the primitive's history. This should always be a consideration if it is important to preserve the solid primitives that were used to construct a 3D solid model. Instead of editing the primitive subobjects at their subobject level, it may be better to add or remove material with a Boolean operation, thus preserving the solid's history.

NOTE

There are several ways to perform a 3D edit. Keep the following options in mind when working with subobject editing.

- Selecting the **MOVE**, **ROTATE**, or **SCALE** command and picking a subobject will not display a gizmo, regardless of the current gizmo button displayed in the **Selection** panel and the **Show gizmos** flyout.
- Selecting the **3DMOVE**, **3DROTATE**, or **3DSCALE** command and picking a subobject will display the appropriate gizmo, regardless of the current gizmo button displayed in the **Selection** panel and the **Show gizmos** flyout.
- Picking a subobject without having selected a command first will display the gizmo corresponding to the current gizmo button in the **Selection** panel and the **Show gizmos** flyout.

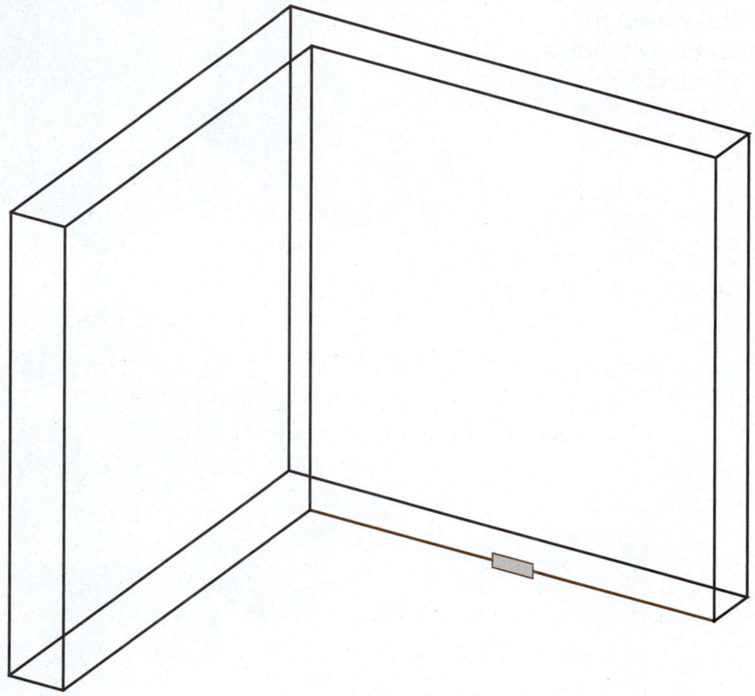

Figure 11-21.
Edge grips are rectangular and displayed in the middle of the edge.

Moving Edges

To move an edge, select it using the [Ctrl] key, as previously discussed. See **Figure 11-22A**. By default, the gizmo corresponding to the gizmo button in the **Selection** panel and the **Show gizmos** flyout appears. If the move gizmo is not current, select the **Move Gizmo** button in the **Selection** panel or the **Show gizmos** flyout. Select the appropriate axis handle and dynamically move the edge or use direct distance entry, **Figure 11-22B**. The move gizmo remains active until the [Esc] key is pressed to deselect the edge.

If you pick an edge grip to turn it hot, the gizmo is bypassed. This places you in the standard grip editing mode. You can stretch, move, rotate, scale, and mirror the edge. In this case, the stretch function works in the same manner as the move gizmo, but less reliably. You must be careful to use either ortho, polar tracking, or direct distance entry, but the possibility for error still exists.

The options available when dynamically moving an edge are similar to those used when moving a face. See **Figure 11-23A**. First, select the edge. Then hover over the base edge grip to display the base grip shortcut menu. Selecting the **Extend Adjacent Faces** option and moving the edge maintains the orientation of the moved edge, but its length is modified. See **Figure 11-23B**. This is because the planes and orientation of adjacent faces are maintained. Selecting the **Move Edge** option and moving the edge maintains the length and orientation of the moved edge. However, the shape and planes of adjacent faces are changed. See **Figure 11-23C**. Selecting the **Allow Triangulation** option and moving the edge maintains the length and orientation of the moved edge. But, if the move alters the planes of adjacent faces, those faces may become *nonplanar.* In other words, the face may now be located on two or more planes. If this happens, adjacent faces are divided into triangles, **Figure 11-23D**. This is visible when the object in **Figure 11-23D** is displayed in two orthographic views. See **Figure 11-24**.

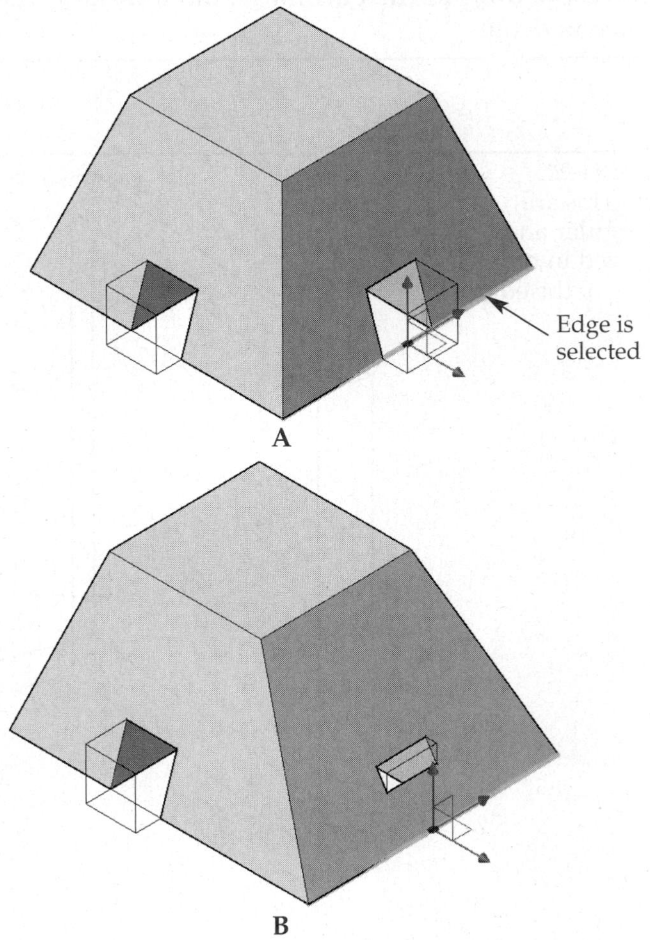

Figure 11-22.
A—Select the edge subobject to be moved. B—The edge is moved. Notice how the size of the primitive used to subtract the cutout is not affected.

The [Ctrl] key is used to access options when dynamically moving an edge. For example, after selecting the edge, pick the gizmo. Then, press and release the [Ctrl] key to cycle through the options.

Figure 11-23.
Moving an edge. A—Selecting the edge and then hovering over the base grip displays the base grip shortcut menu. B—Using the **Extend Adjacent Faces** option maintains the orientation of the moved edge, but its length is modified because the planes of adjacent faces are maintained. C—When using the **Move Edge** option, the edge maintains its length and orientation, but the shape and planes of adjacent faces are changed. D—When using the **Allow Triangulation** option, the adjacent faces may be triangulated.

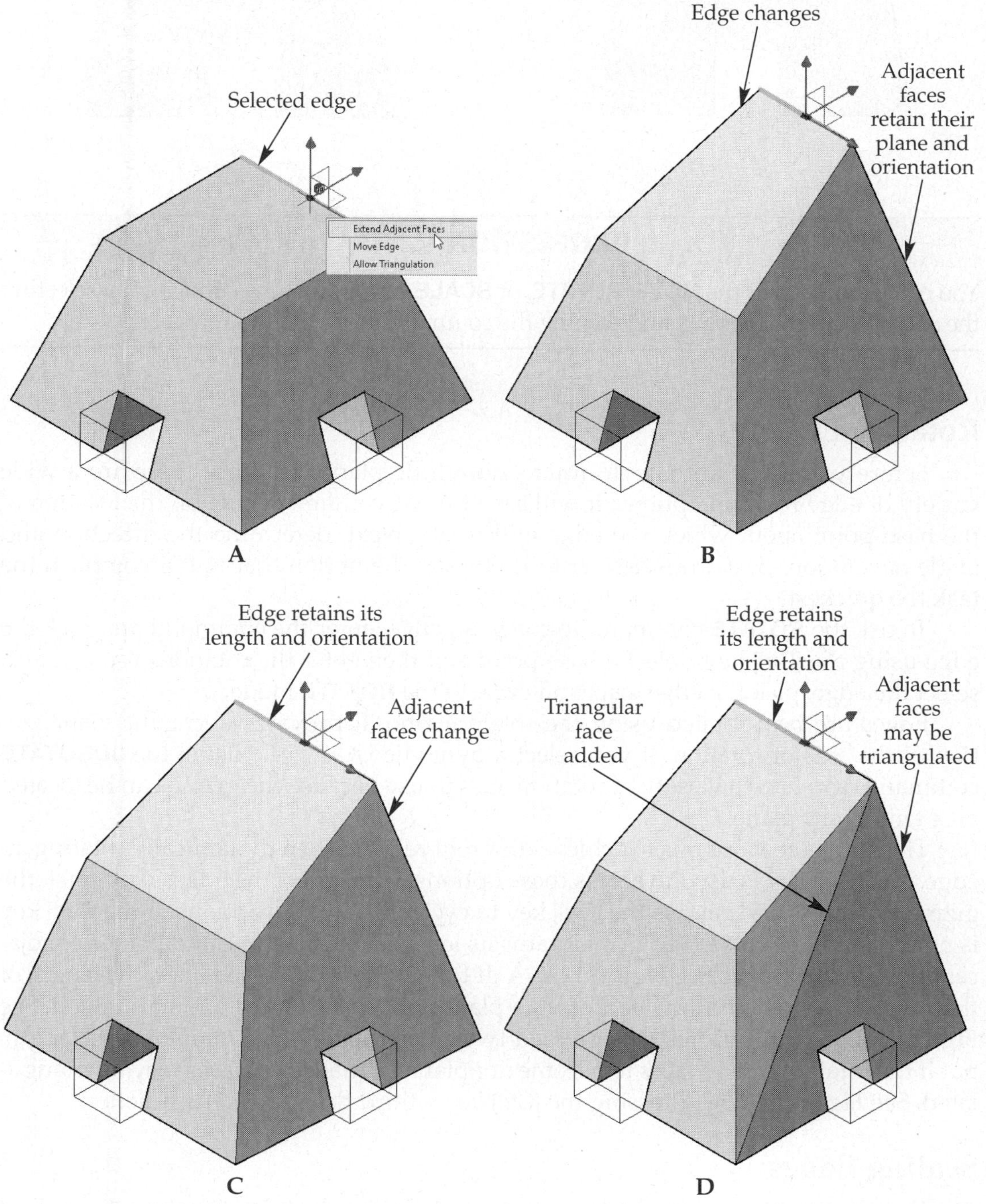

Figure 11-24.
Triangulated faces are clear in plan views. A—Front plan view. B—Side plan view.

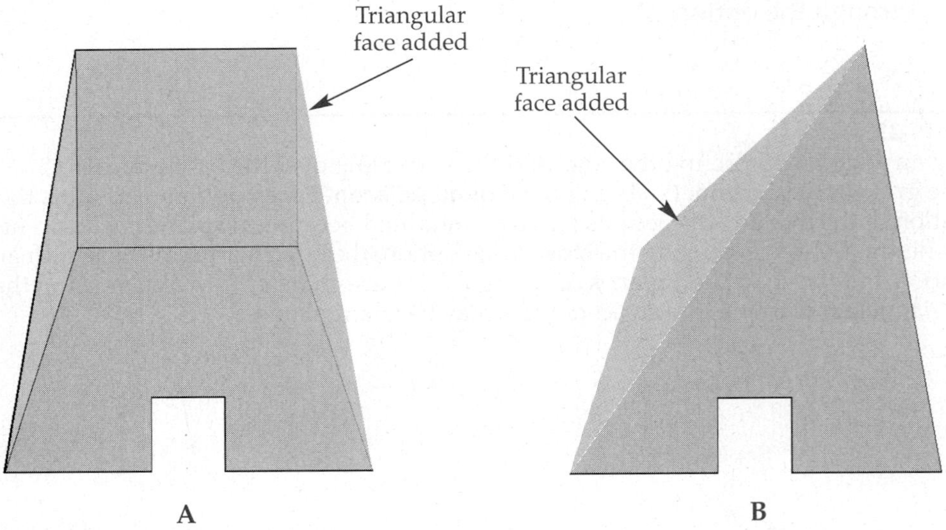

PROFESSIONAL TIP

You can quickly enter the **MOVE**, **ROTATE**, or **SCALE** command for a subobject by selecting the subobject, right-clicking, and picking the command from the shortcut menu.

Rotating Edges

Before you select an edge to rotate, do a little planning. Since there are a wide variety of edge rotation options, it will save time if you first decide on the location of the base point about which the edge will rotate. Next, determine the direction and angle of rotation. Based on these criteria, choose the option that will accomplish the task the quickest.

To use the **ROTATE** command to rotate an edge, enter the command and pick the edge using the [Ctrl] key. Select a base point and then enter the rotation. You can also select the edge, pick the edge grip, and cycle to the **ROTATE** mode.

Edges are best rotated using the rotate gizmo. It provides a graphic visualization of the axis of rotation. If you select a dynamic UCS while using the **3DROTATE** command, you have a variety of rotation axes to use because the gizmo can be located on a temporary plane.

There are a few options to achieve different results when dynamically rotating an edge. The [Ctrl] key is used to access these options. First, select the edge. Then, pick the gizmo and press and release the [Ctrl] key to cycle through the options. If the [Ctrl] key is not pressed, the rotated edge maintains its length, but the shape and planes of adjacent faces are changed. See **Figure 11-25A**. If the [Ctrl] key is pressed once, the length of the rotated edge is modified because the planes of adjacent faces are maintained. See **Figure 11-25B**. If the [Ctrl] key is pressed twice, the rotated edge maintains its length, but if the rotation causes faces to become nonplanar, the adjacent faces may be triangulated. See **Figure 11-25C**. Pressing the [Ctrl] key a third time resets the function.

Scaling Edges

Only linear (straight-line) edges can be scaled. Circular edges, such as the ends of cylinders, can be modified using grips or the **SOLIDEDIT** command. These tools can be used to change the diameter or establish taper angles. See Chapter 12 for a complete discussion of the **SOLIDEDIT** command.

Figure 11-25.
Rotating an edge. A—The [Ctrl] key is not pressed. B—The [Ctrl] key is pressed once. Notice the top edge of the dark face. C—The [Ctrl] key is pressed twice.

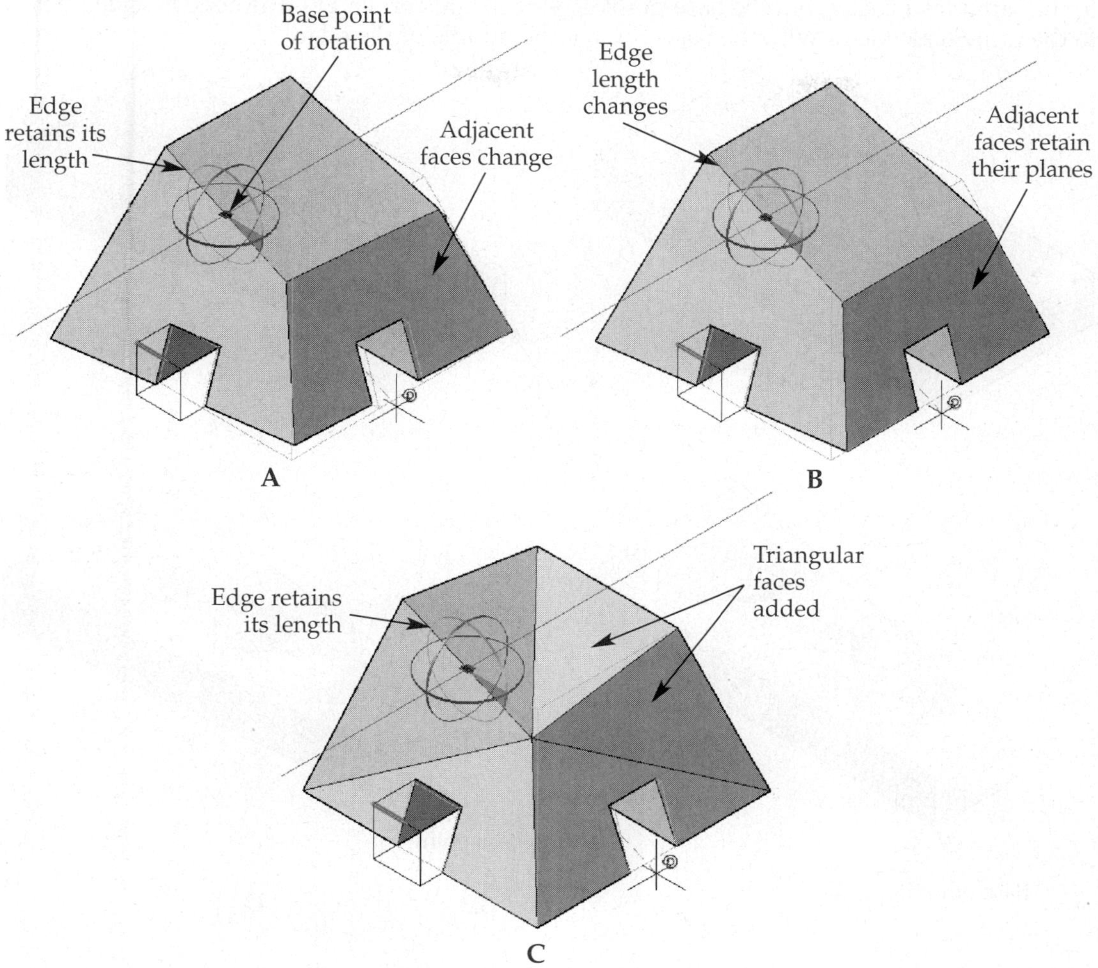

To scale a linear edge, enter the **SCALE** command. Select the edge using the [Ctrl] key. Pick a base point for the operation and enter a scale factor. You can also select the edge, pick the edge grip, and cycle to the **SCALE** mode. However, using the scale gizmo may be the best option.

The direction of the scaled edge is related to the base point you select. The base point remains stationary, while the vertices in either direction are scaled. If you enter the **SCALE** command, you are prompted for the base point. If you select the edge grip, the grip becomes the base point. The differences in opposite end and midpoint scaling of an edge are shown in **Figure 11-26**.

There are a few options to achieve different results when dynamically scaling an edge. The [Ctrl] key is used to access these options. First, select the edge. Then, pick the scale gizmo and press and release the [Ctrl] key to cycle through the options.

If the [Ctrl] key is not pressed, the edge is scaled. The shape and planes of adjacent faces are changed to match the scaled edge. See **Figure 11-27A**.

If the [Ctrl] key is pressed once, the edge is, in effect, not scaled. This is because the planes of adjacent faces are maintained.

If the [Ctrl] key is pressed twice, the edge is scaled, as are edges attached to the modified edge. However, if the scaling causes faces to become nonplanar, they may be triangulated. See **Figure 11-27B**.

Figure 11-26.
The differences in opposite end and midpoint scaling of an edge. A—The original object. B—The edge is scaled down with a base point on the left corner. C—The edge is scaled down to the same scale factor, but the base point is on the right corner. D—The edge is scaled down to the same scale factor with the base point at the middle of the edge.

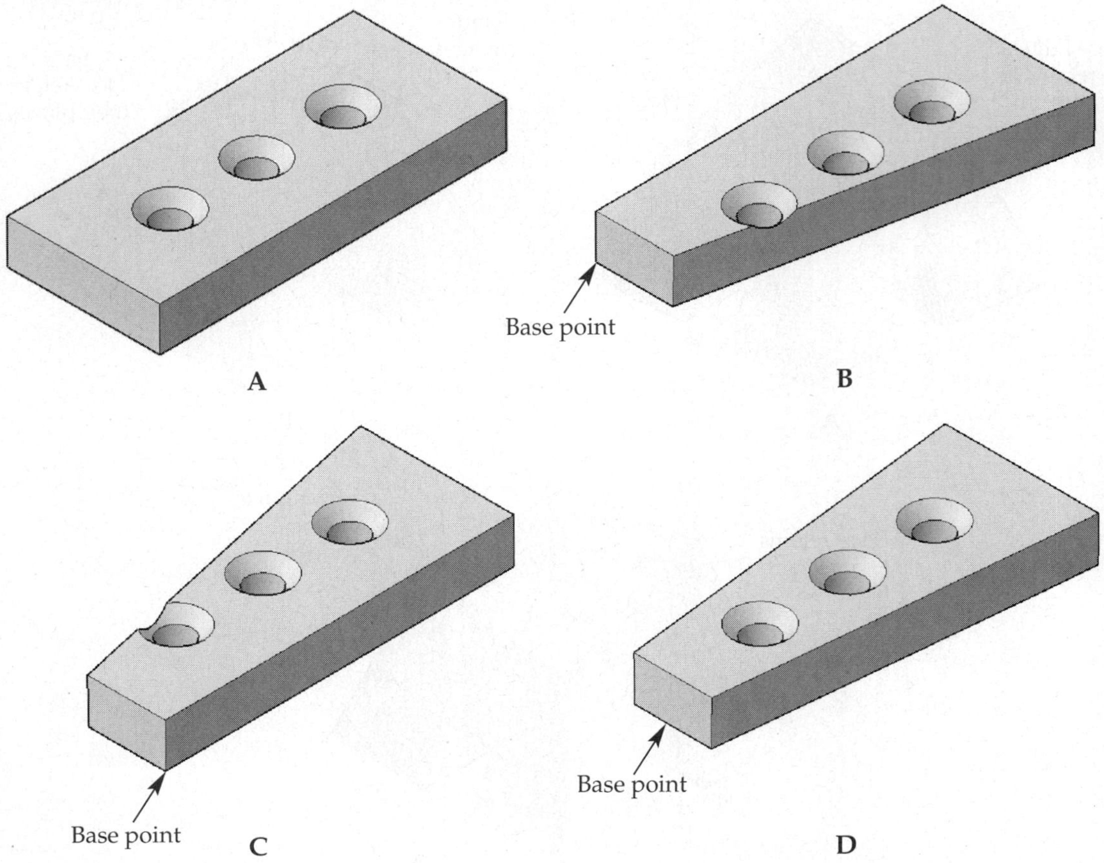

Figure 11-27.
Scaling the front edge with the base point at the middle of the edge. The original object is shown in Figure 11-26A. A—If the [Ctrl] key is not pressed, the edge is scaled and the shape and planes of adjacent faces are changed. B—If the [Ctrl] key is pressed twice, the edge is scaled, as are edges attached to it. If the scaling causes faces to become nonplanar, the adjacent faces may be triangulated.

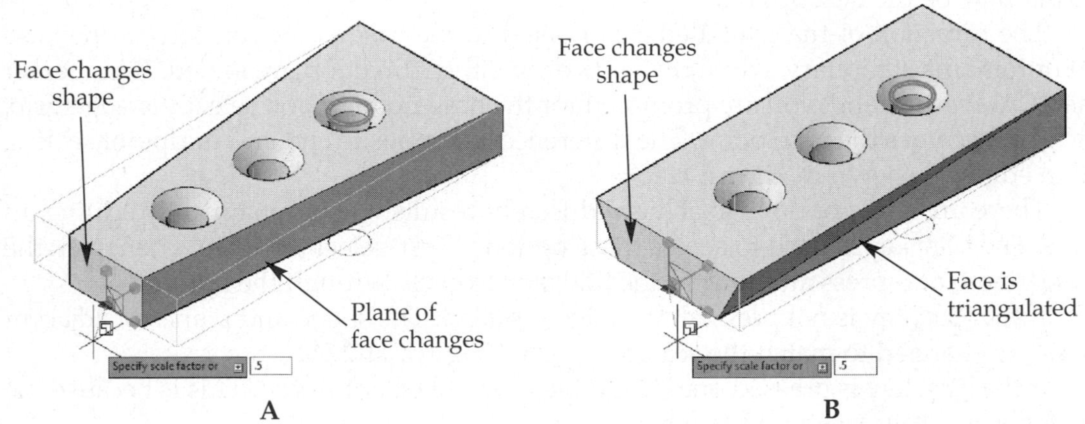

Coloring Edges

To change the color of an edge, use the [Ctrl] key to select the edge. Next, open the **Properties** palette. In the **General** category, pick the drop-down list for the Color property.

Select the desired color or pick Select Color... and select a color from the **Select Color** dialog box. Edges cannot have materials assigned to them.

Deleting Edges

Edges can be deleted in certain situations. In order for an edge to be deleted, it must completely divide two faces that lie on the same plane. If this condition is met, the **ERASE** command or the [Delete] key can be used to remove the edge. The two faces become a single face.

Exercise 11-4 www.g-wlearning.com/CAD/

Complete the exercise on the companion website.

Vertex Subobject Editing

The modification of a single vertex involves moving the vertex and stretching all edges and planar faces attached to it. Vertex grips are circular and located on the vertex, as shown in **Figure 11-28**. Use the vertex subobject filter to assist in selecting vertices. A single vertex cannot be rotated or scaled, but you can select multiple vertices and perform rotating and scaling edits. When editing multiple vertices in this manner, you are, in effect, editing edges.

As with other subobject editing functions performed on a 3D solid primitive, the solid's history is removed when a vertex is modified. The solid can no longer be edited using the primitive grips; only a single base grip is displayed. Further editing of the solid must be done with the **SOLIDEDIT** command, discussed in Chapter 12, or through subobject editing.

Moving Vertices

To move a vertex, select it using the [Ctrl] key. If the move gizmo is not displayed, select it from the gizmo drop-down list in the **Selection** panel or the **Show gizmos** flyout. You can use the move gizmo, the **MOVE** command, or standard grip editing modes to move the vertex. Hovering over the base vertex grip displays a shortcut menu with two move options. See **Figure 11-29A**. The **Move Vertex** option allows the vertex to be moved without triangulating adjacent faces, but the faces may change shape. See **Figure 11-29B**. In some cases, AutoCAD may deem it necessary to triangulate faces. Using the **Allow Triangulation** option results in the triangulation of adjacent faces when the vertex is moved. See **Figure 11-29C**. Pressing the [Ctrl] key during the edit cycles through the move options.

Figure 11-28.
Vertex grips are circular and placed on the vertex.

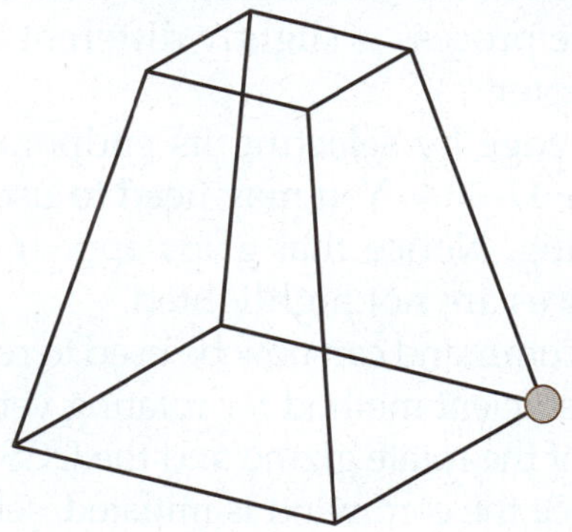

Figure 11-29.
Moving a vertex. A—The original object. B—Using the **Move Vertex** option moves the vertex and changes some of the adjacent faces. B—When using the **Allow Triangulation** option, adjacent faces are triangulated.

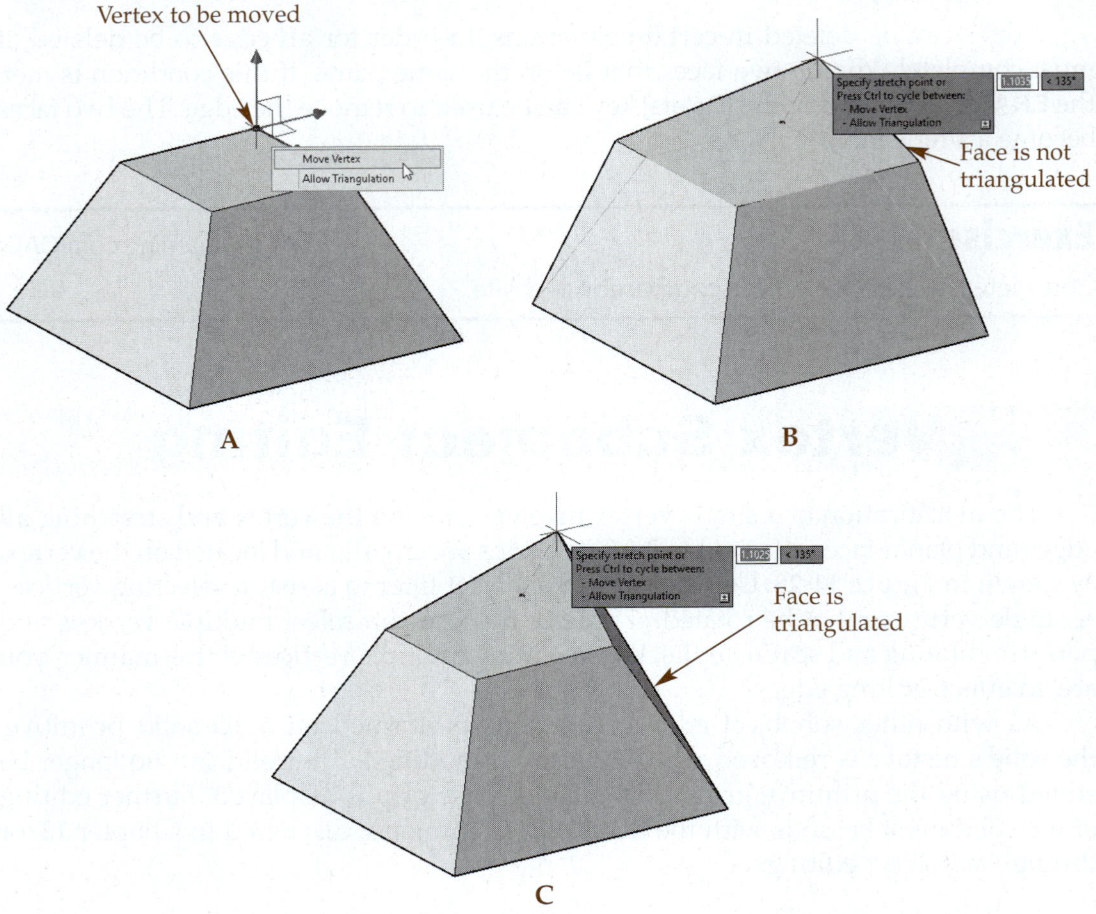

PROFESSIONAL TIP

If you are dragging a vertex and faces become triangulated, you can transparently change your viewpoint to see the effect of the triangulation. Press and hold the [Shift] key. At the same time, press and hold the mouse wheel button. Now, move the mouse to change the viewpoint. This is a transparent instance of the **3DORBIT** command.

Rotating Vertices

As previously stated, a single vertex cannot be rotated or scaled, but two or more vertices can be. Since two vertices define a line, or edge, any edit is an edge modification. However, the process is slightly different from the edge modifications described earlier in this chapter.

To rotate an edge by selecting its endpoints, press the [Ctrl] key and select each vertex. See Figure 11-30A. You may need to use the [Ctrl]+[W] key combination to turn on selection cycling. Notice that grips appear at each selected vertex, but the edges between the vertices are not highlighted.

The **ROTATE** command can now be used to rotate the vertices (if **PICKFIRST** is set to 1). However, a more efficient method for rotating vertices is to use the **3DROTATE** command. The combination of the rotate gizmo and the UCS icon enables you to graphically view the rotation plane. Once the command is initiated, select the vertices and move the base point

Figure 11-30.
To rotate or scale vertices, multiple vertices must be selected. In effect, the edges are modified. A—Vertices are selected to be rotated. B—The rotate gizmo is placed at the base of rotation and the rotation axis is selected. C—The vertices are rotated.

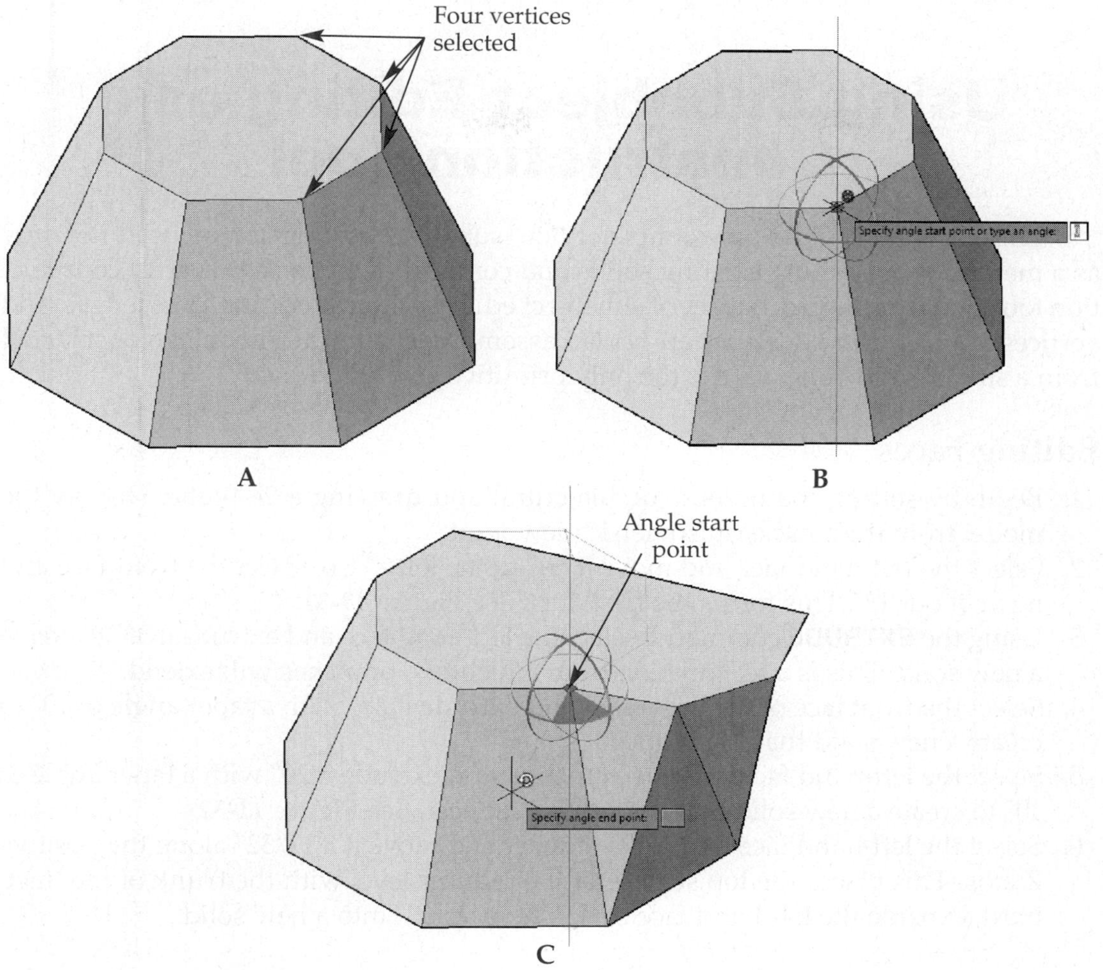

of the rotate gizmo if needed. See **Figure 11-30B**. Then, select the axis of revolution. Finally, pick the angle start point and enter the rotation. See **Figure 11-30C**.

If the [Ctrl] key is not pressed while dynamically rotating the vertices, the area of the selected vertices does not change and adjacent faces are triangulated. This is because the edges of the adjacent faces are attached to the selected vertices, so their edge length changes as the selected edge is rotated. If the [Ctrl] key is pressed once, the adjacent faces are not triangulated unless necessary, but the faces may change shape.

NOTE

If the selected edge does not dynamically rotate at the "angle end point" prompt, then the desired rotation is not possible.

Scaling Vertices

As mentioned earlier, it is not possible to scale a single vertex. However, two or more vertices can be selected for scaling. This, in effect, scales edges. The selection methods are the same as discussed for rotating vertices and the use of the [Ctrl] key while dragging produces the same effects. As the pointer is dragged, the dynamic display of scaled edges may be difficult to visualize. Therefore, it is best to use a scale factor or the **Reference** option to achieve properly scaled edges.

Exercise 11-5

www.g-wlearning.com/CAD/

Complete the exercise on the companion website.

Using Subobject Editing as a Construction Tool

This section provides an example of how subobject editing can be used not only as a method for changing existing solids and composites, but as a powerful construction tool. Some of the procedures of subobject editing, such as editing faces, edges, and vertices, are used to construct an HVAC assembly. The entire model is constructed from a single, solid cube. This is the only primitive you will draw.

Editing Faces

1. Begin by setting the units to architectural and drawing a 24″ cube. Display the model from the southeast isometric viewpoint.
2. Select the left-hand face and move it 60″ to the left. Also, select the front face and move it out 12″. This forms the first duct. See **Figure 11-31**.
3. Using the **EXTRUDE** command, select the left-hand face and extrude it 36″ to create a new solid. This is a tee junction from which two branches will extend.
4. Select the front face of the new solid and extrude it 28″ with a taper angle of 10° to create a new solid that is a reducer.
5. Select the left-hand face of the tee junction and extrude it 20″ with a taper angle of 10° to create a new solid that is a second reducer. See **Figure 11-32**.
6. Select the left-hand face of the 20″ reducer and move it 3 17/32″ along the positive Z axis. This places the top surface of the reducer level with the trunk of the duct. Next, extrude the left-hand face of this reducer 60″ into a new solid.

Figure 11-31.
The left-hand face of the cube is moved 60″. The front face is then moved 12″.

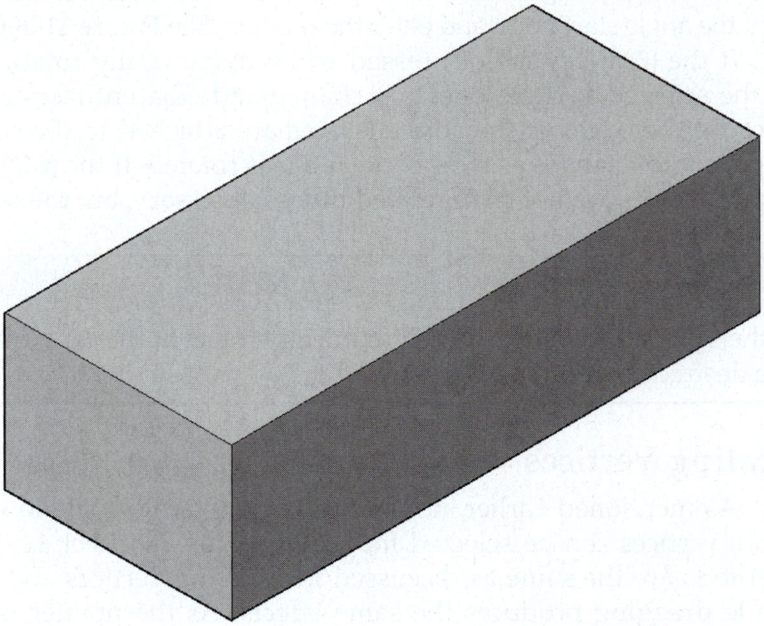

Figure 11-32.
Two reducers are created by extruding faces from the tee junction.

7. Use the **REVOLVE** command to turn the left-hand face of the 60″ extension into a new solid that is a 90° bend. Your drawing should now look like **Figure 11-33**.

Editing Edges and Vertices

1. The bottom surface of the 28″ reducer must be level with the bottom of the tee junction and main trunk. Select the bottom edge of the reducer's front face and move it down 4 15/16″.
2. Select the two top vertices on the 28″ reducer's front face. Move the vertices down (negative Z) 3″.
3. Select the front, rectangular face of the 90° bend and extrude it 72″ into a new solid.
4. Select the front face of the 28″ reducer and extrude it 108″ into a new solid.
5. Select the front face of the new solid created in step 4 and extrude it 26″ to create a new solid that will be a tee junction.
6. Extrude the left-hand face of the tee junction 20″. See **Figure 11-34**.

Figure 11-33.
The left end of the 60″ extrusion is revolved 90° to create an elbow.

Figure 11-34.
The face of the revolved elbow is extruded 72″. The face of the right branch is extruded 108″. The right duct is then extruded 26″ and the left face of that extrusion is extruded 20″.

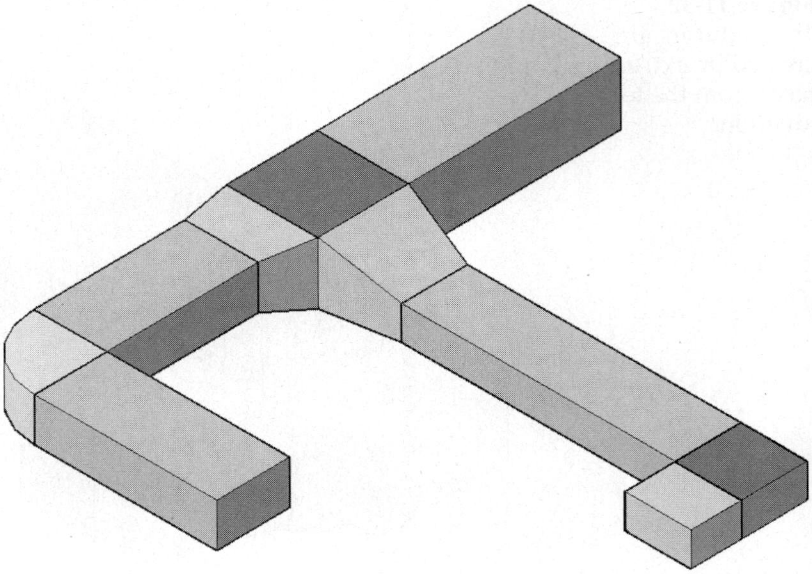

7. Move the top edge of the left-hand face on the 20″ extrusion created in step 6 down 4″.
8. Move each vertical edge of the 20″ extrusion 6″ toward the center of the duct.
9. Mirror a copy of the 20″ extrusion to the opposite side of the tee junction. The completed drawing should look like **Figure 11-35**.

NOTE

For many of the previously discussed subobject editing techniques, you may find it difficult to edit an object after a complex composite solid is created, especially if it has multiple fillets. You may find it necessary to select the subobject, delete it, and start the process over. Or, you may find it necessary to add geometry to the existing solid using the **UNION** command and then edit the new object as needed.

Figure 11-35.
The reducer is mirrored to create the final assembly.

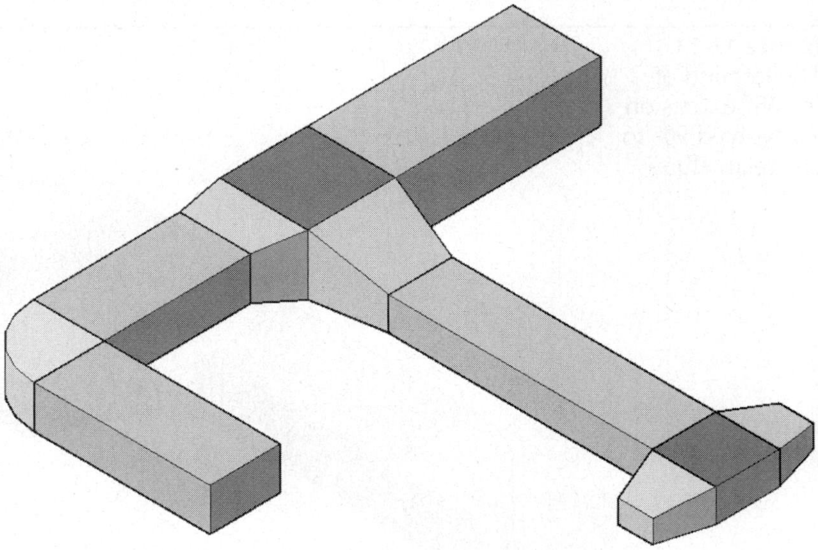

Other Solid Editing Tools

There are other tools that can be used in solid model editing. As you will learn in Chapter 12, the **SOLIDEDIT** command can be used to edit faces, edges, and vertices, much like subobject editing. In addition, you can extrude open or closed geometry with the **PRESSPULL** command, offset edges with the **OFFSETEDGE** command, and explode a solid with the **EXPLODE** command. These methods are discussed in the next sections.

Presspull

PRESSPULL

Ribbon
Home
 > Modeling
Solid
 > Solid

Presspull

Type
PRESSPULL

The **PRESSPULL** command allows you to create extrusions from open or closed 2D profiles, 3D curves, and 3D solid faces. You can extrude open 2D objects such as lines, polylines, splines, or arcs. You can also extrude objects forming closed areas such as circles, closed polylines, and regions. In addition, you can use the **PRESSPULL** command to add to or subtract from an existing solid by selecting and dragging a closed and bounded area of the solid. The extrusion is always applied perpendicular to the plane of the boundary, but can be in the positive or negative direction. When applied to the face of a solid, the **PRESSPULL** command is very similar to the **Extrude Faces** option of the **SOLIDEDIT** command, though dynamic feedback is provided for the extrusion with **PRESSPULL**.

Once the command is initiated, you are prompted to select an object or bounded area to extrude. If you are selecting a closed boundary, pick inside the boundary. Then, drag the boundary to a new location and pick, or, if dynamic input is on, enter the extrusion distance. The **PRESSPULL** command stays active so repetitive selections can occur in a single command sequence.

An example of using the **PRESSPULL** command to add material to or subtract material from a 3D model is shown in **Figure 11-36**. By extruding the cylinder in a positive direction, the new cylinder is automatically unioned to the existing part, **Figure 11-36B**. By extruding the cylinder in a negative direction, the cylinder is subtracted from the wedge, **Figure 11-36C**.

Using the **PRESSPULL** command to create extrusions from open 2D objects is shown in **Figure 11-37**. As is the case when using the **EXTRUDE** command, the resulting object is a surface.

When extruding a 3D face with the **PRESSPULL** command, the extrusion is applied perpendicular to the surface and the shape of adjacent faces is maintained. If the adjacent faces are at an angle to the selected surface, you can maintain the shape of the faces by pressing the [Ctrl] key. You can use this method to maintain the orientation of adjoining surfaces or to create a taper angle. **Figure 11-38A** shows a run of ductwork along the top of walls that are at an angle to each other. The ducts follow the wall angles. A redesign requires the angled ductwork to be expanded. In **Figure 11-38B**, the long duct along the angled wall has been expanded using the **PRESSPULL** command. Then, the adjacent face is selected with the [Ctrl] key to maintain the angle with the face of the long duct. In addition, the face of the next duct at the 90° wall corner is selected and extruded. The result is shown in **Figure 11-38C**. In **Figure 11-38D**, the four front wall ducts have been expanded and unioned to complete the design.

NOTE

The **PRESSPULL** command allows for multiple object selection. To select more than one object, use the **Multiple** option, hold down the [Shift] key, or right-click and select **Multiple** from the context-sensitive shortcut menu.

Figure 11-36.
Using the **PRESSPULL** command. A—Pick inside of a boundary (shown in color) and drag the boundary to a new location. B—The completed operation with a positive distance, resulting in a union. C—The completed operation with a negative distance, resulting in a subtraction.

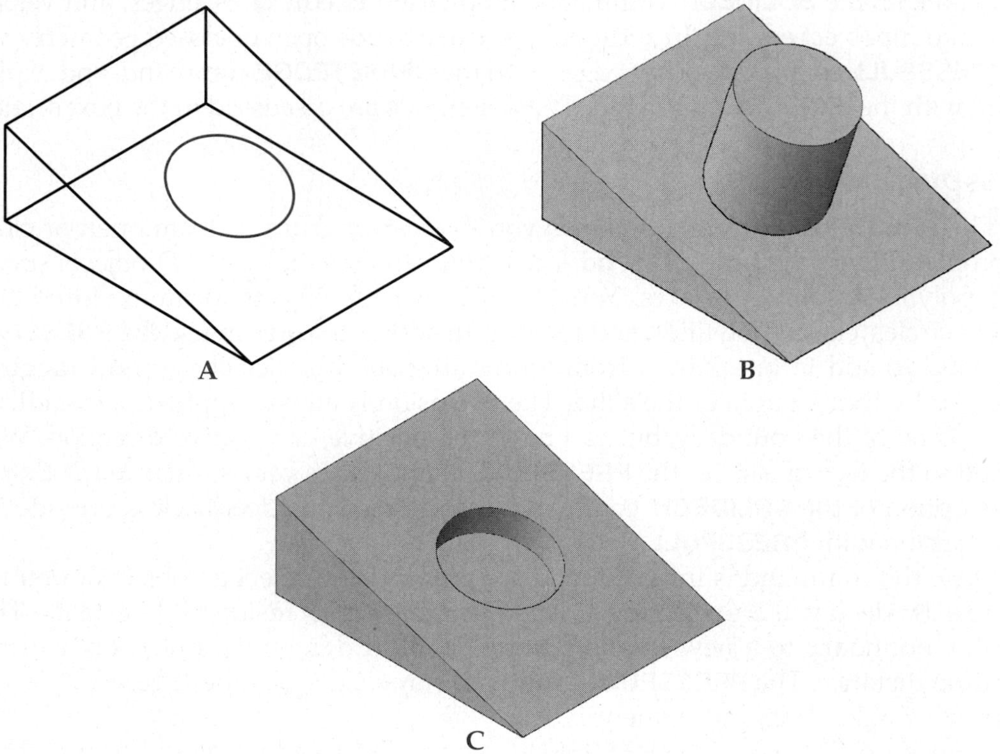

Figure 11-37.
Extruding open 2D objects with the **PRESSPULL** command. The resulting objects are surfaces.

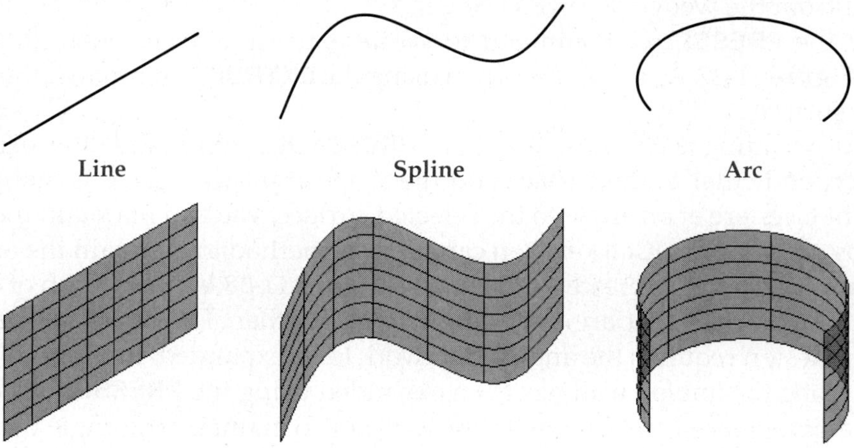

Offsetting Edges

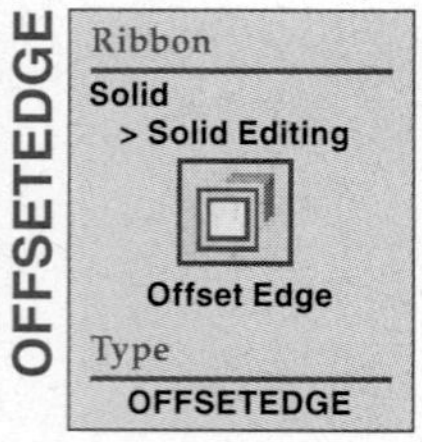

The **OFFSETEDGE** command allows you to offset the edges of a planar face to create a new, closed object on the same plane as the existing face. The new object is a closed polyline or spline that can then be used to create a new 3D object. The planar face you offset can be on a 3D solid or surface model. After entering the **OFFSETEDGE** command, select the face and then pick a point inside or outside of the existing edge. The point you pick determines where the offset is created. You can also use the **Distance** option to create the offset at a specific distance from the existing edge. The **OFFSETEDGE** command allows you to continue selecting faces to create offsets until you press [Esc] or [Enter]. The procedure is similar to that used with the **OFFSET** command.

Figure 11-38.
Using the **PRESSPULL** command to expand a run of ductwork along angled walls. A—The original design. B—After expanding the first face, the adjacent face is selected using the [Ctrl] key. A second face at the corner duct is also selected and extruded. C—The resulting design. D—The front wall ducts are expanded and unioned to complete the design.

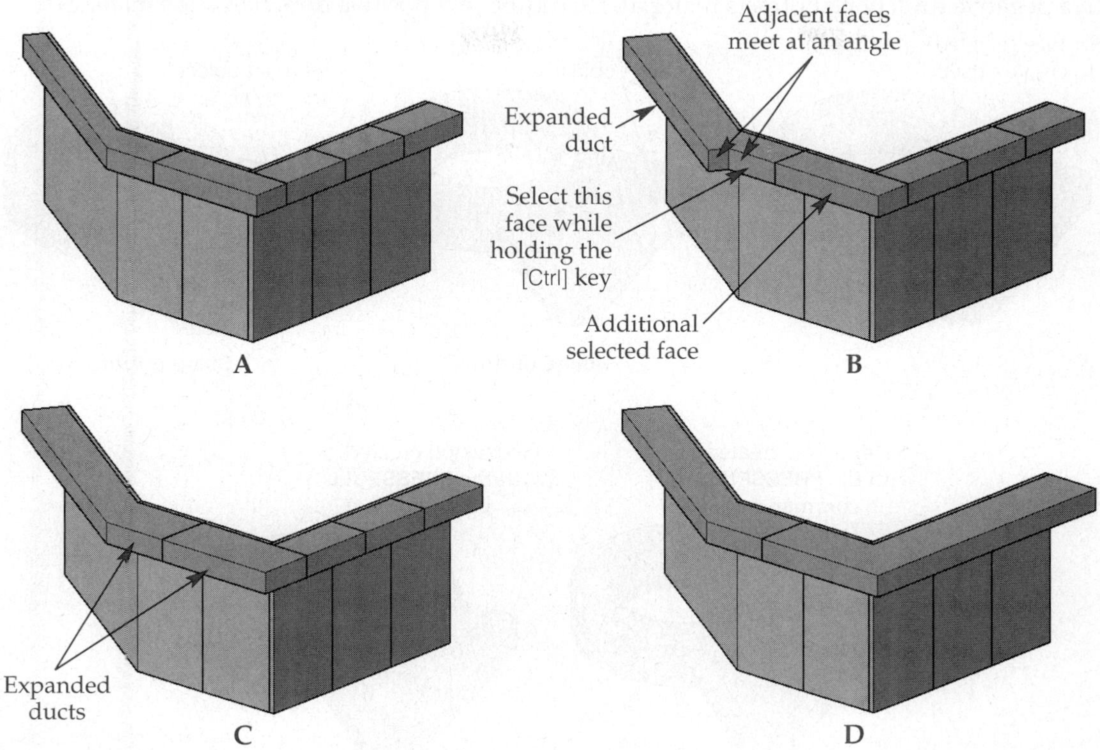

An example of using the **OFFSETEDGE** command is shown in **Figure 11-39**. Before specifying the point through which the offset object is created, you can use the **Corner** option to create the offset object with round or sharp corners. Results after using the **Round** and **Sharp** options are shown in **Figure 11-39B**. In both examples shown, the new offset object is a closed polyline.

Once the offset object is created, you can use the **EXTRUDE** or **PRESSPULL** command to create a new 3D feature. See **Figure 11-39C**. In each example shown, an extrusion is created with the **PRESSPULL** command.

Exercise 11-6

www.g-wlearning.com/CAD/

Complete the exercise on the companion website.

Exploding a Solid

A solid can be exploded. This turns the solid into surfaces and/or regions. Flat surfaces on the solid are turned into regions. Curved surfaces on the solid are turned into surfaces. To explode a solid, select the **EXPLODE** command. Then, pick the solid(s) to explode and press [Enter].

EXPLODE

Ribbon
Home
> Modify
Explode
Type
EXPLODE
X

Exercise 11-7

www.g-wlearning.com/CAD/

Complete the exercise on the companion website.

Figure 11-39.
Using the **OFFSETEDGE** command. A—After entering the command, select a point on the planar face with the edges to offset. B—The offset object can have rounded or sharp corners. In each example shown, a closed polyline is created. C—The new polyline object can be used for modeling purposes. Shown are results after using the **PRESSPULL** command. Extruding in a negative direction subtracts material. Extruding in a positive direction adds material.

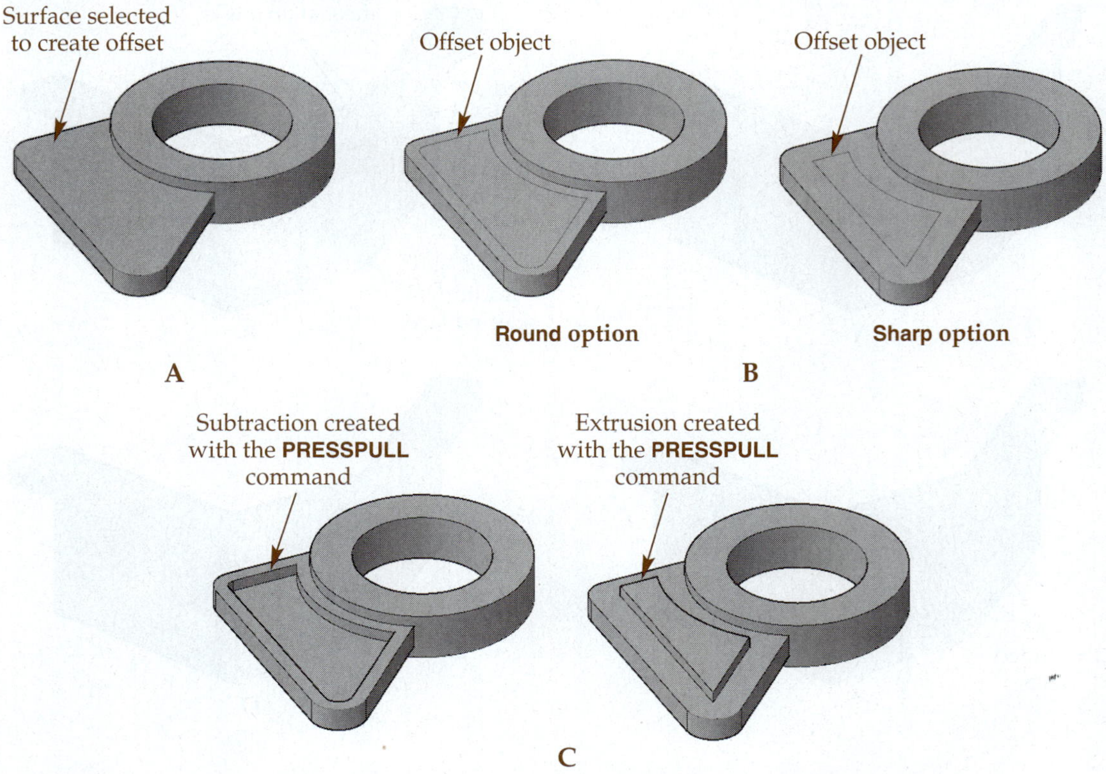

Working with Associative Arrays

A 3D array constructed with the **ARRAYRECT**, **ARRAYPOLAR**, or **ARRAYPATH** command can be created as an associative or non-associative array. Arrays are introduced in Chapter 8. An associative array acts as a single object, much like a block. If you try to use a Boolean operation with an associative array, AutoCAD will not allow the operation to perform. As with a block, an array is a collection of objects that are defined as one object. Think of a 3D associative array as a group of solids in a plastic "wrapper." This type of object cannot be used in a Boolean operation. For example, if you attempt to subtract an associative polar array of cylinders from a plate, AutoCAD will display the following message:

No solids, surfaces, or regions selected.

To use arrayed objects for a Boolean operation, create the array as a non-associative array or explode the array using the **EXPLODE** command. See Figure 11-40. Exploding an array creates individual objects that can then be used for other modeling purposes. Figure 11-40C shows the result after using the **SUBTRACT** command to subtract the individual cylinders from the base plate.

Exercise 11-8

www.g-wlearning.com/CAD/

Complete the exercise on the companion website.

Figure 11-40.
Using arrayed objects in a Boolean operation. A—A base plate with a single cylinder. B—A polar array is created from the cylinder. If the array is created as an associative array, it cannot be subtracted from the base plate in a Boolean operation. C—The result after exploding the array and subtracting the individual cylinders from the base plate. Shown is a shaded view of the model.

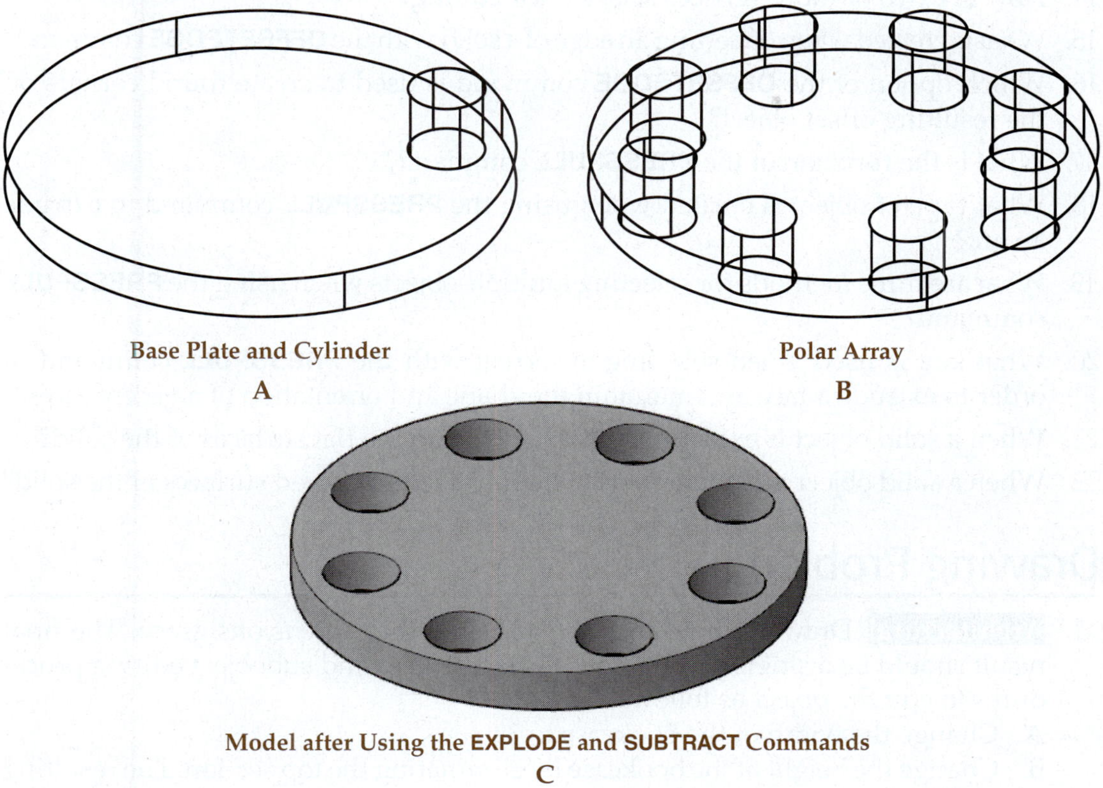

Chapter Review

Answer the following questions using the information in this chapter.

1. How do you select a subobject?
2. How do you deselect a subobject?
3. When grip editing, two types of grips appear on the object. Name the two types of grips.
4. If you have a cylinder primitive with a height of 10 units, but the height requirement has changed to 15 units, explain the procedure for adding 5 units to the cylinder height.
5. How can you change the radius of a fillet or the distances of a chamfer?
6. When moving a face on a solid primitive, how can you accurately control the axis of movement?
7. When moving a face on a solid primitive, which option maintains the planes of adjacent faces while modifying the size of the face?
8. Describe a major difference of function between the **ROTATE** and **3DROTATE** commands.
9. Which system variable enables you to use the **3DROTATE** command in a 3D view even if you select the **ROTATE** command?

10. How does the location and shape of an edge grip differ from a face grip?
11. What is the most efficient tool to use when rotating an edge and how is it displayed?
12. What is the only type of edge that can be scaled?
13. What is the only editing function that can be done when editing a single vertex?
14. How are two or more vertices selected for editing?
15. What is created when offsetting an edge of a solid with the **OFFSETEDGE** command?
16. Which option of the **OFFSETEDGE** command is used to create round corners on the resulting offset object?
17. What is the function of the **PRESSPULL** command?
18. What type of object is created when using the **PRESSPULL** command to extrude an arc?
19. What are three methods for selecting multiple objects when using the **PRESSPULL** command?
20. What key is used when selecting an object with the **PRESSPULL** command in order to extrude a face and maintain the shape and orientation of adjacent faces?
21. When a solid object is exploded, what happens to the flat surfaces of the solid?
22. When a solid object is exploded, what happens to the curved surfaces of the solid?

Drawing Problems

1. **Architectural** Draw the bookcase shown using the dimensions given. The final result should be a single solid object. Then, use grip and subobject editing procedures to edit the object as follows.
   A. Change the width of the bookcase to 3′.
   B. Change the height of the bookcase by eliminating the top section. The resulting height should be 3′-2″.
   C. Save the drawing as P11-1.

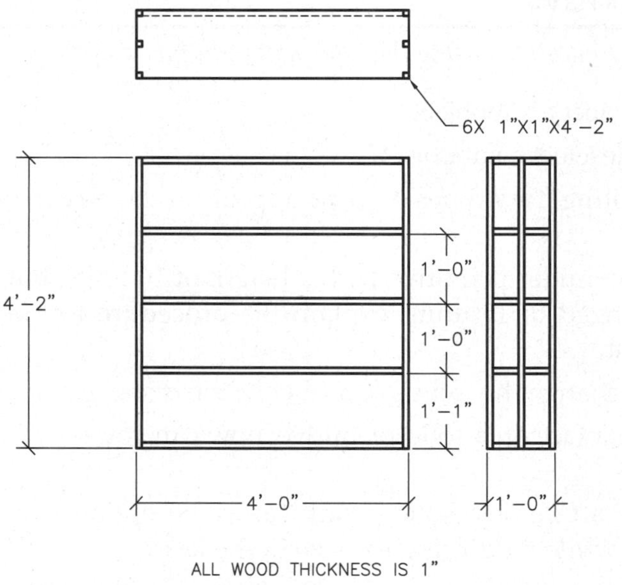

2. **Architectural** Open problem P11-1. Save it as P11-2. Use primitive and subobject editing procedures to create the following edits.
   A. Change the depth of the top of the bookcase to 6-1/2″.
   B. Change the depth of the bottom of the bookcase to 24″.
   C. Reduce the height of the front uprights so they are flush with the top surface of the next lower shelf.
   D. Extend the front of the middle shelf to match the front of the bottom. Add two uprights at the front corners between the bottom and this shelf.
   E. Save the drawing.

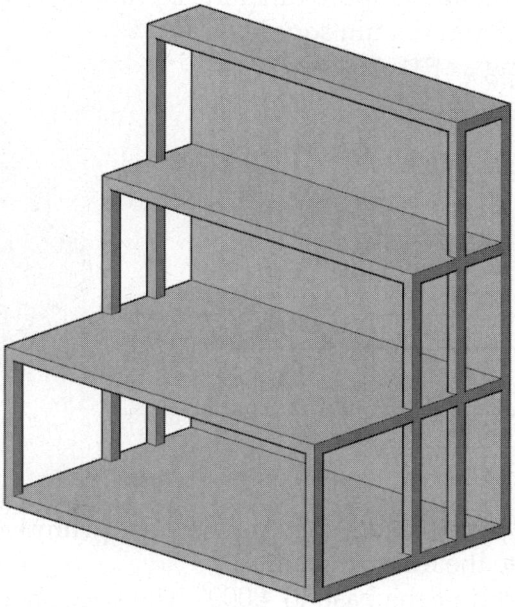

3. **Mechanical** Draw the mounting bracket shown below. Then, use primitive and subobject editing procedures to create the following edits.
   A. Change the 3.00″ dimension to 3.50″.
   B. Change the 2.50″ dimension in the front view to 2.75″.
   C. Change the location of the slot in the auxiliary view from .60″ to .70″ and change the length of the slot to 1.15″.
   D. Change the width of each foot in the top view from 2.00″ to 1.50″. The overall dimension (5.00″) should not change.
   E. Change the angle of the bend from 15° to 45°.
   F. Save the drawing as P11-3.

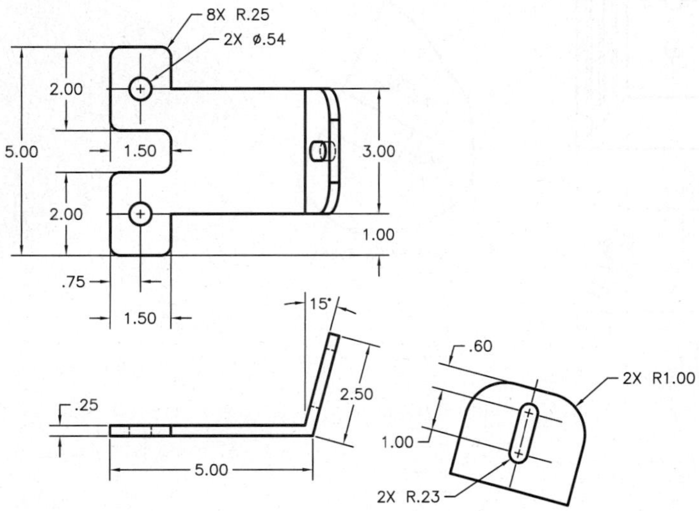

Drawing Problems - Chapter 11

4. **General** Draw as a single composite solid the desk organizer shown in the orthographic views. Then, use primitive and subobject editing procedures to create the following edits. The final object should look like the shaded view.
   A. Change the 3″ height to 3.25″.
   B. Change the 2″ height to 1.85″.
   C. Increase the thickness of the long compartment divider to .5″. The increase in thickness should be evenly applied along the centerline of the divider. Locate three evenly spaced, ∅5/16″ × 1.5″ holes in this divider.
   D. Angle the top face of the rear compartments by 30°. The height of the rear of the organizer should be approximately 4.5″ and all corners on the bottom of the organizer should remain square.
   E. Save the drawing as P11-4.

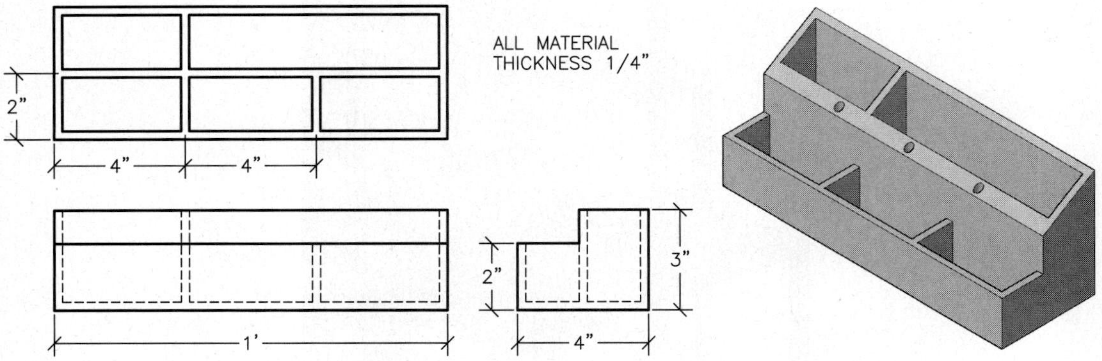

5. **General** Draw the pencil holder shown. Then, use primitive and subobject editing procedures to create the following edits.
   A. Change the depth of the base to 4.000″. The base should be rectangular, not square, and the grooves should become shorter.
   B. Change the height of the top groove from .250″ to .125″.
   C. Change the diameter of two holes from ∅.450″ to ∅.625″.
   D. Change the diameter of the other two holes from ∅.450″ to ∅1.000″.
   E. Rotate the top face 15° away from the side with the grooves. The planes of the adjoining faces should not change. Refer to the shaded view.
   F. Save the drawing as P11-5.

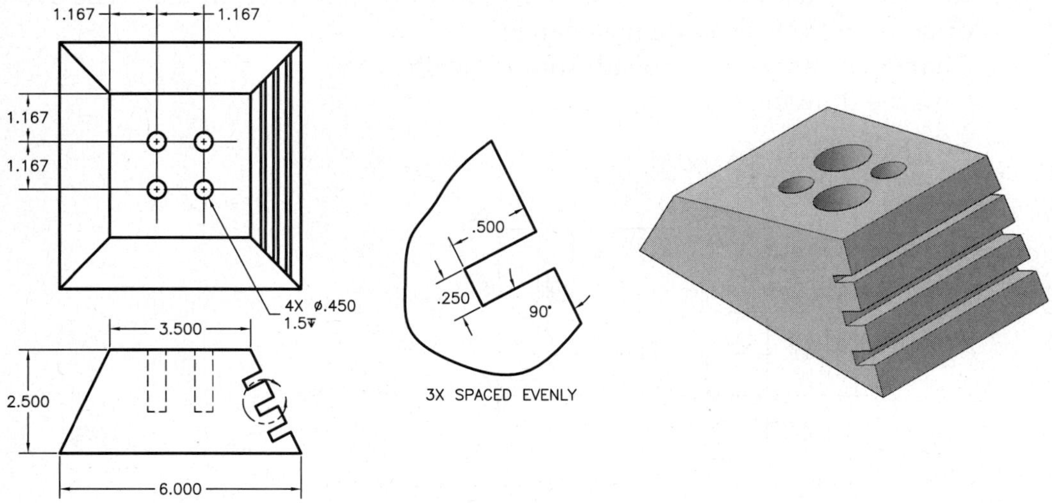

CHAPTER 12

Solid Model Editing

Learning Objectives

After completing this chapter, you will be able to:

- ✓ Change the shape and configuration of solid object faces.
- ✓ Copy and change the color of solid object faces and edges.
- ✓ Break apart a composite solid composed of physically separate entities.
- ✓ Extract edges from a 3D solid using the **XEDGES** command.
- ✓ Use the **SOLIDEDIT** command to construct and edit a solid model.

AutoCAD provides expanded capabilities for editing solid models. As you saw in the previous chapter, grips can be used to edit a solid model. Also, the subobjects that make up a solid, such as faces, edges, and vertices, can be edited. A single command, **SOLIDEDIT**, allows you to edit faces, edges, or the entire body of the solid.

NOTE

Mesh objects cannot be modified using the **SOLIDEDIT** command. The mesh object must be converted to a solid first. If you select a mesh for editing with the **SOLIDEDIT** command, you are given the option of converting it to a solid, as long as the display of the dialog box has not been turned off.

Overview of the SOLIDEDIT Command

The **SOLIDEDIT** command allows you to edit the faces, edges, and body of a solid. Many of the subobject editing functions discussed in Chapter 11 can also be performed with the **SOLIDEDIT** command. The options of the **SOLIDEDIT** command can be accessed in the **Solid Editing** panel on the **Home** or **Solid** tab of the ribbon or by typing SOLIDEDIT. See **Figure 12-1**. The quickest method of entering the command is by using the ribbon. For example, directly select a face editing option from the drop-down list in the **Solid Editing** panel, as shown in **Figure 12-1**.

Figure 12-1.
Accessing the **SOLIDEDIT** command options.

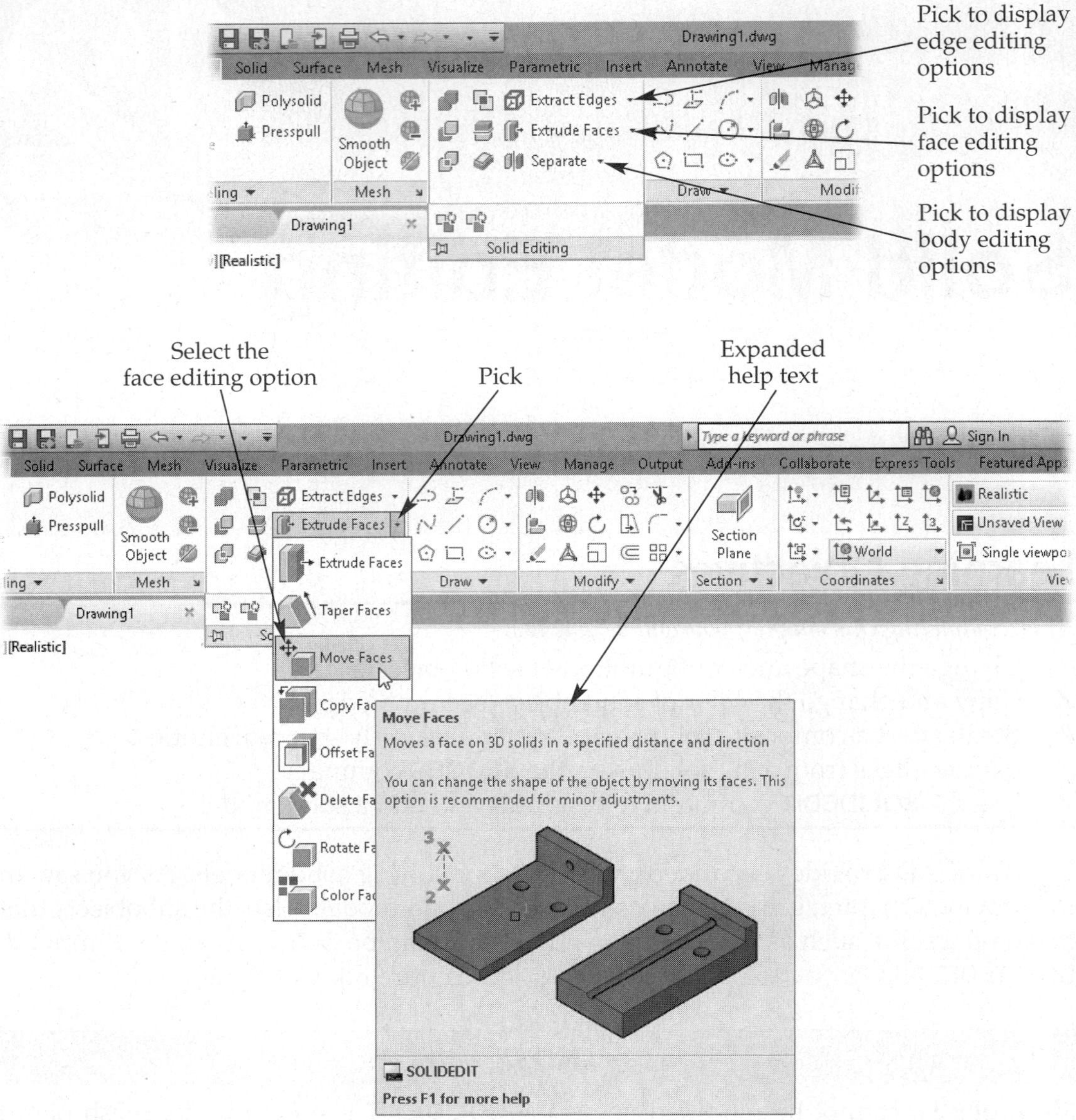

When the **SOLIDEDIT** command is entered at the keyboard, you are first asked to select the component of the solid with which you wish to work. Specify **Face**, **Edge**, or **Body**. The editing options for the selected component are then displayed and are the same as those seen in **Figure 12-1**.

The editing function is directly entered when the option is selected from the ribbon. This is why using the ribbon is the most efficient method of entering the **SOLIDEDIT** command options.

The following sections provide an overview of the solid model editing functions of the **SOLIDEDIT** command. Each option is explained and the results of each are shown. A tutorial later in the chapter illustrates how these options can be used to construct a model.

NOTE

AutoCAD displays a variety of error messages when invalid solid editing operations are attempted. Rather than trying to interpret the wording of these messages, just realize that what you tried to do will not work. Actions that may cause errors include trying to rotate a face into other faces or extruding and tapering an object at too great of an angle. When an error occurs, try the operation again with different parameters or determine a different approach to solving the problem in order to maintain the design intent.

Face Editing

The basic components of a solid are its faces. Many of the **SOLIDEDIT** options are for editing faces. All eight face editing options prompt you to select faces. It is important to make sure you select the correct part of the model for editing. Remember the following three steps when using any of the face editing options.

1. First, select a face to edit. If you pick an edge, AutoCAD selects the two faces that share the edge. If this happens, use the **Remove** option to deselect the unwanted face. A more intuitive approach is to select the open space of the face as if you were touching the side of a part. AutoCAD highlights only that face.
2. Adjust the selection set at the Select faces or [Undo/Remove/ALL]: prompt. The following options are available.
   - **Undo.** Removes the previous selected face(s) from the selection set.
   - **Remove.** Allows you to select faces to remove from the selection set. This is only available when **Add** is current.
   - **All.** Adds all faces on the model to the selection set. This is only available after selecting at least one face. It can also be used to remove all faces if **Remove** is current.
   - **Add.** Allows you to add faces to the selection set. This is only available when **Remove** is current.
3. Press [Enter] to continue with face editing.

Extruding Faces

An extruded face is moved, or stretched, in a selected direction. The extrusion can be straight or have a taper. To extrude a face, select the command and pick the **Face**>**Extrude** option. Remember, the option is directly entered when picking the button in the ribbon. You are then prompted to select the face(s) to extrude. Nonplanar (curved) faces cannot be extruded. As you pick faces, the prompt verifies the number of faces selected. For example, when an edge is selected, the prompt reads 2 faces found. When done selecting faces, press [Enter] to continue.

Ribbon
Home
> Solid Editing
Solid
> Solid Editing
Extrude Faces
Type
SOLIDEDIT

SOLIDEDIT

Next, the height of the extrusion needs to be specified. A positive value adds material to the solid, while a negative value subtracts material from the solid. A taper can also be given.

Specify height of extrusion or [Path]: *(enter the height)*
Specify angle of taper for extrusion <0>: *(enter an angle or accept the default)*
Solid validation started.
Solid validation completed.
Enter a face editing option
[Extrude/Move/Rotate/Offset/Taper/Delete/Copy/coLor/mAterial/Undo/eXit] <eXit>: **X**↵
Solids editing automatic checking: SOLIDCHECK=1
Enter a solids editing option [Face/Edge/Body/Undo/eXit] <eXit>: **X**↵

Figure 12-2 shows an original solid object and the result of extruding the top face with a 0° taper angle and a 30° taper angle. It also shows the original solid object with two adjacent faces extruded with 15° taper angles.

In addition to extruding a face perpendicular to itself, the extruded face can follow a path. Select the **Path** option at the Specify height of extrusion or [Path]: prompt. The path of extrusion can be a line, circle, arc, ellipse, elliptical arc, polyline, or spline. The extrusion height is the exact length of the path. See **Figure 12-3**.

Exercise 12-1

www.g-wlearning.com/CAD/

Complete the exercise on the companion website.

Figure 12-2. Extruding faces on an object. A—The original object. B—The top face is extruded with a 0° taper angle. C—The top face of the original object is extruded with a 30° taper angle. D—The top and right-hand faces of the original object are extruded with 15° taper angles.

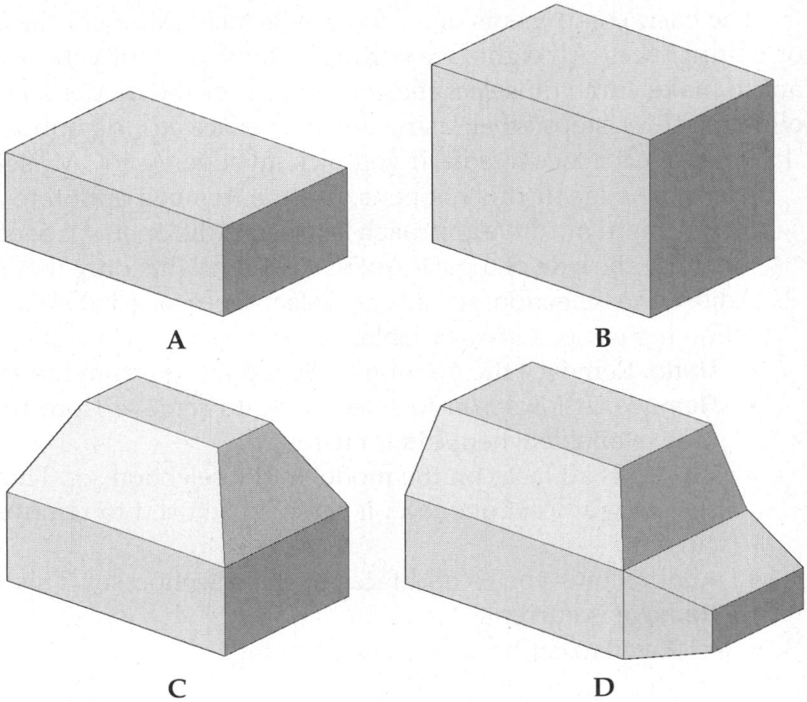

Figure 12-3. The path of extrusion can be a line, circle, arc, ellipse, elliptical arc, polyline, or spline. Here, the paths are shown in color.

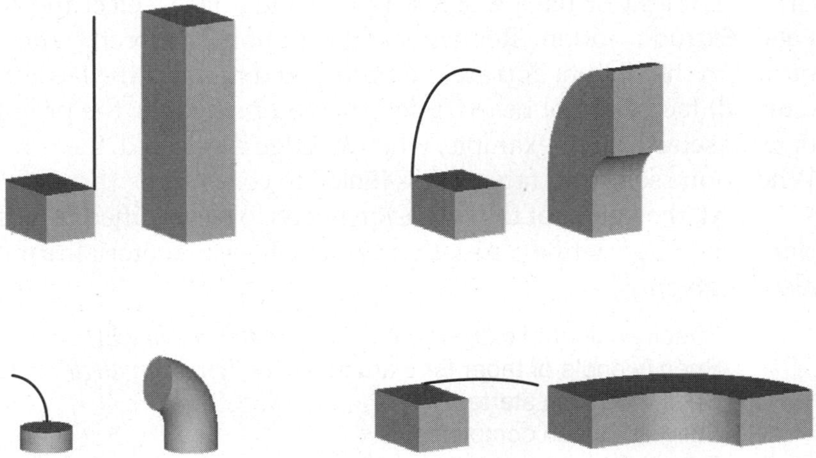

Moving Faces

The **Move** option moves a face in the specified direction and lengthens or shortens the solid object. In another application, a solid model feature (such as a hole) that has been subtracted from an object to create a composite solid can be moved with this option. Object snaps may interfere with the function of this option, so they may need to be toggled off during the operation.

> Ribbon
> Home
> > Solid Editing
> 
> Move Faces
> Type
> SOLIDEDIT
>
> SOLIDEDIT

To move a face, select the command and pick the **Face**>**Move** option. If the button is picked in the ribbon, the option is directly entered. You are then prompted to select the face(s) to move. When done selecting faces, press [Enter] to continue. Next, you are prompted to select a base point of the operation:

Specify a base point or displacement: *(pick a base point)*
Specify a second point of displacement: *(pick a second point or enter coordinates)*
Solid validation started.
Solid validation completed.
Enter a face editing option
[Extrude/Move/Rotate/Offset/Taper/Delete/Copy/coLor/mAterial/Undo/eXit] <eXit>: **X**↵
Solids editing automatic checking: SOLIDCHECK=1
Enter a solids editing option [Face/Edge/Body/Undo/eXit] <eXit>: **X**↵

When adjacent faces are perpendicular, the edited face is moved in a direction so the new position keeps the face parallel to the original. See **Figures 12-4A** and **12-4B**. Faces that are normal to the current UCS can be moved by picking a new location or entering a direct distance. If you are moving a face that is not normal to the current

Figure 12-4.
A—The hole will be moved using the **Move** option of the **SOLIDEDIT** command. B—The hole is moved. C—When the angled face is moved, a portion of it is altered to be coplanar with the vertical face. D—If the angled face is moved more, it becomes completely coplanar to the vertical face. This is a new, single face.

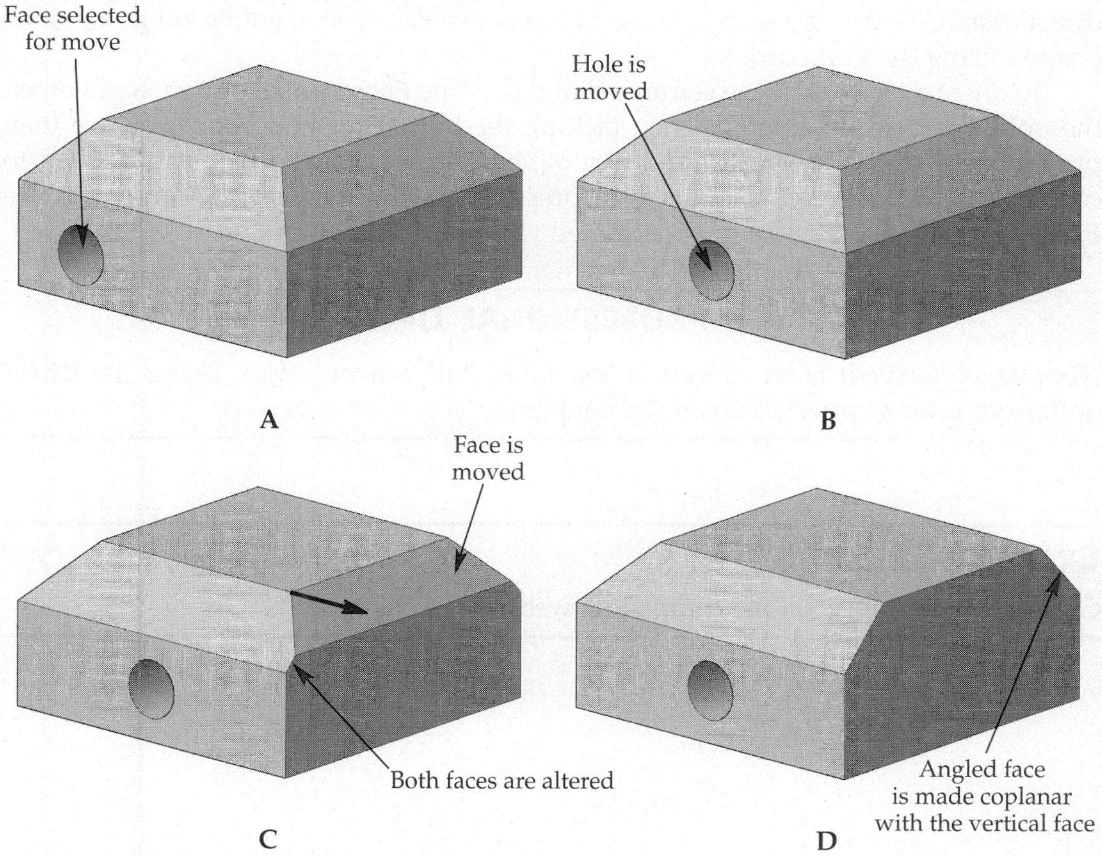

UCS, you can enter coordinates for the second point of displacement, but it may be easier to first use the UCS icon grips or the **Face** option of the **UCS** command to align the UCS with the face to be moved.

When adjacent faces join at angles other than 90°, the moved face will be relocated as previously stated, but only if the movement is less than the dimensional offset of the two faces. For example, in **Figure 12-4B**, the top edge of the angled face is in .5″ from the vertical face. If the angled face is moved outward a distance of less than .5″, it is altered as shown in **Figure 12-4C**. A portion of the angled face becomes coplanar with the vertical face. If the angled face is moved outward a distance greater than .5″, it is altered so that it forms a single plane with the adjacent face. What has actually happened is that the angled face is moved beyond the adjacent face, while remaining parallel to its original position. Thus, in effect, it has disappeared because the adjacent, vertical face cannot be altered. See **Figure 12-4D**. In this example, the angled face was moved .75″. The new vertical face that is created can now be moved.

Exercise 12-2

www.g-wlearning.com/CAD/

Complete the exercise on the companion website.

Offsetting Faces

SOLIDEDIT

Ribbon
Home
> Solid Editing
Solid
> Solid Editing

Offset Faces
Type
SOLIDEDIT

The **Offset** option may seem the same as the **Extrude** option because it moves faces by a specified distance or through a specified point. Unlike the **OFFSET** command in AutoCAD, this option moves all selected faces a specified distance. It is most useful when you wish to change the size of features such as slots, holes, grooves, and notches in solid parts. A positive offset distance increases the size or volume of the solid (adds material). A negative distance decreases the size or volume of the solid (removes material). Therefore, if you wish to make the width of a slot wider, provide a negative offset distance to decrease the size of the solid. Picking points to set the offset distance and direct distance entry are always taken as a positive value, so negative values must be entered using the keyboard.

To offset a face, select the command and pick the **Face>Offset** option. Remember, the option is directly entered when picking the button in the ribbon. You are then prompted to select the face(s) to offset. When done selecting faces, press [Enter] to continue. Next, enter the offset distance, press [Enter], and then exit the command. See **Figure 12-5** for examples of features edited with the **Offset** option.

PROFESSIONAL TIP

Nonplanar (curved) faces cannot be extruded, but can be offset. Using the **Offset** option, you can, in effect, "extrude" a nonplanar face.

Exercise 12-3

www.g-wlearning.com/CAD/

Complete the exercise on the companion website.

Figure 12-5.
Offsetting faces. A—The original objects. The hole is selected to offset. The interior of the L is also selected to offset. B—A positive offset distance increases the size or volume of the solid. C—A negative offset distance decreases the size or volume of the solid.

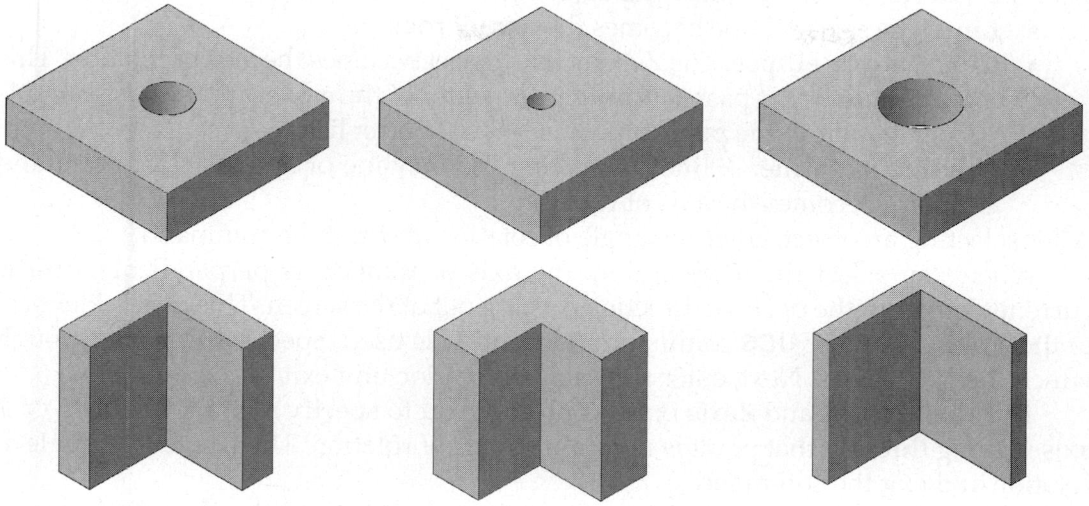

A B C

Deleting Faces

The **Delete** option deletes selected faces. This is a quick way to remove features such as chamfers, fillets, holes, and slots. To delete a solid face, select the command and pick the **Face>Delete** option. If the button is picked in the ribbon, the option is directly entered. You are then prompted to select the face(s) to delete. When done selecting faces, press [Enter] to continue and then exit the command. When a face is deleted, existing faces extend to fill the gap. No additional faces are created. For instance, the inclined surface of a wedge cannot be deleted, as there are no existing faces that can be extended to fill the gap. When deleting the face that is a chamfered or filleted edge, the adjacent edges are extended to fill the gap. See **Figure 12-6**.

Ribbon
Home
> Solid Editing
Delete Faces
Type
SOLIDEDIT

SOLIDEDIT

Rotating Faces

The **Rotate** option rotates a face about a selected axis. To rotate a solid face, select the command and pick the **Face>Rotate** option. Remember, the option is directly entered when picking the button in the ribbon. You are then prompted to select the face(s) to rotate. When done selecting faces, press [Enter] to continue. There are several methods by which a face can be rotated.

Ribbon
Home
> Solid Editing
Rotate Faces
Type
SOLIDEDIT

SOLIDEDIT

Figure 12-6.
Deleting faces. A—The original object with three rounds. B—The faces of two rounds have been deleted.

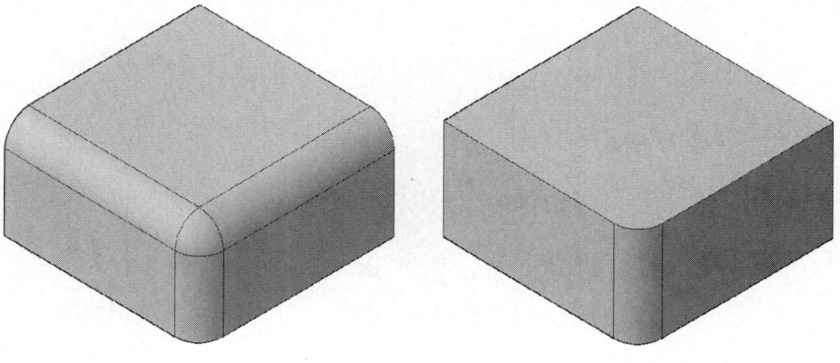

A B

The **2points** option is the default. Pick two points to define the "hinge" about which the face will rotate. Then, provide the rotation angle and exit the command.

The **Axis by object** option allows you to use an existing object to define the axis of rotation. You can select the following objects.

- **Line.** The selected line becomes the axis of rotation.
- **Circle, arc, or ellipse.** The Z axis of the object becomes the axis of rotation. This Z axis is a line that passes through the center of the circle, arc, or ellipse and is perpendicular to the plane on which the 2D object lies.
- **Polyline or spline.** A line connecting the polyline or spline's start point and endpoint becomes the axis of rotation.

After selecting an object, enter the angle of rotation and exit the command.

When you select the **View** option, the axis of rotation is perpendicular to the current view, with the positive direction coming out of the screen. This axis is identical to the Z axis when the **UCS** command **View** option is used. Specify the point through which the axis passes. Next, enter the angle of rotation and exit the command.

The **Xaxis**, **Yaxis**, and **Zaxis** options prompt you to specify a point. The X, Y, or Z axis passing through that point is used as the axis of rotation. Then, enter the angle of rotation and exit the command.

Figure 12-7 provides several examples of rotated faces. Notice how the first and second pick points determine the direction of positive and negative rotation angles.

NOTE

A positive rotation angle moves the face in a clockwise direction looking from the first pick point to the second. Conversely, a negative angle rotates the face counterclockwise. If the rotated face will intersect or otherwise interfere with other faces, an error message indicates that the operation failed or no solution was calculated. In this case, you may wish to try a negative angle if you previously entered a positive one. In addition, you can try using the opposite edge of the face as the axis of rotation by picking the appropriate points.

Figure 12-7.
When rotating faces, the first and second pick points determine the direction of positive and negative rotation angles.

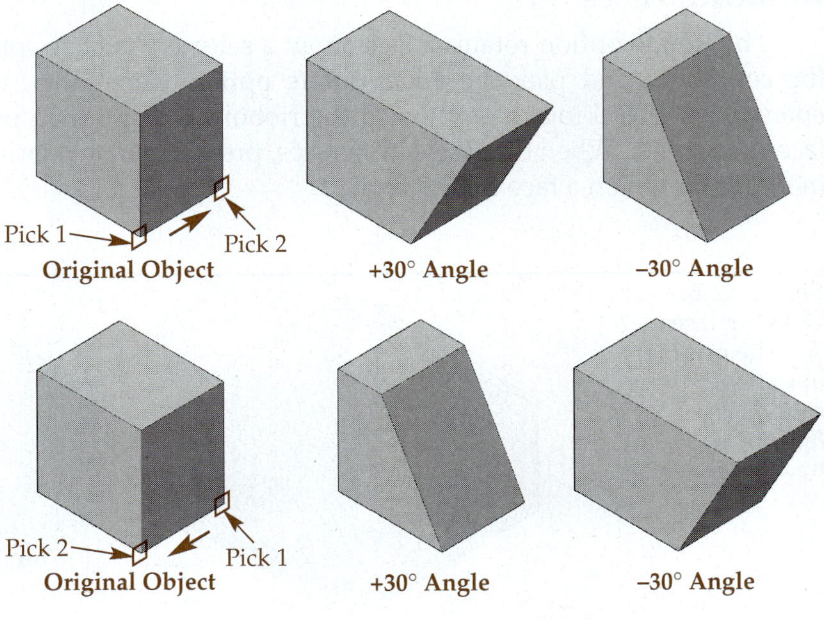

Exercise 12-4

www.g-wlearning.com/CAD/

Complete the exercise on the companion website.

Tapering Faces

The **Taper** option tapers a face at the specified angle, from the first pick point to the second. To taper a solid face, select the command and pick the **Face>Taper** option. If the button is picked in the ribbon, the option is directly entered. You are then prompted to select the face(s) to taper. When done selecting faces, press [Enter] to continue:

> Ribbon
> Home
> > Solid Editing
> Solid
> > Solid Editing
> Taper Faces
> Type
> SOLIDEDIT

Specify the base point: *(pick the base point)*
Specify another point along the axis of tapering: *(pick a point along the taper axis)*
Specify the taper angle: *(enter a taper value)*

Tapers work differently depending on whether the faces being tapered describe the outer boundaries of the solid, a cavity, or a removed portion of the solid. A positive taper angle always removes material. A negative taper angle always adds material. For example, if a positive taper angle is entered for a solid cylinder, the selected object is tapered in on itself from the base point along the axis of tapering, thus removing material. A negative angle tapers the object out away from itself to increase its size along the axis of tapering, thus adding material. See **Figure 12-8.**

On the other hand, if the face (or faces) of a feature such as a hole or slot are tapered, a positive taper angle increases the size of the feature along the axis of tapering. For example, if a round hole is tapered using a positive taper angle, its diameter increases

Figure 12-8.
Tapering faces. A—The original objects. The dark face of the box and the circumference of the cylinder are selected. B—Positive taper angle. C—Negative taper angle.

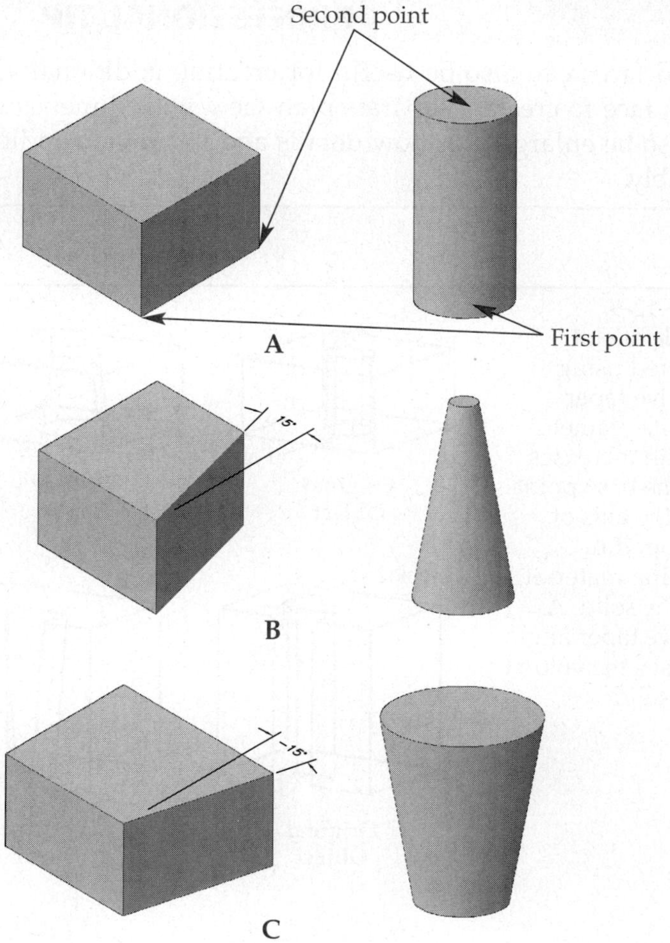

from the base point along the axis of tapering, thus removing material from the solid. Conversely, if the same round hole is tapered using a negative taper angle, its diameter decreases from the base point along the axis of tapering, thus adding material to the solid. **Figure 12-9** shows some examples of tapering features.

Exercise 12-5

www.g-wlearning.com/CAD/

Complete the exercise on the companion website.

Copying Faces

SOLIDEDIT

Ribbon
Home
> Solid Editing

Copy Faces

Type
SOLIDEDIT

The **Copy** option copies a face to the location or coordinates given. The copied face is *not* part of the original solid model. It is actually a region, which can later be extruded, revolved, swept, etc., into a solid. This may be useful when you wish to construct a mating part in an assembly that has the same features on the mating faces or the same outline. This option is quick to use because you can pick a base point on the face, and then enter a single direct distance value for the displacement. Be sure an appropriate UCS is set if you wish to use direct distance entry.

To copy a solid face, select the command and pick the **Face>Copy** option. Remember, the option is directly entered when picking the button in the ribbon. You are then prompted to select the face(s) to copy. When done selecting faces, press [Enter] to continue. You are prompted for a base point for the copy. Pick this point and then pick a second point of displacement or press [Enter] to use the first point as a displacement. See **Figure 12-10** for examples of copied faces.

PROFESSIONAL TIP

Copied faces can also be useful for creating additional views. For example, you can copy a face to create a separate plan view with dimensions and notes. A copied face can also be enlarged to show details and to provide additional notation for design or assembly.

Figure 12-9. If a hole or slot is tapered using a positive taper angle, its diameter or width increases from the base point along the axis of tapering, thus removing material from the solid. A negative taper angle increases the volume of the solid.

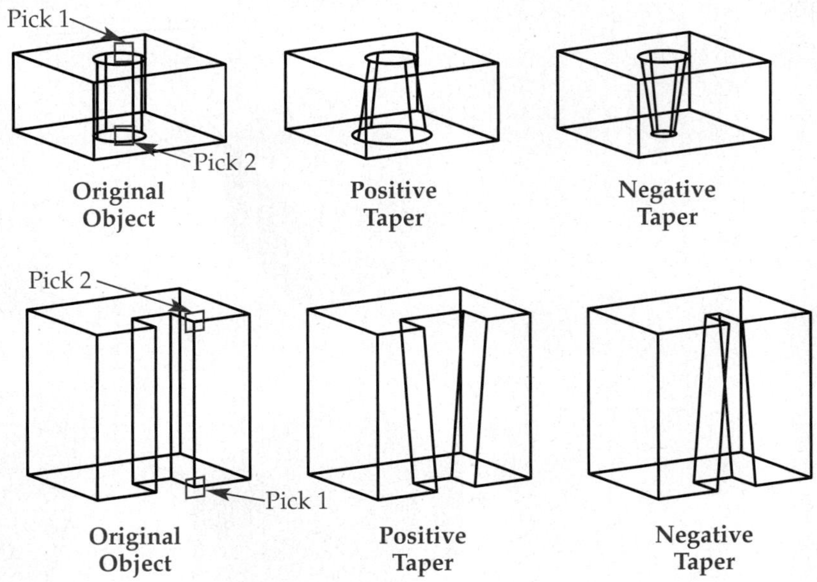

Figure 12-10.
A face can be quickly copied by picking a base point on the face and then entering a direct distance value for the displacement.

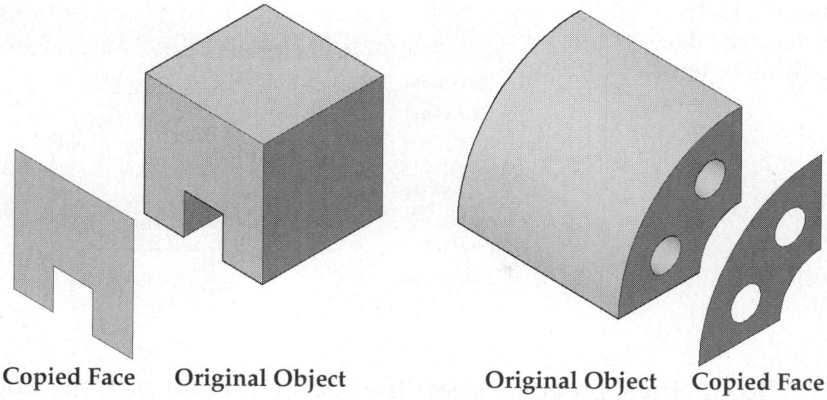

Coloring Faces

You can quickly change a selected face to a different color using the **Color** option. Select the command and pick the **Face>Color** option. If the button is picked in the ribbon, the option is directly entered. You are then prompted to select the face(s) to color. When done selecting faces, press [Enter] to continue. Next, select the desired color from the **Select Color** dialog box that is displayed. Remember, the color of the object (or face) determines the shaded color.

Ribbon
Home
> Solid Editing
Color Faces
Type
SOLIDEDIT

SOLIDEDIT

PROFESSIONAL TIP

Use the **Color** option to enhance features on a solid. For example, if you have a model that has holes or slots, enhance just the color of these features.

NOTE

The **Material** option of the **SOLIDEDIT** command allows you to assign a material to selected faces on a solid. Materials are discussed in Chapter 17.

Exercise 12-6

www.g-wlearning.com/CAD/

Complete the exercise on the companion website.

Edge Editing

Edges can be edited in only two ways with the **SOLIDEDIT** command. They can be copied from the solid. Also, the color of an edge can be changed.

Copying an edge is similar to copying a face. To copy a solid edge, select the command and pick the **Edge>Copy** option. Remember, the option is directly entered when picking the button in the ribbon. See **Figure 12-11**. You are then prompted to select the edge(s) to copy. When done selecting edges, press [Enter] to continue. You are prompted for a base point for the copy. Pick this point and then pick a second point of displacement or press [Enter] to use the first point as a displacement. The edge is copied as a line, arc, circle, ellipse, or spline.

Ribbon
Home
> Solid Editing
Copy Edges
Type
SOLIDEDIT

SOLIDEDIT

Figure 12-11.
Selecting edge-editing options.

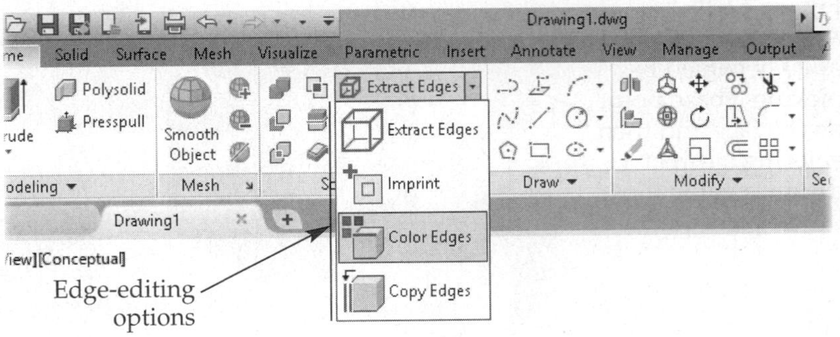

SOLIDEDIT
Ribbon
Home > Solid Editing
Color Edges
Type
SOLIDEDIT

To color a solid edge, select the command and pick the **Edge>Color** option. You are then prompted to select the edge(s) to color. When done selecting edges, press [Enter] to continue. Next, select the desired color from the **Select Color** dialog box that is displayed and pick the **OK** button. The edges are now displayed with the new color. You may need to set a wireframe or hidden visual style current to see the change.

Extracting a Wireframe

XEDGES
Ribbon
Home > Solid Editing
Solid > Solid Editing
Extract Edges
Type
XEDGES

The **XEDGES** command creates copies of, or extracts, all of the edges on a selected solid. You can also extract edges from a surface, mesh, region, or subobject. Once the command is initiated, you are prompted to select objects. Select one or more solids and press [Enter]. The edges are extracted and placed on top of the existing edges. See **Figure 12-12**. The new objects are created on the current layer.

Straight edges and the curved edges where cylindrical surfaces intersect with flat or other cylindrical surfaces are the only edges extracted. Spheres and tori have no edges that can be extracted. The round bases of cylinders and cones are the only edges of those objects that will be extracted.

Figure 12-12.
Extracting edges with the **XEDGES** command. A—The original object. B—The extracted wireframe (edges).

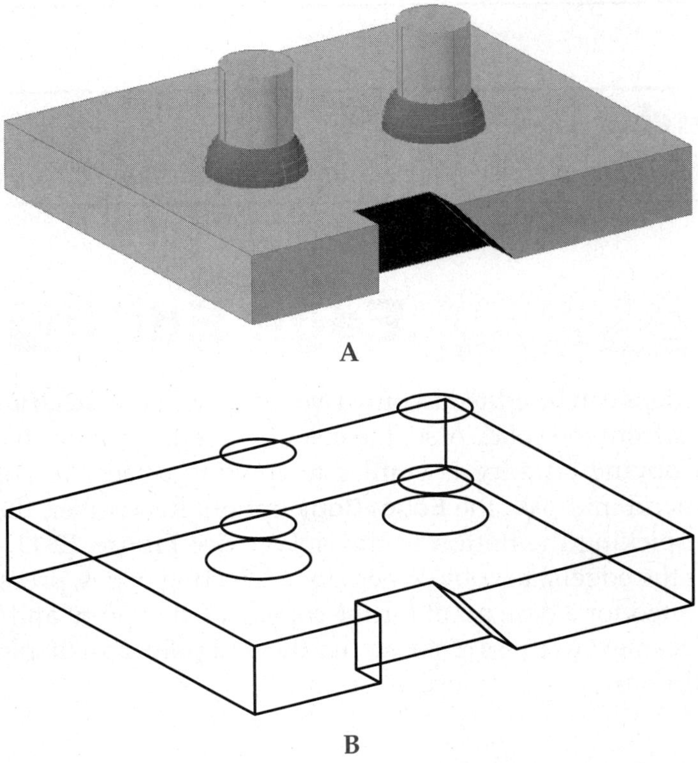

Exercise 12-7

www.g-wlearning.com/CAD/

Complete the exercise on the companion website.

Body Editing

The body editing options of the **SOLIDEDIT** command perform editing operations on the entire body of the solid model. The body options are **Imprint**, **Separate**, **Shell**, **Clean**, and **Check**. The next sections cover these body editing options.

The **Imprint** option is a body editing function. However, since it modifies a body by adding edges, the option is located in the edges drop-down list in the **Solid Editing** panel on the **Home** tab. On the **Solid** tab, it is a separate button in the **Solid Editing** panel. In both places, the **Separate**, **Shell**, **Clean**, and **Check** options are found in the body drop-down list.

Imprint

Arcs, circles, lines, 2D and 3D polylines, ellipses, splines, regions, bodies, and 3D solids can be imprinted onto a solid, if the object intersects the solid. The imprint becomes a face on the surface based on the overlap between the two intersecting objects. Once the imprint has been made, the new face can be modified.

To imprint an object on a solid, select the **SOLIDEDIT** command and pick the **Body>Imprint** option. If IMPRINT is entered at the keyboard or the button is picked on the ribbon, the option is directly entered. Once the option is activated, you are prompted to select the solid. This is the object on which the other objects will be imprinted. Then, select the objects to be imprinted. You have the option of deleting the source objects.

Ribbon
Home
> Solid Editing
Solid
> Solid Editing

Imprint

Type
IMPRINT
SOLIDEDIT

IMPRINT

The imprinted face can be modified using face editing options. **Figure 12-13** illustrates objects imprinted onto a solid model. Two of these are then extruded into the solid to create holes. The **PRESSPULL** command is used on the third object to create a cylindrical feature.

Figure 12-13.
Imprinted objects form new faces that can be modified. A—A solid box with three objects on the plane of the top face. B—The objects are imprinted, then two of the new faces are extruded through the solid and used in subtraction operations. The **PRESSPULL** command is used on the third new face to create the cylindrical feature.

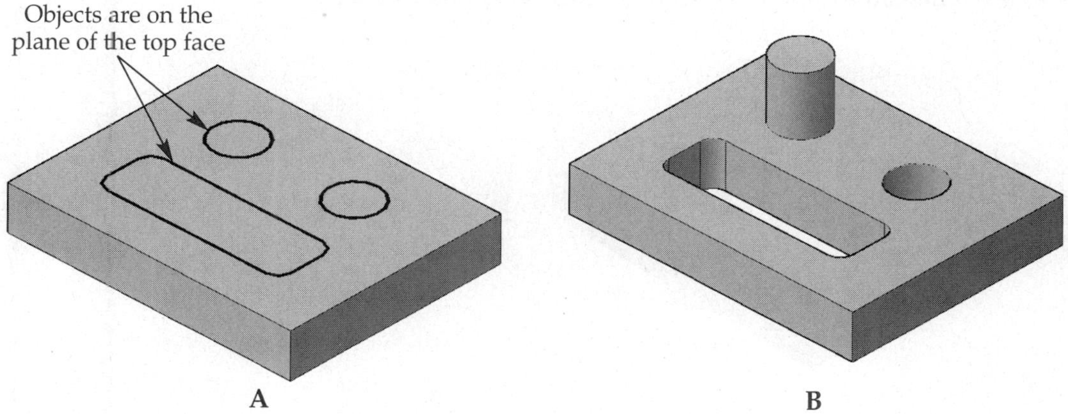

NOTE

Remember that objects are drawn on the XY plane of the current UCS unless you enter a specific Z value. Therefore, before you draw an object to be imprinted onto a solid model, be sure you have set an appropriate UCS for proper placement of the object by using the UCS icon grips or a dynamic UCS. Alternately, you can draw the object on the XY plane and then move the object onto the solid object.

Separate

SOLIDEDIT

Ribbon
Home
> Solid Editing
Solid
> Solid Editing
Separate
Type
SOLIDEDIT

The **Separate** option separates two objects that are both a part of a single solid composite, but appear as separate physical entities. This can happen when modifying solids using the Boolean commands. The **Separate** option may be seldom used, but it has a specific purpose. If you select a solid model and an object physically separate from the model is highlighted, the two objects are parts of the same composite solid. If you wish to work with them as individual solids, they must first be separated.

To separate a solid body, select the command and pick the **Body**>**Separate** option. If the button is picked in the ribbon, the option is directly entered. You are then prompted to select a 3D solid. After you pick the solid, it is automatically separated. No other actions are required and you can exit the command. However, if you select a solid in which the parts are physically joined, AutoCAD indicates this by prompting The selected solid does not have multiple lumps. A "lump" is a physically separate solid entity. In order to separate a solid, the solid must be composed of multiple lumps. See **Figure 12-14.**

Shell

SOLIDEDIT

Ribbon
Home
> Solid Editing
Solid
> Solid Editing
Shell
Type
SOLIDEDIT

A ***shell*** is a solid that has been "hollowed out." The **Shell** option creates a shell of the selected object using a specified offset distance, or thickness. To create a shell of a solid body, select the command and pick the **Body**>**Shell** option. Remember, the option is directly entered when picking the button in the ribbon. You are prompted to select the solid. Only one solid can be selected.

After selecting the solid, you have the opportunity to remove faces. If you do not remove any faces, the new solid object will appear identical to the old solid object when shaded or rendered. The thickness of the shell will not be visible. If you wish to create a hollow object with an opening, select the face to be removed (the opening).

Figure 12-14.
A—After the cylinder is subtracted from the box, the remaining solid is considered one solid.
B—Use the **Separate** option to turn this single solid into two solids.

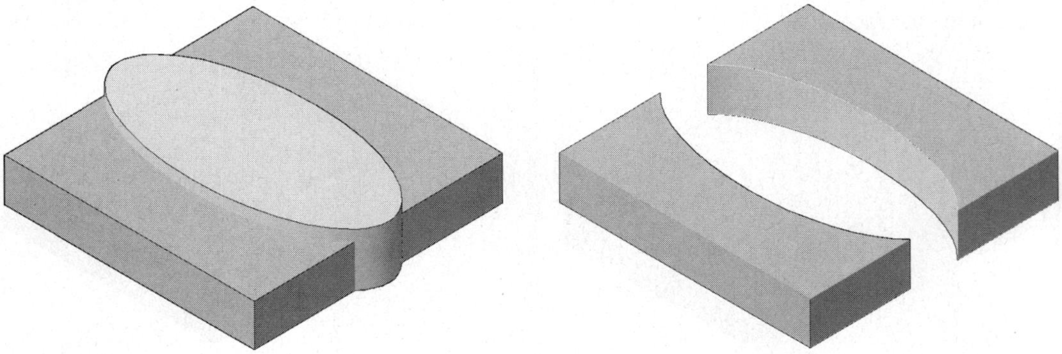

A B

After selecting the object and specifying any faces to be removed, you are prompted to enter the shell offset distance. This is the thickness of the shell. A positive shell offset distance creates a shell on the inside of the solid body. A negative shell offset distance creates a shell on the outside of the solid body. See **Figure 12-15**. If you shell a solid that contains internal features, such as holes, grooves, and slots, a shell of the specified thickness is placed around those features. This is shown in **Figure 12-16**.

If the shell operation is not readily visible in the current view, you can rotate the view by pressing the [Shift] key and pressing the mouse wheel button at the same time to enter the transparent **3DORBIT** command. You can also rotate the view using the view cube, or you can see the results by picking the **X-Ray Effect** button in the **Visual Styles** panel on the **Visualize** tab of the ribbon.

Figure 12-15.
A—The right-front, bottom, and left-back faces (marked here by gray lines) are removed from the shell operation.
B—The resulting object after the shell operation.

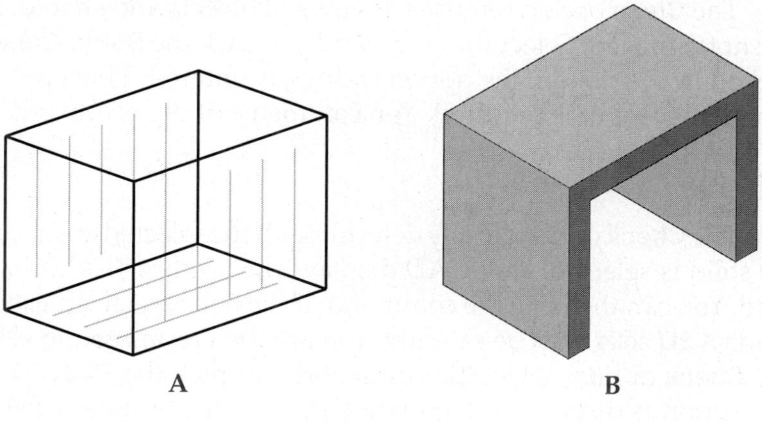

Figure 12-16.
If you shell a solid that contains internal features, such as holes, grooves, and slots, a shell of the specified thickness is also placed around those features.
A—Solid object with holes subtracted.
B—Wireframe display after shelling with a negative offset.
C—The Conceptual visual style is set current.

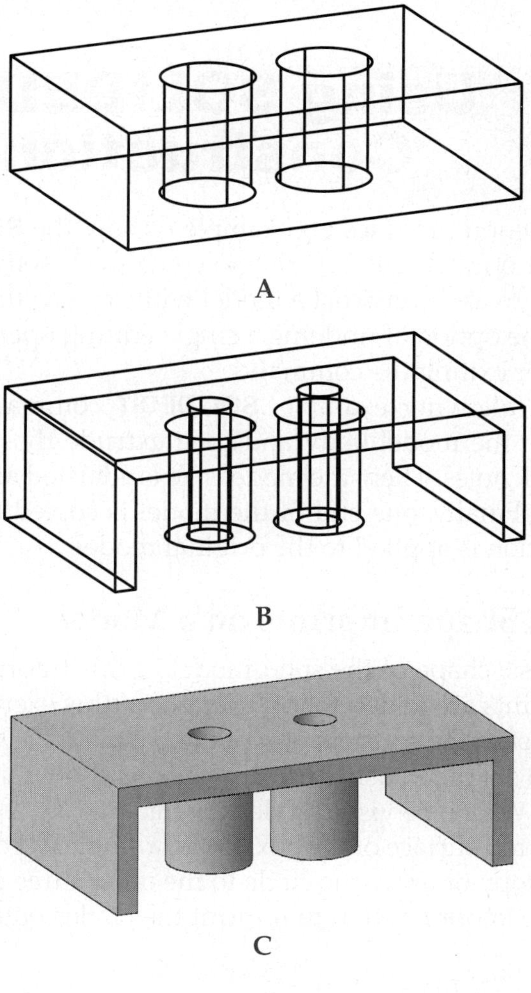

PROFESSIONAL TIP

The **Shell** option of the **SOLIDEDIT** command is very useful in applications such as solid modeling of metal castings or injection-molded plastic parts.

Exercise 12-8

www.g-wlearning.com/CAD/

Complete the exercise on the companion website.

Clean

SOLIDEDIT

Ribbon
Home
> Solid Editing
Solid
> Solid Editing
Clean
Type
SOLIDEDIT

The **Clean** option removes all unused objects and shared edges. Imprinted objects are not removed. Select the command and pick the **Body**>**Clean** option. If the button is picked in the ribbon, the option is directly entered. Then, pick the solid to be cleaned. No further input is required. You can then exit the command.

Check

SOLIDEDIT

Ribbon
Home
> Solid Editing
Solid
> Solid Editing
Check
Type
SOLIDEDIT

The **Check** option simply determines if the selected object is a valid 3D solid. If a true 3D solid is selected, AutoCAD displays the prompt This object is a valid ShapeManager solid. You can then exit the command. If the object selected is not a 3D solid, the prompt reads A 3D solid must be selected. You are then prompted to select a 3D solid. To access the **Check** option, select the command and pick the **Body**>**Check** option. Remember, the option is directly entered when picking the button in the ribbon. Then, select the object to check.

Using SOLIDEDIT as a Construction Tool

This section provides an example of how the **SOLIDEDIT** command options can be used not only to edit, but also to construct a solid model. This approach makes it easy to design and construct a model without selecting a variety of commands. It also gives you the option of undoing a single editing operation or an entire editing session without ever exiting the command.

In the following example, **SOLIDEDIT** command options are used to imprint shapes onto the model body and then extrude those shapes into the body to create countersunk holes. Then, the model size is adjusted and an angle and taper are applied to one end. Finally, one end of the model is copied to construct a mating part and a shell operation is applied to the original model.

Creating Shape Imprints on a Model

The basic shape of the solid model in this tutorial is drawn as a solid box. Then, shape imprints are added to it. Throughout this exercise, you may wish to change the UCS to assist in the construction of the part.

1. Draw a solid box using the dimensions shown in **Figure 12-17**.
2. Set the Wireframe visual style current.
3. On the top surface of the box, locate a single ∅.4 circle using the dimensions given. Then, copy or array the circle to the other three corners as shown in the figure.
4. Use the **Imprint** option to imprint the circles onto the solid box. Delete the source objects.

Figure 12-17.
The initial setup for the tutorial model.

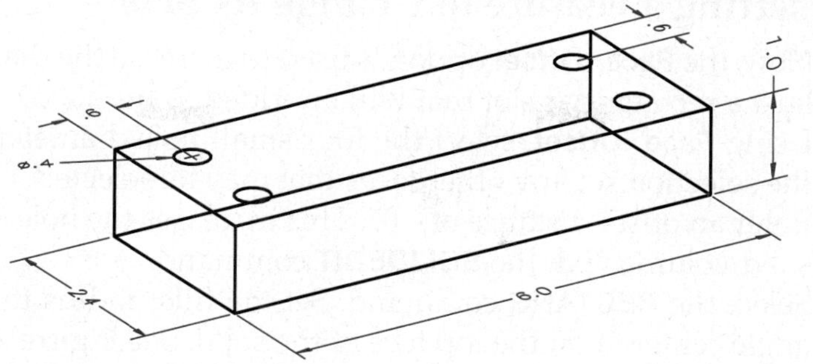

Extruding Imprints to Create Features

The imprinted 2D shapes can now be extruded to create new 3D solid features on the model. Use the **Face>Extrude** option to extrude all four imprinted circles.

1. When you select the edge of the first circle, all features on that face are highlighted, but only the circle you picked and the top face have actually been selected. If you pick inside the circle, only the circle is selected and highlighted. In either case, be sure to also pick the remaining three circles.
2. Remove the top face of the box from the selection set, if needed.
3. The depth of the extrusion is .16 units. Remember to enter –.16 for the extrusion height since the holes remove material. The angle of taper for extrusion should be 35°. Your model should look like **Figure 12-18A**.
4. Extrude the small diameter of the four tapered holes so the holes intersect the bottom of the solid body. Select the holes by picking the small diameter circles. Instead of calculating the distance from the bottom of the chamfer to the bottom surface, you can simply enter a value that is greater than this distance, such as the original thickness of the object. Again, since the goal is to remove material, use a negative value for the height of the extrusion. There is no taper angle. Your model should now look like **Figure 12-18B**.

Moving Faces to Change Model Size

The next step is to use the **Face>Move** option to decrease the length and thickness of the solid body.

1. Select either end face and the two holes nearest to it. Be sure to select the holes *and* the countersinks. Move the two holes and end face two units toward the other end, thus changing the object length to four units.
2. Select the bottom face and move it .5 units up toward the top face, thus changing the thickness to .5 units. See **Figure 12-19**.

Figure 12-18.
A—The imprinted circles are extruded with a taper angle of 35°. B—Holes are created by further extrusion with a taper angle of 0°.

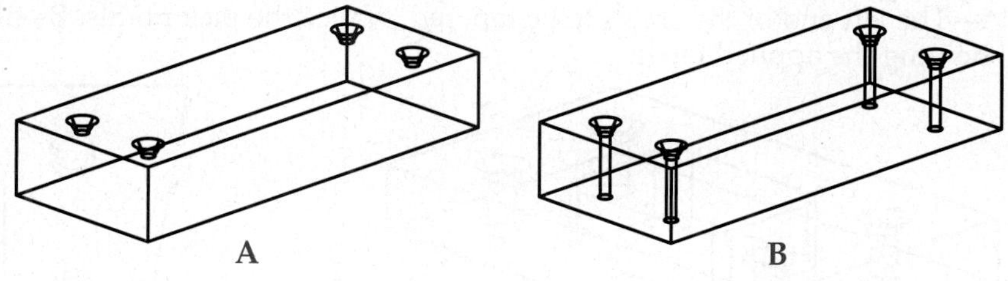

Figure 12-19.
The length of the object is shortened and the height is reduced.

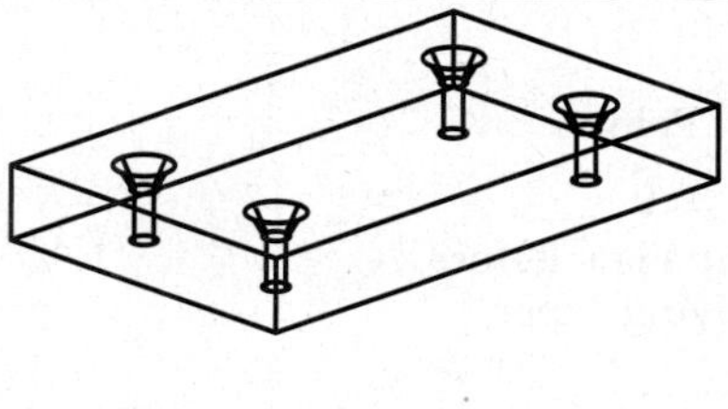

Offsetting a Feature to Change Its Size

Now, the **Face>Offset** option is used to increase the diameter of the four holes and to adjust a rectangular slot that will be added to the solid.

1. Using **Face>Offset**, select the four small hole diameters. Be sure to remove from the selection set any other faces that may be selected.
2. Enter an offset distance of –.05. This increases the hole diameter and decreases the solid volume. Exit the **SOLIDEDIT** command.
3. Select the **RECTANG** command. Set the fillet radius to .4 and draw a 2 × 1.6 rectangle centered on the top face of the solid. See **Figure 12-20A.**
4. Imprint the rectangle on the solid. Delete the source object.
5. Extrude the rectangle completely through the solid (.5 units). Remember to remove from the selection set any other faces that may be selected.
6. Offset the rectangular feature using an offset distance of .2 units. You will need to select all faces of the feature. This decreases the size of the rectangular opening and increases the solid volume. Your drawing should appear as shown in **Figure 12-20B.**

Tapering Faces

One side of the part is to be angled. The **Face>Taper** option is used to taper the left end of the solid.

1. Using **Face>Taper**, pick the face at the left end of the solid.
2. Pick point 1 in **Figure 12-21A** as the base point and point 2 as the second point along the axis of tapering.
3. Enter a value of –10 for the taper angle. This moves the upper-left end away from the solid, creating a tapered end. The result is shown in **Figure 12-21B.**

Figure 12-20.
A—The diameter of the holes is increased and a rectangle is imprinted on the top surface. B—The rectangle is extruded to create a slot.

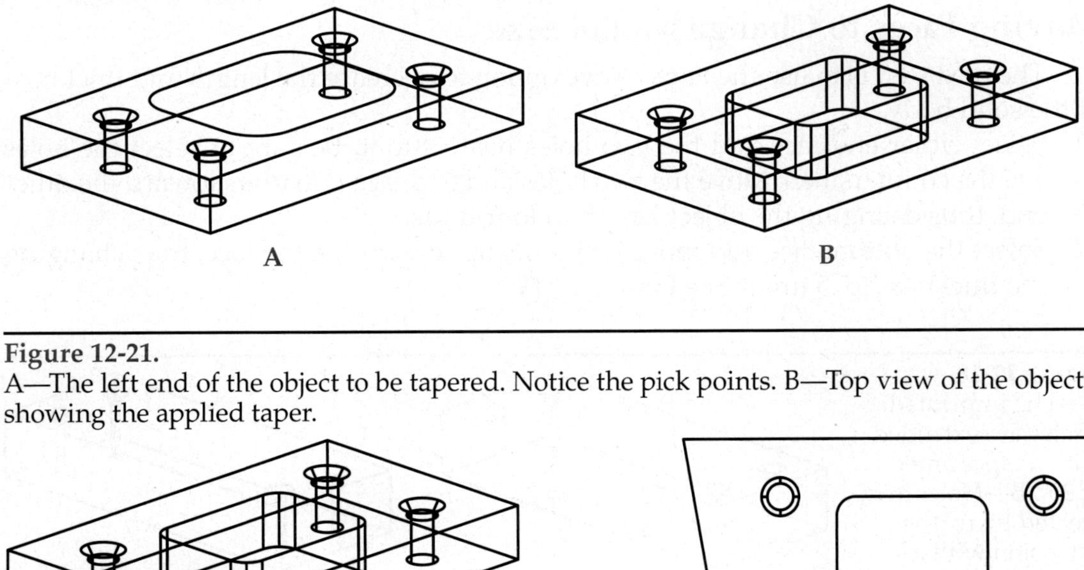

Figure 12-21.
A—The left end of the object to be tapered. Notice the pick points. B—Top view of the object showing the applied taper.

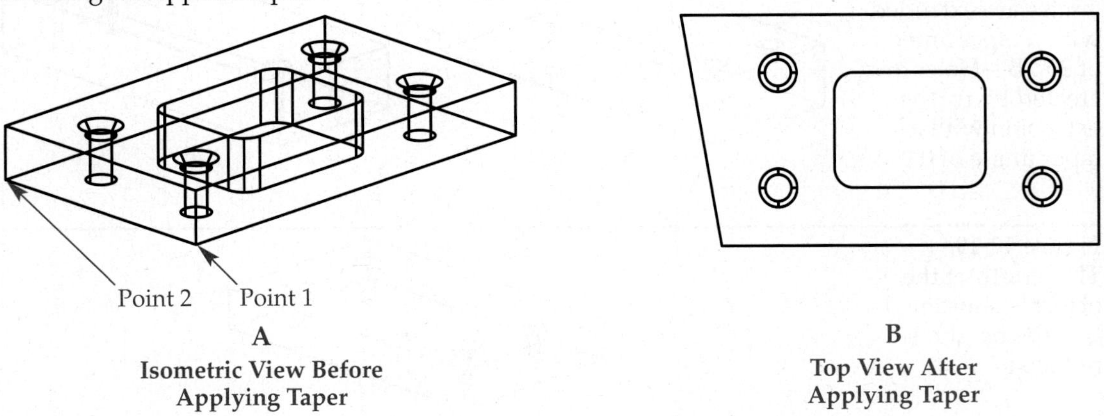

Rotating Faces

Next, use the **Face>Rotate** option to rotate the tapered end of the object. The top edge of the face will be rotated away from the holes, adding volume to the solid.

1. Using **Face>Rotate**, pick the face at the left end of the solid.
2. Pick point 1 in **Figure 12-22A** as the first axis point and point 2 as the second point.
3. Enter a value of **–30** for the rotation angle. This rotates the top edge of the tapered end away from the solid. See **Figure 12-22B**.

Copying Faces

A mating part will now be created. This is done by first copying the face on the tapered end of the part.

1. Using the **Face>Copy** option, pick the angled face on the left end of the solid.
2. Pick one of the corners as a base point and copy the face one unit to the left. This face can now be used to create a new solid. See **Figure 12-23A**.
3. Draw a line four units in length on the negative X axis from the lower-right corner of the copied face. Use the **EXTRUDE** command on the copied face to create a new solid. Select the **Path** option and use the line as the extrusion path. See **Figure 12-23B**. If you do not use the **Path** option, the extrusion is projected perpendicular to the face.

Figure 12-22.
The tapered end of the object is modified by rotating the face. A—Pick the points shown. B—The resulting solid.

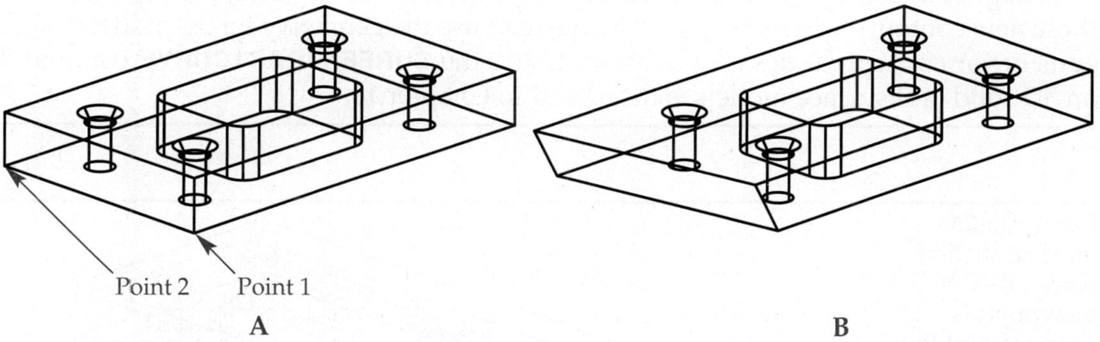

Figure 12-23.
Creating a mating part. A—The angled face is copied. B—The copied face is extruded into a solid.

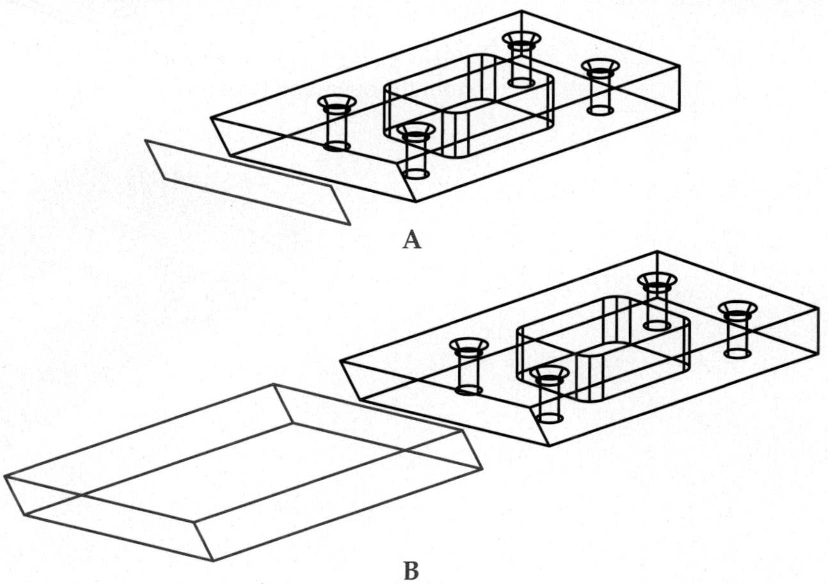

NOTE

The **Face>Extrude** option of the **SOLIDEDIT** command cannot be used to turn a copied face into a solid body.

Creating a Shell

The bottom surface of the original solid will now be shelled out. Keep in mind that features such as the four holes and the rectangular slot will not be cut off by the shell. Instead, a shell will be placed around these features. This becomes clear when the operation is performed.

1. Select the **Shell** option and pick the original solid.
2. Pick the lower-left and lower-right edges of the solid. See **Figure 12-24A**. This removes the two side faces and the bottom face.
3. Enter a shell offset distance of .15 units. The shell is created and the model should appear similar to **Figure 12-24A**.
4. Use the view cube or **3DORBIT** command to view the solid from the bottom. Also, set the Conceptual visual style current. Your model should look like the one shown in **Figure 12-24B**.

NOTE

The **SURFEXTRACTCURVE** command allows you to extract curves such as lines, splines, polylines, or arcs from an existing surface, 3D solid, or face of a 3D solid. This enables the designer to create wireframe geometry that displays the contours of a surface. Once these new contours are created, the designer can use the geometry for 3D model design edits or for conceptual design purposes. Using the **SURFEXTRACTCURVE** command on 3D solid and surface models is discussed in Chapter 10.

Figure 12-24.
A—The shelled object. B—The viewpoint is changed and the Conceptual visual style is set current.

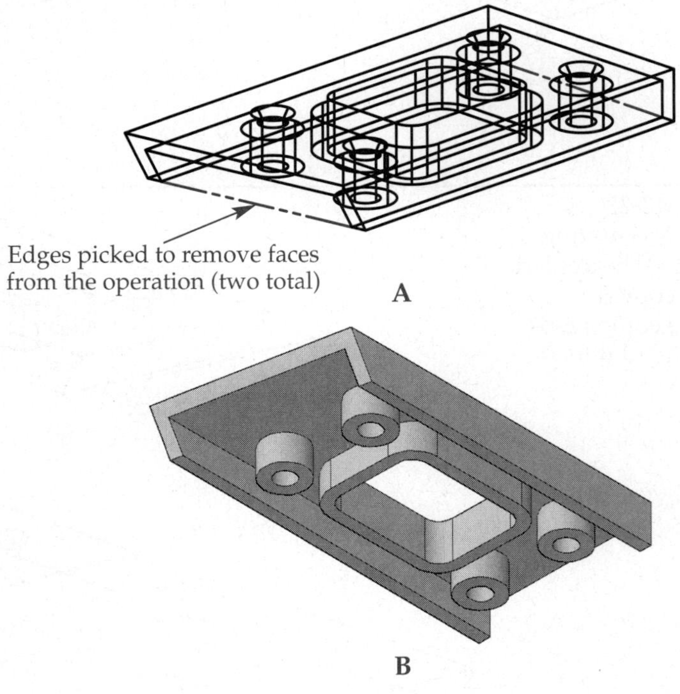

Chapter Review

Answer the following questions using the information in this chapter.

1. What three components of a solid model can be edited using the **SOLIDEDIT** command?
2. When using the **SOLIDEDIT** command, how many faces are highlighted if you pick an edge?
3. How do you deselect a face that is part of the selection set when using the **SOLIDEDIT** command?
4. How can you select a single face?
5. Name the objects that can be used as the path of extrusion when extruding a face.
6. What is one of the most useful aspects of the **Offset Faces** option?
7. How do positive and negative offset distance values affect the volume of the solid?
8. How is a single object, such as a cylinder, affected by entering a positive taper angle when using the **Taper Faces** option?
9. When a shape is imprinted onto a solid body, which component of the solid does the imprinted object become and how can it be used?
10. What is the purpose of the **XEDGES** command?
11. How does the **Shell** option affect a solid that contains internal features such as holes, grooves, and slots?
12. How can you determine if an object is a valid 3D solid?
13. Describe two ways to change the view of your model while you are inside of a command.
14. How can you extrude a face in a straight line, but not perpendicular to the face?

Drawing Problems

1. **General** Complete the tutorial presented in this chapter. Then, perform the following additional edits to the original solid.
   A. Lengthen the right end of the solid by .5 units.
   B. Taper the right end of the solid with the same taper angle used on the left end, but taper it in the opposite direction.
   C. Fillet the two long, top edges of the solid using a fillet radius of .2 units.
   D. Rotate the face at the right end of the solid with the same rotation angle used on the left end, but rotate it in the opposite direction.
   E. Save the drawing as P12-1.

Problems 2–6. Follow the instructions to complete each problem. Do not add dimensions to the models. For objects that have dimensions with tolerances, use the base dimension from which the upper and lower limits are calculated to create the model. Save each drawing as P12-*(problem number).*

2. **Mechanical** Construct the solid part shown using as many **SOLIDEDIT** options as possible. After completing the object, make the following modifications.
   A. Lengthen the 1.250″ diameter feature by .250″.
   B. Change the .750″ diameter hole to .625″ diameter.
   C. Change the thickness of the .250″ thick flange to .375″ (toward the bottom).
   D. Extrude the end of the 1.250″ diameter feature .250″ with a 15° taper inward.
   E. Save the drawing as P12-2.

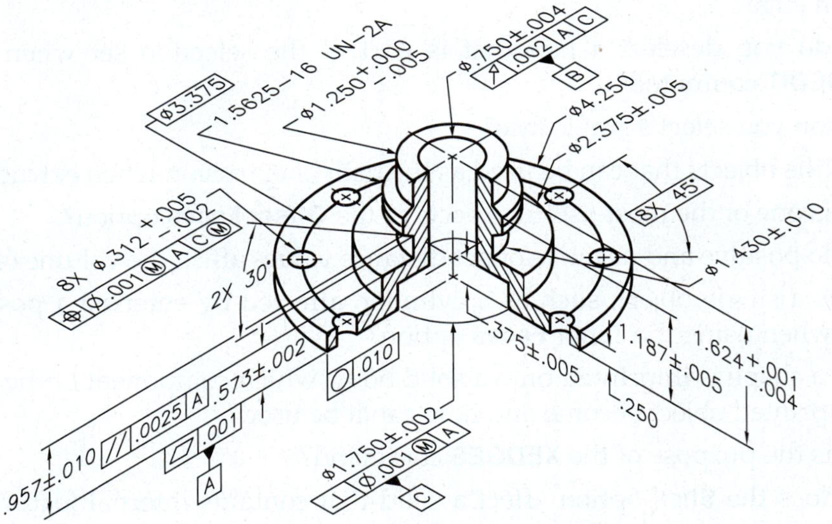

Hub

3. **Mechanical** Construct the solid part shown. Then, perform the following edits on the solid using the **SOLIDEDIT** command.
   A. Change the diameter of the hole to 35.6/35.4.
   B. Add a 5° taper to each inner side of each tooth (the bottom of each tooth should be wider while the top remains the same).
   C. Change the width of the 4.8/4.0 key to 5.8/5.0.
   D. Save the drawing as P12-3.

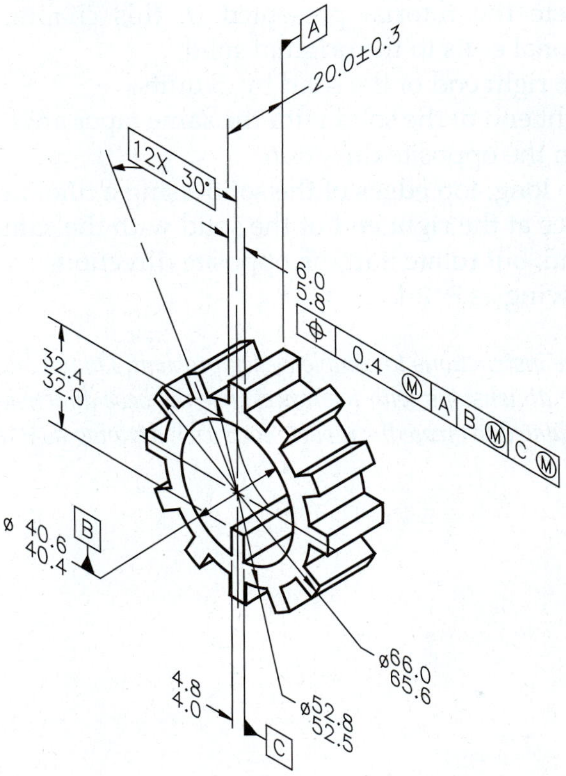

Spline

4. **Mechanical** Construct the solid part shown using as many **SOLIDEDIT** options as possible. Then, perform the following edits on the solid.
   A. Change the depth of the counterbore to 10 mm.
   B. Change the color of all internal surfaces to red.
   C. Save the drawing as P12-4.

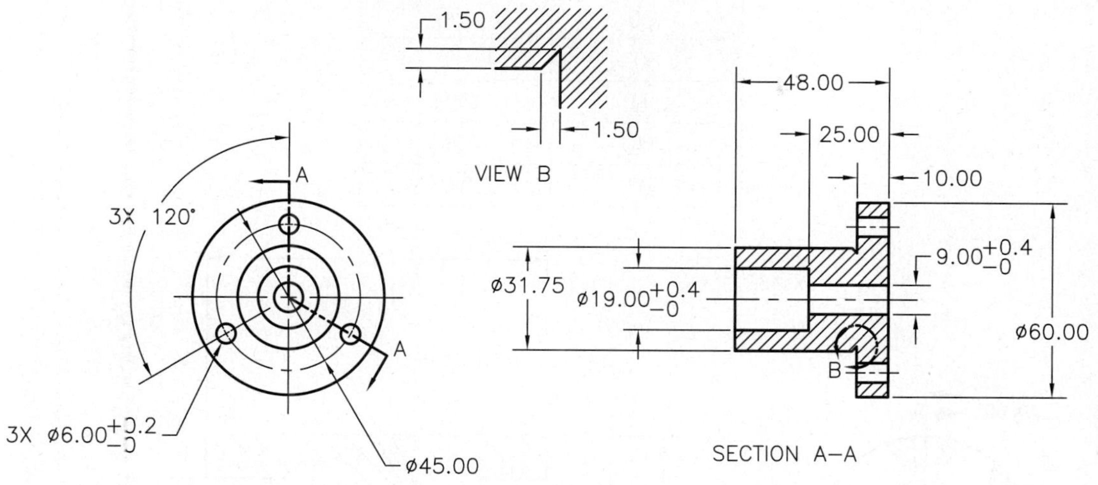

Bushing

5. **Mechanical** Construct the solid part shown using as many **SOLIDEDIT** options as possible. Then, perform the following edits on the solid.
   A. Change the 2.625″ height to 2.325″.
   B. Change the 1.625″ internal diameter to 1.425″.
   C. Taper the outside faces of the .875″ high base at a 5° angle away from the part. Hint: The base cannot be directly tapered.
   D. Save the drawing as P12-5.

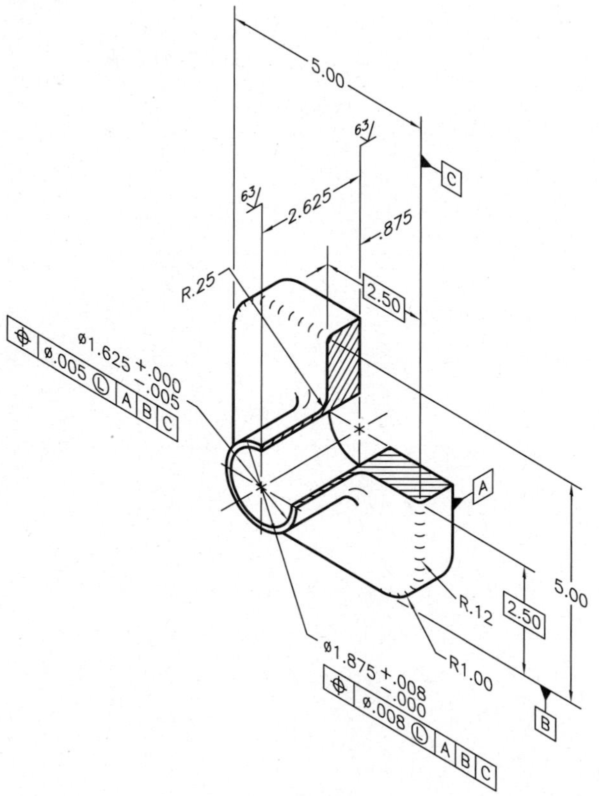

Thrust Washer

6. **Mechanical** Construct the solid part shown using as many **SOLIDEDIT** options as possible. Then, change the dimensions on the model as follows. Save the drawing as P12-6.

Existing	New
100	106
80	82
Ø60	Ø94
Ø40	Ø42
30°	35°

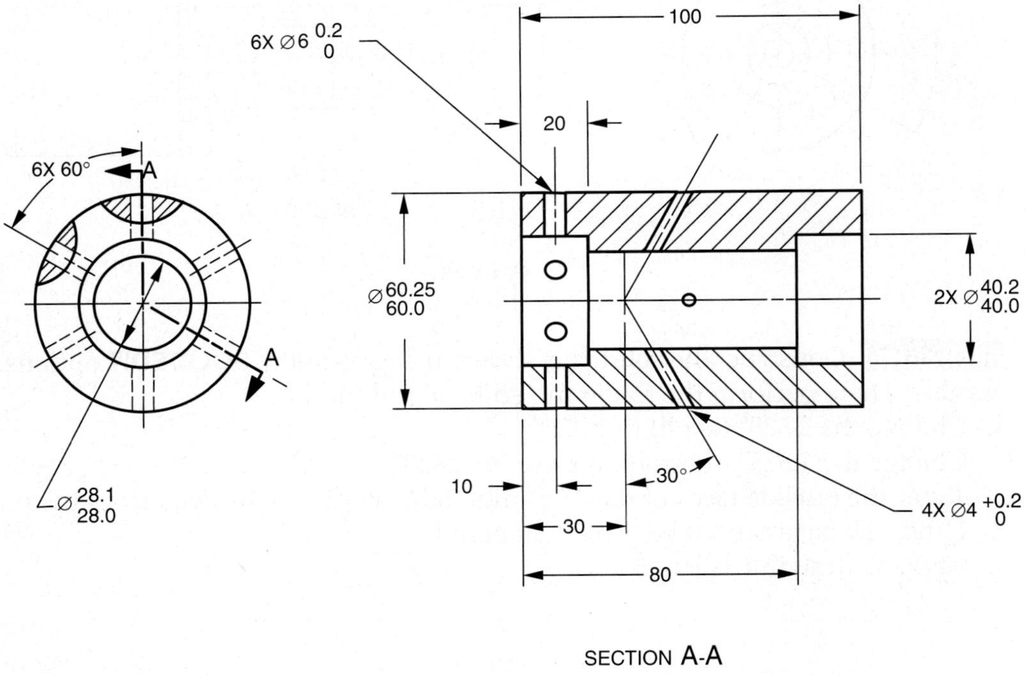

Nozzle

CHAPTER 13

Text and Dimensions in 3D

Learning Objectives

After completing this chapter, you will be able to:

- ✓ Create text with a thickness.
- ✓ Draw text that is plan to the current view.
- ✓ Dimension a 3D drawing.

Creating Text with Thickness

A thickness can be applied to text after it is created. This is done using the **Properties** palette. The thickness setting is located in the **General** section. Once a thickness is applied, the hidden lines can be removed using the **HIDE** command or a shaded visual style. **Figure 13-1** shows six different fonts as they appear after being given a thickness and with the Conceptual visual style set current.

PROPERTIES	
Ribbon	View > Palettes, Properties Palette
Type	PROPERTIES, PR, [Ctrl]+[1]

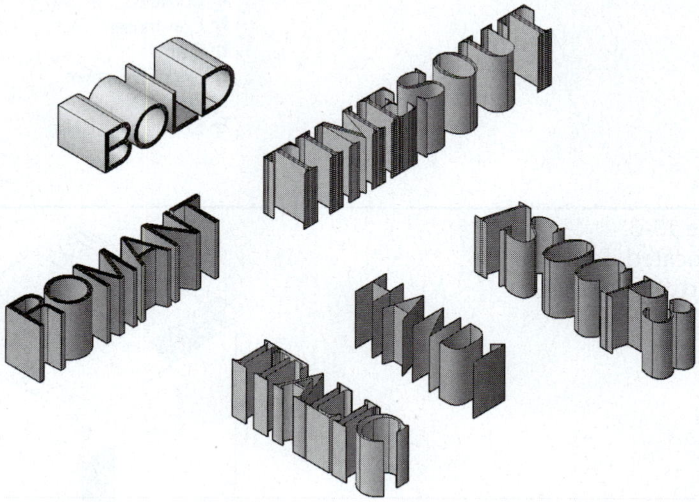

Figure 13-1.
Six different fonts with thickness after the Conceptual visual style is set current.

Only text created using the **TEXT** or **DTEXT** command (a text object) can be assigned thickness. Text created with the **MTEXT** command (an mtext object) cannot have thickness assigned to it. In addition, only AutoCAD SHX fonts can be given thickness. Therefore, when creating a text style to use for 3D purposes, select a text font in the **Text Style** dialog box with a .shx file extension. See **Figure 13-2**. Windows TrueType fonts *cannot* be used to create text with thickness. If a text object assigned a TrueType font is selected, the **Thickness** option will appear in the **Properties** palette. You can assign a thickness value, but it will not be applied to the text.

Text and the UCS

Text is created parallel to the XY plane of the UCS in which it is drawn. Therefore, if you wish to show text appearing on a specific plane, establish a new UCS on that plane before placing the text. You can use the UCS icon grips to create a new UCS. **Figure 13-3** shows several examples of text on different UCS XY planes.

Changing the Orientation of a Text Object

If text is improperly placed or created using the wrong UCS, it can be edited using grips or editing commands. Editing commands and grips are relative to the current UCS. For example, if text is drawn with the WCS current, you can use the **ROTATE** command to change the orientation of the text in the XY plane of the WCS. However, to rotate the text so it tilts up from the XY plane of the WCS, you will need to change the UCS. Rotate the UCS as needed so the Z axis of the new UCS aligns with the axis about which you want to rotate. Then, the **ROTATE** command can be used to rotate the text. See **Figure 13-4**. The **3DROTATE** command can also be used instead of rotating the UCS.

Figure 13-2.
Only AutoCAD SHX fonts can be used for 3D text with thickness.

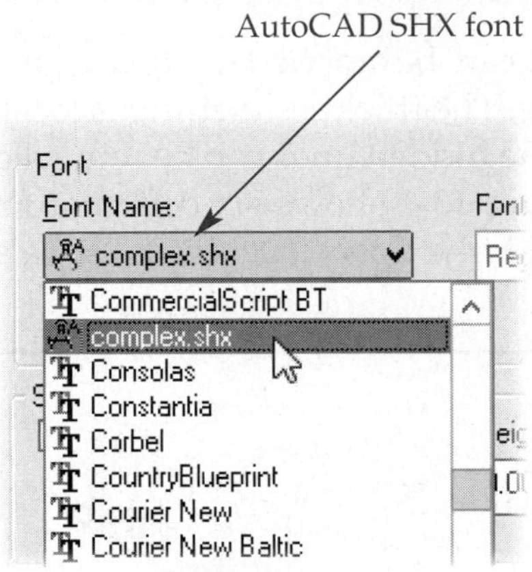

Figure 13-3.
Text located using three different UCSs.

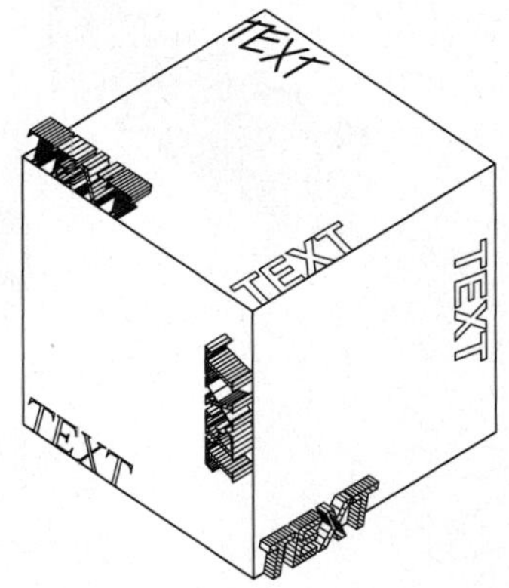

Figure 13-4.
Creating a new UCS to rotate text. A—Pick the UCS icon to display the grips and hover over the X axis grip. Select the **Rotate Around Y Axis** option from the shortcut menu. B—Rotate the UCS so the Z axis of the new UCS aligns with the axis about which you want to rotate. C—The **ROTATE** command is used to rotate the text 90°.

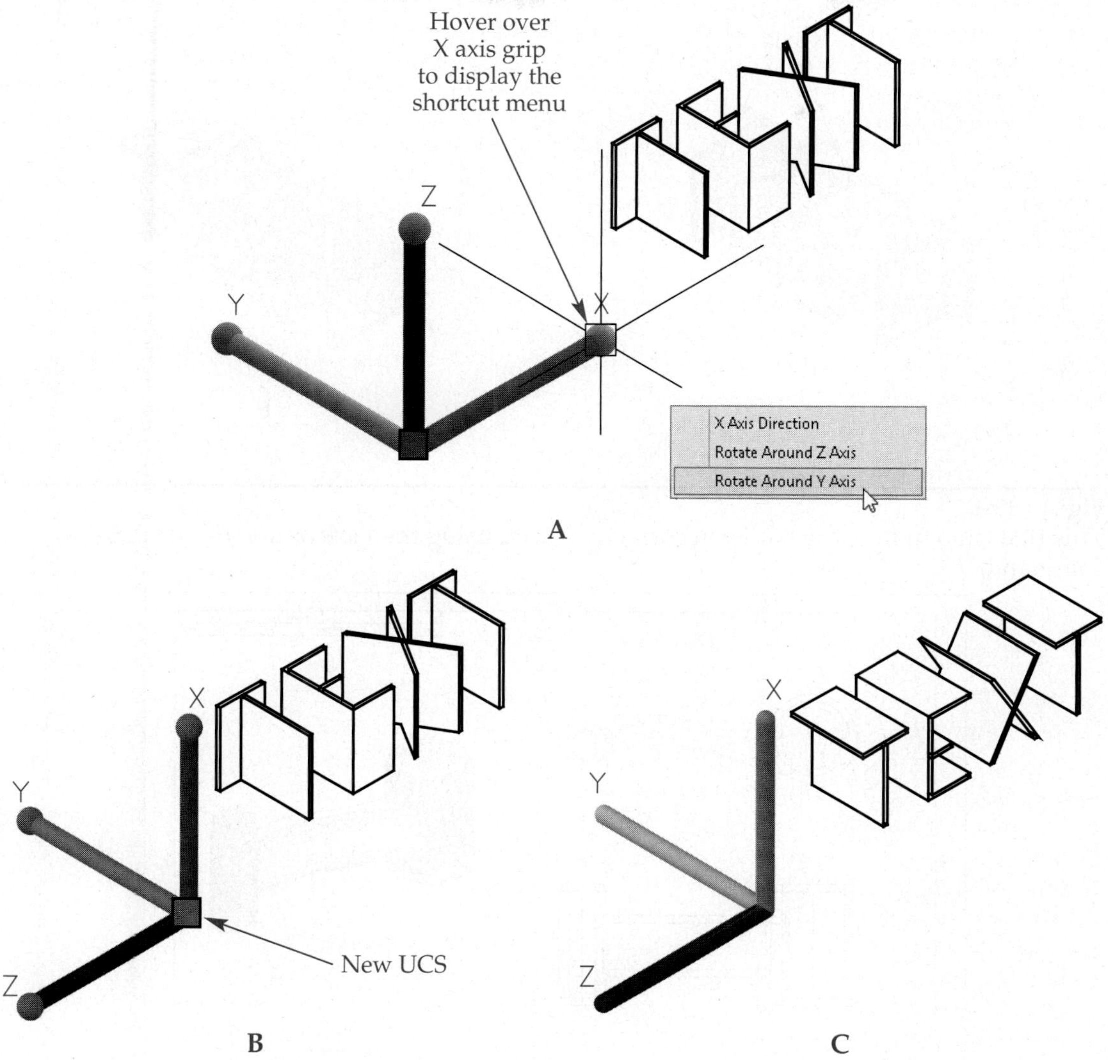

Text objects can also be rotated quickly without first changing the UCS by using the rotate gizmo. With a visual style other than 2D Wireframe set current, simply pick the text. By default, the move gizmo is displayed. Switch to the rotate gizmo, pick the circle about which you want to rotate the object, and then either enter a rotation angle or dynamically move the pointer to achieve the desired result. Using this technique, you can rotate text in any direction regardless of the current UCS. See **Figure 13-5**.

Using the UCS View Option to Create a Title

It is often necessary to create a pictorial view of an object, but with a note or title that is plan to your point of view. For example, you may need to insert the title of a 3D view. See **Figure 13-6**. This is done with the **View** option of the **UCS** command, which is discussed in Chapter 4. With this option, a new UCS is created perpendicular to your viewpoint. However, the view remains unchanged. Inserted text will be horizontal (or vertical) in the current view. Name and save the UCS if you will use it again.

Ribbon
Visualize
> Coordinates
View
Type
UCS

UCS

Figure 13-5.
Using the rotate gizmo to rotate text on any axis regardless of the current UCS. A—Selected text with the rotate gizmo displayed. Notice the circle selected to rotate the text. B—Text is rotated 90° about the X axis.

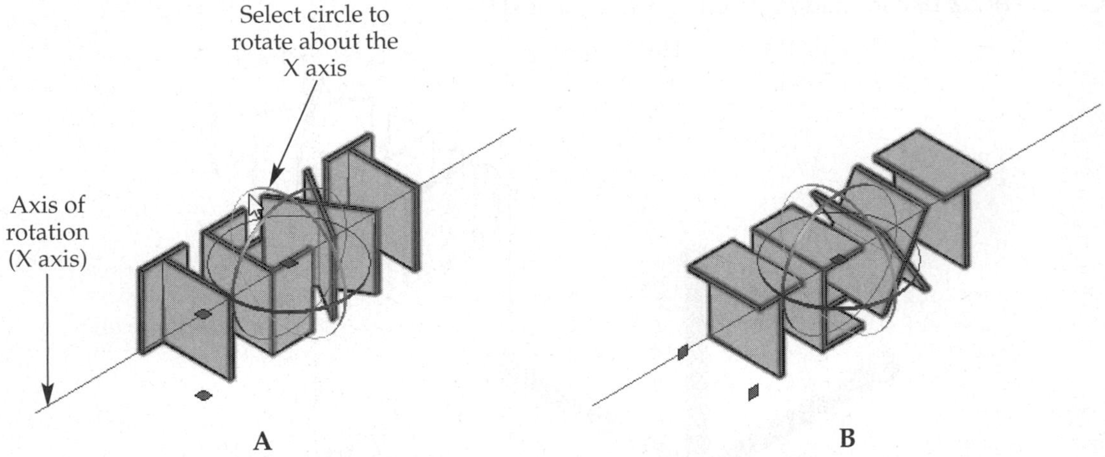

Figure 13-6.
This title (shown in color) has been correctly placed using the **View** option of the **UCS** command.

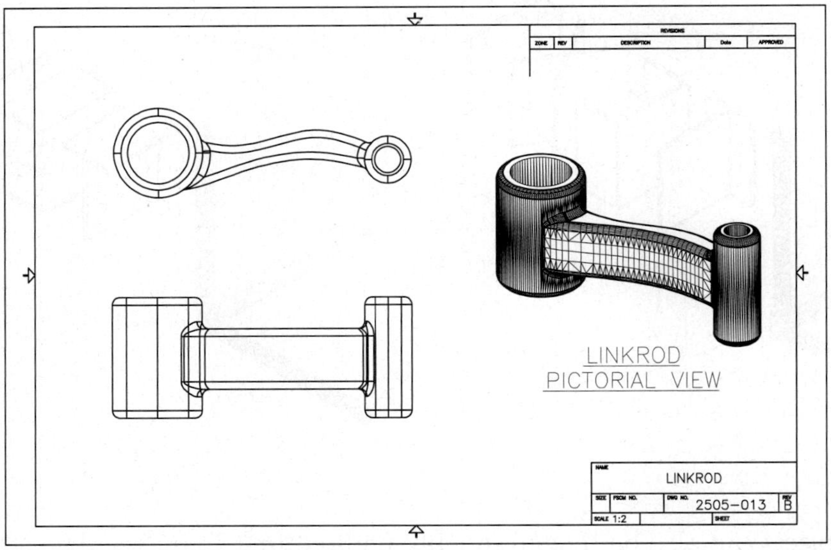

Exercise 13-1

www.g-wlearning.com/CAD/

Complete the exercise on the companion website.

Dimensioning in 3D

Three-dimensional objects are seldom dimensioned for manufacturing, but may be used for assembly. Dimensioned 3D drawings are most often used for some sort of presentation, such as displays, illustrations, parts manuals, or training manuals. All dimensions, including those shown in 3D, must be clear and easy to read. The most important aspect of applying dimensions to a 3D object is planning. That means following a few basic guidelines.

Creating a 3D Dimensioning Template Drawing

If you often create dimensioned 3D drawings, make a template drawing containing a few 3D settings. Starting a drawing based on a template will speed up the dimensioning process because the settings will already be made for you.

- Create named dimension styles with appropriate text heights. See ***AutoCAD and Its Applications—Basics*** for detailed information on dimensioning and dimension styles.
- Establish several named user coordinate systems that match the planes on which dimensions will be placed.
- If the preset isometric viewpoints will not serve your needs, establish and save several 3D viewpoints that can be used for different objects. These viewpoints will allow you to select the display that is best for reading dimensions.
- If 3D dimensioned views are to be used with a multiview 2D drawing, create a paper space drawing layout containing appropriate viewports and the required items listed above.

NOTE

Multiview orthographic and pictorial drawing layouts can be created quickly and efficiently using the **VIEWBASE** command. Creating drawing views from 3D models is discussed in Chapter 14.

Placing Dimensions in the Proper Plane

The location of dimensions and the plane on which they are placed are often a matter of choice. For example, **Figure 13-7** shows several options for placing a thickness dimension on an object. All of these are correct. However, several of the options can be eliminated when other dimensions are added. This illustrates the importance of planning.

The key to good dimensioning in 3D is to avoid overlapping dimension and extension lines in different planes. A freehand sketch can help you plan this. As you lay out the 3D sketch, try to group information items together. Dimensions, notes, and item tags should be grouped so that they are easy to read and understand. This technique is called ***information grouping***.

Figure 13-8A shows the object from **Figure 13-7** fully dimensioned using the aligned technique. Notice that the location dimension for the hole is placed on the top surface. This avoids dimensioning to hidden points. **Figure 13-8B** shows the same object dimensioned using the unilateral technique.

Figure 13-7.
A thickness dimension can be located in many different places. All locations shown here are acceptable.

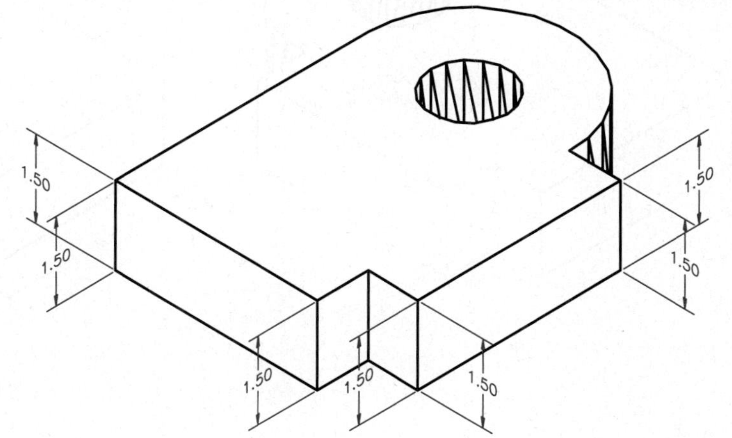

Figure 13-8.
A—An example of a 3D object dimensioned using the aligned technique. B—The object dimensioned with unilateral dimensions.

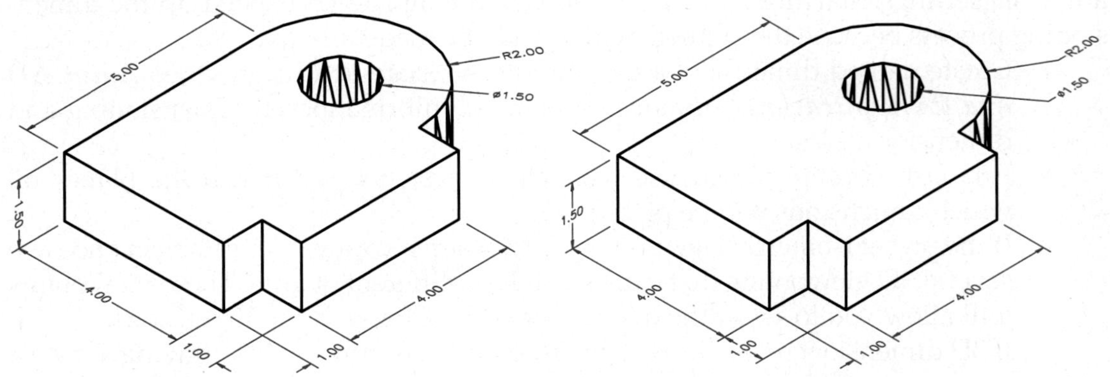

A B

To create dimensions that properly display, it may be necessary to modify the dimension text rotation. The dimension shown in **Figure 13-9A** is inverted because the positive X and Y axes are incorrectly oriented. Using the **Properties** palette, change the text rotation value to 180. The dimension text is then properly displayed, **Figure 13-9B**. Alternately, you can rotate the UCS before drawing the dimension, but this may be more time-consuming.

PROFESSIONAL TIP

Prior to placing dimensions on a 3D drawing, you should determine the purpose of the drawing. For what will it be used? Just as dimensioning a drawing for manufacturing purposes is based on the function of the part, 3D dimensioning is based on the function of the drawing. This determines whether you use chain, datum, arrowless, architectural, or some other style of dimensioning. It also determines how completely the object is dimensioned.

Figure 13-9.
A—This dimension text is inverted. B—The rotation value of the text is changed and the text reads correctly.

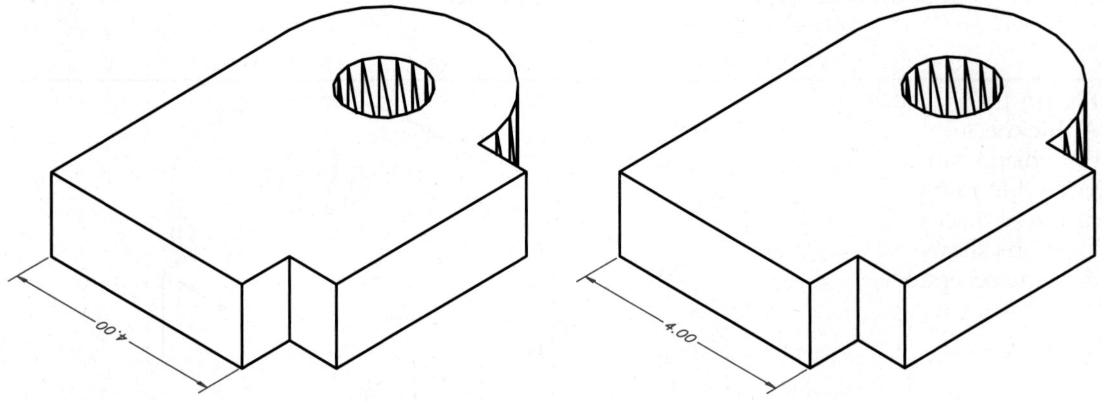

A B

Placing Leaders and Radial Dimensions in 3D

Although standards such as ASME Y14.5 should be followed when possible, the nature of 3D drawing and the requirements of the project may determine how dimensions and leaders are placed. Remember, the most important aspect of dimensioning a 3D drawing is its presentation. Is it easy to read and interpret?

Leaders and radial dimensions can be placed on or perpendicular to the plane of the feature. **Figure 13-10A** shows the placement of leaders and dimensions on the plane of the top surface. **Figure 13-10B** illustrates the placement of leaders and dimensions on two planes that are perpendicular to the top surface of the object. Remember that text, dimensions, and leaders are always created on the XY plane of the current UCS. Therefore, to create the layout in **Figure 13-10B**, you must use more than one UCS.

Figure 13-10.
A—Dimensions placed in the plane of the top surface. B—Dimensions placed using two UCSs that are perpendicular to the top face.

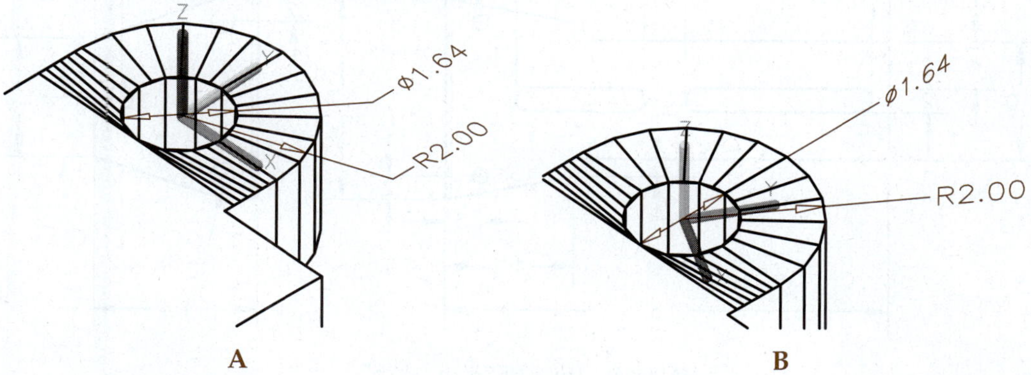

Exercise 13-2

www.g-wlearning.com/CAD/

Complete the exercise on the companion website.

Chapter Review

Answer the following questions using the information in this chapter.

1. How can you create 3D text with thickness?
2. If text is placed using the wrong UCS, how can it be edited to appear on the correct one?
3. How can text be placed horizontally based on your viewpoint if the object is displayed in 3D?
4. Name three items that should be a part of a 3D dimensioning template drawing.
5. What is *information grouping*?

Drawing Problems

For the following problems, create solid models as instructed. If appropriate, apply geometric constraints to the base 2D drawing. If you are using constraints and create profiles that can be converted into regions for solid modeling, apply geometric constraints and save the drawing with a different name than the final solid model.

1. **Mechanical** This is a two-view orthographic drawing of a window valance mounting bracket. Create it as a solid model. Use solid primitives and Boolean commands as needed. Use the dimensions given. Similar holes have the same offset dimensions. Create new UCSs as needed. Display an appropriate pictorial view of the drawing. Then, add dimensions. Finally, add the material note so it is plan to the 3D view. Plot the drawing to scale on a C-size sheet of paper. Save the drawing as P13-1.

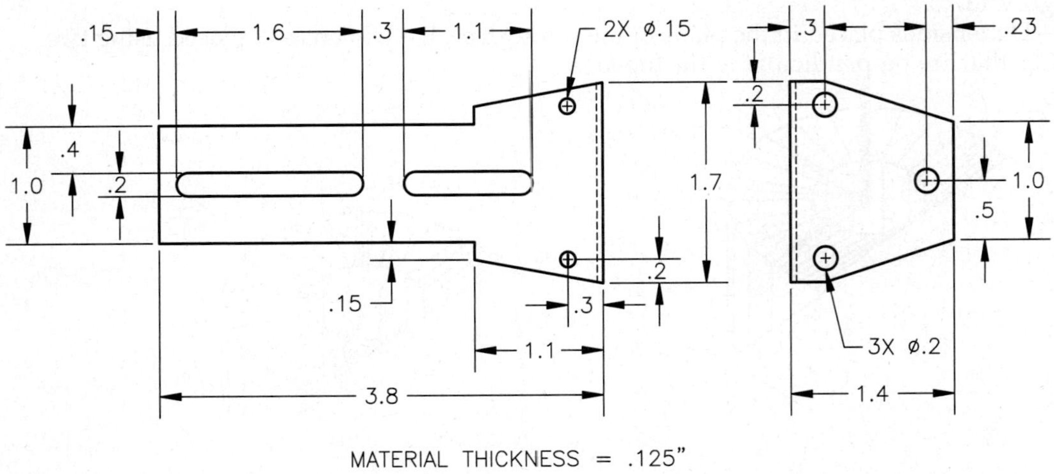

2. **Mechanical** Open P5-3. If you have not completed this problem, construct it using the directions for the problem in Chapter 5. Display the 3D view in a single viewport. Then, add dimensions to the 3D view, as shown in Chapter 5. Plot the drawing on a C-size sheet of paper. Save the drawing as P13-2.

Drawing Problems - Chapter 13

3. **Architectural** Create the end table as a solid model using solid primitives and Boolean commands as needed. The result should be a single object. As a test of your object-editing skills, try drawing the entire model by starting with only a single rectangle. You can copy, resize, extrude, and move objects as you create them from the single rectangle. Use the dimensions given and the following information to construct the model.
   A. Table height is 24″.
   B. Top of bottom shelf is 5″ off the floor.
   C. Table legs must be located no less than 1/2″ from the tabletop edge.
   D. Shelf must be no closer than .75″ from the outside of table legs.
   E. Dimension the table as shown.
   F. Save the drawing as P13-3.

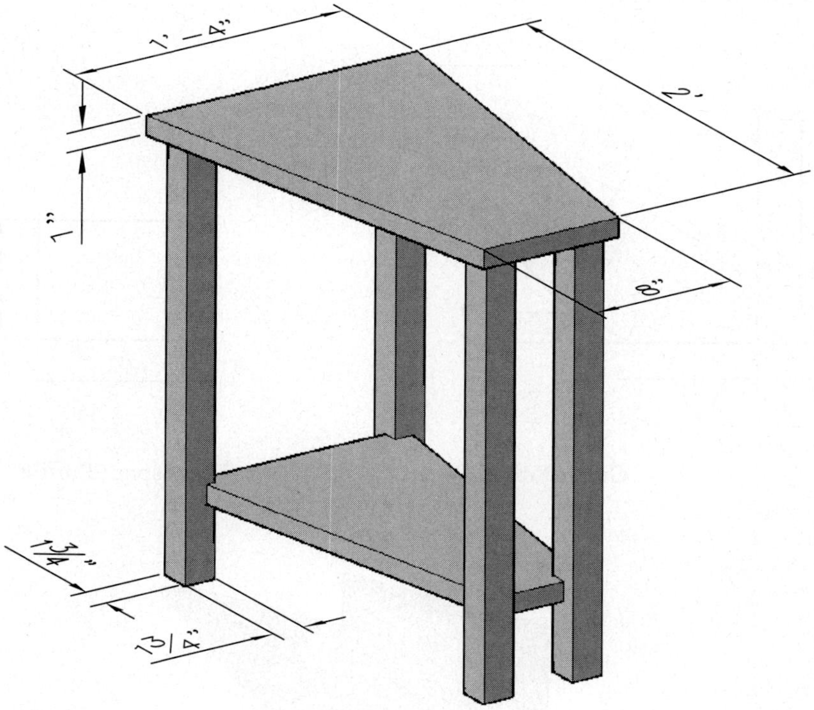

4. **Architectural** Shown are the profiles of a roof gutter (for the collection of rainwater) and a gutter downspout. Draw the profiles in 3D using the dimensions shown. Use the following additional information to construct a 3D model like the one shown in the shaded view.
   A. Offset the gutter profile to create a material thickness of .025″. Be sure to close the ends to create a closed polyline so a 3D solid is created when it is extruded.
   B. Extrude the gutter profile 12″ to create a one-foot section.
   C. Relocate the downspout profile on the underside of the gutter.
   D. Construct an extrusion path for the downspout. Refer to the shaded view, but use your own design.
   E. Extrude the downspout profile along the path.
   F. Dimension the end of the gutter profile in 3D.
   G. Save the drawing as P13-4.

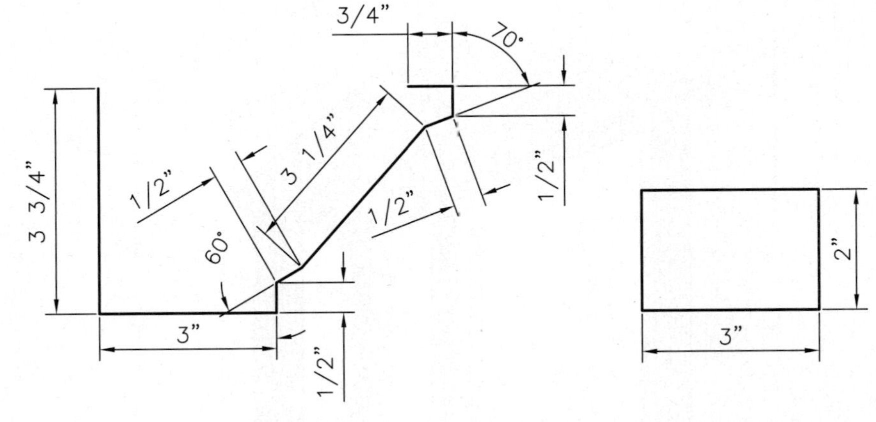

Gutter Profile **Downspout Profile**

A

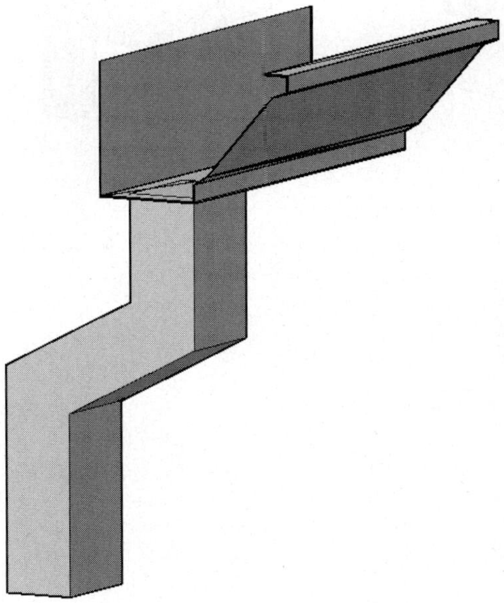

B

5. **Mechanical** Open P5-4. If you have not completed this problem, construct it using the directions for the problem in Chapter 5. Display the 3D view in a single viewport. Then, add dimensions to the 3D view. Plot the drawing on a C-size sheet of paper. Save the drawing as P13-5.

Problems 6 and 7 are mechanical parts. Create a solid model of each part. Dimension each model. Place the title of each model so it is plan to the pictorial view. Plot the finished drawings on B-size paper. Save each drawing as P13-*(problem number).*

6. **Mechanical**

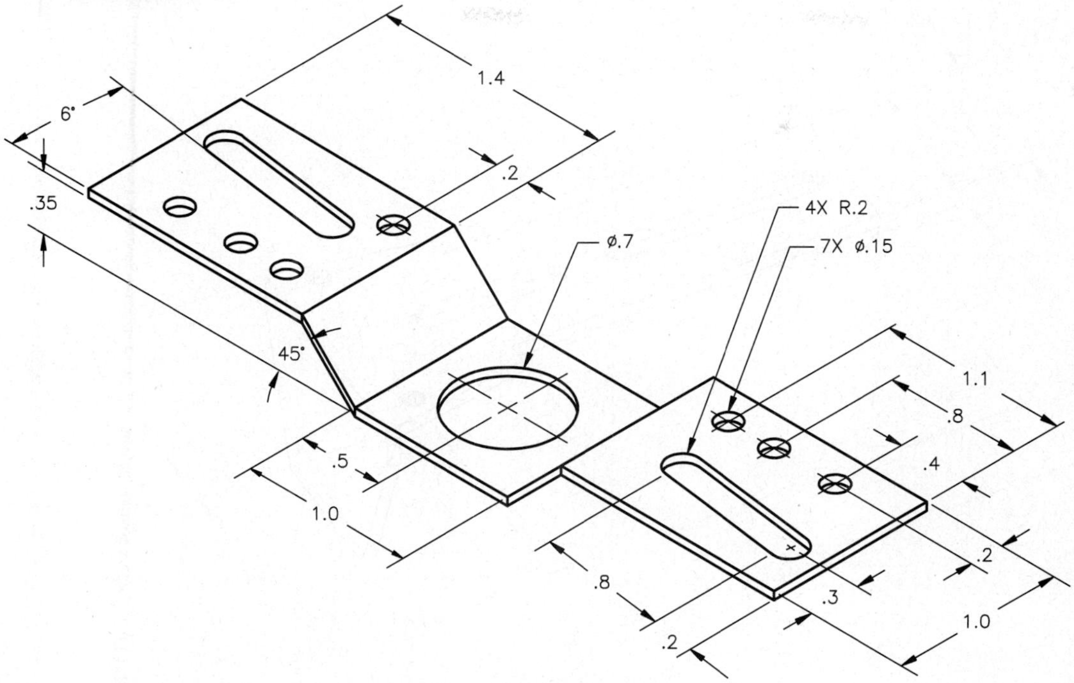

Guide Bracket

7. **Mechanical**

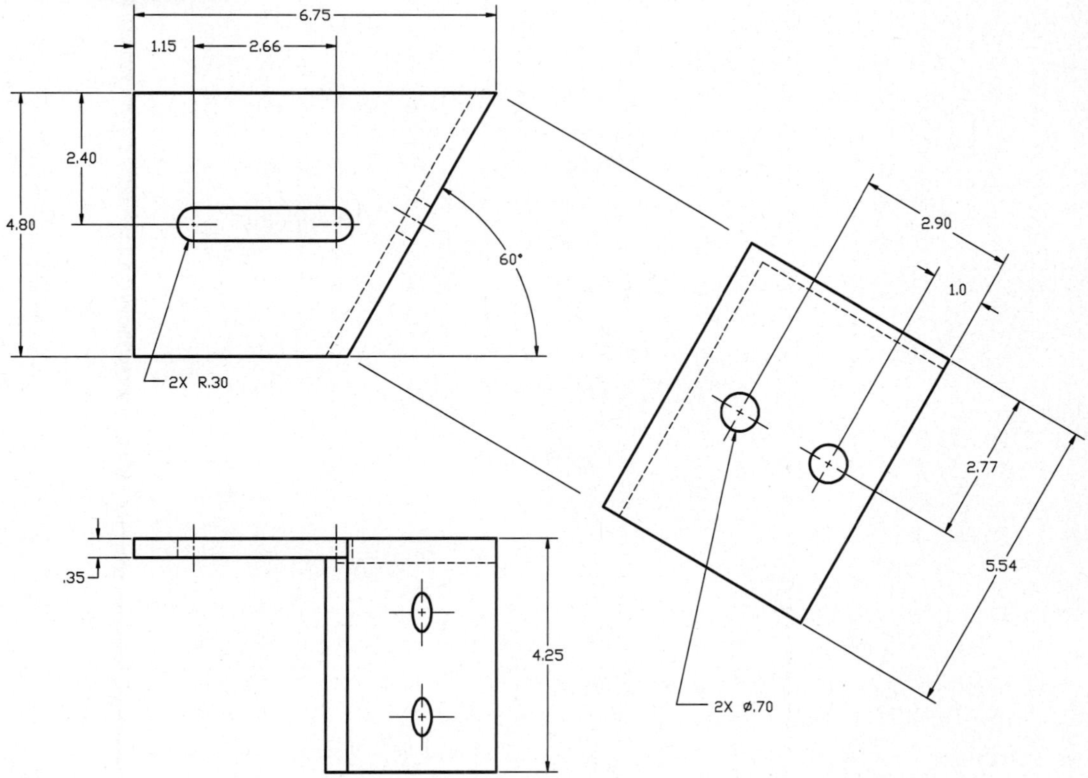

Angle Mount

Three-dimensional models are not typically dimensioned, but there are cases where dimensions may be applied to assist workers on the shop floor. Whenever a 3D model is dimensioned, follow accepted drafting practices when possible, but you may need to violate certain rules in order to clearly describe the part.

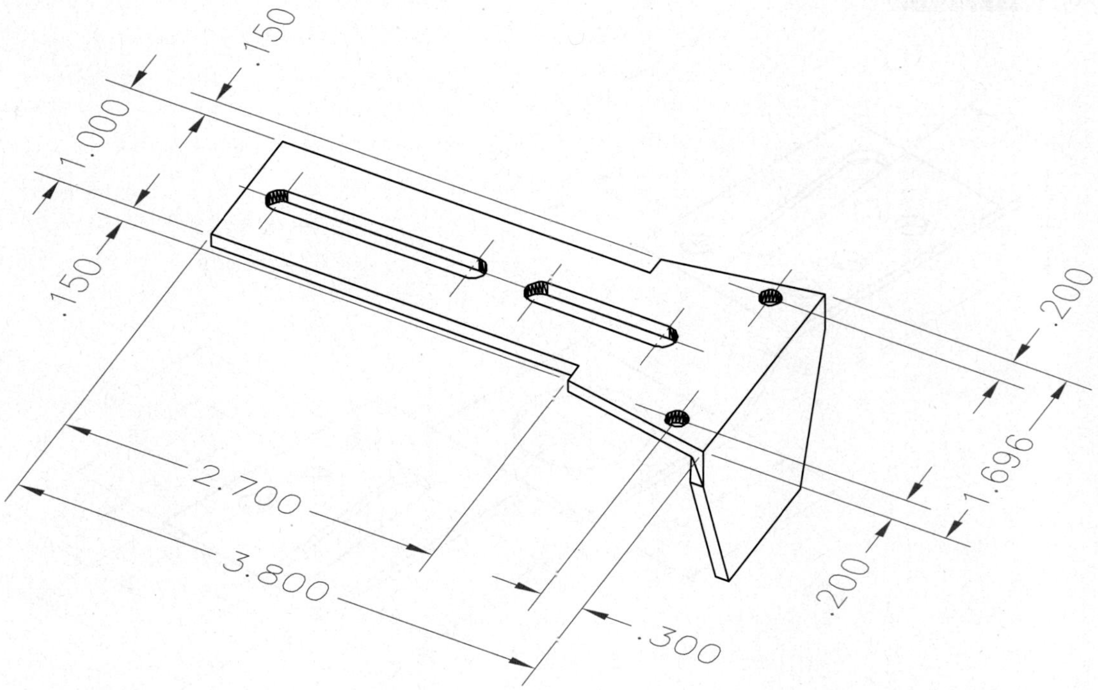

CHAPTER 14

Model Documentation, Analysis, and File Exchange

Learning Objectives

After completing this chapter, you will be able to:

- ✓ Create model documentation using drawing views.
- ✓ Create section views.
- ✓ Create detail views.
- ✓ Create auxiliary views.
- ✓ Construct a 3D section plane through a 3D model.
- ✓ Adjust the size and location of section planes.
- ✓ Create a dynamic section of a 3D model.
- ✓ Construct 2D and 3D section blocks.
- ✓ Analyze solid and surface models.
- ✓ Exchange solid model file data with other programs.
- ✓ Prepare files for 3D printing.

This chapter discusses the tools available in AutoCAD for creating drawing views and showing section views of objects. This chapter also discusses AutoCAD tools for conducting model analysis, importing and exporting model files, and preparing files for 3D printing.

Introduction to Model Documentation

There are multiple techniques available to create drawing views from 3D models in AutoCAD. Using these techniques, orthographic, isometric, section, and auxiliary views can be created in order to produce multiview drawings. The techniques available include the following.

- Use the **VIEWBASE**, **VIEWPROJ**, **VIEWSECTION**, and **VIEWDETAIL** commands to create multiview drawings that are derived from and associated to a 3D solid or surface model. The drawing views are placed in a paper space layout. This is the most efficient method to create multiview drawings in AutoCAD.
- Use the **SECTIONPLANE** command to show the internal features of a 3D solid model. This command can create 2D and 3D section views on an object.

- Use the **SOLVIEW** command to create a multiview drawing from a solid model. This command allows you to create a layout containing orthographic, section, and auxiliary views. The **SOLDRAW** command can then be used to complete profile and section views. The **SOLVIEW** and **SOLDRAW** commands are not discussed in this chapter. The **VIEWBASE**, **VIEWPROJ**, **VIEWSECTION**, and **VIEWDETAIL** commands are typically more useful and provide greater flexibility in editing.

Drawing Views

You can create 2D drawing views from AutoCAD 3D solid or surface models using the **VIEWBASE** command. This command allows you to quickly create a multiview drawing layout. Drawing views created with the **VIEWBASE** command are created in paper space (layout space). After creating a layout of views, you can dimension them using AutoCAD dimensioning commands.

NOTE

The **VIEWBASE** command can be used to create drawing views from AutoCAD 3D models or from Autodesk Inventor® part (IPT), assembly (IAM), or presentation (IPN) files. This chapter covers creating drawing views from AutoCAD 3D models only.

Drawing views created with the **VIEWBASE** command are *associative*. This means that they are linked to the model from which they are created and update to reflect changes to the model geometry. This capability allows you to keep drawing views up-to-date when design changes are required.

When using the **VIEWBASE** command, you create a ***base view*** of the model and then have the option to create additional views that are projected from the base view without exiting the command. You can create both orthographic and isometric views. A base view created with the **VIEWBASE** command is defined as a *parent view*. Views projected from the base view inherit the properties of the base view, such as the drawing scale and display properties, and are placed in orthographic alignment with the base view. If the base view is moved, any projected views are moved with it to maintain the parent-child relationship.

You can also create drawing views from models imported with the **IMPORT** command. The **IMPORT** command is discussed later in this chapter. Drawing views can also be created from an assembly or subassembly to create an assembly drawing. This chapter discusses creation of drawing views from part models.

Drawing View and Layout Setup

As previously mentioned, drawing views created with the **VIEWBASE** command are created in paper space. To activate paper space, pick a layout tab. Then, before creating views, set up the page layout as required. If a default viewport appears in the layout, *delete the viewport*. Then, right-click on the active layout tab and select **Page Setup Manager...** to access the layout settings. Select the desired paper size, drawing orientation, and other settings. If you have a template drawing already created, you can create a new layout based on the template by right-clicking on a layout tab and selecting **From Template....**

VIEWSTD

Type
VIEWSTD

By default, drawing views created with the **VIEWBASE** command are generated using third-angle projection. This is the ANSI standard. The default projection used for drawing views can be set using the **VIEWSTD** command. This command

and other drawing view commands are accessed from the **Layout** tab of the ribbon. See **Figure 14-1**. Pick the dialog box launcher button at the lower-right corner of the **Styles and Standards** panel to initiate the **VIEWSTD** command. This opens the **Drafting Standard** dialog box, **Figure 14-2**.

The options in the **Projection type** area determine the type of projection used for new drawing views. By default, the **Third angle** button is selected. If the **First angle** button is selected, views are created using first-angle projection. Select this option to create views in accordance with ISO standards. The options in the **Thread style** area determine the way threaded ends are shown in section views and the type of circular thread representation used for views showing threaded features. The thread representation options are used when creating views from Autodesk Inventor models or imported models containing threaded features.

The options in the **Shading/Preview** area of the **Drafting Standard** dialog box determine the type of preview that appears when placing a drawing view. When **Shaded** is selected in the **Preview type** drop-down list, a shaded view appears. This is the default option. This option provides a preview image of the view orientation before the view is placed. When the **Bounding box** option is selected, a preview box representing the extents of the view appears. However, no preview image is displayed. The **Shaded view quality** option is set to 100 dpi by default. A higher setting will provide better resolution quality when a shaded preview is displayed. However, for all practical purposes, a setting of 150 dpi or higher does not produce a dramatic difference in resolution quality.

Figure 14-1.
The AutoCAD drawing view commands are accessed from the **Layout** tab of the ribbon.

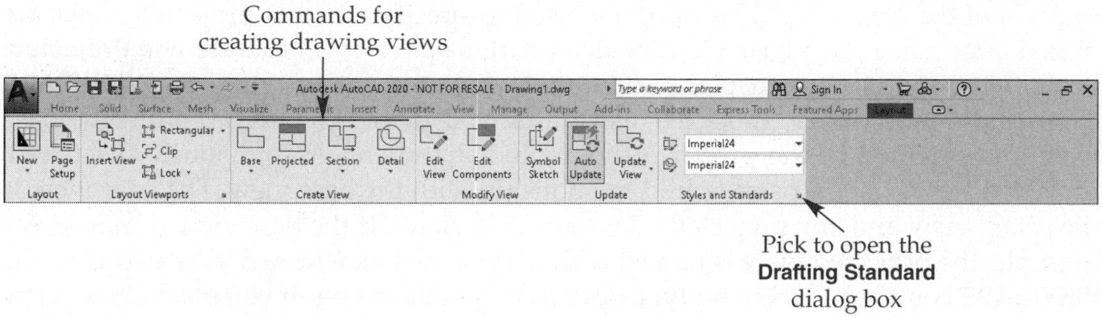

Figure 14-2.
The **Drafting Standard** dialog box.

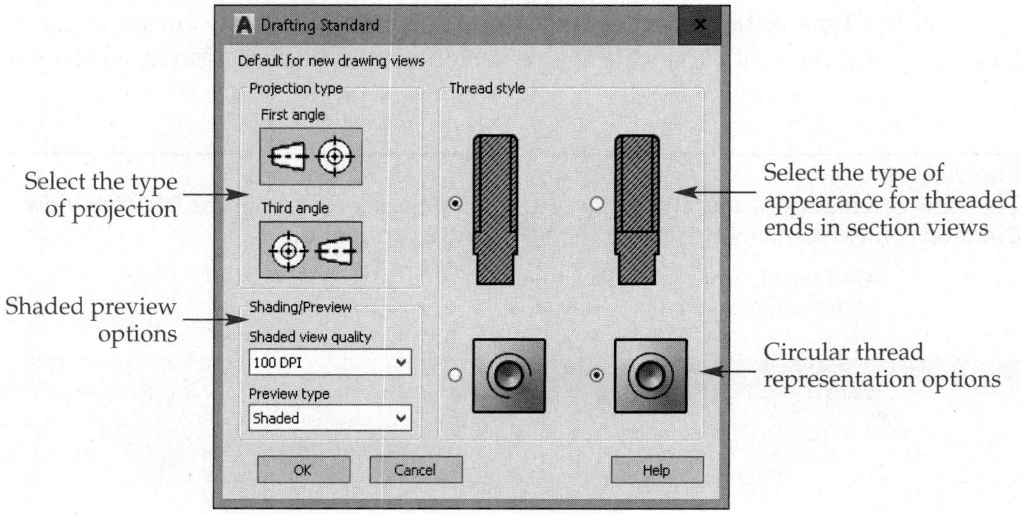

Creating Drawing Views

After setting up a layout and making the appropriate drawing view settings, you are ready to place the base view. As previously discussed, drawing views can be created from a 3D solid or surface model. Drawing views created in the layout are generated from the model you have created in model space. If no model exists in the drawing and you initiate the **VIEWBASE** command, the **Select File** dialog box appears. From this dialog box, you can select a model created in Autodesk Inventor.

VIEWBASE

Ribbon
Layout
> Create View
Base View from Model Space
Type
VIEWBASE

When you select the **VIEWBASE** command from the ribbon, a preview representing a scaled orthographic view of the model appears. This is the ***base view*** of the model. You can then specify the location of the base view or select an option. To select an option, use dynamic input or make a selection from the **Drawing View Creation** contextual ribbon tab. See **Figure 14-3.**

The **Select** option switches you to model space and allows you to add objects to or remove objects from the base view. If necessary, use the appropriate option to select solids or surfaces to add or remove. Use the **Layout** option to return to the layout after making changes. The **Orientation** option can be used to select a different view from the default base view. The **Orientation** options correspond to AutoCAD's six orthographic and four isometric preset views. These views are based on the WCS. By default, the front view is used as the base view by AutoCAD. Depending on the construction of your model, you may want to use a different view, such as the top view, to serve as the base view. In **Figure 14-4**, the top view of the model is selected as the base view. This view will establish the front view on the orthographic drawing. In most cases, the front orthographic view on the drawing describes the most critical contour of the model. You will typically use the front or top view of the 3D model for the front orthographic view on the drawing.

The **Type** option of the **VIEWBASE** command is available before picking the initial location of the base view. This option is used to specify whether projected views are placed after placing the base view. By default, this option is set to **Base and Projected.** With this option, you can place additional, projected views while the **VIEWBASE** command is still active. Projected views that you place are *projected* from the base view and oriented in the proper orthographic alignment. As previously discussed, projected views have a parent-child relationship with the base view. The base view is the *parent* view and any projected views are *child* views. If the base view is moved, for example, the projected view is moved with it. Projected views are not created with the **VIEWBASE** command when the **Type** option is set to **Base only**. If you place a base view in this manner and later decide to create projected views from the base view, you can use the **VIEWPROJ** command. The **VIEWPROJ** command allows you to place projected views from any selected view.

After you pick the location for the base view, you can press [Enter] to place projected views (if the **Type** option is set to **Base and Projected**). Drag the cursor away from the base view and then pick to locate the projected view. The direction in which you move

Figure 14-3.
The view orientation, scale, and other settings can be selected from the **Drawing View Creation** contextual tab when using the **VIEWBASE** command.

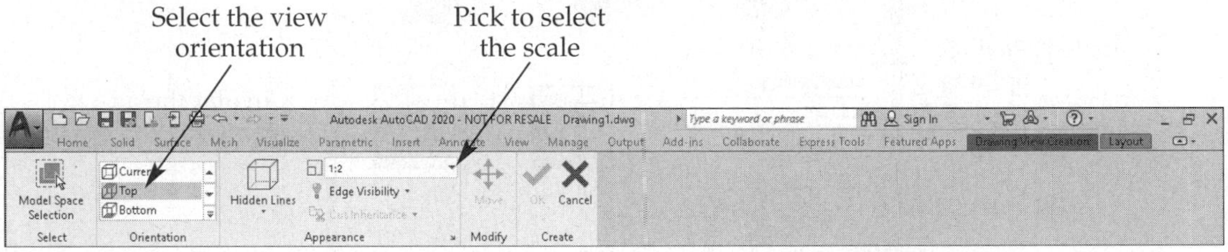

Figure 14-4.
Placing a base view with the **VIEWBASE** command. A—The original model drawn in model space. B—The preview of the base view in paper space. The **Top** view option is selected for the base view because it is the most descriptive view of the model. C—The resulting base view.

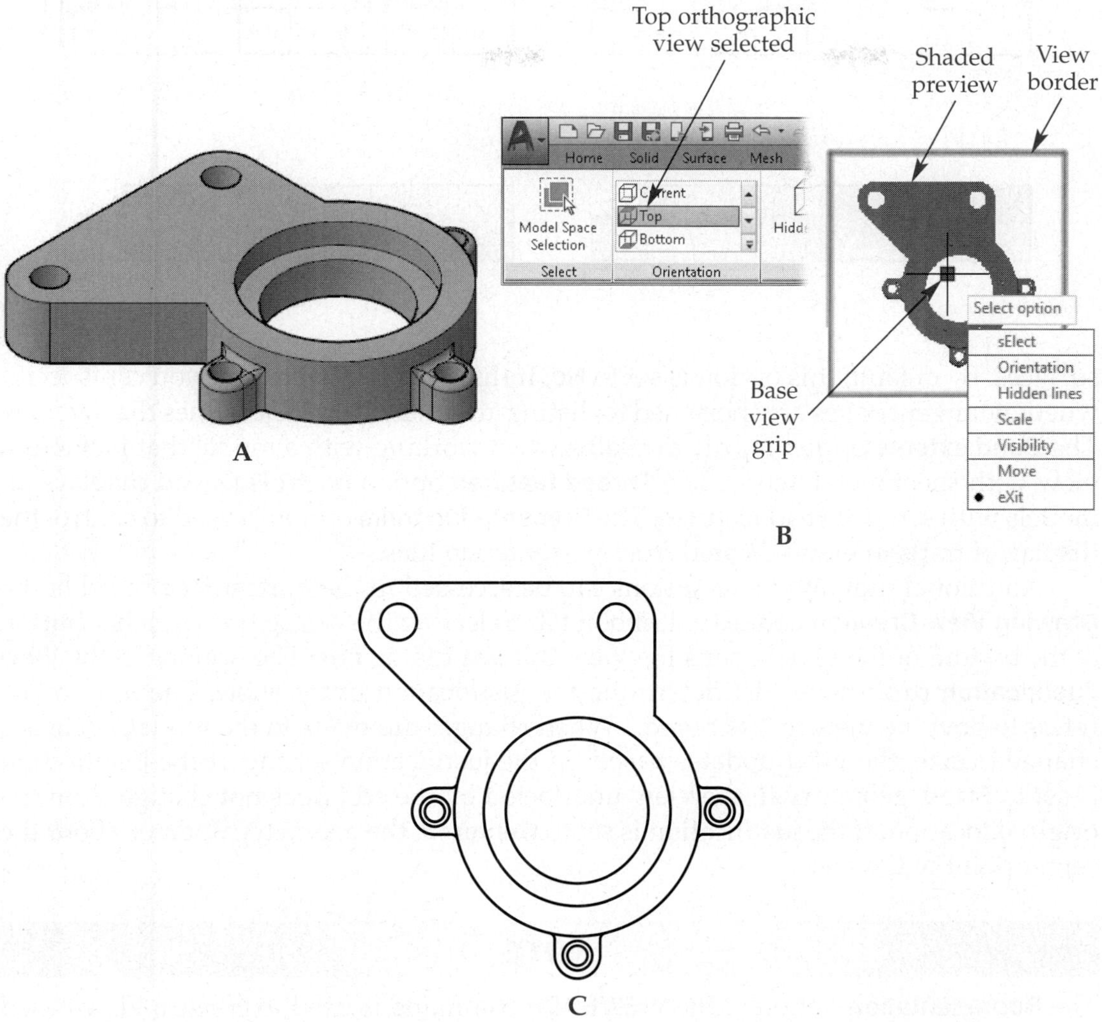

the cursor determines the orientation of the projected view. You can continue placing projected views, or you can press [Enter] to end the command. To end the command without placing projected views, press [Enter] twice.

The display style of the base view can be changed by using the **Hidden Lines** option. The four display style options available are **Visible Lines**, **Visible and Hidden Lines**, **Shaded with Visible Lines**, and **Shaded with Visible and Hidden Lines**. See **Figure 14-5**. These options can be selected using dynamic input or the drop-down list in the **Appearance** panel in the **Drawing View Creation** contextual ribbon tab.

The **Scale** option is used to set the scale of the base view. You can select a scale from the drop-down list in the **Appearance** panel in the **Drawing View Creation** contextual ribbon tab. You can also specify the scale by typing a value.

The **Move** option allows you to move the base view after picking the initial location on screen. When you select the **Move** option, the view is reattached to the cursor so that you can move it to another location. After you pick a new location, the **VIEWBASE** options are again made available.

The **Visibility** option allows you to control the display of drawing geometry in the view. The **Interference edges** option controls whether both object and hidden lines are displayed for interference edges. By default, this option is set to **No**. The **Tangent edges** option controls whether tangential edges are displayed to show the intersection of

Figure 14-5.
Display style options available for drawing views. A—**Visible Lines**. B—**Visible and Hidden Lines**. C—**Shaded with Visible Lines**. D—**Shaded with Visible and Hidden Lines**.

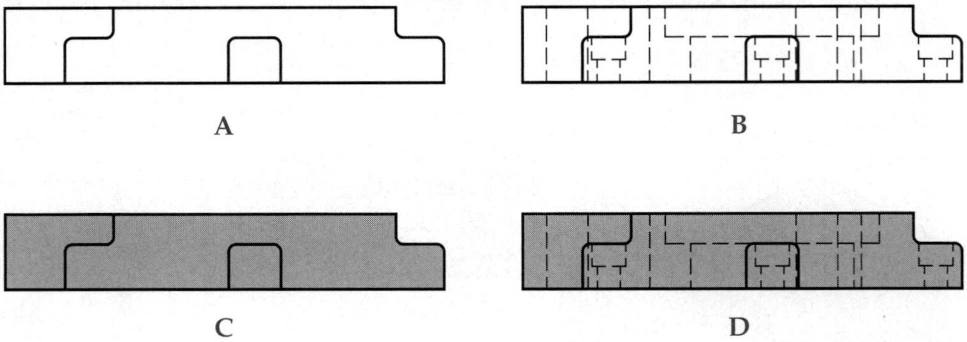

surfaces. By default, this option is set to **No**. If this option is set to **Yes**, you can specify whether tangent edges are shortened to distinguish them from object lines that overlap. The **Bend extents** option is only available when working with a model that includes a view with sheet metal bends. The **Thread features** option controls thread displays on models with screw thread features. The **Presentation trails** option is used to control the display of trails in views created from presentation files.

Additional drawing view options can be accessed in the **Appearance** panel of the **Drawing View Creation** contextual ribbon tab. Selecting the dialog box launcher button at the bottom of the panel opens the **View Options** dialog box. The options in the **View Justification** drop-down list determine the justification of the view. The *justification* refers to how the view is "anchored." When changes are made to the model, such as a change in size, the view updates based on the justification setting. If the justification is set to **Fixed**, geometry in the view unaffected by the edit does not change from the original location. If the justification is set to **Centered**, the geometry updates about the center point of the view.

NOTE

The **Representation** option of the **VIEWBASE** command is used to create a view based on a model representation and is only available when working with models created in Autodesk Inventor.

Working with Drawing Views

Creating a base view creates an AutoCAD ***drawing view*** object. A drawing view has a view border, a base grip, and a parameter grip for changing the scale. The scale and certain other properties of the view can be edited, as discussed later in this chapter. However, the content of the drawing view *cannot* be edited.

When a base view is created, new layers are created by AutoCAD for the drawing view geometry. The layers are created based on support for the type of geometry represented. Object lines (visible lines) in the view are placed on a newly created layer named MD_Visible. Hidden lines in the view are placed on a newly created layer named MD_Hidden, and so on. The drawing view object is placed on the current layer or on the 0 layer. Additional layers may be created by AutoCAD, depending on the type of model, display style used, and edges displayed in the view. However, the layers are only created to organize the drawing view geometry. The layer properties can be modified to change the appearance of the drawing view geometry, but the geometry cannot be otherwise modified.

In **Figure 14-6**, a multiview drawing layout is created by placing three projected views after placing the base view. The projected views include two orthographic views and an isometric view. Notice that wireframe styles are used for the orthographic views and a shaded style is used for the isometric view. Drawing views can be edited after placing them to change properties as needed. In addition, as previously discussed, drawing views update when changes are made to the model. This is discussed in more detail in the next section.

PROFESSIONAL TIP

You can create more than one base view in a layout. Additional base views can be used to create assembly or subassembly drawings.

Updating Drawing Views

Drawing views maintain an associative relationship with the model from which they are created. However, it is important to note that this associative relationship is controlled by the model, not the drawing view. When making design changes, you make changes to the model geometry, *not* the drawing view.

During the design process, it is often necessary to make modifications. If you have created drawing views from a model, and then make modifications to the model, the derived base view and any views projected from the base view are updated automatically by default. This behavior is controlled by the **VIEWUPDATEAUTO** system variable. When this system variable is set to 1 (on), drawing views automatically update when you open a drawing file or activate a layout with drawing views. The

Figure 14-6.
A multiview drawing layout consisting of front, top, right-side, and isometric views. The top view of the model is used as the front view (base view) in the drawing. Notice the different display style options that are used.

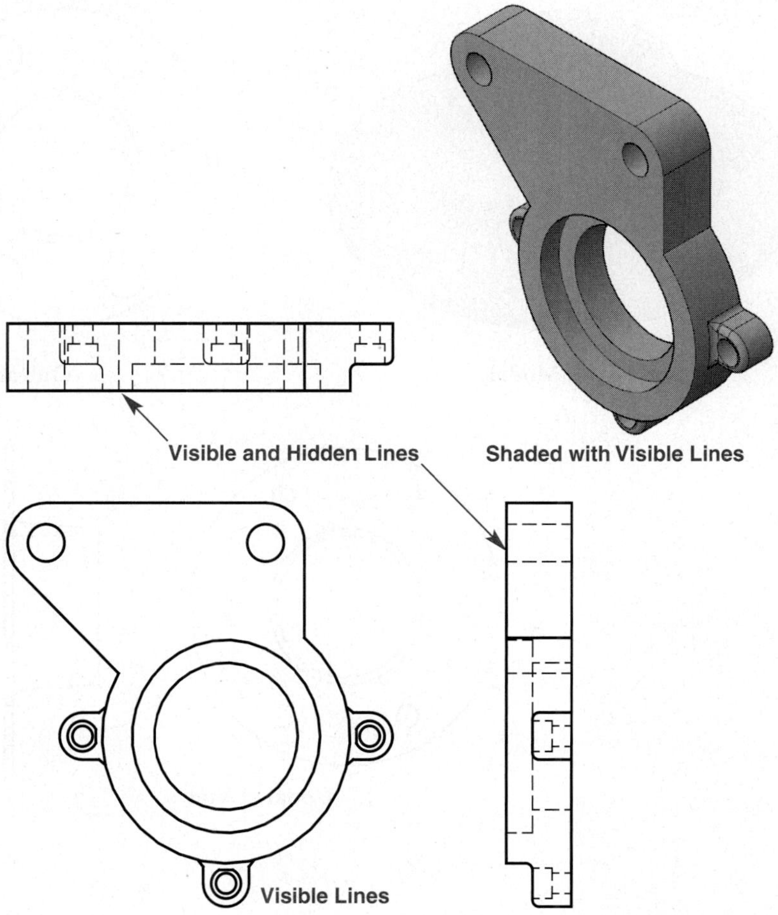

VIEWUPDATEAUTO system variable setting is indicated by the **Auto Update** button located on the **Update** panel of the **Layout** tab on the ribbon. Refer to **Figure 14-1**. With the default setting of 1, the button has a blue background.

Drawing views can be updated manually if needed. For example, if the **VIEWUPDATEAUTO** system variable is set to 0 (off) and changes are made to the model, drawing views are not updated by AutoCAD automatically. If you make changes to a model in model space and then switch to a layout with drawing views of the model, red markers appear to indicate that the views are *out-of-date*. See **Figure 14-7**. A balloon notification also appears on the status bar. You can pick the link in the balloon notification or use one of AutoCAD's view update commands to update the views.

Update all Views

Ribbon
Layout
> Update

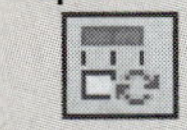

Update all Views

VIEWUPDATE

Ribbon
Layout
> Update

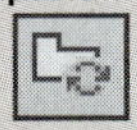

Update View

Type
VIEWUPDATE

In **Figure 14-7A**, the six arrayed holes have been modified by changing the diameter from Ø1.8 to Ø.9. In **Figure 14-7B**, the drawing views appear with markers to indicate they are out-of-date. To update the views, select **Update all Views** from the **Update View** drop-down list in the **Update** panel on the **Layout** tab of the ribbon. After updating the views, the holes are updated to the modified diameter. See **Figure 14-7C**. In this example, all of the views are updated at once. This is the preferred method for updating views. However, you can also update views individually by selecting **Update View** from the **Update View** drop-down list in the **Update** panel on the **Layout** tab of the ribbon. Selecting **Update View** is the same as using the **VIEWUPDATE** command. When using this command, you are prompted to select each view to update.

Figure 14-7.
Updating drawing views after editing the 3D model. A—The holes in the flanged coupler have been edited from Ø1.8 to Ø.9. B—After switching to the layout, markers appear to indicate that the drawing views are out-of-date. C—The views are updated to reflect the changes in the model.

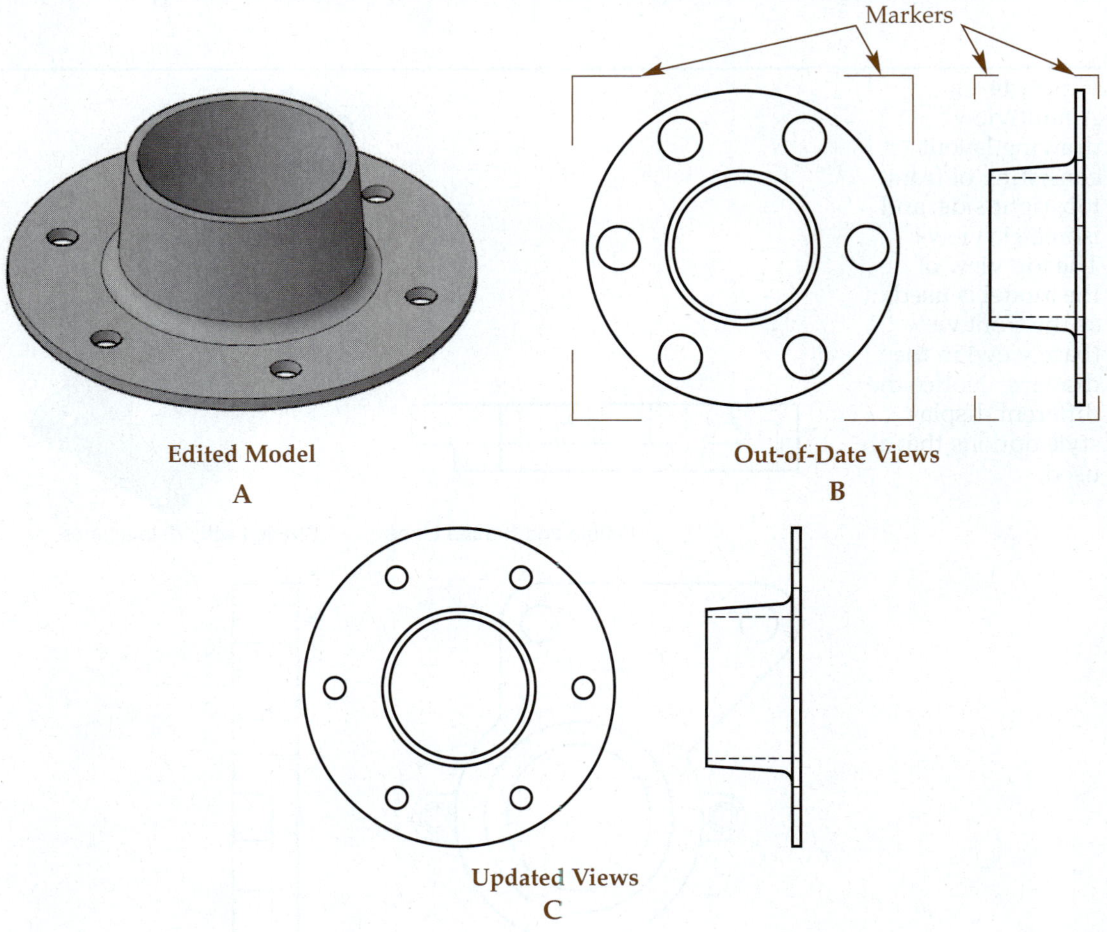

NOTE

Drawing views are different from viewports. You cannot pick inside of the view and edit or modify the drawing geometry by any standard AutoCAD editing technique. Drawing views can only be updated by changing the 3D model geometry.

PROFESSIONAL TIP

In order to modify solid primitive subobjects in a composite solid model, the solid history must be recorded.

Editing Drawing Views

Drawing views can be edited after being created. Like other AutoCAD objects, drawing views can be moved or rotated. In addition, certain properties of drawing views, such as the display style, edge visibility, and scale, can be modified.

The **VIEWEDIT** command can be used to edit the properties of a drawing view. You can quickly initiate this command by double-clicking on a view. The **Drawing View Editor** contextual ribbon tab appears, Figure 14-8. The editing options are similar to the options available when you create a base view.

Ribbon
Layout
> Modify View
Edit View
Type
VIEWEDIT

VIEWEDIT

As previously discussed, projected views inherit the properties of the base view when created. If you select a projected view with the **VIEWEDIT** command, you can change the display style, edge visibility, and scale.

A quick way to change the scale of a view is to use the scale parameter grip. This grip appears when you single-click on a view. Pick on the grip to access a different scale. To move a drawing view, single-click on the view and then pick on the base grip to move the view directly. When moving a parent view, any child views will move accordingly to maintain alignment. When moving a child view, you can move the view, but it cannot be moved out of alignment with the parent view. This applies to orthogonal views. If you move an isometric view, it is not aligned to other views and can be moved freely around the layout.

Picking on a drawing view grip and right-clicking displays a shortcut menu. The **Stretch** option can be used to move the drawing view. The **Rotate** option can be used to rotate the drawing view. You can rotate the view dynamically using the cursor or you can specify a rotation angle. If a drawing view is rotated, any parent-child relationships that exist between the view and other drawing views are broken.

You can break the alignment between parent and child views when moving a child view. To do this, select the drawing view grip and press [Shift] once. The view is free to move to a different location. To reestablish the alignment, press [Shift] again. The alignment is restored and the child view cannot move out of alignment with the parent view.

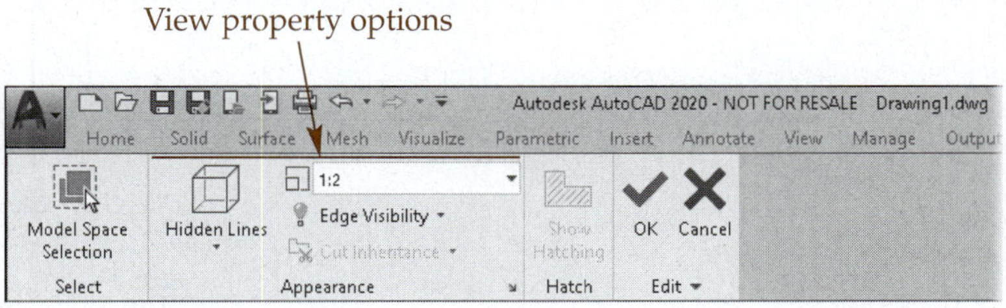

Figure 14-8. The **Drawing View Editor** contextual ribbon tab.

Exercise 14-1

www.g-wlearning.com/CAD/

Complete the exercise on the companion website.

Section Views

A ***section view*** shows the internal features of an object along a section line (cutting plane). A section view is projected from an existing view, such as an orthographic top view. The existing view serves as a parent view. To create a section view, you pick points on the parent view to define the section line (cutting-plane line). You can also select an object, such as a line or polyline, to define the section line. Section views are created using the **VIEWSECTION** command. This command can be used to create full, half, offset, or aligned sections from an AutoCAD 3D model or an Autodesk Inventor model.

Section views created with the **VIEWSECTION** command are created in the same paper space layout as other drawing views. Section views are *associative*. A section view is linked to the parent view that creates the section view. As with other types of drawing views, section views are updated automatically when model changes are made if the **VIEWUPDATEAUTO** system variable is set to 1.

By default, a section identifier is placed with the section line and a section view label is placed with the section view when you create the view. See **Figure 14-9**. The

Figure 14-9.
A section view created with the **VIEWSECTION** command. Shown is a full section created from the top view of the model. A—The model drawing. B—After creating the top view as a base view, a section view is created by drawing the section line through the center of the object. Note the placement of the section identifier and section view label.

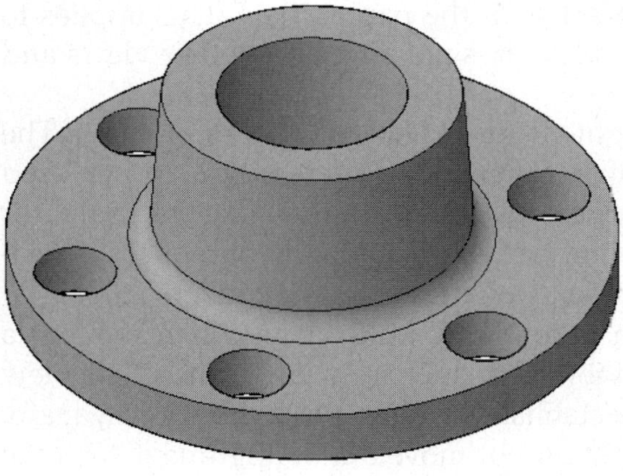

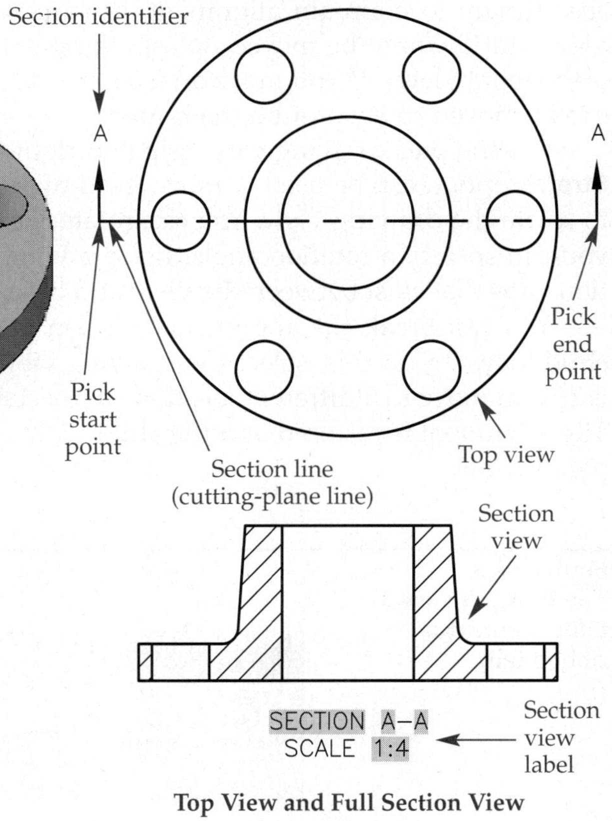

Top View and Full Section View

B

section identifier is automatically incremented when you place additional section views. The text objects used for the section view label contain fields that update according to changes made to the section view. The appearance of elements in the section identifier and section view label is controlled by the section view style. A ***section view style*** defines settings such as the text style and height, direction arrow size and length, and hatch pattern used for sectioning. A section view style is similar to a dimension style and includes similar controls.

Ribbon
Layout
> Styles and Standards
Section View Style
Type
VIEWSECTIONSTYLE

VIEWSECTIONSTYLE

The **VIEWSECTIONSTYLE** command is used to create and modify section view styles. This command accesses the **Section View Style Manager** dialog box, **Figure 14-10**. Picking the **New...** button allows you to create a new section view style. Picking the **Modify...** button opens the **Modify Section View Style** dialog box for the selected style. The tabs in the **New Section View Style** dialog box or the **Modify Section View Style** dialog box are used to make settings for the section identifier, cutting-plane line, section view label, and section line hatching. See **Figure 14-11**. You can apply and modify section view styles during the design process or after a section view is created. Develop standards for section views similar to the standards you develop for text and dimensions. Section view styles should follow company or industry standards.

Once created, section views can be edited by editing the section line and editing properties of the section view, such as the hatch pattern used for the section. The following sections discuss techniques for creating and editing section views.

NOTE

New layers are created by AutoCAD for the drawing view geometry when a section view is created. The section line and section view label are placed on a layer named MD_Annotation. The section pattern is placed on a layer named MD_Hatching.

Full Section

A full section "cuts" the parent view of the object in half. It is created by making a cut completely through the object. Refer to **Figure 14-9**.

Ribbon
Layout
> Create View
Section View Full
Type
VIEWSECTION

VIEWSECTION

To create a full section, select the **VIEWSECTION** command and then select the parent view. The parent view can be a base view or a projected view. Next, select the **Type** option and then select the **Full** option. This option can be accessed directly from the ribbon by selecting the button from the **Section** drop-down list in the **Create View** panel of the **Layout** tab. Once you select the parent view, you are prompted to specify the start point of the section line. Use object snaps and object snap tracking to assist in

Figure 14-10.
The **Section View Style Manager** dialog box is used to create and modify section view styles.

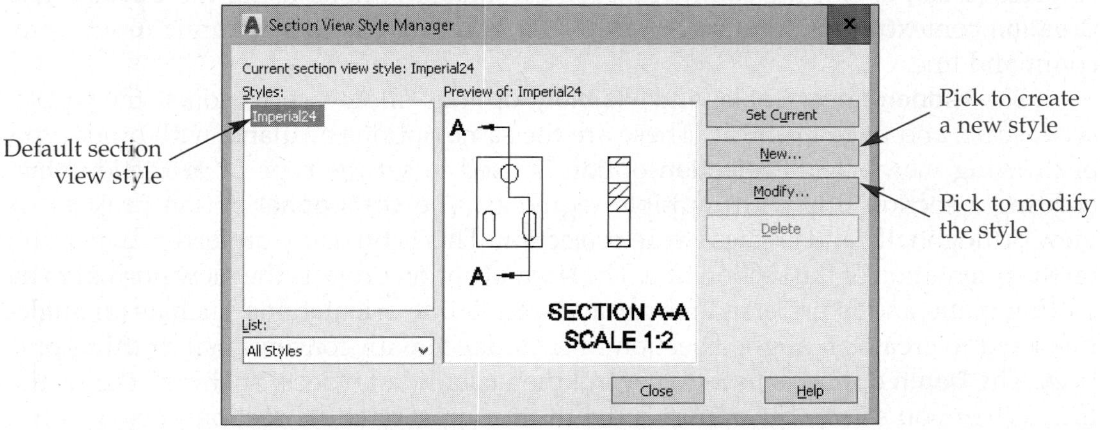

Figure 14-11.
The tabs in the **Modify Section View Style** dialog box are used to modify the appearance of the section identifier, the section line (cutting-plane line), the view label, and the hatch pattern applied to the section view.

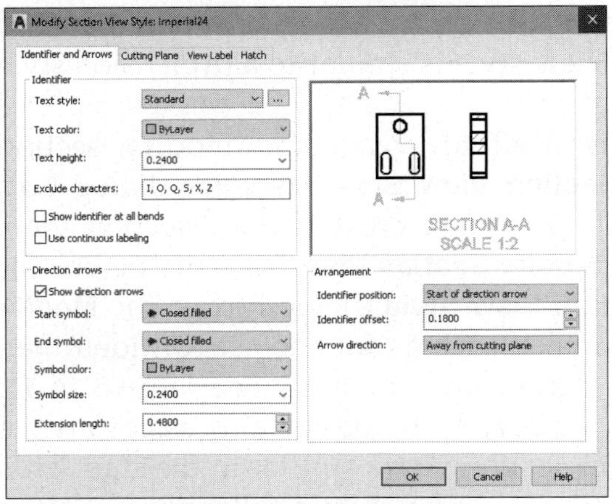

Identifier and Arrows tab

A

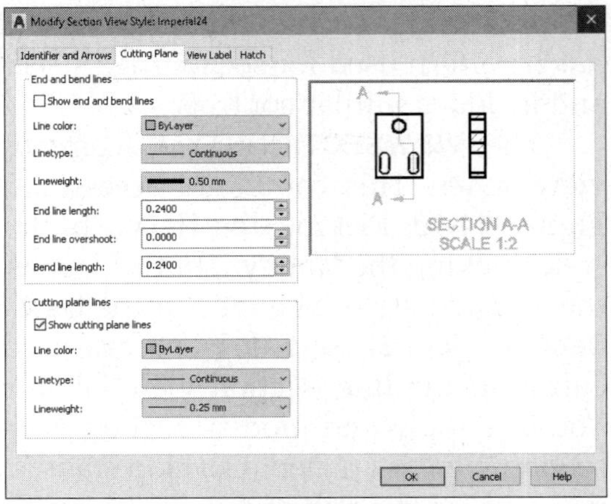

Cutting Plane tab

B

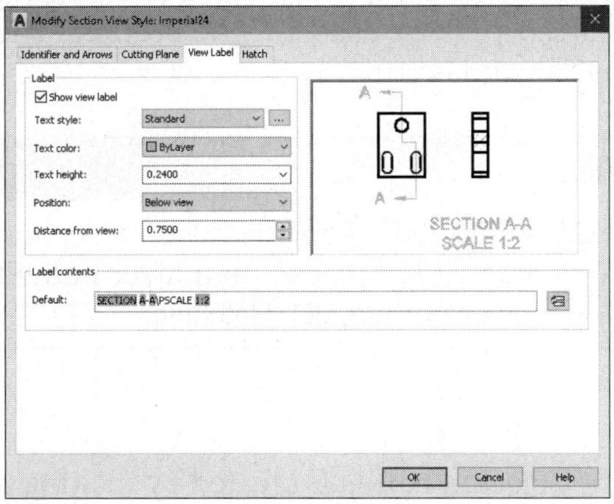

View Label tab

C

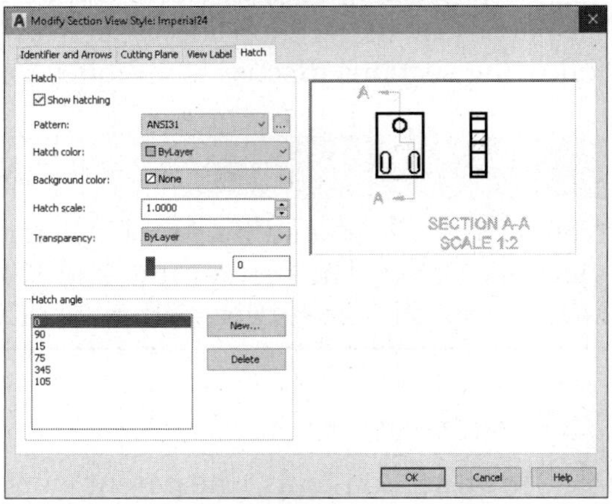

Hatch tab

D

specifying the start point. Then, drag the cursor and pick the end point of the section line. The section line is created and a preview of the section view is aligned perpendicular to the section line. The dragging direction from the section line determines the viewing direction. Move the section view to the desired location and pick. To break the alignment between views, press the [Shift] key once. To restore the alignment, press the [Shift] key again. When you pick a location for the view, you can select an option or press [Enter] to exit the command. You can adjust options using the **Section View Creation** contextual ribbon tab, **Figure 14-12**. You can also use dynamic input or the command line.

The **Hidden lines**, **Scale**, and **Visibility** options allow you to adjust the display style, scale, and edge visibility. These are the same options available with other types of drawing views. The **Projection** option is used to set the type of projection when creating a section line with multiple segments. The **Orthogonal** option projects the view orthogonally and creates a true projection. This is typically preferred, depending on the orientation of the section line. The **Normal** option projects the view normal to the cutting plane and is preferred for certain section line orientations, such as an angled line used to create an aligned section in accordance with conventional drafting practices. The **Depth** option is used to control the visibility of objects "behind" the section line. When you select this option, a depth line appears at the section line. Hovering

Figure 14-12.
The **Section View Creation** contextual ribbon tab.

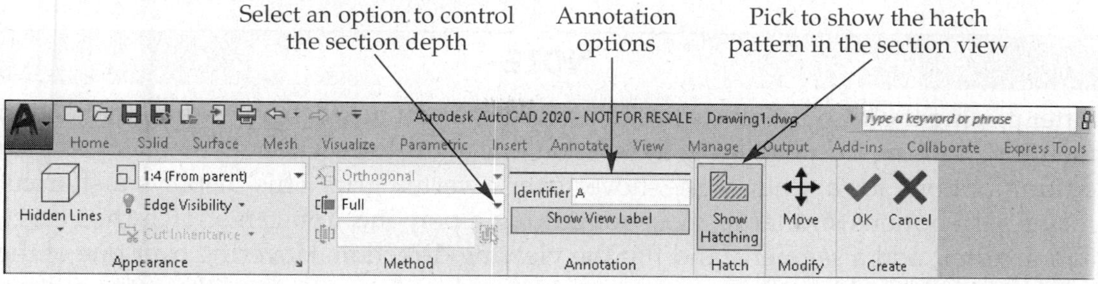

over this line and dragging allows you to set the depth of the section view. Objects that are behind the depth line will not be visible in the section view. Selecting the default **Full** option includes all objects within the section view. Selecting the **Slice** option removes all objects behind the section line, creating a thin representation section view. The **Slice** option may be practical for special section view documentations.

The **Annotation** option allows you to enter the text used for the section identifier and specify whether a view label is shown. As previously discussed, the section identifier is automatically incremented when creating additional section views. The **Hatch** option is used to specify whether a hatch pattern is used for the section view. The **Move** option allows you to adjust the location of the view after selecting the initial position. When using this option, you can press the [Shift] key to break the alignment between views.

Section views can be used to create projected views. In **Figure 14-13**, an isometric view has been created from a full section view to show an isometric representation

Figure 14-13.
An isometric section view is added to the layout by projecting a view from the orthographic section view (front view). The display style of the isometric view is changed to **Shaded with Visible Lines**.

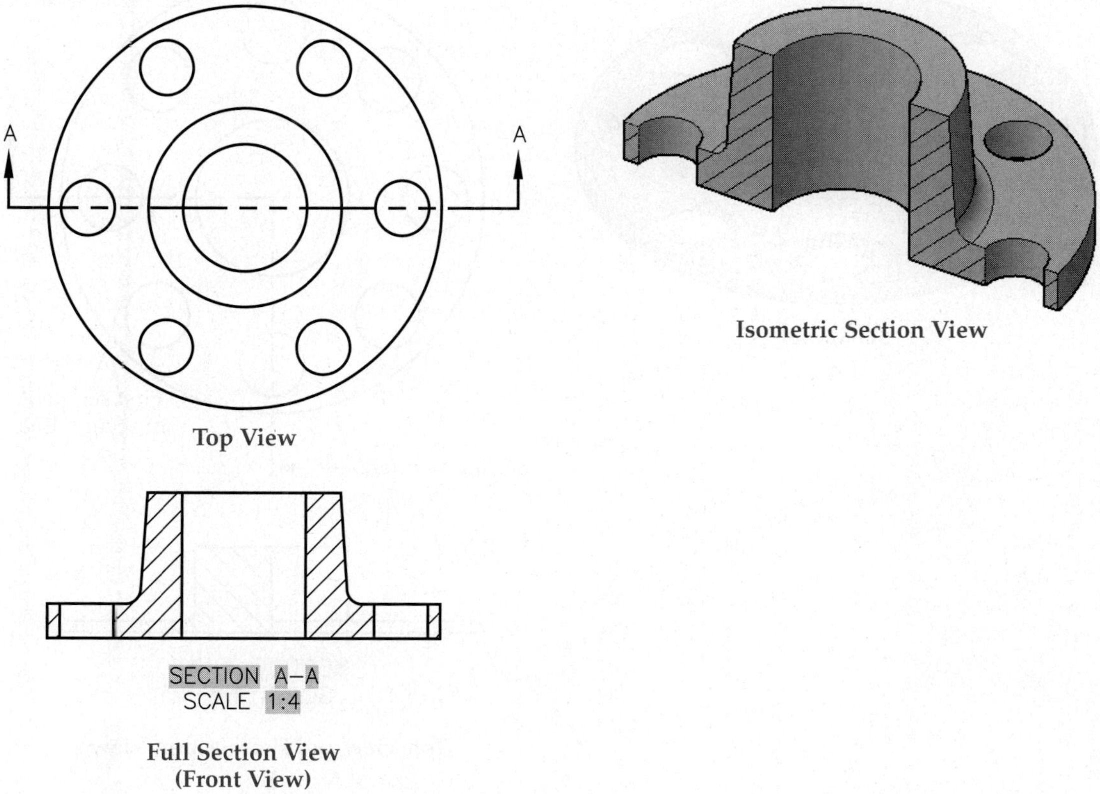

of the "cut." By default, an isometric view projected from a section view shows the section and inherits the display properties of the parent view.

NOTE

After placing the section view, you can use grips to edit the section line and the section identifier. Editing the position of the section line alters the section view. To access grip editing options for a section line, hover over a vertex grip. This displays a shortcut menu with options similar to those for editing a polyline. You can stretch the vertex, add a vertex, add a segment, and flip the viewing direction. Hovering over one of the section identifier grips allows you to move the identifier with or without the section line and reset the identifier to the initial position. Editing section views is discussed in more detail later in this chapter.

Exercise 14-2

www.g-wlearning.com/CAD/

Complete the exercise on the companion website.

Half Section

A half section is half of a full section. It represents one-quarter of the object cut away. See **Figure 14-14**. Half sections are most typically used for symmetrical objects. The half of the object that is not sectioned is usually shown as a solid object with no hidden lines.

Figure 14-14.
Creating a half section view. A—The model drawing. B—After creating the top view as a base view, the half section view is created by drawing a section line consisting of two segments.

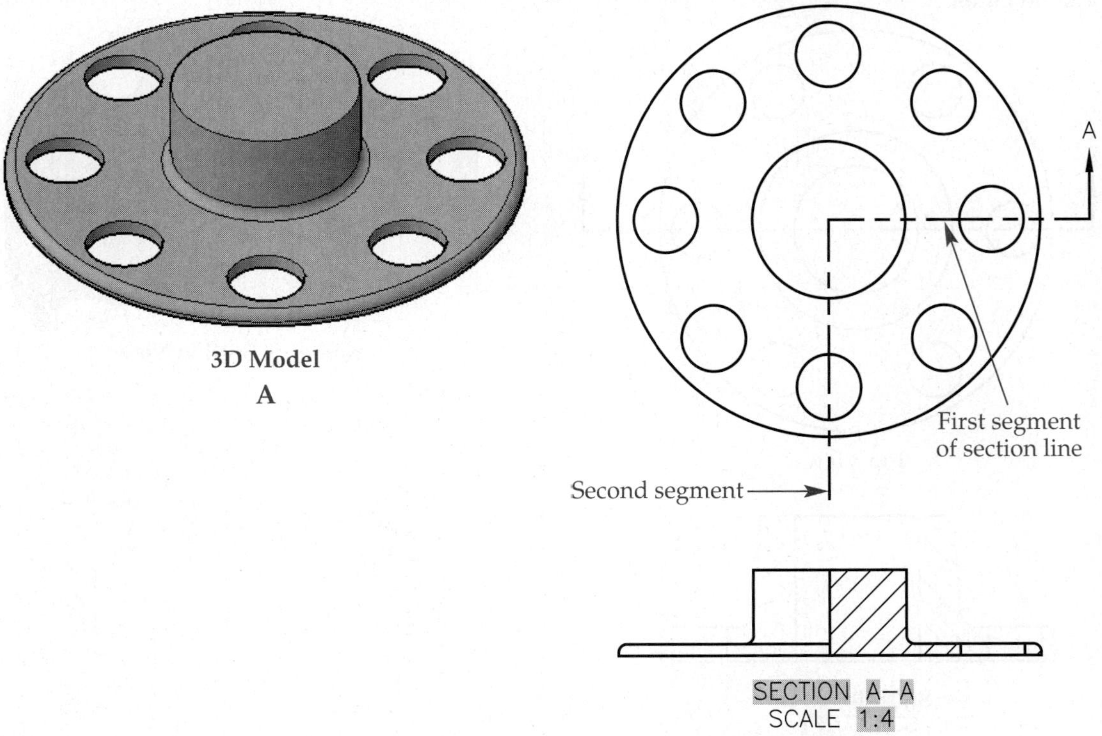

3D Model
A

Top View and Half Section View
B

To create a half section, select the **VIEWSECTION** command and select the parent view. Next, select the **Type** option and then select the **Half** option. The option is accessed directly from the ribbon by selecting the button from the **Section** drop-down list in the **Create View** panel of the **Layout** tab. Three points are required to define the section line. Use object snaps and object snap tracking to assist in specifying each point. If you pick an incorrect point, use the **Undo** option. After drawing the final segment of the section line, pick to locate the view. You can then select an option or press [Enter] to exit the command. You can adjust options using the **Section View Creation** contextual ribbon tab, dynamic input, or the command line. The options are the same as those available when creating a full section.

Ribbon
Layout
> Create View
Section View Half
Type
VIEWSECTION

VIEWSECTION

An example of using the **Depth** option to adjust a half section is shown in **Figure 14-15**. In **Figure 14-15A**, the half section is created with the section line drawn through the middle of the object. In **Figure 14-15B**, the **Depth** option is selected to move the depth line to the back end of the hole feature. In this case, adjusting the depth of the section line helps clarify the interior detail of the part.

Exercise 14-3

www.g-wlearning.com/CAD/

Complete the exercise on the companion website.

Figure 14-15.
Using the **Depth** option to adjust the visibility of objects in a section view. A—A top view and half section view of a model. The **Depth** option is set to **Full**. B—The half section view is adjusted by moving the depth line in the top view. Notice that the back leg no longer appears. The depth line is shown for reference only and does not appear after exiting the command.

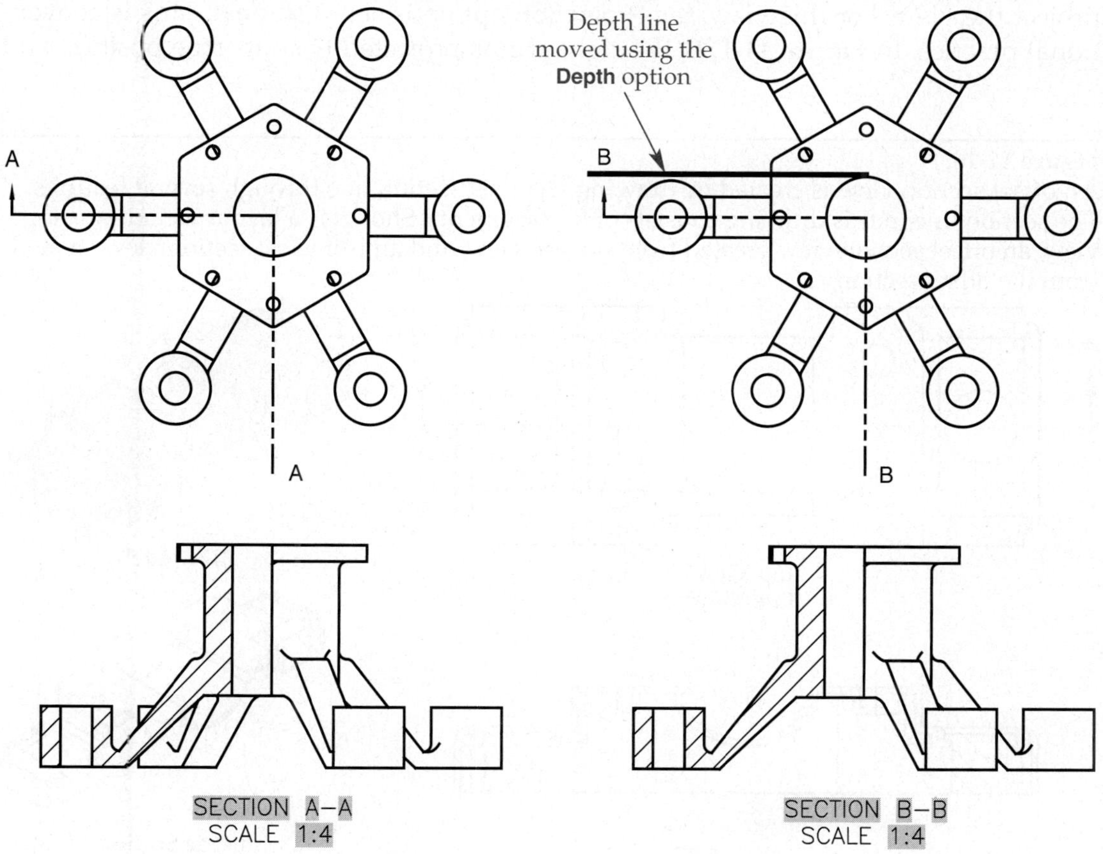

Half Section View
A

Half Section View after Adjusting the Depth
B

Offset Section

An offset section *shifts* (offsets) the section line to pass through certain features of a part or assembly for better clarification of detail. Typically, the section line consists of several segments drawn through features such as holes and bosses. See **Figure 14-16**.

VIEWSECTION

Ribbon
Layout
> Create View
Section View Offset
Type
VIEWSECTION

To create an offset section, select the **VIEWSECTION** command and select the parent view. Next, select the **Type** option and then select the **Offset** option. The option is accessed directly from the ribbon by selecting the button from the **Section** drop-down list in the **Create View** panel of the **Layout** tab. Then, pick the points to define the section line. Select as many points as needed to define the section. Use object snaps and object snap tracking as needed. If you pick an incorrect point, use the **Undo** option. After drawing the final segment of the section line, select the **Done** option. Then, pick to locate the view. You can then select an option or press [Enter] to exit the command. The options are the same as those available when creating a full section.

Exercise 14-4

www.g-wlearning.com/CAD/

Complete the exercise on the companion website.

Aligned Section

An aligned section is made by passing two nonparallel cutting planes through an object. The resulting section view shows features that are oriented at an angle rotated into the same cutting plane. See **Figure 14-17**.

The purpose of an aligned section is to show the true size and shape of a feature. In **Figure 14-17A**, an aligned section is created to show the true size and shape of the right arm. Notice that the right arm is rotated into the center cutting plane to project the view. For this view, the **Projection** option is set to **Normal**. This is conventional practice. In **Figure 14-17B**, the right arm is projected from its true position and

Figure 14-16.
An offset section view is created by drawing an offset section line through several features. The section line in this drawing consists of five segments. Shown is a layout including a top view, an offset section view created from the top view, and an isometric section view created from the offset section.

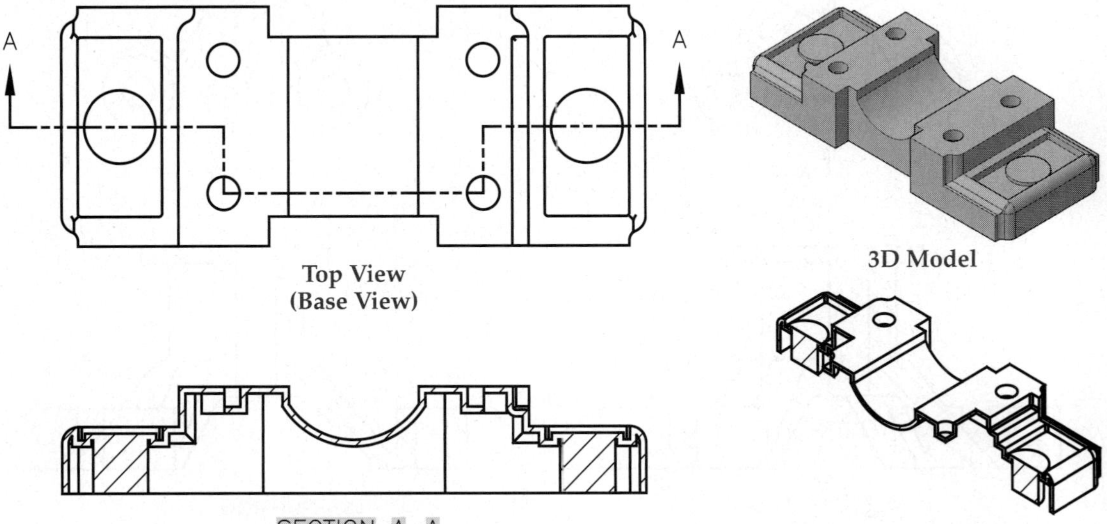

Top View (Base View)

3D Model

Offset Section View (Front View)

Isometric Section View

Figure 14-17.
An aligned section view is created by drawing an angled section line through nonparallel features. A—The **Normal** projection method is used to project the right arm in its true shape and size. Notice that the right arm rotates into the center cutting plane to make the projection. This is conventional practice. B—Using the **Orthogonal** option creates a true projection of the right arm and results in foreshortening in the section view.

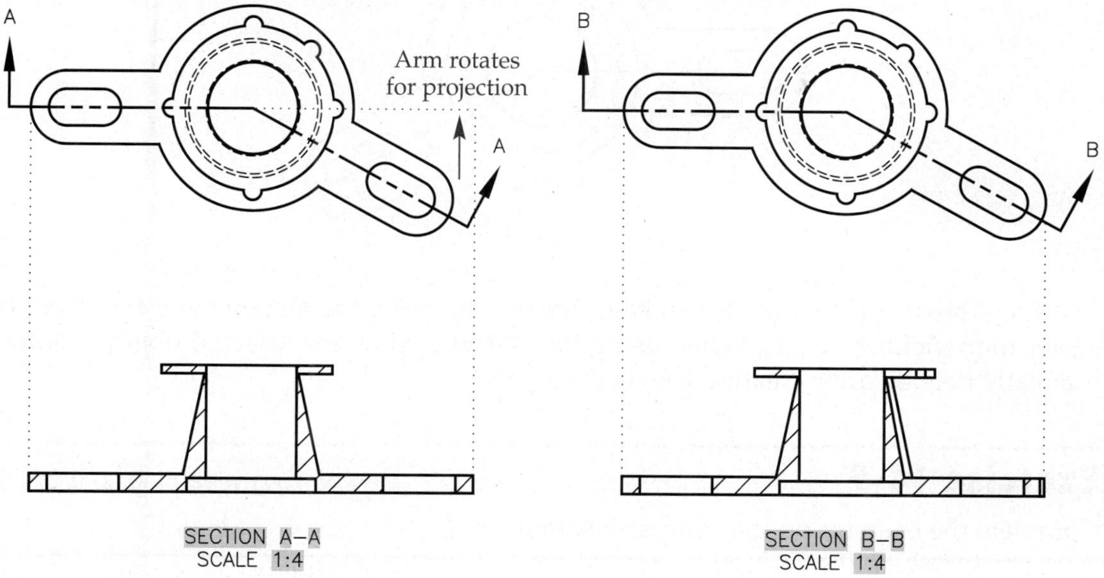

True Size and Shape Projection Conventional Practice (Preferred)

A

True Projection (Avoid)

B

appears foreshortened in the front view. For this view, the **Projection** option is set to **Orthogonal**. In this instance, this is *not* the preferred method to create the view.

To create an aligned section, select the **VIEWSECTION** command and select the parent view. Next, select the **Type** option and then select the **Aligned** option. The option is accessed directly from the ribbon by selecting the button from the **Section** drop-down list in the **Create View** panel of the **Layout** tab. Then, pick the points to define the section line. Use object snaps and object snap tracking as needed. After drawing the final segment of the section line, select the **Done** option. Then, pick to locate the view. You can then select an option or press [Enter] to exit the command. The options are the same as those available when creating a full section.

Ribbon
Layout
> Create View
Section View Aligned
Type
VIEWSECTION

VIEWSECTION

In **Figure 14-17**, notice that the angled segment of the section line extends beyond the lower right arm. To orient the end of the section line in this manner, it may be necessary to edit the section line using grips to stretch the vertex. Another option is to draw the section line first and use the **Object** option of the **VIEWSECTION** command to create the section. This is discussed next.

Creating a Section View from an Object

You can select an object in the paper space layout to use as the section line when creating a section view. This is a useful method when it is difficult to locate points using the **VIEWSECTION** command. To use an object as the section line, select the **VIEWSECTION** command and select the parent view. Next, select the **Type** option and then select the **Object** option. The option is accessed directly from the ribbon by selecting the **From Object** option from the **Section** drop-down list in the **Create View** panel of the **Layout** tab. Then, select the object and press [Enter]. Pick a point to locate the view. The object you select determines the type of section created. In **Figure 14-18**, a polyline is drawn in the desired location prior to accessing the **Object** option of the **VIEWSECTION** command.

Ribbon
Layout
> Create View
Section View Object
Type
VIEWSECTION

VIEWSECTION

Figure 14-18.
A section view can be created from an object using the **Object** option of the **VIEWSECTION** command. The polyline shown here provides an alternate way to create the aligned section for the object shown in Figure 14-17.

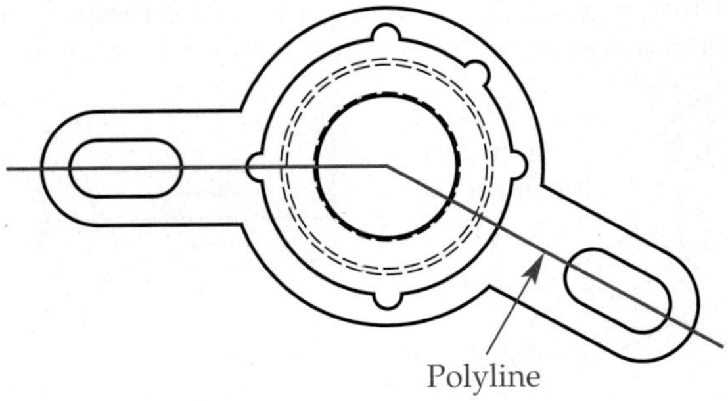

The **Object** option provides an alternate way to create the section view and may be easier than picking points. When using the **Object** option, the selected object is automatically deleted after creating the section view.

Exercise 14-5

www.g-wlearning.com/CAD/

Complete the exercise on the companion website.

Constraining Section Lines

When picking points to define a section line, you have the option to apply constraints to control the location of the line. If **Infer Constraints** is activated, constraints are automatically applied when you pick points. The **Infer Constraints** button on the status bar controls whether constraints are inferred when drawing new objects. The **Infer Constraints** button does not appear on the status bar by default. Select the **Infer Constraints** option in the status bar **Customization** flyout to add the **Infer Constraints** button to the status bar. Inferring constraints provides a way to constrain the geometry of the section line to the geometry in the drawing view. This is similar to applying constraints automatically when drawing objects in model space.

VIEWSYMBOLSKETCH

Ribbon
Layout
> Modify View
View Sketch
Type
VIEWSYMBOL-SKETCH

Constraints can also be applied to the section line after creating the section view. The **VIEWSYMBOLSKETCH** command is used to constrain a section line using geometric and dimensional constraints. After accessing this command, select the section line. The **Parametric** tab is displayed and symbol sketch mode is activated. In this mode, objects other than the drawing view geometry and section line are faded. A polyline replaces the section line. You can use the appropriate tools to constrain the section line to the drawing view geometry. You can also add construction geometry if needed to apply constraints. Geometry that is added is only displayed in symbol sketch mode and does not display when you exit the command. When finished adding constraints, pick the **Finish Symbol Sketch** button in the **Exit** panel of the **Parametric** tab. When prompted, specify whether to save or discard changes made in symbol sketch mode.

PROFESSIONAL TIP

Constrain the section line if future design modifications change the drawing view geometry size or location. For example, apply a fix constraint to constrain the section line to the center of the sectioned object.

Excluding Components from Sectioning

Certain features in section views, such as fasteners, are not shown sectioned. For example, components such as screws, pins, and thin-walled objects in an assembly are shown without section lines. This practice conforms with drafting standards. When creating a section view from a parent view that includes items such as fasteners and shafts, you can use the **VIEWCOMPONENT** command to control how sectioning is applied.

To exclude items from sectioning, access the **VIEWCOMPONENT** command. Select the components in the parent view and then select the **None** option. If you select the **Section** option, the components are included in sectioning. Selecting the **Slice** option creates a thin section at the section line and removes the visibility of objects behind the section line.

Ribbon
Layout
> Modify View

Edit Components
Type
VIEWCOMPONENT

VIEWCOMPONENT

Editing Section Views

The **VIEWEDIT** command, introduced earlier in this chapter, can be used to edit the properties of a section view. You can quickly initiate this command by double-clicking on a section view. This displays the **Section View Editor** contextual ribbon tab. Many of the same options used when creating a section view are available in this tab. Additional options may be available depending on the type of view selected.

The **Cut Inheritance** option is available when a view created from a section view inherits the section cut. An example of this is an isometric view projected from a section view. By default, the isometric view shows the section cut with the sectioned portion hatched. To remove the section cut from the isometric view, expand the **Cut Inheritance** drop-down list in the ribbon and uncheck the **Section cut** option. This resets the view to an isometric view without sectioning.

The **Defer Updates** option, available in the **Edit** panel of the **Section View Editor** contextual ribbon tab, allows you to make edits without previewing the changes dynamically. The effects of changes you make are not displayed until you exit the command by picking the **OK** button in the **Edit** panel. By default, the **Defer Updates** option is inactive. For very large drawings, activating this option may save time by preventing dynamic previews from occurring. For very small drawings, deferring updates provides only a marginal benefit.

As previously discussed, you can edit a section view by using grips to edit the location of the section line. If a section line is constrained, press the [Shift] key to relax constraints and break the alignment with the constrained position.

NOTE

The hatch pattern applied to the section view can be edited in the same manner as a hatch pattern applied in a 2D drawing. Double-click on the hatch to display the **Drawing View Hatch Editor** contextual ribbon tab. You can also hover over the center grip of the hatch pattern to access grip editing options.

Exercise 14-6

www.g-wlearning.com/CAD/

Complete the exercise on the companion website.

Detail Views

A *detail view* shows a selected portion of a view to clarify model details. A detail view is projected from a parent view and is typically shown at a larger scale. As with other types of projected views, the detail view is linked to the parent view.

Detail views are created using the **VIEWDETAIL** command. A detail view is created by drawing a circular or rectangular boundary to define the extents of the view. You can create detail views from an AutoCAD 3D model or an Autodesk Inventor model.

Detail views are associative. As with other types of drawing views, detail views are updated automatically when model changes are made if the **VIEWUPDATEAUTO** system variable is set to 1.

Examples of detail views are shown in **Figure 14-19**. Creating a detail view places a *detail boundary* representing the "cutout" area in the parent view. The detail boundary includes a detail identifier. The detail view is placed in another location on the drawing and includes a view label. The detail identification is automatically incremented when you place additional detail views. The appearance of elements in the detail identifier and detail view label is controlled by the detail view style. A *detail view style* defines settings such as the text style and height, symbol size, and detail boundary appearance.

VIEWDETAILSTYLE

Ribbon
Layout
> Styles and Standards
Detail View Style
Type
VIEWDETAILSTYLE

The **VIEWDETAILSTYLE** command is used to create and modify detail view styles. This command accesses the **Detail View Style Manager** dialog box, **Figure 14-20**. Picking the **New...** button allows you to create a new detail view style. Picking the **Modify...**

Figure 14-19.
A layout of drawing views of a chair including the top, front, and side views and two detail views. The detail views have a larger scale and show details of the seat and leg. Each detail view is drawn with a circular boundary.

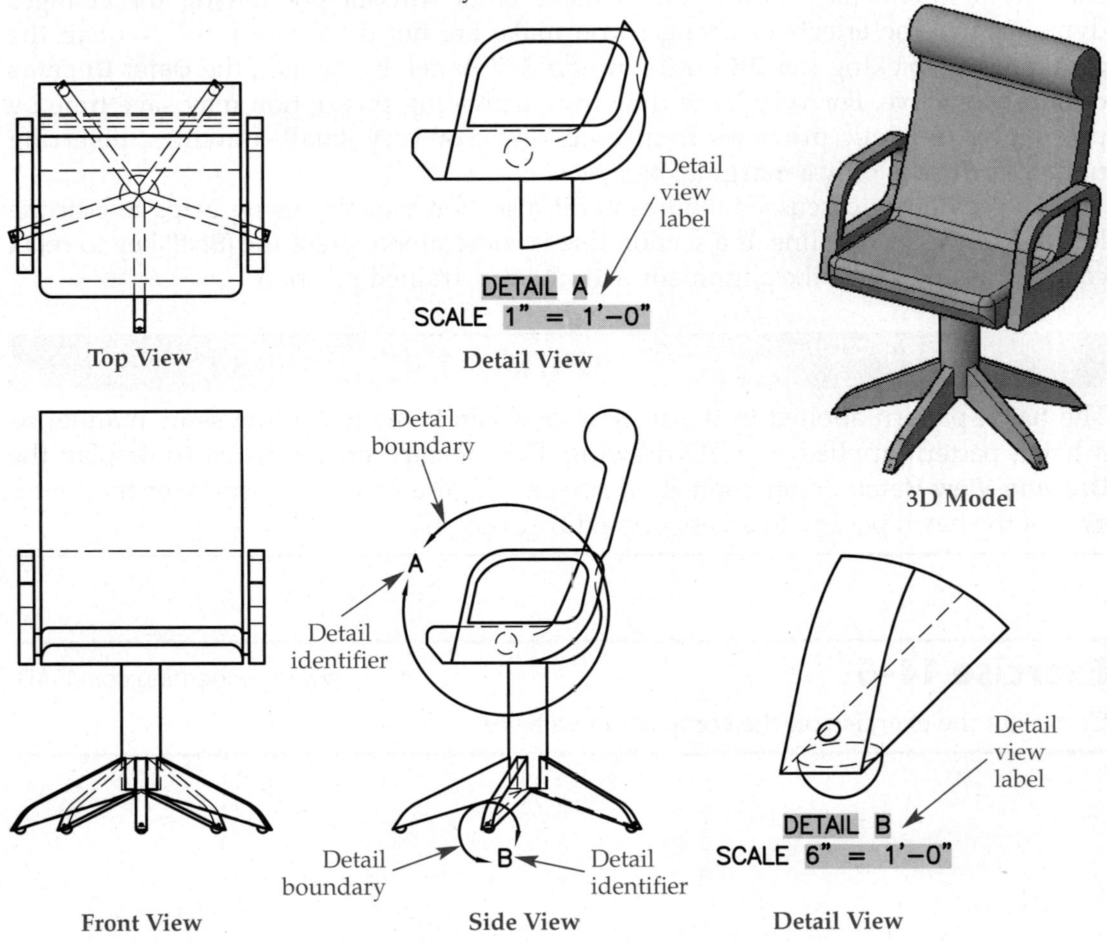

button opens the **Modify Detail View Style** dialog box for the selected style. The tabs in the **New Detail View Style** dialog box or the **Modify Detail View Style** dialog box are used to make settings for the detail identifier, detail boundary, and detail view label. These tabs are shown in **Figure 14-20**. As with other types of styles, develop standards for detail views in accordance with company or industry standards.

Creating a Detail View

Ribbon
Layout
> Create View
Detail View
Circular
Type
VIEWDETAIL

VIEWDETAIL

Ribbon
Layout
> Create View
Detail View
Rectangular
Type
VIEWDETAIL

VIEWDETAIL

Creating a detail view is similar to creating a section view. To create a detail view, select the **VIEWDETAIL** command and then select the parent view. The default method for creating the view is to create a circular detail boundary. Refer to **Figure 14-19**. This is the preferred display for most detail views. You can change the boundary type to rectangular using the **Boundary** option. With the **Rectangular** option, a rectangular detail boundary is drawn and the detail view has a rectangular outline. See **Figure 14-21**. Selecting one of the options from the **Detail** drop-down list in the **Create View** panel of the **Layout** tab on the ribbon begins the command and sets the appropriate boundary type. If you are creating a circular detail boundary, select the parent view and then pick a point to specify the center of the view. At the next prompt, drag the cursor or enter a value to set the size of the boundary. Then, pick a point to locate the view. A

Figure 14-20.
The **Detail View Style Manager** dialog box is used to create or modify detail view styles. A—Pick the **New...** button to create a new detail view style or pick the **Modify...** button to modify an existing style. B—The **Identifier** tab. C—The **Detail Boundary** tab. D—The **View Label** tab.

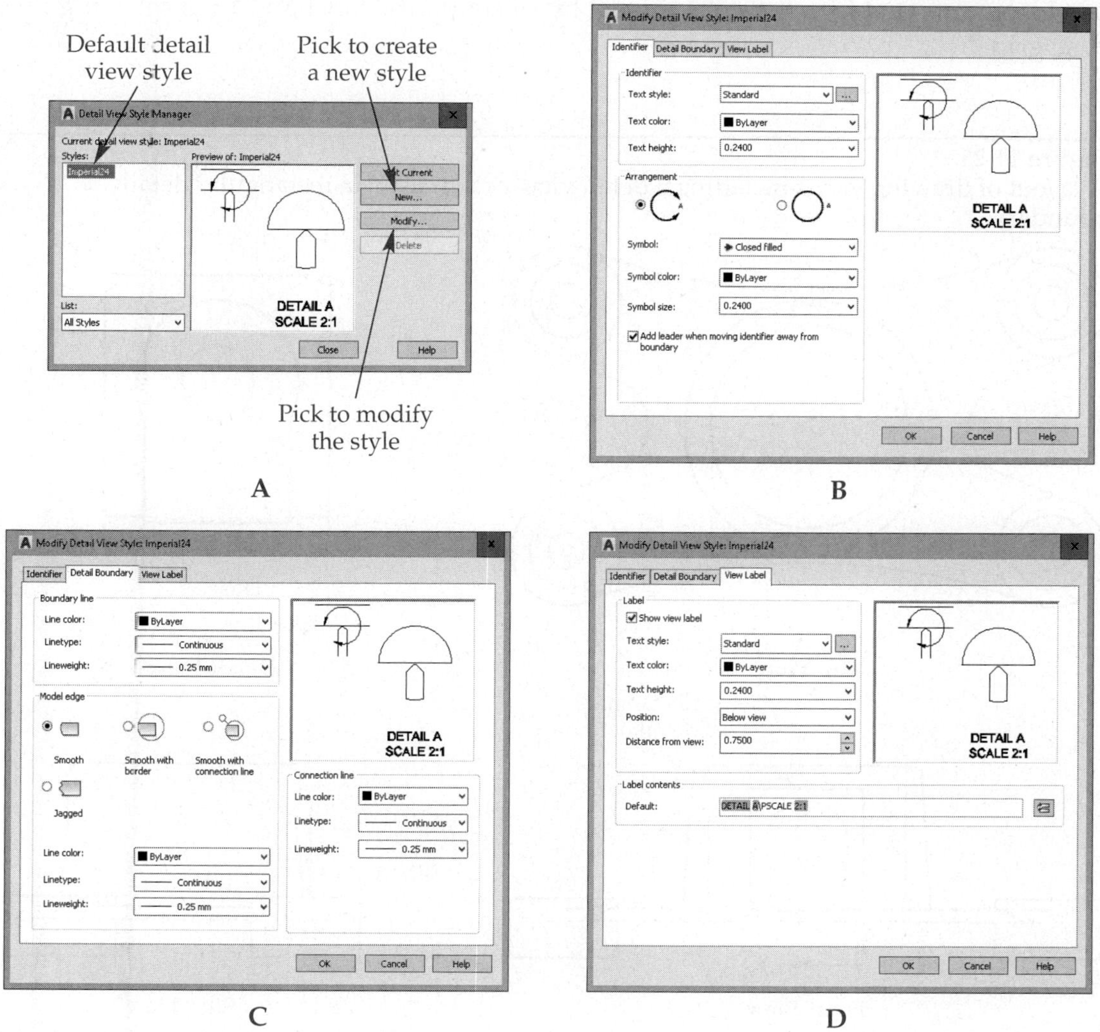

rectangular detail boundary is created in the same manner. After locating the view, you can select an option or press [Enter] to exit the command. You can adjust options using the **Detail View Creation** contextual ribbon tab. You can also use dynamic input or the command line.

The **Hidden Lines**, **Scale**, **Visibility**, and **Move** options are the same as those previously discussed for section views. The **Model Edge** option is used to adjust the edges of the detail view and set border display and leader options. See **Figure 14-22**. The **Smooth** option creates a smooth edge for the view. This is the default option. The **Smooth with Border** option creates a smooth edge for the view and draws a circular or rectangular border, depending on the type of boundary specified. The **Smooth with Connection Line** option creates a smooth edge, draws a circular or rectangular border, and attaches a leader from the detail symbol in the parent view to the detail view. The **Jagged** option creates the view with a jagged edge. With this option, no border is displayed and the view does not have a leader attached. The **Annotation** option allows you to adjust the view identifier and specify whether a view label is shown.

Once created, detail views can be edited by editing the detail boundary or detail identifier. To edit the detail boundary, select the boundary and hover over one of the four boundary grips to display a shortcut menu. The options allow you to stretch the boundary and change the boundary type to circular or rectangular. Hovering over the detail identifier grip allows you to move the identifier or reset the identifier to the initial position.

You can also edit a detail view by using the **VIEWEDIT** command or by selecting the view to display grips. A detail view has a base grip providing access to the standard grip editing options and a parameter grip for changing the scale. The detail view label is an mtext object. It can be moved by selecting the label and then selecting the base grip.

Figure 14-21.
A layout of drawing views including a detail view drawn using a rectangular detail boundary.

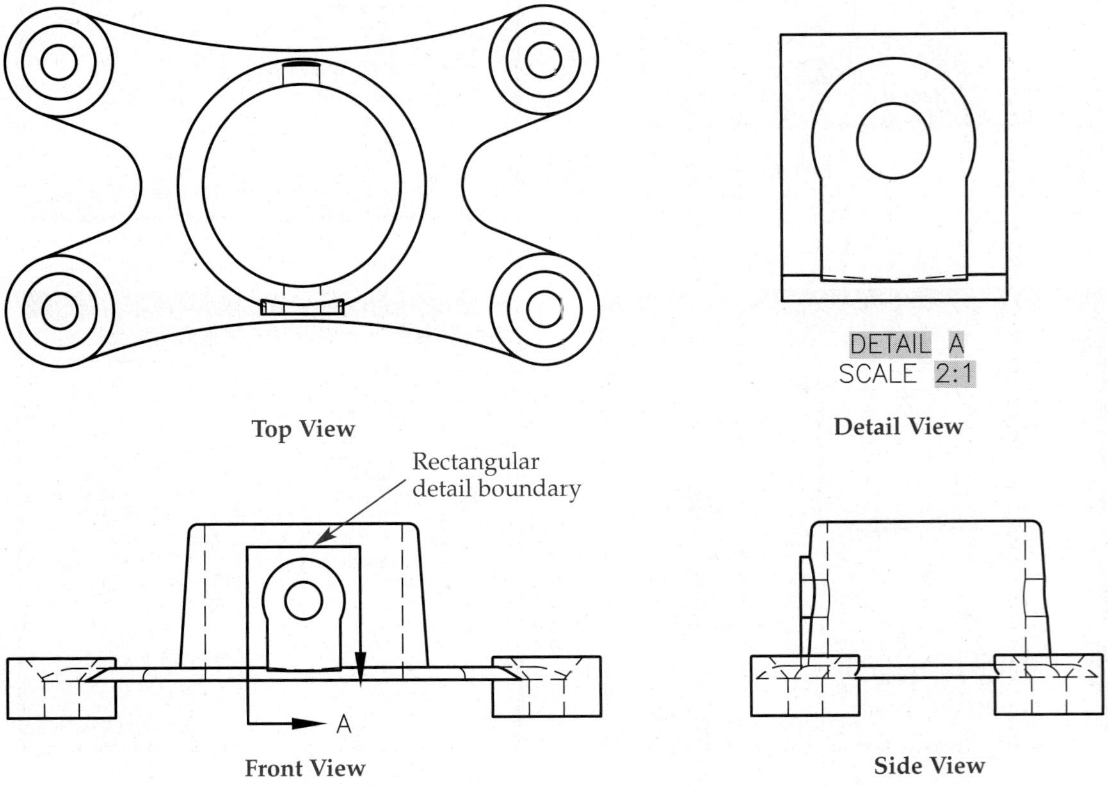

Figure 14-22.
The **Model Edge** option of the **VIEWDETAIL** command is used to adjust the model edges in the detail view. Included are settings for creating a border and a connection leader. A—**Smooth** option. B—**Smooth with Border** option. C—**Jagged** option. D—**Smooth with Connection Line** option.

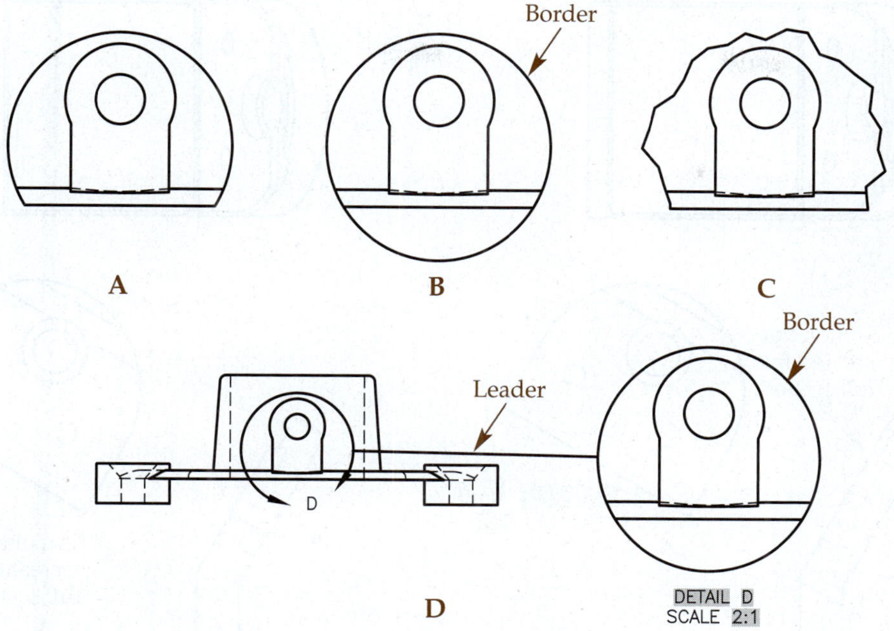

NOTE

The **Properties** palette can also be used to make modifications to the detail boundary and identifier. Deleting the detail boundary deletes the detail view.

Exercise 14-7

www.g-wlearning.com/CAD/

Complete the exercise on the companion website.

Auxiliary Views

An undocumented feature in AutoCAD is the ability to create an auxiliary view by using a section view. Often, a multiview drawing contains inclined surfaces that do not describe the true size or shape of features in a regular orthographic view. To establish an auxiliary view, you can draw a full section line using the **Full** option of the **VIEWSECTION** command. When specifying the section line, pick two points on the inclined surface. If needed, draw a parallel construction line across the inclined surface and use it to create the section line. The auxiliary view plane is oriented parallel to the inclined edge of the surface and the view is created perpendicular to the surface. To remove the display of the section line and view label from the drawing, freeze the MD_Annotation layer. See **Figure 14-23**. Notice that since the section line does not intersect the object, no hatch pattern is created.

Exercise 14-8

www.g-wlearning.com/CAD/

Complete the exercise on the companion website.

Figure 14-23.
An auxiliary view can be created to show the true shape of an inclined surface by drawing a full section view. A—A full section view is generated from the front view of the object by drawing a section line on the inclined surface. B—The completed drawing after freezing the MD_Annotation layer.

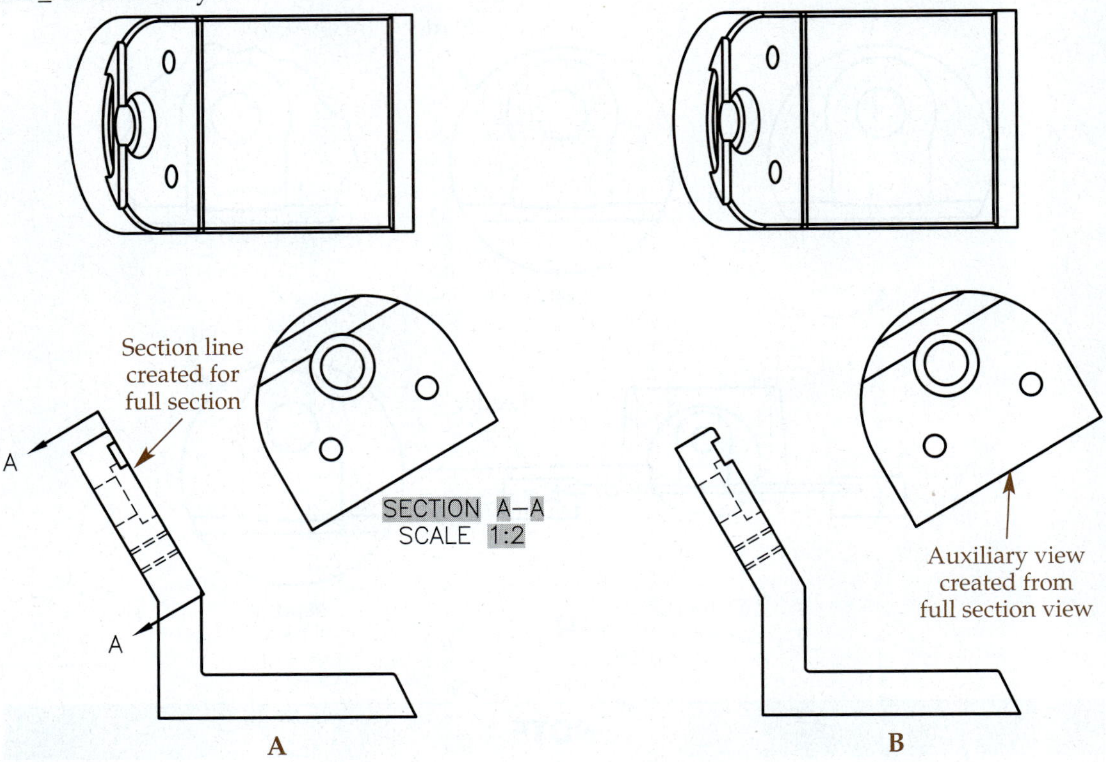

Dimensioning Drawing Views

After placing drawing views on a paper space layout, you can dimension each view as needed in the layout. See **Figure 14-24**. When dimensioning drawing views, make sure that the **DIMASSOC** system variable is set to 2 so that associative dimensions are created. Associative dimensions are associated to the dimensions of the model and update when the physical model changes. However, depending on edits to the model geometry, some dimensions can become disassociated.

Associative dimensions added to a drawing view are attached to the underlying object geometry. As the size or location of an object changes, the associative dimensions automatically adjust to their new size or location. Associative dimensions can become disassociated when a 3D model is modified or updated and the dimensions describing the underlying geometry lose their attached location. The **Annotation Monitor** is used to monitor and notify you of any changes in the associativity of dimensions placed in drawing views. This feature can be turned on by picking the **Annotation Monitor** button on the status bar. Picking this button displays the **Annotation Monitor** icon to the right of the button. See **Figure 14-25A**. The **Annotation Monitor** is turned off by default, but it is automatically activated when a model is edited and dimensions are updated. The status of the **Annotation Monitor** is controlled by the **ANNOMONITOR** system variable.

The **Annotation Monitor** icon provides feedback regarding the state of associative dimensions. If any associative dimensions become disassociated, the icon turns red and a balloon notification appears. See **Figure 14-25B**. In addition, yellow alert icons appear in the layout next to the disassociated dimensions. You can click the balloon notification link to delete the dimensions all at once, or you can pick on individual alert icons in the layout to update the dimensions.

Figure 14-24.
A layout of dimensioned drawing views including a top view and a half section view used for the front view.

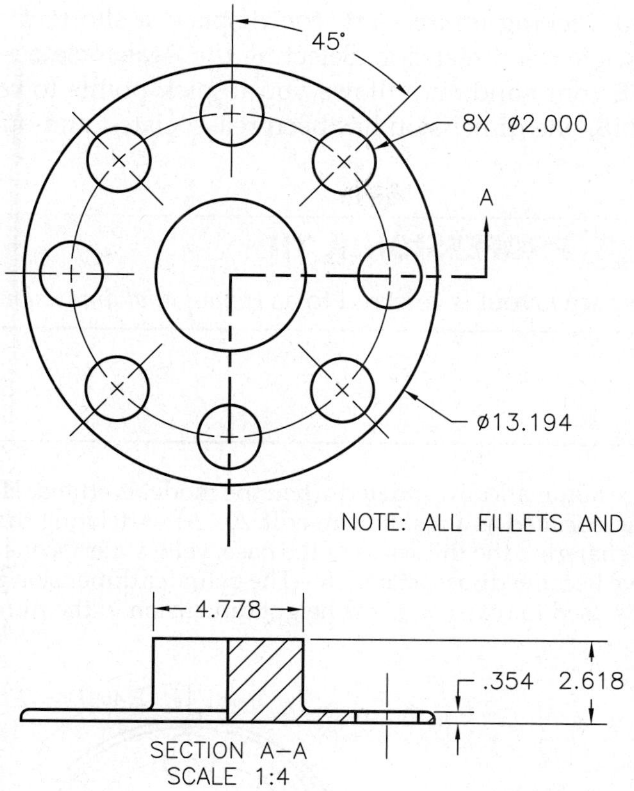

Figure 14-25.
The **Annotation Monitor** is used to monitor the status of associative dimensions placed in drawing views. A—Picking the **Annotation Monitor** button on the status bar turns the feature on and displays the **Annotation Monitor** icon as shown. B—The **Annotation Monitor** icon turns red and a balloon notification appears to indicate dimensions have become disassociated as a result of a model edit.

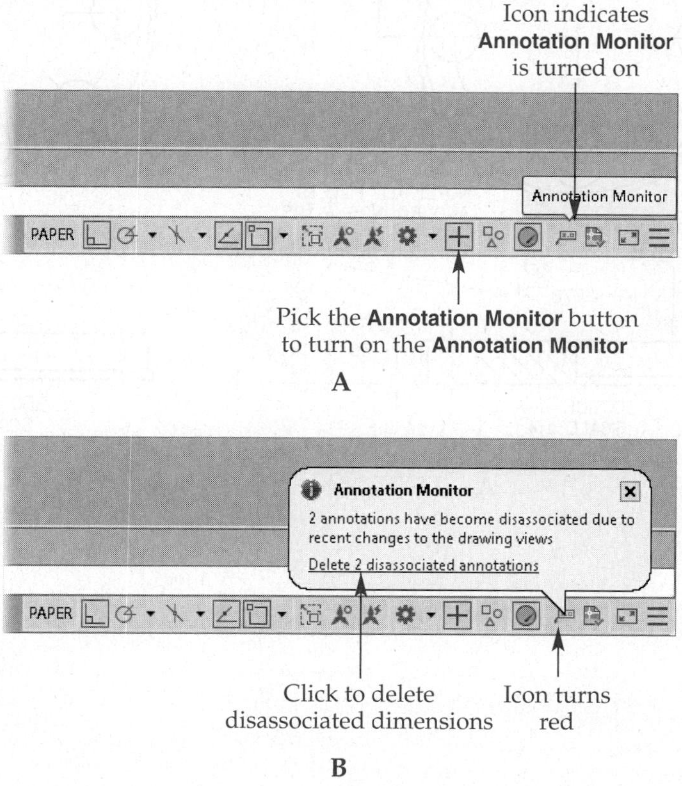

In **Figure 14-26A**, the model has been edited by deleting the center cylinder in **Figure 14-24** and changing the thickness of the base. The drawing views have been automatically updated and yellow alert icons appear to indicate the dimensions that have become disassociated. Picking on an alert icon displays a shortcut menu with options to reassociate or delete the dimension. Selecting the **Reassociate** option initiates the **DIMREASSOCIATE** command and allows you to pick points to redefine the dimension. In **Figure 14-26B**, one dimension has been reassociated and one has been deleted.

PROFESSIONAL TIP

Dimensioning in a paper space layout is referred to as *transspatial dimensioning*.

Figure 14-26.
By default, drawing views are automatically updated when the model is edited. However, dimensions may become disassociated as a result of an edit. A—After deleting the cylinder in the center of the part and changing the thickness of the base, yellow alert icons appear to indicate dimensions that have become disassociated. B—The cylinder dimension is deleted and the **Reassociate** option is used to reassociate the height dimension in the front view.

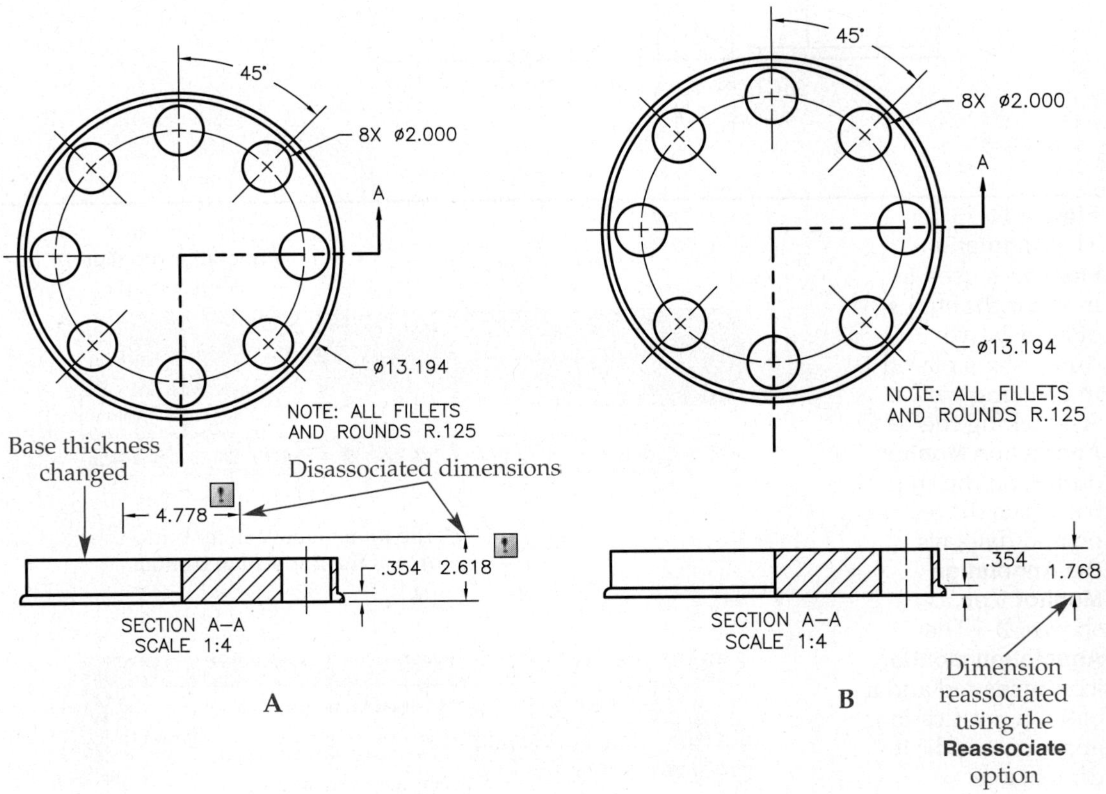

Using the EXPORTLAYOUT Command

You can use the **EXPORTLAYOUT** command to export a layout containing drawing views into a new drawing (DWG) file. However, this technique will break the associativity between the 3D model and the drawing views. The advantage to using this technique is that the drawing geometry can be edited in the same way as any other AutoCAD 2D geometry in model space. The disadvantage is that the 2D geometry has lost any associativity back to the 3D model. When using the **EXPORTLAYOUT** command, you save the exported layout as a DWG file. The drawing views become blocks in the new drawing file.

NOTE

The **FLATSHOT** command can also be used to create a multiview orthographic drawing from a 3D model. However, the resulting drawing views do not have associative properties. The **FLATSHOT** command creates a flat projection of the 3D objects in the drawing from the current viewpoint. The view is created in model space. The view that is created is composed of 2D geometry and is projected onto the XY plane of the current UCS.

Creating Section Planes

The **SECTIONPLANE** command offers a powerful visualization and display tool. It allows you to construct a section plane, known as an AutoCAD ***section object***, that can then be used as a cutting plane to cut through a 3D model. Section planes can be used to cut through solids, surfaces, meshes, regions, and point clouds. Working with point clouds is discussed in Chapter 15.

Once the section object is drawn, it can be moved to any location, jogs can be added to it, and it can be rendered "live" so that internal features and sectioned material are dynamically visible as the cutting plane is moved. The section object stores object property information. A variety of settings allow you to customize the appearance of section features. Additionally, you can generate 2D sections/elevations or 3D sections that can be inserted into the drawing as a block.

Once the **SECTIONPLANE** command is initiated, you are prompted to select a face or the first point on the section object. The **Draw section** option allows you to define a section object by picking multiple points. The **Orthographic** option allows you to locate a section object by specifying an orthographic plane. The **Type** option is used to set the initial display state for the section object when it is created. These options and additional sectioning options are discussed in the following sections.

SECTIONPLANE
Ribbon
Home > Section
Solid > Section
Section Plane
Type
SECTIONPLANE SPLANE

Picking a Face to Construct a Section Plane

The simplest way to create a section plane is to pick a flat face on the 3D object. Access the **SECTIONPLANE** command, move the pointer to the face you wish to select, and then pick it. A transparent section object is placed on the selected face and the model is cut at the plane. See **Figure 14-27**. The section plane can now be moved to create a section anywhere along the 3D model.

The automatic cut that appears on the model when creating a section object is the default display and is called *live sectioning*. Live sectioning is discussed later in this chapter. To turn live sectioning on or off, select the section object and pick the **Live Section** button in the **Section Plane** contextual ribbon tab. See **Figure 14-28**. You can also turn live sectioning on or off by selecting the section object, right-clicking, and

Figure 14-27.
Creating a section object on a face. A—The object before the face is selected. B—The section object is created.

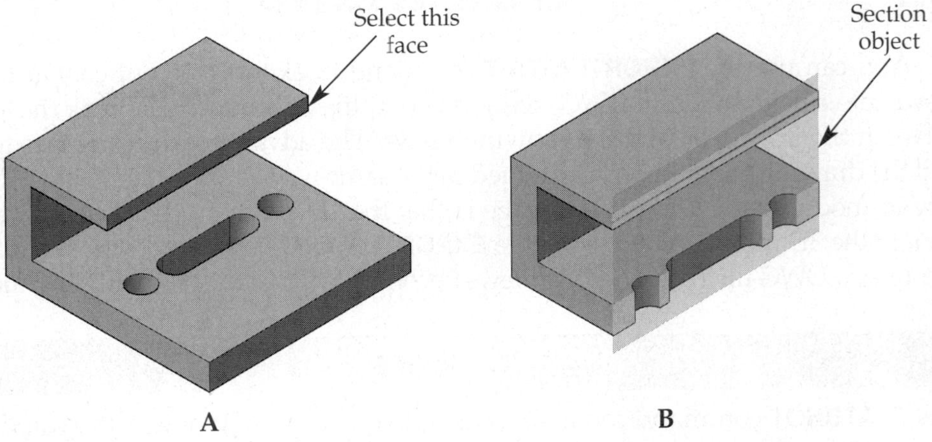

Figure 14-28.
The **Section Plane** contextual ribbon tab contains display settings and editing options for section objects.

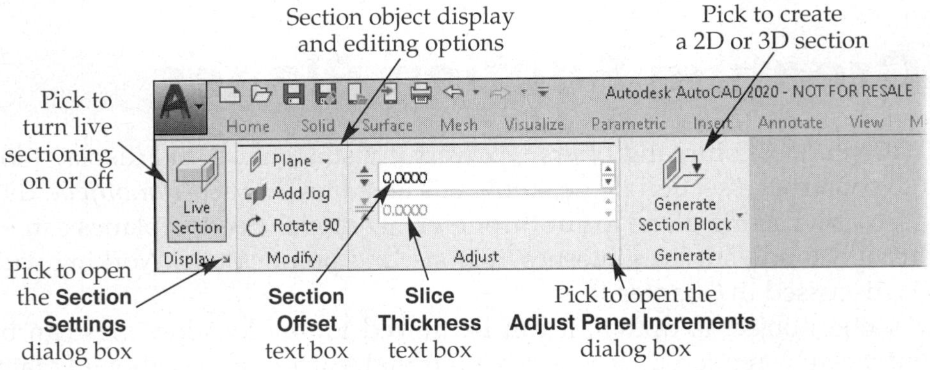

selecting **Activate live sectioning** from the shortcut menu. The **Section Plane** contextual ribbon tab and the shortcut menu contain options for editing section objects and generating section views, as discussed later in this chapter.

Picking Two Points to Construct a Section Plane

A second method for defining a section plane is to pick two points through which the section object passes. The section object is perpendicular to the XY plane of the current UCS. When the command is initiated, pick the first point, which cannot be on a face. See P1 in Figure 14-29. It may be best to turn off the dynamic UCS function or you could end up picking a face as the first point instead of a point. After picking the first point, move the pointer and notice that the section plane rotates about the first point. Next, pick the second point (P2) to define a line that cuts through the model. After the second point is picked, the section object is created. The section plane extends just beyond the edges of the model.

Picking Multiple Points to Construct a Section Plane

The previous method accepts only two points to construct a single section plane. Using the **Draw section** option, you can specify multiple points in order to create section plane *jogs*. A section object drawn in this manner can represent an offset or aligned section plane. As discussed earlier in this chapter, offset and aligned sections are used for objects containing features lying in different planes.

Figure 14-29.
Creating a section object by selecting two points.

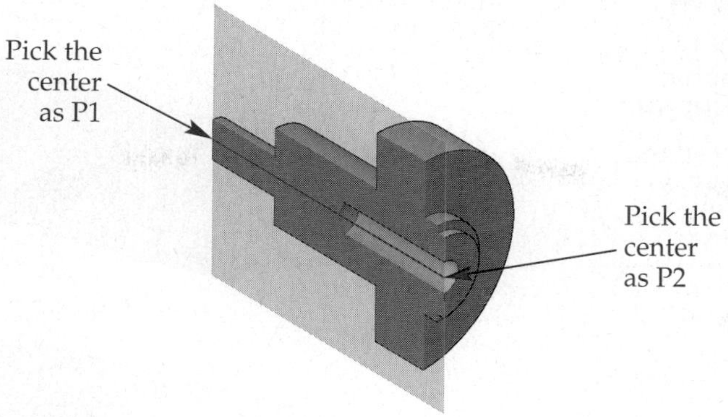

Once the command is initiated, select the **Draw section** option. Pick the start point, using object snaps if necessary. See **Figure 14-30**. Continue picking points as needed. After picking the last point to define the section plane, press [Enter]. You are then prompted to specify a point in the direction of the section view. This point is on the opposite side of the section object from the viewer. Pick a point on the model using object snaps if necessary. The section plane is created.

Exercise 14-9

www.g-wlearning.com/CAD/

Complete the exercise on the companion website.

Creating Orthographic Section Planes

The **Orthographic** option enables you to quickly place a section plane through the front, back, top, bottom, left, or right side of the object. See **Figure 14-31**. The origin is the center point of all objects in the model. Once the command is initiated, select the **Orthographic** option. Then, specify which orthographic plane you want to use as the section plane. The section object is created and all objects in the drawing are affected by it. You can move the section plane by selecting a grip and dragging, as discussed later in this chapter.

Figure 14-30.
Using the **Draw section** option of the **SECTIONPLANE** command to create a section object with multiple segments.

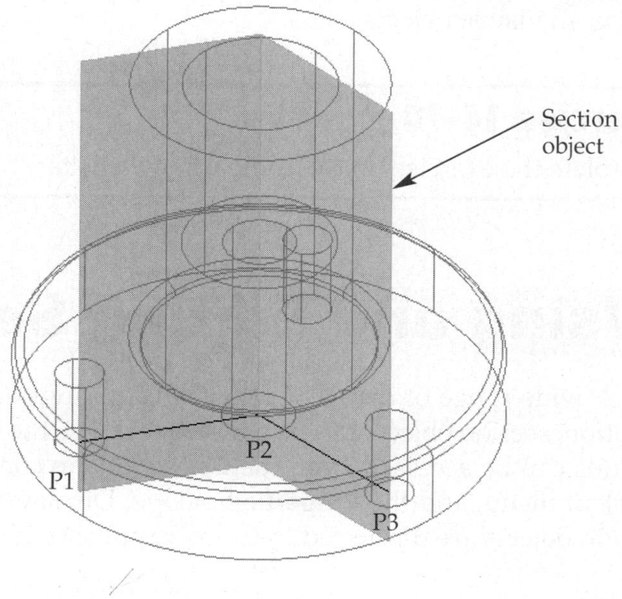

Figure 14-31.
Examples of orthographic **SECTIONPLANE** options. A—Top. B—Bottom. C—Left. D—Right.

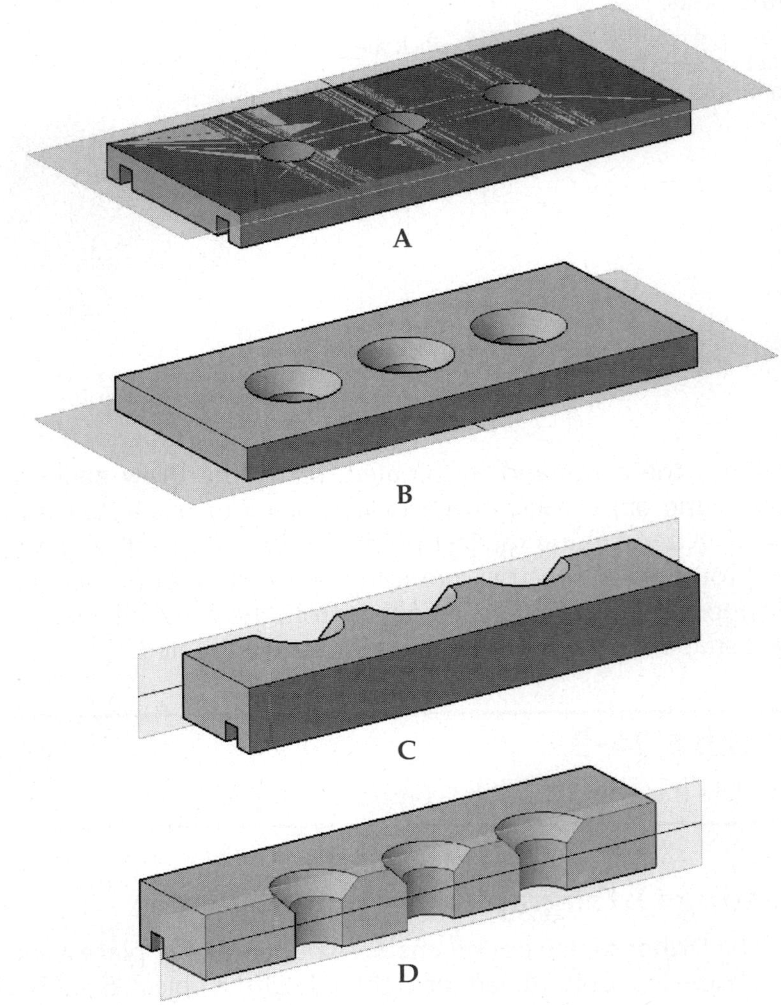

You may encounter a situation in which there is more than one solid on the screen and you want to use the **Orthographic** option to create a section object based on just one object. In this case, create a new layer, move objects you do not want to section to this layer, and then freeze the layer. The section object will be created based on the object that is visible. However, if the section plane passes through objects on frozen layers, those objects will be sectioned when the layers are thawed. A section plane affects all visible objects.

Exercise 14-10

www.g-wlearning.com/CAD/

Complete the exercise on the companion website.

Using and Editing Section Planes

A wide range of display options and settings are available for section objects. In addition, section objects can be edited to achieve the desired model display. Sectioning options can be accessed from the **Section Plane** contextual ribbon tab, the right-click shortcut menu, and the **Properties** palette. Display and editing functions applying to section objects are discussed in the following sections.

Section Object States

There are four possible states for the section object created by the **SECTIONPLANE** command—section plane, section slice, section boundary, and section volume. See **Figure 14-32**. The section object can be changed from one state to another. As previously discussed, the **Type** option can be used to set the section object state when using the **SECTIONPLANE** command. Depending on which state is active, the section object produces different results on the 3D model. To change the section object state, select the section object and then select an option from the **Section Type** drop-down list in the **Modify** panel in the **Section Plane** contextual ribbon tab. You can also pick the triangular menu grip on the section object to access the options, as discussed later in this chapter.

When the section object is created with the **Type** option set to **Plane**, the object is in the ***section plane state***. This is the default setting. A transparent plane is displayed on each segment of the section object. The transparent plane is called the ***section plane indicator***. A line called the ***section line*** connects the pick points (or the edges of the section object). The section plane extends infinitely in the section object's Z direction and along the direction of the object segment (unless connected to other segments). A section object with the section plane state active is shown in **Figure 14-32B**.

The ***section slice state*** creates a thin cut (slice) through the 3D model. See **Figure 14-32C**. A 2D box enclosing the sectioned object represents the thickness of the slice. To set the section slice state when creating the section object, set the **Type** option to **Slice**. To change the thickness of the slice, select the section object and change the value in the **Slice Thickness** text box in the **Adjust** panel in the **Section Plane** contextual ribbon tab.

When the section object is in the ***section boundary state***, a transparent plane is displayed on each segment of the section object. A 2D box extends to the XY-plane boundaries of the section object. See **Figure 14-32D**. The sectioned object fits inside of this footprint. The section plane extends infinitely in the section object's Z direction. To set the section boundary state when creating the section object, set the **Type** option to **Boundary**.

Figure 14-32.
Section object states. A—The original object. B—Section plane. C—Section slice. D—Section boundary. E—Section volume.

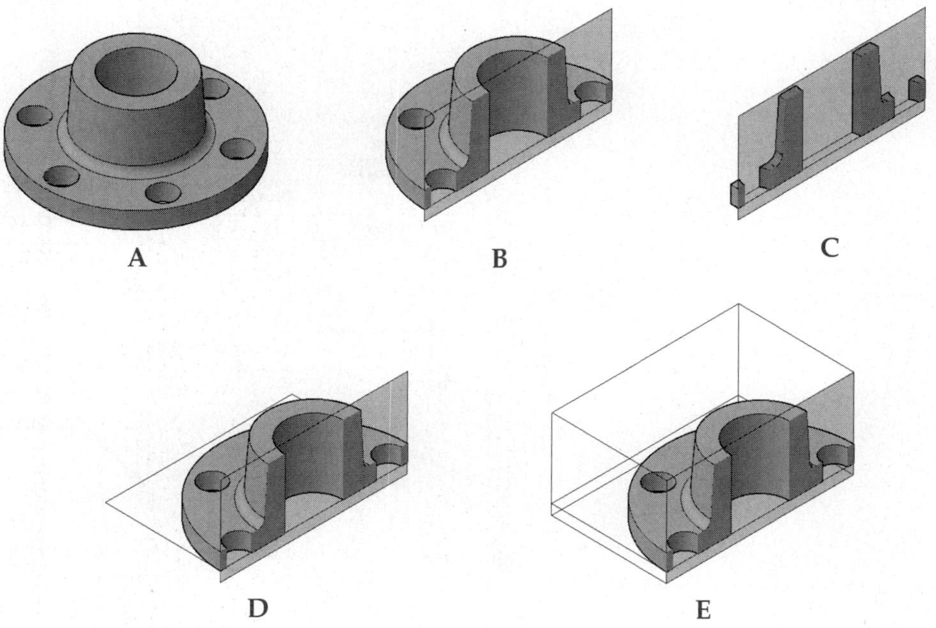

When the section object is in the ***section volume state***, a transparent plane is displayed on each segment of the section object. In addition, a 3D box extends to the XYZ boundaries of the section object. The sectioned object fits inside of this box. See **Figure 14-32E**. To set the section volume state when creating the section object, set the **Type** option to **Volume**.

NOTE

A section object cannot contain jogs when using the slice section state. If the **Type** option is set to **Slice** when you create the section object, the **Draw section** option is unavailable. Select a face, pick two points, or use the **Orthographic** option to complete the command. If a section object containing jog segments is changed to the slice section state, AutoCAD displays an alert that the jogs will be removed. Use caution if you decide to proceed, as the results may be different from what you expect.

Section Object Properties

Once the section object is created, the properties of the section object can be changed. The **Section Object** category in the **Properties** palette contains properties specific to the section object. See **Figure 14-33**. These properties are described in the next sections.

Name

The default name of the first section object is Section Plane(1). Subsequent section planes are sequentially numbered, such as Section Plane(2), Section Plane(3), and so on. It may be beneficial to rename section objects so the name is representative of the section. For example, Front Half Section is much more descriptive than Section Plane(1). To rename a section object, select the Name property. Then, type a new name in the text box.

Figure 14-33.
The properties of a section object can be changed in the **Properties** palette.

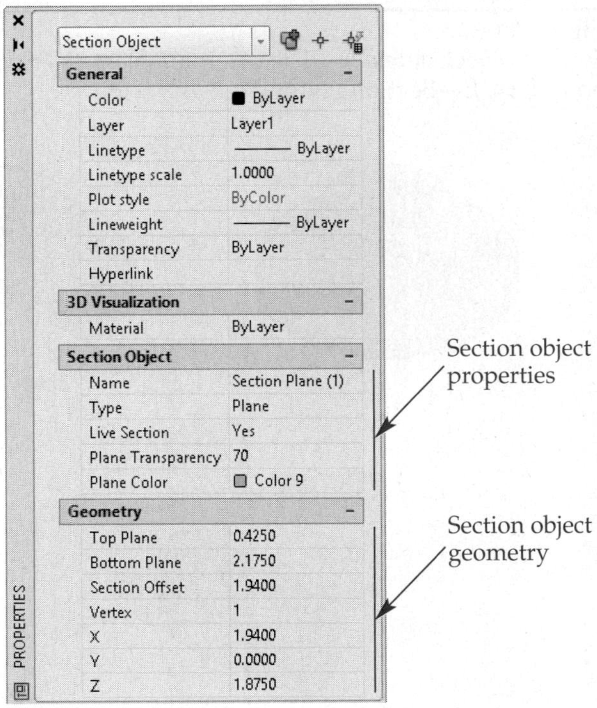

Type

As discussed earlier, the section object is in one of four states. The four states are section plane, section slice, section boundary, and section volume. To change the state of the section object, select the Type property. Then, pick the state in the drop-down list.

Live Section

Live sectioning is a tool that allows you to dynamically view the internal features of a model as the section object is moved. This tool is discussed later in the chapter. To turn live sectioning on or off, select the Live Section property. Then, pick either Yes (on) or No (off) in the drop-down list. This is the same as turning live sectioning on or off using the ribbon or the shortcut menu.

Plane Transparency

The Plane Transparency property determines the opacity of the plane for the section object. The property value can range from 1 to 100. The lower the value, the more opaque the section plane object. See **Figure 14-34**.

Plane Color

The plane of the section object can be set to any color available in the **Select Color** dialog box. To change the color, select the Plane Color property and then select a color from the drop-down list. To select a color in the **Select Color** dialog box, pick the Select Color… entry in the drop-down list. This property only affects the plane of the section object, not the lines defining the boundary, volume, or section line. The color of these lines is controlled by the Color property in the **General** category.

Section Object Grips

When the translucent planes of a section object or the lines representing the section object state are picked, grips are displayed. The specific grips displayed are related to the current section object state. The types of grips are:

- Base grip.
- Menu grip.
- Direction grip.
- Second grip.
- Arrow grips.
- Segment end grips.

Base Grip

The *base grip* appears at the first point picked when defining the section object. See **Figure 14-35**. It is the grip about which the section object can be rotated and scaled. The section object can also be moved using this grip.

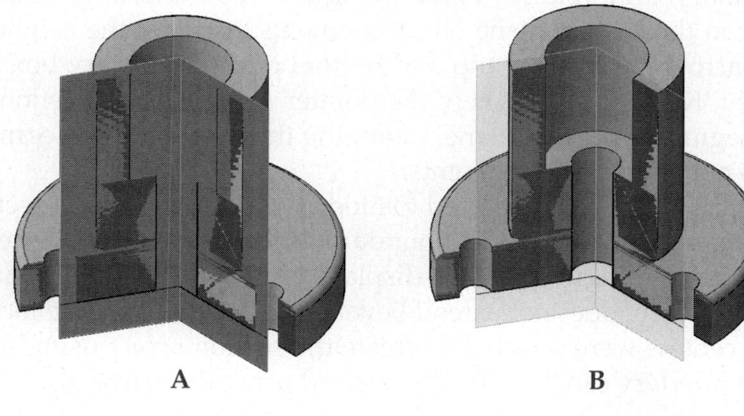

Figure 14-34. A—The Plane Transparency property of the section plane object is set to 1 (or 1% transparent). B—The Plane Transparency property of the section plane object is set to 85 (or 85% transparent).

Figure 14-35.
The types of grips displayed on a section object.

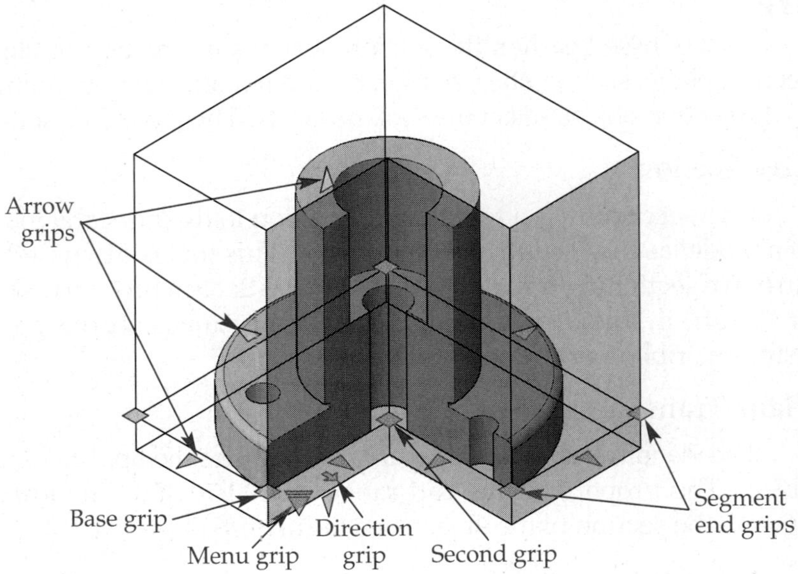

Figure 14-36.
The section state menu is used to change section object states.

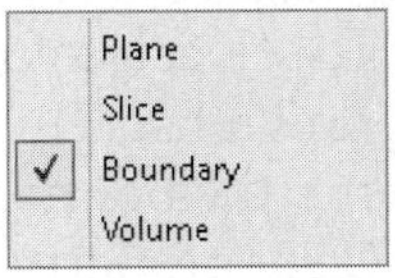

Menu Grip

The ***menu grip*** is always next to the base grip. Refer to **Figure 14-35**. Picking this grip displays the section state menu. See **Figure 14-36**. To switch the section object between states, pick the grip and then select the state from the menu.

Direction Grip

The ***direction grip*** controls the direction in which the section is viewed. Refer to **Figure 14-35**. Pick the grip to flip the view 180°.

Second Grip

The ***second grip*** appears at the second point picked when defining the section object. Refer to **Figure 14-35**. The section object can be rotated and stretched about the base grip using the second grip.

Arrow Grips

Arrow grips are located on all of the lines that represent the section plane, slice, boundary, and volume. Refer to **Figure 14-35**. These grips are used to lengthen or shorten the section plane object segments or adjust the height of the section volume. The arrow grips at the top and bottom of the boundary box are used to change the height. Regardless of where the pointer is moved, the section object only extends in the segment's current plane. Changing the length of one segment of the section plane does not affect other segments.

Grip editing can be used to modify the boundary of a section object when it does not extend past the model boundary. In **Figure 14-37A**, the section object created for the model in **Figure 14-30** is displayed in the section boundary state. Notice that the section object does not extend beyond the boundary of the model. This is because the hole centers were selected to create the section object using the **Draw section** option. The boundary can be quickly corrected using the arrow grips. Notice in **Figure 14-37B**

Figure 14-37.
A—The section object created for the model in Figure 14-30 displayed in the section boundary state. Notice that the section plane does not extend past the model boundary. B—The arrow grip is being used to extend the left side of the section plane past the boundary of the solid model. C—The edited section object. The right side can be corrected in the same manner.

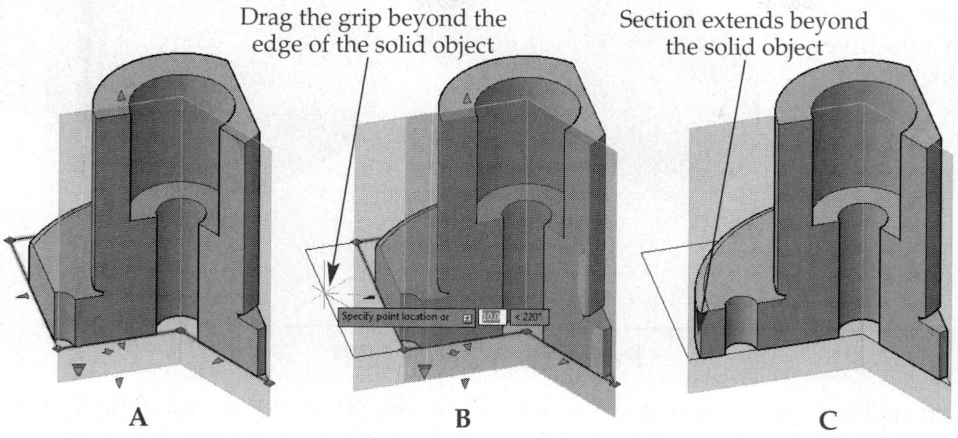

that the arrow grip is being used to extend the left side of the section plane past the boundary of the solid model. This allows any subsequent section views to display the entire object rather than just a portion of it. The right side of the section plane can be extended in the same manner using the opposite arrow grip.

The arrow grips located on the line segments of the section plane move the position of the section plane. As a segment of the section plane is moved, it maintains its angular relationship and connection to any adjacent section plane segment.

Segment End Grips

The ***segment end grips*** are located at the end of each line segment defining the section object state. Refer to **Figure 14-35**. The number of displayed segment end grips depends on whether the section object is in the section plane, section slice, section boundary, or section volume state. These grips provide access to the standard grip editing options of stretch, move, copy, rotate, scale, and mirror. If the rotate option is used, the section plane is rotated about the selected segment end grip. Moving a segment end grip can change the angle between section plane segments.

Adjusting Section Objects

Additional editing functions for section objects are available in the **Section Plane** contextual ribbon tab. Refer to **Figure 14-28**. If you need to move a section object to show a different cut in the model, you can offset the section object by a specified distance. Select the section object and enter a positive or negative offset value in the **Section Offset** text box in the **Adjust** panel. The section object is moved perpendicular to the section plane in a positive or negative direction relative to the WCS. See **Figure 14-38**. Picking the up or down arrow next to the **Section Offset** text box offsets the section plane by an incremental value in a positive or negative direction. The incremental value can be adjusted by picking the dialog box launcher button at the bottom of the **Adjust** panel to access the **Adjust Panel Increments** dialog box.

A section object can also be rotated to show a different cut in the model. To rotate a section object in 90° increments, select the **Rotate 90** option in the **Modify** panel. The section object rotates 90° about the section line, section boundary, or section volume, depending on the current section state. This provides an alternative method to using grips to rotate the section object.

Figure 14-38.
Offsetting a section object. A—The original section object. B—The section object is offset in a positive direction.

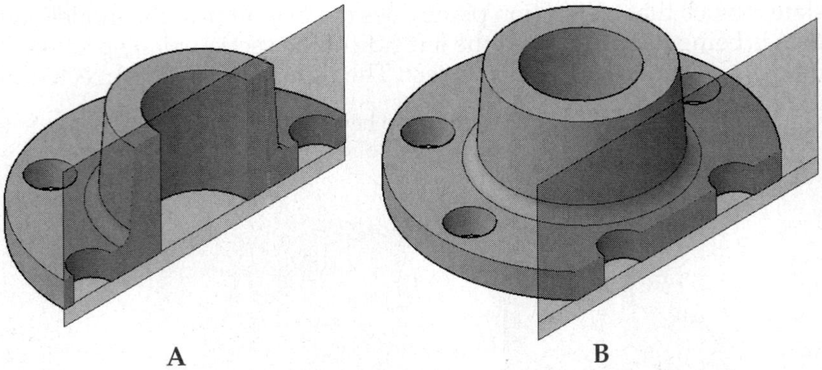

A B

PROFESSIONAL TIP

Standard editing commands can be accessed by selecting the section object and right-clicking to display the shortcut menu. Select **Move**, **Scale**, or **Rotate** from the shortcut menu to adjust the section object location as needed.

Adding Jogs to a Section

SECTIONPLANEJOG

Ribbon
Home
> Section
Solid
> Section

Section Plane Jog

Type
SECTIONPLANEJOG

You can quickly add a jog, or offset, to an existing section object. First, select the section object. Then, select the **Add Jog** option in the **Modify** panel in the **Section Plane** contextual ribbon tab. You can also right-click to display the shortcut menu and select **Add jog to section**. You are then prompted:

Specify a point on the section line to add jog:

Select a point directly on the section line. If any object snap is active, the **Nearest** object snap is temporarily turned on to ensure you pick the line. Once you pick, the jog is automatically added perpendicular to the line segment. See **Figure 14-39A**.

It is not critical that you pick the exact location on the line where you want the jog to occur. Remember, grips allow you to easily adjust the section plane location. Notice in **Figure 14-39B** that the second jog barely cuts through the first hole. The intention is to run the section plane through the middle of the hole. To fix this, drag the arrow grip so the section plane segment is in the desired location, **Figure 14-39C**.

Exercise 14-11

www.g-wlearning.com/CAD/

Complete the exercise on the companion website.

Live Sectioning

LIVESECTION

Ribbon
Home
> Section
Solid
> Section

Live Section

Type
LIVESECTION

Live sectioning is a tool that allows you to view the internal features of models that are cut by the section plane of the section object. The view is dynamically updated as the section object is moved. This tool is used to visualize internal features and for establishing section locations from which 2D and 3D section views can be created.

Live sectioning is automatically turned on when a section object is created. Live sectioning can be turned on and off for individual section objects, but only one section object can be "live" at any given time.

To turn live sectioning on or off, select the section object. Then, select the **Live Section** option in the **Display** panel in the **Section Plane** contextual ribbon tab. You can also enter the **LIVESECTION** command and select the section object to toggle the on/off

Figure 14-39.
Adding a jog to a section object. A—Pick a point on the section line to add a jog. B—The jog is added, but it is not in the proper location. C—Using the arrow grip, the jog is moved to the proper location.

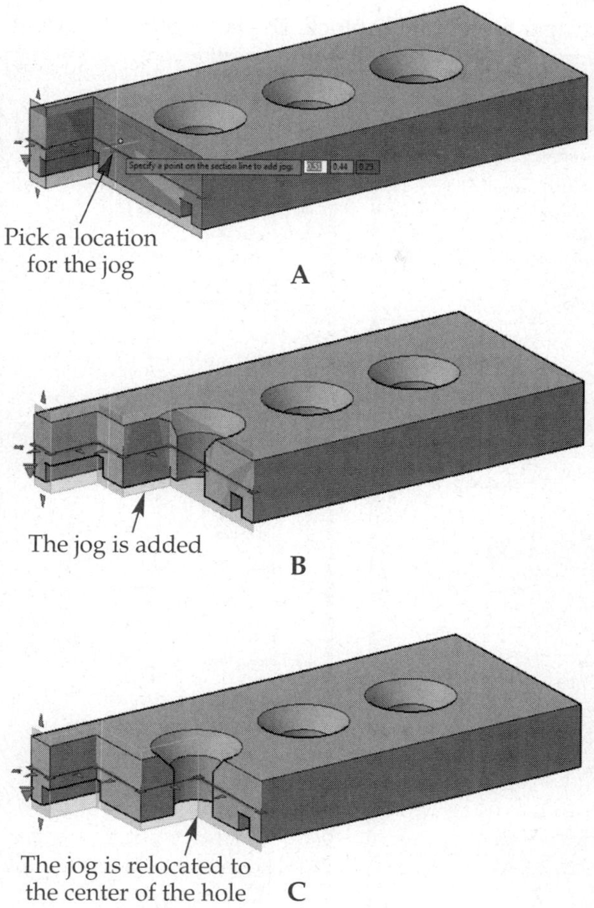

setting. When live sectioning is turned on, the material behind the viewing direction of the section plane is removed. The cross section of the 3D object is shown in gray (by default) and the internal shape of the 3D object is visible.

A wide variety of options allow you to change the appearance of not only the live sectioning display, but also of 2D and 3D section blocks that can be created from the sectioned display. These settings are found in the **Section Settings** dialog box. See **Figure 14-40**. To open this dialog box, select the section object and pick the dialog box launcher button in the lower-right corner of the **Display** panel in the **Section Plane** contextual ribbon tab. You can also pick the dialog box launcher button in the lower-right corner of the **Section** panel in the **Home** or **Solid** tab of the ribbon.

To change the settings for live sectioning, pick the **Live Section settings** radio button at the top of the **Section Settings** dialog box. The categories displayed in the dialog box contain properties related to live sectioning. Settings for 2D and 3D sections and elevations are discussed later in this chapter.

The three categories for live sectioning are **Intersection Boundary**, **Intersection Fill**, and **Cut-away Geometry**. To display a brief description of any property, hover the cursor over the property in the **Section Settings** dialog box. The description is displayed in a tooltip. A check box at the bottom of the **Section Settings** dialog box allows you to apply the properties to all section objects or to just the selected section object.

PROFESSIONAL TIP

Live sectioning can be quickly turned on or off by double-clicking on the section plane object.

Figure 14-40.
Section settings. A—For a 2D block. B—For a 3D block. C—For live sectioning.

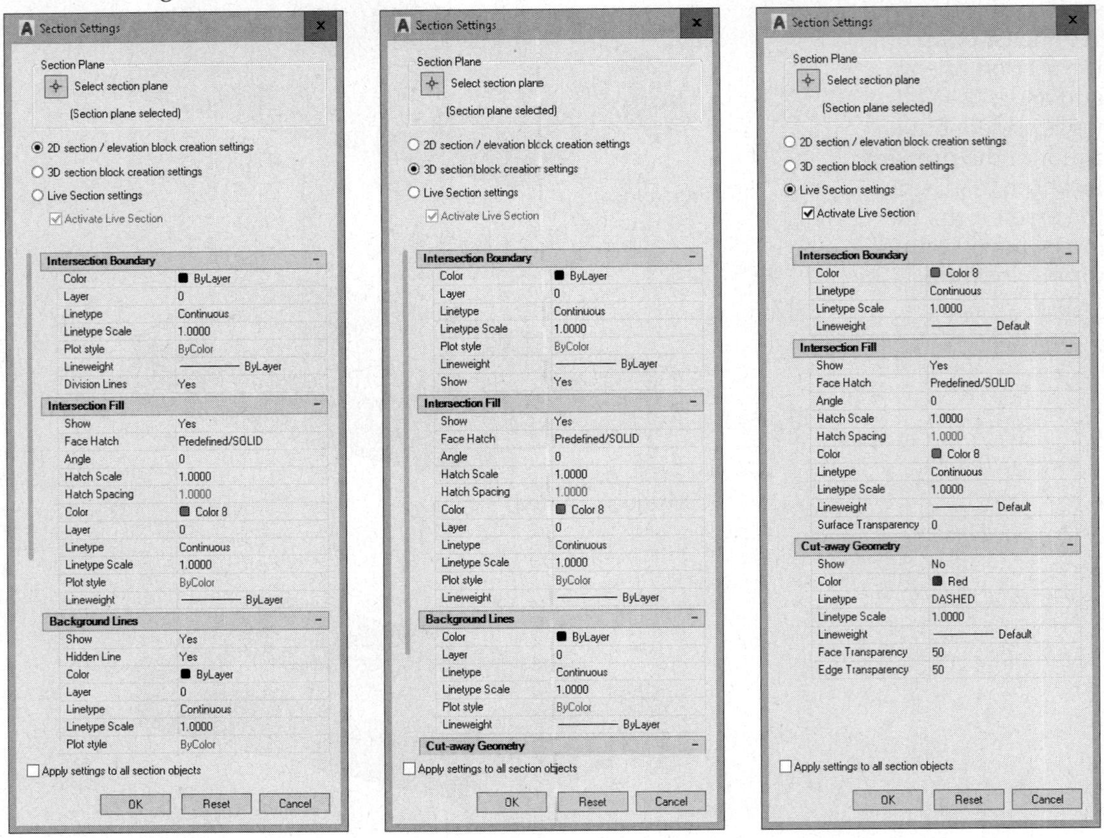

A B C

Intersection Boundary

The intersection boundary is where the model is intersected by the section object. It is represented by line segments. You can set the color, linetype, linetype scale, and line weight of the intersection boundary lines.

Intersection Fill

The intersection fill is the material visible on the model surface where the section object cuts. It is displayed as a solid fill, by default. Any hatch pattern available in AutoCAD can be used as the intersection fill. See **Figure 14-41A**. The angle, hatch scale, hatch spacing, and color can be set. In addition, the linetype, linetype scale, and line weight can be changed. The fill pattern can even be set to be transparent.

Cutaway Geometry

The cutaway geometry is the part of the model removed by the live sectioning. By default, this geometry is not displayed. Changing the Show property to Yes displays the geometry. See **Figure 14-41B**. You can set the color, linetype, linetype scale, and line weight of the lines representing the cutaway geometry. In addition, the Face Transparency and Edge Transparency properties allow you to create a see-through effect, as seen in **Figure 14-42**. Each of these two properties is set to 50 by default.

PROFESSIONAL TIP

You can also display the cutaway geometry without using the **Section Settings** dialog box. Select the section object, right-click, and pick **Show cut-away geometry** in the shortcut menu. Live sectioning must be on.

Figure 14-41.
Live sectioning settings. A—The intersection fill can be displayed as a hatch pattern in any specified color. B—The cutaway geometry removed by the live sectioning is displayed.

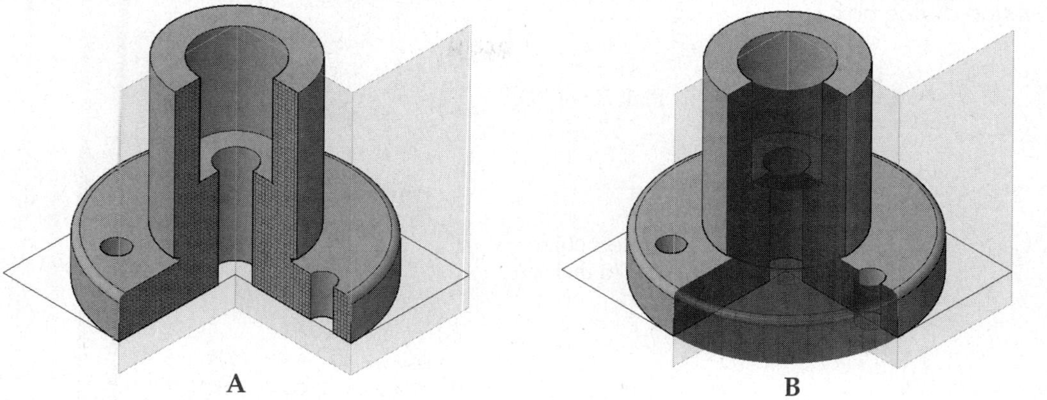

A B

Figure 14-42.
The cutaway geometry is displayed with 100% transparent faces and solid black lines.

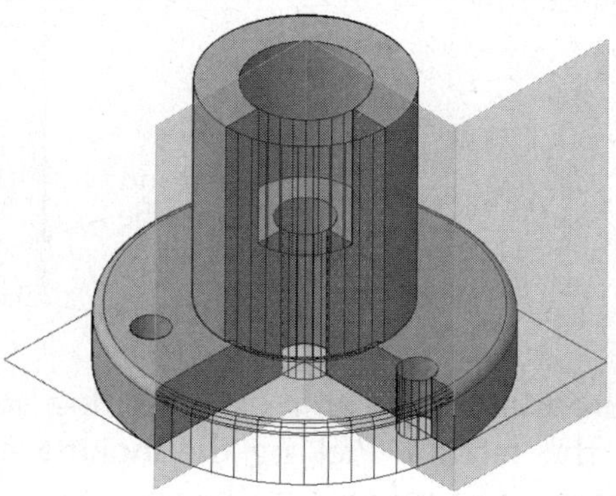

Exercise 14-12

www.g-wlearning.com/CAD/

Complete the exercise on the companion website.

Generating 2D and 3D Sections and Elevations

The **SECTIONPLANETOBLOCK** command provides a fast and efficient method of creating sections. The sections can be either 2D or 3D and are inserted as blocks. Not only can the sections be displayed on the current drawing, they can also be exported as a file that can then be used in any other drawing or document for display, technical drawing, or manufacturing purposes.

Creating Sections

Ribbon
Home
> Section
Solid
> Section
Generate Section to Block
Type
SECTIONPLANE-TOBLOCK
GENERATESECTION

SECTIONPLANETOBLOCK

To create a section, enter the **SECTIONPLANETOBLOCK** command. You can also select the section object and then select the **Generate Section Block** option in the **Generate** panel in the **Section Plane** contextual ribbon tab. The **Generate Section/Elevation** dialog box is displayed. See **Figure 14-43**. In this dialog box, you can specify whether the section will be 2D or 3D, select what is included in the section, and specify a destination for the section. To expand the dialog box, pick the **Show details** button, which looks like a down arrow.

To create a 2D section, pick the **2D Section/Elevation** radio button in the **2D/3D** area of the dialog box. A 2D section is projected onto the section plane, but is placed flat on the XY plane of the current UCS. To create a 3D section, pick the **3D Section** radio button. A 3D section is placed so its surfaces are parallel to the corresponding cut surfaces on the 3D object.

Figure 14-43.
The expanded **Generate Section/ Elevation** dialog box.

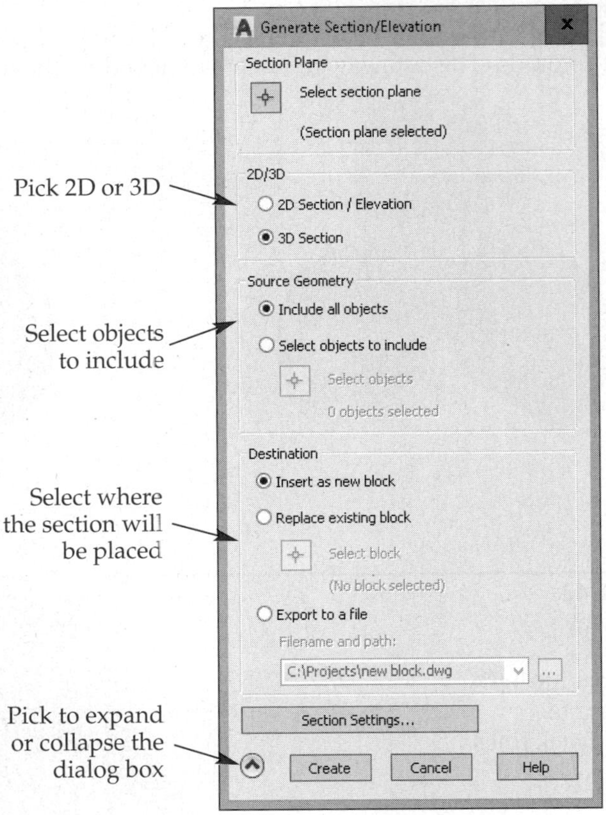

In the **Source Geometry** area of the dialog box, you can specify which geometry is included in the section. Picking the **Include all objects** radio button includes all 3D solids, surfaces, and regions in the section. To limit the section to certain objects, pick the **Select objects to include** radio button. Then, pick the **Select objects** button, select the objects on-screen, and press [Enter]. The number of selected objects is then displayed in the dialog box.

The **Destination** area of the dialog box is where you specify how the section will be placed. To place the section into the current drawing, pick the **Insert as new block** radio button. To update an existing section block, pick the **Replace existing block** radio button. Then, pick the **Select block** button, select the block on-screen, and press [Enter]. You will need to do this if the section object is changed. To save the section to a file for use in other drawings, pick the **Export to a file** radio button. Then, enter a path and file name in the text box.

Once all settings have been made, pick the **Create** button. The section is attached to the cursor and can be placed like a regular block. See **Figure 14-44**. Additionally, the options available are the same as if a regular block is being inserted. Once the block is inserted, it can be moved, rotated, and scaled as needed.

Section Settings

Picking the **Section Settings...** button at the bottom of the **Generate Section/ Elevation** dialog box opens the **Section Settings** dialog box discussed earlier. Using this dialog box, you can adjust all of the properties associated with the type of section being created. Depending on whether the **2D Section/Elevation** or **3D Section** radio button is selected in the **Generate Section/Elevation** dialog box, the appropriate categories and properties are displayed in the **Section Settings** dialog box. Refer to **Figure 14-40**.

The categories discussed earlier in relation to the **Live Section settings** radio button are available. Also, two additional categories are displayed for 2D and 3D sections:

- **Background Lines.** Available for 2D and 3D sections.
- **Curve Tangency Lines.** Only available for 2D sections.

Figure 14-44.
Inserting a 2D section block.

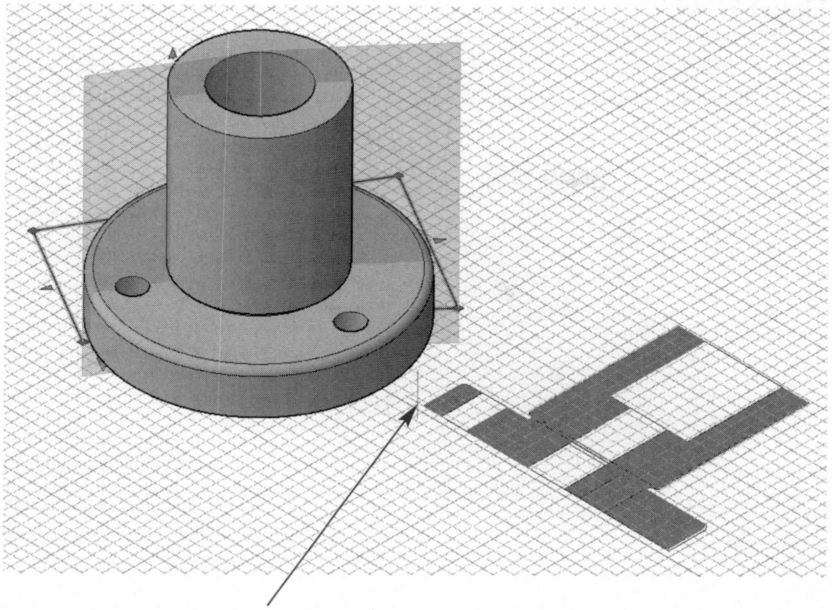

These categories are discussed below. Examples of 2D and 3D sections inserted as blocks in the drawing are shown in **Figure 14-45**. Notice how properties can be set to show cutaway geometry in a different color and to change the section pattern, color, and linetype scale.

Background Lines. The properties in the **Background Lines** category provide control over the appearance of all lines that are not on the section plane. You can choose to have visible background lines, hidden background lines, or both displayed. They can be emphasized with color, linetype, or line weight. The layer, linetype scale, and plot style can also be changed. These settings are applied to both visible and hidden background lines.

Curve Tangency Lines. The properties in the **Curve Tangency Lines** category apply to lines of tangency behind the section plane. For example, the object shown in **Figure 14-45** has a round on the top of the base. This results in a line of tangency behind the section plane where the round meets the vertical edge. You can have these lines displayed or suppressed. In general, lines of tangency are not shown in a section view. If you choose to display these lines, you can set the color, layer, linetype, linetype scale, plot style, and line weight of the lines.

NOTE

When a 3D section is created, you must turn off live sectioning to see the complete sectioned object in the block. With live sectioning on, only the cut surfaces appear in the block.

Updating the Section View

Once the section view is created, it is not automatically updated if the section object is changed. To update the section view, select the section object (not the block) and select the **Generate Section Block** option from the **Generate** panel in the **Section Plane** contextual ribbon tab. Then, in the **Destination** area of the **Generate Section/Elevation** dialog box, pick the **Replace existing block** radio button. If necessary, pick the **Select block** button and select the section block in the drawing. If you want to change the appearance of the section view, pick the **Section Settings...** button and adjust the properties as needed. Finally, pick the **Create** button in the **Generate Section/Elevation** dialog box to update the section block.

Figure 14-45.
A—The section object is created. B—A 2D section block is inserted into the drawing and the view is made plan to the block. C—A 3D section block is inserted into the drawing. Notice how the hatch pattern is displayed. D—The 3D section block is updated and now the cutaway geometry is displayed.

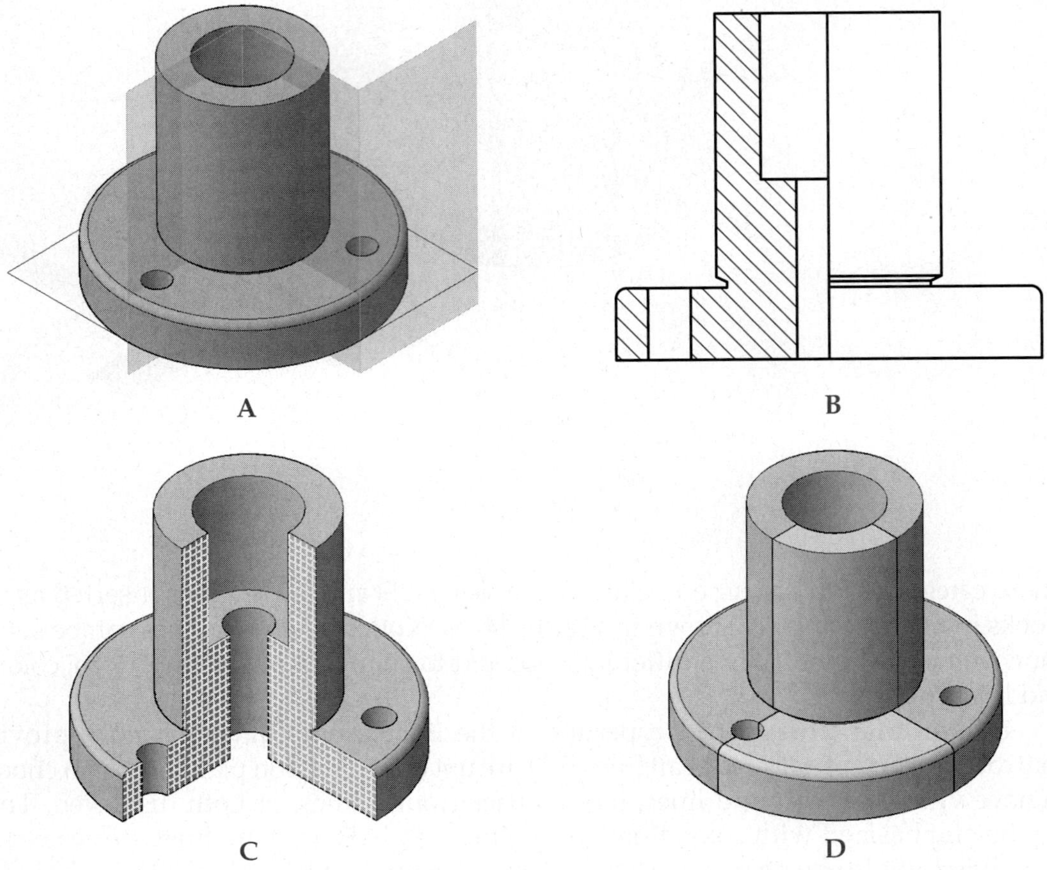

Exercise 14-13

www.g-wlearning.com/CAD/

Complete the exercise on the companion website.

Model Analysis

AutoCAD provides a number of tools for analyzing solid and surface models. Model analysis is conducted to evaluate various data and determine whether design changes are needed prior to manufacturing. Model analysis tools available in AutoCAD are discussed in the following sections.

Solid Model Analysis

MASSPROP

Type
MASSPROP

The **MASSPROP** command allows you to analyze a solid model for its physical properties. The data obtained from **MASSPROP** can be retained for reference by saving the data to a file. The default file name is the drawing name. The file is an ASCII text file with a .mpr (mass properties) extension. The analysis can be used for third-party applications to produce finite element analysis, material lists, or other testing studies.

Once the command is initiated, you are prompted to select objects. Pick the objects for which you want the mass properties displayed and press [Enter]. AutoCAD analyzes the model and displays the results in the **AutoCAD Text Window**. See **Figure 14-46**. The following properties are listed.

Figure 14-46.
The **MASSPROP** command displays a list of solid properties in the **AutoCAD Text Window**.

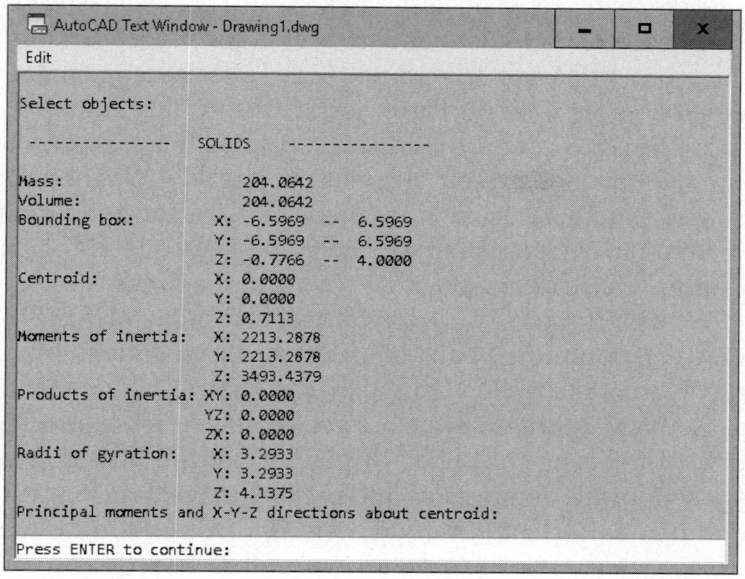

- **Mass.** A measure of the inertia of a solid. In other words, the more mass an object has, the more inertia it has. Note: Mass is *not* a unit of measurement of inertia.
- **Volume.** The amount of 3D space the solid occupies.
- **Bounding box.** The dimensions of a 3D box that fully encloses the solid.
- **Centroid.** A point in 3D space that represents the geometric center of the mass.
- **Moments of inertia.** A solid's resistance when rotating about a given axis.
- **Products of inertia.** A solid's resistance when rotating about two axes at a time.
- **Radii of gyration.** Similar to moments of inertia. Specified as a radius about an axis.
- **Principal moments and X-Y-Z directions about centroid.** The axes about which the moments of inertia are the highest and lowest.

PROFESSIONAL TIP

Advanced applications of solid model design and analysis are possible with Autodesk Inventor software. This product allows you to create parametric designs and assign a wide variety of physical materials to the solid model.

Surface Continuity Analysis

Surface continuity describes the type of transition formed between adjoining surfaces in a model. Surface continuity is discussed in Chapter 10. Depending on the type of model you are working with and manufacturing requirements, you may need to modify the curvature of the model to create a more smooth transition between surfaces. For example, if you are working with a surface model and continuity settings are available, you may need to adjust the settings to produce a more smooth contour.

In AutoCAD, ***zebra analysis*** is used to graphically check for surface continuity. Zebra analysis is also referred to as ***zebra striping***. The **ANALYSISZEBRA** command is used to conduct zebra analysis. When using this command, AutoCAD projects a map of zebra stripes onto the model. The flow of stripes from one surface to another is then checked to determine the surface continuity. Areas where the stripes line up indicate where the model has surface continuity.

The purpose of zebra analysis is to simply check for the quality of continuity or *flow* between surfaces. It does not tell you how accurate the surface is or the quality of the surface constructed. However, good surface continuity is important in many modeling applications. You will typically want the model to have a smooth surface shape and avoid sharp changes in curvature.

Before using zebra analysis, make sure a 3D visual style is set current. Then, initiate the **ANALYSISZEBRA** command and select the model surfaces to analyze. You can select one or more surfaces or solids. Press [Enter] after selecting the surfaces. In **Figure 14-47**, the three surfaces of the air duct are selected. The top and bottom surfaces of the model are loft surfaces. The center surface is a blend surface. In this part, smooth flow design is important. Notice the flow of the zebra stripes in **Figure 14-47B**. The stripes touch but do not align consistently. In this version of the model, the start and end continuity of the blend surface are both set to G0 (positional continuity). In **Figure 14-47C**, changing the continuity to G2 (curvature) creates a smoother flow between the surfaces.

ANALYSISZEBRA

Ribbon
Surface
> Analysis
Analysis Zebra
Type
ANALYSISZEBRA

To turn off the zebra striping display, use the **ANALYSISZEBRA** command. Select the **Turn off** option. You can also select **Analysis Options** from the **Analysis** panel in the **Surface** tab of the ribbon to display the **Analysis Options** dialog box. In the **Zebra** tab, pick the **Clear Zebra Analysis** button to remove zebra striping. The **Zebra** tab contains options for setting the stripe display, the stripe direction, and the colors and thickness used for the zebra striping. This dialog box is also used to set options for surface curvature analysis and draft analysis, as discussed in the next sections.

Surface Curvature Analysis

Once a model is designed, it is then analyzed to confirm the quality of the design before the actual part is manufactured. When working with models containing curved surfaces, you design with the intent that the model stays within specific curvature ranges for manufacturing purposes. If you design a model that falls outside specific design criteria, problems occur and the model cannot be manufactured. ***Surface curvature analysis*** assists in determining the overall smoothness of a 3D model design.

In AutoCAD, surface curvature analysis is performed with the **ANALYSISCURVATURE** command. When using this command, AutoCAD applies a color gradient to the surfaces of the model. This is known as ***color mapping***. The model is then graphically analyzed. Different gradient colors indicate areas of high and low curvature and abrupt changes in curvature.

When you use the **ANALYSISCURVATURE** command, AutoCAD uses the settings specified in the **Curvature** tab of the **Analysis Options** dialog box, **Figure 14-48**. To display this dialog box, use the **ANALYSISOPTIONS** command. The options in the

ANALYSISOPTIONS

Ribbon
Surface
> Analysis
Analysis Options
Type
ANALYSISOPTIONS

Figure 14-47.
Zebra analysis. A—The original air duct model consists of two loft surfaces connected by a blend surface. B—The three surfaces are selected with the **ANALYSISZEBRA** command. Notice that the zebra stripes do not line up. In this version of the model, the start and end surface continuity of the blend surface are both set to G0. C—Adjusting the start and end surface continuity of the blend surface to G2 produces smoother continuity.

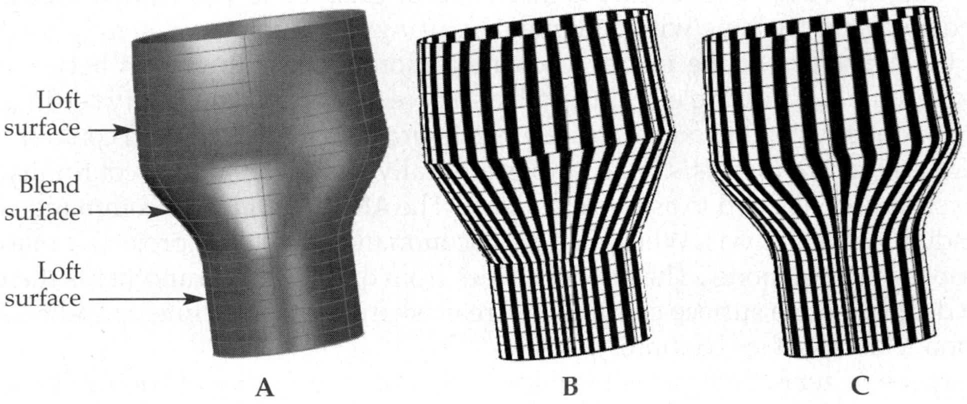

Figure 14-48.
The **Curvature** tab of the **Analysis Options** dialog box.

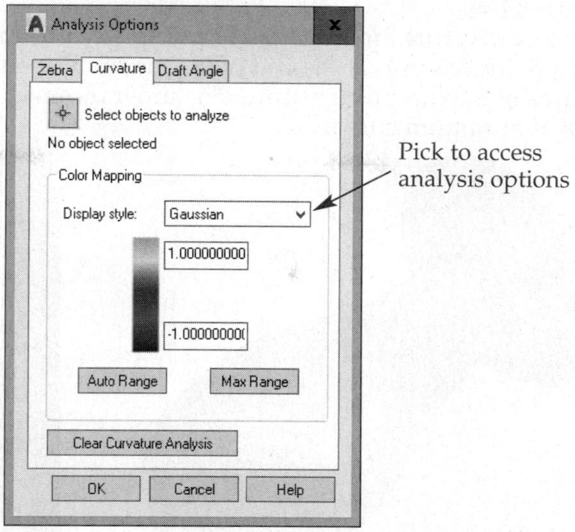

Display style: drop-down list in the **Color Mapping** area determine how the curvature is analyzed. The options are described as follows:

- **Gaussian.** This is the default option. AutoCAD analyzes areas of high and low curvature in the model. The color red is assigned to areas with positive curvature. A positive Gaussian value indicates that the surface has a bowl or spherical shape. The color green is assigned to areas with zero-value curvature (flat surface areas). Cylindrical and conical surfaces are examples of surfaces with zero-value Gaussian curvature. The color blue is assigned to areas with negative curvature. A negative Gaussian value indicates that the surface has a saddle or hyperbolic shape. The **Gaussian** option is suitable for general purposes and works well with swept objects.
- **Mean.** AutoCAD analyzes the mean curvature of values along the U and V directions of the surface. This option is useful when checking for quick or sharp changes in the shape of the surface.
- **Max radius.** AutoCAD analyzes the maximum curvature of values along the U and V directions of the surface. This option is useful for determining how flat or curved specific surfaces are.
- **Min radius.** AutoCAD analyzes the minimum curvature of values along the U and V directions of the surface. This option is useful for checking surfaces with small radius bends.

The values in the text boxes at the ends of the color gradient bar in the **Color Mapping** area define the curvature range from minimum to maximum curvature. The default values are 1.0 and –1.0. The values are used to define the acceptable curvature range. AutoCAD displays the corresponding color in the color gradient bar when a curvature value is reached. Maximum values are indicated in green and minimum values are indicated in blue. The resulting display can vary depending on the range of curvature values entered.

The **Auto Range** and **Max Range** buttons are used to calculate ranges of curvature values of selected objects. When the **Auto Range** button is picked, the calculation is based on 80% of the values in the curvature range. This can provide a starting point for setting the minimum and maximum curvature values. When the **Max Range** button is picked, the calculation is based on the maximum range of values.

Ribbon
Surface
> Analysis

Analysis Curvature
Type
ANALYSIS-CURVATURE

ANALYSISCURVATURE

To select surfaces to analyze, pick the **Select objects to analyze** button in the **Curvature** tab if the **Analysis Options** dialog box is open. Otherwise, use the **ANALYSISCURVATURE** command. Before using this command, make sure that a 3D visual style is set current. Once you initiate the command, select the surfaces and press [Enter]. You can select one or more surfaces or solids. If needed, access the **Analysis Options** dialog box to set the curvature range.

In **Figure 14-49**, a model created from a loft surface is analyzed using the **Min radius** option. In **Figure 14-49B**, the features at the end of the model are checked. In the

Figure 14-49.
Surface curvature analysis. A—The original loft surface. B—The **Min radius** option is used to analyze the features at the end of the model. The outer "ring" is shaded green. This indicates an area of maximum curvature. The inner recessed surface is shaded blue. This indicates an area of minimum curvature.

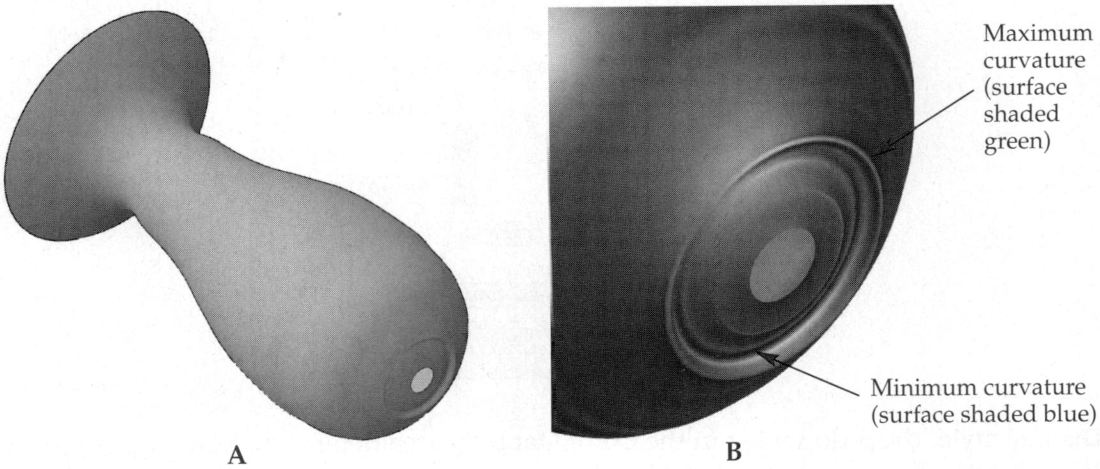

resulting shaded display, the outer portion of the ring-shaped bend is shaded green. This indicates an area of maximum curvature. The recessed surface is shaded blue. This indicates an area of minimum curvature.

To turn off the curvature analysis display, use the **ANALYSISCURVATURE** command. Select the **Turn off** option. You can also pick the **Clear Curvature Analysis** button in the **Analysis Options** dialog box.

Draft Analysis

When designing a part that will be removed from a mold, a ***draft angle*** is designed on the part to ensure that the part can be pulled from the mold. ***Draft analysis*** is used to evaluate a solid or surface model for adequate draft. In AutoCAD, draft analysis is performed with the **ANALYSISDRAFT** command. The procedure is similar to that used with surface curvature analysis. When you select a model to analyze, AutoCAD maps a color gradient to indicate the angle variation of the model. Green colors are mapped to areas where the draft is at the highest angle specified, and blue colors are mapped to areas where the draft is at the lowest angle specified.

When using the **ANALYSISDRAFT** command, the high and low draft angles are specified in the **Draft Angle** tab of the **Analysis Options** dialog box. Use the **ANALYSISOPTIONS** command to open this dialog box, as previously discussed. Then, in the **Draft Angle** tab, use the **Angle:** text boxes to specify the highest and lowest draft angles allowed in the model. The default values are 3.0 and –3.0.

ANALYSISDRAFT
Ribbon
Surface
> Analysis

Analysis Draft
Type
ANALYSISDRAFT

To select a model to analyze, pick the **Select objects to analyze** button in the **Draft Angle** tab if the **Analysis Options** dialog box is open. Otherwise, use the **ANALYSISDRAFT** command. Before using this command, make sure that a 3D visual style is set current. Once you initiate the command, select the model and press [Enter].

In **Figure 14-50**, a computer mouse is analyzed for adequate draft. The model was created as a mesh, converted to a surface, and then converted to a solid. Draft analysis is used to determine if a 4° draft angle can be used for the mold. The model after using the **ANALYSISDRAFT** command is shown in **Figure 14-50B**. The top of the model is shaded green and is within the parameters. However, there are two areas of concern along the side and bottom (where surfaces are shaded blue). This indicates that changes to the design are in order.

To turn off the draft analysis display, use the **ANALYSISDRAFT** command. Select the **Turn off** option. You can also pick the **Clear Draft Angle Analysis** button in the **Analysis Options** dialog box.

Figure 14-50.
Draft analysis. A—The original solid model. B—The resulting analysis based on angle values of 4.0 and –4.0. The top of the model is within the draft parameters. Areas along the side and bottom are shaded blue to indicate the draft is at the lowest angle specified.

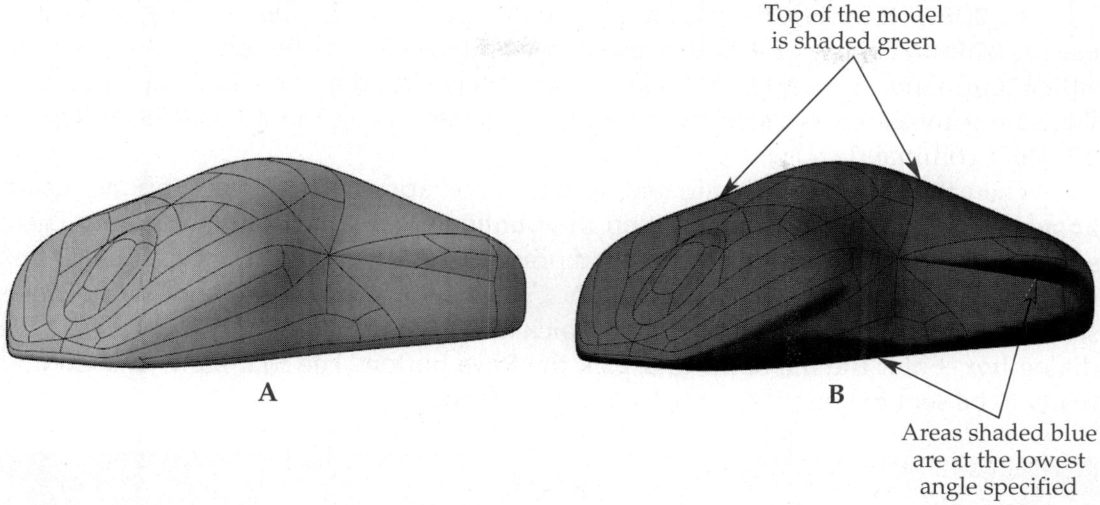

Solid Model File Exchange

AutoCAD drawing files can be converted to files that can be used for testing and analysis. Use the **ACISOUT** or **EXPORT** command to create a file with a .sat extension. These files can be imported into AutoCAD with the **ACISIN** or **IMPORT** command.

Solids can also be exported for use with 3D printing. You can use the **3DPRINTSERVICE** or **3DPRINT** command to prepare files for 3D printing.

Importing and Exporting Solid Model Files

A solid model is frequently used with analysis and testing software or in the manufacture of a part. The **ACISOUT** and **EXPORT** commands allow you to create a type of file that can be used for these purposes. Once the **ACISOUT** command is initiated, you are prompted to select objects. After selecting objects and pressing [Enter], a standard save dialog box is displayed. Enter a name for the SAT file and pick **Save**. When using the **EXPORT** command, the standard save dialog box appears first. After entering a file name and selecting a file type (SAT), you are then prompted to select objects.

ACISOUT
Application Menu
Export
> Other Formats
Type
ACISOUT
EXPORT
EXP

An SAT file can be imported into AutoCAD and automatically converted into a drawing file using the **ACISIN** or **IMPORT** command. Once either command is initiated, a standard open dialog box appears. If you are using the **ACISIN** command, locate the file and pick the **Open** button. If you are using the **IMPORT** command, change the file type to SAT first and then locate the file to open.

ACISIN
Ribbon
Insert
> Import

Import
Type
ACISIN
IMPORT
IMP

Additive Manufacturing

Additive manufacturing is a process used to build a physical model from 3D model file data. ***Stereolithography (SLA)*** is an additive manufacturing process that creates various plastic 3D model prototypes using a computer-generated solid model, a laser, and a vat of liquid polymer. This technology is also referred to as ***rapid prototyping*** or ***3D printing***. Some additive manufacturing processes, such as fused-deposition modeling (FDM), add material in "layers" from a filament that is extruded through a heated nozzle. Other additive manufacturing processes are selective laser sintering (SLS), 3D printing (3DP), multi-jet modeling (MJM), and electron beam melting (EBM). Using one of these processes, a prototype 3D model can be designed and formed in a short amount of time without using standard subtractive manufacturing processes.

Most CAD software today can create a stereolithography file (STL file). AutoCAD can export a drawing file to the STL format, but *cannot* import STL files.

3DPRINTSERVICE

Application Menu
Publish
Send to 3D Print Service
Type
3DPRINTSERVICE

Creating STL Files

The **3DPRINTSERVICE** command is used to create an STL file. An alternative is to use the **3DPRINT** command. Solids and watertight meshes can be selected for use with either command. A ***watertight mesh*** is completely closed and contains no openings. Watertight meshes are converted to 3D solids when using the **3DPRINTSERVICE** or **3DPRINT** command.

3DPRINT

Application Menu
Print
3D Print
Type
3DPRINT
3DP
3DPLOT
RAPIDPROTOTYPE

When the command is initiated, a dialog box appears with two options: **Learn about preparing a 3D model for printing** and **Continue**. Select the **Continue** option. Then, select the solids or watertight meshes and press [Enter]. The **3D Print Options** dialog box is displayed, **Figure 14-51**. This dialog box allows you to change the output dimensions or select other objects. When done, pick the **OK** button to display a standard save dialog box. Enter the file name and pick the **Save** button. The STL file is created and ready to be sent to your 3D printer to create the model.

NOTE

When using the **3DPRINTSERVICE** or **3DPRINT** command, the **FACETRES** system variable is automatically set to 10 for the creation of an STL file. The system variable is reset after the command is completed.

Exercise 14-14

www.g-wlearning.com/CAD/

Complete the exercise on the companion website.

Figure 14-51.
The **3D Print Options** dialog box.

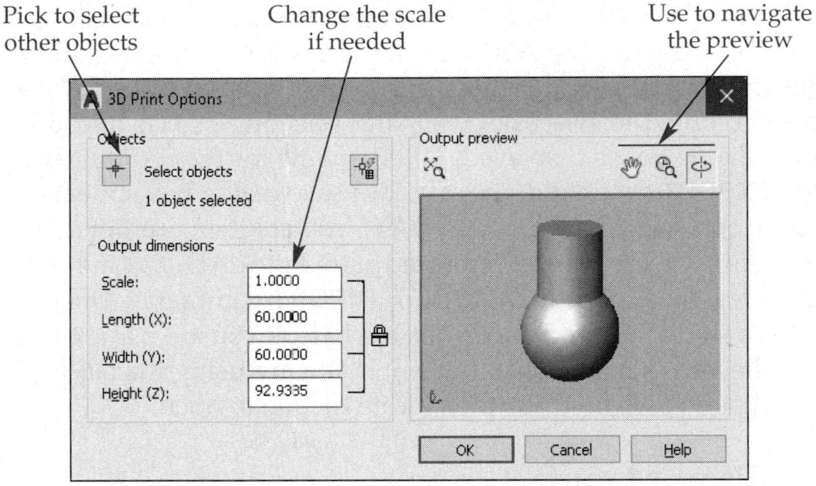

Chapter Review

Answer the following questions using the information in this chapter.

1. Which option of the **VIEWBASE** command is used to change the default view when placing the base view?
2. Explain how to change the default projection used by AutoCAD when placing drawing views.
3. Explain two ways to place a projected view.
4. Describe two ways to locate the section line when creating a section view with the **VIEWSECTION** command.
5. When creating a section view with the **VIEWSECTION** command, what controls the appearance of elements in the section identifier and section view label?
6. What command is used to edit a section view?
7. What are the two types of detail boundaries that can be used when creating a detail view?
8. What option of the **VIEWDETAIL** command allows you to create a smooth edge for the detail view and include a border?
9. To what value should the **DIMASSOC** system variable be set in order to dimension drawing views with associative dimensions?
10. What does the **SECTIONPLANE** command create?
11. How is the **Face** option of the **SECTIONPLANE** command used?
12. Which option of the **SECTIONPLANE** command is used to create sections with jogs?
13. Which section object grips are used to accomplish the following tasks?
    A. Change the section object state.
    B. Lengthen or shorten the section object segment.
    C. Rotate the section view 180°.
14. How is live sectioning turned on or off?
15. Which category in the **Section Settings** dialog box provides control over the material that is removed by the section object?
16. What are the two types of section view blocks that can be created from a section object?
17. What is the function of the **MASSPROP** command?
18. What is the extension of the ASCII file that can be created by **MASSPROP**?
19. What command is used to graphically analyze surface continuity in a model?
20. Briefly explain how to set values defining the acceptable curvature range when using the **ANALYSISCURVATURE** command.
21. Which commands export and import solid models?
22. Which type of file has an .stl extension?
23. What is the purpose of the **3DPRINTSERVICE** command?

Drawing Problems

1. **Mechanical** In this problem, you will create a multiview layout of the base bracket by placing drawing views with the **VIEWBASE** command. Set up a page layout using the ANSI B (17.00 × 11.00 Inches) paper size. Select the appropriate scale for the orthographic views and the isometric view. Edit the views as needed to look like the layout shown.
   A. Construct the model shown using the dimensions provided.
   B. Create the front, top, right-side, and isometric views shown.
   C. The isometric view should be a shaded view.
   D. Dimension the drawing in the layout using the dimensions given. Make sure to use associative dimensions.
   E. Save the drawing as P14-1.

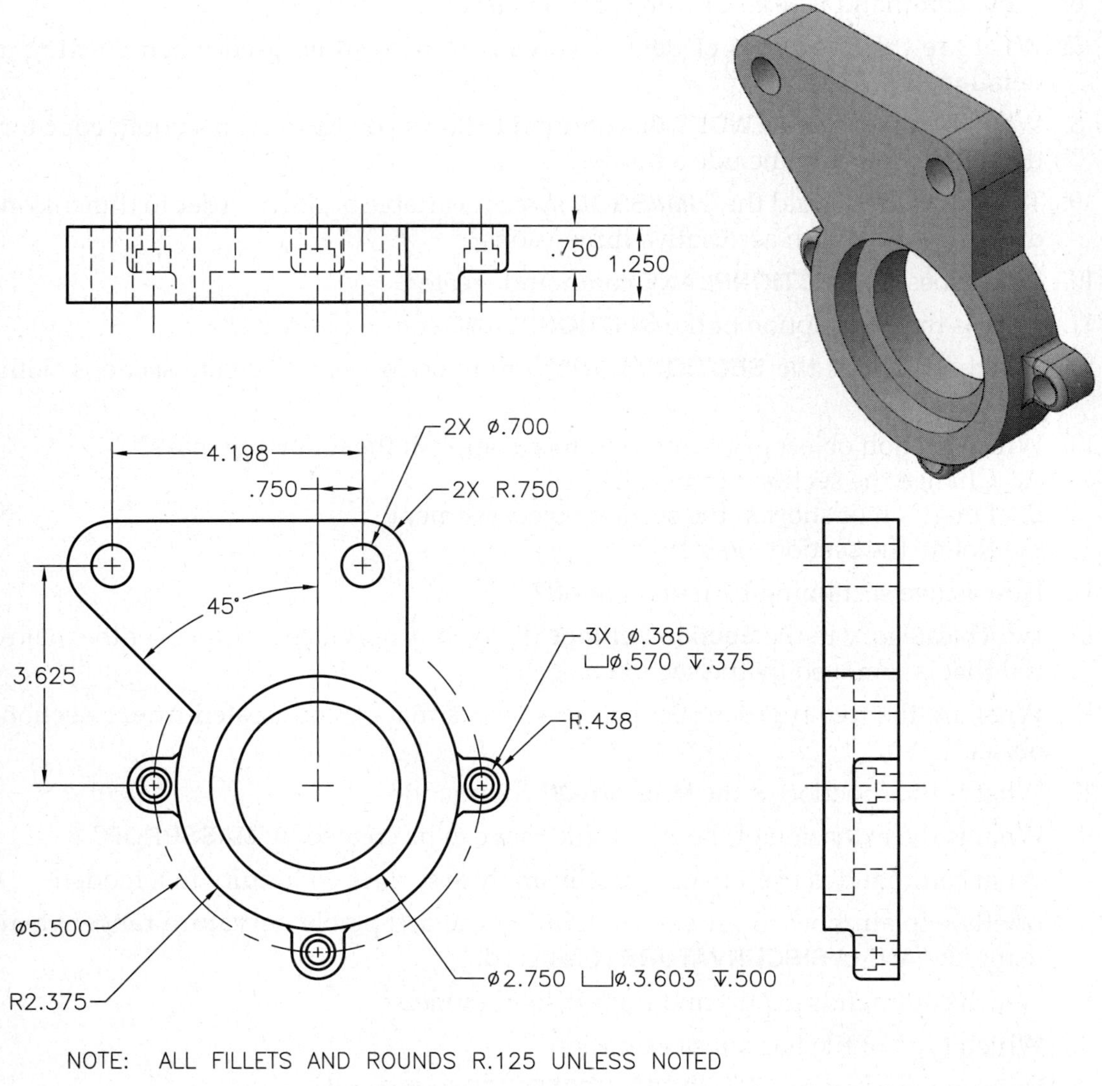

2. **Mechanical** In this problem, you will create a multiview layout of the flanged coupler by placing drawing views with the **VIEWBASE** command. Set up a page layout using the ANSI B (17.00 × 11.00 Inches) paper size. Select the appropriate scale for the orthographic views and the isometric view. Edit the views as needed to look like the layout shown.
   A. Construct the model shown using the dimensions provided.
   B. Create the front, right-side, and isometric views shown.
   C. The isometric view should be a shaded view.
   D. Dimension the drawing in the layout using the dimensions given. Make sure to use associative dimensions.
   E. Save the drawing as P14-2.

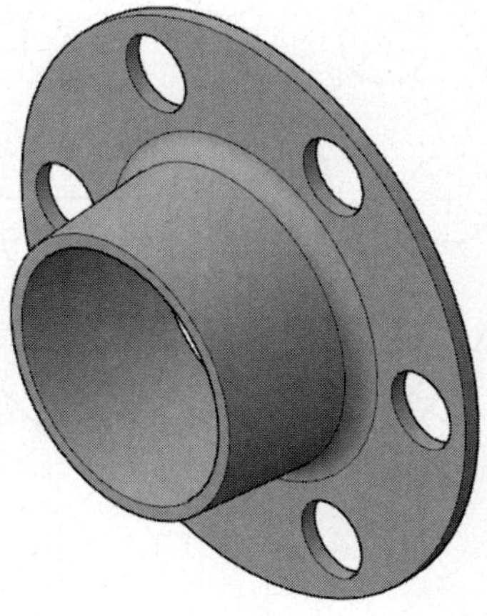

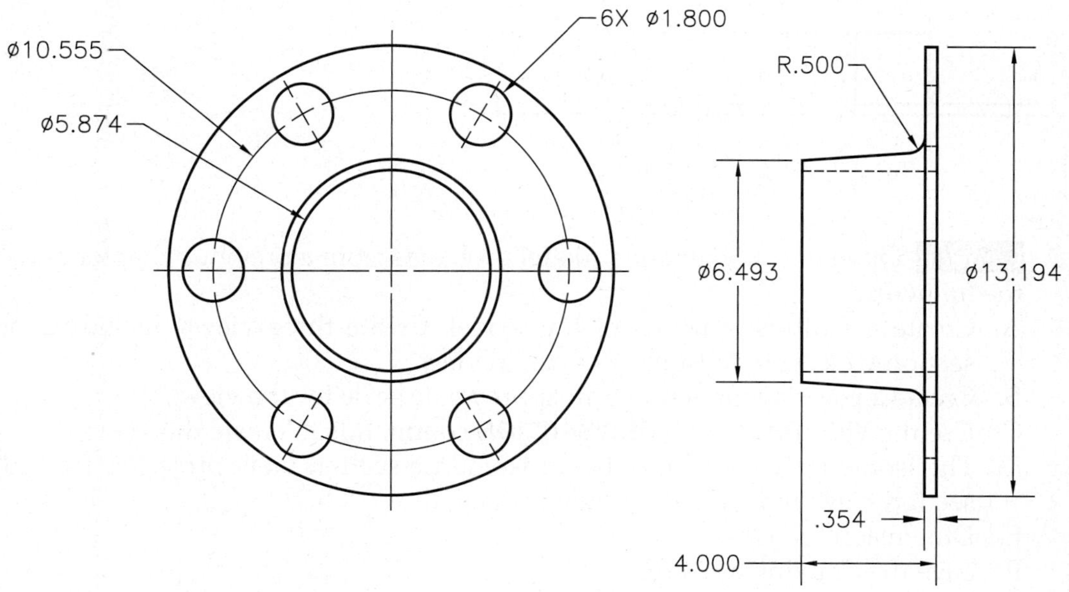

3. **General** Open one of your solid model problems from a previous chapter and do the following.
   A. Create a multiview layout of the model. Use three orthographic views and one isometric view.
   B. Create a page layout and use an appropriate scale for the views.
   C. Use the **VIEWBASE** command to create the views.
   D. The isometric view should be shaded.
   E. Dimension the orthographic views.
   F. Use the **3DPRINTSERVICE** command to export an STL file for 3D printing and name it P14-3.
   G. Save the drawing as P14-3.

Drawing Problems - Chapter 14

4. **Mechanical** In this problem, you will complete a drawing you created in Exercise 14-3 by adding dimensions to the views.
   A. Open drawing EX14-3.
   B. Referring to the layout shown, dimension the top and front views. Make sure to use associative dimensions.
   C. Add the note and the centerlines shown.
   D. Save the drawing as P14-4.

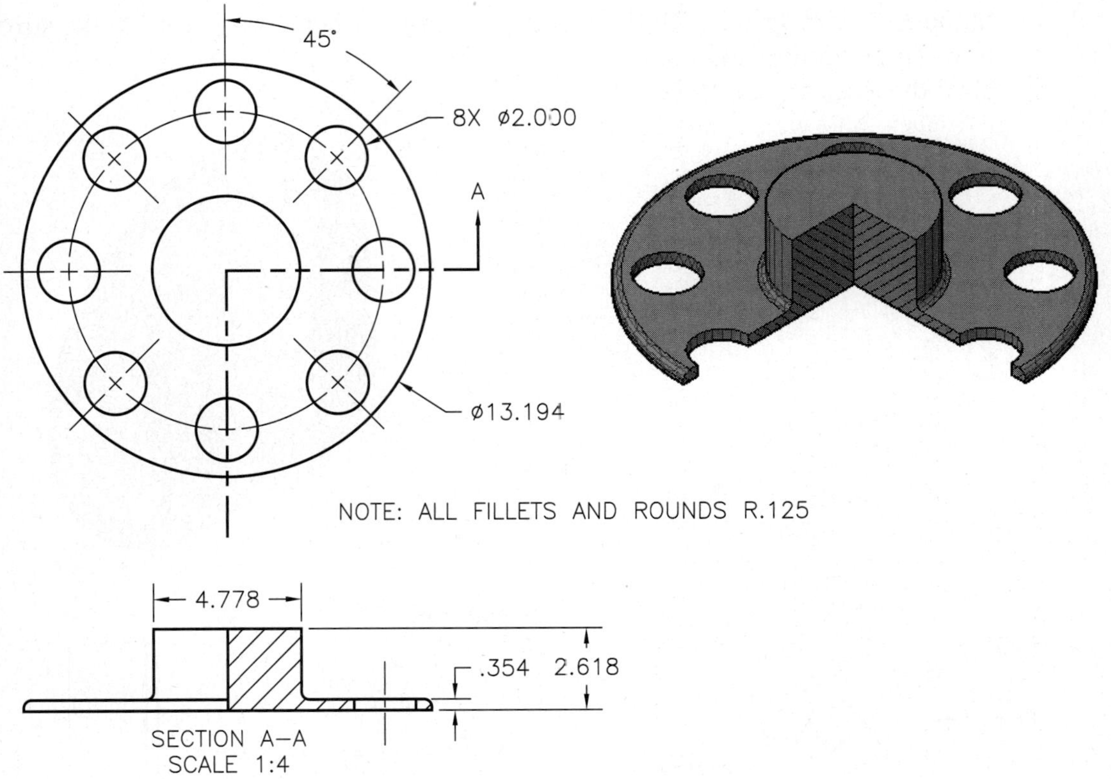

5. **General** Open one of your solid model problems from a previous chapter and do the following.
   A. Create a multiview layout of the model. Create three views, including one section view, and an isometric section view.
   B. Create a page layout and use an appropriate scale for the views.
   C. Use the **VIEWBASE** and **VIEWSECTION** commands to create the views.
   D. The isometric view should be an isometric section view projected from the section view and should be shown shaded.
   E. Dimension the views.
   F. Save the drawing as P14-5.

6. **General** Open one of your solid model problems from a previous chapter and do the following.
   A. Construct a section through the model using the **Draw section** option of the **SECTIONPLANE** command. Cut through as many features as possible.
   B. Display cutaway geometry with a 50% transparency.
   C. Display section lines using an appropriate hatch pattern.
   D. Generate a 3D section block that displays the cutaway geometry in a color of your choice.
   E. Create a layout with a viewport for the 3D block displayed at half the size of the original model.
   F. Save the drawing as P14-6.

7. **General** Choose five solid model problems from previous chapters and copy them to a new folder. Then, do the following.
   A. Open the first drawing. Export it as an SAT file.
   B. Do the same for the remaining four files.
   C. Compare the sizes of the SAT files with the DWG files. Compare the combined sizes of both types of files.
   D. Begin a new drawing and import one of the SAT files.

8. **Mechanical** Draw the object shown as a solid model. Only half of the object is shown; draw the complete object. Do not dimension the object. Then, do the following.
   A. Construct a section plane that creates a full section, as shown.
   B. Display the intersection fill as an ANSI hatch pattern.
   C. Activate live sectioning and view the cutaway geometry with a high level of transparency.
   D. Save the drawing as P14-8.

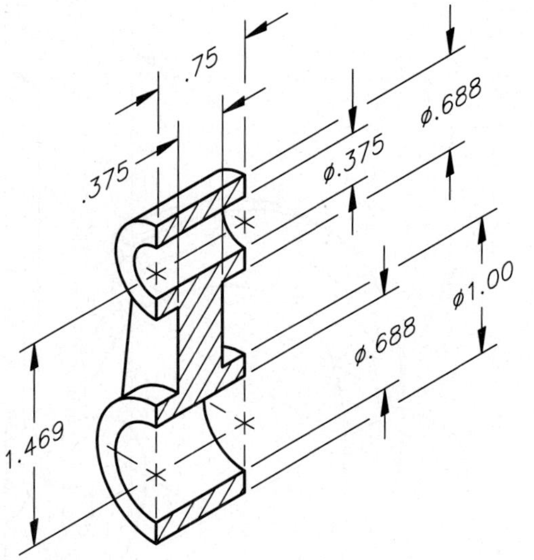

9. **Mechanical** Draw the object shown as a solid model. Do not dimension the object. Then, do the following.
   A. Construct a section object that creates a full section along the centerline of the hole.
   B. Generate a 2D section and display it on the drawing at half the size of the original. Specify section settings as desired.
   C. Generate a 3D section and display it on the drawing at half the size of the original. Do not display cutaway geometry. Specify section settings as desired.
   D. Activate live sectioning. Do not display the cutaway geometry.
   E. On the original solid model, display the intersection fill as an ANSI hatch pattern.
   F. Save the drawing as P14-9.

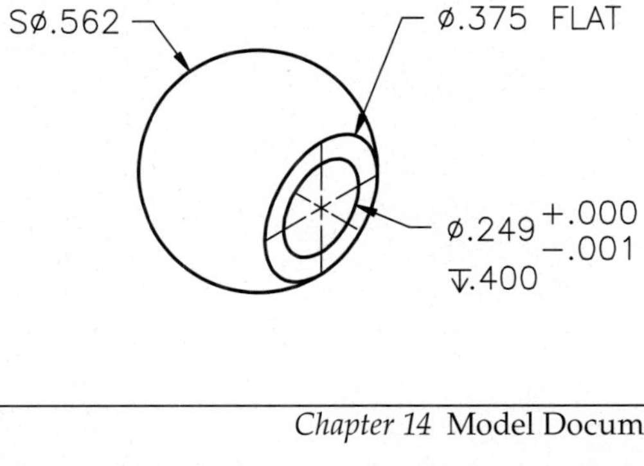

Drawing Problems - Chapter 14

10. **Mechanical** Draw the object shown as a solid model. Only half of the object is shown; draw the complete object. Do not dimension the object. Then, do the following.

A. Construct a section plane that creates a half section.
B. Display the intersection fill as an ANSI hatch pattern.
C. Activate live sectioning and view the cutaway geometry with a low level of transparency.
D. Generate a 2D section and display it on the drawing at full size. Specify section settings as desired.
E. Generate a 3D section and display it on the drawing at half the size of the original. Do not display cutaway geometry. Specify section settings as desired.
F. Save the drawing as P14-10.

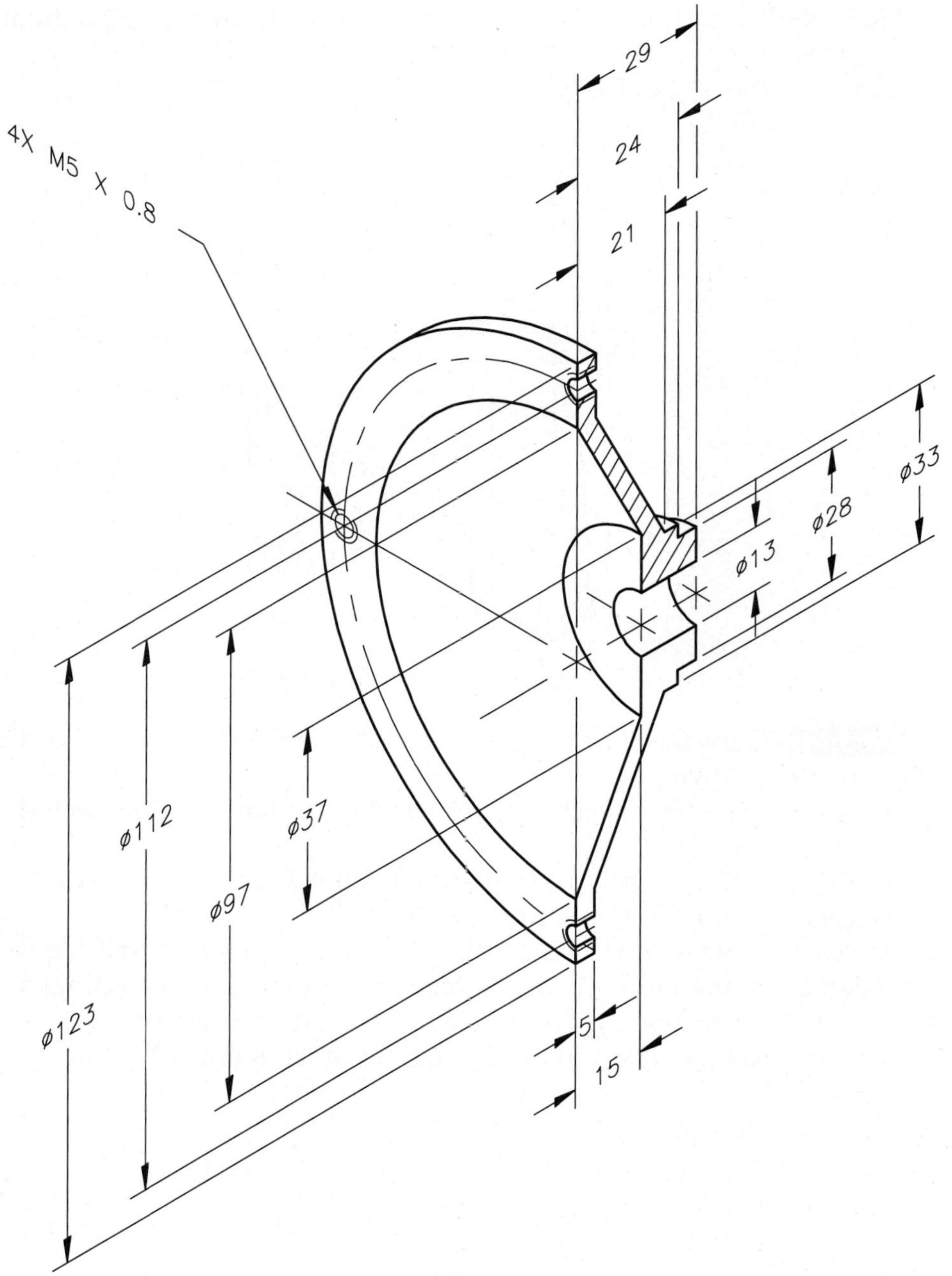

11. **Mechanical** Draw the object shown as a solid model. Use your own dimensions. Then, do the following.
   A. Construct a section plane that creates an offset section. The section should pass through the center of two holes in the base and through the large central hole.
   B. Display the intersection fill as an ANSI hatch pattern.
   C. Activate live sectioning and view the cutaway geometry with a low level of transparency in the color red.
   D. Generate a 3D section of the sectioned solid model and save it as a block.
   E. Create a two-view orthographic layout. Use the orthographic presets to create a top view and a front view. Use an appropriate scale to plot on a B-size sheet.
   F. Create a third floating viewport and insert the 3D section block scaled to half the size of the drawing.
   G. Save the drawing as P14-11.

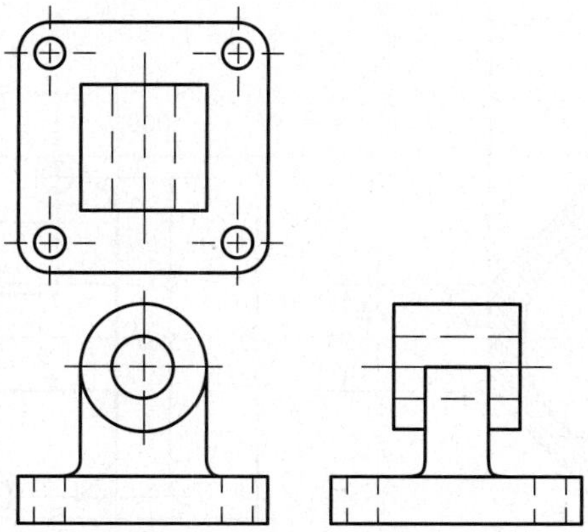

Drawing Problems - Chapter 14

12. **Mechanical** Draw the object shown as a solid model. Do not dimension the object. Then, do the following.
   A. Construct a section plane that creates a half section.
   B. Display the intersection fill as an ANSI hatch pattern.
   C. Activate live sectioning and view the cutaway geometry with a low level of transparency in the color red.
   D. Alter the section plane to create the section shown.
   E. Generate a 3D section and save it as a block.
   F. Create a two-view orthographic layout. Use the orthographic presets to create a top view and a front view. Use an appropriate scale to plot on an A-size sheet.
   G. Create a third floating viewport and insert the 3D section block scaled to half the size of the drawing.
   H. Save the drawing as P14-12.

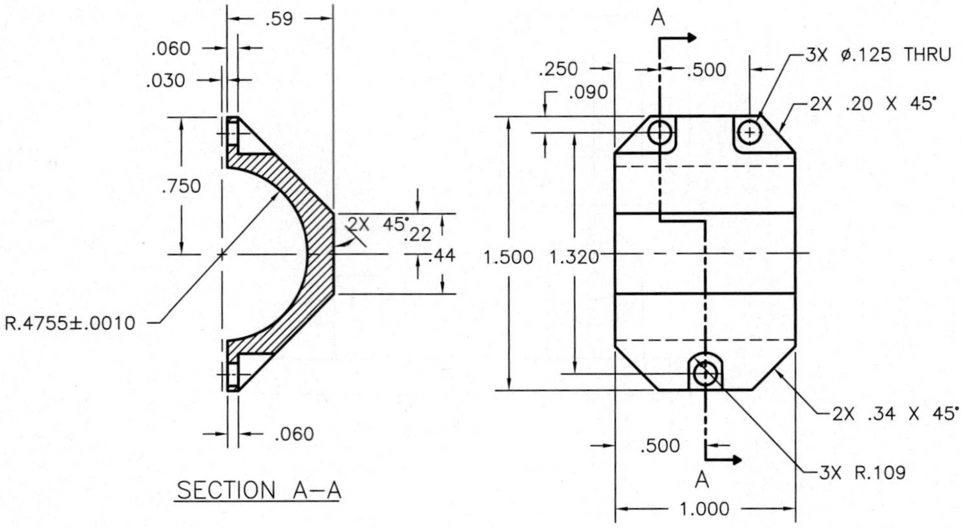

CHAPTER 15
Point Clouds

Learning Objectives

After completing this chapter, you will be able to:

- ✓ Identify applications for point clouds in 3D design and model creation.
- ✓ Use Autodesk ReCap Pro to import, register, and index scan files.
- ✓ Clean up and adjust the display of point clouds.
- ✓ Attach a point cloud to an AutoCAD drawing file.
- ✓ Extract 2D geometry from a point cloud.

A ***point cloud*** is a digital representation of an existing object that is constructed of large numbers of points in 3D space. A point cloud provides accurate 3D data that can be used for design and modeling purposes in a variety of applications, including building construction, surveying, manufacturing, accident and crime scene reconstruction, gaming, and computer animation.

AutoCAD provides tools for working with point clouds to help in 3D visualization, design, and model creation. You can use point clouds to:

- Visualize as-built conditions.
- Create new vector-based drawings using the point cloud data as a reference.
- Apply color mappings and transparency for improved visualization.
- Crop regions to show what is needed or not needed for different projects.
- Generate new geometry using extraction commands.

Working with point clouds in AutoCAD begins with converting raw data from ***scan files*** into a supported point cloud file format. The resulting point cloud file can then be attached to a drawing file for reference purposes. Working with a point cloud is much like working with an external reference file.

The process of importing and converting point cloud data is referred to as ***reality capture***. The following sections provide an introduction to point clouds and the tools available for working with point clouds in AutoCAD.

Laser Scanning

A ***laser scanner*** is a device used to capture data points from an existing structure or product and create ***scan files***. Laser scanning is used by engineers, designers, and animators. The scanning project is planned so that an entire building, facility, mechanical part, assembly, or geographic feature can be scanned with a single scan or multiple overlapping scans. The scanner can be mounted on a tripod or desktop, or held by hand. The scanner rapidly records millions of data points to create the scan file. Multiple scan files may be created, depending on the size of the project. Each data point in a scan file has XYZ values, but the resulting data is different from vector data, which has from-to XYZ values.

Raw scan files are produced in different file formats. Some of the more common scan file formats, such as Faro (.fls and .fws), Leica (.ptg, .pts, and .ptx), and Lidar (.las), are associated with different laser scanner manufacturers. In order to be used in AutoCAD, raw scan files must first be converted to a point cloud file. Point cloud files are created using Autodesk ReCap Pro.

Autodesk ReCap Pro

Autodesk ReCap Pro is a reality capture software program that indexes and registers point clouds. It is an application program that is installed separately from AutoCAD. Autodesk ReCap Pro provides the ability to process extremely large files containing raw scanned data. It contains tools for enhancing and modifying point cloud data for use in design work. Files created in Autodesk ReCap Pro can be used in AutoCAD, Autodesk Inventor, Autodesk Revit, and other Autodesk software, and can be shared online.

Autodesk ReCap Pro converts raw scan files to point cloud files through a process called ***indexing***. Indexed files are saved as reality capture scan (RCS) files. In a typical workflow, an individual RCS file or a set of multiple RCS files is associated with a reality capture project (RCP) file. An RCP file is used to organize information from scan files in a point cloud project. An RCP file references the individual RCS files, but does not contain the files. Either an RCS file or RCP file can then be attached to an AutoCAD drawing.

The process of ***registering*** scan files merges multiple scans into one coordinate system. Registering matches the 0,0,0 coordinate point of multiple scans to create a new merged point location for all scans. Registered scan files in a project are commonly described as a registered data set.

The process of working with scan files in Autodesk ReCap Pro consists of the following steps.

1. Create a new project (RCP file).
2. Import scan files.
3. Register the scan files, if necessary. If the scanning camera was relocated during the scanning process, the 0,0,0 coordinate location of each scan will be different. It will be necessary to register the scans to merge them into one coordinate system.
4. Index the scan files.
5. Clean up the files by removing points, creating scan regions, and changing the display colors.

NOTE

Autodesk ReCap Photo, a companion application to Autodesk ReCap Pro, is designed for photo-based projects. Autodesk ReCap Photo is used to process digital photos taken with a camera or aerial photos taken by an unmanned aerial vehicle (UAV), commonly referred to as a drone. Autodesk ReCap Photo can be used to convert data to an RCS file or a reality capture mesh (RCM) file.

Using Autodesk ReCap Pro

Autodesk ReCap Pro provides an interface for working with scan files and point cloud data. When you open Autodesk ReCap Pro, the home screen appears, **Figure 15-1**. The initial tools that are available allow you to start a new project, open an existing project, set system preferences, and access the help system. Any recent projects appear as thumbnails along the bottom of the home screen.

To start a new scan project, select **New project** at the top of the home screen and then select **Import point cloud**. You are prompted to enter a name for the project and assign a folder location for files associated with the project. Enter the new project name and select the folder where the files will be saved. Then, select **Proceed**. The **Import files** screen is displayed and prompts you to select the scan files to import. You can select individual files, select a folder containing files, or drag the files to the screen. If you choose to select individual files or a folder containing files, use the **Import Point Clouds** dialog box to select the files. You can select from a variety of file formats. Select one or more files and pick the **Open** button. The **Scan settings** screen appears, allowing you to adjust the scan settings used for the point cloud. The scan settings are discussed next.

Scan Settings

The point cloud scan settings allow you to control the filtering and size of the point cloud, the resolution, and the coordinate system. Filtering and clipping options are available when the **Scan settings** screen initially appears and the **Filtering** button is selected.

Figure 15-1.
The Autodesk ReCap Pro home screen. To start a new scan project, select **New project**.

See **Figure 15-2A**. Set the type of filtering by selecting one of the **Filter scans** options. The **Minimal** option preserves most points in the point cloud. The **Standard** option removes weak points. The **Aggressive** option removes points that are not associated with an object or surface. The clipping options define the clipping range and clipping intensity. The clipping range sliders set the minimum and maximum distances from the scanner at which points are preserved. The clipping intensity sliders set the minimum and maximum intensity values at which points are preserved. Points that are too weak or too intense are not included.

Advanced scan settings are accessed by selecting the **Advanced** button. See **Figure 15-2B**. Picking the **Decimation** button activates or disables the decimation grid

Figure 15-2.
Adjusting point cloud scan settings on the **Scan settings** screen. A—The **Scan settings** screen initially appears with the **Filtering** button selected, allowing you to set filtering and clipping options. B—Select the **Advanced** button to access settings for the decimation grid, coordinate system, and axis orientation.

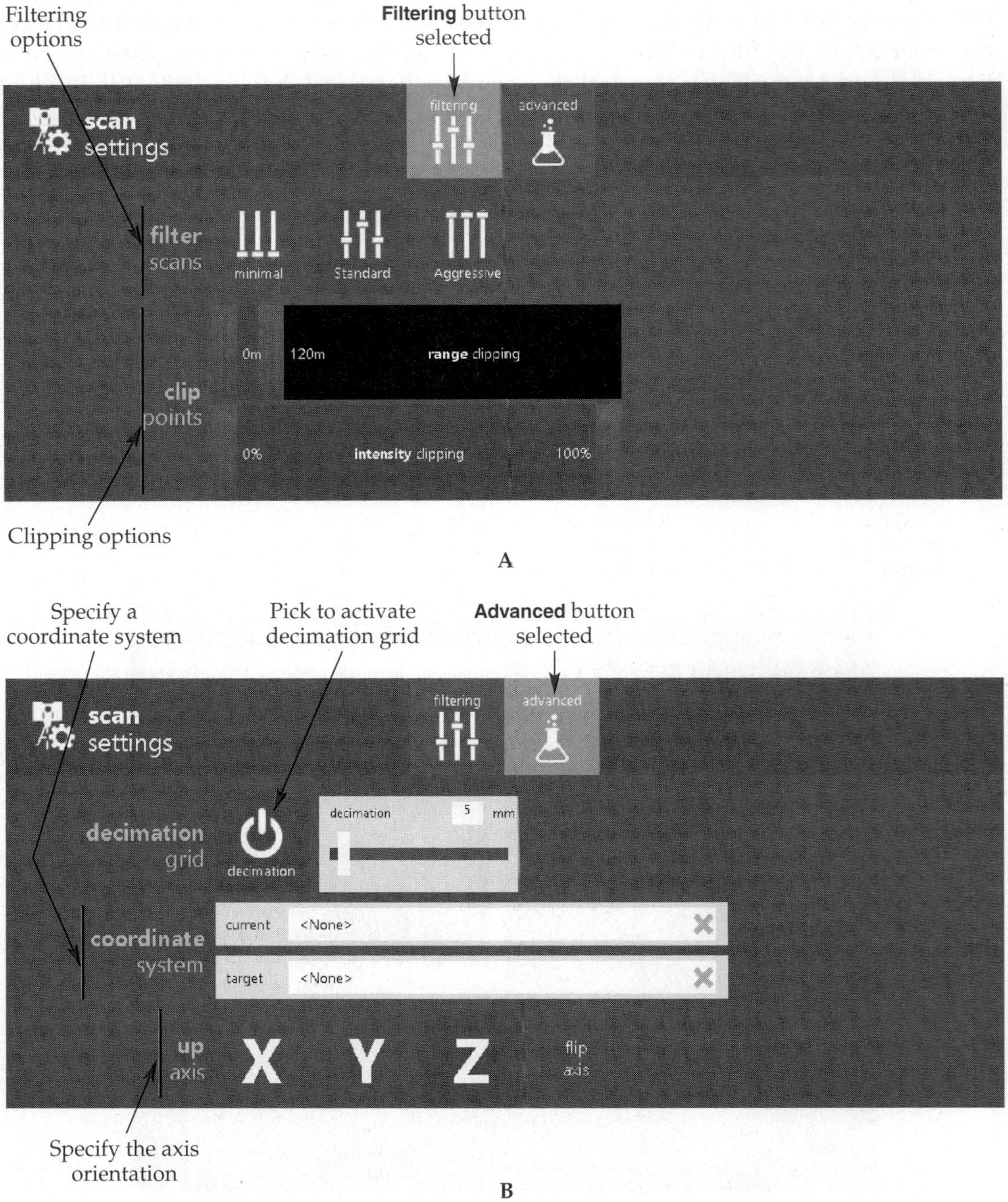

B

setting, which allows you to adjust the resolution, or density, of the point cloud. The value defines the minimum cubic volume of a single point. Drag the slider or enter a value in the text box to adjust the setting. A smaller value produces more points and a higher resolution, resulting in a larger file size. A higher value produces fewer points and a lower resolution, resulting in a smaller file size.

To align the point cloud with a coordinate system, use the **Coordinate system** text boxes. If a coordinate system has been assigned, it is listed in the **Current** text box. Use the **Target** text box to convert to a different coordinate system, if needed. The **Up axis** buttons are used to specify the axis orientation of the model. Pick the **X**, **Y**, or **Z** button to specify which axis is used to set the "up" direction of the model. The Z axis is the default "up" axis. Picking the **Flip axis** button flips the "up" axis orientation.

Once you are done adjusting settings, pick **Import files**. Importing begins and an icon representing each file appears near the top of the **Import files** screen. See **Figure 15-3**. Each icon has a circular progress bar indicating the import status. You can continue working while the files are importing. If you have selected multiple files, you can choose to register the scan files or proceed to indexing the scan files, as discussed next.

Registering Scans

As discussed earlier, registration is the process of aligning scan files so that different files in the project use the same coordinate system. This provides a way to accurately link together scans that are generated from different scanner positions. When scan files are importing, you can choose to skip registration, register the scan files manually, or have the program register the scan files automatically. To access these options, hover the cursor over the **Auto register scans** button on the **Import files** screen. If you choose to register the scan files manually, registration mode is activated and you are prompted to select a scan file to serve as the base file. Then, you select three matching pairs of points between the base file and another scan file.

Figure 15-3.
The **Import files** screen. Each file selected for importing is represented with an icon and the import status is indicated by a circular progress bar.

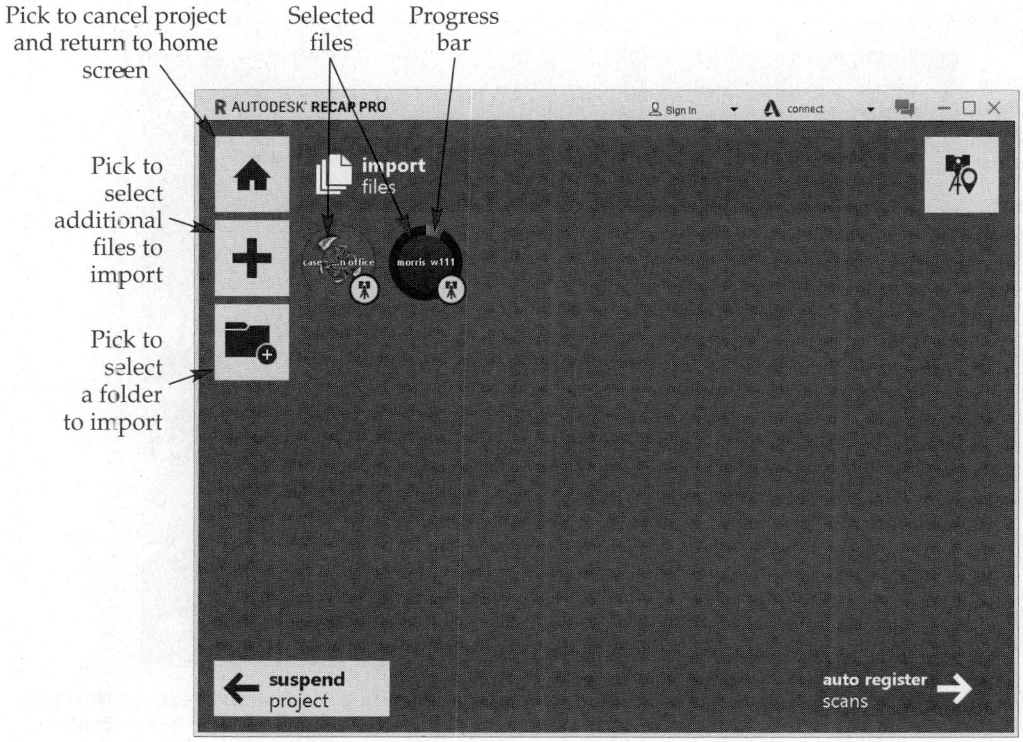

After selecting the points, a preview of the alignment appears and you are prompted to approve the registration. The process is repeated to register the remaining scan files.

If you do not want to register the scan files, you can skip registration. This allows you to start indexing files without performing registration. If you decide to have the program register the scan files automatically, registration is completed once the scan files are imported. During automatic registration, the scan files are processed so that they align properly when indexed together. Once registration is completed, a report is generated and you have the opportunity to preview the alignment. If the results are sufficient, you can proceed to indexing the scans.

Indexing Scans

Select **Index scans** on the **Register scans** screen to begin file indexing. An icon representing each file appears on the **Index scans** screen and the indexing progress is reported at the bottom of the screen. Once the files are indexed, select **Launch project** to begin working with the point cloud.

NOTE

Depending on your computer's memory and speed, indexing may take some time to complete.

Project Screen

The **Project** screen appears when you launch or open a project. It allows you to make edits and changes to the display. See **Figure 15-4**. For example, you can control the point display, view the point cloud in different colors and lighting, create scan regions, and use standard navigation tools to adjust the view. After making changes, you can then save the project.

Figure 15-4.
The **Project** screen appears when you launch a project or open an existing project. Shown is a point cloud of the interior of a college building. *(Pittsburg State University School of Construction)*

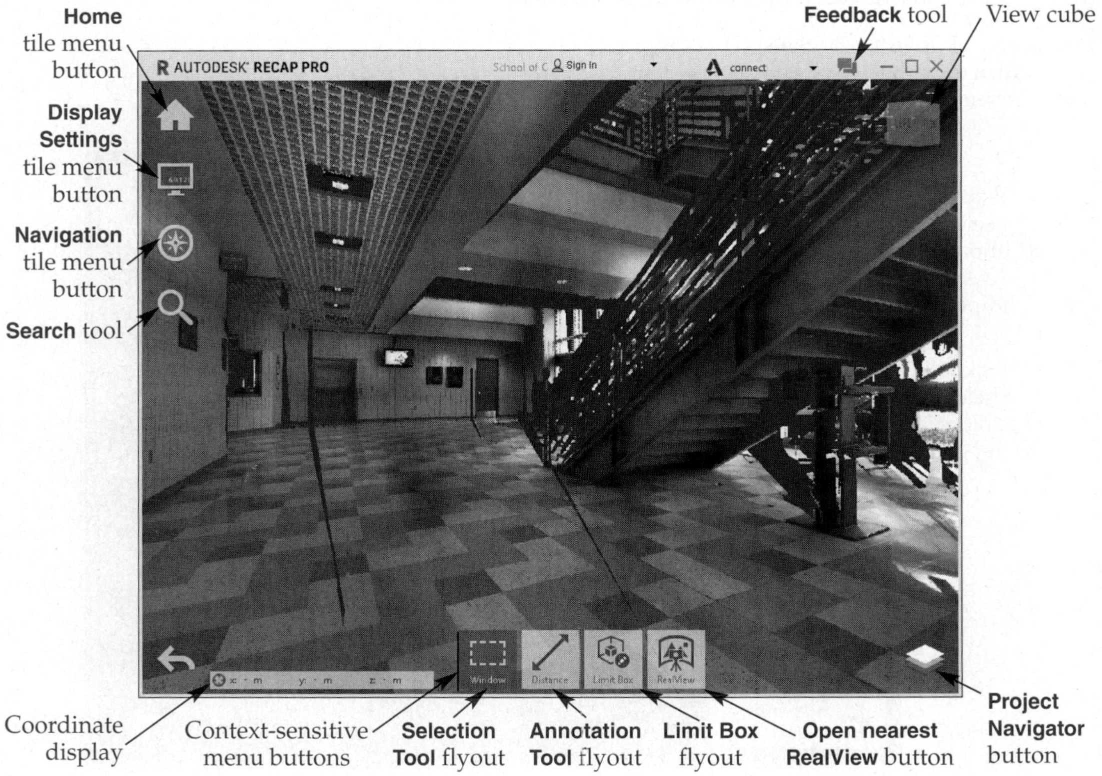

The left side of the **Project** screen includes four expandable tile menus and the **Search** tool button. To expand a tile menu, hover over the tile menu button. The expanded menus are shown in **Figure 15-5**. Many of the buttons in the tile menus display a flyout when you hover over the button. The **Home** tile menu contains tools for managing the project. The **Display Settings** tile menu contains tools for changing the color display, applying lighting, changing the point size, and toggling the display of navigation tools. The **Navigation** tile menu contains the standard navigation tools, such as zoom, pan, and orbit. The view cube, which is displayed by default in the upper-right corner of the **Project** screen, can also be used to adjust the view.

The **Search** tool is used to search for items within the project. This allows you to enter search terms to locate and edit specific items, such as scan regions or view states, which are discussed later in this section.

The **Project** screen contains additional tools at the bottom of the screen. The context-sensitive menu in the middle displays tools appropriate for the current mode, view, or selection. The **Selection Tool** flyout contains tools for making a window, fence, or plane selection in the point cloud. You can use these tools to delete points or create a clipping boundary. The **Annotation Tool** flyout contains tools for measuring a linear distance, measuring an angular value, and inserting a note at a selected point. The **Limit Box** flyout contains tools for configuring a limit box. A ***limit box*** is a cropped area defining the width, length, and depth of a three-dimensional portion of the point cloud. Picking the **Open nearest RealView** button displays a panoramic view of the point cloud. See **Figure 15-6**.

Figure 15-5.
Tile menus that appear on the **Project** screen. Hovering over a tile menu button displays the menu. A—**Home** tile menu. B—**Display Settings** tile menu. C—**Navigation** tile menu.

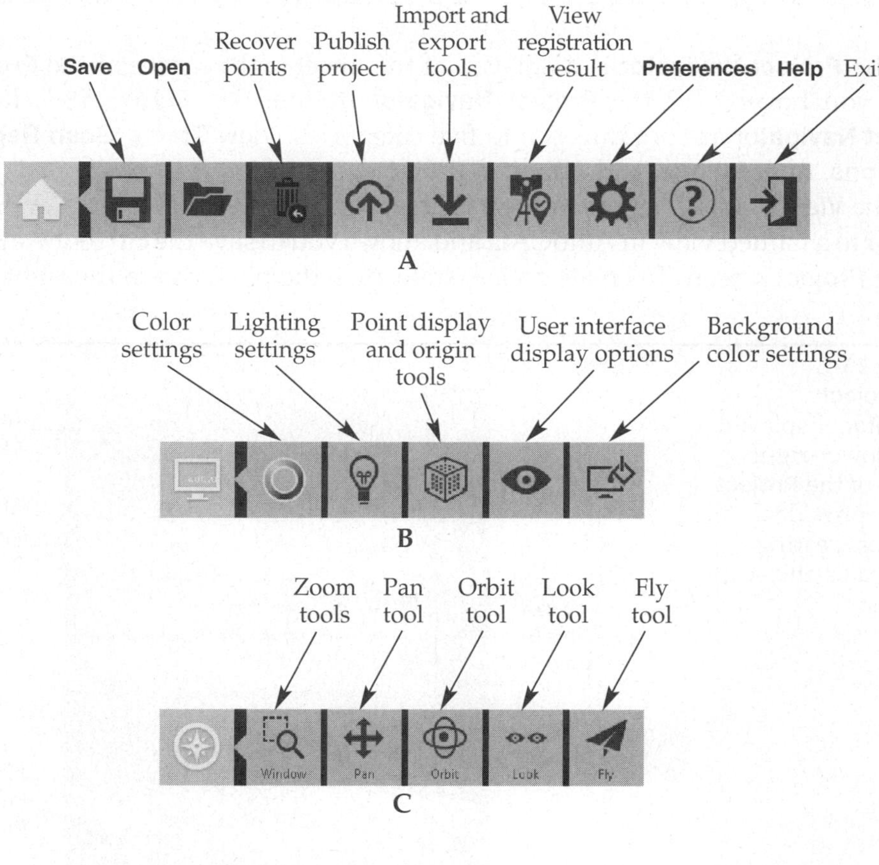

Figure 15-6.
Displaying a panoramic view of the point cloud shown in Figure 15-4. Notice how the view is rotated to show more of the supporting structure of the stairs. *(Pittsburg State University School of Construction)*

The **Project Navigator** is displayed at the lower-right corner of the **Project** screen when you hover over the **Project Navigator** button. See **Figure 15-7**. Items in the **Project Navigator** are organized into five categories: **View States**, **Scan Regions**, **Scan Locations**, **Annotations**, and **Attached Projects**.

The **View States** category is used to create and display view states. A ***view state*** is similar to a named view in AutoCAD and allows you to save the current view displayed on the **Project** screen. To create a view state, pick the plus icon to the right of the **View**

Figure 15-7.
The **Project Navigator**, displayed at the lower-right corner of the **Project** screen, provides tools for creating view states and scan regions.

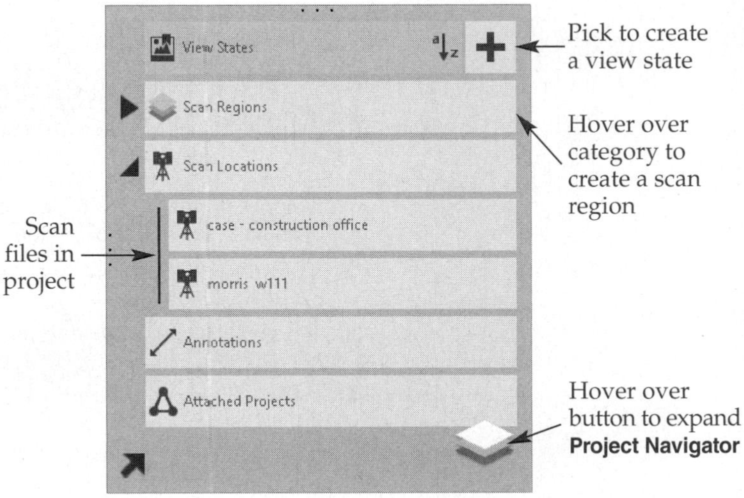

States category. Enter a name for the new view state and press [Enter]. You can then pick on the name of the view state to restore it after making display changes. Once defined, view states can be used to create an animated presentation by playing them as a sequence of keyframes. The **Scan Regions** category is used to create and display scan regions. A ***scan region*** is a portion of the point cloud representing a three-dimensional area. Before creating a scan region, define a limit box to represent the extents. Then, pick the plus icon to the right of the **Scan Regions** category. Enter a name for the new scan region and press [Enter]. When you hover over the name of the scan region, the corresponding portion of the point cloud is highlighted. The **Scan Regions** category also contains tools for exporting scan regions and importing scan region data.

The **Scan Locations** category lists all scan files used in the project. The **Annotations** category lists annotations created in the project. If an Autodesk Navisworks model has been attached to the scan project, it is listed in the **Attached Projects** category.

Once you are done making changes, pick the **Save** button in the **Home** menu to save the project. The point cloud is now ready to be inserted in AutoCAD.

Exercise 15-1

www.g-wlearning.com/CAD/

Complete the exercise on the companion website.

Inserting a Point Cloud File

Inserting a point cloud file into an AutoCAD drawing file allows you to view the point cloud model and use it for reference or design purposes. Use the **POINTCLOUDATTACH** command to attach a point cloud file. In the file selection dialog box that appears, navigate to the point cloud file and select **Open**. You can attach an RCS or RCP file. After selecting a file, the **Attach Point Cloud** dialog box appears with a preview of the point cloud model. Use this dialog box to specify insertion options. The options are similar to those available when attaching an external reference. Specify the path type, insertion point, scale, and rotation. Check the **Use geographic location** check box if geographic data is available from the point cloud file and you want to insert the point cloud using geographic coordinates in the drawing file. Check the **Lock point cloud** check box if you want to lock the position of the point cloud and prevent it from being moved or rotated. Check the **Zoom to point cloud** check box to automatically zoom to the point cloud. This is the default option. Selecting the **Show Details** button displays additional information about the point cloud, including the number of points.

Type
POINTCLOUD-ATTACH
PCATTACH

POINTCLOUDATTACH

NOTE

The maximum number of points displayed for point clouds is controlled by the **Maximum point cloud points per drawing** value in the **3D Modeling** tab of the **Options** dialog box. The maximum value is 25 million. Adjust the initial setting if needed before accessing the **POINTCLOUDATTACH** command. Set the maximum value to 25 million to get the largest density possible for the point cloud.

Exercise 15-2

www.g-wlearning.com/CAD/

Complete the exercise on the companion website.

Point Cloud Display Options

After the point cloud is attached, you can adjust the viewpoint and display as needed. To access point cloud display options, select the point cloud to access the **Point Cloud** contextual ribbon tab. See **Figure 15-8**. The **Display**, **Visualization**, **Cropping**, and **Options** panels provide display options and tools. The options in the **Extract** and **Section** panels are used to extract 2D drawing geometry and create section objects, as discussed later in this chapter.

The tools in the **Display** panel allow you to adjust the point size and density. Use the slider bars to adjust the values, if needed. The navigation tools in the **Display** panel can be used to manipulate the viewpoint or change to a parallel or perspective projection.

The options in the **Visualization** panel allow you to adjust the display color, transparency, and lighting. The options in the **Stylization** drop-down menu are used to set the color style. The **Scan Colors** option uses the original scan colors. The **Object Color** option uses the color assigned to the point cloud object. The **Normal** option assigns colors based on the normal direction of each point. The colors used are based on values assigned to the X, Y, and Z axes. The **Intensity** option assigns colors based on point intensity values. The **Elevation** option assigns colors based on the Z coordinate values of points. The **Classification** option assigns colors based on point classifications.

Figure 15-8.
The **Point Cloud** contextual ribbon tab. *(Model courtesy of Laser Design and GKS Services)*

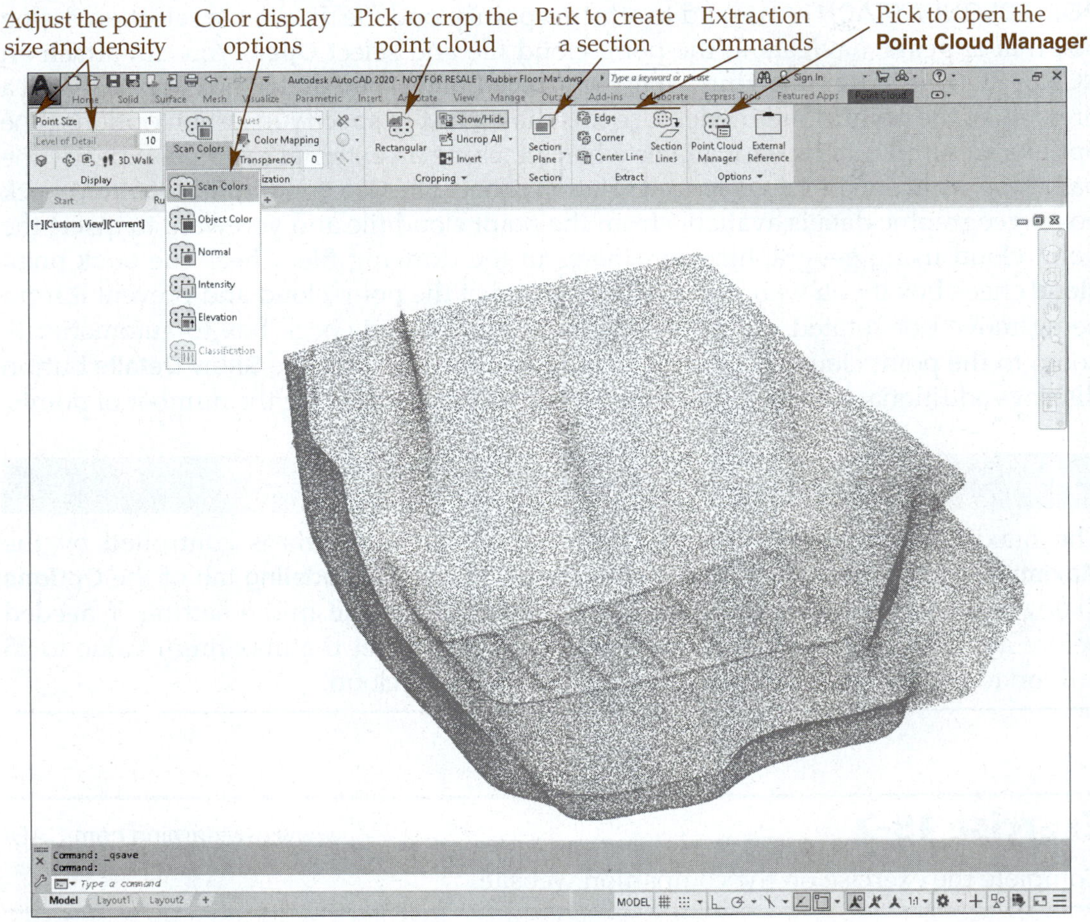

This option is available if classifications are defined in the scan file. Picking the **Color Mapping** button in the **Visualization** panel opens the **Point Cloud Color Map** dialog box, which can be used to customize the **Intensity**, **Elevation**, and **Classification** color styles. Color schemes available for the **Intensity**, **Elevation**, and **Classification** color styles can be selected from the **Color Schemes** drop-down list.

The **Transparency** option is used to apply transparency to the point cloud. Use the slider or the text box to adjust the value. The value can range from 0 to 90. As with other transparency settings in AutoCAD, the higher the transparency value, the greater the transparency.

The **Lighting** drop-down list can be used to adjust the lighting applied to the scene. The default option is **No Lighting**. If you select the **Single Sided** or **Double Sided** option, you can set the shading and light source using the **Shading** and **Light Source** drop-down lists. Lighting is discussed in detail in Chapter 18.

The options in the **Cropping** panel allow you to crop the point cloud using a rectangular, polygonal, or circular boundary. Selecting one of the options from the **Crop** drop-down list initiates the **POINTCLOUDCROP** command and sets the type of boundary used for cropping. Cropping enables you to show relevant areas without displaying the entire point cloud. See **Figure 15-9**. When defining the cropping boundary, you can specify whether to keep the points inside or outside the crop. Once the crop is defined, you can invert, hide, or remove the crop. You can also save the cropped view as a crop state. To save a crop state, expand the **Cropping** panel and pick the **New Crop State** button to activate the **POINTCLOUDCROPSTATE** command. When prompted, enter a name for the crop state. The **Crop State** drop-down list in the expanded **Cropping** panel lists all crop states for the point cloud and can be used to restore a previously created crop state. To delete a crop state, use the **POINTCLOUDCROPSTATE** command.

Ribbon
Point Cloud
> Cropping
Crop
Type
POINTCLOUD-CROP

POINTCLOUDCROP

Ribbon
Point Cloud
> Cropping
New Crop State
Type
POINTCLOUD-CROPSTATE

POINTCLOUDCROPSTATE

Figure 15-9.
Cropping a point cloud. A—A portion of the interior of a building has been cropped to show a wall corner. B—Further cropping is used to isolate a portion of the end wall. *(Pittsburg State University School of Construction)*

A B

NOTE

The projection must be set to parallel in order to crop a point cloud.

POINTCLOUDMANAGER

Ribbon
Point Cloud > Options

Point Cloud Manager

Type
POINTCLOUD-MANAGER

Selecting the **Point Cloud Manager** button in the **Options** panel displays the **Point Cloud Manager**, which provides information about all point clouds attached to the drawing. See **Figure 15-10**. Any scans and scan regions associated with the point cloud are listed. You can control the display of each item by picking the **On/Off** button to the right of the item. To isolate the display of a scan region, right-click on the name and select **Isolate** from the shortcut menu. This displays the scan region and hides other items. The **On** and **Off** options can also be selected from the shortcut menu.

By default, the point cloud is framed by a bounding box. You can control the display of the bounding box by selecting an option from the **Show Bounding Box** drop-down list, which is located in the expanded **Options** panel. Selecting the **External Reference** button in the **Options** panel displays the **External References** palette.

Exercise 15-3

www.g-wlearning.com/CAD/

Complete the exercise on the companion website.

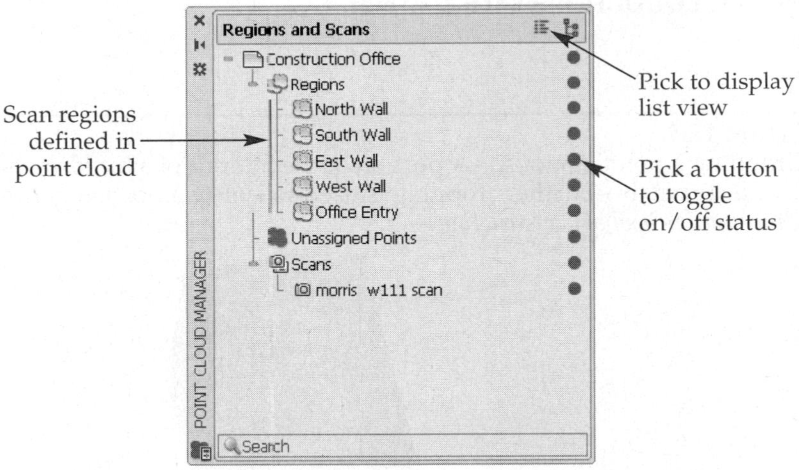

Figure 15-10.
The **Point Cloud Manager** with the default tree view mode active.

Working with Point Clouds

As you have seen, point clouds provide a powerful visualization tool for use in AutoCAD modeling applications. However, point clouds can also be used as a reference in creating new drawing geometry. For example, you can create new 2D geometry based on the existing features in a point cloud, such as the walls and floors in a building. If you are using the 2D geometry to create profile shapes, you can then construct 3D models by extruding or applying other 3D modeling techniques. Several methods for creating new drawing geometry based on point clouds are available in AutoCAD. You can use an existing surface in a point cloud as a drawing surface to construct 2D or 3D objects. In addition, you can extract 2D geometry from point clouds by using extraction commands. Another technique is to create a section object that cuts through a point cloud and then extract 2D geometry from the resulting section boundary. These methods are discussed in the following sections.

NOTE

Hardware acceleration must be enabled in order to work with point clouds. For optimal results, make sure your computer's graphics card is updated with the most current drivers installed.

Point Cloud Drawing Aids and Construction Tools

As with other types of 3D model surfaces, you can orient the UCS to a planar face in a point cloud for construction purposes. Use the **Object** option of the **UCS** command to orient the UCS to a surface in the point cloud. Once the UCS is properly oriented, use standard 2D and 3D drawing commands and object snaps to create new geometry.

AutoCAD provides object snaps for accurately selecting points in a point cloud. Point cloud object snaps assist in creating new geometry and taking measurements of existing features. The available point cloud object snaps are **Node, Intersection, Edge, Corner, Nearest to plane, Perpendicular to plane, Perpendicular to edge**, and **Centerline**. These are set in the **3D Object Snap** tab of the **Drafting Settings** dialog box, introduced in Chapter 1.

NOTE

In order to use some of the point cloud object snaps available in AutoCAD, the point cloud must contain ***segmentation data***. Segmentation data is obtained by identifying groups of neighboring points that make up the same planar or non-planar surface. Segmentation data is processed when scan files are indexed in Autodesk ReCap Pro. Accurate processing of segmentation data is dependent on the type of laser scanner used and the work that is scanned. To determine whether the point cloud contains segmentation data, select the point cloud, open the **Properties** palette, and check the status of the Segmentation property in the **Misc** category.

Another way to orient the UCS to a planar surface on a point cloud is to use a dynamic UCS. Dynamic UCS behavior when detecting a point cloud surface is similar to that when working with solid models. Hover the cursor over the appropriate surface to align the XY plane with the surface and then draw the new object. When the command is completed, the UCS reverts to its previous location. In order for the dynamic UCS function to work correctly, make sure to turn off 3D object snaps before hovering over the desired surface.

Extracting Geometry from Point Clouds

You can generate new 2D geometry from point clouds by using extraction commands to copy or "extract" objects from existing surfaces. You can extract edges of planar segments, corners of planar segments, and centerlines of cylindrical segments. The new geometry is created on the current layer and can be modified in the same way as other 2D objects. The **PCEXTRACTEDGE, PCEXTRACTCORNER**, and **PCEXTRACTCENTERLINE** commands are used to extract geometry from point clouds. These commands are accessed in the **Extract** panel in the **Point Cloud** contextual ribbon tab. Refer to **Figure 15-8**.

PCEXTRACTEDGE

Ribbon
Point Cloud
> Extract
Extract Edge
Type
PCEXTRACTEDGE

PCEXTRACTCORNER

Ribbon
Point Cloud
> Extract
Extract Corner
Type
PCEXTRACT-CORNER

PCEXTRACTCENTERLINE

Ribbon
Point Cloud
> Extract
Extract Centerline
Type
PCEXTRACT-CENTERLINE

PCEXTRACTSECTION

Ribbon
Point Cloud
> Extract
Section Lines
Type
PCEXTRACT-SECTION

The **PCEXTRACTEDGE** command is used to extract an edge where two planar surfaces intersect. Access the command and hover the cursor over the first planar surface. When the surface becomes highlighted, select the surface. Then, hover the cursor over a second planar surface that is adjacent to the first and select the surface. The edge at the intersection of the two surfaces is extracted as a line.

The **PCEXTRACTCORNER** command is used to extract a point where three planar surfaces intersect. Access the command and hover the cursor over the first planar surface. When the surface becomes highlighted, select the surface. Then, hover the cursor over a second planar surface that is adjacent to the first and select the surface. In the same fashion, select a third planar surface that is adjacent to the first two surfaces. A point object is generated at the intersection of the three surfaces.

The **PCEXTRACTCENTERLINE** command is used to extract a line along the center axis of a cylindrical object. Access the command and hover the cursor over a cylindrical surface. When the surface becomes highlighted, select the surface. A line coinciding with the center axis of the surface is extracted.

Point Cloud Sections

Section objects, introduced in Chapter 14, can be applied to point clouds by using the **SECTIONPLANE** command. The **Section** panel in the **Point Cloud** contextual ribbon tab provides access to the orthographic section plane options. Picking the **2 Point** option activates the **SECTIONPLANE** command.

Once the section plane is created, you can use it to extract the lines forming the section of the point cloud. This is an alternative to using other point cloud extraction commands. To extract section lines, make sure that live sectioning is turned on for the section object. Then, select the point cloud and access the **PCEXTRACTSECTION** command. The **Extract Section Lines from Point Cloud** dialog box appears, **Figure 15-11.** This dialog box is used to specify settings for the extracted geometry. Selecting the **Entire cross section** option in the **Extract** area extracts geometry from the area in which the section plane cuts the point cloud. Selecting the **Perimeter only** option extracts geometry along the section plane perimeter. The options in the **Output geometry** area are used to set the layer and color assigned to the geometry and specify whether lines or polylines are extracted. If you are extracting polylines, you can set the resulting polyline width. The **Preview result** check box located at the bottom of the dialog box

Figure 15-11.
The **Extract Section Lines from Point Cloud** dialog box.

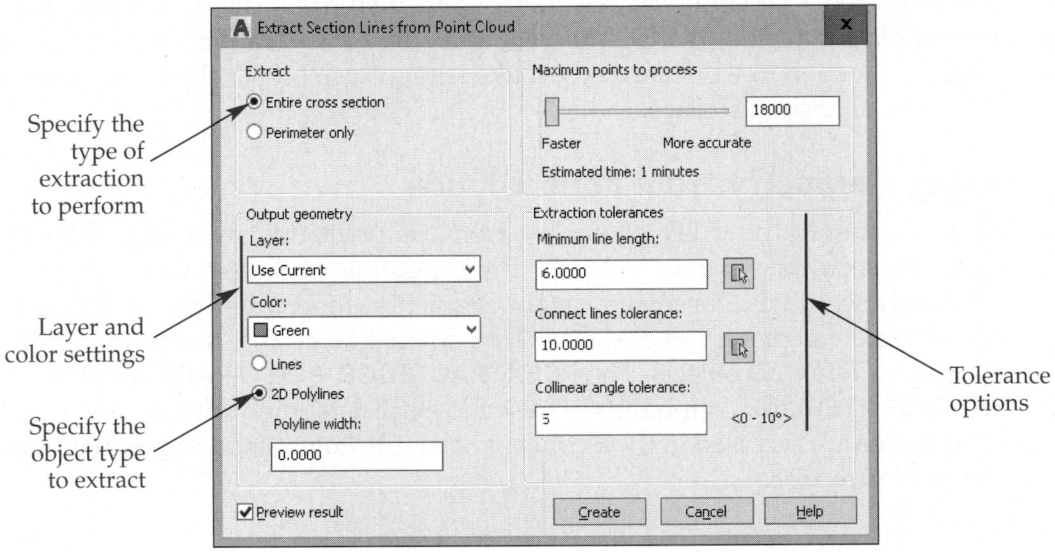

is checked by default. This allows you to preview the extraction results and adjust settings before completing the command.

The slider in the **Maximum points to process** area is used to specify the number of points processed. A higher value increases the accuracy of the extracted geometry, but also increases processing time. At higher settings, the extraction may take a very long time to process, depending on your computer. A time estimate appears below the slider and updates as you move the slider.

The **Extraction tolerances** area contains tolerance settings that specify how the geometry is generated. The **Minimum line length:** value sets the minimum length of extracted objects. Enter a value or pick the button next to the text box to set the length by picking two points in the drawing. The **Connect lines tolerance:** value sets the distance below which neighboring line and arc segments are joined. Enter a value or pick the button next to the text box to set the distance by picking two points in the drawing. The **Collinear angle tolerance:** setting controls the number of object segments generated based on the angle between neighboring segments. The higher the setting, the fewer segments generated. The setting can range from 0 to 10.

After making the appropriate settings, pick **Create** to close the dialog box and begin processing. Once processing is completed, a preview of the extracted geometry is displayed and you are prompted to accept the results. Select **Accept** to complete the operation. Selecting the **Settings** option redisplays the **Extract Section Lines from Point Cloud** dialog box, where you can change the settings to produce different results. Selecting the **Undo** option deletes the extracted geometry and allows you to continue the command. You can use the **Settings** option to adjust the extraction settings or exit the command.

In **Figure 15-12A**, the point cloud has been sectioned using an orthographic section plane. The results after using the **PCEXTRACTSECTION** command are shown in **Figure 15-12B**. In this example, polylines are extracted along the entire cross section. The resulting polylines can be edited by using grips or standard editing commands, such as **PEDIT** and **JOIN**.

Exercise 15-4

www.g-wlearning.com/CAD/

Complete the exercise on the companion website.

Figure 15-12.
Extracting section lines from a point cloud. A—A point cloud sectioned with an orthographic section object using the **Top** option. B—The drawing after extracting polylines from the point cloud using the **Entire cross section** option. The point cloud is shown with transparency applied for illustration purposes only. *(Model courtesy of Laser Design and GKS Services)*

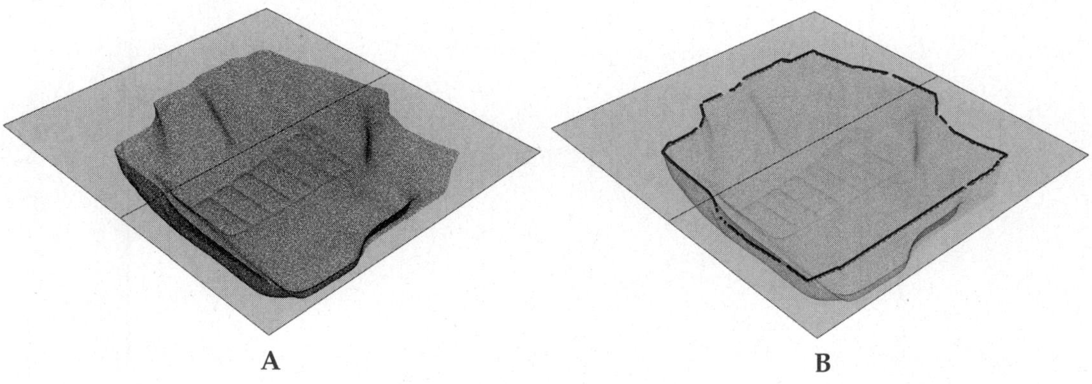

Chapter Review

Answer the following questions using the information in this chapter.

1. What is a *point cloud*?
2. Name at least three applications for which a point cloud can provide accurate data for design and modeling purposes.
3. In Autodesk ReCap Pro, what is the name of the process in which scan files are converted to point cloud files?
4. In Autodesk ReCap Pro, what is the name of the process in which multiple scans are merged into one coordinate system?
5. What is a *limit box*?
6. What is a *scan region*?
7. What command is used to attach a point cloud in AutoCAD?
8. Explain how to adjust the point size and density of a point cloud after it has been inserted into an AutoCAD drawing file.
9. What command is used to extract a line object from an edge in a point cloud where two planar surfaces intersect?
10. What command is used to extract a point object from a point cloud where three planar surfaces intersect?

Drawing Problems

1. **Mechanical** In this problem, you will add model geometry to a point cloud of a floor mat design. Refer to the model shown.
   A. Copy the drawing Floor Mat.dwg available on the companion website to a folder of your choice. Copy the file Floor Mat.rcs to the same folder location. Open the drawing Floor Mat.dwg.
   B. Set the point cloud transparency to 35.
   C. Turn on 3D object snaps. Set the **Node** point cloud object snap as the running object snap mode.
   D. Set the Nibs layer current. Draw three small cylinders in the first bottom section of the floor mat as shown. The cylinders are ∅20″ and 5″ in height. These cylinders will be located at the bottom of the floor mat to act as carpet nibs. (The top surface of each cylinder should be underneath the floor mat.)
   E. Copy the cylinders to the locations shown. There should be 18 total cylinders.
   F. Save the drawing as P15-1.

(Model courtesy of Laser Design and GKS Services)

Drawing Problems - Chapter 15

2. **Architectural** In this problem, you will add overhead speakers to a point cloud of an amphitheater stage design. Refer to the model shown. The point cloud is shown with transparency applied for illustration purposes only.

A. Copy the drawing Amphitheater Stage.dwg available on the companion website to a folder of your choice. Copy the file Amphitheater Stage.rcs to the same folder location. Open the drawing Amphitheater Stage.dwg.

B. Use the **Move and Align** UCS icon origin grip menu option to move the UCS to the front face of the amphitheater roof. Use the **Nearest plane of point cloud** 3D object snap to select the face.

C. Set the Speaker_Support layer current. Using the **Nearest to edge of point cloud** 3D object snap and the **Axis endpoint** option of the **CYLINDER** command, draw a 4-unit cylinder from the front bottom edge of the right side roof to the front bottom edge of the left side roof on the opposite end.

D. Use the **3 Point** UCS icon shortcut menu option to align the UCS with the right end of the cylinder. The X axis should be aligned with the center axis. Refer to the model shown.

E. Set the Speakers layer current. Insert one instance of the Roof Speaker block. Use the **Center of face** 3D object snap to center the block along the cylindrical support.

F. Use the **Array** option of the **COPY** command to make 5 copies of the speaker to the right and left. Spacing is 50 units.

G. Save the drawing as P15-2.

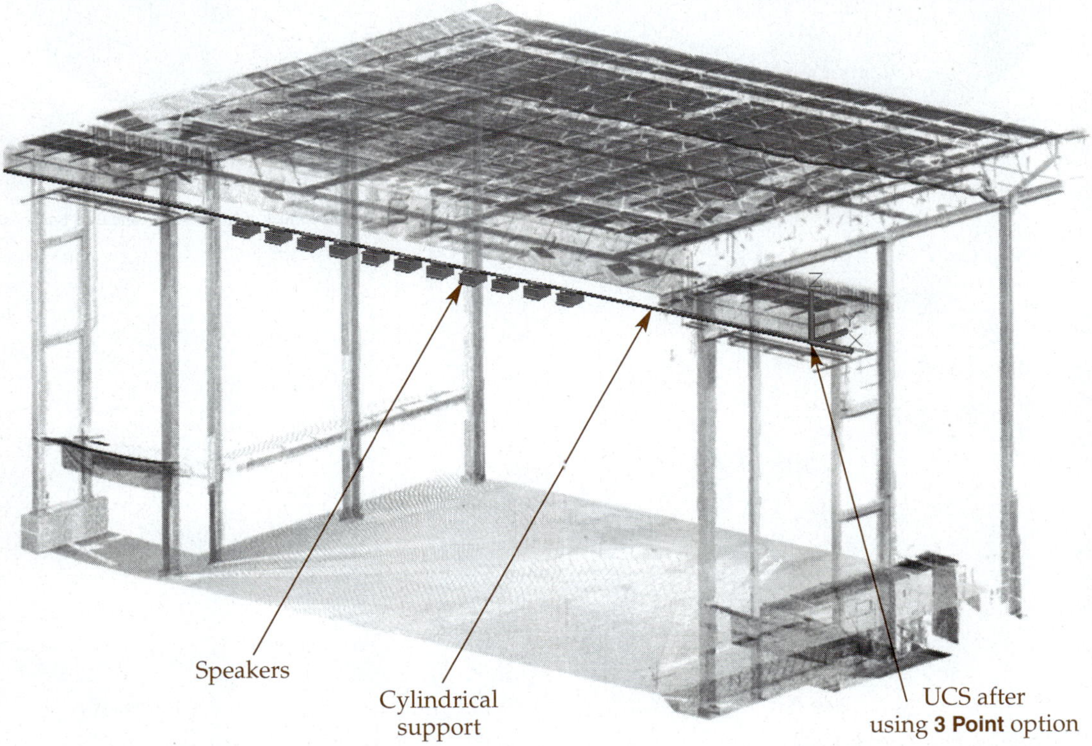

(Red Rocks Amphitheatre, City & County of Denver)

CHAPTER 16

Visual Style Settings and Basic Rendering

Learning Objectives

After completing this chapter, you will be able to:

✓ Control the display of solid models.
✓ Describe the **Visual Styles Manager** palette.
✓ Change the settings for visual styles.
✓ Create custom visual styles.
✓ Export visual styles to a tool palette.
✓ Render a scene using sunlight.
✓ Save a rendered image from the **Render** window.

There are many options available in AutoCAD for controlling the display of models. These settings help with your visualization while you are working on a model and may also impact printed output. In Chapter 1, you were introduced to the default visual styles. In this chapter, you will learn about all visual style settings and how to redefine the visual style. You will also learn how to create your own visual style. Finally, this chapter provides an introduction to lights and rendering.

Controlling Solid Model Display

AutoCAD solid models can be displayed in wireframe form, with hidden lines removed, or in shaded form by using a shaded visual style. The 2D Wireframe visual style is the default display when a drawing is started based on the acad.dwt template. Wireframe displays are the quickest to display. When a drawing is started based on the acad3D.dwt or acadiso3D.dwt template, the default display is the Realistic visual style, which is a shaded display. The following sections introduce display options available in wireframe, hidden, and shaded displays.

NOTE

Many graphics effects in AutoCAD, including the appearance of lines and curves, view transitions, material displays, and rendering, are enhanced by hardware acceleration. Hardware acceleration is enabled by default if your computer video card and graphics support the system. Hardware acceleration can be turned on or off in the **Graphics Performance** dialog box, which is accessed by using the **GRAPHICSCONFIG** command or by right-clicking on the **Hardware Acceleration** button on the status bar and selecting **Graphics Performance...**.

Isolines

ISOLINES

Type
ISOLINES

The appearance of curved surfaces on a solid model is controlled by the **ISOLINES** system variable. ***Isolines*** represent the edges and curved surfaces of a solid model. This setting does *not* affect the final rendered model. However, if the Show property in the **Edge Settings** category in the **Visual Styles Manager** palette is set to Isolines, isolines are displayed when the visual style is set current. The default **ISOLINES** value is four. It can have a value from zero to 2047. All solid objects with curved surfaces are affected by changes to the **ISOLINES** value, as are all visual styles set to display isolines. **Figure 16-1** illustrates the difference between **ISOLINES** settings of four and 12.

The setting of the **ISOLINES** system variable can be changed in the **Visual Styles Manager** palette. You can also type **ISOLINES** and enter a new value, or you can change the **Contour lines per surface** setting in the **Display resolution** area of the **Display** tab in the **Options** dialog box. See **Figure 16-2**. The settings in the **Display** tab in the **Options** dialog box can also be used to control the **DISPSILH** and **FACETRES** system variables, discussed next.

Creating a Display Silhouette

DISPSILH

Type
DISPSILH

In the 2D Wireframe visual style, a model can appear smooth with only a silhouette displayed, similar to the Hidden visual style. This is controlled by the **DISPSILH** (display silhouette) system variable. The **DISPSILH** system variable has two values, 0 (off) and 1 (on). **Figure 16-3** shows solids with **DISPSILH** set to 0 and 1 after setting the 2D Wireframe visual style current and then using **HIDE**.

The setting can be changed by typing **DISPSILH** and entering a new value. You can also set the variable using the **Draw true silhouettes for solids and surfaces** check box in the **Display performance** area of the **Display** tab in the **Options** dialog box. Refer to **Figure 16-2**. A preferred technique is to set **ISOLINES** to 0 and **DISPSILH** to 1 to

Figure 16-1.
Isolines define curved surfaces.
A—**ISOLINES** = 4.
B—**ISOLINES** = 12.

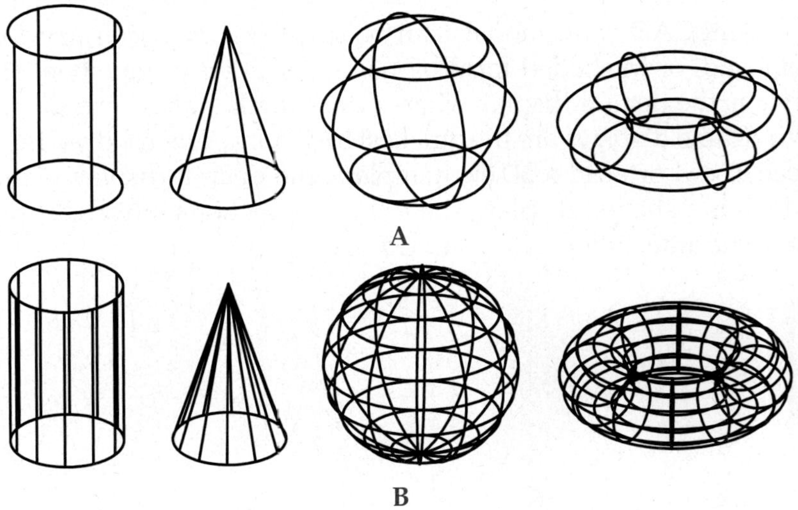

Figure 16-2.
The **ISOLINES**, **FACETRES**, and **DISPSILH** values can be set in the **Options** dialog box.

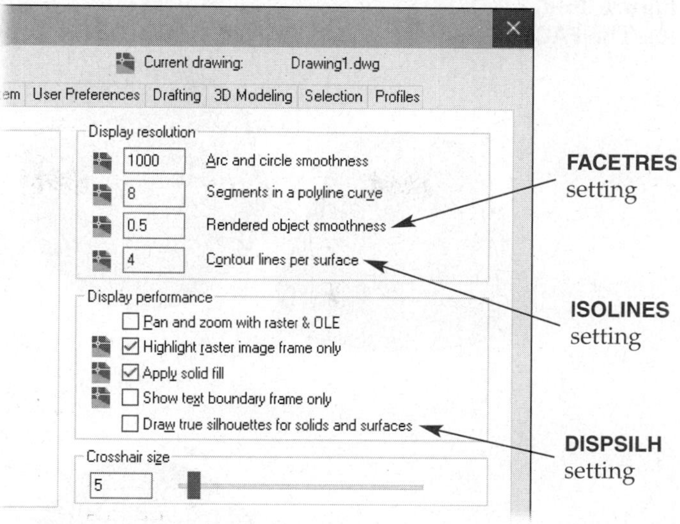

create a true display of the silhouette edge of a 3D model. For the 2D Wireframe visual style, the **DISPSILH** variable can also be set in the **Visual Styles Manager** palette. The variable is controlled by the Draw true silhouettes property in the **2D Wireframe options** category. Setting this property to Yes turns on silhouettes.

Controlling Surface Smoothness

The smoothness of curved surfaces in hidden, shaded, and rendered displays is controlled by the **FACETRES** system variable. This variable determines the number of polygon faces applied to the model. The value can range from .01 to 10.0 and the default value is .5. This system variable can be changed by typing **FACETRES** or by changing the **Rendered object smoothness** setting in the **Display resolution** area of the **Display** tab in the **Options** dialog box. Refer to **Figure 16-2**. For the 2D Wireframe visual style, the variable can be set in the **Visual Styles Manager** palette. The Solid smoothness property in the **Display resolution** category controls the variable. **Figure 16-4** shows the effect of two different **FACETRES** settings.

Type
FACETRES

FACETRES

Figure 16-3.
A—The **HIDE** command used when **DISPSILH** is set to 0. Objects are displayed faceted.
B—The **HIDE** command used when **DISPSILH** is set to 1. Facets are eliminated and only the silhouette is displayed.

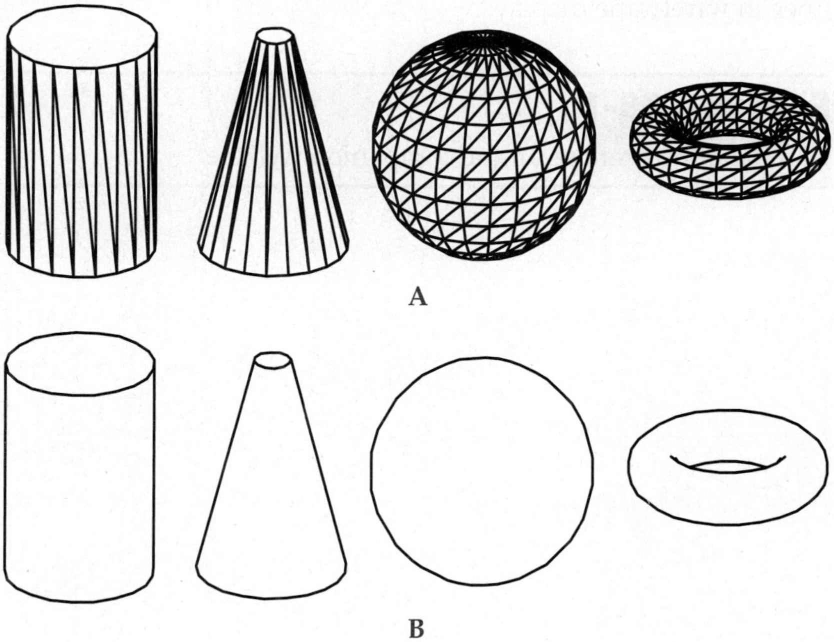

Figure 16-4.
A—The **FACETRES** setting is .5. B—The **FACETRES** setting is 5.0.

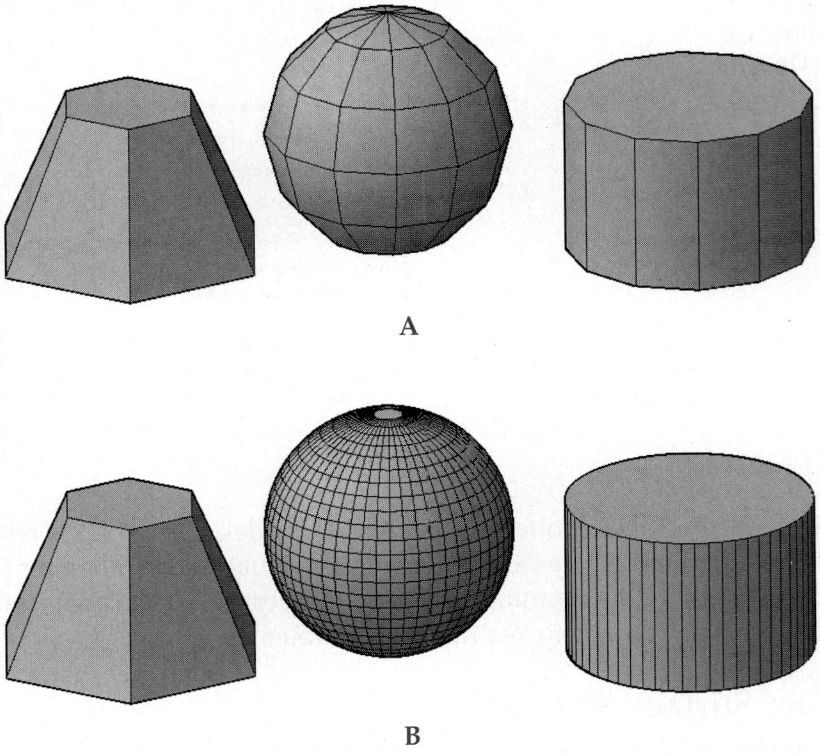

Controlling Line Display

Type
LINESMOOTHING

LINESMOOTHING

The smoothness of 2D lines displayed in the 2D Wireframe visual style is controlled by the **LINESMOOTHING** system variable. By default, this system variable is set to 1 (on). At this setting, antialiasing is turned on. ***Aliasing*** refers to the jagged or "stair-stepped" appearance of pixels defining diagonal or curved edges at a lower resolution. ***Antialiasing*** is a method used to lessen this effect by shading adjacent pixels. Turning on the **LINESMOOTHING** system variable will improve the appearance of 2D object lines in wireframe displays.

Exercise 16-1

www.g-wlearning.com/CAD/

Complete the exercise on the companion website.

Overview of the Visual Styles Manager

The **Visual Styles Manager** palette provides access to all of the visual style settings. This palette is a floating window similar to the **Properties** palette. See **Figure 16-5**. Changes made in the **Visual Styles Manager** redefine the visual style in the current drawing.

> **Ribbon**
> **Visualize**
> **> Visual Styles**
> **> Visual Style Manager...**
> **Type**
> **VISUALSTYLES**
> **VSM**

VISUALSTYLES

At the top of the **Visual Styles Manager** are image tiles for the defined visual styles. See **Figure 16-6**. The default visual styles are 2D Wireframe, Conceptual, Hidden, Realistic, Shaded, Shaded with Edges, Shades of Gray, Sketchy, Wireframe, and X-Ray. User-defined visual styles also appear as image tiles. The image on the tile is a preview of the visual style settings. Selecting an image tile provides access to the properties of the visual style in the palette below. The name of the currently selected visual style appears below the image tiles and the corresponding image tile is surrounded by a yellow border.

To set a different visual style current using the **Visual Styles Manager**, double-click on the image tile. You can also select the image tile and pick the **Apply Selected Visual Style to Current Viewport** button immediately below the image tiles. An icon consisting of a small image of the button is displayed in the image tile of the visual style that is current in the active viewport, as shown in **Figure 16-6**. A drawing icon appears in the image tile if the visual style is current in a viewport that is not active. The AutoCAD icon appears in the image tiles of the default visual styles.

You can also set a different visual style current by selecting the style from the **Visual Style Controls** flyout in the viewport controls located in the upper-left corner of the drawing window. See **Figure 16-7**. Selecting **Visual Styles Manager...** in the **Visual Style Controls** flyout opens the **Visual Styles Manager**.

Figure 16-5.
The **Visual Styles Manager** palette.

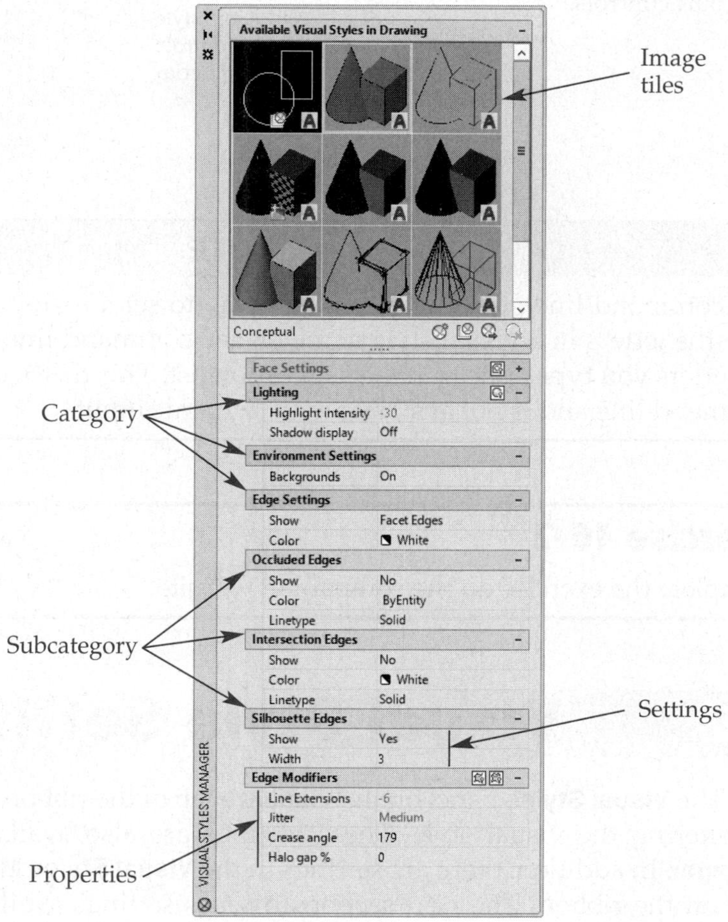

Figure 16-6.
The image tiles correspond to the visual styles. The image on the tile is a preview of the visual style's settings.

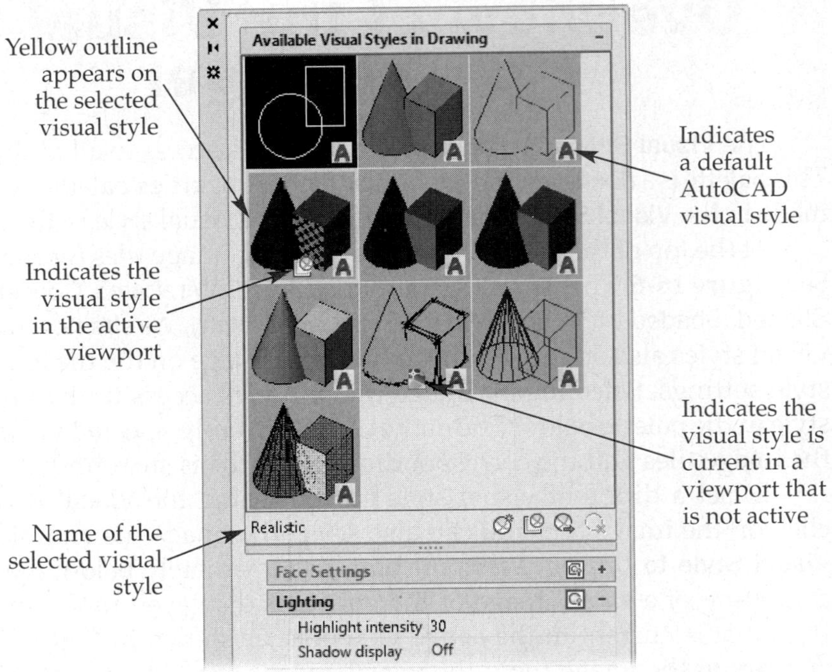

Figure 16-7.
A visual style can be set current by selecting it from the **Visual Style Controls** flyout in the viewport controls.

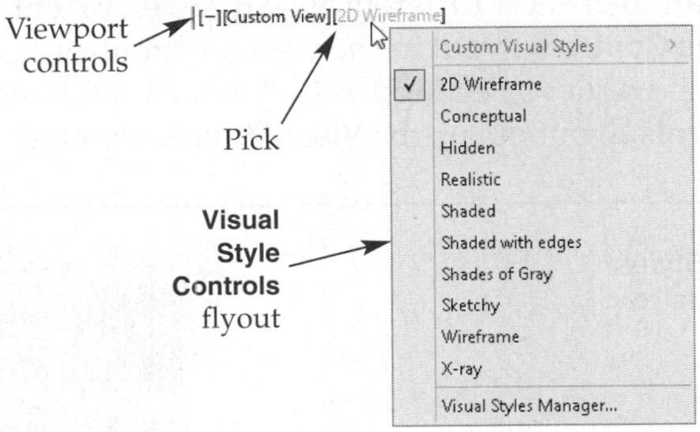

NOTE

The command line offers an additional way to set a visual style current. When you type the letters in a visual style name on the command line, visual styles that match the letters you type appear in the suggestion list. This method can only be used on the command line and is not available with dynamic input.

Exercise 16-2

www.g-wlearning.com/CAD/

Complete the exercise on the companion website.

Visual Style Settings

The **Visual Styles** panel on the **Visualize** tab of the ribbon provides several settings for altering the visual style. These settings are also available in the **Visual Styles Manager**. In addition, there are settings in the **Visual Styles Manager** that are not available on the ribbon. The next sections discuss settings available in the **Visual Styles Manager** for the default visual styles.

Changing any setting in the **Visual Styles Manager** *redefines* the visual style in the current drawing. Changes made using the ribbon are temporary and are not kept when a different visual style is set current. This is important to remember.

2D Wireframe

When the 2D Wireframe visual style is set current, lines and curves are used to show the edges of 3D objects. Assigned linetypes and lineweights are displayed. All edges are visible as if the object is constructed of pieces of wire soldered together at the intersections (thus, the name *wireframe*). Either the 2D or 3D wireframe UCS icon is displayed. OLE objects will display normally as long as they are parallel to the viewing plane. If they are not parallel, only the OLE frame is visible. In addition, the drawing window display changes to the 2D model space context and parallel projection. For the 2D Wireframe visual style, the **Visual Styles Manager** displays the following categories. See **Figure 16-8**.

- **2D Wireframe Options**
- **2D Hide—Occluded Lines**
- **2D Hide—Intersection Edges**
- **2D Hide—Miscellaneous**
- **Display Resolution**

2D Wireframe Options

The Contour lines property controls the **ISOLINES** system variable. The setting is 4 by default. Isolines are suppressed when the **HIDE** command is used with the 2D Wireframe visual style set current. The Draw true silhouettes property controls the **DISPSILH** system variable. It is set to No by default, which is equivalent to a **DISPSILH** setting of 0.

2D Hide—Occluded Lines

The Color property in this category controls the **OBSCUREDCOLOR** system variable. This property determines the color of *occluded* lines (obscured lines hidden from the view) when the **HIDE** command is used. The default setting is ByEntity. This means that, when displayed, obscured lines are shown in the same color as the object.

The Linetype property controls the **OBSCUREDLTYPE** system variable. This property determines whether obscured lines are displayed and in which linetype they are displayed. The default setting is Off, which means that obscured lines are not displayed when the **HIDE** command is used. Selecting a linetype allows you to have obscured lines displayed instead of removed from the view. The available linetypes are: Solid,

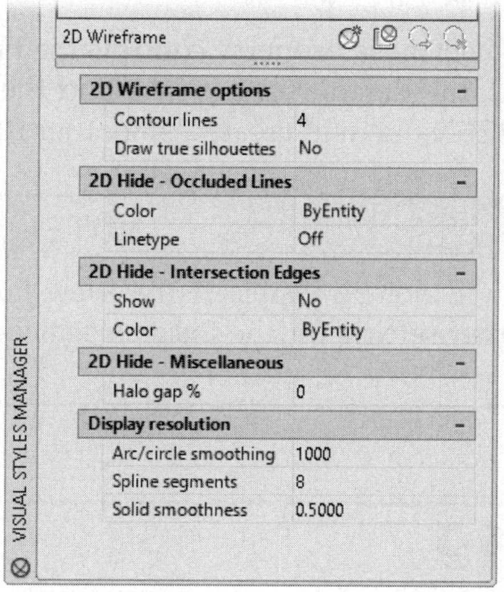

Figure 16-8. The categories and properties available for the 2D Wireframe visual style.

Dashed, Dotted, Short Dash, Medium Dash, Long Dash, Double Short Dash, Double Medium Dash, Double Long Dash, Medium Long Dash, and Sparse Dot. When a linetype is selected from the drop-down list, obscured lines are displayed in that linetype after the **HIDE** command is used.

NOTE

The linetypes available in the Linetype property drop-down list are not the same as the linetypes loaded into the **Linetype Manager** dialog box. They are independent of the zoom level, which means the dash size will stay the same when zooming in and out.

2D Hide—Intersection Edges

This category is used to toggle the display of polylines at the intersection of 3D surfaces and set the color of the lines. The Show property controls the **INTERSECTIONDISPLAY** system variable. This property determines whether polylines are displayed at the intersection of non-unioned 3D surfaces. The default setting is No, which means that polylines are not displayed when the **HIDE** command is used.

The Color property in this category controls the **INTERSECTIONCOLOR** system variable. This property determines the color of the polylines displayed at intersection edges. By default, the setting is ByEntity. This means that, when displayed, the polylines at intersection edges are shown in the same color as the object.

2D Hide—Miscellaneous

The Halo gap % property controls the **HALOGAP** system variable. This property determines the gap that is displayed where one object partially obscures another (between the foreground edge and where the background edge starts to show). The default setting is 0 and the value can range from 0 to 100. The value refers to a percentage of one unit. The gap is only displayed when the **HIDE** command is used. It is not affected by the zoom level.

Display Resolution

The Arc/circle smoothing property controls the zoom percentage set by the **VIEWRES** command. This determines the resolution of circles and arcs when hardware acceleration is disabled. The value can range from 1 to 20,000. The higher the value, the higher the resolution of circles and arcs.

The Spline segments property controls the **SPLINESEGS** system variable. This property determines the number of line segments in a spline-fit polyline. The value can range from –32,768 to 32,767.

The Solid smoothness property controls the **FACETRES** system variable. As previously discussed, the default setting is .5 and the value can range from .01 to 10.0. A higher **FACETRES** value will create a smoother finish on 3D printed models.

NOTE

Polygon faces will not be visible if the Draw true silhouettes property is set to Yes. However, a higher setting for the Solid smoothness property will make curved edges smoother.

Exercise 16-3

www.g-wlearning.com/CAD/

Complete the exercise on the companion website.

Conceptual

When the Conceptual visual style is set current, objects are smoothed and shaded. The shading is a transition from cool to warm colors. The transitional colors help highlight details. The previous projection is retained and that context is set current.

NOTE

The categories and properties for the shaded visual styles (those other than 2D Wireframe) are the same. These settings are discussed in the Common Visual Style Settings section later in this chapter.

Hidden

The Hidden visual style removes obscured lines from your view and makes 3D objects appear solid. The previous projection is retained and that context is set current. The benefit of using Hidden is that you get sufficient 3D display, but it does not push the graphics system too hard. Objects are not shaded or colored. This is very useful when working on complex drawings and/or using a slow computer.

Realistic

As with the Conceptual visual style, the objects have smoothing, highlights, and shading applied to them when the Realistic visual style is set current. In addition, if materials are applied to the objects, the materials are displayed. The previous projection is retained and that context is set current. This visual style is good for adjusting textures and patterns on materials and for a final look at the scene before rendering.

Shaded

This style is very similar to Realistic, except the lighting is smooth instead of smoothest and textures are turned off.

Shaded with Edges

This style is the same as Shaded, except isolines are displayed with the default number of 4 and silhouette edges are displayed with a width setting of 3.

Shades of Gray

This style is similar to the other shaded styles, except object colors are displayed as monochrome. Materials and textures are turned off. Facet edges are turned on, as are silhouette edges. This style is good for adjusting lights and shadows. It is much easier to see what is happening with lights when all objects are the same color.

Sketchy

The Sketchy visual style has most of the same settings as the Hidden visual style. However, objects look like they are hand-sketched. Line extensions and jitter are turned on, as are silhouette edges.

Wireframe

The Wireframe visual style is almost identical in appearance to the 2D Wireframe visual style. The main differences are that the 3D mode UCS icon is displayed and OLE objects display normally regardless of the viewing angle. The previous projection is retained and that context is set current. When working in 3D, a wireframe view is sometimes necessary to select objects normally hidden from your view.

X-Ray

This visual style is very similar to the Shaded with Edges visual style, except the Opacity property is set to 50%. This makes objects appear somewhat transparent.

PROFESSIONAL TIP

Depending on the visual style chosen, drawing window colors for individual elements such as the background, grid lines, crosshairs, etc., may need to be adjusted to make it easier to see and work with the objects in the drawing. Access the **Options** dialog box and select the **Colors...** button in the **Display** tab to change these settings in the **Drawing Window Colors** dialog box. Colors used with 2D model space, 3D parallel projection, and 3D perspective projection are controlled separately.

Common Visual Style Settings

With the exception of 2D Wireframe, all of the visual styles share similar categories and settings in the **Visual Styles Manager**. The visual styles have **Face Settings**, **Lighting**, **Environment Settings**, and **Edge Settings** categories. See Figure 16-9. These categories and the properties available in them are discussed in the next sections.

Face Settings

VSFACESTYLE

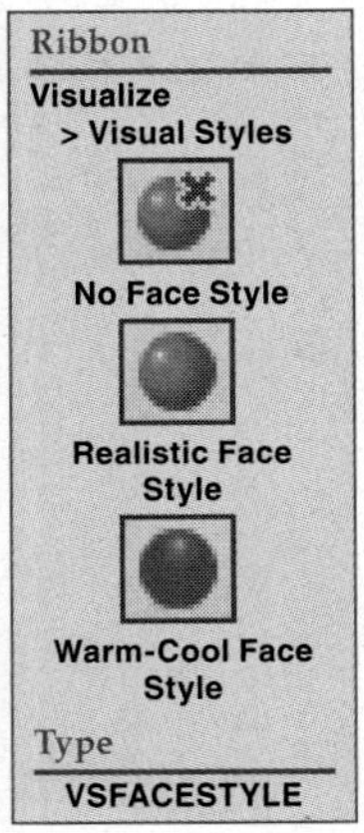

The Face style property controls the **VSFACESTYLE** system variable. The options have slightly different names, but are the same as the **No Face Style**, **Realistic Face Style**, and **Warm-Cool Face Style** options available by selecting the buttons in the **Visual Styles** panel on the **Visualize** tab of the ribbon. Remember, though, selecting a button on the ribbon does not redefine the visual style in the current drawing. Rather, it is a temporary change to the viewport display.

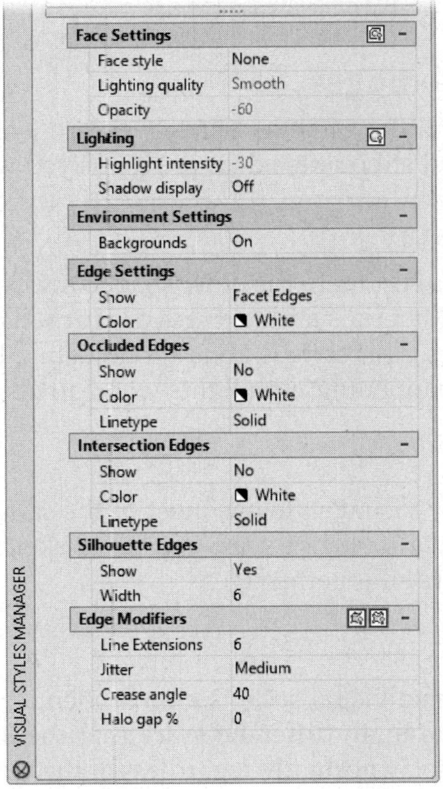

Figure 16-9.
These categories and properties are similar for all of the visual styles except 2D Wireframe.

The Lighting quality property controls the **VSLIGHTINGQUALITY** system variable. This property determines whether curved surfaces are displayed smooth or as a series of flat faces. Lighting quality is unavailable if the Face style property is set to None.

The Color property controls the **VSFACECOLORMODE** system variable. This property determines how color is applied to the faces of an object. The effects can be temporarily applied by picking a button in the face colors flyout in the **Visual Styles** panel on the **Visualize** tab of the ribbon. The choices are:

- Normal. The object color is applied to faces.
- Monochrome. One color is applied to all faces. This also displays and enables the Monochrome color property.
- Tint. A combination of the object color and a specified color is applied to faces. This also displays and enables the Tint color property.
- Desaturate. The object color is applied to faces, but the saturation of the color is reduced by 30%.

VSFACECOLORMODE

Ribbon
Visualize
> Visual Styles
Normal
Monochrome
Tint
Desaturate
Type
VSFACECOLORMODE

The Monochrome color and Tint color properties control the **VSMONOCOLOR** system variable. This system variable determines the color that is applied when the face Color property is set to Monochrome or Tint.

The Opacity property controls the **VSFACEOPACITY** system variable. This property determines how transparent or opaque faces are in the viewport. The value can range from –100 to 100. When the setting is 1, the faces are completely transparent. When the setting is 100, the faces are completely opaque. Settings below 0 set the value, but turn off the effect. To quickly turn the effect on or off, pick the **Opacity** button on the **Face Settings** category title bar. See **Figure 16-10**. This changes the value from negative to positive, or vice versa. This property cannot be changed if the Face style property is set to None.

Opacity may also be controlled by picking the **X-Ray Effect** button in the **Visual Styles** panel on the **Visualize** tab of the ribbon. The adjacent **Opacity** slider sets the value. However, remember, picking a button or making settings on the ribbon is only temporary. The change is not saved to the visual style.

The Material display property controls the **VSMATERIALMODE** system variable. When set to Off, objects display in their assigned color. When the setting is changed to Materials, the objects display the color of the material, but not the textures. When the setting is changed to Materials and textures, full materials are displayed.

CAUTION

Displaying materials and textures on 3D objects in a complex drawing will slow system performance. Set the Material display property to Materials and textures only when it is absolutely necessary.

Figure 16-10.
Picking the **Opacity** button on the **Face Settings** category title bar turns the effect of opacity on or off.

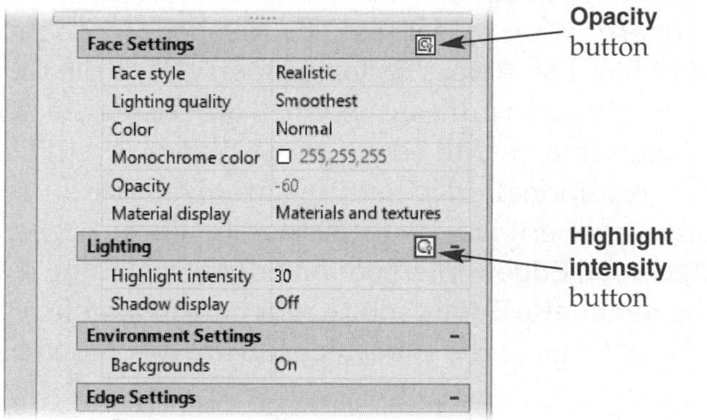

Lighting

The Highlight intensity property controls the **VSFACEHIGHLIGHT** system variable. This property determines the size of the highlight on faces to which no material is assigned. A small highlight on an object makes it look smooth and hard. A large highlight on an object makes it look rough or soft. The initial value is 30 or –30; the value can range from –100 to 100. The higher the setting is above 0, the larger the highlight. Settings below 0 set the value, but turn off the effect. To quickly turn the effect on or off, pick the **Highlight intensity** button on the **Lighting** category title bar. Refer to **Figure 16-10**. This changes the value from negative to positive or vice versa. This property cannot be changed if the Face style property is set to None.

VSSHADOWS

Ribbon
Visualize
> Lights
Shadows Off
Ground Shadows
Full Shadows
Type
VSSHADOWS

The Shadow display property controls the **VSSHADOWS** system variable. This property controls if and how shadows are cast when the visual style is set current. The options have slightly different names, but are the same as those available by selecting a button in the shadows flyout in the **Lights** panel on the **Visualize** tab of the ribbon. Remember, however, a setting made using the ribbon is temporary and does not redefine the visual style.

If the property is set to Ground shadow, objects cast shadows on the ground, but not onto other objects. The "ground" is the XY plane of the WCS. The Mapped Object shadows setting, which corresponds to the **Full Shadows** button on the ribbon, only works if lights have been placed in the scene and hardware acceleration is enabled.

Environment Settings

The Backgrounds property controls the **VSBACKGROUNDS** system variable. This property determines whether the preselected background is displayed in the viewport. Backgrounds can be assigned to a view when a named view is created or edited. After the view is created, restore the view to display the background.

Edge Settings

Properties in the **Edge Settings** category determine the appearance of model edges. This category includes the **Occluded Edges**, **Intersection Edges**, **Silhouette Edges**, and **Edge Modifiers** subcategories. Refer to **Figure 16-5**.

VSEDGES

Ribbon
Visualize
> Visual Styles
No Edges
Isolines
Facet Edges
Type
VSEDGES

The Show property controls the **VSEDGES** system variable. This property determines how edges on solid objects are represented when the visual style is set current. The options are the same as those available by selecting a button in the edge flyout in the **Visual Styles** panel on the **Visualize** tab of the ribbon. The settings of Isolines and Facet Edges determine how edges and curved surfaces are displayed on a 3D model. Setting this property to None turns off isolines and facets and displays no edges. If the Face style property is set to None, the Show property cannot be set to None.

The Color property in the **Edge Settings** category controls the **VSEDGECOLOR** system variable. This property determines the color of all edges on objects in the drawing. It is disabled when the Show property is set to None.

The Number of lines and Always on top properties are displayed when the Show property is set to Isolines. The Number of lines property controls the **ISOLINES** system variable. The Always on top property controls the **VSISOONTOP** system variable. This property determines if isolines are displayed when the model is shaded or hidden. When set to Yes, all edges are displayed, even if they are physically hidden from view.

Additional edge settings are available in the **Edge Settings** category when the Show property is set to Isolines or Facet Edges. These settings are available in the **Occluded Edges**, **Intersection Edges**, and **Edge Modifiers** subcategories. The settings in the **Silhouette Edges** subcategory are available regardless of the Show property setting. The settings in the subcategories are discussed as follows.

Figure 16-11.
A—Occluded lines are not shown. B—The Show property is set to Yes and occluded lines are shown.

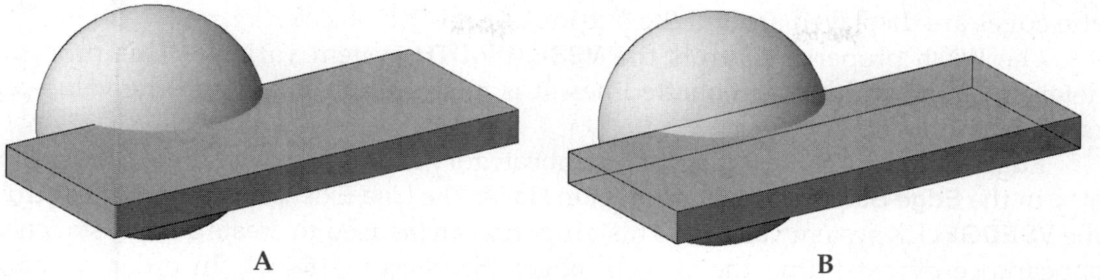

Occluded Edges Subcategory. This subcategory is not available if the Show property in the **Edge Settings** category is set to None. The Show property in this subcategory controls the **VSOCCLUDEDEDGES** system variable. This property determines whether or not occluded edges are displayed in a hidden or shaded view. See **Figure 16-11.**

The Color property in this subcategory controls the **VSOCCLUDEDCOLOR** system variable. The Linetype property controls the **VSOCCLUDEDLTYPE** system variable. These properties are similar to the properties for occluded lines discussed earlier in this chapter in the 2D Hide—Occluded Lines section.

Intersection Edges Subcategory. This subcategory is not available if the Show property in the **Edge Settings** category is set to None. The Show property in this subcategory controls the **VSINTERSECTIONEDGES** system variable. This property determines whether lines are displayed where one 3D object intersects another 3D object. See **Figure 16-12.**

The Color property in this subcategory controls the **VSINTERSECTIONCOLOR** system variable. The Linetype property controls the **VSINTERSECTIONLTYPE** system variable. These properties are similar to the properties for intersection lines discussed earlier in this chapter in the 2D Hide—Intersection Edges section.

PROFESSIONAL TIP

Setting the intersection edges Color property to a color that contrasts with the objects in your model is a good way to quickly check for interference between 3D objects.

Figure 16-12.
A—A line does not appear where these two objects intersect. B—The Show property is set to Yes and a line appears at the intersection.

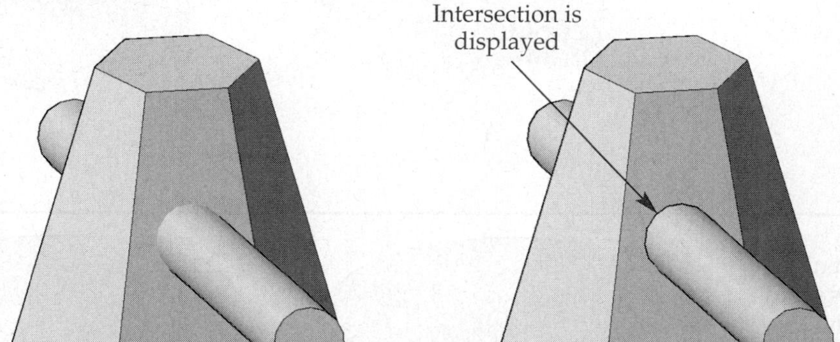

Silhouette Edges Subcategory. This subcategory is available for each setting for the Show property in the **Edge Settings** category. The Show property in this subcategory controls the **VSSILHEDGES** system variable. It determines whether or not silhouette edges are displayed around the outside edges of all objects.

The Width property controls the **VSSILHWIDTH** system variable. This property determines the width of silhouette lines. It is measured in pixels and the value can range from 1 to 25.

Edge Modifiers Subcategory. This subcategory is not displayed if the Show property in the **Edge Settings** category is set to None. The Line Extensions property controls the **VSEDGELEX** system variable. This property can be used to create a hand-sketched appearance by extending the ends of edges. See **Figure 16-13A**. In order to make changes to this property, the **Line Extensions edges** button must be on in the **Edge Modifiers** subcategory title bar. The value for the Line Extensions property can range from –100 to 100, which is the number of pixels. The higher the setting, the longer the extension. A negative value sets the extension length, but turns off the property. Picking the button makes the value positive and applies the effect (or makes the value negative and turns off the effect).

The Jitter property controls the **VSEDGEJITTER** system variable. Jitter makes edges of objects look as if they were sketched with a pencil. See **Figure 16-13B**. In order to make changes to this property, the **Jitter edges** button must be on in the **Edge Modifiers** subcategory title bar. There are four settings from which to select: Off, Low, Medium, and High. The number of sketched lines increases at each higher setting.

When the Show property in the **Edge Settings** category is set to Facet Edges, the Crease angle and Halo gap % properties are displayed in the **Edge Modifiers** subcategory. The Crease angle property controls the **VSEDGESMOOTH** system variable. This property determines how facet edges are displayed based on the angle between adjacent facets. See **Figure 16-14**. The value can range from 0 to 180. This is the number of degrees between facets at which a line is displayed. The Halo gap % property is similar to the setting discussed earlier in the 2D Hide—Miscellaneous section; however, this property controls the **VSHALOGAP** system variable.

Figure 16-13.
A—Line extensions have been turned on for this visual style. B—Jitter has been turned on for this visual style.

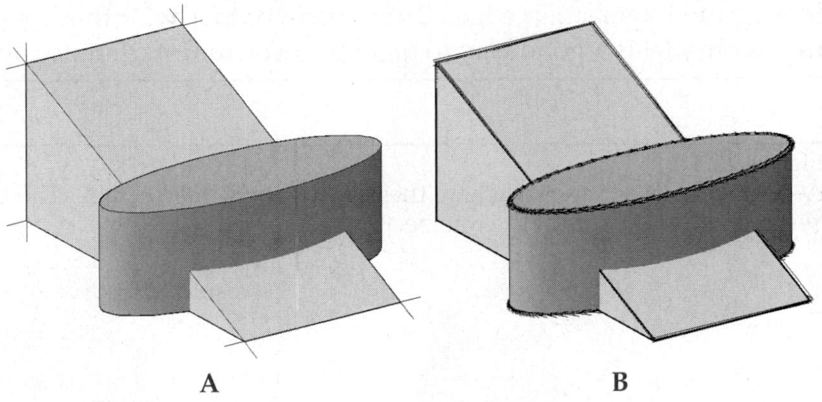

Figure 16-14.
A—The current visual style is set to Conceptual and the Crease angle property is set to 179. B—The Crease angle property is set to 10. Notice the edges between facets.

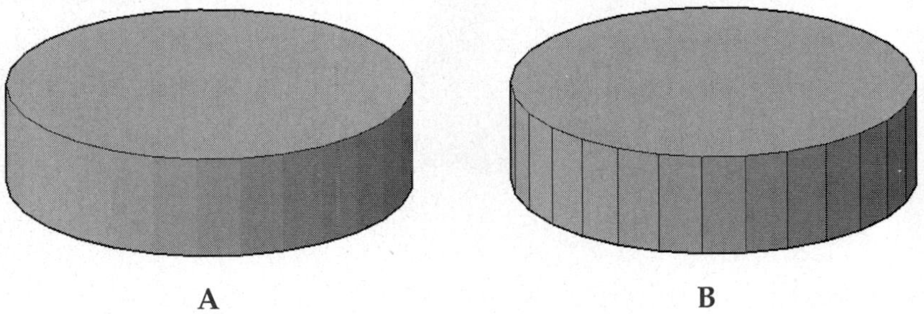

Creating Your Own Visual Style

As you saw in the previous sections, you can customize the default AutoCAD visual styles. However, you may also want to create a number of different visual styles to quickly change the display of the scene. Custom visual styles are easy to create.

To create a custom visual style, open the **Visual Styles Manager**. Then, pick the **Create New Visual Style** button below the image tiles. You can also right-click in the image tile area and select **Create New Visual Style...** from the shortcut menu. In the **Create New Visual Style** dialog box that appears, type a name for the new style and give it a description. See **Figure 16-15**. Then, pick the **OK** button to create the new visual style.

An image tile is created for the new visual style. The name and description of the visual style appear as help text when the cursor is over the image tile. Select the image tile to display the default properties for the new visual style. Then, change the settings as needed to meet your requirements.

Custom visual styles are only saved in the current drawing. They are not automatically available in other drawings. To use the new visual styles in any drawing, they must be exported to a tool palette. This is discussed in the next section.

PROFESSIONAL TIP

To return one of AutoCAD's visual styles to its default settings, right-click on the image tile in the **Visual Styles Manager** and select **Reset to default** from the shortcut menu.

Exercise 16-4

www.g-wlearning.com/CAD/

Complete the exercise on the companion website.

Steps for Exporting Visual Styles to a Tool Palette

To have custom visual styles available in other drawings, export them to a tool palette. Use the following procedure.

1. Create and customize a visual style as described in the previous section.
2. Open the **Tool Palettes** window.
3. Right-click on the **Tool Palettes** title bar and pick **New Palette** from the shortcut menu.
4. Type the name of the new palette, such as My Visual Styles, in the text box that appears. See **Figure 16-16A**.
5. The new palette is added and active. You are ready to export your custom visual styles into it.
6. Select the image tile of the visual style in the **Visual Styles Manager**. Remember, a yellow border appears around the selected image tile.
7. Pick the **Export the Selected Visual Style to the Tool Palette** button below the image tiles in the **Visual Styles Manager**. You can also right-click on the image tile and select **Export to Active Tool Palette** from the shortcut menu.

TOOLPALETTES

Ribbon
View
> Palettes
Tool Palettes
Type
TOOLPALETTES
TP
[Ctrl]+[3]

Figure 16-15.
Creating a new visual style.

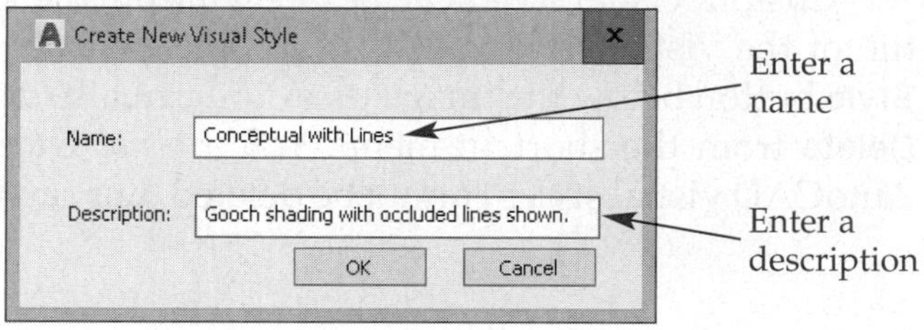

Figure 16-16.
A—Creating a new tool palette on which to place visual style tools.
B—A visual style has been copied to the tool palette as a tool.

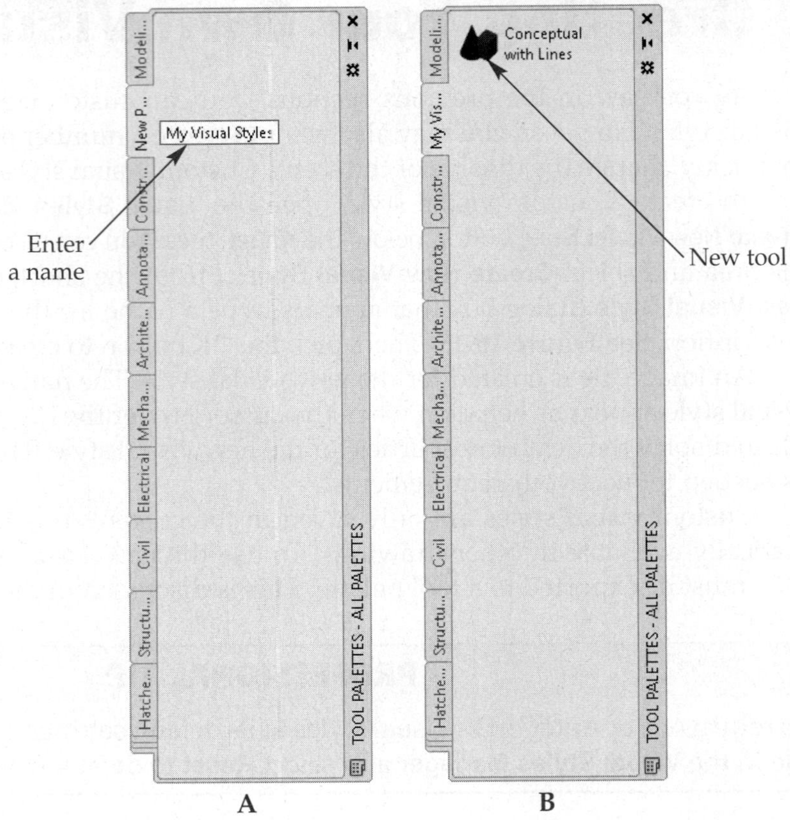

A new tool now appears in the palette with the same image, name, and description as the visual style in the **Visual Styles Manager**. See **Figure 16-16B**. Selecting the tool applies the visual style to the current viewport. You can also right-click on the tool to display a shortcut menu. Using this menu, you can apply the visual style to the current viewport, apply it to all viewports, or add the visual style to the current drawing. The shortcut menu also allows you to rename the tool, access the properties of the visual style, and delete the visual style from the palette.

PROFESSIONAL TIP

A visual style can be added as a tool on a tool palette by dragging its image tile from the **Visual Styles Manager** and dropping it onto the tool palette.

Exercise 16-5 www.g-wlearning.com/CAD/

Complete the exercise on the companion website.

Deleting Visual Styles from the Visual Styles Manager

Custom visual styles can be deleted from the **Visual Styles Manager**. Pick the image tile of the visual style you want to delete. Then, pick the **Delete the Selected Visual Style** button below the image tiles. You can also right-click on the image tile and select **Delete** from the shortcut menu. You are *not* warned about the deletion. The default AutoCAD visual styles cannot be deleted, nor can a visual style that is currently in use.

Plotting Visual Styles

A visual style not only affects the on-screen display, it also affects plots. To plot objects with a specific visual style, use the following guidelines.

Plotting a Visual Style from Model Space

Open the **Plot** dialog box and expand it by picking the **More Options** (>) button. Then, select the desired display from the **Shade plot** drop-down list in the **Shaded viewport options** area. If the desired visual style is current in the viewport, you can also select As displayed from the drop-down list. Finally, plot the drawing.

Plotting a Visual Style from Layout (Paper) Space

When plotting from layout (paper) space, the shade plot properties of the viewports govern how the viewport is plotted. The viewports can be set to plot visual styles in three different ways.

In the first method, select the viewport in layout space and right-click to display the shortcut menu. Pick **Shade plot** to display the cascading menu. Then, select the appropriate visual style.

In the second method, use the **Properties** palette to set the Shade plot property of the viewport. To do this, select the viewport in layout space and open the **Properties** palette. Pick the Shade plot property in the **Misc** category and change the setting to the desired option. The Shade plot property is also available in the **Quick Properties** palette.

In the third method, use the **Visual styles** suboption of the **Shadeplot** option of the **MVIEW** command. When prompted to select objects, pick the border of the viewport. Do not pick the objects in the viewport.

NOTE

The visual style of the viewport may also be selected when you use the **VPORTS** command to create a viewport configuration in the **Viewports** dialog box. Select the viewport in the **Preview** area of the dialog box. Then, pick the visual style desired from the **Visual Style:** drop-down list at the bottom of the dialog box. The **VPORTS** command can be used in model space or layout (paper) space.

Introduction to Rendering

Visual styles provide a way to plot your 3D scene to paper or a file, but control over the appearance is limited to the visual style settings. In Chapter 1, you were briefly introduced to the **RENDER** command. The **RENDER** command allows you to create photorealistic images with complete control over the scene. In this chapter, you will be introduced to AutoCAD's rendering and lighting tools. Materials, lights, and more advanced rendering features are discussed in later chapters.

RENDER

Ribbon
Visualize > Render
Render to Size
Type
RENDER
RR

When you render a scene, you are making a realistic image of your design that can be printed, displayed on a web page, or used in a presentation. To create an attractive rendering, you have to figure out what view you want to display, where the lights should be placed, what types of materials need to be applied to the 3D objects, and the kind of output that is needed. This section shows you how to create a quick rendering of your scene.

Introduction to Lights

Lights provide the illumination to a scene and are essential for rendering. There are three types of lighting in AutoCAD—default lighting, sunlight, and user-created lighting. AutoCAD automatically creates one or two default light sources in every scene. These lights ensure that all surfaces on the model are illuminated and visible. The types of lighting are discussed in more detail in Chapter 18.

A scene can be rendered with the default lights, but the results are usually not adequate to produce a photorealistic image. Normally, user-created lighting or sunlight is added to the scene and adjusted to obtain the desired results. When a light is added to a scene, the default lights must be turned off. The first time you add a light, you receive a warning to this effect (unless the warning has been disabled).

In this section, you will learn how to add sunlight to the scene. Chapter 18 provides detailed information on lighting. Sunlight is produced by an automated distant light. Sunlight can be turned on by picking the **Sun Status** button in the **Sun & Location** panel on the **Visualize** tab of the ribbon. The button background is blue when sunlight is on. See **Figure 16-17**.

SUNSTATUS

Ribbon
Visualize
> Sun & Location
Sun Status
Type
SUNSTATUS

If the **Default Lighting** button is on in the **Lights** panel when the **Sun Status** button is turned on, a warning dialog box is displayed. This dialog box gives you choices to either turn off default lighting or keep it on. You cannot see the effects of sunlight with default lighting turned on, so it is recommended to turn it off.

Another warning dialog box is displayed informing you that sunlight requires a different exposure setting from other lights. Different types of lighting require different exposure settings and AutoCAD is recommending you to adjust the exposure settings before switching to sunlight. Exposure settings are discussed in Chapter 19. For now, choose to keep exposure settings the way they are.

Once sunlight is turned on, you can make adjustments to visualize different effects of the sunlight settings on the objects in your scene. Shadows will display in the viewport only if the current visual style is set to display shadows or if shadows are turned on by selecting one of the options in the shadows flyout in the **Lights** panel in the **Visualize** tab of the ribbon. Full shadows will be shown in the rendering.

The **Date** and **Time** sliders in the **Sun & Location** panel on the **Visualize** tab of the ribbon are active when sunlight is turned on. You can drag the sliders to adjust the date and time. The current date and time are displayed on the right-hand end of the sliders. As you drag the sliders, the lighting and shadows in the scene change to reflect the settings.

Figure 16-17.
The **Lights** and **Sun & Location** panels on the **Visualize** tab of the ribbon. A—Turn off default lighting to see the effects of sunlight in the scene. B—Use the **Date** and **Time** sliders in the **Sun & Location** panel to adjust the date and time.

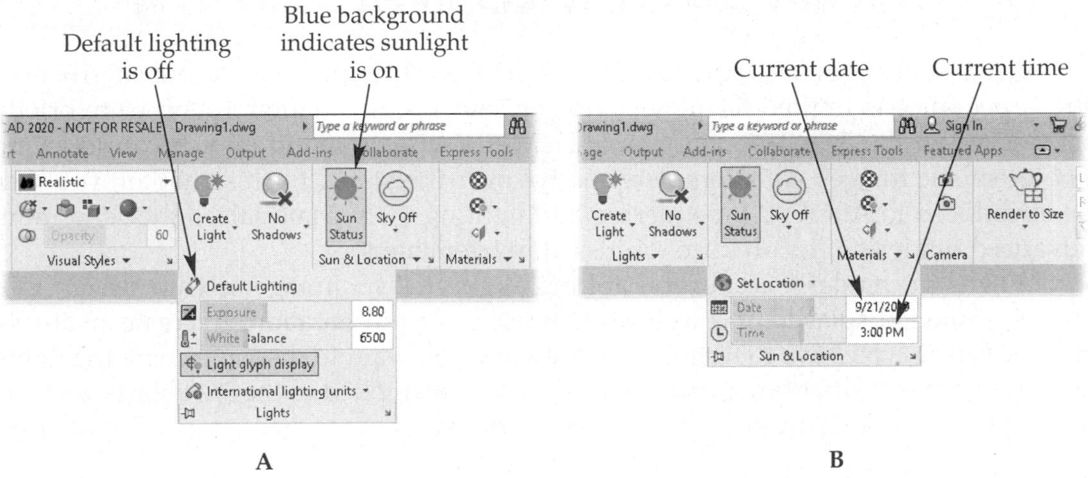

Turning on **Sky Background and Illumination** adds more realism to the scene by filling in the shadow areas with indirect light. See **Figure 16-18**. Picking the flyout under the **Sky Off** button provides access to this setting. The projection must be set to perspective to access the sky settings.

NOTE

If the scene is washed-out with light after turning on sunlight, you can adjust the exposure by using the **Light Exposure** slider in the **Lights** panel in the **Visualize** tab of the ribbon. A higher exposure setting will darken the scene.

Rendering the Scene

The **Render** panel on the **Visualize** tab of the ribbon is shown in **Figure 16-19**. If you pick the **Render to Size** button, the **Render** window appears and AutoCAD immediately starts rendering the viewport. The rendering is processed over a number of levels, with each level improving the quality of the image. A progress bar at the bottom of the window indicates which level is being processed and its percentage of completion. The **Render** window is explained more in the next section.

The **Render to Size** flyout contains options for rendering the scene at different resolutions. The default size is 800 × 600 pixels. Larger resolutions will result in longer rendering times.

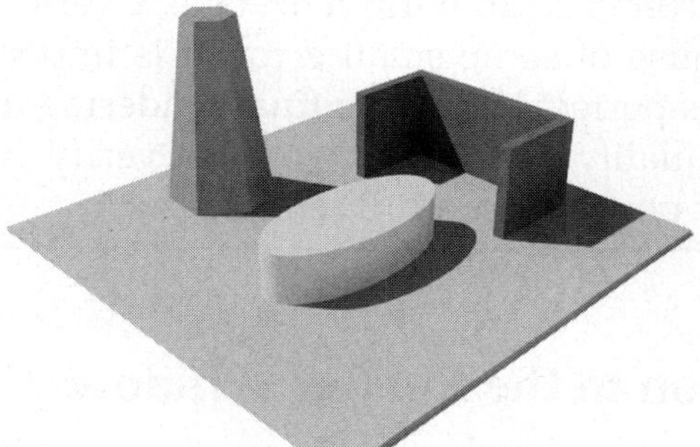

Figure 16-18.
A scene rendered with sunlight. In addition, sky background and illumination has been turned on. Notice that objects in the scene cast and receive shadows.

Figure 16-19.
The **Render** panel on the **Visualize** tab of the ribbon.

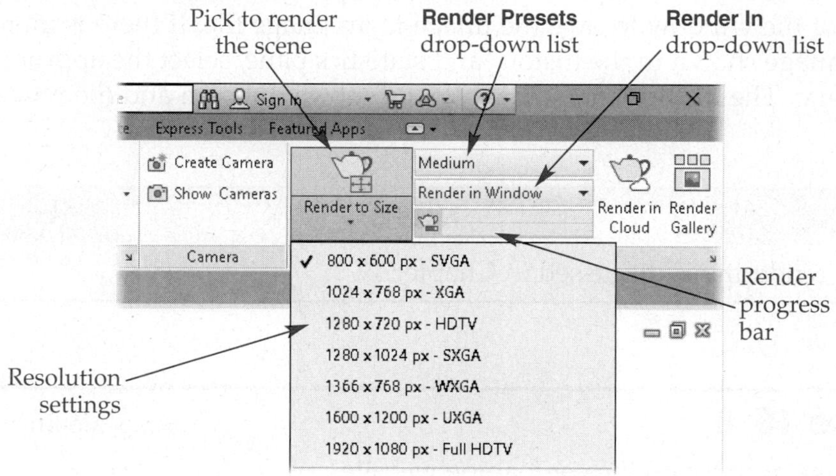

Also in the **Render** panel, you will find the **Render Presets** drop-down list. This is located in the upper-right corner of the panel. This drop-down list gives you a selection of rendering presets based on image quality. The options are:

- Low
- Medium
- High
- Coffee-Break Quality
- Lunch Quality
- Overnight Quality

The Low, Medium, and High presets apply 1, 5, and 10 levels, respectively, of processing during the rendering. The more levels, the longer the rendering takes to complete. The other three presets produce much better quality and are classified by the amount of time the rendering takes to complete. The Coffee-Break Quality preset takes 10 minutes, the Lunch Quality preset takes one hour, and the Overnight Quality preset takes 12 hours, producing the highest-quality rendering.

The options in the **Render In** drop-down list allow you to render the scene to the **Render** window (discussed in the next section) or directly to the viewport. When rendering to the viewport, you can render the entire viewport or a portion of it. If you render with the **Render in Viewport** option, the rendered scene appears in the viewport. The **Render in Region** option prompts you to pick two points of a crop window defining an area of the scene to render. Panning or zooming in the viewport after rendering will cause the rendering to disappear, but this temporary rendering may be enough to identify problem areas and make changes before the final rendering.

PROFESSIONAL TIP

Rendering a complex drawing may take a very long time and you do not want to repeat it because of some small error. It is important to make sure that everything in the scene is perfect before the final rendering. By rendering a cropped region and using lower-quality renderings, you can verify the appearance of any questionable areas without performing a full rendering.

Introduction to the Render Window

The **Render** window has two areas. See **Figure 16-20**. The image pane is where the rendering appears. Immediately below it, the history and statistics pane shows a list of all of the images rendered during the drawing session, with the most recent at the top.

You can zoom into the image in the image pane for detailed inspection. Use the mouse scroll wheel or the zoom buttons at the top of the window. Pick the save button at the top of the window to save the image to an image file. If there is more than one rendered image shown in the history and statistics pane, select the appropriate image before saving. The history and statistics pane shows the path and file name for saved images.

NOTE

Advanced rendering is discussed in Chapter 19.

Exercise 16-6 www.g-wlearning.com/CAD/

Complete the exercise on the companion website.

Figure 16-20.
The **Render** window.

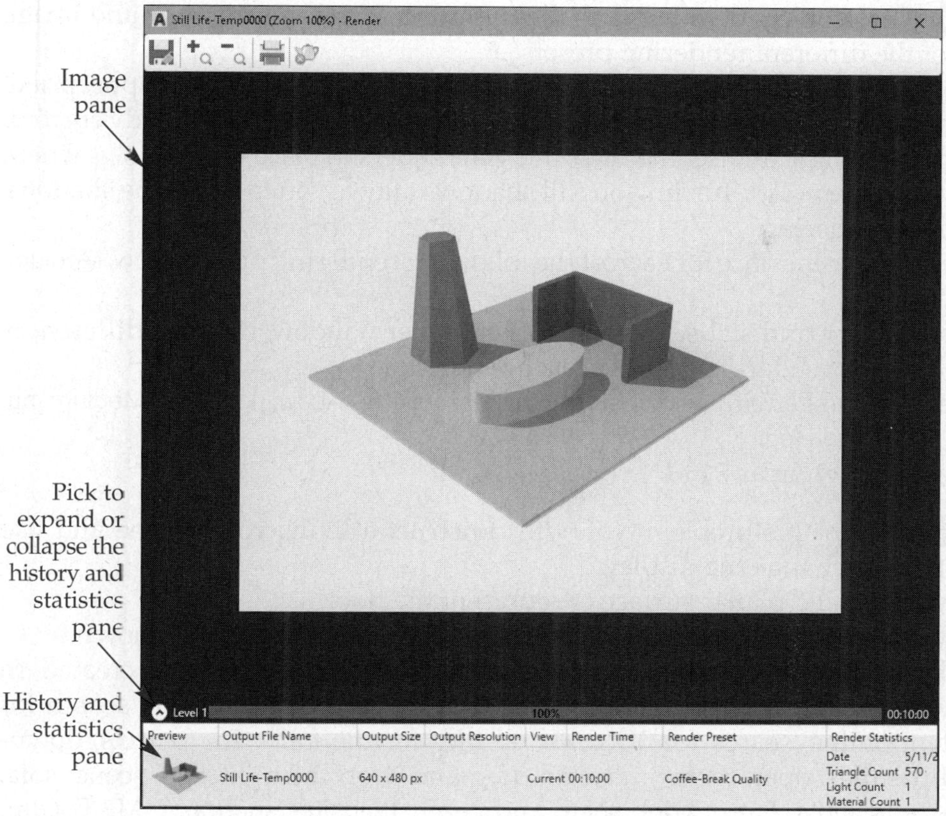

Chapter Review

Answer the following questions using the information in this chapter.

1. Which AutoCAD system variable is used to control the number of isolines displayed on a solid model?
2. What is the **Visual Styles Manager**?
3. Name the 10 default AutoCAD visual styles that can be edited in the **Visual Styles Manager**.
4. Describe the difference between setting the Lighting quality property to Smooth and Faceted.
5. What does the Desaturate setting of the Color property do?
6. What has to be added to a scene before full shadows are displayed?
7. If you want to make your scene look hand sketched, but the Line Extensions and Jitter properties are not available, what other setting(s) do you have to change?
8. How do you set a visual style to display silhouette edges?
9. List the four settings for the Jitter property.
10. What is an *intersection edge*?
11. How do you make your own visual styles available in other drawings?
12. Which visual styles cannot be deleted?
13. How can you turn on sunlight?
14. Explain the function of the **Render in Region** option in the **Render** panel on the ribbon.
15. Name the two main areas of the **Render** window.
16. How can you save a rendered image in the **Render** window?

Drawing Problems

1. **General** This problem demonstrates the differences in rendering time and image quality of the different rendering presets.
   A. Open any 3D drawing from a previous chapter and display an appropriate isometric view. Change the projection to perspective, if it is not already current.
   B. Turn on sunlight and set the **Date** and **Time** sliders to place the shadows where you want them. Tip: Turning on full shadows allows you to locate the shadows without rendering.
   C. Render the scene once for each of the following rendering presets: Low, Medium, and High.
   D. In the history and statistics pane of the **Render** window, note the differences between the rendering times for each rendering.
   E. Save each image with a corresponding name: P16-1-Low.jpg, P16-1-Medium.jpg, and P16-1-High.jpg.
   F. Save the drawing as P16-1.

2. **Architectural** In this problem, you will construct a living room scene and use visual styles to change the display.
   A. Draw a 12′ × 12′ planar surface to represent the floor.
   B. Draw two boxes to represent walls, 12′ × 4″ × 8′. Position them as shown.
   C. Add the furniture shown to the scene. Use objects you have created in previous chapters or search for blocks of furniture online. If you search for content online, search for 3D drawing files for the following objects, download the files, open them, and then drag and drop them into the scene: sofa, table, end table, lamp, book shelf, and chair. Websites such as CAD Forum, CAD Blocks Free, and BIM Object provide free 3D blocks and drawing files for use in AutoCAD. The blocks you use do not have to be exactly the same as shown here and may need to be scaled up or down. Complete the scene by positioning the furniture as shown.
   D. To create a more realistic scene, change the projection to perspective and move your viewing angle to a more natural height.
   E. Apply each of the default visual styles to the viewport and plot each one. Use the **As displayed** option in the **Plot** dialog box. Note the differences in each one.
   F. Save the drawing as P16-2.

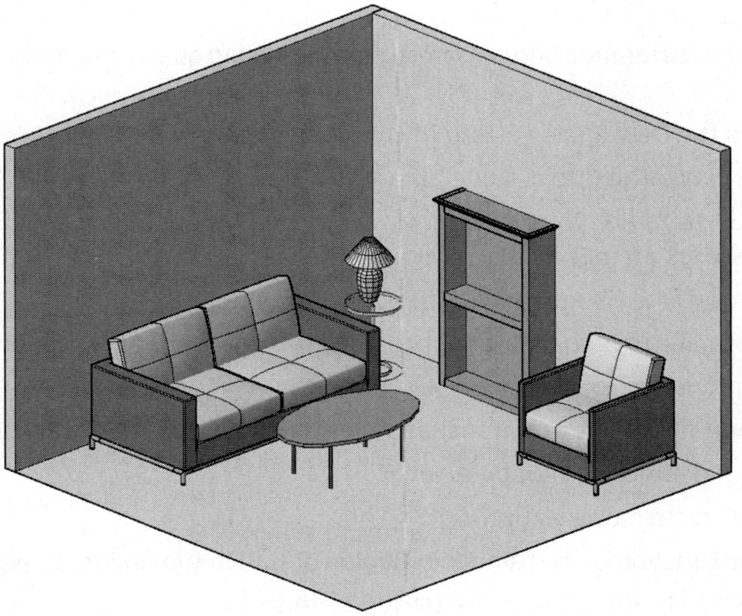

3. **Architectural** In this problem, you will create a new visual style to display the scene as if it is hand sketched.
   A. Open drawing P16-2.
   B. Create a new visual style named Hand Sketched with a description of Displays objects as sketched.
   C. Change the Line Extensions and Jitter settings to make the scene look as shown.
   D. Change any other settings you like.
   E. Plot the scene. Select the new visual style in the **Shade plot** drop-down list.
   F. Export the new visual style to a tool palette so that it can be used in other drawings.
   G. Save the drawing as P16-3.

4. **Mechanical** In this problem, you will create a realistic image of the car fender that you created in Chapter 7.
   A. Open drawing P7-4.
   B. Freeze any layers needed so that only the fender is displayed. Display the fender in the color you want it to be.
   C. Draw a planar surface to represent the ground.
   D. Set the Realistic visual style current. Then, turn on the highlight intensity and full shadows. Also, set the Show property in the **Edge Settings** category to None.
   E. Turn on sunlight and sky background and illumination. Adjust the **Date** and **Time** sliders to make the shadows look as shown.
   F. Render the scene and save it as a JPEG image. Name the file P16-4.jpg.
   G. Save the drawing as P16-4.

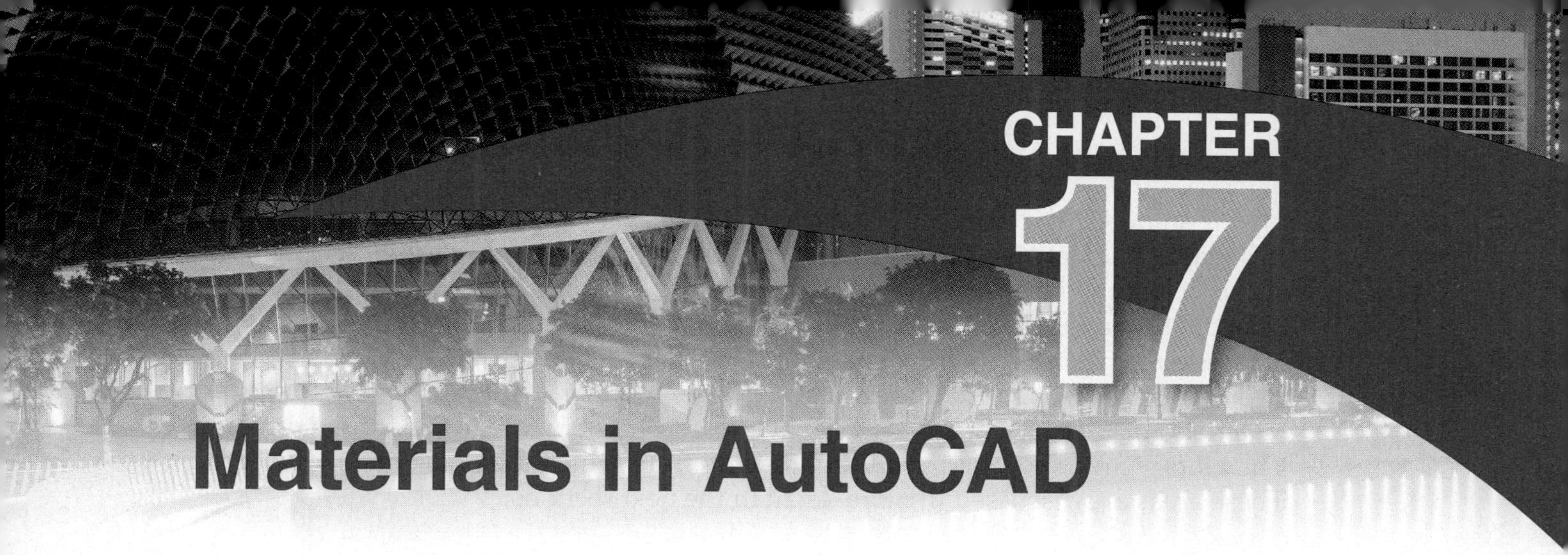

CHAPTER 17

Materials in AutoCAD

Learning Objectives

After completing this chapter, you will be able to:

- ✓ Attach materials to the objects in a drawing.
- ✓ Change the properties of existing materials.
- ✓ Create new materials.

A ***material*** is simply an image stretched over an object to make it appear as though the object is made out of wood, marble, glass, brick, or various other materials. AutoCAD provides an assortment of materials that can be used in your drawings to create a realistic scene. The materials are grouped into categories to make them easier to find.

Materials are easy to attach. They can be dragged and dropped onto the objects, attached to all selected objects, and even attached based on the object's layer. Once the material is attached, you can adjust how the material is *mapped* to the object. If the current visual style is set to display materials in the viewport, you can immediately see the effects on the object. The properties of a material can also be changed to make it look shinier, softer, smoother, rougher, and so on. When you finally render the scene, you will see the full effect of the materials.

The ***materials library*** is the location where all materials are stored. When you install AutoCAD, the Autodesk Material Library is installed. The images in the Base Resolution Image Library are low resolution (512 × 512) and are used with AutoCAD materials. The Medium Resolution Image Library contains medium-resolution images (1024 × 1024) that are good for close-up work or large-scale model rendering. The Medium Resolution Image Library is an additional software option that you must install after installing AutoCAD. When you attempt to render, you may be asked if you want to download the Medium Resolution Image Library. Just follow the prompts to accomplish this task.

Materials Browser

MATBROWSEROPEN

Ribbon
Visualize
> Materials
Materials Browser
Type
MATBROWSER-OPEN
RMAT
MAT

In AutoCAD, the **Materials Browser** palette is used to manage the materials library. In this chapter, the **Materials Browser** palette is referred to as the ***materials browser***. The materials browser provides access to all materials that are available in the Autodesk libraries and from other sources. The **Materials** panel in the **Visualize** tab of the ribbon contains buttons for accessing the materials browser and the materials editor. See **Figure 17-1**.

There are two main areas in the materials browser, **Figure 17-2**. The **Document Materials** area contains materials that have been selected for use in the current drawing. The **Library** area shows all available libraries from which materials may be selected. These areas are discussed in more detail in the next sections.

Figure 17-1.
The **Materials Browser** button is located on the **Materials** panel in the **Visualize** tab of the ribbon. The materials editor is displayed by picking the dialog box launcher on the panel.

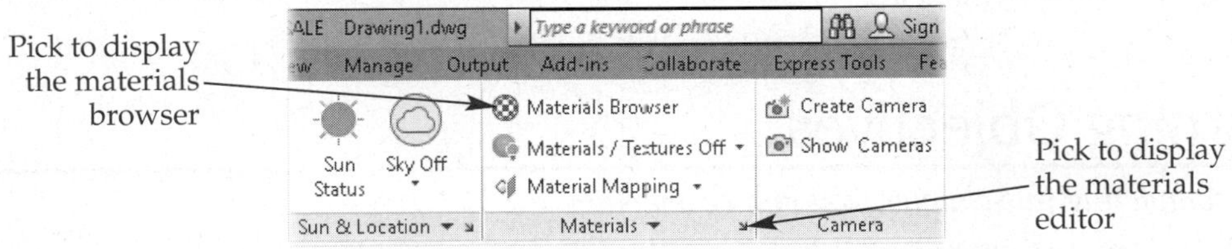

Figure 17-2.
The materials browser contains all of the available materials.

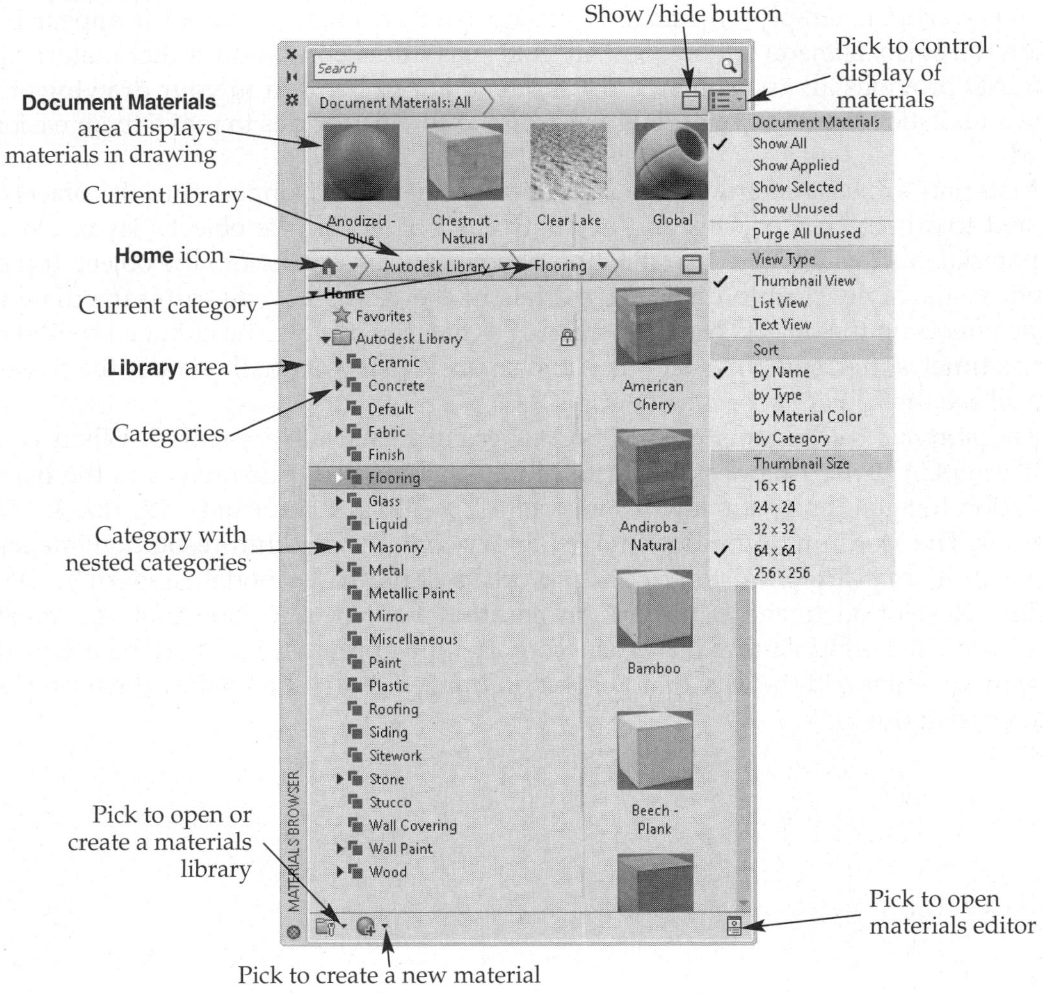

Document Materials Area

Materials are added to the **Document Materials** area from the **Library** area. Hovering over a material swatch in the **Library** area displays two options. See **Figure 17-3**. Picking **Adds material to document** adds the material to the **Document Materials** area. Picking **Adds material to document and displays in editor** adds the material to the **Document Materials** area and opens it in the materials editor (discussed later). After you add a material to the document, a swatch for the material appears in the **Document Materials** area.

Picking the button on the right end of the **Document Materials** title bar opens a drop-down menu. The options in this menu allow you to control which materials are displayed in the **Document Materials** area. You can show all materials, only the materials that are applied to objects in the drawing, selected materials, or unused materials. Materials may be sorted by name, type, color, or category. They can be displayed in thumbnail, list, or text view, and the size of thumbnails can be changed. Refer to **Figure 17-2**. Picking the show/hide button on the **Document Materials** title bar hides the display of the **Library** area. Pick the button again to expand the **Library** area.

Right-clicking on a material in the **Document Materials** area displays a shortcut menu. There are several options available in the shortcut menu:

- **Assign to Selection.** This is active if one or more objects are selected in the scene. The material will be applied to the selected object(s).
- **Select Objects Applied To.** All objects in the drawing that have the material applied to them are selected in the drawing window.
- **Edit.** This displays the materials editor for editing the selected material. The materials editor is discussed later in this chapter.
- **Duplicate.** A copy of the material is added to the drawing. A new material is added with the same name as the original material, but with a sequential number added to the name.
- **Rename.** This option is used to change the name of the material. It is a good idea to rename all of your materials to meaningful names.
- **Delete.** This option removes the material from the drawing. If it is currently being used in the drawing, a warning alerts you to this. Continuing with the deletion removes the material from the objects in the drawing as well as from the **Document Materials** area. It is not, however, removed from the library.
- **Add to.** This option displays a cascading menu with two choices. The material may be added to the Favorites library in the **Library** area. This is excellent for organizing materials that you frequently use and want to make available for quick access in other drawings. You can also add the material to the active tool palette.
- **Purge All Unused.** This option removes from the drawing all materials not currently assigned to an object.

Figure 17-3.
Hovering over a material in the **Library** area displays two buttons that allow you to add the material to the document. Picking the button on the right adds the material to the document and displays it in the materials editor.

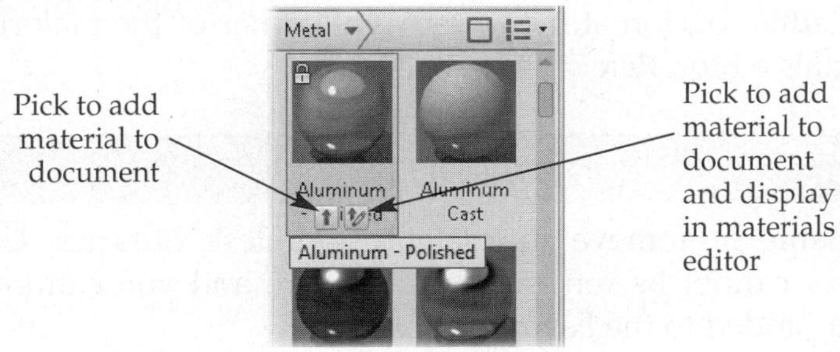

Library Area

The **Library** area of the materials browser contains the open library files. By default, two libraries exist when AutoCAD is installed: Favorites and Autodesk Library. The name of the current library is displayed on the **Library** area title bar. You can select a different library from the drop-down list on the left end of the title bar. Picking the **Home** icon next to the drop-down list deselects the current library and displays the available libraries in the right-hand column of the **Library** area.

The Autodesk Library is composed of the materials installed when AutoCAD was installed and cannot be edited. Favorites can be used to collect and organize commonly used materials, as well as user defined materials. Favorites is empty by default, but you may add your own categories and materials to it. Custom materials are only available in the drawing in which they were created unless saved to Favorites or another user defined materials library. Custom materials that you create in a drawing but do not save to a materials library are called *embedded materials*.

To expand a library in the tree view, pick the triangle next to the library name. Libraries can have categories within them. If the category contains nested categories, a triangle appears next to the name. Refer to **Figure 17-2**. Pick the triangle to expand the item. Categories in the Autodesk Library are organized by material type. Categories in Favorites and user defined libraries must be created by the user. As you make selections in the tree view, the **Library** area title bar updates to display the name of the current item.

Materials are displayed in the right-hand column and change to reflect whatever category is selected. Picking the button on the right end of the **Library** area title bar opens a drop-down menu similar to the drop-down menu in the **Document Materials** area. The options in the **Library** section allow you to select which library is displayed. The view type, sort, and thumbnail size options are the same as those previously discussed for materials displayed in the **Document Materials** area. Picking the show/hide button on the **Library** area title bar hides the tree view and changes the display to a single window displaying materials.

The libraries that you assemble can be saved externally and opened in other drawings. The button with the folder icon at the lower-left corner of the materials browser provides options for managing materials libraries and categories:

- **Open Existing Library.** Allows you to select a library file (*.adsklib) and display it in the materials browser.
- **Create New Library.** Used to create a new library you assemble in the materials browser. Library files are saved with the .adsklib file extension.
- **Remove Library.** Used to delete libraries from the materials browser.
- **Create Category.** Allows you to add categories to your library.
- **Delete Category.** Used to remove categories.
- **Rename.** Allows you to rename libraries and categories.

The **Creates a new material in the document** button at the bottom of the materials browser provides a useful way to create your own materials based on the standard Autodesk material types. Selecting **Concrete**, **Metal**, **Plastic**, etc. from the drop-down list creates a material with the default settings for that material type. The materials editor is opened for you to modify the new material to your liking. Picking the **Opens/Closes material editor** button at the lower-right corner of the materials browser also opens the materials editor. Refer to **Figure 17-2**.

NOTE

You cannot rename or remove the default Autodesk libraries. Categories in the Autodesk Library cannot be renamed or removed, and you cannot add categories. Categories can be added to the Favorites library.

Exercise 17-1

www.g-wlearning.com/CAD/

Complete the exercise on the companion website.

Applying and Removing Materials

To attach a material to an object in the drawing, you can drag the material from the materials browser and drop it onto an object. To apply a material only to a face on an object, hold the [Ctrl] key and pick the face. To apply a different material to an object, simply drag the new material to the object and pick.

Ribbon
Visualize
> Materials
Attach By Layer
Type
MATERIALATTACH

MATERIALATTACH

You can use the **MATERIALATTACH** command to assign materials to the layers in your drawing. Once a material is assigned to a layer, any object on that layer is displayed in the material, as long as the object's material property is set to ByLayer. When objects are created in AutoCAD, the default "material" assigned to them is ByLayer. If your objects are organized on layers, this is the easiest way to attach materials. You can override the layer material by applying a material to individual objects.

Figure 17-4 shows the **Material Attachment Options** dialog box displayed by the **MATERIALATTACH** command. The list on the left side of the dialog box shows the materials loaded into the drawing. The right side of the dialog box shows the layers in the drawing and the material attached to each layer. When no material is attached to a layer, the material is listed as Global. The Global material is a "blank" material in every drawing. To attach a material to a layer, drag the material from the list on the left and drop it onto the layer name on the right. To remove a material from a layer, pick the **X** button next to the material name on the right side of the dialog box.

Ribbon
Visualize
> Materials
Remove Materials

Remove Materials

The **Remove Materials** button on the **Materials** panel in the **Visualize** tab of the ribbon allows you to quickly set an object's material back to ByLayer. When the button is picked, a paintbrush selection cursor is displayed. If you select an object that has a material specifically attached to it, the material is removed. Selecting an object already set to ByLayer has no effect.

Figure 17-4.
Attaching materials to layers.

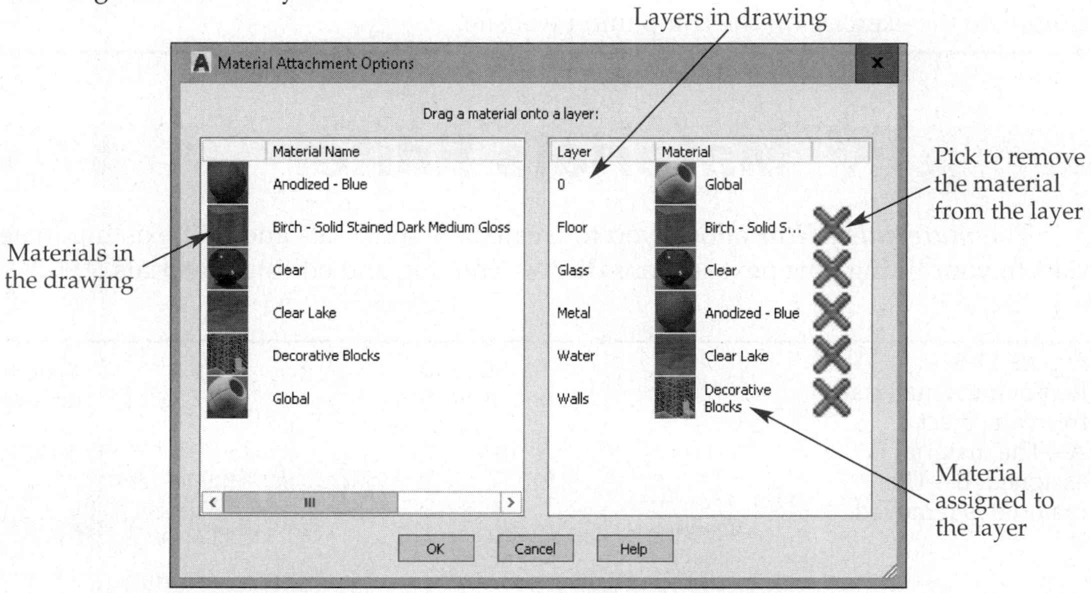

A material can also be removed using the **Properties** palette. To remove a material from an object or subobject, simply change its Material property in the **3D Visualization** category to Global. If a material has not been assigned to the object's layer, the property can also be set to ByLayer. See **Figure 17-5**.

PROFESSIONAL TIP

A material can also be applied to an object by selecting the object in the drawing window, right-clicking on the material swatch in the materials browser, and selecting **Assign to Selection**. A material can be loaded into the drawing without attaching it to an object by picking the material and dragging and dropping it into a blank area of the drawing. This makes the material available in the drawing.

Exercise 17-2

www.g-wlearning.com/CAD/

Complete the exercise on the companion website.

Material Display Options

VSMATERIALMODE

Ribbon
Visualize > Materials

Materials/Textures Off

Materials On/ Textures Off

Materials/Textures On

Type
VSMATERIALMODE

As you learned in the previous chapter, visual styles control how materials are displayed in the viewport. The Material display property of a visual style can be set to display materials and textures, materials only, or neither materials nor textures. The **Materials** panel on the **Visualize** tab of the ribbon has three buttons in a flyout that correspond to, but override, this property setting:

- **Materials/Textures Off.** Objects are displayed in their assigned colors.
- **Materials On/Textures Off.** Objects are displayed in the basic color of the material, but no other material details are displayed.
- **Materials/Textures On.** Objects are displayed with the effects of all material properties visible.

Exercise 17-3

www.g-wlearning.com/CAD/

Complete the exercise on the companion website.

Materials Editor

The ***materials editor*** allows you to create new materials and edit existing materials to your liking. The next sections discuss creating and editing materials.

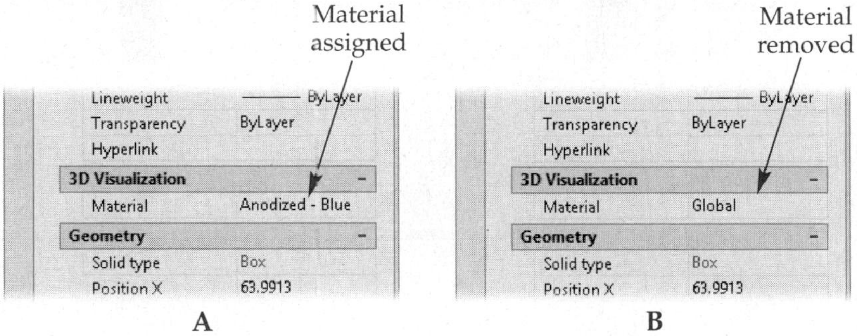

Figure 17-5. Removing a material from an object. A—The material is assigned. B—The material is removed.

Type
MATEDITOROPEN

MATEDITOROPEN

The **Materials Editor** palette is displayed by picking the dialog box launcher button at the lower-right corner of the **Materials** panel in the **Visualize** tab of the ribbon. See **Figure 17-6.** You can also type the **MATEDITOROPEN** command, double-click on a material swatch in the materials browser, or pick the **Opens/Closes material editor** button in the lower-right corner of the browser. In addition, you can hover over a material swatch in the **Document Materials** area to display a small pencil icon. Picking this icon opens the materials editor.

A preview of the material appears at the top of the materials editor. Picking the drop-down menu button at the lower-right corner of the material swatch displays three menus. The options in these menus allow you to change the geometry, type of lighting, and quality used for the preview.

The **Scene** menu provides 13 shapes from which to select: **Sphere**, **Cube**, **Cylinder**, **Canvas**, **Plane**, **Object**, **Vase**, **Draped Fabric**, **Glass Curtain Wall**, **Walls**, **Pool of Liquid**, **Utility**, and **Torus**. The **Environment** menu provides four basic lighting options that allow you to preview a material in a setting similar to where you may be using it. The **Grid Light** option simulates overhead neutral-color lighting that you might find in an office environment. The **Plaza** option simulates an outdoor lighting environment with a bluish tint and strong shadows. The **Snowfield** option produces soft, even light that appears to be coming from everywhere. The **Warm Light** option produces a yellowish cast similar to incandescent indoor lighting.

The options in the **Render Settings** menu set the quality of the preview image. The **Quick** option provides a good quality preview, but the environment settings are poorly shown. The **Draft** option displays the environment nicely and will take a few seconds to update (depending on material complexity and computer speed). The **Production** option provides the highest quality, but the resulting display may take a few minutes to update. The updates occur whenever you change a material property.

Figure 17-6.
The materials editor. A—A preview of the material appears at the top of the palette. Picking the drop-down menu button at the lower-left corner of the palette displays the drop-down menu shown in B. B—This drop-down menu provides options for creating a new material.

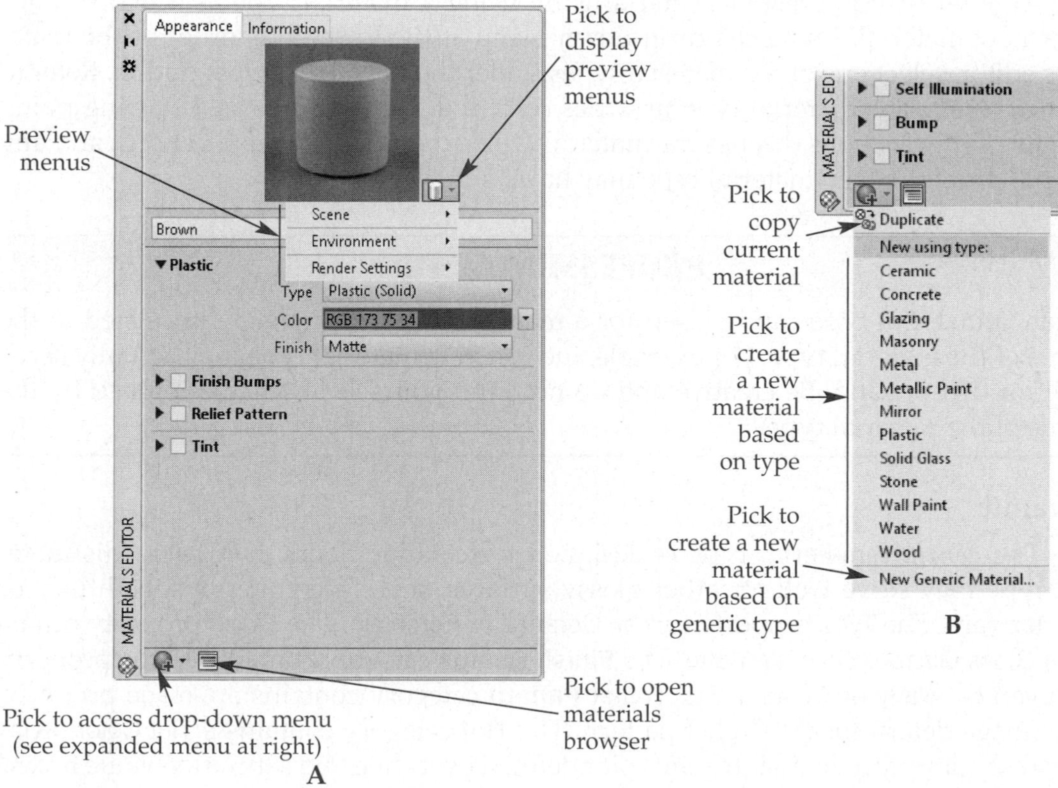

The name of the material can be changed by editing it in the text box located below the material preview. You can also rename the material using the **Name** text box in the **Information** tab.

Materials created or edited in the materials editor are then added to the **Document Materials** area of the materials browser. The materials browser can be opened from the materials editor by picking the **Opens/Closes material browser** button at the bottom of the **Appearance** tab. Refer to Figure 17-6A.

Creating and Modifying Materials

There are three different approaches to creating new materials. You can duplicate an existing material. You can start with an existing material type as a template. Finally, you can start with a generic material, which is like starting from scratch. To access these options, use the drop-down menu at the lower-left corner of the materials editor.

New Material from an Existing Material

By far the easiest way to create your own material is to start with an existing material, create a duplicate, and make any needed modifications. Look through the materials library in the materials browser to find a material that is close to what you want and add the material to the **Document Materials** area. Hover over the material swatch in the **Document Materials** area and select the pencil icon to launch the materials editor.

The material selected in the **Document Materials** area of the materials browser is displayed in the materials editor. Pick **Duplicate** from the drop-down menu at the lower-left corner of the materials editor. Refer to Figure 17-6B. This creates a duplicate material with the same name and a sequential number. Change the name and the properties to your liking (discussed later) and close the materials editor. The new material is in the **Document Materials** area of the materials browser ready to use.

New Material Using an Existing Material Type

Another way to create a material is by using a material type as a template for your new material. Using the drop-down menu at the lower-left corner of the materials editor, select one of the material types under the **New using type:** option. Refer to Figure 17-6B. The material type provides certain default settings as a starting point. The following sections discuss the material types, their possible applications, and any special properties the material type may have.

PROFESSIONAL TIP

Often, a material type can be used for a material that is completely unrelated to the name of the material type. For example, the concrete material type may actually serve well for dirt or sand. Be creative and do not limit yourself to what is implied by the name of the material type.

Ceramic

The *ceramic* material type is designed for ceramic floors. However, this material type may serve well for other glossy surfaces, such as countertops, bathtubs, or dinnerware. The Type property can be Ceramic or Porcelain. The Finish property can be High Gloss/Glazed, Satin, or Matte. The **Finish Bumps** category contains a Type property that can be Wavy or Custom. The **Relief Pattern** category contains an Image property. The image determines the relief pattern. The **Tint** category contains a Tint Color property that allows you to assign a tint color defined by its hue and saturation value mixed with white. This property is also available in the other material types.

Concrete

The *concrete* material type works well for concrete floors, walls, and sidewalks. This material type can be used for anything constructed of concrete, such as an in-ground swimming pool. The Sealant property can be None, Epoxy, or Acrylic. The **Finish Bumps** category contains a Type property that can be Broom Straight, Broom Curved, Smooth, Polished, or Stamped/Custom. The **Weathering** category contains a Type property that can be Automatic or Custom-Image (based on a selected image).

Glazing

The glass in windows is called glazing. Therefore, the *glazing* material type is designed mostly for use on windows or thin glass objects. The solid glass material type is designed for thicker glass. The Color property can be Clear, Green, Gray, Blue, Blue-green, Bronze, or Custom. The Reflectance value determines how reflective the material is. The Sheets of Glass property simulates the effect of multiple panes of glass.

Masonry

Masonry includes brick walls, cobblestone, tile flooring, and so on. The *masonry* material type is designed for use on these objects. The Type property can be CMU (concrete masonry unit) or Masonry. The Finish property can be Glossy, Matte, or Unfinished. The **Relief Pattern** category contains an Image property. The image determines the relief pattern.

Metal

The *metal* material type works well for mechanical parts and other objects made from different types of metals. This material type primarily is for raw, or unfinished, metal. The Type property can be Aluminum, Anodized Aluminum, Chrome, Copper, Brass, Bronze, Stainless Steel, or Zinc. The Finish property can be Polished, Semi-polished, Satin, or Brushed. The **Relief Pattern** category contains a Type property that can be Knurl, Diamond Plate, Checker Plate, or Custom-Image (based on a selected image). The **Cutouts** category contains a Type property that can be Staggered Circles, Straight Circles, Squares, Grecian, Cloverleaf, Hexagon, or Custom. Depending on which type of metal is selected, additional properties may be available for editing.

Metallic Paint

The *metallic paint* material type is for objects made of metal, but with a finish applied. This includes objects like car parts, lawn furniture, and kitchen appliances. The **Flecks** category contains Color and Size properties. The **Pearl** category includes a Type property that can be Chromatic or Second Color. The **Top Coat** category includes a Type property that can be Car Paint, Chrome, Matte, or Custom. The **Top Coat** category also includes a Finish property that can be Smooth or Orange Peel.

Mirror

The *mirror* material type is designed for very reflective objects, such as mirrors. It can also be used for water, glass, or any object that should have a high reflectivity. The Color property determines the color of the mirror. As discussed earlier, the **Tint** category is also available.

Plastic

The *plastic* material type is designed for use on plastic objects. The plastic can be opaque, translucent, glossy, or textured. The Type property can be Plastic (Solid), Plastic (Transparent), or Vinyl. The Finish property can be Polished, Glossy, or Matte. The **Finish Bumps** category contains an Image property, as does the **Relief Pattern** category. The two images do not have to be the same.

Solid Glass

The ***solid glass*** material type is intended for thick glass objects or a volume of liquid, such as a glass of water. The Color property can be Clear, Green, Gray, Blue, Blue-green, Bronze, or Custom. The Reflectance property determines the degree of reflectivity of the material. The Refraction property can be Air, Water, Alcohol, Quartz, Glass, Diamond, or Custom. The Roughness property determines the polish on the material. The **Relief Pattern** category contains a Type property that can be Rippled, Wavy, or Custom.

Stone

The ***stone*** material type works well for stone walls, stone walkways, and marble countertops. The **Stone** category contains an Image property, which is the image applied to the material. This can be a selected image or a specified texture. The **Stone** category also contains a Finish property that can be Polished, Glossy, Matte, or Unfinished. The **Finish Bumps** category contains a Type property that can be Polished Granite, Stone Wall, Glossy Marble, or Custom. The **Relief Pattern** category contains an Image property. This can be a selected image or a specified texture.

Wall Paint

The ***wall paint*** material type works well for interior or exterior painted walls and other objects. The Finish property can be Flat/Matte, Eggshell, Platinum, Pearl, Semi-gloss, or Gloss. The Application property can be Roller, Brush, or Spray.

Water

The ***water*** material type is designed for any liquid. Pools, reflecting ponds, rivers, lakes, and oceans are examples of where this material type may be used. The Type property can be Swimming Pool, Generic Reflecting Pool, Generic Stream/River, Generic Pond/Lake, or Generic Sea/Ocean. The Color property (when available) can be Tropical, Algae/Green, Murky/Brown, Generic Reflecting Pool, Generic Stream/River, Generic Pond/Lake, Generic Sea/Ocean, or Custom. The Wave Height property sets the amplitude of ripples in the liquid.

Wood

The ***wood*** material type works well for various finishes used for flooring, furniture, wood trim, and so on. The **Wood** category contains an Image property, which is the image applied to the material. This can be a selected image or a specified texture. When the **Stain** check box is checked, the color of stain can be set. The **Wood** category also contains a Finish property, which can be Unfinished, Glossy Varnish, Semi-gloss Varnish, or Satin Varnish. The Used For property can be Flooring or Furniture. The **Relief Pattern** category contains a Type property that can be Based on Wood Grain or Custom (which is based on a selected image).

Generic

The generic material type is the "blank canvas" material. All properties are available to create any material needed. This material type is used to create a material from scratch, as discussed in the next section.

PROFESSIONAL TIP

AutoCAD provides fantastic-looking materials to dress up the scene and make it look real. However, after you get comfortable with creating materials, start a library of your own materials. If your project is presented to a customer along with projects from competitors, and your competitors are using standard AutoCAD materials, your project will stand out from the crowd.

Exercise 17-4

www.g-wlearning.com/CAD/

Complete the exercise on the companion website.

New Material from Scratch

Creating a material from scratch gives you complete control over the material. Although there are many options to consider, creating a material from scratch may be the only way to get the exact material you need to complete your scene. To start creating a material from scratch, select the generic material type by selecting **New Generic Material...** from the drop-down menu at the lower-left corner of the materials editor. Refer to **Figure 17-6B**. The **New Generic Material...** option is also available from the materials browser. The properties and settings are discussed in the next sections.

NOTE

New materials are placed in the **Document Materials** area and are only available in the current drawing unless added to the Favorites library or a user defined library.

Generic Category

The ***diffuse color*** is the color of the object in lighted areas, or the perceived color of the material. See **Figure 17-7**. It is the predominant color you see when you look at the object. The Color property sets the diffuse color of the material. In AutoCAD, the ambient and specular colors are determined by the diffuse color.

The ***ambient color*** is the color of the object where light does not directly provide illumination. It can be thought of as the color of an object in shadows. In nature, shadows cast by an object typically contain some of the ambient color.

A full-color image is provided on the companion website. www.g-wlearning.com/CAD/

Figure 17-7.
The three colors of a material are illustrated here. In AutoCAD, the Color property sets the diffuse color. The ambient and specular colors are based on the diffuse color.

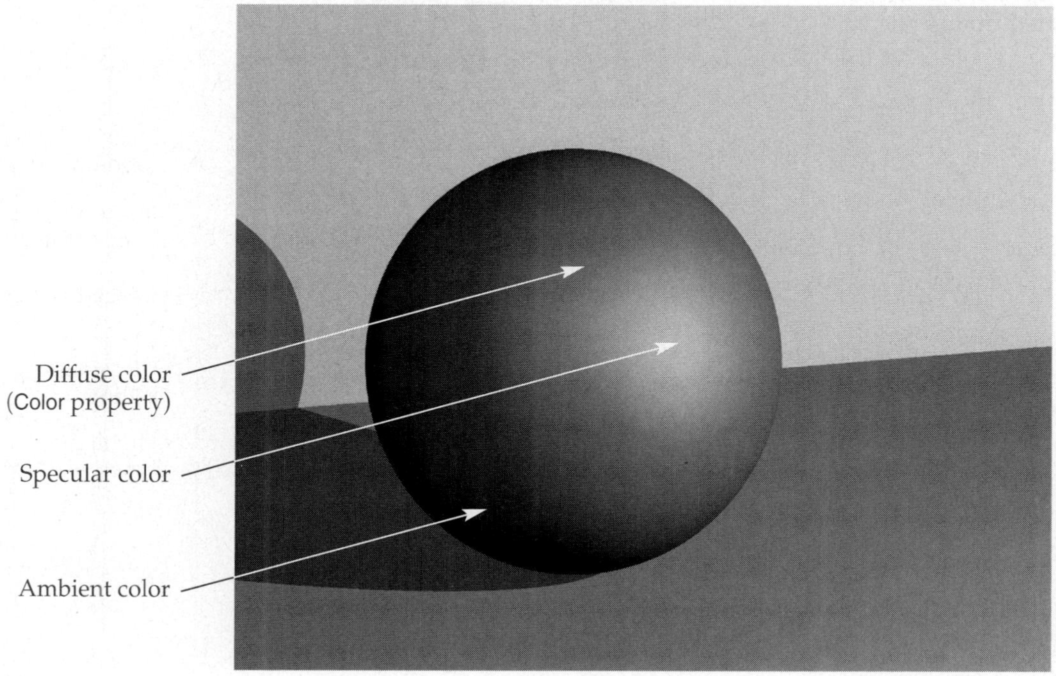

The *specular color* is the color of the highlight (the shiny spot). It is typically white or a light color. The amount of specular color shown is determined by the glossiness and reflectivity of the material and the intensity of lighting in the scene.

You have two options for controlling the Color property of the material. By default, the material color is dark gray (RGB 80 80 80). Pick the current color setting in the edit box to display the **Select Color** dialog box and choose the color you want. See **Figure 17-8**. Picking the button to the right of the color edit box displays a drop-down list. If you select **Color by Object** in the drop-down list, the base color of the object is used as the diffuse color.

The Image property allows you to apply an image or texture to control the appearance of the material. Picking in the image preview area displays a standard open dialog box. When you select an image, it is applied to the material and is displayed in the material preview area, **Figure 17-9**. Also, the texture editor is displayed with the

A full-color image is provided on the companion website. www.g-wlearning.com/CAD/

Figure 17-8.
Setting a color for a material.

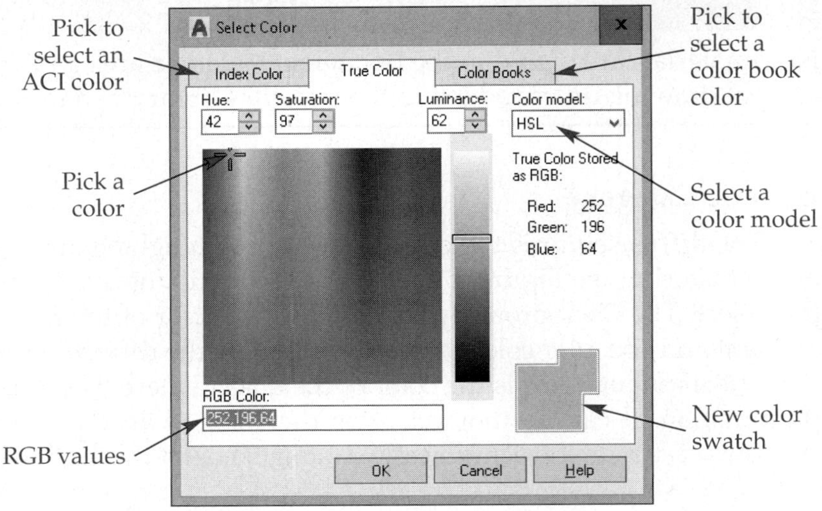

A full-color image is provided on the companion website. www.g-wlearning.com/CAD/

Figure 17-9.
The property settings in the **Generic** category of a material created from scratch.

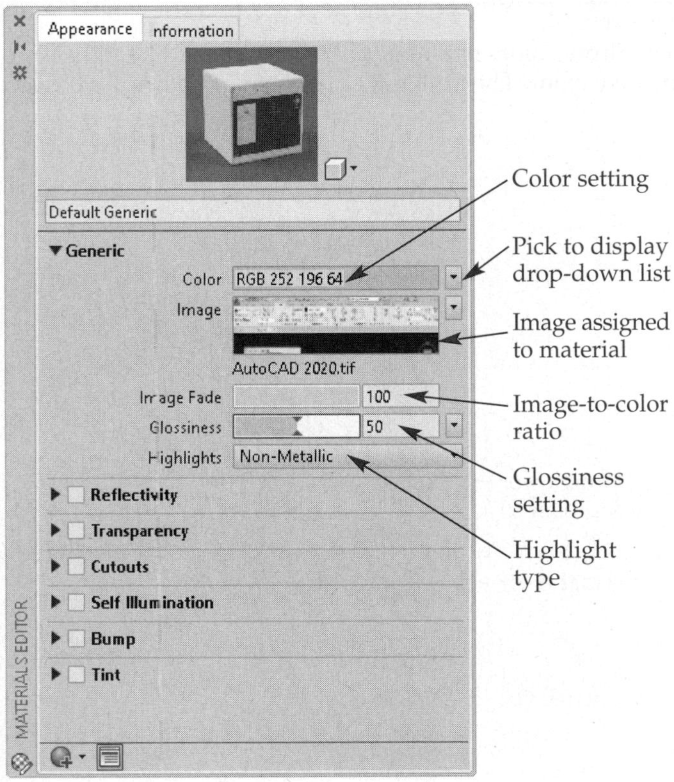

image loaded and ready for editing, if necessary. The button to the right of the image preview area is used to apply a texture instead of an image. Textures and texture editing are discussed later in this chapter.

When an image, such as a bitmap image or digital photograph, is applied, the material color is replaced with the image. **Figure 17-10** shows an object with a computer screen image applied to the material attached to it. The Image Fade property controls the ratio between the image and the object color. When set to 100, the image completely replaces the object color.

The Glossiness property controls how shiny the material appears, **Figure 17-11**. It is a measure of the surface roughness of a material. A setting of 100 specifies a very shiny material, such as a smooth surface. A setting of 0 specifies a matte (dull) material. The button to the right of the setting allows you to add a texture or image to the Glossiness property. The pattern is applied to the glossiness effect. **Reflectivity** must be turned on for the glossiness effect to be seen. Refer to the next section.

The Highlights property determines how the shiny areas are created. Choices for this property are Metallic or Non-metallic. Highlights are brighter when Metallic is selected.

A full-color image is provided on the companion website. www.g-wlearning.com/CAD/

Figure 17-10.
A material with an image applied instead of a color to simulate a tablet screen. A—The screen object's material does not have an image applied. B—A bitmap image has been applied to the image component of the screen object's material. C—The finished tablet.

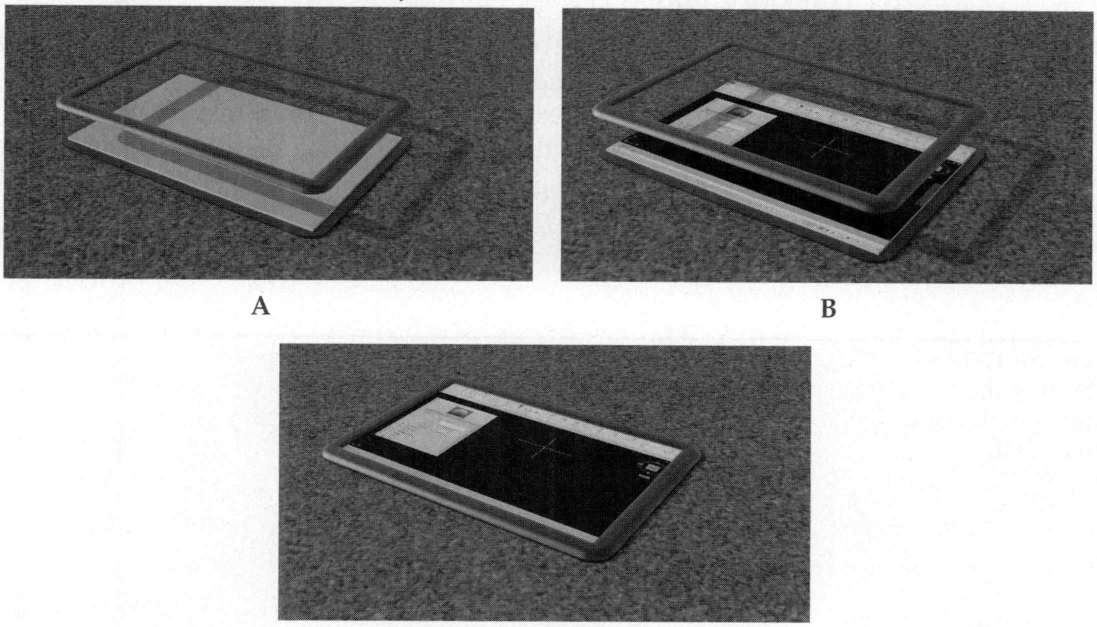

A full-color image is provided on the companion website. www.g-wlearning.com/CAD/

Figure 17-11.
Three different glossiness settings are illustrated here.

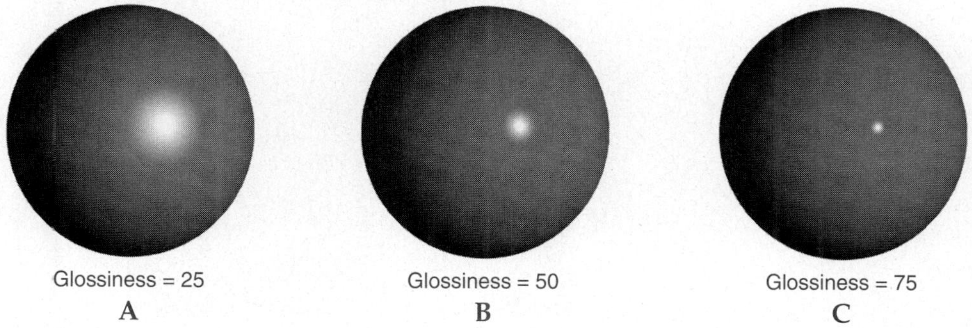

PROFESSIONAL TIP

Highlights can be seen everywhere. Look around you right now at edges and inclined surfaces. The diffuse color of the surface typically has little to do with the color of the highlight. The color of the light source usually determines the predominant highlight color. Most highlights are white or near white because most light sources are white or nearly white. However, highlights in the interior of a home may have a yellow cast to them because incandescent light bulbs generally cast yellow light. Compact fluorescent lights cast a slightly different color, although many are designed to cast the same color as an incandescent bulb. Outside with a clear sky and bright sun, highlights may have a slight blue cast. Keep these points in mind when creating your own materials. These small details are what make a scene realistic.

Reflectivity Category

Reflectivity is a measure of how much light is bounced off the surface. There are two basic property settings for reflectivity, Direct and Oblique. See **Figure 17-12**. The Direct property controls how much light is reflected back for surfaces that are more or less facing the camera. The Oblique property controls how much light is reflected back when the surface is at an angle to the camera.

Each property has a slider/text box that controls the amount of reflectivity. No reflections are created with a setting of 0. The maximum reflections are created with a setting of 100. Object color, lighting, surroundings, and other factors also determine just how reflective a material appears in the scene.

The buttons to the right of the text boxes allow you to add a texture or image to control the reflection. The white areas of the texture or image have a reflection. The black areas do not have a reflection. The degree of reflectivity varies for gray areas and the grayscale values of colors.

Figure 17-12.
Setting the reflectivity for a material.

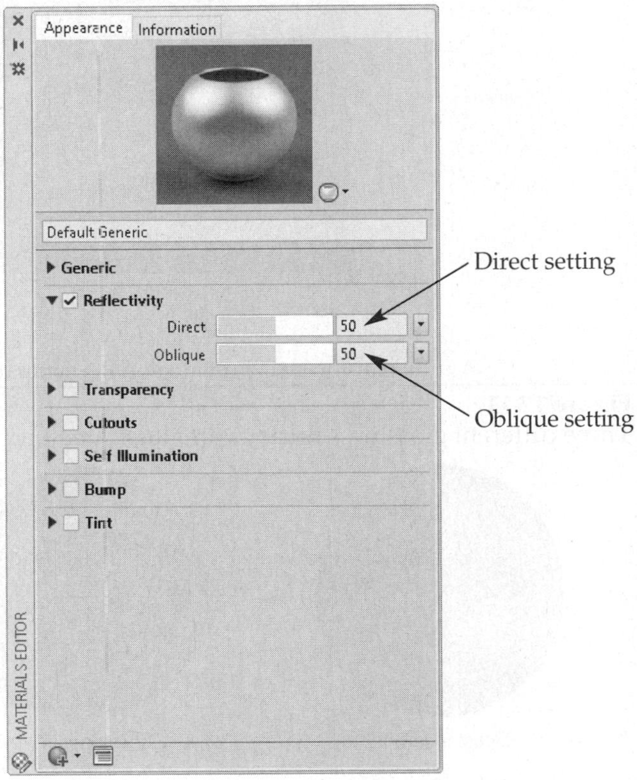

Transparency Category

Transparency is a measure of how much light the material allows to pass through it. Glass, water, crystal, and some plastics, along with other materials, are nearly completely transparent. **Figure 17-13** shows an example of using a transparent material to show the internal workings of a mechanical assembly. The **Transparency** category contains a number of properties that combine to create any transparent or semitransparent material, **Figure 17-14**.

The amount of light that passes through the material is controlled by the Amount property. When this property is set to 0, the material is opaque. A setting of 100 creates a completely transparent material.

A full-color image is provided on the companion website. www.g-wlearning.com/CAD/

Figure 17-13.
The material used for the housing on this mechanism has a transparency setting of about 50.

Figure 17-14.
Setting transparency for a material.

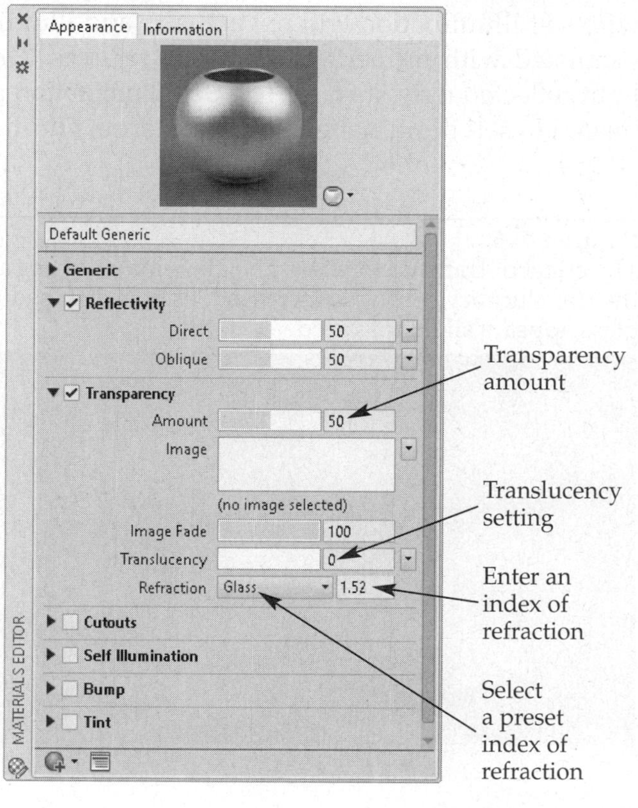

The Image property is used to add a transparency map to the material. White areas in the image or texture are transparent. Black areas are opaque. All other colors produce varying degrees of transparency based on their grayscale values. The Amount property is applied to the transparency map. Maps are discussed later in this chapter.

The Image Fade property determines how much of an impact the transparency map has on the transparency of the material. A setting of 100 means the transparency is completely see-through based on the transparency map. As the setting is decreased, a higher percentage of transparency is determined by the Amount property.

Translucency is a quality of transparent and semitransparent materials that causes light to be diffused (scattered) as it passes through the material. See **Figure 17-15**. This makes any object with the material applied to it appear as if it is being illuminated from within, or glowing. The thicker the material, the more pronounced the effect. When the Translucency property setting is 0, light appears to travel through the material, lighting the opposite side. A setting of 100 creates a material similar in appearance to frosted glass.

Refraction, sometimes specified as the ***index of refraction (IOR)***, is a measure of how much light is bent (refracted) as it passes through transparent or semitransparent materials. Refraction is what causes objects to appear distorted when viewed through a bottle or glass of water. See **Figure 17-16**. The Refraction property sets the IOR. The higher the value, the more that light is bent as it passes through the material. The IOR of water is 1.3333. You can enter a value in the text box or select a preset IOR by picking the name that is displayed to the left of the text box.

Cutouts Category

The Cutouts property allows you to select an image or texture to use for a pattern of cutouts (holes), **Figure 17-17**. Black areas in the image will appear to be see-through, as if there is no object in those areas. White areas in the image have the normal material colors. This is similar to using a transparency map, but without the other transparency settings. **Figure 17-18** shows an example of a cutout map applied to a material.

Self Illumination Category

Self illumination is an effect of a material producing illumination. See **Figure 17-19**. For example, the surface of a neon tube glows. However, in AutoCAD, a material with self illumination will not actually add illumination to a scene. This effect can be simulated with properly placed light sources. ***Luminance*** is defined as the value of light reflected off a surface. The **Self Illumination** category contains several properties related to self illumination and luminance, **Figure 17-20**.

A full-color image is provided on the companion website. www.g-wlearning.com/CAD/

Figure 17-15.
The effect of translucency. A—The glass material has a translucency setting of zero. B—When the translucency setting is increased, light is diffused within the material. Notice how the glass appears slightly frosted.

A

B

A full-color image is provided on the companion website. www.g-wlearning.com/CAD/

Figure 17-16.
Note the effect of refraction on the grout lines in the tile floor behind this drinking glass. The default refraction setting for glass is 1.52.

Figure 17-17.
Adding a cutout map to a material.

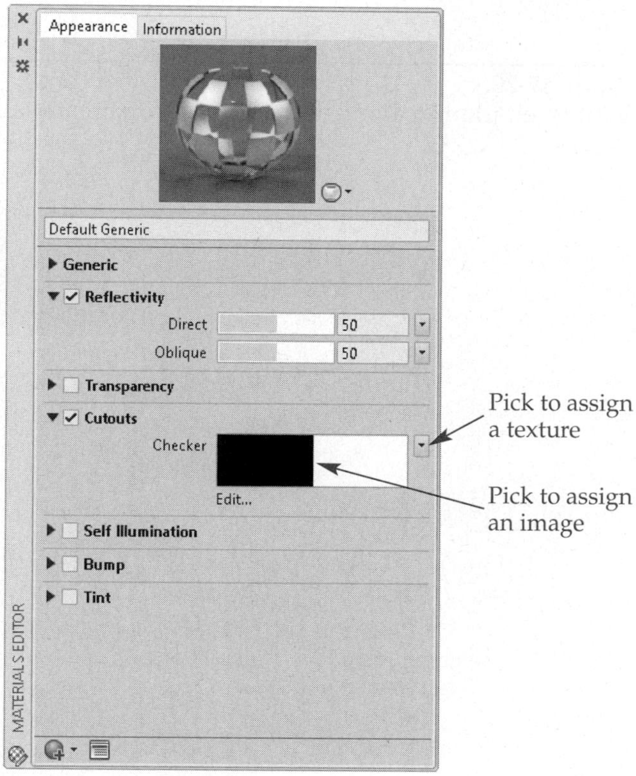

A full-color image is provided on the companion website. www.g-wlearning.com/CAD/

Figure 17-18.
The effect of applying an image to the Cutouts property. A—This black and white image will be used as the cutout map. B—The material on the plane is completely opaque. C—When the cutout map is applied to the material, the dark areas of the map produce transparent areas on the object.

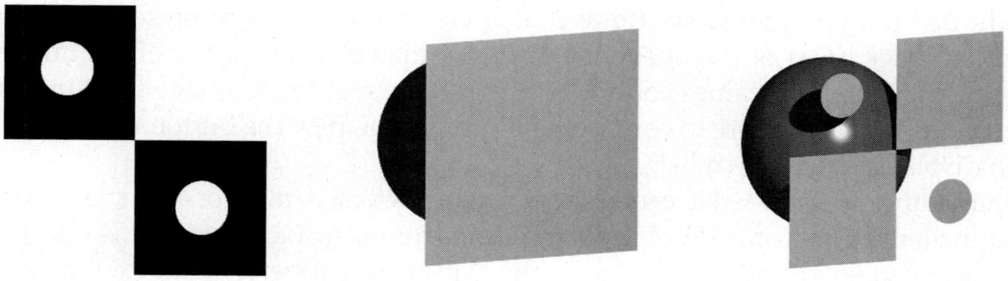

A B C

Figure 17-19.
The effect of self illumination/luminance. A—The globe of this light bulb does not have any self illumination. B—Self illumination is applied to the globe material.

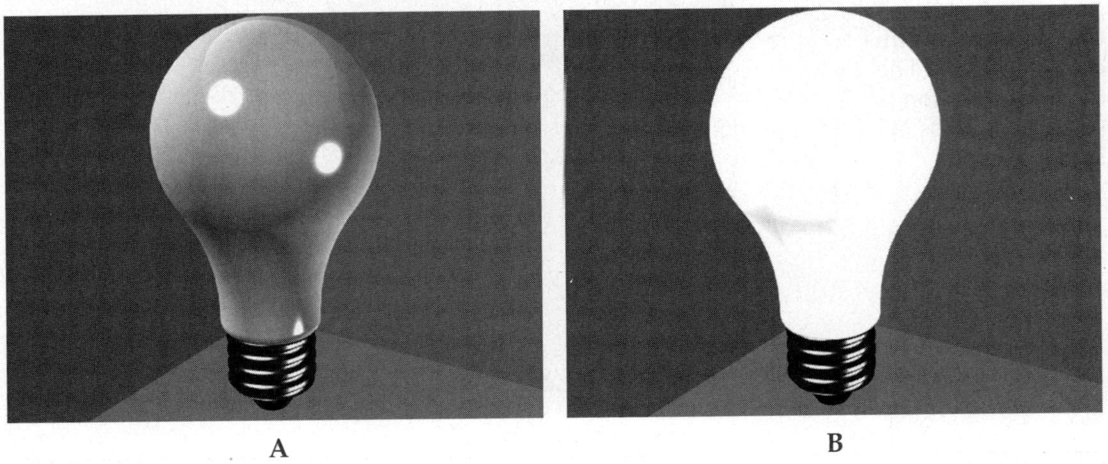

Figure 17-20.
Setting self illumination and luminance for a material.

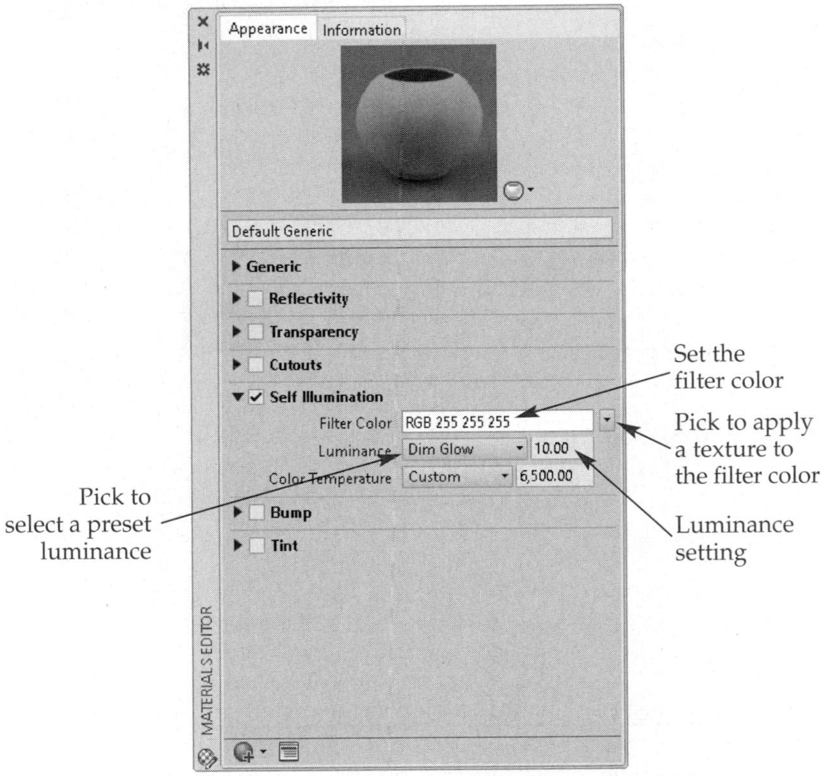

The Filter Color property controls the color of the self illumination effect. Pick in the edit box to open the **Select Color** dialog box. An image may be selected instead of a color. Black areas of the image are not illuminated. White areas of the image are illuminated. Grayscale values control how much the rest of the image illuminates the material. To apply a texture to control the illumination, pick the button next to the edit box to display a drop-down list.

Luminance is expressed in candelas per square meter (cd/m^2). For example, 1 cd/m^2 is the equivalent of one candela of light radiating from a surface area that is one square meter. You can enter a specific value in the Luminance property text box. Or, you can pick the name displayed next to the text box to display a drop-down list with choices for typical materials. Some of these choices include Dim Glow, LED Panel, and Cell Phone Screen.

The Color Temperature property determines the warmth or coolness value of the color. The value is expressed as degrees Kelvin. Candles and incandescent bulbs are warm. Fluorescent lights and TV screens are cool. The drop-down list gives you choices for typical objects with different color temperatures, but you can enter a value directly in the text box.

Bump Category

The **Bump** category contains settings for making some areas of the material appear raised and other areas depressed, **Figure 17-21**. The image or texture used for this effect is called a ***bump map***. The black, white, and grayscale values of the map are used to determine raised and depressed areas. Dark areas of the map appear raised and light areas appear depressed.

For example, to show the texture of a brick wall, you could physically model the grooves into the wall. This would take a lot of time to model and would immensely increase the rendering time because of the increased complexity of the geometry. Using the properties in the **Bump** category is an easier and more efficient way to accomplish the same task. **Figure 17-22** shows a bump map used to represent an embossed stamp on a metal case.

To apply an image as a bump map, pick the Image property swatch to display a standard open dialog box. To apply a texture as a bump map, pick the button next to the edit box to display a drop-down list.

The Amount property determines the relative height of the bump pattern. A setting of 0 results in a flat material, or no bumps. A setting of 1000 creates the maximum difference between low and high areas of the pattern.

Figure 17-21.
Making bump settings for a material.

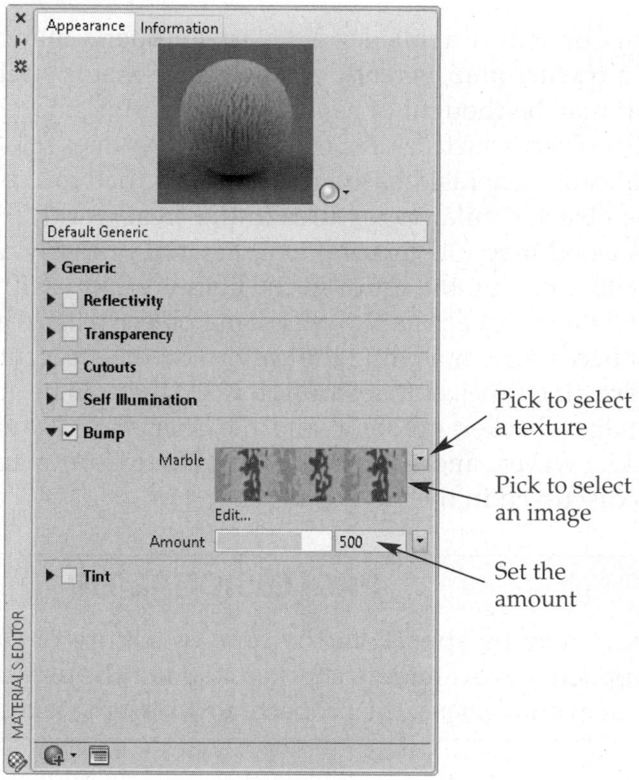

Figure 17-22.
The effect of a bump map. A—This image will be used as the bump map. B—When applied to the material, the bump map simulates an embossed stamp on the metal case.

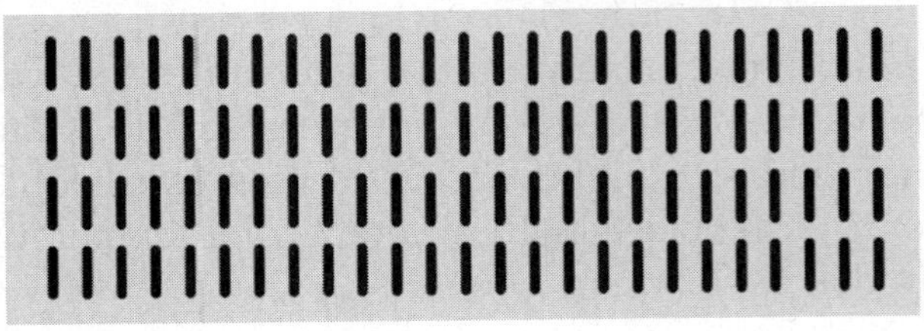

A

B

Maps

As discussed in previous sections, images and textures may be applied to materials to enhance the color, glossiness, reflectivity, transparency, translucency, cutouts, self illumination, and bumpiness. The applied image or texture is called a ***map***. A material that has a map applied to at least one of its properties is called a ***mapped material***.

An image or texture applied to a material property may be 2D or 3D. A ***2D texture***, also called a ***texture map***, is really just an image that is stretched over the surfaces of the object. It may be thought of as a fixed set of pixels, kind of like a mosaic pattern, applied to the object's surfaces. A ***3D texture***, sometimes referred to as a *procedural map*, is mathematically generated based on the colors that you select. This texture extends through the object, similar to textures in the "real world." For example, if you attach AutoCAD's wood map (3D texture) to a material property and assign the material to an object with a cutout, the grain in the cutout will match the grain on the exterior. An image (2D texture) applied to this same object will "reapply" itself to the cutout surface, not necessarily matching the pattern on the adjacent surfaces.

There are nine types of maps in AutoCAD that can be applied to material properties. The image, checker, gradient, and tiles maps are 2D texture maps. The marble, noise, speckle, waves, and wood maps are 3D texture maps. Each map has unique settings, as discussed in the next sections.

PROFESSIONAL TIP

The term *map* may be applied to the type of texture or the property to which the texture is applied. For example, a checker map may be used as a bump map. *Checker* is the type of map and *bump* is the property to which the texture is applied.

Texture Editor

The *texture editor* is used to adjust a map once it has been applied to a material property. It is automatically displayed once a map is assigned. **Figure 17-23** shows the texture editor with an image map displayed for editing. A preview of the map appears at the top of the texture editor. The triangle in the lower-right corner of the preview is used to resize the preview. Below the map preview, the type of map is indicated.

Maps rarely appear on the object in the correct position, scale, or angle. For example, if you are using an image map of a logo to be applied to a box for a packaging design, it may not be positioned in the center of the box by default. The map may be rotated in the wrong direction or it may be too large. The texture editor is where adjustments to the map are made.

The bottom of the texture editor contains the settings for the map. The properties that are available depend on the type of map being edited. As shown in **Figure 17-23**, the **Transforms**, **Position**, **Scale**, and **Repeat** categories are available and automatically expanded when the texture editor is opened for an image map. The next sections discuss the properties for each map type.

Image

Applying an *image map* is straightforward. There are several file types that can be applied as a map. When you specify Image as the map for a property, a standard open dialog box is displayed. Navigate to the image file and open it.

In the texture editor, the preview area shows the image with dimensions for the size of an individual tile. If the Sample Size property in the **Scale** category is changed, the preview dimensions reflect the change. Directly below the preview are the image file name and a slider/text box for adjusting the image brightness. Refer to **Figure 17-23**. To select a different image, pick the name to display a standard open dialog box. A brightness value of 100 means the image is at its full brightness. A value of 0 results in the image being all black. The check box below the slider is used to invert the image colors. This produces an effect similar to a photographic negative

Figure 17-23.
The texture editor is used to adjust a map once it has been applied to a material property. Shown are the properties of an image map.

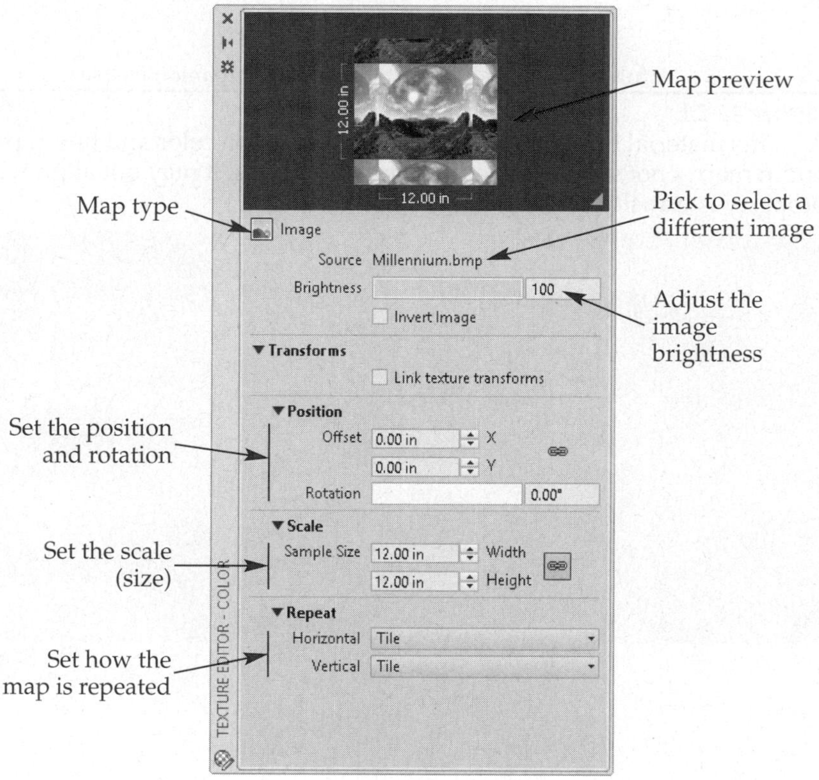

from an old film camera but is most useful for reversing the effects of a bump, cutout, or transparency map.

In the **Transforms** category is the **Link texture transforms** check box. It is very important to check this check box if you are using the same map for different properties in the material and need them to be synchronized in appearance. For example, to make a realistic tile floor, an image of tiles is applied to the color property. The same image map is also used as a bump map to make the grout look recessed. If the scaling is changed for the bump map, but the maps are not synchronized, the grout colors and indentations may not match. See **Figure 17-24**. By linking the texture transforms, all transform settings are the same. Transforms are not linked by default and if you turn on linking after you have made changes, the properties may not be synchronized. Linking must be turned on for *each* material map or it will not work.

The properties in the **Position** category control the location of the map on the object and its rotation. The Offset property moves the image in the X and Y directions. The link button to the right of these text boxes locks the X and Y values together. It is off by default. The Rotation property allows you to rotate the image on the material.

The properties in the **Scale** category control the size of the image. The Width and Height properties are locked together by default. This is important to maintain a proportional ***aspect ratio*** for the image. The properties can be unlocked, but be aware that entering different values for width and height will stretch and distort the image.

The **Repeat** category is where you can set the image to tile or not tile. ***Tiling*** means that the image repeats as many times as it takes to cover the object. If tiling is turned off, there will only be one image on the object. For example, if you are using an image for a label on a box, tiling should be turned off. Otherwise, the box will be completely covered with labels.

Checker

A ***checker map*** creates a two-color checkerboard pattern. By default, the colors are black and white, but different colors or images can be used as well. This map type can be used for checkerboard pattern floor materials. However, by changing various properties, you can simulate many different effects and materials.

A full-color image is provided on the companion website. www.g-wlearning.com/CAD/

Figure 17-24.
A—This material has an image map assigned to the color and bump properties. B—If the bump map is not synchronized to transformations, it may not align with the grout lines if a property is transformed.

A

B

Figure 17-25.
Adjusting the properties of a checker map.

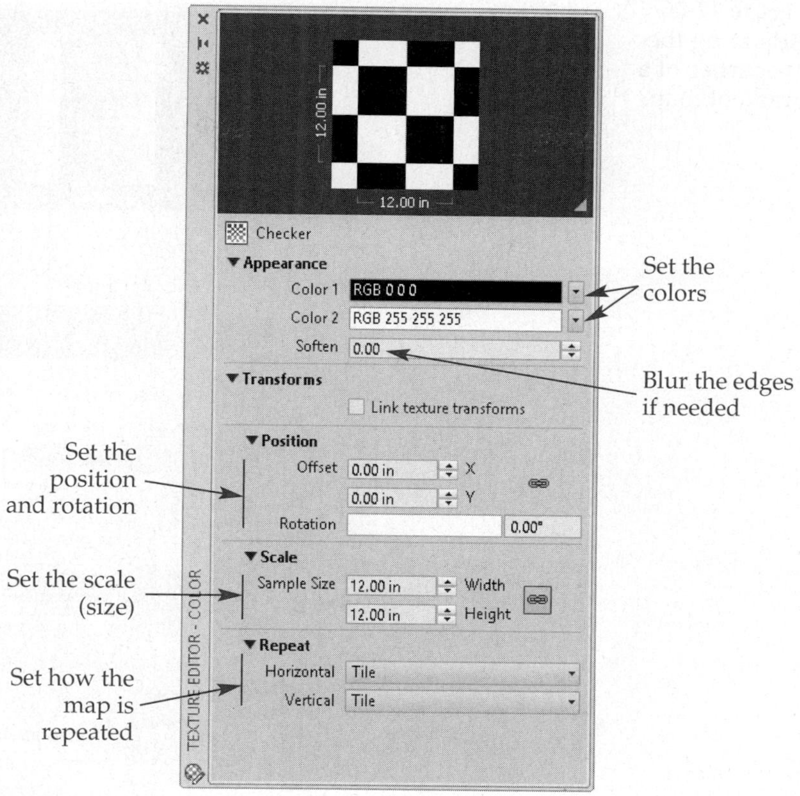

When you specify Checker as the map for a property, the texture editor displays the properties for the map, **Figure 17-25**. In the **Appearance** category, the Color 1 and Color 2 properties set the color of the checkers. The button to the right of each property allows you to specify an image or texture for the checker. For example, you may add a texture map to the Color 1 property and a noise map to the Color 2 property. The possibilities are endless.

The Soften property is used to blur the edges between the checkers. To change the setting, enter a value in the text box or use the up and down arrows. A value of 0.00 creates sharp edges between the checkers. The maximum setting of 5.00 produces edges that are very blurred.

The properties in the **Transforms**, **Position**, **Scale**, and **Repeat** categories are the same as described for an image map. Refer to the Image section for details on these properties.

Gradient

A ***gradient map*** is a texture map that allows you to create a material blending colors in different patterns. This is similar to creating a gradient using the **HATCH** command's **Gradient** option. When you specify Gradient as the map for a property, the texture editor displays the properties for the map, **Figure 17-26**.

The gradient is represented in the **Appearance** category with three nodes at the bottom edge of the ramp. Each node represents a different color in the ramp. By default, the left node (node 1) is black, the middle node (node 2) is gray, and the right node (node 3) is white. The middle node can be moved left or right, as discussed later in this section, to change where the color transitions.

There must be at least three nodes, but you are not limited to three nodes. Picking anywhere in the ramp creates a new node. Selecting a node by picking on it changes the properties directly below the ramp to the settings for that node. The Color property sets the color of the selected node.

Figure 17-26.
Adjusting the properties of a gradient map.

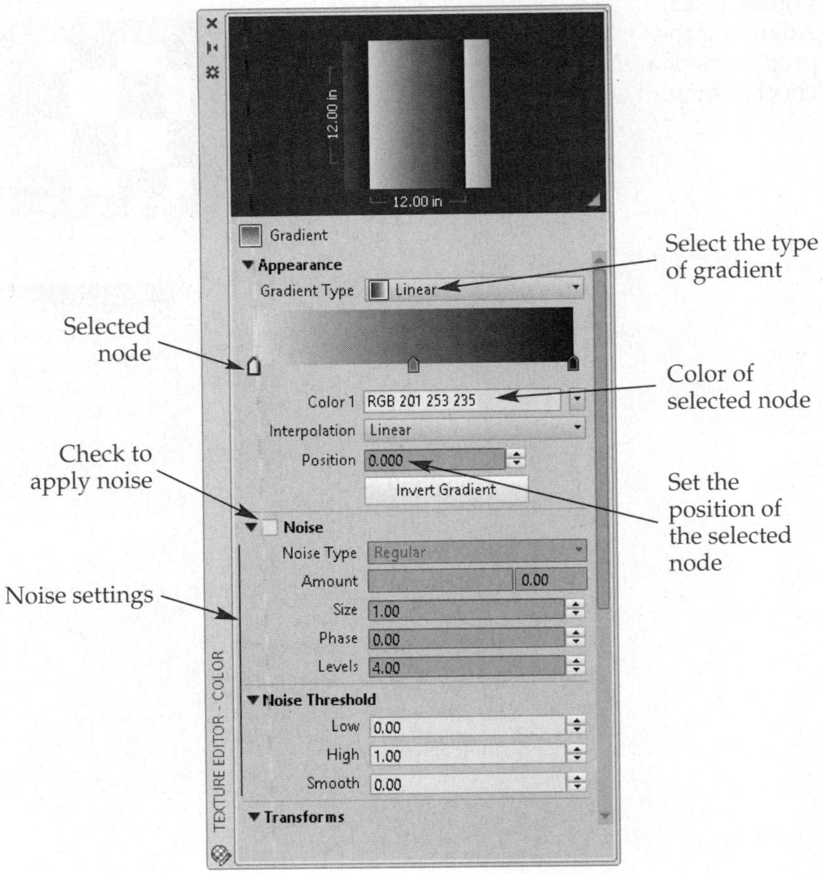

Above the ramp is the **Gradient Type** drop-down list. The setting in the drop-down list controls the pattern of the gradient ramp. See **Figure 17-27**. The default pattern is linear. This results in a typical pattern similar to the ramp display. The options are described as follows:

- **Linear asymmetrical.** This gradient type is similar to the default linear type, but the transition between colors is not symmetrical.
- **Box.** The transition of colors is in the shape of a square.
- **Diagonal.** The transition of colors is linear, but rotated on the surface.
- **Light normal.** The intensity of the light source determines where the transition takes place. The right side of the ramp corresponds to the highest intensity of light and the left side of the ramp is equal to no light.
- **Linear.** This is the default. It is a smooth transition from one node to the next.
- **Camera normal.** The angle between the camera direction and the surface normal controls how the pattern is displayed. The left side of the ramp is 0° between the normal and the camera viewpoint. The right side is 90° between the normal and the camera viewpoint.
- **Pong.** This gradient is a rotated linear transition, similar to the diagonal type, but it pivots about the corner of a box and reverses in the middle of the pattern.
- **Radial.** Colors are arranged in a circular pattern similar to a target.
- **Spiral.** The gradient sweeps about a central point similar to the movement on a radar screen.
- **Sweep.** This gradient is similar to the spiral type, but the center of the sweep is at a corner instead of in the center. Also, the gradient does not repeat like the pong type.
- **Tartan.** This gradient resembles a plaid pattern. It is very similar to the box type.

Figure 17-27.
There are 11 different gradient types available for use in a gradient map. The four shown here are applied to the same object with the same lighting and mapping coordinates. A—Linear asymmetrical. B—Box. C—Diagonal. D—Light normal.

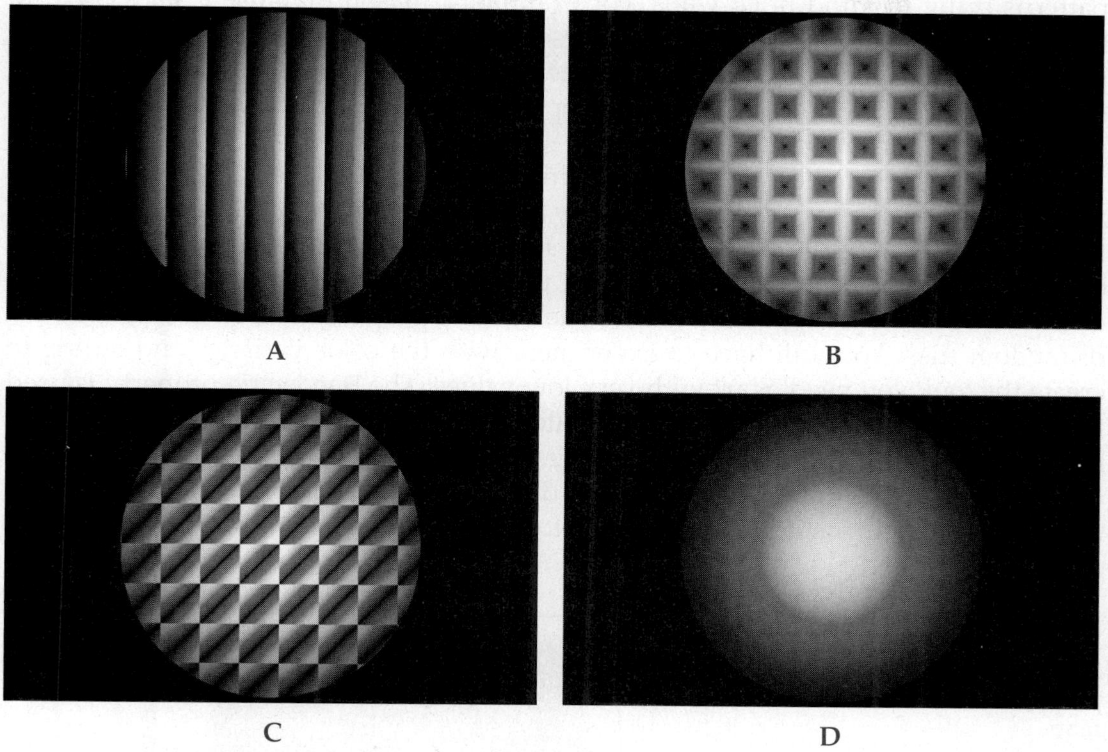

The Interpolation property controls the transition of colors from one node to the next. Transitions are applied to the nodes from left to right, regardless of the node number. The options are:

- **Ease in.** Shifts the transition closer to the node on the right.
- **Ease in out.** Shifts the transition toward the node, but it remains more or less centered on the node.
- **Ease out.** Shifts the transition closer to the node on the left.
- **Linear.** This is the default. The transition is constant from one node to the next.
- **Solid.** No transition between nodes. There is an abrupt change at each node.

The Position property is simply the position of the selected node in the ramp. The node on the left is at the 0 position and the node on the right is at the 1.000 position. Nodes in between will be at varying values between 0 and 1.000. You can also change this value by dragging the nodes left or right.

Picking the **Invert Gradient** button reverses all color values, inverting the gradient pattern. In effect, the ramp is flip-flopped from left to right.

Noise may be added to the gradient map to create an uneven appearance. The properties in the **Noise** and **Noise Threshold** categories are similar to those for the noise map. The noise map is described later in this chapter.

The properties in the **Transforms**, **Position**, **Scale**, and **Repeat** categories are the same as described for an image map. Refer to the Image section for details on these properties.

Tiles

A ***tiles map*** is a pattern of rectangular, colored blocks surrounded by colored grout lines. This may be the most versatile map in the whole collection. Tiles are used to simulate tile floors, ceiling grids, hardwood floors, and many different types of brick walls. When you specify Tiles as the map for a property, the texture editor displays the properties for the map, **Figure 17-28**.

You first need to define the pattern for the map. The **Pattern** category contains properties for defining the pattern. For the Type property, select one of the seven predefined tile patterns or Custom to create your own. The names of the predefined patterns bring to mind brick walls. For example, a mason may use a stack bond to build a brick wall. However, remember these are only *patterns*. You can also use a brick pattern to create tile floors and acoustic ceiling panels. Four of the tile patterns are shown in **Figure 17-29**. The Tile Count property sets how many tiles are in each row and column before the pattern repeats.

As the name implies, the properties in the **Tile Appearance** category determine what the tiles look like. You can choose any color you wish or apply a texture or image to the tiles. Ceramic tile floors look more realistic if each tile is slightly different in color. The Color Variance property can be used to alter the color of random tiles to create a more realistic appearance. The Fade Variance property is used to fade the color of random tiles. You will have to experiment with the color variance and fading to create the look you need. Start with very low values. The Randomize property is used to alter the random color variation in the tiles. This variation is automatically applied, but entering a different random seed changes the pattern. If your scene has more than one object with this material applied to it, duplicate the material and change the seed number of the new material, then apply it to the other object.

Figure 17-28.
Adjusting the properties of a tiles map.

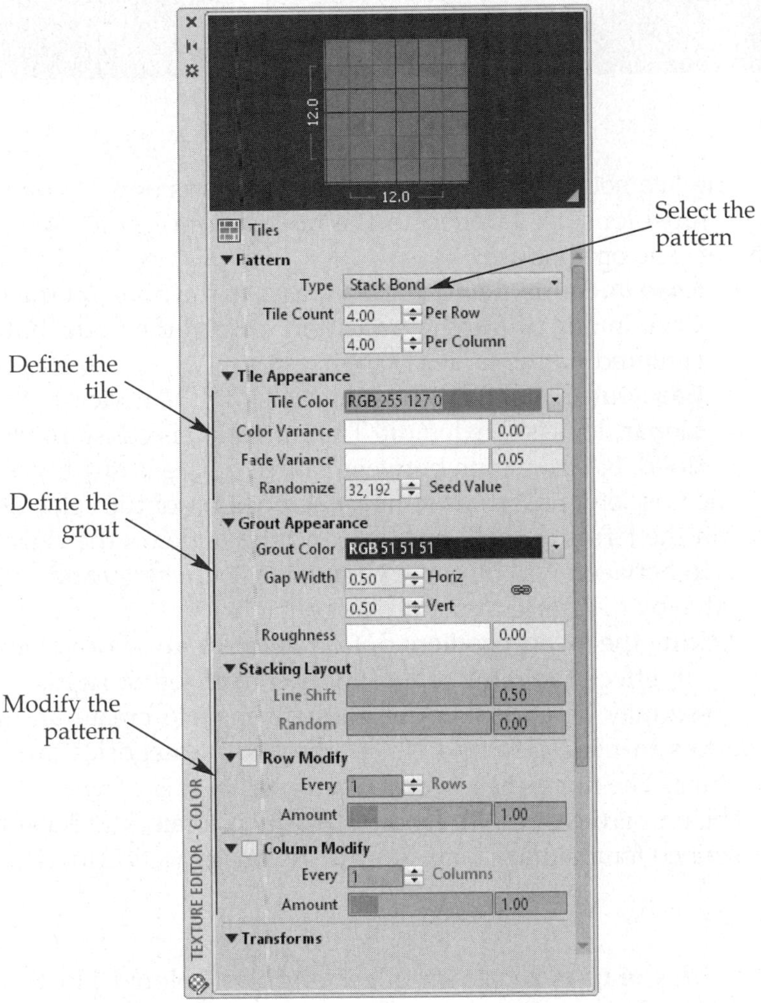

Figure 17-29.
A tiles map can have a custom pattern or predefined pattern. Four of the predefined patterns are shown here.
A—Running bond.
B—English bond.
C—Stack bond.
D—Fine running bond.

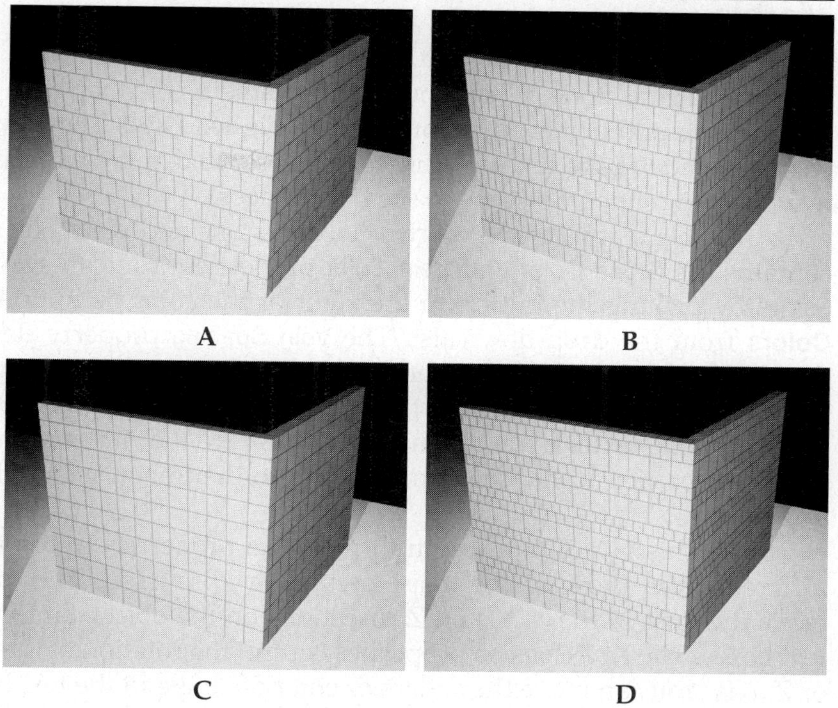

The properties in the **Grout Appearance** category control what the grout looks like. The grout is the line between the tiles. The Grout Color property is set to dark gray by default, but any color, image, or texture may be used. The Gap Width property determines how wide the grout lines are in relation to the tiles. There are horizontal and vertical settings. In some cases, such as for a hardwood floor material, you will have to set the scale differently on the horizontal and vertical axes to make the gap thicker in one direction.

The properties in the **Stacking Layout** category are only available for a custom pattern. The Line shift property changes the location of the vertical grout lines in every other row to create an alternate pattern of tiles. The default value is 0.50 and the range is from 0.00 to 100.00. The Random property randomly moves the same lines. This works nicely for hardwood floor materials. The default value is 0.00 and the range is 0.00 to 100.00.

The properties in the **Row Modify** and **Column Modify** categories are available with all tile pattern types, but may be disabled by default. To enable the settings, pick the check box by the category name. The settings in these areas allow you to change the number of grout lines in the horizontal and vertical directions to create your own pattern. The two Every properties determine which rows and columns will be changed. When set to 0, no changes take place in the row or column. When set to 1, every row or column will be changed. When set to 2, every other row or column will be changed, and so on. The value must be a whole number. The Amount property controls the size of the tiles in the row or column. A setting of 1 means that the tiles remain their original size. A setting of 0.50 makes the tiles one-half of their original size, a setting of 2 makes the tiles twice their original size, and so on. A setting of 0.00 completely turns off the row or column and the underlying material color shows through.

The properties in the **Transforms**, **Position**, **Scale**, and **Repeat** categories are the same as described for an image map. Refer to the Image section for details on these properties.

Marble

A ***marble map*** is a 3D map based on the colors and values you set. It is used to simulate natural stone. When you specify Marble as the map for a property, the texture editor displays the properties for the map, **Figure 17-30**. The viewport may not reflect changes made in the texture editor, even if set to display materials and textures. You may have to render the scene to see the changes.

A marble map is based on two colors—stone and vein. The **Appearance** category contains the Stone Color and Vein Color properties. You can swap the vein and stone colors by picking the button to the right of the color definition and selecting **Swap Colors** from the drop-down list. The Vein Spacing property determines the relative distance between each vein in the marble. The Vein Width property determines the relative width of each vein. Each of these settings can range from 0.00 to 100.00.

The **Link texture transforms** check box in the **Transforms** category works as described earlier for an image map. Refer to the Image section for details on texture transforms.

Since this is a 3D (procedural) map, the properties in the **Position** category are different from those of the maps previously discussed. The three Offset properties move the map in the X, Y, and Z directions on the object. Simply enter a value in the text boxes. The XYZ Rotation properties control the rotation of the map around the X, Y, or Z axis. You can move the sliders or enter an angle in the text boxes.

NOTE

The mathematical calculations that create a 3D (procedural) map are based on the world coordinate system. If you move or rotate the object, a different result is produced.

Figure 17-30.
Adjusting the properties of a marble map.

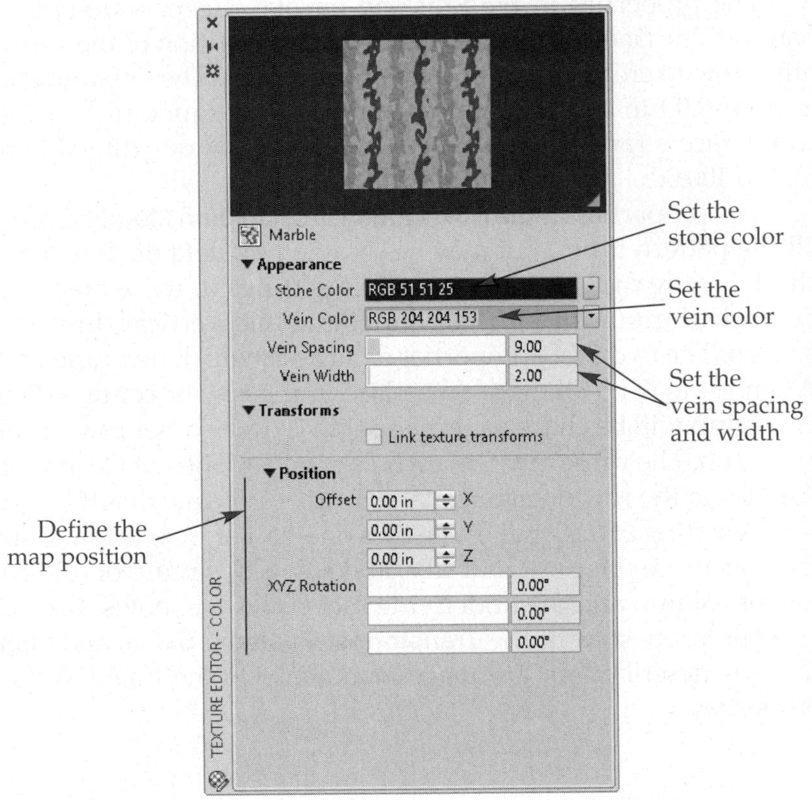

Noise

A ***noise map*** is a 3D map based on a random pattern of two colors used to create an uneven appearance on the material. It is most often used to simulate materials such as concrete, soil, asphalt, grass, and so on. When you specify Noise as the map for a property, the texture editor displays the properties for the map, **Figure 17-31**.

The properties in the **Appearance** area control how the noise looks. First, you need to select the type of noise. The options for the Noise Type property are:

- **Regular.** This is "plain" noise and useful for most applications.
- **Fractal.** This creates the noise pattern using a fractal algorithm. When this is selected, the Levels property in the **Noise Threshold** category is enabled.
- **Turbulence.** This is similar to fractal noise, except that it creates fault lines.

The Size property controls the size scale of the noise. The larger the value, the larger the size of the noise. The value can range from 0.00 to 1 billion. The Color 1 and Color 2 properties control the color of the pattern of noise. You can assign a color, image, or texture to the property. To swap the color definitions, pick the button next to the properties and select **Swap Colors** from the drop-down list.

The properties in the **Noise Threshold** category are used to fine-tune the noise effect. The properties in this category are:

- **Low.** The closer this setting is to 1.00, the more dominant color 1 is. The default setting is 0.00 and it can range from 0.00 to 1.00.
- **High.** The closer this setting is to 0.00, the more dominant color 2 is. The default setting is 1.00 and it can range from 0.00 to 1.00.
- **Levels.** Sets the energy amount for the fractal and turbulence types. Lower values make the fractal noise appear blurry and the turbulence lines more defined. The default setting is 3.00 and it can range from 1.00 upward.
- **Phase.** Randomly changes the noise pattern with each value. This allows you to have materials with the same noise map settings look slightly different. You should have different patterns on different materials. This adds a level of realism to your scene.

Figure 17-31.
Adjusting the properties of a noise map.

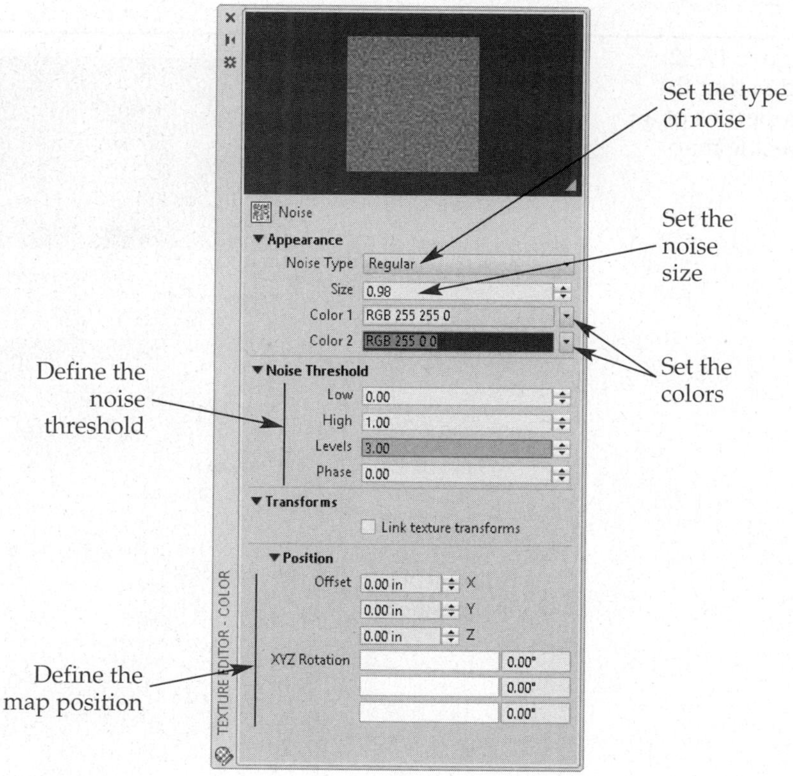

The properties in the **Transforms** and **Position** categories are the same as described for a marble map. Refer to the Marble section for details on these properties.

Speckle

A ***speckle map*** is a 3D map based on a random pattern of dots created from two colors. This map is very useful for textured walls, sand, granite, and so on. When you specify Speckle as the map for a property, the texture editor displays the properties for the map, **Figure 17-32**.

The settings for a speckle map are very simple. In the **Appearance** category, pick colors for the Color 1 and Color 2 properties. You cannot use maps, only colors. To swap the color definitions, pick the button next to the properties and select **Swap Colors** from the drop-down list. The Size property controls the size of the speckles.

The properties in the **Transforms** and **Position** categories are the same as described for a marble map. Refer to the Marble section for details on these properties.

Waves

A ***waves map*** is a 3D map in a pattern of concentric circles. Imagine dropping two or three stones into a pool of water and watching the ripples intersect with each other. A number of wave centers are randomly generated and a pattern created by the overlapping waves is the result. As the name implies, the waves map is usually used to simulate water. When you specify Waves as the map for a property, the texture editor displays the properties for the map, **Figure 17-33**.

In the **Appearance** category, pick colors for the Color 1 and Color 2 properties. You cannot use maps, only colors. To swap the color definitions, pick the button next to the properties and select **Swap Colors** from the drop-down list. The Distribution property can be set to 2D or 3D. This setting determines how the wave centers are distributed on the object. Selecting 3D means that the wave centers are randomly distributed over the surface of an imaginary sphere. This distribution affects all sides of an object. On the other hand, selecting 2D means that the wave centers are distributed on the XY plane. This is much better for nearly flat surfaces, such as the surface of a pond or lake.

Figure 17-32.
Adjusting the properties of a speckle map.

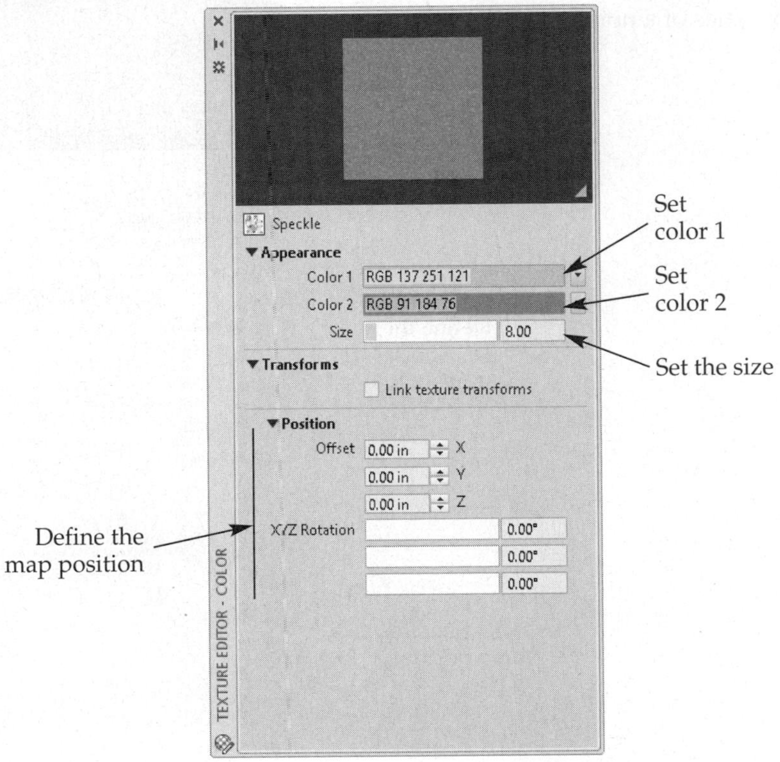

Figure 17-33.
Adjusting the properties of a waves map.

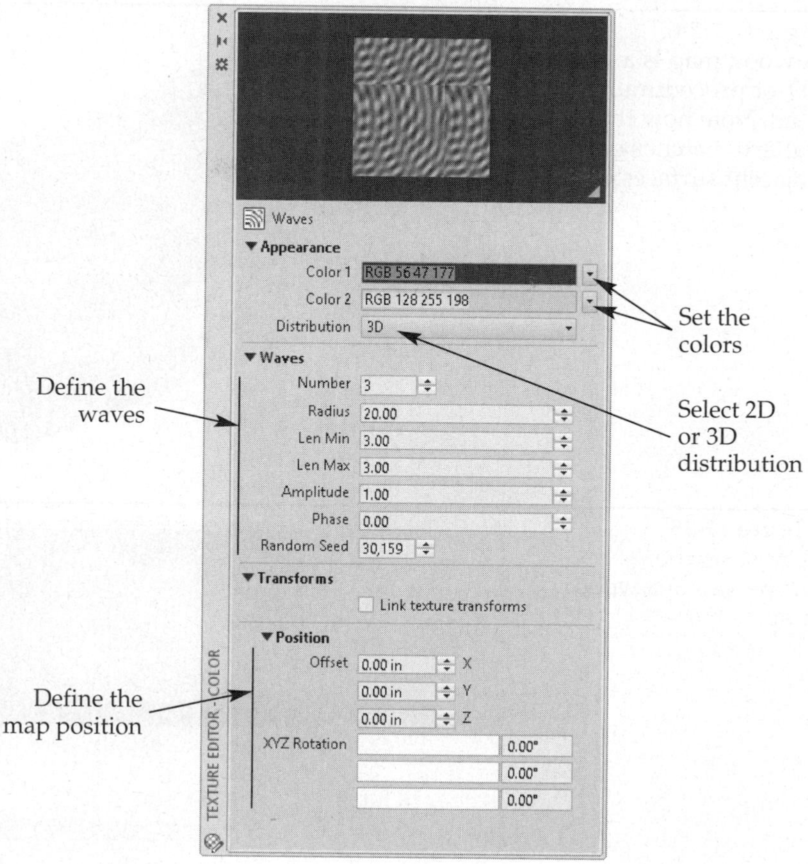

The properties in the **Waves** category define the pattern of waves. The Number property sets the number of wave centers that are generating the waves. The Radius property sets the radius of the circle or sphere from which the waves originate. The Len Min and Len Max properties define the minimum and maximum interval for each wave. The Amplitude property can be thought of as the "power" of the wave. The default value is 1.00, but the value can range from 0.00 to 10000.00. A value less than 1.00 makes color 1 more dominant. For a value greater than 1.00, color 2 is more dominant. The Phase property is used to shift the pattern and the Random Seed property is used to redistribute the wave centers.

The properties in the **Transforms** and **Position** categories are the same as described for a marble map. Refer to the Marble section for details on these properties.

PROFESSIONAL TIP

Remember, you can change the material swatch geometry in the materials editor, such as to a cube, sphere, or cylinder. Some maps, like a waves map, are easier to understand when displayed on a cube.

Wood

A ***wood map*** is a 3D map that generates a wood grain based on the colors and values you select. See **Figure 17-34**. When you specify Wood as the map for a property, the texture editor displays the properties for the map, **Figure 17-35**.

A wood map is based on two colors. The Color 1 and Color 2 properties in the **Appearance** category are used to specify these colors, usually one dark and one light color. To swap the color definitions, pick the button next to the properties and select

Figure 17-34.
A wood map is a 3D, or procedural, map. Note how the pattern matches on adjacent surfaces.

Figure 17-35.
Adjusting the properties of a wood map.

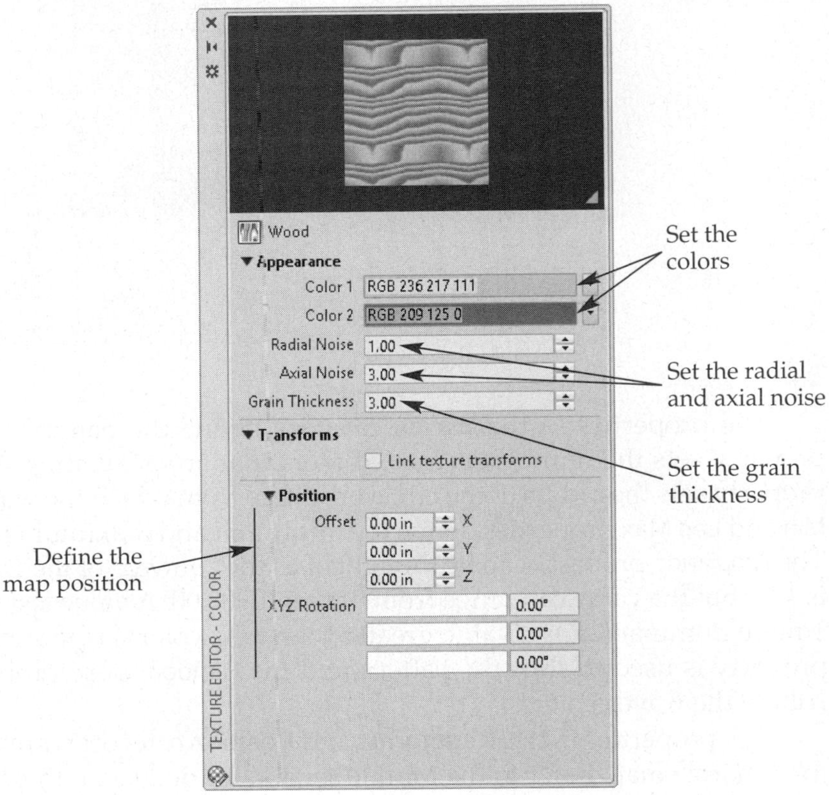

Swap Colors from the drop-down list. The Radial Noise property determines the waviness of the wood's rings. The rings are found by cutting a tree crosswise. The Axial Noise property determines the waviness of the length of the tree trunk. The Grain Thickness property determines the relative width of the grain.

The properties in the **Transforms** and **Position** categories are the same as described for a marble map. Refer to the Marble section for details on these properties.

Exercise 17-5

www.g-wlearning.com/CAD/

Complete the exercise on the companion website.

Adjusting Material Maps

Simply applying a map to a material property rarely results in a realistic scene when the scene is rendered. The maps usually need to be adjusted to produce the desired results. Maps can be adjusted at the material level or the object level. A combination of these two adjustments is usually required to produce a photorealistic rendering.

Material-Level Adjustments

Material-level adjustments involve changing the properties of the map in the material definition. Map properties are discussed in previous sections. Sometimes, these adjustments may be enough to get the materials looking the way you want them.

Other adjustments become necessary when the same material is applied to more than one object in the same scene. If you make changes at the material level, they affect all objects with that same material. If you need different objects to have different settings, then you will have to make object-level adjustments.

Object-Level Adjustments

Material mapping refers to specifying how a mapped material is applied to an object. When a mapped material is attached to an object, a default set of mapping coordinates, or simply *default mapping*, is used to apply the map to the object.

AutoCAD allows you to adjust mapping at the object level for 2D-mapped (texture-mapped) materials. The **MATERIALMAP** command applies a grip tool, or ***gizmo***, based on one of four mapping types: planar, box, spherical, or cylindrical. See **Figure 17-36**. The colored edge represents the start and end of the map. For best results, select the mapping type based on the general shape of the object to which mapping is applied. Do not be afraid to experiment with other mapping types. Any mapping type can be used on any object, regardless of the object's shape. However, only one mapping type can be applied to an object at any given time.

MATERIALMAP
Ribbon
Visualize
> Materials
Planar
Box
Cylindrical
Spherical
Type
MATERIALMAP

After one of the mapping types is selected, you are prompted to select the faces or objects. You can select multiple objects or faces. After making a selection, the gizmo is placed on the selection set. The command remains active for you to adjust the mapping or enter an option.

Drag the grips on the gizmo to stretch or scale the material. The effects of editing a color map are dynamically displayed if the current visual style is set to display materials and textures. Otherwise, exit the command and render the scene to see the effect of the edit. To readjust the mapping, select the same mapping type and pick the object again. The gizmo is displayed in the same location as before.

The **Move** and **Rotate** options of the command toggle between the move and rotate gizmos. Using the gizmos, you can move and rotate the map on the object. The **Reset**

Figure 17-36.
These are the four material map gizmos. From left to right: planar, box, spherical, and cylindrical. The colored edge represents the start and end of the map.

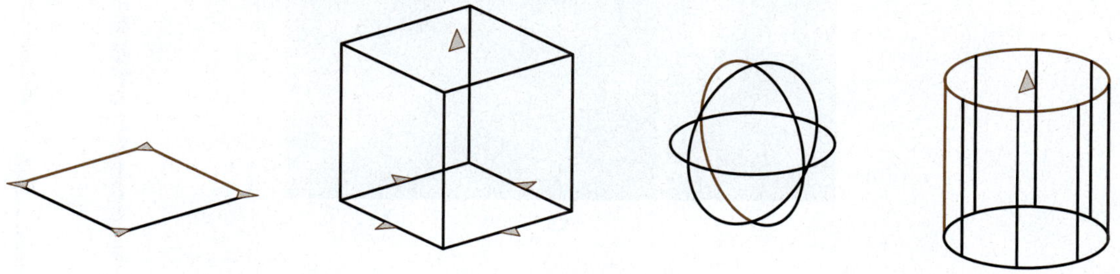

option of the command restores the default mapping to the object. The **Switch mapping mode** option allows you to change between the four types of mapping.

If the command is typed, there is an additional option. The **Copy mapping to** option provides a quick and easy way to apply the changes made to the current object to other objects in the scene. Enter this option, select the face or object to copy from, and then select the faces or objects to copy to. This option is also available if the **Switch mapping mode** option is entered.

For example, look at **Figure 17-37A**. The grain on the stair risers is running vertically when it should run horizontally. First, apply a planar map gizmo to the bottom riser. Next, rotate the mapping 90°, **Figure 17-37B**. Finally, use the **Copy mapping to** option to copy the mapping to the other risers, **Figure 17-37C**.

NOTE

Mapping coordinates can be applied to 3D-mapped (procedural-mapped) materials, but adjusting the coordinates has no effect. This is because the 3D map is generated from mathematical calculations based on the world coordinate system. A 3D-mapped material must be adjusted at the material level.

Figure 17-37.
Correcting material mapping. A—The grain on the risers runs vertically instead of horizontally. B—Rotating the map with the rotate gizmo. C—The corrected rendering. *(Model courtesy of Arcways, Inc., Neenah, WI)*

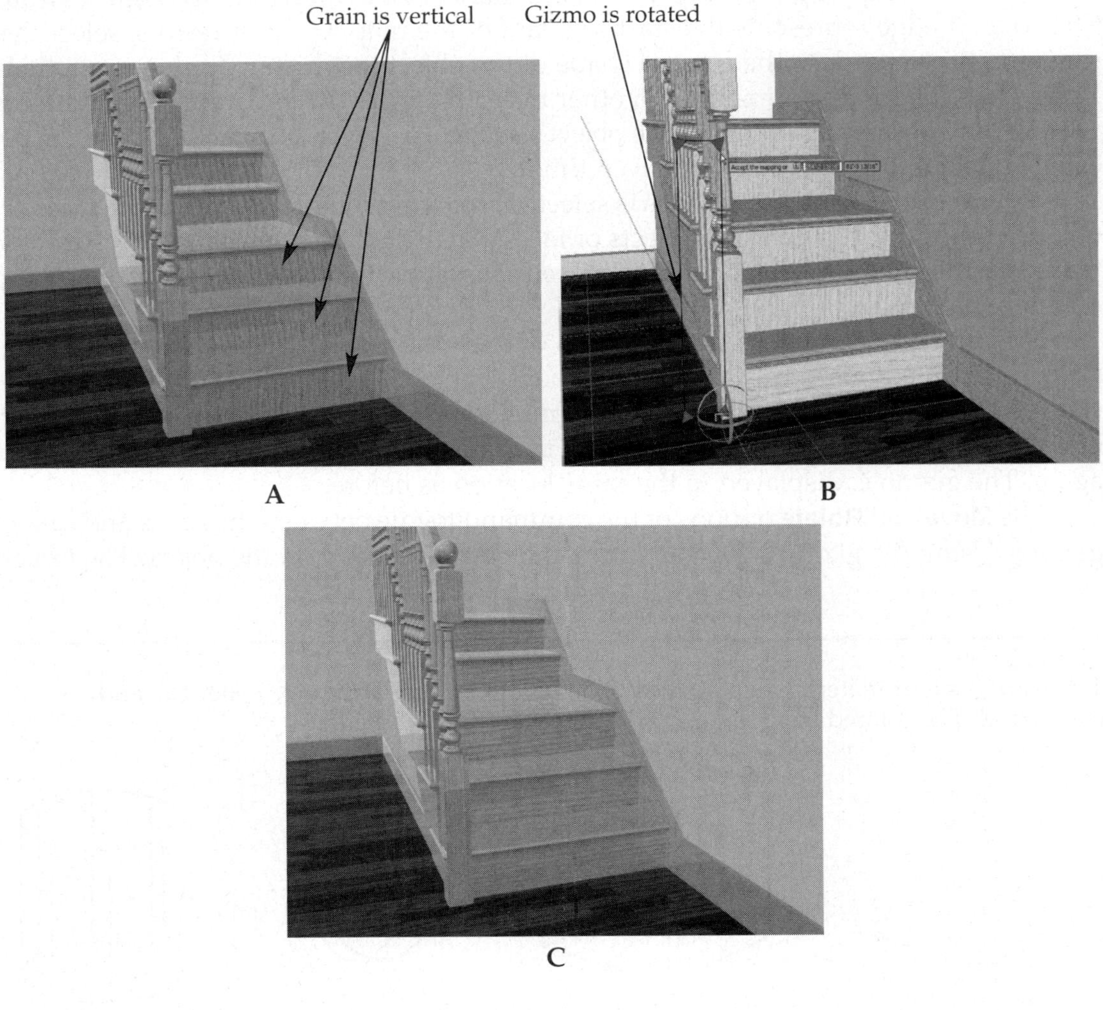

PROFESSIONAL TIP

If you are using a reflectivity, self illumination, or bump map and need to adjust it at the object level, you cannot see the effects of the mapping change in the viewport. Apply the same map as a color map. Also, set the visual style to display materials and textures. Then, adjust the object mapping as needed. The edits are dynamically displayed in the viewport. When the image is in the correct location, remove the color map from the material.

Exercise 17-6

www.g-wlearning.com/CAD/

Complete the exercise on the companion website.

Chapter Review

Answer the following questions using the information in this chapter.

1. Define *material.*
2. Define *materials library.*
3. What is the Favorites section of the **Library** area of the materials browser used for?
4. Describe how to attach a material using the materials browser.
5. How can materials be attached to layers?
6. By default, which material is attached to newly created objects?
7. Which material is used as the base material for creating new materials?
8. Name the 13 shapes that can be used to display the material in the preview in the materials editor.
9. How do you know if a material in the materials browser is being used in the drawing?
10. How can the name of an existing material be changed?
11. Name the 14 basic material types.
12. When creating a material to look like plastic, what is the benefit of using the plastic material type instead of starting from scratch using the generic material?
13. In the **Reflectivity** category of the generic material, there are Direct and Oblique properties. What are these used for?
14. An image mapped to a material property will normally repeat itself to cover the entire object. What is this called and how do you turn it off?
15. When a cutout map is applied to a material, what are the different effects produced by black and white areas in the image?
16. How much illumination does a self illuminated material add to a scene?
17. How is a marble material created?
18. Explain how black and white areas of a map applied to the Transparency property affect the transparency of a material.
19. Explain what the nodes in the ramp of a gradient map are for.
20. Name the four types of mapping available for adjusting texture maps at the object level.

Drawing Problems

1. General In this problem, you will create a scene with basic 3D objects, attach materials to the objects, and adjust the settings of the materials.
   A. Start a new drawing and set the units to architectural.
   B. Draw a 15′ × 15′ planar surface to represent the floor.
   C. Draw two boxes to represent two walls. Make the boxes 15′ × 4″ × 9′. Position them to form a 90° corner. Alternately, draw a polysolid of the same dimensions.
   D. Draw a R2′ × 5′H cone in the center of the room.
   E. Open the materials browser and locate a material similar to the one shown on the floor. Attach this material to the floor.
   F. Locate an appropriate material for the wall and attach it.
   G. Locate an appropriate material for the cone and attach it.
   H. Turn on the sun and adjust the time to create good shadows. Refer to Chapter 16 for an introduction to sun settings.
   I. Render the scene. Save the rendering as P17-1.jpg. Save the drawing as P17-1.

A full-color image is provided on the companion website. www.g-wlearning.com/CAD/

2. **Architectural** In this problem, you will attach materials to the objects in an existing drawing and render the scene.
   A. Open drawing P16-2 from Chapter 16. Save the drawing as P17-2.
   B. Attach the materials of your choice to the objects in the scene. Do not be restricted by the names of the materials. For example, a concrete material may be suitable for items such as foliage or even carpet with a simple color change. Be creative.
   C. If the items in the scene were downloaded from the Internet, they may be blocks with nested layers. Instead of exploding the blocks, use the **MATERIALATTACH** command and attach materials to the layers on which the nested objects reside.
   D. Turn on the sun and adjust the time to create good shadows. Refer to Chapter 16 for an introduction to sun settings.
   E. Render the scene. Save the rendering as P17-2.jpg. Save the drawing.

A full-color image is provided on the companion website. www.g-wlearning.com/CAD/

3. **General** In this problem, you will create custom wood and marble materials.
   A. Start a new drawing and save it as P17-3.
   B. Draw two 5 × 5 × 5 boxes and position them near each other. Using other primitives, cut notches and holes in the boxes. The boxes will be used to test the custom materials.
   C. In the materials editor, create two new materials. Name one Wood-*your initials* and the other Marble-*your initials*.
   D. Attach the wood material to one of the boxes and the marble material to the other box.
   E. Render the scene and make note of the wood grain and marble veins.
   F. Use the texture editor to change the properties of the materials.
   G. Render the scene again and make note of the changes. Using the **Render in Region** button on the **Render** panel in the **Visualize** tab of the ribbon can save time when testing material changes.
   H. When you are satisfied with the materials, save the rendering as P17-3.jpg.
   I. Save the drawing.

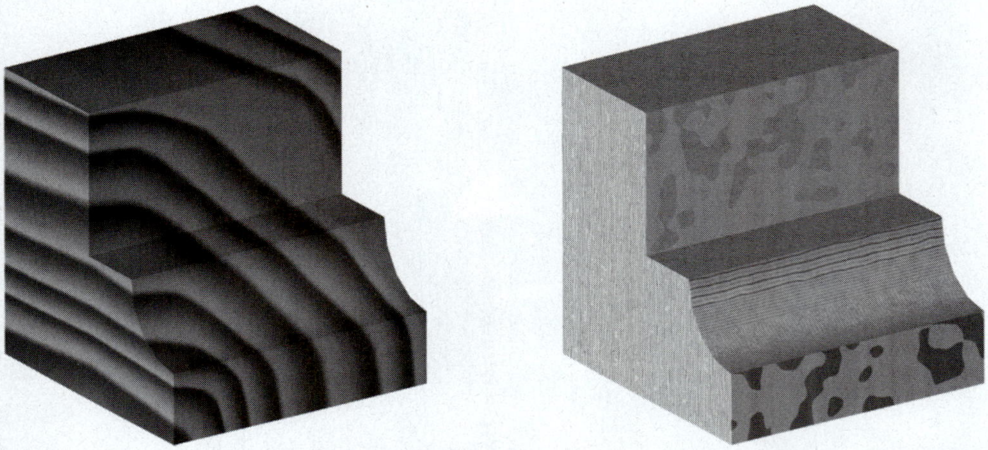

A full-color image is provided on the companion website. www.g-wlearning.com/CAD/

4. **General** In this problem, you will create a bitmap and use it as a transparency map and a bump map.

   A. Draw a rectangle with an array of smaller rectangles inside of it as shown. Sizes are not important and the pattern can be varied if you like.
   B. Display a plan view of the rectangles. Then, copy all of the objects to the Windows Clipboard by pressing [Ctrl]+[C] and selecting the objects.
   C. Launch Windows Paint. Then, paste the objects into the blank file. Notice how the AutoCAD background outside of the large rectangle is also included.
   D. Use the select tool (rectangle) in Paint to draw a window around the large rectangle created in AutoCAD and the smaller rectangles within it. Copy this to the Windows Clipboard by pressing [Ctrl]+[C].
   E. Start a new Paint file without saving the current one and paste the image from the Clipboard into the new blank file. Now, the unwanted AutoCAD background is no longer displayed. If needed, change the small rectangles to black and the lattice to white using the tools in Paint. The colors should be the reverse of what is shown below. Then, save the image file as P17-4.bmp and close Paint.
   F. In AutoCAD, draw a solid box of any size.
   G. Using the materials editor, create a new material.
   H. Assign the P17-4.bmp image file you just created as a transparency map. Adjust the map so that it is scaled to fit to the object.
   I. Render the scene and note the effect.
   J. Turn off the transparency property by unchecking the check box in the category name.
   K. Assign the P17-4.bmp image file as a bump map. Adjust the map so that it is scaled to fit to the object.
   L. Render the scene and note the effect.
   M. Save the drawing as P17-4.

5. General In this problem, you will recreate the scene in the photograph below. The glass is approximately 3″ in diameter at the base, 3 1/4″ in diameter at the top, and 4″ in height. The thickness of the glass is approximately 1/16″ on the sides and 1/2″ at the base. Add the rest of the details needed to complete the scene. Render the scene and make adjustments if needed. Save the drawing as P17-5.

CHAPTER 18
Lighting

Learning Objectives

After completing this chapter, you will be able to:

✓ Describe the types of lighting in AutoCAD.
✓ List the user-created lights available in AutoCAD.
✓ Change the properties of lights.
✓ Add a background to your scene and control its appearance.

In the movie industry, it has often been said that "Lighting is everything." That statement also rings true when creating realistic scenes in AutoCAD. If lights are used incorrectly, the scene will be washed-out with light or too dark to see anything. In Chapter 16, you were introduced to lighting. You learned how to adjust lighting by turning off the default lights and adding sunlight. In this chapter, you will learn all about the lights available in AutoCAD, as well as tips and tricks to help make your scene look its best.

Types of Lights

Ambient light is like natural light just before sunrise. It is the same intensity everywhere. All faces of the object receive the same amount of ambient light. Ambient light cannot create highlights, nor can it be concentrated in one area. AutoCAD does not have an ambient light setting. Instead, it relies on indirect illumination, which is discussed in Chapter 19.

A ***point light*** is like a lightbulb. Light rays from a point light shine out in all directions. A point light can create highlights. The intensity of a point light falls off, or weakens, over distance. A ***target point light*** is the same as a standard point light except that a target is specified. The illumination of the target point light is directed toward the target.

A ***distant light*** is a directed light source with parallel light rays. This acts much like the sun. Rays from a distant light strike all objects in your model on the same side and with the same intensity. The direction and intensity of a distant light can be changed.

A ***spotlight*** is like a distant light, but it projects in a cone shape. Its light rays are not parallel. A spotlight is placed closer to the object than a distant light. Spotlights have a hotspot and a falloff. The light from a standard spotlight is directed toward a target. A ***free spotlight*** is the same as a standard spotlight, but without a target.

A *weblight* is a directed light that represents real-world distribution of light. The illumination is based on photometric data that can be entered for each light. The light from a standard weblight is directed toward a target. A *free weblight* is the same as a standard weblight, but without a target point.

Properties of Lights

Several factors affect how a light illuminates an object. These include the angle of incidence, reflectivity of the object's surface, and the distance that the light is from the object. In addition, the ability to cast shadows is a property of light.

Angle of Incidence

AutoCAD renders the faces of a model based on the angle at which light rays strike the faces. This angle is called the ***angle of incidence***. See **Figure 18-1**. A face that is perpendicular to light rays receives the most light. As the angle of incidence decreases, the amount of light striking the face also decreases.

Reflectivity

The angle at which light rays are reflected off a surface is called the ***angle of reflection***. The angle of reflection is always equal to the angle of incidence. Refer to **Figure 18-1**.

The "brightness" of light reflected from an object is actually the number of light rays that reach your eyes. A surface that reflects a bright light, such as a mirror, is reflecting most of the light rays that strike it. The amount of reflection you see is called the ***highlight***. The highlight is determined by the angle from the viewpoint relative to the angle of incidence. Refer to **Figure 18-1**.

The surface quality of the object affects how light is reflected. A smooth surface has a high specular factor. The ***specular factor*** indicates the number of light rays that have the same angle of reflection. Surfaces that are not smooth have a low specular factor. These surfaces are called ***matte***. Matte surfaces ***diffuse***, or "spread out," the light as it strikes the surface. This means that few of the light rays have the same angle of reflection. **Figure 18-2** illustrates the difference between matte and high specular finishes. Surfaces can also vary in ***roughness***. Roughness is a measure of the polish on a surface. This also affects how diffused the reflected light is.

Figure 18-1.
The amount of reflection, or highlight, you see depends on the angle from which you view the object.

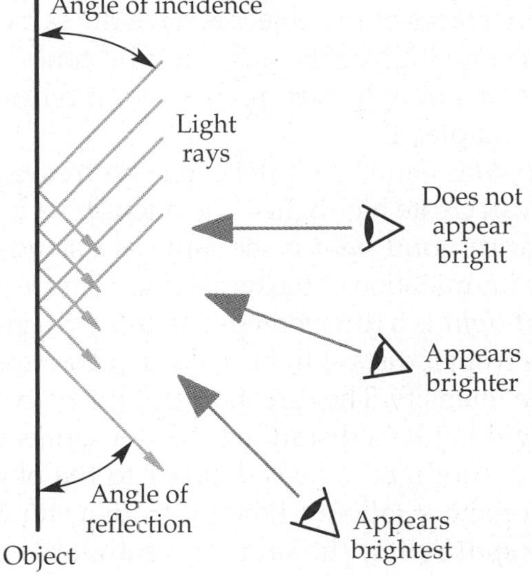

Figure 18-2.
Matte surfaces produce diffuse light. This is also referred to as having a low specular factor. Shiny surfaces evenly reflect light and have a high specular factor.

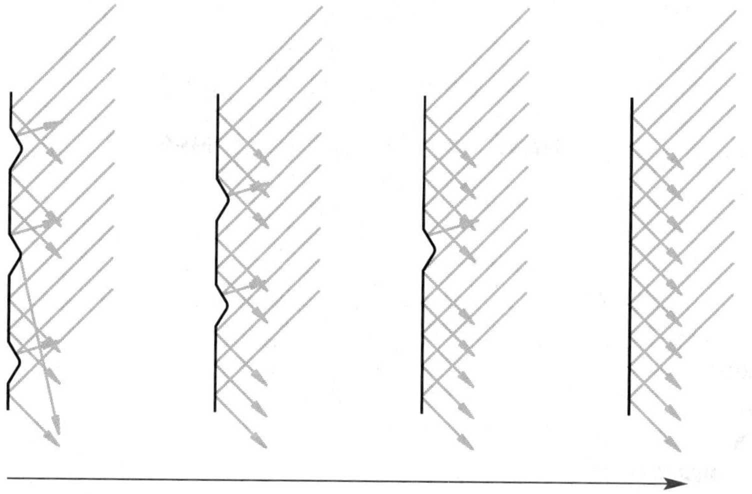

PROFESSIONAL TIP

When rendering metal and other shiny materials, it helps to have a bright object nearby to create some highlights on the object. Adding simple vertical planes to the left and right of the object (out of the camera's view) and applying a self-illuminating material to them will create some nice highlights. Photographers use this trick in the studio.

Hotspot and Falloff

A spotlight produces a cone of light. The ***hotspot*** is the central portion of the cone, where the light is brightest. See **Figure 18-3**. The ***falloff*** is the outer portion of the cone, where the light begins to blend to shadow. The hotspot and falloff of a spotlight are not affected by the distance the light is from an object. Spotlights are the only lights with hotspot and falloff properties.

A full-color image is provided on the companion website. www.g-wlearning.com/CAD/

Figure 18-3.
The hotspot of a spotlight is the area that receives the most light. The smaller cone is the hotspot. The falloff receives light, but less than the hotspot. The larger cone is the falloff.

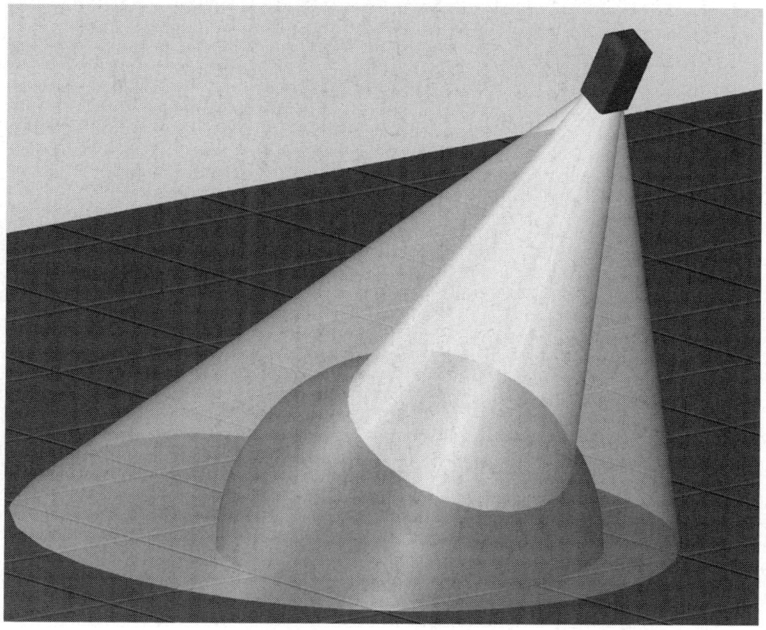

Attenuation

The farther an object is from a point light or spotlight, the less light that reaches the object. See **Figure 18-4**. The intensity of light decreases over distance. This decrease is called ***attenuation***. All lights in AutoCAD, except distant lights, have attenuation. Often, attenuation is called *falloff* or *decay*. However, do not confuse this with the falloff of a spotlight, which is the outer edge of the cone of illumination. All lights that attenuate in AutoCAD are calculated using inverse square attenuation. This is a "real-world" calculation in which the illumination of an object decreases in inverse proportion to the square of the distance. For example, if an object is two units from the light, it receives $(1/2)^2$, or 1/4, of the full light. If the object is four units away, it receives $(1/4)^2$, or 1/16, of the full light.

NOTE

Different types of attenuation other than inverse square attenuation are available when creating point lights and spotlights, as discussed later in this chapter. However, these options are only available to maintain backward compatibility with older releases of AutoCAD and have no effect on the scene.

PROFESSIONAL TIP

The intensity of the sun's rays does not diminish from one point on Earth to another. They are weakened by the angle at which they strike Earth. Therefore, since distant lights are similar to the sun, attenuation is not a factor with distant lights.

A full-color image is provided on the companion website. www.g-wlearning.com/CAD/

Figure 18-4.
Attenuation is the intensity of light decreasing over distance.

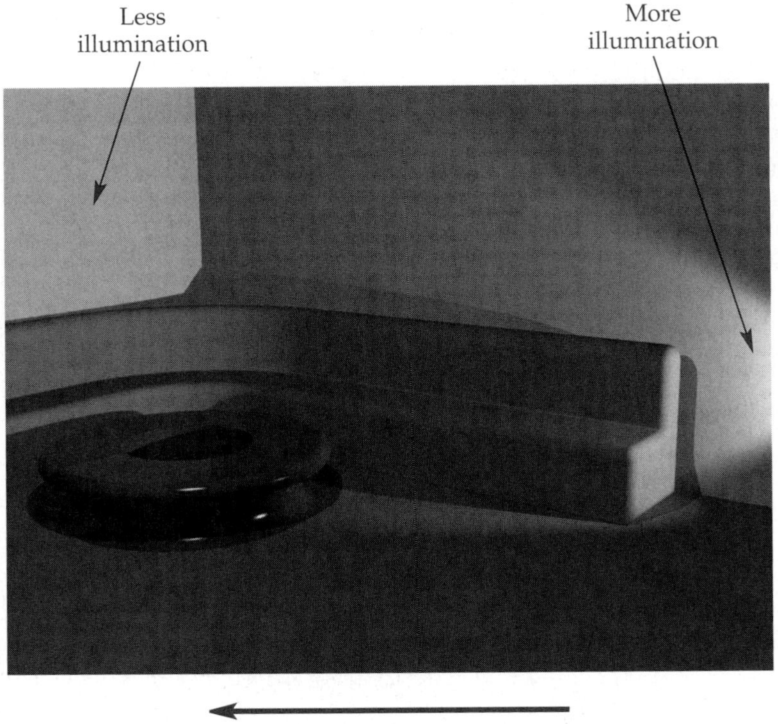

AutoCAD Lights

AutoCAD has three types of lighting: default lighting, sunlight with or without sky illumination, and user-created lighting. ***Default lighting*** is the lighting automatically available in the scene. It is composed of one or two default light sources that appear to be originating from above and to the left of the viewer. As the viewpoint is changed, the light sources follow to maintain the same illumination of the scene. There is no control over default lighting and it must be shut off whenever one of the other types of lighting is used.

As you saw in Chapter 16, ***sunlight*** may be added to any scene. AutoCAD uses a distant light to simulate the parallel rays of the sun. The date and time of day can be adjusted to create different sunlight illumination. ***Sky illumination*** may also be added with sunlight to simulate light bouncing off objects in the scene and particles in the atmosphere. This helps create a more-natural feel.

User-created lighting results when you add AutoCAD light objects to the drawing. There are four types of user-created lights: distant light, weblight, point light, and spotlight. See **Figure 18-5**. A distant light is a directed light source with parallel light rays. A weblight is a directional point light containing light intensity (photometric) data. A point light is like a lightbulb with light rays shining out in all directions. A spotlight is like a distant light, but it projects light in a cone shape instead of having parallel light rays.

When created, point lights, weblights, and spotlights are represented by ***light glyphs***, or icons, in the drawing. To suppress the display of light glyphs, pick the **Light Glyph Display** button in the expanded **Lights** panel on the **Visualize** tab of the ribbon. The button is blue when light glyphs are displayed. The default lights, sun, and distant lights are not represented by glyphs.

Ribbon
Visualize > Lights
Light Glyph Display
Type
LIGHTGLYPH-DISPLAY

LIGHTGLYPHDISPLAY

In this section, you will learn how to add lights. You will also learn how to adjust the various properties of sunlight and AutoCAD light objects. The tools for working with lights can be accessed using the command line, tool palettes, and the **Lights** and **Sun & Location** panels on the **Visualize** tab of the ribbon. See **Figure 18-6**.

So that you will never work with a completely dark scene, default lighting is applied in the viewport and to the rendering if no other lights are added. In order for your lights to be applied, you must switch between default lighting and user lighting. To do this, pick the **Default Lighting** button in the expanded **Lights** panel on the **Visualize** tab of the ribbon. This button toggles the lighting between default lighting and whatever lights are available in the scene. When default lighting is on, the button is blue. If you elected for AutoCAD to do so, the default lighting will be automatically shut off when sunlight is turned on or a user-created light is added to the scene.

A full-color image is provided on the companion website. www.g-wlearning.com/CAD/

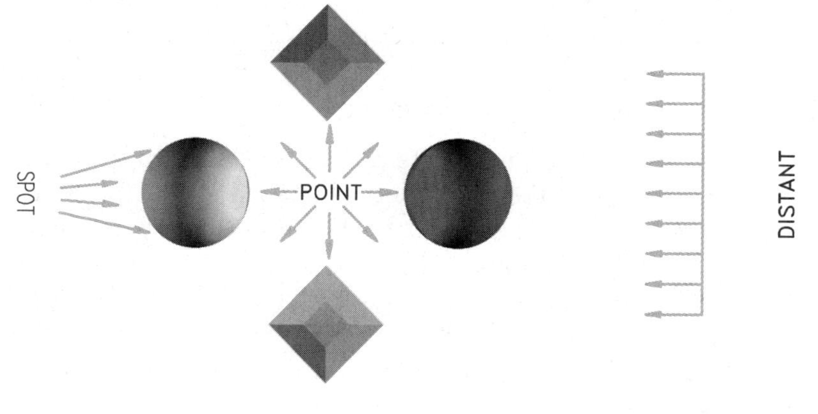

Figure 18-5.
AutoCAD has four types of user-created lights: distant, point, weblight, and spotlight. A weblight is really a targeted point light. It projects in all directions, but may be predominant in one direction.

Figure 18-6.
The tools for adding and controlling lights. A—Tools in the **Lights** panel of the **Visualize** ribbon tab. B—Tools in the **Sun & Location** panel of the **Visualize** ribbon tab.

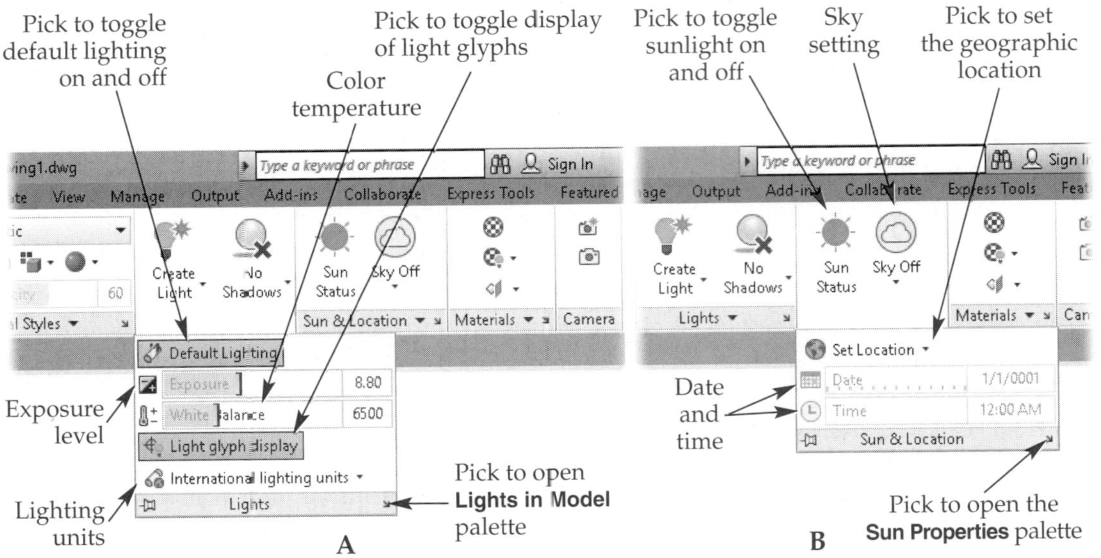

Lighting Units

There are two settings for ***lighting units*** available in AutoCAD: American (foot-candles) and International (lux). Each type of lighting is called photometric lighting. ***Photometric lighting*** is physically correct and attenuates at the square of the distance from the source. For more accuracy, photometric data files can be imported from lighting manufacturers.

The **LIGHTINGUNITS** system variable sets what type of lighting is used. The setting can be changed using the **International lighting units** drop-down list in the expanded **Lights** panel on the **Visualize** tab of the ribbon. Refer to **Figure 18-6**. A setting of 1 for the **LIGHTINGUNITS** system variable specifies American lighting units. A setting of 2 specifies International lighting units. This is the default setting. The only difference between the two settings is that American units are displayed as foot-candles (fc) and International units are displayed as lux (lx).

NOTE

All lights in AutoCAD are calculated using photometric lighting. Non-photometric AutoCAD lighting, also called generic lighting, was available prior to AutoCAD 2016 and was the only type of lighting used prior to AutoCAD 2008. This lighting provided very nice results, but the settings were not based on any real measurements. For compatibility purposes, it is recommended to update non-photometric lights created in previous releases to photometric lights.

Exposure and White Balance

The brightness and color temperature of lighting in the scene can be adjusted using the sliders in the expanded **Lights** panel on the **Visualize** tab of the ribbon. Refer to **Figure 18-6**. These are global controls that provide a quick way to adjust the lighting. The **Light Exposure** slider is used to adjust the brightness. Lower values make the scene brighter and higher values make it darker. The **Light White Balance** slider is used to adjust the color temperature of lighting based on the Kelvin temperature value (k). Higher values make the light warmer and lower values make it cooler. Warmer lights are more yellow and cooler lights appear bluish.

Sunlight and Location

To turn sunlight on or off, pick the **Sun Status** button in the **Sun & Location** panel on the **Visualize** tab of the ribbon. Refer to Figure 18-6. This button is blue when sunlight is on. Sunlight can also be turned on or off in the **Sun Properties** palette, which is discussed in the next section. Sunlight is not represented by a light glyph. The date, time, and geographic location can also be set in the **Sun & Location** panel on the **Visualize** tab of the ribbon. These properties determine how the scene is illuminated by the sun.

Ribbon
Visualize > Sun & Location
Sun Status
Type
SUNSTATUS

SUNSTATUS

To change the current date, drag the **Date** slider left or right. Sunlight must be on for the slider to be enabled. As you drag the slider, the date is displayed on the right-hand end of the slider. The time is changed in the same manner. Drag the **Time** slider left or right. As you drag the slider, the time is displayed on the right-hand end of the slider.

The location of the scene can be set to an actual geographic location. This is important if you want to replicate the lighting and shadows of an actual site. The **Set Location** drop-down list on the **Sun & Location** panel on the **Visualize** tab of the ribbon provides two options for setting the location of the scene: **From Map** and **From File**. The **From Map** option allows you to select a geographic location from the Autodesk online maps service. The **From File** option allows you to specify the location by importing a KML or KMZ file. *KML* stands for Keyhole Markup Language. A KML file contains latitude, longitude, and, sometimes, other data to pinpoint a location on Earth. A KMZ file is a zipped KML file.

Ribbon
Visualize > Sun & Location > Set Location
From Map
Type
GEOGRAPHICLOCATION GEO NORTH NORTHDIR

GEOGRAPHICLOCATION

In order to use the Autodesk online maps service, you must first sign in to or create an Autodesk account. To create an Autodesk account, select the **Sign In** button in the **InfoCenter** and select **Sign In to Autodesk account**. Select **Create Account** to create a user name and password. After creating the account, you can use the Autodesk online maps service.

Select **From Map** from the **Set Location** drop-down list and pick **Yes** when asked if you want to use live map data. The **Geographic Location-Specify Location** dialog box appears and allows you to select a location directly on the map. See Figure 18-7. If necessary, use

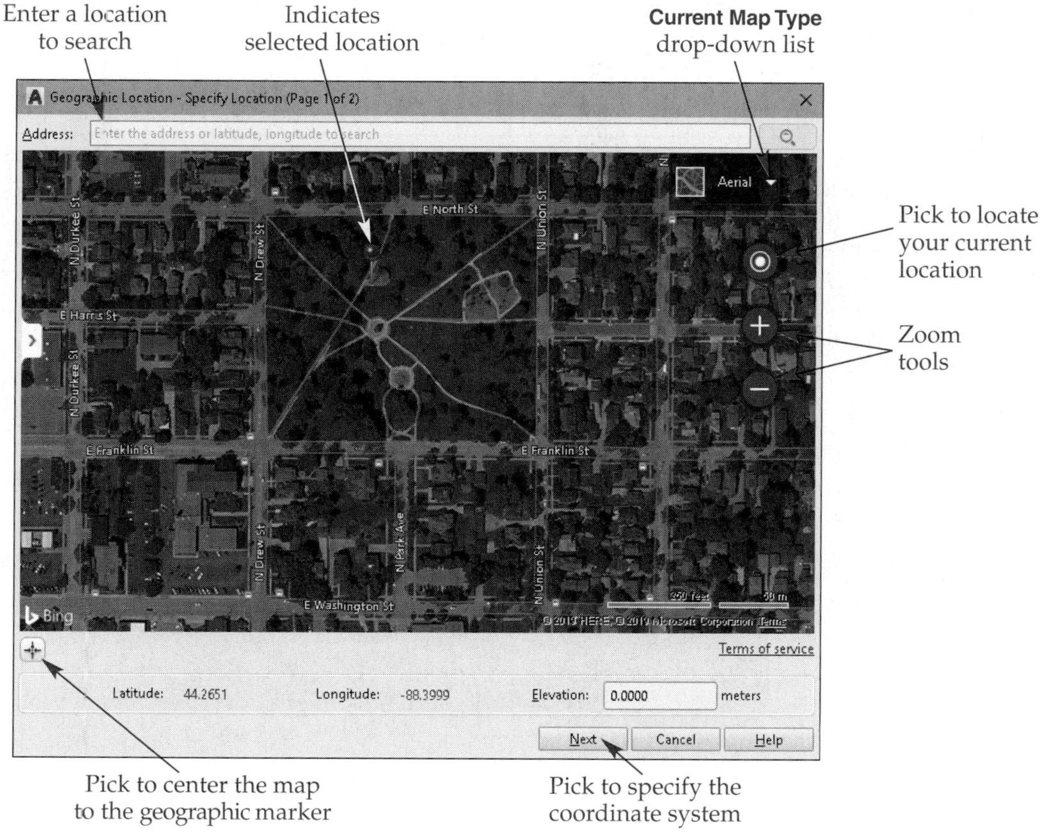

Figure 18-7.
The **Geographic Location-Specify Location** dialog box is used to specify geographic location data. Pick directly on the map to set the location.

the zoom and pan tools to navigate to a specific location. Then, right-click on the location and select **Drop Marker Here**. A geographic marker is placed at the location and the latitude and longitude for the location are saved. Pick **Next** to display the **Geographic Location-Set Coordinate System** dialog box, which allows you to select the coordinate system and time zone. Select the appropriate coordinate system and then pick **Next**. This switches you back to the drawing, where you can specify a point for the geographic location in the drawing. After picking a point, you are prompted to specify the north direction. Pick a point that should be in the north direction, enter an angle, or use the **First point** option to pick two points that define an angle. Once this is set, a geographic marker is added at the selected point and the online map is displayed in the background. The geographic marker looks like a red and white thumbtack. See **Figure 18-8A**.

The **Geolocation** contextual ribbon tab appears once you specify a geographic location. See **Figure 18-8B**. The options in the **Location** panel allow you to edit the location and orientation of the marker or remove the location from the drawing. The **Locate Me** button is available in the **Tools** panel if your operating system is configured with location sensing features. Picking the **Locate Me** button places a marker in the drawing to indicate your current location. The **Mark Position** drop-down list contains options to mark other positions on the map. You can mark a position by specifying the latitude and longitude, picking a point, or specifying your current location. Once you specify a point, you have the option to enter multiline text to label the point location. Type the text and exit the multiline text editor. If a visual style other than 2D Wireframe is current, a yellow position marker is displayed. The position marker includes a point object with a leader attached to the label text.

GEOMAP
Ribbon
Geolocation
> Online Map
Aerial
Type
GEOMAP

The map image drop-down list in the **Online Map** panel provides options for controlling the appearance of the map in the background. A map can be displayed as an aerial image, a road map, or a road map overlaid with an aerial image. The options in the map capture drop-down list in the **Online Map** panel allow you to select a portion of the map and embed it into the drawing. Once you capture a map image, you can display and plot it even if you are no longer connected to the Internet. The **Capture Area** option prompts you to define a rectangular area to capture. The **Capture Viewport** option captures whatever is displayed in the viewport. Selecting one of these options from the ribbon initiates the **GEOMAPIMAGE** command. To use this command, you must first change to a plan view of the WCS.

GEOMAPIMAGE
Ribbon
Geolocation
> Online Map
Capture Area
Type
GEOMAPIMAGE

Figure 18-8.
A—The geographic marker is an icon that represents the point in the drawing selected for the geographic location of the scene. B—The **Geolocation** contextual ribbon tab allows you to adjust the selected location, change the map display, and mark positions on the drawing.

A

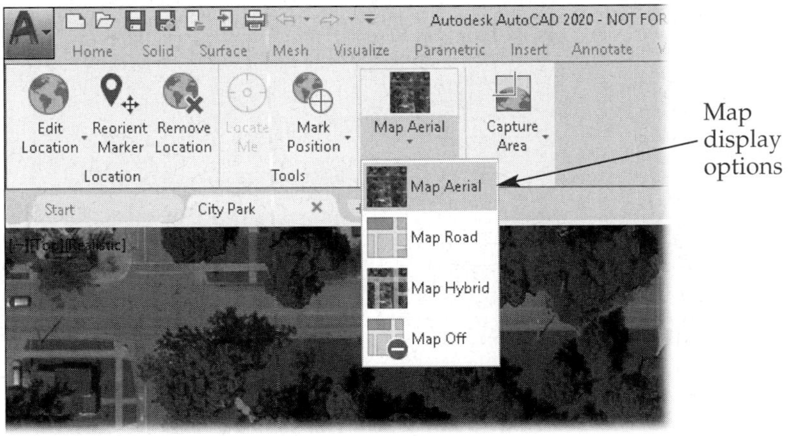

B

Sun Properties Palette

The properties of the sun are set in the **Sun Properties** palette, **Figure 18-9**. The **SUNPROPERTIES** command opens this palette. You can also pick the dialog box launcher button at the lower-right corner of the **Sun & Location** panel on the **Visualize** tab of the ribbon. The sun can be turned on or off using the **Sun Properties** palette. The date and time can also be changed in the palette. Other properties of the sun are only available in the **Sun Properties** palette. The categories in this palette are discussed in the next sections.

Type
SUNPROPERTIES

SUNPROPERTIES

General

The Status property in the **General** category controls turning the sun on or off. The Intensity Factor property determines the brightness of the sun. Setting this property to zero, in effect, turns off sunlight. Increasing the property makes the sunlight brighter. The maximum value for the property is determined by the capabilities of your computer. The Color property is read-only and set to a preselected color based on the geographic location, date, and time.

Sky Properties

The settings in the **Sky Properties** category control the sky. The sky is used in conjunction with sunlight to generate more-realistic lighting in the scene. The Status property determines if the sky effect is off or if both the sky background and illumination are active. In order to set the sky status, the projection must be set to perspective. The Intensity Factor property value is a multiplier for the illumination provided by the sky. The Haze property controls how the sky illumination is diffused. The value can range from 0 to 15. The preset sun color is affected by this value. The Night color property sets the color of the night sky. In a city, the night sky may have an orange tint to it because of the streetlamps in the city. However, in the country, the night sky is nearly black due to the lack of artificial illumination. The default color is true color 1,2,5, which is near black.

Figure 18-9.
The **Sun Properties** palette.

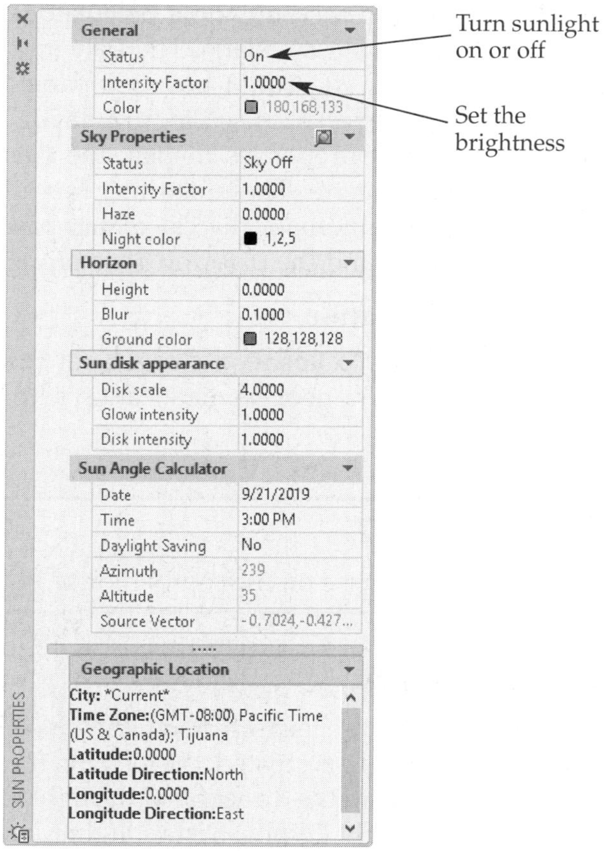

PROFESSIONAL TIP

The sky at night is always lighter than the Earth unless there are other light sources in the scene, such as streetlamps.

NOTE

The Sky Background status property setting in the **Sky Properties** category is only available for backward compatibility with older releases of AutoCAD and has no effect on the scene. The same setting appears in the flyout under the **Sky Off** button in the **Sun & Location** panel on the **Visualize** tab of the ribbon.

The settings in the **Horizon** subcategory control what the horizon looks like and where it is located. Changing the Height property moves the horizon up or down. The default value is 0 and values can range from –10 to 10. The Blur property determines how much the horizon is blurred between the ground and the sky. The default value is 0.1 and values can range from 0 to 10. The Ground color property controls the color of the ground. The default color is true color 128,128,128, which is a medium gray. To see how this affects your scene, rotate your viewpoint to make the horizon visible.

Normally, lights do not appear in the rendered scene at all, only the illumination provided by the lights. The settings in the **Sun disk appearance** subcategory control what the sun looks like in the sky, or on the background. The Disk scale property sets the size of the sun, or solar disk, as it appears on the background. The default value is 4 and values can range from 0 to 25. The Glow intensity property determines the size of the glowing halo around the sun in the sky. The default value is 1 and values can range from 0 to 25. The Disk intensity property controls the brightness of the sun on the background. The default value is 1 and values can range from 0 to 25.

Sun Angle Calculator

The settings in the **Sun Angle Calculator** category determine the angle of the sun in relationship to the XY plane. The Date and Time properties, discussed earlier, can be controlled from the **Sun & Location** panel on the **Visualize** tab of the ribbon. The Daylight Saving property is used to turn daylight saving on or off. The Azimuth, Altitude, and Source Vector properties display the current settings, but are read-only in the **Sun Properties** palette. These settings represent the location of the sun in the sky. Their values are automatically calculated based on the geographic location settings, discussed earlier, and the Date and Time properties.

Geographic Location

The **Geographic Location** category at the bottom of the **Sun Properties** palette displays the current geographic location settings, described earlier. The settings are read-only.

Exercise 18-1

www.g-wlearning.com/CAD/

Complete the exercise on the companion website.

Distant Lights

Distant lights are user-created lights that have parallel light rays. See **Figure 18-10**. When a distant light is created, the location from where the light is originating must be specified along with the direction of the light rays. Distant lights are not represented in the drawing by light glyphs. Distant lights are used to create even, uniform, overhead

Figure 18-10. This example shows the use of a distant light to simulate sunlight shining through a window. Notice how the edges on the shadows of the grill are parallel.

illumination, such as the illumination you would encounter in an office setting. The distance of the objects in the scene to the distant light has no effect on the intensity of the illumination. Distant lights do not attenuate. A distant light can also be used to simulate sunlight without having to set up a time and location. However, the light must be manually moved to change the illumination effect.

NOTE

When creating a distant light, you will receive a warning to the effect that photometric distant lights may overexpose the scene. Your choices are to disable or allow distant lights. If you choose to continue to place the distant light and the scene becomes washed-out with light, adjust the exposure as previously discussed.

Ribbon
Visualize > Lights
Distant
Type
DISTANTLIGHT
LIGHT

DISTANTLIGHT

The **DISTANTLIGHT** or **LIGHT** command is used to create a distant light. You are first prompted to specify the direction from which the light is originating or to enter the **Vector** option. If you pick a point, it is the location of the light. Next, you are prompted to specify the point to which the light is pointing. This is simply the location where the light is aimed.

If you enter the **Vector** option instead of picking a "from" point, you must type the endpoint coordinates (in WCS units) of the direction vector. The light will point from the WCS origin to the entered endpoint. This option is not often used.

After the light location and direction are determined, several other options are available. You can name the light, set the intensity, turn the light on or off, and set the light color.

AutoCAD provides a default name for new lights based on the type of light and a sequential number, such as Distantlight1, Distantlight2, and so on. It is a good idea to provide a meaningful name for a light. This is especially true if there are other lights in the scene. To rename a light, enter the **Name** option. Then, type the name of the light and press [Enter].

To set the brightness of the light, enter the **Intensity factor** option. Then, type a value and press [Enter]. The default value is 1.00. Setting the value to 0.00, in effect, turns off the light. The maximum value depends on the capabilities of your computer.

When a light is created, it is on. To turn the light off, enter the **Status** option. Then, change the setting to Off.

The **Photometry** option controls the luminous qualities of visible light sources. Once you enter the **Photometry** option, you can select one of three options:

- **Intensity**. This is the power of the light source. The value is specified in candelas (cd).
- **Color**. This is the color of the light source. The value can be changed by typing in a name (to get a list of color names, enter ?) or by specifying the Kelvin temperature value.
- **Exit**. This exits the command option.

NOTE

The **Shadow** option is only available for backward compatibility with older releases of AutoCAD. Shadow casting cannot be turned off and shadow-mapped shadows, which have soft edges, are created by AutoCAD by default.

By default, new distant lights cast white light (true color 255,255,255). To change the color of the light, enter the **Filter Color** option. To specify a new true color, simply specify the RGB values and press [Enter]. To specify a color based on hue, saturation, and luminance (HSL), enter the **Hsl** option and specify the values. To enter an AutoCAD color index (ACI) number, enter the **Index color** option and specify the ACI number. To specify a color book color, enter the **Color Book** option and then specify the name of the color book followed by the name of the color.

Once all settings for the distant light have been made, use the **Exit** option to end the command and create the light. Do not press [Esc] to end the command. Doing so actually cancels the command and the light is not created.

PROFESSIONAL TIP

To provide a visual cue to a distant light's location and direction, draw a line with one endpoint at the "from" coordinate and the other endpoint at the "to" coordinate. Then, when creating the distant light, select the endpoints of the line. Since a distant light is not represented by a light glyph, the line serves the purpose of a glyph.

Exercise 18-2

www.g-wlearning.com/CAD/

Complete the exercise on the companion website.

Point Lights

Point lights are user-created lights that have light rays projecting in all directions. See Figure 18-11. When a point light is created, its location must be specified. Since point lights illuminate in all directions, there is no "to" location for a point light. A light glyph represents point lights in the drawing. See Figure 18-12. Point lights have attenuation, which means that the distance of the objects in the scene to the point light affects the intensity of the illumination.

POINTLIGHT

Ribbon
Visualize
> Lights

Point

Type
POINTLIGHT
LIGHT

The **POINTLIGHT** or **LIGHT** command is used to create a point light. You are first prompted to specify the location of the point light. Once the location is established, several options for the light are available. You can name the light, set the intensity, turn the light on or off, adjust the photometry settings, and set the light color. The **Name**, **Intensity Factor**, **Status**, **Photometry**, and **Filter Color** options work the same as the corresponding options for a distant light. However, additional photometry settings are available. The **Intensity** option provides additional options for setting the power of the light source. The **Flux** option can be used to specify the power in lumens (lm). The

Figure 18-11.
A point light is placed inside of the lamp fixture. The light projects in all directions, but the lampshade directs most of the light toward the ceiling.

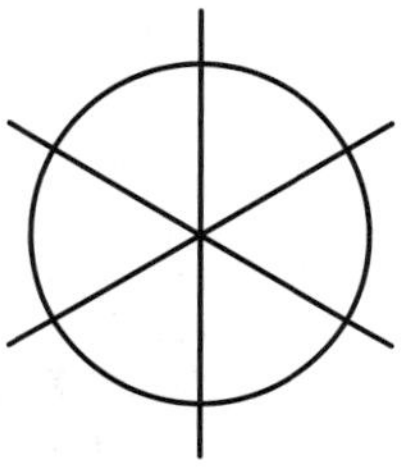

Figure 18-12.
This is the light glyph for a point light.

Illuminance option can be used to specify the illuminance in lux (lx) or foot-candles (fc), depending on the current lighting units. An additional option, **Distance**, is available under **Illuminance**. This defaults to one unit and is used to calculate illuminance.

NOTE

The **Shadow** and **Attenuation** options are only available for backward compatibility with older releases of AutoCAD. Shadow casting cannot be turned off and shadow-mapped shadows are created by AutoCAD, as previously discussed. Inverse square attenuation is used regardless of the attenuation type selected. Setting the attenuation start and end limits has no effect on the scene.

Target Point Lights

The target point light is like a regular point light except that the command starts by asking for a *source* location and a *target* location. The rest of the options are the same. You can type TARGETPOINT or pick the **Targetpoint** option in the **LIGHT** command to create a target light.

PROFESSIONAL TIP

A point light may be used as an incandescent lightbulb, such as the lightbulb in a table lamp. Most of these lightbulbs cast a yellow light. In these cases, you may want to change the color of the light to a light yellow. Compact fluorescent lightbulbs may cast white, light blue, or light yellow light. LED-based lights cast a color based on the color of the LED. Other colors can be used to give the impression of heat or colored lights.

Exercise 18-3

www.g-wlearning.com/CAD/

Complete the exercise on the companion website.

Spotlights

Spotlights are user-created lights that have light rays projecting in a cone shape in one direction. See **Figure 18-13**. When a spotlight is created, the location from where the light is originating must be specified, along with the direction in which the light rays travel. A light glyph represents spotlights in the drawing. See **Figure 18-14**. Spotlights have attenuation, which means that the distance of the objects in the scene to the spotlight affects the intensity of the illumination.

SPOTLIGHT

Ribbon
Visualize
> Lights

Spot

Type
SPOTLIGHT
LIGHT

The **SPOTLIGHT** or **LIGHT** command is used to create a spotlight. You are first prompted to specify the location of the light. This is from where the light rays will originate. Next, you are prompted for the target location. This is simply the location where the light is aimed.

Once the location and target are established, several options for the light are available. The **Name**, **Intensity Factor**, **Status**, **Photometry**, and **Filter Color** options work the same as the corresponding options for a point light. However, a spotlight also has hotspot and falloff settings.

The ***hotspot*** is the inner cone of illumination for a spotlight. Refer to **Figure 18-15**. This is measured in degrees. To set the hotspot, enter the **Hotspot** option and then specify the number of degrees for the hotspot.

A full-color image is provided on the companion website. www.g-wlearning.com/CAD/

Figure 18-13. Three spotlights are used to simulate recessed ceiling lights. Notice how the light from each spotlight projects in a cone.

Figure 18-14. This is the light glyph for a spotlight.

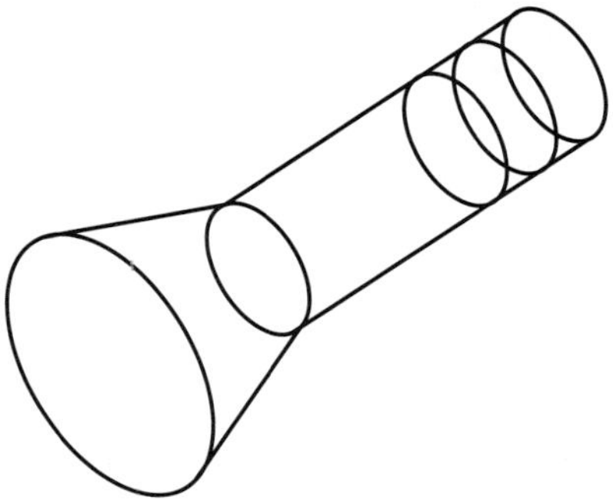

Figure 18-15.
Hotspot and falloff for a spotlight are angular measurements.

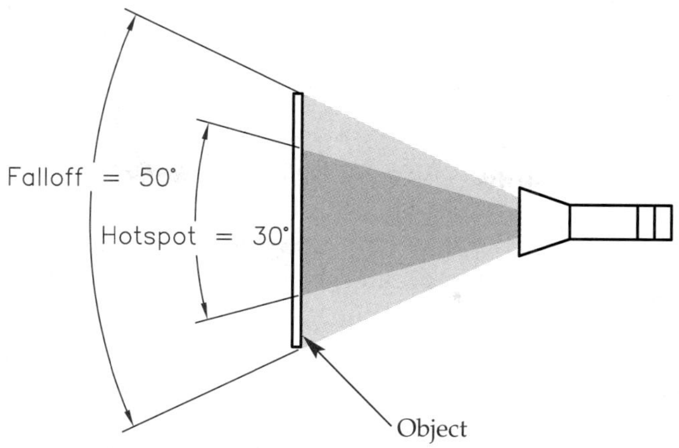

The *falloff*, not to be confused with attenuation, is the outer cone of illumination for a spotlight. Like the hotspot, it is measured in degrees. The falloff value must be greater than or equal to the hotspot value. It cannot be less than the hotspot value. In practice, the falloff value is often much greater than the hotspot value. To set the falloff, enter the **Falloff** option and then specify the number of degrees for the falloff.

NOTE

The **Shadow** and **Attenuation** options are only available for backward compatibility with older releases of AutoCAD. Shadow casting cannot be turned off and shadow-mapped shadows are created by AutoCAD, as previously discussed. Inverse square attenuation is used regardless of the attenuation type selected. Setting the attenuation start and end limits has no effect on the scene.

Once the light is created and you select it in the viewport, grips are displayed. If you hover the cursor over a grip, a tooltip is displayed indicating what the grip will modify. You can use grips to change the location of the spotlight and its target, the hotspot, and the falloff. If you hover over a falloff or hotspot grip, the current angle is displayed in the wireframe cone (if dynamic input is on).

Free Spotlight

A free spotlight is like a standard spotlight except that you do not specify a target, only the light location. The rest of the options are the same. You can type FREESPOT or pick the **Freespot** option in the **LIGHT** command to create a free spotlight. When created, a free spotlight points down the Z axis (from positive to negative) of the current UCS. A free spotlight may be easier to control than a standard spotlight because you do not have to worry about the target point. If you want to change the angle or position of the light, use the **3DMOVE** and **ROTATE3D** commands.

Type
FREESPOT
LIGHT

FREESPOT

PROFESSIONAL TIP

The best way to see how colored lights affect your model is to experiment. Remember, you can render selected areas of the scene. This allows you to see how changes to light intensity and color affect objects without performing a full render.

Exercise 18-4

www.g-wlearning.com/CAD/

Complete the exercise on the companion website.

Weblight

A photometric weblight is really just a targeted point light. The difference is that a weblight provides a more precise representation of the light. Real-world lights appear to evenly illuminate from their source, but, in reality, the shape of the light, the material used in its manufacture, and other factors make all lights distribute their energy in different ways. These data are provided by light manufacturers in the form of light distribution data. Light distribution data can be loaded into the **Photometric Web** subcategory of the **General** category in the **Properties** palette when the light is selected. Select the Web file property, then pick the button to the right and select an IES file. IES stands for Illuminating Engineering Society. The AutoCAD help documentation has additional information on IES files.

Think of the web of a weblight as a spherical cage surrounding the light source. If the light is evenly distributed from its source, the cage is a true sphere. In actuality, a light may emit more light energy in the X direction than in the Z direction. In this case, the cage bulges out further in the X direction. The position of this bulge may be important to the illumination of the scene and you may need to rotate the web to apply more or less light in one direction or another.

WEBLIGHT

Ribbon
Visualize
> Lights
Weblight
Type
WEBLIGHT
LIGHT

The **WEBLIGHT** or **LIGHT** command is used to create a weblight. You are prompted for source and target locations. The **Name**, **Intensity Factor**, **Status**, **Photometry**, **Shadow**, and **Filter Color** options work the same as the corresponding options for the previously discussed lights. However, weblights have an additional **Web** option. When this option is activated, these options are presented:

- **File**. Allows you to select an IES file.
- **X**. Rotates the web around the X axis.
- **Y**. Rotates the web around the Y axis.
- **Z**. Rotates the web around the Z axis.
- **Exit**. Exits the **Web** option.

Point and spotlights can be converted to weblights, and vice versa, using the **Properties** palette. Simply select an existing light and open the **Properties** palette. In the **General** category, the Type property determines whether the light is a point light, spotlight, or weblight. Select the type in the drop-down list. Using the **Properties** palette with lights is discussed in detail later in this chapter.

Free Weblight

FREEWEB

Type
FREEWEB
LIGHT

A free weblight is the same as a standard weblight except there is no target. Only the source location is specified when placing the light. To change the location and direction of the light, use the **ROTATE3D** and **3DMOVE** commands.

Photometric Lights Tool Palette Group

Photometric lights may be easily added to the drawing using the tool palettes in the **Photometric Lights** tool palette group. This palette group contains four palettes: **Fluorescent**, **High Intensity Discharge**, **Incandescent**, and **Low Pressure Sodium**. Lights created with these tools have preset properties for **Intensity Factor** and **Filter Color**. The long fluorescent lights are weblights. The high intensity discharge, low-pressure sodium, and regular incandescent lights are point lights. The incandescent halogen lights are free spotlights. The recessed incandescent lights are a special weblight designed for recessed light fixtures.

Lights in Model and Properties Palettes

The **Lights in Model** palette is extremely useful for controlling the lights in your scene, **Figure 18-16**. Using the **Lights in Model** palette in conjunction with the **Properties** palette, you can manage and edit all of the lights in a scene.

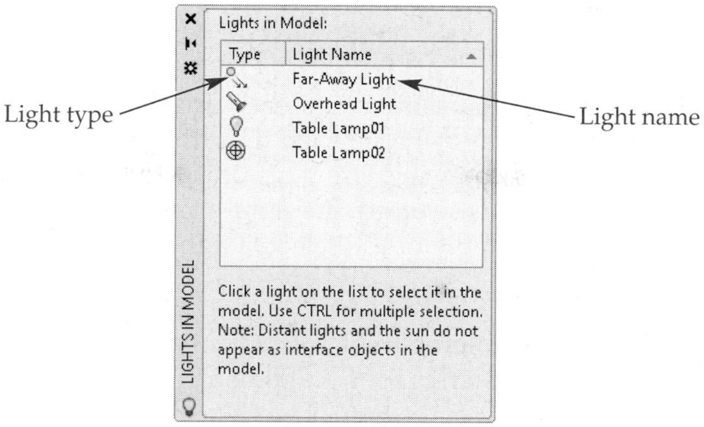

Figure 18-16.
All user-created lights are listed in the **Lights in Model** palette.

Lights in Model Palette

Type
LIGHTLIST

LIGHTLIST

The **LIGHTLIST** command displays the **Lights in Model** palette. This palette can also be displayed by picking the dialog box launcher button at the lower-right corner of the **Lights** panel on the **Visualize** tab of the ribbon. All user-created lights in the scene are displayed in the list. To modify the properties of a light, either double-click on the light name or right-click on it and select **Properties** from the shortcut menu. This opens the **Properties** palette. See Figure 18-17. If the **Properties** palette is already open, you can simply select a light in the **Lights in Model** palette. You can select more than one light by pressing the [Ctrl] key and selecting the names in the **Lights in Model**

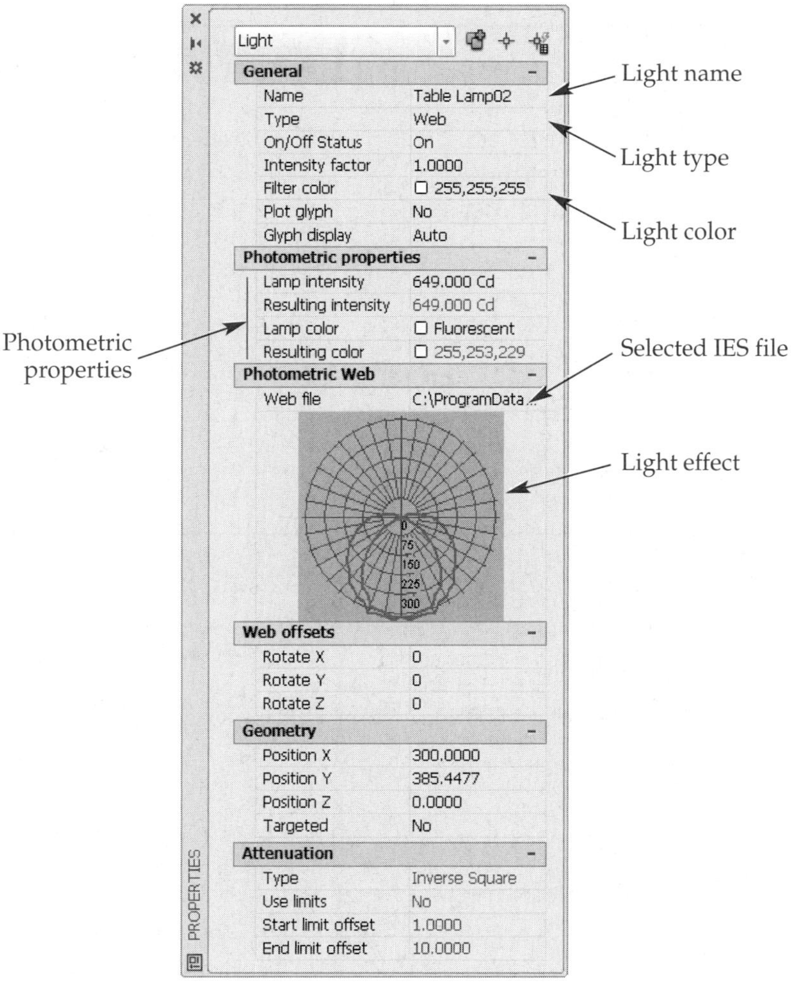

Figure 18-17.
The **Properties** palette with a weblight selected.

palette, which allows you to change all of their settings at the same time. This is an excellent way to make the lighting in your scene uniform or to control a series of lights with a single edit.

A light can be deleted from the scene using the **Lights in Model** palette. To do so, simply right-click on the name of the light and select **Delete Light** from the shortcut menu. The light is removed from the drawing. Using the **UNDO** command restores the light.

If a single light is selected, the right-click menu also contains an option to control the display of the glyph for the selected light. The options are **Auto**, **On**, or **Off**.

Properties Palette

The **Photometric properties** subcategory of the **General** category in the **Properties** palette has special settings for photometric lights. The Lamp intensity property determines the brightness of the light. The value may be expressed in candelas (cd), lumens (lm), or illuminance (lux) values. When you select the Lamp intensity property, a button is displayed to the right of the value. Picking this button opens the **Lamp Intensity** dialog box, **Figure 18-18**. In this dialog box, you can change the illumination units and set the intensity (Lamp intensity property). You can also set an intensity scale factor. This is multiplied by the Lamp intensity property to obtain the actual illumination supplied by the light. The read-only Resulting intensity property in the **Properties** palette displays the result.

The Lamp color property in the **Photometric properties** subcategory controls the color of the light. If you select the property, a button is displayed to the right of the value. Picking this button opens the **Lamp Color** dialog box. See **Figure 18-19**. This dialog box gives you the option to control the color of the light by either standard spectra colors or Kelvin colors. The color selected in the **Filter color:** drop-down list is

Figure 18-18.
The **Lamp Intensity** dialog box is used to set the intensity for a light.

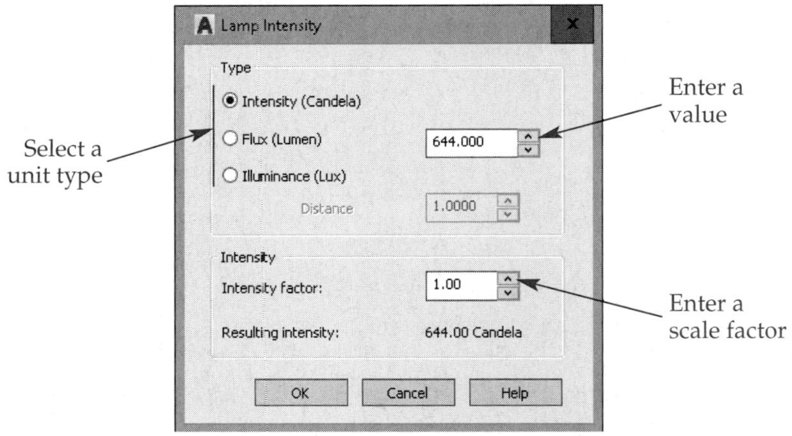

Figure 18-19.
The **Lamp Color** dialog box is used to set the color for the light and a filter color, if needed.

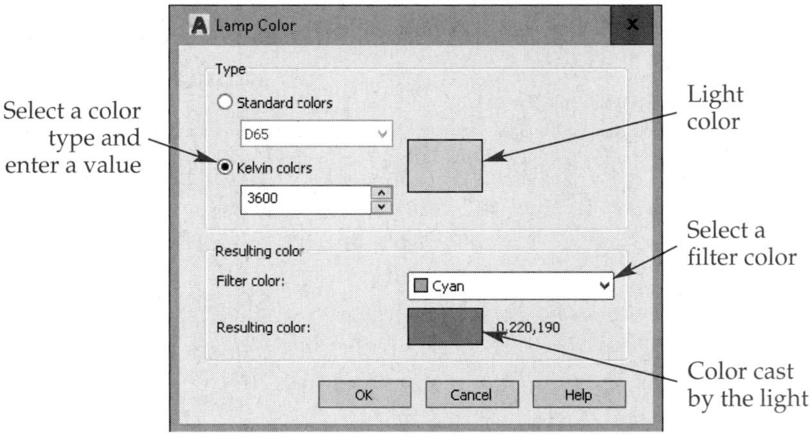

applied to the color of the light. The **Resulting color:** swatch displays the color cast by the light once the filter color is applied. If the filter color is white (255,255,255), then the light color is the color cast by the light.

If a weblight is selected, the **Photometric Web** subcategory in the **General** category is where you can specify an IES file for the light. Select the Web file property, then pick the button to the right and select the IES file. Once the file is selected, its location is displayed in the Web file property. The effect of the data is shown in a graph at the bottom of the **Photometric Web** subcategory. Refer to **Figure 18-17**.

The **Web offsets** subcategory allows you to rotate the web around the X, Y, and Z axes, as discussed earlier in this chapter in the Weblight section.

In the **Geometry** category, you can change the X, Y, and Z coordinates of the light. You can also change the X, Y, and Z coordinates of the light's target. If the light is not targeted, the Target X, Target Y, and Target Z properties are not displayed. To change the light from targeted to free, and vice versa, select Yes or No in the Targeted property drop-down list.

The properties in the **Attenuation** category are the same as those discussed earlier for lights that have attenuation properties. The Type property is set to Inverse Square and is read-only. The attenuation limits properties are read-only and are only for backward compatibility with older releases of AutoCAD.

PROFESSIONAL TIP

It is important to give your lights names that make them easy to identify in a list. If you accept the default names for lights, they will be called Pointlight1, Spotlight5, Distantlight7, Weblight2, etc., making them difficult to identify. Use the **Name** option when creating the light or, after the light is created, the **Properties** palette to change the name of the light.

Determining Proper Light Intensity

Placing lights usually requires adjustments to produce the results you are looking for. In addition, you will also typically spend some time determining the proper light intensity and other settings. As a general guideline, the object nearest to a point light or spotlight should receive the full illumination, or full intensity, of the light. Full intensity of any light that has an attenuation property is a value of one. Remember, as discussed earlier in this chapter, AutoCAD calculates attenuation using the inverse square method. You can calculate the light intensity based on this method to establish an approximate starting point.

For example, suppose you have drawn an object and placed a point light and a spotlight. The point light is 55 units from the object. The spotlight is 43 units from the object. Use the following calculations to determine the intensity settings for the lights based on inverse square attenuation.

Since the point light is 55 units from the object, the object receives $(1/55)^2$, or 1/3025 ($55^2 = 3025$), of the light. Therefore, set the intensity of the point light to 3025 (3025/3025 = 1). Since the spotlight is 43 units from the object, the object receives $(1/43)^2$, or 1/1849 ($43^2 = 1849$), of the spotlight's illumination. Therefore, set the intensity of the spotlight to 1849 (1849/1849 = 1).

However, keep in mind that these settings are merely a starting point. You will likely spend some time adjusting lighting to produce the desired results. In some cases, it may take longer to light the scene than it did to model it.

PROFESSIONAL TIP

If there are no lights in the scene or all lights are turned off, default lighting is activated so that you can still see what you are doing. If you render a scene and the image appears black, lights in the scene may have their intensity set too low or they may be positioned too far away from the objects in the scene.

Shadows

Shadows are critical to the realism of a rendered 3D model. A model without shadows appears obviously fake. On the other hand, a model with realistic materials and shadows may be hard to recognize as computer generated. In AutoCAD, all lights, including default lights, cast shadows. You may not see shadows in the viewport depending on the current visual style settings, but shadows are always calculated in the rendered image. As previously discussed, the **Shadow** option that appears in command prompts when creating lights is only available for backward compatibility with older releases of AutoCAD. All lights display realistic shadows with soft edges.

Adding a Background

A ***background*** is the backdrop for your 3D model. The background can be a solid color, a gradient of colors, an image file, the sun and sky, or the current AutoCAD drawing background color. By default, the background is the drawing background.

BACKGROUND

Type
BACKGROUND

One way to change the background for your drawing is to use the **BACKGROUND** command. This command opens the **Background** dialog box, **Figure 18-20**. The **Type:** drop-down list is used to specify the type of background. The options are None, Solid, Gradient, Image, and Sun & Sky. When the None option is selected, the current AutoCAD viewport color is used. The other background options are discussed in the following sections.

You can also set the drawing background when creating a named view using the **New View/Shot Properties** dialog box. Creating named views is introduced in Chapter 3. A background for a named view is specified using the drop-down list in the **Background** area of the **View Properties** tab in the **New View/Shot Properties** dialog box. The background options correspond to those in the **Background** dialog box.

Figure 18-20.
The **Background** dialog box with the Solid option selected.

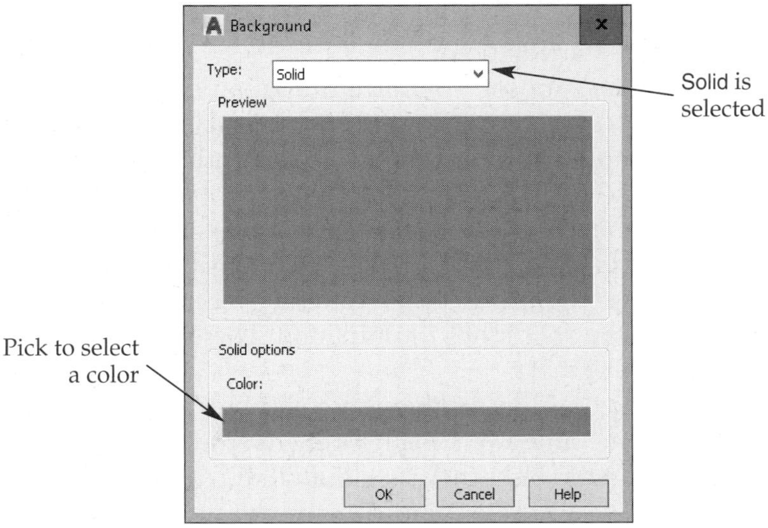

NOTE

An image based lighting map can also be used to set the background of a scene. Image based lighting and other advanced rendering functions are discussed in Chapter 19.

PROFESSIONAL TIP

A named view can be recalled at any time after being created and is useful for making a rendering. Before you create a named view, establish the viewpoint from which you want to see the final rendering. Set the perspective projection current, if desired. These settings are saved with the view.

Solid Backgrounds

The background of the scene can be set to a solid color. Refer to **Figure 18-20**. When the **Type:** drop-down list in the **Background** dialog box is set to Solid, the default color is displayed in the **Preview** area. In the **Solid options** area of the dialog box, pick the horizontal **Color:** bar to open the **Select Color** dialog box. Then, select the background color that you desire. When the **Select Color** dialog box is closed, the color you picked is displayed in the **Preview** area of the **Background** dialog box. Pick the **OK** button to accept your changes and close the **Background** dialog box.

NOTE

Make sure to do a test rendering with the background color you selected. The result may look quite different from your expectations.

Gradient Backgrounds

A gradient background can be composed of two or three colors. If you select Gradient in the **Type:** drop-down list in the **Background** dialog box, the default gradient colors are displayed in the **Preview** area. See **Figure 18-21**. The **Top color:**, **Middle color:**, and **Bottom color:** swatches are displayed on the right-hand side of the **Gradient options** area. Selecting a swatch opens the **Select Color** dialog box for changing the color. To create a two-color gradient composed of the top and bottom colors, uncheck the **Three color** check box. The **Rotation:** text box provides the option of rotating the gradient. Pick the **OK** button to accept your changes and close the **Background** dialog box.

Figure 18-21.
Creating a gradient background.

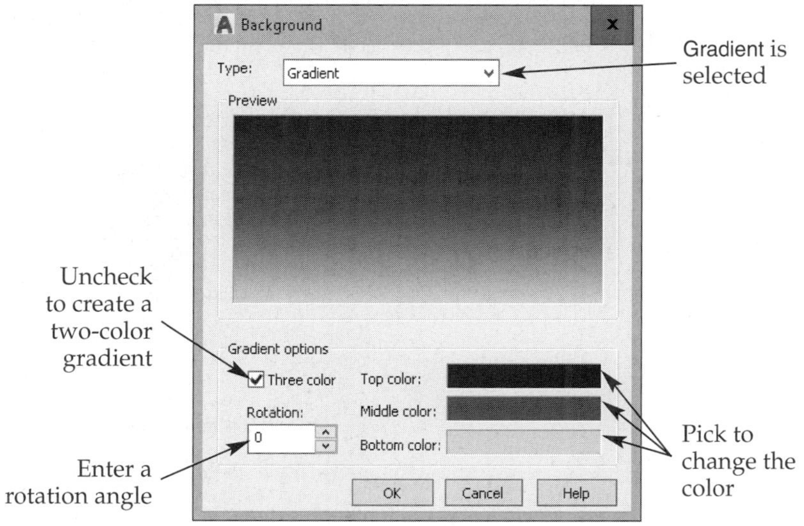

Convincing, clear blue skies can be simulated using a gradient background. Initially, set the **Top**, **Middle**, and **Bottom** color values the same. Then, change the lightness (luminance) in the **True Color** tab in the **Select Color** dialog box. Preview the background and make adjustments as needed.

Using an Image as a Background

An image can be used as a background. This technique can be used to produce realistic or imaginative settings for your models. If you select Image in the **Type:** drop-down list in the **Background** dialog box, a blank image is displayed in the **Preview** area. To locate the image file, pick the **Browse...** button to display a standard open dialog box. These image file types may be used for the background: TGA, BMP, PNG, JFIF (JPEG), TIFF, GIF, and PCX. Locate the image file and select **Open**. See **Figure 18-22**.

Once the image file is selected, it must be adjusted. The **Preview** area of the **Background** dialog box shows the image with a preview of a drawing sheet. This drawing sheet indicates how the image is going to be positioned in the view. Pick the **Adjust Image...** button to open the **Adjust Background Image** dialog box. See **Figure 18-23**.

Figure 18-22.
Setting an image as the background.

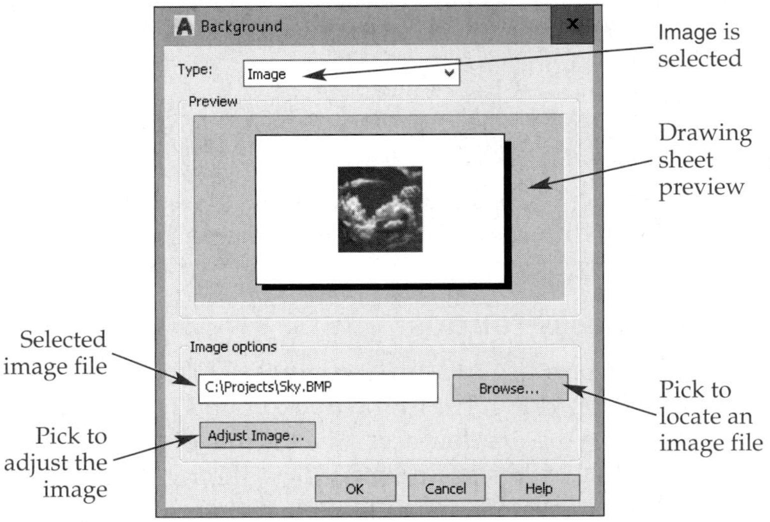

Figure 18-23.
Adjusting the background image.

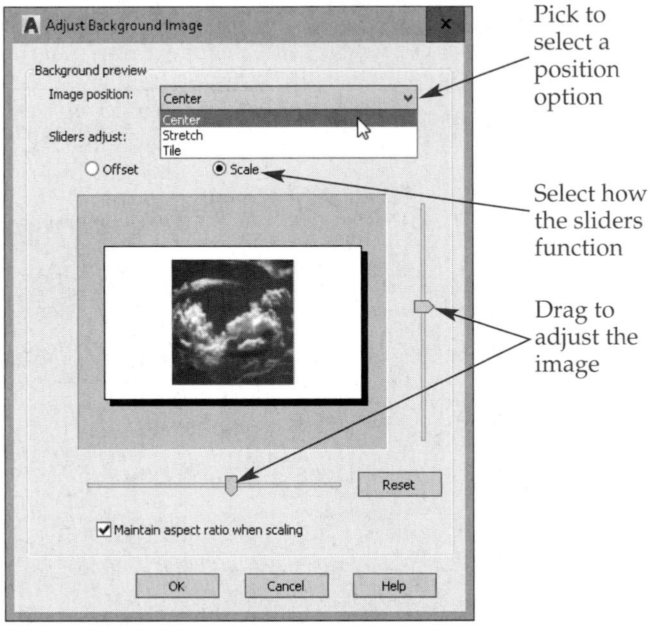

In the **Image position:** drop-down list, pick how the image is applied to the viewport. The Center option centers the image in the view without changing its aspect ratio or scale. The Stretch option centers the image and stretches or shrinks it to fill the entire view. This is one way to plot an image file from AutoCAD. The Tile option keeps the image at its original size and shape, but moves it to the upper-left corner and duplicates it, if needed, to fill the view.

After the image is positioned, use the sliders to adjust it further. The sliders are disabled if Stretch is selected in the **Image position:** drop-down list. The slider function is based on which radio button is picked above the image:

- **Offset**. The sliders move the image in the X or Y direction.
- **Scale**. The sliders scale the image in the X or Y direction. This may distort the image if it is scaled too much in one direction. To prevent distortion, check the **Maintain aspect ratio when scaling** check box at the bottom of the dialog box.

The **Reset** button is located at the bottom-right corner of the preview pane. Picking this button returns the scale and offset settings to their original values.

Once the image is adjusted, pick the **OK** button to close the **Adjust Background Image** dialog box. Then, pick the **OK** button to close the **Background** dialog box.

Sun and Sky

If you select Sun & Sky in the **Type:** drop-down list in the **Background** dialog box, the **Adjust Sun & Sky Background** dialog box is displayed. AutoCAD uses the settings in this dialog box to simulate the sun in the sky. This dialog box has a preview tile at the top and the **General**, **Sky Properties**, **Horizon**, **Sun disk appearance**, **Sun Angle Calculator**, and **Geographic Location** categories. The settings in these categories were discussed earlier in the Sun Properties Palette section in this chapter. Once the sky is set, pick the **OK** button to close the **Adjust Sun & Sky Background** dialog box.

Changing the Background of an Existing View

Ribbon
Visualize
> Named Views
View Manager
Type
VIEW
V

VIEW

If you need to change the background of an existing named view, open the **View Manager** dialog box. Select the view name in the **Views** tree on the left-hand side of the dialog box. Then, in the **General** category in the middle of the dialog box, select the Background override property. See **Figure 18-24**. Next, pick the drop-down list for the property and make a selection. Picking None sets the background to the

Figure 18-24.
Changing the background of an existing, named view.

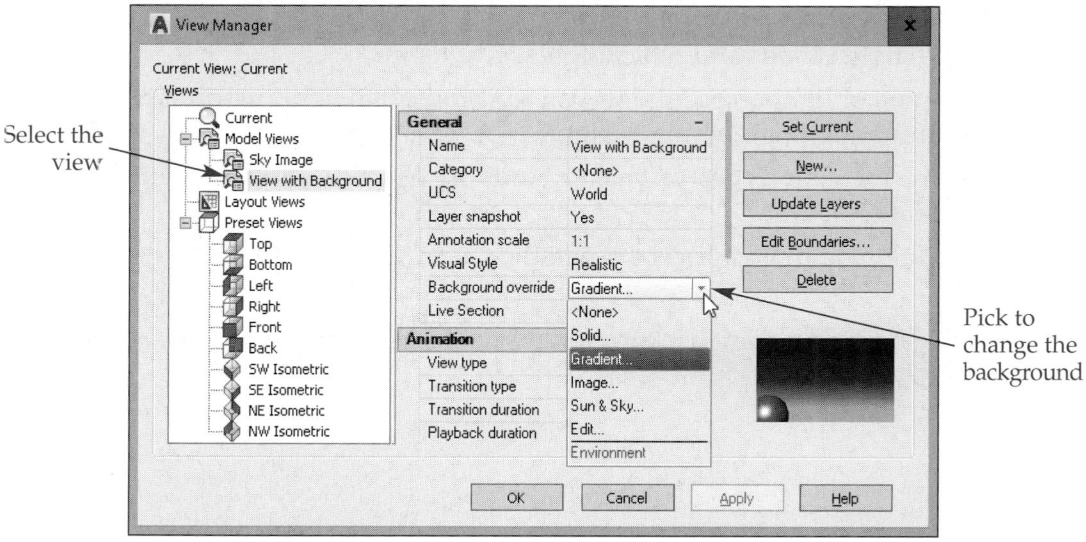

AutoCAD default background. Setting the property to Solid, Gradient, or Image opens the **Background** dialog box, where you can make settings for that type. Selecting Sun & Sky opens the **Adjust Sun & Sky Background** dialog box. Picking Edit opens the **Background** dialog box. Once the background type has been changed or the existing background edited, pick the **OK** button to save the view and close the **View Manager** dialog box.

NOTE

After exiting the **Background** dialog box, you are returned to the **View Manager** dialog box. Picking the **OK** button to exit the **View Manager** dialog box does not necessarily activate the view that you just created or modified. The view must be set current to see the effects of the changes to the background. A view can be set current using the drop-down list in the **Named Views** panel on the **Visualize** tab of the ribbon, the **View Controls** flyout in the viewport controls, or the **View Manager** dialog box.

Exercise 18-5

www.g-wlearning.com/CAD/

Complete the exercise on the companion website.

Chapter Review

Answer the following questions using the information in this chapter.

1. Compare and contrast *ambient light*, *distant lights*, *point lights*, *spotlights*, and *weblights*.
2. Define *angle of incidence*.
3. Define *angle of reflection*.
4. A smooth surface has a(n) _____ specular factor.
5. Describe *hotspot* and *falloff*. Which lights have these properties?
6. What is *attenuation*?
7. What are the three types of lighting in AutoCAD?
8. Briefly describe *photometric lighting*.
9. If you want to use the **GEOMAPIMAGE** command to embed a portion of a geographically located map in the drawing, what view must you set?
10. What are the four types of light objects in AutoCAD?
11. What are light glyphs and which lights have them?
12. Which type of light object allows you to incorporate light distribution data from the Illuminating Engineering Society?
13. What are the four types of backgrounds in AutoCAD, other than the default background?
14. Why would you draw a line between the "from" point and "to" point of a distant light?
15. Describe how a gradient background can be used to represent a clear blue sky.

Drawing Problems

1. **Architectural** In this problem, you will draw some basic 3D shapes to create a building similar to an ancient structure, place lights in the drawing, and render the scene.
   A. Begin a new drawing and set the units to architectural. Save the drawing as P18-1.
   B. Draw a 32′ × 22′ planar surface to represent the floor. Using the materials browser, attach a material of your choice to the floor.
   C. Draw cylinders to represent pillars. Make each ∅2′ × 15′ tall. There are 10 pillars per side. Attach a suitable material to the pillars.
   D. The roof is 32′ × 22′ and 5′ tall at the ridge. Attach an appropriate material.
   E. Create a perspective viewpoint looking into the building.
   F. Turn on sunlight and turn off the default lighting. Set the geographic location to Athens, Greece. Change the date and time to whatever you wish to get the proper lighting and shadows.
   G. Render the scene.
   H. Save the drawing.

A full-color image is provided on the companion website. www.g-wlearning.com/CAD/

2. **Architectural** Using the drawing from Problem 1, you will experiment with different lighting types.
   A. Open drawing P18-1 and save it as P18-2.
   B. Turn off the sun.
   C. Place three point lights inside of the building. Evenly space the lights along the centerline of the ceiling. Adjust the light intensity so that the interior is not washed out. Set the color of the middle light to white. Set the color of the outside lights to red or blue. Render the scene.
   D. Turn off the point lights.
   E. Place two spotlights, one pointing from the front corner to the rear corner and the other pointing between the pillars on the left side of the building. Target them at the floor. Render the scene. Adjust the intensity, hotspot, and falloff as needed.
   F. Turn the point lights back on and render the scene with all five lights active. Adjust the light intensities again if the rendering is too washed out with light.
   G. Save the drawing.

A full-color image is provided on the companion website. www.g-wlearning.com/CAD/

3. **Mechanical** Using a previously created mechanical model, you will apply materials and lights to make it ready for presentation.
   A. Open drawing P6-5 created in Chapter 6. Save it as P18-3.
   B. Draw a planar surface below the flange to represent a tabletop.
   C. Attach an appropriate material, such as a wood or tile material, to the surface.
   D. Open the materials editor and create a new material based on the Metal material type. Experiment with different finish settings until you find one you like. Apply the material to the flange.
   E. Place two spotlights in the drawing and target them at the flange from different angles. Adjust their hotspot, falloff, and intensity to get the proper lighting.
   F. Create a perspective view of the scene. Then, render the scene.
   G. Save the drawing.

A full-color image is provided on the companion website. www.g-wlearning.com/CAD/

Drawing Problems - Chapter 18

4. **Architectural** The building shown will be used to study passive solar heating at different times of the year. Model the building using the overall dimensions given. Use your own dimensions for everything else. The side with the windows should be facing South (–Y in AutoCAD, by default).
   A. Set the geographic location to a city in the northern hemisphere.
   B. Set the date to midsummer and the time to noon.
   C. Turn on sunlight and turn the default lighting off.
   D. Render the scene and note the location of the shadows inside the building.
   E. Change the date to late winter, render the scene again, and note the new location of the shadows. You can easily switch between the rendered images in the **Render** window by selecting each rendering in the history and statistics pane (this is discussed more in the next chapter).
   F. Observing the changes in the shadow locations, what design changes can be made to maximize sun exposure in the cold winter months? What design changes can be made to minimize sun exposure in the heat of summer?
   G. Change the geographic location to somewhere closer to the equator. Then, render the scene in summer and winter. How do the shadows compare to those in the previous location?
   H. Save the drawing as P18-4.

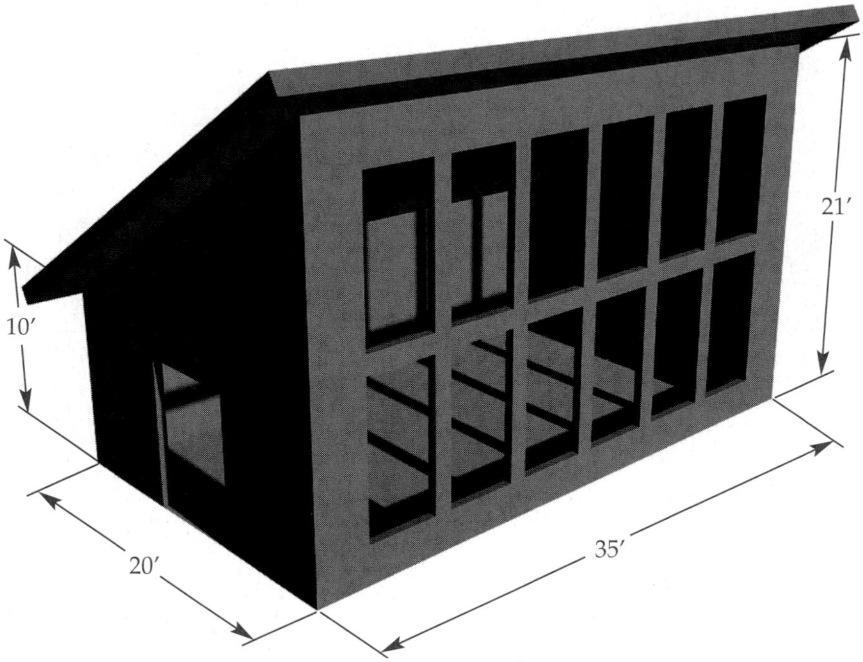

A full-color image is provided on the companion website. www.g-wlearning.com/CAD/

5. **Architectural** In this problem, you will be adding lights to a model and controlling the light properties to create a pleasing scene. Model the courtyard shown. The overall dimensions are 15′ × 16′ × 4′ (wall height). Use your own dimensions for everything else. Add lights as follows.

   A. Turn on the sun and turn off the default lighting. Set the time to late in the day so that the sun is close to the horizon.
   B. In the **Sky Properties** category of the **Sun Properties** palette, select Sky Background and Illumination for the Status property.
   C. Add a point light at the center of each sphere.
   D. Attach a glass material to the spheres that will make them look like they are glowing. There are materials in the Glass category of the Autodesk Library that can be used as a base for creating a new material.
   E. Create a fill light above to illuminate the scene. This can be a point light or spotlight.
   F. Render the scene using the Low preset to see the lighting effects.
   G. Adjust the sun properties to create the look that you want.
   H. Adjust the properties of the point lights and any other lights in the scene. You may have to increase the intensity of the lights to properly illuminate the scene.
   I. When the scene is illuminated the way you want it, render the scene using the Medium or High preset. If time allows, render the scene using the Coffee-Break Quality or Lunch Quality preset.
   J. Save the drawing as P18-5.

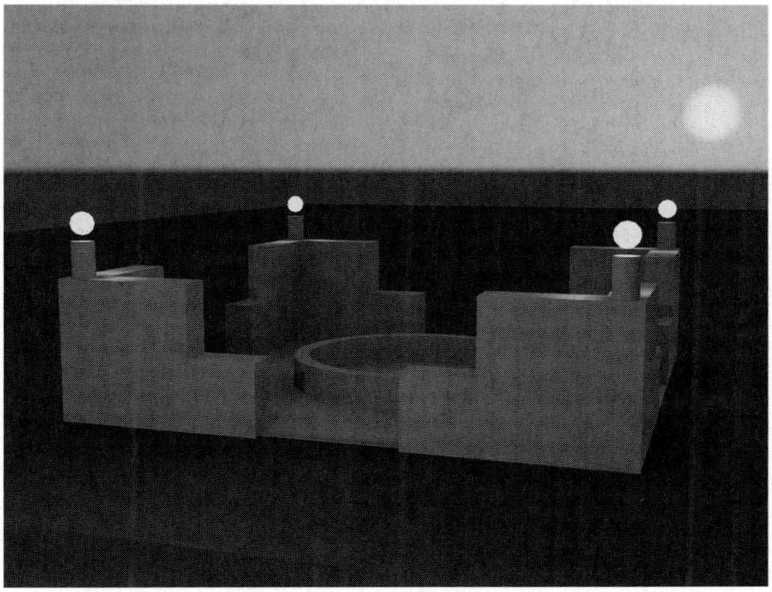

A full-color image is provided on the companion website. www.g-wlearning.com/CAD/

Drawing Problems - Chapter 18

6. **Architectural** In Chapter 8, you completed the kitchen chair model that you started in Chapter 2. In this problem, you will be adding lights to the model and attaching materials to the various components in the chair.
   A. Open P8-8 from Chapter 8.
   B. Create a layer for each of the chair components: seat, legs/crossbars, seatback bow, and seatback spindles.
   C. Draw a planar surface to represent the floor. Place this on its own layer.
   D. Assign materials to each of the layers. You can use materials from the Autodesk Library or create your own materials.
   E. Set the Realistic visual style current.
   F. Adjust material mapping as needed.
   G. Add lighting to the scene.
   H. Render the scene. If you experience problems with the materials, try attaching by object instead of by layer.
   I. Save the drawing as P18-6.

A full-color image is provided on the companion website. www.g-wlearning.com/CAD/

CHAPTER 19

Advanced Rendering

Learning Objectives

After completing this chapter, you will be able to:

- ✓ Make advanced rendering settings.
- ✓ Manage and create render presets.
- ✓ Save a rendering to an image file.
- ✓ Add a render environment to a scene.

In Chapter 16, you learned how to create a view of your scene that is more realistic than a visual style. In that chapter, you used AutoCAD's sunlight feature to create a simple rendering with mostly default settings. In this chapter, you will discover how to make a rendering look truly realistic, or *photorealistic*. Some of the advanced rendering features add significantly to the rendering time and you must learn how to balance the quality of the rendering with an acceptable time frame to get the job done.

Render Window

By default, a drawing is rendered in the **Render** window. The **Render** window allows you to inspect the rendering, save it to a file, compare it with previous renderings, and take note of the statistics. See **Figure 19-1**. There are two main areas of the **Render** window—the image pane and the history and statistics pane.

Image Pane

As AutoCAD processes the scene, the image begins to appear in the image pane in its final form. As discussed in Chapter 16, the rendering is processed over a number of levels. The quality of the image improves as the levels are processed. You can cancel the rendering at any time by pressing [Esc] or picking the **Cancel Rendering** button at the top of the image pane.

Immediately below the image pane is the progress meter. The progress meter displays the progress of the rendering. The current level being processed is shown on the left end of the bar, the percentage of completion is shown in the middle, and the elapsed time is shown on the right end. The rendering is completed when the progress meter reaches 100%.

Figure 19-1.
The **Render** window.

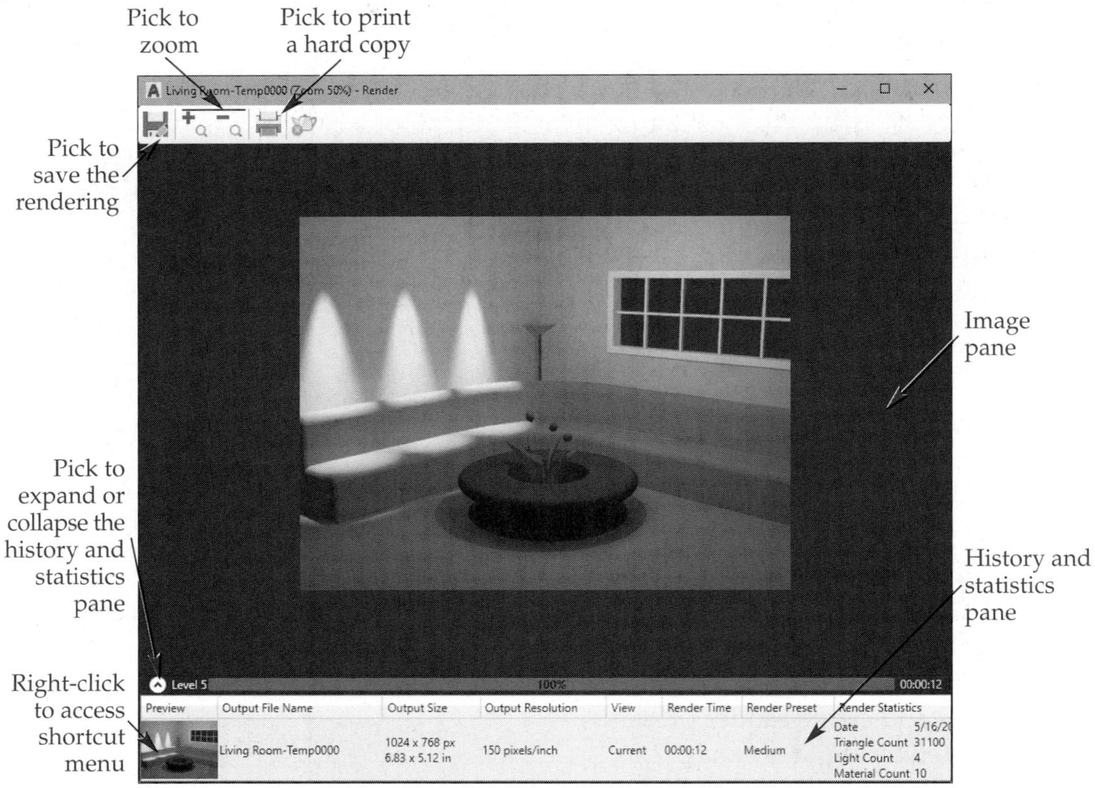

While the rendering is being processed, you can zoom in and out to inspect the image. Use the mouse scroll wheel or the zoom buttons at the top of the window. When the rendering is completed, you can save it to an image file. Pick the save button at the top of the window to display a standard save dialog box. The rendering can be saved to the BMP, TGA, TIF, JPEG, or PNG format. After selecting the file format and picking **Save**, you are prompted to select the quality of the image. The settings are different for each image type.

To print a hard copy of the rendered image, pick the print button at the top of the window. Specify the print device in the **Print** dialog box. This dialog box is similar to the **Plot** dialog box used for printing drawings from model space or paper space.

History and Statistics Pane

The history and statistics pane does not initially appear when the **Render** window is first displayed. To expand or collapse the pane, pick the vertical arrow button shown in **Figure 19-1**. The history and statistics pane contains a list of all of the renderings completed in the drawing in the current drawing session. The items in this list are called ***history entries***. Information and details about each history entry appear under the column headings at the top of the list.

The Preview column displays a thumbnail view of the rendered image. The Output File Name column lists the name of the rendered image. If the rendering has not been saved, the name is displayed in the format FileName-Temp*n* with *n* representing an incremental number. The Output Size column lists the width and height of the image in pixels followed by inches. The Output Resolution column lists the resolution of the image in pixels per inch. The View column lists the name of the view that was rendered. If a named view was not rendered, the listing is Current. The Render Time column lists the amount of time taken to complete the rendering in hours, minutes, and seconds. The Render Preset column lists the name of the render preset used to process the rendering. The Render Statistics column lists the date and time of the rendering and other data, including the triangle count, the number of lights active, and the number of materials used on the objects in the scene.

Selecting one of the history entries redisplays it in the image pane. Right-clicking on a history entry displays a shortcut menu. The options in this menu can be used to save the image, render the image again, and manage the entry. The options in the shortcut menu are:

- **Render Again.** Renders the scene again using the same settings. A new entry is not added to the history pane.
- **Save....** Saves the rendered image to a file using a standard save dialog box. The output file name is changed to the new path and file name.
- **Save Copy....** Saves the rendered image to a new file without changing the current entry.
- **Make Render Settings Current.** Makes all of the rendering settings of the entry the current rendering settings in the drawing. This allows you to render the current scene using the settings of the entry.
- **Remove From the List.** Deletes the entry from the history and statistics pane, but any image files saved from the entry remain.
- **Delete Output File.** Deletes the image file created by saving the entry. The entry remains in the history and statistics pane and any image files that were created as copies are retained.

Exercise 19-1

www.g-wlearning.com/CAD/

Complete the exercise on the companion website.

Render Presets

The quickest and easiest way to control the quality of a rendering is with render presets. AutoCAD provides six standard render presets: Low, Medium, High, Coffee-Break Quality, Lunch Quality, and Overnight Quality.

The Low, Medium, and High presets are based on the number of rendering levels that are processed. See **Figure 19-2**. The higher the number of levels, the better the rendering quality. Higher scene complexity will increase rendering times when using these presets.

- Low. This preset produces the lowest quality, but the quickest rendering time. It applies only 1 rendering level with a raytracing depth of 3 (raytracing is discussed later in this chapter). This is good for a quick view of the rendered scene. Do not expect materials, lights, and shadows to be correctly displayed.
- Medium. This preset will take a little longer than the Low preset but produces better quality. It applies 5 rendering levels with a raytracing depth of 5. Lights, shadows, and materials display nicely, providing a quick, nice-quality rendering.
- High. This preset improves upon the Medium preset. It applies 10 rendering levels with a raytracing depth of 7. This produces the best quality that you can get before going to the timed rendering methods.

The Coffee-Break Quality, Lunch Quality, and Overnight Quality presets are based on the length of time over which rendering occurs. Image quality improves the longer the scene is rendered. Scene complexity does not increase rendering time, but may impact the quality of the rendering.

- Coffee-Break Quality. This preset takes 10 minutes to complete and has a draft rendering accuracy with a raytracing depth of 3. The quality of the rendering may be similar to that produced by the Low preset.

Figure 19-2.
A comparison of renderings made with the first three render presets (Low, Medium, and High) in AutoCAD. A—Low. B—Medium. C—High.

A

B

C

- Lunch Quality. This preset improves upon the Coffee-Break Quality preset. It takes 1 hour to complete and has a high rendering accuracy with a raytracing depth of 5.
- Overnight Quality. This is the highest-quality rendering preset. It takes 12 hours to complete. It applies the highest rendering accuracy with a raytracing depth of 7.

The render presets can be selected in the **Render** panel on the **Visualize** tab of the ribbon or from the drop-down list in the **Render Presets Manager** palette. The **Render Presets Manager** palette also allows you to create and use your own render presets, as discussed in the next section.

PROFESSIONAL TIP

The Low, Medium, and High render presets provide very nice quality renderings in a short period of time. The Coffee-Break Quality, Lunch Quality, and Overnight Quality presets, because of their default settings, sometimes produce worse results than the quicker rendering methods. You will need to experiment with settings on your own computer to determine what works best for your scene.

Render Settings

The **Render Presets Manager** palette is used to control basic render settings. See Figure 19-3. The **RPREF** command opens the palette. The palette can also be displayed by picking the dialog box launcher button at the lower-right corner of the **Render** panel on the **Visualize** tab of the ribbon. The settings in this palette are discussed in the following sections.

Type
RPREF
RENDERPRESETS
RPR

RPREF

Output Settings

The settings at the top of the **Render Presets Manager** palette specify the location where AutoCAD processes the rendering and the output size of the rendered image. The same settings are available in the **Render** panel on the **Visualize** tab of the ribbon. Picking the **Render** button at the top-right corner of the palette renders the scene with the current settings.

There are three settings in the **Render in:** drop-down list. The Window setting renders the scene in the **Render** window. This is the default setting. The Viewport setting renders the scene directly to the viewport. Because the area of the rendering will most likely be larger, expect the rendering to take longer to process. However, it will be easier to see the details of the rendered scene. Zooming or panning will cause the rendering to disappear. You can use the **SAVEIMG** command to save the rendered image to an image file. However, the quality of the result is dependent on the resolution of the monitor, which may produce a low-resolution image. The Region setting also renders the scene directly to the viewport. When using this setting, you are prompted to pick two points defining a rectangular region to render. AutoCAD only processes the objects in this region. This is an excellent way to quickly perform a test rendering and check on lighting, materials, and shadows. As when rendering with

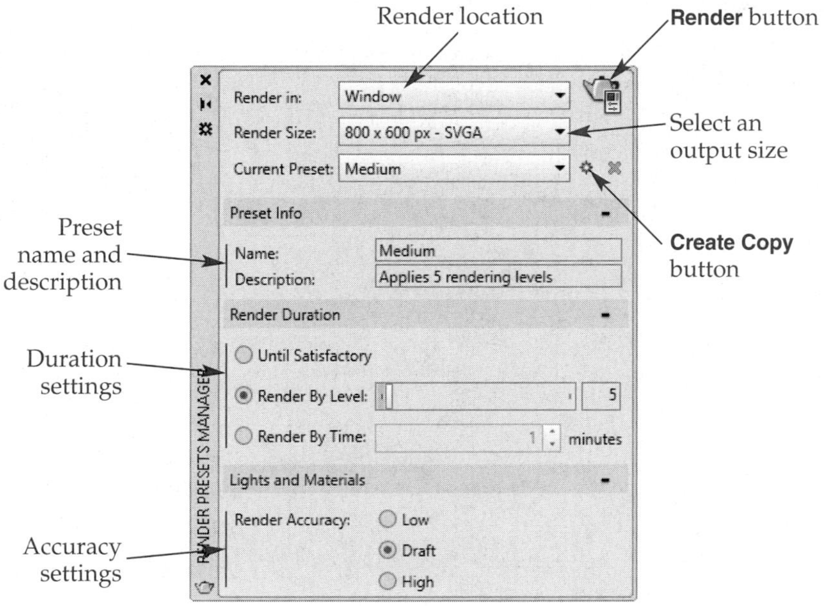

Figure 19-3.
The **Render Presets Manager** palette.

the Viewport setting, the rendering is temporary and will disappear once you zoom or pan in the viewport.

The **Render Size:** drop-down list is available if you are rendering to the **Render** window. The selections in this drop-down list specify the output size of the rendered image in pixels. The seven settings in the top section range from 800 × 600 (SVGA) to 1920 × 1080 (full HDTV). The four settings below the top section are designed for printing the image at 300 dpi (dots per inch) on 8 1/2″ × 11″, 11″ × 17″, ISO A4, and ISO A3 size sheets.

Selecting **More Output Settings...** opens the **Render to Size Output Settings** dialog box, which provides additional settings for file output. See **Figure 19-4**. The **Image Size** drop-down list contains the same output sizes available in the **Render Presets Manager** palette, but you can enter your own width and height using the **Width** and **Height** text boxes. Sizes can be entered in pixels, inches, or centimeters by selecting the appropriate format from the **Units** drop-down list. To maintain the same ratio of width to height when changing the size, check the **Lock Image Aspect** check box. The **Resolution** text box setting determines the image resolution. Higher values produce a larger file size, but improve the quality of the image when printed. The resolution can be set in pixels per inch or centimeter, which can be thought of as dots per inch or centimeter. A good rule of thumb is to set the resolution close to the maximum resolution of the printer that you will be using. The approximate file size resulting from the size and resolution settings is displayed below the **Resolution** text box.

The **Save Image** area is used to save an image file with a specified name and file path. If the **Automatically save rendered image** check box is checked, a file is automatically saved when the rendering is generated. Check the check box and then pick the **Browse...** button to specify a file name and type.

Picking the **Render To Size** button at the bottom of the dialog box renders the scene using the specified output settings. If a custom output size has been defined, it appears in the list of sizes in the **Render Presets Manager** palette and the **Render to Size** drop-down list in the ribbon after the rendering is completed. To save the custom output size and close the **Render to Size Output Settings** dialog box without rendering, pick the **Save** button.

Figure 19-4.
The **Render to Size Output Settings** dialog box.

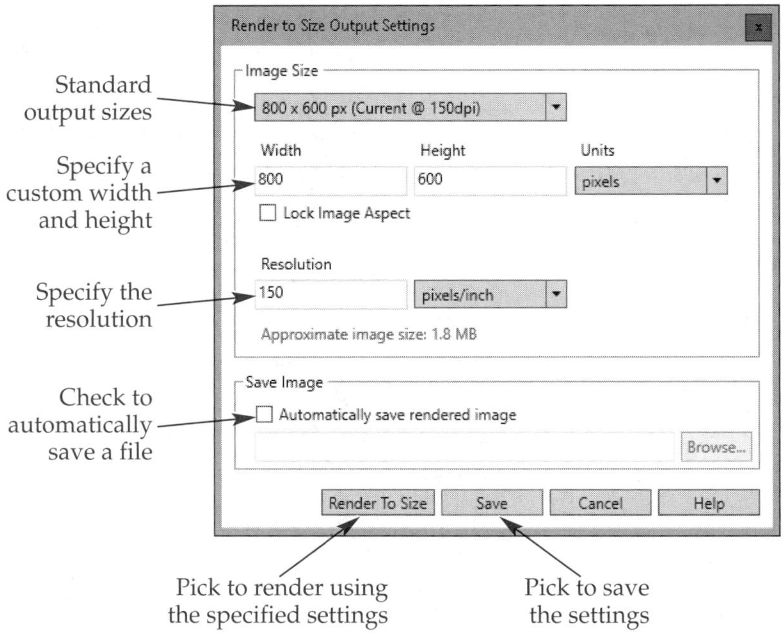

Current Preset and Preset Information

The **Current Preset:** drop-down list in the **Render Presets Manager** palette is used to select one of the six standard render presets or a custom render preset. Picking the **Create Copy** button next to the drop-down list creates a custom preset, as discussed later in this chapter.

The entries in the **Name:** and **Description:** text boxes in the **Preset Info** area correspond to the currently selected preset in the **Current Preset:** drop-down list. These entries are read-only when a standard preset is selected. The **Name:** and **Description:** text box entries can be edited when creating or editing a custom preset.

Render Duration Settings

The settings in the **Render Duration** area allow you to control the duration of the rendering in three different ways. The current setting in this area corresponds to the currently selected preset in the **Current Preset:** drop-down list. If you select a different radio button when a standard preset is current, a custom preset is created. Creating a custom preset is discussed later in this chapter.

The Until Satisfactory setting is used to render without a predefined number of processing levels or set time limit. With this setting, you can simply observe the rendering and then cancel when it reaches a point where the quality is good enough. Picking the **Cancel Rendering** button at the top of the **Render** window ends processing and allows you to save the image to file. The Render By Level: setting is used to set the number of levels that you want AutoCAD to process. Enter a value in the text box or use the slider to set the value. The Render By Time: setting is used to specify the amount of time the rendering takes to complete. Enter the number of minutes in the text box or use the up and down arrows to set the value.

Render Accuracy Settings

The settings in the **Lights and Materials** area are used to specify the quality of the rendering based on a specific lighting model. The current **Render Accuracy:** setting corresponds to the currently selected preset in the **Current Preset:** drop-down list. If you select a different radio button when a standard preset is current, a custom preset is created. Creating a custom preset is discussed in the next section.

The Low setting uses the simplified lighting model. This produces quick results but is not very realistic. Reflections and refractions are limited and global illumination is inactive. Global illumination is discussed later in this chapter.

The Draft setting uses the basic lighting model. Global illumination is active and reflections and refractions are more realistic. Using this setting may be good enough for some situations.

The High setting uses the advanced lighting model. This takes longer to complete, but produces the most realistic results. Global illumination is active and reflections and refractions are calculated at a high level.

Creating a Custom Render Preset

Custom render presets are created in the **Render Presets Manager** palette. The six standard presets can be used as a starting point when creating a custom render preset. Set the appropriate preset current and then pick the **Create Copy** button next to the **Current Preset:** drop-down list. The name of the new preset appears in the **Current Preset:** drop-down list in the format Name-Copy *n* with *n* representing an incremental number. Use the **Name:** and **Description:** text boxes to change the default name and description. Then, adjust the rendering settings as desired. The new custom render preset is available in the **Current Preset:** drop-down list and in the **Render Presets** drop-down list in the **Render** panel on the **Visualize** tab of the ribbon.

Another way to create a custom render preset is to select a standard preset in the **Current Preset:** drop-down list and then change one of the settings in the **Render Duration** or **Lights and Materials** area, as previously discussed. To delete a custom render preset, select the preset in the **Current Preset:** drop-down list and pick the **Delete** button next to the **Create Copy** button. The standard render presets cannot be deleted.

Rendering Fundamentals

Rendering requires significant computer processing resources and much is happening when your computer is processing the scene. As previously discussed, rendering times increase at higher resolution and accuracy settings. The following sections will help you understand the process and troubleshoot problems when they arise.

Sampling

Sampling is a process that tests the scene color at each pixel and then determines what the final color should be. This is most important in transition areas, such as edges of objects or shadows. Increasing the sampling will smooth out the jagged edges and incorrect coloring, but increase rendering time. Increasing the number of samples computed at each pixel means that the color displayed by each pixel will be more accurate. You may also notice thicker lines in the final rendering. Higher render accuracy settings will have higher sampling rates.

Raytracing

Raytracing is a method of calculating reflections, refractions, and shadows by tracing the path of the light rays from the light sources. The render accuracy settings previously described specify the maximum number of times that a ray can be reflected or refracted. See **Figure 19-5**. Try to imagine a light ray bouncing off reflective objects or traveling through transparent objects in your scene. Count how many surfaces the object must contact before it comes to rest. This can help you determine if you will need to increase the render accuracy settings.

Indirect Illumination

Indirect illumination is an AutoCAD mechanism that simulates natural, bounced light. If indirect illumination is not calculated and light does not directly strike an object, the object is black. When rendering without indirect illumination, other lights must be added to the scene to simulate indirect illumination. If you have placed lights in your scene and you render the scene using the Low render preset, you will notice that only the objects directly in the light's path are illuminated. See **Figure 19-6**. This is not how lights work in the real world. Rendering the scene at a higher level of quality will allow AutoCAD to calculate more accurate indirect illumination, resulting in a more realistic scene.

Global Illumination

Global illumination (GI) is an indirect illumination process. Bounced light is simulated by generating photon maps on surfaces in the scene. These maps are created by tracing photons from the light source. Photons bounce around the scene from one object to the next until they finally strike a diffuse surface. When a photon strikes a surface, it is stored in the photon map. This map is calculated for each scene you render. The lights you create and the materials added to the objects in your scene determine the complexity of this map.

A full-color image is provided on the companion website. www.g-wlearning.com/CAD/

Figure 19-5.
A—If the raytracing depth is limited to 3, a light ray will bounce three times and then stop at the next surface. B—If transparent objects exist in the scene, refractions need to be calculated as well as reflections. The light ray will still make it to the back wall as long as the render accuracy settings are high enough. C—Lower render accuracy settings may mean that the light ray will not make it to the back wall.

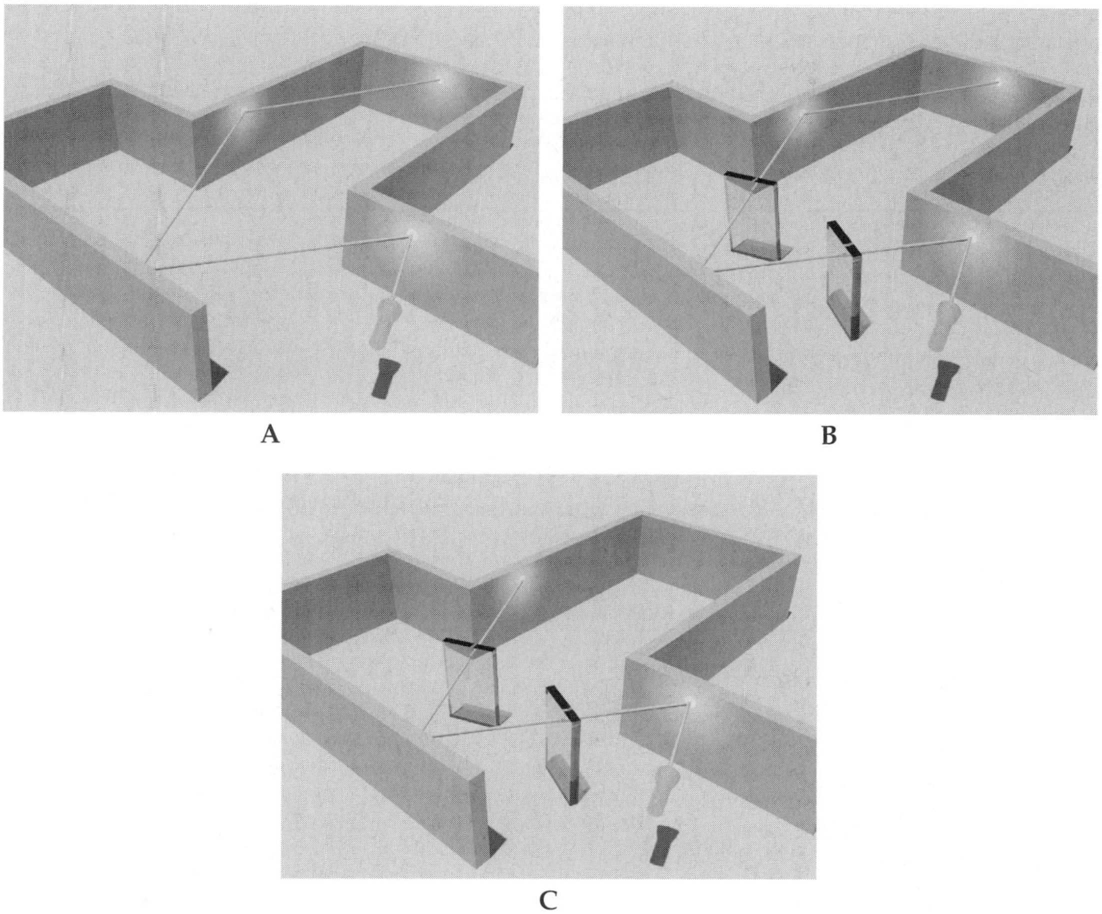

A full-color image is provided on the companion website. www.g-wlearning.com/CAD/

Figure 19-6.
A—This scene has a single point light and a low ceiling above to reflect the light. It has been rendered with the Low render preset. The portion of the floor in the shadow of the cylinder primitive is not in the direct path of light and is not properly illuminated. B—Using a higher render preset activates indirect illumination. Notice that the portion of the floor in the shadow is illuminated. The Medium render preset is used in this example.

Final Gathering

The global illumination process produces a photon map that may have dark and light areas. In other words, it may be splotchy. ***Final gathering*** increases the number of rays in the rendering and cleans up these artifacts. It will also greatly increase rendering time. If you render a scene with one of the higher render presets and watch it all the way through, you may actually see this phase. Final gathering works best with scenes that contain overall diffuse lighting. See **Figure 19-7**.

Exercise 19-2

www.g-wlearning.com/CAD/

Complete the exercise on the companion website.

Render Environment and Exposure

RENDEREXPOSURE

Ribbon
Visualize > Render
Render Environment and Exposure
Type
RENDER-EXPOSURE
RENDER-ENVIRONMENT

The render environment settings allow for the addition of lighting and a background image based on a predefined environment. The **RENDEREXPOSURE** or **RENDERENVIRONMENT** command displays the **Render Environment & Exposure** palette. See **Figure 19-8**. In this palette, you can quickly set up a scene for rendering

A full-color image is provided on the companion website. www.g-wlearning.com/CAD/

Figure 19-7.
A—This scene has a single point light. B—Global illumination is active. Notice the unevenness of the lighting. This can be seen especially on the sofa and in the corner of the walls. C—Final gathering cleans up artifacts and provides a more even illumination.

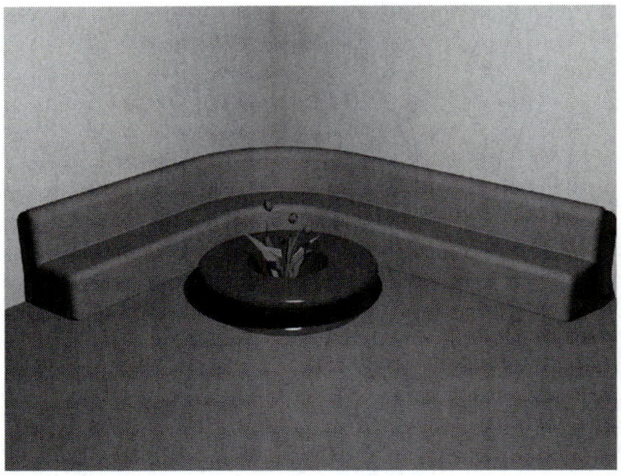

A

B

C

Figure 19-8.
The **Render Environment & Exposure** palette.

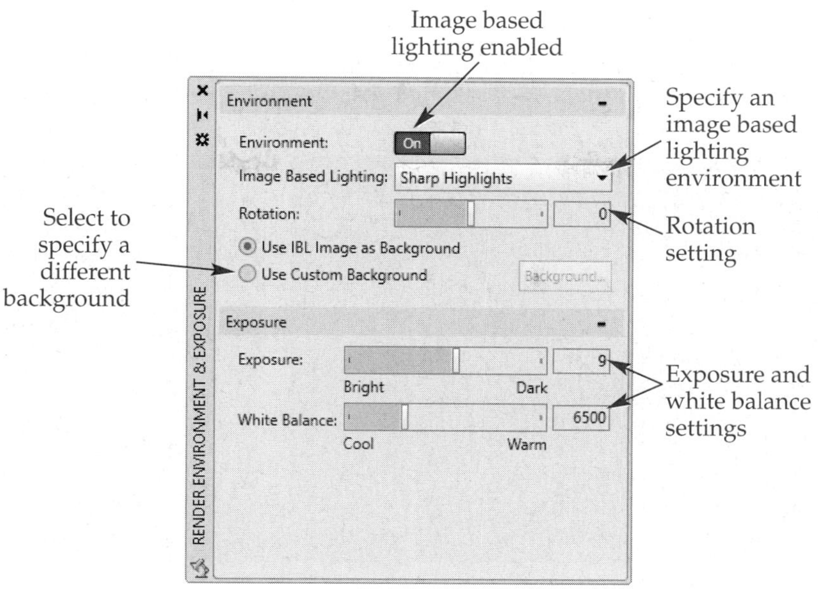

without having to create and position lights or adjust sunlight settings. You can also adjust the exposure settings for lighting in the scene. The **Render Environment & Exposure** palette is divided into two areas: **Environment** and **Exposure**. The settings in these areas are discussed in the following sections.

Environment Settings

To use the render environment settings, the switch at the top of the **Render Environment & Exposure** palette must be set to **On**. When the switch is set to **Off**, AutoCAD uses one of the other types of lighting for the scene (default lighting, sunlight, or user-created lighting, as discussed in Chapter 18).

The settings in the **Environment** area are used to apply image based lighting and a background image to the scene. ***Image based lighting*** simulates a 360° environment around the scene. The lighting used to illuminate the scene is an image lighting map associated with a specific environment, such as an outdoor scene or an interior setting. The environment can be rotated to reposition the light source and change how the scene is illuminated. This provides a way to quickly adjust the lighting to enhance the scene and achieve different results.

The **Image Based Lighting:** drop-down list contains 11 different environments from which to choose. The Gypsum Crater, Dry Lake Bed, Plaza, Snow Field, and Village environments provide a panoramic background image that is applied to the scene when rendered. These environments are designed for exterior scenes and are described below. See **Figure 19-9**.

- Gypsum Crater. In this environment, the ground is covered in grayish rock. There are hills covered with green foliage in the background and a blue sky appears beyond. The lighting is similar to bright sunshine.
- Dry Lake Bed. This is a flat sandy desert environment with bluffs and mountains in the background. The sky is pale blue and the lighting is bright and harsh.
- Plaza. This environment sets the scene in the center of a cobblestone plaza in the late afternoon. Buildings with European architecture surround the plaza. The lighting has a hazy quality to it.
- Snow Field. This environment sets the scene in the middle of a flat snow-covered plane. A road below has been plowed and the sky is mostly overcast with gray wispy clouds in the background. Lighting is even and diffuse.

Figure 19-9.
A comparison of the image based lighting environments for exterior scenes used with an outdoor cooker model. A—Gypsum Crater. B—Dry Lake Bed. C—Plaza. D—Snow Field. E—Village.

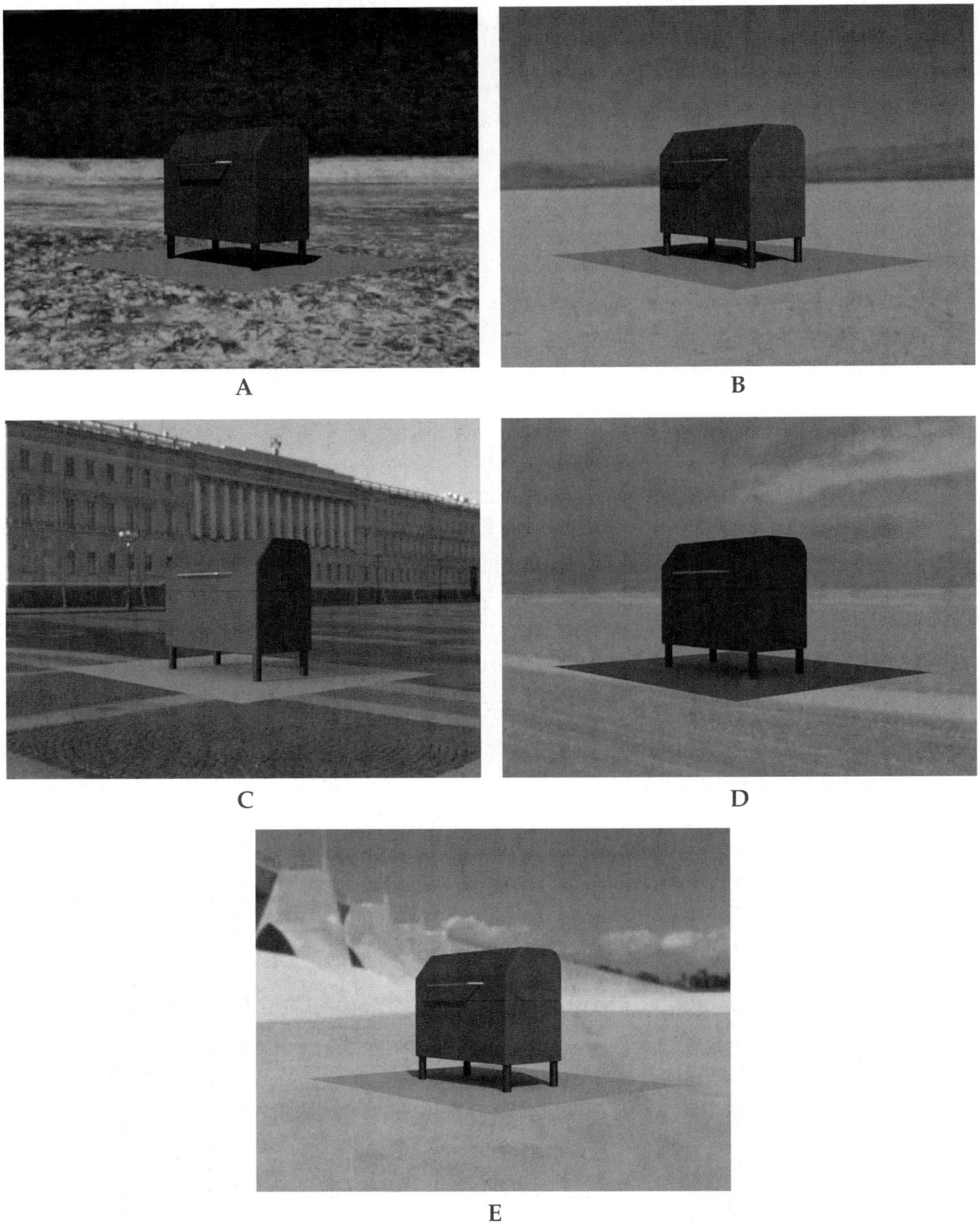

- Village. The name of this environment may be misleading. The scene appears to be at the end of a concrete spillway in bright sunshine. The structures in the background may create an interesting scene.

The Sharp Highlights, Rim Highlights, Grid Lights, Cool Light, Warm Light, and Soft Light environments simulate an indoor setting and have a background that resembles the inside of a room. These environments provide different lighting choices and are very useful for creating presentations of consumer products, mechanical parts, or vehicles. They are described as follows.

- Sharp Highlights. This environment provides a dark and light background with a strong low light from one direction. See **Figure 19-10**.
- Rim Highlights. This environment provides a gray indistinct background with diffuse overhead lighting.
- Grid Lights. In this environment, the scene is positioned inside a room with uniform grid lighting from above.
- Cool Light. This environment is similar to the Grid Lights environment with light panels providing highlights on objects in the scene.
- Warm Light. This environment is the same as the Cool Light environment, but the light is more like incandescent lighting.
- Soft Light. This environment is similar to the Grid Lights environment, but the lighting is dim and diffuse. Very few highlights are produced on objects in the scene.

NOTE

The projection must be set to perspective when using one of the image based lighting environments.

The **Rotation:** setting below the **Image Based Lighting:** drop-down list is used to rotate the environment around your scene. The light source rotates with the background, allowing you to position it to your liking. Use the slider to change the rotation or enter a value in the text box.

A full-color image is provided on the companion website. www.g-wlearning.com/CAD/

Figure 19-10.
A preliminary battery pack design rendered using the Sharp Highlights image based lighting environment.

If the Use IBL Image as Background setting is current, the scene is rendered with the image based lighting background. The Use Custom Background setting allows you to use a background of your choice. You can set the background to a solid, gradient, or image. Pick the **Background...** button to display the **Image Based Lighting Background** dialog box. Select a background type from the **Type:** drop-down list and then make the appropriate settings. The background is displayed in the viewport and is used with the specified image based lighting when the scene is rendered.

Exposure Settings

The Exposure: and White Balance: settings in the **Exposure** area of the **Render Environment & Exposure** palette are used to adjust the brightness and color temperature of lighting in the scene. The same controls are available in the expanded **Lights** panel on the **Visualize** ribbon tab, discussed in Chapter 18. The Exposure: setting is used to brighten or darken the scene and the White Balance: setting is used to make the lighting warmer or cooler. After you specify environment settings and render the scene, you can use the exposure and white balance settings to make fine-tuning adjustments to the lighting. These are global controls that affect the entire scene.

PROFESSIONAL TIP

Renderings can take a long time to process and may require great amounts of computer resources such as memory and powerful graphics hardware. The Autodesk online rendering service is available for generating renderings using online resources. Moving the processing online allows you to save time and free up your computer for other tasks. A user account is required to use the service.

To access the Autodesk online rendering service, pick the **Render in Cloud** button located on the **Render** panel on the **Visualize** tab of the ribbon. Selecting the **Render Gallery** button on the **Render** panel allows you to display your online rendering status and completed renderings.

Exercise 19-3

www.g-wlearning.com/CAD/

Complete the exercise on the companion website.

Chapter Review

Answer the following questions using the information in this chapter.

1. Describe the two panes of the **Render** window.
2. Once a rendering is completed and displayed in the **Render** window, how can it be saved to a file?
3. List the render presets AutoCAD provides.
4. What are the three render output settings available in the **Render in:** drop-down list in the **Render Presets Manager** palette?
5. What dialog box is used to set a custom output size and resolution for the rendering?
6. If you are watching the progress of your rendering in the **Render** window and decide that it looks good enough, how can you cancel processing and save the image?
7. Describe how to create a custom render preset.
8. What is *sampling*?
9. Raytracing calculates shadows, _____, and _____.
10. How does global illumination simulate bounced light?
11. What is the benefit of final gathering?
12. What is *image based lighting*?
13. Name the five image based lighting environments designed for exterior scenes.
14. Name and describe three examples of scenes that could be used with the image based lighting environments designed for interior settings.
15. For what is the Exposure: setting in the **Render Environment & Exposure** palette used?

Drawing Problems

1. **Architectural** Using the drawing from Problem 2 in Chapter 18, you will experiment with different render presets.
   A. Open drawing P18-2 and save it as P19-1. If you did not complete this problem, do so now.
   B. Make sure all of the lights are active and render the scene to the **Render** window using the Medium render preset.
   C. Open the **Render Presets Manager** palette. In the **Lights and Materials** area, change the **Render Accuracy:** setting to Low. Then, render the scene again.
   D. Change the **Render Accuracy:** setting to High and render the scene again. This time, it will probably take longer to render.
   E. Which rendering has the best quality?
   F. Which setting impacted rendering time the most?
   G. Save the last image as P19-1.jpg.
   H. Save the drawing.

2. **General** In this problem, you will experiment with the image based lighting environments designed for indoor scenes.
   A. Start a new drawing and save it as P19-2.
   B. Draw a planar surface that is 50 units × 20 units.
   C. Randomly place various objects (cones, boxes, spheres, etc.) on the plane. Assign a different color or material to each object.
   D. Create a viewpoint that is almost at ground level looking down the length of the plane. Try to get as many objects in the view as possible. Save this as a named view and add a background of some type.
   E. Add a distant light source. Position it and adjust its intensity so that interesting shadows are created in the scene, but the objects are sufficiently illuminated.
   F. Render the scene to see how it looks and adjust the lighting if necessary.
   G. Open the **Materials Browser** and add a different material to each object in the scene. Use different types of materials to see how they are affected by the image based lighting environments.
   H. Render the scene.
   I. Render the scene using each of the Sharp Highlights, Rim Highlights, Grid Lights, Cool Light, Warm Light, and Soft Light image based lighting environments. Determine which one you feel looks the best. Adjust the Exposure: and White Balance: settings in the **Render Environment & Exposure** palette as appropriate.
   J. Render the scene a final time with the render preset of your choice.
   K. Save the image as P19-2.jpg.
   L. Save the drawing.

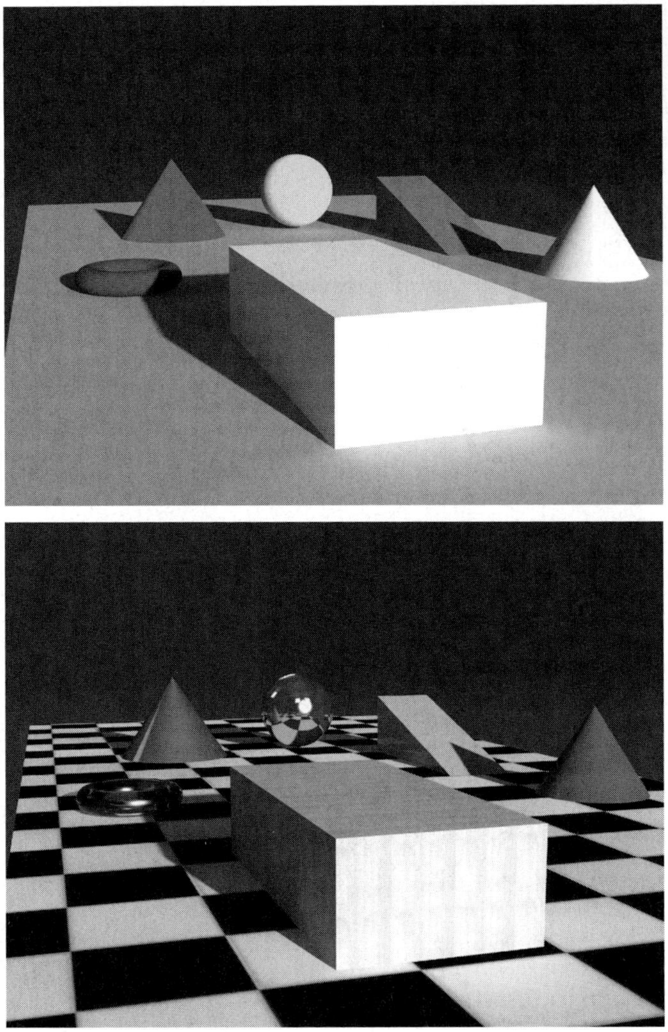

A full-color image is provided on the companion website. www.g-wlearning.com/CAD/

3. **Architectural** In this problem, you will experiment with the image based lighting environments designed for exterior scenes. Open P18-5 created in Chapter 18 and save it as P19-3. If you did not complete this problem, do so now.
   A. Render the scene using the Gypsum Crater, Dry Lake Bed, Plaza, Snow Field, or Village image based lighting environment.
   B. Experiment with different rotation settings until you find one that you like.
   C. Try rendering with the other four exterior image based lighting environments, rotating each one until you decide on the one to use.
   D. Adjust the Exposure: and White Balance: settings in the **Render Environment & Exposure** palette as appropriate.
   E. Render the scene a final time with the render preset of your choice.
   F. Save the image as a file named P19-3.jpg.
   G. Save the drawing.

A full-color image is provided on the companion website. www.g-wlearning.com/CAD/

Drawing Problems - Chapter 19

4. **Architectural** Using the same drawing from Problem 19-3, you will add a custom background to the render environment.
   A. Open drawing P19-3 and save it as P19-4.
   B. Locate an image that you want to use as a background for the scene.
   C. Using the **Render Environment & Exposure** palette, experiment again with the different image based lighting environments. This time, however, pay attention only to the lighting and try to match it to the image you selected.
   D. Select the Use Custom Background setting in the **Render Environment & Exposure** palette and pick the **Background...** button to specify your image as the background.
   E. Once the image is selected, you can adjust it as described in Chapter 18 in the Adding a Background section.
   F. Adjust the Exposure: and White Balance: settings in the **Render Environment & Exposure** palette as appropriate.
   G. Render the scene and save the image to a file named P19-4.jpg.
   H. Save the drawing.

A full-color image is provided on the companion website. www.g-wlearning.com/CAD/

CHAPTER 20

Using Show Motion to View a Model

Learning Objectives

After completing this chapter, you will be able to:

✓ Explain the show motion tool.
✓ Create still shots of 3D models.
✓ Create walk shots of 3D models.
✓ Create cinematic shots of 3D models.
✓ Replay single shots and a sequence of shots.
✓ Change the properties of a shot.

AutoCAD's ***show motion*** tool is a powerful function that allows you to create named views, animated shots, and basic walkthroughs. It can quickly display a variety of named shots. It is also used to create basic animated presentations and displays. This capability is especially useful for animating 3D models that do not require the complexity and detail of fully textured, rendered animations and walkthroughs. Walkthroughs are presented in Chapter 21.

A ***view*** is a single-frame display of a model or drawing from any viewpoint. A ***shot*** is the manner in which the model is put in motion and the way the camera *moves* to that view. Therefore, a single, named view can be modified to create several different shots using camera motion and movement techniques. Using show motion, you can create shots in one of three different formats:

- Still.
- Walk.
- Cinematic.

A ***still shot*** is exactly the same as a named view, but show motion allows you to add transition effects to display it. A ***walk shot*** requires that you use the **Walk** tool to define a camera motion path to create an animated shot. A ***cinematic shot*** is a single view to which you can add camera motion and movement effects to display the view. These shots are all created using the **New View/Shot Properties** dialog box. This is the same dialog box used to create named views.

Understanding Show Motion

Show motion is simply a means for creating, manipulating, and displaying named views. The **NAVSMOTION** command is used for show motion. The process involves the creation of shots using the **ShowMotion** toolbar and the **New View/Shot Properties** dialog box. Once a shot is created, you can give it properties that allow it to be displayed in many different ways. If a saved shot does not display in the manner you desire, it is easily modified.

A ***view category*** is a heading under which different views are filed. It is not necessary to create view categories, especially if you will be making just a few views. On the other hand, if you are working on a complex model and need to create a number of views with a variety of cinematic and motion characteristics, it may be wise to create view categories.

The process of using show motion to create, modify, and display shots begins by first using the **ShowMotion** toolbar. All of your work with shots will be performed using the **New View/Shot Properties** dialog box. Options in this dialog box change based on the type of shot that is selected. These options are discussed in the sections that follow relating to each kind of shot.

PROFESSIONAL TIP

As you first start working with show motion, you may want to create view categories named Still, Cinematic, and Walk. As you create shots, file each under its appropriate category name. This will assist you in seeing how the different shot-creation techniques work. Later, you can apply the view categories to specific components of a complex model. For example, a subassembly of a 3D model may have a view category that contains all three types of shots for use in different types of modeling work or presentations. Creating and using view categories is discussed in detail at the end of this chapter.

Show Motion Toolbar

NAVSMOTION

Type
NAVSMOTION
MOTION
Toolbar
Navigation Bar

ShowMotion

The **ShowMotion** toolbar is displayed at the bottom-center of the screen when the **NAVSMOTION** command is entered. See **Figure 20-1**. The quickest way to display the **ShowMotion** toolbar is to pick the **ShowMotion** button on the navigation bar. The **ShowMotion** toolbar provides controls for creating and manipulating views:

- **Unpin ShowMotion/Pin ShowMotion**
- **Play All**
- **Stop**
- **Turn on Looping**
- **New Shot...**
- **Close ShowMotion**

When the toolbar is pinned, it remains displayed if you execute other commands, minimize the drawing, change ribbon panels, or switch to another software application. The **Unpin ShowMotion** button is used to unpin the toolbar. The **Pin ShowMotion** button is then displayed in its place. If you unpin the toolbar, you must execute the **NAVSMOTION** command each time you wish to use show motion.

Pick the **Play All** button to play all of the shots displayed as thumbnails above the toolbar. The playback of these shots will loop (repeat) if the **Turn on Looping** button is selected. If looping is turned on, a shot, category, or all categories are displayed in a loop whenever the **Play All** button is selected. The **Turn on Looping** button is a toggle. The button image changes to indicate whether looping is turned on.

Figure 20-1.
The **ShowMotion** toolbar is displayed at the bottom of the screen and provides controls for creating and manipulating shots.

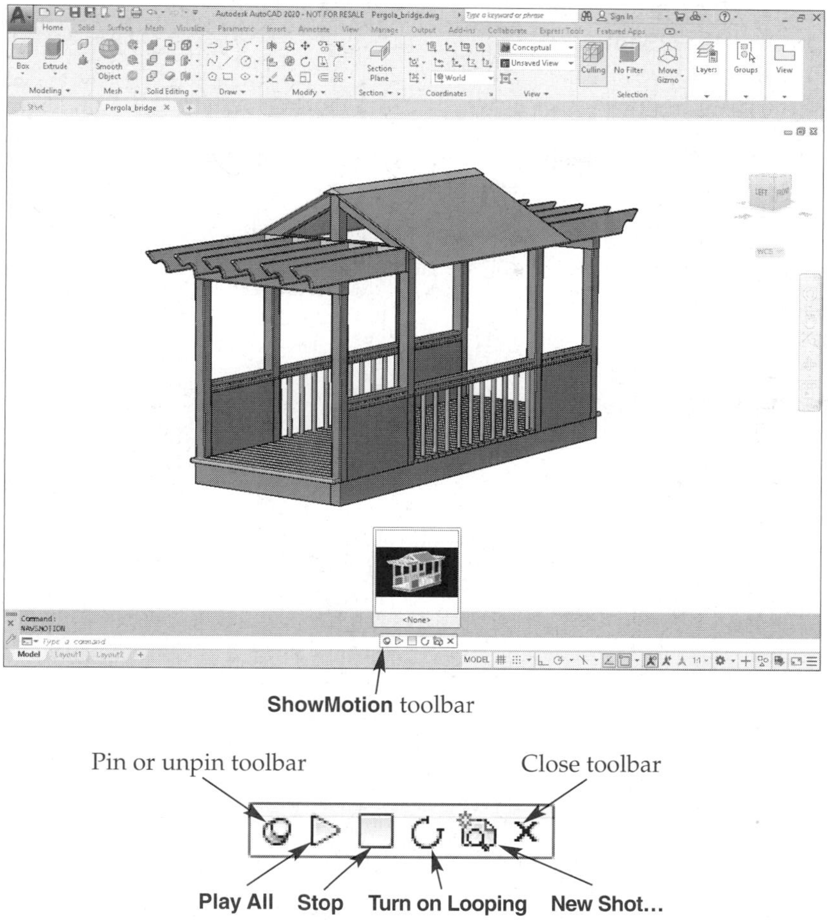

Picking the **New Shot...** button opens the **New View/Shot Properties** dialog box. This dialog box is discussed in the next section. Picking the **Close ShowMotion** button closes the **ShowMotion** toolbar.

New View/Shot Properties Dialog Box

Toolbar
ShowMotion

New Shot...

NEWSHOT

All shot creation takes place inside of the **New View/Shot Properties** dialog box. It is accessed by using the **ShowMotion** toolbar as previously described or by using the **NEWSHOT** command. The dialog box can also be displayed from within the **View Manager** dialog box by picking the **New...** button. The **New View/Shot Properties** dialog box is shown in **Figure 20-2**.

You must supply a shot name and a shot type. A view category is not required, but can help organize shots. This feature is discussed later in the chapter.

The setting in the **Boundary** area determines what is displayed in the shot. The **Current display** radio button is on by default. This means the shot will be composed of what is currently shown in the drawing area. You can adjust the view by picking the **Define window** radio button. This temporarily closes the dialog box so you can draw a rectangular window to define the view. After picking the second corner, you can adjust the view by picking the first and second corners of the window again. Press [Enter] to accept the window and return to the dialog box.

Below the **Boundary** area are two tabs containing options that allow you to define the overall view and then specify the types of movement and motion desired in the shot. These tabs are discussed in the next sections.

Figure 20-2.
The **New View/Shot Properties** dialog box is used to create the three types of shots: still, walk, and cinematic.

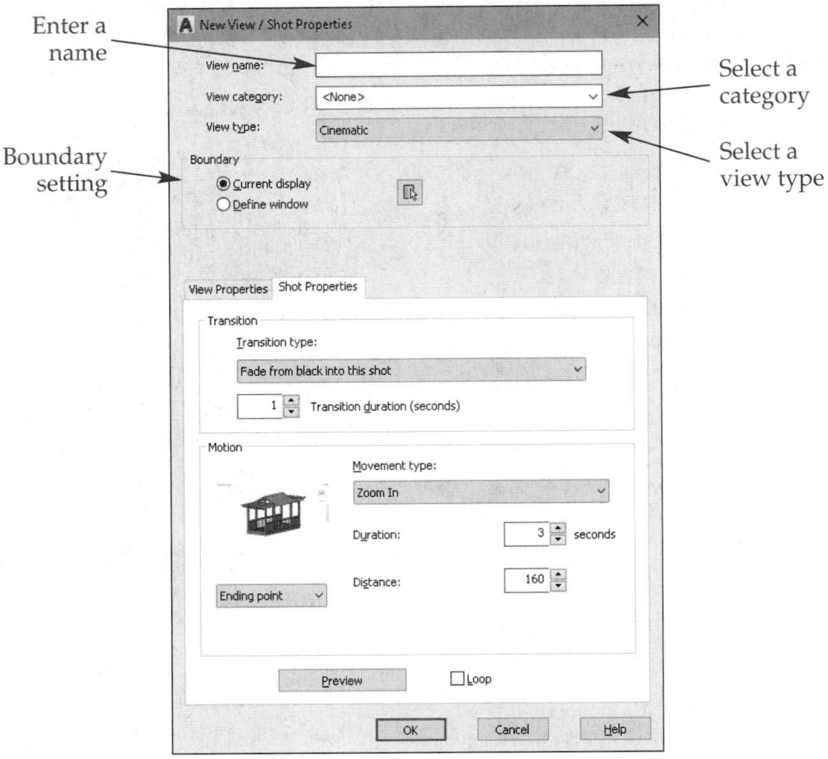

View Properties Tab

The **View Properties** tab of the **New View/Shot Properties** dialog box contains two areas in which you can specify the overall presentation of the model view. See **Figure 20-3**. The overall presentation includes the visual settings and background.

Figure 20-3.
The **View Properties** tab of the **New View/Shot Properties** dialog box.

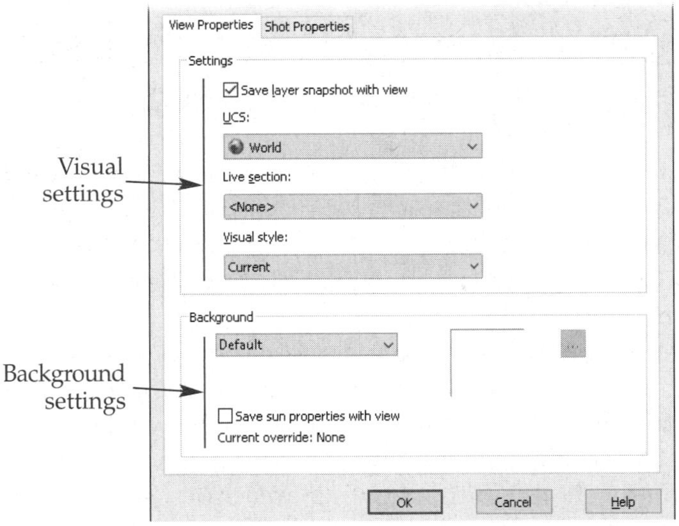

Visual Settings

The options in the **Settings** area of the **View Properties** tab apply to the overall display of the model in the shot. When the **Save layer snapshot with view** check box is checked, all of the current layer visibility settings are saved with the new shot. This is checked by default.

Any UCS currently defined in the drawing can be selected for use with the new shot. Use the **UCS:** drop-down list to select the UCS. When the shot is restored, that UCS is restored, too. If you select <None> in the drop-down list, there is no UCS associated with the shot.

Live sectioning is a tool that allows you to view the internal features of 3D solids that are cut by a section plane object. This feature is covered in detail in Chapter 14. When a section plane object is created, it is given a name. Therefore, if a model contains one or more section plane objects, their names appear in the **Live section:** drop-down list. If a section plane object is selected in this list, the new shot shows the live sectioning for that plane.

The **Visual style:** drop-down list contains all visual styles in the drawing plus the options of Current and <None>. Selecting a visual style from the list results in that style being set current when the shot is played. Selecting Current or <None> will cause the shot to be displayed in the visual style currently displayed on the screen. A detailed discussion of visual styles is provided in Chapter 16.

NOTE

A shot created with live sectioning on will be displayed in that manner even though live sectioning may be currently turned off in the drawing. However, subsequent displays of the model that were created with live sectioning off are shown with it on. Keep this in mind as you develop shots for show motion.

Background Settings

The **Background** area of the **View Properties** tab provides options for changing the background of the new shot. The drop-down list in this area allows you to select a solid, gradient, image, or sun and sky background. You can also choose to retain the default background. If you pick Solid, Gradient, or Image, the **Background** dialog box appears. If you select Sun & Sky, the **Adjust Sun & Sky Background** dialog box appears. These dialog boxes are used to set the background and are discussed in detail in Chapter 18.

The **Save sun properties with view** setting is used to apply the sunlight data to the view. If you are displaying an architectural model using sunlight, you will likely want this check box checked for show motion. Sunlight and geographic location are discussed in detail in Chapter 18.

Shot Properties Tab

Settings in the **Shot Properties** tab of the **New View/Shot Properties** dialog box provide options for controlling the transition and motion of shots. There are numerous movement and motion options in this tab. The specific options available are based on the type of shot selected: still, walk, or cinematic. In addition, the cinematic shot contains a variety of motions that can be applied to the shot and each type of motion contains a number of variables. The options in the **Shot Properties** tab are discussed later in this chapter as they apply to different shots, movements, and motions.

Creating a Still Shot

A still shot is the same as a named view, but with a transition. Open the **New View/Shot Properties** dialog box and enter a name in the **View name:** text box. Next, pick in the **View category:** text box and type a category name or select an existing category using the drop-down list. Remember, it is not necessary to create or select a category at this time. Now, select Still from the **View type:** drop-down list.

NOTE

The Still view type is also used to create named views. A *named view* is a single-frame display of a model from a given viewpoint. Named views can be used for any purpose in AutoCAD, not just for show motion. For example, it is useful to create named views for use in rendering. Named views are saved with the drawing and can be displayed by accessing the drop-down list in the **Named Views** panel on the **Visualize** tab of the ribbon. A named view can be displayed at any time, and can be used for any show motion view.

In the **Shot Properties** tab, select a transition from the **Transition type:** drop-down list. You can select from one of three transitions:

- **Fade from black into this shot.** The screen begins totally black and fades into the current background color.
- **Fade from white into this shot.** The screen begins totally white and fades into the current background color.
- **Cut to shot.** The view is immediately displayed without a transition and the shot movements are applied.

The fade transitions will not function unless hardware acceleration is enabled. This feature is normally activated by default upon software installation. However, you can confirm this by checking the appearance of the **Graphics Performance** tool in the status bar. See **Figure 20-4**. The tool is shaded blue if hardware acceleration is turned on. Hover over the tool to display the tooltip, which should read **Hardware acceleration - On**. If hardware acceleration is not on, right-click on the tool to display the shortcut menu and select **Graphics Performance...** to open the **Graphics Performance** dialog box. Refer to **Figure 20-4B**. In the **Graphics Performance** dialog box, pick the **Hardware Acceleration** button to enable or disable hardware acceleration. Other graphics effects can also be enabled or disabled in this dialog box. See **Figure 20-5**.

After hardware acceleration has been enabled, display the **New View/Shot Properties** dialog box, enter a name, and select a transition. Next, in the **Transition duration (seconds)** text box, enter a length of time over which the transition will occur. If you

Figure 20-4.
Hardware acceleration must be activated for the show motion fade transitions to function.
A—Hover over the **Graphics Performance** tool to confirm that hardware acceleration is turned on.
B—The shortcut menu displayed by right-clicking on the **Graphics Performance** tool.

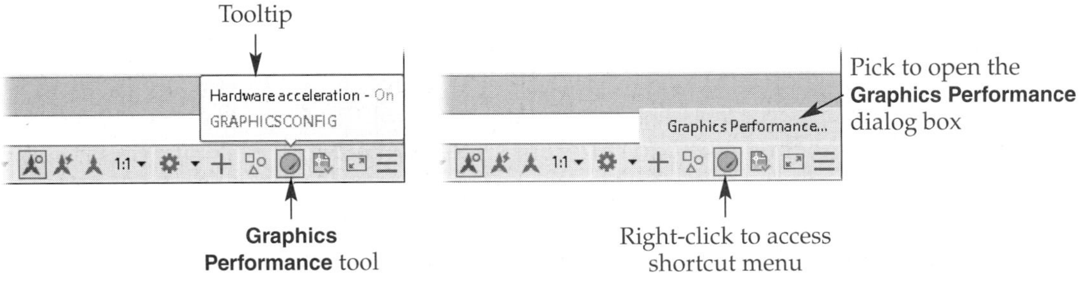

Figure 20-5.
The **Graphics Performance** dialog box.

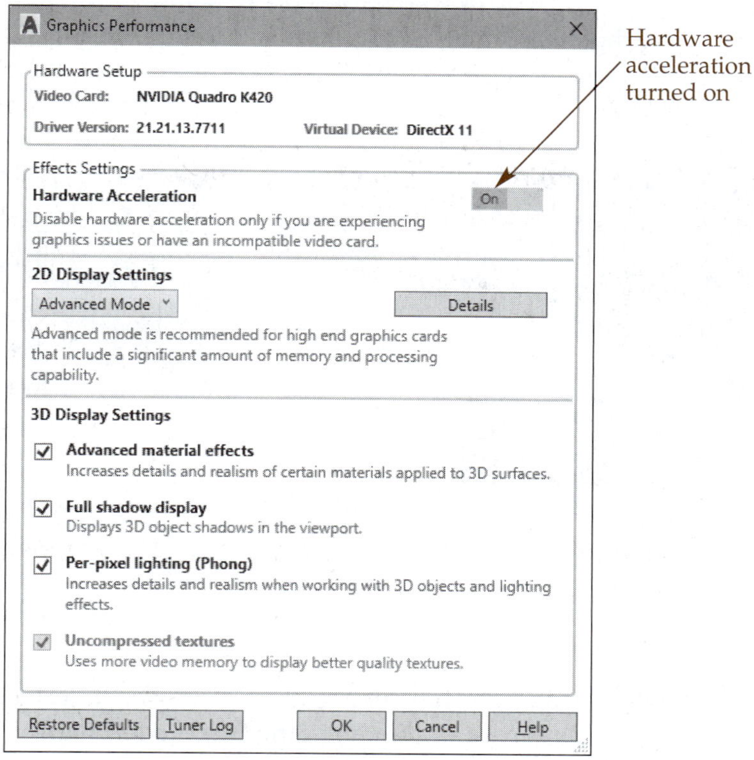

want a fade transition to be complete, be sure to enter a value in the **Duration:** text box that is equal to or greater than the transition duration. Pick the **Preview** button to view the shot, and then edit the transition and motion values as needed.

When finished, pick the **OK** button to close the **New View/Shot Properties** dialog box. The thumbnail image for the new shot is displayed above the **ShowMotion** toolbar. See **Figure 20-6A**. The large thumbnail image represents the view category. Since there was no view category selected for the view, the name <None> is displayed. The small thumbnail image represents the shot just created. Its name may be truncated. Move the cursor into the shot thumbnail image and a large image is displayed. The name of the shot should appear in its entirety. The category thumbnail image is reduced to a small image. See **Figure 20-6B**.

NOTE

If the **Loop** check box at the bottom of the **Shot Properties** tab is checked, the shot will continuously loop through the transition when previewing the playback. This is similar to picking the **Turn on Looping** button in the **ShowMotion** toolbar. All three shot types have this option.

Exercise 20-1

www.g-wlearning.com/CAD/

Complete the exercise on the companion website.

Creating a Walk Shot

To create a walk shot, open the **New View/Shot Properties** dialog box, enter a name, and select a category, if needed. Next, select Recorded Walk from the **View type:** drop-down list. In the **View Properties** tab, set up the view as described for a still shot.

Figure 20-6.
A—The large thumbnail image on the bottom is for the category. The small thumbnail image is for the shot just created. Its name may be truncated. B—Move your cursor into the shot thumbnail image and it converts to a large image. The category thumbnail image is reduced to a small image.

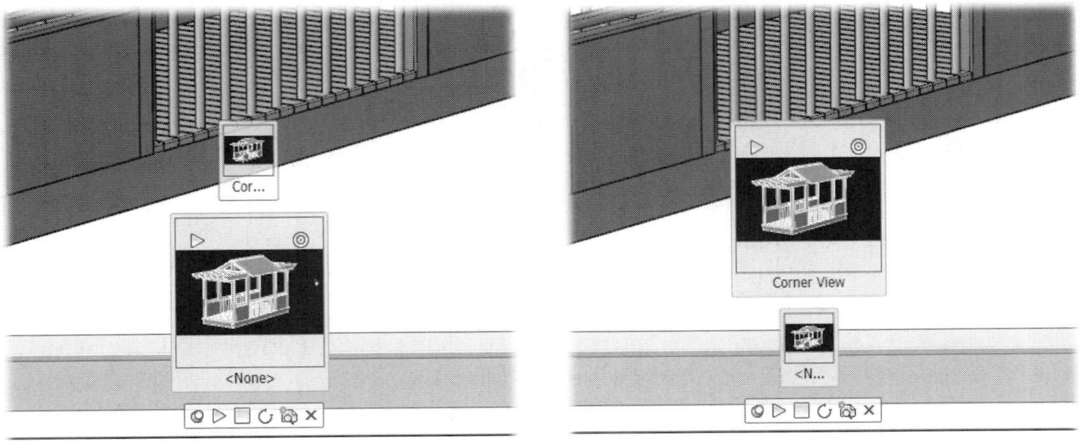

A B

In the **Shot Properties** tab, set up the transition as described for a still shot. The only option in the **Motion** area of the tab is the **Start recording** button. See **Figure 20-7**. The **Duration:** text box is grayed out because the value is based on how long you record the walk. The camera drop-down list below the preview image is also grayed out because the walk begins at the current display and ends at the point where you terminate it.

This type of shot uses the same **Walk** tool found in the steering wheels, discussed in Chapter 3. After picking the **Start recording** button, the **New View/Shot Properties** dialog box is hidden and the **Walk** tool message is displayed. Pick and drag to activate the **Walk** tool. As soon as you pick, the center circle icon is displayed. If needed, you

Figure 20-7.
Creating a walk shot.

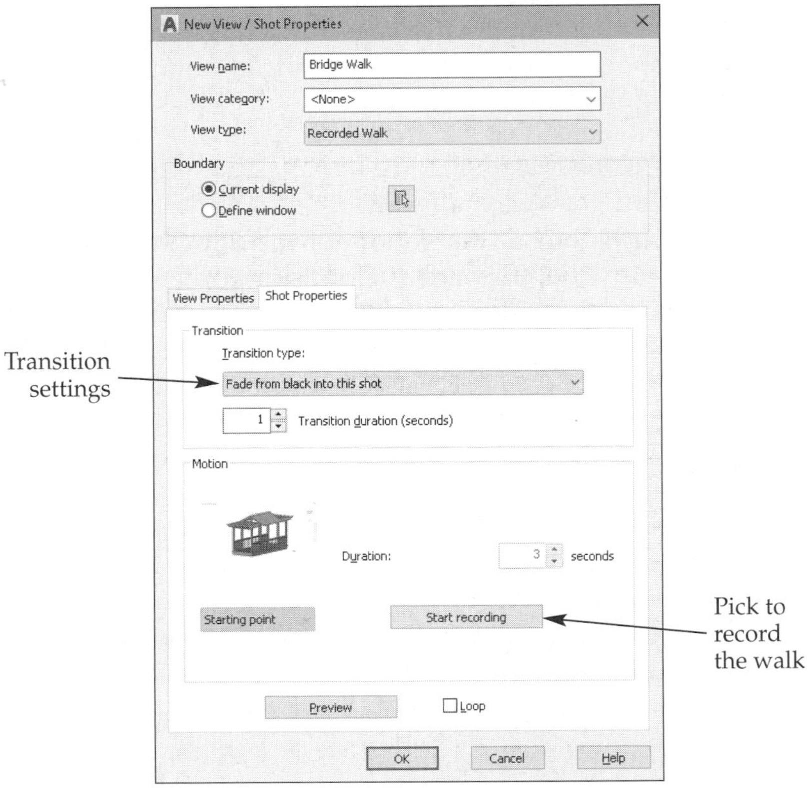

can hold down the [Shift] key to move the view up or down. When the [Shift] key is released (with the mouse button still held down), you can resume walking. The mouse button must be depressed the entire time to record all movements. As soon as you release the button, the recording ends and the dialog box is redisplayed.

Finally, preview the shot. If you need to re-record it, pick the **Start recording** button and begin again. You cannot add to the shot; you must start over. Pick the **OK** button to save the shot and a new thumbnail is displayed in the **ShowMotion** toolbar.

NOTE

The walk shot capability of show motion is limited in its ability to produce a true "walk-through." The path of your "walk" is inexact and it may take several times to create the effect you need. Should you wish to create a genuine walkthrough of an architectural or structural model, it is better to use tools such as **3DWALK** (walkthrough), **3DFLY** (flyby), or **ANIPATH** (motion path animation). These powerful commands are used to create professional walkthroughs and animations and are covered in Chapter 21.

Exercise 20-2

www.g-wlearning.com/CAD/

Complete the exercise on the companion website.

Creating a Cinematic Shot

To create a cinematic shot, open the **New View/Shot Properties** dialog box, enter a name, and select a category, if needed. Next, select Cinematic from the **View type:** drop-down list. In the **View Properties** tab, set up the view as described for a still shot. In the **Shot Properties** tab, pick a transition type and duration as described for a still shot.

The specific options in the **Motion** area of the **Shot Properties** tab are discussed in the next sections, **Figure 20-8**. Pick a type of movement in the **Movement type:** drop-down list. The movement options available will change depending on the type of movement you select. Refer to the chart in **Figure 20-9** as a quick reference for movement options when creating a cinematic shot. Finally, pick the current position of the camera from the camera drop-down list below the preview image.

Pick the **Preview** button to view the shot. Make any changes required before picking the **OK** button to save the shot. Once the **OK** button is picked, the new view is displayed as a small thumbnail image above the large category thumbnail image. See **Figure 20-10**.

Cinematic Basics

The motion and movement options available for a cinematic shot allow you to create a final display that appears to move into position as if the camera is traveling in a path toward the object. The **Motion** area of the **Shot Properties** tab contains a variety of options for creating an array of cinematic shots. Refer to **Figure 20-8**. The options can be confusing unless you understand some basics about essential components of a cinematic shot. The most important aspect is the preview of the current view. All of the motion actions revolve around this view. This image tile is the current position of the camera and is also referred to as the ***key position*** of a shot. This is the position that is displayed when you pick the **Go** button on a thumbnail image above the **ShowMotion** toolbar.

Figure 20-8.
Creating a cinematic shot.

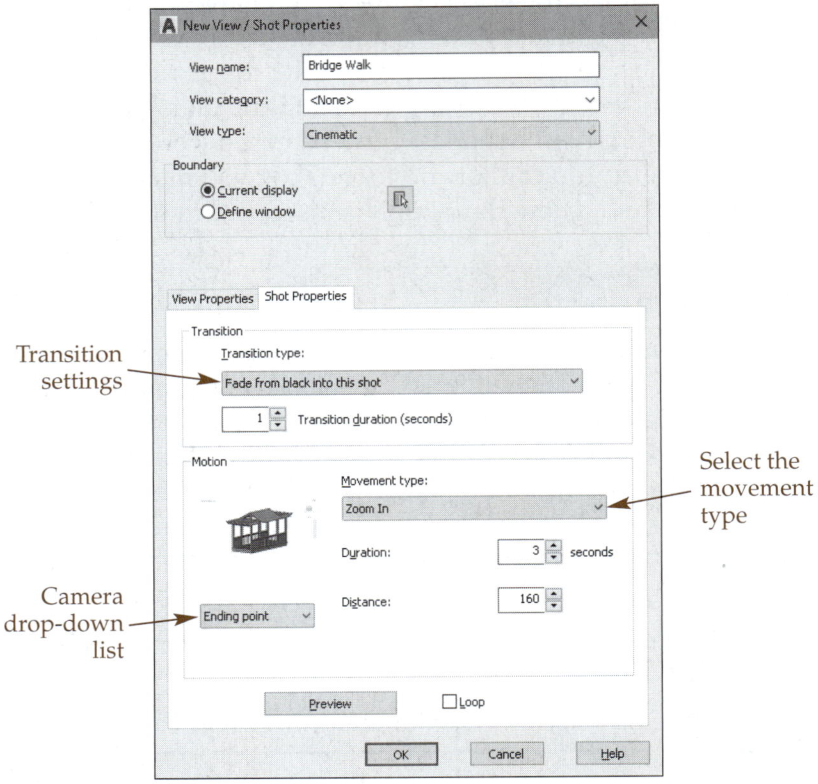

Figure 20-9.
This chart shows the different cinematic movement options based on the movement type selected in the **Movement type:** drop-down list in the **New View/Shot Properties** dialog box.

	Duration	Distance	Look at Camera Point	Distance Up	Distance Back	Distance Down	Distance Forward	Shift Left/Right	Degrees Left/Right	Degrees Up/Down
Zoom In	X	X								
Zoom Out	X	X								
Track Left	X	X	X							
Track Right	X	X	X							
Crane Up	X		X	X	X			X		
Crane Down	X		X			X	X	X		
Look	X								X	X
Orbit	X								X	X

Figure 20-10.
The thumbnail image for the new cinematic shot is displayed above the category thumbnail image.

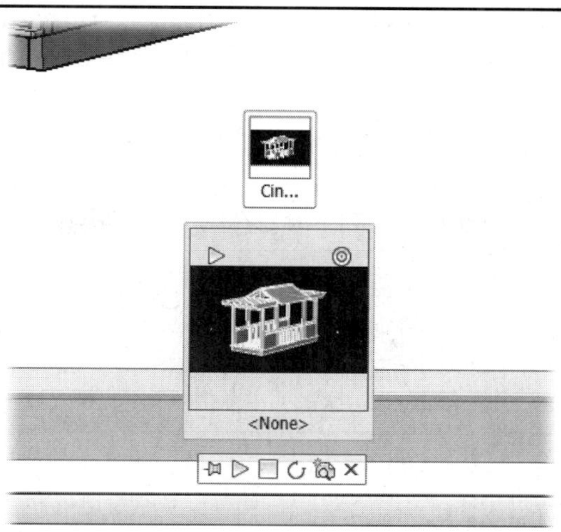

The two elements of a cinematic shot are motion and movement. ***Motion*** relates to the behavior of the object and how it appears to be in motion during the cinematic shot. In addition, it refers to the position of the model at a specified point in the animation. ***Movement*** in a cinematic shot is the manner in which the camera moves in relation to the object.

PROFESSIONAL TIP

If your goal is to create a series of shots that blend together, it is a good idea to first develop a storyboard of the entire sequence. This could be as simple as a few notes indicating how you want the shots to move or even a few sketches noting the required movements and motion values. Planning your shots will save time when you begin creating them in AutoCAD.

Camera Drop-Down List

The camera drop-down list is located below the preview image in the **Motion** area of the **Shot Properties** tab. The preview image represents the position of the camera based on the option selected in the camera drop-down list. The following three options are available in the drop-down list, **Figure 20-11**.

- **Starting point.** The view in the preview image is the display that will be shown at the start of the cinematic shot. All movement options begin at this point.
- **Half-way point.** The view in the preview image is the display that will be shown at the half-way point of the cinematic shot.
- **Ending point.** The view in the preview image is the display that will be shown at the end of the cinematic shot. All movement options take the shot to this point.

Figure 20-11.
Selecting what the key position represents.

Select a camera position

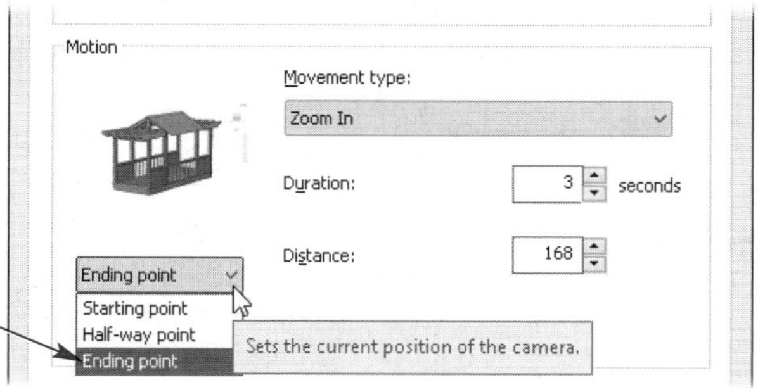

These options, in part, determine how the cinematic shot is created. If Starting point is selected, the cinematic shot begins with the current view. During the animation, the model may move off the screen. If this happens, you may want to select Ending point so the model appears to move into position and stay there.

PROFESSIONAL TIP

The **Preview** button in the **New View/Shot Properties** dialog box is the best way to test any changes you make to the motion options. Each time you make a change, pick the button to see if the effect is what you want. Previewing each change is far more efficient than making several changes before you examine the results.

Movement Type

The **Movement type:** drop-down list in the **Shot Properties** tab of the **New View/ Shot Properties** dialog box is used to select the motion for the cinematic shot. There are eight types of camera movement that can be used with a cinematic shot:

- Zoom in.
- Zoom out.
- Track left.
- Track right.
- Crane up.
- Crane down.
- Look.
- Orbit.

The options available in the **Motion** area of the tab are based on which movement type is selected. The movement types and their options are discussed in the next sections.

Zoom In

When Zoom In is selected in the **Movement type:** drop-down list, the camera appears to zoom into the model in the shot. This movement type has two options, **Figure 20-12**. The value in the **Duration:** text box is the length of time over which the animation is recorded. The value in the **Distance:** text box is the distance the camera travels during the animation. The camera zooms in to cover the distance in the specified duration of time. When Zoom In is selected, the camera drop-down list is automatically set to Ending point. Keep in mind that with this option the current position of the camera represents the final display after the cinematic shot is complete.

Zoom Out

When Zoom out is selected in the **Movement type:** drop-down list, the camera appears to zoom away from the model in the shot. This movement type has **Duration:** and **Distance:** text box options, as described for the Zoom In movement type.

When Zoom out is selected, the camera drop-down list is automatically set to Starting point. With this setting, the camera will zoom out the specified distance and the final image in the shot will be smaller than the preview image. If this is not the effect you want, it may be better to pick Ending point for the current position of the camera. Then, the model will appear to move from behind your view to stop at the current view.

Figure 20-12.
The settings for the Zoom In movement type.

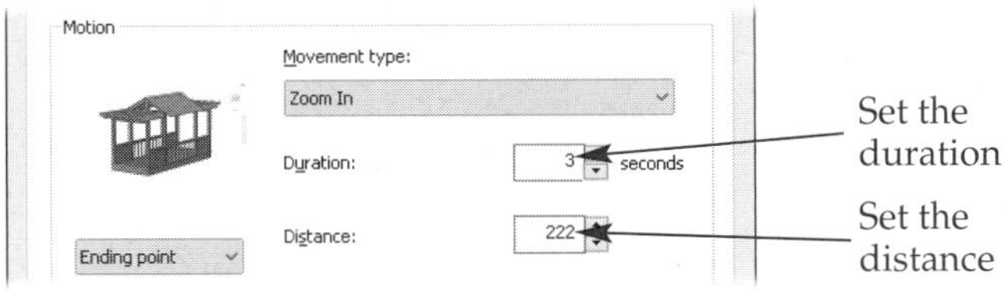

Track Left

When Track left is selected in the **Movement type:** drop-down list, the camera will move from right to left. This results in the view moving from left to right. This movement occurs over the specified distance and duration. The Track left movement type has **Duration:** and **Distance:** text boxes as described previously and the **Always look at camera pivot point** check box, **Figure 20-13**.

If the **Always look at camera pivot point** check box is checked, the center of the view remains stationary. As a result, the view in the shot appears to rotate about this point instead of sliding across the screen. The best way to visualize this motion is to use the **Preview** button with the option checked and then unchecked. This option is available for all "track" and "crane" movement types.

When Track left is selected, the camera drop-down list is automatically set to Half-way point. This means the preview image is the middle point of the animation. It will be displayed at the midpoint of the **Duration:** value.

PROFESSIONAL TIP

If the **Distance:** value is large, the screen may be blank for a few moments until the camera moves enough to bring the model into view. If this is not what you want, decrease the **Distance:** value or check the **Always look at camera pivot point** check box.

Track Right

When Track right is selected in the **Movement type:** drop-down list, the camera will move from left to right. This results in the view moving from right to left. This movement type has **Distance:** and **Duration:** text boxes and the **Always look at camera pivot point** check box. These options function in the same manner as described for the Track left movement type. When Track right is selected, the camera drop-down list is automatically set to Half-way point.

Crane Up

Where the "track" movement types move the view left and right, the "crane" movement types move the view up and down. When Crane up is selected in the **Movement type:** drop-down list, the camera will move from bottom to top and then backward. This results in the view moving from top to bottom and zooming out. This movement type has the **Duration:** text box and the **Always look at camera pivot point** check box described previously. The **Always look at camera pivot point** check box is checked by default.

This movement type has three additional options, **Figure 20-14**. The value in the **Distance Up:** text box is the distance the camera is moved upward. The value in the **Distance Back:** text box is the distance the camera is moved backward. The backward movement is typically short compared to the upward movement.

Figure 20-13.
The settings for the Track left movement type.

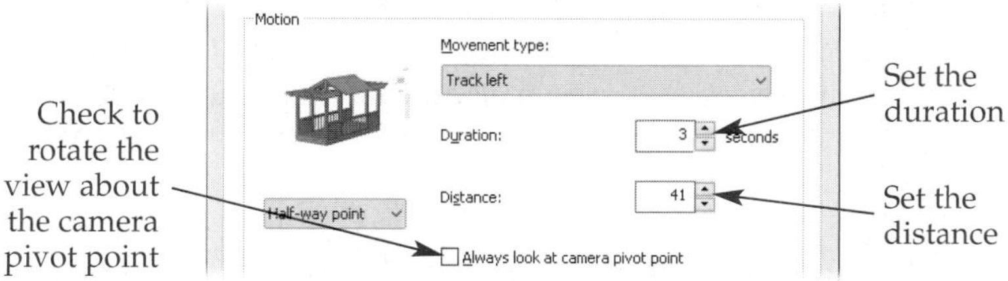

Figure 20-14.
The settings for the Crane up movement type.

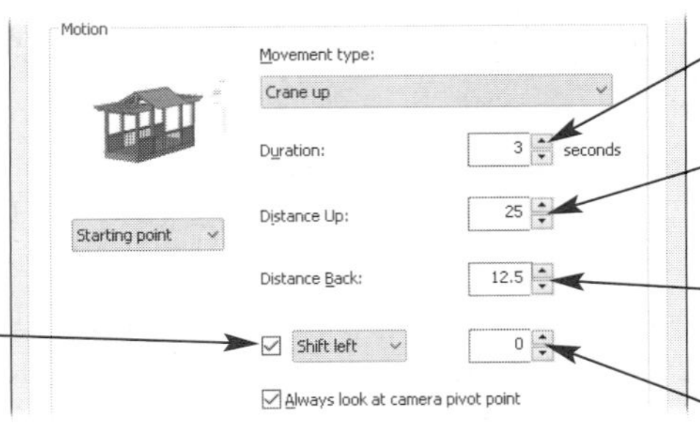

You can add more interest to the motion in the shot by shifting the view left or right. First, enable the shift option by checking the check box below the **Distance Back:** setting. Refer to **Figure 20-14**. Then, select either Shift left or Shift right from the drop-down list. Finally, enter a distance in the text box next to the drop-down list. The camera will be shifted left or right for the distance specified in this text box, resulting in the view shifting in the opposite direction. This shifting option is available for both "crane" motion types.

When Crane up is selected, the camera drop-down list is automatically set to Starting point. This means the preview image shows the beginning of the shot before the movement is applied. If the **Always look at camera pivot point** check box is not checked, and depending on the movement settings, the model may move off the screen in the shot.

Crane Down

When Crane down is selected in the **Movement type:** drop-down list, the camera will move from top to bottom and then forward. This results in the view moving from bottom to top and zooming in. This movement type has the **Duration:** text box and the **Always look at camera pivot point** check box described previously. The **Always look at camera pivot point** check box is checked by default. This movement type also has the left/right shifting option described in the previous section.

Instead of **Distance Up:** and **Distance Back:** settings, this movement type has **Distance Down:** and **Distance Forward:** settings. The value in the **Distance Down:** text box is the distance the camera cranes down in the shot. The value in the **Distance Forward:** text box is the distance the camera is moved forward in the shot.

When Crane down is selected, the camera drop-down list is automatically set to Ending point. This means the preview image shows the end of the shot after the movement is applied. If the **Always look at camera pivot point** check box is not checked, and depending on the movement settings, the model may start off the screen in the shot.

NOTE

All distance values related to the Crane up and Crane down movement types represent how far away from the view the camera must begin before the cinematic shot is started. For example, a large **Distance Back:** value means that the camera may have to begin at a point beyond or even around the view in order to travel the distance back to display the current view (when Ending point is selected in the camera drop-down list). It is always good practice to make a single change to the movements, and then preview the shot. This is especially true for the Crane up and Crane down movement types.

Look

When Look is selected in the **Movement type:** drop-down list, the camera pans based on the values for left/right and up/down to display the view, **Figure 20-15**. For example, if the movement is set to look up 45° and Ending point is selected in the camera drop-down list, then the camera begins at a 45° angle below the view and looks up to display it.

Figure 20-15.
The settings for the Look movement type.

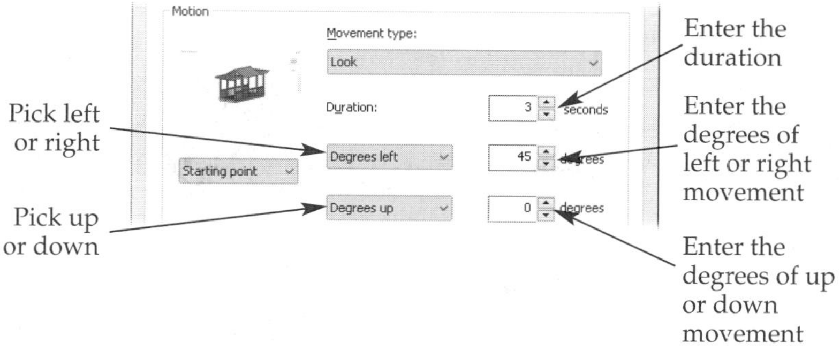

The value in the **Duration:** text box is the length of time over which the animation is recorded. This option is the same as described for the other movement types.

The first drop-down list below the **Duration:** text box is used to specify either left or right movement. Select Degrees left in the drop-down list to have the camera move from right to left, resulting in the view moving from left to right. Select Degrees right to have the camera move from left to right and the view right to left. Next, specify the angular value for this movement in the degrees text box to the right of the drop-down list. For example, suppose you select Degrees left and enter an angle of 15°. In this case, the camera will start 15° to the *right* of the model and rotate to the left in the shot.

The second drop-down list below the **Duration:** text box is used to specify either up or down movement. Select Degrees up in the drop-down list to have the camera move from bottom to top, resulting in the view moving from top to bottom. Select Degrees down to have the camera and view move in the opposite direction. Specify the angular value for this movement in the degrees text box to the right of the drop-down list.

When Look is selected in the **Movement type:** drop-down list, the camera drop-down list is automatically set to Starting point. This means the shot will start with the current view and then apply the movement settings.

Orbit

When Orbit is selected in the **Movement type:** drop-down list, the camera rotates in place based on the values set for left/right and up/down movements. The Orbit movement type has the same options as the Look movement type, as described in the previous section.

When Orbit is selected in the **Movement type:** drop-down list, the camera drop-down list is automatically set to Starting point. This means the shot will start with the current view and then apply the movement settings. Unlike the Look movement type, the view center remains stationary in the shot. As a result, you do not need to worry about the model moving off the screen.

Exercise 20-3 www.g-wlearning.com/CAD/

Complete the exercise on the companion website.

Displaying or Replaying a Shot

As each shot is created, its thumbnail image is placed above the **ShowMotion** toolbar. The shot name is displayed below the thumbnail image. By default, the category thumbnail image is large and each shot thumbnail image is small. If the cursor is moved over one of the shot thumbnail images, all of the shot thumbnail images are enlarged and the category thumbnail image is reduced in size.

Each thumbnail image is composed of an image of the view in the shot, the shot name, and viewing controls. See **Figure 20-16**. The viewing controls are only

Figure 20-16.
Each thumbnail image is composed of the shot image, the shot name, and the viewing controls.

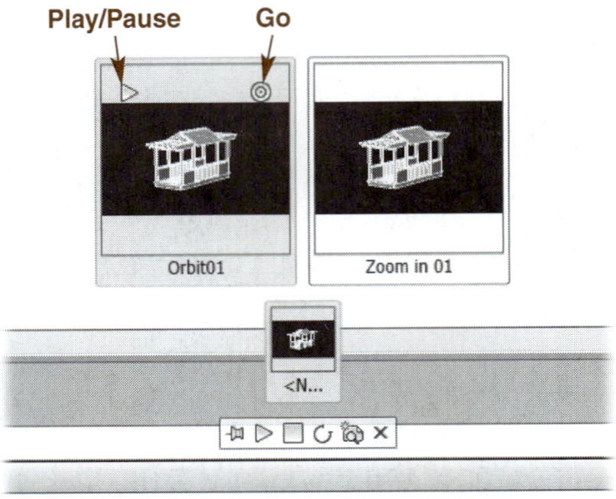

displayed when the cursor is moved over the thumbnail image. The three viewing controls are:

- **Play.** Plays the shot. If looping is enabled, the shot repeats. Otherwise, the shot is played once. When this button is picked, it changes to **Pause**. You can also play the shot by picking anywhere on its thumbnail image. The **VIEWPLAY** command can be used to replay a shot.
- **Pause.** Pauses the shot. When this button is picked, it changes to **Play**. Picking anywhere inside the image or on the **Play** button restarts playing of the shot.
- **Go.** Displays the key position of the view without playing the shot. The **VIEWGO** command also restores a named view.

VIEWPLAY
Type
VIEWPLAY
VPLAY

VIEWGO
Type
VIEWGO
VGO

You can play all of the shots in sequence by picking the **Play** button in the view category thumbnail image or by picking anywhere inside of the view category thumbnail image. Additionally, you can move to the key position of the first shot in a view category by picking the **Go** button on the view category thumbnail image.

Exercise 20-4

www.g-wlearning.com/CAD/

Complete the exercise on the companion website.

Thumbnail Shortcut Menu

Right-clicking on a shot or view category thumbnail image displays the shortcut menu shown in Figure 20-17. This menu provides quick access for modifying and manipulating shots and their thumbnail images.

Picking the **New View/Shot...** entry displays the **New View/Shot Properties** dialog box. Picking the **Properties...** entry initiates the **EDITSHOT** command, which allows you to edit a shot using the **View/Shot Properties** dialog box. This dialog box is the same as the **New View/Shot Properties** dialog box. However, all of the settings of the shot are displayed and can be changed. This option is only available to change the properties of a shot, not a category.

NOTE

If the **View/Shot Properties** dialog box does not open after selecting the **Properties...** shortcut menu option, check the shot name entry at the Enter view name to edit: prompt on the command line. Press [Backspace] once to remove an extra space or retype the shot name and then press [Enter].

Figure 20-17.
Right-click on a shot or view category thumbnail image in the **ShowMotion** toolbar to display this shortcut menu.

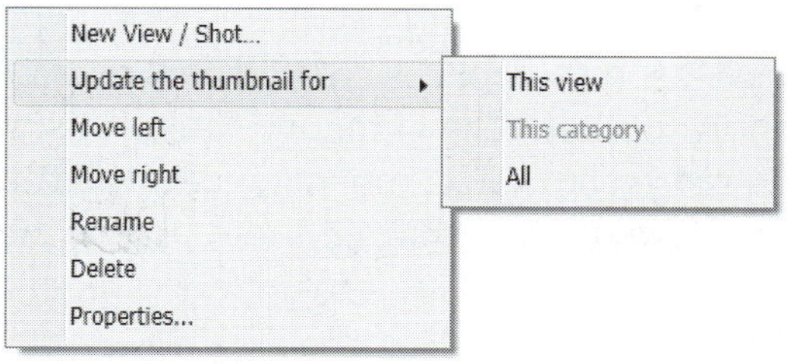

To rename a shot, pick the **Rename** entry in the shortcut menu. The name is highlighted below the thumbnail image. Type the new name and press [Enter]. To delete a shot, select **Delete** from the shortcut menu. There is no warning; the shot is simply deleted. The **UNDO** command can reverse this action.

If there is more than one shot in a category, you can rearrange the order. Right-click on a thumbnail image and pick either **Move left** or **Move right** in the shortcut menu. This moves the shot one step in the selected direction.

When a change is made to the model, it is not automatically reflected in the thumbnail images. If you do not update thumbnail images, they will remain in their original format, regardless of how many changes are made to the model. Selecting **Update the thumbnail for** in the shortcut menu displays a cascading menu with options for updating the thumbnail image:

- **This view.** Updates only the thumbnail image for the shot on which you right-clicked.
- **This category.** Updates all thumbnail images in the category. This option is only enabled when you right-click on a category thumbnail image.
- **All.** This option updates all thumbnail images in all categories and shots.

Creating and Using View Categories

A *view category* is a grouping that can be created in order to separate different types of shots. A view category can also contain shots arranged in a sequence so that they appear connected as they are played together. This is an optional feature, but an efficient method of separating different types of shots or grouping shots to use for a specific purpose. When a new category is created, it is represented by a category thumbnail image displayed above the **ShowMotion** toolbar.

To create a new view category, pick in the **View category:** drop-down list text box in the **New View/Shot Properties** dialog box. Then, enter a name. See Figure 20-18. The name is added to the drop-down list. If you wish to organize shots by categories, be sure to select the view category from the drop-down list before picking the **OK** button to exit the dialog box and create the shot.

You must create a shot to create a view category. If you delete the only shot in a view category, you also delete the category. The view category will no longer be available in the **New View/Shot Properties** dialog box.

Figure 20-18.
To create a new view category, type its name in the **View category:** drop-down list text box.

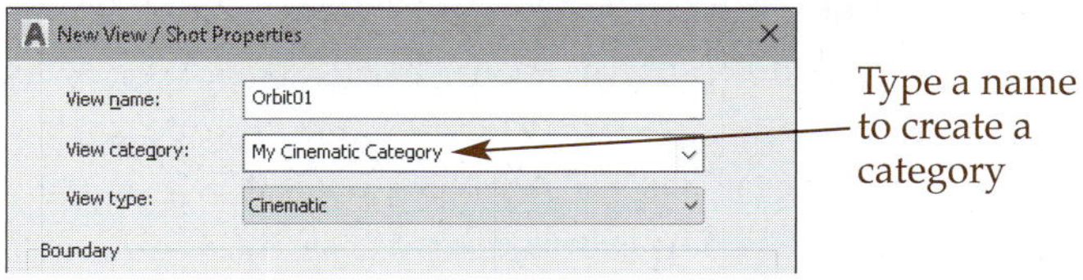

View Category Basics

After a new view category is created, it is represented by a thumbnail image above the **ShowMotion** toolbar. The category name is shown below the thumbnail image. As you create more shots within a category, the thumbnail images for the new shots are placed to the right of existing shots above the category thumbnail image. The thumbnail image for the first shot in a view category is displayed as the view category thumbnail image.

An entire view category and all of the views included in it can be quickly deleted. Simply right-click on the view category thumbnail image and pick **Delete** in the shortcut menu. An alert box appears asking you to confirm the deletion. The shots in a deleted view category cannot be recovered, so be sure there are no shots in the category you wish to save before deleting. However, you can use the **UNDO** command to reverse the deletion.

PROFESSIONAL TIP

Remember, you can change the order of the shots in a view category using the shortcut menu. Picking **Move left** or **Move right** moves the shot one step. Continue moving shots until the order is appropriate.

Playing and Looping Shots in a View Category

A view category is a useful tool for grouping shots to be played together in a sequence. To play the shots in a view category, move the cursor over the view category thumbnail image above the **ShowMotion** toolbar. Then, pick the **Play** button in the thumbnail image. All shots in the category will be played. You can also use the **SEQUENCEPLAY** command to play the shots in a view category.

SEQUENCEPLAY

Type
SEQUENCEPLAY
SPLAY

If you want the shots in a category to run on a continuous loop, pick the **Turn on Looping** button in the **ShowMotion** toolbar. Then, when the **Play** button is picked, the shots in the view category will display until the **Pause** button is picked either on the **ShowMotion** toolbar or in the view category thumbnail image. Pressing the [Esc] key also stops playback. To return to single-play mode, pick the **Turn off Looping** button on the **ShowMotion** toolbar. This button replaces the **Turn on Looping** button.

Changing a Shot's View Category and Properties

If you put a shot in the wrong view category, it is simple to move the shot to a different category. Right-click on the shot thumbnail image and select **Properties...** in the shortcut menu. Select the proper category name in the **View category:** drop-down list in the **View/Shot Properties** dialog box and pick the **OK** button. The shot thumbnail image will move into position above its new view category thumbnail image.

To change shot properties, right-click on the shot thumbnail image and pick **Properties...** in the shortcut menu. All properties of the shot can be changed in the **View/Shot Properties** dialog box. Change the settings as needed. Always preview the shot before you pick the **OK** button to save it.

PROFESSIONAL TIP

Try this procedure for creating a series of shots in a view category that are to be played in sequence.

1. Play each shot to determine where it should be located in the sequence.
2. Move shots left or right in the category as needed to create the proper sequence.
3. Play the category to determine if the shots properly transition from one to the next.
4. If a subsequent shot does not begin where the previous shot ended, right-click on that shot, pick **Properties...**, and make the necessary adjustments in the **View/Shot Properties** dialog box.
5. Play the category again.
6. Repeat the editing process with each shot until the category plays smoothly.

Exercise 20-5

www.g-wlearning.com/CAD/

Complete the exercise on the companion website.

Chapter Review

Answer the following questions using the information in this chapter.

1. For what is the *show motion tool* used?
2. Define *view* as it relates to the show motion tool.
3. Define *shot* as it relates to the show motion tool.
4. List the formats in which a shot can be created.
5. List the six buttons on the **ShowMotion** toolbar.
6. What is *live sectioning* and how can it be included in a shot?
7. Which type of shot is the same as a named view, but with a transition?
8. Which type of shot requires you to navigate through the view as you record the motion?
9. What is the *key position* of a cinematic shot?
10. Define *motion* and *movement* as they relate to a cinematic shot.
11. What is the purpose of the camera drop-down list in the **Shot Properties** tab of the **New View/Shot Properties** or **View/Shot Properties** dialog box?
12. List the eight types of camera movement for a cinematic shot.
13. Briefly describe two ways to play a single shot.
14. What is a *view category*?
15. How is an entire category of shots replayed?

Drawing Problems

1. **General** Open one of your 3D drawings from Chapter 18 or 19. Do the following.
   A. Display a pictorial view of the drawing.
   B. Create a still shot. Use settings of your choice. Create a view category and place the shot in it.
   C. Display a different view of the drawing and create another still shot. Place the shot in the view category you created.
   D. Display a third view of the drawing and create a third still shot. Place the shot in the view category you created.
   E. Play each shot. If necessary, rearrange the shots. Then, play all shots in the category.
   F. Save the drawing as P20-1.

2. **General** Open one of your 3D drawings from Chapter 7 or 8. Do the following.
   A. Display the objects in the Conceptual or Realistic visual style.
   B. Toggle the projection from parallel to perspective.
   C. Create two still shots and four cinematic shots of the model. Use a different motion type with each of the cinematic shots.
   D. Create two new view categories and place one still and two cinematic shots in each view category.
   E. Edit the shots in each view category to create a smooth motion sequence.
   F. Save the drawing as P20-2.

3. **General** Open one of your 3D drawings from Chapter 6. Do the following.
   A. Display a pictorial view.
   B. Create a walk shot. Place it in a view category named Walk Shots.
   C. Play the shot. How does this compare to a cinematic shot as far as ease of creation?
   D. Save the drawing as P20-3.

4. **General** Open drawing P20-3 and do the following.
   A. Create a new cinematic shot using the Orbit motion type. Place it in a view category named Cinematic Shots.
   B. Create another cinematic shot using the Track left or Track right movement type. Check the **Always look at camera pivot point** check box. Place the view in the Cinematic Shots category.
   C. Create a third cinematic shot using the Crane up or Crane down movement type. Make sure the **Always look at camera pivot point** check box is checked. Place the view in the Cinematic Shots category.
   D. Play the Cinematic Shots category. Edit the shots as needed to create a smooth display.
   E. Create three still shots. Place them in a view category named Still Shots. Try creating three different gradient backgrounds to simulate dawn, noon, and dusk.
   F. Play the Still Shots category. Edit the shots as needed to create a smooth display.
   G. Save the drawing as P20-4.

CHAPTER 21

Cameras, Walkthroughs, and Flybys

Learning Objectives

After completing this chapter, you will be able to:

- ✓ Create a camera to define a static 3D view.
- ✓ Activate and adjust front and back clipping planes.
- ✓ Record a walkthrough of a 3D model to a movie file.
- ✓ Record a flyby of a 3D model to a movie file.
- ✓ Create walkthroughs and flybys by following a path.
- ✓ Control the viewpoint, speed, and quality of the animation.

Once you have a 3D design complete, or even while still in the conceptual phase of design, you may want to take a stroll through the model and have a look around. You may also want to strap on some wings and fly over and around the model to see it from above. A ***walkthrough animation*** shows a scene as a person would view it walking through the scene. Walkthroughs are typically used to show the interior of a building, but can be created for exterior scenes as well. A ***flyby animation*** is similar to a walkthrough, except that the person is not bound by gravity. In other words, the scene is viewed as a bird flying above would see it. Flybys often show the exterior of a building.

The **3DWALK** command is used to create a walkthrough by recording views as a camera "walks" through the scene. The **3DFLY** command is very similar, but the movement of the camera is not limited to a single Z value. The **ANIPATH** command allows you to draw a path and link the camera to the path. This chapter discusses these commands and other methods used to create the animation you need. In addition, this chapter discusses creating and using cameras.

Creating Cameras

Cameras are used in AutoCAD to store a viewpoint and easily recall it later when needed for viewing or rendering the scene. After the camera is established, you can zoom, pan, and orbit as needed and then come back to the camera view. It is not necessary to create a camera before using the **3DWALK**, **3DFLY**, and **ANIPATH** commands (discussed later in this chapter) because these commands create their own cameras.

The **CAMERA** command allows you to add a camera to the scene. Cameras are normally placed in the plan view of the scene to make it easy for you to pick where you want to "stand" and where you want to "look." Once the command is selected, you are first prompted to specify the camera location. A camera glyph is placed in the

CAMERA

Ribbon
Visualize > Camera
Create Camera
Type
CAMERA
CAM

scene at the camera location, **Figure 21-1**. Next, you must specify the target location. As you move the cursor before picking the target location, a pyramid-shaped field of view indicates what will be seen in the view. Once you select the target location, the command remains active for you to select an option:

Enter an option [?/Name/LOcation/Height/Target/LEns/Clipping/View/eXit]<eXit>:

The list, or ?, option allows you to list the cameras in the drawing. Select this option and type an asterisk (*) to show all of the cameras in the drawing. You can also enter a name or part of a name and an asterisk. For example, entering HOUSE* will list all of the cameras whose name begins with HOUSE, such as HOUSE_SW, HOUSE_SE, and HOUSE_PLAN.

The **Name** option allows you to change the name of the camera as you create it. If you do not rename the camera, it is given a default, sequential name, such as Camera1, Camera2, Camera3, and so on. It is always a good idea to provide meaningful names for cameras. Names such as Living Room_SW, Corner, or Hallway_Looking East leave no doubt as to what the camera shows. If you choose not to rename the camera at this point, it can be renamed later using the **Properties** palette.

The **Location** option allows you to change the placement of the camera. Enter the option and then specify the new location. You can enter coordinates or pick a location in the drawing.

The **Height** option allows you to change the vertical location of the camera. Enter the option and then enter the height of the camera. The value you enter is the number of units from the current XY plane. If you are placing the camera in a plan view, this option is used to tilt the view up or down from the current XY plane.

The **Target** option allows you to change the placement of the camera target. Enter the option and then specify the new location. You can enter coordinates or pick a location in the drawing.

The **Lens** option allows you to change the focal length of the camera lens. If you change the lens focal length, you are really changing the field of view, or the area of the drawing that the camera covers. The lower the lens focal length, the wider the field of view angle. The focal length is measured in millimeters.

The **Clipping** option is used to turn the front and back clipping planes on or off. These planes are used to limit what is shown in the camera view. Clipping planes are discussed later in this chapter.

The **View** option is used to change the current view to that shown by the camera. This option has two choices—**Yes** or **No**. If you select **Yes**, the active viewport switches to the camera view and the **CAMERA** command ends. If you select **No**, the previous prompt returns.

Once you have made all settings, press [Enter] or select the **Exit** option to end the command. The view (camera) is listed with the other saved views in the drop-down list in the **Named Views** panel on the **Visualize** tab of the ribbon and in the **View Controls** flyout of the viewport controls. Selecting the view makes it the current view

Figure 21-1.
A camera is represented by a glyph. When the camera is selected, the field of view (shown in color) and grips are displayed.

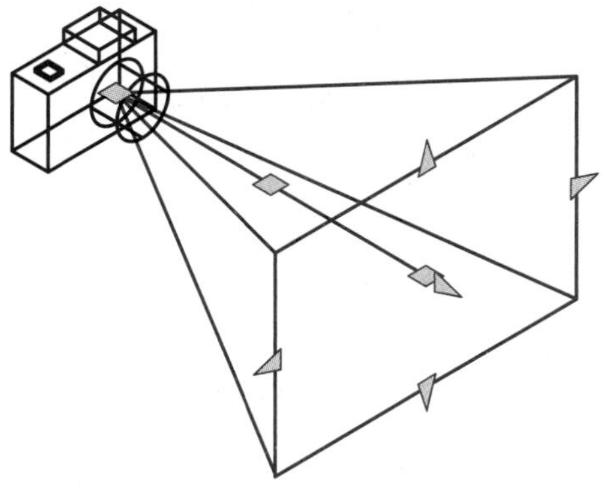

in the active viewport. The view is also listed under the **Model Views** branch in the **View Manager** dialog box. It can be made current by selecting the view, picking the **Set Current** button, and then picking the **OK** button.

PROFESSIONAL TIP

In addition to the camera name, many other camera properties can be changed in the **Properties** palette. The camera and target locations can be changed, the lens focal length and field of view can be adjusted, and the clipping planes can be modified. Also, you can change the roll angle, which is the rotation about a line from the camera to the target, and set the camera glyph to plot.

Camera System Variables

Ribbon
Visualize
> Camera

Camera Display
Type
CAMERADISPLAY

CAMERADISPLAY

The **CAMERADISPLAY** system variable controls the visibility of camera glyphs. When set to 1, camera glyphs are displayed. When set to 0, camera glyphs are not displayed. Creating a camera automatically sets the variable to 1. The **Camera Display** button in the **Camera** panel on the **Visualize** tab of the ribbon toggles the display of camera glyphs off and on.

When creating a camera, if you pick the camera and target locations without using object snaps, you may assume that the camera and target are located on the XY plane (Z coordinate of 0) of the current UCS. This may or may not be true. The **CAMERAHEIGHT** system variable determines the default height of the camera if a Z coordinate is not provided. It is a good idea to set this variable to a typical eye height before placing cameras. There is no corresponding system variable for the target because the target is usually placed by snapping to an object of interest. If X and Y coordinates are entered for the target location, but a Z coordinate is not provided, the Z value is automatically 0. If a camera was previously created in the drawing session and the **Height** option was used, that height value becomes the default camera height.

Cameras Tool Palette

The **Cameras** tool palette provides a quick way to add a camera, but the default tools do not allow for the options described earlier. The **Normal Camera** tool creates a camera with a 50 mm focal length. This camera simulates normal human vision. The **Wide-angle Camera** tool creates a camera with a 35 mm focal length. This type of view is commonly used for scenery or interior views where it is important to show as much as possible with minimal distortion. The **Extreme Wide-angle Camera** tool creates a camera with a 6 mm focal length. This camera produces a fish-eye view, which is very distorted and mainly useful for special effects.

Changing the Camera View

Once the camera is placed, it is easy to manipulate. If you select a camera, the **Camera Preview** window is displayed by default. This window shows the view through the camera, Figure 21-2. The view in the window can be displayed in any of the 3D visual styles or any named visual style. Select the visual style in the drop-down list in the window. If the **Display this window when editing a camera** check box at the bottom of the window is unchecked, the window is not displayed the next time a camera is selected. The next time the drawing is opened, this setting is restored (checked).

When a camera is selected, grips are displayed. Refer to Figure 21-1. If you hover the cursor over a grip, a tooltip appears indicating what the grip will alter. Picking the base grip on the camera allows you to reposition the camera in the scene. If the **Camera Preview** window is open, watch the preview as you move the camera to help

Figure 21-2.
The **Camera Preview** window is displayed, by default, when a camera is selected.

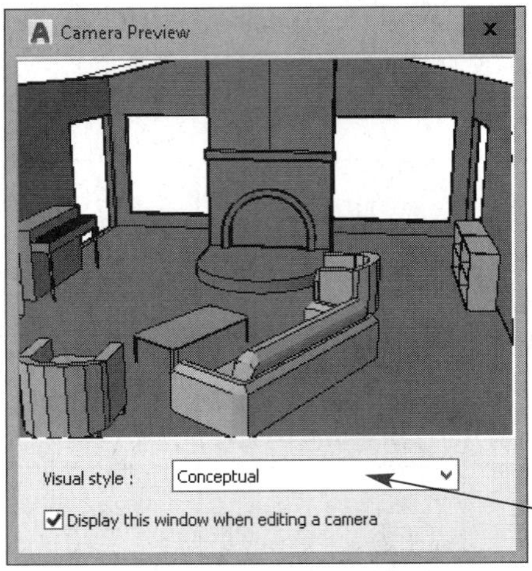

guide you. Selecting the grip on the target allows the target to be repositioned. Again, use the preview in the **Camera Preview** window as a guide. The grip at the midpoint between the camera and target can be used to reposition the camera and target at the same time. If you pick and move one of the arrow grips on the end of the field of view, the lens focal length and field of view are changed.

Camera Clipping Planes

Clipping planes allow you to suppress objects in the foreground or background of your scene. Picture these clipping planes as flat, 2D objects perpendicular to the line of sight that can be moved closer to or farther from the viewer. Only the objects between the front and back clipping planes and within the field of view are seen in the camera view. This is helpful for eliminating walls, roofs, or any other clutter that may take away from the focus of the scene. Clipping planes can be set while creating the camera or later using the **Properties** palette. Clipping planes can be adjusted using grips.

To set the clipping planes while creating the camera, enter the **Clipping** option. You are prompted:

Enable front clipping plane? [Yes/No] <No>:

To enable the front clipping plane, select **Yes**. You are then asked to specify the offset from the target plane. The target plane is described next. Once you enter the offset, or if you answer **No**, you are prompted:

Enable back clipping plane? [Yes/No] <No>:

To enable the back clipping plane, select **Yes** and then specify the offset from the target plane.

The ***target plane*** is the 2D plane that is perpendicular to the line of sight and passing through the camera's target point. Offsets for both front and back clipping planes are from this plane. Positive values place the clipping planes between the camera and the target plane. Negative values place the planes on the opposite side of the target plane from the camera. You can place the clipping planes anywhere in the scene from the camera location to infinity. You cannot, however, place the back clipping plane between the front clipping plane and the camera.

The best way to adjust clipping planes is using the **Properties** palette or grips. Create the camera and then display a plan view of the camera and target (an approximate plan view is okay). Select the camera and open the **Properties** palette. In the **Clipping** category, select the Clipping property. In the property drop-down list, select Front on, Back on, or Front and back on to turn on the appropriate clipping plane(s). Notice that the clipping planes are visible in the viewport, **Figure 21-3**. Next, enter offset values for the Front plane and Back plane properties as appropriate or use grips to set the locations of the clipping planes. By displaying a plan view of the camera and target, you can see where the clipping planes are located and visualize their effect on the scene. If the **Camera Preview** window is open, the clipping is displayed in the preview.

Exercise 21-1

www.g-wlearning.com/CAD/

Complete the exercise on the companion website.

Animation Preparation

The tools presented in this chapter make it easy to lay out a path, plan camera angles, and record the movement of the camera. The resulting animation can be directly output to a number of movie file types that can be shared with others. However, there are some decisions to make first.

It is important to plan exactly what you want to see in the animation. Think like a movie director and plan the "shots." Ask these questions:

- What will be visible from each camera angle?
- Is there a background in place?
- Is the lighting appropriate?
- Will a simple walkthrough suffice or will a flyby be necessary?
- How close is the viewer (camera) going to be to the objects in the scene?

The answers to these questions will help determine the modeling detail required. Do not model anything that will not be seen. Also, do not place detailed materials on objects that are not the focus of the animation. Processing the animation may take a

Figure 21-3.
Adjusting the clipping planes for a camera.

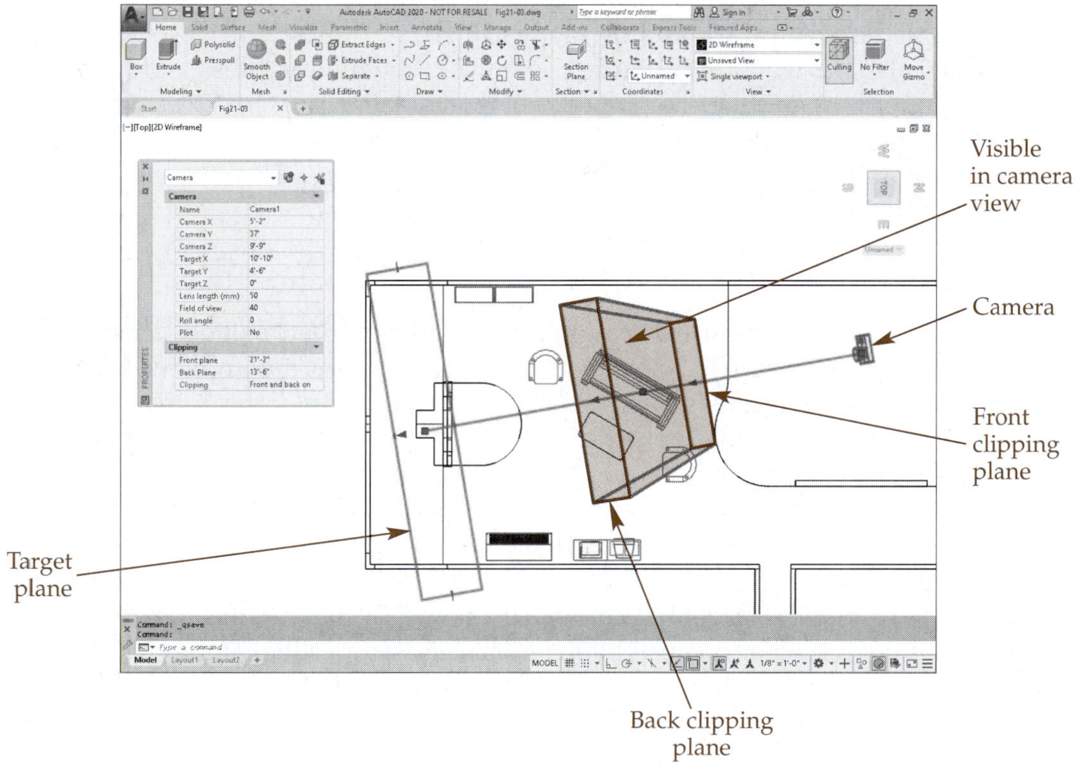

long time. Unnecessary detail may bog down the computer. In addition, walkthroughs and flybys must be created in views with perspective, not parallel, projection.

The "visual quality" of the scene has the biggest impact on the time involved in rendering the animation. An animation can be rendered in any visual style or using any render preset that is available in the drawing. It is a natural tendency to render at the highest level to make the animation look the best. However, a computer animation has a playback rate of 30 frames per second (fps). If a single frame (view) takes three minutes to render, how long will it take to render a 30 second animation? An animation 30 seconds in length has 900 frames (30 fps × 30 seconds). If each frame takes three minutes to render, the entire animation will take 2700 minutes, or 45 hours, to render.

Are you willing to wait two or three days for a 30 second movie? How about your boss or your client? There are trade-offs and concessions to be made. Perform test renderings on static views and note the rendering time. Then, decide on the acceptable level of quality versus rendering time and move ahead with it.

Walking and Flying

The process for creating a walkthrough or a flyby is the same. First, the command is initiated. Then, the movement is defined and recorded. Finally, the recorded movement is saved to an animation file. Note: the **Animations** panel in the **Visualize** tab on the ribbon may not be displayed by default.

When using the **3DWALK** and **3DFLY** commands, you can move through the scene using the arrow keys or the [W], [A], [S], and [D] keys on the keyboard to control your movements. Once either command is initiated, a message appears from the **Help** flyout in the **InfoCenter**, if balloon notifications are turned on. If you expand this message, the key movements are explained. See **Figure 21-4**. To redisplay this message while the command is active, press the [Tab] key.

3DWALK
Ribbon
Visualize
> Animations
Walk
Type
3DWALK
3DW

3DFLY
Ribbon
Visualize
> Animations
Fly
Type
3DFLY

- **Move forward.** Up arrow or [W].
- **Move left.** Left arrow or [A].
- **Move right.** Right arrow or [D].
- **Back up.** Down arrow or [S].
- **Toggle between walk and fly modes.** [F].

You can also navigate through the scene using the mouse. Press and hold the left mouse button and then drag the mouse in the active viewport to "steer" through the scene. With the **3DWALK** command, the camera remains at the same Z value. With the **3DFLY** command, the Z position of the camera can change. The steps for creating a walkthrough or flyby are provided at the end of this section.

Position Locator

When the **3DWALK** or **3DFLY** command is initiated, the **Position Locator** palette appears. See **Figure 21-5**. The preview in this palette shows a plan view of the scene. The purpose of this window is to provide an overview of the scene, in plan, while you develop the animation. It does not need to be displayed to create an animation and can be closed if it takes up too much space or slows down the rendering.

Position and target indicators appear in the plan view to show the location of the camera and its target. The green triangular shape displays the field of view. The ***field of view (FOV)*** is the area within the camera's "vision." The field of view indicator is only displayed when the target indicator is displayed. By default, the position indicator is red. The target indicator is green by default. These properties can be changed in the **General** category at the bottom of the **Position Locator** palette.

You can reposition the camera and the target in the plan view simply by picking and dragging either indicator. The effect of the change is visible in the active viewport.

Figure 21-4.
This message from the **InfoCenter** shows the keys that can be used to navigate through an animation.

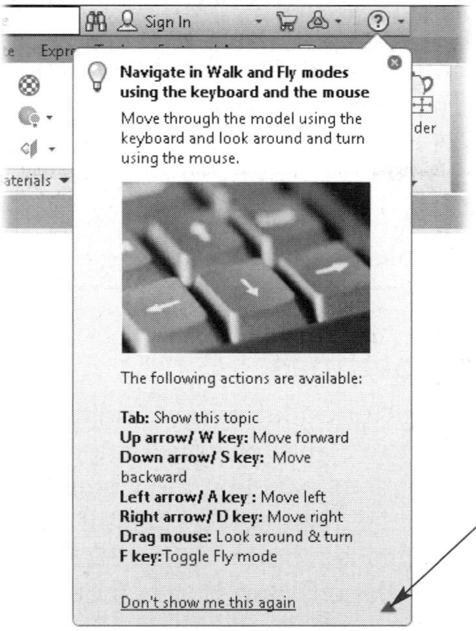

Pick to expand or collapse the message

Moving the position and target indicators closer together reduces the field of view. Picking the field of view lines and dragging moves the position and target indicators at the same time. The Position Z property in the **General** category sets the Z coordinate value for the position indicator. The Target Z property in the **General** category sets the Z coordinate value for the target indicator. The Z coordinate value determines eye level.

In addition to changing the color of the position and target indicators, you can use the properties in the **General** category to modify the display in the **Position Locator** palette. The Position indicator size property determines if both indicators are displayed small, medium, or large. If the Position indicator blink property is set to On, both indicators flash on and off in the preview. The Preview visual style property sets the visual style for the preview. This setting does not affect the current viewport or the animation. The Preview transparency property is set to 50% by default, but can be changed to whatever you want. If the view in the **Position Locator** palette is obscured by something (a roof, perhaps), you may want to set the Preview visual style property to Hidden

Figure 21-5.
The **Position Locator** palette.

Position

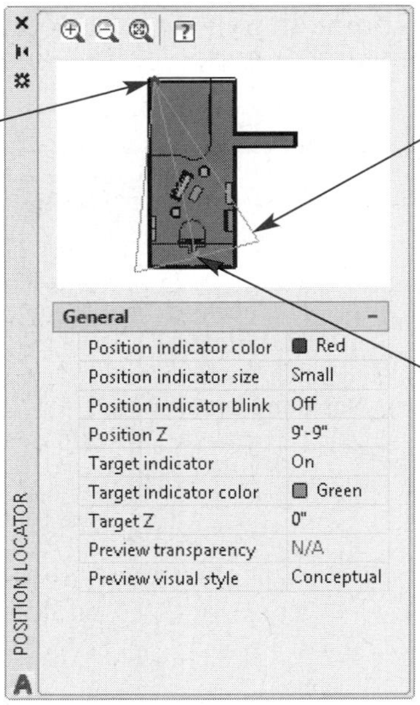

FOV

Target

and the Preview transparency property to 80% or 90%. This will make the objects under the roof visible. If hardware acceleration is on, then the Preview transparency property is disabled. The **Graphics Performance** tool in the status bar indicates the on/off status of hardware acceleration.

Exercise 21-2

www.g-wlearning.com/CAD/

Complete the exercise on the companion website.

Walk and Fly Settings

WALKFLYSETTINGS

Ribbon
Visualize
> Animations

Walk and Fly Settings

Type
WALKFLY-SETTINGS

General settings for walkthroughs and flybys are made in the **Walk and Fly Settings** dialog box. See **Figure 21-6**. Open this dialog box by picking the **Walk and Fly Settings** button in the **Animations** panel on the **Visualize** tab of the ribbon (in the **Walk** flyout). This panel may not be displayed by default. The dialog box can also be displayed by picking the **Walk and Fly...** button in the **3D Modeling** tab of the **Options** dialog box.

The three radio buttons at the top of the dialog box are used to determine when the message shown in **Figure 21-4** is displayed. The check box determines if the **Position Locator** palette is automatically displayed when the **3DWALK** or **3DFLY** command is entered.

The text boxes in the **Current drawing settings** area determine the size of each step and the number of steps per second. The **Walk/fly step size:** setting controls the **STEPSIZE** system variable. This is the number of units that the camera moves in one step. The **Steps per second:** setting controls the **STEPSPERSEC** system variable. This is the number of steps the camera takes each second. Together, these two settings determine how fast the camera moves in the animation.

PROFESSIONAL TIP

You will have to experiment with step size and steps per second values to make an animation that is easy to watch. Start with low numbers (for slow movement) and work your way up. Fast movement is disorienting and makes the viewer feel as if something was not seen, or missed. The viewer should be able to watch at a comfortable pace and get a good look at your design.

To get a feel for the proper speed for a walkthrough, pay attention to the next movie or TV show that you watch. When the director wants you to get a good look at the setting for the scene, the camera very slowly pans around the room. To emphasize distance, the camera slowly zooms in to a target object or person.

Figure 21-6. General settings for the walkthrough or flyby are made in the **Walk and Fly Settings** dialog box.

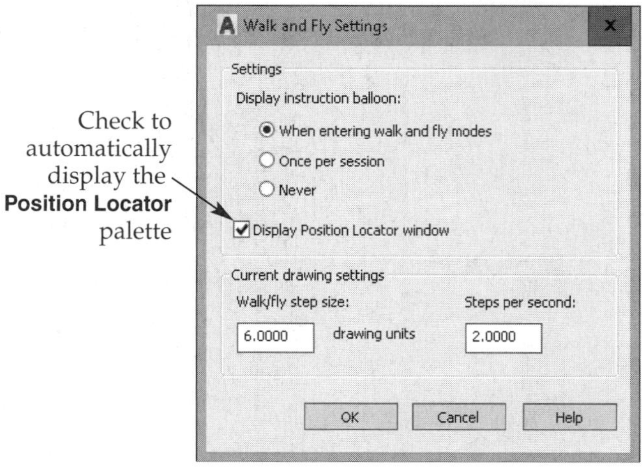

Camera Tools

The expanded **View** panel on the **Home** tab of the ribbon contains some tools for quickly adjusting the camera before starting the animation. See **Figure 21-7**. The **Lens Length** slider controls how much of the scene is seen by the camera. The ***lens length*** refers to the focal length of the camera lens. The higher the number, the closer you are to the subject. The range is from about 1 to 100,000; 50 is a good starting point. There are stops on the slider for standard lens lengths. You can enter a specific value for the lens length or field of view by selecting the text in the slider, typing the value, and pressing [Enter].

Below the **Lens Length** slider are text boxes for the camera and target positions. These can be used to change the X, Y, and Z coordinates of the camera or target before starting the animation.

NOTE

The view of the current viewport is changed when using the camera tools in the expanded **View** panel on the **Home** tab of the ribbon. If you change the current view and want to save it for future use, save it as a new view.

Animation Tools

The **Animations** panel on the **Visualize** tab of the ribbon contains the tools for controlling the recording and playback of the animation. See **Figure 21-8**. Remember, this panel may not be displayed by default. The **Record Animation** button is used to initiate recording of camera movement. After the **3DWALK** or **3DFLY** command is activated, pick the button to start recording. Make sure that you are ready to start moving when you pick the button because recording starts as soon as it is picked.

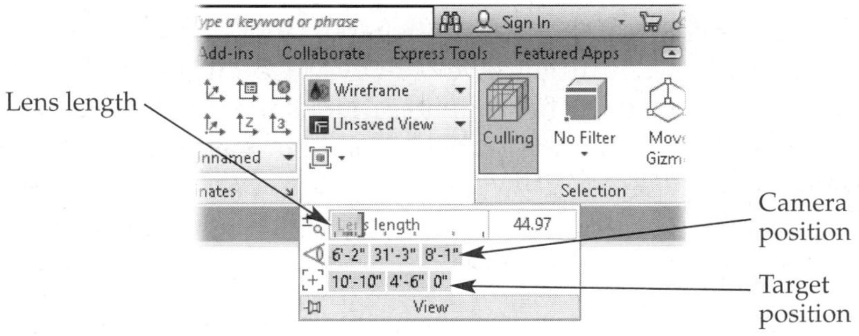

Figure 21-7. Camera tools are located in the expanded **View** panel on the **Home** tab of the ribbon.

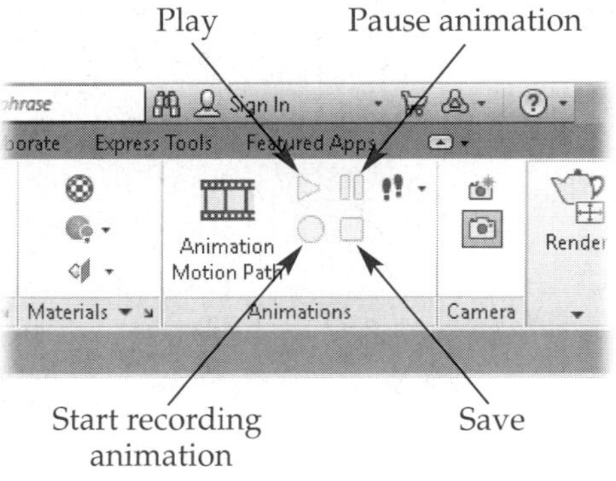

Figure 21-8. The **Animations** panel on the **Visualize** tab of the ribbon is where you can record and play back the walkthrough or flyby animation. This panel may not be displayed by default.

Picking the **Pause Animation** button temporarily stops recording. This allows you to adjust the view without recording the actions. When you are ready to begin recording again, pick the record button to resume.

Picking the **Play Animation** button stops the recording and opens the **Animation Preview** dialog box, where the animation is played. See Figure 21-9. The controls in this dialog box can be used to rewind, pause, and play the animation. The slider can be dragged to preview part of the animation or move to a specific frame. The visual style can also be set using the drop-down list. If the animation is created using a render preset, the file must be played in Windows Media Player or another media player to view the rendered detail. Render presets are available in the **Animation Settings** dialog box, as discussed in the next section.

Picking the **Save Animation** button in the **Animations** panel stops recording and opens the **Save As** dialog box. Name the animation file, navigate to a location, and pick the **Save** button.

CAUTION

While the **3DWALK** or **3DFLY** command is active and the record button is on, you are creating an animation. If you move the camera in the **Position Locator** palette and start re-recording the animation to correct a problem, but do not first exit the current **3DWALK** or **3DFLY** command session, you are adding another segment to the animation you just previewed. To start over, first exit the current command session.

Exercise 21-3

www.g-wlearning.com/CAD/

Complete the exercise on the companion website.

Animation Settings

The **Animation Settings** dialog box may contain the most important settings pertaining to walkthroughs and flybys. See Figure 21-10. These animation settings determine how good the animation looks, how long it is going to take to complete, and how big the file will be. The dialog box is displayed by picking the **Animation settings...** button in the **Save As** dialog box displayed when saving an animation.

Figure 21-9.
The animation is played in the **Animation Preview** dialog box.

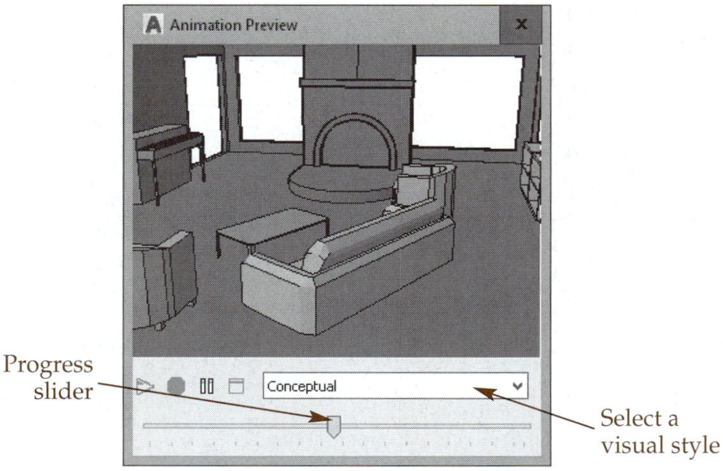

Figure 21-10.
The **Animation Settings** dialog box contains important settings pertaining to walkthroughs and flybys.

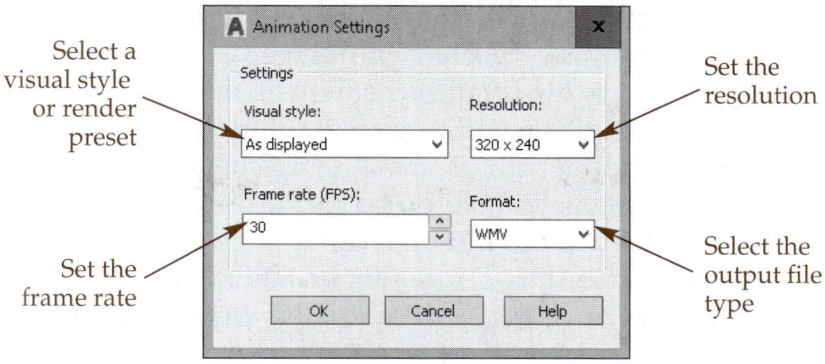

The **Visual style:** drop-down list is used to set the shading level in the animation. The name of this drop-down list is a little misleading because visual styles and render presets are available. Using a higher shading level or render preset will result in a longer rendering time and larger file size. If you have numerous lights casting shadows and detailed materials, settle in for a long wait. A simple, straight-ahead walkthrough of 10 or 15 feet can easily result in 300 frames of animation. If each frame takes about five seconds to render, that equals 1500 seconds, or 25 minutes, to create an animation file that is only 10 seconds long.

The value in the **Frame rate (FPS):** text box sets the number of frames per second for the playback. In other words, this sets the speed of the animation playback. The default is 30 fps, which is a common playback rate for computers.

The **Resolution:** drop-down list offers standard options for resolution, from 160 × 120 to 1024 × 768. These are measured in pixels × pixels. Remember, higher resolutions mean longer processing times and larger file sizes.

The **Format:** drop-down list is used to select the output file type. The file type must be set in this dialog box. It cannot be changed in the **Save As** dialog box. The output file type options are:

- **WMV.** The standard movie file format for Windows Media Player.
- **AVI.** Audio-Video Interleaved is the Windows standard for movie files.
- **MPG.** Moving Picture Experts Group (MPEG) is another very common movie file format.

Depending on the configuration of your computer, you may not have all of these file type options or you may have additional options not listed here.

PROFESSIONAL TIP

Other AutoCAD navigation modes may be used to create a walkthrough or flyby. Any time that you enter constrained, free, or continuous orbit, you can pick the **Record Animation** button to record the movements. After you activate the **3DWALK** or **3DFLY** command, right-click and experiment with some of the other options as you record your animation. You can even combine some of these navigation modes with walking or flying. For example, use **3DFLY** to zoom into a scene, pause the animation, switch to constrained orbit, restart the recording, and slowly circle around your model. Animation settings are also available in this shortcut menu.

Exercise 21-4

www.g-wlearning.com/CAD/

Complete the exercise on the companion website.

Steps to Create a Walkthrough or Flyby

1. Plan your animation. Determine where you are moving from and to, what you are going to be looking at, and what will be the focal point of the scene.
2. Set up a multiple-viewport configuration of three or four viewports.
3. In one of the viewports, create or restore a named view with the appropriate starting viewpoint. Make sure a background is set up, if desired.
4. Start the **3DWALK** or **3DFLY** command. Note in the **Position Locator** palette the camera location, target location, and field of view. Adjust these in the **Position Locator** palette preview area or the expanded **View** panel on the ribbon, if necessary.
5. Position your fingers over the navigation keys on the keyboard.
6. Pick the **Record Animation** button.
7. Practice navigating through the view and then pick the **Play Animation** button to preview your animation. When you are done practicing, make sure to cancel the command and reposition the camera at the starting point.
8. Pick the **Record Animation** button to start over.
9. Start navigating through the view. Try to keep the movements as smooth as possible. Any jerks and shakes will be visible in the animation.
10. When you are done, stop moving forward and then pick the stop (**Save Animation**) button.
11. In the **Save As** dialog box, pick the **Animation settings...** button. In the **Animation Settings** dialog box, select the desired visual style, frame rate, resolution, and output file format. Pick the **OK** button to close the **Animation Settings** dialog box. In the **Save As** dialog box, name and save the file.
12. The **Creating Video** dialog box is displayed as AutoCAD processes the frames, Figure 21-11.
13. When the **Creating Video** dialog box is automatically closed, the animation file is saved and you can view it. Pick the **Play Animation** button and watch the animation in the **Animation Preview** window. You can also locate the file using Windows Explorer. Then, double-click on the file to play the animation in Windows Media Player (or whichever program is associated with the file type).
14. Exit the command. If you are not satisfied with the results and want to try it again, make sure to exit the command before you attempt another walkthrough or flyby.

Figure 21-11.
The **Creating Video** dialog box is displayed as AutoCAD generates the animation.

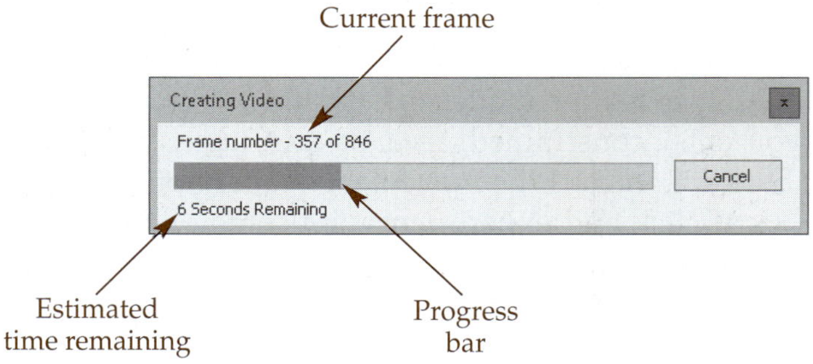

Motion Path Animation

You may have found it difficult to create smooth motion using the keyboard and mouse. Fortunately, AutoCAD provides an easy way to create a nice, smooth animated walkthrough or flyby. This is done using a motion path. A ***motion path*** is simply a straight or curved path along which the camera, target, or both travel during the animation.

One method of using a motion path is to link the camera and target to a single path. The camera and its line of sight then follow the path as a train follows tracks. See **Figure 21-12**.

Another option when using a motion path is to link the camera to a single point in the scene and the target to a path. For example, the target can be set to follow a circle or arc. The camera swivels on the point and "looks at" the path as if it is being rotated on a tripod. See **Figure 21-13**.

A third way to use a motion path is to have the camera follow a path, but have the target locked onto a stationary point. This is similar to riding in a vehicle and watching an object of interest on the side of the road. As the vehicle moves, your gaze remains fixed on the object. See **Figure 21-14**.

Figure 21-12.
A—The camera and target are linked to the same path (shown in color). B—The camera looks straight ahead as it moves along the path.

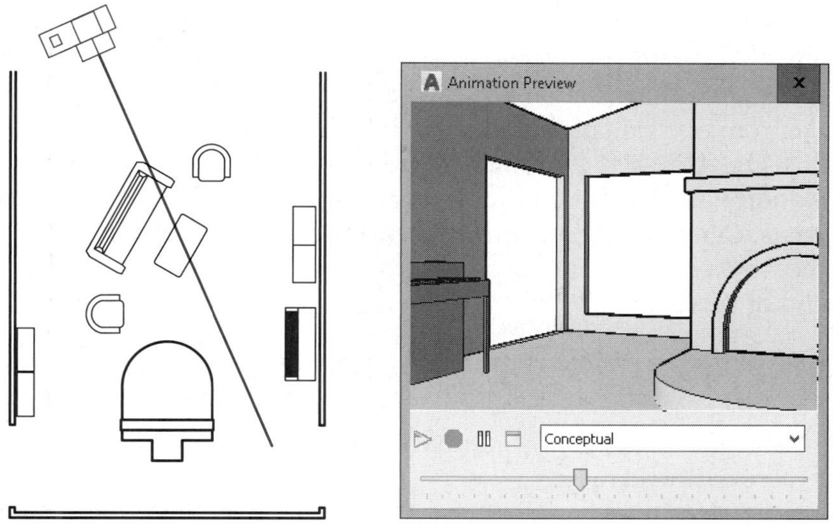

A B

Figure 21-13.
A—The camera is linked to a point so it remains stationary. The target is linked to the circle. B—The camera view rotates around the room as if the camera is on a swivel tripod.

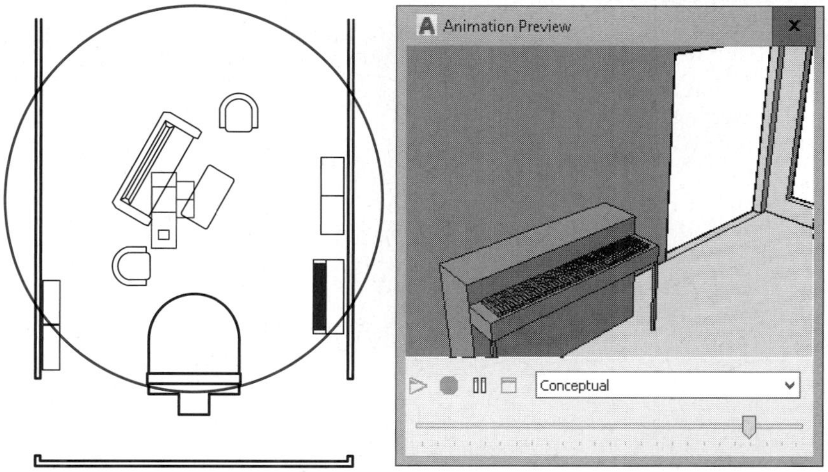

A B

Figure 21-14.
A—The camera is linked to the spline path and the target is linked to the point (shown in color).
B—As the camera moves along the path, it always looks at the point.

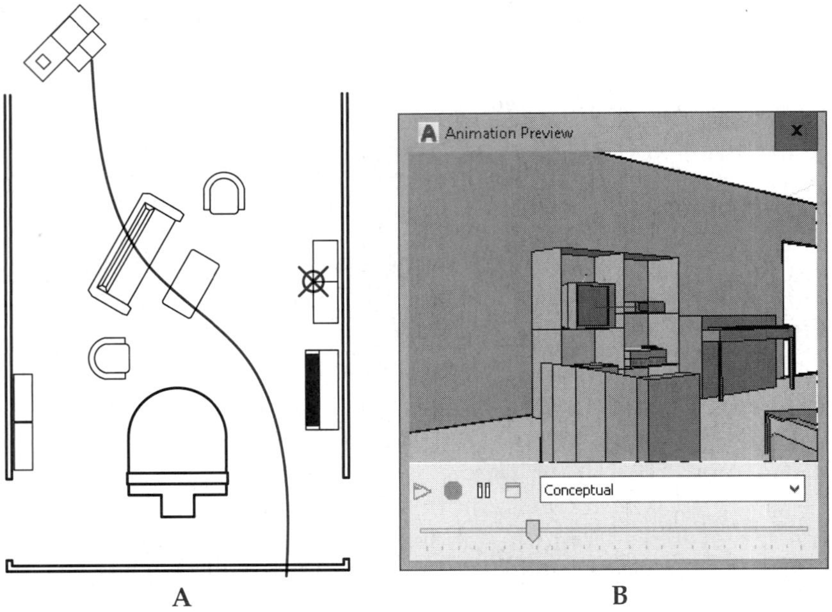

The fourth method of using a motion path is to have both the camera and target follow separate paths. Picture yourself walking into an unfamiliar room. As you walk into the center of the room, your gaze sweeps left and right across the room. In this case, the camera (you) follows a straight-line path and the target (your gaze) follows an arc from one side of the room to the other.

ANIPATH

Ribbon
Visualize
> Animations
Animation Motion Path
Type
ANIPATH

The **ANIPATH** command is used to assign motion paths. The command opens the **Motion Path Animation** dialog box. See **Figure 21-15**. This dialog box has three main areas: **Camera**, **Target**, and **Animation settings**. These areas are described in detail in the next sections. The steps for creating a motion path animation are provided at the end of this section.

NOTE

Selecting a motion path automatically creates a camera. You cannot add a motion path to an existing camera.

Figure 21-15.
The **Motion Path Animation** dialog box is used to create an animation that follows a path.

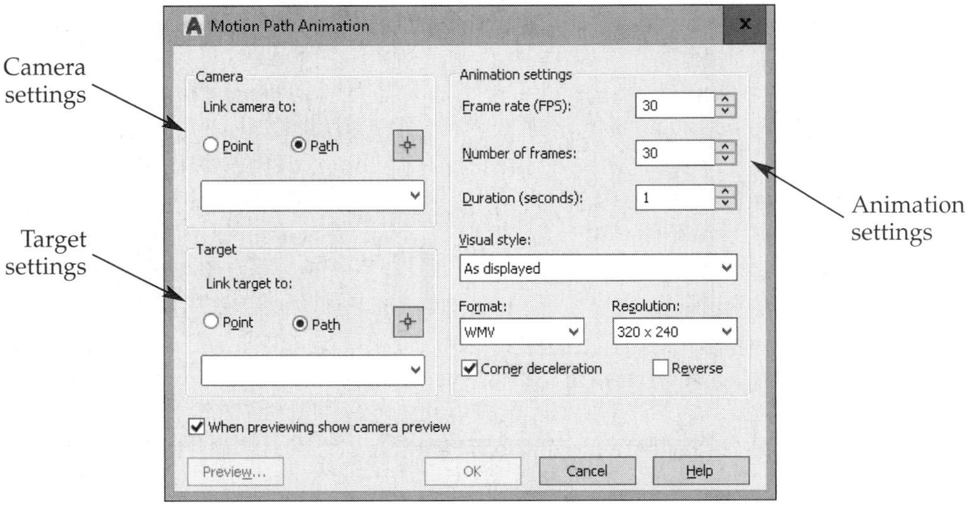

Camera Area

The camera can be linked to a path or a point. To select a path, pick the **Path** radio button and then pick the "select" button next to the radio buttons. The dialog box is temporarily closed for you to select the path in the drawing. The path may be a line, arc, circle, ellipse, elliptical arc, polyline, 3D polyline, spline, or helix, but it must be drawn before the **ANIPATH** command is used. Splines are nice for motion paths because they are smooth and have gradual curves. The camera moves from the first point on the path to the last point on the path, so create paths with this in mind. Once you select the path, you will be prompted to specify a name for the path. If you do not name it, AutoCAD will give it a default, sequential name, such as Path1, Path2, etc.

To select a stationary point, pick the **Point** radio button. Then, pick the "select" button next to the radio buttons. When the dialog box is hidden, specify the location in the drawing. You can use object snaps or enter coordinates. It may be a good idea to have a point drawn and use object snaps to select the point. As when selecting a path, once you select the point, you will be prompted to specify a name for the point.

The camera must be linked to either a path or a point. If neither is selected, the command cannot be completed. If you want the camera to remain stationary as the target moves, select the **Point** radio button and then pick the stationary point in the drawing.

Once a point or path has been selected, it is added to the drop-down list. All named motion paths and selected motion points in the drawing appear in this list. Instead of using the "select" button, you can select the path or point in this drop-down list.

Target Area

The target is the location where the camera points. Like the camera, the target can be linked to a point or a path. To link the target to a path, select the **Path** radio button. Then, pick the "select" button and select the path in the drawing. If the camera is linked to a point, the target must be linked to a path. If the camera is set to follow a path, then you actually have three choices for the target. It can be linked to a path, point, or nothing. To link the target to a point, pick the **Point** radio button. Then, pick the "select" button to select the point in the drawing. The None option, which is selected in the drop-down list, means that the camera will look straight ahead down the path as it moves.

Animation Settings Area

Most of the settings in the **Animation settings** area have the same effect as the corresponding settings in the **Animation Settings** dialog box discussed earlier. However, there are four settings unique to the **Motion Path Animation** dialog box.

The **Number of frames:** text box is used to set the total number of frames in the animation. Remember, a computer has a playback rate of 30 fps. Therefore, if the frame rate is set to 30, set the number of frames to 450 to create an animation that is 15 seconds long ($30 \times 15 = 450$).

The value in the **Duration (seconds):** text box is the total time of the animation. This value is automatically calculated based on the frame rate and number of frames. However, you can enter a duration value. Doing so will automatically change the number of frames based on the frame rate.

By default, the **Corner deceleration** check box is checked. This slows down the movement of the camera and target as they reach corners and curves on the path. If this is unchecked, the camera and target move at the same speed along the entire path, creating very jerky motion on curves and at corners. It is natural to decelerate on curves.

Checking the **Reverse** check box simply switches the starting and ending points of the animation. If the camera (or target) travels from the first endpoint to the second endpoint, checking this check box makes the camera (or target) travel from the second endpoint to the first.

Previewing and Completing the Animation

To preview the animation, pick the **Preview...** button at the bottom of the **Motion Path Animation** dialog box. The camera glyph moves along the path in all viewports. If the **When previewing show camera preview** check box is checked, the **Animation Preview** window is also displayed and shows the animation.

To finish the animation, pick the **OK** button in the **Motion Path Animation** dialog box. The **Save As** dialog box is displayed. Name the file and specify the location. If you need to change the file type, pick the **Animation settings...** button to open the **Animation Settings** dialog box. Change the file type, close the dialog box, and continue with the save.

Steps to Create a Motion Path Animation

1. Plan your animation. Determine where you are moving from and to, what you are going to be looking at, and what will be the focal point of the scene.
2. Draw the paths and points to which the camera and target will be linked. Draw the path in the direction the camera should travel (first point to last point). Do not draw any sharp corners on the paths and make sure that the Z value (height) is correct.
3. Start the **ANIPATH** command.
4. Pick the camera path or point.
5. Pick the target path or point (or None).
6. Adjust the frames per second, number of frames, and duration to set the length and speed of the animation.
7. Select a visual style, the file format, and the resolution.
8. Preview the animation. Adjust settings, if needed.
9. Save the animation to a file.

Exercise 21-5 www.g-wlearning.com/CAD/

Complete the exercise on the companion website.

Chapter Review

Answer the following questions using the information in this chapter.

1. Which system variable controls the display of camera glyphs?
2. Name the three camera tools available on the **Cameras** tool palette and explain the differences between them.
3. When is the **Camera Preview** window displayed, by default?
4. From where is the offset distance for the camera clipping planes measured?
5. What is the difference between the **3DWALK** and **3DFLY** commands?
6. How do you "steer" your movement when creating a walkthrough or flyby animation?
7. What is the *field of view*?
8. What is the purpose of the **Position Indicator** palette?
9. In the **Walk and Fly Settings** dialog box, which settings combine to control the speed of the animation?
10. How do you start recording a walkthrough or flyby?
11. What must be done before correcting a motion error in a walkthrough or flyby?
12. Motion path animation involves linking a camera or target to _____ or _____.
13. Which types of objects may be used as a motion path?
14. If None is selected as the target "path," what does the camera do in the animation?
15. Explain *corner deceleration*.

Drawing Problems

1. **General** In this problem, you will create and manipulate a camera in a drawing from a previous chapter.
   A. Open drawing P19-2 from Chapter 19 and save it as P21-1.
   B. Create at least two viewports and display a plan view in one of them.
   C. Use the **CAMERA** command to create a camera looking at the objects from the southwest quadrant. Change the camera settings as needed to display a pleasing view of the scene.
   D. Name the camera SW View.
   E. Turn on both the front and back clipping planes. Adjust them to eliminate one object in the front and one object in the back.
   F. Open the **Cameras** tool palette and, using the tools in the palette, create three more cameras looking at the scene from various locations. Change their names to Normal, Wide-angle, and Fish-eye to match the type of camera.
   G. Save the drawing.

2. **Architectural** In this problem, you will draw some basic 3D shapes to represent equipment in a small workshop. Then, you will create an animated walkthrough.
   A. Start a new drawing and set the units to architectural. Save it as P21-2.
   B. Draw a planar surface that is 15′ × 30′.
   C. Draw three 9′ tall walls enclosing the two long sides and one short side.
   D. Use boxes and a cylinder to represent equipment and shelves. Refer to the illustration shown. Use your own dimensions.
   E. Use the **3DWALK** command to create an animation of walking into the workshop. Turn and look at the shelves at the end of the animation.
   F. Set the visual style to Conceptual and the resolution to 640 × 480.
   G. Save the animation as P21-2.avi (or the file format of your choice).
   H. Save the drawing.

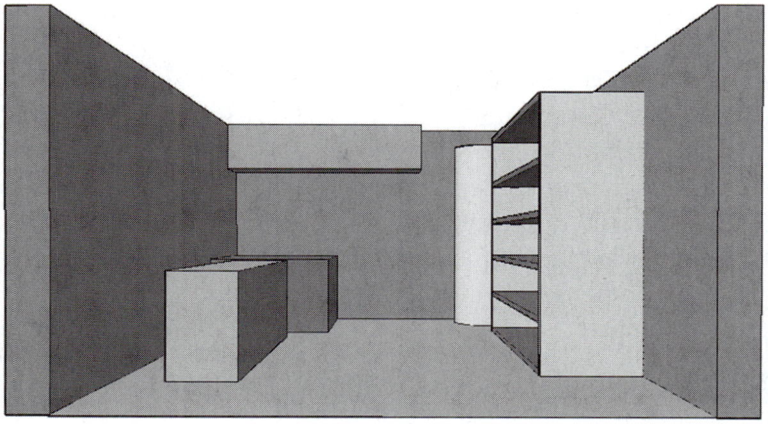

3. **Architectural** In this problem, you will create a motion path animation for the workshop drawn in Problem 21-2.
   A. Open drawing P21-2 and save it as P21-3.
   B. Draw a line and an arc similar to those shown (in color). The dimensions are not important.
   C. Move both objects so they are 4′ off the floor.
   D. Using the **ANIPATH** command, link the camera to the line and the target to the arc. Set the resolution to 640 × 480.
   E. Preview the animation. Adjust the animation settings as necessary. You may need to slow down the animation quite a bit. How do you do this?
   F. Save the animation as P21-3.wmv (or the file format of your choice).
   G. Save the drawing.

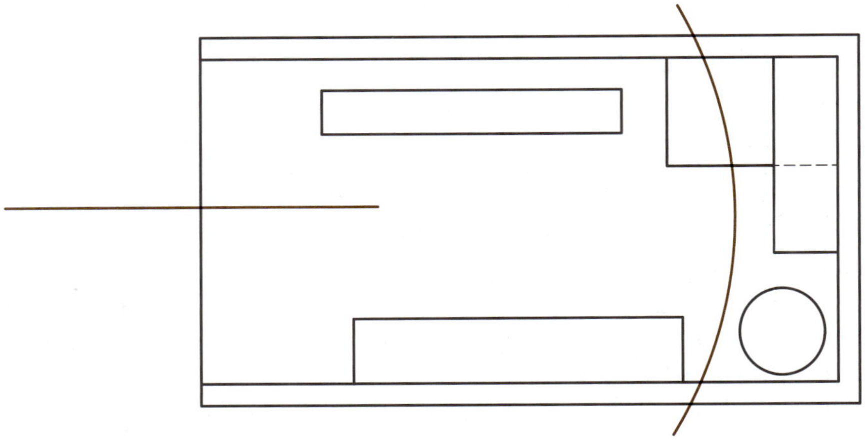

4. **Mechanical** In this problem, you will create a motion path animation for the presentation of a mechanical drawing from a previous chapter.
   A. Open drawing P18-3 and save it as P21-4.
   B. Draw a circle centered on the flange with a radius of 300.
   C. Move the circle 200 units in the Z direction.
   D. Using the **ANIPATH** command, link the camera to the circle and the target to the center of the flange.
   E. Preview the animation and adjust the animation settings as necessary. Due to the materials in the scene, the preview may play slowly, depending on the capabilities of your computer.
   F. Select a visual style that your computer can handle. Set the resolution to 320 × 240.
   G. Save the animation as P21-4.avi (or the file format of your choice).
   H. Save the drawing.

5. **Architectural** In this problem, you will create a flyby of the building that you created in Chapter 18.
   A. Open drawing P18-1 and save it as P21-5.
   B. Create a perspective view of the scene that shows the building from slightly above it. Save the view.
   C. Start the **3DFLY** command. Practice with the movement keys to make sure you know how to fly around the building. Then, cancel the command.
   D. Restore the starting view and select the **3DFLY** command.
   E. Record the flyby and save the animation as P21-5.wmv (or the file format of your choice).
   F. Save the drawing.

INDEX

2D sections, 381–384
2D sketch, as basis for 3D construction, 17–19
2D texture, 460
2D Wireframe visual style, 12–13, 19, 32, 418, 420, 423–424
3DALIGN command, 177–180
3DARRAY command, 185
3D arrays, 185–194
  3D path array, 190–194
  3D polar array, 188–190
  3D rectangular array, 185–187
3D coordinate systems, 1–3, 79–107
3DCORBIT command, 49, 62
3D dimensioning, 334–337
3DEDITBAR command, 259
3D edit bar gizmo, 259
3DFACE command, 214
3DFLY command, 549, 554–558
3DFORBIT command, 49
3DMESH command, 214
3DMOVE command, 176, 182, 280, 287, 495–496
3D object snaps, 16–17
3DORBIT command, 49, 55, 144, 279, 294, 321, 326
3DPOLY command, 84
3D polyline, 84
3DPRINT command, 390
3D printing, 389–390
3D Print Options dialog box, 390
3DPRINTSERVICE command, 390
3DROTATE command, 144, 182–183, 332
  rotating subobjects, 282, 287, 290, 294
3DSCALE command, 287
3D sections, generating, 381–384
3D text, 331–334
3D texture, 460
3D views, 7–11
  isometric and orthographic presets, 9–10
  view cube, 10–11
  viewport controls, 8–9
3DWALK command, 549, 554–560

A

ACISIN command, 389
ACISOUT command, 389
active viewport, 115
additive manufacturing, 389
Adjust Background Image dialog box, 502–503
Adjust Panel Increments dialog box, 377
Adjust Sun & Sky Background dialog box, 503–504, 533
aliasing, 420
ALIGN command, 177, 180
aligned section views, 358–359
aligning, 177–180
ambient color, 451
ambient light, 481
ANALYSISCURVATURE command, 386–388
ANALYSISDRAFT command, 388
ANALYSISOPTIONS command, 386, 388
Analysis Options dialog box, 386–388
ANALYSISZEBRA command, 385–386
angle of incidence, 482
angle of reflection, 482
animation, 549–564
  cameras, 549–553
  motion paths, 561–564
  preparation, 553–554
  walkthroughs and flybys, 554–560
Animation Preview dialog box, 558
Animation Settings dialog box, 558–560, 564
ANIPATH command, 549, 562–564
ANNOMONITOR system variable, 366
Annotation Monitor, 366–367
antialiasing, 420
architectural molding, 206–207
ARRAY command, 185
ARRAYEDIT command, 186
ARRAYPATH command, 185, 190–194
ARRAYPOLAR command, 185, 188–190, 302
ARRAYRECT command, 185–187, 302

arrays, 185–194, 302–303
associative, 302–303
path array, 190–194
polar array, 188–190
rectangular array, 185–187
aspect ratio, 462
associative arrays, 302–303
associative surfaces, 239–241
Attach Point Cloud dialog box, 407
attenuation, 484
Autodesk ReCap Pro, 400–407
indexing, 404
Project screen, 404–407
registering, 403–404
scan settings, 401–403
auxiliary views, 365–366

B

BACKGROUND command, 500
Background dialog box, 500–504, 533
backgrounds, 500–504
changing for named views, 503–504
gradient, 501–502
image, 502–503
solid, 501
sun and sky, 503
banking, 156
blend surfaces, 245–247
Boolean operations, 36
boss, 205–206
box, 24–26
BOX command, 24, 271
bulge magnitude, 246
bump map, 459

C

CAMERA command, 549–550
CAMERADISPLAY system variable, 551
CAMERAHEIGHT system variable, 551
cameras, 549–553
Cameras tool palette, 551
clipping planes, 552–553
views, 551–552
ceramic material type, 448
chamfer, 196
editing, 278
CHAMFER command, 196–197, 205
CHAMFEREDGE command, 196–197, 205
checker map, 462–463
cinematic shots, 537–543
basics, 537–539
movement types, 540–543
clipping planes, 552–553
commands
3DALIGN, 177–180
3DARRAY, 185
3DCORBIT, 49, 62
3DEDITBAR, 259
3DFACE, 214
3DFLY, 549, 554–558
3DFORBIT, 49
3DMESH, 214
3DMOVE, 176, 182, 280, 282, 287, 495–496
3DORBIT, 49, 55, 144, 279, 294, 321, 326
3DPOLY, 84
3DPRINT, 390
3DPRINTSERVICE, 390
3DROTATE, 144, 182–183, 332
3DSCALE, 287
3DWALK, 549, 554–560
ACISIN, 389
ACISOUT, 389
ALIGN, 177, 180
ANALYSISCURVATURE, 386–388
ANALYSISDRAFT, 388
ANALYSISOPTIONS, 386, 388
ANALYSISZEBRA, 385–386
ANIPATH, 549, 562–564
ARRAY, 185
ARRAYEDIT, 186
ARRAYPATH, 185, 190–194
ARRAYPOLAR, 185, 188–190, 302
ARRAYRECT, 185–187, 302
BACKGROUND, 500
BOX, 24, 271
CAMERA, 549–550
CHAMFER, 196–197, 205
CHAMFEREDGE, 196–197, 205
CIRCLE, 42
CONE, 26, 81, 88, 271
CONVTOMESH, 134
CONVTONURBS, 134, 257–258
CONVTOSOLID, 132, 134, 198, 218, 220, 261
CONVTOSURFACE, 132, 134, 218, 220, 260
COPY, 277
CUI, 7
CVHIDE, 258
CVSHOW, 258
CYLINDER, 27, 95, 97, 127, 271
DIMREASSOCIATE, 368
DISTANTLIGHT, 491
DIVIDE, 141
DSETTINGS, 16
DTEXT, 332
EDGESURF, 214
ERASE, 228, 277, 293
EXPLODE, 301–303
EXPORT, 389
EXPORTLAYOUT, 369
EXTRUDE, 18, 42, 90, 92, 126, 238, 256, 325
FILLET, 194–195
FILLETEDGE, 194–196
FLATSHOT, 369
GEOMAPIMAGE, 488
GRAPHICSCONFIG, 418, 420
HELIX, 24, 40
HIDE, 12–13, 15, 49, 331, 423–424

IMPORT, 344, 389
INTERFERE, 38
INTERSECT, 36–38, 144–145, 271
JOIN, 413
LIGHT, 491–495
LIGHTLIST, 497
LINE, 42, 92, 125, 161
LIVESECTION, 378
LOFT, 134, 158–159, 163–164, 238, 240, 249–250, 256, 271
MASSPROP, 384–385
MATEDITOROPEN, 447
MATERIALATTACH, 445
MATERIALMAP, 473
MEASUREGEOM, 42–43
MESH, 214–215
MESHCAP, 229
MESHCOLLAPSE, 229
MESHCREASE, 227
MESHEXTRUDE, 225
MESHMERGE, 230
MESHOPTIONS, 218
MESHPRIMITIVEOPTIONS, 216
MESHREFINE, 223
MESHSMOOTH, 214, 218–219
MESHSMOOTHLESS, 222
MESHSMOOTHMORE, 222
MESHSPIN, 231
MESHSPLIT, 226, 231
MESHUNCREASE, 228
MIRROR, 17, 184, 277
MIRROR3D, 184–185
MOVE, 277, 279–280, 282, 287, 293
MTEXT, 332
MVIEW, 116, 433
NAVBAR, 50
NAVSMOTION, 51, 529
NAVSWHEEL, 63
NAVVCUBE, 10–11, 53, 59
NEWSHOT, 531
NEWVIEW, 51
OFFSETEDGE, 300–302
OPTIONS, 7, 17
PAN, 14, 50
PCEXTRACTCENTERLINE, 411–412
PCEXTRACTCORNER, 411–412
PCEXTRACTEDGE, 411–412
PCEXTRACTSECTION, 412–413
PEDIT, 84, 141, 198, 413
PFACE, 214
PLAN, 10, 20, 54, 63, 106
PLANESURF, 34, 136, 238, 256, 275
PLINE, 42, 84, 92, 141, 143, 161, 198
POINT, 83
POINTCLOUDATTACH, 407
POINTCLOUDCROP, 409
POINTCLOUDCROPSTATE, 409
POINTLIGHT, 492
POLYSOLID, 28–30, 271
PRESSPULL, 18, 42, 299–301, 319
PROJECTGEOMETRY, 255–256
PROPERTIES, 173, 198
PYRAMID, 30, 271
RECTANG, 324
REGEN, 12, 128
REGENALL, 128
REGION, 17, 42, 143
RENDER, 12, 14, 49, 433
RENDERENVIRONMENT, 520
RENDEREXPOSURE, 520
RENDERWINDOW, 14
REVOLVE, 134, 136–138, 140–142, 238, 256, 271, 286, 297
RIBBON, 5
ROTATE, 276–277, 279, 282, 287, 290, 294, 332
ROTATE3D, 495–496
RPREF, 515
RTPAN, 65
RTZOOM, 66
SAVEIMG, 515
SCALE, 277, 279, 287, 291
SECTIONPLANE, 343, 369–373, 412
SECTIONPLANETOBLOCK, 381
SEQUENCEPLAY, 546
SLICE, 199–204
SOLIDEDIT, 174, 198, 271, 278, 290, 293, 299, 307–326
SPHERE, 31, 81, 271
SPLINE, 257
SPOTLIGHT, 494
STRETCH, 277
SUBTRACT, 36–38, 42, 141–142, 271, 273, 302–303
SUNPROPERTIES, 489
SURFBLEND, 245, 249
SURFEXTEND, 245, 253
SURFEXTRACTCURVE, 259, 262–264, 326
SURFFILLET, 245, 249, 251–252
SURFNETWORK, 242
SURFOFFSET, 245, 250–251
SURFPATCH, 245, 247–250
SURFSCULPT, 259, 261
SURFTRIM, 253–255, 264
SWEEP, 40–41, 134–135, 153–156, 238, 256, 271
TEXT, 332
THICKEN, 34, 259–261
TORUS, 33, 271
UCS, 10, 59, 84–88, 91, 93–94, 98–103, 105–106, 127, 411
UCSICON, 3, 87, 89
UCSMAN, 89
UNDO, 11, 58, 498, 545–546
UNION, 36, 38, 144, 194, 252, 271, 273, 286, 298
VIEW, 9, 51, 125
VIEWBASE, 19, 335, 343–344, 346–348
VIEWCOMPONENT, 361
VIEWDETAIL, 343–344, 362–365
VIEWDETAILSTYLE, 362
VIEWEDIT, 351, 364
VIEWGO, 544
VIEWPLAY, 544

commands, *Continued*
VIEWPROJ, 343–344, 346
VIEWRES, 424
VIEWSECTION, 343–344, 352–353, 357–360, 365
VIEWSECTIONSTYLE, 353
VIEWSTD, 344–345
VIEWSYMBOLSKETCH, 360
VIEWUPDATE, 350
VPORTS, 115–116, 118, 121–122, 433
VSCURRENT, 12
WEBLIGHT, 496
WEDGE, 34, 271
WSCURRENT, 5
XEDGES, 318
ZOOM, 14, 50
composite region, 42
composite solid, 35, 174, 271
Boolean operations, 35–36
creating, 35–39
grip editing, 273–275
Conceptual visual style, 12–13, 19, 42, 49–50, 81, 88–89, 125, 176, 331, 425
concrete material type, 449
cone, 26–27
CONE command, 26, 81, 88, 271
context-sensitive panels, 225
continuity, 245
continuous 3D orbit, 62
contour lines, 31
CONVTOMESH command, 134
CONVTONURBS command, 134, 257–258
CONVTOSOLID command, 132, 134, 198, 218, 220, 261
CONVTOSURFACE command, 132, 134, 218, 220, 260
coordinate systems, 1–3, 79–107
cylindrical, 82–83
rectangular, 1–3
spherical, 79–82
COPY command, 277
corner deceleration, 563
corner intersections, 207
counterbore, 204
countersink, 205
crease, 227
Create New Visual Style dialog box, 431
Creating Video dialog box, 560
CUI command, 7
Customization flyout, 95–96, 176, 181, 223, 225, 274, 277–278
Customize User Interface dialog box, 7
CVHIDE command, 258
CVPORT system variable, 120–121
CVSHOW command, 258
CV splines, 257
cylinder, 27–28
CYLINDER command, 27, 95, 97, 127, 271
cylindrical coordinates, 82–83

D

decay, 484. *See* attenuation
default lighting, 485
DELOBJ system variable, 41, 134, 218, 240, 261
detail boundary, 362
detail views, 361–385
Detail View Style Manager dialog box, 362
diffuse, 482
diffuse color, 451
DIMASSOC system variable, 366
dimensioning, 334–337, 366–368
information grouping, 335
DIMREASSOCIATE command, 368
display silhouette, 418–419
DISPSILH system variable, 33, 418–419, 423
distant light, 481, 490–492
DISTANTLIGHT command, 491
DIVIDE command, 141
draft analysis, 388–389
draft angle, 161, 388
Drafting Settings dialog box, 16–17, 95, 411
3D Object Snap tab, 16–17, 411
Drafting Standard dialog box, 345
Drawing Units dialog box, 83
drawing view (object), definition, 348
drawing views, 344–351
creating, 346–349
editing, 351
page layout setup, 344–345
Drawing Window Colors dialog box, 58, 426
DSETTINGS command, 16
DTEXT command, 332
dynamic displays, 54
dynamic input, 22
dynamic UCS function, 96
DYNPICOORDS system variable, 92

E

edges, 287–293, 297–298, 317–318
coloring, 292–293
deleting, 293
editing, 287–293, 297–298, 317–318
moving, 288–289
offsetting, 300–302
rotating, 290
scaling, 290–292
elevations, generating, 381–384
environment settings, 521–523
ERASE command, 228, 277, 293
explicit surface, 239. *See* procedural surface
EXPLODE command, 301–303
EXPORT command, 389
EXPORTLAYOUT command, 369
exposure settings, 524
extracting curves from surfaces, 262–264
Extract Section Lines from Point Cloud dialog box, 412–413

EXTRUDE command, 18, 42, 90, 92, 126, 238, 256, 325
creating features, 141–145
creating models, 131–133, 135–136, 140–141, 143
creating regions, 135
creating surfaces, 138–139
extrusion along a path, 133–134

F

face editing, 279–286, 296–297, 309–317
collapsing, 229
coloring, 283, 317
deleting, 228
extruding, 225–226, 309–310
merging, 230
moving, 280–281, 311–312, 323
offsetting, 312
revolving, 286
rotating, 282–284, 313–314, 325
scaling, 282
spinning, 231
splitting, 226
tapering, 315–316, 324
FACETERDEVNORMAL system variable, 218
FACETERDEVSURFACE system variable, 218
FACETERGRIDRATIO system variable, 218
FACETERMAXEDGELENGTH system variable, 219
FACETERMESHTYPE system variable, 218
FACETRES system variable, 390, 418–420, 424
facets, 215
falloff, 483, 495
features, removing, 204
field of view (FOV), 554
file exchange, 389–390
importing and exporting, 389
fillet, 194
editing, 278
FILLET command, 194–195
FILLETEDGE command, 194–196
FILLETRAD3D system variable, 251
fillet surfaces, 251–252
final gathering, 519
FLATSHOT command, 369
floating viewports, 116
flyby animation, 554–560
creation steps, 560
Position Locator palette, 554–556
Forward tool, 69–70
free spotlight, 481, 495
free weblight, 482, 496
frustum cone, 26
frustum pyramid, 30
Full Navigation wheel, 63–65
full section, 353–356

G

Generate Section/Elevation dialog box, 381–383
generic material type, 450
generic surface, 238
geographic location, 487–488
GEOMAPIMAGE command, 488
gizmos, 181–183, 225, 259. *See also* grips
3D edit bar gizmo, 259
glazing material type, 449
global illumination (GI), 518
gradient background, 501–502
gradient map, 463–465
GRAPHICSCONFIG command, 418, 420
Graphics Performance dialog box, 418, 420, 534–535
grips, 271–275. *See also* gizmos
grip tools, 279. *See* grips, gizmos
GTAUTO system variable, 182
GTDEFAULT system variable, 282
GTLOCATION system variable, 182
guide curves, 163–164

H

half section, 356–357
HALOGAP system variable, 424
helix, 40–41
HELIX command, 24, 40
hidden display, 12
Hidden visual style, 12–13, 418, 425
HIDE command, 12–13, 15, 20, 49, 331, 423–424
highlight, 482
history and statistics pane, **Render** window, 512–513
home view, 53, 58
hotspot, 483, 494

I

image based lighting, 521–524
image map, 461–462
image pane, **Render** window, 511–512
images as backgrounds, 502–503
IMPORT command, 344
Import Point Clouds dialog box, 401
imprint, 319, 322–323
indexing, 400, 404
index of refraction (IOR), 456
indirect illumination, 518
information grouping, 335
interface, 4–7
INTERFERE command, 38
Interference Checking dialog box, 38–39
Interference Settings dialog box, 38
INTERSECT command, 36–38, 144–145, 271
intersection, 35–36
INTERSECTIONCOLOR system variable, 424

INTERSECTIONDISPLAY system variable, 424
isolines, 418
ISOLINES system variable, 418–419, 423, 428
isometric viewpoint presets, 9–10

J

JOIN command, 413

K

key position, 537

L

Lamp Color dialog box, 498
laser scanner, 400
layouts, exporting, 369
layout space, plotting from, 433
lens length, 557
levels, 185
LIGHT command, 491–495
light glyphs, 485
lighting, 485–486
  default, 485
  photometric, 486
  user-created, 485
lighting units, 486
LIGHTINGUNITS system variable, 486
LIGHTLIST command, 497
light properties, 482–484
  angle of incidence, 482
  attenuation, 484
  hotspot and falloff, 483
  reflectivity, 482
lights, 434–435, 481–500
  distant light, 490–492
  exposure and white balance, 486
  free spotlight, 495
  free weblight, 496
  intensity, 499
  lighting units, 486
  point light, 492–493
  properties, 482–484
  shadows, 500
  spotlight, 494–495
  sunlight, 487–488
  target point light, 493
  weblight, 496
Lights in Model palette, 496–498
limit box, 405
LINE command, 42, 92, 125, 161
line display, controlling, 420
LINESMOOTHING system variable, 420
Linetype Manager dialog box, 424
LIVESECTION command, 378
live sectioning, 378–381, 533
loft, 153, 158–164
  controlling with a path, 164
  cross sections, 159–163
  guide curves, 163–164
LOFT command, 134, 158–159, 163–164, 238, 240, 249–250, 256, 271
LOFTNORMALS system variable, 163–164
Loft Settings dialog box, 159–163
Look tool, 69–70
luminance, 456

M

map, 460–475
  adjustments, 473–475
  checker, 462–463
  gradient, 463–465
  image, 461–462
  marble, 468
  noise, 469–470
  speckle, 470
  texture editor, 461
  tiles, 465–467
  waves, 470–471
  wood, 471–472
mapped material, 460
marble map, 468
masonry material type, 449
MASSPROP command, 384–385
MATEDITOROPEN command, 447
MATERIALATTACH command, 445
Material Attachment Options dialog box, 445
material-level map adjustments, 473
MATERIALMAP command, 473
material mapping, 473
materials, 441–475
  applying, 445–446
  creating and modifying, 448–459
  editing, 446–448
  removing, 445–446
materials browser, 442–444
materials editor, 446–448
materials library, 441
material types, 448–450
matte, 482
MAXACTVP system variable, 115
MEASUREGEOM command, 42–43
mesh, 16, 237
MESHCAP command, 229
MESHCOLLAPSE command, 229
MESH command, 214–215
MESHCREASE command, 227
mesh editing, 221–231
  collapsing a face or edge, 229
  creases, applying and removing, 227–228
  extruding a face, 225–226
  gaps, creating and closing, 228–229
  merging faces, 230
  refining, 223
  smoothing, 221–222

spinning faces, 231
splitting a face, 226
MESHEXTRUDE command, 225
MESHMERGE command, 230
mesh modeling, 213–231
converting between object types, 217–221
editing, 221–231
mesh primitives, 216–217
tessellation divisions, 214–215
MESHOPTIONS command, 218
MESHPRIMITIVEOPTIONS command, 216
Mesh Primitive Options dialog box, 215–217, 219
mesh primitives, 216–217
MESHREFINE command, 223
MESHSMOOTH command, 214, 218–219
MESHSMOOTHLESS command, 222
MESHSMOOTHMORE command, 222
MESHSPIN command, 231
MESHSPLIT command, 226, 231
Mesh Tessellation Options dialog box, 218–219
MESHUNCREASE command, 228
metallic paint material type, 449
metal material type, 449
MIRROR command, 17, 184, 277
MIRROR3D command, 184–185
mirror material type, 449
model analysis, 384–389
draft analysis, 388–389
surface continuity, 385–386
surface curvature, 386–388
model documentation, 343–384
auxiliary views, 365–366
detail views, 362–365
dimensioning, 366–368
section views, 352–361
model space viewports, 116
Modify Detail View Style dialog box, 363
Modify Section View Style dialog box, 353–354
motion, 539
motion path animation, 561–564
Motion Path Animation dialog box, 562–564
MOVE command, 277, 279–280, 282, 287, 293
movement, 539
movement types, 540–543
MTEXT command, 332
MVIEW command, 116, 433

N

named views, 51–52
NAVBAR command, 50
navigation bar, 50–51
customizing and repositioning, 75–76
NAVSMOTION command, 51, 529
NAVSWHEEL command, 63
NAVSWHEELMODE system variable, 63
NAVSWHEELOPACITYBIG system variable, 72
NAVSWHEELOPACITYMINI system variable, 72
NAVSWHEELSIZEBIG system variable, 72
NAVSWHEELSIZEMINI system variable, 72
NAVVCUBE command, 10–11, 53, 59
NAVVCUBELOCATION system variable, 60
NAVVCUBEOPACITY system variable, 60
NAVVCUBEORIENT system variable, 61
NAVVCUBESIZE system variable, 60
nested block, 38
network surfaces, 242–244
New Detail View Style dialog box, 363
New Section View Style dialog box, 353
NEWSHOT command, 531
NEWVIEW command, 51
New View/Shot Properties dialog box, 51–52, 500, 529–545
noise map, 469–470
non-uniform rational B-spline, 238. *See* NURBS surface
normal, 160
NURBS surface, 138, 238, 256–259
creating and editing, 257–259

O

OBSCUREDCOLOR system variable, 423
OBSCUREDLTYPE system variable, 423
occluded lines, 423
OFFSETEDGE command, 300–302
offset section views, 358
OPTIONS command, 7, 17
Options dialog box, 7, 17, 58, 97, 244, 407, 418–419, 426, 556
Orbit tool, 54, 68
O-ring, 206
O-ring groove, 206
orthographic displays, 56–57
orthographic section planes, 371–372
orthographic viewpoint presets, 9–10

P

PAN command, 14, 50
Pan tool, 65
paper space, 116, 433
parallel projection, 55
patch surface, 247
path arrays, 190–194
PCEXTRACTCENTERLINE command, 411–412
PCEXTRACTCORNER command, 411–412
PCEXTRACTEDGE command, 411–412
PCEXTRACTSECTION command, 412–413
PDMODE system variable, 83, 141, 143
PEDIT command, 84, 141, 198, 413
perspective projection, 55
photometric lighting, 486
PICKFIRST system variable, 294
pictorial displays, 54–55
planar surface, 34–35
PLAN command, 10, 20, 54, 63, 106
PLANESURF command, 34, 136, 238, 256, 275
plan view, 9

plastic material type, 449
PLINE command, 42, 84, 92, 141, 143, 161, 198
POINTCLOUDATTACH command, 407
Point Cloud Color Map dialog box, 409
POINTCLOUDCROP command, 409
POINTCLOUDCROPSTATE command, 409
point clouds, 399–413
    Autodesk ReCap Pro, 400–407
    display options, 408–410
    extracting geometry, 411–412
    inserting into drawing file, 407
    section objects, 412–413
POINT command, 83
POINTLIGHT command, 492
point lights, 481, 492–493
polar arrays, 188–190
POLYSOLID command, 28–30, 271
polysolids, 28–30
PRESSPULL command, 18, 42, 299–301, 319
PREVIEWCREATIONTRANSPARENCY system variable, 249
primitives, 16, 22–34, 271
Print dialog box, 512
procedural map, 460
procedural surface, 138, 238
projected geometry, trimming surfaces, 254–255
PROJECTGEOMETRY command, 255–256
PROPERTIES command, 173, 198
Properties palette, 20, 34, 132, 157, 163, 173–174, 186, 193, 198, 222, 365, 410–411, 433
PSOLHEIGHT system variable, 28
PSOLWIDTH system variable, 28
pyramid, 30–31
PYRAMID command, 30, 271

Q

Quick Access toolbar, 6, 75

R

rapid prototyping, 389
raytracing, 518
Realistic visual style, 9, 12–13, 19, 425
reality capture, 399
rectangular 3D coordinate system, 1–3
rectangular arrays, 185–187
rectangular coordinate entry, 79
reflectivity, 454
refraction, 456
REGEN command, 12, 128
REGENALL command, 128
region, 42
    calculating area, 42–43
REGION command, 17, 42, 143
registering, 400, 403–404
RENDER command, 12, 14, 49, 433
Render Environment & Exposure palette, 520–524
RENDERENVIRONMENT command, 520
RENDEREXPOSURE command, 520
rendering, 14–15, 433–437, 511–524
    fundamentals, 518–520
    lights, 434–435
    Render window, 511–513
render presets, 513–518
    creating custom, 517–518
    settings, 515–517
Render Presets Manager palette, 515–518
Render to Size Output Settings dialog box, 516
Render window, 14–15, 436–437, 511–513
RENDERWINDOW command, 14
revolutions, 136–145
REVOLVE command, 134, 136–138, 140–142, 238, 256, 271, 286, 297
    creating features, 142–145
    creating regions, 138
    creating surfaces, 138–139
Rewind tool, 67–68
RIBBON command, 5
ribbon panels, 5
right-hand rule of 3D drawing, 3–4
ROTATE command, 276–277, 279, 282, 287, 290, 294, 332
ROTATE3D command, 495–496
roughness, 482
round, 194
RPREF command, 515
RTPAN command, 65
RTZOOM command, 66

S

sampling, 518
SAVEIMG command, 515
saving named UCS, 92–94
SCALE command, 277, 279, 287, 291
scan files, 399–400
scan region, 407
sculpting solids from surfaces, 261–262
section boundary state, 373
section line, 373
section objects
    adjusting, 377
    grips, 375–377
    jogs, 378
    live sectioning, 378–381
    point clouds, 412–413
    properties, 374–375
    states, 373–374
SECTIONPLANE command, 343, 369–373, 412
section plane indicator, 373
section planes, 369–384
    constructing, 369–372
    orthographic, 371–372
section plane state, 373
SECTIONPLANETOBLOCK command, 381
Section Settings dialog box, 379–380
section slice state, 373
section views, 352–361

constraining section lines, 360
creating from an object, 359–360
editing, 361
excluding components, 361
section view style, 353
Section View Style Manager dialog box, 353
section volume state, 374
segmentation data, 411
Select Color dialog box, 452, 458, 501–502
selection cycling, 176–177, 277
SELECTIONCYCLING system variable, 176
self illumination, 456
self intersecting, 33
SEQUENCEPLAY command, 546
shaded display, 12
Shaded visual style, 12–13, 425
Shaded with Edges visual style, 12–13, 425
Shades of Gray visual style, 12–13, 425
shadows, 500
shell, 320–321, 326
shot, 529
Show gizmos flyout, 181, 225, 274, 287–288
SHOWHIST system variable, 174
show motion, 529–547
cinematic shots, 537–543
displaying and replaying shots, 543–544
still shots, 534–535
thumbnail shortcut menu, 544
view categories, 545–547
walk shots, 535–537
ShowMotion toolbar, 51–52, 530–531, 535–537, 543–546
Sketchy visual style, 12–13, 425
sky illumination, 485
SLICE command, 199–204
SMOOTHMESHCONVERT system variable, 220
SMOOTHMESHMAXLEV system variable, 222
SOLDRAW command, 344
solid backgrounds, 501
SOLIDEDIT command, 174, 198, 271, 278, 290, 293, 299
solid extrusion, 131
solid glass material type, 450
SOLIDHIST system variable, 174–176
solid model, 237
solid model display
controlling, 417–420
display silhouette, 418–419
isolines, 418
surface smoothness, 419–420
solid model editing, 307–326
body editing, 319–322
edge editing, 317–318
extracting a wireframe, 318
face editing, 309–317
solid model file exchange, 389–390
solid model history, 174–176
solid modeling, 16, 21–43
composite solids, 35–39
primitives, 22–34
working with regions, 42–43
solid primitives, 22–34
box, 24–26
cone, 26–27
cylinder, 27–28
polysolid, 28–30
pyramid, 30–31
sphere, 31–33
torus, 33
wedge, 33–34
solids
converting, 198–199
exploding, 301–303
grip editing, 271–275
sculpting, 261–262
slicing, 199–204
speckle map, 470
specular color, 452
specular factor, 482
sphere, 31–33
SPHERE command, 31, 81, 271
spherical coordinates, 79–82
SPLINE command, 257
SPLINESEGS system variable, 424
spotface, 204–205
SPOTLIGHT command, 494
spotlights, 481, 494–495. *See also* free spotlight
steering wheels, 63–75
Full Navigation wheel, 63–65
mini wheels, 70–71
Tour Building wheel, 63, 68–70
View Object wheel, 63, 66–68
Steering Wheel Settings dialog box, 68, 71–75
STEPSIZE system variable, 556
STEPSPERSEC system variable, 556
stereolithography (SLA), 389
still shots, 529, 534–535
stone material type, 450
STRETCH command, 277
subdivision surfaces, 213
subobject editing, 204, 271–303
edges, 287–293, 297–298
faces, 279–286, 296–297
fillets and chamfers, 278
grips, 271–275
vertices, 293–295, 297–298
SUBOBJSELECTIONMODE system variable, 224
SUBTRACT command, 36–38, 42, 141–142, 271, 273, 302–303
sun and sky background, 503
sunlight, 434–435, 485
setting by location, 487–488
SUNPROPERTIES command, 489
Sun Properties palette, 489–490
surface associativity, 239–241
SURFACEASSOCIATIVITY system variable, 134, 138–139, 238–240, 256
SURFACEAUTOTRIM system variable, 255
surface continuity, 385–386
surface curvature, 386–388
surface curvature analysis, 386
surface extrusion, 131

surface model, 237
surface modeling, 15, 237–264
  blend surfaces, 245–247
  extending surfaces, 253
  fillet surfaces, 251–252
  network surfaces, 242–244
  NURBS surfaces, 256–259
  offsetting surfaces, 250–251
  patch surfaces, 247–250
  projected geometry, 254–255
  trimming, 253–256
SURFACEMODELINGMODE system variable, 132, 138–139, 238–239, 256–258
surfaces
  converting objects to surfaces, 260
  editing using grips, 275
  extracting curves, 262–264
  trimming, 253–256
  using to sculpt solids, 261–262
surface smoothness, controlling, 419–420
SURFBLEND command, 245, 249
SURFEXTEND command, 245, 253
SURFEXTRACTCURVE command, 259, 262–264, 326
SURFFILLET command, 245, 249, 251–252
SURFNETWORK command, 242
SURFOFFSET command, 245, 250–251
SURFPATCH command, 245, 247–250
SURFSCULPT command, 259, 261
SURFTRIM command, 253–255, 264
SURFU system variable, 34, 242, 244
SURFV system variable, 34, 242
sweep, 153
SWEEP command, 40–41, 134–135, 153–156, 238, 256, 271
swept objects, 153–158
  base point, 155
  creating surfaces and solids, 153–158
  scaling, 156
  twisting, 156–157
system variables
  ANNOMONITOR, 366
  CAMERADISPLAY, 551
  CAMERAHEIGHT, 551
  CVPORT, 120–121
  DELOBJ, 41, 134, 218, 240, 261
  DIMASSOC, 366
  DISPSILH, 33, 418–419, 423
  DYNPICOORDS, 92
  FACETERDEVNORMAL, 218
  FACETERDEVSURFACE, 218
  FACETERGRIDRATIO, 218
  FACETERMAXEDGELENGTH, 219
  FACETERMESHTYPE, 218
  FACETRES, 390, 418–420, 424
  FILLETRAD3D, 251
  GTAUTO, 182
  GTDEFAULT, 282
  GTLOCATION, 182
  HALOGAP, 424
  INTERSECTIONCOLOR, 424
  INTERSECTIONDISPLAY, 424
  ISOLINES, 418–419, 423, 428
  LIGHTINGUNITS, 486
  LINESMOOTHING, 420
  LOFTNORMALS, 163–164
  MAXACTVP, 115
  NAVSWHEELMODE, 63
  NAVSWHEELOPACITYBIG, 72
  NAVSWHEELOPACITYMINI, 72
  NAVSWHEELSIZEBIG, 72
  NAVSWHEELSIZEMINI, 72
  NAVVCUBELOCATION, 60
  NAVVCUBEOPACITY, 60
  NAVVCUBEORIENT, 61
  NAVVCUBESIZE, 60
  OBSCUREDCOLOR, 423
  OBSCUREDLTYPE, 423
  PDMODE, 83
  PICKFIRST, 294
  PREVIEWCREATIONTRANSPARENCY, 249
  PSOLHEIGHT, 28
  PSOLWIDTH, 28
  SELECTIONCYCLING, 176
  SHOWHIST, 174
  SMOOTHMESHCONVERT, 220
  SMOOTHMESHMAXLEV, 222
  SOLIDHIST, 174–176
  SPLINESEGS, 424
  STEPSIZE, 556
  STEPSPERSEC, 556
  SUBOBJSELECTIONMODE, 224
  SURFACEASSOCIATIVITY, 134, 138–139, 238–240, 256
  SURFACEAUTOTRIM, 255
  SURFACEMODELINGMODE, 132, 138–139, 238–239, 256–258
  SURFU, 34, 242, 244
  SURFV, 34, 242
  UCSAXISANG, 106
  UCSBASE, 106
  UCSDETECT, 96, 106
  UCSFOLLOW, 106–107, 126–128
  UCSNAME, 106
  UCSORG, 106
  UCSORTHO, 106
  UCSSELECTMODE, 89, 106
  UCSVIEW, 106
  UCSVP, 106–107, 126–128
  UCSXDIR, 106
  UCSYDIR, 106
  VIEWUPDATEAUTO, 349–350, 352, 362
  VSBACKGROUNDS, 428
  VSEDGECOLOR, 428
  VSEDGEJITTER, 430
  VSEDGELEX, 430
  VSEDGES, 428
  VSEDGESMOOTH, 430
  VSFACECOLORMODE, 427
  VSFACEHIGHLIGHT, 428
  VSFACEOPACITY, 427

system variables, *Continued*
- **VSFACESTYLE**, 426
- **VSHALOGAP**, 430
- **VSINTERSECTIONCOLOR**, 429
- **VSINTERSECTIONEDGES**, 429
- **VSINTERSECTIONLTYPE**, 429
- **VSISOONTOP**, 428
- **VSLIGHTINGQUALITY**, 215, 427
- **VSMATERIALMODE**, 427
- **VSMONOCOLOR**, 427
- **VSOCCLUDEDCOLOR**, 429
- **VSOCCLUDEDEDGES**, 429
- **VSOCCLUDEDLTYPE**, 429
- **VSSHADOWS**, 428
- **VSSILHEDGES**, 430
- **VSSILHWIDTH**, 430

T

target plane, 552
target point light, 481, 493
tessellation divisions, 214–215
tessellation lines, 31. *See* contour lines
TEXT command, 332
text, three-dimensional, 331–334
texture editor, 461
texture map, 460
THICKEN command, 34, 259–261
tiled viewports, 116
tiles map, 465–467
tool palette, copying visual styles to, 431–432
torus, 33
TORUS command, 33, 271
Tour Building wheel, 63, 68–70
translucency, 456
transparency, 455
transspatial dimensioning, 368

U

UCS. *See* user coordinate system
UCSAXISANG system variable, 106
UCSBASE system variable, 106
UCS command, 10, 59, 84–88, 91, 93–94, 98–103, 105–106, 127, 411
UCS Details dialog box, 105
UCSDETECT system variable, 96, 106
UCS dialog box, 85, 89–90, 93–94, 103–105, 107
UCSFOLLOW system variable, 106–107, 126–128
UCS icon, 84, 89
- manipulating, 94–96
- shortcut and grip menus, 85

UCSICON command, 3, 87, 89
UCSMAN command, 89
UCSNAME system variable, 106
UCSORG system variable, 106
UCSORTHO system variable, 106
UCSSELECTMODE system variable, 89, 106
UCS settings, 59
UCSVIEW system variable, 106
UCSVP system variable, 106–107, 126–128
UCSXDIR system variable, 106
UCSYDIR system variable, 106
UNDO command, 11, 58, 498, 545–546
union, 35
UNION command, 36, 38, 144, 194, 252, 271, 273, 286, 298
Up/Down tool, 69–70
user coordinate system, 53, 84–107
- creating from three points, 99
- dynamic, 96–97
- moving and aligning, 86–88, 97–98
- naming and saving, 92–94
- preset orientations, 103
- rotating, 86–88, 90–91, 98–99
- working with the UCS icon, 94–96

user-created lighting, 485

V

vanishing points, 55
vertices, 293–295, 297–298
view, 529
VIEWBASE command, 19, 335, 343–344, 346–348
view category, 530, 545
VIEW command, 9, 51, 125
VIEWCOMPONENT command, 361
View Controls flyout, 8–10, 55
view cube, 10–11, 53–62
- home view, 58
- orthographic displays, 56–57
- pictorial displays, 54–55
- projection, 55–56

View Cube Settings dialog box, 53–54, 59–62
VIEWDETAIL command, 343–344, 362–365
VIEWDETAILSTYLE command, 362
VIEWEDIT command, 351, 364
VIEWGO command, 544
viewing and displaying models, 49–76
- continuous 3D orbit, 62
- named views, 51–52
- navigation bar, 50–51, 75–76
- **PLAN** command, 63
- steering wheels, 63–75

View Manager dialog box, 8–9, 51, 503–504
View Object wheel, 63, 66–68
VIEWPLAY command, 544
viewpoint, 8
viewpoint presets, 9–10
viewport controls, 8–9
Viewport Controls flyout, 8, 10, 50, 53, 63, 75, 118, 121
viewports, 115–128
- adjusting, 123
- configurations, 118–120
- drawing in multiple, 125–128
- joining, 122–123
- making active, 120–121

Viewports dialog box, 118–126, 433
VIEWPROJ command, 343–344, 346
VIEWRES command, 424
VIEWSECTION command, 343–344, 352–353, 357–360, 365
VIEWSECTIONSTYLE command, 353
View/Shot Properties dialog box, 544, 546–547
view state, 406
VIEWSTD command, 344–345
VIEWSYMBOLSKETCH command, 360
VIEWUPDATE command, 350
VIEWUPDATEAUTO system variable, 349–350, 352, 362
Visual Style Controls flyout, 8–9, 12, 14, 19
visual styles, 12–13, 421–433
- 2D Wireframe, 12–13, 19, 32, 418, 420, 423–424
- Conceptual, 12–13, 19, 42, 49–50, 81, 88–89, 125, 176, 331, 425
- creating custom, 431–432
- exporting to a tool palette, 431–432
- Hidden, 12–13, 418, 425
- plotting, 433
- Realistic, 9, 12–13, 19, 425
- Shaded, 12–13, 425
- Shaded with Edges, 12–13, 425
- Shades of Gray, 12–13, 425
- Sketchy, 12–13, 425
- Wireframe, 12–13, 81, 125, 182, 282, 322, 425
- X-ray, 12–13, 87, 94–96, 426

Visual Styles Manager palette, 31–33, 418–419, 421–423, 431–432
VPORTS command, 115–116, 118, 121–122, 433
VSBACKGROUNDS system variable, 428
VSCURRENT command, 12
VSEDGECOLOR system variable, 428
VSEDGEJITTER system variable, 430
VSEDGELEX system variable, 430
VSEDGES system variable, 428
VSEDGESMOOTH system variable, 430
VSFACECOLORMODE system variable, 427
VSFACEHIGHLIGHT system variable, 428
VSFACEOPACITY system variable, 427
VSFACESTYLE system variable, 426
VSHALOGAP system variable, 430
VSINTERSECTIONCOLOR system variable, 429
VSINTERSECTIONEDGES system variable, 429
VSINTERSECTIONLTYPE system variable, 429
VSISOONTOP system variable, 428
VSLIGHTINGQUALITY system variable, 215, 427
VSMATERIALMODE system variable, 427
VSMONOCOLOR system variable, 427
VSOCCLUDEDCOLOR system variable, 429
VSOCCLUDEDEDGES system variable, 429
VSOCCLUDEDLTYPE system variable, 429
VSSHADOWS system variable, 428
VSSILHEDGES system variable, 430
VSSILHWIDTH system variable, 430

W

Walk and Fly Settings dialog box, 556
walk shot, 529, 535–537
Walk tool, 65, 536
walkthrough animation, 554–560
- creation steps, 560
- general settings, 556–559

wall paint material type, 450
water material type, 450
watertight mesh, 390
waves map, 470–471
WCS. *See* world coordinate system
weblight, 482, 496. *See also* free weblight
WEBLIGHT command, 496
wedge, 33–34
WEDGE command, 34, 271
wireframe display, 12
wireframe object, 15, 318
Wireframe visual style, 12–13, 81, 125, 182, 282, 322, 425
wood map, 471–472
wood material type, 450
workspace, 5
world coordinate system (WCS), 59, 63, 79, 84
WSCURRENT command, 5

X

XEDGES command, 318
X-ray visual style, 12–13, 87, 94–96, 426

Z

zebra analysis, 385
zebra striping, 385
ZOOM command, 14, 50
Zoom tool, 66–68

Index - Advanced